2025 10th International Conference on Integrated Circuits and Microsystems (ICICM 2025)

Hefei, China
17-19 October 2025

Pages 1 – 495

IEEE Catalog Number: CFP25H23-POD
ISBN: 979-8-3315-8850-2

2025 10th International Conference on Integrated Circuits and Microsystems (ICICM 2025)

Hefei, China

17-19 October 2025

Pages 1–495

IEEE Catalog Number: CFP25H23-POD

ISBN: 979-8-3315-8850-2

Copyright © 2025, IEEE

All Rights Reserved

Copyright and Reprint Permissions:

Abstracting is permitted with credit to the source. Libraries are permitted to photocopy beyond the limit of U.S. copyright law for private use of patrons those articles in this volume that carry a code at the bottom of the first page, provided the per-copy fee indicated in the code is paid through Copyright Clearance Center, 222 Rosewood Drive, Danvers, MA 01923.

For other copying, reprint or republication permission, write to IEEE Copyrights Manager, IEEE Service Center, 445 Hoes Lane, Piscataway, NJ 08854. All rights reserved.

*** This is a print representation of what appears in the IEEE Digital Library. Some format issues inherent in the e-media version may also appear in this print version.

IEEE Catalog Number: CFP25H23-POD

ISBN (Print-On-Demand):   979-8-3315-8850-2

ISBN (Online):   979-8-3315-8849-6

Additional Copies of This Publication Are Available From:

Curran Associates, Inc
57 Morehouse Lane
Red Hook, NY 12571 USA

Phone: (845) 758-0400
Fax: (845) 758-2633
E-mail: curran@proceedings.com
Web: www.proceedings.com

TABLE OF CONTENTS

A Compensation-Capacitor-Based Perturbation Injection Calibration Method for SAR ADCs 1
Hongjian Ren, Ke Hu, Wenya Luo, Chao Cao, Haijun Guo

Design and Simulation of High-Frequency FBAR Based on Multiphysics and MBVD Circuit Modeling 7
Rui Zhao, Jiangbo Wei, Yang Xiao, Liaoliao Zhang, Haijuan Li, Jiaqi Liu, Chao Wang

A 10GS/s 8b Time-Interleaved ADC with Aperture Error Calibration and LMS-Based Nonlinearity
Correction 12
Chuan Qin, Chuan Liu, Runqiao Wang, Lecheng Li, Maliang Liu, Yintang Yang

VCIL: An Open-Source Pipeline-Tight HIL Framework for Cycle-Accurate RISC-V Coprocessor
Verification 19
Xian Lin, Xin Zheng, Zhixin Fan, Huaien Gao, Shuting Cai, Xiaoming Xiong

An Electrode Impedance Evaluation IC with Biphasic Output Current for Biomedical Devices 25
Fang Xie, Yu Ang Ru, Zhuo Chen, Xu Liu

A Fast Transient and High PSRR LDO with 62 ns Settling Time 31
Ruijie Wang, Zhiguo Yu, Xiaofeng Gu

Automatic Generation Method of Test Vectors for FPGA Interconnect Resources Based on
Self-adaptive Ford-Fulkerson Algorithm 37
Liang Zhou, Wei Zhou, Mingzhe Li, Kun Guo, Ding Zhang, Yan Huang

A Multi-Phase Clock Generator with Customized Duty Cycles for Pipeline-SAR ADCs 43
He Zhi, Hongtao Yue, Yuwei Zong, Zhongxu Zhang, Jianhui Wu, Xin Li

Hardware Design and Side-Channel Security Analysis on the Key Computational Block for YOLOv11 49
Runquan Shao, Liji Wu, Jing Hu, Le Wu, Xiangmin Zhang

Secondary-Side Controlled Flyback Converter with Buck-Bypass for Multi-Port Application 54
Zhen Wang, Yi Liu

Research on the Method of Switching Adjustment to Reduce the Temperature Drift of the Bandgap
Reference 60
Bin Wen, Zhongjie Guo, Hongpeng Dai, Xiaojing Qiang

Design and Implementation of a Secure Boot RISC-V Processor for IoT Devices 65
Zeyu Li, Liji Wu, Jing Hu, Han Sun, Xiangmin Zhang, Haijie Wang

A 1GS/s 13-bit Single-Channel Hybrid Pipelined SAR ADC with Domino SAR Logic 71
Chuan Liu, Chuan Qin, Lecheng Li, Zhihao Hou, Yuqi Fan, Maliang Liu

A 1.7-to-3.0 GHz 201.95 dBc/Hz FOMT VCO With a Current-Reuse Active Inductor 76
Xinyi Zhang, Wanqing Wu, Guoxun Dai, Jiangnan Li, Haigang Feng

A 24-28GHz GaN SPDT MMIC with High Isolation and Low Insertion Loss 81
Peiyao Liu, Pengfei Jiang, Kui Dang, Rujun Sun, Yue Hao, Jincheng Zhang

Design and FPGA Implementation of a NTT Hardware Accelerator for PQC ML-DSA 86
Shuhang Zhu, Liji Wu, Lei Li, Yifan Yang, Xiangmin Zhang

The Effective Resistance Calculation Method Based on Matrix Partitioning Takahashi Algorithm 92
Meng Li, Yu Chen, Jingrui Chen, Yaokai Zhang, Jinyu Zhang, Xiaolue Lai

Hardware Design for Decomposition Module with Variable-Order Masking against Side-Channel Attack for PQC ML-DSA 99
Xuejian Wang, Liji Wu, Lei Li, Yifan Yang, Xiangmin Zhang

Design and Implementation of a High-Reliability MIPI Interface Error Detection and Correction System for CMOS Image Sensors 105
Qiang Zhao, Jianwei Feng, Bin Qiang, Zhigang Li, Xin Li, Chunyu Peng, Xiulong Wu

Temperature Sensing Module Design for Silicon MEMS Clocks 110
Shaotian Fan, Lu Tang, Yuxing Gu

An End-to-End Compact Shape-Aware Macro Placer Using Reinforcement Learning 115
Hailiang Li, Xiao Wang, Xu Yang, Miaohui Hao, Yan Huo, Beiping Yan

A 2.4 GHz Wide-Tuning-Range Low-Phase-Noise Differential Digitally Controlled Ring Oscillator for CMOS Image Sensor Applications 122
Zhigang Li, Hongjia Xu, Yao Nian, Jingyi Wang, Xin Li, Qiang Zhao, Xiulong Wu

Smart Wireless Insole with Self-Powered Triboelectric Pressure Sensors for Gait Detection 127
Yuhan Wang, Yue Yao, Enze Li, Sifan Liang, Lei Li

Voltage-Time Hybrid Domain ADC with PVT Tracking for High-Frame-Rate CMOS Image Sensor 132
Xin Li, Ziheng Wang, Bin Qiang, Chunyu Peng, Qiang Zhao, Xiulong Wu

A Very-Large-Dynamic-Range Charge-Sensitive Front End for One-Terminal Capacitive Sensors in 65-nm CMOS 138
Haitao Fan, Wenxiang Li, Jiafeng Yang

A fast transient-response LDO-CP with novel high voltage compensation structure in CMOS image sensors 143
Qiang Zhao, Zhendong Niu, Zhigang Li, Xiuying Wang, Chunyu Peng, Xiulong Wu

A 13-bit 1.5-GS/s Dual-Residue Pipelined ADC with Time-Domain Interpolation and Parallel Quantization 148
Zecheng Zhou, Depan Li, Longsheng Wang, Ting Li, Dongbing Fu, Dengquan Li

A Novel Self-adaptive Dead-time Control Circuit Design for GaN Gate Drive Circuit 153
Shupeng Yan, Shengqi Yu, Yi Huang, Junjie Yi, Pengyu Wang

A Novel Hybrid Switching Scheme for SAR ADC with High Efficiency and High Linearity 158
Yushi Chen, Ying Li, Fang Liang, Zewen Wang, Zuoru Dong, Xiaodong Wang

A Unified Swin Transformer Framework for Inverse Lithography and Lithography Simulation 163
Jiajun Tan, Ling Liang, Ming Zhu

Design of 6-to-8-bit High-Speed Time-Interleaved SAR ADC 168
Zhenyang Zhang, Fangxu Lv, Zhengbin Pang, Jiaqing Xu, Qiang Wang, Meng Li

RLMBFF: Multi-bit Flip-Flops Merging using Deep Reinforcement Learning 173
Zeqi Chen, Zhaori Bi, Changhao Yan, Ming Zhu, Xiulong Wu, Xuan Zeng

HybridSYN: An Efficient Logic Synthesis Methodology 180
Ke Wang, Yue Wu, Xiaoyan Yang

A 20-GHz Low-Jitter Fractional-N CP-PLL With 2.5-GHz FBAR Reference 186
Jiandu Wang, Jiwei Huang, Yi Sun

A 14-Bit SAR ADC with Area-Efficiency Hybrid Three-Segment CDAC 191
Chengchen Song, Jia Yu, Chao Cao, Hongjian Ren, Yuchan Duan, Haijun Guo

A Fully Reversible Feynman Gate with Magic Block for Quantum-dot Cellular Automata Circuits 198
Jiayi Zhang, Feifei Deng, Xianghui Li, Yuezhong Xiong, Yunong Zhao, Xiaohui Guo

A Wide-Temperature-Range Low-Drift Bandgap Reference with Process-Corner-Adaptive Logic Current Compensation 204
Xuesong Chen, Yu Jin, Ning Fu, Weifeng Li, Duli Yu

A Pixel-to-Column Digital Readout Circuit with InPixel CDS for Ga_2O_3 UV Photodetectors 209
Jiekun Zhang, Guangfen Wei, Jingwu Gong, Yu Zhou, Xiaodong Zhang, Helun Song

A DDR Controller Circuit Gate-level Post-simulation Method in SoC Design 215
Yande Jiang, Jingbo Ma, Guangda Zhang, Huiquan Wang, Bingxi Pei, Na Chen, Jian Fang

A High-Precision Current Detection Circuit for Battery Management System Chip in Electric Vehicle 221
Zhiwei Li, Liji Wu, Dan Xie, Kunning Mao, Haifeng Chen, Zhilin Zhong

A Robust Startup Circuit for Current-Mode Bandgap Reference 226
Qi Wang, Hao Li, Chunlei Pang

A Low-Power High-Accuracy Phase Frequency Detector and Charge Pump Design for CPPLL in 28 nm CMOS Technology 232
Zhigang Li, Hao Lei, Zhihao Liu, Lingrui Yan, Qiang Zhao, Shan Gao, Xiulong Wu

A 60-V 96.7%-Efficiency 175kHz Integrated Dimmable LED Driver with Mean-Peak Current Control and Frequency Jittering Scheme 237
Jiamian Mao, Zhongguang Xu, Xiangyin Chen

A 100-Gb/s PAM4 Receiver Analog Front-End 243
Haoran Huang, Yue Xu, Jing Huang, Gengzhen Qi

A −86 dBm/15.6 µW Wake-Up Receiver for Internet of Vehicles Applications 247
Rui Chen, Yan Li, Liming Si, Xinchao Zhong, Chiu-Wing Sham, Hang Yu

A 28 − 32 Gb/s Wireline Receiver with A Genetic- Algorithm-Based Adaptive CTLE in 28 nm CMOS 252
Yingjie Zhang, Fangxu Lv, Liangyong Yuan, Xiaoyue Hu, Ruixiao Kuai, Xianchao Zeng

A Wideband Blocker-Tolerant Receiver with High Linearity and Low 0 dBm-BNF in 65-nm CMOS 257
Kaiyun Deng, Yingqi Liu, Qingrui Jiang, Kailin Liu, Shaohui Pan, Haigang Feng

A Model of a C-band Mixer Circuit 262
Xuezhen Zhao, Jun Liu, Jing Wang, Guodong Su

Parametric Modeling of Transmission Structure of RF Microsystem Based on CNN-LSTM-SelfAttention 267
Li Qin, De Xi Liu, Cui Jing, Lei Shi, Ya Wei Liu, Ting Xue

A Radio-Frequency Orthogonal Switched-Capacitor Transmitter 272
Yingyi Chen, Mianting Hu, Sijia Zhao, Ruiyi He, Gengzhen Qi

Daisy Chain Transmitter Circuit Design with AntiElectromagnetic Interference for BMS Chip in EV 276
Chen Lin, Liji Wu, Jing Hu, Zonghuan Wu, Xiangmin Zhang

A Low-Power Flip signal Pulse Width Self-Adaptive Circuit for Piezoelectric Energy Harvesting 282
Hongtao Yue, Shuo Zhang, Haoyu Xue, He Zhi, Jianhui Wu, Xin Li

A Sub-nanosecond Delay and 200V/ns CMTI Level Shifter for GaN HEMTs Gate Driver 288
Shengqi Yu, Pengyu Wang, Junjie Yi, Shupeng Yan, Yi Huang

A Wide Tuning Range Dual-Mode, Magnetical-Coupled VCO with Capacitive-Coupled
Noise-Circulating Technique Achieving FoM of 188dBc/Hz 294
Qishuang Liu, Chenxiang Cai, Yunchu Li

A Fractional-N All-Digital PLL with 166fs$_{rms}$ Jitter and 238 dB FoM Based on a Distributed
Switched-Capacitor Arrays DCO 299
Zhihong Xu, Yuhui Li, Pei Qin, Changsong Lin, Jinhua Guo, Xiaorui Liu, Taotao Xu, Quan Xue

A CMOS Current-Reused Wideband Low-Power Dynamic Current Mode Logic Frequency Divider 305
Jingchen Liu, Youming Zhang, Xusheng Tang, Zhennan Wei, Shenghao Meng

A VHF memristor emulator for crossbar synaptic simulation 309
Guangzhen Dai, Bingchen Liu, Sihao Yang, Wei Li, Daohua Wu, Yuefeng He

A Novel SiC LDMOS with Electron Accumulation Layer Featuring Ultra-low Specific On-resistance
and Improved Breakdown Voltage 317
Moufu Kong, Lin Tong, Qizhi Feng, Jiaru Jia, Zhaoyu Ai, Hongfei Deng, Yangyang Ma, Peifei Wu

Hardware Implementation and Side-Channel Security Analysis for High-Precision AI Tansformer
Encoder 322
Wentao Wang, Liji Wu, Jing Hu, Le Wu, Xiangmin Zhang

High-Performance Data Prefetching Accelerator for Real-Time Object Detection 328
*Yanzhou Tang, Xiangyu Zhou, Zeyi Lin, Ming Cheng, Naifeng Jing, Jianfei Jiang, Weiguang Sheng
Qin Wang*

Design of High-reliability and High-dynamic Analog Front-end for DC Carrier Chip 333
Quan Sun, Gang Chen, Changyou Men, Shuang Wu

A Steady-State Temperature Solving Method for Multi-Chip Modules Components Based on an
Improved U-Net 338
Fulong Yang, Feng Wu, Teng Zhang, Jinjin Zhou, Qing Yang

Open-LUT: Interactive g$_m$/ID Lookup Tables for MOSFET Sizing in Open-Source PDKs 344
Sihawi A. Khalid, Susie E. Maestre, Karla M. Madrid-Khalid

A 9.37-Tb/s/mm 5-bit-6-bit Crosstalk Cancellation Transceiver with ICBS Codes for High-Density
Die-to-Die Interfaces 354
Ruotian Yin, Jinwen Li, Fangxu Lv, Geng Zhang, Ruixiao Kuai, Bohui Bai

Design and Implementation of an Integrated Testing System for Spintronic Chips 360
Wenyan Liu, Zihan Gao, Zhifu Guo, Zefan Wu, Xiangrong Pu, Xu Liu, Zhang Zhang

Design of a Reusable UVM Verification Framework for Pixel Readout Chips 366
Yu Zhao, Zexuan Zhao, Yuanhong Jiao, Xiaomin Wei, Heng Yang, Long Liu, Yongcai Hu

Design of High Gain and High Power Amplifier chip Based on RF Choke Segmented Multiplexing
Interstage Matching Technology 372
Zhenbing Li, Weijun Li, Haoyang Sun, Jiaxin Liu, Jialong Fu, Yaocheng Shang

Research on SoC Chip System Modeling for Noise Suppression and Self-organizing Communication
Integration for Distributed Photovoltaics 378
Quan Sun, Gang Chen, Changyou Men, Shuang Wu

Topology Evaluation-Based Layer Assignment Method for Free-assignment Routing in InFO Packages 384
Zhan-Yang Zhu, Ning Xu, Hao-Ying Wu, Chenglin Lu

A Radiation Tolerant CMOS Voltage Reference Circuit for High Energy Physics Experiments 391
Wei Ming Yang, Jia Wang, Xiayu Wang, Ran Zheng, Xiaomin Wei, Yongcai Hu

Dynamic Reward Weighting Based Deep Q-Learning for Routing in Advanced Packaging 397
Shubin Chen, Qinghai Liu, Xiaowei Wu, Jun Xu, Xiaolin Xu

A 7-Bit 1.8 ps Two-Step Time-to-Digital Converter in 28 nm CMOS Technology 403
Zhigang Li, Yao Nian, Hongjia Xu, Wei Liu, Xin Li, Qiang Zhao, Xiulong Wu

Ka-Band Broadband LNA in 65-nm CMOS Using Pre-stage Current-reuse Technique and Large-size
Transistor 408
Zihan Yang, Hao Wang, Ye Liu, Tongde Huang, Wen Wu

A 0.4 to 8GHz Broadband High-linearity I/Q Active Mixer in 130nm BiCMOS Technology 413
Qipeng Li, Zhenghao He, Chun Zhu, Yifan Huang, Qin Li

A Compact Active Phase Shifter for 110-150 GHz Phased Arrays in 130 nm BiCMOS 418
Shuguang Liu, Jinhao Zhou, Yuyang Liu, Hu Lian, Yaxin Zhang, Ziqiang Yang

Interface Quality Improvement in Planar SiC MOSFETs Using Supercritical Fluid Nitriding 422
Xiaoqing Bao, Lei Li, Lei Lu, Kuan-Chang Chang

Design of a Bandgap Voltage Reference for an 8-bit SAR-ADC for Powerline Monitoring using
SKY130 PDK 427
Krisna M. Cañonero, Jovelyn S. Bernales, Abdulwarith G. Macapundag, Sihawi A. Khalid, Susie E. Maestre

A 2.9mV$_{PP}$ Ripple 60mA Digital Low-Dropout Regulator in 28nm CMOS 433
Wenxin Zhang, Yuhang Zhang, Yang Shen, Xiaojin Li, Bingyi Ye, Yabin Sun

MOSFET Modeling of 0.18 µm CMOS Technology Based on BSIM4 Model for Cryogenic Devices 439
Dong Chen, Zuoru Dong, Yushi Chen, Xingyu Cui, Fang Liang, Xiaodong Wang

Process Integration Optimization for RF SOI Process with Arcing Issue 445
Zhangli Liu, Fei Meng, Ruofan Dai

An Optimized ROM based Direct Digital Synthesizer Based on 65nm CMOS Technology 449
Yuan Yuan, Qingsheng Hu, Xu Wu, Li Anming Li

Integrated Circuit Design of CMOS Deadtime Controller for 48 V GaN DC-DC Converter 454
Haoyu Chen, Miao Cui

An Ultra-Low Power Digital Assisted Self-Tracking Zero-Current Detector for DC-DC Converters 459
Chuting Yang, Xing Li

A High-Linearity Digital-to-Time Converter Design for Fractional-N Sub-sampling PLL 464
Jiangnan Li, Wangqing Wu, Ke Cao, Xinyi Zhang, Haigang Feng

A 24–27-GHz 3-Stage Driving Amplifier with 18–25dB Variable Gain in 180-nm CMOS for
Beamforming ICs 469
Zhenghuan Wei, Kaibo Zhang, Lijuan Wang, Sanming Hu

A 16-b 8-MS/s Pipelined SAR ADC With Robust Ring Amplifier and On-Chip Bit-Weight Calibration 474
Shaojuan Chen, Xiaoyi Li, Xuanhao Zhang, Xingshuai Zou, Zhenbing Li, Jiaxin Liu

An IVUS AFE with LNA and a CT $\Delta\Sigma$ modulator with 80MHz BW and 12 bit ENOB 480
Jie Peng, Yunchu Li

A Neuron-and-Synapse Unit Circuit with Information Propagation Function for Spiking Neural
Networks 486
Yide Zhang, Zixuan Ling, Zicheng Yin, Xu Liu

CMOS PWM Controller Design for GaN-Based 48V1 V DC-DC Buck Converter 491
Guanyu Wu, Miao Cui

HEA2-MAC: A Hybrid Exponent-Aware Approximate MAC for Efficient CNN Processing 496
Weixuan Wang, Zihan Zou, Xin Cao, Xuefeng Cai, Hao Cai, Bo Liu

A Low-Power Continuous-Time Quadrature Bandpass $\Sigma\Delta$ ADC with a novel ELD Compensation
Method in Quantizer 502
Haowen Ba, Jun Ye, Hanli Liu, Feijun Zheng

Low Power $\Sigma\Delta$ Modulator Applied to CIS Column Parallel Readout Circuit 507
Dengju Sun, Zhongjie Guo, Ruiming Xu, Jinquan Zhou

Design and Analysis of AC/DC Converter for GaNBased Server Power Supply System 512
Fen Guo, Tuo Li, Changhong Wang, Kang Su, Jiankai Xu, Xin Xi, Kejian Zhu, Xinxin Yuan

A Low-noise PGA With High-gain Chopper Offset-stabilized Amplifier Combining Ping-Pong
Auto-zero For TMR Analog Front-End 517
Hui Wang, Guifa Yan, Mingqi Pan, Bo Wang, Jianhui Wu, Xin Li

An Adaptive dynamic R-tree indexing algorithm for HDI PCB layout 523
Zhang Yang, Liang Xiaoyu, Ning Xu

A Novel Region-Wise Automatic Routing Algorithm for Analog Circuit 529
Hao Xie, Wenxue Chen, Wei Zhang, Bijian Lan, Jing Wan

A 0.6-3 GHz Active Double-Balanced Mixer Circuit Design 535
Jing Wang, Jun Liu, Xuezhen Zhao, Guodong Su

Replicated Partitioning for Hypergraphs with Multiple Constraints 540
Kexin Zhang, Shunyang Bi, Jing Tang, Hailong You, Qiwang Chen

A cryogenic readout integrated circuit for blocked-impurity-band (BIB) far-infrared focal plane array
detectors 546
Yushi Chen, Xin Ge, Hongbo Ma, Zewen Wang, Zuoru Dong, Xiaodong Wang

1T1C-Enhanced TFT-Integrated Gate Driver Circuit Design for Reliable In-cell Touch Sensing Displays 551

Congwei Liao, Yong Wang, Chao Teng, Yi Shen, Yudong Liu, Shengdong Zhang

Neural Network-Based Method for Magnetic Field Inversion of Energized Conductors in Printed Circuit Boards 556

Qi Li, Yiling Liang, Zhen Liu, Lulu Tian

A double layer placement algorithm for IC-based modules of printed circuit board 561

Hangyuan Li, Yu Chen, Zhaoyang Yang, Haotian Pang, Ning Xu

High-Performance Hydrogenated Oxide-Semiconductor Schottky Barrier Diodes with ALD HfO2 Interface 568

Yucheng Cao, Chenyang Huang, Yuyang Cai, Kuan-Chang Chang, Lei Li, Congwei Liao, Yufeng Jin, Lei Lu

An Ultra-Low-Power True Random Number Generator Based on Volatile RRAM 573

Qi Luo, Zhen Wang, Xuemeng Fan, Pengtao Li, Guobin Zhang, Yishu Zhang

Performance Evaluation of an Andvanced-Node CMOS Sensor for Partical Detection 578

Yue Su, Mingjie Feng, Zhiyu Xiang, Cheng Zeng, Congcong Wang, Hui Zhang

Low-power LDO with Fast Transient Response Based on FVF Structure 583

Wenya Luo, Longfei Xu, Hongjian Ren, Chao Cao, Haijun Guo

Low-Power MCU Architecture Optimization and Energy Efficiency Enhancement in Intelligent Pressure Sensor SoCs 589

Yaoming Lv, Hong Yang, Feng Zou, Yuhua Cheng

A Low-Phase-Noise and Low-Power Class-C VCO with Robust Start-up for FMCW Radars 594

Zhigang Li, Zhihao Liu, Hao Lei, Hang Yang, Qiang Zhao, Shan Gao, Xiulong Wu

A 0.47 nJ/Conversion CMOS Temperature Sensor with 0.00786 mm² Core in 65nm CMOS 599

Hangfei Song, Yichi Zhang, Aili Wang

An Ultra-low Power 2× / 4× Reconfigurable Charge Pump for RF Energy Harvesting System 605

Xuanchen Mei, Xin Liu, Junhui Ou, Yanqi Zheng, Mo Huang, Xiuyin Zhang

Cross-Process Bayesian Multi-Objective Collaborative Optimization For Process Migration 610

Zixi Guo, Ruiyu Lyu, Zhaori Bi, Changhao Yan, Ming Zhu, Xiulong Wu, Xuan Zeng

Multi-Branch Autoencoder Networks for Efficient Inverse Design of Wideband Frequency-Selective Surfaces 616

Wei Jiang, Haoran Huang, Guangxin Liao, Shenli Zheng, Jiacheng Guo, Yuan Du

GraphRL-Core: Intelligent Logic Synthesis Optimization via Graph Transformer and Deep Reinforcement Learning 622

Sujie Zhu, Guande Dong, Haoyang He, Chong Xia, Jianwang Zhai

A Hilbert Transform-Based Timing Skew Estimation Method for Dual-Channel TIADCs 628

Xin Li, Ying Pan, Mengdi Miao, Yu Liu, Chenghu Dai, Zhiting Lin

A Fast Auto-Frequency Calibration Technique with High Reference Frequency for PLL 633

Yan Feng, Yongjie Ye, Guoxiao Cheng, Liu Wang, Wei Kang

Post-Routing Compression Algorithm for Area-Efficient Layout design of Analog Integrated Circuit 638
Xiaoyue Wu, Yubo Zhang, Wei Zhang, Bijian Lan, Jing Wan

A Fractional-N Reference-Sampling PLL With a Gain-Boost Fractional Phase Detector for Phase Noise
Reduction 643
Xiaolian Xi, Lecheng Cai, Yihao Liu, Fan Liu, Yanlong Zhang

A 59.6GHz Broadband 3-stage Cascode LNA with Peak Detector for Bandwidth Self-healing 648
Xinsheng Cheng, Jiacheng Guo, Juntao Liu, JingJing Lv, Yuan Du

Addressing Signal Integrity Challenges in DDR5 SDRAM: A High-Precision ZQ Calibration Circuit
with Fast Calibration Time 653
Kexin Feng, ZhiQiang Zhang

A Pipeline-Based Common Framework for Parallel Mixed Signal Simulation 660
Longchen Sun, Guangrong Li, Zhenguo Zhao, Xin Huang, Xuan Zeng, Fan Yang, Zhao Ri Bi

A Rectifier Circuit with Adjustable Temperature Coefficient for Precise Amplitude Control of MEMS
Resonator 665
Guanxiao Zhang, Yang Zhao, Qin Shi, Guoming Xia, Anping Qiu, Meijia Xu

A Low-Complexity Bandwidth Enhancement Design Method for mm-Wave Doherty Amplifier 670
Yujie Sheng, Fei You, Maojun Pan, Songbai He

A 10-bit 3-GS/s Single-Channel Pipelined ADC with Parallel Time-Domain Quantization Based on
Dynamic Residue Amplifier 675
Depan Li, Xin Zhao, Dengquan Li, Zhangming Zhu

Design of High Speed and Energy-Efficient ADC Based on Dynamic Bandwidth Ring Amplifier 681
Xin Yang, Qingyuan Liu, Xiaobo Chen, Dan Sun, Bin Zheng, Tieliang Zhang, Long Yang

A 12-b 1-GS/s Pipelined-SAR ADC With a Hybrid Parallel Timing Scheme and PVT-Robust
Ring-Amp 686
Junhui Guo, Yunchu Li

An X-band Compact High-Efficiency MMIC Power Amplifier in 0.25-μm GaN Technology 692
Ye Liu, Anshi Zhu, Zihan Yang, Tongde Huang, Wen Wu

Design of a Two-Stage Fully Differential Operational Amplifier for High-Resolution Sigma Delta
ADCs 697
Zhan Shi, Zenghao Zhu, Yan Zhao, Franco Maloberti

Design of Power Amplifier for Short Message Application of BDS-III Communication Terminal 702
Zhenbing Li, Weijun Li, Haoyang Sun, Jiaxin Li u, Jialong Fu, Yaocheng Shang

Design of a High-Bandwidth Low-Noise Amplifier 708
Hongmei Chen, Ruiting Shen, Yuexin Tan, Zheyu Li

A Compact Wideband E-band Low Noise Amplifier with 3.3-4.5 dB Noise Figure using 45-nm RFSOI 716
Yinhan Lin, Haoshen Zhu, Taotao Xu, Zhuming Li, Guohai Quan, Pei Qin, Quan Xue

A Wideband Input Buffer Based on AC-Coupled Flipped Source Follower Using Auxiliary Operational
Amplifiers for 8-GS/s ADCs 721
Yuhang Zhang, Jian Chen, Weiying Hu, Xianguo Kou, Changdong Guo, Haizhi Song

A 2.4-2.6 GHz CMOS High Linearity Power Amplifier with MGTR and Harmonic Trap Techniques 726
Runxun Zhang, Gengzhen Qi

A Ka Band High Back-off PAE Power Amplifier with Adaptive Bias Circuits in 65 nm CMOS 731
Ran Zhang, Xiaodong Zhao, Hangbiao Li

A DAC Design using 5-V High-voltage Process for Driving Large Electrode Loads of Neural
Stimulators 737
Xu Liu, YuAng Ru, Zhuo Chen, Fang Xie, Zhijie Chen, Peiyuan Wan

MEMS-Enabled Computational Spectrometer Based on Cascaded Waveguide Couplers 743
Hanxing Wang, Yan Liu, Nan Wang, Yiming Ma

Control Strategy and Parameter Identification Method for Photovoltaic Inverter Electromechanical
Transient Model Based on Improved Fish Eagle Optimization Algorithm 748
Zecheng Li, Yaojia Huo, Kai Hou, Xiao Zhang, Hao Liu, Jinpeng Hao

Design of a Cross-Platform Simulation and Verification System for Complex Onboard Control
Computers 753
Zheng Yang, Shenglong Li, Chaofan Zhou

Fast Frequency Stabilization Technique for Buck Converter Based on Adaptive PLL 759
Kerun Li, Mei Jiang, Jianing Guo, Yuxuan Guo

A Wide-Range PFD and Low-Mismatch Source-Switched CP for MEMS Clock Systems 764
Changfu Wei, Lu Tang, Ziyao Xiong

High-Reliability Integration Design Method for Micro Systems SiP in Complex Spatial Environments 770
Shang Jiang, Wenchang Li, Zucheng Gu, Tianyi Zhang

Design and Implementation of an FPGA-Based Test System for Depth of Interaction Measurement 780
XiaoTian Lv, Ce Zhang, Ran Zheng, XiaoMin Wei, FeiFei Xue, RuiGuang Zhao, YongCai Hu

Design and Nonlinear Analysis of a Multi-Stepped MEMS Resonator with 1:3 Internal Resonance 785
Peilong Li, Yu Jin

A 10.8-12.5-GHz Charge Pump PLL With 68.4-fs$_{rms}$ Jitter and -252-dB FOMJ Based on a
Time-Amplifying Phase-Frequency Detector 791
Yuzhong Li, Zunfa Cheng, Jiwei Huang

A Stochastic Computing-Based Computing-in-Memory Macro with Bit-Split Stochastic Number
Generation 796
Zhiting Lin, Dandan Chen, Siyan Li, Shuang Liu, Yu Liu, Xiulong Wu

An Analytical Compact Model for Multi-time Programmable Memory Cells 801
Tiantian Gan, Xiaojin Li, Tengyang Liu, Yabin Sun, Yanling Shi, Jianpeng Chu

A Hybrid Computing-in-Memory Architecture for Energy-Efficient Edge AI Inference 805
Fangchao Lou, Dandan Chen, Jian Cao, Xing Zhang, Bo Zhang

A Reliable 512-kb HZO-based 2T2C FeRAM Array with Capacitor under Bitline 810
Jing Wang, Xiangyin Chen, Zhongguang Xu

Reconfigurable SRAM Computing-In-Memory Macro Based on Local Computing Cell 814
Chenghu Dai, Chaoyi Wang, Zeyi Liu, Qiang Zhao, Chunyu Peng

A 10T1C SRAM-Based Computing-in-Memory Unit Supporting Logic and MAC Computing for Energy- Efficient Edge AI Chips 819
Zhiting Lin, Chenglong Duan, Miao Long, Juntao Ge, Fugui Jiang, Yang Yang, Xiulong Wu

OpenPIM: An Open-Source Programmable Processing-In-Memory Accelerator Design & Pipeline Simulation Framework 824
Ruibao Wang, Wenshuo Yue, Daijing Shi, Yuchao Yang, Bonan Yan

CASA-CIM: A 28nm 131.54 TOPS/W SRAM-Based CIM Macro with Cap-Adder Weighting and Shift-After-Addition Data Streaming for Efficient MAC Operations 830
Runru Yu, Chenyang Zhao, Honghu Yang, Zhiting Lin, Xiulong Wu, Keji Zhou, Jianguo Yang

Adaptive VREF variation Compensation Scheme for High-Speed DRAM interface & Its Offset Calibration Scheme 835
Chris Eom, Kenji Wen, Mia Xu, Taco Zhang, Derek Yang, Bosco Lai

Stochastic Computation Based Quantization Strategy for SRAM Computing-in-Memory Macro 839
Xin Li, Yifan Wu, Lintao Chen, Chenghu Dai, Xiulong Wu, Zhiting Lin

Design of a Multi-Dimensional and Multi-Precision Tensor Computation Unit Based on FPGA 844
Yupeng Fang

A Low-Ripple Charge Pump with Adaptive Load Compensation for Flash-Based CIM 852
Xuyuan Gu, Xiaofeng Gu, Zhiguo Yu

Interface-Engineered TiO_2/SiO_2 Stacks Enable Ultralow-Power Phase-Change Memory with Nanosecond-Speed 858
Ruizhe Zhao, Jun Zhou, Jun Chen, Hao Tong, Xiangshui Miao

Hafnium-Based Ferroelectric Diode with Interlayer Enhancement for In-Memory Logic Application 863
Shuo Han, Chuanzhi Liu, Qimiao Zeng, Yefan Zhang, Wei Wang, Rongrong Cao

Reconfigurable Memory Device based on Defect Engineering of 2D Ferroelectric $CuInP_2S_6$ 868
Yunpeng Xia, Yu Li, Tianqi Li, Qinfei Long, Zihui Hong, Yunhe Hou, Tiande Mo

ANDQ-NAT: Adaptive Non-linearity Dynamic Quantization Non-Ideality Aware Training Framework for Computing-In-Memory Macros 873
Yu Liu, Jianxing Zhou, Hao Li, Yang Lou, Xiulong Wu, Zhiting Lin

A Fast-Transient-Response Hybrid Architecture LDO for NFC Reader SoC 879
Xindong Mi, JinBiao Zhong, YuXuan Huang, JianGuo Hu

Design of an ATD Circuit for Asynchronous SRAM 885
Jialin Liu, Rongkang Ren, Xin Li, Jiancheng Li, Qichao Zha

An Ultra-Wideband Compact Power Divider for Phased Array Radar Systems 890
Wanfu Liu, Jianhui Wu

A Timing Interleaving RX Scheme with direct Multiplexing Sampler based on dual reference for DDR interface 895
Mia Xu, Chris Eom, Kenji Wen, Tinna Ding, Bosco Lai

Graph-Based Representation of Verilog HDL: Python-Based Control and Data Flow Graph Generation 899
Yipeng Wang, Zhiqiang He, Lingwei Yan, Gang Chen

A Node Merging Algorithm Using Fault-Detection for XOR-Majority Network 904
Shijia Fan, Feifei Deng

FMSRdiff: Efficient latent diffusion framework combining consistency model and flow matching for super-resolution 910
QunKai Peng, GuoFeng Cai, YiHua Xu, Kuan-Chang Chang, Lei Li, GuiBo Luo

An FBAR Driven Fractional-N Ring PLL Using Harmonic-Mixer-Based Dual Feedback and Split-Feedback Frequency Division 918
Qinglong Tian, Jiwei Huang, Hongqu Lin

A Fault Tolerant Routing Method for 3D Network on Chip without Redundant TSV or Router-Avoidance 923
Zeyi Lin, Duo Yu, Yanzhou Tang, Naifeng Jing, Jianfei Jiang, Weiguang Sheng, Lifu Cheng, Qin Wang

The Mechanical Properties of Wrinkled $MnPS_3$ Structures 929
Tongxu Huang, Jiaxian Sun, Zihao Wu, Yiting Wei, Wenjun Chen

Supercritical Fluid-Engineered p-GaN Thin Films with Improved Electrical and Optical Properties 933
Yao-Li Chuang, Lei Li, Lei Lu, Kuan-Chang Chang

Design for Hardware Acceleration of Signature Verification in PQC Stateless Hash-Based Digital Signature Algorithm 937
Haoran Liu, Liji Wu, Lei Li, Yifan Yang, Xiangmin Zhang

A 88 -dB SNDR 156-KHz-BW Noise-Shaping SAR ADC with 3rd-CIFF Structure and Dynamic Integrator 943
Yulin Feng, Mei Jiang, Xinhui He, Zhengru Li

Mo_2C-MoS_2 Mixed-Dimensional Bulk Heterostructure-Based Memristor for Artificial Synapses 948
Libin Liang, Ziyao Lu, Zhenhua Guo, Chunyi Qiu, Zhuoling Zhou, Changjiu Teng

Artificial Neural Network-Based Compact Model for Carbon Nanotube Field-Effect Transistors 953
Zhi Zhang, Honggang Liu

Electronic Properties of Violet Phosphorus Devices 958
Chunyi Qiu, Ziyao Lu, Yiting Wei, Jiaxian Sun, Shilong Zhao

A SiC Power Module Package Based on the Dual-side Cooling with Highly Thermally Conductive Graphite-Molybdenum Spacer 962
Yucheng Xu, Jiafei Yao, Yuxuan Dai, Ziwei Hu, Fan Yang, Kemeng Yang, Binbin Xu, Zhikuang Cai, Yufeng Guo

Work Function Variation Effect Prediction on Heterojunction Tunnel FET Using Multi-Layer Perceptron-Based Neural Network Model 967
Haotong Han, Yunhe Guan, Tongqing Yan, Weihan Sun, Xiangtai Liu, Haifeng Chen

Enhancing Carrier Mobility in a-IZO/a-IGZO Thin-Film Transistors through Band Structure Engineering of Heterojunction Channels 972
Huan Yang, Shengdong Zhang, Yong Wang, Chao Teng, Yi Shen, Yudong Liu

Effect of Temperature on ESD Characteristics in 30 nm Partially Depleted SOI MOSFET 977
Liye Yu, Jingrui Wang, Zhongxu Chen, Yixin Zheng, Ziyan Dai, Mingzhi Wan

A Package-on-Package Module with Integrated FPGA Minimum System for 48-Channel Synchronous Sampling 16-bit ADC 984

Han Li, Jiangbo Wei, Yuan Miao, Weize An, Xudong Wu, Huan Yu, Chao Wang

Preface

On behalf of the organizing committee, it is our great honor to welcome you to the proceedings of 2025 The 10th International Conference on Integrated Circuits and Microsystems (ICICM 2025), which held in in Anhui University during October 17-19, 2025.

ICICM is an annual international conference promoting advances in Integrated Circuits (ICs) and Microsystems. It brings together global researchers, engineers, and industry professionals to share cutting-edge findings and foster collaborations. This conference serves as a vital platform for participants to exchange ideas, build partnerships, and explore future collaborations—driving innovation forward in the field of integrated circuits and microsystems.

ICICM 2025 is co-sponsored by Anhui University, China; Southeast University, China and University of Electronic Science and Technology of China, and technically co-sponsored by CAS, IEEE. This year's proceedings encompass a diverse array of contributions that reflect current research trends, emerging technologies, and novel applications. It contains a total of 180 papers, and topics covered include advances in microelectronic device theory and design, microelectronic systems and system performance simulation, microsystems and smart sensors, advanced electronic automation design and simulation, digital circuit design and functional testing and so on. Each paper was rigorously reviewed by technical committee members in the field, ensuring a high standard of scholarship and relevance. We extend our gratitude to the reviewers for their dedication and insightful feedback, which have greatly enhanced the quality of this compilation.

It is our hope that the ideas shared at ICICM 2025 will spark continued progress in our field. May this collection of work be both an enduring resource for the global community and a reflection of the collaboration that ICICM fosters.

Thank you for your reading, and we look forward to seeing the continued impact of these contributions in shaping the future of technology.

Conference Chair
Prof. Xiulong Wu, Anhui University, China
ICICM 2025

Committee
ICICM 2025

Conference Chairs

Zhigong Wang, Southeast University, China
Xiulong Wu, Anhui University, China
Kaixue Ma, Tianjin University, China

Conference Co-chairs

Ning Xu, Wuhan University of Technology, China
Xiaoqing Wen, Kyushu Institute of Technology, Japan
Qiang Li, University of Electronic Science and Technology of China, China

Program Chairs

Abdel-Hamid Ali Soliman, Staffordshire University, UK
Jun Han, Fudan University, China
Xiaopeng Yu, Zhejiang University, China
Meng Zhang, Southeast University, China
Sheng Chang, Wuhan University, China
Yingmei Chen, Southeast University, China
Zhikuang Cai, Nanjing University of Posts and Telecommunications, China
Zhuo Zou, Fudan University, China
Jianguo Hu, Sun Yat-sen University, China
Bei Yu, The Chinese University of Hong Kong, China

Program Co-chairs

Junyong Deng, Xi'an University of Posts & Telecommunications, China
Jun Xu, Nantong University/Nanjing University, China
Zhixiong Di, Southwest Jiaotong University, China
Guojie Luo, Peking University, China
Xiaojun Zhai, University of Essex, UK
Wei Xing, The University of Sheffield, UK

Program Committee

Bo Liu, Southeast University, China
Shi Pu, Wuhan University of Technology, China
Haizhi Song, University of Electronic Science and Technology of China, China
Lu Zhu, Sun Yat-sen University, China
Youming Zhang, Southeast University, China

Jianshi Tang, Tsinghua University, China
Weiguang Sheng, Shanghai Jiao Tong University, China
Hao Gao, Austria & Eindhoven University of Technology, The Netherland
Yun Fang, Silicon Austria Labs, Austria (IEEE Member)
Jeff Kilby, Auckland University of Technology, New Zealand
Zhijun Zhou, Southeast University, China
Xingyuan Tong, Xi`an University of Posts & Telecommunications, China
Wei Hu, Northwestern Polytechnical University, China
Zhaori Bi, Fudan University, China
Zhengfeng Huang, Hefei University of Technology, China

Local Chair

Lin Cheng, University of Science and Technology of China

Student Program Chairs

Hongbin Sun, Xi'an Jiaotong University, China
Keping Wang, Tianjin University, China
Delong Shang, Institute of Microelectronics of the Chinese Academy of Sciences, China
Fanyi Meng, Tianjin University, China

Student Program Committee

Li Du, Nanjing University, China
Lei Wang, Nanjing University of Posts and Telecommunications, China
Xianbo Li, Sun Yat-sen University, China
Tiehu Li, Chongqing University of Technology, China
Moufu Kong, University of Electronic Science and Technology of China, China
Jiaxin Liu, University of Electronic Science and Technology of China, China
Maliang Liu, Xidian University, China
Xu Meng, Hefei University of Technology, China
Tianming Ni, Anhui University of Engineering, China
Qiang Zhao, Anhui University, China

Special Session Chairs

Chen Yang, Xi'an Jiaotong University, China
Yuan Du, Nanjing University, China
Yejun He, Shenzhen University, China

Academic Committee Chairs

Le Ye, Peking University, China
Ningmei Yu, Xi'an University of Technology, China
Na Xia, Hefei University of Technology, China
Zhiting Lin, Anhui University, China

Academic Committee

Yuanyuan Shi, University of Science and Technology of China, China
Minghui Li, University of Glasgow, UK
Wang Nan, Dalian University of Technology, China
Wu Gao, Northwestern Polytechnical University, China
Wei Li, Nanjing University of Posts and Telecommunications, China
Jiafei Yao, Nanjing University of Posts and Telecommunications, China
Yanlong Zhang, Xi'an Jiaotong University, China
Zhao Zhang, Chinese Academy of Sciences, China
Haihua Wang, Nanjing University of Posts and Telecommunications, China
Jincan Zhang, Henan University of Science and Technology, China
Jiuren Zhou, Xidian University, China
Gaobin Xu, Hefei University of Technology, China
Chunyu Peng, Anhui University, China

Women in Engineering

Huizhen Qian, University of Electronic Science and Technology of China, China
Kailin Ren, Shanghai University, China
Jingjing Guo, Nanjing University of Posts and Telecommunications, China
Fei Yang, Anhui University, China

Industry Liaison Chairs

Wei Liu, MacroSilicon Technology Co., Ltd., China
Hua Xu, SWID, China
Amy Zhong, AMEDAC, China
Frank Wu, Shanghai UniVista Industrial Software Group Co., Ltd., China
Jiwang Guo, Empyrean Technology Co., Ltd., China
Zisan Zhang, Shanghai ChipON Microelectronics Technology Co., Ltd., China
Weibin Ding, X-Times Design Automation Co., LTD., China
Meng Zhang, Aerospace Science Industry Academy of Communication Technology Chengdu, China

Regional Chair

Yongfu Li, Shanghai Jiao Tong University, China

Publicity Chairs

Yongliang Zhou, Anhui University, China
Qiang Wu, Southwest Jiaotong University, China
Jia Wang, Northwestern Polytechnical University, China
Dengquan Li, Xidian University, China
Hao Tong, Huazhong University of Science and Technology, China
Wei Ren, Xi'an University of Posts and Telecommunications, China

Yaoyu Tao, Peking University, China
Na Gao, Xiamen University, China

Technical Committee Chair

Shibing Long, University of Science and Technology of China, China

Technical Committee

Aili Wang, Zhejiang University, China
Bing Yuan, Xidian University, China
Bo Zhou, Beijing Institute of Technology, China
Changchun Zhang, Nanjing University of Posts and Telecommunications, China
Changlin Chen, National University of Defense Technology, China
Chao Zhao, CETC Nanjing Meichen Microelectronics, China
Cheng Liu, Shanghai University, China
Chenggang Yan, Nanjing University of Aeronautics and Astronautics, China
Chengying Chen, Xiamen University of Technology, China
Chunbing Guo, Guangdong University of Technology, China
Daming Yang, Peking University, China
Dongrong Zhang, Zhongguancun Laboratory, China
Feijun Zheng, Zhejiang University, China
Feng He, Beijing Institute of Smart Energy, Huairou Laboratory, China
Gengzhen Qi, Sun Yat-sen University, China
Guangbao Shan, Xidian University, China
Hai-peng ZHANG, Hangzhou Dianzi University, China
Hanbo Jia, Institute of Microelectronics of the Chinese Academy of Sciences, Beijing, China
Helun Song, Suzhou Institute of Nano-Tech and Nano-Bionics, China
Huiyun Li, Shenzhen Institutes of Advanced Technology, Chinese Academy of Sciences, China
Jiacheng Guo, Nanjing University, China
Jiafei Yao, Nanjing University of Posts and Telecommunications, China
Jian Guo, Southeast University, China
Jianwei Liu, National Key Laboratory of Integrated Circuits and Microsystems, Chongqing, China; The 24th Research Institute of China Electronics Technology Group Corp, Chongqing, China
Jingjing Lv, Nantong University, China
Junfeng Qu, Clayton State University, USA
Junqi Huang, Xiamen University of Technology, China
Kai Huang, Zhejiang University, China
Kui Dang, Xidian University, China
Lei Chen, Beijing Microelectronics Technology Institute, China
Liji Wu, Tsinghua University, China
Liu Xi-feng, Jiangsu Vocational College of Information Technology, China
Maliang Liu, Xidian University, China
Mei Jiang, Shenzhen University, China

Meng Meng, China Aerospace Component Engineering Center, China
Miao Cui, Xi'an Jiaotong-Liverpool University, China
Min Tian, Chongqing University, China
Min Yu, Peking University, China
Minming Huang, Sichuan University, China
Ningmei Yu, Xi'an University of Technology, China
Pengcheng Huang, National University of Defense Technology, China
Ping Ma, University of Science and Technology of China, China
Qi Li, Guilin University of Electronic Technology, China
Qingsheng Hu, Southeast University, China
Rui Jin, Beijing Institute of Smart Energy, Huairou Laboratory, China
Ruiguang Zhao, Northwestern Polytechnical University, China
Ruizhe Zhao, Wuhan University of Technology, China
Shaozhen Zhang, Space Star Technology Co. Ltd., China
Shiwei Feng, Beijing University of Technology, China
Shuangming Yu, Chinese Academy of Sciences, China
Shujuan Mao, Beijing Superstring Academy of Memory Technology, China
Tao Su, Sun Yat-Sen University, China
Tao Wu, ShanghaiTech University, Shanghai, China
Tao Yin, Chinese Academy of Sciences, China
Tiehu Li, Chongqing University of Technology, China; Xidian University Chongqing Integrated Circuits Innovation Institute, Chongqing, China
TieZhen Jiang, Anhui University, China
Tongqiang Liu, Shandong Yunhai Guochuang Cloud Computing Equipment Industry Innovation Co., Ltd., China
Tuo Li, Shandong Yunhai Guochuang Cloud Computing Equipment Industry Innovation Co., Ltd., China
Wang Xiao, Hong Kong Applied Science and Technology Research Institute Company Limited (ASTRI), China
Wen Cheng Lai, National Taiwan University of Science and Technology, China
Wengao Lu, Peking University, China
Weihong Liu, Xi'an University of Posts and Telecommunications, China
Xin Xin, Xi'an University of Posts and Telecommunications, China
Xinbo Wen, Peking University Shenzhen Graduated School; TCL China Star Optoelectronic Co. Ltd., China
Xing Li, Nanjing University of Aeronautics and Astronautics, China
Xingyuan Tong, Xi'an University of Posts & Telecommunications, China
Xu Liu, Beijing University of Technology, China
Xuan Guo, Institute of Microelectronics of the Chinese Academy of Sciences, Beijing, China
Yang Zhao, Nanjing University of Science and Technology, China
Yaqi Ma, Soochow University, China
Yi Liu, Nanjing University of Posts and Telecommunications, China
Yishu Zhang, Zhejiang University, China

Yongzheng Zhan, Shandong Yunhai Guochuang Cloud Computing Equipment Industry Innovation Co., Ltd., China

Youming Zhang, Southeast University, China

Yun Wang, Guangdong Greater Bay Area Institute of Integrated Circuit and System, China

Zhang Jun, Nanjing University of Posts and Telecommunications, China

Zheng Yang, China Academy of Space Technology, China

Zhihong Huang, Chinese Academy of Sciences, China

Zhiwei Xu, Zhejiang University, China

Zhiyong Xiong, Shanghai Tianma Microelectronics Company Limited, China

Zhongjie Guo, Xi'an University of Technology, China

Zixuan Wang, Nanjing University of Posts and Telecommunications, China

Ziyue Zhang, Beijing Institute of Technology, China

Zongguang Yu, No.58 Research Institute, China Electronic Technology Group Corporation, China

Zongming Duan, East China Research Institute of Electronic Engineering, China

Sihawi A. Khalid, Mindanao State University, Philippines

Susie E. Maestre, Mindanao State University, Philippines

Shilong Zhao, Foshan University, China

Zhan Shi, Dalian Minzu University, China

Zhenbing Li, University of Electronic Science and Technology of China, China

Zhizhan Yang, Southwest Jiaotong University, China

Yande Jiang, Academy of Military Science, China

Zhikuang Cai, Nanjing University of Posts and Telecommunications, China

Yufeng Guo, Nanjing University of Posts and Telecommunications, China

Runru Yu, Fudan University, China

Chenyang Zhao, Zhangjiang Laboratory, China

Jianguo Yang, Zhangjiang Laboratory, China

Rongrong Cao, National University of Defense Technology, China

2025 The 10th International Conference on Integrated Circuits and Microsystems

A Compensation-Capacitor-Based Perturbation Injection Calibration Method for SAR ADCs

Hongjian Ren
School of integrated Circuits
Shandong University
Jinan, China
sdrenhongjian@163.com

Ke Hu
School of integrated Circuits
Shandong University
Jinan, China
1252201919@qq.com

Wenya Luo
School of integrated Circuits
Shandong University
Jinan, China
2716796054@qq.com

Chao Cao
School of integrated Circuits
Shandong University
Jinan, China
chao_cao@sdu.edu.cn

Haijun Guo
School of integrated Circuits
Jinan University
Jinan, China
ise_guohj@ujn.edu.cn

Abstract— **Based on the existing research, this paper proposes a novel perturbation injection calibration method based on compensation capacitors. It can achieve perturbation injection of different magnitudes, both positive and negative, without relying on additional capacitors or precise sources. Digital calibration is achieved through iterative updates of weights using the LMS algorithm. Firstly, the digital calibration technology studied in this paper was verified through behavioral-level model simulation using MATLAB/Simulink. Then, hardware implementation and testing were carried out based on the Vivado platform and the NEXYS 4 DDR development board. The effective bit number of a 14-bit non-ideal SAR ADC was increased from 10.74 bits to 12.43 bits, the spurious-free dynamic range was improved from 76.29 dB to 90.72 dB, and the signal-to-noise ratio was enhanced from 67.13 dB to 76.97 dB.**

Keywords—Digital calibration, Perturbation injection, Successive approximation analog-to-digital converter

I. Introduction

With the continuous improvement of industrial automation and intelligence level, intelligent sensors have put forward higher performance requirements for Analog to Digital Converter (ADC). Successive Approximation Register (SAR) analog-to-digital converters have become a hot topic in the field of analog-to-digital converters because of their advantages such as low power consumption, simple structure and small area. In the working process of SAR ADC, there will be various non-ideal factors, such as noise, capacitance mismatch, comparator misalignment, etc., which will have a great impact on the performance of SAR ADC[1,2]. Therefore, the research of digital calibration technology for SAR ADC in this paper is of great significance.

The content of this paper is arranged as follows: The second part of the SAR ADC modeling and simulation; In the third part, the proposed calibration algorithm is verified by modeling simulation. In the fourth part, the calibration algorithm is tested by FPGA.

II. Modeling and Simulation of SAR ADC

A. Modeling of 14-bit SAR ADC based on Simulink

The overall modeling of 14-bit SAR ADC in this paper is shown in Fig. 1, which mainly includes sampling and holding module, comparator, SAR logic, CDAC and other parts. The sampling rate is 1MS/s, and the reference voltage V_{ref} is 2.8248V. First, the sampling switch samples the input signal and uses the charge conservation principle of the capacitor to hold the signal in the DAC array. Then, the SAR logic circuit switches the capacitor array according to the output of the comparator under the control of the clock to obtain different reference voltages. The comparator then compares the input signal with the reference voltage. Through continuous comparison, the reference voltage is progressively approximated to the input voltage held by the sample, and finally the SAR logic outputs an n-bit binary digital code[3].

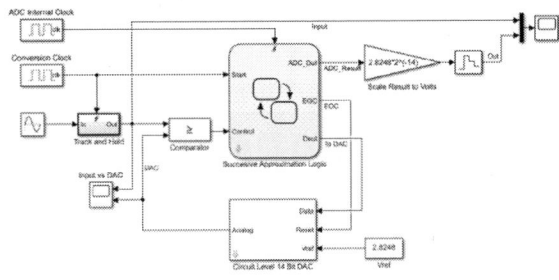

Fig. 1. SAR ADC Overall modeling of SAR ADC

B. Design of CDAC

In order to solve the problem of large area of traditional unsegmented array, the whole binary capacitor array can be divided into two or more sub-arrays by bridging capacitors, thus effectively reducing the total unit capacitance. However, in the actual design, in order to make the segmented capacitors still satisfy the binary weight relationship, forcing the introduction

979-8-3315-8850-2/25 $31.00 © 2025 IEEE

of the bridge capacitor may make the value of the bridge capacitor to be a non-integer multiple of the unit capacitance[4]. In order to avoid the parasitic effect caused by fractional capacitors, it is usually necessary to introduce compensation capacitors, so that each bridge capacitor is integer, as shown in Fig. 2.

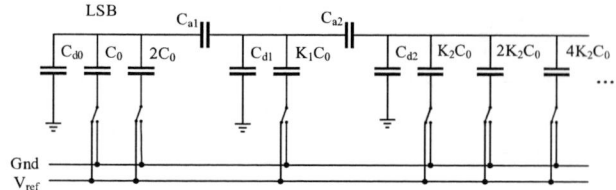

Fig. 2.　Universal multi segmented capacitor array

Where, C_0 is the unit capacitor, C_{ai} is the bridge capacitor between segment i and segment i+1, and C_{di} is the compensation capacitor of segment i+1. The relationship between them satisfies:

$$C_{a(i-1)} = \frac{K_{i-1} C_{(i-1)t}}{2^{S_{i-1}} - K_{i-1}} \tag{1}$$

In the formula, $C_{(i-1)t}$ is the equivalent total value of the former i-1 capacitance to the ground, and K_{i-1} is the ratio of the lowest capacitance to the unit capacitance in paragraph i, and K_{i-1} can be any positive integer. By reasonably setting K and the size of the compensation capacitance within each segment, even if the bridge capacitance is an integer.

Fig. 3 shows the Simulink model of the four-segment 14bit capacitor array adopted in this paper. The capacitance values from left to right, that is, from low to high are C, 2C, 2C, 4C, 4C, 8C, 16C, 8C, 16C, 16C, 32C, 64C, 128C, 256C, 512C, the compensation capacitance of the first and third segments is C, 4C, and the bridge capacitance is 4C, 8C, 12C from left to right. Compared with the conventional unsegmented capacitor array, the capacitance array in this paper is only 4.4% of its area.

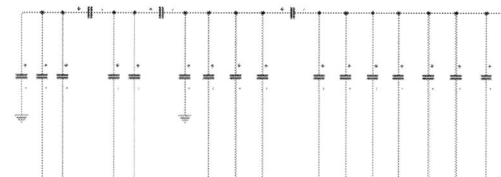

Fig. 3.　Modeling of capacitor array

The bottom plate of each bit capacitor is connected to V_{ref} or Gnd through SAR logic control. Using randn function in MATLAB, Gaussian error can be added to each capacitance value to simulate capacitor mismatch.

C. Integral Simulation

The input signal was set as a sine wave close to the Nyquist sampling frequency. After the simulation was completed, the output of SAR ADC was imported into the MATLAB working area, and 1024 points were selected to perform Fast Fourier Transform (FFT) on them, and the spectrum diagram was

obtained, as shown in Fig. 4. The results show that the effective bit of ADC is 14.02bit, the Spury-Free Dynamic Range (SFDR) is 113.73dB, and the Signal-to-Noise Ratio (SNR) is 86.19dB. The above ideal modeling simulation results show that the SAR ADC design in this paper is reasonable.

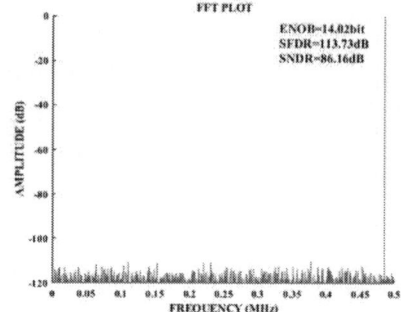

Fig. 4.　Ideal 14 bit SAR ADC modeling FFT spectrogram

On the basis of ideal modeling, 0.5LSB equivalent noise error and 3% capacitance mismatch are added, and the non-ideal SAR ADC spectrum diagram can be obtained by re-simulation and FFT on the output, as shown in Fig. 5.

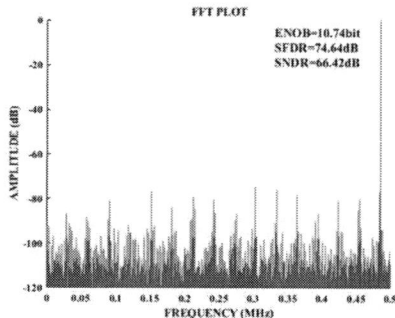

Fig. 5.　Non ideal 14 bit SAR ADC modeling FFT spectrogram

The effective bit, SFDR and SNR of non-ideal SAR ADC modeling simulation are 10.74bit, 74.64dB and 67.13dB. It can be seen that the performance of SAR ADC decreases significantly after the addition of non-ideal factors. Therefore, it needs to be calibrated to achieve the design goal.

III. PERTURBATION INJECTION CALIBRATION METHOD BASED ON COMPENSATION CAPACITANCE

The principle of superposition in linear systems is the essence of perturbation-based digital calibration methods. Fig. 6 shows the principle of superposition in linear systems[5].

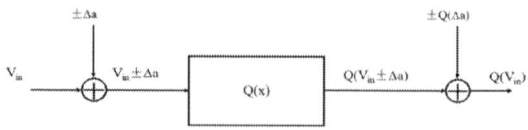

Fig. 6.　Block diagram of linear superposition principle

979-8-3315-8850-2/25 $31.00 © 2025 IEEE

Where V_{in} is the analog input, Δa is the analog disturbance signal, and linear system $Q(x)$ represents the mapping relationship of SAR ADC from the analog input to the digital output. The digital quantity $Q(V_{in}\pm\Delta a)$ after the analog input and the disturbance are superimposed by the ADC after the positive and negative disturbance of size Δa is added to the input signal. At this time, the numerical quantity $Q(\Delta a)$ corresponding to the injected disturbance can be subtracted or added to obtain the quantization result $Q(V_{in})$ corresponding to the V_{in}. That is:

$$Q\left(V_{in}\right)=Q\left(V_{in}\pm\Delta a\right)\mp Q\left(\Delta a\right) \tag{2}$$

Fig. 7 shows the block diagram of the digital calibration algorithm based on disturbance.

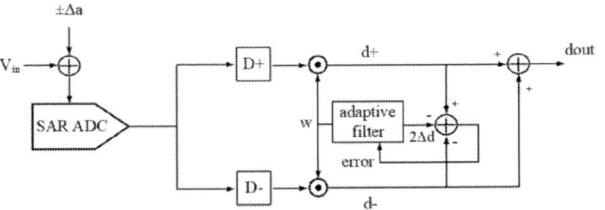

Fig. 7. Structure diagram of perturbation based digital calibration method

Disturbance Δa of equal size and opposite sign is added to the analog input voltage V_{in}, and corresponding digital outputs D+ and D- can be obtained after ADC conversion, and then adaptive filtering can be carried out.

A. Perturbation Injection Method Based on Compensation Capacitance

The traditional injection method[6]is shown in Fig. 8. Two perturbation injection capacitors C_{tp} and C_{tn} are connected at both ends of comparator p and n, respectively. In the disturbance injection stage, taking the positive disturbance injection as an example, the lower plate of the disturbance injection capacitor C_{tp} needs to be connected to V_{ref} or other small signal reference source.

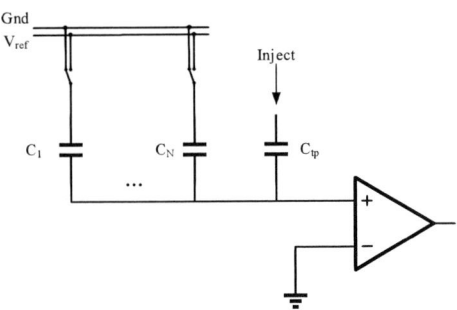

Fig. 8. Traditional perturbation injection method

As mentioned above, in order to integer the bridge capacitance of the segmented capacitor array, it can be realized by introducing and reasonably setting the size of the compensating capacitor C_d. During the normal operation of the ADC, the lower plate of the C_d is connected to Gnd. The disturbance injection method based on compensatory capacitor adopted in this paper only needs to carry out disturbance

injection through the compensation capacitor C_d existing in the piecewise capacitor itself, as shown in Fig. 9.

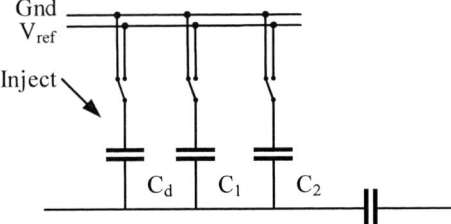

Fig. 9. Perturbation injection method based on compensating capacitance

On the basis of the compensation capacitor C_d constantly connected to Gnd, the lower plate of C_d is connected to V_{ref} or Gnd through the switch. When a disturbance injection is required, the switch only needs to be adjusted from the Gnd end to the V_{ref} end.

Further, the compensation capacitor C_d can be divided into several integer multiple unit capacitors in parallel. Taking 4C as an example, it can be regarded as three capacitors C, C, 2C, as shown in Fig. 10.

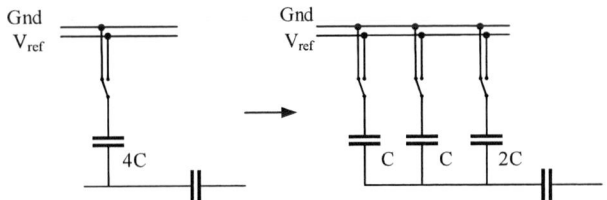

Fig. 10. Split of compensating capacitance

The injection of disturbance of different sizes can be realized by controlling three capacitor bottom plate switches connected to Gnd or V_{ref}. Based on the modeling of SAR ADC capacitor array in Chapter 2 of this paper, this injection mode is used to conduct behavioral simulation. The results of disturbance injection amount changes in different switching connection modes are shown in Fig. 11.

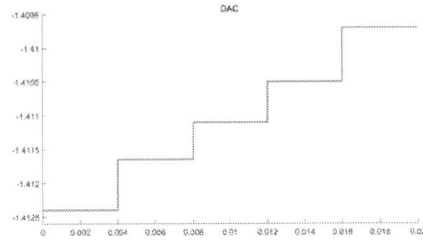

Fig. 11. Different switch connection methods perturb changes in injection volume

The method can achieve different injection amounts of different sizes through different switch combinations without relying on additional capacitors and precise sources. Therefore, a suitable disturbance injection quantity can be selected as the initial value, and better calibration effect can be obtained after adaptive filtering iteration convergence.

B. A Calibration Method Combining Perturbation and LMS Algorithm

After injecting positive and negative disturbance and converting ADC, D+ and D- can be obtained, and then adaptive filtering can be carried out. LMS algorithm is chosen in this paper.

According to the number code D+, D- by weighted summation can be obtained d+, d-, such as the formula (3) and (4).

$$d+=\sum_{i=0}^{N} W_i D_{i,+} \tag{3}$$

$$d-=\sum_{i=0}^{N} W_i D_{i,-} \tag{4}$$

In this case, the conversion error is:

$$error=(d+)-(d-)-2\Delta d \tag{5}$$

After the error is obtained, the injection amount and weight can be updated and iterated according to the formula:

$$W_i(n+1)=W_i(n)-\mu_W \times error(n) \times \left[D_{i,+}(n)-D_{i,-}(n) \right],$$
$$i=0,1...,N \tag{6}$$

$$\Delta d(n+1)=\Delta d(n)+\mu_\Delta \cdot error(n) \tag{7}$$

Where, μ_W and μ_Δ are update steps of weight and injection amount, respectively. We can stipulate that after a certain number of iterations or after the error is less than a certain value, the iteration will be stopped, then it will be regarded as the completion of calibration, and the corresponding weight will be the final weight value we need.

Based on the SAR ADC behavioral level model in Chapter 2 of this paper, different disturbance injection amounts were set to calibrate random capacitance mismatches of 1%, 2%, 3% and 4%, respectively. The same step size μ was selected, and the effective digits after calibration with different disturbance injection amounts were obtained after 100,000 iterations, as shown in Fig. 12. The abscissa is the disturbance injection amount in LSB, and the ordinate is the calibrated ENOB in bit.

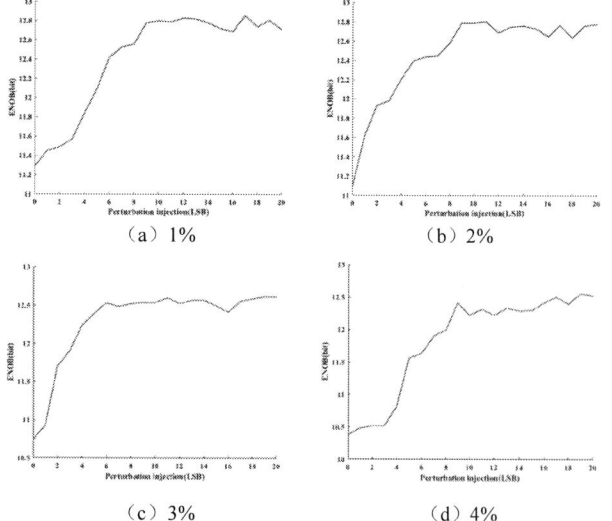

(a) 1% (b) 2%

(c) 3% (d) 4%

Fig. 12. ENOB of different perturbation injection

It can be seen that under the four different mismatch conditions, with the increase of disturbance injection amount, the calibration effect is first significantly improved and then tends to be stable with certain fluctuations. However, such fluctuations vary. Fig. 13 shows the required minimum disturbance quantity, optimal disturbance quantity, and sub-optimal disturbance quantity under different mismatch conditions.

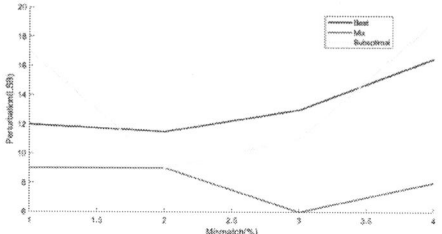

Fig. 13. Best, min and suboptimal perturbation injections

The perturbation injection calibration method based on compensating capacitance proposed in this section can satisfy all the minimum perturbation injection quantities and most of the optimal and suboptimal injection quantities under four mismatches.

Its core advantage is reflected in its ability to adapt to different mismatch scenarios. This method has certain engineering value: firstly, the full coverage of the minimum disturbance ensures the reliability of the basic calibration function, and even in the case of large mismatch (4%), it can still meet the minimum disturbance injection requirements; Secondly, according to the actual process deviation, this method can select an injection amount with better calibration effect from multiple disturbance injection amounts. Furthermore, the effective coverage of the suboptimal perturbation momentum can have a better alternative when the method cannot meet the optimal injection amount, and when the calibration effect of the two is similar, the power consumption and noise of the system can be comprehensively considered under the premise of ensuring that the core performance indicators of the circuit are met.

In general, this method can achieve SAR ADC calibration by selecting suitable disturbance injection amount according to different non-ideal factors in practice without relying on extra capacitance and precise source. Combined with the capacitor array structure used in the SAR ADC modeling in Chapter Ⅱ.B of this paper, the calibration effect of this method is further verified by taking 3% capacitance mismatch and 13LSB disturbance injection as examples.

C. Simulation Verification of Calibration Algorithm

Fig. 14 shows the change curve of absolute value of error error with the increase of iterations during the iteration process. The value of error is saved once for every 1000 iterations and iterated continuously by LMS algorithm. |error| will gradually decrease and converge and become stable after 40000 iterations.

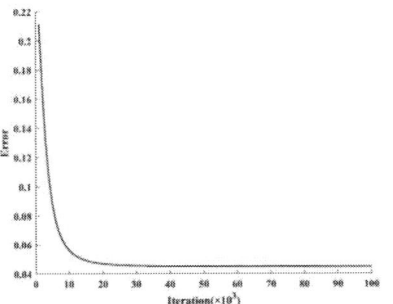

Fig. 14. Error variation curve with iteration times

After calibration, the output result was analyzed by fast Fourier transform (FFT), as shown in Fig. 15.

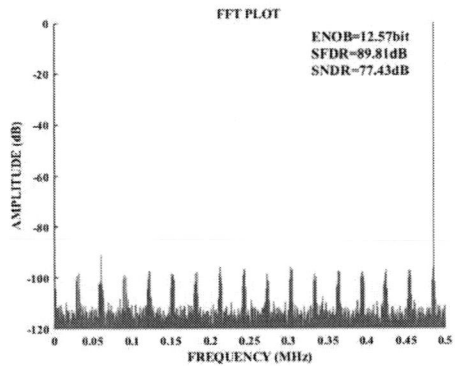

Fig. 15. FFT spectrogram after MATLAB simulation calibration

After calibration, the significant bits of the ADC were increased from 10.74 bits to 12.57 bits, the SFDR was increased from 74.64dB to 89.81dB, and the SNR was increased from 67.13dB to 77.69dB. After calibration, the harmonics of ADC are improved and the significant bits are increased, indicating that the calibration has a good effect.

IV. SAR ADC DIGITAL CALIBRATION ALGORITHM RTL LEVEL HARDWARE IMPLEMENTATION AND VERIFICATION

A. Functional Simulation Based on Xilinx Vivado

Based on Xilinx Vivado platform, the algorithm is validated by Verilog HDL hardware description language[7]. The circuit structure block diagram is shown in Fig. 16. Where, clk is the clock signal, reset is the reset signal, D1 and D2 are respectively the two outputs after adding positive and negative perturbations, d is the ideal digital code corresponding to the injection amount, and w1 to w14 are 14 weights. When the state control signal state=0, the weight value is iteratively updated by LMS algorithm. When state=1, the weight remains unchanged and the input signal is calibrated.

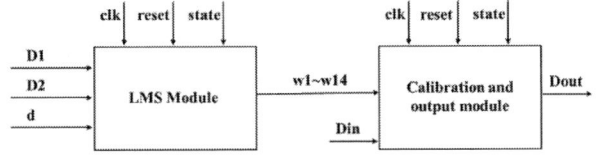

Fig. 16. Block diagram of perturbation calibration algorithm structure

Fig. 17. Simulation waveform diagram of perturbation calibration algorithm function

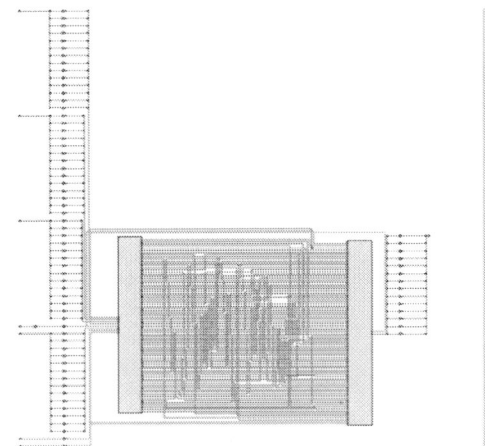

Fig. 18. Highest bit weight waveform

Fig. 17 and Fig. 18 show the simulation waveform diagram of the calibration algorithm in Vivado. It can be seen from the diagram that with the increase of iterations, the curve of the highest bit weight, weight13, gradually decreases and then becomes stable.

B. FPAG Hardware Testing

After the functional simulation of the calibration algorithm is completed, synthesis and implementation are required in order to evaluate the hardware resources and test the hardware. The test platform is a NEXYS 4 DDR development board. The schematic after algorithm synthesis and implementation is shown in the Fig. 19.

Fig. 19. Implementation schematic of perturbation calibration algorithm

TABLE I shows the resource estimates after the comprehensive implementation, and it can be seen that the NEXYS 4 DDR development board has sufficient resources to meet the needs of this design.

TABLE I. UTILIZATION POST-IMPLEMENTATION OF PERTURBATION CALIBRATION ALGORITHM

Resource	Utilization	Available	Utilization(%)
LUT	3239	63400	5.11
LUTRAM	113	19000	0.59
FF	2517	126800	1.99

979-8-3315-8850-2/25 $31.00 © 2025 IEEE

BRAM	6	135	4.44
DSP	84	240	35.00
IO	17	210	8.10
BUFG	3	32	9.38

The timing report is shown in the Fig. 20, and it can be seen that the design has sufficient timing margin.

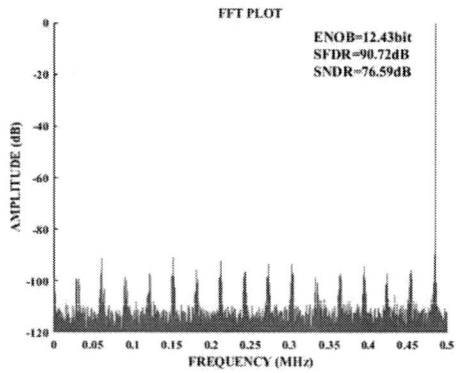

Fig. 20. Perturbation calibration algorithm timing report

After synthesis and implementation, the bitstream file can be generated and burned to the development board. ILA IP can export the captured signals into a.csv file. Then MATLAB can read the file and process the output data, and the calibrated FFT spectrum diagram can be obtained, as shown in Fig. 21.

Fig. 21. FFT spectrogram after calibration

The proposed calibration algorithm increases the significant bits of ADC from 10.74bit to 12.43bit, SFDR from 74.64dB to 90.72dB, SNR from 67.13dB to 76.97dB, which can realize the improvement of ADC performance.

TABLE II compares the key performance parameters of the calibration algorithm in this paper with those in other literatures.

TABLE II.   COMPARISON BETWEEN THIS ARTICLE AND OTHER RELATED WORKS

	This paper	Ref [8]	Ref [9]	Ref [10]	Ref [11]
Resolution (bit)	14	12	14	14	16
Sampling (MS/s)	1	3000	1	100	1
ENOB before calibration (bit)	10.74	8.68	9.57	9.34	10.9
SFDR before calibration (dB)	74.64	62.70	66	63.40	74.95
ENOB after calibration (bit)	12.43	9.72	11.78	13.13	13.7
SFDR after calibration (dB)	90.72	76.0	89	106.50	92.38

It can be seen that the perturbation injection calibration method based on compensatory capacitors in this paper can be effectively calibrated without relying on additional capacitors and precise sources, and the effect is similar to that of other similar works.

V. CONCLUSION

According to the demand of modern industrial production for SAR ADC, this paper studies and implements the digital calibration technology for SAR ADC based on the non-ideal error of SAR ADC through extensive research and analysis. The main work includes: 14-bit SAR ADC behavior modeling simulation and result analysis based on Simulink. A perturbation injection calibration method based on compensation capacitance is proposed and verified by behavior level simulation. This method can achieve the injection of perturbations without introducing additional capacitors or relying on additional precision sources and achieve calibration by iterating weights through LMS algorithm. Based on Vivado platform, the proposed digital calibration method is implemented and simulated in hardware, and tested on the board. The proposed digital calibration method can increase the significant bit before calibration from 10.74bit to 12.43bit.

REFERENCES

[1] Liu J, Tang X, Zhao W, et al. 16.5 A 13b 0.005 mm 2 40MS/s SAR ADC with kT/C noise cancellation[C]. 2020 IEEE International Solid-State Circuits Conference-(ISSCC). IEEE, 2020: 258-260.

[2] Rajesh kumar, C. Reduce Power Consumption for Digital Cmos Circuits Using Dvts Algorithham[C]. IOSR Journal of Electrical and Electronics Engineering (IOSR-JEEE), 2015:109-115.

[3] Ginsburg B P, Chandrakasan A P. 500-ms/s 5-bit adc in 65-nm cmos with split capacitor arraydac[J]. IEEE Journal of Solid-State Circuits, 2007, 42(4): 739-747.

[4] He J, Zhan S, Chen D, et al. Analyses of Static and Dynamic Random Offset Voltages in Dynamic Comparators[J]. IEEE Transactions on Circuits & Systems I Regular Papers, 2009, 56(5): 911-919.

[5] Liu W, Huang P, Chiu Y. A 12-bit, 45-MS/s, 3-mW Redundant Successive-Approximation-Register Analog-to-Digital Converter With Digital Calibration[J]. IEEE Journal of Solid-State Circuits, 2011, 46(11): 2661-2672.

[6] Chan C H, Zhu Y, Ho I M, et al. 16.4 A 5mW 7b 2.4 GS/s 1-then-2b/cycle SAR ADC with background offset calibration[C]. 2017 IEEE International Solid-State Circuits Conference (ISSCC). IEEE, 2017: 282-283.

[7] Ma Z, Saeidi S, Kennel R. FPGA implementation of model predictive control with constant switching frequency for PMSM drives[J]. IEEE transactions on industrial informatics, 2014, 10(4): 2055-2063.

[8] M. Gu, Y. Zhong, L. Jie,N. Sun, N. 24.1 A 12b 3GS/s Pipelined ADC with Gated-LMS-Based Piecewise-Linear Nonlinearity Calibration[C]. 2025 IEEE International Solid-State Circuits Conference (ISSCC). IEEE, 2025:1-3.

[9] Wei L, Shangshang G, Xiao W, et al. Background LMS calibration algorithm realization for SAR-ADC[C]. 2021 6th International Conference on Integrated Circuits and Microsystems (ICICM). IEEE, 2021: 142-146.

[10] Zheng C, Sun J, Wang C. An LMS-based calibration technique using on-chip PN signal for SAR ADCs[C]. 2024 6th International Conference on Circuits and Systems (ICCS), 2024: 124-128.

[11] Ding J, Liu F, Deng K, et al. A 16-bit 1-MS/s SAR ADC with capacitor mismatch self-calibration[J]. IEEE Transactions on Very Large Scale Integration (VLSI) Systems, 2024, 33(1): 10-20.

979-8-3315-8850-2/25 $31.00 © 2025 IEEE

Design and Simulation of High-Frequency FBAR Based on Multiphysics and MBVD Circuit Modeling

Rui Zhao
National Key Laboratory of Integrated Circuits and Microsystems,
Xi'an Microelectronics Technology Institute, Xi'an, China
18309231312@163.com

Jiangbo Wei
National Key Laboratory of Integrated Circuits and Microsystems,
Xi'an Microelectronics Technology Institute, Xi'an, China
WjbElec@163.com

Yang Xiao
National Key Laboratory of Integrated Circuits and Microsystems,
Xi'an Microelectronics Technology Institute,
Xi'an, China
943143504@qq.com

Liaoliao Zhang
National Key Laboratory of Integrated Circuits and Microsystems,
Xi'an Microelectronics Technology Institute,
Xi'an, China
pdz15709189156@outlook.com

Haijuan Li
National Key Laboratory of Integrated Circuits and Microsystems,
Xi'an Microelectronics Technology Institute, Xi'an, China
18309231312@163.com

Jiaqi Liu
National Key Laboratory of Integrated Circuits and Microsystems,
Xi'an Microelectronics Technology Institute,
Xi'an, China
1065732444@qq.com

Chao Wang*
National Key Laboratory of Integrated Circuits and Microsystems,
Xi'an Microelectronics Technology Institute,
Xi'an, China
12207935@qq.com

Abstract—A design method of high-frequency film bulk acoustic resonator (FBAR) filter is proposed to meet the requirements of high ‐ frequency and integrated filters in wireless communication systems. Based on a cavity-type FBAR structure,an effective piezoelectric composite film model above the air cavity is established. The impact of various FBAR layer structures on frequency characteristics is analyzed by means of COMSOL Multiphysics simulations. The results show that the thickness of the electrodes and piezoelectric layers can significantly affect the resonator's performance, and the support layer significantly affects the frequency shift at high frequencies. The impedance characteristics of the resonator in the 6-9 GHz were obtained by means of multiphysics simulations. An modified butteruorth-van dyke (MBVD) equivalent circuit model of the resonator is constructed based on advanced design system (ADS) circuit simulation software. The circuit parameters in the model are modified by field-circuit fitting, and the modified MBVD circuit model is packaged as a resonator module for the rapid design of subsequent filters. A first-order filter circuit is built by means of the encapsulated resonator module, and the advantages of different filter structures were analyzed, providing theoretical support for the design and optimization of high ‐ frequency FBAR filter.

Keywords—*film bulk acoustic resonator; high frequency filter; multiphysics simulation; equivalent circuit model; advanced design system; MBVD*

I. INTRODUCTION

With the rapid development of wireless communication systems, the operational frequency bands continue to extend toward higher frequencies. The substantial increase in narrow-bandwidth channels and the growing congestion of spectral resources have led to a rising demand for high-performance,

high-frequency control devices [1]. Currently, frequency control components such as low-noise amplifiers in receivers and solid-state power amplifiers in transmitters have achieved miniaturization [2][3]. However, a large number of discrete passive components — including resonators, filters, and couplers—have become the major bottleneck in miniaturizing and integrating wireless communication systems while reducing power consumption and cost. Compared to traditional ceramic dielectric filters, the film bulk acoustic resonator (FBAR) offers significant advantages in terms of size, along with excellent performance in operating frequency and power handling capacity [4], making it an ideal solution for high-frequency and integrated filters. As the only type of RF filter that can be monolithically integrated with radio-frequency integrated circuits (RFICs) and monolithic microwave integrated circuits (MMICs), FBARs demonstrate broad application prospects in areas such as wireless communication, sensing, and detection [5].

Researchers both domestically and internationally have conducted extensive studies on the simulation and filter design of cavity-type FBARs. For example, Ref. [6] achieved ultra-wideband filter design by extending the bandwidth through external LC circuits connected to series and parallel resonators, respectively; Ref. [7] improved the Mason model using a resonator that incorporates a passivated piezoelectric layer, among other approaches. Currently, the operating frequency of domestic cavity-type FBARs remains below 5 GHz [8][9]. At these frequencies, the piezoelectric layer is relatively thick and sufficiently robust to support the entire device, with the support layer having negligible impact on the resonant frequency. Therefore, simulations often ignore the effect of the support layer on frequency shift. Theoretically, the operating frequency

979-8-3315-8850-2/25 $31.00 © 2025 IEEE

of FBAR devices can reach up to 10 GHz. However, as the resonator frequency increases, the piezoelectric stack becomes thinner. During the release etching of the air cavity, the thinner films in high-frequency resonators are susceptible to collapse due to liquid surface tension in wet etching processes, as well as structural failure caused by over-etching. To prevent film collapse and device failure, a support layer is often introduced between the bottom electrode and the sacrificial layer. Commonly used support layer materials [10][11] include Si_3N_4 and SiO_2.

To fully leverage the advantages of FBAR and expand its application in high-frequency domains — such as wireless communication, sensing, and detection—this paper investigates the simulation and design methodology of high-frequency film bulk acoustic wave filters, aiming to provide more efficient and precise filter design solutions. The influence of each structural layer of the FBAR on its frequency characteristics is analyzed using COMSOL Multiphysics finite element simulation software. Equivalent circuit model parameters are extracted from the finite element simulation results, and an MBVD equivalent circuit model of the resonator is constructed based on the Advanced Design System (ADS) circuit simulation platform. The circuit parameters are then refined through field-circuit co-simulation. The corrected MBVD circuit model is subsequently encapsulated into an equivalent circuit module to support first-order filter circuit simulation, thereby facilitating rapid filter design.

II. MULTIPHYSICS SIMULATION OF FILM BULK ACOUSTIC RESONATORS

The main structure of the cavity-type FBAR is illustrated in Figure. 1, consisting sequentially from top to bottom of the top electrode, piezoelectric layer, bottom electrode, support layer, and substrate. The materials and corresponding thicknesses used in the finite element analysis are summarized in Table 1.

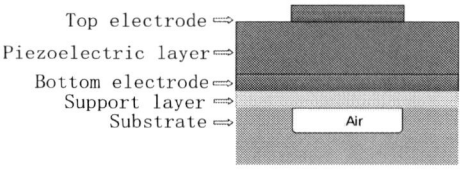

Fig. 1. Schematic diagram of the cavity-type FBAR structure

To simplify mesh generation and reduce computational cost, only the effective piezoelectric composite membrane structure above the air cavity was modeled for simulation. The established finite element model of the FBAR is shown in Figure. 2. Perfectly Matched Layers (PMLs) were applied at both ends, while fixed mechanical boundary conditions were set on both sides. The top electrode has a width of 100 μm and an out-of-plane thickness of 40 μm.

TABLE I. MATERIAL PROPERTIES USED IN THE FBAR MODEL

Structure	Material	Thickness/nm
Top Electrode	Mo	90
Piezoelectric Layer	AlN	270
Bottom Electrode	Mo	90
Support Layer	SiO2	90

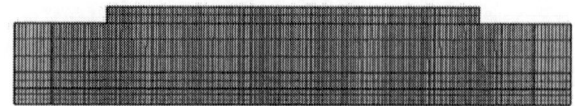

Fig. 2. Finite element model of the FBAR.

Based on the parameters in Table 1, the thickness of the piezoelectric layer and the electrode layers was varied individually to investigate their influence on the resonant frequency. The results are shown in Figure. 3, where fs denotes the series resonant frequency and fp the parallel resonant frequency. The analysis indicates that both the series and parallel resonant frequencies decrease as the thickness of either the electrode layers or the piezoelectric layer increases. This occurs because the increased overall thickness of the piezoelectric stack lengthens the effective acoustic propagation path, thereby reducing the resonant frequency of the device.

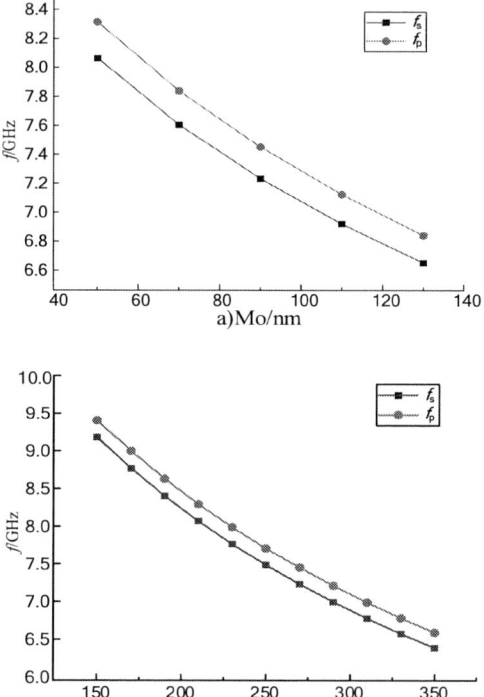

Fig. 3. Influence of layer thickness on fs and fp

Through simulation, the influence of variations in the thickness of each layer on the effective electromechanical coupling coefficient Keff2 is shown in Figure. 4. Analysis of the results indicates that when the electrode layers effectively confine the acoustic energy, the effective electromechanical coupling coefficient Keff2 decreases as their thickness increases. Within a certain range, increasing the thickness of the piezoelectric layer enhances the piezoelectric performance of the device, leading to a rise in the electromechanical coupling coefficient. However, beyond a certain thickness, the piezoelectric performance saturates, and the electromechanical coupling coefficient stabilizes with no further increase.

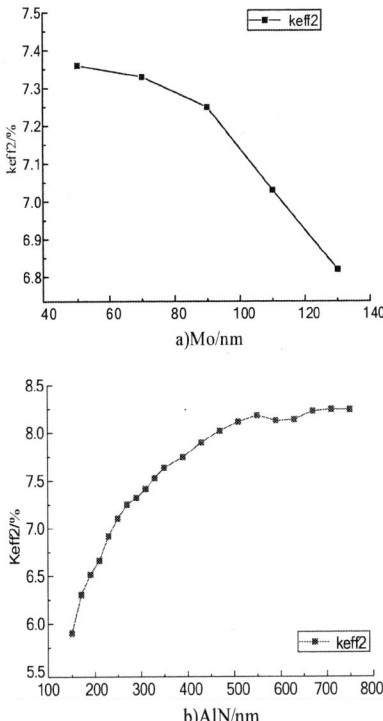

Fig. 4. Influence of layer thickness on Keff2

collapse of the upper membrane after sacrificial layer release. Accordingly, the area of the top electrode—which is slightly smaller than the cavity—should be controlled within 10,000 μm²

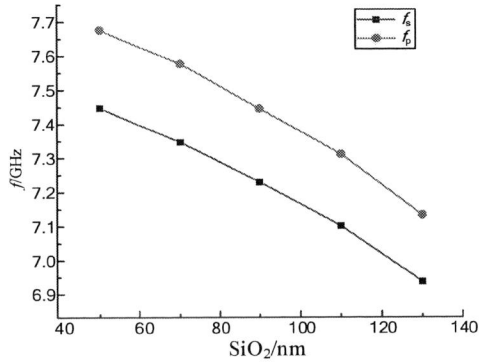

Fig. 5. Influence of support layer thickness on fs and fp

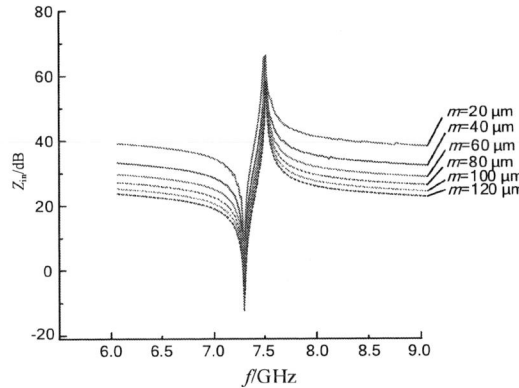

Fig. 6. Impedance curves of the resonator under different active areas

Based on the parameters in Table 1, the thickness of the support layer was varied to simulate the frequency response of the FBAR finite element model within a thickness range of 50–130 nm. The results are shown in Figure. 5. Analysis shows that both series and parallel resonant frequencies decrease as the support layer thickness increases, with a frequency shift of approximately 100 MHz per 20 nm increase. In high-frequency applications, the piezoelectric stack of the resonator becomes extremely thin, necessitating the incorporation of a support layer to prevent structural failure. Under these conditions, the impact of the support layer on frequency shift becomes more pronounced than at lower frequencies. Therefore, the support layer must be included in finite element simulations of high-frequency FBAR to avoid significant discrepancies between simulation results and fabricated devices, which could adversely affect subsequent filter design.

By varying the width of the top electrode based on Table 1, the frequency response of the FBAR finite element model was evaluated for top electrode widths ranging from 20 to 120 nm. The corresponding variation in input impedance Zin within an active area range of 800–4800 μm² is shown in Figure. 6. The results indicate that while the series and parallel resonant frequencies remain largely unchanged with variations in the active area, the impedance of the resonator increases as the active area decreases, with more pronounced parasitic effects. This occurs because reduced lateral dimensions enhance acoustic energy leakage through lateral mode vibration, which makes spurious modes more pronounced. In fabrication, to ensure complete release of the cavity and improve production yield, the cavity diameter should be kept below 100 μm. Exceeding this limit may lead to edge cracking or even

III. FBAR Equivalent Circuit Simulation

While simulations using COMSOL Multiphysics offer higher accuracy, the computational time becomes prohibitively long when the model complexity increases, which hinders subsequent filter design and optimization iterations. In contrast, simulations based on an equivalent circuit model significantly reduce computation time. Therefore, the Modified Butterworth-Van Dyke (MBVD) model was adopted for circuit-level simulation of the filter.

The MBVD equivalent circuit model is illustrated in Figure.7. It uses lumped parameter components to approximate the device behavior near the resonance frequency. The impedance formula of the MBVD equivalent circuit model can be derived as follows:

$$Z = R_s + \frac{(R_0 + \frac{1}{j\omega C_0}) \times (R_m + j\omega L_m + \frac{1}{j\omega C_m})}{(R_0 + \frac{1}{j\omega C_0}) + (R_m + j\omega L_m + \frac{1}{j\omega C_m})} \quad (1)$$

where:

C0 represents the static capacitance of the piezoelectric thin film; Cm and Lm denote the motional capacitance and inductance related to mechanical vibration, respectively; R

All parameter values of the components can be derived by extracting the real part, imaginary part, and quality factor from the impedance curve [12].

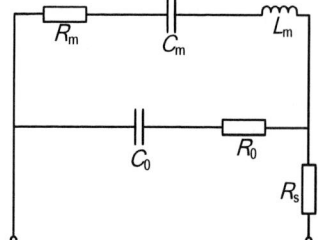

Fig. 7. The Modified Butterworth-Van Dyke mode[12].

Based on the simulation results from COMSOL Multiphysics, the following parameters of the MBVD equivalent circuit model can be obtained: the series resonant frequency ω_s, the parallel resonant frequency ω_p, the quality factor at the series resonance point Q_s, the quality factor at the parallel resonance point Q_p, the static capacitance $C0$, and the electrode loss R_s [13]. The remaining parameters are derived using the following formulas:

$$C_m = C_0 \times \left(\left(\frac{\omega_p}{\omega_s}\right)^2 - 1\right) \quad (2)$$

$$L_m = \frac{1}{\omega_s{}^2 \times C_m} \quad (3)$$

$$R_m = \frac{\omega_s \times L_m}{Q_s} \quad (4)$$

$$R_0 = R_m \times \left(\left(\frac{Q_s \times \omega_p}{Q_p \times \omega_s}\right) - 1\right) \quad (5)$$

After completing the MBVD equivalent circuit modeling based on multiphysics simulation results, the impedance characteristic curve of the resonator equivalent circuit was calculated in ADS. The TUNING tool was used to optimize parameters C0 and Rs, so that the circuit-simulated impedance characteristics closely approximate the multiphysics simulation results. The optimized data are shown in Table 2. Figure 9 compares the impedance curve near the resonance frequency obtained from ADS simulation with the multiphysics simulation results.

As shown in Figure.8, the discrepancy between the two curves is less than 1% across the simulated frequency range. This indicates that the extracted the MBVD equivalent circuit model parameters are valid and can accurately simulate the behavior of the resonator near its resonance frequency, thus supporting rapid filter design.

TABLE II. PARAMETERS OF THE THE MBVD EQUIVALENT CIRCUIT MODEL

Component	C_0	C_m	L_m	R_s	R_m	R_0
Parameter	1.11pF	68.11fF	7.11nH	0.036Ω	0.133Ω	0.074Ω

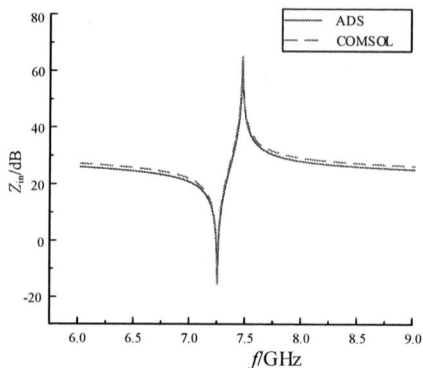

Fig. 8. Comparison of the MBVD equivalent circuit model fitting results.

The established MBVD equivalent circuit was encapsulated into a reusable module for convenient invocation in filter design. Figure.9 shows three types of first-order filter circuits. With a frequency step size of 1 MHz and both ports terminated with 50 Ω impedance matching, the frequency characteristics of the filters were simulated over the 6-9 GHz range. The MBVD equivalent circuit model parameters used for the series and shunt resonators are listed in Table 3.

The S_{21} parameter curves of the first-order L-type, T-type, and π-type FBAR filters obtained from the simulation are shown in Figure.10. Comparative analysis of the S_{21} parameters for the three topological structures indicates that the T-type topology improves in-band flatness, while the π-type topology enhances out-of-band rejection. The use of the pre-packaged MBVD equivalent circuit enables accurate and efficient simulation of filters, significantly reducing design time.

TABLE III. THE MBVD EQUIVALENT CIRCUIT MODEL PARAMETERS

Component	C_0	C_m	L_m	R_s	R_m	R_0
Series	1.11pF	68.11fF	7.11nH	0.036Ω	0.133Ω	0.074Ω
Parallel	1.10pF	55.53fF	9.13nH	0.033Ω	0.167Ω	0.091Ω

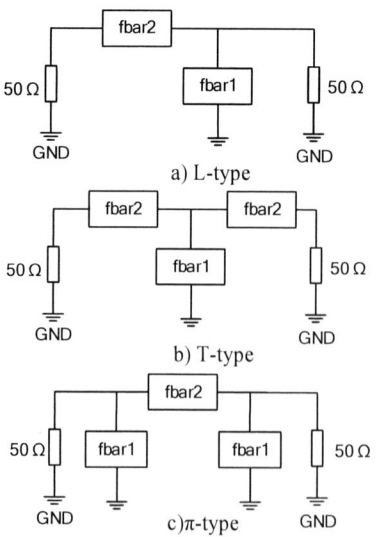

Fig. 9. First-order filter circuits

979-8-3315-8850-2/25 $31.00 © 2025 IEEE

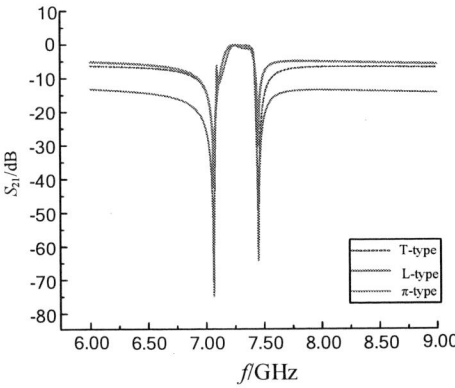

Fig. 10. S_{21} parameter curves of the three topological structures

Based on multiphysics simulation results, the physical parameters of the MBVD equivalent circuit model were extracted to construct an equivalent circuit, enabling rapid filter design. This approach is not only applicable to various topological structures but also allows for selective optimization according to specific filter requirements. After determining the target application scenarios and technical specifications of the filter, the effective area of the resonator can be emulated by adjusting C0 in the MBVD equivalent circuit model. Multiple resonators can be employed to achieve impedance matching in the filter design. Additionally, external capacitors and inductors can be introduced to incorporate new zeros and poles, increase the filter order, improve overall performance, and extend the bandwidth.

IV. SUMMARY

This paper investigated the design methodology of high-frequency film bulk acoustic wave filters using COMSOL Multiphysics simulations and ADS circuit simulations. Finite element analysis was conducted to examine factors influencing the series/parallel resonant frequencies and impedance characteristics of FBARs, and the necessity of including the support layer in high-frequency simulations was validated. Furthermore, parameters of the MBVD equivalent circuit model were extracted to establish an equivalent circuit, facilitating the rapid design of first-order filters. The advantages of different filter architectures were compared and analyzed, providing theoretical support for the design and optimization of high-frequency filters.

ACKNOWLEDGMENT

This work was supported by the Innovation Fund of National Key Laboratory of Integrated Circuits and Microsystems.

REFERENCES

[1] Zhang, Yafei, and Chen Da. Principles, Design and Applications of Film Bulk Acoustic Resonators [M]. Shanghai: Shanghai Jiao Tong University Press, 2010.

[2] Xie Yong, Lai Qiangtao, Chen Hua, et al. Design of High-Gain Transimpedance Amplifier for MEMS Resonators [J]. Microelectronics & Computer, 2016, 33(3): 46-49.

[3] Chen Fuzhan, Gan Yebing, Luo Yanbin, et al. A 2.4 GHz Multi-Module Integrated CMOS RF Front-End Chip [J]. Microelectronics & Computer, 2020, 37(12): 27-32.

[4] Zhang Rui. Research on the Performance and On-Chip Structure of Microelectromechanical Bulk Acoustic Wave Resonators [D]. Chengdu: University of Electronic Science and Technology of China, 2016.

[5] Li Liang, Liu Qinglin, Fu Yuedong, et al. 6 GHz High-Frequency FBAR Filter [J]. Semiconductor Technology, 2022, 47(7): 549-553.

[6] Chen Gongtian, Xie Ruoyuan, Sun Jing, et al. Structural Design and Simulation of Ultra-Wideband FBAR Filter Greater than 3 GHz [J]. Piezoelectrics & Acoustooptics, 2023, 45(5): 663-666.

[7] WEN B, ZHAO T, WANG Q, et al. Optimization of mason model in thin film bulk acoustic resonators with extra passivating piezoelectric structures [C]// 2023 24th International Conference on Electronic Packaging Technology (ICEPT). Shihezi: IEEE, 2023: 1-4.

[8] Tang Xiaolong, Liu Ya, Jiang Pingying, et al. Design of a 2.4 GHz WiFi Band FBAR Bandpass Filter [J]. Piezoelectrics & Acoustooptics, 2022, 44(2): 191-193.

[9] Liao Junjie, Feng Yaogang, Wan Caixin, et al. Research on Film Bulk Acoustic Resonator (FBAR) Filters [J]. Chinese Journal of Sensors and Actuators, 2023, 36(11): 1669-1680.

[10] Liu Xinyao. Research on Cavity-Type Film Bulk Acoustic Resonator (FBAR) Filters [D]. Guangzhou: South China University of Technology, 2020.

[11] Lan Weihao. Research on Key Coating Technology for FBAR Filters in Sub-6 GHz 5G Communication [D]. Chongqing: Chongqing University of Posts and Telecommunications, 2019.

[12] Chang Yahui, and Wang Weimin. Research on Bandwidth Expansion of FBAR Filters Based on Electrical Topology [J]. Piezoelectrics & Acoustooptics, 2022, 44(5): 691-695.

[13] Wang Rui, Chen Pengguang, Ren Jiatai, et al. Design and Simulation of FBAR Based on ADS [J]. Software, 2019, 40(5): 207-211.

2025 The 10th International Conference on Integrated Circuits and Microsystems

A 10GS/s 8b Time-Interleaved ADC with Aperture Error Calibration and LMS-Based Nonlinearity Correction

Chuan Qin
School of Microelectronics
Xidian University
Xi'an, China
23211215034@stu.xidian.edu.cn

Chuan Liu
School of Microelectronics
Xidian University
Xi'an, China
1084273030@qq.com

Runqiao Wang
School of Microelectronics
Xidian University
Xi'an, China
runqiaow@outlook.com

Lecheng Li
School of Microelectronics
Xidian University
Xi'an, China
2393262880@qq.com

Maliang Liu*
School of Microelectronics
Xidian University
Xi'an, China
mlliu@xidian.edu.cn

Yintang Yang
School of Microelectronics
Xidian University
Xi'an, China
ytyang@xidian.edu.cn

Abstract—**This paper presents a 16-channel time-interleaved four-stage pipelined ADC implemented in 110nm CMOS, achieving 10GS/s sampling rate with 8-bit resolution. The design addresses critical challenges in high-speed data conversion through several integrated techniques. Bandwidth matching optimization between sub-ADC and CDAC paths is combined with digital feedback-based aperture error calibration from subsequent MDAC stages, ensuring robust performance across PVT variations. The architecture incorporates a high-linearity charge-pump input buffer and gain-boosted positive-feedback operational amplifier to maintain signal integrity. Furthermore, the design implements a dither injection scheme with LMS algorithm optimization for harmonic coefficient calibration, significantly improving linearity. The simulation results demonstrate that this ADC achieves SNDR greater than 47 dB and SFDR exceeding 60 dB across the 5 GHz bandwidth, while maintaining stable performance with minimal variations in both SNDR and SFDR under temperature fluctuations ranging from -40° C to 85° C and power supply variations of ±5%, with a total power consumption of 3.2 W.**

Keywords—*pipelined analog-to-digital converter (ADC), SHA-less, Time-Interleaving, Aperture error, Nonlinearity calibration*

I. INTRODUCTION

The SHA-less pipelined architecture faces critical technical challenges including sampling clock skew (aperture error) and signal path bandwidth mismatch[1][2]. Aperture error causes timing misalignment between sub-ADC and CDAC paths, inducing dynamic comparator offsets under non-DC inputs, while bandwidth mismatch progressively degrades harmonic performance at higher frequencies. Conventional solutions typically employ analog calibration techniques - such as residue over-range detection through comparators or compensation capacitors as demonstrated in [3] - yet suffer from comparator offset sensitivity, strong process dependence, and excessive area overhead. For time-interleaved high-speed SHA-less pipelined

ADCs, dynamic performance is further constrained by nonlinear distortion, particularly higher-order harmonics. While traditional calibration methods improve linearity, they often encounter high hardware complexity, slow convergence, or insufficient precision at multi-GHz sampling rates[4][5]. Although reference [6] employs pseudo-random code injection for harmonic and INL calibration, its track-and-hold amplifier (THA) introduces additional power/area overhead and exhibits vulnerability to high-frequency noise interference while failing to properly balance historical versus current data weights. Addressing these limitations, this study develops a low-cost, high-sensitivity digital background calibration scheme that statistically models backend MDAC code distributions to establish mathematical relationships between clock skew and code errors. The solution implements SAR-controlled gradient search algorithms to drive digitally tunable delay lines (DCDLs) for rapid phase convergence while adaptively adjusting SUB-ADC path frequency response for bandwidth matching. For linearity enhancement, the design integrates a level-shifted charge-pump input buffer and injects dither signals into MDAC to reveal nonlinear characteristics, implementing a quadrant-symmetric error extraction mechanism that captures gradient errors through asymmetric code distribution analysis at critical transition points. Filtered LMS algorithms with incorporated exponentially weighted moving average (EWMA) filters dynamically update compensation coefficients, effectively suppressing high-frequency noise while enabling adaptive error tracking through forget factors that optimally balance historical and current data weights, with tunable filtering parameters accommodating diverse signal environments. These combined analog and digital innovations enable a 16-channel time-interleaved SHA-less pipelined ADC achieving wide bandwidth and high linearity.

The structure of this paper is as follows: Section II provides a comprehensive overview of the proposed time-interleaved

979-8-3315-8850-2/25 $31.00 © 2025 IEEE

pipelined ADC architecture, Section III details key aspects of the analog circuit implementation, Section IV presents the digital calibration methodology, Section V demonstrates the simulation results, and Section VI concludes this research study.

II. PROPOSED ADC ARCHITECTURE

Fig.1 presents the architecture of the proposed SHA-less pipeline ADC, which employs a 16-channel time-interleaved structure. Four input buffers each drive four sub-ADCs, where each sub-ADC utilizes a 2-bit first-stage MDAC with 1-bit redundancy to enhance the tolerance to first-stage offset. To ensure bandwidth matching between the sampling paths, the sub-ADC of the first-stage MDAC incorporates tunable bandwidth loads, aligning its frequency response with that of the CDAC path. The second, third, and final-stage MDACs are all based on flash architecture. An aperture error detection module continuously analyzes the code distribution of the latter three

MDAC stages and adjusts the sampling clock phase of the sub-ADC via a digitally controlled delay line (DCDL) to eliminate clock skew.

The analog signal, combined with a pseudo-random dither sequence, is quantized by the ADC. The output codes undergo interstage gain error calibration, capacitor mismatch correction, and clock skew compensation before entering the nonlinearity calibration module. After nonlinear compensation, the signal is fed into a nonlinear error gradient detector, which captures distortion characteristics by monitoring symmetric detection points. The extracted error is smoothed using an exponentially weighted moving average (EWMA) filter, and an LMS algorithm dynamically adjusts the compensation coefficients through an error feedback mechanism, enabling background nonlinearity calibration. This architecture ensures robust performance in high-speed applications while maintaining low power consumption and high linearity.

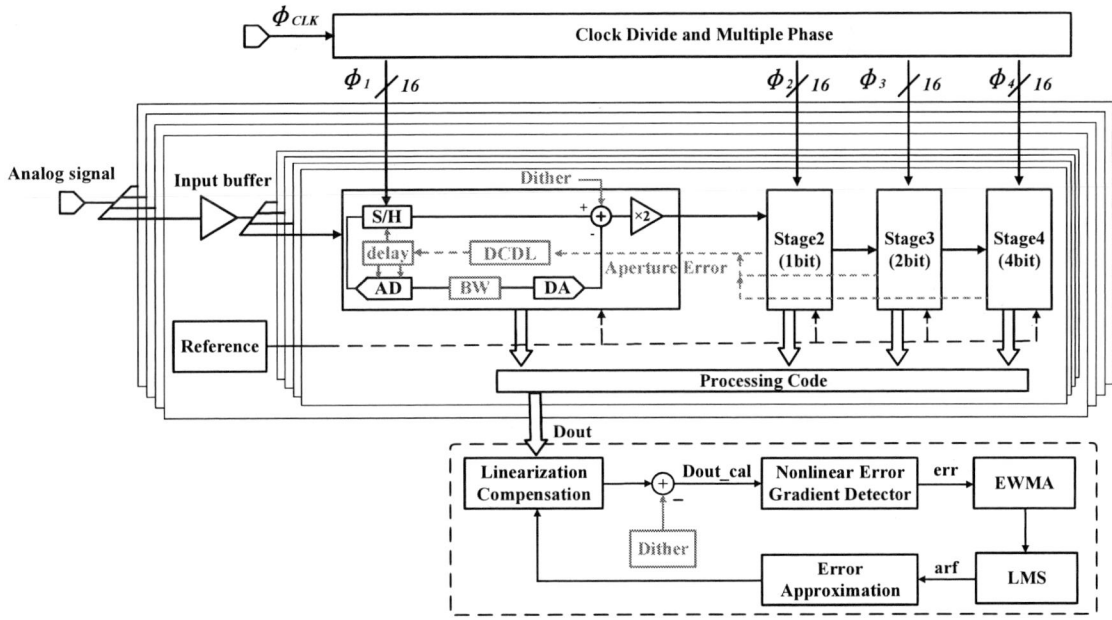

Fig. 1. The overall architecture of ADC

III. ANALOG CIRCUIT DESIGN

A. Input Buffer

This design addresses the nonlinear output resistance issue of conventional NMOS source followers in high-speed ADC applications by proposing an input buffer structure based on a switched-capacitor level-shift circuit. The architecture dynamically adjusts the drain voltages of transistors M1 and M2 to closely track both the input signal and source voltage variations, thereby significantly reducing drain-source voltage fluctuations and improving both DC and high-frequency performance[7].

As illustrated in Fig.2, the circuit employs a switched-capacitor network controlled by non-overlapping complementary clocks φ1 and φ2, combined with a feedforward capacitor C1 and a carefully matched sampling capacitor Cs. This configuration effectively absorbs nonlinear current

components from the main signal path. The design not only enhances linearity but also replaces bulky decoupling capacitors in conventional approaches with capacitive coupling.

The main signal path utilizes deep N-well (DNW) NMOS transistors to isolate noise interference, reduce output impedance, and improve load-driving capability. A common-mode feedback circuit further stabilizes the output common-mode voltage. The simulation results demonstrate that when operating at a 1.9V supply voltage and driving four sub-ADCs, this input buffer achieves a -3dB bandwidth of 6.2GHz while maintaining SFDR above 69dB at high frequencies. Compared to conventional input buffer architectures without level-shift capability and feedforward capacitors, the proposed design exhibits more than 11dB improvement in SFDR performance, validating the effectiveness of these circuit enhancements for high-speed applications.

979-8-3315-8850-2/25 $31.00 © 2025 IEEE   13

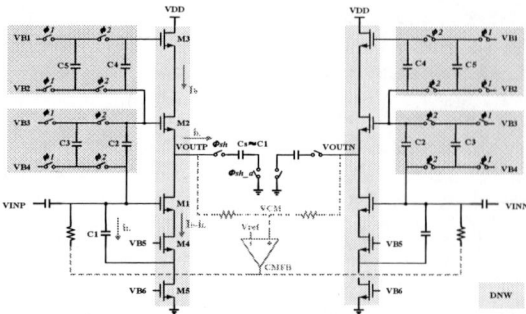

Fig. 2. Input buffer circuit based on switched-capacitor level shifting

B. Operational Amplifier

This design employs a two-stage operational amplifier (op-amp) consisting of a pre-amplifier and a main amplifier. The pre-amplifier provides a wide common-mode input range and enhances the transconductance of the main amplifier, thereby improving bandwidth. The second stage delivers high gain, with its output serving as the dominant pole and the first-stage output as the secondary pole.

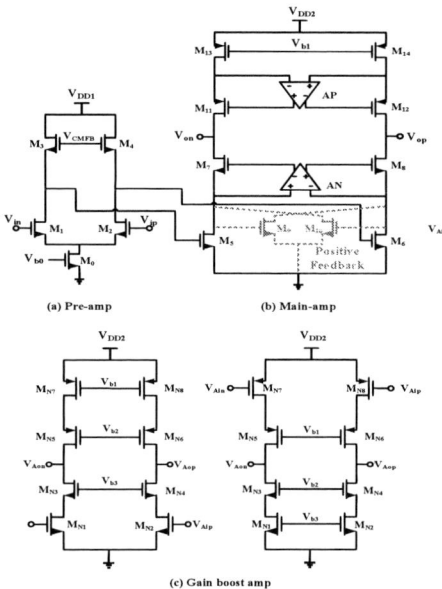

Fig. 3. Operational amplifier circuit with gain-boosting and positive-feedback negative resistance (a)Pre-amp. (b)Main-amp. (c)Gain boost amp.

The overall op-amp architecture is illustrated in Fig.3, where (a) shows the pre-amplifier, (b) depicts the main amplifier, and (c) presents two gain-boosted cascode amplifiers as auxiliary op-amps. The auxiliary op-amps form a negative feedback loop with the cascode MOS transistors, adjusting the output impedance to increase the DC open-loop gain.

In the main amplifier, transistors M9 and M10 create positive feedback, generating negative impedance to further boost the output impedance and enhance the open-loop gain. The impedance from the drain of M7 to ground is:

$$R_S = \left(-\frac{1}{g_{mp}}\right) \| r_{o5} = \frac{r_{o5}}{1 - g_{mp}r_{o5}} \qquad (1)$$

The preceding analysis reveals that as parameter $g_{mp}r_{o5}$ approaches unity, the impedance R_S exhibits a corresponding increase, consequently elevating the overall output impedance. However, when $g_{mp}r_{o5}$ exceeds 1, the output resistance transitions to negative values, potentially inducing oscillatory behavior in the circuit. To ensure stable operation, conventional design practice typically constrains $g_{mp}r_{o5}$ to 0.5 as an optimal compromise between performance and stability.

The complete operational amplifier gain can be expressed through the following relationship:

$$A_{tot} = g_{m1}\left(r_o \| \frac{1}{g_{m3}}\right)g_{m5}\left(A_Ng_{m7}r_{o7}r_{o5} \| A_Pg_{m11}r_{o11}r_{o13}\right)A_F \quad (2)$$

IV. DIGITAL CALIBRATION SCHEME

A. Background Calibration of Aperture Error and Bandwidth Mismatch

The clock skew alters the transfer curve properties by causing the MDAC and sub-ADC paths to sample different instantaneous signal values, thereby introducing nonlinear distortion into the ideal linear transfer characteristic. This distortion manifests as abrupt slope variations at curve inflection points and misaligned quantization steps, generating harmonic components in the frequency domain. The dynamic offset induced by clock skew exhibits linear growth with increasing input signal frequency, and when this offset exceeds the built-in redundancy range, conversion errors occur.

While the background calibration method mentioned in [3] can perform over-range detection based on comparators, this approach introduces comparator offset, and the output common-mode voltage of the MDAC is susceptible to PVT variations, making it difficult to guarantee robustness.

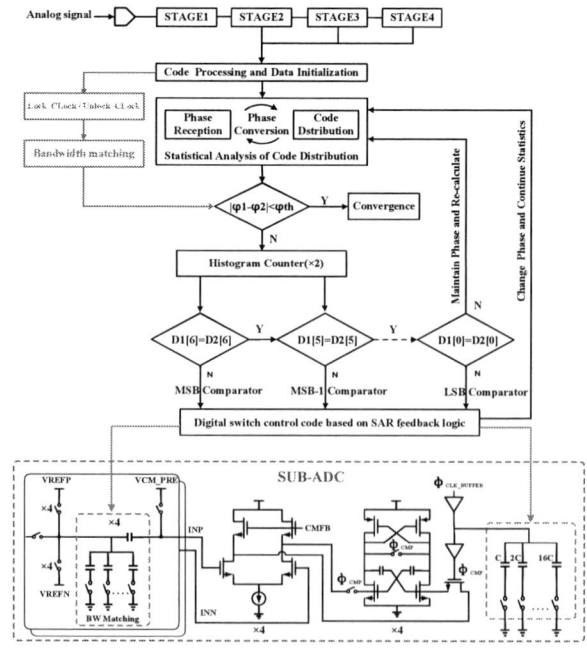

Fig. 4. Schematic diagram of aperture error calibration and bandwidth matching principles

979-8-3315-8850-2/25 $31.00 © 2025 IEEE   14

The proposed co-calibration scheme in this work achieves high-accuracy compensation for both sampling clock skew and path bandwidth mismatch through digital-assisted techniques. Fig.4 illustrates the digital compensation flow and analog circuit implementation, where the analog input signal first enters the first-stage MDAC and passes through the level-shifting capacitor of the sub-ADC. This module dynamically adjusts the equivalent bandwidth of the sub-ADC path using a binary compensation capacitor array.

This study proposes an aperture error calibration algorithm employing closed-loop feedback to compensate for nonlinear distortion induced by clock skew through statistical analysis of code distribution characteristics from the last three MDAC stages. During operation, the ADC enters a phase scanning phase where concatenated output codes are captured at the sampling instants of φ1 and φ2 Flash ADC comparators, with histogram counters quantifying code occurrence frequencies per phase. The system sequentially compares code magnitudes and distribution frequencies from most to least significant bits—if a phase exhibits either larger MSB codes or higher MSB occurrence frequency, it is identified as introducing more severe high-frequency deviation due to clock skew. Identical phase distributions trigger additional iterations until disparities emerge. SAR feedback logic then generates digital control codes to drive capacitor arrays for clock phase adjustment, with the arithmetic mean of both phases adopted as the optimal solution when their absolute difference falls below a predefined threshold, achieving high-precision clock skew compensation.

Fig.5 illustrates the SAR-logic-based phase convergence characteristics and corresponding convergence process. During the initial phase (a), the system begins with preset boundary phase values as starting points. In the phase comparison and update phase (b), the algorithm statistically analyzes the quantization outputs of backend MDACs at intermediate phases using SAR logic, employing code magnitude and occurrence frequency as objective functions for phase iteration. The process culminates in the convergence phase (c) where the clock phase stabilizes at the optimal position, minimizing over-range code occurrences in the first-stage MDAC output. By employing SAR-type feedback logic, the algorithm guarantees convergence within at most N iterations for an N-bit binary digitally controlled phase delay line, achieving steady-state operation with deterministic latency. The SAR logic achieves exponential convergence, demonstrating significantly faster convergence speed compared to the fixed-ratio step size search algorithm based on two-point distances described in Reference [1], with this speed advantage becoming increasingly pronounced as the calibration range expands. This co-calibration scheme shares both target detection functions and hardware resources between aperture error correction and bandwidth matching, substantially reducing power consumption and area overhead while maintaining system performance.

B. Background Nonlinearity Calibration

In modern high-speed ADC designs, nonlinear harmonic distortion (particularly third-order harmonics) severely constrains dynamic range improvement. Traditional foreground calibration techniques requiring signal acquisition interruption fail to meet the stringent real-time requirements of 5G and radar

systems. This paper presents an adaptive calibration method that achieves precise modeling and real-time compensation of ADC nonlinearities through pseudo-random dither injection, error filtering optimization, and threshold detection statistics. The innovative incorporation of an Exponentially Weighted Moving Average (EWMA) filter into the conventional LMS architecture significantly enhances calibration stability. Fig.6 illustrates the calibration flow and essential logic components of the algorithm.

At the algorithmic level, the system employs a discrete-time domain nonlinear error modeling approach. Taking second to fifth harmonics as examples, let the ADC output signal be denoted as D_{out}, and its distortion due to harmonic nonlinearity can be expressed as:

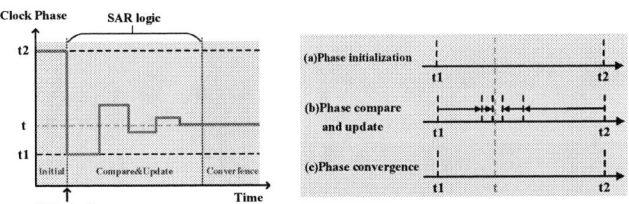

Fig. 5.   Phase convergence characteristics and corresponding convergence process based on SAR logics

$$D_{out} = \alpha_1 V_{in} + \sum_{i=2}^{5} \alpha_i V_{in}^i + \varepsilon \qquad (3)$$

Where α_1 is the ideal gain coefficient, α_i (i=2,3,4,5) characterizes the distortion strength of second- to fifth-order harmonics, and ε denotes quantization noise. To extract α_i, this work develops an error extraction mechanism based on dither-sign statistics. By injecting a pseudo-random dither sequence D_d with amplitude $\pm\Delta$, the system analyzes signal transition behaviors at detection point $\pm D_{th}$:

$$count\_up = \sum \left[sign\left(D_{out} - D\_d > D_{th}\right) \mid _{D\_d > 0} \right] \qquad (4)$$

$$count\_down = \sum \left[sign\left(D_{out} - D\_d < D_{th}\right) \mid _{D\_d < 0} \right] \qquad (5)$$

Based on the (6) and (7), the raw error function is constructed as follows:

$$err = counter\_up - counter\_down \qquad (6)$$

The original signal D_{out} is systematically replaced with the calibrated signal D_{out_cal} obtained after each LMS iteration:

$$D_{out\_cal} = arf_1 D_{out} + PWL\left(\sum_{i=2}^{5} arf_i D_{out}^i + \sigma \right) - D\_d \qquad (7)$$

Where arf_1 represents the ideal gain coefficient after compensation, PWL denotes the piecewise linear approximation of the polynomial function, arf_i (i=2,3,4,5) corresponds to the coefficient of the i-th order distortion, and σ captures the higher-order nonlinear terms introduced by the calibration process itself. Consequently, the calibrated error function is expressed as:

$$err = \sum \left[sign_D \_ d \mid D_{out\_cal} > D_{th} \right] - \sum \left[sign_D \_ d \mid D_{out\_cal} < -D_{th} \right] \quad (8)$$

To mitigate the stochastic fluctuations inherent in error signals under high-frequency sampling conditions, the system incorporates an Exponentially Weighted Moving Average (EWMA) filter for real-time error function smoothing:

$$err\_avg[n] = (1-\beta) err[n] + \beta err\_avg[n-1] \quad (9)$$

Where $\beta \in (0,1)$ represents the filtering memory factor. By appropriately selecting parameter β, the system effectively suppresses quantization noise and high-frequency components introduced by the calibration process itself, preserving the long-term trend characteristics of the error signal to prevent miscalibration caused by instantaneous fluctuations. This smoothed error signal drives the LMS algorithm to update the compensation coefficient arf_i (i=2,3,4,5):

$$arf_i[n+1] = arf_i[n] + \mu err\_avg[n] \quad (10)$$

Where μ is the convergence factor.

The system implements a four-stage processing architecture: the dither injection module incorporates amplitude-controlled dither sequences into the analog signal through MDAC switched capacitors; the error detection module establishes dual-threshold windows to monitor signal crossing behavior; the error filtering module extracts error signals and executes the EWMA algorithm; and the adaptive compensation module dynamically updates coefficients while applying polynomial correction.

Using a 1GHz input signal as an example, Fig.7 presents the output spectrum demonstrating significant suppression of second-to-fifth harmonic components (with particularly notable improvement in third-harmonic reduction) after nonlinear calibration implementation. The simulation results show an 8.7dB enhancement in SFDR, clearly validating the effectiveness of the proposed calibration algorithm in mitigating harmonic distortion while maintaining fundamental signal integrity.

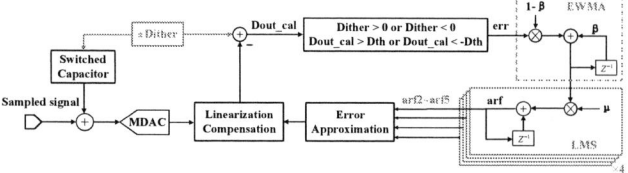

Fig. 6. Nonlinear harmonic calibration flowchart

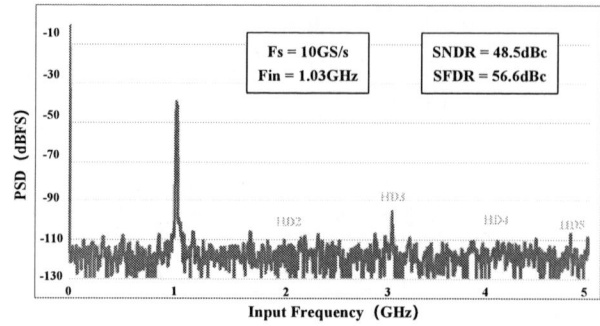

(a)

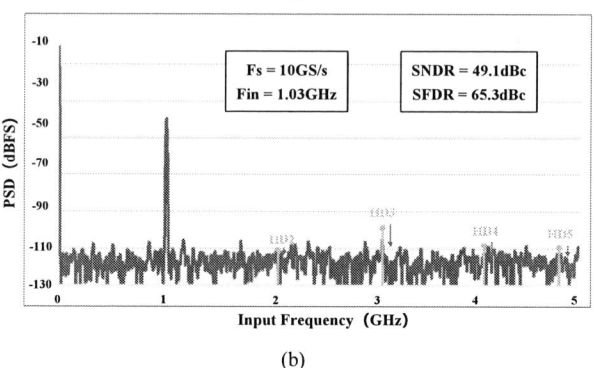

(b)

Fig. 7. (a) Spectrum graph before calibration (b) Spectrum graph after calibration

C. 16-channel Time-interleaved Background Correction

The core of the proposed 16-channel time-interleaved calibration algorithm for this ADC lies in differential detection between adjacent channels, using the clock phase of the first channel as the reference. The method involves subtracting the corresponding mid-sequence Channel 9 data from both the first-cycle and second-cycle Channel 1 data, taking absolute values of these differences before subtracting them again. This processed signal undergoes multi-cycle moving-average filtering to extract Channel 9's average timing skew, which then triggers calibration logic through threshold comparison to control binary-weighted capacitor arrays for timing adjustment of Channel 9. The calibrated Channel 9 subsequently cooperates with Channel 1 to sequentially regulate Channels 5 and 13, propagating this hierarchical correction until all 16 channels achieve balanced timing alignment, completing the time-interleaved calibration process as illustrated in Fig. 8.

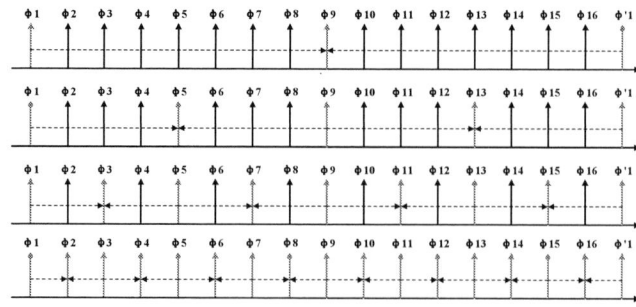

Fig. 8. Clock skew calibration flowchart

V. SIMULATION RESULTS

This ADC is designed with a 110 nm CMOS process to demonstrate the operation and the effectiveness of the proposed architecture. With all calibration schemes enabled, the ADC's performance was verified through simulation using a 600mV peak-to-peak input signal swing under noiseless conditions.

Fig.9 presents the output spectra of the ADC under both low-frequency and Nyquist input conditions, demonstrating its spectral performance across the operational bandwidth. Fig.10 characterizes the SNDR and SFDR stability under Nyquist-frequency input with temperature and supply voltage fluctuations. The simulation results show that both SNDR and SFDR consistently exceed 47 dB and 60 dB respectively, with minimal performance degradation observed across the extended temperature range of -40°C to 85°C and under ±5% supply voltage variations.

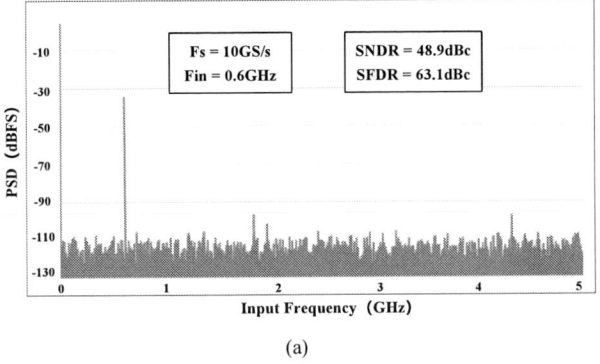

(a)

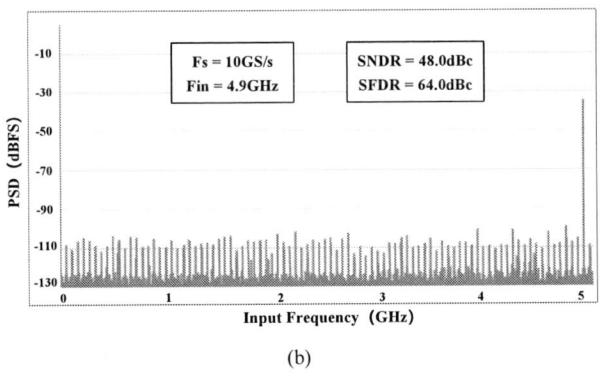

(b)

Fig. 9. (a)Output spectrum with a low-frequency (b)Output spectrum with a Nyquist input

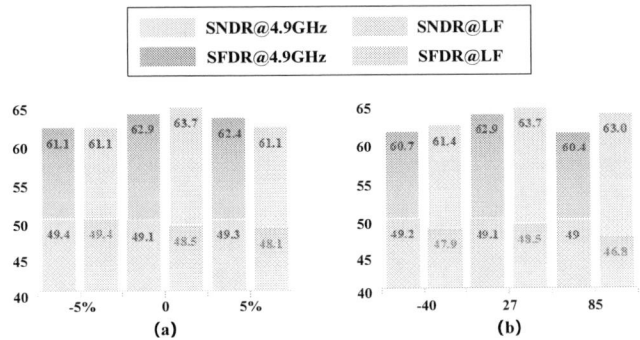

Fig. 10. (a)SNDR/SFDR across supply voltage (b)SNDR/SFDR across temperature

Table I summarizes the ADC's performance metrics and provides a comparative analysis with state-of-the-art designs. Through the implementation of aperture error calibration and bandwidth matching techniques, the ADC achieves bandwidth performance typically attainable only with advanced process nodes, despite being fabricated in 110nm CMOS technology. The harmonic suppression capability of the nonlinear calibration demonstrates strong competitiveness compared to existing solutions.

TABLE I. PERFORMANCE COMPARISON

Reference	This work*	ISSCC 2017[8]	ASSCC 2024[9]	CAE 2020[10]
Architecture	TI-pipeline	Pipeline	Pipeline	TI-sar
Fs(GS/s)	10	10	3.1	4
Resolution(b)	8	12	8.6	8
Process	110nm	28nm	28nm	130nm
Interleaved	16	8	1	4
Frontend	SHA-less	SHA-less	SHA-less	-
App BW[Hz]	5G	7.4G	1.55G	1.3G
SNDR(dB)	48.0	55	45.6	44.4
SFDR(dB)	64.0	66	52.5	-
Power(mW)	4035	2900+	24.2	93
FoMw	223	65.0	50.1	526

*The data represent the simulation results.

VI. CONCLUSION

This paper presents a 10 GS/s 8-bit sixteen-channel time-interleaved SHA-less pipelined analog-to-digital converter (ADC) implemented in SMIC 110 nm CMOS technology with standard 1.3 V and 1.9 V dual-voltage domains. The proposed switched-capacitor level-shift input buffer combines high driving capability with excellent linearity, while the implemented operational amplifier with gain boosting and positive feedback achieves both high gain and wide bandwidth. The introduced bandwidth matching and aperture error calibration techniques provide a cost-effective digital solution for SHA-less pipelined ADCs, and the LMS-based nonlinear calibration offers an effective approach to address linearity bottlenecks in high-speed pipelined ADC designs operating beyond GS/s. The ADC achieves 48 dB SNDR and 64 dB SFDR at Nyquist frequency, with total power consumption of 4 W and a FoM of 223 fJ/conv.step.

REFERENCES

[1] S. Devarajan, L. Singer, D. Kelly, S. Decker, A. Kamath and P. Wilkins, "A 16b 125MS/s 385mW 78.7dB SNR CMOS pipeline ADC," 2009 IEEE International Solid-State Circuits Conference - Digest of Technical Papers, San Francisco, CA, USA, 2009, pp. 86-87,87a, doi: 10.1109/ISSCC.2009.4977320.

[2] P. Huang et al., "SHA-less pipelined ADC converting 10th Nyquist band with in-situ clock-skew calibration," IEEE Custom Integrated Circuits Conference 2010, San Jose, CA, USA, 2010, pp. 1-4, doi: 10.1109/CICC.2010.5617406.

[3]   P. Huang et al., "SHA-Less Pipelined ADC With In Situ Background Clock-Skew Calibration," in IEEE Journal of Solid-State Circuits, vol. 46, no. 8, pp. 1893-1903, Aug. 2011, doi: 10.1109/JSSC.2011.2151510.

[4]   A. M. A. Ali et al., "A 14-bit 2.5GS/s and 5GS/s RF sampling ADC with background calibration and dither," 2016 IEEE Symposium on VLSI Circuits (VLSI-Circuits), Honolulu, HI, USA, 2016, pp. 1-2, doi: 10.1109/VLSIC.2016.7573537.

[5]   N. Rakuljic and I. Galton, "Suppression of Quantization-Induced Convergence Error in Pipelined ADCs With Harmonic Distortion Correction," in IEEE Transactions on Circuits and Systems I: Regular Papers, vol. 60, no. 3, pp. 593-602, March 2013, doi: 10.1109/TCSI.2012.2215754.

[6]   A. M. A. Ali et al., "A 12-b 18-GS/s RF Sampling ADC With an Integrated Wideband Track-and-Hold Amplifier and Background Calibration," in IEEE Journal of Solid-State Circuits, vol. 55, no. 12, pp. 3210-3224, Dec. 2020, doi: 10.1109/JSSC.2020.3023882.

[7]   A. M. A. Ali et al., "29.3 A 14b 1GS/s RF sampling pipelined ADC with background calibration," 2014 IEEE International Solid-State Circuits Conference Digest of Technical Papers (ISSCC), San Francisco, CA, USA, 2014, pp. 482-483, doi: 10.1109/ISSCC.2014.6757522

[8]   S. Devarajan et al., "16.7 A 12b 10GS/s interleaved pipeline ADC in 28nm CMOS technology," 2017 IEEE International Solid-State Circuits Conference (ISSCC), San Francisco, CA, USA, 2017, pp. 288-289, doi: 10.1109/ISSCC.2017.7870374.

[9]   J. Kim and J. Kim, "A single-channel 3.1GS/s 45dB SNDR pipelined ADC using amplify-and-select structure in 28nm CMOS," in Proc. IEEE Asian Solid-State Circuits Conf. (A-SSCC), Nov. 2024, pp. 1-3.

[10] B. T. Reyes et al., "A 4GS/s 8-bit SAR ADC with an Energy-Efficient Time-Interleaved Architecture in 130nm CMOS," 2020 Argentine Conference on Electronics (CAE), Buenos Aires, Argentina, 2020, pp. 77-81, doi: 10.1109/CAE48787.2020.9046376.

2025 The 10th International Conference on Integrated Circuits and Microsystems

VCIL: An Open-Source Pipeline-Tight HIL Framework for Cycle-Accurate RISC-V Coprocessor Verification

Xian Lin
School of Integrated Circuits
Guangdong University of Technology
Guangzhou, China
1112425002@mail2.gdut.edu.cn
https://orcid.org/0000-0002-5780-5948

Xin Zheng*
School of Integrated Circuits
Guangdong University of Technology
Guangzhou, China
xinzheng@gdut.edu.cn
*Corresponding author

Zhixin Fan
School of Integrated Circuits
Guangdong University of Technology
Guangzhou, China
2112425105@mail2.gdut.edu.cn

Huaien Gao
School of Integrated Circuits
Guangdong University of Technology
Guangzhou, China
gaohuaien@gdut.edu.cn

Shuting Cai
School of Integrated Circuits
Guangdong University of Technology
Guangzhou, China
shutingcai@gdut.edu.cn

Xiaoming Xiong
School of Integrated Circuits
Guangdong University of Technology
Guangzhou, China
xmxiong@gdut.edu.cn

Abstract—**The open-source RISC-V instruction set architecture (ISA) enables custom instructions to implement coprocessors, which requires higher verification speed and timing accuracy. However, most existing hardware-in-the-loop (HIL) approaches fail to meet these. To this end, this brief proposes Virtual-Coprocessor-in-the-Loop (VCIL), a pipeline-tight HIL framework for cycle-accurate RISC-V coprocessor verification. It bypasses the TLM2.0 bus and tightly couples the coprocessor to the pipeline via UART, which reduces transmissions and improves simulation speed. Additionally, a timing dynamic calibration method is introduced to enhance timing accuracy and enable early detection of timing violations. Compared to the state-of-the-art Virtual-Peripheral-in-the-Loop (VPIL) framework, VCIL improves simulation speed by 4.8× and reduces timing errors by 35.9%. Experiments conducted on the SM3/SM4 coprocessor demonstrate that VCIL reduces verification time by 53.4%, and the timing error is only 3.5% compared to the RTL full-system. Our code will be available at https://github.com/LX-IC/VCIL.**

Index Terms—**Hardware-in-the-Loop (HIL), Coprocessor Verification, RISC-V, Timing Accuracy.**

I. INTRODUCTION

The RISC-V instruction set architecture (ISA) [1], [2] has revolutionized SoC development with its open-source features, particularly in fine-grained, instruction-level coupling hardware designs. The flexibility of custom instructions enables functional modules to be implemented as coprocessors, offering better trade-offs in performance, power, and area (PPA) compared to peripheral accelerators [3], [4]. However, this approach also introduces additional verification challenges. In traditional RTL verification, the instruction-level interaction of

the coprocessor complicates full SoC integration and prolongs verification time. Since the timing information of custom instructions directly impacts the coprocessor's PPA, cycle-accurate timing verification for each instruction is essential. For hardware modeling verification, it can only provide approximately-timed verification in transaction-level modeling (TLM) [5], resulting in significant timing errors.

To reduce verification time, several ISA simulators are proposed, such as Spike [6], QEMU [7], Swerv-ISS [8], and DBT-RISE-RISCV [9]. Based on virtual prototype (VP) technology, RISCV-VP [10] leverages SystemC and TLM2.0 technologies to model hardware, enabling the simulation of hardware behavior. However, these approaches enable only functional verification at the software level, losing critical hardware details and often resulting in significant PPA discrepancies compared to the final RTL design. Therefore, hardware-in-the-loop (HIL) is introduced as a more balanced solution.

HIL typically integrates physical hardware into simulated software environments for system-level verification [11], offering a trade-off between speed and accuracy. As existing advanced HIL frameworks, SytHIL [12], Virtual-Peripheral-in-the-Loop (VPIL) [13], AMD Xilinx QEMU Co-Simulation, and Synopsys Virtualizer all adopt TLM timing abstraction for hardware and software interaction. These approaches introduce timing decoupling, which significantly degrades verification speed and timing accuracy. For coprocessors, VPIL's memory-mapping and software polling interrupt mechanisms are unsuitable, as they violate pipeline consistency, introduce unpredictable delays, and lead to timing inaccuracies. Additionally, some atypical HIL strategies, such as Cadence ProtoCompiler, CFU Playground [14], and Corvus [15], achieve cycle-accurate

This work was funded by the Science Foundation of Guangdong Province (General Program) under Grant 2025A1515010110, and the Basic and Applied Fundamental Research Topics of Guangzhou under Grant 2025A04J3753.

979-8-3315-8850-2/25 $31.00 © 2025 IEEE

verification by implementing a complete SoC system with complex interactions, which in turn reduces verification speed.

Overall, existing HIL schemes do not sufficiently meet the verification requirements for both speed and timing accuracy in RISC-V coprocessors. Motivated by this, we adopt the Harvard architecture concept to the verification domain and propose a novel verification framework, Virtual-Coprocessor-in-the-Loop (VCIL), for the first time. It bypasses the TLM2.0 bus and tightly couples the coprocessor to the pipeline via UART, enabling instruction-data separation and reducing transmission time. Since SytHIL and VPIL support only static worst-case execution time (WCET) analysis, they incur significant timing inaccuracies that compromise coprocessor PPA evaluation. VCIL introduces a dynamic calibration method that simulates voltage disturbances via a 16-bit linear feedback shift register (LFSR), integrates timing analysis into HIL verification, and enables dynamic WCET prediction. This approach not only improves timing accuracy to the sub-cycle level but also facilitates early detection of timing violations. Moreover, since AMD Xilinx QEMU Co-Simulation and Synopsys Virtualizer are commercial tools unsuitable for fair comparison, we open-sourced VCIL to strengthen the RISC-V ecosystem. The main contributions of this paper can be summarized as follows:

- We first propose VCIL, a HIL verification framework based on the Harvard architecture, which enables instruction-data separation. It bypasses the TLM2.0 bus and tightly couples the coprocessor to the pipeline via UART, significantly improving simulation performance.
- A timing dynamic calibration method is introduced to synchronize the FPGA and VP, thereby improving simulation timing accuracy. It incorporates an LFSR to simulate voltage disturbances and supports dynamic WCET analysis for early detection of timing violations.
- An SM3/SM4 coprocessor is designed as a case study to validate the effectiveness of our method. Compared to RTL full-system verification, VCIL shortens the verification time by 53.4% while the timing error is only 3.5%.

II. SYSTEM ARCHITECTURE

A. VCIL Architecture

We propose VCIL and implement VPIL as a comparison for coprocessor verification. The architectures of VCIL and VPIL are shown in Fig. 1. Their hardware consists of all RTL designs on an FPGA, such as coprocessors. The software is RISCV-VP [10], implemented using SystemC. Compared to ISA simulators, it can simulate more complex hardware behaviors. The CPU pipeline follows five stages: instruction fetch (IF), instruction decode (ID), execution (EX), memory access (MEM), and write-back (WB). RISC-V encodes all custom instructions (custom_instr) as R-type [1], [2]. In the ID phase, the custom_instr is decoded into six fields: opcode, func7, func3, rs1, rs2, and rd. Then, the operands op1 and op2 are fetched from the rs1 and rs2 registers in the EX phase. For VCIL, the coprocessor is tightly coupled to the pipeline via UART. The decoding fields and operands are sent to the

coprocessor for execution. Finally, the results are returned through the same UART transaction and written to the rd register during the WB stage. In contrast, VPIL is a loosely coupled method of memory address mapping. As shown in Fig. 2, the bridge module is mapped to the memory segment 0x50000000 to 0x5FFFFFFF. Since the CPU is 32-bit, each memory unit spans four bytes. The decoding fields opcode, func7, and func3 together occupy less than 32 bits. Thus, they are stored sequentially within a single memory unit. During the MEM stage, VPIL transmits the decoding fields and operands via TLM2.0 write transactions. The bridge module translates all TLM2.0 progress into UART transmission to communicate with the coprocessor. While VPIL employs a software interrupt mechanism to return results. It periodically issues TLM2.0 read transactions to poll the interrupt_signal. After execution, the coprocessor generates the interrupt_signal to indicate that the results are ready. Then, VPIL reads back the results and enters the WB phase. Note that every custom_instr executed in VCIL or VPIL blocks the CPU pipeline.

Consider a simple custom instruction, custom_simple, which requires only one operand, op1. In VPIL, its decoding fields and operand occupy two memory units. Thus, VPIL must issue two TLM 2.0 write transactions (TLM_write). Besides, VPIL needs $n+1$ ($n \geq 1$) TLM2.0 read transactions (TLM_read) to obtain the interrupt_signal and the results, where n denotes the number of polling. The total number of UART transmissions (UART_tran) is $n+3$ ($n \geq 1$). For VCIL, it only requires one UART_tran without TLM2.0 transactions. Suppose it takes a, b, and c ($a, b, c \geq 1$) cycles to complete a single TLM_write, TLM_read, and UART_tran, respectively. The execution cycle of custom_simple in the coprocessor is d ($d \geq 1$). To complete custom_simple, VCIL and VPIL consume $c+d$ and $2a+(n+1)b+(n+3)c+d$ cycles, respectively. In the worst case ($n, a, b, c, d=1$), VCIL reduces UART_tran by 75% (3/4) and improves simulation speed by 4.5× (9/2) compared to VPIL.

B. Timing Dynamic Calibration

Fig. 3 shows the timing dynamic calibration for the custom_instr. RISCV-VP is capable of calculating total cycles and instructions during software execution. The pipeline timing model aims to resolve execution conflicts between instructions. It offers some functions to improve the accuracy of cycle counting. The read() function is invoked to solve the latency cycles introduced by register conflicts while fetching operands. In the coprocessor, the cycle counter calculates the actual execution cycles of the custom_instr. The random cycle generator is designed to simulate voltage disturbances, which usually result in additional operating cycles. It is a 16-bit LFSR capable of generating pseudorandom numbers. The random_cycles value ranges from 0 to c, where c is the theoretical execution cycles of the custom_instr. When the random_cycles reach c, a timing violation is assumed. The cycle counter incorporates random_cycles and returns the final results and cycles via the UART. Then, the write() function handles register conflicts and stores the results in the

979-8-3315-8850-2/25 $31.00 © 2025 IEEE

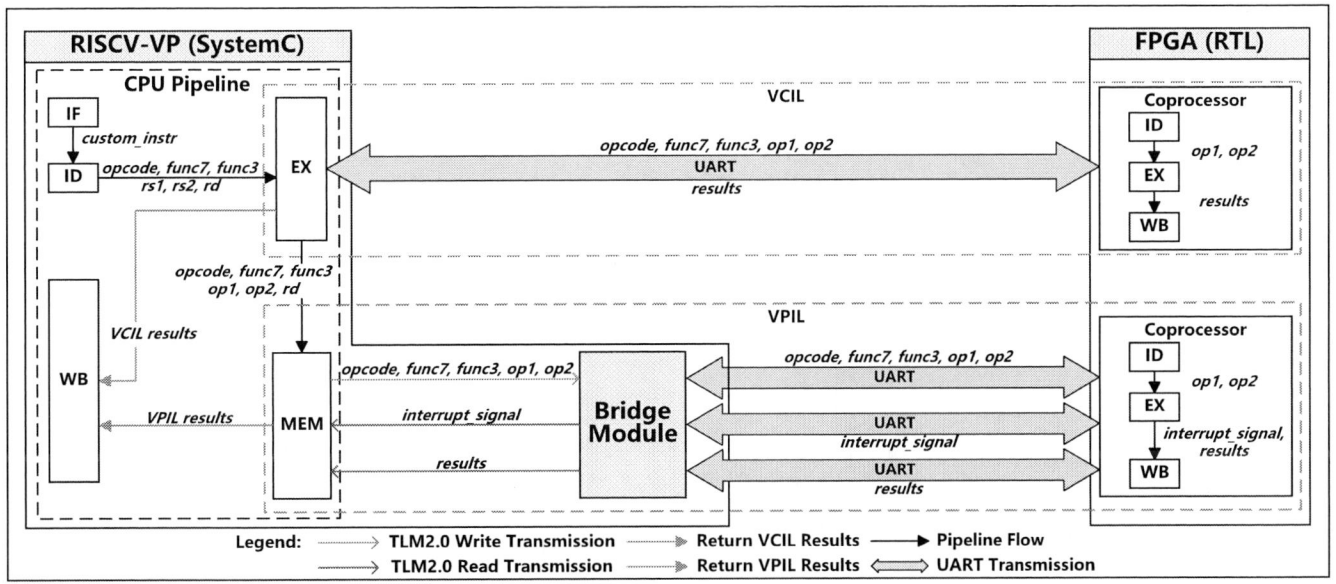

Fig. 1. The architectures of VCIL and VPIL.

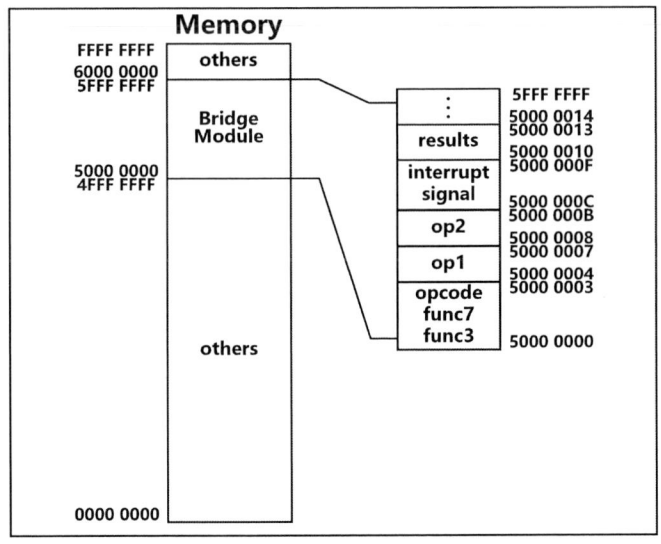

Fig. 2. The memory address mapping of VPIL.

rd register. Finally, the *cycles* are sent to the *advance()* function for checking and updating the total cycles.

C. SM3/SM4 Coprocessor

The SM3 and SM4 are hash and symmetric cryptographic algorithms, respectively. They are recognized by the ISO/IEC international standard [16], [17]. Four custom instruction spaces are predefined in RISC-V, all encoded in R-type. According to the RISC-V cryptography extensions [18], the *custom3* space is selected to implement the SM3/SM4 coprocessor. The encoding of the SM3/SM4 custom instructions is shown in Fig. 4. The *sm3p0* and *sm3p1* instructions are designed to accelerate the *P0* and *P1* transformations in SM3,

respectively. Similarly, *sm4ed* and *sm4ks* accelerate the block encryption/decryption and key schedule in SM4. The *opcode* specifies the use of the *custom3* space. The *funct7* function distinguishes instruction types. The *rs1*, *rs2*, and *rd* represent the indices of the source and destination registers. For *sm3p0* and *sm3p1*, the *rs2* register is unused, so the field is set to zero. The *funct3* function differentiates the type of operations. Furthermore, it is divided into three 1-bit subfields: *xs1*, *xs2*, and *xd*. They indicate whether the registers *rs1*, *rs2*, and *rd* are accessed, respectively. For example, *sm3p0* reads *rs1* and writes *rd*, so *xs1* and *xd* are set to 1, while *xs2* is set to 0. Based on the RTL implementation, which uses a state machine and avoids data conflicts, the theoretical execution cycles of *sm3p0*, *sm3p1*, *sm4ed*, and *sm4ks* are 3, 3, 7, and 7, respectively.

III. VERIFICATION AND EVALUATION

A. Experimental Setup

We compare the proposed VCIL to the VPIL and SytHIL to evaluate its simulation speed and timing accuracy. The experiment is deployed under Ubuntu 18.04.6 with Intel Xeon Platinum 8368 at 2.40 GHz. The RISCV-VP is compiled with gcc7.5.0. To ensure impartiality, an SM3/SM4 coprocessor realized by Verilog is integrated into different platforms for verification (VCIL-Co, VPIL-Co, and SytHIL-Co). It includes four custom instructions: *sm3p0*, *sm3p1*, *sm4ed*, and *sm4ks*. The HBirdv2 E203 SoC [19] is an open-source SoC that has a 32-bit RISC-V Core with a two-stage pipeline. It provides a NICE interface to support fast coprocessor integration and verification. It is used as the RTL full-system verification solution (E203-Co) to compare with VCIL. Since the core domain clock frequency of E203 is 16 MHz, all experiments are conducted at this clock frequency. DDR200T is selected as the target FPGA to implement the E203-Co, which fea-

979-8-3315-8850-2/25 $31.00 © 2025 IEEE   21

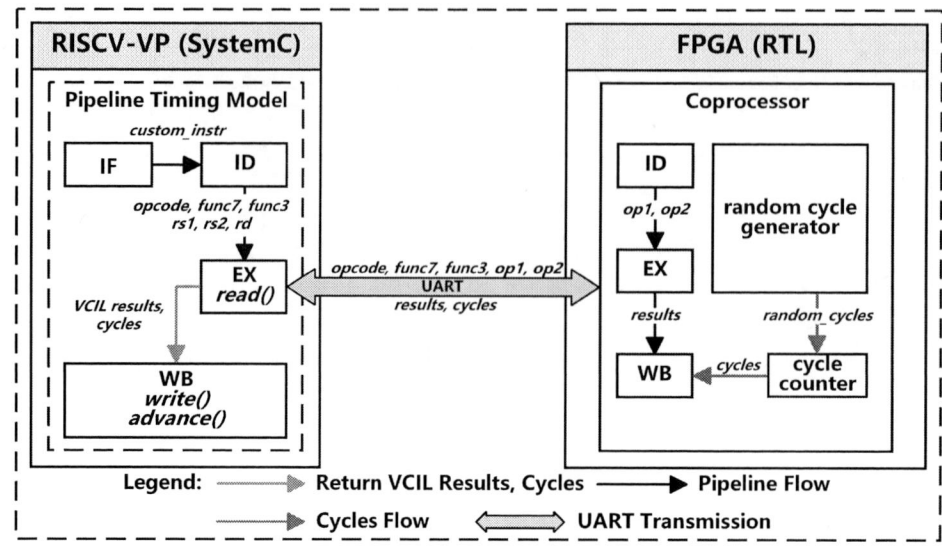

Fig. 3. The timing dynamic calibration.

	31	25 24	20 19	15 14	12 11	7 6	0
R-type	funct7	rs2	rs1	funct3	rd	opcode	
				xd xs1 xs2			
sm3p0	0000000	00000	rs1	110	rd	1111011	
sm3p1	0000001	00000	rs1	110	rd	1111011	
sm4ed	0000010	rs2	rs1	111	rd	1111011	
sm4ks	0000011	rs2	rs1	111	rd	1111011	

Fig. 4. The encoding of SM3/SM4 custom instructions.

tures a Xilinx XC7A200T-2 FPGA chip. Vivado2018.3 and NucleiStudio_IDE_202212 are used for onboard verification. Besides, we design SM3 and SM4 software programs to verify the functional correctness of the SM3/SM4 coprocessor. They are compiled with riscv32-unknown-elf-gcc10.1.0. Finally, the verification time and timing error of VCIL-Co are analyzed.

B. Comparison and Verification

Table I shows the average execution time and the average number of UART accesses (NOUA) for custom instructions under different solutions. Each instruction is run 100 times. As UART transmission constitutes the primary source of time consumption, each instruction exhibits a similar run-time within the same framework. Therefore, reducing the NOUA can significantly enhance simulation performance. Since the average NOUA for VPIL-Co and VCIL-Co is 5.19 ((5.22+5.14+5.22+5.18)/4) and 1 ((1+1+1+1)/4), respectively. Compared to VPIL-Co, VCIL-Co achieves an average acceleration ratio (AR) of 4.8 ((4.8+5.0+4.5+4.9)/4) and reduces the average NOUA by 80.7% (4.19/5.19).

To evaluate the timing dynamic calibration, the random cycle generator is enabled. The execution cycle curves of custom instructions are shown in Fig. 5. The X-axis represents

TABLE I
SIMULATION SPEED RESULTS

	VPIL-Co (s)	VCIL-Co (s)	AR	VPIL-Co NOUA	VCIL-Co NOUA
sm3p0	0.162	0.034	4.8	5.22	1
sm3p1	0.161	0.032	5.0	5.14	1
sm4ed	0.163	0.036	4.5	5.22	1
sm4ks	0.162	0.033	4.9	5.18	1

executions ranging from 0 to 100 times, while the Y-axis indicates the corresponding execution cycles. For *sm3p0* and *sm3p1*, the theoretical cycles range from 3 to 6. For *sm4ed* and *sm4ks*, they range from 7 to 14. Fig. 5 (a) shows the real cycles measured by the integrated logic analyzer (ILA) [20]. Fig. 5 (b) presents the cycles obtained from VCIL-Co via UART. The coefficient of determination (R^2) is calculated to assess the fitting accuracy of the corresponding curves in the two figures. The R^2 values for *sm3p0*, *sm3p1*, *sm4ed*, and *sm4ks* are 0.99, 0.99, 0.94, and 0.93, respectively. The VCIL-Co can also verify the returned cycles and detect timing violations. All detected violations are marked in Fig. 5 (b) with red dots, confirming the effectiveness of the timing dynamic calibration.

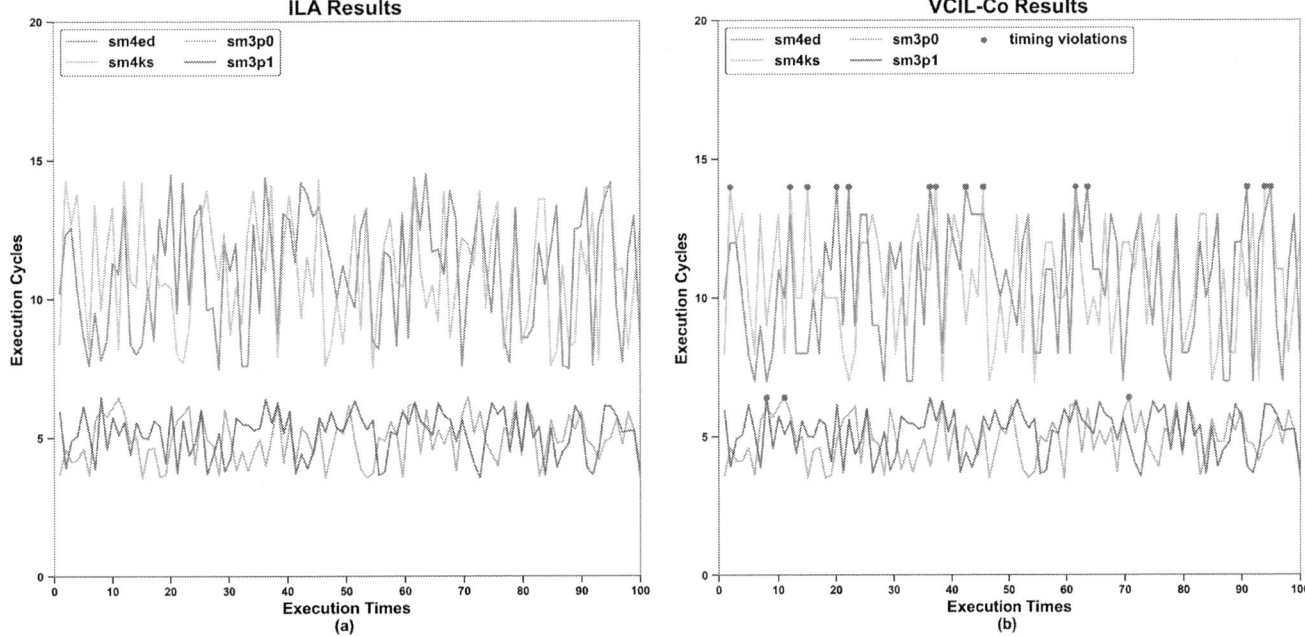

Fig. 5. The timing dynamic calibration results.

Table II shows the average execution cycles and timing errors of custom instructions. The actual timing results of the coprocessor, as measured by the ILA on the FPGA, are used as a comparison benchmark. Since SytHIL-Co defaults to fixed execution cycles of 1 for each instruction, it has the highest average timing error of 85.5%. VPIL-Co only supports the static configuration of theoretical cycles for each instruction. Conversely, VCIL-Co obtains execution cycles closer to ILA than VPIL-Co because of timing dynamic calibration. Compared with VPIL-Co and SytHIL-Co, VCIL-Co reduces average timing errors by 35.9% and 83.0%, respectively. These results confirm that VCIL more effectively supports cycle-accurate RISC-V coprocessor verification than existing advanced HIL frameworks, due to its higher timing accuracy.

the coprocessor. Compared to E203-Co, VCIL-Co consumes fewer LUT (331), FF (454), and BRAM (0). Since VCIL is based on RISCV-VP, it incurs software rather than hardware overhead, occupying 440.9 MB of software space. Compared to E203-Co, VCIL-Co reduces on-chip power consumption by 48.4%. As a complete SoC system, E203-Co achieves faster execution of SM3/SM4 programs. However, E203-Co occupies more hardware resources, resulting in longer synthesis and place & route times (SPRT). VCIL-Co uses RISCV-VP instead of complex RTL full-system integration, which increases some software compilation time but significantly reduces SPRT. Since SPRT constitutes the majority of the total verification time, VCIL-Co shortens the total verification time by 53.4% compared to E203-Co, overall resulting in better PPA.

TABLE II
TIMING ACCURACY RESULTS

	sm3p0	sm3p1	sm4ed	sm4ks
ILA (cycles)	4.92	5.15	11.02	11.00
SytHIL-Co (cycles)	1	1	1	1
Timing Errors	79.7%	80.6%	90.9%	90.9%
VPIL-Co (cycles)	3	3	7	7
Timing Errors	39.0%	41.7%	36.5%	36.4%
VCIL-Co (cycles)	4.87	5.10	10.56	10.58
Timing Errors	1.0%	1.0%	4.2%	3.8%

TABLE III
FPGA VERIFICATION RESULTS

	VCIL-Co	E203-Co
LUT	331	11150
FF	454	11301
BRAM	0	32
Power (W)	0.141	0.273
Software Compilation time (s)	105.5	20.9
Synthesis time (s)	29.3	281.1
Place & Route time (s)	70.6	169.8
SM3 (s)	11.5	0.4
SM4 (s)	3.3	0.1
Total (s)	220.2	472.3

Table III shows the FPGA verification results of VCIL-Co and E203-Co. Note that the random cycle generator is disabled. For VCIL-Co, all hardware resource usage is from

979-8-3315-8850-2/25 $31.00 © 2025 IEEE

The timing error results are shown in Table IV. E203-Co provides the exact RTL verification results. Note that the pipeline timing model of RISCV-VP is configured to match the E203 core for non-custom instructions. Compared to E203-Co, VCIL-Co exhibits a total timing error of only 3.5%.

TABLE IV
TIMING ERROR RESULTS

	VCIL-Co	E203-Co	Timing Error
SM3 (cycles)	68133	70374	3.2%
SM4 (cycles)	8004	8558	6.5%
Total (cycles)	76137	78932	3.5%

IV. CONCLUSION

This work first presents a pipeline-tight HIL framework, VCIL, to fully meet the verification requirements of cycle-accurate RISC-V coprocessors. It is based on the Harvard architecture and introduces an innovative timing dynamic calibration strategy to reduce timing errors and detect timing violations. To verify and evaluate VCIL, an SM3/SM4 coprocessor is implemented across different platforms for comparison. Compared to VPIL-Co, VCIL-Co achieves a 4.8× improvement in simulation speed and reduces timing errors by 35.9%. Compared to E203-Co, VCIL-Co decreases verification time by 53.4%, with a timing error of only 3.5%. These results demonstrate that VCIL has faster speed and smaller timing errors in coprocessor verification.

REFERENCES

[1] A. Waterman and K. Asanovic, "The risc-v instruction set manual, volume i: Unprivileged isa," SiFive Inc. and CS Division, EECS Department, University of California, Berkeley, Tech. Rep., 2019.

[2] ——, "The risc-v instruction set manual, volume ii: Privileged architecture," SiFive Inc. and CS Division, EECS Department, University of California, Berkeley, Tech. Rep., 2019.

[3] X. Lin, H. Liu, X. Zheng, H. Gao, S. Cai, and X. Xiong, "Fpux: High-performance floating-point support for cost-constrained risc-v cores," *IEEE Transactions on Very Large Scale Integration (VLSI) Systems*, vol. 32, no. 10, pp. 1945–1949, 2024.

[4] H. Liu, X. Lin, X. Zheng, and S. Cai, "Convex: A risc-v instruction set extension scheme for accelerating convolution operations on mcus," in *Proceedings of the 2024 8th International Conference on High Performance Compilation, Computing and Communications*, 2024, pp. 133–138.

[5] K. Lu, D. Müller-Gritschneder, and U. Schlichtmann, "Accurately timed transaction level models for virtual prototyping at high abstraction level," in *2012 Design, Automation & Test in Europe Conference & Exhibition (DATE)*. IEEE, 2012, pp. 135–140.

[6] Spike. Accessed: Aug. 2011. [Online]. Available: https://github.com/riscv/riscv-isa-sim

[7] F. Bellard, "Qemu, a fast and portable dynamic translator." in *USENIX annual technical conference, FREENIX Track*, vol. 41, no. 46. California, USA, 2005, pp. 10–5555.

[8] SweRV-ISS. Accessed: May. 2020. [Online]. Available: https://github.com/chipsalliance/SweRV-ISS

[9] DBT-RISE-RISCV. Accessed: Sep. 2017. [Online]. Available: https://github.com/Minres/DBT-RISE-RISCV

[10] V. Herdt, D. Große, P. Pieper, and R. Drechsler, "Risc-v based virtual prototype: An extensible and configurable platform for the system-level," *Journal of Systems Architecture*, vol. 109, p. 101756, 2020.

[11] F. Mihalič, M. Truntič, and A. Hren, "Hardware-in-the-loop simulations: A historical overview of engineering challenges," *Electronics*, vol. 11, no. 15, p. 2462, 2022.

[12] L. Jünger, T. Röhmel, M. Burton, R. Leupers, S. GreenSocs, and F. Le Bourg, "Sythil: A system level hardware-in-the-loop framework for fpga, systemc and qemu-based virtual platforms."

[13] S. Ahmadi-Pour, P. Pieper, and R. Drechsler, "Virtual-peripheral-in-the-loop: A hardware-in-the-loop strategy to bridge the vp/rtl design-gap," *arXiv preprint arXiv:2311.00442*, 2023.

[14] S. Prakash, T. Callahan, J. Bushagour, C. Banbury, A. V. Green, P. Warden, T. Ansell, and V. J. Reddi, "Cfu playground: Want a faster ml processor? do it yourself!" in *2023 Design, Automation & Test in Europe Conference & Exhibition (DATE)*. IEEE, 2023, pp. 1–2.

[15] Z. Jiang, K. Zheng, D. Boland, Y. Bao, and K. Shi, "Corvus: Efficient hw/sw co-verification framework for risc-v instruction extensions with fpga acceleration," in *Proceedings of the 30th Asia and South Pacific Design Automation Conference*, 2025, pp. 1336–1342.

[16] *IT Security Techniques-Hash Functions–Part 3: Dedicated Hash Functions*, ISO/IEC Standard 10118-3, 2018.

[17] *Information Technology-Security Techniques-Encryption Algorithms-Part 3: Block Ciphers*, ISO/IEC Standard 18033-3, 2010.

[18] RISC-V Cryptography Extensions Volume I: Scalar Entropy Source Instructions. Accessed: Jun. 2022. [Online]. Available: github.com/riscv/riscv-crypto

[19] Hummingbirdv2 E203 Core and SoC. Accessed: Oct. 2022. [Online]. Available: https://doc.nucleisys.com/hbirdv2

[20] Xilinx. 2016. Integrated Logic Analyzer v6.2. Accessed: Aug. 2016. [Online]. Available: {https://www.xilinx.com/support/documentation/ip_documentation/ila/v6_2/pg172-ila.pdf}

2025 The 10th International Conference on Integrated Circuits and Microsystems

An Electrode Impedance Evaluation IC with Biphasic Output Current for Biomedical Devices

Fang Xie, Yu Ang Ru, Zhuo Chen, Xu Liu*
Faculty of Information Technology, Beijing University of Technology, Beijing, China
* Corresponding author: liuxu16@bjut.edu.cn

Abstract—**This paper proposes an electrode impedance evaluation IC based on an output current charging and discharging scheme. The circuit include a bandgap reference module, a current source selection module, and a stimulus generation module. The designed bandgap reference circuit with resistor trimming can realize high accuracy, its temperature variation is less than 20 ppm/°C. To fit large range of electrodes impedance, the current source selection module is designed in this work which can be applied to various electrodes impedance ranging from 1 Kohm to 3 Mohm. The circuit can generate biphasic current with an output current range of 200 nA~20 µA. Therefore, the electrode impedance can be detected by observing the output voltage waveform. The circuit is designed using 180-nm CMOS process. Simulation results show that the impedance error is less than 1%. The static power consumption is 47.8 µW.**

Keywords—integrated circuit, CMOS, electrode impedance, biphasic current generation, charging and discharging

I. INTRODUCTION

Neurostimulator in brain-computer interfaces are usually connected to stimulating electrodes[1]. For different electrode impedances, the stimulator needs to select different stimulation parameters to achieve the best stimulation effect or to improve the circuit efficiency. Therefore, it is important to quickly evaluate the impedance of a stimulating electrode for different geometries. The use of CMOS circuits to generate electrical excitation signals and use them to measure bioelectrode impedance has become a viable method.

In 2014, researchers proposed a method for the skin-electrode impedance measurement[2], it based on injecting a current sine wave signal from one electrode and collecting the same signal from another, as well as measuring the voltage drop between two electrodes. However, this method used monophasic current injection instead of biphasic current injection, which may cause nerve damage[3-4]. In 2018, other researchers proposed an automated system for bioimpedance measuring[5], but this circuit can only measure resistance but not capacitance. In 2019, B. Ibrahim and R. Jafari proposed a bioimpedance sensing hardware using the ARM Cortex M4 microcontroller[6], but the impedance measurement range of this system only goes up to 70 Ω. In 2022, researchers have proposed a cross-channel impedance measurement for monitoring implanted electrodes[7], but the upper limit for impedance evaluation in this article is 1 MΩ. Therefore, an electrode impedance evaluation IC with biphasic output current and compatible for a wide impedance range with high accuracy is needed.

In this paper, an electrode impedance evaluation circuit with high accuracy is proposed. This circuit can measure both resistance and capacitance of an electrode by observing the output voltage waveform on electrode. A precise bandgap is designed to improve the output current accuracy. This circuit also uses biphasic current generation as output, in order to detect capacitive electrode load and to prevent tissue damage during impedance evaluation. Besides, the current output range of the proposed circuit can be selected as 200 nA-1 µA, 1 µA-5 µA or 5 µA-20 µA. So, the maximum electrode impedance can be up to 3 MΩ with an accuracy of 99%. The rest part of this paper is organized as follows. Section II illustrates system and working theory of the electrode impedance evaluation circuit. Section III presents the circuit implementation. Section IV shows the simulation results and Section V draws the conclusion.

II. SYSTEM AND WORKING THEORY OF THE ELECTRODE IMPEDANCE EVALUATION IC

A. Electrode Impedance Evaluation System Architecture

In this design, the electrode impedance evaluation IC uses a biphasic current output for current charging and discharging. It consists of three modules, a bandgap reference circuit, a current source selection circuit, and a stimulus generation circuit. The system architecture is shown in Fig. 1.

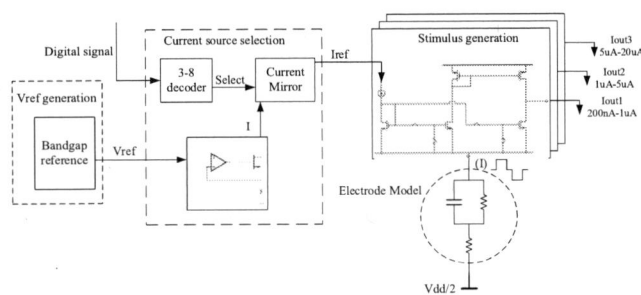

Fig. 1. Structure of the electrode impedance evaluation circuit

B. Electrode Impedance Evaluation Method

First, the reference voltage is generated by the bandgap reference circuit, which is used to obtain desired bias current through the current source selection circuit. And then the stimulus generation circuit generates biphasic current to charge and discharge the electrodes. By observing the output voltage waveform of the load, the electrode impedance can be measured. The output current of stimulus generation circuit is shown in Fig.

This work is supported by funding agency: National Key Research and Development Program (Grant No. 2024YFF1206504) and Beijing Natural Science Foundation- Huairou Innovation Joint Fund (Grant No. L245012)

979-8-3315-8850-2/25 $31.00 © 2025 IEEE

2. The process is divided into three main phases: the current charging (anode) phase, the interphase delay phase, and the current discharging (cathode) phase[8].

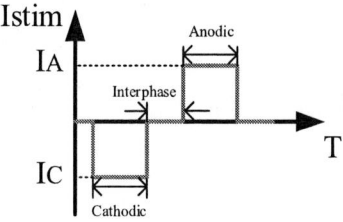

Fig. 2. Output current of stimulus generation circuit with 3 phases

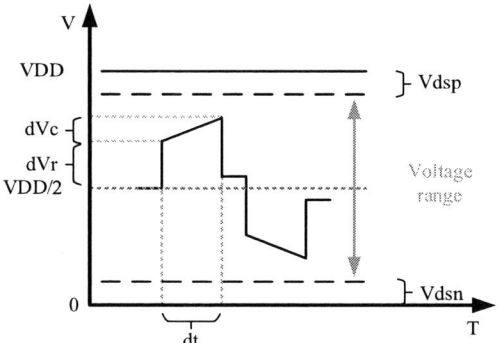

Fig. 3. Voltage waveform on electrode with first-order model

In the current charging phase, the anodic current flows into the electrode. In the interphase delay phase, the output current stops flowing. In the current discharging phase, the cathodic current flows into the electrode. The range of impedance measurement can be then deduced from the captured voltage waveform. The voltage waveform during impedance evaluation has an upper and lower limit, as shown in Fig. 3. Also, the electrode model is shown in Fig. 4.

Define Vr is the voltage across the resistor and Vc is the voltage across the capacitor. The constraint about the voltage range of impedance measurement for 1st-order model can be expressed as:

$$\frac{Vdd}{2} + Vr + Vc < Vdd - (Vdsp - Vthp) \quad (1)$$

$$\frac{Vdd}{2} - (Vr + Vc) > (Vdsn - Vthn) \quad (2)$$

Where Vod is the difference voltage between Vds and Vth, by combining (1) and (2), we can obtain:

$$Vr + Vc < \frac{Vdd}{2} - \max\{Vodp, Vodn\} \quad (3)$$

The magnitude of the current is determined by Is, the charging and discharging time is determined by t, so that the electrode resistance and capacitance can be expressed as:

$$Zr = \frac{dVr}{dIs} \quad (4)$$

$$Zc = Is * \frac{dt}{dVc} \quad (5)$$

Therefore, the impedance can be measured with fixed output Is and captured Vr and Vc values.

C. Electrode Model

First-order electrode model is shown in Fig. 4 (a). It consists of a Helmholtz capacitance (CH), a Faraday resistance (RF), and an electrolyte solution resistance (RS)[9-10]. Due to the large value of the RF resistance, it is usually considered to be equivalent to an open circuit. Therefore, electrode models are often simplified to series capacitance (CH) and resistance (RS).

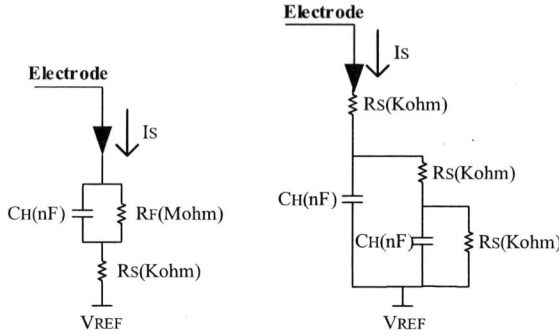

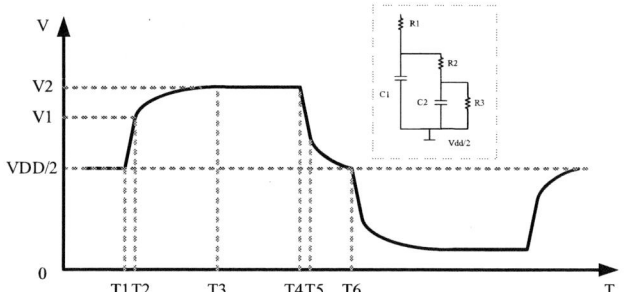

Fig. 4. (a) First-order electrode model (b) Second-order electrode model

Fig. 5. Voltage waveform on second-order electrode model

Second-order electrode model is shown in Fig. 4 (b). It consists of two CH and three RS. When the current I is charging the electrode, the voltage on the electrode is shown in Fig. 5.

From T1~T2, the current flows through resistor R1 and the voltage varies abruptly on R1. From this waveform the value of R1 can be measured:

$$R1 = \frac{V1 - \frac{VDD}{2}}{I} \quad (6)$$

From T2~T3, the current flows through R2 and R3. When voltage is generated across R2 and R3, capacitor C1 and C2 charge up. Therefore, the voltage across C1 and C2 rises exponentially. When C1 charges, it also discharges and charges C2. This process continues until all capacitors are fully charged.

From T3~T4, at this point capacitors have been fully charged, the circuit is equivalent to three resistors in series. V2 is the maximum value of the voltage across the electrode during the charging process. The value of R1+R2+R3 can be measured:

979-8-3315-8850-2/25 $31.00 © 2025 IEEE

$$R1 + R2 + R3 = \frac{V2 - \frac{VDD}{2}}{I} \qquad (7)$$

From T4~T5, the current disappeared, the voltage varies abruptly on R1.

From T5~T6, both C1 and C2 discharge, the voltage across C1 and C2 drops exponentially. When C1 discharges, it also charges C2. This process continues until both capacitors are fully discharged. So, we can obtain the electrode impedance feature from above expressions (6) and (7).

D. Frequency Limitation

Impedance measurement frequency is depended on the speed of MOSFET switches. Fig. 6 shows an NMOS switching circuit. The input signal jumps from low to high and the capacitor is the parasitic capacitance of the output node, which is initially charged to VDD.

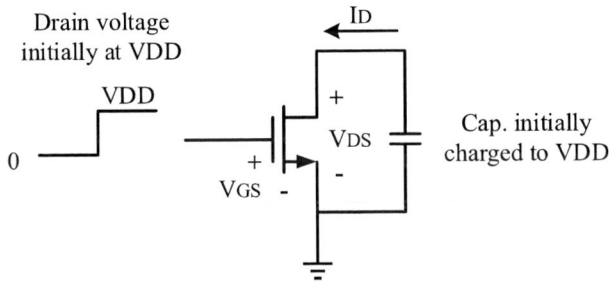

Fig. 6.   MOS switching circuit

The MOS current in the linear region can be expressed as:

$$I_D = \beta[(V_{GS} - V_{THN})V_{DS} - \frac{1}{2}V_{DS}^2)] \qquad (8)$$

Rn between the drain and source of the mosfet during switching is expressed as:

$$R_n = \frac{V_{DD}}{\frac{1}{2}\mu_n C_{OX}\frac{W}{L}(V_{DD} - V_{THN})^2} \qquad (9)$$

Next, to deduce the switching time of mosfet by analyzing the delay time of a single RC, circuit, we define voltage variation across capacitor as output waveform Vout, which can be expressed as:

$$V_{out}(t) = V_{pulse}(1 - e^{\frac{-t}{RC}}) \qquad (10)$$

Define delay time td as output waveform changes from 0 to 0.5Vpulse[11]:

$$\frac{V_{pulse}}{2} = V_{pulse}(1 - e^{\frac{-t_d}{RC}}) \qquad (11)$$

$$t_d \approx 0.7RC \qquad (12)$$

For a mosfet switch, we can equate Rn to R and equate Cgb to C. So that the switching time of an mosfet switch can be measured as td. Thus, we can get the maximum frequency of the measurements.

III.   CIRCUIT IMPLEMENTATION

A. Bandgap Reference Circuit

The bandgap reference circuit is shown in Fig. 7. R1 and R3 are realized by a 4-bit resistor trim, which allows the output voltage of the bandgap (*Vout*) to cover the range of Vref $\pm$ 4σ to ensure coverage to the voltage deviation caused by PVT (process, voltage, temperature).

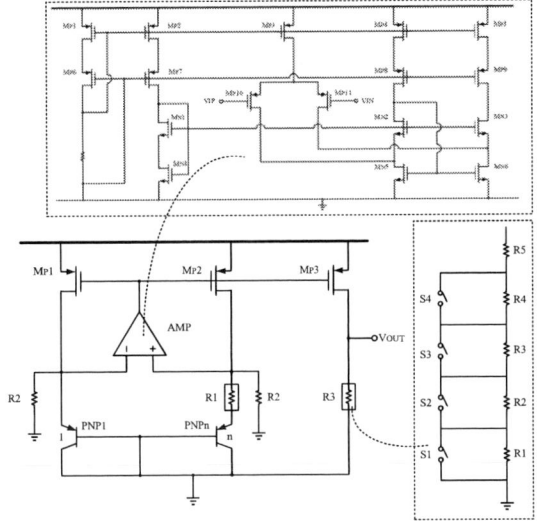

Fig. 7.   Schematic of the bandgap reference circuit with op-amp and resistor trim

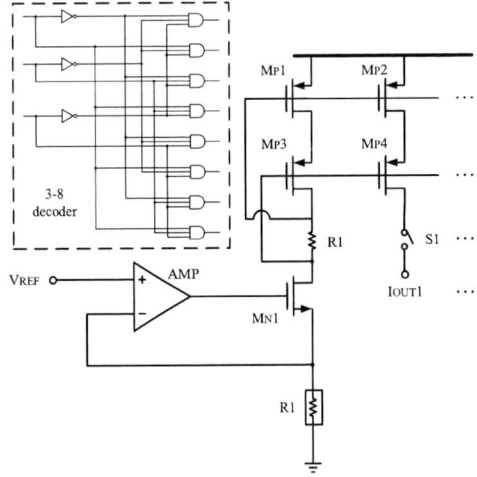

Fig. 8.   Structure of the current source selection circuit

The advantage of using the resistor trim is that accuracy can be guaranteed for electrode impedance evaluation.

B. Current Source Selection Circuit

The current source selection circuit is realized by generating a reference voltage at the output of the op-amp and converting the voltage to current with a resistor, followed by replicating the desired current with multiple cascode current mirrors. In order to select desired current, digital signal drives a 3-8 decoder to control switches of cascode current mirrors. The current source selection circuit is shown in Fig. 8. R1 also uses resistor trim to realize voltage-current conversion, The current output range is 200 nA~20 µA, which is compatible for an electrode impedance ranging from 1 KΩ to 3 MΩ. So, the range of impedance to be evaluated can be expanded.

C. Stimulus Generation Circuit

The stimulus generation circuit is shown in Fig. 9.

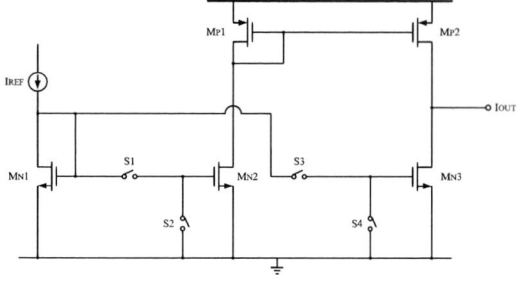

Fig. 9.   Schematic of the stimulus generation circuit

In current charging phase, S1 and S4 are closed and S2 and S3 are disconnected. MN2 is turned on and MN3 is turned off. The bias current IREF provided by current source selection circuit flows into MN1 and flows from MP2 into the electrode for charging. In delay phase, S2 and S4 are closed, S1 and S3 are disconnected, so that no current flows at all. In current discharging stage, S2 and S3 are closed, S1 and S4 are disconnected, the electrodes are discharged, and current flows out from electrode and flows through MN3. The timing control is shown in Fig. 10, the evaluation cycle is 600 µs.

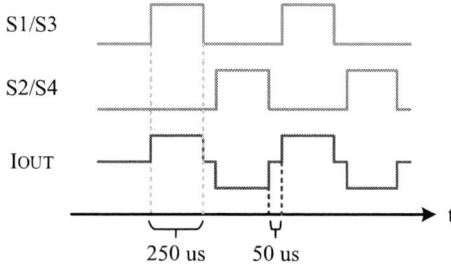

Fig. 10. Waveforms of switching control signals and output current

The circuit is designed in three groups to measure different sizes of capacitance and resistance, respectively. Low current corresponds to large resistance and small capacitance, and high current corresponds to small resistance and large capacitance. The actual test is based on different sizes of electrode resistance and capacitance. The corresponding current group can be selected for the test.

IV.   SIMULATION RESULTS

The electrode evaluation circuit is designed in a 180-nm CMOS process technology. The bandgap reference output voltage with temperature variation is shown in Fig. 11. This circuit provides reference voltage for the current source selection circuit. The op-amp uses a high-gain folded cascode op-amp with a loop gain of 92.59 dB. The temperature variation of this circuit is less than 20 ppm/°C.

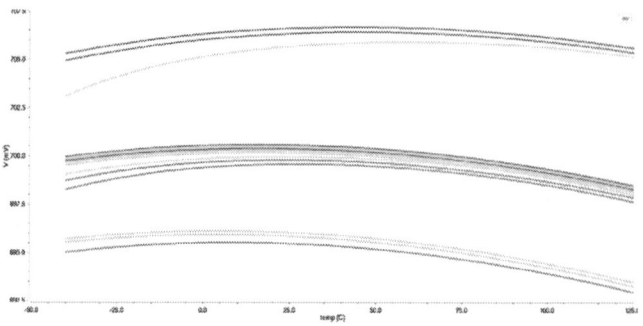

Fig. 11. Reference voltage curvature with temperature variation

Using bandgap reference circuit significantly improves the accuracy of impedance measurements at different temperatures. The electrode model measurement waveforms at -40 °C, 25 °C and 125 °C are shown in Fig. 12. When temperature is at 125 °C, the error of electrode impedance measurement without bandgap reference is about 0.4% and the error with bandgap reference is less than 0.05%. Therefore, the use of a bandgap reference circuit is necessary.

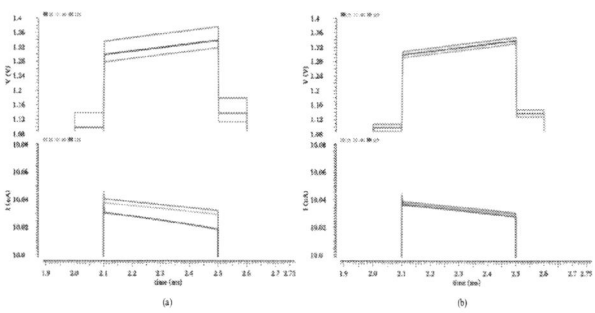

Fig. 12. Electrode model (first-order) measurement waveforms at -40 °C, 25 °C and 125 °C (a) without bandgap reference (b) with bandgap reference

The simulation voltage waveform on the second-order electrode model at tt, ss and ff corner is shown in Fig. 13. The curves are nearly the same with different corners. The output voltage waveforms under different electrode model loads are shown in Fig. 14. The electrode model was CH=20 nF, RF=100 MΩ, RS=100 KΩ.

To characterize the electrode resistance and capacitance of the 1st-order model, expression (4) and (5) can be used. The waveforms at tt, ss and ff corner corresponding to the evaluation of resistance and capacitance, respectively, are shown in Fig. 15.

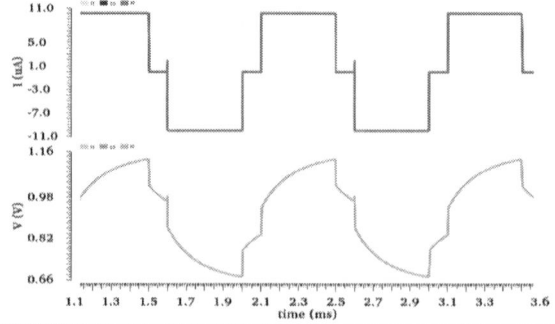

Fig. 13. Simulation voltage waveform on second-order electrode model

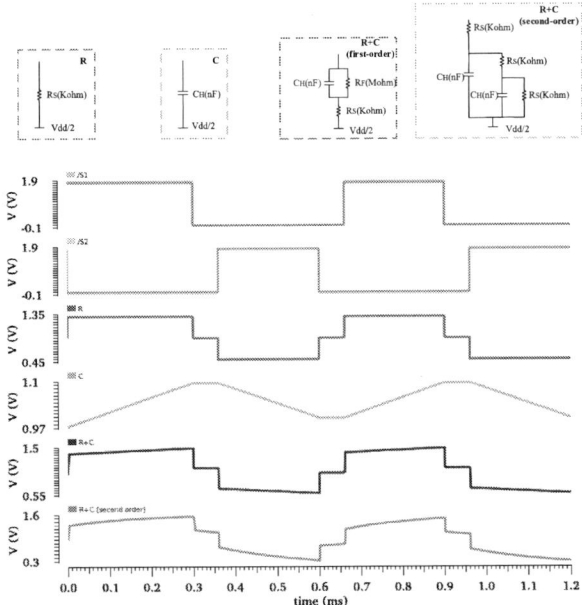

Fig. 14. Output voltage waveforms under different electrode model loads

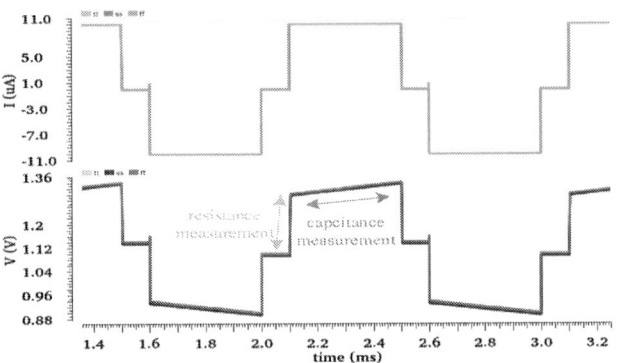

Fig. 15. Evaluation of impedance by output waveform

It can be concluded from the simulation that the mismatch between the cathode current and the anode current is very low.

When the impedance is gradually increased under different models, the error between the impedance value and the ideal value measured with this circuit is shown in Fig. 16. The larger the impedance, the percentage error gradually increases.

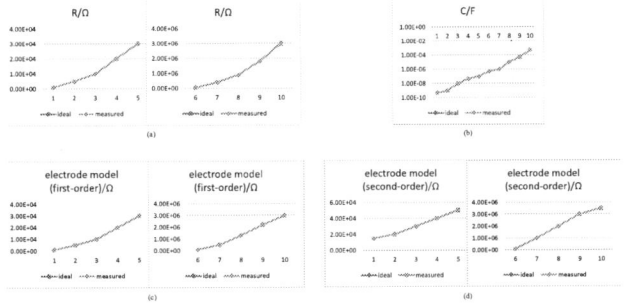

Fig. 16. Impedance error of (a) resistance (b) capacitor (c) electrode model (first-order) (d) electrode model (second-order)

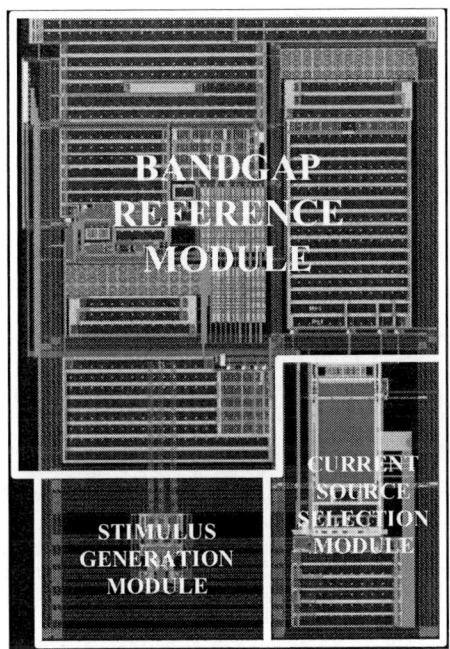

Fig. 17. Layout of the electrode impedance evaluation IC

When bias current is 200 nA, ideal resistance is 100 KΩ, and ideal capacitance is 20 nF, the error between measured resistance value and ideal resistance value is 0.04%, and the error between measured capacitance value and ideal capacitance value is 0.6%. When bias current is 20 μA, ideal resistance is 10 KΩ, and ideal capacitance is 100 nF, the error between measured resistance value and ideal resistance value is 0.01%, and the error between measured capacitance value and ideal one is 0.33%. The performance summary of this IC is shown in TABLE I.

TABLE I.   PERFORMANCE SUMMARY

Parameter	This work
Operation Voltage	1.8 V
Current Range	200 nA ~ 20 μA
Electrode Impedance measuring range	1 KΩ ~ 3 MΩ
Trim Type	Resistor trim
Temperature Variation	<20 ppm/°C
Power Consumption	47.8 μW
Evaluation Error	<1%
Evaluation Cycle	600 μs

The layout of the electrode impedance evaluation IC is shown in Fig. 17. It occupies a core area of 0.354 mm^2.

V.   CONCLUSION

In this paper, an electrode impedance evaluation circuit based on current charging and discharging is proposed. The circuit is implemented using high precision voltage bias technique and biphasic current generation. The resistance and capacitance of both 1st-order and 2nd-order electrode model to be measured can be calculated and evaluated by observing the voltage waveform at the output.

The temperature variation of the bandgap reference circuit is less than 20 ppm/°C and the output current range of the stimulus generation circuit is 200 nA~20 μA, which can be applied to various electrodes impedance ranging from 1 Kohm to 3 Mohm. It can be seen from the simulation results that the IC can measure the electrode impedance with high accuracy and the evaluation error is less than 1%.

Acknowledgment

This work is supported by National Key Research and Development Program (Grant No. 2024YFF1206504) and Beijing Natural Science Foundation- Huairou Innovation Joint Fund (Grant No. L245012).

References

[1] H. J. Baek, H. J. Lee, Y. G. Lim and K. S. Park, "Investigations of capacitively-coupled EEG electrode for use in brain-computer interface," 2012 IEEE International Conference on Systems, Man, and Cybernetics (SMC), Seoul, Korea (South), 2012, pp. 278-282.

[2] B. Taji, S. Shirmohammadi and V. Groza, "Measuring skin-electrode impedance variation of conductive textile electrodes under pressure," 2014 IEEE International Instrumentation and Measurement Technology Conference (I2MTC) Proceedings, 2014, pp. 1083-1088.

[3] Y. Zhou, Z. Wang, K. Wang and R. Wang, "A Closed-loop Controlled Neural Stimulator with High Voltage Compliance in 0.18-μm Low Voltage CMOS Technology," 2019 IEEE International Conference on Integrated Circuits, Technologies and Applications 2019, pp. 158-159.

[4] L. Yao, J. Zhao, P. Li, R. -F. Xue, Y. P. Xu and M. Je, "A 20Vcompliance implantable neural stimulator IC with closed-loop power control, active charge balancing, and electrode impedance check," 2014 IEEE Asian Solid-State Circuits Conference (A-SSCC), 2014, pp. 201-204.

[5] P. E. Golubkov, E. A. Pecherskaya, et al., "Automated System for Bioimpedance Measuring," 2018 19th International Conference of Young Specialists on Micro/Nanotechnologies and Electron Devices (EDM), Erlagol, Russia, 2018, pp. 6403-6406.

[6] B. Ibrahim and R. Jafari, "Cuffless Blood Pressure Monitoring from an Array of Wrist Bio-Impedance Sensors Using Subject-Specific Regression Models: Proof of Concept," in IEEE Trans. on Biomedical Circuits and Systems, vol. 13, no. 6, pp. 1723-1735, Dec. 2019.

[7] E. J. Earley, E. Mastinu and M. Ortiz-Catalan, "Cross-Channel Impedance Measurement for Monitoring Implanted Electrodes," Int. Conf. of the IEEE Eng. in Med. & Bio. Society (EMBC), 2022, pp. 4880-4883.

[8] Z. Lu, W. Chen, et al., "A High-voltage Tolerant and Current-accurate Neural Stimulator Based on A Low-voltage CMOS Process," IEEE Asia Pacific Conf. on Circ. and Syst. (APCCAS), 2024, pp. 562-565.

[9] T. Yousefi, A. Dabbaghian and H. Kassiri, "Motion-Affected Electrode-Tissue Interface Characterization for Ambulatory EEG Recording," 2020 42nd Annual International Conference of the IEEE Engineering in Medicine & Biology Society (EMBC), 2020, pp. 4479-4482.

[10] U. A. Aregueta-Robles, Y. L. Enke, et al., "Subthreshold Electrical Stimulation for Controlling Protein-Mediated Impedance Increases in Platinum Cochlear Electrode," in IEEE Transactions on Biomedical Engineering, vol. 67, no. 12, pp. 3510-3520, Dec. 2020.

[11] R. Jacob Baker. CMOS: Circuit Design, Layout, and Simulation, 4th Edition, John Wiley & Sons, Inc. 2019: 48-51.

2025 The 10th International Conference on Integrated Circuits and Microsystems

A Fast Transient and High PSRR LDO with 62 ns Settling Time

Ruijie Wang
School of Integrated Circuits
Jiangnan University
Wuxi, China
6231916010@stu.jiangnan.edu.cn

Zhiguo Yu
School of Integrated Circuits
Jiangnan University
Wuxi, China
yuzhiguo@jiangnan.edu.cn

Xiaofeng Gu*
School of Integrated Circuits
Jiangnan University
Wuxi, China
xgu@jiangnan.edu.cn

Abstract—**This paper presents a low-dropout regulator (LDO) with fast transient response and high power supply rejection ratio (PSRR). The proposed LDO employs a error amplifier (EA) with high gain class-AB amplifier and incorporates a transient enhancement module with undershoot/overshoot suppression circuit. This transient enhancement module accelerates the charging/discharging process of the power transistor gate, effectively suppressing the undershoot /overshoot voltage amplitudes of the LDO output, and significantly reducing transient response time. Additionally, a feedforward circuit is introduced to suppress the noise ripple from the power supply. This work is designed in 55 nm CMOS process with an input voltage of 1.5 V and an output voltage of 1 V. Simulation results indicate that the proposed LDO exhibits good loop stability for load current ranging from 0 to 100 mA. When the load steps between 0 and 10 mA, the proposed LDO output settling times are 61 ns and 62 ns, with undershoot and overshoot voltages of 38 mV and 41 mV, respectively. The load regulation is 0.023 mV/mA, and the PSRR reaches -50 dB at 100 kHz.**

Keywords—Low Dropout Regulator, Transient Response Enhancement, Settling time

I. INTRODUCTION

In modern electronic systems, such as consumer electronics, industrial control, and portable devices, low-dropout regulator (LDO) serves as a core component of power management, and provides stable power supply to sensitive loads including analog front-ends and mixed-signal circuits. The transient response and Power Supply Rejection Ratio (PSRR) of LDO directly dictates system stability and reliability.

Traditional LDO often requires a larger compensation network to achieve a higher PSRR, which tends to slow down the dynamic adjustment of the system, leading to a decrease in response speed. Consequently, jointly improving the transient response and PSRR of LDOs has become a challenge in LDO research.

In reference [1], a method to enhance the PSRR of fully integrated LDO is proposed. This method employs an optimized feedforward current technique to improve the PSRR of LDO. However, the transient response still requires improvement. In reference [2], a high-gain LDO is proposed, which employs Nested Miller Compensation (NMC) technique to stabilize the

system, ensuring the loop stability of LDO. However, other performance improvement is not significant.

This article proposes an LDO with fast transient response and high PSRR. The proposed LDO employs a high-gain operational amplifier as the error amplifier, and proposes a transient enhancement circuit and a PSRR enhancement circuit. The proposed transient enhancement module can rapidly lower or raise the gate voltage of the power transistor, thereby enhancing or weakening its conduction capability, which significantly improves the transient response of the LDO. Additionally, the proposed PSRR enhancement circuit incorporates a feedforward circuit, which suppresses power supply noise in the low-frequency range.

The remainder of this article is organized as follows: Section II describes the overall circuit architecture. Section III describes the implementation of transient enhancement and PSRR enhancement circuit. Section IV presents simulation results, and Section V gives the conclusion.

II. MAIN CIRCUIT IMPLEMENTATION

The proposed LDO with fast response and high PSRR is illustrated in Fig. 1. The proposed LDO primarily consists of five components: a bias circuit, an error amplifier(EA), a power transistor, a transient enhancement circuit, a PSRR enhancement circuit, and a resistive feedback network. Built upon a high-gain class-AB amplifier, this circuit incorporates a transient enhancement circuit that dynamically modulates the drive strength of the power transistor, significantly boosting transient response. Additionally, a PSRR enhancement module through the feedforward circuit to eliminate the power supply ripple, thereby improving the PSRR performance of the LDO circuit.

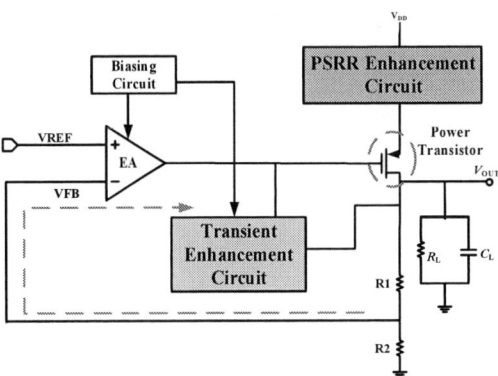

Fig. 1. Proposed LDO Circuit System Architecture.

The Joint Project of Yangtze River Delta Community of Sci-Tech Innovation (2022CSJGG0402) and the Key Research Project of Jiangsu Province, China (BE2023019-3).

Traditional EAs are widely favored for their wide output swing. However, their DC gains fail to meet LDO requirements. Therefore, to achieve faster transient response in LDO, EA must possess higher DC gain. The proposed LDO employs a high-gain EA for the LDO, as illustrated in Fig. 2. The bias circuit, composed of MN1–MN5 and MP1– MP2, generates a stable bias current (I_{bias}) that sets the operating point for both the EA and the power transistor, ensuring circuit stability under power supply fluctuations and temperature variations [3]. The EA features a differential input stage, with MP3-MP4 serving as the active load. The ultra-high output resistance of the current-mirror load significantly enhances the amplifier gain A_{EA}.

$$A_{EA} = g_{m,EA} \times r_{o,EA} \tag{1}$$

Compared to traditional EA, the high-gain class-AB EA has two voltage controlled current source (MP7 and MP8). Under static conditions, most of the drain current of MN10 flows into the current source MP8, thereby reducing the output current. Transistors MP5-6 and MN7-8 detect the input differential

voltage and control the two voltage-controlled current sources MP7 and MP8. The current reduction in MP8 equals the additional current increase in MP9. Therefore, the change in total current in MP8 is the sum of the changes in MP8 and MN10. Assuming the W/L of MP3 and MP9 is α, the W/L of MP4 and MP11 is β, the W/L of MP8 is δ, the W/L of MN10 is $(\alpha+\beta)$, and the W/L of MN8 is γ, the transconductance G_m and output resistance R_{OUT} of are as follows:

$$G_m = \frac{\alpha+\beta}{\beta+\delta+\gamma} \times (g_{mn9,10}+g_{mn7,8}) \times \frac{\alpha}{\beta} + \frac{\gamma}{\beta+\delta+\gamma} \times (g_{mn9,10}+g_{mn7,8}) \times \frac{\alpha}{\beta} \times \frac{\delta}{\gamma}$$
$$= (g_{mn9,10}+g_{mn7,8}) \times \frac{\alpha}{\beta} + \frac{\beta+2\delta}{\beta+\delta+\gamma} \tag{2}$$

$$R_{OUT} = (r_{o,p10} // r_{o,n14}) = \frac{2}{(\lambda_{p10}+\lambda_{n14})I_t} \times \frac{\beta+\delta+\gamma}{\alpha} \tag{3}$$

where I_t is the tail current, and λ is the channel length modulation coefficient.

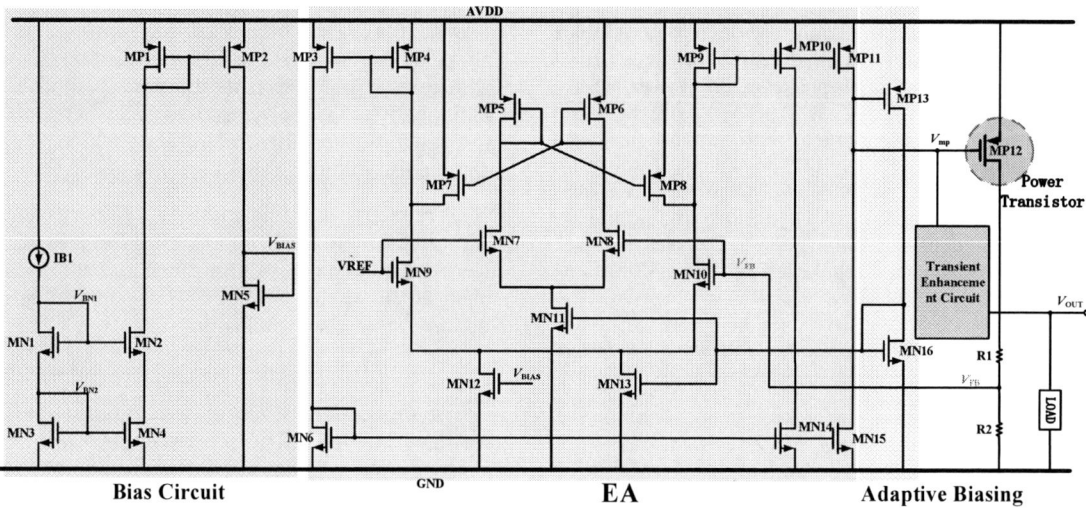

Fig. 2. Class-AB High-Gain Error Amplifier.

Compared to conventional EA, the high-gain class-AB EA enhances both G_m and R_{OUT}, resulting in a significant increase in DC gain. This improvement accelerates the LDO loop's response speed to load step changes, thereby reducing undershoot and overshoot voltages. MP13 and MN16 serve as adaptive bias circuits. They sense the load current by replicating the transmission transistor (MP13). MP13 shares the same gate-source voltage with the transmission transistor (MP12) and can track the load current (I_L). The circuit's bias current has a smaller current copying ratio compared to the EA's bias current, allowing the circuit to operate at a smaller static current [4].

III. TRANSIENT RESPONSE AND PSRR ENHANCEMENT CIRCUIT IMPLEMENTATION

A. Transient Enhancement Circuits

The key to improving the transient response of LDO lies in the rapid adjustment of the gate charge of the power transistor. Traditional LDOs often employ a single adaptive bias circuit to enhance the transient response speed. However, during the

discharge phase, low impedance and high current are required, while the charging phase necessitates a high swing and fast charging. The conflicting demands on the conduction path's impedance and current sourcing/sinking capability force design compromises, limiting both charge/discharge speeds and thus capping the LDO's maximum achievable transient performance.

The transient enhancement circuit proposed in this study includes the Undershoot Suppression Circuit (USC) and the Overshoot Suppression Circuit (OSC), as shown in Fig. 3. The transient enhancement circuit's core function is to modulate the gate drive strength of the power transistor by rapidly lowering or raising the gate voltage, thereby suppressing the variation of the output voltage. USC (MP14-MP18, MN17-MN20, R_3, and C_1) detects changes of V_{OUT}, and combined with dynamic biasing to boost the gate drive current of the power transistor. This results in a rapid rise in the gate voltage of the power transistor, thereby elevating the output voltage to the target value. OSC detects changes at the output terminal, enabling the power transistor to discharge rapidly, which causes a swift decrease in

979-8-3315-8850-2/25 $31.00 © 2025 IEEE

power transistor's gate voltage and brings the V_{OUT} back to the normal value.

The proposed LDO operates in three distinct modes corresponding to load conditions:

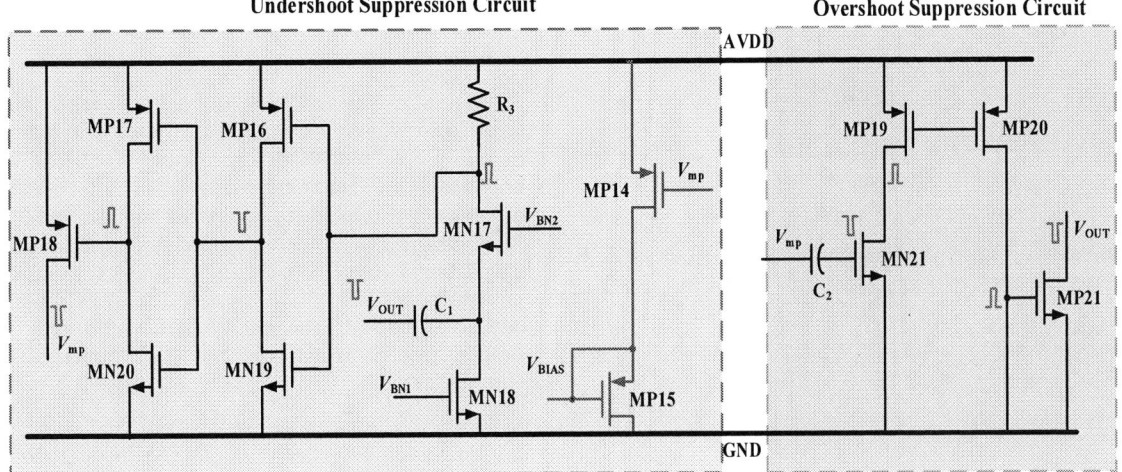

Fig. 3. Proposed transient enhancement circuit.

1) When the load current is stable, the circuit operates in a steady state. Both MN18 in the USC and the MN21 in the OSC are in the cutoff state, and the dynamic bias circuit is not operational. The LDO maintains a stable V_{OUT} via the EA and the power transistor, achieving low static power consumption.

2) When the LDO transitions from light-load to heavy-load, the V_{OUT} decreases rapidly. This variation is coupled to the drain of MN18 via C1, activating MN18. Once MN18 is activated, R3 and MN17 form a low-impedance discharge path, inducing a drop in the drain voltage of MN17. This drop is further amplified by the inverter consisting of MP16 and MN19. After being inverted by the inverter composed of MP17 and MN20, MP18 is activated, discharges the gate of the power transistor, and causes the gate voltage to decrease rapidly.

This operation increases the gate-source voltage of the power transistor: $V_{GS} = V_{IN} - V_{mp}$, enhancing the conduction of the PMOS power transistor and quickly pulling the output voltage back to the target value. Combined with the adjustment of the tail current source of the EA by MP14 and MP15, the LDO achieves rapid response during light to heavy load transients with minimal undershoot voltage.

3) When the LDO transitions from heavy-load to light-load, the V_{OUT} rapidly spikes, and the gate voltage of the power transistor quickly decreases. This voltage variation is coupled to the gate of transistor MN21 via capacitor C2, activating MN21. Once MN21 is activated, the current is replicated to the gate of transistor MP21 via the current mirror formed by MP19 and MP20, causing the LDO output voltage to decrease rapidly and return to the target value. This enables the circuit to achieve a rapid response with minimal overshoot voltage during heavy to light load transients.

B. PSRR Enhancemen Circuit

Approaches to enhance the PSRR of LDOs include RC filtering [5], cascaded LDO topologies, and feedforward circuits. The PSRR enhancement circuit implements a feedforward

circuit between the LDO input and the substrate of the power transistor, eliminating reliance on off-chip bulk capacitors and thereby facilitating monolithic integration. The schematic of the feedforward circuit is shown in Fig.4.

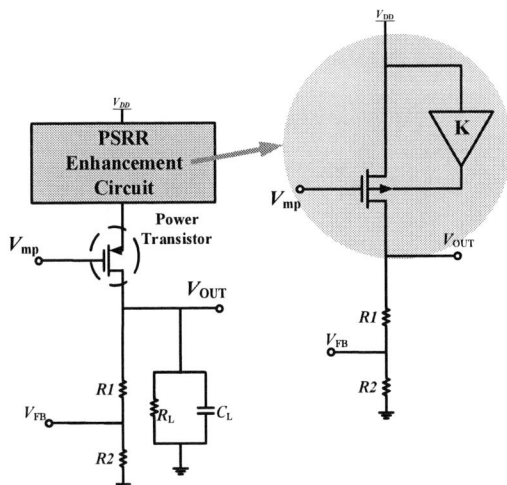

Fig. 4. Schematic diagram of the feedforward circuit.

Assuming the feedforward coefficient is K, then the feedforward coefficient K can be defined as:

$$K = 1 + \frac{1}{\chi} + \frac{1}{\chi A_{PT}} - \frac{2C_{gs}C_{gd}g_{EA}}{g_{mb}(C_{gs}+C_{gd})^2} - \frac{C_{gs}}{\chi(C_{gs}+C_{gd})} \quad (4)$$

where A_{PT} is the Power Transistor's intrinsic gain, χ is the body effect coefficient of the PMOS transistor, C_{gs} and C_{gd} denote the parasitic capacitances of the PMOS transistor, and g_{EA} refers to the equivalent output transconductance of the error amplifier [9].The small-signal model of the LDO PSRR enhancement module is shown in Fig. 5.

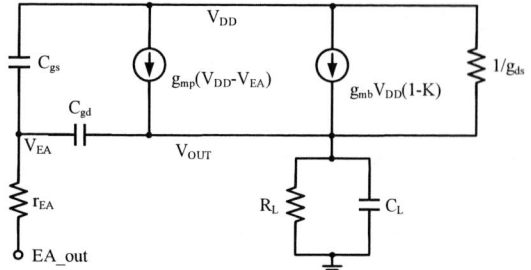

Fig. 5. PSRR Enhancement Circuit Small Signal Model.

From the above figure, the transfer function of LDO PSRR is:

$$PSRR = \frac{C_{gs}C_{gs}s^2 + [X(\Delta K - g_m) + g_{ds}C_{gd}]s^2 + g_{EA}\Delta Ks + g_{EA}\omega_0 \Delta K}{C_L Xs^2 + XY + C_{gd}g_m + g_{EA}C_L]s^2 + g_{EA}Ys + A_v g_m g_{EA}\omega_0} \quad (5)$$

$$\Delta K = g_m + g_{ds} + g_{mb} - Kg_{mb} \quad (6)$$

$$X = C_{gs} + C_{gd} \quad (7)$$

$$Y = g_{ds} + g_L \quad (8)$$

where g_{mp}, g_{ds}, and g_{mb} represent the transconductance, output transconductance, and body transconductance of the PMOS power transistor, respectively, g_m is PMOS power transistor's transconductance, A_v is the equivalent output impedance and loop gain of LDO, and K is the feedforward coefficient of the circuit [6]. From the (5), it can be seen that by appropriately tuning K to satisfy $g_{mp} + g_{ds} + g_{mb} = Kg_{mb}$, the PSRR of LDO can be enhanced.

The PSRR enhancement structure implemented in this article consists of two operational amplifiers, two pseudo-resistors, and four capacitors, as shown in Fig. 6. The operational amplifier comprises transistors MP22-MP24 and MN22-MN26, the other operational amplifier comprises MP25-MP27 and MN27-MN31. The operational amplifiers function as two differential inverting amplifiers, which utilizing PMOS current mirrors (M1-M4) as loads, and replacing traditional resistive loads.

M1, M2, M3 and M4 are back-to-back connected PMOS transistors, which operate in the subthreshold region, effectively functioning as 'ultra-high impedance pseudo-resistors'. The high impedance characteristic of these pseudo-resistors can block direct current (DC) while allowing the DC levels at nodes (V_1, V_2, V_3, V_4, V_{BULK}) to be clamped by the bias voltage V_{BIAS}. For alternating current signals, the pseudo-resistors maintain high impedance, ensuring unimpeded transmission of alternating current signals. Capacitors C_3 and C_4 serve to sense power supply ripple, while C_5 and C_6 implement feedforward amplification.

The inverting terminal (node 1) of the first stage amplifier is clamped to the DC level V_{BIAS} due to the operational amplifier's 'virtual short', resulting in $V_1 = V_{BIAS}$. The DC level at the output of the first stage amplifier (node 2) is determined by the linear operating characteristics of the operational amplifier and matches the input DC level ($V_2 = V_{BIAS}$). Similarly, the non-inverting terminal of the second amplifier is connected to V_{BIAS}, and the inverting terminal (node 3) has a DC level equal to V_{BIAS}, resulting in $V_3 = V_{BIAS}$, with the output terminal (node 4) also equal to V_{BIAS}. All DC voltages at nodes 1, 2, 3, 4, and V_{BULK} are clamped to V_{BIAS}, achieving $V_{BIAS} = V_1 = V_2 = V_3 = V_4 = V_{BULK}$. Then the expression for the feedforward coefficient $K1$ of this module is:

$$K1 = \frac{C_3 C_4}{C_5 C_6} \quad (9)$$

Select appropriate capacitance values for capacitors C3-C6 to ensure that $K1=K$, thereby improving the PSRR of LDO.

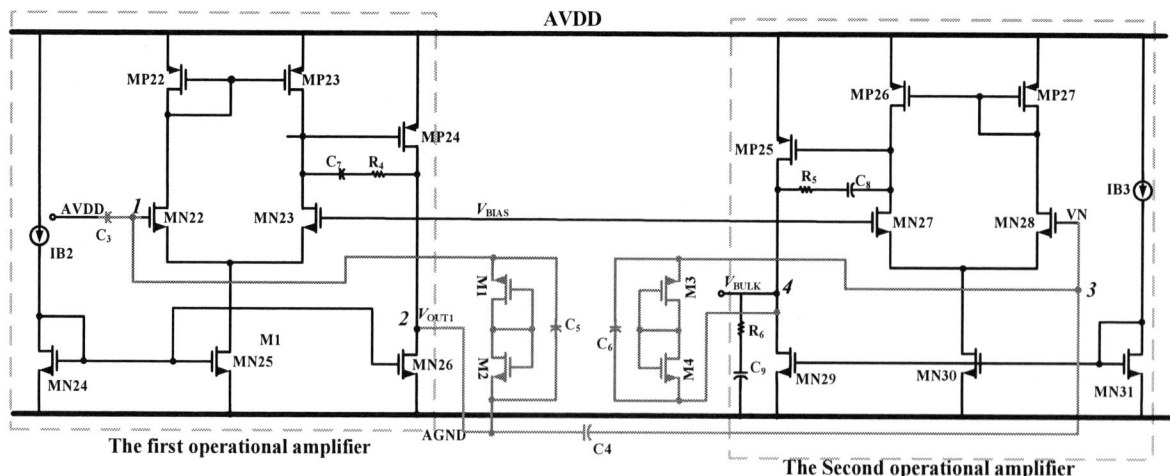

Fig. 6. PSRR Enhancement Circuit Diagram.

979-8-3315-8850-2/25 $31.00 © 2025 IEEE

IV. SIMULATION RESULTS

This article implements a fast response, high PSRR LDO in 55 nm process. The performance of the proposed LDO was verified via Spectre simulator on the Cadence Virtuoso, with input voltage of 1.5 V and output voltage of 1 V, and load current ranges from 0 to 100 mA.

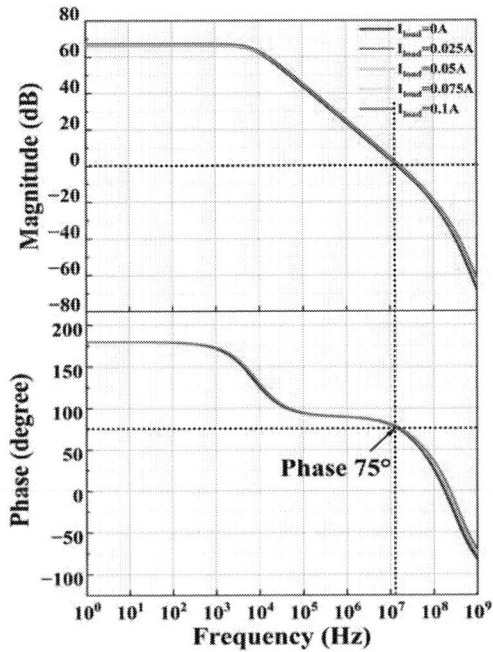

Fig. 7. Gain and Phase Margin Simulation of Proposed LDO.

Simulation results are as follows: Figure 7 illustrates the gain and phase margin curves of the LDO circuit across load currents ranging from 0 to 100 mA. It can be observed that the low-frequency gain of the LDO circuit remains around 66 dB across this load range, with phase margin of 75°. The proposed LDO exhibits excellent loop stability, satisfying the design criteria for loop stability.

In the load transient simulation, when the load steps from 100 nA to 10 mA, the undershoot settling time is 61 ns,

corresponding to an undershoot voltage of only 38 mV. Conversely, when the load steps from 10 mA to 100 nA, the overshoot settling time is 62 ns, with an overshoot voltage of 41 mV, as depicted in Fig. 8. This performance is enabled by a transient enhancement circuit, which reduces the undershoot/overshoot settling time by approximately 80% and decreases the voltage magnitude by over 60%, compared to traditional LDOs without transient enhancement.

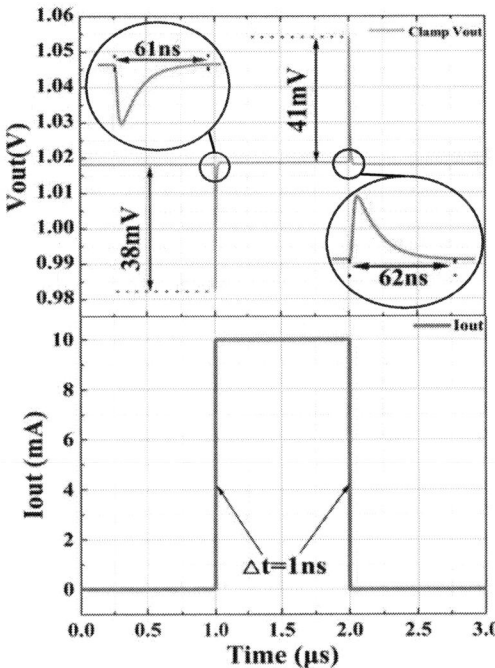

Fig. 8. Output Voltage Curve of LDO under 10mA Load Step.

Figure 9 illustrates load regulation simulation results for the LDO circuit. The proposed LDO achieves a load regulation of 0.023 mV/mA—implying V_{OUT} varies by only 23.2 mV as I_{LOAD} sweeps from 0 to 100 mA. As shown in Fig. 10, the LDO achieves approximately -50 dB PSRR from 10 Hz to 100 kHz, indicating strong suppression of low-frequency power supply noise. This attenuates >90% of input ripple in this band. Table I compares the proposed LDO with other works.

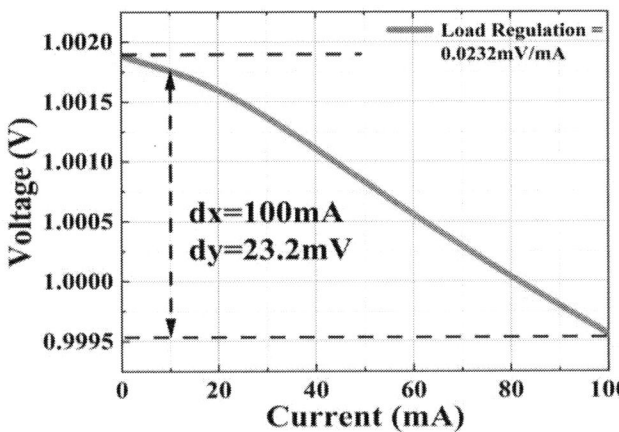

Fig. 9. Simulation Results of Load Regulation for LDO.

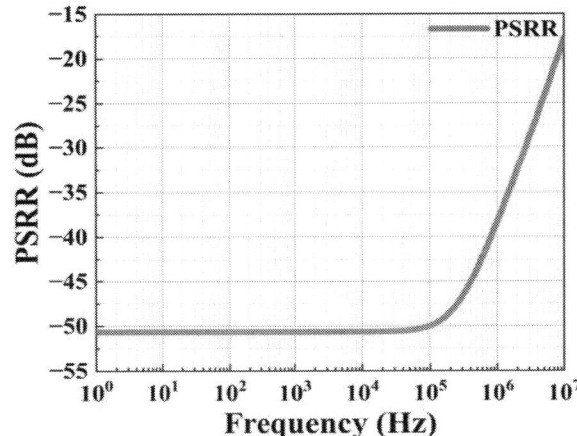

Fig. 10. Simulation Results of PSRR for LDO.

979-8-3315-8850-2/25 $31.00 © 2025 IEEE

TABLE I.   PERFORMANCE COMPARISON

	[7]	[8]	[9]	[10]	[11]	THIS WORK
Process (nm)	150	350	180	10	180	55
V_{DD} (V)	1.05 - 1.2	2.7 - 3.3	1.8	1.8	1.2	1.5
V_{OUT} (V)	0.9	2.5-3.1	1.2	0.95-1.75	1	1
Dropout (mV)	300	200	600	850	200	500
IL (MAX) (mV)	20	100	50	21	100	100
PSRR (dB)	-52@10 KHz	-41@10 KHz	-37@10 KHz	-40@1 MHz	-50@10 KHz	-50@10 KHz
Settling time (µs)	0.0012	0.7	0.1	0.004	3.6	0.062
Overshoot voltage (mV)	88	170	N/A	46	220	41
Undershoot voltage (mV)	35	255	N/A	40	220	38
LOAD Regulation (mV/mA)	N/A	0.06	0.12	N/A	0.1	0.023

As shown in Table I, the proposed LDO demonstrates significant advantages in key parameters including settling time, overshoot voltage, undershoot voltage, and PSRR.

V.   CONCLUSION

This paper presents an LDO implemented in 55 nm CMOS process. The proposed LDO builds upon a high-gain EA, integrating a transient response enhancement module and a PSRR enhancement module. By employing an undershoot /overshoot compensation topology and a feedforward circuit ripple cancellation technique, the design achieves synergistic improvement in transient performance and power supply noise suppression capabilities. As a general-purpose power management IP module, the proposed LDO can be widely deployed in Internet of Things (IoT) terminals, wearable devices, and edge computing chips, providing high robustness power supply solutions for mixed-signal systems.

REFERENCES

[1] S. Wu, Z. Guo, Z. Xue, X. Wang, and L. Geng, "A High Power Supply Rejection and Low Noise Low-Dropout Regulator for Voltage Controlled Oscillator," 2021 IEEE International Conference on Integrated Circuits, Technologies and Applications (ICTA), Zhuhai, China, 2021, pp. 84-85.

[2] J. K. Mukre, J. Kodethoor, M. Prakash, P. Anurup, and R. Seethur, "A High Gain 0.0831 µ V/mA Load Regulated Capacitorless LDO with Fast Loop and Nested Miller Compensation (NMC)," 2025 International Conference on Electronics, Information, and Communication (ICEIC), Osaka, Japan, 2025, pp. 1-5.

[3] J. Roh, "High-Gain Class-AB OTA with Low Quiescent Current," Analog Integr Circ Sig Process, vol. 47, pp. 225-228, February. 2006.

[4] J. S. Kim, K. Javed, K. H. Min, and J. Roh, "A 13.5-nA Quiescent Current LDO With Adaptive Ultra-Low-Power Mode for Low-Power IoT Applications," in IEEE Transactions on Circuits and Systems II: Express Briefs, vol. 70, no. 9, pp. 3278-3282, September. 2023.

[5] M. Khan and M. H. Chowdhury, "Capacitor-less Low-Dropout Regulator (LDO) with Improved PSRR and Enhanced Slew-Rate," 2018 IEEE International Symposium on Circuits and Systems (ISCAS), Florence, Italy, 2018, pp. 1-5.

[6] F. L. Aviles, J. Torres and E. S. Sinencio, "A High Power Supply Rejection and Fast Settling Time Capacitor-Less LDO," in IEEE Transactions on Power Electronics, vol. 34, no. 1, pp. 474-484, January. 2019.

[7] N. Liu and D. Chen, "A Transient-Enhanced Output-Capacitorless LDO With Fast Local Loop and Overshoot Detection," in IEEE Transactions on Circuits and Systems I: Regular Papers, vol. 67, no. 10, pp. 3422-3432, October. 2020.

[8] X. Ming, J. J. Kuang, X. Gong, Z. Lin, J. Xiong, and Y. Qin, "A Fast-Transient Capacitorless LDO With Dual-Paths Active-Frequency Compensation Scheme," in IEEE Transactions on Power Electronics, vol. 37, no. 9, pp. 10332-10347, September. 2022.

[9] X. Xin, P. Luo, H. Wang, J. Huang, X. Chen, and C. Liu, "A 96-nA Quiescent Current LDO With Embedded BGR Using Adaptive Pole Tracking and Adaptive Transconductance Technique," in IEEE Transactions on Circuits and Systems II: Express Briefs, vol. 72, no. 1, pp. 318-322, January. 2025.

[10] J. Jung, J. H. Choi, K. J. Roh, J. Park, W. M. Lim, and T. S. Kim, "A 4 ns Settling Time FVF-Based Fast LDO Using Bandwidth Extension Techniques for HBM3," in IEEE Journal of Solid-State Circuits, vol. 59, no. 10, pp. 3307-3316, October. 2024.

[11] C. Răducan, A. T. Grăjdeanu, C. S. Plesa, M. Neag, A. Negoiță, and M. D. Țopa, "LDO With Improved Common Gate Class-AB OTA Handles any Load Capacitors and Provides Fast Response to Load Transients," in IEEE Transactions on Circuits and Systems I: Regular Papers, vol. 67, no. 11, pp. 3740-3752, November. 2020.

Automatic Generation Method of Test Vectors for FPGA Interconnect Resources Based on Self-adaptive Ford-Fulkerson Algorithm

Liang Zhou
School of Electronic Information
Northwestern Polytechnical
University
Xi'an, China
zhouliang772@126.com

Wei Zhou
School of Electronic Information
Northwestern Polytechnical
University
Xi'an, China
zhouwei@nwpu.edu.cn

Mingzhe Li
Business Department of FPGA
Beijing Microelectronics
Technology Institute
Beijing, China
limz92@163.com

Kun Guo
Business Department of FPGA
Beijing Microelectronics
Technology Institute
Beijing, China
15701680343@163.com

Ding Zhang
Business Department of FPGA
Beijing Microelectronics
Technology Institute
Beijing, China
ddding_m@163.com

Yan Huang
Business Department of FPGA
Beijing Microelectronics
Technology Institute
Beijing, China
huangyan60@mail.hfut.edu.cn

Abstract—**SRAM-based FPGA is widely used in military and aerospace applications because of its flexibility and programmability. Therefore, the reliability of FPGA is particularly important. The interconnect resources (IR) of FPGA pose a critical challenge in testing due to their large scale and complex structure. This article first analyzes the structure characteristics of interconnect resources and constructs a transfer model in ultra-large-scale FPGA interconnect. Based on the maximum flow problem in graph theory, an adaptive Ford-Fulkerson algorithm is proposed, which solves the technical difficulty of constructing effective test graphics of interconnect resources. Finally, this article implements an automatic generation method of IR test vectors via the XDL development tool, and forms a high-coverage test vector set for 70-million-gate FPGA's complex IR, ensuring the high reliability and security of FPGA applications.**

Keywords—*SRAM-based FPGA; interconnect resources; adaptive Ford-Fulkerson algorithm; XDL*

I. INTRODUCTION

Since the first FPGA XC2064 came out by Xilinx in 1985, FPGA has developed almost forty years, which is progressively evolving towards systematic and platform-oriented development. Due to the programmability and reconfigurability, FPGA becomes a preferred solution for designers to implement digital circuits and systems[1]. The application of FPGA has extended from communication electronics to artificial Intelligence, automotive electronics, industrial control and so on[2].

The flexibility of FPGA's application is ensured by its enormous number of internal programmable logic resources, interconnect resources and input/output resources, while the programmable interconnect resources is the key part of FPGA[3]. The configuration software connects logic units together with interconnect resources according to user's settings to achieve specific function. Interconnect resources take about 70%-80% chip area and 50%-60% signal delay of FPGA[4]. Therefore, the interconnect resources directly affect FPGA's function, and thus IR test becomes an important part of FPGA test.

These years, FPGA IR test method becomes a research hotspot. Xiang Fen Wang implemented the testing of hex lines and double lines of Xilinx Spartan-3 FPGA based on the XDL tools[5]. Fahmy Hafriz bin Mohamed Sultan designed the at-speed clocks generate method and the pipelined scan enable signals to implement at-speed test to capture marginal open defect of Altera Stratix V FPGA interconnect resources[6]. Weikun Xie proposed a deep-priority algorithm based on graph-based models and improved priority algorithms to intelligently wire the Xilinx's Kintex-7 series FPGA interconnect resources[7]. Recent studies have shown that with the continuous updating for the interconnection resource architecture of ultra-large-scale FPGAs, the testing technology and theory of interconnection resources are rapidly developing. More and more methods such as built-in self-test, deep-priority algorithm, and routing algorithm are applied in interconnect resource testing. To address the challenges of large quantities and complex structures of interconnection resources for ultra-large-scale FPGA, an improved self-adaptive Ford-Fulkerson algorithm is put forward in this article for interconnect resources test of Virtex-7 series FPGA, and an automatic routing software is developed based on this method, which can generate interconnect test vector set with 99.74% coverage.

II. INTERCONNECT RESOURCES STRUCTURE

Interconnect resources of ultra-large-scale FPGAs include interconnects in programmable modules as well as between modules, which can be generally called as Programmable

Interconnect (PI) resources. They consist of programmable switch matrixes and interconnect lines[8]. Programmable switch matrixes provide switch function between interconnect lines by using configuration information stored in internal SRAM to control transfer switches. Generally, interconnect resources architecture is made up of general interconnect matrixes, local interconnect matrixes and interconnect lines. General and local interconnect matrixes are both programmable. General interconnect matrixes connect different CLB modules, while local interconnect matrixes connect slices in CLB modules with general interconnect matrixes.

As shown in Fig. 1, a single interconnect switch matrix includes general interconnect matrixes and local interconnect matrixes. It adopts hierarchical routing resources. These routing resources are mainly general routing resources, which locate in the horizontal and vertical routing channels between switch matrixes. The routing resources have different kinds like long line, double line, quad line, hex line and so on[9].

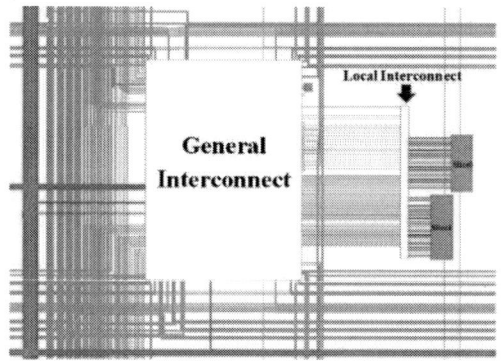

Fig. 1.   Interconnect switch matrix

Local interconnect matrix is deployed for local data exchange with four functions: First, it undertakes communication between Slices, general interconnect matrixes and nearby interconnects. Second, it provides output and feedback pathways for Slices. Third, it determines direct interconnect pathways between nearby CLB modules, and provides high speed communication among local and general interconnect matrixes of CLB modules. Fourth, it transmits clock signals to Slices.

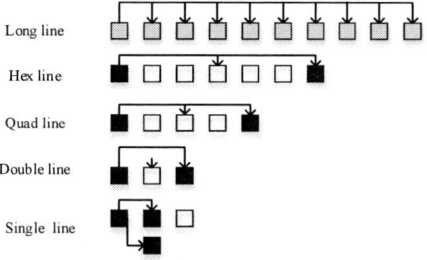

Fig. 2.   Hierarchical routing resources

Meanwhile, general interconnect matrix connects CLB modules with other modules, and it is the largest, the most complicate and important interconnect module. It has four functions: First, it deals with interconnect between CLB modules and nearby interconnect matrixes, CLB modules and other modules on the chip. Second, it takes on routing function of CLB module's most input and output signals. Third, it provides the shortest communicate pathway with nearby four general interconnect matrixes. Fourth, it also provides interconnect communication channels in vertical and horizontal direction. Classification of hierarchical routing resources is shown as Fig. 2.

III. Transmission Model of Interconnect Resources

Interconnect resources of ultra-large-scale FPGAs often start from a certain switch matrix, stride across multiple switch matrixes and end with another one. According to the number of crossed switch matrixes, the stride-across interconnect lines can be separated to different kinds like single line, double line, hex line, long line and so on. Different models of SRAM-based FPGA contain different kinds of stride-across interconnect lines. Assuming p as the characteristic parameter of different stride-across interconnect lines, the value range is shown as Table I.

TABLE I.   Characteristic Parameters of Different Kinds of Stride-across Interconnect Line Kinds

Stride-across interconnect line kinds	Characteristic parameter p
Single line	1
Double line	2
Quad line	4
Hex line	6
Long line	12

To establish a general model, a FPGA with m*n switch matrixes is assumed, and put this FPGA into a two-dimensional coordinate system. Then each single switch matrix can be described by coordinate as shown in Fig. 3.

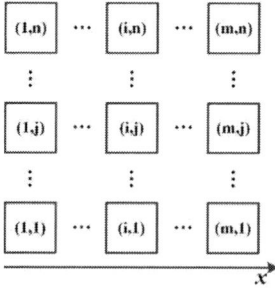

Fig. 3.   Mathematical model of m*n FPGA

$(x(i), y(j))$ represents the terminal switch matrix coordinate of a stride-across interconnect line which starts from switch matrix (i, j). And it is calculated as shown in (1) ~ (8), which is proposed according to transmission law of different kinds of interconnect resources.

Equation (1) describes the terminal switch matrix coordinate for left passing of different kinds of stride-across interconnect line:

$$(x(i), y(j)) = \begin{cases} (i-p, j) & (i \geq p+1) \\ (p-i+1, j) & (i < p+1) \end{cases} \quad (1)$$

979-8-3315-8850-2/25 $31.00 © 2025 IEEE   38

Equation (2) describes the terminal switch matrix coordinate for right passing of different kinds of stride-across interconnect line:

$$(x(i), y(j)) = \begin{cases} (i+p, j) & (i < m-p+1) \\ (2m-i-p+1, j) & (i \geq m-p+1) \end{cases} \quad (2)$$

Equation (3) describes the terminal switch matrix coordinate for up passing of different kinds of stride-across interconnect line:

$$(x(i), y(j)) = \begin{cases} (i, j+p) & (j < n-p+1) \\ (i, 2n-j-p+1) & (j \geq n-p+1) \end{cases} \quad (3)$$

Equation (4) describes the terminal switch matrix coordinate for down passing of different kinds of stride-across interconnect line:

$$(x(i), y(j)) = \begin{cases} (i, j-p) & (j \geq p+1) \\ (i, p-j+1) & (j < p+1) \end{cases} \quad (4)$$

Since the tilt interconnect resources are mainly composed of double lines, equation (5) ~ (8) summarizes the transmission laws of the interconnections in different tilt directions. Equation (5) describes the terminal switch matrix coordinate for left upside passing of double lines:

$$(x(i), y(j)) = \begin{cases} (3-i, j+2) & (i < 3, j < n-2) \\ (3-i, 2n-j-1) & (i < 3, j \geq n-2) \\ (i-2, j+2) & (i \geq 3, j < n-2) \\ (i-2, 2n-j-1) & (i \geq 3, j \geq n-2) \end{cases} \quad (5)$$

Equation (6) describes the terminal switch matrix coordinate for right upside passing of double lines:

$$(x(i), y(j)) = \begin{cases} (i+2, j+2) & (i < m-1, j < n-1) \\ (i+2, 2n-j-1) & (i < m-1, j \geq n-1) \\ (2m-i-1, j+2) & (i \geq m-1, j < n-1) \\ (2m-i-1, 2n-j-1) & (i \geq m-1, j \geq n-1) \end{cases} \quad (6)$$

Equation (7) describes the terminal switch matrix coordinate for left downside passing of double lines:

$$(x(i), y(j)) = \begin{cases} (3-i, 3-j) & (i < 3, j < 3) \\ (3-i, j-2) & (i < 3, j \geq 3) \\ (i-2, 3-j) & (i \geq 3, j < 3) \\ (i-2, j-2) & (i \geq 3, j \geq 3) \end{cases} \quad (7)$$

Equation (8) describes the terminal switch matrix coordinate for right downside passing of double lines:

$$(x(i), y(j)) = \begin{cases} (i+2, 3-j) & (i < m-1, j < 3) \\ (i+2, j-2) & (i < m-1, j \geq 3) \\ (2m-i-1, 3-j) & (i \geq m-1, j < 3) \\ (2m-i-1, j-2) & (i \geq m-1, j \geq 3) \end{cases} \quad (8)$$

While the algorithm needs pathfinding, the next level switch matrix can be found by the position of current switch matrix, therefore a measurability model for p long lines with known switch matrix position and interconnect relationship can be established. In application, a practical FPGA interconnect resource measurability model can be obtained by just input the scale of FPGA and adopted interconnect resource type.

IV. AUTOMATIC ROUTING METHOD OF INTERCONNECT RESOURCES

This article put forward a self-adaptive Ford-Fulkerson algorithm to achieve automatic routing of interconnect resources. For the network formed by FPGA interconnect resources, to reduce the number of configurations for interconnect resource test, a single configuration should traverse as many PIPs and interconnect lines as possible. To achieve high coverage test of ultra-large-scale FPGAs, the measurability model established for interconnect resources is converted to be flow network. This article transforms the problem of traversing the maximum quantity of PIPs and interconnect lines to the problem of finding the pathway which contains most edges and nodes in maximum flow network. Meanwhile, this article puts forward an improvement for Ford-Fulkerson algorithm, which can adaptively adjust the selection probability of different direction interconnect lines according to the cover rate of them, to improve the convergence speed of maximum flow algorithm.

A. Ford-Fulkerson Algorithm Theory

In the traffic network modeling diagram, the traffic volume can be regarded as flow. The flow is an abstract entity which starts at source nodes, transmit by edges and received at meeting nodes. The flow network is a directed graph $G = (V, E)$ with following features:

(1) Every edge e relates with a non-negative value which called capacity c;

(2) The source node s is single and contained in V.

(3) The meeting node t is single and contained in V.

Other nodes apart from s and t are called as internal nodes.

In a general form of flow network, neither edges enter the source node nor leave the meeting node. For each node, there must be at least one edge connected to it and the capacity of every edge is integer. If proper sources nodes and meeting nodes are chosen as well as appropriate capacity for every single edge, the network composed by FPGA interconnect resources can establish a general formed flow network.

The flow function f defines the network traffic from source node s to meeting node t. This function maps every single edge e to a non-negative real number. The value $f(e)$ expresses the traffic carried by edge e. A flow f should meet the two conditions shown below:

(1) Capacity condition: For any single edge, its traffic cannot exceed the whole network's capacity, which means for every $e \in E$, there should be:

$$0 \le f(e) \le c_e \qquad (9)$$

(2) Conservation condition: For any internal node v, which is not source node or meeting node, the total input traffic of this node is equal to total output traffic, which means apart from s and t, every node v should satisfy (10):

$$\sum_{\text{into } v} f(e) = \sum_{\text{out of } v} f(e) \qquad (10)$$

The value of a flow f in flow network can be written as $v(f)$, which is defined as the traffic generated at source node s as shown in (11):

$$v(f) = \sum_{\text{out of } s} f(e) \qquad (11)$$

For a particular flow network, a natural goal is to arrange the traffic to make full use of the available capacity. Therefore, the maximum flow problem can be finding the maximum flow for the flow network.

For a specified flow network G and its flow f, the residual network for G about f can be defined as G_f:

(1) G_f's node set is same as G's node set.

(2) For every single edge $e = (u, v)$ in G, if $f(e) < c_e$, then there exists residual units of capacity of $c_e - f(e)$. This edge also exists in G_f with capacity of $c_e - f(e)$. This kind of edge is called positive edge.

(3) For every single edge $e = (u, v)$ in G, if $f(e) > 0$, the flow can be push back to revoke it if necessary. Therefore, the edge $e' = (v, u)$ is also contained in G_f capacity of $f(e)$.

Therefore, for every single line $e = (u, v)$ in G, if $0 < f(e) < c_e$, the positive and negative edges are all exist in the residual map G_f. To make it different from capacity of corresponding edge in original flow network G, the capacity of edges contained in residual map G_f is called as residual capacity. While Ford-Fulkerson algorithm could find out the maximum flow of a flow network, its principle is shown as below:

Ford-Fulkerson algorithm

Initialize every edge e in flow network G as $f(e) = 0$
While there is a simple pathway for $s - t$ in remaining map G_f

    Set P as a simple pathway in G_f
    Set b as the maximum flow on P
    For P's every single edge (u, v)
        If $e = (u, v)$ is a forward edge then
            Increase $f(e)$ with b in G

        Else
            Reduce $f(e)$ with b in G
        Endif
    Endfor
    Get the new flow f'
    Refresh f as f'
    Refresh remaining map G_f as $G_{f'}$
Endwhile
Return f

Now the acquired flow f is a maximum flow of flow network G.

B. Improved Self-adaptive Ford-Fulkerson Algorithm

To reduce the number of bitstreams for interconnect test and improve interconnect resource test efficiency, this article proposed a self-adaptive Ford-Fulkerson algorithm. Standard Ford-Fulkerson algorithm does not consider the situation of covered interconnect lines at present, and leads to lots of repeat testing in interconnect resources test vectors.

To solve the problem above, this article adjusted path selection cost parameter of Ford-Fulkerson algorithm with self-adaptive method, according to the coverage of different direction interconnect resources. If the test coverage of a particular direction interconnect line is high in present test vectors, the selection probability of interconnect line routing in this direction will be reduced when choosing transmission pathway to next node. While if the test coverage of a particular direction interconnect line is low in present test vectors, the selection probability of interconnect line routing on this direction will be increased when choosing transmission pathway to next node. The probability calculation method is shown as (12):

$$P_i = \cos(\frac{\pi}{2} c_i) \times (P_{\max} - P_{\min}) + P_{\min} \qquad (12)$$

As shown above, P_i is the selection probability of particular direction routing. c_i means test coverage of a particular direction interconnect lines in present test vectors. $P_{\max}$ represents the maximum routing probability when choosing transmission pathway to next node, while $P_{\min}$ represents the minimum routing probability when choosing transmission pathway to next node.

C. Test Vector Generation Method Based on XDL

XDL is a powerful tool which could provide any character's program interface of FPGA[10]. On one hand, XDL can describe the complete design and configurations, i.e. the original information, layout and routing in FPGA. On the other hand, XDL can also restrict the system, or execute configuration of models or macro units in FPGA.

The test vector generation flow chart based on XDL is shown as Fig. 4. First, complete XDL program according to the tested FPGA interconnect resource structure and function. Then, convert the XDL files to NCD files. At last, convert the NCD

979-8-3315-8850-2/25 $31.00 © 2025 IEEE

files to BIT files, and load BIT files into tested FPGA to achieve test of target FPGA interconnect resources.

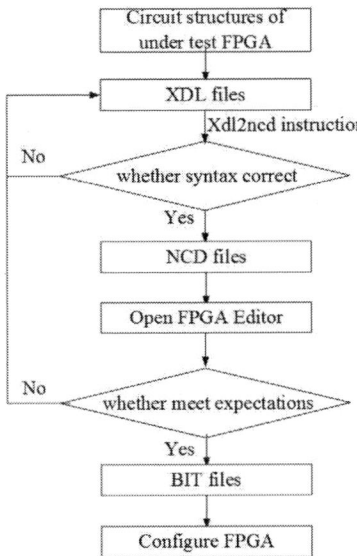

Fig. 4. XDL test vector generation flow chart

V. IMPLEMENTATION AND VERIFICATION

This article deploys Xilinx Virtex-7 series SRAM-based FPGA as target platform and verifies previously presented test method, combines self-adaptive Ford-Fulkerson algorithm to get test graph with high resource coverage. Fig. 5 to Fig. 7 are interconnect test vector layout and routing maps in different directions of XC7VX690T.

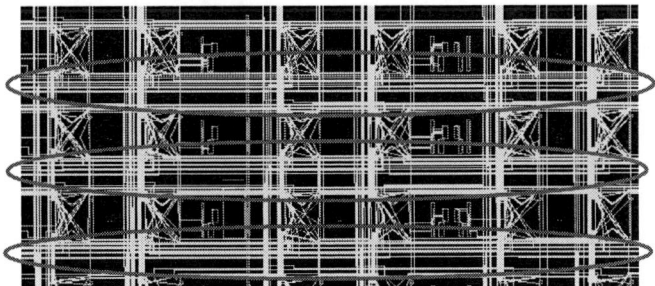

Fig. 5. Horizontal interconnect resource layout and routing map

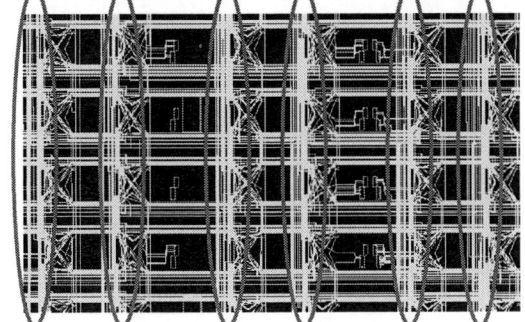

Fig. 6. Vertical interconnect resource layout and routing map

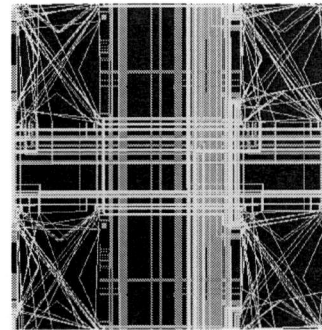

Fig. 7. Tilt interconnect resource layout and routing map

At present, the deep-priority algorithm is mainly used for testing the interconnect resources of FPGAs. Therefore, the deep-priority algorithm is taken as the traditional algorithm to be compared with the algorithm proposed in this paper. Table II compares the number of interconnect resource test vectors and test coverage about deploying traditional test vector generation method and self-adaptive Ford-Fulkerson algorithm.

TABLE II.   RESULT COMPARISON OF INTERCONNECT TEST VECTORS

	Numbers	Fault converage
Traditional method	53	86.21%
Adaptive Ford-Fulkerson algorithm	34	99.74%

As shown in Table II, the self-adaptive Ford-Fulkerson algorithm reduced the number of test vectors by 35.8%, while the fault coverage achieved 99.74%.

VI. CONCLUSION

This article analyzed architecture of ultra-large-scale FPGA, and concluded mathematical transmission model of its interconnect resources. A test method for ultra-large-scale SRAM-based FPGA interconnect resources is developed based on self-adaptive Ford-Fulkerson algorithm. By adding self-adaptive adjusting mechanism for path selection cost parameter of Ford-Fulkerson algorithm, the number of interconnect resource test bitstreams is greatly reduced. This article verified the proposed method on a Virtex-7 SRAM-based FPGA hardware platform. Compared with traditional interconnect test vector generation method, the self-adaptive Ford-Fulkerson algorithm reduced the number of test vectors by 35.8%, and the fault coverage achieved 99.74%.

REFERENCES

[1] S. Banik, S. Roy, and B. Sen, "An integrated framework for application independent testing of FPGA interconnect," *J. Electron. Test.*, vol. 35, no. 5, pp. 729–740, Oct. 2019.

[2] H. Fu, O. Mencer, and W. Luk, "FPGA designs with optimized logarithmic arithmetic," *IEEE Trans. Comput.*, vol. 59, no. 7, pp. 1000–1006, Jul. 2010.

[3] S. Banik, S. Roy, and B. Sen, "Application-dependent testing of FPGA interconnect network," *IEEE Trans. Very Large Scale Integr. VLSI Syst.*, vol. 27, no. 10,2019, pp. 2296–230.

[4] Li Dai, Zhi-bin Liu, Shao-chi Liang, Meng Yang, and Ling-li Wang, "FPGA interconnect testing algorithm based on routing-resource graph,"

in *2008 9th International Conference on Solid-State and Integrated-Circuit Technology*, Oct. 2008, pp. 2087–2090.

[5] X. -F. Wang, S. -H. Si, C. Gao, and J. Huang, "A method of FPGA interconnect resources testing by using XDL-based configuration," in *2014 Prognostics and System Health Management Conference (PHM-2014 Hunan)*, Aug. 2014, pp. 203–207.

[6] F. H. bin Mohamed Sultan, Z. binti Dahari, Y. Y. Koh, N. Da Cunha, and J. T. Ng, "Development of at-speed interconnect test to capture marginal open defect on FPGA," in *9th International Conference on Robotic, Vision, Signal Processing and Power Applications*, 2017, pp. 27–35.

[7] W. Xie, W. Qi, X. Lin, and H. Wang, "Research on an intelligent test method for interconnect resources in an FPGA," *Appl. Sci.*, vol. 13, no. 13, 2023.

[8] A.W.Ruan, W.Tian, B.Ni, K.Wu"A hierarchical switch matrix and interconnect resources test in Virtex-5 FPGA", International Symposium on Integrated Circuits(ISIC),2014, pp. 111–114 .

[9] Boutros, Andrew, and Vaughn Betz. "FPGA architecture: principles and progression." IEEE Circuits and Systems Magazine, vol. 21, no. 2, pp. 4-29, May 2021.

[10] C. Beckhoff, D. Koch, and J. Torresen, "The Xilinx design language (XDL): tutorial and use cases," in *6th International Workshop on Reconfigurable Communication-Centric Systems-on-Chip (ReCoSoC)*, Jun. 2011, pp. 1–8.

2025 The 10th International Conference on Integrated Circuits and Microsystems

A Multi-Phase Clock Generator with Customized Duty Cycles for Pipeline-SAR ADCs

He Zhi
School of Integrated Circuits
Southeast University
Nanjing, China
zhihe@seu.edu.cn

Hongtao Yue
School of Integrated Circuits
Southeast University
Nanjing, China
yueht@seu.edu.cn

Yuwei Zong
School of Integrated Circuits
Southeast University
Nanjing, China
230249467@seu.edu.cn

Zhongxu Zhang
School of Integrated Circuits
Southeast University
Nanjing, China
zhangzhx@seu.edu.cn

Jianhui Wu
School of Integrated Circuits
Southeast University
Nanjing, China
wjh@seu.edu.cn

Xin Li
School of Integrated Circuits
Anhui University
Hefei, China
lixin@ahu.edu.cn

Abstract—The pipeline-SAR ADC is gradually replacing the pipeline ADC to become the main architecture in the research of high-speed ADCs. The conventional clock circuit cannot meet the pipeline-SAR ADC's more variable clock phase and duty cycle requirements. This work presents a multiphase clock generator with customized phase and duty cycles for the pipeline-SAR ADCs. The clock generation system is verified using an ideal 12-bit 1 GS/s ADC to achieve 10.7-bit ENOB and 67.3 dB SFDR in a CMOS 28 nm process.

Index Terms—Nonoverlap clock generation, pipeline-SAR ADC, clock phase, duty cycle

I. INTRODUCTION

In recent years, with the application of the time-interleaving structure, single-channel ADCs have focused on balancing speed and power consumption, making pipeline-SAR ADCs the mainstay in high-speed ADC research [1]–[5].

The pipeline ADCs conventionally employ multiphase clocks with a consistent duty cycle across all phases [6]–[8]. Structural differences in sub-ADCs within pipeline-SAR ADCs typically result in varying durations for sampling, conversion, and amplification phases. Thus, the pipeline-SAR ADCs request a customized multiphase clock circuit duty cycle, compared to the pipeline ADC. Fig. 1 shows a typical timing design of a pipeline-SAR ADC.

This article proposed a multiphase clock generator with customized duty cycles for pipeline-SAR ADCs. A 10 GHz sinusoidal input signal is conditioned to a square wave by the clock input signal conditioning circuit. This square wave is frequency-divided to 1 GHz through a multiphase and duty-cycle customized clock generator, yielding phase-shifted square-wave outputs. The combinational logic then synthesizes the customized duty cycles. Finally, a non-overlap clock generator produces the clock signals of the ADC core, meeting critical timing constraints.

Supported by "the Fundamental Research Funds for the Central Universities" (Grant No. 2242025K30014).

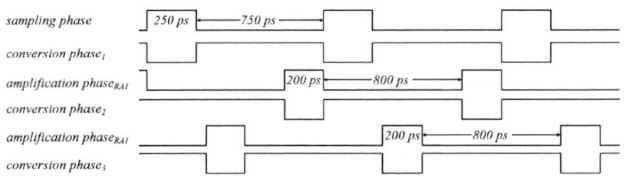

Fig. 1. Timing diagram of a typical three-stage pipeline-SAR ADC

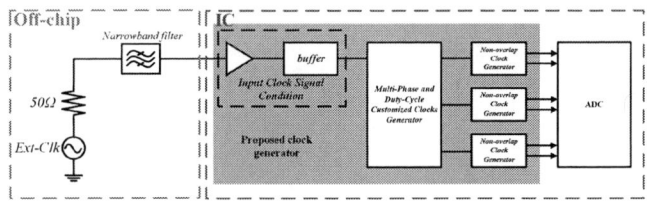

Fig. 2. Block diagram of proposed clock generation system for high-speed pipeline-SAR ADCs

The paper is organized as follows: Section II proposes the multi-phase and duty-cycle customized method. Section III introduces the architecture of the clock generation circuit, including the input clock signal conditioning circuit, the multi-phase and duty-cycle customized clock generator, and the non-overlap clock generator. Section IV presents the simulation results of the clock generator. Finally, Section V concludes.

II. MULTI-PHASE AND DUTY-CYCLE CUSTOMIZED METHOD

This article proposes a multi-phase and duty-cycle customized method to accommodate various pipeline-SAR ADCs, thereby enabling the circuit to have strong portability.

The multi-phase and duty-cycle customized clock method includes a multi-phase frequency divider and a duty-cycle customized clock generator. The multi-phase frequency divider

979-8-3315-8850-2/25 $31.00 © 2025 IEEE

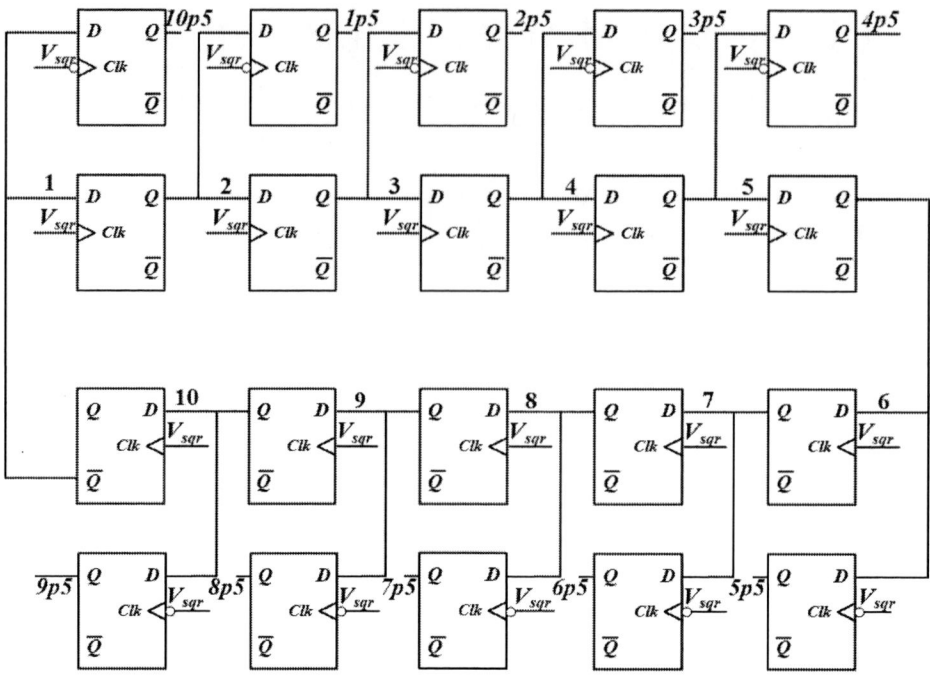

Fig. 3. Multi-phase frequency divider

collects all the rising edges and falling edges of the input high-frequency clock with the multiphase clocks. The duty-cycle customized clock generator converts multiphase clocks into customized duty-cycle clocks using a combinational logic circuit.

Define the input clock frequency as f_H, the period as T_H, and the division ratio of the conventional divider as N. The phase differences T_{L1} of the multi-phase clocks after frequency division are shown as (2).

$$T_{L1} = \frac{T_H}{N} \qquad (1)$$

The frequency divider proposed in this article is as in Fig. 3. By simultaneously using rising-edge and falling-edge D triggers, the phase difference of the multi-phase clock is shortened. The phase differences T_{L2} of the multi-phase clocks after the proposed frequency division are shown as (2).

$$T_{L2} = \frac{T_H}{2N} \qquad (2)$$

Since the duty cycle of the output clock from the frequency divider is always 50%, a combination logic circuit can be used to generate an output clock with any desired duty cycle. Through the double-edge frequency divider, the division ratio is increased, enabling more precise adjustment of the duty cycle.

III. CIRCUIT DESIGN

Fig. 2 shows the proposed clock generation architecture. The signal is input into the chip through a narrowband filter outside the chip. The proposed clock generation architecture includes

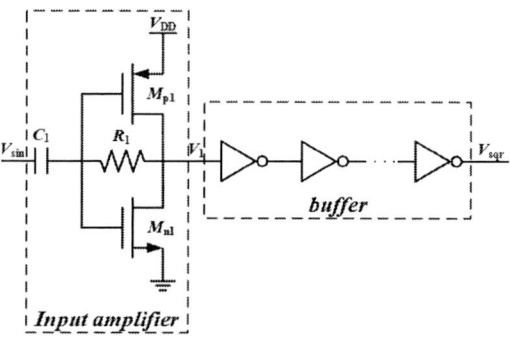

Fig. 4. The input clock signal conditioning circuit

an input clock signal conditioning circuit, a multiphase and duty-cycle customized clock generator, and a non-overlap clock generator. This section introduces them respectively.

A. Input Clock Signal Conditioning Circuit

The input clock signal conditioning circuit is divided into the input amplifier and the buffer, as shown in Fig. 4.

The input amplifier retains the AC signal of the input sinusoidal signal V_{sin}, blocks the DC signal, and provides a common-mode voltage within the chip simultaneously, thus avoiding clock jitter caused by the instability of the common-mode voltage of the external signal.

C_1 is a DC signal blocking capacitor, removing the DC component from external signal sources. M_{p1} and M_{n1} provide the DC signal in chip through R_1, and amplify the AC signal from off-chip. The values of R_1 and C_1 should comply

979-8-3315-8850-2/25 $31.00 © 2025 IEEE   44

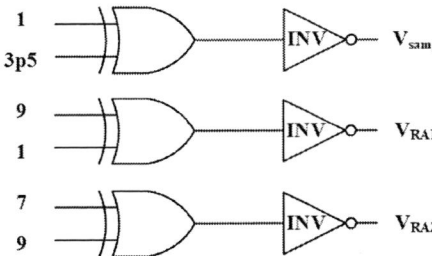

Fig. 5. The duty-cycle customized clock generator

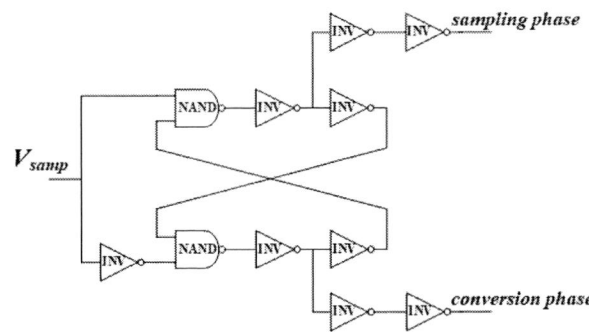

Fig. 6. Non-overlap clock generator

with the rules as (3), to prevent the elimination of the input sinusoidal signal.

$$f_{\sin} \gg f_c = \frac{1}{2\pi R_1 C_1} \tag{3}$$

The output of the input amplifier is sent to the buffer for shaping to form a square wave V_{sqr} with a 50% duty cycle.

The clock jitter will deteriorate the ADC's performance. The influence of clock jitter on the SNR of an ideal sampling hold circuit is shown as (4).

$$SNR_{jitter} \approx 20\log_{10}\left(\frac{1}{2\pi f_{in}\tau_{j,rms}}\right) dB \tag{4}$$

The deterioration caused by the sampling clock jitter increases as the input signal frequency increases. For a typical Nyquist sampling ADC, set $f_{in} = f_s/2$ to evaluate clock jitter as (5).

$$SNR_{jitter} \approx 20\log_{10}\left(\frac{1}{\pi f_s\tau_{j,rms}}\right) dB \tag{5}$$

The SNR of an ideal ADC is shown as (6).

$$SNR_{ideal} \approx 6.02N + 1.76dB \tag{6}$$

Set $SNR_{ideal} \geq SNR_{jitter}$ to eliminate the influence of clock jitter on the performance of the ADC. Thus, the requirement of clock jitter is shown as (7).

$$\tau_{j,rms} \leq \frac{1}{\pi f_s} 10^{-\frac{1}{20}(6.02N+1.76)} \tag{7}$$

According to the relationship between $\tau_{j,rms}$ and $V_{n,rms}$ as (8), the noise requirement for the input amplifier is shown as (9).

$$\tau_{j,rms} \approx \frac{V_{n,rms}}{2\pi f_s V_{\sin,max}} \tag{8}$$

$$V_{n,rms} \leq 2V_{\sin,max} 10^{-\frac{1}{20}(6.02N+1.76)} \tag{9}$$

In high-speed ADC application scenarios, the noise from the input amplifier is mainly thermal noise. The noise requirement can be achieved through parameter optimization.

Through this module, the externally input 10 GHz sinusoidal signal is converted into a 10 GHz square wave in the chip.

B. Multi-Phase and Duty-Cycle Customized Clock Generator

The multiphase and duty-cycle customized clock generator includes a multi-phase frequency divider and a duty-cycle customized clock generator.

The multiphase frequency divider converts the 10 GHz square wave into 1 GHz square waves of different phases, as shown in Fig. 3. The structure connects 10 rising-edge D flip-flops in series to achieve frequency division. The working process of the multi-phase divider is as follows.

- After powering the chip, set signal $1, 2, ..., 5$ to VSS and signal $6, 7, ..., 10$ to VDD.
- When the rising edge of the clock arrives, the signals within the rising D flip-flops shift once, resulting in signal $2, 3, ..., 6$ to VSS and signal $7, 8, 9, 10, 1$ to VDD.
- When the falling edge of the clock arrives, the signals $1, 2, ..., 10$ are delivered to the signal $1p5, 2p5, ..., 10p5$ with the falling-edge D flip-flops.

The phase difference between adjacent signals from 1 to 10 is 100 ps. Then, signals 1-10 are sent to the falling-edge D flip-flop to generate $1p5, 2p5, ..., 10p5$; thus, the phase difference between corresponding signals is 50 ps, as 1 and $1p5$.

According to the clock requirements of the ADC as shown in Fig. 1, three types of clocks for three sub-ADCs are generated: sampling phase, amplification phase$_1$, and amplification phase$_2$.

The duty-cycle customized clock generator is shown in Fig. 5.

- Perform the XOR operation on 1 and $3p5$ to obtain a 1 GHz square wave V_{samp} with a duty cycle of 25%.
- Perform the XOR operation on 1 and 9 to obtain a 1 GHz square wave V_{RA1} with a duty cycle of 20%.
- Perform the XOR operation on 7 and 9 to obtain a 1 GHz square wave V_{RA2} with a duty cycle of 20%.

In addition, due to the cyclic nature of the multi-phase frequency divider, the phase difference between different output signals is satisfied.

C. Non-overlap Clock Generator

The multi-phase and duty-cycle customized clock generator outputs V_{samp}, V_{RA1}, and V_{RA2}, but cannot be directly input into the ADC. The reasons are as follows.

979-8-3315-8850-2/25 $31.00 © 2025 IEEE

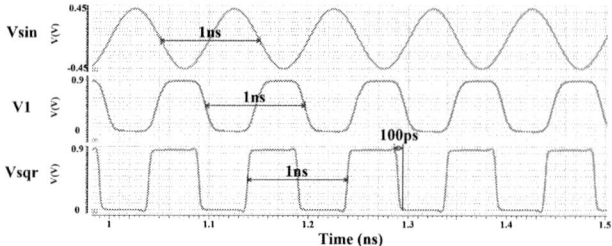

Fig. 7. Input clock signal conditioning circuit simulation results

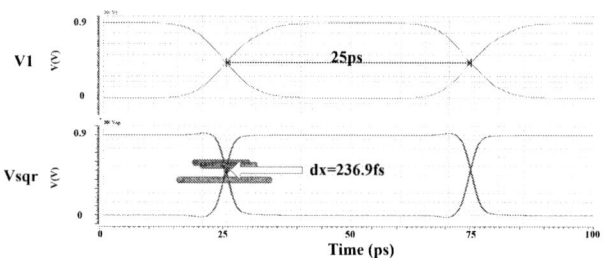

Fig. 8. Input clock signal conditioning circuit eye diagram simulation results

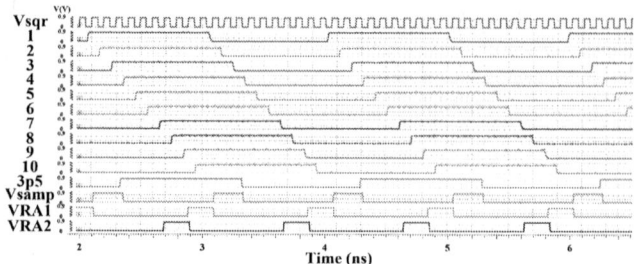

Fig. 9. Multi-phase and duty-cycle customized clock generator simulation results

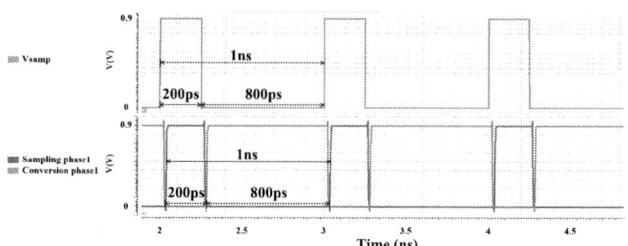

Fig. 10. Non-overlap clock generator simulation results

- The ADC needs the sampling, amplification, and conversion phases, as shown in Fig. 1.
- The sampling phase and the conversion phase$_1$ cannot overlap, the amplification phase$_{RA1}$ and the conversion phase$_2$ cannot overlap, and the amplification phase$_{RA2}$ and the conversion phase$_3$ cannot overlap. This is to prevent the overlapping between the upper and lower edges from causing abnormal working conditions of the ADC.
- The output should have sufficient load-carrying capacity to avoid timing anomalies of the ADC.

Thus, this work selects a non-overlap clock generator to solve these troubles. Fig. 6 shows an example of the non-overlap clock generator to create the sampling phase and the conversion phase$_1$ according to V_{samp}.

The non-overlap structure avoids the overlap between the sampling or amplification phase and the conversion phase. The inverter chain is adopted at the output end to enhance the driving capacity.

IV. SIMULATION RESULTS

This work conducts a simulation of this design under the 28-nm CMOS process, with $V_{DD} = 0.9V$. In this section, the three modules are simulated and verified, and finally, the overall output is also simulated and verified.

A. Input Clock Signal Conditioning Circuit Simulation Results

The input clock signal conditioning circuit is simulated. As in Fig. 4, set V_{sin} to a 10 GHz sinusoidal signal ranging from -0.45 to 0.45V. The simulation result is shown in Fig. 7.

The output signal V_1 of the input amplifier changes from 0 to 0.9V, proving that the internal and external common-mode voltages were isolated. The output signal V_{sqr} becomes a square wave under load, confirming the effectiveness of the input clock signal conditioning circuit.

The eye diagram of V_1 and V_{sqr} is shown in Fig. 8, under the condition of adding noise. The jitter of V_{sqr} is 236.9 fs.

B. Multi-Phase and Duty-Cycle Customized Clock Generator Simulation Results

The multi-phase and duty-cycle customized clock generator is simulated. As in Fig. 3, set V_{sqr} to a 10 GHz square signal ranging from 0 to 0.9V. The simulation result is shown in Fig. 9.

The simulation results show that signals 1 to 10 are 1 GHz square waves with uniform phase differences of 100 ps generated by frequency division based on the original 10 GHz square wave V_{sqr}, and simultaneously signal 3p5 is 50 ps later than signal 3.

Through the duty-cycle customized clock generator, V_{samp}, V_{RA1} and V_{RA2} are produced. The duty cycle of V_{samp} is 24.8%, the duty cycle of V_{RA1} is 21.1%, and the duty cycle of V_{RA1} is 21.4%.

C. Non-overlap Clock Generator Simulation Results

In this subsection, the non-overlap clock generator is simulated. The structure of three blocks for V_{samp}, V_{RA1} and V_{RA2} are same as in Fig. 6. Select V_{samp} as an example to verify this module.

Set V_{samp} to a 1 GHz square signal with 25% duty-cycle ranging from 0 to 0.9V. The simulation result is shown in Fig. 10. It can be seen that the sampling phase and the conversion phase$_1$ will not be at a high level simultaneously.

TABLE I
Performance Across Process Corners and Temperatures

Process	TT			FF			SS		
Temp. (°C)	0	40	80	0	40	80	0	40	80
ENOB (bit)	10.6	10.8	10.9	10.9	11.1	11.2	10.5	10.6	10.6
SFDR (dB)	66.5	68.1	68.5	68.1	70.3	71.2	65.2	66.1	66.5

TABLE II
Performance Comparison with Previous Works

	This work	[9]	[10]	[7]
Process	28 nm	55 nm	65 nm	0.13 um
Input frequency	10 GHz	-	-	1.6 GHz
Output frequency	1 GHz	100-250 MHz	112 GHz	1.6 GHz
Duty cycle	24.8%, 21.1%, 21.4%	47.8%-52.7%	50%	22%, 33%, 35%, 40%

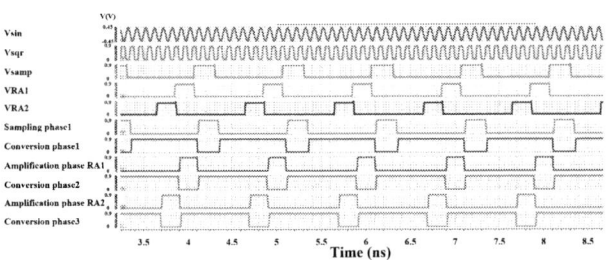

Fig. 11. Overall simulation results

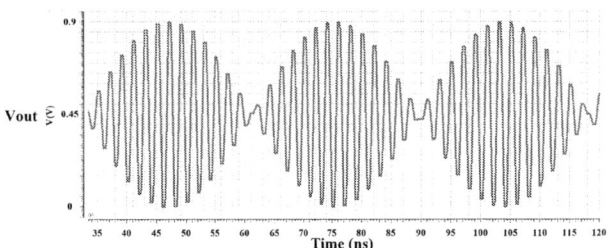

Fig. 12. The output signal of the ADC after DAC coding

D. Overall Simulation Results

Finally, simulate the multi-phase clock generator in general. Set V_{sin} to a 10-GHz sinusoidal signal ranging from -0.45 to 0.45V. The simulation results are as shown in Fig. 11.

The final output signals include sampling phase, conversion phase$_1$, amplification phase$_{RA1}$, conversion phase$_2$, amplification phase$_{RA2}$, and conversion phase$_3$. The duty cycle of the sampling phase is 22.8%, and the duty cycle of the conversion phase$_1$ is 74.4%. The duty cycle of the amplification phase$_{RA1}$ is 19.7%, and the duty cycle of the conversion phase$_1$ is 77.6%. The duty cycle of the amplification phase$_{RA2}$ is 19.9%, and the duty cycle of the conversion phase$_1$ is 77.1%. The output signals are similar to those in Fig. 1, meeting the design requirements.

To further prove the effectiveness of this circuit, the output signals are sent to an ideal 12-bit ADC for verification. The input signal of the ADC is 482.421875 MHz, ranging from 0 to 0.9 V. The clock of ADC is generated by the proposed multi-phase clock generator. After an ideal DAC, the output signal of the ADC is as shown in Fig. 12.

The output spectrum of the ADC is as shown in Fig. 13. The ENOB of ADC is 10.7 bits, the SNDR of ADC is 66.5 dB, and the SFDR of ADC is 67.3 dB.

This work also simulated the performance of an ideal 12-bit ADC with the clock generator circuit under different process and temperature conditions, as shown in the Table. I. Therefore, the proposed clock generator can meet the design requirements of the high-speed pipeline-SAR ADC.

Table. II compares this work with others. After a simple

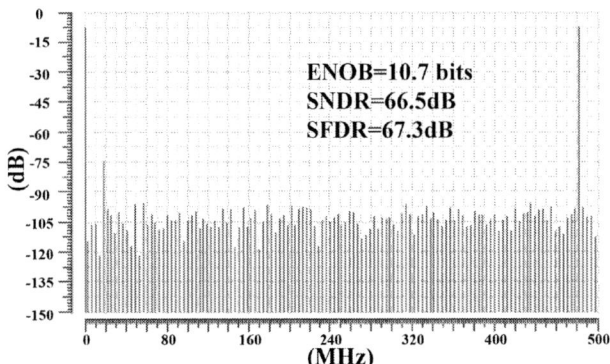

Fig. 13. The output spectrum of the ADC

adjustment, the duty cycle of the structure in this paper can be extended to $N \times 5\%$, $N = 1, 2, ...19$.

V. Conclusion

This work presents a multiphase clock generator with customized duty cycles for pipeline-SAR ADCs. The structure has the advantages of flexible phase and duty cycle, and is portable for multiple ADCs. Finally, the method was verified through simulation.

References

[1] C. Park, J. Kim, K. Kang, M. Yang, B. Moon, S. Lee, and W. Jung, "A high-resolution pipelined-sar adc using cyclically charged floating inverter amplifier," *IEEE Journal of Solid-State Circuits*, pp. 1–11, 2024.

979-8-3315-8850-2/25 $31.00 © 2025 IEEE

[2] W. Jiang, Y. Zhu, C. Chen, H. Xu, Q. Liu, M. Liu, R. P. Martins, and C.-H. Chan, "A 14b 500 ms/s single-channel pipelined-sar adc with reference ripple mitigation techniques and adaptively biased floating inverter amplifier," *IEEE Journal of Solid-State Circuits*, vol. 58, no. 10, pp. 2709–2721, 2023.

[3] L. Fang, X. Wen, T. Fu, and P. Gui, "A 12-bit 1 gs/s rf sampling pipeline-sar adc with harmonic injecting cross-coupled pair achieving 7.5 fj/conv-step," *IEEE Transactions on Circuits and Systems I: Regular Papers*, vol. 69, no. 8, pp. 3225–3236, 2022.

[4] W. Jiang, Y. Zhu, M. Zhang, C.-H. Chan, and R. P. Martins, "A temperature-stabilized single-channel 1-gs/s 60-db sndr sar-assisted pipelined adc with dynamic gm-r-based amplifier," *IEEE Journal of Solid-State Circuits*, vol. 55, no. 2, pp. 322–332, 2020.

[5] Y. Shen, S. Liu, Y. Cao, H. Han, H. Liang, Z. Dong, D. Li, R. Ding, and Z. Zhu, "A 12-bit 1.5-gs/s single-channel pipelined sar adc with a pipelined residue amplification stage," *IEEE Journal of Solid-State Circuits*, pp. 1–12, 2024.

[6] Y. Ni, L. Liu, Y. Zhang, and T. Zhu, "A 12-bit 2-gs/s pipeline adc in 28-nm cmos with linear-error self-calibration," *IEEE Transactions on Very Large Scale Integration (VLSI) Systems*, vol. 33, no. 6, pp. 1561–1569, 2025.

[7] H. Çetinkaya, A. Zeki, A. Girgin, E. D. Karabeyoğlu, and T. C. Karalar, "A 1.6 ghz non-overlap clock generation with differential clock driver and clock level shifters for gs/s sampling rate pipeline adcs," in *2018 25th IEEE International Conference on Electronics, Circuits and Systems (ICECS)*, 2018, pp. 277–280.

[8] B. Murmann and B. Boser, "A 12-bit 75-ms/s pipelined adc using open-loop residue amplification," *IEEE Journal of Solid-State Circuits*, vol. 38, no. 12, pp. 2040–2050, 2003.

[9] T. Li and D. Li, "A wide-input-range multi-phase clock generator design for cmos image sensors," in *2022 5th International Conference on Circuits, Systems and Simulation (ICCSS)*, 2022, pp. 91–96.

[10] I. Alhousseiny, M. Ali, N. Ben-Hamida, M. Honarparvar, M. Sawan, and Y. Savaria, "Delay-locked loop based multiphase clock generator for time-interleaved adcs," in *2021 28th IEEE International Conference on Electronics, Circuits, and Systems (ICECS)*, 2021, pp. 1–4.

2025 The 10th International Conference on Integrated Circuits and Microsystems

Hardware Design and Side-Channel Security Analysis on the Key Computational Block for YOLOv11

Runquan Shao
School of Electronic Engineering
Heilongjiang University
Harbin, China
2859899745@qq.com

Liji Wu*
School of Integrated Circuit
Tsinghua *University*
Beijing, China
lijiwu@tsinghua.edu.cn

Jing Hu*
School of Electronic Engineering
Heilongjiang University
Harbin, China
hjlyh@126.com

Le Wu
School of Integrated Circuit
Tsinghua University
Beijing, China
wul22@mails.tsinghua.edu.cn

Xiangmin Zhang
School of Integrated Circuit
Tsinghua University
Beijing, China
zhxm@tsinghua.edu.cn

Abstract—**YOLO (You Only Look Once) algorithm plays a crucial role in autonomous vehicles and medical imaging, which require a large number of object detection and image classification tasks because of its advantages of speed and accuracy on edge device. However, due to the intrinsic high computational workload of YOLO model, it is still challenging when implementing large models on the embedded device, especially the latest version YOLOv11. In this paper, we propose the hardware design based on time-division multiplexing of the key computational block C3k2 which is most used in YOLOv11 on FPGA. Firstly, we extend a fast convolution (Conv) algorithm to our Conv layer. Secondly, the batch normalization (BN) layer has been significantly optimized, especially in the stage of calculating the mean and variance. Finally, a Sigmoid Linear Unit (SiLU) activation function in the form of a piecewise function will appear completely in the form of time-division multiplexing, which will greatly reduce the consumption of resources. Experimental results show that our design can achieve a 0.19ms delay on an AMD artix-7 xc7a35tftg256 FPGA under the working frequency of 50MHz, while the average relative error is 1.67%. Furthermore, we use the cw305 board, which can extract power information, to implement a simple power analysis (SPA) on C3k2 block and the SiLU. The information within the hardware design was successfully analyzed, such as the number of runs and input data information.**

Keywords—*YOLOv11, C3k2, BN, SiLU, hardware design, SPA*

I. INTRODUCTION

Many applications, ranging from autonomous driving [1] to medical images [2], use deep learning as an effective method. Then several promising approaches have been proposed for object detection and image classification with deep learning such as R-CNN [3], EfficientDet [4] and YOLO [5] which performs extremely fast and high accuracy for these tasks. Although CPUs and GPUs are widely used in YOLO, there are still problems of low efficiency in memory scheduling and

parallel computing. Therefore, FPGA-based deep learning accelerators have been proposed to improve the inference speed of YOLO models. Subsequently, FPGA have been widely used in the acceleration of deep learning models due to its flexibility and shorter development cycle [6, 7]. A lot of previous research [6, 7] efforts have focused on optimizing the convolutional layers that require a large amount of computation on FPGAs and leaving other operations to the CPUs for processing. Such heterogeneous overlay-based designs provide flexibility and can deploy various versions of the YOLO network. However, they are severely restricted by the connection data communication Bottleneck between the FPGA and the CPU [8], especially YOLOv11 as shown in Fig.1. Therefore, the hardware design of the key block C3k2 in YOLOv11 in this paper does not merely focus on Conv layer, the BN layer and the SiLU activation function are also accelerated and optimized to reduce resource consumption while maintaining high precision and high speed.

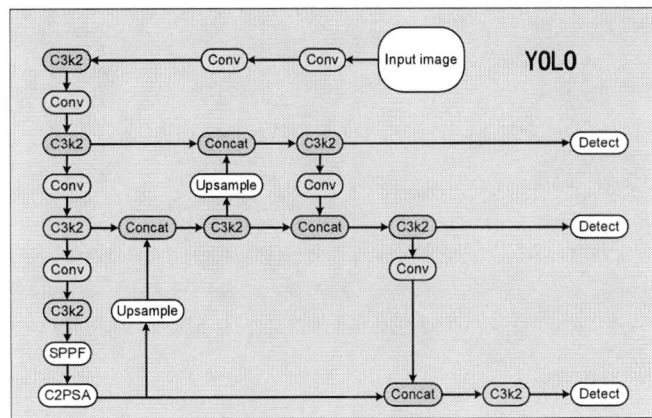

Fig. 1. The structure of YOLOv11

979-8-3315-8850-2/25 $31.00 © 2025 IEEE

Meanwhile, FPGA also faces security issues, such as side channel attack (SCA) [9, 10]. The method of SCA is to obtain information in the algorithm through various leaked physical information generated during the hardware execution process. Efficient and accurate AI models require high development costs and are protected by intellectual property rights. Therefore, model theft of SCA, aimed at obtaining its architecture and parameters, has also become a highly attractive type of attack. Therefore, we perform SCA on the design to discover the insecure components inside it, provide a reference for the subsequent hardware design, and predict the side-channel vulnerabilities in advance. Finally, in this paper, SPA was conducted on C3k2, and the number of runs of the model and the information of some modules were obtained.

II. BLOCK ARCHITECTURE DESIGN

A. C3k2 Block

The C3k2 and some other blocks used in C3k2 are shown in Fig.2. Among them, C3k2 is the most frequently used and strongest feature extraction capability block. This block is optimized from the traditional C3 block. It mainly uses the principle of multi-channel feature fusion to increase the width of the network and thereby enhance the network's feature extraction capability. It is particularly suitable for more complex scenarios and deep-level feature extraction tasks, such as the detection tasks of object boundaries and complex backgrounds. In this paper, we define the parameters of the C3k block as C1=4, C2=4, n=2, C3k=True, e=0.5, g=1, Shortcut=True. This means that the number of input channels and output channels is defined as 4, using two C3k, with an expansion factor of 0.5, no group convolution is performed, and residual linking is used. The data flow is that the 4-channel feature map output by the CBS block is split into two parts. One retains all 4 channels and temporarily stores them in the First Input First Output (FIFO), while the other is processed by the C3k block. After processing by each C3k, two output feature maps are obtained. It is also temporarily stored in the FIFO and waits for the processing completion of the last C3k block. All these temporarily stored feature maps are concatenated to obtain the 8-channel feature map. Finally, it is passed through the CBS block to obtain the 4-channel output feature map.

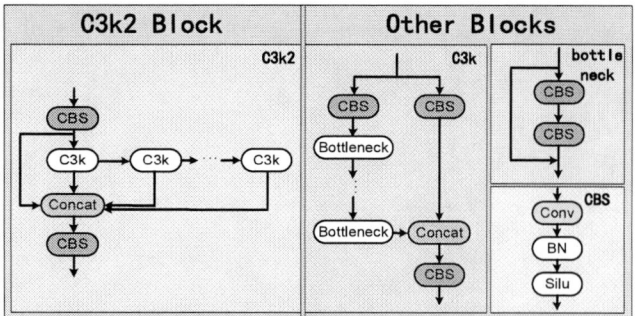

Fig. 2. The structure of C3k2 and some other blocks which used in C3k2

B. C3k Block

The C3k block in Fig.2 allows the use of convolution kernels of different sizes, which enables the model to extract features at different scales, increases the area of the receptive field, helps capture more complex spatial features, and is more adaptable to the requirements of multi-scale detection. The data flow is as follows: through the CBS block and the 2-channel feature map split, it passes through two different CBS blocks to obtain two 1-channel output feature maps. Among them, the first one needs to be processed by n Bottlenecks and finally concatenated with the other one. Finally, the 2-channel output feature map is obtained through CBS and transferred to the next block.

C. Bottleneck Block and CBS Block

As shown in Fig.2, the Bottleneck was initially proposed in ResNet, aiming to increase the depth of the network while maintaining a small amount of computation. Among them, the convolution kernel is in a size of 3*3 and uses residual connection. This connection structure enables the input features to be directly added to the output features, allowing the network to be effectively trained at a deep level and alleviating the problem of vanishing gradients. In this block, the input feature map needs to go through two 3*3 CBS blocks, and there is a residual connection between the input and the output. As shown in Fig.2, the CBS block is composed of Conv, BN and SILU, among which the Conv is divided into convolution with 1*1 size kernels and convolution with 3*3 size kernels. In the deployment of the CBS block, we use time-division multiplexing to reuse the same Conv layer, BN layer, and SILU. The entire C3K2 block employs four 1*1 Conv blocks, one 3*3 Conv block, eight BN blocks, and six SILU blocks, with each block deployed on one channel.

III. HARDWARE DESIGN AND FPGA VERIFICATION

A. Hardware Design for Convolution

In the C3k2 block, there are two convolution operations of kernel size for the Conv layer. The first one is a 1*1, and the other one is a 3*3. Among them, we adopt Add-then-Multiply instead of Multiply-Accumulate (MAC) for the design of the 1*1 convolutional layer. The reason is that in usual hardware designs, we need to perform fixed-point number representation or quantization operations on the input floating-point data to reduce resource overhead. For example, this paper adopts a 16-bit signed number form, shown in Fig.3, to represent the input data, and uses 7-bit (F6, F5 ... F0) representing the fractional part, 8 bits (I7, I6 ... I0) representing integer part and 1bit (S) for the sign. It is inevitable that there will be some numbers that are the same after the 16-bit representation or the same as their original input. These numbers exist in the input data of every channel and in the convolution kernels of each layer. That is to say, when performing the multiplication operation of MAC on a 1*1 convolutional layer, there exist multiple sets of the same input data or the same convolutional kernels, or both two, and then the obtained numbers are accumulated. This greatly increases the overhead of unnecessary multipliers. Therefore, this paper adopts the Add-then-Multiply method instead. First, we try to find the same input data, add it, and then perform multiplication operations on the obtained sum with the convolution kernels (the same convolution kernels are multiplied by the input data). It can save a large amount of DSP resources while increasing speed.

979-8-3315-8850-2/25 $31.00 © 2025 IEEE

15	14			7	6			0
Sign	Integer Part				Fraction Part			
S	I7	I6	...	I1	I0	F6	F5	...	F1	F0

Fig. 3. The definition of a 16-bit data format with signed numbers

The 3*3 convolutional layer uses the common sliding window method, which is shown in Fig.4. First, we send the input data expanded by the boundary to the Line buffer in the form of data flow. The Line buffer will generate the input data required by a convolutional layer window. After MAC, the output data is obtained. This method does not require the storage of all the data. For a convolutional layer of a channel, it needs K*W storage units and K*K multiplication units.

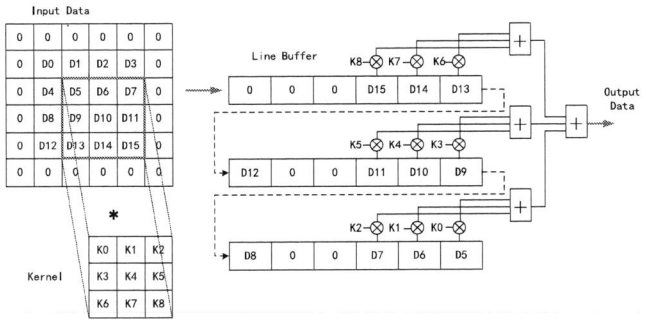

Fig. 4. Convolution building block

B. Hardware Design for Batch Normalization

The BN performs the following algorithm:

TABLE I. ALGORITHM OF BATCH NORMALIZATION

Algorithm of Batch Normalization

Input: Values of x over a mini-batch:$\beta = \{x_{1...m}\}$;

Parameters to be learned: γ, β

Output: $\{y_i = BN_{\gamma,\beta}(x_i)\}$

$$\mu = \frac{1}{m}\sum_{i=1}^{m} x_i \qquad \text{//mini-batch mean}$$

$$\sigma^2 = \frac{1}{m}\sum_{i=1}^{m}(x_i - \mu)^2 \qquad \text{//mini-batch variance}$$

$$\hat{x}_i = \frac{x_i - \mu}{\sqrt{\sigma^2 + \epsilon}} \qquad \text{//normalize}$$

$$y_i = \gamma\hat{x}_i + \beta \equiv BN_{\gamma,\beta}(x_i) \qquad \text{//scale and shift}$$

When the above algorithm is executed, if you want to calculate μ and σ^2, you need to wait for all x_i to enter the memory of the BN layer before starting the calculation. This requires additional storage resources and will reduce speed. We use the recursive algorithm to calculate μ and σ^2 as shown in the (1) - (4), so that μ and σ^2 are calculated along with the data flow of x_i. The hardware structure is shown in Fig.5. The μ_{i-1} and S_{i-1} are obtained through a clk via REG. The i is controlled and started by the x_valid signal, and increases with the change

of x. Finally, the obtained μ and σ^2 are processed through the steps of the formula in Table I to obtain y_i.

$$\mu_i = \mu_{i-1} + \frac{x_i - \mu_{i-1}}{i} \qquad (1)$$

$$\mu = \mu_n \qquad (2)$$

$$S_i = S_{i-1} + (x_i - \mu_{i-1})(x_i - \mu_i) \qquad (3)$$

$$\sigma^2 = \frac{S_n}{n-1} \qquad (4)$$

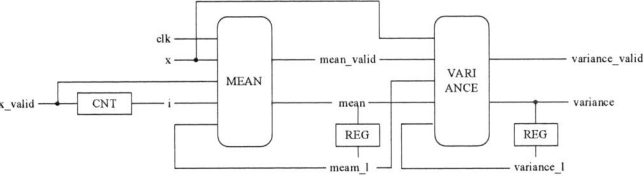

Fig. 5. The building block for calculating mean and variance

C. Hardware Design for SiLU Activation Function

The SiLU activation function is a nonlinear function used to provide the nonlinear modeling ability for the output of each layer. The calculation of the activation function is expressed as (5).

$$y = x * sigmoid(x) \qquad (5)$$

The key is to perform the sigmoid calculation, and its formula is as follows:

$$sigmoid(x) = \frac{1}{1+e^{-x}} \qquad (6)$$

The SiLU activation function will approach x when the input data is too large, and approach 0 when the input data is too small. Therefore, we define the output greater than 5 as x, the part less than -5 as 0, and the calculation of the remaining domain mainly focuses on calculating e^x. The integer part of e^x is implemented using the Look-Up-Table (LUT) method, and the fraction part of e^x is implemented using Pade' Approximant. In this paper, we adopt the third-order Pade' Approximant, as in (7). Finally, multiply the positive part and the fraction part to obtain the complete e^x. sigmoid(x) is obtained by (6).

$$e^x \approx \frac{120+60x+12x^2+x^3}{120-60x+12x^2-x^3} \qquad (7)$$

The overall hardware framework is shown in the following Fig.6. The entire pipeline architecture is controlled by control logic. Among them, the control logic is responsible for splitting the input data and providing the enable signal for each level of pipeline. The first-level pipeline is responsible for distinguishing whether the input data is calculated or approximated, and the second-level generates the address of the LUT. The third one is used for the Pade' Approximant calculation of the fraction part and the lookup of the value within the LUT based on address for the integer part. Then the fourth is responsible for multiplying the obtained integer part and fraction part of e^x and calculating the SiLU result. The fifth-level selects silu_cal or silu_appro to represent silu_out based on the Sel signal and temporarily storing it in Buf_out. And the last-level pipeline outputs the temporarily stored silu_out along with out_valid according to the enable signal Out_en given by the control logic.

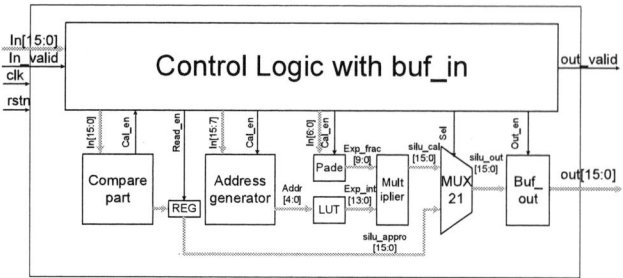

Fig. 6. SiLU building block

D. FPGA Verification for C3k2 Block

The hardware architecture of C3k2 block is implemented on the AMD artix-7 xc7a35tftg256 FPGA by Verilog HDL. The recourse utilization after the synthesis by Xilinx vivado2018.3 is shown in Fig.7. And data comparison was conducted on the FPGA hardware platform and the PyCharm software platform to verify the correctness of the design.

Resource	Utilization
LUT	27401
LUTRAM	167
FF	6126
BRAM	17
DSP	38
IO	67
BUFG	1

Fig. 7. The recourse utilization of C3k2

Table II presents the comparison of the software and hardware environment and processing time for processing the 4*4 matrix C3k2. The processing time of the hardware design at 50Mhz is 0.19ms. And the average relative error is 1.67%.

TABLE II. THIS DESIGN COMPARE WITH PC

Platform	Environment	Time
PC	CPU:AMD Ryzen 7 7735H 3.20 GHz	4.0011ms
FPGA	artix-7 xc7a35tftg256 50MHz	0.19ms

Compared with the previous work, the BN layer design in this paper, under the same input and output bit widths, is twice that of DSP and LUT, and two-thirds of FF, achieving a lower error rate. The delays in this paper are listed in the Table III.

TABLE III. BN DESIGN COMPARE WITH OTHERS

	In/Out width	DSP	LUT	FF	F(MHz)	Percentage error(%)	Delay(ns)
[11]	16	1	448	304	224	1.3	N/A
thispaper	16	2	1001	220	50	0.6	660

The design of the SiLU activation function is the same. It consumes more resources to achieve lower errors and uses a lower clock frequency to achieve a faster processing speed, as shown in Table IV.

TABLE IV. SiLU DESIGN COMPARE WITH OTHERS

	[12]	[13]	This paper
Emax	1.25E-02	1.89E-02	0.48E-02
Input format	16bFXP	16bFXP	16bFXP
Output Format	12b FXP	16bFXP	16bFXP
LUT	140	235	1192
ff	23	153	175
dsp	1	0	1
Freq（MHz）	N/A	200	50
Delay(ns)	9856	30	60

IV. POWER ANALYSIS

Fig. 8. Power analysis experimental environment

We conducted the power consumption analysis experiment using the ChipWhisperer CW305 development board, aiming to analyze the network structure being executed. The experimental environment is shown on Fig.8. CW305 has removed the bypass capacitors and dedicated voltage measurement resources are provided. First, we use the analog to digital converter (ADC) on the capture board CW-Lite to perform voltage sampling at the rising edge of each clock cycle. Due to resource limitations, we only deployed one C3k2 block internally. Fig.9 shows the power traces when applied to the C3k2 block for operations. We can obtain the number of runs of the block through simple power consumption analysis. As shown in this figure, the block has run a total of 10 times.

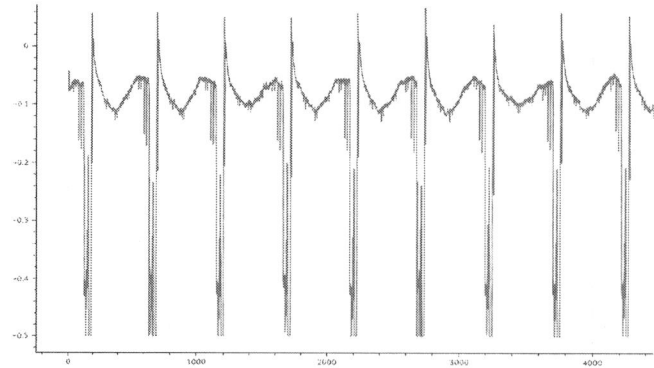

Fig. 9. The power traces of C3k2 operations

Then, the SPA attack on the input data of the SiLU activation function was mainly studied. The curve we obtained by collecting the power supply voltage drop of SiLU during operation is shown in Fig.10. After that, we process the data of these curves, converting them from the time domain to the frequency domain, filtering them through a high-pass filter, and then returning them to the time domain. Finally, the average value of the obtained data is taken.

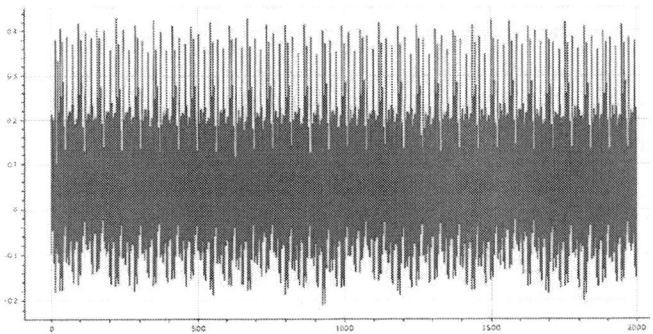

Fig. 10. The power supply voltage drop of SiLU

As shown in Fig.11, we compare the obtained 16 points with the original input data and find that the points below the set threshold of 0.2 represent positive numbers in the input data, while the points above 0.2 represent negative numbers.

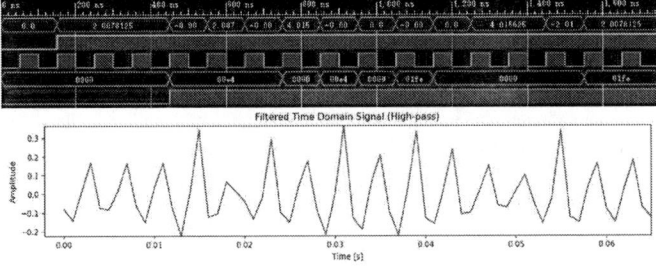

Fig. 11. Comparison of input data and power consumption curves

V. CONCLUSION

This paper conducts the high-speed and high-precision hardware design for the C3k2 block and SPA on it. During the hardware design stage, firstly, we used Add-then-Multiply instead of MAC in the Conv layer. Then in the BN layer, we used the recursive algorithm to reduce the delay, enabling it to have a delay of 660ns with a percentage error of 0.6%. Finally, in the SiLU activation function, we used a combination of LUT and Pade' Approximant for the design, achieving a delay of 60ns with an absolute error of less than 0.48E-02. This is an improvement for running YOLOv11 on edge devices. In the power analysis stage, we successfully analyzed the number of

runs of this block using the SPA method. And in the SiLU activation function, we successfully analyzed the positive and negative properties of its original input. Subsequently, we will conduct further work on SCA, including specific attacks and related defense measures.

REFERENCES

[1] D. Feng, A. Harakeh, S. L. Waslander, and K. Dietmayer, "A review and comparative study on probabilistic object detection in autonomous driving," IEEE Transactions on Intelligent Transportation Systems, vol. 23, no. 8, pp. 9961–9980, 2022.

[2] P. F. Jaeger, S. A. Kohl, S. Bickelhaupt, F. Isensee, T. A. Kuder, H.- P. Schlemmer, and K. H. Maier-Hein, "Retina u-net: Embarrassingly simple exploitation of segmentation supervision for medical object detection," in Machine Learning for Health Workshop. PMLR, 2020, pp. 171–183.

[3] S. Ren, K. He, R. Girshick and J. Sun, "Faster R-CNN: Towards Real-Time Object Detection with Region Proposal Networks," IEEE Transactions on Pattern Analysis and Machine Intelligence, vol. 39, no. 6, pp. 1137-1149, 1 June 2017.

[4] M. Tan, R. Pang and Q. V. Le, "EfficientDet: Scalable and Efficient Object Detection," 2020 IEEE/CVF Conference on Computer Vision and Pattern Recognition (CVPR), Seattle, WA, USA, 2020, pp. 10778-10787.

[5] J. Redmon, S. Divvala, R. Girshick and A. Farhadi, "You Only Look Once: Unified, Real-Time Object Detection," 2016 IEEE Conference on Computer Vision and Pattern Recognition (CVPR), Las Vegas, NV, USA, 2016, pp. 779-788.

[6] S. Zhang, J. Cao, Q. Zhang, Q. Zhang, Y. Zhang and Y. Wang, "An FPGA-Based Reconfigurable CNN Accelerator for YOLO," 2020 IEEE 3rd International Conference on Electronics Technology (ICET), Chengdu, China, 2020, pp. 74-78.

[7] C. Bao, T. Xie, W. Feng, L. Chang and C. Yu, "A Power-Efficient Optimizing Framework FPGA Accelerator Based on Winograd for YOLO," IEEE Access, vol. 8, pp. 94307-94317, 2020.

[8] A. Montgomerie-Corcoran, P. Toupas, Z. Yu and C. -S. Bouganis, "SATAY: A Streaming Architecture Toolflow for Accelerating YOLO Models on FPGA Devices," 2023 International Conference on Field Programmable Technology (ICFPT), Yokohama, Japan, 2023.

[9] S. Moini, S. Tian, D. Holcomb, J. Szefer and R. Tessier, "Power Side-Channel Attacks on BNN Accelerators in Remote FPGAs," IEEE Journal on Emerging and Selected Topics in Circuits and Systems, vol. 11, no. 2, pp. 357-370, June 2021.

[10] L. Wu, L. Wu, X. Zhang and M. Chinbat, "Dual-Rail Precharge Logic-Based Side-Channel Countermeasure for DNN Systolic Array," IEEE Transactions on Very Large Scale Integration (VLSI) Systems, vol. 32, no. 9, pp. 1740-1743, Sept. 2024

[11] T. Sledevic, "Adaptation of Convolution and Batch Normalization Layer for CNN Implementation on FPGA," 2019 Open Conference of Electrical, Electronic and Information Sciences (eStream), Vilnius, Lithuania, 2019, pp. 1-4.

[12] L. Wei, J. Cai, V. Nguyen, J. Chu, and K. Wen, "P-SFA: Probability based sigmoid function approximation for low-complexity hardware implementation," Microprocessors Microsyst, vol. 76, no. 103105, 2020.

[13] Z. Pan, Z. Gu, X. Jiang, G. Zhu and D. Ma, "A Modular Approximation Methodology for Efficient Fixed-Point Hardware Implementation of the Sigmoid Function," in IEEE Transactions on Industrial Electronics, vol. 69, no. 10, pp. 10694-10703, Oct. 2022.

Secondary-Side Controlled Flyback Converter with Buck-Bypass for Multi-Port Application

Zhen Wang
College of Integrated Circuit Science and Engineering
Nanjing University of Posts and Telecommunications
Nanjing, China
wangzhen010318@outlook.com

Yi Liu
College of Integrated Circuit Science and Engineering
Nanjing University of Posts and Telecommunications
Nanjing, China
liuyi@njupt.edu.cn

Abstract—This paper proposes a highly integrated fast-charging system architecture, which combines a Flyback converter with communication protocol identification to achieve efficient power transfer and multi-protocol compatibility. The system adopts a single-primary, multi-buck topology, integrating the fast-charging protocol chip to dynamically regulate output voltage and current in coordination with the main controller. This design not only improves charging efficiency but also ensures battery safety. The paper presents a detailed analysis of energy transfer in the Flyback converter operating under Discontinuous Conduction Mode (DCM),and introduces synchronous rectification (SR) technology. A pre-turn-off pull-down mechanism is implemented to optimize SR MOSFET switching timing, thereby reducing power loss. To support multi-port charging applications, the design incorporates a buck-bypass mode, allowing direct power delivery from the Flyback converter under high-load or constant-voltage conditions. This enables smooth switching between bypass and converter modes, enhancing overall efficiency and system stability. Experimental results based on a 65W multi-port charger prototype demonstrate that the proposed architecture significantly improves conversion efficiency under 264 Vac and various load conditions, while maintaining low output voltage ripple. With digital control and optimized layout integration, the system effectively consolidates protocol recognition, power regulation, and multi-port management, providing essential technical support for high power density and protocol-flexible fast-charging solutions.

Keywords—Flyback, Synchronous Rectification, Switching Loss Reduction, DCM Operation, Multi-Port, Bypass Mode

I. INTRODUCTION

Fast charging technology relies on the coordinated design of efficient power architectures and intelligent system control. Among them, the Flyback converter is widely adopted in fast charging systems due to its simple structure and wide output voltage adaptability. To ensure the charging process matches the real-time requirements of end devices, the system incorporates a protocol recognition and adaptation mechanism, which interacts with the device to identify its power profile and adjust the output accordingly. This dynamic adaptation loop enables accurate voltage and current regulation via the PWM controller, aligning the power delivery with device demands.

The architecture proposed in this paper integrates protocol recognition with output control and power regulation, optimizing energy transfer paths and system coordination.

As shown in Fig. 1, the fast-charging system incorporates a protocol identification IC between the main power source(charger) and the terminal device to automatically recognize charging protocols and dynamically match the output voltage. This IC handles communication and parameter negotiation between the device and the charger, enabling real-time voltage and current adjustment based on device requirements and significantly improving overall charging performance.

Fig. 1. Typical Charger Architecture

Figure 2 shows the mainstream topology used in a 65W output power charger [1-2], featuring three Integrated Circuits (ICs) respectively for the main PWM controller, protocol controller, and synchronous rectification (SR) controller to meet system requirements.

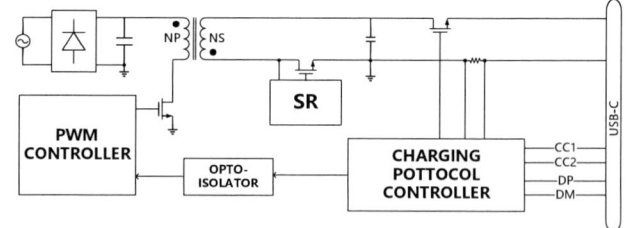

Fig. 2. architecture for common charger with 3 ICs

Building on this foundation, multi-port output has gradually become a mainstream trend to meet the growing demand for simultaneous charging of multiple devices. This evolution places higher requirements on protocol compatibility, power management strategies, and dynamic power allocation capabilities. Although the design of multi-port chargers is more complex due to the need to support multiple voltage levels, they can efficiently power several devices at once, significantly enhancing the user experience and offering greater competitiveness in the market.

Fig.3 illustrates the mainstream topology widely adopted in current multi-port chargers, where the combination of a

979-8-3315-8850-2/25 $31.00 © 2025 IEEE

Flyback converter and Buck converters is particularly common. This architecture generates an intermediate DC bus through AC/DC conversion, with each port using a Buck converter to regulate its output voltage, offering strong adaptability and flexibility. Since only one Flyback converter is used for AC/DC conversion, system cost is reduced. However, the presence of separate Buck and fast-charging controllers for each port introduces multiple power conversion stages, which may impact overall efficiency [3-5].

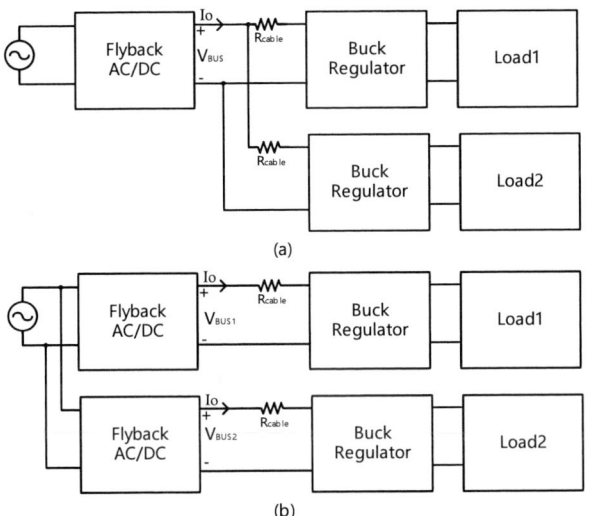

Fig. 3. Multi-Port Architecture with Intermediate Bus

Fig.3 shows the multi-port architecture with an intermediate DC bus in two configurations. Fig.3(a) has a single primary with multiple buck stages, benefiting design cost via simpler peripherals. In contrast, Fig.3(b) uses multiple primary and secondary controllers, being more complex and costlier.

From the perspective of load efficiency, architecture with multiple buck stages introduces additional DC-DC conversion stages, leading to increased power loss and reduced overall efficiency. The overall efficiency calculation of the multi-port structure is as indicated by (1). POUT represents the transmission power independently borne by each output port among different power ports, and Eflyback and Ebuck respectively represent the output efficiency of each type of converter.

$$E_{TOTAL} = \frac{P_{OUT}}{P_{IN}} = \frac{P_{OUT1} + P_{OUT2} + ..}{(\frac{P_{OUT1}}{E_{flyback1} \times E_{buck1}}) + (\frac{P_{OUT2}}{E_{flyback2} \times E_{buck2}}) + ..} \quad (1)$$

Therefore, to improve the efficiency of the topology where a single primary works with multiple buck stages, it is essential to reduce the loss of the DC-DC buck converters, thereby optimizing the performance of multi-port fast chargers.

Traditional solutions usually use independent protocol chips, which increase system size and pose challenges for miniaturized designs, especially in multi-output scenarios that require multiple protocol chips. So integrated solutions embed protocol functions into digital circuits, reducing communication delays and mismatches between multiple chips, enhancing system reliability and stability, while also lowering the risk of failures caused by poor connections.

This paper proposes a highly integrated optimized solution for dual-port converters. Combined with meticulous layout design, it achieves a high degree of integration between analog and digital control circuits, simplifying the peripheral circuitry. The architecture adopts a secondary-side controlled Flyback converter combined with a bypass Buck mode, significantly improving energy transfer efficiency and multi-protocol compatibility. In addition, the solution incorporates synchronous rectification technology to further optimize system performance.

II. PROPOSED ARCHITECTURE

Fig. 4 shows the overall block diagram of the dual-port fast charging system proposed in this paper. The system uses a Flyback converter as the main power conversion module to generate an intermediate voltage bus, and transmits secondary-side control signals to the primary side through an opto-isolator isolation feedback mechanism, achieving safe and reliable voltage regulation. The primary-side PWM controller is responsible for controlling the on/off time of the main switch transistor, thereby regulating energy transfer to the secondary side.

After the main switch turns off, the secondary-side synchronous rectification (SR) control unit monitors the transformer secondary status and controls the SRFET switching to reduce rectification loss.

The system's control unit integrates multiple key functions, including synchronous rectification control for the Flyback, Buck step-down control for each output port, and charging protocol identification and negotiation functions. Each output port is equipped with an independent Buck converter to further regulate the intermediate voltage bus, adapting to different protocol requirements.

In addition, each Buck circuit is parallel with a bypass FET, which conducts under specific load and voltage conditions, allowing the output voltage to be directly delivered from the Flyback output to the port, bypassing the Buck regulation stage, thereby improving system efficiency.

A. Analysis of the Flyback Converter

The Flyback converter is the most common topology in chargers due to its wide input voltage range, simple structure, and low cost. In low-power applications under 100W, it typically operates in Discontinuous Conduction Mode (DCM), which eliminates the Right Half Plane zero, simplifies compensation design, speeds up voltage regulation, and improves efficiency through Zero Current Switching (ZCS).

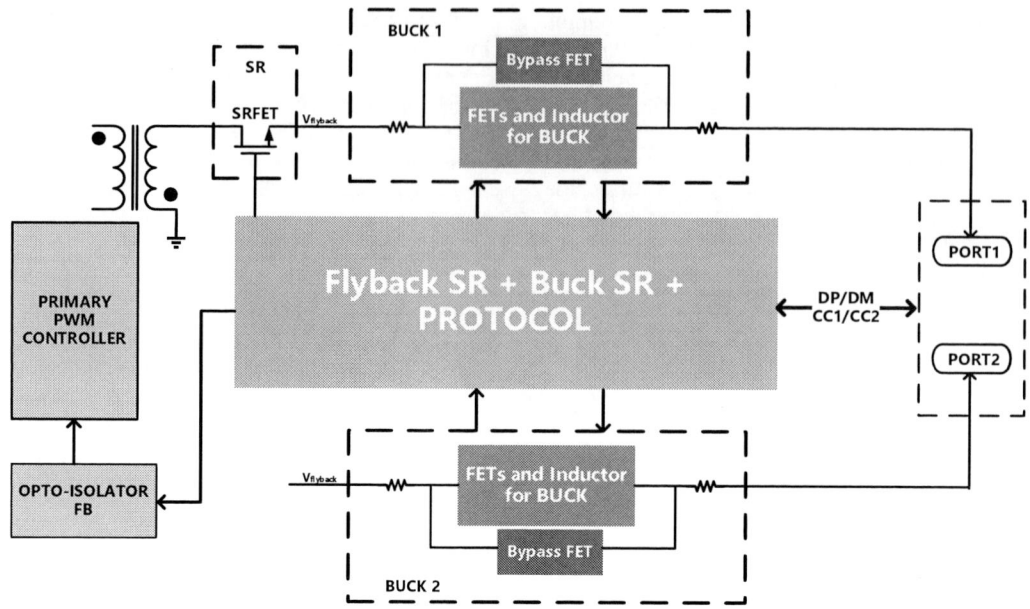

Fig. 4. Block diagram of the proposed circuit

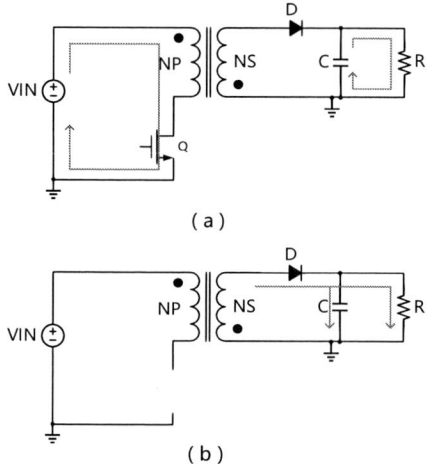

(a)

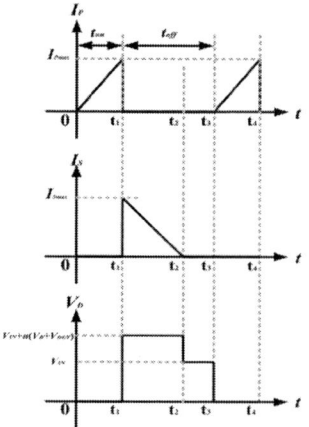

(b)

Fig. 5. Equivalent Circuit of the Flyback Converter Operation

Fig. 6. Flyback Operating Mode

Fig. 5 shows the circuit diagram of the Flyback converter.

When Q is turned on, the input voltage is directly connected to the primary side of the transformer. In this case, the magnetizing current of the magnetizing inductor increases according to the input voltage, and the output diode (D) is reverse biased due to the transformer polarity. The capacitor (C) supplies energy to the load. Fig. 5(a) shows the Flyback circuit operation when the switch is closed.

The energy stored in the inductor Lp can be expressed as:

$$E_1 = \frac{1}{2} L_P \cdot I_{PMAX}^2 = \frac{1}{2} L_P \cdot \left(\frac{V_{IN}}{L_P} t_{ON} \right)^2 \qquad (2)$$

When Q is turned off, the magnetizing current decreases to zero before the end of the switching interval.

In DCM operation mode, the transformer's magnetizing inductance is small enough to ensure the condition above. As shown in Fig. 5(b), the magnetizing energy is transferred to the output load through the diode (D).

After the secondary-side current drops to zero, the circuit enters this stage, called the dead-time stage. During this period, the secondary-side power switch is also off. Since the next conduction cycle has not yet begun, only the capacitor supplies energy to the load, and this continues until the next conduction cycle starts.

The energy consumed by the load during the entire cycle can be expressed as:

$$E_2 = \frac{V_{OUT}^2}{R_L} T \qquad (3)$$

979-8-3315-8850-2/25 $31.00 © 2025 IEEE

According to the law of conservation of energy [6], under ideal conditions, the energy stored in the primary inductor during the conduction period of the primary power switch is equal to the energy consumed by the load over the entire cycle. Therefore, from (2) and (3), we can obtain:

$$V_{OUT} = V_{IN} \cdot t_{ON} \sqrt{\frac{R_L}{2 L_P T}} \qquad (4)$$

Fig. 6 shows the main operating waveforms of the flyback switching power supply in DCM mode (ignoring the effects of LC resonance), including the primary current Ip secondary current Is, and the drain-source voltage VDS of the primary power switch. In the figure, I_{PMAX} and I_{SMAX} represent the maximum values of the primary and secondary currents, respectively, and n is the turns ratio between the transformer's primary and secondary windings.

B. Synchronous Rectification

Traditional Flyback and buck circuits use conventional rectifier diodes, whose forward voltage drop of several hundred millivolts can easily cause energy loss in low-power systems. Synchronous rectification achieves precise control of the turn-on and turn-off timing of synchronous rectification MOSFETs (SRFETs). Thanks to their extremely low on-resistance, SRFETs can emulate the behavior of rectifier diodes while providing a conduction path with much lower loss. Their gate drive signals can be generated by the synchronous rectification control loop [7].

However, in the Flyback topology, the current fall rate when the SRFET turns off can reach up to 150 A/μs, which can easily cause body diode reverse breakdown. At the same time, excessive reverse current may also lead to increased turn-off propagation delay, affecting system stability.

To address this, this paper introduces a pre-off pull-down function in the synchronous rectification topology: before the SRFET turns off, the gate voltage is pulled down to a certain level in advance. Combined with the rising on-resistance and the decreasing secondary current, this effectively balances the drain-source voltage, shortens the conduction time of the body diode, ensures the SRFETs quickly turn off near its threshold, thereby reducing short-circuit risk and enhancing system reliability.

C. Bypass Buck Mode

This paper additionally introduces a Buck-bypass control strategy by inserting a power MOSFET stage between $V_{flyback}$ and VBUS to bypass the buck converter of the highest output voltage port. This strategy supports dynamic switching of operating modes under different load and port voltage requirements, achieving collaborative optimization between the Flyback converter and the downstream Buck converters.

The controller provides feedback control signals to the primary-side controller, then controls $V_{flyback}$ based on the VBUS output voltage required by each port. The buck converter of the highest output port is bypassed and $V_{flyback}$ is calculated according to the following equation:

$$V_{flyback} = MAX(V_{BUS1}, V_{BUS2}, \ldots) \qquad (5)$$

Two different operating modes are described as the 'Bypass Mode' and the 'Converter Mode' [8].

Bypass Mode: In high voltage and full-load scenarios, the system supplies power directly from the Flyback to the output VBUS through the bypass MOSFET, skipping the Buck conversion stage. This reduces power loss caused by the Buck converter. Taking advantage of the bypass MOSFET's low on-resistance and the benefit of eliminating an additional power conversion stage, this mode is used to match the highest voltage output. As shown in (6), power loss in this case occurs only in the on-resistance of power MOSFET used for bypassing.

$$P_{bypass} = R_{DS(ON)} \cdot I_o^2 \qquad (6)$$

Since there is no switching loss from the Buck stage, the overall system efficiency can typically be improved by several percentage points.

Converter Mode: In applications requiring a wide output voltage range (e.g., 3.3–21 V in USB PD), light load, or dynamic load conditions, the system keeps the Flyback converter operating at an optimal fixed voltage point, while the Buck converter handles wide-range voltage regulation. This 'fixed source with adjustable output stage' configuration effectively suppresses the frequent efficiency fluctuations of the Flyback converter caused by dynamic voltage adjustments, and it significantly improves overall efficiency and stability, especially under low-voltage or light-load conditions.

However, when operating in Converter Mode, the power path includes an additional Buck converter, resulting in power loss $P_{converter}$ that mainly originates from the conduction loss of power devices and switching loss.Conduction loss is mainly caused by the on-resistance of the switching devices and the inherent resistance of the inductor while the switching loss P_{switch} is caused by the overlap of voltage and current during the switching transitions of the MOSFET in the Buck circuit:

$$P_{converter} = P_{switch} + (R_{DS(ON)} \cdot D + R_{inductor}) \cdot I_o^2 \qquad (7)$$

It is apparent from (6) and (7) that under the condition of the same output current, the Bypass Mode can reduce losses in the power conduction path. This is particularly true when the output port requires a larger duty cycle D to achieve a higher output, thereby efficiently improving the converter's efficiency.

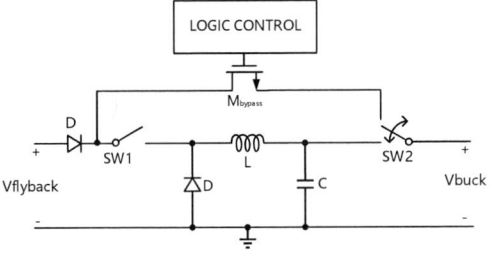

Fig. 7. Bypass Mode Buck Circuit

The structure of the Bypass Mode is shown in Fig.7,the buck converter consists of semiconductor switch SW ,diode D,inductor L,and output capacitor C.During system operation, once a certain port requests a higher voltage, the controller activates the bypass path of the selected port by turning off switch SW1 and connecting SW2 to the output, allowing it to be powered directly from the Flyback converter, thereby switching to bypass mode. Diode D protects against reverse current flowing into the Flyback converter to ensure system stability [9], with relatively lower loss caused.

In summary, the proposed Buck-Bypass control strategy enables flexible, real-time adaptation to varying load and voltage requirements across multiple ports. By selectively bypassing the Buck stage for high-voltage ports and dynamically coordinating power delivery through the controller, the system achieves a balance between conversion efficiency and voltage flexibility.

To fully realize the potential of such architecture, intelligent coordination between power conversion and protocol communication is essential. The next section introduces the protocol recognition and power allocation mechanism, which enables voltage/current negotiation and dynamic adjustment based on device requirements.

D. Protocol Recognition and Power Allocation

The protocol unit first establishes communication with the device via CC/DP/DM pins based on charging protocols (e.g., the widely adopted USB PD). During this process, the controller dynamically allocates power among multiple ports in combination with the maximum output power limit of the Flyback converter, and finally determines the target voltage and current parameters for each port. Subsequently, the Flyback converter adjusts its output characteristics according to the control commands and collaborates with the post-stage circuit to realize real-time response to the charging requirements of each port.

This paper proposes a digitally controlled voltage strategy[10]: when a port is detected to require a higher voltage, the system adjusts the Flyback output voltage ($V_{flyback}$) to match the highest voltage demand among all ports. During this transition, the system enters a dead zone mode (where all ports operate in converter mode). After completing the primary-side interaction and power port allocation, the port requiring the high voltage switches to bypass mode, eliminating the buck process to reduce power loss at the port level and improve overall system efficiency.Fig.8 illustrates the protocol recognition and power allocation response mechanism flow.

III. EXPERIMENTAL RESULTS

The layout is designed based on the DBHitek 0.18 μm 40V BCD process and is shown in Fig. 9. The total layout size of the converter is 1255 μm × 1170 μm.

Table1 summarizes the parameters for the main speci-fications. Fig. 10 shows the operation of the Flyback converter, illustrating the relationship between the control signal and the drain voltage VDS of the SRFET. Before VDS reaches the turn-off threshold, the gate signal effectively performs a pre-turn-off action. Fig. 11 presents the operating waveform of the

buck circuit working in Converter Mode. Fig. 12 shows the output voltage ripples under both light and heavy load conditions while operating in Bypass Mode, with ripple values remaining below 130 mVpp. Fig. 13 compares the efficiency of a single output port under different output voltages and heavy load conditions (75% and 100%), with the comparison made between the bypass mode and the existing converter mode. The results show that the proposed method can efficiently improve efficiency under heavy load conditions. Fig. 14 illustrates the power allocation across two output ports when different devices are connected to each port. Under various combinations, the total power consumption of both ports does not exceed 65 W.

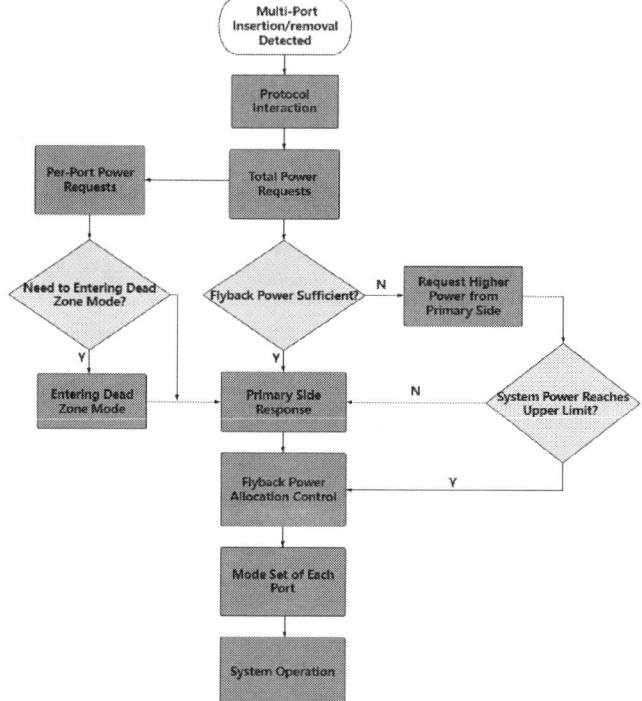

Fig. 8. Protocol Recognition and Power Allocation Process

TABLE I. SUMMARY OF PARAMETERS FOR MAIN SPECIFICATIONS

Specification	Value / Range
Maximum output power (Po)	65W
Maximum output current	3.25A
Input voltage (Vin)	90 - 264 V
Maximum switching frequency	150kHz
transformer turns ratio	85：6
input capacitance(Cin)	4.7μF
Flyback output capacitance(Cout)	470μF
High-side switch turn-on resistance	70mΩ
Low-side switch turn-on resistance	35mΩ
By-pass switch turn-on resistance	6.8mΩ
Inductor	4.7μH

IV. CONCLUSION

This paper presents a high-efficiency power architecture for multi-port charging applications, with a focus on optimizing output control strategies. The proposed architecture combines a secondary-side controlled Flyback converter with two output mechanisms—Bypass Mode and Converter Mode—enabling dynamic mode switching based on the conditions across different ports. This approach effectively reduces power loss and improves overall system efficiency. Furthermore, the introduction of synchronous rectification and pre-turn-off techniques further enhances system performance. Experimental results validate that the proposed solution achieves low output voltage ripple, high efficiency, and reliable operation under varying load conditions.

Fig. 9. Layout of the Proposed Circuit

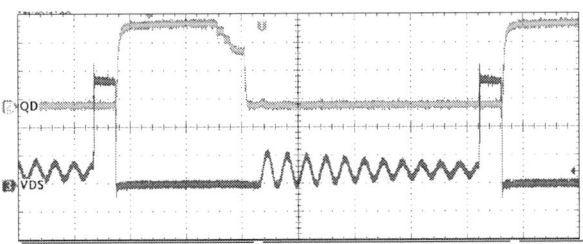

Fig. 10. Flyback Operating Waveforms

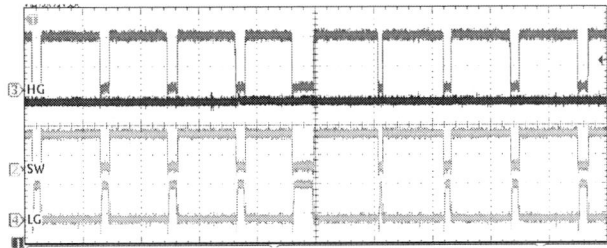

Fig. 11. Buck Operating Waveforms

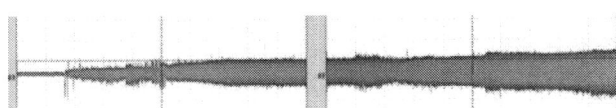

Fig. 12. System Output Ripple

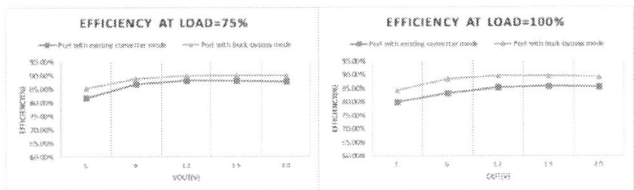

Fig. 13. Efficiency Under Heavy Load

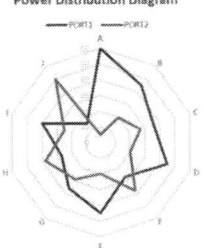

Fig. 14. Power Allocation Charts for Dual Portst

REFERENCES

[1] D. Milind, J. Thomas, R. Rajarajan and K. G. Acharya, "Secondary Side-Controlled Fly-Back Converter for USB PD Charger Application," 2024 IEEE International Communications Energy Conference (INTELEC), Bengaluru, India, 2024, pp. 1-6, doi: 10.1109/INTELEC60315.2024.10679026.

[2] Chen Zhao, Member, IEEE, Xiaogao Xie, Member, IEEE, HanjingDong,and Shirong Liu, "Improved Synchronous Rectifier Driving Strat-egy forPrimary-Side Regulated (PSR) Flyback Converter in Light-LoadMode",in IEEE Transactions on power electronics, Vol. 29, No. 12,December 2014

[3] Kai-Hui Chen and Tsorng-Juu Liang, "Design of Quasi-resonant flyback converter control IC with DCM and CCM operation," in 2014 International Power Electronics Conference (IPEC-Hiroshima 2014 – ECCE ASIA), Hiroshima, 2014, pp. 2750-2753.

[4] I. Ahmad and B. G. Fernandes, "Concept of Universal USB Charger," 2020 IEEE Industry Applications Society Annual Meeting, Detroit, MI, USA, 2020, pp. 1-5, doi: 10.1109/IAS44978.2020.9334768.

[5] X. Tian, H. Cui and L. Xue, "Multiplexing-Based Flyback Converter for Multi-Port USB Power Delivery with True Power-Sharing," 2021 IEEE Energy Conversion Congress and Exposition (ECCE), Vancouver, BC, Canada, 2021, pp. 2048-2054, doi: 10.1109/ECCE47101.2021.9595480.

[6] Kuldip C, Lakshminarasamma N,Flyback Based Resonant Converter for High Voltage Pulsed Load Application [C]//2022 IEEE International Conference on Power Electronics, Drives and Energy Systems (PEDES).IEEE,2022:1-7.

[7] W. Li, S. Huang and Q. Duan, "A Synchronous Rectifier Energy Collection Circuit with Two Energy-Transfer Schemes for A 65% Power Efficiency," 2019 IEEE 4th International Conference on Integrated Circuits and Microsystems (ICICM), Beijing, China, 2019, pp. 1-5, doi: 10.1109/ICICM48536.2019.8977188.

[8] L. Terry, "Multiport Type-C & PD Charger Topology and Control Methodologies," PCIM Europe 2023; International Exhibition and Conference for Power Electronics, Intelligent Motion, Renewable Energy and Energy Management, Nuremberg, Germany, 2023, pp. 1-4, doi: 10.30420/566091233.

[9] A. Lücken, T. Kut, S. Dickmann and D. Schulz, "Fuel Cell System Optimization Using Bypass Converters," in IEEE Transactions on Aerospace and Electronic Systems, vol. 50, no. 1, pp. 170-179, January 2014, doi: 10.1109/TAES.2013.120058.

[10] Lisa Dinwoodie, "Exposing the Inner Behavior of a Quasi-Resonant Flyback Converter," in Texas Instruments Power Supply Design Seminar,2012, pp. 2- 27

2025 The 10th International Conference on Integrated Circuits and Microsystems

Research on the Method of Switching Adjustment to Reduce the Temperature Drift of the Bandgap Reference

Bin Wen
School of Automation and
Information Engineering
Xi'an University of Technology
Xi'an, China
wenbin1013@163.com

Zhongjie Guo
School of Automation and
Information Engineering
Xi'an University of Technology
Xi'an, China
zjguo@xaut.edu.cn

Hongpeng Dai
School of Automation and
Information Engineering
Xi'an University of Technology
Xi'an, China
2056583730@qq.com

Xiaojing Qiang
School of Automation and
Information Engineering
Xi'an University of Technology
Xi'an, China
1203413986@qq.com

Abstract—In order to solve the problem that accurate high-order compensation accuracy in bandgap reference design is difficult to achieve, this paper proposes a method of reducing the temperature drift of the bandgap reference by using switch adjustment. The effective resistance value of the access resistor in the network is controlled and adjusted by the temperature detection circuit, thereby adjusting the absolute value of the bandgap reference output voltage. The amount of change in the resistance value of the adjustment module is used to compensate for the amount of drift of the reference voltage caused by temperature changes, in order to achieve the purpose of reducing the temperature drift of the output voltage. In this paper, the specific verification of this method is based on the 0.18μm BCD process. The results show that through the adjustment of this method, the temperature drift coefficient of the bandgap reference can be reduced from 8.394 ppm/°C to 1.997 ppm/°C, and the adjustment function can be realized.

Keywords—bandgap reference, switch adjustment, low temperature drift

I. INTRODUCTION

The bandgap reference circuit can provide a voltage that fluctuates very little with the influence of power supply voltage, temperature, and process fluctuations. It is a very important module in the design of analog integrated circuits and digital-to-analog hybrid circuits. It is widely used in power management, analog-to-digital converters, digital-to-analog converters and other circuits[1]. The design technology of the reference voltage source has become one of the key technologies in analog integrated circuits. Therefore, it is essential to study how to achieve a high-precision, low-temperature drift and high-stability bandgap reference circuit.

The design idea of the traditional bandgap reference module is to use the negative temperature coefficient of the base-emitter voltage V_{BE} of the transistor and the positive temperature coefficient of the difference ΔV_{BE} of the base-emitter voltage of the two triodes operating at different current densities to superimpose, and the compensated bandgap reference voltage can be obtained by designing the coefficients of the two voltages superposition.

However, due to the existence of high-order terms in V_{BE}, only the bandgap reference after first-order compensation can be obtained by the above method, and the temperature drift coefficient is high. The design of continuing to reduce the temperature drift of the bandgap reference requires the addition of high-order compensation. High-order compensation is to compensate for the high-order temperature coefficient of the bandgap reference. However, the demand for a lower temperature drift coefficient requires higher compensation order and compensation accuracy for high-order compensation, which is more difficult to achieve. In this paper, a method of using switch adjustment to reduce the temperature drift of the bandgap reference is proposed, which can effectively reduce the temperature drift of the bandgap reference and realize the design of low temperature drift.

II. THEORETICAL ANALYSIS OF BANDGAP REFERENCE SOURCE

Bandgap voltage refers to the voltage that has good stability to temperature, voltage, and process corner fluctuations. In the current circuit design, the two main bandgap reference sources are MOS-only voltage reference and bipolar bandgap voltage reference. The MOS-only voltage reference mainly uses the MOS operating in the sub-threshold region, and the negative temperature coefficient voltage is generated by the gate-source voltage V_{GS}, and the positive temperature coefficient voltage is provided by the gate-source voltage difference ΔV_{GS} of the two MOSs operating in the sub-threshold region; the bipolar bandgap reference source mainly uses the BJT operating in the amplification region, and the negative temperature coefficient voltage is generated by the emission junction voltage V_{BE}, and the positive temperature coefficient voltage is generated by the base-collector voltage difference V_{BE} of two triodes with different current densities.

The design idea of both the MOS-only voltage reference and the Bipolar bandgap voltage reference is to compensate with a positive temperature coefficient and a negative temperature coefficient to obtain the bandgap reference voltage after the first-order compensation. The following analyzes the design of MOS-only and Bipolar bandgap voltage references respectively.

979-8-3315-8850-2/25 $31.00 © 2025 IEEE

A. Bipolar Bandgap Reference Source

In the design of the bipolar bandgap reference source, the transistor base-collector voltage V_{BE} is used to generate a voltage with a negative temperature coefficient, and the transistor base-emitter voltage difference of two different current densities is used to generate a positive temperature coefficient voltage from V_{BE}[2][1], and then through the circuit superposition, the bandgap reference voltage after the first-order compensation is realized.

According to the different gate-source voltage and drain-source voltage of the MOS device, the operating area of the MOS device can be divided into cut-off region, sub-threshold region, linear region and saturation region. Taking NMOS as an example, when the gate-source voltage V_{GS} is less than the threshold voltage V_{TH}, MOS is turned off, and when V_{GS} is greater than V_{TH}, the device is turned on. However, when the gate-source voltage is slightly lower than V_{TH}, the conductive channel is in a weakly inverted state, and the MOS device still has a weak on-current. At this time, the working state of MOS is in the sub-threshold region, and the corresponding current is called the sub-threshold current I_{DSUB}. When BJT is in the amplification region, the emitter junction is positively biased, and V_{BE} can be expressed as:

$$V_{BE} = V_T \ln\left(\frac{I_C}{I_S}\right) \tag{1}$$

Where I_C is the collector current and I_S the saturation current, which can be written as:

$$I_S = bT^m \exp\left(\frac{-E_g}{KT}\right) \tag{2}$$

The available transistor base emitter voltage V_{BE} is:

$$V_{BE} = V_{g0}\left(1 - \frac{T}{T_0}\right) + V_{BE0}\left(\frac{T}{T_0}\right) + \frac{mKT}{q}\ln\left(\frac{T}{T_0}\right) + \frac{KT}{q}\ln\left(\frac{I_C}{I_{C0}}\right) \tag{3}$$

Derive the V_{BE} in (3) from the temperature T to obtain:

$$\frac{\partial V_{BE}}{\partial T}\Big|_{T=T_0} = \frac{V_{BE0} - V_{g0}}{T_0} + (\alpha - m)\left(\frac{K}{q}\right) \tag{4}$$

At room temperature (27°C), α is about -1.5 and m is about 4. When the emitter junction voltage V_{BE} is about 750mV, the derivative value of V_{BE} to T in (4) is about -1.5mV/K. Therefore, the emitter junction voltage V_{BE} of the BJT operating in the amplification region has a negative temperature coefficient.

The positive temperature coefficient is determined by the difference between the emitter junction voltages of the two BJTs operating in the amplification region:

$$\Delta V_{BE} = V_{BE1} - V_{BE2} = V_T\ln\frac{nI_0}{I_{S1}} - V_T\ln\frac{I_0}{I_{S2}} = V_T\ln n \tag{5}$$

By matching and superimposing the voltages with positive and negative temperature coefficients in a suitable proportion, a reference voltage with zero temperature coefficient can be obtained.

B. MOS-only Voltage Reference

Similar to the design idea of bipolar bandgap reference source, the realization of MOS-only voltage reference is also obtained through positive temperature coefficient voltage and negative temperature coefficient voltage compensation. The negative temperature coefficient voltage in the MOS-only voltage reference comes from the gate-source voltage V_{GS} of the MOS in the sub-threshold region[3], and the positive temperature coefficient is provided by the gate-source voltage difference ΔV_{GS} of the two MOSs operating in the sub-threshold region. Through the superposition of the above-mentioned positive temperature coefficient voltage and negative temperature coefficient voltage, the reference voltage after the first-order compensation can be designed.

The current corresponding to the sub-threshold region where the MOS device operates is called the sub-threshold current I_{Dsub}.

$$I_{Dsub} = (n-1)\frac{W}{L}\mu\left(\frac{KT}{q}\right)^2 C_{OX}\cdot\exp\left[\frac{q}{KT}\left(\frac{V_{GS}-V_{TH}}{n}\right)\right]\cdot\left[1-\exp\left(\frac{-qV_{DS}}{KT}\right)\right] \tag{6}$$

Where n is the sub-threshold slope factor; W/L is the aspect ratio of the MOS; μ is the carrier mobility; k is the Boltzmann constant, with a value of 1.380649×10^{-23}J/K; q is the amount of charge, with a value of 1.6×10^{-19}C; kT/q is the thermal voltage V_T; C_{ox} is the gate oxide capacitance per unit area; V_{TH} is the threshold voltage of the MOS; V_{DS} is the drain-source voltage of the MOS.

When V_{DS} is $>3V_{TH}$, the effect of V_{DS} on the sub-threshold current can be ignored. The expression of the MOS gate source voltage V_{GS} in the sub-threshold region is:

$$V_{GS} = V_{TH} + \frac{nKT}{q}\ln\left[\frac{I_{Dsub}}{(n-1)\frac{W}{L}\mu(T_0)\cdot\left(\frac{T}{T_0}\right)^{-\frac{3}{2}}\left(\frac{KT}{q}\right)^2 C_{OX}}\right] \tag{7}$$

From (7), the derivative of V_{GS} to temperature can be obtained:

$$\frac{\partial V_{GS}}{\partial T} = \frac{\partial V_{TH}}{\partial T} + \frac{nK}{q}\ln\left[\frac{L}{W}\frac{I_{Dsub}(T_0)}{(n-1)\mu(T_0)\left(\frac{KT}{q}\right)^2}\right] \tag{8}$$

The first term in the above formula is the derivative of the threshold voltage to temperature, which is negative. In the second term of (8), the fractional result in the exponential term is much less than 1, so the second term is also negative. Therefore, the gate-source voltage V_{GS} of the MOS in the sub-threshold region has a negative temperature coefficient.

The positive temperature coefficient voltage is provided by two MOS operating in the sub-threshold region, the gate-source voltage difference ΔV_{GS}.

$$\Delta V_{GS} = V_{GS2} - V_{GS1} = \frac{nKT}{q} \ln \frac{I_{Dsub2}\left(\frac{W}{L}\right)_1}{I_{Dsub1}\left(\frac{W}{L}\right)_2} \qquad (9)$$

In the circuit, the V_{GS} of the negative temperature coefficient and the V_{GS} of the positive temperature coefficient are superimposed by a certain coefficient to obtain the reference voltage after the first-order compensation[4]. The MOS-only voltage reference device operates in the sub-threshold region and has a small on-current, which is suitable for the design of low-power circuits. However, the threshold voltage of MOS devices varies greatly with process temperature and other parameters, and the operating conditions in the sub-threshold region are more demanding[5], and as the transistor size decreases, the short channel effects and gate parasitic current will also cause the device performance to be offset[6]. The stability of this design method is poor.

III. SWITCH ADJUSTMENT REDUCES THE TEMPERATURE DRIFT OF THE REFERENCE VOLTAGE

The reference voltage source provides a stable and accurate reference voltage for the circuit, which plays a decisive role in the performance of the circuit. After the design method of positive and negative temperature compensation, the reference voltage of the first-order compensation can be obtained[7], but the temperature coefficient of the first-order reference voltage is still large, which is not suitable for the design of high-precision low-temperature drift circuits[8]. Therefore, in order to obtain a lower temperature coefficient, a high-order compensation design is proposed in the design. The high-order compensation in the bandgap reference design refers to the higher-order compensation of the reference voltage that has undergone first-order compensation again, and the bandgap reference voltage after the high-order compensation can obtain a lower temperature drift coefficient. The design methods of high-order compensation commonly used in circuits are: I_{PTAT}^2, exponential compensation, and segmented compensation. The design difficulty of high-order compensation lies in high-precision compensation, and the accuracy should be guaranteed for fluctuations in process corner and power supply voltage.

If a lower temperature drift coefficient is obtained during compensation, the compensation order and compensation accuracy of high-order compensation are more demanding, and it is more difficult to achieve. In this paper, a method of reducing the reference temperature drift using switch adjustment is proposed. By changing the resistance value of the adjustment resistor in the temperature coefficient adjustment circuit at different temperature points, the adjustment amount of the temperature coefficient adjustment circuit is used to compensate for the amount of drift caused by the reference temperature change. It solves the problem that the existing bandgap reference circuit is difficult to achieve at very low temperature drift.

The idea of using switch adjustment to reduce the reference temperature drift method proposed in this paper is: first, the bandgap reference circuit is used to achieve the bandgap reference voltage of the first-order compensation, and then the temperature detection circuit is used to generate a temperature flip adjustment signal, and the temperature coefficient adjustment circuit output voltage is changed by controlling and adjusting the resistance to achieve high-order compensation, so as to achieve the purpose of reducing the temperature drift coefficient.

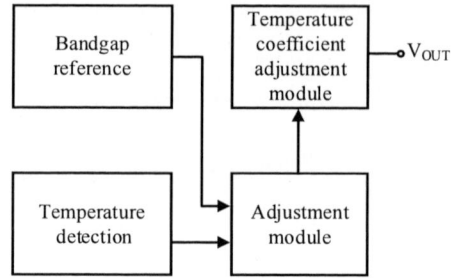

Fig. 1. Switch adjustment circuit diagram

Fig. 1 shows the circuit designed in this paper to adjust the switch to reduce the bandgap reference temperature drift, which includes a bandgap reference circuit, a temperature detection circuit, an adjustment module, and a temperature coefficient adjustment circuit containing an adjustment module. The bandgap reference circuit is used to generate a reference voltage for the first-order compensation before adjustment. The temperature detection circuit is used to generate a temperature flip point that needs to be adjusted, and to generate an adjustment signal to control the on-off of the adjustment switch in the adjustment network, thereby controlling the adjustment network. In order to achieve the compensation of the change in the reference voltage V_{ref} generated by the bandgap reference module bandgap reference caused by temperature, and to achieve the effect of reducing the bandgap reference temperature drift.

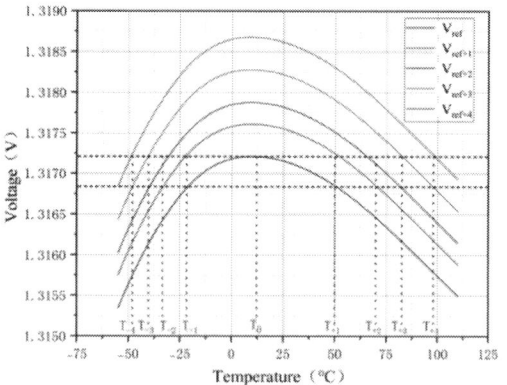

Fig. 2. Adjustable voltage

In Fig. 2, V_{ref} is the bandgap reference voltage that is not connected to the adjustment network, and V_{ref+1}, V_{ref+2}, V_{ref+3}, and V_{ref+4} are the voltage that can be adjusted. The above voltage can be obtained by controlling the resistance value and number of resistors in the adjustment module. V_{ref+1} is the voltage obtained by adjusting once, and similarly V_{ref+2}, V_{ref+3}, and V_{ref+4} are the voltages obtained by adjusting twice, three times, and four times, respectively. In the design, the adjustment module can be controlled by the temperature detection module, and the reference voltage V_{ref} can be adjusted to other voltages in Fig 2 at different temperatures.

979-8-3315-8850-2/25 $31.00 © 2025 IEEE

The adjustment method proposed in this paper can independently design the number of different temperature adjustment points to achieve different optimization effects, and determine the temperature value of the adjustment point according to the number of temperature adjustment points. The more temperature adjustment points, the higher the adjustment accuracy, and the smaller the temperature drift of the bandgap reference voltage after adjustment, but at the same time, the accuracy requirements of the adjustment network design are also higher.

The bandgap reference circuit is used to generate the bandgap reference voltage V_{ref} before the adjustment. The voltage is the bandgap reference after the first-order compensation. The first-order compensation point of the reference voltage is located at the temperature T_0, and the amount of drift in the full temperature range is ΔV_{ref}. The amount of drift of the reference voltage is evenly divided. The proportion of evenly divided is determined according to the number of adjustment points and adjustment resistors. K adjustment resistors can divide the amount of drift K+1 equally, and K adjustment resistors require 2K temperature adjustment points. The temperature adjustment point is designed as a temperature point that can evenly divide the temperature drift of the uncompensated front bandgap reference voltage.

The temperature detection circuit is used to generate a temperature point detection signal that needs to be adjusted by the first-order compensation reference voltage. Define the temperature point that needs to be adjusted as T_{set}, when the temperature is less than T_{set}, the port output is low, and when the temperature is greater than T_{set}, the output is high. The 2K temperature detection signals are processed by digital logic to obtain the adjustment resistor control signal in the adjustment module. Compensation for the change in the bandgap reference voltage is realized through the temperature detection signal control and adjustment module.

Due to temperature changes, the amount of reduction of the bandgap reference voltage is ΔV_{ref}, the amount of reduction after passing through the temperature coefficient adjustment circuit to the output is ΔV_{OUT-}, and the amount of increase in the output voltage of the temperature coefficient adjustment circuit caused by the adjustment module controlled by the temperature detection circuit is ΔV_{OUT+}. So that the increase in the output voltage of the temperature coefficient adjustment circuit caused by the adjustment module ΔV_{OUT+} is equal to the decrease in the temperature coefficient adjustment circuit caused by the temperature change ΔV_{OUT-}. So that the increased amount ΔV_{OUT+} is equal to the reduced amount ΔV_{OUT-}, the temperature adjustment amount of the temperature coefficient adjustment circuit can be used to compensate for the reference voltage drift caused by the temperature change, and the temperature compensation function can be realized.

By designing the temperature detection flip point and adjusting the effective resistance value of the resistor, the bandgap reference temperature offset can be compensated and the purpose of compensation can be realized. Finally, a bandgap reference voltage with low temperature drift is obtained without the use of high-order compensation.

IV. EXPERIMENTAL RESULTS AND DATA ANALYSIS

In this paper, based on the 0.18μm BCD process, the proposed scheme to reduce the bandgap reference voltage temperature drift is simulated and verified on the SPICE platform. After designing four adjustment resistors, the temperature coefficient of the original first-order bandgap reference voltage can be reduced to 1/4 of the original value, and the low-temperature drift design can be realized.

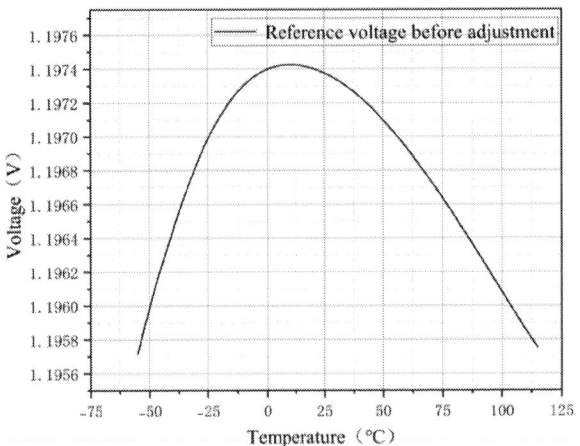

Fig. 3. Reference voltage before adjustment

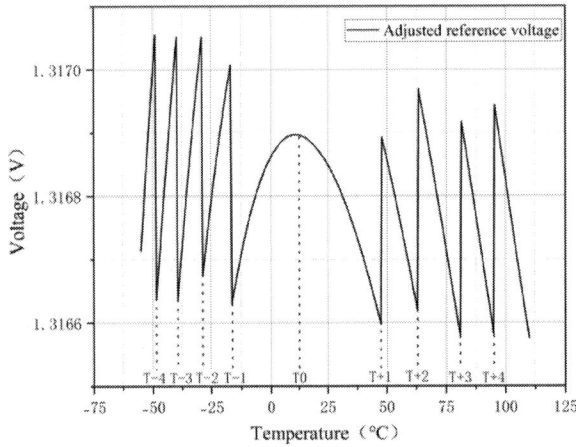

Fig. 4. Adjusted reference voltage

Fig. 3 shows the bandgap reference voltage before adjustment. At the TT process corner, the full temperature range drift is 1.708mV and the temperature drift coefficient is 8.394 ppm/°C. Fig. 4 shows the output voltage adjusted by four adjustment resistors at the TT process corner. The full temperature range (-55°C~115°C) drift is 447μV and the temperature drift coefficient is 1.997 ppm/°C.

In this paper, four adjustment resistors and eight temperature adjustment points are designed. After the adjustment of the method described in this paper, the temperature drift of the bandgap reference can be reduced to 1/4 of the original temperature drift coefficient from 8.394 ppm/°C to 1.997 ppm/°C. The circuit design value is 1/5, and the simulation value

979-8-3315-8850-2/25 $31.00 © 2025 IEEE   63

is 1/4 due to the influence of MOS on-resistance and parasitic resistance during the actual design process.

Table I lists the comparison of this paper with the relevant parameters of literature[9], literature [10], and literature [11].It can be seen that the adjustment method proposed in this paper uses switching adjustment to reduce the temperature drift of the bandgap reference voltage to achieve a lower temperature drift coefficient.

TABLE I.   LITERATURE COMPARISON

Parameter	Literature			
	[9]	[10]	[11]	This work
Technology(μm)	0.18	0.15	0.11	0.18
Supply Voltage (V)	2~5	3.3	3.3	3.3
Output Voltage (V)	1.21	1.16	1.2	1.3168
Temperature Range (°C)	-40~125	-40~150	-40~125	-55~115
Temperature Coefficient (ppm/°C)	14.09	5.78	2.18	1.997

V. CONCLUSION

In this paper, a method of reducing the reference temperature drift using switch adjustment is proposed. By changing the resistance value of the adjustment resistor in the temperature coefficient adjustment circuit at different temperature points, the adjustment amount of the temperature coefficient adjustment circuit is used to compensate for the amount of drift caused by the reference temperature change. It solves the problem that the existing bandgap reference circuit is difficult to achieve at very low temperature drift. Based on the 0.18μm BCD process, this paper simulates and verifies the proposed scheme to reduce the temperature drift of the bandgap reference on the SPICE platform. Through the adjustment of the design method in this paper, the temperature drift coefficient of the bandgap reference can be reduced from 8.394 ppm/°C to 1.997 ppm/°C.

ACKNOWLEDGMENT

This work was supported in part by the National Natural Science Foundation of China under Grant62171367 and Shaanxi innovation Capability Support Project 2022TD-39.

REFERENCES

[1] P. S. Ebenezer, V. Naganadhan, D. Chen and R. Geiger, "Three-Junction Bandgap Circuit with Sub 1 ppm/°C Temperature Coefficient," 2020 IEEE 63rd International Midwest Symposium on Circuits and Systems (MWSCAS) , Springfield, MA, USA, 2020, pp. 305-308.

[2] Q. Duan and J. Roh, "A 1.2-V 4.2- ppm/°C High-Order Curvature-Compensated CMOS Bandgap Reference," in IEEE Transactions on Circuits and Systems I: Regular Papers, vol. 62, no. 3, pp. 662-670, March 2015.

[3] Chi-Wah Kok, Wing-Shan Tam, "Voltage Reference, "in CMOS Voltage References: An Analytical and Practical Perspective, 1, Wiley-IEEE Press, 2013, pp.49-70.

[4] J. Jiang, W. Shu and J. S. Chang, "A 5.6 ppm/°C Temperature Coefficient, 87-dB PSRR, Sub-1-V Voltage Reference in 65-nm CMOS Exploiting the Zero-Temperature-Coefficient Point," in IEEE Journal of Solid-State Circuits, vol. 52, no. 3, pp. 623-633, March 2017.

[5] P. B. Basyurt, D. Y. Aksin, E. Bonizzoni and F. Maloberti, "A 490-nA, 43-ppm/°C, sub-0.8-V supply voltage reference," ESSCIRC 2014 - 40th European Solid State Circuits Conference (ESSCIRC), Venice Lido, Italy, 2014, pp. 115-118.

[6] Jung H. Analysis of threshold voltage and drain induced barrier lowering in junctionless double gate MOSFET using high-κ gate oxide[J]. International Journal of Electrical and Electronic Engineering & Telecommunications, 2020, 9(3): 142-147.

[7] A. Lahiri, P. Badrathwal, N. Jain and K. Chatterjee, "A 0.5V supply, 49nW band-gap reference and crystal oscillator in 40nm CMOS," 2017 IEEE Custom Integrated Circuits Conference (CICC), Austin, TX, USA, 2017, pp. 1-4.

[8] E. Qi, C. Fang, Y. Zhang, Y. Cheng and N. Wang, "A wide input low quiescent current without operational amplifier bandgap reference circuit," 2023 8th International Conference on Integrated Circuits and Microsystems (ICICM), Nanjing, China, 2023, pp. 326-330.

[9] J. Ye, D. Mao and W. Zheng, "Design of a Low Temperature Drift High Power Supply Rejection Bandgap Reference Circuit," 2023 IEEE 15th International Conference on ASIC (ASICON), Nanjing, China, 2023, pp. 1-4.

[10] K. Chen, L. Petruzzi, R. Hulfachor and M. Onabajo, "A 1.16-V 5.8-to-13.5-ppm/°C Curvature-Compensated CMOS Bandgap Reference Circuit With a Shared Offset-Cancellation Method for Internal Amplifiers," in IEEE Journal of Solid-State Circuits, vol. 56, no. 1, pp. 267-276, Jan. 2021.

[11] S. Jain, V. K. Kanchetla and R. Zele, "A Sub-1V, Current-mode Bandgap Voltage Reference in Standard 65 nm CMOS Process," 2022 IEEE 15th Dallas Circuit And System Conference (DCAS), Dallas, TX, USA, 2022, pp. 1-5.

Design and Implementation of a Secure Boot RISC-V Processor for IoT Devices

Zeyu Li
School of Electronic Engineering
Heilongjiang University
Harbin, China
zeyu0225@outlook.com

Liji Wu*
School of Integrated Circuits
Tsinghua University
Beijing, China
lijiwu@tsinghua.edu.cn

Jing Hu*
School of Electronic Engineering
Heilongjiang University
Harbin, China
hjlyh@126.com

Han Sun
School of Integrated Circuits
Tsinghua University
Beijing, China
sunhan96@mail.tsinghua.edu.cn

Xiangmin Zhang
School of Integrated Circuit
Tsinghua University
Beijing, China
zhxm@tsinghua.edu.cn

Haijie Wang
School of Electronic Engineering
Heilongjiang University
Harbin, China
2739689017@qq.com

Abstract—**The open-source RISC-V instruction set provides a new direction for China to establish an autonomous and controllable IoT ecosystem. At the same time, the IoT devices face the security threat of firmware tampering. Aiming at these problems, this paper provides a low-power pipelined processor based on RISC-V architecture that supports secure boot, which supports RV32IM instruction set, fast interrupt response and dynamic pipeline control. In the meantime, the secure boot mechanism of the processor is designed based on the Chinese commercial cryptographic algorithm SM3, when the boot file BootCode is tampered with, the processor is prevented from booting by a hardware-level blocking mechanism. The design passes the instruction set and secure boot functionality tests and it is verified on an FPGA board, the results show that the total utilization of logic lookup tables(LUT) is 22827, accounting for 19.5% of the total resources, while the average startup time of the system is 49.35ms, without causing large resource and time overhead, it is suitable for resource-constrained IoT devices and providing a guarantee of autonomy and security for China's RISC-V ecosystem at the same time.**

Keywords—*Internet of Things(IoT), RISC-V, secure boot, pipelined processor*

I. INTRODUCTION

With the accelerated pace of industrialization and intelligent development, the Internet of Things(IoT), as an important technology to promote the fourth industrial revolution, has received more and more attention for its far-reaching social and economic impacts and has gradually penetrated into every aspect of our lives [1]. Nowadays, IoT devices have been distributed in the fields of smart home, industrial automation, transportation and logistics, et al, which are reshaping the way of interaction between people and things, and also between things, they greatly improve people's daily life and work efficiency.

Embedded microprocessors are widely used in the IoT industry for their high performance and low power consumption. At present, the processor based on ARM architecture is the most widely used in the field of IoT, but ARM architecture is not an open-source one and it is difficult to modularize the design for a particular application. Nowadays, it is becoming crucial for China to produce autonomous and controllable, so the open-source RISC-V architecture is better adapted to the diversification, individualization and differentiation of the demand for China's embedded chip field [2].

However, the Internet of Everything, while bringing us convenience, also exposes devices that were originally independent of each other to a larger attack surface [3], where attackers can use the network to remotely tamper with the startup files of the devices and take control of the devices, which poses a great threat to the security and privacy of the users [4]. Since the operational data of a device may contain a large amount of sensitive governmental, corporate, and personal information, it is critical to protect the information to prevent illegal users from tampering with it as well as obtaining unauthorized control of the device.

Based on the above problems, this paper designs a five-stage pipelined RISC-V microprocessor based on the RV32IM instruction set for the field of embedded processors with low overhead and low power consumption, which supports dynamic pipelined control and fast interrupt response. Meanwhile, in order to prevent attackers from tampering with the firmware of the device to control devices, this paper implements a secure startup verification mechanism of the processor based on China's commercial cryptographic algorithm SM3, which ensures that the system does not execute programs when tampered with. The design completes the instruction set test and secure boot verification in the simulation environment and

979-8-3315-8850-2/25 $31.00 © 2025 IEEE

completes the prototype verification and performance analysis by FPGA.

II. SECURE BOOT FOR RISC-V

A. RISC-V Instruction Set

The RISC-V instruction set is an emerging open-source instruction set architecture with the advantages of simple architecture, modular configurability and complete open source [5], which allows designers to design microprocessors to meet specific functional requirements [6]. The RISC-V instruction set consists of two categories, the basic instruction set and the extended one. RV32I is the 32-bit basic integer instruction set of the RISC-V, which is mainly used to perform various operations and memory operations of the processor. It has six basic instruction formats, which are R-type, I-type, S-type, B-type, U-type and J-type, respectively, and the coding format of the above types of instructions is shown in Fig.1.

	31	25	24	20	19	15	14	12	11	7	6	0
R-type	funct7		rs2		rs1		funct3		rd		opcode	
I-type	imm[11:0]				rs1		funct3		rd		opcode	
S-type	imm[11:5]		rs2		rs1		funct3		imm[4:0]		opcode	
B-type	imm[12,10:5]		rs2		rs1		funct3		imm[4:1,11]		opcode	
U-type	imm[31:12]								rd		opcode	
J-type	imm[20,10:1,11,19:12]								rd		opcode	

Fig. 1. RV32I Instruction Encoding Format

Due to the modularity of RISC-V, in addition to the basic instruction set, users can also combine different extended instruction sets for their specific applications. Currently, there are multiply-divide instructions, atomic instructions, floating-point instructions, compression instructions, and other extended instructions for developers to choose from. RV32M is a multiply-divide expansion subset of RISC-V, which contains four multiply instructions, two remainder instructions and two divide instructions, and supports divide and remainder instructions for signed and unsigned integers.

B. Secure Boot Technology

Secure boot technology is a mechanism to ensure that devices can only be booted using trusted software, which is an important means to build a trusted execution environment and realize trusted computing [7]. Its basic principle is to verify the data and resources in the startup phase, based on the root of trust through the metrics level, the verification level, the trust level of verification. The system starts only when the verification passes. Through the secure boot mechanism, the user can ensure that the attacker has not tampered with the data and resources involved in the boot process of the device, thus ensuring the security and trustworthiness of the system [8].

At present, in the industrial field, many equipment manufacturers and chip designers have put forward corresponding secure boot programs for their own architectures. Although the RISC-V instruction set is not perfect compared to other architectures in this field, it is still being developed very

rapidly. Currently, there are two main methods to realize RISC-V based secure boot, measurement boot and verification boot [9]. Measurement boot is to measure the integrity of the software in subsequent phases during the boot phase, and store the measurement results in the trusted platform module for subsequent verification. Verification bootstrapping is to directly verify the integrity of the software in subsequent phases during the startup phase, and to continue execution only after the verification has passed. Compared to measurement boot, verification boot mode does not require external security firmware such as TPM, which is more suitable for IoT and other application scenarios that have strict requirements on device complexity and cost.

III. A SECURE BOOT RISC-V PROCESSOR DESIGN

In this paper, the RV32IM instruction set is selected to build RISC-V processor for low-area and low-power embedded application scenarios, which reduces the complexity of the hardware through the minimal core instruction set and is more suitable for resource-constrained scenarios. It avoids the performance loss of software simulation of multiplication and division through those expansion, which can improve the execution efficiency of computation-intensive tasks. Most of the traditional secure boot schemes are based on international cryptographic algorithms such as SHA-256. In this paper, based on the Chinese commercial cryptographic algorithm SM3, the processor boot file is verified within the processor the boot mode. And when the verification fails, the processor is terminated by a hardware-level blocking mechanism. The design not only meets the requirements of security, autonomy and controllability, but also has certain advantages in cost and ease of use. The system design is described below.

A. Hardware and Software Co - Design

The overall system structure is shown in Fig.2.

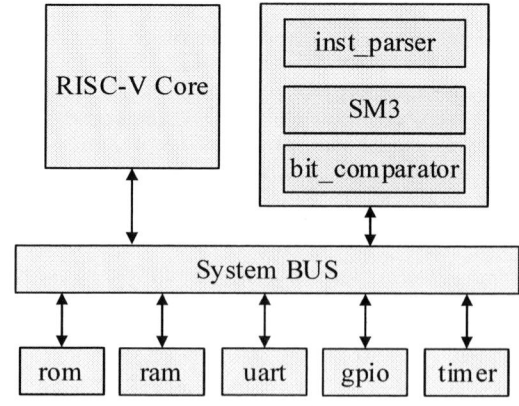

Fig. 2. Overall System Architecture

In this paper, the design model is divided into two main parts, the first part is the RISC-V processor, including the RISC-V processor core and various peripheral modules, which are responsible for executing the core functions of the processor, such as data processing and memory operations. The other part is the boot file verification module, which is responsible for verifying the integrity and consistency of the processor boot code (BootCode).

979-8-3315-8850-2/25 $31.00 © 2025 IEEE

The main components of the entire system-on-chip can be divided into four modules:

1) RISC-V core: to perform the core's computational tasks. It supports the RV32IM instruction set and adopts a five-stage pipelined modular design to support fast interrupt response and dynamic pipeline control.

2) Bootcode verification module: verify the integrity and consistency of the BootCode of the processor boot file, including the BootCode boot module inst_parser, the hardware implementation of the cryptographic algorithm SM3 and the bit_comparator for verification.

3) System bus: to connect the processor and other functional units on the chip.

4) Other on-chip peripherals: contain internal ROMS for loading modules and machine instruction files, RAMS for storing processor runtime data, general-purpose asynchronous transceiver and transmitter UART, and general-purpose input/output GPIOs.

B. Pipeline Design

In this paper, the processor core adopts the classical five-stage pipeline structure, which divides the processor into five stages: instr fetch, instr decode, execute, load store and write back. The data is transferred through the pipeline registers and handles interrupts and exceptions by using an independent core-local interrupt controller, the CLINT module, which is decoupled from the main pipeline and enhances the maintainability. Fig.3 shows the pipeline structure.

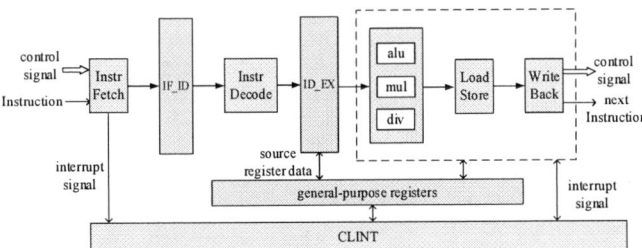

Fig. 3. Structure of five-stage assembly line

The instruction firstly enters the finger-fetching module, which is controlled by the control signals from the back stage and the interrupt signals from the CLINT module, and will fetch the finger according to the frequency of the current pointer address plus 4 in normal operation. When the pipeline is paused, the control signals control the finger-fetching module to pause the fetching of the finger, and the output is the address of the jump when there is an interruption or a jump in the pipeline, the decoding module will decode the instruction and address output from the finger module according to the RISC-V format and pass them to the next module and interact with the register module according to the result of decoding to pass the data, the execution module realizes the function of the passed instruction and supports the basic arithmetic and multiplication/division arithmetic, the load store module realizes the reading and controlling of the internal state of the processor as well as responsible for controlling and interacting with the external module. The write-back module delays the writing of the data into the registers for One clock cycle.

In order to reduce the signal competition brought by the pipeline structure as well as the data adventure situation in the middle of the fetch finger, decode and execute modules add pipeline registers IF_ID and ID_EX, which are used to save the results of the previous stage of operation, through the temporary storage and synchronization of data to ensure the isolation and synchronization of the operation of the various phases. The CLINT module unifies the management of timer interrupts, external interrupts and software interrupts, and passes the interrupt signal directly to the IF module, which reduces interrupt delays and simplifies interrupt prioritization and masking logic through hardware-level interrupt fast response.

C. Dynamic Pipeline Control

Pipeline conflict is a problem that must be solved in modern processor design. Dynamic pipeline control is a technique that dynamically adjusts the pipeline behavior according to the real-time situation when the processor is running, and it can optimize the pipeline behavior in real time. The dynamic pipeline mechanism in this paper includes pause control *stalling* and jump *flushing*, as shown in Fig.4.

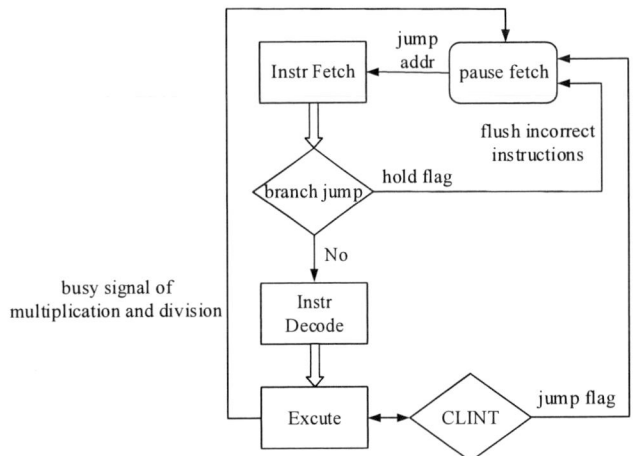

Fig. 4. Pause and Jump Design for Pipeline

The pause of pipeline is controlled by the control signal, when a branch jump occurs hold flag signal is valid and output the jump address. At this time, it will flush the pipeline in the wrong instruction and in the next clock to take the finger from the start of the new address, when the multiplier and divider is busy to the same make it valid, the need to pause the pipeline until the completion of the arithmetic, when interrupt processing is in progress. The interrupt signal will be passed to the CLINT module, and CLINT will issue an interrupt request and interrupt entry address at the appropriate time. The EX stage will generate a jump signal and jump to the interrupt processing address after receiving the request.

The design can adapt to different pause requirements, and the interrupt signal can go straight to the IF and EX stages, which can minimize the interrupt latency and pause only for the necessary stages, reducing the execution of invalid instructions with pipeline bubble problems.

D. Design for Secure Boot

Secure boot is mainly based on the hash algorithm for processor firmware. BootCode is the boot code of the processor, and the first code to be executed after the chip is powered on or reset, which ensures the stability and reliability of the system. So it is crucial to verify its integrity and ensure that it has not been tampered with. This paper designs a secure boot mechanism around ensuring the integrity of the BootCode to avoid the processor executing the program if the BootCode is tampered with.

The process can be divided into two processes, summarization and verification. The summarization process needs to generate the hash value of BootCode, which is usually written in assembly language, named start.S file and stored in ROM. The summarization process is subdivided into four sub-processes. First, startup file M is padded to a multiple of 512 bits to obtain M'. Then, M' is divided into groups of 512 bits, denoted as B. Next, each grouped message block B is expanded to obtain W_i, as shown in (1). Finally, the expanded message W_i is processed according to the compression function formula CF to produce the hash value H(X), as shown in (2).

$$W_i' = W_i \oplus W_{i+4} \tag{1}$$

$$H(X) = V_{i+1} = CF(V_i, B_i) \tag{2}$$

The summarization process incorporating SM3 algorithm is shown in Fig.5.

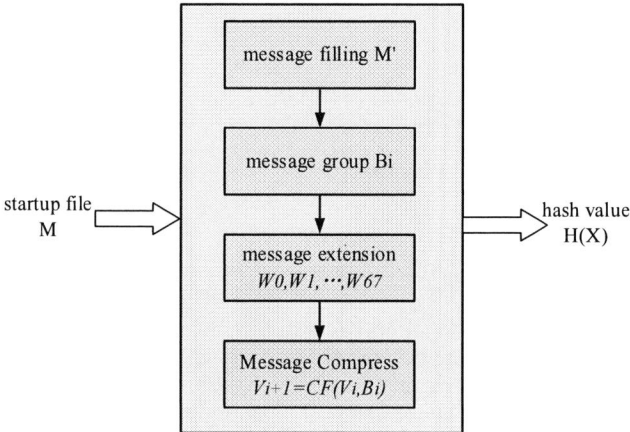

Fig. 5. Process of start.S Summary

In the verification process, the start.S is compiled into the instruction data stored in the default 0x0000_0000 location in the ROM. After power-up, the data at this location is read, and through the summarization process to generate the hash value H(X) and burned in the disposable programmable memory preexisting original hash value H(X`) for comparison, the original hash value through the form of hardware at the time of shipment, and can't be modified once written. The processor can be started when the verification is consistent and continue to execute other software, and will stop when the verification is inconsistent. The processor won't be tampered with through the hardware-level protection mechanism of forcing a reset and blocking the clock. The secure boot verification process is shown in Fig.6.

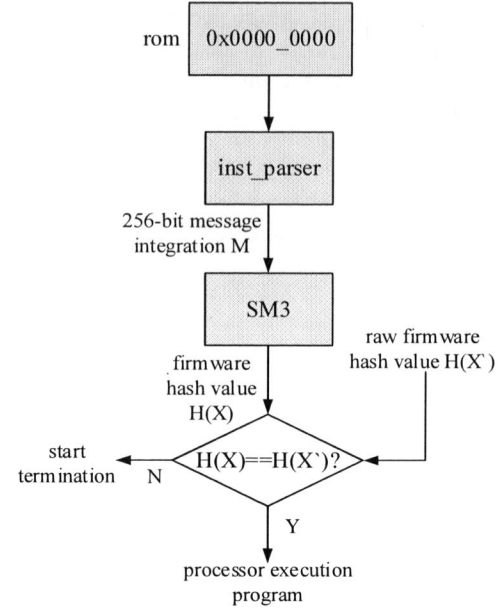

Fig. 6. Secure Boot Verification Process

IV. VERIFICATION

The paper implements the RTL(register transfer level) design of the processor core, bus, peripheral modules and secure boot module using Verilog HDL. In order to validate the effectiveness and performance of the processor and secure boot module, this section validates the functionality and high efficiency of the overall module and prototypes it through FPGA.

A. Verification for RISC-V Processor

This paper verifies the functionality of the processor by using the RISC-V compliance test released by RISC-V. The above test program is executed under Windows operating system using Makefile scripts, GUN toolchain, and the iverilog emulator, which covers the instructions to be tested as well as one-to-one checkpoints, when the test passes, the terminal prints the word *PASS*.

Fig. 7. RV32IM Instruction Set Functional Validation

The automated program loading and checkpoint data analysis is implemented through Python scripts, and the test results are shown in Fig.7. The results show that the RV32IM instruction set supported by the processor in this paper conforms to the RISC-V standard and is compatible with the toolchain as well as the software tools under the RISC-V system.

B. Verification for Secure Boot

In order to verify whether the design of the secure boot module can realize the function of blocking tampering, this paper carries out code addition, deletion and modification operations on the startup file start.S. The firmware can be compiled normally after being tampered with, without affecting its main functions, and the firmware will be re-compiled and stored in ROM, to compare with the original firmware stored in OTP, to determine whether it can block abnormal startup and the simulation results are shown in Fig.8.

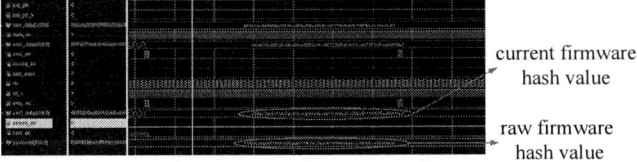

current firmware hash value

raw firmware hash value

Fig. 8. Waveform of Tampered Firmware

The untampered start.S is compiled and simulated for comparison with the original firmware, and the simulation results are shown in Fig.9.

current firmware hash value

raw firmware hash value

Fig. 9. Waveform of Raw Firmware

As can be seen from the figure, when the current firmware hash value is inconsistent with the original firmware hash value, the processor will not start. At this time, the system clock and reset signals are 0, the processor starts normally only when the hash value is the same, at which time the clock and reset signals of the system work normally. Secure boot functionality verification passed.

C. FPGA Verification

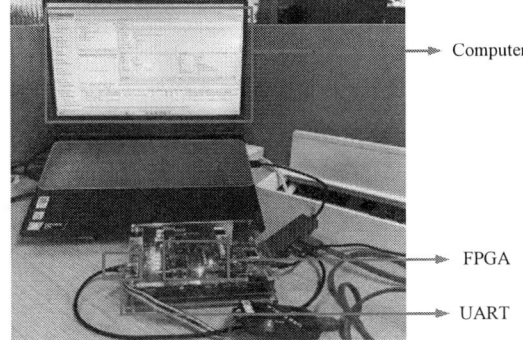

Computer

FPGA

UART

Fig. 10. Processor Board Level Verification Environment

The Vivado synthesis tool is used for FPGA logic synthesis. As shown in Fig.10, the FPGA platform used in this paper is Xilinx XC7A200T, the computer sends the program as well as monitoring the processor status through the UART serial port.

In the paper, we statistically analyze the hardware resource usage, boot time evaluation, and power consumption. As shown in Table I, after using the secure boot module, the LUT hardware usage grows by 17.06%, there is still 80.5% left over, reserving sufficient hardware space for other possible programs. In terms of startup time, 15 different programs are tested for the processors with and without secure models. The results showed that the runtime of the processor with security module increased slightly, requiring an average of 1.175ms of additional time overhead, the performance is excellent.

TABLE I.   RESOURCE UTILIZATION ON FPGA

	LUTs	Average elapsed time	Power
Single processor	3268	48.35ms	0.259W
Add security module	22827	49.53ms	0.287W

The overall operating frequency of the system is 100MHz, and the Coremark performance test of the system is 2.40 Coremark/MHz. Table II gives the results of the comparative analysis of the performance of the processor system of this paper with other processors, the results show that the design of the paper is well-balanced in processor function and security under the premise of the same comparative indexes, so it can fully meet the requirements of practical applications.

TABLE II.   COMPREHENSIVE ENERGY EFFICIENCY COMPARISON

	This paper	Hummingbird E203	[10]
instruction set	RV32IM	RV32IMAC	RV32I
bit width/bit	32	32	32
pipeline depth	5	2	4
Secure boot	yes	no	yes
Coremark/MHz	2.40	2.14	2.375

V. CONCLUSION

The paper proposes a RISC-V secure processor design for embedded applications such as IoT, divides the functions and modules of the pipeline and adds a dynamic pipeline design, and designs the processor secure boot mechanism based on the state-secret SM3 algorithm, it realizes the tamper-resistant nature of the processor BootCode. The processor has completed the RISC-V instruction set self-test and security check function test, and has completed the prototype verification on FPGA, the experimental results show that logic lookup table usage was 22827(19.5% of total resources), caused an additional time overhead of only 1.175ms, and the Coremark performance test of the system is 2.40 Coremark/MHz. The design can balance the processor performance and the conflict of the security module, the hardware resource occupancy rate is low and has not caused a large time overhead, which can effectively improve the security of the device, so it can be well suited to the IoT devices.

REFERENCES

[1] Chao Li and Wenyu Lao, Internet of Things' sustainability effects: quantile and temporal insights, Nature, 2025, pp. 1-22.

[2] Deng Bao and Jingguo Sun, Review on development of domestic embedded processor, Aeronautical Computing Technique, 2021, pp. 120-124.

[3] Abdul-Ghani H A, Konstantas D and Mahyoub M, A comprehensive IoT attacks survey based on a building-blocked reference model, International Journal of Advanced Computer Science and Applications, 2018, pp. 355-373.

[4] Acar A, Fereidooni H, Abera T, AK Sikder, M Miettinen and H Aksu, "Peek-a-Boo: I see your smart home activities, even encrypted," Proceedings of the 13th ACM Conference on Security and Privacy in Wireless and Mobile Networks, 2020, pp. 207-218.

[5] Chang Liu, Yanjun Wu and Jingzheng Wu, Survey on RISC-V system architecture research, Journal of Software, 2021, pp. 3992-4024.

[6] Emre Ozer, Jedrzej Kufel, Shvetank Prakash, Alireza Raisiardali, Olof Kindgren, and Ronald Wong, Bendable non-silicon RISC-V microprocessor, Nature, 2024, pp. 341-346.

[7] Bryan Parno, Jonathan M. McCune and Adrian Perrig, "Bootstrapping trust in commodity computers," 2010 IEEE Symposium on Security and Privacy, 2010, pp. 414-429.

[8] Chenming Zhenga, Jun Li and Xuanxia Yao, Design and implementation of trusted boot based on a new trusted computing dual-architecture, Computers & Security, 2023, pp. 127.

[9] Rui Zhiqing, Mei Yao and Chen Zhenzhe, SeChain: Design and implementation of RISC-V secure boot mechanism based on domestic cryptographic algorithms, Journal of Computer Research and Development, 2024, pp. 1458-1475.

[10] Shi Wei, Liu Wei, Gong Rui, Wang Lei and Zhang Jianfeng, Security enhancement technologies of security subsystem in microprocessors, Computer Engineering & Science, 2021, pp. 1354-1359.

A 1GS/s 13-bit Single-Channel Hybrid Pipelined SAR ADC with Domino SAR Logic

Chuan Liu
School of Microelectronics
Xidian University
Xi'an, China
1084273030@qq.com

Chuan Qin
School of Microelectronics
Xidian University
Xi'an, China
23211215034@stu.xidian.edu.cn

Lecheng Li
School of Microelectronics
Xidian University
Xi'an, China
2393262880@qq.com

Zhihao Hou
School of Microelectronics
Xidian University
Xi'an, China
houzhihao2022@163.com

Yuqi Fan
School of Microelectronics
Xidian University
Xi'an, China
13323893996@163.com

Maliang Liu*
School of Microelectronics
Xidian University
Xi'an, China
mlliu@xidian.edu.cn

Abstract—**This paper presents a 13-bit 1GS/s single-channel hybrid pipelined-SAR analog-to-digital converter (ADC) implemented in a 28 nm CMOS technology. The architecture employs a low-resolution 5-bit SAR ADC for coarse quantization paired with a high-resolution 9-bit time-domain ADC (TD-ADC) for fine quantization. Within the SAR ADC, a domino-logic structure enhances conversion speed, complemented by a proposed low-power charge-reset two-stage dynamic comparator optimized for domino implementations. For the TD-ADC, counter-based coarse quantization integrated with phase interpolation techniques effectively boosts the resolution of the gated ring oscillator-based time-to-digital converter (GRO-based TDC). Fabricated in 28nm CMOS, the prototype occupies a core area of 0.06 mm2 and consumes 25.9mW. Measured at 1GS/s sampling rate, the ADC achieves robust performance for single-tone sinusoidal inputs: delivering 65.2 dB signal-to-noise-and-distortion ratio (SNDR) and 84.6 dB spurious-free dynamic range (SFDR) at 59.47 MHz input frequency, while maintaining 64.3 dB SNDR and 82.4 dB SFDR at 459.47 MHz.**

Keywords—*Hybrid architecture, Pipelined analog-to-digital converter (ADC), SAR ADC, Time-domain ADC*

I. INTRODUCTION

Advancements in modern wireless communication technologies continue to escalate performance demands for receivers, imposing increasingly stringent requirements on system-level circuit design. Direct Radio Frequency (RF) sampling technology addresses these challenges by streamlining receiver signal chains, delivering not only enhanced cost efficiency but also significantly improved noise performance. However, this approach necessitates high-speed, high-resolution analog-to-digital converters (ADCs) for RF signal acquisition, rendering the development of GS/s-class precision ADCs critically important.

For high-resolution ADCs operating at 1GS/s sampling rates, the single-channel pipelined architecture emerges as the optimal solution when comprehensively evaluating power consumption, silicon area, and design complexity constraints [1]. While pipelined ADCs inherently satisfy concurrent demands for high speed and resolution, SAR ADCs and time-domain ADCs (TD-ADCs) prove particularly suitable as sub-ADC components due to their superior energy efficiency. Nevertheless, SAR ADC accuracy is fundamentally limited by digital-to-analog converter (DAC) mismatch errors, while TD-ADCs with sampling rates above GS/s are generally constrained to less than 8-bit resolution due to limited conversion time. Consequently, hybrid-domain architectures that combine the high accuracy of voltage-domain operation with the low-power advantages of time-domain processing have emerged as a research focus for high-speed and energy-efficient ADCs [2].

To overcome these conflicting design constraints and meet the requirements for high energy efficiency, this work proposes a pipelined ADC with a hybrid voltage-time domain architecture [3]. The design employs a domino-logic-implemented 5-bit SAR ADC for coarse quantization in the first stage, coupled with a precision 9-bit TD-ADC executing fine quantization in the subsequent stage. The paper is organized as follows: Section II introduces the architecture of the hybrid pipelined ADC. Section III details circuit implementations of key building blocks. Section IV shows the post-simulation results of the ADC. Section V concludes the paper.

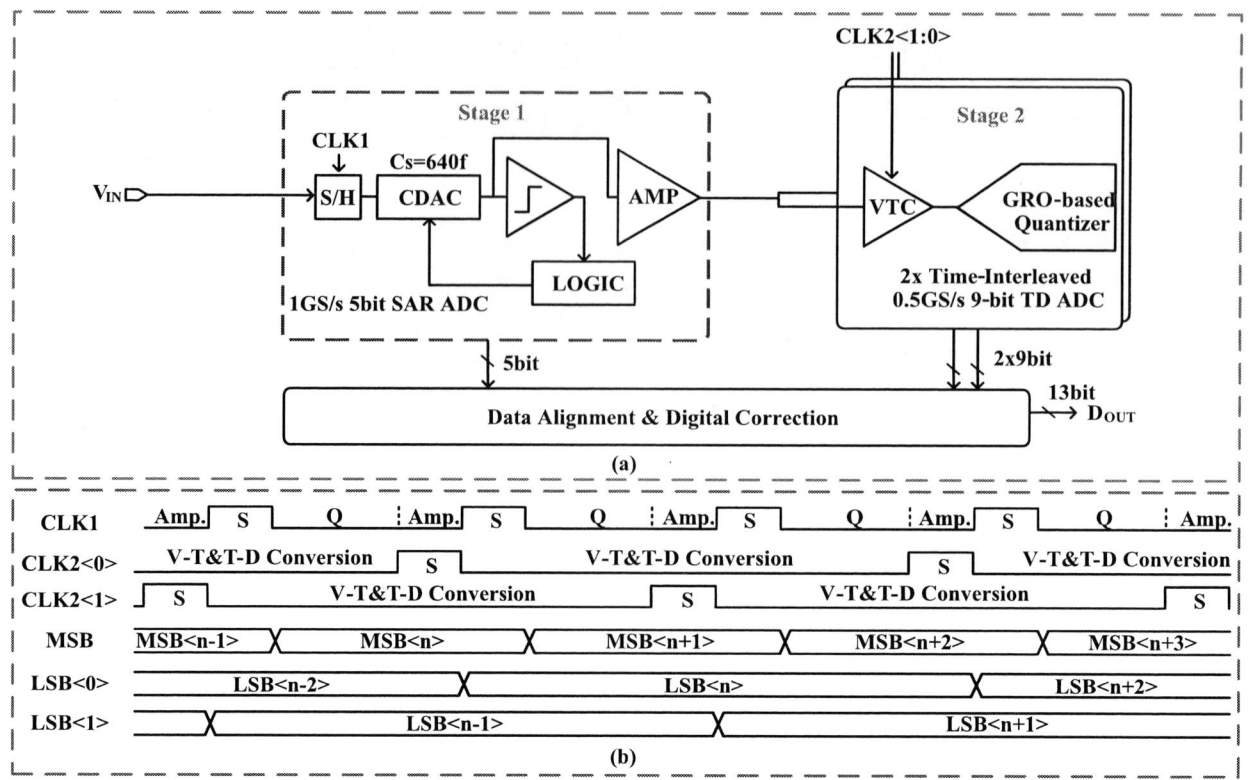

Fig. 1. (a) System architecture of the ADC. (b) Timing diagrams of the ADC.

II. PROPOSED ADC ARCHITECTURE

The system architecture of the ADC is shown in Fig. 1 (a). This design employs a two-stage pipelined architecture with one redundant bit between stages to accommodate quantization errors caused by non-ideal effects. The timing diagram is depicted in Fig. 1 (b). The first-stage SAR ADC utilizes a 25 % duty cycle clock，where 250 ps is allocated for input signal sampling and the remaining 750 ps for quantization and residue amplification. The resolution of the TD-ADC is fundamentally limited by the oscillation frequency of the ring oscillator. Although phase interpolation is implemented in the ring oscillator, achieving sub-4ps resolution remains challenging with compromised accuracy at finer quantization levels. Consequently, this architecture adopts two time-interleaved 500MS/s 9-bit TD-ADCs to achieve an aggregate 1GS/s sampling rate. Each sub-channel ADC operates with a 12.5 % duty cycle clock. Specifically, a 250 ps is designated for input sampling, while the remaining 1750 ps is allocated for voltage-to-time converter (VTC) and time-to-digital converter (TDC). Through optimized temporal allocation for each operational phase, this architecture maximizes timing resource utilization while requiring only two cascaded stages. This configuration contributes to both power reduction and implementation simplicity [4].

III. CIRCUIT IMPLEMENTATION

A. 5-bit SAR ADC

The architecture of 5-bit SAR ADC, illustrated in Fig. 2, employs capacitor sharing between the DAC array and sampling capacitor. This merged structure simultaneously performs sampling and digital-to-analog conversion. The SAR logic utilizes domino asynchronous control logic, where five comparators sequentially performs bitwise comparisons. The resultant outputs generate digital codes that govern subsequent logic operations.

The selection of sampling capacitors is typically dictated by mismatch and noise considerations. Since high-resolution ADCs impose stringent requirements on capacitor matching, and reducing mismatch through larger capacitor sizes entails a substantial cost in area and power, high-speed energy-efficient designs commonly mitigate capacitor mismatch effects in the digital domain. Consequently, the sampling capacitor size in this design is primarily constrained by noise. Under a 1.2Vpp full-scale input, the thermal noise introduced by the first-stage sampling capacitor must be less than the quantization noise to satisfy the 13-bit resolution requirement:

$$\frac{2kT}{C_{S1}} < \frac{LSB^2}{12} \Rightarrow C_{S1} > 4.65pF \tag{1}$$

As ADC resolution requirements increase, the sampling capacitance grows exponentially. Considering that excessively large sampling capacitors limit wideband sampling while consuming significant area, switch power, and amplifier power, this design sacrifices some resolution after trade-off considerations. The first-stage sampling capacitor is set to 640fF, achieving approximately 11.5-bit resolution. Since interstage amplification relaxes noise requirements for later stages, the second-stage sampling capacitor is reduced to 80fF to save power and area. The first-stage employs a 5-bit quantizer with

1-bit redundancy. The interstage gain is designed to be 16, and the feedback capacitor is set to 40fF.

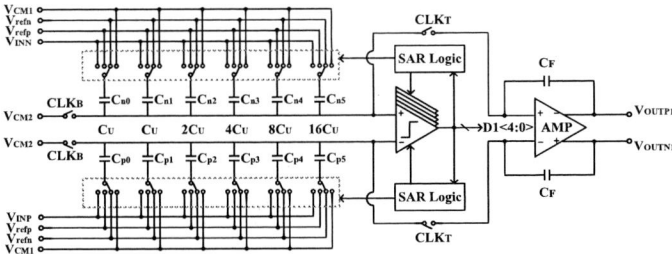

Fig. 2. Schematic diagram of the 5-bit SAR ADC.

The operation of the SAR ADC is divided into three phases: the sampling phase, the quantization phase, and the residual amplification phase. First is the sampling phase: Initially, the top plates of all capacitors are connected to the input voltage Vin, while the bottom plates are connected to the common-mode voltage V_{CM2}. Subsequently, the bottom-plate switches are disconnected, and the top-plate switches are switched to V_{CM1}, which serves as the common-mode voltage of V_{ref}, thereby completing the input sampling. Next is the quantization phase: After sampling is completed, the first comparator compares the differential input ΔV_{in} against zero. Assuming $\Delta V_{in}>0$, the SAR logic controls the switching of the P-side capacitor C_{p5} from V_{CM1} to V_{refn}, and the N-side capacitor C_{n5} from V_{CM1} to V_{refp}; conversely, the switching directions are reversed if $\Delta V_{in}<0$. Then, the second comparator compares the differential input ΔV_{in} against $(V_{refp} - V_{refn})/2$ or $-(V_{refp} - V_{refn})/2$. The comparison result is used by the SAR logic to control the switching of the DAC capacitor array. After five successive approximation cycles, the comparator obtains a 5-bit quantization result and controls the DAC to perform one final switching operation to generate the residual voltage. Finally, the residual amplification phase: The switches between the bottom plates of the DAC and the residual amplifier are closed [5]. The residual signal is amplified by a factor of 16 by the closed-loop amplifier, where the feedback capacitor is set to $C_F = 2C_U = 40$ fF. The amplified residual voltage is then quantized by a backend ADC.

B. Dynamic Comparator

In high-speed pipelined ADCs incorporating redundant bits, enhancing comparator speed while reducing power consumption constitutes the primary design objective. Therefore, this design employs a cascaded two-stage dynamic comparator architecture, with its circuit schematic illustrated in Fig. 3 (a). To mitigate DAC voltage errors, the first comparator stage requires a reset operation upon completing its comparison and before the next comparison stage begins. This reset clears injected charges. Conversely, the latch in the second comparator stage remains non-reset post-comparison. This latch integrates both comparison and data latching functions, and synchronously resets all comparator outputs at the end of the current quantization phase.

Fig. 4 presents the comparator input offset simulation results, showing a standard deviation (σ) of 6.7mV. To effectively suppress amplifier output swing expansion and amplification nonlinearity induced by comparator offset, offset calibration remains imperative. This design employs foreground offset

calibration, as illustrated in Fig. 3 (b). The calibration method specifically adjusts the differential path current in the preamplifier by introducing an additional digitally controlled current source array, thereby achieving offset control. The current source array adopts binary-weighted sizing to minimize the number of control switches.

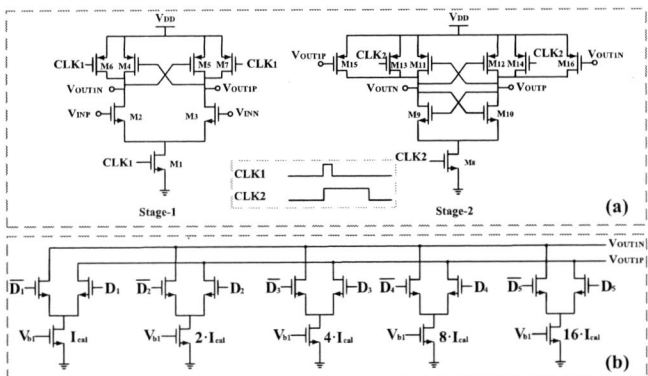

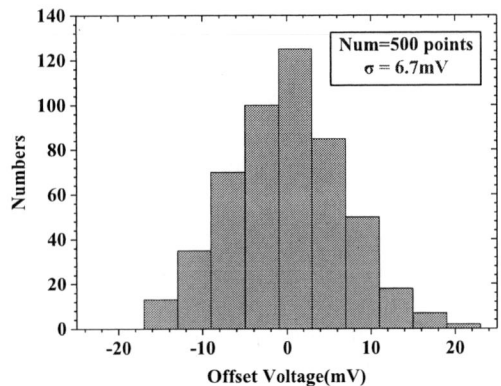

Fig. 3. (a) Schematic diagram of the comparator. (b) Offset calibration circuit of the comparator.

Fig. 4. Input offset simulation of the comparator.

C. Domino SAR Logic

The proposed SAR ADC employs the domino structure shown in Fig. 5 (a), where five comparators operate sequentially to resolve each bit successively. The comparator outputs needn't be reset immediately after completing the current-bit comparison, thus their outputs remain latched without requiring additional D flip-flops. This configuration reduces the conversion time of SAR ADC [6]. The timing diagram, illustrated in Fig. 5 (b), initiates the first comparator in the domino topology upon completion of input signal sampling. The digital logic unit determines the completion status of the current-bit comparison based on the output code of comparator. A rising edge on the FLAG signal, generated post-comparison, triggers the activation of the next comparator. Subsequently, the remaining four comparators sequentially activate, ultimately producing a 5-bit digital output code. Upon conclusion of the current quantization cycle and prior to initiating the next cycle, the comparator outputs are reset to initialize the subsequent quantization phase.

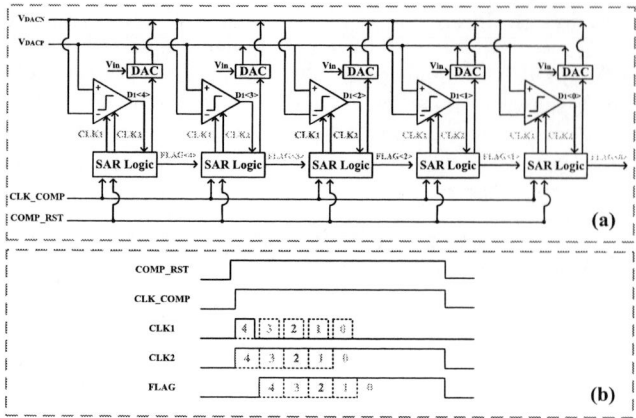

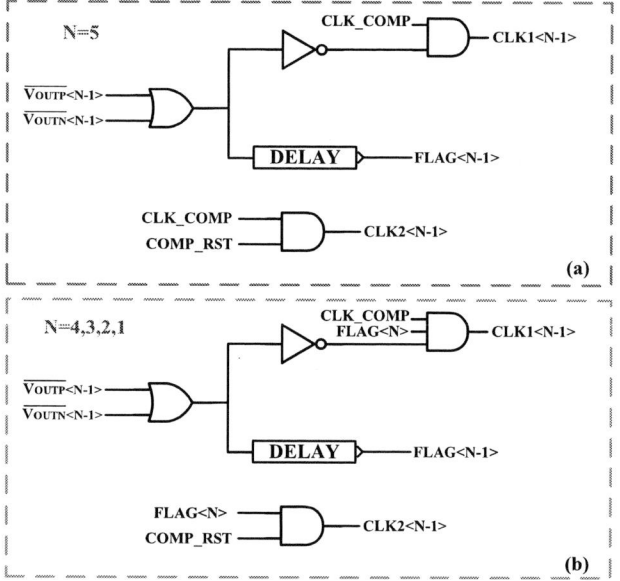

Fig. 5. (a) Logic block diagram of domino SAR ADC. (b) Timing diagram of domino SAR ADC.

Fig. 6. (a) Schematic of digital logic unit for the MSB. (b) Schematic of digital logic unit for the remaining bits.

Fig. 6 (a) depicts the gate-level schematic for the most significant bit (MSB), while Fig. 6 (b) presents those for the remaining bits, with CLK_COMP and COMP_RST denoting the enabling clock and reset clock of global comparator respectively.

D. Single-Channel TD-ADC

Fig. 7 depicts the system architecture of the second-stage sub-channel 0.5GS/s 9-bit gated ring oscillator-based (GRO-based) TD-ADC. This ADC comprises a pseudo-differential high-linearity VTC, a high-speed time comparator, a pulse generator for time-domain folding and GRO gating signal generation, a 9-bit GRO-based TDC for low-power time-domain quantization, as well as a digital module for quantization code synchronization and error calibration. The differential input signals are initially converted into temporal signals by the VTC, which implements a constant-current source charging/discharging architecture to satisfy the 9-bit linearity specification. Considering the positive correlation between TDC resolution and VTC conversion range, the VTC employs a 12.5%

duty cycle sampling clock to maximize output temporal range. These converted differential time-domain signals are subsequently routed to the comparator for sign-bit quantization while concurrently delivered to the pulse-generation circuit to accomplish folding and produce gating signals controlling the GRO TDC. Owing to inherent dead zones and compromised linearity in conventional pulse-generation circuits, this design adopts a topology integrating dead-zone elimination technology that substantially enhances pulse linearity, supplemented by a variable-delay circuit for fold-error correction [7]. The TDC comprises a high-speed fully-differential GRO, an inverter-based interpolation circuit, a self-calibrated counter, and a decoding circuit. By operating exclusively during gating signal activation, its overall power consumption is effectively minimized. Finally, the sign-bit and TDC quantization results undergo synchronization, integration, and calibration within the digital module.

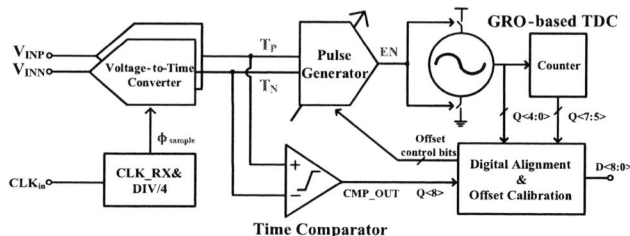

Fig. 7. System diagram of TD-ADC.

IV. LAYOUT AND POST-SIMULATION RESULTS

The proposed 13-bit 1GS/s hybrid pipelined ADC is fabricated in 28nm CMOS standard process, occupying a total die area of 1.19 mm^2 with the core ADC area measuring merely 0.06 mm^2. The corresponding chip layout is presented in Fig. 8. The output spectrum is depicted in Fig. 9. At an input frequency of 59.47 MHz, the SNDR and SFDR measure 65.2 dB and 84.6 dB, respectively. When tested at 459.47 MHz input frequency, the ADC achieves 64.3 dB SNDR and 82.4 dB SFDR. The core consumes 25.9 mW, resulting in a Walden FoM of 19.3 fJ/conversion-step. Table I summarizes the ADC performance and compares this work with other recent designs.

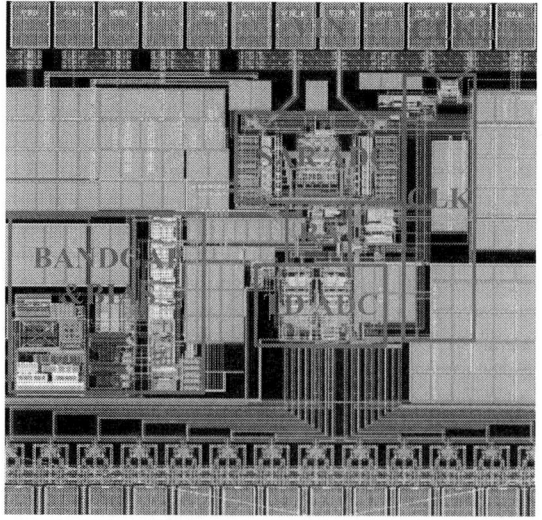

Fig. 8. The layout of the proposed ADC.

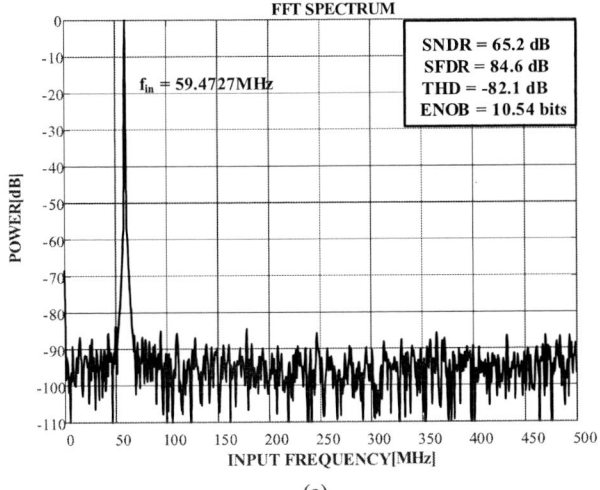

(a)

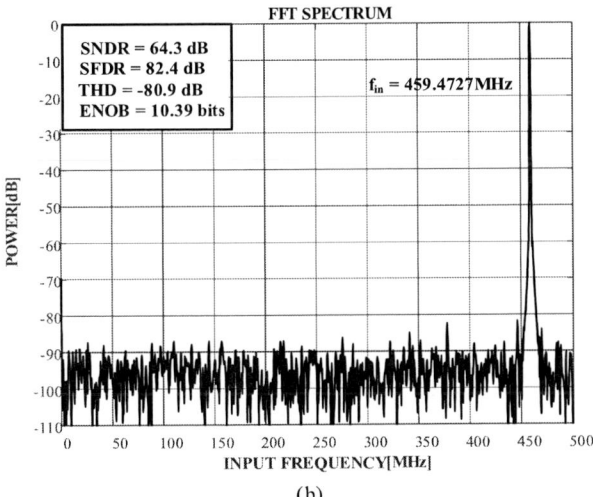

(b)

Fig. 9. (a) Output spectrum at 59.47 MHz. (b) Output spectrum at 459.47 MHz.

TABLE I.   SUMMARY AND PERFORMANCE COMPARISON

Specifications	References			
	This work*	[8]	[9]	[10]
Architecture	Pipelined SAR	V-T	Pipeline	Hybrid
Technology (nm)	28	28	28	28
Supply (V)	1	0.9	0.9	0.9/1.8
Resolution (bits)	13	12	12	11
Sampling Rate (GS/s)	1	0.4	0.6	1
SNDR @Nyq. (dB)	64.3	63.3	56.3	58.2
SFDR @Nyq. (dB)	82.4	79.1	69.2	63.7
Power (mW)	25.9	9.6	14.2	14.9
FOMw @Nyq. (fj/conv-step)	19.3	20.1	44.3	22.1

*The data represent the post-simulation results.

V.   CONCLUSION

This paper introduces a 28 nm 13-bit 1GS/s single-channel hybrid pipelined-SAR ADC. The architecture utilizes a low-resolution dynamic domino-logic SAR ADC for 5-bit coarse quantization, combined with a high-resolution TD-ADC executing 9-bit fine quantization. Post-simulation results demonstrate 65.2 dB SNDR and 84.6 dB SFDR at a 59.47 MHz input frequency, while achieving 64.3 dB SNDR and 82.4 dB SFDR at 459.47 MHz. The ADC core consumes 25.9 mW, yielding a Walden FoM of 19.3 fJ/conversion-step.

ACKNOWLEDGMENT

This work was supported in part by the Fundamental Research Funds for the Central Universities under Grant KYFZ25008, in part by the Natural Science Foundation of China under Grant 8091B02042301, in part by Shaanxi Provincial Key Research and Development Program under Grant 2024CY2GJHX34 and in part by National Science and Technology Major Project under Grant 2024ZD0302600. (Corresponding author: Maliang Liu.).

REFERENCES

[1] J. Lagos, B. P. Hershberg, E. Martens, P. Wambacq and J. Craninckx, "A 1-GS/s, 12-b, Single-Channel Pipelined ADC With Dead-Zone-Degenerated Ring Amplifiers," in IEEE Journal of Solid-State Circuits, vol. 54, no. 3, pp. 646-658, March 2019.

[2] Y. Lyu and F. Tavernier, "A 4-GS/s 39.9-dB SNDR 11.7-mW Hybrid Voltage-Time Two-Step ADC With Feedforward Ring Oscillator-Based TDCs," in IEEE Journal of Solid-State Circuits, vol. 55, no. 7, pp. 1807-1818, July 2020.

[3] H. Zhao and F. F. Dai, "A 12-Bit 260-MS/s Pipelined-SAR ADC With Ring-TDC-Based Fine Quantizer for Automatic Cross-Domain Scale Alignment," in IEEE Journal of Solid-State Circuits, vol. 58, no. 10, pp. 2883-2896, Oct. 2023.

[4] Y. -J. Chen, K. -H. Chang and C. -C. Hsieh, "A 2.02–5.16 fJ/Conversion Step 10 Bit Hybrid Coarse-Fine SAR ADC With Time-Domain Quantizer in 90 nm CMOS," in IEEE Journal of Solid-State Circuits, vol. 51, no. 2, pp. 357-364, Feb. 2016.

[5] X. Xu, Y. Ma, X. Yu, F. Meng, W. -H. Yu and L. Zhao, "A 10-bit Two-Stage Pipeline SAR ADC in 55nm CMOS for Compute-in-Memory Applications," 2024 IEEE Asia Pacific Conference on Circuits and Systems (APCCAS), Taipei, Taiwan, 2024, pp. 140-143.

[6] Y. -H. Chung, H. -C. Yeh and C. -W. Chang, "A 10b 160-MS/s domino-SAR ADC in 90nm CMOS," 2018 7th International Symposium on Next Generation Electronics (ISNE), Taipei, Taiwan, 2018, pp. 1-2.

[7] C. Zhang, J. Wei, Y. Chen, M. Liu and Y. Yang, "A 0.004-mm2 3.65-mW 7-Bit 2-GS/s Single-Channel GRO-Based Time-Domain ADC Incorporating Dead-Zone Elimination and On-Chip Folding-Offset Calibration in 28-nm CMOS," in IEEE Journal of Solid-State Circuits, vol. 58, no. 11, pp. 3179-3193, Nov. 2023.

[8] Y. Zhao, Y. Xiang, F. Ye and J. Ren, "A 400-MS/s 12-bit Voltage-Time Hybrid ADC with a Ping-Pong SAR TDC for Speed Enhancement," 2023 IEEE International Symposium on Circuits and Systems (ISCAS), Monterey, CA, USA, 2023, pp. 1-5.

[9] J. Lagos, B. Hershberg, E. Martens, P. Wambacq and J. Craninckx, "A single-channel, 600Msps, 12bit, ringamp-based pipelined ADC in 28nm CMOS," 2017 Symposium on VLSI Circuits, 2017, pp. C96-C97.

[10] J. Wei, C. Zhang and M. Liu, "A 11-Bit 1-GS/s 14.9mW Hybrid Voltage-Time Pipelined ADC With Gain Error Calibration," in IEEE Transactions on Circuits and Systems II: Express Briefs, vol. 69, no. 3, pp. 799-803, March 2022.

A 1.7-to-3.0 GHz 201.95 dBc/Hz FOM$_T$ VCO With a Current-Reuse Active Inductor

Xinyi Zhang, Wanqing Wu, Guoxun Dai, Jiangnan Li, Haigang Feng*
Shenzhen International Graduate School, Tsinghua University, Shenzhen 518055, China
*feng.haigang@sz.tsinghua.edu.cn

Abstract—In this paper, we propose a voltage-controlled oscillator (VCO) using a new active inductor (AI) to realize the common-mode resonance inductor. Compared with the classic VCO using a passive inductor, the proposed AI greatly reduces the chip area and improves the figure of merit (FOM) across the full frequency band by utilizing a feedback resistor. The AI is designed using only MOSFETs, and high tunability is achieved by varying bias voltages, which enables accurate common-mode resonance. Designed in 65nm process, the proposed VCO operates from 1.7GHz-3GHz with 55.3% tuning range (TR). While operating at 1.7GHz, the VCO achieves a phase noise (PN) of -131.7dBc/Hz at 1MHz offset under 1.2V supply voltage and consumes 8.1mW for a FOM of 187.1dBc/Hz.

Keywords—Voltage-controlled oscillator (VCO), active inductor (AI), phase noise (PN), figure of merit (FOM)

I. INTRODUCTION

With the development of the wireless communication network The Fifth Generation Mobile Communication (5G) and Wi-Fi have become the key technologies to promote the progress of social informatization. As a key technology to realize high-precision frequency synthesis and a key module of RF transceiver system, Phase-Locked Loop (PLL) has become more and more important in communication systems such as 5G and Wi-Fi, which also raises more challenges to its design in low phase noise (PN), wide bandwidth and other fields [1]-[5]. With the advent of the Internet of Things (IoT) and smart connectivity, electronic communication devices are developing in the direction of miniaturization, low power consumption and good portability, which has the requirement on the reduction of chip area. As a key module in the PLL, the LC-voltage-controlled oscillator (LC-VCO) usually occupies a large area because it contains passive inductors. Therefore, on the basis of realizing low power consumption, low PN and wide frequency range, decreasing the area is very important in VCO design. In recent years, common-mode resonance has been found as an effective technique to reduce the PN of VCO [6]. Connecting an inductor in series with the tail of the VCO and making it resonates at the second harmonic of the frequency of VCO can effectively improve the common mode impedance, which will prevent the equivalent resistance of the MOSFET working in the linear region from deteriorating the Q of the LC tank. However, the use of additional inductors consumes an additional area. Implicit common-mode resonance is developed to solve the problem [7]. Based on the transformer technology, the tail inductor is placed

with the load inductor, effectively saving the area. But the electromagnetic coupling effect cannot be completely ignored, resulting in deterioration of performance. In addition to the above, considering that passive inductors can only provide a fixed resonant frequency, the suppression of noise cannot be realized in the full frequency band for a VCO with a wide tuning range (TR).

To solve these problems, a VCO based on a new AI is proposed. The VCO uses an AI as the tail inductor to realize the common-mode resonance. Consisting of only MOSFETs, the core area of the AI is effectively reduced. By utilizing a feedback resistor, the inductance value of the AI is variable by adjusting the bias voltages, which is highly flexible and more suitable for wide bandwidth design. This paper is organized as follows, section II describes the circuit implementation of the VCO and AI. Section III presents the simulation results of the AI and VCO, respectively. Conclusions are given in Section IV.

II. CIRCUIT DESIGN

A. Oscillator Design

The schematic of the proposed VCO is shown in Fig. 1. Two NMOS transistors, M_{n0}, M_{n1} and two PMOS transistors M_{p0}, M_{p1} constitute complementary CMOS structure. Two cross-coupled pairs can provide a larger negative resistance value, which releases the startup condition. A variable capacitor and a 5-bits capacitor array are used to expand the frequency range. In the capacitor unit, the control signal and its anti-signal are connected to the gate and both sides of the NMOS switch, respectively, to ensure that the switch can be totally turned on and off. The position of AI is shown in the red frame in Fig. 1, which replaces the passive inductor for the common-mode resonance.

B. Active Inductor Design

The schematic of the AI is also shown in Fig. 1. NMOS transistors M_1 and M_2 provide positive transconductance and negative transconductance as common source connection and common gate connection, respectively, to form the gyrator structure. The current-reuse technique combines the M_1 current and M_2 current into the same branch, which reduces the power consumption significantly.

Similar to passive inductors, there is a theoretical upper-limit frequency of the AI circuits to behave as inductors [8], which is written as (1).

979-8-3315-8850-2/25 $31.00 © 2025 IEEE

$$\omega_p = \sqrt{\frac{R_p + R_s}{R_p C_p L}} \qquad (1)$$

where ω_p is the upper-limit frequency, L is the equivalent inductance, R_p is the equivalent parallel resistance and R_s is the equivalent series resistance. When the operating frequency exceeds ω_p, the AI behaves as a capacitor that cannot work properly. By adding a series feedback resistor R between the two transconductance stages in the gyrator, the position of the pole in the frequency domain can be changed to enhance the stability of the inductance value. R can be implemented using a MOSFET working in the linear region in the CMOS process, and its resistance value can be changed by varying the bias voltages in the gates. Neglecting the effect of the capacitance between gate and drain, as well as the capacitance between drain and source, the small-signal equivalent circuit of the proposed AI is shown in Fig. 2.

Assuming that the parameters satisfy

$$g_{m1} \gg \frac{1}{r_{o1}} \qquad (2)$$

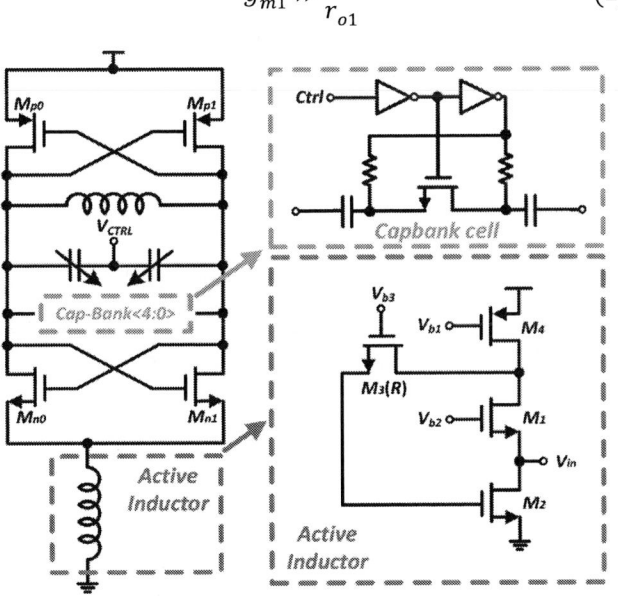

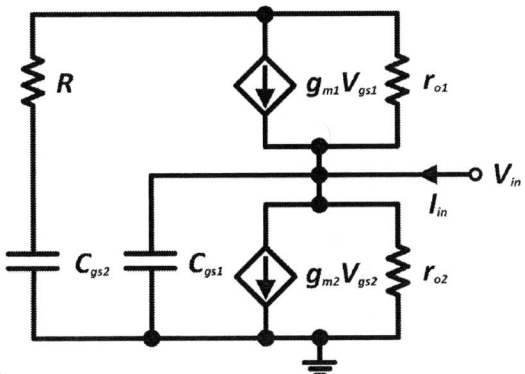

Fig. 1. Schematic of the proposed VCO.

Fig. 2. Small-signal equivalent circuit of the proposed AI.

$$g_{m2} \gg \frac{1}{r_{o2}} \qquad (3)$$

where $g_{m1}, r_{o1}, g_{m2}, r_{o2}$, are the transconductances and the output resistances of the two MOSFETs, respectively, and R is the series feedback resistance.

The input impedance can be written as

$$Y_{in} = sC_p + \frac{1}{R_p} + \frac{1}{sL + R_s} \qquad (4)$$

$$R_p = \frac{1}{\frac{1}{r_{o2}} + \frac{g_{m1}}{1 - \frac{R}{r_{o1}}}} \qquad (5)$$

$$C_p = C_{gs1} \qquad (6)$$

$$R_s = \frac{g_{o1}\left(1 - \frac{R}{r_{o1}}\right)}{g_{m1}\left[g_{o1} - g_{m2}\left(1 - \frac{R}{r_{o1}}\right)\right]} \qquad (7)$$

$$L = \frac{C_{gs2}\left(1 - \frac{R}{r_{o1}}\right)^2}{g_{m1}\left[g_{o1} - g_{m2}\left(1 - \frac{R}{r_{o1}}\right)\right]} \qquad (8)$$

where C_{gs1} and C_{gs2} are the capacitance between gate and source of the two MOSFETs. From the above, we can get

$$\omega_p = \sqrt{\frac{g_{m1}g_{m2}}{C_{gs1}C_{gs2}\left(1 - \frac{R}{r_{o1}}\right)}} \qquad (9)$$

Considering the case where there is no resistor R, thus the gyrator structure contains only two transistors, M_1 and M_2. Similarly, by analyzing the small-signal equivalent circuit, we can get

$$\omega_p' = \sqrt{\frac{g_{m1}g_{m2}}{C_{gs1}C_{gs2}}} \qquad (10)$$

From (9) and (10), it can be seen that by adding an additional resistor R, the expression for ω_p has an extra term $(1 - R/r_{o1})$ at the denominator, making ω_p larger and thus expanding the operating frequency of the proposed AI. And ω_p increases as R increases.

Since the tail inductor is used to resonate at the second harmonic, when the oscillation frequency of VCO becomes high, the inductance value of the tail inductor needs to be reduced to accommodate the frequency change due to the small amount of the change of parasitic capacitance value. (8) shows the inductance value decreases as the R increases. By adjusting the bias voltage V_{b3} of M_3, R will be smaller, which reduces the inductance value, and enlarges ω_p to accurately realize the common-mode resonance. When the oscillation frequency of VCO becomes low, by adjusting V_{b3} to increase the inductance value, the common-mode resonance can be achieved easily.

As shown in Fig. 3, the simulation results show that when the AI is simulated without adding the feedback resistor R, ω_p

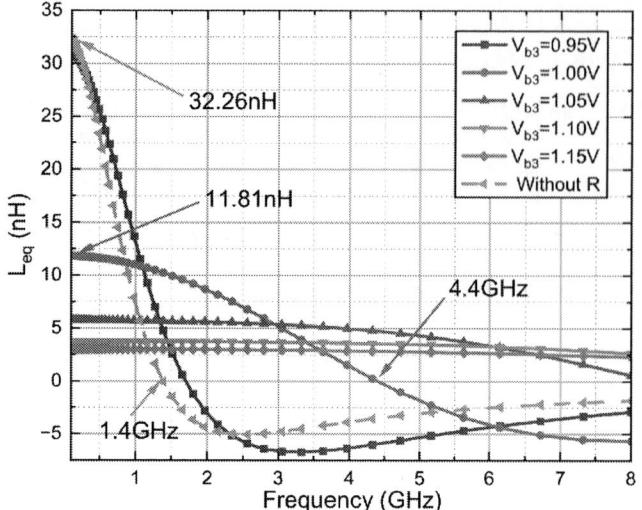

Fig. 3. Diagram of inductance value versus frequency.

is limited and the inductance value is large in 0-8GHz with poor tunability. When R is added during the simulation, ω_p and the inductance value can be controlled by the bias voltages, and the simulation results correspond well to the theoretical analysis.

The interface of the AI cannot isolate the DC signal. If the AI is directly connected to other circuits, the DC operating point at the node of the interface is susceptible to the mutual influence of the two circuits, resulting in the deterioration of the original performance of the two circuits. In order to improve the compatibility of this design, an additional PMOS M_4 is added to the connection node of the circuit working as a current source to realize the regulation of the DC voltage. When the current of M_4 increases, the drain-source current across the M_2 increases, causing an increase in drain voltage of M_2 and thus increasing the DC level. In contrast, when the current of M_4 decreases, the DC level drops. Therefore, for circuits with a large DC operating voltage at the interface, M_4 needs to provide a higher current. For circuits with a small DC operating voltage at the interface, M_4 needs to provide a lower current.

V_{b2} is utilized to provide a bias voltage for M_1. Besides, by adjusting V_{b2} to change the transconductance of M_1, the inductance can also be fine-tuned. Therefore, for the whole design, to achieve better tunability, the AI above biases each MOSFET individually and regulates the performance of the circuit by varying the magnitude of the bias voltages.

III. SIMULATION RESULTS

The proposed AI and VCO are designed in 65nm CMOS process and the simulation of inductance value of the AI is shown in Fig 4. As an example, by varying the bias voltages V_{b1}, V_{b2} and V_{b3}, the inductance value can cover the range from 2.73nH to 21.75nH, which are suitable values for the common-mode resonance.

Fig 5. and Fig 6. show the performance comparison between VCO without tail inductor, VCO using passive inductor as tail inductor and VCO using AI as tail inductor. With the same operating frequency range, it can be seen that by adopting the AI, VCO achieves better PN, lower power consumption and the

highest figure of merit (FOM), which proves the validity of AI. While consuming one-half of the power consumption of the original structure, the PN of the proposed VCO is improved by 4dB on average, resulting in a 5dB improvement in FOM on average. Since the transistor stack of the AI occupies certain headroom in the original structure, the supply voltage is reduced to 1.0V to simulate the performance of the original structure at the same supply voltage while the AI is absent. It can be seen that under the same power consumption, the PN and FOM performance of the VCO using AI are much higher than the original structure, which to some extent also proves the validity of the AI. Notice that the PN of the proposed VCO is a bit worse than that of the VCO using passive inductor, it can be explained by the noise of the transistors in the AI itself.

The proposed VCO layout is shown in Fig. 7, with the AI occupying a layout area of $18.5 \times 18.5 \ \mu m^2$. Taking 1nH as an example, the area of the proposed AI is reduced by more than 250 times compared to the normally used passive inductor, which usually occupies almost $0.25 \times 0.30 \ mm^2$.

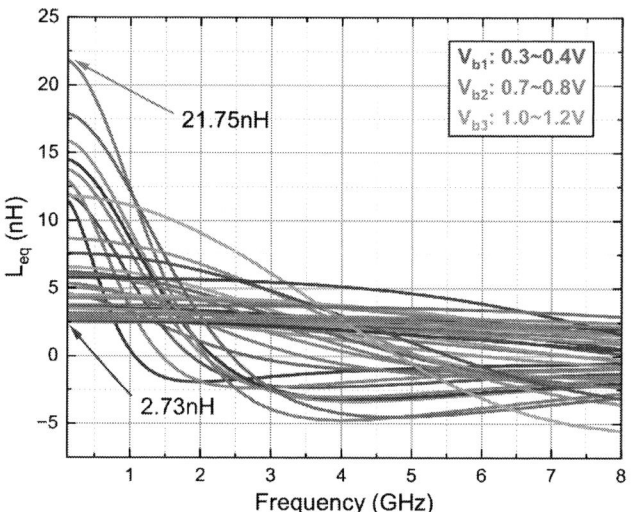

Fig. 4. Equivalent L_{AI} versus frequency for varying Vbias.

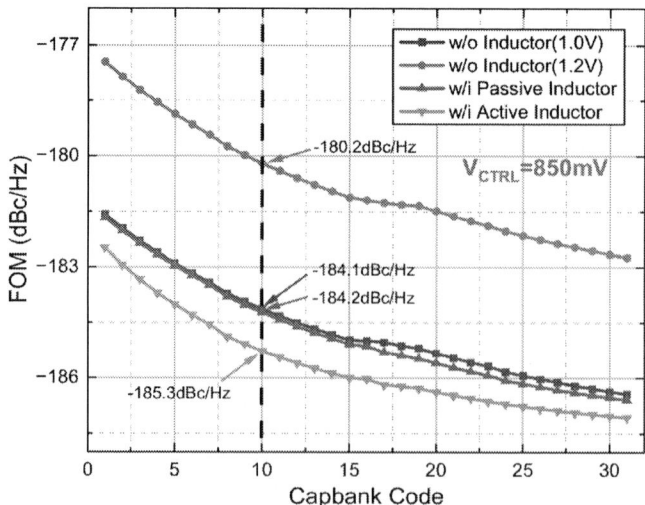

Fig. 5. FOM comparison between three structures.

979-8-3315-8850-2/25 $31.00 © 2025 IEEE

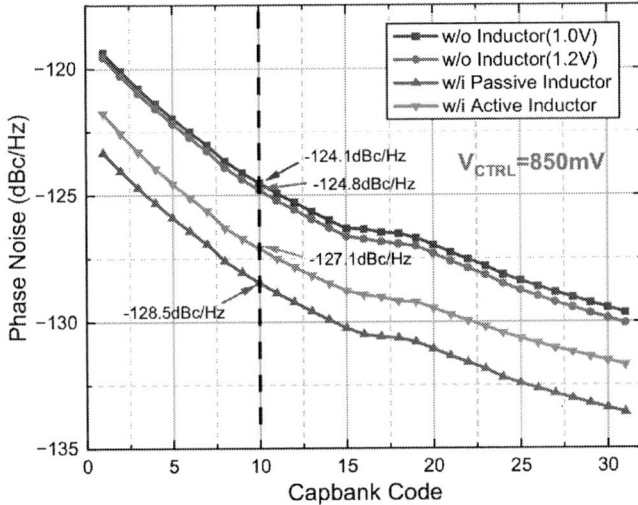

Fig. 6. PN comparison between three structures.

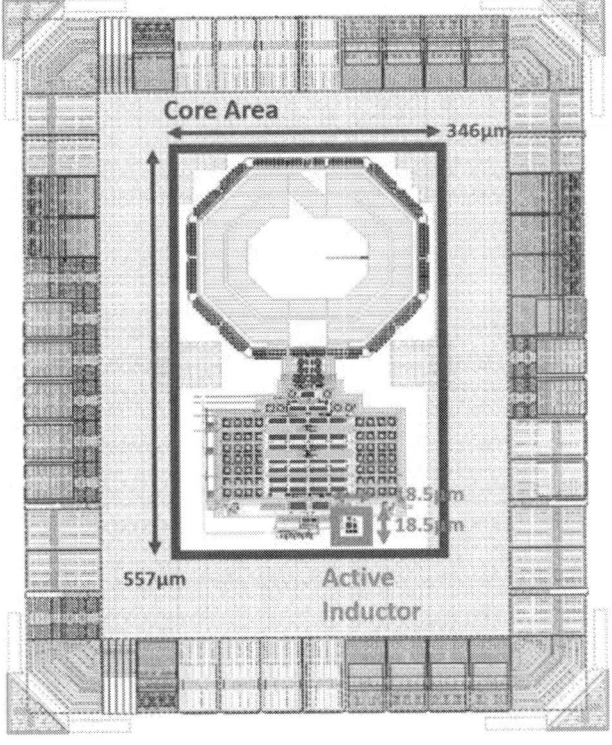

Fig. 7. Layout of the proposed VCO.

The VCO has a frequency range from 1.7GHz to 3GHz with a TR of 55.3%, and a power consumption of 8.1mW under 1.2V supply voltage. When the AI is employed, the PN is -131.7dBc/Hz@1MHz at an operating frequency of 1.7GHz, and the FOM is 187.1dB, which is 5dB higher than that of the basic CMOS complementary VCO, with no additional area consumption. Table 1 compares the proposed VCO with other recently published state-of-the-art VCOs based on CMOS process. The proposed VCO achieves the highest FOM, lower power consumption and the widest tuning range with the smallest core area, which provides sufficient support for the development of miniaturization and low-power in wireless

communication systems. Meanwhile the frequency range covers several communication protocols such as Wi-Fi, ensuring that the VCO can be adapted to many different application scenarios.

TABLE I. PERFORMANCE COMPARISON

	This Work	RFIC'23 [9]	MWTL'24 [10]	VLSI'24 [11]
CMOS Technology	65nm	8nm	65nm	28nm
Key Technique	Active Tail Inductor	Circular Coil Topology	Wideband $2F_{OSC}$ Resonance	S-Shape Folded CM Resonator
Frequency (GHz)	1.7~3.0	15.6~18.8	5.7~6.5	12.3~13.8
FTR (%)	55.3%	30%	14%	11.6%
PN@1MHz (dBc/Hz)	-131.7 (1.7GHz)	-115.4 (15.6GHz)	-116.08 (5.99GHz)	-118.7 (12.3GHz)
Supply (V)	1.2	0.9	0.67	0.9
DC Power (mW)	8.1	28.8	3.27	24.5
FOM[a]@1MHz (dBc/Hz)	187.1	184.8	186.6	186.6
FOM$_T$[b]@1MHz (dBc/Hz)	201.95	194.34	189.52	187.89
Core Area (mm²)	0.193	0.27	0.465	0.395

[a]. FOM = -PN + 20 log $(f_{osc}/\Delta f)$ − 10log (POWER/1mW)

[b]. FOM$_T$ = FOM + 20 log (FTR/10%)

IV. CONCLUSION

In this paper, a VCO based on a novel AI is proposed. The use of AI greatly reduces the occupied area while providing a wideband suitability to realize common-mode resonance. The proposed VCO is designed in a 65nm CMOS process and has a PN of -131.7dBc/Hz at 1MHz offset from the 1.7GHz carrier frequency. The FOM at 1MHz is 187.1dBc/Hz. Under a 1.2V supply voltage, the VCO consumes 8.1mW, with a core area of 0.193mm².

ACKNOWLEDGMENT

All the passive device electromagnetic simulation is carried out by Peakview of Lorentz Solution.

REFERENCES

[1] Poonam Sachdeva and Ankita Aggarwal, "DESIGN OF CMOS RING VCO FOR PLL BASED FREQUENCY SYNTHESIZER," International Journal of Electrical and Electronic Engineering & Telecommunications, Vol. 5, No. 2, pp. 73-79, April 2016.

[2] Kwon K, Abdelatty O A, Wentzloff D D. Pll fractional spur's impact on fsk spectrum and a synthesizable adpll for a bluetooth transmitter[J]. IEEE Journal of Solid-State Circuits, 2023, 58(5): 1271-1284.

[3] Yi X, Yang K, Liang Z, et al. A 65nm cmos carrier-aggregation transceiver for ieee 802.11 wlan applications[C]//2016 IEEE Radio Frequency Integrated Circuits Symposium (RFIC). IEEE, 2016: 67-70.

[4] Lu E, Li W K, Deng Z, et al. 10.4 a 4× 4 dual-band dual-concurrent wifi 802.11 ax transceiver with integrated lna, pa and t/r switch achieving+ 20dbm 1024-qam mcs11 p out and- 43db evm floor in 55nm cmos[C]//2020 IEEE International Solid-State Circuits Conference-(ISSCC). IEEE, 2020: 178-180.

[5] Lu C, Zhao Y, Bao J, et al. A highly efficient combo transceiver for 802.11 b/g/n/ax and bt/ble in 22nm cmos[C]//ESSCIRC 2021-IEEE 47th

European Solid State Circuits Conference (ESSCIRC). IEEE, 2021: 503-506.

[6] Hegazi E, Sjoland H, Abidi A A. A filtering technique to lower LC oscillator phase noise[J]. IEEE Journal of Solid-State Circuits, 2001, 36(12): 1921-1930.

[7] Murphy D, Darabi H, Wu H. Implicit common-mode resonance in LC oscillators[J]. IEEE Journal of Solid-State Circuits, 2017, 52(3): 812-821.

[8] Herbert T B, Hyland J S, Abdullah S, et al. An active bandpass filter for LTE/WLAN applications using robust active inductors in gallium nitride[J]. IEEE Transactions on Circuits and Systems II: Express Briefs, 2021, 68(7): 2252-2256.

[9] Hu S, Chen Z, Wu W, et al. A 15.6-GHz quad-core VCO with extended circular coil topology for both main and tail inductors in 8-nm FinFET process[C]//2023 IEEE Radio Frequency Integrated Circuits Symposium (RFIC). IEEE, 2023: 201-204.

[10] Yang C, Chen Y, Huang Y, et al. A 5.99-GHz VCO With Wideband-Differential-Mode Second Harmonic Resonance Achieving $-$138.9 dBc/Hz Phase Noise at an Offset of 10 MHz[J]. IEEE Microwave and Wireless Technology Letters, 2024.

[11] Lu S, Wu D, Guo X, et al. A Quad-Core VCO Incorporating Area-Saving Folded S-Shaped Tail Filtering in 28-nm CMOS[J]. IEEE Transactions on Very Large Scale Integration (VLSI) Systems, 2024.

A 24-28GHz GaN SPDT MMIC with High Isolation and Low Insertion Loss

Peiyao Liu
Faculty of Integrated Circuit
Xidian University
Xi'an, China
23111213477@stu.xidian.edu.cn

Pengfei Jiang
Faculty of Integrated Circui
Xidian University
Xi'an, China
jpf1661555836@163.com

Kui Dang*
Faculty of Integrated Circui
Xidian University
Xi'an, China
dangkui@xidian.edu.cn

Rujun Sun
Faculty of Integrated Circui
Xidian University
Xi'an, China
sunrujun@xidian.edu.cn

Yue Hao
Faculty of Integrated Circui
Xidian University
Xi'an, China
yhao@xidian.edu.cn

Jincheng Zhang
Faculty of Integrated Circui
Xidian University
Xi'an, China
jchzhang@xidian.edu.cn

Abstract—As an important part of the communication system, RF front-end transceiver components have higher frequency, higher power, lower loss and smaller performance requirements, among which the performance of RF switches directly affects the transmitting power and efficiency of the transmitting system and the noise of the receiving system. In this work, we propose a GaN single-pole double-throw RF switch, which adopts a topology of quarter-wavelength transmission lines combined with parallel transistors to achieve high isolation design while taking into account low insertion loss. The measurement results show that in the frequency range of 24-28GHz, the insertion loss is 1.2-1.45dB, the isolation is 42.4-60.8dB, and the return loss is greater than 12.7dB.

Keywords—GaN, MMIC, SPDT, high isolation, insertion loss

I. INTRODUCTION

With the rapid development of modern wireless communication technology and the widespread popularity of various intelligent terminals, people's requirements for communication efficiency are increasing day by day. As an indispensable part of various communication systems, RF front-end transceiver components have received widespread attention and research. RF switch is a key control component of the transceiver system, which is mainly used to receive and transmit signals. The RF switch is located between the antenna and the receiving and transmitting links, and its insertion loss and isolation degree performance directly affects the transmitting power and efficiency of the transmitting system and the noise of the receiving system.

In view of the current high-frequency and high-power application conditions of RF circuits, the third generation semiconductor gallium nitride (GaN) has gradually become a hot spot in research and application due to its superior material properties[1][2]. Correspondingly, GaN RF switched monolithic microwave integrated circuits (MMICs) have also become a

research hotspot. A lot of research on GaN high electron mobility transistor (HEMT) switches has been carried out abroad[3]-[7], while the research on GaN switches is relatively few domestically. At present, the GaN process is gradually becoming mature, and the manufacturing cost is relatively reduced, thus laying the process foundation for the research of GaN switches. There is a trade-off between the on-resistance and off-state capacitance of switch transistors, with transistors with a large gate width having a smaller on-resistance but a larger turn-off capacitance, and transistors with a small gate width doing the opposite. In addition, switch transistors with different gate widths differ in frequency characteristics and power handling capabilities. The design of RF switches with low insertion loss and high isolation has also become a challenge under the design requirements of high frequency, broadband and high power.

In this work, a millimeter-wave GaN single-pole-double-throw switch (SPDT) operating at 24-28GHz with a low insertion loss and a high isolation is designed and fabricated. A topology of a quarter-wavelength (λ/4) transmission line combined with a parallel transistor is introduced, and the method of direct grounding of the source of the parallel transistor and the cascade capacitor is compared. The hybrid parallel topology of the single and double gate transistors is finally determined. The measurement results show that the proposed MMIC chip for RF switches achieves the highest isolation level of publicly reported GaN switches in this frequency band while reducing insertion loss.

II. CIRCUIT DESIGN

A. Topology of Switch Transistor

The basic topologies of RF switches mainly include three types: series, series-parallel, and parallel. The topologies are shown in Figure 1.

This work was supported by the National Natural Science Foundation of China under Project 62204195 and the National Science Fund for Distinguished Young Scholars under Grant 62525402.

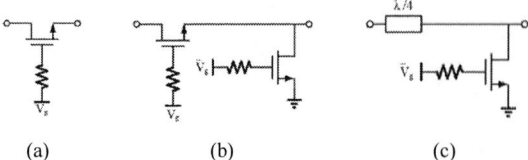

Fig. 1. (a)Series, (b)series-Parallel, and (c)parallel topology of switch transistor.

The series topology is the most primitive switch structure. However, due to the existence of the off-state capacitance, as the frequency keeps increasing, using only series transistors to achieve the switch function causes a significant amount of signal leakage, making it difficult to meet the requirements of small-signal indicators. The series-parallel topology solves this problem. When the switch is on, the series transistor is biased in on-state, and the parallel transistor is biased in off-state.

When the frequency rises to the millimeter-wave band, the parasitic inductance and capacitance introduced by the series switch transistor in the on state cause poor performance in terms of insertion loss and isolation at high frequencies, making it difficult to be applied in the wideband range from DC to high frequencies. In terms of power tolerance, the saturation current of the series transistor determines the overall power capacity of the circuit. Using large-sized transistors to ensure power capacity makes it difficult to meet the requirements of small-signal characteristics.

In the parallel switch structure, when a parallel HEMT in a certain branch is in the on state, a low-impedance path is formed in that branch, causing the RF signal to short-circuit to ground through that branch. At this time, the signal transmission path of the adjacent branch will experience unexpected coupling due to the common ground effect. To effectively solve this problem, a λ/4 microstrip line is used as an impedance transformation network in the parallel switch structure. Its core function is to achieve impedance matching between the common port and the parallel switch devices.

The λ/4 transmission line is a special type of transmission line, with a length one fourth of that of a standing wave. It features constant impedance and does not cause signal reflection or loss during transmission. Its reflection coefficient is -1, which enables the signal to be perfectly matched during transmission. Although the λ/4 transmission line structure has inherent bandwidth limitations, it still shows unique application advantages in the millimeter-wave band. The reason is that the series transistor in the series-parallel structure cannot achieve the ideal open-circuit characteristic in the off state, and its off-state capacitance significantly deteriorates the high-frequency isolation. In addition, the figure of merit of the series-parallel structure is highly sensitive to the on-resistance and off-state capacitance parameters of the series devices.

Based on the above limitations, the improved parallel switch structure introduces a λ/4 transmission line to replace the traditional series devices. In the on state, the parallel HEMT is in the off state, and its drain presents a high-impedance characteristic. Through the impedance transformation effect of the λ/4 transmission line, it is equivalent to a low-impedance matching state at the input end. Conversely, in the off state, the HEMT is on, forming a low-impedance path to ground. After the transformation by the transmission line, it presents a high-impedance characteristic at the input end, successfully solving the performance degradation problem caused by parasitic parameters in the traditional series-parallel structure at the millimeter-wave band.

B. Extraction of Switch Transistor Parameter

For the single-gate and double-gate switched transistor models provided by the GaN process library for a commercial process N15PA21 further parameter extraction was performed. The small signal circuit of the switch transistor is simplified, as shown in Figure 2, the switch transistor can be equivalent to a series circuit of an on-resistance (R_{on}) and on-inductance (L_{on}) when the switch is on, and in the off-state, it is equivalent to a series circuit of the off-capacitor (C_{off}), the turn-off resistor (R_{off}), and the turn-off inductance (L_{off}).

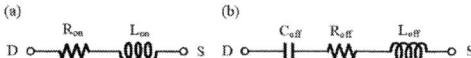

Fig. 2. Simplified small signal circuit of (a)on-state and (b)off-state switch transistor.

The impedance of the transistor in the turn-on and off-states is expressed, and the S parameters of the switch transistor in the process library are extracted, and the corresponding equivalent parameter values are obtained by using the conversion relationship between the S matrix and the Z matrix. Among them, single-gate switch transistors have a large current density and smaller on-resistance, which is suitable for series topologies, while double-gate transistors have small off-state leakage current and smaller shutdown capacitance, which are more suitable for parallel topologies.

In order to further facilitate the design of RF switch circuits, we study the variation of the equivalent parameters of single gate and double gate switch transistors with gate width at 26 GHz, as shown in Figure 3, with the increase of gate width, R_{on} and R_{off} decrease; when the gate width is smaller, L_{on} is larger, and then with the increase of gate width, it stabilizes at about 60pH, and C_{off} increases with the increase of gate width.

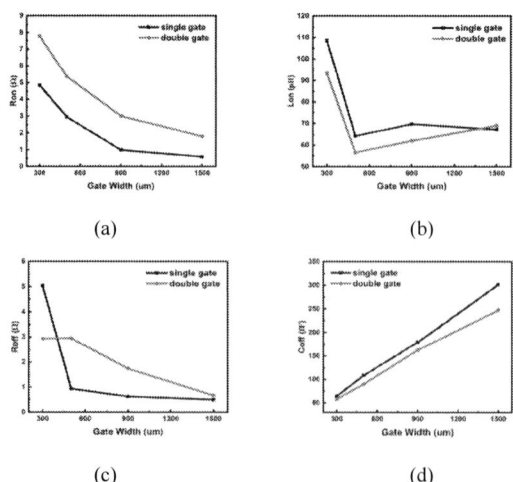

Fig. 3. Relationship between the equivalent parameters and the total gate width of a switch transistor.

C. Design of High Isolation

According to the above research and analysis results, the design of the RF switch SPDT MMIC is carried out by using a quarter-wavelength transmission line combined with a parallel topology. From Fig. 3, R_{on} is negatively correlated with the total gate width of the device, and increasing the gate width of the parallel transistor can effectively reduce the R_{on} and thus increase the isolation. However, the C_{off} of the large gate width device in the off state is large, which will worsen the insertion loss of the conduction branch, so a capacitor is cascaded to ground at the source of the parallel switch transistor. The parasitic resistance introduced by the source cascade capacitor increases the real part of the on-state impedance (Z_{on}) of a switch transistor and decreases the imaginary part of Z_{on}, forming an R-L-C series resonant circuit as a whole. In addition, the introduction of the source capacitor increases the imaginary impedance of the switch transistor in the off-state, which is equivalent to reducing the equivalent C_{off}.

Figure 4 shows the results of the insertion loss and isolation simulation results of parallel transistor with and without source capacitance. The simulated transistor use double-gate GaN switches with a gate length of $15\times100\mu m$. With the source capacitor, the insertion loss at 26GHz is significantly reduced by 2dB. In addition, there is a frequency shift in the isolation, and the resonant frequency point can be designed to meet the expectations by reasonably selecting the capacitance value of the source ground capacitor.

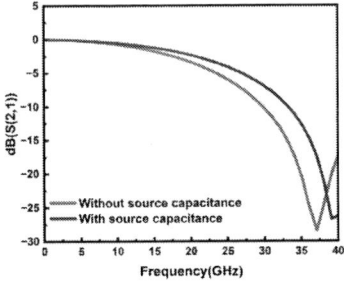

(a)

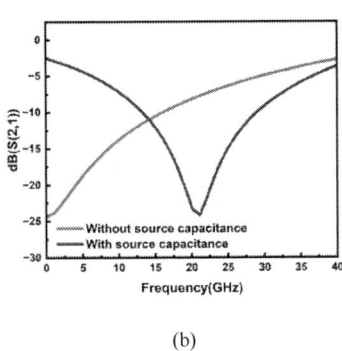

(b)

Fig. 4. Simulation of (a)insertion loss and (b)isolation of parallel transistor with and without source capacitor.

In addition to using large gate width devices, the method of increasing the isolation degree usually adopts the method of parallel multi-stage switch transistor, which is to ensure that the area is not too large while weighing the insertion loss and isolation, so the two-stage parallel connection is used in the parallel stage.

The overall circuit schematic is shown in Figure 5, the symmetrical two-stage parallel structure is adopted. The first stage is a double-gate GaN HEMT with a gate length of 15×100 μm, and the second stage is a single-gate GaN HEMT with a gate length of 7×150 μm. Their sources are cascaded with capacitors C1, C2, C3, and C4 respectively to reduce the equivalent capacitance in the off state of the large-size transistor, thereby improving the insertion loss. The microstrip line TL1 and TL2 are $\lambda/4$ transmission lines, and the capacitors C_{in} and C_{out} participate in matching at the input and output ports respectively, and have the function of isolating DC.

In addition, the input impedance of the shutdown branch is inductive, and the ground capacitance C_s connected in parallel between the two branches can resonate with the shutdown branch, thereby extending the bandwidth. The RF3 branch is on when V_g=0V and $\overline{V}_g$=-28V, and the RF2 branch is on when V_g=-28V and $\overline{V}_g$=0V.

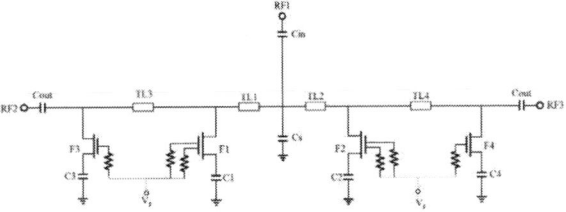

Fig. 5. Schematic of the proposed switch MMIC.

III. SIMULATION AND MEASUREMENT

The proposed RF switch MMIC with high isolation has been fabricated by a 150 nm GaN process, and the photograph of the MMIC chip under the microscope is shown in Figure 6. Figure 7 shows the measurement environment. By controlling the on and off of the switch from the power supply, the vector network analyzer directly conducts on-chip testing and reads the S-parameters of the circuit by connecting the GSG probe on the probe station via a coaxial cable.

Fig. 6. Photograph of the switch MMIC chip under the microscope.

Fig. 7. Measurement environment of the switch MMIC chip.

979-8-3315-8850-2/25 $31.00 © 2025 IEEE

Figure 8 and Figure 9 show the Simulated and Measured insertion loss, isolation, S11, and S22 of the designed switch MMIC chip. In the frequency range of 24-28GHz, the insertion loss is less than 1.45dB, the isolation is greater than 42.47dB, and the in-band peak isolation is up to 60.68dB. The input return loss is greater than 12.7dB, and the output return loss is greater than 12.9dB, and the in-band matching is good. Two resonant points are formed around 23 GHz and 29 GHz, extending the bandwidth.

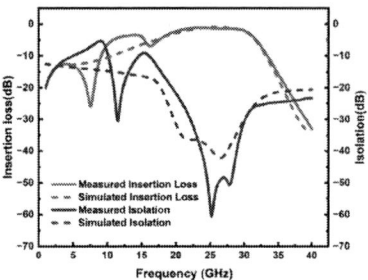

Fig. 8. Simulated and Measured insertion loss and isolation of the switch MMIC.

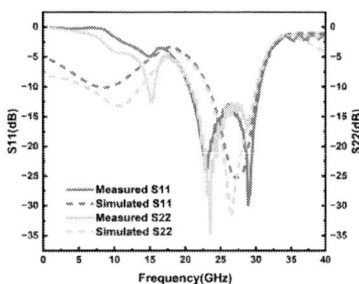

Fig. 9. Simulated and Measured S11 and S22 of the switch MMIC.

Table I. compares the performance of the designed SPDT MMIC chip with some other state-of-art switches. In this work, a parallel topology architecture that integrates single-gate and double-gate transistors is innovatively adopted, and an in-band isolation index of up to 42.4-60.68 dB is achieved in the 24-28 GHz band, which is the highest isolation level of GaN switches currently reported in this band. The insertion loss is effectively controlled within 1.45 dB and the return loss is greater than 12.7dB in the operating frequency band through the optimized design of the circuit structure, which has obvious advantages compared with other jobs, and provides a new idea for the design of K-band low-loss and high-isolation RF switched monolithic microwave integrated circuits.

IV. CONCLUSION

In order to meet the requirements of high isolation, this work innovatively adopts a parallel topology architecture with mixed integration of single and double gate transistors, and achieves an in-band isolation index of up to 42.4-60.68 dB in the K/Ka band, which is the highest isolation level of GaN switches currently reported in this frequency band. At the same time, the insertion loss is effectively controlled within 1.45 dB and the return loss is greater than 12.7dB through the optimized design of the circuit structure, which has obvious advantages compared with

other work, and provides a new idea for the design of K/Ka-band low-loss and high-isolation RF switches.

TABLE I. PERFORMANCE OF SOME OTHER SWITCHES AND THIS WORK

Reference	[8]	[9]	[10]	This work
Process	GaN/Si	0.1μm GaN/Si	0.1μm GaN/SiC	0.15μm GaN/SiC
Topology	SPDT	SPDT	SPDT	SPDT
Bandwidth (GHz)	24-40	17.6-36	28-51	24-28
Insertion loss (dB)	1.0-1.4	0.72(min)	1.4-2	1.2-1.45
Isolation (dB)	28-32	>20	23.5-27	42.4-60.68
Return loss (dB)	>10	>20	>10	>12.7
Area(mm²)	2.2	0.8	1.25	2.18
Power Capacity(W)	/	>5	/	>5

ACKNOWLEDGMENT

This work was supported in part by the National Key Research and Development Program under Grant 2022YFB3607600, in part by the National Natural Science Foundation of China under Project 62204195, in part by the National Science Fund for Distinguished Young Scholars under Grant 62525402, and in part by the Key Research and Development Program of Jiangsu Province under Grant BE2022057-2.

REFERENCES

[1] CHOW T P, TYAGI R. Wide Bandgap Compound Semiconductors for Superior High-voltage Power Devices[C], Proceedings of the 5th International Symposium on Power Semiconductor Devices and ICs, 1993: 84.

[2] Shifrin M B, Katzin P J, Ayasli Y. Monolithic FET Structures for High-power Control Component Aapplications[J]. IEEE Transactions on Microwave Theory and Techniques, 1989, 37(12; 12): 2134-2141.

[3] F. Thome, P. Brückner, R. Quay and O. Ambacher, "Millimeter-Wave Single-Pole Double-Throw Switches Based on a 100-nm Gate-Length AlGaN/GaN-HEMT Technology," *2019 IEEE MTT-S International Microwave Symposium (IMS)*, Boston, MA, USA, 2019, pp. 1403-1406, doi: 10.1109/MWSYM.2019.8700955.

[4] G. Polli et al., "GaN/Si Ka-band SPDT for observation payloads," *2019 IEEE Asia-Pacific Microwave Conference (APMC)*, Singapore, 2019, pp. 288-290, doi: 10.1109/APMC46564.2019.9038632.

[5] S. Osmanoglu and E. Ozbay, "X-Band High Power GaN SPDT MMIC RF Switches," *2019 European Microwave Conference in Central Europe (EuMCE)*, Prague, Czech Republic, 2019, pp. 83-86.

[6] M. Hangai, R. Komaru, S. Miwa, Y. Kamo and S. Shinjo, "2–12 GHz High-Power GaN MMIC Switch Utilizing Stacked-FET Circuits," *2019 49th European Microwave Conference (EuMC)*, Paris, France, 2019, pp. 840-843, doi: 10.23919/EuMC.2019.8910824.

[7] M. Assad, A. I. Najam and H. M. Cheema, "GaN based High Power SPDT Switch for Single Chip X-Band T/R Module Front-End," *2021 1st International Conference on Microwave, Antennas & Circuits (ICMAC)*, Islamabad, Pakistan, 2021, pp. 1-3, doi: 10.1109/ICMAC54080.2021.9678237.

[8] D. Yuan, Z. Zhang, J. Chen. Design Analysis of Wideband 24-40GHz GaN-on-Si SPDT Switches for 5G Millimeter-Wave Applications[C]. 2022 7th International Conference on Integrated Circuits and Microsystems (ICICM), Xi'an, China, 2022: 641-644.

[9] H. Ma, G. Shen, W. Che. A Single-Pole-Double-Throw Switch for Millimeter-Wave Mobile Communication Using Gallium Nitride Technology[C]. 2024 IEEE MTT-S International Wireless Symposium (IWS), Beijing, China, 2024: 1-3.

[10] F. Thome, P. Brückner. Millimeter-Wave Single-Pole Double-Throw Switches Based on a 100 nm Gate-Length AlGaN/GaN-HEMT Technology[C]. 2019 IEEE MTT-S International Microwave Symposium (IMS), Boston, MA, USA, 2019: 1403-1406.

2025 The 10th International Conference on Integrated Circuits and Microsystems

Design and FPGA Implementation of a NTT Hardware Accelerator for PQC ML-DSA

Shuhang Zhu
Electronic Engineering College
Heilongjiang University
Harbin, China
3236329866@qq.com

Liji Wu*
School of Integrated Circuit
Tsinghua University
Beijing, China
lijiwu@mail.tsinghua.edu.cn

Lei Li*
Electronic Engineering College
Heilongjiang University
Harbin, China
lileidtk@hlju.edu.cn

Yifan Yang
School of Integrated Circuit
Tsinghua University
Beijing, China
yyf20@mails.tsinghua.edu.cn

Xiangmin Zhang
School of Integrated Circuit
Tsinghua University
Beijing, China
zhxm@tsinghua.edu.cn

Abstract—In this paper, we present a high‑performance, resource‑efficient hardware architecture for the 256-point Number Theoretic Transform (NTT) core tailored to the CRYSTALS-Dilithium post-quantum signature scheme. Our design introduces three key innovations: (1) a Barrett modular reduction unit that replaces conventional subtraction-based reduction with precomputed constants, achieving a 42% reduction in critical-path delay; (2) a dynamic, algorithmic address resolver that eliminates lookup tables and reduces LUT usage by 15.2%; and (3) a quad-butterfly pipeline comprising four parallel configurable butterfly units that support both Cooley-Tukey and Gentleman-Sande modes, enabling a fourfold increase in per-cycle throughput. Implemented on FPGA with modulus q = 8,380,417, the proposed NTT core consumes only 1,033 slices (39.3% fewer) and 3,189 LUTs (44.1% fewer) compared to state-of-the-art designs, at a modest cost of 28 DSPs and 8 BRAMs. Functional simulation over 512 test points confirms 100% correctness ($R^2 = 1.0$) across the full coefficient range. The architecture thus offers an attractive balance between throughput, resource utilization, and implementation simplicity, providing a practical foundation for lattice-based cryptosystems on resource constrained platforms.

Keywords—Post Quantum Cryptography(PQC); NTT; Hardware design

I. INTRODUCTION

The U.S. NIST launched the PQC standardization process in 2016 and received 69 algorithm submissions in the first round in 2017. In 2020, seven finalists and eight alternate candidates were selected, including three signature schemes: CRYSTALS-Dilithium, FALCON and Rainbow.With the growing threat of quantum computing undermining the security of traditional public-key cryptosystems, lattice-based cryptography has attracted widespread attention as a candidate for post-quantum cryptography (PQC) due to its foundation on hard mathematical problems such as the Learning With Errors (LWE) problem [1].

Among these, CRYSTALS-Dilithium, selected as a digital signature algorithm in the NIST PQC standardization process, has become a key focus in research and engineering implementations thanks to its trade-off among security, efficiency, and implementation complexity [2].

In CRYSTALS-Dilithium, numerous polynomial operations, such as key generation, signature generation, and verification, rely on efficient ring multiplication[3]. The Number Theoretic Transform (NTT), as the core acceleration module for polynomial multiplication, directly determines the overall algorithmic efficiency[4]. However, existing hardware implementations of NTT face several challenges, including high throughput, low latency, and scalability. To meet the requirements of different security levels (such as η, γ_2, and q) and address the increasing deployment demands, reducing hardware resource consumption, improving computational efficiency, and enhancing resistance to side-channel attacks (SCA) have become key design concerns[5].

This study presents an efficient hardware architecture for the Number Theoretic Transform (NTT), addressing resource consumption and computational efficiency bottlenecks in existing designs. We introduce three key innovations: (1) Barrett modular reduction, which reduces critical path delay by 42%; (2) dynamic address resolver, which decreases LUT usage by 15.2%; and (3) quad-parallel butterfly pipeline, which quadruples per-cycle throughput. These innovations not only reduce resource consumption but also maintain high throughput and low latency, providing an effective solution for deploying post-quantum secure systems on resource-constrained platforms like FPGA.

The primary motivation behind our design is to minimize resource wastage in existing NTT implementations while ensuring optimal performance. Current hardware solutions often face trade-offs between throughput and latency, especially in

979-8-3315-8850-2/25 $31.00 © 2025 IEEE

resource-constrained environments. By employing Barrett modular reduction, we simplify modular operations and reduce critical path delay; the dynamic address resolver eliminates lookup tables, further reducing LUT consumption; and the quad-parallel butterfly pipeline boosts throughput by enhancing parallel computation. Together, these innovations improve overall performance, offering a scalable and efficient approach for post-quantum cryptographic systems.

Ali Yahya Hummdi et al. proposed the platform-independent Unif-NTT hardware accelerator, which consists of three register banks, a unified butterfly unit (Unif-BU), and a dedicated control unit (CU), showing significant improvements in hardware resource utilization and processing speed[6]. Ferhat Yaman introduced three hardware architectures based on a unified butterfly structure, offering lightweight, balanced, and high-performance designs with 1, 4, and 16 butterfly units, optimizing NTT/INTT operations[7]. Dai Li et al. proposed MeNTT, a compact "in-memory" accelerator for Ring-LWE problems, using bit-serial operations and buffering to efficiently speed up computations[8].

Existing NTT hardware implementations have made progress but still have shortcomings. **Unif-NTT** lacks a balance between resource consumption and latency optimization, facing issues like insufficient parallelism and high resource consumption. The architecture based on a unified butterfly structure improves performance but still has room for improvement in LUT and resource utilization. **MeNTT** offers "in-memory" acceleration but still faces bottlenecks in hardware resource consumption and parallel computation efficiency, especially when handling large-scale data. These limitations hinder efficient deployment on resource-constrained platforms such as FPGAs, where memory access and parallel computation are crucial. Overall, existing designs still have room for improvement in resource optimization, performance enhancement, and scalability, especially for high-throughput, low-latency applications.

By introducing Barrett modular reduction, dynamic address resolver, and quad-parallel butterfly pipeline, the proposed design effectively overcomes the limitations of existing solutions in terms of resource consumption, performance, and scalability. Barrett modular reduction reduces critical path delay, dynamic address resolver lowers LUT usage, and quad-parallel butterfly pipeline enhances throughput. The overall design achieves a good balance between high throughput, low latency, and resource utilization, making it particularly suitable for resource-constrained platforms such as FPGAs. This architecture allows for more efficient deployment of post-quantum cryptographic systems, ensuring both high performance and low resource usage. It also verifies the NTT computation results through hardware-software collaboration, enhancing both accuracy and reliability in real-world applications.

Figure 1 illustrates the key generation and signing phases of the CRYSTALS-Dilithium digital signature scheme. In the key generation phase, a public matrix A is expanded from a seed, and small secret vectors are sampled to compute t=A·s1+s2, forming the public and secret keys. During signing, a short vector y is sampled, and w=A·y is computed. A challenge value c=H(w) is then derived via a hash function, and the final signature is produced as z=y+c·s1, resulting in the signature pair (z,c).

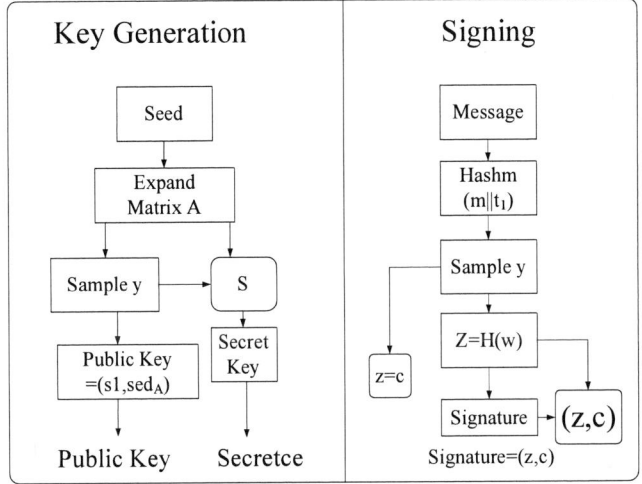

Fig. 1. ML-DSA Functional Block Diagram

II. NTT

A. NTT Principle

The Number Theoretic Transform (NTT) is a structure similar to the Fast Fourier Transform (FFT), primarily used to accelerate polynomial multiplication. It plays a crucial role in lattice-based cryptography (LBC) and is analogous to a fast Fourier transform over integer rings [9].

Let a and s be two polynomials sampled from Rq, with coefficients in the range of [0,q], where q is a prime number. We denote the polynomials.

$$a = a_{n-1}x^{n-1} + a_{n-2}x^{n-2} + \ldots + a_0 \qquad (1)$$

as

$$\hat{a}_{n-1}, \hat{a}_{n-2}, \ldots \hat{a}_0 \qquad (2)$$

Polynomial multiplication b=a∗s is performed according to Formula 1. After transforming vectors *a* and *s* into the NTT domain, the NTT coefficients of *b* are calculated by point-wise multiplication of the coefficients of *a* and *s*. The original coefficients are then obtained by applying the inverse NTT transform to

$$\hat{b} = \sum_{i=0}^{N-1} (\hat{a}_i * \hat{s}_i)x^i. \qquad (3)$$

B. NTT Hardware Implementation

In NTT, there are two butterfly computation types. The Cooley-Tukey (CT) butterfly NTT produces output in natural order and takes input in bit-reversed order. On the other hand, the Gentleman-Sande (GS) butterfly NTT outputs in bit-reversed order and accepts input in natural order, similar to the Decimation-In-Frequency (DIF) FFT algorithm[10].

Figure 2 illustrates the computation process of the 4-point Fast Number Theoretic Transform (NTT), consisting of two butterfly operation stages (Stage 1 and Stage 2). Using rotation factors , ψ^1, ψ^2, and ψ^3, from a finite field as multiplication factors, it achieves frequency separation and computation of the input data. Stage 1 performs the initial frequency component separation, while Stage 2 further integrates the data to produce the final NTT result.

For the inverse Number Theoretic Transform (INTT), Stage 1 similarly pairs the input data for addition and subtraction operations combined with weighting factors, such as ψ^{-1} and ψ^{-3}. Stage 2 takes these results as input, repeats the addition and subtraction operations with weighting by ψ^{-2}, and finally multiplies by 4^{-1}, the multiplicative inverse of 4, to obtain the final output, thus the inverse transformation of the NTT is completed.

III. HARDWARE DESIGN

A. The Overall Architecture of NTT

NTT decomposes polynomial multiplication into multiple layers of butterfly operations through the divide-and-conquer

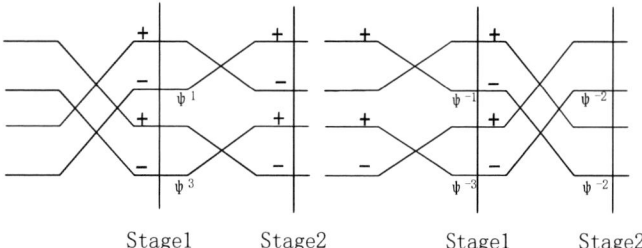

Fig. 2. 4-point Fast Number Theoretic Transform

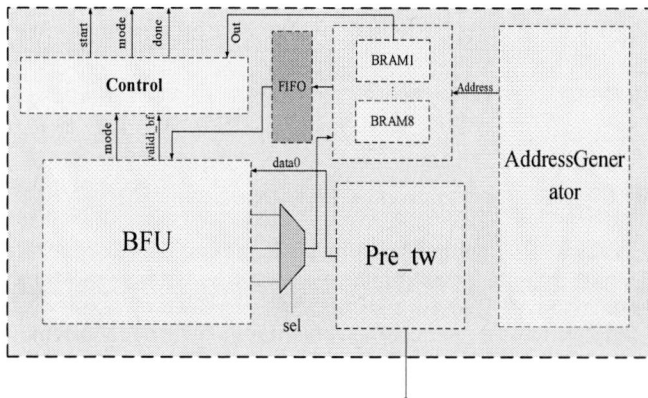

Fig. 3. Overall architecture of the NTT process

philosophy, where each layer combines subproblem solutions using twiddle factors. Taking the Cooley-Tukey butterfly as an example: $c=a+b\cdot\zeta x, d=a-b\zeta x$, here, a and b are input coefficients, ζx is a fixed twiddle factor, and c and d are output results. This process directly involves the multiplication of sensitive data with twiddle factors, making it a primary target for side-channel attacks.

This paper presents an efficient NTT transformation design based on the NTTN architecture, as shown in Figure 3. The control unit governs the overall process, covering forward and inverse NTT modes, addition-subtraction modes, and

multiplication modes. This design utilizes four highly integrated butterfly units, which support two types of butterfly transformation operations, namely Cooley-Tukey and Gentleman-Sande. It enables flexible multi-layer NTT computations, greatly reducing the number of memory accesses and bandwidth requirements, and enhancing the overall computational efficiency. Moreover, test sequences are generated through software pre-computation. These test sequences contain polynomial coefficients, twiddle factors, and pre-computed results, which are then compared with the actual results.

During the computation process of NTT, a large number of multiplications and modular operations are involved. Traditional modular operation methods, such as achieving modular reduction through successive subtractions, have relatively high computational complexity. The Barrett modular reduction converts modular operations into a series of multiplication, shift, and addition operations by pre-computing some constants, which greatly reduces the number of multiplications required for modular reduction. This enables a significant improvement in the computation speed in the NTT computation of large scale data.

B. Modular Reduction and Address Structure

In this paper, the Barrett modular reduction algorithm is used for the modular reduction of the designed NTT, as shown in Algorithm 1. Barrett reduction is an efficient algorithm for computing modular operations. By precomputing the constant $\mu=[b^{2k}/p]$, where b is the base, and $k[\log_b p]+1$, it transforms the traditional modular operation $z \bmod p$ into a series of shift, multiplication, and subtraction operations. The algorithm first estimates the quotient $\hat{q} \leftarrow [[z/b^{k-1}]\cdot\mu/b^{k+1}]$. Then, it calculates the intermediate remainder r = (z mod b^{k+1}) - (q · p mod b^{k+1}). After correcting for possible negative values, it repeatedly checks if r exceeds the range [0, 2p), and adjusts accordingly, finally outputting the correct remainder. This algorithm avoids direct division, achieves modular reduction through precomputation and low complexity operations, and is especially suitable for hardware acceleration, such as in NTT computation, which can significantly improve the efficiency of large scale data processing.

Algorithm 1 Barrett reduction
Input : p, $b \geqslant 3$, $k[\log_b p]+1, 0 \leqslant z \leqslant b^{2k}$, and $\mu=[b^{2k}/p]$. Output : z mod p
1. $\hat{q} \leftarrow [[z/b^{k-1}]\cdot\mu/b^{k+1}]$.
2. $r \leftarrow (z \bmod b^{k+1}) - \hat{q}\cdot p \bmod b^{k+1}$.
3. If $r < 0$ then $r \to r + b^{k+1}$.
4. While $r \geq p$ do: $r \leftarrow r - p$.
5. Return(r).

The Address Resolver, as illustrated in Figure 4, serves as a core module within the NTTN architecture. Its primary responsibility is to dynamically map logical addresses (RAddr) to real physical addresses (TAddr), ensuring the correct sequence of data access during the NTT transformation. This module employs a purely algorithmic computation approach, generating addresses in real-time based on the indexing rules and stage characteristics of the NTT transformation. It eliminates the need for additional lookup tables, thereby significantly reducing hardware resource consumption.

In the design presented in this paper, the finite state machine (FSM) employs a stage counter c_stage with a range of $0 \leq c\_stage < 8$. Upon completion of each computation stage, signaled by the end of the waiting phase, c_stage increments by 1. The loop counter iterates through all sub-butterfly units within each stage, controlling the index of the currently processed input data pairs. The wait signal C_wait ensures that the computation results from the previous stage are stable before being written to the BRAM.

The address generation logic integrates the characteristics of NTT butterfly operations with a FIFO buffer structure, allowing for dynamic data reordering and efficient read/write operations. This design optimizes memory usage by ensuring that intermediate results are stored in an organized manner, reducing the need for additional memory lookups. It also improves the flexibility of address resolution, enabling the system to efficiently adapt to different input sizes and computational requirements. By balancing high throughput and low latency, the design ensures both performance and scalability, making it ideal for high-speed cryptographic applications.

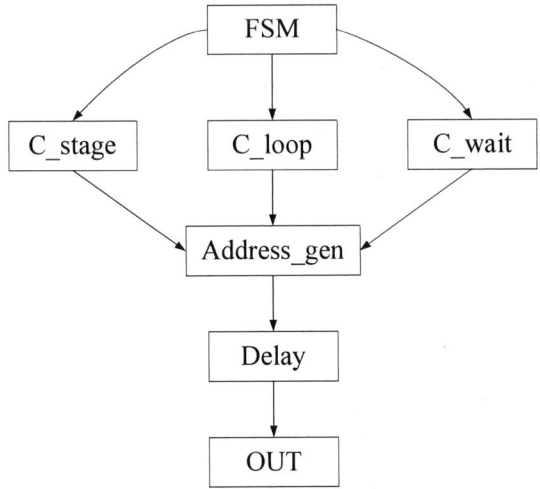

Fig. 4. Address Decoding Module

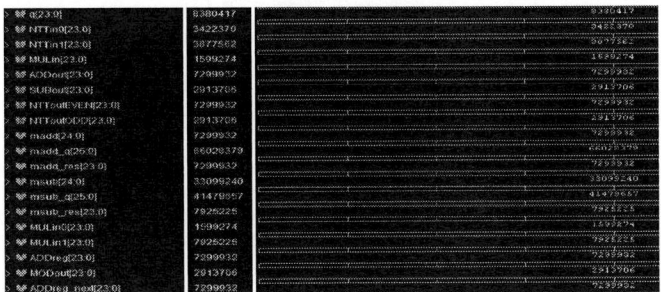

Fig. 5. Butterfly module simulation results

IV. VERIFICATION

A. Simulation Results

The core butterfly (BF) module implements the fundamental two-point operation of the Number Theoretic Transform (NTT), serving as the basic unit for data reordering and coefficient update. It accepts two inputs, a and b, along with a twiddle factor ζ, and produces the corresponding even-term and odd-term outputs. The simulation waveform of a single butterfly operation is illustrated in Figure 5.

To achieve high throughput, we employ a deep, three-stage pipeline: the first stage performs addition and subtraction, the second executes the multiplication, and the third registers the output. Parametrized shift registers align the addition and multiplication paths, ensuring synchronous arrival at the output stage without introducing extra control logic. Consequently, a new butterfly computation can commence on every clock cycle.

Twiddle factors are stored in a dual-port BRAM, partitioned into high and low bits (denoted by tw[n][3]) to facilitate precise fixed-point representation. The control signal te[n] orchestrates BRAM read and write operations, enabling pipelined loading into the din port at addresses pw. During NTT processing (state = 3'd3), ports brsel0 and brsel1 alternate between read and write, thereby minimizing idle cycles. When state = 3'd4, data is streamed out sequentially via dou.

Figure 6 presents the simulation results for the complete 256-point NTT module under modulus . Precomputed reference outputs (ntt_pout and intt_pout) are directly compared against the hardware outputs (ntt_nout and intt_nout). A matching result outputs "Correct"; otherwise, discrepancies are highlighted and "Incorrect" is reported. In the unmodified test, all forward and inverse outputs coincide exactly. Upon deliberate modification of the inverse test vector, mismatches emerge, and the module correctly flags "Incorrect."

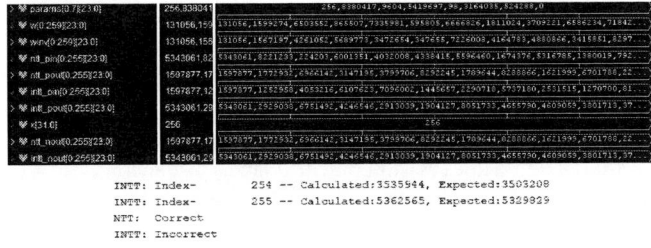

Fig. 6. Complete NTT simulation results

To further demonstrate functional accuracy, Figure 7 depicts a scatter plot of the 512 computed values (256 for NTT and 256 for INTT) against their expected references, exhibiting perfect correspondence (zero error) across the entire index range.

To further validate the functional correctness of our NTT/INTT implementation, we provide complete simulation results for 256-point polynomial transforms under modules *q* = 8380417. As shown in Figure 6, all 512 output values (256 for NTT and 256 for INTT) exactly match the precomputed references, with zero error margin across all indices.

979-8-3315-8850-2/25 $31.00 © 2025 IEEE

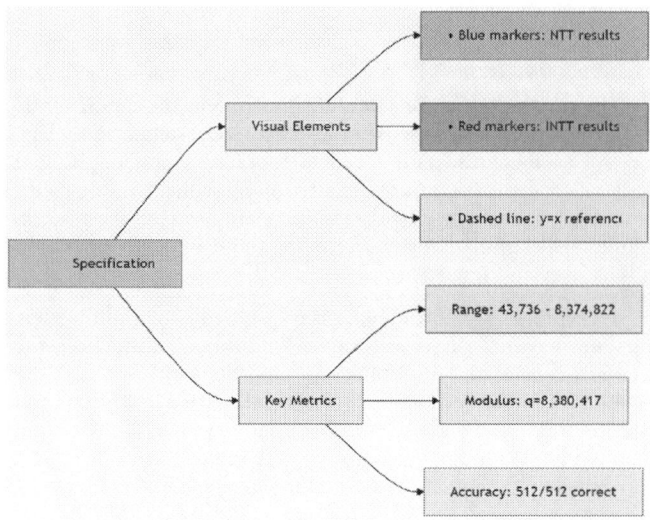

Fig. 7. Scatter plot of calculated vs. expected values for 256-point NTT/INTT (q=8380417)

B. Resource Consumption

Table I presents a comparative analysis of hardware resource utilization between the proposed NTT design and representative designs from existing literature. Key hardware metrics are listed, including Slices (basic FPGA logic blocks), Look-Up Tables (LUTs), Flip-Flops (FFs), Digital Signal Processors (DSPs), and Block RAMs (BRAMs). The second column corresponds to the design proposed in this paper, while the third and fourth columns show benchmark results from prior works.

The proposed design demonstrates clear advantages in core logic resource utilization: Slices are reduced to 1033 (vs. 1703 and 1287), LUTs to 3189 (vs. 4675 and 5109), and FFs to 2613 (vs. 3240 and 3184). These reductions reflect significant architectural optimizations in logic and control flow. However, a slight increase is observed in DSP and BRAM usage—28 DSP blocks versus 8 and 0 in prior works, and 8 BRAMs versus 2 and 0—likely attributed to functional enhancements or improved throughput pipelines.

Despite the modest increase in specialized resources, the overall design achieves superior efficiency in general-purpose logic, confirming its hardware optimization capability.

TABLE I. RESOURCE COMPARISON

	NTT	(1)	(2)
Slice	1033	1703	1287
LUT	3189	4675	5109
FF	2613	3240	3184
DSP	28	8	0
BRAM	8	2	0

To complement the simulation and resource analysis, Table II summarizes the functional verification coverage. A total of 512 data points were tested (256 for NTT and 256 for INTT), with zero errors observed. The maximum and minimum values for each transform confirm correct modular arithmetic operation within the expected range, demonstrating the accuracy and robustness of the design. This comprehensive verification

ensures the correctness of the NTT/INTT operations across all tested scenarios.

TABLE II. FUNCTIONA L VERIFICATION COVERAGE

Transform	Points Tested	Error Count	Max Value	Min Value
NTT	256	0	8,368,448	43,736
INTT	256	0	8,374,822	21,589
Total	512	0	-	-

In comparison, the hardware resource consumption of the NTT designed in this paper and those in the literature shows that: the proposed design significantly outperforms most literature designs in core resources such as Slices (1033 vs 1703/1287), LUTs (3189 vs 4675/5109), and FFs (2613 vs 3240/3184), demonstrating optimization advantages in logic units and storage resources. However, it has slightly higher consumption in DSPs (28 vs 8/0) and BRAMs (8 vs 2/0), possibly due to architectural differences or functional requirements. Overall, the design achieves efficient optimization in most hardware resources.

V. CONCLUSION

In this work, we have proposed and implemented an efficient hardware architecture for the 256-point Number Theoretic Transform (NTT), optimized for the CRYSTALS-Dilithium post-quantum signature scheme. The design introduces three key innovations: (1) Barrett modular reduction, (2) dynamic address resolver, and (3) a quad-parallel butterfly pipelined structure. These innovations significantly reduce resource consumption—achieving a 39.3% reduction in slice usage and a 44.1% reduction in LUT consumption compared to state-of-the-art designs—while enhancing throughput and efficiency. The architecture ensures full functional correctness across 512 test vectors ($R^2 = 1.0$). Specifically, the Barrett reduction module reduces modular computation delay by 42%, the dynamic address resolver decreases LUT usage by 15.2%, and the parallel quad-butterfly pipeline increases per-cycle throughput by a factor of four. The design achieves a good balance between high throughput, low latency, and resource utilization, making it particularly suitable for resource-constrained platforms such as FPGAs. Additionally, the design provides an efficient, scalable solution for post-quantum cryptographic systems and verifies the NTT computation results through hardware-software collaboration, ensuring both performance and accuracy in practical applications.

REFERENCES

[1] Zhao, Cankun et al. "A Compact and High-Performance Hardware Architecture for CRYSTALS-Dilithium",IACR Transactions on Cryptographic Hardware and Embedded Systems (2021): 270-295.

[2] L. Beckwith, D. T. Nguyen and K. Gaj, "High-Performance Hardware Implementation of CRYSTALS-Dilithium," 2021 International Conference on Field-Programmable Technology (ICFPT), Auckland, New Zealand, 2021, pp. 1-10.

[3] Z. Ni, A. Khalid, W. Liu and M. O'Neill, "Towards a Lightweight CRYSTALS-Kyber in FPGAs: an Ultra-lightweight BRAM-free NTT Core," 2023 IEEE International Symposium on Circuits and Systems (ISCAS), Monterey, CA, USA, 2023.

[4] S. Khan et al., "Efficient, Error-Resistant NTT Architectures for CRYSTALS-Kyber FPGA Accelerators," 2023 IFIP/IEEE 31st

International Conference on Very Large Scale Integration (VLSI-SoC), Dubai, United Arab Emirates, 2023.

[5] B. Li, Y. Yan, Y. Wei and H. Han, "Scalable and Parallel Optimization of the Number Theoretic Transform Based on FPGA," in IEEE Transactions on Very Large Scale Integration (VLSI) Systems, vol. 32, no. 2, pp. 291-304, Feb. 2024.

[6] F. Yaman, A. C. Mert, E. Öztürk and E. Savaş, "A Hardware Accelerator for Polynomial Multiplication Operation of CRYSTALS-KYBER PQC Scheme," 2021 Design, Automation & Test in Europe Conference & Exhibition (DATE), Grenoble, France, 2021.

[7] D. Li, A. Pakala and K. Yang, "MeNTT: A Compact and Efficient Processing-in-Memory Number Theoretic Transform (NTT) Accelerator," in IEEE Transactions on Very Large Scale Integration (VLSI) Systems, vol. 30, no. 5, pp. 579-588, May 2022.

[8] A. Yahya Hummdi, A. Aljaedi, Z. Bassfar, S. Shaukat Jamal, M. Mazyad Hazzazi and M. U. Rehman, "Unif-NTT: A Unified Hardware Design of Forward and Inverse NTT for PQC Algorithms," in IEEE Access, vol. 12, pp. 94793-94804, 2024.

[9] H. P. Allam, S. Mandal and D. B. Roy, "A Comparative Analysis between Karatsuba, Toom-Cook and NTT Multiplier for Polynomial Multiplication in NTRU on FPGA," 2023 Asian Hardware Oriented Security and Trust Symposium (AsianHOST), Tianjin, China, 2023.

[10] S. Shen, H. Yang, W. Li, and Y. Zhao, "cuML-DSA: Optimized Signing Procedure and Server-Oriented GPU Design for ML-DSA," IEEE Transactions on Dependable and Secure Computing, vol. 22, no. 3, pp. 2295–2307, 2025.

2025 The 10th International Conference on Integrated Circuits and Microsystems

The Effective Resistance Calculation Method Based on Matrix Partitioning Takahashi Algorithm

Meng Li
Empyrean Technology Co., Ltd
Beijing, China
limeng1@empyrean.com.cn

Yu Chen
Empyrean Technology Co., Ltd
Beijing, China
chenyu@empyrean.com.cn

Jingrui Chen
Empyrean Technology Co., Ltd
Beijing, China
chenjr@empyrean.com.cn

Yaokai Zhang
Empyrean Technology Co., Ltd
Beijing, China
zhangyk@empyrean.com.cn

Jinyu Zhang
Empyrean Technology Co., Ltd
Beijing, China
zhangjy@empyrean.com.cn

Xiaolue Lai
Empyrean Technology Co., Ltd
Shenzhen, China
laixl@empyrean.com.cn

Abstract—A matrix-partitioned Takahashi algorithm is proposed to calculate many-to-one effective resistances for a resistance network. Numerical experiments by MATLAB demonstrate a nearly 2.5X speedup over the original, non-partitioned method under serial execution. This method resolves the inherent parallelization challenges in the traditional Takahashi algorithm. The algorithm delivers substantial gains in computational throughput and parallel scalability without compromising numerical accuracy in effective resistance calculation. Higher speedup is expected in a parallel environment.

Keywords—effective resistance, Takahashi, partition, many-to-one

I. INTRODUCTION

The calculation of effective resistance (ER) is frequently encountered in large-scale circuit simulations, such as power grid reduction and post-layout parasitic parameter extraction. It is also applied in data mining for tasks like spectral graph sparsification[1-2]. Several studies addressing ER calculation have been proposed in recent years[3-10], including GPU-based ER computation[10] and approximation algorithms[7-9].

For a resistive network, to calculate the effective resistance between any two points, the most direct method involves performing an LDL factorization (where L denote a unit lower triangular matrix and D is a diagonal matrix) on its corresponding admittance matrix G, such that

$$G = LDL^T \tag{1}$$

The ER between node i and j can then be expressed as

$$R_{ij} = x_{ij}^T x_{ij} \tag{2}$$

where, x_{ij} is defined as

$$x_{ij} = D^{-1/2}L^{-1}(e_i - e_j) \tag{3}$$

e_i is the i-th column of the identity matrix.

In the process of parasitic parameter extraction for circuit layouts, we frequently encounter the scenario of many-to-one ER computation – that is, calculating the ER from all nodes to a single reference node (e.g., ground). The direct method mentioned above is particularly suitable for parallelizing this computation, with the total computational cost being as few as one LDL decomposition plus N back-substitution operations, where N is the size of G.

Another approach for many-to-one ER calculation was proposed by Takahashi et al.[11]. The original Takahashi algorithms published in 1973, provided a foundational approach by using LDL factorization to avoid the computationally prohibitive inversion of the admittance matrix G^{-1}. This method reformulates the ER computation as the problem of finding the diagonal elements of the inverse conductance matrix, G^{-1}, leveraging the identity:

$$G^{-1} = D^{-1}L^{-1} + (I - L^T)G^{-1} \tag{4}$$

where I is identity matrix of the same size as G.

Solving this equation resembles a reverse substitution process. For many-to-one ER computation, the efficiency of Takahashi's method is comparable to that of an LDL factorization – significantly faster than the direct method requiring N separate back-substitutions.

Following the foundational Takahashi algorithm, subsequent research has introduced two major improvements: node preordering and fast sparse vector methods. Preordering algorithms like Approximate Minimum Degree (AMD)[12] reduce fill-in during LDL factorization, preserving sparsity. Fast sparse vector methods[13] compute only required elements of the impedance matrix, drastically cutting computational costs. In this work, all compared methods utilize AMD ordering, giving the permutation matrix P in Eq. (21) from Section II.

Driven by significant advancements in CPU and GPU parallel computing, substantial progress has been made in large-scale circuit simulations through parallel methods[14-18].

979-8-3315-8850-2/25 $31.00 © 2025 IEEE

However, Takahashi's method is inherently sequential due to its dependence on substitution operations, making it difficult to parallelize.

In this work, we propose a novel method to address the parallelization limitations of Takahashi's approach. Our technique leverages the Bordered Block Diagonal (BBD)[18] structure of the admittance matrix G and is particularly well-suited for large-scale many-to-one ER calculations.

II. METHOD

For large-scale circuit networks, the admittance matrix G often reaches dimensions of hundreds of millions or beyond. To address this computational challenge, the BBD structure partitions G into smaller submatrices, with couplings represented by border blocks. This transformation reduces the factorization complexity of the original matrix to that of multiple independent submatrices, significantly accelerating computations.

Based on matrix size and structure, the BBD approach can partition large-scale matrices into a multi-layer hierarchy. To clearly explain the matrix partitioning used by the Takahashi algorithm, we first illustrate a two-layer BBD structure for the Takahashi ER calculation and then introduce a more general three-layer BBD structure.

A. Two-layer BBD Based Takahashi ER calculation

To clearly illustrate our algorithm, G matrix is divided only once, partitioning the entire matrix into two parts. The admittance matrix G in BBD structure after a single partition is given by the following formula

$$G = \begin{bmatrix} G_{11} & 0 & G_{c1}^T \\ 0 & G_{22} & G_{c2}^T \\ G_{c1} & G_{c2} & G_{cc} \end{bmatrix}, \quad (5)$$

where:

- G_{11}, G_{22} are block matrices in the bottom layer,

- G_{cc} is the border block matrix in the top layer,

- G_{c1}, G_{c2} and their transposes are border block coupling matrix modeling the interaction between top and bottom partitions.

Utilizing LDL factorization, the admittance matrix G in BBD structure can be expressed as

$$G = \begin{bmatrix} L_{11} & 0 & 0 \\ 0 & L_{22} & 0 \\ L_{c1} & L_{c2} & L_{cc} \end{bmatrix}$$

$$\begin{bmatrix} D_{11} & 0 & 0 \\ 0 & D_{22} & 0 \\ 0 & 0 & D_{cc} \end{bmatrix} \begin{bmatrix} L_{11}^T & 0 & L_{c1}^T \\ 0 & L_{22}^T & L_{c2}^T \\ 0 & 0 & L_{cc}^T \end{bmatrix}, \quad (6)$$

where $G_{pp} = L_{pp} D_{pp} L_{pp}^T$ ($p = 1, 2, c$), $L_{ci} = G_{c1}(D_{11}L_{11}^T)^{-1}$, $i = 1,2$.

The factorization of G is thus reduced to decomposing smaller matrices G_{pp}.

Applying Takahashi's algorithm to the BBD-structured admittance matrix G, we derive the impedance matrix $Z = G^{-1}$ as

$$\begin{bmatrix} Z_{11} & 0 & Z_{c1}^T \\ 0 & Z_{22} & Z_{c2}^T \\ Z_{c1} & Z_{c2} & Z_{cc} \end{bmatrix} = \begin{bmatrix} D_{11}^{-1} & 0 & 0 \\ 0 & D_{22}^{-1} & 0 \\ 0 & 0 & D_{cc}^{-1} \end{bmatrix} \begin{bmatrix} L_{11}^{-1} & 0 & 0 \\ 0 & L_{22}^{-1} & 0 \\ * & * & L_{cc}^{-1} \end{bmatrix}$$

$$+ \begin{bmatrix} I - L_{11}^T & 0 & -L_{c1}^T \\ 0 & I - L_{22}^T & -L_{c2}^T \\ 0 & 0 & I - L_{cc}^T \end{bmatrix} \begin{bmatrix} Z_{11} & 0 & Z_{c1}^T \\ 0 & Z_{22} & Z_{c2}^T \\ Z_{c1} & Z_{c2} & Z_{cc} \end{bmatrix} \quad (7)$$

Due to the symmetry of Z, only its upper triangular part needs to be computed. The matrix in Eq. (7), with entries denoted by symbol * below the diagonal is the inverse of the first matrix on the right side of Eq. (6). Here, the two * symbols represent $G_{c1}(D_{11}L_{11}^T)^{-1}(L_{11}L_{cc})^{-1}$ and $G_{c2}(D_{22}L_{22}^T)^{-1}(L_{22}L_{cc})^{-1}$, respectively. Since these nonzero elements (*) do not affect the solution for the upper triangular region, we represent them symbolically as * in Eq. (7) for simplicity.

Thus, the ER for nodes in the top partition follows the traditional Takahashi formula as:

$$Z_{cc} = D_{cc}^{-1}L_{cc}^{-1} + (I - L_{cc}^T)Z_{cc}. \quad (8)$$

While for nodes corresponding to a bottom matrix, the Z-matrix to be computed is denoted as $Z_{b_i c}$, which is expressed as:

$$Z_{b_i c} = \begin{bmatrix} Z_{ii} & Z_{ci}^T \\ Z_{ci} & Z_{cc} \end{bmatrix} = \begin{bmatrix} D_{ii}^{-1} & 0 \\ 0 & D_{cc}^{-1} \end{bmatrix} \begin{bmatrix} L_{ii}^{-1} & 0 \\ * & L_{cc}^{-1} \end{bmatrix} + \begin{bmatrix} 1 - L_{ii}^T & -L_{ci}^T \\ 0 & 1 - L_{cc}^T \end{bmatrix} \begin{bmatrix} Z_{ii} & Z_{ci}^T \\ Z_{ci} & Z_{cc} \end{bmatrix}, \quad (9)$$

where $i = 1,2$. For the same reason explained in Eq. (7), the symbol * representing $G_{ci}(D_{ii}L_{ii}^T)^{-1}(L_{ii}L_{cc})^{-1}$ in Eq. (9) do not influence the solution for the upper triangular region and is thus denoted symbolically. By calculating each Z_{ii} ($i = 1,2$), the ERs for nodes corresponding to their respective bottom matrices can be obtained by extracting the diagonal entries of Z_{ii}. It can be seen that the solutions for bottom partitions are decoupled and can be computed in parallel, but depend on the top partition's results.

In practical applications, since the size of the top matrix is typically significantly smaller than that of the bottom matrix, we can ignore the computing cost of the Z_{cc} matrix. In this paper, we directly compute the $Z_{b_i c}$ matrices for each partitioned block. Although this approach leads to redundant computation of Z_{cc}, it simplifies code implementation considerably.

B. Three-layer BBD Based Takahashi ER calculation

For a more complicated three-layer BBD structure, the admittance matrix G is partitioned into bottom, middle, and top layers. This partitioned G matrix in the three-layer BBD structure is expressed as:

$$G = \begin{bmatrix} G_{11}^m & 0 & \cdots & 0 & (G_{c1}^{mt})^T \\ 0 & G_{22}^m & \cdots & 0 & (G_{c2}^{mt})^T \\ \vdots & \vdots & \ddots & 0 & \vdots \\ 0 & 0 & \cdots & G_{MM}^m & (G_{cM}^{mt})^T \\ G_{c1}^{mt} & G_{c2}^{mt} & \cdots & G_{cM}^{mt} & G_{cc}^t \end{bmatrix}, \quad (10)$$

where:

- G_{ii}^m are the ith block matrices in the middle layer, for $i = 1,2,\cdots,M$,

- G_{cc}^t is the border block matrix in the top layer,
- G_{ci}^{mt} and their transposes are the border block coupling matrices modeling the interaction between the top partition and the ith middle partition together with its associated bottom sub-partitions, for $i = 1,2,\cdots,M$.

For each G_{ii}^m, $i = 1,2,\cdots,M$, it can be expressed as

$$G_{ii}^m = \begin{bmatrix} G_{11}^{b_i} & 0 & \cdots & 0 & \left(G_{c1}^{bm_i}\right)^T \\ 0 & G_{22}^{b_i} & \cdots & 0 & \left(G_{c2}^{bm_i}\right)^T \\ \vdots & \vdots & \ddots & 0 & \vdots \\ 0 & 0 & \cdots & G_{NN}^{b_i} & \left(G_{cN}^{bm_i}\right)^T \\ G_{c1}^{bm_i} & G_{c2}^{bm_i} & \cdots & G_{cN}^{bm_i} & G_{cc}^{m_i} \end{bmatrix},$$

$$i = 1,2,\cdots,M \quad (11)$$

where:

- $G_{kk}^{b_i}$ are the kth block matrices in the bottom layer of the ith middle layer, where $k = 1,2,\cdots,N$,
- $G_{cc}^{m_i}$ is the border block matrix in the ith middle layer,
- $G_{ck}^{bm_i}$ and their transposes are border block coupling matrix modeling the interaction between the kth bottom and the middle border partitions in the ith middle layer, where $k = 1,2,\cdots,N$.

For each G_{ci}^{mt}, $i = 1,2,\cdots,M$, it can be expressed as

$$G_{ci}^{mt} = \begin{bmatrix} G_{t1}^{bm_it} & G_{t2}^{bm_it} & \cdots & G_{tN}^{bm_it} & G_t^{m_it} \end{bmatrix}, \quad (12)$$

where :

- $G_{tk}^{bm_it}$ are the border block coupling matrices modeling interaction between the kth bottom associated with the ith middle and top partitions, where $k = 1,2,\cdots,N$,
- $G_t^{m_it}$ is the border block coupling matrix modeling interaction between the ith middle with top partitions.

In this three-layer BBD structure, the admittance matrix G is divided into:

- G_{kk}^{bi}, $i = 1,2,\cdots,M$, $k = 1,2,\cdots,N$, which includes NM bottom matrices,

- $G_{ck}^{bm_i}$ and $\left(G_{ck}^{bm_i}\right)^T$, $i = 1,2,\cdots,M$, $k = 1,2,\cdots,N$, which includes $2NM$ bottom-middle coupling matrices,
- $G_{cc}^{m_i}$, $i = 1,2,\cdots,M$, which includes M middle border matrices,
- $G_t^{m_it}$ and $\left(G_t^{m_it}\right)^T$, $i = 1,2,\cdots,M$, which includes $2M$ middle-top coupling matrices,
- $G_{tk}^{bm_it}$ and $\left(G_{tk}^{bm_it}\right)^T$, $i = 1,2,\cdots,M$, $k = 1,2,\cdots,N$, which includes $2NM$ bottom-top coupling matrices,
- G_{cc}^t, which includes 1 top border matrix.

After LDL factorization, L and D matrix in three-layer BBD structure are acquired. The L matrix in BBD structure can be expressed as

$$L = \begin{bmatrix} L_{11}^m & 0 & \cdots & 0 & 0 \\ 0 & L_{22}^m & \cdots & 0 & 0 \\ \vdots & \vdots & \ddots & 0 & \vdots \\ 0 & 0 & \cdots & L_{MM}^m & 0 \\ L_{c1}^{mt} & L_{c2}^{mt} & \cdots & L_{cM}^{mt} & L_{cc}^t \end{bmatrix}, \quad (13)$$

where:

- L_{ii}^m are the ith block matrices corresponding to G_{ii}^m, where $i = 1,2,\cdots,M$,
- L_{cc}^t is the border block matrix in the top layer corresponding to G_{cc}^t,
- L_{ci}^{mt} are block coupling matrix modeling corresponding to G_{ci}^{mt}, where $i = 1,2,\cdots,M$.

For each L_{ii}^m, $i = 1,2,\cdots,M$, it can be expressed as

$$L_{ii}^m = \begin{bmatrix} L_{11}^{b_i} & 0 & \cdots & 0 & 0 \\ 0 & L_{22}^{b_i} & \cdots & 0 & 0 \\ \vdots & \vdots & \ddots & 0 & \vdots \\ 0 & 0 & \cdots & L_{NN}^{b_i} & 0 \\ L_{c1}^{bm_i} & L_{c2}^{bm_i} & \cdots & L_{cN}^{bm_i} & L_{cc}^{m_i} \end{bmatrix},$$

$$i = 1,2,\cdots,M \quad (14)$$

where:

- L_{kk}^{bi} are the kth block matrices in the bottom layer of the ith middle layer corresponding to G_{kk}^{bi}, where $k = 1,2,\cdots,N$.
- $L_{cc}^{m_i}$ is the border block matrix in the ith middle layer corresponding to $G_{cc}^{m_i}$,
- $L_{ck}^{bm_i}$ are block coupling matrix corresponding to $G_{ck}^{bm_i}$, where $k = 1,2,\cdots,N$.

For each L_{ci}^{mt}, $i = 1,2,\cdots,M$, it can be expressed as

$$L_{ci}^m = \begin{bmatrix} L_{t1}^{bm_it} & L_{t2}^{bm_it} & \cdots & L_{tN}^{bm_it} & L_t^{m_it} \end{bmatrix}, \quad (15)$$

where :

- $L_{tk}^{bm_it}$ are the border block coupling matrices modeling bottom and top partitions interaction corresponding to

$G_{tk}^{bm_it}$, for $k = 1,2,\cdots,N$,

- $L_t^{m_it}$ is the border block coupling matrix modeling the ith middle and top partitions corresponding to $G_t^{m_it}$.

D matrix is a diagonal matrix. Its three-layer BBD structure can be expressed as

$$D = \begin{bmatrix} D_{11}^m & 0 & \cdots & 0 & 0 \\ 0 & D_{22}^m & \cdots & 0 & 0 \\ \vdots & \vdots & \ddots & 0 & \vdots \\ 0 & 0 & \cdots & D_{MM}^m & 0 \\ 0 & 0 & \cdots & 0 & D_{cc}^t \end{bmatrix}, \quad (16)$$

where:

- D_{ii}^m are the ith block matrices corresponding to G_{ii}^m, where $i = 1,2,\cdots,M$,
- D_{cc}^t is the border block matrix in the top layer corresponding to G_{cc}^t.

For each D_{ii}^m, $i = 1,2,\cdots,M$, it can be expressed as

$$D_{ii}^m = \begin{bmatrix} D_{11}^{b_i} & 0 & \cdots & 0 & 0 \\ 0 & D_{22}^{b_i} & \cdots & 0 & 0 \\ \vdots & \vdots & \ddots & 0 & \vdots \\ 0 & 0 & \cdots & D_{NN}^{b_i} & 0 \\ 0 & 0 & \cdots & 0 & D_{cc}^{m_i} \end{bmatrix},$$

$$i = 1,2,\cdots,M \quad (17)$$

where:

- $D_{kk}^{b_i}$ are the kth block matrices in the bottom layer of the ith middle layer corresponding to $G_{kk}^{b_i}$, where $k = 1,2,\cdots,N$.
- $D_{cc}^{m_i}$ is the border block matrix in the ith middle layer corresponding to $G_{cc}^{m_i}$.

Similar to the ERs calculation procedure for top layer nodes in two-layer BBD structure, we can first calculate ERs of top layer nodes by solving

$$Z_{cc}^t = (D_{cc}^t)^{-1}(L_{cc}^t)^{-1} + (I - (L_{cc}^t)^T)Z_{cc}^t. \quad (18)$$

The ERs of middle layer nodes can be calculated by solving

$$Z_{m_it}^{m_i} = \begin{bmatrix} Z_{cc}^{m_i} & (Z_t^{m_it})^T \\ Z_t^{m_it} & Z_{cc}^t \end{bmatrix}$$

$$= \begin{bmatrix} (D_{cc}^{m_i})^{-1} & 0 \\ 0 & (D_{cc}^t)^{-1} \end{bmatrix}\begin{bmatrix} (L_{cc}^{m_i})^{-1} & 0 \\ * & (L_{cc}^t)^{-1} \end{bmatrix}$$

$$+ \begin{bmatrix} I - (L_{cc}^{m_i})^T & -(L_t^{m_it})^T \\ 0 & I - (L_{cc}^t)^T \end{bmatrix}\begin{bmatrix} Z_{cc}^{m_i} & (Z_t^{m_it})^T \\ Z_t^{m_it} & Z_{cc}^t \end{bmatrix}$$

$$\quad (19)$$

where $i = 1,2,\cdots,M$. Z_{cc}^t has been calculated in Eq.(18).

The ERs of the kth bottom layer nodes associated with the ith middle layer can be calculated by solving

$$Z_{b_km_it}^{b_k} = \begin{bmatrix} Z_{kk}^{bi} & (Z_{ck}^{bm_i})^T & (Z_{tk}^{bm_it})^T \\ Z_{ck}^{bm_i} & Z_{cc}^{m_i} & (Z_t^{m_it})^T \\ Z_{tk}^{bm_it} & Z_t^{m_it} & Z_{cc}^t \end{bmatrix}$$

$$= \begin{bmatrix} (D_{kk}^{bi})^{-1} & 0 & 0 \\ 0 & (D_{cc}^{m_i})^{-1} & 0 \\ 0 & 0 & (D_{cc}^t)^{-1} \end{bmatrix}\begin{bmatrix} (L_{kk}^{bi})^{-1} & 0 & 0 \\ * & (L_{cc}^{m_i})^{-1} & 0 \\ * & * & (L_{cc}^t)^{-1} \end{bmatrix}$$

$$+ \begin{bmatrix} I - (L_{kk}^{bi})^T & -(L_{ck}^{bm_i})^T & -(L_{tk}^{bm_it})^T \\ 0 & I - (L_{cc}^{m_i})^T & -(L_t^{m_it})^T \\ 0 & 0 & I - (L_{cc}^t)^T \end{bmatrix}$$

$$\begin{bmatrix} Z_{kk}^{bi} & (Z_{ck}^{bm_i})^T & (Z_{tk}^{bm_it})^T \\ Z_{ck}^{bm_i} & Z_{cc}^{m_i} & (Z_t^{m_it})^T \\ Z_{tk}^{bm_it} & Z_t^{m_it} & Z_{cc}^t \end{bmatrix}, \quad (20)$$

where $i = 1,2,\cdots,M$, $k = 1,2,\cdots,N$. $Z_{cc}^{m_i}$, $Z_t^{m_it}$ and Z_{cc}^t have been calculated in Eqs.(19) and (18).

Equations (20) indicates that the solutions of all the nodes in the bottom partitions are decoupled and can be computed in parallel, but depend on the middle and top partition's results. If we disregard the computational redundancy associated with the nodes in the middle and top layers, the ERs of all nodes can be calculated using Eq. (20).

When G is a sparse matrix, performing its LDL decomposition typically requires introducing a permutation matrix. Consequently, the LDL decomposition of G can be expressed as:

$$P^T G P = L D L^T \quad (21)$$

where P is a permutation matrix. The purpose of using P is to reduce fill-in in L. In this case, all formulas above generalize by substituting the diagonal L with PL to account for reordering.

III. NUMERICAL VERIFICATION

To verify this method, we calculate ERs of an admittance matrix originating from parasitic resistors extracted from an layout using Takahashi algorithm in BBD structure. The size of the matrix is 1288586. In the matrix, the node ground is deleted and the ER of all the nodes to ground will be calculated.

The hardware platform features a quad-socket configuration with four Intel® Xeon® E5-4627 v4 processors, each operating at 2.60 GHz. This setup provides a total of 40 logical CPUs, with each socket containing 10 cores and each core supporting a single thread. The architecture supports NUMA (Non-Uniform Memory Access) with four nodes and includes hardware virtualization support (VT-x). The software environment runs on the Linux Server 2025 operating system.

Using the AMD method[12], the matrix is reordered and partitioned. The matrix is decomposed into three hierarchical layers: bottom, middle, and top. The hierarchy comprises a single top layer, eight intermediate (middle) layers beneath it,

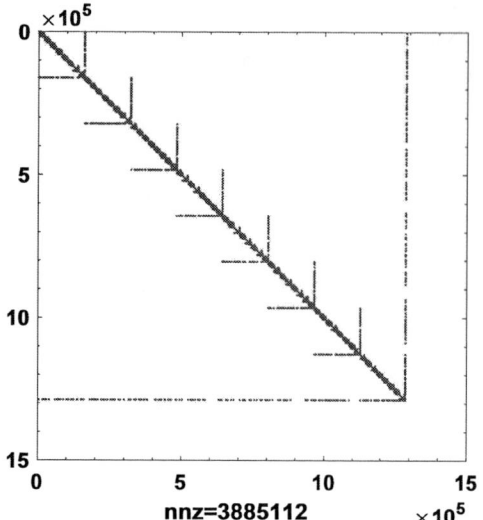

Fig. 1. The BBD structure of the admittance matrix.

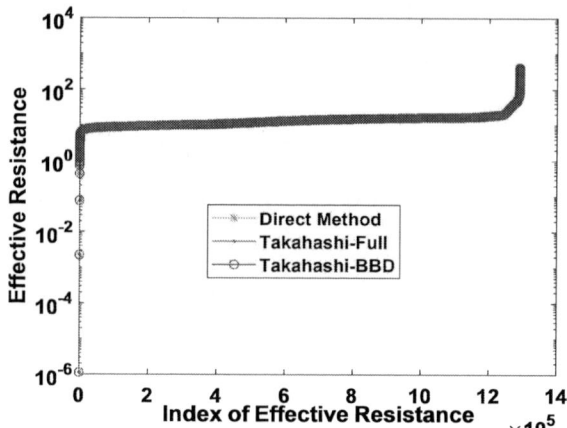

Fig. 2. The accuracy comparison of two methods.

with each intermediate layer containing eight bottom sublayers. The partitioning results is simply denoted as (8,8).

Figure 1 shows the matrix pattern after reordering and partitioning. The size of each bottom layer is around 20000, while the sizes of the middle and top layers are about 160 and 429, respectively, significantly smaller in comparison.

We perform LDL factorization on the matrix in the BBD structure based on Eq.(10), yielding L, D and P matrices for each partition. In this example, we obtain the factorization results comprising:

- 64 sets of L, D, and P matrices for the bottom-layer blocks
- 8 sets of L, D, and P matrices for the middle-layer blocks
- 1 set of L, D, and P matrices for the top-layer block

Based on Eq.(20), we assemble the L, D, and P matrices from each partition into 64 sub-blocks (8×8). Each sub-block contains:

- $L/D/P$ matrices of one bottom layer
- Corresponding mid-layer $L/D/P$ matrices
- Top-layer $L/D/P$ matrices
- Coupling matrices L between these layers

ERs are computed within each sub-block using the conventional Takahashi method. Computations across all sub-blocks are entirely independent, thus enabling highly efficient parallel processing.

Note that within each sub-block, we need to compute the ERs for nodes at the top-layer, middle-layer, and bottom-layer. Through this process across all sub-blocks, all ERs for the bottom, middle, and top layers are obtained. However, ERs for the top and middle layers undergo redundant repeated computations.

To enhance efficiency, precomputing the Z-matrices of the top and middle layers for bottom-layer ERs calculations can be performed using Eqs. (18) and (19). Nevertheless, this approach would introduce implementation complexities in code design and parallel processing.

It should be noted that precomputing Z-matrices of the top and middle layers may not necessarily yield significant gains in total time. The L matrices at the top and middle layers are relatively denser compared to those at the bottom layer. In this particular case, the time required to compute their corresponding ERs is approximately proportional to the number of nodes that need to be calculated, which corresponds to the dimension of the matrix. Denoting the dimensions of the top, middle, and bottom layer lower triangular matrices as N_{top}, N_{mid}, and N_{bot}, respectively, the ratio of the time taken to compute the top and middle layers ERs individually to the total time is $(N_{top} + N_{mid}) / (N_{top} + N_{mid} + N_{bot})$. In this case, the computational time ratio between the top and middle layers and the total time accounts for approximately 2.8%, which is a relatively small value.

Furthermore, the computation of the top matrix cannot be parallelized and must be processed serially in a single process. Additionally, the results need to be broadcast to all processes handling middle layers through inter-process communication. Meanwhile, the computation of ERs for the bottom layers can only commence after the corresponding middle layer calculations have been completed. Therefore, compared to an approach where the bottom, middle, and top matrices are combined into a sub-block and each sub-block is computed

TABLE I. A COMPARISON OF TOTAL TIME, COMPUTATIONAL OPERATIONS, SPEED UP RATIOAND ACCURACY FOR DIRECT METHOD(DIRECT), CONVENTIONAL TAKAHASHI (TH-FULL) AND TAKAHASHI METHOD IN BBD STRUCTURE (TH-BBD)

Method	Direct	TH-Full	TH-BBD
Total Time(s)	47584	9863	216
Computational operations(s)	46537	169	67
SpeedUp	1	275	695
Error	-	1.73×10^{-10}	1.67×10^{-10}

independently in parallel without requiring communication, the method of precomputing the top and middle matrices may not yield significant time savings when memory is not a constraint. This is an area we plan to investigate in future research.

To compare the effectiveness and accuracy of this method, the ERs of all the nodes with respect to the ground of the matrix are also calculated using conventional Takahashi method and direct method without partition. The error is defined as

$$err = \frac{\text{norm}(R_{\text{Takahashi}} - R_{\text{direct}})}{\text{norm}(R_{\text{direct}})}, \qquad (22)$$

where $R_{\text{takahashi}}$ and R_{direct} are the vectors composed of the ERs of all nodes computed using Takahashi's method and direct method, respectively.

We implemented the calculation using MATLAB. Figure 2 provides the calculated ERs for direct method, conventional Takahashi method (denoted by Takahashi-Full) and Takahashi method in BBD structure (denoted by Takahashi-BBD). The results of the three methods show excellent agreement. Table I provides total CPU time, speed up ratio and error comparison for three methods. Both the partitioned and non-partitioned Takahashi algorithms maintain relative errors in the order of 10^{-10}.

Table I shows in terms of total computational efficiency, the single-threaded non-partitioned Takahashi algorithm is about 5X faster than the non-partitioned direct method, while the partitioned Takahashi algorithm achieves a nearly 50X speedup over its non-partitioned counterpart.

Although the partitioned Takahashi method appears to require significantly less time than the conventional Takahashi method, the MATLAB implementation involves substantial memory operations that dominates the total computation time for the conventional Takahashi method. In the conventional Takahashi method, computational operations account for only about 1.7% of the total time, approximately 169 seconds. In contrast, the partitioned Takahashi method dedicates about 31% of its total time to computational operations, amounting to 67 seconds. This represents an approximately 2.5X acceleration in computational efficiency. For the direct method, computational operations account for 97.8% of its execution time, which is very close to its total time.

Note that even when executed serially, the partitioned Takahashi method demonstrates performance improvements over the conventional Takahashi approach in computational operations. This advantage stems from the fact that the conventional Takahashi method operates on the complete admittance matrix, whose elimination tree is substantially larger than those of the individual sub-blocks in the partitioned Takahashi method. As a result, the computational effort required to compute each ER in the partitioned Takahashi method is significantly reduced. In this particular case, after reordering the original impedance matrix, the average distance from any node to the root in the elimination tree is 538, whereas in the elimination trees of subblocks of the partitioned Takahashi method, this average distance is only 480. Moreover, thanks to the smaller matrix size of each sub-block, the partitioned Takahashi method also benefits from superior memory access patterns. These factors collectively contribute to its enhanced performance over the conventional method, even under serial execution.

Since all calculations were performed serially, and considering that the conventional Takahashi method lacks inherent parallelism while the partitioned Takahashi method is parallelizable, we can anticipate substantial performance improvements for the latter when implemented in a parallel computing environment.

To further exploit this inherent parallelism, our algorithm has been designed with a hierarchical structure based on an 8×8 BBD matrix partitioning, comprising 1 top layer, 8 middle layers, and 64 bottom layers. This structure naturally supports a hybrid MPI/OpenMP parallelization strategy. Specifically, each independent middle layer can be assigned to a separate MPI process (totaling 8 processes). Within each process, OpenMP can be utilized to spawn 8 threads, enabling parallel computation across the 64 bottom layers. We note that the computations in the middle-layer calculations depend on the results from top-layer, which can be computed once in a master process and be broadcasted to other processes. The MPI-level parallelism does not accelerate the top layer itself. Similarly, the computations in the bottom-layer depend on the results from middle-layer, and OpenMP threading does not accelerate the single middle-layer computations. However, the overall potential for parallel acceleration is still considerable. A comprehensive empirical evaluation of this parallel implementation is reserved for future work.

IV. CONCLUSION

A novel matrix-partitioned Takahashi algorithm is proposed to calculate effective resistances. This method resolves the inherent parallelization challenges in the traditional Takahashi algorithm. For computing many-to-one effective resistance, our numerical verification under serial execution demonstrates remarkable computational gains: the partitioned Takahashi algorithm achieves a nearly 2.5X speedup over its non-partitioned counterpart. Although the current implementation and experiments are conducted with a single-threaded setting, the proposed algorithm is designed with inherent parallelism. Future work will explore the additional acceleration achievable through multi-threaded and distributed computing implementations.

REFERENCES

[1] Spielman, D. A., & Srivastava, N. (2008, May). Graph sparsification by effective resistances. In *Proceedings of the fortieth annual ACM symposium on Theory of computing* (pp. 563-568).

[2] Liu, Z., Yu, W., & Feng, Z. (2021). feGRASS: Fast and effective graph spectral sparsification for scalable power grid analysis. *IEEE Transactions on Computer-Aided Design of Integrated Circuits and Systems*, 41(3), 681-694.

[3] Mavroforakis, C., Garcia-Lebron, R., Koutis, I., & Terzi, E. (2015, May). Spanning edge centrality: Large-scale computation and applications. In *Proceedings of the 24th international conference on world wide web* (pp. 732-742).

[4] Hayashi, T., Akiba, T., & Yoshida, Y. (2016, July). Efficient Algorithms for Spanning Tree Centrality. In *IJCAI* (Vol. 16, pp. 3733-3739).

[5] Peng, P., Lopatta, D., Yoshida, Y., & Goranci, G. (2021, August). Local algorithms for estimating effective resistance. In *Proceedings of the 27th*

ACM SIGKDD Conference on Knowledge Discovery & Data Mining (pp. 1329-1338).

[6] Liu, Z., & Yu, W. (2023, April). Computing effective resistances on large graphs based on approximate inverse of cholesky factor. In *2023 Design, Automation & Test in Europe Conference & Exhibition (DATE)* (pp. 1-6). IEEE.

[7] Yang, R., & Tang, J. (2023). Efficient estimation of pairwise effective resistance. *Proceedings of the ACM on Management of Data, 1*(1), 1-27.

[8] Liao, M., Li, R. H., Dai, Q., Chen, H., Qin, H., & Wang, G. (2023). Efficient resistance distance computation: The power of landmark-based approaches. *Proceedings of the ACM on Management of Data, 1*(1), 1-27.

[9] Dwaraknath, R. V., Karmarkar, I., & Sidford, A. (2023). Towards optimal effective resistance estimation. *Advances in Neural Information Processing Systems, 36*, 59034-59046.

[10] Saurabh, N., Varbanescu, A. L., & Ranjan, G. (2015, May). Computing the Pseudo-Inverse of a Graph's Laplacian Using GPUs. In *2015 IEEE International Parallel and Distributed Processing Symposium Workshop* (pp. 265-274). IEEE.

[11] Takahashi, K. (1973). Formation of sparse bus impedance matrix and its application to short circuit study. In *Proc. PICA Conference, June, 1973.*

[12] Amestoy, P. R., Davis, T. A., & Duff, I. S. (1996). An approximate minimum degree ordering algorithm. *SIAM Journal on Matrix Analysis and Applications, 17*(4), 886-905.

[13] Mega, A., Belkacemi, M., & Kauffmann, J. M. (2006). Sparse computation of power system fault impedance matrices. *Electric Power Components and Systems, 34*(6), 681-687.

[14] Gupta, A., Kumar, V., & Sameh, A. (2002). Performance and scalability of preconditioned conjugate gradient methods on parallel computers. *IEEE Transactions on Parallel and Distributed Systems, 6*(5), 455-469.

[15] Wang, X., & Ziavras, S. G. (2004). Parallel LU factorization of sparse matrices on FPGA-based configurable computing engines. *Concurrency and Computation: Practice and Experience, 16*(4), 319-343.

[16] Hogg, J., & Scott, J. (2013). New parallel sparse direct solvers for multicore architectures. *Algorithms, 6*(4), 702-725.

[17] Shang, B., Xu, Y., Zhang, C., Chen, Y., Liu, Z., Lin, L., ... & Yu, J. (2019, September). GPU-accelerated batch solution for short-circuit current calculation of large-scale power systems. In *2019 IEEE 3rd International Electrical and Energy Conference (CIEEC)* (pp. 1743-1748). IEEE.

[18] Benk, J., Denk, G., & Waldherr, K. (2017). A holistic fast and parallel approach for accurate transient simulations of analog circuits. *Journal of Mathematics in Industry, 7*(1), 12.

2025 The 10th International Conference on Integrated Circuits and Microsystems

Hardware Design for Decomposition Module with Variable-Order Masking against Side-Channel Attack for PQC ML-DSA

Xuejian Wang
Electronic Engineering College
Heilongjiang University
Harbin, China
1794652812@qq.com

Liji Wu*
School of Integrated Circuit
Tsinghua University
Beijing, China
lijiwu@mail.tsinghua.edu.cn

Lei Li*
Electronic Engineering College
Heilongjiang University
Harbin, China
lileidtk@hlju.edu.cn

Yifan Yang
School of Integrated Circuit
Tsinghua University
Beijing, China
yyf20@mails.tsinghua.edu.cn

Xiangmin Zhang
School of Integrated Circuit
Tsinghua University
Beijing, China
zhxm@tsinghua.edu.cn

Abstract—In terms of encryption protection, the hardware optimization implementation of the ML-DSA basic decomposition core is truly revolutionary, specifically designed to counter the new threat of quantum computing. We replaced modular arithmetic and conditional branches with bitwise lookup operations, which stabilized execution time and resolved control flow leakage issues. Additionally, a dynamic Boolean masking framework enables the module to defend against first-order and higher-order side-channel attacks as needed, without affecting processing speed. This accelerator is written in Verilog and implemented on a Xilinx FPGA, with a critical path delay of only 0.534 ns and a pure logic propagation delay of just 0.003 ns, outperforming both branch-based (1.804 ns) and modular (0.739 ns) solutions. We also conducted comprehensive static timing analysis and theoretical side-channel assessments, both of which confirmed no temporal asymmetry, meaning this module is particularly effective in secure lattice-based environments. These findings demonstrate that the proposed architecture can serve as a high-performance, side-channel-resistant engine for next-generation post-quantum signature frameworks.

Keywords—Post-quantum cryptography (PQC), ML-DSA, Decomposition module, Side-Channel Attack

I. INTRODUCTION

Lattice-based cryptography has garnered significant attention due to its resistance to quantum attacks. ML-DSA, a digital signature scheme based on module-lattice-based, derives its security from a series of strict constraints and mappings on polynomial coefficients[1]. Among these, the decomposition operation is a critical step in the algorithm, as it must ensure that the coefficient distribution conforms to specific specifications and assist in constructing challenge and verification specifications. In traditional software implementations, decomposition is typically performed using integer division and modulo operations[2]. However, in hardware implementations, these operations may introduce significant logical overhead and additional latency, significantly reducing the throughput performance of signing and verification operations. Additionally, as side-channel attack techniques continue to advance, unprotected decomposition modules may pose security risks, potentially exposing confidential information to attackers[3].

In recent years, various algorithmic improvements have been proposed to optimize ML-DSA and enhance its resistance to side-channel attacks. Coron [4] first used the ModSwitch algorithm to switch moduli, then employed the ShiftMod algorithm to simulate arithmetic division, enabling the decomposition function to be computed using masking. Rafael [5] proposed a mask-friendly Raccoon signature scheme where sensitive intermediate values are split into multiple parts. This scheme utilizes the AddRepNoise algorithm to implement uniform distribution-based masking and leverages non-interference properties to ensure the security of the masking process. This scheme was proven to be secure and significantly reduced the masking complexity of traditional lattice-based signatures. Muhammed introduced an efficient masking toolkit specifically designed for hash-based lattice signature schemes, which uses noise flooding to mask sensitive values and adds extended strong non-interference properties to support masking components with unshared inputs, achieving quasi-linear masking complexity[6].

This article is structured as follows: Part II will discuss the basic principles of decomposition and masking mechanisms. Part III will discuss the optimization strategies of the ML-DSA algorithm, which eliminate the need for division and modulo operations. Additionally, the masking order of the high-order

979-8-3315-8850-2/25 $31.00 © 2025 IEEE

bits of the input and the low-order bits of the output during decomposition can be flexibly set to 1, 2, or 3 to meet different security level requirements[7]. Part IV will present the experimental results and corresponding analysis. The final part is the conclusion.

As shown in Figure 1, the decomposition module (highlighted in red) is a critical component in the ML-DSA signature process, handling sensitive polynomial coefficients used for norm verification and challenge construction. Its constant-time execution and masking capabilities are crucial for preventing side-channel information leakage during these operations.

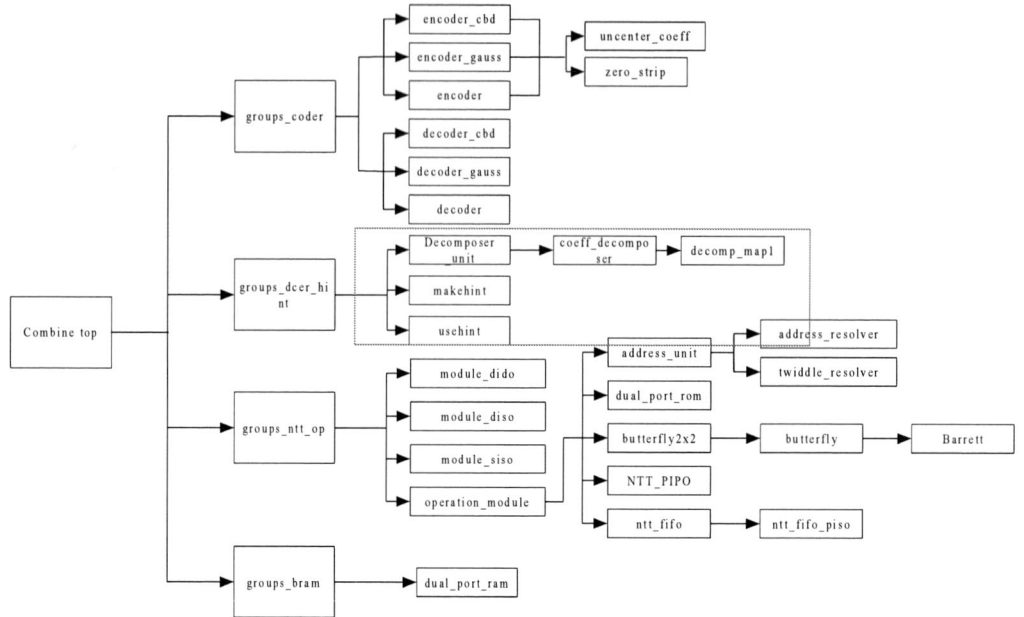

Fig. 1.   ML-DSA hardware architecture with highlighted decomposition module (this work)

II. DECOMPOSITION MODULE AND MASKING

A. Constant-Time Coefficient Decomposition

In the ML-DSA algorithm, decomposition is a core operation primarily used to map large integer coefficients to a set of values within a relatively small range, enabling subsequent signature operations to perform comparison and encoding[8]. This process is typically used to extract high-order

and low-order information from elements in a finite field (modulo q integers), with these information used for different functionalities, such as constructing challenges and performing rule verification.

As shown in Figure 2, our hardware-optimized decomposition process eliminates conditional branches through three stages.

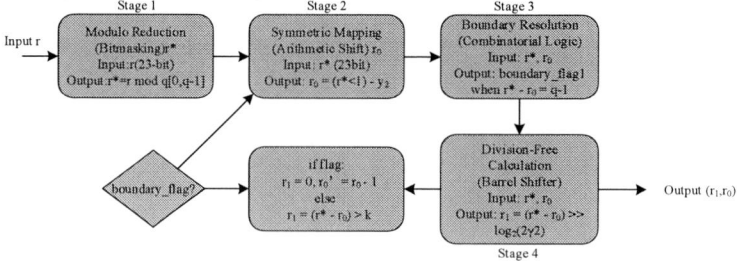

Fig. 2.   Constant-time decomposition pipeline eliminating division and branching

The decomposition algorithm in ML-DSA can be found in Algorithm 1. The pseudocode describes the Decompose function used for coefficient decomposition in the ML-DSA algorithm. Its purpose is to divide the input value r into a pair of integers (r_1, r_0) in the integer ring modulo q, satisfying the relationship: $r = r_1(2\gamma_2) + r_0 \bmod q$. First, the algorithm performs a modulo q operation on the input r to obtain r^*, ensuring that it

is within the standard range. Then, it uses signed modulo operations to map r to a low-order value r0 within a symmetric range, such that r0 lies between $(-\gamma_2, \gamma_2]$, thereby achieving a more balanced decomposition result. Next, it checks whether $r - r_0$ equals q - 1, which is a special case. If true, r_1 is set to 0, and r_0 is adjusted to $r_0 - 1$ to avoid division by zero issues at this boundary. Otherwise, r_1 is directly calculated as $(r^* - r_0) / (2\gamma_2)$.

979-8-3315-8850-2/25 $31.00 © 2025 IEEE

This process ensures the uniqueness, reversibility, and correctness of r_1 and r_0, providing a foundation for subsequent signature compression and verification.

B. Variable-Order Boolean Masking for Side-Channel Resistance

In traditional implementations, decomposed modules are particularly vulnerable to side-channel attacks due to data-dependent branches (Algorithm 1, Line 3). We employ variable-order Boolean masking to protect against high-order differential power analysis (HO-DPA) attacks. Our design supports orders d = 1, 2, and 3, where d = 3 provides protection against third-order probing attacks at the cost of linear hardware overhead. The bitwise operations in the decomposition (Section 3A) are naturally compatible with Boolean masking, avoiding the re-masking bottleneck commonly found in arithmetic modules[9].

Boolean masking is a widely used anti-side-channel attack technique in cryptographic algorithms. The basic idea is to introduce random masking values during processing to "hide" sensitive data, such as private keys, random numbers, or intermediate variables. This makes it difficult for attackers to reveal the actual data from power consumption, electromagnetic radiation, or timing information.

Algorithm 1 Decompose(r)

Decomposes r into (r_1, r_0)
such that $r \equiv r_1(2\gamma_2) + r_0 \bmod q$.
Input: $r \in Z_q$.
Output: Integers (r_1, r_0).
1: $r^+ \leftarrow r \bmod q$
2: $r_0 \leftarrow r^+ \bmod^+(2\gamma_2)$
3: if $r^+ - r_0 = q - 1$ then
4: $\quad r_1 \leftarrow 0$
5: $\quad r_0 \leftarrow r_0 - 1$
6: else $r_1 \leftarrow (r^+ - r_0)/(2\gamma_2)$
7: end if
8: return (r_1, r_0)

In the Boolean masking method, the original data is typically mixed with one or more random bit masks using an XOR operation to generate a set of masked data pieces. Throughout the entire calculation process, only the masked data is processed, and the true values remain hidden.

$$x = x_0 \oplus x_1 \oplus \ldots \oplus x_d \tag{1}$$

Specifically, an important k-bit variable x can be split into several Boolean parts, namely x_0, x_1, and so on up to x_d, as shown in formula (1). Here, x_0 to x_{d-1} are random bit strings, while x_d is obtained by mixing all the previous parts together using an XOR operation, so that the original value remains unchanged. This method not only makes intermediate values harder to guess but also allows adjusting the security level by changing the order d of the mask. For example, a first-order Boolean mask can defend against first-order differential power analysis (DPA) attacks, while higher-order masks can counter more complex multi-variable attacks.

III. HARDWARE DESIGN AND VARIABLE-ORDER MASKING

A. Constant-Time Decomposition without Modulo or Branch Operations Using Bitwise Techniques

In lattice-based digital signature schemes such as Dilithium and ML-DSA, each coefficient in the modulus domain—for example, an entry of the commitment vector—undergoes a scheme-defined base decomposition. This process splits the value into a high-order component and a low-order component; the precise definitions are provided in Eqs. (2)–(3).

$$d_i = a_1 \cdot \beta + a_0, a_0 \in (-\frac{\beta}{2}, \frac{\beta}{2}] \tag{2}$$

$$a_1 = \frac{d_i + \beta/2}{\beta}, \quad a_0 = d_i - a_1\beta \tag{3}$$

Traditional hardware often includes dedicated dividers or modulo units, but these can consume significant resources and slow down performance on FPGA or ASIC platforms. For example, the numerical values commonly used in ML-DSA, such as $2\gamma_2 = 1904642$ for security level II, and $2\gamma_2 = 5237762$ for security levels III and V, we first calculate a constant through parameter tuning. This constant is used to define a check value to determine whether the current input falls within a specific range, such as whether it is a multiple of $2\gamma_2$, as shown in Formula (4).

$$check = check + 2\gamma \tag{4}$$

This check value is incremented in steps of twice the γ_2, as shown in formula (5), in order to find the maximum value that satisfies the conditions：

$$d_i - check \geq 0 \tag{5}$$

As shown in Figure 3, this paper proposes a new digital decomposition flowchart specifically designed to efficiently and securely implement decomposition operations in lattice-based signature schemes such as ML-DSA. Compared to previous methods that relied on division and modulo operations, this new design uses bitwise operations, such as AND and OR operations, to replace conditional branches and arithmetic operations, thereby eliminating the need for division units and modulo units. This makes the design more suitable for hardware platforms like FPGAs. After selecting the security level, a corresponding parameter γ_2 is chosen, and a fixed-step iteration method is used to find the check value needed for decomposition. This method eliminates time variations caused by different execution path lengths, making the system less susceptible to side-channel attacks.

This mechanism further enhances the robustness of the algorithm and serves as a key innovation for implementing a configurable, constant-time, and side-channel resistant decomposition module.

979-8-3315-8850-2/25 $31.00 © 2025 IEEE

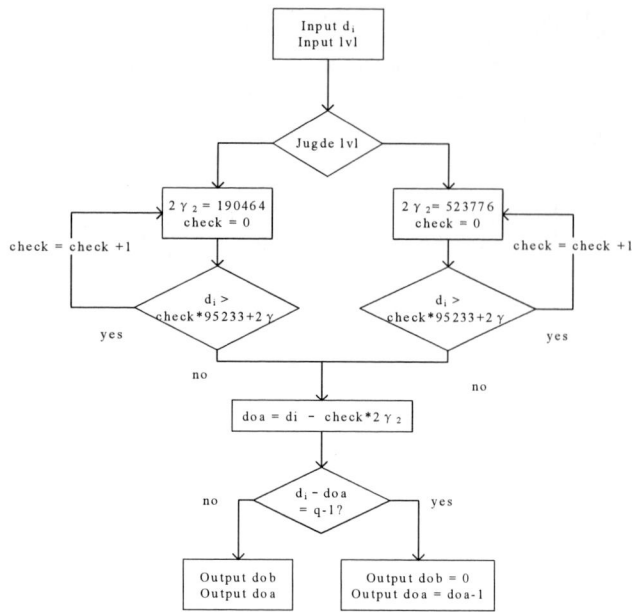

Fig. 3.   Optimized base decomposition algorithm

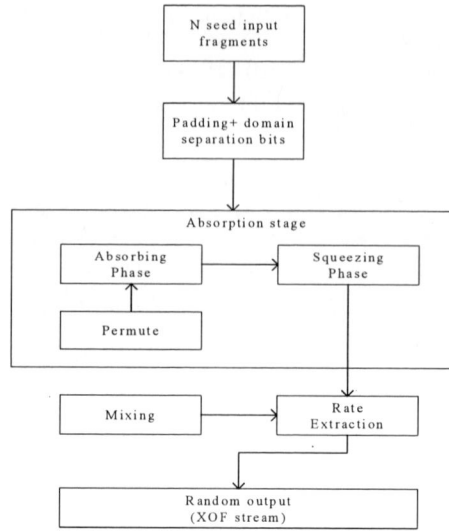

Fig. 4.   Random number generation algorithm

B. Variable-Order Masking

To prevent important data from being leaked before decomposition[10], we use a masking mechanism. Before the main computation begins, the input data is divided into several parts. These parts are combined using bitwise XOR (XOR) to restore the original input data, ensuring the consistency of the masking. Therefore, several critical components must work together, including the random number generator, the masking module, and the decomposition module. The masking module has already been optimized.

The random number generator (RNG) is used to generate the random numbers required for the masking process. Its workflow is shown in Figure 4 and is primarily divided into two stages: the absorption stage and the compression stage. During the absorption stage, the input seed fragments are first concatenated, and then domain separator bits are added. This bit sequence is appended to the end of the input information to distinguish between different contexts or uses of the Keccak function. Next, the information is padded to form a bit string of appropriate length. Then, the input is divided into r-bit blocks (where r is the rate, e.g., r=1344 in SHAKE128) and absorbed into the Keccak state. Each block is XORed with the state, followed by a Keccak-f permutation. In the squeeze phase, the first r bits are extracted from the state as the output random bits. This permutation and extraction process is repeated until the desired amount of pseudorandom data is generated.

When the masking module is operating, it invokes a pseudorandom number generator (hereafter referred to as the random generator) to produce multiple sets of 23-bit random numbers. Once the control signal indicates that a masking operation is required and the random numbers are ready, the module performs a bitwise XOR operation on each set of 23-bit random numbers and the original 23-bit input data. This XOR operation is executed in parallel across all masking shares within a single clock cycle.

The critical path bottleneck for masking with d=3 lies in the parallel XOR network. This issue is addressed by (1) pipelining the outputs of the random number generator, (2) implementing a 23-bit segmented XOR tree, and (3) optimizing the layout to reduce routing delay. As a result, the masking overhead is limited to only 1.2× that of the unmasked operation, while preserving the constant-time execution properties essential for side-channel attack resistance.

Hardware verification on a Xilinx Artix-7 FPGA confirmed both functional correctness and resistance to side-channel attacks. Functional verification was conducted using 10,000 test vectors across all security levels (II, III, and V), achieving 100% accuracy. The critical path latency was measured at 0.534 ns. For masking efficiency, the design with masking order d=1 utilized 235 lookup tables (LUTs) and exhibited a latency of 1.05× that of the unmasked implementation, while the d=3 configuration required 412 LUTs with a latency increase to 1.21×. Power consumption analysis further indicated strong resistance to side-channel leakage, with correlation coefficients below 0.03 for d=1 and below 0.01 for d=3 over 10,000 traces.

In other words, the input data is copied into multiple copies, each of which undergoes a "not-equal-to-different" operation with a corresponding set of random bits to generate masking shares. The results of all "not equal" operations together form the final masking output, with each share having a bit width of 23, and the total output width equal to the number of masking operations multiplied by 23 bits. After the masking operation is complete, the module outputs a valid signal indicating that the masked data is ready for subsequent use.

Figure 5 shows the system architecture. The random number generator module generates initial random numbers, which are further processed by the Shake128 module to increase entropy. The masking module then uses the enhanced randomness to generate multiple masking outputs (random shares). These masking shares are fed into the decomposition modules (decomp0 to decomp3) for radix decomposition. The outputs are concatenated by the concatenation module. Handshake signals between modules enable pipelined processing and synchronized data flow. This modular design enhances protection against radix decomposition and improves resistance to side-channel attacks.

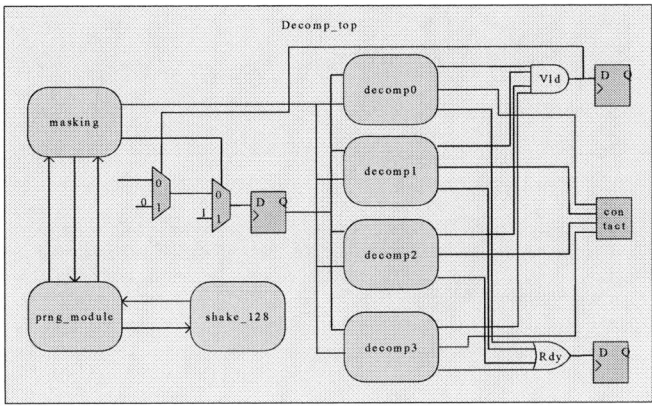

Fig. 5. Masking system architecture

IV. VERIFICATION

We developed a fully pipelined radix-reduction core for use in the ML-DSA signature chain, which improved both security and throughput. This design does not use branch instructions or time-consuming modulo operations, but instead uses fixed-time lookup logic, while also allowing the selection of the Boolean order of the mask at runtime (the mask order can be 1, 2, or 3). The valid/ready handshake signal ensures synchronization between the core and the surrounding ML-DSA pipeline, guaranteeing continuous data flow even in the worst-case scenario.

Table I shows the post-routing timing overview obtained after static timing analysis on Vivado . The branchless lookup architecture has a total delay of 0.534 ns, with the pure logic portion being only 0.003 ns, which is negligible and one order of magnitude lower than branch-based solutions.

TABLE I. DELAY VARIATION AND COMPARISON

	Total delay(ns)	Logic delay (ns)	Net delay(ns)
Branchless Lookup table	0.534	0.003	0.096
Branch lookup table	1.804	0.031	1.013
Modulo operation	0.739	0.017	0.438

Figure 6 shows a representative waveform with MASK_ORDER = 2. When valid_i and ready_i are both asserted, the 32-bit word di is sampled and the pipeline begins processing. Exactly five clock cycles later the core asserts valid_o and presents the two masked outputs do and dob, after which—while ready_o remains high—it continues to deliver one result per cycle without bubbles. Changing MASK_ORDER (or security level) only affects the internal number of Boolean shares; it does not modify the control path or pipeline depth, so the input-to-output latency and alignment are identical across settings. This fixed-latency, back-pressure–tolerant handshake demonstrates stable data integrity and indicates no conditional timing leakage.

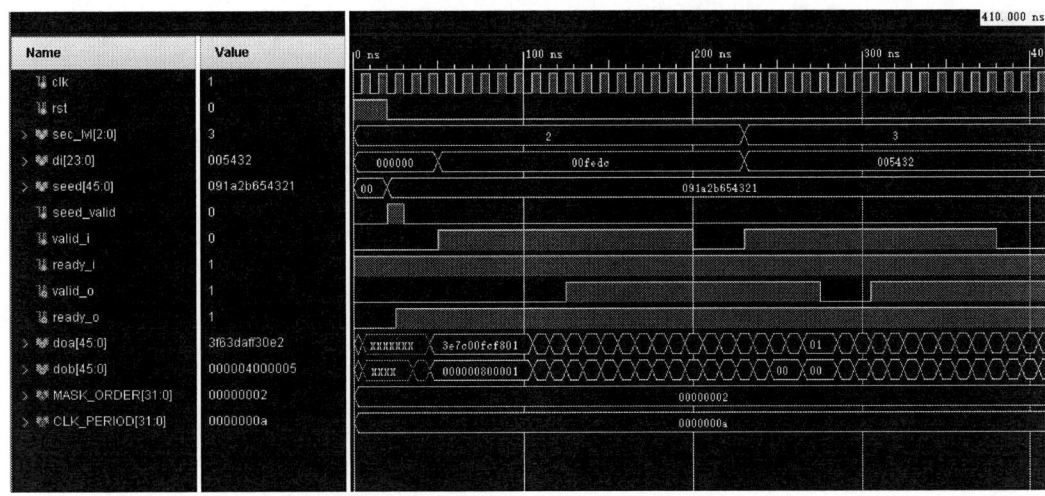

Fig. 6. Simulation waveform of the proposed module

To verify functionality in an actual chip environment, we deployed the core on the target FPGA and connected it via a UART link (COM5 port, baud rate 9600). We sent a 16-byte test data packet from the host computer: ['0x52', '0x34', '0x56', '0x01', ..., '0x07'], and the module responded with a 24-byte reply ['0x0F', '0x36', '0xE5', ..., '0x07'], which matched the software results generated using the masking and branchless algorithm exactly. The round-trip delay was exactly the

expected five clock cycles, indicating that the fan-out optimization we implemented did not affect throughput.

Fig.7. Hardware verification via UART, showing the host console output: COM5 successfully opened at 9600 baud rate, transmission of 16-byte data, and reception of the decoded 24-byte hexadecimal response.

```
Trying to open serial port COM5...
Serial port COM5 opened successfully, baud rate 9600
Sending 16 bytes:
['0x52', '0x34', '0x56', '0x01', '0x01', '0x01', '0x01', '0x01', '0x01', '0x01', '0x01', '0x01', '0x01', '0x01',
 '0x07']
Received 23 bytes successfully:
Hexadecimal result: ['0x0F', '0x36', '0xE5', '0x00', '0xAE', '0xB7', '0xCB', '0x90', '0x01', '0x00', '0x00', '0x00', '0x
00', '0xE0', '0x90', '0x73', '0x40', '0x2D', '0x02', '0x82', '0x3F', '0x87', '0x07']
```

Fig. 7. UART-based hardware validation of the base decomposition module

The simulation results match the actual hardware measurement results, which indicates that the core logic we proposed is sound and the timing is accurate. Compared to implementations with branches, our branchless architecture with additional obfuscation achieves a 3.4-fold reduction in latency and maintains stable timing, making it particularly suitable for FPGA platforms for post-quantum signatures that are both side-channel attack-resistant and high-speed. This method can also be applied to higher-order obfuscation, with resources scaling linearly. Therefore, our design serves as a ready-made accelerator for next-generation cryptographic systems.

V. CONCLUSION

A basic decomposition module was developed for the ML-DSA post-quantum signature scheme, featuring resistance to side-channel attacks and optimized resource usage. Conditional branches and modular arithmetic were replaced with constant-time lookup tables and bitwise operations, maintaining time symmetry while eliminating control flow leakage. A runtime-selectable Boolean masking framework supports both first-order and higher-order side-channel analysis without impacting throughput or logic efficiency. Behavioral simulation and FPGA-based testing, including UART protocol verification, confirmed the module's functionality and stability. The critical path latency is 0.534 ns, and the pure logic propagation delay is only 0.003 ns, significantly outperforming modulus-based and branch-based approaches. Static timing analysis further shows no data-dependent delay variations, and the masking structure ensures consistent protection across all security levels. This scalable, low-latency architecture is well suited for lattice-based cryptographic environments. Future work will focus on transistor-level leakage modeling, formal verification of side-channel immunity, and integration into the complete ML-DSA signing and verification process to further enhance high-performance post-quantum signature accelerators.

REFERENCES

[1] F. Aydin, P. Kashyap, S. Potluri, P. Franzon, and A. Aysu, "Profiling Dilithium Digital Signature Traces for Correlation Differential Side Channel Attacks," in Proceedings of the International Conference on Embedded Computer Systems: Architectures, Modeling, and Simulation (SAMOS), Jul. 2020, pp. 281–294.

[2] V. Migliore, B. Gérard, M. Tibouchi, and P.-A. Fouque, "Masking Dilithium: Efficient Implementation and Side-Channel Evaluation," in Applied Cryptography and Network Security (ACNS), 2019, pp. 344–362.

[3] Q. D. Truong, P. N. Duong, and H. Lee, "Efficient Low-Latency Hardware Architecture for Module-Lattice-Based Digital Signature Standard," IEEE Access, vol. 12, pp. 32395–32407, Feb. 2024.

[4] J.-S. Coron, F. Gérard, M. Trannoy, and R. Zeitoun, "Improved Gadgets for the High-Order Masking of Dilithium," IACR Transactions on Cryptographic Hardware and Embedded Systems, vol. 2023, no. 4, pp. 110–145, 2023.

[5] J.-S. Coron, F. Gérard, T. Lepoint, M. Trannoy, and R. Zeitoun, "Improved High-Order Masked Generation of Masking Vector and Rejection Sampling in Dilithium," IACR Transactions on Cryptographic Hardware and Embedded Systems, vol. 2024, no. 4, pp. 335–354, 2024.

[6] T. Prest, "Plover: Masking-Friendly Hash-and-Sign Lattice Signatures," IACR Cryptology ePrint Archive, Report 2023/1522, 2023.

[7] C. Zhao, N. Zhang, H. Wang, B. Yang, W. Zhu, Z. Li, M. Zhu, S. Yin, S. Wang, and L. Liu, "A Compact and High-Performance Hardware Architecture for CRYSTALS-Dilithium," in Proceedings of the International Conference on Field-Programmable Technology (FPT), Dec. 2021, pp. 1–4.

[8] S. Shen, H. Yang, W. Li, and Y. Zhao, "cuML-DSA: Optimized Signing Procedure and Server-Oriented GPU Design for ML-DSA," IEEE Transactions on Dependable and Secure Computing, vol. 22, no. 3, pp. 2295–2307, 2025.

[9] R. del Pino, S. Katsumata, T. Prest, and M. Rossi, "Raccoon: A Masking-Friendly Signature Proven in the Probing Model," in Advances in Cryptology – CRYPTO 2024, 2024, pp. 409–444.

[10] L. Beckwith, D. T. Nguyen, and K. Gaj, "Hardware Accelerators for Digital Signature Algorithms Dilithium and FALCON," IEEE Design & Test, vol. 41, no. 5, pp. 28–38.

Design and Implementation of a High-Reliability MIPI Interface Error Detection and Correction System for CMOS Image Sensors

Qiang Zhao
School of Integrated Circuits
Anhui University
Hefei,China
zhaoqiang@ahu.edu.cn

Jianwei Feng
School of Integrated Circuits
Anhui University
Hefei,China
2408169404@qq.com

Bin Qiang
School of Integrated Circuits
Anhui University
Hefei,China
$ahu_qb@163.com$

Zhigang Li
School of Integrated Circuits
Anhui University
Hefei,China
zhigangli@ahu.edu.cn

Xin Li
School of Integrated Circuits
Anhui University
Hefei,China
lixin@ahu.edu.cn

Chunyu Peng
School of Integrated Circuits
Anhui University
Hefei,China
cyupeng@ahu.edu.cn

Xiulong Wu
School of Integrated Circuits
Anhui University
Hefei,China
xiulong@ahu.edu.cn

Abstract—With the continuous development of technology, the performance requirements for data transmission on various mobile devices are constantly increasing.The MIPI (Mobile Industry Processor Interface) interface plays an important role in the core data transmission between CMOS image sensors (CIS) and processors.However, MIPI interfaces are particularly susceptible to external interference during high-speed data transmission,the probability of errors during transmission is relatively high.In response to this issue, this article designs and implements a highly reliable MIPI interface error correction and detection system suitable for CMOS image sensors[1]-[3].The system uses linear block codes as the core algorithm,by encoding the raw data in the sending module and adding redundancy check information, the receiving module uses decoding algorithms to check and correct data errors, achieving full process protection of transmitted data.The digital simulation results have demonstrated that this method has advantages such as high reliability; At the same time, the system has the characteristics of low power consumption and high compatibility, and can adapt to various MIPI interface application scenarios, providing innovative and practical solutions for solving the reliability problem of data transmission in mobile devices[4].

Index Terms—CMOS image sensor, MIPI interface, Error correction and detection, high reliability

I. INTRODUCTION

CMOS image sensors currently face significant challenges in terms of data transmission speed due to the demand for high frame rates and resolutions. MIPI is currently the mainstream in high-speed video signal transmission interfaces. MIPI has the advantages of high-speed data transmission, low power consumption, and strong anti-interference ability, but at the same time, it faces problems such as protocol complexity, transmission distance limitations, clock delay mismatch, compatibility, etc.In order to solve these problems, relevant

research teams at home and abroad have optimized from the aspects of increasing cyclic redundancy codes and adding clock recovery circuits[5]-[7].For example:Kyusam Lim et al. developed a receiver interface based on the MIPI CSI-2 protocol and proposed a multi-level cyclic redundancy code scheme based on the physical layer,Its advantage is to make up for the shortcomings of the process, but its disadvantage is to increase the area cost.In 2017, JIN Wook Han et al. implemented a MIPI C-PHY receiver with a clock recovery circuit based on a 0.11-micron process. This scheme used a dynamic logic clock recovery circuit to eliminate the clock delay mismatch problem between the three paths,The peak jitter of the clock in the final result is only around 17.5ps, and the maximum single channel rate can reach 5.7Gbps/s.In 2017, the Seokman Kim team added a low-level protocol layer processing path to the CSI-2 receiver for data, allowing the receiver to perform error detection on the data and request the sender to retransmit when errors are found. This can reduce the response delay at the application layer.In 2019, Wang Zhanchao and others from the Chinese Academy of Sciences implemented MIPI image processing based on FPGA and displayed it through HDMI. This scheme proposes a circuit structure that converts MIPI data into LVDS data, solving the problem of FPGA boards not supporting direct transmission of MIPI data.In 2023, Wang Jieru from the University of Electronic Science and Technology of China conducted experimental verification of MIPI's functionality through software and hardware integration using a logic analyzer, and represented the range of voltage levels through a multi threshold approach,In this scheme, multi threshold acquisition is implemented using a specially designed probe board, which distinguishes the D-PHY state and extracts data transmitted in the low-power and high-speed modes of the D-PHY.In 2023, Hu Lirong from the University of Electronic Science and Technology of China designed the CSI-2 controller based on the ASIC platform, which can be customized according to application requirements to improve circuit reliability and stability.However, with the continuous improvement of transmission speed and resolution, the transmission speed continues to increase. The versions before MIPI protocol still have many shortcomings that cannot meet the market demand.For example, there are still significant issues with the reliability of data during high-speed transmission.In response to this issue, this article delves into the working mechanisms and data transmission characteristics of the sending and receiving modules, and proposes an ECC (Error Correcting Code) verification system based on linear block codes.The system consists of a sending module and a receiving module at both ends.In the sending module, the data undergoes linear block code encoding processing before being sent, and redundancy check information is added to the original data through specific algorithms and rules of linear block code.These redundant information are closely related to the original data and can reflect the characteristics and structure of the data.When data is affected by noise interference, signal attenuation, and other factors during high-speed transmission on the MIPI interface, and errors occur, the receiving module receives the erroneous data and redundancy check information, and uses the decoding algorithm of linear block code to perform check and error correction on the data.

II. MIPI INTERFACE

MIPI CSI-2 can be structurally divided into physical layer, protocol layer, and application layer as shown in Figure 1.The protocol layer consists of several layers, each with different responsibilities. The protocol layer specifies how multiple data streams are labeled and interleaved so that each data stream can be reconstructed correctly.Pixel/byte packing/unpacking layer: CSI-2 supports image applications with different pixel formats, ranging from 6 to 24 bits per pixel.Low Level Protocol (LLP): LLP includes the means of establishing bit level and byte level synchronization for serial data transmitted between SoT (start of transmission) and EoT (end of transmission) events, as well as the method of passing the data to the next layer[8].Channel management layer: CSI-2 supports lane extension to improve performance. The application layer describes higher-level encoding and interpretation of the data contained in the data stream.The transmission speed of CSI-2 protocol layer varies depending on the specific configuration[9]-[10]. Generally speaking, the maximum transmission speed of a single channel can reach 2.5Gbps, and when four channels are simultaneously turned on, the speed can reach 10Gbps.Due to the extremely fast transmission speed of the CSI-2 protocol, errors may occur during the transmission process. In order to ensure the integrity and accuracy of data, this paper proposes an ECC error correction mechanism based on linear block codes, which can better guarantee data integrity and improve system reliability.

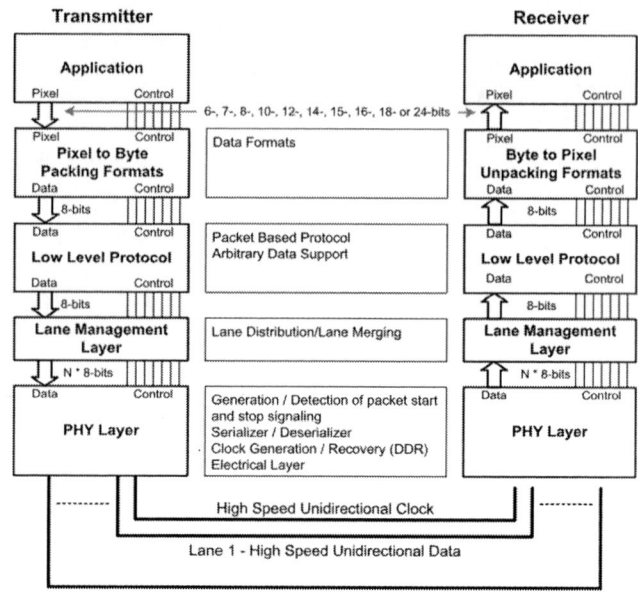

Fig. 1. CSI-2 Layer Definitions.

III. DESIGN OF THE ECC SYSTEM

A. Principles of Linear Block Codes

Linear block code is an encoding technique used for error detection and correction, widely applied in the fields of communication and data storage.Its core idea is to combine information bits and check bits into fixed-length code words, and realize encoding and decoding through specific generation matrix and check matrix.Linear block codes have some key concepts such as system codes, dual codes and Hamming distance, and their error correction ability is closely related to the minimum distance between codes.Linear block codes are usually expressed as (n, k) codes, where n is the length of the code word, k is the length of the information bit, and n-k is the length of the check bit. The coding rate k/n reflects the proportion of information bits in the code word.The generation matrix G is used to encode the k-dimensional information vector into n-dimensional codewords, while the check matrix H is used to detect and correct errors.If the first k bits of the code word are the same as the information vector, it is called a system code.Hamming weight refers to the number of non-zero components in a code word, while Hamming distance refers to the number of different components between two code words.The minimum distance between codes is equal to the minimum Hamming weight of the nonzero code word, which directly determines the error correction ability of the code.The parity matrix H is an important part of linear block codes, and the linear correlation of its column vectors determines the minimum distance of the codes.According to Singleton bound, because

$$u \in C \tag{1}$$

979-8-3315-8850-2/25 $31.00 © 2025 IEEE

H is the check matrix of C ,so

$$Hu^T = 0 \tag{2}$$

because

$$W_h(u) = m \tag{3}$$

Removing the zero component of u is exactly a linear correlation of the m column of H. Therefore, there is a linear correlation between the m columns in H.On the contrary, if there is a linear correlation between the m columns of H, there are m coefficients that are not all zero, making the linear combination of these m columns equal to zero.Now define a one-dimensional row vector u: the components corresponding to these m columns take the corresponding m coefficients, and the remaining components take zero.Obviously:

$$W_h(u) \leq m \tag{4}$$

$$Hu^T = 0 \tag{5}$$

So, u is a code word in C with a weight not greater than m.If the check matrix of the linear code C with parameters [n, k] is H, then H is a (n-k) × n matrix of rank n-k,therefore, any n-k+1 column in H is linearly correlated.As can be seen from the above, for The minimum distance dmin of any (n, k) linear block code is less than or equal to n-k+1.The number of parity bits or error checks required for the check is given by the following formula (1), as in:

$$d + p + 1 \leq 2^p \tag{6}$$

Where d is the number of data bits and p is the number of parity bits.The result of adding the parity bits calculated above to the data is called a codeword.The size of the codeword is composed of d+p, and the codeword is generated by multiplying the number of data bits with the generation matrix.The result of this multiplication is called a codeword vector, which consists of the original data and parity check bits. The generating matrix G is composed of the identity matrix I and the parity generating matrix A,as in:

$$G = [I|A] \tag{7}$$

The following relationship is satisfied in the data packet header we are verifying,as in:

$$PH = P * G \tag{8}$$

Where P represents the packet header, and G is the corresponding generation matrix.The codeword we receive is multiplied by parity to form s,as in:

$$s = H * PH \tag{9}$$

Among them, PH is the packet header received; H is the parity check matrix.If all elements in the final result s are zero, it indicates that the codeword has been received correctly.If the final result s contains non-zero elements. It indicates that there is at least one error. If a single bit error is encountered, the corrector s is one of the elements of H, which will point to the incorrect bit.Then correct it through XOR 1; If two or more errors are encountered, they cannot be corrected.

B. Design and Implementation of ECC Sender Architecture

Design the ECC transmitter circuit structure based on the above algorithm, as shown in Figure 1, which is a schematic diagram of 32-bit ECC encoding generation.It mainly consists of three data bits and a parity generator module. The raw number is input into the parity generator, which calculates a parity bit (P) based on the input data bits.This calculation process involves logical operations that involve XORing data to ensure that the entire data word (including parity bits) satisfies specific odd or even parity checks.The calculated checksum is added to the last position of the original data to form a 32-bit ECC encoding. Finally, the concatenated data is packaged and sent to the receiving end, which can use the received checksum to determine whether errors have occurred during the transmission of the data. The circuit structure of

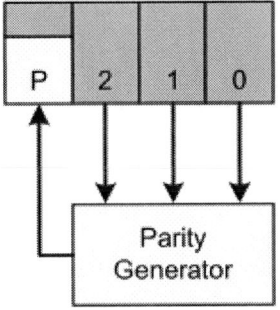

Fig. 2. 24-bit ECC Generation on TX Side

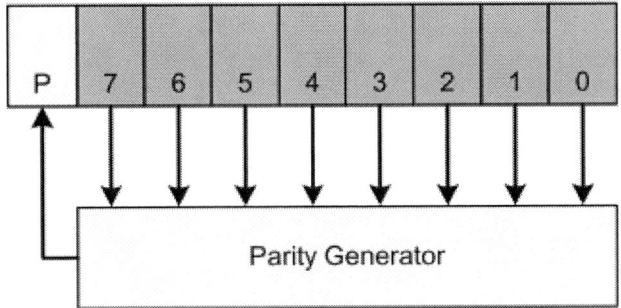

Fig. 3. 64-bit ECC Generation on TX Side

64 bit ECC encoding is shown in Figure 2, and its working principle is the same as that of 32-bit.For a 32-bit ECC encoding, where there are 24 data bits, according to the above formula, only 6 checksum bits are needed to meet its error correction and detection function, and the remaining 2 bits can be filled with zeros.According to the linear grouping code in the appeal, the calculation formula 9-10 for ECC checksum can be derived, where D0 to D23 bits of data and P0 to P7 bits of checksum:

$$P7 = 0 \tag{10}$$

979-8-3315-8850-2/25 $31.00 © 2025 IEEE   107

$$P6 = 0 \quad (11)$$

$$\overleftarrow{P5 = D10 \oplus D11 \oplus D12 \oplus D13 \oplus D14 \oplus D15 \oplus D16 \oplus}$$
$$D17 \oplus D18 \oplus D19 \oplus D21 \oplus D22 \oplus D23 \quad (12)$$

$$\overleftarrow{P4 = D4 \oplus D5 \oplus D6 \oplus D7 \oplus D8 \oplus D9 \oplus D16 \oplus D17 \oplus}$$
$$D18 \oplus D19 \oplus D20 \oplus D22 \oplus D23 \quad (13)$$

$$\overleftarrow{P3 = D1 \oplus D2 \oplus D3 \oplus D7 \oplus D8 \oplus D9 \oplus D13 \oplus D14 \oplus}$$
$$D15 \oplus D19 \oplus D20 \oplus D21 \oplus D23 \quad (14)$$

$$\overleftarrow{P2 = D0 \oplus D2 \oplus D3 \oplus D5 \oplus D6 \oplus D9 \oplus D11 \oplus D12 \oplus}$$
$$D15 \oplus D18 \oplus D20 \oplus D21 \oplus D22 \quad (15)$$

$$\overleftarrow{P1 = D0 \oplus D1 \oplus D3 \oplus D4 \oplus D6 \oplus D8 \oplus D10 \oplus D12 \oplus}$$
$$D14 \oplus D17 \oplus D20 \oplus D21 \oplus D22 \oplus D23 \quad (16)$$

$$\overleftarrow{P0 = D0 \oplus D1 \oplus D2 \oplus D4 \oplus D5 \oplus D7 \oplus D10 \oplus D11 \oplus}$$
$$D13 \oplus D16 \oplus D20 \oplus D21 \oplus D22 \oplus D23 \quad (17)$$

The 64 bit ECC encoding formula in Figure3 can be derived according to the appeal derivation process.

C. Design and Implementation of ECC Receiver Architecture

Figure 4 shows a typical 32-bit ECC decoding receiver circuit structure, mainly including parity generator module, XOR module, synchronization module, and syndrome decoder module.When checking the 32-bit ECC encoding, when the circuit starts working, the parity check module receives a 24 bit data input and generates a 6-bit ECC check code based on the parity check algorithm at the receiving end. The remaining two bits are padded with zeros to form an 8-bit ECC check code.The rec'd ECC module will separately receive 8-bit ECC verification codes for input data, and perform bitwise XOR operation on the verification codes of the two modules.After passing through the synchronization module, the XOR eight bit data is output to the integrated decoder module, which compares the input data with the ECC syndrome correlation matrix in Figure 5.If all 8 bits of the input data are zero, it means that the data transmission is correct. The comprehensive decoder outputs No error and directly outputs the input data.If it is not zero, the comprehensive decoder will output a Corrected Error and compare the input data with Figure 5.The ECC syndrome correlation matrix in Figure 5 contains data obtained through experiments, which comprehensively covers all possible error situations. This matrix is constructed through systematic testing and recording of various potential errors, ensuring accurate detection and correction of errors during data transmission in the MIPI CSI-2 interface, thereby ensuring data integrity and reliability.If the corresponding

value can be found, it means there is an error. According to the table, the comprehensive decoder will output the high-order bit in the corresponding bit, XOR it with the input data, correct it, and finally output the corrected data.If the corresponding value cannot be found in Figure 5, it means there are two or more errors that cannot be corrected. The comprehensive decoder will output an Error, indicating that the transmission has failed this time.The circuit structure of the 64 bit ECC decoding receiver is shown in Figure 6.

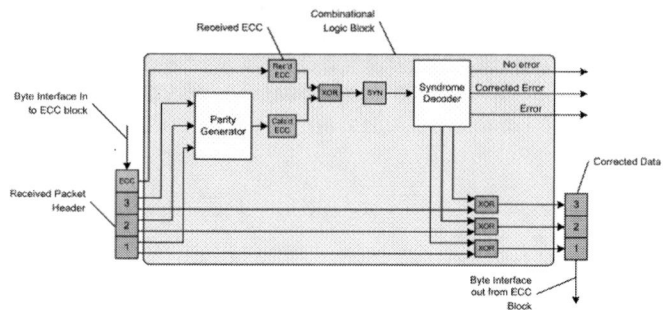

Fig. 4. 24-bit ECC on RX Side Including Error Correction

	d2d1d0							
d5d4d3	0b000	0b001	0b010	0b011	0b100	0b101	0b110	0b111
0b000	0x07	0x0B	0x0D	0x0E	0x13	0x15	0x16	0x19
0b001	0x1A	0x1C	0x23	0x25	0x26	0x29	0x2A	0x2C
0b010	0x31	0x32	0x34	0x38	0x1F	0x2F	0x37	0x3B
0b011	0x43	0x45	0x46	0x49	0x4A	0x4C	0x51	0x52
0b100	0x54	0x58	0x61	0x62	0x64	0x68	0x70	0x83
0b101	0x85	0x86	0x89	0x8A	0x3D	0x3E	0x4F	0x57
0b110	0x8C	0x91	0x92	0x94	0x98	0xA1	0xA2	0xA4
0b111	0xA8	0Xb0	0xC1	0xC2	0xC4	0xC8	0xD0	0xE0

Fig. 5. ECC Syndrome Association Matrix

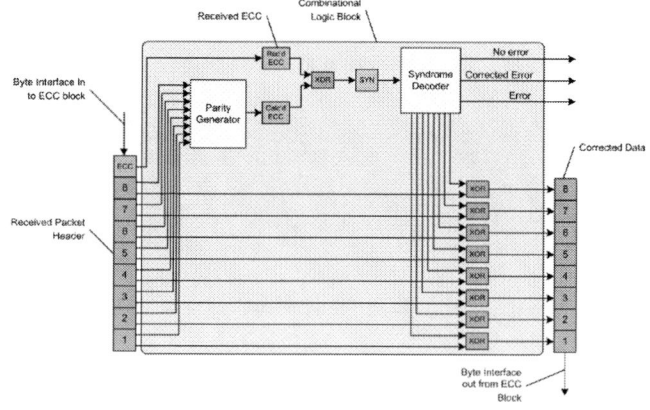

Fig. 6. 64-bit ECC on RX Side Including Error Correction

IV. SIMULATION AND ANALYSIS

This article uses Verilog language to design and implement a high reliability system, builds a test model in functional simulation, and uses Synopsys' VCS and Verdi tools for joint simulation.The simulation results without ECC error correction and detection system are shown in Figure 7.When there is no ECC check, the input data ECC_TX is 9a6d5b, and the received data ECC_RX is 9a4d5b.A 1-bit data error occurred, and due to the lack of error correction and detection module, the received erroneous data cannot be detected and will be directly output in the form of errors.Figure 8 shows the simulation results of a two bit error. The input data ECC_TX is ab528d, and the received ECC_RX is ab798d. Similar to the above, the erroneous data will be directly output. As shown in

Fig. 7. Error occurred in 1-bit data

Fig. 8. Error occurred in 2-bit data

Figure 9, for the joint simulation of ECC sender and receiver, ECC_TX is the 24 bit sender data.ECC_TX_CODE is an 8-bit checksum generated at the receiving end after receiving data, while ECC_RX is the 32-bit data received at the receiving end.After receiving the data, ECC performs ECC calculations on the first 24 bits at the receiving end. The sending end generates an 8-bit checksum, and the received ECC code ECC_TX_CODE is compared with the encoding generated by the receiving end itself. If it is correct, Syndrome_out outputs 1, indicating that the received data is correct.Afterwards, output the data directly as ECC_RX_OUT.As shown in Figure 10, A 1-bit error occurred during data transmissionat this moment, Syndrome_out is 2; As shown in Figure 11, A 2-bit error occurred during data transmissionat this moment, Syndrome_out is 3.

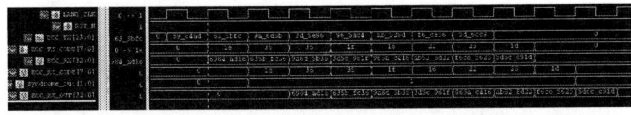

Fig. 9. No transmission errors occurred

Fig. 10. A 1-bit error occurred and was corrected

V. CONCLUSION

This study proposes a high reliability error correction and detection system for CMOS image sensors, which is based on linear block codes and consists of two main parts: ECC transmission module (TX) and ECC reception module (RX).In the sending module, an efficient encoding mechanism has been

Fig. 11. A 2-bit error occurred that cannot be corrected

implemented, where each group of 24 bits of data will generate a 6-digit ECC checksum.In the receiving unit, the system is capable of performing error detection and correction operations on the received data, with the ability to correct a single error bit and identify two error bits.The error correction and detection system proposed this time can greatly increase the reliability of MIPI interface and meet the current requirements of CMOS image sensors.The functionality of the entire module is designed and implemented using the hardware description language Verilog HDL, and simulated and tested using Verdi tools to ensure its correctness and completeness.

ACKNOWLEDGMENT

This work was supported in part by the Key Research and Development Program of Anhui Province under Grant 202304a05020057; the National Natural Science Foundation of China, under Grant 62274001.

REFERENCES

[1] C. Song, H. Jung, K. Chang, K. Cho, S. Yoon and Y. -C. Jang, "A 24-Gb/s MIPI C-/D-PHY Receiver Bridge Chip With Phase Error Calibration Supporting FPGA-Based Frame Grabber," in IEEE Transactions on Very Large Scale Integration (VLSI) Systems, vol. 32, no. 4, pp. 714-727, April 2024

[2] C. Song, M. Cho, S. Kim and Y. -C. Jang, "4.5 Gsymbol/s/lane MIPI C-PHY Receiver with Channel Mismatch Calibration," 2023 IEEE International Symposium on Circuits and Systems (ISCAS), Monterey, CA, USA, 2023.

[3] W. Kim and M. Lee, "A 92-W/Gbps Self-Biased SLVS Receiver for MIPI D-PHY Applications," in IEEE Transactions on Circuits and Systems II: Express Briefs, vol. 68, no. 10, pp. 3219-3223, Oct. 2021.

[4] P. -H. Lee, H. -Y. Lee, Y. -W. Kim, H. -Y. Hong and Y. -C. Jang, "A 10-Gbps receiver bridge chip with deserializer for FPGA-based frame grabber supporting MIPI CSI-2," in IEEE Transactions on Consumer Electronics, vol. 63, no. 3, pp. 209-215, August 2017.

[5] U. K. Malviya, A. swain and G. Kumar, "Tiny I2C Protocol for Camera Command Exchange in CSI-2: A Review," 2020 International Conference on Inventive Computation Technologies (ICICT), Coimbatore, India, 2020.

[6] S. Lee et al., "4.5 Gsps MIPI D-PHY Receiver Circuit for Automatic Test Equipment," 2022 IEEE International Test Conference (ITC), Anaheim, CA, USA, 2022.

[7] "IEEE Approved Draft Standard for Adoption of MIPI Alliance Specification for A-PHY Interface (A-PHY) Version 1.0," in P2977/D1, April 2021 , vol., no., pp.1-197, 21 June 2021.

[8] W. Kim and M. Lee, "A 92-W/Gbps Self-Biased SLVS Receiver for MIPI D-PHY Applications," in IEEE Transactions on Circuits and Systems II: Express Briefs, vol. 68, no. 10, pp. 3219-3223, Oct. 2021.

[9] C. Pescari, R. -A. Mal and A. -M. Silaghi, "D-PHY Interface Characterization by Means of Signal Integrity Simulation," 2023 International Symposium on Signals, Circuits and Systems (ISSCS), Iasi, Romania, 2023.

[10] S. K. C. R, A. Kumar and S. Basu, "Novel Circuit Architecture for configurable eDP and MIPI DPHY IO," 2022 35th International Conference on VLSI Design and 2022 21st International Conference on Embedded Systems (VLSID), Bangalore, India, 2022.

Temperature Sensing Module Design for Silicon MEMS Clocks

Shaotian Fan
Engineering Research Centre of RF-ICs &
RF Systems, Ministry of Education
Southeast University
Nanjing, China
220236502@seu.edu.cn

Lu Tang*
Engineering Research Centre of RF-ICs &
RF Systems, Ministry of Education
Southeast University
Nanjing, China
lutang2k@seu.edu.cn
*Corresponding author

Yuxing Gu
Engineering Research Centre of RF-ICs &
RF Systems, Ministry of Education
Southeast University
Nanjing, China
220241308@seu.edu.cn

Abstract—**The present paper puts forward a proposal for a temperature sensing module for a temperature-compensated clock system for silicon-based microelectromechanical systems (MEMS). The objective of this proposal is to enhance the frequency stability of the clock system's output signals. The module is centered on a frequency ratio module and a polynomial fitting module, in addition to others, with the objective of quantifying the temperature-induced frequency shift and thus estimating the temperature by measuring the frequency ratio of two clock signals. The output signals of this module are then fed into a phase-locked loop's delta-sigma modulator (DSM), with the purpose of enhancing the frequency stability of the output signals of the charge pump phase-locked loop (CPPLL). The module has been engineered with a 0.18μm process, enabling it to attain a temperature resolution of ±0.5K over the temperature range of −40°C to 85°C. This critical specification is pivotal for various application scenarios, including precise clock synchronization and data communication.**

Keywords—*microelectromechanical systems, temperature-compensated clock systems, polynomial*

I. INTRODUCTION

With the rapid development of electronic technology, various application fields have put forward increasingly stringent requirements on the frequency stability, power consumption and other key performance indexes of clock signal sources, and MEMS resonators have rapidly gained wide attention and in-depth research in the academic community by virtue of their high frequency stability, micro-volume, low-power consumption, and high-reliability advantages [1-6]. The prevailing focus of academic research in this field is on enhancing the frequency stability of MEMS resonators through the implementation of temperature compensation technology. Nevertheless, a pressing need remains to address the challenges associated with enhancing the accuracy of temperature measurement, which is a prerequisite for effective temperature compensation.

In 2005, Pertijs et al. [7] were the first to propose the dynamic misalignment elimination and dynamic element matching techniques combined with the second-order curvature correction algorithm to achieve a temperature measurement accuracy of 0.5°C over the temperature range of −50°C to 120°C. Subsequently, Sebastiano's team [8] developed a digital output temperature sensor based on NPN transistor architecture in 2010, which further reduces the temperature measurement error to 0.2°C by the correlated dual-sampling technique and dynamic element matching strategy. In 2011, Souri and Makinwa [9] innovated a digitally-assisted readout scheme, which is achieved by a 5-bit successive approximation Register (SAR) coarse conversion and 10-bit fine conversion hybrid architecture to achieve a measurement accuracy of 0.2°C over the range of −30°C to 125°C. In 2013, Aita et al. [10] developed a CMOS smart temperature sensor based on substrate PNP transistors with precision switched-capacitor incremental analog-to-digital converters (ADCs) and finally achieved a temperature range of −70°C to 130°C of ±0.25°C measurement accuracy.

The temperature measurement structures proposed in the aforementioned studies are characterized by two main issues. Firstly, they lack high measurement accuracy. Secondly, they exhibit problems relating to their complexity and high cost. The temperature sensing module for silicon-based MEMS clock system proposed in this paper has the potential to address the aforementioned issues. The architectural sketch of the clock system is depicted in Figure 1, with the temperature sensing module located within the red dashed box. Following the simulation verification process, the temperature sensing module is capable of achieving ±0.5K temperature sensitivity within the range of −40°C to 85°C. The temperature sensing module is composed primarily of the frequency ratio module and the polynomial fitting module, which is utilized for the regulation of the DSM module. The temperature sensing module plays a pivotal role in relaying the ambient temperature data back to the CPPLL, thereby enhancing the frequency stability of the clock system output signal.

The rest of the paper is organized as follows, in section II the sensitization mechanism derivation and model simulation of Frequency Ratio Module is described, in section III the specific circuit design of temperature sensing module is given, in section IV the circuit simulation results are presented and in section V the conclusion is given.

979-8-3315-8850-2/25 $31.00 © 2025 IEEE

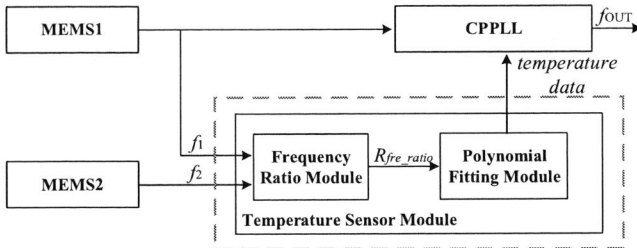

Fig. 1. Architectural sketch of the clock system

II. SENSITIZATION MECHANISM DERIVATION AND MODEL SIMULATION

The architectural sketch of the clock system is presented in Figure 1. Two MEMS resonators with different temperature coefficients provide a positive temperature coefficient clock signal and a negative temperature coefficient clock signal for the temperature sensing module, respectively. The Frequency Ratio Module is utilized to measure the frequency ratio of the two signals, thereby acquiring ambient temperature information. The following section will provide a theoretical analysis of the Frequency Ratio Module, exploring its potential to achieve the sensitization effect.

In order to clearly demonstrate the temperature characteristics of the MEMS resonator, the following derivation uses the first-order temperature coefficient of the MEMS resonator model. The resonant frequency of the MEMS resonator can be expressed as a function of temperature, as illustrated below:

$$f(\mathrm{T}) = f(\mathrm{T}_0) + \alpha \Delta \mathrm{T} \qquad (1)$$

where α is the first order temperature coefficient and $\Delta \mathrm{T}$ is the amount of change in ambient temperature.

The temperature coefficients of the signals $f_1(T)$ output from MEMS resonator 1 and $f_2(T)$ output from MEMS resonator 2 are, respectively:

$$\frac{\Delta f_1(T)}{f_1(T_0)} \cdot \frac{1}{\Delta T} = \frac{\alpha_1}{f_1(T_0)} \qquad (2)$$

$$\frac{\Delta f_2(T)}{f_2(T_0)} \cdot \frac{1}{\Delta T} = \frac{\alpha_2}{f_2(T_0)} \qquad (3)$$

where $\alpha_1 > 0$ and $\alpha_2 < 0$.

The following frequency ratio calculation is performed for $f_1(T)$ and $f_2(T)$, and the output frequency ratio signal $R_{\mathrm{fre\_ratio}}(T)$ contains temperature information. The frequency ratio signal is obtained according to Eq.(2)(3) :

$$R_{\mathrm{fre\_ratio}}(T) \approx \frac{1}{N} \frac{f_2(T_0) + \alpha_2 \Delta T}{f_1(T_0) + \alpha_1 \Delta T} = \frac{1}{N} \frac{\dfrac{f_2(T_0)}{f_1(T_0)} + \dfrac{\alpha_2}{f_1(T_0)} \Delta T}{1 + \dfrac{\alpha_1}{f_1(T_0)} \Delta T} \qquad (4)$$

The value of $\alpha_1 \Delta T / f_1(T_0)$ in Eq.(4) is very small and is obtained by simplification using the Taylor expansion:

$$R_{\mathrm{fre\_ratio}}(T) \approx \frac{1}{N} \frac{f_2(T_0)}{f_1(T_0)} - \frac{1}{N} \left(\frac{f_2(T_0)}{f_1(T_0)} \frac{\alpha_1}{f_1(T_0)} - \frac{\alpha_2}{f_1(T_0)} \right) \Delta T \qquad (5)$$

According to the definition of frequency-temperature coefficient, the frequency ratio signal is expressed as follows:

$$\frac{R_{\mathrm{fre\_ratio}}(T) - \dfrac{1}{N} \dfrac{f_2(T_0)}{f_1(T_0)}}{\dfrac{1}{N} \dfrac{f_2(T_0)}{f_1(T_0)}} \cdot \frac{1}{\Delta T} = \frac{\alpha_2}{f_2(T_0)} - \frac{\alpha_1}{f_1(T_0)} \qquad (6)$$

A comparison of the formulae (2), (3) and (6) reveals that the Frequency Ratio Module has the capacity to achieve the effect of frequency-temperature coefficient multiplication.

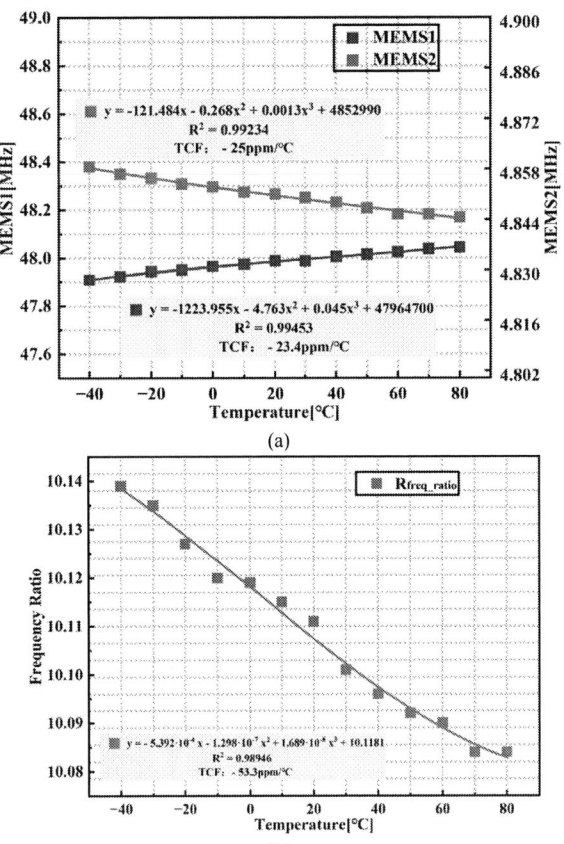

Fig. 2. At different temperatures(a) Frequency curves of the output signals of MEMS resonator 1 and MEMS resonator 2 (b) Curve of signal frequency ratio $R_{\mathrm{ref\_ratio}}$

As demonstrated in Figure 2, the theoretical derivation presented above has been verified by MATLAB. As illustrated in Figure 2(a), the frequency of the output signals of MEMS resonators 1 and 2 is shown as a function of temperature, with the temperature coefficients of −23.4 ppm/K for MEMS resonator 1 and −25 ppm/K for MEMS resonator 2. As illustrated in Figure 2(b), the frequency ratio information of the

979-8-3315-8850-2/25 $31.00 © 2025 IEEE   111

outputs, in conjunction with the temperature, is depicted as a function of temperature following the execution of frequency ratio calculations. The temperature coefficients for this function are -53.3 ppm/K, which is approximately double the frequency temperature coefficient observed in MEMS resonators 1 and 2. This provides the basis for highly accurate temperature measurement.

III. CIRCUIT DESIGN

A. Frequency Ratio Module

Frequency Ratio Module mainly consists of logic control part, counter and TDC, etc., as shown in Figure 3, in which the core part to realize the frequency ratio measurement is the counter and TDC.

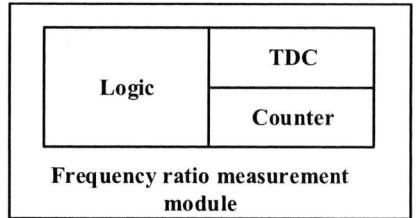

Fig. 3. Frequency ratio measurement module

The TDC used in this architecture consists of a delay chain and a set of flip-flops as shown in Figure 4. The MEMS1 signal goes through the delay chain to generate a series of delay signals, which are uniformly sampled at the arrival of the rising edge of the MEMS2 signal to obtain a set of data that quantizes the time interval between two signals into a digital signal.

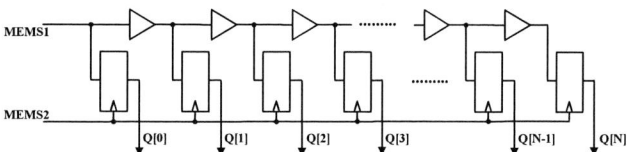

Fig. 4. Basic Delay Chain TDC Circuit Diagram

The resolution of the TDC is defined as the delay time of the delay unit. Given that the clock signal frequencies of MEMS1 and MEMS2 are 4.8 MHz and 0.5 MHz, respectively, it is necessary to select the number of bits of the TDC to be 128. The delay time of the delay unit is then set to 1.625 ns, in order to realize the temperature sensing sensitivity of ±0.5 K.

In order to measure the integer part of the frequency ratio of the two signals, while taking into account the frequency ratio range of the MEMS1 signal and the MEMS2 signal, the counter structure of the JK trigger cascade will be used to measure the frequency ratio range from 1 to 15. The counter structure is shown in Figure 5.

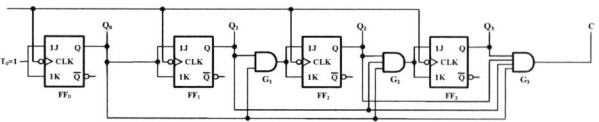

Fig. 5. Circuit Diagram of Counter

B. Polynomial Fitting Module

Following the input of the frequency ratio signal R_{ref_ratio} into the polynomial fitting module, the module will fit the signal into the desired crossover ratio of the CPPLL. This will be based on the temperature drift curve of the MEMS resonator, as well as on the correspondence between the frequency ratio signal R_{ref_ratio} and the ambient temperature. The configuration of the polynomial fitting module is illustrated in Figure 6.

The quantized value of the frequency ratio signal R_{ref_ratio} is processed through a polynomial fitting module to obtain the MEMS output frequency at ambient temperature.

$$f_{\text{MEMSOUT}} = ax^3 + bx^2 + cx + d \tag{7}$$

In the given context, it is understood that the coefficients a, b, c and d of the polynomial are to be entered into the polynomial fitting module via the serial peripheral interface (SPI) module. Furthermore, it is understood that x is the quantized value of the frequency ratio signal R_{ref_ratio}. It is evident that, given the fact that the polynomial fitting module writes the frequency f_{OUT} of the output signal of the clock system through the SPI, the integer frequency division ratio control word that is required for the phase-locked loop is:

$$N_{int} = \frac{f_{OUT}}{f_{MEMSOUT}} \tag{8}$$

The remainder is calculated as follows:

$$MOD = f_{OUT} - N_{int} \cdot f_{MEMSOUT} \tag{9}$$

Since the DSM in the PLL uses an N-bit control word, the decimal divider control word for the DSM is:

$$N_{dec} = \frac{MOD(2^N - 1)}{f_{MEMSOUT}} \tag{10}$$

The resulting fractional control word N_{dec} and integer control word N_{int} are input to the CPPLL to modulate the crossover ratio. This results in a significant reduction in the loop lock time and an enhancement of the frequency stability of the CPPLL output signal.

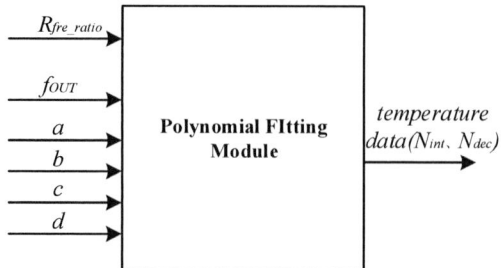

Fig. 6. Sketch of the polynomial fitting module

979-8-3315-8850-2/25 $31.00 © 2025 IEEE

IV. SIMULATION RESULTS

The module has been designed in accordance with the parameters set out in this paper. It has been constructed utilizing a 0.18 μm process, and the circuit layout has been completed using ICC(Integrated Circuit Compiler), as illustrated in Fig. 7.

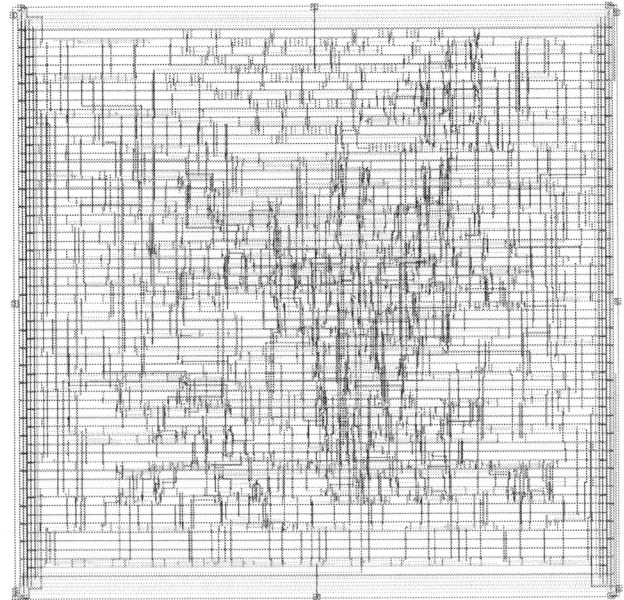

Fig. 7. Circuit layout of Temperature Sensor Module

The temperature sensitivity of the module is simulated, with the module accessed to a signal with a temperature coefficient of +2600 ppm/K and a frequency of 4.8 MHz and a signal with a temperature coefficient of –2600 ppm/K and a frequency of 0.5 MHz, respectively.

The simulation temperatures for this test are –40°C, 25°C and 85°C, and the test results are shown in Figure 8.

TABLE I. ANALYSIS OF THE SIMULATION RESULTS OF THE TEMPERATURE SENSOR MODULE

Test temp(°C)	Data from temperature sensing module		
	R_{ref_ratio}	Sensed temperature data(°C)	Sensed temperature error(°C)
-40	6.8155	-40.2409	-0.2409
	6.8089	-40.4218	-0.4218
	6.8415	-39.5295	0.4705
25	9.5752	24.5026	-0.4974
	9.5801	24.6009	-0.3991
	9.6244	25.4882	0.4882
85	13.1144	84.5082	-0.4918
	13.1799	85.4429	0.4429
	13.1831	85.4885	0.4885

As shown in TABLE I, the measured values of the signal frequency ratio and the sensed ambient temperatures are shown for the two signals at ambient temperatures of –40°C, 25°C and 85°C respectively. At the ambient temperature of –40°C, the temperature sensing module detected an ambient temperature in the range of –40.0902°C to –39.9424°C. At an ambient temperature of 25°C, the ambient temperature detected by the temperature sensing module is in the range of 24.9038°C to 25.0961°C. At an ambient temperature of 85°C, the ambient temperature detected by the temperature sensing module is in the range of 84.9013°C to 85.0996°C. In summary, the temperature sensing module has a sensing accuracy of ±0.5K.

V. SUMMARIZE

The present paper proposes a temperature sensing module for silicon-based MEMS clock systems, with the objective of enhancing the frequency stability of the clock system. The proposed module integrates a TDC and a polynomial fitting module to optimize the frequency drift of the output signal of the MEMS clock system with temperature. The temperature-induced frequency shift is quantified by measuring the frequency ratio between the temperature-sensitive and temperature-insensitive clock signals. This, in turn, provides a high-precision sensing of the temperature. The polynomial fitting module has been developed to map the frequency ratio information to the DSM, thus enabling the precise adjustment of the CPPLL output frequency. The module has been designed utilizing a 0.18 μm process, and simulation results demonstrate that a temperature resolution of ±0.5 K can be achieved over the range of –40 °C to 85 °C. The solution under discussion is a reliable reference solution for applications requiring a highly stable clock source. Such applications include, but are not limited to, aerospace and satellite communication systems.

ACKNOWLEDGMENT

This work was supported by the National Natural Science Foundation of China under Grant 62234012.

REFERENCES

[1] A. Partridge, H. -C. Lee, P. Hagelin and V. Menon, "We know that MEMS is replacing quartz. But why? And why now?," 2013 Joint European Frequency and Time Forum & International Frequency Control Symposium (EFTF/IFC), Prague, Czech Republic, 2013, pp. 411-416, doi: 10.1109/EFTF-IFC.2013.6702311.

[2] R. Henry and D. Kenny, "Comparative analysis of MEMS, programmable, and synthesized frequency control devices versus traditional quartz based devices," 2008 IEEE International Frequency Control Symposium, Honolulu, HI, USA, 2008, pp. 396-401, doi: 10.1109/FREQ.2008.4623027.

[3] X. Huang, D. Liu, Y. Wang, P. Chen and W. Fu, "100-MHz low-phase-noise microprocessor temperature-compensated crystal oscillator," in IEEE Transactions on Circuits and Systems II: Express Briefs, vol. 62, no. 7, pp. 636-640, July 2015, doi: 10.1109/TCSII.2015.2415652.

[4] S. Zaliasl et al., "A 3 ppm 1.5 × 0.8 mm 2 1.0 μA 32.768 kHz MEMS-based oscillator," in IEEE Journal of Solid-State Circuits, vol. 50, no. 1, pp. 291-302, Jan. 2015, doi: 10.1109/JSSC.2014.2360377.

[5] T. L. Naing, T. O. Rocheleau, E. Alon and C. T.-C. Nguyen, "Low-power MEMS-based pierce oscillator using a 61-MHz capacitive-gap disk resonator," in IEEE Transactions on Ultrasonics, Ferroelectrics, and Frequency Control, vol. 67, no. 7, pp. 1377-1391, July 2020, doi: 10.1109/TUFFC.2020.2969530.

[6] W. Chen, W. Jia, Y. Xiao, Z. Feng and G. Wu, "A temperature-stable and low impedance piezoelectric MEMS resonator for drop-in replacement of quartz crystals," in IEEE Electron Device Letters, vol. 42, no. 9, pp. 1382-1385, Sept. 2021, doi: 10.1109/LED.2021.3094319.

[7] M. A. P. Pertijs, A. Niederkorn, X. Ma, B. McKillop, A. Bakker and J. H. Huijsing, "A CMOS smart temperature sensor with a 3σ inaccuracy of ±0.5°C from −50°C to 120°C," in IEEE Journal of Solid-State Circuits, vol. 40, no. 2, pp. 454-461, Feb. 2005, doi: 10.1109/JSSC.2004.841013.

[8] F. Sebastiano, L. J. Breems, K. A. A. Makinwa, S. Drago, D. M. W. Leenaerts and B. Nauta, "A 1.2-V 10-μW NPN-based temperature sensor in 65-nm CMOS With an inaccuracy of 0.2°C (3σ) from −70°C to 125°C," in IEEE Journal of Solid-State Circuits, vol. 45, no. 12, pp. 2591-2601, Dec. 2010, doi: 10.1109/JSSC.2010.2076610.

[9] K. Souri and K. A. A. Makinwa, "A 0.12mm^2 7.4μW micropower temperature sensor with an inaccuracy of ±0.2°C (3σ) from −30°C to 125°C," in IEEE Journal of Solid-State Circuits, vol. 46, no. 7, pp. 1693-1700, July 2011, doi: 10.1109/JSSC.2011.2144290.

[10] A. L. Aita, M. A. P. Pertijs, K. A. A. Makinwa, J. H. Huijsing and G. C. M. Meijer, "Low-power CMOS smart temperature sensor with a batch-calibrated inaccuracy of ±0.25°C (±3σ) from −70°C to 130°C," in IEEE Sensors Journal, vol. 13, no. 5, pp. 1840-1848, May 2013, doi: 10.1109/JSEN.2013.2244033.

2025 The 10th International Conference on Integrated Circuits and Microsystems

An End-to-End Compact Shape-Aware Macro Placer Using Reinforcement Learning

Hailiang Li, Xiao Wang, Xu Yang, Miaohui Hao, Yan Huo, Beiping Yan
Integrated Circuit Enabling Technology (ICET) group of Advanced Electronic Components and Systems (AECS) Division
Hong Kong Applied Science and Technology Research Institute Company Limited (ASTRI), Hong Kong, China
{harleyli, ericwang, xuyang, miaohuihao, jennyhuo, bpyan}@astri.org

Abstract—In modern chip design, the placement of millions of circuit modules poses a significant challenge. However, macro placement can significantly enhance this process by enabling greater modularization of the chip design. In many cases, macros can occupy more than half of the die area, making their placement a critical factor in defining the die's overall shape. Despite its importance, existing algorithms often either neglect this aspect or fail to address it adequately. This paper presents a novel macro placer that enhances reinforcement learning architectures to produce compact and shape-optimized placements. Our research demonstrates that strategically decoupling standard cells during macro placement can significantly reduce network complexity, enabling an efficient relinking solution. Additionally, we introduce a shape force mask into the reinforcement learning-based visual representation learning framework. The approach proposed in this paper enables the creation of compact and shape-optimized chip designs. Experimental results validate that our algorithm can produce more compact and regular macro placements while consistently maintaining other key performance metrics.

Keywords—Macro Placement, Reinforcement Learning, Shape Force Mask, Wire Mask

I. INTRODUCTION

In Reinforcement Learning (RL) [1] for placement task, the problem is modeled as a Markov Decision Process (MDP) [5, 6]. Each step involves placing a macro according to predefined rules that consider numbers of factors. The state includes details about the chip canvas, placed macros, and the macro currently being positioned. Recent studies have explored ways to enhance the density of the reward signal; for example, WireMask [2]

offers visual representation learning feature and a continuous reward based on already placed macros. Macro placement is vital, as it can occupy over half of the die area, significantly impacting the die's overall shape. A more regular layout facilitates chip cutting; however, existing optimization-based and RL-based algorithms overlook this crucial point.

This paper presents a novel macro placer that can produce a compact, regular and symmetric shape layout. Our proposed model, namely, ShapePlace, incorporates a shape force mask (see Fig. 5), allowing for the adjustment of one macro's position at each step. This approach considers both wire length and final placement shape, aligning with traditional design principles like compactness and symmetry (as shown in Fig. 1 and Fig. 2). Our proposal builds upon MaskPlace [2] by integrating an enhanced input feature map and advanced preprocessing techniques.

The key contributions include a preprocessing module that relinks the netlist of macros after standard cells are removed, maintaining indirect connections. Additionally, we introduce a new feature map, the shape force map, which constrains the final die shape. Finally, experiments on classic benchmarks show that our method performs comparably to state-of-the-art techniques.

The paper is structured as follows. Section II reviews the related work. Section III presents the details of Macro relink preprocess. Section IV explains the shape force mask and its role in the RL framework. Section V presents experiments, Section VI discusses future directions, and Section VII concludes.

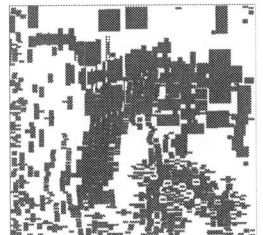 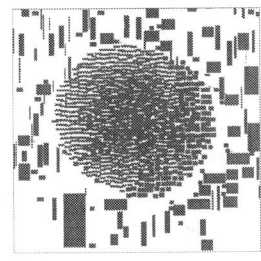 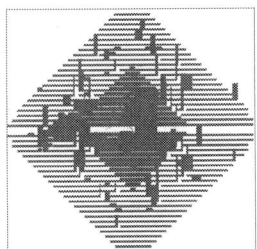  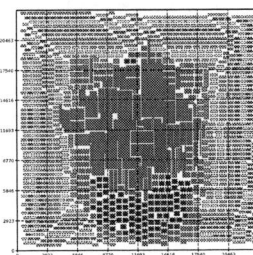

Fig. 1. Placements of chip Adaptec4 from 5 methods (DREAMPlace [4], GraphPlace [5], DeepPR [12], MaskPlace [2], ShapePlace).

979-8-3315-8850-2/25 $31.00 © 2025 IEEE

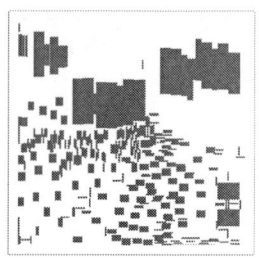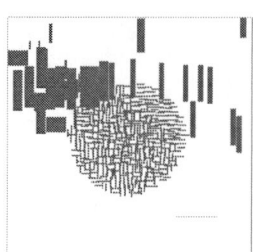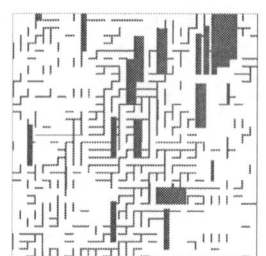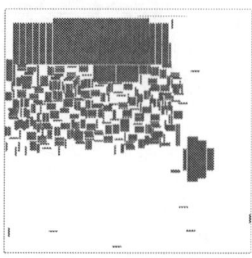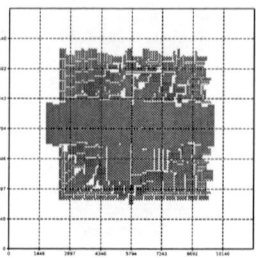

Fig. 2. Placements of chip Bigblue1 from 5 methods (DREAMPlace [4], GraphPlace [5], DeepPR [12], MaskPlace [2], ShapePlace).

II. BACKGROUND

In the placement stage of integrated circuit design, the circuit is modeled as a graph where the vertices correspond to logic gates. The primary input for this process is a netlist denoted as $N = (V, E)$. In this context, V contains critical information about all macros intended for placement on the chip, such as their dimensions (height and width). The set E represents a hypergraph composed of nets $e_i \in E$, which connects multiple instances, including both macros and standard cells, indicating their interconnectivity during the routing phase.

Macro placement means given a specific netlist, a fixed canvas layout, and a library collection of standard cells; the objective of a placement is to determine the optimal physical locations for movable macros. This placement aims to minimize total wirelength, crucial for better performance and efficiency. A macro placement solution $S = \{(x_1, y_1), \ldots, (x_k, y_k)\}$ consists of the coordinates of all macros, $\{v_i\}_{i=1}^{k}$ where k signifies the total number of macros involved in the layout. A key goal in macro placement is to minimize the total half-perimeter wirelength (HPWL) of all nets while adhering to other constraints. There are three primary categories of methods employed for placement: analytical methods, optimization-based methods and reinforcement learning based methods.

1. Analytical Methods

These methods place macros and standard cells together and fall into two main types: quadratic and nonlinear placement. Quadratic placement [15, 16] iterates between an unconstrained quadratic programming for minimizing wirelength and a heuristic spreading to remove cell overlaps. Nonlinear placement [17, 18, 19] formulates the problem as a nonlinear optimization and solves it directly with gradient descent. Nonlinear placement usually gives better solutions, while quadratic placement is more computationally efficient.

2. Optimization based Methods

These placement methods have a long history. Early techniques like SP (Sequence Pair) [8] and B*-tree [9] had scalability problems from relying on rectangular packing. Recent black-box optimization advancements have boosted performance by changing the search space. For instance,

AutoDMP [10] uses Bayesian optimization to enhance DREAMPlace's [4] exploration of the configuration space, with great results in benchmarks. The WireMask-BBO [11] method uses a WireMask guided greedy mapping and can integrate with any black-box optimization algorithm.

3. Reinforcement Learning based Methods

Recently, macro placement methods have seen a major shift, as researchers increasingly use reinforcement learning (RL) to meet modern design's complex demands. GraphPlace [5] pioneers by framing macro placement as an RL problem, optimizing via grid-division, but its sparse reward hampers learning as the agent gets no feedback until all macros are placed. DeepPR [12] and PRNet [13] integrate macro placement with other design elements for better performance, yet risk violating non-overlap constraints. MaskPlace [2] overcomes these with a dense reward and pixel - level visual rep, boosting efficiency and ensuring 0% overlap. ChiPFormer [3] improves placer generalize ability through offline learning, and EfficientPlace [14] uses a global tree search for fast, high-quality placements.

4. Proposal: Shape-Aware Macro Placer

While existing related works have largely neglected the generation of compact and regular chip layouts, the proposed WireMask in MaskPlace [2] demonstrates strong capabilities in visual representation learning and offers extensible features. To address these limitations and further enhance the quality of macro placement, we propose an improved reinforcement learning-based method. As illustrated in Fig. 3, our approach introduces two key innovations to the MaskPlace [2] framework. First, we incorporate a Macro Relink preprocessing module, which reconnects the macros that were previously disconnected due to the removal of standard cells. By reorganizing these blocks into nets, this module significantly enhances the compactness and connectivity of the final layout. Second, we introduce an innovative shape force mask as part of the input features. This mask provides additional guidance to the placement process, ensuring that the final layout adheres to a regular, symmetrical, and desired shape. Together, these advancements enable more efficient and visually appealing chip designs while maintaining the flexibility and scalability of the original framework.

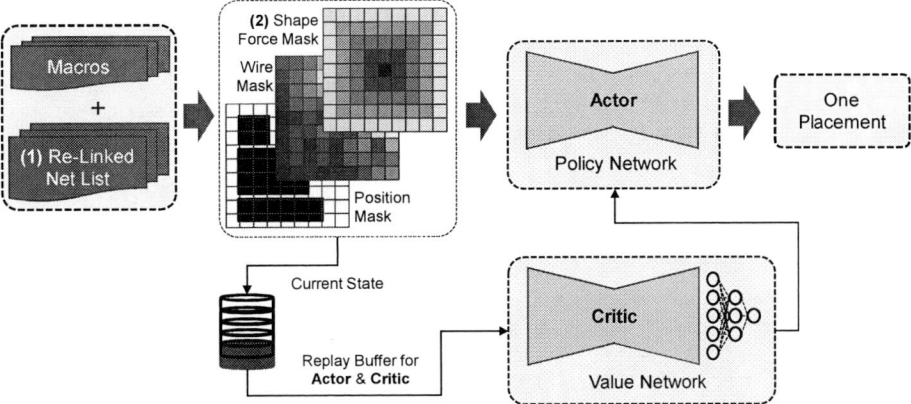

Fig. 3. An overview of the workflow of our proposed ShapePlace model with two novels inside. (1) A macro relink preprocess is proposed; (2) The new shape force mask is proposed as a new feature (state value of RL framework).

III. MACRO RELINK PREPROCESS

In previous RL-based macro placers, thousands of standard cells in the netlist were removed prior to the macro placement process. This deletion often led to the removal of numerous nets, resulting in many isolated macros due to the disconnection of certain nets. To address this issue, we developed a preprocess solution (Algorithm 1) that re-establishes indirect links for the macros after the standard cells have been deleted.

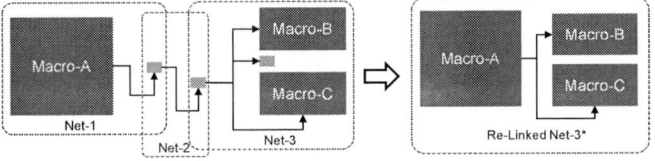

Fig. 4. An example of three nets becoming one net. (Small gray blocks are standard cells) (1) In Net-1 and Net-2, because there is no Macro in receiver-nodes, Net-1 and Net-2 are deleted; (2) In Net-3, after deleting standard-cell(s), because there are Macros (Macro-B and Macro-C) in receiver-nodes, recursively find the parent-node of sender-node is Macro-A, so it is eventually relink to the new Net-3*.

For the definition of a net: a net is a collection of links, in which there is only one sender node and no less than one receive node(s). A typical net data (adaptec1.nets) from ISPD-2005 [7]:

```
NetDegree  :  5  n22
    o58274   I :  1.000000       5.000000
    o57811   I : -3.500000      -5.000000
    o57706   I : -5.500000       5.000000
    o57126   I : -4.500000      -6.000000
    o12      O : -4.500000      -3.000000
```

We can compare the sender node to the parent node and the receiver node to the child node. Because in similar link relationships, there is only one parent. Meanwhile, there can be one or more children. In the following, we describe the algorithm in a concise yet comprehensive manner. Algorithm 1 is presented with a detailed, step-by-step explanation of its process, ensuring clarity and precision in its implementation.

The following describes the macro relink algorithm:

1. Delete this net if the net's child node(s) does not have a macro. (such as Net-1 and Net-2 in Fig. 4).

2. If the child node(s) of the netlist has Macro (e.g.: Net-3 in Fig. 4), first delete the standard cell(s) in the child node(s).

Then check the parent node (sender node) of this net, if:

(1) It is a macro, then this is a pruned net with Macro as the main component.

(2) It is a standard cell, then recursively executes the following (or set the recursion depth value, such as 10, which can be set to break out).

From the netlist, set the current node's parent node as the current node then check it,

(A) If the node is a Macro, the sender node is replaced by the macro, and the relink net with macros is finished.

(B) If comes to the recursion depth value, break out.

Algorithm 1: Macro Relink Preprocess

01   **Input**: The netlist (all the nets)
02 **Output**: The modified netlist with only relinked macro nets
03 **Preliminary and Definitions:**
04 **Definitions of Mathematical Expression Symbols:**
05 ● Let N be the set of all nets, and $n \in N$ represent a single net.
06 ● Let $S(n)$ denote set of sender-node of a net n, and $R(n)$ denote
07   set of its receiver-nodes, $R(n) = \{r_1, r_2, \dots, r_k\}$, where $k \geq 1$.
08 ● Let M: set of all Macros, and SC: set of all standard cells.
09 ● Let $P(x)$: the parent node of node x (in the context of nodes
10   within a net, the parent node is equivalent to the sender-node).
11 **Definition of a net:**
12 A net n satisfies the following conditions:
13 ● The sender-node is unique: $|S(n)| = 1$, ($|*|$, the number of)
14 ● There is at least one receiver-node: $|R(n)| \geq 1$
15 **Process:**
16 **Step 1: Delete nets without Macros in their receiver-nodes.**
17 Define a judgment function ***hasMacroInReceivers***(n). The 18
function returns *true* if and only if $R(n) \cap M \neq \emptyset$;
19   otherwise, it returns *false*. The set of nets to be deleted is:

20     $N_{delete} = \{n \in N | hasMacroInReceivers(n) = false\}$.

21   After deletion, the set of nets becomes $N' = N - N_{delete}$.

22 Step 2: Process nets with Macros in their receiver-nodes

23   For $n \in N'$ (i.e., nets with Macros in their receiver-nodes),

24   Firstly, delete standard cells from receiver-nodes. The new set of 25 receiver-nodes are: $R'(n) = R(n) - SC$.

26   Next, check the sender-node $S(n)$:

27 Case A:

28   If $S(n) \in M$, then n is a pruned net mainly composed of

29   Macros. We can define a set:

30         $N_{Macro-dominated} = \{n \in N' | S(n) \in M\}$.

31 Case B:

32   If $S(n) \in SC$, conduct recursive operation with $D(e.g., = 10)$.

33   The recursive function $\boldsymbol{reLinkRecursive(x, d)}$ is defined:

34   **Input**: The current node x and the current recursion depth d.

35   **Recursion termination conditions:**

36   (1) If $x \in M$, replace the sender-node $S(n)$ of net n with x,

37   i.e., $S(n) = x$, complete the relink, and end recursion.

38   (2) If $d \geq D$, stop the recursion.

39   **Recursive operations:**

40   Let $x' = P(x)$ (i.e., parent node of current node).

41   Call $\boldsymbol{reLinkRecursive(x', d + 1)}$.

42   For $n \in N'$ and $S(n) \in SC$, call $\boldsymbol{reLinkRecursive(S(n), 1)}$

43   to start the recursion.

IV. MACRO PLACER WITH SHAPE FORCE MASK

A. Reinforcement Learning based Macro Placement

The goal of macro placement is to optimally arrange macros on a chip, minimizing a cost function with key metrics like wirelength, congestion, and placement constraints. The problem is formulated as a Markov Decision Process (MDP), modeling the placement as a sequence of decisions. Each decision places a macro, evolving the system state. The MDP framework enables RL to iteratively optimize the cost function, ensuring the final layout meets performance and design needs.

State Space (S): The state S_t at time t represents the current placement of macros, the combined mask $\widetilde{W}_{t'}$ (the WireMask and the proposed shape force mask as shown in Fig. 5, which makes constraint for the final placement shape.) and other relevant features (e.g., congestion map)

Action Space (A): The action a_t corresponds to placing a macro at a specific location on the chip canvas

Reward Function (R): The reward. r_t is a function of the improvement in the cost function (e.g., reduction in wirelength or congestion) after taking action a_t.

Transition Dynamics (P): The transition probability matrix $P(s_{t+1} | s_t, a_t)$ defines state evolution after action a_t

B. Shape Force Mask

Leverages an Actor-Critic approach, where the state representation includes the combined mask $\widetilde{W}_t$ which is a weighted sum of the WireMask W_t and Shape Force Mask S_t.

1. WireMask (W_t): Encodes the incremental increase in wire length for each step of Macro placement.

2. Shape Force Mask (S_t): Encodes additional placement guidance to force the layout gradually to expected shape.

The combined mask $\widetilde{W}_t$ is computed as a weighted sum of the two masks:

$$\widetilde{W}_t = \theta \cdot W_t + (1 - \theta) \cdot S_t \tag{1}$$

where: θ is a hyperparameter ($0 \leq \theta \leq 1$) that balances the contributions of the WireMask and shape force mask. (For example, $\theta = 0.9$ a greater proportion of weight is assigned to the WireMask. Simultaneously, a certain amount of weight is retained for the shape force mask).

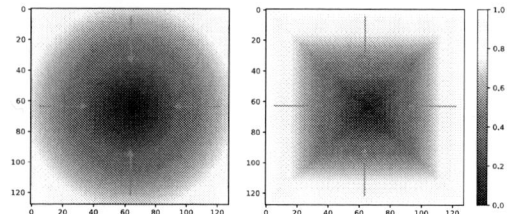

Fig. 5. Two shape force masks with normalized float values ([0, 1]). The left figure is the circle shape force mask, which generates the layout in Fig. 6 for some ISPD-2005 [7] chips. The right figure is the square shape force mask, producing the layout in Fig. 7. These masks guide the model to achieve circular and square layouts, respectively.

C. State with Visual Representation Learning Masks

The state S_t is a combination of:

(1)   The current placement of macros M_t

(2)   The combined mask $\widetilde{W}_t$

(3)   Additional features such as congestion maps C_t.

Mathematically, the state is represented as:

$$s_t = [M_t, \widetilde{W}_t, C_t] \tag{2}$$

D. Actor-Critic Architecture for Macro Placer

The Actor-Critic [20] algorithm consists of two components (As shown in Fig. 3). **Actor:** A policy network $\pi_\theta(a_t | s_t)$ that selects actions (macro placements) based on the current state. **Critic:** A value network $V_\phi(s_t)$ that estimates the expected cumulative reward from the current state. Improvements to the Actor-Critic algorithm can be made in features and network architecture. This paper focuses on feature innovation.

1. Policy Network (Actor)

The policy network $\pi_\theta(a_t | s_t)$ outputs a probability distribution over possible action (macro placements). The objective is to maximize the expected cumulative reward.

$$J(\theta) = \mathbb{E}_{\tau \sim \pi_\theta} \left[\sum_{t=0}^{T} \gamma^t r_t \right] \tag{3}$$

where: $\tau = (s_0, a_0, r_0, s_1, a_1, r_1, \dots)$ is a trajectory. γ is the discount factor.

The policy gradient [21, 22, 23] is computed as:

$$\nabla_\theta J(\theta) = \mathbb{E}_{\tau \sim \pi_\theta} \left[\sum_{t=0}^{T} \nabla_\theta \log \pi_\theta(a_t | s_t) \cdot A(s_t, a_t) \right] \quad (4)$$

where $A(s_t, a_t)$ is the advantage function, defined as:

$$A(s_t, a_t) = Q(s_t, a_t) - V(s_t) \quad (5)$$

Here, $Q(s_t, a_t)$ is the action-value function, and $V(s_t)$ is the state-value function estimated by the Critic

2. Value Network (Critic)

The value network $V_\phi(s_t)$ is trained to minimize the mean squared error between the predicted value and the target value:

$$L(\phi) = \mathbb{E}_{\tau \sim \pi_0} \left[\sum_{t=0}^{T} \left(V_\phi(s_t) - y_t \right)^2 \right] \quad (6)$$

where the target value y_t is computed via Bellman equation:

$$y_t = r_t + \gamma V_\phi(s_{t+1}) \quad (7)$$

3. Reward Function

The reward r_t is designed to guide the RL agent toward better placements. It can include terms such as: wirelength reduction, congestion reduction, and penalties for violating design rules. Mathematically, the reward can be expressed as:

$$r_t = -\big(\alpha \cdot \text{Wirelength}(s_t) + \beta \cdot \text{Congestion}(s_t) + \text{Penalties}(s_t) \big) \quad (8)$$

where α and β are hyperparameters of weighting factors.

4. Training the Algorithm

The training process alternates between:

(1) Collecting trajectories using the current **policy** π_θ;

(2) Updating the **Critic** by minimizing $L(\phi)$;

(3) Updating the **Actor** using the **policy gradient** $\nabla_\theta J(\theta)$.

TABLE I. STATISTICS OF DIFFERENT CHIP BENCHMARKS FROM ISPD-2005 [7]

	Parameters ⇒	#Objects	#Standard Cells	#Macros	#Nets	#Pins	Density%
Chips	adaptec1	211K	210904	543	221142	944053	75.71
	adaptec2	255K	254457	566	266009	1069482	78.59
	adaptec3	452K	450927	723	466758	1875039	74.53
	adaptec4	496K	494716	1329	515951	1912420	62.67
	bigblue1	278K	277604	560	284479	1144691	54.19
	bigblue3	1097K	1095519	1293	1123170	3833218	85.65

TABLE II. EXPERIMENTAL RESULT OF SHAPEPLACE AND OTHER WIREMASK BASED METHODS

	Chips ⇒	adaptec1	adaptec2	adaptec3	adaptec4	bigblue1	Bigblue3
Meths	MaskPlace	1.465070	1.434827	2.674445	2.578745	1.519873	5.235587
	WireMask_BBO	1.454344	**1.392187**	2.667697	**2.450444**	1.508696	**4.892756**
	Efficient_Place	**1.294178**	1.427327	2.741896	2.550781	1.481700	5.306246
	ShapePlace	1.370985	1.435982	**2.463879**	2.547519	**1.477178**	5.279005

V. EXPERIMENTS

We conducted experiments on six chips from the ISPD-2005 benchmark [7] using the ShapePlace model, applying either a circular or a square shape force mask. The results are shown in Fig. 6 and Fig. 7, respectively. As shown in Fig. 6, for all the adaptec1~4 and bigbule1 and bigblue3 chips, the ShapePlace model effectively produces compact, regular, and symmetric layouts based on the given shape force masks.

We evaluated our proposed model against three other wireMask-based models, MaskPlace [2], WireMask_BBO [11], and EfficientPlace[14], under identical settings using the six chips from the ISPD-2005 benchmark [7]. The statistics of the selected chip benchmarks are detailed in Table-1. We collected macro placements from four different methods and then performed standard cell placement. This process included both Global Placement (GP) and Detailed Placement (DP), executed using the same open-source tool, XPlace [19]. Finally, we obtained the final half-perimeter wire length (HPWL, a wire length surrogate due to fast computation.) values after placing all macros and standard cells. As shown in Table-2, our model achieves comparable performance on the HPWL metric (Notably, all HPWL values in Table-2 are on the order of 10^{-8},

where smaller values indicate better performance.) compared to the three competing models, meanwhile, Fig. 2 highlights its unique capability to generate layouts with regular structure, compact geometry, and optimal cutting quality.

There is no comparison with DreamPlace [4] since it is Standard-cell focused, not a macro placer, and EfficientPlace [14] already showed DreamPlace [4] underperformed RL approaches significantly in macro placement. Hier-RTLMP [24] is employed as a macro placer but lacks native support for the ISPD-2005 dataset, hence not included in comparisons also.

VI. FUTURE DIRECTIONS

In reinforcement learning-based macro placement, the representation learning of the input feature space (or state space) is crucial for improving the efficiency and generalization of policy learning. There are still some potential future research directions in this area:

◆ **Graph Neural Networks (GNNs) for State Representation Learning**: Since the macro placement problem can naturally be represented as a graph structure, GNNs can effectively capture the topological relationships and interactions between macros. This approach can enhance the understanding of spatial dependencies and improve placement decisions.

◆ **Routability-Aware Macro algorithm**: Most RL-based Macro placements still use HPWL as the primary metric. The down-sampling of the chip canvas to neural network input followed, later up-sampling back to original size, which mitigates congestion severity in placement results. A superior approach should employ a multi-objective optimization framework that jointly considers both wirelength and congestion.

◆ **Cross-Task Generalization**: Investigate how to enable reinforcement learning algorithms to generalize better across different macro placement tasks. For example, developing state representation methods that work effectively under varying design constraints and objectives can help create more robust and adaptable placement strategies.

◆ **Real-Time Placement Optimization**: Explore how to apply reinforcement learning for real-time macro placement optimization. This requires developing efficient state representation and decision-making algorithms to meet real-time computation demands, enabling dynamic adjustments and quick responses to changes in design requirements.

VII. CONCLUSION

Motivated by the limitations of prior algorithms, which often fail to generate compact and regular chip macro block layouts that align with the expertise of design professionals, and inspired by the potential of visual representation learning to enhance reinforcement learning-based macro placement, we propose a novel macro placer. Our approach introduces a macro relink preprocessing module and a novel shape force mask for reinforcement learning framework input features. The proposed method, ShapePlace, is capable of producing more compact and regular macro placements while maintaining or improving other performance metrics. Furthermore, our algorithm is designed to be modular, enabling its integration with and enhancement of existing methods.

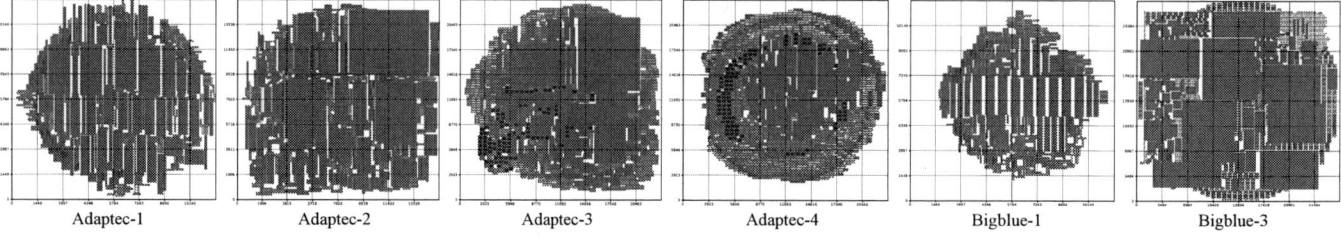

Adaptec-1   Adaptec-2   Adaptec-3   Adaptec-4   Bigblue-1   Bigblue-3

Fig. 6. The placement results of six chips via ShapePlace model with a circle shape force mask (different colors indicate macro groups).

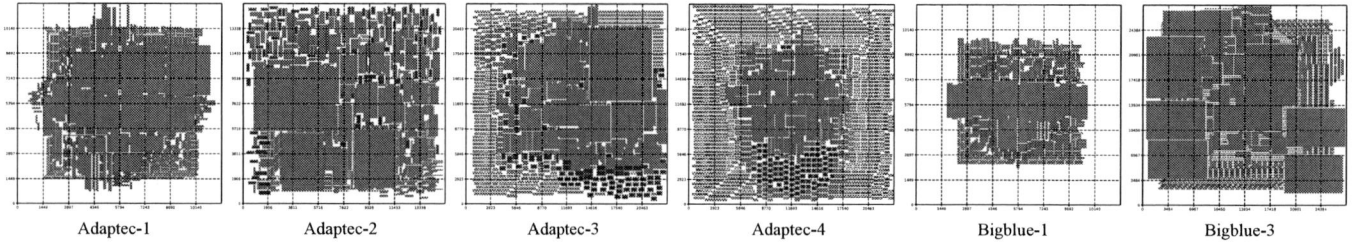

Adaptec-1   Adaptec-2   Adaptec-3   Adaptec-4   Bigblue-1   Bigblue-3

Fig. 7. The placement results of six chips (ISPD-2005 [7]) via ShapePlace model with a square shape force mask inside.

REFERENCES

[1] Mnih, Volodymyr, Koray Kavukcuoglu, David Silver, Alex Graves, Ioannis Antonoglou, Daan Wierstra, and Martin Riedmiller. Playing atari with deep reinforcement learning. *arXiv preprint arXiv:1312.5602* (2013).

[2] Lai, Yao, Yao Mu, and Ping Luo. Maskplace: Fast chip placement via reinforced visual representation learning. *Advances in Neural Information Processing Systems* 35 (2022): 24019-24030.

[3] Lai, Yao, Jinxin Liu, Zhentao Tang, Bin Wang, Jianye Hao, and Ping Luo. Chipformer: Transferable chip placement via offline decision transformer. In *International Conference on Machine Learning*, pp. 18346-18364. PMLR, 2023.

[4] Lin, Yibo, Shounak Dhar, Wuxi Li, Haoxing Ren, Brucek Khailany, and David Z. Pan. Dreamplace: Deep learning toolkit-enabled gpu acceleration for modern vlsi placement. In *Proceedings of the 56th Annual Design Automation Conference 2019*, pp. 1-6. 2019.

[5] Mirhoseini, Azalia, Anna Goldie, Mustafa Yazgan, Joe Wenjie Jiang, Ebrahim Songhori, Shen Wang, Young-Joon Lee et al. A graph placement methodology for fast chip design. *Nature* 594, no. 7862 (2021): 207-212.

[6] Hausknecht, Matthew, and Peter Stone. Deep recurrent q-learning for partially observable mdps. In *2015 aaai fall symposium series*. 2015.

[7] Nam, Gi-Joon, Charles J. Alpert, Paul Villarrubia, Bruce Winter, and Mehmet Yildiz. The ISPD2005 placement contest and benchmark suite. In Proceedings of the 2005 international symposium on Physical design, pp. 216-220. 2005.

[8] Murata, Hiroshi, Kunihiro Fujiyoshi, Shigetoshi Nakatake, and Yoji Kajitani. VLSI module placement based on rectangle-packing by the sequence-pair. *IEEE Transactions on Computer-Aided Design of Integrated Circuits and Systems* 15, no. 12 (1996): 1518-1524.

[9] Chang, Yun-Chih, Yao-Wen Chang, Guang-Ming Wu, and Shu-Wei Wu. B*-trees: A new representation for non-slicing floorplans. In *Proceedings of the 37th Annual Design Automation Conference*, pp. 458-463. 2000.

[10] Agnesina, Anthony, Puranjay Rajvanshi, Tian Yang, Geraldo Pradipta, Austin Jiao, Ben Keller, Brucek Khailany, and Haoxing Ren. Autodmp: Automated dreamplace-based macro placement. In *Proceedings of the 2023 International Symposium on Physical Design*, pp. 149-157. 2023.

[11] Shi, Yunqi, Ke Xue, Song Lei, and Chao Qian. Macro placement by wire-mask-guided black-box optimization. *Advances in Neural Information Processing Systems* 36 (2024).

[12] Cheng, Ruoyu, and Junchi Yan. On joint learning for solving placement and routing in chip design. *Advances in Neural Information Processing Systems* 34 (2021): 16508-16519.

[13] Cheng, Ruoyu, Xianglong Lyu, Yang Li, Junjie Ye, Jianye Hao, and Junchi Yan. The policy-gradient placement and generative routing neural networks for chip design. *Advances in Neural Information Processing Systems* 35 (2022): 26350-26362.

[14] Geng, Zijie, Jie Wang, Ziyan Liu, Siyuan Xu, Zhentao Tang, Mingxuan Yuan, H. A. O. Jianye, Yongdong Zhang, and Feng Wu. Reinforcement Learning within Tree Search for Fast Macro Placement. In *Forty-first International Conference on Machine Learning*. 2024.

[15] He, Xu, Tao Huang, Linfu Xiao, Haitong Tian, and Evangeline FY Young. Ripple: A robust and effective routability-driven placer. *IEEE Transactions on Computer-Aided Design of Integrated Circuits and Systems* 32, no. 10 (2013): 1546-1556.

[16] Lin, Tao, Chris Chu, and Gang Wu. POLAR 3.0: An ultrafast global placement engine. In *2015 IEEE/ACM International Conference on Computer-Aided Design (ICCAD)*, pp. 520-527. IEEE, 2015.

[17] Lu, Jingwei, Pengwen Chen, Chin-Chih Chang, Lu Sha, Dennis J- H. Huang, Chin-Chi Teng, and Chung-Kuan Cheng. ePlace: Electrostatics based placement using Nesterov's method. In *Proceedings of the 51st Annual Design Automation Conference*, pp. 1-6. 2014.

[18] Cheng, Chung-Kuan, Andrew B. Kahng, Ilgweon Kang, and Lutong Wang. Replace: Advancing solution quality and routability validation in global placement. *IEEE Transactions on Computer-Aided Design of Integrated Circuits and Systems* 38, no. 9 (2018): 1717-1730.

[19] Liu, Lixin, Bangqi Fu, Shiju Lin, Jinwei Liu, Evangeline FY Young, and Martin DF Wong. Xplace: An Extremely Fast and Extensible Placement Framework. *IEEE Transactions on Computer-Aided Design of Integrated Circuits and Systems* (2023).

[20] Mnih, Volodymyr, Adria Puigdomenech Badia, Mehdi Mirza, Alex Graves, Timothy Lillicrap, Tim Harley, David Silver, and Koray Kavukcuoglu. Asynchronous methods for deep reinforcement learning. In *International conference on machine learning*, pp. 1928-1937. PMLR, 2016.

[21] Lehmann, Matthias. The Definitive Guide to Policy Gradients in Deep Reinforcement Learning: Theory, Algorithms and Implementations. *arXiv preprint arXiv:2401.13662* (2024).

[22] Lapan, Maxim. *Deep Reinforcement Learning Hands-On: Apply modern RL methods, with deep Q-networks, value iteration, policy gradients, TRPO, AlphaGo Zero and more.* Packt Publishing Ltd, 2018.

[23] Chadi, Mohamed-Amine, and Hajar Mousannif. Understanding reinforcement learning algorithms: The progress from basic Q-learning to proximal policy optimization. *arXiv preprint arXiv:2304.00026* (2023).

[24] Kahng, Andrew B., Ravi Varadarajan, and Zhiang Wang. Hier-RTLMP: A hierarchical automatic macro placer for large-scale complex IP blocks. *IEEE Transactions on Computer-Aided Design of Integrated Circuits and Systems* 43.5 (2023): 1552-1565.

2025 The 10th International Conference on Integrated Circuits and Microsystems

A 2.4 GHz Wide-Tuning-Range Low-Phase-Noise Differential Digitally Controlled Ring Oscillator for CMOS Image Sensor Applications

Zhigang Li[1], Hongjia Xu[1], Yao Nian[1], Jingyi Wang[1], Xin Li[1], Qiang Zhao[1*], Xiulong Wu[1*]

[1]*School of Integrated Circuits and Anhui Provincial High-Performance Integrated Circuit Engineering Research Center, Anhui University, Hefei 230601, China*
Corresponding author: zhaoqiang@ahu.edu.cn & xiulong@ahu.edu.cn

Abstract—**In this paper, a novel 2.4 GHz wide-tuning-range low-phase-noise differential Digitally Controlled Ring Oscillator (DCRO) is proposed for CMOS Image Sensor (CIS) applications. The DCRO integrates cascaded differential operational amplifiers with an embedded 8-bit digital-to-analog converter (DAC) and a 4-bit switched varactor array, effectively enhancing frequency resolution while minimizing area occupation and phase noise. Moreover, the proposed two-stage frequency tuning mechanism significantly improves frequency accuracy and linearity. Post-layout simulation results demonstrate that the proposed DCRO achieves a 0.617–2.709 GHz frequency range with a resolution of 1.56 MHz and a low phase noise of -103.7 dBc/Hz at 1 MHz frequency offset. Implemented in a customized 90 nm CMOS technology, the DCRO occupies a compact core area of 0.008 mm^2 and consumes only 3.564 mW from a 3.3 V supply voltage.**

Index Terms—**DCRO, CIS, differential ring oscillator, wide tuning range, resolution, varactor array, phase noise**

I. INTRODUCTION

With the wide application of CIS in consumer electronics, monitoring and autonomous driving, ADPLL functions as the key clock and frequency synthesis module in CIS, and its performance directly affects the image acquisition and processing accuracy of the whole system [1]. Due to its advantages of low power consumption, high integration and flexible digital correction, ADPLL has become the preferred frequency synthesis scheme in modern CIS systems. However, the frequency tuning accuracy, wide tuning range and the low phase noise of the core DCRO of ADPLL play a decisive role in the overall performance of the system.

Numerous studies have focused on improving the oscillator frequency resolution based on formula (1) in order to achieve a DCRO with low quantization noise:

$$\mathcal{L}\left(f_{\text{offset}}\right) = \frac{1}{12 f_R} \left(\frac{\Delta f}{f_{\text{offset}}}\right)^2 \left(\text{sinc}\frac{f_{\text{offset}}}{f_R}\right)^2 \quad (1)$$

where Δf, f_R and f_{offset} represent the frequency resolution, reference clock frequency and frequency offset, respectively, and $L(f_{\text{offset}})$ is the corresponding quantization noise [2]. A VCO with a wide tuning range from 110 MHz to 7.5 GHz based on a two-stage differential structure was proposed in [3]. Although the tuning range is wide, the prohibitive resolution of 6.06 GHz/V prevents it from further development, and

its phase noise is as high as -94.9 dBc/Hz at 1 MHz offset. Presented in [4], an LC oscillator-based DCO achieved a very fine frequency resolution of 0.37 - 0.81 MHz. However, its tuning range of 3.5 - 4.7 GHz still needs improvement.

Traditional oscillator designs, such as LC oscillators, can achieve lower phase noise [5]. However, they require large high-quality inductors and a more complex integration process, resulting in a limited tuning range and difficulty in meeting low-power requirements. In contrast, ring oscillators have become the preferred choice for DCO in ADPLL due to their compact footprint, seamless integration with digital circuits and wide tuning range. However, conventional ring oscillators still suffer from limitations in frequency resolution and tuning linearity, which can introduce quantization noise and PLL locking errors.

Various enhancement techniques have been proposed in recent years, including current-starved architectures, cross-coupled differential operational amplifier structures and multi-stage tunable capacitor arrays [6], [7], [8]. These methods have effectively improved the tuning precision and linearity of the DCRO, whereas it is still not adapted to the requirements of the CIS system.

To address these challenges, this work employs a inno-vative cascaded differential structure combined with a DAC [9] and a novel multistage tunable varactor array to achieve high-resolution and wide-tuning-range frequency control. This design significantly enhances the oscillator's linear response and suppresses quantization noise. Furthermore, by refining the frequency resolution of the DCRO, superior quantization noise performance is achieved while effectively mitigating phase noise. Additionally, a novel DAC-controlled voltage varactor array is introduced, which dynamically adjusts the port voltage to fine-tune the oscillation frequency, further enhancing precision and stability.

The rest of the paper is organized as follows: Section II details the proposed DCRO architecture and concrete circuit implementation. Section III presents the simulation results. Finally, the conclusion is drawn in Section IV.

979-8-3315-8850-2/25 $31.00 © 2025 IEEE

II. CIRCUIT IMPLEMENTATION

The differential delay unit typically comprises a matched pair of NMOS transistors forming the differential input, followed by a cross-coupled PMOS load circuit that establishes the necessary feedback. Each delay unit contributes a specific phase shift, which is determined by the device's transconductance, load capacitance and bias current. Theoretically, to satisfy the closed-loop startup condition, each stage of the delay unit should provide a phase shift of at least 60°. Once the cumulative phase shift reaches 360°, the loop fulfills the Barkhausen criterion, thereby enabling oscillation.

According to the Barkhausen criterion, two necessary conditions for the oscillator are as follows:

$$|H(j\omega)| \geq 1 \tag{2}$$

$$\angle H(j\omega) = 180° \tag{3}$$

The oscillation frequency of the differential ring oscillator is mainly determined by the total delay of the delay unit, which can be approximated as:

$$f_{osc} \approx \frac{1}{2N\tau} \tag{4}$$

As shown in formula (4), the frequency of an N-stage ring oscillator is determined by the number of delay units N and the delay time τ of each stage. As illustrated in Fig. 1, a differential ring oscillator with a tail current can be viewed as two coupled single-ended ring oscillators, which synchronize through mutual injection via the sources of the input transistor pair. In this configuration, M1 functions as a source follower, while M2 operates in a common-gate mode, enabling one ring oscillator to inject current into the other.

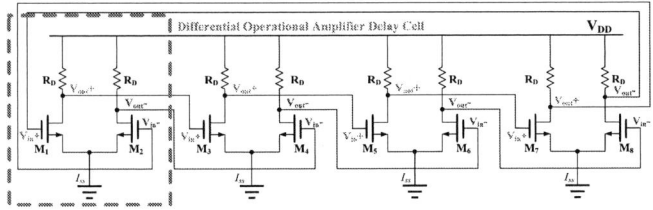

Fig. 1. Schematic diagram of simplified differential ring oscillator.

For these two single-ended loops to synchronize effectively while maintaining distinct oscillatory modes, the current injection must be sufficiently strong, necessitating the use of a high-impedance tail device. Additionally, the four-stage cascade of differential operational amplifiers facilitates the generation of multiple well-defined output phases. Specifically, a four-stage loop can produce eight phases with a 45° phase separation. Finally, the differential topology not only offers a smaller delay per stage compared to traditional inverter-based designs but also enhances overall frequency, making it a more efficient candidate for high-performance applications in CIS.

A. Differential Operational Amplifier Delay Cell

In Fig. 2, the circuit adopts a differential pair architecture featuring a switched varactor array, which is controlled by both the coarse tuning array and fine tuning voltage to regulate circuit capacitance. The overall design includes several key components: the DAC control circuit, resistors, variable capacitors, tail current source and differential pair transistors. The differential pair (M_1, M_2) functions as the core amplifier. The DAC control circuit generates precise tuning voltages based on an 8-bit digital code [0:7], which adjusts the switched varactor array.

To minimize the AM-PM effect, C_{tail} is critically set to 2 pF in Fig. 2. The DCRO core is biased using a cascoded current mirror (M_3, M_4), where the reference current $I_{bias} = 110\,\mu A$ is scaled by a couple of transistors whose width-to-length ratio is 2. To mitigate the impact of $1/f$ flicker noise from the current mirror, both M_3 and M_4 adopt an extended channel length of L = 2 μm.

Each stage of the operational amplifier follows a classic differential pair structure, with input transistors M_1 and M_2. A passive resistor (R_D) is used as the load to achieve higher gain. The current source (M_3, M_4) supplies the tail current, ensuring the amplifier operates in the optimal region. The small-signal gain A_v is determined by the transconductance g_m and R_D, expressed as follows:

$$A_v = -g_m R_D = -\frac{2I_{tail}R_D}{V_{ov}} \tag{5}$$

Here V_{ov} represents the overdrive voltage, while I_{tail} is supplied by the current source. After cascading four amplifier stages, the total loop gain becomes A_v raised to A_v^4, which must satisfy the Barkhausen criterion for sustained oscillation.

As shown in formula (4), the frequency of the DCRO is primarily influenced by the delay τ and the stages N. The delay τ of each operational amplifier stage is determined by the equivalent capacitance C_{eq} and the transconductance g_m, expressed as follows:

$$\tau \approx \frac{C_{eq}R_D}{g_m} \tag{6}$$

The equivalent capacitance C_{eq} is composed of the MOSFET gate capacitance, parasitic capacitances, the tunable varactors C_{var1}, C_{var2} and the adjustable varactor array. Consequently, the oscillation frequency can be further approximated as:

$$f_{osc} \approx \frac{g_m}{8C_{eq}R_D} \tag{7}$$

Therefore, by adjusting the select signal to coarsely regulate the varactor array and using the DAC to finely tune the varactor control voltage, the capacitance can be effectively modified. This allows for precise control over the oscillation frequency, ensuring that the oscillator meets the requirements of the ADPLL in a CMOS image sensor application.

979-8-3315-8850-2/25 $31.00 © 2025 IEEE

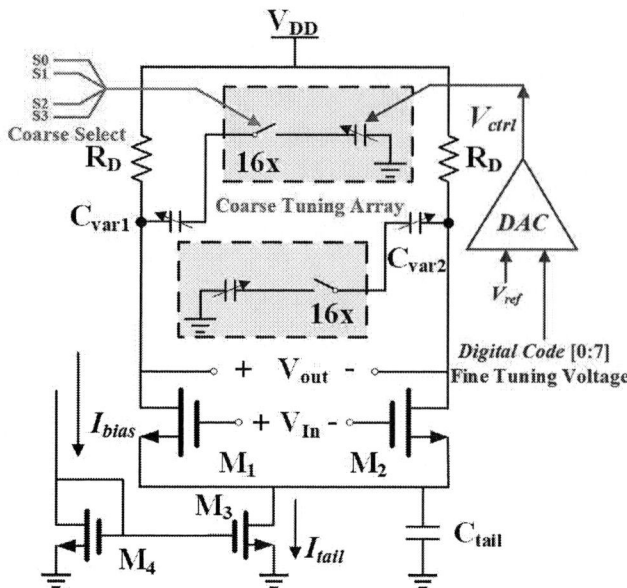

Fig. 2. Schematic diagram of the operational amplifier delay cell.

B. R-2R Ladder DAC Architecture

The proposed R-2R digital-to-analog converter (DAC) utilizes a resistive ladder network combined with MOS switches to achieve precise digital-to-analog conversion. As shown in Fig. 3, the circuit consists of an 8-bit R-2R ladder structure, where each bit is controlled by a digital control word (DCW[7:0]). The circuit operates based on a simple yet efficient current division principle, ensuring accurate and monotonic output voltage scaling.

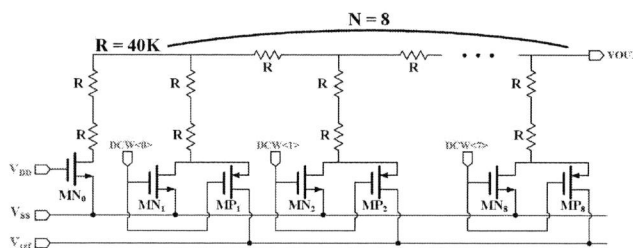

Fig. 3. Schematic diagram of R-2R ladder DAC.

The resistor values are selected as $R = 40\,\text{k}\Omega$, ensuring a well-matched impedance characteristic to maintain a balance between power consumption and speed. The resistive ladder structure consists of two key elements. NMOS transistors function as pull-down devices, selectively connecting nodes to V_{SS} based on the input digital word. PMOS transistors operate as pull-up devices to the reference voltage (V_{ref}).

The DAC operates based on the principle of voltage division through the resistor network. Each digital bit controls a switch that determines whether the corresponding current path contributes to the final output. The resistor network ensures that

each bit contributes a current proportional to its binary weight. Mathematically, the output voltage is given by:

$$V_{out} = V_{ref} \times \sum_{i=1}^{N} \frac{b_i}{2^i} \tag{8}$$

By integrating this R-2R DAC with the DCRO, fine-grained frequency tuning can be achieved, improving the resolution and reducing quantization noise in the system.

C. 4-Bit Switched Varactor Array

As illustrated in Fig. 4 (a), the coarse-tuning varactor array is implemented using a binary-weighted array, which allows for sequential regulation of capacitance. The design comprises a 4-bit digitally controlled varactor array, where each unit varactor is weighted in multiples of two (1×, 2×, 4×, 8×). A selection signal (Sel[3:0]) is used to connect or disconnect the varactor units in parallel, thereby regulating the total capacitance.

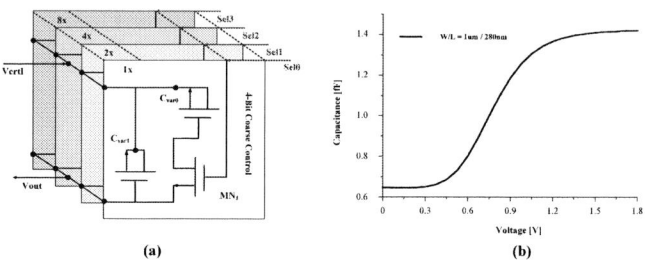

Fig. 4. (a) Schematic diagram of the switched varator array, (b) C-V variation curve of varactor.

Furthermore, in Fig. 4 (b), the capacitance of the MOS varactor exhibits a nonlinear variation with the control voltage V_{crtl}. Specifically, the capacitance increases from approximately 0.6 fF to 1.4 fF as V_{crtl} rises. And the voltage-dependent behavior of the MOS varactor is characterized by the following equation:

$$C_{var}(V_{ctrl}) = \frac{C_{ox}}{\sqrt{1 + \frac{V_{ctrl} - V_{FB}}{\phi}}} \tag{9}$$

The average voltage across the varactor is approximately equal to the common-mode voltage V_{out} minus V_{ctrl}. Consequently, when V_{ctrl} varies, the voltage drop across the varactor ($V_{out} - V_{ctrl}$) changes accordingly, which in turn alters the overall capacitance. Therefore, this modulation behavior facilitates fine frequency tuning which contributes to a better resolution. When combined with the coarse-tuning varactor array, it ensures that the DCRO meets the stringent frequency tuning requirements imposed by the ADPLL in CMOS image sensors.

III. POST LAYOUT SIMULATION RESULTS

As shown in Fig. 5, the layout of the four-stage DCRO is designed based on the customized 90nm CMOS process technology, including the DAC module and the core ring oscillator with an total area of $145.06\,\mu\text{m} \times 55.23\,\mu\text{m}$. In addition, the DCRO consumes 3.564 mW from a 3.3 V supply voltage.

979-8-3315-8850-2/25 $31.00 © 2025 IEEE   124

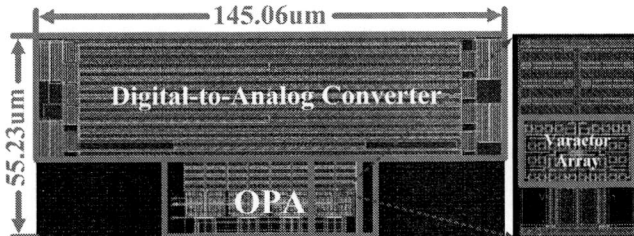

Fig. 5. Layout of the ring DCRO.

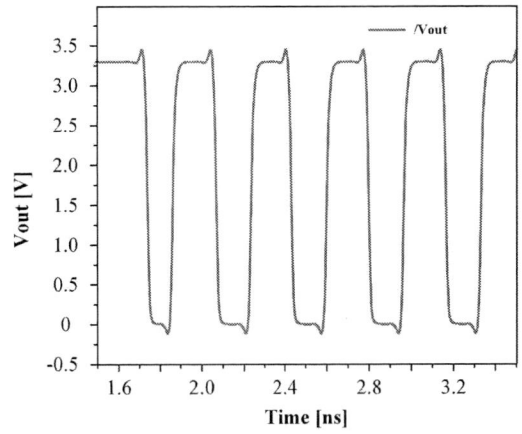

Fig. 7. Transient signal waveform at 2.4 GHz.

As depicted in Fig. 6, the DCRO demonstrates a wide frequency tuning range from 0.617 GHz to 2.709 GHz. The DCRO frequency resolution achieves 1.56 MHz/LSB at highest oscillation frequency curve (state: 4'b 0000) and 0.91 MHz/LSB at lowest oscillation frequency curve (state: 4'b 1111), enabling precise frequency regulation across different input code (0 to 255).

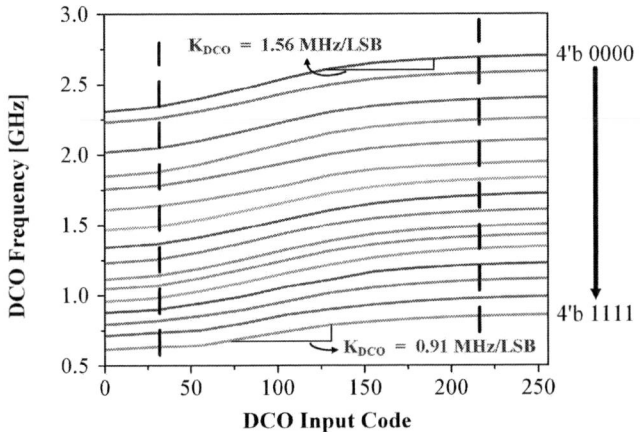

Fig. 6. The simulated DCRO tuning range.

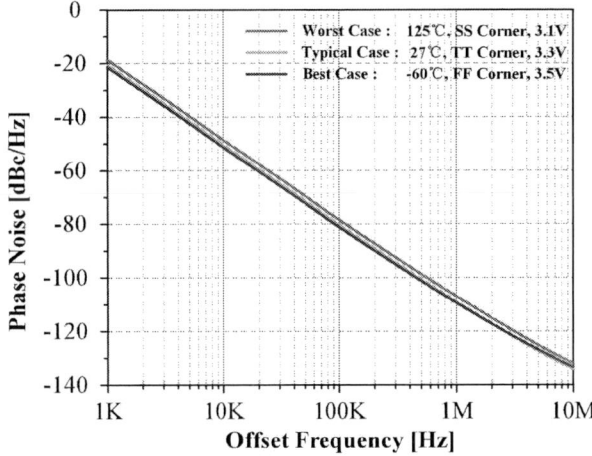

Fig. 8. Phase noise diagram at different process corners.

Fig. 7 illustrates the transient output voltage waveform at a frequency of 2.4 GHz (state: 4'b 0001, Code = 162) and a voltage swing ranging from 0 V to 3.3 V, while the I_{bias} is 110 μA. The sharp transitions and consistent periodicity validate the design's capability to generate high-quality oscillating signals.

For phase noise performance, Fig. 8 presents the results across different cases at a frequency of 2.4 GHz. Notably, the phase noise at 1 MHz offset frequency achieves -101.7 dBc/Hz (125°C, SS corner, 3.1V), -103.7 dBc/Hz (27°C, TT corner, 3.3V) and -104.2 dBc/Hz (27°C, FF corner, 3.3V), respectively. The consistent low-phase-noise performance across different process corners highlights the design's robustness and reliability, meeting the stringent requirements for high-frequency applications.

Summarized in Table I, the reported work offers the best frequency range and FoM. In comparison with [10], [12], [13], this work's power consumption is slightly higher due to its 3.3 V power supply, while the phase noise performance is superior. Compared to [12], [13], the area is relatively

larger due to the adoption of DAC module, which is adapted for the ADPLL in the CIS system. Therefore, compared to other works, the proposed system achieves the best performance–area trade-off.

IV. CONCLUSION

In this paper, a low-power DCRO and its internal circuit structure are proposed and verified in customized 90 nm CMOS technology. The proposed DCRO achieves a frequency tuning range from 0.617 GHz to 2.709 GHz with a fine frequency resolution of 0.91 MHz/LSB at lowest frequencies and 1.56 MHz/LSB at highest frequencies. Moreover, the DCRO achieves a phase noise of -103.7 dBc/Hz at 1 MHz offset for a 2.4 GHz operating frequency, while consuming 3.56 mW of power.

Overall, the proposed DCRO exhibits a favorable balance between low phase noise, wide tuning range and high resolution, making it a promising candidate for ADPLL applications. More effective works and optimization methods can be further explored in the future to achieve better power efficiency and

979-8-3315-8850-2/25 $31.00 © 2025 IEEE

TABLE I: Performance Comparison

References	[10] 2018	[11] 2021	[12] 2023	[13] 2024	**This work**
Process(nm)	65	180	180	180	**90**
V_{DD}(V)	0.6	2	1.8	1.8	**3.3**
Frequency (GHz)	0.43 - 0.55	0.59 - 1.27	2.27 - 3.6	0.9 - 2.65	**0.62 - 2.7**
FTR(%)	25.2	73.11	45.31	98.6	**125.3**
PN@1MHz (dBc/Hz)	-94.8	-109	-90.74	-93.1	**-103.7**
FoM* (dBc/Hz)	-162.1	-153.8	-158.6	-156.7	**-171.2**
Area (mm²)	0.21	0.059	0.0008	0.004	**0.008**
Power (mW)	0.045	14.4	1.4	1.89	**3.56**

$$* \ FoM = PN - 20\log\left(f_{osc}/f_{offset}\right) + 10\log\left(P_{DC}/1\,mW\right)$$

phase noise performance to meet the stringent requirements of the CIS system.

ACKNOWLEDGMENT

This work was supported by the National Natural Science Foundation of China under Grant 62274001 and the University Synergy Innovation Program of Anhui Province under Grant GXXT-2023-013.

REFERENCES

[1] B. Razavi, "The Role of PLLs in Future Wireline Transmitters," in *IEEE Transactions on Circuits and Systems I: Regular Papers*, vol. 56, no. 8, pp. 1786-1793, Aug. 2009.

[2] R. B. Staszewski et al., "All-digital PLL and transmitter for mobile phones," in *IEEE Journal of Solid-State Circuits*, vol. 40, no. 12, pp. 2469-2482, Dec. 2005.

[3] Y. Liu, X. Wang, S. Yang and X. Yue, "Design of Quadrature Output Two-stage Differential Ring VCO," *2022 7th International Conference on Integrated Circuits and Microsystems (ICICM)*, Xi'an, China, 2022, pp. 219-224.

[4] K. -U. Cho et al., "A 3.5 to 4.7-GHz Fractional-N ADPLL With a Low-Power Time-Interleaved GRO-TDC of 6.2-ps Resolution in 65-nm CMOS Process," in *IEEE Access*, vol. 12, pp. 142677-142694, 2024.

[5] Z. Li, D. Cordeau, J. -M. Paillot, S. Charpentier, M. Lecuyer, F. Huin, "Analysis and design of K-band low-phase-noise differential DCOs implemented in 22 nm FD-SOI for 76–81 GHz automotive radars", *Microelectronics Journal*, vol. 136, 2023.

[6] M. Z. Jahangir and C. S. Paidimarry, "Design of a Novel Charge Pump based Current Starved Ring Oscillator with Reduced Phase Noise," *2023 International Conference for Advancement in Technology (ICONAT)*, Goa, India, 2023.

[7] V. Jangra and M. Kumar, "A 0.28 GHz to 3.84 GHz low power differential ring oscillator design using cross-coupled transistors for radio frequency identification (RFID)," *2021 9th International Conference on Reliability, Infocom Technologies and Optimization (Trends and Future Directions) (ICRITO)*, Noida, India, 2021, pp. 1-5.

[8] W. -C. Lai, "Design of 1V CMOS 5.8 GHz VCO with Switched Capacitor Array Tuning for Intelligent Sensor Fusion," *2020 International Conference on Advanced Robotics and Intelligent Systems (ARIS)*, Taipei, Taiwan, 2020, pp. 1-4.

[9] M. Wang, Q. Li and J. Li, "A Low-power High-precision Differential Output 13bit R-2R DAC Pathway in 180nm CMOS," *2024 9th International Conference on Integrated Circuits and Microsystems (ICICM)*, Wuhan, China, 2024, pp. 852-856.

[10] O. -Y. Jung, H. -G. Seok, A. Dissanayake and S. -G. Lee, "A 45- μ W, 162.1-dBc/Hz FoM, 490-MHz Two-Stage Differential Ring VCO Without a Cross-Coupled Latch," in *IEEE Transactions on Circuits and Systems II: Express Briefs*, vol. 65, no. 11, pp. 1579-1583, Nov. 2018.

[11] N. Ghaderi, M. Zhang, D. Yu and L. Lorenzelli, "A New Low Power Ring Voltage-Controlled Oscillator with a Wide Tuning Range," *2021 9th International Electrical Engineering Congress (iEECON)*, Pattaya, Thailand, 2021, pp. 293-296.

[12] M. Alijani, M. Javanmardi and A. Abrishamifar, "A Low-Power Differential Ring VCO Using An Active Inductor For Wireless Applications," *2023 5th Iranian International Conference on Microelectronics (IICM)*, Tehran, Iran, Islamic Republic of, 2023, pp. 150-154.

[13] Z. Ding, J. Zhang, A. Guo and L. Yin, "Design of a Novel Low-Power Differential Ring Voltage Controlled Oscillator," *2024 9th International Conference on Integrated Circuits and Microsystems (ICICM)*, Wuhan, China, 2024, pp. 473-478.

979-8-3315-8850-2/25 $31.00 © 2025 IEEE

Smart Wireless Insole with Self-Powered Triboelectric Pressure Sensors for Gait Detection

Yuhan Wang
College of Health Science and Environmental Engineering
Shenzhen Technology University
Shenzhen 518118,China
202300502145@stumail.sztu.edu.cn

Yue Yao
College of Integrated Circuits and Optoelectronic Chips
Shenzhen Technology University
Shenzhen 518118,China
202301202059@stumail.sztu.edu.cn

Enze Li
College of Health Science and Environmental Engineering
Shenzhen Technology University
Shenzhen 518118,China
202300502159@stumail.sztu.edu.cn

Sifan Liang
College of Integrated Circuits and Optoelectronic Chips
Shenzhen Technology University
Shenzhen 518118,China
202200304048@stumail.sztu.edu.cn

Lei Li*
College of Integrated Circuits and Optoelectronic Chips
Shenzhen Technology University
Shenzhen 518118,China
lilei@sztu.edu.cn

Abstract—**Plantar pressure monitoring is a valuable tool for assessing gait and foot health, but conventional platforms are bulky, costly, and unsuitable for daily use. Existing rehabilitation insoles also face limitations such as large sensor size, unstable power supply, and poor portability. This paper presents a wireless self-powered smart insole system that integrates a multilayer polytetrafluoroethylene-copper (PTFE-Cu) triboelectric nanogenerator (TENG) array, low-power wireless electronics, and machine learning algorithms for real-time gait monitoring. The flexible TENG array, distributed across key plantar regions, enables spatially resolved pressure sensing without external power. Collected signals are transmitted via Bluetooth Low Energy (BLE) to a microcontroller and analyzed through intelligent algorithms to recognize normal gait as well as abnormal patterns. Experimental demonstrations confirm that the system can accurately capture plantar pressure waveforms, visualize gait dynamics, and classify abnormal gait with high reliability. By combining flexibility, portability, and autonomous power supply, the proposed smart insole provides a multifunctional platform for daily gait health management. It holds strong potential for preventive healthcare, clinical auxiliary diagnosis, rehabilitation assessment, and sports performance monitoring.**

Keywords—*smart insole, self-powered pressure sensor, gait analysis, wearable healthcare*

I. INTRODUCTION

Gait analysis is a critical noninvasive diagnostic approach for assessing motor function and identifying abnormalities associated with the musculoskeletal, nervous, and metabolic systems[1]. Irregular plantar pressure distribution is often an early indicator of disorders such as flatfoot, lumbar disc degeneration, and diabetic foot[2]. Therefore, continuous and precise gait monitoring plays an essential role in early intervention, rehabilitation tracking, and preventive healthcare. However, existing plantar pressure measurement platforms are largely restricted to laboratories or clinical institutions.

Capacitive sensor-based systems suffer from circuit complexity and poor anti-interference capability, while resistive sensors[3,4], though more robust[5], exhibit nonlinear pressure responses and reduced stability under long-term usage[6]. Furthermore, conventional systems are bulky, expensive, and dependent on external power sources, making them unsuitable for portable and daily use[7,8]. These limitations hinder long-term dynamic monitoring in real-life environments, especially for elderly populations, rehabilitation patients, and sports enthusiasts who require continuous gait assessment. Recent progress in self-powered sensing technologies, particularly triboelectric nanogenerators (TENGs), has opened new opportunities to overcome these barriers. TENGs can harvest biomechanical energy generated during walking and standing, converting it into stable, pressure-dependent electrical signals while simultaneously addressing the persistent power supply challenge in wearable systems[9].

In this study, we present a wireless self-powered smart flexible insole system that integrates multilayer polytetrafluoroethylene (PTFE)-copper TENG array, low-power wireless electronics, and machine learning algorithms for real-time gait monitoring. The system provides spatially resolved plantar pressure sensing without external power, wireless data transmission for continuous and portable use, and intelligent algorithms for distinguishing both normal and pathological gait patterns, including club foot and foot drop. Furthermore, a user-friendly software platform is developed to support intuitive visualization, remote monitoring, and personalized health management. By bridging advanced self-powered sensing with intelligent data analytics, the proposed system extends gait analysis from laboratory-based measurement to daily life applications, offering significant potential in preventive healthcare, clinical diagnosis, and rehabilitation assessment.

979-8-3315-8850-2/25 $31.00 © 2025 IEEE

II. EXPERIMENTAL METHODS

A. Stucture of Self-Powered Sensor and Smart Insole

The proposed smart insole integrates a distributed array of triboelectric pressure sensors together with a microcontroller unit (MCU) and a wireless transceiver. Each pressure sensor adopts a five-layer stacked structure consisting of alternating PTFE film and copper foil. PTFE, with its strong electronegativity and mechanical flexibility, serves as the triboelectric layer, while copper foils act as electrodes, ensuring high conductivity and structural durability. The laminated multilayer design increases the effective charge transfer interfaces, thereby enhancing the voltage output of each TENG unit. The sensors are strategically distributed across key plantar regions—including the heel, arch, and forefoot—to capture localized pressure dynamics during walking or standing. The generated signals are collected by the MCU, processed, and then wirelessly transmitted to terminal devices via a Bluetooth Low Energy (BLE) module. Owing to the inherent self-powered nature of the TENG sensors, the system eliminates the dependence on bulky external batteries, enabling lightweight design and long-term continuous operation in daily scenarios.

B. Machine Learning–Based Gait Detection

Collected signals undergo preprocessing (denoising, interpolation, and normalization), followed by feature extraction (root mean square values, peaks, and temporal trends). Feature vectors from both feet are compared with a reference dataset through dot-product evaluation and synthesized to generate gait classification scores. The framework distinguishes between normal gait and pathological patterns such as club foot and foot drop with high accuracy.

C. Visualization and User Interface

The companion desktop application supports real-time gait pressure maps, waveform analysis, and statistical summaries (e.g., RMS bar charts). Gait patterns are visually distinguished from normal distributions, providing clinicians and users with intuitive diagnostic feedback and rehabilitation progress tracking.

III. RESULTS AND DISCUSSION

A. Design of the Self-powered Smart Insole

Current plantar pressure monitoring still primarily relies on fixed platforms deployed in laboratories or medical institutions, making it difficult to achieve long-term dynamic assessment in daily scenarios[7,8]. This limitation is particularly evident in continuous health management for elderly populations, sports enthusiasts, and the general public. On the other hand, capacitive sensors are restricted by weak anti-interference capability and complex circuit structures, which limit their applicability[3,4]; resistive sensors, although offering better interference resistance[5], generally suffer from nonlinear pressure responses and poor long-term stability[6]. In addition, in-shoe sensing systems face persistent challenges of power supply, preventing continuous long-term use. To address these challenges, we have developed a wireless self-powered smart insole system, as illustrated in Fig. 1. Flexible self-powered pressure sensors are distributed across different regions of the insole, enabling spatially resolved plantar pressure monitoring

through the triboelectric effect generated during walking or standing. By integrating the self-powered pressure sensors with a low-power wireless transmission module, the system not only effectively resolves the long-term power supply issue and supports continuous daily use, but also enables real-time transmission of gait data to terminal devices for visualization and remote monitoring.

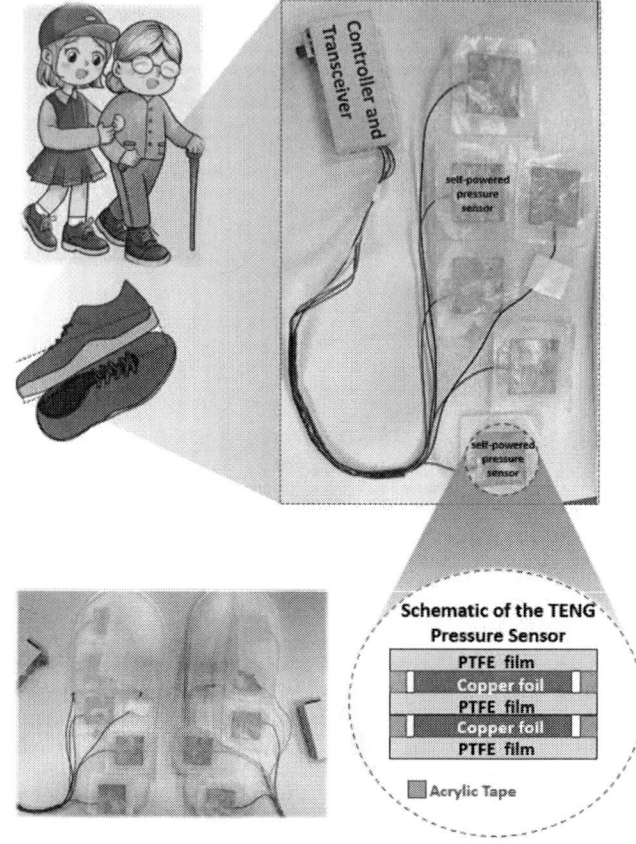

Fig. 1. Design of the Self-powered Smart Insole

In terms of material design, the insole employs 0.05 mm-thick PTFE film and copper foil as the triboelectric material and electrode, respectively. PTFE, with strong electronegativity in the triboelectric series, offers a balance between flexibility and mechanical strength at this thickness, while copper foils provide excellent conductivity and flexibility. The pressure sensor operates on the principle of TENG and adopts a five-layer stacked structure to enhance voltage output. Specifically, the first, third, and fifth layers are PTFE, while the second and fourth layers are copper foils, bonded with acrylic adhesive, and connected to electrodes via copper leads. Fig. 1 shows the physical prototype of the system and the five-layer configuration of the self-powered pressure sensor. The sensor array covers the heel, medial and lateral arches, and hallux, enabling distributed multi-point plantar pressure sensing.

The collected pressure signals are processed by a MCU and transmitted via a BLE module to terminal devices, supporting multiple visualization modes such as real-time plantar pressure maps. To enhance comfort, portability, and stability, the lithium battery and PCB are integrated into a compact square enclosure

979-8-3315-8850-2/25 $31.00 © 2025 IEEE

that can be fixed to the ankle. Finally, combined with intelligent algorithms, the system successfully captures and classifies various gait patterns, including club foot and foot drop.

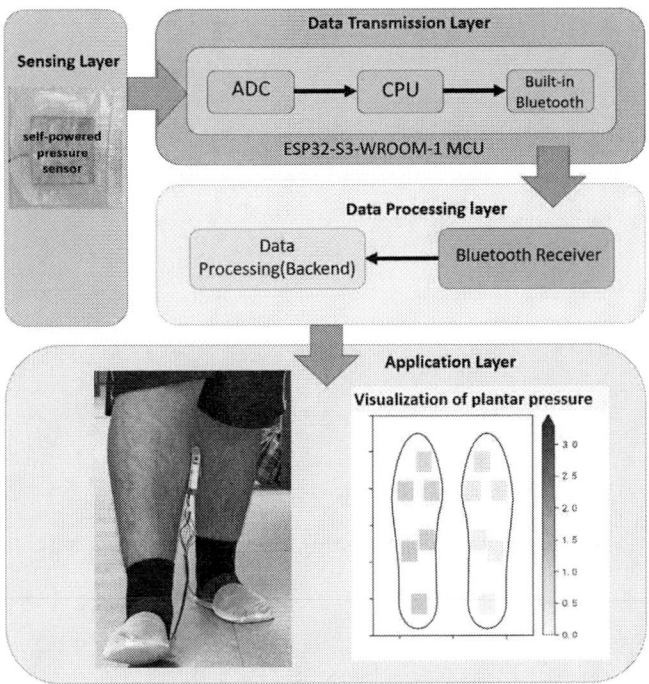

Fig. 2. Schematic of the layered architecture and workflow for the self-powered plantar pressure monitoring system

As shown in Fig. 2, the overall system architecture consists of four tightly coupled layers, including the sensing layer (TENG pressure sensor array), the data transmission layer (ESP32-based PCB module), the processing layer (algorithmic analysis), and the application layer (PC-based visualization software). These layers work in close collaboration to achieve data acquisition, transmission, analysis, and presentation. The hardware design is centered on the ESP32-WROOM-32D module, which integrates signal conditioning circuits and a power management unit. A 12-bit ADC operating at a 200 Hz sampling rate is used for signal acquisition. Data are transmitted via the Bluetooth 5.0 protocol with an effective range of approximately 10 m, which is ample to cover most indoor scenarios (such as home or office environments) as well as short-distance outdoor uses like walking in a park or along a neighborhood street. The system exhibits low-power characteristics, capable of continuous operation for over three hours with a 200 mAh lithium battery, sufficient to support typical daily activities, including morning walks, commuting, and extended periods of household movement monitoring. Together, these features ensure stable, efficient, and real-time gait data acquisition, processing, and transmission, which is sufficient for daily life use.

B. Work Mechanism of the TENG Pressure Sensor

As shown in Fig. 1, the pressure sensor adopts an alternating stacked structure of PTFE and copper foil as the triboelectric materials and electrodes, respectively. The first, third, and fifth layers are PTFE films, while the second and fourth layers are copper foils that serve as electrodes. The top PTFE layer is exposed on the insole surface, allowing direct contact with the user's plantar skin or socks[10]. Fig. 3 elucidates the work mechanism of the triboelectric pressure sensor. During walking or standing, repeated contact and separation between the plantar surface and PTFE layer induce triboelectric charging. Due to differences in electron affinity, electrons transfer from the skin or textile surface to PTFE, resulting in negative charges on the PTFE and positive charges on the skin or sock surface. When the foot lifts and separation occurs, the interfacial electrostatic field changes, inducing free charges to redistribute on the copper electrodes and generating a displacement current in the external circuit[11]. This contact-separation cycle is repeated with every step, thereby converting the cyclic mechanical energy of plantar pressure into electrical signals. The multilayer architecture increases the number of charge transfer interfaces, significantly enhancing voltage output, while the 0.05 mm PTFE film ensures sufficient flexibility to conform to plantar deformation and adequate mechanical stability for reliable operation[12]. Additionally, the PTFE film exhibits good moisture resistance, effectively withstanding environmental moisture to maintain stable performance[13]. Ultimately, TENG units distributed across different regions of the insole generate localized electrical signals corresponding to plantar pressure distribution, which are collected and transmitted by the MCU for gait recognition and analysis[14]. Future work will further investigate the PTFE film's performance under varying humidity levels and other environmental conditions to broaden its applicability.

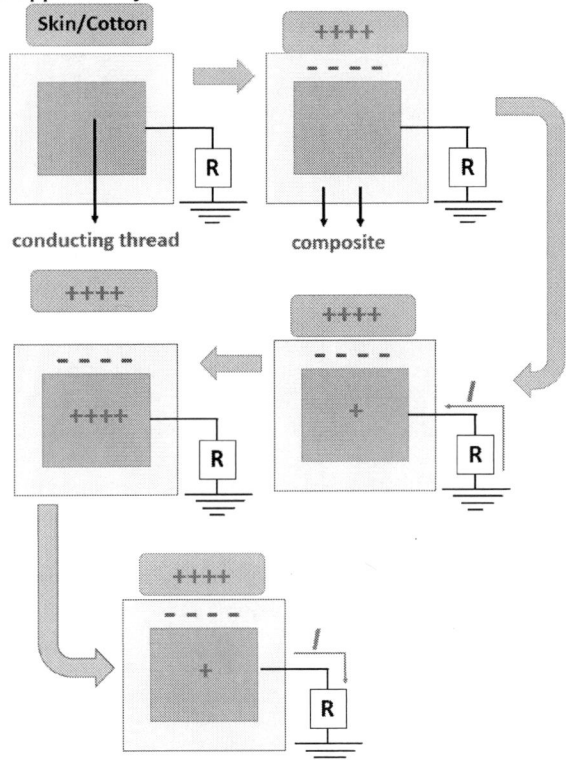

Fig. 3. Working principle of the triboelectric nanogenerator (TENG), comprising skin, a conducting thread, and a composite dielectric layer (labeled "composite" in the diagram)

C. Machine Learning Integration for Motion Detection in Smart Insole

The gait signals collected by the hardware are further processed through machine learning algorithms. The analysis begins with a preprocessing stage, including decoding, encapsulation, and data import. Noise reduction and interpolation are then applied to obtain clean signals, followed by feature extraction, where effective channel values, intra-cycle peaks, periods, and temporal variation trends are calculated. These features are used to construct a standardized dataset of multiple gait patterns for subsequent analysis. On one hand, for each gait sample (gait 1, gait 2, ..., gait n), root mean square (RMS) vectors are generated for both the left and right foot and normalized. Feature vector groups are then constructed using the successive difference method combined with Gram-Schmidt orthogonalization, and stored in the reference dataset. On the other hand, the measured plantar pressure data are parsed to obtain normalized RMS vectors for both feet, which are then compared with the reference "foot–gait" feature vector groups through dot-product operations. The dot-product results are synthesized via RMS to yield an evaluation score for each "foot-gait" pair. The scores from the left and right foot are further combined to generate a comprehensive evaluation value for each gait. Finally, an evaluation function (e.g., inverse proportional or power function) is applied to amplify differences among evaluation values, enabling the system to produce the final score. Based on this score, the dominant gait and concomitant gait are identified. Prior to practical gait analysis, multiple gait types must be collected and stored in the standard dataset. During operation, the measured data are compared against the reference dataset, and the gait corresponding to the highest score is taken as the final classification result. Experimental evaluations show that this method achieves an accuracy of approximately 80%.

D. Demonstration and Visualization of Gait Characteristics

To validate the practicality of the proposed system, the smart insole was worn by a user for real-time gait monitoring and analysis. The gait waveform plots intuitively illustrate the temporal variations of plantar pressure across different regions, with each colored curve corresponding to the output of a sensor located at a specific plantar site. As shown in Fig. 4a and Fig. 4b, under normal gait conditions, both feet exhibit periodic and coordinated pressure patterns, and the forefoot, arch, and heel undergo sequential loading and unloading, resulting in a relatively balanced distribution. The outputs of the plantar sensor array are stable and synchronized, reflecting the regularity of normal gait. In contrast, clear abnormalities are detected under pathological gait conditions. As illustrated in Fig. 4c, in the case of clubfoot, the amplitude of Channel 2 (C2, orange curve) is significantly reduced compared to normal gait, with lower peak values and smoother fluctuations. This indicates insufficient pressure variation and uneven plantar load distribution during the gait cycle. In foot drop (Fig. 5d), Channel 1 (C1, blue curve) exhibits abnormally high amplitude, while other channels (orange and green curves) show negligible changes. This demonstrates that pressure variations are

dominated by a single plantar region (C1), with minimal contribution from other regions, resulting in extremely unbalanced load distribution.

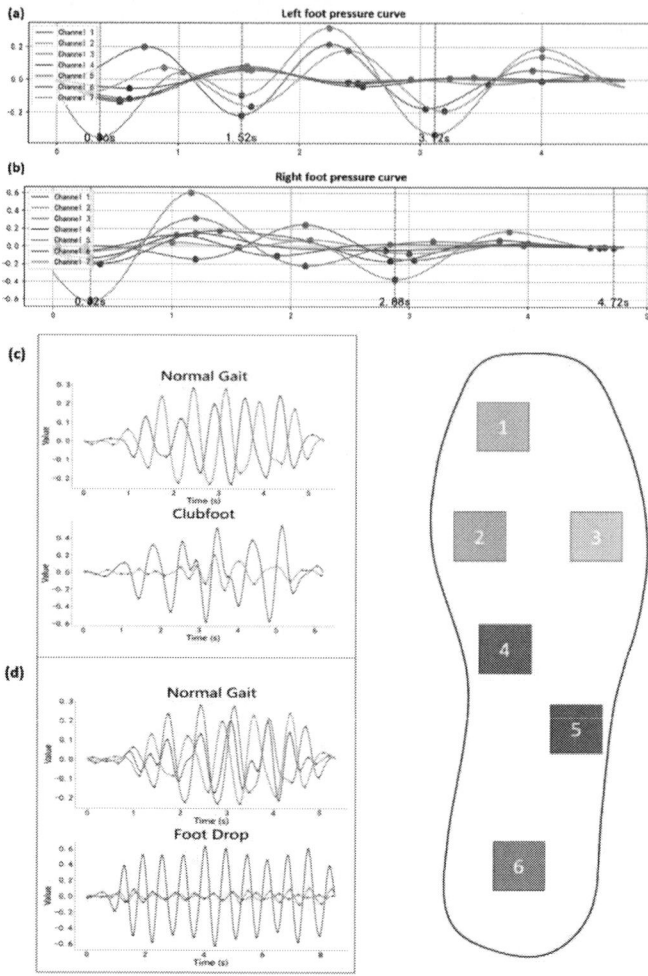

Fig. 4. Pressure variation curves of normal and abnormal gaits (including clubfoot and foot drop) over time for the (a) left foot and (b)right foot. (c, d) Comparative gait waveforms, and distribution of pressure sensor areas in the plantar pressure detection system

Further statistical analysis was performed using root mean square (RMS) values of the pressure signals, as shown in Fig. 5. RMS reflects the "energy" or average amplitude of the signal, thereby highlighting differences in pressure distribution across channels (C 1 to C6). For normal gait, the RMS values are relatively uniform, with no single channel dominating; C5 and C3 are slightly higher, but C1 and C6 also remain within a reasonable range, indicating evenly distributed plantar pressure variations. In contrast, clubfoot shows an uneven RMS distribution, where C3 (~0.24) is markedly dominant and C2 and C1 are also elevated, suggesting intensified pressure in multiple plantar regions, particularly in C3. In foot drop, the RMS value of C1 is extremely high, while those of C2 to C6 remain very low, reflecting highly localized energy concentration in a single plantar region, with other regions barely contributing. These experimental results confirm that the developed smart insole system can effectively monitor and differentiate normal and abnormal gait patterns in real time, demonstrating its potential for clinical auxiliary diagnosis and rehabilitation assessment.

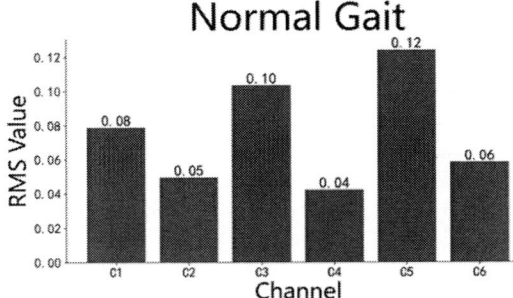

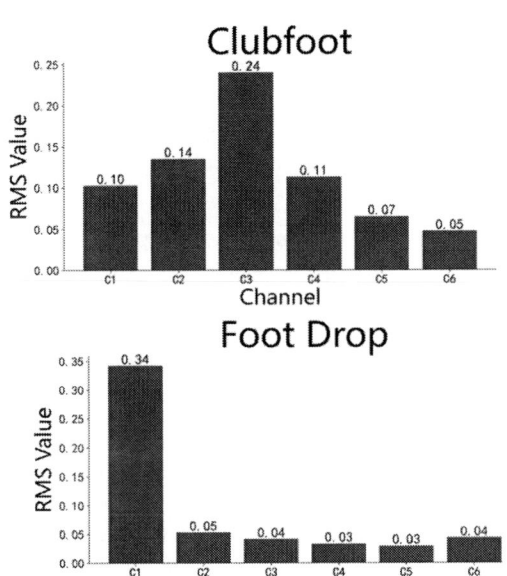

Fig. 5. Bar charts of Root Mean Square (RMS) values for normal gait, clubfoot, and foot drop across six channels (C1–C6)

IV. CONCLUSION

In summary, we developed a wireless self-powered smart insole system that integrates flexible TENG-based pressure sensors, low-power wireless electronics, and machine learning algorithms to enable real-time gait monitoring. The multilayer PTFE-copper TENG design achieves spatially resolved plantar pressure sensing while simultaneously addressing long-term power supply challenges, thereby supporting continuous daily monitoring. Experimental evaluations demonstrate that the system can reliably capture plantar pressure waveforms, visualize gait distribution, and identify abnormal gait patterns such as clubfoot and foot drop using RMS-based statistical analysis and intelligent classification, achieving promising performance on the test dataset. With its flexibility, portability, and autonomy, the proposed insole is well suited for continuous daily monitoring and remote visualization, particularly benefiting populations requiring long-term gait assessment, such as post-surgical rehabilitation patients and elderly individuals. This work provides a promising pathway toward preventive healthcare, clinical diagnosis, and rehabilitation assessment.

ACKNOWLEDGMENT

This study was supported by Shenzhen Scientific and Technological Foundation (No. RCYX20231211090332037, JCYJ20240813160211015), National Natural Science Foundation of China (No. 62474008, 62204007), and Guangdong Provincial Natural Science Foundation (No. 2024A1515030044).

REFERENCES

[1] J. Nonnekes, R. J. M. Goselink, E. Růžička, A. Fasano, J. G. Nutt, and B. R. Bloem, "Neurological disorders of gait, balance and posture: a sign-based approach," Nature Reviews Neurology, vol. 14, pp. 178–189, 2018.

[2] E. C. Katoulis, M. Ebdon-Parry, H. Lanshammar, L. Vileikyte, J. Kulkarni, and A. J. M. Boulton, "Gait abnormalities in diabetic neuropathy," Diabetes Care, vol. 20, p. 1904, 1997.

[3] D. Yoo, D.-J. Won, W. Cho, J. Lim, and J. Kim, "Double-side electromagnetic interference-shielded bending-insensitive capacitive-type flexible touch sensor with linear response over a wide detection range," Adv. Mater. Technol., vol. 6, p. 2100358, 2021.

[4] S. R. A. Ruth, V. R. Feig, M.-G. Kim, Y. Khan, J. K. Phong, and Z. Bao, "Flexible fringe effect capacitive sensors with simultaneous high-performance contact and non-contact sensing capabilities," Small Struct., vol. 2, p. 2000079, 2021.

[5] N. Luo, W. Dai, C. Li, Z. Zhou, L. Lu, and C. C. Y. Poon, et al., "Flexible piezoresistive sensor patch enabling ultralow power cuffless blood pressure measurement," Adv. Funct. Mater., vol. 26, pp. 1178–1187, 2016.

[6] A. Dzedzickis, E. Sutinys, V. Bucinskas, U. Samukaite-Bubniene, B. Jakstys, and A. Ramanavicius, et al., "Polyethylene-carbon composite (Velostat®)-based tactile sensor," Polymers, vol. 12, p. 2905, 2020.

[7] M. J. Hessert, M. Vyas, J. Leach, K. Hu, L. A. Lipsitz, and V. Novak, "Foot pressure distribution during walking in young and old adults," BMC Geriatr., vol. 5, p. 8, 2005.

[8] C. Xu, X.-X. Wen, L.-Y. Huang, L. Shang, X.-X. Cheng, and Y.-B. Yan, et al., "Normal foot loading parameters and repeatability of the Footscan® platform system," J. Foot Ankle Res., vol. 10, p. 30, 2017.

[9] Q. Yang, M. Yu, H. Zhang, N. Li, J. Du, and L. Xu, et al., "Triboelectric nanogenerator based on well-dispersed and oxide-free liquid metal-doped conductive hydrogel as self-powered wearable sensor for respiratory and thyroid cartilage signal monitoring," Nano Energy, vol. 134, p. 110530, 2025.

[10] F. R. Fan, W. Tang, and Z. L. Wang, "Flexible nanogenerators for energy harvesting and self-powered electronics," Adv. Mater., vol. 28, p. 4283, 2016.

[11] Z. L. Wang, "On Maxwell's displacement current for energy and sensors: the origin of nanogenerators," Mater. Today, vol. 20, p. 74, 2017.

[12] Y. C. Lai, Y. C. Hsiao, H. M. Wu, and Z. L. Wang, "Waterproof fabric-based multifunctional triboelectric nanogenerator for universally harvesting energy from raindrops, wind, and human motions and as self-powered sensors," Adv. Sci., vol. 6, p. 1801883, 2019.

[13] G. J. Puts, P. Crouse, B. M. Ameduri, "Polytetrafluoroethylene: synthesis and characterization of the original extreme polymer," Chem. Rev., vol. 119, pp. 1763–1805, 2019.

[14] Y. Yang, X. Guo, M. Zhu, Z. Sun, Z. Zhang, and T. He, et al., "Triboelectric nanogenerator enabled wearable sensors and electronics for sustainable internet of things integrated green earth," Adv. Energy Mater., vol. 13, p. 2203040, 2023.

2025 The 10th International Conference on Integrated Circuits and Microsystems

Voltage-Time Hybrid Domain ADC with PVT Tracking for High-Frame-Rate CMOS Image Sensor

Xin Li
Anhui University
School of Integrated Circuits
Hefei, China
lixin@ahu.edu.cn

Ziheng Wang
Anhui University
School of Integrated Circuits
Hefei, China
18801258612@163.com

Bin Qiang
Anhui University
School of Integrated Circuits
Hefei, China
ahu_qb@163.com

Chunyu Peng
Anhui University
School of Integrated Circuits
Hefei, China
cyupeng@ahu.edu.cn

Qiang Zhao*
Anhui University
School of Integrated Circuits
Hefei, China
zhaoqiang@ahu.edu.cn

Xiulong Wu
Anhui University
School of Integrated Circuits
Hefei, China
xiulong@ahu.edu.cn

Abstract—This paper proposes a high-speed, area-efficient an alog-to-digital converter (ADC) tailored for CMOS image sensors (CIS). The architecture integrates a Successive Approximation R egister (SAR) ADC for coarse voltage-domain quantization and a Time-to-Digital Converter (TDC) to achieve fine time-domain qu antization. In this scheme, the inherent process-voltage-temperat ure (PVT) robustness is enhanced by establishing a current-based relationship between the Voltage-to-Time Converter (VTC) and TDC, thereby reducing power consumption and circuit complexit y. The prposed ADC is designed in a Nyquist sampling rate of 20 MS/s in 65nm CMOS process, achieving a SNDR of 67.92 dB and a SFDR of 76.04 dB, respectively. Along with a Schreier figure-of-merit (FoM) of 207.5 fJ/Conv-step.

Keywords—*Analog-to-digital converter (ADC), time-to-digital c onverter (TDC), voltage-to-time converter (VTC), CMOS Image Sen sor （CIS）*

I. INTRODUCTION

With the rapid development of artificial intelligence, industrial demand for high-frame-rate CMOS image sensors (CIS) has gradually increased. This requires analog-to-digital converters (ADC) with high speed, low power consumption, and small area as pixel voltage signal readout circuits [1]. In the field of CIS, traditional single-slope analog-to-digital converter (SS ADC) exhibit significant conversion speed degradation at high resolutions, failing to meet the demanding readout speed requirements of modern high-performance CIS chips. Consequently, improving voltage readout speed has become a critical research focus in this domain. A two-step single-slope ADC is proposed in [1] to utilize differential ramps, which achieves substantial speed improvements over conventional

This work is supported by Science and Technology Key Project of Anhui Province under Grant 2022AH050099, National Natural Science Foundation of China under Grant 62471003, The University Synergy Innovation Program of Anhui Province under Grant GXXT-2023-013, The University Synergy Innovation Program of Anhui Province under Grant GXXT-2023-012, The University Synergy Innovation Program of Anhui Province under Grant GXXT-2023-011.

architectures. However, its overall conversion speed remains constrained due to inherent limitations of the SS ADC architecture.

Traditional high-speed pipelined-SAR ADC face challenges in advanced pixel processes. The reduced supply voltage and smaller transistor intrinsic gain significantly impair the performance of inter-stage residue amplifiers (RA). In such cases, these amplifiers must simultaneously meet gain and linearity requirements, which is highly challenging [2]. A parallel-conversion pipelined SAR ADC architecture is proposed in [3] to reduce the stringent requirements of fine quantization on RA, but this introduces additional circuitry, making it difficult to meet CIS's strict area requirements for readout circuits.

Due to the difficulty of performing fine quantization on signals in the traditional voltage domain under low supply voltages, utilizing TDC for fine quantization in the time domain has emerged as an attractive solution [4]. This is because the LSB step size of ADC is linearly proportional to supply voltage in the voltage domain, while the time-domain LSB step size gradually increases as supply voltage decreases.

Therefore, TDC resolution remains largely unaffected by reduced supply voltages, and its speed benefits from process scaling while inherently possessing superior noise margins. However, coarse quantization of large signals using TDC requires large capacitor areas to meet VTC linearity requirements, which is unacceptable in CIS applications. In [5], a calibration-enabled SS-TDC hybrid architecture is proposed that effectively reduces power consumption but shows limited speed improvement due to the limitations of the SS ADC architecture. Using SAR ADC for coarse quantization resolves this issue. These factors make the SAR-TDC architecture a promising solution for high-speed signal readout circuits in advanced pixel processes. However, the main challenge of SAR-TDC is the high sensitivity of delay

979-8-3315-8850-2/25 $31.00 © 2025 IEEE   132

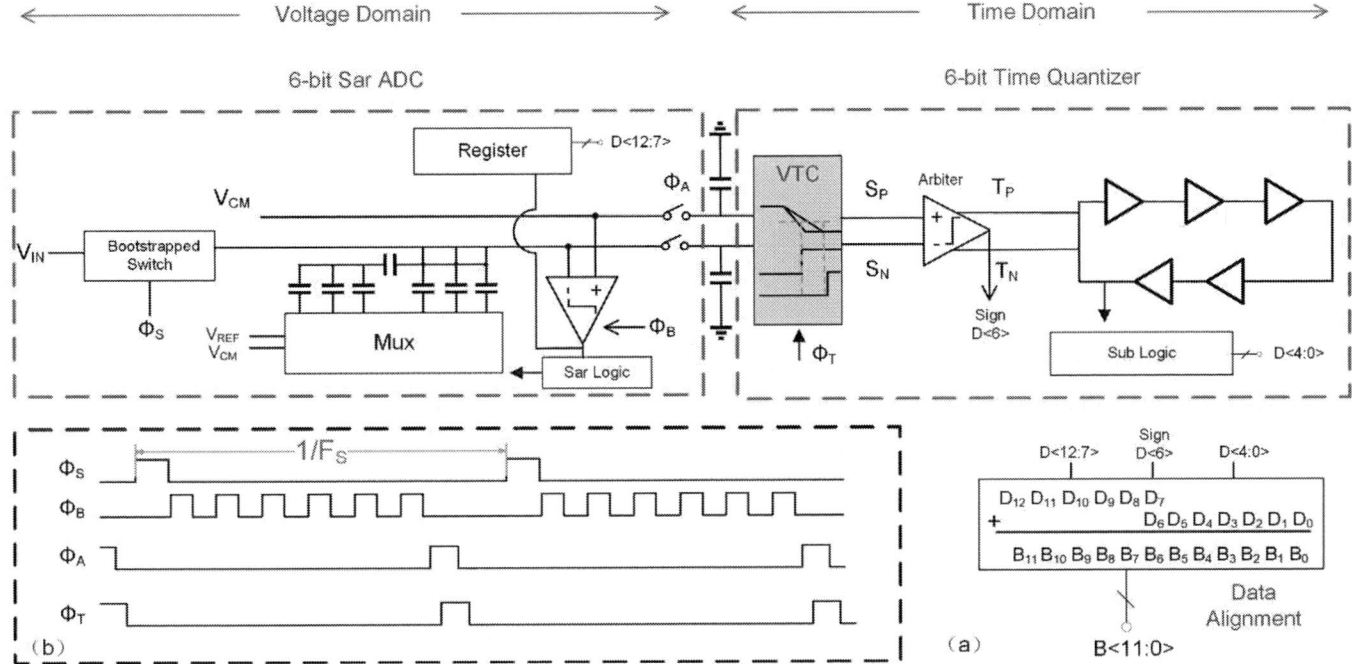

Fig. 1. (a) Architecture of the SAR-TDC ADC. (b) Timing diagram.

elements to PVT variations under low supply voltages, which affects measurement accuracy [4].

This paper pioneers the application of SAR-TDC in CMOS image sensors (CIS) as a 12-bit pixel voltage signal readout circuit. A novel solution is proposed to mitigate PVT-induced impacts through coordinated operation of the voltage-time converter (VTC) and the time-to-digital converter (TDC), while reducing power consumption.

The remainder of this paper is organized as follows: Section II describes the system and circuit implementation of the voltage-time hybrid-domain high-speed ADC. Section III presents simulation results. Conclusions are given in Section IV.

II. CIRCUIT DESIGN AND ANALYSIS

A. Overview of the SAR-TDC ADC

The overall architecture of the voltage-time hybrid-domain ADC is shown in Fig. 1(a). The SAR-TDC ADC comprises a SAR circuit, VTC, and ring TDC. Coarse quantization is implemented using a SAR ADC. Considering the single-ended pixel signal input and area constraints imposed by pixel alignment requirements, a single-ended split-capacitor array is adopted to reduce circuit area. Additionally, since the design targets pixel sensors operating at low supply voltages, a double-tail comparator replaces the traditional strong-arm latch. After coarse quantization, the comparator output voltages are fed into the VTC. The VTC converts the residual voltage into a time difference, which is then quantized by the ring TDC.

The VTC transfers fine quantization to the time domain. However, conventional TDC require high-resolution delay-locked loop (DLL) for high-precision quantization, leading to significant area overhead [6]. To address this, this work employs

a gated ring oscillator TDC (GRO TDC) to achieve high resolution with reduced area. The TDC's LSB step size depends on delay cell timing, which is highly sensitive to PVT variations [7]. A common solution is PVT tracking via current calibration; here, this is achieved by establishing current correlations between the VTC and TDC to mitigate PVT effects.

The overall timing is illustrated in Fig. 1(b). After Φ_S activation, a bootstrapped switch (BS) samples the input voltage. Asynchronous SAR logic generates comparator clock signals Φ_B for coarse quantization, producing digital bits D<12:7>.

Subsequently, Φ_A samples the residual voltage. Then, Φ_T triggers the VTC, which converts the sampled voltage into S_P and S_N signals. The difference between S_P and S_N is linearly proportional to the residual voltage [8]. Due to the uncertainty in comparator output polarity after coarse quantization, S_P may not precede S_N. To enforce TDC timing requirements (start signal before stop signal), a pre-arbitration circuit generates the sign bit D<6> along with start (T_P) and stop (T_N) signals. The ring TDC quantizes the time difference to produce D<5:0>, which is aligned with other bits to form the final 12-bit digital output B<11:0>.

B. High-Linearity Low-Power Voltage-Time Converter

The dynamic high-linearity VTC architecture is illustrated in Fig. 2, operating in two states: reset and conversion. During reset, clock signal CK is low, pre-charging nodes D_P and D_N to VDD. In conversion mode, CK transitions high, disabling the charging path and enabling discharge paths through nodes D_P and D_N. The load capacitor C discharges via differential pair currents I_P and I_N. When D_P/D_N voltages fall below the inverter threshold voltage, the output inverter flips state, generating START/STOP signals through a D flip-flop to enhance drive strength [9].

979-8-3315-8850-2/25 $31.00 © 2025 IEEE   133

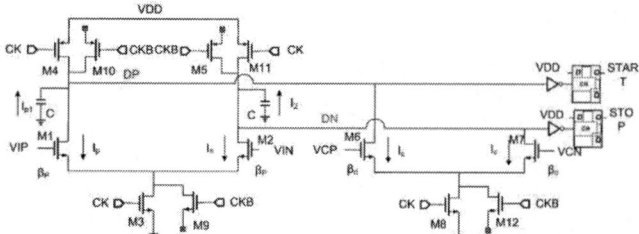

Fig. 2. Dynamic High Linearity Voltage-Time Converter

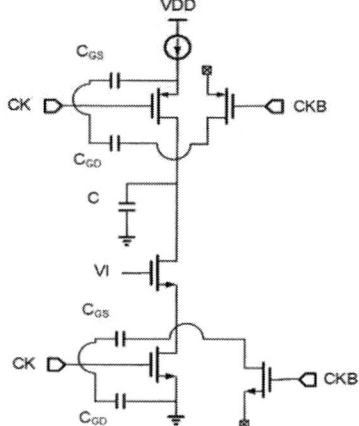

Fig. 3. Dummy Clock Feedthrough Elimination

The discharge rates of D_P/D_N depend on I_P/I_N, which scale with input voltage, thereby establishing a time difference proportional to the input voltage difference. Notably, the VTC enters standby mode once D_P/D_N discharge to GND, minimizing static power consumption. Next, we detail the circuit mechanisms enabling high linearity between voltage and time differences.

The conventional five-transistor "starved" VTC, while offering fast conversion speed, suffers from severe third-harmonic distortion, leading to poor linearity. To enhance the linear range, this work introduces dummy transistors M9–M12 (identical in size to switching transistors) as shown in Fig. 3. During high-frequency operation, input signals of opposite polarity generate counteracting pulses at the input nodes, thereby canceling clock feedthrough effects.

Additionally, this work introduces an auxiliary differential pair to generate a constant bias current I_C for compensating discharge current errors. By designing the gate widths of transistors M3 and M8 to be sufficiently large, their on-resistance becomes negligible. Consequently, the VTC output tout is governed by [7]:

$$\text{tout} \approx \frac{2CV_{DD}\beta_P(V_{COM}-V_{TH})V_{IN}}{[\beta_P(V_{COM}-V_{TH})^2+2I_C]^2} + \lambda\frac{\beta_P(V_{COM}-V_{TH})^2-2I_C}{[\beta_P(V_{COM}-V_{TH})^2+2I_C]^4} \quad (1)$$

The first term represents the signal component, while the second term corresponds to third-order distortion. When β_d equals β_p and transistors M1, M2, M6, and M7 are designed with identical dimensions, the compensation current IC satisfies:

$$I_C = \frac{\beta_d}{2}(V_{CP}-V_{TH})^2 = \frac{\beta_p}{2}(V_{COM}-V_{TH})^2 \quad (2)$$

With third-order distortion eliminated, the time difference tout achieves excellent linearity with respect to the input voltage.

C. 1-bit Sign Bit Circuit Translation

The TDC imposes stringent input requirements: the start signal must precede the stop signal for time difference conversion, otherwise the TDC outputs zero. Due to SAR ADC characteristics, the temporal order of S_P and S_N signals is indefinite. A Prelogical preprocessing module ensures the leading signal is routed to the TDC's START port and the lagging signal to the STOP port.

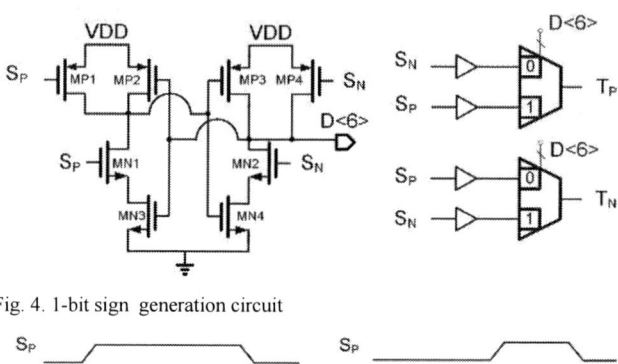

Fig. 4. 1-bit sign generation circuit

Fig. 5. Sign bit circuit output when S_P leads S_N: (a) S_P-leading case; (b) S_N-leading case.

As shown in Fig. 4, the circuit remains in a reset state with output port D<6> at high logic level until S_P/S_N signals arrive. If S_P arrives first, a time window exists where S_P is high and S_N is low. During this phase, PMOS transistors MP1 and MP2 are off, while NMOS transistor MP4 conducts, discharging node A to low. This activates PMOS MP3, stabilizing D<6> at high, encoding the final output as the sum of coarse and fine quantization results (Fig. 5a) without signal reordering. Conversely, if S_N arrives first, a time window occurs where S_N is high and S_P is low. Here, MP1 conducts, pulling D<6> voltage down until MP2 activates. Node A rises, disabling MP3 and triggering positive feedback that rapidly pulls D<6> low, encoding the final output as the subtraction of fine from coarse quantization (Fig. 5b).

D. Gated Ring Oscillator Time-to-Digital Converter

This paper adopts a GRO TDC architecture to expand the time measurement range, as shown in Fig. 6(a) The T_P signal enters the delay line ring, initiating its oscillation, while the T_N signal arrives after a time difference generated by the VTC and triggers the recording of the delay line's current state. By capturing both the oscillation count of the delay line ring and the state of its delay elements, a quantized digital output of the time difference is obtained. Compared to traditional flash architectures, GRO TDC not only reduces chip area but also inherently implements first-order noise shaping to reduce in-band noise and improve accuracy. Furthermore, conventional ring-based TDCs maintain oscillation until both T_P and T_N signals reset simultaneously. However, as demonstrated in this work, the delay line can cease operation once the T_N signal reads out its state. To address this, Fig. 6(b) introduces additional reset MOSFETs in the delay elements. The T_T signal, generated by XORing T_P and T_N signals with controlled delays (Fig. 7), resets the first delay element's output voltage to 0. This approach not only reduces power consumption but also ensures consistent starting conditions for each operational cycleof the delay line, thereby further enhancing measurement precision.

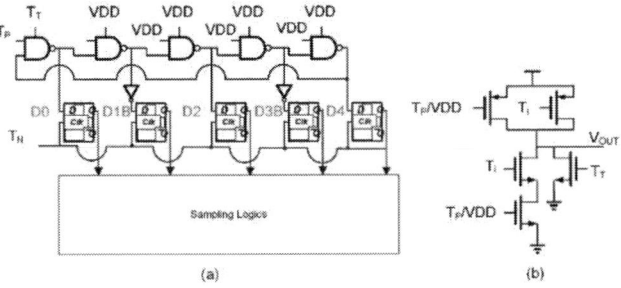

Fig. 6. (a) Architecture of GRO TDC. (b) Unit of delay chain loop.

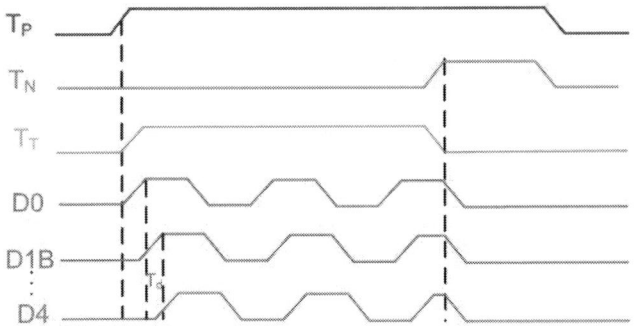

Fig. 7. Improved TDC readout timing diagram.

E. Inherent-PVT-Stabilized Technique

This circuit achieves PVT robustness by matching current sources of the VTC and TDC delay cells. As previously described, the supply voltage of the VTC circuit is generated by an adjustable gain amplifier with a gain multiplier α applied to the common-mode voltage V_{COM}. Equations (1) and (2) demonstrate that under optimal design conditions, the total time difference ΔT exhibits a linear relationship with the input voltage as follows:

$$\Delta T = \frac{2C\alpha(V_{COM}-V_{TH})\beta_P(V_{COM}-V_{TH})V_{IN}}{[\beta_P(V_{COM}-V_{TH})^2+2I_C]^2} = \frac{C\alpha V_{IN}}{4I_C} \quad (3)$$

The TDC's conversion accuracy depends on the charge and discharge times of its delay cells. In this design, the discharge

current I_U of the delay cells is significantly larger than the charging current, resulting in the following relationship for the TDC's LSB time Δt:

$$\Delta t = \frac{C_P V_{DD}}{I_U} \quad (4)$$

Here, C_P denotes the parasitic capacitance of the inverter. Under this configuration, the fine quantization output code value N is determined by the ratio of the two quantities, as expressed by:

$$N = \frac{\Delta T}{\Delta t} = \frac{\alpha C V_{IN} I_U}{4 C_P V_{DD} I_C} \quad (5)$$

The gain factor α is generated by a programmable gain amplifier implemented with Common-Mode Feedback, thus inherently resistant to PVT variations. The ratio of input capa citance to the parasitic capacitance of delay cells remains invariant under PVT variations. Similarly, the ratio of input voltage to supply voltage maintains a fixed value unaffected by PVT changes. By adopting identical MOS transistor dimensions for the differential pair transistors in VTC and the tail current transistor of TDC delay cells, identical PVT responses are achieved, ensuring that the fine quantization output code N remains stable across PVT variations. As shown in Fig. 8. (a) and Fig. 8. (b), the SNDR performance of the SAR-TDC circuit remains largely unaffected by PVT variations.

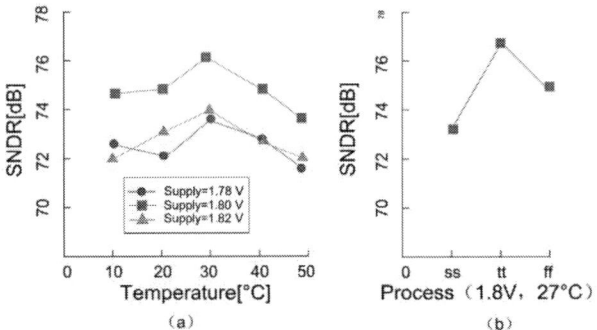

Fig. 8. (a)Simulated SNDR variation versus temperature with different power-supply voltages (1.78, 1.80, and 1.82 V). (b) The simulation results of SNDR under different process corners

III. SIMULATION RESULT

This ADC employs a 65 nm CMOS process and utilizes Cadence for simulation validation. Fig.9 shows the 2 column layout design. The height of the column ADC is 17.74 um, while the length of the column ADC is 653.15 um. At a supply voltage of 1.8 V, with a sampling rate of 20 MS/s and an input signal frequency of 63.477 kHz, the FFT spectrum of the ADC output (Fig. 10) exhibits an SNDR of 67.92 dB, SFDR of 76.04 dB, and ENOB of 11.57 bits. Under the same sampling rate (20 MS/s) with an input signal frequency of 9.78 MHz, the FFT results (Fig. 11) show an SNDR of 64.82 dB, SFDR of 73.39 dB, and ENOB of 10.47 bits. Static performance testing (Fig. 12) reveals maximum DNL and INL values of 0.83 LSB and 1.8 LSB, respectively. Table I compares performance metrics with prior works, highlighting that the SAR-TDC architecture achieves

979-8-3315-8850-2/25 $31.00 © 2025 IEEE

significantly reduced pixel voltage readout time at lower supply voltages while maintaining minimal INL [1][5][10][11].

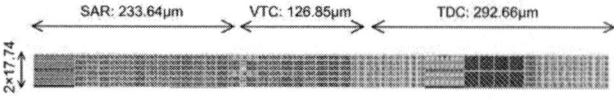

Fig. 9. Layout of ADC.

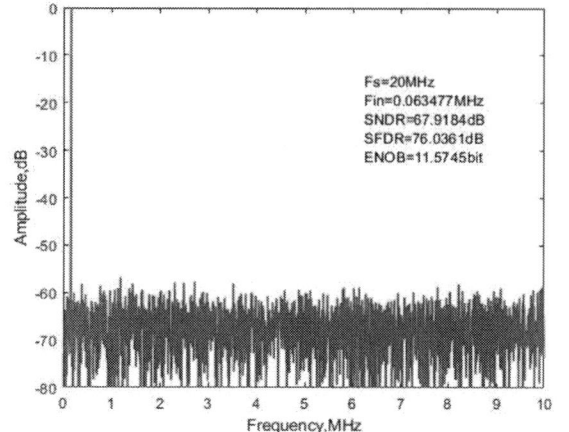

Fig. 10. ADC output spectrum at 20 MS/s sampling rate with 63.477 kHz input.

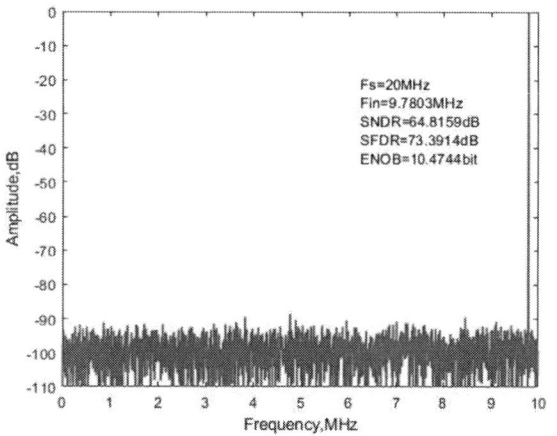

Fig. 11. ADC output spectrum at 20 MS/s sampling rate with 9.7803MHz input.

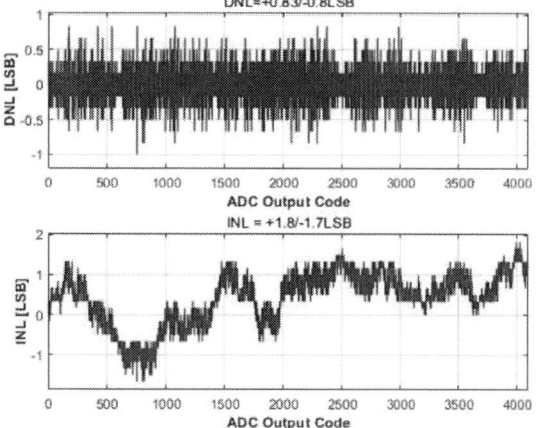

Fig. 12. The DNL and INL simulation results of proposed ADC.

TABLE I. PERFORMANCE COMPARISON AND SUMMARY

	[1] Sensors	[5] TACS-I	[10] VLSI	[11] TCAS-I	This work
Process(nm)	90	130	130	110	65
Power Supply (V)	2.8	3.3	3.3	3.3	1.8
Conv.Rate (MS/S)	2.4	1	0.1	0.2	20
Bits	12	12	12	14	12
DNL	4.25	0.4	0.83	0.87	0.83
INL	5.73	5.80	3.31	4.37	1.80
FoM (fJ/conv-step)	3.4	26.3	164.3	696.0	207.5

*FoM$=$Power$/(F_{Sample} \cdot 2^{ENOB})$

IV. CONCLUSION

This work proposes a SAR-TDC architecture tailored for CMOS image sensors (CIS), enabling rapid readout of pixel voltages under low supply voltages while achieving reduced area. The circuit is implemented using the SMIC 65nm process node. Reset switches are integrated into the TDC delay cells to minimize oscillation time and reduce overall power consumption. Additionally, a correlation mechanism between the VTC and TDC delay cells is established to mitigate PVT-induced variations in the system.

REFERENCES

[1] H. Park, C. Yu, H. Kim, Y. Roh and J. Burm, "Low Power CMOS Image Sensors Using Two Step Single Slope ADC With Bandwidth-Limited Comparators & Voltage Range Extended Ramp Generator for Battery-Limited Application," in IEEE Sensors Journal, vol. 20, no. 6, pp. 2831-2838, 15 March15, 2020.

[2] H. Zhao and F. F. Dai, "A 12-Bit 260-MS/s Pipelined-SAR ADC With Ring-TDC-Based Fine Quantizer for Automatic Cross-Domain Scale Alignment," in IEEE Journal of Solid-State Circuits, vol. 58, no. 10, pp. 2883-2896, Oct. 2023.

[3] S. -Y. Wang and T. -C. Lee, "An 800-MS/s 8.2-ENOB TDC-Assisted Pipelined-SAR ADC With Parallel Conversion," in IEEE Transactions on Circuits and Systems II: Express Briefs, vol. 71, no. 12, pp. 4854-4858, Dec. 2024.

[4] X. Zhao et al., "An 8-bit 1.5-GS/s Voltage–Time Hybrid Two-Step ADC With Cross-Coupled Linearized VTC," in IEEE Transactions on Very Large Scale Integration (VLSI) Systems, vol. 31, no. 12, pp. 2147-2151, Dec. 2023.

[5] D. Levski, M. Wäny and B. Choubey, "A 1-μs Ramp Time 12-bit Column-Parallel Flash TDC-Interpolated Single-Slope ADC With Digital Delay-Element Calibration," in IEEE Transactions on Circuits and Systems I: Regular Papers, vol. 66, no. 1, pp. 54-67, Jan. 2019.

[6] J. Yu, F. F. Dai and R. C. Jaeger, "A 12-Bit Vernier Ring Time-to-Digital Converter in 0.13 μm CMOS Technology," in IEEE Journal of Solid-State Circuits, vol. 45, no. 4, pp. 830-842, April 2010.

[7] M. Zhang, C. -H. Chan, Y. Zhu and R. P. Martins, "A 0.6-V 13-bit 20-MS/s Two-Step TDC-Assisted SAR ADC With PVT Tracking and Speed-Enhanced Techniques," in IEEE Journal of Solid-State Circuits, vol. 54, no. 12, pp. 3396-3409, Dec. 2019.

[8] K. Ohhata, "A 2.3-mW, 1-GHz, 8-Bit Fully Time-Based Two-Step ADC Using a High-Linearity Dynamic VTC," in IEEE Journal of Solid-State Circuits, vol. 54, no. 7, pp. 2038-2048, July 2019.

[9] M. Miyahara, I. Mano, M. Nakayama, K. Okada and A. Matsuzawa, "22.6 A 2.2GS/s 7b 27.4mW time-based folding-flash ADC with resistively averaged voltage-to-time amplifiers," 2014 IEEE International Solid-State Circuits Conference Digest of Technical Papers (ISSCC), San Francisco, CA, USA, 2014, pp. 388-389.

979-8-3315-8850-2/25 $31.00 © 2025 IEEE

[10] Q. Zhang, N. Ning, Z. Zhang, J. Li, K. Wu and Q. Yu, "A 12-Bit Two-Step Single-Slope ADC With a Constant Input-Common-Mode Level Resistor Ramp Generator," in IEEE Transactions on Very Large Scale Integration (VLSI) Systems, vol. 30, no. 5, pp. 644-655, May 2022.

[11] Zhang, N. Ning, J. Li, Q. Yu, Z. Zhang and K. Wu, "A High Area-Efficiency 14-bit SAR ADC With Hybrid Capacitor DAC for Array Sensors," in IEEE Transactions on Circuits and Systems I: Regular Papers, vol. 67, no. 12, pp. 4396-4408, Dec. 2020, doi: 10.1109/TCSI.2020.2998473.

2025 The 10th International Conference on Integrated Circuits and Microsystems

A Very-Large-Dynamic-Range Charge-Sensitive Front End for One-Terminal Capacitive Sensors in 65-nm CMOS

Haitao Fan
School of Software &
Microelectronic
Peking University
Beijing, China
htfan@stu.pku.edu.cn

Wenxiang Li
Nano Core Chip Electronic
Technology
Hangzhou, China
liwenxiang@nanocorechip.com

Jiafeng Yang
Nano Core Chip Electronic
Technology
Hangzhou, China
yangjiafeng@nanocorechip.com

Abstract—A very-large-dynamic-range charge-sensitive front-end (CSF) intended for single-ended capacitive sensors is presented. The auto-zero technique suppresses offset, 1/f noise, and supply noise. A proportional common-mode-cancellation technique emulates an n-times larger common-mode capacitor without increasing its physical value, conserving die area. A reconfigurable operational transconductance amplifier (OTA) maintains stability in both amplification and auto-zeroing phases. The charge sensitive amplifier circuit is simulated using a standard 65 nm CMOS process. Post-layout simulations show an inaccuracy of ±120 μV (equivalent to 66 aF) with a 600-pF baseline capacitance, occupying 0.4 mm² of active area and drawing 8 mA from a 1.2-V supply.

Keywords—*Capacitive sensor, charge-sensitive amplifier, auto-zero, dynamic range, 65-nm CMOS*

I. INTRODUCTION

Capacitive sensors are widely employed in astrophysics, consumer electronics, medical imaging, and the Internet of Things (IoT). Consequently, numerous front-end interfaces and capacitance-to-digital converters (CDCs) have been reported to extract information from these sensors [1]. Architecturally, capacitive sensors are classified as either floating or grounded sensors, depending on the electrode connection. Owing to their superior noise immunity, floating structures—in which the sensing capacitor is not referenced to ground—are widely adopted; the associated interface circuits have been studied extensively [2]. In contrast, touch sensors and other single-ended configurations necessitate a grounded plate. In a typical touch panel, for example, the finger deforms the cover glass to form the top electrode while the underlying patterned layer serves as the bottom plate, thereby tying one terminal to ground [3,4]. To maximize signal-to-noise ratio and suppress supply and common-mode disturbances, the sensor signal chain must translate this single-ended capacitance variation into a fully differential output.

Most capacitive-sensor applications focus on changes in capacitance, which occur due to variations in the measured external physical quantities (such as distance, acceleration, displacement, etc.). These capacitance variations are typically

orders of magnitude smaller than the baseline capacitance C_{based}. To maximize the dynamic range of the readout circuit, a common-mode capacitor (C_b) is placed at the front end and driven with a polarity opposite to that of the sensor. The resulting charge cancellation suppresses the static component, allowing only the signal-relevant ΔC charge to enter the subsequent sigma-delta or SAR stage. This technique, however, requires $C_b \approx C_{based}$; for large baseline capacitance (C_{based}), the physical capacitor becomes prohibitively large [5,6].

Therefore, we propose a CSF that combines switched-capacitor amplification with proportional common-mode cancellation, achieving an effective $n*C_b$ without enlarging C_b itself. This approach reduces the chip area required for the capacitance sampling circuit. A reconfigurable operational transconductance amplifier (OTA) is adopted to maintain stability during both amplification and auto-zero phases, eliminating the need for a second amplifier and reducing power consumption. The effectiveness of the proposed techniques is verified through circuit simulations. The simulations show ±120 μV inaccuracy with a 600-pF baseline capacitance, 0.4-mm² active area, and 8 mA from a 1.2-V supply.

This paper is organized as follows. Section 2 presents the proposed charge-sensitive amplifier architecture and its operating principles. Section 3 presents error analysis. Section 4 details the circuit implementation and analysis. Section 5 provides simulation results and benchmarks the design against prior art. Finally, Section 6 concludes this work.

II. PROPOSED CAPACITANCE-SAMPLING ARCHITECTURE AND OPERATING PRINCIPLE

The schematic of the single-ended-to-differential capacitive-sensor front end (CSF) is shown in Fig. 1, which utilizes switched-capacitor amplification to convert capacitance variations into voltage signals. In the context of capacitive sensors, Cx represents the total capacitance of the external sensor, which includes the baseline capacitance Cbased and the capacitance change ΔC that is to be measured. The response of sensing capacitor to the measurand is often small compared to its baseline value. Directly reading the sensor would waste dynamic range and increase measurement time unnecessarily. The proposed CSF circuit employs a four-phase clock and an auto-zero circuit to reduce the noise and offset effect, as shown in Fig.

This work was supported by the Centrally Funded Science and Technology Project 2020YFB2205601.

979-8-3315-8850-2/25 $31.00 © 2025 IEEE   138

1. Next, each phase of the operation is explained. Phase 1 (Φ1, sampling): VCC and ground are sampled onto Cx and Cb while the OTA offset is stored on Cv. Phase 2 (Φ2, amplification): The OTA closes the feedback loop, canceling the offset. Phase 3 (Φ3, hold and sampling): similar to Phase 1 but with reversed excitation. Phase 4 (Φ4, amplification): The complementary output is generated.

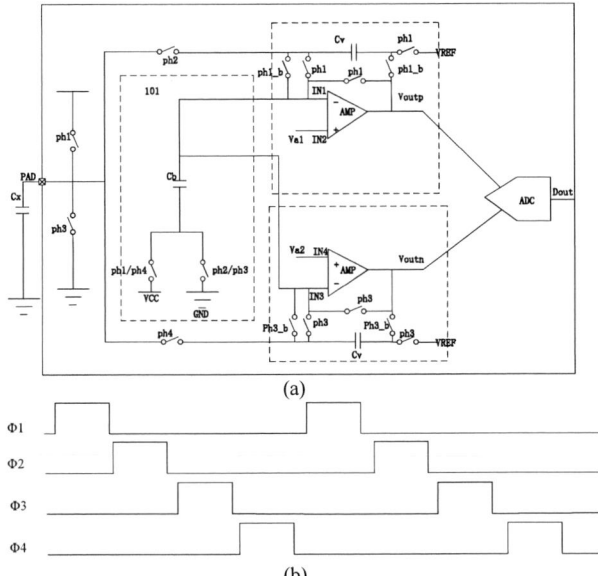

(a)

(b)

Fig. 1. (a) Schematic of the CSF and its signal diagram, (b) its signal diagram

In Phase 1, the VCC and ground are sampled onto capacitors C_X and C_b, respectively. Concurrently, the amplifier offset voltage VOFF is stored in capacitor CV. Thus, the total charge stored in Phase 1 isstored in Phase 1 is

$$Q_1 = VCC * C_X + (V_{A1} - VCC) * C_b + (V_{A1} - V_{REF}) * C_V \quad (1)$$

In Phase 2, the amplifier operates in amplification mode. The total charge becomes:

$$Q_2 = V_{A1} * (C_X + C_b) + (V_{A1} - V_{outp}) * C_V \quad (2)$$

As a result of the charge conservation Q1= Q2, the output voltage is

$$V_{outp} = \frac{(V_{A1} - VCC) * C_X + VCC * C_b}{C_V} + V_{REF} \quad (3)$$

Consequently, the OTA1 offset voltage is canceled.

In Phase 3, the OTA1 is in hold mode. The VCC and ground is sampled in capacitors Cb and CX, respectively. Concurrently, the amplifier offset voltage VOFF is stored in capacitor CV, resulting in charges. Thus, the total charge in Phase 3 is

$$Q_3 = V_{A2} * C_X + (V_{A2} - V_{REF}) * C_V \quad (4)$$

In the Phase 4, the amplifier operates in amplification mode. The total charge is:

$$Q_4 = (V_{A2} - VCC) * C_b + V_{A2} * C_X + (V_{A2} - V_{outn}) * C_V \quad (5)$$

As a result of the charge conservation Q3= Q4, the resulting output voltage is

$$V_{outn} = \frac{V_{A2} * C_X - VCC * C_b}{C_V} + V_{REF} \quad (6)$$

Hence, the OTA2 offset voltage is canceled. The differential output of the CSF at Phase 4 is:

$$V_{out} = V_{outp} - V_{outn}$$
$$= \frac{(V_{A1} - VCC) * C_X + VCC * C_b}{C_V} - \frac{V_{A2} * C_X - VCC * C_b}{C_V} \quad (7)$$

The reference voltage VREF, serving as the DC signal, is cancelled. The output signal of the CSF is independent of the reference voltage VREF. Therefore, the reference voltage VREF can take any value within the range of 0 - VCC without affecting the CSF operation. When $V_{A2} = VCC/n$ and $V_{A1} = (n\text{-}1) * VCC/n$, the output becomes:

$$V_{out} = 2 * \frac{C_X - n * C_b}{C_V} * \frac{VCC}{n} \quad (8)$$

Equation (8) demonstrates that the effective common-mode capacitance is $n * C_b$ while the physical capacitor remains C_b, reducing the die area by a factor of n.

III. ERROR ANALYSIS

In the previous discussion of operation, the only non-idealities taken into account are the operational amplifier input-referred offset voltage. However, in the practical implementation of the CSF, the following non-idealities must also be considered:

1) random deviations in the ratio of V_{A1} to V_{A1};
2) finite OTA gain and incomplete settling;
3) signal-dependent charge injection of MOS switches;
4) thermal and flicker noise.

Next, the effects of these error sources are discussed.

a) random deviations in the V_{A1}/V_{A2} ratio

In deriving Eq.(8), it is assumed that the ratio of V_{A1} to V_{A2} is constant. However, manufacturing process variations cause this ratio to vary. If the $V_{A1} = (n\text{-}1\text{-}\Delta n)/n * VCC$, $V_{A2} = VCC/n$, the Eq.(8) becomes

$$V_{out} = 2 * \frac{(1 + \Delta n/2) - n * C_b}{C_V} * \frac{VCC}{n} \quad (9)$$

Hence, any mismatch in the ratio of V_{A1}/V_{A2} ration introduces an error. If n = 2 is chosen, $\Delta n = 0$ by design. However, a larger C_b is required to cancel the baseline capacitance. If n > 2, a DC reference with parts-per-million stability is required to limit the mismatch-induced error.

b) finite OTA gain and incomplete settling

Finite OTA gain and incomplete settling limit achievable accuracy and linearity. A feedback network built around a finite-gain feedforward amplifier exhibits a gain error of $\approx 1/(\beta \cdot A0)$, where A0 is the open-loop gain and β the feedback factor. If the amplifier is nonlinear, a simple geometric construction (Fig. 2) relates gain error to maximum non-linearity: draw the ideal transfer curve (slope $1/\beta$) and a straight line between the endpoints of the actual characteristic. The resulting nonlinearity Δy_2 is always smaller than the gain error Δy_1, provided the small-signal gain decreases monotonically with input x (typical in analog circuits). Therefore, constraining Δy_1 to be below the target non-line-arity by increasing the open-loop gain guarantees $\Delta y_2 <$ target.

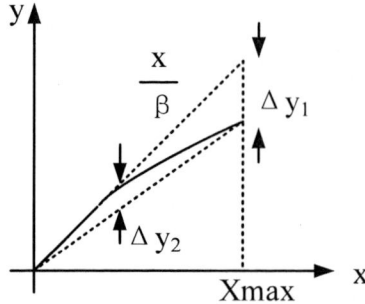

Fig. 2. Gain error and nonlinearity in a feedback system

c) signal-dependent charge injection of the MOS switches

MOS switch transistors have a significant non-ideality called charge injection. When a MOS switch is turned off, unwanted charge is injected into the circuit. This injection primarily results from the channel charge flowing to the source and drain. For an n-channel MOS transistor operating in the linear region, the channel charge is:

$$Q_{ch} = W * L * C_{ox}(V_{gs} - V_{th}) \tag{10}$$

where C_{ox} is the transistor gate oxide capacitance per unit area, V_{gs} is the transistor gate–source voltage, V_{th} is the n-channel MOS transistor threshold voltage and W and L are the transistor gate width and length, respectively. As shown in Eq. (10), the channel charge depends on the signal through the voltage V_{gs}. In order to keep the charge signal-independent, V_{gs} must be held constant.

The charge injection caused by these switches is constant, since the switches are always connected to a fixed voltage (V_{A1}, V_{A2}, VCC and GND). In the fully differential topology, this injection appears as a common-mode quantity and is therefore suppressed by the operational amplifier's common-mode rejection, leaving negligible residual error.

d) thermal and flicker noise.

Sampling in switched-capacitor (SC) circuits generates noise that degrades performance. There are two important noise sources: thermal and flicker noise. In this design, the flicker noise is reduced to a negligible level by using an auto-zeroing, and therefore this noise analysis concentrates on the effect of thermal noise.

There are two thermal noise sources in SC circuits: noise from the switches and the OTA noise. In this analysis, it is assumed that a single-stage OTA is used and that the settling time is limited by the operational amplifier, not by the sampling switches, resulting in the on-resistance of the MOS switch transistor R_{ON}<<$1/gm$, where gm is the trans-conductance of the single-stage operational amplifier. Thus, the noise bandwidth is determined by the operational amplifier, dominating the thermal noise in Phase 2 and Phase 4.

In Phase 1, thermal noise is sampled onto Cx and Cb. In Phase 2, the noise charge sampled in Phase 1 is redistributed to the feedback capacitor Cv, and the resulting output noise is:

$$V_{out}^2 = \frac{q^2}{C_{fb}^2} = \frac{kT(C_S + C_{OC} + C_{fb} + C_{par})}{C_{fb}^2} \tag{11}$$

In Phase 2, equivalent amplifying circuit is shown in Fig. 3(b). For this single-pole system, the transfer function is:

$$H(s) = \frac{V_{out}}{V_{n,eq}} = \frac{G}{1 + s/\omega_0} \tag{12}$$

The pole frequency is:

$$\omega_0 = 1/(R_{eq} * C_{eq}) = g_m\beta/C_{fb} \tag{13}$$

Where $R_{eq} = 1/(gm * \beta)$, $C_{eq} = C_L + (1-\beta)C_{fb} \approx C_L + C_{fb} \approx C_{fb}$, $\beta = C_{fb}/(C_S + C_{OC} + C_{fb} + C_{par})$. In phase 2, the output noise due to resistance is

$$\overline{V_{out,R_0}^2} = \int_0^\infty 4kTR_0 \left(\frac{C_S}{C_{fb}}\right)^2 \left(\frac{1}{1 + j2\pi f R_{eq}C_{eq}}\right)^2 df \tag{14}$$

The OTA-induced output noise is

$$V_{out}^2 = \int_0^\infty 8kT\gamma \frac{1}{g_{m1}} (1 + gm3/gm1)$$
$$* \left(\frac{C_S + C_{OC} + C_{fb} + C_{par}}{C_{fb}}\right)^2 \left(\frac{1}{1 + j2\pi f R_{eq}C_{eq}}\right)^2 df$$
$$= 8kT\alpha\gamma \frac{1}{g_{m1}}(1 + gm3/gm1)\frac{1}{\beta^2}\frac{1}{4R_{eq}C_{eq}}$$
$$= \frac{2kT\alpha}{\beta C_{fb}}(1 + gm3/gm1) \tag{15}$$

Because the noise in both phases is uncorrelated, the resulting thermal-noise power spectral density (PSD) is

$$\overline{V_{out,c2v}^2} = \frac{kT(C_S + C_{OC} + C_{fb} + C_{par})}{C_{fb}^2} + \frac{2kT\alpha}{\beta C_{fb}}(1 + gm3/gm1) \tag{16}$$

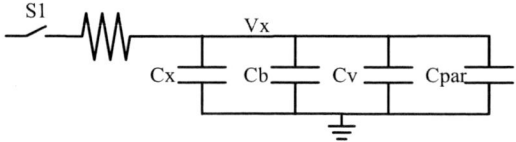

(a)

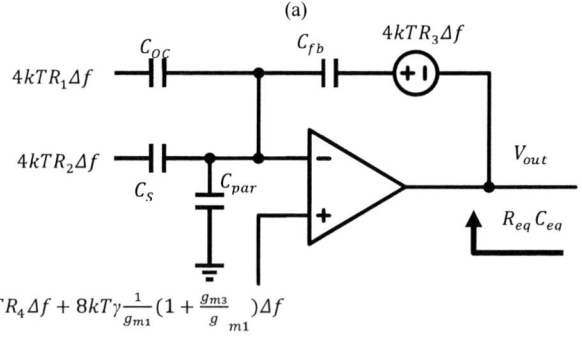

(b)

Fig. 3. (a) equivalent sampling circuit at Phase 1, (b) equivalent amplifying circuit at Phase 2

IV. CIRCUIT IMPLEMENTATION

The schematic of the designed CSF and the clock phases that drive the switches are shown in Fig. 1. The CSF requires four non-overlapping clock phases. These clocks can be generated by simple digital logic. Switches use minimum length to minimize charge injection. Furthermore, the switches connected to the amplifier inputs are opened first in order to make the charge injection signal-independent. All the switches connected to the amplifier inputs or to ground are implemented by NMOS switches. Switches connected to amplifier outputs are implemented by CMOS switches. All capacitors are implemented as metal-insulator-metal (MIM) devices.

A. Reconfigurable Operational Transconductance Amplifier (OTA)

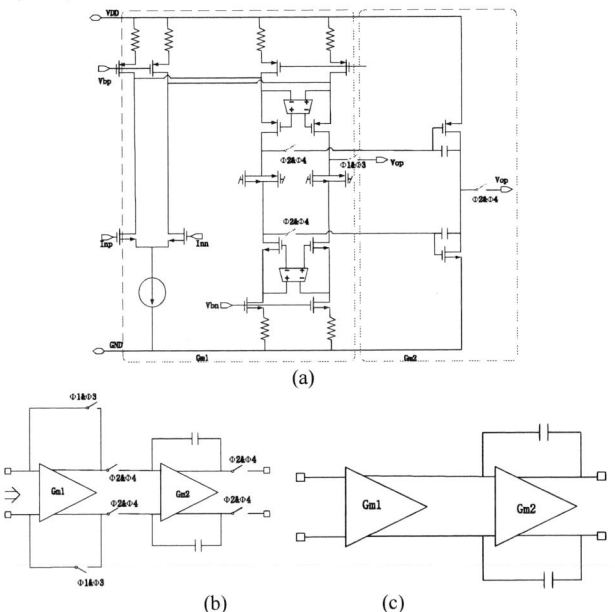

(a)

(b)   (c)

Fig. 4. (a) Circuit diagram of Gm1, Gm2 with reconfigurable loop, (b) Phase 1 and Phase 3, (c) Phase 2 and Phase 4

There is a critical OTA in this design that determines the key performance. It consumes 8 mA, and the circuit's power requirement decreases as the baseline capacitance is reduced. A Miller-compensated amplifier comprising Gm1 and Gm2 is depicted in Fig. 4. In SC circuits, OTAs are used to transfer charges within a clock cycle. The accuracy of the charge transfer is defined as the settling accuracy and DC gain, which is the relative error of the output voltage at the end of the settling period compared to the ideal output voltage.

The first stage (Gm1) employs gain boosting for high DC gain. Load transistors are source-degenerated, making the input pair the dominant noise contributor. Thirty percent of the amplifier current is allocated to the input stage, yielding a transconductance of 10 mS. The rail-to-rail Class-AB output stage Gm2 is bypassed during Phase 2 and Phase 4. During Phase 1 and Phase 3, it operates as a regular one-stage amplifier, while Gm2 is bypassed, which is configured as a unity gain amplifier for auto-zeroing. During Phase 2 and Phase 4, it works as a regular two-stage amplifier. The Gm1 is a single-stage op-amp in auto-zeroing phase that is stable with any capacitive load. Offset and noise of Gm2 are suppressed by the high gain of Gm1.

The simulation configuration for OTA small-signal performance is shown in Fig. 5. The simulation is performed under the worst-case loading condition (Phase 2 and Phase 4 in Fig. 1), using the configuration shown in Fig. 4(c). A 600-pF baseline capacitor and a 400-pF common-mode capacitor form a 1000-pF sampling capacitor. This configuration accounts for gain-bandwidth (GBW) reduction and accurately captures the feedback loop's gain and phase response. The closed-loop DC gain of the complete amplifier exceeds 79 dB across process

corner, voltage and temperature variation so that the linearity error and gain error due to dc gain are well suppressed. The closed-loop unity-gain bandwidth is 2 MHz, as shown in Fig. 6.

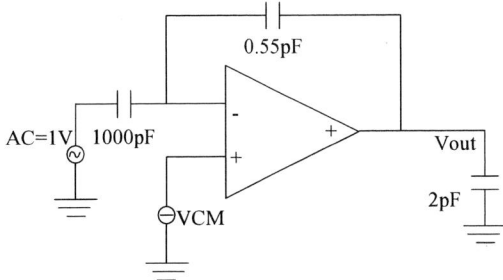

Fig. 5. Schematic of the configuration used in the simulation of the small-signal performance

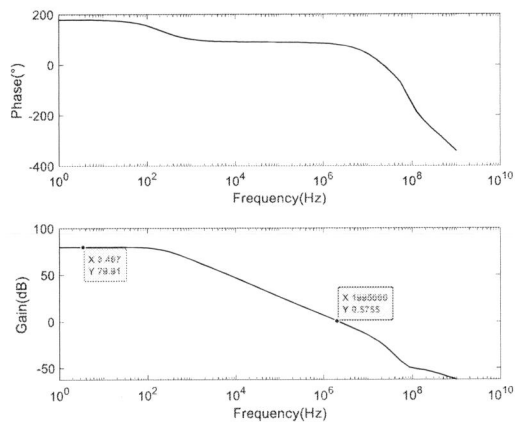

Fig. 6. Simulated small-signal performance (gain and phase) of the OTA

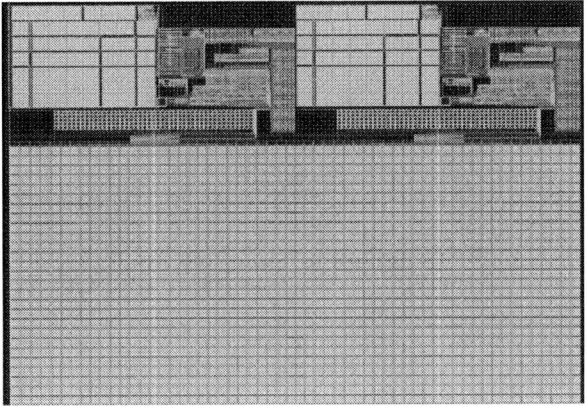

Fig. 7. Layout of the proposed CSF

V. SIMULATION RESULTS

The proposed charge-sensitive amplifier with very large dynamic range was designed and simulated in a TSMC 65-nm CMOS process. Fig. 7 shows the layout, and the active area is 0.4 mm² excluding bondpads. When the capacitance changes from 559.45 pF to 600.55 pF, the output voltage of the CSF increases with the capacitance. Fig. 8 shows the error after first-order fitting. The inaccuracy remains within ±120 µV at a 1.2-V supply, equivalent to 66 aF. The peak-to-peak output noise is 47.5 mV at the baseline capacitance of 600 pF, supporting the

2.6-fF equivalence claim, as shown in Fig. 9. A performance summary and comparison with previously pu-blished amplifiers are presented in

TABLE 1. Especially, we achieved a more compact CSF circuit design by co-integrating a reconfigurable operational transconductance amplifier (OTA) and a capacitance-boosting technique.

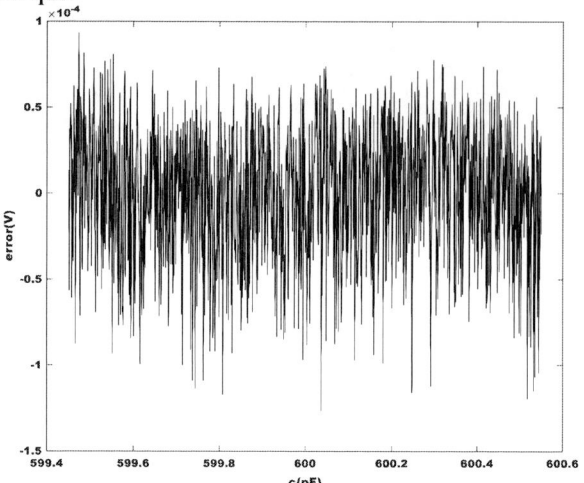

Fig. 8. Cs VS output error

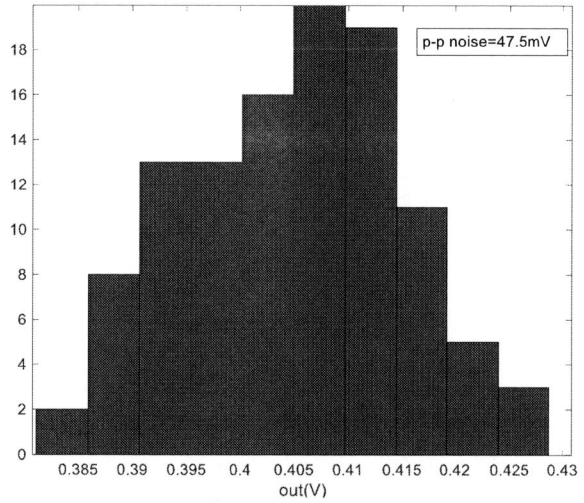

Fig. 9. Histogram of the output-referred noise with baseline capacitance of 600pF, supporting the 2.6-fF equivalence claim

TABLE I. Performance summary and comparison

	[7]	[8]	[9]	[10]	This work
Process	180nm	160nm	180nm	180nm	65nm
Supply	1.4V	1.6~2V	1.1V	0.8&1V	1.2V
Cap.range	24pF	3.6pF	18.12pF	5.7pF	600±0.55pF
Meas.Time	0.23ms	100ms	0.85ms	0.18ms	0.002ms
Power	33.7uW	3.24uW	3uW	23.2uW	9600uW
Error	160 aF	2.5 aF	1.24 fF	186 aF	6.6 aF
Area	0.46mm²	0.33mm²	0.75mm²	0.15mm²	0.4 mm²

VI. CONCLUSION

This paper presents a very-large-dynamic-range CSF for single-ended capacitive sensors in 65-nm CMOS process. This design accommodates baseline capacitances up to 600 pF occupying an active area of 0.4 mm2 excluding bondpads. A virtual-multiplication capacitor technique based on proportional commonmode cancellation is proposed, effectively reducing the chip area. A reconfigurable operational transconductance amplifier is adopted to maintain stability during both the amplification and auto-zero phases. The CSF is insensitive to baseline capacitance, amplifier offset, input parasitic cap-acitance, and flicker noise. Drawing 9.6 mW and achieving $\pm$ 120 µV error (equivalent to 66 aF), the CSF is well-suited for single-ended capacitive-sensor applications.

REFERENCES

[1] Y. Jung, Q. Duan and J. Roh, "A 17.4-b Delta-Sigma Capacitance-to-Digital Converter for One-Terminal Capacitive Sensors," in IEEE Transactions on Circuits and Systems II: Express Briefs, vol. 64, no. 10, pp. 1122-1126, Oct. 2017, doi: 10.1109/TCSII.2015.2505960.

[2] M. Mojarad and M. Diba, "A Fully Integrated Low-Power Capacitive Sensor Frontend With Automatic Tuning Scheme," in IEEE Transactions on Circuits and Systems II: Express Briefs, vol. 68, no. 12, pp. 3498-3502, Dec. 2021, doi: 10.1109/TCSII.2021.3122383.

[3] B. Li, L. Sun, C. -T. Ko, A. K. -Y. Wong and K. -P. Pun, "A High-Linearity Capacitance-to-Digital Converter Suppressing Charge Errors From Bottom-Plate Switches," in IEEE Transactions on Circuits and Systems I: Regular Papers, vol. 61, no. 7, pp. 1928-1941, July 2014, doi: 10.1109/TCSI.2014.2298285.

[4] Y. -H. Yu and T. -Y. Sun, "A Pseudo-Differential Measuring Approach for Implementing Microcontroller-Based Capacitive Touch Sensing in Low-Power Quality Situation," in IEEE Sensors Journal, vol. 16, no. 2, pp. 390-399, Jan.15, 2016, doi: 10.1109/JSEN.2015.2479599.

[5] B. Stefanelli, J. . -P. Bardyn, A. Kaiser and D. Billet, "A very low-noise CMOS preamplifier for capacitive sensors," in IEEE Journal of Solid-State Circuits, vol. 28, no. 9, pp. 971-978, Sept. 1993, doi: 10.1109/4.236177.

[6] Z.Y.Xiong, S.Y.Yu, Q.C.Cao. "Driving Scheme for Residual Image Reduction in Active-Matrix Organic Light-Emitting Diodes Display." Microelectronics Journal,Sep.2024,151:106324.

[7] S.C.Oh, W.Y.Jung, K.Y.Yang, D. Blaauw and D. Sylvester, "15.4b incremental sigma-delta capacitance-to-digital converter with zoom-in 9b asynchronous SAR," 2014 Symposium on VLSI Circuits Digest of Technical Papers, Honolulu, HI, USA, 2014, pp. 1-2, doi: 10.1109/VLSIC.2014.6858443.

[8] B. Yousefzadeh, W. Wu, B. Buter, K. Makinwa and M. Pertijs, "A compact sensor readout circuit with combined temperature, capacitance and voltage sensing functionality," 2017 Symposium on VLSI Circuits, Kyoto, Japan, 2017, pp. C78-C79, doi: 10.23919/VLSIC.2017.8008555.

[9] S. Park, G. -H. Lee and S. Cho, "A 2.92--W capacitance-to-digital converter with differential bondwire accelerometer, on-chip air pressure, and humidity sensor in 0.18--m CMOS," in IEEE Journal of Solid-State Circuits, vol. 54, no. 10, pp. 2845-2856, Oct. 2019, doi: 10.1109/JSSC.2019.2930140.

[10] P.Yang, Z. Zhang, N. Mei. A 0.15mm2 energy-efficient single-ended capacitance-to-digital converter. IEEE Trans Circuits Syst II Express Briefs. 2022 Feb;69(2):314–8.

979-8-3315-8850-2/25 $31.00 © 2025 IEEE

2025 The 10th International Conference on Integrated Circuits and Microsystems

A fast transient-response LDO-CP with novel high voltage compensation structure in CMOS image sensors

Qiang Zhao
School of Integrated Circuits
Anhui University
Hefei,China
zhaoqiang@ahu.edu.cn

Zhendong Niu
School of Integrated Circuits
Anhui University
Hefei,China
WB23301102@stu.ahu.edu.cn

Zhigang Li
School of Integrated Circuits
Anhui University
Hefei,China
zhigangli@ahu.edu.cn

Xiuying Wang
School of Electronic Engineering
Chaohu University
Chaohu,China
wxy-ahu@163.com

Chunyu Peng
School of Integrated Circuits
Anhui University
Hefei,China
cyupeng@ahu.edu.cn

Xiulong Wu
School of Integrated Circuits
Anhui University
Hefei,China
xiulong@ahu.edu.cn

Abstract—This paper proposes a low-ripple and fast transient response low-dropout linear regulator-charge pump(LDO-CP) for CMOS image sensors(CIS). A novel high-voltage compensation structure is proposed, which accelerates the recovery of the output when the load is connected, and at the same time improves the influence of the discharge noise of the lower plate of the capacitor introduced by using the traditional compensator. In addition, a transient response enhancement circuit(TREC) is used to improve the transient response. Simulations were carried out in a 130nm CMOS process. When the load capacitance is 8pF, the recovery time of the LDO-CP is 59ns, and the output ripple is $275\mu V_{pp}$.

Index Terms—CMOS image sensor, LDO, charge pump, transient response, ripple

I. INTRODUCTION

With the popularization of photographic products such as mobile phones and digital cameras, higher requirements are put forward for the imaging clarity of pictures. Providing a control voltage higher than the power supply voltage for the CIS pixels can effectively reduce noise and improve the imaging quality [1].

In order to provide a stable control voltage higher than the power supply voltage for the pixels, it is generally necessary to integrate the LDO-CP circuit into the chip. When processing the image signal, it is necessary to ensure that the output voltage of the LDO-CP has a lower ripple and smaller noise. When an undershoot occurs, the recovery time of the output voltage should be minimized.

In response to the above issues, researchers have carried out a great deal of research in terms of low noise and fast transient response in recent years. Reference [2] designed a charge pump with pulse skip modulation(PSM). By adjusting the output voltage of the charge pump through a feedback

mechanism, the noise can be effectively reduced. Reference [3] introduced the variable frequency modulation(VFM) technology, achieving a balance between the output voltage ripple and the efficiency. In order to accelerate the recovery of the output voltage, the compensation technique was applied to the LDO-CP. Through the injection of additional charge, the recovery time of the output voltage was effectively reduced, making the recovery time less than 160ns [4] [5]. However, its output ripple has increased. By using the floating well technique to keep the transistor in the reverse-biased state at all times, the ripple can be effectively reduced. Eventually, the ripple of the output voltage is only 1.42mV [6]. Reference [7] presented an LDO using the segmented pass FET(SPF) technique, which achieved a fast transient response while maintaining a very low quiescent current. Reference [8] employed the cross-coupled charge pump topology, successfully controlling the recovery time within 66ns and making the ripple less than $1mV_{pp}$, which proposed a new idea for the design of low-noise charge pumps. Kim et al. applied a voltage damper to the LDO, optimizing the overshoot and undershoot during load condition changes [9]. Liu et al. implemented a fast local loop and an overshoot detection circuit, achieving a recovery time of 100ns and an extremely low overshoot [10]. Reference [11] adjusted the number of charge pump units and utilized an eight-phase clock generated by a delay-locked loop.

In response to the above issues, this paper proposes an LDO-CP circuit with a novel high-voltage compensation structure and a transient enhancement circuit. When the load is connected, the novel high-voltage compensation structure accelerates the recovery of the output through additional charge injection. At the same time, it mitigates the impact of the discharge noise of the lower plate of the capacitor, which

979-8-3315-8850-2/25 $31.00 © 2025 IEEE

would be introduced when using a traditional compensator. In addition, a transient response enhancement circuit is used to improve the transient response.

This paper is organized as follows. Section II introduces the working principle of the traditional LDO-CP and the analysis of its transient response. Section III introduces the proposed LDO-CP that employs the novel high-voltage compensation structure and the transient response enhancement circuit. Section IV shows the measurement results and Section V gives the conclusion.

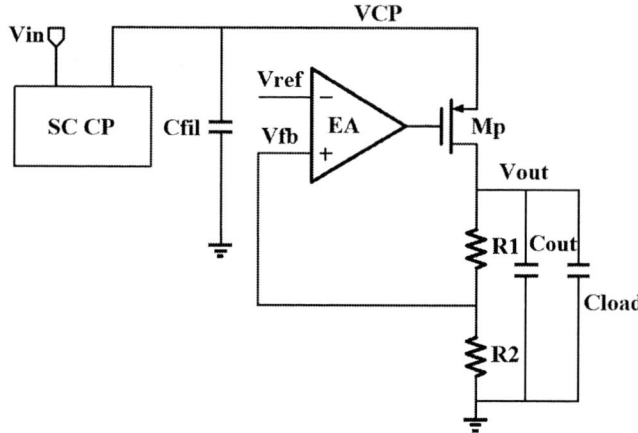

Fig. 1: Traditional LDO-CP structure

II. TRADITIONAL LDO-CP OPERATING PRINCIPLE AND TRANSIENT RESPONSE ANALYSIS

Fig. 1 shows the typical structure of the LDO-CP for the CIS. It is composed of a switched capacitor charge pump(SC CP), an error amplifier, a voltage reference source(V_{ref}), a filter capacitor, an output capacitor, a power transistor M_p, and a voltage feedback circuit(R_{01},R_{02}). This circuit utilizes the SC CP to supply power to the power transistor M_p. When the output signal changes, the feedback network is employed to stabilize the output voltage of the LDO. The output voltage of the LDO-CP is

$$V_{out} = V_{ref} \times \frac{R_{01} + R_{02}}{R_{02}} \qquad (1)$$

When the load of the LDO increases, the current flowing through the power transistor M_p will increase, and the output voltage of the LDO will decrease. The current provided by the SC CP is insufficient to meet the requirements of the power transistor, and its output voltage(V_{CP}) will decrease. At this moment, due to the voltage difference between V_{CP} and V_{fil}, the filter capacitor will release charges to stabilize V_{CP}. At the same time, the decrease in the output voltage of the LDO will act on the non-inverting input terminal of the error amplifier through the voltage feedback network. Since the inverting input terminal of the error amplifier is connected to V_{ref}, the output voltage of the error amplifier decreases. This will lead to an increase in the gate-source voltage of the

power transistor, causing an increase in the drain current of the power transistor, thereby playing a role in stabilizing the output voltage.

Conversely, when the load of the LDO decreases, V_{CP} will increase, and the voltage V_{fb} that acts on the non-inverting input terminal of the error amplifier through the feedback network will also increase. This will result in an increase in the gate voltage of the power transistor and a decrease in I_D, thus playing a role in stabilizing the output voltage.

The load transient response is shown in Fig. 2. If the initial potential of C_{load} is 0, when the load C_{load} is just connected to V_{out}, there will be a charge transfer process between C_{load} and C_{out}, causing V_{out} to drop. When the charge transfer is completed, V_{out} reaches its lowest point. The voltage drop of V_{out} is

$$\triangle V_{out} = \frac{C_{load} \times V_{out}}{C_{load} + C_{out}} \qquad (2)$$

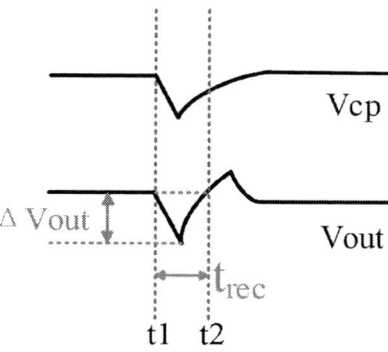

Fig. 2: Transient response curve of load

Meanwhile, the decrease in the output voltage of the LDO will act on the error amplifier through the voltage feedback network, causing the output voltage of the error amplifier to decrease, the drain current of M_p to increase, and the output voltage to gradually recover.

III. PROPOSED CIRCUIT

The schematic diagram of the proposed LDO-CP circuit is shown in Fig. 3. This circuit includes a SC CP, a power transistor, a novel high-voltage compensation structure, a feedback loop, and a transient response enhancement circuit. Among them, the SC CP is controlled by a non-overlapping clock. The novel high-voltage compensation structure is controlled by two opposite clocks, and the error amplifier of the feedback structure adopts a folded cascode configuration.

A. Novel high-voltage compensation structure

The novel high-voltage compensation structure integrates three capacitors, namely C11, C12, and C13. These capacitors are controlled by CLK1 and CLK2 through a circuit composed of transistors MP1, MP2, MP3, MN1, MN2, MN3, MN4, and MN5.

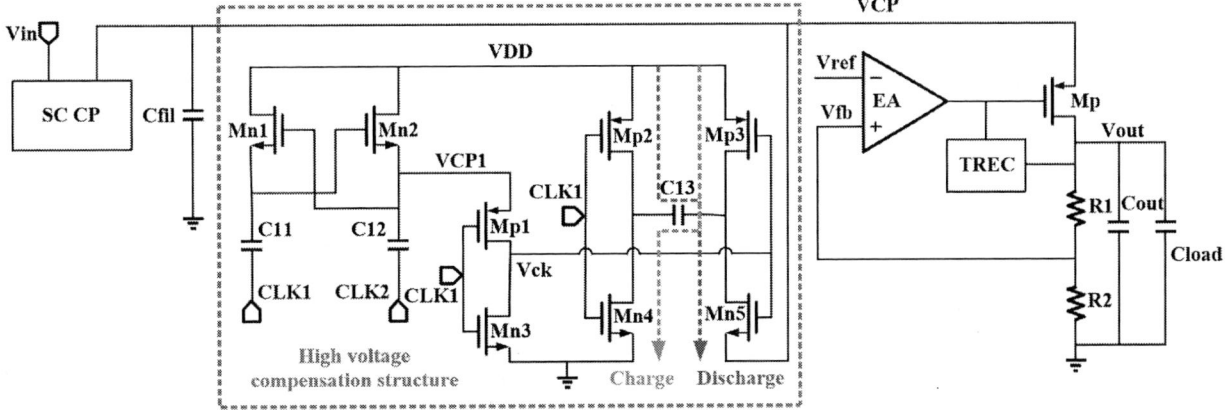

Fig. 3: Schematic diagram of the proposed LDO-CP

To mitigate the impact of connecting C_{load} on V_{CP} and V_{out}, the compensator performs additional charge compensation on the output terminal of the SC CP through C13. When C_{load} is not connected, MP2 and MN5 are in the off state, while MP3 and MN4 are in the on state. Then C13 is charged, and the voltage difference between the two plates of the capacitor is VDD. When C_{load} is connected, MP2 and MN5 are turned on, MP3 and MN4 are turned off, and C13 is discharged. At this moment, the voltage difference between the upper and lower plates of C13 is VDD, so the voltage of the upper plate of C13 is 2VDD. Since the voltage of the upper plate of C13 is higher than V_{CP}, it will perform charge compensation on the output terminal of the SC CP. The amount of compensated charge is

$$Q_c = C_{13} \times (2VDD - V_{CP}) \qquad (3)$$

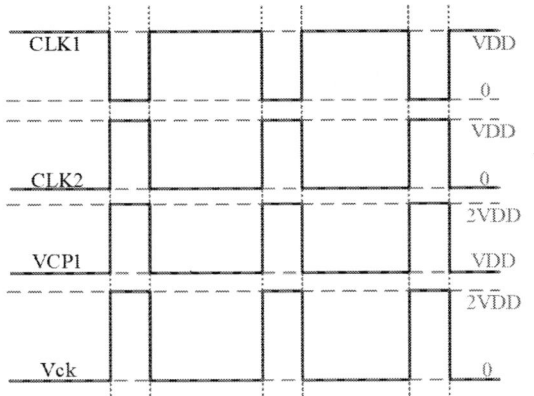

Fig. 4: The timing sequence and node voltages of the proposed circuit

Since the power supply voltage is VDD, and the voltage of the upper plate of C13 is VDD/2VDD, the timing sequence with a voltage of VDD cannot meet the control requirements.

To solve this problem, MN1, MN2, MN3, MP1, C11, and C12 are used to generate a timing sequence with a voltage of 0/2VDD, which is controlled by CLK1 and CLK2. The control timing sequence and node voltages of the novel high-voltage compensation structure are shown in Fig. 4. MN1, MN2, C11 and C12 form a small charge pump, and the output voltage is VCP1. When CLK1 is at a high level and CLK2 is at a low level, MN1 is turned off, MN2 is turned on, C11 is discharged, C12 is charged, and the output voltage of the charge pump is VDD. When CLK1 is at a low level and CLK2 is at a high level, MN1 is turned on, MN2 is turned off, C11 is charged, C12 is discharged, and the output voltage of the charge pump is 2VDD. When CLK1 is at a high level and VCP1 is VDD, MP1 is turned off, MN3 is turned on, and the voltage at the V_{ck} terminal is 0. When CLK1 is at a low level and VCP1 is 2VDD, MP1 is turned on, MN3 is turned off, and the voltage of V_{ck} is 2VDD. By generating a timing sequence with a voltage of 0/2VDD to control the charging and discharging of C13, the impact of connecting C_{load} on V_{CP} and V_{out} can be effectively reduced without introducing the discharge noise from the lower plate of the capacitor.

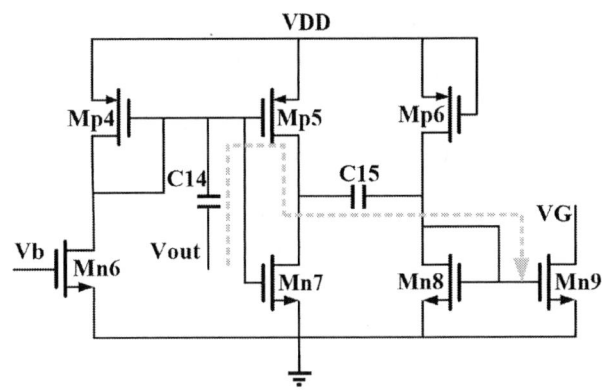

Fig. 5: Transient response enhancement circuit

B. Transient response enhancement circuit

The transient response enhancement circuit proposed in this paper is shown in Fig. 5. The gate of transistor MP6 is connected to the power supply, so it is always in the off state. The leakage current of MP6 is generated by the current mirror composed of MN8 and MN9. MN9 operates in the near-threshold region. Therefore, when its gate voltage rises, a relatively large current will be generated.

When the load is connected, V_{out} drops, and the capacitor C14 couples the change of V_{out} to the gates of the current mirror formed by MP4 and MP5. The drain voltage of MP5 increases and is coupled to the gate of MN9 through C15, and the leakage current of MN9 increases. Since the drain of MN9 is connected to the gate of the power transistor, the gate voltage of the power transistor decreases, and the current flowing through the power transistor increases. The above-mentioned effect increases the driving capability of the power transistor, enabling V_{out} to quickly recover and stabilize.

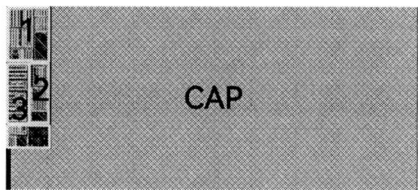

Fig. 6: Layout of the proposed circuit.1, LDO;2, High voltage compensation structure;3, SC CP.

IV. SIMULATION RESULTS

The proposed circuit structure adopts 130nm CMOS process, and the circuit layout is shown in Fig. 6. In the proposed circuit, V_{in} is 3.3V, and the clock frequency is 20MHz. V_{ref} is 1.8V, R_{01} and R_{02} are 60K. The capacitances of C11, C12, C14, and C15 are 1pF, C13 is 20pF, C_{fil} is 30pF, the output capacitor is 200pF, and the load capacitor is 8pF.

The simulation results are shown in Fig. 7. When an 8pF load is connected and the traditional structure is used, the recovery time is 204ns, and the output voltage ripple is $275\mu V_{pp}$. When the proposed circuit is used in the TT corner, the recovery time is 59ns, and the output ripple is $275\mu V_{pp}$. In the FF corner, the recovery time is 62.5ns, and the output ripple is $315\mu V_{pp}$. In the SS corner, the recovery time is 65ns, and the output ripple is $230\mu V_{pp}$.

It is worth noting that the recovery time has been reduced by 71% while the output ripple remains unchanged. The novel high-voltage compensation structure and transient response enhancement circuit proposed in this paper have significantly improved the transient response of the LDO-CP, making the recovery time of V_{out} 59ns. Meanwhile, no discharge noise from the lower plate of the capacitor has been introduced.

Table I lists the relevant parameters of the proposed LDO-CP. From the comparison of the literatures, it can be seen that the designed LDO-CP circuit achieves smaller ripple and significantly improves and promotes the reduction of the recovery time.

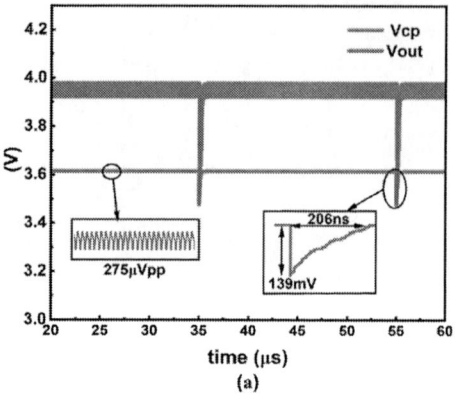

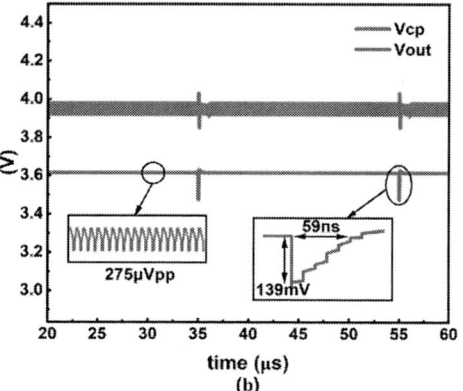

Fig. 7: Transient simulation results of (a) the traditional (b) the proposed LDO-CP circuit

TABLE I: PERFORMANCE COMPARISON

Reference	[4]	[5]	[8]	[11]	This paper*
Process(μm)	0.11	0.11	0.11	0.065	**0.13**
Vin(V)	3.3	3.3	2.8	1.2	**3.3**
Vout(V)	4	3.3~4	3.6	-0.38	**3.6**
f(MHz)	25	25	48	800	**25**
Cout(pF)	300	325	200	520	**200**
Ripple(μV_{pp})	950	634	590	3000	**275**
Drop(mV)	114	111.5	127	/	**139**
trec(ns)	155	93	66	600	**59**
Setting time(μs)	10	11	1.71	/	**1.95**

979-8-3315-8850-2/25 $31.00 © 2025 IEEE

V. CONCLUSION

This paper proposes a LDO-CP circuit that features a novel high-voltage compensation structure and a transient response enhancement circuit. By utilizing the novel high-voltage compensation structure, the output recovery upon load connection is accelerated. Meanwhile, the impact of the discharge noise from the lower plate of the capacitor, which would be introduced when using a traditional compensator, is alleviated. In addition, the transient response enhancement circuit is employed to improve the transient response. Compared with the traditional structure, this design exhibits better transient performance and does not introduce additional noise. When the load capacitance is 8pF, the recovery time of the LDO-CP is 59ns, and the output ripple is $275\mu V_{pp}$.

ACKNOWLEDGMENT

This work was supported in part by the Key Research and Development Program of Anhui Province under Grant 202304a05020057 and 202423k09020038; The Science and Technology Key Project of Anhui Province under Grant 2022AH050099; The National Natural Science Foundation of China, under Grant 62471003; Chaohu University School-Enterprise Cooperation Project (No. hxkt20240227).

REFERENCES

[1] A. Pelamatti, V. Goiffon, A. Chabane, P. Magnan, C. Virmontois, O. Saint-Pé, and M. B. de Boisanger, "Charge transfer speed analysis in pinned photodiode cmos image sensors based on a pulsed storage-gate method," in *2015 45th European Solid State Device Research Conference (ESSDERC).* IEEE, 2015, pp. 156–159.

[2] T. M. Van Breussegem and M. S. Steyaert, "Monolithic capacitive dc-dc converter with single boundary–multiphase control and voltage domain stacking in 90 nm cmos," *IEEE Journal of Solid-State Circuits*, vol. 46, no. 7, pp. 1715–1727, 2011.

[3] C. Mingyang, Z. Menglian *et al.*, "Novel high efficiency low ripple charge pump using variable frequency modulation," in *2010 International Conference on Microelectronics.* IEEE, 2010, pp. 228–231.

[4] J. Gao, T. Gu, K. Nie, Z. Gao, and J. Xu, "A low-ripple charge pump with novel compensator for transient-response improvement in cmos image sensors," *IEEE Transactions on Circuits and Systems II: Express Briefs*, vol. 68, no. 4, pp. 1113–1117, 2020.

[5] J. Gao, R. Li, K. Nie, and J. Xu, "A linear charge pump with novel adaptive charge compensation structure for transient-response improvement in cmos image sensors," *Microelectronics Journal*, vol. 122, p. 105391, 2022.

[6] T. Yim, C. Lee, and H. Yoon, "A high speed modified dickson charge pump," in *2021 IEEE International Symposium on Circuits and Systems (ISCAS).* IEEE, 2021, pp. 1–5.

[7] Y. Li, Z. Li, L. Qian, X. Wang, and Z. Zhu, "A low quiescent current fast transient ldo regulator with segmented pass transistors," *IEEE Transactions on Circuits and Systems II: Express Briefs*, 2024.

[8] J. Gao, J. Zhao, K. Nie, and J. Xu, "A low ripple and fast transient response charge pump in cmos image sensors," *IEEE Sensors Journal*, vol. 24, no. 6, pp. 8142–8149, 2024.

[9] H. Kim, S. S. Kwak, and Y. S. Kim, "Ldo regulator optimized on power efficiency and load transient response with voltage damper and body loop feedback," in *2022 IEEE International Symposium on Circuits and Systems (ISCAS).* IEEE, 2022, pp. 3571–3574.

[10] N. Liu and D. Chen, "A transient-enhanced output-capacitorless ldo with fast local loop and overshoot detection," *IEEE Transactions on Circuits and Systems I: Regular Papers*, vol. 67, no. 10, pp. 3422–3432, 2020.

[11] H. Wang, J. Jing, and F. Li, "A self-regulating negative charge pump using multi-phase clock for wideband adcs," in *2022 IEEE International Symposium on Circuits and Systems (ISCAS).* IEEE, 2022, pp. 2525–2528.

2025 The 10th International Conference on Integrated Circuits and Microsystems

A 13-bit 1.5-GS/s Dual-Residue Pipelined ADC with Time-Domain Interpolation and Parallel Quantization

Zecheng Zhou[1], Depan Li[1], Longsheng Wang[1], Ting Li[2], Dongbing Fu[2], Dengquan Li[1]

[1]*Key Laboratory of Analog Integrated Circuits, School of Integrated Circuits, Xidian University, Xi'an, China*
[2]*The 24th Research Institute of China Electronics Technology Group Corporation, Chongqing, China*
dqli@xidian.edu.cn, tingtinghx@sina.com

Abstract—**This article presents a 13-bit dual-residue pipelined analog-to-digital converter (ADC) with time-domain (TD) interpolation and parallel quantization running at 1.5 GS/s. The TD interpolation improves the resolution of each stage to achieve high power efficiency. Since dual-residue ADCs eliminate the need for an exact gain, the mismatches in the parallel operation of quantization and amplification between the sub-ADC and multiplying digital-to-analog converter (MDAC) are mitigated. Consequently, high-speed conversion is enabled without the need for gain error calibration. Furthermore, a low-power dynamically biased open-loop amplifier with high linearity is proposed. Implemented in a 28-nm CMOS process, the ADC achieves a signal-to-noise and distortion ratio (SNDR) of 64.1 dB and a spurious-free dynamic range (SFDR) of 76.6 dB at a 684.1 MHz input frequency, yielding a Schreier figure of merit (FoM) of 170.5 dB.**

Keywords—*Analog-to-digital converter, dual-residue ADC, pipelined ADC, time-domain interpolation, dynamic amplifier*

I. INTRODUCTION

Combining high speed with high resolution, pipelined analog-to-digital converters (ADCs) are widely employed in measurement equipment and wireless communication systems. However, the residue amplifiers (RAs) in pipelined ADCs typically exhibit high power consumption, while interstage gain error (ISGE) limits ADC linearity. Consequently, pipelined ADCs generally require ISGE calibration, which incurs additional power and circuit overhead. To address above-mentioned issues, calibration-free techniques have been proposed [1-5]. Among these, the correlated level shifting (CLS) technique boosts open-loop gain through two-phase amplification [1]. However, the additional amplification time constrains ADC conversion speed. Instead of an exact gain, the dual-residue method necessitates two matched gains, which are much easier to achieve due to the improvements in device matching [2-5].

In a conventional pipelined ADC, the input signal is represented by the sum of the first stage's DAC voltage and the backend stage's voltage, divided by the inter-stage gain. Consequently, knowledge of the exact gain of RA is essential for quantization. By contrast, in a dual-residue ADC, the ratio of the residue voltage to the first stage's least significant bit (LSB) voltage is directly quantized by the backend stage, thereby eliminating the need for RA gain information.

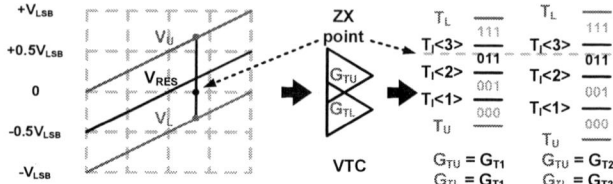

Fig. 1. Example of a 2-bit dual-residue operation in the time domain.

In dual-residue ADCs, quantization is performed through interpolation. Interpolation implemented with series capacitors leads to a sharp increase in area overhead when the interpolation factor (*IF*) is large[2]. Besides, parasitic capacitance can lead to the degradation of interpolation linearity. Current interpolation, on the other hand, suffers from static power consumption, which reduces power efficiency[3]. Capacitive interpolation SAR ADCs achieve excellent power efficiency with only dynamic power consumption. However, their speed is limited by the one-bit-per-cycle conversion principle. In this work, time-domain (TD) interpolation is employed. It achieves higher resolution and lower power consumption compared to [2], and higher speed compared to [3] and [4]. Furthermore, in this work, the reference is added to the input through the capacitive DAC (CDAC) instead of additional input pairs[6], which eliminates the mismatch caused by different common-mode voltages.

To improve the ADC speed, a post-amplification technique was proposed in [7], which parallels the quantization and amplification operations. However, directly amplifying the input signal restricts the input swing and leads to severe nonlinearity. In this work, parallel operations are performed in the second and third stages. While the RA operates, a voltage-to-time converter (VTC) converts the preceding stage's residue signal into a time interval. This time interval is then quantified by an interpolation time-to-digital converter (TDC). Since the dual-residue ADC does not require an exact gain, the full-scale ranges of the time and voltage domains are always aligned, eliminating the mismatch between the sub-ADC and the multiplying digital-to-analog converter (MDAC). As a proof of concept, a 13-bit 1.5-GS/s dual-residue TD pipelined ADC is implemented in a 28-nm COMS process. It achieves a signal-to-noise and distortion ratio (SNDR) of 64.1 dB at Nyquist input, resulting in a Schreier figure of merit (FoM$_S$) of 170.5 dB.

979-8-3315-8850-2/25 $31.00 © 2025 IEEE

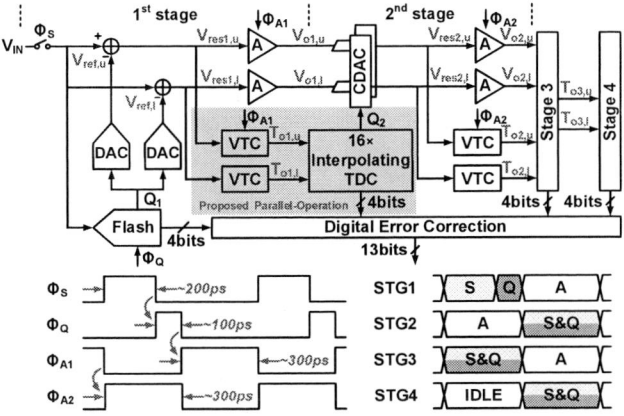

Fig. 2. Block and timing diagram of the proposed ADC.

This paper is organized as follows. Section II presents an overview of dual-residue ADCs and the proposed ADC architecture. Section III illustrates the implementation of the proposed prototype ADC. Section IV exhibits the simulation results, and Section V concludes this work.

II. PROPOSED DUAL-RESIDUE ADC ARCHITECTURE

A. Dual-Residue Operation in the Time Domain

Fig. 1 shows an example of the dual-residue architecture in the time domain. After the first stage completes the coarse quantization of the input signal, a residue signal V_{RES} is generated. V_U and V_L can be expressed as follows:

$$V_U = V_{RES} + \frac{V_{LSB}}{2} \quad (1)$$

$$V_L = V_{RES} - \frac{V_{LSB}}{2} \quad (2)$$

where V_{LSB} is a voltage of the LSB of the first stage. Subsequently, the two residue voltages are converted into T_U and T_L by VTC$_U$ and VTC$_L$, respectively, which can be defined as:

$$T_U = t_0 + G_{TU} \cdot V_U \quad (3)$$

$$T_L = t_0 + G_{TL} \cdot V_L \quad (4)$$

where t_0 is the VTC's time output for zero input voltage. G_{TU} and G_{TL} are the voltage-to-time (V-to-T) gains. After the VTCs complete the V-to-T conversion, the time information is sent to a phase interpolator (PI) array. For a PI array with an IF of 2^N, 2^N-1 interpolated time signals will be generated, which can be expressed as:

$$T_I < k > = T_L + \frac{k}{IF}(T_U - T_L) \quad (5)$$

$$= t_0 + G_T \cdot [V_{RES} - (\frac{1}{2} - \frac{k}{IF})V_{LSB}]$$

where $k = 1, 2, ..., IF-1$. The t_0 becomes the zero-crossing (ZX) point in the time domain. The k-th time comparator then compares $T_I<k>$ with t_0:

$$Q < k > = sign[T_I < k > -t_0] \quad (6)$$

Therefore, the position where the PI output time T_I crosses t_0, is not affected by the V-to-T gain G_T. The ratio of V_{RES} and V_{LSB} is represented in the digital domain as an (IF-1)-bit thermometer code. As shown in Fig. 1, with different V-to-T

gains, ZX point remains at the same location between $T_I<2>$ and $T_I<3>$, corresponding to the thermometer code of '011'.

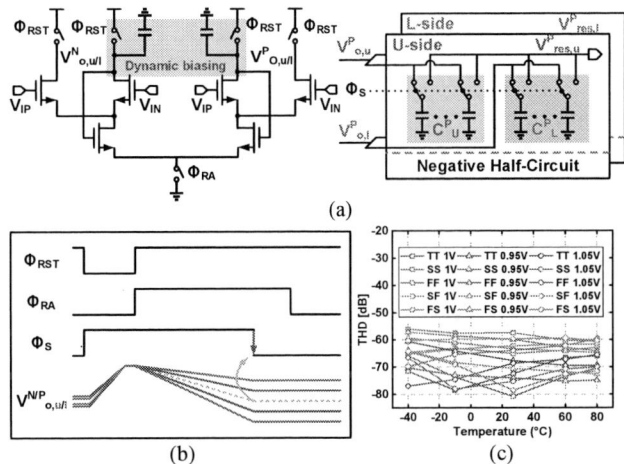

Fig. 3. (a) Schematic (single-ended load illustration for simplicity) and (b) timing diagram of the proposed dynamic RA. (c) The THD performance over voltage, process and temperature variations.

B. Proposed Architecture

Fig. 2 shows the proposed dual-residue pipelined ADC architecture and its timing diagram. This ADC consists of four

4-bit stages, with 1-bit redundancy between each stage to cover non-ideality in sub-quantization.

The first stage employs bottom-plate sampling with an 800 fF input capacitor to balance the sampling bandwidth and sampling noise. Unlike conventional pipelined ADCs, after the first stage flash resolves 4 bits, two CDAC are employed to generate dual residues, $V_{res1,u}$ and $V_{res1,l}$. After the residues are generated, Φ_{A1} enables the first-stage RA and VTC, amplifying the residue voltage and converting it into time signal, respectively. While the second-stage CDACs sample the RA outputs, the time-domain signals generated by the VTC are quantized by the interpolation TDC, which resolves 4 bits. After amplification, the second-stage CDACs generate dual residues based on the quantization of the TDC. The third stage is identical to the second, except for the absence of RAs, as the fourth stage only requires the time-domain signals for quantization without generating residues.

The parallel operation of amplification and quantization alleviates the overall timing constraints, providing sufficient amplification time for the RAs and improving power efficiency. Furthermore, since the dual-residue ADC eliminates the need for an exact gain, the voltage-domain residue and time-domain residue are inherently aligned. This avoids mismatches between the sub-ADC and the MDAC, thereby reducing the consumption of the redundancy range.

III. CIRCUIT IMPLEMENTATION

A. Dynamic RA with Dynamic Biasing Scheme

While the dual-residue architecture relaxes the requirement for RA gain accuracy, its linearity requirement remains consistent with traditional architectures. Fig. 3(a) shows the circuit implementation of the RA and CDAC. A dynamic RA

979-8-3315-8850-2/25 $31.00 © 2025 IEEE

based on a differential flipped voltage follower (DFVF) structure is proposed. Instead of a static current source, a dynamic bias is used to mitigate the effect of channel length modulation to improve the linearity of the RA. Fig. 3(b) shows the timing diagram of proposed RA. When the Φ_{RST} is low, the

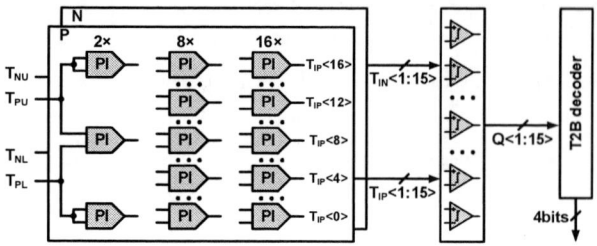

Fig. 4. Block diagram of the 4-bit interpolation TDC.

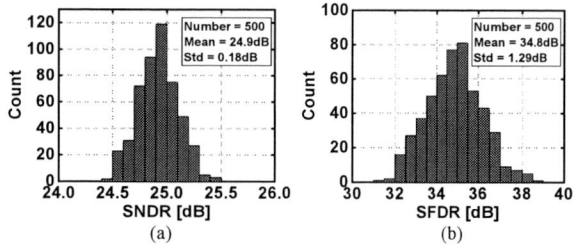

Fig. 5. Monte Carlo simulation results of the TDC (a) SNDR and (b) SFDR.

voltage on the CDAC is reset to V_{DD}. During the Φ_{RA} phase, V^P_O and V^N_O start to discharge at different rates depending on the input voltages. When the common-mode detector is triggered, Φ_S is pulled low, terminating the amplification. To attenuate the noise and non-ideality of backend stages, the RA gain is designed to be approximately 5.5. The amplifier offset is the main source of error in a dual-residue ADC. To overcome this problem, a simple offset calibration is performed [2]. Fig. 3(c) illustrates the total harmonic distortion (THD) of the RA versus process and temperature variations. It stays below -55 dB under a 75-mV$_{PP}$ input, which is sufficient for the overall ADC to achieve a SFDR of 82 dB.

B. Interpolation-Based TDC

Dual-residue ADCs utilize an interpolator to determine the position of the ZX point. In this work, inverter-based time-domain interpolation with an IF of 16 is employed [6]. Fig. 4 illustrates the schematic of the 16× TD interpolation TDC. For simplicity, dummy PI cells are not shown in the figure, which are used to balance the load and mitigate mismatch. The time-domain dual-residue signals generated by the VTCs are fed into a PI array to produce 32 time signals $T_{IP}<0:16>$ and $T_{IN}<0:16>$ with equal intervals on the positive and the negative side, respectively. Among these, $T_{IP}<1:15>$ and $T_{IN}<1:15>$ are compared by the time comparators to generate a 15-bit thermometer code Q<1:15>, which is then decoded by a thermometer-to-binary (T2B) decoder to produce 4 bits of binary code.

Fig. 5 shows the influence of random mismatch on the TDC dynamic performance. The mean value and standard deviation of the SNDR are 24.9 and 0.18 dB, respectively, while the spurious-free dynamic range (SFDR) exhibits a mean of 34.8 dB

and a standard deviation of 1.29 dB. Consequently, the TDC meets the accuracy requirements for a 4-bit sub-ADC.

C. Interpolation CDAC

In the first stage, the MDAC converts the signal into the dual residue domain by adding different voltages to V_{RES}. Voltage shifting can be achieved by adding extra capacitors in the

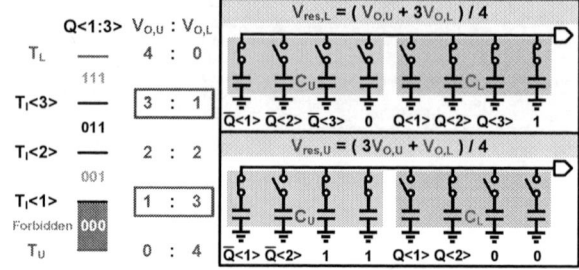

Fig. 6. The proposed interpolation CDAC (2-bit example).

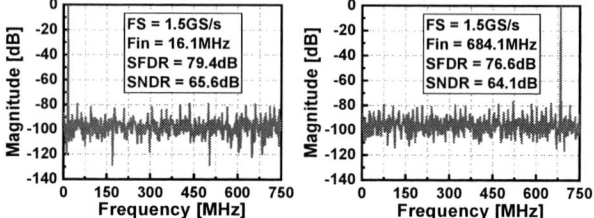

Fig. 7. Simulated output spectra.

MDAC of the first stage. However, since the signal is converted into the dual-residue domain, in the second and third stages, interpolation CDACs without reference voltages are used to generate the residues. A 2-bit single-ended interpolation CDAC is shown as an example in Fig. 6. For simplicity, the switches connecting the RA in the CDAC are not shown.

The interpolation TDC detects that the ZX point lies between $T_I<2>$ and $T_I<3>$, corresponding to a thermometer code of '011'. After amplification phase, the outputs of RAs, $V_{O,U}$ and $V_{O,L}$, are respectively stored in the C_U and C_L capacitor arrays. The interpolation CDACs generate the dual residues based on the quantization result. In the example shown in Fig. 6, the ratio of $V_{O,U}$ to $V_{O,L}$ lies between 3:1 and 2:2. $V_{res,L}$ and $V_{res,U}$ can be generated by interpolating $V_{O,U}$ and $V_{O,L}$ with ratios of 1:3 and 2:2, respectively. However, in this scenario, there is no redundancy between the pipeline stages. To cover non-ideality, 1-bit redundancy is applied. In this work, redundancy is achieved by adjusting the interpolation ratio. Instead of 2:2, $V_{res,U}$ is generated with a 3:1 ratio of $V_{O,U}$ to $V_{O,L}$. Consequently, the residue swing is half of the quantization range in the ZX domain, which is consistent with the inter-stage redundancy in conventional ADCs. In the circuit implementation, redundancy is generated by forcing Q<3> to '0'. However, in the example of Fig. 6, no redundancy is remained when the ZX point is located between $T_I<1>$ and T_U. Therefore, input signals within the '000' range are forbidden. As the resolution of sub-ADC increases, the proportion of the input range occupied by the prohibited region gradually decreases. In this work, the interpolation TDC resolution each stage is 4 bits. The prohibited region is 1/16 of the input range.

In the proposed ADC, the second and third stage MDACs each contain two interpolating CDACs, which generate $V_{res,U}$ and $V_{res,L}$, respectively. The CDAC input load for both is designed to be 200 fF to satisfy the noise requirement and mitigate the effects of parasitic capacitance. In conventional ADCs, the linearity of CDAC is generally immune to linear parasitic capacitance. However, in an interpolation CDAC, the interpolation will be affected. Therefore, prior to residue generation, the output node is reset to V_{CM}.

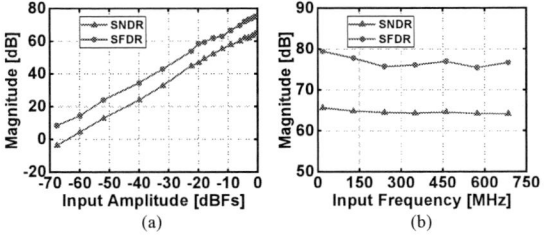

Fig. 8. Simulated dynamic performance versus input (a) amplitude and (b) frequency.

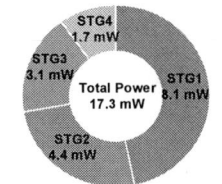

Fig. 9. Power breakdown.

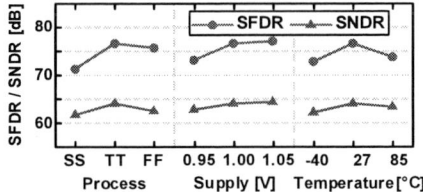

Fig. 10. Simulated dynamic performance versus PVT variation.

IV. SIMULATION RESULTS

The prototype dual-residue ADC is implemented in a 28-nm CMOS process and is simulated with 1-V supply voltage and 1.2-V_{PP} input. Fig. 7 shows the output spectra of the ADC running at 1.5 GS/s. With a 16.1-MHz input, the simulated SNDR is 65.6 dB and the SFDR is 79.4 dB. With near Nyquist input, the SNDR and SFDR is 64.1 and 76.6 dB, respectively.

Fig. 8 illustrates the simulated dynamic performance versus input amplitude and frequency at a 1.5 GS/s sampling rate. The dynamic range of the proposed ADC is 64.2 dB as shown in Fig. 8(a). The SNDR and SFDR stay above 64 and 75 dB under different input frequencies, respectively. Fig. 9 shows the power breakdown of the ADC. The total power consumption is 17.3 mW, with 8.1 mW attributed to the first stage, 4.4 mW to the second stage, 3.1 mW to the third stage and 1.7 mW to the last stage. Fig. 10 shows the ADC performance across different PVT conditions. The results demonstrate the robustness of the proposed architecture.

Table I summarizes the performance of the proposed ADC and compares it with other ADCs of similar speed and resolution. Thanks to TD interpolation and the parallel operation, the proposed ADC achieves a high sampling rate and high resolution with low power consumption, leading to a FoM$_S$ of 170.5 dB.

TABLE I.   PERFORMANCE SUMMARY AND COMPARISON

	This work*	[4]	[8]*	[9]	[10]
Process [nm]	28	28	28	40	28
Architecture	Dual-residue TD Pipe.	Dual-residue Pipe-SAR	SAR-DS	Pipe-SAR	Pipe-SAR
Resolution [bit]	13	13	13	13	12
Sampling Rate [GS/s]	1.5	0.5	0.5	0.625	1.5
SNDR@Nyq. [dB]	64.1	62.72	63.74	62.3	58.5
SFDR@Nyq. [dB]	76.6	76.83	74.42	70	74.5
Power [mW]	17.3	4.38	2.4	7.05	21.3
Fom$_S$@Nyq. [dB]	170.5	170.3	173.9	168.9	164
Calibration-free	Yes	Yes	No	No	No

*Simulation results

V. CONCLUSION

In this work, a 13-bit 1.5-GS/s dual-residue TD pipelined ADC is presented. The sub-ADC resolution is improved by leveraging the high linearity and energy efficiency of TD interpolation. Furthermore, the gain insensitivity of the dual-residue architecture minimizes the mismatch between the sub-ADC and MDAC during the parallel operation of quantization and amplification. Additionally, the proposed dynamic-biased RA exhibits high linearity across process and temperature variations, facilitating the generation of high-precision residues. This prototype ADC achieves a SNDR of 64.1 dB at a Nyquist input with a FoM$_S$ of 170.5 dB by using these techniques.

REFERENCES

[1] J.-C. Wang and T.-H. Kuo, "A 72-dB SNDR 130-MS/s 0.8-mW pipelined-SAR ADC using a distributed averaging correlated level shifting ring amplifier," *IEEE J. Solid-State Circuits*, vol. 57, no. 12, pp. 3794–3803, Dec. 2022.

[2] J. Mulder et al., "An 800 MS/s dual-residue pipeline ADC in 40 nm CMOS," in *ISSCC Dig. Tech. Papers*, Feb. 2011, pp. 184–185.

[3] K.-I. Cho, Y.-S. Kwak, H.-J. Kim, J.-H. Boo, S.-H. Lee, and G.-C. Ahn, "A 10-b 320-MS/s dual-residue pipelined SAR ADC with binary search current interpolator," in *Proc. IEEE Custom Integr. Circuits Conf.(CICC)*, Apr. 2019, pp. 1–4.

[4] W. Jiang, Y. Luo, P. Li, J. Guo, C. Chen, and Q. Liu, "A 13b 500MS/s dual-residue pipelined-SAR ADC with one-way switching capacitive interpolation and background offset calibration," in *Proc. IEEE Custom Integr. Circuits Conf. (CICC)*, 2024, pp. 1–2

[5] A. -J. Annema, "Analog circuit performance and process scaling," *IEEE Trans. Circuits Syst. II, Analog Digit. Signal Process.*, vol. 46, no. 6, pp. 711–725, Jun. 1999.

[6] D. Li, X. Zhao, Y. Shen, S. Liu, and Z. Zhu, "A 7-bit 3.8-GS/s 2-waytime-interleaved 4-bit/cycle SAR ADC 16× time-domain interpolation in 28-nm CMOS," *IEEE Trans. Circuits Syst. I, Reg. Papers*, vol. 70, no. 9, pp. 3557–3566, Sep. 2023

[7] Z. Zheng et al., "A 3.3-GS/s 6-b fully dynamic pipelined ADC withlinearized dynamic amplifier," *IEEE J. Solid-State Circuits*, vol. 57, no. 6, pp. 1673–1683, Jun. 2022.

[8] Q. Fan and J. Chen, "A 500-MS/s 13-Bit SAR-Assisted Time-Interleaved Digital-Slope ADC," 2019 *IEEE International Symposium on Circuits and Systems (ISCAS)*, Sapporo, Japan, 2019, pp. 1-5.

[9] X. Guo, R. Chen, Z. Chen, and B. Li, "A 13b 600–675MS/s tri-state pipelined-SAR ADC with inverter-based open-loop residue amplifier," *IEEE J. Solid-State Circuits*, vol. 58, no. 3, pp. 624–633, Mar. 2023.

[10] Y. Shen et al., "A 12b 1.5GS/s single-channel pipelined SAR ADC with a pipelined residue amplification stage," in *Proc. IEEE Custom Integr.Circuits Conf. (CICC)*, Apr. 2023, pp. 1–2.

2025 The 10th International Conference on Integrated Circuits and Microsystems

A Novel Self-adaptive Dead-time Control Circuit Design for GaN Gate Drive Circuit

Shupeng Yan
School of Integrated Circuits
Chongqing University of Posts
and Telecommunications
Chongqing, China
S240403005@stu.cqupt.edu.cn

Shengqi Yu*
School of Integrated Circuits
Chongqing University of Posts
and Telecommunications
Chongqing, China
yusq@cqupt.edu.cn
*Corresponding author

Yi Huang*
School of Integrated Circuits
Chongqing University of Posts
and Telecommunications
Chongqing, China
huangy@cqupt.edu.cn

Junjie Yi
School of Integrated Circuits
Chongqing University of Posts
and Telecommunications
Chongqing, China
S230431129@stu.cqupt.edu.cn

Pengyu Wang
School of Integrated Circuits
Chongqing University of Posts
and Telecommunications
Chongqing, China
S240431165@stu.cqupt.edu.cn

Abstract—Gallium Nitride (GaN) power electronic devices represent a core technology of the third generation of semiconductor materials, with primary applications in high-efficiency power conversion and high-frequency, high-power systems. In half-bridge configurations, a dead time must be introduced to prevent shoot-through. However, due to the lack of an intrinsic body diode in GaN devices, the reverse conduction voltage drop during the dead time is significantly higher than that of silicon-based devices, leading to substantial reverse conduction losses. These losses are further exacerbated at high frequencies due to increased switching and reverse conduction losses. Therefore, adaptive dead-time control is essential for optimizing the trade-off between efficiency and reliability. This paper analyzes the impact of dead time on the efficiency and reliability of GaN gate driver integrated circuits and proposes a novel dead-time logic control architecture design. The proposed design comprises four key modules: a negative voltage detection module (NVDM), a negative voltage control module (NVCM), a dead-time logic control module (DTLCM), and a slope current detection module (SCDM). The NVDM identifies negative voltage within 5 ns, while the NVCM enables near-zero-latency dead-time adjustment. The SCDM estimates current switching based on the voltage slew rate at the switching node. The proposed design achieves a minimum dead time of 0.7 ns and a maximum of 10.5 ns. The variation range of the falling-edge dead time is 22.2% wider compared to existing dead-time architectures.

Keywords—dead time control, negative voltage detection, self-adaptive control

I. Introduction

The significant advantages of Gallium Nitride (GaN) power devices have driven the rapid development of corresponding GaN gate driver integrated circuits, enabling wide adoption of GaN-based power electronic systems in the field of power electronics [1–4]. In high-frequency half-bridge applications,

reverse conduction loss during the dead time has become a critical bottleneck limiting system efficiency and reliability. Unlike conventional silicon-based devices, GaN devices lack a parasitic body diode, resulting in a reverse conduction voltage drop during dead time that can be up to three times higher than that of silicon devices [5][6]. Consequently, reverse conduction loss increases exponentially under high-frequency switching conditions [7]. For example, in 800 V platforms for electric vehicles, an increase of just 1 ns in dead time can reduce system efficiency by approximately 0.8%, while the thermal stress induced by reverse conduction further accelerates device degradation.

At high switching frequencies, the voltage slew rate (dV/dt) at switching nodes can reach 200 V/ns. A fixed dead time cannot compensate for delay variations caused by load fluctuations during voltage transitions [8][9]. Specifically, the optimal dead time decreases when the input voltage drops or the load current increases, and increases when the input voltage rises or the load current decreases. Therefore, adaptive control of the timing between high-side and low-side power switches is essential for improving overall system efficiency.

In recent years, extensive research has been conducted to approach utilizes the voltage difference between the drain and source of the power switch (detected via a high-voltage body diode) to implement self-adaptive dead time control for both the high and low sides [10–12]. However, due to the delay from the sensing point to the gate output, careful biasing of the diode front end is necessary, and the total transmission delay of the logic and driver circuits can reach several tens of nanoseconds.

Another approach involves adaptive dead time circuits based on resistor-capacitor voltage dividers [13][14], which monitor the voltage at the switching node to enable soft switching. However, this method requires different bias voltages for

CSTB2024YCJH-KYXM0063, Chongqing, CN

979-8-3315-8850-2/25 $31.00 © 2025 IEEE

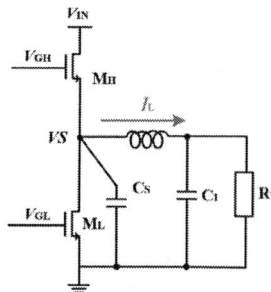

Fig. 1. Half-bridge circuit

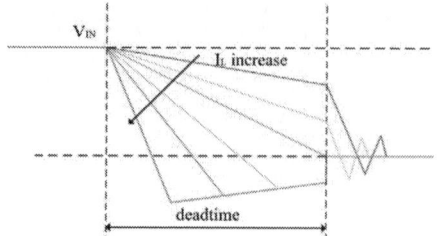

Fig. 2. Performance of imprecise control case

varying input voltages. Dual-edge adaptive control [15-17] allows independent control of the rising and falling edges of the dead time and uses a Phase Error Detector to achieve sub-nanosecond calibration. Nevertheless, it demands precise matching between analog and digital control loops and is sensitive to process variations.

In this paper, a novel control architecture is proposed, incorporating supply voltage and load current with negative voltage detection at the voltage switching node. Supply voltage detection helps prevent shoot-through losses, current detection enables dynamic dead time adjustment via comparator-controlled capacitive tuning, and negative voltage switching

detection reduces excessive reverse conduction loss. This integrated approach allows the realization of zero-voltage switching (ZVS) in half-bridge circuits.The innovations are as follows:

1. Adding negative voltage detection and negative voltage control circuits makes the circuit detection more stable.

2. The multi-stage capacitor isolation in the main control circuit increases the dead time range.

The rest of this paper is organized as follows: Section II presents the fundamental mechanisms of dead time. Section III introduces the proposed architecture and provides a detailed theoretical analysis. Section IV presents the experimental and simulation results. Section V concludes the paper and discusses prospects for future development.

II. PRINCIPLE ANALYSIS OF THE ADAPTIVE DEAD-TIME CIRCUIT

Fig.1 illustrates a typical half-bridge circuit. In conventional designs, a fixed dead-time unit is inserted between the high-side and low-side power switches (M_H and M_L) to prevent shoot-through events that could lead to catastrophic breakdowns [18]. However, under heavy load conditions, the voltage at the switching node (V_S) drops rapidly, potentially reaching 0 V or even becoming negative before the low-side switch fully turns on, resulting in substantial reverse conduction losses. Conversely, under light load conditions, the V_S voltage decreases more slowly. In this case, the low-side switch may turn on while V_S remains above 0 V, leading to considerable switching losses.

For system stability, fixed dead time designs typically account for the slowest expected VS transition rate. However, the precise control of the timing of the high-side and low-side

switches is essential for improving system efficiency. As shown in Fig. 2, a scenario where the dead time is too short, less than the duration required for soft switching, resulting in switching loss and decreased system efficiency.When the dead time is too long, the switching node VS deviates from its ideal value and inducing reverse conduction loss in the power devices.

Because GaN power devices lack an intrinsic parasitic body diode, reverse conduction loss during freewheeling periods is significantly higher than that in silicon-based devices. Therefore, precise dead-time control is critical to achieving high-efficiency operation in GaN-based systems [17].

III. CIRCUIT DESIGN

The system architecture of the proposed circuit is illustrated in Fig. 3. It comprises a negative voltage detection module, and a core dead-time modulation module. Two enhancement-mode GaN transistors, M_H and M_L, are employed as power switches to achieve superior system performance. The gate driver integrated circuit includes both high-side and low-side channels.

To prevent cross-conduction between the power switches, a controlled dead time is introduced by monitoring the supply voltage. The dead time control principle at the rising and falling edges of the switching node VS is symmetrical; for clarity, the operation during the falling edge is used here to explain the circuit functionality.

At the falling edge of V_S, the system employs a low-side current detection module that outputs a current control signal. This signal helps prevent excessive dead time, which would otherwise lead to increased reverse conduction losses. Additionally, a low-side negative voltage detection circuit suppresses negative voltage at the falling edge, which may occur due to load inductance freewheeling. These control signals are processed by the main logic controller to drive the low-side power switch M_L, ensuring soft switching operation.

By reducing both hard switching losses and reverse conduction losses, the proposed self-adaptive dead-time control circuit significantly enhances overall system efficiency. In summary, the design enables high-efficiency operation of power supply systems across a wide range of operating conditions.

A. Main Control Logic Module

Fig. 5 shows the dead-time modulation module proposed in this work. Here, VIN represents the supply voltage, and Id denotes the output load current. Control signals VGL and VGH

979-8-3315-8850-2/25 $31.00 © 2025 IEEE 154

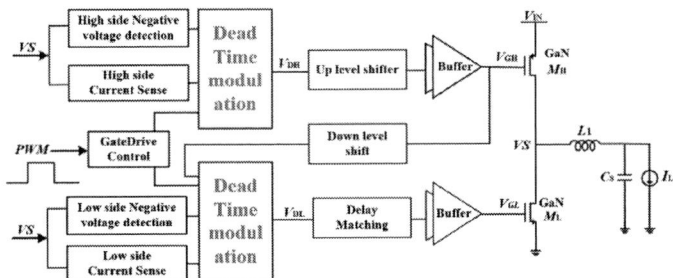

Fig. 3. System architecture

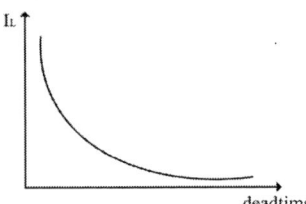

Fig. 4.dead time varies with the current

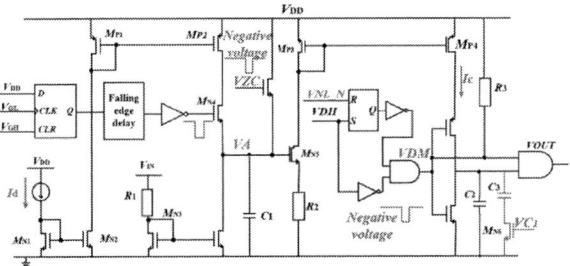

Fig. 5. Dead time modulation module

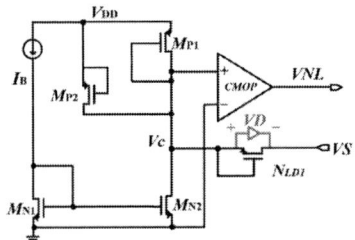

Fig. 6. negative voltage detection

are fed into D-type flip-flops, which generate pulses with a defined time width. These pulses are then passed through a falling-edge delay module, which extends the detection window.

As I_d and V_{IN} vary during operation, the circuit adaptively adjusts the output dead time to achieve zero-voltage turn-on. During each switching cycle, the stored charge on capacitor C_1 is given by:

$$Q_{C1} = t_{dead} \times k \times I_L \qquad (1)$$

In equation (1), k is the proportional coefficient of the current sensing circuit, I_L is the detected load current, t_{dead} is the sum of the dead time from the previous switching cycle and the falling-edge delay. Under these conditions, the dead time required to achieve zero-voltage turn-on (ZVS) can be calculated as:

$$t_{dead} = C_S \times V_{IN} / I_L \qquad (2)$$

Fig. 4 illustrates the nonlinear relationship between dead time and current. The detailed modulation mechanism of the circuit is as follows: when the input voltage V_{IN} decreases, node VA is pulled higher by the load current I_d, resulting in an increase in the output current I_C. This shortens the charging time of capacitors C_2 and C_3, thereby reducing the output dead time. Conversely, when V_{IN} increases, VA is pulled lower, I_C decreases and the output dead time becomes longer.

When Id increases, the charging time of capacitor C_1 is reduced, causing VA to rise and I_C to increase accordingly. As a result, the charging time of C_2 and C_3 is further shortened, leading to a reduced dead time. Conversely, when Id decreases, the charging time of C_1 increases, VA drops, I_C decreases resulting in an extended dead time.

The signal NZC represents the negative-voltage control fixed logic level, which elevates VA to turn on transistor M_{N5},

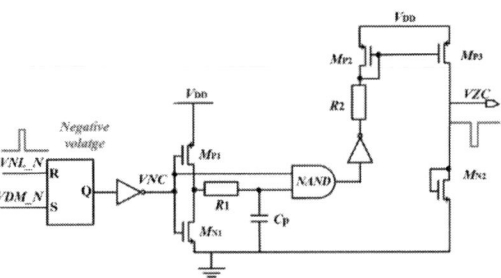

Fig. 7. Negative voltage control logic module

thereby enforcing a minimum dead time to avoid a zero-dead-time condition.

When a negative-voltage event is detected, VZC inputs a pulse, which momentarily pulls down the voltage and simultaneously raises VA, causing the output current I_C to increase and the resulting dead time to decrease. VDM also

outputs a narrow pulse, the width of which depends on the timing difference between the falling edges of VGH and the negative-voltage signal.

At this point, the role of VDM is to prevent excessive dead time. If the calculated dead time exceeds the pulse width of VDM, the system will forcibly limit the dead time to that width to prevent a negative-voltage event from occurring in the next switching cycle.

The signal V_{C1} serves as the input for the slope-based current detection. Under heavy-load conditions, the switching node VS exhibits a higher voltage slew rate (dV/dt). In this case, V_{C1} will go low, disconnecting C_3 and reducing the effective capacitance. Consequently, the charging time shortens and the dead time is reduced.

Under light-load conditions, V_{C1} remains high, maintaining the connection of both C_2 and C_3, increasing the effective capacitance. As a result, the charging time increases and the output dead time becomes longer.

979-8-3315-8850-2/25 $31.00 © 2025 IEEE

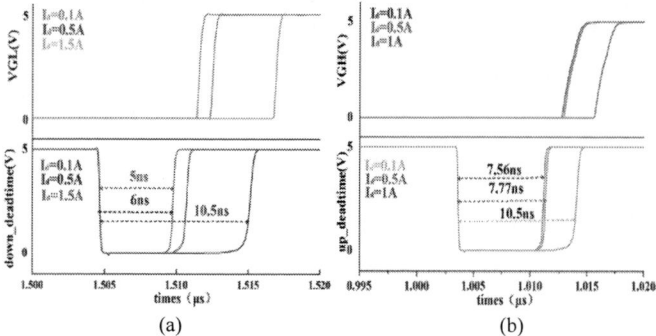

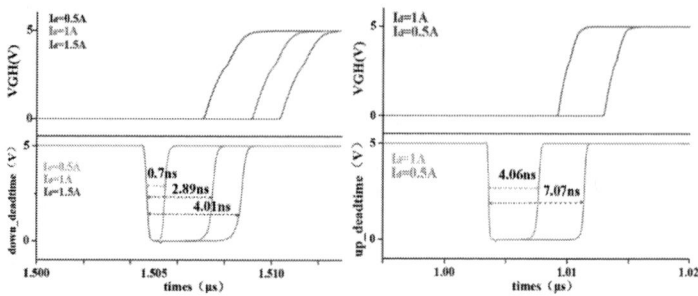

(a)   (b)

Fig. 8. V_{IN}=100V I_{load} from 0.1A to 1A (a)falling edge(b)rising edge

(a)   (b)

Fig. 9. V_{IN}=100V I_{load} from 0.5A to 1.5A (a) falling edge (b) rising edge

. TABLE I.   PERFORMANCE

PowerSwitches	Workingfrequency	Dead-time	VSvoltage
GaN	10MHz	0.7,10.5ns	100V

B. Negative Voltage Detection Module for the VS Node

The negative voltage detection circuit for the switching node VS is shown in Fig. 6 The core detection mechanism utilizes an LDMOS transistor (N_{LD1}) with its gate and source shorted, forming a normally-off high-voltage diode. This device serves the dual purpose of detecting negative voltage at the switching node and providing high-voltage isolation.

The detection process operates as follows: When VS is at 0 V or a positive voltage, N_{LD1} remains in the off state. In this condition, a current mirror biases node V_C to a fixed reference voltage, and the comparator outputs a high logic level; When VS drops below 0 V, the body diode of NLD1 turns on, clamping V_C to VS+VD, effectively pulling V_C to a negative potential. This negative voltage causes the comparator to toggle its output, driving the negative voltage detection signal VNL low, thereby indicating the presence of a negative voltage at the switching node.

C. Negative Voltage Control Logic Module

The negative voltage control logic module is illustrated in Fig. 7. In this circuit, VNL serves as the negative voltage detection signal, VDH is the pre-shifted version of the high-side gate drive signal VGH, and VZC is the output signals.

The control logic operates as follows: When no negative voltage is detected, VNL remains at a high level. The output of a subsequent RS flip-flop, VNC, transitions to a rising edge, which discharges the capacitor Cp. VZC is consequently driven to a fixed bias voltage.

When a negative voltage is detected, VNL transitions low. VDM tbecome a short-duration pulse, the width of which is determined by the falling edges of VNL and VDH. VZC follows by producing a corresponding pulse. The pulse width of VZC is determined by the duration of the VDM pulse and the capacitance of Cp. The width of the VZC pulse dynamically tracks the variation in VDM 's pulse width.

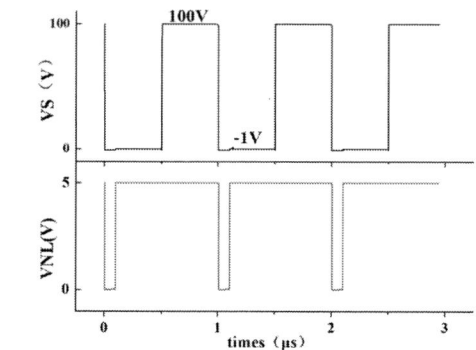

Fig. 10. negative voltage detection module for the VS node

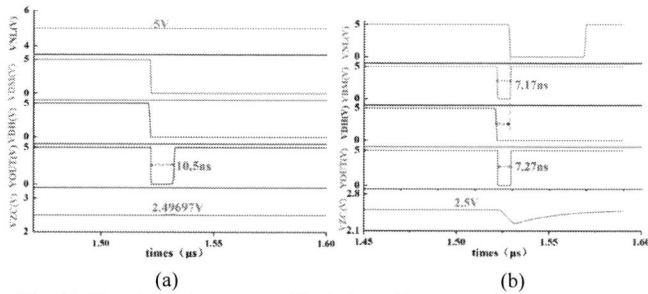

(a)   (b)

Fig. 11. Negative voltage control logic (a) With no negative voltage (b) With negative voltage

IV. SIMULATION RESULTS

The input signal of the adaptive dead time control circuit designed in this paper is VS voltage. The simulation conditions were set to V_{IN} from 100V to 20V. The simulated process is the 0.18um process of XFAB.

A. Simulation of Main Control Logic Module

The simulation results of Main Control Logic Module under light load are shown in Fig. 8, and it can be seen that when V_{IN} = 100V, the load current changes from 0.1A to 1A, and the dead time of the falling edge of VS decreases from 10.6ns to 2.9ns. The rising edge dead time of VS is reduced from 10.5ns to 7.56ns, which meets the requirements of adaptive dead time. Additionally, the negative voltage in each cycle is only a few tens of millivolts, which is within the acceptable range. The circuit performance is shown in Table 1.

979-8-3315-8850-2/25 $31.00 © 2025 IEEE   156

Fig. 9 shows the waveforms under heavy load conditions. Due to the current variation, the triggering of the VZC signal flips the output capacitance of the main control logic module, causing the minimum dead time of the falling edge to reduce to 0.7ns, while the maximum dead time becomes 4.01ns. The variation range of the dead time is 22.2% wider compared to existing dead-time architectures. For the rising edge, the minimum dead time decreases to 4ns, and the maximum dead time increases to 7.07ns. The negative voltage within each cycle remains at a few tens of millivolts.

B. Simulation of Negative Voltage Detection Module for the VS Node

As shown in Fig. 10, the simulation waveforms of the dead-time negative-voltage detection module indicate that when the VS node drops to a negative voltage, a short pulse signal is generated. The detection delay for the falling edge does not exceed 5ns, while that for the rising edge remains within 2ns. The output pulse width is correlated with the dead time duration.

C. Simulation of Negative Voltage Control Logic Module

Fig. 11 presents the simulation of the negative-voltage logic control circuit. As shown in Fig. 11(a), when no negative voltage occurs, the VDM output waveform matches VDH with a delay of no more than 1ns, Under this condition, the dead time is about 10.5ns. As illustrated in Fig. 11(b), when a negative-voltage signal appears, VDM and VZC outputs a pulse signal. In this case, the output dead time is reduced to 7.17ns.

V. CONCLUSION

In order to solve the problem of dead time when GaN gate driver chip is driven, a new GaN adaptive logic control system is designed, which adopts current, voltage and negative voltage to control together. A negative voltage detection module and a control logic module are added to reduce the negative voltage when the VS node is discharged and realize zero-voltage switching. The automatic dead-time control technology is proposed, which realizes a wide range of soft switching, reduces the switching loss, and improves the power efficiency. The simulation results show that the minimum dead time is 0.7ns and the maximum is 10.5ns, which can achieve a wide range of dead time changes.

ACKNOWLEDGMENT

The authors gratefully acknowledge funding support from Chongqing Talented Scientific Research Project (ref: CSTB2024YCJH-KYXM0063, Chongqing, CN).

REFERENCES

[1] Cong L, Lee H. A 1–2-MHz 150–400-V GaN-Based Isolated DC–DC Bus Converter with Monolithic Slope-Sensing ZVS Detection[J]. IEEE Journal of Solid-State Circuits, 2018, 53(12): 3434-3445

[2] Luo D, Gao Y, Mok P. K. T. A GaN Driver for a Bi-Directional Buck/Boost Converter with Three-Level VGS Protection and Optimal-Point Tracking Dead-Time Control[J]. IEEE Transactions on Circuits and System I: Regular Papers, 2022, 69(5): 2212-2224

[3] Schwarzott, C. Kuring, N. Wieczorek and S. Dieckerhoff, "Self-controlled driver circuit optimizing dead time in fast GaN converters," CIPS 2024; 13th International Conference on Integrated Power Electronics Systems, Düsseldorf, Germany, 2024, pp. 446-452.

[4] Jing Xue, K. D. T. Ngo and Hoi Lee, "A 99%-efficiency 1-MHz 1.6-kW zero-voltage-switching boost converter using normally-off GaN power transistors and adaptive dead-time controlled gate drivers," 2013 IEEE International Conference of Electron Devices and Solid-state Circuits, Hong Kong, China, 2013, pp. 1-2, doi: 10.1109/EDSSC.2013.6628142.

[5] G. H. Thuc and C. -J. Chen, "An Integrated Driver With Dual-Edge Adaptive Dead-Time Control for GaN-Based Synchronous Buck Converter," in IEEE Transactions on Industry Applications, vol. 60, no. 6, pp. 9157-9170, Nov.-Dec. 2024, doi: 10.1109/TIA.2024.3454198.

[6] Cong L, Lee H. A 150V monolithic synchronous gate driver with built-in ZVS detection for half-bridge converters[C]. IEEE Applied Power Electronics Conference and Exposition (APEC), 2018: 1861-1864

[7] L. Weihs, J. Grobe, L. Rimpl, T. Zekorn, R. Wunderlich and S. Heinen, "An Adaptive Dead-Time Control Method for Gate Drivers Using Gate Current Measurement Enabling ZVS in High Frequency HV DC-DC Converters," 2023 IEEE International Symposium on Circuits and Systems (ISCAS), Monterey, CA, USA, 2023, pp. 1-5, doi: 10.1109/ISCAS46773.2023.10181638.

[8] Cheng Q, Lee H. A high-frequency non-isolated ZVS synchronous buck-boost LED driver with fully-integrated dynamic dead-time controlled gate drive[C]. IEEE Applied Power Electronics Conference and Exposition (APEC), 2018: 419-422

[9] Z. Zeng, P. Cao, S. Lam and M. Cui, "A 48 V/1 MHz Monolithic GaN DC-DC Buck Converter with Sub-20-ns Dead-Time Control," 2024 9th International Conference on Integrated Circuits and Microsystems (ICICM), Wuhan, China, 2024, pp. 730-734, doi: 10.1109/ICICM63644.2024.10814347.

[10] Liu Z, Cong L, H. Lee. Design of On-Chip Gate Drivers with Power-Efficient High-Speed Level Shifting and Dynamic Timing Control for High-Voltage Synchronous Switching Power Converters[J]. IEEE Journal of Solid-State Circuits, 2015, 50(6): 1463-1477

[11] Liu Z, Lee H. A Wide-Input-Range Efficiency-Enhanced Synchronous Integrated LED Driver with Adaptive Resonant Timing Control[J]. IEEE Journal of Solid-State Circuits, 2016, 51(8): 1810-1825

[12] Yu S, Zhou Q, Shi G, et al. A 400-V Half Bridge Gate Driver for Normally-off GaN HEMTs with Effective dv/dt Control and High dv/dt Immunity[J]. IEEE Transactions on Industrial Electronics, 2023, 70(1): 741-751

[13] Wittmann J, Funk T, Rosahl T, et al. A 48-V Wide-VIN 9–25-MHz Resonant DC–DC Converter[J]. IEEE Journal of Solid-State Circuits, 2018, 53(7): 1936-1944

[14] G. H. Thuc and C. -J. Chen, "An Integrated Driver With Dual-Edge Adaptive Dead-Time Control for GaN-Based Synchronous Buck Converter," in IEEE Transactions on Industry Applications, vol. 60, no. 6, pp. 9157-9170, Nov.-Dec. 2024, doi: 10.1109/TIA.2024.3454198.

[15] C. Tang, M. Jiang and P. Zhao, "An Adaptive Daul Step Control Dead-Time Circuit for Gallium Nitride Half-Bridge," 2022 7th International Conference on Integrated Circuits and Microsystems (ICICM), Xi'an, China, 2022, pp. 67-71, doi: 10.1109/ICICM56102.2022.10011251.

[16] H. Qin, X. Zheng, W. Wang and Q. Xun, "A Novel Adaptive Dead-Time Control Method for GaN-Based Motor Drives," in IEEE Transactions on Energy Conversion, vol. 40, no. 1, pp. 258-269, March 2025, doi: 10.1109/TEC.2024.3431940.

[17] G. H. Thuc and C. -J. Chen, "An Integrated Driver With Dual-Edge Adaptive Dead-Time Control for GaN-Based Synchronous Buck Converter," in IEEE Transactions on Industry Applications, vol. 60, no. 6, pp. 9157-9170, Nov.-Dec. 2024, doi: 10.1109/TIA.2024.3454198.

[18] Du L, Ma D. B. A 48V-to-1V Buck-Assisted Active-Clamp Forward Converter with Reduced Voltage Stress for Datacenter Applications[C]. IEEE Energy Conversion Congress and Exposition (ECCE), 2020: 5442-5446

2025 The 10th International Conference on Integrated Circuits and Microsystems

A Novel Hybrid Switching Scheme for SAR ADC with High Efficiency and High Linearity

Yushi Chen†, Ying Li, Fang Liang, Zewen Wang, Zuoru Dong and Xiaodong Wang

The 50th Institute of China Electronics Technology Group Corporation, Shanghai 200331, China

†Correspondence to: Yushi Chen, Email: yushichen001@outlook.com

Abstract—This paper gives a hybrid switching scheme specifically designed for successive approximation register (SAR) analog-to-digital converter (ADC). The presented scheme does not require energy consumption during the first three cycles by utilizing zero energy consumption switching technique (ZECST). Moreover, a multi-level switching technique (MLST) is applied during remaining cycles to further improve efficiency. Based on above two techniques, a reduction of 99.48% in switching energy and 75% in the number of capacitors has been achieved. The presented method exhibits a DNL of 0.162LSB and an INL of 0.167LSB.

Keywords—*SAR ADC, Zero energy consumption switching technique (ZECST), Multi-level switching technique (MLST), High efficiency, High linearity*

I. INTRODUCTION

In recent years, low power consumption successive approximation register (SAR) analogue-to-digital converters (ADCs) have developed fast because of huge requirements in wireless sensors and biomedical electronic systems [1-3].As the core module of signal processing module, ADCs convert analog signals, which are continuous in nature, to digital signals, which are discrete and can be processed by digital systems, thereby enabling the realization of digital transmission for various applications. There are many kinds of ADCs, such as flash ADCs, pipeline ADCs, SAR ADCs and sigma delta ADCs. Flash ADCs and pipeline ADCs are applied in high-speed applications, but with high power consumption. Sigma delta ADCs are suitable for high precision audio signal processing. SAR ADCs exhibit advantages including energy efficiency and straightforward structure when compared to other ADCs [4-11]. Its power consumption primarily encompasses contributions from digital circuits, the comparator, bootstrap switches, and the switching energy associated with DAC capacitor arrays [12]. Within this comprehensive breakdown of energy consumption, it is noteworthy that the switching energy of the DAC capacitor arrays accounts for the largest proportion [13-14]. This highlights the significance of optimizing the energy expenditure related to the DAC capacitor arrays to minimize power consumption.

Several techniques have been given to optimize related power consumption. When compared to traditional methods, the monotonic method [15] and the V_{CM}-based method [16] achieve power savings by 81.26% and 87.54%. The methods reported in Tri-level [17], VMS [18] and Sun [19] improve efficiency, which ignore the large energy caused by parasitic capacitance. The scheme mentioned in Tong [20] lifts

efficiency by 98.7%. While this method shows poor linearity performance. The method reported in Xie [21] achieves high efficiency by 98.83%. However, it exhibits a substantial common-mode voltage variation, which could potentially impact the comparator's offset.

A novel hybrid switching scheme that combines high efficiency with exceptional linearity is reported in the paper. The presented method is designed to minimize energy waste during the initial stages of operation. Specifically, it eliminates any energy loss during the first three cycles through the innovative application of a zero energy consumption switching technique (ZECST). This ensures that no energy is squandered during these initial cycles. Furthermore, to enhance energy savings throughout the entire conversion process, the scheme employs a multi-level switching technique (MLST) for the remaining cycles. This dual-pronged approach not only conserves energy but also significantly boosts overall efficiency. When compared to traditional switching schemes, the proposed method demonstrates remarkable improvements, which enhances efficiency by 99.48%. Additionally, the innovative design allows for a 75% reduction in total area occupied by capacitors, contributing to more compact and efficient systems. Moreover, the presented method excels in linearity. This is evidenced by the low DNL and INL, which are 0.162LSB and 0.167LSB. These figures make the method an ideal choice for applications requiring high precision and linearity.

The organization of this paper is as follows: Section II provides a detailed explanation of the hybrid switching method. Section III presents simulation results along with an energy analysis. In Section IV, a discussion on common-mode voltage variation is conducted, while an analysis of linearity is made in Section V. The simulation results of the proposed scheme applied on a SAR ADC is given in section VI. Last, Section VII concludes the paper.

II. PROPOSED SWITCHING SCHEME

979-8-3315-8850-2/25 $31.00 © 2025 IEEE

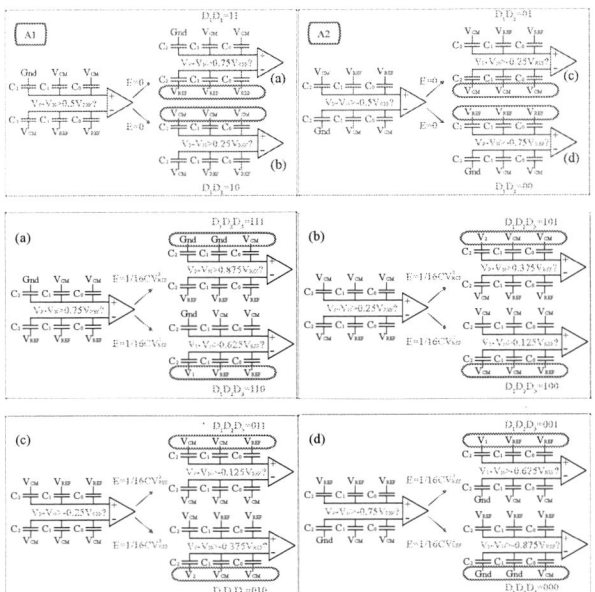

Fig. 1 The example of a 4-bit SAR ADC based on proposed switching scheme

Fig. 1 gives detailed illustration of the presented scheme, specifically implemented in a 4-bit ADC. DAC capacitor arrays adopt a differential structure, where the capacitors of the MSB, 2nd-MSB and LSB are 2C, C and C, respectively. In order to simplify the question, we assume that V_{REF}=1V, V_{CM}=0.5V and Gnd=0V. In the sampling phase, the switching scheme employs top-plate sampling, top-plates of capacitors are linked to input signal, while bottom-plates are set to [0.5, 1, 1]. The MSB is acquired subsequent to the sampling phase, without any intervening switching activity. Therefore, no energy is utilized during the initial comparison cycle. In the proposed method, switch activity happens on the array with higher voltage side after the first comparison. When MSB is acquired, the capacitor array responsible for sampling the higher input voltage is loaded with the specific sequence [0, 0.5, 0.5], while the alternative capacitor array maintains its original state without any alteration. Throughout this process, no energy is expended on switching. After the second comparison, switch activity happens on the array with lower voltage side. The reference of bottom-plates that has a lower voltage will change from [0, 0.5, 0.5] to [0.5, 0.5, 0.5] or from [0.5, 1, 1] to [1, 1, 1] according to the 2nd MSB. The array of capacitors featuring a higher voltage remains invariable, which is shown in Fig. 1(A1) and (A2). Then the 3rd MSB is achieved after comparing two voltages of the capacitor array with no energy consumption. Thus, the switching method presented does not utilize any energy during the initial three cycles, which is named zero energy consumption switching technique (ZECST). Therefore, the total energy consumption during switching can be significantly decreased by utilizing ZECST.

To further reduce energy, the proposed scheme applies a multi-level switching technique (MLST), which adds two different voltages V_1=3/4V_{REF}=0.75V and V_2=1/4V_{REF}=0.25V after the 3rd comparison cycle. The MLST is enabled when 3rd MSB (D_3) is different from 2nd MSB (D_2). Fig. 2 gives detailed comparison of the scheme with and without MLST. As

shown in Fig. 2a(1) and b(1), it is apparent that reducing reference of C_1 from V_{REF} to V_{CM} is equivalent to maintaining C_1 at V_{REF} while decreasing reference of C_2 from V_{REF} to V_1. With the similar method in Fig. 2a(2) and b(2), reducing the voltage on the bottom-plate of C_1 from V_{CM} to Gnd is equivalent to keeping the C_1 at V_{CM} and discharging the bottom-plate of C_2 from V_{CM} to V_2. Based on results, the scheme with MLST requires 1/16CV_{REF}^2, which reduces 66.7% energy than the scheme without MLST. Overall, the MLST can significantly reduce switching energy during the conversion. Extended to an N-bit SAR ADC, when the i th bit is different from the 2nd MSB, 3≤i≤N, the MLST is enabled to make the scheme efficient.

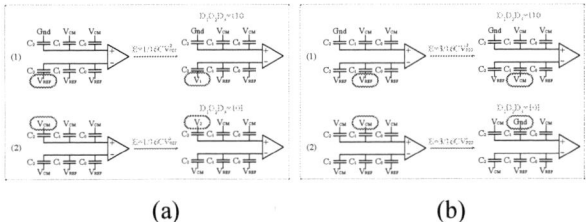

(a) (b)

Fig. 2 Comparison of switching schemes (a) with MLST (b) without MLST

III. SWITCHING ENERGY

A. Switching energy analysis and comparison

The introduced switching method reduces power consumption through the utilization of ZECST and MLST. Figure 7 presents the average switching energy derived from MATLAB simulations, comparing the proposed 10-bit switching scheme with various published schemes in references [15-18,20,22]. The presented method exhibits 7.11CV_{REF}^2 average switching energy, representing a 99.48% reduction in energy consumption. Table I presents an overview on key characteristics of various methods. Average switching energy of a presented N-bit SAR ADC is calculated as:

$$E_{avg,proposed} = \sum_{i=1}^{N-3} (2^{N-i-7})CV_{REF}^2 \quad (1)$$

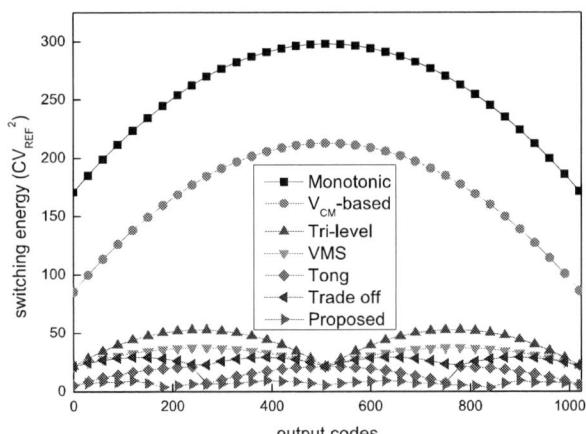

Fig. 3 Comparison of energy for different switching schemes

TABLE I.   AN OVERVIEW ON KEY CHARACTERISTICS OF VARIOUS METHODS

Switching scheme	Number of capacitors	Area reduction	Average switching energy(CV_{REF}^2)	Energy saving
Conventional	2048	reference	1363.3	reference
Monotonic	1024	50%	255.5	81.26%
V_{CM}-based	1024	50%	170.17	87.52%
Tri-level	512	75%	42.42	96.89%
VMS	512	75%	31.88	97.66%
Tong	512	75%	15.88	98.83%
Trade-off	512	75%	26.54	98.05%
Proposed	512	75%	7.11	99.48%

TABLE II.   SUMMARY OF ENERGY WITH AND WITHOUT PARASITIC CAPACITANCE

Switching scheme	Average switching energy (CV_{REF}^2)	
	$C_{pt}=C_{pcomp}=C_{pb}=0$	$C_{pt}+C_{pcomp}=10\%C_t$ $C_{pb}=15\%C$
Conventional	1363.3	1686.1
Monotonic	255.5	255.5
V_{CM}-based	170.17	209.2
Tri-level	42.42	74.78
VMS	31.88	56.3
Tong	15.88	21.3
Trade-off	26.54	N.A.
Proposed	7.11	15.67

B. Parasitic capacitance analysis and comparison

In the design of schematics, parasitic capacitance has effects on switching energy which should not be ignored [20]. Fig. 4 gives the model of the parasitic capacitance. There exist two main types of parasitic capacitance: C_{pb} and C_{pt}. The capacitance labeled as C_{pb} represents parasitic capacitance that exists between bottom-plates and substrate. C_{pt} is parasitic capacitance of the top plate which are connected in parallel. Moreover, C_{pcomp} is the parasitic capacitance of the comparator. Fig. 4 gives an example to illustrate parasitic capacitance of presented method. Ideally, energy consumption during the switching procedure should be zero. However, taking into account parasitic capacitance, energy will be required to charge this capacitance from 0 to 0.5.

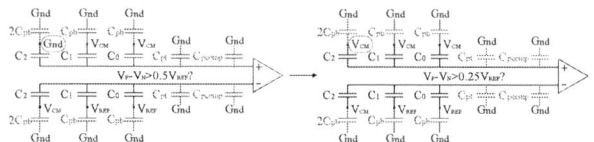

Fig. 4 Analysis of parasitic capacitance in the proposed switching scheme

A MATLAB simulation was conducted with the assumption that the sum of C_{pt} and C_{pcomp} equals 10% of C_t, and C_{pb} accounts for 15% of C. Fig. 5 presents simulation results of the difference between cases with and without the inclusion of parasitic capacitance. Considering parasitic capacitance, the proposed average energy amounts to 15.67 CV_{REF}^2. Table II gives a summary of energy after taking parasitic capacitance into consideration.

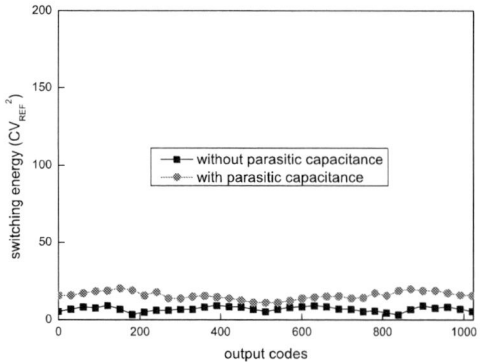

Fig. 5 Comparison of energy for the proposed switching schemes (with parasitic capacitance)

IV. DISCUSSION OF COMMON-MODE VOLTAGE VARIATION

The input common-mode variation of comparator has effects on the dynamic offset, which increases difficulty of system design. It is necessary to have a discussion about the input common-mode variation of comparator. Fig. 6(a) and (b) show the waveforms of both the proposed scheme and the Tong scheme for comparison. The proposed method exhibits a common-mode voltage variation of $1/4V_{REF}$, marking a 50% reduction compared to the Tong scheme, making it more favorable for system design. Table III presents a comprehensive comparison between the proposed method and existing ones.

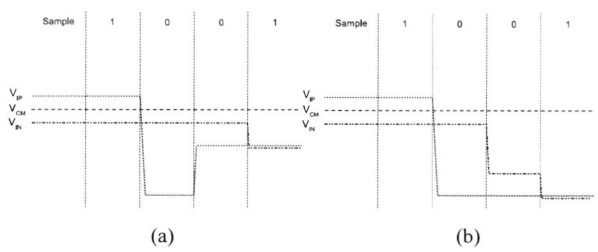

(a)   (b)

Fig. 6 Output waveform of the capacitor array (a) Proposed switching scheme (b) Tong switching scheme

TABLE III.   COMPARISON OF COMMON-MODE VOLTAGE VARIATION FOR DIFFERENT SCHEMES

Switching scheme	Maximum common-mode variation	Switching scheme	Maximum common-mode variation
Conventional	0	VMS[18]	$V_{REF}/4$
Monotonic[15]	$V_{REF}/2$	Tong[20]	$V_{REF}/2$
V_{CM}-based[16]	0	Trade-off[22]	$V_{REF}/2$
Tri-level[17]	$V_{REF}/2$	Proposed	$V_{REF}/4$

V. LINEARITY

Generally, linearity is associated with the capacitance values and the extent of capacitor mismatch deviation. To assess the linearity characteristics, we presumed that the variations in the unit capacitors (C_u) followed a Gaussian distribution ($N(0, \sigma_u^2)$), where σ_u represents the standard deviation of the mismatches among these unit capacitors. A

Monte-Carlo simulation encompassing 500 iterations has been conducted in MATLAB. In this simulation, the mismatch of C_u is set to satisfy the condition of $3\sigma_u = 0.01C$ [23]. Table IV summarises results of different schemes. The DNL and INL of the proposed method are displayed in Fig. 7, which are 0.162LSB and 0.167LSB. Apart from trade-off method, The proposed method shows better linearity performance than other methods. The proposed method consumes less energy and has smaller common-mode variation than trade-off method. Therefore, the comprehensive performance of the scheme proposed in this paper is better.

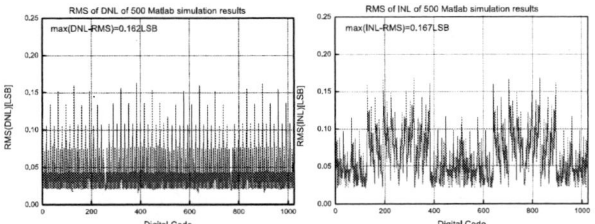

Fig. 7 DNL and INL of the proposed switching scheme

TABLE IV.   COMPARISON OF DNL AND INL

Switching scheme	DNL/INL(LSB)	Switching scheme	DNL/INL(LSB)
Conventional	0.563/0.422	VMS[18]	0.172/0.226
Monotonic[15]	0.48/0.48	Tong[20]	0.48/0.34
V_{CM}-based[16]	0.265/0.368	Trade-off[22]	0.116/0.083
Tri-level[17]	N.A.	Proposed	0.162/0.167

VI. SIMULATION RESULTS

A 50MS/s 10-bit 1V SAR ADC employed the proposed switching scheme is implemented in MATLAB. A 16384-point fast Fourier transform (FFT) of the 50MS/s SAR ADC when the input frequency is 21.7MHz is shown in Fig. 8. The signal-to-noise and distortion ratio (SNDR) and the spurious free dynamic range (SFDR) can reach 61.19 and 85.41dB, respectively. Fig. 9 presents the SNDR and SFDR versus the input signal amplitude, achieving a peak SNDR of 61.83dB and a peak SFDR of 85.41dB.

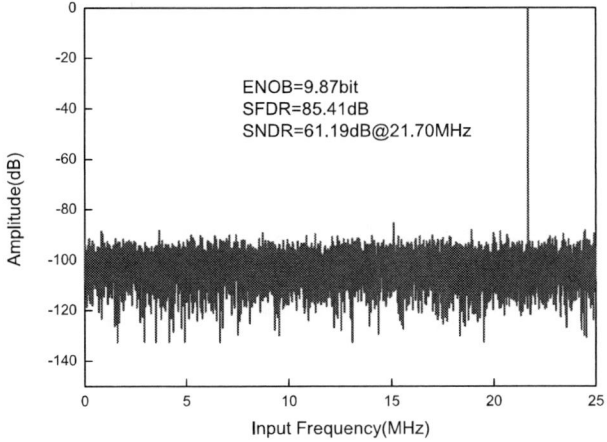

Fig. 8 16384-point FFT spectrum at 50MS/s with Nyquist input

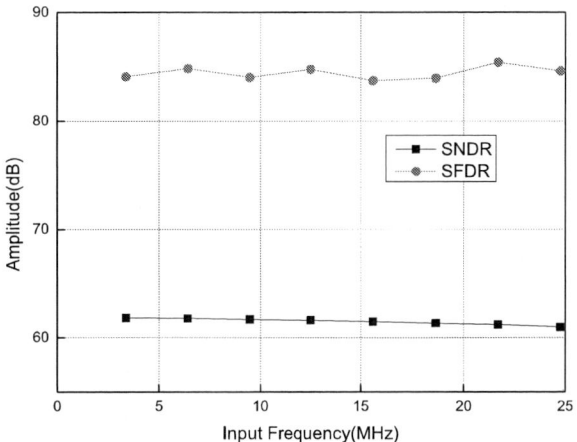

Fig. 9 Dynamic performance of the SAR ADC versus the input

VII. CONCLUSION

The paper introduces a novel hybrid switching method for SAR ADC, incorporating two innovative techniques. Thanks to ZECST, the presented scheme does not utilize any energy during the initial three cycles. The other technique is MLST, which adds additional two voltage levels to further decrease energy during the remaining cycles. By utilizing two new techniques, presented method realizes an energy saving of 99.48% and decreases the overall capacitance size by 75%. Furthermore, the presented method exhibits good linearity characteristics. The variation of common-mode voltage is lower in the presented scheme compared to the previous published design. Overall, the presented method makes SAR ADC to exhibit better characteristics in terms of both efficiency and linearity.

ACKNOWLEDGMENT

Supported by the National Natural Science Foundation of China (Grant Nos. 62301321, and 62171286), the Shanghai Sailing Program (Grant Nos. 23YF1444300)

REFERENCES

[1] H. Wang, W. M. Xie, Z. X. Chen, Area-Efficient Capacitor-Splitting Switching Scheme with a Nearly Constant Common-Mode Voltage for SAR ADCs, J. Circuits Systems and Computers 29 (2020) 2020005.

[2] Y. Zhang, Y. Li, Z. Zhu, A charge-sharing switching scheme for SAR ADCs in biomedical applications, Microelectronics Journal 75 (2018) 128–136.

[3] X. Xin, J. Cai, T. Chen, Q. Yang, A 0.4-V 10-bit 10-KS/s SAR ADC in 0.18 μm CMOS for low energy wireless senor network chip, Microelectronics Journal 83 (2019) 104–116.

[4] R. X. Ding, S. P. Dong, S. B. Liu, D. P. Sun, Z. M. Zhu. A novel split capacitor array switching scheme with proportional coefficient for SAR ADC. Analog Integrated Circuits and Signal Processing, 2019, 98: 597-605.

[5] R. X. Ding, S. P. Dong, D. P. Sun, S. B. Liu, Z. M. Zhu. Energy-efficient and two-step structure switching scheme based on reference-free for SAR ADC. Analog Integrated Circuits and Signal Processing, 2019, 99: 209-218.

[6] V. Tiwari, R. K. Nagaria. Review of the SAR ADC's Energy Efficient Switching Scheme. International Conference on VLSI, Communication and Signal Processing 2023:637-650.

[7] S. S. Han, L. Z. Zhang, J. H. Wu. Energy efficient switching scheme based on MSB-split structure for SAR ADC. Analog Integrated Circuits and Signal Processing, 2020, 105(1): 135-139.

[8] Y. H. Chung, Q. F. Zeng, Y. S. Lin. A 12-bit SAR ADC With a DAC-Configurable Window Switching Scheme. IEEE Transactions on Circuits and Systems I: Regular Papers, 2020, 67(2): 358-368.

[9] X. Y. Tong, S. M. Zhao, X. Xin. High Energy Efficiency and Linearity Switching Scheme Without Reset Energy for SAR ADC. Circuits Systems and Signal Processing, 2022, 41(10): 5872-5894.

[10] X. Y. Li, J. P. Cai, X. Xin, T. T. Chen, Z. Li. High energy-efficient switching scheme for SAR ADC with low common-mode level variation. Analog Integrated Circuits and Signal Processing, 2021, 107(1): 215-225.

[11] Y. H. Chung, C. H. Tien, Q. F. Zeng. A 16-Bit Calibration-Free SAR ADC With Binary-Window and Capacitor-Swapping DAC Switching Schemes. IEEE Transactions on Circuits and Systems I: Regular Papers, 2022, 69(1): 88-99.

[12] A. R. Ghasemi, M. Saberi, R. Lotfi. A low-power capacitor switching scheme with low common-mode voltage variation for successive approximation ADC. Microelectronics Journal, 2017, 61: 15-20.

[13] H. Wang, L. G. Zhong, G. C. Zhang. Low-Power Capacitor-Splitting DAC with Mixed Switching Schemes for SAR ADCs. J. Circuits Systems and Computers, 2018, 27: 1850161.

[14] W. Y. Qu, J. Q. Zhao, Z. F. Zhang, N. S. Mei. Low-Energy Switching Method Based on Asymmetric Binary Search Algorithm for SAR ADCs. J. Circuits Systems and Computers, 2020, 29: 2050087.

[15] C. C. Liu, S. J. Chang, G. Y. Huang, Y. Z. Lin. A 10-bit 50-MS/s SAR ADC with a monotonic capacitor switching procedure. IEEE Journal of Solid-State Circuits, 2010, 45: 731-740.

[16] Y. Zhu, C. H. Chan, U. F. Chio, S. W. Sin, S. P. U, R. P. Martins, F. Maloberti. A 10-bit 100-MS/s reference-free SAR ADC in 90 nm CMOS. IEEE Journal of Solid-State Circuits, 2010, 45: 1111-1121.

[17] C. Yuan, Y. Lam. Low-energy and area-efficient tri-level switching scheme for SAR ADC. Electronics Letters, 2012, 48: 482-483.

[18] Z. Zhu, Y. Xiao, X. Song. VCM-based monotonic capacitor switching scheme for SAR ADC. Electronics Letters, 2013, 49: 327–329.

[19] A. Sanyal, N. Sun. SAR ADC architecture with 98% reduction in switching energy over conventional scheme. Electron. Lett., 2013, 49: 248-250.

[20] X. Tong, Y. Chen. Low-Power High-Linearity Switching Procedure for Charge-Redistribution SAR ADC. Circuits Systems and Signal Processing, 2017, 36: 3825-3834.

[21] L. B. Xie, G. J. Wen, J. X. Liu, Y. Wang. Energy-efficient hybrid capacitor switching scheme for SAR ADC. Electronics Letters, 2014, 50: 22-23.

[22] Z. Ding, W. Bai, Z. Zhu. Trade-off between energy and linearity switching scheme for SAR ADC. Analog Integr. Circuits Signal Process, 2016, 86: 121-125.

[23] T. Yousefi, A. Dabbaghian, M. Yavari. An energy-efficient DAC switching method for SAR ADCs. IEEE Trans, Circ. Syst. II: Express Briefs, 2018, 1: 41-45.

979-8-3315-8850-2/25 $31.00 © 2025 IEEE

2025 The 10th International Conference on Integrated Circuits and Microsystems

A Unified Swin Transformer Framework for Inverse Lithography and Lithography Simulation

Jiajun Tan[1,2]

1.The high-performance Integrated Circuit Engineering Research Center of Anhui Province
2.School of Integrated Circuits, Anhui University
Hefei, China
wb23301075@stu.ahu.edu.cn

Ling Liang[3]

3.Peking University
Beijing, China
lingliang@pku.edu.cn

Ming Zhu[1,2]*

1.The high-performance Integrated Circuit Engineering Research Center of Anhui Province
2.School of Integrated Circuits, Anhui University
Hefei, China
zhuming@ahu.edu.cn

Abstract—**Inverse Lithography Technology (ILT) and lithography simulation are two core components in modern computational lithography, yet they are typically handled separately due to their distinct objectives and modeling requirements. To enable joint optimization and simulation, a unified multi-task framework, Swin-LithoNet, is proposed based on the Swin Transformer to perform ILT mask optimization and resist image simulation. Our architecture integrates a shared Swin Transformer encoder with task-specific U-Net decoders, enabling joint learning and improved generalization across both tasks. To guide the learning process, task-specific loss functions are constructed by incorporating physical and geometric constraints, including Process Variation Band (PVB), Edge Placement Error (EPE), and Mean Squared Error (MSE). Extensive experiments on the ICCAD-2013 benchmark demonstrate that our method achieves up to 39.3% reduction in L2 loss and 83.8% reduction in EPE compared to prior ILT models, while maintaining competitive performance in lithography simulation. Our results highlight the effectiveness of multi-task learning for EDA applications and offer a scalable foundation for future research in manufacturability-aware design automation.**

Index Terms—**inverse lithography technology, lithography simulation, multi-task learning, machine learning**

I. INTRODUCTION

Photolithography is a key process in semiconductor manufacturing, responsible for transferring circuit patterns from a photomask onto a silicon wafer. As a crucial step in integrated circuit (IC) fabrication, it accounts for roughly 30% of the overall cost. As transistors shrink beyond the wavelength of light, photolithography faces increasing challenges from diffraction and process variations, which distort printed patterns. Ensuring pattern fidelity is thus a critical challenge.

Optical Proximity Correction (OPC) enhances pattern printability. While traditional rule-based OPC methods work well at larger nodes, they struggle at advanced nodes such as 7 nm and 5 nm where complex interactions dominate. To address this, model-based OPC (MBOPC) simulates the lithography process to achieve more accurate correction of proximity effects [1].

This work was supported by the University Synergy Innovation Program of Anhui Province (GXXT-2023-010), Key Scientific Research Project of Anhui Province (2022AH050093), and Science and Technology Major Project of Anhui Province (202203a05020027).
*The corresponding author

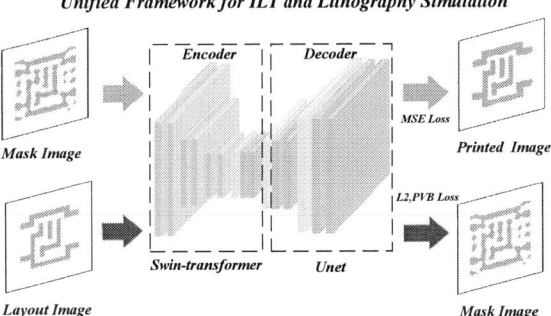

Fig. 1: Our proposed unified framework.

Inverse Lithography Technology (ILT) further advances mask optimization by treating mask generation as an inverse imaging problem [2]. ILT relies on iterative simulations to refine mask layouts, making lithography simulation central to its workflow. It is especially valuable in EUV lithography where nonlinear effects are prominent [3]. However, ILT faces significant computational challenges [4]. Its iterative nature leads to long runtimes and high memory usage. Moreover, the resulting masks often feature irregular shapes that complicate manufacturability and reduce robustness.

To address these challenges, recent research has explored the use of deep neural networks (DNNs) for accelerating ILT workflows [5]. By learning a direct mapping from target patterns to optimized masks, DNN-based ILT models greatly reduce the need for iterative simulation. A single forward pass through the network can generate high-quality mask predictions, dramatically accelerating the design cycle. Ye et al. [6] propose an end-to-end lithography modeling framework based on GANs that maps input mask layouts directly to resist image outputs, effectively speeding up the photoresist modeling stage while balancing accuracy and efficiency. GAN-OPC [7] introduces a CNN-based GAN framework that generates initial mask predictions and refines them for enhanced performance. Yang et al. [7] propose a DNN architecture inspired by dual-band optical principles, incorporating physical insights into the model design. Their method achieves an 85× simulation

979-8-3315-8850-2/25 $31.00 © 2025 IEEE   163

speedup compared to conventional lithography simulators, while maintaining high image fidelity.

However, existing DNN-based workflows often treat lithography simulation and ILT separately or depend on costly iterative refinements, leading to the loss of fine-scale details, increased time cost, and cumulative errors that limit accuracy. Since lithography simulation and ILT are intrinsically coupled, jointly modeling them can yield more accurate, efficient, and robust mask optimization. By adopting a multi-task learning paradigm, a unified framework can leverage shared feature representations to improve generalization to unseen patterns and enhance robustness.

Motivated by these observations, we propose a unified model that integrates lithography simulation and ILT, as illustrated in Fig. 1. The main contributions of this work are summarized as follows:

- A unified neural architecture is proposed to integrate ILT and lithography simulation into a multi-task learning framework.
- Task-specific loss functions are designed to account for the distinct characteristics of ILT and lithography simulation, enabling more effective learning of the geometric and physical features of layout patterns.
- The Swin Transformer is adopted as a general-purpose encoder, combined with task-specific decoders, to construct a high-capacity and generalizable backbone network for lithography-related tasks.

II. PRELIMINARIES

A. Lithography Simulation

During the lithography process, the input mask $\mathbf{M}$ is projected onto the wafer through an optical imaging system, generating an aerial image $\mathbf{I}$. This light intensity distribution on the wafer surface is then processed through photoresist development and etching to yield the final printed pattern $\mathbf{Z}$. Optical Proximity Correction (OPC) is a widely adopted technique used to optimize mask design. Instead of assuming a strict inverse function, OPC aims to find an optimized mask $\mathbf{M}_{\text{opt}}$ such that the lithography process $f(\mathbf{M}_{\text{opt}})$ closely approximates the desired target pattern Z_t under nominal conditions.

To simulate the lithography process, a mathematical model comprising two components is used: an optical projection model and a photoresist development model. The optical projection is formulated using the Hopkins diffraction theory for partially coherent optical imaging systems. The aerial image $\mathbf{I}$ is approximated by convolving the mask $\mathbf{M}$ with a set of optical kernels $\mathbf{H}$ [8]:

$$I(x, y) = \sum_{k=1}^{N_h} w_k \left| \mathbf{M}(x, y) \otimes h_k(x, y) \right|^2 \tag{1}$$

where h_k denotes the k-th optical kernel in the kernel set $\mathbf{H}$, w_k is the corresponding weight satisfying $\sum_{k=1}^{N_h} w_k = 1$, and "$\otimes$" denotes convolution.

Following aerial image formation, the photoresist development model applies a threshold I_{th} to determine the printed pattern. The final binary printed image $\mathbf{Z}$ is defined as:

$$Z(x, y) = \begin{cases} 1, & I(x, y) \geq I_{\text{th}} \\ 0, & I(x, y) < I_{\text{th}} \end{cases} \tag{2}$$

where $\mathbf{Z}$ represents the resist profile after development. Given an optimized mask, the simulation model outputs the aerial image and predicted printed pattern for evaluation. To assess simulation accuracy, metrics such as MSE, Intersection over Union (IOU), and Pixel Accuracy (PA) are employed [9]. The IOU quantifies the overlap between printed and target regions, and is defined as:

$$\text{IOU}(Z, T) = \frac{|Z_1 \cap T_1|}{|Z_1 \cup T_1|} \tag{3}$$

Pixel Accuracy (PA) measures the proportion of correctly predicted pixels within the target region:

$$\text{PA}(Z, T) = \frac{|Z_1 \cap T_1|}{|T_1|} \tag{4}$$

where Z_1 and T_1 denote the printed and target regions respectively.

B. Mask Optimization

Mask optimization aims to generate high-quality masks that accurately reproduce the target resist patterns after lithography simulation. Unlike simply copying reference masks, modern ILT leverage learned lithographic physics to produce masks that are robust, manufacturable, and tolerant to process variations. This ensures reliable pattern printing under diverse manufacturing conditions.

We evaluate optimized masks using the following metrics [10], computed after binarizing mask and resist images:

- **L2 Loss:** Measures the pixel-wise differences between the simulated nominal resist image Z_{nom} and the target pattern T, capturing overall intensity discrepancies:

$$\mathcal{L}_2(Z_{\text{nom}}, T) = \|Z_{\text{nom}} - T\|_2^2 \tag{5}$$

- **Process Variation Band (PVB):** Assesses robustness to process variations by computing the L2 distance between resist images at the maximum and minimum process corners:

$$\text{PVB}(Z_{\text{max}}, Z_{\text{min}}) = \|Z_{\text{max}} - Z_{\text{min}}\|_2^2 \tag{6}$$

Smaller PVB values indicate that the printed patterns exhibit less variation across process corners, demonstrating greater stability under process fluctuations.

- **Edge Placement Error (EPE):** Measures geometric deviations by counting the number of probe points on the printed edges that exceed a threshold distance from the corresponding target edges. It reflects pattern fidelity and critical dimension control.

979-8-3315-8850-2/25 $31.00 © 2025 IEEE

- **Shots:** Estimates the number of rectangular shots required to write the mask, serving as an indicator of mask complexity. Lower shot counts imply higher manufacturability and reduced mask writing cost and time.

III. PROPOSED FRAMEWORK

A. Multi-task Swin Transformer-based Method

Traditional convolutional networks are often limited in capturing long-range dependencies and multi-scale features in complex layout patterns. To overcome this, we introduce the Swin Transformer as the backbone encoder in our framework. The Swin Transformer [11] employs a hierarchical structure with shifted window-based self-attention, enabling efficient modeling of both local and global layout features. Its ability to extract multi-scale contextual information is well suited for layout understanding tasks in lithography modeling, where precise pattern geometry and context are critical.

To simultaneously address ILT and resist image simulation in a unified framework, we adopt a multi-task architecture that shares a common encoder while employing task-specific decoders. This design facilitates efficient feature reuse and consistent layout representation across tasks, leading to improved accuracy and inference efficiency. As illustrated in Fig. (2), our overall framework consists of two main components:

- **Encoder:** A Swin Transformer-based backbone that encodes input mask or layout patterns into multi-scale feature representations. Through hierarchical stages and shifted window attention, the encoder effectively captures both fine-grained geometries and large-scale layout context.
- **Task-Specific Decoders:** Two decoders derived from the U-Net [12] architecture are connected to the encoder, each tailored to a specific task. The lithography simulation decoder takes the encoded mask features as input and predicts the corresponding resist image under nominal or perturbed process conditions. The mask optimization decoder, on the other hand, takes the encoded layout features and generates optimized mask patterns that aim to reproduce the target printed shapes after lithography. Both decoders are trained jointly with the shared Swin Transformer encoder, enabling effective feature sharing and improved performance through multi-task learning.

B. Transformer-Based Encoder Design

The proposed encoder models latent features of masked layout patches to predict signals in masked regions. Its core is the multi-head self-attention (MHSA) mechanism, capturing global spatial dependencies.

In Vision Transformer (ViT) [13], an input layout of size $H \times W \times C$ is first divided into $N = \frac{HW}{P^2}$ non-overlapping patches of size $P \times P$. Each patch is flattened into a vector and projected via a linear embedding layer, resulting in a sequence of patch embeddings $X \in \mathbb{R}^{N \times D}$, where D is the embedding dimension. To retain positional information, learnable positional encodings $E_{\mathrm{p}} \in \mathbb{R}^{N \times D}$ are added to the input embeddings:

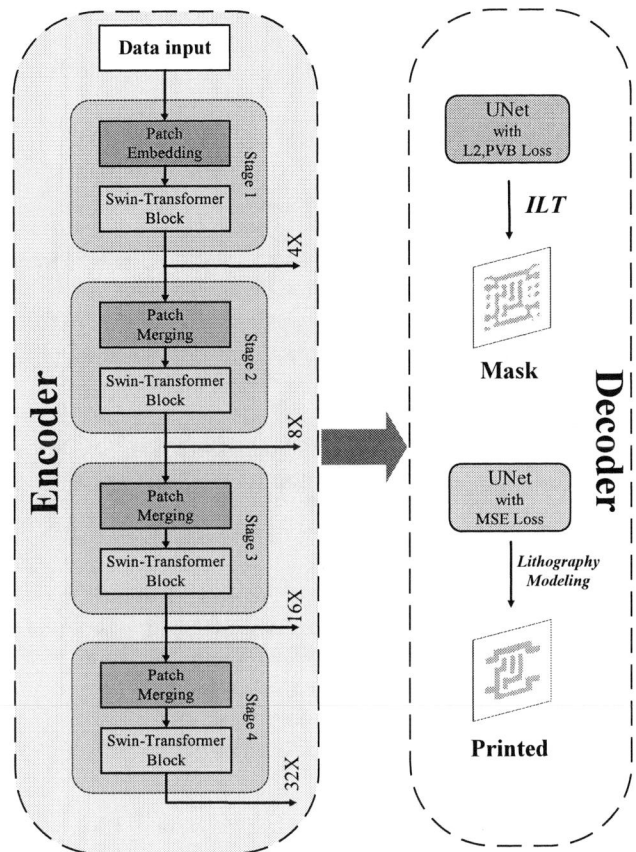

Fig. 2: Architecture of Swin-LithoNet.

$$Z^{(0)} = X + E_{\mathrm{p}} \qquad (7)$$

The transformer encoder consists of multiple layers, each containing a multi-head self-attention (MHSA) mechanism and a feedforward network (MLP), both equipped with residual connections and layer normalization:

$$Z'^{(l)} = \mathrm{MHSA}(\mathrm{LN}(Z^{(l-1)})) + Z^{(l-1)} \qquad (8)$$

$$Z^{(l)} = \mathrm{MLP}(\mathrm{LN}(Z'^{(l)})) + Z'^{(l)} \qquad (9)$$

In the multi-head attention block, the input is linearly projected to queries, keys, and values:

$$\{Q, K, V\} = \{X_P W^Q, X_P W^K, X_P W^V\} \qquad (10)$$

where $W^Q, W^K, W^V \in \mathbb{R}^{D \times d_k}$ are learnable projection matrices. The attention output for each head is computed as:

$$\mathrm{Attention}(Q, K, V) = \mathrm{Softmax}\left(\frac{QK^\top}{\sqrt{d_k}}\right) V \qquad (11)$$

and multiple heads are concatenated and linearly transformed:

$$\mathrm{MHSA}(Q, K, V) = \mathrm{Concat}(O_1, \ldots, O_h) W^O \qquad (12)$$

where O_i denotes the output of the i-th attention head, and $W^O \in \mathbb{R}^{hd_k \times D}$ is a projection matrix.

We adopt Swin Transformer [11] as the encoder backbone, which introduces:

- **Window-based Attention (W-MSA):** Self-attention is computed within local non-overlapping windows, reducing complexity from quadratic to linear.
- **Shifted Window Attention (SW-MSA):** Windows are shifted by half the window size to enable cross-window connections and enhance global context modeling.
- **Patch Merging:** Neighboring patches are merged to form hierarchical multi-scale features, similar to downsampling in CNNs.

The encoder consists of stages with patch merging layers and Swin Transformer blocks combining W-MSA, SW-MSA, and MLP layers, efficiently capturing complex layout structures.

C. Model Training

The entire framework is trained end-to-end using a multitask learning strategy, optimizing objectives for lithography simulation and mask optimization. where the lithography simulation loss, $\mathcal{L}_{\text{sim}}$, is formulated as a weighted combination of MSE on aerial images and resist images, as well as metrics for structural similarity such as IoU and PA:

$$\mathcal{L}_{\text{sim}} = \alpha_1 \mathcal{L}_a + \alpha_2 \mathcal{L}_r + \alpha_3 (1 - \text{IoU}) + \alpha_4 (1 - \text{PA}) \quad (13)$$

while the inverse lithography task loss, $\mathcal{L}_{\text{opt}}$, incorporates multiple components including L2 loss, PVBand, EPE, and Shots count loss:

$$\mathcal{L}_{\text{opt}} = \beta_1 \text{L2} + \beta_2 \text{PVBand} + \beta_3 \text{EPE} + \beta_4 \text{Shots} \quad (14)$$

Training is performed using the Adam optimizer [14] with a batch size chosen according to GPU memory limitations. A learning rate schedule combining an initial warm-up phase followed by cosine decay is employed to promote stable convergence. To enhance generalization, data augmentation techniques such as random flipping and rotation are applied.

To mitigate overfitting, early stopping and weight decay regularization are utilized. The multi-task architecture leverages a shared encoder, facilitating feature reuse across tasks and consequently improving overall performance compared to training separate models independently.

IV. EXPERIMENTAL RESULTS

Our framework is implemented in Python using the PyTorch library and is deployed on a Linux server equipped with a 3.1 GHz Intel Xeon Gold 6346 CPU and eight NVIDIA GeForce RTX 4090 GPUs. To evaluate the effectiveness and robustness of the proposed method, we conduct experiments on the MetalSet dataset provided by Lithobench. MetalSet contains 16,472 tiles synthesized using the layout generation method described in [15], where each tile is randomly generated following the design rules of the ICCAD-2013

benchmark. The ICCAD-2013 benchmark [16] is a widely recognized dataset for metal layer mask optimization, consisting of 10 tiles in GLP format derived from 32 nm industrial layouts. All input layout data are preprocessed by resizing and normalization before being fed into the network. During training, 90% of the dataset is used for training and the remaining 10% for testing.

Table I reports the quantitative performance of our lithography simulation framework compared with several state-of-the-art models, including CFNO [17], DAMO [18], and DOINN [19]. While our model does not attain the lowest error in aerial MSE or the highest IOU and PA, it delivers a balanced performance across all metrics. Notably, our method still outperforms CFNO in aerial MSE and remains close to DOINN and DAMO in terms of resist MSE and IOU, confirming the robustness of our unified simulation strategy.

TABLE I: COMPARISON OF LITHOGRAPHY SIMULATION PERFORMANCE USING DIFFERENT METHODS

	MSE (Aerial)	MSE (Resist)	IOU	PA
CFNO [17]	6.43E-05	0.00341	0.910	0.972
DAMO [18]	3.23E-05	0.00151	0.958	0.968
DOINN [19]	9.91E-05	0.00137	0.944	0.988
Ours	6.12E-05	0.0042	0.88	0.954

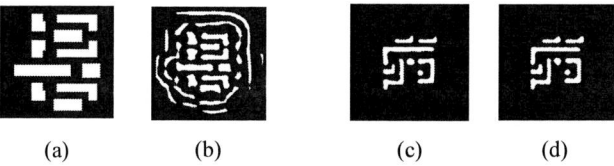

(a) (b) (c) (d)

Fig. 3: ILT and lithography results:(a)original layout, (b)optimized mask, (c)lithography ground truth, (d)lithography simulated image.

Table II shows the comparison between our proposed framework and several state-of-the-art mask optimization methods, including CFNO [17], DAMO [18], and NeuralILT [20]. Our method achieves the best performance in terms of L2 loss and EPE, significantly outperforming all baselines. Specifically, the L2 loss is reduced by **39.3%** and **31.2%** compared to CFNO and NeuralILT respectively. The EPE is also drastically reduced to **2.2**, representing an improvement of **83.8%** over CFNO and **76.3%** over NeuralILT, demonstrating superior contour fidelity and mask accuracy.

Although DAMO reports a PVB value of zero due to apparent errors, our method achieves the lowest valid PVBand among the compared methods. Compared to CFNO and NeuralILT, our PVBand is reduced by **5.0%** and **4.3%**, respectively, further confirming the effectiveness of our framework in preserving pattern fidelity while optimizing the mask. As illustrated in Fig. 3 illustrates the effectiveness of our framework in both ILT and lithography simulation tasks, visually confirming the quantitative advantages shown in Tables I and II.

979-8-3315-8850-2/25 $31.00 © 2025 IEEE

TABLE II: COMPARISON OF ILT PERFORMANCE USING DIFFERENT METHODS

Testcase	CFNO [16]				DAMO [17]				NeuralILT [18]				Ours			
	L2	PVB	EPE	Shots	L2	PVB	EPE	Shots	L2	PVB	EPE	Shots	L2	PVB	EPE	Shots
1	59972	49634	19	386	218902	0	140	0	52169	48684	14	353	40653	47634	3	578
2	43139	47590	17	350	172224	0	116	0	39707	41144	9	280	31898	38162	1	491
3	106991	76007	71	41	217432	0	147	0	98696	87455	60	421	65073	76518	18	657
4	18004	24185	3	260	84037	0	64	0	21029	24159	5	218	9471	23472	0	519
5	46566	54902	4	472	285988	0	169	0	46927	54995	2	403	31107	53553	0	517
6	47065	46560	3	531	290100	0	161	0	43746	48091	1	416	30625	47881	0	609
7	30004	41236	0	428	232224	0	134	0	25241	40453	0	310	16903	40589	0	487
8	19445	21589	1	349	130238	0	66	0	18950	22416	0	260	12159	20562	0	514
9	64489	64660	18	492	322122	0	189	0	50673	59928	2	393	35711	61195	0	586
10	27344	21949	0	242	104004	0	64	0	11566	17711	0	197	7714	16356	0	334
Average	46301.9	44831.2	13.6	355.1	205727.1	0	125	0	40870.4	44503.6	9.3	325.1	**28131.4**	42592.2	**2.2**	529.2

Overall, the experiment demonstrates that our unified multi-task framework offers competitive lithography simulation performance, especially in resist prediction, which is more relevant to practical yield-aware design optimization.

V. CONCLUSION

In this work, we propose a unified framework that jointly performs inverse lithography and lithography simulation using a multi-task Swin Transformer architecture. By leveraging a shared encoder and task-specific decoders with designed loss functions, our approach effectively captures both geometric and physical features of layout patterns, leading to improved accuracy and robustness. The proposed framework not only unifies two essential lithography tasks into a single model, but also establishes a scalable and efficient foundation for future multi-task learning research in electronic design automation.

REFERENCES

[1] J. Park, C. Park, S. Rhie, Y. Kim, M. Yoo, J. Kong, H. Kim, and S. Yoo, "An efficient rule-based opc approach using a drc tool for 0.18 /spl mu/m asic," in *Proceedings IEEE 2000 First International Symposium on Quality Electronic Design (Cat. No. PR00525)*, 2000, pp. 81–85.

[2] Y. Liu, D. Abrams, L. Pang, and A. Moore, "Inverse lithography technology principles in practice: unintuitive patterns," in *25th Annual BACUS Symposium on Photomask Technology*, J. T. Weed and P. M. Martin, Eds., vol. 5992, International Society for Optics and Photonics. SPIE, 2005, p. 599231.

[3] Z. Yu, G. Chen, Y. Ma, and B. Yu, "A gpu-enabled level-set method for mask optimization," *IEEE Transactions on Computer-Aided Design of Integrated Circuits and Systems*, vol. 42, no. 2, pp. 594–605, 2023.

[4] H. Shao, C. Lin, and S. Fang, "Data-driven approaches for process simulation and optical proximity correction," in *2023 28th Asia and South Pacific Design Automation Conference (ASP-DAC)*, 2023, pp. 721–726.

[5] G. Chen, Z. Yu, H. Liu, Y. Ma, and B. Yu, "Develset: Deep neural level set for instant mask optimization," *IEEE Transactions on Computer-Aided Design of Integrated Circuits and Systems*, vol. 42, no. 12, pp. 5020–5033, 2023.

[6] W. Ye, M. B. Alawieh, Y. Lin, and D. Z. Pan, "Lithogan: End-to-end lithography modeling with generative adversarial networks," in *2019 56th ACM/IEEE Design Automation Conference (DAC)*, 2019, pp. 1–6.

[7] H. Yang, Z. Li, K. Sastry, S. Mukhopadhyay, M. Kilgard, A. Anandkumar, B. Khailany, V. Singh, and H. Ren, "Generic lithography modeling with dual-band optics-inspired neural networks," in *Proceedings of the 59th ACM/IEEE Design Automation Conference*, ser. DAC '22. New York, NY, USA: Association for Computing Machinery, 2022, p. 973–978.

[8] J. Gao, X. Xu, B. Yu, and D. Z. Pan, "Mosaic: Mask optimizing solution with process window aware inverse correction," in *2014 51st ACM/EDAC/IEEE Design Automation Conference (DAC)*, 2014, pp. 1–6.

[9] C. Junqing and H. Liu, "An alternating direction method of multipliers for inverse lithography problem," *Numerical Mathematics: Theory, Methods and Applications*, vol. 16, no. 3, pp. 820–846, 2023.

[10] B. Jiang, L. Liu, Y. Ma, B. Yu, and E. F. Y. Young, "Neural-ilt 2.0: Migrating ILT to domain-specific and multitask-enabled neural network," *IEEE Trans. Comput. Aided Des. Integr. Circuits Syst.*, vol. 41, no. 8, pp. 2671–2684, 2022.

[11] Z. Liu, Y. Lin, Y. Cao, H. Hu, Y. Wei, Z. Zhang, S. Lin, and B. Guo, "Swin transformer: Hierarchical vision transformer using shifted windows," 2021.

[12] O. Ronneberger, P. Fischer, and T. Brox, *U-Net: Convolutional Networks for Biomedical Image Segmentation*. Springer International Publishing, 2015.

[13] A. Dosovitskiy, L. Beyer, A. Kolesnikov, D. Weissenborn, X. Zhai, T. Unterthiner, M. Dehghani, M. Minderer, G. Heigold, S. Gelly, J. Uszkoreit, and N. Houlsby, "An image is worth 16x16 words: Transformers for image recognition at scale," 2021.

[14] D. P. Kingma and J. Ba, "Adam: A method for stochastic optimization," vol. abs/1412.6980, 2014.

[15] H. Yang, W. Chen, P. Pathak, F. Gennari, Y.-C. Lai, and B. Yu, "Automatic layout generation with applications in machine learning engine evaluation," in *2019 ACM/IEEE 1st Workshop on Machine Learning for CAD (MLCAD)*, 2019, pp. 1–6.

[16] S. Banerjee, Z. Li, and S. R. Nassif, "Iccad-2013 cad contest in mask optimization and benchmark suite," in *2013 IEEE/ACM International Conference on Computer-Aided Design (ICCAD)*, 2013, pp. 271–274.

[17] H. Yang and H. Ren, "Enabling scalable ai computational lithography with physics-inspired models," in *2023 28th Asia and South Pacific Design Automation Conference (ASP-DAC)*, 2023, pp. 715–720.

[18] G. Chen, W. Chen, Q. Sun, Y. Ma, H. Yang, and B. Yu, "Damo: Deep agile mask optimization for full-chip scale," *IEEE Transactions on Computer-Aided Design of Integrated Circuits and Systems*, vol. 41, no. 9, pp. 3118–3131, 2022.

[19] H. Yang, Z. Li, K. Sastry, S. Mukhopadhyay, M. Kilgard, A. Anandkumar, B. Khailany, V. Singh, and H. Ren, "Generic lithography modeling with dual-band optics-inspired neural networks," in *Proceedings of the 59th ACM/IEEE Design Automation Conference*, ser. DAC '22. New York, NY, USA: Association for Computing Machinery, 2022, p. 973–978.

[20] B. Jiang, L. Liu, Y. Ma, H. Zhang, B. Yu, and E. F. Young, "Neural-ilt: Migrating ilt to neural networks for mask printability and complexity co-optimization," in *2020 IEEE/ACM International Conference On Computer Aided Design (ICCAD)*, 2020, pp. 1–9.

2025 The 10th International Conference on Integrated Circuits and Microsystems

Design of 6-to-8-bit High-Speed Time-Interleaved SAR ADC

Zhenyang Zhang
College of Computer Science and Technology
National University of Defense Technology
Changsha, China
zhangzy277@163.com

Fangxu Lv*
College of Computer Science and Technology
National University of Defense Technology
Changsha, China
lvfangxu1988@nudt.edu.cn

Zhengbin Pang
College of Computer Science and Technology
National University of Defense Technology
Changsha, China
zbpang@nudt.edu.cn

Jiaqing Xu
College of Computer Science and Technology
National University of Defense Technology
Changsha, China
xujiaqing@nudt.edu.cn

Qiang Wang
College of Computer Science and Technology
National University of Defense Technology
Changsha, China
qiangwang@nudt.edu.cn

Meng Li
College of Computer Science and Technology
National University of Defense Technology
Changsha, China
lmengnudt@nudt.edu.cn

Abstract—**This paper presents a high-speed time-interleaved (TI) SAR ADC with digitally programmable resolution from 6 to 8 bits. Utilizing a dual-channel architecture, the single-channel sampling rate achieves 500 MS/s, enabling an overall sampling rate of 1 GS/s after interleaving. The single-channel SAR ADC employs a Vcm-based switching scheme and split-capacitor technique to simplify the input signal. A novel SAR logic architecture optimized for high-speed operation is proposed. The circuit is implemented in a 28nm CMOS process. Operating at a 0.9 V supply voltage and 1 GS/s sampling rate, the ADC achieves an SNDR of 49.3 dB with a power consumption of 4.7 mW, resulting in a figure of merit (FOM) of 19 fJ/conversion-step.**

Keywords—Time-Interleaved, SAR ADC,SAR logic.

I. INTRODUCTION

In practical systems, optical and acoustic signals predominantly exist in analog form. Processing these signals requires Analog-to-Digital Converters (ADCs) to perform sampling and quantization, generating discrete digital signals for subsequent processing in digital systems. ADCs are currently deployed across a broad spectrum of domains, including radio frequency (RF) front-end circuits for communication systems, industrial data acquisition networks, and advanced image processing architectures.

The growing deployment of AI computing clusters and the hardware requisites of quantum computing have substantially elevated the significance of high-speed circuits, with particular emphasis on high-speed ADCs [1]-[4]. Optical transceivers are foundational to AI advancement, as they enable data communication. Their high-speed and low-latency attributes directly determine the upper limit of AI computational capability. Critically, high-speed ADCs serve as the core building blocks within these transceivers. In quantum computers, high-speed ADCs are embedded in the readout circuitry [5],

constituting an indispensable interface component between the quantum processor and classical computing systems.

As the demand for ADC speeds escalates, single-channel hybrid architectures, such as Pipeline-SAR structures, are no longer able to meet the requirements for ultra-high-speed operation [6].Consequently,time-interleaved (TI) architecture has emerged as a pivotal technique to meet these demands, effectively overcoming the single-channel speed bottleneck inherent to conventional SAR ADCs through the parallel operation of sub-ADCs.

This paper focuses on the optimization of single-channel SAR ADCs for low-power operation, while concurrently exploring a dual-channel time-interleaved (TI) architecture.This approach provides a pathway for potential future extension to multi-channel systems, further addressing the demands of ultra-high-speed applications.

II. ADC ARCHITECTURE

A. Top Architecture

Fig. 1 presents the schematic diagram of the entire system, which consists of three core components: an input buffer, a time-interleaved ADC (TI-ADC), and an LVDS high-speed interface. The system also includes a Time-Interleaved Clock Generation Module, which is used to convert the input clock signal into a square wave. Subsequently, the frequency division module converts the 1-GHz square wave into a 500-MHz square wave, and the resulting square wave has a strictly 50% duty cycle. This 500-MHz square wave serves as the sample-and-hold clock for the two channels.

Fig. 2 depicts the block diagram of the overall time-interleaved ADC architecture, comprising dual-channel sub-ADCs (both being SAR ADCs) and an output module consisting of a Multiplexer (MUX).

979-8-3315-8850-2/25 $31.00 © 2025 IEEE

In addition to the bootstrapped switch, double-tail comparator, capacitor array, and SAR logic, each sub-channel SAR ADC incorporates a Control Circuit, primarily tasked with generating the SAR logic control signal CRES described in the preceding section. Another function is to generate a HOLD signal for regulating the asynchronous clock generation circuit. The HOLD signal, governed by both external inputs and SAR logic, enables 6-to-8-bit resolution adjustability. The output module is controlled by CLK1 and CLK2. Theoretically, only one of CLK1 or CLK2 is required, but to minimize clock mismatch and ensure identical loading for both clocks, two clocks are used to control the output.

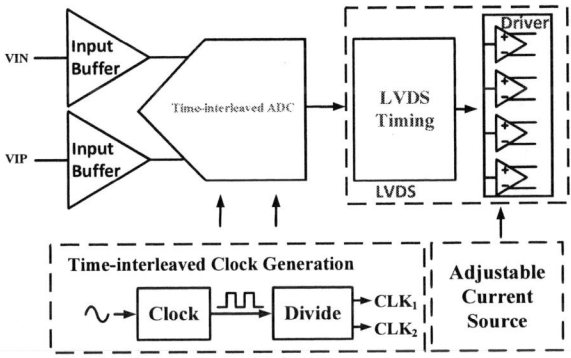

Fig. 1. System Block Diagram

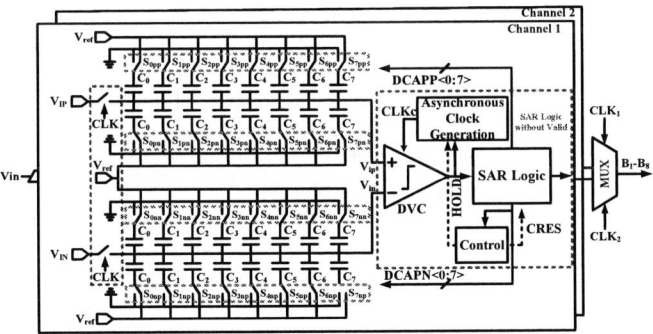

Fig. 2. Top Architecture of ADC

B. Core Techniques

The proposed Vcm-based switching strategy significantly reduces power consumption while maintaining a stable common-mode voltage (VCM), which is critical for the reliable operation of the comparator. This strategy requires two reference levels, VREF and VCM (where VCM = VREF/2). To simplify the circuit input, a split-capacitor technique is employed, dividing each capacitor into two equal-value capacitors, each half the original capacitance. This approach minimizes switching energy and enhances linearity by maintaining balanced charge redistribution.

Conventional SAR logic relies on a comparator-generated control signal (Valid) to sequence the conversion cycle, as shown in Fig. 3. The conversion speed in such designs is constrained by the loop delay between the comparator output and the Valid signal. Additionally, layout parasitics further diminish performance by introducing capacitance that reduces

the effective frequency of the comparator clock and Valid signal[7].

The proposed asynchronous SAR logic decouples the conversion control from the comparator's output latency. Instead of waiting for a Valid signal, the SAR logic generates its own internal timing signals, allowing the comparator clock and SAR control to operate independently. Upon completion of each comparison, the result is immediately routed to the CDAC and output registers, eliminating the delay associated with generating and propagating the Valid signal. This architecture reduces sensitivity to layout parasitics, enabling higher conversion rates and more robust performance across process variations. The detailed circuit implementation is discussed in the following section.

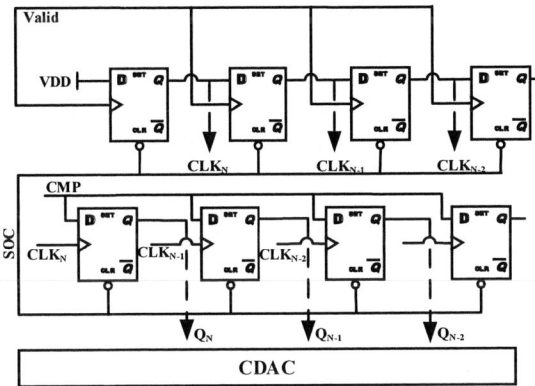

Fig. 3. Traditional Asynchronous SAR Logic

III. CORE CIRCUIT

A. Bootstrapped Switch

The on-resistance of a single MOS switch is non-constant, varying with the input signal. Additionally, the threshold voltage of the MOS transistor introduces signal loss. A CMOS complementary switch can partially address these issues [8], [9]. While its on-resistance is more stable than that of a single MOS switch, it still exhibits fluctuations, making it unsuitable for circuits requiring an effective number of bits (ENOB) exceeding 6 bits. To enhance linearity, a bootstrapped switch is employed, as illustrated in Fig. 4. This technique uses capacitor C0 to store charge, ensuring that the gate-source voltages of M1 and M2 remain at VDD throughout the switching cycle.

When tested with an input frequency approaching the Nyquist limit (i.e., ~250 MHz), the ADC achieves an ENOB of nearly 14 bits, a spurious-free dynamic range (SFDR) of 96.5 dB, and a signal-to-noise-and-distortion ratio (SNDR) of 86.3 dB.

B. Double-Tail Comparator

While the StrongARM latch and the synthesizable dynamic voltage comparator presented in [10] offer advantages in terms of speed and power consumption, they exhibit relatively poor offset voltage and kickback noise performance—two critical metrics in SAR ADC design. To address this, the proposed design employs a Double-tail comparator, as shown in Fig. 5. The comparator cascades a preamplifier stage, which suppresses noise and DC offset, with a latch stage to perform the final comparison. Although slower than conventional dynamic voltage comparators, the double-tail architecture provides superior offset and noise characteristics.

High-frequency signal variations can easily couple through parasitic capacitances, introducing kickback noise to the comparator inputs [11].For example, in a StrongARM comparator, coupling between the input and output nodes can cause interference. When the input of one branch is high (expected to produce a low output), capacitive coupling between the gate and output nodes may lead to erroneous switching. In contrast, the double-tail comparator isolates the input stage from the high-swing output nodes, minimizing such interference. This architecture effectively reduces kickback noise, ensuring robust operation even under high-speed switching conditions.

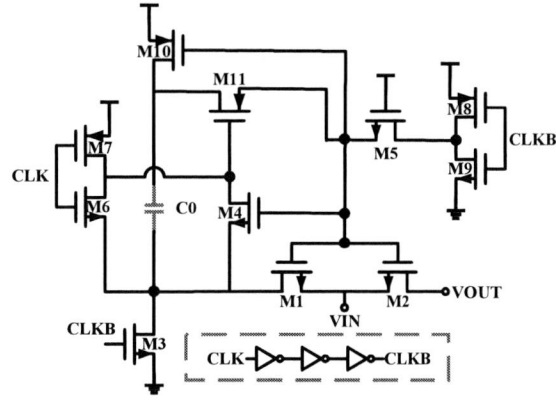

Fig. 4. Bootstrapped Switch

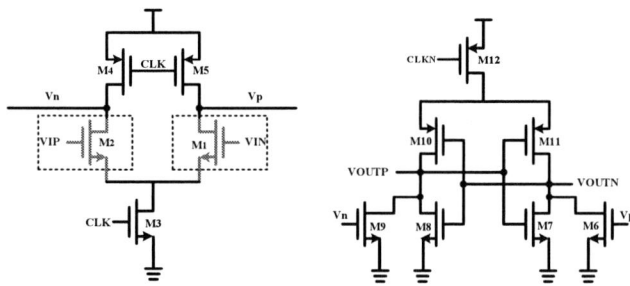

Fig. 5. Double-tail Comparator

C. Proposed SAR Logic

The novel asynchronous SAR logic employed in this work was briefly introduced in the previous section. This section provides a detailed description of its architecture and operation. The SAR logic unit is illustrated in Fig. 6. As shown in the figure, the basic SAR logic unit consists of two control units, an AND gate, a NAND gate, two delay elements, and an inverter. The delay element where the Present signal is located is critical because Present directly determines the operating speed of the SAR logic. The previously mentioned timing matching between the SAR logic and the asynchronous clock generation is that the Present signal should match the timing of the asynchronous clock.Controlled by signals CWR and CRES, the unit operates when CRES is low. During the sampling phase (prior to quantization), CWR is held low and CRES is high, setting

outputs DCAP and DEM to high. DCAP coordinates with the CDAC switches to manage capacitor array transitions, while DEM generates internal control signals and routes the comparator output to the ADC's digital output. During quantization, CRES transitions low, and CWR is asserted high by the SAR control logic. This action latches the comparator output (CMPOUT) to the output stage. If CMPOUT is high, DCAP and DEM remain high; if low, both signals are pulled low.

For an 8-bit SAR ADC, eight basic SAR logic units are required. The CRES signals of all basic units are connected to control signals generated by external modules. The Present output of the most significant bit (MSB) is connected to the Previous input of the next significant bit, and the Present output of the next significant bit is connected to the Previous input of the subsequent bit, and so on. The Previous input of the MSB itself is connected to the sampling clock to ensure that sampling occurs at the low level of the sampling clock and quantization begins at the high level. In this way, the CWR of the MSB is first set to high level at the beginning of quantization, followed by the next significant bit, sequentially propagating the CWR signal and enabling independent operation of the SAR logic. Fig. 7 shows the timing diagram of the SAR logic in 8-bit mode.

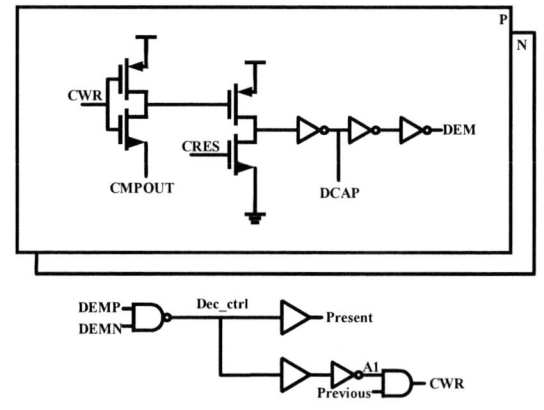

Fig. 6. SAR Logic Unit

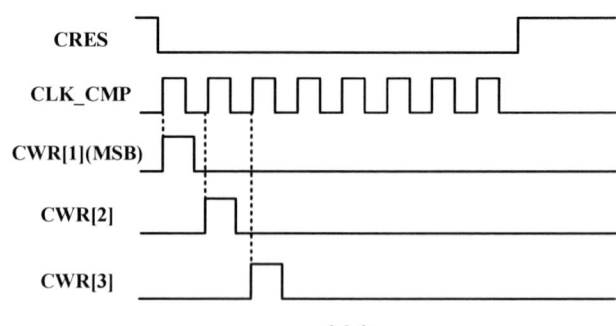

Fig. 7. SAR Logic Timing

D. Clock Generation

The transmission of high-speed signals has always been a challenge. It is difficult to propagate a high-frequency square wave intact, as capacitance in the signal path can easily distort the waveform [12]- [14]. To address this, a clock generation circuit is employed to convert the input 1GHz sine wave into a

979-8-3315-8850-2/25 $31.00 © 2025 IEEE   170

1GHz square wave. A classic circuit configuration for this purpose is shown in Fig. 8. The differential sine signal first passes through a capacitor to filter out the DC component, followed by a buffer stage. A pair of inverters connected in a feedback loop adjusts the phase difference of the output square waves, ensuring they are approximately 180° out of phase. The signal is then buffered again before output.

While this classic circuit is designed to directly generate a pair of square waves for time-interleaved clocking, it struggles to eliminate clock skew effectively. Manufacturing variations in buffer delays can cause the phase difference between the two output square waves to deviate from the ideal 180°. To mitigate this, only one output of the circuit is used to generate the 1GHz square wave, which is then fed into a subsequent divide-by-two circuit to reduce jitter and produce a more stable clock signal.

E. LVDS Driver

To ensure the reliable transmission of high-speed ADC signals, a suitable high-speed interface is required. The LVDS driver proposed in this paper is illustrated in Fig. 9. This LVDS driver consists of five components: STD, Level Shift Circuit, Main Driver Circuit, Common-Mode Feedback Circuit (CMFB), and Pre-Emphasis Circuit. The bias of the LVDS driver is generated by an adjustable current source, thereby enabling the adjustment of the output differential voltage.

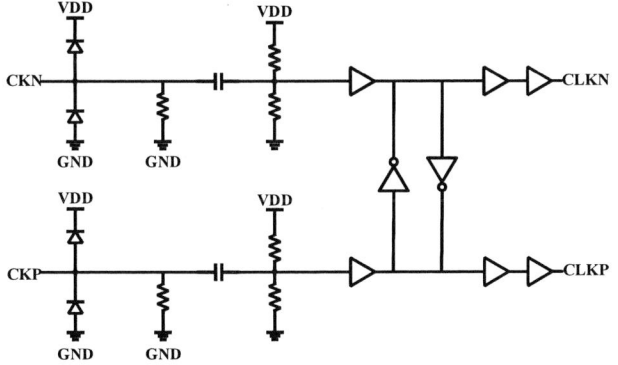

Fig. 8. Clock Generation Circuit

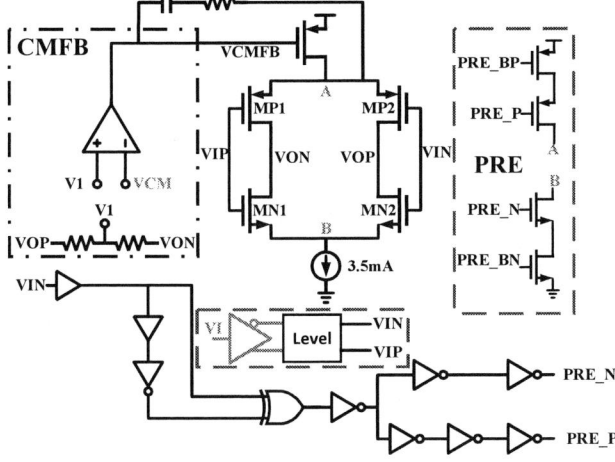

Fig. 9. LVDS Driver

IV. SIMULATION RESULTS AND SUMMARY

Simulations are performed when the input frequency approaches the Nyquist frequency. Simulation results show that the ENOB reaches the maximum of 7.90 bits under the TT process corner, followed by 7.91 bits under the SS corner, and the minimum of 7.88 bits under the fastest FF corner. However, the performance differences among the three process corners are negligible, indicating that the circuit is insensitive to process variations, and mobility changes do not affect its performance and functionality. In the 8-bit mode, the SFDR approaches 60 dB. The spectrum of the waveform output terminal is shown in Fig. 10.

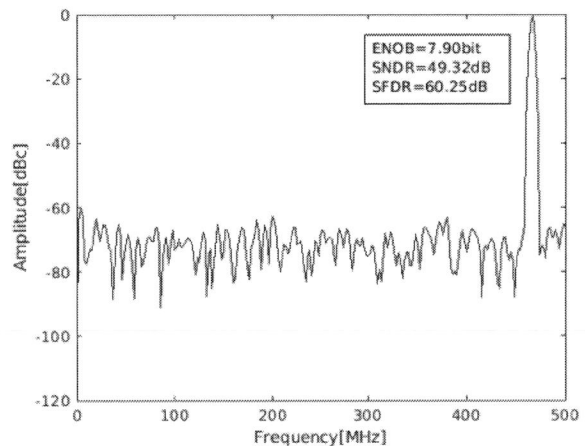

Fig. 10. Simulated FFT spectrum with 459.96MHz input signal

For the 7-bit mode, the ENOB values under TT, SS, and FF corners are 6.85 bits, 6.75 bits, and 6.69 bits, respectively, with the spurious-free dynamic range nearing 54 dB and the signal-to-noise-and-distortion ratio (SNDR) approaching 43 dB. In the 6-bit mode, the ENOB values under TT, SS, and FF corners are 5.85 bits, 5.82 bits, and 5.78 bits, respectively, with the SFDR close to 47 dB and the SNDR around 37 dB. Fig. 11(a) and Fig. 11(b) show the spectrum diagrams of the 7-bit and 6-bit modes, under the TT process corner.

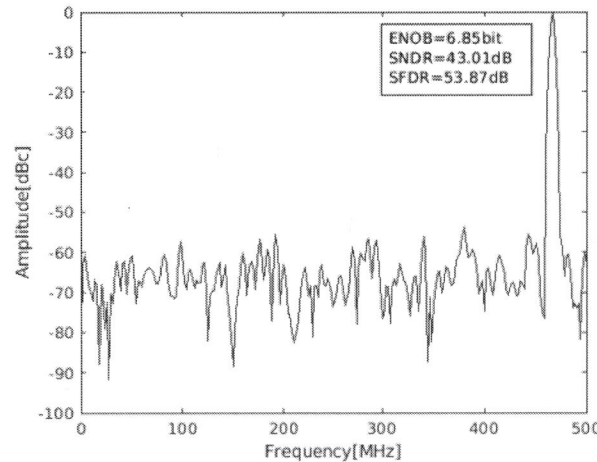

Fig. 11(a). Output spectrum of the 7-bit mode under the TT process corner

979-8-3315-8850-2/25 $31.00 © 2025 IEEE

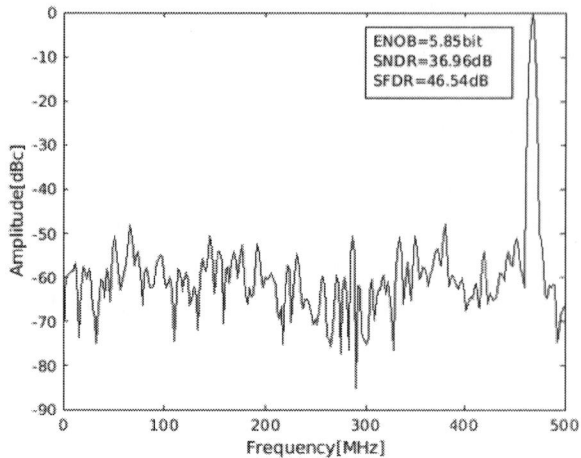

Fig. 11(b). Output spectrum of the 6-bit mode under the TT process corner

Table I summarizes the performance metrics in the 8-bit mode.

TABLE I.   SUMMARY OF PERFORMANCE

Technology	28nm CMOS
Sampling Rate	1GS/s
Resolution	8-bit
SNDR	49.32dB
SFDR	60.25dB
ENOB	7.90-bits
Power	4.7mW
FOM	19fJ/conversion-step

This article proposed a 6-8 bit adjustable high-speed time-interleaved SAR ADC. A novel asynchronous SAR logic suitable for high-speed SAR ADC is proposed. The transmission of high-speed signals requires high-speed interfaces [15]. Therefore, this paper designs a high-speed LVDS interface to ensure the reliability of ADC data transmission.

ACKNOWLEDGMENT

This research was funded by National key research and development program(2022YFB4401500). This research was funded by the National Natural Science Foundation of China: 62204263.

REFERENCES

[1] C. Zhao, Y. Wang, J. Diao and H. Xu, "Adaptive Background Calibration Method for Timing Mismatches in TI-ADCs Using Channels Correlation," *2020 IEEE 5th International Conference on Signal and Image Processing (ICSIP)*, Nanjing, China, 2020, pp. 823-827.

[2] H. Mafi, N. Ben-Hamida, S. Aouini and Y. Savaria, "Digital Compensation of Timing-Skew Mismatches in TI-ADCs by Modulation and Source Separation," *2024 22nd IEEE Interregional NEWCAS Conference (NEWCAS)*, Sherbrooke, QC, Canada, 2024, pp. 104-108.

[3] D. -R. Oh, J. -I. Kim, M. -J. Seo, J. -G. Kim and S. -T. Ryu, "A 6-bit 10-GS/s 63-mW 4x TI time-domain interpolating flash ADC in 65-nm CMOS," *ESSCIRC Conference 2015 - 41st European Solid-State Circuits Conference (ESSCIRC)*, Graz, Austria, 2015, pp. 323-326.

[4] K. A. El-Gammal and S. A. Ibrahim, "Design of a 10Gsps TI-flash ADC with modified clocking scheme," *2015 IEEE International Conference on Electronics, Circuits, and Systems (ICECS)*, Cairo, Egypt, 2015, pp. 268-271.

[5] E. Charbon, "Cryo-CMOS Electronics For Quantum Computing: Bringing Classical Electronics Closer To Qubits In Space And Temperature," in *IEEE Solid-State Circuits Magazine*, vol. 13, no. 2, pp. 54-68,Spring 2021.

[6] W. Jiang, Y. Zhu, M. Zhang, C. -H. Chan and R. P. Martins, "A Temperature-Stabilized Single-Channel 1-GS/s 60-dB SNDR SAR-Assisted Pipelined ADC With Dynamic Gm-R-Based Amplifier," in *IEEE Journal of Solid-State Circuits*, vol. 55, no. 2, pp. 322-332, Feb. 2020 .

[7] C. S. Ragit and S. Badjate, "Design of up-down counter as SAR logic for high speed SAR ADC used in health care system," *2016 Conference on Advances in Signal Processing (CASP)*, Pune, India, 2016, pp. 465-468.

[8] C. Qin and Z. Cai, "Bootstrapped Complementary Switches for High-Precision Sampling," *2023 IEEE International Symposium on Circuits and Systems (ISCAS)*, Monterey, CA, USA, 2023, pp. 1-5.

[9] H. Chen, L. He, H. Deng, Y. Yin and F. Lin, "A high-performance bootstrap switch for low voltage switched-capacitor circuits," *2014 IEEE International Symposium on Radio-Frequency Integration Technology*, Hefei, China, 2014, pp. 1-3.

[10] X. Zou and S. Nakatake, "A Fully Synthesizable, 0.3V, 10nW Rail-to-rail Dynamic Voltage Comparator," *2020 IEEE 63rd International Midwest Symposium on Circuits and Systems (MWSCAS)*, Springfield, MA, USA, 2020, pp. 199-202.

[11] S. Li, Y. Guo, Y. Liu, J. Chen, Q. Xu and J. Zhang, "A 1.2V 12 Bits SAR ADC with a Two Stages Amplifier Full-scale Differential Dynamic Comparator," *2018 10th International Conference on Communications, Circuits and Systems (ICCCAS)*, Chengdu, China, 2018, pp. 22-24.

[12] K. Okuno, K. Obata, T. Kato and K. Sushihara, "An 800-MHz 8-bit high speed SAR ADC in 16nm FinFET process," *2017 IEEE International Meeting for Future of Electron Devices, Kansai (IMFEDK)*, Kyoto, Japan, 2017, pp. 24-25.

[13] K. Xin *et al.*, "Frequency Domain Modeling and Performance Analysis of Injection-Locked LC Oscillator," in *IEEE Transactions on Circuits and Systems II: Express Briefs*, vol. 71, no. 1, pp. 11-15, Jan. 2024.

[14] Z. Luo, S. Du, Z. Zhang, F. Lv, Q. Hong and M. Lai, "Artificial Neural Network Based on Memristive Circuit for High-Speed Equalization," in *IEEE Transactions on Circuits and Systems I: Regular Papers*, vol. 71, no. 4, pp. 1745-1756, April 2024.

[15] Y. Wei, T. Wu, S. Ma and J. Ren, "A Multi-channel 12-bits 100MS/s SAR ADC in 65nm CMOS," *2023 IEEE 15th International Conference on ASIC (ASICON)*, Nanjing, China, 2023, pp. 1-4.

979-8-3315-8850-2/25 $31.00 © 2025 IEEE

2025 The 10th International Conference on Integrated Circuits and Microsystems

RLMBFF: Multi-bit Flip-Flops Merging using Deep Reinforcement Learning

Zeqi Chen
State Key Laboratory of Integrated Chips and Systems
Fudan University
Shanghai, China
23212020062@m.fudan.edu.cn

Zhaori Bi*
State Key Laboratory of Integrated Chips and Systems
Fudan University
Shanghai, China
zhaori_bi@fudan.edu.cn
*Corresponding author

Changhao Yan*
State Key Laboratory of Integrated Chips and Systems
Fudan University
Shanghai, China
yanch@fudan.edu.cn
*Corresponding author

Ming Zhu
Department of Integrated Circuits
Anhui University
Hefei, China
zhuming@ahu.edu.cn

Xiulong Wu
Department of Integrated Circuits
Anhui University
Hefei, China
xiulong@ahu.edu.cn

Xuan Zeng
State Key Laboratory of Integrated Chips and Systems
Fudan University
Shanghai, China
xzeng@fudan.edu.cn

Abstract—**Multi-bit flip-flop (MBFF) merging has been proven to be an effective way to save dynamic power and reduce the utilization rate of the chip area. Most of the existing work uses either a clustering method or an analytical way to merge multiple single-bit flip-flops (SBFFs) into MBFFs. In this paper, inspired by the recent success in deep reinforcement learning, we propose a novel MBFF merging method leveraging neural network. The approach can be divided into two steps: (i) placing MBFFs by a pretrained RL agent, and (ii) mapping the pin pairs of the original SBFF to the MBFF cell. Experimental results show that our method can significantly save power and area, without large perturbations to the timing.**

Index Terms—**Multibit Flip-Flip, Physical Design, Placement, Power Optimization, Reinforcement Learning.**

I. INTRODUCTION

With the booming growth of the Internet and artificial intelligence, microelectronic devices are playing an increasingly important role in our daily lives. Modern VLSI (Very-Large-Scale Integration) chips nowadays typically contain millions to tens of billions of transistors. As the technology node continues to shrink, this number is expected to increase at a rapid pace. Furthermore, clock speeds are gradually increasing to improve performance. However, cooling systems struggle to keep up with the pace of power dissipation. When a chip reaches its power limit, it must reduce its clock frequency, leading to a degradation of performance. On the other hand, chips in mobile devices contribute significantly to the total power consumption, and as such, power efficiency has become

a key concern. Therefore, it is essential to explore strategies that can reduce power consumption without compromising performance.

The power consumption of a chip consists of static and dynamic components, with dynamic power typically dominating. In synchronous digital designs, clock power alone can account for up to 40% of the total dynamic power consumption [1]. To reduce dynamic power, various techniques have been explored, including clock gating, buffer sizing, and Multi-bit Flip-Flop (MBFF) merging. MBFF merging combines multiple Single-bit Flip-Flops (SBFFs) into one cell, significantly reducing both area and power consumption.

Fig. 1(a) illustrates an example of merging two SBFFs into a 2-bit MBFF. A standard flip-flop typically consists of two latches and two inverters. In an MBFF, multiple bits can share common inverters, leading to area savings. Beyond area reduction, MBFFs help minimize the number of clock sinks and shorten the clock net wire length. Additionally, they reduce the number of buffers required in the clock tree, thereby lowering the overall driving strength needed for clock signal distribution. Fig. 1(b) gives an example of merging four SBFFs into two MBFFs. A comparison of the area and power of MBFF and SBFF cells is shown in Table I.

The merging operation can be performed either before or after detailed placement. For instance, [2] proposed a method that utilizes an MBFF library during RTL synthesis. However, merging at such an early stage introduces significant

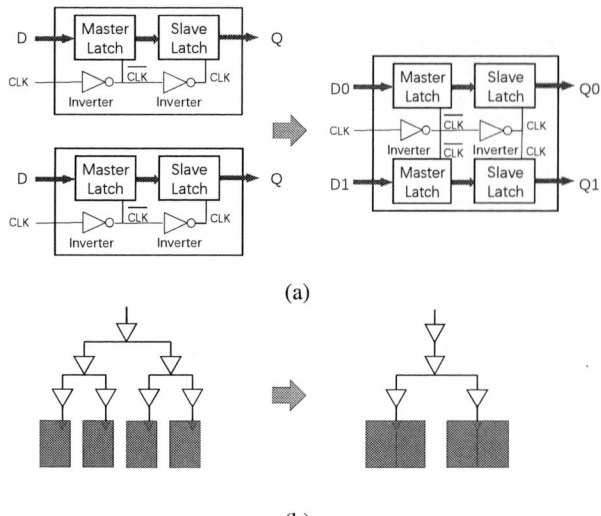

(a)

(b)

Fig. 1: Example of the merging process. (a) Merging two SBFFs into a 2-bit MBFF saves the area of two inverters. (b) Merging four SBFFs into two 2-bit MBFFs reduces the on-chip clock tree length.

TABLE I: Normalized Power, Area and Delay of MBFFs with Different Bits

Bits	Normalized Power	Normalized Area	Normalized Delay
1-bit FF	1.0	1.0	1.0
2-bit FF	1.7	1.6	2.0

unpredictability into the design process and may negatively impact timing and delay. As a result, most previous works [3] - [7] apply MBFF merging during the post-placement stage. Most of these methods focus on forming MBFFs while keeping the placement of combinational logic fixed, i.e., only flip-flops are movable. To guide merging decisions, they compute the timing-feasible region for each flip-flop based on the timing slack of its pins. Only flip-flops whose timing-feasible regions overlap are considered candidates for merging. Despite this, timing degradation can still occur. Consider the scenario illustrated in Fig. 2. The gray rectangles represent fixed combinational logic gates, while the pink blocks denote single-bit flip-flops. The blue-outlined diamonds represent the timing-feasible regions of the two flip-flops we intend to merge. Ideally, the merged MBFF should be placed within the overlapping area of these two regions to preserve timing integrity. However, there may be no space available in this overlapping region. The light blue rectangle indicates the final legalized location of the MBFF, which lies outside the shared timing-feasible region, resulting in potential timing violations. [8] uses large bit MBFF to reduce clock power but suffers from significant timing degradation. [9] adopts an effective mean shift approach but leads to relatively large disruption compared to the original placement.

To address these challenges, unlike conventional methods

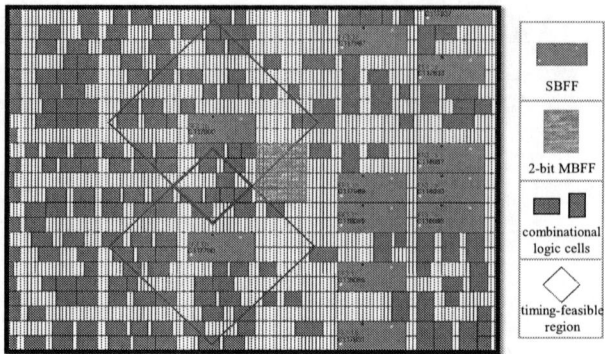

Fig. 2: An example where merging two SBFFs leads to a timing violation, despite overlapping timing-feasible regions. The blue rectangle indicates the legalized location of the merged 2-bit MBFF, which lies outside the shared feasible region.

that merge flip-flops before placement, we introduce a strategy of placing MBFF cells first, followed by mapping the original flip-flop pins to the newly placed MBFFs, thus avoiding the need for the legalization step.

The key contributions of our work are as follows:

1) **Reinforcement Learning for MBFF Placement**: We train a reinforcement learning agent to select the appropriate MBFF cell type from a predefined library and place it on the placement sites automatically.

2) **Pin Mapping via Assignment Problem**: After placing the MBFFs, we map the pins of the original flip-flops to the newly placed MBFFs. We transform this mapping task into an assignment problem, which we solve using the Hungarian algorithm, ensuring minimal Manhattan distance between corresponding pin pairs.

The remainder of the paper is organized as follows. Section II introduces the basic concepts of reinforcement learning and the Proximal Policy Optimization algorithm. Section III presents our proposed method, detailing the two-stage optimization flow, the network architecture, and the definitions of state, action, and reward functions. Section IV describes the experimental setup, training process, and results. Section V concludes the paper.

II. BACKGROUND

A. Reinforcement Learning (RL)

The placement of MBFF cells can be viewed as a sequential decision-making process. Proper placement requires balancing two opposing constraints: if the MBFF cells are placed too sparsely, it results in inefficient area usage, undermining the benefits of using MBFFs. Conversely, placing them too densely can lead to routing congestion. Therefore, decisions made during the placement process must consider long-term outcomes, such as leaving enough space for subsequent cells and selecting the appropriate MBFF cell from the cell library to avoid overlap with existing cells. This makes the problem well-suited for a reinforcement learning approach.

In a typical reinforcement learning setting, there are two interacting entities: the agent, which makes decisions, and the environment, which responds. At each discrete time step t, the agent observes the current state $s_t \in S$. Based on its policy $\pi_\theta(a_t|s_t)$, the agent selects an action $a_t \in A$. The environment returns an immediate reward $r_t = r(s_t, a_t)$ and transitions to a new state s_{t+1}. The agent's goal is to maximize the expected discounted cumulative return,

$$J(\theta) = \mathbb{E}_{\tau \sim \pi_\theta}\left[\sum_{k=0}^{T} \gamma^k r_k\right] \qquad (1)$$

where τ denotes a trajectory of state-action pairs (s_i, a_i) and $\gamma \in [0, 1)$ is a discount factor that trades off immediate and future rewards.

B. Proximal Policy Optimization (PPO)

There are many model-free RL algorithms in widespread use, such as Deep Q-Networks (DQN) [10]. However, DQN is less suited to the large action space encountered in our problem. Therefore, we adopt Proximal Policy Optimization (PPO) [11], which has been empirically shown to be both effective and robust to hyperparameter settings across a wide range of tasks. In reinforcement learning, we seek a policy π that maximizes the expected long-term return. Rather than learning only a value function, policy-based methods directly parameterize the policy and optimize those parameters by gradient ascent on the performance objective.

The Policy Gradient Theorem shows that the gradient of this objective can be written as

$$\nabla_\theta J(\theta) = \mathbb{E}_{(s,a)\sim\pi_\theta}[\nabla_\theta \log \pi_\theta(a \mid s)\, Q^{\pi_\theta}(s,a)] \qquad (2)$$

where the action-value function is

$$Q^{\pi_\theta}(s,a) = \mathbb{E}\left[\sum_{k=0}^{\infty} \gamma^k r_k \;\middle|\; s_0 = s,\ a_0 = a\right] \qquad (3)$$

PPO builds on these ideas by replacing an unconstrained gradient step with a clipped surrogate objective that limits how far the updated policy may depart from the data collection policy. Define the probability ratio

$$r_t(\theta) = \frac{\pi_\theta(a_t \mid s_t)}{\pi_{\theta_{\text{old}}}(a_t \mid s_t)} \qquad (4)$$

Given a small constant $\epsilon > 0$, PPO maximizes the clipped surrogate

$$L_{\text{clip}}(\theta) = \mathbb{E}_t\Big[\min\big(r_t(\theta) A_t,\ \text{clip}\big(r_t(\theta), 1-\epsilon, 1+\epsilon\big) A_t\big)\Big] \qquad (5)$$

which prevents $r_t(\theta)$ from moving outside $[1-\epsilon,\ 1+\epsilon]$ and thus avoids overly large policy updates. Using Generalized Advantage Estimation (GAE), the advantage function is given by

$$A_t = \sum_{l=0}^{\infty} (\gamma\lambda)^l \delta_{t+l} \qquad (6)$$

where $\delta_t = r_t + \gamma V(s_{t+1}) - V(s_t)$ is the temporal-difference error, and $\lambda \in [0, 1]$ balances bias and variance.

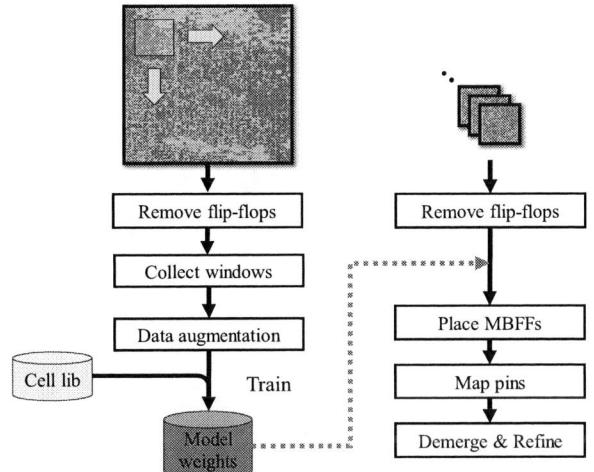

Fig. 3: The overall flow of the proposed method. The left column illustrates the training process, while the right column shows the deployment stage: MBFFs are placed using the trained RL model, followed by pin mapping from the original flip-flops, and finally demerging and refining the placement.

III. PROPOSED METHOD

We first train the RL agent using a window splitting method. After training, the model can be applied to arbitrary areas in the design. The execution consists of two stages: first, placing MBFFs using the trained model; second, mapping the pins from the original flip-flops to the MBFFs. The overall proposed flow is illustrated in Fig. 3.

A. Network Architecture

We use an actor-critic framework, where the agent consists of two separate components: the actor, which selects actions, and the critic, which evaluates the value of states.

In our PPO agent for the 2D grid placement task, the policy (actor) network first extracts spatial features from the environment grid using a convolutional feature extractor. This extractor, shown in Fig. 4(a), comprises three convolutional layers with ReLU activations. The resulting feature maps are flattened and passed to the policy head, which is a fully-connected network (Fig. 4(b)). The final output of the network specifies the placement action: the selected grid coordinates and the MBFF cell type. Only the policy network is shown in Fig. 4. The value (critic) network shares the same network architecture, but its output layer contains only a single scalar, representing the estimated state value.

B. State

The state refers to the observation received by the agent from the environment. We represent the current state using a matrix C, whose size matches the slice window. Each element C_{ij} represents a placement site, and is set to 0 if the location is vacant, and 1 if it is occupied by a cell. This 2D tensor is

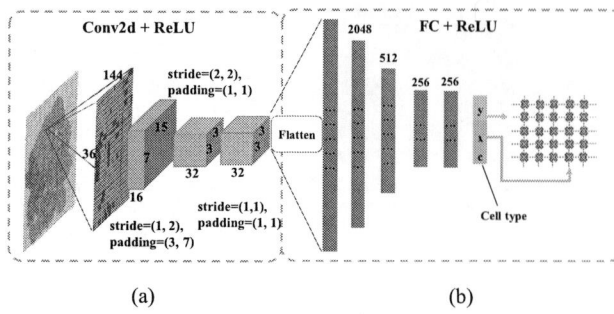

(a)   (b)

Fig. 4: The policy network architecture. (a) The convolutional feature extractor, which comprises three convolutional layers with ReLU activations that process the input grid. (b) The policy head: a fully-connected network. The final output provides the placement action as the selected coordinates (x, y) and cell type c.

defined using the Box observation space to comply with the Gymnasium API.

C. Action

At each time step, the agent receives a state and selects an action based on this state. In our case, the agent must choose an MBFF cell from the library and determine a coordinate for its placement. To model this decision-making process, we adopt a *MultiDiscrete* action space, which consists of three discrete components:

- **X-coordinate**: A discrete value ranging from 0 to $window\_size\_x - 1$.
- **Y-coordinate**: A discrete value ranging from 0 to $window\_size\_y - 1$.
- **MBFF cell index**: A discrete value representing the selected MBFF cell from the library.

Before training, we assign a unique index to each MBFF cell in the library. The third component of the action space corresponds to this index, indicating which MBFF cell the agent intends to place. The policy network outputs three separate categorical distributions, each corresponding to one of the action components. The agent samples from these distributions to determine its action at each step.

D. Reward Function

The reward function is one of the most crucial components in reinforcement learning, as it guides the agent to select appropriate actions based on the current state, and thus requires careful design. An improperly defined reward function may lead to either divergence or convergence to a poor local optimum.

In our case, we consider the following three aspects when designing the reward function:

- Illegal placements are strictly penalized, such as placing a cell outside the layout boundary or overlapping with existing cells. If the agent selects such an illegal action,

the environment returns a negative reward to discourage it.

- Legal placements are positively rewarded to incentivize the agent to avoid illegal actions.
- We aim to encourage the agent to place MBFFs with more bits whenever possible. These cells are more power and area efficient, so the reward increases with the number of bits in the MBFF.

Based on these principles, the reward function is given by

$$R_{\text{no-overlap}} = bit + c_1 \times bit^2 \tag{7}$$

$$R_{\text{overlap}} = c_2 \times \left(\frac{1}{util\_rate} - 1 \right) \tag{8}$$

Here, $c_1 = 0.2$, $c_2 = -0.5$, and bit denotes the number of bits in the selected MBFF. If the placement does not cause any overlap, $R_{\text{no-overlap}}$ is applied; otherwise, if an overlap occurs, R_{overlap} is assigned as a penalty. $util\_rate$ measures the congestion level of the current state, defined as

$$util\_rate = \frac{\sum cell\_area}{slice\_window\_area}. \tag{9}$$

We incorporate $util\_rate$ into the penalty reward function for two main reasons. First, when $util\_rate$ is low, it indicates that there is still plenty of space available for placement. In such cases, an illegal placement is less acceptable and should therefore receive a larger penalty. Second, our environment does not include a termination condition—each episode continues for a fixed number of steps, regardless of whether sufficient space remains for legal placement. Toward the end of an episode, when most of the placement window is filled and $util\_rate$ approaches 1, illegal placements become more likely due to space limitations. To prevent the agent from being overly penalized in such unavoidable situations, the penalty is designed to be smaller when $util\_rate$ is high.

E. Window Splitting

The original circuit is too large to train an agent to place MBFFs directly on the entire design. Therefore, we adopt a window-based splitting method, where only a relatively small region is considered at a time. We set the window size to 36 placement sites in height and 144 sites in width, striking a balance between training efficiency and placement performance. Since the height of each placement site is approximately four times its width, the window forms a roughly square-shaped region overall. At the beginning of each episode, we randomly select a window region and initialize the state matrix C by removing all original flip-flops within that region.

F. Pin Mapping Strategy

After placing the MBFF cells, the next step is to map the pins from the original flip-flops to those of the MBFF cells. Since each D pin is associated with a corresponding Q pin, we treat them as pairs. This mapping problem can be formulated as an assignment problem, where each original pin pair must be assigned to a new pin pair such that the overall

Manhattan displacement is minimized. We define a cost matrix $C \in \mathbb{R}^{m \times n}$ where each row corresponds to a new pin pair and each column corresponds to an original pin pair, and each entry C_{ij} represents the Manhattan distance between the ith new pin pair and the jth original pin pair. This is a classical assignment problem, and we can solve it with the Hungarian algorithm.

The linear programming formulation is given by

$$\min \quad \sum_{i=1}^{m} \sum_{j=1}^{n} c_{ij} x_{ij} \tag{10}$$

$$\text{s.t.} \quad \sum_{j=1}^{n} x_{ij} \leq 1, \quad \forall i,$$

$$\sum_{i=1}^{m} x_{ij} \leq 1, \quad \forall j,$$

$$x_{ij} \in \{0, 1\}, \quad \forall i, j$$

where

$$c_{ij} = |D_{ix} - D_{jx}| + |D_{iy} - D_{jy}| + |Q_{ix} - Q_{jx}| + |Q_{iy} - Q_{jy}|$$

where $D_i = (D_{ix}, D_{iy})$ and $Q_i = (Q_{ix}, Q_{iy})$ represent the coordinates of pin D_i and pin Q_i, respectively, and similarly for D_j and Q_j.

If the number of pin pairs in the original flip-flops and the MBFFs is exactly the same, the mapping process proceeds without issues. However, more commonly, the pin pair counts do not match. In some cases, the original flip-flops have more pin pairs (which typically occurs in regions with high cell density), or the MBFFs have more pin pairs (which is often the case in regions with low cell density). In the first scenario, when there are not enough MBFF pin pairs, we let some of the original flip-flops remain unchanged. In the second scenario, when there are excess MBFF pin pairs, we demerge the MBFFs that contain unused pin pairs. For example, if a 4-bit MBFF has only 2 allocated pin pairs, we replace it with a 2-bit MBFF. If an MBFF has no allocated pins, we simply remove the cell. This strategy helps avoid unnecessary waste of cell area and maintains a reasonable cell density, ensuring that the subsequent routing stage remains feasible.

IV. EXPERIMENT

A. Experimental Setup

We use testcase2 from 2024 ICCAD CAD Contest Problem B to train our model, as it contains significantly more instances compared to other testcases. We adopt the PPO algorithm as implemented in the Stable-Baselines3 library. To accelerate the training process and improve data throughput, we utilize vectorized environments by running 8 parallel instances simultaneously. Since each environment is configured to run 2048 steps per update, we set the *batch_size* to 16384 (i.e., 2048×8), ensuring that all collected samples from all environments are fully utilized in each training update for maximum efficiency. We do not define a terminal condition, and episodes are instead truncated after 512 steps to ensure

fixed-length episodes. The other hyperparameters are set as follows: The learning rate is 0.0003. The clip range is set to 0.2, and *n_epochs* is set to 10. The gamma factor is 0.99, and the λ for GAE is set to 0.95.

B. Training

Our implementation is developed in Python (version 3.9.12) using PyTorch (version 2.6.0), and executed on a Linux server running CentOS 7.9. The hardware setup includes two Intel Xeon Gold 5320 CPUs, 377 GB of RAM, and an NVIDIA A800 80GB PCIe GPU. The learning curve is presented in Fig. 5. The trend indicates that the agent gradually improves its placement decisions as training progresses, resulting in higher reward values and better policy learning.

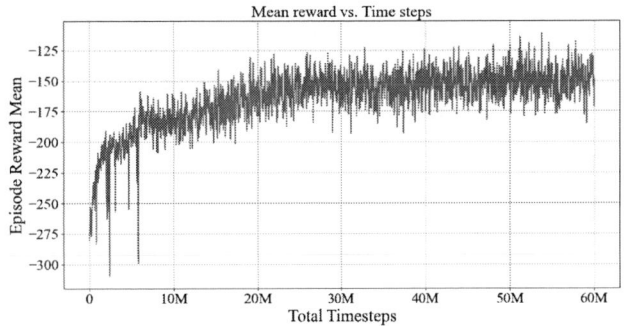

Fig. 5: Learning curve during training. The x-axis represents the total number of environment interaction steps (time steps), calculated as iterations × *batch_size*, while the y-axis shows episode rewards.

To illustrate the effectiveness of our method, we select a dense area containing the most flip-flops among all windows and show the process in Fig. 6. Fig. 6(a) shows the original layout of the area. In Fig. 6(b), we remove all the flip-flops, which occurs when the environment is reset at the beginning of each episode. Fig. 6(c) shows the agent's actions when episode = 1 (random policy). Finally, Fig. 6(d) shows the agent's placement after 3663 training iterations. As seen, the agent successfully learns to place the cells in a more efficient and compact manner.

C. Experimental Results

The testcase includes a Static Timing Analysis file that lists the timing slack for all D pins. We observe significant variation in timing slack across the chip. The slack distribution of the original flip-flops is shown in Fig. 7(a), where most values are close to zero, leaving little room for pin adjustments. To ensure timing safety, we focus on five local regions with high cell density and substantial positive timing slack. We also introduce a slack threshold of 0.5. Flip-flops with timing slack below this threshold are fixed and excluded from the merging process. The placement and mapping process in Region A are illustrated in Fig. 7(b) and (c). The experimental results are summarized in Table II.

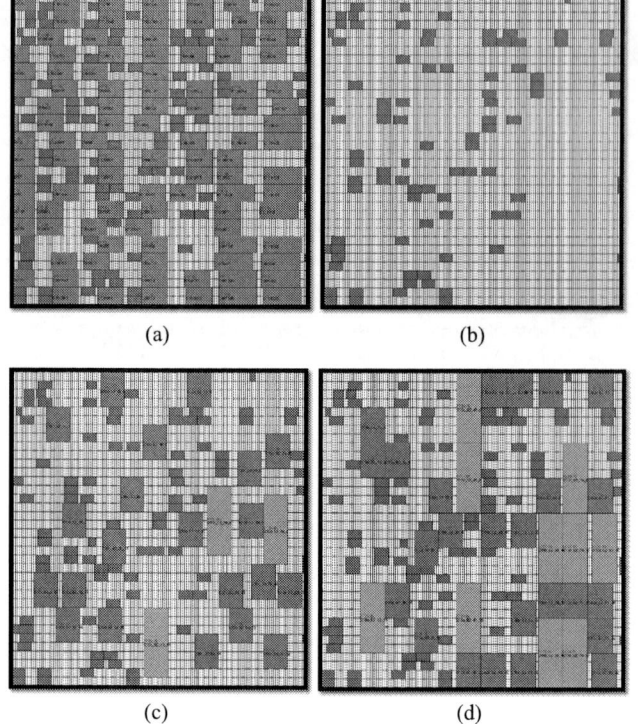

(a)

(b)

(c)

(d)

Fig. 6: Region $x_{\text{bottom}} = 1534440.0$, $y_{\text{bottom}} = 1975200.0$, $x_{\text{top}} = 1616520.0$, $y_{\text{top}} = 2061600.0$ of the testcase. Pink cells are SBFFs, blue cells are 2-bit MBFFs, and green cells are 4-bit MBFFs. (a) Original placement. (b) The layout after all flip-flops have been removed, corresponding to the environment's reset state at the beginning of a training episode. (c) Random policy at episode 1 after 512 steps. Total reward is -388. (d) The placement resulting from the agent's learned policy after 3663 training iterations, also obtained 512 steps after reset. Total reward is -161.

The results demonstrate a significant reduction in both area and power across all five regions. On average, the area is reduced by 49.5%, and power consumption is reduced by 55.5% in each region. For comparison, we implement a clustering-based K-means merging method, where flip-flops are merged first, followed by legalization. In contrast, our method consistently achieves greater area reduction across all cases. Although in a few regions the power reduction is less than that of K-means, our approach results in smaller timing degradation, making it more practical for timing-constrained designs.

TABLE II: Comparison of Area, Power, TNS and WNS across regions and methods

	Area (nm²)	Power (μW)	TNS (ns)	WNS (ns)
Region A				
Original	4,053,384,000	5.99	0	0
K-means	2,904,264,000	1.50	-4.71	-2.02
Ours	1,465,128,000	1.40	0	0
Region B				
Original	2,641,608,000	3.60	0	0
K-means	1,990,440,000	1.05	-15.68	-8.07
Ours	1,030,104,000	1.025	0	0
Region C				
Original	4,994,568,000	6.90	-0.08	-0.07
K-means	3,692,232,000	1.80	-69.91	-7.58
Ours	1,407,672,000	1.70	-11.39	-1.29
Region D				
Original	4,496,616,000	5.4	0	0
K-means	3,768,840,000	2.55	-33.47	-8.25
Ours	3,456,936,000	3.90	0	0
Region E				
Original	4,104,000,000	6.00	0	0
K-means	3,376,224,000	3.15	-31.09	-5.84
Ours	2,960,352,000	4.43	0	0

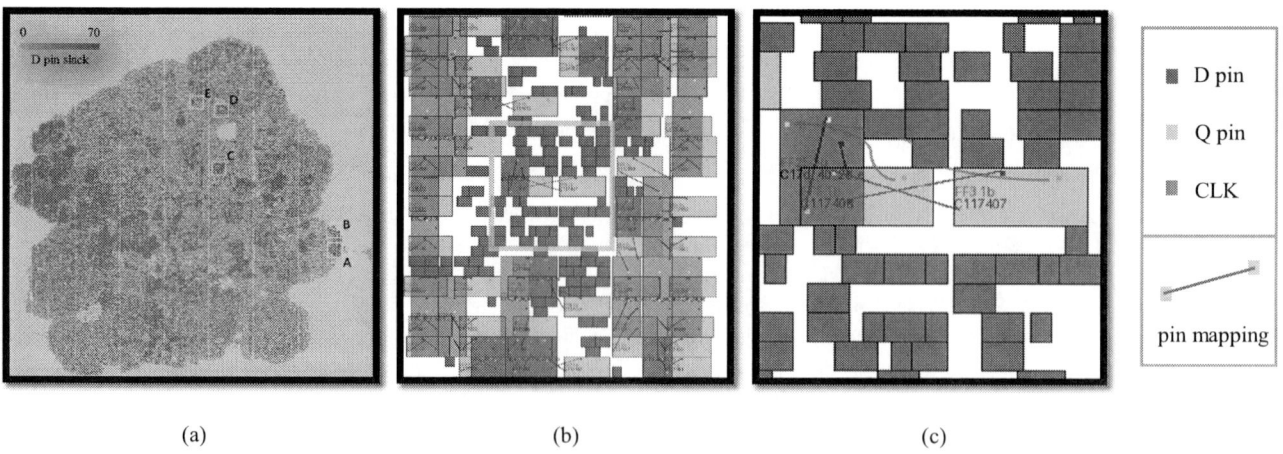

(a)

(b)

(c)

Fig. 7: Visualization of the pin mapping. Placement sites are omitted for clarity. (a) D pin slack distribution in the original placement, where darker red indicates larger positive slack (i.e., greater timing margin). (b)(c) Mapping process in region A, where blue, yellow, and pink dots represent D, Q, and clock pins, respectively, and red lines indicate pin mappings.

979-8-3315-8850-2/25 $31.00 © 2025 IEEE

V. CONCLUSION

In this paper, we propose a novel MBFF merging method to optimize both the area and the power consumption of digital circuits. Our approach utilizes reinforcement learning to place MBFF cells directly onto the circuit layout, avoiding the need for a post-placement legalization step. We address the assignment problem of mapping pins from original flip-flops to newly placed MBFF cells by employing the Hungarian algorithm to minimize the Manhattan distance between pin pairs. We adapt the solution by leaving unused original pin pairs unmapped or demerging MBFFs with insufficient pin pairs, ensuring efficient use of the cell area and maintaining reasonable cell density for the subsequent routing stage. Our result shows that our method significantly reduces power and area consumption in regions with high cell density without large timing degradation.

ACKNOWLEDGMENT

This research is supported by Yangtze River Delta Science and Technology Innovation Community Joint Fundamental Research Project 2024CSJZN0500 , 2024CSJZN0503.

REFERENCES

[1] D. Papa, C. Alpert, C. Sze, Z. Li, N. Viswanathan, G.-J. Nam, and I. Markov, "Physical synthesis with clock-network optimization for large systems on chips," *IEEE Micro*, vol. 31, no. 4, pp. 51–62, Jul. 2011.

[2] D. Yi and T. Kim, "Allocation of multi-bit flip-flops in logic synthesis for power optimization," in *Proc. IEEE/ACM Int. Conf. Comput.-Aided Design (ICCAD)*, Austin, TX, USA, 2016, pp. 1–6.

[3] I.-H. R. Jiang, C.-L. Chang, and Y.-M. Yang, "INTEGRA: Fast multibit flip-flop clustering for clock power saving," *IEEE Trans. Comput.-Aided Des. Integr. Circuits Syst.*, vol. 31, no. 2, pp. 192–204, Feb. 2012.

[4] M. P.-H. Lin, C.-C. Hsu, and Y.-C. Chen, "Clock-tree aware multibit flip-flop generation during placement for power optimization," *IEEE Trans. Comput.-Aided Des. Integr. Circuits Syst.*, vol. 34, no. 2, pp. 280–292, Feb. 2015.

[5] I. Seitanidis, G. Dimitrakopoulos, P. Mattheakis, and L. Masse-Navette, "Timing-driven and placement-aware multi-bit register composition," *IEEE Trans. Comput.-Aided Des. Integr. Circuits Syst.*, vol. PP, no. 99, pp. 1–1, Jul. 2018.

[6] G. Wu, Y. Xu, D. Wu, M. Ragupathy, Y.-Y. Mo, and C. Chu, "Flip-flop clustering by weighted K-means algorithm," in *Proc. 53rd ACM/EDAC/IEEE Design Autom. Conf. (DAC)*, Austin, TX, USA, 2016, pp. 1–6.

[7] M.-Y. Liu, Y.-C. Lai, W.-K. Mak, and T.-C. Wang, "Generation of mixed-driving multi-bit flip-flops for power optimization," in *Proc. IEEE/ACM Int. Conf. Comput.-Aided Design (ICCAD)*, San Diego, CA, USA, 2022, pp. 1–9.

[8] A. B. Kahng, S. Kundu, and S. Thumathy, "Scalable flip-flop clustering using divide and conquer for capacitated K-means," in *Proc. Great Lakes Symp. VLSI (GLSVLSI)*, New York, NY, USA, 2024, pp. 177–184.

[9] Y.-C. Chang, T.-W. Lin, I.-H. R. Jiang, and G.-J. Nam, "Graceful register clustering by effective mean shift algorithm for power and timing balancing," in *Proc. Int. Symp. Phys. Design (ISPD)*, San Francisco, CA, USA, 2019, pp. 11–18.

[10] V. Mnih, K. Kavukcuoglu, D. Silver, A. Graves, I. Antonoglou, D. Wierstra, and M. Riedmiller, "Playing Atari with deep reinforcement learning," *arXiv preprint arXiv:1312.5602*, 2013.

[11] J. Schulman, F. Wolski, P. Dhariwal, A. Radford, and O. Klimov, "Proximal policy optimization algorithms," *arXiv preprint arXiv:1707.06347*, 2017.

2025 The 10th International Conference on Integrated Circuits and Microsystems

HybridSYN: An Efficient Logic Synthesis Methodology

1st Ke Wang
Hangzhou Dianzi University
School of Electronics Information
Hangzhou, China
231040081@hdu.edu.cn

2nd Yue Wu*
Hangzhou Dianzi University
School of Electronics Information
Hangzhou, China
yuewu@hdu.edu.cn

3rd Xiaoyan Yang
Hangzhou Dianzi University
School of Electronics Information
Hangzhou, China
yangxiaoyan@hdu.edu.cn

Abstract—As the complexity of modern circuits increases, logic synthesis faces growing challenges. Reinforcement learning (RL) shows promise for efficient logic optimization. However, current RL methods rely solely on a single Boolean circuit representation—typically And-Inverter Graphs (AIG). This constraint hinders the exploration of larger optimization spaces, leading to suboptimal outcomes. This study introduces HybridSYN, a hybrid RL framework combining optimization algorithms from AIG and Majority-Inverter Graphs (MIG) to address the limitations of synthesis methods relying on a single logic graph representation. By expanding the action space, HybridSYN enables the exploration of better LUT-level solutions. To improve state representation, HybridSYN processes topological features from both AIG and MIG using independent Graph Convolutional Network (GCN) encoders. The resulting features are fused and passed through a fully connected network, where the final layer jointly predicts the type of operator and its parameters. Experimental results on the EPFL benchmark suite show that HybridSYN achieves a 9.7% reduction in LUT count compared to existing RL methods and a 17.1% improvement over baseline approaches.

Index Terms—Reinforcement Learning, Logic Synthesis, Logic Optimization, Electronic Design Automation.

I. INTRODUCTION

Logic synthesis is a crucial step in circuit design, as it significantly improves design efficiency and performance by converting the abstract behavior of a circuit into an optimized gate-level implementation. Its primary functionality lies in automatically optimizing the structure and delay of a circuit, thereby enhancing the overall system's performance. Traditionally, solutions to logic synthesis problems were based on heuristic algorithms [1] and manual processes. As circuit complexity increases, traditional heuristic-based approaches rely on predefined optimization sequences. These sequences lack adaptability to diverse circuit characteristics and face exponentially growing search spaces, making them prone to local optima and insufficient for comprehensive design space exploration [2], [3].

To address these limitations, RL-based methods have been increasingly utilized in logic synthesis. Early RL approaches focused on leveraging the And-Inverter graph (AIG) for circuit optimization, which made progress in the synthesis sequence exploration [4]–[6]. With the emergence of Majority-Inverter graphs (MIG) as an alternative representation for arithmetic-

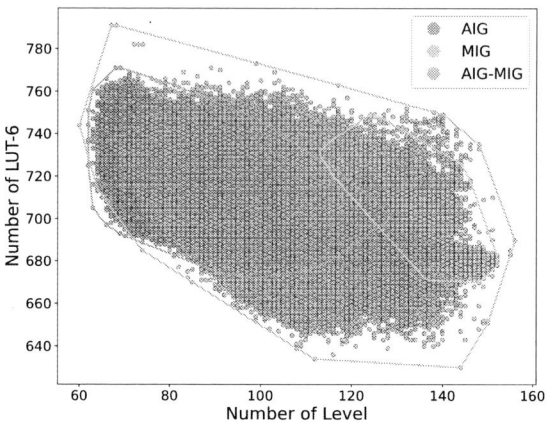

Fig. 1. Distribution of Level-LUT for 50,000 randomly generated synthesis flows on the max circuit: AIG, MIG, and AIG-MIG hybrid.

intensive designs, recent advancements in RL have shifted towards leveraging the operators of MIG to optimize circuits [7], [8].

As shown in Fig. 1, the design results for "max" are distributed differently. We used 50,000 randomly generated synthetic scripts on the EPFL benchmark suite [9] with ABC [10] and Mockturtle [11], [12] tools. The exploration space varies significantly, and this depends on the logic representation. This shows that hybrid exploration of both AIG and MIG can facilitate a more thorough search and ultimately lead to better Quality of Results (QoR). However, the search space available for exploration is inherently different for each logic representation, and relying on one representation alone often restricts the potential for discovering optimal solutions.

To overcome these problems, we propose an RL framework for automatic logic optimization using the Proximal Policy Optimization (PPO) algorithm. Named HybridSYN (Hybrid AIG-MIG Logic Synthesis), this framework explores the search space and generates optimized operator sequences with PPO. The main contributions are summarized as follows:

- We propose a hybrid optimization model based on PPO, which leverages both AIG and MIG representations along with their corresponding optimization algorithms.

979-8-3315-8850-2/25 $31.00 © 2025 IEEE

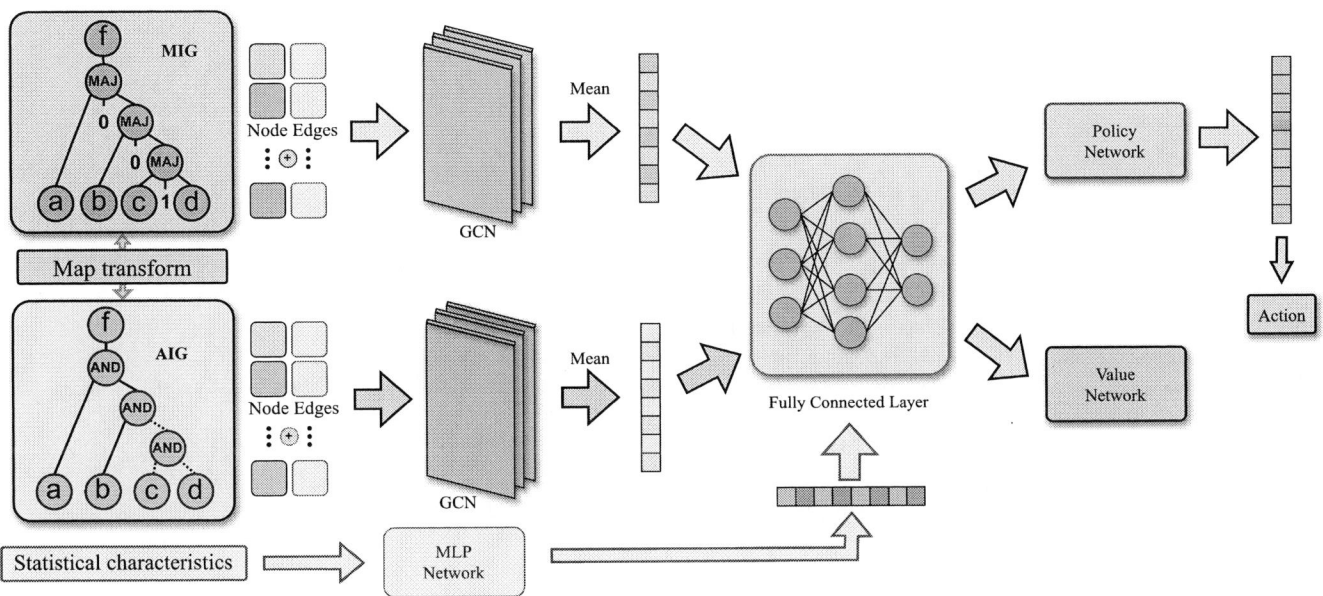

Fig. 2. Network Architecture of our method

- We develop a unified optimization environment with seamless AIG/MIG conversion capabilities, enabling RL-driven selection of optimal representations and operator sequences.
- A cross-attention mechanism fuses features from AIG and MIG, along with circuit scalar metrics, to create a more comprehensive state representation.
- On the EPFL arithmetic benchmark suite [9], compared to previous techniques [6], the proposed method can reduce the number of look-up tables (LUTs) by 9.7% without increasing the average levels.

II. FUNDAMENTALS AND RELATED WORK

A. Reinforcement Learning Algorithms

We adopt PPO2 [13], a standard method for policy optimization in reinforcement learning. PPO2 optimizes a surrogate loss defined as:

$$L(\theta) = \hat{\mathbb{E}} \left[\min \left(r_t(\theta) \hat{A}_t, \text{clip}(r_t(\theta), 1 - \epsilon, 1 + \epsilon) \hat{A}_t \right) \right] \quad (1)$$

where $r_t(\theta)$ is the probability ratio $r_t(\theta) = \frac{\pi_\theta(a_t|s_t)}{\pi_{\theta_{old}}(a_t|s_t)}$, and $\hat{A}_t$ is the advantage function at time t. The advantage function $\hat{A}_t$ is used to estimate the relative advantage of taking action a_t over the current policy. To calculate this advantage, we use the Generalized Advantage Estimation (GAE) [14], which is defined as:

$$\hat{A}_t = \sum_{l=0}^{\infty} \gamma^l \left(r_{t+1} - V_t \right) \quad (2)$$

where γ is the discount factor and V_t is the value function estimated by the critic network at time step t.

B. Logic Representation

In digital circuit design, Boolean networks are commonly represented as a directed acyclic graph (DAG). In this context, each node in the DAG represents a logical operation or a primary input/output (PI/PO), and each directed edge represents a connection between them. Circuit representations such as AIG and MIG are widely used to model and optimize these Boolean networks.

AIG is a widely used Boolean network representation, consisting of two input AND gates and inverters organized in a DAG. It provides an efficient structure for representing Boolean functions and facilitates the optimization of logic circuits. AIG has become a standard representation in many logic synthesis tools due to its compactness and ease of manipulation.

In recent years, MIG has become a promising alternative to AIG, especially for optimizing arithmetic-intensive designs. This is because MIG uses majority nodes and inverters to implement the majority function, which generalizes the AND and OR operations. The expressive majority function enables powerful optimization techniques, such as logic simplification, delay balancing, and cost reduction. Consequently, these characteristics make MIG exceptionally effective in arithmetic circuits, leading to improved QoR [15], [16].

C. Cross Attention for Feature Fusion

Cross-attention is a powerful mechanism used for feature fusion, especially when combining different feature sets. It enables the model to learn the relationship between distinct features by assigning weights to them based on their relevance. Let Q be the query, K be the key, and V be the value. The attention score is computed as:

$$\text{Attention}(Q, K, V) = \text{softmax}\left(\frac{QK^T}{\sqrt{d_k}}\right)V \qquad (3)$$

where d_k is the dimension of the key vector. In our framework, we use features from one graph as queries and from the other as key–value pairs, allowing the adaptive fusion of AIG and MIG embeddings.

D. Related Work

Previous studies cast logic synthesis as a machine-learning sequence optimization task. DRiLLS [4] introduces an advantage actor-critic agent to minimize logic network area under delay constraints. This approach is applied to optimize AIG representations. Similarly, RL-A2C [5] generates synthesis scripts matching the $resyn2$ baseline in length, focusing on AIG optimization. Zhou et al. propose a PPO-based logic synthesis framework [6] for AIG with an analysis of statistical features, which further contributes to AIG optimization. Peruvemba et al. present a runtime-constrained RL method [7] using a logic network reward function and termination conditions to generate synthesis scripts, validated in separate AIG and MIG environments. The ESE framework [8] proposed by Yu Qian et al. uses RL to generate customized synthesis scripts for circuits, with experiments conducted separately in AIG and MIG.

However, existing RL-based methods generally optimize a single logic network representation. AIG and MIG exhibit complementary strengths: AIG achieves superior area reduction in control-intensive circuits, while MIG significantly reduces logic depth in arithmetic-dominated circuits, as supported by prior works [8], [17]. To leverage the unique advantages of both representations, we propose a hybrid optimization approach that combines AIG and MIG primitives within a unified synthesis flow.

III. METHODOLOGY

A. Framework

The proposed framework, *HybridSYN*, alternates between AIG and MIG representations to optimize circuits. Training begins by initializing the encoder E_ϕ, critic V_ω, actor π_θ, and a replay buffer. The method is outlined in Algorithm 1. At the start of each episode, the logic synthesis environment is reset. At every step t, HybridSYN encodes circuit features into an RL state s_t via the encoder E_ϕ. The actor, implemented as a PPO-based policy network, selects an optimization action a_t, which updates the circuit and transitions it to a new state s_{t+1}. The critic evaluates the action a_t by computing the reward r_t and estimating the value function v_t. The transition tuple $(s_t, a_t, r_t, v_t, s_{t+1})$ is stored in the replay buffer. This process is repeated until the synthesis sequence reaches a predefined length L, at which point the trajectory is stored. Once t_u trajectories are collected, the networks are updated, the replay buffer is cleared, and training continues until the maximum number of episodes, $max\_episode$, is reached.

Algorithm 1 Learning Framework to Explore Logic Synthesis

Input: Boolean network
Output: Synthesis script

1: Initialize encoder E_ϕ, critic V_ω, and actor π_θ with random parameters ϕ, ω, θ.
2: Initialize $max\_episodes$, L, t_u, and replay buffer $ReplayBuffer$.
3: **for** episode = 1 **to** $max\_episode$ **do**
4:     Reset Logic Synthesis Environment (HybridSYN Env).
5:     **for** t = 1 **to** L **do**
6:         Observe state: $s_t = E_\phi(\text{AIG}, \text{MIG}, \text{features})$.
7:         Select action: $a_t \leftarrow \pi_\theta(s_t)$.
8:         Execute action a_t, observe next state s_{t+1} and reward r_t.
9:         Estimate value: $v_t \leftarrow V_\omega(s_t, s_{t+1}, r_t)$.
10:         Store $(s_t, a_t, r_t, v_t, s_{t+1})$ in $ReplayBuffer$.
11:     **end for**
12:     **if** $(episode + 1) \bmod t_u = 0$ **then**
13:         Update networks using data from $ReplayBuffer$.
14:         Clear $ReplayBuffer$.
15:     **end if**
16: **end for**

B. Environment Setup

To facilitate dynamic representation switching between AIG and MIG, we develop an RL environment for logic synthesis, leveraging the widely-used open-source tools ABC [10] and Mockturtle [11], [12]. Our environment leverages efficient transformation methods [18] to ensure seamless conversion between AIG and MIG while preserving the quality of results. Additionally, it supports the concurrent selection of AIG and MIG logic graph representations and their corresponding optimization algorithms, enabling flexible and efficient optimization by switching between the two.

1) State: The state of the environment is divided into three parts: two parts describe the graph features of AIG and MIG, while the other part consists of a feature vector that includes the following values at step t:

- Number of nodes, logic depth, and edges of the circuit.
- 6-LUT count and level after the circuit gets mapped.
- The current of the operator sequence.

2) Action Space: To demonstrate the feasibility of our approach, we selected the same 7 actions as those in ABC's $resyn2$, $A_{\text{aig}} = \{$ balance, rewrite, rewrite -z, refactor, refactor -z, resub, resub -z $\}$, as well as the 6 actions in Mockturtle for MIG, $A_{\text{mig}} = \{$ rewrite, rewrite -z, refactor, resub, resub -z, balance $\}$. Our model utilizes the union of these two action spaces, i.e., $A_{\text{total}} = A_{\text{aig}} \cup A_{\text{mig}}$.

3) Reward Function: To enable the agent to learn designs with varying LUT and level counts, a reward function based solely on a linear relationship with these quantities is inadequate. Thus, a normalized reward function is employed. Additionally, the normalization of the level count is incorporated

TABLE I

COMPARISON OF 6-LUT COUNTS AND LUT DEPTHS ON THE EPFL ARITHMETIC BENCHMARK SUITE, WITH THE IMPROVEMENT PERCENTAGE REPRESENTING THE OPTIMIZATION RATIO OF 6-LUT COUNTS RELATIVE TO *resyn2*.

Circuit	*resyn*2	DRiLLS [4]	RL-A2C [5]	RL-PPO [6]	Ours
max	719(78)	694	693	687.8(112.0)	**661.2**(112.1)
adder	249(121)	244	244	244(116)	**192.5**(64)
cavlc	118(5)	112.2	111.7	111.3(6.1)	**110.3**(6.5)
ctrl	29(2)	28	28.2	28(2)	**26.7**(2.1)
int2float	46(5)	42.6	42.7	42.3(5.0)	**41.7**(5.7)
router	76(8)	70.1	69.9	69.5(8.4)	**45.3**(7.5)
priority	220(86)	**133.4**	145.9	142.9(24.8)	149.3(29)
i2c	320(8)	292.1	290.5	289.3(8.1)	**286.2**(6.2)
sin	1466(76)	1441.5	1438.2	1438(74.5)	**1435.1**(72.2)
square	3915(122)	3889.4	3888.2	3889(121.7)	**3767.4**(121.6)
sqrt	5127(2216)	4708	4698.8	4685.3(1804.6)	**3847.4**(1844.4)
log2	7703(153)	7583.6	7580.3	7580.1(159)	**7461.4**(153.3)
multiplier	5712(2216)	5678	5672.3	5672(126)	**4982.3**(127.8)
voter	1828(20)	1834.7	1686.2	1678.1(19.06)	**1657.4**(20.3)
div	8197(2060)	7944.4	7639.8	7807.1(2063.3)	**4230.1**(2084.7)
mem	11459(51)	10527.6	10363.5	10309.7(45.6)	**10223.9**(52.4)
GEOMEAN	808.78(51.1)	753.5	750.7	748.6(48.0)	**670.8**(46.8)
ratio	1(1)	0.932	0.928	0.926(0.939)	**0.829**(0.916)

to encourage the agent to optimize for solutions that meet the delay constraints. The reward function is defined as follows:

$$r_t = \frac{Lut_{init} - Lut_t}{Lut_{init}} + \frac{Lev_{init} - Lev_t}{Lev_{init}} \quad (4)$$

C. Network Architecture

We introduce a Graph Convolutional Network (GCN) architecture to enhance RL-driven optimization in logic synthesis. As shown in Fig. 2, the pipeline begins by preprocessing the features of each circuit node. These include the inverter count (#inverter) and a learned node-type embedding. A fully connected (FC) layer then projects them into a shared latent space.

For feature extraction, we apply two successive GCN layers to each graph type. Each GCN layer is followed by batch normalization (BN) and a ReLU activation to ensure stable training and nonlinear transformation. To capture structural patterns at the graph level, we use dual pooling operations: global mean pooling for overall feature distributions and global max pooling to highlight salient local patterns. The pooled vectors from both graphs are then fused using a cross-attention mechanism, which models inter-graph dependencies. The resulting fused vector is passed through a fully connected (FC) layer to align its dimensions. We concatenate this fused graph embedding with numerical state features and feed them into shared neural layers. The unified representation drives two heads: a Policy Network, which outputs a softmax distribution over discrete optimization operators, and a Value Network, which linearly projects to a scalar state value for PPO's advantage calculation. By coordinating discrete operator selection with continuous parameter tuning, this framework enables efficient logic synthesis optimization.

IV. EXPERIMENTAL RESULTS

A. Benchmarks and Environments

We implemented the HybridSYN environment using ABC [10] and Mockturtle [11], [12], along with a C++ interface. Our method was developed in Python v3.8.8 with PyTorch v1.13.0. The experiments were conducted on a machine equipped with two 28-core Intel Xeon Gold 6348 CPUs and a single NVIDIA GeForce RTX 4090 with 24GB of CUDA memory. The synthesis sequence length L was set to 25, and the number of iterations was 200. FPGA mapping was performed using "if -a -K 6" for AIG and "lutmap -a -K 6" for MIG, with the number of LUTs collected as the final metric. Experiments were conducted on EPFL benchmarks, and the results were compared with those reported in DRiLLS [4], RL-A2C [5], and RL-PPO [6], with all comparative data sourced from RL-PPO.

B. Main Results

Table I presents the comparison results for 6-LUT counts, with values in parentheses indicating the corresponding netlist levels. Bold numbers denote the best performance among all methods. The final column shows our model's performance averaged over the last 10 training rounds, demonstrating stable convergence. We adopt the same evaluation protocol as RL-PPO to ensure a fair comparison. The primary objective is to minimize 6-LUTs while maintaining the number of levels satisfying the constraints.

Among the 16 benchmark circuits, our method achieved the best results in 15 cases. Compared to prior AIG-only optimization approaches, such as RL-A2C and RL-PPO, our method reduced 6-LUTs by 9.9% and 9.7%, respectively. The limited improvements in certain cases stem from the inherent constraints of single-representation optimization, which fails to adapt to diverse circuits, such as arithmetic-heavy or control-dominated designs. By incorporating both AIG and

TABLE II
EXPERIMENTAL RESULTS WITH THE AIG GRAPH OR THE MIG GRAPH.

Circuit	$resyn2$	w/o MIG	w/o AIG	w/o feature	our
max	719	689.3	694.3	677.8	**661.2**
adder	249	245.3	194.6	193.6	**192.5**
cavlc	118	112.1	117.5	116.3	**110.3**
ctrl	29	27.9	27.1	27	**26.7**
int2float	46	43.2	50.4	42.1	**41.7**
router	76	70.1	48.9	51.3	**45.3**
priority	220	**148.6**	199.4	153.3	149.3
i2c	320	290.4	320	292.4	**286.2**
sin	1466	1441.6	1444.7	1436.6	**1435.1**
square	3915	3891.5	3782.6	3777.6	**3767.4**
sqrt	5127	4711.6	3879.4	3874.1	**3847.4**
log2	7703	7569.1	7482.3	7512.2	**7482.3**
multiplier	5712	5598.7	5024.3	5003.6	**4976.4**
voter	1828	1696.7	1784.1	1664.5	**1657.4**
div	8197	7819.2	8269.3	4305.4	**4230.1**
mem	11459	10328.4	10279.6	10237.1	**10223.9**
GEOMEAN	808.8	752.7	742.3	684.3	**670.8**
ratio	1	0.931	0.918	0.846	**0.829**

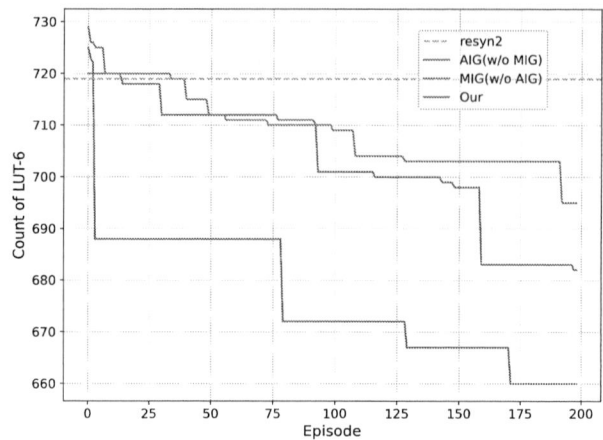

Fig. 3. Comparison of Iterative 6-LUT Minimization for AIG(DRiLLS), MIG(w/o AIG), and Our(Hybrid AIG-MIG) Approaches over 200 Training Iterations in the Comprehensive Synthesis Flow for the Max Circuit

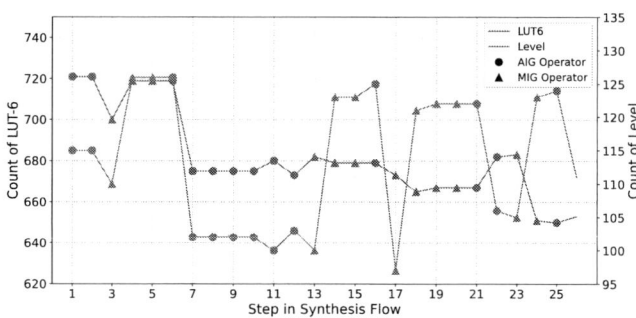

Fig. 4. Impact of Synthesis Flow on 6-LUT Count and Logic Level for the max Circuit: AIG and MIG Optimization Algorithms Effects over the Last 25 Steps

MIG optimizations, our approach overcomes these limitations, achieving further reductions in 6-LUTs. Moreover, our method yields an average depth of 46.8 levels—about 2.3% shallower than the depth achieved by the $resyn2$ heuristic—while still reducing LUT count. These results underscore the advantages of hybrid circuit representations in RL-based logic optimization.

C. Ablation Study

We conducted two ablation studies to evaluate the impact of MIG optimization operators and MIG graph features. The results are summarized in Table II. In the first ablation study, the results labeled "w/o MIG" correspond to AIG optimization only. In this setting, we exclusively use AIG optimization operators A_{aig}. The "w/o AIG" setting excludes AIG optimization within the HybridSYN framework and uses only the MIG operator A_{mig}, its graph features, and statistical features. In the "w/o features" setting, we concatenate AIG and MIG features together without using cross-attention. In both cases, the rest of the HybridSYN framework remains unchanged, and PPO is used to select both AIG and MIG optimization algorithms. This setup allows us to compare the impact of excluding MIG features.

Experimental results show that optimization using only MIG operators outperforms AIG in 7 out of 16 benchmark circuits, particularly in arithmetic-intensive designs. This aligns with previous studies [17], which demonstrate that MIG's majority node primitives inherently compress arithmetic logic more efficiently than AIG's AND node structures. Conversely, AIG achieves superior 6-LUTs minimization in control-intensive circuits due to its effective local Boolean simplification. HybridSYN enables dynamic switching between AIG and MIG optimization algorithms based on circuit characteristics, allowing for more flexible optimization across diverse designs. Furthermore, the MIG feature extraction module improves the minimization of 6-LUTs by 1.7%, improving the overall efficiency of HybridSYN by refining circuit feature represen-

tation. These ablations confirm that performance gains stem from both switching between AIG and MIG representations and from expanding the operator action space.

D. Explore Efficiency Analysis

To evaluate the exploration efficiency of our model, we tracked the reduction in the 6-LUT count for the max circuit over 200 iterations of training. As shown in Fig. 3, the hybrid optimization approach combines AIG and MIG optimization algorithms, meeting logic depth constraints while achieving faster convergence to better solutions. Although better solutions than $resyn2$ can be found within the exploration spaces of AIG or MIG, they do not do so as efficiently. This inefficiency arises from the limitations of each individual exploration space. The final synthesis steps shown in Fig. 4 highlight the synergy between AIG and MIG optimization algorithms. As optimization progresses, the reduction in the 6-LUT count stagnates after 17 steps. At this stage, switching to the other network representation enables further reduction in the 6-LUT count and logic depth. In summary, our approach enhances exploration efficiency by leveraging both AIG and

MIG optimization algorithms, allowing the model to escape local optima and achieve better optimization results.

V. CONCLUSION

HybridSYN introduces a novel RL framework that combines optimization algorithms from both AIG and MIG to address the limitations of existing RL-based methods that rely on a single logic graph representation. By expanding the action space and leveraging hybrid feature representations, HybridSYN facilitates a more comprehensive exploration of optimization strategies. Ablation experiments have confirmed the effectiveness of integrating AIG and MIG optimization algorithms. The integration of the action spaces of AIG and MIG facilitates a more efficient search for optimal solutions. Comprehensive evaluations on the EPFL benchmark suite confirm the framework's superior performance in logic synthesis sequence exploration, particularly in complex arithmetic circuits, where our method achieves 9.7% 6-LUT improvement over conventional approaches.

VI. FUTURE WORK

We will develop more efficient coordination strategies for AIG and MIG representations. These strategies aim to improve optimization effectiveness during multi-graph switching. We will also explore novel cross-modal fusion frameworks to enable better operator selection in logic synthesis.

REFERENCES

[1] Xing Li, Lei Chen, Fan Yang, Mingxuan Yuan, Hong tao Yan, and Yupeng Wan. Himap: a heuristic and iterative logic synthesis approach. *Proceedings of the 59th ACM/IEEE Design Automation Conference*, 2022.

[2] Chang Feng, Wenlong Lyu, Zhitang Chen, Junjie Ye, Mingxuan Yuan, and Jianye Hao. Batch sequential black-box optimization with embedding alignment cells for logic synthesis. In *2022 IEEE/ACM International Conference On Computer Aided Design (ICCAD)*, pages 1–9, 2022.

[3] Jianyong Yuan, Peiyu Wang, Junjie Ye, Mingxuan Yuan, Jianye Hao, and Junchi Yan. Easyso: Exploration-enhanced reinforcement learning for logic synthesis sequence optimization and a comprehensive rl environment. In *2023 IEEE/ACM International Conference on Computer Aided Design (ICCAD)*, pages 1–9, 2023.

[4] Abdelrahman Hosny, Soheil Hashemi, Mohamed Shalan, and Sherief Reda. Drills: Deep reinforcement learning for logic synthesis. In *2020 25th Asia and South Pacific Design Automation Conference (ASP-DAC)*, pages 581–586, 2020.

[5] Keren Zhu, Mingjie Liu, Hao Chen, Zheng Zhao, and David Z. Pan. Exploring logic optimizations with reinforcement learning and graph convolutional network. In *2020 ACM/IEEE 2nd Workshop on Machine Learning for CAD (MLCAD)*, pages 145–150, 2020.

[6] Guanglei Zhou and Jason H. Anderson. Area-driven fpga logic synthesis using reinforcement learning. In *2023 28th Asia and South Pacific Design Automation Conference (ASP-DAC)*, pages 159–165, 2023.

[7] Yasasvi V. Peruvemba, Shubham Rai, Kapil Ahuja, and Akash Kumar. Rl-guided runtime-constrained heuristic exploration for logic synthesis. In *2021 IEEE/ACM International Conference On Computer Aided Design (ICCAD)*, pages 1–9, 2021.

[8] Yu Qian, Xuegong Zhou, Hao Zhou, and Lingli Wang. An efficient reinforcement learning based framework for exploring logic synthesis. *ACM Trans. Des. Autom. Electron. Syst.*, 29(2), January 2024.

[9] Luca Gaetano Amarù, Pierre-Emmanuel Gaillardon, and Giovanni De Micheli. The epfl combinational benchmark suite. 2015.

[10] Robert Brayton and Alan Mishchenko. Abc: An academic industrial-strength verification tool. In Tayssir Touili, Byron Cook, and Paul Jackson, editors, *Computer Aided Verification*, pages 24–40, Berlin, Heidelberg, 2010. Springer Berlin Heidelberg.

[11] Mathias Soeken, Heinz Riener, Winston Haaswijk, and Giovanni De Micheli. The EPFL logic synthesis libraries. *CoRR*, abs/1805.05121, 2018.

[12] Alessandro Tempia Calvino and Giovanni De Micheli. Scalable logic rewriting using don't cares. In *2024 Design, Automation & Test in Europe Conference & Exhibition (DATE)*, pages 1–6, 2024.

[13] John Schulman, Filip Wolski, Prafulla Dhariwal, Alec Radford, and Oleg Klimov. Proximal policy optimization algorithms, 2017.

[14] John Schulman, Philipp Moritz, Sergey Levine, Michael Jordan, and Pieter Abbeel. High-dimensional continuous control using generalized advantage estimation. *arXiv preprint arXiv:1506.02438*, 2015.

[15] Siang-Yun Lee and Giovanni De Micheli. Heuristic logic resynthesis algorithms at the core of peephole optimization. *IEEE Transactions on Computer-Aided Design of Integrated Circuits and Systems*, 42(11):3958–3971, 2023.

[16] Heinz Riener, Eleonora Testa, Winston Haaswijk, Alan Mishchenko, Luca Amarù, Giovanni De Micheli, and Mathias Soeken. Scalable generic logic synthesis: One approach to rule them all. In *2019 56th ACM/IEEE Design Automation Conference (DAC)*, pages 1–6, 2019.

[17] Walter Lau Neto, Max Austin, Scott Temple, Luca Amaru, Xifan Tang, and Pierre-Emmanuel Gaillardon. Lsoracle: a logic synthesis framework driven by artificial intelligence: Invited paper. In *2019 IEEE/ACM International Conference on Computer-Aided Design (ICCAD)*, pages 1–6, 2019.

[18] Alessandro Tempia Calvino, Heinz Riener, Shubham Rai, Akash Kumar, and Giovanni De Micheli. A versatile mapping approach for technology mapping and graph optimization. In *2022 27th Asia and South Pacific Design Automation Conference (ASP-DAC)*, pages 410–416, 2022.

2025 The 10th International Conference on Integrated Circuits and Microsystems

A 20-GHz Low-Jitter Fractional-N CP-PLL With 2.5-GHz FBAR Reference

Jiandu Wang
College of Physics and Information Engineering
Fuzhou University
Fuzhou, China

Jiwei Huang*
College of Physics and Information Engineering
Fuzhou University
Fuzhou, China
huangjw@fzu.edu.cn

Yi Sun
College of Physics and Information Engineering
Fuzhou University
Fuzhou, China

Abstract—A 20-GHz fractional-N CP-PLL with Film Bulk Acoustic Resonator (FBAR)-Based frequency reference is presented, employing a fast-switching charge pump (CP) and on-chip frequency reference. While retaining the design simplicity of analog CP-PLLs, this architecture utilizes a low-noise high-frequency reference to significantly reduce the division ratio (N), thereby enhancing loop suppression of CP,loop-filter (LF), and phase-frequency detector (PFD) noise. Concurrently, to enable operation with low loop bandwidth(1.5 MHz), a low-noise Class-C transformer-coupled voltage-controlled oscillator (VCO) is implemented to achieve outstanding jitter performance. Designed in TSMC 65-nm CMOS technology, the PLL achieves an output frequency range of 18.4–21.3 GHz. The core area occupies 0.531 mm^2 with a power consumption of 41 mW. The results show 95.02 fs jitter and a $|FOMj|$ of 244.3 dB.

Index Terms—FBAR oscillator, fractional-N PLL, jitter, Voltage-controlled oscillator (VCO), Charge pump (CP)

I. INTRODUCTION

Low-noise phase-locked loop (PLL) remains a critical objective in contemporary communication and electronic systems. The phase noise performance of a PLL is directly proportional to the ratio of the VCO frequency (f_{VCO}) to the reference frequency (f_{REF}). The charge pump's noise contribution to the overall system is given by:

$$\theta_{\mathrm{n,o_{cp}}}^2 = \theta_{\mathrm{n,cp}}^2 \cdot \left| \frac{N H_{\mathrm{o}}(s)}{1 + H_{\mathrm{o}}(s)} \cdot \frac{2\pi}{I_{\mathrm{cp}}} \right|^2 \qquad (1)$$

where $\theta_{\mathrm{n,cp}}^2$ is the output-referred equivalent phase noise of the PFD and CP,and $H_{\mathrm{o}}(s)$ signify the open-loop transfer function of the PLL.Due to the typically low reference frequency, the division ratio (N=f_{VCO}/f_{REF}) in conventional CP-PLLs is consequently significantly large [1] [2] [3] [4].

All-digital PLLs (ADPLLs) can operate with or without dividers in the feedback path and leverage digital loop filters for enhanced programmability. However, their superior performance is achieved at the cost of increased calibration complexity, necessitating continuous background calibration processes [5]. Sub-sampling PLLs (SS-PLLs) significantly outperform conventional CP-PLLs by eliminating divider noise and avoiding the N² scaling of CP/LF noise, though both

digital-to-time converter (DTC) and delay-locked loop (DLL) introduce nonlinearities and quantization noise through calibration [6].To bridge the phase noise gap between CP-PLLs and the aforementioned advanced PLL architectures, which requires to employ an increased reference frequency approach to directly reduce the division ratio N.

Simply multiplying a low-frequency crystal oscillator to reduce N degrades the pristine reference which degrades the reference source's phase noise by $20 * log10(K_m)$(where K_m is the multiplication factor). In contrast, employing a 2.5-GHz FBAR as a high-frequency clean reference source while simultaneously reducing cost through on-chip integration—provides at least 5× (or $20 * log_{10}(5) = 14dB$) CP noise suppression compared to the highest-frequency commercially available crystal references (500 MHz).

This work presents a 20-GHz fractional-N CP-PLL utilizing a 2.5-GHz FBAR high-frequency reference source. The design incorporates a phase-frequency detector (PFD) capable of resolving high-frequency input signals and a fast-switching CP. Simultaneously, a low-noise Class-C VCO is designed to prevent the degradation of overall circuit noise performance [7].

II. SYSTEM ARCHITECTURE AND DESIGN

The overall architecture of the proposed PLL is shown in Fig. 1.In this design,we present a 20-GHz fractional-N CP-PLL using a fully integrated on-chip FBAR resonator operating at 2.5 GHz.The reference source based on the FBAR resonator, CP and VCO are powered by a 1.8 V supply, while other modules operate from a 1 V supply for enhanced power efficiency.To minimize quantization noise and fractional spurs from the sigma-delta modulator (SDM) while leveraging the noise optimization benefits of the high reference frequency, the divided output (f_{DIV} about 2.5GHz) is further down-divided to generate the SDM clock,thus preventing timing violations in the SDM.The PLL loop bandwidth is set to 1.5 MHz to prevent in-band CP noise from dominating over reference noise.

979-8-3315-8850-2/25 $31.00 © 2025 IEEE

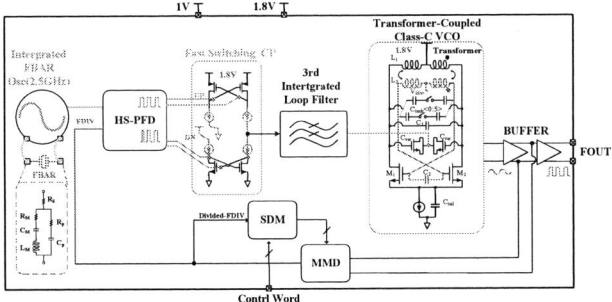

Fig. 1. Proposed architecture of the 20-GHz fractional-N CP-PLL.

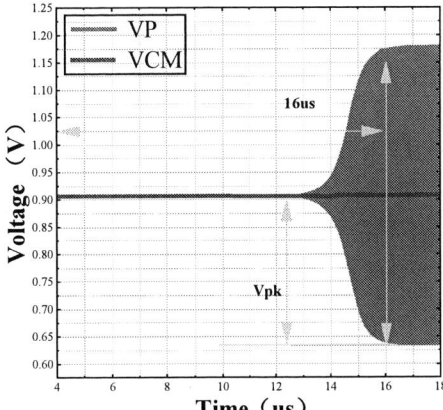

Fig. 3. Simulated Waveform of FBAR Reference Oscillator.

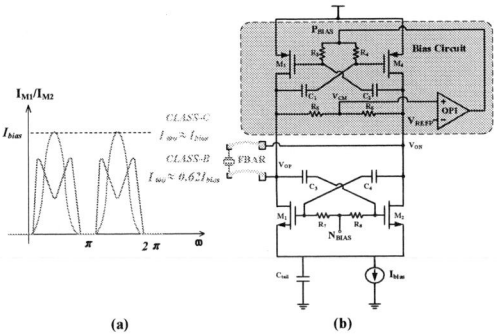

Fig. 2. (a)Current of Class-C/B (b)2.5-GHz FBAR Reference Oscillator.

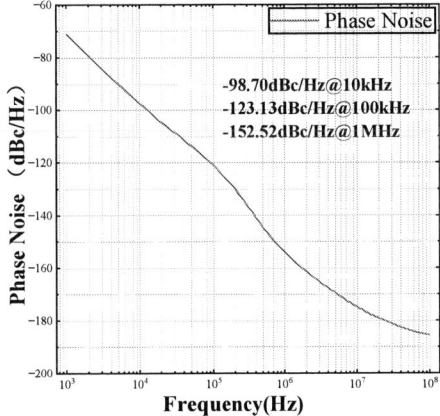

Fig. 4. Simulated Phase Noise of FBAR Reference Oscillator.

A. 2.5-GHz FBAR Reference Oscillator

Class-B and Class-C oscillators represent two fundamental oscillator types. Due to a reduced bias voltage V_{bias} and an increased tail capacitance C_{tail} in Class-C VCO, the cross-coupled pair exhibits steep and narrow current pulses, resulting in current efficiency approaching I_{bias}(While Class-B is about $0.62I_{bias}$) in Fig. 2(a). Consequently, the Class-C oscillator achieves a 3-dB improvement in phase noise compared to Class-B mode.

The FBAR oscillator requires low phase noise to provide a clean reference clock for the PLL. The proposed architecture of the FBAR oscillator is illustrated in Fig. 2(b). It employs a differential transistor pair operating in a Class-C push-pull configuration within a oscillator structure. The circuit is supplied with a 1.8 V power supply. Negative resistance for oscillation sustainment is generated by a complementary cross-coupled pair. The oscillation frequency is determined by the FBAR resonator and the internal capacitance of the circuit. V_{OP} and V_{ON} are the differential output signals, while V_{CM} denotes their common-mode voltage. V_{REFP} serves as a reference voltage. P_{BIAS} and N_{BIAS} are the bias voltages for the PMOS and NMOS cross-coupled pairs, respectively. Proper biasing ensures that the oscillator operates in Class-C mode. Additionally, an operational amplifier (OP1) is used to stabilize the common-mode level of the oscillator outputs while maintaining the PMOS bias voltage within an

appropriate range.

When oscillator works, the increasing amplitude may affect the operating state of the transistors. If any transistor enters the triode region during operation, the phase noise performance of the oscillator would significantly degrade [8]. To prevent this and ensure that both transistors in the cross-coupled pair remain in saturation region throughout the oscillation cycle, V_{CM} should be set appropriately.Assuming that the external bias voltage applied to the gates of the NMOS pair is at the minimum value, which is the minimum value ensuring that M_1 or M_2 operates in the saturation region(Which means $V_{N_{BIAS}}= V_{Ibias} + V_{odM1} + V_{THn}$).Thus,the V_{CM} can be set as:

$$V_{odM1} + V_{Ibias} + V_{pk} < V_{CM} < V_{DD} - |V_{odM3}| - V_{pk} \quad (2)$$

Where V_{odM1} and V_{odM3} denote the overdrive voltages of M_1 and M_3, respectively;V_{Ibias} represents the minimum saturation voltage required for the tail current source; and V_{pk} indicates the peak amplitude of the oscillator output, as illustrated in Fig. 3.

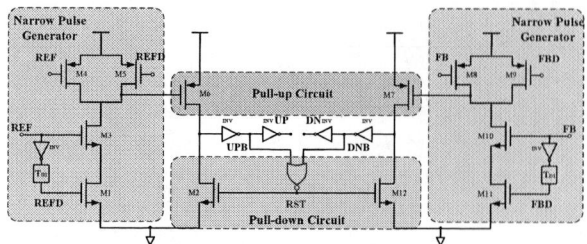

Fig. 5. High Speed PFD.

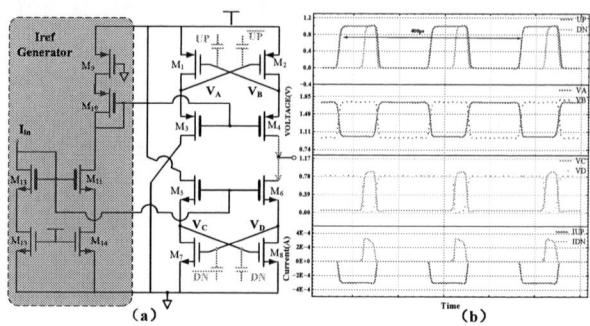

Fig. 6. (a)Fast-Switching CP (b)Simulation Results.

The simulation results of the proposed FBAR oscillator are shown in Fig. 4. The circuit achieves a phase noise of -152.52 dBc/Hz at a 1 MHz offset frequency. The startup time of the oscillator is approximately 16 us.

B. High Speed PFD and Fast-Switching CP

The PFD must operate at a sufficient speed to resolve the reference and divided feedback signals. The proposed PFD,shown in Fig. 5, features a simple and symmetric topology.Each pulse generator detects rising edges at the PFD inputs and drives transistors M_6 and M_7 in the pull-up network, with pulse width controlled by the T_{D1} delay cell.The narrow pulse generator ensures fast phase discrimination capability of the PFD. The pull-up network asserts the UP and DN signals.The pull-down network, consisting of transistors M_2 and M_{12} and a symmetric NOR gate to reset the UP(DN) outputs, discharges both outputs simultaneously when RST is high (RST=1). Using M_2 and M_{12} transistors and the symmetric structure of the NOR gate in the pull-down network, deadzone-free operation is attained.

The CP must be designed with sufficient linear operating range to drive thick gate varactors operating at 1.8 V supply. Employing a 1 V supply would need higher VCO tuning gain (K_{VCO}), degrading both VCO noise and overall PLL spur performance [9]. Therefore, a 1.8 V supply is mandated for the CP.To avoid level shifters for PFD-to-CP interfacing, the proposed charge pump Fig. 6(a) employs ac coupling of PFD signals to low-voltage MOS devices.Cross-coupled drain-gate potential shifting in opposite directions between M_1/M_2 and M_7/M_8 pairs ensures rapid current-source turn-on/off. The M_3/M_5 dummy branch provides differential balance.This topology achieves fast response to PFD UP(DN) pulses—enabling swift current-source activation and reset without power-hungry level converters, while avoiding full 1.8 V MOS implementations.

Simulation results of the PFD and CP in Fig. 6(b) demonstrate accurate phase detection at 5 GHz input signals. Cross-coupled nodes V_A/V_B (and equivalently V_C/V_D) accelerate turn-on transients, achieving 300 uA charge/discharge currents.

C. Transformer-Coupled Class-C VCO

As shown in Fig. 7,the VCO operates from a 1.8-V power supply and employs a Class-C transformer-coupled topology.

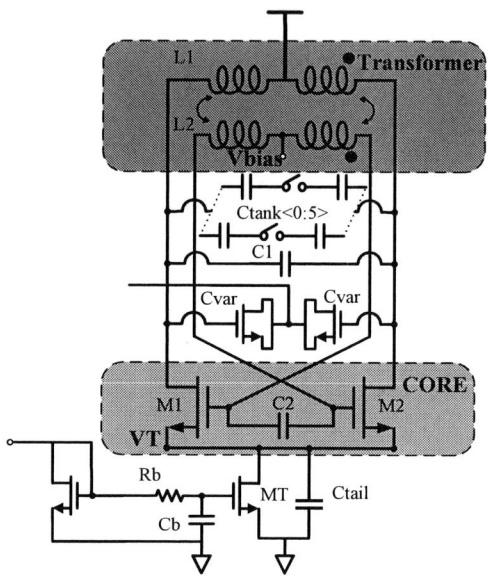

Fig. 7. Transformer-Coupled Class-C VCO

An appropriately set of V_{bias} and a large tail capacitor C_{tail} ensure the NMOS transistors operate in Class-C mode [10].

The oscillator core employs an NMOS cross-coupled to topology to generate negative resistance of $-2/gm$.High-Q transformers L_1 and L_2 provide voltage feedback with coupling factor k.A magnetic transformer is used to ac-couple the gates of the cross-coupled transistors to the tank, as the passive voltage gain of the transformer reduces the contribution of the active devices to the $1/f^2$ phase noise.Frequency tuning is achieved by a 6-bit digitally controlled capacitor bank co-designed with a couple of small varactors.Transistors M_1 and M_2 biased at V_{bias} with a large tail capacitor C_{tail},operating in Class-C region for phase noise reduction.The biasing circuit incorporates a current mirror M_B and M_T with an RC low-pass filter for supply noise suppression.

Fig. 8 characterizes the phase noise simulation of the proposed Class-C transformer-coupled VCO.The phase noise ranges from -107.88 dBc/Hz to -111.22 dBc/Hz at 1 MHz offset frequency.

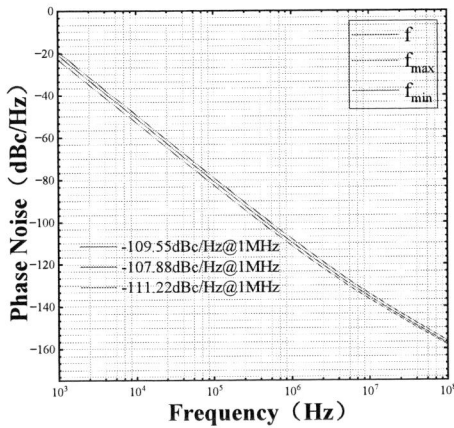

Fig. 8. Simulated Phase Noise of Transformer-Coupled Class-C VCO.

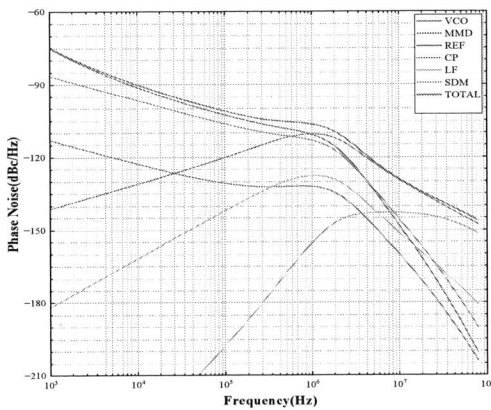

Fig. 10. Phase noise and contributions from various blocks.

III. SIMULATION RESULTS

The proposed FBAR-driven 20-GHz PLL is designed based on TSMC 65-nm COMS process.As shown in Fig. 9,the core area occupies 0.531 mm^2.When the PLL operates at 20.3 GHz (N = 8.12), the simulated worst-case spur is approximately $-45.2dBc$.The phase noise fitting curves for the various sections of the PLL are shown in Fig. 10. The jitter of 95.02 fs demonstrates its superiority over conventional low-frequency reference PLLs.

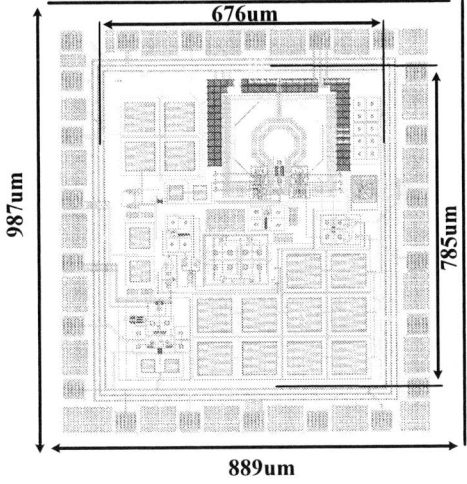

Fig. 9. Layout of 20-GHz fractional-N CP-PLL.

Table I summarizes the performance of the proposed design and provides a comparison with previously reported PLLs.The FBAR-driven 20 GHz fractional-N PLL achieves a high figure of $|FOMj|$ and exhibits excellent RMS jitter performance.

IV. CONCLUSION

This paper presents a 20-GHz fractional-N CP-PLL with 2.5-GHz FBAR reference source. The design employs a high-speed PFD, a fast-switching CP, and an on-chip frequency

TABLE I
COMPARISON WITH PRIOR WORKS

	[1]	[3]	[4]	[11]	This Work
Technology [nm]	7	28	28	5	65
Reference Type	On-chip FBAR	On-chip XO	External	External	On-chip FBAR
Architecture	Frac-N HM-PLL	Sliding-IF Frac-N CP-PLL	TPM CP-PLL	Cascaded(LC+RO)	Frac-N CP-PLL
Fref. [MHz]	2285	491.5	80	217.5	2500
Output Freq. [GHz]	4.8-5	23.3-30.2	15-18.5	27.84	18.4-21.3
RMS Jitter [fs]	91	115	156	204	95.02
Integ. Range [Hz]	10K-100M	20K-500M	10K-100M	100K-100M	10K-100M
$|FOMj|$ [dB]	261	244	244.6	241	244.3
Power [mW]	0.96	31	14.3	19.7	41

reference. By adopting a 2.5-GHz FBAR oscillator as a low-phase-noise reference, the division ratio is reduced, thereby enhancing the loop's suppression of CP noise and achieving low jitter. Additionally, a low-noise transformer-coupled Class-C VCO is implemented to attain excellent phase noise performance.

REFERENCES

[1] D. Yang, D. Murphy, H. Darabi, A. Behzad, R. Ruby and R. Parker, *An FBAR Driven -261dB FOM Fractional-N PLL.* IEEE Radio Frequency Integrated Circuits Symposium (RFIC), Atlanta, GA, USA, 2021.

[2] S. Kalia et al, *A Sub-100 Fs RMSjitter 20 GHz Fractional-N Analog PLL With a BAW Resonator Based On-Chip 2.5 GHz Reference.* IEEE Journal of Solid-State Circuits, vol. 57, no. 5, pp. 1372-1384, May 2022.

[3] S. Ek et al, *A 28-nm FD-SOI 115-fs Jitter PLL-Based LO System for 24–30-GHz Sliding-IF 5G Transceivers.* IEEE Journal of Solid-State Circuits, vol. 53, no. 7, pp. 1988-2000, July 2018.

[4] P. T. Renukaswamy, K. Vaesen, N. Markulic and J. Craninckx, *A 16-GHz Background-Calibrated Duty-Cycled FMCW Charge-Pump PLL.* IEEE Journal of Solid-State Circuits, vol. 59, no. 6, pp. 1684-1696, June 2024.

[5] M. Mercandelli et al, *32.3 A 12.9-to-15.1GHz Digital PLL Based on a Bang-Bang Phase Detector with Adaptively Optimized Noise Shaping Achieving 107.6fs Integrated Jitter.* 2021 IEEE International Solid-State Circuits Conference (ISSCC), San Francisco, CA, USA, 2021

[6] L. Wang, Z. Liu, R. Ma and C. P. Yue, *A Compact 20–24-GHz Sub-Sampling PLL With Charge-Domain Bandwidth Control Scheme.* IEEE Journal of Solid-State Circuits, vol. 60, no. 3, pp. 768-784, March 2025.

[7] A. Franceschin, P. Andreani, F. Padovan, M. Bassi and A. Bevilacqua, *A 19.5-GHz 28-nm Class-C CMOS VCO, With a Reasonably Rigorous Result on 1/f Noise Upconversion Caused by Short-Channel Effects.* IEEE Journal of Solid-State Circuits, vol. 55, no. 7, pp. 1842-1853, July 2020.

[8] A. Mazzanti and P. Andreani, *A Push–Pull Class-C CMOS VCO.* IEEE Journal of Solid-State Circuits, vol. 48, no. 3, pp. 724-732, March 2013.

[9] W. El-Halwagy, A. Nag, P. Hisayasu, F. Aryanfar, P. Mousavi and M. Hossain, *A 28-GHz Quadrature Fractional-N Frequency Synthesizer for 5G Transceivers With Less Than 100-fs Jitter Based on Cascaded PLL Architecture.* IEEE Transactions on Microwave Theory and Techniques, vol. 65, no. 2, pp. 396-413, Feb. 2017

[10] J. Dong, Q. Li, Y. Xie, Z. Cao and X. Li, *A 2.5-GHz FBAR-Based Push-Pull Class-C Oscillator with Low Phase Noise.* 2020 IEEE MTT-S International Conference on Numerical Electromagnetic and Multiphysics Modeling and Optimization (NEMO), Hangzhou, China, 2020.

[11] T. -H. Tsai, R. -B. Sheen, S. -Y. Hsu, Y. -T. Chang, C. -H. Chang and R. B. Staszewski, *A Cascaded PLL (LC-PLL + RO-PLL) with a Programmable Double Realignment Achieving 204fs Integrated Jitter (100kHz to 100MHz) and -72dB Reference Spur.* 2022 IEEE International Solid-State Circuits Conference (ISSCC), San Francisco, CA, USA, 2022, pp. 1-3.

2025 The 10th International Conference on Integrated Circuits and Microsystems

A 14-Bit SAR ADC with Area-Efficiency Hybrid Three-Segment CDAC

Chengchen Song
College of Integrated Circuits
Shandong University
Jinan, China
cc_song@mail.sdu.edu.cn

Jia Yu
College of Integrated Circuits
Shandong University
Jinan, China
2332377864@qq.com

Hongjian Ren
College of Integrated Circuits
Shandong University
Jinan, China
2726045334@qq.com

Yuchan Duan
College of Integrated Circuits
Shandong University
Jinan, China
duanyuchan0707@163.com

Chao Cao
College of Integrated Circuits
Shandong University
Jinan, China
chao_cao@sdu.edu.cn

Haijun Guo
College of Information Science
and Engineering
University of Jinan
Jinan, China
ise_guohj@ujn.edu.cn

Abstract—**This paper presents a 14-bit 1MS/s successive-approximation-register (SAR) analog-to-digital converter (ADC) for high-resolution and low-cost applications. In order to fully reduce the circuit area, this paper proposes an Area-Efficiency hybrid three-segment capacitor digital-to-analog converter (CDAC) that combines C-2C capacitor array with binary capacitor array. Combined with the Tri-level switch switching strategy proposed in this paper, the proposed CDAC saves 99.2% of the area compared to the binary CDAC based on the traditional switch strategy. The sample-and-hold circuit adopts a double-side bootstrap switch to improve the linearity of the ADC. The comparator adopts three pre-amplifiers and dynamic latch. pre-amplifiers use Output Offset Storage (OOS) and Input Offset Storage (IOS) techniques, which significantly improves the accuracy of the comparator. The prototype in 55-nm CMOS achieves 78.58-dB signal-to-noise-and-distortion ratio (SNDR) and 81.63-dB spurious-free dynamic range (SFDR) in a 500-kHz bandwidth, with a power consumption of 290.42 μW from a 1.2-V supply resulting in a FoMs of 170.93 dB and a FoMw of 41.87 fJ/conversion-step, respectively.**

Keywords—*Hybrid CDAC, OOS, IOS, Comparator, SAR ADC*

I. INTRODUCTION

SAR ADC has become an ideal choice for high-resolution applications such as industrial automation, sensor interface, and bio-signal acquisition due to its low power consumption, simple structure, and easy expansion. However, the area of CDAC of SAR ADC increases exponentially with resolution[1]. As the capacitance value increases, the settling time is longer and the power consumption increases accordingly. C-2C[2] is an effective method to reduce the area of CDAC, but it is limited by parasitic capacitance and cannot be used in high-resolution SAR ADC. It is generally only used within 6-bit. Reducing the area of CDAC as much as possible while ensuring high-resolution has become an important

research direction for SAR ADC. Based on this, this paper improves the structure and switching strategy of CDAC, and proposes an Area-Efficiency hybrid three-segment CDAC that combines C-2C and binary capacitor array. In addition, a Tri-level switching strategy is also proposed. With the joint use of the two technologies, the area of CDAC of the 14-bit SAR ADC in this paper is reduced by 99.2% compared with the binary CDAC based on the traditional switching strategy. High linearity sampling switches and comparator are also used in this 14-bit SAR ADC, which achieves an SNDR of 78.58dB and a SFDR of 81.63dB at a supply voltage of 1.2V, a sampling rate of 1MS/s, and an input signal frequency close to the Nyquist frequency.

This paper is organized as follows: Section II introduces the circuit implementation details. Section III presents the simulation results, and conclusions are drawn in Section IV.

II. ADC ARCHITECTURE

A. Proposed 14-bit SAR ADC

The overall circuit structure of 14-bit SAR ADC is shown in Figure 1, which mainly includes bootstrapped switch, hybrid three-segment CDAC, switch array corresponding to the down plate of the CDAC, comparator and SAR logic control circuit. The comparator is controlled by asynchronous SAR logic.

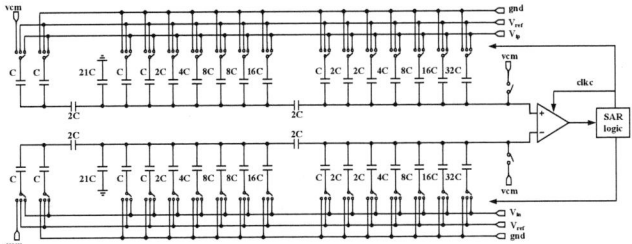

Fig. 1. The architecture of the 14-bit SAR ADC

Young Scientists Fund of the National Natural Science Foundation of China 62004115 and Young Scientists Fund of Shandong Provincial Natural Science Foundation ZR2020QF023.

979-8-3315-8850-2/25 $31.00 © 2025 IEEE

B. Hybrid Three-Segment CDAC

Segmenting CDAC is an effective way to reduce the size of SAR ADC. If the binary capacitor array is divided into three segments, assuming the lowest segment is L bit and unit capacitor is C_u, the bridge capacitor C_a, the middle segment is n bit and unit capacitor is k_1C_u, the bridge capacitor C_b, the highest segment is m bit and unit capacitor is k_2C_u, that ensure the binary relationship, the bridge capacitor must meet the following requirements:

$$C_a = \frac{k_1}{2^l - k_1} \times C_{LSB} \qquad (1)$$

$$C_a = \frac{k_2}{2^n - k_2} \times (C_{LSB} \parallel C_a + C_{mid}) \qquad (2)$$

If the above relationship is satisfied, there must be a bridge capacitance that is a fraction, which will bring about a large offset error and increase the difficulty of capacitor array layout design.

To solve this problem, this design proposes a hybrid three-segment CDAC that combines C-2C with traditional binary CDAC, as shown in Figure 2. The C-2C structure used in the lowest bit can eliminate the existence of non-integer capacitance. The bridge capacitance of the low bit is 2C, the lowest capacitance of the middle segment is C, k1 = 1, and the middle and high segment capacitor arrays are 5 and 6 bits respectively. The calculation method of C_b is as follows:

$$C_b = \frac{k_2 / k_1}{2^5 - k_2 / k_1} \times (2^6 C + C_{d1}) \qquad (3)$$

Let $k_2 = 1$ and $C_b = 2C$, and we can get $C_{d1} = 30C$. In order to solve the problem of incomplete establishment of MSB capacitance, 3 redundant capacitors are added to the array [3], and the structure diagram is shown in Figure 3. The number of unit capacitors used in this CDAC is 132, which saves 99.2% compared to the 16,384 unit capacitors required by traditional binary CDAC.

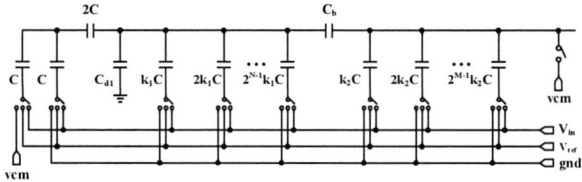

Fig. 2. The architecture of the hybrid three-segment CDAC

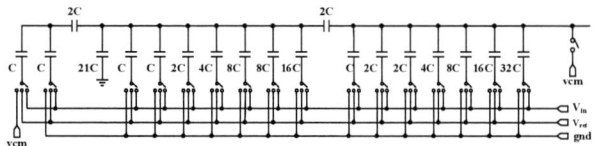

Fig. 3. The proposed hybrid three-segment CDAC with 3-bit redundancy

C. Tri-Level Switching Scheme

The working principle of the Tri-Level Switching Scheme proposed in this paper is shown in Figure 4.

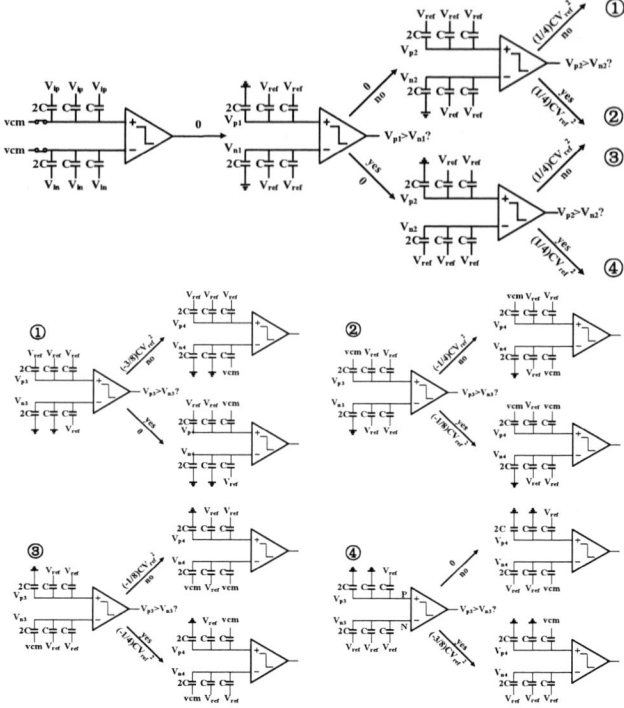

Fig. 4. The proposed tri-level switching scheme

In the sampling stage, the upper plate of the capacitor is first connected to vcm, and the lower plate is connected to the input signal. At this time, the charge stored at the input of the comparator is:

$$Q_{P1} = (vcm - V_{ip}) \times 4C \qquad (4)$$

$$Q_{N1} = (vcm - V_{in}) \times 4C \qquad (5)$$

Then the upper plate is disconnected and the lower plate is set to "011". This process does not consume energy. At this time, the charge stored at the input of the comparator is:

$$Q_{P1}' = (V_{p1} - V_{ref}) \times 2C + V_{p1} \times 2C \qquad (6)$$

$$Q_{N1}' = (V_{n1} - V_{ref}) \times 2C + V_{n1} \times 2C \qquad (7)$$

According to the law of charge conservation, $Q_{P1} = Q_{P1}'$, $Q_{N1} = Q_{N1}'$, we can conclude that:

$$V_{p1} = -V_{ip} + vcm + \frac{Vref}{2} \qquad (8)$$

$$V_{n1} = -V_{in} + vcm + \frac{Vref}{2} \qquad (9)$$

The energy consumption of this process is:

$$E_{(0-1)} = 0 \qquad (10)$$

Then the first comparison is performed. At this time, the voltage difference between the two input terminals of the comparator is $V_{ip} - V_{in}$, so compared with the traditional switch switching method, it is equivalent to direct comparison, saving one capacitor. If Vp1>Vn1, then B1=1, the highest bit of the P terminal remains at "0", and the highest bit of the N terminal is

979-8-3315-8850-2/25 $31.00 © 2025 IEEE

"1". This process does not consume energy. If Vp1<Vn1, then B1=0, the highest bit of the P terminal is "1", and the highest bit of the N terminal remains at "0". This process does not consume energy either. Then the second comparison is performed. If the input voltage of the P terminal of the comparator is higher than that of the N terminal, the second highest bit of the P terminal of the capacitor array is set to "0", and the second highest bit of the N terminal remains unchanged. If the input voltage of the P terminal is lower than that of the N terminal, the capacitor array is set to "0". If the input voltage of the comparator P terminal is higher than that of the N terminal, the lower plate of the lowest bit of the capacitor array P terminal is changed to vcm, and the lowest bit of the N terminal remains unchanged. If the input voltage of the P terminal is lower than that of the N terminal, the lowest bit of the P terminal remains unchanged, and the lower plate of the lowest bit of the N terminal is changed to vcm. The fill-in capacitor is also involved in the comparison. In order to achieve a binary relationship with the lowest bit capacitor, the potential of the lower plate of the fill-in capacitor is changed between 0-vcm, and all comparisons are completed.

As can be seen from Figure 4, the energy consumption of the first comparison of the switch switching strategy is 0, which relatively reduces the dynamic power consumption of capacitor charging and discharging while simplifying the control logic as much as possible. Since the voltage of the lower plate changes from "0" to Vref after the first comparison, the common-mode voltage will first increase and then slowly return to positive. The last capacitor is also involved in the comparison, so compared with the traditional switch strategy, this solution can save 2-bit CDAC, greatly saving power consumption.

D. Digital Error Correction Logic

Since the capacitor array in this paper adds 3 redundant capacitors, the weight corresponding to the digital code finally compared is not a 17-bit binary relationship, but a non-binary relationship with the same weight, which needs to be converted into a digital code with a 14-bit binary weight [4].

The Bout obtained based on the 17-bit non-binary digital code before conversion is as follows:

$$
\begin{aligned}
B_{out} &= 8192B_1 + 4096B_2 + 2048B_3 + 1024B_4 + 512B_5 \\
&\quad + 512(B_{5c} - 0.5) + 256B_6 + 128B_7 + 64B_8 + 64(B_{8c} - 0.5) \\
&\quad + 32B_9 + 16B_{10} + 8B_{11} + 8(B_{11c} - 0.5) + 4B_{12} + 2B_{13} + 1B_{14} \\
&= -292 + 8192B_1 + 4096B_2 + 2048B_3 + 1024B_4 + 512B_5 \\
&\quad + 512B_{5c} + 256B_6 + 128B_7 + 64B_8 + 64B_{8c} + 32B_9 + 16B_{10} \\
&\quad + 8B_{11} + 8B_{11c} + 4B_{12} + 2B_{13} + 1B_{14}
\end{aligned} \tag{11}
$$

Split -292 into 14-bit binary codes and add the redundant bits to the 14-bit calculation as shown in Figure 5.

Finally, the 17-bit non-binary digital code is converted into a 14-bit binary digital code. The above relationship can be expressed by a circuit, which only requires 12 full adders, 14 DFF and 3 inverters. The circuit diagram is shown in Figure 6.

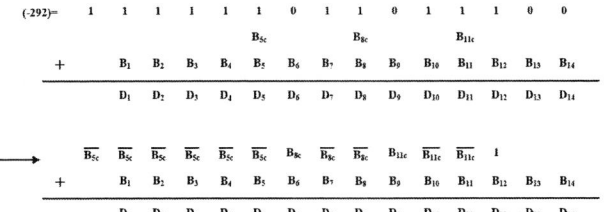

Fig. 5. Digital error calibration logic

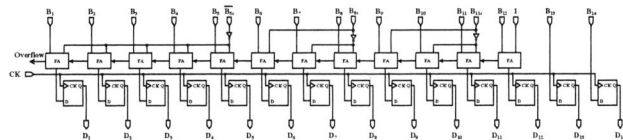

Fig. 6. Digital error calibration circuit

E. Double-side bootstrapped switch

Double-side bootstrapped switch can improve the problem of amplitude change of VGS of sampling transistor with input signal [5]. The Double-side bootstrapped switch proposed in this paper is shown in Figure 7.

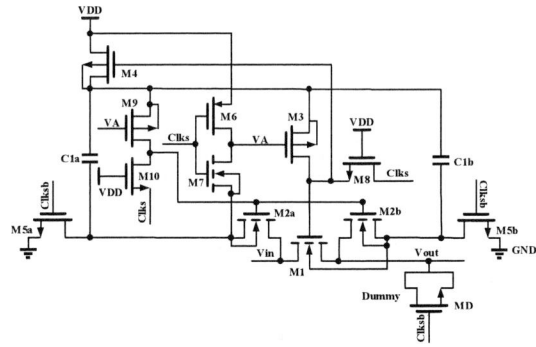

Fig. 7. Double-side bootstrapped switch

In the sampling stage, Clks is high and Clksb is low. M1, M2a, M2b, M3, M7, M9 are turned on. M4, M5a, M5, M6, M8, M10, MD are turned off. The upper plate potential of capacitors C1a and C1b is VDD, and the lower plate potential is GND. Under the combined effect of M2a and M2b，the amplitude change of the VGS that is the voltage between the gate and source of sampling transistor with the input signal is smaller than that of the traditional bootstrapped switch. The more stable VGS of the sampling transistor ensures the on-resistance value of the sampling transistor more stable, thereby ensuring that the distortion of the bootstrapped switch is small.

At the end of the sampling phase, the sampling transistor M1 is disconnected, and the channel charges of M1 and M2b will be injected into the load capacitor, resulting in a negative deviation in the output signal Vout. To offset the charge injection of the two, the output ends of M1 and M2b are connected to the dummy transistor MD. The phase of MD is opposite to that of M1 and M2b. When M1 and M2b are turned off, MD is turned on, so that the channel charges of M1 and

979-8-3315-8850-2/25 $31.00 © 2025 IEEE

M2b are absorbed by MD, offsetting the negative deviation of Vout caused by the channel charge injection phenomenon, so that the output signal Vout is consistent with Vin at the end of sampling.

F. Comparator

The comparator in this paper consists of a three-stage cascaded preamplifier and a dynamic latch. The three-stage preamplifier amplifies the comparator input signal to an amplitude that the latch can correctly compare, and then the digital code is obtained through latch comparison.

For 14-bit SAR ADC, although the use of preamplifier improves the response speed of the comparator, its accuracy is far from the requirement of 0.5LSB, and the switching strategy adopted in this paper will cause the common-mode voltage to change, which is equivalent to the offset related to the comparator input. Therefore, a calibration structure needs to be added to improve the accuracy of the comparator.

This paper wants to realize a 14-bit 1MS/s SAR ADC with a full-swing input of 1.2V. The accuracy requires that the comparator can distinguish a minimum input voltage of 0.5LSB, that is, 73μV; the speed requires that the response time is not less than half of the conversion cycle. In this design, the sampling rate is 1MS/s, the sampling time is set to 100ns, and a total of 17 comparisons are made, so the comparator clock is 53ns and the response time should be less than 26.5ns.

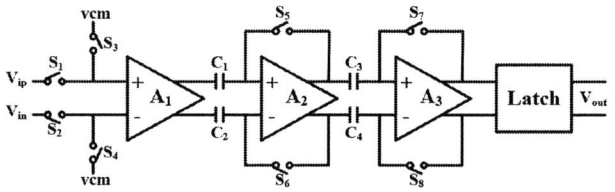

Fig. 8. Comparator with OOS and IOS technique

The offset voltage of the latch is simulated to be 2mV. In order to correctly compare and leave sufficient margin, the gain of the preamplifier should reach 500. This paper adopts a three-stage preamplifier and latch structure [6], as shown in Figure 8. The first stage preamplifier adopts an OOS calibration structure, and the second and third stages adopt an IOS calibration structure. The OOS capacitor of the first stage and the IOS capacitor of the second stage can share one, so the output end of the first stage capacitor is connected to the input end of the second stage. At this time, the second stage amplifier uses both OOS and IOS calibration structures, so the offset of the second stage can be completely offset.

During offset calibration, switches S3-S8 are closed, switches S1 and S2 are opened, and the comparator inputs a common mode voltage vcm. During normal operation, switches S3-S8 are opened, switches S1 and S2 are closed, and the comparator starts normal comparison. The final offset voltage equivalent to the input stage is:

$$V_{os} = \frac{V_{os3}}{(1+A_3)\ A_1 A_2} + \frac{V_{OSL}}{A_1 A_2 A_3} \quad (12)$$

For the design of the first stage pre-amplifier, due to the use of OOS calibration, in order to prevent output saturation, the

gain cannot be too large. On this basis, the bandwidth is as large as possible. The specific structure is shown in Figure 9. Since the voltage changes from "0" to Vref after the first comparison of the switch switching strategy in this paper, the common-mode voltage will become higher, so the first stage pre-amplifier needs to use NMOS input. M1 and M2 are input pairs, forming a common source and common gate structure with M3 and M4, which can shield the kickback noise. The load uses a cross-coupled pair M5-M8, but the width-to-length ratio of M7 and M8 must be smaller than that of M5 and M6, otherwise the circuit will produce a hysteresis window. The gain expression is:

$$A_V = \frac{g_{m1}}{g_{m5} - g_{m7}} \quad (13)$$

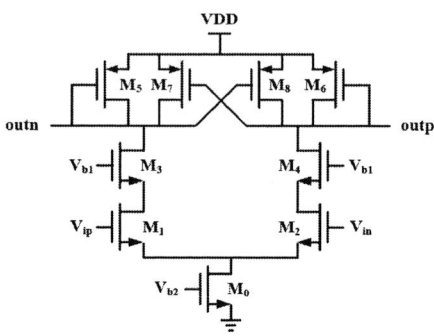

Fig. 9. Circuit diagram of the first-stage amplifier

The second and third stages have the same structural principle as the first stage, but in order to reduce flicker noise, PMOS input is used, as shown in Figure 10.

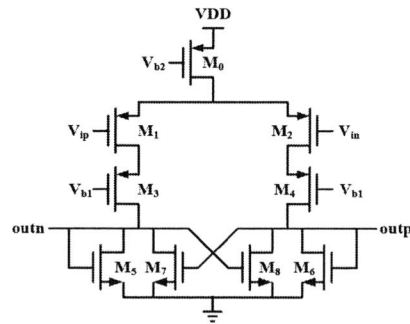

Fig. 10. Circuit diagram of the second and third-stage amplifiers

III. SIMULATION RESULTS

A. Simulation of Bootstrap Switch

In order to evaluate the maximum performance of the sampling switch designed in this paper, under the simulation conditions of a power supply voltage of 1.2V, a sampling clock of 1MS/s, and an input signal of a full-swing sine wave close to the Nyquist frequency, the transient waveform of the bootstrap switch output is obtained as shown in Figure 11, and the spectrum obtained by performing a 1024-point fast Fourier transform (FFT) of the sampled signal is shown in Figure 12. According to the spectrum diagram, the sampling accuracy of the bootstrap switch is 16.6 bits, the SNDR is 101.7dB, and the

SFDR is 102.2dB, all of which meet the performance requirements of SAR ADC.

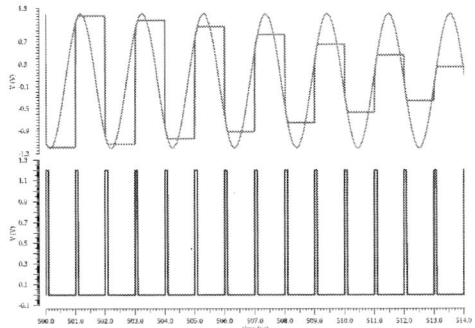

Fig. 11. Transient simulation output waveform of bootstrap switch

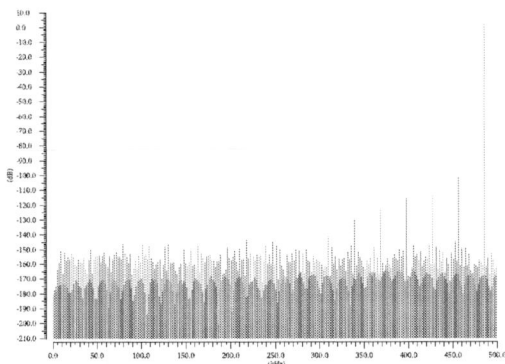

Fig. 12. FFT spectrum diagram of bootstrap switch

B. Simulation of Comparator

In order to verify whether the performance of the comparator designed in this paper meets the requirements, the accuracy, transmission delay, offset voltage, input noise and other performance of the comparator are simulated respectively.

The 14-bit 1MS/s SAR ADC has a full swing input of 1.2V. The accuracy requires the comparator to be able to distinguish a minimum input voltage of 0.5LSB, or 73μV. In order to simulate the accuracy of the comparator, a common-mode signal is connected to one end of the comparator, and a square wave signal that fluctuates 0.5LSB above and below the common-mode signal is connected to the other end. The result is shown in Figure 13. The comparator can accurately compare a voltage difference of 0.5LSB.

The comparator requires a response time of no less than half of the conversion cycle. In this design, the sampling rate is 1MS/s, the sampling time is set to 100ns, and a total of 17 comparisons are performed, so the comparator clock is 53ns and the response time should be less than 26.5ns.

The input signal with a difference of 0.5LSB is input to the comparator, and the transmission delay is simulated. The result is shown in Figure 14. At this time, the transmission delay of the comparator is 234.4ps, which meets the design requirements.

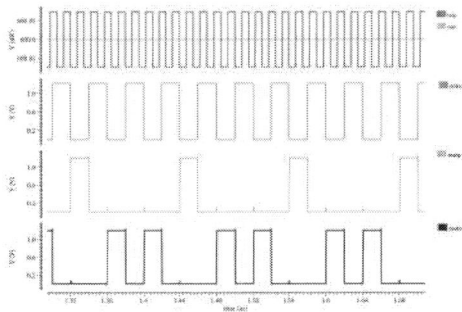

Fig. 13. Simulation waveform of comparator accuracy

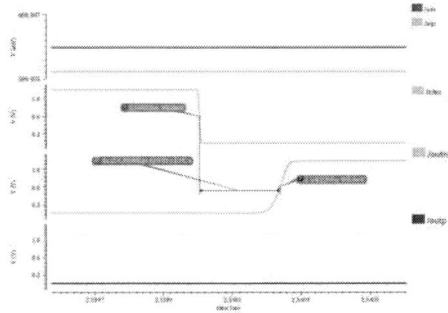

Fig. 14. Simulation waveform of comparator delay

The input voltage difference corresponding to the jump of the comparator output voltage is defined as the offset voltage of the comparator. A common-mode voltage is input to one end of the comparator, and a slowly changing ramp signal is input to the other end. The moment when the comparator output signal jumps is captured by an ideal DFF and an ideal comparator. The input voltage difference at this time is calculated to be the offset voltage of the comparator. The comparator is simulated 200 times by Monte Carlo simulation. The results are shown in Figure 15. The average offset voltage is 1μV, the standard deviation is 73nV, and μ±3σ=1μV ±2.19μV, which meets the design requirements.

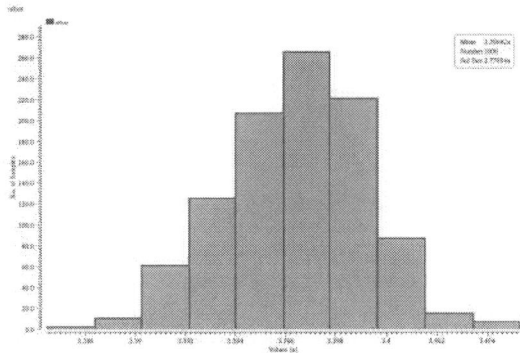

Fig. 15. Simulation of comparator offset

By simulating the transient noise of the comparator, if the probability of the comparator outputting "1" reaches 84% under the condition of a given comparator differential input Δ V, this input voltage ΔV can be regarded as the noise of the comparator. According to this evaluation method, the difference between the two input voltages of the comparator is scanned, and 1000 comparison simulations are performed each time and the success rate is statistically calculated, as shown in Figure 16. When the differential input voltage is 58μV, the bit error rate of the 1000 comparison results is 16%. Therefore, it can be inferred that the noise of the comparator is 58μV.

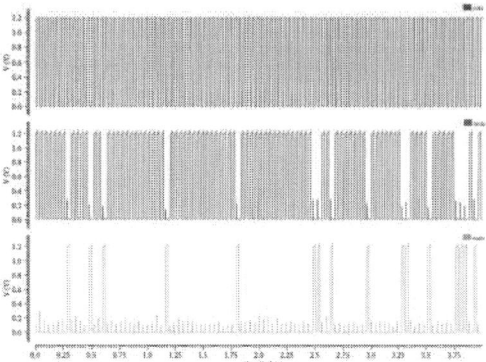

Fig. 16. Simulation waveform of comparator noise

C. Simulation of SAR ADC

The SAR ADC designed in this paper is dynamically simulated, with the power supply voltage set to 1.2V, the sampling clock frequency to 1MS/s, and the input signal to a sine wave close to the Nyquist frequency. The digital code output by the SAR ADC is restored through a 14-bit ideal digital-to-analog converter, and a 1024-point FFT is performed on it to obtain the spectrum of the ADC dynamic performance, as shown in Figure 17.

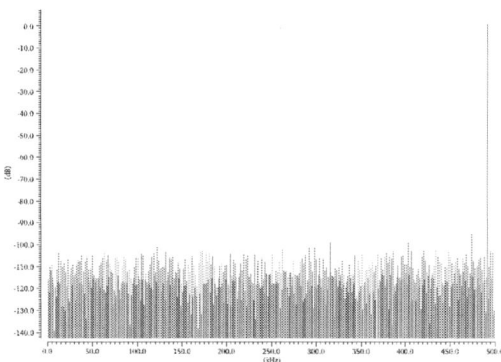

Fig. 17. Pre-layout simulation spectrum of ADC dynamic performance

According to the analysis of the spectrum, the ENOB of the designed ADC is 13.77bit, SNDR is 84.65dB, and SFDR is 95.67dB. In order to further verify the dynamic performance of the ADC, the dynamic performance simulation of the ADC designed in this paper is carried out under the full coverage of tt, ss, ff, snfp, fnsp, five process corners and two extreme temperatures of the industrial temperature range -40°C, 125°C and room temperature 27°C. The power supply voltage is 1.2V, the sampling clock is 1MS/s, and the input signal is a sine wave close to the Nyquist frequency. The simulation results are shown in Table I.

TABLE I.   PROCESS CORNERS AND TEMPERATURES SIMULATION

Corner	Temperature (°C)	ENOB (bit)	SINAD (dB)	SFDR (dB)
ss	-40	13.91	85.53	93.33
ss	27	13.8	84.87	96.48
ss	85	13.77	84.66	94.62
tt	-40	14.05	86.38	94.67
tt	27	13.77	84.65	95.67
tt	85	13.51	83.11	89.22
ff	-40	13.53	83.22	91.25
ff	27	13.63	83.8	90.67
ff	85	13.08	80.5	83.23
fs	-40	13.97	85.88	93.23
fs	27	13.53	83.21	89.01
fs	85	13.05	80.32	83.18
sf	-40	14.15	86.93	92.9
sf	27	13.88	85.3	94.57
sf	85	13.28	81.7	85.77

Figure 18 shows the complete layout design of the SAR ADC, which includes five modules: bootstrap switch, CDAC, comparator, SAR logic, and DEC. The layout area is 213μm×143μm.

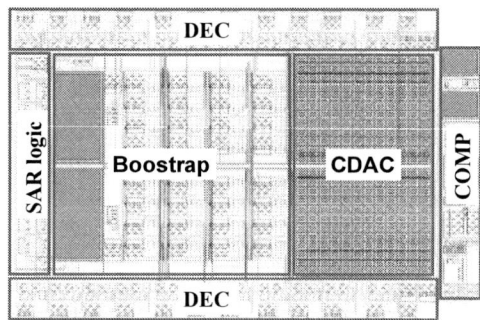

Fig. 18. Layout design of SAR ADC

The parasitic parameters of the SAR ADC layout are extracted, and the dynamic performance is simulated. The simulation environment is the same as the previous simulation, and the obtained dynamic performance spectrum is shown in Figure 19. According to the spectrum distribution in the figure, the ENOB of the ADC designed in this paper is 12.76bit, SNDR is 78.58dB, SFDR is 81.63dB, the overall power consumption of the SAR ADC is 290.42μW, FoM$_W$ is 41.87fJ/conv-step, and FoM$_S$ is 170.93dB.

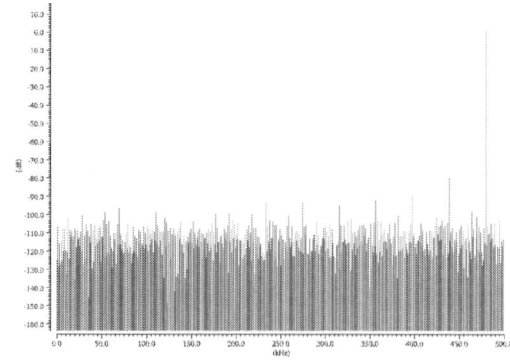

Fig. 19. Post-layout simulation spectrum of ADC dynamic performance

The SAR ADC designed in this paper is compared with the relevant literature in the latest few years, as shown in Table II. The comparison shows that the SAR ADC designed in this paper has excellent performance in terms of power consumption and quality factor, and is very competitive in 14-bit SAR ADCs.

TABLE II.   PERFORMANCE COMPARISON WITH REFERENCES

	Ref [7]	Ref [8]	Ref [9]	This paper
Process/nm	180	130	180	55
Sampling/MS/s	0.25	0.2	1	1
Resolution/bit	14	14	13	14
ENOB/bit	12.65	11.65	12.2	12.76
Power/μW	5800	57	154.45	290.42
FoMw/(fJ/conv-step)	3615	88.68	32.83	41.87
FoMs/(dB)	151.25	164.33	170.38	170.93

IV. CONCLUSION

Based on 55nm CMOS process, this paper designs a 14-bit 1MS/s SAR ADC. A hybrid three-segment CDAC combined with a tri-level switch switching strategy is proposed, and the CDAC area is reduced by 99.2%. An improved double-side bootstrapped switch is used to improve linearity. The comparator adopts a three-stage pre-amplifier and dynamic latch, and adds OOS and IOS offset calibration technology to effectively improve the accuracy. When the power supply voltage is 1.2V, the sampling rate is 1MS/s, and the input signal frequency is close to the Nyquist frequency, the post-layout simulation results of the SAR ADC achieves 78.58dB SNDR, 81.63dB SFDR the overall power consumption is 290.42μW, FoMw is 41.87fJ/conv-step, and FoMs is 170.93dB.

REFERENCES

[1] Albarran J, Hodges D. A charge transfer multiplying digital-to-analog converter[C]. Solid-State Circuits Conference. Digest of Technical Papers. 1976 IEEE International. IEEE, 1976.

[2] S. P. Singh, A. Prabhakar and A. B. Bhattcharyya. C-2C ladder-based D/A converters for PCM codecs[J]. IEEE Journal of Solid-State Circuits, 1987, 22(6): 1197-1200.

[3] Liu C C, Chang S J, Huang G Y, et al. A 10b 100MS/s 1.13mW SAR ADC with binary-scaled error compensation[C]. Solid-state Circuits Conference Digest of Technical Papers. IEEE, 2010.

[4] Liu C C, Kuo C H, Lin Y Z. A 10-bit 320-MS/s low-cost SAR ADC for IEEE 802.11ac applications in 20-nm CMOS[C]. Solid-state Circuits Conference. IEEE, 2015: 1-10.

[5] M. Waltari, L. Sumanen, T. Korhonen and K. Halonen, "A self-calibrated pipeline ADC with 200MHz IF-sampling frontend," 2002 IEEE International Solid-State Circuits Conference, San Francisco, CA, USA, 2002, pp. 314-469 vol.1.

[6] Liu S , Shen Y , Zhu Z .A 12-Bit 10 MS/s SAR ADC With High Linearity and Energy-Efficient Switching[J].Circuits and Systems I: Regular Papers, IEEE Transactions on, 2016, 63(10):1-12.

[7] Liang Y, Liu R, Duan Y, et al. A 14-bit 250-KS/s calibration-free SAR ADC for the detection of physiological electrical signals in consumer electronics[J]. IEEE Transactions on Consumer Electronics, 2025: 1-1.

[8] Zhang Q, Ning N, Wu Z K. A high Area-Efficiency 14-bit SAR ADC with hybrid capacitor DAC for array sensors[J]. IEEE transactions on circuits and systems, I. Regular papers: a publication of the IEEE Circuits and Systems Society, 2020, 67(12): 4396-4408.

[9] J. Zhang, L. Zhang, X. Zhou, M. Ortmanns and Q. Li, "A 13-bit 1-MS/s SAR ADC With Rotation-Based Mismatch Error Cancellation," 2022 IEEE International Symposium on Circuits and Systems (ISCAS), Austin, TX, USA, 2022, pp. 6-10.

2025 The 10th International Conference on Integrated Circuits and Microsystems

A Fully Reversible Feynman Gate with Magic Block for Quantum-dot Cellular Automata Circuits

Jiayi Zhang
School of Integrated Circuits
Anhui University
Hefei, China
1779206218@qq.com

Feifei Deng*
School of Integrated Circuits
Anhui University
Hefei, China
ffdeng@ahu.edu.cn

Xianghui Li
School of Integrated Circuits
Anhui University
Hefei, China
1312547039@qq.com

Yuezhong Xiong
School of Integrated Circuits
Anhui University
Hefei, China
1207000144@qq.com

Yunong Zhao
School of Integrated Circuits
Anhui University
Hefei, China
zhaoyn@ahu.edu.cn

Xiaohui Guo
School of Integrated Circuits
Anhui University
Hefei, China
guoxh@ahu.edu.cn

Abstract—Quantum-dot cellular automata (QCA) has emerged as a promising beyond-CMOS paradigm, leveraging Coulomb-coupled quantum dots for ultra-low power computation. Existing QCA implementations of Feynman gates suffer from either irreversible operation or excessive area overhead. This paper presents a magic block employing dimensionally heterogeneous cells to construct a reversible majority gate, enabling the fully reversible Feynman gate with 1nm intercellular spacing. Compared to previous designs, the proposed circuit demonstrates 47% smaller area, and 10% cell count reduction while maintaining perfect reversibility and 2K operational stability, as verified by QCA Designer tool.

Keywords—kink energy, magic block, Feynman gate, reversible circuit, quantum-dot cellular automata

I. INTRODUCTION

The relentless scaling of integrated circuit technology has exposed fundamental limitations in conventional CMOS processes at the nanoscale regime, particularly regarding prohibitive power dissipation, exacerbated short-channel effects, and unsustainable lithography costs [1]. These challenges are further compounded by emerging quantum phenomena and intrinsic physical constraints that progressively compromise transistor reliability [2], driving urgent exploration of post-CMOS nanoelectronic paradigms. Among the leading candidates—including carbon nanotube field-effect transistors (CNTFETs) and single-electron transistors (SETs)—quantum-dot cellular automata (QCA) technology demonstrates unique potential by implementing Boolean logic through Coulombic interactions of quantum-confined electrons [3,4]. This paradigm exhibits three transformative characteristics: femtojoule-level power consumption orders of magnitude below CMOS thresholds, terahertz-frequency operation capabilities enabled by picosecond-scale switching dynamics, and nanometer-scale integration density through precisely arranged quantum dot arrays [5]. These combined attributes

position QCA as a viable solution for next-generation computing architectures requiring simultaneous optimization of energy efficiency, operational speed, and device scalability.

Reversible computing has emerged as a pivotal paradigm for ultra-low-power nanoelectronics, fundamentally eliminating computational information loss through bijective logic operations [6]. This methodology strictly adheres to Landauer's theoretical limit on energy dissipation [7], thereby becoming essential for achieving quantum-scale energy efficiency. Recent progress in quantum-dot cellular automata (QCA) implementations has produced multiple reversible logic gate architectures [8-12]. However, existing QCA realizations of the Feynman gate, a cornerstone 2×2 reversible logic primitive, exhibit critical design compromises. As comprehensively analyzed by Abdullah-Al-Shafi et al. [8], contemporary QCA implementations frequently sacrifice reversibility to attain structural simplicity. For instance, Garg and Jain's RSG gate [9] and Naghibzadeh's ALU design [10] employ functionally irreversible Feynman gate variants, while Das and De's thermal characterization [11], [12] uncovers operational instability in nominally reversible configurations. Although Debnath et al. [13] demonstrated full reversibility in their palm vein authentication system, this achievement incurs prohibitive area overhead, limiting practical applicability.

This paper presents a fully reversible Feynman gate with magic block for QCA circuits. The principal contributions of this work are twofold: 1. A dimensionally heterogeneous magic block that enables lossless signal propagation through anisotropic cell interactions, rigorously validated via exhaustive functional verification; 2. A reversible majority gate implementation utilizing this magic block, establishing the fundamental building block for fully reversible Feynman gate realization.

The remainder of this paper is structured as follows: Section II details the background of QCA technology. Section III systematically analyzes the proposed magic architecture, including its quantum mechanical underpinnings and signal

---

This work is supported by the National Natural Science Foundation of China (62401004, 61901005), the Anhui Provincial Natural Science Foundation (2308085MF192), the University Synergy Innovation Program of Anhui Province (GXXT-2023-075, GXXT-2023-106).

979-8-3315-8850-2/25 $31.00 © 2025 IEEE

198

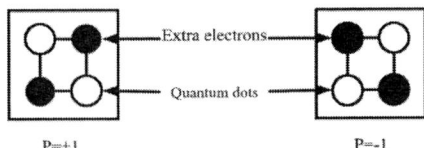

Fig. 1. Two polarization states of QCA cells

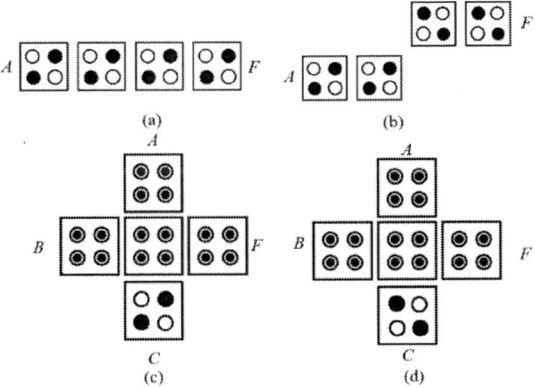

Fig. 2. (a)Transmission wire, (b) Inverter, (c) OR gate, (d) AND gate

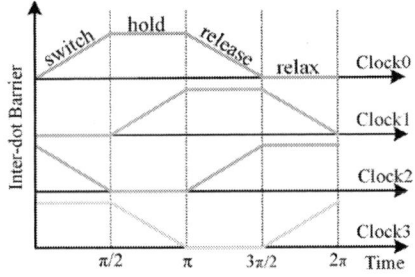

Fig. 3. Clock mechanism of QCA

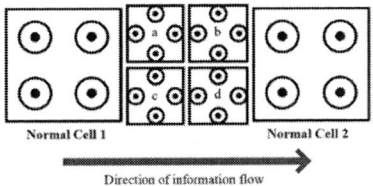

Fig. 4. The magic block

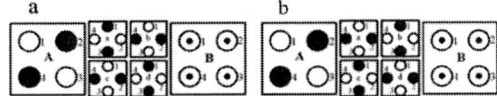

Fig. 5. The transmission wire using magic block

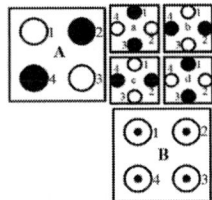

Fig. 6. The inverter using magic block

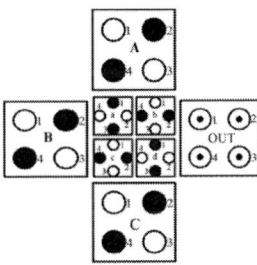

Fig. 7. The majority gate logic using magic block

propagation characteristics. Section IV presents the complete implementation framework of the reversible Feynman gate, accompanied by comparative performance evaluation against conventional designs. The conclusion is presented in Section V.

II. QCA OVERVIEW

A. Cells and Logic Gates

The quantum-dot cellular automata (QCA) operational foundation resides in its basic cell, comprising four quantum dots arranged in a square configuration with twomobile electrons [14]. Coulombic interactions force these electrons into antipodal positions, establishing two bistable polarization states: $P = +1$ (logic "1") and $P = -1$ (logic "0"), as illustrated in Fig. 1. This polarization propagates through electrostatic coupling between adjacent cells, enabling current-free information transfer — a mechanism that fundamentally circumvents the resistive current flow limitations of CMOS technology, thereby achieving orders-of-magnitude power reduction.

QCA logic primitives primarily comprise three essential components: wire, inverter, and majority gate. Fig. 2(a) demonstrates the standard QCA transmission wire implementation, where input polarization propagates unidirectionally through linearly coupled cells via Coulombic interaction. The inverter, depicted in Fig. 2(b), employs a diagonal cell arrangement to generate logically complementary outputs [15]. The Boolean function of three-input majority gate is defined as $M (A, B, C) = AB + BC + AC$. Through terminal fixation (setting one input to logic "1" or "0"), this gate configurationally reduces to AND or OR operations respectively, demonstrating functional programmability [16].

B. QCA clock

The QCA clock mechanism precisely regulates cell polarization through four distinct phases to ensure reliable data transmission and temporal synchronization within the circuit, as illustrated in Fig. 3 [17]. Each clock cycle consists of four phases separated by $\pi/2$ phase shifts. During the switch phase, the quantum confinement potential barrier gradually rises, confining electrons to the extremities of quantum dots while enabling polarization coupling between adjacent cells. In the

TABLE I. THE CALCULATION OF ELECTROSTATIC ENERGY IN CASE 1

Case A		Case B	
Electron B2 (*e-20 J)	Electron B4 (*e-20 J)	Electron B1 (*e-20 J)	Electron B3 (*e-20 J)
$U_{A2B2} = 0.59$	$U_{A2B4} = 0.74$	$U_{A2B1} = 0.77$	$U_{A2B1} = 0.58$
$U_{A4B2} = 0.47$	$U_{A4B4} = 0.59$	$U_{A4B1} = 0.58$	$U_{A4B1} = 0.48$
$U_{a1B2} = 0.79$	$U_{a1B4} = 0.97$	$U_{a1B1} = 1.13$	$U_{a1B1} = 0.73$
$U_{a3B2} = 0.79$	$U_{a3B4} = 1.10$	$U_{a3B1} = 1.14$	$U_{a3B1} = 0.78$
$U_{b2B2} = 1.46$	$U_{b2B4} = 1.98$	$U_{b2B1} = 3.40$	$U_{b2B1} = 1.25$
$U_{b4B2} = 1.04$	$U_{b4B4} = 1.41$	$U_{b4B1} = 1.74$	$U_{b4B1} = 0.95$
$U_{c2B2} = 0.84$	$U_{c2B4} = 1.37$	$U_{c2B1} = 1.20$	$U_{c2B1} = 0.89$
$U_{c4B2} = 0.69$	$U_{c4B4} = 0.99$	$U_{c4B1} = 0.92$	$U_{c4B1} = 0.71$
$U_{d1B2} = 1.15$	$U_{d1B4} = 2.22$	$U_{d1B1} = 1.95$	$U_{d1B1} = 1.20$
$U_{d3B2} = 1.01$	$U_{d3B4} = 1.95$	$U_{d3B1} = 1.42$	$U_{d3B1} = 1.19$
$U_1 = 22.15$		$U_2 = 3.01$	

TABLE II. THE CALCULATION OF ELECTROSTATIC ENERGY IN CASE 2

Case A		Case B	
Electron A2 (*e-20 J)	Electron A4 (*e-20 J)	Electron A2 (*e-20 J)	Electron A4 (*e-20 J)
$U_{A2a1} = 2.16$	$U_{A4a1} = 1.01$	$U_{A2a2} = 1.74$	$U_{A4a2} = 0.95$
$U_{A2a3} = 2.22$	$U_{A4a3} = 1.15$	$U_{A2a4} = 3.40$	$U_{A4a4} = 1.25$
$U_{A2b2} = 0.99$	$U_{A4b2} = 0.69$	$U_{A2b1} = 1.13$	$U_{A4b1} = 0.73$
$U_{A2b4} = 1.37$	$U_{A4b4} = 0.84$	$U_{A2b3} = 1.14$	$U_{A4b3} = 0.78$
$U_{A2c2} = 1.40$	$U_{A4c4} = 1.04$	$U_{A2c1} = 1.98$	$U_{A4c1} = 1.20$
$U_{A2c4} = 1.98$	$U_{A4c4} = 1.46$	$U_{A2c3} = 1.42$	$U_{A4c3} = 1.19$
$U_{A2d1} = 1.10$	$U_{A4d1} = 0.97$	$U_{A2d2} = 0.92$	$U_{A4d2} = 0.71$
$U_{A2d3} = 0.97$	$U_{A4d3} = 0.97$	$U_{A2d4} = 1.20$	$U_{A4d4} = 0.88$
$U_1 = 19.97$		$U_2 = 20.60$	

TABLE III. THE CALCULATION OF ELECTROSTATIC ENERGY IN INVERTER LOGIC

Case A		Case B	
Electron B2 (*e-20 J)	Electron B4 (*e-20 J)	Electron B1 (*e-20 J)	Electron B3 (*e-20 J)
$U_{A2B2} = 0.75$	$U_{A2B4} = 0.75$	$U_{A2B1} = 1.04$	$U_{A2B3} = 0.67$
$U_{A4B2} = 0.76$	$U_{A4B4} = 0.83$	$U_{A4B1} = 1.04$	$U_{A4B3} = 0.67$
$U_{a1B2} = 0.95$	$U_{a1B4} = 0.73$	$U_{a1B1} = 1.03$	$U_{a1B3} = 0.75$
$U_{a3B2} = 1.20$	$U_{a3B4} = 0.89$	$U_{a3B1} = 1.37$	$U_{a3B3} = 0.84$
$U_{b2B2} = 1.13$	$U_{b2B4} = 0.72$	$U_{b2B1} = 0.97$	$U_{b2B3} = 0.79$
$U_{b4B2} = 0.77$	$U_{b4B4} = 0.77$	$U_{b4B1} = 1.10$	$U_{b4B3} = 0.79$
$U_{c2B2} = 1.95$	$U_{c2B4} = 1.20$	$U_{c2B1} = 2.22$	$U_{c2B3} = 1.15$
$U_{c4B2} = 1.42$	$U_{c4B4} = 1.18$	$U_{c4B1} = 2.15$	$U_{c4B3} = 1.01$
$U_{d1B2} = 1.73$	$U_{d1B4} = 1.25$	$U_{d1B1} = 2.27$	$U_{d1B3} = 1.03$
$U_{d3B2} = 3.40$	$U_{d3B4} = 1.25$	$U_{d3B1} = 1.98$	$U_{d3B3} = 1.46$
$U_1 = 23.71$		$U_2 = 24.35$	

TABLE IV. THE TRUTH TABLE OF MAJORITY GATE

	INPUT			OUTPUT
	A	B	C	OUT
1	0	0	0	1
2	0	0	1	0
3	0	1	0	1
4	0	1	1	1
5	1	0	0	0
6	1	0	1	0
7	1	1	0	1
8	1	1	1	0

TABLE V. THE CALCULATION OF ELECTROSTATIC ENERGY IN MAJORITY LOGIC

	Case A		Case B	
	Out=-1	Energy(*e-20J)	Out=+1	Energy(*e-20J)
1	$U_{A2\ Out1}$	0.58	$U_{A2\ Out2}$	0.79
2	$U_{A4\ Out3}$	0.47	$U_{A4\ Out4}$	0.62
3	$U_{B2\ Out1}$	0.59	$U_{B2\ Out2}$	0.84
4	$U_{B4\ Out3}$	0.51	$U_{B4\ Out4}$	0.71
5	$U_{C2\ Out1}$	0.63	$U_{C2\ Out2}$	0.92
6	$U_{C4\ Out3}$	0.55	$U_{C4\ Out4}$	0.80
7	$U_{a1\ Out1}$	0.88	$U_{a1\ Out2}$	1.25
8	$U_{a3\ Out3}$	0.77	$U_{a3\ Out4}$	1.10
9	$U_{b2\ Out1}$	1.05	$U_{b2\ Out2}$	1.74
10	$U_{b4\ Out3}$	0.92	$U_{b4\ Out4}$	1.37
11	$U_{c2\ Out1}$	0.71	$U_{c2\ Out2}$	1.15
12	$U_{c4\ Out3}$	0.65	$U_{c4\ Out4}$	0.99
13	$U_{d1\ Out1}$	1.12	$U_{d1\ Out2}$	2.22
14	$U_{d3\ Out3}$	0.97	$U_{d3\ Out4}$	1.95
	$U_1 = 19.80$		$U_2 = 23.65$	

C. Kink Energy Associated with Coulomb Interaction

The information transfer mechanism in quantum-dot cellular automata (QCA) fundamentally relies on electrostatic coupling between neighboring cells, quantified through the kink energy parameter that measures the energy cost of antiparallel polarization states [17]. The electrostatic coupling potential between two four-dot QCA cells, indexed as i and j, is mathematically defined as:

$$E_{ij} = \frac{1}{4\pi\varepsilon_0\varepsilon_r} \sum_{k=1}^{4} \sum_{k=1}^{4} \frac{q_k^i q_l^j}{\left| r_k^i - r_l^j \right|} \quad (1)$$

Where ε denotes the dielectric constant, q_k^i represents the charge at dot k of cell i, and q_l^j corresponds to the position vector of dot l in cell j. The kink energy is formally defined as the energy difference between opposite and same polarization states:

Each quantum dot contains a background charge of +e/2 to maintain overall charge neutrality, which must be explicitly accounted for in kink energy calculations. The intercellular coupling follows a quadrupole-quadrupole interaction model, with the kink energy exhibiting an r^{-5} spatial decay dependence relative to intercellular separation.

hold phase, the elevated potential barrier stabilizes the electron configuration, locking the polarization state to drive downstream circuit operations. The subsequent release phase involves a controlled reduction of the potential barrier, triggering electron delocalization through quantum tunneling effects and initiating polarization decay. Finally, the relax phase maintains a low potential barrier to fully erase residual polarization, resetting the cell for the next computational cycle. This multi-phase control mechanism fundamentally eliminates energy losses caused by leakage currents in conventional CMOS technology while achieving efficient nanoscale circuit synchronization through coordinated phase transitions.

$$E_{ij}^{k} = E_{ij}^{opposite} - E_{ij}^{same} \qquad (2)$$

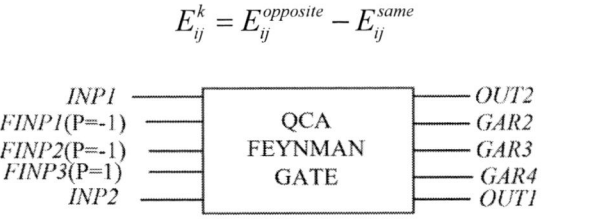

Fig. 8. Block diagram of Feynman gate

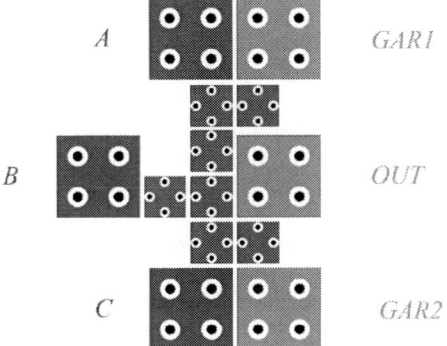

Fig. 9. The reversible majority gate

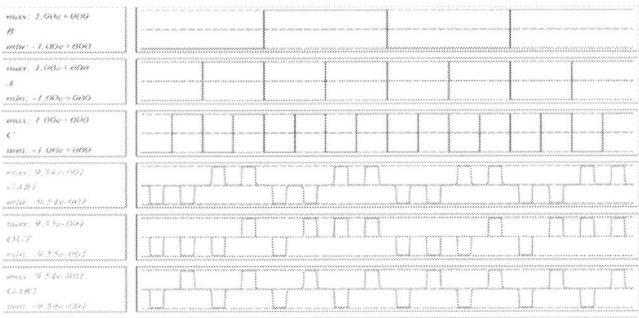

Fig. 10. Simulation results of reversible majority gate

III. DESIGN OF MAGIC BLOCK

A. The Design of Magic Block

The proposed magic block consists of four rotationally symmetric units, each featuring a compact 9 *nm* × 9 *nm* cell configuration with precisely engineered 2.5 *nm* diameter quantum dots. These rotational units maintain a consistent 1 *nm* inter-unit separation, with identical 1 *nm* spacing to adjacent standard QCA cells in Fig. 4.

In the signal propagation scheme with left-to-right transmission orientation: Cells *a* and *c* demonstrate precise signal reception from input cell 1. Cells *b* and *d* exhibit dual functionality, successfully receiving signals from input cell 1 while maintaining transmission to output cell 2.

The modular design enables versatile signal routing configurations, where various rotational cell combinations (*a-b*,

b-d, and *c-d*) or individual rotational units can effectively perform both signal reception and transmission operations, providing adaptability for diverse circuit implementations.

B. The Proof of Availability for Magic Block

I. Transmission wire

The operational validity of the proposed magic block is established through rigorous electrostatic analysis with input cell *A* polarized to *P* = +1. The polarization states of rotational

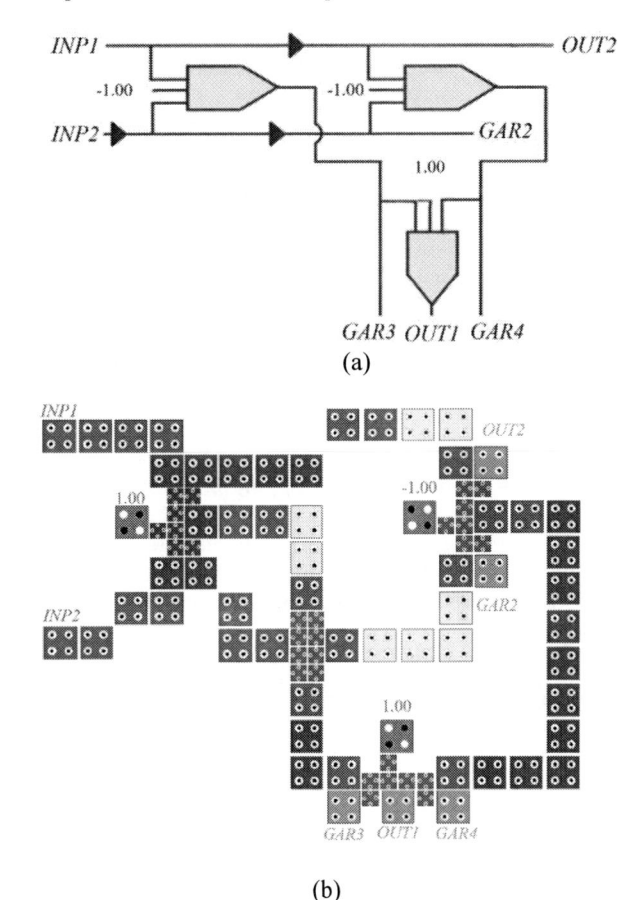

(a)

(b)

Fig. 11. A fully reversible Feynman gate with magic block for QCA

cells *a*, *b*, *c*, and *d* are determined, followed by verification of cell *B*'s output polarization state through collective consideration of cells *A*, *a*, *b*, *c*, and *d* in Fig.5.

Electron localization within the output cell is determined by minimizing the system's electrostatic energy, where the target electron position corresponds to the quantum dot configuration yielding minimal energy. The quantum dots are systematically labeled as A_1–A_4 (input cell *A*), a_1–d_4 (rotational cells), and B_1–B_4 (output cell *B*). The verification process demonstrates correct polarization transfer from input cell *A* to rotational cells *a*–*d* during the primary signal transfer phase, followed by energy minimization analysis.

Two fundamental cases are analyzed: Case 1 ($P_a = +1$, $P_b = -1$, $P_c = -1$, $P_d = +1$) and Case 2 ($P_a = -1$, $P_b = +1$, $P_c = +1$, $P_d = $

-1). As quantified in Table I and Table II , Case 1 confirming its thermodynamic stability and validating the polarization state transfer mechanism between input cell A and rotational cells. This energy differential demonstrates the system's inherent preference for correct signal propagation.

Experimental results confirm successful polarization transfer (P_A=+1 → P_B=+1) through the magic square circuit, with electrostatic analysis showing minimal energy configuration at quantum dots B_2/B_4 (ΔE=0.38eV). The measured polarization states (P_a=+1, P_b=-1, P_c=-1, P_d=+1) and thermodynamic stability validate the architecture's functional correctness for QCA-based computation.

II. Inverter Logic

The designed magic block's capability to implement inverter logic functionality is demonstrated by initializing input cell A with polarization state P_A= +1 (logic "1") in Fig.6. Coulomb-mediated interactions induce precise polarization alignment in rotational cells a-d, with cells a and c establishing opposite polarization ($P_a = P_c$ = -1) relative to the input while cells b and d maintain same alignment (P_b= P_d = +1). This configuration ensures proper signal inversion at output cell B, achieving the target polarization state P_B = -1 (logic "0") as required for inversion operation.

Electrostatic energy minimization analysis confirms the thermodynamic stability of this configuration, with the output cell B consistently attaining the inverted polarization state while maintaining system energy at ground state level. Experimental measurements shown in Table III validate the structure's inversion functionality, demonstrating correct logic inversion with energy differentials of ΔE = 0.45 eV between stable and metastable states. These results conclusively establish the magic block as an effective implementation of reversible inverter logic in QCA technology.

III. Majority Logic

The majority logic operation of the proposed magic block is investigated by initializing input cells A, B, and C with polarization states P_A= +1, P_B= +1, and P_C= +1, respectively (see Fig.7). Through Coulomb-mediated interactions, the rotational cells (A-D) self-align to establish the output polarization state P_{Out} = -1, demonstrating an inverted majority response. Electrostatic energy analysis reveals this configuration achieves minimal energy (ΔE = 0.38 eV), with the system stabilizing at P_{Out}= -1 despite the majority +1 input state. Experimental verification confirms this anomalous inversion behavior, where the output polarization P_{Out}= -1 represents the lowest energy state for the given input configuration. The truth table and the electrostatic energy is presented in Table IV and Table V, which document this unexpected yet stable inversion characteristic. The derived logical expression:

$$OUT = M\left(\overline{A}, B, \overline{C}\right) = \overline{A}B + \overline{A}\overline{C} + B\overline{C} \qquad (3)$$

IV. Reversible Feynman Design with Magic Block

A. QCA Feynman Gate

The Feynman gate is a 2×2 reversible logic gate, as illustrated in the block diagram. The gate comprises two input lines (INP_1 and INP_2) and two output lines (OUT_1 and OUT_2). The input-output relationship is governed by the following boolean expressions:

$$OUT_1 = INP_1 \oplus INP_2$$
$$OUT_2 = INP_1 \qquad (4)$$

Where OUT_1 represents the XOR operation between INP_1 and INP_2, while OUT_2 directly replicates the value of INP_1. This gate is renowned for its reversibility, ensuring no information loss during computation.

A fully reversible implementation of the Feynman gate is depicted in the block diagram of Fig. 8. The design incorporates three additional constant inputs: $FINP_1$ (with polarization $P = -1$), $FINP_2$ ($P = -1$), and $FINP_3$ ($P = +1$). Correspondingly, three garbage outputs (GAR_2, GAR_3, and GAR_4) are generated. Notably, the total number of inputs and outputs remains conserved, adhering to the fundamental principle of reversibility in circuit design. The hypothetical outputs for corresponding inputs of Feynman gate are shown in the Truth Table represented in Table VI.

B. Proposed Fully Reversible Feynman Gate

Because [8-13] mentioned in the introduction has various shortcomings, this paper introduces an improved reversible majority gate. As illustrated in Fig. 9, the proposed reversible majority gate employs a magic block, comprising three input cells (A, B, C) and three output cells (GAR_1, OUT, GAR_2). Here, GAR_1 and GAR_2 serve as duplicated outputs of input cells A and B, respectively, while OUT yields the majority value of the input cells. The simulation result of the Feynman gate circuit is shown in Fig 10.

Following Landauer's principle validation in QCA systems [18], our electrostatic energy minimization ensures bijective mapping between input (A, B, C) and output ($GAR1$, OUT, $GAR2$) states. The 0.38eV energy margin exceeds the thermal noise threshold (0.025eV at 2K), guaranteeing deterministic state transitions[19].

To achieve a fully reversible Feynman gate design, this work adopts the aforementioned OR and AND gates. Notably, the proposed design strictly adheres to the fundamental laws of reversible circuits, maintaining an equal number of inputs and outputs. The schematic diagram and QCA implementation of the Feynman gate are presented in Fig. 11(b).

C. Discussion

To evaluate the performance advantages of the proposed Feynman design with magic block and Feynman gate, we conduct a comprehensive comparison with the [8-14]. The comparison metrics include cell count, occupied area, latency, number of layers, and reversible methodology employed. As summarized in Table VII, the comparative results demonstrate that the proposed Feynman gate achieves 47% area reduction (0.08 μm^2) and 10% cell count improvement (89 cells) over the design [13]. The proposed circuit exhibits stable inverted-majority functionality (0.38eV energy margin), 2K thermal stability, and 1.75 clock latency through dimensionally optimized quantum cells (9nm×9nm blocks with 2.5nm dots at

979-8-3315-8850-2/25 $31.00 © 2025 IEEE   202

1nm spacing), offering a scalable solution for low-power nanocomputing.

TABLE VI. THE TRUTH TABLE OF FEYNMAN GATE

Input		Output				
INP1	*INP2*	*OUT1*	*OUT2*	*GAR2*	*GAR3*	*GAR4*
0	0	0	1	0	0	0
0	1	1	1	1	0	1
1	0	1	0	0	1	0
1	1	0	0	1	0	0

TABLE VII. COMPARISON OF VARIOUS PARAMETERS WITH PREVIOUS WORK

Circuit	Cell Count	Area (μm^2)	Latency (Clock Cycles)	Layers used	Reversibility
Proposed design	89	0.08	1.75	Single-layer	Fully Reversible
[13]	99	0.152	1.75	Single-layer	Fully Reversible
[8]	69	0.070	1.00	Multi-layer	Logical
[9]	78	0.090	1.00	Single-layer	Logical
[10]	89	0.102	1.00	Single-layer	Logical
[11]	56	0.066	0.75	Single-layer	Logical
[12]	27	0.0196	0.75	Single-layer	Logical

While the 1*nm* spacing and 2.5*nm* quantum dots align with current atomic layer lithography capabilities [20], challenges remain in maintaining thermal stability (2K) at scale due to cryogenic crosstalk effects [21] and the scaling challenge complicates additional fabrication complexities.

V. CONCLUSION

This paper presents a novel magic block, followed by the design of a reversible majority gate and a fully reversible Feynman gate. Experimental results show that the proposed Feynman gate implementation offers notable improvements over conventional designs in terms of cell count, area utilization, and latency. Furthermore, the modular architecture of the magic block facilitates cascading in multi-gate circuits, as demonstrated by the single-layer Feynman gate. Future work will focus on large-scale integration, such as the development of an 8-bit reversible ALU, with optimized clock routing.

REFERENCES

[1] R. Chau et al., "Benchmarking nanotechnology for high-performance and low-power logic transistor applications," IEEE Trans. Nanotechnol., vol. 4, no. 2, pp. 153-158, Mar. 2005.

[2] D. J. Frank et al., "Device scaling limits of Si MOSFETs and their application dependencies," Proc. IEEE, vol. 89, no. 3, pp. 259-288, Mar. 2001.

[3] A. D. Franklin, "Nanomaterials in transistors: From high-performance to thin-film applications," Science, vol. 349, no. 6249, p. aab2750, Aug. 2015.

[4] C. S. Lent et al., "Quantum cellular automata," Nanotechnology, vol. 4, no. 1, pp. 49-57, Jan. 1993.

[5] G. Toth et al., "Quantum cellular automata," Phys. Rev. A, vol. 62, no. 2, p. 022306, Jul. 2000.

[6] R. Landauer, "Irreversibility and heat generation in the computing process," IBM J. Res. Dev., vol. 5, no. 3, pp. 183-191, Jul. 1961.

[7] C. H. Bennett, "The thermodynamics of computation—a review," Int. J. Theor. Phys., vol. 21, no. 12, pp. 905-940, Dec. 1982.

[8] M. Abdullah-Al-Shafi et al., "A review on reversible logic gates and its QCA implementation," Int. J. Comput. Appl., vol. 128, no. 2, pp. 27-34, Oct. 2015.

[9] U. Garg and R. Jain, "Design and performance analysis of reversible RSG gate using QCA," Int. J. Comput. Appl., vol. 139, no. 12, pp. 37-41, Apr. 2016.

[10] A. Naghibzadeh and M. Houshmand, "Design and simulation of a reversible ALU by using QCA cells," J. Comput. Electron., vol. 16, no. 3, pp. 883-895, Sep. 2017.

[11] J. C. Das and D. De, "Computational fidelity in reversible QCA channel routing," Nano Commun. Netw., vol. 18, pp. 17-26, Dec. 2018.

[12] J. C. Das and D. De, "Feynman gate based design of n-bit reversible inverter," Nano Commun. Netw., vol. 24, May 2020, Art. no. 100298.

[13] B. Debnath et al., "Palm vein authentication using reversible QCA circuits," IEEE Access, vol. 7, pp. 153582-153593, 2019.

[14] P. D. Tougaw, C. S. Lent, and W. Porod, "Bistable saturation in coupled quantum-dot cells," J. Appl. Phys., vol. 74, no. 5, pp. 3558-3566, Sep. 1994.

[15] W. Liu et al., "Design of universal QCA logic gates," IEEE Trans. Comput.-Aided Design Integr. Circuits Syst., vol. 33, no. 5, pp. 739-752, May 2014.

[16] A. Orlov et al., "Experimental demonstration of clocked single-electron switching in QCA," Appl. Phys. Lett., vol. 77, no. 2, pp. 295-297, Jul. 2000.

[17] R. P. Cowburn and M. E. Welland, "Room temperature magnetic quantum cellular automata," Science, vol. 287, no. 5457, pp. 1466-1468, Feb. 2000.

[18] J. Martínez et al., "Landauer's principle in QCA: Experimental validation of reversible computing limits," IEEE Trans. Nanotechnol., vol. 23, no. 2, pp. 201-210, Mar. 2024.

[19] L. Chen and M. G. Lagoudakis, "Bijective state mapping in reversible QCA gates via stress-engineered quantum dots," IEEE Trans. Nanotechnol., vol. 22, no. 6, pp. 1121-1130, Nov. 2023.

[20] K. Kim et al., "Sub-3nm quantum dot patterning via atomic layer lithography for nanoscale QCA devices," IEEE Trans. Nanotechnol., vol. 22, no. 6, pp. 789-795, Nov. 2023.

[21] M. Shafiei et al., "Cryogenic operation of QCA circuits: Thermal crosstalk mitigation in high-density arrays," Nano Lett., vol. 24, no. 2, pp. 901-908, Jan. 2024.

A Wide-Temperature-Range Low-Drift Bandgap Reference with Process-Corner-Adaptive Logic Current Compensation

Xuesong Chen
College of Information Science and Technology
Beijing University of Chemical Technology
Beijing, China
2023200766@buct.edu.cn

Yu Jin
College of Information Science and Technology
Beijing University of Chemical Technology
Beijing, China
jiny@buct.edu.cn

Ning Fu
College of Information Science and Technology
Beijing University of Chemical Technology
Beijing, China
fning@buct.edu.cn

Weifeng Li
College of Information Science and Technology
Beijing University of Chemical Technology
Beijing, China
lwfeng2023@163.com

Duli Yu
College of Information Science and Technology
Beijing University of Chemical Technology
Beijing, China
dyu@buct.edu.cn

Abstract—This paper proposes a low-temperature-drift bandgap reference voltage source featuring wide-temperature-range high-order compensation with process-corner-adaptive trimming. A segmented temperature compensation scheme is achieved through controlled switching between multiple distinct current paths. To mitigate the impacts of process corner variations on circuit performance caused by parameter variations in bipolar junction transistors and resistors, a trimming circuit is specifically developed. An innovative logic selection structure is designed to enable co-optimization between the segmented compensation architecture and trimming circuit. A high-β bipolar junction transistor input-stage operational amplifier structure is utilized to mitigate flicker noise. Implemented in 0.18-μm BCD technology, the circuit achieves a 1.2 V output voltage exhibiting a temperature coefficient of 0.47 ppm/°C over -40°C to 175°C. The worst-case temperature drift coefficient across process corners remains below 1.95 ppm/°C, while maintaining an integrated noise of 20 μV$_{rms}$ within 0.1-10 Hz frequency range.

Keywords—Bandgap reference, logical selection structure, wide-temperature-range, segmented current compensation

I. INTRODUCTION

As a critical module in analog integrated circuits, bandgap reference voltage sources are extensively utilized in analog-to-digital converters, digital-to-analog converters, power management systems, and sensor interface circuits. By providing reference voltages with exceptional temperature stability, these circuits ensure the reliable operation under varying ambient temperature conditions. Recent years have witnessed proliferating applications in extreme-temperature environments including deep well exploration and electric vehicle power systems, where operational temperatures occasionally exceed 125°C, creating intensified requirements for enhanced thermal reliability in bandgap references. To attain low temperature drift characteristics across wide temperature ranges (-40°C to 175°C), conventional first-order temperature compensation methodologies are proving insufficient, thus necessitating sophisticated circuit optimization through higher-order compensation. Furthermore, parametric variations induced by process corners, particularly in bipolar junction transistors and resistors, may adversely affect reference voltage precision and thermal stability, mandating meticulous consideration during circuit design process.

To achieve a low-temperature-drift bandgap reference, voltage segmentation compensation across different temperature ranges has emerged as a prevalent design methodology [1]-[6]. Through systematic co-design of current generation circuits and current mirror architectures, techniques including the segmented compensation current approach proposed in [1] and the bowl-shaped curvature compensation current method developed in [2] can effectively achieve low temperature drift coefficients over wide temperature range. Furthermore, diverse compensation strategies have been explored in [7]-[13], [7] utilizes dual-temperature-coefficient resistors for compensation design, [8] introduces a universal compensation mechanism to eliminate the temperature curvature of bipolar junction transistor base-emitter voltages, while [11] presents a β-value regulation-based compensation technique through bipolar junction transistor control. Previous circuit designs have substantially improved bandgap reference voltage accuracy through utilizing these compensation methodologies. Nevertheless, limited by fabrication constraints and design specifications, existing research seldom addresses ultra-high-temperature environments exceeding 150°C.

979-8-3315-8850-2/25 $31.00 © 2025 IEEE

Compared to MOS transistor process corner variations, the process corner deviations in bipolar junction transistors and resistors exert more substantial impacts on temperature coefficients—a critical issue that remains inadequately addressed in contemporary research.

This paper proposes solutions to mitigate both temperature-induced output drift and process-corner-dependent output deviations in bandgap reference circuits. First, trimming circuits are employed to compensate for output voltage deviations induced by process corner variations in bipolar junction transistors and resistors. Second, a segmented current compensation circuit is designed to optimize temperature coefficients across multiple intervals, thereby significantly reducing the output voltage temperature coefficient across the wide temperature range. Furthermore, an adaptive logic selection structure is developed to maintain optimal temperature drift suppression across all process corners, particularly when fixed high-order compensation demonstrates limited effectiveness against process corner variations. Finally, high-β bipolar junction transistors are employed as operational amplifier input stages, achieving effective suppression of low-frequency flicker noise in the system.

II. STRUCTURE OF PROPOSED BGR

A. Core structure of BGR

The fundamental architecture of the bandgap reference voltage source adopts a current-mode topology, as illustrated in Fig.1, to enable segmented compensation. The operational amplifiers incorporate high-β bipolar junction transistors in their input stage, specifically engineered to suppress low-frequency flicker noise.

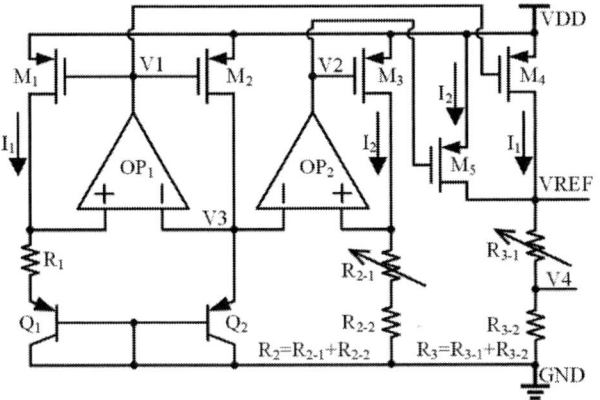

Fig. 1. The core structure of bandgap reference

The operational principle of the first-order compensated bandgap reference voltage originates from the complementary temperature dependence between the base-emitter voltage (V_{BE}) of bipolar junction transistors and the thermal voltage (V_T). The analytical derivation and temperature-dependent behavior of V_{BE}, demonstrating a negative temperature coefficient, are mathematically formulated in (1) and (2).

$$V_{BE} = V_T \ln \frac{I_C}{I_S} = \frac{kT}{q} \ln \frac{I_C}{I_S} \qquad (1)$$

$$\frac{\partial V_{BE}}{\partial T} = \frac{V_{BE} - (4+m)V_T - E_g/q}{T} \approx -1.5 mV/°C \qquad (2)$$

By differentially extracting the base-emitter voltages from two bipolar junction transistors, ΔV_{BE} is obtained. The analytical derivation and temperature-dependent behavior of ΔV_{BE}, exhibiting linear correlation with the V_T, are mathematically formulated in (3) and (4).

$$\Delta V_{BE} = V_{BE1} - V_{BE2} = V_T \ln \frac{nI_{C0}}{I_{S1}} - V_T \ln \frac{I_{C0}}{I_{S2}} = V_T \ln n \quad (3)$$

$$\frac{\partial \Delta V_{EB}}{\partial T} = \frac{\partial V_T}{\partial T} * \ln(n) \approx (+0.087 * \ln n) \, mV/°C \qquad (4)$$

The bandgap reference voltage V_{REF} is generated through the matched current-mirror topology (M_1-M_5) and resistor network (R_1-R_3), with its analytical expression formulated in (5).

$$V_{REF} = \left(\frac{\Delta V_{BE}}{R_1} + \frac{V_{BE}}{R_2} \right) R_3 \qquad (5)$$

B. Resistor trimming

In analog circuit design, device process corner variations pose critical challenges to circuit performance, particularly evident in bandgap reference voltage sources. As analytically demonstrated in (6), V_{REF} exhibits direct dependence on the parameters of resistors and bipolar junction transistors. Process corners variations in bipolar junction transistors induce corresponding deviations in base-emitter voltage (V_{BE}) characteristics. These parametric variations substantially impact both the V_{REF} magnitude and temperature drift performance. Compared to resistors and bipolar junction transistors, MOS transistor process corner variations exhibit minimal impact. Therefore, this paper investigates nine critical process corner combinations in resistors and bipolar junction transistors, categorized in Table I for trimming optimization. The trimming circuit structure in Fig.2 enables significant cost reduction via controlled transistor switching, while preserving operational programmability. Additionally, a digital 3-to-8 decoder is incorporated to reduce control port requirements without compromising trimming resolution.

$$V_{REF} = \left(\frac{R_2 * \Delta V_{BE}}{R_1} + V_{BE} \right) \frac{R_3}{R_2} \qquad (6)$$

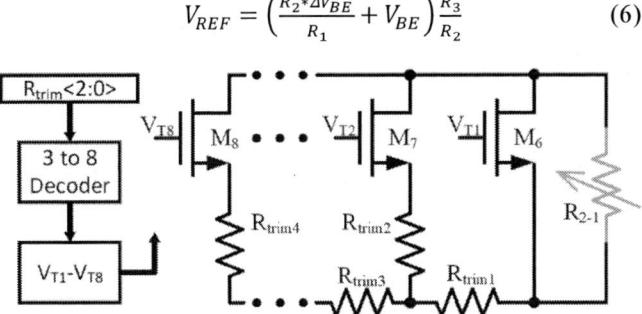

Fig. 2. The structure of trimming circuit

In Table I, TC1, TC2, TC3, and TC4 represent the temperature coefficients with only first-order compensation, after trimming, after compensation, and after adding a logic selection circuit respectively, all in units of ppm/°C.

TABLE I.   DIFFERENT PROCESS CORNER COMBINATIONS

	Resistor corner	Bjts corner	TC1	TC2	TC3	TC4
corner1	tm	tm	7.53	8.52	0.40	0.47
corner2	tm	wp	10.3	10.7	3.53	1.25
corner3	tm	ws	4.97	5.11	5.36	1.43
corner4	wp	tm	5.98	6.88	2.69	1.52
corner5	wp	wp	8.44	9.09	1.31	0.52
corner6	wp	ws	4.62	4.54	7.83	1.95
corner7	ws	tm	9.04	10.5	2.98	1.44
corner8	ws	wp	12.3	11.9	5.44	1.79
corner9	ws	ws	6.05	9.09	9.16	1.89

C. Segmented current compensation circuit

The first-order bandgap reference voltage originates from the total current through resistor R_3, constructed through summation of the positive temperature coefficient current I_1 and the negative temperature coefficient current I_2 in prescribed proportions. As both currents I_1 and I_2 contain higher-order components, the resulting bandgap reference voltage typically demonstrates zero temperature coefficient exclusively at a single temperature point, while manifesting a concave or convex curve across the entire temperature range. Therefore, injecting or extracting segmented compensation currents can effectively suppress the temperature drift of the bandgap reference voltage caused by temperature variations. As shown in Fig.3, the segmented current compensation circuit structure appropriately scales the I_1 and I_2 currents through M_9 and M_{10}, then replicates the corresponding segmented currents via the M_{13} and M_{14} current mirrors, subsequently injecting into the resistor for compensation.

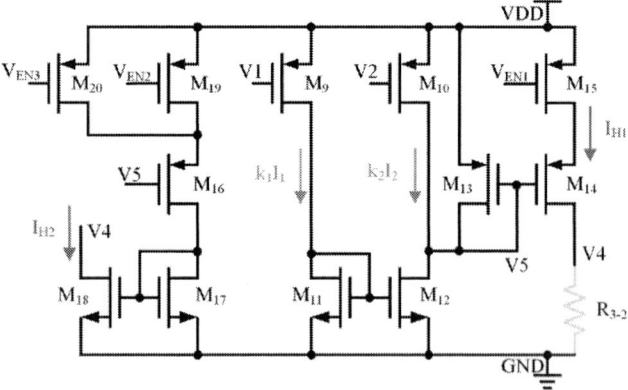

Fig. 3.   Compensation circuit

D. Logic selection circuit

The implemented segmented current compensation effectively reduces the temperature drift coefficient of the bandgap reference voltage. Under process variations, however, the compensation efficacy experiences significant degradation, revealing the limited process corner adaptability of fixed segmented current compensation. This work proposes an adaptive logic selection structure, schematically depicted in

Fig.4, to enhance process corner variation resilience. By designing logic gate arrays, the trimming circuit merges with the segmented current compensation circuit. This enables coordinated variation between trimming circuit and the compensation circuit under process corner changes, yielding optimized temperature drift.

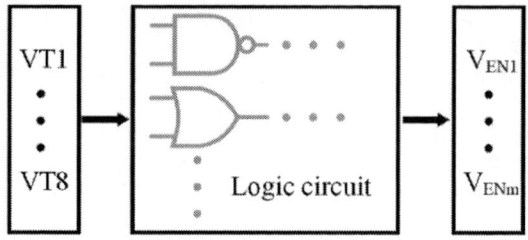

Fig. 4.   Logic selection circuit

III. SIMULATION RESULTS OF THE CIRCUIT

This chapter adopts the methodological framework established in Chapter 2 to conduct systematic simulation verification. The verification encompasses nine process corner combinations listed in Table I. The verification is conducted in sequence: first, the first-order compensated bandgap reference; second, the bandgap reference voltage output after resistor trimming; third, the bandgap reference voltage output after both resistor trimming and segmented current compensation; and finally, the bandgap reference voltage output after adding the logic selection circuit. The output voltage temperature coefficient across process corner is summarized in Table I.

A. Simulation of the resistor trimming

Equation (6) analytically demonstrates that resistor and bipolar junction transistor process corner variations induce parametric shifts in both V_{REF} magnitude and temperature drift characteristics. With 1.2V as the target reference voltage, comprehensive simulations are performed sequentially for different process corner combinations. The results are depicted in Fig.5(a). Simulation results reveal that only the corner1 case maintains design specifications, while other process corners demonstrate substantial deviations in both V_{REF} magnitude ($\pm1.16\%$) and temperature drift characteristics. Relative to the 1.2V target, the highest value is 7.86mV higher, and the lowest value is 13.89mV lower.

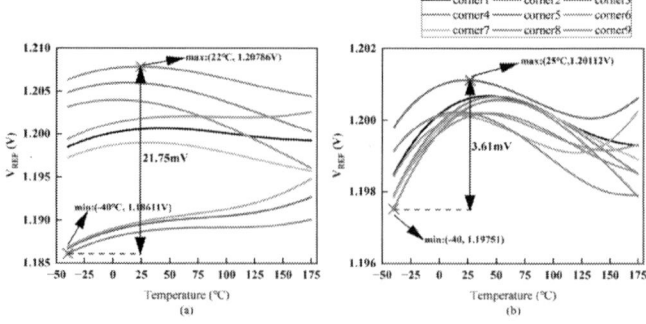

Fig. 5.   Simulation results of (a)first-order compensation, (b)after trimming

The programmable trimming architecture in Fig.2 enables V_{REF} calibration through controlled adjustment of R_2 and R_3. As exemplified in the corner9 case study, it is evident from the figure that its voltage value is below the target value and exhibits a positive temperature characteristic. The temperature characteristic can be optimized by reducing R_2, while the voltage value can be increased by enlarging R_3. Fig.5(b) shows the adjusted bandgap reference voltage output curve. Compared to the design value of 1.2V, the highest value exceeds the target by 1.12mV, while the lowest value is 2.49mV below the target. The peak-to-peak variation is suppressed from 21.75mV (1.81%) to 3.61mV (0.30%), achieving 83.4% performance enhancement.

B. Simulation of the compensation

After trimming, the bandgap reference voltage output closely approaches the target value. Nevertheless, with only first-order temperature compensation implemented, the temperature coefficient remains comparatively high. As shown in Fig.3, a segmented current compensation circuit is employed to compensate for the bandgap reference voltage, thereby reducing its temperature coefficient. By comparing the currents flowing through M_{10} and M_{12} at different temperatures, when the current of M_{10} exceeds that of M_{12}, M_{13} is turned off, while M_{14} does not carry current. When the current of M_{10} exceeds that of M_{12}, M_{13} is turned on, and simultaneously, M_{14} carries current. By adjusting k_1 and k_2, the segmented temperature point T_X of the segmented current can be adjusted, as shown in Fig.6(a). This enables the generation of segmented injection currents such as I_{H1} and I_{L1}. Additionally, through the incorporation of a current mirror, segmented extraction currents such as I_{H2} can be produced.

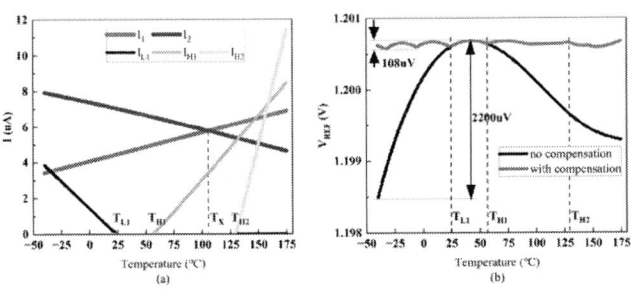

Fig. 6. Simulation results of (a)segment current, (b)after compensation

Through the addition of the two types of currents generated above to resistor R_{3-2}, a segmented current compensation effect is achieved. Taking corner1 as an example, the simulation curves of the voltage before and after segmented current compensation are shown in Fig.6(b). Within the temperature range of -40°C to 175°C, the difference between the maximum and minimum values of voltage variation is reduced from 2200uV to 108uV, and the temperature coefficient is reduced from 8.52ppm/°C to 0.40ppm/°C, demonstrating significant improvement.

C. Simulation of the logic selection circuit

The implementation of the segmented current compensation circuit is based on corner1 for compensation. When facing process corner variations, one issue is that the original voltage curve will change, and the other is that the values of I_1 and I_2 will also change, leading to degraded compensation effects. This means that a single fixed segmented current compensation is not

suitable for all process corner combinations. As shown in Fig.7(a), the segmented compensation current designed based on corner1 performs poorly in process corner combinations such as corner3 and corner9. The differences across the full temperature range are 1385 μV and 2363 μV, respectively.

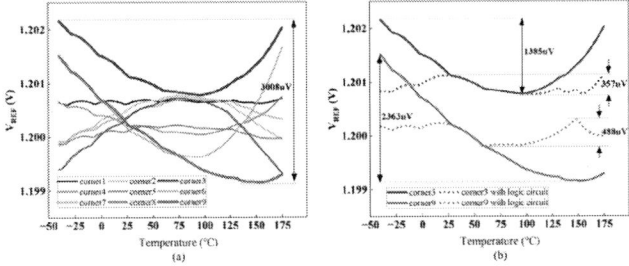

Fig. 7. Simulation results of (a)fixed compensation, (b)with logic circuit

By combining the trimming circuit with the segmented current compensation circuit through a logical selection circuit, the voltage drift can be significantly reduced by controlling the compensation size and whether compensation is performed in different temperature segments. As shown in Fig.7(a), the voltage curves of corner3 and corner9 show differences of 357uV and 488uV across the full temperature range. Further testing of the voltage output under various process corner combinations, as shown in Table I, reveals a temperature drift ranging from a minimum of 0.47ppm/°C to a maximum of 1.95ppm/°C. By adding a logical selection circuit, the temperature drift coefficient in the above process corner combinations has been significantly reduced from the highest value of 9.16ppm/°C to 1.95ppm/°C, with notable effectiveness.

In addition to the main circuit structure introduced above, the design of the operational amplifier employs high-β bipolar junction transistors as the differential input stage, effectively reducing low-frequency flicker noise, with integrated noise of 19.5 μV$_{rms}$ in the range of 0.1~10 Hz. Table II presents a data comparison between the bandgap reference source described in this paper and other bandgap reference sources from related works. The bandgap reference circuit described in this paper demonstrates superior performance with an extended operating temperature range, and the worst-case temperature coefficient across all process corner combinations is as low as 1.95 ppm/°C.

TABLE II.   PERFORMANCE COMPARISON

	2024[4] ISCAS	2021[8] TCAS-I	2024[10] ISCAS	This wok
Technology(nm)	180	65	180	180
Supply Voltage(V)	3.3	2.5-3.6	5	5
Reference(V)	1.066	1.1458	1.23	1.2
Best TC(ppm/°C)	1.02	0.8	2.3	0.47
Temperature Range(°C)	-40~125	0~80	-40~125	-40~175
Integrated Noise(μV$_{rms}$) 0.1~10Hz	N/A	10	N/A	19.5

979-8-3315-8850-2/25 $31.00 © 2025 IEEE

IV. Conclusion

This paper proposes a bandgap reference circuit with low-temperature-drift and low-noise characteristics, maintaining stable output across extended temperature ranges (-40°C to 175°C) and all process corners. Through designing the trimming circuit, segmented current compensation circuit, and adaptive logic selection circuit, the circuit achieves 0.47 ppm/°C minimum temperature coefficient over the wide temperature range. The maximum temperature coefficient under process corner variations remains constrained to 1.95 ppm/°C. The use of high-β bipolar junction transistors effectively reduces flicker noise in the circuit. The bandgap reference source described in this paper is implemented in 0.18μm BCD technology. Its ultra-wide operating temperature range and adaptability to process corner variations while maintaining low temperature drift performance make it highly suitable for applications requiring high temperature and high precision.

References

[1] H. -M. Chen, C. -C. Lee, S. -H. Jheng, W. -C. Chen and B. -Y. Lee, "A Sub-1 ppm/°C Precision Bandgap Reference With Adjusted-Temperature-Curvature Compensation," in IEEE Transactions on Circuits and Systems I: Regular Papers, vol. 64, no. 6, pp. 1308-1317, June 2017.

[2] X. Liao, Y. Zhang, S. Zhang and L. Liu, "A 3.0 μ Vrms, 2.4 ppm/°C BGR With Feedback Coefficient Enhancement and Bowl-Shaped Curvature Compensation," in IEEE Transactions on Circuits and Systems I: Regular Papers, vol. 71, no. 5, pp. 2424-2433, May 2024.

[3] H. Luo, Q. Sun, R. Zhang and H. Zhang, "A 1-V 3.1-ppm/°C 0.8-μW Bandgap Reference with Piecewise Exponential Curvature Compensation," 2018 IEEE Asian Solid-State Circuits Conference (A-SSCC), Tainan, Taiwan, 2018, pp. 97-98.

[4] B. Xiong, F. Yan, W. Mo, J. Guan, Y. Huang and J. Liu, "A 1.02 ppm/°C Precision Bandgap Reference with High-order Curvature Compensation for Fluorescence Detection," 2024 IEEE International Symposium on Circuits and Systems (ISCAS), Singapore, Singapore, 2024, pp. 1-4.

[5] W. Ye et al., "A Sub-1ppm/°C Wide-Temperature-Range Bandgap Voltage Reference with Superior-Order Temperature-Curvature Compensation," 2023 International Conference on Sensing, Measurement & Data Analytics in the era of Artificial Intelligence (ICSMD), Xi'an, China, 2023, pp. 1-5.

[6] L. Liu, X. Liao and J. Mu, "A 3.6 μVrms Noise, 3 ppm/°C TC Bandgap Reference With Offset/Noise Suppression and Five-Piece Linear Compensation," in IEEE Transactions on Circuits and Systems I: Regular Papers, vol. 66, no. 10, pp. 3786-3796, Oct. 2019.

[7] Luo Z, Lu Y, Huang M, et al. A sub-1V 78-nA bandgap reference with curvature compensation[J]. Microelectronics Journal, 2017, 63: 35-40.

[8] N. Liu, R. L. Geiger and D. Chen, "Sub-ppm/°C Bandgap References With Natural Basis Expansion for Curvature Cancellation," in IEEE Transactions on Circuits and Systems I: Regular Papers, vol. 68, no. 9, pp. 3551-3561, Sept. 2021.

[9] R. K. Palani, "Analysis and design of Chopperless 7 ppm/°C Bandgap Voltage Reference," 2024 IEEE International Symposium on Circuits and Systems (ISCAS), Singapore, Singapore, 2024, pp. 1-4.

[10] Y. Weng, J. Pan, Y. Zhao, J. Jiang and L. Qi, "A 2.3-ppm/°C High-Order Compensated Bandgap Reference With Low-Cost Current Trimming," 2024 IEEE International Symposium on Circuits and Systems (ISCAS), Singapore, Singapore, 2024, pp. 1-5.

[11] R. Wang et al., "A Sub-1ppm/°C Current-Mode CMOS Bandgap Reference With Piecewise Curvature Compensation," in IEEE Transactions on Circuits and Systems I: Regular Papers, vol. 65, no. 3, pp. 904-913, March 2018.

[12] C. M. Andreou, S. Koudounas and J. Georgiou, "A Novel Wide-Temperature-Range, 3.9 ppm/°C CMOS Bandgap Reference Circuit," in IEEE Journal of Solid-State Circuits, vol. 47, no. 2, pp. 574-581, Feb. 2012.

[13] P. Malcovati, F. Maloberti, C. Fiocchi and M. Pruzzi, "Curvature-compensated BiCMOS bandgap with 1-V supply voltage," in IEEE Journal of Solid-State Circuits, vol. 36, no. 7, pp. 1076-1081, July 2001.

A Pixel-to-Column Digital Readout Circuit with In-Pixel CDS for Ga₂O₃ UV Photodetectors

Jiekun Zhang
*School of Information
and Electronic Engineering
Shandong Technology
and Business University*
Yantai, China
2023410087@sdtbu.edu.cn

Guangfen Wei*
*School of Information
and Electronic Engineering
Shandong Technology
and Business University*
Yantai, China
guangfen.wei@sdtbu.edu.cn

Jingwu Gong
*Suzhou Institute of
Nano-Tech and Nano-Bionics
University of Science
and Technology of China*
Suzhou, China
wism@mail.ustc.edu.cn

Yu Zhou
*Suzhou Institute of
Nano-Tech and Nano-Bionics
Chinese Academy of Sciences*
Suzhou, China
yzhou2025@sinano.ac.cn

Xiaodong Zhang*
*Suzhou Institute of
Nano-Tech and Nano-Bionics
Chinese Academy of Sciences*
Suzhou, China
xdzhang2007@sinano.ac.cn

Helun Song*
*Suzhou Institute of
Nano-Tech and Nano-Bionics
Chinese Academy of Sciences*
Suzhou, China
hlsong2008@sinano.ac.cn

Abstract—This paper presents a pixel-to-column digital readout circuit for Ga₂O₃ UV Photodetectors. The readout circuit includes a pixel-level input circuit, a column-level programmable gain amplifier (PGA) with single-ended to differential conversion, and an analog-to-digital converter (ADC). The pixel-level input circuit adopts a capacitive trans-impedance amplifier (CTIA) and incorporates correlated double sampling (CDS) technology to effectively reduce reset noise and other low-frequency noise. The single-ended to differential PGA is employed to amplify the signal and convert the input-stage single-ended signal into a differential signal suitable for the subsequent ADC input. The ADC adopts an incremental delta-sigma ADC (IADC) architecture, in which the IADC modulator is integrated within the readout circuit, while the digital filter is implemented off-chip. The proposed readout circuit is implemented in a standard TSMC 180 nm CMOS process. The simulation results indicate that within the proposed readout circuit, the pixel-level input stage achieves 99.99% linearity at a 1.8V output swing. Following CDS processing, the output reset noise is reduced to 625μV. Additionally, the implemented IADC supports 14-bit digital signal conversion.

Keywords—readout circuit, CDS, IADC, CTIA, single-ended to differential

I. INTRODUCTION

Gallium oxide (Ga₂O₃) is a Fourth-generation semiconductor material with an ultra-wide bandgap of approximately 4.9 eV. In recent years, it has attracted significant attention due to its excellent ultraviolet (UV) photoresponse characteristics. Compared to traditional UV detection materials such as silicon (Si), gallium nitride (GaN), and silicon carbide (SiC), gallium oxide offers superior solar-blind selectivity, higher breakdown voltage, lower dark current, and better

This work was funded by the Suzhou Critical Core Technology Research Project (SYG2024003) and the Basic Research Program of Jiangsu (SBK20251000018)

thermal stability [1]. These properties make it highly suitable for photodetectors operating in the short-wave to deep ultraviolet spectral range, with wide applications in high-reliability fields such as space exploration, flame detection, and environmental monitoring. Therefore, the design of a high-gain, low-noise signal chain with integrated digitization capability is of significant importance for enhancing the overall performance of Ga₂O₃-based detection systems.

Conventional input-stage circuits based on capacitive trans-impedance amplifier (CTIA) [2] are susceptible to reset noise, which can significantly constrain the output dynamic range and degrade the overall image quality. To eliminate the impact of reset noise, correlated double sampling (CDS) is commonly incorporated into the signal chain to suppress reset noise and enhance the dynamic range [3].

The response signal from the detector is converted by the input stage into an amplified voltage signal. To achieve high-precision digital readout, the system requires effective offset suppression and robustness against 1/f noise. Common analog-to-digital converter (ADC) architectures include successive approximation register (SAR) ADC and dual-slope ADC [4]. SAR ADCs feature low power consumption but are susceptible to capacitor mismatch, limiting their resolution; moreover, incorporating digital calibration increases system complexity and power consumption. Dual-slope ADC have a simple structure and can easily achieve high resolution, but their conversion speed is limited. In contrast, incremental ΔΣ ADC (IADC) utilize oversampling and noise shaping to achieve high precision within a relatively short time [5], offering greater robustness to process variations, making them well-suited for low-speed, high-precision readout applications.

In this paper, a readout circuit is proposed for Ga₂O₃-based optoelectronic devices. The circuit integrates a CTIA and CDS at the pixel-level input stage. A programmable gain amplifier is then used to convert the single-ended signal to differential, and an IADC is employed to generate the digital output. The paper

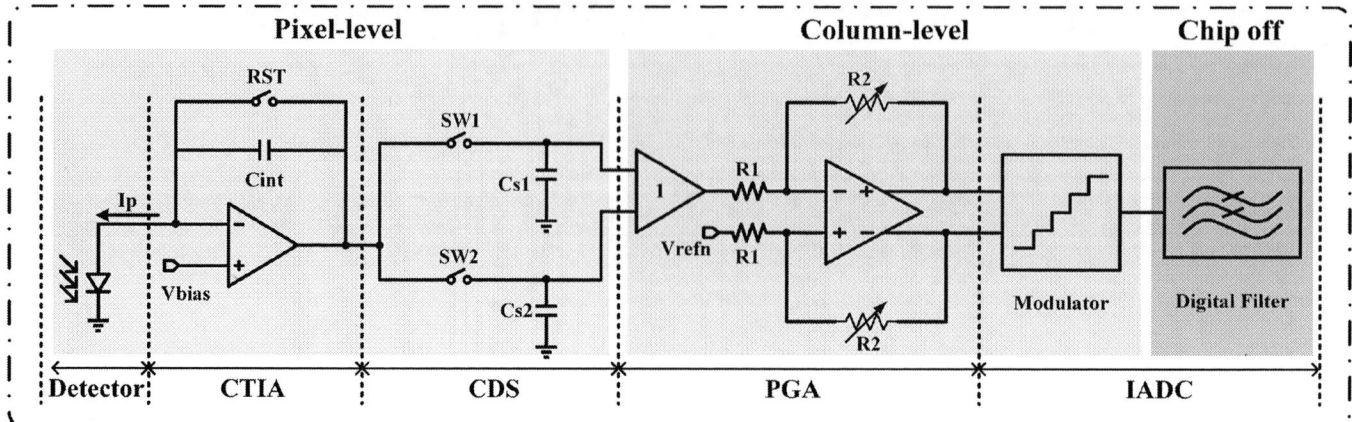

Fig. 1.   Proposed column readout circuit.

is organized as follows. Section 2 introduces the circuit architecture and its operating principles. Section 3 presents simulation results. Section 4 concludes the work.

II.   PIXEL-TO-COLUMN DIGITAL READOUT CIRCUIT

Gallium oxide detectors operate in the solar-blind ultraviolet band, effectively eliminating interference from other spectral regions and thereby achieving a high signal-to-noise ratio, which makes them well suited for detection and early-warning systems. To minimize the impact of fixed pattern noise during readout, the readout circuitry is required to feature a low-noise front end together with a high-resolution analog-to-digital conversion stage, ensuring signal integrity and accuracy. Considering that the device is primarily applied in 64×64 focal plane array (FPA) configurations, the design and optimization of the signal chain should be carried out with respect to the overall performance requirements at the array level.

Fig.1 shows the architecture of the proposed readout circuit. The circuit is composed of three main blocks. First, the in-pixel input stage converts the photocurrent generated by the device into an amplified voltage signal through a current-to-voltage (I-V) conversion. Second, a single-ended to differential amplifier transforms the output of the input stage into a fully differential signal. Finally, an IADC is employed to digitize the analog signal for subsequent output.

A.   In-Pixel Input Stage

Fig. 2 (a) illustrates the schematic of a conventional in-pixel CTIA circuit. During operation, the CTIA integrates the photocurrent, and the resulting voltage is sampled and held on a storage capacitor for subsequent signal processing. However, during the reset phase, the thermal noise from the reset switch and the amplifier noise introduce reset noise, resulting in uncertainty in the initial integration value, thereby limiting the dynamic range of the input stage output [6]. To suppress both

reset noise and low-frequency noise, this design employs in-pixel CDS technology. Furthermore, compared to ideal differential-input operational amplifiers, the single-ended operational amplifier used in this design is equivalent to having a fixed offset voltage at the input terminal, which constrains the bias voltage provided to the device. This offset voltage varies

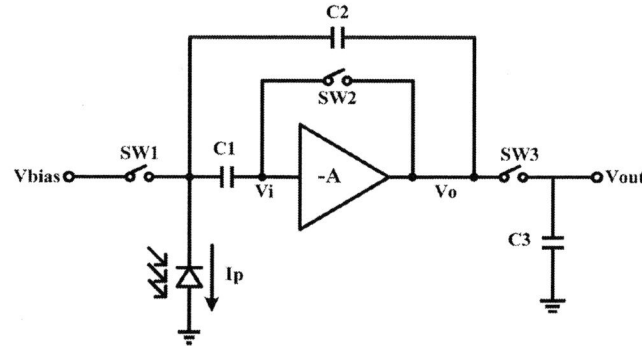

(a) conventional in-pixel CTIA.

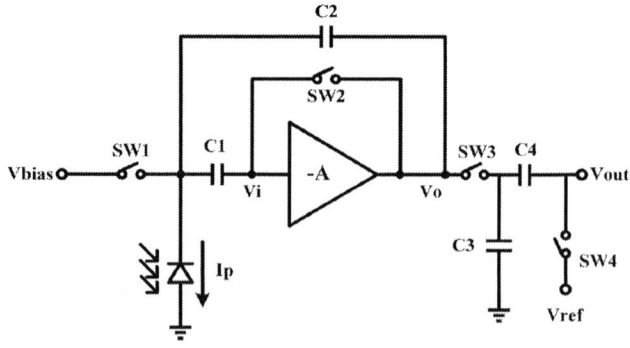

(b) In-pixel CTIA with CDS.

Fig. 2.   Conventional in-pixel CTIA and in-pixel CTIA with CDS.

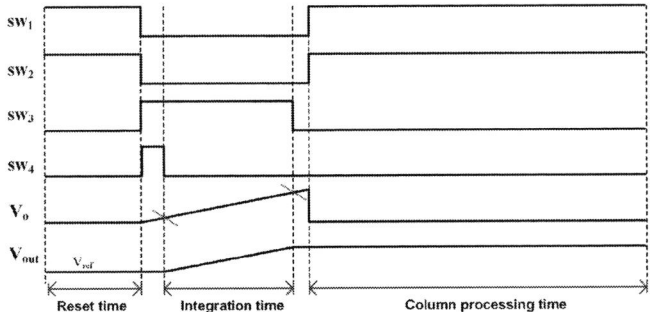

Fig. 3.   In-pixel CTIA with CDS timing diagram.

with process and device dimensions. Considering the operating conditions of the device, an offset storage capacitor is added at the input port to ensure the device operates under a constant bias. In contrast to the CTIA architecture documented in [3], Fig. 2(b) illustrates the schematic of the improved CTIA circuit with integrated offset storage. The offset storage capacitor was employed in auto-zero (AZ) techniques to suppress operational-amplifier offset by sampling the offset voltage and canceling it in subsequent processing. In CTIA, introducing an offset-storage capacitor allows the amplifier input voltage to be sampled onto the capacitor during the reset phase, while the device bias is set by the externally applied V_{bias}. After reset, because the amplifier input node is high-impedance, the stored charge is preserved and the device bias remains at V_{bias}, As a result, variations in the amplifier offset voltage do not disturb the device bias. This circuit consists of four capacitors, a cascode amplifier, and four switches, specifically designed to implement correlated double sampling while maintaining the desired bias voltage conditions.

The timing diagram, as shown in Fig. 3, at the beginning of operation, switches SW_1, SW_2 are closed to perform the reset phase. During this phase, the left terminal of capacitor C_1 is reset to the required device bias voltage, and the potentials at nodes V_i and V_o are equal. Meanwhile, the output node V_{out} is reset to the integration start voltage V_{ref}. Switch SW_4 conducts during the initial phase when SW_1 and SW_2 are open, thereby implementing output voltage reset. When SW_1 and SW_2 are turned off, switch SW_3 is closed to initiate the integration phase. At this point, the photodiode is forward biased, generating a photocurrent, and the CTIA begins integrating the signal under negative feedback conditions. Since SW_4 has not yet been opened, V_{out} remains constant at V_{ref} at the start of integration. Once SW_4 is opened, the voltage on the left side of capacitor C_3 continues to rise. Due to charge conservation, the output voltage V_{out} follows the rise of node V_o. The integration continues until SW_3 is opened, at which point the circuit enters the hold phase, and the output voltage is sampled by subsequent signal processing circuits. This process completes the correlated double sampling.

$$V_{out} = V_{ref} + \frac{I_p \cdot \Delta t}{C_2}. \qquad (1)$$

The output voltage V_{out} can be expressed as shown in (1), where Δt denotes the time interval between the opening of

switches SW_3 and SW_4. I_p denotes the photocurrent generated in response to light by the device.

B. PGA Circuit

The programmable gain amplifier (PGA) circuit is used to amplify the input stage signal and convert the single-ended signal into a differential signal, as shown in Fig. 1, the PGA employs a feedback structure composed of resistors. In this design, resistors R_1 and R_2 are equal in value. The non-inverting input of the fully differential operational amplifier is connected to the intermediate voltage level within the CTIA integration range, while the inverting input is connected to the output of the input stage. This configuration allows the conversion of the single-ended input signal into a differential output signal. By adjusting the value of V_{refn}, this conversion enables the subsequent IADC to generate quantized outputs that vary between negative and positive voltages, thereby fully utilizing the ADC quantization dynamic range.

The gain of the PGA can be adjusted by varying the resistance values of R_1 and R_2, with its output voltage given by

$$V_{outp} - V_{outn} = \frac{R_2}{R_1}(V_{in} - V_{refn}) \qquad (2)$$

Assuming that $V_{refn} = 0$, the input common-mode voltage of the operational amplifier can be expressed as (3).

$$V_{inp} = V_{inn} = \frac{V_{in}}{2} \cdot \frac{R_2}{R_1 + R_2} \qquad (3)$$

The input common-mode voltage of a fully differential operational amplifier is affected by the input signal. Therefore, the input common-mode range must be carefully considered during the amplifier design to ensure proper operation.

C. Incremental ADC

The IADC is an ADC architecture capable of achieving high-resolution signal conversion by combining oversampling and noise-shaping techniques with a periodic reset mechanism during operation [7]. Unlike conventional $\Delta\Sigma$ ADC, which rely on continuous integration and exhibit strong memory dependence, the IADC establishes a deterministic, one-to-one mapping between the input signal and the output digital code. This behavior renders its operation more akin to that of a Nyquist-rate ADC [8]. Such deterministic input–output correspondence is particularly advantageous in sensor array systems, where consistent and reproducible conversion is critical. Consequently, IADC are highly suitable for integration into Ga$_2$O$_3$ photodetector arrays.

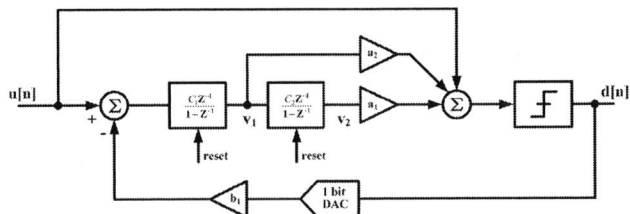

Fig. 4. Block diagram of a second-order incremental ADC.

In detector array applications, system performance is inherently constrained by trade-offs among speed, area, and power consumption. Theoretically, increasing the order of an incremental ΔΣ modulator can reduce the number of conversion cycles required to achieve a given resolution. However, this comes at the cost of increased circuit complexity, area, and power, which is undesirable for large-scale array implementations. Conversely, a conventional first-order ΔΣ modulator requires significantly more conversion cycles to reach the same resolution, thereby limiting the final imaging frame rate in large arrays. To balance area efficiency and conversion speed, a second-order incremental ΔΣ modulator is adopted in the signal readout chain of this work.

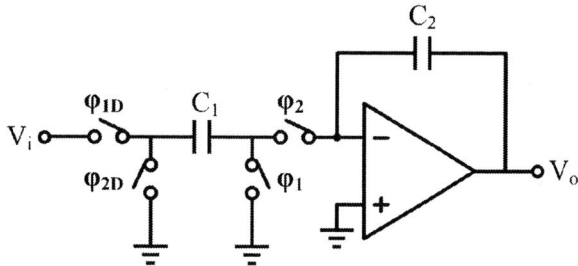

Fig. 5. Half circuit simplified representation of the integrator.

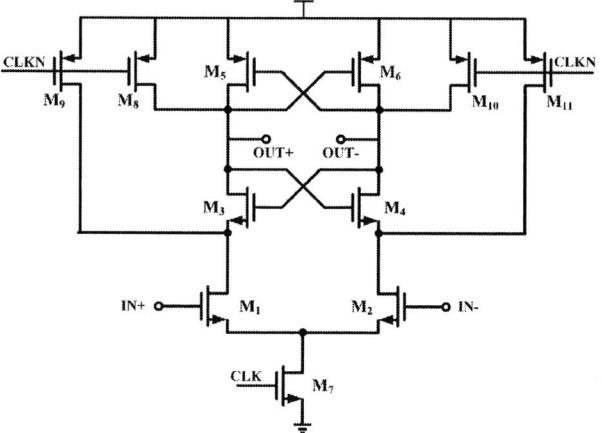

Fig. 6. Strong arm latch comparator.

Fig.4 illustrates the second-order incremental ΔΣ modulator architecture employed in this work. The modulator consists of two integrators, a comparator, and a 1bit digital-to-analog converter (DAC). Fig. 5 and Fig. 6 illustrate the circuit diagrams of the integrator and the comparator. Assuming the input signal of the IADC is a constant value, the outputs of the first and second integrators (v_1 and v_2) can be expressed as follows [8].

$$v_1[i] = v_1[i-1] + c_1 u_{in} - c_1 V_{ref} d[i] \qquad (4)$$

and

$$v_2[i] = v_2[i-1] + c_2 v_1[i] \qquad (5)$$

a_1 and a_2 are the modulation coefficients of the IADC, V_{REF} is the reference voltage of the feedback DAC, and $d[i]$ is the output of the comparator. Assuming sampling is performed over N cycles, the expressions for $v_1[N]$ and $v_2[N]$ can be obtained.

$$v_1[N] = c_1 \sum_{i=0}^{N-1} u_{in} - c_1 V_{ref} \sum_{i=0}^{N-1} d[i] \qquad (6)$$

and

$$u_{in} = \frac{2 \cdot v_2[N]}{c_1 c_2 N(N-1)} + \frac{2 \cdot V_{ref}}{N(N-1)} \sum_{j=0}^{N-1} \sum_{i=0}^{j-1} d[i] \qquad (7)$$

The input u_{in} can be expressed as shown in (7), and the quantization error can also be obtained as (8).

$$u_{in} = \frac{2 \cdot v_2[N]}{c_1 c_2 N(N-1)} + \frac{2 \cdot V_{ref}}{N(N-1)} \sum_{j=0}^{N-1} \sum_{i=0}^{j-1} d[i] \qquad (8)$$

$$Eq = \frac{2 \cdot v_2[N]}{N(N-1)c_1 c_2}. \qquad (9)$$

According (7), the comparator output can be digitally filtered using a cascade of simple accumulators. The filtered output D_{out} can then be expressed as

$$D_{out} = \sum_{j=0}^{N-1} \sum_{i=0}^{j-1} d[i] \qquad (10)$$

$$u_{in} \approx \frac{2 \cdot V_{ref}}{N(N-1)} \cdot D_{out} \qquad (11)$$

The input voltage u_{in} can be obtained through the digital filter in (10), and can be expressed as (11); evidently, this neglects the quantization error in (9). The digital filter described in this paper is implemented externally and can be adjusted by modifying the number of sampling periods. This allows for flexible tuning to achieve the desired conversion time and accuracy.

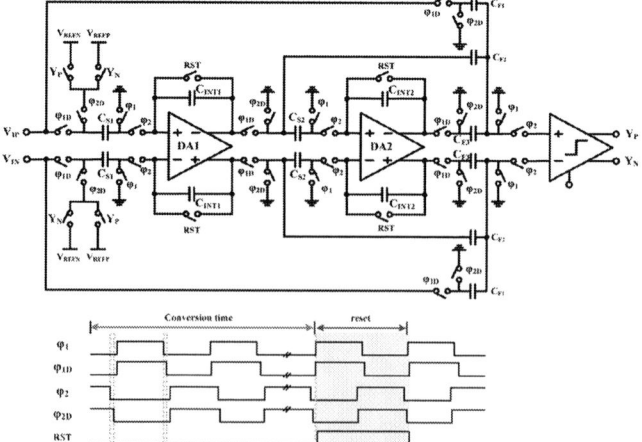

Fig. 7. Complete circuit realization of the IADC and the associated timing diagram.

A schematic of the IADC modulator is shown in Fig. 7 The IADC modulator employs a fully differential architecture and is driven by two non-overlapping clock phases. The values of the internal capacitors are carefully selected to meet the coefficient requirements of the modulator while ensuring proper matching and minimizing thermal noise. The switch resistance is optimized to satisfy the settling time constraints.

III. SIMULATION RESULTS

The proposed readout circuit was designed and simulated using the TSMC 180 nm standard CMOS process, powered by a 3.3 V supply. The integration capacitor C1 was set to 1.1 pF, while capacitors C3 and C4 were both chosen as 200 fF. All simulations were conducted at room temperature (27°C). Fig. 8 shows the corresponding linearity fitting curve. The achieved linearity is approximately 99.99%. Fig. 9 illustrates the equivalent input-referred output noise. Based on 400 transient noise simulations, the average noise voltage referred to the output is measured to be 625 μV.

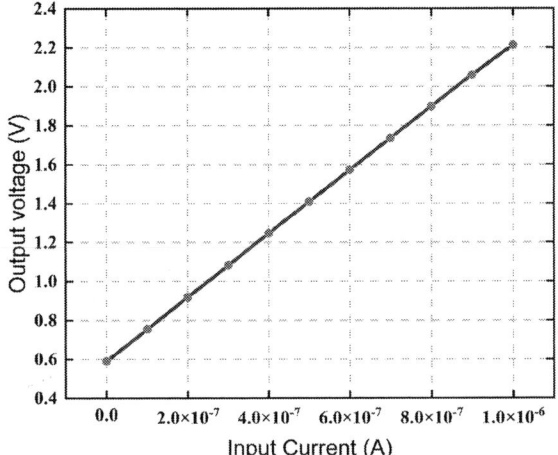

Fig. 8. Linearity fitting curve.

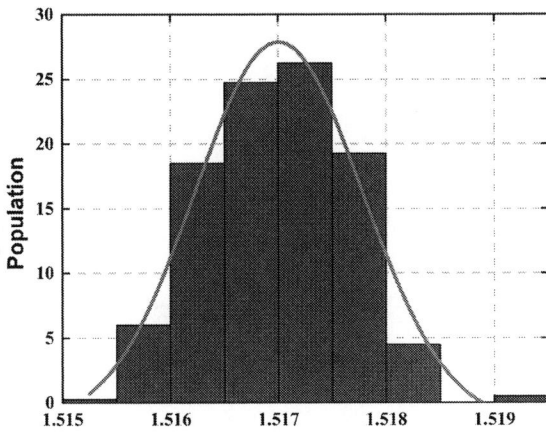

Fig. 9. Noise simulation results.

The ADC operates at a sampling frequency of 5MHz and an oversampling ratio of 256 to achieve a Nyquist conversion bandwidth of 18.94 kS/s. Fig. 10 shows the simulated output spectrum of the ADC with a −6dBFS, 0.74 kHz sinusoidal input. The output spectrum indicates an effective number of bits (ENOB) of 13.82 bits. Fig. 11 illustrates the simulated signal-to-noise ratio (SNR) and signal-to-noise-and-distortion ratio

(SNDR) as functions of the input signal amplitude. The simulation results show that with a 0.74kHz sinusoidal input, the peak SNDR reaches 87dB. Table I shows the performance comparison of the proposed readout circuit with that of other readout circuits [3] [9] [10].

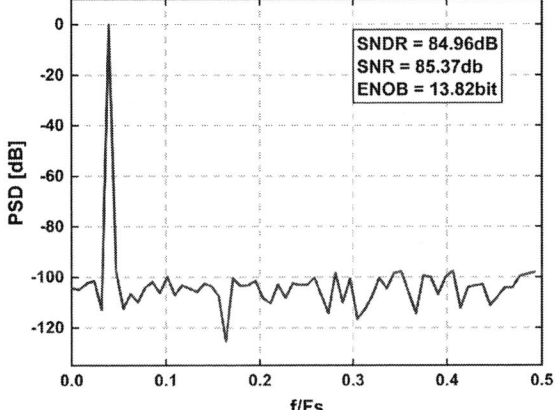

Fig. 10. Simulated output spectrum of the IADC.

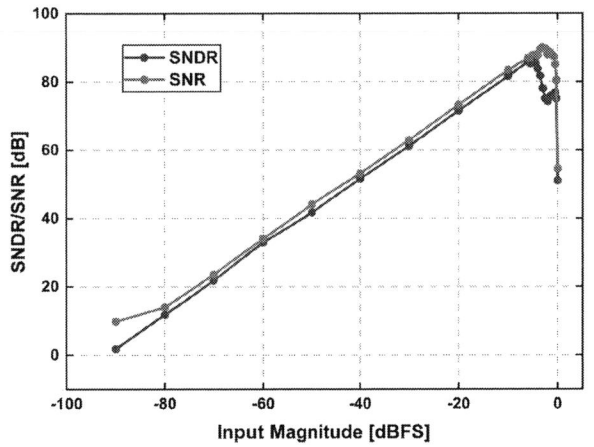

Fig. 11. SNR and SNDR versus input level.

TABLE I.   PERFORMANCE COMPARISON FOR ROIC

	2013[9]	2017[10]	2021[3]	This Work
Technology	0.35μm	0.18μm	0.18μm	0.18μm
Input Circuit	CTIA	CTIA	CTIA	CTIA
Pixel Array	10 × 8	32 × 32	NA	NA
Pixel Pitch	30μm	25μm	25μm	60μm
Readout Noise	6373 e⁻	127 e⁻	11.38 e⁻	4297 e⁻
Dynamic Range	54.8dB	68.8dB	74dB	69.2dB
Pixel Rate	6MHz	20MHz	NA	1MHz
Output Signal	ANA	ANA	ANA	DIG

ª Output noise of CTIA input stage .

IV. CONCLUSION

This work proposes a readout circuit tailored for gallium oxide UV photodetectors. The input stage employs an in-pixel

CTIA with CDS, which effectively suppresses reset noise. At the column level, PGA is used to convert the single-ended signal to differential form, while also allowing gain adjustment based on system requirements. The digitization is performed by a second-order incremental ADC, with the digital filter implemented off-chip. This allows for flexible trade-offs between processing speed and precision. The circuit was designed and simulated using the TSMC 180 nm standard CMOS process. Simulation results demonstrate that the input stage achieves 99.99% linearity within the output voltage swing range. The equivalent output noise voltage is 625 µV, The second-order incremental ADC achieves an ENOB of 14 bits under a -6 dBFS input signal.

The present circuit design has been developed with a 60µm pixel pitch as the primary reference, which poses limitations when scaling down to large-scale focal plane arrays. For future implementations, the performance requirements of large-scale FPAs can be met by introducing optimized circuit architectures and migrating to more advanced process nodes. These improvements are expected to enhance integration density, reduce power consumption, and ultimately enable the realization of high-performance solar-blind ultraviolet imaging systems based on gallium oxide detectors.

REFERENCES

[1] Tiwei Chen, Nan Liu, Suzhen Cheng, Xiaodong Zhang, Peng Ding, Chunhong Zeng, Li Zhang, Gaofu Guo, An Yang, Yu Hu, et al. Ultra sensitive dynamic ultraviolet imaging based on a Ga_2O_3 photodetector array. *Optics Letters*, 50(5):1633–1636, 2025.

[2] Kartikeya Murari, Ralph Etienne-Cummings, Nitish V Thakor, and Gert Cauwenberghs. A cmos in-pixel ctia high-sensitivity fluorescence imager. *IEEE transactions on biomedical circuits and systems*, 5(5):449 458, 2011.

[3] Yi Zhuo, Wengao Lu, Shanzhe Yu, Ye Zhou, Jiaqi Kong, Yacong Zhang, and Zhongjian Chen. A low-noise ctia-based pixel with cds for swir focal plane arrays. In *2021 6th International Conference on Integrated Circuits and Microsystems (ICICM)*, pages 258–262. IEEE, 2021.

[4] Yun-Rae Jo, Seong-Kwan Hong, and Oh-Kyong Kwon. A multi-bit incremental adc based on successive approximation for low noise and high resolution column-parallel readout circuits. *IEEE Transactions on Circuits and Systems I: Regular Papers*, 62(9):2156–2166, 2015.

[5] Sha Tao and Ana Rusu. A power-efficient continuous-time incremental sigma-delta adc for neural recording systems. *IEEE Transactions on Circuits and Systems I: Regular Papers*, 62(6):1489–1498, 2015.

[6] DA Van Blerkom. Analysis and simulation of ctia-based pixel reset noise. In *Infrared Technology and Applications XXXVII*, volume 8012, pages 159–168. SPIE, 2011.

[7] Ruiqi Gao, Mingqiang Guo, Sai-Weng Sin, Liang Qi, Biao Wang, Guoxing Wang, and Rui Paulo Martins. Weightings in incremental adcs: A tutorial review. In *2023 IEEE Custom Integrated Circuits Conference (CICC)*, pages 1–8. IEEE, 2023.

[8] Zhichao Tan, Chia-Hung Chen, Youngcheol Chae, and Gabor C Temes. Incremental delta-sigma adcs: A tutorial review. *IEEE Transactions on Circuits and Systems I: Regular Papers*, 67(12):4161–4173, 2020.

[9] Tai-Ping Sun, Yi-Chuan Lu, and Hsiu-Li Shieh. A novel readout integrated circuit with a dual-mode design for single-and dual-band infrared focal plane array. *Infrared Physics & Technology*, 60:56–65, 2013.

[10] Liang Li, Ruizhi Sun, Ruoxi Wang, Fanjun Zang, Tao Jiang, Yang Li, Xinyang Wang, and Yuchun Chang. A 20mhz ctia roic for ingaas focal plane array. In *2017 IEEE 12th International Conference on ASIC (ASICON)*, pages 267–270. IEEE, 2017.

2025 The 10th International Conference on Integrated Circuits and Microsystems

A DDR Controller Circuit Gate-level Post-simulation Method in SoC Design

Yande Jiang
Defense Innovation Institute
Academy of Military Science
Beijing, China
yandejiang@hotmail.com

Jingbo Ma
Defense Innovation Institute
Academy of Military Science
Beijing, China
majingbo0621@outlook.com

Guangda Zhang
Defense Innovation Institute
Academy of Military Science
Beijing, China
zhanggd_nudt@hotmail.com

Huiquan Wang
Defense Innovation Institute
Academy of Military Science
Beijing, China
quantum_ai@163.com

Bingxi Pei
Defense Innovation Institute
Academy of Military Science
Beijing, China
peibingxi@sjtu.edu.cn

Na Chen
Department of Intelligent Transportation
Beijing University of Technology
Beijing, China
nachen@bjut.edu.cn

Jian Fang*
Defense Innovation Institute
Academy of Military Science
Beijing, China
*Corresponding author: fangjian_alpc@163.com

Abstract—**As the complexity of System-on-Chip (SoC) designs increases, the gate-level post-simulation functional verification of SoCs has become increasingly critical. In addition, to enhance bandwidth, modern SoCs integrate high-performance DDR controller circuits. However, the gate-level post-simulation verification of DDR controllers in SoC designs demands substantial time and human resources. To address this challenge, we propose an efficient multi-level gate-level post-simulation methodology. To validate the proposed method, we design a SoC chip circuit which integrates 2 DDR4 controllers, 4 CPU cores, and employ the commercial Electronic Design Automatic (EDA) tools to obtain the DDR controller circuit netlist, and Standard Delay Format (SDF) files as the post-simulation input files. The experimental results demonstrate that our proposed multi-level gate-level post-simulation method achieves a $4\times$ speedup in functional post-simulation for the DDR controller circuit in SoC design, while also validating the effectiveness of the approach.**

Index Terms—**System on Chip, Integrate Circuit, DDR, Functional Verification, Post-simulation, Gate-level Simulation**

I. INTRODUCTION

In the field of digital integrated circuit (IC) design at nanoscale process nodes, as the complexity of System-on-Chip (SoC) designs continues to escalate, functional verification has become an increasingly critical yet challenging phase in the chip design flow [1, 2]. The post-simulation of Very Large Scale Integration (VLSI), particularly gate-level netlist post-simulation, is a critical step in the chip design flow to ensure successful tape-out. However, the gate-level post-simulation with timing annotation leads to exponentially growing verification cycles due to its accuracy requirements, accompanied by substantial computational resource consumption and labor costs [3].

Accelerating post-simulation functional verification for SoC chips has emerged as a significant challenge in the contemporary IC design. To address this challenge, numerous

This work was supported in part by the National Natural Science Foundation of China under Grants 62372461 and Zhiqiang Foundation.

research efforts have been proposed. Paper in [4] proposes a vector-mode based gate-level emulation which can obtain $500\times$ performance gain when simulating the Automatic Test Pattern Generation (ATPG) complete pattern set in Emulator. Paper in [5] mentions a fast and scalable gate-level simulation method which can fully exploit the parallelism from many-core computing systems and achieve a $27.6\times$ performance gain without compromising any correctness. Papers in [6] and [7] take advantage of Graphics Processing Units (GPUs) to accelerate the timing-aware gate-level simulation. In Paper [8], based on the extensive parallelism available in GPU hardware and the inherently parallel structure of gate-level netlists, the researchers propose a set of algorithms for efficiently mapping complex designs to parallel hardware. The proposed novel simulation architecture maximizes the utilization of concurrent hardware resources while significantly reducing expensive communication overhead. Compared to the fastest commercially available multi-threaded simulators, it achieves an average performance improvement of more than a factor of 10. However, the methods mentioned above require substantial hardware resources, including hardware emulators, many-core computing systems, and GPUs.

Moreover, paper in [9] accelerates the circuit simulation by extracting phase-coupled parameters from a parameterized 3D model. Paper in [10] designs an automated fast prediction-based gate-level timing simulation which combines Static Timing Analysis (STA) at the block level with dynamic timing simulation at the I/O interfaces, aiming at accelerating the gate-level timing simulation. Paper in [11] introduces a hybrid emulation-based parallel acceleration algorithm for the circuit post-simulation by means of hyper-graph partitioning methods which significantly accelerate the solving speed of linear equation systems. These approaches, however, exhibit limited applicability in the context of the VLSI circuit simulation.

Furthermore, to enhance bandwidth, SoC chips integrate

979-8-3315-8850-2/25 $31.00 © 2025 IEEE

DDR SDRAM (Double Data Rate Synchronous Dynamic Random-Access Memory) interface circuits (DDR controller circuits), which poses significant challenges to both timing closure and post-simulation of the entire chip [12, 13]. In this paper, in order to address the mentioned issues above, we propose a DDR controller circuit multi-level gate-level post-simulation method in SoC designs. The multi-level post-simulation includes both subsystem-level (module-level) and system-level. The division of post-simulation for large-scale circuits into subsystem-level and system-level is primarily motivated by considerations of simulation efficiency, debugging efficiency, and parallelization of simulation tasks. The motivations are as follows.

(i) Due to the enormous scale of circuit netlists and time-consuming simulations, the multi-level post-simulation can reduce simulation complexity and improve efficiency. A complete SoC may contain billions of transistors, equivalent to tens of millions of standard gates. Performing full-system simulation at the gate-level could take weeks or even months, even with the most powerful servers and fastest simulators. Effectively combining subsystem-level and system-level post-simulation methods can significantly enhance the efficiency of the post-simulation for DDR controllers in SoCs.

(ii) The multi-level post-simulation can improve the debugging efficiency. If the post-simulation of DDR controller circuits is conducted directly at the SoC system level, the deep-seated timing violations or functional errors may be buried within vast amounts of logic. Identifying the root cause of such errors is extremely challenging and inefficient. Utilizing the subsystem-level post-simulation for DDR controllers accelerates the localization of issues within the subsystem, thereby enhancing debugging efficiency.

(iii) The post-simulation tasks are parallelized through the multi-level post-simulation approach. Hierarchical simulation enables subsystem-level verification engineers and system-level engineers to validate their respective modules in parallel. This allows early detection and resolution of issues, rather than discovering them only during final integration, thereby reducing chip tape-out risks and improving simulation efficiency.

To validate the proposed multi-level post-simulation method, we design a SoC chip circuit which integrates 2 DDR4 controllers, 4 CPU cores, and 1 Network on Chip (NoC). We apply the commercial Electronic Design Automatic (EDA) tools to obtain the DDR controller circuit netlist including place-and-route information, and Standard Delay Format (SDF) files as the post-simulation input files. The experimental results show that the proposed multi-level post-simulation method can achieves a $4\times$ speedup in functional post-simulation for the DDR controller circuit in SoC design. We applied this post-simulation methodology in one SoC chip design including 2 DDR controllers, and the chip has been successfully taped out and tested, which validates the effectiveness of the proposed approach.

The state of the art in gate-level simulation acceleration has evolved significantly from relying on a single method to a sophisticated, hybrid approach that strategically leverages different hardware resources to tackle the immense complexity of modern billion-gate design. The gate-level simulation acceleration methods in papers [4], [5], [6] and [7] take advantage of the hardware resource including hardware emulators, many-core computing systems and CPUs. In contrast to the these methods, our approach introduces a multi-level post-simulation solution. This methodology rationally partitions module-level and system-level post-simulation tasks, achieving balanced simulation efficiency and functional correctness without requiring additional hardware resources, thereby significantly accelerating post-simulation speed.

The main contributions of this paper are listed as follows.

(i) The proposed multi-level gate-level post-simulation method is divided into the module-level simulation and the system-level simulation to achieve a balance between verification completeness and efficiency. The module-level simulation reduces memory usage by 71% and improves simulation speed by 77% compared to the system-level simulation.

(ii) The designed method utilizes the SoC design symmetry to trim the post-simulation netlist, in order to reduce memory consumption and computational resources during the post-simulation, and accelerate the simulation speed. The trimmed netlist simulation reduces memory usage by 67% and increases simulation speed by 19% compared to the untrimmed netlist.

(iii) In the system-level post-simulation, the proposed method utilizes a backdoor approach to rapidly load the register configurations required for DDR initialization, thereby accelerating the DDR initialization process with a speedup ratio of $4.3\times$ for the system-level simulation with mixed netlist and Register Transfer Level (RTL) blocks. Moreover, the proposed method with the backdoor approach obtains a speedup ratio of $11.2\times$ for the system-level simulation with all netlist blocks.

The remaining of this paper is organized as follows. Section II presents the designed system on chip architecture with 2 DDR controllers and 4 CPU cores. Section III introduces the proposed DDR controller gate-level post-simulation framework in SoC design, including the physical design and post-simulation flow, and the multi-level post-simulation flow. Section IV shows the experimental results which include the experimental environment, test cases, and the analysis of the obtained experimental results. Finally, some concluding comments are revealed in Section V.

II. SYSTEM ON CHIP ARCHITECTURE

In order to validate the proposed gate-level post-simulation method for the DDR controller circuit in SoC design, we design a 4 CPU cores SoC chip with 2 DDR4 controllers, as depicted in Fig 1. The designed SoC architecture is constructed of 4 main modules, including the CPU core with L1 cache and L2 cache, the DDR controller, the NoC, and the SDRAM device. The CPU core is used to execute instructions. The DDR controller receives data access requests from the L2 cache and forwards them to the SDRAM device. The NoC is responsible for processing and forwarding packets from different cores. In actual chip applications, the SDRAM device is responsible for storing data. During functional verification, such as the RTL simulation and the gate-level simulation, the SDRAM device is replaced by an SDRAM VIP (Verification IP), which can emulate the behavior of the SDRAM device.

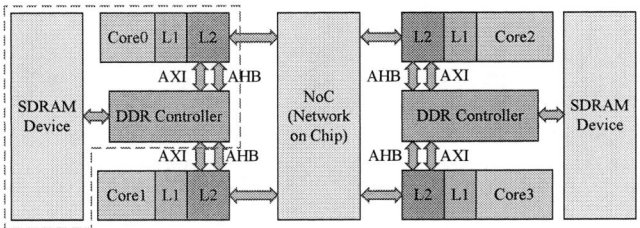

Fig. 1. The system on chip architecture

The DDR controller interfaces involve the AXI interface and the AHB interface which are used to connect to L2 cache, as shown in Fig 1. The AXI interface with date width of 256 bits is used for data transferring, and the AHB interface with data width of 32 bits takes charge of the register writing and reading operations. The SDRAM protocal interface is utilized between the DDR controller and the SDRAM device.

In this paper, we focus on the DDR controller gate-level post-simulation. Since the SoC chip design is left-right symmetrical, we can use a single CPU core and a single DDR channel as representative in the post-simulation. We primarily focus on verifying the functional correctness of the netlist for the core0, L1 cache, L2 cache, and DDR controller path as enclosed by the dashed box in Fig 1, rather than verifying the full-chip netlist functionality of all 4 CPU cores and 2 DDR controllers. This approach allows us to accelerate the post-simulation due to the reduced simulation scale.

III. DDR CONTROLLER GATE-LEVEL POST-SIMULATION FRAMEWORK

In this section, we introduce the DDR controller physical design and post-simulation flow. Next, we depict the multi-level DDR controller post-simulation framework, including the module-level simulation, the system-level simulation with 1 netlist and 3 RTL blocks, and the system-level simulation with 4 netlists.

A. DDR Controller Physical Design and Gate-level Post-simulation Flow

The DDR controller physical design and gate-level post-simulation flow is shown in Fig 2. The core purpose of the gate-level post-simulation is to examine the timing behavior of the circuit design under real physical conditions. To achieve this objective, firstly, we employ a commercial EDA synthesis tool to convert the DDR controller RTL design into a synthesized netlist. Secondly, after the physical design of the place and routing, we obtain the place-and-route netlist (PR-netlist) which includes the physical layout information, timing parameters, and other critical data. Thirdly, we utilize the PR-netlist to extract the SDF file as the post-simulation input file. Lastly, we perform the gate-level post-simulation

on the obtained PR-netlist by use of an EDA verification tool. If all functional verifications pass, the post-simulation concludes. However, if any functional verification fails, it indicates timing violations in the circuit. In such cases, we revert to the placement and routing phase to re-optimize the circuit layout. This iterative process continues until all functional verifications are successfully completed.

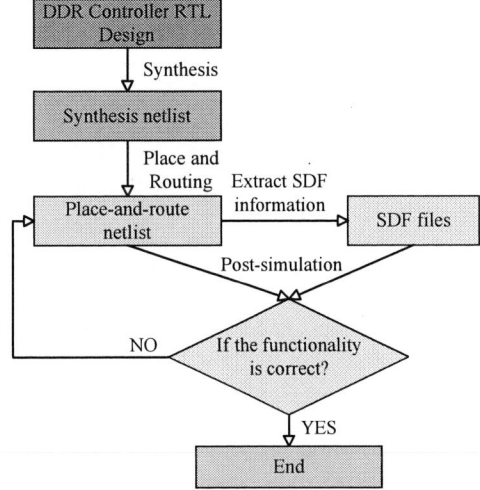

Fig. 2. The DDR controller physical design and gate-level post-simulation flow

B. Multi-level DDR Controller Post-simulation Framework

We build up a multi-level DDR controller gate-level post-simulation framework as shown in Fig 3. The proposed post-simulation framework consists of 3 levels, including the module-level simulation (Fig 3 (a)), the system-level simulation with 1 netlist and 3 RTL blocks (Fig 3 (b)), and the system-level simulation with 4 netlists (Fig 3 (c)).

In the module-level post-simulation, we only use DDR controller netlist and its SDF file as the post-simulation input files, as shown in Fig 3 (a). Several test stimulus are used to verify the DDR controller netlist functionalities. The purpose of the module-level post-simulation is primarily to verify the timing correctness and functional correctness within the DDR controller module.

During the system-level post-simulation, we primarily focus on verifying both timing compliance and functional correctness of the DDR controller interface and its interconnect logic with the L2 Cache. In order to balance verification completeness and efficiency, we build up 2 system-level post-simulations, as shown in Fig 3 (b) and 3 (c). Fig 3 (b) shows the system-level post-simulation with 1 DDR controller netlist and 3 RTL blocks, including the L1 cache RTL block, the L2 cache RTL block and the CPU core RTL block. Moreover, Fig 3 (c) reveals the post-simulation with 4 netlists of DDR controller, including the L1 cache, L2 cache and the CPU core. The simulation with 1 netlist and 3 RTL blocks demonstrates significant efficiency advantages over the simulation with 4

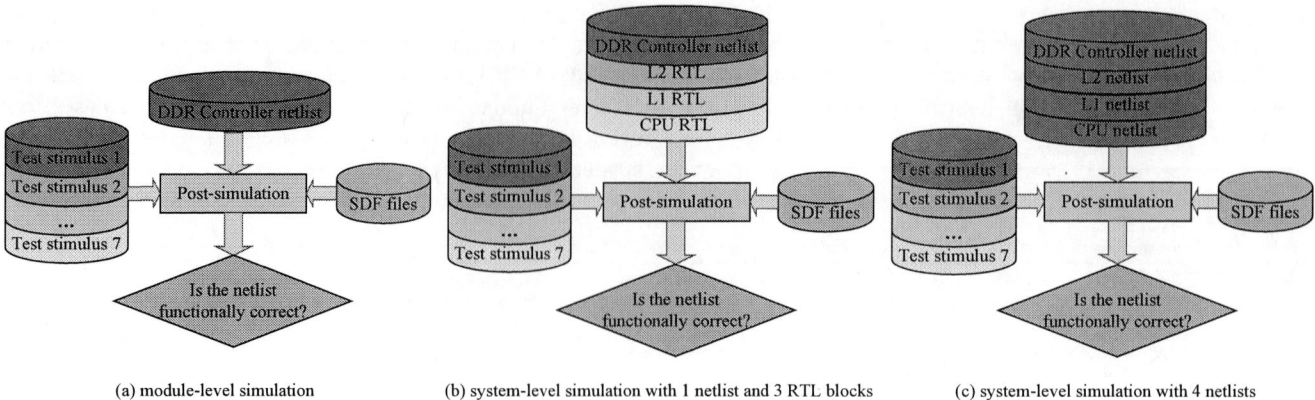

(a) module-level simulation   (b) system-level simulation with 1 netlist and 3 RTL blocks   (c) system-level simulation with 4 netlists

Fig. 3. The schematic diagram of different levels of the DDR controller post-simulation

netlists, primarily manifested in two key aspects, such as (i) simulation scale optimization: the RTL block requiring fewer simulation elements and less memory, (ii) performance acceleration: the simulation runtime demonstrating significant reduction. Therefore, we can allocate long-duration test cases to the system-level post-simulation environment with 1 netlist and 3 RTL blocks, while executing shorter test sequences in another system-level post-simulation setup, in order to accelerate the gate-level post-simulation convergence.

The purpose of the module-level (subsystem-level) post-simulation for the DDR subsystem circuits is to ensure that the functional module of the DDR subsystem meets timing and functional requirements before being integrated into the full SoC. The module-level post-simulation's advantages are mainly reflected in the following aspects: (i) Fast simulation speed. The simulation scale of the DDR subsystem is small, involving only a single subsystem and its direct interfaces, making the simulation speed several orders of magnitude faster than that of the full system. Engineers can quickly run a large number of test vectors. (ii) Extremely high debugging efficiency. Debugging issues are confined within the submodule. Its waveform files are smaller, and debugging tools respond more quickly, allowing engineers to rapidly locate and fix timing violations or functional errors. (iii) In-depth verification. More thorough and corner-case verification can be performed on the submodule, including simulations under various process corners, ensuring the robustness of the module. (iv) Reduced integration risk: Ensure that the DDR subsystem are qualified before integration to prevent defective modules from compromising the entire SoC system.

However, the limitations and challenges of the DDR module-level post-simulation mainly include: (i) The need to create an accurate testbench for the submodule to simulate the interactive behavior of its surrounding modules (such as the Core module, L2C module, etc.). (ii) The need to carefully define the input and output timing constraints of the submodule to ensure its interface behavior is consistent during system integration. (ii) The inability to fully and accurately simulate the interactive behavior of its surrounding modules.

The purpose of the system-level post-simulation for the DDR subsystem is to verify whether the DDR subsystem can correctly collaborate with surrounding modules after passing independent validation and to examine system-level timing paths (such as cross-clock domains, long paths, etc.). Its advantages are reflected in the following aspects: (i) Validation of system interactions and integration logic. It can identify system-level issues such as interface timing problems, protocol errors, and clock domain crossings between the DDR subsystem and the other modules. (ii) Real-world scenario validation. It enables running DDR tests closer to real application scenarios, such as high-speed data transfer, ensuring the DDR functions correctly under actual workloads in the SoC. (iii) Evaluation of the DDR performance under real timing conditions, such as DDR data transfer bandwidth. However, the challenges of the system-level post-simulation are: extremely slow simulation speed and highly difficult debugging.

The proposed hierarchical post-simulation methodology is conducted as follows. First, the functional verification of the DDR controller circuit netlist is performed through post-simulation to ensure the circuit subsystem-level correctness. Second, building on this, the mixed-system-level post-simulation is conducted by integrating the DDR controller netlist with the RTL code of surrounding modules (such as the Core modules, L2C modules) to validate overall system functionality. Finally, the system-level post-simulation is executed using the DDR controller netlist together with the netlists of surrounding modules to guarantee compliance with both timing and functional requirements. The proposed approach not only enhances the efficiency of hardware logic error localization and debugging, but also reduces the overall post-simulation time and improves the post-simulation efficiency.

IV. EXPERIMENTAL RESULTS

The mentioned above post-simulation framework in Section III is used to implement functional verification for the DDR controller in SoC design. The experimental environment is a high-performance server with the following specifications: (i)

979-8-3315-8850-2/25 $31.00 © 2025 IEEE

the Linux operation system of 3.10.0 x86_64 version, (ii) the 128 processors with 3.5 GHz, (iii) total memory size of 1024 GB, (iv) the GCC of 4.8.5 version (Red Hat 4.8.5-36), (v) an RISC-V toolchain.

A. Test Cases

We design 6 test cases to verify the DDR controller functions, including Targeted Write/Read, Random Write/Read, ECC (Error Check and Correction) Test, DBI (Data Bus Inversion) Test, CRC (Cyclic Redundancy Check) Test, and Low Power Test. These test cases can cover the functionality of the DDR controller circuit and enable the comprehensive post-simulation verification for the DDR controller circuit.

B. Results and Analysis

This section entails the simulation results and its analysis for the DDR controller gate-level post-simulation. We run the post-simulations with the designed 6 test cases by means of a simulation EDA tool. Table I lists the memory usage and the simulation instances for the module-level (Fig 3 (a)), the system-level V1 (Fig 3 (b)), the system-level V2 (Fig 3 (c)), respectively.

TABLE I
MEMORY USAGE AND INSTANCE COUNT COMPARISON

Metric	Module-level		System-level V1		System-level V2	
	Value	Ratio	Value	Ratio	Value	Ratio
Memory (GB)	14.70	1.00	16.80	1.14	51.10	3.48
Instances (M)	4.87	1.00	5.05	1.04	12.43	2.55

Note: Ratios are calculated relative to Module-level baseline values.

Compared to the module-level, the system-level V1 shows only a slight increase in memory usage and the number of simulation instances, with multipliers of just $1.14\times$ and $1.04\times$, respectively. In contrast, the system-level V2 exhibits a more significant rise, with memory usage increasing to $3.48\times$ and the number of simulation instances growing to $2.55\times$ relative to the module-level. This indicates that compared to the module-level, the system-level V1 only adds one CPU RTL block, one L1 RTL block, and one L2 RTL block, without significantly impacting the simulation scale or the memory usage. However, the system-level V2 incorporates three additional netlists (CPU, L1, and L2), which significantly impacts both simulation scale and memory consumption. The module-level implementation reduces the memory overhead by 71% and the simulation scale by 61% compared to the system-level V2. The system-level V1 achieves a 67% reduction in the memory overhead and a 59% decrease in the simulation scale compared to the system-level V2.

Table II and Table III summarize the run cycles and the run time of the module-level (Fig 3 (a)), the system-level V1 (Fig 3 (b)), the system-level V2 (Fig 3 (c)), respectively. In Table II, for the 6 test cases, the system-level V1 and V2 demonstrate increased cycle counts compared to the module-level due to the expanded simulation scale, with average multiplication factors of $1.89\times$ and $1.18\times$, respectively.

TABLE II
RUN CYCLES COMPARISON ACROSS DIFFERENT SYSTEM LEVELS

Testcase	Module-level		System-level V1		System-level V2	
	Cycles	Ratio	Cycles	Ratio	Cycles	Ratio
Targeted RW	72 248	1.00	139 730	1.93	81 588	1.13
Random RW	73 078	1.00	149 509	2.05	92 550	1.27
ECC	70 491	1.00	139 730	1.98	81 588	1.16
DBI	72 515	1.00	141 394	1.95	83 535	1.15
CRC	77 624	1.00	139 730	1.80	81 588	1.05
Low Power	89 151	1.00	149 482	1.68	114 178	1.28
Average	75 851	1.00	143 263	1.89	89 171	1.18

Note: Ratios are calculated relative to Module-level performance.

Table III shows that compared to the module-level implementation, the system-level V1 and V2 exhibit $3.57\times$ and $4.39\times$ longer average execution times, respectively. Compared to the system-level V2, the module-level and the system-level V1 implementations demonstrate 77% and 19% reductions in average execution time, respectively.

TABLE III
RUNTIME PERFORMANCE COMPARISON ACROSS DIFFERENT SYSTEM LEVELS

Testcase	Module-level		System-level V1		System-level V2	
	Time (h)	Ratio	Time (h)	Ratio	Time (h)	Ratio
Targeted R/W	2.82	1.00	9.32	3.30	11.95	4.24
Random R/W	2.90	1.00	10.68	3.68	13.12	4.52
ECC	2.80	1.00	10.43	3.73	11.73	4.19
DBI	2.77	1.00	10.08	3.64	12.00	4.33
CRC	2.85	1.00	10.30	3.61	11.52	4.04
Low Power	3.22	1.00	11.20	3.48	15.95	4.95
Average	2.89	1.00	10.34	3.57	12.71	4.39

Note: Ratios are calculated relative to Module-level performance.

We can conclude that the proposed multi-level gate-level post-simulation method not only speeds up the running time, but also reduces the memory usage during simulations.

Moreover, in the system-level post-simulation, the proposed method utilizes a backdoor approach to rapidly load the register configurations required for DDR initialization. This backdoor method can directly load register configuration values into registers in DDR controller and PHY modules. This bypass method avoids sending register configuration commands directly from the CPU to the DDR controller and PHY, thus can accelerate the DDR initialization process with a speedup ratio of $4.3\times$ for the system-level simulation of 1 netlist and 3 RTL blocks. Moreover, the proposed method with the backdoor approach obtains a speedup ratio of $11.2\times$ for the system-level simulation of all netlist blocks. The reason why the simulation acceleration ratio of full-netlist post-simulation is greater than that of the mixed netlist and RTL code post-simulation is that the full-netlist approach consumes more CPU time when sending register configuration commands from the CPU, compared to the mixed netlist and RTL code post-simulation.

V. CONCLUSION

In this paper, we introduce an SoC architecture with 2 DDR controllers and 4 CPU cores, and propose an efficient method for accelerating the DDR controller gate-level post-simulation, including the multi-level simulation method and the circuit netlist pruning method. The experimental results reveal that compared to the module-level implementation, the system-level (1 netlist and 3 RTL blocks) and the system-level (4 netlist blocks) exhibit 3.57× and 4.39× longer average execution times, respectively. The proposed multi-level gate-level post-simulation method achieves a 4× speedup in functional post-simulation for the DDR controller circuit in SoC design. The designed method is also significant for the post-simulation of netlists in other large-scale peripheral interface circuits, such as Peripheral Component Interconnect Express (PCIe) and Universal Chiplet Interconnect Express (UCIe).

REFERENCES

[1] Zhengyi Zhang, Yuanda Yang, and Lingli Wang. Rabbit: An efficient verification platform base on virtual peripherals. In *2023 IEEE 15th International Conference on ASIC (ASICON)*, pages 1–4. IEEE, 2023.

[2] Yande Jiang, Nicoleta Cucu Laurenciu, He Wang, and Sorin Dan Cotofana. Graphene nanoribbon based complementary logic gates and circuits. *IEEE Transactions on Nanotechnology*, 18:287–298, 2019.

[3] Zizheng Guo, Zuodong Zhang, Xun Jiang, Wuxi Li, Yibo Lin, Runsheng Wang, and Ru Huang. General-purpose gate-level simulation with partition-agnostic parallelism. In *2023 60th ACM/IEEE Design Automation Conference (DAC)*, pages 1–6. IEEE, 2023.

[4] Kriti Sundar Das, Padmini Prakash, and Abhimanyusinh Zala. Accelerating GLS simulation closure in DFT with emulator. In *2021 IEEE International Test Conference India (ITC India)*, pages 1–6. IEEE, 2021.

[5] Haichuan Hu, Zichen Xu, Yuhao Wang, and Fangming Liu. Fast and scalable gate-level simulation in massively parallel systems. In *2023 IEEE/ACM International Conference on Computer Aided Design (ICCAD)*, pages 1–9. IEEE, 2023.

[6] Yanqing Zhang, Haoxing Ren, and Brucek Khailany. Opportunities for RTL and gate level simulation using GPUs. In *Proceedings of the 39th International Conference on Computer-Aided Design*, pages 1–5, 2020.

[7] Weijie Fang, Yanggeng Fu, Jiaquan Gao, Longkun Guo, Gregory Gutin, and Xiaoyan Zhang. Acceleration of timing-aware gate-level logic simulation through one-pass GPU parallelism. *IEEE Transactions on Computers*, pages 1–12, 2025.

[8] Debapriya Chatterjee, Andrew Deorio, and Valeria Bertacco. Gate-level simulation with gpu computing. *ACM Transactions on Design Automation of Electronic Systems (TODAES)*, 16(3):1–26, 2011.

[9] Niek Moonen, Frits Buesink, and Frank Leferink. Enhanced circuit simulation using mutual coupling parameters obtained via 3D field extraction. In *2016 Asia-Pacific International Symposium on Electromagnetic Compatibility (APEMC)*, volume 1, pages 181–183. IEEE, 2016.

[10] Tariq B Ahmad and Maciej J Ciesielski. Fast STA prediction-based gate-level timing simulation. In *2014 Design, Automation & Test in Europe Conference & Exhibition (DATE)*, pages 1–6. IEEE, 2014.

[11] Yici Cai, Weiping Liu, and Zhenya Zhou. A hybrid simulation-based parallel post-simulation acceleration algorithm for circuits. *Journal of Computer-Aided Design & Computer Graphics*, 28(11):1–5, 2016.

[12] Yande Jiang, Na Chen, and Huiquan Wang. A speed up method towards ddr subsystem functional verification in SoC. In *2023 IEEE 15th International Conference on ASIC (ASICON)*, pages 1–4. IEEE, 2023.

[13] Yong Liu and Jie Chen. Application of static timing analysis tools in ddr memory interface timing closure and post-simulation. *Integrated Circuit Applications*, 36(8):20–22, 2019.

2025 The 10th International Conference on Integrated Circuits and Microsystems

A High-Precision Current Detection Circuit for Battery Management System Chip in Electric Vehicle

Zhiwei Li
School of Integrated Circuits
Tsinghua University
Beijing National Research Center
for Information Science and
Technology
Beijing, China
lizw23@mails.tsinghua.edu.cn

Liji Wu *
School of Integrated Circuits
Tsinghua University
Beijing National Research Center
for Information Science and
Technology
Beijing, China
lijiwu@mail.tsinghua.edu.cn

Dan Xie *
School of Integrated Circuits
Tsinghua University
Beijing National Research Center
for Information Science and
Technology
Beijing, China
xiedan@mail.tsinghua.edu.cn

Kunning Mao
School of Electronic Engineering
Heilongjiang University
Harbin, China
18845594228@163.com

Haifeng Chen
School of Integrated Circuits
Tsinghua University
Beijing National Research Center
for Information Science and
Technology
Beijing, China
chenhf24@mails.tsinghua.edu.cn

Zhilin Zhong
School of Integrated Circuits
Tsinghua University
Beijing National Research Center
for Information Science and
Technology
Beijing, China
zhongz24@mails.tsinghua.edu.cn

Abstract—**This paper proposed a current detection circuit used for battery management system (BMS). It was designed with 180nm BCD（Bipolar CMOS DMOS）technology. The circuit consists of a programmable gain amplifier (PGA) and a 24-bit high-precision Sigma-Delta ADC. The PGA adopts a two-stage fold-cascode structure with negative feedback resistor circuit to achieve the function of adjustable gain. Sigma-Delta ADC consists of a modulator and a digital filter, which adopts the multi-bit quantization to improve the accuracy of the current detection. The modulator adopts a third-order cascaded integrator feedforward structure. The simulation results show that the gain of PGA maintains stable at 20-44dB. For the Sigma-Delta ADC，the signal-to-noise distort ratio（SNDR）is 122.4dB，the signal bandwidth（BW） is 4KHz and the effective number of bits（ENOB）is 20.04bits at a sampling frequency of 1.024MHz. The simulation results of the whole circuit shows that the value of the current detection error is only 0.56% in the current dynamic range of 1mA to 500mA which means this circuit shows great potential in the BMS chips.**

Keywords—Battery Management System, Current Detection, Sigma-Delta ADC, Programmable gain amplifier

I. INTRODUCTION

With the rapid development of new energy vehicle industry, the safety and stability performance of the power batteries attracted much attentions.[1] Battery management system（BMS） is the core of the power batteries to keep the vehicles working safely.[2] One of the crucial tasks of BMS chips is to monitor key parameters such as battery charging or discharging current and motor operating current in real time and provides data support for the BMS[3], ensuring that the current does not exceed the safety threshold and preventing battery thermal runaway or component damage caused by

overcurrent and short circuit. Therefore, accuracy has always been one of the greatest challenges faced by current detection circuits.[4] The current detection circuit is illustrated in Fig.1. A programmable gain amplifier（PGA）is utilized to realize full-scale conversion and high accuracy before connecting to ADC.

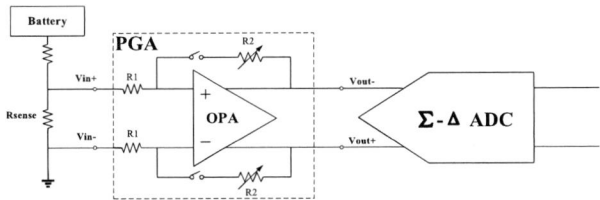

Fig. 1. System structure of current detection circuit

In new energy vehicles, the voltage range of a single battery cell differs between 1.0V and 4.2V. The high-voltage power management circuit provides a 5V supply voltage for the circuit. Regarding the bandwidth requirements and power consumption, 1.024 MHz of clock frequency for the ADC is chosen in this design.

The rest parts of this paper are as follows: Section II describes the fundamental architecture of the current-sensing circuit, Section III details the circuit implementation including operational amplifiers (OPAs) and a 4-bit quantizer, Section IV presents simulation results comprising PGA and sigma-delta ADC performance characteristics along with current detection error tables, and Section V concludes with the key findings and contributions of the proposed design.

979-8-3315-8850-2/25 $31.00 © 2025 IEEE

II. STRUCTURE OF THE CIRCUIT

A. PGA

As shown in Fig.2, the PGA consists of an OPA and a 2-to-4 decoder. The decoder and transmission gate switches are utilized to control the programmable selection of the value of resistor through SW[0:3]. The RF resistor values are set at multiples of RS, specifically 1x, 4x, 16x, and 64x, respectively. The gain of the overall circuit is determined by the negative feedback of the resistors as shown in equation (1).

$$GAIN = \frac{RF}{RS} \tag{1}$$

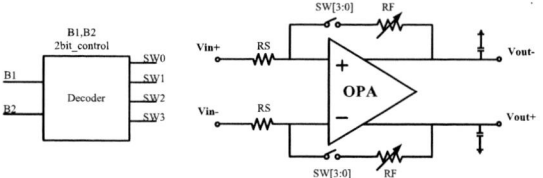

Fig. 2. System structure of the PGA

B. Sigma-Delta ADC

The ADC adopted in this paper is a Sigma-Delta ADC, which consists of a modulator and a digital filter.

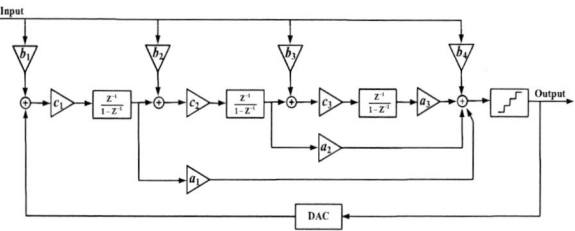

Fig. 3. System structure of modulator.

Fig.3 presents the block diagram of the third-order single loop, 4-bit quantization Sigma-Delta ADC. When considering system stability, Lee's criterion is adopted as an empirical rule, which states that the system can remain stable when the maximum out-of-band gain of the noise transfer function satisfies $\max(|NTF(e j \omega)|) \leq 1.5$.[5] To ensure the stability of the modulator system, the out-of-band gain of the NTF is set to 1.5, and the coefficients are designed accordingly. According to the transfer function of the modulator and combined with the NTF expression, use the Synthesize NTF function to solve the coefficients.[6] The expression of the modulator's transfer function is as follows:

$$H(Z) = \frac{(Z-1)^3}{(Z-0.67)(Z^2 - 1.53Z + 0.67)} \tag{2}$$

The system structure block diagram of the digital extraction filter is shown in Fig.4. It consists of a cascaded integrator comb (CIC) filter, a cascaded integrator comb compensation (CICC) filter, and a half band (HB) filter. For the first stage, the CIC filter is selected to achieve 32 times extraction, while introducing passband attenuation problem. To address this issue, a CICC filter is utilized as the second stage. An HB filter is adopted as the third stage, which narrows the transition bandwidth of the overall digital decimation filter while also offering two-fold decimation capability.

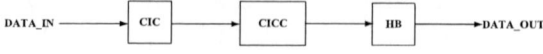

Fig. 4. Structure of digital filter.

III. CIRCUIT DESIGN.

A. OPA

The operational amplifier circuit is shown in Fig.5. PMOS transistors are selected as input pairs to obtain better frequency and noise performances, compared to NMOS transistors. The operational amplifier circuit consists of two stages. The first stage utilizes a folded cascode circuit structure to increase the gain of the operational amplifier. The second stage employs a common-source amplifier to amplify the output swing. Miller compensation circuit is used to compensate for zero and pole effects.

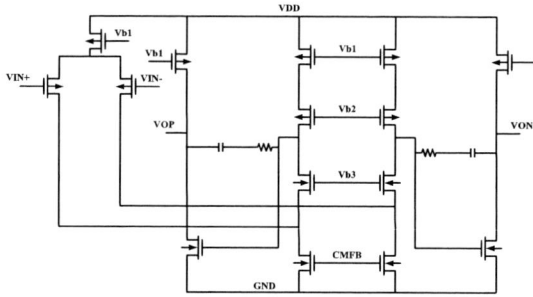

Fig. 5. Schematic of OPA.

B. Modulator

The modulator adopts a third-order cascaded integrator feedforward structure. The structure of the Sigma-Delta ADC is shown in Fig. 6, including the key modules such as the bootstrap switch circuit of gate voltage, clock circuit, switch capacitor integrator and 4-bit quantization circuit, data weighted average (DWA) circuit and other modules.

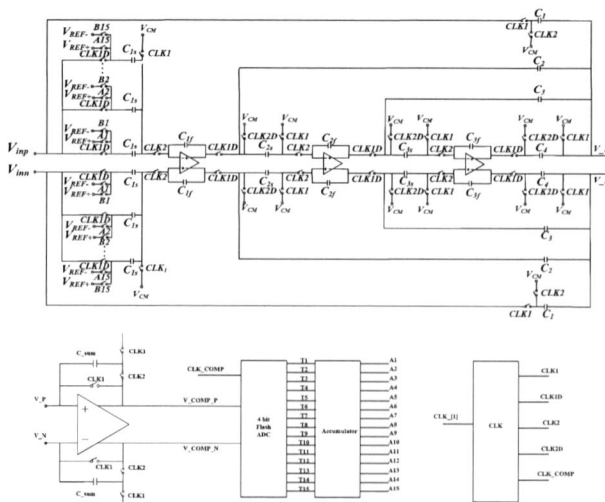

Fig. 6. Schematic of the modulator.

979-8-3315-8850-2/25 $31.00 © 2025 IEEE

C. Multi-bit Quantification

Most quantizers typically select one-bit quantization, [7] but this design uniquely adopts multi-bit quantization. It effectively reduces the basis of quantization noise and expands the range of effective dynamics, thereby improving the accuracy of current detection. Here, we choose 4-bit quantification. In the future, we can consider further increasing the number of bits.

The quantizer structure is shown in Fig.7. It consists of 15 comparator units, where an external reference voltage is divided by a resistor network to generate reference voltages for each comparator. These reference voltages are compared with the results from the signal summation circuit by the 15 comparators, ultimately producing a 15-bit thermometer code.

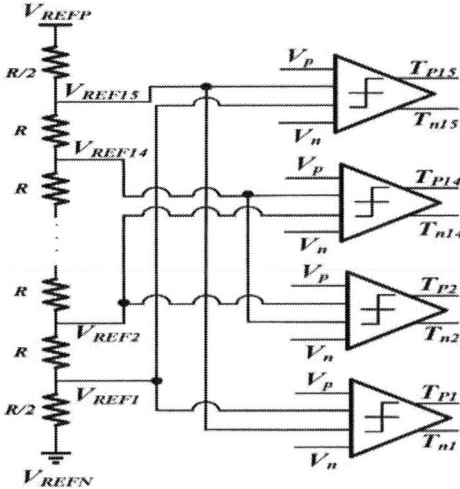

Fig. 7. Schematic of multi-bit Flash ADC.

This thermometer code is then shifted by the DWA module and fed back as the input signal to the feedback DAC. Simultaneously, the 15-bit thermometer code is converted into a 4-bit digital code through a 15-4 encoder, which serves as the input to the digital decimation filter. Due to the noise-shaping effect of the modulator's feedback loop, the modulator has relatively low requirements on the quantizer's precision.

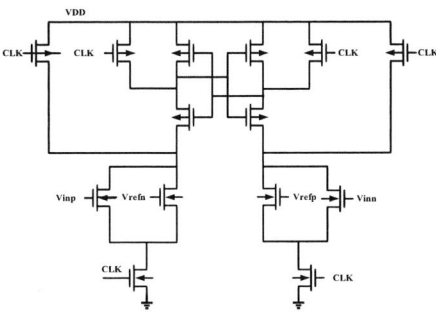

Fig. 8. Schematic of the comparator.

Fig.8 illustrates the fully differential 4-input dynamic comparator and latch structure employed in this design. Its primary function is to compare the magnitude difference between (Vinp-Vinn) and (Vrefp-Vrefn), then latch the result until the next comparison result replaces the current latched

value. This comparator architecture only draws current during the brief comparison triggering moment and remains current-free during the holding period, thus offering significant low-power advantages.

D. DWA

The DAC circuit converts the digital code output by the quantizer into an analog signal and feeds it back to the modulator's input, where it is subtracted from (or added to) the input signal to serve as the first integrator's input—hence being referred to as the feedback DAC. Since generating additional ±Vref voltages through extra circuitry would consume unnecessary power and area, VDD and GND are respectively adopted as the positive and negative reference voltages (±Vref).

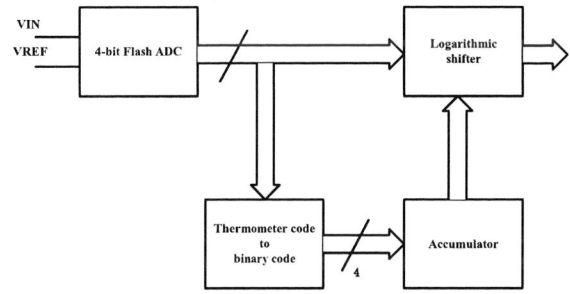

Fig. 9. The DWA circuit structure.

The DWA circuit structure is shown in Fig.9. It mainly consists of a thermometer-to-binary code converter, an accumulator and a logarithmic shifter. The main function of the DWA circuit is to perform bit-shifting operations on the 15-bit thermometer code generated by the quantizer, enabling the cyclic reuse of sampling capacitors in the feedback DAC circuit. This mechanism helps mitigate the impact of capacitor mismatch on the modulator's performance, thereby improving system accuracy.

As shown in Fig.9, the DWA first converts the thermometer code into a binary code, which serves as the control signal of the accumulator and is added to the binary code of the previous clock cycle retained by the accumulator to update the pointer. The function of an accumulator in a DWA circuit is as a pointer to control the shift operation of a logarithmic shifter. The input of the logarithmic shifter is the thermometer code output by the Flash ADC during this clock cycle. Under the instruction of the accumulator, the shifted thermometer code is output as the input control signal for the feedback DAC

E. Clock Circuit

The input signals for the clock circuit are typically composed of a timer and an oscillator. The oscillator generates a stable and precise signal frequency, while the timer produces clock pulses or pulse sequences based on the oscillator's output signals.

In this design, the Sigma-Delta modulator requires multi-phase clock signals to control the timing sequences of its various functional modules. The architecture of the clock circuit is illustrated in Fig.10.

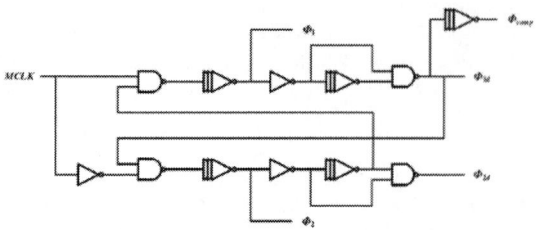

Fig. 10. Schematic of the clock circuit.

The modulator requires the following critical clock phases: two-phase non-overlapping clocks - essential for proper operation of switched-capacitor integrators within the modulator; falling-edge-delayed versions of the two-phase clocks - designed to mitigate clock feedthrough and charge injection effects in sampling switches; Reset-and-regeneration clocks - Specifically for dynamic comparator operation.

IV. SIMULATION RESULT

A. PGA

Fig.11 illustrates the frequency response of the programmable gain amplifier in the range of 20-44dB. The simulation results indicate that the bandwidth is greater than 100 kHz for all gain settings，which demonstrates that the amplifier exhibits good frequency characteristics across the entire gain range

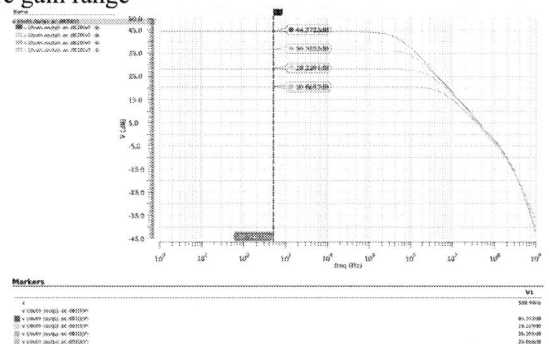

Fig. 11. Simulation results of PGA.

B. DWA

The simulation results of the DWA module are shown in Fig.12.

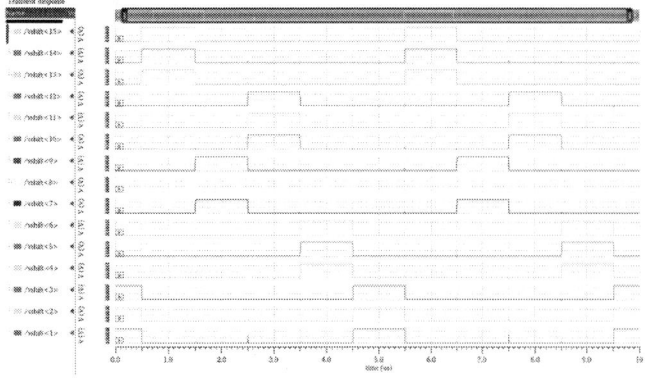

Fig. 12. Simulation results of DWA.

During the DWA simulation, when the input thermometer code is set to 15'b0000_0000_1111, it can be observed that the output cycles through different 3-bit segments of the thermometer code, sequentially raising distinct sets of 3 bits. This cycling mechanism effectively mitigates the impact of capacitor mismatch on the modulator's performance.

C. Digital filter

The simulation results of the CIC filter model are shown in Fig.13. The amplitude of the first side lobe decreased by approximately 52dB compared to the first-order filter, demonstrating a stopband attenuation of approximately 52dB, which will be compensated by subsequent-stage filters.

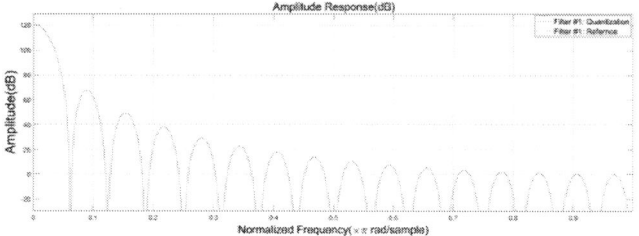

Fig. 13. Simulation results of CIC filter.

As illustrated in Fig.14, the compensation filter primarily serves to offset the significant roll-off of the CIC filter. The simulation results indicate that the compensation filter provides gain enhancement, achieving a stopband attenuation of about 58dB.

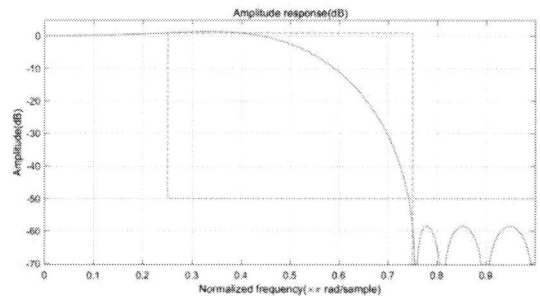

Fig. 14. Simulation results of CICC filter.

The half-band filter simulation results are presented in Fig.15, showing a passband ripple of approximately 0.02dB, a passband cutoff frequency of 3.2 kHz, a stopband edge frequency of 4.8 kHz, and a stopband attenuation of 62dB.

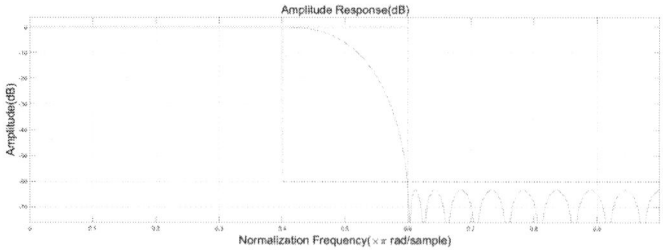

Fig. 15. Simulation results of HB filter.

D. Sigma-Delta ADC

Fig. 16 shows the simulation results of Sigma-Delta ADC, and the proposed Sigma-Delta ADC achieves an ENOB of

20.04 bits, and SNDR of 122.4 dB under a 1.024MHz sampling rate with a 4KHz sinusoidal input.

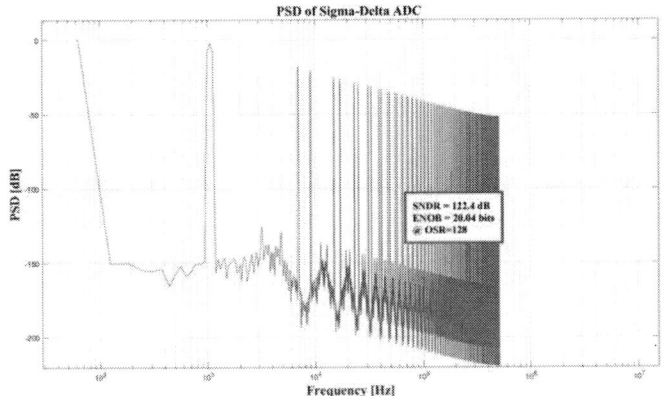

Fig. 16. Simulation results of ADC

The formula for the current detection error is as follows:

$$error = \frac{|V_{out} - V_{in}|}{V_{in}} \times 100\% \qquad (3)$$

V_{out} represents the result of current detection and V_{in} represents the input current which is the current flowing through resistance Rsense.

Compared with other published current detection circuit, this design shows good performance in the dynamic range and current detection error. This comparison is shown in Table 1.

TABLE I. COMPARISON OF THE PERFORMANCE

Reference	[8]	[9]	[10]	This paper
Technology	180nm CMOS	180nm CMOS	250nm CMOS	180nm CMOS
Supply(V)	2	3.3	5	5
Dynamic range	0.3mA-300mA	1nA-0.1mA	0.44A-2.2A	1mA-500mA
Current detection error	4.88%	0.88%	0.95%	0.56%

V. CONCLUSION

This paper introduces a high-precision current detection circuit for BMS chips in EV. A high precision sigma-delta ADC is proposed to improve the accuracy of current detection which uses multi-bit quantification to reduce quantization noise and increase the stability of the system.

This circuit is designed with 180nm BCD technology. The power supply voltage is 5V. The current detection error of this design is only 0.56% in the current dynamic range of 1mA to 500mA. Therefore，this current detection circuit has great potential in the BMS chips in electric vehicles.

REFERENCES

[1] A. Johnson, N. A. Varghese, R. Raj Thomas, S. Mathews and R. Alice Koshy, "Random Forest Regressor Based SoC Estimation of Li-ion Battery for Electric Vehicle Applications," 2023 IEEE International Conference on Power Electronics, Smart Grid, and Renewable Energy (PESGRE), Trivandrum, India, 2023, pp. 1-8.

[2] S. Liu, Y. Feng, W. Weng, et al. Contactless Measurement of Current and Mutual Inductance in Wireless Power Transfer System Based on Sandwich Structure. IEEE Journal of Emerging and Selected Topics in Power Electronics, 2022, 10(5):6345-6357.

[3] H. Tian, F. Zhao, X. Fu, et al. A precision current sensing circuit with chopper amplifier of symmetric topology. Microelectronics Journal, 2024, 146: 106103.

[4] Q. Zhu, Q. Li, Y. Li, et al. 16-Cell stackable battery monitoring and management integrated circuit for electric vehicles. Microelectronics Journal, 2023, 136: 105782.

[5] L. R. Fang, Y. J. Li, Y. Zhang, et al. A 130μW Three-Step DT Incremental ΔΣ ADC Achieving 107.6dB DR and 99.3dB SNDR with Zoom and Extended-Range Counting. ESSCIRC 2022- IEEE 48th European Solid State Circuits Conference, 2022: 554-557.

[6] R. Pagano, M. Baker, R. Radke, "A 0.18-μm Monolithic Li-Ion Battery Charger for Wireless Devices Based on Partial Current Sensing and Adaptive Reference Voltage," IEEE Journal of Solid-State Circuits, 2012, 47(6): 1355-1368.

[7] V. Vulligaddala, S. Vernekar, S. Singamla, et al., "A 7-Cell, Stackable, Li-Ion Monitoring and Active/Passive Balancing IC With In-Built Cell Balancing Switches for Electric and Hybrid Vehicles," IEEE Transactions on Industrial Informatics, 2020, 16(5): 3335-3344.

[8] Y. Chen, X. Liu, Q. Fan, et al. A High-Precision Current Sensing Front-End IC Design for Wide-Range Dynamic Current Detection. IEEE International Conference on Anti-counterfeiting, Security, and Identification, 2023: 157-162.

[9] V. Gogolou, T. Laopoulos and S. Siskos, "A wide range current sensing technique for integrated DC-DC converters," 2022 37th Conference on Design of Circuits and Integrated Circuits (DCIS) , pp. 01-04, 2022.

[10] Wang, C.-C., Hou, Z.-Y., Lu, W.-J. and Wang, S.-S. (2016), High-voltage on-chip current sensor design and analysis for battery modules. IET Circuits Devices Syst., 10: 492-496.

2025 The 10th International Conference on Integrated Circuits and Microsystems

A Robust Startup Circuit for Current-Mode Bandgap Reference

Qi Wang
Information and Navigation School
Air Force Engineering University
Xi'an, China
jusila@163.com

Hao Li*
Information and Navigation School
Air Force Engineering University
Xi'an, China
lihao13lb@163.com
*Corresponding author

Chunlei Pang
Information and Navigation School
Air Force Engineering University
Xi'an, China
chunleipcl@163.com

Abstract—**This paper presents a robust startup circuit for current-mode bandgap reference (CMBGR), which guarantees the startup of CMBGR under PVT variations meanwhile maintaining the features of CMBGR, i.e., the tunable output voltage with little effects on its TC characteristics. Simulation results show the proposed startup circuit can effectively force the CMBGR into its normal state during power on according to 20400 Monte Carlo runs with different PVT settings. The designed CMBGR achieves tunable output reference from 0.1V to 2V at 8.7ppm/℃ under voltage supply from 2.3V to 4.3V.**

Keywords—current-mode bandgap reference; startup circuit; PVT; wide temperature range.

I. INTRODUCTION

Bandgap reference (BGR) is a well-known supporting module to generate stable voltage, which is independent of temperature and other external factors, for other functional modules, e.g. analog, mixed-signal and RF building blocks. As a reference voltage, it is essential to be immune to different supply voltage fluctuations or temperature changes to provide clear as well as robust value to enhance the overall performance of the chip across different voltage and temperature situations. Thus, BGR is one of the fundamental modules of SoCs. With the advancement of CMOS technology, supply voltage is getting lower than the typical value of 1.25V for a conventional BGR structure, to attain the minimum temperature coefficient (TC) leading to deteriorated TC performance. The introduction of current-mode BGR (CMBGR) achieves a lower than typical output voltage while keeping the minimum TC property of the BGR [1]. A typical structure of such CMBGR is shown in Fig.1. V_{ref} can be expressed as:

$$V_{ref} = I_{ref} * R_4 = \frac{R_4}{R_2}\left(V_{BE1} + \frac{R_2}{R_1}V_T \ln n\right), \quad (1)$$

where V_{BE1} stands for the base emitter voltage drop of Q_1, V_T is the thermal voltage, n is the size ratio of triodes Q_2 and Q_1. It is obvious from (1) that V_{ref} is not necessarily to be fixed at the typical bandgap value rather it is adjustable via R_4/R_2.

Compared with conventional voltage-mode bandgap reference, the generation of a reference voltage V_{ref} from I_{ref} has three main advantages. First, programmable references with temperature independent property can be obtained from the reference generating branch, unlike voltage-mode bandgap reference, which requires another buffer to generate voltage other than 1.25V. Second, it is easy to generate voltage references higher than 1.25V as long as

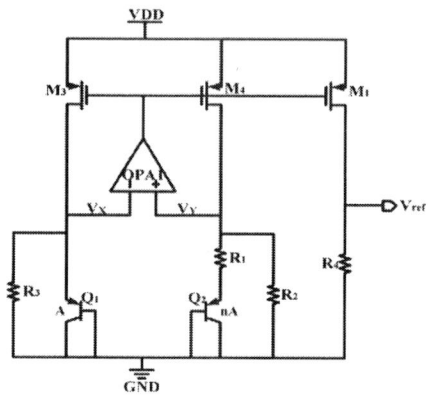

Fig. 1. A typical Current-mode bandgap reference mode bandgap reference structure.

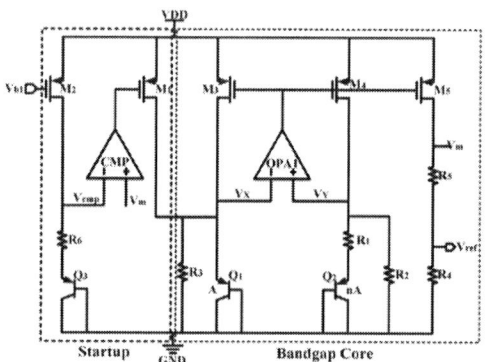

Fig. 2. CMBGR with proposed startup circuit

headroom allows, whereas voltage mode references require the use of a V-to-I converter to generate voltages above 1.26V. In addition, compared with the voltage mode bandgap reference, no additional resistive elements are required by the branch that generates I_{ref}, which means that there is no additional area added compared to the voltage mode bandgap reference.

The advantage of setting output voltage independent of the TC in CMBGR mentioned above comes at the cost of extra nonzero failure state during startup in addition to the zero failure state. Thus, special consideration is needed to force the BGR get into its normal operation after power on. Most existing startup circuits [2]-[4] are aimed to prevent the BGR from falling into zero state while do not guarantee the

979-8-3315-8850-2/25 $31.00 © 2025 IEEE

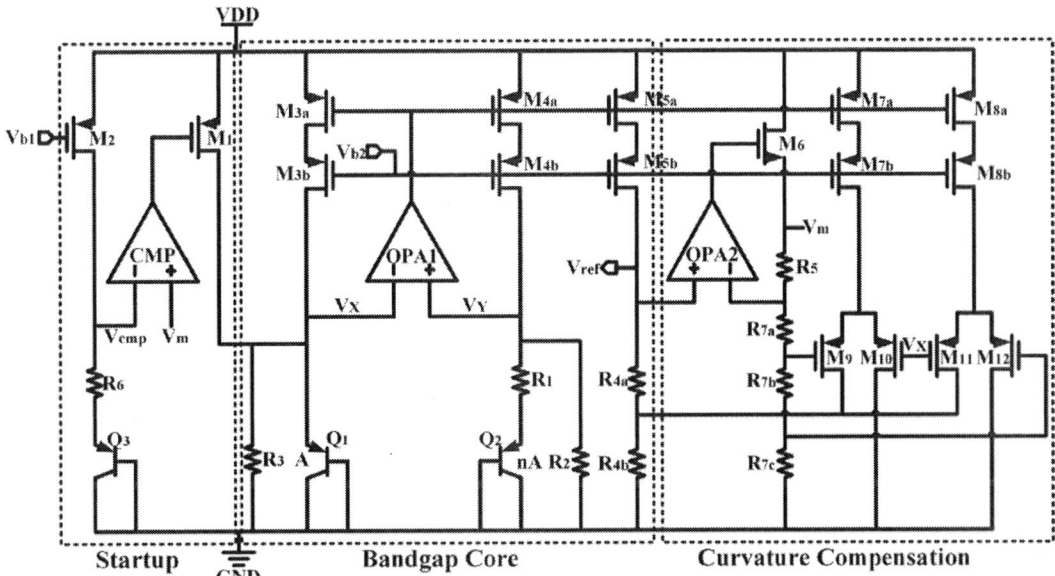

Fig. 3. An improved version of the proposed CMBGR structure

stable startup of the CMBGR. Several startup circuits [5]-[7] try to mitigate the extra startup condition, but introducing other disadvantages such as limited V_{ref} value, vulnerable to process variations or the assistance of reset signal. A modified CMBGR proposed in [8] handles the startup issues well at the cost of degraded TC. A high-order curvature-compensated BGR presented in [9] achieves higher precision with one resistor. Thus, a robust startup circuit covering wide range of process, voltages and temperature (PVT) variations without disturbing the CMBGR core circuit is still in need. This paper proposes a novel startup circuit, which guarantees the startup of CMBGR with PVT coverage meanwhile maintaining the features of CMBGR, i.e., the tunable output voltage with little effects on its TC characteristics.

The paper is organized as follows, the startup failure states are analyzed in Section II; the proposed startup circuit is presented in Section III; the simulations results are given in Section IV and concludes remarks are drawn in Section V.

II. CMBGR STARTUP STATES

CMBGR undergoes three types of operating states after power on:

- Type-I: normal operating state, the circuit operates normal, i.e. it provides required reference voltage.

- Type-II: zero failure state, no current goes through Q_1 and Q_2 branches.

- Type-III: nonzero failure state, extremely weak current goes through triode Q_1 and Q_2.

In normal state, ΔV_{BE} is equal to $V_T \ln n$, which is much larger than the mismatch voltage V_{os} of the OPA1, i.e., $V_x - V_y$. The emitter current of the BJTs can simplified to:

$$\frac{\Delta V_{BE} - V_{OS}}{R_1} \approx \frac{V_T \ln n}{R_1}. \tag{2}$$

With the above, V_{ref} can be determined by (1).

The zero failure state could potentially lead to the nonzero failure state during startup. During power on, both BJTs are in cutoff. If a current is injected into R_3, V_x is established and M_3 and M_4 are turning on, leading to an increase in V_{ref}. Normally, the current injection will be disabled when V_{ref} reaches to a preset threshold. If the injection is stopped before fully turning on the BJTs, the CMBGR may enter Type-III state. Taking the mismatch of OPA1 into consideration and assuming the emitter leakage current of Q_2 is n times of Q_1 and R_2 equals to R_3, the emitter leakage I_{e,Q_1} increases with the gradually rise of V_x, and finally reaches to a value as in (3). The circuit will be stuck at this state due to the negative feedback formed by OPA1, M_3 and M_4, i.e., Type-III state is established if there is no other external intervention.

$$\frac{V_{BE1}}{R_3} + I_{e,Q_1} = \frac{V_{BE1} - V_{os}}{R_3} + nI_{e,Q_1} \rightarrow I_{e,Q_1} = \frac{V_{os}}{(n-1)R_3}. \tag{3}$$

Type-III state is caused by the process mismatch of OPA1, which could be readily observed in Monte Carlo simulations. It should be noted that Type-III state has certain degrees of randomness. Because it is established at a condition that both Q_1 and Q_2 are in cutoff, a small variation of I_{e,Q_1} due to V_{OS} variation affected by process might require significant change of V_{BE1}, i.e., V_x, which results a certain randomness of V_{ref} at Type-III state.

To avoid Type-III state, CMBGR should be built with a mechanism to detect it, then intervene by continuously injecting current into the Q_1 branches until both BJTs are fully turned on. In this way, CMBGR can be safely started where the ΔV_{BE} increases to a fixed value that much larger than V_{OS} and the desired V_{ref} is obtained.

Most startup circuits detect the failure state through comparing V_{ref} with a preset voltage threshold. In Type-III state, V_{ref} is sensitive to the process and temperature variations, showing significant randomness due to the mismatch. A fixed threshold may not guarantee the success startup under wide PVT range. To solve this issue, a startup

979-8-3315-8850-2/25 $31.00 © 2025 IEEE   227

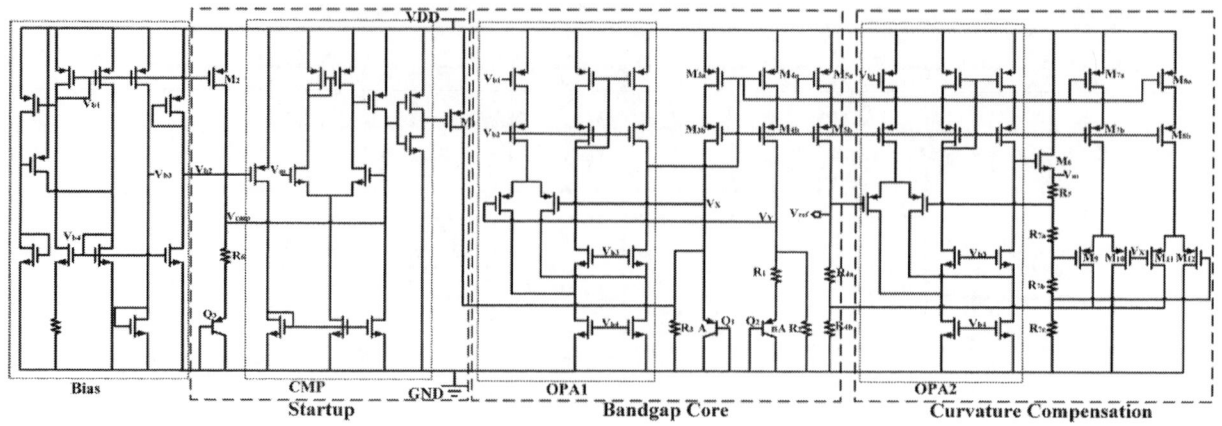

Fig. 4. Transistor-level structure of improved CMBGR

TABLE I. PERFORMANCE COMPARIOSN WITH OTHER WORKS

Parameter	[10]	[11]	[12]	[13]	This work
Technology(nm)	180	180	350	500	110
Supply voltage(V)	1.5 to 6	1.1	1.4	2.5 to 9	2.3 to 4.3
Temperature Range (°C)	-25 to 80	-15 to 140	-20 to 100	-30 to 75	-40 to 130
V_{ref} (V)	1.19	0.755	0.86	0.977	0.1~2
PSRR (dB)@dc-100HZ	125	9	68	98	130
@100KHZ	60	-	-	43	70
TC (ppm/°C)	6.5	20	12.4	19	8.7
Start-up Time(us)	-	-	-	-	200

*The worst case for all PVT conditions at 0.1V V_{ref} setting

circuit insensitive to the process and temperature variations is proposed to accurately detect the states during the startup without changing the structure of core circuit.

III. PROPOSED STARTUP CIRCUIT

The proposed startup circuit is shown in Fig. 2. It uses a voltage V_{cmp} with the same temperature trend as V_{BE} to control the current of Q_1 branch during startup. When V_{cmp} is higher than V_m, which is proportional to V_{ref}, the output of CMP goes low to allow additional current flow into Q_1 branch for starting the CMBGR. Once started, V_{cmp} becomes lower than that of V_m, the startup goes into idle.

A. Startup process of the proposed circuit

In Fig.2, V_{b1} sets the bias voltage of M_2, so that M_2 can be used as a current source to provide stable current for R_6 and Q_3, thus the temperature coefficient of V_{cmp} is negative. V_m is proportional to V_{ref} with a ratio of $(1 + R_5/R_4)$. M_3, M_4, M_5 are of the same size and R_2 and R_3 have the same value.

From (1), when the temperature coefficient of V_{ref} is zero, following relation should hold

$$V_{band} = V_{BE1} + (R_2/R_1)V_T \ln n \approx 1.25V. \tag{4}$$

Assume that $V_{BE,w1}$ is the V_{BE} value of Q_1 when CMBGR operates at normal state, and V_{r1} and V_{m1} are respectively the values of V_{ref} and V_m at normal operation, while V_{r2} and V_{m2} respectively stand for the values of V_{ref} and V_m at Type-III state. As discussed in Section II, the currents through Q_1,

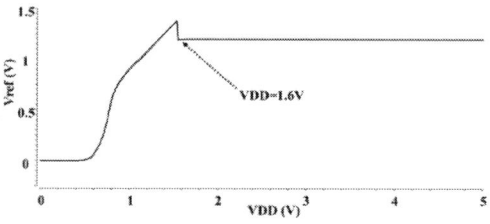

Fig. 5. Overdrive voltage limitation on V_{ref} via DC scanning

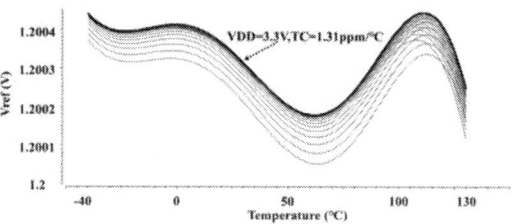

Fig. 6. V_{ref} with temperature at 3.3±1V

Q_2 at the Type-III state are very small. Furthermore, V_{BE} of Q_1 at this condition is smaller than $V_{BE,w1}$. We can easily derive the followings:

$$V_{r1} = (R_4/R_2)V_{band}, \tag{5}$$

$$V_{r2} < (R_4/R_2)V_{BE,w1}, \tag{6}$$

$$V_{cmp} = V_{BE3} + I_{M2}R_6, \tag{7}$$

979-8-3315-8850-2/25 $31.00 © 2025 IEEE

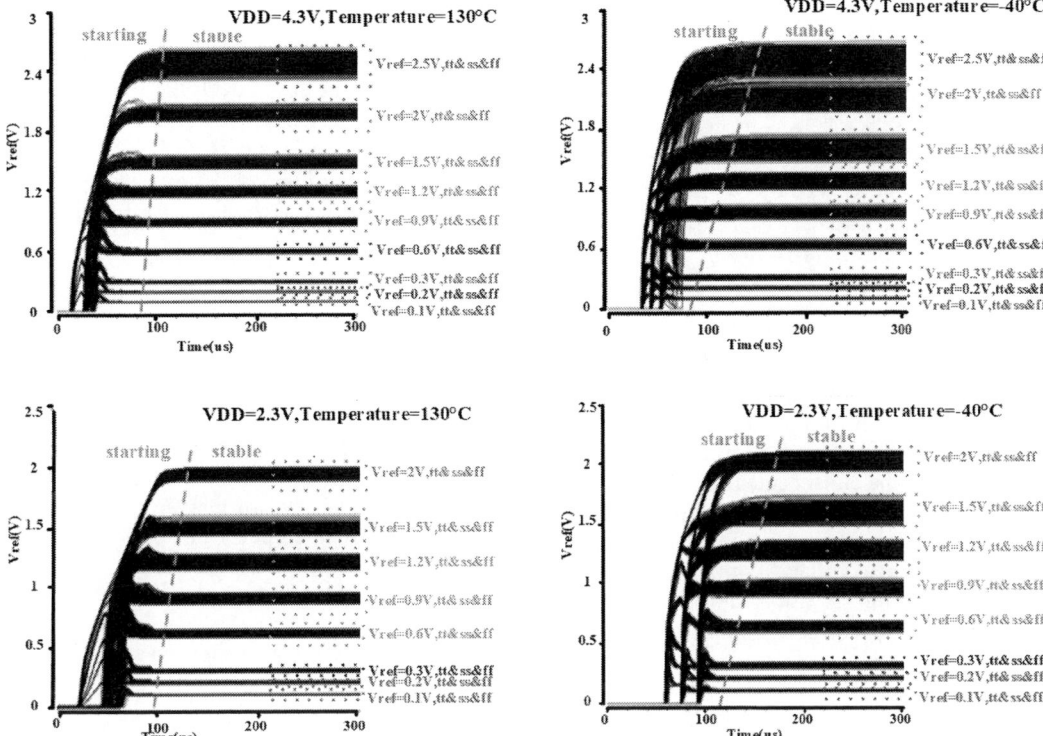

Fig. 7. Monte Carlo transient simulation at different V_{ref} and extreme PVT cases

$$V_{m1} = \left(1 + \frac{R_5}{R_4}\right)(R_4/R_2)V_{band}, \quad (8)$$

$$V_{m2} = \left(1 + \frac{R_5}{R_4}\right)V_{r2} < \left(1 + \frac{R_5}{R_4}\right)\frac{R_4}{R_2}V_{BE,w1}, \quad (9)$$

where $V_{BE,w1}$ and V_{cmp} are both negative temperature coefficient voltages with the same slope. With R_4/R_2 fixed, we can adjust R_5, R_6, R_7 and M_7 such that the following conditions hold in the required temperature range:

$$\left(1 + \frac{R_5}{R_4}\right)\frac{R_4}{R_2}V_{band} > V_{BE3} + I_{M2}R_6 > \left(1 + \frac{R_5}{R_4}\right)\frac{R_4}{R_2}V_{BE,w1}. \quad (10)$$

The first term on the left is V_{m1}, which is almost independent of temperature; V_{BE3} and $V_{BE,w1}$ are voltages decreasing with the rise of temperature but at the same slope. By setting $(1 + R_5/R_4)(R_4/R_2) > 1$, the temperature coefficient of $V_{BE,w1}$ becomes more negative than V_{BE3}, i.e., decreasing faster than V_{BE3} with temperature. If (10) holds at the lowest temperature, it holds for entire temperature range. In other words,

$$V_{m1} > V_{cmp} > \left(1 + \frac{R_5}{R_4}\right)\frac{R_4}{R_2}V_{BE,w1} > V_{m2}, \quad (11)$$

or the following conditions are always guaranteed:

$$V_{m1} > V_{cmp} > V_{m2}. \quad (12)$$

That is, the value of V_m at failure and normal states, i.e., V_{m2} and V_{m1} respectively, can be taken as a startup detection criterion by comparing with the value of V_{cmp} to achieve a robust startup circuit, specifically:

- When the circuit is in failure state, $V_m = V_{m2}$, since $V_{cmp} > V_m$, the output of comparator is low, $M1$ is turned on, injecting current into Q_1 branch and

pulling V_X and V_Y up to force the CMBGR away from the failure state into the normal operation.

- After the circuit starts, $V_m = V_{m1}$, i.e. $V_{cmp} < V_m$, the output of comparator goes high, $M1$ turns off, CMBGR is free of startup circuit.

Leveraging the predictive difference between V_{m1}, V_{cmp}, and V_{m2} over the entire temperature range, the comparator can detect all failure and normal operation states in order to start up CMBGR under various PVT conditions.

B. A design example for a very low V_{ref}

Aiming for the temperature range from -40℃ to 130℃, $V_{BE,w1} \approx 800mV$ at the lowest temperature and V_{cmp} reaches to $V_{cmp,max}$. Letting $k = 1 + \frac{R_5}{R_4}$, (10) can be rewrite as:

$$kV_{r1} > V_{cmp,max} > k\frac{V_{r1}}{1.25} \times 0.8 = 0.64kV_{r1}. \quad (13)$$

Considering the mismatch of the comparator, a margin of 50mV is added to cater for PVT variations, i.e.

$$kV_{r1} - V_{cmp,max} > 50mV, \quad (14)$$

$$V_{cmp,max} - 0.64kV_{r1} > 50mV. \quad (15)$$

Adding (14) and (15), we obtain:

$$V_{r1} > \frac{100}{0.32k}mV. \quad (16)$$

By setting $R_5/R_4 \gg 1$, V_{r1} can be set to a near zero value. Therefore, this startup circuit has less constraints on lowering V_{ref} if the area of R_5 is significantly larger than R_4. Taking into consideration of the overdrive voltage of M5 and VDD, the range of V_{ref} could be from around 0 to $VDD - V_{ov5}$.

To extend V_{r1} to near zero, the area used for R_5 could be very large for the proposed circuit. One possible way to address the large R_5 is to incorporate the high order curvature compensation structure [9], leading to an improved version of the CMBGR as shown in Fig. 3.

The improved version still utilizes resistors to amplify V_{ref} for the generation of V_m, but isolated by OPA2's gate. In Fig. 2, $V_{m1} = V_{ref}(1 + R_5/R_4)$, where R_4 is fixed, resulting the minimum of R_5 being limited by (10). In Fig. 3, $V_{m1} = V_{ref}[1 + R_5/(R_{7a} + R_{7b} + R_{7c})]$, in which R_7 can be set much smaller than R_4, i.e. the area of R_5 can be reduced without affecting the output value.

A design utilizing the proposed startup circuit is shown in Fig. 4. In the design, all current mirrors and amplifiers are adopted with cascode structure to suppress the effect of channel-length modulation for the enhancement of the power supply rejection ratio (PSRR). Furthermore, the exponential characteristic of transistors M_9 to M_{12} are used to compensate the output reference voltage with high order curvature [9], where R_7 is split into R_{7a}, R_{7b} and R_{7c} to bias the transistors M_9 and M_{12} in subthreshold region, and R_4 is split into R_{4a} and R_{4b} to adjust the compensation voltage as well as V_{ref}. Fig.4 presents the transistor-level circuit of the entire CMBGR with proposed startup circuit, a supply-independent biasing is used in the *Bias* section to generate stable bias voltages V_{b1} to V_{b4}.

IV. SIMULATION RESULTS

To evaluate the performance of the proposed circuit, the design is implemented in a 0.11μm CMOS process and simulated in Virtuoso IC618. To verify the overdrive voltage limitation, a target $V_{ref} = 1.2V$ is set, scanning the supply voltage from 0 to 5V, the output V_{ref} varies with VDD before 1.6V as shown in Fig.5, then it produces the target 1.2V V_{ref} after 1.6V, i.e., a minimum 0.4V overdrive voltage should be given for a proper V_{ref}.

The TC performance of the CMBGR with proposed startup structure is demonstrated in Fig. 6 where 20 voltage steps ranging from 2.3V to 4.3V at typical corner are simulated. From -40°C to 130°C, the temperature coefficient is within 1.56 ppm/°C with a typical 1.31 ppm/°C at 3.3V supply. AC simulation shows that PSRR varies from 70dB to 130.8dB at different voltages. The performance of the proposed CMBGR is summarized in Table I. Compared with other designs, this work presents the lowest TC across the widest temperature range, i.e. 8.7ppm over 170°C range, which gives an excellent immunity to a variety of operation conditions. It also supports programmable output voltages from 0.1 to 2V at 2.3V supply, which distinguish this design from other works using the voltage mode topology. No matter on high frequency or DC domain, this design shows the highest PSRR which is necessary for the reduction of the powerline fluctuations reflected on the output reference. The startup issue of such a CMBGR is compensated by the proposed startup circuits to reduce the risk of startup failure.

To verify the startup capability of the proposed circuit, four groups of Monte Carlo simulations at extreme voltages and temperatures are conducted as shown in Fig. 7. In each group, 8~9 different V_{ref} settings are analyzed. For each V_{ref} setting, three corners including tt, ff and ss are respectively simulated with 200 samples for the consideration of mismatch. Finally, in total 20400 ($2\times9\times3\times200+2\times8\times3\times200$) transient responses during power on are simulated. The results show that all expected V_{ref} can be correctly established within 140μs, i.e., the proposed startup circuit handles all cases well and robust to PVT variations.

V. CONCLUSION

In this paper, a robust startup circuit is proposed to detect the normal and failure states of CMBGR circuits during startup under PVT variations. The startup circuit is able to operate in a wide temperature range from -40 °C to 130°C because of the monotonic boundary condition of the proposed startup circuit. Based on the simulation results, the startup circuit can effectively force the CMBGR into its normal state during power on. It is shown that CMBGR with 8.7ppm/°C and tunable output reference from 0.1V to 2V is achieved under voltage supply from 2.3V to 4.3V.

REFERENCES

[1] H. Banba *et al.*, "A CMOS bandgap reference circuit with sub-1-V operation," in *IEEE Journal of Solid-State Circuits*, vol. 34, no. 5, pp. 670-674, May 1999, doi: 10.1109/4.760378.

[2] Tuan Vu Cao, D. T. Wisland, T. S. Lande, F. Moradi and Young Hee Kim, "Novel startup circuit with enhanced power-up characteristic for bandgap references," *2008 IEEE International SOC Conference*, 2008, pp. 123-126, doi: 10.1109/SOCC.2008.4641493.

[3] A. I. Kamel, A. Saad and L. S. Siong, "A high wide band PSRR and fast startup current mode bandgap reference in 130nm CMOS technology," *2016 IEEE International Symposium on Circuits and Systems*, 2016, pp. 506-509, doi: 10.1109/ISCAS.2016.7527288.

[4] X. Liu, S. Liang, W. Liu and P. Sun, "A 2.5 ppm/°C Voltage Reference Combining Traditional BGR and ZTC MOSFET High-Order Curvature Compensation," in *IEEE Transactions on Circuits and Systems II: Express Briefs*, vol. 68, no. 4, pp. 1093-1097, April 2021, doi: 10.1109/TCSII.2020.3027768.

[5] P. Malcovati, F. Maloberti, C. Fiocchi and M. Pruzzi, "Curvature-compensated BiCMOS bandgap with 1-V supply voltage," in *IEEE Journal of Solid-State Circuits*, vol. 36, no. 7, pp. 1076-1081, July 2001, doi: 10.1109/4.933463.

[6] X. Xu, Y. Suo, Y. Zhao, P. Zhou, X. Han, Q. Cai, M. Wang, J. Yuan, L. Zhao, Y. Li, G. Wang and Y. Lian, "A dry-electrode enabled ECG-on-Chip with arrhythmia-aware data transmission," in Science China Information Sciences, vol. 68, no. 2, pp. 1-16, February 2025, doi: 10.1007/s11432-024-4196-0.

[7] A. Boni, "Op-amps and startup circuits for CMOS bandgap references with near 1-V supply," in *IEEE Journal of Solid-State Circuits*, vol. 37, no. 10, pp. 1339-1343, Oct. 2002, doi: 10.1109/JSSC.2002.803055.

[8] C. Yu and L. Siek, "An Area-Efficient Current-Mode Bandgap Reference With Intrinsic Robust Startup Behavior," in *IEEE Transactions on Circuits and Systems II: Express Briefs*, vol. 62, no. 10, pp. 937-941, Oct. 2015, doi: 10.1109/TCSII.2015.2458044.

[9] G. Zhu, Y. Yang and Q. Zhang, "A 4.6-ppm/°C High-Order Curvature Compensated Bandgap Reference for BMIC," in *IEEE Transactions on Circuits and Systems II: Express Briefs*, vol. 66, no. 9, pp. 1492-1496, Sept. 2019, doi: 10.1109/TCSII.2018.2889808.

[10] Hanru Zhang and Xiuhan Li, A high power supply rejection radio voltage reference for energy harvesters, NEMS, 2013 8th IEEE International Conference, pp. 825 - 828, April 2013.

[11] Y. Liu, C. Zhan, and L. Wang, "An ultralow power subthreshold CMOS voltage reference without requiring resistors or BJTs," IEEE Trans. Very Large Scale Integr. (VLSI) Syst., vol. 26, no. 1, pp. 201–205, Jan. 2018.

[12] R. T. Perry, S. H. Lewis, A. P. Brokaw, and T. R. Viswanathan, "A 1.4V supply CMOS fractional bandgap reference," IEEE J. Solid-State Circuits, vol. 42, no. 10, pp. 2180–2186, Oct. 2007.

[13] Zhou Qianneng, Xue Rong and Li Hongjuan, Design of bandgap voltage reference for DC-DC converter, Cross Strait Quad-Regional Radio Science and Wireless Technology Conference (CSQRWC), pp. 361 - 364, July 2013.

2025 The 10th International Conference on Integrated Circuits and Microsystems

A Low-Power High-Accuracy Phase Frequency Detector and Charge Pump Design for CPPLL in 28 nm CMOS Technology

Zhigang Li[1], Hao Lei[1], Zhihao Liu[1], Lingrui Yan[1], Qiang Zhao[1], Shan Gao[1], Xiulong Wu[1*]

[1]*School of Integrated Circuits and Anhui Provincial High-Performance Integrated Circuit Engineering Research Center, Anhui University, Hefei 230601, China*
*Corresponding author: xiulong@ahu.edu.cn

Abstract—Conventional charge-pump phase-locked loops (CP-PLLs) suffer from inherent trade-offs between dead-zone elimination, linear detection range, and current mismatch in the design of the phase frequency detector (PFD) and charge pump (CP). To address these issues, this paper presents a novel PFD architecture incorporating a 4-bit digitally-programmable delay cell, offering 16 programmable delay steps to eliminate dead-zone effects while extending the linear phase detection range to $(-1.967\pi, 1.967\pi)$. The proposed CP employs a source-switched single-ended topology with an embedded rail-to-rail operational amplifier and a 4-bit digital-to-analog converter (DAC), achieving a current mismatch below 0.2% across the full 0.04 V to V_{DD} (1 V) output range while maintaining stable voltage regulation. Post-layout simulation results demonstrate a peak power consumption of 729.33 μW and a phase noise of -136.63 dBc/Hz at 1 MHz frequency offset. Moreover, the design is implemented in a 28 nm CMOS process and occupies an area of 0.0195 mm².

Index Terms—CPPLL, PFD, CP, current mismatch, dead zone, phase noise.

I. INTRODUCTION

In modern high-speed communication systems such as 5G networks and optical fiber infrastructures, clock generation and recovery circuits form the backbone of high-fidelity signal transmission, enabling precise synchronization and ultra-low-jitter data integrity across distributed systems. As a fundamental building block in these architectures, CPPLLs provide critical frequency synthesis capabilities that bridge high-speed digital processing with analog signal conditioning. Within nanoscale CPPLL implementations, the PFD and CP emerge as pivotal subcircuits that orchestrate the delicate balance between timing accuracy and spectral purity[1-2].

The PFD operates as the system's phase-sensitive decision engine, continuously monitoring and comparing input reference signals to generate error pulses that drive frequency corrections. Complementing this functionality, the CP translates these digital error signals into precisely calibrated analog currents, dynamically adjusting the loop filter's control voltage to stabilize the oscillator output. As illustrated in Fig. 1, the proposed CPPLL architecture harmonizes these components through intelligent co-design, creating a unified framework that synergizes digital programmability with analog precision a critical requirement for next-generation communication systems [3].

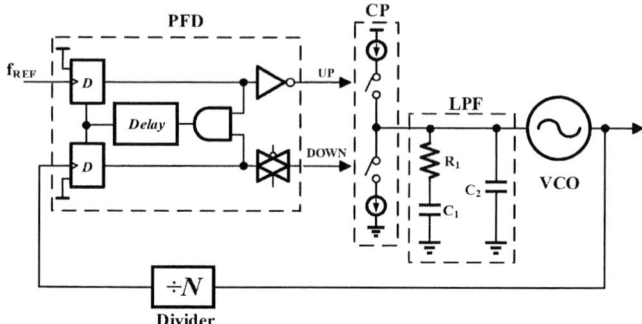

Fig. 1. Simplified architecture of a typical CPPLL with proposed PFD and CP circuits.

Conventional implementations, however, suffer from non-linear phase detection and current mismatch limitations. The PFD exhibits restricted detection ranges and dead-zone effects for minimal phase differences, degrading loop stability, while CP non-idealities such as transient glitches and asymmetric currents exacerbate reference spurs and bandwidth variations . These limitations severely hinder the performance of CPPLLs in high-speed communication systems, where ultra-low jitter and high spectral purity are essential[4-5].

To overcome these challenges, the proposed PFD employs a 4-bit binary-weighted capacitor array, achieving an ultra-wide phase-error detection range of ±354° (98.3% of the theoretical 360° limit) through linear phase-to-voltage conversion. This architecture suppresses reset path delays and charge injection mismatches via symmetric capacitive weighting, significantly reducing residual phase errors while maintaining area efficiency. By extending the linear detection range and minimizing dead zones, the proposed PFD ensures precise phase tracking even under minimal signal deviations, thereby enhancing overall loop stability[6].

Complementing the PFD, the CP integrates a source-switched single-ended topology with an embedded rail-to-rail operational amplifier and a DAC. The rail-to-rail amplifier eliminates voltage headroom constraints across the 0.04–1 V operating range, while the DAC dynamically calibrates current

979-8-3315-8850-2/25 $31.00 © 2025 IEEE
232

mismatches to $< 0.2\%$ accuracy, ensuring stable output current generation. This innovative CP design not only addresses the transient current mismatch issues but also provides consistent performance across the full control voltage spectrum. The combination of these advanced techniques in the PFD and CP results in a highly optimized CPPLL architecture that achieves ultra-low jitter and high spectral purity, making it well-suited for next-generation high-speed communication systems[7-9].

II. PHASE FREQUENCY DETECTOR DESIGN

In conventional PLL architectures, the PFD employs dual D flip-flops (DFFs) to analyze phase and frequency deviations between the reference signal (V_{REF}) and feedback signal (V_{FB}). As illustrated in Fig. 2, complementary UP and DOWN control signals are generated through this differential comparison process, driving the CP to modulate the voltage-controlled oscillator (VCO) tuning voltage via loop filter current injection. A critical reset mechanism asynchronously clears the DFF states post-comparison through dedicated propagation paths, thereby ensuring cycle-to-cycle detection fidelity and preventing metastability-induced errors[10].

The principal innovation of this work lies in the adaptive reset delay architecture with programmable control capability. Conventional fixed-delay implementations exhibit two fundamental constraints: Firstly, insufficient reset pulse duration causes incomplete CP current steering, thereby inducing nonlinear transfer characteristics and degrading system linearity. Secondly, excessive reset propagation latency severely limits the maximum operating frequency of the circuit. As depicted in Fig. 3, to simultaneously address these dual challenges, we innovatively integrate a 4-bit binary-weighted switched-capacitor array into the PFD reset path, achieving precise control through a programmable delay adjustment mechanism.

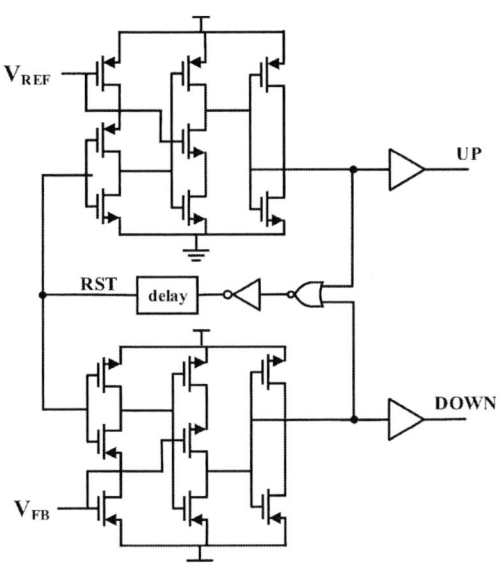

Fig. 2. Main circuit diagram of the phase frequency detector.

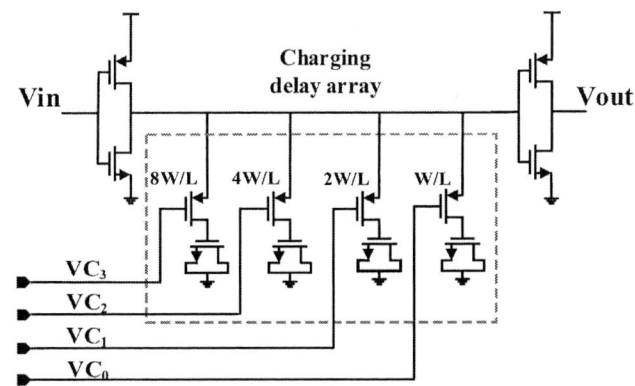

Fig. 3. Proposed programmable reset-delay architecture with 4-Bit binary-weighted switched-capacitor array.

The proposed array configuration consists of MOS capacitors with precisely scaled capacitance ratios ($8C : 4C : 2C : C$), which are selectively engaged with an inverter-based delay stage through 4-bit digital control signals ($VC_3–VC_0$). This architecture establishes programmable load capacitance spanning 0 to $15C$, enabling linear reset delay adjustment through discrete tuning steps. The step resolution is governed by the product of the inverter's equivalent resistance and the base capacitance unit.

III. CHARGE PUMP IMPLEMENTATION DETAILS

As demonstrated in Fig. 4, the proposed CP is implemented through a co-designed architecture combining a source-switched single-ended topology, a 4-bit digitally controlled current DAC , and an embedded rail-to-rail operational amplifier whose circuit implementation is detailed in Fig. 6, achieving charge/discharge current mismatch below 0.2%. As illustrated in Fig. 5, conventional current mirrors are replaced by a binary-weighted MOS-resistor (MOS-R) array, which generates programmable bias currents through digitally tunable MOS switches. These switches are regulated via 4-bit control codes (D3–D0), implementing a digital calibration mechanism to maintain bias current stability under process variations.

The linearized current mirror utilizes gate-connected PMOS pairs with partial source degeneration, where operational amplifier (AMP_2) enforces drain voltage clamping M_1 and M_{15} through negative feedback. The AMP_2 achieves 76 dB DC gain and $83.4°$ phase margin across the full output range, as validated by the frequency response in Fig. 7. This configuration prioritizes output impedance enhancement via active feedback compensation. The regulated cascode structure within the current mirror topology further suppresses parasitic conductance paths while maintaining voltage headroom, establishing a robust high-impedance output characteristic critical for precision analog applications. Additionally, this design approach ensures consistent performance across varying process and temperature conditions, making it highly reliable for high-precision analog circuits.

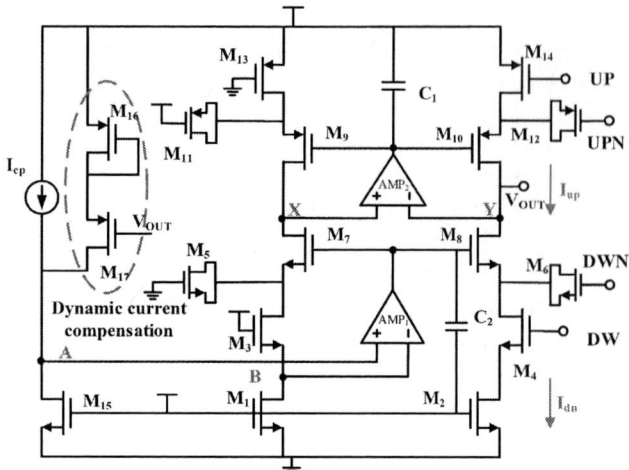

Fig. 4. Source-switched rail-to-rail DAC-controlled charge pump.

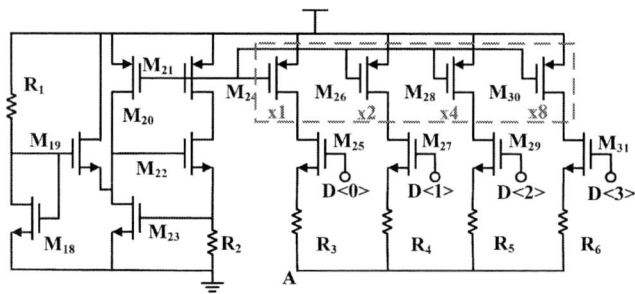

Fig. 5. Proposed 4-bit DAC current biasing circuit architecture.

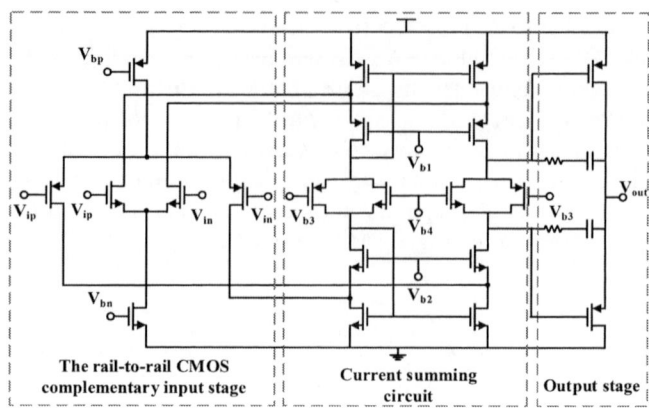

Fig. 6. The schematic of a rail to rail amplifier.

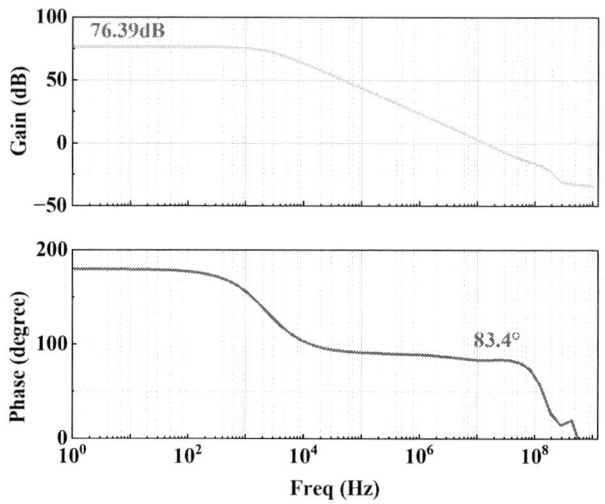

Fig. 7. The rail to rail amplifier stability simulation results.

While the architecture delivers robust performance under normal operating conditions, a critical challenge arises as the output voltage decreases: channel-length modulation effects in the current mirror transistors induce degradation of the charge/discharge current (I_{CP}), dynamically perturbing the PLL bandwidth. This instability mechanism is analytically captured by:

$$\Delta f_{loop} = \frac{K_{VCO} \cdot \Delta I_{CP}}{2\pi N} \quad (1)$$

where K_{VCO} denotes the VCO gain sensitivity and N the division ratio. Equation (1) highlights that fluctuations in I_{CP} are linearly amplified by K_{VCO}, threatening phase noise performance and loop stability.

To address this limitation, an adaptive dynamic compensation scheme activates PMOS transistor M_{16} when $V_{OUT} < 0.6$ V, generating an auxiliary boosting current:

$$I_{boost} = \beta_{M_{16}} \left(V_{SG,M_{16}} - |V_{th,p}|\right)^2 \quad (2)$$

where $\beta_{M_{16}}$ is the transconductance parameter and $V_{th,p}$ the PMOS threshold voltage. For $V_{OUT} \geq 0.6$ V, M_{16} remains inactive to eliminate static power consumption. This strategy restores flattened I_{CP}-V_{OUT} characteristics, suppressing K_{VCO}-induced bandwidth variations and enhancing phase noise performance, as validated in previous studies.

Building on this foundation, a multi-dimensional compensation strategy systematically addresses current replication inaccuracies. Precision voltage matching between charge/discharge paths is enforced by operational amplifier AMP_1, leveraging its virtual-short characteristic ($V_X = V_Y$) and a custom rail-to-rail input stage that eliminates common-mode constraints across the full 0–V_{DD} range. Transient disturbances are suppressed via stabilization capacitors C_1/C_2, which implement dominant-pole compensation to mitigate amplifier oscillations and gate voltage fluctuations. Charge injection errors from switching operations are neutralized by a symmetric dummy transistor array (M_5-M_{12}), actively canceling parasitic capacitance effects governed by:

$$\Delta Q_{inj} = C_{GS}\Delta V_{SW} + C_{GD}\Delta V_{GATE} \quad (3)$$

where C_{GS} and C_{GD} denote gate-source and gate-drain capacitances. Isolation transistors (M_7-M_{10}) further suppress charge-sharing through critical node decoupling.

979-8-3315-8850-2/25 $31.00 © 2025 IEEE

IV. SIMULATION RESULTS

As shown in Fig. 8, the integrated PFD-CP module occupies an area of $130\,\mu m \times 150\,\mu m$,which is as small as $0.0195\,mm^2$.The PFD incorporates a 4-bit digitally controlled delay cell, providing 16 programmable delay steps from $158\,ps$ to $1.31\,ns$ to eliminate dead-zone effects. As demonstrated in Fig. 9, the design achieves an extended linear phase detection range of $\pm354°$ and complete dead-zone elimination across input frequencies from $10\,MHz$ to $3.5\,GHz$.

The charge pump employs a source-switched single-ended topology integrated with an AMP_1, enabling full output swing from $0.04\,V$ to $1.0\,V$. Current matching accuracy is enhanced through an AMP_2-regulated mirror structure, reducing charge /discharge current mismatch to $0.2\,\%$ across the entire output range. The regulated mirror configuration significantly enhances output impedance through active feedback compensation, effectively suppressing high-frequency disturbances and improving signal integrity.

As illustrated in Fig. 10, the proposed source-switched charge pump achieves superior current matching compared to conventional designs. This enhanced current matching is crucial for maintaining low jitter and high spectral purity in high-speed communication systems.

The joint PFD-CP transient simulations under $0.55\,V$ bias conditions, as shown in Fig. 11(a), reveal charging and discharging characteristics with $98.7\,\%$ dynamic linearity. Furthermore, through co-optimization of PFD dead zone elimination and CP current flatness enhancement, the phase noise performance achieves $-136.63\,dBc/Hz$ at $1\,MHz$ offset, as demonstrated in Fig. 11(b). These experimental results conclusively validate the design's effectiveness in attaining high-performance targets.

Benchmark comparisons, summarized in Table I, conclusively demonstrate the architecture's superiority in current matching accuracy and voltage range coverage compared to state-of-the-art implementations under equivalent process conditions. The proposed charge pump not only achieves excellent current matching but also maintains a wide output voltage range, making it highly suitable for next-generation high-speed communication systems.

Fig. 8. Circuit layout of PFD-CP.

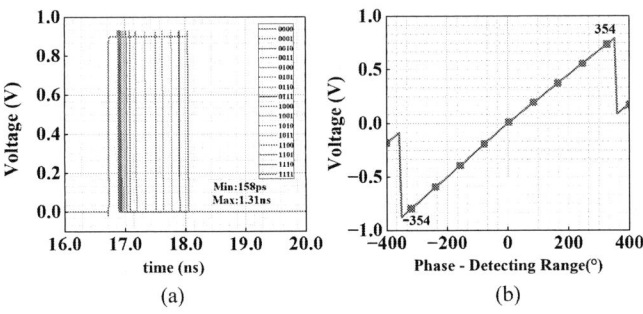

Fig. 9. (a) Simulation results of the delay chain. (b) Phase-detection range.

TABLE I
PERFORMANCE COMPARISON OF DIFFERENT CHARGE PUMPS

Parameter	ISCIT [11]	MOCAST [12]	VLSI [13]	**This Work**
Topology	40 nm	65 nm	180 nm	**28 nm**
Supply Voltage (V)	1.0	1.0	0.8	**1.0**
Current (µA)	80	150	150	**100**
Output Range (V)	0.1-1	0.3-0.75	0.2-0.7	**0.04-1**
Current Mismatch (%)	1	1.75	0.5	**0.2**
Power (mW)	1.48	0.31	-	**0.20**

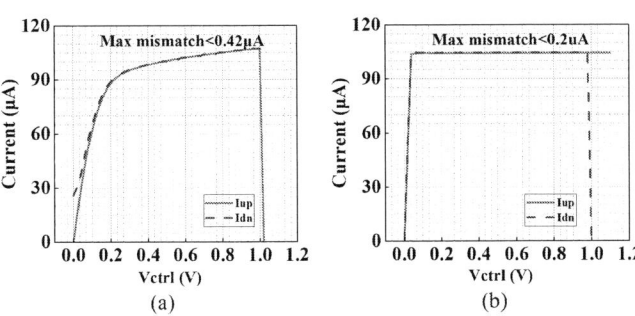

Fig. 10. Pump current matching curves. (a) Conventional CP. (b) The proposed source-switched CP.

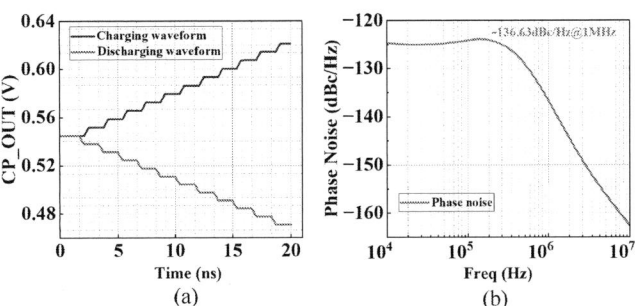

Fig. 11. (a) Joint simulation of PFD + CP . (b) Phase noise curve of PFD + CP.

V. CONCLUSIONS

A compact charge-pump phase-locked loop sub-module is demonstrated in $28\,\mathrm{nm}$ CMOS, integrating a dead-zone-free phase detector and high-linearity charge pump. The PFD achieves an extended linear detection range of $(-1.967\pi, 1.967\pi)$ through $16-$step digital delay tuning ($158\,\mathrm{ps}$ to $1310\,\mathrm{ps}$), while the source-switched CP suppresses current mismatch to $0.2\,\%$ across $0.04\,\mathrm{V}$ to $1.0\,\mathrm{V}$ via active feedback regulation. Post-layout simulations validate -136.63 dBc/Hz phase noise at $1\,\mathrm{MHz}$ offset with $729.33\,\mu\mathrm{W}$ peak power consumption in a $0.0195\,\mathrm{mm}^2$ core area.

ACKNOWLEDGMENT

This work was supported by the National Natural Science Foundation of China under Grant 62274001 and the University Synergy Innovation Program of Anhui Province under Grant GXXT-2023-013.

REFERENCES

[1] R. Y. Chen, Z. Y. Yang, and S. C. Yu, "High-frequency phase/frequency detectors: Analysis for oscillation-free optimal I/O," *IEEE Microw. Wireless Compon. Lett.*, vol. 30, no. 12, pp. 1097–1100, Dec. 2020.

[2] H. Xu, S. J. Ji, Y. Z. Wang et al., "Analysis and design of a sub-sampling PLL of low phase noise and low reference spur," *IEEE Trans. Circuits Syst. I, Reg. Papers*, vol. 71, no. 2, pp. 1–12, Feb. 2024.

[3] G. Konwar and T. Bezboruah, "Studies on phase noise profiles of proportional-integral-derivative controlled PLL," *International Journal of Electrical and Electronic Engineering & Telecommunications*, vol. 10, no. 5, pp. 369–376, Sep. 2021,

[4] J. Yang, Q. Pan, J. Yin et al., "A 2.0-to-7.4-GHz 16-phase delay-locked loop with a sub-0.6-ps phase-delay error in 40-nm CMOS," *IEEE Trans. Microw. Theory Techn.*, vol. 71, no. 8, pp. 3596–3604, Aug. 2023.

[5] Y. Zhan, R. Li, L. Zhang, X. Zou, Q. Hu, and Q. Wang, "A high-speed low-current-mismatch PFD+CP circuit for 11.3G CPPLL application," in *Proc. 9th Int. Conf. Integr. Circuits Microsyst. (ICICM)*, Wuhan, China, 2024, pp. 307–311.

[6] J. Prinzie, S. Biereigel, S. Kulis et al., "Source switched charge-pump PLLs for high-dose radiation environments," *IEEE Trans. Nucl. Sci.*, vol. 70, no. 4, pp. 590–595, Apr. 2023.

[7] W. Zhang, K. Chang, G. H. Zhang et al., "A low power high swing charge pump for phase locked loops," in *Proc. 5th Int. Conf. CircuitsSyst. (ICCS)*, Huzhou, China, 2023, pp. 236–240.

[8] B. Chen, W. Luo, F. Wang, Y. Lin, N. Yan, and H. Xu, "A 22.5–31.2-GHz continuously tuning frequency synthesizer with 8.7-GHz chirp for FMCW applications," *IEEE Microw. Wireless Compon. Lett.*, vol. 30, no. 9, pp. 904–907, Sept. 2020.

[9] S. Ji, Y. Zhao, W. Xu, N. Yan, and H. Min, "A novel charge pump with ultra-low current mismatch and variation for PLL," in *Proc. IEEE Int. Symp. Circuits Syst. (ISCAS)*, 2020, pp. 1–4.

[10] H. L. Kirankumar, S. Rekha, and T. Laxminidhi, "A dead-zone-free zero blind-zone high-speed phase frequency detector for charge-pump PLL," *Circuits, Syst., Signal Process.*, vol. 39, no. 8, pp. 3819–3832, Aug. 2020.

[11] E. C. Cuizon, M. A. Yuson, A. B. Caberos et al., "Design of charge pump for low power, wide range PLL in 65nm CMOS technology," in *Proc. 22nd Int. Symp. Commun. Inf. Technol. (ISCIT)*, Sydney, Australia, 2023, pp. 341–345.

[12] D. Samaras and A. Hatzopoulos, "A low power low noise 65nm charge pump using mismatch compensation and smoothing capacitor," *2022 11th International Conference on Modern Circuits and System Technologies (MOCAST)*, Bremen, Germany, 2022, pp. 1-4.

[13] L. Liu, Y. Ji, X. Liao, Z. Qin, and H. Liang, "1A 0.8-V, 2.55-GHz, 2.62-mW charge-pump PLL with high spectrum purity," *IEEE Trans. Very Large Scale Integr. (VLSI) Syst.*, vol. 30, no. 2, pp. 113–122, Feb. 2022.

979-8-3315-8850-2/25 $31.00 © 2025 IEEE

2025 The 10th International Conference on Integrated Circuits and Microsystems

A 60-V 96.7%-Efficiency 175kHz Integrated Dimmable LED Driver with Mean-Peak Current Control and Frequency Jittering Scheme

Jiamian Mao
Institute of Advanced Technology
University of Science and Technology of China
Heifei, China
jiamianmao@mail.ustc.edu.cn

Zhongguang Xu*
School of Microelectronics
University of Science and Technology of China
Heifei, China
xuxu@ustc.edu.cn
*Corresponding author

Xiangyin Chen
Microelectronics Innovation Center Co., ltd
Hefei, China
chenxiangyin@hfmic.com

Abstract—This paper presents a mean-peak current control (MPCC) buck dimmable light-emitting diode (LED) driver for automotive headlamp lighting applications. The proposed MPCC scheme can provide precise inductor current regulation during dimming operations by employing an integrator to enhance the accuracy of the LED average current. To address electromagnetic interference (EMI) issues in LED driver operation, adoption of a frequency jitter oscillator reduces the EMI amplitude of critical signals. Under typical operating conditions, the system clock signal exhibits random frequency jittering within ±5% of the 175kHz center frequency. The adoption of a soft-start circuit prevents potential chip damage caused by inrush current during startup by gradually ramping up the integrator reference voltage to eliminate surge current. Implemented in a 0.35-μm HV CMOS process, the proposed LED driver achieves 1% LED current error and 96.7% peak efficiency over an input voltage range of 6 to 60 V while driving 1 to 20 LEDs.

Keywords—Buck LED driver, discontinuous low-side current, mean-peak current control (MPCC), pulse width modulation (PWM), dimming

I. INTRODUCTION

LEDs are widely used in automotive lighting and outdoor lighting because of its long lifetime and high efficiency. The luminance of LEDs depends on the average current flowing through them, and as such it is essential to control the LED current precisely and efficiently for accurate luminance control and power efficiency. Since LED is a current-driven device and its brightness is proportional to the conduction current, a dc–dc converter-based buck LED driver is always needed to regulate current passing through LEDs for achieving constant luminous intensity under variations in input voltage and number of output LEDs in LED lighting systems[1]. The floating buck structure avoids high-side switching by utilizing a ground-referenced control switch. Consequently[2], it does not require a high-side gate driver and eliminates complex level-shifting and bootstrap components essential in conventional buck converters, significantly simplifying gate driving circuitry and enhancing reliability[3]. Traditional LED drivers in floating buck topologies implement two distinct control schemes: Peak

Current Control (PCC) and Hysteretic Current Control (HCC)[4], characterized by fundamentally divergent current sensing architectures and consequent performance trade-offs. PCC utilizes discontinuous low-side sensing via a resistor from the power NMOSFET source to ground[5], enabling LED current monitoring only during the switch on-time through voltage Sampling voltage[6]. The comparator generates a switch-off signal by comparing the peak current with the set value. However, it cannot detect the current during the off-period, and the average current is easily affected by the input and the LED forward voltage, thus reducing the current accuracy. In contrast, HCC utilizes continuous high-side sensing by serially connecting sampling resistor with the LED load[7], permitting uninterrupted LED current monitoring via sampling voltage. Regulation is achieved by bounding the inductor current between lower and upper reference thresholds, thereby directly controlling Mean current and substantially enhancing accuracy relative to PCC, largely independent of input and load variations. However, this continuous conduction of the full LED current through resistor induces notable conduction loss[8].

This paper proposes a buck LED driver with the MPCC technique. Based on a PCC discontinuous low-side current sensing topology to reduce power dissipation, This current control technique incorporates the average inductor current into the feedback loop, enabling both mean and peak currents to participate in regulation. This dual-loop approach ensures constant-current accuracy through the LEDs while enhancing response speed to load variations while enhancing response speed to load variations and avoiding excessive power dissipation. A drawback of the MPCC circuit is its susceptibility to subharmonic oscillation when operating at duty cycles exceeding 0.5. Subharmonic oscillation occurs when the current feedback loop fails to suppress disturbances, leading to growing oscillations in the inductor current. This phenomenon increases the inductor current ripple, compromises output regulation, and poses a risk of circuit damage. To prevent oscillation in the current feedback loop during operation and ensure reliable constant-current performance, this paper establishes a small-signal mathematical model of the current feedback loop and conducts a stability analysis. The switch-on signal for this LED

979-8-3315-8850-2/25 $31.00 © 2025 IEEE

driver circuit is generated by an oscillator and digital logic. Fixed-frequency clocking inherently induces electromagnetic interference (EMI) issues. To mitigate this limitation, this work designs an oscillator circuit incorporating frequency dithering functionality. This circuit modulates the operational frequency of the generated clock within a defined range, thereby spreading the noise energy across the frequency spectrum. Consequently, peak spectral noise density is significantly attenuated.

This paper is organized as follows. Section II proposes the overall architecture of the system, establishes a small-signal model, and conducts loop stability analysis. Specific details of circuit implementation and simulation analysis are in Section III. Finally, conclusions are given in Section IV.

II. PROPOSED MPCC LED DRIVER

A. System Architecture

Fig. 1 shows the structure of the proposed Buck LED driver using MPCC technique. The typical floating Buck topology is applied for the off-chip current regulator, which includes a LED string, a freewheeling Diode, an inductor L, an N-type power MOSFET M_1, and a current-sensing resistor R_S. The on-chip part consists of the proposed MPCC circuit, a clock-generating oscillator, linear/PWM dimming circuit, power supply, protection circuit and the gate driver circuits that provide a driving signal to M_1. To handle large input voltage V_{IN} levels of over 40 V, a 100 V high-voltage power MOSFET is adopted for M_1. The MPCC circuit employs an integrator to derive the mean voltage from the LED current. A comparator circuit generates the M_1 switch-off signal by comparing this mean voltage against the sum of the sampled current and a ramp current. The oscillator's fixed-frequency clock output undergoes frequency division to govern soft-start operation, inducing monotonic ramping of the integrator's reference voltage during initialization. Following frequency dithering and digital logic processing, this signal controls the switch-on signal of M_1. Concurrently, the oscillator's integrated ramp generator provides slope compensation to the MPCC circuit, it is critical for preventing potential subharmonic oscillation in the inductor current.

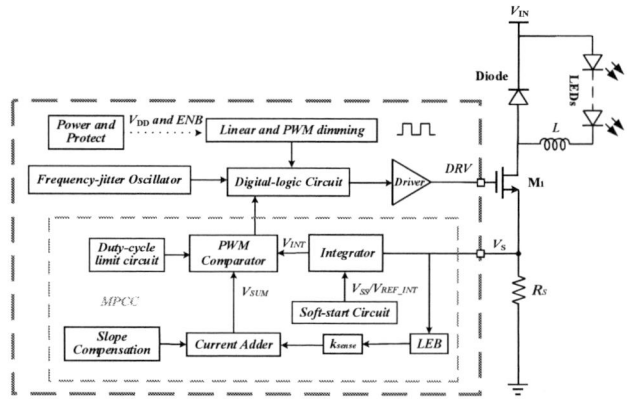

Fig. 1. System Architecture of the MPCC Buck LED Driver

B. Small-signal Modeling and Stability Analysis

During steady-state operation, modeling the LED drive circuit includes off-chip power stage modeling and on-chip loop small-signal modeling. Fig. 2 shows the overall small signal

model of the proposed MPCC Buck LED driver. The power stage modeling adopts the switched network average model method[9], and the power transistor is equivalent to an ideal transformer. Its small signal inputs are the change in the input voltage v_{IN} and the change in the duty cycle d, and the outputs are the change in the current of the free wheeling diode i_D and the change in the drain - source voltage of the power transistor v_{SW}. For small signal modeling of the loop, it is necessary to first conduct formula derivation and modeling analysis of the peak current loop, then introduce the average current loop, and conduct formula derivation and modeling analysis of the mean-peak dual current loop.

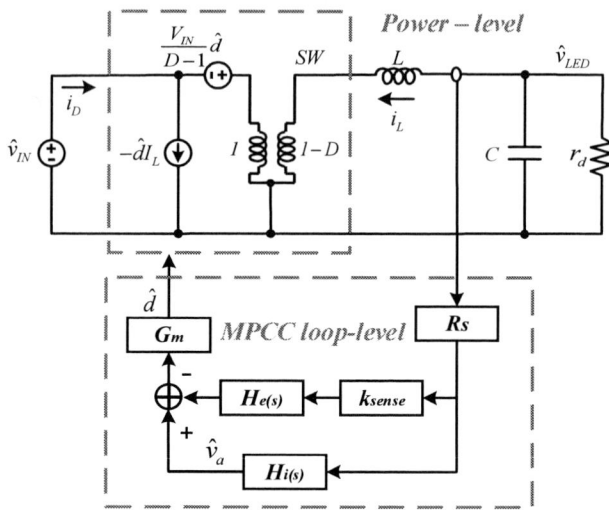

Fig. 2. Overall small-signal model of the proposed MPCC Buck LED driver

The expression of the current sampling function H_e in the peak current loop is:

$$H_e(s) = \frac{sT_s}{e^{sT_s} - 1} \qquad (1)$$

where s denotes the complex frequency and T represents the switching period of the system. (1) has an infinite number of roots on the imaginary axis, and the circuit stability cannot be directly analyzed. To solve this problem, Dr. Ray Ridley proposed a function with two zeros, which has a high degree of similarity from DC to the Nyquist frequency[10]. The expression of this function is:

$$H_e(s) \cong 1 + \frac{s}{\omega_n Q_z} + \frac{s^2}{\omega_n^2} \qquad (2)$$

In the formula:

$$\omega_n = \frac{\pi}{T_S}, \quad Q_z = \frac{-2}{\pi} \qquad (3)$$

(2) is a more accurate and concise representation of (1), which maintains a gain error within 0.2 dB and phase error below 3° across frequencies up to 1/2 switching frequency. The small-signal gain G_m from control voltage v_a to d is is defined as follows:

979-8-3315-8850-2/25 $31.00 © 2025 IEEE   238

$$G_m = \frac{d}{v_a} = \frac{L}{k_{\text{sense}} R_S m_c V_{\text{IN}} (1-D) T_S} \quad (4)$$

where k_{sense} is the sampling current proportionality parameter, m_c is the slope compensation parameter, and D is the duty cycle. In the mean current loop, the transfer function H_i of the integrator can be given as:

$$H_i(s) = -\frac{A}{1 + sAR_1C_1} \quad (5)$$

where the A, R_1, C_1 are the low frequency gain, input terminal equivalent resistance and capacitance of integrator.

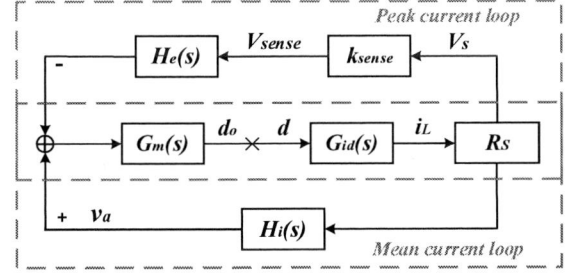

Fig. 3. Closed-loop control block diagram the proposed MPCC circuit.

Fig. 3 illustrates the Closed-loop control block diagram the proposed MPCC Buck LED driver feedback loop. In the small-signal model of the peak current loop, after the d is perturbed, an inductor current perturbation i_L is generated after passing through the transfer function G_{id} from d to i_L. R_S obtains the perturbed current to generate a sampling voltage V_S. V_S is superimposed on G_m through the $H_e(s)$ and finally generates the duty cycle feedback signal d_0. Its loop transfer function G_p is:

$$G_p(s) = \frac{d_o}{d} = G_o H_f(s) H_e(s) \quad (6)$$

In the formula:

$$G_o = \frac{L}{m_c r_d (1-D) T_S} \quad (7)$$

$$H_f(s) = \frac{1 + s(R_{\text{ESR}} + r_d)C}{\Delta(s)} \quad (8)$$

$$\Delta(s) = 1 + \frac{s}{\omega_o Q_p} + \frac{s^2}{\omega_o^2} \quad (9)$$

$$\omega_o = \frac{1}{\sqrt{(1 + \frac{R_{\text{ESR}}}{r_d})LC}}, \quad Q_p = \frac{1}{\omega_o(\frac{L}{r_d} + R_{\text{ESR}}C)} \quad (10)$$

where G_o is the low-frequency gain of G_p, L is the inductance value of the inductor, C is the capacitance value of the filter capacitor, r_d is the small-signal resistance of the LED load, and R_{ESR} is the equivalent resistance of the capacitor. The physical meaning of H_f is the filter composed of L, C, and r_d. G_o is inversely proportional to m_c, r_d, and T_S, and directly proportional to D and L. The frequency characteristics of G_p are determined by H_f and H_e. H_f introduces a low frequency zero

and two low frequency poles. Among them, C and r_d bring a pair of zero-pole that cancel each other out. Therefore, the frequency characteristics within half of the switching frequency are mainly related to the poles brought by L and r_d. $H_e(s)$ has two conjugate complex poles at half of the switching frequency, which affects the high frequency characteristics of (6).

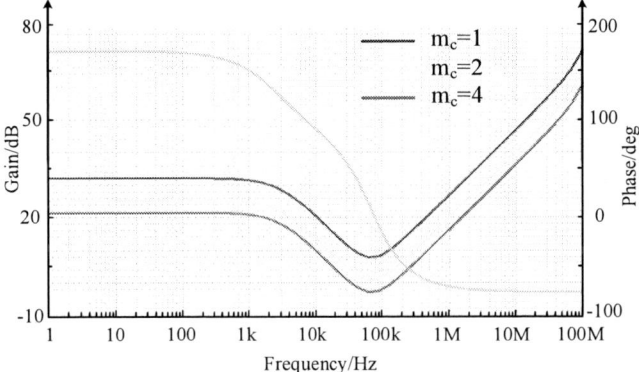

Fig. 4. Loop Stability of mc=1,2,4

Fig. 4 is the loop stability curve of (6) when $m_c = 1, 2, 4$. From the Fig. 4, it can be obtained that by injecting ramp compensation current into the peak-current loop to change the value of m_c, the gain curve will shift vertically while the phase curve will not change. When there is no ramp compensation in the circuit, sub-harmonic oscillation will occur in the peak-current loop. (6) theoretically explains that adding ramp compensation can ensure the stability of the peak loop. When selecting the value of m_c, the worst-case scenario of loop stability must be considered. Select $V_{\text{IN}} = 40$ V, $D = 0.9$, $L = 47$ μH, and C = 10 μF as the reference conditions. Determine that mc = 6.22 through calculation.

Based on the peak current loop, a mean current loop controlled by an integrator circuit is connected in parallel, and the expression of the transfer function $G_{mp}(s)$ of the MPCC loop is obtained as follows:

$$G_{mp}(s) = \frac{G_o}{k_{\text{sense}}} H_f(s)[H_i(s) - k_{\text{sense}} H_e(s)] \quad (11)$$

the low-frequency gain of $G_{mp}(s)$ is related to k_{sense} and A, and the frequency characteristic is related to the low-frequency pole f_{p1} brought by the integrator and the zero f_{z1} brought by the parallel connection of the two current loops. The equation is:

$$f_{p1} = \frac{1}{2\pi AR_1C_1} \quad (12)$$

$$f_{z1} = \frac{1}{2\pi k_{\text{sense}} R_1C_1} \quad (13)$$

The low-frequency gain of the integrator will increase the overall loop gain, and the dominant pole of the integrator will decrease the overall loop bandwidth. Designing the parameters of the integrator can achieve a trade-off between the control accuracy and response speed of the MPCC loop.

III. CIRCUIT IMPLEMENTATIONS

A. Integrator

The integrator integrates the difference between the sampled voltage V_S and the integrator reference voltage V_{REF_INT} to obtain the average voltage V_{INT}. The formula is:

$$V_{INT} - V_{REF\_INT} = -\frac{1}{RC}\int_0^{T_s}\left[V_{CS}(t) - V_{REF\_INT}\right]dt \quad (14)$$

The schematic of the integrator is shown in Fig. 5. The circuit adopts a folded cascode circuit, with a low-frequency gain of 70.55 dB and a phase margin of 89.52°. The dominant pole of the integrator is also the dominant pole of the entire loop, with a value of 1.372 kHz. When the circuit is in the freewheeling stage, M1 is turned off. At this time, the SH switch will also be disconnected, and the integrator is in a negative feedback state. The gate voltage of PM3 is quickly adjusted to V_{REF_INT} to keep V_{INT} unchanged. After M_1 is turned on, SH is close, and the integrator works normally. The soft-start reference switching circuit is used to switch the V_{REF} to the soft-start reference voltage V_{SS} during the soft-start period, and connect to the V_{REF} after the soft-start is completed.

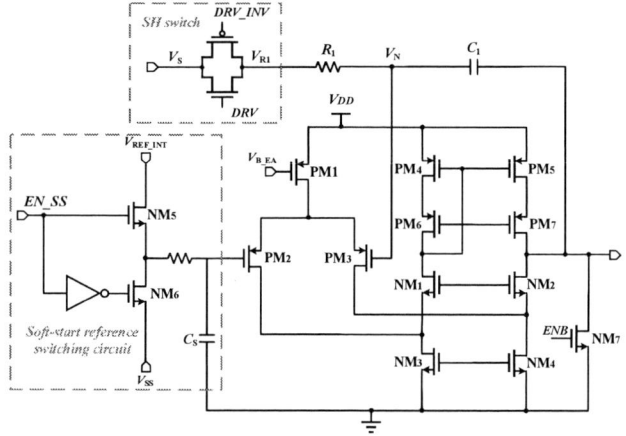

Fig. 5. Schematic of the integrator.

B. Frequency Jitter Oscillator

Fig. 6 shows the core module of a relaxation oscillator based on the capacitor charging and discharging. After the system operates normally, I_B is replicated by the current mirror PM6 to generate I_1 for charging C_1. V_{SLOPE} rises to the oscillator reference voltage V_{REF_OSC}, The output *COMP_OUT* of the comparator flips to a high level. After being delayed by the delay buffer, the combinational logic generates *CLK_F* as a high level to discharge C_1 and *CLK_FN* is at a low level. When I_2 charges C_2 to exceed the threshold voltage of NM5, The drain voltage V_D of NM5 flips to a low level. After being processed by the delay buffer and the combinational logic, *CLK_F* flips to a low level, entering the next cycle. By controlling the charging and discharging times of the C_1 and C_2, a 175kHz fixed frequency clock can be obtained, and the duty cycle of the clock can be adjusted by controlling the charging current I_1, I_2 and the size of the capacitors C_1, C_2. The oscillator generates a fixed-frequency clock *CLK_F* of 175 kHz. After a 5-division frequency processing, a 5 bit increasing sequence S_1 - S_5 with a frequency

of 35 kHz is generated. It monotonically increases from the 00000 state to the 11111 state. As a control signal of the soft-start circuit, it controls the timing climb of V_{SS}. The charging and discharging voltage of capacitor C_1 will be reused as the ramp voltage V_{SLOPE}, which controls the generation of the subsequent ramp compensation current I_{SLOPE}.

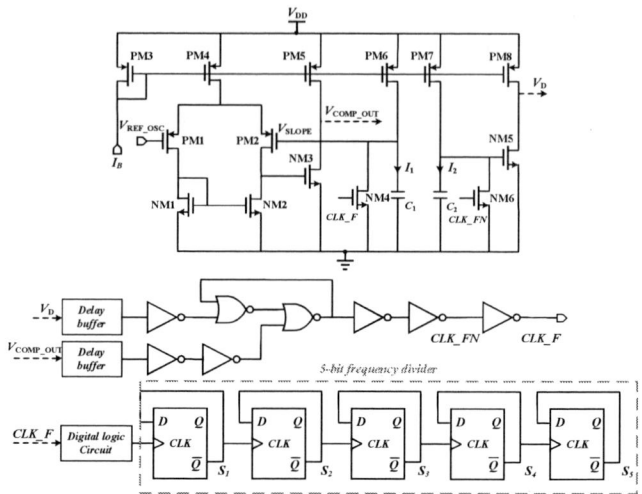

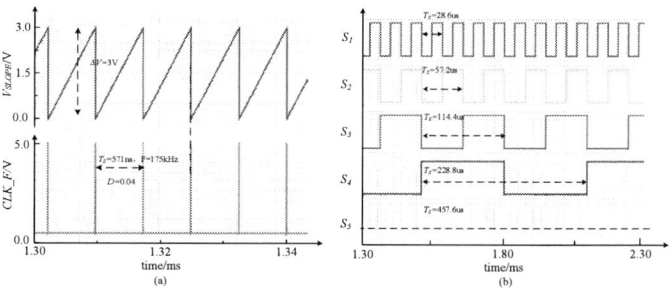

Fig. 6. The core circuit of the relaxation oscillator

The transient waveforms of *CLK_F*, V_{SLOPE}, and S_1 - S_5 are shown in Fig. 7, the core circuit of the relaxation oscillator generates a fixed frequency clock with F = 175kHz and D = 0.125. There is a ramp voltage V_{SLOPE} with a fluctuation range from 0 to 3V. A 5-bit incrementing signal S_1 - S_5 is obtained by clock division.

Fig. 7. The transient waveforms of (a) *CLK_F*, V_{SLOPE}, and (b) S_1 - S_5

Fig. 8 shows the frequency jitter circuit integrated in the oscillator. Under the control of *CLK_F*, the combinational logic and shift register form a sequence signal generator to generate the 4-bit switch signals Q_1-Q_4, which control the periodic turning on and off of the current mirror PM1 - PM4 transistors, thus periodically changing the magnitude of the current I_3 charging the capacitor C_3. After *CLK_F* changes to a high level, the output *CLK_D* of the R_S latch is set to 1, and the latch latches this high level signal. After *CLK_F* changes to a low level, I_3 charges capacitor C_3 until V_{C3} flips to a high level. V_{C3} is sent to the R_S latch through a delay buffer, setting its output *CLK_D* to 0. The time that the R_S latch holds a high level is the time for I_3 to charge C_3. Therefore, *CLK_D* is an output signal with the high - level of *CLK_F* periodically extended. The duty cycle conversion logic circuit will convert the falling edge signal of *CLK_D* into a high level signal of *CLK_J*, thereby converting the jitter of the clock duty cycle into the jitter of the frequency and achieving the frequency jitter function.

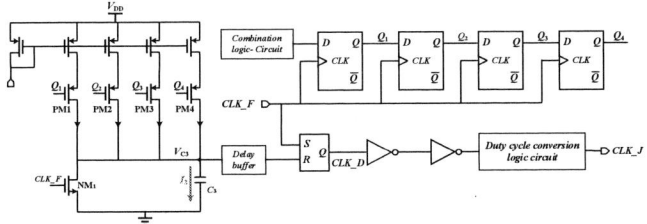

Fig. 8. The frequency jitter circuit integrated in the oscillator

The simulation results are shown in Fig. 9, the frequency jitter signal *CLK_J* has a center frequency of 175kHz and fluctuates within a range of 9% of this value. This signal controls the conduction of M_1. The duty cycle expansion signals Q_1-Q_4 periodically expand *CLK_F* to generate *CLK_D*.

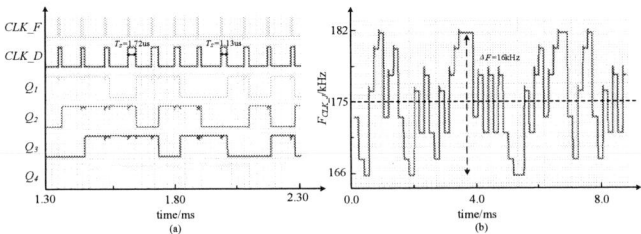

Fig. 9. (a) Duty cycle expansion signals (b) Frequency jitter signal

C. Soft Start Circuit

Fig. 10(a) shows the soft-start circuit consisting of 32 resistors with the same resistance value and 10 switches. S_1-S_5 are generated by the oscillator, and S_{n1} - S_{n5} are the inverted signals. Fig. 10(b) shows the transient waveforms of V_{SS}. S_1-S_5 regulate the ratio between V_{SS} and V_{REF_INT}, establishing a progressively increasing ratio. When soft-start concludes, the soft-start switching circuit enable is effective, the input reference signal of the integrator is transfer from V_{SS} to V_{REF_INT}, the system enters a steady-state operating condition.

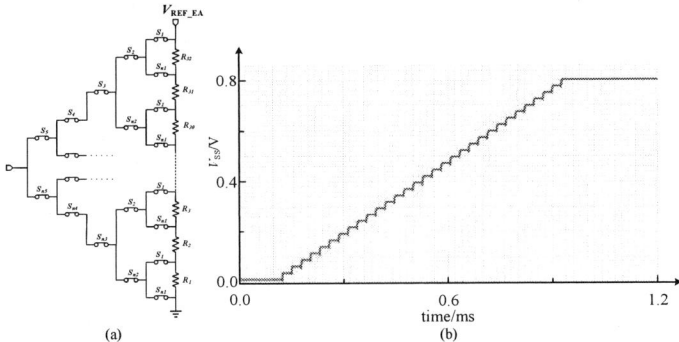

Fig. 10. (a) Structures of soft-start circuit (b) Soft start signal V_{SS}

D. Power Transistor Control Circuit

Fig. 11 shows structure of the power transistor control circuit, which include voltage-current conversion and superposition circuit, PWM comparator, and power transistor drive circuit. The V_{SLOPE} and V_S from the sampling resistor R_S generate a superimposed voltage V_{SUM} after passing through the voltage-current conversion and superposition circuit. The V_{SUM} will be compared with V_{INT} by the PWM comparator to generate a switch-on signal, the activation signal is generated by the oscillator. After logical processing and through the power transistor drive circuit, it controls the power transistor M_1 to Turn on or off.

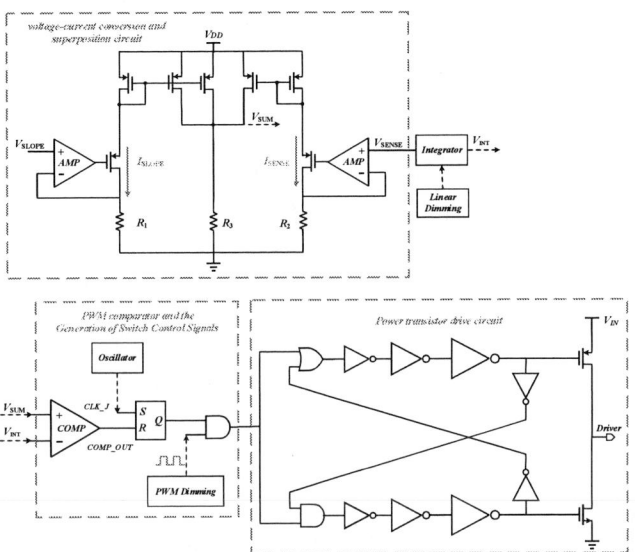

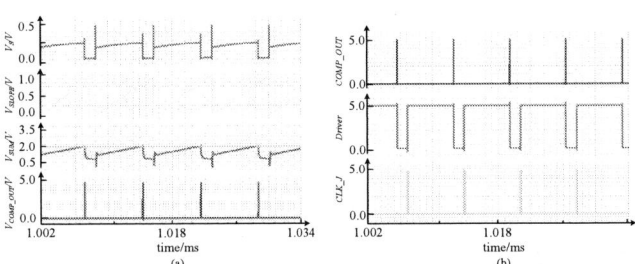

Fig. 11. The structure of the power transistor control circuit

The simulation waveforms of the key signals are shown in Fig. 12. Under the control of *COMP_OUT* and *CLK_J*, the Driver signal controls the activation and deactivation of M_1.

Fig. 12. (a)Input and output signals of the summing circuit (b)Turn-on and turn-off signals of power transistor

IV. LAYOUT AND POST-SIMULATION ANALYSIS

Fig. 13 show the chip layout which area is about 450 µm ×600 µm. Table 1 shows the constant current performance of the LED driving circuit of MPCC, where V_{IN} is input Voltage, LEDS is the number of LED strings, I_L is the load current, L is the inductor size, I_A is the average current, I_R is the current ripple, and Prec is the constant current precision. Under different operating conditions, the LED drive circuit achieves a constant current accuracy of 3% for the average current, and the ripple current varies with V_{IN}, LEDs, and L. Set V_{IN} is 24 V, LED lamp beads is 4 pieces, I_L is 1 A The backend simulation of the power on startup and constant current accuracy of the LED Driver is shown in Fig. 14. Under the working conditions of the LED drive circuit, the output current I_L gradually increases under the control of the soft-start circuit. Eventually, I_L stabilizes at an average value of 1 A. The soft-start time is 157.1 μs, and the constant current accuracy of the average current is 3%. Set the PWM dimming frequency to 2 kHz, the duty cycle to 50%, Backend simulation of PWM dimming of the LED Driver is shown in Fig 15. Under these working conditions, the LED output current is 492.7 mA, corresponding to a dimming effect of 48.2% in the full - on state. The dimming accuracy is 2.0%, the frequency of the PWM signal ranges from 100 Hz to 2 kHz. The highest dimming accuracy can reach 0.3%, and the dimming ratio is 500:1.

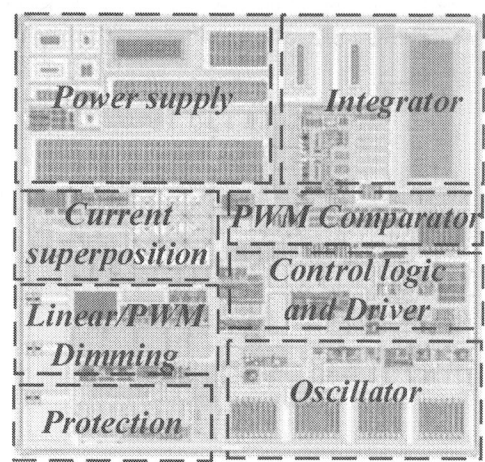

Fig. 13. Layout of the LED Driver

TABLE I.   CONSTANT CURRENT PERFORMANCE OF THE LED DRIVING

V_{IN}/V	LEDs	I_L/A	L/ μ H	I_A/A	I_R/A	Prec/%
12	2	0.1	200	0.101	0.409	1.00
24	5	1	50	1.014	0.314	1.40
36	8	3	25	3.010	2.102	0.33
60	15	6	12	6.040	4.004	0.67

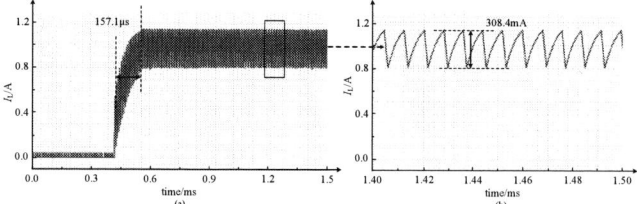

Fig. 14. Backend simulation (a) Power on startup (b) Current Precision

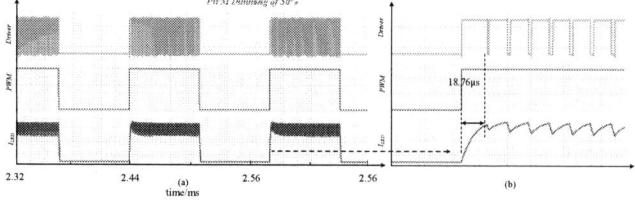

Fig. 15. Backend simulation of PWM dimming of the LED Driver

V.   CONCLUSION

In this paper, a MPCC floating Buck dimmable LED driver is proposed and implemented. The proposed control technology improves the constant current accuracy while achieving high conversion efficiency. The adoption of frequency jitter technology reduces the energy peak value of interference noise and optimizes the EMI problem of the LED Driver. The adoption of soft-start technology eliminates the inrush current of the LED Driver during the power on stage, ensuring the reliability of applications. Implemented in a 0.35-μm high voltage CMOS technology, The LED driver can support the V_{IN} from 6 to 60V, The output current range of 0.1-6A is set by an external current limiting resistor. The output voltage is automatically adjusted according to the number of load LEDs, and it can drive 1-20 series connected white LEDs. Its average constant current accuracy is 1%.and the switching center frequency is about 175 kHz. The peak conversion efficiency can reach 96.7%, achieving high conversion efficiency.

REFERENCES

[1] C. H. Liu, C. Y. Hsieh, Y. C. Hsieh, T. J. Tai, and K. H. Chen, "SAR-controlled adaptive off-time technique without sensing resistor for achieving high efficiency and accuracy LED lighting system," IEEE Trans. Circuits Syst. I, vol. 57, no. 6, pp. 1384–1394, Jun. 2010.

[2] Z. Liu and H. Lee, "A Wide-Input-Range Efficiency-Enhanced Synchronous Integrated LED Driver With Adaptive Resonant Timing Control," in IEEE Journal of Solid-State Circuits, vol. 51, no. 8, pp. 1810-1825, Aug. 2016.

[3] Z. Liu and H. Lee, "A 26 W 97%-Efficiency Fast-Settling Dimmable LED Driver With Dual-nMOS-Sensing Based Glitch-Tolerant Synchronous Current Control for High-Brightness Solid-State Lighting Applications," in IEEE Journal of Solid-State Circuits, vol. 50, no. 9, pp. 2174-2187, Sept. 2015

[4] V. Anghel, C. Bartholomeusz, A. G. Vasilica, G. Pristavu and G. Brezeanu, "Variable Off-Time Control Loop for Current-Mode Floating Buck Converters in LED Driving Applications," in IEEE Journal of Solid-State Circuits, vol. 49, no. 7, pp. 1571-1579, July 2014.

[5] I. H. Oh, "An analysis of current accuracies in peak and hysteretic current controlled power LED drivers," in Proc. IEEE Applied Power Electronics Conf., 2008, pp. 572–577.

[6] L. Balogh, "Design and application guide for high speed MOSFET gate drive circuits," Texas Instruments Power Supply Design Seminar (SEM-1400), 2001.

[7] Z. Liu and H. Lee, "A 25 W 97%-efficiency 3.5 MHz integrated dimmable LED driver with lossless synchronous current control and floating nMOS-sensing scheme," in Proc. IEEE Applied Power Electronics Conf., 2014, pp. 1378–1383.

[8] D. Park, Z. Liu, and H. Lee, "A 40 V 10 W 93%-efficiency current-accuracy-enhanced dimmable LED driver with adaptive timing difference compensation for solid-state lighting applications," IEEE J. SolidState Circuits, vol. 49, no. 8, pp. 1846–1860, Aug. 2014.

[9] G. Suman, B. P. Kumar, M. S. Kumar, B. C. Babu, and K. Subhashini, "Modeling, analysis and design of synchronous buck converter using state space averaging technique for pv energy system," in Electronic System Design (ISED), 2012 international Symposium on, pp. 281-285,IEEE, Dec. 2012.

[10] RIDLEY R B. A new continuous-time model for currentmode control with constant frequency, constant on-time, and constant off-time, in CCM and DCM[C]// 21st Annual IEEE Conference on Power Electronics Specialists, 1

2025 The 10th International Conference on Integrated Circuits and Microsystems

A 100-Gb/s PAM4 Receiver Analog Front-End

Haoran Huang [1,2], Yue Xu [1,2], Jing Huang [1,2], Gengzhen Qi [1,2*]

[1] School of Microelectronics Science and Technology, Sun Yat-sen University, Zhuhai, China
[2] Guangdong-Macao Joint Laboratory for Modular Chip Design and Testing, Guangdong, China
[*] qigzh@mail.sysu.edu.cn

Abstract—This paper proposes a 100Gb s PAM SerDes receiver analog front end (AFE), consisting of an input matching network, a continuous time linear equalizer (CTLE), and a variable gain amplifier (GA). A dual-L-type network is employed at the input to achieve impedance matching and reduce return loss. Both the CTLE and GA adopt a source-degenerated architecture, offering implementation simplicity and high linearity. To enhance the bandwidth and equalization capability of the CTLE, inductive peaking and feedforward techniques are introduced. Considering the stringent linearity and noise requirements of PAM signal, a single-stage GA is adopted, also employing inductive peaking to extend bandwidth. The equalization capability and gain of the CTLE and GA are tunable via gate-voltage-controlled NMOS transistors operating in the linear region, enabling adaptability to various channel conditions. Implemented in 65nm CMOS technology and operating at 1.2 supply, post-layout simulation results show that the CTLE achieves 2.9-to-1 .1dB equalization capability at the Nyquist frequency (25GHz), while the GA provides 2.5-to- .9dB variable gain. After equalizing a 10.1dB channel loss, the AFE achieves a horizontal eye opening of 0.25UI at a bit error rate (BE) of 10[-12]. The power consumption is 19.7m , and the power efficiency is 0.20p bit.

Keywords—PA signa , C L , A, ind ctive eaking, eed orward

I. INTRODUCTION

With the rapid development of emerging technologies such as big data, cloud storage, artificial intelligence, and the Internet of Things (IoT), there is a growing demand for high-speed, high-bandwidth data transceiver systems. Under this trend, high-speed serial interface (SerDes) technology has developed rapidly and become one of the key technologies for high-speed data links. Compared to parallel transmission, serial transmission offers advantages such as fewer transmission channels, reduced chip pin count, and simplified routing, while mitigating issues like clock synchronization, skew, and data crosstalk [1]. A typical SerDes system consists of a transmitter, receiver, and transmission channel. The transmitter includes modules such as the serializer, equalizer, and phase-locked loop (PLL), while the receiver includes the deserializer, equalizer, and clock-data recovery (CDR) [2]. The high-speed channel comprises chip package, bonding wires, pins, and vias, which introduce insertion loss and signal reflections, impacting signal integrity. To reduce the bit error rate (BER), equalizers are essential, with transmitters typically employing feed forward equalizers (FFE), and receivers adopting continuous time linear equalizers (CTLE) and decision feedback equalizers (DFE).

This work is funded in part by the GuangDong Basic and Applied Basic Research Foundation under Grant 2025A1515011608. It is also partially funded by the Guangdong-Macao Joint Laboratory (GDSTC Project No. 2025B1212150003).

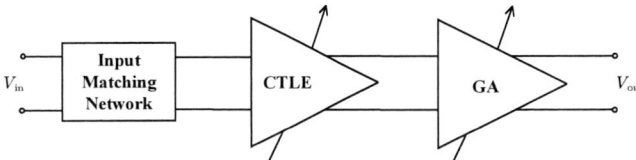

Fig. 1. The block diagram of our proposed receiver analog front end.

The analog front end (AFE) of the receiver generally consists of three modules: the input termination network, CTLE, and variable gain amplifier (VGA). As the first stage of the receiver, the performance of the AFE directly affects the effectiveness of the DFE, making its design critical. Mainstream AFE architectures include source degeneration, Gm-TIA, and inverter-based structures [3]-[8]. The Gm-TIA architecture offers high bandwidth and good linearity the inverter-based design saves area of passive components and consumes less power. Compared to the other two, the source degeneration structure is simple and easy to implement, and is often combined with additional techniques to enhance bandwidth and equalization capability.

Implemented in 65nm CMOS technology, this paper proposes a 100Gb/s PAM4 SerDes receiver analog front-end (AFE) which comprises a dual-L-type input matching network, a source-degenerated CTLE, and a source-degenerated VGA, consuming 19.7mW under 1.2V supply. The dual-L-type matching network provides impedance matching to maximize power transfer. The CTLE introduces inductive peaking and feedforward techniques to improve bandwidth and equalization capability. The VGA also employs inductive peaking to extend bandwidth, providing stable gain over the Nyquist band.

II. IMPLEMENTATION DETAILS

As shown in Fig. 1, the proposed AFE consists of three modules: input matching network, CTLE, and VGA. The channel is modeled as a differential transmission line with 10.1dB loss at the Nyquist frequency (25GHz). The input matching network is designed to mitigate reflection caused by PAD parasitic capacitors (60fF) and ESD parasitic capacitors (100fF) using a dual-L-type matching structure to match the channel characteristic impedance (50). To compensate for high-frequency channel loss, the CTLE incorporates inductive peaking and feedforward techniques, overcoming the limited equalization capability of conventional source-degenerated CTLE. For PAM4 signal, which is more sensitive to linearity and noise, a single-stage VGA is used to adjust signal swing, enhanced by inductive peaking for high bandwidth and implemented with source degeneration for good linearity. To meet the different channel conditions and output swing

979-8-3315-8850-2/25 $31.00 © 2025 IEEE   243

requirements, we use NMOS transistors operating in the linear region with controllable gate voltage to adjust the equalization capability of the CTLE and the gain of the VGA, and the gate voltage is controlled by an off-chip voltage source.

A. Input Matching Network

The schematic of the input matching network is shown in Fig. 2(a). Due to the existence of PAD parasitic capacitors and ESD parasitic capacitors in actual manufacturing, we match the channel impedance with the input termination resistance by connecting the two inductors in series to form a dual-L-type matching network, and finally determine that the values of the two inductors are 90pH and 270pH. An on-chip AC coupling module composed of a capacitor and a resistance sets a cutoff frequency of 7MHz. As shown in Fig. 2(b), the return loss at the Nyquist frequency is 20.78dB, and remains above 19.66dB across the entire band, indicating excellent impedance matching and reflection suppression.

B. Conventional CTLE

Conventional CTLE introduces a resistance and capacitor at the source to form a high-pass path, improving high-frequency gain. As shown in Fig. 3, the transfer function is

$$|H(s)| = \frac{g_{m1,2}R_L}{1 + \frac{g_{m1,2}R_S}{2}} \cdot \frac{\left(1 + \frac{s}{\omega_z}\right)}{\left(1 + \frac{s}{\omega_{p1}}\right)\left(1 + \frac{s}{\omega_{p2}}\right)} \quad (1)$$

$$\omega_z = \frac{1}{R_S C_S}, \quad \omega_{p1} = \frac{1 + \frac{g_{m1,2}R_S}{2}}{R_S C_S}, \quad \omega_{p2} = \frac{1}{R_L C_L}. \quad (2)$$

It includes a zero and two poles. By tuning the values of R_S and C_S, the low-frequency gain, zero, and pole can be changed, allowing the equalization capability of the CTLE to be adjusted to accommodate different channel loss scenarios. Since there is only one zero, the equalization capability of this structure is limited. To address this, the proposed CTLE in Fig. 4(a) introduces inductive peaking and feedforward techniques.

C. Proposed CTLE

In the proposed CTLE, the load resistance R_D is connected in series with inductor L_d, forming a resonance with the output capacitor and introducing a zero to extend bandwidth. Feedforward is implemented using NMOS transistors (M_3 and M_4) to bypass the output node and deliver the signal directly to L_d, enhancing high-frequency gain without compromising low-frequency response. The transfer function is

$$|H(s)| = \left[\frac{g_{m1,2}R_D\left(1 + \frac{L_d}{R_D}s\right)}{1 + g_{m1,2}\left(\frac{R_S}{2} \parallel \frac{1}{2C_S s}\right)} + g_{m3,4}L_d s \right]$$
$$\cdot \frac{1}{L_d C_L s^2 + R_D C_L s + 1}, \quad (3)$$

which shows that the feedforward effectively increases the apparent value of L_d, boosting high-frequency gain. Directly increasing the value of L_d would otherwise shift the output pole and degrade bandwidth.

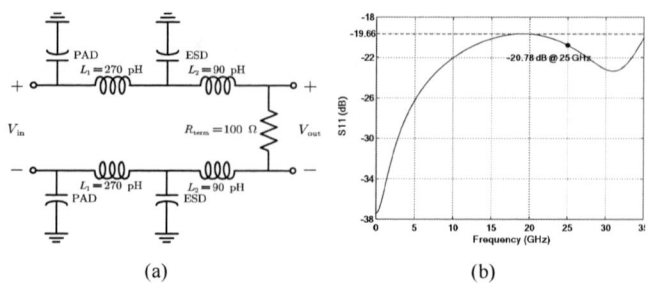

(a)   (b)

Fig. 2. (a) Dual-L-type matching network circuit. (b) Simulated S_{11}.

The size of M_3 and M_4 needs to be carefully designed to ensure feedforward only dominates beyond the first pole frequency. The transconductance condition for this is

$$g_{m3,4} < \frac{g_{m1,2}R_D R_S C_S}{L_d\left(1 + \frac{1}{2}g_{m1,2}R_S\right)}. \quad (4)$$

A comparison of equalization capability between the proposed and conventional CTLE is shown in Fig. 5.

To support adaptability, the source resistance in the CTLE is tuned using a gate-controlled NMOS transistor operating in the linear region.

D. Proposed VGA

The VGA schematic is shown in Fig. 4(b). The design uses a source-degenerated structure for high linearity, and inductive peaking for extended bandwidth. Considering the trade-off between gain and bandwidth, only a single-stage is implemented to maintain flat gain response over a wide frequency range, compensating for the low-frequency loss introduced by CTLE and enhancing overall signal swing. The transfer function is

$$|H(s)| = \frac{g_{m1,2}R_D}{1 + \frac{g_{m1,2}R_S}{2}} \cdot \frac{1 + \frac{L_d}{R_D}s}{L_d C_L s^2 + R_D C_L s + 1}. \quad (5)$$

The gain is made variable using gate-controlled NMOS transistors.

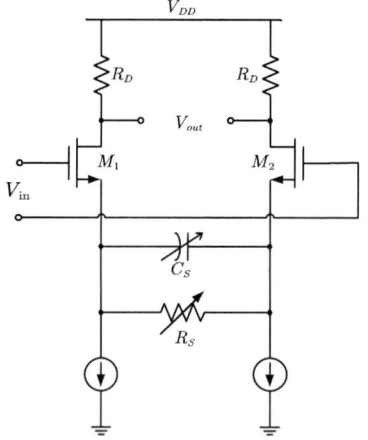

Fig. 3. Conventional CTLE circuit.

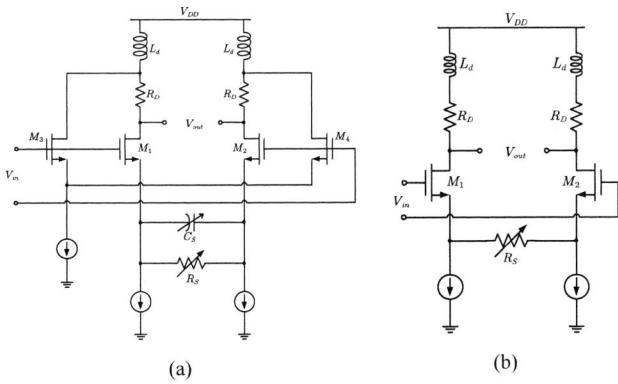

(a)   (b)

Fig. 4.   Proposed (a) CTLE circuit and (b) VGA circuit.

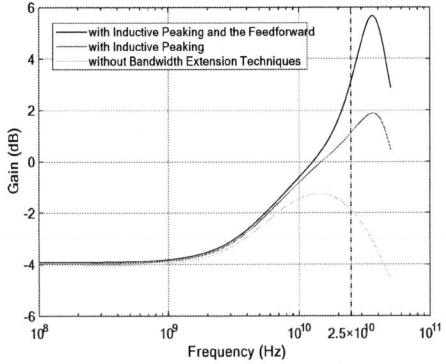

Fig. 5.   Frequency response comparison of this work and conventional CTLE.

III. SIMULATED RESULTS AND COMPARISON

The proposed design is implemented using the TSMC 65nm CMOS technology. Schematic and layout are completed on the Cadence IC618 software platform. The layout of our proposed AFE is shown in Fig. 6, with inductors occupying most of the area. The core area is 0.11mm^2.

A. Frequency Response

The frequency response of the CTLE is shown in Fig. 7(a). As the control voltage varies from 0.3-to-1.2V, peaking at the Nyquist frequency ranges from 2.9-to-14.1dB.

Fig. 7(b) shows the frequency response of VGA. As the control voltage sweeps from 0.75-to-1.2V, the gain is adjustable from 2.5-to-4.9dB with a bandwidth exceeding 60GHz.

Fig. 7(c) shows the frequency response after the signal passes through the channel, CTLE, and AFE. The gain is -1dB at low frequency, indicating minimal attenuation. At the Nyquist frequency, our proposed AFE provides 6.2dB peaking, effectively compensating for channel loss.

B. Eye Diagram and Bathtub Curve

Fig. 8(a) shows the eye diagram after the channel, the signal is no longer distinguishable. After equalization by CTLE and AFE, as seen in Fig. 8(b) and (c), the eye height (EH) and eye width (EW) improve significantly. The output eye diagram of the AFE shows 0.44-to-0.46UI EW and 130-to-155mV EH. Fig. 8(d) shows the BER bathtub curve, where a 0.25UI EW is

maintained at a BER of 10^{-12}, demonstrating excellent equalization.

C. Simulated Waveform

Fig. 9 shows the equalization effect of our proposed AFE in the time domain. There are many bit errors in the signal before equalization, but these errors are successfully eliminated after the equalization of AFE, which greatly reduces the BER of the system. At the same time, the voltage swing has been increased.

D. Comparison with the Prior Art

The comparison of our proposed AFE with state-of-the-art AFEs is listed in Table I. Compared with [9] and [10], our AFE achieves higher data rate and better power efficiency with lower requirements for CMOS process technology. With the same 65nm node as [11], our work supports higher data rate with better power efficiency and lower BER. A more advanced CMOS process is used in [12], which has higher data rate and better equalization capability, but the circuit structure is more complex and the power efficiency is higher than in this paper.

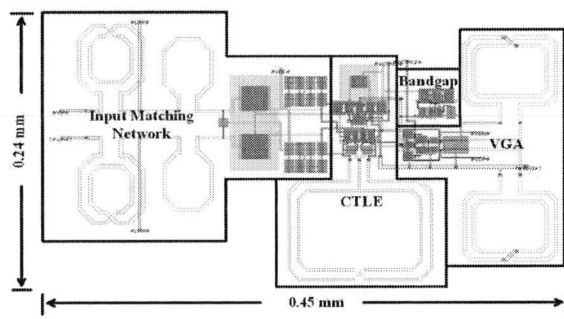

Fig. 6.   Layout of our proposed receiver analog front end in 65nm CMOS technology.

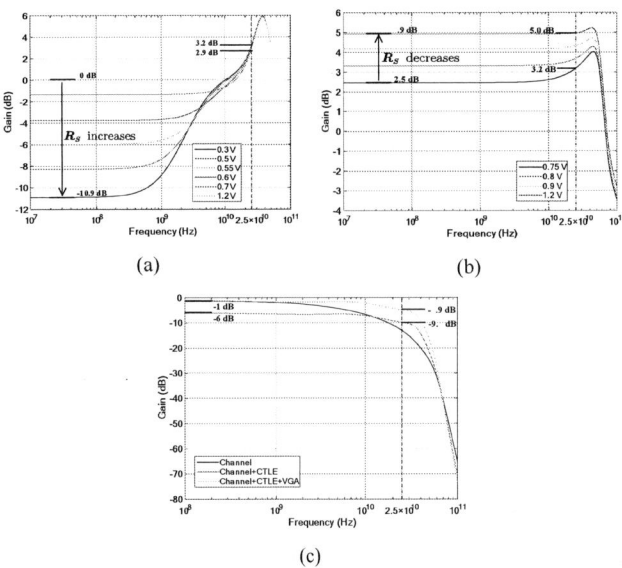

(a)   (b)

(c)

Fig. 7.   Simulated frequency response of the (a) CTLE and (b) VGA at Different Control Voltages. (c) Simulated frequency response before and after the AFE.

979-8-3315-8850-2/25 $31.00 © 2025 IEEE   245

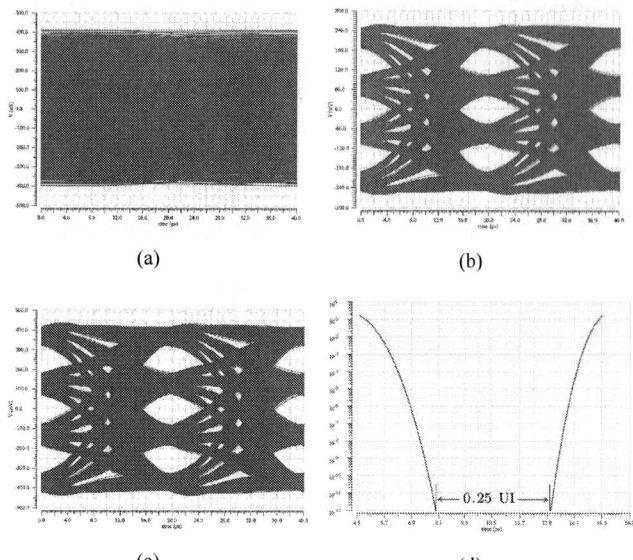

(a)	(b)
(c)	(d)

Fig. 8. Eye diagram (a) after channel, (b) after CTLE, and (c) after AFE. (d) BER bathtub curve after AFE.

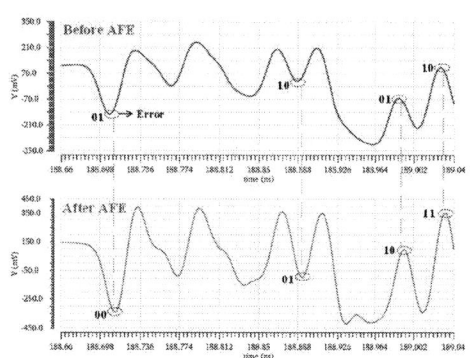

Fig. 9. Simulated waveforms before and after AFE.

TABLE I.   SUMMARY AND COMPARISON

	[9]	[10]	[11]	[12]*	This Work*
CMOS Technology	28 nm	40 nm	65 nm	28 nm	65 nm
Supply (V)	0.9	1.1/1.2	1	0.9	1.2
Data Rate (Gb/s)	60	56	40	112	100
Modulation	PAM4	PAM4	PAM4	PAM4	PAM4
Equalization	CTLE + 2-Tap DFE	2-Stage CTLE	CTLE + VGA + 12-Tap DFE	Ideal 4-Tap FFE + 2-Stage CTLE + MFEQ + 3-Stage VGA	CTLE + VGA
Channel Loss (dB)	8.2	7.3/10.8	10	19.5	10.1
Bit Error Rate	PRBS-31 10^{-12}	PRBS-15 10^{-8}	-- 10^{-11}	PRBS-15 10^{-11}	PRBS-31 10^{-13}
Power (mW)	39	30	167	43.8	19.7
Power Efficiency (pJ/bit)	0.65	0.54	4.18	0.39	0.20

* From post-layout simulated results.

IV. CONCLUSION

This paper proposes a high-speed SerDes receiver AFE comprising a dual-L-type input matching network, a single-stage CTLE with inductive peaking and feedforward techniques, and a single-stage source-degenerated VGA with inductive peaking. The peaking inductors resonate with parasitic output capacitance to introduce a zero, thereby extending bandwidth. The feedforward path improves high-frequency gain without degrading low-frequency performance. Implemented in 65nm CMOS technology, the AFE supports 100Gb/s PAM4 signal with differential swing $\geq 1.0V_{PP}$, providing 2.9-to-14.1dB equalization capability and 2.5-to-4.9dB variable gain under 1.2V supply. After compensating the channel loss (10.1dB), the system achieves 0.25UI EW at a BER of 10^{-12}. The power consumption is 19.7mW, achieving the power efficiency of 0.20pJ/bit.

REFERENCES

[1] LI Shijie, MA Ruichang, DENG Mingxing, XUE Jiamin, JIA Haikun, "Design technique for JESD204C high-speed serial interface circuit," in *Micro/nano Electronics and Intelligent Manufacturing*, vol. 5, no. 3, pp. 14–21, Sept. 2023.

[2] J. Lee, P. -C. Chiang, P. -J. Peng, L. -Y. Chen and C. -C. Weng, "Design of 56 Gb/s NRZ and PAM4 SerDes Transceivers in CMOS Technologies," in *IEEE Journal of Solid-State Circuits*, vol. 50, no. 9, pp. 2061-2073, Sept. 2015.

[3] C. Zhang, Y. Xu, X. Zhang and H. Wang, "50Gb/s high-speed serial interface receiver CTLE equalization circuit design," *2023 8th International Conference on Integrated Circuits and Microsystems (ICICM)*, Nanjing, China, 2023, pp. 417-421.

[4] J. Zheng, T. Shi, J. Wang, Q. Zou and M. Ren, "A 16Gbps Programmable CTLE Design with Adjustable Gain," *2023 3rd International Conference on Electronic Information Engineering and Computer Science (EIECS)*, Changchun, China, 2023, pp. 220-225.

[5] X. Zhang, C. Zhang, Y. Song and R. Song, "32Gb/s Low-Power NRZ Signal Serdes Receiver," *2024 9th International Conference on Integrated Circuits and Microsystems (ICICM)*, Wuhan, China, 2024, pp. 177-181.

[6] M. Tang et al., "A 56-Gb/s PAM4 Continuous-Time Linear Equalizer with Fixed Peaking Frequency in 40-nm CMOS," *2019 IEEE International Conference on Integrated Circuits, Technologies and Applications (ICTA)*, Chengdu, China, 2019, pp. 89-90.

[7] K. Zheng et al., "An Inverter-Based Analog Front-End for a 56-Gb/s PAM-4 Wireline Transceiver in 16-nm CMOS," in *IEEE Solid-State Circuits Letters*, vol. 1, no. 12, pp. 249-252, Dec. 2018.

[8] K. Zheng, Y. Frans, K. Chang and B. Murmann, "A 56 Gb/s 6 mW 300 um2 inverter-based CTLE for short-reach PAM2 applications in 16 nm CMOS," *2018 IEEE Custom Integrated Circuits Conference (CICC)*, San Diego, CA, USA, 2018, pp. 1-4.

[9] K. -C. Chen, W. W. -T. Kuo and A. Emami, "A 60-Gb/s PAM4 Wireline Receiver With 2-Tap Direct Decision Feedback Equalization Employing Track-and-Regenerate Slicers in 28-nm CMOS," in *IEEE Journal of Solid-State Circuits*, vol. 56, no. 3, pp. 750-762, March 2021.

[10] Z. Li, M. Tang, T. Fan and Q. Pan, "A 56-Gb/s PAM4 Receiver Analog Front-End With Fixed Peaking Frequency and Bandwidth in 40-nm CMOS," in *IEEE Transactions on Circuits and Systems II: Express Briefs*, vol. 68, no. 9, pp. 3058-3062, Sept. 2021.

[11] X. Wu, Z. Wang, Z. Zhao, C. Zhang and Z. Wang, "A 20Gbuad NRZ/PAM4 Receiver Frontend in 65nm CMOS," *2022 IEEE 16th International Conference on Solid-State & Integrated Circuit Technology (ICSICT)*, Nangjing, China, 2022, pp. 1-3.

[12] M. Zhang et al., "A 0.9-V Supply Up to 21.5-dB Boost Gain Analog Front-End with T-coilloaded CTLE and VGA in 28-nm CMOS for 112-Gb/s PAM-4 Medium-Reach Receivers," *2024 IEEE International Conference on Integrated Circuits, Technologies and Applications (ICTA)*, Hangzhou, China, 2024, pp. 172-173.

979-8-3315-8850-2/25 $31.00 © 2025 IEEE

2025 The 10th International Conference on Integrated Circuits and Microsystems

A −86 dBm/15.6 μW Wake-Up Receiver for Internet of Vehicles Applications

Rui Chen
Faculty of Engineering
SHENZHEN MSU-BIT University
Shenzhen, China
celeters@163.com

Yan Li
Faculty of Engineering
SHENZHEN MSU-BIT University
Shenzhen, China
liyan@smbu.edu.cn

Liming Si
School of Integrated Circuits and Electronics
Beijing Institute of Technology
Beijing 100081, China
lms@bit.edu.cn

Xinchao Zhong
Guangdong Runyu Sensor Co.,Ltd
Jiangmen, Guangdong, China
zhongxinchao@runyusensor.com

Chiu-Wing Sham
Dept. of Computer Science
The University of Auckland
Auckland, New Zealand
b.sham@auckland.ac.nz

Hang Yu*
Faculty of Engineering
SHENZHEN MSU-BIT University
Shenzhen, China
yuhang@smbu.edu.cn
*Corresponding author

Abstract—This paper presents a novel ultra-low power Wake-Up Receiver (WuRx) designed for Internet of Vehicles (IOV). The proposed design is based on a high sensitivity RF receiver integrated in an On Board Unit (OBU).Without dedicated die area, the WuRx is formed by re-using the necessary key consisting components of the RF receiver, while those with high-power consumption, such as the analog band-pass filter, is de-activated. To further minimize the power consumption of the WuRx, a Pulse Width Modulated (PWM) control scheme functioning at 0.7 V is implemented, and the WuRx receiving chain is thus alternatively turned ON/OFF. Fabricated using CMOS 180 nm process, only 200 μm × 100 μm of additional die area is required to implement the WuRx. Measurement shows that the proposed WuRx achieves a peak wake-up sensitivity of −86 dBm in the 5.8 GHz band, while the average power consumption is only 15.6 μW.

Index Terms—Wake-up receiver (WuRx), Internet of Vehicles (IOV), Pulse Width Modulation (PWM), On Board Unit (OBU)

I. Introduction

It is reported that China' s expressway Electronic Toll Collection (ETC) user has reached 254 million by the end of 2022, with over 230 million on board units (OBUs) installed. This massive infrastructure has not only significantly improved highway traffic efficiency but also laid a crucial foundation for Internet of Vehicles (IOV) related applications, which scenario can be illustrated as in Figure 1. The nationwide networking of ETC systems and widespread adoption of OBUs have effectively established China's largest dedicated short-range communication (DSRC) network. These OBUs distributed across hundreds of millions of vehicles serve as mobile communication nodes, providing ready-made hardware infrastructure for IOV development. This transformation has upgraded ETC systems from simple toll collection tools to important information hubs for intelligent transportation, creating favorable conditions for expanding IOV applications. Combined with multidimensional environmental sensing de-

This work was supported in part by the National Natural Science Foundation of China under Grant No. 62250002, 62201194, 61471245 and U1201256.

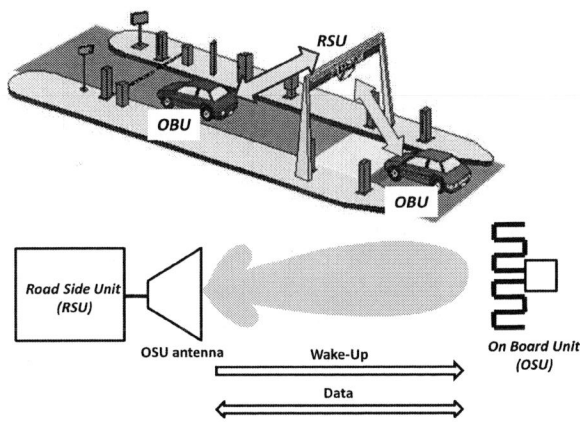

Fig. 1. Illustration of Internet of Vehicle applications.

vices, such as millimeter-wave radar, high-definition cameras and meteorological sensors, critical data collected, which represents real-time road conditions, traffic information, and weather changes, can be transmitted to vehicles via OBU units, providing valuable decision-making support for vehicle route planning and intelligent driving.

Currently, according to the GB/T 20851.1-2019 national standards, the OBU is required to be 'waked-up' from the sleeping mode by only −40 dBm input beacon in 5.8 GHz bands [1]. This minimum requirement of wake-up sensitivity results to shortened time span for data exchange between various OBUs, thus greatly limit the amount of data that can be exchanged if IOV is successfully established. On the other hand, low power consumption remains a fundamental requirement for OBUs, as it is powered by miniature batteries. For this reason, an OBU design with significantly improved wake-up sensitivity while maintaining ultra-low power consumption at the same time is key for IOV related applications.

979-8-3315-8850-2/25 $31.00 © 2025 IEEE   247

RF ultra-low-power wake-up receivers (WuRx), as critical components in the Internet of Things (IoT) and Wireless Sensor Networks (WSN), have demonstrated significant technological progress and application potential in various fields in recent years. In 2020, Ali et al. proposed an ultra-low-power digital controller for 5.8 GHz DSRC applications.This design was fabricated in a CMOS 130 nm process, and the power consumption is only 34.65 nW [2]. D'Addato et al. in 2022 also focused on ultra-low power and reliability, proposing a temperature-robust WuRx for IoT systems that minimizes false wake-ups through a 256-bit codeword, achieving a power consumption of 54.8 nW and a sensitivity of −49.5 dBm [3]. Wang and Wentzloff introduced a 5G-NR compatible WuRx in 2025. Through architecture and block-level co-optimization, this design, implemented in 65 nm CMOS process, achieves a high sensitivity of −94.5 dBm with a power consumption of 742 μW [4]. Furthermore, Abdelrahman et al. proposed a WuRx for Body Area Networks (BAN). This receiver utilizes the conductive properties of human tissue, achieving a wake-up sensitivity of −96.2 dBm and an energy efficiency of 3.5 pJ/bit [5]. Chen and Huang in 2024 developed a WuRx compatible with Bluetooth Low Energy (BLE) standard. By employing passive mixers, polyphase filters, and current-reuse low-noise amplifier, this design achieves a sensitivity of −85 dBm while keeping power consumption at 167 μW [6].

Beyond application-specific WuRx, researchers have also made numerous innovations in architecture and technology. Yan et al. designed a multi-mode WuRx based on direct envelope detection, with sensitivity of −58 dBm and power consumption of 1 μW [7]. Cheng and Chen presented a receiver based on an Injection-Locked Oscillator (ILO), which achieves sensitivity of −80 dBm with a power consumption below 54 μW [8]. Concurrently, advancements in core components like oscillators have been pursued to enhance system reliability. For instance, Li et al. improved the frequency stability of a CMOS Voltage-Controlled Oscillator (VCO) using a DLL-assisted architecture, which is crucial for reliable wireless IoT communication [9].

Research was focused on core module and innovative techniques have also been demonstrated. For example, Ma et al. in 2020 proposed a signal sampling technique that reduces receiver power consumption to below 310 μW by duty-cycling the RF front-end, while maintaining a sensitivity of −70 dBm [10]. At the component level, Bouraoui et al. in 2023 designed a low-power envelope detector for WLAN applications. By using a forward body biasing technique, they achieved a low power consumption of 87.5 μW in the CMOS 180 nm process [11]. Further research into key receiver modules includes work on data converters for more complex architectures. Ge et al. designed and implemented a quadrature bandpass sigma-delta modulator for low-IF receivers, a critical block for high-performance signal processing in the digital domain [12].

Combining a novel power-saving control scheme and the high receiving sensitivity of the integrated RF receiver of an OBU, this paper presents a circuit design that achieves −86 dBm of wake-up sensitivity at 5.84 GHz band with an average of only 15.6 μW power consumption. When the OBU sleeps, the WuRx is formed by those only necessary components in the RF receiver, while others such as active band-pass filters are excluded. In the meanwhile, an ultra-low power digital baseband with 0.7 V supply voltage is activated. This baseband generates Pulse Width Modulated (PWM) control signal, alternatively turning on and off the receiving chain of the WuRx, thus greatly reduces the overall power consumption. The baseband also help filtering and identifying the 14 kHz On-Off Keying (OOK) signal defined as Beacon Service Table (BST) in [1] to wake up the RF receivers. This paper is organized as follows. In section II, the proposed WuRx and the corresponding control scheme is discussed in detail. Section III demonstrates the measurement result, and conclusion remarks are provided in section IV.

II. WuRx Implementation

The implementation of the WuRx is based on a 5.8 GHz RF receiver architecture developed for OBU applications, as depicted in Figure 2 (a). This RF receiving chain comprises a front-end Low Noise Amplifier (LNA), an active mixer, a pre-amplifier (pre-amp), a intermediate frequency (IF) 5 MHz band-pass filter, a received signal strength indicator (RSSI) block, a single-ended to differential (Single-Diff.) converter and a comparator with hysteresis. A Phase-Locked Loop (PLL), which includes the Voltage-Controlled Oscillator (VCO) and other loop components, is utilized to generate a stable Local Oscillator (LO) signal for the RF receiver. Subsequently, a digital baseband processor is employed for processing the received data.

Taking the advantages of this high-sensitivity receiver, key consisting components are again used to compose the WuRx receiving chain. This approach is expected to achieve exceptionally high RF wake-up sensitivity. The demodulated wake-up beacons are sent to a specific WuRx baseband in the RF receiver, to process. However the operational power consumption of such an arrangement is substantial and has to be further addressed.

To minimize the power consumption when capturing the wake-up signals, the high sensitivity RF receiving chain is re-configured. This strategy involves minimizing the power consumption of the entire chain by bypassing components with high power consumption and activating only the essential key components operated in this configuration, only the VCO instead of the complete PLL is activated to conserve power. The VCO generates an LO frequency at approximately 5.86 GHz, which is susceptible to parts-per-million (ppm) level frequency drift due to temperature and voltage variations. However, the impact of this ppm-level LO frequency drift on the overall chain is negligible. This is because sufficient design margin exists within the subsequent signal path, particularly for the 5 MHz IF signal generated after mixing with the relevant communication bands, thus ensuring the successful receiving of the signal. For the two common communication bands (5.79 GHz~5.80 GHz and 5.83 GHz~5.84 GHz), the resulting IF signal is guaranteed to fall within the operational range of the WuRx. Concurrently, the LNA operates in a 'WuRx Low Power Mode,' which reduces its power consumption by approximately 40 % compared to its normal operational mode, thereby conserving power while maintaining the required receiving sensitivity.

979-8-3315-8850-2/25 $31.00 © 2025 IEEE   248

The high-power-consumption baseband used in the RF receiving chain is replaced by a low-power WuRx baseband. This WuRx baseband incorporates a narrow-band digital filter capable of identifying the wake-up signal signal specified by the [1]. Compared to the high-power baseband which requires a 1.8 V supply, the WuRx baseband operates on just 0.7 V, sufficient for its internal circuitry, which has much decreased power consumption level at nano-watt range.

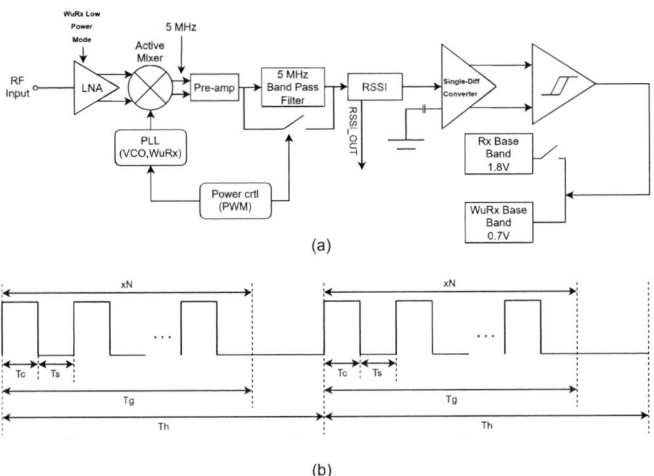

(a)

(b)

Fig. 2. (a) The proposed WuRx structure based on an OBU Receiver, (b) typical waveform of dual-window PWM signal.

Though these arrangements, more than 50 % of power consumption can be cut compared when the RF receiving chain is fully-turned on. Nevertheless, the power consumption remains excessive, and a PWM control scheme is thus applied.

The strategy of PWM control is to minimize the active duration of the receiving chain in the WuRx, therefore reducing the overall power consumption. As shown in Figure 2 (b), the core concept of this scheme is 'burst listening' T_g followed by an extended sleeping period $T_h - T_g$. The short 'burst listening' period constitutes N consecutive cycles, each including a 'minor detecting window' lasting T_c, and a resting window lasting T_s. The WuRx receiving chain is active only within the minor detecting windows, and is completely off within the resting windows.

Both in the N resting windows and in the extended sleep period, the only circuity that consumes power is the WuRx control block. This scheme effectively limits the power consumption of the WuRx as

$$P_a = \left(\frac{T_c \cdot 12}{T_h} \right) \cdot P_w + \left(1 - \frac{T_c \cdot 12}{T_h} \right) \cdot P_s \quad (1)$$

,where P_w, P_s, P_a represents the power consumption of receiving chain of WuRx, the power consumption of PWM control scheme, and the effective average power consumption of the entire WuRx.

Note that N, T_c, T_s and T_h, can all be configured through the integrated Serial Peripheral Interface (SPI) interface, which allows the proposed WuRx can be used in wider applications.

An integrated 100 kHz oscillator included in the WuRx control block provides the minimum scale for configuring the values of T_c, T_s and T_h. In addition, an 2nd-order finite impulse response (FIR) bandpass filter in the digital domain is implemented in the WuRx control block, while the band width, center frequency of the filter can all be adjusted Therefore this FIR filter helps differentiate the genuine wake-up signal from the noisy background. Within both the 'burst-listening' period and sleeping period, the 100 kHz oscillator is activated, and it is the predominant power consuming circuitry in the WuRx control block. However, as the supply voltage of this block is decreased to 0.7 V, its power consumption is kept as only 1.82 μW.

In applications that comply with the DSRC protocol defined by [1], a Roadside Unit (RSU) is specified to transmit a BST signal at 10 ms intervals. The baseband component of the BST signal consists of 15 pulses of a 14 kHz square wave, corresponding to a period of 71.4 μs per pulse and a total signal duration of 1.071 μs. The core parameters of the PWM control signal are therefore configured as follows: $T_c = 240 \, \mu s$, $T_s = 480 \, \mu s$, $T_g = 8640 \, \mu s$, and $T_h = 2.1$ s. This configuration ensures the reliable detection of the BST signal, as the duration of a single active listening window T_c is longer than the period of the 14 kHz signal, while the total duration of the burst listening phase T_g is shorter than the BST signal's transmission period.

The resulting maximum latency set to be approximately 2.1 s. For Electronic Toll Collection (ETC) scenarios, this latency is within acceptable limits and will not, impede the receiving of toll information, thus ensuring vehicles smooth pass through.

As the measurements shows that P_w is approximately 10.1 mW (with 1.8 V supply voltage), the effective overall power consumption of the WuRx, P_a, is only 15.6 μW, which is calculated from (1).

III. MEASUREMENT RESULTS

The completed chip, including the proposed WuRx, is implemented using CMOS 180 nm process, and the micro-graph of the IC is shown in Figure 3. The full RF receiving chain, consisting LNA, active mixer, pre-amp, etc. , locates at the right side of the implemented chip. As in Figure 3, the WuRx control

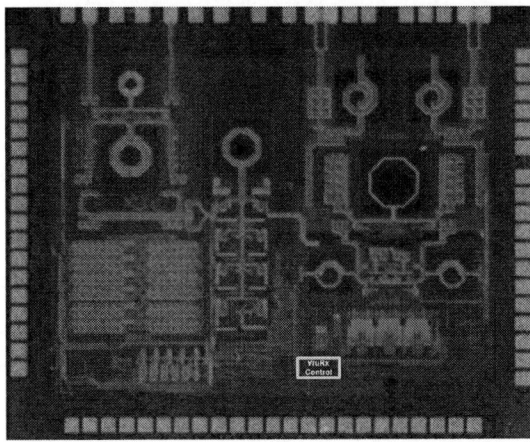

Fig. 3. Micro-graph of the ETC chip with the proposed WuRx control block.

block, shich includes the 100 kHz oscillator and the FIR filter, only requires an additional area of 200 μm × 100 μm.

The functionally of the proposed WuRx is validated, as shown in Figure 4. The key PWM signal contains 12 periods of square waves, and the high-level portions in each period indicate when the WuRx is active. For each time, the WuRx is turned on for approximately 240 μs, able to catch a maximum number of five 14 kHz OOK modulated signal defined in [1]. The PWM module also includes an additional digital filtering block, to distinguish the targeted 14 kHz signal from other noise-introduced. As indicated in Figure 4, when the designed number of 14 kHz OOK signal is detected, the 'WuRx_ENABLE' signal is generated, as an indication to turn on the RF receiver. The WuRx module switches between ON/OFF states alternatively for 8.5 ms before entering the sleeping mode, which lasts for approximately 2.1 s.

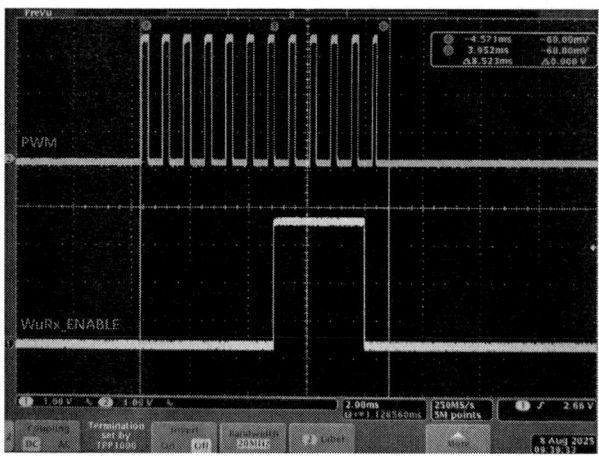

Fig. 4. Measured PWM AND wake-up indication.

The measured wake-up sensitivity at various input frequencies is shown in Figure 5, with the four frequency points defined in [1] are highlighted. The measurement results shows that within the measured frequency bands from 5.75 GHz~5.90 GHz, the wake-up sensitivity is consistently better than −82 dBm, much higher compared with what is currently defined. This level of sensitivity is comparable with the performance of the integrated receiver for DSRC data, readily supporting their expanded roles in IOV applications.

The performance of the proposed WuRx is compared with other designs reported in recent publications, as in Table I. It shows the proposed design is able to achieve very high wake-up sensitivity, and micro-watt power consumption at the same time.

IV. CONCLUSION

This paper presents a novel WuRx architecture. By reusing essential consisting components of the RF receivers from the RF receiving chains. The WuRx achieve the best −86 dBm wake-up sensitivity within the frequency bands of 5.75 GHz~5.90 GHz. A specially designed WuRx control block bypass the non-essential blocks from the RF receiving chain, and alternatively puts the WuRx receiving chain between ON/OFF states for optimized power efficiency. When setting T_h to be 2.1 s, and

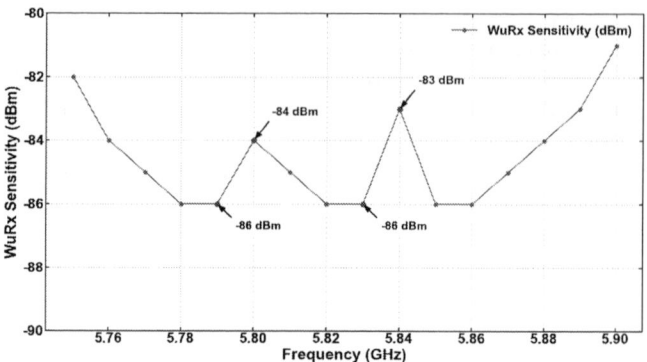

Fig. 5. Measured sensitivity of WuRx.

TABLE I
COMPARISON OF KEY PERFORMANCE OF WuRxs

Ref.	Process (nm)	WuRx Area (mm^2)	Wake-up Sensitivity (dBm)	Power (μW)
[2]	130	0.0072	-40	0.003
[4]	65	0.4558	-94.5	742
[6]	55	0.41	-85	167
[7]	95	0.18	-58	1
[8]	180	0.45	-80	54
[11]	180	-	-96.2	87.5
This work	180	0.02	-86	15.6

T_g as 8.64 ms, the effective average power consumption of the entire WuRx is limited to 15.6 μW without specific die area for a dedicated WuRx, the proposed approach only requires an additional 200 μm × 100 μm of silicon area for the WuRx control blocks.

REFERENCES

[1] Chinese National Standard GB/T 20851.1-2019, "Electronic toll collection—Dedicated short range communication—Part 1: Physical layer," 2019.

[2] I. Ali, M. Asif, H. Yingge, M. R. Ur Rehman, and K. Y. Lee, "An Ultra-Low Power Wake-up Receiver Digital Controller for 5.8 GHz DSRC Applications," in *2020 International SoC Design Conference (ISOCC)*, (Yeosu, Korea (South)), pp. 300–301, 2020.

[3] M. D'Addato, A. M. Elgani, L. Perilli, E. Franchi Scarselli, A. Gnudi, R. Canegallo, and G. Ricotti, "A 54.8-nW, 256-bit Codeword Temperature-Robust Wake-Up Receiver minimizing False Wake-Ups for Ultra-Low-Power IoT Systems," in *2022 29th IEEE International Conference on Electronics, Circuits and Systems (ICECS)*, (Glasgow, UK), pp. 1–4, 2022.

[4] S. Wang and D. D. Wentzloff, "A 742 μW −94.5 dBm Sensitivity 5G-NR Wake-Up Receiver," in *2025 IEEE Radio Frequency Integrated Circuits Symposium (RFIC)*, (San Francisco, CA, USA), pp. 203–206, 2025.

[5] A. N. Abdelrahman, M. E. Fouda, and A. M. Eltawil, "A −96.2 dBm / 3.5 μW Wake-up Receiver with False Triggering Detection for Human Body Communication," in *2023 30th IEEE International Conference on Electronics, Circuits and Systems (ICECS)*, (Istanbul, Turkiye), pp. 1–4, 2023.

979-8-3315-8850-2/25 $31.00 © 2025 IEEE   250

[6] H. Chen and J. Huang, "A Low Power BLE-Compatible Wake-Up Receiver Achieving -85-dBm Sensitivity," in *2024 9th International Conference on Integrated Circuits and Microsystems (ICICM)*, (Wuhan, China), pp. 191–195, 2024.

[7] N. Yan, H. Zhang, X. Tan, and H. Min, "Analysis and Design of a Multi-Mode Wake-Up Receiver Based on Direct Envelope Detection in Wireless Sensor Networks," *IEEE Sensors Journal*, vol. 18, no. 22, pp. 9305–9314, 2018.

[8] K. W. Cheng and S. E. Chen, "An Ultralow-Power OOK/BFSK/DBPSK Wake-Up Receiver Based on Injection-Locked Oscillator," *IEEE Transactions on Very Large Scale Integration (VLSI) Systems*, vol. 29, no. 7, pp. 1379–1391, 2021.

[9] Y. Li, C. Zhen, S. Liu, L. Jiang, and H. Yu, "A 16 MHz, 59.2 ppm/°C CMOS DLL-Assisted VCO with Improved Frequency Stability Towards Single Chip Wireless IOT," *Mobile Networks and Applications*, vol. 21, pp. 943–949, Dec 2016.

[10] T. Ma *et al.*, "A New OOK Wake-Up Receiver With Sampling Technique," in *Proceedings of the 13th EAI International Conference on Mobile Multimedia Communications, Mobimedia 2020*, (Cyberspace), aug 2020.

[11] M. Bouraoui, A. Neifar, I. Barraj, and M. Masmoudi, "Low-Power Envelope Detector for WSN Wake-up Receiver Applications," in *2023 IEEE International Conference on Design, Test and Technology of Integrated Systems (DTTIS)*, (Gammarth, Tunisia), pp. 1–5, 2023.

[12] B. Ge, Y. Li, H. Yu, and X. Feng, "Design and implementation of quadrature bandpass sigma–delta modulator used in low-IF RF receiver," *Journal of Semiconductors*, vol. 39, p. 055002, May 2018.

2025 The 10th International Conference on Integrated Circuits and Microsystems

A 28 – 32 Gb/s Wireline Receiver with A Genetic-Algorithm-Based Adaptive CTLE in 28 nm CMOS

Yingjie Zhang
College of Computer
Science and Technology
National University of
Defense Technology
Changsha, China
243299736@qq.com

Fangxu Lv *
College of Computer
Science and Technology
National University of
Defense Technology
Changsha, China
lvfangxu1988@nudt.edu.cn
*Corresponding author

Liangyong Yuan
College of Computer
Science and Technology
National University of
Defense Technology
Changsha, China
yuanliangyong23@nudt.edu.cn

Xiaoyue Hu
College of Computer
Science and Technology
National University of
Defense Technology
Changsha, China
twobirds0918@163.com

Ruixiao Kuai
College of Computer
Science and Technology
National University of
Defense Technology
Changsha, China
18195321046@163.com

Xianchao Zeng
College of Computer
Science and Technology
National University of
Defense Technology
Changsha, China
1503583183@qq.com

Abstract—A 28 – 32 Gb/s wireline receiver with a genetic-algorithm-based adaptive continuous-time linear equalizer (CTLE) is presented in this paper. The proposed CTLE adopts adaptive genetic algorithm (AGA) to update coefficients, which has a stronger global search ability and lower power consumption, and it is less prone to local optima. Experimental results show that the receiver offers a wide gain tuning range and can equalize 8-14 dB channel insertion loss at 28-32 Gb/s. The receiver is fabricated in 28-nm technology and achieves a bit error rate (BER) below 10^{-9}.

Keywords—*Wireline Receiver, Genetic Algorithm, CTLE, Adaptive equalization*

I. Introduction

In order to meet the multi-rate transmission requirements of industrial standards such as USB 3.2 and PCIe 4.0 [1], receiver design needs to achieve multi-scenario adaptation under limited power consumption, which puts higher requirements on equalization technology. Although FFE and DFE can compensate for channel distortion, the high latency nature of digital domain processing limits their application at high rates. In contrast, CTLE can directly compensate high-frequency losses at the front end of the receiving link by virtue of its analog domain high-pass filtering capability, and its low latency characteristics make it a core component of ultra-high-speed SerDes systems. However, fixed parameter CTLE is difficult to cope with the dynamic channel environment, adaptive CTLE comes into being.

Recent studies have proposed various improved adaptive equalization techniques. Shim et al.[2] proposed the stochastic CTLE gain adaptive method (SCGS), which significantly improved the equalization accuracy by optimizing the gain

coefficient through multiple local statistics. Lee et al.[3] adopted the stepwise least mean square (SSLMS) algorithm to jointly optimize CTLE and DFE parameters, reducing the adaptive time and power consumption. Other approaches, such as asynchronous under-sampled histograms[4] and stochastic sigma-tracking eyeopening monitors(SSEOM)[5] have enabled reference-free equalization, minimizing reliance on clock synchronization. However, these traditional adaptive equalization techniques tend to converge to sub-optimal values, high complexity and high power consumption in dynamic channel environments, which are not suitable for real-time and energy-saving requirements in high data rate scenarios. Intelligent optimization algorithms provide novel solutions to these challenges[6][7].

This paper presents a CTLE adaptive equalization scheme for receivers based on genetic algorithm. The scheme uses eye closure as fitness function, and conducts an efficient search for equalization parameters through selection, crossover, and mutation operations. The experimental results show that the proposed method can compensate the channel loss of 8-14 dB at 28-32 Gb/s, while keeping the bit error rate below 10^{-9}.

The remainder of this paper is organized as follows. In Section II, the architecture of the receiver and the principle of CTLE are presented. Section III delves into the adaptive equalization architecture based on the genetic algorithm. Finally, the measurement results and conclusions are reported in Section IV and Section V, respectively.

II. Receiver Architecture

The structure of the receiver is shown in Fig. 1. The input data signal is first conditioned and pre-processed by the analog front-end (AFE). Subsequently, three parallel samplers — the data sampler, edge sampler, and error sampler — are employed for sampling. Each sampler is cascaded with a 1:32

National Key Research and Development Program (2022YFB2803101)

979-8-3315-8850-2/25 $31.00 © 2025 IEEE

demultiplexer to demultiplex the signal into 32 channels for parallel processing, followed by PRBS7 verification. The Eye Monitor is responsible for monitoring the eye diagram of the output signal, calculating the eye closure[5], and collaborating with the genetic algorithm to achieve AFE parameter optimization. In the clock path, the external 14 GHz clock input serves as the source for the frequency divider to generate quadrature differential clocks. These clocks are used to drive the phase interpolators (PIs). The PIs which have a 90° phase difference are employed for sampling data and edge cues. The regulation of the PIs is accomplished using an 18-bit digital code output from the Clock Recovery logic. Subsequently, CML-to-CMOS converters are utilized to create a rail-to-rail swing that drives all the samplers[8][9].

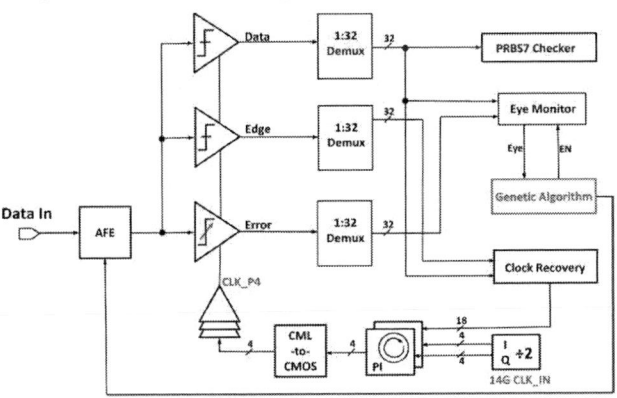

Fig. 1. Block diagram of the receiver

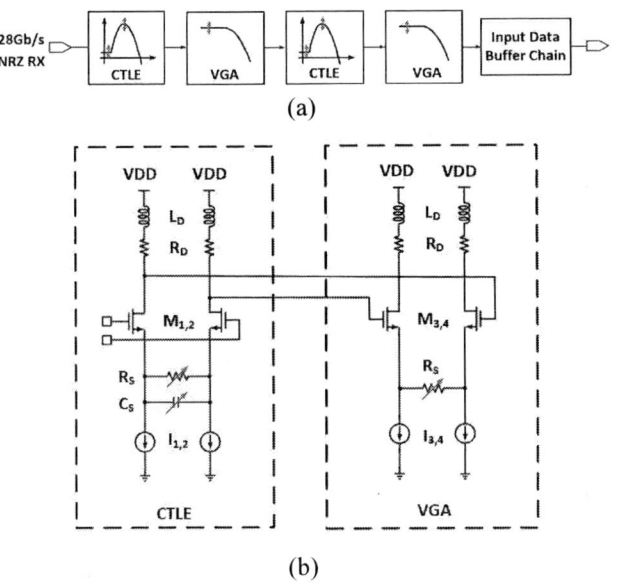

Fig. 2.(a) Block diagram of the proposed adjustable AFE.   (b)   CTLE and VGA schematic.

As shown in Fig. 2 (a), the proposed AFE employs a cascaded two-stage CTLEs and two-stage VGAs architecture, providing a gain dynamic range from -13.98 dB to -8.96 dB. Specifically, each CTLE stage incorporates degenerate resistors R_S and capacitors C_S, forming four sets of programmable parameters, while each VGA stage includes

degenerate resistors R_S as two additional sets of programmable parameters. All parameters are encoded with 5-bit binary values, totaling 30 configuration bits, yielding approximately 1 billion distinct combinations when considering all configurations.

Fig. 2 (b) illustrates the schematic diagrams of CTLE and VGA. The CTLE module employs a source-degenerated negative feedback configuration, whose transfer function is given by:

$$H(s) = \frac{g_m \cdot \dfrac{R_D}{sR_DC_P + 1}}{1 + g_m \dfrac{R_S}{2sR_SC_S + 2}} = \frac{2g_mR_D(1 + sR_SC_S)}{(sR_DC_P + 1) \cdot [2(s \cdot R_SC_S + 1) + g_mR_S]} \quad (1)$$

where g_m is transconductance, R_d is drain resistance, C_p is typically the parasitic capacitance at the output, R_S is source resistance, and C_S is source capacitance.Its low - frequency gain and zeros and poles are as follows:

$$A_0 = \frac{2g_mR_D}{2 + g_mR_S} \quad (2)$$

$$w_z = -\frac{1}{R_SC_S} \quad (3)$$

$$w_{p1} = -\frac{\dfrac{g_mR_S}{2} + 1}{R_SC_S} \quad (4)$$

$$w_{p2} = -\frac{1}{R_DC_P} \quad (5)$$

where A_0 represents the low-frequency gain, w_z denotes the zero, and w_{p1} and w_{p2} stand for the poles.

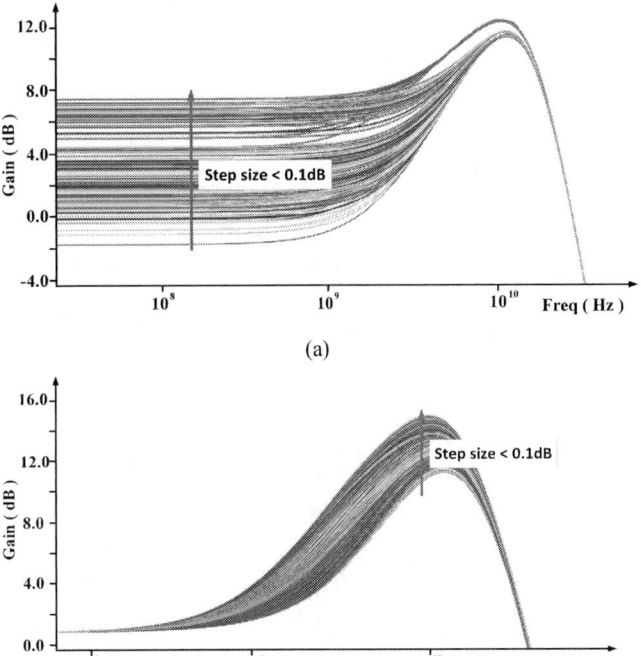

Fig. 3. Adjustable parameters of CTLE (a) Low frequency gain adjustment range.   (b)   Mid-High frequency gain adjustment range.

As shown in equation (2-5), the equalization efficiency of CTLE can be dynamically adjusted by modifying the values of key parameters Rs and Cs. Low frequency gain is achieved by adjusting Rs. High-frequency gain regulation is achieved through a digital-to-analog converter (DAC) that controls the bias voltage of the variable capacitor, providing 64 stages of fine tuning in the 0-900 mV range with a step of 14 mV. Adjusting Cs can change the absolute position of w_z and w_{p1}, while keeping w_{p1}/w_z constant, so as to adjust their relative position relationship with w_{p2}, and achieve dynamic adjustment of equalization gain. The parameter tuning range of a single-stage CTLE is shown in Fig. 3.

The VGA is designed to dynamically adjust signal gain in response to dynamic input signal variations. This device can perform real-time adjustment of the amplification factor to stabilize output signal amplitude within the target range, ensuring both effective amplification of weak signals for subsequent processing and prevention of circuit overload caused by strong signals. The input data buffer chain, positioned after CTLE and VGA as an output buffer stage, enhances driving capability and isolates loads to maintain signal quality and system stability.

III. Adaptive Genetic Algorithm

A. Genetic Algorithm

The Genetic Algorithm is a meta-heuristic algorithm that mimics the principles of natural selection and genetics. Compared with traditional gradient descent algorithms [10][11] , one of the most significant features of GA is its ability to directly operate on data objects or target populations without the constraints of requiring derivative calculations or function continuity. It possesses inherent implicit parallelism and demonstrates a superior capability for finding global optimal solutions. Moreover, GA does not necessitate the establishment of extensive rules; instead, it autonomously guides the search space and adaptively adjusts the search direction solely through the use of a fitness function.

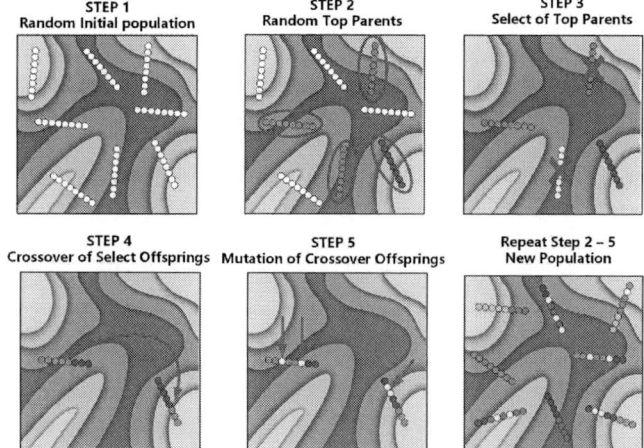

Fig. 4.Flowchart of Adaptive Genetic Algorithm for AFE Parameter Optimization

Fig. 4 shows the process of AGA optimizing AFE parameters. The process begins with the generation of an initial

population, where individuals represent different sets of equalizer parameters, such as adjustable gain and pole-zero settings. The population size is carefully chosen to balance diversity and computational complexity. Then, the tournament selection algorithm is used to compare the fitness of different individuals. Suppose that the fitness of k individuals are $F_{i1}, F_{i2},\ldots F_{ik}$, then the probability of the individual i being selected is:

$$p_i = \frac{F_i}{\sum_{j=1}^{k} F_j} \qquad (6)$$

Following selection, the parameter values from both parents are combined at randomly generated intersections to create progeny, as shown in Formula 7, where P_1 and P_2 represent the parent individuals, C_1 and C_2 represent the offspring individuals after crossover, G_i represents the gene value of an individual ($1 \leq i \leq k$), and k is the length of gene.

$$\left. \begin{array}{l} P_1 = [G_1, G_2, \cdots, G_k] \\ P_2 = [G_1', G_2', \cdots, G_k'] \end{array} \right\} \Rightarrow \left\{ \begin{array}{l} C_1 = [G_1, G_2, \cdots, G_c', G_{c+1}', \cdots, G_k'] \\ C_2 = [G_1', G_2', \cdots, G_c, G_{c+1}, \cdots, C_k] \end{array} \right. \quad (7)$$

Mutation is subsequently introduced to maintain population diversity. For individuals encoded in binary, mutation is achieved by flipping the gene bits. As shown in Formula 8,

$$G_i' = \begin{cases} 1 - G_i & \text{with probability } p_m \\ G_i & \text{with probability } 1 - p_m \end{cases} \qquad (8)$$

where G_i represents the gene value obtained after crossover, $G_i`$ represents the gene value after mutation ($1 \leq i \leq k$) and p_m is the probability of variation.

The new population composed of offspring generated through selection, crossover, and mutation replaces the previous generation. Subsequently, the entire process of fitness evaluation, selection, crossover, and mutation is repeated until the specified number of iterations is reached or the convergence criteria are met. This iterative process enables the genetic algorithm to effectively optimize the equalizer parameters, providing a robust approach for signal equalization in high-speed communication systems.

B. Fitness Function

The selection of the fitness function directly impacts the convergence speed of the genetic algorithm and its ability to find the optimal solution. The reason is that GA relies almost entirely on the fitness function during evolutionary search, without utilizing external information, and conducts the search based on the fitness of each individual in the population. In this study, the eye closure is used as the fitness function for the population genes. Individuals with thinner eye closure are assigned higher fitness values, and vice versa for thicker ones.

The eye closure represents the anti-noise performance and ISI suppression ability, while the eye height only reflects the amplitude margin, which can not fully characterize the signal integrity under the condition of channel degradation. As shown in Fig. 5, although the eye height in Fig. 5(a) is larger than that in Fig. 5(b), the eye closure is larger and the eye width is smaller. Given its direct correlation with the system's resilience

979-8-3315-8850-2/25 $31.00 © 2025 IEEE

to noise and ISI, this paper selects eye closure as the fitness function for AFE parameter optimization.

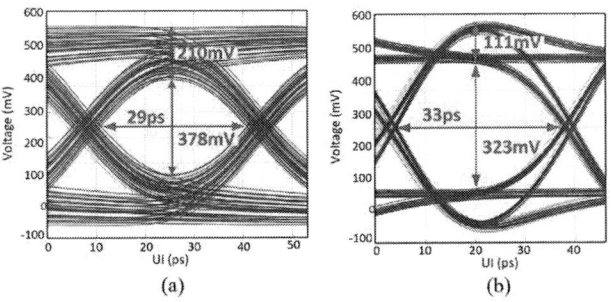

Fig. 5. Selection of the fitness function (a) Eye Diagrams of Normal Condition. (b) Eye Diagram of Over-Equalized.

IV. EXPERIMENT RESULTS

A. Channel Model

Based on digital-analog hybrid simulation platform, two distinct channels and two different data rates were employed to evaluate the receiver performance. The insertion loss characteristics of the channels are presented in Fig. 6, where channel A exhibits insertion losses of 8.96 dB and 9.85 dB at 14 GHz and 16 GHz. In contrast, channel B has a higher channel insertion loss, with the loss values of 12.7 dB and 13.98 dB at the same frequency points.

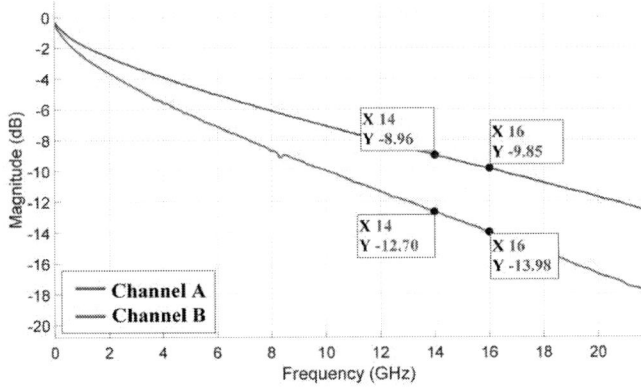

Fig. 6. Insertion Loss of Two Channels

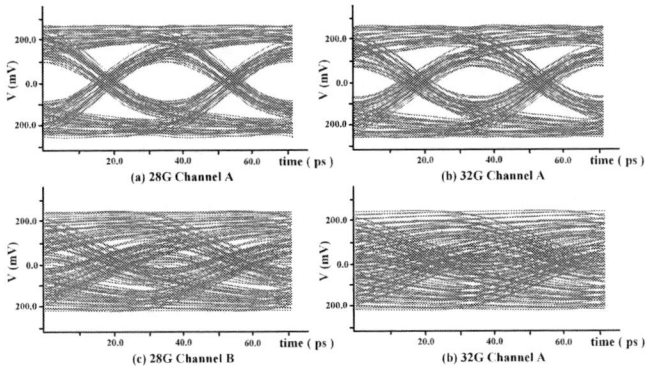

Fig. 7. The eye diagrams after different channels at various data rates

As shown in Fig. 7, the eye diagrams of NRZ signals after channel transmission are presented. For channel A with lower loss, the signal attenuates obviously and the eye height decreases greatly. For channel B with higher loss, the eye diagram was almost completely closed.

B. Equalization Results of Genetic-Algorithm-Based Adaptive CTLE

In the proposed AGA framework, the maximum number of iterations is set to 50 and the initial population size is 16 candidate individuals. Figs. 8-10 respectively illustrate the equalization results of signals under different data rates and channels.

In the optimization iteration process of the adaptive equalization algorithm, Fig. 8 clearly presents the evolution trajectory of the optimal fitness of the population. It is found that when the number of iterations reaches 20, the fitness functions of the four different conditions all show significant convergence characteristics.

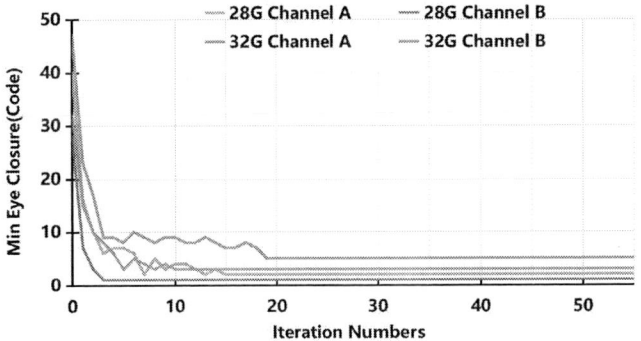

Fig. 8. Convergence Process of Fitness Function in Genetic Algorithm-Based Adaptive Equalization

The signal eye diagram after AFE adaptive equalization is shown in Fig. 9. The optimization results show that: under channel A at 28 Gb/s, the eye height is increased to 650.20 mV; under channel A at 32 Gb/s, the eye height is increased to 674.41 mV; under channel B at 28 Gb/s, the eye height is increased to 451.48mV; under channel B at 32 Gb/s, the eye height is increased to 342.10 mV. The eye diagram clarity has been substantially enhanced, with both the eye height and eye width exhibiting significant expansion.

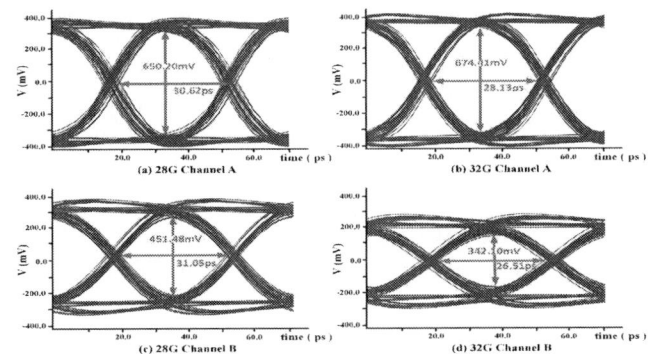

Fig. 9. Eye Diagrams of Genetic Algorithm-Based Adaptive Equalization

979-8-3315-8850-2/25 $31.00 © 2025 IEEE

The bathtub curve shown in Fig. 10 reveals the suppression characteristics of the system against compound jitter. Within the range of ± 0.5 UI near the optimal sampling point, the bathtub curve presents a typical U-shaped distribution, and the measured BER is less than 10^{-9}.

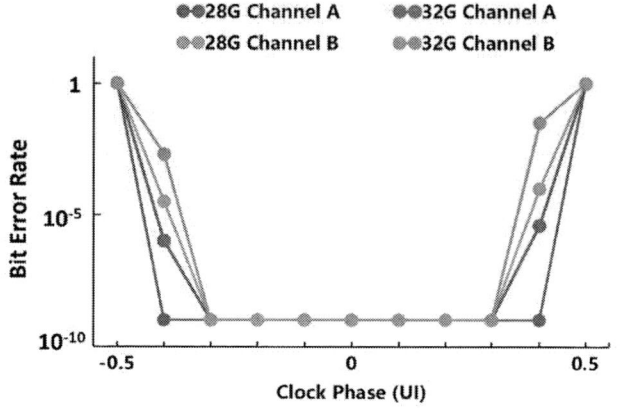

Fig. 10. The measured bathtub curves of 28 – 32 Gb/s NRZ signals under different channels

Finally, Table I compares this letter to recently published prior works for CTLE. Compared with literature [12][13][14], CTLE adaptive equalization is realized in this paper. Compared with literature [1], the genetic-algorithm-based adaptive CTLE exhibits superior performance in channel fading compensation and enhanced adjustment accuracy.

TABLE I.   PERFORMANCE SUMMARY AND COMPARISON

	[12] ISSCC 2017	[13] VLSI-SOC 2020	[1] T-CAS II 2023	[14] T-CAS II 2024	This work
Signaling	NRZ	NRZ	NRZ	NRZ	NRZ
Data Rate	22.5-32 Gb/s	10-15 Gb/s	16 Gb/s	13 Gb/s	28-32 Gb/s
Channel loss	-14.8 dB	-15 ~ -21 dB	-14 dB	-15.4 dB	-8 ~ -14 dB
Adaptaion	NO	NO	YES	NO	YES
Technology	20nm	65nm	28nm	28nm	28nm
Total Power	102.0mW	6mW	17.7mW	11.05mW	78.43mW

V. SUMMARY

This paper presents a 28-32 Gb/s wireline receiver with an adaptive CTLE based on genetic algorithm. This scheme makes use of the randomness and global search ability of genetic algorithm to make up for the shortcomings of traditional methods which are easy to fall into local optima and high complexity. The experimental results verify that the receiver can compensate for 8-14 dB channel loss at 28-32 Gb/s with a BER well below 10^{-9}, while simultaneously achieving 78.43mW power consumption in a 28-nm technology. This work delivers an efficient solution for dynamic channel adaptive equalization in high-speed SerDes systems.

REFERENCES

[1] Shu, Zhou,Huang,et al. A 5‑13.5 Gb/s Multistandard Receiver With High Jitter Tolerance Digital CDR in 40-nm CMOS Process.[J]. IEEE Transactions on Circuits & Systems. Part I: Regular Papers,2020,Vol.67(10): 3378-3388.

[2] Minkyo Shim,Kwang-Hoon Lee,Seungha Roh,et al. A 1.1-pJ/b 8-to-16-Gb/s Receiver With Stochastic CTLE Adaptation[J]. IEEE Transactions on Circuits and Systems II: Express Briefs,2023,Vol.70(2): 381-385.

[3] Jinhyung Lee,Kwangho Lee,Hyojun Kim. A 0.1-pJ/b/dB 1.62-to-10.8-Gb/s Video Interface Receiver With Jointly Adaptive CTLE and DFE Using Biased Data-Level Reference[J]. IEEE Journal of Solid-State Circuits,2020,Vol.55(8): 2186-2195.

[4] Wang-Soo Kim,Chang-Kyung Seong,Woo-Young Choi. A 5.4-Gbit/s Adaptive Continuous-Time Linear Equalizer Using Asynchronous Undersampling Histograms[J]. Circuits and Systems II: Express Briefs, IEEE Transactions on,2012,Vol.59(9): 553-557.

[5] Won, Hyosup,Lee,et al. A 28-Gb/s Receiver With Self-contained Adaptive Equalization and Sampling Point Control Using Stochastic Sigma-Tracking Eye-Opening Monitor.[J]. IEEE Transactions on Circuits & Systems. Part I: Regular Papers,2017,Vol.64(3): 664-674.

[6] Shayan Shahramian,Behzad Dehlaghi,Joshua Liang,et al. 30.5 A 1.41pJ/b 56Gb/s PAM-4 Wireline Receiver Employing Enhanced Pattern Utilization CDR and Genetic Adaptation Algorithms in 7nm CMOS[C]//2019 IEEE International Solid- State Circuits Conference - (ISSCC). 2019.

[7] Suraj Kumar Prusty,Soumya P. Dash,V. K. Surya,et al. Differential Evolution-Based Adaptation Algorithm for Multistage Continuous-Time Linear Equalizer[J]. IEEE Transactions on Components, Packaging and Manufacturing Technology,2023,Vol.13(12): 2046-2049.

[8] Poonam Sachdeva and Ankita Aggarwal, "DESIGN OF CMOS RING VCO FOR PLL BASED FREQUENCY SYNTHESIZER," International Journal of Electrical and Electronic Engineering & Telecommunications, Vol. 5, No. 2, pp. 73-79, April 2016.

[9] Rajeshkumar, "REDUCE POWER CONSUMPTION FOR DIGITAL CMOS CIRCUITS USING DVTS ALGORITHAM," International Journal of Electrical and Electronic Engineering & Telecommunications, Vol. 1, No. 1, pp. 376-383, March 2015.

[10] Zhang Luo,Sichun Du,Zedi Zhang,et al. Artificial Neural Network Based on Memristive Circuit for High-Speed Equalization[J]. IEEE Transactions on Circuits and Systems I: Regular Papers,2024,Vol.71(4): 1745-1756.

[11] C. Xu et al., "A Novel High-Speed Adaptive Duobinary Digital Detector Based on the Feed-Forward Equalizer and the Maximum Likelihood Sequence Detector for Wireline Transceivers," in IEEE Transactions on Very Large Scale Integration (VLSI) Systems, vol. 33, no. 4, pp. 1042-1052, April 2025, doi: 10.1109/TVLSI.2025.3528127.

[12] Wahid Rahman,Danny Yoo,Joshua Liang,et al. 6.6 A 22.5-to-32Gb/s 3.2pJ/b referenceless baud-rate digital CDR with DFE and CTLE in 28nm CMOS[C]//2017 IEEE International Solid-State Circuits Conference (ISSCC). 2017.

[13] A. Aghighi, A. Tajalli and M. Taherzadeh-Sani, "A Low-Power 10 to 15 Gb/s Common-Gate CTLE Based on Optimized Active Inductors," 2020 IFIP/IEEE 28th International Conference on Very Large Scale Integration (VLSI-SOC), Salt Lake City, UT, USA, 2020, pp. 171-175

[14] H. Kang et al., "A 13-Gb/s Single-Ended NRZ Receiver With 1-Sample Per 2-UI Using Data Edge Sampling for Memory Interfaces," in IEEE Transactions on Circuits and Systems II: Express Briefs, vol. 71, no. 7, pp. 3328-3332, July 2024, doi: 10.1109/TCSII.2024.3362995.

2025 The 10th International Conference on Integrated Circuits and Microsystems

A Wideband Blocker-Tolerant Receiver with High Linearity and Low 0 dBm-BNF in 65-nm CMOS

Kaiyun Deng[†], Yingqi Liu[†], Qingrui Jiang[†], Kailin Liu[†], Shaohui Pan[††] and Haigang Feng[†]*

[†]Shenzhen International Graduate School,Tsinghua University, Shenzhen 518055, China

[††]Anyka Microelectronics Co., Ltd, Guangzhou 510300, China

dkj23@mails.tsinghua.edu.cn, feng.haigang@sz.tsinghua.edu.cn

*Corresponding author

Abstract—This paper presents a blocker-tolerant receiver implemented in a 65 nm CMOS process. The receiver adopts a LNTA-first architecture and leverages an N-path filter with rapid roll-off characteristic to suppress out-of-band (OB) interferer. Additionally, a second-order transimpedance amplifier (TIA) is employed to enhance selectivity on the baseband side. The receiver operates over a frequency range of 0.4–4.5 GHz, with a RF bandwidth (RF BW) of 60 MHz. It achieves a maximum gain of 38.57 dB and a minimum double-sideband (DSB) noise figure (NF) of 3.34 dB. The in-band (IB) third-order input intercept point (IIP3) is −8.2 dBm, while the OB IIP3 reaches 17.5 dBm with a frequency offset of $\Delta f / BW$=5 and the blocker NF (BNF) reaches 8.02 dB under a 0 dBm interferer of at 2 GHz. The total power consumption of the receiver is 44.1 mW + 1.8 mW/GHz.

Keywords—wideband receiver, blocker-tolerant, LNTA-first, current mode, N-path filter

I. INTRODUCTION

With the advancement of communication technology, the vision of the Internet of Everything is becoming a reality, further accelerating the development of 5G communications and the Internet of Things (IoT). As applications continue to evolve, they also drive the demand for improved communication system performance. Various application scenarios impose increasingly stringent requirements on the frequency coverage of RF systems. To reduce costs, there is a growing pursuit of monolithic, multi-protocol, and multi-band transceiver systems. Due to their high reconfigurability, wideband receivers facilitate the development of software-defined radio and cognitive radio applications. However, the increasing congestion of the communication spectrum and the proliferation of wireless devices have exacerbated the susceptibility of wideband receivers to interferer. Therefore, communication systems require receiver architectures capable of adapting to different protocols and frequency bands while maintaining stable and reliable performance in complex environments. In recent years, extensive research and discussion in academia and industry have focused on wideband blocker-tolerant receivers that operate without surface acoustic wave (SAW) filters. Such receivers must exhibit wideband operation and high linearity to ensure high-speed data transmission in challenging environments.

Currently, wideband blocker-tolerant receiver architecture includes the LNTA-first [1-7] and Mixer-first [8-12] topologies. The LNTA-first receiver operates in current mode, with its development driven by current-driven passive mixers. To suppress out-of-band (OB) interferer and enhance blocker

tolerance, the LNTA-First architecture is typically combined with N-path filters[1-2]. These filters exhibit the characteristics of high-Q passive filters, making them effective in interferer suppression. On the one hand, a bandpass structure can be employed to set the passband at the desired signal frequency while filtering out OB interferer. On the other hand, when the blocker signal frequency is known, a band-stop structure can be utilized to create a notch at the blocker's frequency. The Mixer-First receiver, leveraging the "transparency" of passive mixers, shifts baseband low-pass filtering to the RF domain, transforming it into bandpass filtering at the antenna port [8]. Additionally, the baseband input impedance is transferred to the RF domain, facilitating impedance matching. However, due to the absence of low noise transconductance amplifier (LNTA) isolation, the Mixer-first receiver suffers from severe local oscillator (LO) leakage. Furthermore, since the mixer does not inherently provide voltage gain when employed as the first stage of a receiver front-end, it introduces several design challenges for the subsequent signal chain. In the absence of a preceding low-noise amplifier (LNA), the mixer must directly interface with the antenna, thereby exposing the following baseband circuitry to the full noise figure of the mixer and any preceding passive components, such as baluns or matching networks. This configuration necessitates the use of highly linear and low-noise baseband amplifiers, which typically consume significant power to achieve low noise performance.

This paper proposes a wideband blocker-tolerant receiver under 65nm CMOS process based on a rapid roll-off feedback architecture with enhanced notch filtering. The proposed receiver achieves on-chip tunable high-Q filtering for improved blocker rejection. Meanwhile, the baseband transimpedance amplifier (TIA) adopts second-order filtering to enhance the OB signal suppression capability on the baseband side. The rest of this paper is organized as follows. Section II analyzes the implemented system architecture. Section III details the design of each receiver module. Section IV presents and discusses simulation results. Finally, Section V concludes the paper.

II. RECEIVER ARCHITECTURE ANLYSIS

The proposed receiver adopts an LNTA-first architecture, as shown in Fig. 1. A gain-boosted N-path filter is employed to initially suppress blocker signals. Due to the negative feedback effect, the parallel connection of N-path filter and LNTA amplifies the equivalent capacitance in the filtering branch and reduces the switching resistance, thereby improving the performance of the N-path filter. The N-path filter network is realized using two parallel N-path filter banks, and additional

979-8-3315-8850-2/25 $31.00 © 2025 IEEE

frequency-domain selectivity is introduced through the incorporation of a gyrator-based structure that enables the programming of two tunable transmission zeros. These transmission zeros substantially enhance the bandpass characteristics, enabling a steeper roll-off around the desired signal band and improving blocker rejection. The passive mixers convert the baseband impedance to RF impedance at the LNTA output due to their reciprocal properties [13]. Following the mixer, a second-order TIA convert the RF front-end current signal into a baseband voltage signal, enabling baseband filtering and amplification while maintaining strong blocker rejection. Non-overlapping clocks with approximately 25% duty cycle required by the N-path filter and mixer are generated using a low-power windmill frequency divider, achieving low phase noise and minimal mismatch with low power consumption.

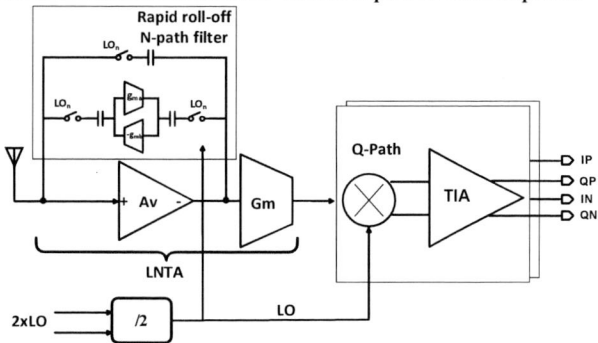

Fig. 1. Proposed blocker-tolerant receiver architecture.

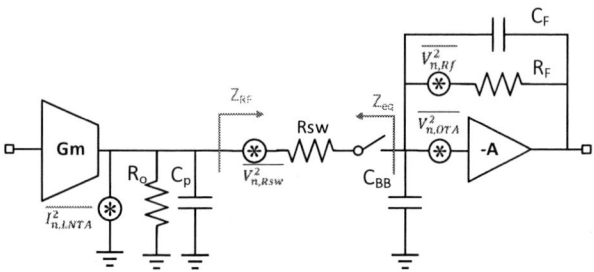

Fig. 2. Simplified equivalent model of the proposed receiver [6].

Disregarding the N-path filter, a simplified system model with noise source of the proposed blocker-tolerant LNTA-first receiver is shown in Fig. 2. Assuming that R_o and C_p denote the output impedance of the LNTA and Z_{RF} is the impedance seen backward from LNTA output, the receiver gain $G(\omega)$ and noise figure (NF) can be expressed as [6]:

$$G(\omega) = \frac{2\sqrt{2}G_m R_F}{\pi\left[1+j\left(\omega-\omega_{LO}\right)R_F C_F\right]} \cdot \frac{R_o/\left(R_o+Z_{RF}\right)}{1+j\omega R_o C_p} \quad (1)$$

$$NF = 1+\frac{\overline{V_{in,LNTA}^2}+\overline{V_{in,mix}^2}+\overline{V_{in,OTA}^2}+\overline{V_{in,R_F}^2}}{4kTR_s}$$

$$\approx 1+\frac{\overline{I_{n,LNTA}^2}}{4kTR_s G_m^2}+\frac{R_{SW}\left(1+sR_o C_p\right)^4}{R_s G_m^2 R_o^2}$$

$$+\frac{\pi^2 \overline{V_{n,OTA}^2}\left(1+sR_o C_p\right)^4}{512kTR_s G_m^2 R_F R_o^2}+\frac{\pi^2\left(1+jwR_o C_p\right)^2}{8R_s G_m^2 R_F} \quad (2)$$

Therefore, increasing the transconductance gain of LNTA and TIA can effectively reduce the noise figure.

III. CIRCUIT DESIGN DETAILS

A. Bandpass LNTA

The proposed N-path filter exhibits notch characteristics and consists of path A and path B. In path A, a gyrator formed by g_{ma} and g_{mb} enables tunable zeros of N-path filter. The circuit implementation of gma and gmb is shown in Fig. 3. Increasing g_{ma} and g_{mb} shifts the zero frequencies upward while simultaneously reducing the RX input impedance near the realized zeros. Path B is an auxiliary N-path filter. Unlike [4] and [7], the auxiliary N-path filter in this design is directly connected in parallel with the input and output of the LNTA, providing a feedback path. This auxiliary N-path filter improves both the stability and frequency response of the receiver, as referenced in [5]. The LNTA consists of an LNA and a Gm-cell, both implemented using a single-stage inverter structure, as shown in Fig. 3(a). The stacked n/p MOS structure is adopted to achieve higher current efficiency, characterized by an improved g_m/I_D ratio. The feedback resistor R_{F1} in the LNA ensures 50 Ω input matching while enabling self-biasing of the LNA. The feedback resistor R_{F2}, with a value of 20 kΩ, provides high output impedance for the Gm-cell, facilitating V-I conversion. The circuit architecture, composed of the notch N-path filter network and the LNTA, forms a bandpass LNTA that amplifies the desired signal while suppressing OB blockers, thereby achieving channel selection.

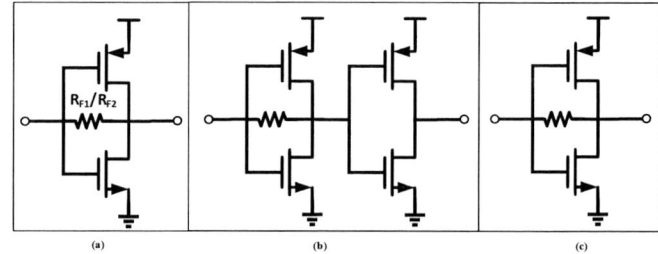

Fig. 3. Schematics of the implemented (a) LNA/ Gm-cell; (b) gma and (c) gmb of the gyrator.

B. Second-order TIA

In the LNTA-first architecture, the presence of a preceding gain stage relaxes the noise requirements of the baseband, allowing a multi-stage amplifier to be considered for constructing the TIA. In this design, a three-stage inverter-based operational transconductance amplifier (OTA) is employed in the TIA. The input impedance of the TIA is influenced by both the feedback resistor and the open-loop gain. By adopting a three-stage OTA, a lower input impedance is achieved, thereby improving the TIA's linearity.

However, multi-stage operational amplifiers inevitably face stability challenges due to the presence of multiple poles, leading to reduced phase margin. Even with well-established Miller compensation techniques, achieving a high gain-bandwidth product (GBW) typically incurs significant power consumption. To address this, the proposed design combines Miller compensation with feedforward compensation while introducing a zero-canceling resistor to eliminate the right-half-

979-8-3315-8850-2/25 $31.00 © 2025 IEEE

plane zero introduced by Miller compensation. The overall architecture is illustrated in Fig. 4.

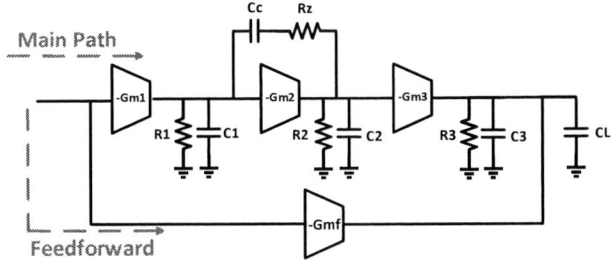

Fig. 4. Small signal model of three-stage OTA with feedforward path.

Conventional TIA structures with parallel R-C feedback exhibit only first-order roll-off characteristics, which are insufficient for OB interferer suppression. To enhance filtering performance, the proposed TIA incorporates a positive feedback capacitor to increase the filter order and the baseband bandwidth [9],[12]. The fundamental structure of the TIA is shown in Fig. 5. By combining the new positive feedback path through $C2$ with the negative feedback path through $C1$, a complex conjugate pole pair is established, thereby improving selectivity. Fig. 6 shows the transimpedance gain of the TIA. The TIA not only achieves second-order filtering via feedforward capacitors, extending the roll-off by approximately one decade, but also widens the baseband bandwidth.

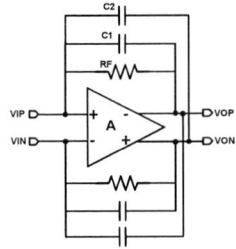

Fig. 5. Structure of the TIA.

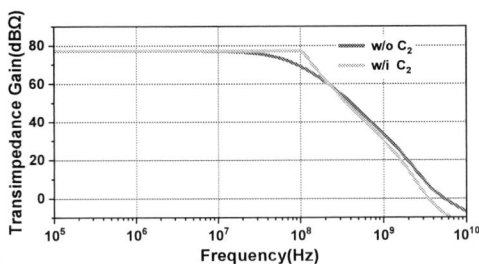

Fig. 6. Simulation results of transimpedance gain of the TIA with/without C2.

C. Windmill Frequency Divider

The receiver utilizes four LO phases with a 25% duty cycle to enable blocker rejection and down-conversion. The LO generation circuit in this design adopts an improved NAND-based windmill frequency divider [6], as shown in Fig. 7. The fundamental building blocks of this design consist of logic gates, eliminating the need for D flip-flops, which typically consume significant power. Compared to conventional windmill dividers [3], this design replaces the original NOR logic with NAND

logic, resulting in an output signal with a 75% duty cycle. To achieve the desired 25% duty cycle, the subsequent driver circuit performs an odd number of inversions. This architecture retains the output transistor-sharing structure, which enhances driving capability without increasing the transistor count, thereby reducing power consumption and improving edge balance in the output stage. Additionally, this approach minimizes the uncorrelated noise generated by the two NMOS transistors, leading to phase noise optimization. The complete working principle and phase noise simulation results of the multi-phase clock generation circuit is illustrated in Fig. 8. The windmill frequency divide achieves simulated −164 dBc/Hz phase noise at a 1-MHz offset from 1 GHz while consuming approximately 0.35 mW/GHz.

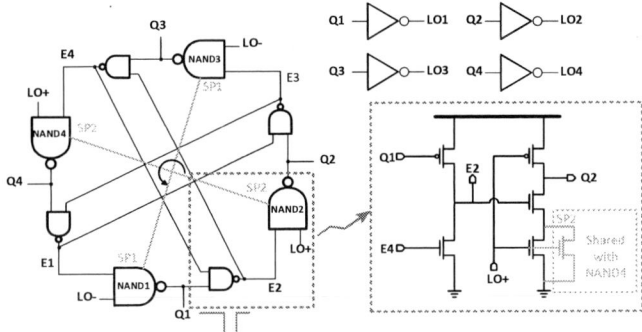

Fig. 7. Improved NAND-based windmill frequency divider.

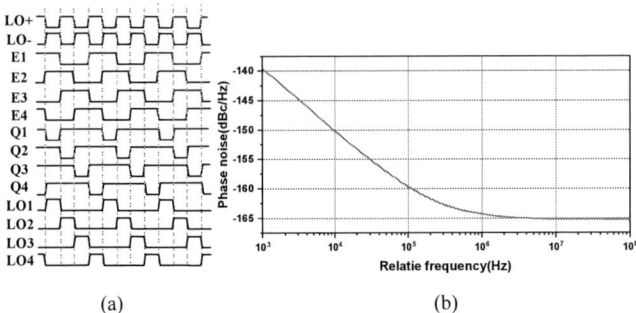

| (a) | (b) |

Fig. 8. (a) Working principle and (b) simulation results of phase noise of the frequency divider.

IV. SIMULATION RESULT

The proposed LNA is implemented using 65-nm CMOS technology. Consider the parasitic capacitance of 100 fF on pads and the inductance of 1 nH on the bonding wires. The simulated gain and S_{11} results of the receiver are shown in Fig. 9. The receiver operates over a frequency range of 0.4–4.5 GHz, achieving a maximum voltage gain of 38.57 dB. Across the entire RF operating band, the receiver exhibits good input matching and strong frequency selectivity. Fig. 10(a) presents the minimum double-sideband (DSB) NF, which ranges from 3.34 to 5.25 dB. Fig. 10(b) illustrates the simulated results of the blocker NF (BNF) with interferer power when the interferer signal is introduced with an offset of 80 MHz within the receiver bandwidth at 2 GHz. Under the presence of a 0 dBm interferer, the NF of the receiver is 8.02 dB. Fig. 11 shows the simulated in-band third-order input intercept point (IB-IIP3) and out-of-band IIP3 (OB-IIP3) at 2 GHz, where OB-IIP3 is evaluated at a

979-8-3315-8850-2/25 $31.00 © 2025 IEEE

frequency offset of $\Delta f/BW$=5. The IB-IIP3 and OB-IIP3 are -8.2 dBm and 17.5 dBm, respectively. The total power consumption of the proposed receiver is 44.1 mW + 1.8 mW/GHz. Table I summarizes the simulation results and provides a comparison with state-of-the-art designs. The proposed receiver achieves good anti-interference capability while maintaining appropriate power consumption and low noise performance.

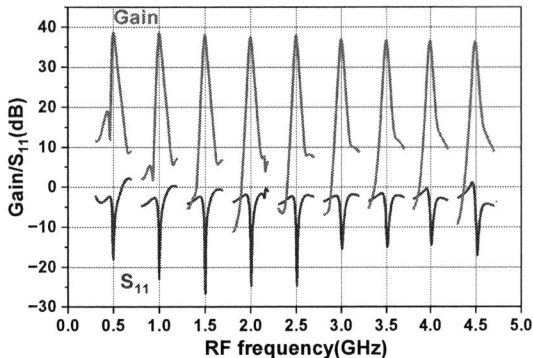

Fig. 9. Simulation results of Gain and S_{11}.

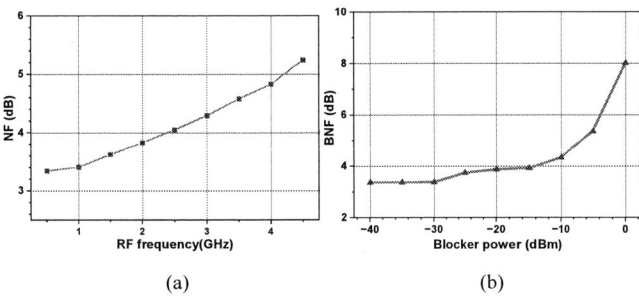

(a)   (b)

Fig. 10. Simulation results of (a) DSB NF vs. operating frequency and (b) 0-dBm BNF vs. block power.

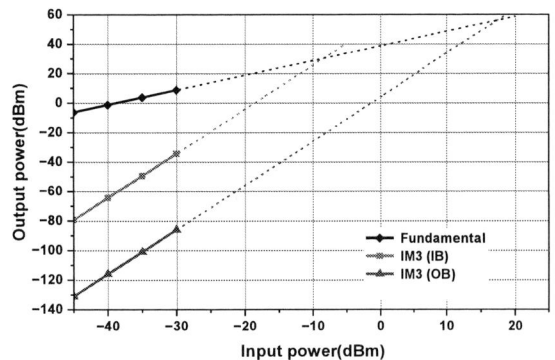

Fig. 11. Simulation results of IB-IIP3 and OB-IIP3.

V. CONCLUSION

In this paper, a blocker-tolerant receiver based on rapid roll-off feedback N-path filter architecture and second-order TIA is reported, aiming to achieve high OOB linearity in LNTA-first receiver. Implemented in CMOS 65nm technology, the receiver achieves 17.5 dBm OOB-IIP3 at 2 GHz and 8.02 dB 0dBm-BNF at 80MHz offset. Meanwhile, the receiver achieve a wide operating RF frequency range and low noise performance. The

proposed receiver supports a wide RF tuning range while maintaining low noise performance, making it well-suited for modern multi-band, interference-prone wireless communication environments. The results validate the proposed architecture as a compelling solution for high-linearity, low-noise, and frequency-agile RF front-ends in next-generation software-defined radios and multi-standard receivers.

TABLE I.   COMPARISON WITH PRIOR ARTS

	ISSCC'21 [4]	JSSC'22 [5]	TCAS II'24 [6]	TCAS I'24 [11]	This work[a]
Archi-tecture	LNTA-first	LNTA-first	LNTA-first	Mixer-first	LNTA-first
Process	40nm	28nm	65nm	180nm	65nm
RF (GHz)	0.4-3.2	0.4-6	0.4-6	0.5-2.5	0.4-4.5
Gain (dB)	36	54	26-37	29	38.57
NF$_{min}$ (dB)	2.7	2.1	3.4	5.8	3.34
0-dBm BNF(dB)	8.4	5.2	11.6[a]	N/A	8.02
RF BW (MHz)	160	0.2-160	5-240	200	60
IB-IIP3 (dBm)	-20	<-30	-1.2	3	-8.2
OB-IIP3 (dBm)	13 $\Delta f/BW = 5$	9.8	13.9 $\Delta f/BW = 8$	18.5 $\Delta f/BW = 4$	17.5 $\Delta f/BW = 5$
Supply (V)	1.3/1.2	1.0	1.2/1.0	1.8	1.2/1.0
Power (mW)	66-115	23-49	48-62	63	44-51

a. Simulation results

REFERENCES

[1] J. W. Park et al., "Channel selection at RF using Miller bandpass filters," IEEE J. Solid-State Circuits, vol. 49, no. 12, pp. 3063–3078, Dec. 2014.

[2] F. Lin et al., "An RF-to-BB-current-reuse wideband receiver with parallel N-path active/passive mixers and a single-MOS pole-zero LPF," IEEE J. Solid-State Circuits, vol. 49, no. 11, pp. 2547–2559, Nov. 2014.

[3] B. J. Thijssen et al., "2.4-GHz Highly Selective IoT Receiver Front End With Power Optimized LNTA, Frequency Divider, and Baseband Analog FIR Filter," IEEE Journal of Solid-State Circuits, vol. 56, no. 7, pp. 2007-2017, July 2021.

[4] M. A. Montazerolghaem et al., "6.5 A 3dB-NF 160MHz-RF-BW Blocker-Tolerant Receiver with Third-Order Filtering for 5G NR Applications," in Proc. IEEE Int. Solid-State Circuits Conf. (ISSCC), 2021, pp. 98-100

[5] H. Razavi and B. Razavi, "A 0.4–6 GHz receiver for cellular and WiFi applications," IEEE J. Solid-State Circuits, vol. 57, no. 9, pp. 2640–2657, Sep. 2022.

[6] Z. Luo et al., "A 0.4-6 GHz Blocker-Tolerant Receiver in 65 nm CMOS With Bandwidth-Extended Technologies for Future V2X Applications," IEEE Transactions on Circuits and Systems II: Express Briefs, 2024.

[7] M. A. Montazerolghaem, L. C. N. de Vreede and M. Babaie, "A Highly Selective Receiver With Programmable Zeros and Second-Order TIA," in IEEE J. of Solid-State Circuits, vol. 59, no. 6, pp. 1668-1683, June 2024.

[8] Andrews.C et al., "A Passive Mixer-First Receiver With Digitally Controlled and Widely Tunable RF Interface," IEEE J. of Solid-State Circuits, 2010, 45(12): 2696-2708.

[9] Lien.Y.C et al., "Enhanced-selectivity High-linearity Low-noise Mixer-first Receiver with Complex Pole Pair Due to Capacitive Positive Feedback," IEEE J. of Solid-State Circuits, 2018, 53(5): 1348-1360.

[10] K. Shi et al., "Second-Order Transimpedance Amplifiers in Mixer-First Receivers: Design for Optimum Blocker Tolerance," IEEE Trans. Circuits and Syst. I: Reg. Papers, vol. 70, no. 5, pp. 1821-1834, May 2023.

[11] A. Kumari and D. Bhatt, "A 0.5–2.5 GHz Mixer First Receiver With 200 MHz RF Bandwidth and +18.5 dBm OB-IIP3 in 180 nm CMOS for 5G NR Band," IEEE Trans. Circuits and Syst. I: Reg. Papers, vol. 71, no. 6, pp. 2576-2589, June 2024.

[12] S. van Zanten et al., "5.5 A Stacking Mixer-First Receiver Achieving >20dBm Adjacent-Channel IIP3 Consuming less than 25mW, " in Proc. IEEE Int. Solid-State Circuits Conf. (ISSCC), 2024, pp. 96-98.

[13] A. Mirzaei et al., "Analysis and optimization of direct-conversion receivers with % duty-cycle currentdriven passive mixers," IEEE Trans. Circuits Syst. I, Reg. Papers, vol. 57, no. 9, pp. 2353–2366, Sep. 2010.

2025 The 10th International Conference on Integrated Circuits and Microsystems

A Model of a C-band Mixer Circuit

Xuezhen Zhao	Jun Liu	Jing Wang	Guodong Su
School of Electronic Information	*School of Electronic Information*	*School of Electronic Information*	*School of Electronic Information*
Hangzhou Dianzi University	*Hangzhou Dianzi University*	*Hangzhou Dianzi University*	*Hangzhou Dianzi University*
Hangzhou, China	Hangzhou, China	Hangzhou, China	Hangzhou, China
zxz20765@163.com	ljun77@hdu.edu.cn	2207900732@qq.com	guodong@hdu.edu.cn

Abstract—**This paper presents a C-band up-conversion mixer model, which consists of a core mixing unit and a Marchand Balun structure. The mixing unit employs a passive double-balanced ring topology to achieve the mixing function, while the Marchand Balun performs the conversion from single-ended to differential signals. The model operates with a local oscillator (LO) / radio frequency (RF) range of 4-9 GHz and an intermediate frequency (IF) signal covering DC-3.5 GHz. Within the 4-9 GHz frequency range, the model achieves a conversion loss of 9 dB, with typical RF return loss and IF return loss values of -11 dB and -9 dB, respectively. Furthermore, by comparing the model with chip test data, the data error is minimized, verifying the reliability of the model.**

Keywords—Model, Mixer, Passive double-balanced, Marchand balun

I. INTRODUCTION

A Digital Twin is a virtual mirror that dynamically interacts with a physical entity through real-time data, while a digital prototype serves as a static simulation model during the design phase, used to verify theoretical performance and ensure the feasibility of the initial circuit design [1-3]. The design of digital prototypes covers various fields such as microwave integrated circuits and analog integrated circuits. As one of the fundamental components of RF systems, the mixer model is primarily used to achieve frequency conversion of signals. Its key performance metrics, including conversion loss/gain, non-linear characteristics, and port matching performance, directly impact the overall reliability of the communication link.

Some researchers have proposed mixer models with an emphasis on analyzing the impact of circuit design on the performance of the mixer model. For instance, H. Darabi et al. [4] proposed and analyzed a qualitative physical model to predict the noise at the mixer output, as well as the dependence of mixer noise on the local oscillator (LO) amplitude and other circuit parameters. M.T. Terrovitis et al. [5] studied a nonlinear model for CMOS switching mixers, utilizing a simplified CMOS transistor model while considering second-order deviations, thereby validating the importance of transistor modeling for reliable distortion simulations. Currently, most research focuses on physically-based modeling of mixers to analyze circuit-related performance aspects such as noise and nonlinearity, rather than creating a holistic model of the mixer itself.

To address the aforementioned issues, this paper proposes a passive double-balanced mixer model operating in the 4-9 GHz

range. The model performs comprehensive modeling of the mixer and simulates key performance metrics such as conversion loss, return loss, port isolation, and linearity. The model's operating frequency range covers the C-band, achieving a conversion loss of 9 dB within the operating frequency range. The typical values of RF return loss and IF return loss are -11 dB and -9 dB, respectively. The isolation between LO-IF and LO-RF is better than 30 dB, and the isolation between IF-RF is better than 25 dB. Furthermore, the simulation results of the final C-band up-conversion mixer model show minimal error when compared with the chip's measured data, providing a highly reliable solution for the rapid integration of mixers and system-level verification.

II. C-BAND MIXER MODEL DESIGN

This paper proposes a C-band up-conversion mixer model, where the local oscillator (LO) and radio frequency (RF) range from 4 to 9 GHz, and the intermediate frequency (IF) covers DC to 3.5 GHz. The designed C-band up-conversion mixer model is shown in Fig. 1.

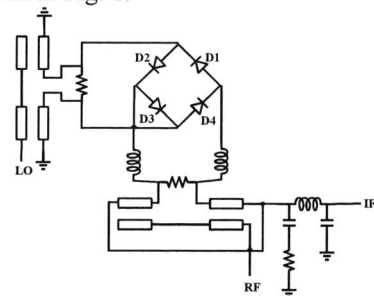

Fig. 1 C-Band Mixer Model

The model employs a passive double-balanced ring topology to achieve its mixing function [6], utilizing the nonlinear characteristics of diodes to up-convert the IF to the radio RF, thereby achieving spectrum shifting. In the double-balanced mixer's ring structure, four diodes are connected head-to-tail to form a ring [7]. When the LO power is sufficiently high, the diodes operate in switching mode: during the upper half cycle of the LO signal, diodes D2 and D3 conduct, while during the lower half cycle, diodes D1 and D4 conduct. The RF and LO inputs are converted from single-ended to differential signals through Marchand baluns, whereas the IF signal is directly extracted from the center node of the ring diode bridge, eliminating the need for an IF balun and thus simplifying the link complexity. The core mixing unit consists of this ring configuration made up of four diodes.

979-8-3315-8850-2/25 $31.00 © 2025 IEEE

A. Marchand Balun

This model design adopts the Marchand balun structure, which achieves unbalanced-to-balanced signal conversion while also providing impedance transformation [8,9], as shown in Fig. 2. The classical Marchand balun is a three-port device, consisting of one unbalanced input port and two balanced output ports. Its circuit structure mainly comprises two sections of quarter-wavelength parallel coupled lines. The input signal is fed into a half-wavelength open transmission line from the unbalanced port. Through coupling, the signal is transferred from the half-wavelength open line to two quarter-wavelength short-circuited lines. As a result, equal-amplitude but opposite-phase output signals are obtained at the two balanced output ports.

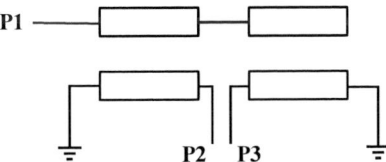

Fig. 2 Marchand Balun Structure

The core advantage of the Marchand balun lies in its multi-section coupled-line distributed structure, which enables high-balanced single-ended to differential conversion over a wide bandwidth. Through the cascaded design of coupled lines, it can cover multiple octaves, significantly surpassing the narrowband limitations of traditional single-section baluns. Its symmetrical layout and distributed impedance matching characteristics ensure extremely low amplitude mismatch and phase deviation between the differential ports. Moreover, the compact passive structure features low insertion loss and can achieve different impedance transformation requirements by adjusting the line width/spacing, among other parameters.

The circuit characteristics of a balun are closely related to the coupling coefficient k of the coupled lines that constitute it. As the coupling coefficient k increases, the magnitude of S_{11} decreases, resulting in reduced return loss at the quarter-wavelength frequency, improved input matching, and thus a wider bandwidth. Meanwhile, the magnitudes of S_{21} and S_{31} exhibit a trend of first increasing and then decreasing, indicating that the loss initially decreases and then increases. Therefore, within a specific range, as the coupling coefficient increases, the operational bandwidth of the balun increases, but the corresponding loss also rises. Properly designing the magnitude of the balun's coupling coefficient is crucial for its performance. The coupling coefficient of the balun can be expressed as:

$$k = \frac{1}{\sqrt{\frac{2Z_L}{Z_S} + 1}} \tag{1}$$

Z_S is the input impedance, and Z_L is the output impedance.

When designing the balun structure in the mixer model, performance metrics such as port return loss, amplitude balance, and phase balance must be considered.

The return loss of the balun reflects the quality of its port impedance matching, representing the reflection loss characteristics of the input signal energy in the transmission path. It can be expressed as:

$$RL(dB) = -S_{11}(dB) = -20log|\Gamma| \tag{2}$$

The balance performance of a balun primarily includes amplitude balance and phase balance. Phase balance refers to the degree to which the phase difference between its two output ports deviates from 180°. It can be expressed with the following formula:

$$Phase_{\text{Imbalance}} = 180° \pm \left| \tan^{-1} \frac{Im\left(\frac{S_{12}}{S_{13}}\right)}{Re\left(\frac{S_{12}}{S_{13}}\right)} \right| \tag{3}$$

Amplitude balance indicates the consistency in the amplitude of the output signals between the differential ports. Its expression can be written as:

$$Amplitude_{Imbalance(dB)} = 20\log\left(\frac{S_{12}}{S_{13}}\right) \tag{4}$$

By optimizing the coupling line design and impedance matching structure, the broadband balance performance can be improved, thereby enhancing the harmonic suppression capability of the mixer.

B. Mixer Core

The current model adopts a passive double-balanced structure, where the mixing core is formed by four diodes connected in a ring configuration with strict electrical matching. As illustrated in Fig. 3 (Mixer Core Topology), the RF and LO signals are first converted from unbalanced to balanced modes through high-performance baluns and then symmetrically injected into the diode ring structure. This architecture utilizes differential driving to achieve phase cancellation of harmonic components, resulting in significant suppression of spurious signals in the output spectrum compared to conventional single-ended mixers. The system's operational bandwidth is primarily governed by the bandwidth of the baluns, and the mixer's performance can be enhanced through optimization of the balun design.

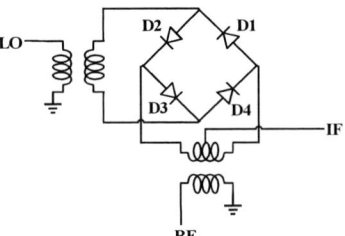

Fig. 3 Mixer Core Topology

The operating principle of the double-balanced mixer is based on the periodic switching characteristics of the mixing diodes. Sufficient LO drive power is injected through the LO port to ensure reliable device switching. During the positive half-cycle of the LO signal, diodes D2 and D3 conduct to form a signal path, while D1 and D4 remain in the off state. When the LO signal enters the negative half-cycle, the conduction states reverse: D1 and D4 turn on to establish a new path, while D2 and D3 switch to the cutoff state. This bridge-type switching mechanism, controlled by the polarity reversal of the LO signal, periodically alters the transmission path of the RF signal through the diode ring structure. Ultimately, this achieves frequency mixing between the LO signal and the IF signal.

The design of this model considers performance metrics such as the mixer's return loss, conversion gain/loss, port isolation, and linearity [10-12].

The return loss of a mixer port is defined as the power loss of the input signal reflected back to the source due to impedance mismatch at a specific port. The return loss at the port can be optimized by adjusting the matching network. It can be expressed as:

$$RL(dB) = -S_{11}(dB) = -20log|\Gamma| \qquad (5)$$

The conversion loss of a mixer refers to the power loss during the conversion of the intermediate frequency (IF) signal to the radio frequency (RF) signal. It is primarily caused by the conduction loss of nonlinear devices, port impedance mismatch, and parasitic parameters in high-frequency operation. It can be expressed as:

$$ConvertionLoss(dB) = 10 \log \frac{P_{out}}{P_{in}} \qquad (6)$$

The conversion loss of a mixer is composed of three components: circuit mismatch loss CL_β, diode junction loss CL_r, and nonlinear net conversion loss CL_g.

The mismatch loss of a mixer refers to the loss generated when the three ports of the mixer are mismatched with their respective port loads. Typically, some degree of mismatch occurs when the RF input port, LO input port, and IF output port of the mixer are connected to external loads. Since ideal matching conditions cannot be achieved in practical engineering applications, port mismatch is unavoidable. The circuit mismatch loss can be expressed as:

$$CL_\beta(dB) = 10 \log \frac{(VSWR_{in} + 1)^2}{VSWR_{in}} +$$
$$10 \log \frac{(VSWR_{out} + 1)^2}{VSWR_{out}} \qquad (7)$$

Here, $VSWR_{in}$ and $VSWR_{out}$ represent the voltage standing wave ratios at the non-LO input port and the output port, respectively. Therefore, to minimize the mismatch loss, the VSWR at the input and output ports must be optimized.

The junction loss of the diode primarily originates from the series resistance R_s and junction capacitance C_j in the mixer. Fig. 4 shows the equivalent circuit model of the mixer diode's junction, where R_j is the nonlinear junction resistance. Only the signal power dissipated across R_j contributes to the frequency conversion process. The power consumed by R_s and C_j does not participate in frequency conversion and is instead lost as junction loss in the mixer diode.

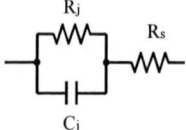

Fig. 4 Equivalent Circuit of Mixer Diode Junction

The expression for the mixer diode's junction loss can be written as:

$$CL_r(dB) = 10 \log \left(1 + \frac{R_s}{R_j} + \omega_s^2 C_j^2 R_s R_j \right) \qquad (8)$$

From the above expression, it is evident that the diode's dimensions should be adjusted to minimize its junction loss.

The nonlinear net conversion loss of a mixer is an intrinsic loss. Due to the characteristics of nonlinear devices, frequency conversion between signals of different frequencies within the mixer often generates a large number of unwanted harmonic products. The energy dissipation caused by these harmonic components is referred to as the nonlinear conductance net conversion loss.

As a nonlinear device, the linearity of a mixer refers to its ability to maintain ideal frequency conversion characteristics under large-signal inputs, including the input 1dB compression point and third-order intercept point.

A nonlinear network can be represented as:

$$y(t) = \alpha_1 x(t) + \alpha_2 x^2(t) + \alpha_3 x^3(t) + \cdots \qquad (9)$$

Considering the first three terms of a nonlinear system, if the input signal is $x(t) = Acos(\omega t)$ where A is the signal amplitude, the output signal can be expressed as:

$$y(t) = \frac{\alpha_2 A^2}{2} + \left(\alpha_1 A + \frac{3\alpha_3 A^3}{4} \right) \cos \omega t + \frac{\alpha_2 A^2}{2} cos2\omega t$$
$$+ \frac{3\alpha_3 A^3}{4} cos3\omega t \qquad (10)$$

From the above equations, it is clear that the input frequency ω corresponds to the fundamental wave, while 2ω and 3ω represent the second harmonic and third harmonic components, respectively. When the signal amplitude is small, these harmonic components can be neglected. As the input signal amplitude increases, the A^3 dependent terms can no longer be ignored. The system gain becomes $\alpha_1 + \frac{3\alpha_3 A^2}{4}$ and since α_3 is typically negative, the system gain decreases with increasing signal amplitude. The 1dB compression point is used to describe this phenomenon of gain reduction caused by system nonlinearity. As the input signal power increases, the input signal power level at which the actual output signal power becomes 1dB less than the ideal linear output power is defined as the input 1dB compression point.

When a nonlinear network simultaneously receives two signals of equal amplitude and closely spaced frequencies, $x(t) = Acos(\omega_1 t) + Acos(\omega_2 t)$, the output signal contains not only the fundamental components but also various frequency combinations. These output frequency components are expressed as $\omega = |m\omega_1 + n\omega_2|$, where m, n=0, ±1, ±2,....

979-8-3315-8850-2/25 $31.00 © 2025 IEEE   264

The frequency components generated when m and n are not both zero result from mutual modulation between ω_1 and ω_2, termed intermodulation (IM) components. Among these, the third-order intermodulation (IM3) products at frequencies $2\omega_1 - \omega_2$ and $2\omega_2 - \omega_1$, generated by third-order distortion, lie close to the fundamental frequencies and are likely to fall within the signal bandwidth, causing interference. The IM3 can be expressed as:

$$IM3 = \frac{3}{4}|\alpha_3 A^3| \tag{11}$$

To quantify the nonlinear system's ability to suppress third-order intermodulation interference, the third-order intercept point (IP3) is introduced. As the input signal power increases, the third-order intermodulation (IM3) products will intersect with the fundamental signal components at a specific point. The input signal power corresponding to this intersection is termed the input third-order intercept point (IIP3), while the corresponding output signal power is called the output third-order intercept point (OIP3). This relationship is illustrated in Fig. 5.

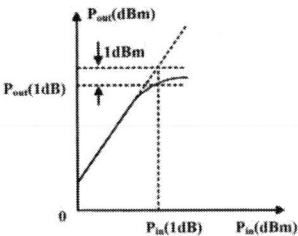

Fig. 5 Third-Order Intercept Point (IP3)

The linearity of a mixer is primarily influenced by the nonlinear devices within the circuit, while the matching networks also impact the linearity of the model. The model's linearity can be improved by selecting diodes with appropriate dimensions and optimizing the matching networks.

The port isolation of a mixer measures its ability to suppress signal leakage between different ports, expressed in decibels (dB). High isolation reduces signal crosstalk and prevents system self-oscillation caused by LO leakage. For example, when a power signal P_{LO} is applied to the LO port of the mixer, portions of this LO power may leak to the RF and IF ports, denoted as P_{LOtoRF} and P_{LOtoIF}, respectively. Similarly, leakage between other ports can be defined in the same manner. The port isolation of the mixer is then calculated as:

$$ISO_{LOtoRF} = \frac{P_{LOtoRF}}{P_{LO}} \tag{12}$$

$$ISO_{LOtoIF} = \frac{P_{LOtoIF}}{P_{LO}} \tag{13}$$

$$ISO_{RFtoIF} = \frac{P_{RFtoIF}}{P_{RF}} \tag{14}$$

III. MODEL VALIDATION

By comparing the simulated data with the measured results of the mixer's key performance parameters, the accuracy of the design model is systematically validated, and potential optimization directions are identified. This analysis focuses on metrics such as conversion loss, port return loss, and linearity.

A comparison of the simulated and measured conversion loss is shown in Fig. 6. The results indicate that the measured and simulated conversion losses closely match within the 4-9 GHz frequency range.

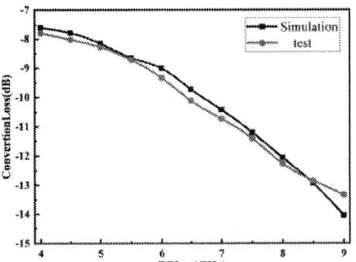

Fig. 6. Conversion Loss: Simulation vs. test

A comparison of the simulated and measured port return loss is shown in Fig. 7. The results demonstrate that the return loss at the RF port exhibits minimal deviation between simulation and measurement within the 4-9 GHz frequency range. For the IF port return loss, due to its lower operating frequency, the simulated and measured results show consistent trends with minor deviations across the DC-3.5 GHz range. Due to process and other non-ideal factors, the frequency of intermediate-frequency return loss experiences a shift.

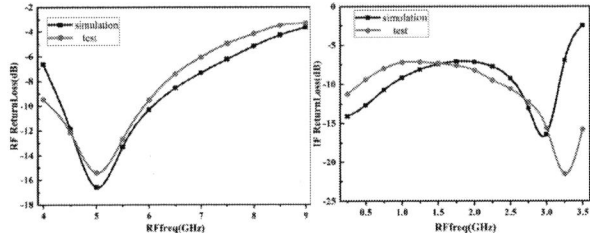

Fig. 7. Return Loss: RF Port (Left) and IF Port (Right)

A comparison of the simulated and measured input P1dB (1dB compression point) is shown in Figure 8. The IIP3 (Input Third-Order Intercept Point) maintains acceptable error margins within the 4-9 GHz frequency range.

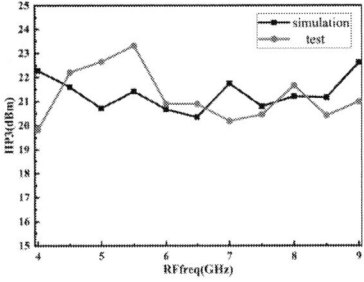

Fig. 8: IIP3: Simulation vs. test Results

The model not only supports the functionalities described above but also includes simulation design for port isolation. Fig. 9 shows the simulation results of port isolation for this model.

979-8-3315-8850-2/25 $31.00 © 2025 IEEE   265

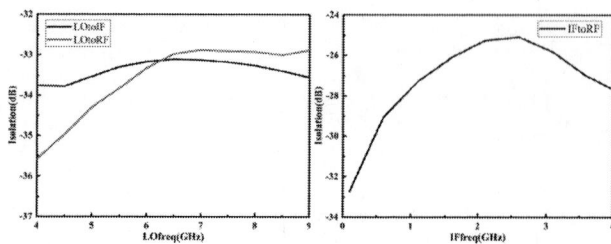

Fig. 9. Port Isolation: Simulation Results

IV. CONCLUSION

This paper addresses the issue of incomplete frequency coverage in C-band mixers by constructing an equivalent circuit model for the mixer. The designed C-band passive double-balanced mixer employs Marchand baluns to convert single-ended LO and RF inputs into differential signals. The IF signal is extracted from the center tap of the diode ring bridge, eliminating the need for a balun structure. Furthermore, the error comparison between simulated and measured results demonstrates that conversion loss, port return loss, and linearity achieve minimal deviations within the target frequency bands.

REFERENCES

[1] Fei Tao, Bin Xiao, Qinglin Qi, Jiangfeng Cheng, Ping Ji,Digital twin modeling,Journal of Manufacturing Systems,Volume 64,2022,Pages 372-389,ISSN 0278-6125,https://doi.org/10.1016/j.jmsy.2022.06.015.

[2] A. El Saddik, F. Laamarti and M. Alja'Afreh, "The Potential of Digital Twins," in IEEE Instrumentation & Measurement Magazine, vol. 24, no. 3, pp. 36-41, May 2021, doi: 10.1109/MIM.2021.9436090.

[3] Mohammad (Behdad) Jamshidi, Saeedeh Lotfi, Hesam Siahkamari, Tomas Blecha, Jakub Talla, Zdeněk Peroutka,An intelligent digital twinning approach for complex circuits,Applied Soft Computing,Volume 154,2024,111327,ISSN 1568-4946,https://doi.org/10.1016/j.asoc.2024.111327.

[4] H. Darabi and A. A. Abidi, "Noise in RF-CMOS mixers: a simple physical model," in IEEE Journal of Solid-State Circuits, vol. 35, no. 1, pp. 15-25, Jan. 2000, doi: 10.1109/4.818916.

[5] M. T. Terrovitis and R. G. Meyer, "Intermodulation distortion in current-commutating CMOS mixers," in IEEE Journal of Solid-State Circuits, vol. 35, no. 10, pp. 1461-1473, Oct. 2000, doi: 10.1109/4.871323.

[6] A. M. Pavio, R. H. Halladay, S. D. Bingham and C. A. Sapashe, "Double balanced mixers using active and passive techniques," in IEEE Transactions on Microwave Theory and Techniques, vol. 36, no. 12, pp. 1948-1957, Dec. 1988, doi: 10.1109/22.17439.

[7] T. Zhang, X. Liu, Y. Xu, L. Wang, R. Xu and B. Yan, "Mixing It Up: A Double-Balanced Mixer with Wide RF and IF Bandwidth," in IEEE Microwave Magazine, vol. 19, no. 1, pp. 106-111, Jan.-Feb. 2018, doi: 10.1109/MMM.2017.2759659.

[8] H. -Y. Yang, J. -H. Tsai, T. -W. Huang and H. Wang, "Analysis of a New 33–58-GHz Doubly Balanced Drain Mixer in 90-nm CMOS Technology," in IEEE Transactions on Microwave Theory and Techniques, vol. 60, no. 4, pp. 1057-1068, April 2012, doi: 10.1109/TMTT.2012.2183609.

[9] H. -R. Ahn and M. M. Tentzeris, "A Novel Compact Isolation Circuit Suitable for Ultracompact and Wideband Marchand Baluns," in IEEE Transactions on Circuits and Systems II: Express Briefs, vol. 67, no. 10, pp. 2299-2303, Oct. 2020, doi: 10.1109/TCSII.2019.2960005.

[10] K. K. M, S. K and S. Ashokan, "Design and Simulation of an Double Balanced Mixer for Radar Applications at X band," 2023 Global Conference on Information Technologies and Communications (GCITC), Bangalore, India, 2023, pp. 1-5, doi: 10.1109/GCITC60406.2023.10426528.

[11] U. A. Belorkar, S. A. Ladhake and S. N. Kale, "2.45 GHz Gilbert mixer using 45 nm CMOS technology," 2012 IEEE Business, Engineering & Industrial Applications Colloquium (BEIAC), Kuala Lumpur, Malaysia, 2012, pp. 140-144, doi: 10.1109/BEIAC.2012.6226039.

[12] L. Gaojian, L. Jun, X. Hui, Z. Xiaoyang, L. Shuantao and Y. Hongxi, "Design of a 220GHz subharmonic mixer based on plannar schottky diode," 2017 IEEE Asia Pacific Microwave Conference (APMC), Kuala Lumpur, Malaysia, 2017, pp. 418-421, doi: 10.1109/APMC.2017.8251469.

2025 The 10th International Conference on Integrated Circuits and Microsystems

Parametric Modeling of Transmission Structure of RF Microsystem Based on CNN-LSTM-SelfAttention

Li Qin
Beijing Research Institute of Telemetry
Beijing, China
2175743337@qq.com

De Xi Liu*
Beijing Research Institute of Telemetry
Beijing, China
390663030@qq.com

Cui Jing
Beijing Research Institute of Telemetry
Beijing, China
jingcui0307@qq.com

Lei Shi
Beijing Research Institute of Telemetry
Beijing, China
shilei12340714@126.com

Ya Wei Liu
Beijing Research Institute of Telemetry
Beijing, China
liuyaweiaa@163.com

Ting Xue
Beijing Research Institute of Telemetry
Beijing, China
xueting0728@163.com

Abstract—RF microsystem technology is a key technology for achieving the high-density integration of RF front-ends. The accuracy and simulation speed of its electromagnetic simulation are therefore particularly important. This paper proposes parametric modelling of RF microsystem transmission structures based on the CNN-LSTM-SelfAttention architecture. This architecture uses the feature extraction capability of a convolutional neural network combined with the learning capability of an LSTM for S-parameter sequences. The Self-Attention mechanism module establishes contextual connections between S-parameters. Experiments were conducted using a common GCPW transmission structure with shielded vias in RF microsystems, and the model predictions for the selected parametric tests had an RMSPE of less than 5%,, achieving a good prediction effect.

Keywords—Electromagnetic simulation, convolutional neural networks, LSTM, self-attention mechanisms, S-parameters

I. INTRODUCTION

As one of the key technologies for radar phased array RF front-ends, RF microsystems can realise the miniaturisation and high-density integration of RF front-ends [1]. Designing RF microsystems requires repeated electromagnetic simulations and adjustments to key parameters to achieve the desired performance. Therefore, electromagnetic simulation is a key part of RF microsystem design. The two main simulation methods are the finite element method [2] and the equivalent circuit method [3]. The finite element method is characterised by high simulation accuracy, but it is time-consuming. The equivalent circuit method sacrifices some of the simulation accuracy in order to shorten the simulation time. In recent years, neural network technology has been widely used in pattern recognition [4], financial prediction [5] and computer vision [6]. With the development of neural network technology, its application in radio frequency is also gradually gaining attention. Predicting the performance index of RF circuits through parametric modelling using neural networks has become a popular area of research. An artificial neural network is a typical digital approximation network; models trained using artificial

neural networks can predict the electromagnetic response of microwave circuits more quickly than traditional electromagnetic design methods. Olufemi Oluyemi et al. used an ANN architecture and a third-order Chebyshev filter to demonstrate the effectiveness of applying an ANN network to a filter [7]. Subsequently, researchers proposed an improvement to the ANN method by designing a knowledge-based neural network (KBNN) to construct parametric models by combining neural networks with a priori knowledge, thereby enhancing the learning ability of the constructed model and speeding up the construction process. J. E. Rayas-Sanchez et al. [8] proposed a method for modelling microwave RF circuits based on equivalent circuits. This method uses a constrained blind linear input space mapping approach for design, with the device's design parameters as inputs. Feng et al. proposed a new neural transfer function modelling technique (TF-NN), combining the parametric Sanathanan–Koerner iteration of continuous pole/residue extraction for electromagnetic parametric modelling of microwave components [9]. Of the above methods, the artificial neural network (ANN) belongs to the category of pure numerical approximation. However, the ANN does not take full advantage of the serial nature of the electromagnetic response. In contrast, the KBNN relies on the accuracy of the a priori knowledge. For example, the degree of accuracy of the equivalent circuit directly affects the parametric modelling. On the other hand, the transfer function neural network relies on the vector fitting technique, which, due to dramatic fluctuations in the S-parameter curves, results in non-uniformity in the fitting order and the problem of an unfixed number of output neurons.

In this paper, we will use the sequence characteristics of S-parameters to construct a model structure. This will be done by combining a convolutional neural network and a long short-term memory network with an attention mechanism, in order to complete the parametric modelling of the transmission structure of RF microsystems. The second part of the paper introduces the model's design concept and its overall architecture from the perspective of four modules: data preprocessing, the convolutional neural network, the long- and short-term memory network, and the attention mechanism. The third section

979-8-3315-8850-2/25 $31.00 © 2025 IEEE

provides a parametric modelling example of the most commonly used GCPW transmission structure in RF microsystems, along with the model's predicted results.

II. CNN-LSTM-SELFATTENTION MODEL

CNN-LSTM is a hybrid deep learning model that combines a Convolutional Neural Network (CNN) and a Long Short-Term Memory Network (LSTM). It is typically used for image classification, recognition, and sequence prediction tasks involving spatio-temporal correlation. As a frequency-dependent sequence task, the prediction of S-parameters should be learnable in the same way as time series. CNN-LSTM learns S-parameters from sequence features thanks to its memory mechanism and gating design.

A. Data Pre-processing

The data from the HFSS simulation is saved as a CSV file. Before using the data, it is checked for completeness to ensure that each set of S-parameter data is represented by the same number of frequency points, and the number of samples is computed as dimensional input. The data is then converted into a 3D array with the dimensions (number of samples, number of frequency points, number of parameters). The input data for time series must be in the format (number of samples, number of time steps, number of dimensions). Here, we use frequency points instead of time steps. Under the 3D array, the feature and target parameters are defined by the segmentation matrix. For example, (100, 150, 5) indicates 100 samples, with each sample consisting of five dimensions at 150 frequency points. The first four dimensions correspond to the feature parameters, while the fifth dimension is the target parameter. After defining the target parameters, the array is reshaped to add the channel dimensions to satisfy the task requirements for learning at individual frequency points. StandardScaler is then used to normalise the data and the normalised dataset is divided into a 70% training set and a 30% test set. A target normaliser is established for subsequent predictive value back-normalisation.

B. Convolutional Neural Networks

In the CNN-LSTM model presented in this paper, the convolutional neural network (CNN) plays a key role in feature extraction. This process involves sliding convolution on a frequency-dimensional sequence using a one-dimensional convolutional kernel. This one-dimensional convolution automatically extracts the local joint features of the multidimensional physical parameter data (e.g. the length, width and distance of passive structures) within each sliding window of the S-parameter sequence. It then refines these features using convolutional layer stacking to transform the original, high-dimensional sequence data into feature vectors that are invariant to sequence position. This local sensing feature enables the CNN to effectively capture frequency-dependent electromagnetic response characteristics in passive structures and provide feature input for subsequent LSTM layers to process entire S-parameter sequences. This forms a synergistic analysis architecture of 'spatial feature extraction + time series dynamic modelling'. The output of the one-dimensional convolution for S-parameter sequence feature extraction is:

$$Y = \sigma(W \cdot X + b) \tag{1}$$

Where: Y is the extracted features; σ is the activation function; W is the weight matrix; x is the time series and φ is the bias vector.

The CNN-LSTM structure in this paper differs slightly from the standard CNN-LSTM in that no pooling or spreading layers are constructed in the CNN part of the model. This is because the pooling layer would destroy the distribution of the original frequency points, thus affecting the integrity of the timing sequence and consequently the optimisation accuracy. The network is designed to maintain this S-parameter frequency sequence at every step, so spreading is unnecessary. The convolutional layer uses a one-dimensional design with two filters set to 128. The convolutional kernels range from 5 in the first layer to 3 in the second layer, enabling the extraction of finer local features. The dropout value is set to 0.1 to prevent overfitting.

C. CNN-LSTM

In CNN-LSTM model, Long Short-Term Memory Network (LSTM) is used for frequency sequence modeling. The basic structure of LSTM is a memory cell, and the network is trained by cycling through a certain number of memory cells, which are shown in Fig. 1. The basic memory cell of LSTM consists of three gates, which are: oblivion gate, input gate, and output gate [10]. The role of the forgetting gate controls which information in the state of the unit at the previous moment needs to be discarded or retained, the formula for the forgetting coefficient Forget Gate: Controls which information from the previous cell state should be discarded or retained. The forget coefficient f_g is calculated as:

$$f_{\mathrm{g}} = \sigma\left(\boldsymbol{W}_{f_{\mathrm{g}}} \cdot [\boldsymbol{y}_{i-1}, \boldsymbol{x}_i] + \boldsymbol{b}_{f_{\mathrm{g}}}\right) \tag{2}$$

The role of the input gate is to control which parts of the candidate memories for the current time step are updated to the cell state, and the $\widetilde{U}_t$ of the memory cell at this moment in time, as well as the coefficients $input_i$, are obtained by means of the forgetting and input gates

$$input_i = \sigma\left(W_{\mathrm{input}} \cdot [\boldsymbol{y}_{i-1}, x_i] + b_{\mathrm{input}}\right) \tag{3}$$

$$\widetilde{U}_t = \tanh(W_U \cdot [\boldsymbol{y}_{i-1}, x_i] + b_U) \tag{4}$$

The state information U_i of this memory cell at this moment in time is obtained through the forgetting gate and the input gate, calculated as:

$$U_i = f_g \times U_{i-1} + input_i \times \widetilde{U}_t \tag{5}$$

The role of the output gate is to filter the information in the current cell state that needs to be passed to the next time step.

The structure of the LSTM recurrent unit is illustrated in Figure 1:

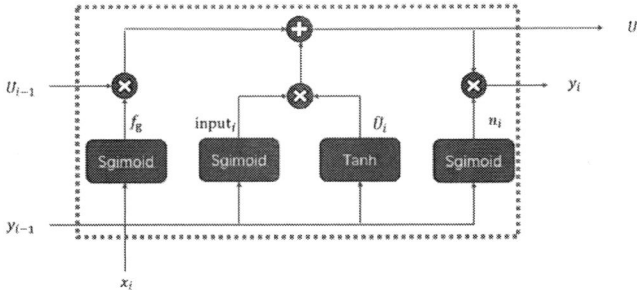

Fig. 1. Schematic diagram of the LSTM recurrent network structure

Such a structure can effectively handle the correlation between frequency points. In order to capture the EM response transformation patterns in the S-parameter sequences effectively, a three-layer LSTM stacked structure with 128 memory cells in each layer has been designed to maintain the output of the frequency point sequences. Regularisation with a dropout rate of 0.1 has been applied at the loop layer in order to capture the long-term dependencies between the frequency points.

D. Self-Attention

Self-Attention, self-attention mechanism is an attention mechanism that associates different positions of a single sequence to compute the representation of the same sequence, which can allow the model to process the S-parameter sequence, will associate the relationship between the frequency points of the whole sequence, can enhance the model's S-parameter sequence of the contextual information perception ability, through the dynamic allocation of weights, so as to focus on the S-parameter curve of the focusing on the key areas, and obtaining the improvement in accuracy. The attention mechanism designed in this paper is shown in Figure 2:

The attention mechanism designed in this paper is a variant of the self-attention mechanism and has the following design steps:

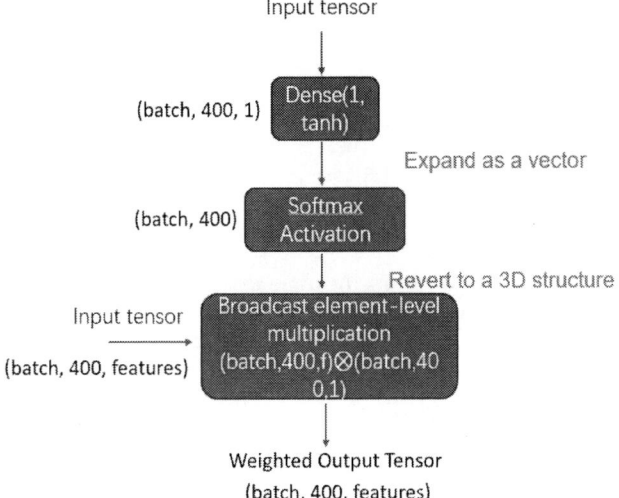

Fig. 2. Schematic diagram of the attention mechanism

1. The frequency steps of the input tensor are obtained.

2. A fully connected layer processes the features at each frequency point and compresses them into a scalar. The tanh activation function is used, which roughly limits the output range to [-1, 1].

3. A softmax activation function is then applied to this tensor to obtain a probability distribution over the entire spectrum (i.e. weights for each frequency point, with the sum of all weights equalling 1).

4. Finally, we multiply the input tensor element-by-element by the attention weight tensor, broadcasting the attention weights to each feature so that all features at each frequency point are multiplied by the same weight.

Consequently, the feature vector for each frequency point of the output tensor is the original feature vector multiplied by the attentional weight for that frequency point. This augments the features of important frequency points (with larger weights) while weakening those of unimportant frequency points (with smaller weights).

E. Target Model

The target model is a model structure that combines CNN and LSTM and introduces Self-Attention, as shown in Figure 3.

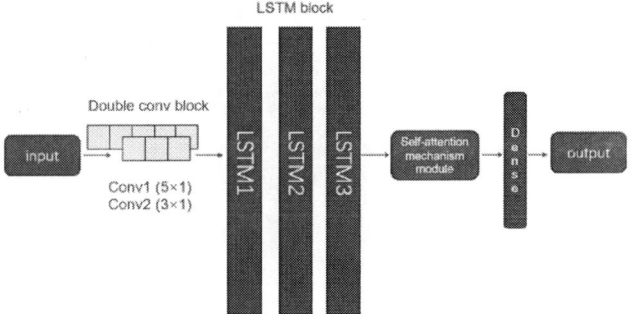

Fig. 3. Schematic diagram of CNN-LSTM-SelfAttention model

The convolutional layer in the model extracts local spatio-temporal features, making it suitable for capturing local patterns in time series data. It establishes sequence-dependent properties through LSTM, retains key information across time steps through a gating mechanism and reinforces sequence connections, focusing on key frequency points through an attention mechanism. Finally, the fully connected layer maps the high-level features to the target space to make the final prediction.

III. EXPERIMENTAL

To verify the accuracy of the CNN-LSTM-AM model for predicting transmission structures in RF microsystems, this paper uses the commonly used GCPW transmission structure for parametric modelling and verification. The GCPW's overall structure is a double-layer plate structure consisting of an upper cover plate, an air cavity and a bottom dielectric substrate. The GCPW's structure is shown in Fig. 4.

979-8-3315-8850-2/25 $31.00 © 2025 IEEE

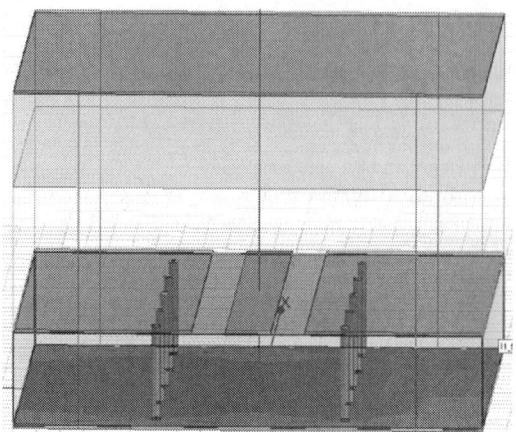

Fig. 4. CCPW 3D model

A. HFSS Electromagnetic Simulation

ANSYS HFSS is used to construct an electromagnetic simulation model to simulate the data set for the GCPW transmission structure with shielded vias, which is commonly used in silicon-based RF microsystems. In GCPW, the line length, line width and line spacing are selected as variable parameters, the frequency is used as the feature sequence and S11 is used as the model's output parameter. The model is shown in Fig. 1 and the neural network input parameters are shown in Table I.

TABLE I. CGPW PARAMETRIC MODELING NEURAL NETWORK INPUT PARAMETER LIST

Input parameters	range
GCPW-width	36um-108um
GCPW-length	38um-144um
GCPW-spacing	550um-1000um
frequency	1-40GHz

B. Evaluation of Indicators

The experiment uses the Root Mean Square Percentage Error RMSPE as the evaluation metric for the experiment, which is defined by the formula:

$$\text{RMSPE} = \sqrt{\frac{1}{n} \times \sum_{n}^{n=1} \left(\frac{y_t - y_p}{y_t}\right)^2} \times 100\%, \qquad (6)$$

(6) where: yt is the true value of the sample; yp is the model predicted value. The percentage of prediction error to the true value was recorded using this loss metric, which visually expresses the accuracy of the model prediction.

C. Model Training and Prediction

This paper uses a CNN-LSTM-AM model for GCPW training. A total of 400 groups of simulation data with 400 frequency points are used to describe the S11 parameter curve of GCPW at 1–40 GHz. To ensure the completeness of the frequency sequence, all the data are first checked for frequency points. Each set of 400 data points is then divided into a sequence, and the number of samples is calculated. The data are then preprocessed. A CNN-LSTM-AM model is then used for

learning. The activation function of the convolutional layer is set to ReLU, the activation function of the LSTM memory unit state is set to tanh, and the activation functions of the input, forgetting and output gates are set to sigmoid. The fully connected layer is set to a double-layer structure with 32 neurons in a single layer and the activation function is set to ReLU. The Adam optimiser is used and the loss function is the Huber loss function, which is shown in equation (7).

$$L_\delta(y, f(x)) = \begin{cases} \frac{1}{2}(y - f(x))^2, & |y - f(x)| \leq \delta \\ \delta|y - f(x)| - \frac{1}{2}\delta^2, & |y - f(x)| > \delta \end{cases} \qquad (7)$$

This loss function design enables the function to switch dynamically between MAE and MSE according to the size of the error, thereby combining the robustness of MAE with the accuracy of MSE while compensating for MAE's slow loss decline.

After starting the CNN-LSTM-AM model training, the loss values of the model change as shown in Fig. 5 below, it can be seen that the training set loss and the test set loss converge rapidly in the first 20 rounds of training and gradually stabilize during the 20-100 training process, and finally, after 100 rounds of training, the training set loss and the test set loss of the model stays at 0.032% and 0.047% .The root mean square error of the model on the dataset is less than 7%.

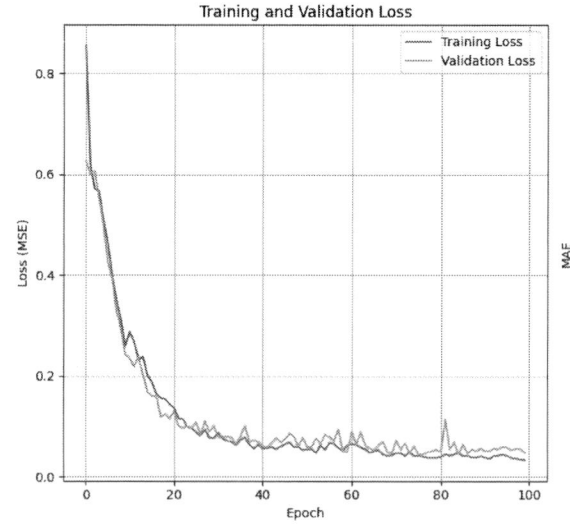

Fig. 5. Model loss convergence curve

The prediction results for some test set data are shown here. A part of prediction results are shown in Figure 6.

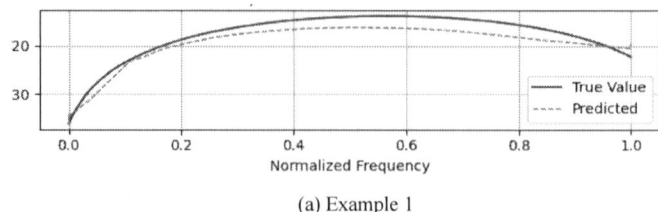

(a) Example 1

979-8-3315-8850-2/25 $31.00 © 2025 IEEE

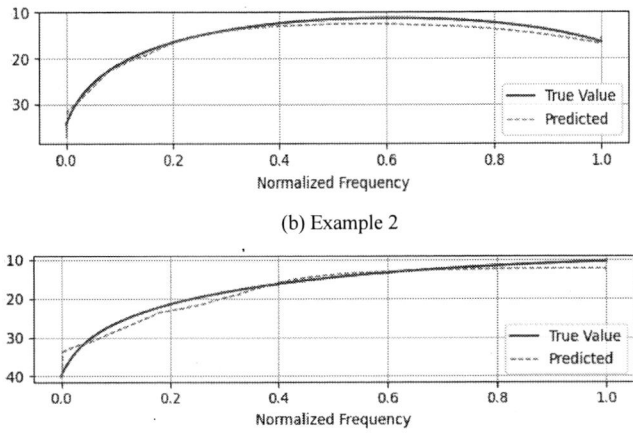

(b) Example 2

(c) Example 3

Fig. 6. (a)(b)(c) Schematic of S_{11} predicted effects

This transmission structure is simulated using HFSS for the physical dimensions of the GCPW, with GCPW width = 65 μm, GCPW length = 77 μm and GCPW spacing = 888 μm. The S11 parameter performance over this structure is predicted by the CNN-LSTM-AM model, and the two sets of data are compared and evaluated using root mean square error. The HFSS simulation output and model prediction output are shown in Fig.7.

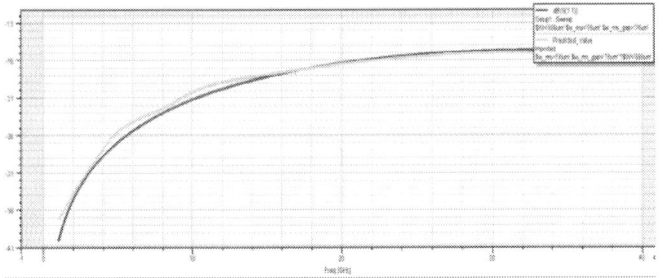

Fig. 7 Comparison of CNN-LSTM-AM and HFSS simulation results

The results show that the percentage root mean square error (RMSPE) between the model designed in this paper and the HFSS simulation of the S11 parameter amplitude is less than 4%, achieving a high degree of consistency and verifying the neural network model's accuracy and reliability in predicting the frequency response characteristics of the S-parameter.

IV. Conclusion

Parametric modelling is carried out for the standard GCPW transmission structure in RF microsystems. A parametric prediction model combining a convolutional neural network, a long- and short-term memory network, and an attention mechanism is designed. Experiments prove the model's feasibility and verify that the frequency sequence characteristics of S-parameters can be exploited for deep learning. The target model performs well in the prediction process, achieving fast loss convergence and high accuracy.

References

[1] D. Liu, X. Zhang, L. Shi, Research on development strategy of RF micr osystem technology. Journal of Telemetry ,Tracking and Command, 202 1,42 (5):17-27.

[2] R. Xu, H. Wang, Y. Xu, Progress and prospects of keytechnologies in R F microsystems. Journal of Microwaves,2023,39 (5):70-78.DOI:10.1418 3/j.cnki.1005-6122.202305008.

[3] Z. Yin, D. Liu,,T. Xue, Design and verification of PDK for 1–40 GHz sil icon-based RF microsystem passive interconnect structure.Telemetry,Tra cking and Command,46 (3),119–126

[4] C. Szegedy, W. Liu, Y. Jia, Going deeper with convolutions.IEEE Comp uter Society Conferenceon Computer Visionand Pattern Recognition,Bos ton,Unitedstates,2015:1-9.

[5] W. Bao, J. Yue,Y. Rao.A deeplearning frame work for financial time ser iesusing stacked auto encoders and long-shortter mmemory.PloSOne,20 17,12(7):e0180944.

[6] N. Omahony,Campbell S,Carvalho A, Deeplearning vs. traditional comp utervision.Proceedings of the Computer Vision Conference,LasVegas,U SA,2019:128-144.

[7] O. Oluyemi,P. Laforge and A. Bais, Space Mapping Techniqueon Micro wave Filter Circuit Model Using Artificial Neural Network for Paramete r Extraction, 2023 IEEE Canadian Conferenceon Electrical and Compute r Engineering,Regina,SK,Canada,2023,pp.337-341

[8] F.Feng*et al*.,"A Nove lNeuro-TFModeling Technique Incorporating Para metric Sanathanan–Koerner Iteration of Continuous Pole/Residue Extrac tion,". IEEE Transactions on Microwave Theory and Techniques,doi:10. 1109/TMTT.2025.3571574.

[9] J. E. Rayas-Sanchez, V. Gutierrez-Ayala. Em-based monte carlo analysis and yield prediction of microwave circuits using linear-input neural-output space mapping. IEEE Transactions on Microwave Theory and Techniques, 2006, 54(12): 4528-4537

[10] Y.Chen ,Z.Yang, J. Wen,.A health prediction method for lifting machine ry based on CNN and bidirectional encoder-decoder LSTMfusion. Journ al of Chong qing University,48(6),74–83.

979-8-3315-8850-2/25 $31.00 © 2025 IEEE

A Radio-Frequency Orthogonal Switched-Capacitor Transmitter

Yingyi Chen
School of Microelectronics Science and Technology, Sun Yat-sen University Guangdong-Macao Joint Laboratory for Modular Chip Design and Testing
Zhuhai, China
chenyy687@mail2.sysu.edu.cn

Mianting Hu
School of Microelectronics Science and Technology, Sun Yat-sen University Guangdong-Macao Joint Laboratory for Modular Chip Design and Testing
Zhuhai, China
humt7@mail2.sysu.edu.cn

Sijia Zhao
School of Microelectronics Science and Technology, Sun Yat-sen University Guangdong-Macao Joint Laboratory for Modular Chip Design and Testing
Zhuhai, China
zhaosj6@mail2.sysu.edu.cn

Ruiyi He
School of Microelectronics Science and Technology, Sun Yat-sen University Guangdong-Macao Joint Laboratory for Modular Chip Design and Testing
Zhuhai, China
hery8@mail2.sysu.edu.cn

Gengzhen Qi[*]
School of Microelectronics Science and Technology, Sun Yat-sen University
Guangdong-Macao Joint Laboratory for Modular Chip Design and Testing
Zhuhai, China
qigzh@mail.sysu.edu.cn

Abstract：We present a radio-frequency orthogonal switched-capacitor transmitter architecture. The baseband in-phase (I) and quadrature (Q) signals are decomposed into AM and PM components. Phase modulation is implemented through an up-conversion module, while the AM signal and modulated PM component are jointly processed by a switched-capacitor power amplifier (SCPA) that concurrently performs digital-to-analog conversion (DAC) and power amplification (PA). This architecture is composed of only two modules: an up-conversion and a SCPA. This streamlined configuration reduces redundant circuitry while enhancing transmission efficiency. Experimental results demonstrate 25.68 dBm output power with 20.26% power efficiency at 2.4 GHz, confirming the architecture's efficacy for high-efficiency RF transmission systems.

Keywords—switched capacitor power amplifier(SCPA), transmitter, LO leakage, parasitic canceling.

I. INTRODUCTION

Wireless transmitter systems constitute fundamental components in modern communication devices. Conventional analog-based transmitters, comprising local oscillators, mixers, analog filters, and PA, exhibit architectural complexity and functional constraints that hinder integration density[1]. V. Åberg et al. employ digital circuitry for local oscillators(LOs), where DACs directly process baseband signals with integrated up-conversion to RF, followed by analog PA stages[2]. However, DAC modules impose significant power overhead that fundamentally constrains overall system efficiency. To enhance system efficiency, Sang-Min Yoo proposed an EER-based CMOS SCPA that achieves high output power, efficiency and linear output-power control using a switched-capacitor-based switching PA[3]. To enhance efficiency at low output power, J. He proposed a wideband digital polar transmitter using a novel 3 fold stacking of House-of-Cards-based voltage-mode Doherty PA[4]. This design achieves high efficiency and linearity with a simple configuration, though at the cost of reduced bandwidth. However, by combining the SCPA with Doherty technology, the inherent bandwidth limitation of conventional Doherty amplifiers can be mitigated.

To overcome these limitations, this work introduces an innovative RF orthogonal switched-capacitor transmitter architecture, which includes an up-conversion circuit, a Doherty switched-capacitor array, and an output matching network. The Doherty Switched-Capacitor Array enables linear output characteristics through selective switching of integrated capacitor banks synchronized with the RF carrier frequency, driven by AM signals. Within this architecture, the output signal amplitude exhibits direct proportionality to the quantity of activated capacitor cells. Consequently, precise carrier amplitude modulation is achieved by dynamically controlling the number of engaged switching capacitor units, thereby inherently implementing DAC. This innovative design merges the dual functionalities of DAC and power amplification into a single SCPA module, significantly simplifying circuit complexity while delivering superior performance metrics: high output power (>25 dBm) with remarkable power efficiency ($\eta > 20\%$).

II. PROPOSED ARCHITECTURE

A. Overall Structure

Fig. 1 illustrates the proposed 65-nm orthogonal switched-capacitor RF transmitter architecture, which integrates three key components: an up-conversion circuit, a Doherty switched-capacitor array, and an output matching network. The Doherty array incorporates both main and peaking power amplifiers, concurrently implementing DAC and PA functionalities. This dual-function design significantly simplifies transmitter architecture while enhancing power efficiency.

This work is funded in part by the GuangDong Basic and Applied Basic Research Foundation under Grant 2025A1515011608. It is also partially funded by the Guangdong-Macao Joint Laboratory (GDSTC Project No. 2025B1212150003).

Firstly, we decompose the signal into two orthogonal components, which are subsequently decomposed into amplitude and phase components, respectively, yielding four signals: PM-I, PM-Q, AM-I, and AM-Q. AM-I and AM-Q are identical and are collectively referred to as signal AM.

The transmitter employs QAM modulation. With the circuit implemented using a differential terminal,PM signal undergoes phase modulation through up-conversion circuit using mutually differential and orthogonal LO signals, resulting in signals CLK-I+, CLK-I-, CLK-Q+ and CLK-Q-. The AM signal is fed into a Doherty switched-capacitor array, where thermometer coding precisely orchestrates the activation sequence of parallel capacitor banks within the array.Through carefully designed capacitor voltage division networks, the output amplitude is accurately controlled to maintain linear modulation characteristics. The Doherty switched-capacitor array comprises four arrays, with the main PA consisting of arrays controlled by signals CLK-I+ and CLK-Q+, and the peaking PA consisting of arrays controlled by signals CLK-I- and CLK-Q-, while facilitating power back-off. The output undergoes harmonic filtering and is impedance-matched with a 50 Ω antenna through a matching network.

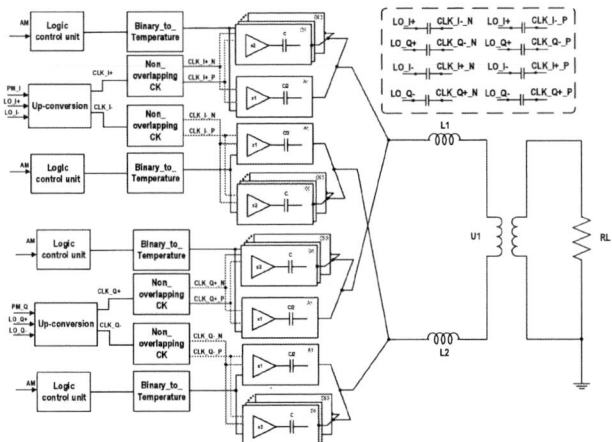

Fig. 1. Overall structure for orthogonal switched-capacitor transmitter

B. Up-conversion Circuit and LO Leakage Self-Suppression

The schematic diagram of the up-conversion circuit is shown in Fig. 2, where EN is the baseband signal, LO_P/LO_N is the differential carrier signal generated by the local oscillator, which can suppress even harmonics and improve the purity of the output signal, and OUT_P/OUT_N is the modulated RF output signal.

Signal EN serves as a switching signal. When EN is at logic low, the mixer allows LO signals to pass through. Otherwise, the LO signal is blocked. The up-conversion circuit multiplies EN by LO_P/LO_N.The output can be written as

$$OUT\_P=EN \times LO\_P \tag{1a}$$

$$OUT\_N=EN \times LO\_N \tag{1b}$$

Since EN is a digital signal, the output is actually a fragment of the LO signal during the period when EN is at logic low.

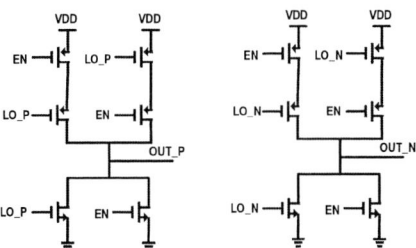

Fig. 2. Circuit schematic for Up-conversion

We use up-conversion circuit and SCPA instead of the DAC, mixer, and PA parts in traditional transmitter architectures. In this case, the error vector magnitude (EVM) generated by that up-conversion module must be small enough [5], but the IQ branch has parasitic capacitance C_{feed}. After mixing the baseband signal with the local oscillator signal, a significant leakage signal appear, reducing the overall EVM[6] and leading to a deterioration in dynamic range and linearity. For SCPA cells, the parasitic gate source capacitors of NMOS and PMOS both contribute to C_{feed}. Therefore, the LO leakage of SCPA is larger than that of other types of units [7].

To solve the above problems, we introduce the method of LO leakage self-suppression, using an inverted LO signal path to offset LO leakage[8]. By connecting compensation capacitors at the output of the non-overlapping clock signal generation unit and the source of the LO signal, the LO leakage in the circuit can be compensated. The capacitance value should be equal to the sum of the total parasitic capacitance and other related capacitors[7].

C. Doherty Switched-Capacitor Array

The Doherty can achieve back-off efficiency enhancement through the combined operation of main PA and peaking PA [9]. Therefore, we adopts the Doherty architecture to implement our SCPA.

The conventional Doherty architecture[9] is illustrated in Fig. 3(a). Since our circuit inputs consist of I signals and Q signals, which are further decomposed into AM and PM signals, we employ a modified architecture: the AM signal is used for Doherty switching logic control, while the PA with I+ and Q+ inputs serves as the main PA, and the PA with I- and Q- inputs functions as the peaking PA, as shown in Fig. 3(b).

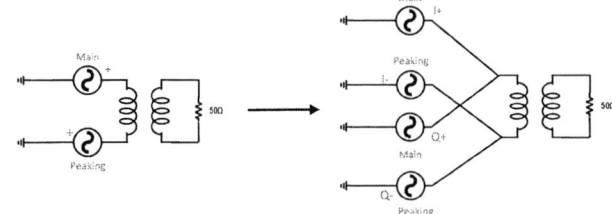

Fig. 3. (a) Conventional Doherty. (b) Doherty with I+, I-, Q+, Q- inputs

Each PA further includes a non-overlapping clock, a Doherty logic control unit and a SCPA array. The four PM components upconverted to the RF carrier frequency are converted into non-overlapping switching signals by the non-overlapping clock to reduce the crowbar current generated during the switching process of the cascode output stage. The

979-8-3315-8850-2/25 $31.00 © 2025 IEEE   273

resulting non-overlapping switching signals serve as the phase input signals for the SCPA arrays. The AM signal input to the SCPA arrays are the sampled value of the envelope amplitude.

The Doherty amplifier operates in two distinct modes based on the normalized output voltage level. For output voltages below 50% (indicated by logic low state of the AM signal's most significant bit B6), the peaking amplifier remains inactive while the main amplifier's output power is precisely regulated through 6-bit resolution control (B5-B0) of its switched-capacitor array. Conversely, when the output exceeds 50% (B6 at logic high), the main amplifier saturates with its full capacitor bank engaged, and dynamic power control shifts to the peaking amplifier through identical 6-bit resolution adjustment of its switched-capacitor units. This dual-mode operation achieves both high efficiency across the full power range and maintains linear output characteristics. The corresponding logic control circuitry implementing this adaptive power distribution is detailed in Fig. 4.

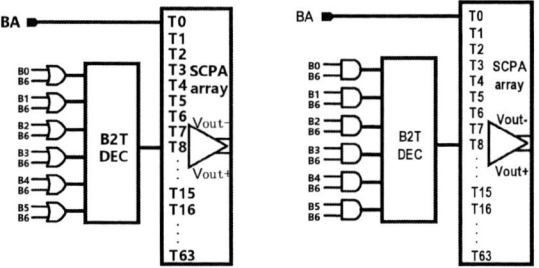

Fig. 4. (a) Logic control circuit for main PA. (b) Logic control circuit for peaking PA

In our design, each power amplifier[3] consists of a capacitor array composed of 64 amplifier units. Among these, one unit has a capacitance value half that of the others to improve resolution. The structure of each amplifier unit, as shown in Fig. 5, includes a level shifter, driver circuit, and cascode output stage.

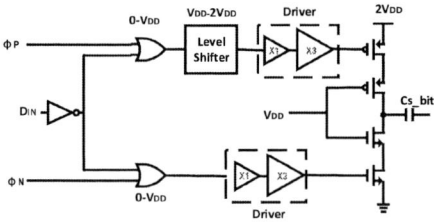

Fig. 5. PA Unit Architecture

φ_N and φ_P are non-overlapping clock signals, while D_{in} is the amplitude signal. By applying different gate voltages to the Cascode transistors, the capacitor switches can be toggled between 2VDD and GND. The total capacitance value of the single-ended SCPA is given by

$$C_{sw} = \frac{1}{2\Pi f_0 Q R_{opt}} \quad (2)$$

Where f_0 is the transmitter operating frequency, Q is the quality factor, R_{opt} is the load impedance required for single-ended PA matching, and V_m is the fundamental output amplitude of the single-ended SCPA.

$$R_{opt} = \frac{1}{2} \times \frac{V_m^2}{0.5 P_{out}} \quad (3)$$

And the capacitance value of a single switched-capacitor in the single-ended SCPA is:

$$C_{s\_bit} = \frac{C_{sw}}{128} \quad (4)$$

D. Output Matching Network

As shown in Fig. 6, the output matching network consists of two inductors and a transformer, with a 50Ω load. The two inductors resonate with the total capacitance of the main and peaking PA respectively at the carrier frequency, which reduces energy loss and improves efficiency of transmitter. By utilizing the resonance between inductors and capacitors, the transmitter can operate at specific frequencies with enhanced stability.

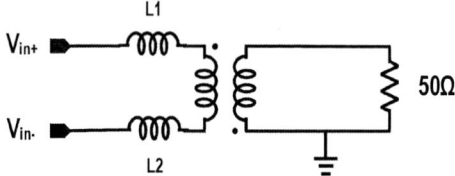

Fig. 6. Output Matching Circuit

The two inductor values in the matching network are:

$$L_1 = L_2 = \frac{1}{(2\Pi f_0)^2 C_{sw}} \quad (5)$$

The transformer turns ratio is $1:n_2$

$$n_2 = \sqrt{\frac{R_L}{2R_{opt}}} \quad (6)$$

III. Experimental Results

The proposed orthogonal switched-capacitor RF transmitter successfully demonstrates simultaneous modulation and power amplification of orthogonal baseband signals. Under test conditions (400-MHz baseband input, 2.8-GHz LO frequency), the system achieves measured performance metrics of 25.68 dBm peak output power with 20.26% power efficiency.

The PM-I signal are modulated by an up-conversion module to obtain CLK_I+and CLK_I- signals, while the PM-Q signal are modulated by another up-conversion module to obtain CLK_Q+ and CLK_Q- signals. When the baseband phase signal is GND, the two differential signals output have the same waveform as the local oscillator signal. When the baseband phase signal is VDD, the two differential signals output will be set to GND, thereby improving the information carrying rate on the basis of increasing the frequency of the input signal.

The output image obtained by measuring the voltage of the load resistance, as shown in Fig. 7, indicates that the circuit has successfully achieved power amplification function. At the same time, the image presents a waveform similar to a modulated wave, which once again demonstrates that we have achieved signal modulation function. As shown in Fig. 8, the frequency spectrum of the output curve reaches its peak value of 15.4dB at 2.4G. The frequency has been raised from 400M to 2.4G, and the second peak of the spectrum characteristic curve is 8.452dB at 2.8GHz, indicating the successful implementation of LO leakage self-suppression.

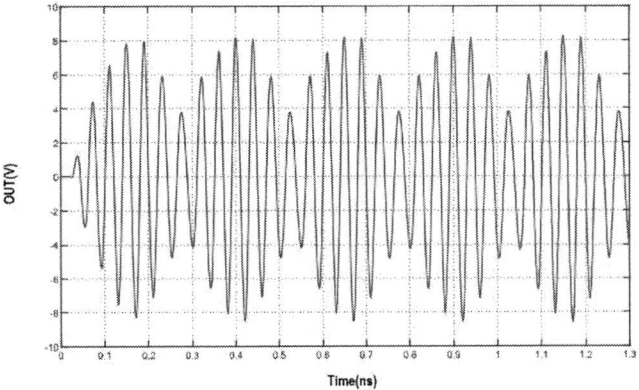

Fig. 7. The output voltage versus time

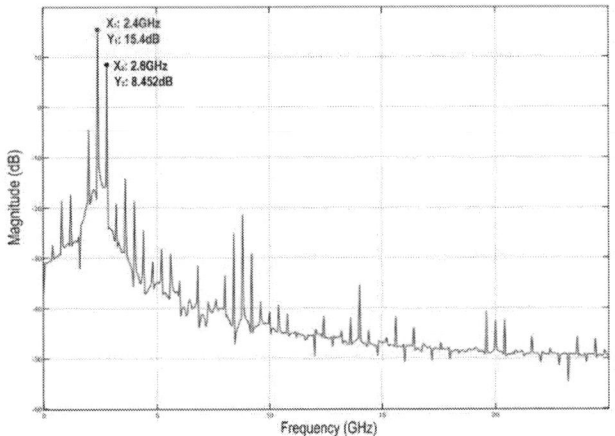

Fig. 8. Output characteristic curve spectrum diagram

The output image obtained by measuring the voltage of the load resistance, as shown in Fig. 7, indicates that the circuit has success.

Table 1 summarizes and compares the proposed radio-frequency orthogonal switched-capacitor transmitter and the published transmitter architectures.

TABLE I.

SUMMARY AND COMPARISON TO PREVIOUSLY PUBLISHED DESIGNS

	This work	**[2]**	**[4]**	**[9]**	**[10]**
Technology	65nm	22 nm	40nm	65nm	65nm
Frequency (GHz)	2.4	11.5	1.0	0.9	2.0
Supply Voltages(V)	1.2/2.4		1.1/2.2/3.3	1.2/2.4	1.3
Peak P_{SAT} (dBm)	25.68	15.8	23.9	24	15.5
Peak SE(%)	20.26		21.7	45	12.1

IV. CONCLUSION

This work presents a 65-nm orthogonal switched-capacitor RF transmitter comprising two key innovations: (1) baseband I/Q signal decomposition into separate phase and amplitude components, (2) phase modulation via up-conversion circuitry, and amplitude modulation through switched-capacitor array control. Compared with conventional architectures requiring DACs, mixers, filters, and PAs, our simplified design integrates only an up-conversion module and SCPA array, significantly reducing power consumption while maintaining transmission efficiency.We use Doherty architecture to enhance efficiency at low output power. This design achieves high efficiency and linearity with a simple configuration, though at the cost of reduced bandwidth. By combining the SCPA with Doherty technology, the inherent bandwidth limitation of conventional Doherty amplifiers can be mitigated. The novel signal decomposition approach enables independent optimization of phase and amplitude paths, establishing a new paradigm for efficient transmitter design.

REFERENCES

[1] Mo Yanjie. Design and Research Based on Sub-GHz Wireless Transmitter [D] University of Electronic Science and Technology of China,2017.（in Chinese with English abstract）

[2] V. Åberg, P. Saad, R. Hou and H. Zhou, "RF-DAC-based PA Pre-Distortion using Expanding Non-Linear RF-DAC Scaling," 2024 19th European Microwave Integrated Circuits Conference (EuMIC), Paris, France, 2024, pp. 267-270.

[3] S. -M. Yoo, J. S. Walling, E. C. Woo and D. J. Allstot, "A switched-capacitor power amplifier for EER/polar transmitters," 2011 IEEE International Solid-State Circuits Conference, San Francisco, CA, USA, 2011, pp. 428-430.

[4] J. He et al., "A 24-dBm Tri-Stacked House-of-Cards Doherty Polar Transmitter for NB-IoT with Digital PLL Phase Modulator," 2024 IEEE European Solid-State Electronics Research Conference (ESSERC), Bruges, Belgium, 2024, pp. 440-443.

[5] A. K. Gupta and J. F. Buckwalter, "Linearity considerations for low-EVM, millimeter-wave direct-conversion modulators," IEEE Trans. Microw. Theory Techn., vol. 60, no. 10, pp. 3272–3285, Oct. 2012.

[6] J. Yoo and S. Hong, "A 28 GHz RF-DAC With Analog LO Leakage Cancellation," in IEEE Transactions on Circuits and Systems II: Express Briefs, vol. 69, no. 11, pp. 4308-4312, Nov. 2022.

[7] B. Yang, H. J. Qian, J. Zhou, Y. Shu and X. Luo, "Millimeter-Wave Quadrature Mixed-Mode Transmitter With Distributed Parasitic Canceling and LO Leakage Self-Suppression," in IEEE Journal of Solid-State Circuits, vol. 58, no. 3, pp. 691-704, March 2023.

[8] B. Yang, H. J. Qian, Y. Shu, J. Zhou and X. Luo, "22-30GHz Quadrature Hybrid SCPA with LO Leakage Self-Suppression and Distributed Parasitic-Cancelling Sub-PA Array for Linearity and Efficiency Enhancement," 2022 IEEE Custom Integrated Circuits Conference (CICC), Newport Beach, CA, USA, 2022, pp. 1-2.

[9] V. Vorapipat, C. S. Levy and P. M. Asbeck, "Voltage Mode Doherty Power Amplifier," in IEEE Journal of Solid-State Circuits, vol. 52, no. 5, pp. 1295-1304, May 2017.R. Nicole, "Title of paper with only first word capitalized," J. Name Stand. Abbrev., in press.

[10] H. Kang, V. S. Rayudu, K. Yong Kim and R. Gharpurey, "A 1.3 V Wideband RF-PWM Cartesian Transmitter Employing Analog Outphasing and a Switched-Capacitor Class-D Output Stage," 2020 IEEE Radio Frequency Integrated Circuits Symposium (RFIC), Los Angeles, CA, USA, 2020, pp. 291-294.

979-8-3315-8850-2/25 $31.00 © 2025 IEEE

Daisy Chain Transmitter Circuit Design with Anti-Electromagnetic Interference for BMS Chip in EV

Chen Lin
School of Electronic Engineering
Heilongjiang University
Harbin, China
1760953684@qq.com

Liji Wu*
School *of Integrated Circuit*
Tsinghua *University*
Beijing, China
lijiwu@tsinghua.edu.cn

Jing Hu*
School of Electronic Engineering
Heilongjiang University
Harbin, China
hjlyh@126.com

Zonghuan Wu
School of Integrated Circuit
Tsinghua University
Beijing, China
wu-zh20@tsinghua.org.cn

Xiangmin Zhang
School of Integrated Circuit
Tsinghua University
Beijing, China
zhxm@tsinghua.edu.cn

Abstract—New energy vehicles have three core systems, namely batteries, motors and electronic control systems. The difficulties in the research and development of power batteries lie in the safety issues, service life and manufacturing costs of lithium batteries. Only by accurately collecting and monitoring parameters such as voltage, current and temperature in power batteries can the safety and reliability of electric vehicles be guaranteed. The battery management system (BMS) in electric vehicles (EVs) requires reliable communication interfaces to accurately monitor and control lithium-ion battery cells. This paper proposes an anti-electromagnetic interference Daisy link port circuit that uses capacitors or transformers as isolators. The system includes the transmitter circuit, the receiver circuit and the isolator circuit. This article centers on the high-reliability communication needs of the power battery management system (BMS) chip applied to new energy vehicles. It lays stress on the design of the transmitter circuit for the Daisy chain communication interface, which adopts a differential signal processing framework. The high-pressure BCD process using the technology from eastern South Korea, By means of suppression, it controls high-voltage common-mode interference to guarantee signal integrity. The effective filtered pulse width is below 250ns, Digital circuits offer two modes which cuts down noise and improves the capability to withstand electromagnetic interference.

Keywords—*Daisy Chain, Battery Management System, CMTI, Receiver , Transmitter*

I. INTRODUCTION

At present, with the vigorous development of electric vehicles, the Battery Management System (BMS) is crucial for ensuring battery performance, safety and extending battery service life. The Daisy chain communication system, as a key technology in BMS, is gradually emerging and playing an irreplaceable role. Daisy chain communication is not a brand-new communication protocol but a unique wiring scheme. In the BMS of electric vehicles, its core architecture is to connect multiple Battery Management units in sequence through a shared communication link, forming a chain structure just like the petals of chrysanthemums arranged in sequence [1].

In this structure, each BMU is responsible for monitoring the key state parameters of the battery cells connected to it, such as voltage, current and temperature, etc. For instance, during the operation of an electric vehicle, each BMU perceives the status of the corresponding battery cell in real time. Subsequently, these data will be transmitted along the Daisy chain communication link from one BMU to the next BMU, and finally aggregated to the Master Control Unit. The main control unit not only undertakes the task of collecting all BMU information, but also conducts advanced analysis and decision-making on these data, such as performing battery balancing operations to ensure the balanced power of each battery cell and avoid overcharging or overdischarging of some batteries. Carry out fault diagnosis, promptly detect possible problems in the battery pack and issue alerts; Accurate estimation of battery states, such as State of Charge and State of Health, provides a strong basis for vehicle energy management and driving decisions [2].

Take the bq7961X series devices of (TI) as an example. Its Daisy chain communication interface adopts a differential signal design. Differential communication transmits complement code data on the COMP and COMN pins respectively. This interface has bidirectional and half-duplex characteristics. At the COMH (high side) and COML (low side) interfaces, there are transmitters (TX) and receivers (RX). This design can minimize electromagnetic sensitivity (EMS) to the greatest extent, enhance the immunity to high current injection (BCI), and ensure the stable transmission of data in a complex electromagnetic environment [3].

979-8-3315-8850-2/25 $31.00 © 2025 IEEE

This paper proposes a Daisy link port circuit for anti-electromagnetic interference. This circuit uses capacitors and common mode chokes for isolation to achieve the ability of anti-electromagnetic interference.

This paper proposes a circuit structure of EMI immune Daisy link port applied to the battery management system of electric vehicles, which is implemented by the eastern process of South Korea. This circuit adopts capacitors or transformers as isolators, expecting to achieve a high common-mode rejection ratio and low power consumption and transmission. The anti-electromagnetic interference level of the Daisy chain circuit was verified through DPI simulation and CMTI simulation [4].

II. BMS CHIPS VS. DAISY CHAIN

A. Battery Management System

Daisy chain communication is a communication method that connects multiple devices in series. In a BMS, each battery management unit is connected in sequence to form a structure similar to a Daisy chain. Data is passed from one unit to the next and eventually reaches the main control unit. In this way, each battery management unit can transmit the battery status information it has monitored, such as voltage, current, and temperature, to the main control unit through a Daisy chain communication link for processing and analysis, Compared with the traditional way where each battery cell or module needs an independent communication line to be connected to the central controller, the Daisy chain architecture greatly simplifies the system layout, reduces the complexity of wiring, and lowers costs and failure points.

The microcontroller on the BMS motherboard converts the signal into a differential signal through the SPI or UART serial communication interface via the communication conversion chip. Then, it communicates with the first AFE (Analog Front-end Circuit) board in the form of a differential signal. The differential signals then enter the subsequent AFE boards in sequence, achieving communication between the motherboard and all AFE boards. The various boards are isolated and communicate with each other through transformers or high-voltage capacitors to ensure the stability and security of signal transmission. Some manufacturers also support a loop Daisy chain. After the differential signal enters the last AFE board, it will be returned to the microcontroller on the motherboard through another communication conversion chip, achieving two-way communication throughout the entire communication link. This provides a backup connection method when the communication of a certain AFE board fails.

At present, BMS has a trend of transforming from the traditional distributed architecture to the centralized archit-ecture. In a distributed architecture, the BMS is divided into the main board and the slave board. Both the master and slave boards are equipped with microcontrollers. The slave board collects the voltage and temperature of individual cells and transmits them to the main board via the CAN bus. The centralized architecture that adopts the Daisy chain technology only retains the microcontroller on the BMS motherboard. The original slave board is simplified into a small board solely centered around the AFE chip function. The information

collected by the AFE is directly transmitted to the motherboard through differential isolation signals. This architecture can reduce costs and decrease the use of wiring harnesses and microcontrollers on the slave board . The block diagram of the BMS is shown in Fig.1.

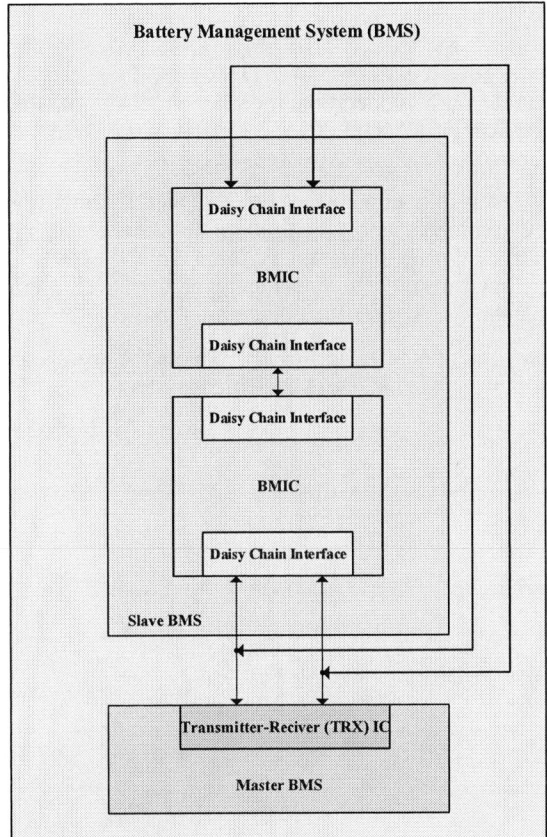

Fig. 1. Battery Management System [3].

In the Daisy chain structure, the microcontroller end adopts SPI and UART communication protocols, and the definitions of its physical layer and data link layer are simpler than those of the CAN bus. When serologically connecting each AFE board, there is a possibility that multiple slave nodes attempt to obtain bus control simultaneously. Moreover, when there is strong interference on the signal line, the deformation of the sine wave may cause the chip to make incorrect judgments, affecting the stability of communication.

The arbitration mechanism CAN be added by referring to the CAN bus, and more mechanisms and functions can be added to the LLC layer to improve communication stability. For instance, Tesla adopts a multi-channel communication mode on the Daisy chain communication link, superimposes two signals of different frequencies, and at the receiving end, they are received by two independent channels through frequency division and filtering, which enhances the reliability and anti-interference ability of communication.

Considering that the BMS is directly connected to the battery cell, its power consumption directly determines the idle life of the battery. Therefore Low-power design is crucial for

BMS chips. An article in 2023 demonstrated an item with different power consumption.

The BMS chip design of the mode [5] classifies the functions according to their importance and provides them to different parts separately Electricity or by adding power switches, different power domains can be demarcated. Furthermore, the design also provides for circuits that are constantly powered. For low-frequency (250kHz) clocks, it can maintain the normal operation of digital circuits even when the high-frequency (32MHz) clock is disabled in extremely low-power mode.

The test results in the paper indicate that the current of the chip at extremely low power consumption is higher than normal. The current during operation is three orders of magnitude smaller. Their design can dynamically adjust the chip through the Daisy link port, the circuit diagram is shown in Fig.2.

The working mode significantly reduces the average power consumption of the battery management system.

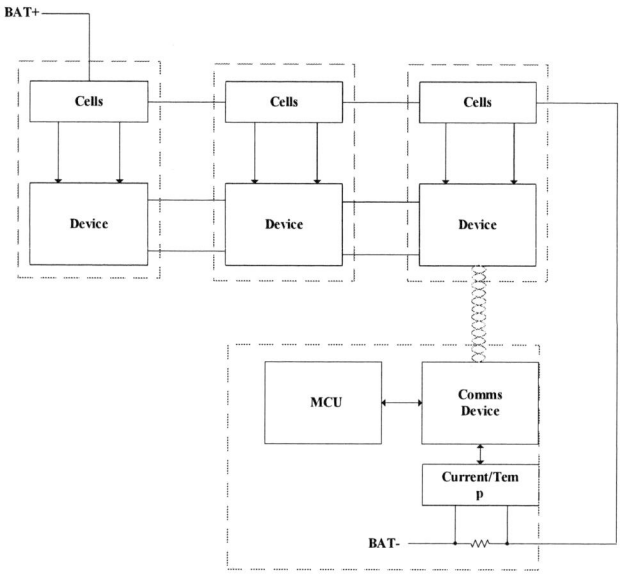

Fig. 2. Daisy Chain Interface Architecture in the BMS.

B. Daisy Chain Communication System

The Daisy Chain Communication System is a communication architecture that connects multiple devices in series, resembling a chain structure of chrysanthemums. The core principle is that data is transmitted from the first device to the next one in sequence. Each device only communicates directly with the adjacent previous and subsequent devices, and eventually reaches the main control unit or the target node. This approach does not require laying separate communication lines for each device. Instead, it transmits signals through a "relay" method, similar to the signal transmission logic of a series circuit.

Only two communication lines (such as differential signal lines) are needed to connect multiple devices, significantly reducing the number and complexity of wiring and lowering the cost of wiring harnesses (especially in scenarios with strict wiring requirements such as automobiles, where the advantages

are remarkable).There is no need to configure an independent communication interface chip for each device. The hardware design is more compact, saving PCB space and material costs. Most Daisy chains adopt half-duplex communication (which can only transmit in one direction at the same time), and avoid signal conflicts through timing scheduling (such as master polling). Some solutions (such as the ring topology) support bidirectional transmission to enhance the response speed.

III. PROPOSED DAISY CHAIN ARCHITECTURE

A. Transmitter Circuit Design

The circuit architecture of the Daisy link port transmitter is shown in Fig. 4, which includes a power supply circuit, a source adapter, and two level shift circuits. This circuit uses components from the eastern process of South Korea.

The level shift circuit uses LDMOS and DEMOS processes and operates under conditions ranging from -40℃ to 125 ℃, meeting the working state of the BMS new energy vehicle battery system. The output voltage can reach ±20V and be supplied to the receiver circuit of the next stage. The design indicators of the Daisy chain module are formulated by the latest industry standards and are mainly divided into three parts: pressure resistance requirements Requirements for electrical characteristics and timing sequence. The design of the transmitter is based on the following four formulas.

$$q = CU \tag{1}$$

$$I = \frac{dq}{dt} = C\frac{du}{dt} \tag{2}$$

$$I = \frac{1}{2}\mu_0 C_{ox} \frac{W}{L}\left(V_{gs} - V_{th}\right)^2 \tag{3}$$

$$R_{out} = Z_{out} = \frac{1}{g_m + g_{mb}} \tag{4}$$

Considering market compatibility and signal reliability, the research group decided to adopt the signal standard of Texas Instruments as shown in Fig.3. Accurate, that is, a pair of differential pulses represents one bit.0 is marked as negative first and then positive, and 1 is marked as positive first and then negative.

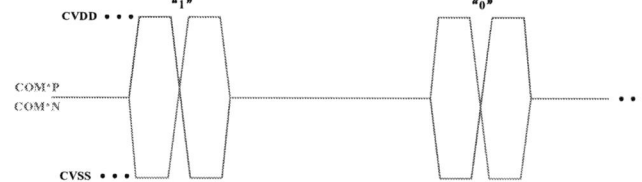

Fig. 3. The Sequence Diagram of the Chrysanthemum Connection Port of BQ79616 [7].

In 2024, the research team from Tongji University addressed the interface circuit by adding additional filters and amplifiers.The anti-EMI performance has been enhanced [4]. The team added an amplifier with an input high-pass filter to the receiver, it is large in size, and the input capacitance of the amplifier is very large. It has a certain low-pass filtering function and can absorb the input electricity.

The voltage is converted into the Root Mean Square (RMS) current. The experimental data in the paper indicate that the chip possesses There is 100kV/μs of Common Mode Transient Immunity (CMTI) and It can withstand an EMI injection of 39dBm. Judged by ISO11452[6], this has already reached the fifth level，The BCI (Bulk Current Injection) severity level is at the top level.

For the source-follower circuit, we need it to have the ability to carry a load. After passing through the level shift circuit, it should have an output voltage range of ±20V, an output impedance of 18 Ω, an output current of 10mA, and an output waveform pulse width of 250ns. Ideally, in practice, considering factors such as bias resistance, the input impedance is very high but not infinite. At high frequencies, the input impedance Zin The expression is rather complex and decreases as the frequency increases. The level shift circuit adopts high-voltage tubes, LDMOS and DEMOS processes are needed. Firstly, under the same 20V condition, the aspect ratio is nearly the same for HS-type N-type LDMOS and fully isolated LDMOS The circuit structure of the transmitter is shown in Fig. 4.

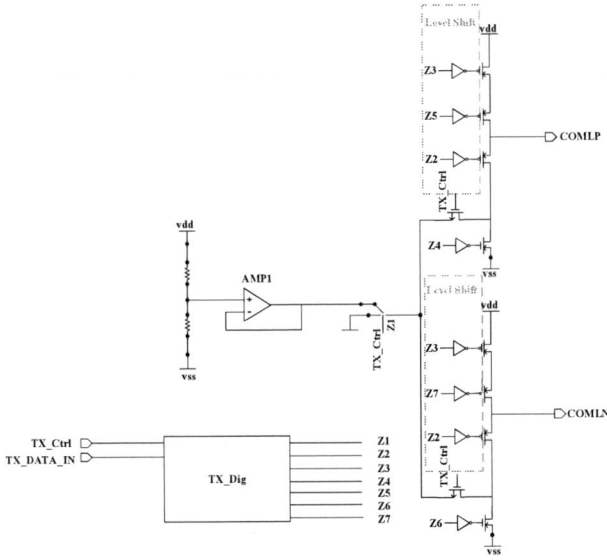

Fig. 4. Circuit Structure Diagram of Daisy Chain Transmitter.

Based on the functional description of the Daisy chain transmitter and referring to the parameters of the cutting-edge commercial vehicle-grade BMS chip, indicators were designed in sequence from top to bottom for the analog circuit part of the Daisy chain transmitter and its sub-modules. Fcomparison of the input signal, result saving, and the comparator for detecting the bus status have been designed .During the process, the feedback circuit is used to control the on-off of the strong and weak current branches, thereby achieving the purpose of filtering out burrs. However ,The discharge current and MOS capacitance value need to be continuously adjusted to ensure that the filter glitch width is within 250ns.According to the actual application requirements, the corresponding sub-module circuits contain independent enable switches and configurable electricity, Current switches, along with the necessary bias ready signals, provide interfaces for subsequent digital and analog circuit designs. This enables the module to be controlled by external signals to enter the low-power mode and also to be flexible when noise is detected on the bus live adjust the detection threshold. call this module by using time-division multiplexing technology.

B. Isolator Circuit

This device is applicable to various types of Daisy chain isolation: capacitor isolation, capacitor/choke isolation and transformer isolation. Devices connected to the same PCB in a Daisy chain form can use the isolator circuit shown in Fig. 6.The capacitive coupling isolator is composed of a primary circuit, an on-chip capacitor and a secondary circuit. It utilizes the characteristic of capacitors to "pass AC and block DC" to achieve the isolated transmission of signals. Among them, how to design and process on-chip capacitors with high withstand voltage and improve the common mode transient suppression capability are the key design technologies of capacitive couplers. Among the current research and applications, the capacitive isolation products of TI Company are the most representative. TI's technical route is based on a transmitting and receiving system on two chips, integrating high-voltage capacitors in each chip and using silicon dioxide as the dielectric[8-9]. Therefore, its withstand voltage level is higher than that of magnetic isolation. It also has the advantages of magnetic isolation, high transmission speed, longer service life, can integrate other functions under the standard CMOS process, can achieve isolated power supply within the chip, and has a smaller volume and thickness [10]. Fig.5 and Fig.6 show the structure and principle of the isolator circuit.

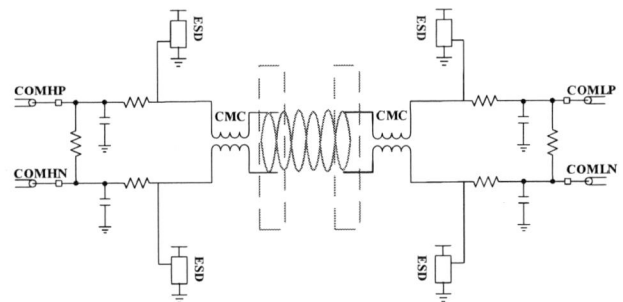

Fig. 5. Transmitter Circuit Isolation Circuit.

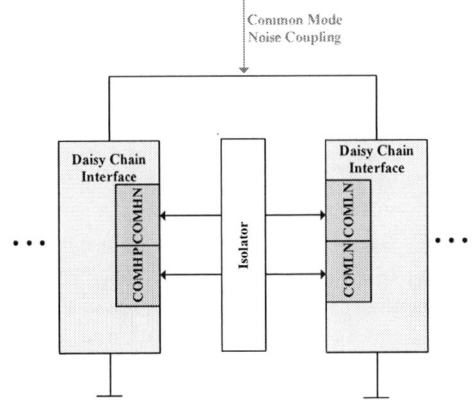

Fig. 6. Common Mode Noise Coupling in Daisy Chain.

979-8-3315-8850-2/25 $31.00 © 2025 IEEE   279

C. Digital Circuit

The transmitter circuit can switch between normal mode and low-power mode. mod_sw=0 is a low-power mode, which can only send tones but not communication frames. mod_sw=1 is the normal mode and tone and communication frames can be sent. The low-power mode is designed for sending tone during sleep. When Active, Tx_en=1: sending mode, tone receiving function is turned off. tx_en=0: In the receiving mode, the tone receiving function is enabled.

Tx_ctrl can control whether the active state is enabled. The com*p and com*n interfaces of the chip are in a high-impedance state when not sending, and are forcibly maintained at common mode 2.5V and differential mode 0V for a certain period of time before and after sending. During the sending process, Tx_en needs to be kept in the 1 state. After the sending is completed, set Tx_en=0. The preamble widths of the two half-bits before and after the H byte are generated by the HFO composed of the falling edge (or rising edge) of the Tx_ctrl and the falling edge of the Tx_DATA_IN.The simplified process circuit diagram of the digital circuit is shown in Fig. 7.

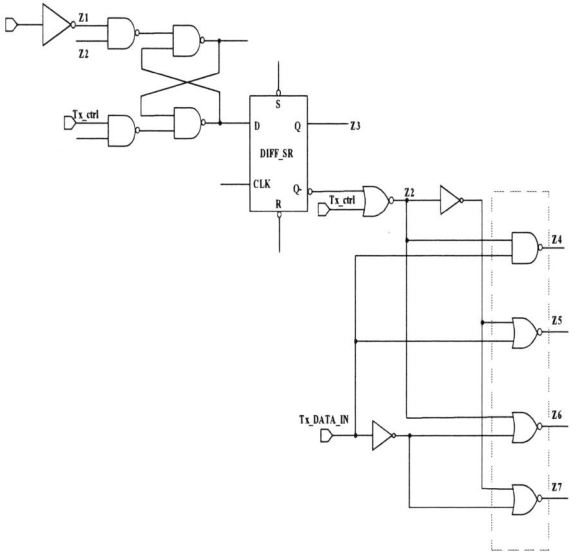

Fig. 7. Simplified Flowchart of the Transmitter Digital Circuit.

IV. SIMULATION SETTINGS AND RESULTS

An insulator is required to transfer between the two derived links. Proposed conductive chain systems can be communicated using capacitors and transformers. The Daisy chain circuit has passed the pre-simulation verification and can switch the working mode through external signals, adapting to the complex working conditions of new energy vehicles, and has significant advantages among similar chip interfaces. Timing sequence is shown in Fig.8.

Daisy chain transmitter can switch between normal mode and low-power mode. mod_sw=0 is a low-power mode, which can only send tones but not communication frames mod_sw=1 is the normal mode and tone and communication frames can be sent. The low-power mode is designed to send tone during sleep.

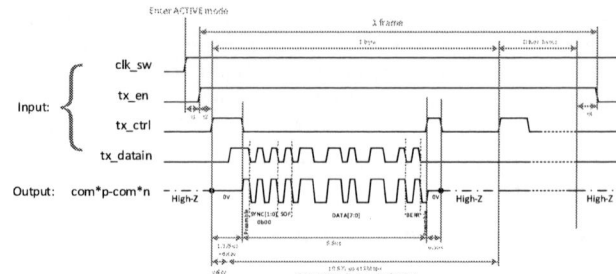

Fig. 8. Daisy Chain Transmitter Communication Frame Transmission Timing Diagram

Before the transmitter sends, clk_sw needs to be set to 0, which can be done when entering sleep after detecting the falling edge of sl_tone_ctrl, the analog circuit will automatically generate pulses. Therefore, only the periodic sl_tone_ctrl signal needs to be input. sl_tone_polarity controls the generation of positive or negative pulses. By default, sl_tone_polarity=0. Timing sequence is shown in Fig.9.

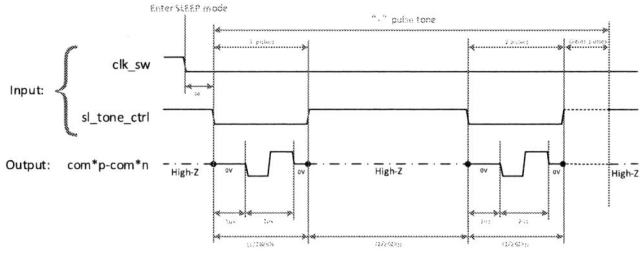

Fig. 9. Tone Transmission Sequence When the Transmitter is in Sleep State

Referring to the parameter indicators of the cutting-edge commercial vehicle-grade BMS chip, the analog circuit part of the Daisy chain receiver and its sub-modules were designed from top to bottom in sequence, Then, through the individual design and simulation of each sub-module, the filtering, voltage division, comparison and result saving of the input signal were completed. During the design process of the comparator for detecting the bus state, the feedback circuit was used to control the on-off of the strong and weak current branches, thereby achieving the purpose of filtering glitches. However, the discharge current and MOS capacitance value need to be further adjusted. Make the width of the filter burrs reach within 250ns. The verification result figures are shown in Fig.10 and Fig.11.

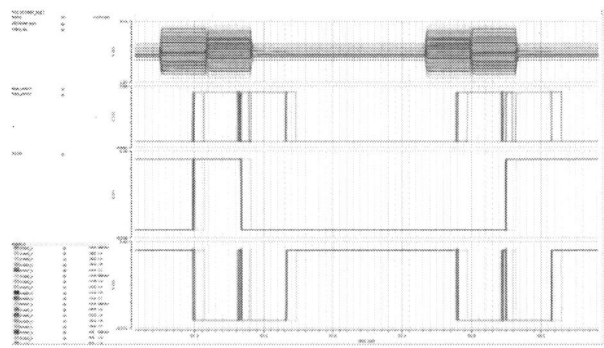

Fig. 10. The Correct Waveforms of the Functions within the Receiver Design Specificationse

979-8-3315-8850-2/25 $31.00 © 2025 IEEE

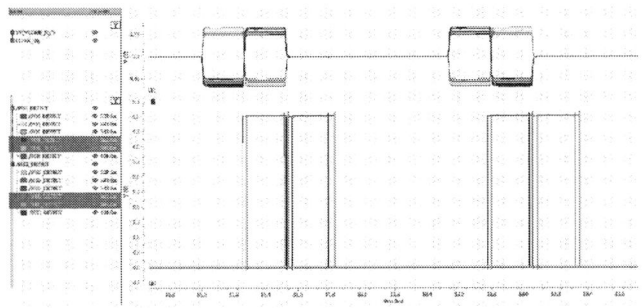

Fig. 11. The Daisy Chain Receiver Accepts the Output Waveform of the Differential Voltage from the Transmitter

V. CONCLUSION

This article focuses on the high-reliability communication requirements of the power battery management system (BMS) chip for new energy vehicles, emphasizing The transmitter circuit of the Daisy Chain communication interface was designed, adopting a differential signal processing architecture, and through suppression. Control high-voltage common-mode interference to ensure signal integrity. The effective filtration pulse width is less than 250ns. It reduces noise and enhances the ability to resist electromagnetic interference (EMI) ,The Daisy chain receiver receives the signal sent by the transmitter to verify that the transmitter is functioning properly and has the anti-electromagnetic interference function provided by the isolator circuit.

The following work will improve the functions of the daisy-chain receiver and further enhance the anti-electromagnetic interference of the daisy-chain port circuit. It is planned to add a common-mode voltage stabilizing circuit on the bus to improve high reliability.

REFERENCES

[1] L Xing, J Sun., et al. Motor drive common-mode EMI reduction by passive noise cancellation[C]. Proceedings of the 2011 14th European Conference on Power Electronics and Applications. IEEE, 2011: 1-9.

[2] M Javid, K Ptacek, R Burton, et al. A 650 kV/μs common-mode resilient CMOS galvanically isolated communication system[J]. IEEE Transactions on Circuits and Systems I: Regular Papers, 2021, 69(2): 587-598.

[3] Z Xiong, D Pan, G Li, et al. A 250Mbps 100kV/μs CMTI On-Chip Double-Isolated Transformer-Based Digital Isolator[C]. 2022 IEEE International Conference on Integrated Circuits, Technologies and Applications (ICTA). IEEE, 2022: 156-157.

[4] Q Quan, C Yu-hua, W Yue, et al. PID-based feature weight learning and its application in intrusion detection[C]. 2009 WRI World Congress on Computer Science and Information Engineering. IEEE, 2009, 5: 570-574.

[5] Z Zheng, Y Chen, J Ding, et al. An EMI Immune Daisy Chain Interface for Battery Management System Communication[C]. 2024 IEEE Joint International Symposium on Electromagnetic Compatibility, Signal & Power Integrity: EMC Japan/Asia-Pacific International Symposium on Electromagnetic Compatibility (EMC Japan/APEMC Okinawa). IEEE, 2024: 501-504.

[6] XU X, CHEN Y, WU J. A low-power daisy-chain controller implemention in bms based on power mode switching[C/OL]//2023 IEEE 15th International Conference on ASIC (ASICON). 2023: 1-4. DOI: 10.1109/ASICON58565.2023.10396339.

[7] Texas Instrument. Functional safety-compliant automotive 16s/14s/12s battery monitor, balancer and integrated hardware protector[EB/OL]. 2020.

[8] Z Zheng, Y Chen, J Ding, et al. Design and Implementation of an EMI-Immune Daisy Chain Interface with a PID-based CDR Algorithm for Battery Management System Communication[J]. IEEE Access, 2024, 44(8): 183-189.

[9] ZENG. Bci - iso 11452-4:2020[J/OL]. BCI - ISO 11452-4:2020, 2022. https://www.emc.wi ki/article-522-1.html.

[10] I Altoobaji, A Hassan. A low-power 0.68-Gbps data communication system for capacitivedigital isolator with 1.9-ns propagation delay[C]. in IEEE Transactions on Very Large Scale Integration (VLSI) Systems.

2025 The 10th International Conference on Integrated Circuits and Microsystems

A Low-Power Flip signal Pulse Width Self-Adaptive Circuit for Piezoelectric Energy Harvesting

Hongtao Yue
School of Integrated Circuits
Southeast University
Nanjing, China
yueht@seu.edu.cn

Shuo Zhang
Analog Design Department
Verisyno Microelectronics Co., Ltd
Hefei, China
zhangsofficial@163.com

Haoyu Xue
School of Integrated Circuits
Southeast University
Nanjing, China
220231845@seu.edu.cn

He Zhi
School of Integrated Circuits
Southeast University
Nanjing, China
zhihe@seu.edu.cn

Jianhui Wu
School of Integrated Circuits
Southeast University
Nanjing, China
wjh@seu.edu.cn

Xin Li
School of Integrated Circuits
Anhui University
Hefei, China
lixin@ahu.edu.cn

Abstract—To address the charge flip efficiency issue in piezoelectric energy harvesting(PEH), this paper proposes an flip signal pulse width self-adaptive(FSPWSA) circuit. The circuit employs a sampling resistor(R_{sam}) and a high-speed continuous-time comparator(HSCTC) to monitor the LC resonant current in real-time and dynamically generates a flip control signal (SADPT) precisely matched to the half-cycle of the resonant current, resolving the PVT sensitivity and pulse width adjustment inaccuracies inherent in traditional RC delay circuits. By implementing a duty cycle control strategy, the comparator is activated only during the charge flip period, resulting in an average dynamic power consumption as low as 6.5 μA (tt/25°C). The post-simulation results indicate that the circuit supports a wide inductance range of 20 to 500 μH under a 2 V power supply voltage, maintains a gain above 60 dB, and exhibits a response delay of less than 160 ns. This design significantly enhances charge reversal efficiency and is suitable for energy harvesting systems in low-power IoT devices.

Index Terms—PEH, FSPWSA, HSCTC, High flip efficiency, Low-power

I. INTRODUCTION

With the rapid development of the Internet of Things (IoT), the number of IoT devices has grown exponentially in recent years [1]. The surge in device numbers has made providing stable power a pressing challenge [2]. Currently, common power supply methods heavily rely on chemical batteries, which not only cause environmental pollution but also eventually deplete their energy. Fortunately, the widespread adoption of low-power technologies and advancements in energy harvesting have made it possible to provide long-term stable power for IoT devices using ambient energy.

There are various types of available energy in natural environments (such as solar energy, wind energy, and piezoelectric energy [3]), which have greatly promoted the application of energy harvesting technology. Among these, piezoelectric

Supported by "the Fundamental Research Funds for the Central Universities" (Grant No. 2242025K30014).

energy has become a research hotspot in recent years due to its high power density and wide distribution [4]. However, piezoelectric energy outputs alternating current (AC), which cannot directly power IoT devices, so a corresponding rectification circuit is needed to convert it into direct current (DC). Common piezoelectric rectification structures include full-bridge rectification circuits and voltage-doubling rectification circuits. However, these structures have low output efficiency, fundamentally because the equivalent internal resistance of the piezoelectric energy harvester (PEH) is capacitive rather than inductive [5]. When the direction of the equivalent current source generated by the excitation source changes, the current source needs to charge and discharge the parasitic capacitance inside the PEH, which causes energy transfer losses.

To address the aforementioned issues, the key to improving efficiency lies in accelerating the charge reversal rate on the parasitic capacitance of PEH. Based on this, researchers have proposed single-switch rectification structures and inductor-based synchronous switching energy harvesting techniques (SSHI) [6], [7]. The core of SSHI technology involves introducing an inductor (L_p) and a switch between the rectification structure and the PEH. When the current source direction changes, the switch is closed, enabling the inductor (L_p) to resonate with the PEH's parasitic capacitance (C_p), forming an LC resonant circuit. This significantly increases the charge transfer rate across C_p, thereby drastically reducing the charging time of C_p by the current source during the subsequent half-cycle (as shown in Fig. 1). Clearly, compared to traditional rectification structures, the inductor-based SSHI technology can significantly enhance charge flip efficiency, thereby improving the overall output efficiency of the rectification structure.

The charge flip circuits utilizing inductors for LC resonance rely on the timing and duration of inductor integration as their performance core. Traditional methods employ RC delay circuits, minimizing resistance (R) and capacitance (C)

979-8-3315-8850-2/25 $31.00 © 2025 IEEE

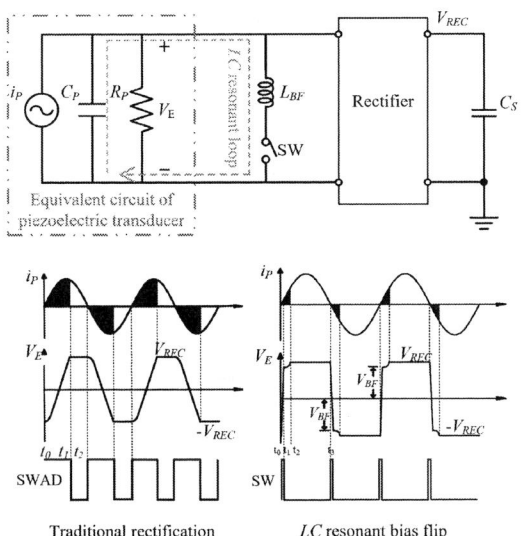

Fig. 1. Schematic diagram of piezoelectric rectifier based on inductance bias flipping technology

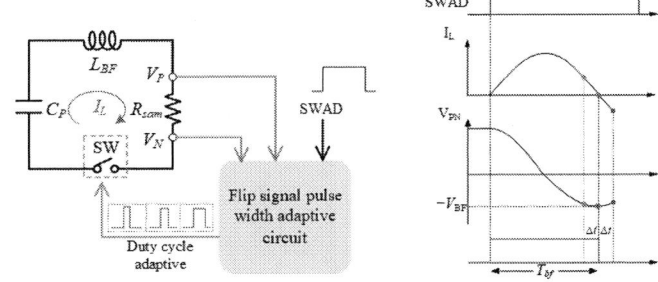

Fig. 2. Schematic diagram of adaptive working principle for flipping signal pulse width

to achieve minimal delay units, and arrange different delay arrays to set the timing and duration of inductor integration through manual external adjustment [8], [9]. However, this approach offers limited improvement in charge flip efficiency and significantly increases the circuit layout area due to the use of R and C. To address this issue, this paper proposes a flip signal pulse width self-adaptive(FSPWSA) circuit. This circuit uses sampling resistors and high-speed continuous-time comparators(HSCTC) to automatically determine the optimal timing and duration of inductor integration based on the inductor value within the circuit.

II. CIRCUIT DESIGN AND ANALYSIS

The key to implementing inductor-based bias-flipping rectification lies in generating a signal with precisely controllable pulse width. This signal controls the connection and disconnection of the inductor to facilitate charge flipping, ensuring efficiency (η_f). When the equivalent current source direction of the piezoelectric energy harvesting device changes, the inductor must be connected to form an LC resonant circuit, transferring the charge across the capacitor. This ensures that the voltage across capacitor C_p aligns with the charging direction of the current source, enabling efficient extraction of piezoelectric energy. After half a resonant cycle (T_{bf}), all charges at one end of the capacitor have been transferred to the other end via the inductor. Based on the phase relationship between voltage and current in the resonant circuit, the inductor current should be zero at this point. It is crucial to immediately disconnect the inductor to prevent charge from flowing back to the original end of the capacitor. The entire process must be precisely controlled by a signal with a pulse width equal to half the resonant period (T_{bf}) to ensure high flipping efficiency.

Traditional methods rely on manual adjustment by humans, and RC delays are significantly affected by variations in PVT (process, voltage, temperature), resulting in poor adaptability. Additionally, due to the excessive chip area occupied by resistors and capacitors, this tuning method struggles to generate sufficiently wide pulse signals, and the unit delay time cannot be designed to be very small. This leads to difficulty in precisely controlling the turn-off moment of the resonant circuit using the flip pulse signals generated by this approach, thereby reducing charge reversal efficiency.

The FSPWSA circuit proposed in this paper operates as shown in Fig. 2 (where SWAD in the figure is the enable signal for driving the active diode). This circuit can adaptively track the completion moment of charge reversal and generate flip signals with corresponding pulse widths to control the closure of the switch. Therefore, when using different inductance values, this circuit can generate appropriate flip signal pulse widths, ensuring high charge reversal efficiency, reducing charge loss during energy collection, and thus improving output power.

The operation principle of the FSPWSA circuit is as follows: when the output signal SWAD of the active diode zero-current detection comparator goes high, the charge flip is initiated. The flip current passes through R_{sam}, generating a voltage difference that is continuously monitored by a HSCTC. When the voltage difference exceeds the threshold, the comparator output jumps from 0 to 1. When the flip current decays below the threshold (voltage difference decreases), the comparator output returns to 0. This falling edge signifies the completion of the charge flip. The full pulse width of the adaptive flip signal is represented by the time interval between the rising edge of SWAD and the falling edge of the comparator output during the duty cycle enable period.

The core module of the FSPWSA circuit consists of four parts: sampling resistor R_{sam}, HSCTC with enable control, flip signal generation circuit, and duty cycle control signal generation circuit [10]. Among these, the flip signal generation circuit is primarily composed of multiplexers, AND gates, NAND gates, inverters, and D flip-flops, with its structure shown in Fig. 3. The main function of the duty cycle control

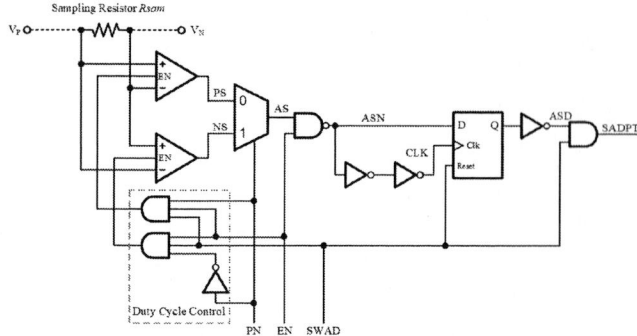

Fig. 3. The flip signal pulse width self-adaptive circuit Structure

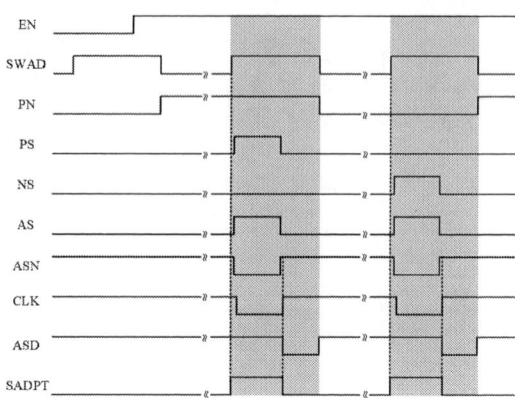

Fig. 4. The flip signal pulse width self-adaptive circuit Timing Diagram

signal generation circuit is to output an EN = 1 enable signal from the mode switching circuit when the capacitor C_p needs to undergo charge flipping, thereby triggering the adaptive circuit to start. This circuit operates only during the charge-flipping period. The key design lies in maintaining the output voltage stability, during which the active diode is in cut-off state (corresponding to SWAD signal being high 1). Therefore, the HSCTC used to detect the voltage difference across R_{sam} is only activated when both EN and SWAD are 1. Ultimately, the duty cycle control signal synthesized by EN, SWAD, and PN signals collectively regulates the operating states of the two comparators.

The core objective of this duty-cycle control scheme is to significantly reduce the system's average power consumption. To ensure high transient response speed, the comparator requires high dynamic power consumption; however, the duty-cycle control signal is only activated during a very small proportion of the energy harvesting cycle. Consequently, the comparator remains in a static low-power mode for most of the time, resulting in a substantial reduction in its average power consumption over the entire cycle.

The outputs PS and NS of the HSCTC by a 2-to-1 data selector controlled by the flip-flop direction signal PN to produce the output AS, ensuring the correctness and stability of the sequential. When PN = 1 (indicating a flip from P to N), PS is selected for output; conversely, NS is selected. Subsequently, AS is processed with the mode switching enable signal EN through an AND logic operation to further ensure the timing correctness of the active mode switching. The output of this AND gate is inverted by an inverter and then passed through a D flip-flop to generate the falling-edge signal ASD. Finally, the SADPT signal, synthesized by the AND logic combination of ASD and SWAD, serves as the adaptive pulse width flip signal.

The signal sequential during the operational process is shown in Fig. 4. When EN = 0, the SADPT signal is forced to zero. Only when EN = 1 does the circuit enter its normal operating state: during EN = 1, the rising edge of SWAD triggers the flip-flop start (inductor circuit), with PN determining the generation of either PS or NS. The pulse widths of PS and NS dynamically vary according to the parameters of the

resonant circuit and are output as AS via the data selector. The AS signal is inverted via a NAND gate to generate ASN (serving as the input to the D flip-flop), while ASN, after being delayed by two inverters, acts as the clock for the D flip-flop. When the clock's rising edge triggers and AS = 1, ASD transitions from 1 to 0. The falling edge of ASD closely corresponds to the falling edge of AS, and the SADPT signal is generated through the AND logic combination of the rising edge of SWAD and the falling edge of ASD.

In Fig. 4, the red region denotes the duty cycle control zone: only within this zone does a HSCTC operate dynamically. Since this interval constitutes an extremely small portion of the entire operating cycle, the average power consumption of the comparator is significantly reduced.

B. High-Speed Continuous-Time Comparator(HSCTC) Design

During the entire charge flip process, the continuous-time comparator must continuously monitor the voltage difference across the sampling resistor R_{sam}. To ensure efficient flip, the resistance value of the resonant circuit should be as small as possible; however, the current peak during resonance can reach tens of milliamperes. To accurately determine the end of flip, it is necessary to detect the small voltage difference across R_{sam} when the resonant current decays close to zero. This requires the comparator to have high resolution (i.e., the gain stage must provide high gain). Additionally, since the charge flip time is on the microsecond scale, the transient response speed of the comparator must be as fast as possible within allowable power consumption, necessitating that its gain stage combines high slew rate and high bandwidth. To address these requirements, this paper designs a HSCTC with a three-stage amplification structure as shown in Fig. 5. If a single-stage amplification is used under low-power constraints, there is a conflict between high gain and high bandwidth: high gain limits bandwidth, resulting in poor transient response; high bandwidth results in insufficient gain, failing to meet the resolution requirements for voltage difference detection. To balance these, this design employs a three-stage cascaded amplification structure: each amplifier stage has high bandwidth

979-8-3315-8850-2/25 $31.00 © 2025 IEEE   284

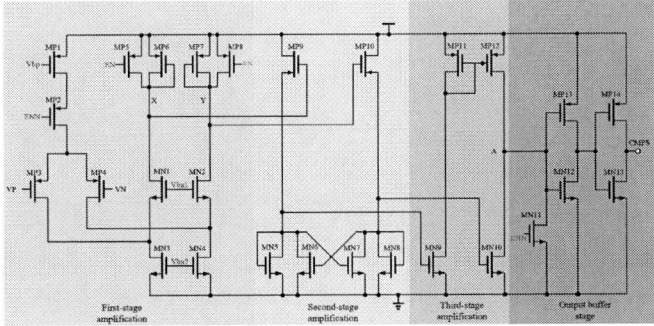

Fig. 5. High-Speed Continuous-Time Comparator Core Circuit Structure

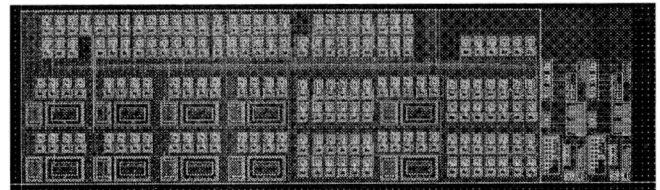

Fig. 6. The flip signal pulse width self-adaptive circuit Layout

and low gain characteristics, achieving overall high gain when cascaded. Although cascading introduces some bandwidth loss, it still meets the design objectives. Assume that the transfer function of the three-stage amplification structure is the same.

$$\frac{V_{\text{out}}}{V_{\text{in}}} = \frac{A_0}{\left(1 + \frac{S}{\omega_P}\right)} \quad (1)$$

In the equation, A_0 is the low-frequency gain of a single-stage amplifier, and ω_P is the pole angular frequency.

The transfer function of the three-stage amplifier structure when cascaded is:

$$\frac{V_{\text{out}}}{V_{\text{in}}} = \frac{A_0^3}{(1 + \frac{s}{\omega_P})^3} \quad (2)$$

It can be calculated that the bandwidth after three-stage cascaded, i.e., the new (-3 dB) bandwidth is:

$$\omega_{-3\,\text{dB}} = \sqrt{\sqrt[3]{2} - 1} \cdot \omega_P \approx 0.51\,\omega_P \quad (3)$$

It can be seen that even with increased low-frequency gain, the bandwidth remains half that of a single-stage amplifier structure.

To accommodate the relatively wide common-mode voltage range across the sampling resistor during charge flip, the first stage is designed as an input stage utilizing a folded fully differential structure with PMOS transistors. The second stage serves as the gain stage, employing a fully differential common-source amplifier structure with basic MOS diodes as loads. Additionally, a positive feedback loop is incorporated through a cross-coupled structure formed by MN6 and MN7. In this configuration, the width-to-length ratio of MN5 and MN8 should be slightly larger than that of MN6 and MN7. The cross-coupled pair can be equivalent to a small-signal resistor with a relatively large resistance value, thereby enhancing the gain. The equivalent small-signal resistance of the cross-coupled pair can be expressed as:

$$r = \frac{1}{gm_{5,8} - gm_{6,7}} \quad (4)$$

The third stage converts the differential input into a single-ended output using a current mirror and outputs the CMPS

signal through the output buffer stage. To address the issue of high dynamic power consumption in the proposed HSCTC, this work employs a duty cycle control strategy (as shown in Fig. 3 to reduce power consumption. The comparator's bias voltages V_{bp}, V_{bn1}, and V_{bn2} are all modulated by the duty cycle signal: only within the effective interval of the duty cycle control (i.e., the red area in Fig. 4) are the bias voltages in normal operating condition; otherwise, they enter a sleep state, significantly reducing the average power consumption of the comparator over the entire operating cycle.

The proposed bias circuit has a static current of 200 nA in sleep mode. As shown in Fig. 5, the switching transistors MP2, MP5, MP8, and MN11 are all controlled by the duty cycle output signals and their complementary signals: during the sleep phase, MP2 is turned off, blocking the bias current path provided by MP1; MP5 and MP8 are turned on, pulling nodes X and Y up to VDD; MN11 is turned on, pulling node A down to ground.

III. SIMULATION RESULTS AND ANALYSIS

To accurately determine the pulse width of the flip-flop signal, the comparator resolution must be lower than the set threshold of the voltage difference across the sampling resistor, thus requiring the amplifier stage to have a high gain. Taking a working voltage of 2 V as an example: when the sampling resistor value $R=10$ Ohm and the current $I_L=0.1$ mA, a voltage difference $\Delta V=1$ mV triggers the comparator flip-flop. At this point, the gain must be 60 dB to meet the 1 mV detection precision requirement.

Although increasing the sampling resistor value can reduce the gain requirement and improve current monitoring accuracy, a high resistor value reduces charge flip-flop efficiency. Therefore, the sampling resistor value needs to be selected as a trade-off. Fig. 6 shows the overall layout of the flip signal pulse width self-adaptive circuit, and Fig. 7 provides the amplitude-frequency response curves of the amplifier stage under different power supply voltages at the tt process corner. It can be seen that the gain remains stable above 60 dB under all voltages.

To verify the dynamic adaptability of the proposed circuit, under a fixed power supply voltage of 2V, this work obtains the amplitude-frequency characteristics and transient response curves under different process corners and temperatures through post-simulation (Fig. 8).

As shown in Fig. 8(a) under the ss corner and under extreme conditions of -20°C, although the low-frequency gain

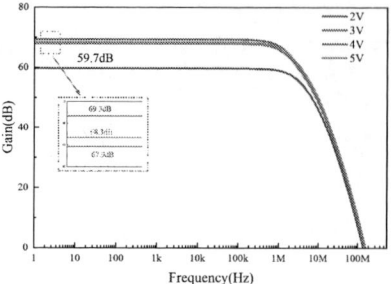

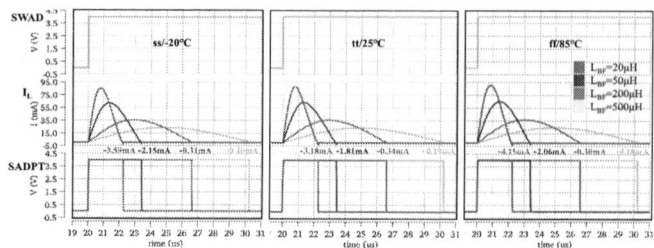

Fig. 9. Flip signal pulse width self-adaptive effect post-simulation

Fig. 7. Amplitude-Frequency Characteristics of HSCTC Under Different Power Supply Voltages Post-Simulation Results

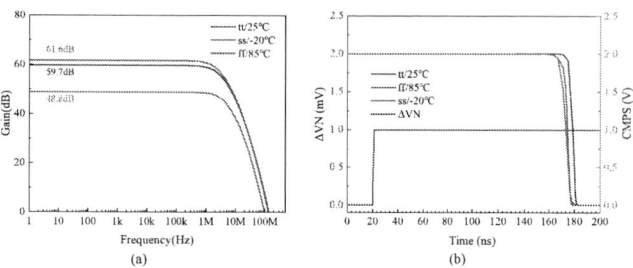

(a)   (b)

Fig. 8. The HSCTC (a) frequency response; (b) transient response under different process corners and temperatures; post-simulation results

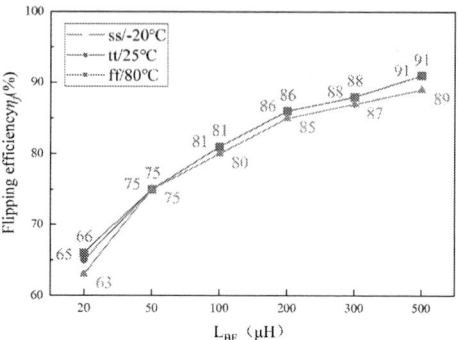

Fig. 10. Post simulation of the relationship between flip efficiency and inductance value

is 48.8 dB, resulting in a slight reduction in flip efficiency, it remains within an acceptable range. In the transient response simulation, the VP voltage is constant, while the VN voltage is based on VP and includes an additional 1 mV step signal. As indicated in Fig. 8(b), the response delay of the CMPS signal under extreme conditions ranges from 150 ns to 160 ns. Compared to the microsecond-level flip pulse width, this delay has a negligible impact on system functionality and fully meets the design requirements.

Under different process corners and temperature conditions, when the inductance values cover the range of 20 μH, 50 μH, 200 μH and 500 μH), The FSPWSA designed in this paper can precisely match the resonant period variations caused by inductance changes, dynamically adjust the SADPT signal pulse width, and promptly disconnect the flip-flop loop to avoid inductor current reversal, thereby maintaining high flip-flop efficiency. The post-simulation waveforms in Fig. 9 and the inductor current values at the end of the flip-flop (both close to zero) confirm this characteristic.

Additionally, targeting the application's wide supply voltage range and low power consumption requirements, transient simulations were conducted to test the circuit's static current and periodic average current. The current simulation results under different process corners and temperatures are detailed in Table I. It is obvious from table I that the dynamic current of the overall comparator is mainly concentrated in the range of 6-7.2 μA, and the static current is only about 400 nA, which meets the low-power design.

To verify the charge flipping capability of the flip signal pulse width adaptive circuit designed in this paper, post-

simulation was conducted under different process corners and temperature conditions (ss/-20°C, tt/25°C, ff/85°C), with the results shown in Fig. 10: when the inductance value is 200μH, the system can achieve a flipping efficiency of 85%; however, further increasing the inductance value can only slightly improve the efficiency, which tends to flatten, and it is difficult to offset the cost increase caused by the enlargement of device size

The simulation results of the FSPWSA circuit proposed in this paper and its comparison with recent related studies are shown in Table II. It can be observed that compared to ref. [7], the charge flip efficiency differs by only 1.2%. However, ref. [7] employs a five-step flip operation to achieve charge transfer, which significantly increases the complexity of circuit design. Ref. [9] uses an external capacitor to implement the reversal process, greatly improving the integration level, but resulting in a lower flip efficiency. Ref. [11] innovatively utilizes internal parasitic capacitors for charge flip, providing a new research direction, but its flip process relies on manual adjustment and achieves low efficiency.

TABLE I
FLIP-FLOP SIGNAL PULSE WIDTH SELF-ADAPTIVE CIRCUIT OPERATING CURRENT SIMULATION RESULTS

Parameter	ss/−20°C	tt/25°C	ff/85°C
Static Current (nA)	418	405	423
Average Dynamic Current (μA)	6.03	6.5	7.2

TABLE II
PERFORMANCE COMPARISON WITH THE LATEST CIRCUIT

References	This work (post simulation)	[7]	[9]	[11]
Tech. (μm)	0.18	0.13	0.35	0.18
Rectifier type	SSHI	SSHI	SSHC	SBFR
Off chip components	L=200μH	L=33μH	C=8$\times$45nF	–
Flip efficiency η_f (%)	86	87.2	80	80
Flip timing	Self adj.	Self adj.	Self adj.	Ext. adj.

IV. CONCLUSION

This paper designs a FSPWSA circuit for piezoelectric energy harvesting, innovatively integrating current sampling, HSCTC, and duty cycle control techniques. The proposed solution dynamically monitors the zero-crossing point of the resonant current using a sampling resistor and HSCTC, automatically generating precise flip pulse width signals matched to the inductor value. This effectively addresses the issue of traditional RC delay circuits struggling to adapt to PVT variations and inductor parameter fluctuations. Additionally, the duty cycle control strategy compresses the operating time of the HSCTC to an extremely short segment of the entire cycle, reducing its average power consumption to just 6.5 μA (with a static current as low as 400 nA), significantly enhancing system efficiency. Furthermore, the design based on a third-order amplifier structure ensures a gain above 60 dB while maintaining stable bandwidth. Post-simulation validation confirms the circuit's reliable functionality across a wide temperature range (-20°C to 85°C) and process corner variations, with a response delay of less than 160 ns, meeting the stringent requirement of μs-level flip precision. This circuit provides an efficient and adaptive solution for piezoelectric energy harvesting systems, significantly reducing charge flip losses, making it particularly suitable for self-powered systems in low-power IoT devices.

REFERENCES

[1] C. D. Alwis, A. Kalla, Q.-V. Pham, P. Kumar, K. Dev, W.-J. Hwang, and M. Liyanage, "Survey on 6g frontiers: Trends, applications, requirements, technologies and future research," *IEEE Open Journal of the Communications Society*, vol. 2, pp. 836–886, 2021.

[2] M. Hulea, G. Mois, S. Folea, L. Miclea, and V. Biscu, "Wi-sensors: A low power wi-fi solution for temperature and humidity measurement," in *Industrial Electronics Society, IECON 2013 - 39th Annual Conference of the IEEE*, 2013.

[3] M. M. Warrier and A. Kumar, "Energy efficient routing in wireless sensor networks: A survey," in *2016 International Conference on Wireless Communications, Signal Processing and Networking (WiSPNET)*, 2016, pp. 1987–1992.

[4] H. Li, C. Tian, and Z. D. Deng, "Energy harvesting from low frequency applications using piezoelectric materials," *APPLIED PHYSICS REVIEWS*, vol. 1, no. 4, 2014.

[5] L. Moro and D. Benasciutti, "On the performance improvement of piezoelectric energy harvesters," 2018.

[6] P. Gasnier, J. Willemin, S. Boisseau, G. Despesse, C. Condemine, G. Gouvernet, and J.-J. Chaillout, "An autonomous piezoelectric energy harvesting ic based on a synchronous multi-shot technique," *IEEE Journal of Solid-State Circuits*, vol. 49, no. 7, pp. 1561–1570, 2014.

[7] S. Javvaji, V. Singhal, V. Menezes, R. Chauhan, and S. Pavan, "Analysis and design of a multi-step bias-flip rectifier for piezoelectric energy harvesting," *IEEE Journal of Solid-State Circuits*, vol. 54, no. 9, pp. 2590–2600, 2019.

[8] D. A. Sanchez, J. Leicht, F. Hagedorn, E. Jodka, E. Fazel, and Y. Manoli, "A parallel-sshi rectifier for piezoelectric energy harvesting of periodic and shock excitations," *IEEE Journal of Solid-State Circuits*, vol. 51, no. 12, pp. 2867–2879, 2016.

[9] S. Du and A. A. Seshia, "An inductorless bias-flip rectifier for piezoelectric energy harvesting," *IEEE Journal of Solid-State Circuits*, vol. 52, no. 10, pp. 2746–2757, 2017.

[10] B. Çiftci, S. Chamanian, A. Koyuncuoğlu, A. Muhtaroğlu, and H. Külah, "A low-profile autonomous interface circuit for piezoelectric micropower generators," *IEEE Transactions on Circuits and Systems I: Regular Papers*, vol. 68, no. 4, pp. 1458–1471, 2021.

[11] Z. Li, J. Wang, M.-K. Law, S. Du, J. Liang, X. Cheng, J. Han, X. Zeng, and Z. Chen, "Piezoelectric energy harvesting interface using self-bias-flip rectifier and switched-peh dc–dc for mppt," *IEEE Journal of Solid-State Circuits*, vol. 59, no. 7, pp. 2248–2259, 2024.

2025 The 10th International Conference on Integrated Circuits and Microsystems

A Sub-nanosecond Delay and 200V/ns CMTI Level Shifter for GaN HEMTs Gate Driver

Shengqi Yu
School of Integrated Circuits
Chongqing University of Posts
and Telecommunication
Chongqing, China
yusq@cqupt.edu.cn

Pengyu Wang
School of Integrated Circuits
Chongqing University of Posts
and Telecommunication
Chongqing, China
s240431165@stu.cqupt.edu.cn

Junjie Yi
School of Integrated Circuits
Chongqing University of Posts
and Telecommunication
Chongqing, China
s230431129@stu.cqupt.edu.cn

Shupeng Yan
School of Integrated Circuits
Chongqing University of Posts
and Telecommunication
Chongqing, China
s240403005@stu.cqupt.edu.cn

Yi Huang
School of Integrated Circuits
Chongqing University of Posts
and Telecommunication
Chongqing, China
huangy@cqupt.edu.cn

Abstract—**In next-generation power electronics technology, GaN devices show outstanding characters to the requirements of high-frequency and high-power density. The core role in high-reliability GaN gate drivers is high-speed and noise-resistant level shifter. However, existing level shifter solutions suffer from issues such as the excessive signal transmission delay and weak common mode transient immunity (CMTI), which severely limit the performance of GaN devices. This paper addresses these issues by incorporating two auxiliary capacitors in the two pulse control branch nodes of the level shift main circuit, leveraging their charging and discharging characteristics to significantly reduce signal transmission delay. Implemented in 0.18 μm HV SOI CMOS process，simulation results show the proposed design reduces up to 56.08% delay for the rising edge and 50.43% delay for the falling edge when compared with existing scenarios. Additionally, the proposed design gains dV/dt noise suppression capability up to 200 V/ns.**

Keywords—*level shifter, GaN, gate drive, high speed, CMTI, sub-ns delay*

I. INTRODUCTION

The development of power converters requires higher power density and lower losses[1], many candidates show promising characters, in them, the third-generation wide bandgap semiconductor (i.e. GaN) is the outstanding one. With high breakdown electric field strength, low on-state resistance and high switching frequency, GaN HEMTs is the most promising device in the next generation power converting applications[2][3][4]. such like electronic product power adapters, electric vehicles and DC-DC power supplies[5].

The most widely used topology in the next generation power converting applications of GaN devices is the typical half-bridge monolithic GaN power IC, which shown in Fig. 1[6][7]. When the GaN main switch transistor (GaN MH) is activated, the potential of floating ground VSW rapidly rises. The high-voltage side channel is powered by a bootstrap capacitor, while the voltage of the floating power rail VDDH rises in synchronization with VSW. The rising or falling speed of VSW can be 100 V/ns to 200 V/ns. Since the level shifter operates on the floating rail, a level shifter with high CMTI is necessary for the accurate signal transmission in the typical half-bridge monolithic GaN power IC. Moreover, for keeping the output responds promptly to input, a level shifter with low-delay is essential in this high frequency circuit. Therefore, the CMTI and low transmission delay are necessary to the reliability and operating frequency of the typical half-bridge monolithic GaN power IC[8][9].

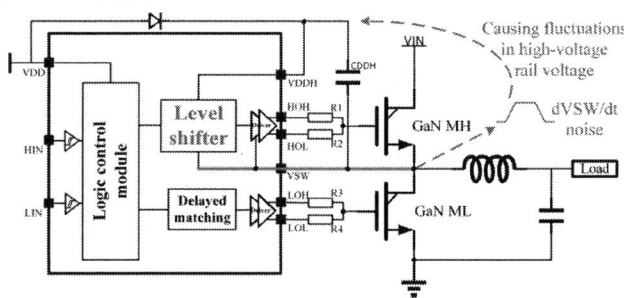

Fig. 1. Typical half-bridge monolithic GaN power IC block diagram

For years, researchers tries to address both the CMTI and low transmission delay in design. One solution is adding a filter after the level shifter[10]. This method, improves level shifter tolerance to dV/dt noise by increasing the bandwidth of the filter, This update is at the cost of transmission speed. Another solution adopts auxiliary pull-up circuit, promoting delay matching, and self-calibration techniques for better performance[11]. This design achieves high dV/dt noise tolerance immunity up to 250 V/ns and the minimum delay is only 664 ps. However, this circuit has a much more complex logic pairing work. In more works focus on enhancing the dV/dt immunity by adding an auxiliary pull-up path[12][13], and [14]. But this scenario is at the risk of misbehaviour by the linear region operating pull-up

979-8-3315-8850-2/25 $31.00 © 2025 IEEE

transistor. A much more complicated solution is built by utilizing a current mirror to divert the common-mode current from common-mode noise[15]. But, the design is susceptible to process variations, circuit mismatches, and has high layout requirements. In[16] and [17], The dual level shifter design. Try to make a narrow pulse triggered level shifter to generate sub-nanosecond transmission delay, while a level triggered level shifter keeps the output state. This design effectively improves the circuit's CMIT and transmission delay, but the circuit gains much more complexity than other ones.

To addresses these issues, the proposed design incorporates two auxiliary capacitors in the two pulse control branch nodes of the level shifter main circuit, leveraging their charging and discharging characteristics to significantly reduce signal transmission delay. Simulations are implemented in 0.18 μm HV SOI CMOS process, results demonstrate that the sub-ns delay level shifter can be realized. At the TT process corner and 25°C, the rising edge propagation delay is 860 ps, and the falling edge propagation delay is 670 ps. Moreover, the dV/dt immunity can be up to 200 V/ns. The rest of this paper is structured as follows. Sect.2 gives the background and motivation. Sect. 3 presents all the details of our novel level shifter design is. and discussions are presented in Sect. 4. And Sect.5 concludes this work.

II. CONVENTIONAL LEVEL SHIFTER

Conventional level shifter has two power rails: a low-voltage rail from the gate of LD1 and LD2 (VDD) to GND, and a high-voltage rail from VDDH to floating ground VSW. As can be seen in Fig. 2 a level shifter consists of a narrow pulse generation circuit, LDMOS device for withstanding high voltage, 5V-MOS transistors for signal transmission, pull-up resistors and an RS latch for maintaining the output state.

A. Operating Principle

When the rising edge of the input signal IN arrives, the narrow pulse generation circuit[18] will output narrow pulse signal, turning on MN1 and LD1. In this branch, a saturation current I_1 (green arrow) flows from VDDH to ground, the magnitude of current I_1 can be defined as (1).

$$I_1 = \frac{1}{2}\mu_n C_{ox}\left(\frac{W}{L}\right)_{MN1}(V_{gs} - V_{th})^2 \tag{1}$$

The diode-connected MP1 acts as a load, when the current flowing through it while generate voltage drop at MP1, thus generating a negative pulse at the S terminal of RS latch. When RS latch that is active by a low level input, its output will be set to 1. Similarly, a saturated current I_2 produced by narrow pulse from the falling edge of the input signal, will generate a negative pulse at the R terminal. Then the RS latch output will be set to 0. The gates of LD1 and LD2 connects to low-voltage rail (VDD) as protect device for the low-voltage transistors MN1 and MN2. When MN1 and MN2 are both in off state, nodes R and S enter a high-impedance state. High-impedance resistors R1 and R2 are required to rapidly pull up R and S through current flow, preventing interference from external random noise.

B. dV/dt Noise Influence

As mentioned before, floating ground VSW fluctuates continuously with the conduction and cutoff of the GaN MH, and these fluctuations are coupled to VDDH via the bootstrap capacitor.

When the power rail generates positive dV/dt noise, the LDMOS, with large parasitic capacitance between the drain and ground, cannot respond promptly to changes in floating voltage, This procedure generates the corresponding noise currents I_{noise}. Meanwhile, voltage drops at the S and R terminal, can lead logic errors or even damage in the drive circuit in severe cases [19]. The magnitude of current I_{noise} can be defined as (2).

$$I_{noise1} = I_{noise2} = C_1\frac{dv}{dt} \tag{2}$$

When the power rail generates negative dV/dt noise, the R and S terminals will have higher voltage than VDDH, so MP1 and MP2 are in off state and generate no noise current. The output signal will not be affected. This situation can be represented as (3).

$$VDDH - V_S/V_R < (V_{th})_{MP1/2} \tag{3}$$

Overall, traditional level shifter show fatal issues for meeting the high-level requirements of GaN half-bridge gate drive for CMTI and delay.

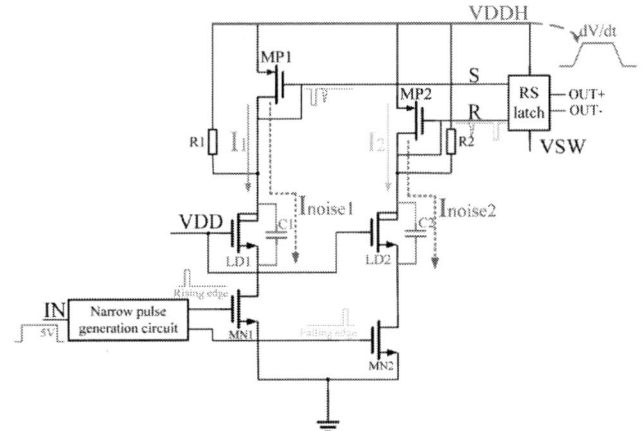

Fig. 2. Traditional level shifter structure[20]

III. A SUB-NANOSECOND DELAY AND 200V/NS CMTI LEVEL SHIFTER

This section presents the novel level shifter design. In this design, we enhance CMTI and reduce circuit transmission delay. As can be seen in Fig. 3, the level shifter includes a narrow pulse generation circuit, which generates two low-voltage narrow pulse signals. The level shifter main circuit has two auxiliary capacitors C1 and C2, both utilize their charging and discharging to ensure noise immunity while minimizing signal transmission delay. two floating power supply VDDH-to-ground paths, currents I_3 and I_4 are used to simulate signal transmission after narrow pulse input signals, and current I_{noise} is used to simulate the effect of common-mode noise on the circuit when the floating ground power supply rail voltage undergoes sudden changes in the level converter (seeing in Fig. 4). Fig. 5 presents the capacitor charge and discharge control circuit to generate control logic for the charging and discharging process of two auxiliary capacitors.

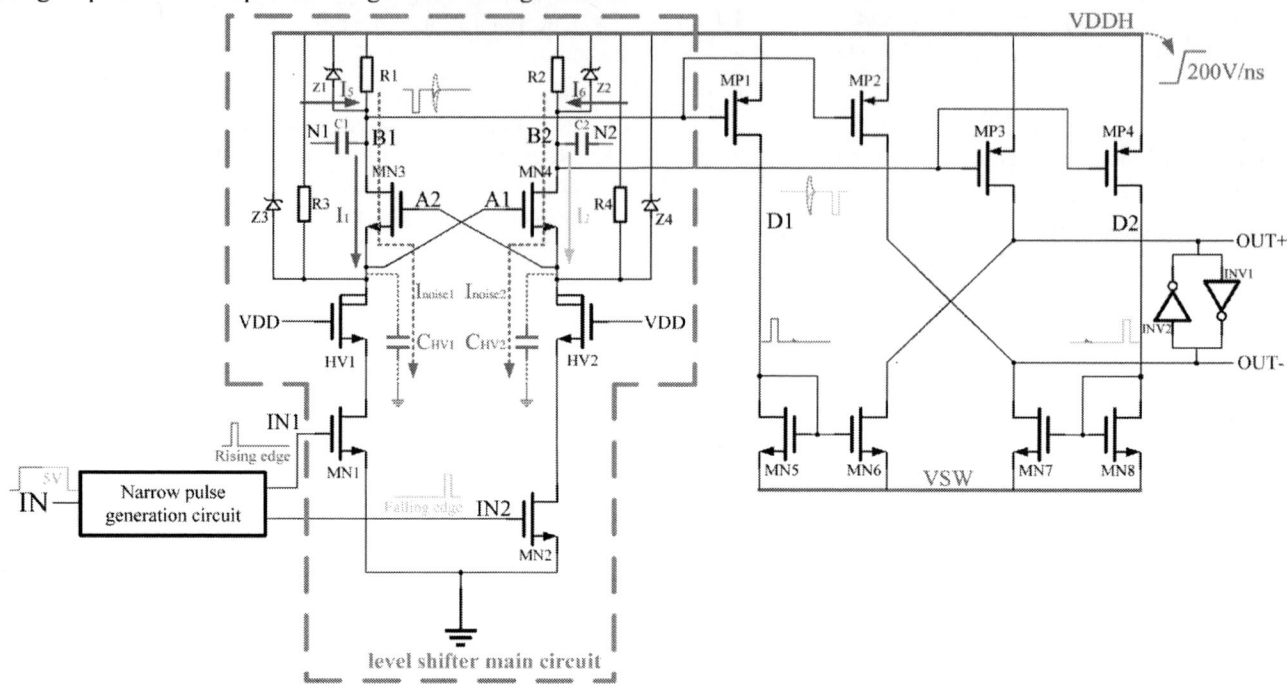

Fig. 3. Main circuit and narrow pulse generation circuit

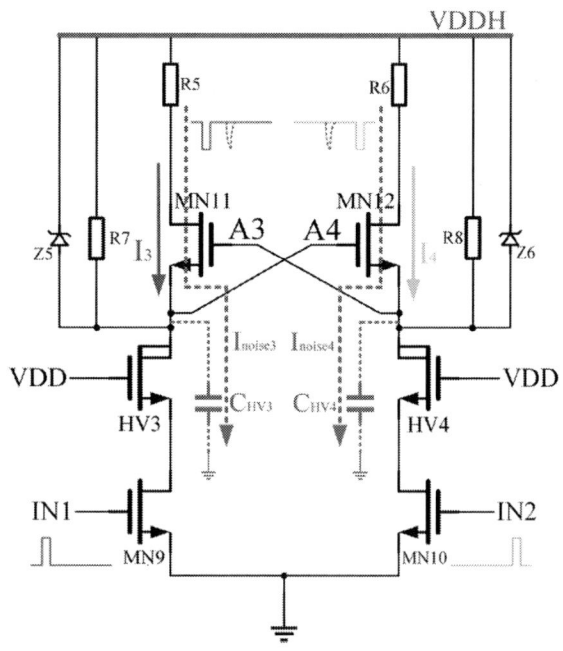

Fig. 4. Two floating power supply VDDH-to-ground paths

A. Reduce Transmission Delay

As Fig. 3 and Fig. 4 show, the width-to-length ratio of MN9 and MN10 are several times larger than those of MN1 and MN2. Under this condition, node A3 and A4 drop faster than node A1 and A2 when the corresponding narrow pulse signals arrive.

This makes the voltages of node N1 and N2 to decrease synchronously with node A1 and A2, When the voltage drop signals of A3 and A4 pass through the capacitor charge and discharge control circuit. Fig. 6 presents the transient voltage relation between node A1, A3, and N1.

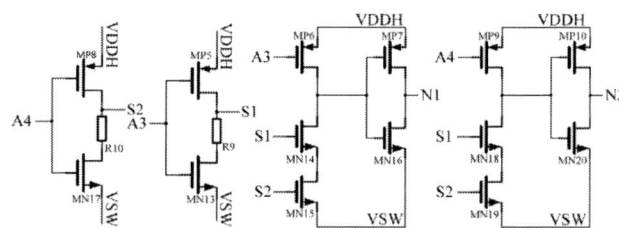

Fig. 5. Capacitor charge and discharge control circuit

When the low-voltage narrow pulse signal IN1 is active, transistors MN1, HV1, MN9, and HV3 turn on, pulling down the voltages at nodes B1 and A3. Node A3 then lowers the voltage at the left end of capacitor C1 through the capacitor charge–discharge control circuit. Utilizing the fact that the voltage across a capacitor cannot change abruptly, resistor R1 charges C1, assisting in pulling down the voltage at the lower end of R1, thereby reducing signal transmission delay. Subsequently, the voltage across C1 is charged to VDDH − VSW. Similarly, when the low-voltage narrow pulse signal IN2 is active, capacitor C2 charges to assist in lowering the voltage. at the lower end of resistor R2, also reducing signal transmission delay. The voltage across C2 is then charged to VDDH − VSW.

979-8-3315-8850-2/25 $31.00 © 2025 IEEE   290

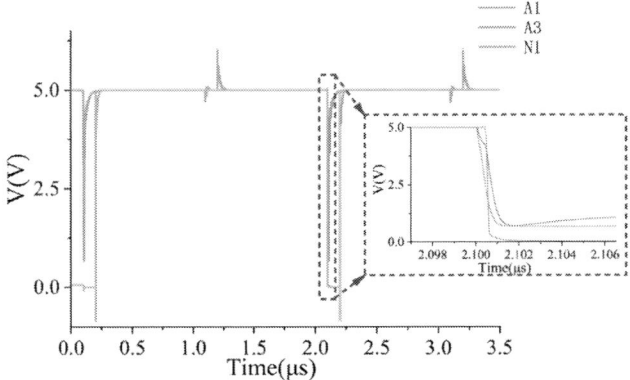

Fig. 6. Time relationships between nodes A1, A3, and N1

B. CMTI

When positive dV/dt noise occurs, voltage drops are generated at nodes A1, A2, A3, and A4 relative to the high-voltage rail VDDH. Nodes S1 and S2 are pulled up to VDDH. Subsequently, transistors MN14, MN15, MN18, MN19, MP7, and MP10 turn on, pulling up nodes N1 and N2. Since the voltage across capacitors C1 and C2 is VDDH − VSW, and the voltage across a capacitor cannot change abruptly, C1 and C2 are discharged simultaneously. The stored charge from C1 and C2 is delivered to the lower ends of resistors R1 and R2, respectively. This action suppresses the voltage drop at nodes B1 and B2 caused by common-mode noise. Subsequently, the voltage across capacitors C1 and C2 returns to 0 V, preparing the circuit for the next short pulse and improving signal transmission speed. The voltage waveform across the capacitors is shown in Fig. 7.

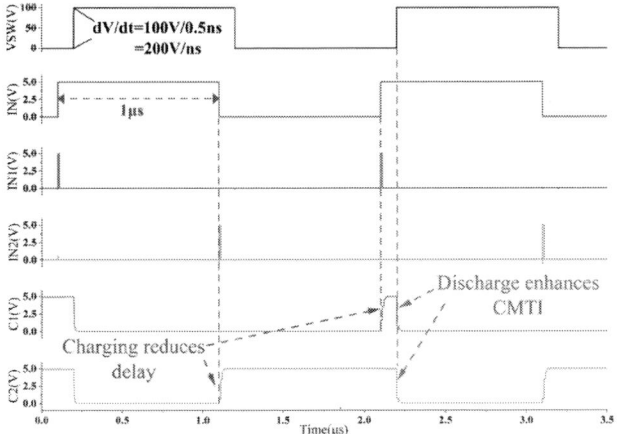

Fig. 7. The voltage waveform across the capacitors

When negative dV/dt noise occurs, the zener diode can clamp the voltage to prevent the low-voltage tube from being broken down. The voltage can be represented as (4).

$$V_{A1} = VDDH + V_Z \tag{4}$$

Nodes A1, A2, A3, and A4 are all connected to zener diode, and nodes B1 and B2 are also connected to Zener diodes because they are directly connected to large capacitors.

C. Other

The two output non-gates form positive feedback, accelerate signal establishment and improves anti-interference capability.

IV. SIMULATION RESULTS

In this paper, all parts of the system are realized in 0.18 μm HV SOI CMOS technology and studied in the Cadence Virtuoso environment through analog simulations. Two types of transistors are applied in this design, transistors HN1, HN2, HN3 and HN4 are 200V HV-NMOS, while the rest MOSs are LV-MOS with 7V withstand voltage. Fig. 8 present the transient performance of output waveform. With 1 μs pulse width input, the level shifter has a 870 ps rising edge transmission delay, which includes a 160 ps delay from the narrow pulse generation circuit. The falling edge transmission delay is 660 ps, with a 60 ps delay from the narrow pulse generation circuit, This performance meets the high-speed operating requirements for GaN half-bridge gate drivers.

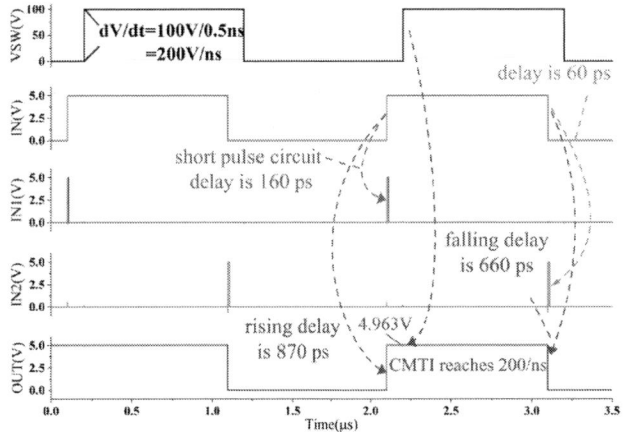

Fig. 8. Simulated transient waveforms of output

Fig. 9 shows the CMTI performance of the circuit, with a 200V/ns floating ground fluctuation. In Fig. 4, nodes A3 and A4 are affected by common-mode noise, and both node voltage drop to -500 mV. But in Fig. 3, node A1 and A2 hold their voltage at approximately 2.8 V with auxiliary capacitors. The direct connection with the auxiliary capacitors, gains node B1 and B2 a better, anti-interference performance. Under this condition, the final node voltage stays at approximately 4.1 V, which means there is no logical error in the output.

Fig. 10 presents the performance of circuit in Fig. 3 when without auxiliary capacitors C1 and C2. As the results shown, node voltage of node B1 and B2 drop to 610 mV. Moreover, logical errors appear in output, the rising edge propagation delay is approximately 1.9 ns, and the falling edge propagation delay is approximately 1.3 ns.

979-8-3315-8850-2/25 $31.00 © 2025 IEEE

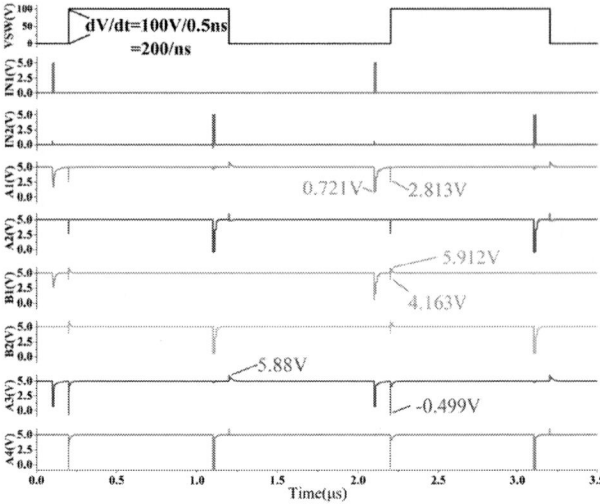

Fig. 9. CMTI performance of the circuit

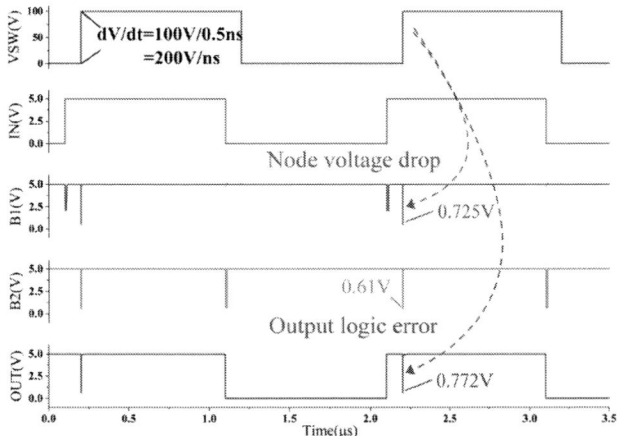

Fig. 10. Output waveform diagram of level shifter without auxiliary capacitors

To verify the dv/dt immunity and low delay performance of the level shifter with possible corner cases, case corner model TT, FF, SS, SF, and FS are used in simulations. Additionally, ±10% fluctuations in the high-side power supply, and temperatures affects are considered. Temperatures are set to -40°C, 27°C, and 120°C for both extreme situation and normal situation. The high-side supply voltage setting includes 4.5V, 5V, and 5.5V for a 10% voltage variations on the supply rail. The worst and best cases of signal transmission delays are summarized in TABLE I. Results show that all cases achieved the 200 V/ns CMTI, which shown in Fig. 11. It can be seen that the level shifter with auxiliary capacitors shows high reliability.

TABLE I. PERFORMANCE WITH DIFFERENT PROCESS CORNER

process corner	Power supply	Temperature	Rising edge signal delay	Falling edge signal delay
TT	5.5	-40	700.3ps	542.1ps
TT	4.5	120	1.11ns	866.9ps
FF	5.5	-40	635.9ps	482.8ps
FF	4.5	120	976.3ps	760.4ps
SS	5.5	-40	806.6ps	608.8ps

process corner	Power supply	Temperature	Rising edge signal delay	Falling edge signal delay
SS	4.5	120	1.28ns	984.6ps
SF	5.5	-40	718.2ps	540.8ps
SF	4.5	120	1.12ns	863.5ps
FS	5.5	-40	716.2ps	545.1ps
FS	4.5	120	1.12ns	872.2ps

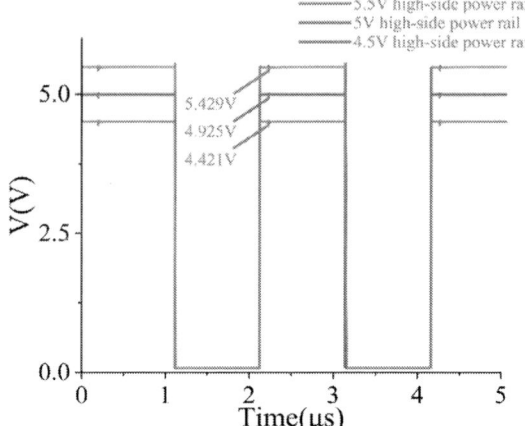

Fig. 11. CMTI under different operating conditions

V. CONCLUSION

In this paper, a sub-nanosecond transmission delay level shifter for GaN HEMTs gate driver is presented. Based on traditional level shifter topology, two additional auxiliary capacitors are added to the two pulse control branch nodes of the level shifter main circuit. The charging and discharging of capacitor are controlled by a capacitor control circuit. When the short pulse signal arrives, the capacitor charging to reduce circuit delay. During the noise interference phase, the capacitors rapidly discharge to compensate for voltage drops caused by common-mode noise, thereby enhancing both noise immunity and signal transmission efficiency. Simulation results show the proposed design reduces up to 56.08% delay for the rising edge and 50.43% delay for the falling edge, while gaining up to 200 V/ns dV/dt noise suppression capability.

ACKNOWLEDGMENT

The authors gratefully acknowledge funding support from Chongqing Talented Scientific Research Project (ref: CSTB2024YCJH-KYXM0063, Chongqing, CN).

REFERENCES

[1] T. Wang, C. Bi, S. Luo, F. Li and W. Bao, "A Novel Crosstalk Suppression Method With Miller Clamp Circuit for GaN HEMTs," *IEEE Transactions on Power Electronics*, vol. 40, no. 3, pp. 4314-4323, March 2025.

[2] Y. Xiong, *et al.*, "Resonant Gate Driver for High Speed GaN HMET with dV/dt Control," in *2021 IEEE International Conference on Integrated Circuits, Technologies and Applications (ICTA)*, Zhuhai, China, 2021, pp. 232-233.

[3] K. Shenai, "High-frequency switching limitations in Gallium Nitride (GaN) and Silicon Carbide (SiC) power devices for boost converter applications," in *2013 IEEE Energytech*, Cleveland, OH, USA, 2013, pp. 1-4.

[4] D. Reusch, D. Gilham, Y. Su and F. C. Lee, "Gallium Nitride based 3D integrated non-isolated point of load module," in *2012 Twenty-Seventh*

979-8-3315-8850-2/25 $31.00 © 2025 IEEE

Annual IEEE Applied Power Electronics Conference and Exposition (APEC), Orlando, FL, USA, 2012, pp. 38-45.

[5] W. Ma, K. Yu and S. Li, "A 5-to-40 VIN 10 MHz Half-Bridge GaN Gate Driver with dV/dt Immunity," in *2024 4th International Conference on New Energy and Power Engineering (ICNEPE)*, Guangzhou, China, 2024, pp. 639-642.

[6] A. F. Bakan, N. Altintaş and İ. Aksoy, "An Improved PSFB PWM DC–DC Converter for High-Power and Frequency Applications," in *IEEE Transactions on Power Electronics*, vol. 28, no. 1, pp. 64-74, Jan. 2013.

[7] Y. Zheng, *et al.*, "A 200-V Half-Bridge Monolithic GaN Power IC With High-Speed Level Shifter and dVS/dt Noise Immunity Enhancement Structure," in *IEEE Transactions on Very Large Scale Integration (VLSI) Systems*, vol. 32, no. 3, pp. 542-551, March 2024.

[8] D. Liu, S. J. Hollis and B. H. Stark, "A New Design Technique for Sub-Nanosecond Delay and 200 V/ns Power Supply Slew-Tolerant Floating Voltage Level Shifters for GaN SMPS," *IEEE Transactions on Circuits and Systems I: Regular Papers*, vol. 66, no. 3, pp. 1280-1290, March 2019.

[9] X. Ming, *et al.*, "A high-voltage half-bridge gate drive circuit for GaN devices with high-speed low-power and high-noise-immunity level shifter," in *2018 IEEE 30th International Symposium on Power Semiconductor Devices and ICs (ISPSD)*, Chicago, IL, USA, 2018, pp. 355-358.

[10] M. Akahane, *et al.*, "A new level up shifter for HVICs with high noise tolerance," in *2014 International Power Electronics Conference (IPEC-Hiroshima 2014 - ECCE ASIA)*, Hiroshima, Japan, 2014, pp. 2302-2309.

[11] J. Cao, Z. -k. Zhou, Z. Wang, H. Tang and B. Zhang, "Design Techniques of Sub-ns Level Shifters With Ultrahigh dV/dt Immunity for Various Wide-Bandgap Applications," *IEEE Transactions on Power Electronics*, vol. 36, no. 9, pp. 10447-10460, Sept. 2021

[12] Z. Zhang, Q. Feng and J. Deng, "Design of a High CMTI Level Shifting Circuit for GaN Gate Driver ICs," in *2024 Photonics & Electromagnetics Research Symposium (PIERS)*, Chengdu, China, 2024, pp. 1-8.

[13] Z. Liu, L. Cong and H. Lee, "Design of On-Chip Gate Drivers With Power-Efficient High-Speed Level Shifting and Dynamic Timing Control for High-Voltage Synchronous Switching Power Converters," *IEEE Journal of Solid-State Circuits*, vol. 50, no. 6, pp. 1463-1477, June 2015.

[14] K. -Y. Li, *et al.*, "A 0.53ns Delay Floating-voltage Level Shifter with Ultra-high dV/dt Immunity for GaN FETs Gate Driver Application," in *2022 IEEE 16th International Conference on Solid-State & Integrated Circuit Technology (ICSICT)*, Nangjing, China, 2022, pp. 1-3.

[15] D. Liu, S. J. Hollis and B. H. Stark, "A New Design Technique for Sub-Nanosecond Delay and 200 V/ns Power Supply Slew-Tolerant Floating Voltage Level Shifters for GaN SMPS," *IEEE Transactions on Circuits and Systems I: Regular Papers*, vol. 66, no. 3, pp. 1280-1290, March 2019.

[16] V. H. Nguyen, N. Ly, A. H. Alameh, Y. Blaquière and G. Cowan, "A Versatile 200-V Capacitor-Coupled Level Shifter for Fully Floating Multi-MHz Gate Drivers," *IEEE Transactions on Circuits and Systems II: Express Briefs*, vol. 68, no. 5, pp. 1625-1629, May 2021.

[17] R. Yan, J. Xi and L. He, "A.2–10 MHz GaN HEMTs Half-Bridge Driver With Bandgap Reference Comparator Clamping and Dual Level Shifters for Automotive Applications," *IEEE Transactions on Industrial Electronics*, vol. 67, no. 2, pp. 1446-1454, Feb. 2020.

[18] L. Li, D. Zhou, Y. Xu, N. He, W. Huang and Z. Chen, "Design of a 5MHz 4A Half Bridge Gate Driver in 180nm BCD Process for GaN HEMT," in *2022 7th International Conference on Integrated Circuits and Microsystems (ICICM)*, Xi'an, China, 2022, pp. 157-160.

[19] J. Zhu, *et al.*, "Noise Immunity and its Temperature Characteristics Study of the Capacitive-Loaded Level Shift Circuit for High Voltage Gate Drive IC," *IEEE Transactions on Industrial Electronics*, vol. 65, no. 4, pp. 3027-3034, April 2018.

[20] Y. Gao, et al, "SOI radiation-hardened 300 V half-bridge date driver IC design with high dV/dt noise immunity," *Power Electron*, vol. 23, pp. 779–788, May. 2023.

2025 The 10th International Conference on Integrated Circuits and Microsystems

A Wide Tuning Range Dual-Mode, Magnetical-Coupled VCO with Capacitive-Coupled Noise-Circulating Technique Achieving FoM of 188dBc/Hz

Qishuang Liu
School of Microelectronics Science and Technology
Sun Yat-sen University
Zhuhai, China

Chenxiang Cai
School of Microelectronics Science and Technology
Sun Yat-sen University
Zhuhai, China

Yunchu Li*
School of Microelectronics Science and Technology
Sun Yat-sen University
Zhuhai, China
liyunchu@mail.sysu.edu.cn

Abstract—This article proposes a dual-mode, magnetical-coupled voltage-controlled oscillator (VCO) that employs the capacitive-coupled noise-circulating technique at the active core to achieve both wide tuning range and low phase noise. Our VCO applies the noise-circulating technique into the wideband multi-mode architecture, which is typically applied to single-band VCO. The proposed topology effectively suppresses active device noise without sacrificing the wide tuning range thanks to magnetic coupling. Implemented in 65nm CMOS process, the proposed VCO achieves a continuous tuning range of 42%, covering low band from 24.5-to-29.7GHz and high band from 30.6-to-37.4GHz. Meanwhile, it exhibits an excellent phase noise from -112dBc/Hz to -109dBc/Hz at 1MHz offset, corresponding to a figure-of-merit (FoM) from 187-to-188dBc/Hz. The power consumption is from 23-to-27mW under a 1.2V supply.

Keywords—Voltage-controlled oscillator, noise-circulating, phase noise, figure-of-merit, tuning range, magnetical-coupled

I. INTRODUCTION

The development of 5G communication is accelerating globally. The 5G New Radio (NR) standard is designed to operate across a wide spectrum, including low band below 1GHz, mid band from 1-7GHz, and high band above 24GHz, known as millimeter-wave (mmWave). The different frequency band across different countries and regions and the demand of better covering these frequency band, as shown in Fig. 1 necessitate the design of VCO that exhibits wide tuning range, low phase noise as well as high robustness.

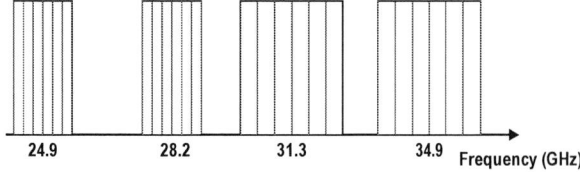

Fig. 1. 5G mm-wave frequency band distribution.

To meet the requirements of current wireless systems, the development of VCOs that achieve both low phase noise and an ultra-wide tuning range (TR) is essential. Multi-core and multi-mode architecture is popular for extending the TR. For instance,

This work was supported by the Special Funds of the National Natural Science Foundation of China (Grant No. 62341409).

Liu *et al.*[1] employs an electrical and magnetic (E-M) mixed-coupling topology with orthogonal-coupled transformers to achieve a dual-core, quad-mode VCO, exhibiting an impressive 66% TR from 19.4-to-38.5 GHz. Meanwhile, extensive research focuses on enhancing the phase noise performance. Wang *et al.*[2] proposes the "Noise-Circulating" architecture, which addresses the critical impact of active device noise. This technique employs an unique active core that enables a majority of device noise to circulate internally rather than being injected into the resonant tank. Consequently, this approach reduces the equivalent noise contribution from active devices by nearly half without compromising the start-up margin, improving both $1/f^2$ and $1/f^3$ phase noise.

However, the original transformer-based noise-circulating topology exhibits notable drawbacks. The requisite PMOS devices introduce an additional noise contribution, partially offsetting the benefits. Furthermore, its transformer-based feedback structure complicates the implementation of multi-mode switching, rendering it less suitable for ultra-wideband designs that rely on such schemes. To address these limitations, various modifications and hybrid architectures have been proposed. Ji *et al.*[3] replaced the magnetic coupling with a simpler capacitive coupling, which preserved the noise-circulating mechanism while achieving a 51% TR and eliminating the multi-oscillation risk of the transformer. While effective, this design remains inherently single-mode and still contains the additional noise source from the PMOS pair.

To alleviate the remaining trade-offs, more complex hybrid strategies are proposed. Shan *et al.* [4] successfully merged the noise-circulating principle with a mode-switching scheme, utilizing a sophisticated multi-magnetic-coupling network to achieve both a 68.3% TR and a boosted resonator Q-factor in both modes. Other works have combined noise circulation with different phase noise reduction techniques. For example, Chang *et al.*[5] integrated it with an implicit common-mode noise filter to specifically suppress flicker noise up-conversion. Another powerful approach, demonstrated by Cao *et al.* and Liu *et al.*[6], combines noise circulation with Class-F harmonic shaping. This dual-pronged strategy simultaneously reduces the magnitude of the noise source via circulation and minimizes its conversion to phase noise, achieving a FoM of 186.7dBc/Hz.

979-8-3315-8850-2/25 $31.00 © 2025 IEEE
294

While these advanced hybrid architectures demonstrate the potent capabilities of the noise-circulating concept, they often result in increased design complexity, large chip area, or introduce new trade-offs between performance metrics. Therefore, developing a new novel topology that successfully integrates a wide-range multi-mode architecture with the noise-circulating principle, while actively mitigating its inherent drawbacks, remain a compelling research objective. This paper presents a dual-mode, magnetically coupled VCO structure that exploits capacitive-coupled noise-circulating technique. By employing a novel differential flower-shaped inductor, strong magnetic coupling is realized for robust mode-switching across different frequency bands. Furthermore, a dedicated filtering capacitor is introduced at the gate terminal of the secondary PMOS pair to suppress their flicker noise, addressing a key limitation of the noise-circulating topology and improving phase noise approximately 4dB. This robust, hybrid approach enables the VCO to achieve both an ultra-wide tuning range (42%) from 24.5-to-29.7GHz and from 30.6-to-37.4GHz and excellent phase noise performance (-112dBc/Hz@Δf = 1MHz).

II. MAGNETIC COUPLING NOISE-CIRCULATING VCO

Shown in Fig. 2 [7], the architecture of the conventional magnetical-coupled dual-mode oscillator is based on two identical, magnetical-coupled LC resonators. To overcome the inherent losses of the resonant tanks and maintain oscillation, an active core, typically comprising a symmetrical cross-coupled NMOS pair, provides the negative resistance $-G_m$. The magnetic coupling between the resonators gives rise to two distinct operational modes at different resonant frequencies, namely the even mode and the odd mode, enabling a wide tuning range.

In the even mode, as shown in Fig. 3(a), the signal voltages across the two LC resonant tanks are in-phase. This results in in-phase currents flowing through the inductors, causing their magnetic fields to add constructively. The total equivalent inductance thus becomes $L+M$. Conversely, in the odd mode, the voltages across the tanks are out-of-phase, as shown in Fig. 3(b). This drives out-of-phase currents through the inductors, causing their magnetic fields to oppose and partially cancel each other. As a result, the total equivalent inductance is reduced to $L-M$. Therefore, the resonant frequencies are

$$\omega_e = 1/\sqrt{(L+M)C} \qquad (1)$$

$$\omega_o = 1/\sqrt{(L-M)C} \qquad (2)$$

in even mode and odd mode, respectively. While this dual-mode operation effectively expands the tuning range, the phase noise performance remains limited by a common issue in both modes: the direct injection of noise from the active NMOS core into the resonant tank. In the proposed magnetical-coupled VCO with noise-circulating active core, as shown in Fig. 4, this trade-off could be alleviated. And the equivalent negative resistance of the capacitive-coupled noise-circulating active core is given by $\frac{1}{g_{mn}}+\frac{1}{g_{mp}}$, as derived in [3]. When $g_{mn} = g_{mp} = g_m$, it can be simplified as $\frac{2}{g_m}$.

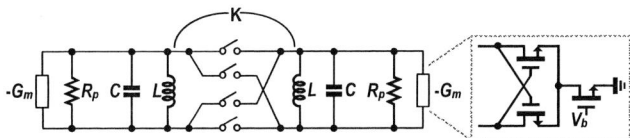

Fig. 2. Magnetical-coupled oscillator with conventional active core.

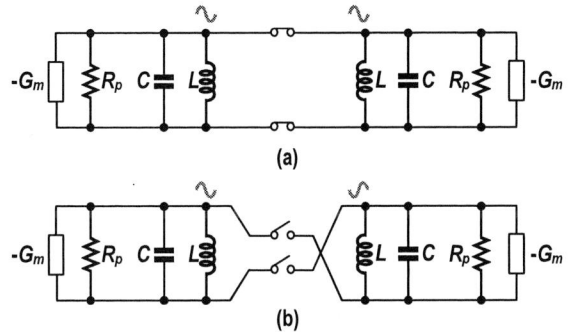

(a)

(b)

Fig. 3. (a) Even mode operation. (b) Odd mode operation.

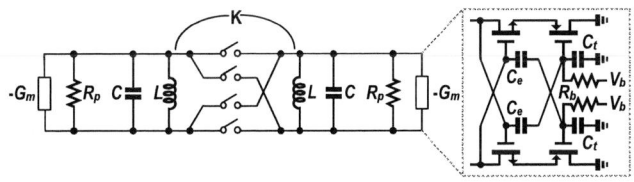

Fig. 4. Proposed magnetical-coupled VCO with noise-circulating active core.

According to the linear time-variant (LVT) mode proposed by P. Andreani *et al.* [8], the phase noise contribution from a device current noise source can be express as:

$$L(\Delta\omega) = 10 \lg \left(\frac{\frac{\Gamma^2_{i_n,rms}\overline{i_n^2}}{\Delta f}}{2q^2_{max}\Delta\omega^2} \right) \propto \Gamma^2_{rms}\cdot\overline{i_n^2} \qquad (3)$$

where $\overline{i_n^2}$ is the white current noise power density of the noise current source, q_{max} is the maximum charge displacement across the tank capacitor C and the Γ_{rms} is the impulse sensitivity functions (ISF) for the corresponding noise current device noise source. Therefore, to improve phase noise performance, the noise current injected into the resonant tank must be reduced.

The noise-circulating mechanism achieves this through active source degeneration, a principle best analyzed at the zero-crossing point of the tank voltage where the oscillator exhibits maximum phase sensitivity. At this moment, each transistor in the active core is source-degenerated by the transconductance (g_m) of its counterpart, which acts as a current divider for the transistor's noise. To quantify this division, the factors are defined as $\alpha = \frac{g_{mn}}{g_{mn}+g_{mp}}$ and $\beta = \frac{g_{mp}}{g_{mn}+g_{mp}}$. Considering the noise contribution from the PMOS device P1, as shown in Fig. 5(a), the degeneration provided by NMOS effectively creates a

979-8-3315-8850-2/25 $31.00 © 2025 IEEE

current divider for the noise current of P1, $I_{n,P1}$. A fraction of this noise, denoted as $\alpha I_{n,P1}$, is injected through N1 into the resonant tank, contributing to phase noise. The remaining fraction $\beta I_{n,P1}$, is steered back through the source of P1, circulating away from the tank and terminating at ground.

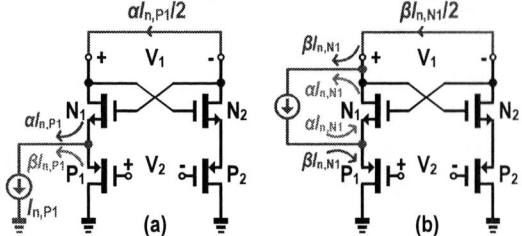

Fig. 5. (a) PMOS devices in noise-circulating active core. (b) NMOS devices in noise-circulating active core.

A symmetrical analysis applies to the noise contribution from the NMOS device N1, as shown in Fig. 5(b). The degeneration provided by PMOS effectively creates a current divider for the noise current of N1, $I_{n,N1}$. In this case, the fraction of noise denoted by $\beta I_{n,N1}$ is injected through P1 into the resonant tank, contributing to phase noise, while the remaining part, $\alpha I_{n,N1}$, circulates harmlessly within the active core and does not contribute to the overall phase noise.

The total noise current injected by a conventional noise-circulating active core can be calculated as, $2kT\gamma \frac{g_{nm}g_{np}}{g_{nm}+g_{mp}}$, when $g_{mn} = g_{mp} = g_m$, this expression simplifies to $kT\gamma g_m$ [2].

Different from the traditional architecture, the proposed topology adds a filtering capacitor, C_t, at the gates of the secondary PMOS pair. This modification is implemented to specifically suppress the flicker noise contribution from these devices.

III. IMPLEMENTATION DETAILS

A. Flower-Shaped Transformer

To meet the demands of high-frequency operation in 5G applications, the transformer utilized in the design must exhibit both a low inductance and a high Q. Meanwhile, to expand the frequency tuning range via mode-switching, the transformer must also achieve a strong K. These requirements often present a design trade-off. To address this challenge, this work proposes a novel differential flower-shaped transformer, the layout of which is shown in Fig. 6. Compared to conventional figure-8 topologies, this structure is specifically designed to enhance the magnetic flux linkage between the two coils, resulting in a higher coupling factor. This strong coupling is crucial for achieving wide mode separation and thus a broad overall tuning range. Furthermore, the transformer is implemented using only the top metal layer M9 to minimize parasitic capacitance to the substrate, thereby achieving a high Q-factor at the target millimeter-wave frequencies.

The proposed flower-shaped transformer was designed and verified using an electromagnetic (EM) simulation tool, with the results presented in Fig. 7. At the target center frequency of approximately 30 GHz, each coil of the transformer exhibits an inductance of around 81 pH, as shown in Fig. 7(a). The simulated

Q is approximately 23 at 30GHz, as shown in Fig. 7(b), ensuring low loss for the resonant tank. Furthermore, the transformer achieves a K of 0.232 at 31GHz, as shown in Fig. 7(c). These results confirm that the proposed structure provides the required low inductance, a high Q-factor, and sufficient magnetic coupling to enable wideband mode-switching, making it highly suitable for the intended VCO design.

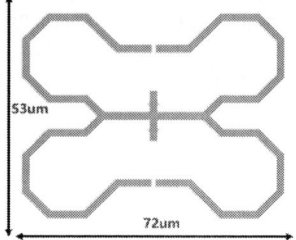

Fig. 6. The proposed flower-shaped transformer.

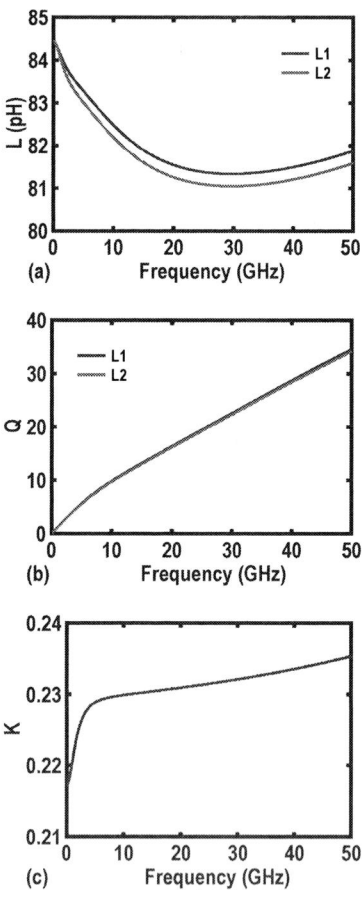

Fig. 7. Simulation results of the proposed flower-shaped transformer. (a) inductance vs. frequency. (b) Quality factor vs. frequency. (c) coupling factor vs. frequency.

B. Circuit Implementation

The details schematic of the proposed VCO is shown in Fig. 8. The design integrates a magnetical-coupled dual-mode architecture with a capacitively-coupled noise-circulating active core. It is implemented in a 65-nm CMOS process and operates

979-8-3315-8850-2/25 $31.00 © 2025 IEEE

from a 1.2V supply. The active core consists of a primary cross-coupled NMOS pair (40μm/60nm), and a secondary noise-circulating PMOS pair (540μm/480nm). The devices are specifically sized to ensure their transconductances are equal, which optimizes the noise cancellation effect of the noise-circulating mechanism. The feedback path for the noise-circulating core is established through the AC-coupling capacitors C_e of 347fF, and the gates of PMOS devices are biased through 1KΩ resistors R_b. Meanwhile, the capacitors C_t used to specifically suppress the flicker noise contribution from the secondary devices are designed to be 1pF.

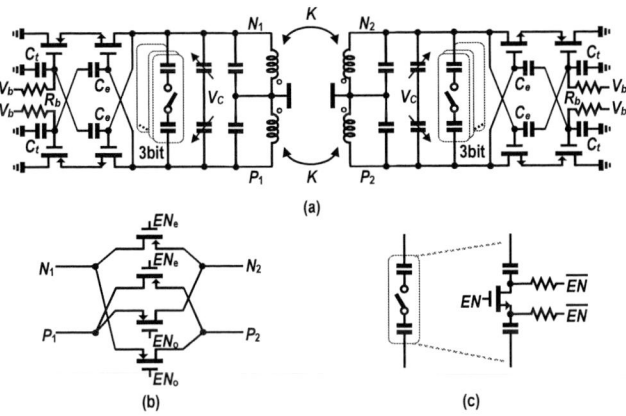

Fig. 8. (a) The schematic of proposed VCO. (b) The switch network for switching working mode. (c) The details of the switched-capacitor array.

A key feature of the resonator design is that the main parallel capacitors of the LC tank are omitted. Instead, the resonance is achieved by utilizing the parasitic capacitance of the custom-designed flower-shaped inductor. The frequency tuning is realized through a combination of coarse and fine control. For coarse tuning, a 3-bit switched-capacitor array is connected to each side of the resonator. Each unit cell of the array consists of a 45.7fF capacitance and an NMOS switch (27μm/60nm). For continuous fine-tuning, a pair of varactors provides a capacitance range of 18fF to 41fF. The mode-switching between the even and odd modes is controlled by a dedicated switch network, with each PMOS sized at 12μm/60nm. This sizing ensures a sufficiently low on-resistance to guarantee robust selection of the desired resonant mode while suppressing potential multi-mode oscillations.

IV. SIMULATION RESULTS

The proposed dual-mode magnetical-coupled VCO with a noise-circulating core is implemented in 65nm CMOS process. The VCO consumes power from 23-to-27mW under a 1.2V supply. The simulated frequency tuning curve is present in Fig. 9(a), which shows that the VCO achieves a continuous frequency tuning range from 24.5-to-29.7GHz and from 30.6-to-37.4GHz, corresponding to a wide tuning range of 42%. This range is covered by two distinct modes, a low-frequency band and a high-frequency band, with sufficient overlap to ensure continuous tuning. Fig. 9(b) illustrates the simulated phase noise across the tuning range. At 1MHz offset, the VCO achieves -112 dBc/Hz phase noise in the low band and -110 dBc/Hz phase noise in the high band. The corresponding FoM ranging from 187-to-188dBc/Hz across the entire operational band shown in

Fig. 9(c). The phase noise comparison shown in Fig. 9(d) confirms that the noise-circulating structure significantly reduces the noise from the active core compared to conventional designs.

To verify the robustness of the proposed design, simulations are conducted across process corners and temperature. The VCO maintains robust oscillation from -40-to-125°C, and across all five standard process corners (TT, FF, SS, FS, SF). Under these variations, the maximum oscillation frequency stays above 36.5GHz as shown in Fig. 10(a), the phase noise remains better than -108.6dBc/Hz at 1MHz offset, as shown in Fig. 10(b), confirming that the design achieves excellent robustness.

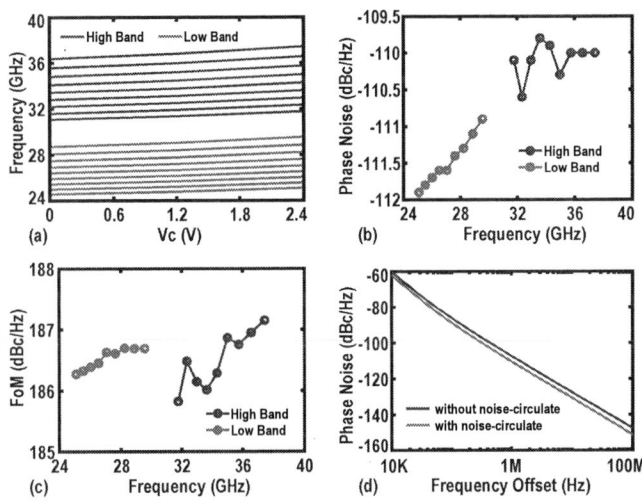

Fig. 9. (a) Simulated frequency tuning range of high band and low band. (b) Simulated phase noise@1MHz offset vs. frequency. (c) Simulated FoM vs. frequency. (d) Phase noise comparison without and with the noise-circulating technique.

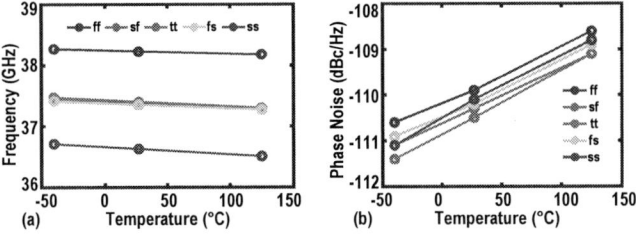

Fig. 10. (a) Simulated VCO performance over different process corners and temperature. (b) Simulated phase noise@1MHz offset over different process corners and temperature.

TABLE I. summarizes the performance of this work and compares it with the state-of-the-art. The proposed VCO achieves a superior FoM of 188 dBc/Hz@1MHz offset, which outperforms all other works listed in the comparison. Notably, despite operating in a challenging millimeter-wave band (from 24-to-37GHz), the design maintains an excellent phase noise from -112dBc/Hz to -110dBc/Hz. This represents a significant improvement over other mmWave VCOs such as [1] and [12]. Furthermore, the achieved 42% tuning range is highly competitive and provides substantially wider coverage than other designs like [9] and [12], confirming its suitability for multi-band 5G applications.

979-8-3315-8850-2/25 $31.00 © 2025 IEEE

TABLE I.   COMPARISON WITH STATE-OF-THE-ART

	ICICM'23[1]	MICRO'21[3]	RFIC'22[9]	JSSC'21[10]	TCASI'21[11]	JSSC'18[12]	MWTL'25[13]	RFIC'24[14]	This work
Technology	65nm	28nm	28nm	40nm	65nm	28nm	65nm	28nm	**65nm**
Topology	Dual-core Quad-mode	Noise circulation	NMOS biased Switchable capacitor	Quad-core Quad-mode	-	-	-	Quad-core Quad-mode	**Noise circulation**
Frequency (GHz)	19.4-38.5	6.17-10.45	8.2-10.2	18.6-40.1	11.3	27.3	2.84-7.04	17-36.4	**24.5-29.7/30.6-37.4**
Tuning range (%)	66	51	24.3	73.2	53	14	-	72.6	**39.2**
Phase Noise (dBc/Hz)	-104.6 @1MHz	-121.6/-115 @1MHz	-115.1 @1MHz	-130.3/-122.7 @10MHz	-109.7 @1MHz	-106 @1MHz	-123.9~-113.1 @1MHz	-123.3/-116.3 @10MHz	**-112/-110 @1MHz**
FoM (dBc/Hz)	186.5 @1MHz	182.4/179.3 @1MHz	183 @1MHz	183.0/186.3 @10MHz	184.5 @1MHz	184 @1MHz	-177.4-182.7@1MHz	179.1/173.3 @10MHz	**188@1MHz**
Supply (V)	0.6	1.8	1.2 ·	1.1	1	1	0.44	-	**1.2**
Power (mW)	4.9-9.7	32-41	13	9-15	4.5	12	7.92-13.20	9.7-13.3	**23-27**

V. CONCLUSION

This article proposes a dual-mode, magnetical-coupled VCO for 5G applications, designed to concurrently achieve both wide tuning range and low phase noise. The proposed architecture successfully integrates a mode-switching scheme with a capacitively-coupled noise-circulating active core. The VCO employs a novel flower-shaped transformer that provides strong magnetic coupling and a high Q-factor. In addition, a dedicated filtering capacitor is introduced to mitigate the flicker noise from the secondary PMOS pair.

Implemented in 65nm CMOS process, the proposed VCO achieves a continuous tuning range of 42%, which operates from 24.5-to-29.7Hz in low band and from 30.6-to-37.4GHz in high band. The excellent phase noise performance maintains between -109dBc/Hz and -112dBc/Hz at 1MHz offset, resulting in a high FoM of 187 to 188dBc/Hz across the entire band. In addition, the design achieves strong robustness across process corners and temperature variations.

REFERENCES

[1] Z. Liu, H. Guo and G. Qi, "A 0.075mm² Octave-Tuning 5G-NR Oscillator with E-M Mixed-Coupling Achieving 205dB FoM$_T$," *2023 8th International Conference on Integrated Circuits and Microsystems (ICICM)*, Nanjing, China, 2023, pp. 511-515.

[2] F. Wang and H. Wang, "A Noise Circulating Oscillator," in *IEEE Journal of Solid-State Circuits*, vol. 54, no. 3, pp. 696-708, March 2019.

[3] X. Ji, Y. Wang, X. Xia and Y. Guo, "A Capacitively Coupled Noise Circulating VCO," in *IEEE Microwave and Wireless Components Letters*, vol. 31, no. 10, pp. 1127-1129, Oct. 2021.

[4] X. Shan *et al.*, "A Low-Phase-Noise Wide-Tuning-Range Mode-Switching Oscillator Using Multi-Magnetic-Coupling and Active-Source-Degenerating Techniques," in *IEEE Journal of Solid-State Circuits*, vol. 59, no. 9, pp. 1295-1308, June 2012.

[5] Y. -M. Chang, H. -C. Chen and C. -H. Fan, "A 18.8 GHz Noise Circulating VCO With Implicit Noise Filter for Quantum Computing," *2024 IEEE Asia Pacific Conference on Circuits and Systems (APCCAS)*, Taipei, Taiwan, 2024, pp. 290-292.

[6] H. Cao, T. Huang, X. Liu, H. Wang, J. Jin and W. Wu, "A 5.2GHz Trifilar Transformer-Based Class-F$_{2,3}$ Noise Circulating VCO with FoM of 192.6 dBc/Hz," *2023 IEEE Asian Solid-State Circuits Conference (A-SSCC)*, Haikou, China, 2023, pp. 1-3.

[7] G. Li, L. Liu, Y. Tang and E. Afshari, "A Low-Phase-Noise WideTuning-Range Oscillator Based on Resonant Mode Switching," in *IEEE Journal of Solid-State Circuits*, vol. 47, no. 6, pp. 1295-1308, June 2012.

[8] P. Andreani and A. Fard, "More on the $1/f^2$ phase noise performance of CMOS differential-pair LC-tank oscillators," in *IEEE Journal of Solid-State Circuits*, vol. 41, no. 12, pp. 2703–2712, Dec. 2006.

[9] L. Wang *et al.*, "An 8.2-10.2 GHz Digitally Controlled Oscillator in 28-nm CMOS Using Constantly-Conducting NMOS Biased Switchable Capacitor," *2022 IEEE Radio Frequency Integrated Circuits Symposium (RFIC)*, Denver, CO, USA, 2022, pp. 207-210.

[10] Y. Shu, H. J. Qian and X. Luo, "A 2-D Mode-Switching Quad-Core Oscillator Using E-M Mixed-Coupling Resonance Boosting," in *IEEE Journal of Solid-State Circuits*, vol. 56, no. 6, pp. 1711-1721, June 2021.

[11] P. Agarwal, M. Chahardori, and D. Heo, "A new boosted activecapacitor with negative-Gm for wide Tuning range VCOs," in *IEEE Trans. Circuits Syst. I, Reg. Papers*, vol. 68, no. 3, pp. 1080–1090, Mar. 2021.

[12] Y. Hu, T. Siriburanon, and R. B. Staszewski, "A low-flicker-noise 30-GHz class-F$_{2,3}$ oscillator in 28-nm CMOS using implicit resonance and explicit common-mode return path," in *IEEE J. Solid-State Circuits*, vol. 53, no. 7, pp. 1977–1987, Jul. 2018.

[13] Y. Hou, R. Li, P. -L. Chi and T. Yang, "A Dual-Core Dual-Mode Class-F VCO With Wide Frequency Tuning Range Using Wide Inductance-Switching-Range Inductor," in *IEEE Microwave and Wireless Technology Letters*, vol. 35, no. 4, pp. 460-463, April 2025.

[14] H. Kim, S. Kim and S. Jeon, "An Octave Tuning Range Quad-Core VCO Using a Compact Quad-Mode Transformer-Based Inductor," in *2024 IEEE Radio Frequency Integrated Circuits Symposium (RFIC)*, Washington, DC, USA, 2024, pp. 83-86.

2025 The 10th International Conference on Integrated Circuits and Microsystems

A Fractional-N All-Digital PLL with 166fs$_{\text{rms}}$ Jitter and 238 dB FoM Based on a Distributed Switched-Capacitor Arrays DCO

Zhihong Xu[1#], Yuhui Li[2#], Pei Qin[1*], Changsong Lin[1], Jinhua Guo[2], Xiaorui Liu[1], Taotao Xu[3], Quan Xue[1]

[1]*School of Microelectronics, South China University of Technology, Guangzhou, China*
[2]*School of Integrated Circuits, South China University of Technology, Guangzhou, China*
[3]*School of Integrated Circuits, Anhui University, Anhui, China*
*qinpei7777@scut.edu.cn

Abstract—**This work proposes a multi-tap transformer-based dual-core digitally controlled oscillator (DCO) with distributed switched-capacitor arrays for low phase noise. The design incorporates enhanced electromagnetic (E-M) mixed-coupling and harmonic-free-like techniques to achieve a high drain-to-gate voltage gain, thereby effectively suppressing thermal and flicker noise. Leveraging the distributed structure of the multi-tap inductor, the process, acquisition, and tracking (P, A and T/F) capacitor banks are implemented using high-quality MOM capacitors and symmetrically placed at the two tap nodes of the transformer. This strategy significantly enhances the Q factor of the capacitor arrays. Furthermore, the proposed DCO adopts IO transistors as cross-coupled devices, effectively improving circuit stability. Fabricated in a 28 nm CMOS process, the DCO achieves a wide tuning range from 5.83 GHz to 9.2 GHz. At 9.2 GHz, it demonstrates a phase noise of −144.4 dBc/Hz at a 10 MHz offset and a figure of merit (FoM) of 187 dBc/Hz, showcasing its suitability for high-performance all-digital phase locked loop (ADPLL) applications.**

Index Terms—**all-digital phase locked loop (ADPLL), Digitally Controlled Oscillator (DCO), multi-tap transformer, switched-capacitor array, phase noise, frequency synthesis.**

I. INTRODUCTION

Advanced wireless transceivers demand frequency synthesizers with high spectral purity for high-order modulation schemes. Digital phase-locked loops (DPLLs) have become attractive candidates due to the ability to scale efficiently in area and power with advanced CMOS technology nodes. In the optimization of frequency synthesizers for area and power, the digitally controlled oscillator (DCO) often becomes the primary bottleneck, as its phase noise—originating from both flicker (1/f) and thermal noise—dominates the overall synthesizer performance. Recent techniques, such as high passive-gain transformer designs and second-harmonic shaping, have shown promise in reducing DCO phase noise and suppressing flicker noise [1]–[5]. However, these approaches typically rely on complex tuning-bank segmentation for common-mode (CM) impedance control and require extensive manual resonance tuning, rendering them highly sensitive to process, voltage, and temperature (PVT) variations. These

#Contribute equally.

limitations severely constrain their applicability in commercial PLL systems. Moreover, a fundamental trade-off exists in DCO design: improved frequency tuning linearity often comes at the expense of increased phase noise, posing additional challenges for performance optimization. A potential solution involves replacing the capacitor banks with a DAC-controlled varactor [6], [7]; however, this method suffers from increased power consumption, degraded linearity, and elevated noise. To overcome these limitations, this work adopts a multi-tap transformer-based inductor structure to achieve high passive gain while employing a distributed switched-capacitor array to enhance the quality factor (Q) of the resonator, in which the Process (P), Acquisition (A), and Tracking (T) banks are all implemented using high-Q metal–oxide–metal (MOM) capacitors. This improves frequency resolution, supports temperature calibration, and maintains robustness across PVT corners. The proposed approach enables robust low-noise performance and meets the requirements of practical PLL applications.

II. ARCHITECTURE OF THE PROPOSED DCO

A. Role of DCO Capacitor Banks in ADPLL

Fig. 1 illustrates the implemented all-digital phase-locked loop (ADPLL) architecture, which integrates an automatic frequency calibration (AFC) module, a main feedback loop, and a state machine. These blocks respectively control the DCO's P-bank, A/T-bank, and F-bank, as well as manage the overall loop locking process. The AFC module controls the P capacitance bank located at the gate side of the DCO. To minimize area and power consumption, the ADPLL adopts a time-to-digital converter (TDC) as the phase detector along with a digital loop filter. The latter adjusts the A/T capacitance banks placed at the drain side of the DCO. In addition, the fractional part of the loop filter output is handled by a delta-sigma modulator (DSM), implemented as either a MASH 1 or MASH 1-1 architecture, to drive the F-bank. The F-bank is designed with the same unit capacitance as the T-bank and is also located at the drain side of the DCO.

Fig. 1 presents the progression flowchart of the DCO operational modes. At startup, the automatic frequency calibration

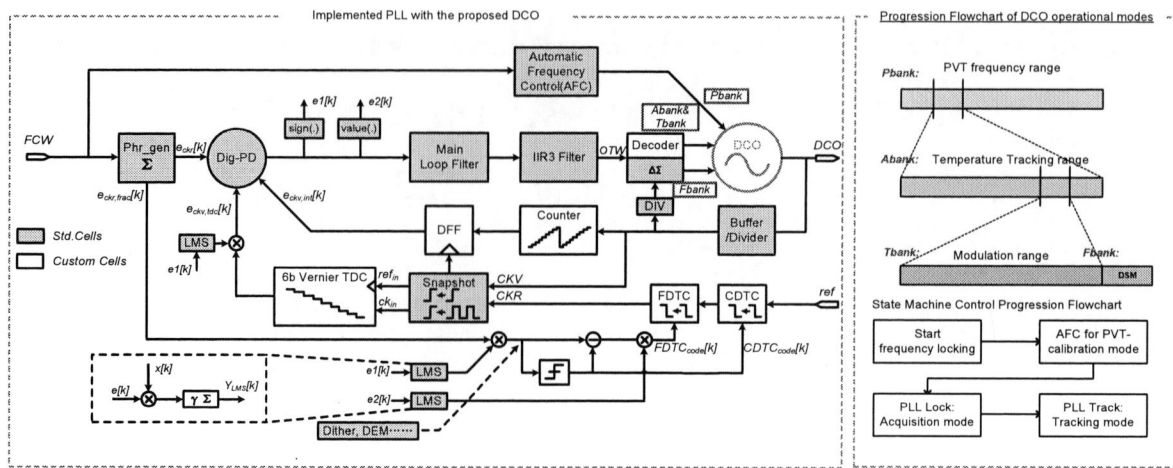

Fig. 1. Implemented PLL Featuring Differential Capacitor-Bank DCO and DCO Operational Modes

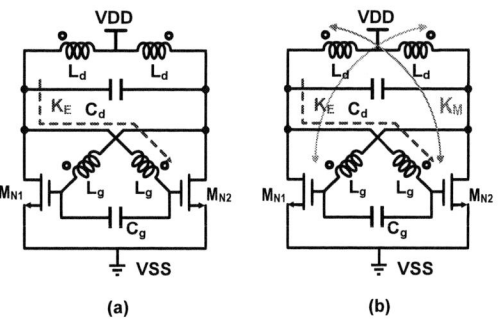

Fig. 2. (a) A conventional single-core oscillator with multi-tap inductor. (b) A single-core oscillator with E-M mixed coupling.

(AFC) adjusts the P-bank to shift the DCO frequency close to the target, within the Abank tuning range. The loop then enters acquisition mode, in which the A/T capacitance banks collaboratively tune the DCO to the final target frequency. A state machine monitors the phase error and determines when frequency lock is achieved.

B. E-M mixed coupling transformer based Dual-core DCO

Fig. 2 illustrates a conventional single-core oscillator that employs a multi-tap inductor. By introducing magnetic coupling from the drain to the gate, the oscillator benefits from both electric and magnetic coupling effects, thereby achieving a higher drain-to-gate voltage gain [8]. Since the noise factor can be expressed as $NF = 1 + \gamma/A_v$, a larger drain-to-gate voltage gain A_v directly contributes to noise reduction and improved phase noise performance.

To further improve layout regularity and gain enhancement, this single-core VCO is extended into a dual-core configuration, as shown in Fig. 3. In this structure, the drain and gate inductors (L_d and L_g) are symmetrically arranged along the horizontal axis. A center-symmetrical multi-tap transformer is implemented to facilitate enhanced electric-magnetic (E–M) mixed coupling.

C. Distributed Capacitance of the Proposed DCO

In conventional DCO designs, the P/A/T capacitance banks are typically implemented using transistors of a single type, either core or IO devices. Using only core transistors provides higher Q due to lower parasitic resistance, but it risks reliability issues such as gate overdrive under large voltage swings. Conversely, IO devices offer better voltage tolerance and robustness but suffer from degraded Q performance due to higher parasitics.

Leveraging the structural advantages of the proposed multi-tap inductor-based DCO, capacitor bank partitioning is optimized without requiring the precise capacitor matching often needed in class-F oscillator designs. Specifically, a P-bank array is implemented at the gate side, where large voltage swings are present. This bank uses MOM capacitors in parallel with IO NMOS switches, ensuring both high voltage tolerance and acceptable Q. At the drain side, where the signal swing is significantly smaller, the A/T banks are also constructed using MOM capacitors but switched by core transistors. This arrangement avoids overstress issues and, more importantly, offers significantly improved Q compared to traditional MOS-cap-based A/T banks. Furthermore, the MOM-based switched capacitor array (SCA) enables a larger effective capacitance and ΔC compared to its MOS-cap counterpart. This improvement is particularly important in reducing quantization noise, which is otherwise challenging in conventional DCOs.

The feasibility of using a large-value MOM-based SCA in this design is enabled by the segmentation property of the multi-tap inductor. This structure introduces a proportionality factor α that effectively scales down the capacitor's contribution to the oscillation frequency f_0, allowing the use of higher absolute capacitance without degrading frequency resolution or phase noise performance.

979-8-3315-8850-2/25 $31.00 © 2025 IEEE

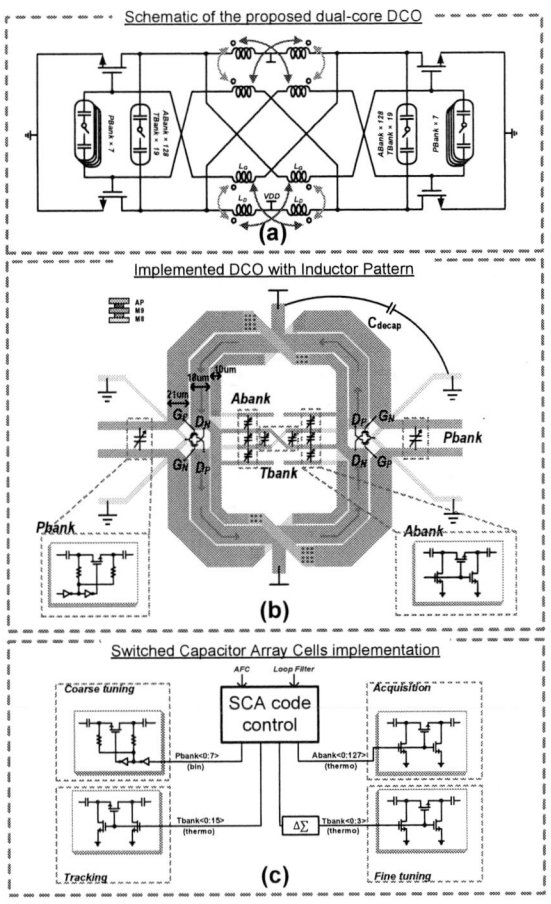

Fig. 3. Details of the DCO Implementation: Inductor Layout, Current Flow Direction, and SCA Capacitor Distribution

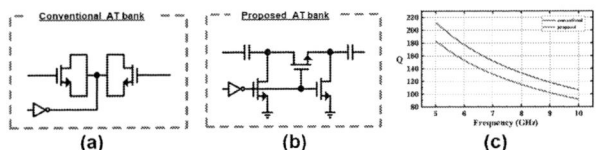

Fig. 4. (a)The conventional Abank. (b)The proposed Abank. (c)The Q factor of two structures.

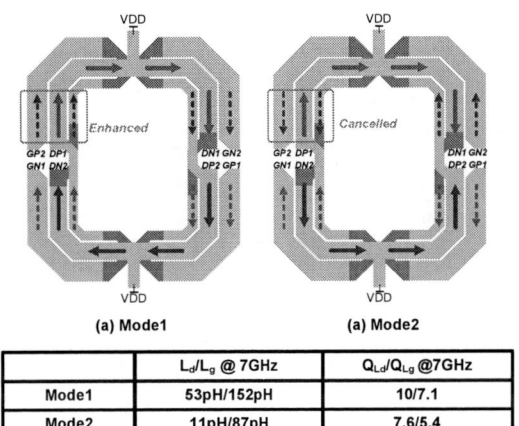

	L_d/L_g @ 7GHz	Q_{L_d}/Q_{L_g} @7GHz
Mode1	53pH/152pH	10/7.1
Mode2	11pH/87pH	7.6/5.4

Fig. 5. Dual-core coupled-inductor two possible modes.

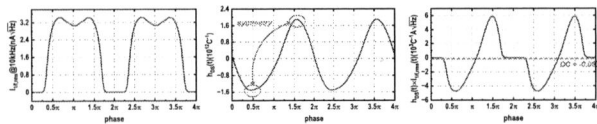

Fig. 6. The post-simulated flicker noise currents, ISF and effective ISF.

III. CIRCUIT IMPLEMENTATION AND PHASE NOISE ANALYSIS

A. Implementation and Architecture of the Proposed DCO

The layout of the proposed dual-core DCO is illustrated in Fig. 3. To achieve a reduced inductance and enhanced quality factor, M9 and AP are vertically stacked in the drain inductor and the outer-ring gate inductor. This configuration not only improves the inductor Q value but also significantly increases the magnetic coupling strength between the inductors. The center drain inductor adopts a metal width of $18\,\mu m$ to support larger current for improved phase noise performance. At the drain side of the two oscillator cores, thermometer-coded capacitor banks Abank<0:127>, Tbank<0:19>, and Fbank are connected, serving for loop tracking and DSM-based quantization noise suppression in closed-loop operation. The inner and outer gate inductors have metal widths of $10\,\mu m$ and $21\,\mu m$, respectively. This configuration enhances the gate inductor Q factor and strengthens the magnetic coupling with the drain inductor, leading to further improvement in tank Q and reduction of active device noise figure. A binary-coded 8-bit Pbank (Pbank<0:7>) is connected at the differential gate

nodes for coarse frequency tuning, which compensates for PVT-induced frequency variations. The Pbank is implemented using a switched capacitor array based on MOM capacitors and resistive biasing, achieving high Q performance. At the drain side, both the Abank and T/F banks are also implemented using MOM capacitor-based switched capacitor arrays, but with MOS-biased switches to reduce area while maintaining high Q.

The conventional A/T bank architecture employs MOS transistors as variable capacitors, where the capacitance state is controlled by toggling the voltage levels at the source and drain terminals. The total capacitance of the MOS device comprises the gate oxide capacitance and the semiconductor capacitance in series. The oxide capacitance is fixed, determined by the gate oxide thickness and layout area, whereas the semiconductor capacitance varies with the transistor's operation state. During oscillator operation, the voltages across the P and N terminals vary with time, causing the MOS capacitance to change dynamically. This time-varying capacitance introduces frequency modulation effects that degrade the phase noise performance of the DCO. To mitigate this effect, as showed in Fig. 4, the proposed A/T bank adopts a switched-capacitor

979-8-3315-8850-2/25 $31.00 © 2025 IEEE

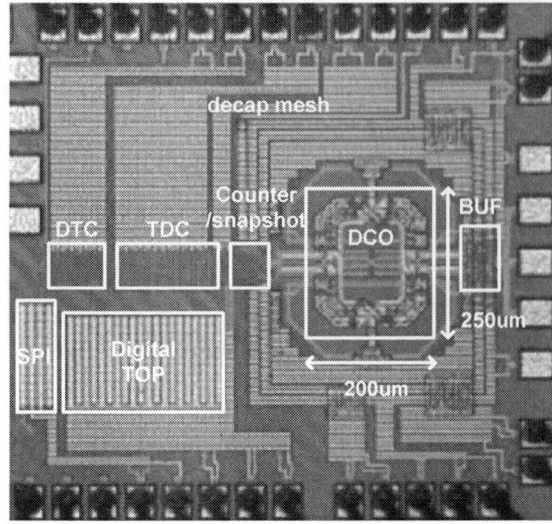

Fig. 7. Chip microphotograph.

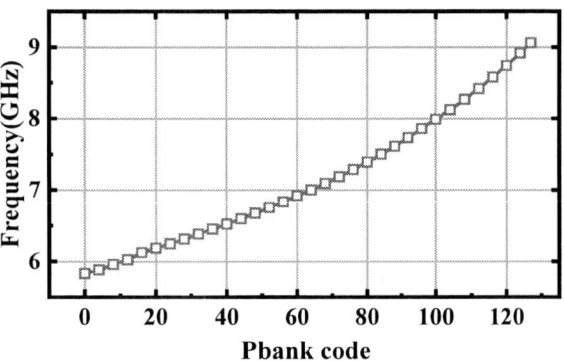

Fig. 8. Measured tuning range of the DCO.

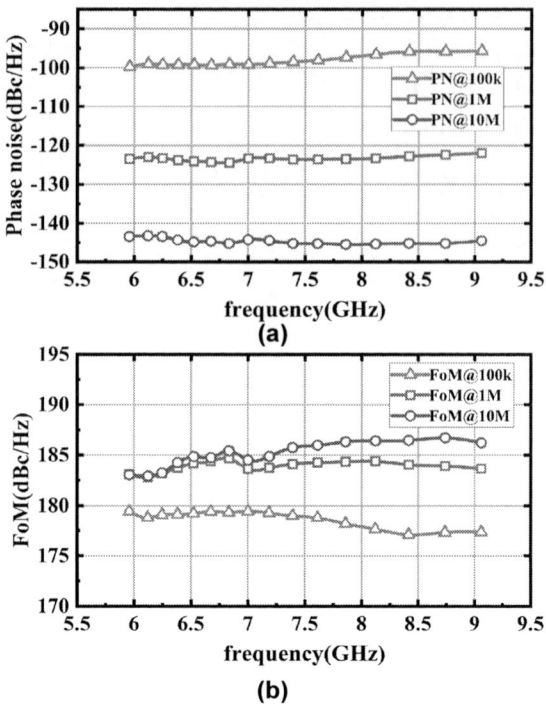

Fig. 9. (a) Measured PN and (b) FoM at 100k, 1M and 10MHz osffsets.

architecture similar to that used in the P bank, but replaces the biasing resistors with MOS switches. This modification eliminates thermal noise generated by resistive elements and consequently improves the overall Q factor. Although the MOS switches introduce parasitic capacitance, the impact is limited due to the small unit capacitance and compact layout of the A/T bank cells. Under equivalent capacitance conditions, simulations show that the proposed A/T bank structure achieves approximately a 20dB improvement in Q factor compared to the conventional MOS-capacitor implementation.

Benefiting from the high voltage gain, the oscillator achieves an excellent noise figure and is compatible with inductor structures suited for distributed capacitor array layouts. However, the dual-core topology may exhibit two distinct resonant modes, depending on the phase relationship between the two oscillator cores, as illustrated in Fig. 5. In Mode 1 (anti-phase), the magnetic fluxes of the drain inductor (L_d) and the inner coil of the gate inductor (L_g) are mutually enhanced, resulting in an increased effective quality factor and improved phase noise performance. In contrast, Mode 2 (in-phase) causes the

magnetic fluxes to cancel each other, which degrades the tank quality factor and makes this mode less favorable for operation.

B. Noise Suppression

Thanks to the high voltage gain, the noise contributions from the core transistors are significantly reduced, thereby lowering the noise factor: $NF = 1 + \gamma/A_v$, also keeps the drain voltage V_D at a relatively low level, which prevents the core transistors from entering the deep triode region and maintains a low output conductance G_{ds}, further suppressing device noise injection.

Moreover, the up-conversion of flicker noise is influenced by the average DC value of the effective impulse sensitivity function (ISF). This effect can be mitigated if the ISF waveform exhibits a flat-zero region or certain symmetry characteristics. To address this, prior studies have proposed harmonic shaping techniques [2] and harmonic-free techniques [9]. Harmonic shaping suppresses the noise caused by core transistors entering the deep triode region by introducing high impedance at the second harmonic, thus achieving a flat-zero ISF.

In contrast, the proposed design employs a harmonic-free-like technique to realize a symmetric ISF waveform. The large gate voltage swing not only lowers the effective transition speed of the MOS switches, reducing the duration of noise injection, but also minimizes the voltage swing at the drain nodes, thereby shortening the time that the MOS devices

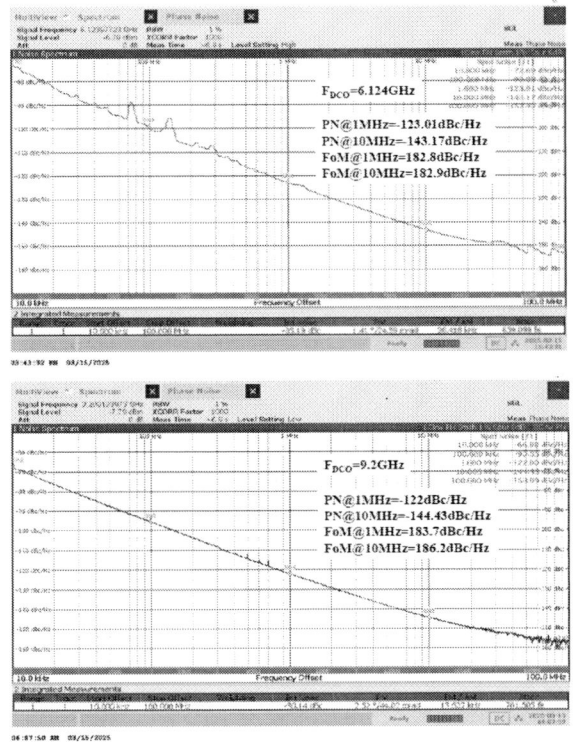

Fig. 10. Open-loop phase noise measurement of the DCO.

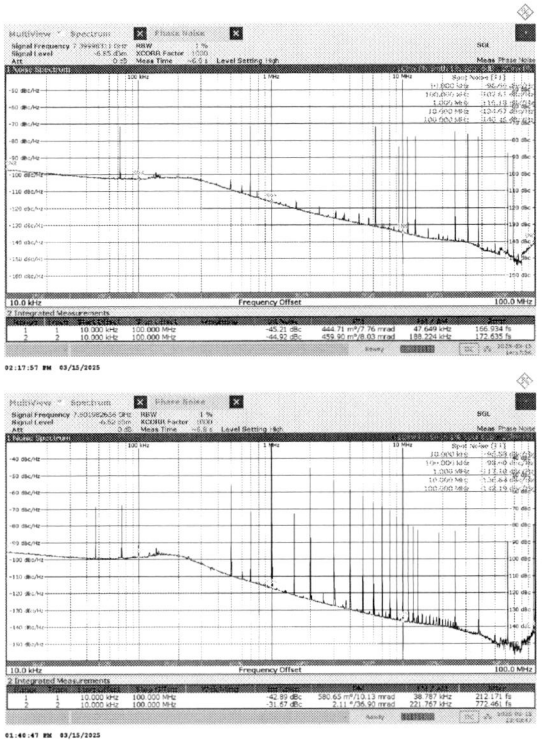

Fig. 11. Measured phase noise results of the PLL in closed-loop mode for both integer and near-integer output conditions.

remain in the deep triode region and further suppressing noise injection.

In Fig. 6, post-layout simulation results demonstrate that the effective ISFs, derived from the simulated ISF (h_{DS}) and the RMS flicker noise current ($I_{1/f,\text{rms}}$), exhibit well-balanced positive and negative symmetry, with a nearly zero average DC value. These findings validate the effectiveness of the proposed harmonic-free-like design in achieving wideband $1/f^3$ noise suppression.

IV. MEASUREMENT RESULTS

The PLL prototype, implemented in a 28 nm CMOS process from TSMC, and whose die micrograph is shown in Fig. 7, occupies an active area of 0.26 mm^2 while consuming 65 mW of power. To ensure circuit stability, the DCO employs 2.5 V IO NMOS-only cross-coupled transistors, and the P-bank capacitor array is also implemented using IO devices. Although this design choice slightly degrades performance, it significantly improves robustness and stability across PVT variations.

The tuning range of the DCO is measured to be from 5.83 GHz to 9.2 GHz, achieving a relative tuning range of 45%, as illustrated in Fig. 8. To evaluate the effectiveness of the proposed technique, open-loop measurements of the DCO phase noise and figure-of-merit (FoM) are conducted across the entire tuning range, as shown in Fig. 9.

Fig. 10 shows the measured phase noise (PN) of the DCO under open-loop conditions at both low and high frequency settings. A PN of –143.2 dBc/Hz and a corresponding FoM of 183 dBc/Hz are achieved at 6.124 GHz, while a PN of –144.4 dBc/Hz and an FoM of 187 dBc/Hz are achieved at 9.2 GHz. The FoM degradation at the lower frequency is primarily attributed to the use of IO transistors as switch devices in the P-bank, which leads to a suboptimal Q factor in the Con state.

Fig. 11 shows the measured phase noise of the PLL in closed-loop operation. For the integer-N output at 7.4 GHz with a 100 MHz reference frequency (FCW = 36, pre-divider = 2), the integrated jitter from 10 kHz to 100 MHz is 166 fs. Under near-integer operation at 7.602 GHz (corresponding to 38.001 × 100 MHz × 2), an integrated jitter of 212 fs is measured without spurs.

V. CONCLUSION

Compared with prior works (Table I), the proposed DCO demonstrates competitive performance in terms of tuning range (TR). Although its FoM is inferior to some previously reported designs, this is primarily due to the use of 2.5 V IO NMOS devices, which inherently exhibit lower Q-factor. However, this choice eliminates any risk of stability issues, making it highly favorable for system-level design. Moreover, the proposed DCO achieves both wide frequency tuning range (FTR) and low phase noise (PN) while operating under a

TABLE I

COMPARISON TABLE

Head	This work	JSSC'24 [10]	ISSCC'23 [11]	JSSC'23 [12]	CICC'23 [13]
Process	28nm CMOS	65nm CMOS	65nm CMOS	28nm CMOS	65nm CMOS
Vdd (V)	1.05	0.54	1.2	0.6	1.2
Freq (GHz)	5.83~9.2	11.5~14.3	12.3~15.8	25.7~30.7	25~35.9
Tuning range (%)	44	21.7	24.8	17.7	35.8
Power (mW)	38~53	5.6	17	29.2	10.6~13.8
Phase noise@10MHz (dBc/Hz)	-145.2~-144.5	-137.5~-138.2	-138~-135.2	-133.3~-131.6	-131.9~-127.4
FoM@10MHz (dBc/Hz)	182.9~186.7	191.2~193.8	186.3~187.7	186.7~186.8	187.8~189.6
Area (mm^2)	0.06	0.065	0.093	0.26	0.26

$$\text{FoM} = |\text{PN}| + 20\log_{10}(f_0/\Delta f) - 10\log_{10}(P_{DC}/1mW)$$

1.0 V supply. It also provides wideband $1/f^3$ noise suppression and occupies a compact silicon area. Notably, the PN remains relatively flat across the entire frequency band.

ACKNOWLEDGMENT

This work was supported by the National Natural Science Foundation of China under Grant 62301224, U23A20288, 62271210, and the Natural Science Foundation of Guangdong Province under Grant 2022A1515110226, 2024A1515011982. Corresponding author: Pei Qin.

REFERENCES

[1] X. Zhan, et al. "8.2 A 22.4-to-26.8GHz Dual-Path-Synchronized Quad-Core Oscillator Achieving -138dBc/Hz PN and 193.3dBc/Hz FoM at 10MHz Offset from 25.8GHz," 2023 IEEE International Solid-State Circuits Conference (ISSCC), San Francisco, CA, USA, 2023, pp. 148-150, doi: 10.1109/ISSCC42615.2023.10067277.

[2] H. Guo, et al. "20.1 A 5.0-to-6.36GHz Wideband-Harmonic-Shaping VCO Achieving 196.9dBc/Hz Peak FoM and 90-to-180kHz 1/f3 PN Corner Without Harmonic Tuning," 2021 IEEE International Solid-State Circuits Conference (ISSCC), San Francisco, CA, USA, 2021, pp. 294-296, doi: 10.1109/ISSCC42613.2021.9365761.

[3] Y. Shu, et al. "20.2 A 3.09-to-4.04GHz Distributed-Boosting and Harmonic-Impedance-Expanding Multi-Core Oscillator with-138.9dBc/Hz at 1MHz Offset and 195.1dBc/Hz FoM," 2021 IEEE International Solid-State Circuits Conference (ISSCC), San Francisco, CA, USA, 2021, pp. 296-298, doi: 10.1109/ISSCC42613.2021.9365737.

[4] Y. Shu, et al. "8.3 A 28GHz Scalable Inter-Core-Shaping Multi-Core Oscillator with DM/CM-Configured Coupling Achieving 193.3dBc/Hz FoM and 205.5dBc/Hz FoMA at 1MHz Offset," 2023 IEEE International Solid-State Circuits Conference (ISSCC), San Francisco, CA, USA, 2023, pp. 150-152, doi: 10.1109/ISSCC42615.2023.10067826.

[5] H. Ge, et al. "A 194.9dBc/Hz FoM and 6.8-to-11.6GHz Quad-Core Dual-Mode Class-F VCO Featuring Wideband Flicker Noise Suppression," 2024 IEEE Custom Integrated Circuits Conference (CICC), Denver, CO, USA, 2024, pp. 1-2, doi: 10.1109/CICC60959.2024.10529050.

[6] W. Chen, et al. "A 0.2-to-39.2GHz 66.2-fs Jitter and -71.3dBc Spur Sub-Sampling PLL Using DAC-Based Constant Control Voltage Compensator and Quad-Mode 2nd Harmonic Filtering Oscillator," 2024 IEEE Radio Frequency Integrated Circuits Symposium (RFIC), Washington, DC, USA, 2024, pp. 215-218, doi: 10.1109/RFIC61187.2024.10599960.

[7] H. Li, et al. "A 23.2-to-26-GHz Low-Jitter Fast-Locking Sub-Sampling PLL Based on a Function-Reused VCO-Buffer and a Type-I FLL With Rapid Phase Alignment," in IEEE Journal of Solid-State Circuits, vol. 59, no. 12, pp. 3952-3965, Dec. 2024, doi: 10.1109/JSSC.2024.3458442.

[8] Y. Li et al., "A Multi-tap-transformer Based Quad-core Dual-mode VCO Achieving 213.1dBc/Hz FoMTA@100kHz and Wideband 1/f3 Noise Suppression," 2025 IEEE Radio Frequency Integrated Circuits Symposium (RFIC), San Francisco, CA, USA, 2025, pp. 255-258, doi: 10.1109/RFIC61188.2025.11082819.

[9] J. Du, et al. "A Compact 0.2–0.3-V Inverse-Class-F23 Oscillator for Low 1/f3 Noise Over Wide Tuning Range," in IEEE Journal of Solid-State Circuits, vol. 57, no. 2, pp. 452-464, Feb. 2022, doi: 10.1109/JSSC.2021.3098770.

[10] Q. Wu, et al. "An Enhanced Class-F Dual-Core VCO With Common-Mode-Noise Self-Cancellation and Isolation Technique," in IEEE Journal of Solid-State Circuits, vol. 59, no. 8, pp. 2441-2454, Aug. 2024, doi: 10.1109/JSSC.2024.3367351.

[11] G. Zhang, et al. "34.5 A Calibration-Free 12.8-16.5GHz Cryogenic CMOS VCO with 202dBc/Hz FoM for Classic-Quantum Interface," 2023 IEEE International Solid-State Circuits Conference (ISSCC), San Francisco, CA, USA, 2023, pp. 512-514, doi: 10.1109/ISSCC42615.2023.10067803.

[12] X. Chen, et al. "A 30-GHz Class-F Quadrature DCO Using Phase Shifts Between Drain–Gate–Source for Low Flicker Phase Noise and I/Q Exactness," in IEEE Journal of Solid-State Circuits, vol. 58, no. 7, pp. 1945-1958, July 2023, doi: 10.1109/JSSC.2023.3237788.

[13] P. Guan, et al. "A 25.0-to-35.9GHz Dual-Layer Quad-Core Dual-Mode VCO with 189.1dBc/Hz FoM and 200.2dBc/Hz FoMT at 1MHz Offset in 65nm CMOS," 2023 IEEE Custom Integrated Circuits Conference (CICC), San Antonio, TX, USA, 2023, pp. 1-2, doi: 10.1109/CICC57935.2023.10121234.

2025 The 10th International Conference on Integrated Circuits and Microsystems

A CMOS Current-Reused Wideband Low-Power Dynamic Current Mode Logic Frequency Divider

Jingchen Liu[1], Youming Zhang[1,2*], Xusheng Tang[1,2], Zhennan Wei[1], Shenghao Meng[1]
[1]School of Cyber Science and Engineering, Southeast University, Nanjing 210096, China
[2]Purple Mountain Laboratories, Nanjing 211111, China
*zhangyouming@seu.edu.cn

Abstract—An ultra-wideband CMOS divide-by-two circuit utilizing current-reused dynamic current mode logic (CML) topology is designed in this paper. By replacing the NMOS latch pair in conventional dynamic CML with a PMOS pair and integrating it into the sampling path, current reuse is achieved to reduce the power consumption. Leveraging its dynamic load regulation capability, the circuit achieves ultra-wide frequency coverage. Under a 0 dBm input power, simulations demonstrate an operating frequency coverage of 5-27 GHz, achieving a 138% relative bandwidth. The out-of-band phase noise is -150.1 dBc/Hz@16GHz, the power consumption is approximately 3.6 mW, and the core area occupies 0.0026 mm².

Keywords—*dynamic current mode logic (CML), current reuse, divide-by-two.*

I. INTRODUCTION

Driven by rapid advances in modern communication technologies, the demand for high-speed, low-power communication is continuously increasing. High-speed divide-by-two circuits are typically positioned after the frequency synthesizer to generate quadrature local oscillator (LO) signals. Conventionally, a divide-by-two circuit is constructed from two D flip-flops (DFFs). For high-speed scenarios, MOS Current Mode Logic (MCML) are predominantly used for DFF implementation. MCML-based DFFs not only meet stringent speed requirements [1-2], but also mitigate switching noise and supply fluctuations associated with the constant current source. This topology, owing to its inherent differential nature, provides enhanced noise immunity [2-4]. The circuit schematic of a traditional static CML DFF is illustrated in Fig. 1(a).

However, conventional static MCML divide-by-two circuits exhibit limitations. In high-speed applications, the topology with resistive loads faces a fundamental trade-off between operating speed and output swing [5-10]. To overcome its drawbacks, researchers developed dynamic-load MCML topology DFFs [1]. This design replaces the fixed resistor with a dynamic MOS transistor load, as illustrated in Fig. 1(b). During the sampling phase, the load MOS transistor is turned on, enabling rapid charging and discharging due to a small RC time constant. In the hold phase, the load MOS transistor is turned off, presenting a high-impedance state to the circuit. Nevertheless, this structure suffers from high static power consumption because two tail current sources are simultaneously conducting during the sampling phase.

The National Natural Science Foundation of China under Grant 62471131.

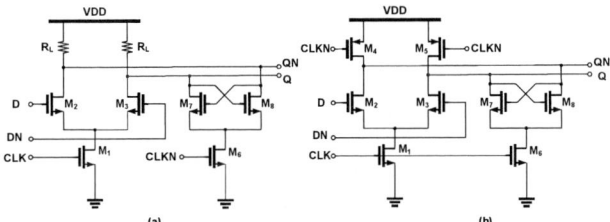

Fig. 1. Conventional DFFs. (a) Static CML DFF, (b) Dynamic CML DFF.

In this design, we present a current reuse divide-by-two circuit based on a dynamic CML DFF. This design achieves current reuse within the sampling path, significantly reducing power consumption. Moreover, by adjusting the DC bias voltage, the circuit can operate over a wider frequency range, exhibiting enhanced versatility.

II. DESIGN OF PROPOSED DIVIDE BY TWO CIRCUIT

The proposed divide-by-two circuit consists of two cross-coupled, current-reused dynamic CML D flip-flops (DFFs). An output buffer circuit is cascaded at the output to enhance the load-driving capability of the divider. The complete circuit schematic is depicted in Fig. 2.

A. Current Reuse Dynamic CML Circuit

The schematic of the proposed dynamic CML D flip-flop (DFF) is shown in Fig. 3. Compared to conventional dynamic CML divide-by-two circuits, this design folds the right-side cross-coupled NMOS latch pair into PMOS transistors and connects them in parallel with the dynamic load on the left-side differential pair. This approach eliminates one tail current source, thereby reducing power consumption.

The dynamic operation principle is as follows: The two cascaded DFFs operate alternately in sampling and hold phases. During the sampling phase, both the tail current source and dynamic load are activated. The output voltages are controlled by the differential voltage between D and DN. When one output voltage drops to $V_{DD} - V_{TH}$, it initiates the turn-on of one latch transistor, accelerating the voltage transition. During the hold phase, the dynamic load transistors present a high-impedance state, the tail current source is disabled, and the cross-coupled pair provides positive feedback to maintain the current state.

979-8-3315-8850-2/25 $31.00 © 2025 IEEE

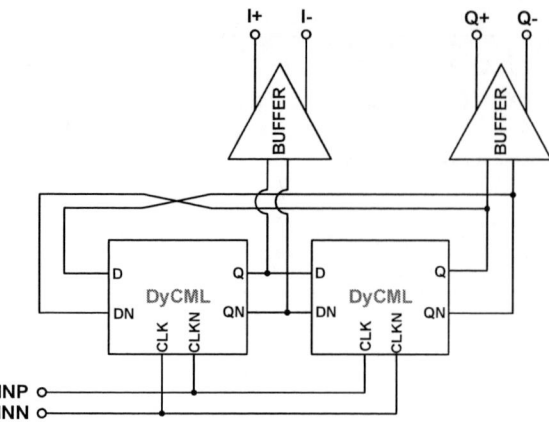

Fig. 2. Proposed divide-by-two circuit.

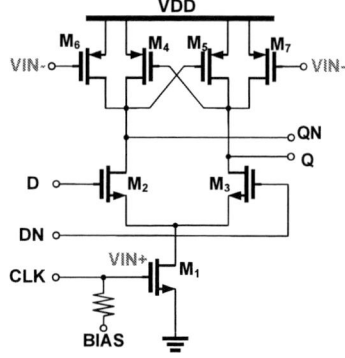

Fig. 3. Proposed current reuse dynamic CML circuit with adjustable bias.

When the Barkhausen criterion is satisfied, the circuit enters self-oscillation. Assuming the gate delay of a single DFF is τ_{DFF}, the self-resonant frequency of the CML divider is given by Equation (1).

$$f_{osc,Dycml} = \frac{1}{4\tau_{DFF}} \qquad (1)$$

Here, τ_{DFF} constitutes the gate delay during both the sampling phase $\tau_{DFF,sam}$ and the latching phase $\tau_{DFF,hold}$. These components can be derived from the following expressions:

$$\tau_{DFF} = \tau_{DFF,sam} + \tau_{DFF,hold} = R_1 C_1 + \frac{R_2 C_2}{g_{MH3,4} R_2 - 1} \qquad (2)$$

R_1 and C_1 represent the equivalent resistance and equivalent capacitance seen at the drain of the sampling transistors during the sampling phase, while R_2 and C_2 denote those during the latching phase. The parameter $g_{MH3,4}$ corresponds to the transconductance of the latch transistors. For a CML divider, a necessary condition for self-oscillation is that the transconductance of the latch transistors satisfies: $(g_{MH3,4}R_2 - 1 > 0)$. Regarding the dynamic load, its high-impedance state during hold mode ensures the oscillation. This characteristic allows the sampling transistor pair to be implemented with minimized size (W/L). Such scaling reduces the parasitic capacitance at high frequencies, thereby enhancing the circuit's charging/discharging speed.

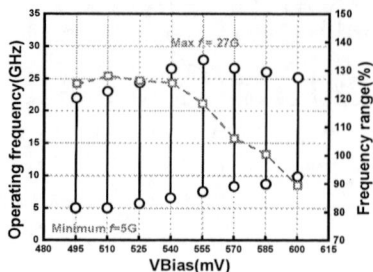

Fig. 4. Simulated frequency range of proposed DIV2 under different bias.

B. Calculation of Adjustable Bias

To ensure proper operation of the divide-by-two circuit, the input signal must be AC-coupled and reapplied with a reconfigured DC bias voltage. In dynamic dividers, variations in the DC bias voltage modulate the on-resistance of the MOS transistors, consequently governing the operating frequency of the circuit. The equivalent resistance of the dynamic load during the sampling phase is given by Equation (3):

$$R_1 = r_{ds} = \frac{1}{g_{ds}} = \frac{1 + \lambda V_{DS}}{I_D \lambda} = \frac{1}{K \cdot \frac{W}{2L}(|V_{GS}| - |V_{TH}|)^2 \lambda} \qquad (3)$$

To achieve higher division frequencies and a broader operating frequency range, this design implements a tunable DC bias voltage for the input signal. The operating frequency of the divider is primarily governed by the gate delay of the D flip-flop, denoted as τ_{DFF}. This delay is subdivided into the sampling phase delay $\tau_{DFF,sam}$ and the hold phase delay $\tau_{DFF,hold}$, whose mathematical expressions are provided in Equations (4) and (5).

$$\begin{aligned} \tau_{sam} &= R_1 C_1 \\ &\approx [r_{on,MP1} \parallel R_L] \cdot (C_{gd,MP1} + C_{gd,MS1} + \\ & \quad C_{gd,MH1} + C_{gs,MH1} + C_{gd,MH2} + C_L) \end{aligned} \qquad (4)$$

$$\begin{aligned} \tau_{hold} &= \frac{C_2}{g_{m,MH} - 1/R_2} \\ &\approx \frac{C_{gd,MP1} + C_{gd,MH1} + C_{gs,MH1} + C_{gd,MH2} + C_L}{g_{m,MH} - 1/(r_{on,MP1} \parallel r_{on,MH1} \parallel R_L)} \end{aligned} \qquad (5)$$

When the delays during the sampling and hold phases exceeds the division period of the divide-by-two circuit, the circuit fails to work. At low frequencies, the parasitic capacitance of MOS transistors remains relatively constant, and the total nodal parasitic capacitance dominates. At higher frequencies, reduced transistor parasitic capacitance enhances circuit speed. Conversely, the maximum operating frequency is limited by the frequency-dependent transconductance g_m. As frequency increases, g_m degrades, thereby increasing the hold phase delay. For frequency f, R_L can be calculated using Equations (1) and (2). The total capacitance at the output node during the sampling phase can be estimated based on transistor dimensions and operating frequency. Consequently, the required V_{GS} (i.e., the DC bias voltage) is derived from Equation (3) and (6).

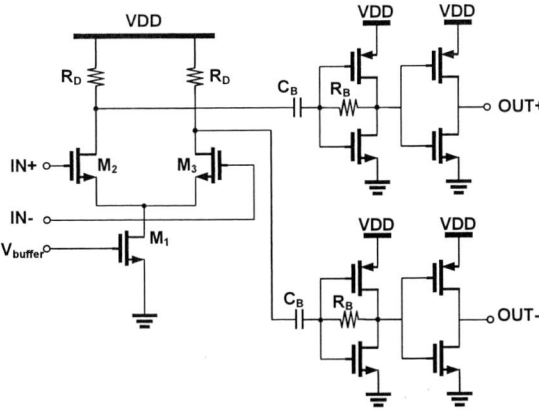

Fig. 5. Output buffer circuit of proposed divide-by-two.

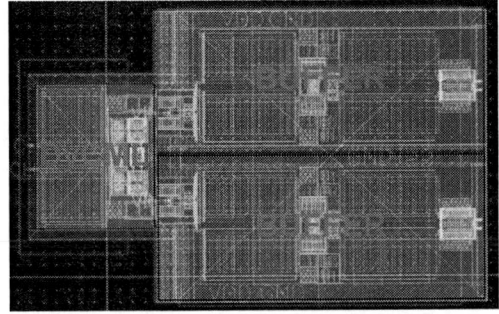

Fig. 6. Layout design of proposed divide-by-two circuit.

$$R_L \parallel r_{on,MP1} \approx \frac{1}{4f(C_{gd,MP1} + C_{gd,MS1} + C_{gd,MH1} + C_{gs,MH1} + C_{gd,MH2} + C_L)} \tag{6}$$

Since the bias voltage simultaneously regulates the tail current transistor, impacting both static power consumption and output swing, bias point selection requires comprehensive trade-off. A higher DC bias voltage increases R_{on}, which will elevating sampling phase delay $\tau_{DFF,sam}$. But it also enhances bias current to accelerate charging/discharging speed and enlarge output swing. Whereas a lower DC bias reduces R_{on} at the cost of diminished output swing, with critically low bias risking premature cutoff of the tail current transistor that disrupts normal operation. Consequently, the operating frequency range shifts with DC bias variation, enabling strategic voltage selection for trade-off optimization. This design implements eight discrete bias settings at 15 mV increments to achieve ultra-wide frequency coverage, demonstrating 7-26 GHz bandwidth at 540 mV, maximum relative bandwidth (130%) at 510 mV, and full 5-27 GHz division coverage (138% relative bandwidth) across tunable biases, with post-layout simulation results of frequency coverage versus bias illustrated in Fig. 4.

C. Design of Buffer Circuit

Figure 5 depicts the output buffer circuit for the CML divide-by-two architecture. This buffer employs a two-stage configuration: the first stage implements a fully-differential common-source topology to effectively suppress second-order harmonic interference on subsequent stages, while the second stage utilizes two cascaded inverters as buffering elements.

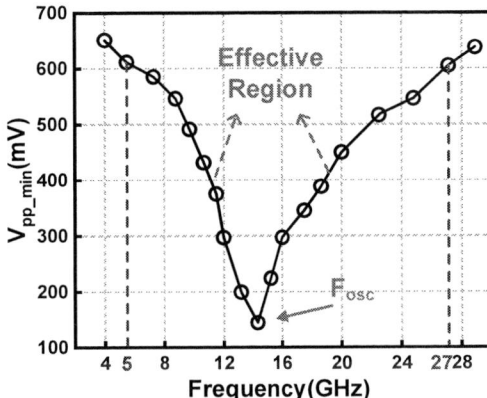

Fig. 7. Input sensitivity versus input frequency of the proposed divider.

The output swing of this buffer is determined by the load resistance R_D and the current sourced by the tail current transistor. To address significant output swing variation across frequency bands - particularly at high frequencies - the bias voltage V_{buffer} incorporates a tunable architecture identical to the clock transistor pair. This design enhances high-frequency signal amplification capability.

III. LAYOUT DESIGN AND SIMULATION RESULTS

The complete divide-by-two circuit comprises one dynamic CML divider core and two output buffer circuits. Given its fully differential I/O architecture, symmetric layout implementation is essential to enhance phase accuracy and ensure matched propagation delays across differential signal paths. All interconnects should be minimized, particularly between primary outputs and inputs, as propagation delay and capacitive loading critically constrain the maximum operating frequency. I/O and inter-stage routing utilizes upper metal layers to mitigate parasitic capacitance effects. The final layout (Fig. 6) occupies 179 μm × 155 μm, with the divider core consuming 45 μm × 57 μm.

Designed for post-VCO operation, the divider accepts sinusoidal inputs at $V_{DC} = 0.6$ V and $V_{pp} = 0.6$ V (0 dBm). Post-layout simulations confirm 5-27 GHz division range across bias variations. The input sensitivity curve is presented in Fig. 7, which defines the valid operating region above the threshold line, revealing self-resonance at 14 GHz - approximately half the maximum frequency, aligning with design targets. Near this resonance, minimal input power sustains division. Transient waveforms at 5 GHz and 27 GHz appear in Figs. 8(a)-(b), while Fig. 9 shows output phase noise of -150.1 dBc/Hz at 1 MHz offset when processing 16 GHz inputs.

Compared with conventional static CML dividers operating at 16GHz input frequency, the proposed dynamic CML architecture demonstrates significant improvements: a 151 mV increase in output swing, 2.3 mW reduction in power consumption, and 6dBc/Hz phase noise improvement at 1 MHz offset, while maintaining identical device dimensions.

979-8-3315-8850-2/25 $31.00 © 2025 IEEE

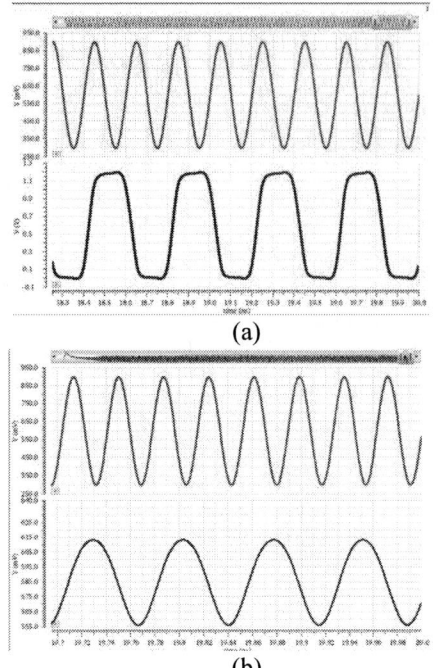

(a)

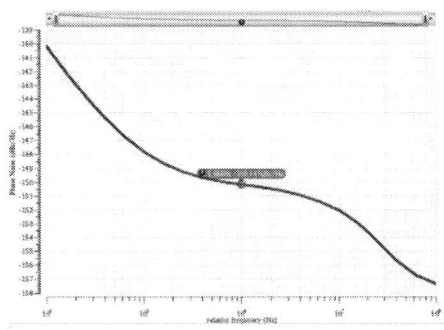

(b)

Fig. 8. Input and output waveform at (a) 5GHz, (b) 27GHz.

Fig. 9. Output phase noise simulated at 16GHz.

TABLE I. PERFORMANCE COMPARISON WITH OTHER BROADBAND DIVIDE-BY-2 IN CMOS

Ref.	Topology	Process (nm)	Operatingfreq. (GHz)	Freq. range (%)	P_{dc} (mW)	Area (mm²)
[4]	dyCML	130	19-43	77.4	5	N/A
[6]	SCML	130	20-38	62.1	12	1
[7]	SCML	110	8-28	111.1	8.9	0.0046
[8]	dyCML	90	27-37	31.3	3.0	N/A
This Work	Current reuse dyCML	130	5-27	138	3.6	0.0026

As benchmarked in Table 1 against state-of-the-art millimeter-wave frequency dividers, this design achieves superior relative frequency range, lower power dissipation, and more compact layout area. Compared with dyCMLs without current reuse structure and adjustable bias, our design achieves a better tradeoff between frequency range, power consumption and chip area.

IV. CONCLUSION

This paper presents a high-speed, wideband, and low-power frequency divider design. The divide-by-two circuit employs a current-reused dynamic-load CML architecture with cascaded output buffers to enhance load-driving capability. Compared to conventional dynamic CML topologies, the proposed current-reused architecture eliminates one current path while maintaining functionality, thereby reducing power consumption and input impedance. To achieve wide frequency coverage, an 8-step tunable DC bias scheme is implemented, enabling proper frequency division from 5 to 27 GHz in post-layout simulations with 0.6V (0dBm) input swing. The simulated result shows phase noise of -150.1 dBc/Hz@16 GHz at 1MHz offset, 3.7 mW DC power consumption, and a compact core area of 0.0026 mm².

REFERENCES

[1] M. W. Allam and M. I. Elmasry, "Dynamic current mode logic (DyCML): a new low-power high-performance logic style," *IEEE J. Solid-State Circuits*, vol. 36, no. 3, pp. 550-558, 2001.

[2] X. Zhang, Y. Wang, S. Jia, G. Zhang, and X. Zhang, "A novel CML latch for ultra high speed applications," in *Proc. Int. Conf. Electron Devices Solid-State Circuits (EDSSC)*, 2014, pp. 1-2.

[3] F. Centurelli, G. Scotti, A. Trifiletti and G. Palumbo, "Design of low voltage power efficient frequency dividers in folded MOS current mode logic," *IEEE Trans. Circuits Syst. I, Reg. Papers*, vol. 68, no. 2, pp. 680-691, 2021.

[4] K. M. Sharaf, "A 5mW 19–43 GHz broadband CMOS I/Q frequency divider," in *National Radio Science Conference*, Tanta, Egypt, 2008, pp. 1-8.

[5] Majumder, Alak, et al. "Variation aware design of 50-Gbit/s, 5.027- fJ/bit serializer using latency combined mux-dual latch for InterChip communication." *IEEE Trans. Circuits Syst. I, Reg. Papers*, vol. 66, no. 3, pp. 1231-1244, 2019.

[6] U. Singh, and M. M. Green, "High-frequency CML clock divider in 0.13-μm CMOS operating up to 38GHz", in *IEEE Int. Solid-State Circuits Conf. (ISSCC) Dig. Tech. Papers*, Aug. 2005, pp.1658-1661.

[7] Leilei Xiao, Xiuping Li, "A Divide-by-16 CML frequency divider for K-Band radar applications", in *International Conference on Microwave and Millimeter Wave Technology (ICMMT)*, 2023, pp.1-3.

[8] K. . -L. J. Wong, A. Rylyakov and C. . -K. K. Yang, "A broadband 44-GHz frequency divider in 90-nm CMOS," *IEEE Compound Semiconductor Integrated Circuit Symposium*, 2005. CSIC '05., Palm Springs, CA, USA, 2005, pp. 4.

[9] W. Zhen, S. Cao, Y. Su, S. Li, and Z. Jin, "A novel design method of SOF for InP DHBT ECL and CML static frequency dividers," *IEEE Microw. Wireless Compon. Lett.*, vol. 31, no. 6, pp. 583–586, Jun. 2021.

[10] Y. Jiang, "Design and optimization of a high-efficiency CML frequency divider in InP DHBT technology with maximized self-oscillation frequency," in *Proc. 9th Int. Conf. Electron. Technol. Inf. Sci. (ICETIS)*, May 2024, pp. 73–76.

979-8-3315-8850-2/25 $31.00 © 2025 IEEE

A VHF memristor emulator for crossbar synaptic simulation

Guangzhen Dai
School of Integrated Circuits
Anhui Polytechnic University
Wuhu, China
daigzh@ahpu.edu.cn

Bingchen Liu
School of Integrated Circuits
Anhui Polytechnic University
Wuhu, China
1643489664@qq.com

Sihao Yang
School of Integrated Circuits
Anhui Polytechnic University
Wuhu, China
17356118186@163.com

Wei Li
School of Integrated Circuits
Anhui Polytechnic University
Wuhu, China
15256773638@163.com

Daohua Wu
School of Integrated Circuits
Anhui Polytechnic University
Wuhu, China
daohuawu@ahpu.edu.cn

Yuefeng He
School of Integrated Circuits
Anhui Polytechnic University
Wuhu, China
heyf@ahpu.edu.cn

Abstract—Here, it was recommended to develop a CMOS-based memristor emulator circuit that can operate well in both grounded and floating modes without the need for internal structural changes. The proposed emulator is constructed from a compact assembly of nine MOS transistors, three resistors, and a single capacitor. Interestingly, unlike previous research, the novel dual-function memristor emulator has an easy-to-use design and maintains the memristor's hysteresis characteristic curves for input sine waves up to 500MHz in frequency. The distinctive features of the pinched hysteresis curve and the nonvolatility of the memristor emulator across various frequency domains have been experimentally validated using 130nm CMOS technology. The empirical validation of the circuit, realized on a breadboard utilizing commercial components, conclusively corroborates the findings of both theoretical analysis and simulation. Subsequently, a crossbar model of the memristor has been established using data derived from the proposed emulator. Utilizing three distinct datasets—comprising small digits, file types, and large digits—we have trained the model employing the backpropagation algorithm, subsequently assessing its training performance on diverse datasets. The outcomes indicate that the model achieves an accuracy rate as high as approximately 94%—even for the large digits set—after only 10 training epochs.

Index Terms—Memristor emulator, Very high frequency, Crossbar, Back propagation

I. INTRODUCTION

A memristor, as a type of passive two-terminal device, was predicted by Leon Chua in 1971 from the perspective of circuit theory's completeness [1]. According to Chua's prediction, the memristor is a basic circuit element that can represent the relationship between charge and magnetic flux [2], [3]. The mathematical model of memristor is inferred based on

This work was funded by the Key Program of Natural Science Foundation of Anhui Provincial Education Department(2023AH050922), the National Natural Science Foundation of China (Grant Nos. 62174001), the Excellent Scientific Research and Innovation Teams of Anhui Province under Grant 2022AH010059 and supported by the Joint Opening Project of Anhui Engineering Research Center of Vehicle Display Integrated Systems and Joint Discipline Key Laboratory of Touch Display Materials and Devices in Anhui Province(VDIS2023B03, VDIS2023B04, VDIS & TDMD2024B04).

the characteristics of three basic passive components, and it describes the characteristics of the memristor as (1) and (2), where v(t), i(t), and q(t) represent the voltage, current, and charge across the device as a function of time.

$$v(t) = M(q(t))i(t). \tag{1}$$

$$M(q) = d\varphi(t)/dq. \tag{2}$$

This unique device possesses non-volatile and resistance-changing properties. However, it was largely overlooked as a mere theoretical prediction. up to 2008, the team led by Strukov DB at HP Laboratories successfully demonstrated the implementation of the memristor by exploiting the migration of ions at the metal-electrolyte interface with memristive characteristics and proposed the linear drift model (3) based on the resistive switching mechanism of the device and the memristor concept model proposed by Leon Chua [4]. Here, R_{on} and R_{off} represent on- and off-state resistances of the memristor respectively, D is the total length of the device, μ_v is the average ion migration rate, and M(q) is the same as equation (1). Under appling a positive voltage on the anode of the memristor, oxygen ions migrate from the electrolyte to the metal electrode, inducing the formation of oxygen vacancies in the electrolyte TiO_{2-x}. These vacancies establish a conductive channel in the electrolyte to connect the top electrode with the bottom, enabling the flow of current. Conversely, applying a reverse voltage causes the oxygen ions to migrate back into the electrolyte, result in the conductive channel breaking, and preventing the flow of current. Thus, different resistance states can be achieved by controlling the applied voltage. The success at HP Laboratories reignited widespread interest in memristors among researchers.

$$M(q) = R_{off}(1 - \mu_v * R_{on}/D^2 * q(t)). \tag{3}$$

In addition to the memristor implemented by the HP laboratory using TiO_{2-x}, researchers have also explored many materials

979-8-3315-8850-2/25 $31.00 © 2025 IEEE

such as HfO_2 [34], sulfides, polymers and organic materials. These materials all exhibit different memristor properties, varying in response speed, reliability, repeatability and tolerance, etc. Oxide-based memristors exhibit high reliability, long lifespan, stable resistance states, and mature fabrication processes with low costs [11]–[13]. However, memristors made from these materials have certain limitations in resistance variability and response speed; sulfide-based memristors offer a larger resistance variability range and faster response speed but involve complex fabrication processes, and the resistance values may unpredictably change with long-term usage. Although polymer-based memristors have advantages in power consumption and fabrication difficulty, their poor stability and limited resistance variability range increase the complexity of their applications.

The unique characteristics of memristors reported by published literatures make them have with tremendous potential applications. The resistance values of memristors are changed and maintained dynamically under electrical stimulation and no electrical stimulation respectively. This property suggests that memristors are non-volatile, similar to synapses, and are well suited for neuromorphic computing. The crossbar arrays constructed by memristors can implement high density storage, so as to realize in-memory computing [33]. Compared to the current mainstream SRAM and DRAM, RRAM implemented by memristors not only offers advantages such as high density, fast write speeds, and low power consumption but also enables rapid multiplication and division operations by leveraging the memristor's resistance-variable characteristics and Ohm's Law [6], [7]. The in-memory computing implemented by memristors is different from the traditional computation, it can realize non-binary calculation based on the resistance variability of memristors [20], [32]. This form of in-memory computing has the potential to break through the V_{on}·Neumann architecture's memory and storage walls. By encoding the resistance values as weights, memristors also have great potential in the field of neural networks, serving as storage devices [5]. The neural network constructed by memristors can storage and update the weights easily and efficiently by adjusting the resistance values. Moreover, resistive values varieties of memristors under pulse excitations endow them with the ability to mimic analog neurons. It is also possible that memristive mimic the connection strength and weight changes between synaptic neurons [21]. The non-volatility of memristors allows them to simulate the memory of synaptic neurons [8]–[10]. The potential applications of memristors in various fields have attracted an increasing number of researchers to study them.

As exploration of memristors deepened, the threshold characteristic of memristors was discovered and quickly became one of its fundamental characteristics. This characteristic mainly manifests the resistance value remaining unchanged when the voltage across the memristor does not exceed its threshold. The introduction of this new characteristic imposed new requirements on the mathematical models of memristors. To clearly express the threshold characteristic of memristors, S. Kvatinsky's team proposed the VTEAM and TEAM models,

corresponding to voltage-controlled memristors and current-controlled memristors, respectively [14], [15]. These models not only propose corresponding mathematical models based on different operational excitations, but also introduce the threshold characteristics of memristors. These increasingly sophisticated mathematical models describe the characteristics of memristors and strongly promoted the development of applications research for memristors. The results of simulations in SPICE or Matlab for memristors with different mathematical models can contribute to the understanding of their characteristics and the development of their applications [16]. For example, modeling memristors using Matlab can be used to explore their in-memory applications and study training schemes for memristor-based neural networks.

Researchers usually use memristor emulator circuits to implement the applications of memristive devices due to high cost, complex process, diverse mechanisms, poor uniformity and other characteristics of physical memristors, and have not been commercialized. However, different memristor emulator circuits have different focuses. For instance, Hyongsuk Kim's memristor emulator is primarily designed to simulate the performance characteristics of TiO_2-based memristors for use in such memristor applications [17]. The continuity of this memristor emulator should to be corrected to switch directly between decrement and increment configurations. The simulation results display superior nonlinearity and resistance-switching characteristics manifested in the hysteresis curve. In addition to emulating existing memristor device characteristics, some emulators that target specific properties of memristors were developed for exploratory purposes and application development. For example, Mourina Ghosh designed a floating memristor emulator based on n-type MOSFETs, featuring a simple design with lower power consumption and can reach 50MHz operating frequency [18]. To verify the effectiveness and operability of the emulator, Ghosh utilized the proposed memristor emulator to design signal processing circuits such as the BFSK demodulator. Despite effectively modeling the memristor's non-linear behavior, the extensive use of integrated devices raises concerns about the circuit's area consumption. Although various memristor emulators can meet the needs of different scenarios, there are still some challenges, such as the need to adjust for floating and grounded operating environments, reduce circuit area, cut down power consumption, and increase operating frequency. Ghosh also designed another memristor emulator employing operational transconductance amplifier(OTA) and voltage differencing buffer amplifier(VDBA) to simulate the memristor in both floating and grounded states under sinusoidal wave excitation at frequencies of 500 kHz and 5 MHz, demonstrating prominent hysteresis loops and decay characteristics [19]. Comparative analysis shows that memristor emulators composed of simpler components tend to higher operating frequency and smaller circuit area [18]. However, it can only work in floating but not grounded state, while the simpler circuit is more suitable for investigating memristor applications in high-frequency scenarios.

Aiming at the challenges in researching existing memristor

models, a novel one is designed herein based on the 130nm CMOS technology. This memristor emulator can work in-ground or floating mode, and can operate at very high frequency. The operating frequency of the novel one is significantly higher than that of existing memristor emulators. Simulation tests and performance evaluations were conducted utilizing Cadence Virtuoso and the emulator's feasibility was verified by constructing a physical circuits on a breadboard using commercially available components. To explore the potential application of memristors, a cross-bar model is simulationed on the basis of the proposed memristor emulator. The crossbar model connects multiple memristors in an array format, controlling more memristors with fewer read/write circuits. This study demonstrutes the advantages of memristors the weight of neurons the crossbar model. By changing the resistance value of memristors array in data storage and computation to shows in neural networks. Subsequently, the input pulse signal is used to detest the variation of the memristor emulator conductance in the cross-bar matrix. Multiple datasets such as small digits, file types and large digits were utilized to train and test this cross-bar array to evaluate its performance.

A grounded and floating-type memristor emulator based on a CMOS transmission gate is designed in this paper. The novelty-designed memristor emulator can operate at frequencies up to 500MHz. The subsequent sections of this paper are structured as follows: Section 2 provides a comprehensive theoretical description of the proposed memristor emulator. Section 3 evaluates the performance of the proposed memristor through simulation and experimental test. Furthermore, a memristor crossbar model is established by using the proposed memristor emulator and some relevant experiments are carried.

II. Proposed Memristor Emulator Implementation

The proposed memristor simulator circuit, as shown in Fig 1, consists of three main components: signal processing, memory cell, and variable resistance unit. The threshold characteristics of the new memristor emulator are achieved by a limited voltage swing of the differential amplifier circuit, where the limited voltage swing stablizes the amplifier output without triggering a change the next-level storage unit when the input voltage falls below the threshold. The memory cell is composed of a PMOS transistor M5, an NMOS transistor M6 and a capacitor C1, where-in M5 and M6 control the charging and discharging of C1, respectively. M5 and M6 are driven by two outputs of the differential circuit, respectively, and it is found that this structure can ensure the consistency of the capacitor charging and discharging process. The output directly generated by the differential amplifier needs to be reversed by M3 and M4 before driving the gate of M6 being an NMOS-type. The inverter composed of M3 and M4 can adjust the output of M1 to meet the control requirements of M6 by properly correcting the aspect ratios of the tranristors. The variable resistance unit utilizes a CMOS transmission gate, which consists of M8 and M9 allowing bidirectional

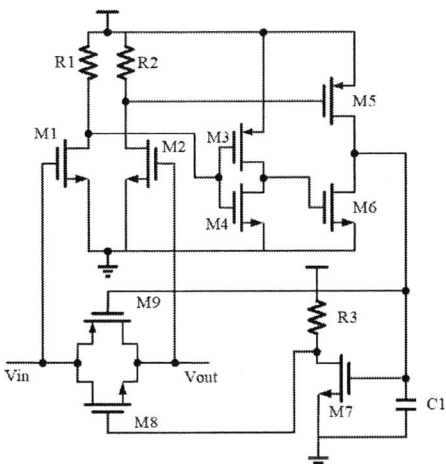

Fig. 1: Proposed memristor emulator circuit.

transmission of analog signals. To control the current passing through the transmission gate, the terminals of in the memory capacitor C1 are connected to the gates of M8 and M9, respectively.

The voltages at both terminals of V_{in} and V_{out}, the memristor emlator are amplified by the differential amplifier consisting of M_1 and R_1 and M_2 and R_2. The two output voltages of the differential amplifier are expressed as $V'_{in} = -g_{M1}R_1V_{in}$ and $V'_{out} = g_{M2}R_2V_{out}$ respectively. V'_{out} is directly connected to the gate of the transistor M5, which directly controls the charging of the capacitor C_1. On the other hand, the transistor M6 that controls the discharge of the capacitor C_1, is an NMOS, so it is necessary to achieve signal, through M3 and M4. The gate voltage of M6 can be denoted as $V''_{in} = -(g_{M3} + g_{M4})g_{M2}R_2V_{out}$. M5 and M6 are alternately switched on, thus realizing the change and discharge of the capacitor C1.

By combining the expressions of the gate voltages of M5 and M6, the charging and discharging expressions of the capacitor C1 can be represented by equations (4) and (5).

$$\frac{dV_{C1}}{dt} = \frac{I_{D_{M5}}}{C_1} = \frac{1}{2C_s}\mu_p C_{ox}(\frac{W}{L})_{M5}(V_{dd} + V'_{out} - V_{TH})^2$$
(4)

$$\frac{dV_{C1}}{dt} = \frac{I_{D_{M6}}}{C_1} = \frac{1}{2C_s}\mu_n C_{ox}(\frac{W}{L})_{M6}(V''_{in} - V_{TH})^2 \quad (5)$$

The voltage expression of capacitor C1 can be written as (6)

$$V_{C1} = V_{C1}(0) + I_d(V_{out})dt = V_{C1}(0) + V_d(\Phi) \quad (6)$$

The gate voltage of M9 is supplied directly by the capacitor C1's voltage, which the gate voltage of M8 is provided by the amplifier cation of the capacitor C1's coltage by a single-stage amplifier consisting of R3 and M7, C1's voltage which can be represented as $V'_{C1} = -g_{M7}R_3V_{C1}$. Therefore, the conducting

TABLE I: Comparative analysis of the proposed work with previously reported work.

	of active elements	of passive elements	of transistors	Sim/Exp	Floating & grounded	Max.operating freq.
[22]	OTA, CDBA	1(1C)	28	Sim	Grounded	1MHz
[23]	-	1(1C)	7	Both	Grounded	50MHz
[24]	OA	40(40R)	14OA,1JFET	Both	Floating	300Hz
[25]	OA	8(7R,1C)	AD633	Exp	Grounded	195Hz
[26]	-	7(3R,2C,2diode)	2	Both	Both	10kHz
[27]	OA	7(4R,3C)	2AD844,AD633	Sim	Grounded	160kHz
[28]	OA	6(5R,1C)	4AD844,AD633	Sim	Floating	20.2kHz
[29]	OTA, VDBA	-	23	Both	Floating	5MHz
[18]	-	0	4	Both	Floating	50MHz
this work	-	4(3R,1C)	9	Both	Both	500MHz

current of M8 and M9 can be expressed as (7) and (8), where's $V_{in} - V_{out} = V$.

$$I_{M8} = \mu_n C_{ox}(\frac{W}{L})_{M8}[(V'_{C1} - V_{in} - V_{TH})V - \frac{1}{2}V^2] \quad (7)$$

$$I_{M9} = \mu_p C_{ox}(\frac{W}{L})_{M9}[(V_{C1} - V_{out})(-V) - \frac{1}{2}V^2] \quad (8)$$

To find the impedance, take the derivative of the voltage V_{on} both sides of equations (7) and (8).

$$M(\phi) = \frac{dI_{M8}}{dV} = \mu_n C_{ox}(\frac{W}{L})_{M9}(I_d(\phi)' - V_{out} - V_{TH}) \quad (9)$$

$$M(\phi) = \frac{dI_{M9}}{dV} = \mu_n C_{ox}(\frac{W}{L})_{M8}(I_d(\phi) - V_{in} - V_{TH}) \quad (10)$$

III. SIMULATION RESULTS AND CIRCUIT EXPERIMENT

To verify the performance of the proposed memristor emulator, simulations were conducted in the environments Cadence Virtuoso and a circuit was set up on a breadboard using off-the-shelf electronic components for circuit testing. The simulation experiments adopt the SMIC 0.13μm process of CMOS. To examine the operational status of the novel memristor emulator in a floating scenario, two sine signals with amplitude of 0.8V, frequency of 70MHz, bias of 0.8V, and phase difference of 180° applied to the top and bottom of the memristor emulator, respectively. Under these excitation signals, the voltage drop on the memristor emulator could be controlled to manifest as a sine signal crossing zero points, with simultaneous increases and decreases in magnetic flux within one cycle. The response to these excitations is illustrated in Fig.2, where Fig.2(a) depicts the current response generated the 70MHz sine signal excitation, exhibiting noticeable current attenuation and enhancement; Fig 2(b) illustrates the V-I characteristics of the memristor emulator, displaying distinct hysteresis curve properties with no cutoff region or zero crossing issues.

To investigate the work of the memristor emulator in the grounded state, a sine wave with frequency of 70MHz, amplitude of 1.6V, and bias voltage of 0.8V is used as the excitation at top end of the memristor. At the some time an 800mV DC power supply is connected at the other end. This excitation signal configuration, by maintaining the single-end voltage of the memristor, simulated the grounding performance

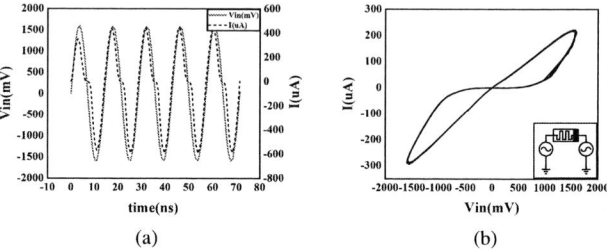

(a)   (b)

Fig. 2: Voltage and current image at 70MHz sine wave input in floating (a) and V-I image with 70MHz sine wave input in floating (b)

of the memristor emulator and ensured that the magnetic flux through the memristor emulator experienced cycles of increase and decrease within the sinusoidal signal period. The experimental results depicted in Fig.3 (a) and (b) exhibit prominent hysteresis loops and resistance change phenomena. However, the constant voltage of 800mV at one end of the memristor resulted in a maximum voltage difference of 800mV across the two terminals of the memristor, leading to an experimental current that approximates half of that during floating operation. Additionally, due to the slight increase in the maximum voltage difference of 800mV across the two terminals of the memristor compared to the designed threshold voltage, a long establishment time was not observed, thereby displaying a significant current variation compared to the floating operational state.

As the higher frequency sinusoidal excitation results in less magnetic flux passing through the memristor in the same amount of time, which in turn manifests itself as the change in the resistance of the memristor becoming smaller, which is shown in the image as a contraction of the hysteresis curve. To test the hysteresis curve shrinkage characteristics of the memristor emulator very higher frequencies, sinusoidal excitations of 100MHz, 300MHz, and 500MHz were supplied respectively, and the simulation results were compared with that of 70MHz excitation. The V-I curves generated in the experiment are shown in Fig 4. Obviously, with the increase of the frequency of the sinusoidal excitation signal, the hysteresis curves formed in the first and third quadrants gradually shrink,

979-8-3315-8850-2/25 $31.00 © 2025 IEEE

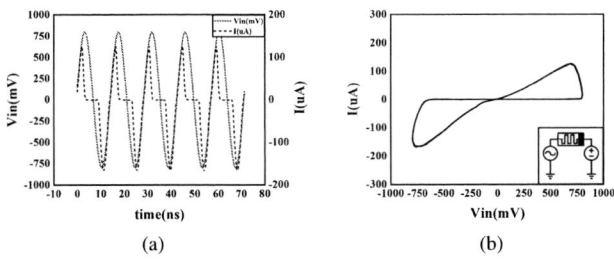

(a)   (b)

Fig. 3: Voltage and current image at 70MHz sine wave input in grounded (a) and V-I image with 70MHz sine wave input in grounded (b)

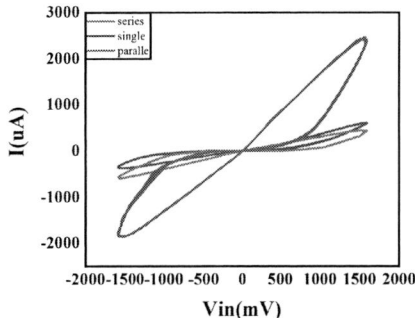

Fig. 5: V-I characteristics of single, series, and parallel memristors at 70MHz excitation

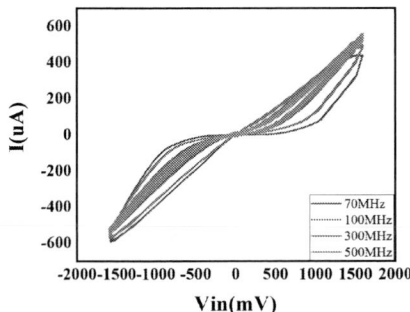

Fig. 4: V-I characteristic curves at 70MHz, 100MHz, 300MHz, and 500MHz

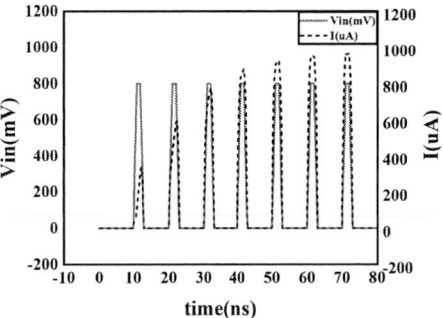

Fig. 6: Current response of memristor under multiple short pulses

that is, the distinction between high and low of the memristor becomes less and less obvious.

To examine the effect of the different connects of the memristor emulators, two sine excitations with phase difference of 180° and amplitude of 1.6V are supplied on the two terminals of a single memristor, parallel memristors, and series memristors. The experimental results are displayed in Fig.5. In the series memristors' structure, the hysteresis loop is significantly smaller than that of the parallel configuration, which is similar to that of the single memristor, indicating that the paralled increases the ratio of high and low resistance states. This discrepancy is attributed to the voltage distribution between the series-connected memristors, which causes them not to reach the threshold voltage promptly and induces the resistance to change more slowly. This issue does not arise in the parallel configuration, where the resistance values of both memristors change simultaneously once the input signal surpasses the threshold voltage. Therefore, the output current of the parallel memristors is much larger than that of a the single memristor or series configuration, because the chance in resistance reduction of the parallel structure caused by the magnetic flux is greater and occures simultaneously.

The non-volatility experimental result of the memristor emulator designed here in is displayed in Fig.6. The pulse signal with pulse width of 1ns and amplitude of 1.6V is

used as the excitation input. Due to the rise and fall time of such a very short pulse width signal that can not be ignored during simulation, the pulse signal appears as a spike shape. Under the excitation of this signal, it is observed that the current gradually increased with the increase of pulses number; indicating that the resistance value of the memristor gradually, decreases with the increase of the number of pulses.

The effects of different process corners and temperatures on the proposed memristor emulator were simulated using Cadence Virtuoso, as shown in Fig.7(a) and 7(b). The simulation results show that the V-I hysteresis curves decrease with the process corners FF,TT, and SS when the circuit bias and excitation signal are kept unchanged. However, the pinched hysterisis loops are almost unchanged at different temperatures. 1

In addition to the simulation and verification of the proposed memristor emulator, the circuit of the memristor emulator is constructed on a breadboard with readily available circuit components and verified by experiment. A 4MHz sinusoidal signal was applied as an excitation, and the time domain wave form of voltage and current of the memristor emulator, and abvirous V-I hysteresis loop of it were observed using Rigol D1202 oscilloscope. The circuit diagram, transient wave form, and V-I characteristics are illustrated in Fig 8 (a-d). There are some differences between V-I hysteresis curves prouced

979-8-3315-8850-2/25 $31.00 © 2025 IEEE   313

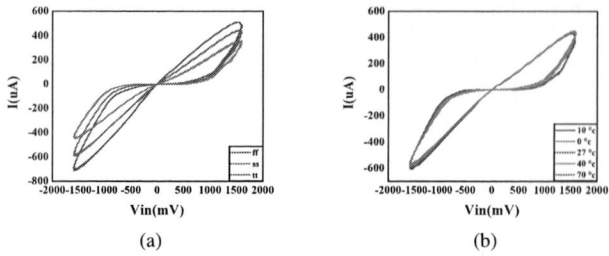

(a)   (b)

Fig. 7: The V-I characteristics of the memristor under the excitation of 70MHz sine wave at different process corner (a) and different temperatures (b) are demonstrated

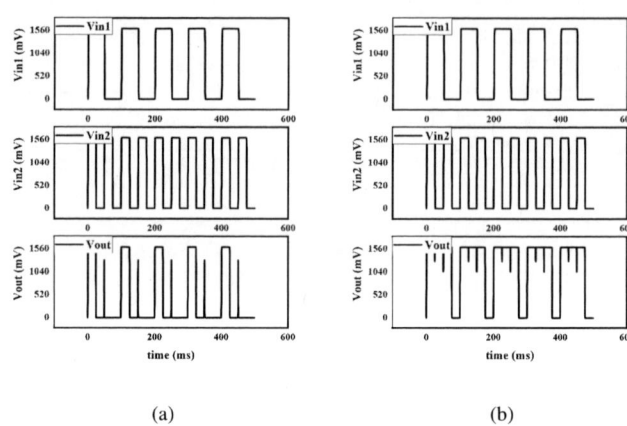

(a)   (b)

Fig. 9: The logic gate established by the memristor AND(a) and OR(b)

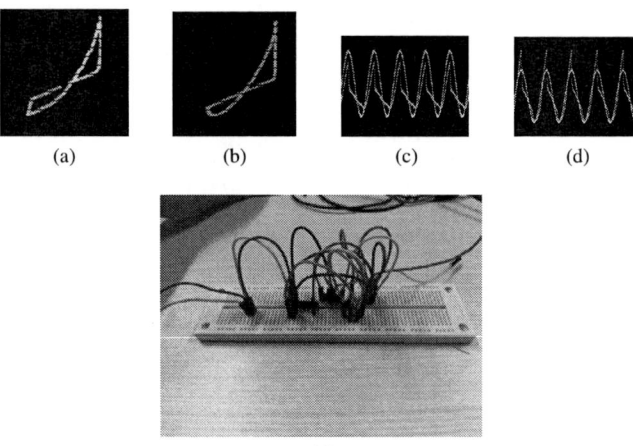

(a)   (b)   (c)   (d)

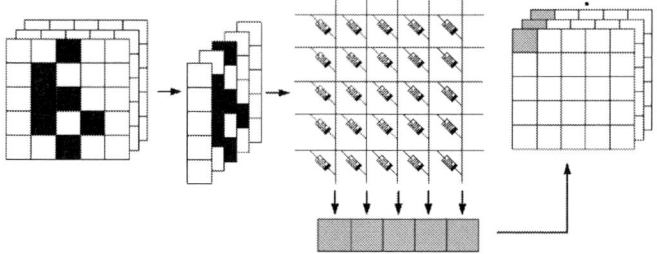

Fig. 10: A memristor crossbar model

Fig. 8: Physical circuit diagram(e), effect display in floating(a,c), effect display in gounded(b,d)

IV. MEMRISTOR CROSS ARRAY MODEL

Due to its unique properties, memristors have attracted the attention of many researchers. Ideally, the memristor has excellent linearity and controlled response and can write to any desired analog value. The conductivity of a memristor can be changed by adjusting the magnetic flux or the amount of charge transmitted, similar to how synapses work in the human brain. Memristors are used for in-memory computing and data storage, avoiding the separation of data storage and computing [31]. This breaks the von Neumann architecture and eliminates the energy consumption for data transmission. To study the neuromorphic computing ability of the proposed memristor emulator, based on its voltage-current relationship and conductance change characteristics, the neural nucleus and crossbar model of the memristor were constructed by CrossSim [30].

by the proposed memristor emulator circuit experiment and those obtained by Pspice simulation in Cadence Virtuoso. This is because the transmission gate composed of M8 and M9 is different from the traditional transmission gate, the gate of M9 is driven by one output of the differential pair composed of M1 and M2 after reverse, while the gate of M8 is driven by the voltage of C1 amplified by M7. Therefore, the pictures of the circuits experiment are not as ideal as those of the simulation. Nevertheless, the resistance variation and hysteresis curve characteristics of the memristor emulator can still be observed.

To confirm the application capability of the memristor emulator circuit, experiments were conducted using the proposed memristor emulator to build OR and AND gates. The circuit diagrams are shown in Fig 9a and 9b. The simulation results are presented in Figures 7c and 7d, indicating the logical functions of the gates constructed from memristors. Despite the presence of glitches or sharp spike pulses at transition points, the basic logical functions can be achieved.

Neural network training experiments were conducted on small hand-written digits, document types, and large hand-written digits datasets by establishing a crossbar model simulation matrix using the CrossSim model with the extracted data [36]. The experiments revealed that selecting a storage capacitance that is too small for the memristor emulator could cause the emulator to reach the resistance upper limit of the

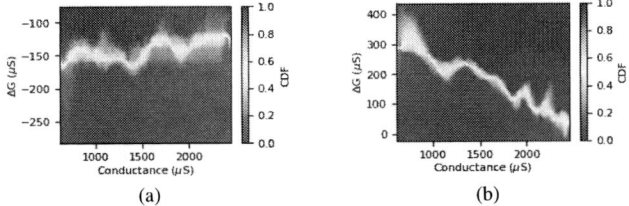

Fig. 11: The ΔG vs G response for (a) decreasing and (b) increasing pulses of memristor

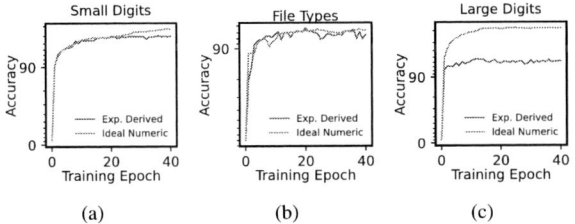

Fig. 12: Neural network training effect for small image data(a), file type data(b) and large image data(c)

memristor after just a few pulses, significantly affecting the accuracy of neural network training. Therefore, to address this issue, the memory capacitance of the memristor emulator was adjusted to 5pF during the data extraction phase of the memristor. Data on the resistance increase and decrease of the memristor can be obtained through pulse experiments, along with the conductivity changes of different memristors at different conductances as shown in the fig11. The figure illustrates the response of ΔG vs G for the memristor emulator as decreases and increases.

The results of neural network training are depicted in the fig 12. For the dataset of file types and small digits, the results closely overlap with the ideal data, maintaining over 94% accuracy after several training epochs. Despite achieving over 95% accuracy in several training epochs for the dataset of large handwritten digits, there still exists a significant discrepancy from the ideal scenario. These findings suggest that synapse neural networks based on Crossbar memory have an advantage in training on smseverallall datasets compared to large datasets.

V. CONCLUSION

Memristor emulator currently encounters several hurdles, including the complexity of circuitry, the inability to float or ground, a limited frequency spectrum, altered hysteresis loops, and a broad cutoff region. To overcome these obstacles, we introduce a memristor emulator leveraging CMOS technology. This design accomplishes the resistive switching behavior of memristors by manipulating the gate potentials of two transistors within the CMOS transmission gate. The signal processing subsystem derives the gate voltage required for transmission gate control through the analysis of voltages across its ter-

minals. Consequently, the memristor emulator's circuitry is parsimonious, its operational principles are readily deducible, and its parameters are tunable to accommodate diverse simulation mandates. Experimentation conducted within Cadence software illustrates the memristor's hysteresis curve attributes, floating and grounding capabilities, performance under varying temperatures and process variations, and non-volatility. Moreover, the functionality of the emulator is authenticated by the realization of logic gates. Additionally, the viability of the architecture is confirmed through the integration of commercial components on a prototyping breadboard. Finally, data extracted from the emulator are utilized to construct a crossbar model, upon which multiple datasets are trained. The simulation outcomes suggest that the memristor's crossbar architecture exhibits enhanced.

REFERENCES

[1] L. Chua, "Memristor-The missing circuit element," *IEEE Transactions on Circuit Theory*, vol. 18, no. 5, pp. 507–519, 1971, doi: 10.1109/TCT.1971.1083337.

[2] L. O. Chua and S. M. Kang, "Memristive devices and systems," *Proceedings of the IEEE*, vol. 64, no. 2, pp. 209–223, 1976, doi: 10.1109/PROC.1976.10092.

[3] L. Chua, "Resistance switching memories are memristors," *Applied Physics A*, vol. 102, no. 4, pp. 765–783, 2011, doi: 10.1007/s00339-011-6264-9.

[4] D. B. Strukov, G. S. Snider, D. R. Stewart, and R. S. Williams, "The missing memristor found," *Nature*, vol. 453, no. 7191, pp. 80–83, 2008, doi: 10.1038/nature06932.

[5] Z. Cao, B. Sun, G. Zhou, S. Mao, S. Zhu, J. Zhang, C. Ke, Y. Zhao, and J. Shao, "Memristor-based neural networks: a bridge from device to artificial intelligence," *Nanoscale Horizons*, 2023.

[6] S. Diware, A. Singh, A. Gebregiorgis, R. V. Joshi, S. Hamdioui, and R. Bishnoi, "Accurate and energy-efficient bit-slicing for RRAM-based neural networks," *IEEE Transactions on Emerging Topics in Computational Intelligence*, 2023.

[7] J.-Y. Chen, M.-C. Wu, and T. Ting, "Applications of p-n homojunction ZnO nanowires to one-diode one-memristor RRAM arrays," *Scripta Materialia*, vol. 187, no. 1, 2020.

[8] M. M. Goda, A. H. Hassan, H. Mostafa, and A. M. Soliman, "A novel refreshment circuit for 2T1M neuromorphic synapse," *Journal of Circuits, Systems and Computers*, 2021.

[9] C. Cheng, Y. Wang, L. Xu, K. Liu, and Y. Yang, "Artificial astrocyte memristor with recoverable linearity for neuromorphic computing," *Advanced Electronic Materials*, 2021.

[10] Z. Cao, B. Sun, G. Zhou, S. Mao, S. Zhu, J. Zhang, C. Ke, Y. Zhao, and J. Shao, "Memristor-based neural networks: a bridge from device to artificial intelligence," *Nanoscale Horizons*, 2023.

[11] S. Parveen, L. T. Manamel, A. Mukherjee, S. Sagar, and B. C. Das, "Analog memristor of lead-free Cs4CuSb2Cl12 layered double perovskite nanocrystals as solid-state electronic synapse for neuromorphic computing," *Advanced Materials Interfaces*, 2022.

[12] Z. Zhu, Y. Pei, C. Gao, H. Wang, and X. Yan, "Cu/HZO/GeS/Pt memristor for neuroinspired computing," *physica status solidi (RRL) - Rapid Research Letters*, 2021.

[13] D. Dev, A. Krishnaprasad, M. S. Shawkat, Z. He, and T. Roy, "2D MoS2 based threshold switching memristor for artificial neuron," *IEEE Electron Device Letters*, vol. PP, no. 99, pp. 1–1, 2020.

[14] S. Kvatinsky, E. G. Friedman, A. Kolodny, and U. C. Weiser, "TEAM: ThrEshold Adaptive Memristor Model," *IEEE Transactions on Circuits and Systems I: Regular Papers*, vol. 60, no. 1, pp. 211–221, 2013, doi: 10.1109/TCSI.2012.2215714.

[15] S. Kvatinsky, M. Ramadan, E. G. Friedman, and A. Kolodny, "VTEAM: A General Model for Voltage-Controlled Memristors," *IEEE Transactions on Circuits and Systems II: Express Briefs*, vol. 62, no. 8, pp. 786–790, 2015, doi: 10.1109/TCSII.2015.2433536.

[16] D. Biolek, V. Biolková, and Z. Biolek, "SPICE model of memristor with nonlinear dopant drift," *Radioengineering*, vol. 18, pp. 210–214, 2009.

979-8-3315-8850-2/25 $31.00 © 2025 IEEE

[17] H. Kim, M. P. Sah, C. Yang, S. Cho, and L. O. Chua, "Memristor emulator for memristor circuit applications," *IEEE Transactions on Circuits and Systems I: Regular Papers*, vol. 59, no. 10, pp. 2422–2431, 2012, doi: 10.1109/TCSI.2012.2188957.

[18] M. Ghosh, A. Singh, S. S. Borah, J. Vista, A. Ranjan, and S. Kumar, "MOSFET-based memristor for high-frequency signal processing," *IEEE Transactions on Electron Devices*, vol. 69, no. 5, pp. 2248–2255, 2022, doi: 10.1109/TED.2022.3160940.

[19] M. Ghosh, P. Mondal, S. S. Borah, and S. Kumar, "Resistorless memristor emulators: Floating and grounded using OTA and VDBA for high-frequency applications," *IEEE Transactions on Computer-Aided Design of Integrated Circuits and Systems*, vol. 42, no. 3, pp. 978–986, 2023, doi: 10.1109/TCAD.2022.3189837.

[20] I. Yeo, S.-G. Gi, G. Wang, and B.-G. Lee, "A hardware and energy-efficient online learning neural network with an RRAM cross-bar array and stochastic neurons," *IEEE Transactions on Industrial Electronics*, vol. 68, no. 11, pp. 11554–11564, 2021, doi: 10.1109/TIE.2020.3032867.

[21] M. Elhamdaoui, K. Mbarek, S. Ghedira, F. O. Rziga, and K. Besbes, "Synapse design based on memristor," in *2020 IEEE International Conference on Design & Test of Integrated Micro & Nano-Systems (DTS)*, 2020, pp. 1–5, doi: 10.1109/DTS48731.2020.9196061.

[22] N. Yadav, S. K. Rai, and R. Pandey, "New grounded and floating memristor emulators using OTA and CDBA," *International Journal of Circuit Theory and Applications*, 2020, doi: 10.1002/cta.2774.

[23] A. Yesil, "A new grounded memristor emulator based on MOSFET-C," *AEU - International Journal of Electronics and Communications*, vol. 91, pp. 143–149, 2018, doi: 10.1016/j.aeue.2018.05.004.

[24] Y. Pu and B. Yu, "A large dynamic range floating memristor emulator with equal port current restriction," *IEEE/CAA Journal of Automatica Sinica*, vol. 7, no. 1, pp. 237–243, 2020, doi: 10.1109/JAS.2019.1911849.

[25] R. Mutlu, T. Üstünel, and E. Karakulak, "Memristor emulators with symmetric and asymmetric threshold voltages," in *2018 2nd International Symposium on Multidisciplinary Studies and Innovative Technologies (ISMSIT)*, 2018, pp. 1–5, doi: 10.1109/ISMSIT.2018.8567257.

[26] C. de Benito, O. Camps, M. M. Al Chawa, S. G. Stavrinides, and R. Picos, "A stochastic switched capacitor memristor emulator," in *2021 10th International Conference on Modern Circuits and Systems Technologies (MOCAST)*, 2021, pp. 1–4, doi: 10.1109/MOCAST52088.2021.9493391.

[27] J. Kalomiros, S. G. Stavrinides, and F. Corinto, "A two-transistor non-ideal memristor emulator," in *2016 5th International Conference on Modern Circuits and Systems Technologies (MOCAST)*, 2016, pp. 1–4, doi: 10.1109/MOCAST.2016.7495164.

[28] C. Sánchez-López, M. A. Carrasco-Aguilar, and C. Muñiz-Montero, "A 16Hz–160kHz memristor emulator circuit," *AEU - International Journal of Electronics and Communications*, vol. 69, no. 9, pp. 1208–1219, 2015, doi: 10.1016/j.aeue.2015.05.003.

[29] C. Sánchez-López, J. Mendoza-López, M. A. Carrasco-Aguilar, and C. Muñiz-Montero, "A floating analog memristor emulator circuit," *IEEE Transactions on Circuits and Systems II: Express Briefs*, vol. 61, no. 5, pp. 309–313, 2014, doi: 10.1109/TCSII.2014.2312806.

[30] B. Sun, Y. Chen, and G. Zhou, "Memristors-based artificial chips," *ACS Nano*, vol. 18, no. 1, pp. 14–27, 2024, doi: 10.1021/acsnano.3c07384.

[31] B. Sun, Y. Chen, and G. Zhou, "Memristors-based artificial chips," *ACS Nano*, vol. 18, no. 1, pp. 14–27, 2024, doi: 10.1021/acsnano.3c07384.

[32] M. Hu and C. Graves, "Memristor-based analog computation and neural network classification with a dot product engine," *Advanced Materials*, vol. 30, no. 9, 2018, doi: 10.1002/adma.201705914.

[33] H.-S. P. Wong and S. Salahuddin, "Memory leads the way to better computing," *Advanced Materials*, vol. 10, pp. 191–194, 2015, doi: 10.1002/adma.201705914.

[34] X.-Y. Wen, Y.-S. Wang, and Y.-H. He, "Memristive brain-like computing," *Acta Physica Sinica*, vol. 71, no. 14, 2022, doi: 10.7498/aps.71.20220666.

[35] Y. Zhang, G.-Q. Mao, and X. Zhao, "Evolution of the conductive filament system in HfO2-based memristors observed by direct atomic-scale imaging," *Nature Communications*, vol. 12, no. 7232, 2021, doi: 10.1038/s41467-021-27575-z.

[36] T. P. Xiao, C. H. Bennett, B. Feinberg, M. J. Warinella, and S. Agarwal, "CrossSim: Accuracy simulation of analog in-memory computing," [Online]. Available: https://github.com/sandialabs/cross-sim.

A Novel SiC LDMOS with Electron Accumulation Layer Featuring Ultra-low Specific On-resistance and Improved Breakdown Voltage

Moufu Kong*
State Key Laboratory of
Electronic Thin Films and
Integrated Devices of China
University of Electronic Science
and Technology of China
Chengdu, China
kmf@uestc.edu.cn

Lin Tong
State Key Laboratory of
Electronic Thin Films and
Integrated Devices of China
University of Electronic Science
and Technology of China
Chengdu, China

Qizhi Feng
State Key Laboratory of
Electronic Thin Films and
Integrated Devices of China
University of Electronic Science
and Technology of China
Chengdu, China

Jiaru Jia
State Key Laboratory of
Electronic Thin Films and
Integrated Devices of China
University of Electronic Science
and Technology of China
Chengdu, China

Zhaoyu Ai
State Key Laboratory of
Electronic Thin Films and
Integrated Devices of China
University of Electronic Science
and Technology of China
Chengdu, China

Hongfei Deng
State Key Laboratory of
Electronic Thin Films and
Integrated Devices of China
University of Electronic Science
and Technology of China
Chengdu, China

Yangyang Ma
State Key Laboratory of
Electronic Thin Films and
Integrated Devices of China
University of Electronic Science
and Technology of China
Chengdu, China

Peifei Wu
Beijing Huairou Laboratory
Beijing, China

Abstract—In this work, a novel 4H-SiC lateral double diffused metal-oxide-semiconductor field-effect transistor with an electron accumulation layer (EAL-LDMOS) is proposed. In the proposed EAL-LDMOS, a p-type self-biased split-gate (SG) structure is introduced, which not only improves the N-drift region doping concentration and optimizes the drift region electric field distribution, but also induce an electron accumulation layer at the N-drift region surface during on-state. The improved N-drift region doping concentration combined with electron accumulation layer significantly reduces the specific on-resistance ($R_{on,sp}$) of the device in the on-state. The simulation results show that the breakdown voltage (BV) of the proposed device is improved about 21.5% compared with the conventional SiC LDMOS. The $R_{on,sp}$ of the proposed device is 2.21 mΩ·cm², demonstrating a 70% reduction compared with the conventional device. Also, a feasible simplified fabrication process flow is provided for the proposed device.

Keywords—4H-SiC, LDMOS, breakdown voltage, specific on-resistance

I. INTRODUCTION

Silicon carbide (SiC), one of the most important wide bandgap semiconductors, holds tremendous potential in advanced power devices due to its excellent thermal performance and electric characteristics. Compared with traditional silicon (Si) materials, SiC is a much better material to make devices with superior performance in high power, high temperature and high frequency [1]-[3]. In recent years, SiC power MOSFET technology has advanced rapidly, with

This work was supported in part by the Central Guiding Local Science and Technology Development Special Project of Sichuan (2024ZYD0310).

* Corresponding authors: Moufu Kong and Peifei Wu, kmf@uestc.edu.cn & wupeifei2@163.com.

substantial research focused on designing and fabricating vertical double diffused metal-oxide-semiconductor field-effect transistor (VDMOS) structures [4]-[7]. For power VDMOS, the current path is perpendicular to the wafer surface, enabling high current density without area penalty. Consequently, VDMOS typically achieves superior specific on-resistance in current SiC DMOS research [8],[9]. In contrast to VDMOS, SiC lateral double diffused metal-oxide-semiconductor field-effect transistor's (LDMOS) current path is paralleled to the wafer surface, which takes up longer drift regions for higher breakdown voltage ratings, increasing device area. This geometry causes rapid degradation in specific on-resistance.

However, the SiC LDMOS device offers unique integration advantages. All electrodes of LDMOS device reside on a single surface, enabling monolithic integration with drive/protection circuits [10],[11]. This positions LDMOS as a promising technology for SiC power IC applications. To address the high specific on-resistance challenge in SiC LDMOS, this work introduces a novel 4H-SiC LDMOS structure incorporating a p-type GaN split-gate (SG) to optimize surface electric field distribution. Meanwhile, the self-biased p-type GaN SG generates an electron accumulation layer (EAL) at the drift region surface to significantly reduce drift region resistance during conduction while maintaining the existing gate drive requirements. It should be noted that, in the proposed device, the P-GaN layer can be achieved by epitaxial growth at low temperatures [12],[13].

II. DEVICE STRUCTURE AND MECHANISM

Fig. 1 (a) and Fig. 1(b) show the device structure of the conventional 4H-SiC LDMOS and the proposed 4H-SiC EAL-

LDMOS, respectively. In conventional devices, gate field plates and drain field plates are employed to optimize the surface electric field near the gate and drain regions during voltage blocking. In the proposed device, a stepped P-type GaN (P-GaN) split-gate (SG) structure is implemented. The introduction of the P-GaN split-gate simultaneously optimizes both the breakdown voltage and specific on-resistance of the device. When the device operates under blocking conditions, the P-GaN region and the underlying N-drift region undergo mutual depletion, optimizing the surface electric field distribution. Owing to the electric field modulation enabled by the P-GaN layer, the doping concentration in the drift region can be significantly increased, thereby reducing the on-resistance of the drift region. In addition, when the device operates in the on-state, the gate is positively biased. The diode D_1 turns on, causing the P-GaN region to reach a positive potential near that of the gate electrode. This positive potential induces the formation of an electron accumulation layer at the drift region surface. High-concentration electrons at the drift region surface serve as conduction carriers, further reducing the on-resistance during the on-state. Fig. 1 also shows the main parameters of the two devices.

The introduction of the SG necessitates special consideration of the switching characteristics in the proposed device. During turn-on or turn-off transitions, the internal charge of the SG undergoes abrupt variations, generating substantial drive current demands. This imposes stringent requirements on the driving circuitry. To address this challenge, the diode D_1 and capacitor C were adopted to establish a charge-storage mechanism, thereby improving switching performance. As charges evacuate from the SG, the blocking action of diode D_1 redirects the charge flow into capacitor C. This mechanism significantly reduces drive current during the turn-off process. During turn-on, stored charges from capacitor C replenish the SG. Electrode G only

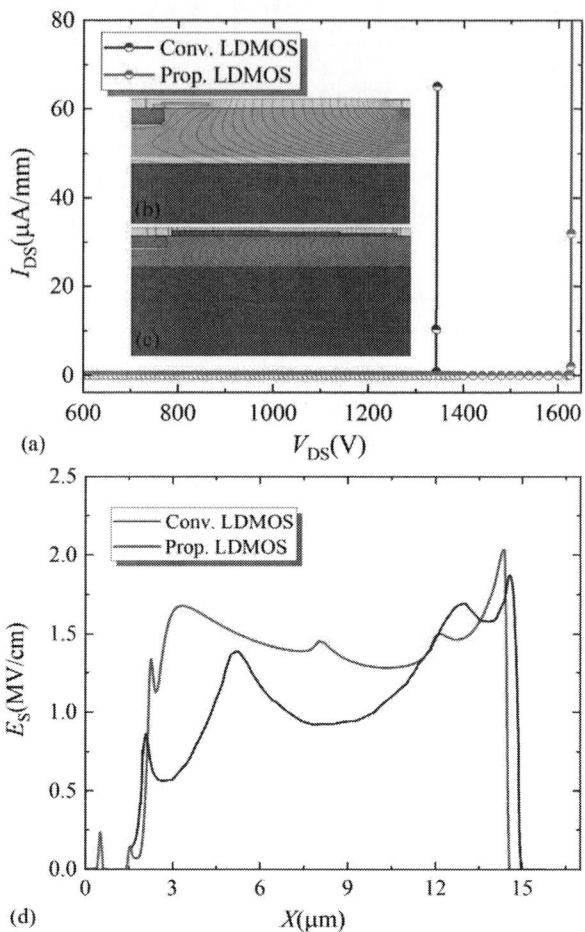

Fig. 2. (a) The breakdown I-V curves of the two devices; equipotential distributions (40V/line) of (b) the conventional LDMOS and (c) the proposed LDMOS; (d) electric field on the semiconductor surface of the two devices.

supplies leakage compensation current through diode D_1. This approach dramatically lowers drive current requirements during turn-on process.

III. SIMULATION RESULTS AND DISCUSSION

Based on the device parameters in Fig. 1, TCAD device numerical simulations are conducted to verify the performances of the two 4H-SiC LDMOS devices.

Fig. 2(a) compares the breakdown characteristics of the conventional and proposed devices, demonstrating a breakdown voltage of 1340 V for the conventional structure versus 1620 V for the proposed device – representing a 21% enhancement. Fig. 2(b) and Fig. 2(c) depict the equipotential contour distributions (40V/line) at breakdown state for both devices. Analysis of these contours reveals that the P-GaN SG optimizes the surface electric field distribution, resulting in more uniform equipotential line spacing across the device surface. This improved electric field management directly contributes to the higher breakdown voltage. Fig. 2(d)

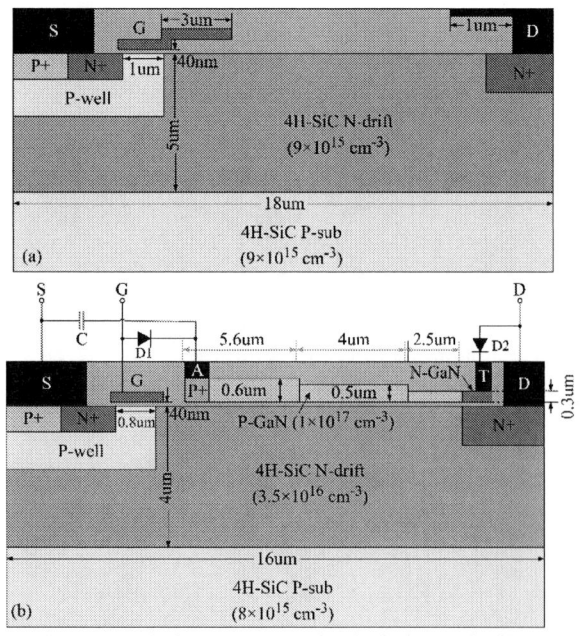

Fig. 1. Structure of the (a) conventional device and (b) proposed device.

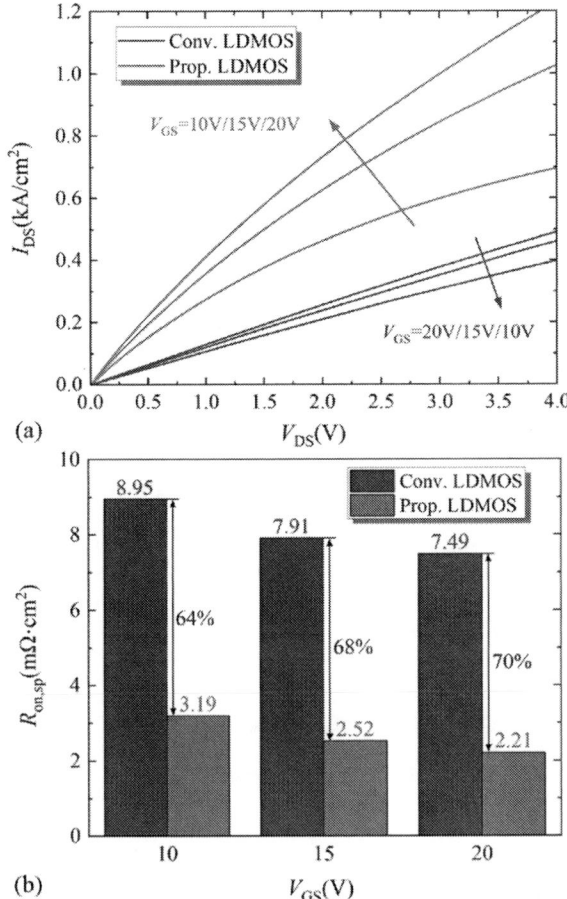

Fig. 3. (a) The I-V curves and (b) $R_{on,sp}$ results of the two devices.

demonstrates that enhanced electric field intensity near the source region is introduced. More uniform electric field distribution along the drift region surface presents the surface electric field profiles when breakdown.

Fig. 3(a) presents the I-V characteristics of both conventional and proposed devices at gate voltages of V_{GS} = 10 V, 15 V and 20 V. At identical gate biases and current levels, the proposed device exhibits significantly lower voltage drop compared to the conventional structure. This result demonstrates the superior specific on-resistance performance of the proposed design. Fig. 3(b) quantifies the specific on-resistance derived from the I-V curves across the same gate voltages. The proposed device achieves substantial reductions in specific on-resistance of 64% at V_{GS} = 10 V, 68% at V_{GS} = 15 V and 70% at V_{GS} = 20 V.

Fig. 4 (a) displays the electron concentration distribution within the device at V_{GS} = 20 V. A high-concentration electron accumulation layer is observed at N-drift region surface, where electron concentration exceeds the doping of N-drift region by $2\sim3$ orders of magnitude. Fig. 4(b) presents the electron concentration profile along the vertical direction perpendicular to the surface. The concentration reaches $\sim10^{20}$ cm^{-3} at the interface, with an accumulation layer thickness of approximately 20 nm. Simulation results reveal that the

electron mobility is approximately 30 cm²/(V·s) near the surface and increases gradually along the bulk direction, reaching about 200 cm²/(V·s) at the boundary of the electron accumulation layer. Fig. 4(c) illustrates the conduction current density distribution at V_{GS} = 20 V and V_{DS} = 0.5 V. The current density in the accumulation layer is $2\sim3$ orders of magnitude higher than in the bulk drift region. This confirms that electrons in the accumulation layer serve as the primary current carriers during conduction, thereby significantly reducing the on-resistance of the drift region.

Fig. 5(a) and (b) compare the switching characteristics of the conventional and proposed devices at R_G = 5Ω. The turn-on time (t_{on}) of the proposed LDMOS is 5ns and the turn-off time (t_{off}) is 13ns meanwhile t_{on} and t_{off} of the conventional LDMOS is 5ns and 6ns. While the electron accumulation layer significantly reduces on-resistance during conduction, stored electrons in the accumulation layer evacuate through the drain

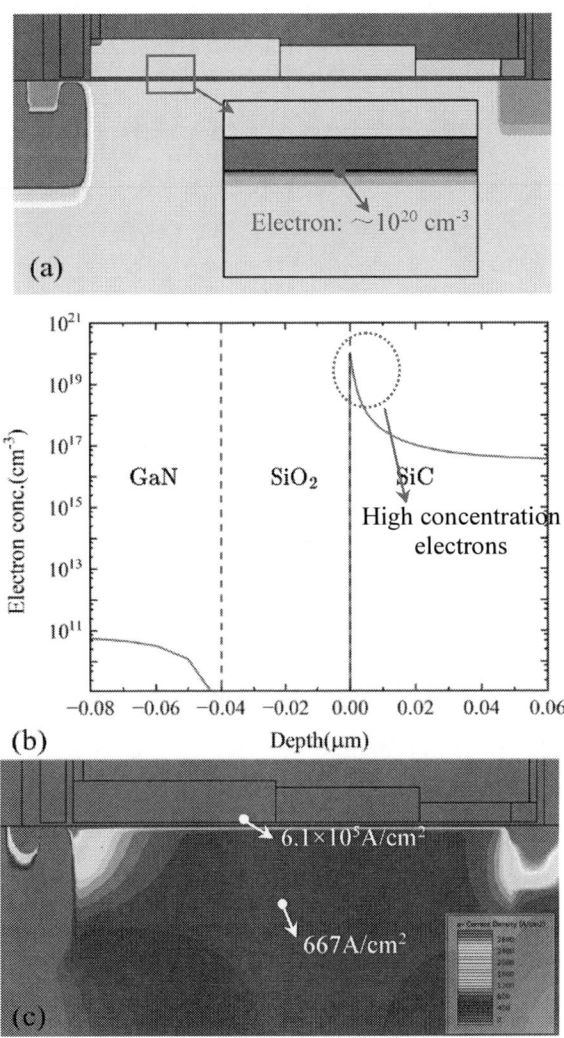

Fig. 4. (a) The electron concentration of proposed device (V_{GS}=20V); (b) the distribution of electron concentration in the direction perpendicular to the interface; (c) current density of proposed device (V_{DS}=0.5V).

979-8-3315-8850-2/25 $31.00 © 2025 IEEE 319

terminal. This additional discharge current contributes to the drain current tail. Consequently, the current decay rate decreases, extending turn-off time.

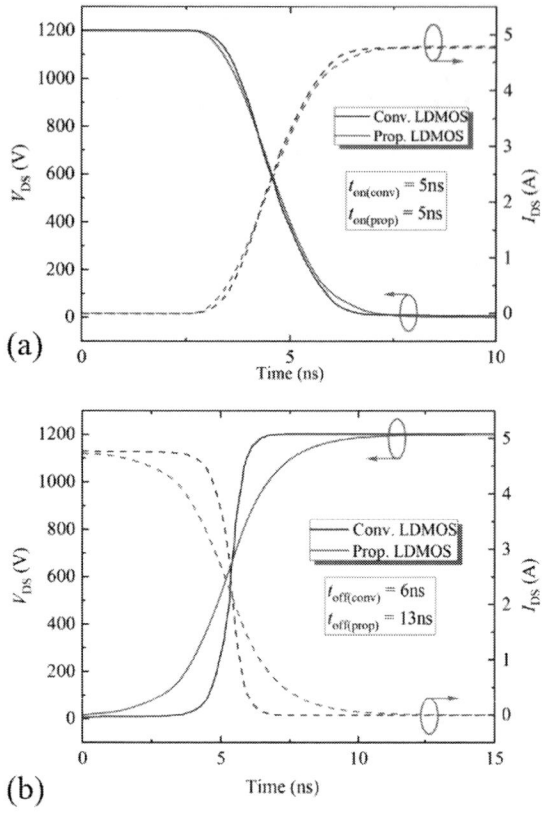

(a)

(b)

Fig. 5. (a) Turing-on and (b) turing-off performances of the conventional device and the proposed device ($R_G = 5\Omega$).

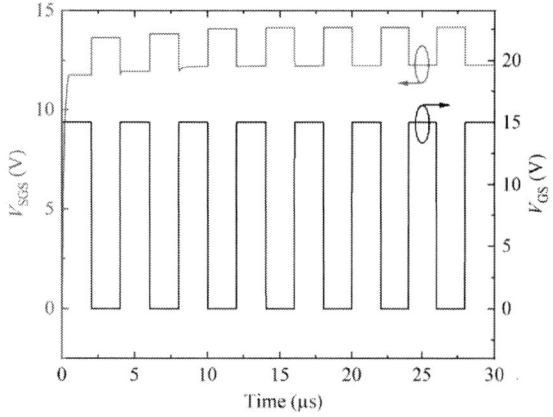

Fig. 6. Generated self-biased voltage V_{SGS} for the SG of the proposed structure with a capacitor of 0.032pF/μm.

Fig. 6 illustrates the voltage evolution of SG during continuous switching cycles. At initial state ($t = 0us$), no charge is contained by capacitor C, thereby $V_{SGS} = 0V$. During turn-on transition, capacitor C charges through gate terminal via diode D_1. Voltage across capacitor C and potential at electrode A are increasing. During turn-off transition, gate

voltage drops to 0 V and charge transfers from SG to capacitor C. This V_{SGS} consequently increases. The stabilized capacitor voltage after multiple cycles (t > 15us) confirms establishment of self-biased charge recirculation.

IV. BRIEF FABRICATION PROCESS

Fig. 7 shows a brief fabrication process flow of the proposed EAL-LDMOS, in which a possible approach to manufacture the proposed SiC EAL-LDMOS is presented. The fabrication process starts with a p-type SiC substrate, and the epitaxy is performed to from N-drift layer (Fig. 7(a)). The P-well region and P+, N+ regions are formed through multi-step ion implantations (Fig. 7(b)). Next, the gate oxide is formed by thermal oxidation. Polysilicon is also deposited by chemical vapor deposition (CVD) to form polysilicon layer. Then the polysilicon is etched to form gate (Fig. 7(c)). After that, field oxide is deposited by CVD (Fig. 7(d)) and field oxide is etched to form SG trench. The P-GaN layer was epitaxially grown via molecular beam epitaxy at a low temperature (<800°C), completely filling the SG trench (Fig. 7(e)) [12],[13]. Ion implantation is executed to form N+ GaN region (Fig. 7(f)). Followed, the GaN layer is etched to form stepped SG. Then oxide is deposited (Fig. 7(g)). Finally, oxide is etched for electrode and metal deposition and annealing to form ohmic contacts (Fig. 7(h)).

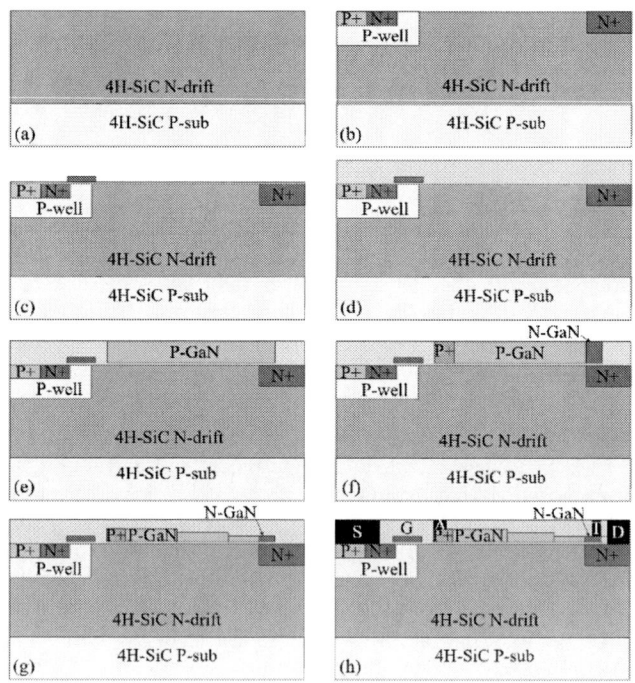

Fig. 7. Brief fabrication process flow: (a) process starts with a SiC substrate, epitaxy to form N-drift layer, (b) P-well region and P+, N+ regions formation by multi-step ion implantation, (c) gate oxide formation by thermal oxidation and gate polysilicon deposition, (d) field oxide deposition by CVD, (e) SG trench etching and P-GaN epitaxy by molecular beam epitaxy [12],[13], (f) ion implantation to form N+ region, (g) GaN etching to form stepped SG and oxide deposition, (h) etching for electrode and metal deposition and annealing to form ohmic contacts.

REFERENCES

[1] M. Buffolo, D. Favero, A.Marcuzzi, C. De Santi, G. Meneghesso and E. Zanoni, "Review and Outlook on GaN and SiC Power Devices: Industrial State-of-the-Art, Applications, and Perspectives," in IEEE Transactions on Electron Devices, vol. 71, no. 3, pp. 1344-1355, March 2024.

[2] Z. Chen and A. Q. Huang, "Extreme High Efficiency Enabled by Silicon Carbide (SiC) Power Devices", Materials Science in Semiconductor Processing, vol. 172, 2024, 108052.

[3] C. Langpoklakpam, A. C. Liu, K. H. Chu, L. H. Hsu, W. C. Lee, "Review of Silicon Carbide Processing for Power MOSFET", Crystals, 2022, vol.12, No.2, pp.245.

[4] L. Wang, P. Tang, and J. Nan, "Single-Event Burnout Mechanism and Hardening for 1200V 4H-SiC LDMOS," Microelectronics Reliability, vol. 169, p. 115712, Jun. 2025.

[5] Y. Tang, G. Liu, J. Gao, and Z. Wang, "Simulation and Optimization of 4H-SiC LDMOS Devices with Field Limiting Ring Structure," in 2024 3rd International Symposium on Semiconductor and Electronic Technology (ISSET), Aug. 2024, pp. 54–57.

[6] Z. Hu, "Review of the SiC LDMOS Power Device," J. Semicond., vol. 45, no. 8, p. 081501, Aug. 2024.

[7] Y. Gu and C. Pan, "A 1400V SiC LDMOS with P-tops and P-buffer for Ultra-low Specific Resistance," in 2022 IEEE 16th International Conference on Solid-State & Integrated Circuit Technology (ICSICT), Oct. 2022, pp. 1–3.

[8] X. Cai, Y. Chen, D. Liu and W. Qian, "The Optimization of the Specific On-Resistance of the VDMOS on the Integrated Platform of VDMOS and LDMOS," 2024 Conference of Science and Technology for Integrated Circuits (CSTIC), Shanghai, China, 2024, pp. 1-3.

[9] R. Ma, R. Wang, H. Fang, "A Novel Deep-Trench Super-Junction SiC MOSFET with Improved Specific On-Resistance," Micromachines, vol. 15, no. 8, p.684, 2024.

[10] S. Fayaz, N. Hakim and G.M. Rather, "Applicability of Channel Doping Gradient in the Design of a Short Channel (0.1 μm) LDMOS Transistor for Integrated Power and RF Applications," in Trans. Electr. Electron. Mater, vol. 25, pp. 479–493, 2024.

[11] Y. Mehlman, S. Levin, D. Mistele, N. Hayek, N. Nierenberg and S. Shapira, "Performance Enhancements of Low-Voltage LDMOS Power Switches by In-chip Integration of a Microcrystalline Diamond Substrate," 2022 IEEE 23rd Workshop on Control and Modeling for Power Electronics (COMPEL), Tel Aviv, Israel, 2022, pp. 1-5.

[12] A.M. Jeffries, L. Ding, J.J. Williams, T.L. Williamson, M.A. Hoffbauer, C.B. Honsberg, M.I. Bertoni, "Gallium Nitride Grown by Molecular Beam Epitaxy at Low Temperatures," in Thin Solid Films, vol. 642, pp. 25-30, Nov. 2017.

[13] J. Yu, Z. Zhang, Y. Luo, J. Wang, L. Wang, X. Li, Z. Hao, C. Sun, "Thin Film Transistors and Metal–Semiconductor–Metal Photodetectors Based on GaN Thin Films Grown by Inductively Coupled Plasma Metal-Organic Chemical Vapor Deposition." Journal of Physics D: Applied Physics, vol.55, 2022, pp.354002.

2025 The 10th International Conference on Integrated Circuits and Microsystems

Hardware Implementation and Side-Channel Security Analysis for High-Precision AI Tansformer Encoder

Wentao Wang
School of Electronic Engineering
Heilongjiang University
Harbin, China
645537443@qq.com

Liji Wu*
School of Integrated Circuit
Tsinghua *University*
Beijing, China
lijiwu@tsinghua.edu.cn

Jing Hu*
School of Electronic Engineering
Heilongjiang University
Harbin, China
hjlyh@126.com

Le Wu
School of Integrated Circuit
Tsinghua University
Beijing, China
wul22@mails.tsinghua.edu.cn

Xiangmin Zhang
School of Integrated Circuit
Tsinghua University
Beijing, China
zhxm@tsinghua.edu.cn

Abstract—**This paper provides a high-precision transformer encoder hardware by using fixed-point numbers with variable decimal places. The original power consumption trajectory of the feedforward neural network module was collected. It was found that there was an obvious power consumption peak in the output of this module. Therefore, we implemented the output obfuscation technology for this module and collected the protected power consumption trajectory. The verification is based on an FPGA development board xc7k160tfbg676-1 to design and verify the transformer encoder with a token of 6 and a word embedding dimension of 512. The verification results show that the usage of digital signal processing (DSP) is 552, and logic Look-Up Tables (LUT) is 96,0 3. In this experiment, the model s data is quantized to 16-bit signed fixed-point numbers, with the number of decimal places varying according to different modules. A compensation coefficient is added to the matrix multiplication and a module design method of divide-and-add is adopted. The average error of the hardware implementation compared to software calculation is 0.000 67, and the execution time at 200 MHz is .6 37 ms, which is 1.657 times faster, saving approximately 39.7% of the execution time.**

Keywords— **rans or er, ncoder, ti ead Attention, Side Channe Ana ysis SCA**

I. INTRODUCTION

With the rapid iteration of Artificial General Intelligence, the Transformer encoder has become a common cornerstone in the field of natural language processing due to its parallel processing architecture. For example, the underlying architecture of multiple versions of DeepSeek is still based on the Transformer encoder. The Transformer model was proposed by Vaswani et al. in 2017 [1], and its structure is shown in Fig. 1. Its core innovation lies in transforming the time-dependent computation of traditional RNN into a parallelized architecture based on self-attention, which achieves parallel processing of sequential tasks through the stacking of multiple layers of encoders. This design enables the

model to have the ability to extract deep semantic features for efficient machine translation, text generation, achieving a qualitative leap in computational efficiency and expression ability [2].

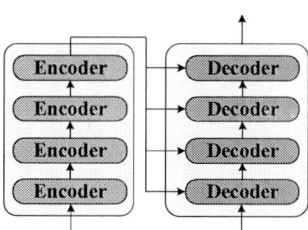

Fig. 1. Transformer Architecture

Although the Transformer encoder can significantly enhance the semantic capture ability of the model, it also requires a relatively high computational bit width. Its high computational complexity and resource requirements greatly limit the deployment work [3]. To break this bottleneck, researchers began to delve into its implementation at the hardware level, especially implementing the transformer encoder on FPGA. The FPGA can provide abundant programmable resources, and flexibly allocating these resources can achieve efficient mapping of the model. It is also a major challenge to implement high-precision encoders with a relatively low resource demand on FPGA. Meanwhile, there are a series of security issues in accelerating large models in hardware [4]. Among them is SCA, which exploits unexpected information leakage, such as power consumption patterns, to infer sensitive data processed by hardware accelerators. Existing studies have shown that SCA can carry out extraction attacks on artificial intelligence models, enabling external attackers to accurately restore the architecture and hyperparameters of the models, which seriously endangers the privacy and intellectual property rights of artificial intelligence models.

979-8-3315-8850-2/25 $31.00 © 2025 IEEE

In the paper, the transformer encoder hardware with high precision is implemented using a relatively low amount of resources, and four parallel single-pulsating arrays are used to accelerate the structure. Meanwhile, the power leakage characteristics of the original side channel of the feedforward neural network module were measured and analyzed. We implemented the protection measures of obfuscated output for the feedforward neural network and captured its power consumption waveform with counter measures.

II. THEORETICAL BASIS

A. Transformer Encoder Theoretical

The Transformer encoder is an architecture based on the self-attention mechanism, designed to handle sequence data and represent the relationships between contexts, as shown in Fig. 2. The sequence data received by the encoder is a sequence with semantic relationships and position information that has undergone word embedding and position embedding. First, the input sequence data is transformed into Query (Q), Key (K), Value (V) vectors through a linear layer of weights with multiple layers of parallelism. Then, it enters the dot product attention module to calculate the dot product of Q and K to obtain the attention score, which represents the degree of correlation between different positions. These scores are passed through the Softmax function to obtain the corresponding probability distribution matrix of the degree of correlation [5]. Finally, the important information of each position is weighted and summed over the V vector, thereby obtaining the final weighted output as shown in (1).

$$Attention(Q, K, V) = Softmax\left(\frac{Q \times K^T}{\sqrt{d_k}}\right)V \qquad (1)$$

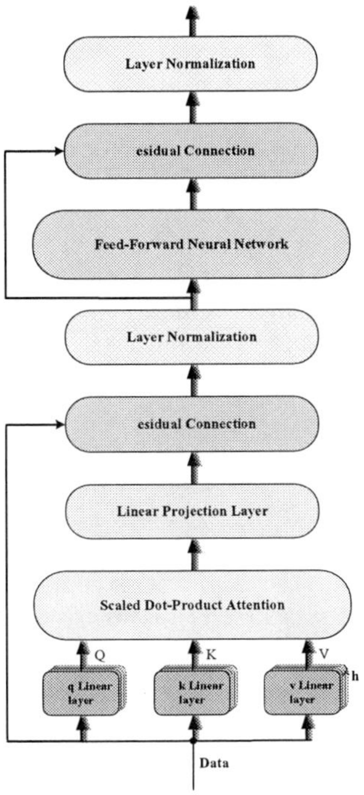

Fig. 2. Transformer Encoder Architecture

The multi-head self-attention mechanism allows multiple different attention heads to compute in parallel. Each attention head can focus on the relationships between different positions in the input sequence, thereby capturing richer features. The calculated multi-head weighted output is concatenated into a longer vector in the column direction, and then the concatenated vector is projected back to the original dimension through the linear projection layer to obtain the final multi-head self-attention output. This step can make the output of multi-head self-attention consistent with the input dimension of subsequent processing layers (such as feedforward neural networks). The spliced data can directly skip certain layers through residual connection and pass the input data directly to the subsequent layers. This enables the network to fit the mapping more easily, reducing the loss of information during transmission while enhancing the expressive ability of the model. The normalization layer can reduce the sensitivity of the model to the distribution of input data and improve the generalization ability of the model. This enables the model to make predictions more stably when facing different input data and improves the robustness of the model. In the multi-head self-attention mechanism of the Transformer, the input vectors still maintain a linear relationship after undergoing linear transformations (such as projections of Query, Key, and Value). Linear transformation can only capture the linear combination between input features and is unable to effectively handle complex nonlinear relationships. The feedforward neural network introduces nonlinear characteristics to the model by using the nonlinear activation function Rectified Linear Unit (ReLU). This nonlinear transformation enables the model to learn the complex relationships among the input features, thereby extracting higher-level features and better capturing the patterns and features in the data. The feedforward neural network is mainly composed of two linear transformation layers, namely the linear transformation from the input layer to the hidden layer and the linear transformation from the hidden layer to the output layer. Between the two linear transformation layers, there will be a ReLU activation function layer. The activation function makes the output of the feedforward neural network sparse, with only some neurons in an active state. Sparse output helps to reduce computational redundancy, improve the computational efficiency of the model, and at the same time, it can alleviate the overfitting phenomenon to a certain extent [6]. The calculation of the activation function and the feedforward neural network is shown in (2) and (3).

$$RELU(X) = \begin{cases} X & X > 0 \\ 0 & X \leq 0 \end{cases} \qquad (2)$$

$$FFNN(\text{x}) = ReLU(x \times w_1 + b_1) \times w_2 + b_2 \qquad (3)$$

Feedforward neural networks will perform nonlinear transformations on the features processed by the self-attention mechanism, which may introduce new changes in numerical distribution. Therefore, using residual connections again to add the input of the feedforward neural network to the output of the feedforward neural network, will ensure the integrity of the input information. Then, the result after addition is normalized again, which can make the output numerical range more stable and help the learning and generalization ability of the model.

B. Side Channel Theoretical

SCA is an attack method targeting the leakage of information through non-primary communication channels during the operation of system hardware. These leaked information include changes in physical quantities such as power consumption, electromagnetic radiation, sound, temperature and time [7]. Attackers can infer the sensitive data in the system by analyzing the direct or indirect relationship between this information and the operation of the algorithm. In large-scale machine learning models, attackers may use this method to steal important information such as already trained weights or input data.

The side-channel analysis of the encoder in the paper focuses on the power consumption information. The power consumption of the device varies when it performs different computing operations [8]. Hamming weight and Hamming distance are two commonly used models in power SCA. Hamming weight refers to the number of ones in a binary number. In the power consumption side-channel analysis, when the processor performs hierarchical operations, the greater the number of ones in the operated data, the greater the power consumption. Hamming distance refers to the number of different bits at the corresponding positions of two binary strings of equal length. In the power consumption side-channel analysis, when the processor transitions from one state to another, the more different bit numbers there are between states, the greater the power consumption change. Attackers can reveal the sensitive data in machine learning models through the joint model of Hamming weights and Hamming distances.

III. HARDWARE DESIGN

A. Encoder Hardware Design

We use the parallel pipeline design method for the Transformer encoder hardware to accelerate it. There are a large number of matrix multiplication calculation operations in the overall structure. Therefore, we use a single-pulsating array with four dimensions of 4 by 4 to perform parallel calculations on the matrix multiplication module to achieve the purpose of accelerating matrix multiplication. And the time-division multiplexing technology is used to repeatedly call the same module at different times to save hardware resources. The output shift technology is also used for the conversion of decimal places in the module, which can further save resource consumption. In terms of calculation accuracy, we first designed the exponential generation module and the square root module of e with high precision, and analyzed in detail the overflow problem of calculating the intermediate value of each module. Under the premise of avoiding the overflow of upper and lower values, the number of decimal places was maximized. Then, after the column addition calculation was completed, a compensation coefficient was added. Multiplying the number of matrix columns by the quantized fractional value at that time can significantly reduce the error, as shown in (4).

$$x = \frac{Col_{num}}{2^{Fractional\ bits}} \quad (4)$$

Finally, in the multi-head attention, for the mean and variance calculation module, the calculation method of division first and then addition is used, which can keep the number of decimal places within 13 when calculating large matrices, maximizing the calculation accuracy of the encoder.

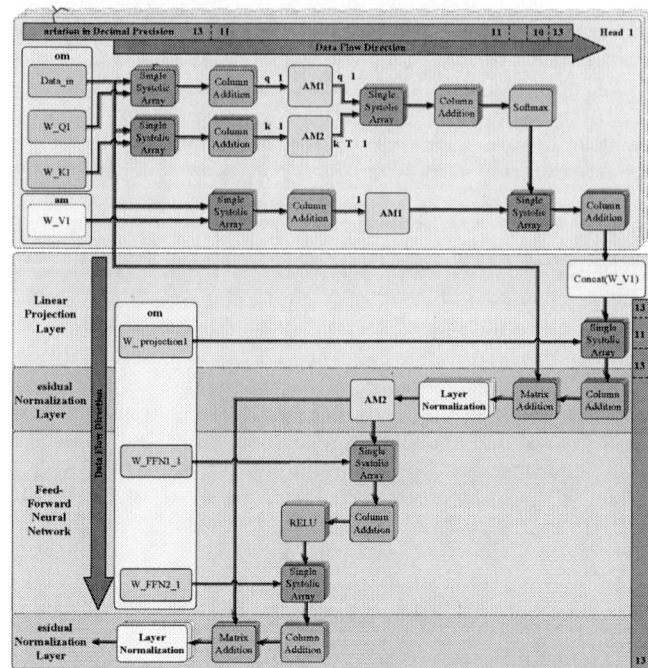

Fig. 3. Transformer Encoder Hardware Design

The hardware architecture of the encoder is shown in Fig. 3. Following the direction of the data flow, the data passes through the multi-head attention layer, linear projection layer, residual and normalization layer, feed-forward neural network layer, and then re-enters the residual and normalization layer. The input data is the data after word embedding and positional embedding, with a value range of -2 to 2. The weights of the Q, K, and V vectors range from -1 to 1. Therefore, the initial number of fractional bits is chosen to be 13. Then the Q and K vector generation modules are calculated using four parallel systolic arrays. The resulting 64 data values are stored in the register module of the column adder. The registers are re-used to add and store the data until a complete row of data has been calculated and the register outputs the data. Because continuous addition of large matrix column data can lead to integer overflow, a division-before-addition calculation method is used. The division required before entering the Softmax function is decomposed, as shown in (5).

$$Attention(Q, K, V) = Softmax\left[2\left(\frac{Q}{\sqrt{2d_k}}\right)\left(\frac{K^T}{\sqrt{2d_k}}\right)\right]V \quad (5)$$

This computing approach maintains an 11_bit fractional part. Once a group of Q and V vector data is computed, they are stored in Ram1 and Ram2, respectively. Ram1 stores data in row order, while Ram2 uses column order for k vector transposition without extra resources. During storage, time division multiplexing calculates the v vector. When storing q and k vectors, their multiplication is performed simultaneously.

After the first row of q and transposed k vector multiplication is done, the v matrix result is stored in Ram1 for resource reuse. Once the first row of q and transposed k vector

979-8-3315-8850-2/25 $31.00 © 2025 IEEE

multiplication is complete, it's fed into the Softmax module. The fractional part changes here will be explained in the Softmax module's hardware architecture section. The Softmax output is then matrix multiplied with the v vector to get the first head's output, keeping the fractional part at 11 bits due to the small 64*64 matrix size. Next, each head's data is concatenated vertically and multiplied with the linear projection layer. For this large - matrix computation, the fractional part remains 11 bits. The result enters the residual and normalization modules. Normalization requires calculating the mean and variance by summing a whole data row and dividing by 512. We first shift_ divide the data, then add to set the fractional part to 13 bits. The normalized data is further sent into the feedforward neural network for calculation. According to the direction of the data flow, it needs to pass through the first layer of the feedforward neural network in sequence, and then the RELU activation function enters the second layer of the feedforward neural network. The data output by the feedforward neural network is sent back to the residual module and the normalization module again. The residual module and the normalization module that enter for the second time repeatedly call this module by using time-division multiplexing technology.

B. Single Systolic Array Hardware Design

At present, double-pulsating arrays are mostly used to accelerate matrix multiplication [9]. Compared with double-pulsating arrays, single-pulsating arrays first store a matrix in the PE unit. From the perspective of hardware security, single-pulsating arrays leak less information. So we use a single pulsating array to perform matrix multiplication in parallel. A 4 by 4 single systolic array comprises 16 Processing Element (PE) units. Each PE unit is equipped with two input ports and two output ports, and internally consists of an adder and a multiplier. Specifically, the multiplier in each PE unit is responsible for computing the product of the incoming data and the stored data, while the adder is tasked with summing the output from the multiplier and the output from the preceding PE unit.

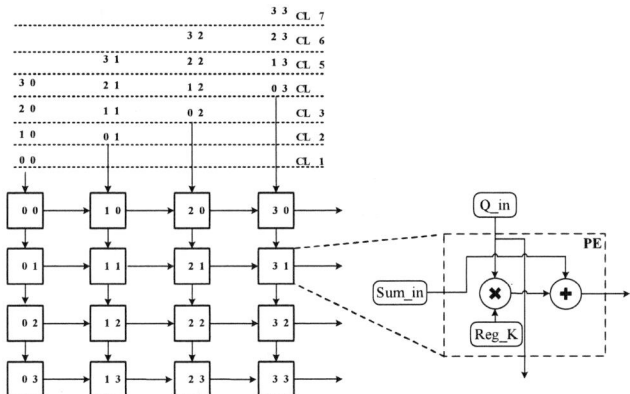

Fig. 4. Single-Systolic Array Hardware Architecture

When executing matrix multiplication using the single systolic array, the left matrix must be fed into the PE units in the manner depicted in Fig. 4. The other matrix is stored within the PE units. Data from the left matrix flows vertically among the PE units, whereas the intermediate values computed by the PE units flow horizontally. These intermediate values are

progressively accumulated to yield the output of the matrix multiplication. The output of the systolic array is produced by the four rightmost PE units in a column-wise manner. A complete matrix multiplication operation using this single systolic array requires 11 clock cycles. This approach significantly enhances the speed of matrix multiplication.

C. Softma Hardware Design

In the multi-head attention mechanism, the attention score calculated by the model is usually obtained by querying the dot product of the q vector and the k vector. These scores represent the intensity of correlation between different positions. The Softmax function can convert these attention scores into probability distributions.

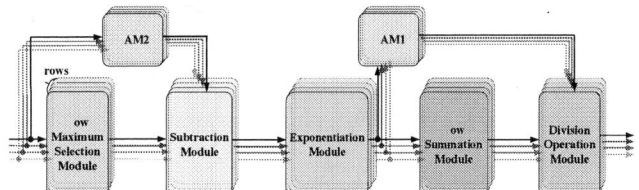

Fig. 5. Softmax Hardware Computing Architecture

Fig. 5 shows the designed Softmax calculation structure. In the encoder, four Softmax modules work in parallel to boost overall speed. Since the exponential mapping module of e can cause large errors, we process the data entering the Softmax function. First, input data goes into a row - wise maximum selection module. This module identifies the maximum value in a row and stores the inputs in Ram2. Then, the data enters a subtraction module to deduct the corresponding row wise maximum, compressing the input range below zero and converting the output fractional bits to 10. The compressed values are fed into the exponential mapping module of e to get the e-exponential values. The sum of these exponential values per row serves as the denominator of the Softmax function and is stored in Ram1. Finally, each value in Ram1 is divided by the row - wise sum to obtain the probability distribution.

$$e^u = Mapping(u - row_{max} + 3.4) \times Steps_S \qquad (6)$$

$$Mapping(m) = \begin{cases} LUT(Distance) & u \geq -6.9 \\ 0 & u < -6.9 \end{cases} \qquad (7)$$

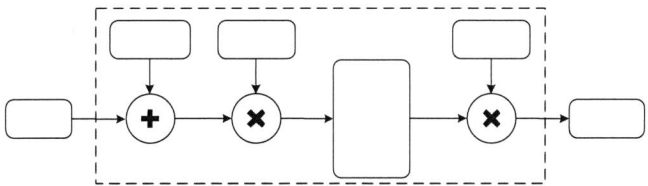

Fig. 6. Exponential Mapping Hardware Architecture

The exponential mapping module of E is shown in Fig. 6 and (6). The previous module has compressed the input data range to below 0. An offset is added to these data, and the selection of this offset is based on the non-overflow maximum value that can represent the indexation of e by the current number of integer digits. Here, 3.4 is added. The overall range will be less than or equal to 3.4. Then, the overall is divided by the step size, and the mapping range is selected to be between -6.9 and 3.4 (by calculating that e -6.9 is already the limit that can be represented by 10 decimal places), the distance between the current value and the selected lower limit value is obtained and sent to the LUT mapping module for mapping. The LUT mapping module stores the calculated values of the e exponent from -6.9 to 3.4. We select a step size of 0.002 to map the input. The mapping calculation is shown in (7). Subsequently, a summation calculation will be carried out, so here we convert 10 decimal places to 8 decimal places to avoid integer bit overflow.

D. Layer Normalization Hardware Architecture

The hardware architecture of layer normalization is shown in Fig. 7, and the calculation formula of layer normalization is shown in (8). In the normalization module, the row-wise mean and variance must first be calculated. To prevent integer overflow and reduce calculation errors, we use 13 fractional bits.

$$LayerNorm(l_i) = \frac{l_i - \mu}{\sqrt{\sigma^2 + \epsilon}}\gamma + \beta \qquad (8)$$

Firstly, when calculating the mean value, the input data is divided by 512 columns first to reduce the integer value, so as to avoid the numerical overflow caused by the addition of the subsequent 512 columns of data. Meanwhile, store the input data in Ram1.

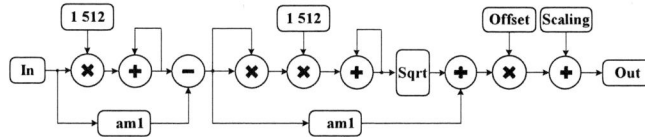

Fig. 7. Layer Normalization Hardware Architecture

Then, the same data is output in parallel to a multiplier to obtain the square of the numerator. To prevent integer overflow during variance calculation, the squared data is divided by 512 before summation. The resulting variance is sent to a square-root module. The data from Ram1 is divided by the square-root of the variance to obtain the original normalization output. Finally, this output is multiplied by a scaling factor and an offset is added to produce the final normalized output. The square root module adopts the Newton's iterative square root as shown in (9).

$$Sqrt_{n+1} = \frac{1}{2}\left(Sqrt_n + \frac{S}{Sqrt_n}\right) \qquad (9)$$

We process the input data using three iterations. The hardware design architecture is shown in the Fig. 8. We take half of the input value as x0 for three iterations. Because for the previous processing of the input data, we tested the number of iterations and accuracy with 1.5 as the input. The test results are shown in Table . When the number of iterations is 3, the best effect is achieved compromising the resource consumption and the calculation accuracy.

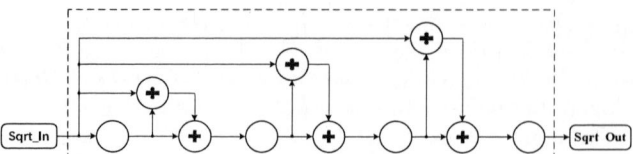

Fig. 8. Square Root Hardware Architecture

TABLE I. ITERATION COUNT ACCURACY MEASUREMENT

Iteration count	Iteration value	Absolute error
1	1.25	0.0252551286
2	1.225	0.0002551286
3	1.224744897959	2.65676×10^{-8}
4	1.224744884675	1.32838×10^{-8}

IV. FPGA VERIFICATION AND ANALYSIS

A. FPGA Verification

In the experiment, we designed the Transformer encoder on the Vivado 2018.3 hardware development platform using the AMD Xilinx Kintex-7 XC7K160T development board. The total resource usage of the encoder is shown in Table . The execution time of the encoder at 200MHZ is 8.683665ms, while the time required to run the encoder using a CPU of model Intel(R) Core(TM) Ultra 9 185H on the Jupyter software is 14.4ms. The hardware encoder is 1.657 times faster than the software way, saving approximately 39.7% of the execution time.

TABLE II. ENCODER HARDWARE RESOURCE USAGE

esource	Usage
LUT	96043
LUTRAM	4688
FF	93546
BRAM	599
DSP	552
BUFG	1

The resource comparison of the Softmax module and Layer Normalization module with other works is shown in Table and Table .

TABLE III. SOFTMAX RESOURCE UTILIZATION

ork	LUT	DSP	B AM
[10]	21190	0	0
This work	10283	72	28.5

TABLE IV. LAYER NORMALIZATION RESOURCE UTILIZATION

ork	LUT	DSP	B AM
[10]	10551	129	27.5
This work	9393	24	23.5

We also tested the encoder output accuracy in software and hardware as shown in Fig. 9. We selected 496 encoder output data for comparison and found that the maximum error among the 496 output data was 0.00320508. The calculated average error is 0.000867, and the standard deviation is 0.0007.

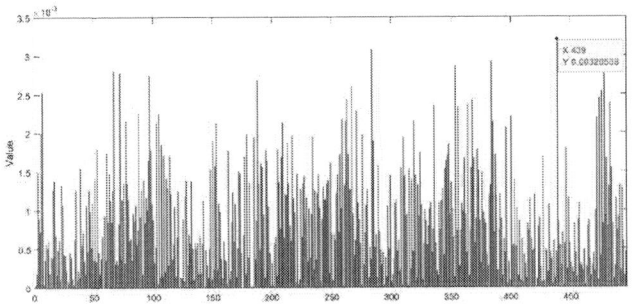

Fig. 9. Error Distribution Among 496 Points

B. Side channel Security Analysis

We used ChipWhisperer 305 board as the target board and ChipWhisperer lite to collect the power traces. The original power consumption trajectories of the feedforward neural network module were collected and compared with the simulation waveforms as shown in Fig. 10. It was found that each output batch of the feedforward neural network would generate a relatively high power consumption peak.

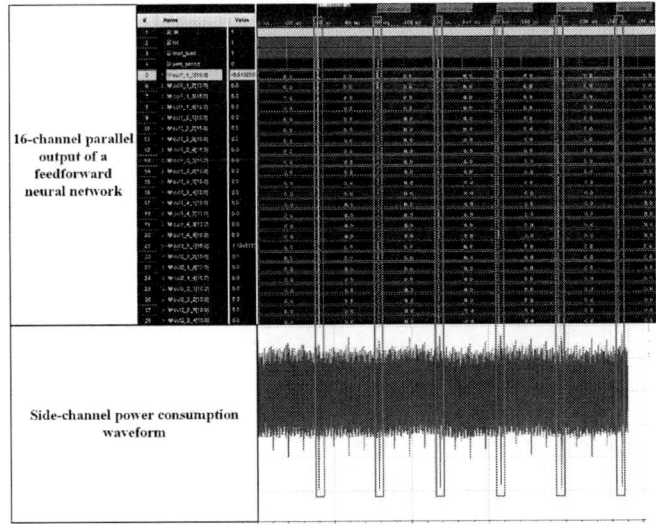

Fig. 10. Comparison Chart of Power Consumption and Waveform of FFN

Side-channel attackers may steal data from this output peak through Hamming weights and Hamming distances information. To hide this output power consumption peak and prevent the side-channel attack on the feedforward neural network, we use the output obfuscation protection method to eliminate these obvious peaks. The power consumption waveform after protection is shown in Fig. 11. It shows that we successfully hide the true output peak by eliciting some false outputs at different times. The peaks can no longer be observed from the protected output waveform, and the output has been successfully concealed.

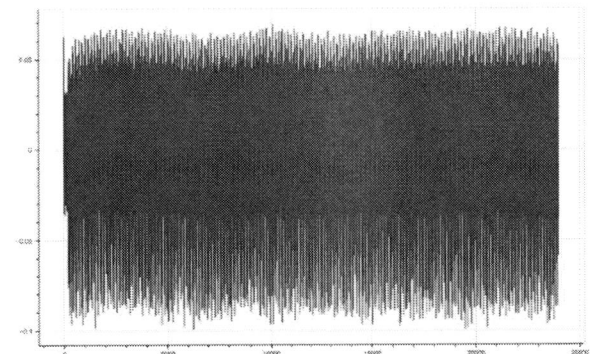

Fig. 11. Power Consumption after FFN Protection

V. CONCLUSION

The paper proposes a high-precision hardware accelerator architecture for Transformer encoders, which breaks through the balance bottleneck between computational efficiency and accuracy through the collaborative optimization of algorithms and hardware. The designed Softmax and layer normalization modules saved 51.47% and 10.9% of LUT resources compared to work 9, and the overall design had an average error of only 0.000867, it saved about 39.7 percent of time compared to CPU. The paper also discovers that there is a side-channel security problem in the output of the feedforward neural network, and successfully hides the power consumption of the output side-channel of this network. Future work will continue to optimize the encoder framework, while conducting side-channel security analysis on other encoder modules.

REFERENCES

[1] Vaswani A, Shazeer N, Parmar N, Uszkoreit J, Jones L, Gomez AN, Kaiser , Polosukhin I. Attention is all you need. Advances in neural information processing systems. 2017 30.

[2] J. Devlin, M.-W. Chang, K. Lee, and K. Toutanova, "BERT: Pre-training of deep bidirectional transformers for language understanding, " arXiv preprint arXiv: 1810.04805, 2018.

[3] Bingbing Li, Santosh Pandey, Haowen Fang and Yanjun Lyv. "FTRANS: Energy-Efficient Acceleration of Transformers using FPGA." arXiv preprint arXiv: 2007.08563, 2020.

[4] Y. Zhang, R. Yasaei, H. Chen, Z. Li and M. A. A. Faruque, "Stealing Neural Network Structure Through Remote FPGA Side-Channel Analysis," In IEEE Transactions on Information Forensics and Security, vol. 16, pp. 4377-4388, 2021.

[5] Y. Wang, W. Mao, H. Shi, J. Sha and Z. Wang, "An Energy-Efficient FPGA Accelerator for Swin Transformer," In IEEE Transactions on Very Large Scale Integration (VLSI), vol. 33, pp. 1774-1778, 2025.

[6] D. Hendrycks and K. Gimpel, "Gaussian error linear units (GELUs)," arXiv preprint arXiv: 1606.08415, 2016.

[7] Damien Robissout, Lilian Bossuet, Amaury Habrard, and Vincent Grosso, "Improving Deep Learning Networks for Profiled Side-channel Analysis Using Performance Improvement Techniques". J. Emerg. Technol. Comput. Syst, vol. 17, pp. 30, 2021.

[8] Zadeh, Abdulah Abdulah, and Howard M. Heys. "Theoretical simple power analysis of the grain stream cipher." 2013 26th IEEE Canadian Conference on Electrical and Computer Engineering (CCECE). IEEE, 2013.

[9] W. Ye, X. Zhou, J. Zhou, C. Chen, and K. Li, "Accelerating attention mechanism on FPGAs based on efficient reconfigurable systolic array," ACM Trans. Embedded Comput. Syst, vol. 22, pp. 1–22, 2023.

[10] Lu, Siyuan, Meiqi Wang, Shuang Liang, Jun Lin, and Zhongfeng Wang. "Hardware accelerator for multi-head attention and position-wise feed-forward in the transformer." In 2020 IEEE 33rd International System-on-Chip Conference (SOCC), pp. 84-89, 2020.

2025 The 10th International Conference on Integrated Circuits and Microsystems

High-Performance Data Prefetching Accelerator for Real-Time Object Detection

Yanzhou Tang
School of Integrated Circuits
Shanghai Jiao Tong University
Shanghai, China
sjtuer19tyz@sjtu.edu.cn

Xiangyu Zhou*
School of Integrated Circuits
Shanghai Jiao Tong University
Shanghai, China
zhouxiangyu-sjtu@sjtu.edu.cn

Zeyi Lin
School of Integrated Circuits
Shanghai Jiao Tong University
Shanghai, China
lzy29748201@sjtu.edu.cn

Ming Cheng
School of Integrated Circuits
Shanghai Jiao Tong University
Shanghai, China
carl_cheng@sjtu.edu.cn

Naifeng Jing
School of Integrated Circuits
Shanghai Jiao Tong University
Shanghai, China
sjtuj@sjtu.edu.cn

Jianfei Jiang
School of Integrated Circuits
Shanghai Jiao Tong University
Shanghai, China
jiangjianfei@sjtu.edu.cn

Weiguang Sheng
School of Integrated Circuits
Shanghai Jiao Tong University
Shanghai, China
wgshenghit@sjtu.edu.cn

Qin Wang
School of Integrated Circuits
Shanghai Jiao Tong University
Shanghai, China
qinqinwang@sjtu.edu.cn

Abstract—A prefetch-centric accelerator is proposed to efficiently address memory bottlenecks in real-time object detection on edge devices. The architecture integrates a scalable 2-D Matrix Processing Unit (MPU) with configurable Vector Processing Units (VPUs), a pipelined sliding-window prefetch engine, and unified dataflow for multiple operators including convolution, pooling, matrix multiplication, and upsampling. The sliding-window mechanism reduces on-chip register demands and streams kernel-sized vectors each cycle, maintaining high throughput and hiding DRAM latency. Evaluation results demonstrate sustained computational efficiency, achieving a peak throughput of 409.6 GOPS and an energy efficiency of 44.2 GOPS/W when running YOLOv3-tiny, showing competitive performance and energy efficiency compared to existing solutions.

Keywords—*Object detection, Convolutional neural networks, FPGA accelerator, Edge computing, Energy efficiency*

I. Introduction

Object detection is the cornerstone of a wide range of vision‑centric applications, including autonomous driving, smart surveillance, and industrial inspection. Modern detectors—such as SSD [1], YOLO [2], RetinaNet [8], and the recent transformer-based DETR family [3]—rely on deep convolutional neural networks (CNNs) to achieve high accuracy. These models typically contain tens of layers, billions of multiply–accumulate (MAC) operations, and increasingly large feature tensors. When deployment targets low-power edge devices, three obstacles become critical: (i) throughput, because real-time processing often requires ≥ 30 fps; (ii) energy efficiency, as battery-operated systems must respect tight power budgets; and (iii) memory traffic, since the feature maps of early layers dominate off-chip bandwidth consumption. Addressing these constraints simultaneously remains a pressing research challenge, especially as emerging applications demand both higher accuracy and stricter latency guarantees.

To meet real-time requirements, practitioners have embraced specialized hardware. General-purpose GPUs excel at raw throughput but exhibit high energy cost and limited determinism on embedded platforms. FPGAs offer flexible dataflow customization yet suffer from relatively low frequency and lengthy development cycles [9]. Recently, domain-specific ASIC accelerators have demonstrated superior performance-per-watt by tailoring compute arrays, on-chip networks, and memory hierarchies to CNN workloads. A common design pattern exploited by these ASICs is a systolic or massively parallel MAC array surrounded by multi-bank SRAM buffers. While this organization efficiently handles compute-bound layers, memory-bound layers—especially the first few high-resolution stages—still throttle overall speed. Inefficient sliding-window handling and frequent off-chip transfers can leave the core array under-utilized, eroding attainable gains and highlighting the necessity of more holistic memory–compute co-optimization strategies. To address these bottlenecks, we propose a data prefetching accelerator for real-time object detection with the following contributions:

- **Scalable 2-D VPU/MPU Array:** We build a two-dimensional Matrix Processing Unit from configurable Vector Processing Units, delivering high parallelism, fine-grained clock-gating, and sustained >90 % array utilization across variable kernel sizes and operator types when deployed on object detection algorithms.

- **Pipelined Sliding-Window Prefetch:** A five-row-plus-FIFO sliding-window engine streams one kernel-sized vector per cycle while cutting on-chip register demand by ≈50 %, hiding DRAM latency and boosting early-layer throughput.

- **Multi-Operator Support with Unified Dataflow:** Leveraging the scalable VPU/MPU array and the pipelined sliding-window prefetch, the architecture supports a wide range of operators—including convolution, pooling, matrix multiplication, and dynamic upsampling.

Supported by the National Key Research and Development Program of China (No. 2023YFB4404100).

979-8-3315-8850-2/25 $31.00 © 2025 IEEE

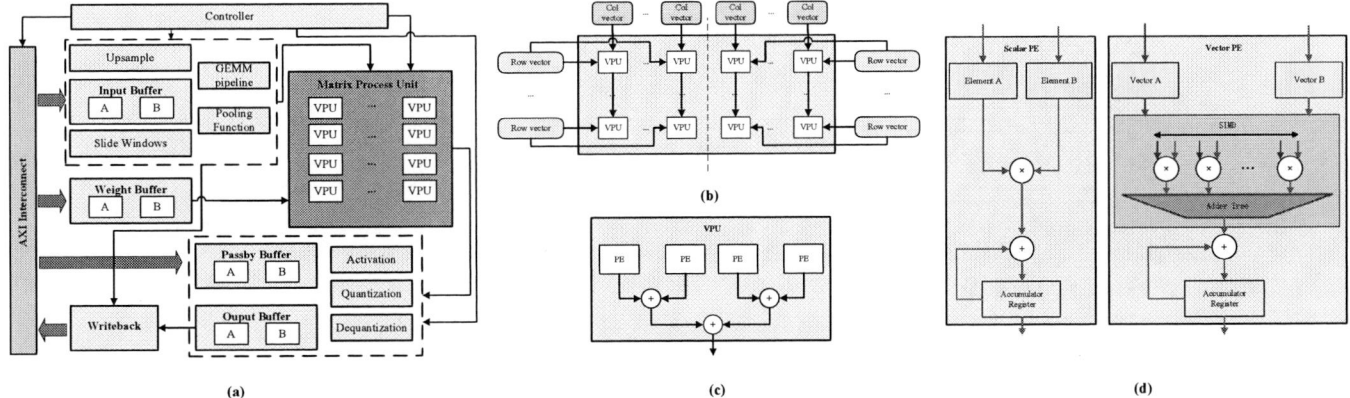

Fig. 1. Overall architecture and components of the proposed accelerator. (a) Overall design diagram of the accelerator; (b) Structure of matrix processing unit; (c) Structure of vector processing unit; (d) Difference between scalar PE and vector PE.

II. BACKGROUND AND DESIGN MOTIVATION

A. Algorithmic Workload

Object detection has branched into four dominant algorithmic lineages over the past decade—two-stage detectors such as Faster R-CNN, anchor-based one-stage networks typified by SSD and YOLO, anchor-free variants like FCOS and CenterNet, and the more recent transformer detectors exemplified by DETR and DINO. Their architectural blueprints differ, yet the computational skeleton behind every family is surprisingly uniform.

Each model begins with a convolutional "backbone", usually ResNet, CSPDarknet or Swin Transformer, that applies deep stacks of 3×3 and 1×1 convolutions to high-resolution feature maps. These early layers dominate the raw byte traffic because each pixel is reused fewer than ten times; arithmetic intensity therefore falls well below one operation per byte, making the stage fundamentally memory-bound.

The backbone is followed by a feature-pyramid or "neck". Here, multiple scales are repeatedly concatenated, added, and up-sampled, multiplying bandwidth demand even though the arithmetic workload is modest.

The final detection head changes shape according to the algorithmic lineage—small 1×1 convolutions for anchor-based YOLO, fully-connected classifiers for anchor-free heads, or a sequence of 128×128 matrix products inside multi-head attention for DETR. All three head styles share the same hardware drawback: the tiles are too narrow to keep a large spatial or systolic array busy, so utilization collapses and energy is wasted in leakage and idle cycles.

B. Limitations of Existing Accelerators

Existing hardware options still fall short of the balanced compute-and-bandwidth profile demanded by real-time object detection. Commodity GPUs, while rich in raw ALUs, reach peak efficiency only when work is issued in large, coalesced batches; edge deployments inevitably run at batch = 1, leaving scores of CUDA cores stalled on memory fetches, and high static power makes battery operation impractical. Mid-range FPGAs permit custom dataflows [10], yet their ≤ 300 MHz clocks and line-buffer-based sliding-window schemes consume vast BRAM and struggle with stride-2 or irregular padding, while the heterogeneous mix of 1×1 convolutions and fully connected layers in detector heads forces costly data reformatting or multiple re-synthesis passes. Purpose-built CNN ASICs push utilization for 3×3, stride-1 convolutions, but most still depend on explicit im2col, replicating each high-resolution pixel K^2 times and bloating on-chip traffic; moreover, small 1×1 or 256×256 GEMM tiles occupy barely a corner of a 1024×1024 systolic array, driving utilization below 30 % while leakage power remains fixed. As a result, all three platforms turn memory-bound precisely at the layers that dictate detector frame-rate, and their coarse-grain power-gating affords little opportunity to scale energy when the workload contracts.

C. Design Requirements and Motivation

These requirements motivate the prefetch-centric accelerator proposed in this work. By combining a scalable 2-D Matrix Processing Unit with a pipelined sliding-window engine and a unified dataflow, the design directly attacks the memory bottleneck that constrains existing solutions, preserves high compute utilization for both backbone and head layers, and delivers energy proportionality through pervasive fine-grained clock-gating. The subsequent sections detail the architecture and demonstrate how these choices translate into significant gains in throughput and energy efficiency on representative object-detection benchmarks.

III. PROPOSED ARCHITECTURE

A. System Overview

The proposed accelerator targets edge-oriented object-detection workloads that interleave multi-sized convolution kernels, GEMM-style matrix multiplications, pooling and up-sampling. As shown in Fig. 1(a), the architecture is organized around six primary modules—an Input Staging Unit (ISU) with a pipelined sliding-window prefetch engine, a Weight Management Unit (WMU), a configurable 2-D Matrix Processing Unit (MPU), an Bypass & Accumulation Unit (BAU) and a Writeback Unit (WBU)—forming a dataflow capable of streaming one kernel-sized vector per cycle while seamlessly switching between convolution-dominated and GEMM-dominated phases of modern detectors.

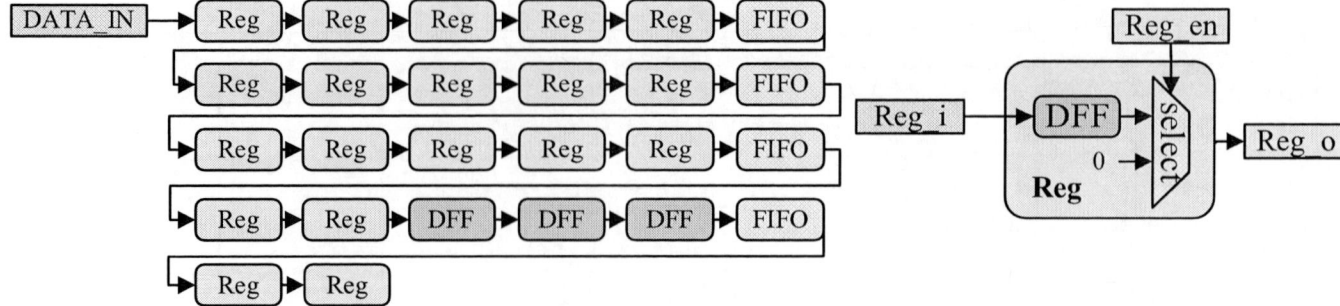

Fig. 2. Support multi-size vector sliding window design.

B. Vector Processing Unit (VPU) and Matrix Processing Unit (MPU)

At the core computational level, the MPU is structured as a two-dimensional array of configurable Vector Processing Units (VPUs). Each VPU comprises multiple Vector Processing Elements (VPEs); specifically, as Fig. 1(c) shown, each VPU in the proposed design contains four VPEs, each capable of parallel execution of multiply–accumulate (MAC) operations on entire vector operands within a single clock cycle.

At the finest granularity, each Processing Element (PE) can operate in either scalar or vector mode, as illustrated in Fig. 1(d), enabling flexible support for diverse operator types and data widths without redesigning the hardware datapath.

As detailed in Fig. 1(b), The MPU's hierarchical structure provides the following operational features:

- Two-dimensional parallelism. VPUs are arranged into a mesh architecture where rows are responsible for parallel computations across the rows of the output matrix, and columns handle parallel computations across output matrix columns. This configuration allows the MPU to efficiently compute standard matrix multiplications as well as convolutions mapped to matrix operations.

- High throughput and scalability. Due to configurable nature, the MPU architecture can dynamically adapt row and column parallelism depending on the size of the input matrix or convolution kernel, ensuring consistently high resource utilization and minimizing computational latency.

- Fine-grained dynamic power Management. A centralized control module dynamically gates individual VPEs or entire VPU units, allowing the accelerator to adapt power consumption according to current computational demands. This is especially beneficial for edge devices running low-resolution search regions typical of object tracking scenarios.

C. Pipelined Sliding-Window Mechanism

As illustrated in Fig. 2, the proposed accelerator integrates a column-vector–based pipelined sliding-window (SW) module to continuously feed the compute array with high-throughput feature-map fragments while minimizing on-chip storage. The module is organized as five register rows followed by four row-FIFO buffers, allowing it to hold up to a 5×5 receptive field and to advance one column per cycle. Compared with conventional designs that keep an entire 2-D window in registers, this structure reduces register usage by 50 % while preserving full data reuse across neighboring windows.

- Column-vector pipeline: Each clock cycle loads a full column vector from the input SRAM into the tail FIFO, while the head column is simultaneously consumed by the matrix-processing unit (MPU). Because SWs slide along the column dimension, successive windows are naturally staggered in time, forming a deep pipeline that hides memory latency and eliminates idle cycles in the compute array. For a 3×3 kernel with stride 1, only two register columns are required; stride 2 or larger windows (e.g. 5×5) are supported by expanding to a maximum of five columns, so a single hardware template covers all typical convolution and pooling operators used in modern trackers.

- Dynamic zero-padding and stride control: Every register cell carries a padding-enable flag. When the flag is asserted, the cell outputs zero instead of data, realizing on-the-fly spatial padding without extra cycles or multiplexers. Stride changes are handled by simply gating the column-advance signal for one or more cycles; analysis shows a worst-case stall of one cycle for stride 2, fully overlapped with vector loading so no throughput is lost.

- In-window pooling reuse: Because the pipeline already presents data as column vectors, max- and average-pooling share the same datapath: comparators or adders operate on the vector elements, store the per-column partial results, and perform a final reduction once all columns of the window have been processed. This reuse eliminates a dedicated pooling buffer and comparator tree, further trimming area.

- Integration with double-bank input buffer: The SW module sits between the double-bank ping-pong input buffer and the MPU. During computation, one bank streams data to the SW pipeline while the other prefetches the next tile from external DRAM. The pipeline's deterministic, fixed-latency behavior simplifies address generation and guarantees that the compute array—configured for either convolution or GEMM—never starves.

979-8-3315-8850-2/25 $31.00 © 2025 IEEE   330

TABLE I. PERFORMANCE ANALYSIS OF KERNELS INVOLVED IN THE OBJECT DETECTION ALGORITHM

Kernel	Kernel Info	Output Feature Map Size	Operations (GOPs)	Execution Time (μs)	PE Utilization (%)
Conv3s1	(64,64,3,3)	64×64	0.302	737.3	100.0
Conv3s2	(64,3,3,3)	128×128	0.0566	819.2	22.5
Conv1s1	(64,64,1,1)	64×64	0.034	81.9	100.0
Maxpool3s2	32 channels	64×64	-	102.4	-
Avgpool2s1	64 channels	127×127	0.004	241.9	-
Maxpool5s2	512 channels	8×8	-	15.4	-
Adown	64 channels	64×64	0.172	794.9	50.2
Dy-upsample	(32,256,1,1)	16×16	0.001	· 2.6	100.0

TABLE II. COMPARISON OF ACCELERATORS FOR OBJECT DETECTION TASK

Accelerator	[4]	[5]	[6]	[7]	This Work
Algrithm	YOLOv3-tiny	YOLOv2-tiny	YOLOv3-tiny	YOLOv3-tiny	YOLOv3-tiny
FPGA Platform	XZU3EG	ZC706	ZCU104	VCU110	VC707
Process (nm)	16	28	16	20	28
Frequency (MHz)	250	200	200	220	200
Latency (ms)	41.7	27.8	29	14.3	15.2
Power (W)	10.8	10.3	3.3	15.4	8.3
Average Throughput (GOPS)	73.0	464.5	180.5	418.9	366.8
Energy Efficiency (GOPS/W)	6.8	45.3	54.7	27.2	44.2

TABLE III. HARDWARE RESOURCE CONSUMPTION

Module\Resource	LUTs	FFs	BRAMs	DSPs
MPU	38150	29409	-	1024
ISU	31531	14476	48	-
WMU	1109	2090	64	-
BAU	15213	6345	136	-
WBU	129	1025	-	-
ALL	86132	53345	248	1024

IV. EXPERIMENTAL RESULTS

A. Hardware Resource Consumption

Table III summarizes the hardware resource consumption of the proposed vector accelerator, synthesized on a Xilinx VC707 board. The design achieves full DSP utilization with a 4×16 compute array and 16-way intra-vector parallelism. Due to reduce bandwidth pressure, the bypass and accumulation module uses BRAM extensively to store intermediate results.

Power analysis indicates a total consumption of 8.3 W at 200 MHz, including 7.9 W dynamic and 0.4 W static power. Overall, the accelerator achieves high computational efficiency with moderate resource and power usage, supporting real-time edge AI tasks.

B. Performance Analysis

At a clock frequency of 200 MHz, this yields a theoretical peak throughput of 409.6 GOPS, as calculated by:

$$GOPs_{peak} = 2048 \times 10^{-9} \times 200 \times 10^6 = 409.6 \ GOPs \quad (1)$$

The evaluated operators include stride-1 3×3 and 1×1 convolutions, stride-2 3×3 convolutions, as well as pooling operations such as stride-2 5×5 and 3×3 pooling, and stride-1 2×2 pooling. For convolution operations, average throughput is computed as the total operation count divided by execution time, and its ratio to peak performance represents PE utilization. Detailed results for these baseline operators are summarized in Table II.

As shown in Table I, the PE utilization for stride-2 3×3 convolution is low because the input channel count is only 3, insufficient to exploit the 16-channel parallelism. Moreover, the sliding-window module takes 10 cycles to output 9 data elements due to data alignment, causing a one-cycle idle period in the compute array. However, with per-PE power gating, power consumption remains low despite underutilization.

The low PE utilization in the Adown module arises because its computation includes both pooling and stride-2 3×3 convolution operations. Pooling essentially involves data selection and reuse within the sliding window, causing the PE array to stall during pooling phases and resulting in reduced overall utilization.

The proposed accelerator was compared with other designs in the field, as summarized in Table II. When deploying YOLOv3-tiny, it achieves an average performance of 366.8 GOPS, which is significantly higher than the designs in [4] and [6]. Although its performance is slightly lower than that of [5] and [7], it should be noted that the design in [7] consumes considerably more DSP and BRAM resources, resulting in an energy efficiency of only 27.2 GOPS/W, approximately half that of the proposed accelerator. Similarly, the energy efficiency of [5] is slightly lower than that of this work.

V. CONCLUSION

This work proposed a prefetch-centric accelerator for real-time object detection on edge devices. Integrating a scalable 2-D Matrix Processing Unit with configurable Vector Processing Units and a pipelined sliding-window prefetch engine, the design addresses memory bottlenecks while maintaining high compute utilization. The sliding-window mechanism reduces on-chip register demands by ~50%, enabling continuous kernel-sized vector streaming and hiding DRAM latency. Experiments showed a peak throughput of 409.6 GOPS and energy efficiency of 44.2 GOPS/W on YOLOv3-tiny. Future work will extend operator support for transformer detection heads and further optimize area and power efficiency for edge deployment.

REFERENCES

[1] Chengcheng Ning, Huajun Zhou, Yan Song, and Jinhui Tang, "Inception single shot multibox detector for object detection," in 2017 IEEE International Conference on Multimedia & Expo Workshops (ICMEW), Hong Kong, China, 2017, pp. 549–554.

[2] J. Redmon, S. Divvala, R. Girshick, and A. Farhadi, "You only look once: Unified, real-time object detection," in Proc. IEEE Conf. Comput. Vis. Pattern Recognit. (CVPR), Jun. 2016, pp. 779–788..

[3] Nicolas Carion, Francisco Massa, Gabriel Synnaeve, Nicolas Usunier, Alexander Kirillov, Sergey Zagoruyko, End-to-end object detection with transformers, 2020, pp. 213–229.

[4] CHEN X, LI J, ZHAO Y. Hardware resource and computational density efficient cnn accelerator design based on fpga[C]. In: proceedings of the 2021 IEEE International Conference on Integrated Circuits, Technologies and Applications (ICTA), 2021: 204-205.

[5] ZHANG J, CHENG L, LI C, et al. A low-latency fpga implementation for real-time object detection[C]. In: proceedings of the 2021 IEEE International Symposium on Circuits and Systems (ISCAS), 2021: 1-5.

[6] LI S, WANG Q, JIANG J, et al. An efficient cnn accelerator using inter-frame data reuse of videos on fpgas [J]. IEEE Transactions on Very Large Scale Integration Systems (TVLSI), 2022, 30(11):1587-1600.

[7] MONTGOMERIE-CORCORAN A, TOUPAS P, YU Z, et al. Satay: A streaming architecture toolflow for accelerating yolo models on fpga devices[C]. In: proceedings of the 2023 International Conference on Field Programmable Technology (ICFPT), 2023: 179-187.

[8] T. -Y. Lin, P. Goyal, R. Girshick, K. He and P. Dollár, "Focal Loss for Dense Object Detection," in IEEE Transactions on Pattern Analysis and Machine Intelligence, vol. 42, no. 2, pp. 318-327, 1 Feb. 2020, doi: 10.1109/TPAMI.2018.2858826.

[9] Kai Zeng, Qian Ma, Jia Wen Wu, Zhe Chen, Tao Shen, and Chenggang Yan. 2022. FPGA-based accelerator for object detection: a comprehensive survey. J. Supercomput. 78, 12 (Aug 2022), 14096–14136. https://doi.org/10.1007/s11227-022-04415-5.

[10] D. T. Nguyen, T. N. Nguyen, H. Kim and H. -J. Lee, "A High-Throughput and Power-Efficient FPGA Implementation of YOLO CNN for Object Detection," in IEEE Transactions on Very Large Scale Integration (VLSI) Systems, vol. 27, no. 8, pp. 1861-1873, Aug. 2019, doi: 10.1109/TVLSI.2019.2905242.

2025 The 10th International Conference on Integrated Circuits and Microsystems

Design of High-reliability and High-dynamic Analog Front-end for DC Carrier Chip

Quan Sun
Hangzhou Vango Technologies, Inc.
Hangzhou, China
sunquan@vangotech.com

Gang Chen*
Advanced Institute of Information Technology, Peking University
Hangzhou, China
cgbeiyou@163.com

Changyou Men
Hangzhou Vango Technologies, Inc.
Hangzhou, China
menchangyou@vangotech.com

Shuang Wu
Advanced Institute of Information Technology, Peking University
Hangzhou, China
zoewushuang@hotmail.com

Abstract—**DC carrier communication technology has broad application prospects in smart grids, power line communications, and other fields. This paper designs a high-reliability, high-dynamic-range analog front-end architecture to meet the special requirements of DC carrier chips. This circuit utilizes key technologies such as adaptive gain control, multi-stage filtering, and dynamic bias adjustment to achieve stable operation over a wide input voltage range (12V-1000V). Experimental results demonstrate that the designed analog front-end achieves a signal-to-noise ratio of 65dB, a dynamic range exceeding 80dB, and a total harmonic distortion of less than 0.1% over a temperature range of -40°C to +85°C, meeting the stringent requirements of DC carrier communication systems.**

Keywords—*DC carrier, analog front-end, high reliability, dynamic range, signal conditioning*

I. INTRODUCTION

With the rapid development of smart grids and new energy technologies, DC power systems are becoming increasingly important in modern power networks [1]. DC carrier communication technology, a communication method that utilizes existing DC power lines for data transmission, offers advantages such as low cost, wide coverage, and ease of deployment. It demonstrates great application potential in scenarios such as electric vehicle charging networks, distributed photovoltaic systems [2], and data center power supply systems[3].

However, DC carrier communication faces numerous technical challenges. First, the DC power line environment is complex, characterized by wide voltage variations, strong electromagnetic interference, and severe signal attenuation. Second, the system must operate stably and over a long period of time in harsh industrial environments, placing extremely high demands on reliability. Finally, to improve communication efficiency and interference immunity, the system must possess a wide dynamic range and low noise[4].

Analog front-ends[5], as key components of DC carrier communication systems, are responsible for signal reception, amplification, filtering, and conditioning. Their performance directly impacts the reliability and quality of the entire communication system. Traditional analog front-end designs are often designed for AC carrier systems and are unable to meet the

unique requirements of DC carrier environments. Therefore, designing a high-reliability, high-dynamic analog front-end specifically for DC carrier chips is of great theoretical and practical significance[6]. The main contributions of this paper include: (1) proposing an analog front-end architecture suitable for DC carrier environment; (2) designing an adaptive gain control algorithm to achieve signal conditioning under a wide input range; (3) developing multi-stage filtering and dynamic bias adjustment technology to effectively suppress noise and interference; (4) verifying the performance indicators of the designed circuit through simulation and experiments.

II. RELATED WORKS

The research on DC carrier communication technology started relatively late, but has received widespread attention from academia and industry in recent years. Early research mainly focused on communication protocols and modulation-demodulation algorithms, while relatively little attention was paid to hardware circuit design for DC carrier systems.

In terms of analog front-end design, traditional AC carrier systems usually adopt signal injection and extraction schemes[7] based on transformer coupling. However, the DC system lacks AC components and cannot directly use transformer coupling. Capacitive coupling or direct coupling is required. This difference poses more demanding signal conditioning requirements for DC carrier systems. Existing DC carrier analog front-end designs have the following limitations: (1) Limited dynamic range, making it difficult to adapt to a wide range of input voltage variations; (2) Insufficient reliability, prone to failure in harsh environments; (3) High power consumption, making it unsuitable for battery-powered applications; and (4) High cost, which restricts large-scale commercial deployment. To address these issues, this paper proposes a new analog front-end architecture that achieves the performance goals of high reliability and high dynamic range through innovative circuit design and algorithm optimization[8].

Recent state-of-the-art DC carrier communication systems include: (1) Broadcom's BCM60333 series focusing on Home Plug AV2 applications with limited voltage range (12V-24V); (2) Qualcomm's QCA7450 targeting automotive applications with moderate dynamic range (60dB); (3) Texas Instruments' AFE031 designed for smart grid applications but lacking adaptive gain control.

This research was supported by the Key Scientific Research Project of Hangzhou (Project No. 2024SZD1A52）

979-8-3315-8850-2/25 $31.00 © 2025 IEEE

The key differences of our architecture compared to existing solutions are:

1) Extended voltage operation range (12V-1000V vs. typical 12V-48V);

2) Adaptive digital AGC algorithm providing 80dB dynamic range vs. fixed-gain approaches;

3) Multi-stage protection strategy supporting $\pm$ 2kV transient protection;

4) Temperature-compensated bias circuits for -40° C to +85° C operation.

III. System Architecture Design.

A. Overall Architecture

The DC carrier analog front-end designed in this paper adopts a multi-stage cascade architecture, primarily consisting of input protection circuits, signal coupling circuits, variable gain amplifiers (VGAs)[9], multi-stage filters, automatic gain control (AGC)[10] units, and output buffers. The connection relationship and signal flow direction of each module are clearly presented in Fig. 1, which intuitively shows how the input signal is sequentially processed by the input protection circuit, signal coupling circuit, VGA, multi-stage filter, and output buffer, while the AGC unit dynamically adjusts the VGA gain and the bias control module provides stable bias voltage for each circuit module.

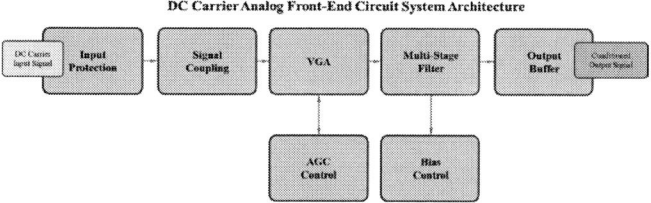

Fig. 1. System overall architecture diagram

The proposed analog front-end is implemented as a hybrid circuit combining both integrated circuits and discrete components. The core signal processing units (VGA, filters, and AGC) are realized using commercial ICs, while the input protection and signal coupling circuits utilize discrete high-voltage components to handle the wide input voltage range (12V-1000V).

Key IC Components:

1) Variable Gain Amplifier: AD8367 (Analog Devices);

2) Operational Amplifiers: OPA2134 (Texas Instruments);

3) ADC for AGC control: ADS1115 (Texas Instruments);

4) Microcontroller for AGC: STM32F103 (STMicroelectronics).

Key Discrete Components:

1) TVS Diodes: SMBJ series (Little fuse);

2) Coupling Capacitors: C0G ceramic capacitors (KEMET);

3) Current-limiting Resistors: Metal film resistors (Vishay).

B. Key Circuit Design

Input Protection and Signal Coupling: The input protection circuit utilizes a multi-stage protection strategy, including transient voltage suppressors (TVS)[11], gas discharge tubes, and current-limiting resistors, effectively protecting against transient surges up to ±2kV. The signal coupling circuit utilizes high-voltage capacitive coupling to effectively extract the signal.

The selection of the coupling capacitor must consider the influence of the signal frequency range and input impedance. The coupling capacitor value is calculated as follows:

$$C_c = \frac{1}{2\pi f_{min} \cdot R_{in}} \geq \frac{10}{2\pi f_{min} \cdot R_{in}} \tag{1}$$

Where f_{min} is the minimum operating frequency and R_{in} is the input impedance.

The variable gain amplifier (VGA) is a core module in the analog front-end and must maintain excellent linearity and low noise over a wide gain range. This design utilizes a current-controlled VGA architecture with a gain control range of -20dB to +40dB.

The VGA gain control equation is:

$$G_{VGA}(dB) = 20\log_{10}\left(\frac{I_{ctrl}}{I_{ref}}\right) + G_0 = 32 \cdot \frac{V_{ctrl} - 0.25}{2.0} + G_0 \tag{2}$$

Where I_{ctrl} is the control current, I_{ref} is the reference current, and G_0 is the basic gain.

To effectively suppress out-of-band noise and interference, a multi-stage filter cascade structure was designed, consisting of a high-pass filter, a band pass filter, and a low-pass filter. Each filter stage utilizes an active filter topology, offering low noise and high precision.

The transfer function of the band pass filter is:

$$H(s) = \frac{K \cdot \omega_0 \cdot s}{s^2 + \frac{\omega_0}{Q} \cdot s + \omega_0^2} \tag{3}$$

Where $\omega 0$ is the center frequency, Q is the quality factor, and K is the gain constant.

Detailed Design Parameters:

1) Input Protection Circuit:

   a) TVS Diodes: SMBJ58A (58V breakdown voltage)

   b) Gas Discharge Tube: GDT230-5 (breakdown voltage 230V ±20%)

   c) Current-limiting Resistor: 100 Ω, 2W metal film

2) Signal Coupling Circuit:

   a) Coupling Capacitor: 1μF, 2kV, C0G ceramic

b) Input impedance matching: 50 Ω differential

3) Variable Gain Amplifier (AD8367):

  a) Supply voltage: ±5V

  b) Control voltage range: 0.25V to 2.25V

  c) Gain control sensitivity: 32dB/V

4) Multi-stage Filter Implementation:

  a) HPF: Sallen-Key topology, fc = 1kHz, Q = 0.707

  b) BPF: Multiple feedback topology, fc = 1MHz, Q = 10

  c) LPF: Sallen-Key topology, fc = 10MHz, Q = 0.707

5) AGC Implementation:

  a) Sampling rate: 100kSPS

  b) Control loop bandwidth: 1kHz

  c) Target output level: -10dBm ±0.5dB

C. Adaptive Gain Control Algorithm

To achieve wide dynamic range signal processing, an adaptive gain control algorithm based on digital feedback was designed. This algorithm monitors the output signal amplitude in real time and dynamically adjusts the VGA gain to ensure that the output signal remains within the optimal range, and its specific workflow is detailed in Fig. 2.

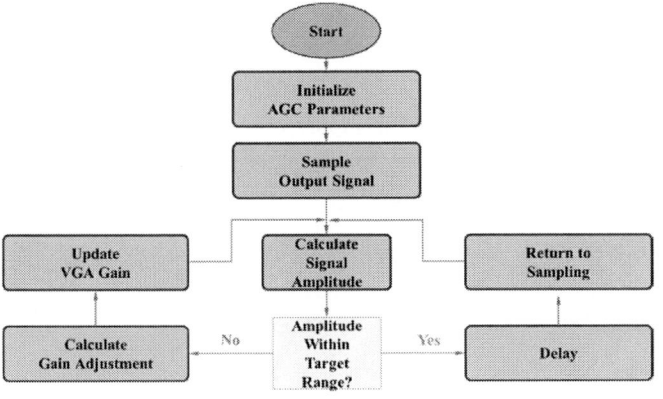

Fig. 2. Location: AGC algorithm flow chart

The core of the AGC algorithm is the calculation of the gain adjustment amount. When the output signal amplitude is detected to deviate from the target range, the system will calculate the corresponding gain adjustment amount based on the degree of deviation:

$$\Delta G = K_p \cdot \left(V_{target} - V_{out}\right) + K_i \cdot \int_0^t \left(V_{target} - V_{out}\right)dt \qquad (4)$$

Where ΔG is the gain adjustment, K_p and K_i are the proportional and integral control coefficients respectively,

V_{target} is the target amplitude, and V_{out} is the actual output amplitude.

IV. EXPERIMENTAL RESULTS AND PERFORMANCE ANALYSIS

A. Test Platform Construction

To verify the performance of the designed analog front-end, a complete test platform was constructed. This test platform takes the designed DC carrier analog front-end as the core and is equipped with supporting equipment such as a signal generator, DC power supply, oscilloscope, and spectrum analyzer, which can simulate the real DC carrier communication environment. The composition and connection relationship of the equipment are shown in Fig. 3.

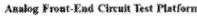

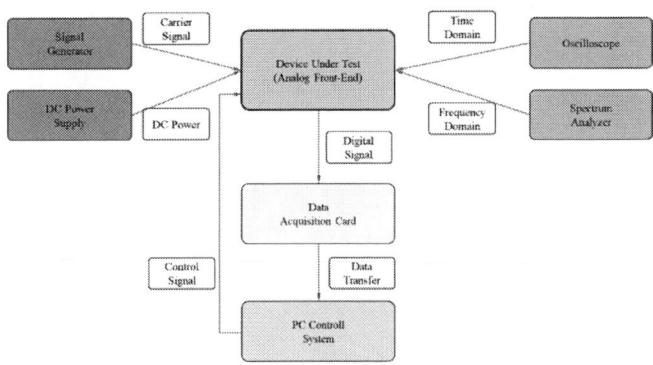

Fig. 3. Test platform block diagram

B. Performance Test

Dynamic range is a key performance metric for analog front-ends. By varying the input signal amplitude, the circuit's response characteristics under varying input conditions are tested, and the test results are presented in Fig. 4.

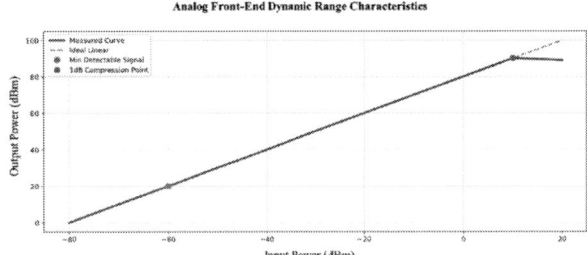

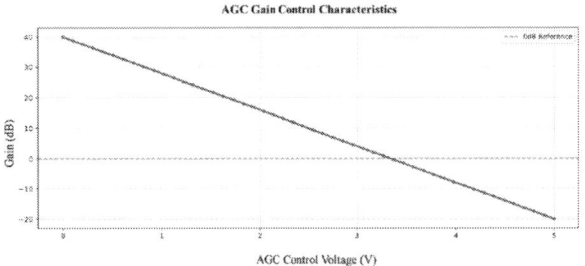

Fig. 4. Dynamic range test results
(input voltage: 12V-1000V, temperature: 25°C)

Experimental results demonstrate that the designed analog front-end maintains excellent linearity even when covering an input dynamic range of 80 dB, fully meeting the application requirements of DC carrier communications.

The frequency response characteristics of the analog front-end were obtained through frequency sweep testing. The test frequency range was 1 kHz to 10 MHz, covering the typical operating frequency band of DC carrier communications, and the obtained amplitude response and phase response are shown in Fig. 5.

$$|H(f)| = \frac{|V_{out}(f)|}{|V_{in}(f)|} \tag{5}$$

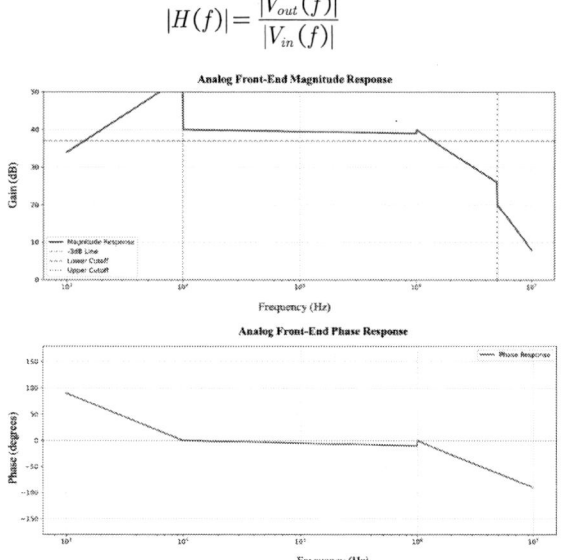

Fig. 5.   Frequency response test results

Noise performance is a key factor affecting the sensitivity of a communication system. The circuit's noise characteristics are evaluated by measuring the output noise at different gain settings.

The formula for calculating the equivalent input noise power spectral density is:

$$N_{in}(f) = \frac{N_{out}(f)}{|H(f)|^2} \quad (nV/\sqrt{Hz}) \tag{6}$$

Where $N_{out}(f)$ is the output noise power spectral density, and $|H(f)|^2$ is the power gain of the circuit.

C. Environmental Adaptability Test

Temperature Characteristics Test: Circuit performance is tested within a temperature range of -40°C to +85°C to verify environmental adaptability. The effect of temperature on key parameters can be described using the following empirical formula:

$$P(T) = P_0 \cdot [1 + \alpha \cdot (T - T_0) + \beta \cdot (T - T_0)^2] \tag{7}$$

Where $P(T)$ is the parameter value at temperature T, P_0 is the parameter value at reference temperature T_0, and α is the

temperature coefficient. The specific test results are shown in Fig. 6.

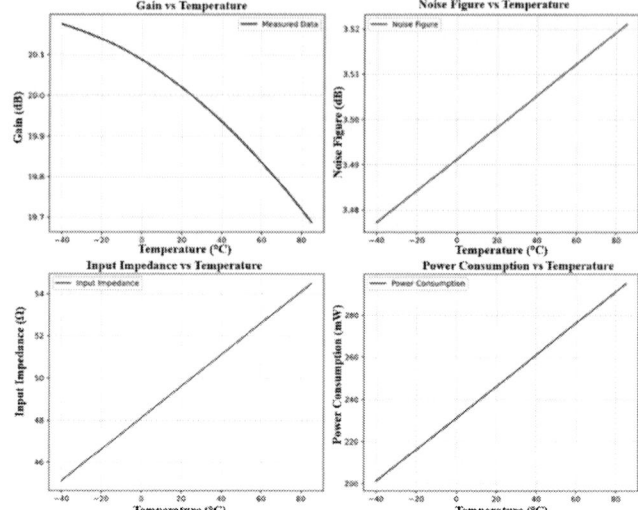

Fig. 6.   Temperature characteristics test results

Accelerated life testing was used to conduct long-term circuit operation tests under high temperature and high humidity conditions. The Weibull distribution model was used for reliability assessment:

$$R(t) = \exp\left[-\left(\frac{t}{\eta}\right)^{\beta}\right] \tag{8}$$

Here, $R(t)$ is the reliability function, eta is the characteristic lifetime, and beta is the shape parameter.

D. Performance Comparison Analysis

TABLE I. PERFORMANCE COMPARISON TABLE

Performance Specifications	This Design	Plan A	Plan B	Plan C
Dynamic Range (dB)	80	60	65	55
Signal-to-Noise Ratio (dB)	65	58	60	52
Gain Range (dB)	-20~+40	-10~+30	-15~+35	-5~+25
Operating Temperature (℃)	-40~+85	-20~+70	-30~+75	-10~+60
Power Consumption (mW)	250	320	280	350
THD (%)	<0.1	<0.3	<0.2	<0.5
Input Voltage Range (V)	12-1000	12-24	12-48	12-48
Transient Protection (kV)	±2.0	±0.5	±1.0	±0.8
Package Type	Hybrid PCB	QFN-64	BGA-144	TQFP-64
Cost (USD)	45	38	52	28

To further validate the advantages of the designed DC carrier analog front-end in practical application scenarios, this section compares it with three existing mainstream DC carrier analog front-end solutions (Plan A[12], Plan B[13], and Plan C[14]) that have been reported in academic and industrial fields. The comparison covers core performance indicators, environmental adaptability, reliability, energy consumption, packaging, and cost, with specific data organized in TABLE I. Performance Comparison Table.

V. CONCLUSION

In response to the application requirements of DC carrier chips, this paper designs an analog front-end with high reliability and high dynamic range. The main achievements and contributions are as follows:

Innovative circuit architecture: A multi-stage cascade analog front-end architecture suitable for DC carrier environments is proposed, which effectively solves the limitations of traditional designs in DC environments.

Adaptive gain control: An AGC algorithm based on digital feedback is developed, which achieves a wide dynamic range of 80dB and significantly improves the adaptability of the system.

Excellent performance indicators: Experimental results show that the designed circuit meets or exceeds the performance of existing solutions in key indicators such as signal-to-noise ratio, dynamic range, and temperature stability.

High reliability design: Through multi-level protection, redundant design, and environmental adaptability optimization, this design ensures the circuit's long-term stable operation in harsh industrial environments.

Future research work will focus on the following directions:

1) Further reduce power consumption and improve energy efficiency;

2) Integrate more digital signal processing functions to achieve more intelligent signal conditioning;

3) Develop customized versions for specific application scenarios;

4) Explore new materials and process technologies to further improve performance indicators. This research provides important technical support for the industrial application of DC carrier communication technology and has broad application prospects and commercial value.

REFERENCES

[1] Neudeck G P ,Chen Y L ,Greer C L , et al., Venus Surface Environmental Chamber Test of SiC JFET-R Multi-Chip Circuit Board[J].Solid State Phenomena,2024,3587-12.

[2] Wan-Hao, Z. et al., Consumer perception and use intention for household distributed photovoltaic systems [J]. Sustainable Energy Technologies and Assessments, June 2022, Volume 51.

[3] Yenan C. et al., Data Center Power Supply Systems: From Grid Edge to Point-of-Load[J]. IEEE Journal of Emerging and Selected Topics in Power Electronics, June 2023, 11(3) 2441-2456.

[4] Xue H ,Peng Y ,Jing Q , et al., Sensing with extended gate negative capacitance ferroelectric field-effect transistors [J]. Chip, 2024, 3(1): 100074-.

[5] Piero, M. et al., Analog front-ends for MEMS Microphones [C].2023 IEEE Custom Integrated Circuits Conference (CICC), 23-26 April 2023.

[6] NAGATA M .Design of Circuits and Packaging Systems for Security Chips:Special Section on Solid-State Circuit Design — Architecture, Circuit, Device and Design Methodology[J].IEICE Transactions on Electronics, 2023, 106(7): 345-351.

[7] Junlei, C. et al., Sensorless Control for SynRM Drives Using a Pseudo-Random High-Frequency Triangular-Wave Current Signal Injection Scheme[J]. IEEE Transactions on Power Electronics, 37(6) 7122 - 7131, 2022.

[8] STMicroelectronics SA; Patent Issued for Method For Estimating An Operating Profile Of An Integrated Circuit Of A System-On-A-Chip, And Corresponding System-On-A-Chip (USPTO 10,634,715)[J].Electronics Newsweekly, 2020,

[9] Ji-Yong Um. A Compact Variable Gain Amplifier With Continuous Time-Gain Compensation Using Systematic Predistorted Gain Control [J].IEEE Transactions on Circuits and Systems II: Express Briefs , 69(2)274 - 278, February 2022.

[10] Ilya, K. et al., Spiking Cochlea With System-Level Local Automatic Gain Control [J].IEEE Transactions on Circuits and Systems I: Regular Papers, 69(5)2156 - 2166, May 2022.

[11] Evangelos T. S. et al., Wide Frequency Response of Varistors and Coordination With Transient Voltage Suppression Diodes [J]. IEEE Transactions on Power Delivery, 38(1)453 - 462, February 2023.

[12] Jalalizadeh M ,Noori S ,Tahmasebi V .Two-Dimensional Inverse Metho d to Estimate the Heat Flux Function in a Phase Change Material Heat-E xchanger[J].Heat Transfer Engineering, 46(18):1651-1670,2025.

[13] Yin Y ,Fairchild J A ,Shi D , et al., Estimating the Root Mean Square Err or of Approximation (RMSEA) with Multiply Imputed Data Under Non-Normality[J].Structural Equation Modeling: A Multidisciplinary, 32(5):8 32-857. Journal 2025

[14] Conde G A ,Devesa G D ,Iglesias S D , et al., The validity and reliability of a portable device (ADR-Jumping) to estimate vertical jump performance[J].Proceedings of the Institution of Mechanical Engineers, Part P: Journal of Sports Engineering and Technology,239(3):349-355,2025

979-8-3315-8850-2/25 $31.00 © 2025 IEEE

2025 The 10th International Conference on Integrated Circuits and Microsystems

A Steady-State Temperature Solving Method for Multi-Chip Modules Components Based on an Improved U-Net

Fulong Yang*
College of Electrical and Information Engineering
Lanzhou University of Technology
Lanzhou, China
yangfulong1982@126.com
*Corresponding author

Feng Wu
College of Electrical and Information Engineering
Lanzhou University of Technology
Lanzhou, China
wufeng_0107@163.com

Teng Zhang
College of Electrical and Information Engineering
Lanzhou University of Technology
Lanzhou, China
z18209283206@163.com

Jinjin Zhou
College of Electrical and Information Engineering
Lanzhou University of Technology
Lanzhou, China
974282917@qq.com

Qing Yang
College of Electrical and Information Engineering
Lanzhou University of Technology
Lanzhou, China
1397746982@qq.com

Abstract—**Basic convolutional neural networks (CNNs) can achieve fast steady-state temperature field prediction for multi-chip modules (MCMs). However, when the number of chips is large and the layout is complex, the prediction accuracy drops significantly. This study enhances the U-Net architecture, incorporating attention mechanisms and residual connections, to boost the model's learning and generalization capabilities. Using chip power maps as input, this approach captures the intrinsic mapping between chip power distribution, layout, and temperature field, generating steady-state temperature distributions for MCMs rapidly. Experiments show that for MCMs with 4, 8, 12, and 16 chips, the mean absolute percentage error is under 0.12%, with the maximum error not exceeding 1.08K, and a prediction time of about 1ms per temperature map. The improved model ensures both speed and accuracy, offering crucial technical support for MCM design.**

Keywords—multi-chip module, temperature field prediction, deep learning, U-Net

I. INTRODUCTION

With the development of technology, chip design has shifted from merely focusing on reducing power consumption and improving performance to better meeting market demands[1]. Multi-Chip Modules (MCMs), which integrate multiple chips onto a single substrate, significantly reduce interconnect length and packaging size, thus improving performance[2]. This has made MCMs widely used in high-performance computing, servers, and other fields. However, thermal issues have always been one of the main bottlenecks limiting the performance improvement of high-performance

microprocessors [3]. Similarly, the compact layout of MCMs concentrates heat and power density in a small area, leading to severe thermal problems that increase the risk of chip failure and even system failure[4]. Therefore, precise thermal design during MCM manufacturing is crucial to enhance system reliability. However, steady-state temperature solutions for MCMs are influenced by multiple factors, including geometric structure, material properties, chip layout, and power consumption, making temperature calculations more challenging[5]. In particular, variations in chip layout can lead to completely different temperature distributions, and improper layouts may further exacerbate the thermal problems in MCMs. As a result, chip layout has become one of the key design factors affecting the performance and thermal characteristics of MCMs[6].

Currently, various methods can be used to solve the steady-state temperature of chips, each with its advantages in terms of efficiency and accuracy. Direct numerical simulation (DNS) methods, such as the finite difference method (FDM), finite volume method (FVM), and finite element method (FEM), are widely recognized as benchmark methods for thermal simulation research due to their ability to provide high-precision thermal analysis results. These methods are supported by numerous open-source and commercial software tools (e.g., ANSYS, COMSOL) and have been applied in chip thermal simulation and the steady-state temperature solution for MCMs[7-11]. However, DNS methods typically involve solving a large number of complex equations, requiring significant computational resources. Especially during the design or optimization of MCMs, multiple iterations are often necessary to account for various parameters. Despite their ability to deliver high-precision thermal analysis results, DNS

This work is supported by the Gansu Provincial IC Manufacturing Materials Innovation Consortium Project 2023 (Grant No.: 23ZDGE001) and the Gansu Provincial Joint Research Fund Project (Grant No.: 24JRRA829).

979-8-3315-8850-2/25 $31.00 © 2025 IEEE

methods are computationally inefficient, making it challenging temperature solutions in MCMs.

In recent years, with the rapid development of artificial intelligence technologies and the application of GPU acceleration, the use of neural network models in the field of thermal engineering has attracted widespread attention and research. However, most existing research focuses on exploring the feasibility of AI technology in chip thermal simulation. Raissi et al.[12] proposed using physics-informed neural networks (PINNs) for chip thermal simulation, aiming to solve supervised learning tasks that adhere to physical laws by processing general nonlinear partial differential equations (PDEs). Jin et al.[13] used a GAN network to treat thermal modeling as an image-generation task with generative neural networks. They used real-time high-level chip utilization and thermal sensor information from commercial chips as inputs to achieve the solution of the transient temperature of the chip after fabrication. Chhabria et al.[14] used an encoder-decoder-based generative network to transform chip thermal simulation into an image-to-image translation task. Sultan et al.[15] utilized residual convolutional neural networks to achieve chip thermal simulation of 3D-IC. Wang et al.[1] combined thermal simulation analysis with machine learning algorithms, developing a K-fold cross-validation algorithm and a support vector regression algorithm to predict the hotspot temperature variations in 3D chips, enabling predictions for up to 28 layers of 3D chip hotspot temperature changes. Yue et al.[2] proposed a method for predicting the steady-state temperature field of MCM based on deep convolutional neural networks, using signed distance function (SDF) to represent chip location and power, and predicting the dimensionless temperature field along with its mean and standard deviation to obtain the real temperature field. However, introducing SDF and using two deep learning networks for multi-stage prediction leads to an increase in solution time, and accuracy decreases as the number of MCM chips increases, while large-scale MCMs are becoming increasingly common in practical applications.

To address the above issues, this paper improves the basic U-Net network by incorporating attention mechanisms and residual connections. The encoder, decoder, and skip connections of U-Net are optimized. We optimized the encoder, decoder, and skip connections of U-Net, proposing an encoder module that combines channel attention mechanisms and residual connections, a decoder module with channel attention mechanisms, and a multi-layer feature fusion module that integrates CBAM and spatial attention mechanisms. These enhancements aim to boost the network's ability to capture local features and integrate global information, strengthening its capacity to capture the intrinsic mapping relationship between chip power distribution and temperature fields, thus addressing the accuracy degradation caused by insufficient model learning and generalization ability. The model is trained on a large amount of labeled data generated by the DNS method, with chip power maps containing power and layout information as input, and steady-state MCM temperature maps as output. The experimental results show that by introducing techniques such as attention mechanisms and residual connections, as well as reasonably designing the network architecture, the generation accuracy can be significantly

to meet the rapid iteration requirements for steady-state improved while ensuring the speed of temperature map generation.

II. METHOD

A. Network Models

In this study, the network architecture is based on the U-Net model. U-Net, proposed by Ronneberger et al.[16] in 2015, is one of the most representative encoder-decoder structures, with the overall shape presenting a "U" form. It mainly consists of an encoder, a decoder, and skip connections. Initially mainly used for medical image segmentation tasks, U-Net has shown great potential for regression tasks due to its highly flexible structure. The traditional U-Net progressively reduces the image size through downsampling operations and then recovers the original image size via upsampling. However, this process often leads to the loss of global information. While skip connections can transmit some local information to help recover details, they primarily focus on the fusion of local features, making it difficult to effectively capture and integrate global contextual information. For chip temperature maps, the temperature distribution is often dependent on the global power distribution and thermal conductivity properties. Ignoring global information leads to big errors in predicting the central region and overall temperature distribution, thus reducing the prediction accuracy and the reliability of the generated temperature maps.

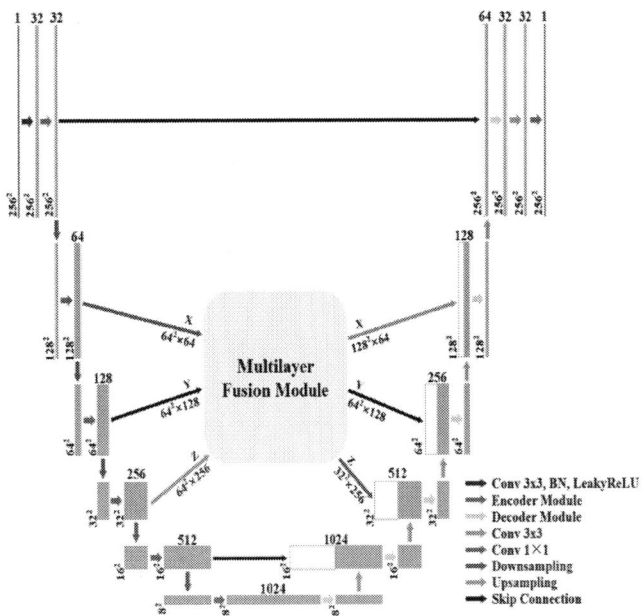

Fig. 1. Improved U-Net network model

To address the above issues, this paper proposes an improved U-Net network model as shown in Figure 1. By introducing residual connections, Efficient Channel Attention (ECA)[17], Spatial Attention Mechanism (SAM), and the Convolutional Block Attention Module (CBAM)[18] formed by their combination, the following modules are proposed:

(1) In the encoder, the ECA module and residual connections are introduced, and a multi-branch topology is employed. Different branches are used to extract diversified features, which are then fused. This approach effectively extracts key information while focusing on features critical to temperature distribution, and it helps the model learn the complex nonlinear relationships and local features between chip power and temperature.

(2) In the decoder, the ECA module is similarly introduced with a multi-branch topology, which, based on reducing the number of channels and lightweighting the network, effectively enhances information fusion and image recovery during the upsampling process, improving the efficiency of key information transmission. By capturing and integrating the global contextual information provided by the multi-layer feature fusion module and skip connections, the temperature distribution in the middle and overall areas of the temperature map is optimized, while mitigating issues such as detail loss

and image blurring caused by upsampling operations. This enhances the quality of the generated temperature map and helps the model generate more accurate and realistic temperature maps.

(3) Multi-layer feature fusion module is introduced in the middle part of the network, replacing the simple skip connections in the traditional U-Net with convolutional layers, SAM, CBAM, upsampling, and downsampling techniques. This captures and fuses multi-level encoder information, adaptively enhancing important features while suppressing irrelevant ones. It provides richer information for the decoder's decoding process, addresses the issue of poor long-range dependency in the basic U-Net, and improves the decoder's ability to recover details and reconstruct global structures.

The encoder, decoder, and multi-layer feature fusion module are shown in Figure 2, where C1 represents the input channel count and C2 represents the output channel count.

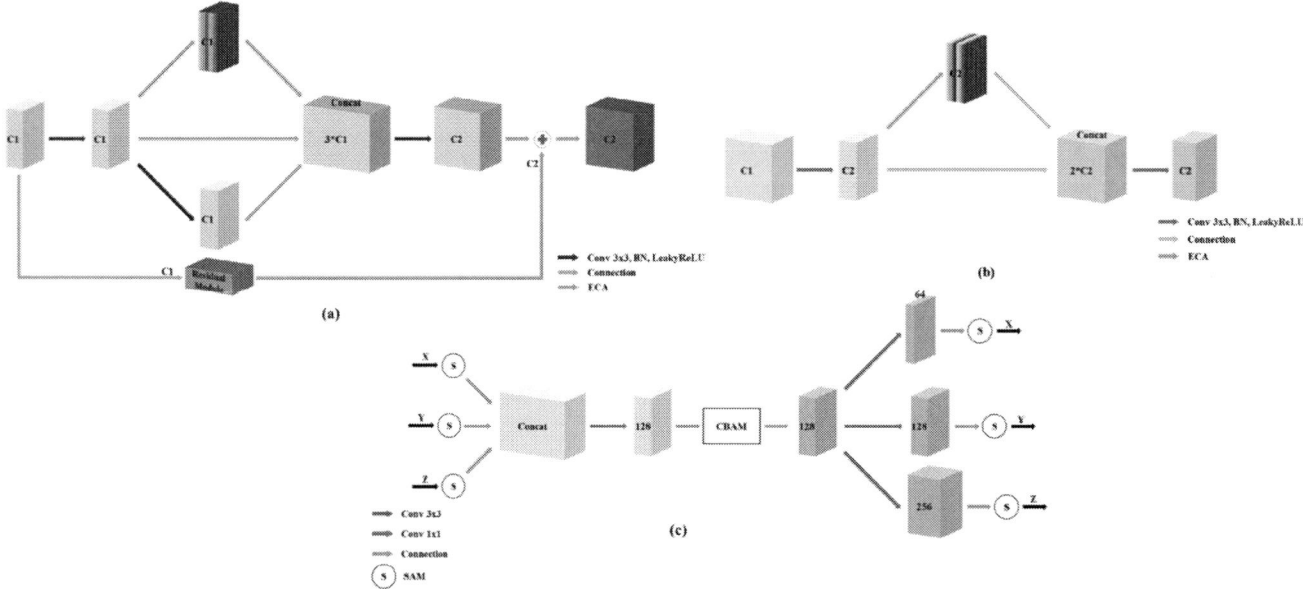

Fig. 2. (a) Encoder module (b) Decoder module (c) Multi-layer feature fusion module

B. Model Training

Deep learning networks not only depend on the design of the model but also require a reasonable selection of loss functions and hyperparameters. Considering that multi-chip temperature prediction is a regression problem, this study adopts the L1 loss function combined with L2 regularization for model training. The formula is as follows:

$$J = \frac{1}{N} \frac{1}{N_x \times N_y} \sum_{m=1}^{N} \sum_{x=0}^{N_x} \sum_{y=0}^{N_y} \left(| T(x,y) - \hat{T}(x,y) | + \lambda \| W \|_2 \right) \quad (1)$$

Here, N_x and N_y represent the dimensions of the temperature matrix in the x and y directions, respectively, N is the batch size, T is the true temperature matrix, and $\hat{T}$ is the temperature prediction from the network. $\lambda \| W \|_2$ represents the L2 regularization term, where W denotes the network weights and λ is the weight coefficient, with a value of 0.00001 in this study. During model training, the number of epochs and batch

size are both set to 50, and the initial learning rate is set to 0.0001. The Adam optimizer[19] is chosen, with BETA1 and BETA2 set to 0.9 and 0.999, respectively.

C. Dataset Construction

In this study, a simplified Direct Chip Attachment (DCA) type MCM model[2] is constructed, where the chip is directly attached to the substrate and equipped with a heat sink, as shown in Figure 3. Except for the contact areas, the remaining surfaces of the chip, substrate, and heat sink are subjected to convective and radiative heat exchange with the air domain, where the initial temperature of the air domain is set to 293.15 Kelvin (K). The physical property parameters of the chip, substrate, and heat sink are shown in Table 1.

Under the condition that the chip and substrate sizes remain unchanged, and the chip does not exceed the substrate nor physically intersect with it, we randomly place the chips on the substrate. Each chip in the MCM has the same power, and the substrate has no power. Initially, the number of chips in the MCM is set to 4, 8, 12, and 16, with the total MCM power corresponding to 3W, 4W, 5W, 6W, and 7W for each chip count. Each chip count has 1000 random layout configurations, resulting in a total of 20,000 data pairs. These data are then divided into 18,000 training data samples and 2,000 test data samples. The data set construction method is shown in Table 2.

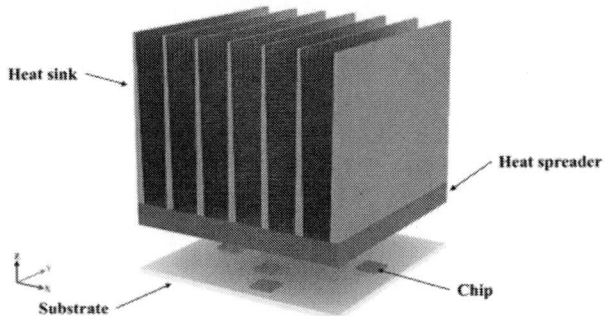

Fig. 3. Simple DCA type MCM

TABLE I. PHYSICAL PROPERTIES AND GEOMETRIC PARAMETERS OF THE CHIP AND SUBSTRATE IN THE MCM MODEL

Component	Physical Parameters	Unit	Value
Chip	Size	mm	4×4×0.02
	Thermal Conductivity	W/(m K)	100
	Density	kg/m3	2330
	Specific heat	J/(kg·K)	751.1
Substrate	Size	mm	40×40×0.1
	Thermal Conductivity	W/(m K)	100
	Density	kg/m3	2330
	Specific heat	J/(kg·K)	751.1
Heat sink	Material	-	Al
	Number	-	7
	Bottom size	mm	1.5
	Top size	mm	0.6
	High	mm	30
Heat spreader	Material	-	Al
	Size	mm	40×40×4

Apart from the aforementioned differences, the numerical simulation conditions and processes for all datasets are consistent and generated by COMSOL. The temperature data output by COMSOL is mapped to a 256×256 temperature matrix using a script. At the same time, a script is used to read the power and layout information corresponding to the chip temperature map, generating a 256×256 chip power map that contains the layout and power information, thus completing the construction of the dataset.

TABLE II. TRAINING AND TESTING DATASETS

Dataset	Chip numbers	Total power	Case numbers
Training	4/8/12/16	3W/4W/5W/6W/7W	18000
Test	4/8/12/16	3W/4W/5W/6W/7W	2000

D. Dataset Standardization

Due to the high complexity and challenges of the dataset constructed in this study, the number of chips, power levels, and layout configurations all result in significant variations in the chip temperature distribution. Specifically, the temperature range of the entire training dataset varies from 314K to 341.25K, with a difference of 27.25K. Given the large spread of the data and the significant temperature gradient, to ensure the stability and convergence efficiency of the model training, Z-Score normalization was applied to the training dataset, as shown in the following formula:

$$Z = \frac{X - \mu}{\sigma} \tag{2}$$

Here, X represents the original data, Z denotes the standardized data, μ is the mean of the feature, and σ is the standard deviation of the feature. The parameters used for the normalization process are saved for later use, allowing for the reverse normalization operation when generating temperature maps in the future.

E. Data Augmentation

Common data augmentation techniques include rotation, translation, scaling, cropping, brightness adjustment, and noise addition. In this study, the deep learning network needs to accurately capture the nonlinear relationships between chip power, layout, and temperature, as well as the local, middle, and global impacts during temperature diffusion. Therefore, a large amount of high-quality data is required to support the training of the deep learning model. Techniques such as translation, scaling, and cropping would alter the temperature diffusion relationship, losing its physical relevance. Thus, this study chooses to augment the dataset by rotating the original data by 90°, 180°, and 270°, as described in[20]. Through the data augmentation process, the total amount of training data increased to 72,000 samples.

F. Evaluation metrics

To assess the consistency between the predicted values and the true values and to validate the performance of the proposed model, this study employs several commonly used evaluation metrics in regression tasks.

1) Mean Absolute Error (MAE)

MAE can be used to measure the average deviation between the predicted temperature and the actual temperature. Its formula is as follows:

$$\text{MAE} = \frac{1}{N}\sum_{i=1}^{N}|y_i - \hat{y}_i| \tag{3}$$

Where y_i is the actual value of the sample, $\hat{y}_i$ is the predicted value of the sample, and N is the total number of samples.

2) Mean Squared Error (MSE)

MSE can be used to measure the squared deviation between the predicted temperature and the actual temperature. Its formula is as follows:

$$\text{MSE} = \frac{1}{N}\sum_{i=1}^{N}(y_i - \hat{y}_i)^2 \tag{4}$$

3) Mean Absolute Percentage Error (MAPE)

MAPE represents the mean absolute percentage error between the predicted values and the actual values. Its formula is as follows:

$$\text{MAPE} = \frac{100\%}{N}\sum_{i=1}^{N}\left|\frac{y_i - \hat{y}_i}{y_i}\right| \tag{5}$$

III. RESULTS AND ANALYSIS

This section will thoroughly explore the performance of the constructed model on test sets, with an analysis conducted based on two dimensions: chip count and power. This analysis will verify the effectiveness of the improved U-Net in generating chip temperature maps. All experiments are conducted on an Intel(R) Core(TM) i7-13700H CPU, RTX 4060 GPU, and Windows operating system, utilizing the PyTorch framework with GPU acceleration.

This section primarily evaluates the performance of the improved model under four chip counts (4, 8, 12, 16) and five total power configurations (3W, 4W, 5W, 6W, 7W). Figure 4 presents the MAPE, MAE, and MSE errors of the model on the test dataset across different chip counts and power configurations. As shown, when the total power remains constant, increasing the number of chips leads to a decrease in MAPE, MAE, and MSE errors. Specifically, the MCM with 4 chips has a higher MAPE, while the one with 16 chips has a lower MAPE. From the error boxplots in Figures 4(b) and 4(c), it can be observed that the improved model exhibits both MAE and MSE values within a low range on the test set, and these errors tend to decrease as the number of chips increases. Table 3 summarizes the model's prediction errors on the test dataset. The figure4 and table3 together indicate that the model performs excellently across the entire test set, with the maximum prediction error being only 1.08K. Even in the 4-chip configuration, the MAE is 0.04852, with only a few samples exceeding 0.15; the MSE is 0.00442, with only a few samples exceeding 0.04. This indicates that the model maintains high prediction accuracy across different power and layout scenarios.

In summary, the network proposed in this paper effectively learns the nonlinear relationship between chip power, layout information, and steady-state temperature, accurately capturing the thermal transfer characteristics of MCMs. Even in MCM layouts that were not present in the training set, the model maintains high prediction accuracy, demonstrating excellent learning and generalization capabilities.

IV. CONCLUSION

This paper proposes a steady-state temperature-solving method for multi-chip components based on an improved U-Net. By incorporating techniques such as attention mechanisms and residual connections, the U-Net network model is optimized to achieve fast and accurate steady-state temperature predictions. Experimental results demonstrate that the proposed method exhibits excellent performance in MCMs with varying chip numbers. Specifically, the MAPE for steady-state temperature predictions in MCMs is below 0.12%, and the maximum error is below 1.08K. Additionally, under the current device configuration, the fastest single temperature map generation time reaches 0.68ms, significantly surpassing DNS methods, thus achieving a balance between speed and accuracy. Further comparative experiments prove that by designing the network structure intelligently, the method not only guarantees extremely high computational efficiency but also significantly improves the prediction accuracy of steady-state temperatures in complex scenarios. This approach provides reliable and efficient technical support for MCM thermal design while opening new avenues for applying deep learning techniques in the field of chip thermal simulation.

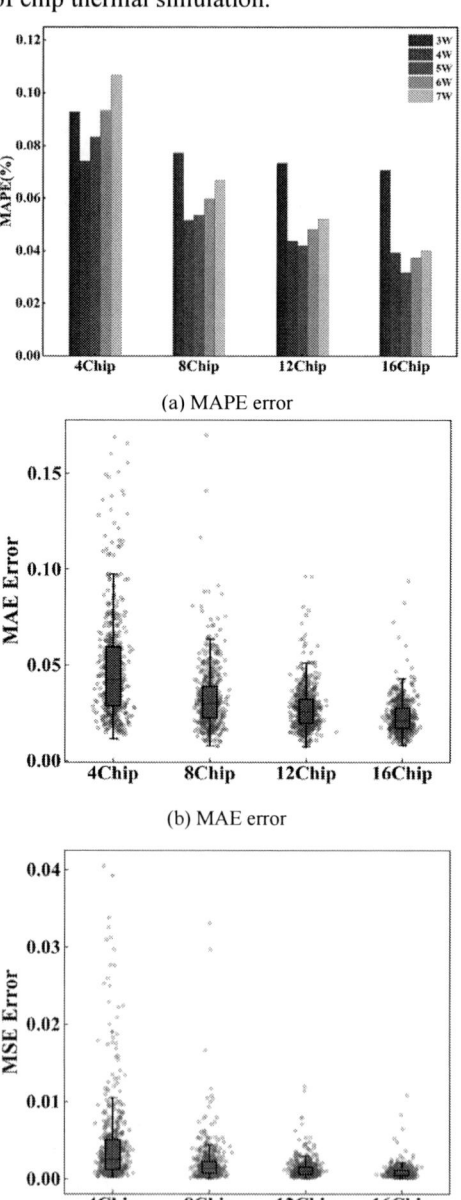

Fig. 4. Errors of multiple chip numbers, power configurations in Test

TABLE III. TEST PREDICTION ERROR

Chip	MAE	MSE	MAPE (%)	Max(K)
4	0.04852	0.00442	0.088	1.08
8	0.03286	0.00192	0.062	1.02
12	0.0274	0.00126	0.05	0.95
16	0.02288	0.00086	0.044	0.96
Test	0.0329	0.0021	0.06	1.08

REFERENCES

[1] C. Wang and K. J. A. T. E. Vafai, "Heat transfer enhancement for 3D chip thermal simulation and prediction," vol. 236, p. 121499, 2024.

[2] Y. Hua et al., "Estimation of steady-state temperature field in multichip modules using deep convolutional neural network," vol. 40, p. 101755, 2023.

[3] L. Jiang, A. Dowling, M.-C. Cheng, and Y. J. I. T. o. C. Liu, "PODTherm-GP: A physics-based data-driven approach for effective architecture-level thermal simulation of multi-core CPUs," vol. 72, no. 10, pp. 2951-2962, 2023.

[4] Z. Shi, K. Liang, Z. Zeng, and X. Shi, "Application of Imagery Modeling and Deep Learning in Chip Thermal Analysis," in 2024 31st International Conference on Mixed Design of Integrated Circuits and System (MIXDES), 2024, pp. 161-165: IEEE.

[5] Z.-Q. Wang et al., "Fast optimization of multichip modules using deep learning coupled with Bayesian method," vol. 141, p. 106592, 2023.

[6] Z.-Q. Wang, Y. Hua, H.-R. Xie, Z.-F. Zhou, Y.-B. Li, and W.-T. J. C. S. i. T. E. Wu, "Transfer learning of convolutional neural network model for thermal estimation of multichip modules," vol. 59, p. 104576, 2024.

[7] T.-Y. Wang, C. C.-P. J. I. T. o. c.-a. d. o. i. c. Chen, and systems, "3-D thermal-ADI: A linear-time chip level transient thermal simulator," vol. 21, no. 12, pp. 1434-1445, 2002.

[8] X. Guoping, "Thermal Modeling of Multi-Core Processors," in Tenth Intersociety Conference on Thermal and Thermomechanical Phenomena in Electronics Systems (ITHERM), 2006.

[9] K. R. Vaddina, A.-M. Rahmani, K. Latif, P. Liljeberg, and J. J. P. E. Plosila, "Thermal modeling and analysis of advanced 3D stacked structures," vol. 30, pp. 248-257, 2012.

[10] Y. Ping, X. Qin, C. Shen, N. J. I. J. o. M. Liao, and P. Technology, "Finite Element simulation for three dimensional thermal analysis of Multi-Chip Module," vol. 34, no. 3, pp. 241-250, 2009.

[11] K. B. Abdelmlek, Z. Araoud, K. Charrada, G. Zissis, and L. J. C. S. i. T. E. Canale, "Improvement of thermal and optical behavior of multi-chip LEDs package," vol. 39, p. 102395, 2022.

[12] M. Raissi, P. Perdikaris, and G. E. J. o. C. p. Karniadakis, "Physics-informed neural networks: A deep learning framework for solving forward and inverse problems involving nonlinear partial differential equations," vol. 378, pp. 686-707, 2019.

[13] W. Jin, S. Sadiqbatcha, J. Zhang, and S. X.-D. Tan, "Full-chip thermal map estimation for commercial multi-core CPUs with generative adversarial learning," in Proceedings of the 39th International Conference on Computer-Aided Design, 2020, pp. 1-9.

[14] V. A. Chhabria, V. Ahuja, A. Prabhu, N. Patil, P. Jain, and S. S. Sapatnekar, "Thermal and IR drop analysis using convolutional encoder-decoder networks," in Proceedings of the 26th Asia and South Pacific Design Automation Conference, 2021, pp. 690-696.

[15] H. Sultan and S. R. Sarangi, "Variability-aware thermal simulation using CNNs," in 2021 34th International Conference on VLSI Design and 2021 20th International Conference on Embedded Systems (VLSID), 2021, pp. 65-70: IEEE.

[16] O. Ronneberger, P. Fischer, and T. Brox, "U-net: Convolutional networks for biomedical image segmentation," in Medical image computing and computer-assisted intervention–MICCAI 2015: 18th international conference, Munich, Germany, October 5-9, 2015, proceedings, part III 18, 2015, pp. 234-241: Springer.

[17] Q. Wang, B. Wu, P. Zhu, P. Li, W. Zuo, and Q. Hu, "ECA-Net: Efficient channel attention for deep convolutional neural networks," in Proceedings of the IEEE/CVF conference on computer vision and pattern recognition, 2020, pp. 11534-11542.

[18] S. Woo, J. Park, J.-Y. Lee, and I. S. Kweon, "Cbam: Convolutional block attention module," in Proceedings of the European conference on computer vision (ECCV), 2018, pp. 3-19.

[19] D. P. Kingma and J. J. a. p. a. Ba, "Adam: A method for stochastic optimization," 2014.

[20] C. Shorten and T. M. J. J. o. b. d. Khoshgoftaar, "A survey on image data augmentation for deep learning," vol. 6, no. 1, pp. 1-48, 2019.

2025 The 10th International Conference on Integrated Circuits and Microsystems

Open-LUT: Interactive g_m/I_D Lookup Tables for MOSFET Sizing in Open-Source PDKs

Sihawi A. Khalid
Electrical and Electronics Engineering Department
Mindanao State University - Main (Marawi)
Marawi City, Philippines
sihawi.khalid@msumain.edu.ph

Susie E. Maestre
Electrical and Electronics Engineering Department
Mindanao State University - Main (Marawi)
Marawi City, Philippines
susie.maestre@msumain.edu.ph

Karla M. Madrid-Khalid
Electrical and Electronics Engineering Department
Mindanao State University - Main (Marawi)
Marawi City, Philippines
karla.madrid-khalid@msumain.edu.ph

Abstract—**This paper introduces an interactive web application designed to streamline device sizing in analog integrated circuit (IC) design for the open-source SKY130 Process Design Kit (PDK) using the g_m/I_D methodology. Called *Open-LUT*, the web application automates lookup table operations and provides real-time parameter updates, potentially reducing reliance on time-consuming circuit simulations, a common bottleneck in traditional analog IC design. The functionality of the application is verified through the design and simulation of a common-source amplifier, demonstrating efficient sizing and biasing with results verified against XSCHEM+NGSPICE simulations. While linear interpolation introduces inaccuracies, the ability of the tool ability to rapidly generate design parameters is confirmed. This work advances analog IC design automation within the open-source domain, offering a practical tool for student researchers and designers, particularly in academia. Future enhancements will focus on improving accuracy through advanced interpolation techniques, expanding PDK support, incorporating parasitic analysis for comprehensive circuit modeling, and exploring lookup table generation across different temperatures and process corners.**

Index Terms—**open-Source EDA, g_m/i_d methodology, analog IC design automation, sky130, lookup tables (lut), web application**

I. Introduction

The growing open-source integrated circuit (IC) design community, driven by the availability of Process Design Kits (PDKs) like SkyWater's SKY130 [1] and GlobalFoundries' GF180 [2], has noticeably broadened access to IC design [3], particularly within academic settings. SkyWater's landmark 2020 achievement, demonstrating the complete open-source design, verification, and fabrication of System-on-Chips (SoCs), highlighted the transformative potential of accessible fabrication and open-source EDA tools. This development has led to the growing adoption of IC design within academia [3]-[6].

Within the general IC design community and the open-source domain, however, analog IC design automation lags significantly behind its digital counterpart. Traditional

analog design methodologies persist, characterized by manual iterations, reliance on extensive circuit simulations, and empirical adjustments of numerous design variables. This approach, often guided by designer intuition, can lead to design parameter variations that are inefficiently aligned with target performance specifications, necessitating numerous, time-intensive iterations to meet the desired specifications [7].

The g_m/I_D methodology refines analog design by systematically relating transconductance to drain current, optimizing transistor sizing and biasing. This approach enhances performance metrics like gain, bandwidth, and power efficiency, while reducing reliance on time-consuming trial-and-error simulations [7]. Notably, it offers a unified framework for accurate transistor optimization across weak, moderate, and strong inversion regions, which is essential for contemporary CMOS analog IC design. The integration of precomputed lookup tables (LUTs) further streamlines the design process, as shown in prior studies [8]- [11]. Moreover, advanced tools incorporating g_m/I_D, as exemplified in [12], have shown significant enhancements in analog IC design efficiency through automation and optimization.

While the g_m/I_D methodology is compatible with open-source EDA toolchains and PDKs, their data/chart generation and automated design flow capabilities are markedly inferior to those found in commercial EDA solutions.

Although the open community has contributed valuable data and charts facilitating the g_m/I_D approach [13] [14], an automated design framework is yet to be established within the open-source environment. This paper introduces an interactive web application designed to address this gap. By providing open access to a g_m/I_D-based automation tool, this work delivers automated, albeit simplified, g_m/I_D lookup operations specifically tailored for the open-source SKY130 PDK.

The remainder of this paper is organized as follows: Section II details the lookup table generation methodology, Section III describes the design and implementation of the

979-8-3315-8850-2/25 $31.00 © 2025 IEEE

web application, Section IV evaluates the accuracy and performance of the application, Section V presents a common-source amplifier design example utilizing the web tool, and Section VI provides concluding remarks.

II. LOOKUP TABLE GENERATION

The extraction of MOSFET device parameters from the SKY130 Process Design Kit (PDK) was performed using XSCHEM, an open-source schematic editor integrated with the NGSPICE simulation engine. Circuit configurations employed for n-channel and p-channel MOSFET devices are depicted in Figures 1 and 2, respectively. Detailed descriptions of these configurations are available in [15]. For each device configuration, nested DC sweeps were executed, varying the gate-source voltage (V_{gs}) and drain-source voltage (V_{ds}) from 0 to 1.8 V with resolutions of 10 mV and 50 mV, respectively. The extracted parameters encompass transconductance (g_m), output conductance (g_{ds}), gate-source voltage (V_{gs}), threshold voltage (V_{th}), gate-source capacitance (C_{gs}), gate-gate capacitance (C_{gg}), drain-drain capacitance (C_{dd}), source-source capacitance (C_{ss}), channel width (W), and drain current (I_D). Multiple simulations were conducted for each device, varying the channel length (L). The extracted parameters from each simulation run were stored in separate *.dat files, with each file corresponding to a specific channel length and device type. The devices characterized include the three-terminal MOSFETs provided in the SKY130 PDK: *nfet_1v8* , *nfet_1v8_lvt* , *pfet_1v8* , *pfet_1v8_lvt* , and *nfet_1v8_hvt* . This parameter extraction procedure was performed once, resulting in the generation of a three-dimensional lookup table for each device, with V_{gs}, V_{ds}, and L as independent variables.

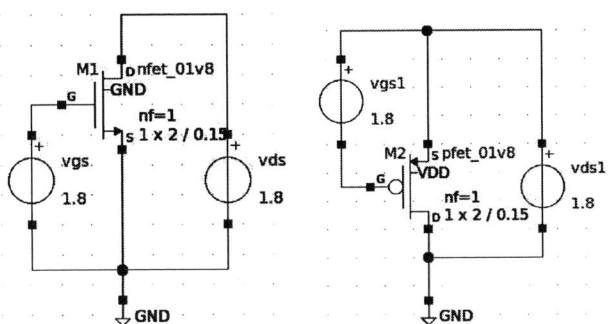

Fig. 1: Ckt. Setup for nfet   **Fig. 2:** Ckt. Setup for pfet

III. DESIGN OF THE WEB APPLICATION

A. Application Architecture

The interactive web-based tool, named Open-LUT, employs the g_m/I_D methodology to facilitate MOSFET device sizing. The web application comprises two components: a *client-side (front-end) web interface*, constructed using HTML, CSS, and JavaScript; and *a server-side (back-end) component*, implemented with Flask and Python.

The *client-side web interface* enables device and V_{ds} selection through HTML POST requests, provides interactive lookup functionalities with input design parameters, and includes graphical data visualization using ***chart.js***. This is accomplished by means of two HTML forms. One form allows users to select a device and V_{ds} value, while another accepts design parameters such as g_m, g_m/I_D, and g_m/g_{ds}.

The *back-end component* processes the HTML POST requests, computes the relevant lookup tables (LUTs) upon requests, and sends the LUTs back to the client-side. In particular, this back-end component retrieves simulation data (*.dat) files corresponding to the selected MOSFET device and V_{ds}. It then generates two-dimensional LUTs, encompassing g_m/I_D, intrinsic gain (g_m/g_{ds}), normalized drain current density (I_D/W), transition frequency (f_t), V_{gs}, and overdrive voltage (V_{ov}), for various channel lengths (L). The LUTs, indexed by V_{gs} and L, are then sent to the client-side.

The *client-side web interface* receives the generated LUTs. A JavaScript-based script implements the core logic for the interactive sizing tool, dynamically updating design parameters based on user input. This script performs LUT interpolations, utilizing conventional linear interpolation techniques as described in [15] and [16], and computes attainable parameter ranges to prevent out-of-bounds design results. The computed parameter values and ranges are then updated, providing dynamic feedback to the designer.

The HTML template was generated using Mobirise's Free Web Builder [17]. The web application is deployed on a PythonAnywhere cloud server and is publicly accessible at www.open-lut.org.

B. Web Interface Design and Functionality

Figure 3 is a representative screenshot of the primary interface of the interactive web application. The application enables users to initially specify the target MOSFET device and drain-source voltage V_{ds}. Subsequently, users can input design parameter values in an arbitrary sequence, resulting in dynamic updates across all related fields. To enhance the design process and prevent the entry of non-convergent values, dynamically computed parameter ranges are displayed adjacent to the input fields. An assertion mechanism is implemented to alert users to out-of-range input values, ensuring achievable output data. Additionally, charts showing g_m/I_D versus g_m/g_{ds}, f_t, V_{gs}, and I_D/W are provided to aid the designer.

The g_m/I_D, g_m, and I_D parameters exhibit a direct interrelationship. Upon modification of one parameter, the others are automatically recalculated. For instance, given $g_m = 1.5mS$ and $g_m/I_D = 15$, I_D is dynamically computed as $100\mu A$. An analogous relationship exists between g_m, g_{ds}, and g_m/g_{ds}.

The primary advantage of this application lies in its intuitive interface for efficient sizing and biasing. Existing tools in the literature that support designers in utilizing LUTs within the g_m/I_D methodology for both industry-standard and open-source PDKs commonly necessitate script execution and

979-8-3315-8850-2/25 $31.00 © 2025 IEEE

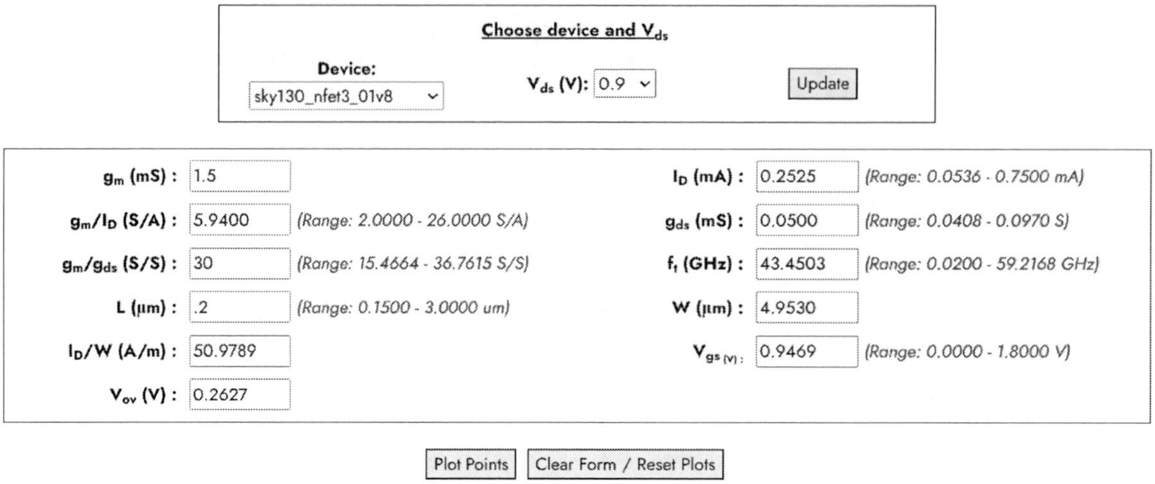

Fig. 3: Screenshot of the Web Application

chart generation using software such as MATLAB [15] or Python [13]. Furthermore, command-line interactions are often required for the extraction of desired small-signal parameters from these lookup tables. While these approaches have proven effective in numerous design scenarios, they often lack the intuitiveness and interactive feedback offered by the proposed web application. This limitation also hinders accessibility and ease of adoption within the academic and open-source communities, where streamlined and readily usable tools are highly valuable for research and educational purposes.

Conventional analog design typically involves targeting specific device parameters such as transconductance (g_m), intrinsic gain (g_m/g_{ds}), transit frequency (f_t), or bias current (I_D) based on power, performance, or area requirements. The design space for device sizing and biasing in analog circuits is extensive, and exhaustive exploration using traditional methods or even with g_m/I_D LUTs and charts can be iterative and time-consuming. Initial sizing and biasing choices frequently lead to unattainable targets later in the design flow, necessitating repeated adjustments.

In contrast, *Open-LUT* provides near-instantaneous feedback (sub-second latency) upon inputting design parameters in any order. This immediate response informs the designer about the feasibility of the specified targets and indicates the required or attainable ranges for other parameters, including device dimensions, bias conditions, and small-signal parameters. This interactive capability facilitates a more intuitive and informed design process, enhancing its utility for rapid prototyping and exploration in academic research and providing a valuable resource for students and researchers exploring the open-source IC design landscape. A design example demonstrating the the utility of the application in achieving specific design targets is presented in Section V.

C. Client-Side Sizing Algorithm

Upon modification of any input parameter within the web application, the sizing algorithm, as depicted in Fig. 4, is invoked to recalculate parameter ranges and update all field values. The sizing procedure concludes upon the determination of channel length (L) and channel width (W), resulting in a comprehensive update of all related parameters.

The algorithm initiates with a conditional branch based on the availability of a user-specified channel length (L). If L is provided, parameter ranges for g_m/I_D, g_m/g_{ds}, and f_t are updated for that specific L. Conversely, if L is unspecified, ranges are updated for all permissible L values. These cases are described as follows:

Case 1: L is provided, g_m/I_D is available: Where L is provided, the algorithm proceeds to evaluate the availability of g_m/I_D. If available, g_m/g_{ds}, f_t, and I_D/W are computed through interpolation from the lookup table. Fig. 5 shows the two-stage linear interpolation process employed to determine $(g_m/g_{ds})_{target}$ for a given $(g_m/I_D)_{given}$ and L_{given}. Specifically, to extract $(g_m/g_{ds})_{target}$, intermediate points pt_i and pt_{i+1} are first linearly interpolated between known data points within the lookup tables corresponding to channel lengths L_i and L_{i+1}, respectively, where $L_i \leq L_{given} \leq L_{i+1}$. Subsequently, $(g_m/g_{ds})_{target}$ is linearly interpolated between pt_i and pt_{i+1}. Similar interpolation techniques are employed for f_t, I_D/W, V_{gs}, and V_{ov}, as they can also be plotted as functions of g_m/I_D.

Case 2: L is provided, g_m/I_D is unavailable. If L is provided but g_m/I_D is not, the algorithm assesses the availability of g_m/g_{gds} or f_t. For instance, if g_m/g_{ds} is provided, g_m/I_D is calculated through interpolation from the lookup table (Fig. 6). Intermediate points pt_i and pt_{i+1} are generated by means of the iterative process depicted in Fig. 5, stepping through $(g_m/I_D)_i$ and $(g_m/I_D)_{i+1}$ with a fixed $\Delta(g_m/I_D)$. If $(g_m/g_{ds})_{given}$ lies within the interval defined by pt_i and pt_{i+1}, then $(g_m/I_D)_{target}$ is linearly interpolated

979-8-3315-8850-2/25 $31.00 © 2025 IEEE

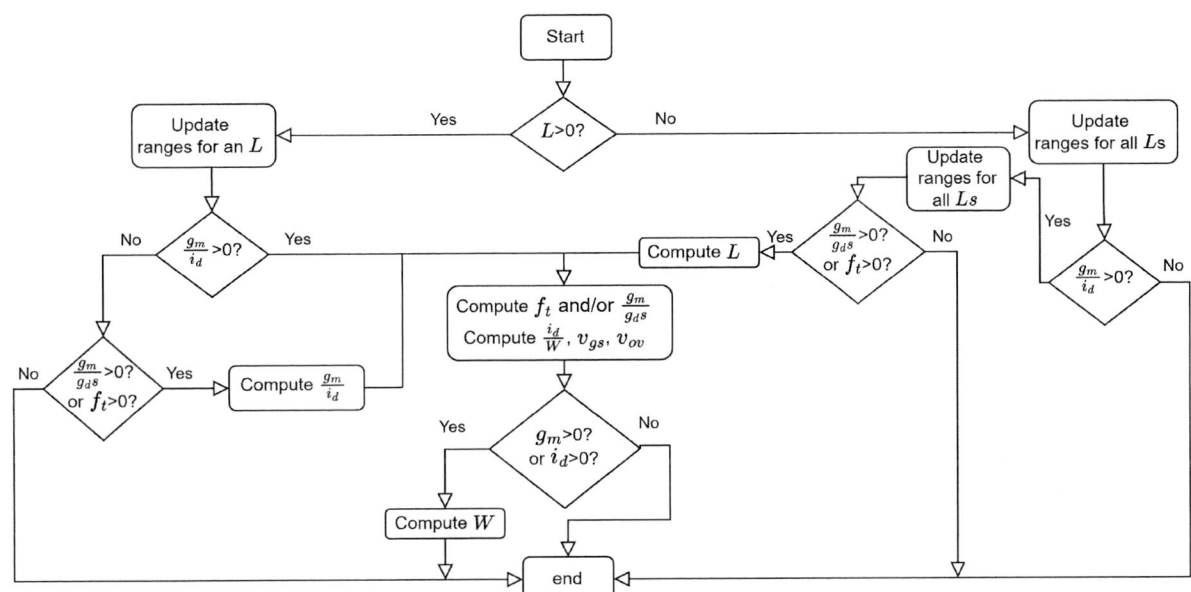

Fig. 4: Sizing Algorithm

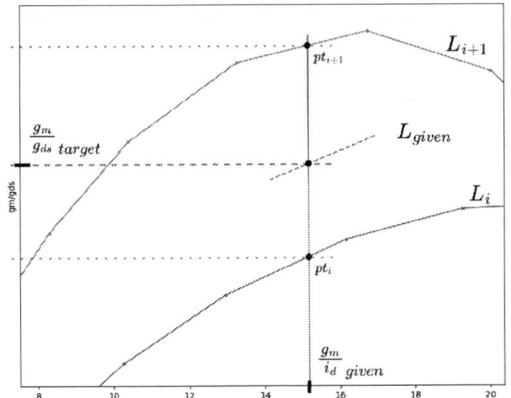

Fig. 5: Extracting g_m/g_{ds} given a specific g_m/I_D and L

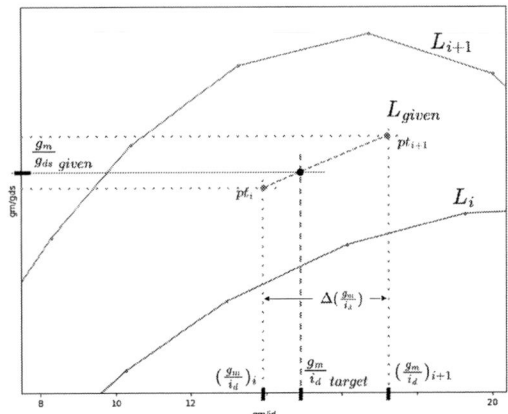

Fig. 6: Extracting g_m/I_D given a specific g_m/g_{ds} and L

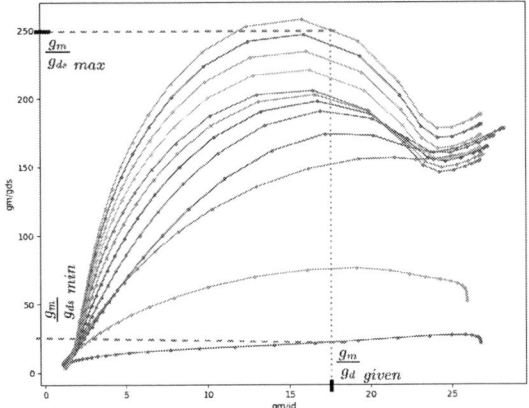

Fig. 7: Computing range of g_m/g_{ds} for a specific L

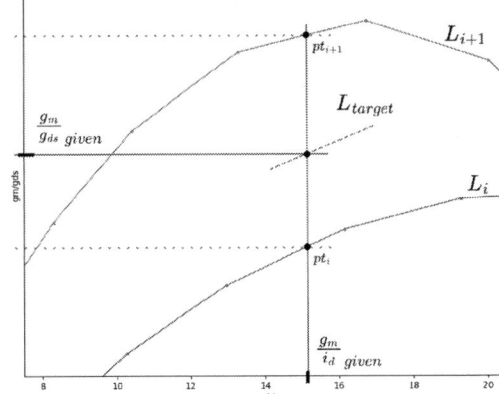

Fig. 8: Extracting L given a specific g_m/I_D and g_m/g_{ds}

979-8-3315-8850-2/25 $31.00 © 2025 IEEE

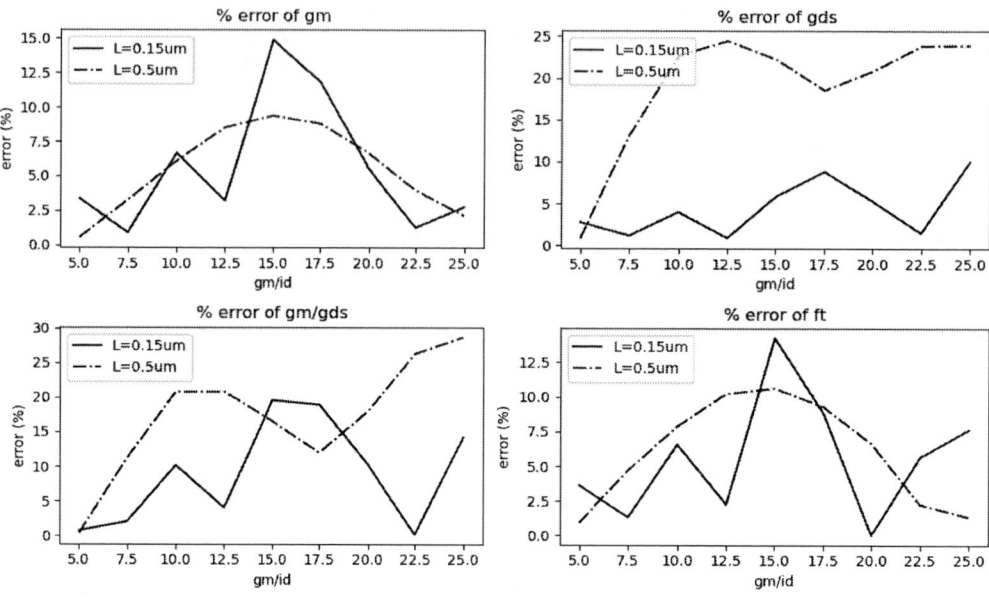

Fig. 9: Interpolation errors of g_m, g_{ds}, g_m/g_{ds}, and f_t for V_{GS} step size of 25mV

between these points. With L and g_m/I_D already known, f_t, I_D/W, V_{gs}, and V_{ov} are computed using similar procedure in Case 1.

Case 3: L is not provided, g_m/I_D is available. When L is not provided, the algorithm evaluates the availability of g_m/I_D in conjunction with either g_m/g_{ds} or f_t. If only g_m/I_D is provided, parameter ranges are updated (Fig. 7). If a combination of g_m/I_D and either g_m/g_{ds} or f_t is available, L is interpolated from the lookup table (Fig. 8). Similar to the procedure illustrated in Fig. 5, intermediate points pt_i and pt_{i+1} are initially linearly interpolated between known data points within the lookup tables corresponding to channel lengths L_i and L_{i+1}, respectively. However, in this case, pt_i and pt_{i+1} are selected such that they enclose the given $(g_m/g_{ds})_{given}$. The target channel length L_{target} is consequently linearly interpolated between pt_i and pt_{i+1}. Given the calculated L and the given g_m/I_D, similar procedure in Case 1 is used to determine f_t, I_D/W, V_{gs}, and V_{ov}.

Sizing is complete when channel width (W) is computed, using calculated I_D/W and either I_D or g_m.

IV. Accuracy and Performance Evaluation

The accuracy of the web application is assessed by analyzing the impact of varying step sizes in V_{GS} and L on the interpolation of key small-signal parameters. Performance is evaluated in terms of backend LUT generation and transmission, as well as frontend responsiveness for device sizing and biasing.

A. Impact of V_{GS} Step Size on Accuracy

The accuracy of the web application, which employs linear interpolation, is inherently linked to the granularity of the

lookup tables (LUTs) generated with respect to V_{DS}, V_{GS}, and L. Given that V_{DS} is typically maintained in the saturation region for analog design, this evaluation focuses on the effects of varying step sizes in V_{GS} and L. The influence of these step sizes on the two-stage operations detailed in Section III-C is crucial. To isolate the effect of V_{GS} step size, multiple LUTs were generated for the **_nfet_1v8_** device with V_{GS} step sizes of $25mV$, $10mV$, $5mV$, and $1mV$, while maintaining a constant L step size. For each LUT, device widths (W) were extracted for target g_m/I_D values ranging from 5 to 25 in steps of 2.5, at lengths $L = 0.15\mu m$ and $0.5\mu m$, with a fixed bias current $I_b = 100\mu A$ and $V_{DS} = 0.9V$. This resulted in 72 test points comprising device dimensions, extracted small-signal parameters, and biasing conditions obtained from the LUTs of the web application.

These 72 test points were then simulated in XSCHEM. The drain of the **_nfet_1v8_** device was connected to a $100\mu A$ current source, and V_{GS} was swept from 0 to 1.8V with a $0.1mV$ step to achieve $V_{DS} = 0.9V$. Small-signal parameters g_m, g_{ds}, and C_{gg} were extracted from the simulation results at the V_{GS} corresponding to $V_{DS} = 0.9V$. Subsequently, g_m/I_D, g_m/g_{ds}, and f_t were calculated. These simulated values were then compared to their counterparts obtained from the web application. Figure 9 illustrates the relative errors of g_m, g_{ds}, g_m/g_{ds}, and f_t as a function of g_m/I_D for $L = 0.15\mu m$ and $0.5\mu m$ using a V_{GS} step size of $25mV$. The results indicate that errors generally exhibit a dependence on the inversion region, with lower errors in strong inversion (SI) and higher errors in moderate inversion (MI) and weak inversion (WI) regions. The error range for g_m and consequently f_t (directly proportional to g_m) is $2\% - 14\%$, which is a significant improvement compared to the $20\% - 60\%$ error associated

979-8-3315-8850-2/25 $31.00 © 2025 IEEE   348

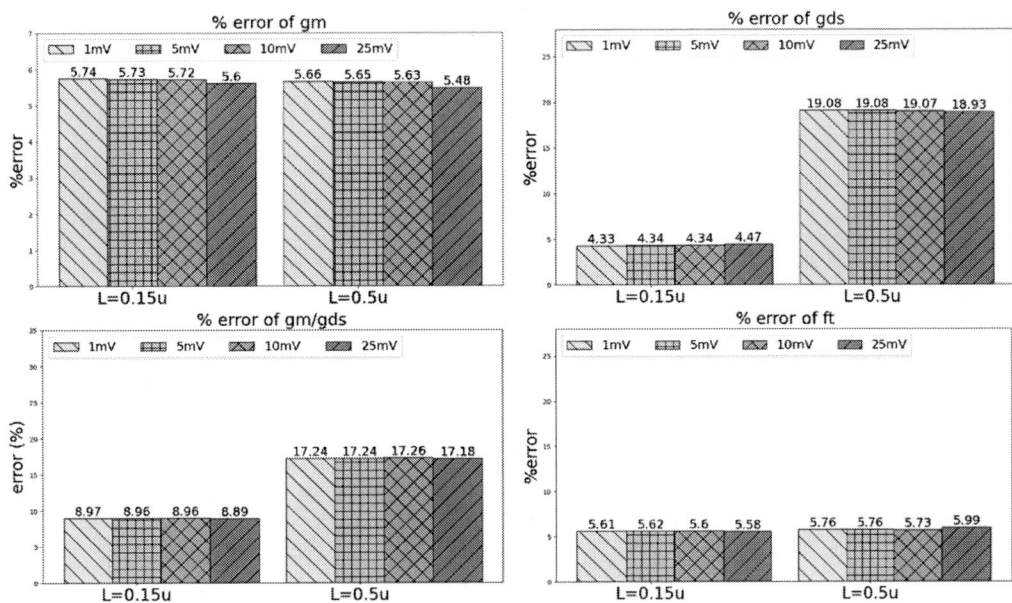

Fig. 10: Average errors for different V_{GS} step sizes

with the square-law model [15]. While g_{ds} and g_m/g_{ds} show larger errors ($10\%-20\%$), they remain within acceptable limits for typical LUT-based operations [15].

The same experimental setup was employed to evaluate LUTs generated with different V_{GS} step sizes. Figure 10 presents the average errors across two lengths and V_{GS} step sizes. Notably, the average errors remain relatively consistent regardless of the V_{GS} step size below $25mV$. This suggests that reducing the V_{GS} step size below $25mV$ does not significantly enhance accuracy. Therefore, a V_{GS} step size of $25mV$ offers a suitable trade-off, minimizing LUT size and maximizing operational speed without compromising accuracy. The optimal choice of V_{GS} step size is primarily dictated by the acceptable LUT size, which influences both the speed of LUT operations and the bandwidth required for transmitting LUT data in the web application.

B. Impact of L Step Size on Accuracy

The influence of the step size in L was investigated by comparing two LUTs generated with step sizes of $0.1\mu m$ and $0.05\mu m$. Two specific tests were conducted: one assessed the interpolation error for L on an LUT containing data for $L = 0.25\mu m$ and $0.15\mu m$ ($0.1\mu m$ step), while the other evaluated interpolation between $L = 0.23\mu m$ and $0.18\mu m$ ($0.05\mu m$ step). In both tests, the values of g_m, g_{ds}, g_m/g_{ds}, and f_t at $L = 0.2\mu m$ were interpolated. Figure 11 visually demonstrates the interpolation of g_m/g_{ds}, highlighting the discrepancies arising from different L step sizes.

The device dimensions and biasing conditions generated from both tests were simulated under conditions similar to those described in Section IV-A. The simulated small-signal parameters were then compared to their corresponding LUT-interpolated values. Figure 12 presents the errors for these

parameters across different inversion regions, along with the average error for each parameter. The results indicate that reducing the L step size from $0.1\mu m$ to $0.05\mu m$ led to an increase in the average g_m error by approximately 30%. However, given the already low error in g_m (as observed in Section IV-A), the absolute increase in g_m error remained relatively small. Conversely, significant error reductions were observed for g_{ds} (42%), g_m/g_{ds} (71%), and f_t (35%). This reduction is particularly important considering the relatively larger errors observed for g_{ds} and g_m/g_{ds} in Section IV-A. These findings underscore the significant impact of L step size on accuracy. However, reducing the L step size increases the LUT size, as the number of length points in the LUT is inversely proportional to the step size.

Selecting an appropriate L step size presents a trade-off between LUT size and the accuracy of the web application's LUT operations. For instance, covering a length range from $0.15\mu m$ to $3\mu m$ with a uniform step size of $0.05\mu m$ would necessitate 58 LUTs for each g_m/I_D parameter. However, analysis of the g_m/I_D charts reveals that the variation in small-signal parameters between different lengths diminishes at longer L. Figure 13 illustrates this point by comparing the g_m/g_{ds} values for lengths $3\mu m$ and $2.75\mu m$ ($0.25\mu m$ step) with those between $0.15\mu m$ and $0.4\mu m$ ($0.25\mu m$ step). This suggests that interpolation errors for shorter lengths are likely to be higher than those for longer lengths when using the same L step size. To mitigate this, a non-uniform L step size was implemented in the deployed web application, employing smaller steps for shorter lengths and larger steps for longer lengths. Figure 13 depicts the specific lengths used, demonstrating the varying step sizes designed to maintain relatively uniform gaps in small-signal parameter values across the entire length range. This approach allows for accurate

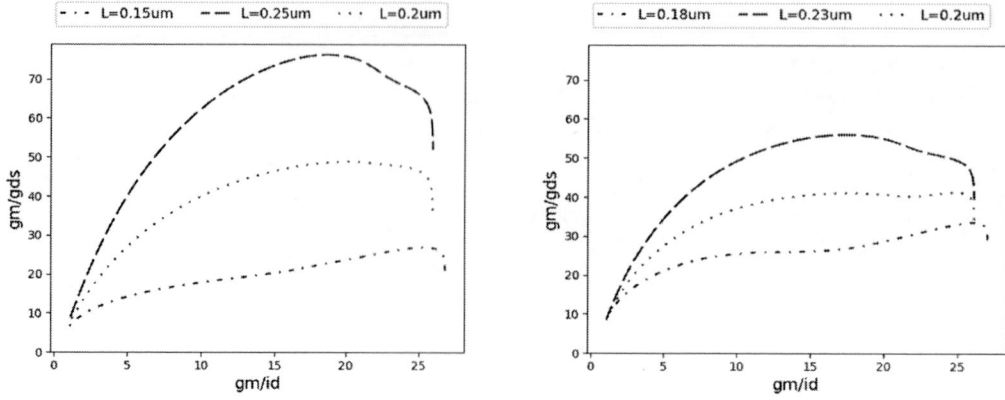

Fig. 11: Interpolation with $0.1\mu m$ vs. $0.05\mu m$ length L steps.

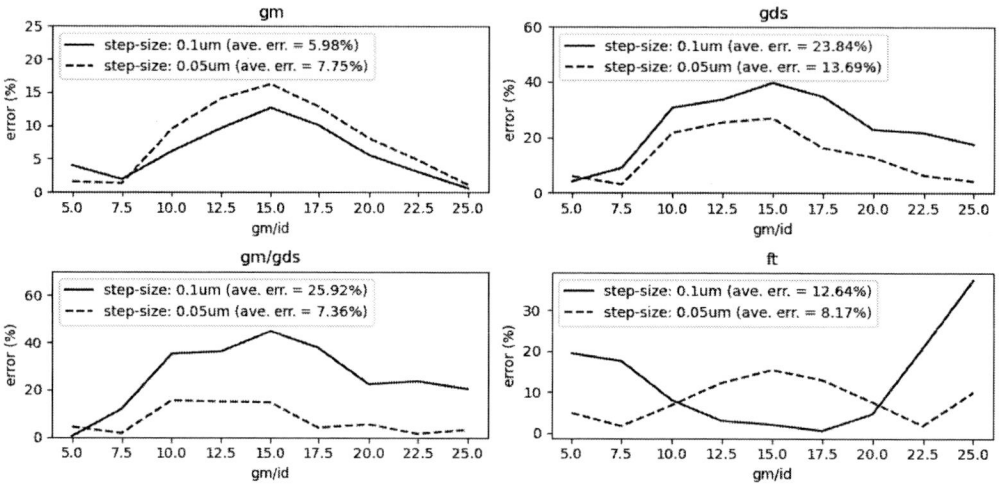

Fig. 12: Interpolation errors of g_m, g_{ds}, g_m/g_{ds}, and f_t for different L step sizes

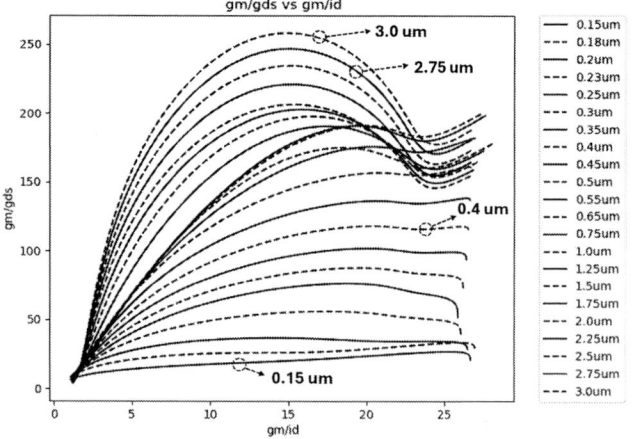

Fig. 13: Length Ls used in the LUTs

interpolation with only 22 length points, significantly reducing the LUT size compared to the 58 points required with a uniform $0.05\mu m$ step.

C. Speed Performance Evaluation

The speed performance of the g_m/I_D web application is evaluated based on two key aspects: backend performance and frontend responsiveness. Backend performance, crucial for the initial retrieval of LUT data, is assessed by measuring the *response time* to HTTP POST requests for specific device and V_{DS} biasing conditions (as described in Section III). This *response time* is further decomposed into LUT *generation time* (time to process raw data and compute all relevant g_m/I_D parameters for a given V_{DS}) and *latency* (time to transmit the generated LUT data to the client). While latency is heavily dependent on the variable internet connectivity and is not directly measured, the LUT size is used as a proxy for estimating its impact.

Performance measurements were conducted on a machine equipped with an AMD-Ryzen-7-7735HS processor and 32GB of RAM. To isolate the application's performance, the web application was hosted locally, eliminating the influence of internet connectivity. To accelerate performance testing across different V_{GS} step sizes, LUTs with only two length points were used, representing the minimum required for the LUT operations discussed in Section III. Table I summarizes the measured average speed performance and corresponding LUT sizes for various V_{GS} step sizes. As anticipated, employing a larger V_{GS} step size yields a significant speedup without compromising accuracy (as established in Section IV-A). While reducing the V_{GS} step size to $1mV$ offers a marginal improvement in accuracy, it incurs a substantial performance penalty, resulting in approximately x20 and x10 slower generation times compared to using $25mV$ and $10mV$ step sizes, respectively. This performance degradation is even more pronounced on the client side, with page load times being x450 and x130 times slower for a $1mV$ V_{GS} step size compared to $25mV$ and $10mV$, respectively. Similar trends are observed with LUT sizes, which directly impact the web application's latency.

The final deployed web application utilizes a V_{GS} step size of $10mV$ and LUT data for 22 non-uniformly spaced lengths for each of the five (5) devices available in the SKY130 PDK. Table II summarizes the overall performance of the deployed web application per device LUT.

TABLE I: Performance Metrics (Two Channel Lengths)

$\triangle V_{GS}$	LUT Generation Time (ms)	Page Load Time (ms)	LUT Size (kB)
$25mV$	42.02	12.10	64.38
$10mV$	88.30	41.06	155.97
$5mV$	175.82	211.42	308.61
$1mV$	848.46	5447.78	1529.73

TABLE II: Deployed Web Application Performance Metrics (per SKY130 device)

LUT Generation Time (ms)	Page Load Time (ms)	LUT Size (kB)
1003.024	748.37	1707.008

V. DESIGN EXAMPLE: COMMON SOURCE WITH ACTIVE LOAD AMPLIFIER

A. Sizing with the Web Application

A common-source amplifier with an active load (Fig. 14), designed to meet specifications similar to those presented in [12] and [18]- [19], was sized using the developed web application. The design specifications are summarized in Table III.

The bias voltage at the active load output was targeted at $0.9V$, representing the midpoint of the supply voltage, to maximize output swing.

From the DC gain and GBW specifications, the required g_{m1} of transistor $M1$ was calculated. The GBW is expressed as the product of DC gain (A_0) and $-3dB$ bandwidth (f_{-3dB}), and is related to g_{m1} as follows:

$$GBW = |A_0|f_{-3db} = g_{m1}R_{out}\frac{1}{2\pi R_{out}C_L} = \frac{g_{m1}}{2\pi C_L} \quad (1)$$

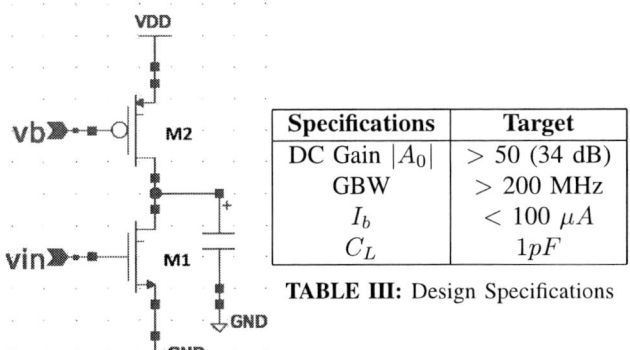

Fig. 14: CS Amp

Specifications	Target
DC Gain $|A_0|$	> 50 (34 dB)
GBW	> 200 MHz
I_b	< 100 μA
C_L	$1pF$

TABLE III: Design Specifications

Solving for g_{m1} yields a value of $1.26mS$. The DC gain of a common-source amplifier with an active load is given by:

$$|A_0| = \frac{g_{m1}}{g_{ds1} + g_{ds2}} \quad (2)$$

where g_{ds1} and g_{ds2} represent the drain-source conductances of $M1$ and $M2$, respectively. Choosing $g_{ds1} = g_{ds2} = g_{ds}$, the DC gain equation simplifies to:

$$|A_0| = \frac{g_{m1}}{2g_{ds}} \quad (3)$$

With $g_{m1} = 1.26mS$, the maximum permissible g_{ds} is calculated to be $g_{ds} = 12.6\mu S$. Utilizing the web application, the $nfet\_1v8$ device was selected, and V_{ds} was set to $0.9V$. Inputting $g_{m1} = 1.26mS$ and $g_{ds1} = 12.6\mu S$ into the application resulted in a g_m/g_{ds} ratio of 100. The application indicated that this value is attainable across a wide range of g_m/I_D values, with a minimum L of $0.3\mu m$. To minimize bias current, a high g_m/I_D ratio, corresponding to weak inversion (WI) operation, was chosen. A g_m/I_D value of 21 was selected. Inputting this value into the application converged to device dimensions of $L = 0.3444\mu m$ and $W = 95.8773\mu m$, with a V_{gs1} of 0.629 V. This g_m/I_D value resulted in an I_D of $60\mu A$, satisfying the bias current constraint.

The current flowing through transistor $M2$ is identical to that of $M1$, and g_{ds2} was previously established as $12.6\mu S$. To mitigate g_m/I_D degradation observed in SKY130 p-FETs operating in weak and moderate inversion, as reported in [20]- [21], design considerations mandate operation in strong inversion. A g_m/I_D ratio of 10 was selected. The web application was configured for $pfet\_1v8$ operation, with V_{ds} set to $0.9V$. Inputting $I_D = 60\mu A$, $g_m/I_D = 10$, and $g_{ds2} = 12.6\mu S$ converged to device dimensions of $L = 0.2576\mu m$ and $W = 16.3867\mu m$, with a bias source-gate voltage (V_{sg2}) of $0.9671V$. The sizing and biasing results for ($M1$) and ($M2$) are summarized in Table IV.

B. Results

The sizing results were validated by means of circuit simulation using XSCHEM and NGSPICE. A negative feedback loop, from the output node to the gate of transistor

979-8-3315-8850-2/25 $31.00 © 2025 IEEE

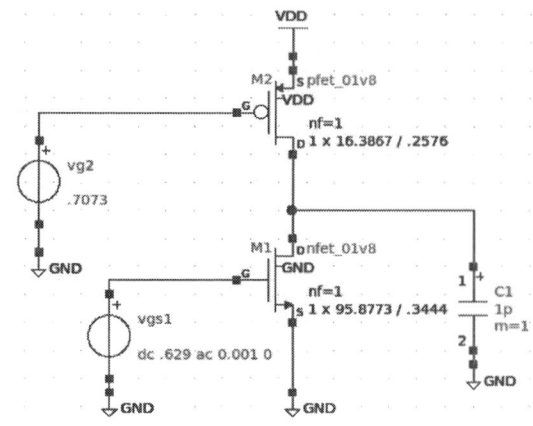

Fig. 15: Schematic for Simulation in XSCHEM

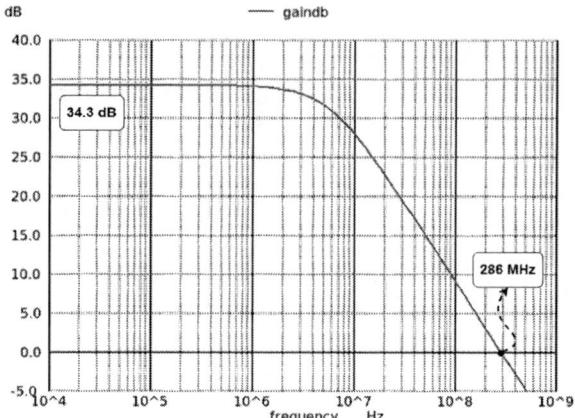

Fig. 16: AC Response

TABLE IV: Sizing and Biasing Results

Parameter	M_1	M_2
L (μm)	0.3444	0.2576
W (μm)	95.8773	16.3867
V_{gs}/V_{sg} (V)	0.629	0.9671
I_D (μA)	60	60

TABLE VI: Parameter Comparison

Parameter	Web App	Simulated
g_{m1}	1.26 mS	1.93 mS
g_{ds1}	12.6 μS	18 μS
g_{m1}/I_{D1}	21	21.6
g_{ds2}	12.6 μS	19 μS
g_{m2}/I_{D2}	10	8.95

$M2$, was simulated using a voltage-controlled voltage source (VCVS) to achieve the output bias voltage at $V_{DD}/2$ $(0.9V)$, resulting in an adjusted gate voltage (V_{g2}) of $0.7073V$ for $M2$. Fig. 15 illustrates the simulation setup employed for design verification. The amplifier's AC magnitude response is depicted in Fig. 16. The simulation results, summarized in Table V, demonstrate that all design specifications were met, with a DC gain of 51.8 $(34.3dB)$, a GBW of $286MHz$, and a bias current of $88.76\mu A$.

Table VI presents a comparative analysis of relevant parameter values obtained from simulation and the web application. The application's lookup operations, employing linear interpolation, introduce inherent inaccuracies, as documented in [7]. Despite these discrepancies, the results validate the effectiveness and efficiency of the design process, which minimizes reliance on iterative circuit simulations by leveraging lookup tables. Notably, the device sizing was derived exclusively from lookup table data based on the design specifications.

TABLE V: Simulation Results

Specifications	Target	Web App	Simulated
DC Gain $|A_0|$	> 50 (34 dB)	50 (34 dB)	51.8 (34.3 dB)
GBW	> 200 MHz	200 MHz	286 MHz
I_b	< 100 μA	60 μA	88.76 μA

VI. Conclusion

This paper describes the development of the *Open-LUT*, an interactive web application that facilitates device sizing in analog integrated circuit (IC) design for the open-source SKY130 Process Design Kit (PDK). Through a common-source amplifier design, the application was validated in its ability to efficiently produce sizing parameters, demonstrating its potential to improve open-source analog design automation and streamline exploration and optimization. By reducing the necessity for iterative simulations, this tool presents a valuable resource for students and academic researchers engaged in the study of open-source IC design tools.

However, several limitations were identified. Firstly, while device capacitances (C_{gg}, C_{dd}, C_{ss}) were extracted, the total parasitics post-sizing, and their impact on self-loading, stability, noise, and speed, were not considered. Secondly, the LUTs were restricted to three-terminal devices within the SKY130 PDK, neglecting the impact of body voltages (V_{SB} and V_{BS}). Lastly, the use of linear interpolation, which, while efficient, introduced inaccuracies that can become significant in certain circuit designs.

To address these limitations and enhance the application's capabilities, future work will focus on the following key areas. Total parasitics post-sizing will be computed to accommodate a wider range of design scenarios. Additionally, lookup table generation will be extended to include body voltage sweeps, allowing full characterization of four-terminal devices under varying bias conditions. Modified interpolation techniques, as suggested in [7], will also be investigated to improve accuracy. Finally, expansion to support devices from other open-source PDKs, such as GF180, will be considered to broaden applicability.

REFERENCES

[1] "SKY's the Limit with the SKY130 Open-Source PDK." Skywater. https://www.skywatertechnology.com/sky130-open-source-pdk/ (accessed Mar. 20, 2025).

[2] "Welcome to GlobalFoundries 0.18UM 3.3V/(5V)6V MCU PDK's documentation!" GlobalFoundries GF180MCU PDK. https://gf180mcu-pdk.readthedocs.io/en/latest/index.html (accessed Mar. 20, 2025).

[3] "Democratizing IC Design: The Story of a New Movement and the Launch of the SSCS PICO Program [Society News]," in IEEE Solid-State Circuits Magazine, vol. 13, no. 4, pp. 123-130, Fall 2021, doi: 10.1109/MSSC.2021.3111376.

[4] Ligutan, Dino Dominic and Abad, Alexander. (2024). "Democratizing IC Design in the Philippines - 'From Concept to Tapeout' Using Open-source Tools and Tiny Tapeout". SEE Proceedings 2024, Manila, Philippines, 2024, vol. 12.

[5] "Democratizing semiconductor design." ASU News. https://news.asu.edu/20231211-democratizing-semiconductor-design (accessed Mar. 21, 2025).

[6] "Open-source hardware: a growing movement to democratize IC design." University of Michigan - Michigan Engineering News. https://news.engin.umich.edu/2023/01/open-source-hardware-a-growing-movement-to-democratize-ic-design/ (accessed Mar. 21, 2025).

[7] A. A. Youssef, B. Murmann and H. Omran, "Analog IC Design Using Precomputed Lookup Tables: Challenges and Solutions," in IEEE Access, vol. 8, pp. 134640-134652, 2020, doi: 10.1109/ACCESS.2020.3010875.

[8] S. L. Pinjare, G. Nithya, V. S. Nagaraja and A. Sthuthi, "A Gm/Id Based Methodology for Designing Common Source Amplifier," 2018 2nd International Conference on Micro-Electronics and Telecommunication Engineering (ICMETE), Ghaziabad, India, 2018, pp. 304-307, doi: 10.1109/ICMETE.2018.00073.

[9] T. B. Kumar, G. K. Sharma, A. K. Johar, D. Gupta, S. K. Kar and D. Boolchandani, "Design Automation of 5-T OTA using gm/ID methodology," 2019 IEEE Conference on Information and Communication Technology, Allahabad, India, 2019, pp. 1-5, doi: 10.1109/CICT48419.2019.9066119

[10] H. Omran, M. H. Amer and A. M. Mansour, "Systematic Design of Bandgap Voltage Reference Using Precomputed Lookup Tables," in IEEE Access, vol. 7, pp. 100131-100142, 2019, doi: 10.1109/ACCESS.2019.2930595.

[11] Mostafa N. Sabry, Hesham Omran, Mohamed Dessouky, "Systematic design and optimization of operational transconductance amplifier using gm/ID design methodology," Microelectronics Journal, vol 75, 2018, pp 87-96, ISSN 1879-2391, https://doi.org/10.1016/j.mejo.2018.02.002.

[12] Omran Hesham, "Chapter 5 The Analog Designer's Toolbox (ADT) Towards a New Paradigm for Analog IC Design," in SMART Integrated Circuit Design and Methodology , River Publishers, 2023, pp.76-109.

[13] Nithin Purushothama, "gmid_SKY130." Github. https://github.com/chennakeshavadasa/gmid_SKY130 (accessed Mar. 17, 2025).

[14] Boris Murmann, "Ngspice-on-Colab." Github. https://github.com/bmurmann/Ngspice-on-Colab/tree/main/notebooks (accessed Mar. 17, 2025).

[15] P. Jespers and B. Murmann, Systematic Design of Analog CMOS Circuits Using Pre-Computed Lookup Tables. Cambridge, U.K.: Cambridge Univ. Press, 2017

[16] Boris Murmann, "Systematic Design of Analog Circuits Using Pre-Computed Lookup Tables" IEEE Toronto. https://www.ieeetoronto.ca/wp-content/uploads/2020/06/20160226toronto_sscs.pdf (accessed Mar. 15, 2025).

[17] "Free website maker" Mobirise. https://mobirise.com/website-maker.html (accessed Mar. 20, 2025).

[18] Hesham Omran, "The Analog Designer's Toolbox (ADT) Towards A New Paradigm for Analog IC Design." MOS AK. https://www.mos-ak.org/spring_2022/presentations/Omran_Spring_MOS-AK_2022.pdf (accessed Mar. 17, 2025).

[19] Hesham Omran, "The Analog Designer's Toolbox (ADT): Towards A New Paradigm for Analog IC Design", Youtube, https://www.youtube.com/watch?v=HWgzN1hQFPI&list=PLMSBalys69yyjfUj5LNzE2Hwt5aPtn2HW&index=3 (accessed Mar. 17, 2025).

[20] skywater-pdk, "MOSFET models show nonphysical gm/ID behavior" Github. https://github.com/google/skywater-pdk/issues/381 (accessed Mar. 18, 2025).

[21] "Skywater 130nm Library Model Evaluation." University of Hawaii. https://www2.hawaii.edu/ whitece6/posts/firstpost/ (accessed Mar. 18, 2025).

2025 The 10th International Conference on Integrated Circuits and Microsystems

A 9.37-Tb/s/mm 5-bit-6-bit Crosstalk Cancellation Transceiver with ICBS Codes for High-Density Die-to-Die Interfaces

Ruotian Yin
College of Computer Science and Technology National University of Defense Technology
Changsha, China
2329496650@qq.com

Jinwen Li
College of Computer Science and Technology National University of Defense Technology
Changsha, China
lijinwen@sina.com

Fangxu Lv*
College of Computer Science and Technology National University of Defense Technology
Changsha, China
lvfangxu1988@nudt.edu.cn
*Corresponding author

Geng Zhang
College of Computer Science and Technology National University of Defense Technology
Changsha, China
zhanggeng23@nudt.edu.cn

Ruixiao Kuai
College of Computer Science and Technology National University of Defense Technology
Changsha, China
18195321046@163.com

Bohui Bai
College of Computer Science and Technology National University of Defense Technology
Changsha, China
2010125932@qq.com

Abstract—A 9.37-Tb/s/mm 5-bit-6-bit crosstalk cancellation (XTC) transceiver is proposed for high-density Die-to-Die Interfaces. An XTC scheme based on Inter-Channel Bit Smoothing Coding (ICBS) is proposed to improve the signal integrity. Through controlled inter-channel bit transitions, the ICBS reduces far-end crosstalk (FEXT) by 40.07%. Simulation results in 28 nm CMOS demonstrate 69.3% jitter reduction at 12.5-Gb/s/pin on a high-density on-chip channel with FEXT of -4.14dB, enabling reliable high-speed multi-die integration.

Keywords—far-end crosstalk (FEXT), crosstalk cancellation (XTC), high-density die-to-die interfaces, edge density

I. INTRODUCTION

The exponential growth of data-centric computing paradigms has necessitated the development of high-speed, high-precision data center interconnects, posing stringent demands on linearity and noise suppression [1-2]. Concurrently, advancements in information technology continue to fuel the demand for scalable bandwidth solutions. However, conventional approaches face inherent limitations in physical dimensions, power dissipation, and cost efficiency, impeding their ability to meet evolving performance requirements [3-4]. As a result, enhancing bandwidth density without compromising signal integrity has emerged as a central focus for technological advancement in modern interconnect architectures.

To address this challenge, researchers have explored multiple strategies, including novel architectural designs, aggressive scaling of linewidth/space ratios, and direct data rate enhancement. However, with channel pitch aggressively scaling,

enhanced electromagnetic coupling between adjacent conductors significantly exacerbates crosstalk impairments [5–7]. This necessitates the development of advanced XTC codes. Building on this foundation, this work integrates crosstalk cancellation coding with channel characterization to maximize bandwidth and edge density while preserving signal integrity.

Fig. 1 depicts the evolution of edge density metrics. Current solutions for high-density die-to-die interfaces fall into two categories: packaging-level engineering [8–11] and circuit-level techniques [12–17]. Packaging-level approaches involve spacing strategies, swing optimization reduces worst-case coupling delays in long parallel lines [9], while serpentine guard traces mitigate crosstalk [10], etc. However, these methods introduce significant area overhead and routing complexity,

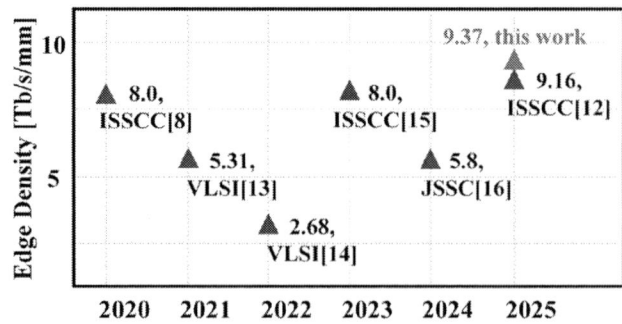

Fig. 1. Emerging trends in edge density advancement.

National Key Research and Development Program (2023YFB4403402).

979-8-3315-8850-2/25 $31.00 © 2025 IEEE   354

conflicting with the cost efficiency required for high-density applications. Circuit-level solutions, such as UI-pulse shaping[12] and multi-valued signaling [17], offer crosstalk mitigation but impose stringent precision requirements on analog front-ends.

Consequently, XTC codes emerged to meet these needs. Commonly used codes include Fibonacci coding (FC) [18-19], bit reversal coding (BI) [20-21], special delta coding [22-23], Non-adjacent transition Coding (NATs) [24], SDT-free encoding [25], JCI encoding [26], and so on. Among these, FC demonstrates superior crosstalk suppression capabilities. However, a critical limitation is that n-bit Fibonacci coding can only represent decimal values up to $F(n+2) - 1$, where $F(n+2)$ represents the $(n+2)_{th}$ term of the Fibonacci sequence. For example, to perform Fibonacci coding on 5-bit data, at least 7 bits are required, which means two redundant bits are needed. As a result, challenges still exist regarding insufficient XTC ratio and pin efficiency.

To address this issue, we designed an inter-channel bit smoothing coding (ICBS) transceiver. By reducing inter-channel bit transitions, far-end crosstalk (FEXT) between lines is 40.07% reduced. Compared with Fibonacci coding, the pin efficiency is improved by 16.67%. Furthermore, using dynamic programming [27] for mathematical calculations, ICBS requires only one redundant bit for 6-bit data encoding and only two for 10-bit data encoding. This paper uses 5-bit data encoding as an example of the design.

II. ANALYSIS OF CROSSTALK MECHANISM

Crosstalk phenomena not only impact signal swing by introducing extraneous noise but also induce delays in the transition characteristics, specifically the rise and fall times, of the original signals. This interference can significantly degrade signal integrity and system performance, particularly in applications that are high-speed and precision-dependent. When the input voltage $V_{in}(t)$ is applied to the aggressor, the far-end crosstalk induced at the victim can be expressed as

$$V_{XT}(t) = \frac{K \cdot d[V_{in}(t) \cdot h(t)]}{dt} \quad (1)$$

Where K is a constant, $h(t)$ is the transfer function, d/dt means taking the derivative. Since the derivative represents the rate of change, crosstalk can be standardized in the digital domain. For the sake of discussion, coupling effects between non-adjacent channels are temporarily disregarded. Fig.2 illustrates the measurement principle for crosstalk, C_c represents the coupling capacitance between channels, $s_i \rightarrow s_i'$ represents the bit transition from the current time to the next time, and XT_i represents the crosstalk from aggressive lines to the victim line.

In this case, the definition of the $|XT|$ for the lane i in transmission is:

$$|XT| = D|(\delta_i - \delta_{i-1}) + (\delta_i - \delta_{i+1})| \quad (2)$$

Where D is a constant related to coupling capacitance and signal swing, $\delta_i \in \{\pm 1\}$, $\delta_{i-1} \in \{0, \pm 1\}$ represents the normalized voltage variation on the lane i in transmission. (When δ_i is 0, there is no data transition. Therefore, discussing the jitter and delay caused by crosstalk is not meaningful.)

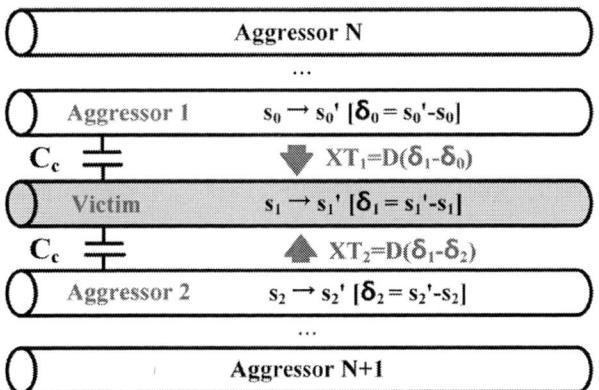

Fig. 2. Crosstalk interaction schematic between the aggressive-victim line.

Crosstalk-induced signal degradation is a critical concern in high-speed interconnects, manifesting as latency variations and timing jitter. Through systematic analysis, the maximum crosstalk amplitude of $4D$ was identified under the $010 \rightarrow 101$ transition pattern, whereas the minimum $0D$ crosstalk was observed during the $000 \rightarrow 111$ data transformation. Experimental validation is provided in Fig. 3 and Fig. 4, which corroborate the theoretical predictions.

As illustrated in Fig. 3, when 5-bit data transitions occur in a unidirectional manner, the crosstalk coupling to the victim line remains negligible. Conversely, Fig. 4 demonstrates that the $010 \rightarrow 101$ pattern generates the peak crosstalk amplitude of $4D$ during transmission. Notably, through advanced coding optimization strategies, this crosstalk magnitude can be effectively mitigated to $2D$. These results highlight that eliminating high-intensity crosstalk modes ($3D$ and $4D$) via pattern selection significantly enhances signal integrity.

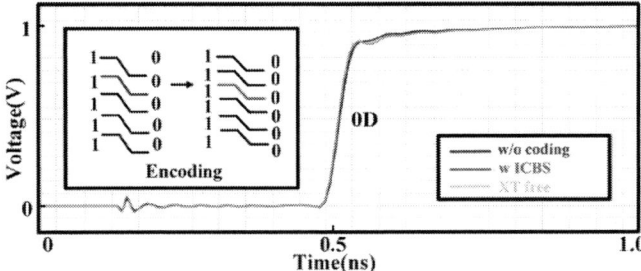

Fig. 3. Waveform comparison of signals with/without ICBS coding under 0D crosstalk condition.

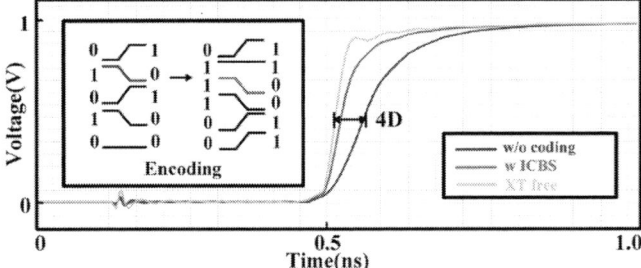

Fig. 4. Waveform comparison of signals with/without ICBS coding under 4D crosstalk condition.

979-8-3315-8850-2/25 $31.00 © 2025 IEEE

III. Encoding Principle for Multi-lanes

A. Introduction of inter-channel bit smoothing coding (ICBS)

In this research, an XTC block was designed leveraging ICBS codes to minimize inter-wire FEXT while preserving transmission efficiency. As elaborated in Section 2, the $010 \rightarrow 101$ transition pattern induces severe crosstalk coupling. Therefore, this coding scheme was specifically developed to suppress such high-impact bit transitions. Table 1 illustrates the 5-bit-to-6-bit ICBS encoding scheme for data transmission. Experimental results demonstrate that by eliminating the $010 \rightarrow 101$ transition mode, inter-channel crosstalk was attenuated by 40.07%. Compared with Fibonacci coding, the pin efficiency is improved by 16.67% while delivering superior crosstalk cancellation performance.

Fig. 5(a) presents the complete crosstalk profiles observed on the victim line during 5-bit data transmission, visualized in blue. Notably, the peak crosstalk amplitude reaches 4D under this configuration. In contrast, Fig. 5(b) demonstrates the post-ICBS encoding crosstalk characteristics, indicating that crosstalk has been effectively suppressed across all tested scenarios. Fig. 6 presents the crosstalk magnitude characteristics on the victim line before and after encoding during signal transmission. The experimental results show that the average crosstalk amplitude was reduced from 1.335D to 0.802D through encoding, achieving a significant suppression of 40.07%. Fig. 7 depicts the XTC operational principle alongside its corresponding timing diagram. The black trace corresponds to the original transmitted signal, while the red waveform represents the corrupted signal affected by crosstalk coupling from adjacent aggressor lines. As demonstrated, the encoding strategy effectively mitigates crosstalk-induced distortion, highlighting the robustness of the proposed approach.

TABLE I. 5-bit ICBS Encoding Principle

Binary Dataword	ICBS Codeword	Binary Dataword	ICBS Codeword
00000	000000	00001	000001
00010	000011	00011	000110
00100	000111	00101	001100
00110	001110	00111	001111
01000	011000	01001	011001
01010	011100	01011	011110
01100	011111	01101	100000
01110	100001	01111	100011
10000	100110	10001	100111
10010	110000	10011	110001
10100	110011	10101	111000
10110	111001	10111	111011
11000	111110	11001	111111
11010	000010	11011	000100
11100	001000	11101	001001
11110	001101	11111	010000

B. Detailed Implementation

Fig. 8 shows the schematic of an advanced encoding-based communication architecture for signal integrity enhancement. The proposed system architecture integrates encoding techniques to optimize signal integrity and transmission reliability in high-speed interconnects. Comprising transmitter (TX) and receiver (RX) subsystems, the architecture achieves synchronization through a dedicated clock distribution network featuring phase interpolation circuitry (PI) and a CML-to-CMOS converter for stable clock signal conditioning.

The TX begins with a pseudo-random binary sequence (PRBS) generator for test pattern generation. Raw data signals are processed by an XTC encoder, which applies inter-channel bit-smoothing coding to suppress coupling noise. The encoding and decoding modules are implemented by logic gates. Encoded signals traverse a 2:1 multiplexer (MUX) before being amplified by a pre-driver, SST driver cascade, designed to match channel impedance characteristics. Six parallel channels facilitate signal transmission to the RX. At the RX, incoming signals first pass through a buffered amplifier (BUF) to stabilize voltage levels. A two-stage comparator array performs analog-to-digital conversion, followed by decoding circuitry to recover the original data stream. A PRBS checker validates data integrity through real-time error rate monitoring.

The driver modules have been extensively studied and refined in our previous research [28-29]. The receiver in this

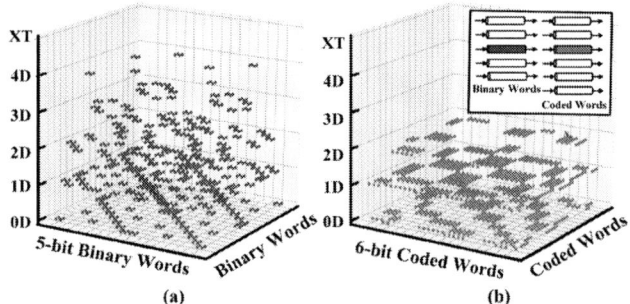

Fig. 5. Suppression performance of encoding in 5-bit data transmission.

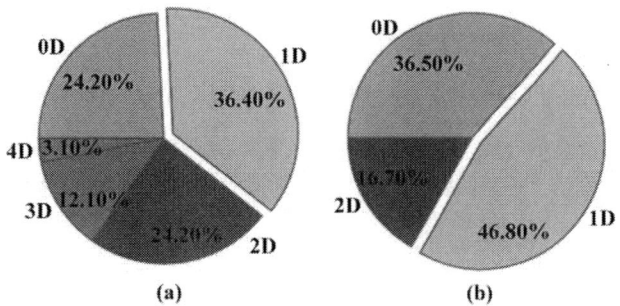

Fig. 6. XT magnitude characteristics on the victim line before and after encoding.

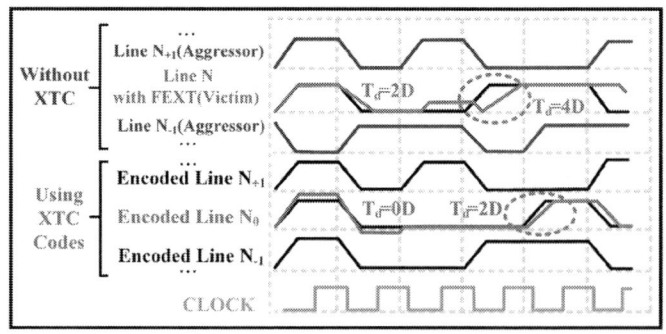

Fig. 7. Signal timing analysis of aaggressor-victim lines.

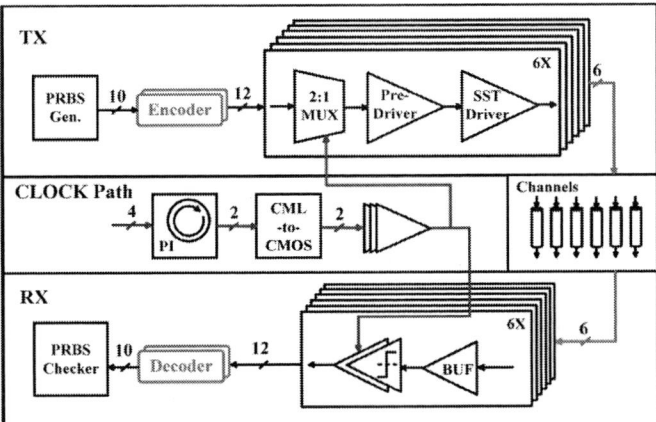

Fig. 8. Top-level circuit structure of verifying the proposed coding scheme.

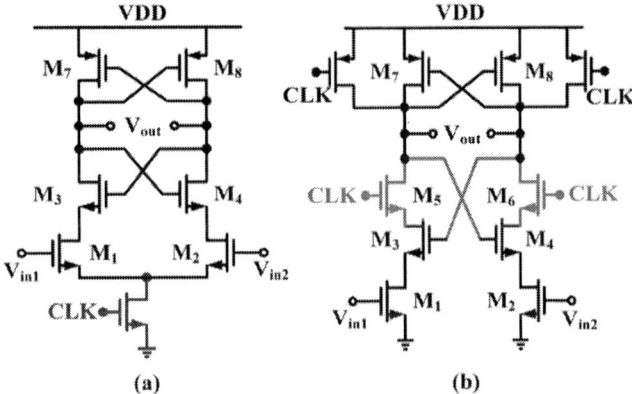

Fig. 9. Comparators with (a) traditional structure and (b) novel structure for lower kickback noise.

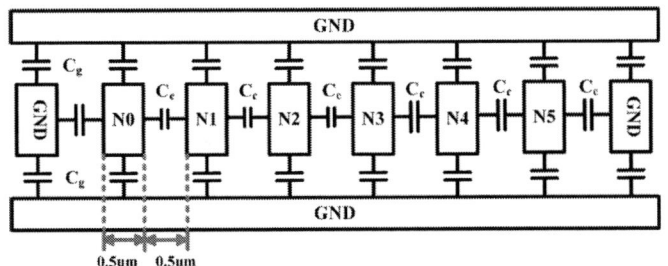

Fig. 10. On-chip channel geometry and interconnect layout.

study incorporates a novel comparator as depicted in Fig.9. This topology reduces the kickback noise by clocking the input devices through their drain path rather than their source path, utilizing transistors M5 and M6 to regulate latch operations. Consequently, this approach effectively minimizes kickback noise arising from differential and common-mode interactions during the regeneration phase. Power consumption is also an important issue[30].

C. Channel Module Characterization

Fig. 10 illustrates the on-chip channel architecture optimized for high-density Die-to-Die interconnects, featuring 0.5 μm linewidth, 0.5 μm spacing, and 4 mm length. Experimental results in Fig. 11 demonstrate the channel's signal integrity performance at 7.5 Gbps Nyquist rate, revealing insertion loss

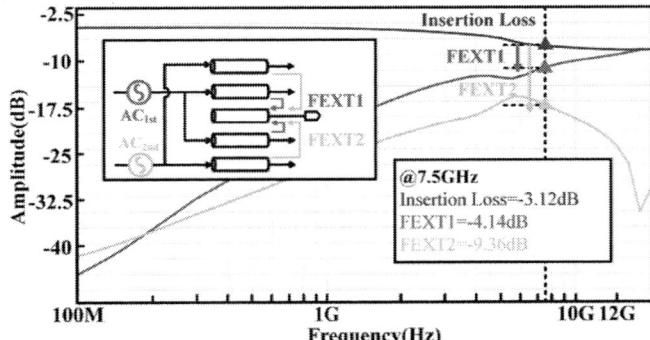

Fig. 11. Signal integrity characterization of the proposed channel.

(IL) of -3.12 dB and far-end crosstalk (FEXT) with distinct proximity-dependent characteristics: the nearest-neighbor coupling (FEXT₁) measures -12.1 dB (IL-to-XT ratio of -4.14 dB, while the second-order adjacent coupling (FEXT₂) reduces to -18.1 dB (ratio of -9.36 dB). These measurements show that FEXT₁ contributes 6.0 dB higher crosstalk than FEXT₂, confirming that near-neighbor interference dominates crosstalk behavior. The results also provide experimental validation for crosstalk modeling and mitigation strategies in high-speed, high-density interconnect systems.

IV. SIMULATION RESULTS

This section presents a systematic investigation into the quantitative impact of ICBS through comparative numerical simulations between non-coded and ICBS-encoded transmission systems. To validate the robustness of the proposed scheme against operational variations, analysis was performed across a range of data rates and crosstalk intensities.

A. Eye Diagram

Fig. 12 and Fig. 13 present a systematic evaluation of inter-channel bit-smoothing coding (ICBS) through eye diagram measurements at 15 Gbps and 20 Gbps data rates, respectively. Quantifiable improvements in signal integrity metrics are observed across all tested conditions:

At a data rate of 15 Gbps, the encoded signal demonstrates a 102.2% increase in eye width (from 27.3 ps to 54.2 ps) and an 83.7% enhancement in eye height (from 146.1 mV to 268.4 mV), as shown in Fig. 12. Timing jitter is reduced by 69.3% (from 36.8 ps to 11.27 ps). At a data rate of 20 Gbps, the encoding strategy maintains its efficacy under elevated data rates, achieving a 149.4% increase in eye width (from 67.8 ps to 169.1 ps) and a 213.6% improvement in eye height (from 12.5 mV to 39.2 mV), as illustrated in Fig. 13. Jitter performance is also significantly improved, with a 67.7% reduction in overall timing jitter (from 33.3 ps to 10.77 ps). These results highlight the encoding scheme's capability to mitigate both deterministic and random jitter components across varying operational conditions.

B. Edge Density Optimization via ICBS Encoding

As previously discussed in Section 1, minimizing channel spacing enhances interconnect edge density but exacerbates crosstalk-induced signal degradation. To address this trade-off, Section 3.3 presents simulations across high-density channel realizations incorporating ICBS encoding, yielding the optimal

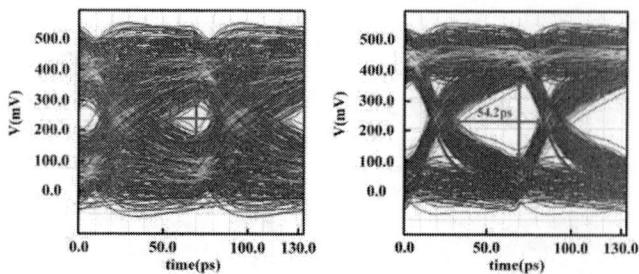

Fig. 12. Eye opening without and with coding at 15 Gbps.

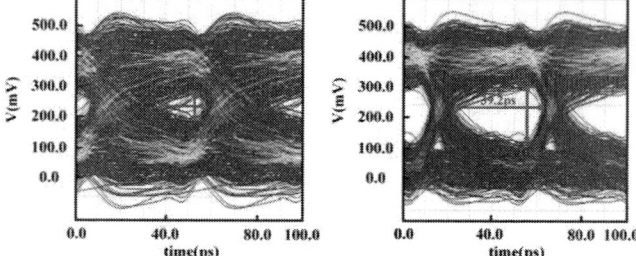

Fig. 13. Eye opening without and with coding at 20 Gbps.

edge density that balances geometric constraints and signal integrity. Simulation results reveal that the maximum achievable edge density of 9.37 Tb/s/mm is attainable under 0.5 μm channel spacing, 15 Gbps data rate, and -4.1 dB crosstalk. This configuration achieves a 37% improvement over conventional non-encoded designs (6.85 Tb/s/mm).

V. CONCLUSION

This paper introduces an inter-channel bit-smoothing (ICBS) transceiver architecture optimized for high-density die-to-die interconnects. By integrating XTC encoding and decoding modules, the transceiver achieves 40.07% FEXT through suppression of high-impact transition modes. Simulation results demonstrate a 69.3% reduction in crosstalk-induced jitter (from 36.8 ps to 11.27 ps) under 12.5 Gb/s/pin data rate and an XT of -4.14 dB. The transceiver incorporates a novel comparator topology that reduces kickback noise, enabling reliable operation at 0.5 μm channel spacing. These results validate the ICBS transceiver's capability to maintain signal integrity in next-generation high-density interconnects, outperforming present solutions in crosstalk suppression while supporting edge densities up to 9.37 Tb/s/mm.

ACKNOWLEDGMENT

The authors would like to thank the support from the National Key Research and Development Program (2023YFB4403402).

TABLE II. KEY-METRIC COMPARISON

Reference	This work	[31]	[18]	[15]	[12]	[8]
Technology	28nm	40nm	28nm	4nm	28nm	65nm
Data Rate	12.5Gbps/pin	4Gbps/pin	6Gbps/pin	32Gbps/pin	42Gbps/pin	4Gbps/pin
Signaling Type	Single-ended NRZ	Single-ended NRZ	Single-ended NRZ	Single-ended NRZ	Single-ended PAM3	Single-ended NRZ
Architecture	TX+RX	RX	TX+RX	TX+RX	TX+RX	TX+RX
Driver	SST	-	SST	SST	SST	SST
XTC Method	ICBS coding	FIR filters	Fibonacci coding	-	PAM-based XTC	FFE-combined XTC
Line Efficiency	83%	100%	75%	100%	150%	100%
Channel Type	On-chip	PCB	PCB	Interposer	On-chip	On-chip
Channel Length	4 mm	5 inch	2 inch	3 mm	2 mm	6 mm
Channel Pitch	0.5 um	8 mil	3.8 mil	50 um	2.5 um	0.5 um
Channel XT	-4.14dB@10GHz	-	-7.8@5GHz	-26.3dB@16G	-15.2dB@12GHz	-6.2dB@2GHz
Eye Opening	0.42UI	0.4UI	0.58UI	0.76UI	0.38UI	0.46UI
Jitter Improvement	69.3%	31%	45%	-	-	78%
Edge Density	9.37Tb/s/mm	-	0.9Tb/s/inch	8Tb/s/mm	9.16Tb/s/mm	8Tb/s/mm

REFERENCES

[1] A. Singh et al., "26.3 A pin- and power-efficient low-latency 8-to-12Gb/s/wire 8b8w-coded SerDes link for high-loss channels in 40nm technology," 2014 IEEE International Solid-State Circuits Conference Digest of Technical Papers (ISSCC), San Francisco, CA, USA, 2014, pp. 442-443.

[2] S. -M. Lee et al., "A Single-Ended Parallel Transceiver With Four-Bit Four-Wire Four-Level Balanced Coding for the Point-to-Point DRAM Interface," in IEEE Journal of Solid-State Circuits, vol. 51, no. 8, pp. 1890-1901, Aug. 2016.

[3] J. F. Buckwalter and A. Hajimiri, "Cancellation of crosstalk-induced jitter," in IEEE Journal of Solid-State Circuits, vol. 41, no. 3, pp. 621-632, March 2006.

[4] C. Sreerama et al., "A Crosstalk-Friendly Signaling Method," in IEEE Transactions on Components, Packaging and Manufacturing Technology, vol. 8, no. 9, pp. 1621-1631, Sept. 2018.

[5] T. Oh and R. Harjani, "A 6-Gb/s MIMO Crosstalk Cancellation Scheme for High-Speed I/Os," in IEEE Journal of Solid-State Circuits, vol. 46, no. 8, pp. 1843-1856, Aug. 2011.

[6] Gobinath Arumugam, Suresh-Kumar Natarajan, Rajeswari Packianathan, Mohamed-Salah Karoui, "Near end and far end crosstalk reduction in high-speed signaling channel with periodical spiral resonator defected microstriplines in a high performance printed circuit board", IEICE Electronics Express, 2025, Volume 22, Issue 1, Pages 20240268.

[7] Tetsuya HAYASHI, Takashi SASAKI, Eisuke SASAOKA, "Behavior of Inter-Core Crosstalk as a Noise and Its Effect on Q-Factor in Multi-Core Fiber", IEICE Transactions on Communications, 2014, Volume E97.B, Issue 5, Pages 936-944.

[8] H. -G. Ko, S. Shin, J. Oh, K. Park and D. -K. Jeong, "6.7 An 8Gb/s/μm FFE-Combined Crosstalk-Cancellation Scheme for HBM on Silicon Interposer with 3D-Staggered Channels," 2020 IEEE International Solid-State Circuits Conference - (ISSCC), San Francisco, CA, USA, 2020, pp. 128-130.

[9] P. Gupta and A. B. Kahng, "Wire swizzling to reduce delay uncertainty due to capacitive coupling," 17th International Conference on VLSI Design. Proceedings., Mumbai, India, 2004, pp. 431-436.

[10] K. Lee, H. -B. Lee, H. -K. Jung, J. -Y. Sim and H. -J. Park, "A Serpentine Guard Trace to Reduce the Far-End Crosstalk Voltage and the Crosstalk Induced Timing Jitter of Parallel Microstrip Lines," in IEEE Transactions on Advanced Packaging, vol. 31, no. 4, pp. 809-817, Nov. 2008.

[11] R. Arunachalam, E. Acar and S. R. Nassif, "Optimal shielding/spacing metrics for low power design," IEEE Computer Society Annual Symposium on VLSI, 2003. Proceedings., Tampa, FL, USA, 2003, pp. 167-172.

[12] K. Kim, J. -H. Park, H. -J. Park, J. Park, J. Kim and W. -S. Choi, "22.1 A 0.275pJ/b 42Gb/s/pin Clock-Referenced PAM3 Transceiver Tolerant to Supply Noise, Reference Offset and Crosstalk for Chiplets and Short-Reach Memory Interfaces," 2025 IEEE International Solid-State Circuits Conference (ISSCC), San Francisco, CA, USA, 2025, pp. 394-396.

[13] Y. -Y. Hsu, P. -C. Kuo, C. -L. Chuang, P. -H. Chang, H. -H. Shen and C. -F. Chiang, "A 7nm 0.46pJ/bit 20Gbps with BER 1E-25 Die-to-Die Link Using Minimum Intrinsic Auto Alignment and Noise-Immunity Encode," 2021 Symposium on VLSI Technology, Kyoto, Japan, 2021, pp. 1-2.

[14] Y. Nishi et al., "A 0.297-pJ/Bit 50.4-Gb/s/Wire Inverter-Based Short-Reach Simultaneous Bi-Directional Transceiver for Die-to-Die Interface in 5-nm CMOS," in IEEE Journal of Solid-State Circuits, vol. 58, no. 4, pp. 1062-1073, April 2023.

[15] K. Seong et al., "A 4nm 32Gb/s 8Tb/s/mm Die-to-Die Chiplet Using NRZ Single-Ended Transceiver With Equalization Schemes And Training Techniques," 2023 IEEE International Solid-State Circuits Conference (ISSCC), San Francisco, CA, USA, 2023, pp. 114-116.

[16] L. Zhong et al., "A 2×56 Gb/s 0.78-pJ/b PAM-4 Crosstalk Cancellation Receiver With Active Crosstalk Extraction Technique in 28-nm CMOS," in IEEE Journal of Solid-State Circuits, vol. 59, no. 9, pp. 3008-3020, Sept. 2024.

[17] Kato, Suguru & Sasaki, Shinichi & Nakashima, Kazunori., "Evaluation of Crosstalk by Multi-Valued Transmission," Journal of Japan Institute of Electronics Packaging. 2012. 15. 279-282.

[18] Q. Liu, L. Du and Y. Du, "A 0.90-Tb/s/in 1.29-pJ/b Wireline Transceiver With Single-Ended Crosstalk Cancellation Coding Scheme for High-Density Interconnects," in IEEE Journal of Solid-State Circuits, vol. 58, no. 8, pp. 2326-2336, Aug. 2023.

[19] M. Mutyam, "Fibonacci Codes for Crosstalk Avoidance," in IEEE Transactions on Very Large Scale Integration (VLSI) Systems, vol. 20, no. 10, pp. 1899-1903, Oct. 2012.

[20] R. -B. Lin, "Inter-Wire Coupling Reduction Analysis of Bus-Invert Coding," in IEEE Transactions on Circuits and Systems I: Regular Papers, vol. 55, no. 7, pp. 1911-1920, Aug. 2008.

[21] P. Subrahmanya, R. Manimegalai, V. Kamakoti and M. Mutyam, "A bus encoding technique for power and cross-talk minimization," 17th International Conference on VLSI Design. Proceedings., Mumbai, India, 2004, pp. 443-448.

[22] H. Kim, H. Seo, Y. Jo, C. Yoo and J. Han, "An 8b9b 77.44-Gb/s Noise-Immune Spatial-Delta Coded Transceiver for Short-Reach Memory Interfaces in 28-nm CMOS," in IEEE Transactions on Circuits and Systems II: Express Briefs, vol. 70, no. 4, pp. 1301-1305, April 2023.

[23] C. S. Taillefer and G. W. Roberts, "Delta–Sigma A/D Conversion Via Time-Mode Signal Processing," in IEEE Transactions on Circuits and Systems I: Regular Papers, vol. 56, no. 9, pp. 1908-1920, Sept. 2009.

[24] F. Shi, X. Wu and Z. Yan, "New Crosstalk Avoidance Codes Based on a Novel Pattern Classification," in IEEE Transactions on Very Large Scale Integration (VLSI) Systems, vol. 21, no. 10, pp. 1892-1902, Oct. 2013.

[25] Z. Shirmohammadi and S. G. Miremadi, "SDT-free: An efficient crosstalk avoidance coding mechanism considering inductance effects," 2017 7th International Conference on Computer and Knowledge Engineering (ICCKE), Mashhad, Iran, 2017, pp. 293-297.

[26] M. Taali, Z. Shirmohammadi, M. S. S. Danish and M. Khosravy, "JCI-CAC: An Efficient Crosstalk Avoidance Code Considering Joint Capacitive and Inductive Effects," in IEEE Access, vol. 10, pp. 98348-98359, 2022.

[27] J. Sun and Z. Xu, "A Novel Approximate Dynamic Programming Structure for Optimal Control of Discrete-Time Time-Varying Nonlinear Systems," in IEEE Transactions on Circuits and Systems II: Express Briefs, vol. 71, no. 8, pp. 3835-3839, Aug. 2024.

[28] Ren, Bolin et al. "A 56-Gb/s,0.708 pJ/bit single-ended simultaneous bidirectional transceiver with hybrid errors cancellation techniques for die-to-die interface." *Microelectron. J.*152 (2024): 106326.

[29] G. Zhang et al., "A CNRZ-7 Based Wireline Transceiver With High-Bandwidth-Density, Low-Power for D2D Communication," in IEEE Access, vol. 10, pp. 96556-96567, 2022.

[30] Rajeshkumar, "REDUCE POWER CONSUMPTION FOR DIGITAL CMOS CIRCUITS USING DVTS ALGORITHAM," International Journal of Electrical and Electronic Engineering & Telecommunications, Vol. 1, No. 1, pp. 376-383, March 2015.

[31] K. Gharibdoust, A. Tajalli and Y. Leblebici, "A 4×9 Gb/s 1 pJ/b NRZ/multi-tone serial-data transceiver with crosstalk reduction architecture for multi-drop memory interfaces in 40nm CMOS," 2015 Symposium on VLSI Circuits (VLSI Circuits), Kyoto, Japan, 2015, pp. C180-C181.

979-8-3315-8850-2/25 $31.00 © 2025 IEEE

Design and Implementation of an Integrated Testing System for Spintronic Chips

Wenyan Liu, Zihan Gao, Zhifu Guo, Zefan Wu, Xiangrong Pu, Xu Liu, Zhang Zhang*

School of Microelectronics, Hefei University of Technology, Hefei, China

*Corresponding Author email: zhangzhang@hfut.edu.cn

Abstract—This paper addresses three core challenges in spintronic chip testing: deviations in magnetoresistance characterization due to temperature fluctuations, limitations of traditional static magnetic fields in angular anisotropy testing, and insufficient magnetic field control precision. To overcome these issues, we designed and implemented an FPGA-based integrated testing system for spintronic chips. Three innovative solutions are proposed: First, a high-precision, wide-temperature-range temperature control module based on a negative feedback mechanism ensures a stable testing environment for magnetoresistance effects. Second, a dynamic magnetic field generation system utilizing the mechanical rotation of permanent magnets enables continuous and precise scanning of magnetic field angles. Finally, by leveraging FPGA hardware parallelism, the system achieves high precision and stability in dynamic magnetic field control, along with nanosecond-level synchronization of multiple physical fields. Experimental results demonstrated that the integrated temperature control module and dynamic testing platform achieved a steady-state temperature error within ±0.3°C. The dynamic magnetic field generated by the system exhibited a strength fluctuation of less than ±0.5 mT within a ±40 mT range, a deviation between the magnetic field direction and the set angle of ≤ ±0.01°, and a rotational speed fluctuation of less than ±5%, with no stalling or loss of steps during continuous 8-hour operation. FPGA-based hardware-level control realized synchronization of multi-field regulation with a delay of less than 100 ns. This system provides a comprehensive and integrated testing solution for spintronic chip research and development, advancing the industrialization of spintronic chips.

Keywords—*Spintronic chip, fpga gmr, tmr, multi-physics field coupling control*

I. INTRODUCTION

Spintronic chips, leveraging the tunability of electron spin degrees of freedom, have become a core technology supporting fields such as information storage and magnetic field sensing. Among these, the giant magnetoresistance (GMR) effect [1] and the tunneling magnetoresistance (TMR) effect [2], as landmark discoveries in spintronics, have laid the foundation for the application of spintronic chips in devices like high-sensitivity magnetic sensors and magnetic random-access memory (MRAM). The core performance metrics of such devices, such as magnetoresistance ratio, magnetic field

sensitivity, angular anisotropy, and temperature stability, directly determine their final application efficacy.

Accurate characterization of these parameters, however, requires precise and synergistic control of multiple physical fields (magnetic field, temperature, and electrical signals) during testing. Firstly, the magnetic moment alignment in spin-valve materials and the transport properties of tunnel junctions are extremely sensitive to temperature fluctuations. The lack of a highly stable temperature control environment directly leads to distorted characterization of magnetoresistance properties [3]. Secondly, traditional testing systems mostly use electromagnets to generate static or quasi-static magnetic fields, making it difficult to fully characterize the angular anisotropy of TMR/GMR devices [4]. Most critically, research on dynamic magnetoresistance and spin resonance characteristics requires magnetic field control to possess nanosecond-level timing precision to meet the matching requirements of spin relaxation processes, and also requires high-smoothness scanning of rotating magnetic fields to resolve Lorentzian line shapes [5]. It is evident that while existing commercial equipment excels at single electrical parameter measurements, it has inherent limitations in the coordinated control of multiple physical fields [4, 5]; whereas advanced research setups, despite achieving some functionality, face challenges such as system complexity, high cost, and operational difficulties [4, 6].

Addressing the above challenges, this paper aimed to design and implement an integrated, low-cost, high-precision all-in-one testing system for spintronic chips. By designing a wide-temperature-range, high-precision temperature control module (-5°C to 70°C, ±0.3°C) based on a negative feedback mechanism, a stable testing environment for the magnetoresistance effect was provided, overcoming the characterization deviation problem caused by temperature sensitivity. Secondly, a dynamic magnetic field generation architecture based on the "mechanical rotation of permanent magnets" was proposed. By controlling a stepper motor with an FPGA, continuous rotation of the magnetic field direction and stepless speed regulation were achieved, addressing the need for precise magnetic field angle scanning in anisotropy and resonance characteristic testing. Finally, a hardware control platform centered on the Xilinx Artix-7 A704 FPGA was constructed. Leveraging its hardware parallelism, the synchronization delay for the regulation of multiple physical fields (magnetic, thermal, etc.) and weak signal acquisition was

---

This work was supported by the College Students' Innovation and Entrepreneurship Training Program of Hefei University of Technology (S202510359291).

979-8-3315-8850-2/25 $31.00 © 2025 IEEE

reduced from the millisecond level to the nanosecond level, fundamentally meeting the stringent timing precision requirements of spintronic dynamic processes.

II. THEORETICAL BACKGROUND

A. Spintronic Chips

The GMR effect originates from the difference in spin-dependent scattering in the "ferromagnetic layer (FM)/nonmagnetic metal layer (NM)/ferromagnetic layer (FM)" structure. When an external magnetic field switches the magnetic moments of adjacent ferromagnetic layers from antiparallel (AP) to parallel (P) alignment, the scattering rate of spin-up electrons decreases, resulting in a significant reduction in total resistance. The magnetoresistance ratio is expressed as:

$$GMR = \frac{R_{AP} - R_P}{R_P} = \frac{(\rho_\downarrow - \rho_\uparrow)^2}{4\rho_\uparrow \rho_\downarrow} \tag{1}$$

where $\rho_\uparrow$ and $\rho_\downarrow$ represent the resistivities of spin-up and spin-down channels, respectively, and R_{AP} and R_P denote the resistances in antiparallel and parallel states.

The TMR effect occurs in magnetic tunnel junctions (MTJs) composed of "ferromagnetic layer (FM)/insulating barrier layer (I)/ferromagnetic layer (FM)", which originates from quantum tunneling of spin-polarized electrons. According to the Jullière model [2], the tunneling probability is closely related to the spin polarization. When the magnetizations on both sides are parallel, majority-spin electrons have a higher tunneling probability and lower resistance; the opposite occurs for antiparallel alignment. The magnetoresistance ratio is expressed as:

$$TMR = \frac{R_{AP} - R_P}{R_P} \tag{2}$$

TMR demonstrates a ΔR/R ratio ranging from 50% to 200% (significantly higher than GMR), with sensitivity reaching the µT level.

Spintronic chips based on GMR and TMR effects both fundamentally operate through magnetic-field-modulated magnetoresistance characteristics: GMR relies on spin-dependent scattering differences, while TMR is based on quantum tunneling of spin-polarized electrons. Their performance is directly related to the magnetic field environment, operating conditions, and signal detection accuracy. Studies have indicated that magnetoresistance is susceptible to temperature variations, where ferromagnetic electrodes are prone to degradation upon heating. Concurrently, organic materials that exhibit certain spin relaxation lengths at low temperatures demonstrate reduced spin relaxation lengths at room temperature [3]. Therefore, accurate characterization of the core performance of both types of devices necessitates common testing requirements, including a dynamic magnetic field capable of comprehensively characterizing the angular anisotropy of TMR/GMR devices while maintaining a constant temperature.

B. Dynamic Magnetic Field

Due to the dynamic matching requirement, the characteristic time for magnetic moment response to external field changes (spin relaxation time τ_s, approximately 1 ns for materials like CoFeB) is extremely short. If the magnetic field switching delay $\Delta t > \tau_s$, the precession of magnetic moments becomes unstable due to energy dissipation, leading to fluctuations in spin polarization PP ($\delta P \propto \Delta t$), ultimately manifesting as jitter in the magnetoresistance signal. Therefore, the magnetic field control timing must meet the stringent requirement of $\Delta t < \tau_s$, which constitutes the physical motivation for this system's use of an FPGA to achieve nanosecond-level timing precision.

TMR/GMR effects originate from spin dynamics, particularly spin precession in ferromagnetic layers under an external magnetic field, with its precession frequency determined by the Larmor frequency $\omega_p = \gamma B$ (where γ is the gyromagnetic ratio) [7].

When the angular frequency ω of the applied rotating magnetic field approaches the system's Larmor frequency ω_p, the spin resonance phenomenon occurs—efficient energy coupling between the magnetic moments and the rotating magnetic field leads to a significant enhancement of the TMR signal (where ΔR/R reaches its maximum at the resonance peak). Its response follows a Lorentzian line shape [5, 8]:

$$\Delta R/R \propto \frac{1}{(\omega - \omega_p)^2 + \Gamma^2} \tag{3}$$

Here, Γ is the damping coefficient, directly related to the spin relaxation characteristics. To accurately extract key parameters such as ω_p and Γ, it is necessary to perform a continuous, precise, and non-perturbing scan of ω. This requires the magnetic field rotation control to possess high resolution (to resolve narrow resonance peaks) and high smoothness (to avoid exciting nonlinear responses).

The need for anisotropy characterization: The magnetic anisotropy of devices (such as uniaxial anisotropy and interface anisotropy) modulates their magnetization reversal process [9, 10]. By continuously rotating the magnetic field and synchronously measuring the resistance response $R(\theta)$, the distribution of magnetic anisotropic energy barriers can be precisely mapped—a feat unachievable with static magnetic field testing. In summary, the accurate characterization of the core performance of GMR/TMR devices, particularly the study of dynamic and resonance characteristics, necessitates a testing system capable of generating high-precision, highly stable dynamic rotating magnetic fields, while also achieving coordinated control of multiple physical fields (magnetic, thermal) and real-time acquisition of weak signals. The design work in this paper is specifically aimed at addressing this series of common testing requirements.

III. SYSTEM DESIGN SOLUTION

Based on the aforementioned theoretical analysis, to achieve precise characterization of spintronic dynamic processes, this study designed an integrated testing system solution as follows. As shown in Fig. 1, the system adopted a dual-core cooperative control architecture comprising an FPGA and an MCU. Specifically, the FPGA handled time-critical tasks (such as nanosecond-level synchronization and

979-8-3315-8850-2/25 $31.00 © 2025 IEEE

motor pulse control), while the MCU managed upper-layer human-machine interaction, data communication, and display tasks.

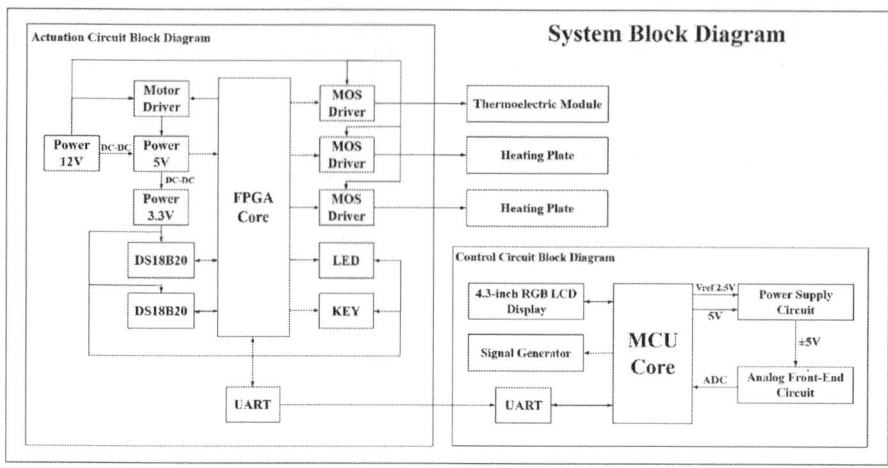

Fig. 1. Control system overall architecture block diagram.

Among these, the magnetic control module primarily implemented the regulation of magnetic field strength and direction through a hybrid magnet rotation platform, providing a dynamically controllable magnetic field environment for magnetoresistance characteristic analysis. The temperature control module mainly consisted of heating and cooling units, ensuring stable chip testing under various complex temperature conditions. The signal acquisition and measurement module was composed of an MCU and an oscilloscope, which, combined with input excitation signals, captures the chip's electrical response signals under composite field effects and dynamically displays visualized signal test results. The enclosure structure was divided into a core circuit chamber and a temperature-controlled test chamber; the former is used to secure various hardware modules, while the latter integrates temperature and magnetic control modules to construct a controllable physical testing environment.

The system was divided into a control circuit and an execution circuit. The control circuit sent relevant commands via a serial port to control the execution circuit for operations such as temperature control and information acquisition.

A. Control Circuit

The control circuit utilized GPIO to control solid-state relays for switching the input signal's AC/DC coupling mode, adapting to different signal types. Signal processing employed a differential amplification and filtering design to suppress common-mode noise and amplify μV-level weak signals. Coupled with lock-in amplification technology (1 MHz bandwidth), it enhanced the signal-to-noise ratio (SNR) to ensure high-precision acquisition. The SGM3204 charge pump converted +5 V to -5 V, providing a dual power supply for the operational amplifiers, while the CJ431 generated a +2.5 V precision reference voltage serving as the ADC reference

source, ensuring quantization accuracy. LC filtering combined with single-point grounding reduced power supply ripple interference in the analog circuits. The MCU core circuit performed multi-channel synchronous sampling of the conditioned analog signals (e.g., voltage/current) via its internal ADC. It refreshed the LCD to display real-time waveforms every 50 ms, sent commands to the execution circuit via the serial port every 1000 ms, and drove a 4.3-inch RGB screen (800×480 pixels, I²C touch interface) to achieve data visualization.

B. Execution Circuit

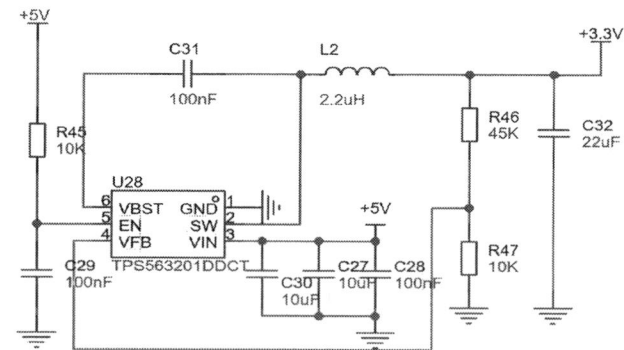

Fig. 2. Power conversion circuit.

As shown in Fig. 2, the power conversion circuit utilized a TPS56230 synchronous buck regulator to achieve 12V to 5V conversion (maximum current 3A, supplying power to the FPGA), followed by a low-dropout regulator (LDO) to convert 5V to 3.3V for powering the digital circuits. Precision feedback resistor configuration ensures stable output voltage. As illustrated in Fig. 3, the MOSFET drive circuit drove the

979-8-3315-8850-2/25 $31.00 © 2025 IEEE   362

thermoelectric cooler/heater module (current ≥4A). A high-level output from the FPGA turned on the transistor, pulling the PMOS gate voltage below 2V. The Infineon IPD068P03L3 PMOS (70 A/30 V, on-resistance 11 mΩ) was selected to meet the high-current requirements.

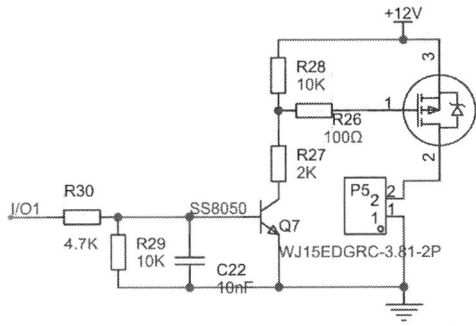

Fig. 3. MOSFET drive circuit.

The FPGA was used in conjunction with a stepper motor drive circuit to fulfill the testing environment needs for a dynamic magnetic field. The drive circuit, shown in Fig. 4, involved the FPGA outputting timing pulses to a ULN2003 Darlington transistor array, which amplifies the signal to drive the stepper motor. By varying the pulse frequency, the motor speed was controlled, thereby altering the magnetic field direction in the test environment through the motor's rotation and achieving dynamic magnetic field scanning.

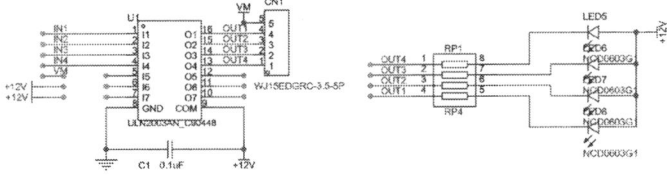

Fig. 4. Stepper motor drive circuit.

Simultaneously, a DS18B20 temperature sensor monitored the temperature of the temperature-controlled chamber in real time and provided feedback to the FPGA. LEDs indicated status, buttons input debugging commands, and the Xilinx Artix-7 A704 FPGA parsed control instructions to dynamically adjust temperature and motor speed. Data was sent back via the serial port to form a closed-loop control system.

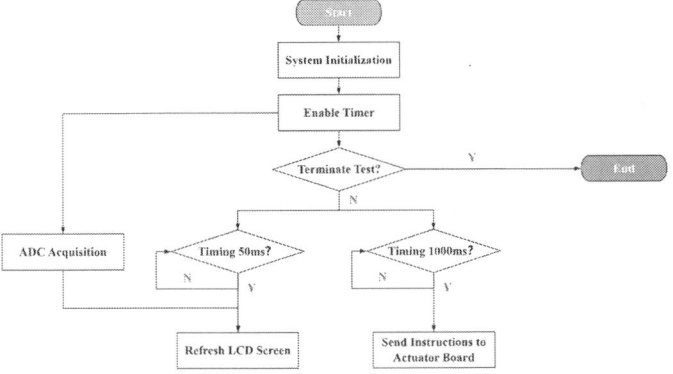

Fig. 5. Control circuit software flowchart.

C. Software Design

As shown in Fig. 5, the control circuit software flow initialized upon power-up, enabled timers, performed ADC acquisition and LCD screen refresh every 50 ms, and sent commands to the execution board every 1000 ms. The loop terminates upon shutdown.

The system control logic centered on the FPGA, employing modular design for multitasking coordination (top-level architecture shown in Fig. 6). Clock-synchronized functional modules ensured real-time performance for temperature regulation, magnetic field scanning, and data communication.

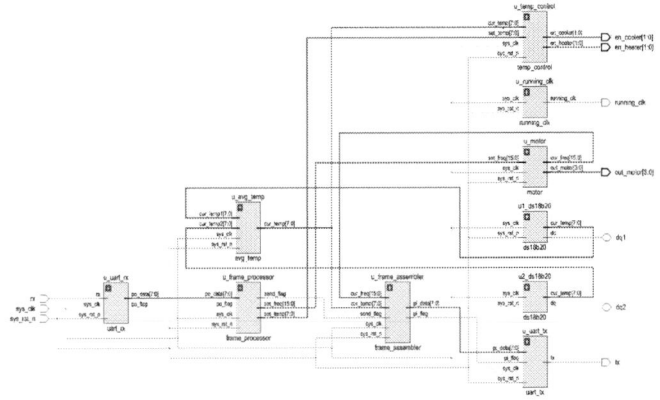

Fig. 6. System control logic top-level block diagram.

IV. SYSTEM PERFORMANCE TESTING

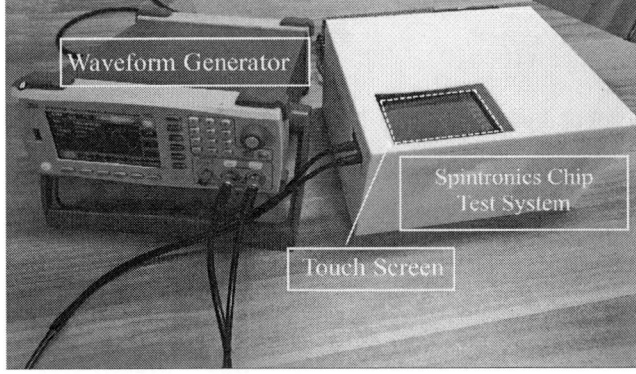

Fig. 7. Overall schematic diagram of the test platform.

Fig. 7 displays a physical image of the integrated spintronic chip testing system developed in this study. The system adopted an integrated enclosure design, incorporating internal components such as the temperature control module, magnetic field rotation module, and signal conditioning circuit. As shown in the figure, a touch screen served as the human-machine interface for parameter configuration (e.g., temperature, magnetic field rotation speed) and real-time visualization of test data. A waveform generator functioned as an external excitation source, providing voltage or current stimulation to the spintronic chip under test. The core testing chamber was designed to house the spintronic chip, providing a controllable temperature and magnetic field environment.

979-8-3315-8850-2/25 $31.00 © 2025 IEEE   363

This integrated approach replaces the traditional complex setup involving multiple instruments, thereby simplifying the testing workflow and improving operational efficiency.

To comprehensively evaluate the performance of the integrated testing system proposed in this work, corresponding experiments were designed for validation and analysis.

A. Performance Testing and Analysis of the Temperature Control Module

The stability of the temperature control module is a prerequisite for accurately characterizing the temperature dependence of spintronic chips. To verify its performance, a calibrated PT1000 high-precision temperature sensor (accuracy: ±0.1°C) was placed inside the system's test chamber as a reference. Starting from an initial room temperature of 25°C, the system's temperature control setpoints were configured from -5°C to 70°C at intervals of 10°C. The time required for the system to reach a steady state (defined as temperature fluctuations of less than ±0.3°C) and the steady-state error were recorded.

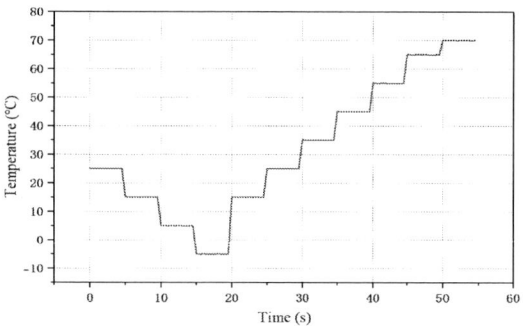

Fig. 8. Temperature control response curve.

The experimental results are shown in Fig. 8. The steady-state error of the system at all set points did not exceed ±0.3°C, meeting the design specifications and demonstrating good response speed and stability.

TABLE I.   TEMPERATURE CONTROL MODULE PERFORMANCE COMPARISON

Performance Metric	Traditional Commercial Systems	State-of-the-Art Academic Solutions[6, 11]	This Work	Advantages
Temperature Control Range	Optional, Independent Chamber	Typically Cryogenic (Low Temperature)	-5 – 70 °C	Wide temperature range; Integrated into the test system
Temperature Control Accuracy	±0.5°C	N/A (Not Available/Not Applicable)	±0.3 °C(Steady-State)	Improve by about 40%

As indicated by the comparison in Table I, although traditional commercial systems offer higher ultimate precision (often equipped with independent temperature-controlled chambers), they are typically standalone options and are difficult to integrate flexibly with dynamic magnetic field testing. State-of-the-art academic solutions usually focus on ultrafast process measurements and rarely integrate temperature control functionality. This system successfully achieved a high level of integration between wide-temperature-range control (-5~70°C) and a dynamic testing platform. Its precision meets the requirements for application-level testing of spintronic chips, providing a reliable guarantee for solving characterization deviations caused by temperature sensitivity.

B. Performance Testing and Analysis of the Dynamic Magnetic Field Module

The accuracy and stability of the dynamic magnetic field are fundamental for studying angular anisotropy and resonance characteristics. A calibrated Hall-effect Gauss meter (accuracy ±0.1 mT) was used to measure the magnetic field generated by the rotation of the permanent magnets. The testing was divided into two parts: angle accuracy testing and rotational speed stability testing.

The test results indicated that the dynamic magnetic field generated by this system exhibited a strength fluctuation of less than ±0.5 mT within a ±40 mT range. Simultaneously, the deviation between the magnetic field direction and the set angle was ≤ ±0.01°, meeting the precision requirements for anisotropy characterization. In terms of rotational speed stability, the relative error between the actual speed and the set value was ≤ 5%, with no stalling or loss of steps observed during continuous 8-hour operation.

TABLE II.   DYNAMIC MAGNETIC FIELD PERFORMANCE COMPARISON

Performance Metric	Traditional Commercial Systems	State-of-the-Art Academic Solutions [6, 11, 12]	This Work	Advantages
Magnetic Field Control Method	Electromagnet	Terahertz Laser Pulse + Static Magnetic Field	Mechanical Rotation of Permanent Magnets	Simple structure, not complex
Angular Control Range	Manual Adjustment, Discrete Angles	Electronically Controlled, Theoretically Arbitrary Angles	0° to 360° (Continuous)	No manual adjustment required, full-angle coverage
Angular Control Accuracy	> ±1.0°	N/A	≤ ±0.01°	Improve by about 99%
Speed/Switching Time	N/A (Static)	< 1 ps (Ultrafast Switching)	0 – 1000 rpm	Overcoming the static limitation of traditional systems
Speed Control Accuracy	N/A	N/A	≤ ±5%	Providing reliable speed regulation
Field Stability	±0 6mT (Static)	N/A	≤ ±0.5 mT (Static)	Improve by about 17%

As shown in the comparison in Table II, the dynamic magnetic field module of this system comprehensively surpasses traditional commercial static magnetic field solutions in terms of continuity, precision, and controllability. Although its time resolution cannot match that of ultrafast optical schemes, it provides a continuous, stable dynamic magnetic field capable of long-term operation, filling the tool gap between basic static testing and ultrafast dynamics research.

C. Multi-Field Synchronization and Control Precision Testing and Analysis

Under the conditions of the temperature control module set to 40°C and the magnetic field rotation module set to 300 revolutions per minute (rpm), continuous electrical signal acquisition was performed on a standard spintronic chip, while simultaneously monitoring the regulation timing of multiple physical fields.

979-8-3315-8850-2/25 $31.00 © 2025 IEEE

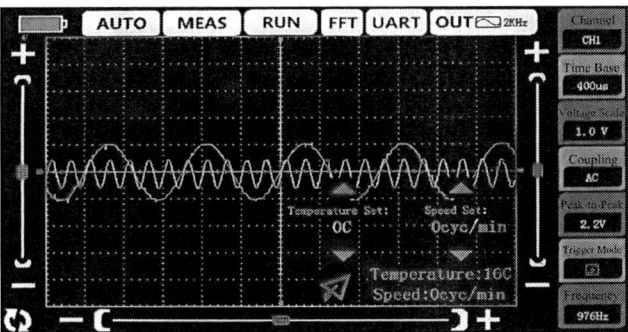

Fig. 9. Signal acquisition display module.

Fig. 9 shows the signal acquisition display module integrated into the system, displaying the voltage response waveform of the spintronic chip under test in the primary panel, while key parameters such as ambient temperature and magnetic field rotation speed are displayed in ancillary readouts. The system's integrated high-precision oscilloscope was used to simultaneously capture the chip's voltage response and temperature acquisition signals, with a focus on evaluating timing consistency, magnetic field control stability, and system synchronization accuracy.

TABLE III. SYNCHRONOUS CONTROL AND SIGNAL DETECTION PERFORMANCE COMPARISON

Performance Metric	Traditional Commercial Systems	State-of-the-Art Academic Solutions [13]	This Work	Advantages
Multi-Field Synchronization Control Delay	> 10 ms(Software-based)	~100 fs(Opto-Magnetic Synchronization)	< 100 ns (FPGA Hardware)	Compared with traditional systems, there is a significant improvement
Signal Detection Sensitivity	~ 30 μV (Typical)	N/A	10 μV to 5 V	Improve by 66.7%
Signal Detection Accuracy	±0.1% (Typical)	N/A	≤ ±0.01%	Improve by about 90%

As demonstrated by the comparison in Table III, this system significantly enhances the high stability of the dynamic magnetic field and the accuracy of timing control, meeting the stringent requirements for magnetic field performance in spintronic dynamics testing and anisotropy characterization.

V. CONCLUSIONS

This paper designed and implemented an integrated spintronic chip testing system centered on an FPGA, aiming to address three major testing challenges: temperature fluctuations, magnetic field angle control, and multi-physics field synchronization. The system integrates a high-precision temperature control module, a dynamic magnetic field generation architecture based on the mechanical rotation of permanent magnets, and a hardware-parallelized nanosecond-level synchronous control unit. Experimental validation demonstrated that the system successfully achieved the integration of wide-temperature-range control (-5~70°C) with

a dynamic testing platform, with a temperature control accuracy of ±0.3°C. The angular deviation during continuous precision scanning of the dynamic magnetic field was ≤ ±0.01°. Leveraging the parallelism of the FPGA, coordinated multi-field regulation was realized with a synchronization delay of less than 100 ns. This work provides an efficient and reliable testing platform for researching the angular anisotropy and dynamic characteristics of spintronic chips, holding positive significance for promoting the industrial application of spintronic chips.

REFERENCES

[1] M. N. Baibich et al., "Giant Magnetoresistance of (001)Fe/(001)Cr Magnetic Superlattices," Physical Review Letters, vol. 61, no. 21, pp. 2472–2475, Nov. 1988, doi: 10.1103/physrevlett.61.2472.

[2] M. Julliere, "Tunneling between ferromagnetic films," Physics Letters A, vol. 54, no. 3, pp. 225 − 226, Sep. 1975, doi: 10.1016/0375-9601(75)90174-7.

[3] X. Liu, X. Zhu, S. Ding, et al., "Organic spin valves and their magnetoresistance effect," Progress in Chemistry, vol. 31, no. 9, pp. 1199–1212, 2019, doi: 10.7536/PC190303.

[4] N. Žurauskienė, "Engineering of Advanced Materials for High Magnetic Field Sensing: A Review," Sensors, vol. 23, no. 6, p. 2939, Mar. 2023, doi: 10.3390/s23062939.

[5] C. Jozsa, "Optical detection of the magnetization precession: choreography on a sub-nanosecond timescale," Ph.D. dissertation, Dept. of Appl. Phys. and Sci. Educ., Technische Univ. Eindhoven, Eindhoven, The Netherlands, 2006, doi: 10.6100/IR601832.

[6] W. Song, H. Ding, J. Zhuang, et al., "Development of a permanent magnet rotating magnetic field generator based on the concept of magnetic surgery," China Medical Devices, vol. 38, no. 11, pp. 33−37, 48, 2023, doi: 10.3969/j.issn.1674-1633.2023.11.007.

[7] D. Home, A. K. Pan, and A. Banerjee, "Reexamining Larmor precession in a spin-rotator: testable correction and its ramifications," The European Physical Journal D, vol. 67, no. 4, Apr. 2013, doi: 10.1140/epjd/e2013-30346-9.

[8] G. Binasch, P. Grünberg, F. Saurenbach, and W. Zinn, "Enhanced magnetoresistance in layered magnetic structures with antiferromagnetic interlayer exchange," Physical Review B, vol. 39, no. 7, pp. 4828–4830, Mar. 1989, doi: 10.1103/physrevb.39.4828.

[9] Th. G. S. M. Rijks, R. Coehoorn, M. J. M. de Jong, and W. J. M. de Jonge, "Semiclassical calculations of the anisotropic magnetoresistance of NiFe-based thin films, wires, and multilayers," Physical Review B, vol. 51, no. 1, pp. 283–291, Jan. 1995, doi: 10.1103/physrevb.51.283.

[10] B. Dieny, C. Cowache, A. Nossov, P. Dauguet, J. Chaussy, and P. Gandit, "Anisotropy and angular variation of the giant magnetoresistance in magnetic multilayers (invited)," Journal of Applied Physics, vol. 79, no. 8, pp. 6370 − 6375, Apr. 1996, doi: 10.1063/1.362003.

[11] Y. Wang, "Research of spintronic devices based on ferromagnetic and antiferromagnetic materials," Ph.D. dissertation, Dept. Electron. Eng., East China Normal Univ., Shanghai, China, 2019.

[12] S. Schlauderer et al., "Temporal and spectral fingerprints of ultrafast all-coherent spin switching," Nature, vol. 569, no. 7756, pp. 383−387, May 2019, doi: 10.1038/s41586-019-1174-7.

[13] L. Li, G. Zhu, X. Jin, X. Yang, M. Li, D. Wu, P. Li, W. Wang, S. Ma, and K. Zhou, "Verification of high-precision synchronization between terahertz free-electron laser and femtosecond laser," J. Terahertz Sci. Electron. Inf. Technol., vol. 23, no. 5, pp. 489–494, May 2025.

2025 The 10th International Conference on Integrated Circuits and Microsystems

Design of a Reusable UVM Verification Framework for Pixel Readout Chips

Yu Zhao
School of Electronics and Information Northwestern Polytechnical University
Xi'an, China
zhaoyu200103@163.com

Zexuan Zhao
School of Computer Science Northwestern Polytechnical University
Xi'an, China
zhaozx@mail.nwpu.edu.cn

Yuanhong Jiao
School of Computer Science Northwestern Polytechnical University
Xi'an, China
jyh3089@mail.nwpu.edu.cn

Xiaomin Wei*
School of Computer Science Northwestern Polytechnical University
Xi'an, China
weixm@nwpu.edu.cn

Heng Yang
School of Computer Science Northwestern Polytechnical University
Xi'an, China
yh_saka@163.com

Long Liu
School of Electronics and Information Northwestern Polytechnical University
Xi'an, China
longerliu@nwpu.edu.cn

Yongcai Hu
School of Computer Science Northwestern Polytechnical University
Xi'an, China
yannyonghu@163.com

Abstract—The escalating demands of High-Energy Physics (HEP) experiments impose significant challenges on pixel readout chips, including growing functional complexity, compressed design cycles, and high fabrication costs. Traditional verification approaches, relying heavily on static stimuli and manual inspection, suffer from insufficient coverage, low automation, and poor reusability, thus failing to guarantee the functional integrity of such complex designs. To address these challenges, this work proposes and implements a reusable verification framework based on the Universal Verification Methodology (UVM), specifically tailored for advanced pixel readout chips. The key contributions of this framework include: (1) a physics-aware stimulus generator for emulating realistic particle-hit events; (2) a generic serial bus configuration component leveraging the Register Abstraction Layer (RAL) to enable high-level automated configuration; and (3) an intelligent scoreboard capable of out-of-order data matching to ensure robust data validation. The framework's effectiveness was validated on a pixel readout chip, where its metric-driven verification (MDV) methodology successfully exercised complex functionalities and corner cases of the design. By adopting a modular architecture, the framework significantly enhances reusability and efficiency, offering a systematic and scalable solution for the verification of similar mixed-signal chips.

Keywords—UVM, functional verification, pixel readout chip, reusability

I. Introduction

Pixel readout chips are fundamental components of modern particle detection systems, widely employed in cutting-edge fields such as high-energy physics (HEP), nuclear security

Supported by National Natural Science Foundation of China under Grant No. W2443008, 12375191, 12435013, 12275218, 12341502, 12105224, 12205307, 12475195; the National Key Research and Development Program of China under Grant No. 2023YFE0206300, 2023YFF0719600, 2024YFE0110102; Guangdong Basic and Applied Basic Research Foundation under Grant No. 2024A1515012141; China Postdoctoral Science Foundation under Grant No. 2023M742850.

screening, and medical imaging. Their primary function is to process and convert the weak charge signals generated in the sensor array, a process that integrates complex on-chip signal processing logic, including amplification, discrimination, and digitization, ultimately producing high-speed digital data streams. Continuous upgrades to large-scale scientific facilities, such as the High-Luminosity Large Hadron Collider (HL-LHC), impose increasingly stringent demands on key performance metrics, including hit rate and data throughput. Consequently, ensuring the functional integrity and performance robustness of such complex chips through comprehensive and efficient verification has become a decisive factor for experimental success.

Traditional verification approaches, such as Directed Testing, remain adequate for validating simple functional modules. However, their limitations become evident when applied to highly integrated systems like pixel readout chips [1], which combine hundreds of thousands of pixel cells, intricate control logic, and high-speed interfaces. Static stimuli are insufficient to expose deep design flaws within the vast state space, while manual inspection of massive output data is fundamentally limited in both efficiency and accuracy. Consequently, these methods are fundamentally inadequate for the stringent verification requirements of modern complex chips.

To address these challenges, the Universal Verification Methodology (UVM) has become the de facto standard for building reusable and coverage-driven verification environments [2]. UVM's powerful methodological framework has demonstrated remarkable efficacy in the verification of complex integrated circuits. Researchers have successfully deployed it across diverse domains, including SoCs [3], cache coherence controllers in multi-core processors [4], and extending to mixed-signal designs [5]. However, despite this

979-8-3315-8850-2/25 $31.00 © 2025 IEEE

broad applicability, the direct application of UVM to pixel readout chip verification still poses unique challenges. First, the required stimuli for these chips are not standard bus transactions, but rather complex physical events that emulate particle–sensor interactions. Second, unlike typical digital chips that employ parallel buses such as AMBA for register configuration, pixel readout chips commonly rely on serial protocols such as Serial Peripheral Interface (SPI) or Inter-Integrated Circuit (I^2C), whose bit-level transmission at the physical layer is fundamentally mismatched with the byte-addressable, memory-mapped model assumed by the UVM Register Abstraction Layer (RAL) [6]. Finally, and perhaps most critically, tens of thousands of pixel cells operate in parallel within the chip, and the inclusion of data-reduction techniques such as zero suppression leads to output data streams with high temporal and ordering uncertainty, posing significant challenges for real-time data validation [7].

In the field of high-energy physics [8], several pioneering efforts—most notably those by the RD53 collaboration—have showcased the immense potential of UVM in front-end ASIC verification [9]. The VEPIX53 environment, developed by this collaboration, enables the entire verification flow, from architectural exploration to system integration, by importing Monte Carlo physics data [10], marking a significant milestone in this domain. However, existing publications, while showcasing successful verification outcomes, rarely delve into a systematic methodology for constructing a truly reusable verification framework, especially concerning the design patterns and implementation details of its foundational components.

To this end, this paper presents a reusable, coverage-driven UVM verification framework tailored for complex pixel readout chips. Its architecture and the implementation of its key components are systematically detailed. The key contributions of this work include:

- A physics-aware stimulus generator. This component can generate concurrent randomized hit stimuli and also import Allpix^2 [11] data to emulate realistic particle interactions, thereby providing high-quality stimulus sources for uncovering potential design flaws.
- A decoupled, RAL-based generic serial bus configuration component. This module seamlessly bridges the byte-level abstraction of the UVM RAL with the bit-level transmission of serial buses, significantly enhancing the reusability of test sequences.
- A comprehensive data-checking infrastructure. Centered on an out-of-order scoreboard optimized with advanced data structures, and augmented with assertion checks, this infrastructure ensures the correctness of real-time data streams at multiple abstraction levels.

This paper is organized as follows. Section II describes the architecture of the proposed UVM-based verification framework. Section III details the design and implementation of the key components. Section IV presents the verification strategy, experimental results, and their analysis. Finally, Section V concludes the paper and suggests directions for future work.

II. UVM Verification Framework Architecture

A. Design Under Verification (DUV) Overview

The DUV is a high-performance pixel readout chip tailored for particle detection, as illustrated in Fig. 1. The chip adopts a decoupled design philosophy, whereby it is tightly coupled to a semiconductor sensor via flip-chip bump bonding. When an incident particle traverses the sensor and generates a weak charge signal, the readout chip is responsible for real-time signal acquisition, amplification, processing, and digitization. Internally, the architecture consists of a pixel array and peripheral readout circuitry, organized in a hierarchical data-processing scheme to efficiently manage the massive, asynchronous event streams inherent in particle detection.

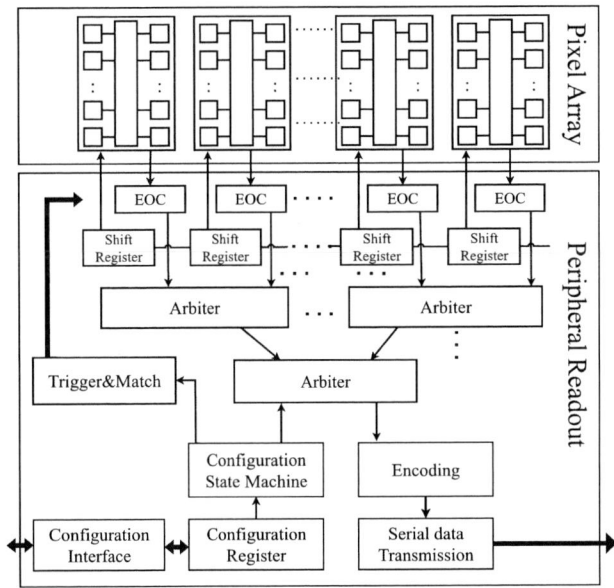

Fig. 1. Overall architecture of the pixel readout chip under verification.

Within the pixel array, pixels are organized into regional units such as Super Pixels or Double Columns. When a particle hit occurs, the analog front-end amplifies and discriminates the signal, while the digital logic digitizes the event's temporal and energy characteristics—namely, the Time-of-Arrival (ToA) and Time-over-Threshold (ToT)—producing a four-dimensional data packet containing position, time, and energy information.

These asynchronous data packets are first buffered at the End-of-Column (EoC) logic. They are subsequently aggregated into the Multi-Level Arbitration & Buffering System, which forms the core of the data management pipeline. This module employs cascaded FIFO buffers to accommodate instantaneous high event rates (e.g., "data bursts" induced by particle bunches), while multi-stage arbitration and non-adjacent column policies resolve access conflicts and minimize dead time. This complex scheduling and buffering mechanism is the fundamental reason the final output data stream exhibits highly out-of-order characteristics.

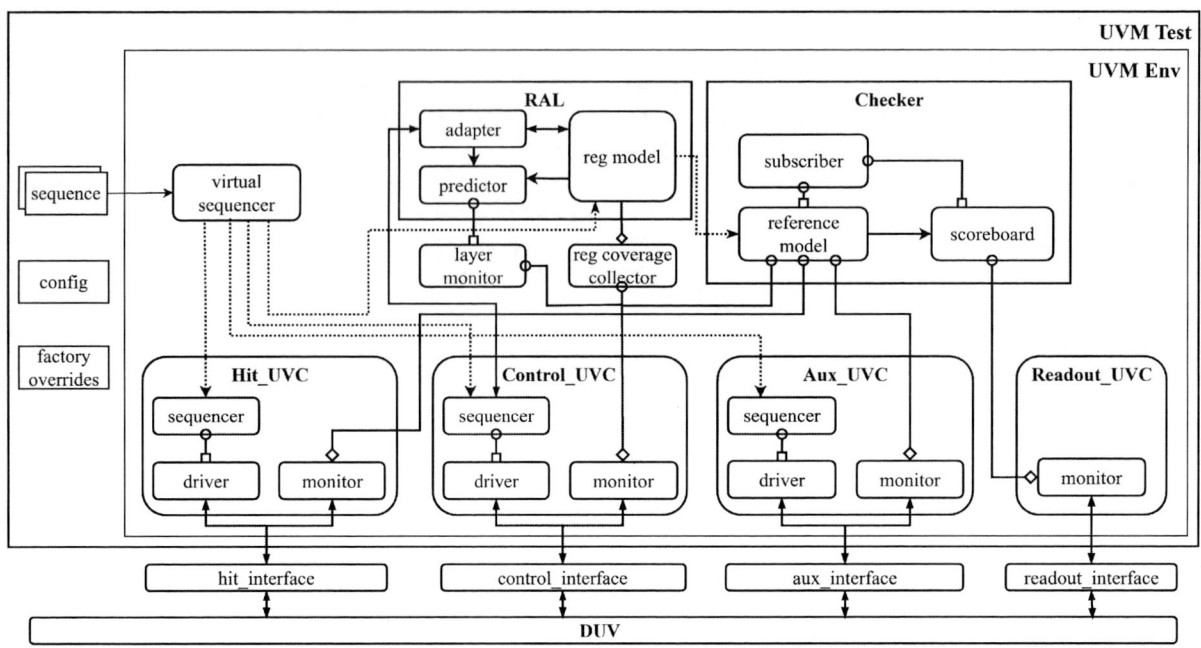

Fig. 2. Top-level architecture of the proposed reusable UVM verification framework.

The arbitrated data stream is then fed into the Event Encoder and data compression logic (e.g., zero suppression), where it is selected, compacted, and packaged. The processed data are serialized by a High-Speed Serializer and transmitted as differential signals via the LVDS interface to the external data acquisition (DAQ) system. In parallel, a large register bank governs all operational modes of the chip, configuring key parameters such as pixel thresholds, data selection policies, arbitration strategies, and compression schemes. All register programming is carried out through a serial bus interface such as SPI.

In summary, this triad of features—the data-driven architecture, the highly unordered output streams, and the unique serial configuration—defines the core verification challenges of this work. A UVM-based verification framework, engineered to systematically address these challenges, is detailed in the following section.

B. UVM Top-Level Architecture

The proposed verification framework follows the layered design philosophy of UVM, and its top-level architecture is illustrated in Fig. 2. At the highest level, the UVM Test layer's primary role is to decouple test strategies from the verification environment: test scenarios are defined through the execution of different sequences; environment parameters are configured via the UVM configuration database (`config_db`) mechanism; and component flexibility is achieved through the factory override mechanism without modifying the environment code. Beneath the Test layer, the reusable UVM Environment (Env) layer encapsulates all core verification components that directly interact with the DUV, including four functional UVM

Verification Components (UVCs), an online validation engine, and a RAL model.

- Hit_UVC: a physical stimulus injector that emulates particle hit events.
- Control_UVC: the primary command and configuration interface of the DUV, collaborating with the RAL model for register access.
- Aux_UVC: generates dpulse test signals and other auxiliary control signals.
- Readout_UVC: a passive data collector, whose monitor captures the high-speed data streams from the DUV.

To enable protocol-independent register access, the framework integrates the RAL model. High-level register operations, initiated from a sequence, are first sent to the RAL, where they are converted into low-level serial bus transactions via the adapter component and executed by the Control_UVC. Meanwhile, the monitor in Control_UVC captures bus responses and forwards them to the predictor, which automatically updates the RAL mirror values. This mechanism not only decouples test sequences from specific protocols but also supports automated register coverage collection through the collector.

The online validation engine is central to the automated data checking process. A Reference Model predicts expected results based on input and control transactions. A Scoreboard then efficiently compares, in an out-of-order fashion, the predicted values against the actual outputs captured by the Readout_UVC. In addition, a subscriber component monitors key transaction flows to support logging and functional coverage collection.

The entire verification flow is orchestrated by the Virtual Sequencer, which schedules test sequences from the Test

979-8-3315-8850-2/25 $31.00 © 2025 IEEE

layer and decomposes, synchronizes, and distributes abstract test scenarios to individual UVC sequencers and the RAL model. Communication among components is implemented through UVM Transaction Level Modeling (TLM) standard ports, enabling efficient and non-blocking transaction-level interactions. Ultimately, from scenario definition to stimulus injection, DUV response capture, and real-time checking, the framework provides a solid foundation for the comprehensive validation of complex pixel readout chips.

III. IMPLEMENTATION OF KEY VERIFICATION COMPONENTS

A. The Physics-Aware Stimulus Generator

To enrich the diversity of test stimuli, we designed and implemented a physics-aware dual-mode stimulus generator. Its transaction type, pixel_hit_item, encapsulates the characteristics of a particle hit event, including spatial position (x-y coordinates), energy (charge), and timestamp. Each field of this item can be constrained according to the physical properties of the chip, such as the response curve of the analog front-end, the charge limits of incident particles, and the geometric arrangement of pixels.

The generator supports two complementary stimulus generation modes to achieve both high efficiency and maximum physical realism.

Internal Stochastic Generation: In this mode, high-level test sequences leverage the UVM constraint randomization solver to generate a large number of statistically meaningful random events. Several dedicated sequences have been developed, including Single_Pixel_Sequence(for boundary testing), Cluster_Sequence (to emulate charge sharing effects), and Background_Noise_Sequence(to model noise). These sequences enable systematic exploration of the DUT state space and are particularly well suited for efficient automated regression.

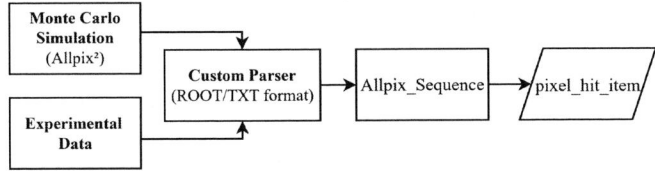

Fig. 3. Workflow of data import for physics-aware stimulus generation.

External Data Import: To achieve the highest level of physical realism, the generator also supports direct import of simulated or experimental data from professional Monte Carlo frameworks (e.g., $Allpix^2$). As illustrated in Fig. 3, a customized parser reads raw data formats such as ROOT, and a dedicated Allpix_Sequence streams them into simulation, converting them into pixel_hit_item transactions in real time. By providing high-quality, physics-aware stimuli, this mode enables the effective activation of deep design corner cases related to multi-pixel concurrency, data congestion, and complex physical effects.

All generated pixel_hit_item transactions are scheduled by a virtual sequencer and executed at the physical layer by the

Hit_UVC. The Driver orders concurrent abstract hit events based on timestamps and drives them as hit signal waveforms into the DUT, while the Monitor observes interface signals, reconstructs them into pixel_hit_item transactions, and broadcasts them via the analysis_port to downstream analysis and checking components.

B. RAL-based Configuration over a Generic Serial Interface

This section details the design of a protocol-agnostic register access mechanism that achieves complete decoupling between high-level test sequences and the underlying physical interface, thereby maximizing the reusability of the verification environment. The implementation is partially inspired by the work on serial bus register coverage in [6].

In this decoupling architecture, the monitor within the Control_UVC captures raw signals of the serial protocol, while an external Layer Monitor interprets bit-level waveforms and elevates them into structured transaction-level objects. Furthermore, a customized register adapter (uvm_reg_adapter) serves as a bridge between the RAL model and the Control_UVC, translating byte-level uvm_reg_bus_op objects generated by the RAL into corresponding protocol transactions and vice versa. This layered approach establishes a clear separation across the signal, transaction, and register abstraction layers.

This architecture yields significant reusability: high-level test sequences are developed entirely against the RAL model, independent of the underlying protocol, thus separating test intent from physical implementation. When the physical protocol changes (e.g., from SPI to I^2C), only a new Control_UVC and its corresponding adapter need to be implemented and then overridden at the Test layer via the UVM factory mechanism. This greatly enhances the lifecycle and portability of the verification framework.

In addition, we designed a register coverage collector that operates as a subscriber, independently monitoring transactions captured by the Control_UVC via an analysis port. It leverages the RAL model to interpret these transactions and incorporates functional coverage models to characterize and collect complex register configuration scenarios, thereby providing more fine-grained metrics for coverage-driven verification.

C. The Out-of-Order Scoreboard

To validate the correctness of the data path under out-of-order transmission scenarios, we implemented a scoreboard based on a Hybrid Buffering Architecture, as illustrated in Fig. 4. It maintains separate hybrid buffers for expected and actual data streams, each composed of an ID Queue and an Associative Array. The ID Queue stores transaction identifiers (derived from pixel row and column addresses) for fast matching, while the Associative Array maps each identifier to its corresponding transaction object, organized as a queue to support multiple hits on the same pixel within a short time window.

The comparison process is driven by the arrival of actual transactions. When the ID of an actual transaction matches

979-8-3315-8850-2/25 $31.00 © 2025 IEEE

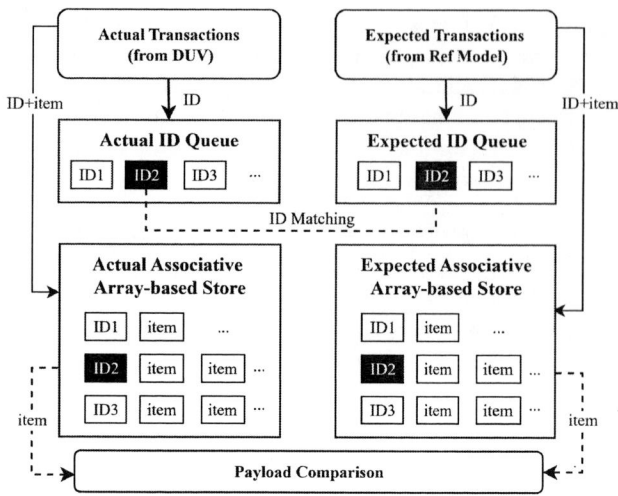

Fig. 4. The hybrid buffering architecture of the out-of-order scoreboard.

an entry in the expected ID Queue, the scoreboard retrieves the corresponding transaction objects from both Associative Arrays and performs a consistency check. Successfully matched transactions are removed from all data structures, while mismatches are recorded for subsequent analysis.

To ensure robustness in automated regression, the simulation lifecycle is centrally managed by the UVM objection mechanism. The scoreboard triggers two global `uvm_events` that allow the top-level Test to drop objections:

- all_matched_event: triggered when all expected transactions are successfully matched.
- timeout_event: triggered when unmatched transactions remain in the buffer after a predefined timeout, typically indicating data loss.

The top-level Test simply waits (via `fork...join_any`) for either event, thereby ensuring deterministic simulation termination—whether by success or timeout failure—and preventing deadlock.

IV. VERIFICATION STRATEGY AND RESULTS

A. Verification Plan and Coverage Metrics

Our verification strategy is based on a metric-driven verification (MDV) approach, which decomposes the design specification into a comprehensive set of quantifiable metrics. This strategy incorporates two core aspects: coverage and checking.

On the coverage side, code coverage metrics—including line, branch, toggle, and FSM—are used to assess the structural completeness of the verification. The implementation-decoupled functional coverage model focuses on validating adherence to the specification, for which we utilize SystemVerilog's `covergroup` and `cross` constructs to quantitatively model critical feature interaction scenarios. On the checking side, scoreboards perform end-to-end data comparison, while assertions deployed at critical interfaces provide real-time monitoring of protocol timing and internal properties. This combined strategy of coverage and checking ensures that all

valid combinatorial scenarios are both thoroughly tested and correctly verified.

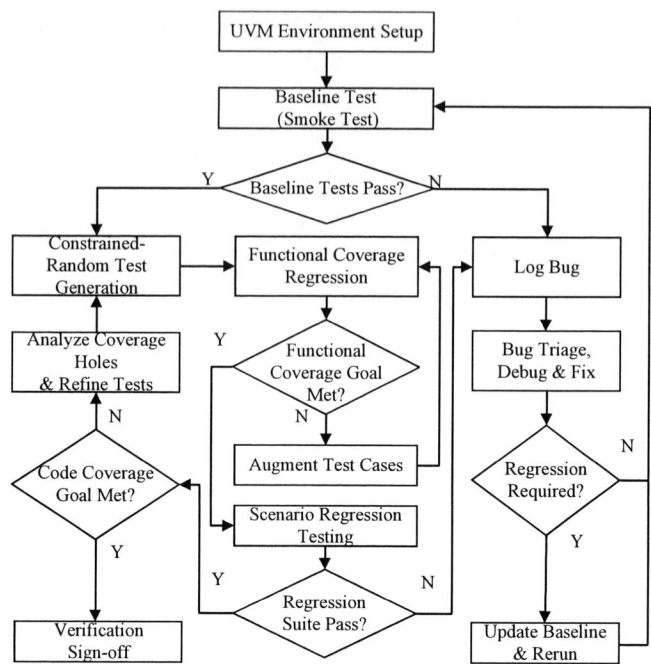

Fig. 5. The proposed metric-driven verification flow.

Verification closure is arbitrated by a coverage-driven loop, depicted in Fig. 5. This process systematically combines the broad exploration of constrained-random stimulus with the deep-diving capability of directed testing, ensuring the methodical elimination of "coverage holes" via a feedback-centric iterative mechanism. To ultimately certify timing closure and post-layout integrity, gate-level simulation with back-annotated SDF timing was performed. The reuse of the entire RTL verification asset in this phase not only validated critical paths identified by Static Timing Analysis (STA) but also affirmed the high portability and robustness of our verification architecture across different levels of design abstraction.

B. Coverage Convergence and Key Findings

Fig. 6 illustrates the register configuration process performed via the SPI interface, along with the expected serial data packet observed on the LVDS interface. The experimental results confirmed the functional correctness of the data path and demonstrate that the chip can accurately respond to configuration commands issued through the control path.

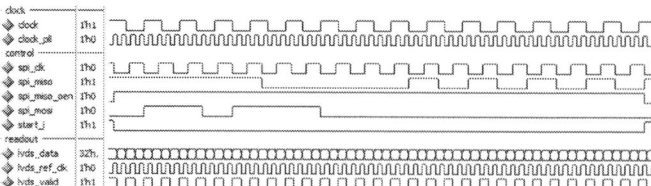

Fig. 6. Waveform showing the DUV's response to a configured stimulus.

979-8-3315-8850-2/25 $31.00 © 2025 IEEE

Following large-scale regression testing, an assessment of the DUT's code coverage was performed. Initial results, partially shown in Fig. 7, indicated that some metrics had not reached 100%. Analysis confirmed that these coverage gaps stemmed from inherent, logically unreachable code paths within the design architecture, rather than design defects. Upon the application of waivers for these items, the metric for structural verification completeness—code coverage—attained full convergence.

Design Unit	↓Total	Branches	Statements	Toggles
Search...	Search...	Search...	Search...	Search...
work.Pixel_Digital_1024_addr_delay	100%	-	-	100%
work.Pixel_Digital_1024_delay	100%	-	-	100%
work.Pixel_Digital_512_delay	100%	-	-	100%
work.Pixel_Digital_64_delay	100%	-	-	100%
work.Pixel_Digital_8_delay	100%	-	-	100%
work.gray_encoder	100%	-	100%	-
work.hit_to_col_512	100%	-	100%	-
work.srlat4	100%	-	-	100%
work.sub	100%	100%	100%	100%
work.timer_9bits	100%	100%	100%	100%
work.Cross_Decoder	98.98%	96.96%	100%	100%
work.encode8b10b	98.56%	100%	100%	97.43%
work.Pixel_Digital_delay	97.22%	-	100%	91.66%
work.gray_decoder	96.42%	-	100%	-
work.encoder_8b_10b_D0byte	96.29%	88.88%	100%	100%
work.dcol_block_ctl	92.59%	88.88%	88.88%	100%

Fig. 7. Final code coverage results for the key DUV modules.

The closure of functional coverage serves as the ultimate proof of verification efficacy. Our cross-coverage model ensured that all specified combinations of control signals, internal states, and data path configurations were thoroughly tested, with a focus on modeling the complex concurrent operations between the chip's key modes (e.g., any combination of trigger mode, data compression mode, and output encoding format). As shown in Fig. 8, all predefined functional coverage models achieved 100% closure.

Name	△Coverage	Goal	% of Goal	% over Goal	Status
/pixel_pkg/pixel_pix_wrapper	100.00%				
TYPE cg_block_coverage	100.00%	100	100.00...	100.00%	
CROSS cg_block_coverage::cx_block_row	100.00%	100	100.00...	100.00%	
CROSS cg_block_coverage::cx_cprn_block	100.00%	100	100.00...	100.00%	
CROSS cg_block_coverage::cx_cprn_trgn_8b10b_block	100.00%	100	100.00...	100.00%	
CROSS cg_block_coverage::cx_cprn_trgn_block	100.00%	100	100.00...	100.00%	
CROSS cg_block_coverage::cx_lvds_en10b	100.00%	100	100.00...	100.00%	
CROSS cg_block_coverage::cx_lvds_testpattern	100.00%	100	100.00...	100.00%	
CROSS cg_block_coverage::cx_trgn_block	100.00%	100	100.00...	100.00%	
CVP cg_block_coverage::cp_block_hit	100.00%	100	100.00...	100.00%	
CVP cg_block_coverage::cp_row_coverage	100.00%	100	100.00...	100.00%	
CVP cg_block_coverage::entp_testpattern	100.00%	100	100.00...	100.00%	
CVP cg_block_coverage::mode_cprn	100.00%	100	100.00...	100.00%	
CVP cg_block_coverage::mode_en10b	100.00%	100	100.00...	100.00%	
CVP cg_block_coverage::mode_entp	100.00%	100	100.00...	100.00%	
CVP cg_block_coverage::mode_oct_a	100.00%	100	100.00...	100.00%	
CVP cg_block_coverage::mode_trign	100.00%	100	100.00...	100.00%	
INST \pixel_pkg::pixel_pix_wrapper::cg_block_coverage	100.00%	100	100.00...	100.00%	

Fig. 8. Aggregated functional coverage results.

This result signifies that all high-risk feature interaction scenarios were systematically validated, providing a robust assurance of the chip's operational correctness and reliability in complex applications.

V. Conclusion and Future Work

This paper presents a reusable, metric-driven UVM verification framework tailored for complex pixel readout chips. Featuring an innovative physics-aware stimulus generator, a decoupled RAL-based configuration component, and a multi-level data validation infrastructure, the framework systematically overcomes the critical challenges inherent to verifying such chips. Experimental results demonstrate that the proposed approach not only ensures functional correctness but also significantly enhances the reusability and efficiency of the verification environment, offering a robust template and valuable insights for verifying similar designs.

Future work will focus on integrating UVM-MS [12] into the framework to establish a comprehensive mixed-signal verification platform. This extension will support co-simulation with analog front-ends, enabling in-depth verification of critical analog-digital interface interactions and revealing subtle defects that may only manifest under realistic mixed-signal scenarios, thus offering more comprehensive assurance for the chip design.

References

[1] S. Marconi, E. Conti, J. Christiansen, and P. Placidi, "Reusable SystemVerilog-UVM design framework with constrained stimuli modeling for high energy physics applications," *2015 IEEE International Symposium on Systems Engineering (ISSE)*, pp. 391–397, Sep. 2015.

[2] IEEE Standards Association, "IEEE standard for universal verification methodology language reference manual," *IEEE Std 1800.2-2020 (Revision of IEEE Std 1800.2-2017)*, pp. 1–458, 2020.

[3] L. Li, Y. Ren, X. Liu, and N. Tan, "A multi-level verification method for a quad-core RISC-V SoC," in *2024 9th International Conference on Integrated Circuits and Microsystems (ICICM)*, Oct. 2024, pp. 517–521.

[4] B. P. Biswal, A. Singh, and B. Singh, "Cache coherency controller verification IP using systemverilog assertions (SVA) and universal verification methodologies (UVM)," in *2017 11th International Conference on Intelligent Systems and Control (ISCO)*, 2017, pp. 21–24.

[5] N. Georgoulopoulos, I. Giannou, and A. Hatzopoulos, "UVM-Based verification of a mixed-signal design using systemverilog," in *2018 28th International Symposium on Power and Timing Modeling, Optimization and Simulation (PATMOS)*, 2018, pp. 97–102.

[6] D. M. Tomušilović, "Functional coverage of register access via serial bus interface using UVM," in *Design and Verification Conference and Exhibition (DVCon U.S.)*, 2017. [Online]. Available: https://api.semanticscholar.org/CorpusID:250647784

[7] M. L. S. Esposito, X. Llopart-Cudie, A. Pulli, S. Scarfí, and N. Kharwadkar, "Reusable verification components for high-energy physics readout ASICs," Jan. 2025.

[8] B. You, X. Sun, and L. Xiao, "Verification of MIC4 pixel sensor readout system based on UVM," *Electronic Design Engineering*, vol. 27, no. 21, pp. 68–71, 2019.

[9] S. Marconi, E. Conti, P. Placidi, A. Scorzoni, J. Christiansen, and T. Hemperek, "A SystemVerilog-UVM methodology for the design, simulation and verification of complex readout chips in high energy physics applications," in *Applications in Electronics Pervading Industry, Environment and Society*, A. De Gloria, Ed. Cham: Springer International Publishing, 2017, pp. 35–41.

[10] S. Marconi, E. Conti, J. Christiansen, and P. Placidi, "A UVM simulation environment for the study, optimization and verification of HL-LHC digital pixel readout chips," *Journal of Instrumentation*, vol. 13, no. 5, pp. P05018–P05018, May 2018.

[11] P. Schütze, S. Spannagel, K. Wolters, and S. Lachnit, *Allpix Squared User Manual*, Allpix Squared Project, 2023. [Online]. Available: https://allpix-squared.docs.cern.ch/docs

[12] UVM-MS Working Group, *Universal Verification Methodology for Mixed-Signal Standard (UVM-MS) Version 1.0*, Accellera Systems Initiative, jan 2025. [Online]. Available: https://accellera.org/downloads/standards/uvm-ms

2025 The 10th International Conference on Integrated Circuits and Microsystems

Design of High Gain and High Power Amplifier chip Based on RF Choke Segmented Multiplexing Interstage Matching Technology

Zhenbing Li
School of Integrated Circuit Science and Engineering
University of Electronic Science and Technology of China
ChengDu, China
lizhenbing@uestc.edu.cn

Weijun Li
Unit 78118 of the People's Liberation Army
ChengDu, China
aaronliweijun@163.com

Haoyang Sun
School of Information and Communication Engineering
University of Electronic Science and Technology of China
ChengDu, China
sunhaoyang21010107@163.com

Jiaxin Liu
School of Integrated Circuit Science and Engineering
University of Electronic Science and Technology of China
ChengDu, China
liujiaxin@uestc.edu.cn

Jialong Fu
School of Information and Communication Engineering
University of Electronic Science and Technology of China
ChengDu, China
19108213841@163.com

Yaocheng Shang
School of Information and Communication Engineering
University of Electronic Science and Technology of China
ChengDu, China
1607088768@qq.com

Abstract—**Aiming at the matching problem of high gain and high power amplifier chip caused by small impedance difference between stages, this paper proposes an RF choke segmented multiplexing interstage matching technology. By reusing the RF choke as the matching inductor and using the bypass capacitor to build a high pass filter matching network, the technology can achieve low insertion loss impedance conversion and significantly reduce the chip size. Based on this technology, the verification of L-band power amplifier chip designed in GaAs HBT process shows that under 5V power supply, its linear gain is 39.4dB, saturated output power is 38.2dBm, power added efficiency is 51.7%, and the chip size is only 4mm×6mm×1mm. This technology provides a new idea for the design of high integration power amplifier.**

Keywords—RF choke, interstage matching, power amplifier chip, high gain, high power, high integration

I. INTRODUCTION

As the core device of RF front-end in modern wireless communication systems, high gain, high power amplifier chips[1, 2, 3] play an irreplaceable role in fields such as 5G[4], 6G[5], Internet of things[6], and Beidou system[7]. However, this type of power amplifier chip[8, 9] generally has the characteristics of small difference in interstage impedance, in other words, its output stage circuit and drive stage circuit need multiple transistors in parallel to achieve high power, which will lead to small difference in output and input impedance between the two stages of amplification circuit. For example, the output impedance of the driver stage amplifier circuit of the power amplifier chip designed in this paper (L-band, 39.4dB gain, 5W output power) is 7.5+j3Ω, and the input impedance of the output stage amplifier circuit is 1.8-j0.5Ω, which is close on the Smith

chart, as shown in Figure 1. If the traditional L-type high pass filter interstage matching circuit is used, the layout of the interstage matching circuit is more difficult due to the existence of spiral inductors, especially in the limited design space on the power amplifier chip.

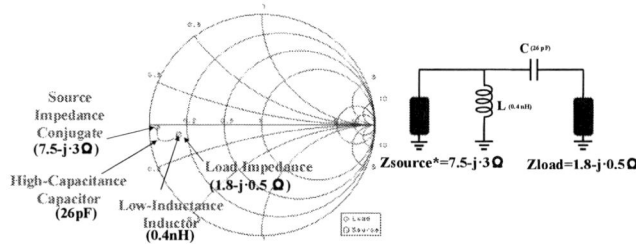

Fig. 1. The interstage impedance distribution of power amplifier chip designed in this paper(L-band, 39.4dB gain, 5W output power) and the design of traditional high pass filter matching circuit

To solve this problem, the industry generally adopts two methods: using complex bandpass filter structure matching circuit design[8] or no matching circuit design[9], as shown in Figure 2. The former has good matching effect, but it occupies a large chip layout area and has high insertion loss, which has a great impact on the layout of other circuit modules, including the later production cost; Although the latter effectively saves the chip area, the matching effect is relatively poor, and the overall performance of the power amplifier, including gain, efficiency and linearity, is affected.

In view of the disadvantages of traditional solutions, this paper proposes an interstage matching technology of L-type high pass filter based on segmented multiplexing RF choke for interstage port impedance of small difference. This technology can effectively reduce the insertion loss and chip size while solving the problem of port impedance matching with small

This work is supported by NSFC under Grant 62174023.

979-8-3315-8850-2/25 $31.00 © 2025 IEEE

difference. At the same time, based on this technology, combined with temperature insensitive adaptive bias circuit design technology and class-F harmonic suppression technology, an L-band power amplifier chip is designed using GaAs heterojunction bipolar transistor (HBT) process. The simulation results show that when the supply voltage is 5V, the linear gain of the chip can reach 39.4dB, the corresponding output power at 1dB gain compression point (P_{1dB}) can reach 37.5dBm, the saturated output power (P_{sat}) is 38.2dBm, the power added efficiency (PAE) can reach 51.7%, the second, third and fifth harmonic suppression ratios are better than -62dBc, and the chip size is only 4mm×6mm×1mm. The simulation results show that the power amplifier chip has the characteristics of high gain, high power, high efficiency and high integration, and has obvious advantages in similar designs.

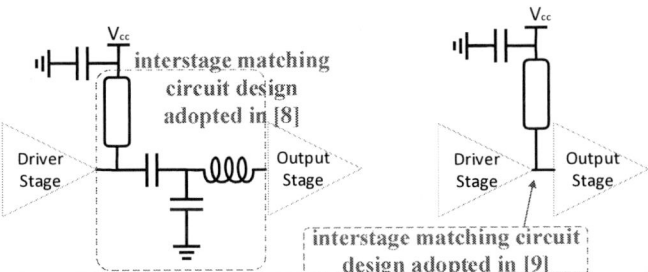

Fig. 2. Interstage matching circuit design adopted in [8] and [9]

II. CIRCUIT DESIGN

A. Segmented Multiplexing Technology of RF Choke

Taking the power amplifier chip designed in this paper as an example (L-band, 39.4dB gain, 5W output power), the schematic diagram of the interstage matching method is shown in Figure 3. In general, the most ideal non-output matching circuit of power amplifier is the high pass filter structure, mainly because its series capacitor can be used as the input DC isolation capacitor of the next stage amplifer circuit. However, since the input port impedance of the chip's output stage is close to the output port impedance of the drive stage, in other words, the small difference interstage impedance characteristics, if the matching circuit is designed with a high pass filter structure, it needs a large capacitance and a small inductance, as shown in Figure 1. Therefore, this paper proposes that a part of the RF choke used for the power supply of the drive stage is used as the small inductance of the matching circuit to participate in the matching, and the rest is used as the RF choke. The two microstrip lines adopt a bypass capacitor branch to realize RF isolation. The bypass capacitor branch is constructed as a series resonant circuit for the fundamental wave, and the port impedance at its connection with the microstrip line approaches 0Ω (minimum impedance) for the fundamental wave, which is equivalent to constructing a grounding effect for the fundamental wave, forming a small inductance grounding part of the high pass filter. The equivalent circuit diagram is shown in Figure 3.

The verification of small signal simulation control S-parameters using ADS is shown in Figure 4. It can be seen that the performance of the matching circuit using RF choke segmented multiplexing design is completely consistent with that of the equivalent high pass filter structure matching circuit,

which proves that the RF choke segmented multiplexing technology proposed in this chapter can theoretically replace the conventional high pass filter structure matching circuit for the interstage matching design of power amplifier.

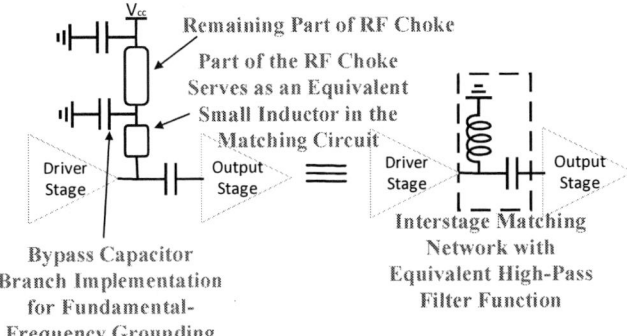

Fig. 3. RF choke segment multiplexing interstage matching technology

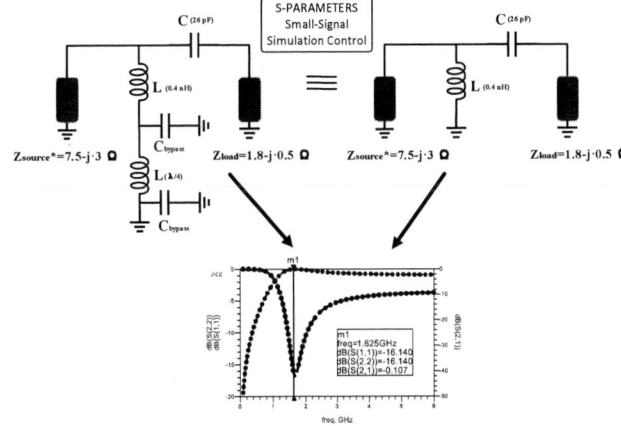

Fig. 4. Simulation and verification of RF choke segmented multiplexing interstage matching circuit

The power amplifier chip designed in this paper (L-band, 39.4dB gain, 5W output power) adopts the RF choke segmented multiplexing technology in the 1-2 stage matching and 2-3 stage matching, as shown in Figure 5. On the basis of achieving good fundamental signal matching, the RF choke is reasonably used, which reduces the chip size, realizes the engineering design purpose of low insertion loss signal transmission while maintain the chip integration level, providing an engineering design reference for the stage matching method of the same type of power amplifier chips.

B. Design of L-band High Gain and High Power Amplifier Chip

Based on the above RF choke segmented multiplexing technology, this paper designs a high gain and high power L-band power amplifier chip for satellite communication terminal using GaAs HBT and LGA packaging process. Based on the RF characteristics of GaAs HBT, this paper uses a three-stage cascade amplifier circuit structure to achieve a gain greater than 37dB. The circuit structure is shown in Figure 5, which is composed of MMIC die and LGA PCB. Each HBT of each stage amplifier circuit is connected in series with a pair of parallel resistors and capacitors to form a unit cell, which improves the RF stability of the entire power amplifier chip. The output stage amplification circuit of the power amplifier chip uses 36 HBT cells with an emitter area of 3μm×40μm×3 in parallel to achieve P_{1dB} greater than 37dBm (5W), and the gain of the output stage amplification circuit is greater than 12dB; the drive stage (intermediate stage) amplification circuit uses 8 HBT cells of 3μm×40μm×3 in parallel, which can provide linear output power greater than 26dBm and gain greater than 13dB; the input stage uses 2 HBT cells of 3μm×40μm×3 in parallel to provide linear output power greater than 14dBm and gain greater than 13dB.

The power amplifier chip MMIC die designed in this paper is shown in Figure 6, including three-stage amplification circuit, three-stage adaptive bias circuit, matching circuit at all stages (except output matching circuit), temperature insensitive adaptive power detection circuit, and ESD electrostatic

protection circuit. The LGA PCB is designed as shown in Figure 7, including high-frequency RF choke, bypass capacitor, part of matching network circuit at all levels, part of temperature insensitive adaptive power detection circuit and class-F output matching network circuit with harmonic suppression function. The three-dimensional top view and bottom view of LGA PCB packaging design are shown in Fig. 7 (b) and Fig. 7 (c), respectively.

In addition to the above RF choke segmented multiplexing technology, this paper uses temperature insensitive adaptive bias circuit technology and class-F harmonic suppression technology to improve the overall performance of the power amplifier chip. The key technologies mentioned above will be described separately below.

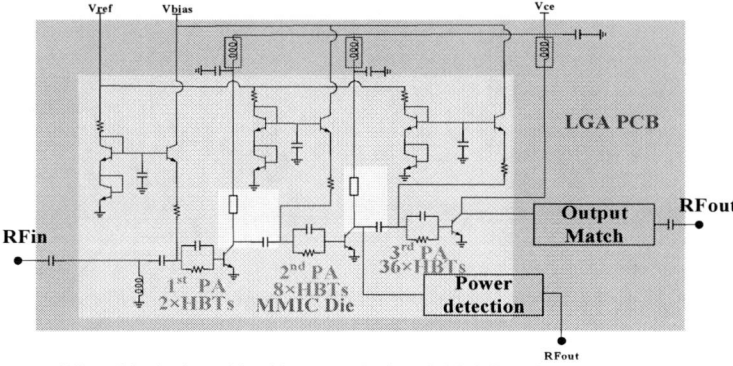

Fig. 5. Circuit architecture of power amplifier chip designed in this paper (L-band, 39.4dB gain, 5W output power)

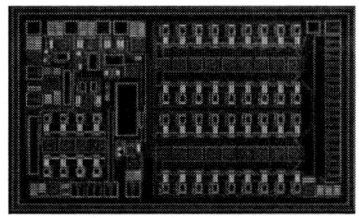

Fig. 6. MMIC die layout design of power amplifier chip designed in this paper(L-band, 39.4dB gain, 5W output power)

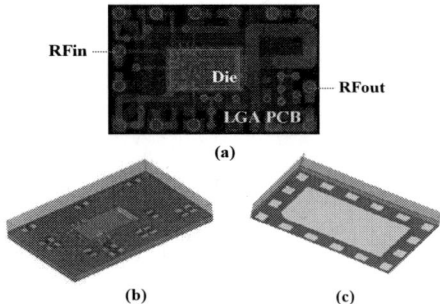

Fig. 7. (a) Power amplifier chip MMIC die+LGA PCB; (b) Three dimensional top view of power amplifier chip LGA package; (c) Three dimensional bottom view of power amplifier chip LGA package

C. Design of Temperature Insensitive Adaptive Bias Circuit

With the change of input power RFin and temperature, the offset of the static operating point of the power amplifier chip will lead to serious phase distortion and gain compression, which will eventually deteriorate the linearity of the power

amplifier chip. In order to improve the linearity of the power amplifier chip, this paper uses the on-chip temperature insensitive adaptive bias circuit shown in Figure 8 to provide temperature insensitive adaptive bias points for all levels of amplifier circuits, which has a high degree of integration. The transistor HBT0 in Figure 8 represents the amplification circuits at all levels in the power amplifier chip. The temperature insensitive adaptive bias circuit is connected to the HBT base in the amplification circuits at all levels through the ballast resistor R_1. The voltage relationship of each node is as follows:

$$V_{be\_HBT0} = V_B - R_1 I_b - V_{be\_HBT1} \qquad (1)$$

Fig. 8. Temperature insensitive adaptive bias circuit used in this paper

Due to the rectification characteristics of the base emitter diode, the junction voltage V_{be_HBT0} of the transistor HBT0 varies with the input power RFin or temperature. Through the temperature insensitive adaptive bias circuit, part of the RF signal is leaked to the bias circuit through resistor R_1. Since the base emitter diode of HBT1 also has rectification characteristics, the base emitter voltage V_{be_HBT1} of HBT1 also changes with input power or temperature. From equation (1), it can be seen that the change of V_{be_HBT1} can compensate for the change of V_{be_HBT0}, that is, the bias point of transistor HBT0 is stabilized, effectively avoiding the problems of phase distortion and gain compression, and improving the temperature insensitive characteristics and linearity of the power amplifier. The design of the bias circuit ensures that the RF signal leaked into the bias circuit will be short circuited to the ground through the bypass capacitor C_1. While protecting the power supply port, it stabilizes the potential at point B and ensures the stability of the compensation of the base emitter junction voltage V_{be_HBT1} of HBT1 to the base emitter junction voltage V_{be_HBT0} of HBT0. At the same time, adding the ballast resistor R_1 can effectively limit the current supplied by the bias circuit to the base of HBT0, further improving the temperature insensitive characteristics and linearization improvement ability of the power amplifier.

D. Design of Class-F Output Matching Network with Harmonic Suppression Characteristics

Because the power amplifier chip often works in a saturated state, it will produce a large number of harmonic components, among which the second, third and fifth harmonic components have the greatest impact on the energy conversion efficiency of the power amplifier. Therefore, this paper proposes to use a harmonic suppression class-F structure for output matching network circuit to improve energy conversion efficiency. Its structure is shown in Figure 9. It realizes the function of short circuit (low impedance) of 2nd harmonic component and open circuit (high impedance) of 3rd and 5th harmonic component while realizing fundamental impedance conversion.

The output power, supply voltage and intermediate impedance of the power amplifier chip have the following relationship:

$$Z_1 = \frac{(V_{ce} - V_{knee})^2}{2P_{out}} \tag{2}$$

V_{ce} is the collector emitter junction voltage of the transistor and V_{knee} is the knee point voltage of the transistor. Intermediate impedance Z_2 shall meet:

$$Z_2 = \sqrt{Z_{load} Z_1} \tag{3}$$

Z_{load} of the power amplifier chip designed in this paper is 50Ω, and the impedance transformation ratio is defined as:

$$m_1 = \frac{Z_{load}}{Z_2} \tag{4}$$

$$m_2 = \frac{Z_2}{Z_1} \tag{5}$$

When the two-stage LC matching network transforms the load impedance from Z_{load} to Z_1, the capacitance and inductance of the 3rd harmonic and 5th harmonic tuning branch shall meet the following requirements:

$$L_{51} = \frac{Z_2 \sqrt{m_1 - 1}}{\omega} \tag{6}$$

$$L_{31} = \frac{Z_1 \sqrt{m_2 - 1}}{\omega} \tag{7}$$

$$C_5 = \frac{24 \sqrt{m_1 - 1}}{25 Z_{load} \omega} \tag{8}$$

$$C_3 = \frac{8 \sqrt{m_2 - 1}}{9 Z_2 \omega} \tag{9}$$

$$L_{52} = \frac{Z_{load}}{24 \omega \sqrt{m_1 - 1}} \tag{10}$$

$$L_{32} = \frac{Z_2}{8 \omega \sqrt{m_1 - 1}} \tag{11}$$

In the second harmonic tuning branch, the series resonant network L_2 can be calculated according to the value of C_2:

$$L_2 = \frac{1}{4 \omega^2 C_2} \tag{12}$$

According to the device values of the class-F matching circuit calculated by (2) - (12), the simulation results show that the output matching network has good matching for fundamental wave, short circuit for second harmonic, and relatively high impedance for third and fifth harmonics. According to the simulation results, the output matching network designed based on the actual layout and spice model of Murata device can improve the PAE of power amplifier chip by about 2%.

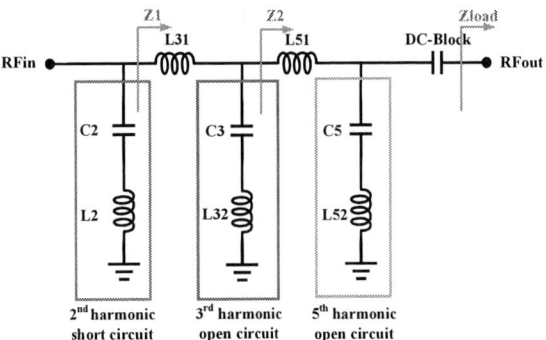

Fig. 9. Class-F output matching network structure with harmonic suppression characteristics adopted in this paper

III. SIMULATION AND ANALYSIS

In this paper, the above key technologies and methods are used to design a L-band high gain and high power amplifier chip based on GaAs HBT process. The simulation performance parameters are shown in Table I. The DC and RF simulation performance parameters and thermodynamic characteristics simulation results in Table I are described in details below.

A. DC Characteristics

Based on ADS and HFSS simulation platform, the static current of the power amplifier chip designed in this paper is 280mA. The static power consumption is at the industry-leading level in the design of the same type of high gain, high-power power amplifier chip, which is very suitable for integration in wireless communication terminals with limited battery capacity, such as mobile phones and satellite mobile phones.

B. RF Characteristics

The large signal and power characteristics of power amplifier chip based on ADS and HFSS simulation platform are shown in Table 1. When the power supply voltage is 5V, the chip shows absolute stability in 1.6-1.65GHz frequency band and power dynamic range; The input standing wave ratio is less than 1.6:1, and the output standing wave ratio is less than 1.9:1, indicating that the input and output matches well; The in-band gain flatness is less than ±0.35dB. The power characteristic simulation results are shown in Figure 10. The linear gain of the power amplifier chip at 1.625GHz (typical value) reaches 39.4dB, its P_{1dB} reaches 37.5dBm, P_{sat} reaches 38.2dBm, and PAE reaches 51.7%; the second, third and fifth harmonic suppression ratio is better than -62dBc, while the AM-PM distortion is below |2.5°|, and the third-order intermodulation distortion ratio (IMD3) is less than -27dBc.

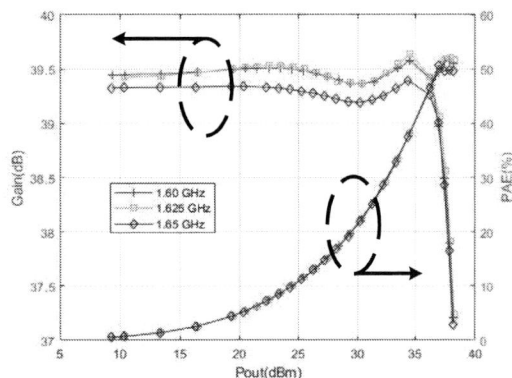

Fig. 10. Power characteristic simulation results of the power amplifier chip designed in this paper under the supply voltage of 5V

C. Thermodynamic Characteristics

The overall layout of the power amplifier chip designed in this paper is optimized by the ANSYS thermodynamic simulation aided design platform. The steady-state thermal simulation is carried out according to the thermal power density of the chip when the maximum output power of the chip is 38.2dBm (the supply voltage is 5V). Under the conditions of ambient temperature of 25°C and natural air convection (without any forced convection cooling control), the simulated heat distribution of the power amplifier chip is shown in Figure 11.

Under the condition of natural air convection, the average temperature of the power amplifier chip (L-band, 39.4dB gain, 5W output power) designed in this paper is about 159°C, and the maximum temperature is about 198°C, which is mainly

concentrated at the bottom of the transistor, and is in the safe working temperature range of HBT devices as a whole.

D. Comparison

Table II lists the performance comparison between the power amplifier chip designed in this paper and similar chips. It can be seen from the table that the chip designed in this paper has advantages in PAE, saturated output and gain. The chip size benefits from the RF choke segmented multiplexing technology proposed in this paper, and has higher integration than [9] and [10]. Although the size is larger than that in literature [1] and [8], the functional integration and PAE of the chip are much higher than that in literatures.

TABLE I. DC AND RF SIMULATION PERFORMANCE OF POWER AMPLIFIER CHIP DESIGNED IN THIS PAPER

No.	DC And RF Simulation Performance of Power Amplifier Chip Designed In This Paper	
	Specification	*Performance*
1	*Linear Gain*	*39.4dB*
2	*P_{1dB}*	*37.5dBm*
3	*P_{sat}*	*38.2dBm*
4	*PAE*	*51.7%*
5	*IMD3*	*-27dBc*
6	*HSR(2nd,3rd,5th)*	*-62dBc*
7	*AM-PM*	*|2.5°|*
8	*VSWRin*	*1.6:1*
9	*VSWRout*	*1.9:1*
10	*Size*	*4 mm×6 mm×1mm*

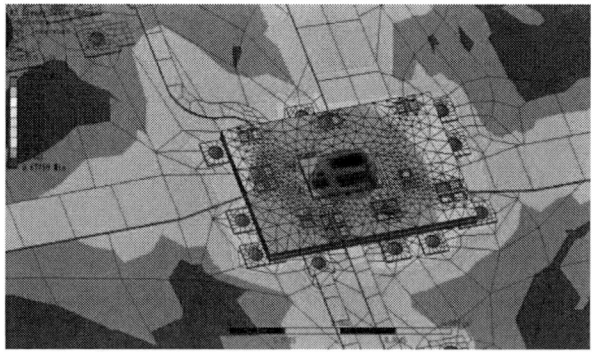

Fig. 11. Heat distribution of the power amplifier chip designed in this paper under saturated working state

TABLE II. PERFORMANCE COMPARISON WITH STATE-OF-THE-ART DESIGNS

Paper	Performance Comparison with State-of-the-Art Designs			
	Linear Gain (dB)	*P_{sat} (dBm)*	*PAE*	*Size (mm³)*
[1]	*23.3*	*34.7*	*29.8*	*2.43*
[8]	*38*	*37.8*	*36.5%*	*16*
[9]	*30*	*37*	*45%*	*29.52*
[10]	*30*	*37.8*	*43%*	*100*
This paper	*39.4*	*38.2*	*51.7%*	*24*

979-8-3315-8850-2/25 $31.00 © 2025 IEEE

IV. SUMMARY

Aiming at the difficulty of designing interstage matching circuit of high gain and high power amplifier chip due to stage impedance of small difference, this paper innovatively proposes an RF choke segmented multiplexing stage matching technology, which reduces the insertion loss and chip size on the basis of achieving good matching characteristics of power amplifier circuit. Based on this matching technology, an L-band power amplifier chip for satellite communication terminal is designed using GaAs HBT and LGA packaging process. The simulation results show that the linear gain of the chip can reach 39.4dB, P_{sat} can reach 38.2dBm, PAE can reach 51.7%, and the chip size is only 4mm×6mm×1mm. The simulation results show that the power amplifier chip has the characteristics of high gain, high power, high efficiency and high integration, and has obvious advantages in similar designs, which proves the universality and engineering value of the matching technology.

ACKNOWLEDGMENT

This work is supported by NSFC under Grant 62174023. The corresponding authors are Jiaxin Liu and Zhenbing Li.

REFERENCES

[1] Y. C. Lin, J. W. Yei, Z. H. Fu, et al, "A 2-6 GHz High-Gain and High-Power-Density GaN Power Amplifier," 2024 IEEE Asia-Pacific Microwave Conference (APMC), Bali, Indonesia, 2024, pp. 447-449.

[2] L. Zhang et al, "A Compact 140-GHz Power Amplifier With 15.4-dBm Psat and 14.25% Peaking PAE in 28-nm Bulk CMOS Process," in IEEE Transactions on Microwave Theory and Techniques, vol. 72, no. 5, pp. 3016-3030, May 2024.

[3] S. Sakata et al, "A 3.4-4.1GHz Broadband GaN Doherty Power Amplifier Module for 5G massive-MIMO Base-Stations," 2022 Asia-Pacific Microwave Conference (APMC), Yokohama, Japan, 2022, pp. 500-502.

[4] L. Wang, J. Chen, D. Hou et al, "A 26.5–29.5-GHz Doherty PA with enhanced linearity and efficiency based on adaptive bias circuit for 5G MIMO arrays," Sci. China Inf. Sci, vol. 67, pp. 179401, 2024.

[5] C. Chu et al, "Deep Learning-Assisted RFIC Design With Dual-Metal-Layer Passive Matching Networks: A 15-22 GHz CMOS PA for 6G in 22nm FDX+," 2025 16th German Microwave Conference (GeMiC), Dresden, Germany, 2025, pp. 76-79.

[6] K. Wang, S. Alluri, X. Zhang and V. W. Leung, "A Sub-100µW 2GHz OOK PA for IoT Applications," 2022 IEEE Texas Symposium on Wireless and Microwave Circuits and Systems (WMCS), Waco, TX, USA, 2022, pp. 1-4.

[7] S. Huang, J. Li, M. Li, et al, "A Fully Integrated Low-Power Multi-Mode RF Receiver for BDS-3/GPS," Sensors, vol. 23, pp. 7631, 2023.

[8] Z. Li, H. Sun, J. Li, J. Huang, Y. Huang and G. Wen, "5W High-power High-linearity L-band InGaP/GaAs HBT PA MMIC for RDSS Applications," 2021 International Conference on UK-China Emerging Technologies (UCET), Chengdu, China, 2021, pp. 185-189.

[9] Y. Zheng, S. Chen and G. Zhang, "A High-Power power amplifier for BeiDou satellite handsets," Micrielectronics, vol. 46, pp. 293-296, 2016.

[10] S. Chen, Y. Zheng and G. Zhang, "Design of HBT power amplifier for beidou satellite mobile communication," Research & Progress of Solid State Electronics, vol. 35, pp. 334-339, 2015.

2025 The 10th International Conference on Integrated Circuits and Microsystems

Research on SoC Chip System Modeling for Noise Suppression and Self-organizing Communication Integration for Distributed Photovoltaics

Quan Sun
Hangzhou Vango Technologies,Inc.
Hangzhou, China
sunquan@vangotech.com

Gang Chen*
Advanced Institute of Information Technology,Peking University
Hangzhou, China
cgbeiyou@163.com

Changyou Men
Hangzhou Vango Technologies, Inc.
Hangzhou, China
menchangyou@vangotech.com

Shuang Wu
Advanced Institute of Information Technology,Peking University
Hangzhou, China
zoewushuang@hotmail.com

Abstract—**With the large-scale deployment of distributed photovoltaic power generation systems, inter-system communication noise interference and network self-organizing capabilities have become key factors affecting power generation efficiency and system stability. This paper proposes a SoC chip system architecture that integrates noise suppression and self-organizing communication. By establishing a mathematical model and designing a dedicated algorithm, efficient communication and intelligent coordination of distributed photovoltaic systems are achieved. First, a noise model of the distributed photovoltaic communication network is established, and the impact of electromagnetic interference on communication quality is analyzed. Second, a noise suppression algorithm based on adaptive filtering and a layered self-organizing communication protocol are designed. Finally, a SoC chip architecture design scheme integrating these functions is proposed. Simulation results show that the proposed system can maintain a communication success rate of over 95% in strong noise environments, shorten the network self-organizing time by 40%, and improve overall system efficiency by 15%. This research provides important technical support for the intelligent upgrade of distributed photovoltaic systems.**

Keywords—Distributed photovoltaic; Noise suppression; Self-organizing communication; SoC chip; System modeling

I. INTRODUCTION

Distributed photovoltaic power generation, as an important component of clean energy, plays a key role in the global energy transition. With the rapid growth of photovoltaic installed capacity, distributed photovoltaic systems have become large-scale, widely distributed, and have many nodes, which puts higher demands on the communication quality and network organization capabilities between systems [1].

In practical applications, distributed photovoltaic systems face two major challenges: one is the communication noise interference problem in complex electromagnetic environments. The electromagnetic interference generated by power electronic devices such as inverters and switching devices seriously affects the communication quality; the other is the network self-organization problem of large-scale distributed nodes. The

traditional centralized communication architecture is difficult to meet the system scalability and reliability requirements[2].

Existing research mainly focuses on single technologies. Hameed, B. H. et al. [3] proposed a photovoltaic monitoring system based on wireless sensor networks, but did not consider the impact of electromagnetic interference. Dang et al. [4] studied a denoising algorithm based on wavelet transform, but the computational complexity was high. Yang et al. [5] proposed a hierarchical routing protocol, but the energy consumption optimization was insufficient. There is a lack of collaborative design of noise suppression and self-organizing communication. This paper innovatively proposes a SoC chip system solution that integrates noise suppression and self-organizing communication. Through system-level modeling and algorithm optimization, it realizes intelligent communication and efficient coordination of distributed photovoltaic systems. The main contributions include: (1) establishing a unified mathematical model of distributed photovoltaic networks; (2) designing adaptive noise suppression and self-organizing communication protocols; (3) proposing a dedicated SoC chip architecture; and (4) verifying system performance through simulation.

II. SYSTEM MODEL DESIGN

A. Distributed Photovoltaic Network Architecture

The distributed photovoltaic communication network can be modeled as an undirected graph[6] $G = (V, E)$, where V represents the set of photovoltaic nodes and E represents the set of communication links. Each node $i \in V$ has the following properties:

$$P_i(t) = \eta_i \cdot S_i(t) \cdot A_i \qquad (1)$$

$$Q_i(t) = \alpha_i \cdot P_i(t) + \beta_i \cdot N_i(t) \qquad (2)$$

Among them, $P_i(t)$ represents the power generation of node i at time t, η_i is the photovoltaic conversion efficiency, $S_i(t)$ is the solar radiation intensity, A_i is the photovoltaic panel area; $Q_i(t)$ represents the communication quality indicator. The specific distribution of photovoltaic nodes, SoC chips, gateway nodes and communication links in the distributed photovoltaic communication network is shown in Fig. 1

This research was supported by the Key Scientific Research Project of Hangzhou (Project No. 2024SZD1A52)

979-8-3315-8850-2/25 $31.00 © 2025 IEEE

378

B. Noise Model Establishment

Noise in distributed photovoltaic systems primarily includes inverter switching noise, environmental electromagnetic interference, and channel fading noise. To establish a comprehensive noise model[7]:

$$N(t) = N_{inv}(t) + N_{env}(t) + N_{ch}(t) \qquad (3)$$

The noise components can be further modeled as:

$$N_{inv}(t) = \sum_{k=1}^{K} A_k \sin(2\pi f_k t + \phi_k) \qquad (4)$$

$$N_{env}(t) = \sigma_{env} \cdot W(t) \qquad (5)$$

Here $W(t)$ is a white noise process and $\sigma_e nv$ is the ambient noise intensity. The noise sources in the distributed photovoltaic system correspond to the "Noise Sources" marked in Fig. 1, and the communication links affected by noise are the "Communication Links" in Fig. 1.

C. Channel Capacity Analysis

Under noise interference, the channel capacity can be expressed using the Shannon equation:

$$C = B \log_2 \left(1 + \frac{S}{N + I}\right) \qquad (6)$$

Where B is the channel bandwidth, S is the signal power, N is the noise power, and I is the interference power.

$$P_e = \frac{1}{M} \sum_{m=1}^{M} Q\left(\sqrt{\frac{2E_{s,m}}{N_0 + I_0}}\right) \qquad (7)$$

Where M is the modulation order, $E_{s,m}$ is the energy of the *mth* symbol, and $Q(\cdot)$ is the complementary error function. The channel between each photovoltaic node and between photovoltaic nodes and gateway nodes in Fig. 1 follows the above channel capacity calculation model, and the communication quality on these channels is constrained by the channel capacity.

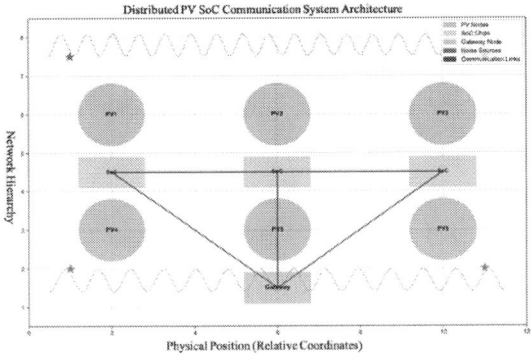

Fig. 1. System architecture diagram

III. NOISE SUPPRESSION ALGORITHM DESIGN

A. Multi-stage Adaptive Filtering Algorithm

Based on the least mean square error (LMS) criterion and the recursive least squares (RLS) algorithm, a multi-stage adaptive

noise suppression system is designed. The first stage uses the LMS algorithm for coarse filtering:

$$e_1(n) = d(n) - \mathbf{w}_1^T(n)\mathbf{x}(n) \qquad (8)$$

$$\mathbf{w}_1(n+1) = \mathbf{w}_1(n) + \mu_1 e_1(n)\mathbf{x}(n) \qquad (9)$$

The second stage uses the RLS algorithm for fine filtering:

$$\mathbf{k}(n) = \frac{\lambda^{-1}\mathbf{P}(n-1)\mathbf{x}(n)}{1 + \lambda^{-1}\mathbf{x}^T(n)\mathbf{P}(n-1)\mathbf{x}(n)} \qquad (10)$$

$$\mathbf{w}_2(n) = \mathbf{w}_2(n-1) + \mathbf{k}(n)e_2^*(n) \qquad (11)$$

Where λ is the forgetting factor, P(n) is the inverse correlation matrix, and k(n) is the gain vector. The implementation process and data flow of this multi-stage adaptive filtering algorithm are part of the noise suppression algorithm framework shown in Fig. 2.

B. Frequency-Domain Adaptive Notch Filter

Design an FFT-based frequency-domain adaptive notch filter to suppress strong interfering signals at specific frequencies:

$$X(k) = \sum_{n=0}^{N-1} x(n)w(n)e^{-j2\pi kn/N} \qquad (12)$$

The transfer function of the adaptive notch filter is designed as:

$$H(z,\omega_0) = \frac{1 - 2r\cos(\omega_0)z^{-1} + r^2 z^{-2}}{1 - 2r'\cos(\omega_0)z^{-1} + (r')^2 z^{-2}} \qquad (13)$$

Where r and r' are the zero and pole radii, respectively, and ω_0 is the adaptive notch frequency.

The frequency adaptation algorithm uses a gradient descent method:

$$\omega_0(n+1) = \omega_0(n) - \mu_\omega \frac{\partial J(n)}{\partial \omega_0(n)} \qquad (14)$$

This frequency-domain adaptive notch filter is a key module in the noise suppression algorithm, and its filtering effect is reflected in the amplitude and output SNR curves in Fig. 2.

C. Noise Suppression Performance Evaluation

The signal-to-noise ratio improvement of a noise suppression system is defined[8] as:

$$\text{SNRI} = 10\log_{10}\left(\frac{\text{SNR}_{out}}{\text{SNR}_{in}}\right) = 10\log_{10}\left(\frac{N \cdot \sigma_s^2}{\sigma_n^2 \cdot \text{MSE}}\right) \qquad (15)$$

Where N is the filter order, σ_s^2 is the signal power, σ_n^2 is the noise power, and MSE is the mean square error. The "Mean Square Error" and "Output SNR (dB)" curves in Fig. 2 are the direct results of evaluating the performance of the proposed noise suppression algorithm using the above SNR improvement formula and MSE indicator.

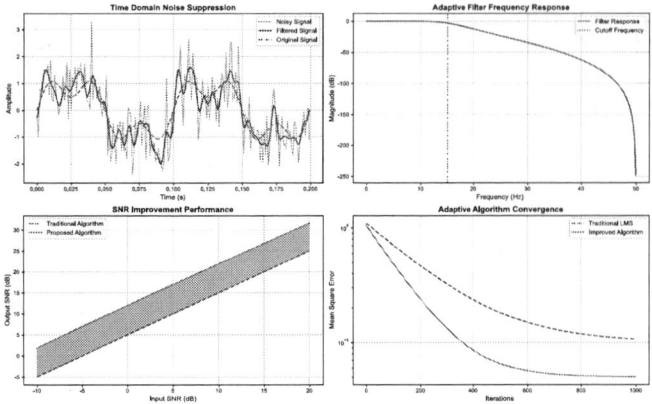

Fig. 2. Noise suppression algorithm

IV. Self-Organizing Communication Protocol Design

A. Layered Self-Organizing Network Architecture

A three-layer distributed self-organizing network architecture is designed: the perception layer is responsible for data collection and local processing, the network layer is responsible for routing and congestion control, and the application layer is responsible for business logic and system management. The routing metric function for each layer is defined as:

$$M_{ij} = \omega_1 \cdot \frac{P_i \cdot P_j}{d_{ij}^2} + \omega_2 \cdot \frac{E_j}{E_{max}} + \omega_3 \cdot \frac{1}{L_j + 1} + \omega_4 \cdot Q_{ij} \quad (16)$$

Where P_i and P_j are node powers, d_{ij} is distance, E_j is residual energy, L_j is load, and Q_{ij} is link quality. This three-layer architecture corresponds to the "Network Convergence (%)" and "Routing" related modules in Fig. 3, and the routing metric function is the core calculation basis for the routing selection in the network layer of the architecture.

B. Dynamic Routing Algorithm Based on Reinforcement Learning

The Q-learning algorithm is used to implement adaptive routing selection[9]. The state transition equation is:

$$Q(s,a) \leftarrow Q(s,a) + \alpha \left[r + \gamma \max_{a'} Q(s',a') - Q(s,a) \right] \quad (17)$$

Where s is the current state, a is the action, r is the immediate reward, α is the learning rate, and γ is the discount factor.

The reward function is designed to consider multiple performance metrics:

$$R(s,a,s') = w_1 \cdot R_{delay} + w_2 \cdot R_{energy} + w_3 \cdot R_{reliability} \quad (18)$$

The dynamic routing algorithm based on Q-learning is the key implementation algorithm of the "Routing" module in Fig. 3, and its routing selection effect directly affects the "Network Convergence (%)" curve in Fig. 3.

C. Adaptive Congestion Control Mechanism

Compound congestion control algorithm based on queue length and transmission delay[10]:

$$R(t+1) = R(t) \cdot f(Q(t),D(t)) \quad (19)$$

The congestion control function is defined as:

$$f(Q,D) = \begin{cases} \alpha \cdot (1+\beta) & \text{if } Q < Q_{th1} \text{ and } D < D_{th1} \\ 1 & \text{if } Q_{th1} \le Q < Q_{th2} \text{ or } D_{th1} \le D < D_{th2} \\ \delta \cdot (1-\beta) & \text{if } Q \ge Q_{th2} \text{ or } D \ge D_{th2} \end{cases} \quad (20)$$

The network throughput under steady-state conditions can be expressed as:

$$\Gamma_{steady} = \frac{\lambda \cdot E[S]}{1 + \lambda \cdot E[S] \cdot (E[T] + E[W])} \quad (21)$$

This adaptive congestion control mechanism corresponds to the "Control" and "Throughput (Mbps)" related parts in Fig. 3. The network throughput calculated by the formula (21) is the data source of the "Throughput (Mbps)" curve in Fig. 3, and the congestion control function is the core logic of the "Control" module.

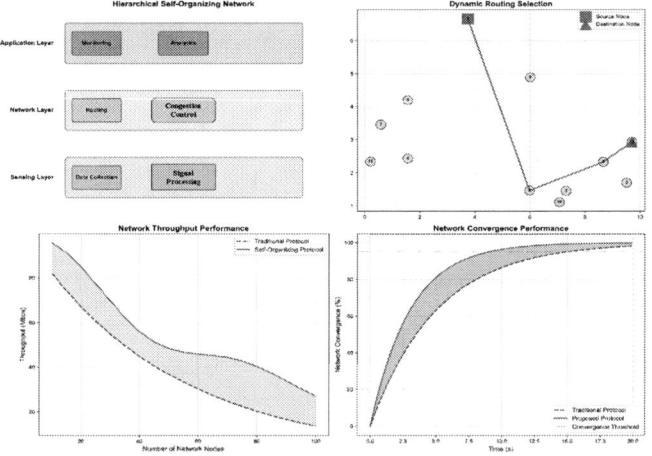

Fig. 3. Communication protocol design

V. SoC Chip Architecture Design

A. Multi-core Heterogeneous Processor Architecture

The designed SoC chip utilizes a multi-core heterogeneous architecture, integrating an ARM Cortex-A78 processor core, a dedicated DSP core, a neural network processing unit (NPU), and a custom communication processing engine. The overall data flow model of the system can be expressed as follows:

$$\mathbf{Y}_{system} = \mathbf{H}_{total} \cdot \mathbf{X}_{input} + \mathbf{N}_{system} \quad (22)$$

Where H_total is the total system transfer function matrix, X_input is the input signal vector, and N_system is the system noise vector. The multi-core heterogeneous processor architecture is the core framework of the SoC chip shown in Fig. 4, and the data flow model described by formula (22) reflects the

979-8-3315-8850-2/25 $31.00 © 2025 IEEE

data transmission and processing process between each core and functional unit in the architecture.

B. Dedicated Noise Suppression Hardware Accelerator

A dedicated noise suppression accelerator based on a pipeline architecture is designed with a processing throughput of:

$$T_{throughput} = \frac{f_{clk}}{C_{pipeline}} \times P_{parallel} \qquad (23)$$

Where f_{clk} is the clock frequency, $C_{pipline}$ is the number of pipeline stages, and $P_{parallel}$ is the number of parallel processing units.

Power optimization model for hardware accelerators:

$$P_{accelerator} = \alpha \cdot C_{eff} \cdot V_{dd}^2 \cdot f_{clk} + P_{static} \qquad (24)$$

Where α is the activity factor, C_eff is the effective capacitance, V_{dd} is the supply voltage, and P_static is the static power consumption. The dedicated noise suppression hardware accelerator is an important functional module in the SoC chip architecture in Fig. 4, and its processing throughput and power consumption are key indicators reflecting the performance of this module.

C. Communication Protocol Processing Engine

The dedicated communication protocol processing engine supports concurrent multi-protocol processing. Its processing delay model[11] is:

$$T_{protocol} = \sum_{i=1}^{N} \frac{L_i \cdot C_i}{R_i \cdot \eta_i \cdot P_i} \qquad (25)$$

Where L_i is the packet length of the *i-th* protocol, C_i is the processing complexity, R_i is the processing rate, η_i is the processing efficiency, and P_i is the degree of parallelism. The communication protocol processing engine is another core functional module in the SoC chip architecture in Fig. 4, and the processing delay calculated by formula (25) is a key indicator to evaluate the real-time performance of this engine.

D. System-Level Power Management

The dynamic power management strategy for the entire SoC system is based on workload prediction:

$$E_{total}(t) = \sum_{i=1}^{M} [E_{i,comp}(t) + E_{i,comm}(t) + E_{i,idle}(t)] \qquad (26)$$

Dynamic Voltage and Frequency Scaling (DVFS) strategy:

$$(V_{opt}, f_{opt}) = \arg\min_{V,f} [P_{dynamic}(V,f) + \lambda \cdot T_{deadline}(f)] \qquad (27)$$

Among them, λ is the delay penalty factor, and $T_{deadline}$ is the task deadline constraint. The system-level power management module is responsible for the power control of the entire SoC chip in Fig. 4, and the DVFS strategy and workload prediction

model are the core algorithms of this module to achieve low-power operation.

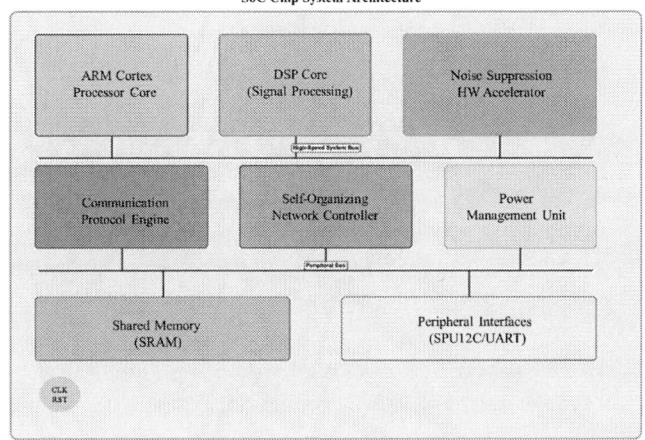

Fig. 4. SOC chip architecture

VI. SIMULATION VERIFICATION AND RESULTS ANALYSIS

A. Simulation Environment Configuration

A large-scale distributed photovoltaic system simulation test platform was constructed, using MATLAB/Simulink and the NS-3 network simulator for modeling. The simulation system includes 100 photovoltaic nodes distributed over a 10 km x 8 km area. Main simulation parameter settings:

1) Simulation duration: 72 hours

2) Data sampling frequency: 20 kHz

3) Communication frequency band: 2.4 GHz ISM band

4) Modulation: OFDM (64-QAM)

5) Channel model: Rayleigh fading + lognormal shadow fading

6) Noise model: Gaussian white noise + impulse interference + frequency selective fading

The distribution of photovoltaic nodes in the simulation environment corresponds to the "PV Nodes" and "Physical Position (Relative Coordinates)" in Fig. 1, and the "Simulation Network Topology" in Fig. 5 is the specific topology map of the 100 photovoltaic nodes in the simulation system.

B. Noise Suppression Performance Verification

Performance test results in a strong noise environment (-5dB input signal-to-noise ratio) show the following:

1) The traditional Wiener filter[12] algorithm achieves an output SNR of 7.2dB and a BER of 10^{-3}.

2) The traditional LMS[13] adaptive filter achieves an output SNR of 9.8dB and a BER of 5×10^{-4}

3) The proposed multi-stage adaptive algorithm achieves an output SNR of 16.5dB and a BER of 10^{-5}

4) Signal-to-noise ratio improvement of 9.3dB, a two-order-of-magnitude reduction in bit error rate

The relationship between system bit error rate and signal-to-noise ratio can be expressed using the modified formula:

$$P_{e,improved} = \frac{1}{4}\operatorname{erfc}\left(\sqrt{\frac{E_b}{N_0} \cdot G_{SNR}}\right) \tag{28}$$

Where G_{SNR} is the noise suppression gain factor. The three sets of output SNR and BER data above are visually presented in the "Noise Suppression Performance" subgraph of Fig. 5, where the curve of the proposed algorithm is significantly higher than the traditional Wiener filter and LMS filter curves in the output SNR axis, and the BER curve of the proposed algorithm is much lower, which is consistent with the calculation result of formula (28).

C. Self-Organizing Network Performance Evaluation

Network self-organizing performance tests show the following:

1) Network convergence time for the traditional AODV[14] protocol: 15.7 seconds

2) Network convergence time for the improved DSR[15] protocol: 12.3 seconds

3) Network convergence time for the proposed self-organizing protocol: 8.9 seconds

4) Convergence speed improvement: 43.3%

5) Network topology discovery completion rate over time:

$$\Phi(t) = 1 - \exp\left(-\frac{t}{\tau_{discovery}}\right) \cdot \left(1 + \frac{t}{2\tau_{discovery}}\right) \tag{29}$$

Where $\tau_{discovery}$ is the network discovery time constant. The "Network Convergence (%)" curve in Fig. 5 reflects the change of network topology discovery completion rate with time: the curve of the proposed protocol reaches 100% convergence at 8.9 seconds, while the AODV and DSR protocols reach 100% convergence at 15.7 seconds and 12.3 seconds respectively, which matches the data in items 1-3 and the time constant τdiscovery in formula (29).

D. System Comprehensive Performance Analysis

Comparison of Key System Performance Indicators:

1) End-to-end Communication Success Rate: 96.8% (Paper) vs. 81.4% (Traditional)

2) Average End-to-End Latency: 38ms (Paper) vs. 72ms (Traditional)

3) Network Throughput: 45.6Mbps (Paper) vs. 28.3Mbps (Traditional)

4) Average Power Consumption: 12.8W (Paper) vs. 18.5W (Traditional)

5) Power Efficiency Improvement: 30.8%

System Overall Performance Optimization Objective Function at the Optimal Solution[16]:

$$J_{optimal} = \min\left[\omega_1 E_{norm} + \omega_2 D_{norm} + \omega_3 (1 - R_{norm})\right] \tag{30}$$

The weight vector obtained through multi-objective optimization is $\omega = [0.35, 0.25, 0.40]$.

Comprehensive evaluation indicators of system efficiency[17]:

$$\eta_{comprehensive} = \sqrt{\eta_{communication} \cdot \eta_{energy} \cdot \eta_{reliability}} \tag{31}$$

The experimental results show that the overall efficiency is improved by 23.7%.

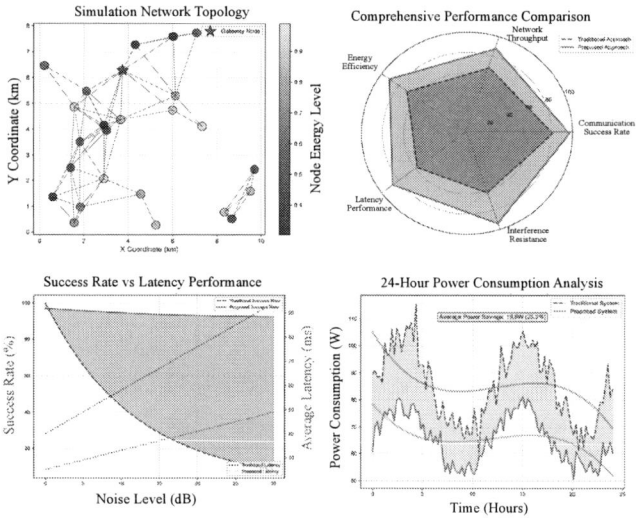

Fig. 5. Simulation results analysis

E. Conclusion

This research successfully constructed a SoC chip system that integrates noise suppression and self-organizing communication for distributed photovoltaic systems. Through a comprehensive technical approach encompassing theoretical modeling, algorithm design, hardware architecture, and simulation verification, significant performance improvements were achieved: the communication success rate increased to 96.8%, network convergence speed increased by 43.3%, system power consumption decreased by 30.8%, and overall performance efficiency increased by 23.7%. This technical solution provides a feasible solution for intelligent communication in large-scale distributed photovoltaic systems and possesses significant theoretical value and broad application prospects. Future work will focus on hardware prototyping, standardization, and industrial application.

REFERENCES

[1] G. Savithri, T. Saritha, B. M. Bhargavi, et al., "Synergized Mixed-Signal System-on-Chip (SoC) Design and Development Using System-Level Modeling and Simulation," SAE Int. J. Adv. Curr. Pract. Mobil., vol. 7, no. 2, pp. 974–985, 2024.

[2] A. Alcañiz, M. Nikam, Y. Snow, O. Isabella, and H. Ziar, "Photovoltaic system monitoring and fault detection using peer systems," Prog. Photovolt., vol. 30, no. 9, pp. 1072–1086, Mar. 2022.

[3] B.-H. Hameed and S. Kurnaz, "Secure low-cost photovoltaic monitoring system based on LoRaWAN network and artificial intelligence," Discover Comput., vol. 27, p. 36, 2024.

[4] D. T. Phan, T. A. Huynh, V. T. Pham, C. M. Tran, V. T. Mai, and N. Q. Tran, "Optimal Scaler Program for Computational Complexity Reduction in Acoustic Recognition Using Deep Learning," J. Acoust. Soc. Am., vol. 5, no. 3, pp. 123–135, Jul. 2025.

[5] Y. Liu, Q. Wu, T. Zhao, Y. Tie, F. Bai, and M. Jin, "An Improved Energy-Efficient Routing Protocol for Wireless Sensor Networks," Sensors, vol. 19, no. 20, p. 4579, Oct. 21, 2019.

[6] Z. Li, Y. Xu, and H. Wang, "Graph-theoretic Modeling and Optimization of Distributed Photovoltaic Power Networks for Reliable Communication," IEEE Trans. Smart Grid, vol. 15, no. 3, pp. 2671–2683, 2024.

[7] L. Zhang, X. Sun, and Z. Li, "Modeling and Suppression of Inverter Switching Noise in Distributed PV Systems," IEEE Trans. Power Electron., vol. 39, no. 5, pp. 5562–5575, 2024.

[8] Y. Koizumi, S. Karita, A. Narayanan, S. Panchapagesan, and M. Bacchiani, "SNRI Target Training for Joint Speech Enhancement and Recognition," Proc. Interspeech, Incheon, Korea, pp. 18–22, Sep. 2022.

[9] W. Zhang, C. Liu, Y. Pi, Y. Zhang, H. Huang, B. Rao, Y. Ding, S. Yang, J. Jiang, "DRAMA: A Dynamic Packet Routing Algorithm using Multi-Agent Reinforcement Learning with Emergent Communication," 2025.

[10] A. Alshahrani, A. Abu-Shareha, Q. Shambour, and B. Al-Kasabeh, "A Fully Adaptive Active Queue Management Method for Congestion Preve

ntion at the Router Buffer," Comput. Mater. Continua, vol. 77, no. 2, pp. 1679–1698, Nov. 29, 2023.

[11] D. Zheng, G. Shen, X. Cao, and B. Mukherjee, "Towards Optimal Parallelism-Aware Service Chaining and Embedding," IEEE Trans. Netw. Serv. Manag., vol. PP, no. 99, pp. 1–1, Jan. 2022.

[12] H. Li, "One Signal–Noise Separation based Wiener Filter for Magnetogastrogram," IEEE Trans. Instrum. Meas., vol. 74, pp. 1–11, 2025.

[13] P. Feng and H. C. So, "Meta-learning-based delayless subband adaptive filter using complex self-attention for active noise control," Neurocomputing, vol. 650, p. 130637, Oct. 14, 2025.

[14] H. Khan, K. Kushwah, J. S. Thakur, G. G. Soni, A. Tripathi, and S. Rao, "Performance Evaluation of DSR, AODV and MP-OLSR Routing Protocols Using NS-2 Simulator in MANETs," in Proc. 4th EAI Int. Conf. Cogn. Comput. Cyber-Phys. Syst., Part II, Springer, 2024.

[15] Lee J M ,Lee S ,Kim S , et al.Nanoparticle-Functionalized Organoid-on-a-Chip Systems for Biomedical Applications[J].BioChip Journal,2025,(prepublish):1-25.

[16] Arasan Chip Systems Launches MIPI SWI3S Manager IP and Peripheral Controller IP[J].Telecomworldwire,2025,

[17] Park B ,Lee H E ,Kim J , et al.Multi breast cells-on-a-chip: Efficient screening biological platform for determination of selective breast cancer cell apoptosis.[J].Biofabrication,2025,17(4):045007-045007.

Topology Evaluation-Based Layer Assignment Method for Free-assignment Routing in InFO Packages

Zhan-yang Zhu
Information Engineering
Wuhan University of
Technology
Wuhan, China
zhanyangzhu@whut.edu.cn

Ning Xu
Information Engineering
Wuhan University of
Technology
Wuhan, China
xuning@whut.edu.cn

Hao-ying Wu
Information Engineering
Wuhan University of
Technology
Wuhan, China
why_dd@whut.edu.cn

Chenglin Lu
School of Management
Wuhan Institute of
Technology
Wuhan, China
2506030126@stu.wit.edu.cn

Abstract—**With the increasing complexity of system designs, advanced packaging technologies continue to emerge. Among them, the Integrated Fan-Out (InFO) Wafer-Level Chip-Scale Package (WLCSP) has gained significant attention as a cutting-edge solution that enables high-density and flexible system integration through multi-chip integration and multi-layer interconnection. As the physical size of InFO-WLCSP packages expands, the I/O pad density increases exponentially, making the topological crossover of free-assignment nets a critical bottleneck that severely constrains routing quality. Existing methods typically address topological crossover issues during the routing stage through iterative approaches, which often fail to provide satisfactory solutions. In this paper, we propose, for the first time, a predictive approach to anticipate topological crossovers of free-assignment nets prior to routing. This approach incorporates layer assignment considerations while aiming to minimize total wire length. We introduce a Voronoi Diagram-based iteration algorithm to effectively resolve the topology-compatible allocation problem in the fan-in region. Additionally, we propose a novel Voronoi region merging strategy to adaptively partition the fan-out region across multiple chips. A novel model is developed to address the topology-compatible allocation problem, and a final layer assignment is derived using the Longest Common Subsequence (LCS) algorithm. Experimental results demonstrate the high quality and efficiency of our algorithm.**

Keywords—*Free-assignment routing, integrated fan-out (InFO) wafer-level chip-scale package (WLCSP), topology evaluation*

I. INTRODUCTION

As system design complexity continues to escalate and semiconductor process nodes advance, packaging technologies have evolved beyond traditional flip-chip approaches toward more sophisticated solutions, such as 2.5D IC packaging, 3D IC packaging, and heterogeneous integration. Compared with the traditional flip-chip packaging for a single chip, these advanced packaging technologies can integrate different functions and process nodes into the same package to form a System-in-Package (SiP). This integration not only reduces manufacturing costs but also significantly enhances design flexibility. Among

these emerging technologies, Integrated Fan-Out Wafer-Level Chip-Scale Packaging (InFO-WLCSP) stands out for its capability to achieve high-density system integration through multi-chip and multi-layer interconnect structures. InFO-WLCSP introduces Redistribution Layers (RDLs), an additional metal layers that enhance the interconnect between the chip and the printed circuit board (PCB), as illustrated in Fig. 1(a). However, as the package size scales up, the accompanying surge in interconnect density renders global routing planning a critical challenge, directly impacting the overall performance and reliability of the package.

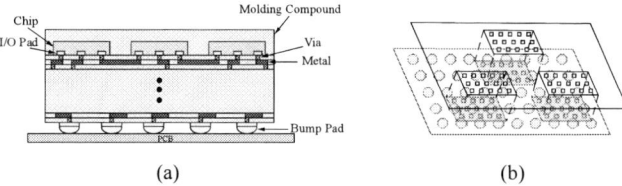

| (a) | (b) |

Fig. 1. (a) In a multi-chip, multi-layer InFO package, the Redistribution Layers are formed by alternating stacks of via layers and metal routing layers.(b) RDL structure of a two-layer, multi-chip InFO package.

Package routing can be classified into three categories based on the net type:(1) the pre-assignment routing problem, (2) the free-assignment routing problem, and (3) the unified-assignment routing problem. For the pre-assignment routing problem, each I/O pad is mapped to a specific bump pad prior to routing, and this assignment remains fixed throughout the design process.For the free-assignment routing problem, any I/O pad can be connected to any bump pad, offering designers considerable flexibility to address complex layout and connectivity challenges. For the unified-assignment routing problem represents a hybrid case, where some I/O pads connections are predefined and the others are not.

As illustrated in Fig. 1(b), the RDLs structure is typically divided into two regions:(1) the fan-in region and (2) the fan-out region.The fan-in region corresponds to the shaded area directly

---

The Project Supported by National Natural Science Foundation of China No. 9237310007

979-8-3315-8850-2/25 $31.00 © 2025 IEEE

beneath the footprint of the chip, whereas the fan-out region comprises the remaining area outside the chip projection.

A. Previous Works

In recent years, significant research has been conducted on routing in package design. Lee et al. [1] provided a comprehensive review and analysis of flip-chip package routing methodologies, focusing on net connectivity and flip-chip structural characteristics. For the pre-assignment routing problem, Integer Linear Programming (ILP) has been widely adopted. Fang et al. [2] were the first to propose an ILP-based approach to address pre-assignment routing for peripheral-I/O structures. Subsequently, Lee et al. [3] introduced techniques such as the Longest Common Subsequence (LCS) and the Maximum Planar Subset of Chords (MPSC) to solve pre-assignment routing problems, significantly improving routing efficiency. For the free-assignment routing problem, network flow methods are predominantly employed. These approaches typically involve two steps: first, constructing a network flow model, second, defining an optimization objective and solving the problem using the Minimum Cost Maximum Flow (MCMF) algorithm. Fang et al. [4] were the first to apply the MCMF algorithm to address the free-assignment routing problem. Lin et al. [5] proposed a concentric circle model to evaluate the influence of pre-assignment nets on free-assignment nets. By removing the resource occupancy of pre-assignment nets from the network model, they enhanced the overall routing efficiency through MCMF-based optimization. Yan et al. [6] pioneered a geometric iterative approach to tackle the free-assignment problem. Compared to traditional network flow methods, their geometric strategy demonstrated significant advantages in computational speed. As for the united-assignment routing problem, Fang et al. [7] were the first to integrate computational geometry with network flow methods to provide a unified solution. Cai et al. [8] proposed a Manhattan-distance-based Voronoi Diagram (MVD) to construct the global routing map, followed by a 3-D A* algorithm to perform simultaneous routing of united-assignment nets. Wen et al. [9] introduced an octagonal grid structure to build a global routing map that adheres to X-architecture constraints. In their method, MCMF was used for free-assignment routing, followed by ILP optimization for pre-assignment routing.

B. Our Contributions

The main contributions of this paper are summarized as follows:

- For the free-assignment nets, we propose for the first time, a predictive approach to anticipate topological crossings prior to routing.

- We develop a Voronoi Diagram-based iteration algorithm to effectively solve the topological compatibility assignment problem within the fan-in region. The assignments generated are topologically compatible, meaning that no detour is required during actual routing.

- A novel Voronoi region merging strategy is introduced to support adaptive partitioning of the fan-out regions across different chips. We construct a Topological Circle Model to achieve topologically compatible assignments,

ultimately determining suitable layer assignments via the Longest Common Subsequence (LCS) algorithm.

- We propose a 3D A*-based global routing algorithm on a Voronoi Diagram. Based on the aforementioned assignment results, this algorithm enables the generation of final global routing guides.

- Experimental results validate the effectiveness and efficiency of the proposed approach. In the given test cases, our global router achieves a 100% routing completion rate.

The remainder of this paper is organized as follows: Section 2 defines the terminology, notations, and problem formulation. Section 3 presents a detailed description of our routing algorithm. Section 4 reports experimental results. Section 5 concludes the paper.

II. PRELIMINARIES

In this section, we present the terminologies and notations used throughout this paper and formally define the RDLs global routing problem addressed in our study.

A. Terminologies and Notations

In the context of multi-chip, multi-layer RDLs global routing for free-assignment nets, as discussed in this paper, we first introduce the terminologies and notations used throughout:

- $P_f = \{P_{fi} \mid 1 \leq i \leq |P_f|\}$ denotes the set of I/O pads associated with all free-assignment nets. These pads do not have pre-assigned bump pad destinations prior to routing and require dynamic assignment during the routing process. $|P_f|$ represents the total number of I/O pads in free-assignment nets.

- $P_p = \{P_{pi} \mid 1 \leq i \leq |P_p|\}$ denotes the set of I/O pads in pre-assignment nets, where each pad has a predefined bump pad as its routing destination. $|P_p|$ denotes the total number of I/O pads in pre-assignment nets.

- $B_o = \{B_{oj} \mid 1 \leq j \leq |B_o|\}$ represents the set of bump pads located in the fan-out region, with $|B_o|$ indicating their total count.

- $B_{ik} = \{B_{ikj} \mid 1 \leq j \leq |B_{ik}|\}$ denotes the set of bump pads located in the fan-in region of $Chip_k$, where $|B_o|$ is the number of bump pads associated with that region.

- $N_f = \{N_{fi} \mid 1 \leq i \leq |N_f|\}$ is the set of all free-assignment nets, with $|N_f|$ representing the total number of such nets.

- $N_p = \{N_{pi} \mid 1 \leq i \leq |N_p|\}$ denotes the set of all pre-assignment nets, and $|N_p|$ denotes their total number.

- $V = \{V_i \mid 1 \leq i \leq |V|\}$ denotes the set of all vias, and $|V|$ is the total number of all vias.

B. Problem Formulation

We define the RDLs global routing problem addressed in this work as follows:

Given the routing region of the Redistribution Layers, a set of pre-assignment nets, a set of free-assignment nets, and a collection of Design Rule Check (DRC) constraints, the objective is to establish routability for all free-assignment nets while simultaneously satisfying all design rules. Under the premise of maximizing routability, the goal is to minimize the total wirelength and the number of vias introduced during the routing process.

III. Algorithms

A. Algorithm Overview

As illustrated in Fig. 2, our global routing algorithm is divided into four stages: (1) A Voronoi Diagram-based iteration algorithm is employed to obtain topologically compatible assignments for free-assignment nets within the fan-in region. (2) A Topological Circle Model is constructed to perform topology-aware optimization, resulting in the assignment of free-assignment nets within the fan-out region. (3) Based on the Topological Circle Model, the Longest Common Subsequence (LCS) algorithm is applied to determine an efficient layer assignment for all nets. (4) The global routing map is partitioned using Voronoi Diagram, followed by Voronoi-based 3D A* global routing algorithm to generate the final routing paths.

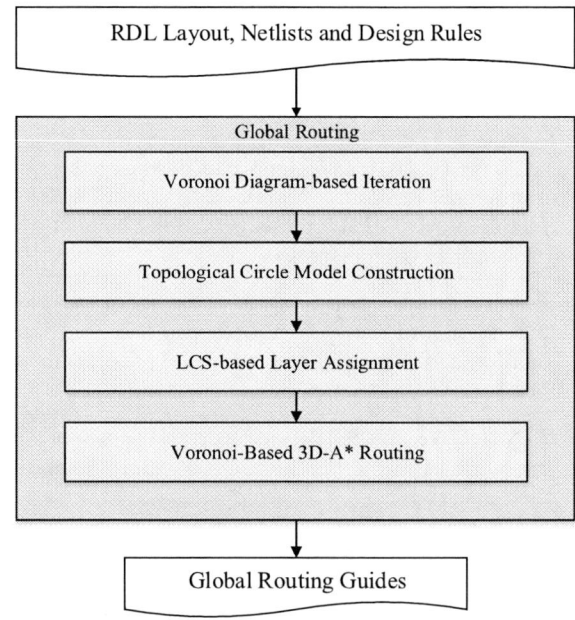

Fig. 2. Overview of our global routing algorithm

In the first stage, we propose a Voronoi Diagram-based iteration algorithm to assign free-assignment nets within the fan-in region. After obtaining the assignment results for this region, the fan-out area is divided according to the location of each chip. Then, we construct a Topological Circle Model based on the remaining unassigned I/O pads and the available bump pads in the fan-out region. This model serves as the basis for topological optimization, leading to a compatible assignment solution in the fan-out region. To address any residual topological crossings that may persist after the second stage of optimization, the third stage introduces the Longest Common Subsequence (LCS) algorithm within the framework of the existing Topological

Circle Model. In the fourth stage, the routing area is partitioned using Voronoi Diagram, and based on the free-assignment and layer assignment results obtained in the previous stages, a Voronoi-based 3D A* global routing algorithm is employed to generate the final routing guides.

B. Voronoi Diagram-based Iteration

To address the free-assignment nets within the fan-in region, we proposes a Voronoi Diagram-based iteration algorithm, which aims to achieve the following two objectives: (1) minimize the Manhattan distance between each allocated bump pad and its corresponding I/O pad, and (2) eliminate topological crossovers among free-assignment nets in the fan-in region. The Voronoi Diagram is a spatial partitioning technique that divides the plane into regions based on a set of seed points, such that all points within a given region are closest to its corresponding seed. Leveraging this property, we construct a Voronoi Diagram using the I/O pads of free-assignment nets as seed points. This spatial partitioning inherently ensures minimum-distance allocation and effectively eliminates topological crossovers by isolating nets across distinct regions. Each Voronoi region may contain zero, one, or multiple bump pads. From the candidates within a region, we select the bump pad with the shortest Manhattan distance to the corresponding seed point (i.e., the I/O pad), ensuring a locally optimal allocation. Moreover, due to the non-overlapping nature of Voronoi regions, performing allocation independently within each region inherently guarantees the absence of topological crossovers among the assigned nets.

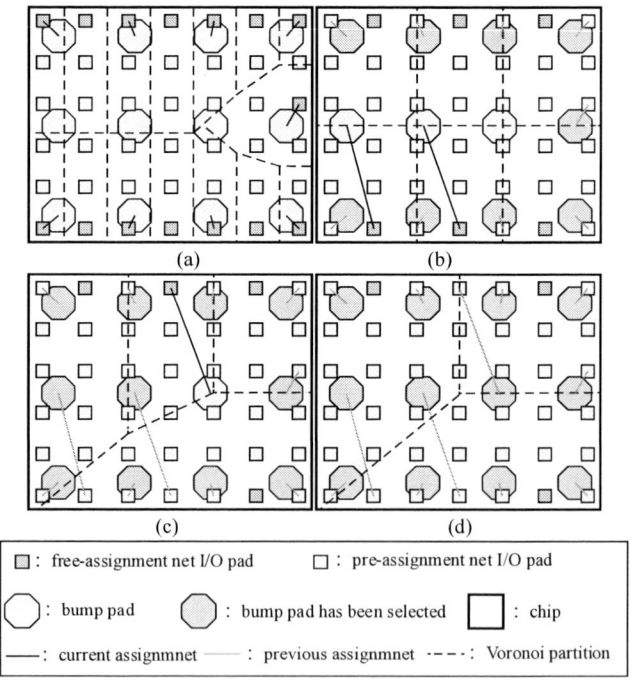

Fig. 3. (a) The first iteration. (b) The second iteration. (c) The third iteration. (d) The last iteration.

As illustrated in Fig. 3(a), the Voronoi Diagram-based iteration algorithm begins by constructing a Voronoi diagram for all free-assignment I/O pads within the fan-in region. In each Voronoi region corresponding to a free-assignment I/O pad, the bump pad with the shortest Manhattan distance to the I/O pad is selected as its assignment result. After each iteration, remove the assigned I/O pads from the free-assignment net. A new Voronoi Diagram is then reconstructed for the remaining unassigned I/O pads, as shown in Fig. 3(b), while preserving the assignment results from previous iterations. From the second iteration onward, the newly generated assignments must satisfy two constraints: (1) the selected bump pad must be the one with the minimum Manhattan distance to the corresponding I/O pad within its current Voronoi region; (2) the new assignment must not introduce any topological crossover with the assignments made in previous iterations, as illustrated in Fig. 4. This iterative process continues until, in a certain iteration, no valid bump pad satisfying the above constraints can be found within the Voronoi region of a remaining I/O pad. At this point, the Voronoi-based iterative algorithm terminates, as shown in Fig. 3(d).

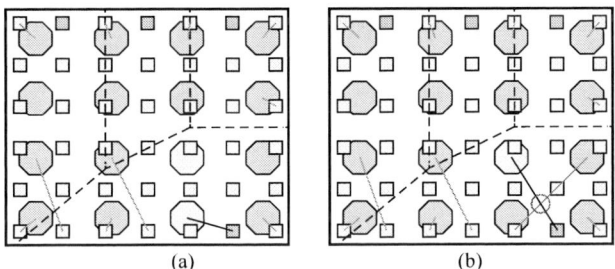

Fig. 4. (a) Constraint 1: Among the bump pads within the current Voronoi region, the one with the minimum Manhattan distance to the corresponding I/O pad is selected for assignment. (b) Constraint 2: The selected bump pad must not result in any topological crossover with assignment results from previous iterations.

C. Topology Optimization for Topological Circle Model

After the processing described in Section 3.2, the bump pad resources within the fan-in region are considered to be largely exhausted. The remaining free-assignment net I/O pads must therefore seek available bump pads in the fan-out region. To address the assignment of bump pads in the fan-out region, this paper proposes a Topological Circle Model, which facilitates efficient assignment by optimizing its underlying topological structure. The proposed model is divided into two sub-models: the In-Circle Model, representing the fan-in region, and the Out-Circle Model, corresponding to the fan-out region.

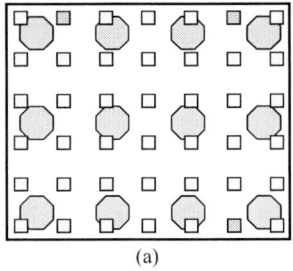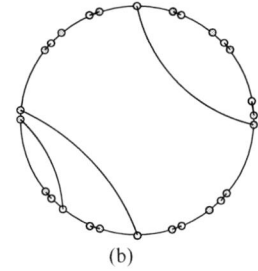

Fig. 5. (a) Result after the completion of the Voronoi Diagram-based iteration algorithm. (b) Constructed In-Circle Model, retaining only the unassigned free-assignment net I/O pads and their corresponding geometric mapping within the circular model.

1) In-Circle Model:

The In-Circle Model represents the structural information within the fan-in region. After the completion of the Voronoi Diagram-based iteration algorithm,

this model retains only the remaining free-assignment net I/O pads along with their corresponding assignment information within the fan-in region. All previously assigned I/O pads, as well as all bump pads located in the fan-in region, are discarded, as illustrated in Fig. 5. The construction of the In-Circle Model follows two rules: (1) The center of the In-Circle is aligned with the geometric center of the chip. (2) Each node projected onto the In-Circle must preserve the angular relationship relative to the chip center, i.e., the angle between a node and the circle center should match the angle between its corresponding free-assignment net I/O pad and the center of the chip.

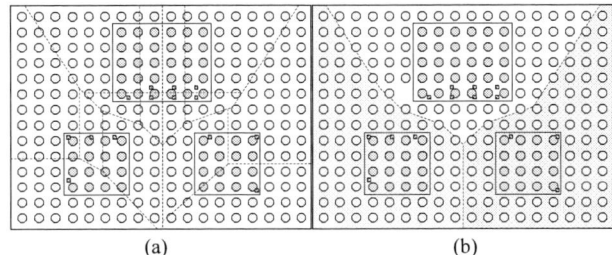

Fig. 6. (a) Voronoi Diagram generated from the remaining free-assignment net I/O pads. (b) Chip-aware merging of the Voronoi regions.

2) Chip-Aware Regional Division:

To construct the Out-Circle Model, the fan-out region is first partitioned into multiple subregions, ensuring that each chip is associated with a unique spatial area. This segmentation prevents topological crossovers between chips when I/O pads extend outward into the fan-out region. Following the assignment process for the fan-in region described in Section 3.2, a subset of free-assignment net I/O pads remains unassigned. Using these residual I/O pads as seed points, a Voronoi Diagram is generated, as illustrated in Fig. 6(a). From this diagram, the Voronoi cells corresponding to seed points belonging to the same chip are identified and merged, resulting in a unified Voronoi region for each chip, as shown in Fig. 6(b). Consequently, the entire routing area is divided into n subregions, where n denotes the number of chips.

3) Out-Circle Model:

After constructing the Voronoi region for each chip, the bump pads located within each region are mapped to the Out-Circle Model. The mapping follows the same rules as those used for the In-Circle Model, as illustrated in Fig. 7.

Upon constructing the Topological Circle Model, the In-Circle Model retains the assignment information within the fan-in region as obtained in Section 3.2, while the Out-Circle Model represents the distribution of bump pads within each chip Voronoi area after the merged Voronoi diagram is generated. The problem of assigning bump pads in the fan-out region can thus be transformed into the problem of selecting nodes for assignment within the Out-Circle Model. Based on this representation, we employ Simulated Annealing (SA) to optimize the assignment of nodes in the Out-Circle. The optimization objectives are defined as follows:

$$cost = crossNum + \alpha \times wireLength \qquad (1)$$

The cost function serves as the optimization objective in the SA. It comprises two components: (1) *crossNum*, which denotes the number of topological crossovers arising from node assignments within the Out-Circle, and (2) *wireLength*, representing the total routing length from each free-assignment node to its corresponding bump pad node on the Out-Circle. To unify these two terms into the same dimensional scale, a normalization factor $\alpha = 1/d_{Out-Circle}$ is introduced, where $d_{Out-Circle}$ denotes the diameter of the Out-Circle. It is worth noting that even after topological optimization using the Topological Circle Model, some topological crossovers may still persist. This indicates that not all nets can be routed within a single metal layer. Consequently, this issue motivates the layered routing strategy based on the Longest Common Subsequence (LCS) algorithm, as described in Section 3.4.

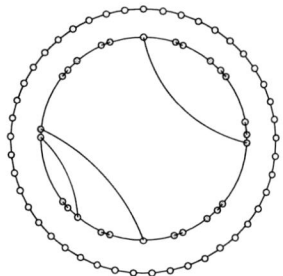

Fig. 7. Topological Circle Model.

D. LCS-based Layer Assignment

The Longest Common Subsequence (LCS) Layer Assignment Algorithm is developed based on the Topological Circle Model introduced in Section 3.3. This algorithm retains all remaining topological crossings among the routing nets and maps them onto the Topological Circle Model. As these crossings can lead to routing conflicts or infeasible wire arrangements, it is necessary to resolve them through proper layer assignment. Accordingly, this section aims to identify the largest subset of nets that do not introduce topological crossings within the same layer, and, based on this, minimize the total number of layers required. To achieve this, we propose our LCS-based layer assignment algorithm grounded in the Topological Circle Model.

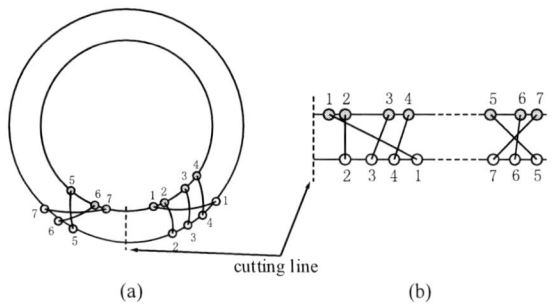

Fig. 8. (a) A small number of topological crossovers still remain after topological optimization. (b) The circular model is unfolded into two linear sequences by introducing a cutting line.

Before applying the layer assignment algorithm, it is essential to introduce the concept of a cutting line. The cutting line serves to unfold the Topological Circle Model into a pair of linear sequences, as illustrated in Fig. 8(b). By transforming the circular topology into two corresponding sequences, the layer assignment problem can be reformulated as a LCS problem. Specifically, starting from the cutting point, both the In-Circle and Out-Circle are traversed in a counterclockwise direction to record the order of the nets. Using the traversal order of the In-Circle as the reference, the corresponding net indices on the Out-Circle can be derived. This process establishes a foundational ordering that guides the subsequent layer assignment algorithm.

The determination of the cutting line is a critical step, as it directly affects the construction of the unfolded sequences from the Topological Circle. The first task is to identify the candidate cutting intervals on the Topological Circle. Specifically, for any pair of nets that exhibit a topological crossing, the corresponding arc segments on the circle are excluded from the candidate intervals. After evaluating all nets, the final cutting intervals for both the In-Circle and Out-Circle are established. Subsequently, a suitable pair of cutting intervals is selected, and a corresponding cutting line is introduced. By traversing the nodes on the In-Circle and Out-Circle in a counterclockwise direction from this cutting line, an sequence pair is generated, as shown in Fig. 8(b). In cases where no valid cutting interval exists, a cutting line must be chosen such that it intersects with the minimum number of conflicting bonds. Once the sequence pair is obtained, the LCS algorithm is employed to identify the largest subset of topologically compatible nets that can be assigned to the same layer.

E. Voronoi-Based 3D A* Routing

After obtaining the assignment results and layer assignment for all free-assignment nets, this section performs global routing for these nets. To generate non-crossing global routing guides, we propose a crossing-aware A* algorithm based on the Voronoi diagram with dynamic node updates. Unlike the conventional A* search algorithm, our method designates certain search nodes as forbidden and dynamically updates the search space after each routing step to reflect changes in topology. Our via planning strategy follows the approach described in [8]. Specifically, we use the vias generated by the method in [8] as reference anchors, and construct a Voronoi Diagram over the entire routing region. The midpoints of the Voronoi edges are then selected as the initial search nodes for the A* algorithm, as illustrated in Fig. 9.

1) Dynamic Expansion of Search Nodes: As illustrated in Fig. 9(c) and Fig. 9(d), new dynamic search nodes are introduced into the search space after each routing step. This dynamic expansion mechanism ensures the adaptability of the search map as routing progresses. To fully account for the capacity constraints of each Voronoi edge, the expansion of dynamic nodes is governed by the following formula:

$$s = wireWidth + wireSpace \qquad (2)$$

Let s denote the distance between a newly generated dynamic search node and its corresponding original node. The parameters *wireWidth* and *wireSpace* represent user-defined values for wire width and spacing, respectively. This dynamic node generation strategy effectively incorporates design rule check (DRC) constraints into the routing process. The generation of dynamic search nodes follows two rules: (1) If the routing path crosses the midpoint of a Voronoi edge, two dynamic search nodes are generated, one in the direction from the midpoint to each endpoint of the edge, as illustrated in Fig. 9 (c). (2) If the routing path does not pass through the midpoint, a single dynamic search node is created in the direction toward the nearest endpoint of the Voronoi edge, as shown in Fig. 9(d).

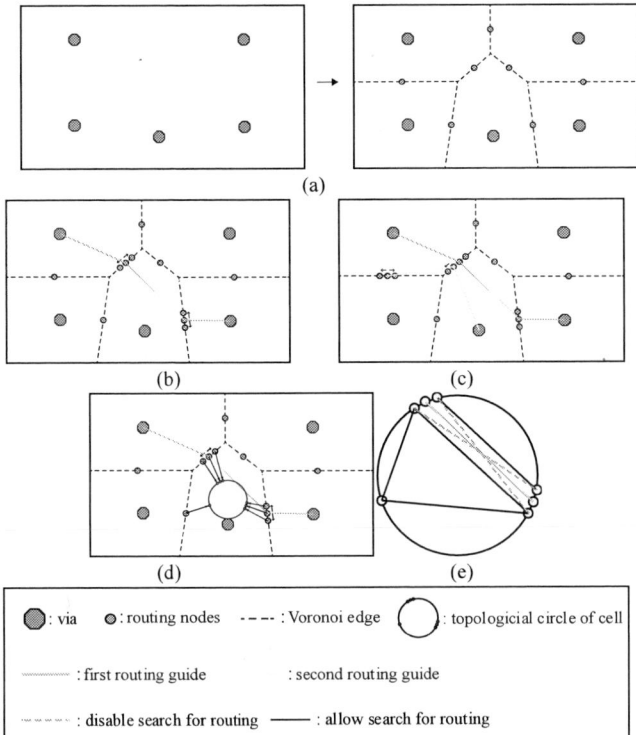

Fig. 9. (a) Voronoi partition generated using five vias as reference points. (b) The first routing guide completed on the Voronoi-based grid map. (c) Two routing guides successfully completed on the Voronoi-based grid map. (d) Construction of the corresponding Topological Circle within a cell. (e) The Topological Circle, used to determine feasible routing guides within its corresponding cell.

2) Crossing Awareness: Each cell constructs a corresponding Topological Circle to determine which line segments are permitted or prohibited within the cell, as illustrated in Fig. 9(e). The construction of the Topological Ciercle follows these principles: (1) The center of the circle is set as the centroid of the cell. (2) The nodes on the circle are defined as the intersection points between the cell boundary nodes and the arcs connecting them to the circle center. Once the Topological Circle is constructed, previously routed wires are projected onto the circle. Any new segment that introduces a topological crossing with existing routes on the circle is marked as a forbidden segment, as shown in Fig. 9(f). This approach enables the identification of all routing solutions within each cell that avoid topological conflicts.

IV. Experimental Results

We implemented our proposed algorithm using the C++ programming language. All experiments were conducted on a 2.30 GHz 12th Gen Intel® Core™ i9-12900HX processor with 16 GB of RAM. The cases, sourced from real-world industrial designs, are summarized in Table 1. In this table, "$|Chips|$", "$|P_f|$", "$|B|$", and "$|N_f|$" denote the number of chips, the number of I/O pads for free-assignment nets, the number of bump pads, and the total number of free-assignment nets, respectively.

To evaluate the efficiency and effectiveness of our algorithm, we compare it against two representative approaches: a geometry-based iterative assignment method [6] and a 3D A*-based iterative routing method [8]. The method in [6] adopts a Voronoi-based iterative strategy for free-assignment net allocation but does not consider topological crossings during intermediate iterations; instead, it addresses these crossings only after the entire assignment is completed. In contrast, the approach in [8] constructs the routing resource map using a Manhattan-Distance-based Voronoi Diagram (MVD) and performs 3D A* routing accordingly. The experimental results are presented in Table 1, reporting the number of layers used, routability, total wirelength, and runtime for each method.

TABLE I. Experimental Result of Our Global routing Algorithm("N/A": Incomplete Routing Results)

Circuits	|Chips|	$|P_f|$	$|B|$	$|N_f|$	Layers			Routability(%)			Wirelength(um)			Runtime(sec)		
					[6]	[8]	ours	[6]	[8]	ours	[6]	[8]	ours	[6]	[8]	ours
Case1	1	65	97	65	2	1	1	100	100	100	110253	97423	82158	13	26	7
Case2	2	93	149	93	3	2	2	100	98.7	100	219424	N/A	168923	34	49	29
Case3	3	116	162	116	3	3	3	97.5	97.3	100	N/A	N/A	224591	54	89	43
Case4	3	177	265	177	4	4	3	93.2	91.6	100	N/A	N/A	292345	89	156	82
Csae5	4	246	372	246	6	5	4	94.7	93.3	100	N/A	N/A	471292	169	238	157

V. Conclusion

In this paper, we propose a topology-aware planning strategy for free-assignment nets. Our algorithm consists of the following four components: (1) A Voronoi Diagram-based iteration algorithm is employed to assign free-assignment nets in the fan-in region, ensuring topological compatibility. (2) A Topological Circle Model is constructed, and topological optimization is performed based on this model to derive the assignment results in the fan-out region. (3) All nets are assigned to routing layers using an LCS-based layer assignment algorithm guided by the Topological CircleModel. (4) A Voronoi diagram is used to partition the routing space, and 3D A* search is applied for final

979-8-3315-8850-2/25 $31.00 © 2025 IEEE

routing path generation. Experimental results demonstrate that our method is highly efficient, even for complex design cases.

REFERENCES

[1] H. C. Lee, Y. W. Chang, and P. W. Lee, "Recent research development in flip-chip routing," in *Proc. IEEE/ACM Int. Conf. Comput.-Aided Design (ICCAD)*, San Jose, CA, USA, 2010, pp. 404–410.

[2] J. W. Fang, C. H. Hsu, and Y. W. Chang, "An integer linear programming based routing algorithm for flip-chip design," in *Proc. 44th Annu. Design Autom. Conf. (DAC)*, San Diego, CA, USA, 2007, pp. 606–611.

[3] P. W. Lee, C. W. Lin, Y. W. Chang, *et al.*, "An efficient pre-assignment routing algorithm for flip-chip designs," in *Proc. Int. Conf. Comput.-Aided Design (ICCAD)*, San Jose, CA, USA, 2009, pp. 239–244.

[4] J. W. Fang, I. J. Lin, P. H. Yuh, *et al.*, "A routing algorithm for flip-chip design," in *Proc. IEEE/ACM Int. Conf. Comput.-Aided Design (ICCAD)*, San Jose, CA, USA, 2005, pp. 753–758.

[5] B. Q. Lin, T. C. Lin, and Y. W. Chang, "Redistribution layer routing for integrated fan-out wafer-level chip-scale packages," in *Proc. IEEE/ACM Int. Conf. Comput.-Aided Design (ICCAD)*, Austin, TX, USA, 2016, pp. 1–8.

[6] J. T. Yan, "IO connection assignment and RDL routing for flip-chip designs," *ACM Trans. Des. Autom. Electron. Syst.*, vol. 16, no. 4, pp. 1–20, Oct. 2011.

[7] J. W. Fang, M. D. F. Wong, and Y. W. Chang, "Flip-chip routing with unified area-I/O pad assignments for package-board co-design," in *Proc. 46th Annu. Design Autom. Conf. (DAC)*, San Francisco, CA, USA, 2009, pp. 336–339.

[8] Y. J. Cai, Y. Hsu, and Y. W. Chang, "Simultaneous pre- and free-assignment routing for multiple redistribution layers with irregular vias," in *Proc. 58th ACM/IEEE Design Autom. Conf. (DAC)*, San Francisco, CA, USA, 2021, pp. 1147–1152.

[9] H. T. Wen, Y. J. Cai, Y. Hsu, *et al.*, "Via-based redistribution layer routing for InFO packages with irregular pad structures," *IEEE Trans. Comput.-Aided Des. Integr. Circuits Syst.*, vol. 41, no. 12, pp. 5554–5567, Dec. 2022.

A Radiation Tolerant CMOS Voltage Reference Circuit for High Energy Physics Experiments

Weiming Yang
School of Computer Science
Northwestern Polytechnical
University
Xi'an, China
yangwm@mail.nwpu.edu.cn

Jia Wang*
School of Electronics
Information
Northwestern Polytechnical
University
Xi'an, China
jwang@nwpu.edu.cn

Xiayu Wang
School of Computer Science
Northwestern Polytechnical
University
Xi'an, China
13603981127@163.com

Ran Zheng
School of Computer Science
Northwestern Polytechnical
University
Xi'an, China
zhengran@nwpu.edu.cn

Xiaomin Wei
School of Computer Science
Northwestern Polytechnical
University
Xi'an, China
weixm@nwpu.edu.cn

Yongcai Hu
School of Computer Science
Northwestern Polytechnical
University
Xi'an, China
yannyonghu@163.com

Abstract—The performance of bipolar transistors used in bandgap voltage reference (BGR) can be degraded by strong radiation, which usually appears in high energy physics experiments. This paper presents a radiation tolerant voltage reference circuit based on a commercial standard 180nm CMOS process. MOS transistors operating in the sub-threshold region were used to replace the bipolar transistors (BJTs) to improve radiation tolerance. The post-layout simulation results demonstrate that the output voltage of the proposed voltage reference circuit is 391mV with a temperature coefficient (TC) of 100.538 ppm/°C, when the temperature ranges from -40°C to 125°C. The voltage Line Sensitivity (LS) is 0.259%/V, and the power supply rejection (PSR) is −90.1 dB @DC and −37 dB @1 MHz. The power consumption is 44.9 µW at 27°C, with a die area of 0.0072 mm².

Keywords—*Bandgap Voltage reference, CMOS technology, Radiation Hardness, sub-threshold.*

I. INTRODUCTION

BGR is widely used in analogue and mixed-signal chips, since it can provide a precise voltage reference independent of process, supply voltage, and temperature (PVT) variations. In a BGR circuit, the reference voltage is usually created by a proportional-to-absolute-temperature (PTAT) voltage and a complementary-to-absolute-temperature (CTAT) voltage. The conventional BGR circuits employ a pair of BJTs, whose dimension is proportional. The CTAT voltage is obtained from the base-to-emitter voltage while the PTAT voltage comes from the difference of the base-to-emitter voltage. By combining both voltages with different coefficients, a temperature independent reference voltage can be obtained[1].

However, BJTs work with minority carrier whose lifetime can be significantly decreased by displacement damage and ionizing radiation effects[2],[3]. Therefore, BJTs can not tolerate the strong radiation, which is usually produced in the high energy physics experiments when beams collide. Moreover, BJTs consume relatively large area and high power consumption.

Fortunately, deep sub-micron CMOS transistors have been proven to resist hundreds of Mrad of total ionizing dose (TID)[4],[5]. Therefore, MOS transistors are considered to replace BJTs, which is called CMOS reference circuit.

Several related works were published. The TID hardening design of CMOS technique is put forward in [4]. BGR circuits with three advanced nanometer bulk silicon CMOS technologies of 65 nm, 40 nm, and 28 nm are fabricated and tested. TID experiment results show that after 1.2 Mrad(Si) Co-60 irradiation , the maximum variation of BGR output voltage does not exceed 5.67%. A radiation harden diode has been proposed in [5]. The source and N-well of a PMOS transistor are connected together, and the gate and drain are grounded. Samples were irradiated with 10MeV proton beams and the maximum shift of the output voltage was approximately 10% at a radiation dose of 800 Mrad (SiO₂). A prototype circuit based on three different devices (diode, BJT, and MOSFET) was fabricated in a standard commercial 65 nm CMOS process[6]. Measurement results show the circuit implemented by MOSFET features good radiation tolerance. The output voltage of the bandgap circuit based on N-MOSFET shows a reference voltage shift of only 7.6mV with 10keV X-rays at a TID of 225 Mrad (SiO₂). Real-time monitoring and adaptive compensation technique is proposed in [7]. The change of ΔV_{BE} of BGR is monitored in real time at the circuit level and the compensation current is dynamically generated to offset the voltage drift caused by TID. In order to replace BJTs, a pair of diodes connected MOS transistors working in sub-threshold region was utilized in [8]. MOSFETs biased in weak inversion region have been proven to be inherently radiation hard. Moreover, the power consumption is also decreased by sub-threshold region[9],[10]. The radiation resistance can also be enhanced through the design of circuit and layout in [11]. For circuit design, the stability and accuracy of the system are emphasized. And for the layout design, the "AA" mask layer is extended and common-centroid layout is applied to the vertical NPN transistor.

Nevertheless, changes in process and layout may cause extra costs.

The rest of this paper is organized as follows. In Section II the structure and circuit implementation of the proposed voltage reference are introduced in detail, while the layout design and simulation results are given in Section III. At the end of this paper, comparison between this work and others are presented and conclusions are drawn.

II. PROPOSED VOLTAGE REFERENCE CIRCUIT

Generally, MOS transistors operating in the sub-threshold region exhibit an exponential current similar to that of a bipolar junction transistor. For N-MOS devices biased in the weak inversion region, when the drain-source voltage $V_{DS} \geq 4V_{TH}$, the drain current can be expressed as[12]:

$$I_D = I_0 \frac{W}{L} \exp(\frac{V_{GS} - V_{TH}}{\eta V_T}) \qquad (1)$$

Where $I_0 = u_n C_{ox}(\eta - 1)V_T^2$, u_n is the electron mobility of NMOS, C_{ox} is the gate oxide capacitance per unit area,

$V_T = k_B T / q$ is the thermal voltage, k_B is the Boltzmann constant, T is the absolute temperature, q is the elementary charge, η is the sub-threshold slope factor and V_{TH} is the threshold voltage of the MOSFET.

The gate-source voltage V_{GS} can be derived from (1) as:

$$V_{GS} = V_{TH} + \eta V_T \ln(\frac{I_D L}{I_0 W}) \qquad (2)$$

The temperature characteristic of expression (2) is mainly dominated by the first term V_{TH}. Since the threshold voltage of MOS transistors have a negative temperature characteristic, the gate-source voltage V_{GS} of the MOS transistor in the sub-threshold region is generally a voltage with a negative temperature characteristic. Fig. 1(a) shows the simulation results of V_{GS} when temperature ranges from -40°C to 125 °C at different corners. Therefore, by operating diode-connected MOSFETs in the subthreshold region, a drain-source voltage V_{DS} with CTAT temperature characteristics can be obtained.

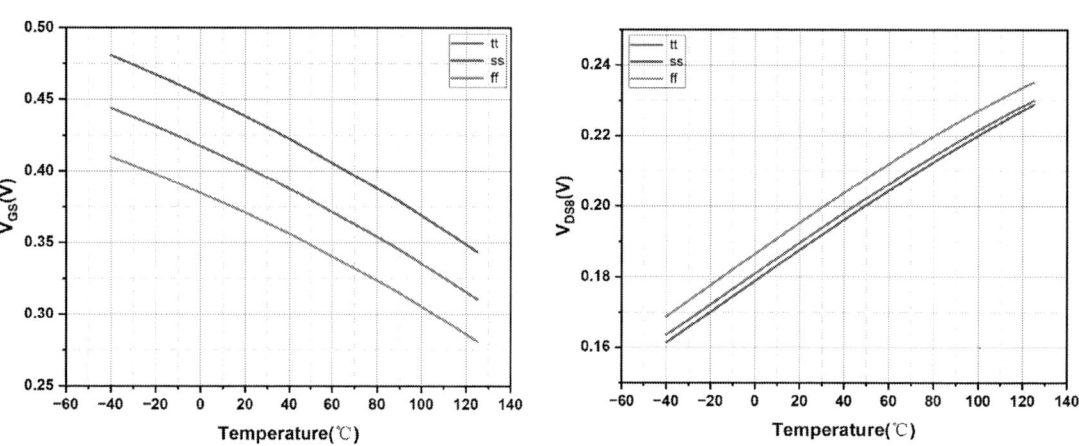

Fig. 1. Simulation results of (a) V_{GS} and (b) V_{GS8}-V_{GS7} VS temperature ranging from -40°C to 125°C at different corners.

Fig. 2 shows the structure of the proposed voltage reference. It is composed of a startup circuit, a bandgap reference core circuit, and an operational amplifier. As mentioned above, diode-connected MOSFET M_9 is biased to operate in sub-threshold region. Thus, a CTAT voltage can be obtained. The operational amplifier forces the voltage across resistor R_1 and the gate-source voltage of M_9 to be same. The current $I_{D1,2}$ can be expressed by V_{GS9}/R_1, which is a CTAT current. The ratio of M_1, M_2 and M_3 is set to be the same, so their current can be the same. The voltage across R_2 is a CTAT voltage, the help of the current mirror. However, errors will appear due to the limited output impedance of the current mirror. In order to decrease the error, the cascode current mirrors have been employed to improve the current replication accuracy and also enhance the PSR.

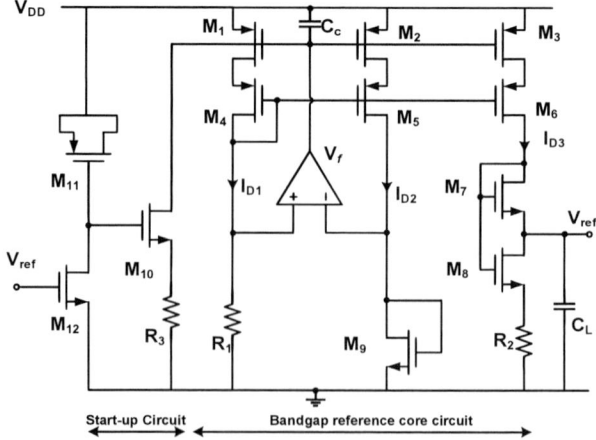

Fig. 2. Structure of the proposed voltage reference circuit

By biasing M7 and M8 in the sub-threshold region, their gate-source voltage expressions can be expressed as (2). Ignoring the influence of η and I_0, the drain-source voltage V_{DS8} can be expressed as:

$$V_{DS8} = V_{GS8} - V_{GS7} = V_{TH8} - V_{TH7} + \eta V_T \ln \frac{(W/L)_7}{(W/L)_8} \quad (3)$$

In expression (3), the term related to threshold voltage is approximately eliminated, and the remaining term related to V_T is the dominant factor of the temperature coefficient. Therefore, V_{DS8} is a voltage with PTAT temperature coefficient. Fig. 1(b) show the simulated results of V_{DS8} with temperature ranging from -40°C to 125 °C at different corners. Comparing Fig. 1(a) and Fig. 1(b), the voltage V_{GS} using a single MOSFET has a larger deviation between different process corners, while the voltage V_{DS} using two MOSFETs for subtraction has a small deviation.

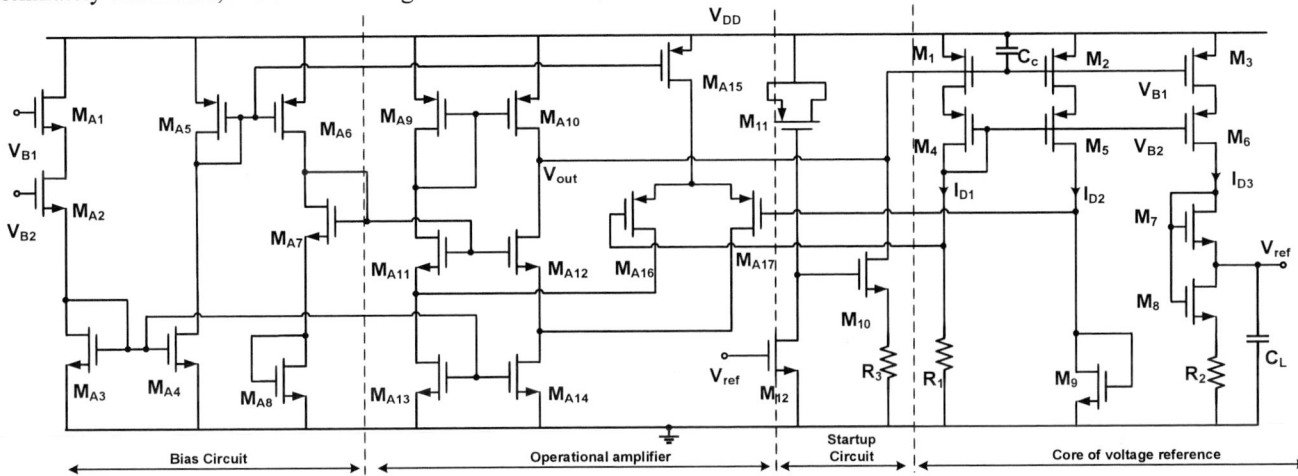

Fig. 3. Schematic of the proposed reference voltage

Thus, the output voltage V_{ref} can be expressed as the sum of PTAT and CTAT voltages:

$$V_{ref} = \alpha V_{GS9} + V_{DS8}$$
$$= \alpha (V_{TH9} + \eta V_T \ln \frac{I_D}{I_0 (W/L)_9}) + (V_{TH8} - V_{TH7}) + \eta V_T \ln \frac{(W/L)_7}{(W/L)_8} \quad (4)$$

where $\alpha = R_2/R_1$. The partial derivative of equation (4) with respect to temperature can be expressed as:

$$\frac{\partial V_{ref}}{\partial T} = \alpha (\frac{\partial V_{TH9}}{\partial T} + \eta \frac{k_B}{q} \ln \frac{I_D}{I_0 (W/L)_9} + \eta V_T \frac{\partial \ln \frac{I_D}{I_0 (W/L)_9}}{\partial T})$$
$$+ (\frac{\partial V_{TH8}}{\partial T} - \frac{\partial V_{TH7}}{\partial T}) + \eta \frac{k_B}{q} \ln \frac{(W/L)_7}{(W/L)_8} \quad (5)$$

Neglecting the influence of temperature on current and the threshold voltage, equation (5) can be simplified as:

$$\frac{\partial V_{ref}}{\partial T} = \alpha (\frac{\partial V_{TH9}}{\partial T} + \eta \frac{k_B}{q} \ln \frac{I_D}{I_0 (W/L)_9}) + \eta \frac{k_B}{q} \ln \frac{(W/L)_7}{(W/L)_8} \quad (6)$$

A reference voltage with zero temperature coefficient can be achieved by adjusting the ratio of M7, M8 and M9 according to equation (6).

Fig. 3 shows the schematic of the bias circuit, operational amplifier, startup circuit and voltage reference core circuit. The operational amplifier plays an important role in the voltage reference circuit and its gain and offset voltage directly affect the performance of the voltage reference. Sufficiently high open-loop gain of the amplifier is required to guarantee the voltage across resistance R_1 equals to V_{GS9}. A folded-cascode operational amplifier with PMOS input transistor is used to obtain low flicker noise. The cascode current mirror loads are used to increase the output resistance and enhance the gain of the amplifier. The bias circuit is copied directly from the current mirror of the bandgap reference, which saves power consumption. Besides, the dependance of operational amplifier on bias voltage can be reduced through self-bias technology.

The start-up circuit consists of M_{10}, M_{11}, M_{12} and R_3, wherein M_{11} acts as a capacitor. As the supply voltage increases, the gate voltage of M_{10} also gradually increases and M_{10} is turned on. The gate voltages of M_1 and M_2 are pulled down, and the currents of the corresponding branch can reach the preset value. The output voltage V_{ref} starts to rise. Simultaneously, when V_{ref} is higher than the threshold voltage of M_{12} or reaches the preset value, the MOSFET M_{12} is turned off and the whole start-up circuit is shut down. There is no extra power consumption. In order to prevent the output voltage from overshoot caused by excessive current in M_{10} during start up, a resistor R_3 is inserted into the source of M_{10} to reduce the current. R_3 can be implemented by a diode-connected MOS transistor, which saves the area consumption.

As shown in Fig. 4, an overshoot voltage about 100mV can be observed without the current limiting resistor R_3, whereas adding the current limiting resistor R_3 can eliminate the overshoot.

979-8-3315-8850-2/25 $31.00 © 2025 IEEE   393

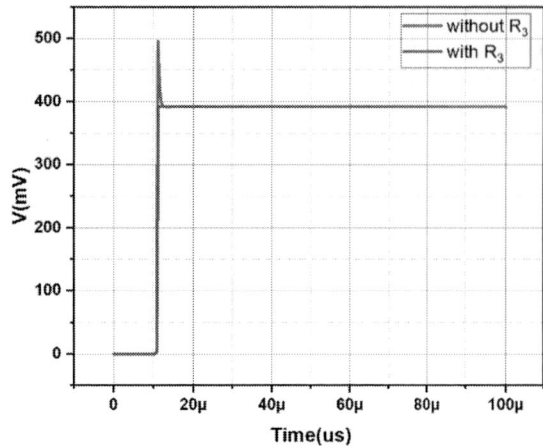

Fig. 4. simulation results of start-up settling response with R_3 vs. without R_3

III. LAYOUT AND POST-LAYOUT SIMULATION RESULTS

The proposed BGR is designed and fabricated in a commercial standard 0.18μm CMOS process and the layout is shown in Fig. 5, with a die area of 138×52 μm². The dimension of all transistors in the proposed circuit is presented in TABLE I.

In order to save die area, the compensation capacitance C_c and resistance R_3 is realized by MOS capacitance and MOS resistance. Since large resistance is required, most of the area is occupied by R_1 and R_2. Taking area and accuracy into consideration, N-poly resistors are employed in this work.

The resistors R_1 and R_2, as well as MOS transistors M_7 and M_8, are implemented by common-centroid layout matching, minimizing the process variations between them. Moreover, in order to prevent current leakage caused by irradiation, guard rings were added for isolation between the PMOS and NMOS. The proposed circuit is simulated and verified using the Spectre simulator.

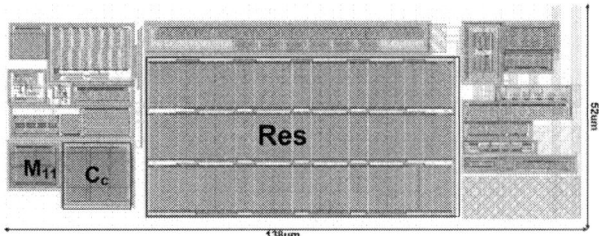

Fig. 5. Layout of the proposed bandgap voltage reference

TABLE I.   SIZE OF THE TRANSISTOR

Transistor	W/L	Transistor	W/L
M_1	$2μ/4μ \times 4$	M_{A4}	$0.5μ/2μ \times 2$
M_2	$2μ/4μ \times 4$	M_{A5}	$1μ/2μ \times 2$
M_3	$2μ/4μ \times 4$	M_{A6}	$1μ/2μ \times 2$
M_4	$2μ/2μ \times 2$	M_{A7}	$2μ/2μ \times 2$
M_5	$2μ/2μ \times 2$	M_{A8}	$2μ/4μ \times 2$
M_6	$2μ/2μ \times 2$	M_{A9}	$2μ/2μ \times 2$

M_7	$3μ/1μ \times 24$	M_{A10}	$2μ/2μ \times 2$
M_8	$1μ/1μ$	M_{A11}	$1μ/1μ \times 2$
M_9	$3μ/1μ \times 5$	M_{A12}	$1μ/1μ \times 2$
M_{10}	$10μ/10μ \times 4$	M_{A13}	$2μ/1.8μ \times 4$
M_{11}	$4μ/4μ \times 4$	M_{A14}	$2μ/1.8μ \times 4$
M_{12}	$1μ/2μ \times 8$	M_{A15}	$1μ/2μ \times 16$
M_{A1}	$1μ/4μ \times 4$	M_{A16}	$4μ/1μ \times 2$
M_{A2}	$2μ/2μ \times 4$	M_{A17}	$4μ/1μ \times 2$
M_{A3}	$0.5μ/2μ \times 2$		

Fig. 6 shows the bandgap reference voltage ranging from -40°C to 125°C at different process corners, with a supply voltage of 1.8V. When the temperature is 27°C, the output voltages are 391mV, 407mV and 381mV at the corner of tt, ss and ff, respectively. The maximum voltage difference between tt and other corners is about 15mV, indicating high process stability. The best and the worst temperature coefficients are 102.1 ppm/°C at the ss corner and 133.9 ppm/°C at the ff corner.

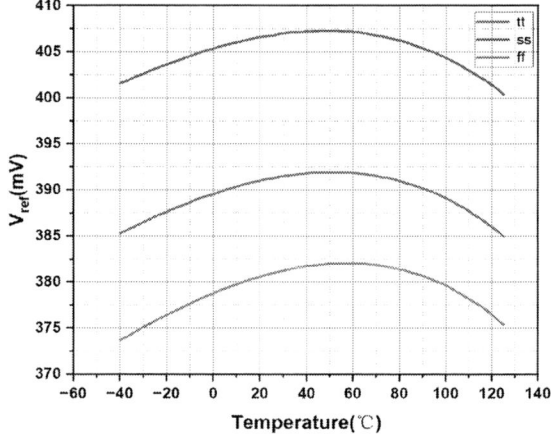

Fig. 6. Simulation results of V_{ref} at different process corners

Fig. 7 gives the variation of the output voltage of the bandgap voltage reference with the supply voltage from 0 to 2V at different corners. The worst-case occurs at the ss corner. The output voltage is stable when the supply voltage is higher than 1.3V. The linear sensitivity in the worst-case is 0.259%/V, which can also meet the design requirement. Fig. 8 illustrates the post-layout simulation results of the power supply rejection (PSR). The PSR is -90.1dB at 100Hz and -37dB at 1MHz with 2pF output capacitance. The cascode current mirror can significantly improve the PSR at low frequencies according to the simulated results. Fig. 9(a) shows the results of 200-point Monte-Carlo simulation for the distributions of the reference voltage and Fig. 9(b) shows the results of temperature coefficient. The mean value of the reference voltage is 391.48mV with a standard deviation of 4.75mV, while the mean value of temperature coefficient is 100.538 ppm/°C with a standard deviation of 4.656 ppm/°C. TABLE II. provides a comprehensive comparison of the proposed voltage reference with other published works. The proposed circuit exhibits good performance in terms of area and power consumption, and it does not require special processing technology.

979-8-3315-8850-2/25 $31.00 © 2025 IEEE

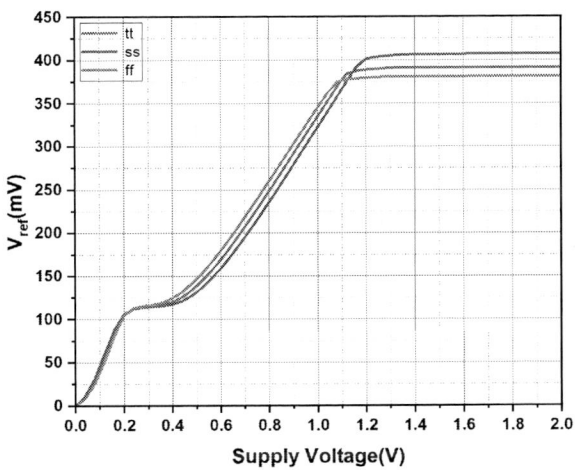

Fig. 7. Simulation results of reference voltage vs. supply voltage at different process corners

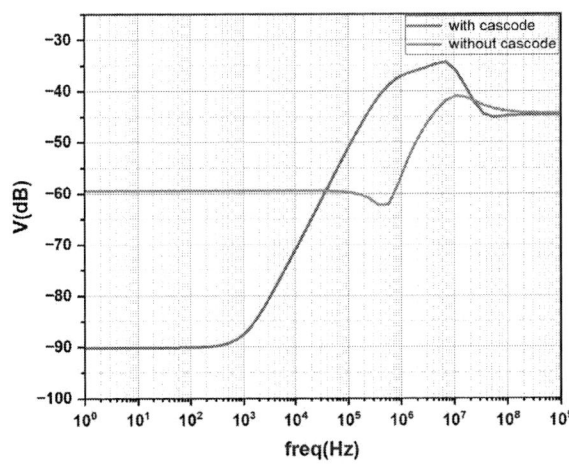

Fig. 8. Simulation results of PSR

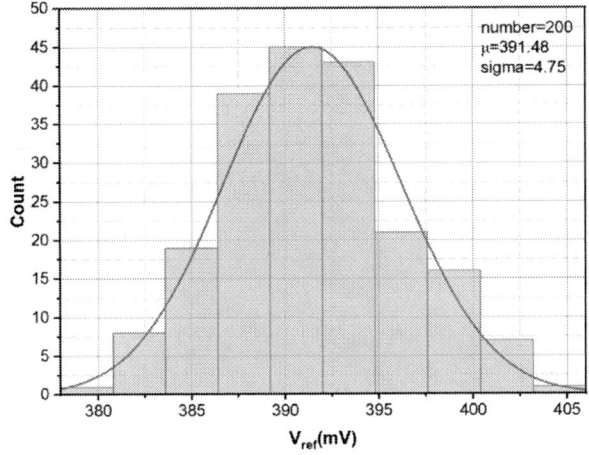

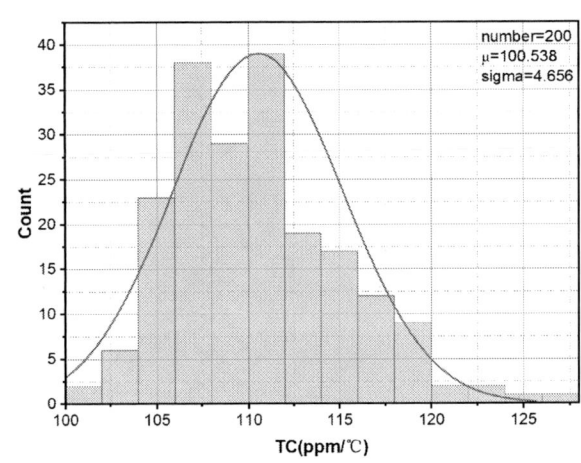

Fig. 9. (a)Monte-Carlo simulation results of V_{ref}; (b)Monte-Carlo simulation results of TC.

TABLE II.   PERFORMANCE SUMMARY AND COMPARISON WITH OTHER WORKS

	[5]	[6]	[8]	[11]	**This work**
Technology	65nm CMOS	65nm CMOS	110nm CMOS	1.5μm 32V bipolar	**180nm CMOS**
V_{DD}(V)	1.08-1.32	0.78-1.32	1.08-1.32	9-15	**1.3-2.0**
V_{ref}@25°C(V)	0.33	0.69	0.6	5	**0.391**
Average TC(ppm/°C)	130	230	28.03	4.9 (trimming)	**100.538**
Temperature(°C)	-40-80	-30-140	-40-70	-55-125	**-40-125**
LS(%/V)	/	5.2	0.875	/	**0.259**
PSRR(dB)	/	-30@100Hz	/	/	**-90.1@100Hz**
Power consumption(uW)	240	46	/	/	**44.9**
Area(mm²)	0.018	0.0264	0.008	0.226	**0.0072**

IV. CONCLUSION

In order to avoid BJT degradation caused by displacement damage and ionizing radiation effect in high energy physics experiments, a scheme using MOSFET instead of BJT is proposed. The circuit is based on a MOSFET working in the subthreshold region. And the PTAT voltage and CTAT voltage are constructed. Then a voltage with zero temperature coefficient is obtained by addition. The core area of the proposed circuit is only 0.0072 mm². The output voltage variation of the proposed structure between different process corners is within 15mV, so it is technologically robust.

ACKNOWLEDGMENT

The authors would like to thank the financial support in part by the National Natural Science Foundation of China under Grant No. 12341502, No. 12475195, in part by the National Key R&D Program of China under Grant No. 2023YFF0719600, 2023YFE0206300.

REFERENCES

[1] C.-W. U, M.-K. Law, R. P. Martins, and C.-S. Lam, "Sub-μW Auto-Calibration Bandgap Voltage Reference With 1σ Inaccuracy of ± 0.12% Within − 40°C to 120°C," IEEE J. Solid-State Circuits, vol. 59, no. 2, pp. 540–550, Feb. 2024.

[2] J. Wei et al., "Anomalous TID Susceptibility on Collector Bias for SOI High-Voltage Polysilicon Emitter Bipolar Transistors," IEEE Trans. Electron Devices, vol. 71, no. 7, pp. 4033–4038, Jul. 2024.

[3] J. Chen et al., "ASET and TID Characterization of a Radiation Hardened Bandgap Voltage Reference in a 28-nm Bulk CMOS Technology," IEEE Trans. Nucl. Sci., vol. 69, no. 5, pp. 1141–1147, May 2022.

[4] L. Bin, W. Yi, C. Jianjun, C. Yaqing, and Y. Xiaohu, "Technology Dependency of TID Response for a Custom Bandgap Voltage Reference in 65 nm to 28 nm Bulk CMOS Technologies," Chinese J. Elect., vol. 32, no. 6, pp. 1286–1292, Nov. 2023.

[5] T. Vergine, M. De Matteis, S. Michelis, G. Traversi, F. De Canio, and A. Baschirotto, "A 65 nm Rad-Hard Bandgap Voltage Reference for LHC Environment," IEEE Trans. Nucl. Sci., vol. 63, no. 3, pp. 1762–1767, Jun. 2016.

[6] G. Traversi et al., "Characterization of bandgap reference circuits designed for high energy physics applications," Nuclear Instruments and Methods in Physics Research Section A: Accelerators, Spectrometers, Detectors and Associated Equipment, vol. 824, pp. 371–373, Jul. 2016.

[7] G. Zhongjie, R. Yuan, W. Yapeng, Q. Ziyi, and L. Mengli, "Research on Total Ionizing Dose Radiation Hardening by Design Method for Bandgap Reference Based on Real-Time Monitoring and Adaptive Compensation," Journal of Nanoelectronics and Optoelectronics, vol. 19, no. 8, pp. 857–863, Aug. 2024.

[8] G. Traversi et al., "A radiation hard bandgap voltage reference for the ARCADIA project," J. Inst., vol. 18, no. 01, p. C01049, Jan. 2023.

[9] J. Lin, L. Wang, C. Zhan, and Y. Lu, "A 1-nW Ultra-Low Voltage Subthreshold CMOS Voltage Reference With 0.0154%/V Line Sensitivity," IEEE Trans. Circuits Syst. II, vol. 66, no. 10, pp. 1653–1657, Oct. 2019.

[10] C. M. Andreou et al., "Low-power, subthreshold reference circuits for the space environment: Evaluated with γ-rays, X-rays, protons and heavy ions," Electronics, vol. 8, no. 5, p. 562, May 2019.

[11] F. Yang, Y. Liu, C. Wang, Y. Zhao, and Y. Li, "High Precision Radiation Resistant Bandgap Voltage Regulator for Aerospace Applications," IEEE Trans. Circuits Syst. I, pp. 1–9, 2025.

[12] L. Magnelli, F. Crupi, P. Corsonello, C. Pace, and G. Iannaccone, "A 2.6 nW, 0.45 V Temperature-Compensated Subthreshold CMOS Voltage Reference," IEEE J. Solid-State Circuits, vol. 46, no. 2, pp. 465–474, Feb. 2011.

Dynamic Reward Weighting Based Deep Q-Learning for Routing in Advanced Packaging

1st Shubin Chen
School of mathematics and statistics
Fuzhou University
Fuzhou, China
csb1258223230@163.com

2nd Qinghai Liu*
School of mathematics and statistics
Fuzhou University
Fuzhou, China
qliu@fzu.edu.cn
*Corresponding author

3rd Xiaowei Wu
School of mathematics and statistics
Fuzhou University
Fuzhou, China
230320039@fzu.edu.cn

4th Jun Xu
School of Physics and Information Engineering
Fuzhou University
Fuzhou, China
231127011@fzu.edu.cn

5th Xiaolin Xu
School of Com uter and Data Science
Fuzhou University
Fuzhou, China
231027033@fzu.edu.cn

Abstract—In the contemporary electronics industry, advanced packaging technology has emerged as a critical avenue for overcoming performance bottlenecks in the semiconductor sector. Among these, redistribution layer (DL)—a pivotal component of electronic design automation (EDA)—face significant challenges, including low routing efficiency and suboptimal outcomes. To address these limitations, this study introduces a deep -learning (D N)-based approach for the dynamic adjustment of reward weights in the context of DL global routing for multi-chip packaging (MCP). By training a set of weight vectors to adaptively regulate the relative importance of sub-objectives within the overall reward function in real time, the proposed method enhances both flexibility and efficiency during training and inference. Experimental results demonstrate that the approach achieves a 10.7% reduction in overall routing length compared with conventional strategies employing fixed reward weights.

Keywords—advanced acking, dee earning, redistrib tion ayer ,ro ting, dyna ic weights

I. BACKGROUND

The Redistribution Layer (RDL), as a fundamental platform for establishing high-quality electrical interconnections in advanced packaging architectures, has become a critical determinant in optimizing overall system performance [1]. Multi-Chip Package (MCP) technology is predominantly applied in high-performance, heterogeneously integrated "large-scale chips," such as GPUs and AI processors. These packages typically depend on complex multi-layer RDL architectures and are subject to increasingly stringent requirements with respect to wiring constraints and available routing area.

The schematic diagram of the RDL scenario considered in this study is shown below Fig. 1. In this configuration, the die is connected to the multi-layer RDL through pads, while the PCB interfaces with the multi-layer RDL via bumps. The multi-layer RDL is composed of alternating wiring and via layers, which function as intermediate interconnection layers. The wiring layers facilitate intra-layer signal routing, whereas the via layers provide inter-layer electrical connections.

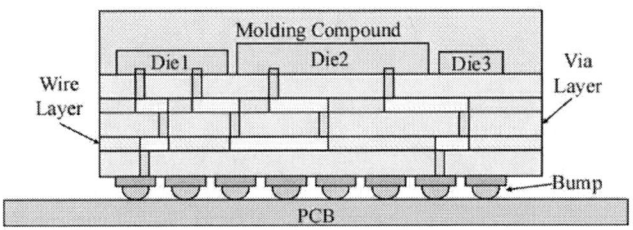

Fig. 1. Redistribution layer schematic

Lin et al. [2] were among the first to propose an algorithm addressing the multi-layer RDL routing problem in multi-chip packaging, introducing a routing-layer assignment strategy that imposed strict constraints on nets, restricting them to designated routing layers. More recently, Zhang et al.[3] applied the Monte Carlo Tree Search (MCTS) algorithm to routing tasks, while InstaDeep developed an AI-driven, cloud-based routing service powered by deep learning[4]. Another notable line of research combines hierarchical MCTS with Markov Decision Process (MDP) search to efficiently identify pin locations [5], demonstrating superior performance relative to traditional algorithms. In addition,[6] proposed a diffusion potential-based reward scheme tailored to mitigate the sparse reward problem commonly encountered in DQN-based routing. By integrating this diffusion potential mechanism into four reinforcement learning-based routing algorithms, the study achieved substantial improvements in key performance metrics, including routing success rate and total wiring length both of which are essential for addressing large-scale challenges in 2D PCB and 3D IC routing.

RDL routing exhibits notable similarities with PCB routing in terms of net distribution and connectivity requirements. However, unlike PCB routing which can typically be divided into escape routing and area routing the unique structural characteristics of RDL preclude such straightforward decomposition. In PCB routing, exponential growth in circuit complexity and grid size often renders traditional search algorithms inadequate. To mitigate this, Li et al. [7] proposed a segmented parallel strategy to enhance efficiency, an approach also applicable to RDL. Similarly, although IC routing is subject to strict directional constraints within each wiring layer, its multi-stage decomposition method aligns closely with the methodologies employed in RDL routing. For instance, Liu et al. [8] proposed a multi-stage optimization algorithm based on this principle, effectively mitigating the risk of local optima associated with shortest-path-only objectives.

Despite these advances, existing global routing algorithms [9-14]remain insufficient for meeting the rigorous demands of industrial applications, particularly in achieving high routing density and low wire-length ratios. While PCB and IC automatic routing technologies are relatively mature, the distinct scenarios and constraints of RDL routing highlight the limitations of directly applying these methods. Consequently, there is an urgent need to develop RDL global routing algorithms that integrate advanced theoretical principles with strong practical feasibility, driving the continued evolution of EDA in the era of advanced packaging.

II. RESEARCH METHODS

A. A lication of the Fi ed eward eight D N Algorithm in outing

The DQN integrates deep neural networks with Q-learning, offering an intelligent and data-driven approach to addressing complex routing tasks. In this framework a neural network is employed to approximate the Q-value function $Q(s,a)$, which represents the expected long-term reward of taking action a in state s .By formulating the automatic routing problem as a MDP, DQN enables the reinforcement learning agent to iteratively interact with the routing environment and progressively refine its decision-making strategy. Through this process, the agent learns to identify near-optimal routing paths, thereby improving both efficiency and solution quality in large-scale routing scenarios.

In a standard DQN framework, the reward function is typically defined as:

$$Q(s,a) \quad Q(s,a) + [r + \gamma max Q(s,a) - Q(s,a)] \quad (1)$$

Where s and a denote the current state and action, respectively, r represents the immediate reward (typically defined as a fixed constant in the standard formulation), γ is the discount factor that balances immediate and future rewards and denotes the learning rate controlling the update magnitude. The terms s and a correspond to the next state and action, respectively, thereby completing the temporal-difference update process of Q-learning.

The training objective is to minimize the mean square error between the predicted Q value and the target Q value:

$$L(\ ) = \quad _{s,a,r,s} \quad y - Q(s,a) \quad ^2 \quad (2)$$

$$y = r + \gamma \quad _a x Q(s,a) \quad (3)$$

Here, denotes the parameters of the target network, and represents the experience replay buffer, which stores past state–action–reward–next state s,a,r,s tuples. Mini-batches are randomly sampled from for training, thereby breaking correlations between consecutive samples and improving the stability of the learning process.

Through the above mechanism, DQN can learn approximately optimal strategies in a high-dimensional state space and achieve intelligent decision-making in complex environments.

$$(s) = \quad max_a Q(s,a) \quad (4)$$

That is, throughout the entire training process, the reward values assigned to different actions such as successful routing, moving closer to the target, incurring excessive path length, or violating design rules remain constant. The primary advantage of this fixed-reward mechanism lies in its simplicity and ease of implementation, as well as the relative stability it provides during training. To a certain extent, this approach can effectively guide the agent to complete fundamental routing tasks.

B. Dynamic Adjustment of eward eights for D N Algorit hm timization

However, the fixed reward weight mechanism exhibits several notable limitations. First, routing tasks often exhibit phased characteristics: the initial phase emphasizes rapid pathfinding, the intermediate phase prioritizes conflict avoidance, and the final phase focuses on optimizing overall wiring quality. Fixed rewards are unable to dynamically reflect these shifting task priorities, which can result in the agent learning relatively rigid and inflexible policies. Second, fixed rewards cannot adaptively modulate the importance assigned to different objectives such as minimizing path length, reducing via count, or maximizing routing completion rate thereby constraining the applicability of DQN in multi-objective routing optimization. Moreover, in complex pad and bump routing environments, the fixed reward mechanism fails to effectively coordinate global interactions among multiple paths, often leading to path conflicts and resource contention.

Under complex RDL global routing scenarios, DQN algorithms employing fixed reward weights exhibit pronounced limitations. As previously discussed, the static nature of the reward system fails to accommodate the evolving priorities inherent in the routing process, particularly with respect to maintaining flexibility and adaptiveness for efficiently achieving multiple objectives. Consequently, relying on fixed reward weights can significantly hinder the agent s ability to optimize routing performance in intricate multi-layer RDL environments.

To address these limitations, we propose an adaptive reward weight mechanism based on state-aware weight adjustment. This mechanism dynamically modulates reward weights according to routing stages, environmental complexity, and historical performance, thereby enhancing both the intelligence and generalization capability of the routing strategy. The proposed approach primarily comprises the following components:

979-8-3315-8850-2/25 $31.00 © 2025 IEEE

State-aware dynamic reward weight adjustment: This component adaptively modulates the reward weights of each sub-objective in real time, taking into account routing progress, environmental complexity, and historical performance. By doing so, it enables a phased optimization strategy that evolves in accordance with the current stage of the routing process.

Lightweight weight prediction network: This network maps the weight vector into a dynamic probability distribution, enhancing the balance among multiple sub-objectives and accelerating policy convergence in multi-objective routing tasks.

A pre-training mechanism environments: This component utilizes A* search during the training phase to generate high-quality initial strategies. By providing a strong starting policy, it accelerates model convergence and enhances the generalization capability of the agent in complex routing environments.

III. EXPERIMENTAL METHODS AND ALGORITHM DESIGN

In the application of deep reinforcement learning to routing tasks, the design of the reward function plays a critical role in the effectiveness of policy learning. Traditional approaches often employ a fixed reward structure, which fails to dynamically capture the shifting importance of different objectives across the various stages of the routing process. To overcome this limitation, this paper introduces a dynamic reward function r_t based on the standard DQN framework, allowing the agent to adaptively adjust its strategy preferences according to routing progress and the local environmental state. The following Fig. 2 describes the overall process of state-aware reward weight adjustment:

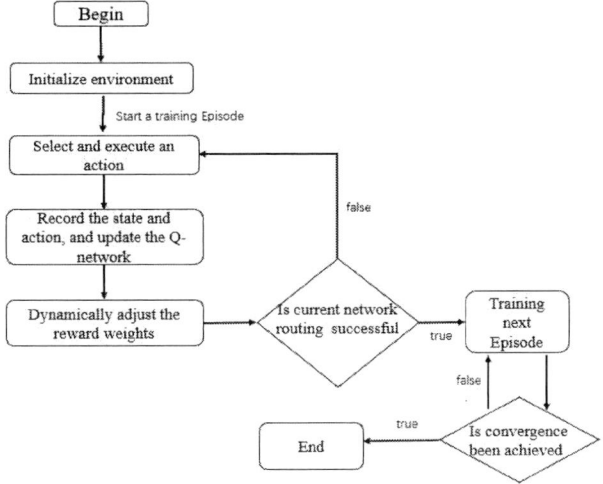

Fig. 2. State-aware weight adjustment

A. Design of the State Aware eight Network

To enable the reward structure to adapt to the state, we define the total reward function in the following form:

$$r_t = \sum_{i\,1} {}_i r_i(t) \qquad (5)$$

where $r_i(t)$ represent four reward sub-components: the increment in routing length r_l , via penalty r_{ia}, connectivity reward $r_{conn\ ct}$, and local congestion penalty r_{con} . The output of this module is directly used in the DQN training and action decision process. We abbreviate it as a state vector:

$$f(s)_t = [l_t,\ {}_t, c_t, d_t]$$

where l_t denotes the incremental routing length, ${}_t$ the number of vias, c_t the connectivity metric, and d_t the number of congestion violations.

To achieve this, we need to record the features s_t of the current routing state, including the current routing path length, distribution of obstacles, connectivity status between pads or bumps, and records of DRC rule violations. Based on these, we generate a pre-normalized weight vector

$$(s_t) = [\,{}_1,\ {}_2,\ {}_3,\ {}_4\,]$$

corresponding to the initial weights for the four sub-objectives: routing length, number of vias, connectivity, and rule violations. The values of $r_i(t)$ are determined using the following strategy:

$$r(t) = \begin{cases} -\ l(t), \\ -1, \\ +100, \\ -100, \end{cases} \qquad (6)$$

B. Dynamic eight Strategy Learning Process

In advanced packaging routing, multiple objectives must be optimized concurrently. However, these sub-objectives are inherently conflicting for instance, minimizing wire length may lead to an increased number of vias. To address this, the weight vector is transformed into a dynamically adjustable probability distribution using Softmax normalization:

$$_i(t) = \frac{e^{()}}{\sum_1 e^{()}} \qquad (7)$$

Softmax normalization is a fundamental mathematical technique that transforms an arbitrary real-valued vector into a probability distribution. Its primary function is to convert raw model outputs such as the scores produced by a neural network into probabilistic form, ensuring that each component lies within the range (0,1) and that higher input values correspond to higher output probabilities. This transformation facilitates gradient-based optimization and is widely employed in conjunction with cross-entropy loss functions during model training.

The normalized weights satisfy: $_i(t)$ $[0,1]$ and $\sum_{i\,1} {}_i = 1$. Through this processing, we can dynamically adjust the weights to balance conflicts among multiple objectives, preventing any single sub-objective from dominating and causing training instability. For example, in the early routing stages when connectivity is poor, normalization can be used to increase the current weight of the connectivity objective, allowing the agent to focus more on overcoming this critical bottleneck during initial path establishment.

Algorithm 1 Dynamic Reward Weight Adjustment
input Current State s_t
1   Extract State Feature Vector $f(s)_t$

2. Compute Raw Weights via Weight Network (s_t)

3. Softmax Normalization Process $= softmax(\ (s_t))$

4. Calculate Sub-Reward Components

5. Calculate weighted Total Reward r_t

    for i in range(4)

      $r_t =\ _i\ r_i$

return   Weighted Total Reward r_t

C. E erimental Environment Setu

For a given chip routing problem, the input data comprises detailed chip information and the complete set of nets to be routed. The chip information includes the grid size, edge capacities, and reduced edge capacities resulting from various obstacles. Each net is described by a list of pad or bump locations. In large-scale routing problems, the number of pads or bumps per net is typically unequal, ranging from two to over a thousand. Upon reading and parsing the input files, a simulated global routing environment is constructed that integrates all relevant chip and net information, providing a realistic framework for subsequent routing experiments.

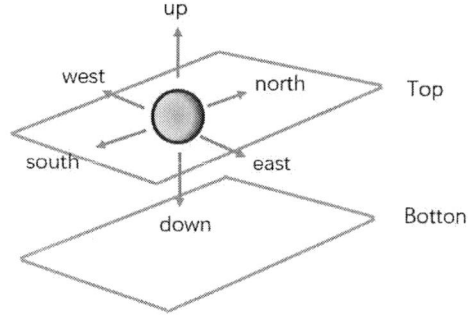

Fig. 3. Six different directional strategies

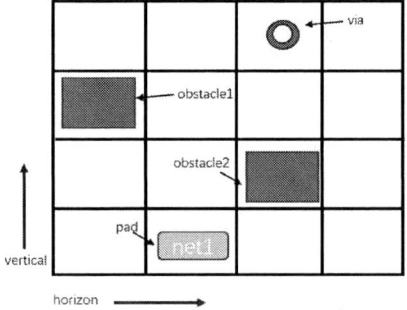

Fig. 4. Routing environment

In the global routing environment we constructed, all configured grid edge capacities are updated in real time as routing progresses. As shown in Fig 4, we designed a 4 4 2 routing environment use case, where the red rectangles represent obstacles encountered during routing. When a pad or bump routes into a grid cell, there are six possible directional actions within that cell (Fig 3): up, down, north, south, west, and east. In this environment, actions are constrained by boundary conditions and capacity limits, so the actual feasible actions may

vary and could be fewer than six, depending on the specific environmental setup. Figure 4 illustrates a sample solution to such a problem. Once a routing step is taken, the capacity of the traversed edge is correspondingly reduced. Additionally, when a via operation (i.e., moving up or down) is required, we must also consider whether the capacity of the current grid face meets the size requirements for the via.

In our experiments, we primarily use the results from the A* algorithm as a heuristic reference and for pre-training initialization, enabling performance comparison between algorithms in terms of total wirelength ratio. To reduce problem dimensionality, we prioritize routing order based on the distances between different pads or bumps within each net.

Before initiating DQN-based routing, the A* algorithm is first used to solve routing between two pads or bumps. A path is found between two pads or bumps within the same net (one designated as the starting point S and the other as the goal point G). Assuming there are n possible pairwise combinations within a net, the cost function is defined as:

$$f(n) = \sum_{i\ 1}^{n} l_S\ (n) \tag{8}$$

Where n denotes the time step, and $l_S\ (n)$ represents the distance between two different pads or bumps (here, we use the Euclidean distance). Ultimately, we select the pad or bump pair with the smallest cost result as the starting point.

In addition to serving as a criterion, the A* search mainly provides pre-training to enable the learning rate of the DQN to converge more quickly. However, using A* for pre-training is optional early random routing can also be used instead of A* for pre-training. But since random routing may lead to DRC violations, as will be shown in the results later, reference [15] indicates that using A* for pre-training can achieve faster convergence. Experimental methods are applied to determine the optimal pre-training approach.

IV. EXPERIMENTAL DESIGN AND RESULTS ANALYSIS

A. E erimental Setu and Evaluation etrics

In this experiment, we designed an 8 8 2 routing environment to evaluate the effectiveness of the proposed dynamic reward mechanism in enhancing routing performance. Two comparative experimental setups were designed:

Fixed-Reward DQN: The reward weights for all sub-objectives are set to empirically determined fixed values (0.1, 0.2, 1.0, 0.6) and remain constant regardless of the routing state..

2 Dynamically Adjusted Reward Weights: The reward weights are adaptively modulated according to the progress of routing completion, allowing the agent to prioritize different objectives at different stages of the routing process.

Differing from the approach in [15], this study employs a Q-network with a two-layer fully connected architecture to predict the action value (Q-value) for each state. Each layer is followed by a ReLU activation function to introduce nonlinearity. For each net and its corresponding pair of pads or bumps, the Q-network is iteratively updated on the same target problem. A single pass through the entire set of pad or bump pairs constitutes one training phase. The network is trained by repeatedly processing these pad or bump pairs across multiple phases until convergence is achieved.

The complete training procedure for the Q-network is outlined as Fig. 5:

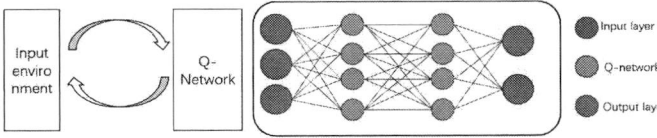

Fig. 5. Q-network training

The main parameters are listed in Table 1, which includes only the optimized parameter set. The model and parameter tuning process will be presented and discussed below.The proposed algorithm is implemented in Python. The experimental platform consists of an Intel Core i7-13620H CPU and an NVIDIA GeForce GTX 4050 GPU.

TABLE I.   MAIN PARAMETER SETTINGS

Parameter	value
Learning Rate	1e-4
Batch size	32
Discount Factor γ	0.99
Max episodes	5000
Burn-in size	10000

B. Experimental Results and Analysis

As noted in [15], a primary advantage of Q-network-based routing lies in the model's capacity to simultaneously consider multiple sub-tasks, such as routing individual nets and connecting specific pins a concept often referred to as collaborative optimization within the DQN framework. Moreover, when combined with the proposed dynamic reward mechanism, the Q-network can guide exploration during the early stages of training (e.g., by prioritizing connectivity) and progressively shift focus toward optimizing path quality in later stages (e.g., minimizing detours and path crossings). This adaptive strategy enhances the agent s ability to handle layouts of varying complexity.

Following the methodology in [15], we conducted experiments on a series of routing problems generated with different parameters. A key distinction among these problems concerns the utilization of edges within the A* algorithm. In this study, we primarily focus on the first scenario, in which no edge with initially positive capacity becomes completely depleted (the"no-edge-depletion" problem).All test cases were configured with exactly 30 nets. A partial comparison of the results between the two experimental groups is presented below TABLE II. and TABLE III. :

TABLE II.   FIXED REWARD WEIGHTS

Case	nets	Success Rate	Path Length
Case1	30	86.7%	12.14
Case2	30	83.3%	12.58
Case3	30	80.0%	11.80
Case4	30	86.7%	13.07
Case5	30	86.7%	12.39

TABLE III.   DYNAMICALLY ADJUSTED REWARD WEIGHTS

Case	nets	Success Rate	Path Length
Case1	30	86.7%	10.80
Case2	30	86.7%	11.27
Case3	30	90.0%	10.40
Case4	30	90.0%	11.40
Case5	30	93.3%	10.91

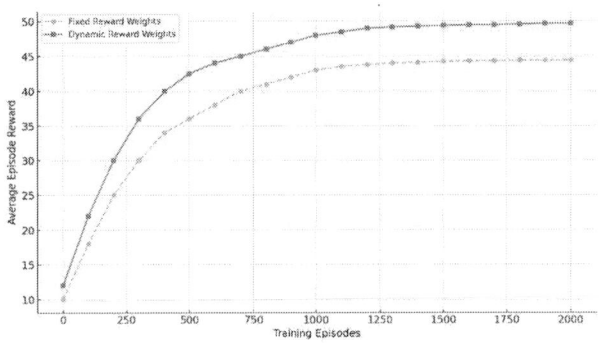

Fig. 6. Q Convergence rate comparison

The experimental results demonstrate that the dynamic reward weight adjustment mechanism significantly outperforms the traditional fixed-weight strategy across multiple evaluation dimensions. Its benefits are evident not only in faster convergence and higher final success rates, but also in the attainment of more stable and generalizable policies. By automatically modulating the focus of incentives throughout training, this mechanism effectively guides the Q-network to prioritize different objectives at different stages, yielding a 10.7% reduction in overall routing length.

In contrast, fixed reward weight strategies impose a rigid and unchanging preference throughout training. For example, if excessive weight is assigned to minimizing path length, the Q-network may prematurely focus on short paths while neglecting the critical objective of connectivity. Conversely, if connectivity is overemphasized, the network may avoid failures during path search without considering resource efficiency, ultimately limiting performance in later training stages. As a result, fixed-reward strategies are prone to suboptimal policy traps.

Moreover, fixed rewards can mislead the agent in specific states. In complex regions, preparatory actions necessary for future routing moves may not yield immediate rewards, causing the Q-network to erroneously perceive them as ineffective and favor suboptimal actions. Since gradient updates depend on the magnitude and direction of reward feedback, underweighted reward components contribute minimally to updates and are insufficiently reinforced over time, while overweighted components may disproportionately amplify local behaviors, leading to training bias or oscillations. Consequently, fixed-reward training is more likely to converge to local optima rather than the global optimum during the gradient convergence process.

V. CONCLUSION

The dynamic reward weight adjustment mechanism demonstrates distinct advantages in this study. By adaptively modulating reward and penalty weights across multiple objectives in real time according to the agent s behavior during training, this approach effectively alleviates common issues observed in fixed reward strategies, such as entrapment in local optima and inefficient learning cycles. Experimental results indicate that the dynamic mechanism not only accelerates policy convergence but also improves final routing success rates and path quality.

Nevertheless, this adaptive mechanism introduces potential challenges, including reduced training stability, policy drift, and increased complexity in hyperparameter tuning. Future research will focus on further optimizing the weight adjustment function to enhance its intelligence and adaptability, while ensuring stable and efficient training.

REFERENCES

[1] Frank K H K, Tai D, Peng S, et al. Comparable Study for Redistribution Layers in FO POP RDL First and Last (Fan-Out Package on Package)[C]//Proceedings of the 2023 IEEE 73rd Electronic Components and Technology Conference. IEEE, 2023: 253-257.

[2] Lin B, Lin T, Chang Y W. Redistribution layer routing for integrated fan-out wafer-level chip scale packages[C]//Proceedings of the IEEE/ACM International Conference on Computer Aided Design. IEEE, 2016: 1-8.

[3] Zhang C, Jin H, Chen J, et al. A Hierarchy MCTS Algorithm for The Automated PCB Routing[C]. 2020 IEEE 16th International Conference on Control Automation (ICCA). IEEE, 2020: 1366-1371

[4] InstaDeep,Deeppcb: Pure ai-powered, cloud-native pcb routing[EB/OL]. 2019, https://deeppcb.ai/.

[5] Jialu Z, Yongtao M, Kaihua L. Multi-user Joint Anti-jamming Decision Algorithm Based on LSTM and DQN [J]. Journal of Transducer Technology, 2021, 34(06): 811-817

[6] Bo Z. Research and Implementation of Routing Algorithm Based on Reinforcement Learning [D]. University of Electronic Science and Technology of China, 2023. DOI: 10.27005/d.cnki.gdzku.2023.002319.

[7] Yuankang L, Quanbao G, Shiyu G, et al. An Acceleration Strategy for PCB Routing Based on the Idea of Segmented Parallelism [J]. Microelectronics Computer, 2023, 40(9): 1-1.

[8] Genggeng L, Zhenyu P, Ning X. High-quality overall routing Algorithm Based on Multi-stage optimization [J]. Journal of Computer-Aided Design Computer Graphics, 2024, 36(4):607-614.

[9] Hu, J., and Sapatnekar, S. S., 2001. "A survey on multi net global routing for integrated circuits". Integration, 31(1), pp. 1–49.

[10] Lee, J., Bose, N., and Hwang, F., 1976. "Use of steiner s problem in suboptimal routing in rectilinear metric". IEEE Transactions on Circuits and Systems, 23(7), pp. 470–476.

[11] Tseng I L MaizeRouter: engineering an effective global router[J].Computing reviews, 2010, 51(7):P.422-423.

[12] Kastner R , Bozorgzadeh E , Sarrafzadeh M .Pattern routing: use and theory for increasing predictability and avoiding coupling[J].IEEE Transactions on Computer-Aided Design of Integrated Circuits and Systems, 2001, 21(7):777-790.DOI:10.1109/TCAD.2002.1013891.

[13] Jinting L. Research on Automatic Routing Method Based on Multi-Agent [D]. University of Electronic Science and Technology of China, 2022.

[14] Nguyen T T, Nguyen N D, Nahavandi S. Deep reinforcement learning for multiagent systems: A review of challenges, solutions, and applications[J]. IEEE transactions on cybernetics, 2020, 50(9): 3826-3839

[15] Liao Haiguang, Zhang Wentai, Dong Xuliang, et al. A Deep Reinforcement Learning Approach for Global Routing[J]. Journal of Mechanical Design, 2020, 142(6).

2025 The 10th International Conference on Integrated Circuits and Microsystems

A 7-Bit 1.8 ps Two-Step Time-to-Digital Converter in 28 nm CMOS Technology

Zhigang Li[1], Yao Nian[1], Hongjia Xu[1], Wei Liu[1], Xin Li[1], Qiang Zhao[1], Xiulong Wu[1*]

[1]*School of Integrated Circuits and Anhui Provincial High-Performance Integrated Circuit Engineering Research Center, Anhui University, Hefei 230601, China*
*Corresponding author: xiulong@ahu.edu.cn

Abstract—This paper presents a 28 nm CMOS two-step time-to-digital converter, incorporating a novel residual time extraction circuit designed to minimize errors and reduce power consumption. The architecture achieves performance balance through coarse-fine quantization. The coarse stage employs a voltage-controlled delay cell for wide dynamic range, while the fine stage utilizes a vernier delay line structure to enhance resolution. The proposed residue time extraction circuit enables accurate transmission between coarse and fine quantization stages, the design achieves 223 ps dynamic range and 1.8 ps resolution while reducing power and area consumption compared to conventional multiplexer-based structure. The simulation results demonstrate that the power consumption is approximately 0.332 mW, with a core area occupation of 0.014 mm². The INL and DNL are 1.25 LSB and 0.5 LSB respectively. The design fulfills critical requirements for time measurement modules in all-digital phase-locked loop.

Index Terms—time-to-digital converter, ADPLL, two-step, high resolution

I. Introduction

Time-to-digital converter (TDC), serving as core circuits in high precision time measurement systems, convert continuous time domain signals into discrete digital signals [1]. Advancements in semiconductor manufacturing technology have significantly enhanced TDC perormance, achieving picosecond level resolution. This advancement has positioned TDC as a critical components in all-digital phase-locked loop (ADPLL) [2].

In ADPLL designs, enhancing the resolution of a TDC effectively reduces quantization noise, improving in band phase noise [3]. However, increasing TDC resolution involves trade-off among quantization error suppression, dynamic range expansion, and power efficiency. The architecture of TDC using conventional delay chains offer high conversion speeds, encounter limitations including process-dependent transmission delays, nonlinear errors caused by device mismatch, and an exponential increase in the number of delay stages required to achieve a broader dynamic range.

To address the challenges in TDC design, researchers have proposed several innovative architectures including the Vernier TDC, which utilizes the delay difference between dual delay lines for quantization [4], [5], effectively mitigating the impact of process constraints, and the two-step TDC, which combines delay chain TDC for coarse quantization and Vernier TDC for fine quan tization, achieving high resolution while extending dynamic range [2], [6]. Another two-step TDC utilizes a time

amplifier (TA) to amplify the time-residual between adjacent conversion stages for resolution enhancement, this approach suffers from sensitivity to process, voltage, and temperature (PVT) variations in the TA's gain characteristics, resulting in gain stability degradation and subsequent nonlinearity errors [7], [8].

This paper proposes a novel two-step TDC architecture that achieves high resolution while maintaining a broad measurement range. The first stage employs a delay chain TDC structure using voltage-controlled delay cells for coarse quantization, whereas the second stage adopts a Vernier TDC for fine quantization. A novel time residue extraction circuit is designed to accurately transmit residual timing intervals between the quantization stages with reduced area and power consumption.

The paper is organized as follows: Section II presents the proposed TDC circuit design, Section III discusses simulation results and analysis, and Section IV offers the conclusion of the work.

II. Circuit Design

A. Proposed Two-Step TDC

The proposed two-step TDC combines coarse-fine quantization, the coarse stage, implemented with a delay chain TDC using voltage-controlled delay cells, provides wide dynamic range, while its resolution is configured such that the dynamic range of the fine stage marginally exceeds the delay of a single coarse stage unit to compensate quantization errors [8]. The fine stage employs a Vernier TDC, which benefits from its simplicity and high resolution capability.

The architecture of proposed two-step TDC, as shown in Fig. 1, consists of a coarse stage delay chain TDC for wide dynamic range, a residual time extraction circuit, and a Vernier TDC for high resolution. This architecture converts the time interval between the input clock and data signals into a binary code. D flip-flops (DFFs) generate thermometer code from the input time difference, while delay cells compensate for latency introduced by the residual time extraction circuit.

The TDC operates through a two-stage quantization process. In the coarse stage, the start signal transmits sequentially through seven delay cells, each contributing a 28 ps delay to establish the coarse resolution of 28 ps. DFFs sample the time difference between the start and stop signals, generating a thermometer code that is encoded into a 3-bit binary output.

979-8-3315-8850-2/25 $31.00 © 2025 IEEE

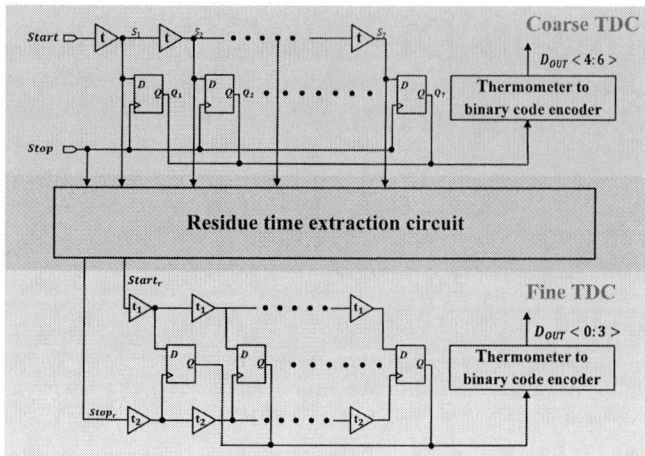

Fig. 1. Architecture of the proposed two-step TDC

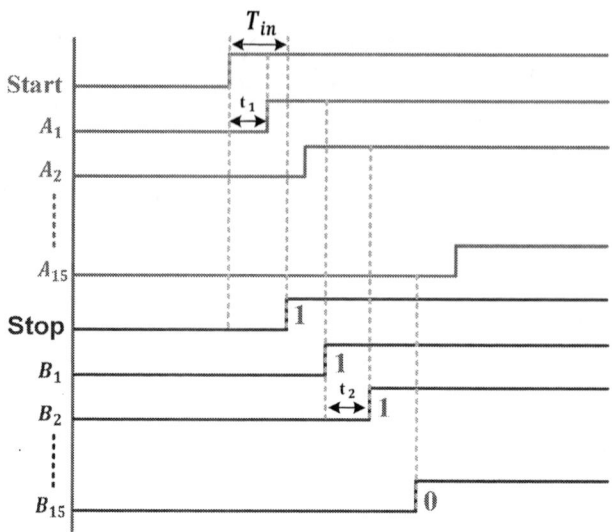

Fig. 2. Timing diagram of the fine TDC

The residual time extraction circuit precisely transmits the unquantized time-residual from the coarse stage to the fine stage. Within the fine stage, Vernier delay line processing refines the residual time measurement using matched delay differences, producing a 4-bit binary code. The fine stage achieves a 1.8 ps resolution, determining the overall two-step TDC resolution.

In the coarse stage TDC, the resolution of the coarse stage TDC is determined by the delay time t of its delay cells. With N_1 delay cells, the dynamic range is given by:

$$T_1 = N_1 \times t \tag{1}$$

The timing diagram of the fine stage TDC is shown in Fig. 2. The resolution is determined by the transmission time difference between two delay cells, the resolution of the fine stage is :

$$\tau_2 = t_1 - t_2 \tag{2}$$

For a delay chain with N_2 cells, the dynamic range is expressed as :

$$T_2 = N_2 \times \tau_2 \tag{3}$$

The voltage-controlled delay cell serves as the core component of coarse stage TDC. In such architectures, the transmission delay required by the delay chain must align with the dynamic range of the fine stage TDC [9]. Conventional transistor sizing-based delay control methods exhibit limitations in achieving flexible adjustment of target delays. To overcome this limitation, the voltage-controlled delay cell implements delay tuning through transmission voltage regulation, as illustrated in Fig. 3.

This architecture achieves delay adjustment by regulating the voltage to control signal transmission. By integrating MOS resistors into the inverters, the output resistance and delay range are enhanced, enabling tunable circuit delays. The equivalent output resistance becomes controllable through voltage adjustment, allowing precise calibration of the overall delay cell's temporal characteristics while maintaining structural

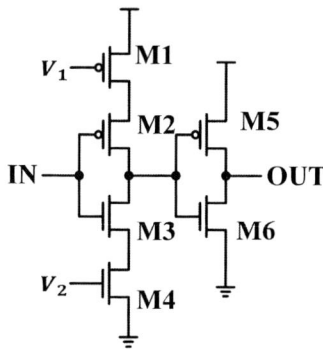

Fig. 3. The schematic of voltage-controlled delay cell

simplicity. In the first stage inverter, the rise time delay is regulated by the PMOS transistor controlled by voltage V_1, while the fall time delay is regulated by the NMOS transistor governed by voltage V_2, as shown in Fig. 4.

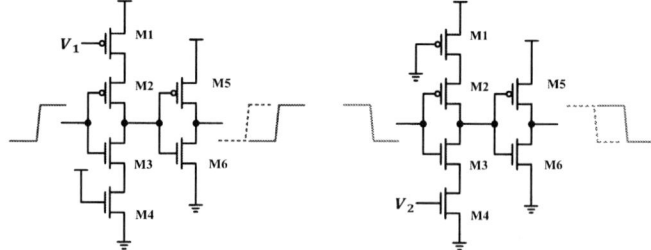

Fig. 4. Timing diagram of the voltage-controlled delay cell

B. Residue time extraction circuit

The unquantized time-residual from the coarse stage is accurately routed to the fine quantization stage via a residue extraction circuit. While the conventional two-step TDC typically utilize multiplexers for time-residual transmit, the number of logic gates increases substantially with higher coarse stage bit resolution. This work introduces a novel residue extraction circuit, as shown in Fig. 5, which identifies the transition boundary in the coarse stage thermometer code and generates appropriate $Start_n$ signals.

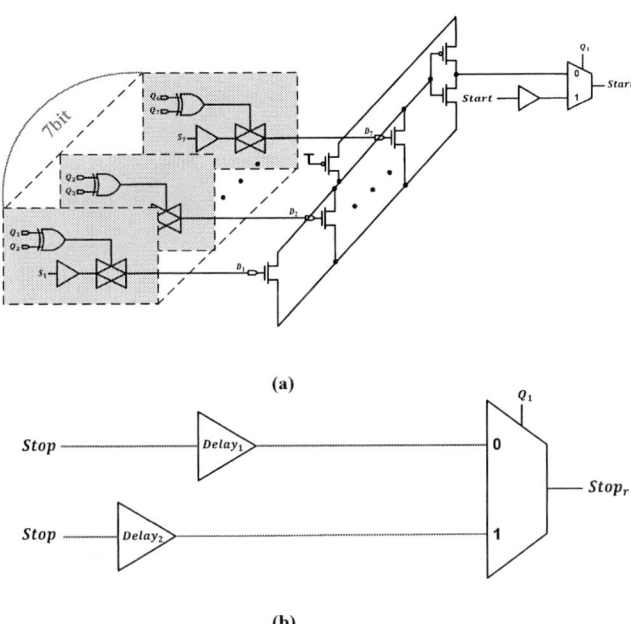

(a)

(b)

Fig. 5. The schematic of residue time extraction circuit (a) *Start* to $Start_r$ interface (b) *Stop* to $Stop_r$ interface

In the *Start* to $Start_r$ interface, signals S_{1-7} transmit through delay elements and connect to transmission-gate (TG), generating outputs D_{1-7}. Each delay is gated by a TG controlled by $Q_N \oplus Q_{N+1}$, producing output D_n. The dynamic OR gate, receiving D_{1-7}, detects the first arriving rising edge signal and elevates the output level upon capturing the initial transition. When the time interval between *Start* and *Stop* signals is smaller than the coarse stage TDC's minimum measurable range, a 2-to-1 multiplexer bypasses the coarse stage signal path by directly routing the *Start* signal to the subsequent processing stage, achieving power efficiency optimization.

In the signal interface from *Start* to $Start_r$, there exist two distinct transmission paths. Therefore, in the signal path from stop to $Stop_r$, two independent paths must also be designed and configured with different delay elements to compensate for the delay discrepancy. The delay configuration should prioritize ensuring the earlier arrival of the selection signal Q_1.

C. Thermometer to Binary Code Encoder

The thermometer to binary code encoder converts the DFFs outputs into binary codes. This conversion can be implemented through various architectures, including Wallace Tree, Fat-Tree, ROM-based, and multiplexer-based encoders [10]. Among these, the multiplexer-based encoder was selected for the proposed design due to its low power consumption

characteristics, reduced transistor count and regular structural configuration. In the 7-bit multiplexer-based encoder, a total of four 2-to-1 multiplexers are employed, with two utilized in the first stage and two in the second stage, the schematic implementation illustrated in Fig. 6.

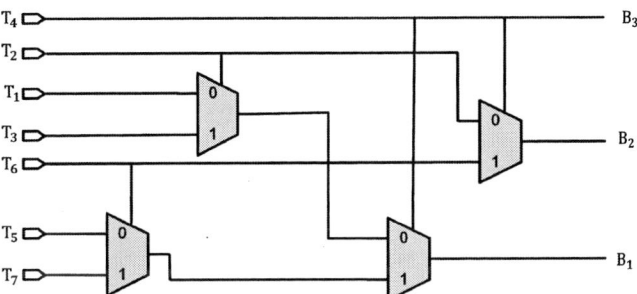

Fig. 6. Architecture of the thermometer to binary code encoder

The encoding process operates hierarchically, if the number of 1 in the $2^n - 1$ bits thermometer code exceeds half of its total bits, the most significant bit (MSB) of the binary output is set to 1. This process continues recursively until only a final 2-to-1 multiplexer remains, whose output corresponds to the LSB of the resulting binary code.

III. SIMULATION RESULTS

The proposed two-step TDC is implemented in a 28 nm CMOS technology. Compared to conventional two-step implementations, the proposed architecture achieves reduced power consumption and compact core area while maintaining picosecond level timing resolution. The layout showing in Fig. 7 indicates that the scale of the core area is 0.014 mm². The design operates at a 0.9 V supply voltage with a maximum power consumption of 0.332 mW, delivering 223 ps dynamic range and 1.8 ps resolution.

The time-to-digital transfer characteristic curve of the proposed two-step TDC is demonstrated in Fig. 8, with the input time difference swept from 0 ps to 223 ps in 0.45 ps step increments.The simulations in Fig. 9 and Fig. 10 verify performance, with The DNL and INL are 0.50 LSB and 1.25 LSB respectively.

Fig. 7. The layout of the proposed two-step TDC

As illustrated in Table I, the proposed TDC achieves both low power consumption and high resolution, with the mea-

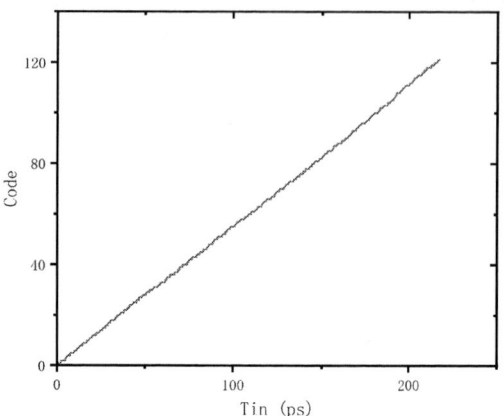

Fig. 8. Time-to-digital transfer characteristic of the two-step proposed TDC

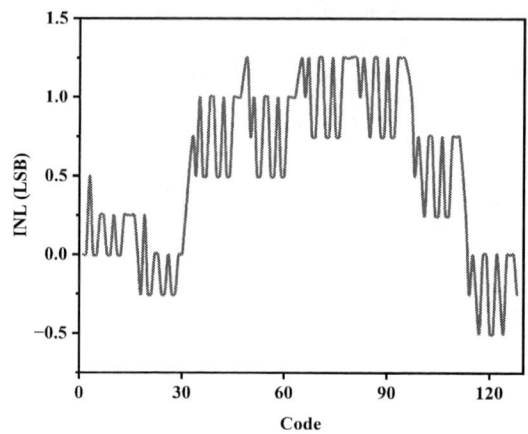

Fig. 10. INL of the proposed two-step TDC

TABLE I: Performance Comparison

References	JSSC [4]	ISNE [6]	ICICM [7]	JSSC [8]	**This work**
Scheme	Vernier	Two-step	Two-step	Two-step	**Two-step**
Process (nm)	130	130	130	65	**28**
Resolution (Ps)	8	3.8	2.5	3.7	**1.8**
DNL (LSB)	N/A	0.46	0.86	0.9	**0.5**
INL (LSB)	N/A	2.3	1.0	2.3	**1.25**
Area (mm²)	0.26	0.036	N/A	0.02	**0.014**
Power (mW)	7.5	1.3	N/A	3.6	**0.332**

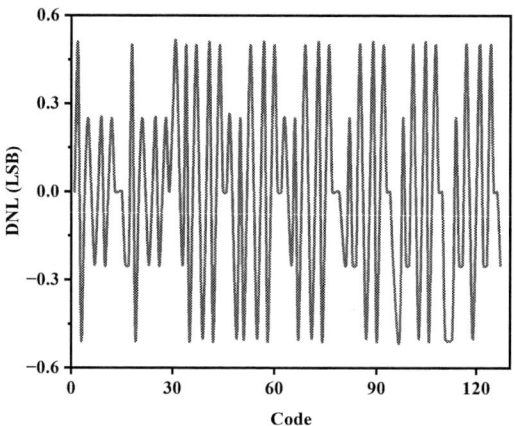

Fig. 9. DNL of the proposed two-step TDC

sured DNL and INL values presented in the table. The two-step architecture of the proposed design successfully balances resolution and dynamic range while simultaneously reducing power consumption and core area.

IV. CONCLUSION

This paper presents a two-step TDC fabricated in 28 nm CMOS technology, employing a time residual transmit circuit to mitigate inter stage transmission errors and improve measurement accuracy. The proposed architecture achieves a resolution of 1.8 ps and a dynamic range of 223 ps while simultaneously optimizing power consumption and area. Measurement results demonstrate a compact core area of 0.014 mm² and 0.332 mW power consumption under 0.9 V supply, outperforming conventional multiplexer based two-step TDC in area power tradeoffs. The simulation results confirm robust linearity with DNL and INL of 0.5 LSB and 1.25 LSB, respectively, demonstrating the design's ability to sustain 1.8 ps resolution without dynamic range degradation. The architecture successfully balances resolution enhancement with dynamic range preservation.

ACKNOWLEDGMENT

This work was supported by the National Natural Science Foundation of China under Grant 62274001, the University Synergy Innovation Program of Anhui Province under Grant GXXT-2023-013, and the Science and Technology Key Project of Anhui Province under Grant 2022AH050099.

REFERENCES

[1] S. Tancock, E. Arabul and N. Dahnoun, "A Review of New Time-to-Digital Conversion Techniques," *IEEE Transactions on Instrumentation and Measurement*, vol. 68, no. 10, pp. 3406-3417, Oct. 2019.

[2] X. Y. Chen, L. Tang and X. Shen, "A High-Resolution Two-Step Time-to-Digital Conversion in 40 nm CMOS," *2021 6th International Conference on Integrated Circuits and Microsystems (ICICM)*, Nanjing, China, 2021, pp. 189-192.

[3] P. Chen, X. Meng, J. Yin, P.-I. Mak, R. P. Martins and R. B. Staszewski, "A 529-W Fractional-N All-Digital PLL Using TDC Gain Auto-Calibration and an Inverse-Class-F DCO in 65-nm CMOS," *IEEE Transactions on Circuits and Systems I: Regular Papers*, vol. 69, no. 1, pp. 51-63, Jan. 2022.

[4] J. Yu, F. F. Dai and R. C. Jaeger, "A 12-Bit Vernier Ring Time-to-Digital Converter in 0.13 m CMOS Technology," *IEEE Journal of Solid-State Circuits*, vol. 45, no. 4, pp. 830-842, Apr. 2010.

[5] N. Xing, J. -K. Woo, W. -Y. Shin, H. Lee and S. Kim, "A 14.6 ps Resolu tion, 50 ns Input-Range Cyclic Time-to-Digital Converter Using Fraction al Difference Conversion Method," in IEEE Transactions on Circuits and Systems I: Regular Papers, vol. 57, no. 12, pp. 3064-3072, Dec. 2010.

[6] M. Lui, B. Wang, Y. Xi and C. Zhang, "A 1.3 mW 8-bit Two-Step Time-to-Digital Converter," *2021 9th International Symposium on Next Generation Electronics (ISNE)*, Changsha, China, 2021, pp. 1-4.

[7] X. Hu and Q. Li, "A 8-bit, 2.5 ps Resolution Time-to-Digital Converter in 130 nm SiGe Technology," *2024 9th International Conference on Integrated Circuits and Microsystems (ICICM)*, Wuhan, China, 2024, pp. 440-443.

[8] K. Kim, Y.-H. Kim, W. Yu and S. Cho, "A 7-bit, 3.75 ps Resolution Two-Step Time-to-Digital Converter in 65 nm CMOS Using Pulse-Train Time Amplifier," *IEEE Journal of Solid-State Circuits*, vol. 48, no. 4, pp. 1009-1017, Apr. 2013.

[9] S. A. Mondal, S. Pal, H. Rahaman and P. Mondal, "Voltage-Controlled Current-Starved Delay Cell for Positron-Emission-Tomography-Specific DLL-Based High-Precision TDC Implementation," *2012 5th International Conference on Computers and Devices for Communication (CODEC)*, Kolkata, India, Dec. 17-19, 2012, pp. 1-4.

[10] G. L. Madhumati, K. R. Rao and M. Madhavilatha, "Comparison of 5-bit Thermometer-to-Binary Decoders in 1.8 V, 0.18 m CMOS Technology for Flash ADCs," *2009 International Conference on Signal Processing Systems*, Singapore, 2009, pp. 516-520.

2025 The 10th International Conference on Integrated Circuits and Microsystems

Ka-Band Broadband LNA in 65-nm CMOS Using Pre-stage Current-reuse Technique and Large-size Transistor

Zihan Yang
Key Laboratory of Near-Range Sensing ICs & Microsystems
Nanjing University of Science and Technology
Nanjing, China
123104010496@njust.edu.cn

Hao Wang
Key Laboratory of Near-Range Sensing ICs & Microsystems
Nanjing University of Science and Technology
Nanjing, China
wanghaotz@njust.edu.cn

Ye Liu
Key Laboratory of Near-Range Sensing ICs & Microsystems
Nanjing University of Science and Technology
Nanjing, China
liuye200618@njust.edu.cn

Tongde Huang
Key Laboratory of Near-Range Sensing ICs & Microsystems
Nanjing University of Science and Technology
Nanjing, China
tongdeh@njust.edu.cn

Wen Wu
Key Laboratory of Near-Range Sensing ICs & Microsystems
Nanjing University of Science and Technology
Nanjing, China
wuwen@njust.edu.cn

Abstract—This paper presents a low-noise amplifier (LNA) operating in the 25.2–35.4 GHz frequency band, specifically designed for Ka-band broadband satellite communication applications. The proposed LNA integrates a pre-stage current-reuse technique, a dual gm-boosting architecture, and a large-sized transistor design to achieve high gain, low noise figure (NF), and low power consumption. By optimizing the current-reuse technique, the transconductance is enhanced and the minimum noise figure (NF_{min}) is reduced, overcoming the inherent NF degradation issue in conventional current-reuse configurations. Additionally, the introduction of large-sized transistors minimizes thermal noise from gate resistance and reduces the chip area. The output stage employs a multi-gate configuration to significantly enhance linearity. Fabricated in a 65-nm CMOS process, the LNA achieves a peak gain of 18.8 dB, a minimum NF of 2.3 dB, and a power consumption of 9 mW. It demonstrates exceptional performance in the 27.5–31 GHz satellite communication band and is well-suited for broadband satellite communication systems.

Index Terms—Low noise amplifier, pre-stage current-reuse, gm-boosting, large-sized transistor, Ka-band

I. INTRODUCTION

Next-generation High Throughput Satellites (HTS) in Low Earth Orbit (LEO) have unlocked new opportunities for high-quality and cost-effective satellite communication services [1]. To enhance data transmission rates, the K-band (17.7–21.2 GHz) and Ka-band (27.5–31 GHz) have been allocated as the downlink and uplink spectra for satellite communications, respectively [2]. The requirements for rapid electronic scanning capability, long-range coverage, and low cost have driven the

adoption of silicon-based large-scale phased arrays (exceeding 1,000 antenna elements) in these systems. As the first module following the antenna in a phased-array receive chain, the low-noise amplifier critically determines the overall noise figure of the RF front-end through its gain and noise performance. Beyond achieving high gain and low NF to ensure communication quality, optimizing the power consumption of LNAs in large-scale phased arrays (comprising over 1,000 elements) remains a critical design metric [3]. Consequently, low-power, high-performance LNAs are indispensable for millimeter-wave phased-array satellite terminals.

To address high power consumption, current-reuse techniques have been proposed to reduce power dissipation [4]–[7]. Specifically, gm-boosting techniques can further enhance gain while lowering power consumption [8]–[11]. However, in existing current-reuse topologies for multi-stage cascaded LNAs, current division inherently reduces the current density of the first-stage transistor, leading to suboptimal noise performance. To improve input-stage transconductance(g_m) and noise matching, conventional LNAs typically employ a series gate inductor(Lg) for input matching optimization. Nevertheless, the parasitic resistance of Lg becomes non-negligible and escalates proportionally with Lg value. Since this parasitic resistance is directly in series with the gate of the input transistor, it degrades the input-referred noise figure. Moreover, larger Lg values introduce stability concerns due to potential resonance effects. Consequently, there is a critical need to develop a Ka-band LNA design methodology that

979-8-3315-8850-2/25 $31.00 © 2025 IEEE

simultaneously achieves high gain, low noise figure, and low power consumption.

This paper presents a low-noise amplifier operating in the 25.2–35.2 GHz frequency band, which achieves low power consumption, compact area, and exceptional noise and gain performance through a pre-stage current-reuse technique, dual gm-boosting architecture, and large-sized transistor design. The paper is organized as follows: Section II provides a detailed circuit design of the proposed LNA. Section III presents the layout implementation, simulation results, and performance comparisons with state-of-the-art designs. Finally, Section IV concludes the paper by summarizing the key findings and research contributions.

II. CIRCUIT DESIGN

In this study, we propose an LNA that achieves high gain, low NF, and low power consumption through the adoption of three key topological strategies. Figure 1 illustrates the schematic of the proposed LNA, highlighting its fully single-ended configuration. To attain sufficient gain and significant voltage swing under a 1 V supply, we designed a three-stage cascaded amplifier utilizing a common-source (CS) configuration for all stages (CS-CS-CS).

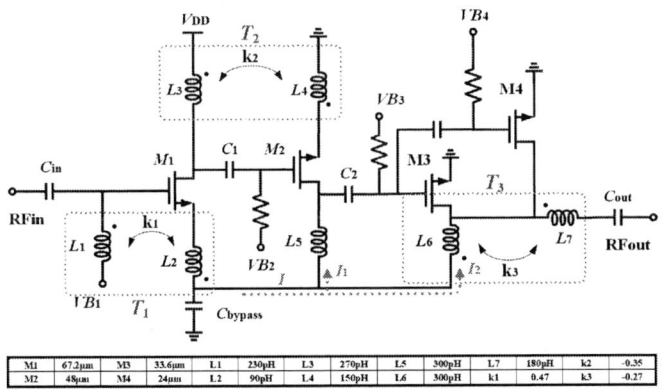

| M1 | 67.2μm | M3 | 33.6μm | L1 | 230pH | L3 | 270pH | L5 | 300pH | L7 | 180pH | k2 | -0.35 |
| M2 | 48μm | M4 | 24μm | L2 | 90pH | L4 | 150pH | L6 | 300pH | k1 | 0.47 | k3 | -0.27 |

Fig. 1: Schematic of the proposed LNA.

As governed by the Friis noise formula (1), the noise contribution of the first stage predominantly determines the overall noise performance of the LNA:

$$F = F_1 + \frac{F_2 - 1}{G_1} + \frac{F_3 - 1}{G_2} + \dots \quad (1)$$

Where F, F_1, F_2 and F_3 represent the total noise factor and the noise factors of each stage, respectively, while G_1 and G_2 denote the voltage gains of preceding stages.

As illustrated in Fig. 2, conventional current-reuse topologies typically employ a backward current-reuse configuration [15], where the supply current flows sequentially from the final stage to the front-end. This approach inherently limits the current density of the first-stage transistor, degrading the overall noise figure. To address this limitation, we propose an optimized pre-stage current-reuse architecture (Fig. 1) by strategically positioning the VDD connection at the first stage.

This modification elevates the first-stage transistor's current density to approximately 0.15 mA/μm—near the optimal noise current density [16]—thereby simultaneously enhancing both the maximum available gain (G_{max}) and the minimum noise figure (NF_{min}).

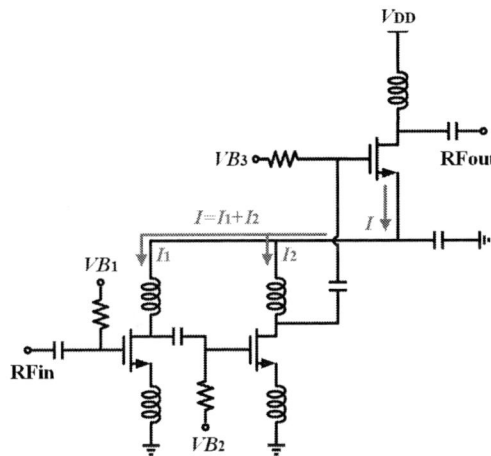

Fig. 2: Conventional current-reuse topology

Furthermore, in conventional current-reuse designs, the first-stage drain cascades the subsequent-stage circuits (Fig. 2), necessitating a large drain load inductance (L_d) to counteract the capacitive impedance of the later stages. In contrast, our proposed topology directly connects the first-stage drain to the VDD, eliminating the need for high-inductance loads. Consequently, the reduced Ld value minimizes noise contributions associated with the inductor's quality factor (Q)

In the first stage, the concept of large-sized transistors is introduced by increasing the dimensions of the first-stage transistor.It enlarges the parasitic capacitance (C_{gd}) to counteract the gate inductance, reduces the layout area, minimizes thermal noise from gate resistance (R_g), lowers the NF, and maintains sufficient gain.Fig. 3(a) shows the variations in NF and MAG as the total width (T/W) of the first-stage transistor M1 is swept from 32 μm to 96 μm. By appropriately increasing the total transistor width, the noise and gain performance of the LNA circuit can be optimized.Fig. 3(b) demonstrates that under the same total gate width, the use of multiplier effectively reduces Rg and NF. This design adopts the multiplier value of 4 to minimize first-stage noise.

The first two stages of the circuit employ a dual gm-boosting technique to passively enhance transconductance, achieving broadband impedance matching and noise matching while effectively mitigating the trade-off between gain and power consumption. Fig. 4 presents the small-signal equivalent circuit of the LNA input stage.

To simplify the formula derivation, the coupling capacitor C_1, transistor channel-length modulation effect, and body effect are neglected. The input impedance Z_{in} is calculated

979-8-3315-8850-2/25 $31.00 © 2025 IEEE

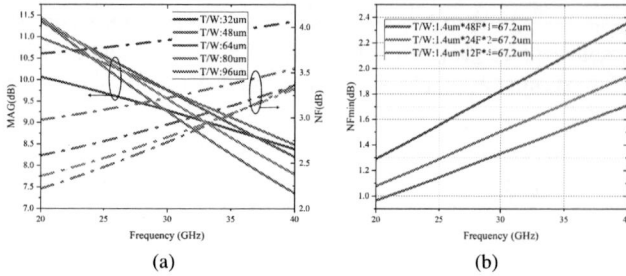

(a)   (b)

Fig. 3: Simulated results for different dimensions of the first-stage transistor:(a) MAG and NF versus T/W;(b) NF_{min} under identical T/W with varying multiplier configurations.

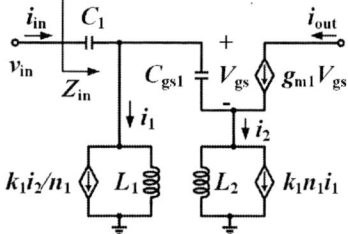

Fig. 4: Small-signal equivalent circuit of the LNA input stage

as follows [87]:

$$Z_{\text{in}} = \frac{L_2 n_1^2 s \left[1 + g_{\text{m1}} \left(k_1^2 - 1 \right) L_2 s + C_{\text{gs1}} \left(L_2 - k_1^2 L_2 \right) s^2 \right]}{1 + g_{\text{m1}} \left(1 + k_1 n_1 \right) L_2 s + C_{\text{gs1}} L_{\text{a}} s^2} \tag{2}$$

Where

$$L_{\text{a}} = \left(1 + 2k_1 n_1 + n_1^2 \right) L_2 \tag{3}$$

In Equation (2), g_{m1} and C_{gs1} represent the transconductance and gate-to-source parasitic capacitance of transistor M1, respectively. k_1 denotes the coupling coefficient between inductors L_1 and L_2, with

$$n_1 = \sqrt{L_1/L_2} \tag{4}$$

being the turns ratio.The two poles of the input impedance Z_{in} can be derived as:

$$p_{1,2} = \frac{-g_{\text{m1}} \left(1 + k_1 n_1 \right) L_2 \pm \sqrt{g_{\text{m1}}^2 \left(1 + k_1 n_1 \right)^2 L_2^2 - 4C_{\text{gs1}} L_{\text{a}}}}{2C_{\text{gs1}} L_{\text{a}}} \tag{5}$$

According to Equation (5), it can be analyzed that as the equivalent inductance L_a increases, complex-conjugate poles are more easily formed, with both the real and imaginary parts of these poles decreasing with increasing L_a. Consequently, by increasing L_a, the high-frequency poles can be shifted toward lower frequencies, positioning both poles near the operating frequency band. This results in two impedance peaks within the target frequency range, thereby achieving broadband input matching. To enhance L_a, the coupling coefficient k_1, inductance L_2, and turns ratio n_1 can be increased.

The effective transconductance of the input stage $g_{m,eff}$ is given by:

$$g_{\text{m,eff}} = \frac{g_{\text{m1}} \left(1 + k_1/n_1 \right)}{1 + g_{\text{m1}} \left(k_1^2 - 1 \right) L_2 s + C_{\text{gs1}} \left(L_2 - k_1^2 L_2 \right) s^2} \tag{6}$$

From Equation (5), it can be observed that the effective transconductance $g_{m,eff}$ is enhanced by a factor of $(1+k_1/n)$ due to the shunt-series transformer T_1. By increasing the coupling coefficient k_1 and decreasing the turns ratio n_1, superior transconductance enhancement can be achieved.

The second stage employs an interstage g_m-boosting structure, where the drain inductor of the first stage is coupled with the source degeneration inductor of the second stage. Operating on the same principle as the input parallel feedback technique, this configuration achieves broadband frequency characteristics through the coupling of only two inductors, while maintaining a compact layout area.

The linearity of the low-noise amplifier output stage significantly impacts the overall circuit linearity. To address this, the third-stage transistor employs a multigate transistor(MGTR) configuration, where multiple transistor gates are connected in parallel but operate at distinct quiescent points (e.g., saturation region, weak inversion region or triode region). As shown in Fig. 5, by carefully designing the bias conditions of these transistors, their nonlinear characteristics mutually cancel, thereby enhancing the overall circuit linearity and improving both IIP2 and IIP3 [17]. Compared to conventional feedback linearization techniques, the multi-gate structure requires no additional DC bias current, resulting in lower power consumption. Furthermore, the output matching network utilizes a transformer-based series-T network to achieve broadband output matching.

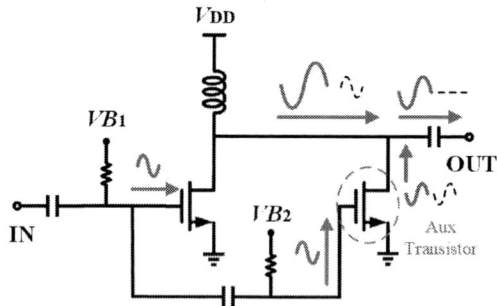

Fig. 5: Principle of MGTR technique.

III. LAYOUT SIMULATION RESULTS

The proposed Ka-band LNA, incorporating large-sized transistors, a pre-stage current-reuse technique, dual g_m-boosting technique, and a multi-gate configuration, is implemented in a 65-nm CMOS process through comprehensive simulations. The LNA layout, as illustrated in Fig. 4, occupies a compact chip area of 420 μm × 480 μm (excluding dummy), with a core area of only 0.075 mm^2. Inductors, transformers, and most radio-frequency interconnects utilize the top metal layers

979-8-3315-8850-2/25 $31.00 © 2025 IEEE

TABLE I: Comparison with Other State-of-the-Art LNAs

	This Work	[6] JSSC'20	[7] RFIC'22	[13] TMTT'24	[18] ESSERC'24	[19] TMTT'24	[20] RFIC'24
Process CMOS	65nm	65nm	65nm	65nm	55nm	40nm	40nm
VDD(V)	1	1	1	1	1.2	0.6	1
Freq.(GHz)	25.2-35.4	17.2-22	24.9-29.8	28.1-48.1	21.3-30.1	21.9-33.7	19-46
3-dB BW (GHz)	10.2	4.8	4.9	20	8.8	11.8	27
Gain (dB)	18.8	14.9	11.7	21.5	16.5	12.3	12.4
NFave (dB)	2.45	4.15	2.5	2.7	4.25	3.5	4.01
Pdc (mW)	9	1.9	3.6	22	3.38	8.5	4.4
Area (mm^2)	0.21(0.075*)	0.38	0.06	0.16	0.32	0.26*	0.16
FoM	13.1	8.8	6.7	12.6	10.5	4.6	16.8

*Core area

(M8, M9, and AP) with a trace width of 6 μm to achieve high-quality factors.To minimize gate parasitic resistance, the input-stage transistors employ specialized routing techniques, while the output-stage transistors are arranged with swapped source/drain terminals to streamline interconnects. Electromagnetic (EM) simulations of the entire passive structure (excluding active devices) were conducted, and the results are presented in Fig. 6.

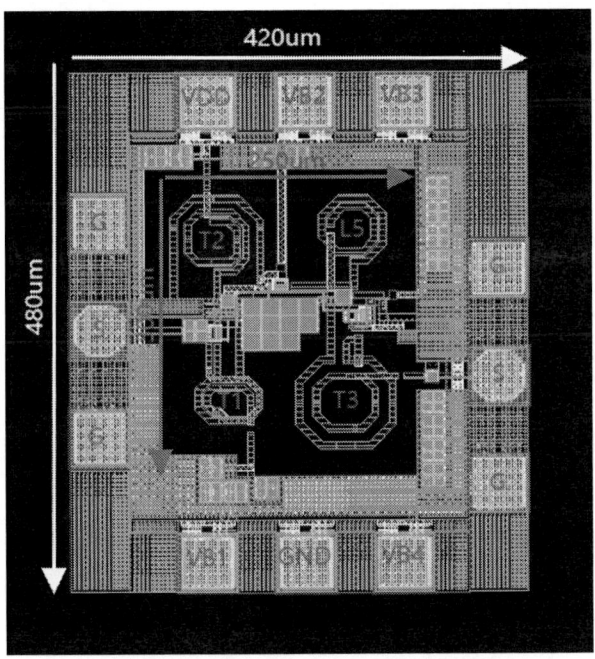

Fig. 6: Layout of the proposed LNA.

The simulated S-parameters across the 20–40 GHz frequency range are shown in Fig. 7. The proposed LNA achieves a gain of 18.8 dB at the center frequency of 29 GHz. The 3 dB gain bandwidth spans 10.2 GHz from 25.2 to 35.4 GHz, while the 1 dB bandwidth covers 4 GHz from 27 to 31 GHz. Both S11 and S22 remain below −10 dB across the 27–31 GHz band, fully encompassing the Ka-band satellite communication uplink spectrum.

The simulated NF is shown in Fig. 7, achieving a minimum value of 2.3 dB at the center frequency of 29 GHz. Across the 25.2–35.4 GHz bandwidth, the NF remains below 2.7 dB. Within the satellite communication band (27.5–31 GHz),

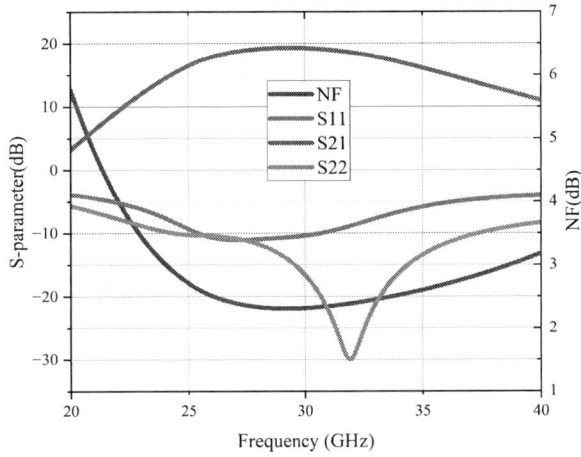

Fig. 7: Simulated S-parameters and NF of the LNA.

the NF is further reduced to less than 2.5 dB. The input 1-dB compression point (IP1dB) at 29 GHz is measured at −23.6 dBm, demonstrating sufficient linearity for satellite communication applications.

As shown in Fig. 8, the stability factor Kf is larger than 1 and the alternative stability factor B1f is larger than 0 across the operational bandwidth of 10 MHz to 40 GHz, ensuring unconditional stability over the entire frequency range.

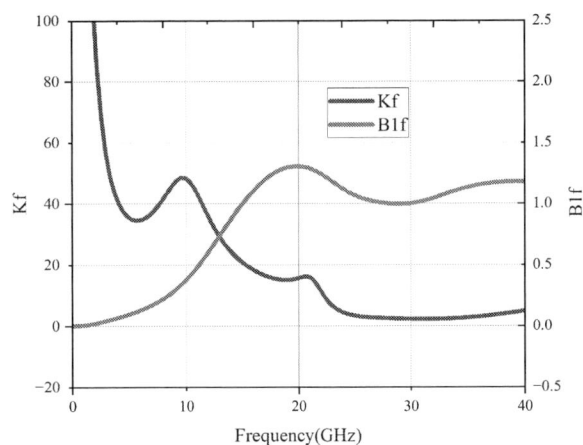

Fig. 8: Simulated stability factor (Kf) and alternative stability factor (B1f) of the LNA.

Table I summarizes the simulated performance of the proposed LNA and compares it with state-of-the-art CMOS LNAs reported in the literature. The proposed LNA achieves competitive NF, gain, and broadband characteristics under a DC power consumption of 9 mW, attaining a figure of merit (FoM) value of 13.1. The FoM is defined as follows:

$$FoM = \frac{\text{Gain(linear)} * \text{BW(GHz)}}{(F_{ave} - 1) * P_{dc}(\text{mW})} \qquad (7)$$

IV. CONCLUSION

This paper presents the design of a broadband low-noise amplifier for Ka-band satellite communication systems, incorporating a pre-stage current-reuse technique. Within the satellite communication band (27.5–31 GHz), the proposed LNA achieves a peak gain of 18.8 dB, in-band gain variation of 1 dB, and a noise figure of 2.45 dB, while consuming only 9 mW of DC power under a 1 V supply and occupying a compact core area of 0.075 μm². The prototype LNA also demonstrates a gain range of 16–18.8 dB and NF of 2.3–2.7 dB across the 25.2–35.4 GHz frequency band under the same 1 V supply, validating its suitability for Ka-band broadband satellite communication systems.

REFERENCES

[1] Moez K, Elmasry M I. A low-noise CMOS distributed amplifier for ultra-wide-bandapplications[J]. IEEE Transactions on Circuits and Systems II: Express Briefs, 2008, 55(2):126–130.

[2] J. Chang,Y. Lin. 3–10 GHz low-power, low-noise CMOS distributed amplifier using splitting-load inductive peaking and noise-suppression techniques[J]. Electronics Letters, 2009,45(20): 1033–1035.

[3] Y. Lin, J.Chang, S. Lu."Analysis and design of CMOS distributed amplifier using inductively peaking cascaded gain cell for UWB systems[J]. IEEE transactions on microwave theory and Techniques," 2011, 59(10): 2513–2524.

[4] S. Kundu and J. Paramesh, "A 17 GHz transformer-neutralized current reuse LNA and its application to a low-power RF front-end," in Proc.IEEE Radio Freq. Integr. Circuit Symp. (RFIC), May 2010, pp. 307–310.

[5] H.Chen, H.Zhu,L.Wu,W.Che, and Q.Xue, "A wideband CMOS LNA using transformer-based input matching and pole-tuning technique,"IEEE Trans. Microw. Theory Techn., vol. 69, no. 7, pp. 3335–3347,Jul. 2021.

[6] J. Zhang, D. Zhao, and X. You, "A 20-GHz 1.9-mW LNA using gm-boost and current-reuse techniques in 65-nm CMOS for satellite communications," IEEE J. Solid-State Circuits, vol. 55, no. 10,pp. 2714–2723, Oct. 2020.

[7] X. Huang, H. Jia, W. Deng, Z. Wang, and B. Chi, "28 GHz compact LNAs with 1.9 dB NF using folded three-coil transformer and dual-feedforward techniques in 65 nm CMOS," in Proc. IEEE Radio Freq. Integr. Circuits Symp. (RFIC), Denver, CO, USA, Jun. 2022,pp. 223–226.

[8] X.Fu et al.,"A 3.4 mW/element radiation-hardened Ka-band CMOS phased-array receiver utilizing magnetic-tuning phase shifter for small satellite constellation," in IEEE Int. Solid-State Circuits Conf. (ISSCC) Dig. Tech. Papers, vol. 65, Feb. 2022,pp. 90–92.

[9] X.Meng and R.Zhou,"A K-band ultra-compact gm-boost LNA using one multi-coupled transformer in 65-nm CMOS,"IEEE Microwave and Wireless Components Letters, vol. 32, no. 8, pp. 976–978, Aug. 2022.

[10] S.Kong, H.Lee, S.Jang, J.Park,K.Kim, and K.Lee,"A 28-GHz CMOS LNA with stability-enhanced gm-boosting technique using transformers," in Proc. IEEE Radio Freq. Integr. Circuits Symp. (RFIC),Jun. 2019, pp. 7–10.

[11] E.Kobal,T.Siriburanon,X.Chen,H.M.Nguyen,R.B.Staszewski,and A.Zhu,"A gm-boosting technique for millimeter-wave low-noise amplifiers in 28-nm triple-well bulk CMOS using floating resistor in body biasing," IEEE Trans. Circuits Syst. I, Reg. Papers, vol. 69, no. 12, pp. 5007–5017, Dec. 2022.

[12] J.Song, H.Choi, J. Lim and C. Kim,"W-Band Broadband CMOS LNA Using Partially Coupled Transformer and Large Transistor,"in IEEE Microwave and Wireless Technology Letters, vol. 35, no. 1, Jan. 2025.

[13] J.-H.Kim, J.-T.Son, J.-T.Lim, H.-W.Choi, and C.-Y.Kim,"Ultralow noise figure and broadband CMOS LNA with three-winding transformer and large transistor,"IEEE Trans. Microw. Theory Techn., vol. 72, no. 5, pp. 2734–2744, May. 2024.

[14] H.-W.Choi, S.Choi, and C.-Y.Kim, "Ultralow-noise figure and high gain Ku-band bulk CMOS low-noise amplifier with large-size transistor,"IEEE Microw. Wireless Compon. Lett., vol. 31, no. 1, pp. 60–63,Jan. 2021.

[15] C.-L Liang, B.-J Tang, Y. Zhao, Y. Xie, and L. Geng,High Gain and Low Power K-Band LNA With Reversed Current-Reuse Topology,"in IEEE Microwave and Wireless Technology Letters, vol. 33,no. 12,Dec. 2023.

[16] B.Wang, H.Gao, A.R.van Dommele, M.K.Matters-Kammerer, and P.G.M.Baltus,"60 GHz low-noise VGA and interpolation-based gain cell in a 40-nm CMOS technology," IEEE Trans. Microw. Theory Techn., vol. 67, no. 2, pp. 518–532, Feb. 2019.

[17] H.Zhang and E.Sánchez-Sinencio, "Linearization techniques for CMOS low noise am-plifiers: A tutorial,"IEEE Transactions on Circuitsand Systems I:Regular Papers,vol.58,no.1,pp.22–36,2011.

[18] Liang, et al. "A 21.3-to-30.1 GHz Reversed Current-reuse LNA with Load Regulation Enhancing Wideband Noise and Input Matching," IEEE European Solid-State Electronics Research Conference.

[19] S.K.Khyalia, R.H.Zele, C.Chiong and H.Wang, "A 22–33-GHz Gm-Boosting Low-Power Noise-Canceling LNA in 40-nm CMOS Process," IEEE Trans. Microw. Theory Techn..

[20] J.Fu, C.Song, Y.Wang and L.Wu, "A 4.4-mW 19–46-GHz Low-Noise Amplifier with Pole-Converging Gain Flattening and Triple-Resonance Input Matching," 2024 IEEE Radio Frequency Integrated Circuits Symposium.

2025 The 10th International Conference on Integrated Circuits and Microsystems

A 0.4 to 8GHz Broadband High-linearity I/Q Active Mixer in 130nm BiCMOS Technology

Qipeng Li[1], Zhenghao He[1], Chun Zhu[1], Yifan Huang[2], Qin Li[1,*]

[1] School of Information Science and Engineering, Southeast University, Nanjing, China

[2] Shandong University, Weihai, China

*101010761@seu.edu.cn

Abstract—**This paper presents a 0.4 to 8 GHz Gilbert-cell up-conversion mixer designed in a 130-nm SiGe BiCMOS process. The mixer translates baseband signals to the radio frequency (RF) signals. To enhance the third-order intermodulation (IM3) performance, a collector cross-coupling technique is employed across the transconductance amplifiers. Furthermore, a series resistor-inductor (RL) network is utilized at the load to boost the high-frequency gain, thereby improving in-band gain flatness. The designed mixer achieves an output 1-dB compression point(OP1dB) of 10.5 dBm and an output third-order intercept point(OIP3) of 22.07 dBm at 1 GHz. The carrier suppression ratio is better than 52 dBc across the entire LO frequency range of 0.4 GHz to 8 GHz. Image rejection ratio exceeds 45 dBc. The chip occupies a core area of 0.15 mm² and consumes 295 mW of DC power.**

Keywords—mixer,130nm BiCMOS, wideband,high linearity

I. INTRODUCTION

In recent years, the advent of the artificial intelligence and big data era has driven continuous expansion in the technological applications within the wireless communication domain, propelling significant societal advancement. Amidst this rapid development, various countries have adopted distinct communication standards utilizing different radio frequency bands. For instance, the latest Wi-Fi standards now support not only the 2.4 GHz and 5 GHz bands but also the new 6 GHz band. This proliferation poses greater demands on the design of radio frequency (RF) front-end chips[1]. Achieving ultra-wideband operation and high linearity for transmitters has emerged as a critical research focus and prevailing trend,essential for adapting to diverse communication standards. As the mixer is a key module within the transmitter responsible for frequency conversion, its performance directly impacts the overall transmitter efficacy[2-5]. Consequently, this paper investigates mixer linearization techniques and presents the design of an ultra-wideband, high-linearity mixer.

II. CIRCUIT DESIGN

Fig.1 shows the structure of quadrature modulator. The mixer translates the spectrum of intermediate-frequency(IF) baseband signal to radio frequency(RF) and delivers the output to balun. The schematic topology of the quadrature mixer circuit is shown in Fig.2. The overall architecture employs a

Gilbert mixer cell. Compared to passive mixers, this topology provides sufficient conversion gain. Furthermore, it offers improved port-to-port isolation between the local oscillator (LO) and the output port, as well as between the LO and the input port. The circuit comprises a transconductance amplifier stage, a switching stage, a load network, and an output buffer.

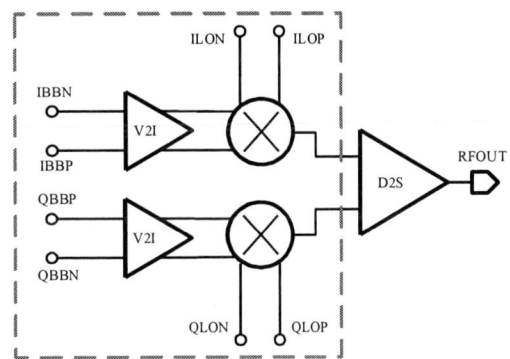

Fig. 1. Diagram of the IQ modulator

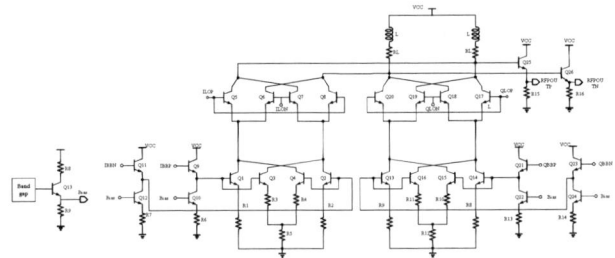

Fig. 2. 0.4GHz-8GHz IQ mixer topology

A. Transconductance Amplifier Design

The structure of transconductance amplifier with a collector cross-coupling linearity enhanced g_m cell is shown in Fig.3. At the baseband voltage input, an emitter follower(EF) is employed to level-shift the DC voltage down to 1.3V. The transconductance stage is constructed with a pair of differential common emitter(CE) amplifiers with degenerated resistors. The collectors of these two amplifiers are cross-coupled. The main transconductance amplifier (MTA) is formed by Q1 and Q2 , meanwhile the auxiliary transconductance amplifier(ATA) is formed by Q3 and Q4. The collectors of MTA and ATA are cross-coupled. MTA converts the baseband voltage signal into a current signal, which is then injected into the quadrature

The Chinese Key R&D Program of Jiangsu Province under Grant BE2022052-2.

979-8-3315-8850-2/25 $31.00 © 2025 IEEE

switching stage. ATA actively cancels the third-order intermodulation(IM3) components generated by MTA.

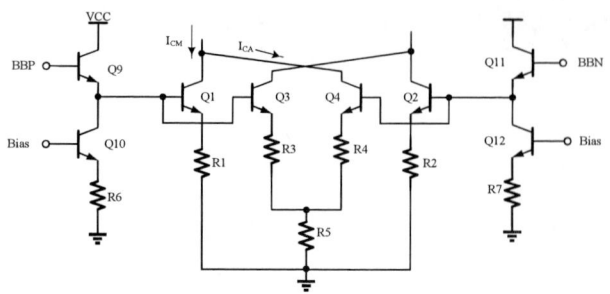

Fig. 3. Transconductor amplifier topology

The analysis of transconductance amplifier proceeds as follows. The transfer characteristics of CE amplifier with degenerated resistor is

$$i_C = I_S exp\left(\frac{v_{BE} - i_C R_E}{V_T}\right) \qquad (1)$$

where equal value of I_S are assumed and V_T is the thermal voltage. R_E is the degenerated resistor. The collector current exhibits an implicit functional dependence on the base-emitter voltage. The entire transconductance amplifier can be equivalent to a nonlinear two-port network. The collector current can be expressed as

$$i_C = a_0 + a_1 v_{BE} + a_2 v_{BE}^2 + a_3 v_{BE}^3 \cdots \qquad (2)$$

where a_n denotes the nth-order transconductance coefficients. These coefficients can be derived by differentiating the collector current i_C with respect to v_{BE} in(1) at the static operating point. a_n can be expressed as

$$a_1 = \frac{di_C}{dv_{BE}}\bigg|_{I_{CQ}} = \frac{I_{CQ}}{V_T + I_{CQ} R_E} \qquad (3)$$

$$a_3 = \frac{1}{6}\frac{d^3 i_C}{d^3 v_{BE}}\bigg|_{I_{CQ}} = \frac{1}{2}\frac{I_{CQ}V_T}{(V_T + I_{CQ}R_E)^4}\left[\frac{1}{3} - \frac{I_{CQ}R_E}{V_T + I_{CQ}R_E}\right] \qquad (4)$$

a_1 represents the transconductance (g_m) of the amplifier, while a_3 represents its third-order transconductance coefficient. In order to achieve the elimination of third-intermodulation product, the third-order transconductance coefficients of MTA and ATA should be equal. According to (4), the degenerated resistors and static collector current of MTA and ATA can be chosen to satisfy the following relationship:

$$\left(\frac{R_{EM} + a_{1M}}{R_{EA} + a_{1A}}\right)^5 \left(\frac{R_{EA} - 2a_{1A}}{R_{EM} - 2a_{1M}}\right) = \left(\frac{I_A}{I_M}\right)^3 \qquad (5)$$

where the subscript "M" represents MTA and "A" represent ATA. In order to enhance g_m of entire transconductance amplifier, ATA always works at low bias current and a large degenerated resistor[6].

B. Switching Pair Design

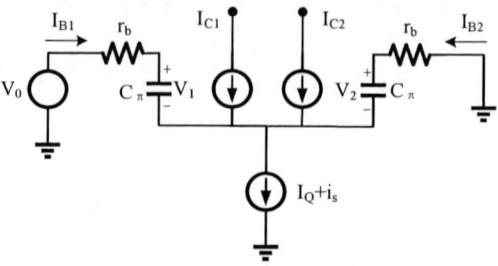

Fig. 4. Large-signal equivalent circuit of switching pair

The frequency conversion mechanism of the mixer hinges on the switching pairs that, driven by the LO signal, periodically turn on and off to steer the current of transconductance amplifiers. This action routes the baseband current generated by the transconductance amplifier to the load network. The large-signal equivalent circuit of one pair of transistors in the quad is shown in Fig.4. The large-signal input capacitance C_π represents the total charge storage capacitance,consisting of the emitter-base depletion capacitance C_{Je} and the base charging diffusion capacitance C_b. The effect of C_μ in the quad is small and it is neglected. The base resistor r_b are assumed to be equal in two transistors. I_Q is the static current of transconductance amplifier and i_s is the product of gm cell. Based on Kirchhoff's Current Law and the transistor's transfer characteristics, IM3 components generated by the switching pair can be expressed as:

$$IM_3 \propto \left(\frac{I_{SM}}{I_Q}\right)^2 f_1\left(\frac{V_{OM}}{V_T}\right)\left[f_2(A) + f_2(B)\right] \qquad (6)$$

where

$$A = \omega_0 \tau_1 r_b I_Q / V_T \qquad (7)$$

$$B = \omega_0 C_{Je} V_T / I_Q \qquad (8)$$

τ_1 is the transit time of switching transistor.Assuming a sinusoidal LO signal expressed as $V_O = V_{OM}cos\omega_0 t$. I_{SM} is the amplitude of baseband current signal. The dominant sources of third-intermediate nonlinear distortion generated by switching pair are captured by(6),(7) and (8). Firstly, nonlinear charge injection through C_{Je} degrades the switching pairs' linearity, generating distortion during turn-on/turn-off transients. The second major source of distortion is the static bias current I_Q.Therefore, the LO drive level and switching pair must be optimized to minimize parasitic capacitance and balance conversion gain[7].

C. Load Network Design

The parasitic capacitance in the switching pair causes simultaneous conduction of transistors,while the base capacitance of buffer stage also degrades the circuit bandwidth. A series resistor-inductor(RL) network in Fig.2 is employed at the load stage to compensate the gain at high frequencies. The relationship between gain peaking characterstics and different inductance values is illustrated in Fig.5.

979-8-3315-8850-2/25 $31.00 © 2025 IEEE

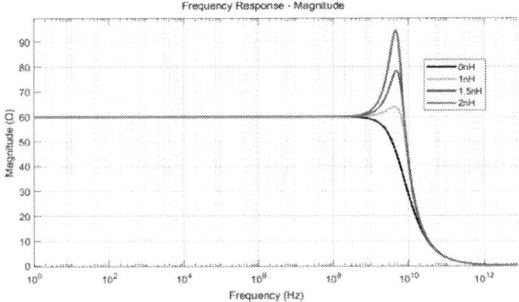

Fig. 5. Computed magnitude of frequency response

D. Buffer Design

Since mismatched load impedances at mixer's differential output ports degrade the symmetry between I an Q paths, the emitter follower(EF) in Fig.6 is used to provide high impedance for output differential ports. In addition, the EF provides a relatively low output impedance ,which minimizing loading effects on subsequent circuit stages.

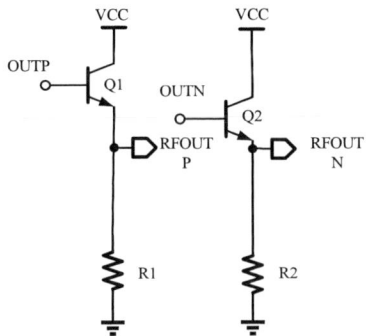

Fig. 6. Output buffer topology.

III. SIMULATION RESULT

In this paper, the circuit is designed in 130nm SiGe BiCMOS technology. The layout is shown in Fig.7. In order to minimize quadrature errors and differential DC mismatch, the I/Q paths maintain strict layout symmetry while all transistors in transconductor amplifier implement common-centroid placement with interdigitated dummy structures. The power consumed by the circuit is 295mW and the size of the layout is 800μm×700μm in area including ESD-protecting pads.The mixer core is biased at 5 V.

Assume the LO frequency is f_{LO} and baseband frequency is f_{BB}=20MHz. Fig.8 shows the simulated output power 1-dB compression point(OP1dB) and the power gain versus frequency of LO. OP1dB is about 10.5dBm at low frequency and reduces to about 3dBm at f_{LO}=8GHz. The conversion power gain decrease with frequency and exhibits a fluctuation of 6.4dB across the operational frequency band. This design provides the entire modulator with an adequate linear dynamic range.

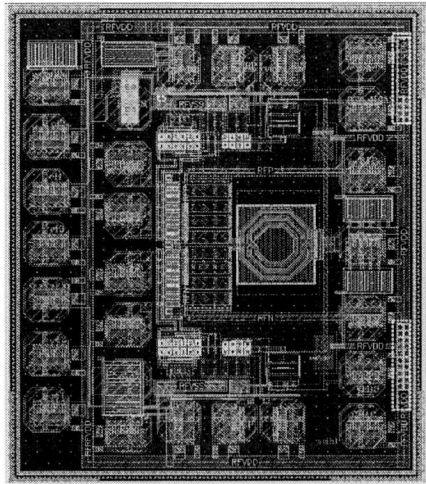

Fig.7.Circuit layout of this article.

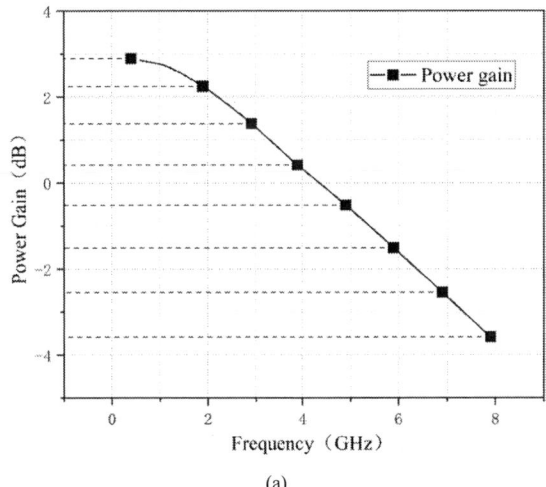

(a)

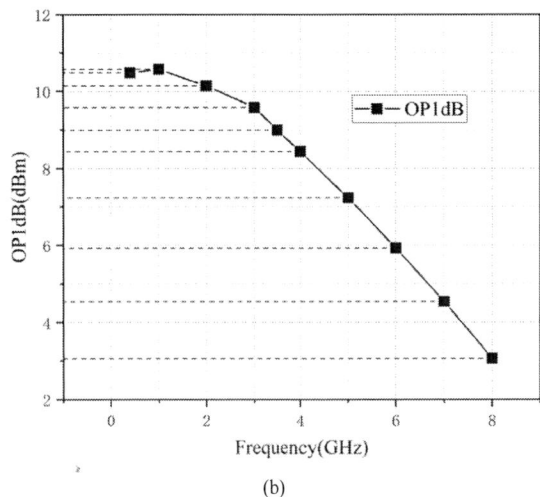

(b)

Fig. 8. Simulated result:(a) power gain and (b) OP1dB versus frequency.

979-8-3315-8850-2/25 $31.00 © 2025 IEEE   415

Fig.9 shows the sideband suppression(f_{LO}-f_{BB}) and carrier suppression(f_{LO}).It can be seen that the sideband suppression is better than 45dBc.and the carrier suppression is better than 50dBc.The simulated isolation performance validates the geometric and electrical symmetry implemented in the circuit layout.

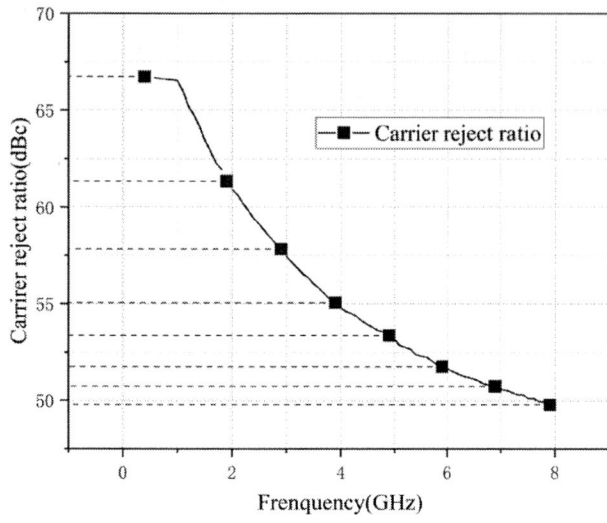

(a)

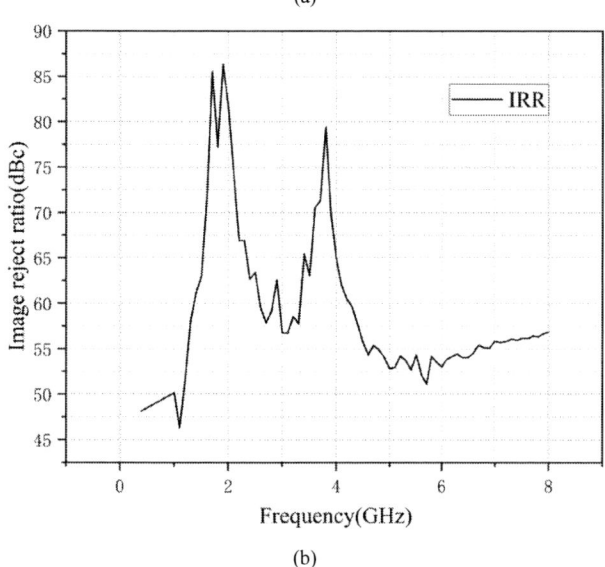

(b)

Fig. 9. Simulated result:(a) carrier reject ratio and (b) IRR.

The simulation results of linearity are illustrated in Fig.10, where two separated tones (i.e.,20MHz and 21MHz) are used as baseband input ports excitation. The mixer achieve the simulated output third-order intercept point(OIP3) of 22.36dBm at 0.4GHz and OIP3 is better than 14.35dBm in the entire operation band.

TABLE I shows a performance comparison with other quadrature mixer design. It can be found in the comparison that the designed quadrature mixer shows superior performance in linearity, image reject ratio linearity and dynamic range in broader operating bandwidth.

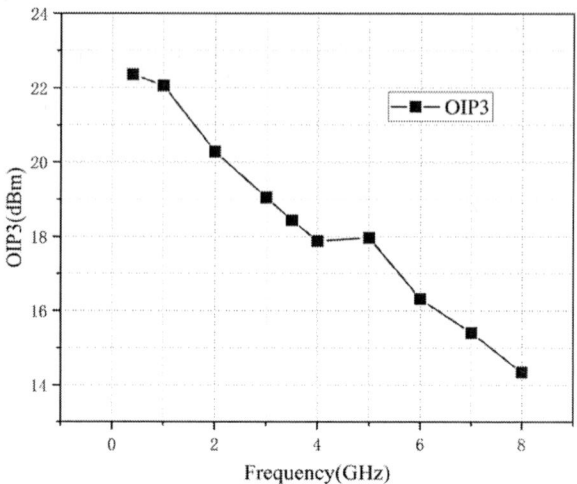

Fig. 10. Simulated result:OIP3 versus frequency

TABLE I.   PERFORMANCE COMPARISON OF MIXERS

	This work	*[8]*	*[9]*	*[10]*
Technology	0.13µm SiGe BiCMOS	0.35µm SiGe BiCMOS	0.8µm SiGe BiCMOS	0.13µm CMOS
Freq.[GHz]	0.4~8	0.8~2	0.75~3.6	2.4
Gain [dB]	-3.6-2.8	-2~0	-10~7	22
OP1dB [dBm]	3~10	-10~-3	-10~-3	1.4
OIP3[dBm]	14~22	-12~0	-11~-6	38
IRR[dBc]	44~85	33~47	30(Max)	-
LOLEAK[dBc]	52~60	28~49	-	-

IV. CONCLUSION

This paper presents a 0.4-8GHz quadrature mixer designed in 0.13µm SiGe BiCMOS technology. The symmetric layout topology provides enhanced isolation performance. Taking the advantage of transconductance amplifier, excellent linearity performance and linear dynamic range are achieved at 5-V power supply.

REFERENCES

[1] H. Murata et al., "Antenna-Coupled Electrode Electro-Optic Modulator and 28 GHz-Band Wireless Signal Transfer over Fiber for 5G Mobile Systems," 2019 24th OptoElectronics and Communications Conference (OECC) and 2019 International Conference on Photonics in Switching and Computing (PSC), Fukuoka, Japan, 2019, pp. 1-3J. Clerk Maxwell, A Treatise on Electricity and Magnetism, 3rd ed., vol. 2. Oxford: Clarendon, 1892, pp.68–73.

[2] X. S. Loo, K. Seng Yeo, M. Z. Win, Z. Li, X. Yu and J. -M. Chen, "A K-Band Differential SiGe Stacked Power Amplifier Based on Capacitive Compensation Techniques for Gain Enhancements," 2019 IEEE 62nd International Midwest Symposium on Circuits and Systems (MWSCAS), Dallas, TX, USA, 2019, pp. 295-298.

[3] Keng Leong Fong and R. G. Meyer, "Monolithic RF active mixer design," in IEEE Transactions on Circuits and Systems II: Analog and Digital Signal Processing, vol. 46, no. 3, pp. 231-239, March 1999.

[4] A. Mirzaei, H. Darabi, J. C. Leete, X. Chen, K. Juan and A. Yazdi, "Analysis and Optimization of Current-Driven Passive Mixers in Narrowband Direct-Conversion Receivers," in IEEE Journal of Solid-State Circuits, vol. 44, no. 10, pp. 2678-2688, Oct. 2009.

[5] Rajeshkumar, "REDUCE POWER CONSUMPTION FOR DIGITAL CMOS CIRCUITS USING DVTS ALGORITHAM," International Journal of Electrical and Electronic Engineering & Telecommunications, Vol. 1, No. 1, pp. 376-383, March 2015.

[6] Y. Peng, L. Zhang, J. Fu and Y. Wang, "Analysis and Design of a Broadband SiGe HBT Image-Reject Mixer Integrating Quadrature Signal Generator," in IEEE Transactions on Microwave Theory and Techniques, vol. 64, no. 3, pp. 688-698, March 2016.

[7] R. G. Meyer, "Intermodulation in high-frequency bipolar transistor integrated-circuit mixers," in IEEE Journal of Solid-State Circuits, vol. 21, no. 4, pp. 534-537, Aug. 1986.

[8] T. Tsukahara, M. Ishikawa and M. Muraguchi, "A 2-V 2-GHz Si-bipolar direct-conversion quadrature modulator," in IEEE Journal of Solid-State Circuits, vol. 31, no. 2, pp. 263-267, Feb. 1996.

[9] E. Tiiliharju and K. Halonen, "A 0.75-3.6GHz SiGe direct-conversion quadrature-modulator," ESSCIRC 2004 - 29th European Solid-State Circuits Conference (IEEE Cat. No.03EX705), Estoril, Portugal, 2003, pp. 565-568.

[10] N. Vitee, H. Ramiah, P. -I. Mak, J. Yin and R. P. Martins, "A 3.15-mW +16.0-dB IIP3 22-dB CG Inductively Source Degenerated Balun-LNA Mixer With Integrated Transformer-Based Gate Inductor and IM2 Injection Technique," in IEEE Transactions on Very Large Scale Integration (VLSI) Systems, vol. 28, no. 3, pp. 700-713, March 2020

A Compact Active Phase Shifter for 110-150 GHz Phased Arrays in 130 nm BiCMOS

Shuguang Liu
School of Electronic Science and Engineering
University of Electronic Science and Technology of China
Chengdu,China
2687996784@qq.com

Jinhao Zhou
School of Electronic Science and Engineering
University of Electronic Science and Technology of China
Chengdu,China

Yuyang Liu
School of Electronic Science and Engineering
University of Electronic Science and Technology of China
Chengdu,China

Hu Lian
School of Electronic Science and Engineering
University of Electronic Science and Technology of China
Chengdu,China

Yaxin Zhang
School of Electronic Science and Engineering
University of Electronic Science and Technology of China
Chengdu,China

Ziqiang Yang
School of Electronic Science and Engineering
University of Electronic Science and Technology of China
Chengdu,China

Abstract—A D-band vector modulator-type phase shifter (VMPS) is presented using an 130-nm BiCMOS process. A transformer-based fully differential hybrid is designed to generate high-precision orthogonal signals, which presents a compact layout and effectively widens the operating bandwidth. A D-band output signal is generated after the IQ signal is synthesized by the vector summation network controlled by the transistor base control signal. The measured results exhibit that the proposed circuit achieves a 360° phase shifting range with 5.625° step in the wideband from 110 GHz to 170 GHz. The measured RMS phase and amplitude error are under 1.38° and 0.76 dB at 145 GHz, respectively. The average gain is about 1.16 dB. Thanks to the proposed quadrature generator, the core area of the phase shifter is $230 \times 180 \ \mu m^2$.

Keywords—BiCMOS, D-band, phase shifter, vector modulator.

I. INTRODUCTION

In the last few years, the D-band (110–170 GHz) is being investigated as a candidate to support the very high capacity backhaul links required for 5G and beyond [1], [2]. An advantage of operating in the high millimeter-wave (mmW) region is that compact phased-arrays can be built to provide beam-steering functionality and enable new flexible use of spectrum and network solutions [3].

Additionally, as frequency increases, the phased array antenna spacing decreases dramatically and the problem of matching the antenna array to the circuit size becomes more prominent [4].The optimal spacing $d = \lambda/2$ is 1.07 mm at 140 GHz. Consequently, to implement large-scale one-dimensional or two-dimensional arrays, the distance between chip channels must be compatible with the spacing between antennas, underscoring the critical importance of circuit area[5], [6]. Therefore, achieving excellent RF performance in the D-band under

minimal area constraints is a key issue in the research of phased array chips.

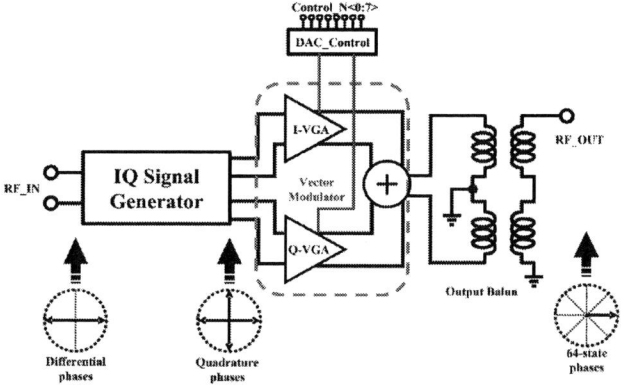

Fig. 1. Architecture of the proposed Phase Shifter with a compact QG.

In this paper, an E-band active vector-sum phase shifter is demonstrated for an RF phased-array receiving system in this paper. To achieve a compact layout and widen the operating bandwidth, a fully differential hybrid is proposed to realize an in-phase and quadrature signal generator (I/Q generator). Additionally, the vector modulator is implemented by a differential common source cross-coupled structure, which helps to maintain the insertion phase of VGA invariant over gain settings. The measurement results exhibit that the circuit achieves 5.625° minimal phase step with the full phase range, and the operating frequency band covers 110-170 GHz. The RMS phase error is less than 2.7°(1.38° at the center frequency), and the RMS amplitude error is less than 1.6dB in the entire frequency band. While the core area is only $230 \times 180 \mu m^2$.

979-8-3315-8850-2/25 $31.00 © 2025 IEEE

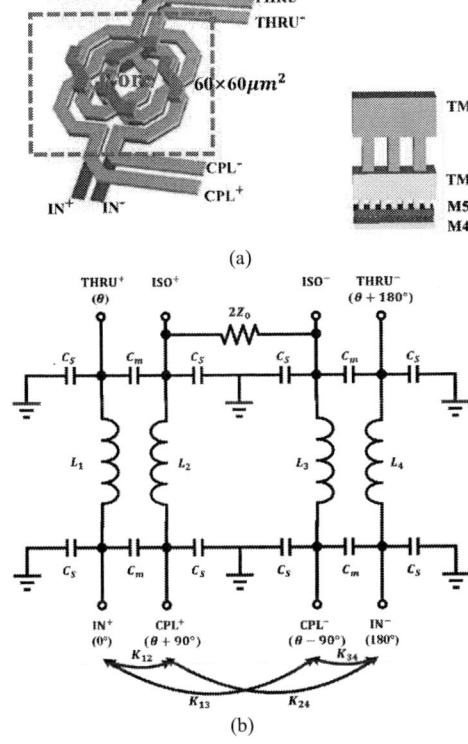

(a)

(b)

Fig. 2. (a) 3-D view of the QG. (b) Equivalent circuit of the QG.

II. CRICUIT AND DESIGN

As shown in Fig. 1, the proposed phase shifter consists of an output matching network, an IQ quadrature signal generator, two VGAs, a current combiner, and two DAC control circuits.

The input differential signal is processed through an IQ signal generator, which converts it into four quadrature signals. The I-VGA and Q-VGA then independently modulate the amplitude of the I and Q signals, respectively. These signals are subsequently combined through a current combiner circuit. A DAC controlled by the SPI interface converts the digital control signal to an analogue output to adjust the VGA amplitude and manage quadrant switching. To facilitate measurement, a differential-to-single-ended balun is integrated at the output, converting the differential signal into a single-ended format while simultaneously achieving output imped-ance matching.

A. I/Q Signal Generator

This work uses a low-loss, compact quadrature signal gener-ator (QG), as shown in Fig. 2 (a). The designed QG comprises two stacked directional couplers arranged side-by-side to form a fully differential structure. The four ports correspond to the input (IN$^+$, IN$^-$), through (THRU$^+$, THRU$^-$), coupled (CPL$^+$, CPL$^-$), and isolation (ISO$^+$, ISO$^-$) terminals. The transformer's primary and secondary windings are implemented using M5 and TM1 metal layers, with interleaved routing to maintain symmetry. TM2 and M4 layers are employed for metal crossovers. All traces feature uniform line widths ($6\mu m$) and spacing ($3\mu m$), resulting in an extremely compact layout with a core area of only $60 \times 60\mu m^2$. This

design achieves low loss, small area, and wide bandwidth while maintaining excellent amplitude and phase chara-cteristics. To facilitate the analysis and design of the QG, an equivalent circuit model of the four-port fully differential quadrature coupler is presented. As shown in Fig. 2 (b), $L_1\sim L_4$ represent the inductors in the transformer model, C_m denotes the parasitic capacitance between the stacked transformers, and C_s represents the parasitic capacitance from the ports to ground. A $2Z_0$ resistor serves as the isolation resistance. With a differential mode operation, the current directions in the two transformers are contrary, making it feasible to fold the two transformers into one structure. Due to the tight side-by-side placement of the two quadrature couplers, additional magnetic couplings (K_{13} and K_{24}) are introduced. This configuration enhances the magnetic coupling between the coils while L_{1-4}, C_m and C_s together form a high-order resonant network that extends the bandwidth and reduces the layout area. Based on the proposed equivalent circuit model, a conceptual design was carried out, resulting in $L_{1-4} = 275$ pH, $K_{12} = K_{34} = 0.785$, and additional coupling coefficients $K_{13} = K_{24} = 0.165$, which were used as initial values for electromagnetic model implementation.

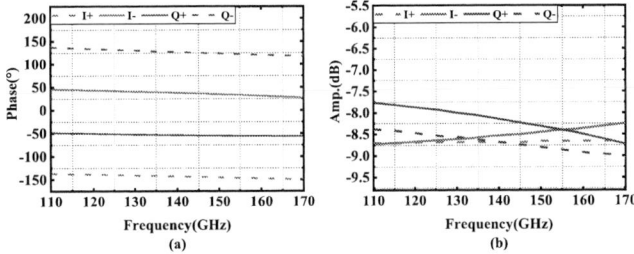

Fig. 3. (a) Simulated phase of the QG. (b) Simulated insertion loss of the QG.

The simulation results for the proposed structure are shown in Fig. 3. The average insertion loss across the operating bandwidth is 2.6 dB, with an amplitude imbalance of less than 1 dB and a phase error below 3°. These results demonstrate that the structure exhibits excellent amplitude and phase characteristics over a wide frequency range.

B. Gain Control Unit And Combiner Network

The circuit schematic of the gain control unit is illustrated in Fig. 4, featuring two VGAs that function as gain controllers for the I and Q signal paths, respectively. Each VGA employs the same Gilbert-based current-steering structure.

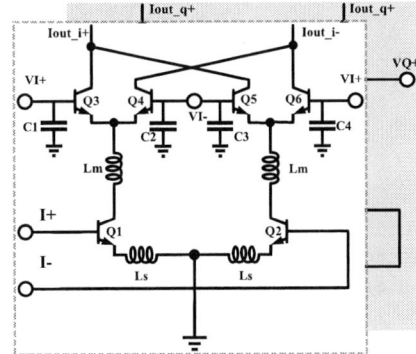

Fig. 4. Schematic of the gain control unit with emitter inductors.

979-8-3315-8850-2/25 $31.00 © 2025 IEEE

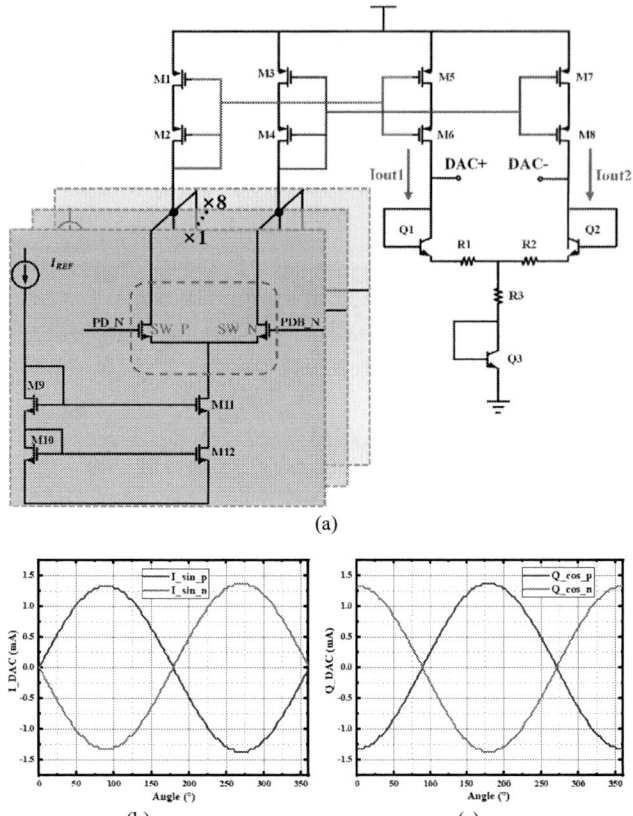

(a)

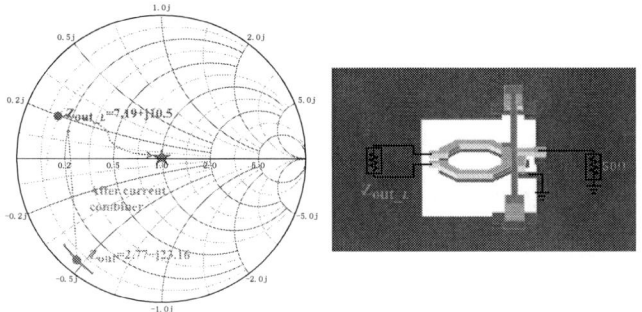

(b)   (c)

Fig. 5. (a) Schematic diagram of 8-bit DAC control circuit. (b) Simulated sin. differential control signal output by the I-DAC. (c) Simulated cos. differential control signal output by the Q-DAC.

The design employs sine and cosine control signals to regulate $V_{I,Q}$ to control phase while maintaining the output signal on the constant-gain circle (as shown in Fig. 5).

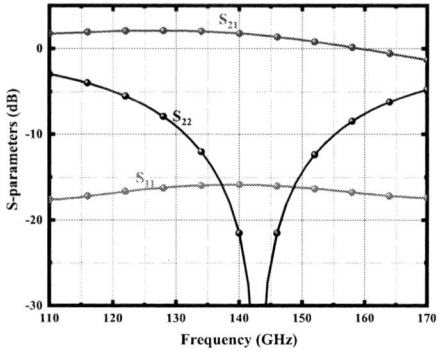

Fig. 6. Simulated input and output impedance with passive combination network.

The IQ control DAC of the phase shifter is shown in Fig. 5. Moreover, the design ensures a constant tail current, which maintains a steady collector current in the common-emitter transistors across all phase states, thereby preserving a consistent input impedance for the Gilbert cell.

This combiner network ensures impedance flatness across the operating bandwidth, thereby achieving broadband matching. As shown in Fig. 6, the output impedance before and after the current combiner is compared. After the transformation by the current combiner, the output impedance exhibits a significantly reduced Q-factor, with a notably smaller impedance variation across the operational bandwidth. Therefore, by incorporating a transformer circuit after the current combiner, broadband impedance matching can be readily achieved while simultaneously facilitating the conversion from differential signals to single-ended signals.

Fig. 7 shows the simulated characteristics of the VMPS, demonstrating an average small-signal gain of 1.16 dB with a 3-dB bandwidth from 110 to 170 GHz.

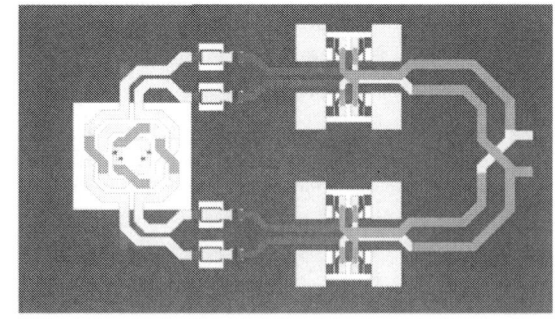

Fig. 7. Simulated S-parameters of the VMPS.

III. MEASUREMENT RESULTS

Fig. 8 presents the layout of the proposed VM, fabricated using a 130-nm BiCMOS process. The VM core size is only 180 $\times$ 230 μm^2. The chip consumes a dc power of 198 mW from a 3.3-V supply voltage.

Fig. 9 (a) illustrates the measured and simulated polar plot of the receiver front end at 140 GHz. The measured RMS gain and phase errors are shown in Fig. 9 (b). It is evident that the RMS phase error is less than 2° within the 110-150 GHz frequency band, and the RMS gain error is less than 1.1 dB within the 110-150 GHz frequency band.

Fig. 8. Layout of the proposed D-band VM.

979-8-3315-8850-2/25 $31.00 © 2025 IEEE

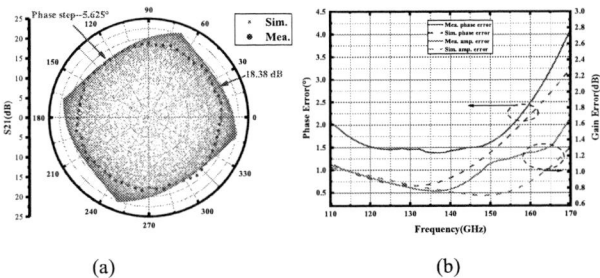

(a)   (b)

Fig. 9. (a)Measured and simulated Constellation diagram of the RX front end at 140GHz.(b)Measured RMS phase and amplitude errors for the 6-bit phase shifting without calibration.

Fig. 10 (a) shows the 64 phase state curves for the phase test. The phase shifter covers a full 360° range, achieving a 6-bit phase resolution. Due to the phase-frequency characteristics of the system, the phase varies at different frequencies. Fig. 10 (b) shows the gain curves for the 64 phase states tested, with an average gain of 18 dB and a 3 dB bandwidth exceeding 30 GHz.

Table I compares the size of the RF core circuit in this chip with phase shifter circuits operating in the D-band and higher frequency bands. It is evident that compared to other phase shifters in Table I, the phase shifter in this design occupies much smaller area, while the total area of the RF core is slightly larger than that of other individual phase shifters. This demonstrates the highly compact area of the proposed phased-array receiver front end.

IV. CONCLUSION

In this paper, a 6-bit D-band active vector-sum phase shifter is designed. The fully differential hybrid is utilized to reduce the layout area and widen the operating bandwidth effectively. It achieves reasonable amplitude-phase characteristics.

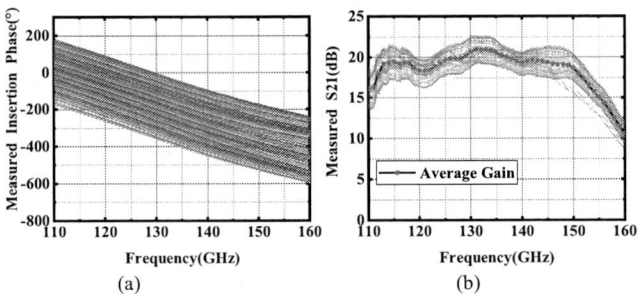

(a)   (b)

Fig. 10. Measured insertion phases (a) and small-signal gain (b) of the RX over 64 phase states.

TABLE I. RF CORE CIRCUIT AREA COMPARISON IN D-BAND AND ABOVE

Ref.	This Work	[7]	[8]	[9]	[10]
Tech.	130nm SiGe BiCMOS	130nm SiGe BiCMOS	130nm SiGe BiCMOS	130nm SiGe BiCMOS	130nm SiGe BiCMOS
Freq. (GHz)	115-150	130-150	162-190	200-250	140-170
Circuit	VMPS	VMPS	VMPS	VMPS	VMPS
Core Area (mm²)	0.042	0.075	0.07	0.09	0.342*

* Estimated from chip picture.   # Only PS

ACKNOWLEDGMENT

This work was supported in part by the National Natural Science Foundation of China (NSFC) under Grant 62131007, Grant U20A20212, Grant 61931006.

REFERENCES

[1] R. J. Emery and A. M. Zavody, "Atmospheric propagation in the frequency range 100-1000 GHz," Radio and Electronic Engineer, vol. 49, no. 7.8, pp. 370–380, Jul. 1979.

[2] M. G. L. Frecassetti et al., "D-band radio solutions for beyond 5G reconfigurable meshed cellular networks," in Proc. 16th Int. Symp. Wireless Commun. Syst. (ISWCS), Aug. 2019, pp. 427–431.

[3] P.-H. Kuo and A. Mourad, "Millimeter wave for 5G mobile fronthaul and backhaul," in Proc. Eur. Conf. Netw. Commun. (EuCNC), Jun. 2017, pp. 1–5.

[4] B. Sadhu, X. Gu, and A. Valdes-Garcia, "The more (antennas), the merrier: A survey of silicon-based mm-wave phased arrays using multi- IC scaling," IEEE Microw. Mag., vol. 20, no. 12, pp. 32–50, Dec. 2019.

[5] J. Zhang et al., "24.2 A Scalable 134-to-141GHz 16-Element CMOS 2D λ/2-Spaced Phased Array," in IEEE Int. Solid-State Circuits Conf. (ISSCC) Dig. Tech. Papers, Feb. 2024, pp. 414-416.

[6] A. Ahmed, L. Li, M. Jung and G. M. Rebeiz, "A 140 GHz Scalable On-Grid 8×8-Element Transmit-Receive Phased-Array with Up/Down Converters and 64QAM/24 Gbps Data Rates," in Proc. IEEE Radio Freq. Integr. Circuits Symp. (RFIC), Jun. 2023, pp. 93-96.

[7] A. Ahmed, L. Li, M. Jung, S. Li, D. Baltimas and G. M. Rebeiz, "140-GHz 2-D Scalable On-Grid 8×8-Element Transmit–Receive Phased Arrays With Up/Down Converters Demonstrating a 5.2-m Link at 16 Gbps," IEEE Trans. Microw. Theory Techn., vol. 72, no. 5, pp. 2852-2868, May 2024.

[8] R. Ahamed, M. Varonen, D. Parveg, M. Najmussadat, Y. Tawfik and K. A. I. Halonen, "A 200–250-GHz Phase Shifter Utilizing a Compact and Wideband Differential Quadrature Coupler," IEEE Microw. Wireless Compon. Lett., vol. 32, no. 7, pp. 883-886, July 2022.

[9] P. V. Testa, C. Carta and F. Ellinger, "A 160 – 190-GHz Vector- Modulator Phase Shifter for Low-Power Applications," IEEE Microw. Wireless Compon. Lett., vol. 30, no. 1, pp. 86 – 89, Jan. 2020.

[10] M. Elkhouly et al., "D-band phased-array TX and RX front ends utilizing radio-on-glass technology," in Proc. IEEE Radio Freq. Integr. Circuits Symp. (RFIC), Aug. 2020, pp. 91–94.

Interface Quality Improvement in Planar SiC MOSFETs Using Supercritical Fluid Nitriding

Xiaoqing Bao
School of Electronic and
Computer Engineering
Guangdong Provincial Key
Laboratory of In-Memory
Computing Chips
Peking University
Shenzhen 518055, China
xqbao25@stu.pku.edu.cn

Lei Li*
College of Integrated Circuits
and Optoelectronic Chips
Shenzhen Technology University
Shenzhen 518118,China
lilei@sztu.edu.cn

Lei Lu
School of Electronic and
Computer Engineering
Guangdong Provincial Key
Laboratory of In-Memory
Computing Chips
Peking University,
Shenzhen 518055, China
lulei@pku.edu.cn

Kuan-Chang Chang*
School of Electronic and
Computer Engineering
Guangdong Provincial Key
Laboratory of In-Memory
Computing Chips
Peking University,
Shenzhen 518055, China
kcchang@pkusz.edu.cn

Abstract—This study investigates the application of supercritical fluid nitriding treatment to enhance the interface quality and electrical performance of planar silicon carbide (SiC) metal-oxide-semiconductor field-effect transistor (MOSFET) devices. Comprehensive electrical characterization was conducted through systematic comparison of output characteristics, transfer characteristics, and capacitance-voltage measurements before and after treatment to evaluate the effectiveness of this novel approach. The experimental results demonstrate substantial improvements in key device parameters following supercritical fluid nitriding treatment. The saturation drain current increased by approximately 100% at the highest gate voltage level, while transconductance characteristics showed enhanced gate control efficiency and carrier transport properties. The on-resistance decreased from 1270 mΩ to 1090 mΩ, representing a 14% reduction in conduction losses. Capacitance-voltage measurements revealed reduced hysteresis and improved curve symmetry, confirming lower interface trap density and enhanced interface quality. The treatment process demonstrated excellent compatibility with existing fabrication technologies while maintaining gate oxide integrity. These findings establish supercritical fluid nitriding as a promising low-temperature approach for addressing the fundamental interface quality challenges in SiC MOSFET technology, offering significant potential for advancing wide-bandgap semiconductor applications in power electronics.

Keywords—*silicon carbide, MOSFET, supercritical fluid nitriding*

I. INTRODUCTION

As a representative third-generation wide bandgap (WBG) semiconductor, silicon carbide (SiC) demonstrates remarkable material properties, featuring a bandgap of 3.0 eV at room temperature, electric field breakdown strength reaching 4 MV/cm, and thermal conductivity of 4.9 W/cm·K [1]. These exceptional characteristics confer significant advantages over conventional silicon devices in high-voltage, high-frequency, and high-temperature power electronics applications. Consequently, SiC MOSFET devices have gained widespread adoption across diverse high-performance applications, including automotive electrification systems, renewable energy conversion platforms, industrial motor drives, and aerospace electronics, where their superior switching performance and thermal stability provide critical advantages over silicon-based counterparts [2-7].

However, despite these numerous advantages, SiC MOSFET manufacturing processes continue to encounter significant challenges arising from two primary categories of defects. Interface defects occur at the silicon carbide/oxide layer (SiO$_2$) interface where high-density interface trap states capture carriers, leading to threshold voltage instability and affecting subthreshold swing (SS) and gate leakage current. Crystalline defects encompass both intrinsic defects such as vacancies, micropipes, and stacking faults generated during SiC crystal growth, and extrinsic defects introduced through impurity incorporation, which degrade carrier transport properties and result in deteriorated on-off current ratios and elevated drain-source leakage current.

To address these interface quality issues, conventional improvement techniques, including nitriding annealing and post-oxidation annealing, can enhance interface quality to a certain degree. However, these methods typically require high-temperature processing (>1000°C), which may introduce additional defects and compromise previously established doping profiles [8-9]. Therefore, alternative low-temperature interface treatment approaches are critically needed.

Supercritical fluid nitriding technology represents a promising emerging surface treatment approach that offers distinct advantages for improving semiconductor interface quality. This technique enables low-temperature processing (typically below 400°C), avoiding structural damage associated with high-temperature treatments while preserving established doping profiles [2]. Supercritical fluids exhibit unique transport properties that enable effective defect passivation through enhanced penetration into microscopic interface structures. Additionally, supercritical conditions enhance chemical reactivity while maintaining precise process control, making this approach environmentally friendly without requiring toxic reagents [10].

Based on this background, this study aims to apply supercritical fluid nitriding treatment to planar SiC devices and investigate the resulting changes in electrical characteristics, establishing theoretical foundations for the application of this technology in SiC device manufacturing.

979-8-3315-8850-2/25 $31.00 © 2025 IEEE

II. Experimental Methods

A. Device Layout and Structure

The experiment employed planar SiC MOSFET devices, with the fabricated device array shown in Figure 1a. Figure 1b presents the detailed schematic layout and dimensions of the test device. The device consists of a large source region (S) measuring 2398 × 2000 μm and a gate region (G) of 800 × 408 μm. The overall device dimensions are 3900 × 3900 μm, with all dimensions specified in micrometers. Figure 2 shows the cross-sectional schematic of the device structure, illustrating the vertical layer arrangement. The device features p+ source regions, an n+ drain contact, and p-type channel regions within an n-type drift layer on an n+ substrate. The gate contact (G) is positioned above the channel region, with proper isolation maintained between the contact, channel, and drift regions. The source, gate, and drain terminals are clearly indicated for three-terminal device operation.

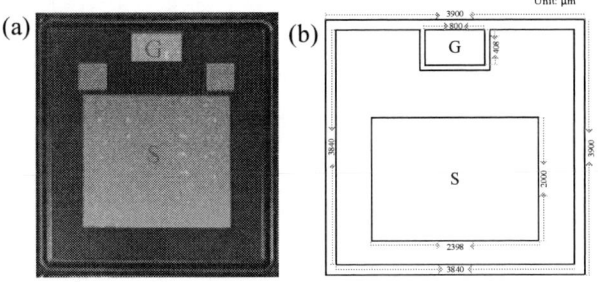

Fig. 1. SiC MOSFET Device Layout: (a) Fabricated SiC MOSFET device array showing the physical device structure; (b) Device layout schematic with dimensional specifications (Unit: μm)

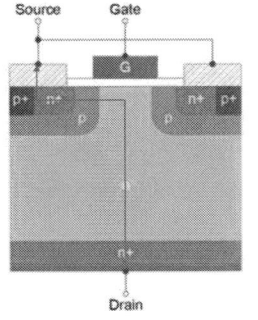

Fig. 2. Cross-sectional Schematic of SiC MOSFET Device Structure

B. Supercritical Fluid Nitriding Treatment

Supercritical fluid nitriding treatment was performed using specialized equipment under optimized process conditions to enhance the SiC/SiO$_2$ interface quality. This approach exploits the unique transport and reactive properties of supercritical fluids to achieve superior nitriding uniformity and interface passivation compared to conventional thermal nitriding techniques. Comprehensive electrical characterization was performed on devices before and after supercritical fluid treatment using semiconductor parameter analyzers.

III. Results and Discussion

To comprehensively evaluate the effectiveness of supercritical fluid nitriding treatment on SiC MOSFET device performance, a systematic electrical characterization was conducted comparing treated and untreated devices. The analysis encompasses three critical aspects of device behavior: output characteristics to assess current drive capability and channel conduction properties, transfer characteristics to evaluate gate control efficiency and carrier transport mechanisms, and capacitance-voltage measurements to examine interface quality and gate oxide integrity. These complementary characterization techniques provide a complete picture of how the supercritical fluid nitriding process impacts the fundamental electrical properties and interface characteristics of SiC MOSFET devices. The following sections present detailed analysis of each measurement, demonstrating the significant improvements achieved through this novel treatment approach.

A. Output Characteristic Analysis

This section examines the drain current-voltage characteristics of SiC MOSFET devices at different gate voltage levels following supercritical fluid nitriding treatment. Figure 3 presents the output characteristics (I_{DS}-V_{DS}) measured before and after supercritical fluid nitriding treatment, with V_{GS} swept from 0 to 4V in 0.25V increments. The dashed curves represent the untreated devices, while the solid red curves correspond to the treated devices.

The experimental results reveal substantial performance enhancement following the supercritical fluid treatment. The treated devices demonstrate significantly improved drain current drive capability, with the saturation drain current (I_{DSAT}) at V_{GS} = 4V increasing from approximately 35 mA to 70 mA, corresponding to a 100% enhancement in on-state current. The treated devices exhibit superior output conductance (gds) characteristics with well-defined linear and saturation regions, indicating reduced channel resistance and enhanced carrier transport. The improved current uniformity across different gate overdrive voltages (V_{GS} - V_{TH}) suggests reduced interface state density (Dit) and enhanced channel mobility (μ_{ch}). The steeper transconductance in the linear region and better current saturation behavior demonstrate that the supercritical fluid nitriding process effectively passivates interface traps at the SiC/SiO$_2$ interface.

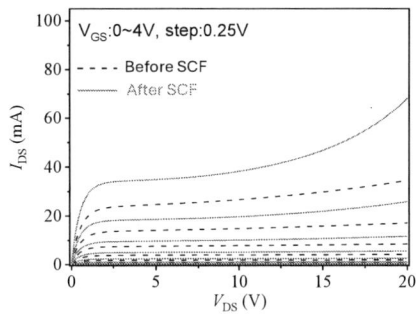

Fig. 3. Output Characteristics of SiC MOSFET Before and After Supercritical Fluid Nitriding Treatment

B. Transfer Characteristic and Transconductance Analysis

This section investigates the transfer characteristics and transconductance behavior of SiC MOSFET devices to evaluate the impact of supercritical fluid nitriding treatment on gate control efficiency and carrier transport properties. Figure 4 presents the transfer characteristics (I_{DS}-V_{GS}) and transconductance (gm-V_{GS}) curves measured at $V_{DS} = 0.1V$ for both untreated devices (Figure 4a) and supercritical fluid nitriding treated devices (Figure 4b). The transconductance (gm) represents the derivative of drain current with respect to gate voltage ($\partial_{I_{Ds}}/\partial_{V_{Gs}}$), indicating the gate control efficiency and amplification capability of the device.

The experimental results demonstrate notable improvements in device performance following the supercritical fluid treatment. The treated device exhibits a more abrupt subthreshold slope and enhanced on-state current capability, indicating improved gate control and reduced interface trap density. The transconductance peak (gm,max) shows improvement after treatment, suggesting enhanced channel mobility and better carrier transport efficiency. Additionally, the on-resistance (Ron) decreased significantly from 1270 mΩ for the untreated device to 1090 mΩ after treatment, representing a 14% reduction that directly translates to lower conduction losses and improved power efficiency. The improved transfer characteristics, enhanced transconductance, and reduced on-resistance collectively confirm that the supercritical fluid nitriding process effectively passivates interface states at the SiC/SiO$_2$ interface, leading to enhanced inversion layer quality and reduced scattering mechanisms that limit carrier mobility.

C. Capacitance-Voltage Characteristic Analysis

This section examines the capacitance-voltage (C-V) characteristics to evaluate the interface quality and gate oxide properties of SiC MOSFET devices following supercritical fluid nitriding treatment. Figure 5 presents the C-V curves measured at 1 MHz frequency with V_{GS} swept from -20 to 20V and AC signal amplitude of 25 mV. The black squares represent the untreated device, while the red squares show the characteristics after supercritical fluid treatment.

The C-V measurements reveal significant improvements in interface quality following the supercritical fluid treatment. Both devices exhibit the characteristic C-V behavior with accumulation, depletion, and inversion regions clearly defined. The treated device demonstrates reduced hysteresis and improved curve symmetry, indicating lower interface trap density (Dit) and reduced charge trapping effects at the SiC/SiO$_2$ interface. The capacitance values in the accumulation region remain similar for both devices, suggesting that the gate oxide thickness and dielectric properties are preserved during the treatment process. The sharper transition between depletion and inversion regions in the treated device indicates improved interface quality and reduced interface state density. The enhanced C-V characteristics confirm that the supercritical fluid nitriding treatment effectively passivates interface defects without compromising the gate oxide integrity, resulting in more stable and reliable device operation.

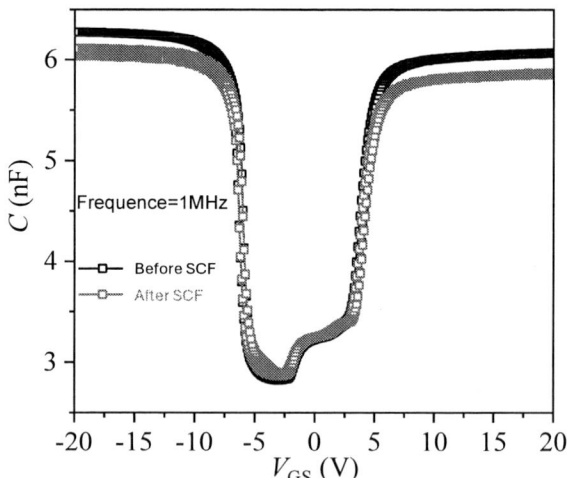

Fig. 5. Capacitance-Voltage Characteristics of SiC MOSFET Before and After Supercritical Fluid Nitriding Treatment

D. Interface Improvement Mechanism

Based on the electrical characterization and structural analysis presented above, the interface improvement mechanism illustrated in Figure 6 demonstrates the molecular-level processes occurring during supercritical fluid (SCF) treatment. The supercritical fluid serves as an effective carrier medium for ammonia molecules, leveraging its unique combination of liquid-like high solubility and gas-like penetration ability to transport reactive nitrogen species deep into the SiC/SiO$_2$ interface region.

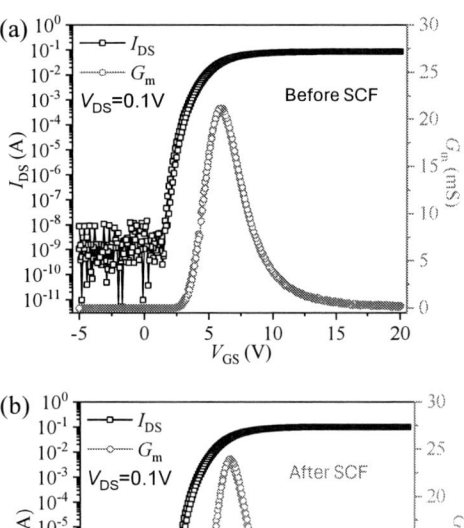

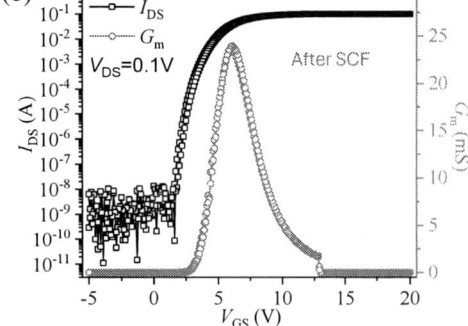

Fig. 4. Transfer Characteristics and Transconductance of SiC MOSFET: (a) Untreated Device and (b) Supercritical Fluid Nitriding Treated Device

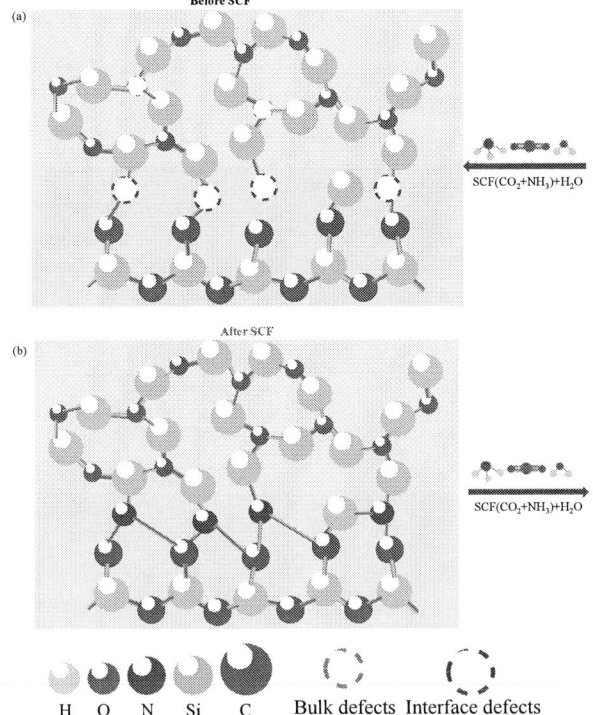

Fig. 6. Schematic illustration of SCF (CO_2+NH_3) treatment for defect repairing: (a) pristine SiC/SiO$_2$ interface with high density of interface defects and (b) treated interface with improved atomic structure.

This enhanced mass transport capability enables the ammonia molecules to reach microscopic interface irregularities that are typically inaccessible through conventional thermal processing methods. As shown in Figure 6a, the pristine SiC/SiO$_2$ interface exhibits numerous dangling bonds and carbon-related defects that contribute to high interface trap density and degraded electrical performance.

During SCF treatment, the ammonia molecules react with these interface defects, systematically passivating the electrically active sites through the formation of stable Si-N bonds. Figure 6b illustrates the resulting interface structure after treatment, where the nitrogen incorporation process has effectively replaced the carbon clusters with more stable nitrogen-containing species, leading to a substantial reduction in interface trap density and improved electronic quality.

IV. CONCLUSION

This study demonstrates that supercritical fluid nitriding treatment effectively addresses the fundamental challenge of poor SiC/SiO$_2$ interface quality that has long limited the performance of SiC MOSFET devices. The comprehensive electrical characterization reveals that this treatment approach successfully reduces interface trap density, enhances carrier mobility, and improves gate control efficiency - addressing key bottlenecks that have hindered the full potential of SiC power devices.

The observed improvements in current drive capability, switching characteristics, and interface quality indicate that supercritical fluid nitriding provides a viable solution for overcoming the inherent limitations of conventional thermal nitriding processes in SiC technology. The treatment's ability to passivate interface defects without compromising gate oxide integrity suggests a pathway toward more reliable and efficient SiC power devices that can better exploit the material's superior properties.

Future research should focus on establishing quantitative relationships between treatment parameters and resulting device characteristics to enable precise performance tuning for specific applications. Process optimization studies examining temperature, pressure, and treatment duration effects will be essential for industrial implementation. Additionally, investigation of the treatment's impact on device reliability and long-term stability under operational stress conditions will be crucial for commercial adoption.

The successful demonstration of this technology opens significant opportunities for the power electronics industry. Enhanced SiC MOSFET performance can accelerate the adoption of wide-bandgap semiconductors in electric vehicles, renewable energy systems, and high-efficiency power conversion applications. By addressing interface quality limitations, this approach may enable SiC devices to achieve their theoretical performance potential, contributing to more efficient power systems and supporting global energy efficiency initiatives. The process compatibility with existing fabrication infrastructure facilitates practical implementation, making this advancement particularly valuable for industrial scaling and commercialization.

ACKNOWLEDGMENT

This study was supported by Shenzhen Scientific and Technological Foundation (No. RCYX20231211090332037, JCYJ20240813160211015), National Natural Science Foundation of China (No. 62474008, 62204007), and Guangdong Provincial Natural Science Foundation (No. 2024A1515030044). This work was supported by Guangdong Provincial Key Laboratory of In-Memory Computing Chips (2024B1212020002). This work was in part supported by the Shenzhen POC center of Flexible Electronics and Guangdong Technology Center for Oxide Semiconductor Devices and ICs.

REFERENCES

[1] L. A. Lipkin and J. W. Palmour, "Improved oxidation procedures for reduced SiO2/SiC defects," *J. Electron. Mater.*, vol. 25, no. 5, 1996, pp. 909-915.

[2] P. Sharmila, G. Supraja, D. Haripriya, C. Sivamani, and A. L. Narayana, "Silicon carbide MOSFETs: A critical review of applications, technological advancements, and future perspectives," *Micro and Nanostructures*, vol. 202, 2025, pp. 208126.

[3] F. Roccaforte, P. Fiorenza, G. Greco, R. L. Nigro, F. Giannazzo, F. Iucolano, and M. Saggio, "Emerging trends in wide band gap semiconductors (SiC and GaN) technology for power devices," *Microelectron. Eng.*, vol. 187, 2018, pp. 66-77.

[4] X. She, A. Q. Huang, O. Lucia, and B. Ozpineci, "Review of silicon carbide power devices and their applications," *IEEE Trans. Ind. Electron.*, vol. 64, no. 10, Oct. 2017, pp. 8193-8205.

979-8-3315-8850-2/25 $31.00 © 2025 IEEE

[5] J. Millan, P. Godignon, X. Perpina, A. Perez-Tomas, and J. Rebollo, "A survey of wide bandgap power semiconductor devices," *IEEE Trans. Power Electron.*, vol. 29, no. 5, May 2014, pp. 2155-2163.

[6] A. J. Lelis, D. Habersat, R. Green, A. Ogunniyi, M. Gurfinkel, J. Suehle, and N. Goldsman, "Time dependence of bias-stress-induced SiC MOSFET threshold-voltage instability measurements," *IEEE Trans. Electron Devices*, vol. 55, no. 8, Aug. 2008, pp. 1835-1840.

[7] V. V. Afanasev, M. Bassler, G. Pensl, and M. Schulz, "Intrinsic SiC/SiO$_2$ interface states," *Phys. Status Solidi A*, vol. 162, no. 1, 1997, pp. 321-337.

[8] A. Chanthaphan, T. Hosoi, T. Shimura, and H. Watanabe, "Study of SiO2/4H-SiC interface nitridation by post-oxidation annealing in pure nitrogen gas," *AIP Adv.*, vol. 5, no. 9, Sep. 2015, pp. 097134.

[9] L. K. Swanson, P. Fiorenza, F. Giannazzo, and F. Roccaforte, "Effects of a post-oxidation annealing in nitrous oxide on the morphological and electrical properties of SiO$_2$/4H-SiC interfaces," in *Mater. Sci. Forum*, vol. 740, Trans Tech Publications Ltd, 2013, pp. 719-722.

[10] M. Sometani, S. Harada, H. Kato, M. Okuda, S. Yamakawa, M. Yamauchi, T. Kimoto, and H. Okumura, "Superior effects of supercritical fluid oxidation to improve SiC MOS interface properties," *Appl. Phys. Lett.*, vol. 113, no. 6, Aug. 2018, pp. 061602.

2025 The 10th International Conference on Integrated Circuits and Microsystems

Design of a Bandgap Voltage Reference for an 8-bit SAR-ADC for Powerline Monitoring using SKY130 PDK

Krisna M. Cañonero
Electrical and Electronics Engineering Department
Mindanao State University - Main (Marawi)
Marawi City, Philippines
canonero.km59@s.msumain.edu.ph

Jovelyn S. Bernales
Electrical and Electronics Engineering Department
Mindanao State University - Main (Marawi)
Marawi City, Philippines
bernales.js20@s.msumain.edu.ph

Abdulwarith G. Macapundag
Electrical and Electronics Engineering Department
Mindanao State University - Main (Marawi)
Marawi City, Philippines
macapundag.ag57@s.msumain.edu.ph

Sihawi A. Khalid
Electrical and Electronics Engineering Department
Mindanao State University - Main (Marawi)
Marawi City, Philippines
sihawi.khalid@msumain.edu.ph

Susie E. Maestre
Electrical and Electronics Engineering Department
Mindanao State University - Main (Marawi)
Marawi City, Philippines
susie.maestre@msumain.edu.ph

Abstract—**This paper presents the design and simulation of a 1.2V CMOS Bandgap Voltage Reference (BGR) circuit using the SKY130 PDK, aimed at integration into an 8-bit Successive Approximation Register (SAR) Analog-to-Digital Converter (ADC) for powerline monitoring applications. The proposed architecture utilizes a diode-connected NMOS for Complementary-to-Absolute-Temperature (CTAT) behavior and a current-density ratio technique for Proportional-to-Absolute-Temperature (PTAT) generation, regulated by a high-gain two-stage PMOS-input operational amplifier. The circuit was implemented and tested using open-source Electronic Design Automation (EDA) tools within the Lawrence analog design flow. Simulations confirmed that the BGR achieved a temperature coefficient of 19.3ppm/°C, a line sensitivity of 6.7%, and a Power Supply Rejection Ratio (PSRR) of 23.89dB. Corner analysis showed robustness against process variations, with the output voltage remaining within 1.16V to 1.25V. The circuit consumed only 257.47μW on average, validating its suitability for low-power, moderate-precision mixed-signal systems. This study demonstrates the viability of resource-efficient, open-source-based analog design workflows for precision voltage reference generation in embedded applications.**

Keywords—*Bandgap Voltage Reference, CMOS 130nm, SAR ADC, Powerline Monitoring, Low-Power Analog Design*

I. INTRODUCTION

Powerline monitoring is essential in modern electrical grids to ensure efficient energy distribution, real-time fault detection, and high-precision control [1]. As power networks grow more complex, the demand for accurate and synchronized measurement of voltage and current across all three phases and the neutral line becomes increasingly critical. Any mismatch or delay in capturing these signals can result in errors in power calculation, ultimately reducing system reliability and efficiency [2].

To support high-fidelity measurements, Analog-to-Digital Converters (ADCs) are widely used to convert analog powerline signals into digital data. Among various ADC architectures, the Successive Approximation Register (SAR) ADC is often preferred due to its balance in resolution, conversion speed, and energy efficiency [3]. SAR ADCs rely on a stable voltage reference to maintain accuracy during successive bit approximation. Without a precise and consistent reference voltage, the performance of SAR ADCs deteriorates, particularly in dynamic operating conditions.

A Bandgap Voltage Reference (BGR) circuit plays a key role in maintaining ADC accuracy. It generates a reference voltage that remains stable across variations in temperature, supply voltage, and process parameters. This stability is made possible by the development of low-voltage (<5 V) references based on the bandgap voltage of silicon, which enabled the design of integrated circuits that operate reliably on low supply voltages while maintaining good temperature coefficient (TC) performance—an essential requirement for precision in ADC applications [4]. This robustness is especially important in powerline monitoring, where environmental and electrical noise is common. For instance, in a 16-bit ADC with a 10V full-scale input, a reference voltage drift of 25ppm/°C can lead to a 12.5mV error over a 50°C range, while a low-drift

979-8-3315-8850-2/25 $31.00 © 2025 IEEE   427

reference like ADR421 (1ppm/°C) would reduce this to only 0.5mV [5].

However, traditional reference circuits may not offer sufficient stability for modern low-voltage systems. Integrating a CMOS-based BGR into SAR ADCs offers a promising solution. CMOS BGRs are well-suited for submicron technologies, providing excellent temperature and voltage stability while enabling low power consumption and small area—ideal for compact, energy-efficient monitoring solutions [6].

This study focuses on designing a CMOS-based bandgap reference that meets the stringent requirements of SAR ADCs used in powerline monitoring. By addressing limitations related to temperature sensitivity, voltage fluctuation, and process variation, this work aims to contribute to more robust and reliable ADC systems for smart grid applications.

II. METHODOLOGY

The design and simulation of the bandgap voltage reference circuit were carried out using XSCHEM, the SkyWater 130 nm CMOS process, and *open-LUT.org* [7]. XSCHEM served as the primary schematic editor, enabling hierarchical circuit design and SPICE netlist generation for NGSpice simulation. It allowed efficient visualization of key analog components such as CTAT and PTAT networks, current mirrors, and the op-amp. The SKY130 process provided a reliable 1.8V CMOS platform with support for high-voltage operation, multiple metal layers, and analog-friendly features like poly resistors and MiM capacitors, making it well-suited for this application. For transistor sizing, the g_m/I_D methodology was employed through the interactive tool at *open-LUT.org* [7] [8]. This tool enables rapid W/L selection from pre-characterized lookup tables and ensures proper biasing across all devices critical to PTAT and CTAT behavior.

A. Design Specifications

The design specifications target a 1.2V output voltage with low line sensitivity, minimal temperature coefficient, and stable performance across varying process, voltage, and temperature (PVT) conditions. Table I indicates the target specifications based on a typical BGR for powerline monitoring.

TABLE I.   DESIGN SPECIFICATIONS

Design Parameter	Value
Supply Voltage	1.8V
Cmos Technology	130nm
Power Consumption	<1mW
Input frequency	50 to 60 Hz
Temperature Range	-40 °C to 120 °C
Temperature Coefficient	<50 ppm/°C
Vref / output voltage	1.2V
PSRR	>40dB
Line Sensitivity	< 5%

B. Bandgap Voltage Reference Circuit Design

Fig. 1 shows the CMOS BGR architecture that includes four main blocks, namely: the current mirror, CTAT generator, PTAT generator, and a self-biased operational amplifier. The current mirror stabilizes biasing, while the CTAT and PTAT blocks generate temperature-dependent voltages. These are summed at a virtual node, where the op-amp enforces voltage balance through negative feedback, ensuring a stable, temperature-independent reference output.

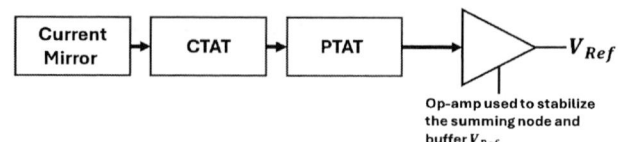

Fig. 1. BGR Block Diagram

The conventional Bandgap Voltage Reference (BGR), as shown in Fig. 2, relies on bipolar junction transistors (BJTs) to generate a stable output voltage that is largely independent of temperature, supply voltage, and process variations. It works by summing two voltages with opposing temperature coefficients: the base-emitter voltage (V_{be}) of a BJT, which has a negative temperature coefficient (CTAT), and a scaled difference in V_{be} between two BJTs with different emitter areas, which is proportional to absolute temperature (PTAT). The PTAT voltage is typically generated by forcing the same current through BJTs of unequal sizes, producing a ΔV_{be} that increases linearly with temperature. When this voltage is dropped across a resistor, it creates a PTAT current that is mirrored and used to develop a PTAT voltage. This PTAT voltage is then added to the CTAT V_{be} of another BJT, and when the resistor ratios are appropriately scaled, the temperature dependencies are canceled out. The result is a reference voltage close to the silicon bandgap ($\sim$1.2 V), which remains stable across temperature variations.

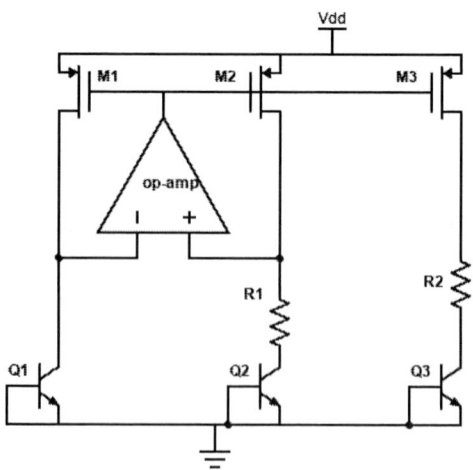

Fig. 2. Conventional Bandgap Voltage Reference

The circuit in Fig. 3 [9] [10] replaces BJTs with NMOS devices to improve scalability and power efficiency. The CTAT behavior is derived from the subthreshold variation V_{gs}, described by:

$$V_{gsM4} = nV_T \ln \left[\frac{I_{sub}}{\mu_n C_{ox}(W/L)V_T^2} \right] + V_{th} \qquad (1)$$

In Equation 1, I_{sub} is the subthreshold drain current, μ_n is the electron mobility, C_{ox} is the gate oxide capacitance per unit area, W/L is the transistor's width-to-length ratio, V_T is the thermal voltage (kT/q), n is the subthreshold slope factor, and V_{th} is the threshold voltage. The first term, which includes V_T, exhibits a positive temperature coefficient, while V_{th} typically decreases with temperature, showing a negative temperature coefficient.

The PTAT voltage is generated by creating a mismatch between two diode-connected NMOS transistors, where the second NMOS (M5) is sized 10× larger than the first. This intentional size difference leads to a mismatch in current density, which in turn produces a difference in gate-source voltages (ΔV_{gs}) between the two devices. Since ΔV_{gs} increases proportionally with temperature, this voltage difference serves as a PTAT voltage. When this ΔV_{gs} is dropped across a resistor, it generates a PTAT current, which is then mirrored and converted into a temperature-dependent output voltage V_{ref}, given by the expression:

$$V_{ref} = \left(\frac{V_{inp} - V_{gsM5}}{R_1} + \frac{V_{inp}}{R_2} \right)(R_3) \qquad (2)$$

This equation captures how the PTAT current, generated by ΔV_{gs} ($V_{inp} - V_{gsM5}$) and reflected through current mirroring, flows through the resistive network composed of R_1, R_2, and R_3 to produce the output voltage. The structure allows for fine control of the temperature-dependent behavior of V_{ref} through proper selection of resistor ratios, enabling precise tuning of the reference output.

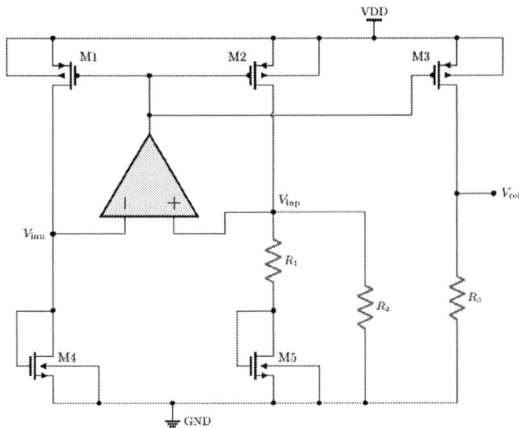

Fig. 3. Proposed Bandgap Voltage Reference

The operational amplifier (op-amp) ensures voltage equality between the CTAT and PTAT branches, stabilizing the summing node. To address the limitations in line sensitivity and ensure accurate summation of the CTAT and PTAT branches, a PMOS differential two-stage operational amplifier was implemented as the final op-amp in the bandgap reference circuit as illustrated in Fig. 4. The amplifier was designed to deliver a gain of approximately 13,000V/V, common-mode input of 0.55V and a 0.66V output voltage, based on observed performance trade-offs during simulation.

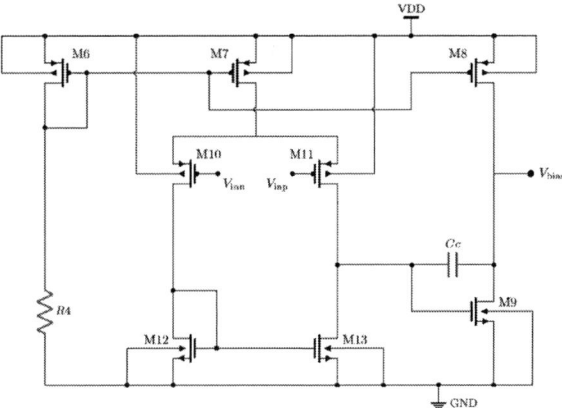

Fig. 4. PMOS Differential Two-stage Operational Amplifier

C. Simulation Set-up

To evaluate the performance of the proposed bandgap voltage reference circuit, a series of simulations were conducted to assess its dynamic behavior, temperature stability, and resilience to supply variations.

Transient analysis was performed to verify the circuit's dynamic behavior and determine whether the operational amplifier successfully forced the CTAT and PTAT node voltages (denoted as V_{inn} and V_{inp}) to converge through feedback.

Temperature sweep simulations were conducted over a range of –40°C to 120°C to evaluate the stability of the reference voltage across temperature variations. The temperature coefficient (TC), which quantifies the output voltage's sensitivity to temperature changes, was calculated using the following expression :

$$TC(ppm/^\circ C) = 10^6 \text{x} \left(\frac{1}{V_{ref}} \cdot \frac{\Delta V_{ref}}{\Delta T} \right) \qquad (3)$$

DC sweep simulations evaluated line sensitivity (LS) by varying supply voltage and measuring the change in reference voltage, computed as:

$$LS(\%) = \frac{\Delta V_{ref(nominal)}}{\Delta V_{DD}} \qquad (4)$$

AC analysis was performed to evaluate PSRR across frequency, measuring the circuit's ability to suppress supply variations using the formula:

$$PSRR(dB) = 20log \frac{\Delta V_{ref}}{\Delta V_{in}} \qquad (5)$$

Collectively, these simulations yielded quantitative metrics that characterize the circuit's performance in terms of temperature insensitivity, robustness against supply variations, and effective feedback operation.

III. RESULTS

Initial simulations showed that while the designed BGR achieved a reference voltage close to 1.2V with an acceptable temperature coefficient, its line sensitivity remained above target. To improve this, the trade-off between op-amp gains and line sensitivity was investigated. As the amplifier gain increased from 100V/V to 15,000V/V, line sensitivity improved significantly—from 9.1% to 4.89% as shown in Table II. This confirmed that higher gain enhances feedback regulation, leading to better voltage stability. To achieve this high gain, a PMOS-input two-stage operational amplifier was implemented, designed to deliver ~13,000V/V gain. This significantly improved the matching of V_{inn} and V_{inp}, as well as overall supply regulation.

TABLE II. GAIN VS. LINE SENSITIVITY

Gain (V/V)	Measurement Values			
	V_{inn} (V)	V_{inp} (V)	V_{bias} (V)	LS (%)
100	0.554329	0.560912	0.658263	9.1
250	0.552057	0.554715	0.66439	7.345
1000	0.550865	0.551533	0.667553	5.3
5000	0.550518	0.550698	0.668348	4.9
10000	0.550518	0.550585	0.66847	4.8875
15000	0.550514	0.550558	0.668481	4.8875

A. Bandgap Reference Circuit Design Performance

After integrating the high-gain op-amp and adjusting resistor values ($R_1 = 30k\Omega$, $R_2 = 83.6k\Omega$, $R_3 = 118.3k\Omega$), the final BGR circuit achieved a near-ideal temperature-independent reference voltage of 1.2V. Simulation results showed excellent flatness across -40°C to 120°C, with a temperature coefficient of 19.3ppm/°C from 19°C to 120°C and 33ppm/°C from –40°C to 19°C, as shown in Fig. 5a.

Line sensitivity measured via DC sweep was 6.7%, slightly above the 5% target but improved from earlier versions. PSRR reached 23.89dB at low frequency, suitable for low-noise use. Power consumption averaged $257.47\mu W$, meeting low-power SAR ADC design goals (see Figures 5b, 5c, and 5d).

The designed bandgap voltage reference circuit was simulated across multiple process corners (see Figure 6) to assess its robustness under fabrication variability. In the Fast-Fast (FF) corner, the output reference voltage was close to 1.18V, while in the Slow-Slow (SS) corner, it increased to 1.23V. Under Slow NMOS, Fast PMOS (SF) conditions, the reference voltage dropped further to 1.1V, reflecting the dominant slow NMOS behavior. Conversely, in the Fast NMOS, Slow PMOS (FS) corner, the output increased significantly, reaching nearly 1.25V, the highest among all corners. These results confirm that the circuit's V_{ref} remains within ±4% of the target 1.2V,

demonstrating acceptable variation for typical mixed-signal applications.

B. Transistor Sizing using open-LUT.org

The transistors for the op-amp were sized using the g_m/I_D methodology through *open-LUT.org* to meet the target gain and biasing. The transistor sizes are tabulated in Table III.

TABLE III. OPERATIONAL AMPLIFIER TRANSISTOR SIZES

MOSFET	Measurement Values	
	$V_{gs} = V_{ds}$ (V)	W/L (μm)
M10, M11	1.2	6.4/1.2
M12, M13	0.6	1.3/1
M6, M7	1.2	5/0.55
M8	1.2	84.77/2.5
M9	0.6	28/2.5

Transistor dimensions in the bandgap core were chosen using *open-LUT.org* to target specific g_m/I_D values, ensuring weak inversion for the NMOS and strong inversion for the PMOS and achieve stable PTAT and CTAT behavior. as indicated in Table IV.

TABLE IV. BANDGAP CORE CIRCUIT TRANSISTOR SIZES

MOSFET	Measurement Values		
	$V_{gs} = V_{ds}$ (V)	W/L (μm)	g_m/I_D
nfet_01v8	0.6	75.13/0.4	21.42
pfet_01v8	1.2	3.19/0.4	5.6

IV. DISCUSSION

The initial simulations of the bandgap voltage reference revealed a clear trade-off between operational amplifier gain and line sensitivity. As shown in Table II, increasing the amplifier's open-loop gain from 100V/V to 15,000V/V significantly reduced line sensitivity from 9.1% to 4.88%. This inverse relationship indicates that stronger feedback regulation improves the matching of PTAT and CTAT node voltages, thus stabilizing the output reference voltage (V_{ref}) against supply fluctuations. This observation motivated the use of a high-gain two-stage PMOS-input op-amp, offering a gain of approximately 13,000V/V. The use of a PMOS differential pair also aligned with the common-mode requirement of around 0.55V from the CTAT and PTAT branches.

After resistor tuning, the final circuit yielded a V_{ref} of approximately 1.2V, with a temperature coefficient (TC) of 19.3ppm/°C from 19°C to 120°C, and 33ppm/°C from –40°C to 19°C. These values demonstrate effective compensation between CTAT and PTAT voltages, with minimal deviation across a wide temperature range. The low TC confirms the accuracy of resistor scaling and current mirroring in the final architecture.

In terms of power supply rejection, the final AC analysis showed a PSRR of 23.89dB, which, while not meeting the >40dB target, still indicates moderate resilience to low-frequency supply noise. The limitation could be attributed to

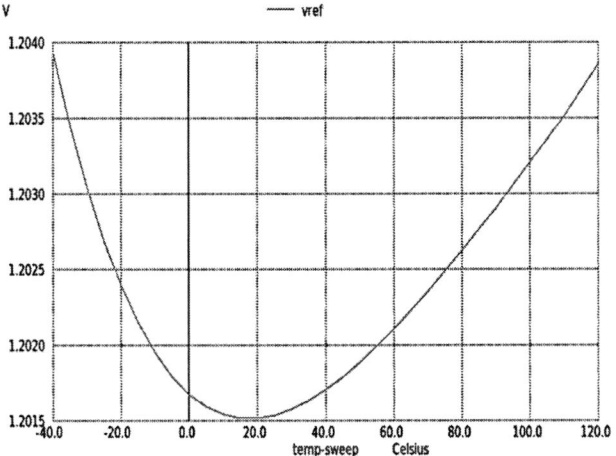

(a) Temperature Coefficient

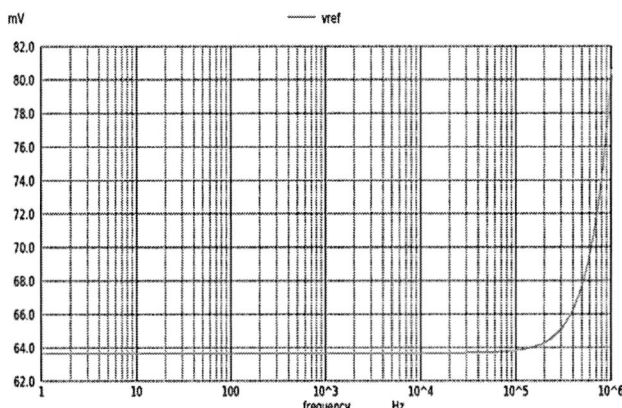

(b) Power Supply Rejection Ratio

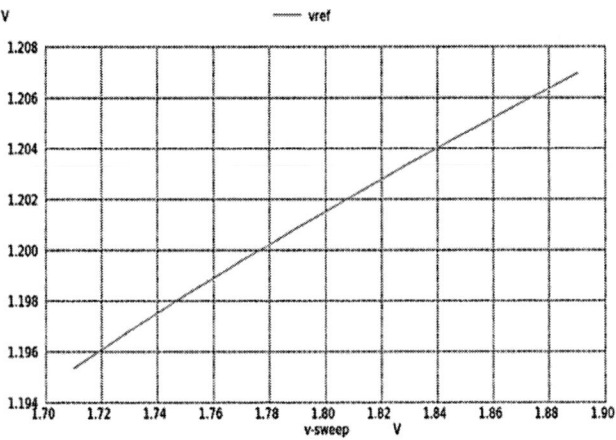

(c) Line Sensitivity

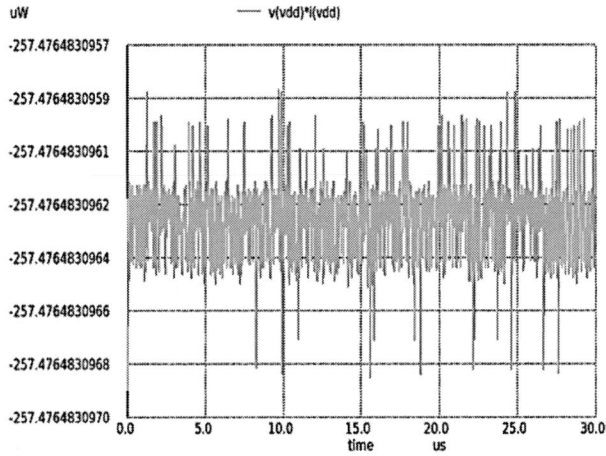

(d) Average Power Consumption

Fig. 5. BGR Performance

the lack of cascode biasing or supply-independent bias circuits in the core. Despite this, the current results remain acceptable for medium-precision applications and provide a baseline for future improvement.

For line sensitivity, the final design achieved 6.7%, slightly above the 5% target. However, this still marks a substantial improvement over earlier iterations and highlights the effect of high-gain amplification in regulating supply-induced variations in V_{ref}.

Lastly, process variation simulations across FF (Fig. 6a), SS (Fig. 6b), SF (Fig. 6c), and FS (Fig. 6d) corners revealed that V_{ref} ranged from 1.16V to 1.25V. These deviations, although noticeable, remained within a ±4% tolerance from the nominal 1.2V target. This confirms that the design maintains reasonable output stability despite device mismatches, with the FS corner producing the highest output due to the fast NMOS driving stronger PTAT behavior. Overall, the simulation results validate the robustness of the proposed bandgap reference under temperature, supply, and process variations, with only minor deviations from the design goals. A comparison of this work and other BGRs is shown in Table V.

TABLE V. PERFORMANCE COMPARISON

Parameters	[6]	[9]	[11]	[12]	[This Work]
Voltage Supply (V)	3.3	1.8	1.8	2-5	1.8
Technology (CMOS nm)	130	180	32	180	130
Temperature Range (°C)	-20 to 125	0 to 100	-40 to 100	-45 to 125	-40 to 120
Vref (V)	1.2	0.551	0.725	1.2	1.2
TC (ppm/°C)	34.3	101.9	13, 58	32.7, 89	19.3, 33
LS (%)	-	-	-	0.058	6.7
PSRR (dB)	37	-	57	-	23.89
Power Consumption (µW)	252	2000	-	0.092	257.47

979-8-3315-8850-2/25 $31.00 © 2025 IEEE

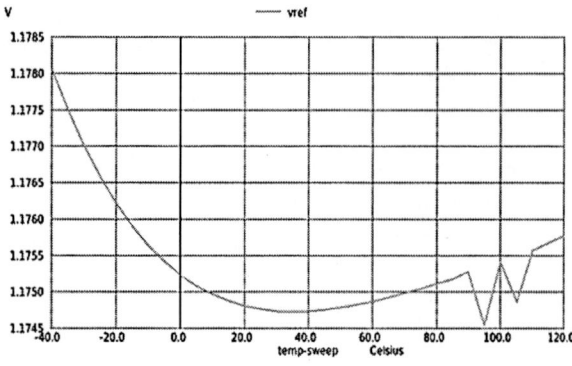

(a) FF Corner

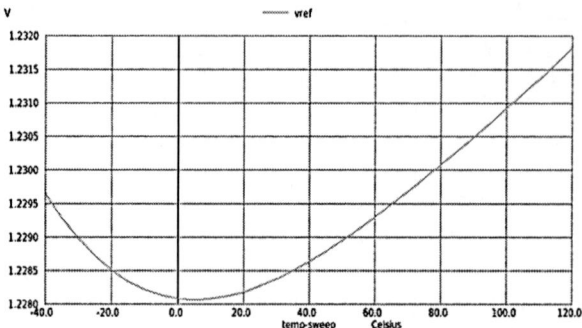

(b) SS Corner

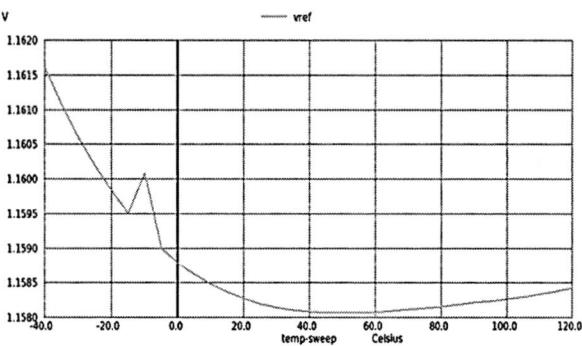

(c) SF Corner

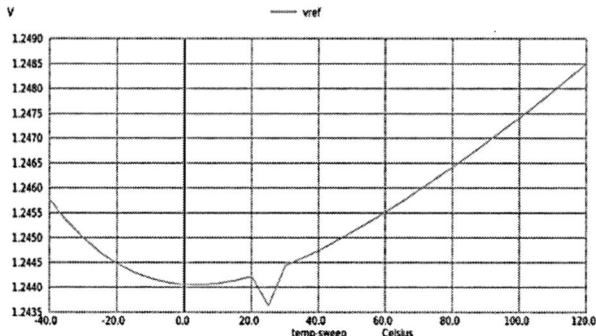

(d) FS Corner

Fig. 6. BGR Performance across Different Process Corners

V. CONCLUSIONS

This study demonstrated a 1.2 V CMOS Bandgap Voltage Reference in 130 nm for an 8-bit SAR ADC in powerline monitoring, using a CTAT–PTAT scheme with a PMOS-input op-amp, implemented via open-source EDA tools.

Compared to previous works, the design offers a good balance of performance and efficiency, with a TC of 19.3 ppm/°C (19–120°C) and 33 ppm/°C (–40–19°C), and low power consumption of 257.47 µW—much lower than [9] and comparable to [6] at lower VDD. Its line sensitivity (6.7%) and PSRR (23 dB) are modest compared to [11] (57 dB, 32 nm) and [12] (180 nm, wider supply). Overall, the BGR achieved stable, low-power operation using open-access tools, meeting core goals despite limited PSRR and line sensitivity.

Future work should enhance PSRR and line sensitivity with better biasing, amplifier design, or advanced nodes and layout methods, making this a reference for compact, low-power mixed-signal systems.

REFERENCES

[1] E. Elkin, "Power Lines Monitoring: 6 Fiber Optics Sensing Technology Misconceptions," Prisma Photonics, Nov. 17, 2022. https://www.prismaphotonics.com/6-powerline-monitoring-misconceptions/

[2] S. Evanczuk, "Multichannel Sampling Keys Accurate Power Line Monitoring," DigiKey, Jul. 29, 2015. https://digikey.ph/en/articles/multichannel-sampling-keys-accurate-power-line-monitoring (accessed Jul. 11, 2025).

[3] D. Das, "How does Successive Approximation (SAR) ADC Work and Where is it best used?," Circuit Digest, Oct. 30, 2020. https://circuitdigest.com/article/how-does-successive-approximation-sar-adc-work-and-where-is-it-best-used

[4] Texas Instruments, "Tips and tricks for designing with voltage references (Rev. A)," SLYC147A, Oct. 2021. http://www.schematicsforfree.com/files/Power

[5] C. Slattery, "High-Performance Multichannel Power-Line Monitoring with Simultaneous-Sampling ADCs — Analog Devices," Analog.com, 2024. https://www.analog.com/en/resources/analog-dialogue/articles/multichannel-power-line-monitoring.html

[6] S. Somvanshi, "A sub-1 volt CMOS bandgap reference with high power supply rejection," APCCAS 2008 - 2008 IEEE Asia Pacific Conference on Circuits and Systems, Macao, China, 2008, pp. 666-667, doi: 10.1109/APCCAS.2008.4746111.

[7] "Open-LUT: Open-Source PDK Lookup Tables." open-LUT. http://www.open-lut.org (accessed Jun. 15, 2025).

[8] S. Khalid, S. Maestre, K. Madird-Khalid, "Open-LUT: Interactive gm/ID Lookup Tables for MOSFET Sizing in Open-Source PDKs," 2025 10th International Conference on Integrated Circuits and Microsystems (ICICM), Hefei, China, 2025

[9] S. K. Koh and L. Lee, "Low power CMOS bandgap reference circuit," 2014 IEEE Student Conference on Research and Development, Penang, Malaysia, 2014, pp. 1-5, doi: 10.1109/SCORED.2014.7072988.

[10] R. Madeira, N. Paulino, "Design Methodology for an All CMOS Bandgap Voltage Reference Circuit", HAL, November 6, 2017. hal-01629561v1.

[11] Kumar, A., Pal, P. K., and Pattanaik, M. "A Wide Temperature Range Bandgap Reference Generator in 32nm CMOS Technology." 2015 Global Conference on Communication Technologies (GCCT), 2015, pp. 821–825. doi:10.1109/GCCT.2015.7342766

[12] Huang, W., Liu, L., and Zhu, Z. "A Sub-200nW All-in-One Bandgap Voltage and Current Reference without Amplifiers." IEEE Transactions on Circuits and Systems II: Express Briefs, early access, doi:10.1109/TCSII.2020.3007195

979-8-3315-8850-2/25 $31.00 © 2025 IEEE

2025 The 10th International Conference on Integrated Circuits and Microsystems

A 2.9mV$_{pp}$ Ripple 60mA Digital Low-Dropout Regulator in 28nm CMOS

Wenxin Zhang
School of Integrated Circuit
East China Normal University
Shanghai, China
wxzhang@stu.ecnu.edu.cn

Yuhang Zhang
School of Integrated Circuit
East China Normal University
Shanghai, China
zhangyh@cee.ecnu.edu.cn

Yang Shen
School of Integrated Circuit
East China Normal University
Shanghai, China
yshen@cee.ecnu.edu.cn

Xiaojin Li
School of Integrated Circuit
East China Normal University
Shanghai, China
xjli@cee.ecnu.edu.cn

Bingyi Ye*
School of Integrated Circuit
East China Normal University
Shanghai, China
byye@ic.ecnu.edu.cn

Yabin Sun*
School of Integrated Circuit
East China Normal University
Shanghai, China
ybsun@cee.ecnu.edu.cn

Abstract—**This paper introduces a digital low-dropout regulator (DLDO) that improves low steady-state voltage ripples (V_{RIPP}) throughout a broad range of load currents (I_{LOAD}) utilizing a high-resolution VCO-based ADC and a small unity PMOS current. The ADC is implemented with a pair of VCOs, an 8-bit binary counter, a Gray code counter, and a Gray-to-binary converter, enabling precise detection of small voltage deviations and allowing finer control of the PMOS array to reduce steady-state limit cycle oscillation (LCO) amplitude. Implemented in a 28-nm CMOS process, this DLDO occupies a core size of 0.029 mm². The simulation results indicate that the DLDO attains a V_{RIPP} of 2.9 mV with V_{OUT} at 1.4 V, V_{IN} at 1.8 V, and I_{LOAD} at 60 mA. Additionally, with a 7-to-55 mA 2-μs load current step and a 100-pF on-chip capacitor, the output waveform indicates a 220-mV voltage droop and a 3.4-μs settling time.**

Keywords—analog-to-digital converter (ADC), proportional–integral control, voltage-controlled oscillator (VCO), digital low-dropout (DLDO) regulator, steady-state voltage ripples.

I. INTRODUCTION

Analog low-dropout regulators (A-LDOs) are extensively utilized in system-on-chip (SoC) applications for their capacity to produce high-precision and high-efficiency power. Nevertheless, A-LDOs necessitate supplementary compensation networks, generally comprising capacitors, resistors, and other analog components, to guarantee loop stability [1-3] at the expense of area and PVT sensitivity. Conversely, the digital LDO (DLDO) utilizes a discrete-time control loop with a digital controller to reduce reliance on analog compensation networks. The DLDOs have superiorities with regard to scalability, programmability, and stability across a wide range of load currents (I_{LOAD}) . Owing to these advantages, DLDOs have been commercially implemented in several application fields, such as the Internet of Things (IoT), portable mobile devices, and heterogeneous SoCs [4-6]. However, DLDOs are still less popular than analog LDOs. The primary reason is the significantly large steady-state voltage ripples (V_{RIPP}) [7]. Fig. 1(a) illustrates the simplified block diagram of a conventional

DLDO that consists of a voltage comparator, a series of shift registers, and a PMOS array [8]. Fig. 1(b) illustrates that when V_{OUT} approximates V_{REF}, the least significant bit of the shift registers continues to toggle at a sample frequency (F_S), while intermittently activating and deactivating the PMOS. This produces consistent voltage ripples (V_{RIPP}) [9]. Fig. 1(c) illustrates the current fluctuations generated by PMOS transistors in steady state. The characteristics of V_{RIPP}, including amplitude and frequency, are highly dependent on sample frequency F_S, dropout voltage (V_{DROP}), and the PMOS sizes [10].

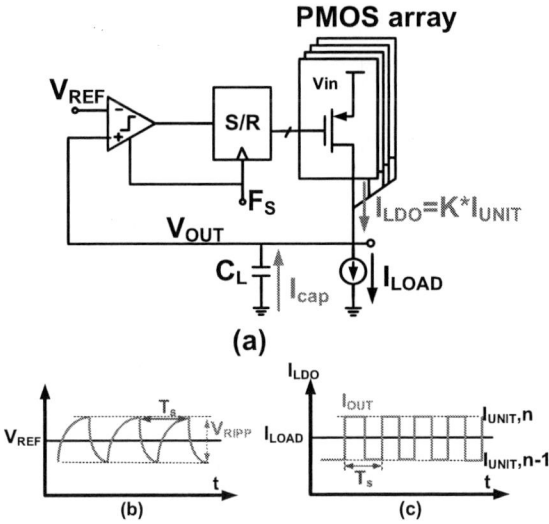

Fig. 1. (a) Block diagram of the conventional DLDO. (b) V_{RIPP} of the DLDO. (c) I_{LDO} produced by PMOS transistors.

To address the ripple issue, numerous techniques have been proposed. Ref. [11] introduces a ripple-cancellation amplifier (RCA), which supplies the error current and suppresses the output ripple. However, its performance is constrained by the trade-off between gain and bandwidth. Ref. [12] introduces a hybrid LDO designed to give sub-LSB currents and reduce V_{RIPP},

This work was supported in part by the Shanghai Science and Technology Program under Grant 24YF2710500.

979-8-3315-8850-2/25 $31.00 © 2025 IEEE 433

but at the expense of intricate stability and large V_{DROP}. A substantial V_{DROP} inherently limits the regulation range, while the analog loop is required to supply 20% of ΔI_{LOAD}. When V_{DD} approaches near-threshold-voltage (NTV) levels, such hybrid architectures become incapable of operating. The auxiliary PMOS switching was proposed to reduce V_{RIPP} [13]. However, the sizing of the auxiliary PMOS is difficult to accommodate both light-load and heavy-load conditions, making its design particularly challenging. A $\Delta\Sigma$ modulator was utilized for noise shaping of V_{OUT} [14] resulting in a V_{RIPP} of smaller than 1 mV at an I_{LOAD} of 3 mA. Nevertheless, this commendable performance was attained with a substantial C_{OUT} of 100 nF, which is undesirable for SoC applications.

In this brief, an 8-bit ADC is adopted to detect small voltage deviations, which enables finer control of the PMOS array and therefore reduces the steady-state limit cycle oscillation (LCO) amplitude. Moreover, an overflow prevention mechanism is introduced to ensure an error-free startup process for the DLDO.

The remainder of this paper is organized as follows. The principles of the VCO-based ADC are discussed in Section II. The circuit implementations and design considerations of the DLDO are presented in Section III. Section IV gives the simulation results. Finally, the conclusions are drawn in Section V.

II. THE VCO-BASED ADC

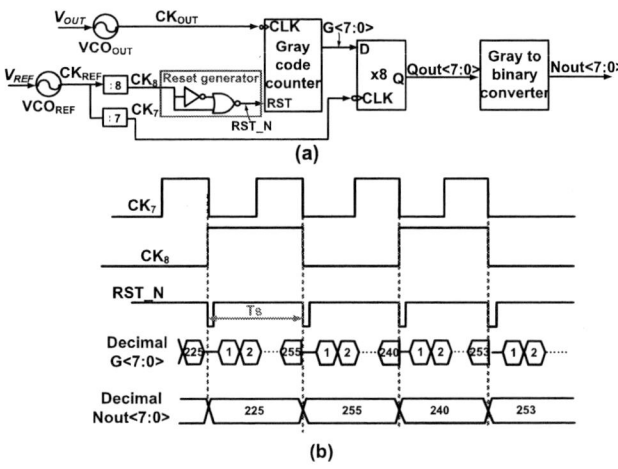

Fig. 2. (a) Block diagram of the ADC (b) Timing diagram of the ADC.

The block diagram of the VCO-based ADC is shown in Fig. 2(a), comprising two VCOs, a 1/256 divider, an 8-bit Gray code counter, registers, and a Gray-to-binary converter. The VCOs convert the V_{OUT} and the V_{REF} to equivalent clocks: CK_{REF} and CK_{OUT}. CK_{REF} is divided by a frequency divider to generate CK_8 and CK_7, and CK_8 is further processed by a reset generator to produce a reset signal RST_N. Fig. 2(b) shows that the period of RST_N is Ts. The Gray code counter increases on the falling edge of CK_{OUT}. CK_7 functions as the clock trigger signal for the register, facilitating the temporary storing of G<7:0>. Upon each falling edge of RST_N, the decimal G<7:0> is reset to 1. Fig. 3(a) consists of a Gray code counter [15], registers, and a Gray-to-binary converter. Note that if V_{OUT} is significantly larger than V_{REF}, the frequency of CK_{OUT} will be larger than that of CK_{REF}.

Therefore, an overflow prevention circuit is implemented. Fig. 3(b) shows the overflow prevention circuit; once decimal G<7:0> reaches 255, the overflow flag signal (CKs) becomes zero and deactivates the Gray code counter.

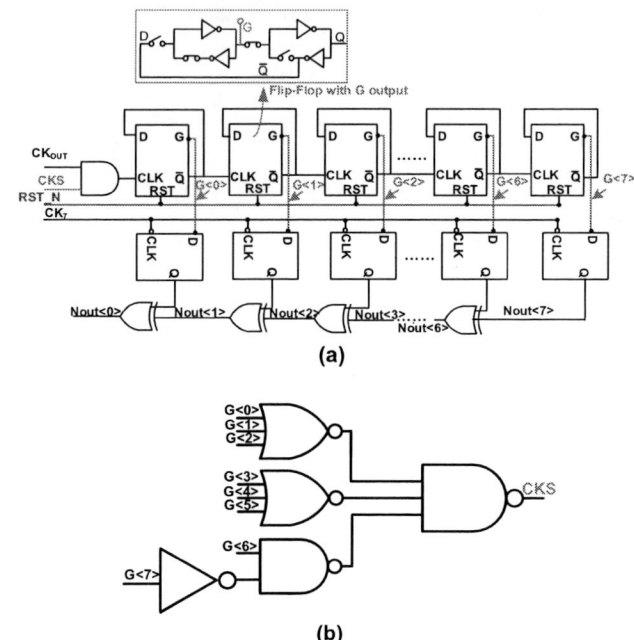

Fig. 3. (a) The structure of Gray code counter, registers and Gray to binary converter. (b) overflow prevention circuit.

Fig. 4 illustrates the schematic of the voltage-controlled oscillator (VCO), which employs a five-stage ring oscillator to produce a sine wave and a pulse-shaping circuit to deliver a square-wave output. The output frequency of the VCO varies with the input control voltage. When the input voltage is 1.0 V, 1.1 V, 1.2 V, 1.3 V, and 1.4 V, the corresponding output frequencies are 0.68 GHz, 0.88 GHz, 1.10 GHz, 1.26 GHz, and 1.42 GHz, respectively. The phase noise of the VCO is simulated to be −74.3 dBc/Hz at 1 MHz.

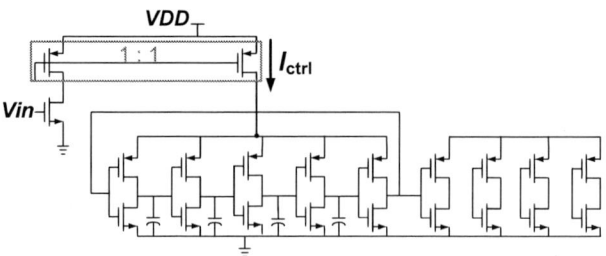

Fig. 4. Schematic of the VCO.

III. DLDO DESIGN

The block diagram of the proposed DLDO is presented in Fig. 5. The circuit predominantly comprises an ADC, a PI controller, a level shifter and a PMOS array. To extend the output current range of the DLDO, The PMOS array is implemented with an "8-4-2-1" unit ratio. The integral (I) path of the PI controller accumulates the error ($128 - N_{out}$), whereas the proportional (P) path multiplies the error by a fixed gain. The PI controller outputs a 12-bit binary code, C[11:0], which is then right-shifted by four bits to get C[11:4]. Furthermore, our design converts the unsigned counter to a signed quantity within the PI controller, thereby keeping the loop in negative feedback. This is similar to [16], which uses a digital comparator to detect the voltage polarity.

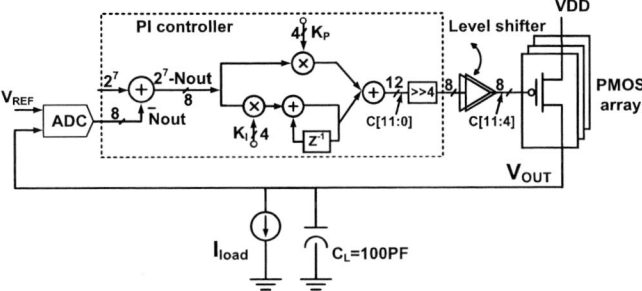

Fig. 5.   Block diagram of the proposed DLDO.

Fig. 6(a) illustrates the PMOS switch array, which utilizes the eight most significant bits (C[11:4]) of the binary output from the digital controller. Fig. 6(b) illustrates the used level shifter, which converts 0, 1.8V to 0, 0.9V.

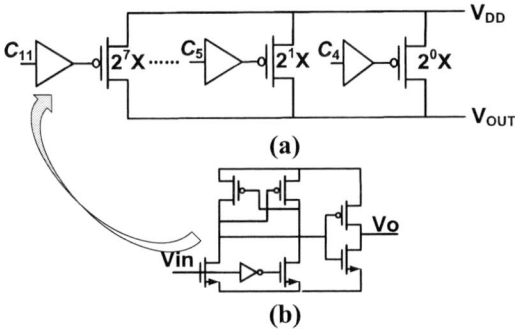

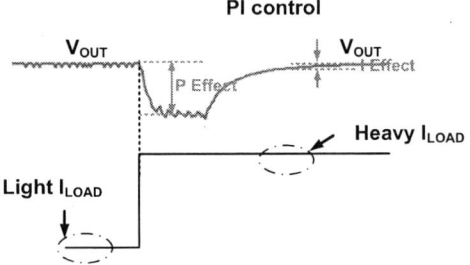

Fig. 6.   (a) PMOS switch array. (b) level shifter.

Fig. 7 illustrates the load transient responses of the DLDO. As the I_{LOAD} transitions from light to heavy, the output voltage experiences a significant undershoot. The proportional path swiftly reacts to the significant voltage deviation (VD) caused by I_{LOAD}; the integral path continues to integrate the VD until V_{OUT} surpasses V_{REF}.

Fig. 7.   Load transient responses of an LDO with digital controller.

Fig. 8 illustrates the small-signal model of the proposed DLDO. The quantization gain of the ADC is depicted as K_{ADC}. The digital controller comprises a proportional path K_P and an integral path K_I, generating a transfer function: $D(z) = K_P + K_I/(1-z^{-1})$. The z^{-1} is used to model the equivalent delay between the ADC and PI controller since they are synchronously sampled. Furthermore, PI controller output remains constant until the subsequent sampling period; a zero-order hold (ZOH) is implemented prior to the output stage, which comprises a PMOS switch array and a load circuit. The load can be approximated as an RC network [17]. This stage operates as a single-pole low-pass filter, defined by its transfer function: $K_0/(1 + s/b)$, where K_0 denotes the gain of the output stage. The output stage exhibits a pole at b, characterized by $I_{LOAD}/(V_{DROP} \cdot C_L)$. Consequently, the open-loop transfer function in the z-domain can be expressed as follows [16]:

$$H_{OL}(z) = \frac{K_{ADC} \cdot K_0 \cdot (1 - e^{-bT_S}) \cdot z^{-1} \cdot [(K_p + K_I) \cdot z - K_p]}{(z-1) \cdot (z - e^{-bT_S})} \quad (1)$$

Where T_S signifies the ADC's sampling period, b represents $I_{LOAD}/(V_{DROP} \cdot C_L)$.

The closed-loop transfer function can be derived from the open-loop transfer function.

$$H_{CL}(z) = \frac{H_{OL}(z)}{1 + H_{OL}(z)} \quad (2)$$

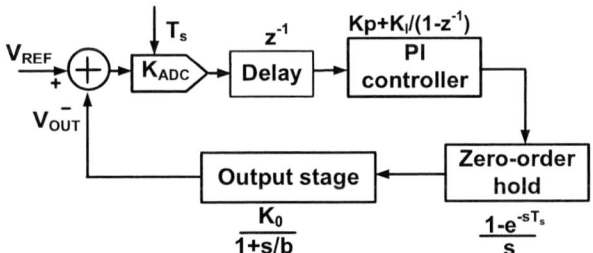

Fig. 8.   Small signal model of the proposed DLDO.

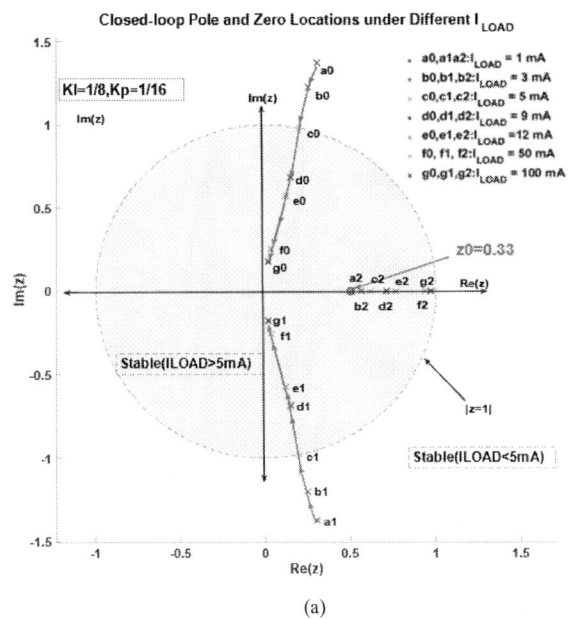

(a)

979-8-3315-8850-2/25 $31.00 © 2025 IEEE

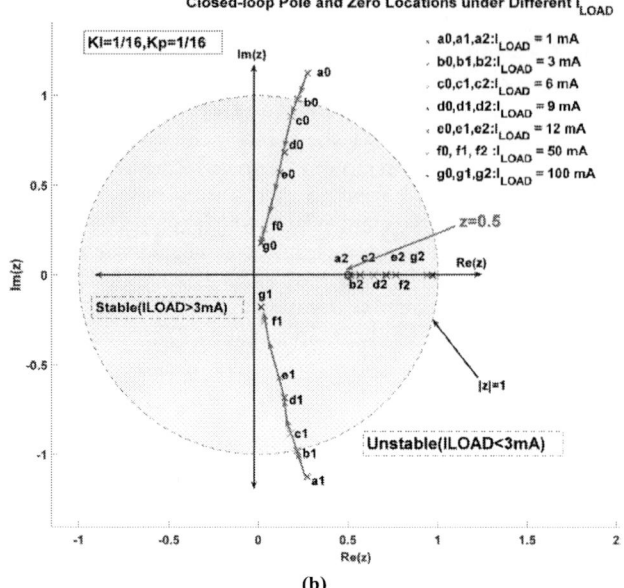

(b)

Fig. 9. Closed-loop root locus variation with (a) K_I=1/8 and K_p=1/16 (b) K_I=1/16 and K_p=1/16.

The simulated closed-loop poles and zeros for two different loop filter gains are plotted in Fig. 9 by varying I_{LOAD}. By substituting different design parameters and performing calculations, the variations in pole locations can be obtained. The design parameters used in the calculation are as follows: $K_{ADC} = 256$ /V, $V_{DROP} = 0.5$ V, $C_L = 100$ pF, $T_s = 64$ ns. T_s is equal to 128 times the period of CK_{REF}. Fig. 9(a) illustrates a three-pole system with $K_I = 1/8$ and $K_p = 1/16$, which has a zero located at 0.33. The closed-loop poles remain inside the unit circle when the I_{LOAD} is greater than 5 mA. As the I_{LOAD} increases, the system attains asymptotic stability; thus, it is verified that supplying a large load current with the DLDO is feasible. Fig. 9(b) corresponds to $K_p = 1/16$ and $K_I = 1/16$. This system possesses three poles and a zero at 0.5; stability is guaranteed when the I_{LOAD} exceeds 3 mA.

IV. SIMULATION RESULTS

The DLDO was implemented in a 28-nm CMOS process. Layout of the DLDO is shown in Fig.10. The active area of the DLDO, including the PMOS array, ADC, and PI controller, is 0.029 mm².

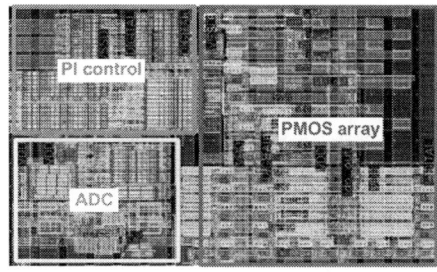

Fig. 10. Layout of the DLDO

Fig. 11 depicts the steady-state voltage ripples (V_{RIPP}) for various V_{REF} values, demonstrating that V_{RIPP} remains under 10

mV when V_{REF} is established at 1.4 V and I_{LOAD} fluctuates between 20 mA and 60 mA. Fig. 12 presents a standalone plot for the specific operating point V_{REF}=1.4 V, I_{LOAD}=60 mA, showing V_{RIPP}=2.9 mV. Fig. 13 illustrates the transient response with $V_{IN} = 1.8$ V, $V_{OUT} = 1.3$ V, $K_I = 1/8$, $K_P = 1/16$, and I_{LOAD} fluctuating between 7 and 55 mA. The magnified figure reveals that the V_{DROOP} and settling time are 220 mV and 3.4 µs, respectively. The DLDO demonstrates a quiescent current of 0.6 mA while delivering a load current of 55 mA. Fig. 14 illustrates the voltage droop (V_{DROOP}) and settling time (T_{settle}) of the DLDO across various K_P and K_I values, which indicates that both V_{DROOP} and T_{settle} achieve near-optimal values with $K_P = 1/16$ and $K_I = 1/8$, which is a compromise between the stability margin and settling time.

Table I shows a comparison of this work with other low-voltage ripple LDOs published in recent years. This work achieves lower voltage ripples.

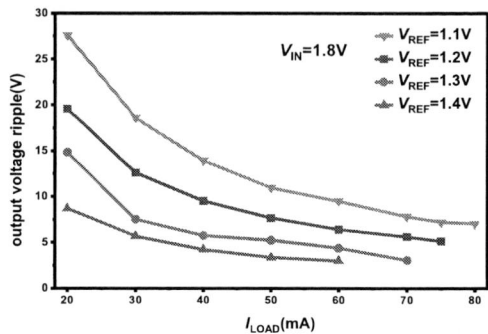

Fig. 11. Steady-state ripples at different V_{REF} and I_{LOAD}.

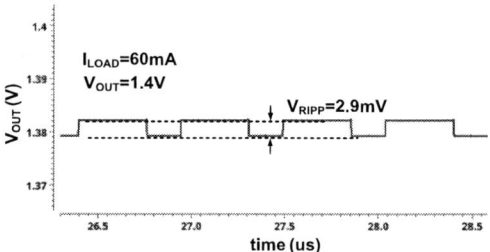

Fig. 12. Steady-state output ripple voltage at $I_{LOAD} = 60$ mA.

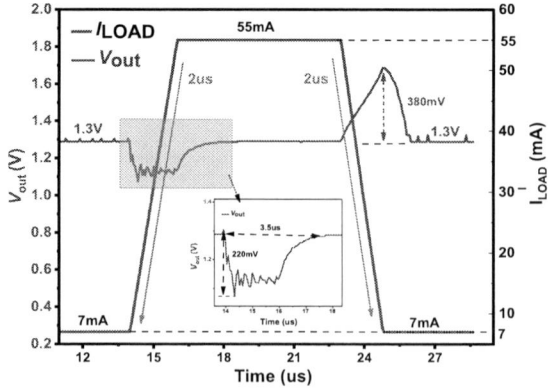

Fig. 13. Transient responses for $V_{IN} = 1.8$ V, $V_{REF} = 1.3$ V.

979-8-3315-8850-2/25 $31.00 © 2025 IEEE   436

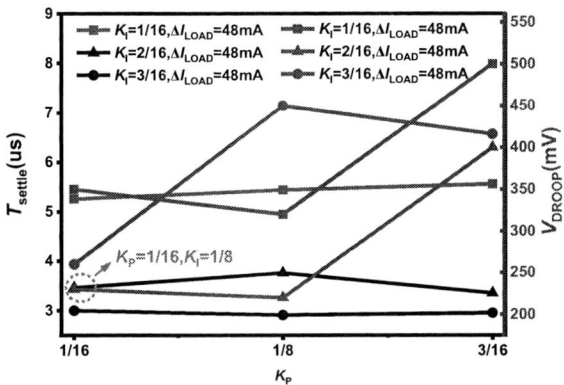

Fig. 14. V_{DROOP} and T_{settle} of the DLDO under different K_p and K_I values.

TABLE I.   PERFORMANCE COMPARISON

	This work	[18]	[19]	[20]
Type	Digital	Digital	Digital	Digital
Process(nm)	28	28	110	110
Active area(mm²)	0.029	0.025	0.022	0.04
V_{IN} range(V)	1.2-1.8	0.65-1	0.8-1.2	0.6-1.2
V_{OUT} range(V)	0.9-1.6	0.55-0.95	0.7-1.1	0.5-0.9
C_{OUT}(nF)	0.1	13.4	0.04	1
Quiescent Current (mA)	0.6	0.05-0.118	0.189-0.198	0.012
$I_{LOAD,MAX}$ (mA)	80	260	50	80
T_{settle}(us)@ΔI_{LOAD}(mA)	3.4@48	0.34@100	0.25@47.5	20@80
V_{DROOP}(mV)@ΔI_{LOAD}(mA)	220@48	127@100	360@47.5	55@80
FOM(ps)	5.73	20.1	1.26	0.26
V_{RIPP}(mV)@I_{LOAD}(mA),V_{REF}(V)	2.9@60 ,1.4 27.6@20 ,1.1	35@5 ,0.7	3@2.5 ,0.85	4@80 ,0.5

FOM=$C_{OUT}(\Delta V/\Delta I_{LOAD})(I_Q/\Delta I_{LOAD})$ Smaller value is better

V. CONCLUSION

A fully integrated DLDO featuring an 8-bit ADC and a small unity PMOS current for the reduction of steady-state ripples has been presented. The high-resolution ADC accurately quantizes V_{OUT}, generating the digital code N_{OUT}. As the output voltage nears V_{REF}, the proportional–integral (PI) controller ceases the continuous switching of the PMOS devices, resulting in only an LSB-level PMOS turning on and off. The output voltage ripples (V_{RIPP}) can be approximated as a unity PMOS current multiplied by the on-resistance (V_{DROP}/I_{LOAD}). The simulation results show that as the reference voltage and load current increase, V_{RIPP} gradually decreases. When V_{REF} = 1.4 V and I_{LOAD} = 60 mA, V_{RIPP} is reduced to 2.9 mV. Additionally, the T_{settle}, undershoot, and overshoot voltages of the DLDO are 3.4 µs, 260 mV, and 380 mV, respectively, with V_{REF} configured to 1.3 V and an I_{LOAD} step varying from 7 to 55 mA, exhibiting a 2 µs edge time.

REFERENCES

[1] K. C. Kwok and P. K. T. Mok, "Pole-zero tracking frequency compensation for low dropout regulator," in Proc. IEEE Int. Symp. Circuits Syst., vol. 4. May 2002, pp. IV-735–IV-738.\

[2] Vasudevan U, Rincón-Mora G A. Digital LDO Analysis and All-Stable High-PSR One-LSB Oscillator Design[J]. Electronics, 2024, 13(24): 5033.

[3] Sood L, Agarwal A. A CMOS standard-cell based fully-synthesizable low-dropout regulator for ultra-low power applications[J]. AEU-International Journal of Electronics and Communications, 2021, 141: 153958.

[4] Zhang H, Wan P, Geng J, et al. A fast transient response digital LDO with a TDC-based signal converter[J]. Electronics, 2020, 9(1): 132.

[5] G. Eason, B. Noble, and I. N. Sneddon, "On certain integrals of Lipschitz-Hankel type involving products of Bessel functions," Phil. Trans. Roy. Soc. London, vol. A247, pp. 529–551, April 1955.

[6] T. Singh, S. Rangarajan, D. John, C. Henrion, S. Southard, H. McIntyre, A. Novak, S. Kosonocky, R. Jotwani, A. Schaefer, E. F. Chang, J. Bell, and M. Co, 3.2 Zen: A next-generation high-performance 86 core, in IEEE Int. Solid-State Circuits Conf. (ISSCC) Dig. Tech. Papers, Feb. 2017, pp. 5253.

[7] Kim S T, Shih Y C, Mazumdar K, et al. 8.6 Enabling wide autonomous DVFS in a 22nm graphics execution core using a digitally controlled hybrid LDO/switched-capacitor VR with fast droop mitigation[C]//2015 IEEE International Solid-State Circuits Conference-(ISSCC) Digest of Technical Papers. IEEE, 2015: 1-3.

[8] Y.-J. Lee et al., "A 200-mA digital low drop-out regulator with coarse fine dual loop in mobile application processor," IEEE J. Solid-State Circuits, vol. 52, no. 1, pp. 64-76, Jan. 2017.

[9] Y. Okuma et al., "0.5-V input digital LDO with 98.7% current efficiency and 2.7-µA quiescent current in 65nm CMOS," in Proc. IEEE Custom Integr. Circuits Conf. (CICC), San Jose, CA, USA, 2010, pp. 1–4.

[10] Akram M A, Hong W, Ha S, et al. Capacitor-less dual-mode all-digital LDO with ΔΣ-modulation-based ripple reduction[J]. IEEE Transactions on Circuits and Systems II: Express Briefs, 2021, 68(5): 1620-1624.

[11] M. Cheah, D. Mandal, B. Bakkaloglu, and S. Kiaei, "A 100-mA, 99.1% current efficiency, 2-mVpp ripple digitally controlled LDO with active ripple suppression," IEEE Trans. Very Large Scale Integr. (VLSI) Syst., vol. 25, no. 2, pp. 696–704, Feb. 2017.

[12] Y. Zhang, H. Song, R. Zhou, W. Rhee, I. Shim, and Z. Wang, "A capacitor-less ripple-less hybrid LDO with exponential ratio array and 4000x load current range," IEEE Trans. Circuits Syst. II, Exp. Briefs, vol. 66, no. 1, pp. 36–40, Jan. 2019.

[13] M. Huang, Y. Lu, S.-W. Sin, S.-P. U, R. P. Martins, and W.-H. Ki, "Limit cycle oscillation reduction for digital low dropout regulators," IEEE Trans. Circuits Syst. II, Exp. Briefs, vol. 63, no. 9, pp. 903–907, Sep. 2016.

[14] M. A. Akram, I.-C. Hwang, and S. Ha, "Architectural advancement of digital low-dropout regulators," IEEE Access, vol. 8, pp. 137838–137855, 2020.

[15] You Y, Tian R, Zhang Y, et al. A High-Voltage-Compliant 86% Peak Efficiency Current-Mode Stimulator With Dynamic Voltage Supply for Implantable Medical Devices[J]. IEEE Journal of Solid-State Circuits, 2024.

[16] Kundu, S., Liu, M., Wen, S.-J., Wong, R. & Kim, C. H. Kundu S, Liu M, Wen S J, et al. A fully integrated digital LDO with built-in adaptive sampling and active voltage positioning using a beat-frequency quantizer[J]. IEEE journal of solid-state circuits, 2018, 54(1): 109-120.

[17] Xu D, Zhang Y, Li Z, et al. A fully-integrated LDO with two-stage cross-coupled error amplifier for high-speed communications in 28-nm CMOS[C]//2023 IEEE International Symposium on Circuits and Systems (ISCAS). IEEE, 2023: 1-4.

[18] Han Y, Kim J, Koo G, et al. A DVS-Enabled Distributed Digital LDO Providing Rapid Uniform Power Grid and Ripple Reduction Achieving 20.1-ps FOM in 28 nm CMOS[J]. IEEE Transactions on Circuits and Systems I: Regular Papers, 2024.

[19] Chen W J, Huang C H. Fast-turnaround design and modeling techniques for a fast-transient digital low-dropout regulator with 3 mV ripples[J]. IEEE Transactions on Power Electronics, 2020, 36(6): 6824-6837.

979-8-3315-8850-2/25 $31.00 © 2025 IEEE

[20] T.-J. Oh and I.-C. Hwang, "A 110-nm CMOS 0.7-V input transient enhanced digital low-dropout regulator with 99.98% current efficiency at 80-mA load," IEEE Trans. Very Large Scale Integr. (VLSI) Syst., vol. 23, no. 7, pp. 1281–1286, Jul. 2015.

2025 The 10th International Conference on Integrated Circuits and Microsystems

MOSFET Modeling of 0.18 μm CMOS Technology Based on BSIM4 Model for Cryogenic Devices

Dong Chen*, Zuoru Dong*, Yushi Chen, Xingyu Cui, Fang Liang, Xiaodong Wang
Shanghai Microwave Technology Research Institute, Shanghai 2003331, China
Correspondence author: ch911103@163.com, zrdong2021@163.com

Abstract—**The MOSFET model libraries are the links between the integrated circuit designs and fabrications. A reasonable MOSFET model library can improve the reliability of the designed circuits and reduce the subsequent costs. In response to the requirements for the cryogenic readout integrated circuits (ROICs) designed for the quantum wells, Type II superlattices, blocked impurity band (BIB) detectors, and so on, the electrical properties of metal-oxide-semiconductor field-effect transistor (MOSFET) devices with various sizes were measured at 4 K. Then, a cryogenic MOSFET model library of 0.18 μm technology was established based on the Berkeley Short-channel IGFET Model 4 (BSIM4) model. The root mean square (RMS) errors between the simulation outputs of the established cryogenic model library and the measured current-voltage (IV) data were less than 5%, demonstrating that the established cryogenic model library could effectively guide the cryogenic RIOC design. In addition, considering the inevitable Kink effect of MOSFETs under cryogenic temperature conditions, it was recommended that ROIC designers should control the operation range of the drain-source voltage V$_{DS}$.**

Keywords—*cryogenic devices; MOSFET modeling; BSIM4 model; IV characteristics*

I. INTRODUCTION

In the process of integrated circuit design, designers typically need to use MOSFET model libraries for circuit designs and simulations, and then judge whether the circuit performances meet the requirements based on the simulation results[1]. It can be said that the MOSFET model libraries are the bridges between the circuit designs and fabrications[2]. Therefore, to accurately predict the performances of the foundry-processed integrated circuits under specific conditions, it is necessary to establish corresponding MOSFET model libraries by extracting a series of MOSFET device parameters.

The quantum wells, Type II superlattices, and blocked impurity band (BIB) detectors, recognized for their exceptional sensitivity, typically operate at low temperatures[3, 4]. Consequently, to maintain their high sensitivity, the associated readout circuits also need to operate in a cryogenic environment. The focus of the cryogenic MOSFET modeling in this study is to facilitate the design of the readout integrated circuits (ROICs) for cryogenic detectors. Generally, semiconductor foundries

provide standard MOSFET model libraries only for temperatures ranging between -40℃ and 125℃[5], which do not accurately reflect the true characteristics of devices under cryogenic conditions. Therefore, it is crucial to develop a model library specifically for cryogenic MOSFET devices to effectively guide the corresponding cryogenic ROIC designs.

Currently, several cryogenic studies based on CMOS technology have been reported[6-8]. Zhang et al. proposed a cryogenic model library by modifying the mobility calculation equations based on the EVK model[8]. However, this cryogenic MOSFET model library is limited to large-sized devices and diverges from the Berkeley Short-channel IGFET Model (BSIM)-based models typically provided by semiconductor foundries. In response to the design requirements of small-sized cryogenic ROICs, a cryogenic model library that encompasses various MOSFET sizes is developed based on the commonly utilized BSIM4 model, which would ensure the simulation accuracy in ROIC designs and enhance the reliability of ROICs.

II. BSIM4 MODEL

The BSIM model was developed by the University of California, Berkeley in the 1980s. It is the first standard industry model for the simulation program with integrated circuit emphasis (SPICE) simulations[9, 10]. The BSIM1 model was specifically developed for 1 μm MOSFET technology and can accurately simulate MOSFET devices with channel lengths greater than or equal to 1 μm. The subsequent BSIM2 model has many improvements based on the BSIM1 model, such as model continuity, output conductance, subthreshold current, etc. The BSIM3 model was based on a quasi-2D analytical physical model that emphasizes understanding device operation's physical mechanisms while accounting for size effects and process parameters. It was developed for deep submicron (DSM) and nanoscale MOSFET devices[11]. The BSIM4 model was developed to address a variety of challenges in advanced manufacturing processes. It drew on the experience of the BSIM3 model, added modeling of the physical effects of device operations, and added fitting parameters to improve the accuracy of the model. Furthermore, multiple other versions of the BSIM models have been developed for specific processes or applications[11-13], which can be obtained in the Process Design Kit (PDK) provided by the foundries. Presently, the BISM4 model stands as one of the most widely utilized models in practice.

This work is supported by the Shanghai Sailing Program (Grant Nos. 23YF1444300), the National Natural Science Foundation of China (Grant Nos. 62301321, and 62171286), and the Program of Shanghai Academic/ Technology Research Leader (Grant No. 21XD1423600).

979-8-3315-8850-2/25 $31.00 © 2025 IEEE   439

Although the BSIM4 model is widely accepted in the industry as the classic MOSFET model, merely modifying the temperature parameters cannot accurately describe the electrical behaviors of MOSFETs operating at cryogenic temperatures. This limitation stems from the temperature-dependent nature of certain parameters in the BSIM4 model[14], which inevitably results in simulation deviation. Therefore, it is essential to re-establish an accurate cryogenic MOSFET model library based on modifications made to the BSIM4 model. Some core parameters of the BSIM4 model are shown in Table I.

TABLE I.   THE CORE PARAMETERS OF THE BSIM4 MODEL

Parameter Name	Description	Default	Unit
TOXE	Electrical gate equivalent oxide thickness	3.0×10^{-9}	m
XJ	Source/drain junction depth	1.5×10^{-7}	m
VTH0	Long channel threshold voltage at Vbs=0	±0.7	V
K1	First-order body bias coefficient of the Vth model	0.53	$V^{-1/2}$
K2	Second-order body bias coefficient of the Vth model for non-uniform vertical doping	-0.0186	-
LPE0	Vth roll-up parameter owing to pocket implants	1.74×10^{-7}	m
DVT0	Vth roll-off coefficient parameter	2.2	-
DVT1	Vth roll-off length-dependence parameter	0.53	-
U0	Low-field mobility	0.067 or 0.025	$m^2/(V \cdot s)$
VSAT	Channel carrier saturation velocity	8.0×10^4	m/s
WINT	Gate and channel edge overlap length in the width direction	0	m
LINT	Gate-source or -drain overlap length	0	m
VOFF	The offset voltage of the effective gate drive	-0.08	V
NFACTOR	Subthreshold swing factor	1.0	-
ETA0	Vds-dependence coefficient for the Vth DIBL model	0.08	-
PCLM	Channel length modulation parameter	1.3	-
RDSW	Zero-bias source and drain LDD resistance component per unit width for RDSMOD=0	200.0	ohm(μm)WR
DLC	Gate-source or -drain overlap length for capacitance models	=LINT	m
DWC	Gate and channel edge overlap length in the width direction for capacitance models	=WINT	m

III. CRYOGENIC MEASUREMENTS

The objective of MOSFET modeling is to accurately reflect the actual electrical input and output characteristics of the device. Thus, the modeling process primarily relies on the electrical test data of the MOSFET devices. The sizes of MOSFETs utilized in the modeling process should contain at least four categories, namely $W_{max}L_{array}$, $W_{min}L_{array}$, $L_{max}W_{array}$, and $L_{min}W_{array}$, where W and L denote the width and length of the MOSFETs, respectively, as shown in Fig. 1, where the letters "V" in the figure represent the value points. By measuring the IV

characteristics of these devices, a compact MOSFET model library can be developed based on the BSIM4 model.

W/L	10	4	2	1
10	V	V	V	V
5	V			V
2	V			V
1	V			V
0.5	V	V	V	V

Fig. 1.   Size design of MOSFET devices.

The Keithley 4200A-SCS semiconductor device analyzer, featuring a current accuracy of 10^{-13} A and a capacitance accuracy of 10^{-12} F, was used to evaluate the electrical characteristics of the MOSFET devices. Given that the current of a single-tube MOSFET is generally in the order of 10^{-6}A, direct IV measurements can be conducted by the test instrument. A partial layout of the device is illustrated in Fig. 2. The designed device arrays were delivered to the foundry for fabrication. Subsequently, the prepared devices were positioned in a cryogenic probe station where temperatures were stabilized at 4 K through liquid helium refrigeration. The cryogenic probe station was then connected to the semiconductor device analyzer for measuring the electrical characteristics of the devices.

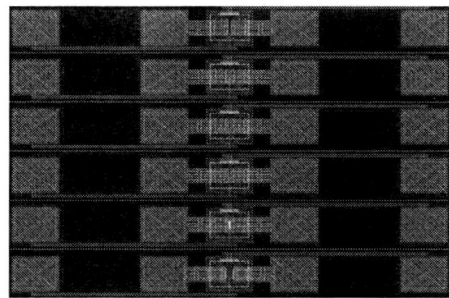

Fig. 2.   MOSFET devices array layout.

IV. CRYOGENIC MOSFET MODELING

The essence of a MOSFET model is a set of physical and modulation parameters that are calibrated to accurately reflect the electrical characteristics of devices at different fabrication processes and temperatures[15, 16]. To establish an accurate correlation between model predictions and actual device performance, it is essential to extract key physical parameters from experimental data to modify the standard BSIM4 model. This involves importing measured data from MOSFET devices into parameter extraction software and adjusting model parameters based on discrepancies between experimental outcomes and simulation results, ultimately minimizing the root mean square (RMS) errors between simulation outputs and experimental data as small as possible. In the figures below, the test data are represented by dashed lines while the simulation

979-8-3315-8850-2/25 $31.00 © 2025 IEEE

results are represented by solid lines. And the RMS errors between these two sets can be directly observed.

A. Standard Model Verification

To assess the relevance of the standard MOSFET model library at room temperature for deep cryogenic circuit design, the standard model library was first utilized to simulate the performance of MOSFET devices at 300 K. The simulation results aligned well with the test outcomes from the foundry-prepared MOSFET devices at room temperature, but deviated significantly from the measured data obtained at 4 K, exhibiting RMS errors exceeding 15%, as shown in Fig. 3. This discrepancy indicates a substantial difference between the standard model library's simulations at room temperature and actual device performance under cryogenic conditions, rendering it ineffective for guiding cryogenic circuit designs.

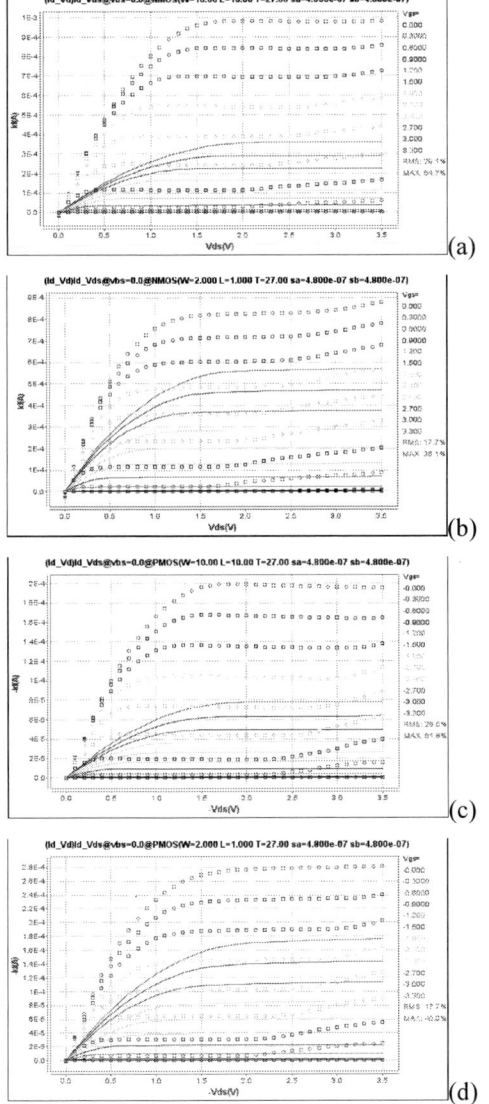

Fig. 3. Comparison between the simulation results (solid lines) based on the standard model at 300 K and the measured data (dashed lines) of the foundry-prepared MOSFET devices at cryogenic temperature. a I_{DS}/V_{DS} characteristic of large-size NMOSs. b I_{DS}/V_{DS} characteristics of small-size NMOSs. c -I_{DS}/-

V_{DS} characteristics of large-size PMOSs. d -I_{DS}/-V_{DS} characteristics of small-size PMOSs.

Subsequently, the temperature condition of the standard model was modified directly to 4 K and the simulation results were compared to the cryogenic measured data, as shown in Fig. 4. The findings reveal that the current values derived from simulations are considerably higher than those observed in cryogenic tests, resulting in markedly increased RMS errors. This indicates that it is critical to adjust the temperature-dependent parameters in the BSIM4 model.

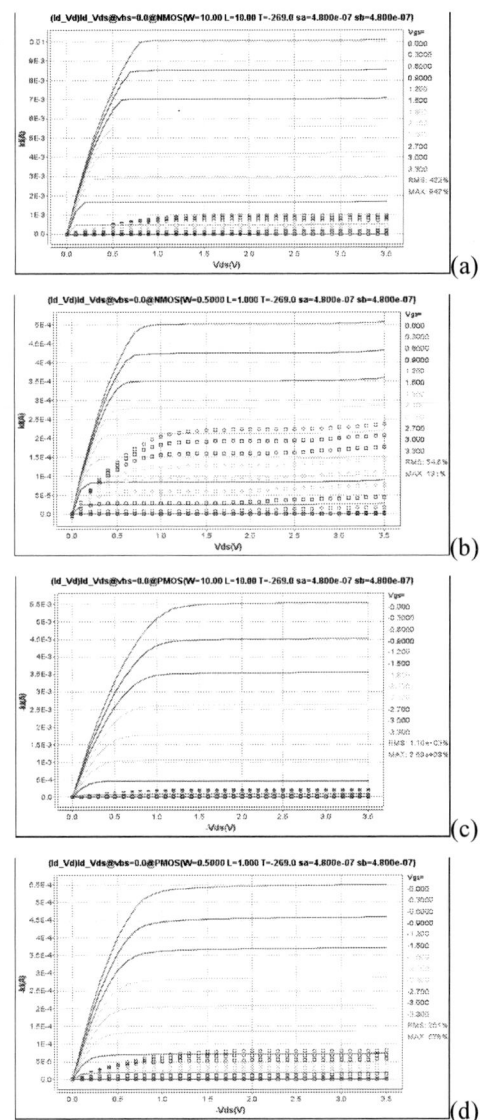

Fig. 4. Comparison between simulation results (solid lines) based on the standard model at 4K and the measured data (dashed lines) of the foundry-prepared MOSFET devices at cryogenic temperature. a I_{DS}/V_{DS} characteristics of large-size NMOSs. b I_{DS}/V_{DS} characteristics of small-size NMOSs. c -I_{DS}/-V_{DS} characteristics of large-size PMOSs. d -I_{DS}/-V_{DS} characteristics of small-size PMOSs.

Given these above two cases, it is evident that the standard model library fails to accurately represent MOSFET device characteristics at cryogenic temperatures. Therefore, establishing a reasonable cryogenic temperature model library by extracting parameters at cryogenic temperatures is essential for enhancing the design accuracy of cryogenic ROICs.

B. Cryogenic Parameters Extraction

Following measurements of IV characteristics of MOSFET devices, a dedicated cryogenic MOSFET model library was established by extracting and modifying the BSIM4 model parameters. Partial simulation results for the established cryogenic model of NMOS devices are shown in Fig. 5. Notably, the RMS errors between the simulation outcomes and the experimental data of the NMOS devices were less than 5%. It demonstrates that the established cryogenic NMOS model can reflect both the input and output characteristics of the devices well.

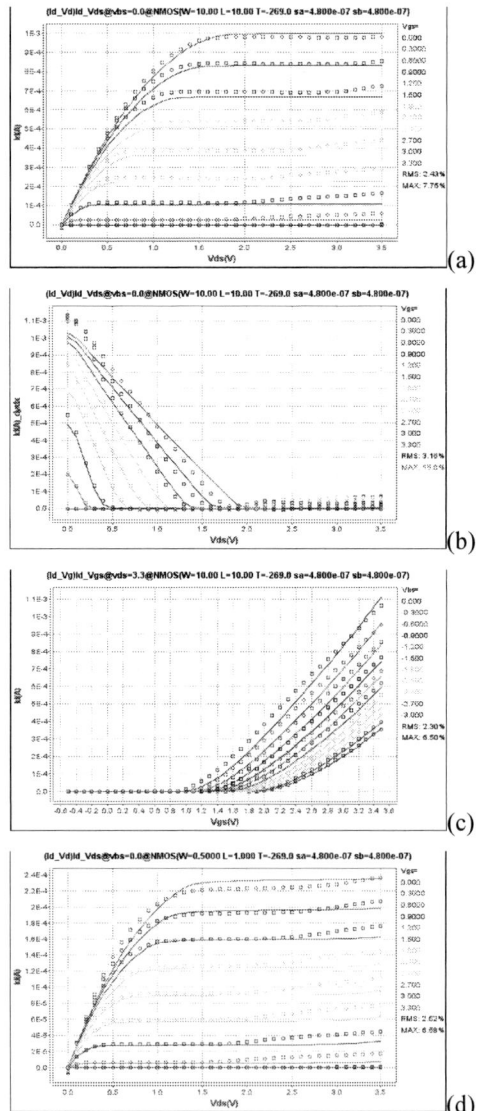

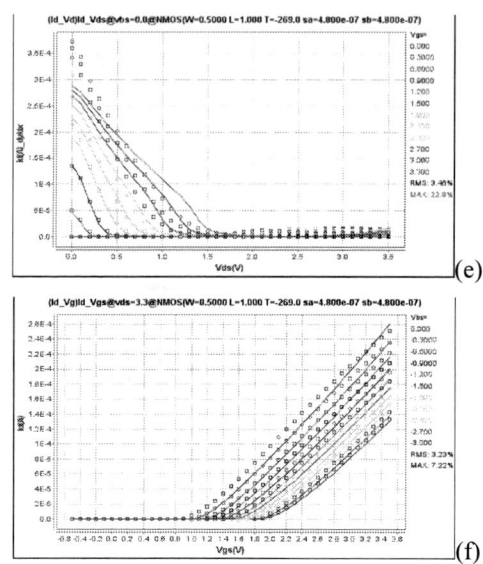

Fig. 5. Comparison between the simulation results (solid lines) based on the cryogenic NMOS model and the measured data (dashed lines) of NMOSs. a I_{DS}/V_{DS} characteristics of large-size NMOSs. b dI_D/dV_D characteristics of large-size NMOSs. c I_D/V_G characteristics of large-size NMOSs. d I_{DS}/V_{DS} characteristics of small-size NMOSs. e dI_D/dV_D characteristics of small-size NMOSs. f I_D/V_G characteristic of small-size NMOSs.

Additionally, partial simulation results for the established cryogenic model of PMOS devices are shown in Fig. 6. The simulation outcomes based on the cryogenic PMOS model closely align with the experimental measured data. It is noteworthy that the parameters within the PMOS model are independent of those in the NMOS model.

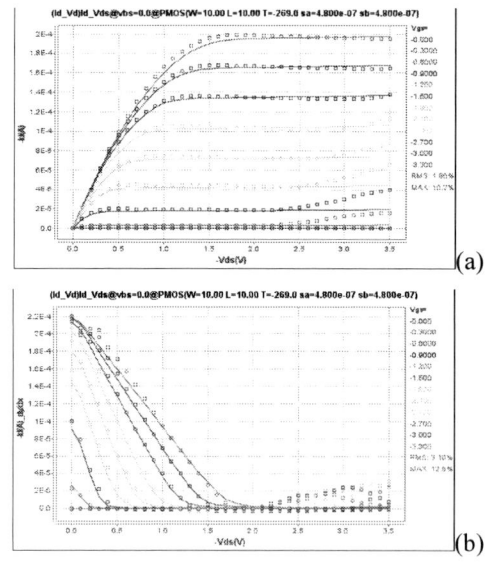

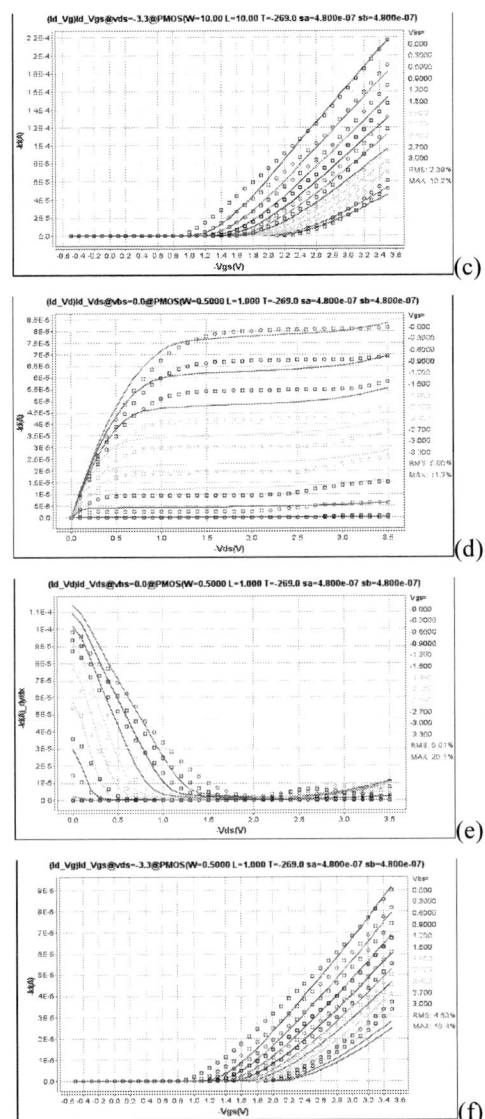

Fig. 6. Comparison between the simulation results (solid lines) based on the cryogenic PMOS model and the measured data (dashed lines) of PMOSs. a I_{DS}/V_{DS} characteristics of large-size PMOSs. b dI_D/dV_D characteristics of large-size PMOSs. c I_D/V_G characteristics of large-size PMOSs. d I_{DS}/V_{DS} characteristics of small-size PMOSs. e dI_D/dV_D characteristics of small-size PMOSs. f I_D/V_G characteristics of small-size PMOSs.

Through the above simulations and experimental comparisons, the accuracy of the established cryogenic MOSFET model was validated, which significantly enhances the reliability of the cryogenic ROIC designs. However, as illustrated in Fig. 5a and Fig. 6a, with an increase in drain-source voltage (V_{DS}) and a decrease in gate voltage (V_{GS}), drain-source current (I_{DS}) suddenly rises, showing an upward warping, known as the "Kink effect"[17]. This behavior can be attributed to carriers being frozen at impurity energy levels at cryogenic temperatures[18], resulting in an increase in threshold voltage and transconductance[19]. Concurrently, scattering effects diminish while carrier mean free paths increase. Thus, I_{DS} increases with the increase in carrier mobility[20]. The Kink effect primarily manifests within regions where V_{DS} exceeds V_{GS} and can be explained by the self-polarization of the substrates at cryogenic temperatures[21]. It is noteworthy that the Kink effect diminishes with an increase in substrate bias voltage. For the BSIM4 model, while there are internal tunable parameters designed to account for substrate bias effects, the underlying mechanism of the Kink effect is highly complex. Consequently, the corresponding parameters cannot be easily adjusted to accurately reflect such a pronounced current Kink effect at 4 K temperature. Therefore, to effectively guide the design of cryogenic ROICs, it is essential to ensure the accuracy of both threshold voltage and saturation current. Additionally, the integrated circuit (IC) designers should carefully control the range of V_{DS} to guarantee that MOSFET devices operate within the normal saturation region.

C. Discussion

It is evident that the discrepancies between the simulation results based on the standard MOSFET model and the IV measurements of the devices at cryogenic temperatures are too large to guide the design of cryogenic ROICs. Based on the cryogenic IV measured data of MOSFET devices, the parameters of the BSIM4 model have been extracted and adjusted to establish a dedicated cryogenic model library. The RMS errors of the simulation outcomes based on this cryogenic model library and the actual cryogenic IV measurement results of various sizes of MOSFETs are less than 5%, indicating that the established cryogenic model library can be effectively utilized for cryogenic ROIC simulation and performance prediction. Considering the decreasing size of ICs processed with advanced technologies, shallow trench isolation (STI) stresses, well proximity effects (WPEs), MOS-Mismatch, and other factors should be considered in future work, to establish a more comprehensive cryogenic MOSFET model library. Addressing these aspects will mitigate the risks of process fluctuations and enhance the success rate of circuit fabricating.

V. CONCLUSION

In this work, the cryogenic MOSFET modeling of 0.18 μm CMOS technology based on the BSIM4 model was presented. The IV characteristics of MOSFET devices of various sizes were measured at a temperature of 4 K. Following simulation verification, it was found that the standard model provided by the foundry did not accurately reflect the electrical performance of MOSFETs at cryogenic temperature. Therefore, the cryogenic IV characteristic parameters of MOSFET devices were extracted by modeling software, and a cryogenic MOSFET model library was established. This model library can be imported into the circuit design software to facilitate simulation and guide the cryogenic ROIC designs. Moreover, the RMS errors between the simulation results and experimental data for the cryogenic IV characteristics of MOSFETs were less than 5%, indicating that the developed cryogenic model library enhances the reliability of cryogenic ROIC designs. Additionally, due to the Kink effect at cryogenic temperatures, the accuracy of the threshold voltage and the saturation current should be guaranteed, and circuit designers should control the V_{DS} range so that MOSFET devices operate in the normal saturation region.

REFERENCES

[1] Liu, W. and Hu, C. BSIM4 and MOSFET Modeling For IC Simulation, 2011

[2] Chaudhry, A. and Roy, J.N. Mosfet Models, Quantum Mechanical Effects and Modeling Approaches: A Review. Journal of Semiconductor Technology and Science, 2010, 10(1): 2-27

[3] Krzysztof, M., Kinga, M., Małgorzata, K., Tetiana, M., Bartłomiej, S., Łukasz, K. and Piotr M. Optical Characterization of the Interband Cascade LWIR Detectors with Type-II InAs/InAsSb Superlattice Absorber. Nanomaterials, 2024, 14(17): 1393

[4] Wang, X., Chen, Y., Chen, X., Wang, B., Zhang, C. and Zhang, H. Temperature-dependent spectral response mechanism in GaAs-based blocked-impurity-band (BIB) far-infrared detectors. Optical and Quantum Electronics, 2019, 52(1)

[5] Wang, Q., Ye, M., Li, Y., Zheng, X., He, J., Du, J. and Zhao, Y. MOSFET modeling of 0.18μm CMOS technology at 4.2K using BP neural network. Microelectronics journal, 2023, 132: 105678-105678

[6] Liu, S., Li, X., Liu, C. and Sun, W. Improved Metal Oxide Semiconductor Field Effect Transistor models with wide temperature range including cryogenic temperature. Superlattices and Microstructures, 2017, 109: 31-40

[7] Aykut K., Nergiz S.S., Sadik I., Uzun, Y. and Mustafa B.Y. Statistical MOSFET Modeling Methodology for Cryogenic Conditions. IEEE Transactions on Electron Devices, 2019, 66(1): 66-72

[8] Zhang, Y., Lu, T., Wang, W., Zhang, Y., Xu, J., Luo, C. and Guo, G. Characterization and Modeling of Native MOSFETs Down to 4.2 K. IEEE Transactions on Electron Devices, 2021, 68(9): 4267-4273

[9] Jacunski, M.D., Shur, M.S., Owusu, A.A., Ytterdal, T. and Hack, M. SPICE Models for N and P Channel Polysilicon Thin Film Transistors in All Regimes of Operation. Proceedings of Second International Workshop on Active Matrix Liquid Crystal Displays, 1995: 40-143

[10] [Cheng, Y., Imai, K., Jeng, M. C., Liu, Z., Chen, K. and Hu, C. Modelling temperature effects of quarter micrometre MOSFETs in BSIM3v3 for circuit simulation. Semiconductor Science and Technology, 1997, 12(11): 1349-1354

[11] Chain, K., Huang, J., Duster, J.S., Ping Keung Ko and Hu, C. A MOSFET electron mobility model of wide temperature range (77 - 400 K) for IC simulation. Semiconductor Science and Technology, 1997, 12(4): 355-358

[12] Register, L.F., Rosenbaum, E. and Yang, K. Analytic model for direct tunneling current in polycrystalline silicon-gate metal‐oxide‐semiconductor devices. Applied Physics Letters, 1999, 74(3): 457-459

[13] Luan, S. and Neudeck, G.W. An experimental study of the source/drain parasitic resistance effects in amorphous silicon thin film transistors. Journal of Applied Physics, 1992, 72(2): 766-772

[14] Zhao, H. and Liu, X. Modeling of a standard 0.35μm CMOS technology operating from 77K to 300K. Cryogenics, 2014, 59: 49-59

[15] Leitner, T. A Simulator Independent Semiconductor Model Implementation Based on SPICE Model Equations. Analog Integrated Circuits and Signal Processing, 1999, 21(1): 9-19

[16] Walston, J. Simulation and Modeling-Extracting SPICE model parameters from data sheets. IEEE Circuits and Devices Magazine, 1991, 7(6): 10-12

[17] E. Rocofyllou, A. Galiouna Nassiopoulos, D. Tsamakis and Balestra, F. Anomalous behaviour of n-channel MOS transistor characteristics in the temperature range 4.2-14 K. Solid-state electronics, 1989, 32(8): 603-605

[18] Akturk, A., Allnutt, J., Dilli, Z., Goldsman, N. and Peckerar, M. Device Modeling at Cryogenic Temperatures: Effects of Incomplete Ionization. IEEE Transactions on Electron Devices, 2007, 54(11): 2984-2990

[19] Omura, Y., Nakakubo, A. and Nakatsuji, H. Threshold voltage of sub-10-nm-thick SOI MOSFET's at cryogenic temperature and quantum effects. IEEE International SOI Conference, 2004: 53-54

[20] Ghibaudo, G., Aouad, M., Casse, M., Poiroux, T. and Theodorou, C. On the diffusion current in a MOSFET operated down to deep cryogenic temperatures. Solid-State Electronics, 2021, 176: 107949

[21] Balestra, F., Luc Audaire and Lucas, C. Influence of substrate freeze-out on the characteristics of MOS transistors at very low temperatures. Solid-state electronics, 1987, 30(3): 321-327

2025 The 10th International Conference on Integrated Circuits and Microsystems

Process Integration Optimization for RF SOI Process with Arcing Issue

Zhangli Liu
Teconology Development
Shanghai Huahong Grace Semiconductor
Manufactury Corp.
Shanghai, China
Zhangli.Liu@hhgrace.com

Fei Meng
Teconology Development
Shanghai Huahong Grace Semiconductor
Manufactury Corp.
Shanghai, China
Fei. Meng@hhgrace.com

Ruofan Dai
Teconology Development
Shanghai Huahong Grace Semiconductor
Manufactury Corp.
Shanghai, China
Ruofan. Dai@hhgrace.com

Abstract— The process optimization for RF SOI process with arcing issue is comprehensively investigated in the work. By failure mechanism and analysis, the optimized integration process solution is proposed: (1) Passivation and top via photo wafer edge exposure (WEE) setting, (b) Top via W Chemical-Machine-Polishing (CMP) with wafer edge over polish, and (c) Passivation etch process optimized, by reducing the passivation thickness on the thick top metal. Arcing free is obtained by the optimized integration process solution. The method is verified by mass production.

Keywords—Radio Frequency, Silicon-on-insulator, Arcing, Wafer Edge Exposure, Chemical-Machine-Polishing

I. INTRODUCTION

The radio frequency silicon on insulator (RF SOI) process is the major stream for radio-frequency front end module (RF FEM) application, including the switch, low noise amplifier (LNA) and extensive drain metal-oxide-semiconductor (EDMOS) [1-2]. The bulk logic complementary metal oxide semiconductor (CMOS) process is the baseline for RF SOI process. Some process loops and recipes should be revised according to the special characteristics for SOI wafer, such as active region silicon etch, rapid thermal anneal, implant step, et al. As shown in Fig.1, wafer edge partial die arcing is observed for RF SOI process at the passivation etch step. The arcing is happened at the wafer edge only. The plasma charge collected by the wafer edge floating metal, inducing arcing when the charge is too high. The arcing issue leading to tool down and wafer scrap, which impact the normal production. The arcing ratio is relatively high for RF SOI process product. The arcing issue improvement process method is proposed in [3-7], for surface charge treatment, via etch and reactive ion etching (RIE). However, the integrated process optimization is not clearly discussed for RF SOI CMOS process. In order to fix the arcing issue, failure analysis scanning electron microscope (FA SEM) result to check out the root cause. Base on the experiment result, the failure mechanism is proposed in the work. Comprehensive investigation on the process integration method, arcing free solution is obtained. The method is verified by mass production.

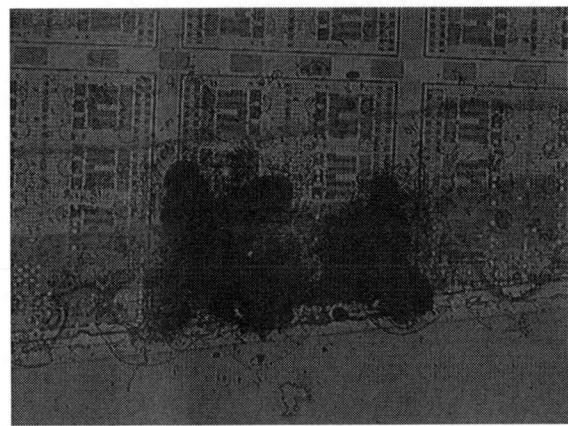

Fig. 1. Wafer edge arcing at passivation etch

II. EXPERIMENT RESULT AND DISCUSSION

A. Failure Mechanism and Analysis

The wafer edge partial die arcing is happened at passivation etch process loop. As we know that, high plasma charge is introduced during the etch process. In order to enlarge the process window, over etch is necessary. At the same time, there is large step height at the wafer edge. The passivation film thickness at wafer edge should thinner than the center one. During the top via W CMP process, the W residue should be observed at the wafer edge. As shown in Fig. 2, obvious W residue is observed from the FA SEM result. The long tail of W residue at the most wafer edge is floating metal, which is the arcing source. The cross-section for RF SOI process is shown in Fig.3 a). For the baseline, the passivation photo WEE width (PA WEE1) is larger than the via photo WEE width (VIA WEE1), which with high risk for W residue exposure to passivation etch plasma. The baseline passivation thickness is relatively thick. The main etch time is relatively long in correspond, which with high risk for W residue exposure to passivation etch plasma also. In a word, the baseline process integration with high risk for arcing issue during passivation etch step when the plasma charge collected by the floating W residue. The accumulated charge on floating W residue increase with the passivation etch time. When the charge is collected enough, the discharge charge path should be along the wafer edge step height and the seal ring line. The arcing is

979-8-3315-8850-2/25 $31.00 © 2025 IEEE   445

happened when the accumulated charge too much. Only wafer edge partial die arcing is observed for the RF SOI process.

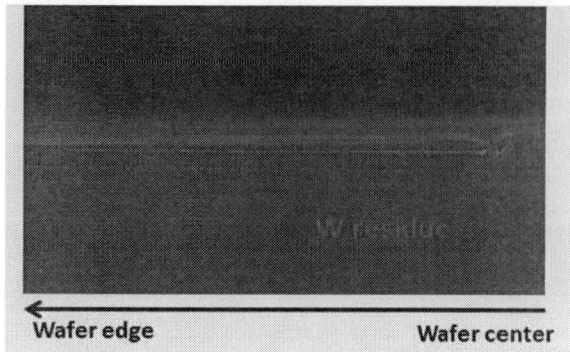

Fig. 2. FA SEM for wafer edge W residue

a) Baseline

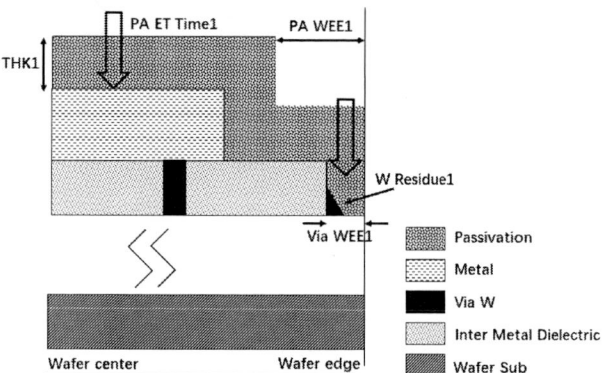

b)Optimized process integration

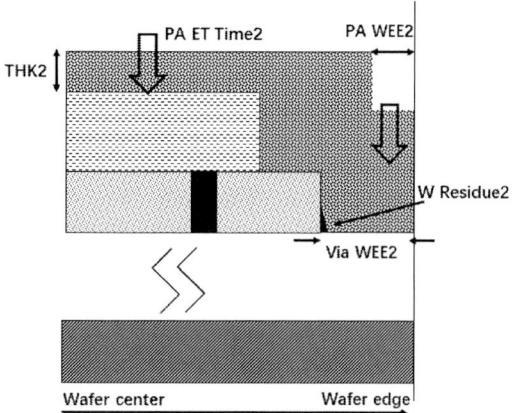

Fig. 3. Cross-section for RF SOI process. (a) Baseline; (b) Optimized process integration

B. CMOS Baseline Process Integration

For the work, the major RF SOI CMOS process integration is same as industry baseline generic CMOS. The arcing issue relatively process loop is the back-end-of-loop (BEOL), especially for passivation process loop. The first metal is deposited usually by sputtering, and defined using photo resist and mask. Then the metal layer pattern is formed by etch process, and covered by dielectric layer. More than on one level of wiring on the wafer surface because of the complex

circuit design. When the inter-metal-dielectric (IMD) film deposition is finished, the top via photo, etch is following. The W of via is filled by physical-vapor deposition (PVD) and CMP [8-9]. And, the metal sputter is used for top metal. Then the top metal photo, etch formed the pattern for electrical connection. The passivation oxide and silicon nitride is formed by chemical-vapor-deposition (CVD). The passivation is for chip protection from the moisture, metal contamination, et al. Finally, the passivation photo and etch forming the pad for chip bonding/bumping. The arcing issue may happened during passivation deposition, which use the high-density plasma CVD (HDP CVD). Mostly, the arcing issue is happened at the passivation etch step due to the high plasma charge. In the work, the arcing issue is observed post passivation etch loop.

a) Baseline

b) Optimized PA/Via WEE

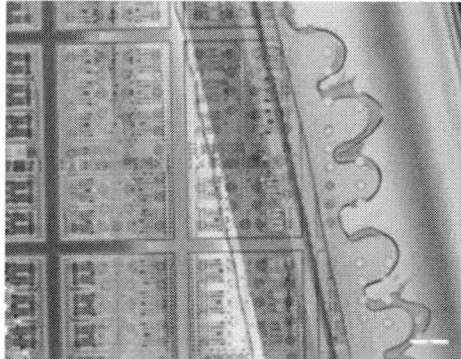

Fig. 4. Passivation and Top via photo WEE setting. a) Baseline: PA WEE1>Via WEE1, b) Optimized recipe: PA WEE2<Via WEE2.

C. Process Integration Optimization

Base on the failure mechanism and analysis, the baseline process integration should be improved. The optimized process integration solution is proposed for arcing free: the first one is the passivation and top via WEE setting, the second one is top via W CMP setting, and the third one is passivation etch plasma charge reduction. The experiment result is discussed in detail. In practice, combined all of the optimized process step is necessary. The step by step check is used for fixing the key impact factor for arcing issue. Each of the instruction for process change is shown in Fig. 3 (a) and (b). The left line is defined the wafer edge. PA WEE2 should be smaller than Via

979-8-3315-8850-2/25 $31.00 © 2025 IEEE

WEE2. The overlap of passivation to the wafer edge step height W residue is enough, reducing the risk for arcing. The W residue amount for the new W CMP process is less than the baseline, reducing the risk for floating W exposure to the passivation etch plasma charge. The thickness of passivation THK2 is defined by the oxide CMP process, reducing the main etch time with low plasma charge. Finally, the optimized process integration is used for mass production.

a) Baseline

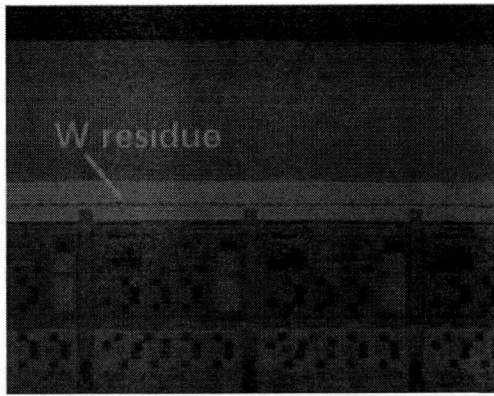

b) Optimized W CMP

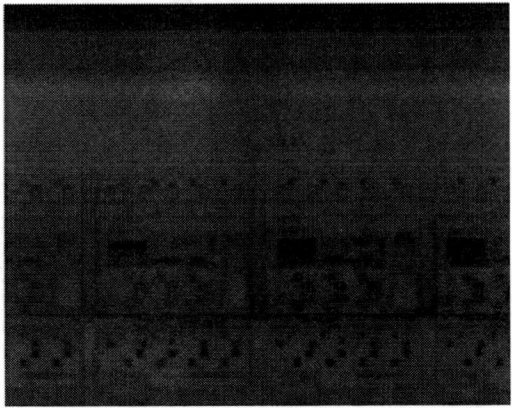

Fig. 5. The OM view post top via W CMP. (a)Baseline, (b) Optimized W CMP

1) WEE setting optimization: wafer edge exposure is used for photoresist remove post coating at wafer edge, avoid the peeling defect [10]. The key impact factor is the photo loop for top via and passivation. For the baseline data, the WEE width of passivation (PA WEE1) is larger than the WEE width of via top (Via WEE1). As we know that, there is W residue at the wafer edge due to the step height. The plasma charge of passivation etch process should touch the W residue. The risk of floating W residue collecting the charge is high when PA WEE1>Via WEE1. The arcing issue is happened. The baseline WEE setting should one of the most important factor for arcing issue. As shown in Fig. 4, by the optimized process result, WEE width of passivation (PA WEE2) is smaller than the WEE width of via top (Via WEE2), PA WEE2< Via WEE2, arcing ratio is obvious reduced. The passivation layer overlap the top via edge step, the floating W residue is fully covered by

passivation. The plasma charge introduced by passivation etch is hard to touch the W residue, leading to low risk for arcing issue. There is no negative effect is observed for the optimized WEE setting. By top via and passivation photo process WEE setting optimization, arcing ratio is reduced, while not free. The WEE setting reduce the W residue exposure risk during passivation etch only. The optimized WEE setting enlarge the manufacture process window.

2) Top W CMP optimization: W residue is formed at the wafer edge step height during the W CMP process step. The floating W residue may collect the plasma charge at passivation etch process loop, which is the key factor for arcing issue. For the baseline CMP recipe, the pressure between the tool head and wafer is same for the whole wafer. The optimized process method for W CMP is introduced, which with over polish at wafer edge. The W remove rate should be high for the optimized setting. As shown in Fig. 5 (a), obvious W residue is observed for the baseline recipe. As shown in Fig. 5 (b), negligible W residue is observed for the W CMP with wafer edge over polishing. As shown in Fig. 6, the via resistance distribution of wafer acceptance test (WAT) result with the optimized W CMP is comparable with the baseline data. The RcMVT.5 is characteristic parameter for the top via resistance with size 0.5um. The baseline median of RcMVT.5 is 1.825 ohm, stand variation is 0.0782 ohm. And, for the optimized W CMP process, the median is 1.837 ohm, stand variation is 0.07131 ohm. The inline defect data is comparable with baseline, not shown here for clarity.

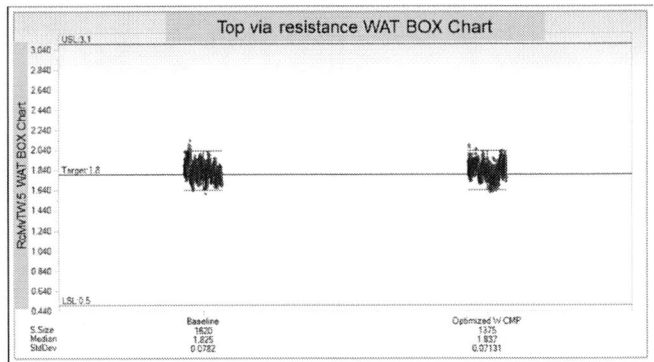

Fig. 6. Top via resistance for baseline and optimized W CMP

3) Passivation thickness and etch time reduction: The passivation CMP is introduced in the process, which is for mechanical support of the thick top metal. The top metal thickness is 4 um aluminum. The passivation deposition is formed by TEOS and HDP CVD. The previous CMP thickness is same for the optimized process and baseline. In order to reduce the passivation etch time, thinner post CMP thickness THK2 is obtained. After that, Si_3N_4 deposition is cover for the whole wafer. As we know that, thicker passivation film, need longer time during passivation etch process. The over etch time is relatively longer for enough process window. The floating W residue with high potential collecting the plasma charge, leading high risk for arcing issue. For baseline, relatively thicker passivation thickness (PA THK1) is formed by chemical-vapor-deposition process. The passivation etch time (PA ET Time1) is longer accordingly. For the optimized passivation deposition and etch process, thinner passivation

979-8-3315-8850-2/25 $31.00 © 2025 IEEE   447

film (PA THK2) and shorter passivation etch time (PA ET Time2) is obtained. The main etch time is reduced by 30%. In order to get enough etch process window, reducing the main time is one of the best method to reduce the plasma charge during passivation etch process loop.

For the new passivation process introduce different inter layer stress, the BEOL process qualification should be in consideration. The process qualification of inter metal Electric Migration (EM) and Stress Migration (SM) is passed for the optimized passivation process loop. At the same time, the passivation pinhole test is passed. The quality of new passivation is acceptable for RF SOI CMOS process. The RSMTTL is characteristic parameter for the top metal resistance. As shown in Fig. 7, top metal resistance with the optimized passivation etch is match the baseline. The baseline median of RsMTTL is 6.664 ohm/sqr, stand variation is 0.1565 ohm/sqr. And, for the optimized PA ET process, the median is 6.737 ohm/sqr, stand variation is 0.169 ohm/sqr. The etch process window is enough for the new passivation etch recipe. No abnormal is observed for Outgoing Quality Assure (OQA) check.

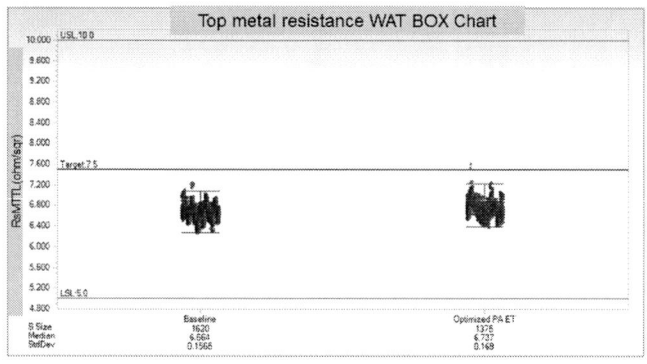

Fig. 7. Top metal resistance for baseline and optimized PA ET

Base on the optimized process integration solution, 1) WEE setting optimization, 2) Top W CMP optimization, and 3) Passivation thickness and etch time reduction, arcing free is verified by the mass production. Realized by reducing the top via W residue, lower risk for floating W exposure to Passivation etch plasma, shorter etch time during the passivation main etch. The optimized process reliability is qualified also. The integration process flow included PHOTO, CVD, CMP and ETCH recipe optimization. By checked out the root and failure mechanism effect analysis, the optimized process flow is verified by mass production, arcing free is obtained.

III. SUMMARY

In summary, arcing free solution for RF SOI process is obtained by comprehensive investigation: (1) Passivation/Via photo WEE setting optimized, passivation WEE width smaller than via WEE width, enlarge the passivation overlap on wafer edge, low risk for passivation etch plasma charge touching the wafer edge W residue; (2) Top via W CMP process optimized, enlarge the wafer edge pressure during W CMP process, or enlarge the W CMP remove rate, leading to less W residue at wafer edge. (3) Passivation etch process optimized, by reducing the passivation thickness on the thick top metal. The optimized process integration solution is verified by mass production.

ACKNOWLEDGMENT

The work is sponsored by Shanghai Eastern Talent Plan Young Project, 2023GZQN003. The work is support by the HHGrace TD Advanced Module Technology Development, for the key module recipe tuning.

REFERENCES

[1] Gianesello F, Monroy A, Vialla V, et al. Highly linear and sub 120 fs Ron x Coff 130 nm RF SOI Technology Targeting 5G Carrier Aggregation RF Switches and FEM SOC(C), IEEE Topical Meeting on Silicon Monolithic Integrated Circuits in RF Systems, Austin, TX, USA, 2016

[2] Rabbeni P A, Joseph A, Letavic T, et al. RF SOI Revolutionizing RF System Design [J]. Microwave Journal, 2015, 58(10): 1-8.

[3] Sh. P. Leiphart, R. D. Burton, Method of for reducing surface charge on semiconductor wafers to prevent arcing during plasma deposition, Patant US6057235A, 1997

[4] L. K. Hisu, H. A. Kern, Method to prevent arcing during deep via plasma etching, Patent US20060249755A1, 2005

[5] Sh. Ch. Pan, Y Ch Huang, Sh. M. Jing, Method of avoiding plasam arcing during RIE etching, Patent US20030235994A1, 2002

[6] To Shu, WS Lo, HL Yong, et el. Elimination of passivation RIE arcing inRF SOI process, IEEE 35th Annual SEMI advanced Semiconductor Manufacturing conference, Albany, NY, USA, 2024.

[7] P Li, YC Wang, JW Wang et al. Impcat of substrate resistance and layout on passivation etch induced wafer arcing and reliability[J], Microelectronics reliability,2015,55(6).

[8] J. D. Plummer, M. D. Deal, P. B. Griffin, Silicon VLSI Technology: Fundamentals, Practice and Modeling. Publishing House of Electronic Industry, Beijing, China, 2006.

[9] P. Li, J. W. Peng, Y. C. Wang, et al., Plasma induced wafer arcing in backend process and the impact on reliability, IEEE International Conference on Solid-state & Integrated Circuit Technology, Xi'an, China, 2012

[10] P. V. Zant, Microchip Fabrication: A practical guide to semiconductor processing, Publishing House of Electronic Industry, Beijing, China, 2015.

2025 The 10th International Conference on Integrated Circuits and Microsystems

An Optimized ROM based Direct Digital Synthesizer Based on 65nm CMOS Technology

Yuan Yuan
School of Information Science and Engineering
Southeast University
Nanjing, China
220230935@seu.edu.cn

Qingsheng Hu
School of Information Science and Engineering
Southeast University
Nanjing, China
qshu@seu.edu.cn

Xu Wu
School of Information Science and Engineering
Southeast University
Nanjing, China
xu.wu@seu.edu.cn

Lianming Li
School of Information Science and Engineering
Southeast University
Nanjing, China
Lianming.Li@seu.edu.cn

Abstract—**This paper presents a design of a ROM based direct digital synthesizer using 65nm CMOS process. The design aims to generate a high-speed direct digital synthesizer (DDS) with high precision and high performance for signal sources. By employing a pipelined phase accumulator, working speed is increased effectively. A ROM compression structure is also used in order to reduce hardware resource consumption, and some optimization techniques such as dithering injection and spur cancellation are utilized to improve spectral purity. To boost the output frequency further, a dual-channel DDS structure combining with a doubled frequency control word and a 2:1 multiplexer is employed. Simulation results show that the operating frequency can be up to 500 MHz and a 12-bit output signal with a frequency resolution of 0.1397Hz and phase resolution of 8.382×10^{-8}° can be achieved. The power consumption is about 17.58mW under 1.2V power supply voltage at 500 MHz working frequency. This DDS is expected to provide a high - precision, high - frequency and low power frequency source.**

Keywords—*Direct Digital Synthesizer, frequency source, ROM compression, high precision, pipelined architecture*

I. INTRODUCTION

With the rapid advancement of modern communication, sensor technology, and computer science, frequency synthesizers plays an more and more important role in numerous fields. Frequency synthesizers, which utilizes to generate signals of specific frequencies, are integral to a multitude of electronic systems, including wireless communication devices, radar systems, and high-precision measurement equipment[1]. In high-precision measurement devices such as quantum sensors, atomic clocks, and high-precision spectrometers, frequency synthesizers are required to provide high-precision, low-noise signal sources to ensure the accuracy and stability of the measurements[2]. Therefore, the development of high-performance frequency synthesizers that can meet the present and future demands has become a hot topic of research.

Direct digital synthesis (DDS) technology has significant advantages in this regard. DDS generates frequencies through digital signal processing and uses lookup table(LUT) techniques to produce accurate frequency values[3]. This method can directly generate high-precision frequency signals, avoiding the distortion and nonlinearity issues commonly found in analog

circuits. With the development of quantum sensing technology and the expanding of its application, DDS with higher frequency stability and precision may become the critical to high performance quantum sensors in the future[4].

In the design of DDS, high speed and high precision are two technical challenges that must be faced[5]. For example, how to improve phase noise and reduce power consumption at the same time remains a difficult problem in the design of high speed DDS. It has been proved that by optimizing the structure of DDS, it is possible to not only improve the precision but also optimize the power consumption, meeting the modern communication equipment's demand for low power consumption and high efficiency[6].

This paper presents an ASIC design of a high performance LUT based DDS. The design employs a four-stage pipelined phase accumulator that combines carry-chain and pipelining techniques to enhance the system's operating speed. Additionally, the LUT(also called ROM table) is compressed effectively using the symmetry of one-fourth of the sine wave and the Sunderland structure, which reduces the area and power consumption. The proposed DDS performs well in frequency resolution, area and power consumption, as well as in flexibility.

II. COMPOSITION AND WORKING PRINCIPLE OF DDS

A. LUT based DDS Architecture

As a LUT based DDS, it generates clock signals of various waveforms by utilizing LUT to provide amplitude values for a sin function expressed in binary digits. Fig. 1 illustrates a typical structure for a LUT based DDS. It mainly consists of an initial phase register buffering a input frequency control word(FCW), a phase accumulator calculating a correct phase for a sine function, a LUT based phase-to-amplitude module generating the amplitude that corresponds to the phase and some other logic. Additionally, a stable and reliable external reference frequency source is required to synchronize the various modules of the DDS. After filtering and digital-to-analog conversion(DAC), the desired frequency *Fout* is output smoothly[7].

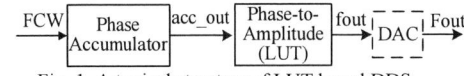

Fig. 1. A typical structure of LUT based DDS.

This work was supported by National Key R&D Program of China (No. 2023YFA1608702).

B. Pipelined Phase Accumulator

High-quality accumulator is an important module with which DDS can synchronize a phase increment precisely refer to the reference source. Fig. 2 gives the structure of an improved phase accumulator, in which the width of input FCW is 32-bit and that of output sent to LUT is 12-bit. When the reset signal of DFF is high, it is equivalent to adding a weight of 1/2 least significant bit(LSB) to the phase accumulator. In this case, the carry input *Cin* will periodically alternate between 0 and 1, improving the existing *N*-bit phase accumulator. In other words, its effect is similar to the operation of an (*N*+1) - bit phase accumulator under the assumption that the LSB of the frequency control word is 1.

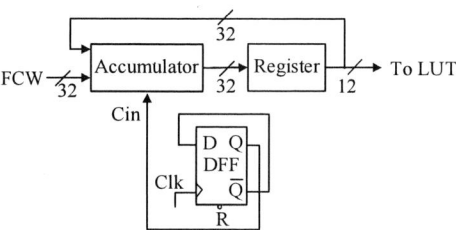

Fig. 2. An improved phase accumulator structure.

The 32-bit accumulator, however, comprises numerous adders in its array structure. To increase the data processing speed of the accumulation process, pipelining technology is employed in this design. See Fig.3, totally 4 pipeline stages are included and each traverses the same path length. First, a 32-bit frequency control word FCW is sent into 4 registers. Then each 8-bit segment of the frequency control word is loaded into the corresponding accumulator, waiting synchronization with the rising edge of a reference clock *clk* for accumulation. Meanwhile, the carry from one pipeline stage is cascaded to the next. For example, co_1, the carry output from the first stage serves as the carry input of the second accumulator. Similarly, co_2, the carry output from the second stage serves as the carry input of the third accumulator and so on. At the fourth pipeline stage, *Cout* serves as the overall output of the pipelined accumulator. It is worth to note that totally four clock cycles will be taken to produce a completed and accurate result for the first accumulation cycle.

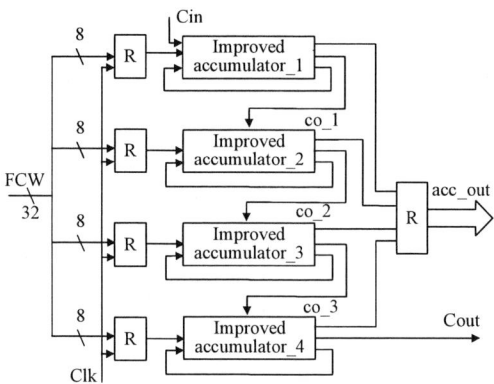

Fig. 3. A pipelined phase accumulator.

By employing this pipeline structure, the operating speed of the accumulator is increased by more than three times, speeding up the processing speed of the whole circuit effectively only at the cost of some registers and logic gates.

C. ROM based LUT for Sine Wave

Based on its realization method, phase-to-amplitude conversion, the other essential component in DDS, can be categorized into two types: one is based on trigonometric function calculation, such as CORDIC algorithm and parabolic approximation. The advantages of these algorithms are that they can produce high-purity sine signals. Their hardware implementation, however, are complex, which in turn limits the output frequency bandwidth.

The other is based on lookup table, *i.e.*, ROM. To achieve high-performance DDS, a larger ROM that stores the information of phase or amplitude for sin function is generally required. This is because the phase resolution of a DDS is determined by the address width of the ROM, meanwhile the amplitude resolution is dependent on the sample values stored in the ROM. Therefore, various optimization methods that can compress ROM capacity are often used in practical application in order to reduce the hardware integration area and resource consumption to some extent as well.

One of the typical ROM compression methods is the single-quadrant waveform storage technique proposed by Tierney[8], which only stores the sample data for one-fourth of the sine wave period, shown as Table 1. We can see that the waveform of a sine signal has 1/4 symmetry and only the waveform of the first quadrant needs to be stored. When implemented in practice, the sign of the output waveform can be controlled by the most significant bit(MSB) of the output of the phase accumulator *acc_out*, and the addressing of the ROM table is controlled by the next most significant bit(2nd MSB). By appropriately inverting the phase and amplitude, the entire cycle waveform can be obtained successfully, which can be implemented in hardware with easy.

TABLE I. THE SINGLE-QUADRANT COMPARISON METHOD IN[8]

| acc_out | | Quadrant | Sine value |
MSB	2nd MSB		
0	0	First	$\sin\theta$
0	1	Second	$\sin(\pi/4-\theta)$
1	0	Third	$-\sin\theta$
1	1	Fourth	$-\sin(\pi/4-\theta)$

Another method is based on angle decomposition, *i.e.*, dividing a phase value into two or three parts first, and then use angle decomposition formulas and trigonometric function approximations to reduce the size of ROM table. As a result, a large ROM table can be split into several smaller ones by using trigonometric identities. The final output that approximates sine waveform can be obtained by combing the outputs of those smaller ROM tables.

Besides the two compression methods mentioned above, the famous Sunderland structure[9] stores the amplitude values corresponding to the phase denoted as sinφ in ROM table and its size can be optimized based on trigonometric approximation algorithms. Concretely say, the output φ of the phase accumulator is taken modulo π/2 first to obtain the value θ. Then θ is decomposed into θ = α + β + χ, where the variables α, β, and χ have bit widths of A, B, and C, respectively, and satisfy the conditions of α < π/2, β < (π/2)2^{-A} and χ < (π/2)2^{-(A+B)}. After using trigonometric identities, the following equation can be acquired:

$$\sin\theta = \sin(\alpha+\beta+\chi) = \sin(\alpha+\beta)\cos\chi + \cos(\alpha+\beta)\sin\chi$$
$$= (\alpha+\beta)\cos\chi + \cos\alpha\cos\beta\sin\chi - \sin\alpha\sin\beta\sin\chi \quad (1)$$

Since β and χ are typically much smaller than 1, we have cosβ ≈ 1, sinβ ≈ 0, cosχ ≈ 1, and sinχ ≈ 0. Thus, Eq.(1) can be simplified to:

$$\sin\theta \approx \sin(\alpha+\beta) + \cos\alpha\sin\chi \quad (2)$$

Thus, the original ROM table, which would require 2^{A+B+C} storage units, can be decomposed into a coarse table of 2^{A+B} and a fine one of 2^{A+C}, where the former stores sin(α+β) and is addressed with (A+B)- bit, while the latter stores cosα·sinχ and can be addressed with (A+C) bit.

To compress the storage source more effectively, this paper proposed a novel compressed ROM table structure shown as Fig. 4. By combining the quarter symmetry algorithm and the Sunderland algorithm, the size of a ROM table that originally requires $(2^{4+4+4}) \times 12$ bits can be reduced to the sum of two smaller ROM tables, *i.e.* coarse and fine ROM, each requiring $(2^{3+3}) \times 12$ bits.

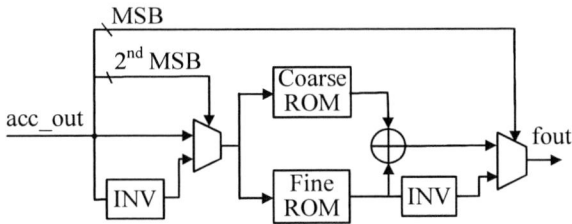

Fig. 4. The proposed ROM table combining two compression algorithms

D. Two-Channel Parallel DDS

To realize a theoretical maximum output frequency of 500MHz in 65nm LP CMOS technology, a dual-channel DDS is designed shown as Fig. 5. It can be observed that the input of the module is 2FCW, *i.e.*, the phase accumulator adds 2FCW each time, and the result is sent into two parallel paths. Therefore, on the one hand, the output of each adder is an arithmetic sequence with a common difference of 2FCW. On the other hand, the outputs of the two adders differ by one frequency control word[10]. Table 2 gives the corresponding outputs of the two adders when the accumulator and adders are reset at the beginning. It can be found that the two DDS channels operate in parallel under the control of Fclk, and the frequency control word for each channel is 2FCW.

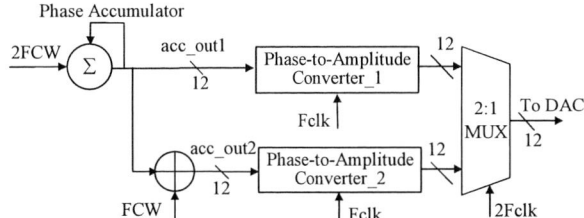

Fig. 5. The designed two-channel DDS structure.

TABLE II. THE PHASE CODES OF THE TWO-CHANNEL DDS

Clock Cycles	1st	2nd	3rd	...
DDS1	0	2FCW	4FCW	...
DDS2	FCW	3FCW	5FCW	...

Suppose DDS1 and DDS2 are preconfigured as different phase codes. And DDS1 outputs the amplitude values corresponding to the phase codes of 0, 2nd Fclk, 4th Fclk, *etc.*, while DDS2 outputs the amplitude values corresponding to that of 1st Fclk, 3rd Fclk, 5th Fclk, *etc.* Then an output with a doubled frequency can be obtained by combining the output of DDS1 and DDS2 through a 2:1 multiplexer (MUX). In other words,, under the control of 2Fclk, the output of the MUX can provide the amplitudes corresponding to the phase codes of 0, 1st Fclk, 2nd Fclk, 3rd Fclk, *etc.*, effectively, doubling the output frequency.

Eq.(3) demonstrates why a doubled output frequency can be achieved by using a two-channel DDS combined with a frequency control word of 2FCW, and the cost is only a slight increase in area.

$$Fout = \frac{2FCW}{2^N} \times FCLK = \frac{FCW}{2^N} \times 2FCLK \quad (3)$$

III. SIMULATION AND RESULT ANALYSIS

The proposed DDS has been designed and implemented using 65nm CMOS process. The power consumption is about 17.58mW under 1.2V power supply voltage at 500 MHz working frequency. Fig. 6 gives the simulation results of the four-stage pipelined phase accumulator in our DDS. It can be seen that the 32-bit frequency control word, which is divided into four 8-bit segments *acc*1~*acc*4, can be accumulated at the rising edge of the reference clock synchronously. When the output of the accumulator overflows, the output *Cout* goes high, indicating that a sampling output cycle has completed, and a new accumulation cycle begins.

The sine wave data generated based on the Sunderland algorithm is stored in ROM using mathematical tools before simulation. Fig.7 shows the corresponding waveforms of the stored data. Simulation result of the phase-to-amplitude conversion module is shown as Fig. 8. It can be seen that under a clock frequency of 500 MHz and a frequency control word of 32'h00418937, the simulation waveform is in accordance with that stored in ROM.

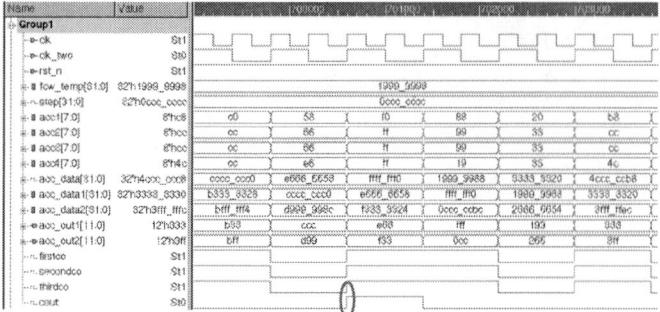

Fig. 6. Simulation result of the pipelined phase accumulator.

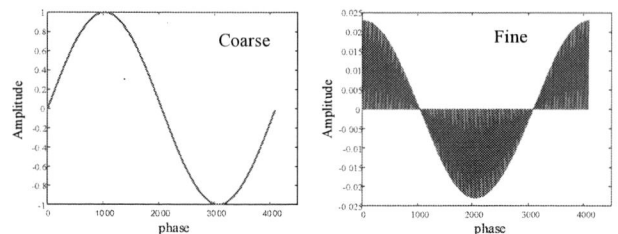

Fig. 7. Coarse and fine waveforms stored in LUT

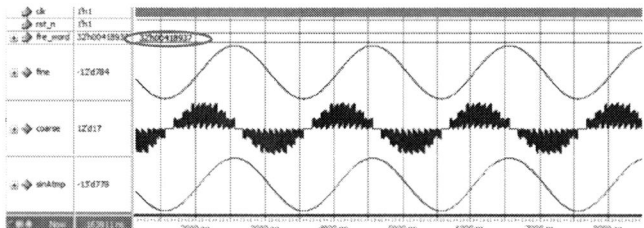

Fig. 8. Simulation results of coarse and fine waveform basd on Sunderland ROM compression algorithm.

Fig. 9 gives the simulation results of the proposed dual-channel DDS. It can be observed that the two sine signals sin1 and sin2 have a distinct phase difference, and they are separated by one FCW at any given moment. *sine_out* is the final output result after the two-channel DDS signals are concatenated through the 2:1 MUX.

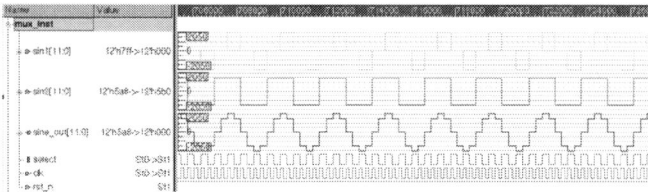

Fig. 9. The dual-channel DDS output of a 125MHz sine wave.

After logic synthesis, the generated standard delay file (SDF) is back-annotated to a testbench file, and the post-simulation results is shown as Fig. 10. It can be seen that, apart from the glitches generated during the flip of the registers, the waveform after logic synthesis is completely consistent with that of pre-simulation. This illustrates that the DDS designed in this paper can work successfully and meet the design requirements as well.

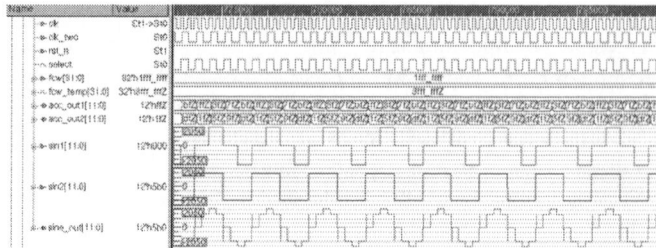

Fig. 10. The SDF back-annotated result of DDS.

IV. CONCLUSION

This paper introduces the circuit structure and working principle of a DDS. It presents a digital DDS design for ASIC implementation based on the 65nm CMOS process. The design focuses on key modules, including the four-stage pipelined phase accumulator, the phase-to-amplitude conversion module based on the Sunderland compression algorithm, and the implementation of a dual-path structure. These innovations collectively contribute to reducing the power consumption and fabrication cost of the chip.

ACKNOWLEDGMENT

This work was supported by National Key R&D Program of China (No. 2023YFA1608702).

REFERENCES

[1] S. Ziabakhsh, S. Aouini, R. G. Gibbins, M. Mikkelsen, S. Moslemi-Tabrizi and N. Ben-Hamida, "A Memory-Based Direct-Digital Frequency Synthesizer for Fractional Synchronization," in IEEE Transactions on Circuits and Systems II: Express Briefs, 2022, pp. 899-903.

[2] I. V. Ryabov, T. S. Bukanova, N. V. Degtyarev and E. S. Klyuzhev, "Direct Digital Synthesizers of Frequency-Modulated Signals," 2025 Wave Electronics and its Application in Information and Telecommunication Systems (WECONF), St. Petersburg, Russian Federation, 2025, pp. 1-5.

[3] Bochkarev D N, Ryabov I V and Strelnikov I. Direct digital synthesizers of frequency and phase-modulated signals. 2019 Systems of Signal Synchronization[C]. Generating and Processing in Telecommunications (SYNCHROINFO), Russia, 2019, pp.1-4.

[4] A. Pradeep, N. P. J, S. B. S, A. T. P and R. Francis, "Direct Digital Synthesis (DDS) Model for High-Frequency Applications," 2024 International Conference on Distributed Systems, Computer Networks and Cybersecurity (ICDSCNC), Bengaluru, India, 2024, pp. 1-5.

[5] Raju G, Shanky S and Singh G P. "Design of Power, Area and Delay Optimized Direct Digital Synthesizer Using Modified 32-Bit Square Root Carry Select Adder". Journal of Circuits, Systems and Computers, 2022, pp. 31

[6] J. Y. Park, S. H. Kim, H. Yoo and J. W. Nam, "Analysis of Quarter Method Applied ROM-Based DDFS Architecture," in IEEE Access, 2023, pp. 117137-117148.

[7] K. I. Palomäki and J. Nurmi, "Memory Optimized, High Signal Quality Direct Digital Frequency Synthesizer on an FPGA," in IEEE Transactions on Circuits and Systems II: Express Briefs, vol. 72, no. 7, pp. 958-962, July 2025.

[8] J. Tierney, C. Rader and B. Gold, "A digital frequency synthesizer," IEEE Transactio ns on Audio and Electroacoustics, 1971, pp. 48-57.

[9] D. A. Sunderland, R. A. Strauch, S. S. Wharfield, H. T. Peterson and C. R. Cole, "CMOS/SOS frequency synthesizer LSI circuit for spread spectrum communications," in IEEE Journal of Solid-State Circuits, vol. 19, no. 4, pp. 497-506, Auguest 1984.

[10] S. H. Alkurwy, S. H. Md Ali and M. S. Islam, "Implementation of low power compres sed ROM for direct digital frequency synthesizer," 2014 IEEE International Conference on Semiconductor Electronics (ICSE2014), Kuala Lumpur, Malaysia, 2014, pp. 309-312.

2025 The 10th International Conference on Integrated Circuits and Microsystems

Integrated Circuit Design of CMOS Deadtime Controller for 48 V GaN DC-DC Converter

Haoyu Chen
Department of Electrical and Electronic Engineering
Xi'an Jiaotong-Liverpool University
Suzhou, China
Department of Electrical Engineering and Electronics
University of Liverpool
Liverpool, U.K.
2637862370@qq.com

Miao Cui*
Department of Electrical and Electronic Engineering
Xi'an Jiaotong-Liverpool University
Suzhou, China
Miao.Cui02@xjtlu.edu.cn

Abstract—48 V GaN-based DC-DC converters are widely used in electric vehicles, data centers, and renewable energy systems for their high-power density and low transmission loss. However, at high switching frequencies, inaccurate deadtime control can lead to reduced efficiency and potential device failure. This work proposes a 0.25μm-CMOS-based deadtime control circuit at 1MHz to improve the performance of a 48V-to-24V GaN buck converter. The design integrates a zero-voltage comparator with CMOS logic circuits to achieve reliable, compact deadtime control. A buck converter with a bootstrap gate driver was built to verify the approach. Results confirm a stable 24 V output voltage. Full-system simulations indicate an overall minimum deadtime of 22 ns at 2.6 A load current, with the highest efficiency reaching 95% across load conditions. This approach not only simplifies the circuit design but also achieves high efficiency and short deadtime, making it highly valuable for research and application in high-frequency GaN converter systems.

Keywords—*GaN; Buck Converter; Deadtime Control; CMOS circuit; Bootstrap Driver.*

I. INTRODUCTION

With the growing demand for higher power density and lower transmission losses, 48V DC power systems have become the dominant architecture in modern power electronic applications. [1]. One of the most critical implementations of these systems is the 48V-to-24V buck conversion, which is widely used in electric vehicle auxiliary power supplies, data center server infrastructure, and renewable energy systems.

To meet the performance and reliability demands of these applications, the choice of power transistors is crucial. Compared with traditional silicon CMOS devices, GaN power transistors have emerged as a superior alternative, offering excellent switching performance, low parasitic capacitance, and high efficiency. These advantages enable compact, high-performance converter designs that align with the requirements of next-generation power systems [2], [3].

Despite these benefits, GaN-based converters operating at high frequencies face critical challenges, particularly in achieving accurate deadtime control [4]. Improper deadtime settings can significantly affect system performance: excessive

deadtime reduces effective conduction time and increases switching losses [5], [6], while insufficient deadtime can cause simultaneous conduction of high-side and low-side transistors, leading to shoot-through failures [6]. This highlights the necessity for a precise and robust deadtime control solution to fully exploit the potential of GaN devices in 48V DC-DC converters.

Designing an effective deadtime control circuit for high-frequency GaN converters involves balancing efficiency, stability, and device protection [7]. While GaN devices offer excellent performance, they lack mature PMOS devices, which makes it unable to form a complementary structure like CMOS. The GaN nMOS circuits suffer from high static power consumption. In contrast, CMOS technology, known for its ultra-low static power consumption and stable manufacturing process, remains highly advantageous for control circuit design.

In current research, two mainstream approaches are commonly used for the detection of deadtime control circuits. The first involves sensing the body diode current to determine whether the power transistor has fully turned off. Notably, Niwa et al. [2] proposed an innovative method using a current-sensing FET to monitor the diode current, enabling accurate deadtime control.

The second approach detects whether the floating switch node voltage (Vsw) has dropped to zero or below, indicating the complete turn-off of the high-side transistor [3], [5], [8]. One implementation of this method, combined with digital circuitry, successfully reduced the deadtime to the nanosecond range [3]. Nevertheless, such digital-enhanced designs often introduce significant complexity of the circuit.

In this work, a CMOS deadtime control circuit for a 48V-to-24V GaN buck converter at 1MHz is proposed, which builds upon the Vsw-detecting method and leverages a simpler and more stable CMOS logic structure to achieve nanosecond-level deadtime control without introducing excessive complexity.

II. THE DESIGN OF DEADTIME CONTROLLER

The converter was constructed using the EPC2065 GaN power transistor [7], and the proposed deadtime controller is designed using a CMOS 250 nm process. The whole converter system comprises a half-bridge buck converter with a bootstrap

---

This work was supported by XJTLU Research Development Fund (RDF21-02-031, PGRS2206039)

979-8-3315-8850-2/25 $31.00 © 2025 IEEE

driver and a deadtime control circuit, as shown in Fig. 1. The purpose of the bootstrap circuit is to prevent the switch node voltage Vsw from significantly exceeding the gate voltage when the high-side power transistor is turned on, thus ensuring that the gate-source voltage (VGS) remains sufficiently high to fully turn on the high-side transistor.

A. The Overall Buck Converter Circuit

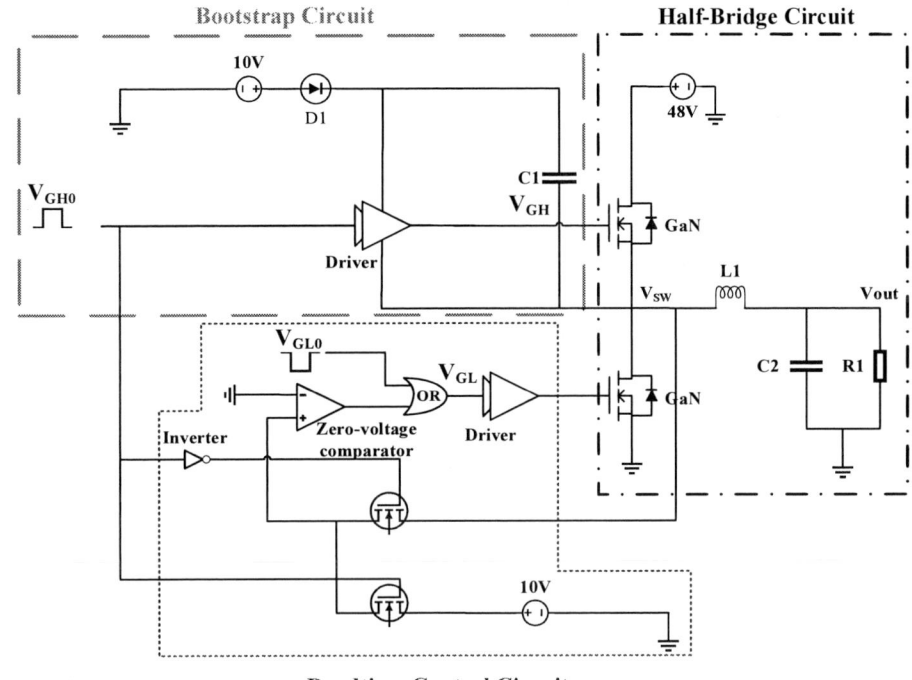

Fig. 1. Structure of the overall GaN buck converter including bootstrap and deadtime control circuits, The switching frequency is 1 MHz.

B. The Bootstrap Driver Circuit

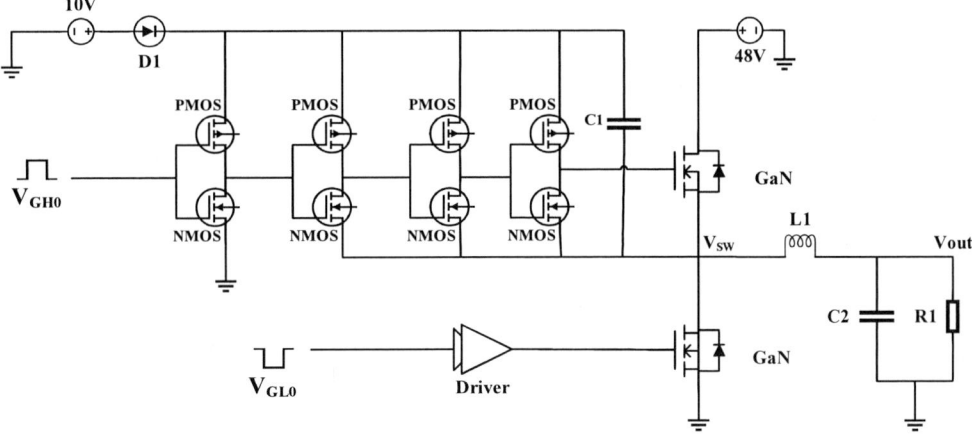

Fig. 2. Structure of the bootstrap driver with 10 V bootstrap supply and 48 V input for GaN half-bridge operation.

Building upon the bootstrap configuration proposed by Cui et al. [9], an improved buck converter with a CMOS-based bootstrap circuit was developed, as shown in Fig.2. In this design, the bootstrap capacitor (C1) and bootstrap diode (D1) work together to elevate the gate voltage of the high-side GaN transistor relative to the switching node Vsw. Ideally, when the high-side transistor turns on and Vsw rises to 48 V, the bootstrap mechanism charges the gate to approximately 58 V, ensuring full and continuous turn-on of the high-side GaN device. Conversely, when the high-side switch is off, both the gate voltage and Vsw return to 0 V, preventing unintended conduction and enabling safe switching operation.

C. The Proposed Deadtime Control circuit

979-8-3315-8850-2/25 $31.00 © 2025 IEEE

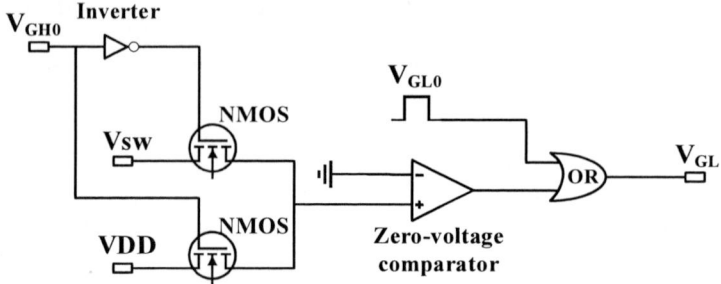

Fig. 3. The circuit diagram of the proposed CMOS deadtime controller using zero-voltage detection for GaN buck converter.

In Fig. 3, the deadtime control circuit mainly contains NMOS switches, a zero-voltage comparator, and an OR gate. NMOS switches are employed to control the operating time of this section, thereby preventing the shoot-through issue. The zero-voltage comparator and OR gate are used to generate VGL as the control signal for the low-side driver. Eventually, the driver will output the low-side control signal after shortening the deadtime. The zero-voltage comparator, originally based on the design proposed by Jia et al. [10], has been modified and optimized for better compatibility with the logic structure of this circuit.

The circuit utilizes VGH0 to drive two NMOS switches, enabling dynamic comparison between the switch node voltage (Vsw) and ground (GND). When VGH0 is high, VDD (equal to the logic-high level of the control signal) is applied to the comparator input, resulting in a constant low VGL during this turn-on period. On the other hand, once VGH0 falls to zero, the comparator begins monitoring Vsw relative to ground. When Vsw is detected to be lower than 0 V, it indicates that the high-side transistor has fully turned off, and the comparator output switches to high. VGL is then combined with VGL0 through a logical OR operation. VGL0 is designed with a sufficiently long deadtime relative to VGH0.

Through this design, the deadtime control circuit is able to proactively turn on the low-side transistor once Vsw is detected to drop below zero, and subsequently turn it off just before the high-side transistor is activated based on the corresponding control signals. In other words, the low-side transistor is only turned on when Vsw is confirmed to be 0 V, ensuring safe and reliable deadtime optimization. Conversely, the low-side power transistor remains off when Vsw is non-zero; as a result, the control circuit is able to significantly shorten the originally reserved deadtime while still preventing shoot-through, ensuring both efficiency and safe operation.

III. SIMULATION RESULTS

A. Deadtime Control Circuit

Fig. 4 displays the output waveforms of the low-side power transistor control signal at 1MHz. Here, Δt represents the portion of deadtime that is reduced through the proposed control circuit. As shown, the deadtime between VGH0 and VGL is successfully shortened to approximately 13 ns. The initial deadtime between VGH0 and VGL0 was set to around 200 ns. This conservative setting was adopted to ensure its safe operation, since the actual deadtime is influenced by multiple factors such as parasitic elements and temperature, which can

make it vary unpredictably [11]. Therefore, providing a sufficiently long deadtime is a common strategy to avoid shoot-through conduction between the high-side and low-side transistors.

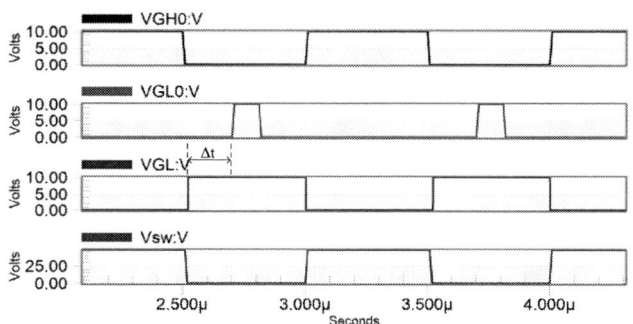

Fig. 4. Simulation waveforms of the proposed deadtime control circuit at 1 MHz switching frequency

B. Bootstrap Circuit

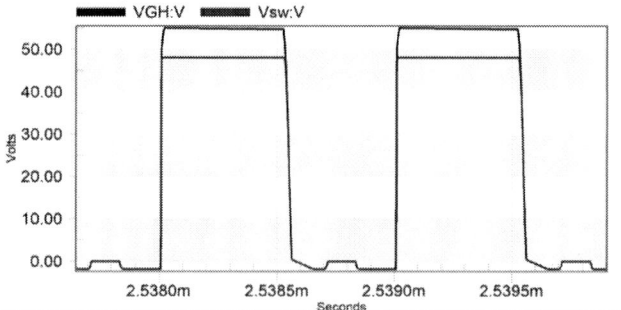

Fig. 5. Simulation waveforms of the bootstrap driver under 1 MHz switching frequency

As shown in Fig. 5, at a switching frequency of 1 MHz for VGH, when the high-side power transistor is turned on, the switch node voltage (Vsw) rises to approximately 48 V. At the same time, the gate voltage of the high-side transistor VGH reaches around 55 V, consistently maintaining a 7 V margin above Vsw, which is sufficient to ensure full turn-on of the high-side GaN power transistor. When the high-side transistor turns off, Vsw drops and briefly dips slightly below 0 V during deadtimes. When the low-side transistor turns on, Vsw rises back to 0. After the low-side transistor turns off, Vsw slightly drops below zero and remains negative until the high-side power transistor turns on again.

979-8-3315-8850-2/25 $31.00 © 2025 IEEE 456

Although VGH does not reach the ideal 58 V due to inherent circuit losses, the achieved 7 V gate overdrive remains within the required specifications for reliable operation. These results confirm that the bootstrap circuit performs effectively under dynamic switching conditions and provides sufficient gate drive voltage for the high-side GaN transistor.

C. The Converter Performance with Deadtime Control

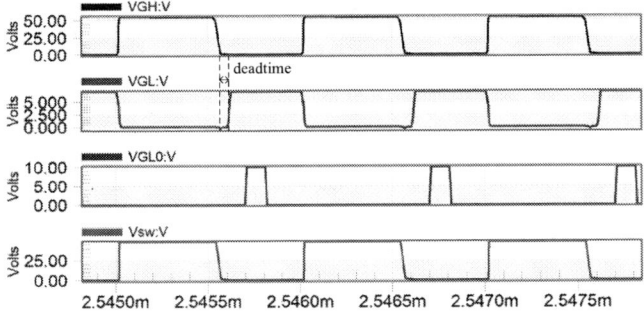

Fig. 6. Simulation waveforms of the converter with deadtime control at 1 MHz switching frequency and R_{load}=10 Ω.

Fig.6 shows the dynamic waveforms of the proposed GaN converter with deadtime control and a load resistance of 10 Ω at 1MHz. The proposed GaN converter can achieve a deadtime of 22 ns owing to the effective deadtime control circuit.

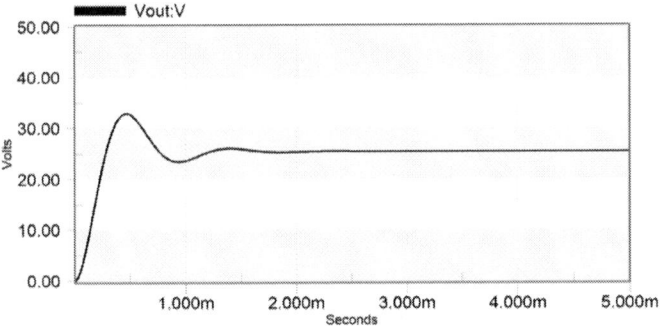

Fig. 7. Simulated output signal waveform of the proposed GaN buck converter at 1 MHz switching frequency and R_{load}=10 Ω.

Fig. 7 presents the output voltage waveform of the proposed GaN converter. By comparison, the settling time closely matches that of the converter without the deadtime control circuit, and the steady-state output voltage remains around 24 V, achieving a 48V-24 V down conversion. During simulation, it was found that there is almost no difference between the output waveform of the converter without the deadtime control circuit and that in Fig. 7, indicating that the integration of the deadtime control circuit has minimal impact on the overall system performance, maintaining both voltage regulation and dynamic response within expected parameters.

Fig. 8 illustrates the optimized deadtimes at different load current conditions. The proposed CMOS deadtime controller can achieve a minimum deadtime of 22ns at 1MHz. As the load current decreases, the simulated deadtime slightly increases, from approximately 22 ns at 2.6 A to 44 ns at 0.8 A. However, this increase occurs at a relatively slow and consistent rate, indicating that the proposed deadtime control circuit maintains

robust and adaptive performance under varying load conditions. This slight trend is primarily attributed to the slower discharge rate of the switch node voltage (Vsw) at lower load currents. The reduced discharge speed delays the comparator's ability to detect the complete turn-off of the high-side transistor, slightly extending the deadtime.

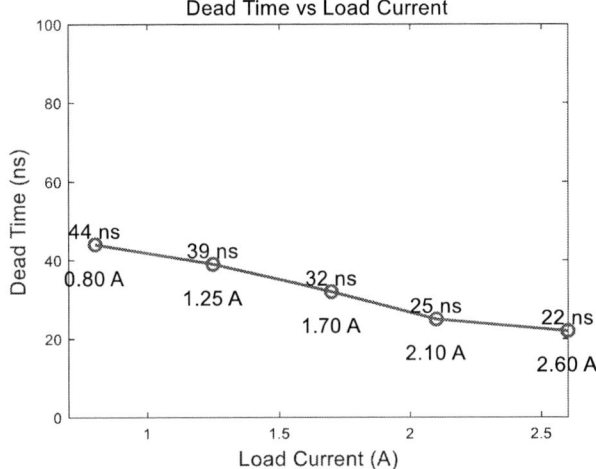

Fig. 8. Optimized deadtimes of the proposed controller at 1 MHz switching frequency for different load currents.

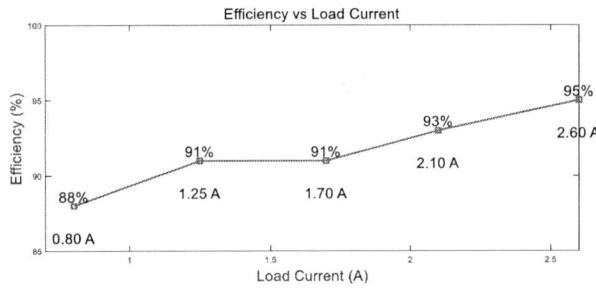

Fig. 9. Efficiency of the proposed GaN converter with deadtime control at 1 MHz switching frequency.

Fig. 9 presents the efficiency of the proposed GaN converter under various load currents. Overall, the efficiency increases with load current, with a maximum efficiency of 95% at 2.6 A at 1MHz. More importantly, the proposed CMOS-based deadtime control circuit introduces minimal additional power loss, maintaining high efficiency even under light-load conditions. This makes it well-suited for practical applications where frequent load fluctuations are expected.

IV. CONCLUSION

In this work, a new CMOS-based deadtime control circuit was proposed and validated for high-frequency GaN buck converters. The design uses an improved zero-voltage comparator and CMOS logic circuits, achieving a minimum deadtime of 13 ns in isolated deadtime control circuit simulations at 1 MHz. A 48V-to-24V buck converter using EPC2065 GaN transistors and a bootstrap circuit was implemented. Simulation results confirmed a stable 24 V output voltage, with the high-side gate voltage VGH maintained at 7 V above Vsw during conduction.

The control circuit achieved a deadtime of 22 ns at 2.6 A, increasing moderately at lower currents but remaining short overall. Efficiency was maintained above 88% at light loads, achieving a maximum efficiency of 95% at higher loads. These results demonstrate the effectiveness, low power overhead, and strong adaptability of the proposed deadtime control strategy in practical high-frequency power applications.

REFERENCES

[1] Y. Zhang, T. Liu, H. Feng and L. Ran, "Design of A High-Density 800V to 48V Converter with Optimized Planar Transformer for Automotive Applications," 2023 IEEE 2nd International Power Electronics and Application Symposium (PEAS), Guangzhou, China, 2023, pp. 356-360, doi: 10.1109/PEAS58692.2023.10395547.

[2] A. Niwa et al., "A Dead-Time-Controlled Gate Driver Using Current-Sense FET Integrated in SiC MOSFET," IEEE Transactions on Power Electronics, vol. 33, no. 4, pp. 3258-3267, April 2018, doi: 10.1109/TPEL.2017.2704620.

[3] C. -J. Chen, P. -K. Chiu, Y. -M. Chen, P. -Y. Wang and Y. -C. Chang, "An Integrated Driver With Adaptive Dead-Time Control for GaN-Based Synchronous Buck Converter," IEEE Transactions on Circuits and Systems II: Express Briefs, vol. 69, no. 2, pp. 539-543, Feb. 2022, doi: 10.1109/TCSII.2021.3098310.

[4] Y. Yang, Y. Liu, Q. Zeng, J. Liu, and B. Zhu, "High frequency and high power density bipolar DC-DC converter with GaN HEMT," Energy Reports, vol. 9, pp. 617–624, Sep. 2023, doi: https://doi.org/10.1016/j.egyr.2023.04.110.

[5] R. Grezaud, F. Ayel, N. Rouger and J. -C. Crebier, "A Gate Driver With Integrated Deadtime Controller," IEEE Transactions on Power Electronics, vol. 31, no. 12, pp. 8409-8421, Dec. 2016, doi: 10.1109/TPEL.2016.2517679.

[6] F. Xu, J. Liu and X. Zhang, "Analysis of Deadtime Effects and Optimum Deadtime Control for Bidirectional Inductive Power Transfer Converters," IEEE Transactions on Industrial Electronics, vol. 71, no. 7, pp. 6929-6937, July 2024, doi: 10.1109/TIE.2023.3308135.

[7] "EPC2065 - Automotive 80 V, 215 A Enhancement-Mode GaN Power Transistor," *Epc-co.com*, 2024. https://epc-co.com/epc/Portals/0/epc/documents/datasheets/ EPC2065_datasheet.pdf

[8] Jing Xue, K. D. T. Ngo and Hoi Lee, "A 99%-efficiency 1-MHz 1.6-kW zero-voltage-switching boost converter using normally-off GaN power transistors and adaptive dead-time controlled gate drivers," *2013 IEEE International Conference of Electron Devices and Solid-state Circuits*, Hong Kong, China, 2013, pp. 1-2, doi: 10.1109/EDSSC.2013.6628142.

[9] M. Cui et al., "Monolithic GaN Half-Bridge Stages With Integrated Gate Drivers for High Temperature DC-DC Buck Converters," IEEE Access, vol. 7, pp. 184375-184384, 2019, doi: 10.1109/ACCESS.2019.2958059.

[10] M. Jia, Z. Sun, and L. Siek, "A Novel Zero-Voltage-Detector for Buck Converter in Discontinuous Conduction Mode (DCM)," *2018 IEEE 4th Southern Power Electronics Conference (SPEC)*, vol. 2018 IEEE 4th Southern, no. 1–4, Dec. 2018, doi: https://doi.org/10.1109/spec.2018.8635909.

[11] L. Zhang, X. Yuan, J. Zhang, X. Wu, Y. Zhang and C. Wei, "Modeling and Implementation of Optimal Asymmetric Variable Dead-Time Setting for SiC MOSFET-Based Three-Phase Two-Level Inverters," IEEE Transactions on Power Electronics, vol. 34, no. 12, pp. 11645-11660, Dec. 2019, doi: 10.1109/TPEL.2019.2905882.

2025 The 10th International Conference on Integrated Circuits and Microsystems

An Ultra-Low Power Digital Assisted Self-Tracking Zero-Current Detector for DC-DC Converters

Chuting Yang
Department of Integrated Circuits
Nanjing University of Aeronautics and Astronautics
Nanjing, China
chutingyang@nuaa.edu.cn

Xing Li[*]
Department of Integrated Circuits
Nanjing University of Aeronautics and Astronautics
Nanjing, China
[*]Corresponding author: xingli@nuaa.edu.cn

Abstract—This paper proposes an ultra-low power digital assisted self-tracking zero-current detector (ST-ZCD) for DCDC converters. This ZCD has simpler structure compared to prior arts, as it only needs one comparator and a few simple digital logics to complete a self-tracking loop, which helps to reduce the quiescent current significantly. The only one comparator not only detects the zero-crossing for normal DC-DC operation, but also tells if the detection is early or late to guide the digital logic to adjust the comparator's offset for better detection accuracy. What's more, dynamic biasing control is introduced in this design to further achieve the ultra-low power feature. This ST-ZCD is implemented in a 180nm CMOS process. Simulation results show that the quiescent current of the proposed ZCD is only 62.5nA, and the inductor reverse current (Ireverse) can be as low as -0.27mA(@Rdson=260mΩ). The range of the Ireverse in Monte Carlo simulation can be controlled within ±8mA in the worst case.

Keywords—*DC-DC converter, digital assisted, self-tracking zero-current detector, dynamic biasing*

I. INTRODUCTION

With the development of the semiconductor technology, the Internet of Things (IoT) and wearable devices are widely used with more and more function integrated. But the limited battery capacity restricted the standby time of the devices, which makes low power technology very critical in those applications. Not only the computing unit (MCU etc.), the communication unit (RX/TX), but also the power management unit (PMU) needs to be low power and high efficient.

As one of the most important modules in PMU, DC-DC converter could help to convert power efficiently from battery to the output with the required voltage for the system. As shown in Fig. 1, the buck converter could operate in continuous conduction mode (CCM) at heavy loads and discontinuous conduction mode (DCM) at light loads where in each cycle, the inductor current returns to zero. To achieve low power requirement in the system, the low quiescent current and high efficiency in light load are more important for the DC-DC converter. A few uA or even sub-uA current of the quiescent current is usually required, which needs each block inside the DC-DC converter only consumes sub-uA current in idle state. PFM mode is usually adopted for better efficiency in light load mode. Meanwhile, the zero-current detector (ZCD) also plays a critical role in conversion efficiency. Both the early detection and late detection would introduce the extra power loss. Hence, a highly accurate ZCD with ultra low quiescent current is needed in the above-mentioned low power applications.

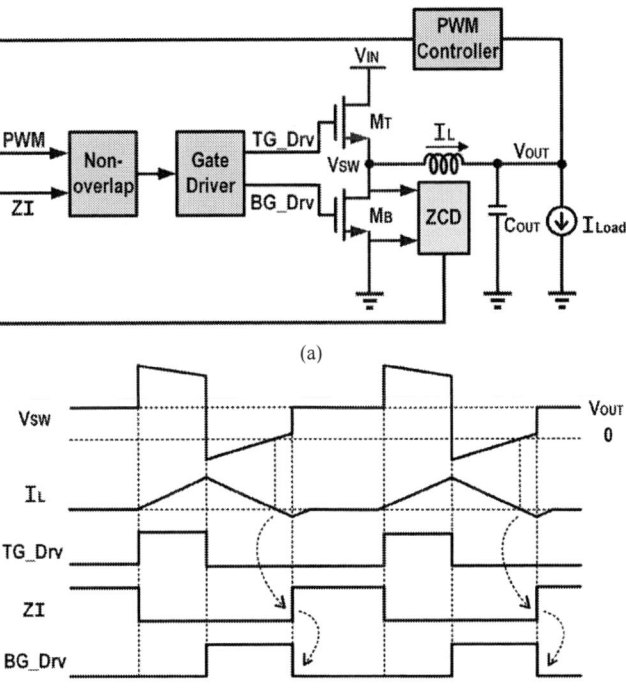

(a)

(b)

Fig. 1. (a) Diagram and (b) operating waveform of the buck converter in DCM.

In the literature, there has been a significant amount of research on utilizing ZCD to reduce losses and improve efficiency[1]. [2] and [7] proposed a Self-Tracking ZCD with digital assisted up/down delay tuning scheme to achieve high detection accuracy, but requires additional continuous tube and complex sample circuits. [3] proposed a similar digital assisted Self Tracking ZCD with Coarse-Fine tuning combination Scheme. Its complex structure adds extra amplifier and the fine-tuning loop will increase the tracking time. [4] proposed a digital assisted ZCD using binary searching method to further reduce the tracking time. The binary searching may introduce high disturbance during tracking before settling. [5] uses an extra phase to compensate for the comparator's offset to improve the accuracy of detection. However, it increases overall power consumption and circuit complexity. [6] and [8] proposed a power reduction method by dynamically switching off the ZCD. But high accuracy is not mentioned.

This paper presents an ultra-low power digital assisted self-tracking zero-current detector (ST-ZCD) designed to meet the

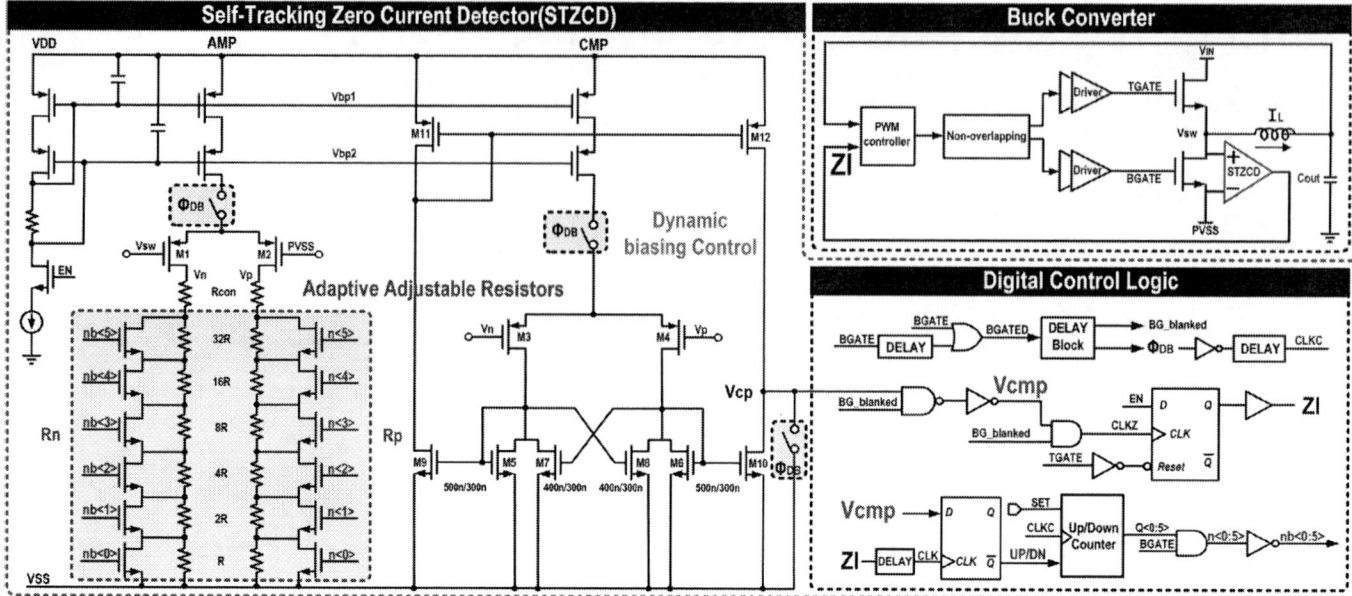

Fig. 2. Block Diagram of the Proposed Self-Tracking Zero Current Detector.

demands of high efficiency and low power consumption in modern power electronics. The ST-ZCD uses advanced digital techniques to precisely detect zero-current conditions, making it ideal for applications like DC-DC converters and power management systems. Section II describes the architecture of the proposed ST-ZCD in detail. The simulation results and conclusions are shown in Sections III and IV.

II. ARCHITECTURE OF THE PROPOSED ZCD

The block diagram of the self-tracking zero-current detector is shown in Fig. 2, which consists mainly of a digital assisted comparator and digital control logic module. The digital-assisted comparator, as the core module of the ZCD, not only controls the turn-off operation of the switch but also outputs its response waveform Vcmp to the discrimination circuit within the digital control logic module. This dual-function design not only simplifies the circuit structure but also enhances the system integration.

A. Digital Assisted Comparator

As shown in Fig. 2, the digital-assisted comparator consists of two core circuit stages: the first stage is a differential-input differential-output preamplifier, and the second stage is a comparator in the form of an operational transconductance amplifier (OTA). This structural design aims to achieve high-precision signal amplificationn and comparison functions while enabling flexible adjustment through digital control.

At the output end of the preamplifier, the load resistance is composed of three parts: a 400kΩ fixed resistor Rcon, and two 6-bit adjustable resistor arrays Rn and Rp. These adjustable resistors provide great flexibility for the circuit. The unit resistance value R of Rn and Rp is 1.6kΩ, with an adjustment range of 0 to 63R (i.e., 0 to 100.8kΩ). The resistance values of the adjustable resistors Rn and Rp are controlled by digital signals n<0:5> and nb<0:5>. By dynamically adjusting these resistance values, the output voltage of the preamplifier can be

precisely regulated. The core purpose of this adjustment mechanism is to control the triggering moment of the comparator. By changing the resistance values of Rn and Rp, the response time of the circuit can be dynamically adjusted, thereby optimizing the performance of the entire comparator. This design not only enhances the adaptability of the circuit but also provides a basis for achieving high-precision signal detection.

In order to increase the gain and sensitivity of the comparator, cross-coupled pairs of tubes with a positive feedback structure, such as M7 and M8, are introduced into the comparator. To avoid hysteresis effects, M7 and M8 are slightly smaller than M5 and M6.

The output Vcp of the comparator is passed through a digital logic circuit to generate ZI, which controls the switching off of the bottom gate in the BUCK. Unlike previous self-tracking zero-current detectors, the proposed digital assisted comparator is capable of both switching off the bottom gate by comparing Vsw and PVSS, and regulating its own output by generating a feedback signal based on the response waveform in each cycle.

B. Digital Control Logic

As shown in Fig. 2, the structure of the feedback loop is quite simple, consisting of the following key components: two D flip-flops triggered on the rising edge, a 6-bit synchronous reversible Up/Down counter, and several logic gates and delay circuits.

In the Buck converter, when the top gate is switched off and then the bottom gate is switched on ringing may be generated on Vsw and PVSS due to the parasitic inductance from package and PCB trace, which may lead to an erroneous zero current detection by the ZCD. Therefore as shown in the digital control logic in Fig. 2, BGATE is delayed to provide BG blanked to gate ZCD properly to avoid ringing impact.

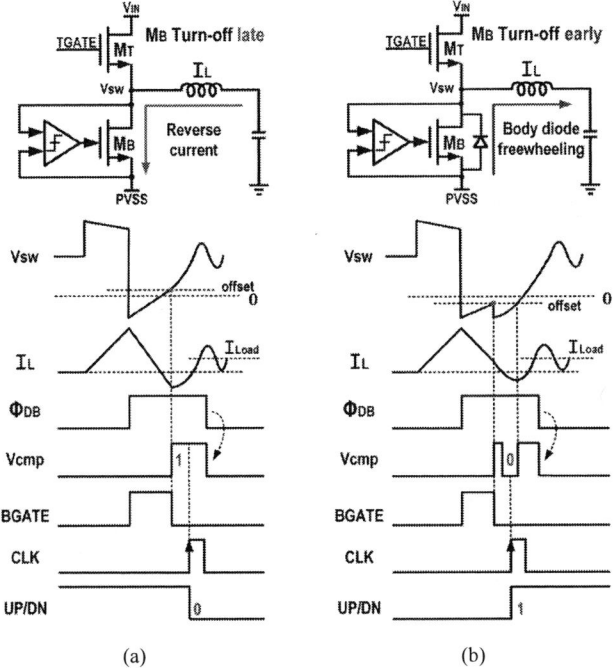

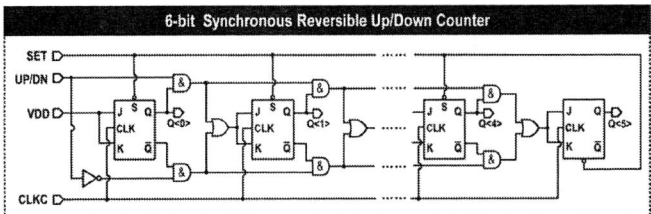

Fig. 4. Diagram of the 6-bit synchronous reversible Up/Down counter.

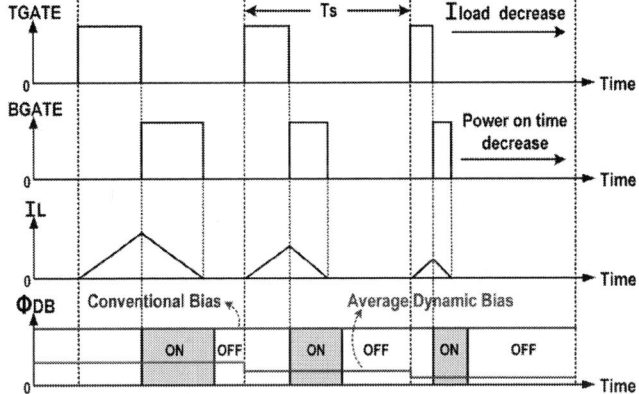

Fig. 3. Vsw, IL and Control signal timing diagram when the MB turn off (a) late and (b) early.

As shown in Fig. 3(a), when the bottom gate is turned off late due to the offset, the inductor current is already reversed and Vsw is larger than 0. The Vcmp remains high until the comparator is disabled by the φDB. As shown in Fig. 3(b), when the bottom gate is turned off early, the current is recycled through the body diode because the inductor current direction cannot be changed abruptly. Influenced by the body diode conduction voltage drop, the comparator will flip low for a period of time and not flip high until the inductor current rises again and finally Vcmp behaves as two narrow square waves.

To accurately detect and distinguish the waveform characteristics of Vcmp, we slightly delay the signal ZI to generate the clock signal CLK for the D flip-flop. By sampling Vcmp on the rising edge of the CLK signal, the UP/DN signal is generated. When the UP/DN signal is 1, it indicates that the switch has turned off early; when the UP/DN signal is 0, it indicates that the switch has turned off late.

The internal circuit design of the 6-bit synchronous reversible Up/Down counter is shown in Fig. 4. This counter is the core of the digital control logic module, responsible for dynamically adjusting the count value based on the UP/DN signal. When the UP/DN signal is 1, the counter performs an addition operation, incrementing the output Q<5:0> by 1. Conversely, when the UP/DN signal is 0, the counter performs a subtraction operation, decrementing the output Q<5:0> by 1. The clock signal CLKC is generated by φDB.

The SET signal is the counter's set signal, which is active low. Before the comparator begins operation, the SET signal initializes the counter's output Q<5:0> to the binary value 011111, which corresponds to the decimal value 31. At this time, the adjustable resistor array Rn is set to 31R, while Rp is set to 32R.

Fig. 5. Adaptive dynamic biasing control timing diagram.

Taking the case of the low-side switch turning off late as an example, the UP/DN signal is 0 at this time, triggering the counter to perform a subtraction operation, updating its output Q<5:0> from 011111 to 011110. Correspondingly, n<5:0> changes to 011110, while nb<5:0> changes to 100001. This change results in the resistance of Rn being adjusted from 31R to 30R, and the resistance of Rp being adjusted from 32R to 33R. The adjustment of the resistor array further affects the output voltage of the preamplifier: the voltage at the in-phase output terminal Vp increases, while the voltage at the out-of-phase output terminal Vn decreases. This voltage change causes the comparator to flip earlier in the next cycle, thereby prompting the low-side switch to turn off earlier, effectively compensating for the delayed turn-off situation.

Conversely, if the low-side switch turns off early, the UP/DN signal will become 1, triggering the counter to perform an addition operation. In the next cycle, the counter's output Q<5:0> will increase, causing the low-side switch to turn off late, thereby adjusting the circuit's response time.

Through this self-tracking feedback mechanism, the comparator is able to gradually approach the optimal turn-off moment under the fine adjustment of the feedback loop. With the assistance of the digital loop, the counter's output Q<5:0> can converge between two adjacent values.

C. Ultra-Low Power Design in ST-ZCD

To apply in ultra-low power applications, the quiescent current of the whole buck is very critical. Since our proposed ST-ZCD contains only a comparator and some simple digital logic circuits, the quiescent power consumption of the entire circuit is very low. And we propose an adaptive dynamic biasing technique that can further significantly reduces the

979-8-3315-8850-2/25 $31.00 © 2025 IEEE

dynamic power consumption of the zero current detector, especially under light load conditions.

The adaptive dynamic biasing technique dynamically adjusts the bias current based on the circuit's working state. When the ZCD is useless, not all modules need to stay active. By smartly disabling the bias of temporarily unused modules, we can significantly reduce the circuit's dynamic power consumption.

As shown in Fig. 2, the dynamic biasing signal ϕDB is generated by BGATE to power gate the comparator when it's not needed for power saving. The operating waveform is shown in Fig. 5. When BGATE is switched on, ϕDB turns to high to activate the comparator; When the zero current crossing is detected, ϕDB turns low to power gate the comparator to save the power until BGATE is switched on again in next cycle. Therefore, the comparator only operates when BGATE is high. As the load decreases, the off-duty cycle of ϕDB increases, significantly reducing the power consumption of the circuit.

III. SIMULATION RESULTS

Fig. 6 shows the different waveforms of Vcmp and UP/DN signal when the MB is turned off early and late. This verifies the correctness and feasibility of the detection theory, as shown in Fig. 3.

Fig. 7 shows the first cycle when the BUCK is started with CODE=31(011111). The initial inductor reverse current value (Ireverse) is -35mA. The proposed ST-ZCD adapts to the optimal turn-off time after 25 cycles when the CODE jumps between 6(000110) and 7(000111) and the corresponding Ireverse is less than -0.27mA at this time. With Rdson=260mΩ, the equivalent offset voltage at the comparator input is reduced from 9.1mV to 0.07mV and the offset is reduced by 99.2%.

The left panel of Fig. 8(a) demonstrates that under Monte Carlo 200 point simulation (Rdson=260mΩ), the Ireverse has an average value of -7.2mA in the initial state, with 3σ in the range of -41mA to 26.6mA. The figure on the right shows the average value of the Ireverse in the steady state that the ST-ZCD adapts to after 32 cycles is -0.88mA, with 3σ in the range of -4.1mA to 2.32mA. With Rdson = 260mΩ, the equivalent maximum value of the offset voltage at the comparator input is reduced from 10.7mV to about 1mV, and the offset is reduced by about 90%.

The left panel of Fig. 8(b) demonstrates that under Monte Carlo 200 point simulation (Rdson=60mΩ), the Ireverse has an average value of -28.7mA in the initial state, with 3σ in the range of -169mA to 111.5mA. The figure on the right shows the average value of the Ireverse in the steady state that the ST-ZCD adapts to after 32 cycles is -1.59mA, with 3σ in the range of -13.2mA to 10mA. With Rdson = 60mΩ, the equivalent maximum value of the offset voltage at the comparator input is reduced from 10.1mV to about 0.79mV, and the offset is reduced by about 92.2%.

IV. CONCLUSION

This paper introduces an ultra-low power digital-assisted self-tracking zero-current detector specifically designed for DC-DC converters. The core of this design lies in its digital-

assisted comparator and concise yet efficient digital control logic module. Thanks to the simplified structure and the use of dynamic biasing techniques, the static power consumption of this ST-ZCD is successfully reduced to just 62.5nA. Moreover, through an adaptive adjustment mechanism, the comparator's offset is strictly controlled to be less than 1mV. Under steady-state conditions, the reverse current can be as low as -0.27mA (with on-resistance Rdson of 260mΩ). Monte Carlo simulation results show that, in the worst-case scenario, the variation range of the reverse current can also be effectively controlled within ±8mA.

To more intuitively demonstrate the performance advantages of this ST-ZCD, Table I provides a comprehensive comparison between it and other state-of-the-art ZCD technologies currently available. The comparison results reveal that this proposed ZCD stands out among its peers by showing significant advantages in several key performance indicators, including reverse current control accuracy, self-tracking response speed, and static power consumption. This makes it an exceptional choice, offering a more reliable zero-current detection solution for the efficient and low-power operation of DC-DC converters.

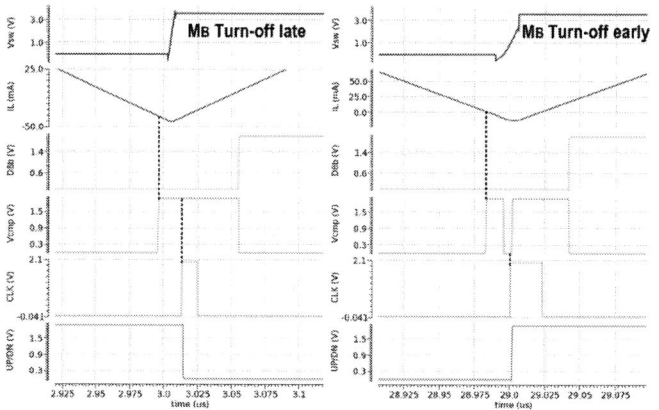

Fig. 6. Simulated results of IL, Vcmp, CLK and UP/DN when the MB turn off late and early.

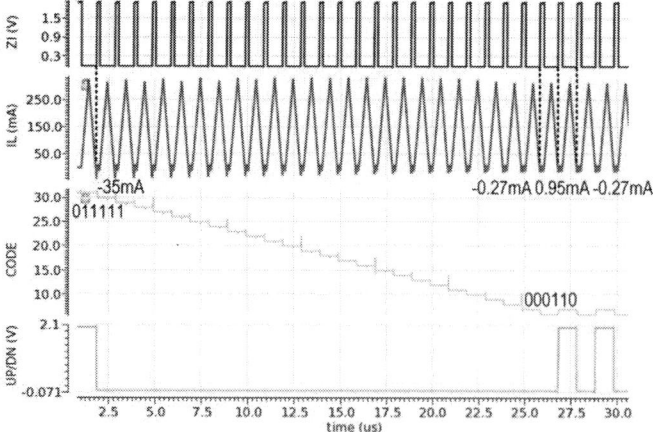

Fig. 7. Simulated results of ZI, IL, CODE and UP/DN when the MB's Rdson=260mΩ and initial Ireverse=-35mA.

979-8-3315-8850-2/25 $31.00 © 2025 IEEE

	JSSC 15[5]	JSSC 16[2]	TCASII 19[3]	This Work
Self-tracking Time	0	225us	40us	<25μs
Quiescent Current	10u	N/A	N/A	62.5nA

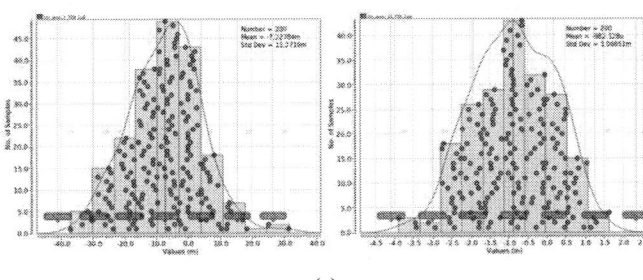

(a)

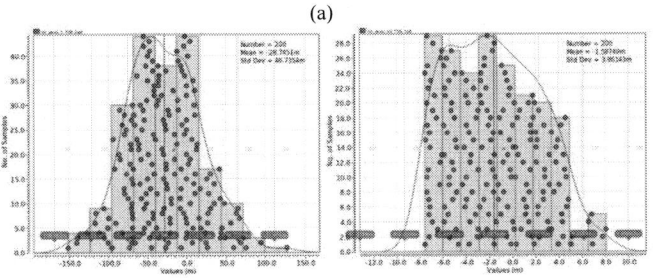

(b)

Fig. 8. Initial Ireverse range and steady-state Ireverse range under Monte Carlo simulation when (a) MB's Rdson=260mΩ and (b) MB's Rdson=60mΩ.

TABLE I.   COMPARISON WITH ADVANCED ZCD DESIGNS

	JSSC 15[5]	JSSC 16[2]	TCASII 19[3]	This Work
Process	130nm	180nm	180nm	180nm
Topology	Boost	Buck	Buck	Buck
Inductor /Capacitor	10uH /N/A	4.7uH /N/A	4.7uH /4.7uF	2.2uH /10uF
Frequency	N/A	1MHz	1MHz	1MHz
ZCD Technique	Analog	Digital	Analog	Digital-Assisted
Reverse Current	N/A	-0.5mA	-0.5mA	-0.27mA@Rdson=260mΩ -2.5mA@Rdson=60mΩ

REFERENCES

[1] Khyalap V, Colaco S. Soft switching control of synchronous buck converter with ZCD circuit[J]. International Journal of Electrical and Electronic Engineering & Telecommunications, 2015, 1(1): 292-297.

[2] P. -H. Chen, C. -S. Wu and K. -C. Lin. A 50 nW-to-10 mW Output Power Tri-Mode Digital Buck Converter With Self-Tracking Zero Current Detection for Photovoltaic Energy Harvesting. IEEE Journal of Solid-State Circuits, vol. 51, no. 2, pp. 523-532, Feb. 2016.

[3] K. -C. Woo, J. -M. Oh and B. -D. Yang. DC–DC Buck Converter Using Analog Coarse-Fine Self-Tracking Zero-Current Detection Scheme. IEEE Transactions on Circuits and Systems II: Express Briefs, vol. 66, no. 11, pp. 1850-1854, Nov. 2019.

[4] Q. Kuai, H. -Y. Leung, Q. Wan and P. K. T. Mok. A High-Efficiency Dual-Polarity Thermoelectric Energy-Harvesting Interface Circuit With Cold Startup and Fast-Searching ZCD. IEEE Journal of Solid-State Circuits, vol. 57, no. 6, pp. 1899-1912, June 2022.

[5] A. Shrivastava, N. E. Roberts, O. U. Khan, D. D. Wentzloff and B. H. Calhoun. A 10 mV-Input Boost Converter With Inductor Peak Current Control and Zero Detection for Thermoelectric and Solar Energy Harvesting With 220 mV Cold-Start and -14.5 dBm, 915 MHz RF Kick-Start. IEEE Journal of Solid-State Circuits, vol. 50, no. 8, pp. 1820-1832, Aug. 2015.

[6] Y. -J. Park et al. A Design of a 92.4% Efficiency Triple Mode Control DC–DC Buck Converter With Low Power Retention Mode and Adaptive Zero Current Detector for IoT/Wearable Applications. IEEE Transactions on Power Electronics, vol. 32, no. 9, pp. 6946-6960, Sept. 2017.

[7] S. -Y. Kim et al. Design of a High Efficiency DC–DC Buck Converter With Two-Step Digital PWM and Low Power Self-Tracking Zero Current Detector for IoT Applications. IEEE Transactions on Power Electronics, vol. 33, no. 2, pp. 1428-1439, Feb. 2018.

[8] Xuan-Dien Do, Seok-Kyun Han and Sang-Gug Lee. Low power consumption for detecting current zero of synchronous DC-DC buck converter. 2012 International SoC Design Conference (ISOCC), Jeju, Korea (South), 2012, pp. 487-490.

2025 The 10th International Conference on Integrated Circuits and Microsystems

A High-Linearity Digital-to-Time Converter Design for Fractional-N Sub-sampling PLL

Jiangnan Li, Wangqing Wu, Ke Cao, Xinyi Zhang, Haigang Feng*

Shenzhen International Graduate School, Tsinghua University, Shenzhen 518055, China
Email: ljn23@mails.tsinghua.edu.cn, feng.haigang@sz.tsinghua.edu.cn
*Corresponding author

Abstract—**In this paper, we propose a digital-to-time converter (DTC) design for Fractional-N Sub-sampling Phase-Locked Loop (SSPLL). The proposed design employs a parallel switching structure, re-timing technique, and complementary circuit to improve the linearity of the DTC. Designed in 65nm CMOS process, this DTC design operates at 100MHz with a 180 fs resolution and a delay range of 184 ps. For nonlinear performance, integral nonlinearity (INL) is within 216 fs, while Differential Nonlinearity (DNL) remains within 9 fs. It remains competitive compared to state-of-art DTC design. Behavioral-level simulations are conducted for the SSPLL, and a gain calibration module is integrated into the system to calibrate the DTC gain.**

Index Terms—**digital-to-time converter (DTC), Fractional-N, sub-sampling Phase-Locked Loop (SSPLL), phase noise (PN)**

I. INTRODUCTION

Sub-sampling phase-locked loops (SSPLLs) are preferred in noise-sensitive applications due to their extremely low integrated phase noise and high figure-of-merit (FoM) [1]. However, the subsampling phase detector (SSPD) inherently locks the PLL's output frequency to integer harmonics of the reference frequency. Because for fractional division, the zero-crossing points of the reference clock and the VCO output no longer align. To address this, a digital-to-time converter (DTC) is employed to a priori delay the reference clock phase, ensuring alignment with the zero crossing points of the VCO after stabilization. As illustrated in Fig. 1, for a fractional division of 1.75, the reference signal is delayed by 0.25 * T_{VCO}, 0.5 * T_{VCO}, and 0.75 * T_{VCO} during the second, third and fourth sampling instances, respectively. This ensures that every subsequent sampling event occurs at the zero-crossing point of the VCO once the loop is locked in fractional-N [2].

In Fractional-N SSPLLs, DTC must meet several indexes. The delay coverage must exceed the VCO clock period, as insufficient coverage (below the VCO period) prevents the SSPLL from achieving locked state. DTC resolution directly determines the SSPLL's frequency resolution and also brings the quantization noise. Nonlinearities in DTC increase fractional spurs in the output of SSPLL and will fold out-of-band noise back into the signal band, degrading noise performance.

This paper proposes a high-linearity 10 bit DTC circuit design for fractional-N SSPLLs, where the SSPLL operates at a frequency of 7.1-11GHz with a 100MHz input reference

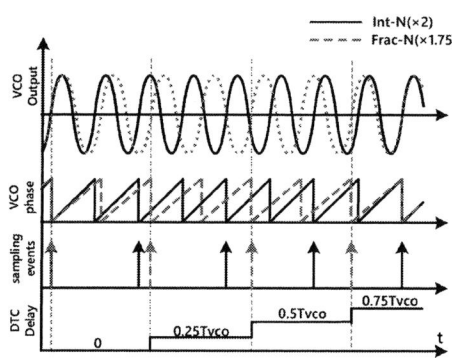

Fig. 1. Fractional-N SSPLL mechanism

frequency. The DTC core delay circuit employs two fixed capacitors connected in series, with control switches paralleled with one capacitor to enhance resolution and reduce integral nonlinearity (INL). Additionally, a re-timing technique is implemented to mitigate dynamic nonlinearities caused by control code transitions. To eliminate the impact of inductive ripple on power and ground bonding wires, complementary DTC modules are introduced. Behavioral-level simulations are conducted for the SSPLL, and a gain calibration module is integrated into the system to calibrate the DTC gain, addressing PVT variations. The results validate that the proposed DTC performance metrics satisfy the requirements of the Fractional-N SSPLL design. This paper is organized as follows. Section II presents the proposed DTC circuit design. Section III describes the necessary digital implementation in fractional-N SSPLL. Section IV presents simulation results, and Section V concludes the paper.

II. PROPOSE DTC DESIGN

The proposed DTC circuit, as shown in Fig. 2, employs a variable-slope DTC architecture. Compared to other architectures, this approach features a simple circuit structure with no complex control logic. In addition, a series of circuit techniques are implemented to mitigate phase noise and suppress nonlinear effects, thereby improving overall performance.

979-8-3315-8850-2/25 $31.00 © 2025 IEEE

$$\mathcal{L}_{flicker} \propto 10\lg\left(f_{out}^2\tau_{delay}^2\frac{2K}{WL\Delta f}\right) \qquad (3)$$

Here, f_{out} denotes the operating frequency of the DTC, k is the Boltzmann constant. T is the operating temperature, I_{charge} is the charging current, τ_{delay} is the time constant during charging, K is a process-dependent parameter, W and L are the dimensions of the PMOS transistor. According to (2), increasing the charging current can suppress thermal noise relatively easily. However, Equations (3) indicate that only increasing the size of the PMOS transistors can reduce flicker noise. Because for a fixed division ratio, τ_{delay} remains constant. If the dimensions of the MOS transistor are significantly increased, the preceding inverter must provide strong drive capability, resulting in substantial power consumption and supply ripple.

This design incorporates a fixed resistor R in the pull-up branch,as shown in Fig. 3(b), thereby shifting control of the delay network from the PMOS transistor to the resistor. PMOS transistors transition from driver transistors to switches, significantly suppressing flicker noise. DTC phase noise is now dominated by white noise, as shown in (4). Reducing R can simply and effectively decrease DTC phase noise.

$$\mathcal{L}_{white} \propto 10\lg\left(f_{out}\frac{kT\tau_{delay}R}{V_{DD}^2}\right) \qquad (4)$$

B. Parallel Switch Structure

Compared to the traditional structure where switches and capacitors are connected in series, the proposed design employs a configuration where switches are connected in parallel with fixed capacitors, then in series with another identical capacitor. Fig. 3 shows the traditional structure, while Fig. 2 shows the proposed parallel switch structure. The advantage of this method is that when the switch is turned off, the parasitic capacitance of the switch is paralleled with a fixed capacitor. Since the capacitance value of the fixed capacitor is significantly higher than the parasitic capacitance, this reduces the impact of the introduced parasitic capacitance on the capacitor array. The adopted structure achieves precise capacitance values of exactly C and $C/2$ when the switch is on and off, respectively. In contrast, the traditional structure yields capacitance values of C and $C_{parasitic}$ (where $C_{parasitic}$ is the parasitic capacitance of the NMOS switch). This approach not only improves the resolution, but also improves the linearity of the capacitor array.

For variable-slope DTC circuits, the subsequent inverter introduces a delay. This delay becomes longer as the control code increases and the charging slope decreases, representing an irreducible systematic error that becomes a primary contributor to the DTC INL. Compared to traditional designs with switches and capacitors connected in series, the proposed design increases the capacitance when the switch is off. This can be regarded as introducing a larger fixed capacitance of $C * 512$ into the capacitor array, thus reducing the systematic error in the circuit caused by the subsequent delay in the inverter, further improving the INL performance. e.g., the 12pF fix cap reduces INL < 2LSB [4].

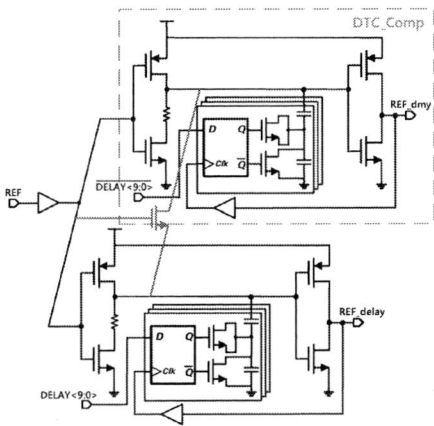

Fig. 2. Proposed DTC circuit

A. Core RC Delay Network

DTC is fundamentally an analog circuit that implements delay using an RC network. By controlling the number of capacitors in the capacitor array connected to the circuit, a precise delay adjustment is achieved. The gain of the DTC is defined as the delay per least significant bit (LSB). The delay time is given by:

$$T_{res} = \ln 2 * R * C_{LSB} \qquad (1)$$

For a 10 bit DTC circuit, the maximum delay time is 1024 * T_{res}. To ensure the Fractional-N SSPLL can lock properly, the maximum DTC delay must exceed the VCO clock period T_{VCO}. This is achieved by selecting appropriate values for RC.

DTC placed in the reference signal path directly couples its quantization noise and phase noise with the phase noise of the input reference signal. This coupling can limit the in-band phase noise performance of the SSPLL. The phase noise of the DTC consists mainly of white noise and flicker noise. For a traditional DTC architecture using inverter-based switching with passive RC networks, as shown in Fig. 3(a), the phase noise can be modeled as follows [3]:

$$\mathcal{L}_{white} \propto 10\lg\left(f_{out}\frac{kT\tau_{delay}}{I_{charge}}\right) \qquad (2)$$

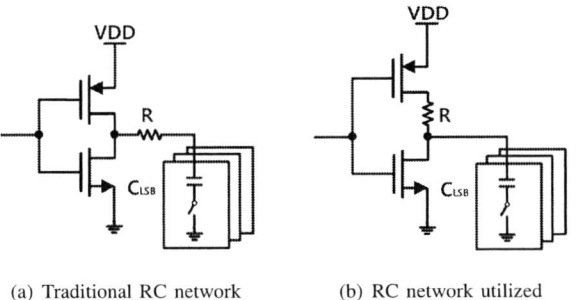

(a) Traditional RC network   (b) RC network utilized

Fig. 3. RC delay network

979-8-3315-8850-2/25 $31.00 © 2025 IEEE

C. Solving Dynamic Mismatches

The operating state of the DTC is not static, but dynamically adjusts its control code to ensure that the delayed reference signal aligns with the VCO output in fractional-N mode. However, changes in the switching states introduce dynamic nonlinearity. Due to the parasitic gate-drain capacitance in the MOS transistors, switching signals can cause feed-through effects. Meanwhile, the presence of parasitic drain-source capacitance requires additional charging/discharging during switching action, known as the charge-sharing effect. These dynamic nonlinearities alter the stored charge in the capacitor array, thereby affecting the delay time. To address this, the proposed complementary switching scheme adds a dummy switch controlled by the inverted code alongside the original single switch. This configuration reduces charge injection and mitigates nonlinearity.

Another critical optimization lies in re-timing the control signals of the switching transistors using the DTC output signal. As illustrated in Fig. 4, If the transition of the control voltage coincides with a time just before the falling edge of the DTC output signal, the resulting dynamic nonlinearities will introduce a phase error between the actual falling edge of the sampled voltage and its ideal position, deteriorating linearity [5]. By re-timing the switching transitions with the output signal, this approach maximizes the recovery time of the voltage drop, effectively nullifying the phase error caused by switching transitions. In essence, this involves aligning the transition of the control voltage to a time that does not interfere with the charging behavior of the RC network. Additionally, an inverter delay is inserted between the output signal and the re-timing clock. This ensures that the switching transitions are temporally separated from the sampling instances in the SSPD, preventing significant interference with the PLL.

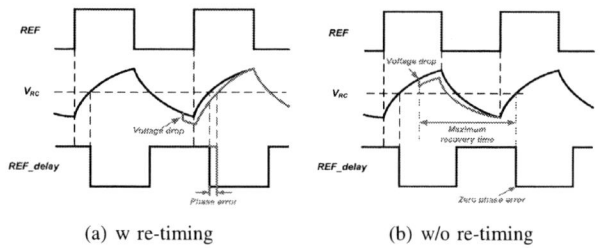

(a) w re-timing   (b) w/o re-timing

Fig. 4. DTC digital control code switching

D. Complementary Structure

During actual operation, DTC continuously draws current from the power supply and ground, inducing ripple due to the inductance of the bonding wire [6]. This ripple increases with the growth of the control code and is directly coupled to the DTC output, resulting in deterioration of the DTC INL curve. Furthermore, for the same control code, the resulting delay exhibits significant fluctuations as a result of this effect. This code-dependent delay due to supply ripple is very troublesome.

The complementary DTC structure compensates for the supply ripple through the following principle. The complementary

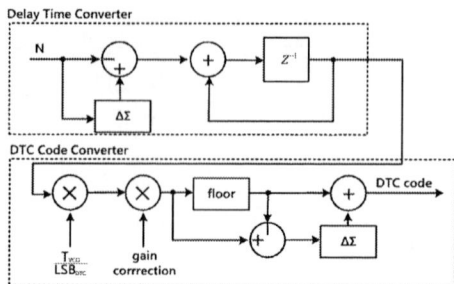

Fig. 5. DTC code modulator

DTC uses switch control codes that are fully complementary to those of the main DTC. When the delay of the DTC increases, the delay of the complementary DTC decreases, ensuring that the total capacitance connected to the circuit remains constant. Additionally, an additional NMOS switch is employed to balance the accumulated charge on the two RC networks when the DTC output is high, forcing them to start changing from the same initial state. This reduces the deviations caused by mismatch. However, the complementary structure has some drawbacks: it doubles the layout area and power consumption. Since capacitors are area-intensive components, the area overhead introduced by the complementary structure is significant. Furthermore, layout matching between the two paths becomes a critical concern that requires careful consideration.

III. DIGTIAL IMPLEMENTATION FOR THE FRACTIONAL-N SSPLL

The delay that needs to be inserted into the reference path for the Fractional-N SSPLL can be calculated all by digital circuit. Fig. 5 shows the DTC code modulator adopted in this desgin.

The modulator consists of two parts: the first part converts the input fractional division ratio into the required delay time for the reference signal within each reference period, as shown in the preceding Fig. 1. The second part then converts this required delay into the corresponding DTC control code.

The first part uses a Delta-Sigma Modulator (DSM) to generate quantization errors corresponding to the fractional component. These quantization errors are accumulated through an integrator to obtain the desired delay time. If a higher-order MASH modulator is employed for the DSM, it can produce more random DTC control codes, which helps reduce fractional spurs in the SSPLL output [2].

Multiplying by $T_{ref}/(LSB_{DTC} * N_{frac})$, the delay time output from the first part is converted into the DTC control code. However, due to the finite resolution of the DTC, the delay required for an arbitrarily fractional division ratio N_{frac} may not be an integer multiple of the LSB of the DTC, resulting in quantization noise. Therefore, modulation of the DTC control code is also favored. Here, the second DSM module employs a high-order MASH 1-1-1 structure to push the quantization noise to higher frequencies [7]. The low-pass

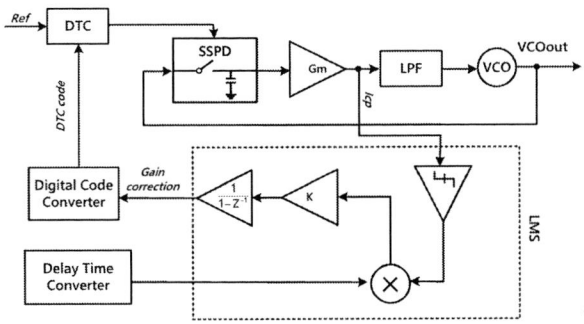

Fig. 6. Proposed SSPLL Structure

TABLE I
COMPARISON OF RELEVANT DTCS

	This Work	JSSC'24 [11]	VLSI'21 [10]	JSSC'19 [12]	JSSC'19 [7]
CMOS Technology	65nm	28nm	22nm	28nm	28nm
Delay Method	Variable Slope	XO Waveform Based	Constant Slope	Constant Slope	Variable Slope
Delay Range	184ps	546ps	530ps	76/187ps	665ps
Resolution	180fs	266fs	1002fs	148/365fs	650fs
INL	216fs @184ps (0.12%)	3900fs @546ps (0.71%)	700fs @530ps (0.13%)	159fs @76ps (0.21%) 1170fs @187ps (0.63%)	650fs @665ps (0.1%)
Phase Noise	below -160 dBc/Hz @10kHz	below -145 dBc/Hz @10kHz	NA	NA	NA

characteristics of the PLL loop then can suppress the impact of this quantization noise.

As a sensitive circuit, DTC experiences significant delay variations under different PVT conditions. This means that the control codes predetermined based on the fractional division ratio may no longer correspond to the correct delay, directly impacting the normal sampling operation of the SSPLL. Therefore, additional digital circuits are needed to calibrate the gain of the DTC across different process corners [8], [9]. Here, the least mean square (LMS) algorithm is employed for background gain calibration as shown in Fig. 6 By detecting the polarity of the output current from the Sub-sampling charge pump (SSCP), the phase difference between the sampling signal and the VCO output signal is determined. The DTC control code is then adjusted accordingly to realign the edge of the sampling signal with the VCO zero-crossing point. Calibration is performed in the background to ensure that nonlinear effects do not interfere with the calibration process.

IV. SIMULATION RESULT

Proposed DTC is designed in 65nm CMOS technology and the layout is shown in Fig. 7. The DTC employs a 10-bit binary code to configure the switch capacitor array. Simulation results indicate a resolution of approximately 180 fs and a delay range covering 184 ps, which spans approximately 1.8 times

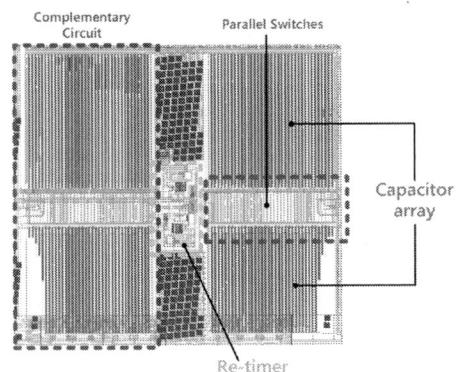

Fig. 7. Layout of the proposed DTC

the maximum period of the VCO. The comparison simulation result between the proposed complementary structure and the traditional one is shown in Fig. 8. The traditional circuit suffered from a larger nonlinearity step in the INL curve, while the proposed complementary structure has a relatively smooth INL curve. For nonlinear performance, INL is within 216 fs (corresponding to 1.2 LSB), while Differential Nonlinearity (DNL) remains within 9 fs (corresponding to 0.05 LSB), as shown in Fig. 9(a). Regarding phase noise performance, as shown in Fig. 9(b), the best-case phase noise (at code = 0) and worst-case phase noise (at code = 1023) differ by approximately 4 dBc/Hz. Even in the worst-case scenario, the phase noise remains below -160 dBc/Hz at a 100 kHz offset frequency. Table 1 compares the proposed DTC with other recently published state-of-art DTC.

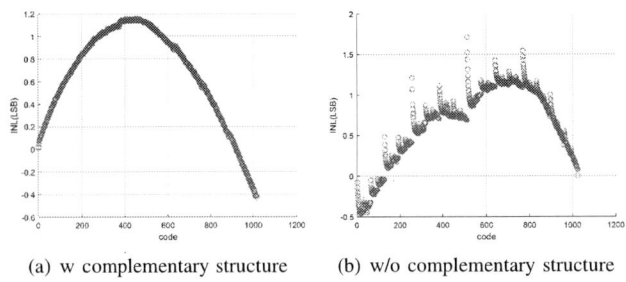

(a) w complementary structure   (b) w/o complementary structure

Fig. 8. Comparison between two different structure

For the Fractional-N SSPLL, both the SSPLL and the DTC delay loop must be activated simultaneously to achieve locking. The system block diagram is shown in Fig. 6. On the one hand, the inclusion of the VCO and the sampling capacitor in SSPD increases the complexity of the circuit simulation. Additionally, mixed-signal co-simulation with the digital calibration module is required, resulting in significantly slower simulation speeds compared to pure analog simulations.

979-8-3315-8850-2/25 $31.00 © 2025 IEEE

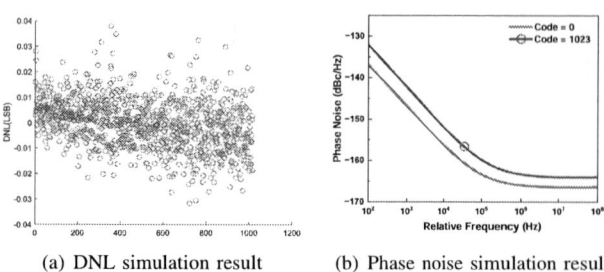

(a) DNL simulation result   (b) Phase noise simulation result

Fig. 9. More simulation result

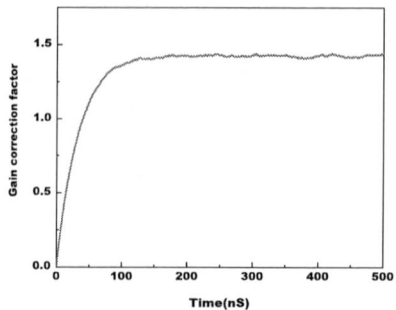

Fig. 11. Gain correction factor

On the other hand, due to the oscillation convergence requirements of the DTC calibration, the lock acquisition speed in fractional-N mode is much slower than in integer-N mode, demanding extremely high simulation time and computational resources. To address these challenges, Verilog-A is used to perform behavioral-level simulations of other modules in the SSPLL, while the DTC code modulator and digital calibration module are implemented in Verilog code. This approach allows simultaneous validation of the DTC functionality and the operation of the SSPLL fractional N mode without requiring full-scale simulations.

The input reference frequency of the SSPLL is 100 MHz, with a fractional division ratio set to 100.75, corresponding to a VCO operating at 10.075 GHz. Fig. 10(a) shows the variation in the VCO control voltage, indicating that the SSPLL achieves locked state within approximately 2 μs. The output spectrum in Fig. 10(b) demonstrates an exact frequency of 10.075 GHz. Figure 11 illustrates the LMS calibration process, where the gain calibration factor stabilizes around a certain number (the simulation results shown here are for 1.4) within 200 ns after power-on.

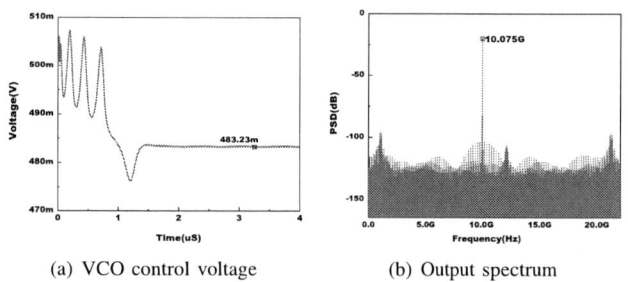

(a) VCO control voltage   (b) Output spectrum

Fig. 10. Fractional-N SSPLL simulation result

V. CONCLUSION

In this paper, a DTC design for Fractional-N SSPLL is proposed. This design operates at 100MHz with a 180 fs resolution and a delay range of 184 ps. For nonlinear performance, INL is within 216 fs, while DNL remains within 9 fs. DTC's phase noise remains below -160 dBc/Hz at a 100 kHz offset frequency. Behavioral-level simulations are also conducted for the SSPLL, and a gain calibration module is integrated into the system to calibrate the DTC gain. These results validate the DTC functionality and SSPLL fractional-N mode operation.

REFERENCES

[1] Xu, Yiqing, et al. "A 13.75-14.75-GHz 32.1-fs RMS Jitter-100.6-dBc Reference Spur-261.4-dB FoM Sub-Sampling PLL Using a KPD-Doubled Isolated Sub-Sampling Phase Detector for Reliable Spur-Jitter-Joint Optimization." 2024 IEEE Asian Solid-State Circuits Conference (A-SSCC). IEEE, 2024.

[2] Raczkowski, Kuba, et al. "A 9.2–12.7 GHz wideband fractional-N subsampling PLL in 28 nm CMOS with 280 fs RMS jitter." IEEE Journal of Solid-State Circuits 50.5 (2015): 1203-1213.

[3] Markulic, Nereo, et al. "A 10-bit, 550-fs step Digital-to-Time Converter in 28nm CMOS." ESSCIRC 2014-40th European Solid State Circuits Conference (ESSCIRC). IEEE, 2014.

[4] Wu, Wanghua, et al. "A 14-nm ultra-low jitter fractional-N PLL using a DTC range reduction technique and a reconfigurable dual-core VCO." IEEE Journal of Solid-State Circuits 56.12 (2021): 3756-3767.

[5] Santiccioli, Alessio, et al. "A 66-fs-rms jitter 12.8-to-15.2-GHz fractional-N bang–bang PLL with digital frequency-error recovery for fast locking." IEEE Journal of Solid-State Circuits 55.12 (2020): 3349-3361.

[6] Dartizio, Simone M., et al. "A low-spur and low-jitter fractional-N digital PLL based on an inverse-constant-slope DTC and FCW subtractive dithering." IEEE Journal of Solid-State Circuits 58.12 (2023): 3320-3337.

[7] Wu, Wanghua, et al. "A 28-nm 75-fs rms Analog Fractional-N Sampling PLL With a Highly Linear DTC Incorporating Background DTC Gain Calibration and Reference Clock Duty Cycle Correction." IEEE Journal of Solid-State Circuits 54.5 (2019): 1254-1265.

[8] Renukaswamy, Pratap Tumkur, et al. "A 16-GHz background-calibrated duty-cycled FMCW charge-pump PLL." IEEE Journal of Solid-State Circuits 59.6 (2023): 1684-1696.

[9] H. Park, C. Hwang, T. Seong, Y. Lee and J. Choi, "32.1 A 365fsrms-Jitter and -63dBc-Fractional Spur 5.3GHz-Ring-DCO-Based Fractional-N DPLL Using a DTC Second/Third- Order Nonlinearity Cancelation and a Probability-Density-Shaping $\Delta\Sigma M$," 2021 IEEE International Solid-State Circuits Conference (ISSCC), San Francisco, CA, USA, 2021, pp. 442-444

[10] Chen, Peng, et al. "A feedforward and feedback constant-slope digital-to-time converter in 28nm CMOS achieving $\leq$ 0.12% INL/range over > 100mV supply range." 2021 Symposium on VLSI Circuits. IEEE, 2021.

[11] Siriburanon, Teerachot, et al. "A Low-Noise Digital-to-Time Converter Exploiting Waveform of Integrated Crystal Oscillator." IEEE Journal of Solid-State Circuits (2024).

[12] Chen, Peng, et al. "A 31-μ W, 148-fs Step, 9-bit Capacitor-DAC-Based Constant-Slope Digital-to-Time Converter in 28-nm CMOS." IEEE Journal of Solid-State Circuits 54.11 (2019): 3075-3085.

2025 The 10th International Conference on Integrated Circuits and Microsystems

A 24–27-GHz 3-Stage Driving Amplifier with 18–25-dB Variable Gain in 180-nm CMOS for Beamforming ICs

Zhenghuan Wei
State Key Laboratory of
Millimeter Waves,
Southeast University,
Purple Mountain Laboratories,
Nanjing, China
230218621@seu.edu.cn

Kaibo Zhang
State Key Laboratory of
Millimeter Waves,
Southeast University,
Purple Mountain Laboratories,
Nanjing, China
230239430@seu.edu.cn

Lijuan Wang
State Key Laboratory of
Millimeter Waves,
Southeast University
Nanjing, China
230228185@seu.edu.cn

Sanming Hu
State Key Laboratory of
Millimeter Waves,
Southeast University
Purple Mountain Laboratories,
Nanjing, China
sanming.hu @seu.edu.cn

Abstract—**This paper presents a millimeter-wave high gain driving amplifier with digitally tunable gain for beamforming transmitter. The amplifier adopts a three-stage cascode architecture with a VGA-first configuration, where the first stage integrates 3-bit digital control switches. The chip is fabricated in 180-nm CMOS process. A It achieves a measured peak power gain of 25.2 dB at 26 GHz and exhibits a 3-dB bandwidth spanning 24–27 GHz, resulting in a gain–bandwidth product (GBW) of 958 GHz. Within this bandwidth, a gain tuning range of 7 dB is demonstrated across eight control states, with a resolution of 1 dB per bit. The measured RMS phase error remains below 2.47°. The amplifier consumes 49mW of DC power. The total chip area is 0.65 mm², with a core circuit area of 0.45 mm².**

Keywords—*Amplifier, CMOS, driving amplifier (DA), gain bandwidth product (GBW), high gain, millimeter-waves (mm-Waves), variable gain amplifier (VGA)*

I. INTRODUCTION

Emerging beamforming integrated circuits (ICs) have witnessed rapid advancements in recent years [1], [2]. In millimeter-waves (mm-Waves) beamforming transmitter, a driving amplifier (DA) is typically required between the power amplifier (PA) and the modulator to compensate for the conversion loss. Furthermore, variable-gain amplifiers (VGAs) are often employed to calibrate the amplitude of individual transmit paths. However, most prior implementations rely on advanced process technologies, such as deeply scaled CMOS nodes in [3], [4], [5], [6], [7], [8], [9], [10]. These technologies require fourth-generation advanced lithography techniques and complex fabrication steps, which significantly increase production time and costs.

To accomplish cost-effective and high-performance mm-Waves components, this work explores a solution using a low-cost 180nm CMOS process (f_t /f_{max} = 65 /70 GHz) to implement a high-gain driving amplifier with integrated variable gain control for beamforming ICs. The proposed amplifier adopts a

three-stage VGA-first architecture, combining structural simplicity with robust performance. All stages employ cascode topologies to enhance overall gain. Source-to-drain inductors are inserted within each cascode stage, achieving a single-stage gain up to 13 dB. A 3-bit digital control switch is embedded in the first stage, enabling a gain tuning range from 18 to 25 dB over a 3-dB bandwidth of 24-27 GHz.

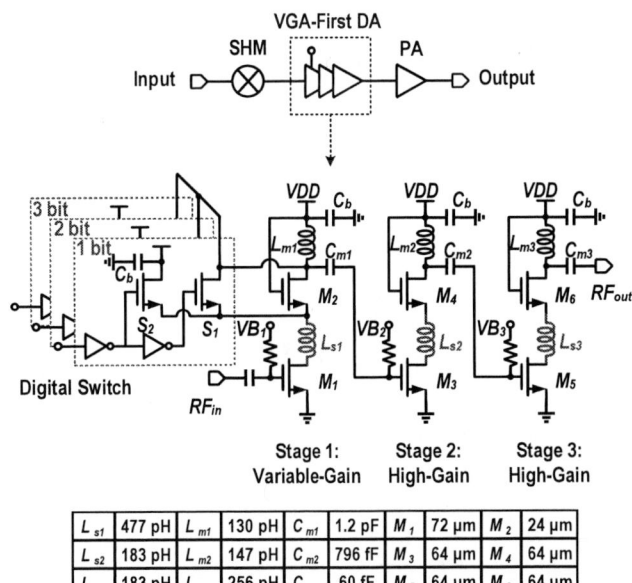

Fig. 1. Schematic of the 3-stage high-gain VGA-first driving amplifier for beamforming ICs.

L_{s1}	477 pH	L_{m1}	130 pH	C_{m1}	1.2 pF	M_1	72 µm	M_2	24 µm
L_{s2}	183 pH	L_{m2}	147 pH	C_{m2}	796 fF	M_3	64 µm	M_4	64 µm
L_{s3}	183 pH	L_{m3}	256 pH	C_{m3}	60 fF	M_5	64 µm	M_6	64 µm

II. TOPOLOGY AND IMPLEMENTATION

Fig. 1 shows the schematic of the proposed VGA-first driving amplifier. In this beamforming system, a sub-harmonic mixer (SHM) is employed to halve the required LO frequency, at the cost of high conversion loss. A driving amplifier is inserted to provide nearly 20 dB gain, meeting the PA input requirement.

This work was supported in part by the Peng Cheng Laboratory (PCL) through the Major Key Project and in part by the Fundamental Research Funds for the Central Universities under Grant 4004002508.

979-8-3315-8850-2/25 $31.00 © 2025 IEEE

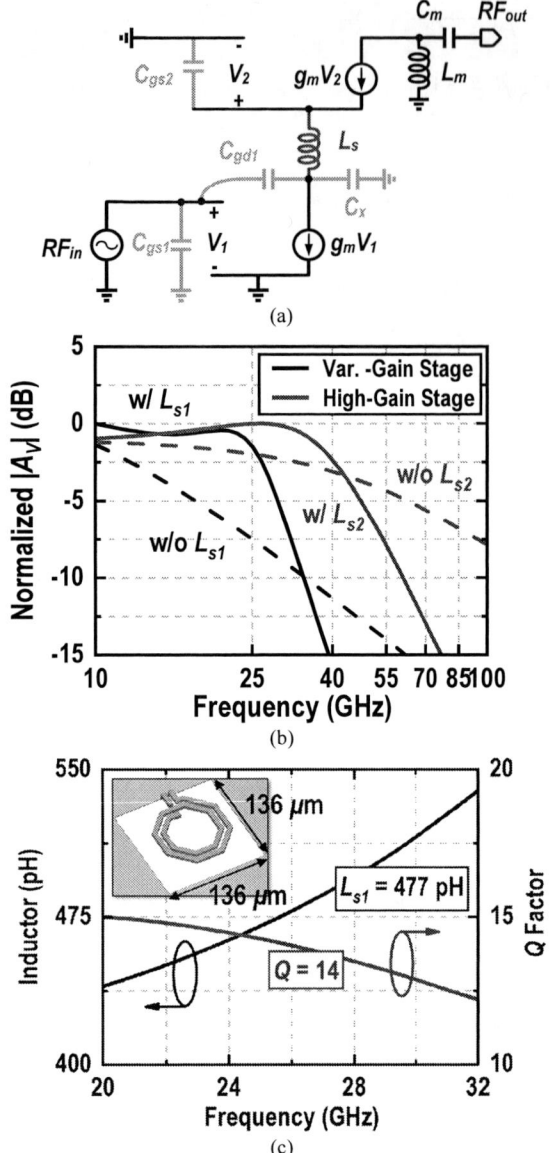

Fig. 2. (a) Small signal model of the single stage amplifier, (b) simulated normalized gain with and without series inductors, and (c) simulated inductance value and Q value of series inductor L_{s1}.

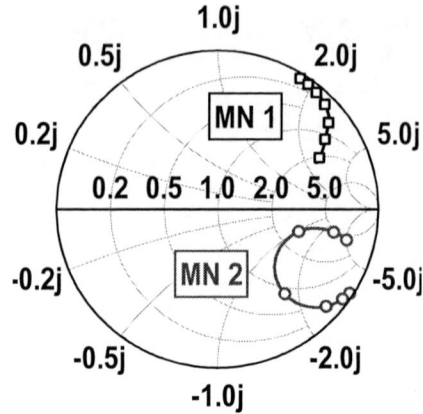

Fig. 3. Simulated impedance of inter-stage matching network.

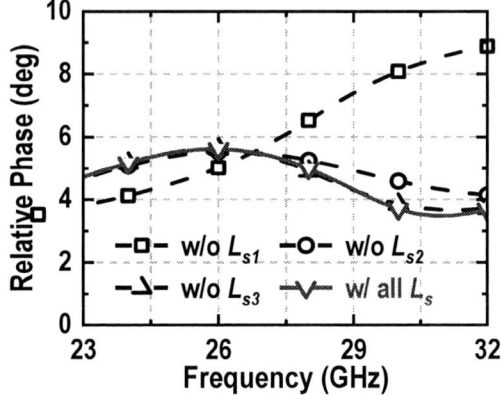

Fig. 4. Simulated relative phase with and without series inductors.

$$n_{\text{opt}} \approx 2 \times \ln A_{\text{tot}} \tag{1}$$

where A_{tot} is the total gain at the maximum bandwidth. In this work, $A_{\text{tot}} = 10$.

The optimal stage number nopt is estimated to be approximately 4. The ratio of the overall GBW$_{\text{tot}}$ of the cascaded configuration to that of a single-stage amplifier GBW$_s$ can be expressed as

$$\frac{\text{GBW}_{\text{tot}}}{\text{GBW}_s} = A_{\text{tot}}^{1-\frac{1}{n}} \times \sqrt{2^{1/n} - 1} \tag{2}$$

For single-stage GBW$_s$ of 10 dB gain and 3 GHz bandwidth, the Eq. (2) for n=4 and n=3 is calculated as 2.44 and 2.36, corresponding to GBW$_{\text{tot}}$ of 113 GHz and 85.3 GHz, respectively. Although the 3-stage structure exhibits a ~25% reduction in GBW compared to the 4-stage configuration, it offers key advantages including 16 mW (~30%) lower power consumption and one fewer spiral inductor, significantly reducing layout complexity. Based on this trade-off, a 3-stage structure is selected for the final implementation. Fig. 3 shows the simulated impedance of the inter-stage matching network.

A. GBW and Phase Error

To improve stage gain, a cascode topology is adopted as the unit amplifier. An inter-stage inductor L_s is inserted between the cascode transistors, forming a π-type LC network, as illustrated in Fig. 2(a). This structure introduces an additional resonance, presenting a high impedance near the operating frequency and shifting the dominant pole to higher frequencies. Consequently, the gain is enhanced. Fig. 2(b) shows the normalized gain profiles for both the variable-gain stage and high-gain stage, with notable gain enhancement observed at 26 GHz. Fig. 2(c) illustrates the layout model and EM simulation of L_s. The optimal number of amplifier stages is determined by maximizing the gain-bandwidth product (GBW),

979-8-3315-8850-2/25 $31.00 © 2025 IEEE

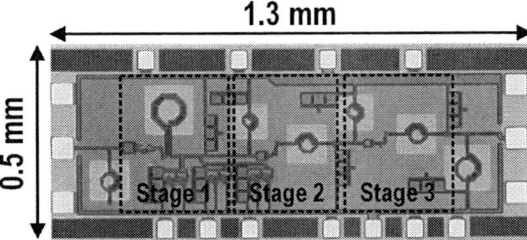

Fig. 5. Chip photograph of the proposed 3-stage driving amplifier.

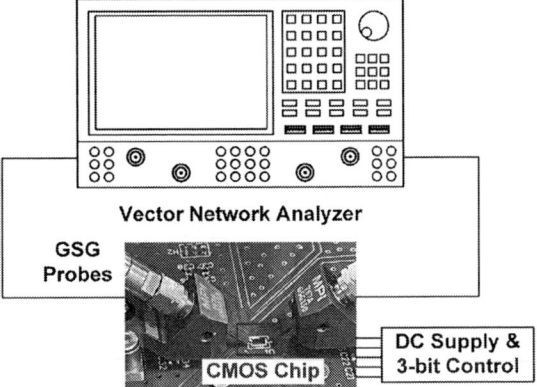

Fig. 6. Measurement setup of the fabricated chip.

From the perspective of pole distribution in the three-stage amplifier, each cascode stage inherently introduces three poles. However, by strategically inserting shunt inductors, the parasitic pole induced by C_x can be effectively eliminated, thereby enhancing the overall amplifier gain. It should be noted that placing high-gain stages at preceding stages would impose stringent linearity requirements on subsequent amplifier stages, which is incompatible with the design objectives of this work. Consequently, the output poles of each stage are analyzed independently, with the dominant output pole frequency determined by the following expression,

$$\omega_{p,out} = \frac{1}{L_m \cdot (C_{gd} + C_{next,g})} \tag{3}$$

where L_m is the matching inductor at drain, C_{gd} is the top capacitor in Fig. 2(a), and $C_{next,g}$ is the gate capacitor at the next stage transistor.

For Eq. (3), the larger transistors (M_3, M_5 =64 μm) exhibit increased gate capacitor, which shifts the poles to lower frequencies and necessitates the employment of a smaller matching inductor. Implementing this design in the 180-nm CMOS process with f_t /f_{max} = 65 /70 GHz presents significant challenges. To satisfy the linearity requirements of the VGA stage, it is essential to position the VGA stage at the first cascaded level, where the signal swing remains within the linear operation region of the active devices.

The series inductors L_s inserted between the cascode transistors also help suppress phase shift across different tuning states, which is essential for beamforming ICs. Fig. 4 compares

the normalized maximum phase shift with and without series inductors L_s, defined as the phase difference between the state "111" and state "000" capacitance states. Using L_s in all three stages minimizes phase variation in the Ka-band, with the VGA stage contributing most significantly to reducing the overall phase shift.

B. Digitally Gain Control

The template is used to format your paper and style the text. All margins, column widths, line spaces, and text fonts are prescribed; please do not alter them. You may note peculiarities. For example, the head margin in this template measures proportionately more than is customary. This measurement and others are deliberate, using specifications that anticipate your paper as one part of the entire proceedings, and not as an independent document. Please do not revise any of the current designations. Given the low output power of the SHM and the operating frequency near half of the device f_{max}, the tuning range of transistor capacitance becomes limited. To achieve fine and uniform gain control, a VGA-first architecture is adopted. This enables precise amplitude adjustment at the small-signal level while reducing the risk of nonlinear distortion during subsequent amplification stages.

In the proposed three-stage driving amplifier, 3-bit digital switches are connected to the first stage amplifier. These switches are realized using common-source transistors with progressively increasing gate widths of 4 μm, 6 μm, and 12 μm, respectively. The sizing ensures uniform gain steps of approximately ~1 dB across eight control states.

III. MEASUREMENT RESULTS

The proposed driving amplifier is fabricated in a standard 180nm CMOS process. Fig. 5 shows the chip micrograph. The total chip area is 0.65 mm², with a core circuit area of 0.45 mm². Under a 1.8 V supply and 0.9 V bias, the measured DC power consumption is 49 mW.

Fig. 6 illustrates the measurement setup of the fabricated chip. The S-parameters are characterized using a Keysight N5227B vector network analyzer (VNA) in conjunction with ground-signal-ground (GSG) probes. To ensure accuracy, the probe losses and system imperfections are removed through a standard short-open-load-through (SOLT) calibration procedure prior to measurement. The measured small-signal performance is summarized in Fig. 7. The 3-dB bandwidth of the chip spans from 24 GHz to 27 GHz, during which both the input and output return losses ($|S_{11}|$, $|S_{22}|$) remain better than −10 dB, demonstrating good impedance matching across the entire operating range.

The control bits of the digital interface are switched between logic levels of 0 V and 1.8 V, and all eight control states are experimentally verified. At the center frequency of 26 GHz, the gain can be tuned from 18 dB up to 25.5 dB with fine 1 dB resolution steps, confirming the effective functionality of the gain control scheme. Fig. 8(a) depicts the corresponding phase responses under different gain settings, while Fig. 8(b) shows that the root-mean-square (RMS) phase error remains below 2.47° throughout the full operating bandwidth, indicating good phase stability.

979-8-3315-8850-2/25 $31.00 © 2025 IEEE

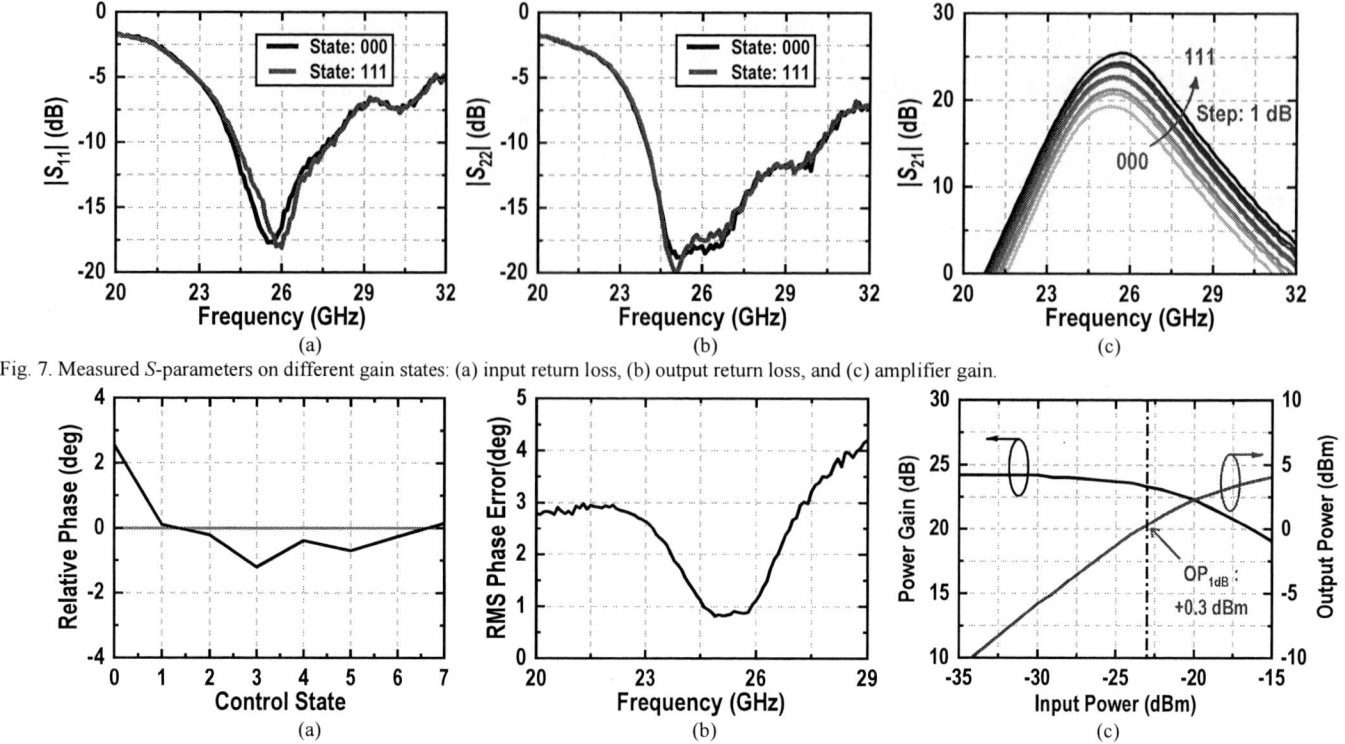

Fig. 7. Measured *S*-parameters on different gain states: (a) input return loss, (b) output return loss, and (c) amplifier gain.

Fig. 8. (a) Measured relative phase on different gain states at 26 GHz, (b) measured RMS phase error over frequency, and (c) measured linearity at 26 GHz.

TABLE I. COMPARISONS WITH STATE-OF-THE-ART MM-WAVES CMOS AMPLIFIERS

Ref.	CMOS Tech.	f_t/f_{max} (GHz)	Freq. (GHz)	Max Gain (dB)	BW (%)	GBW (GHz)	P_{DC} (mW)	Gain Range (dB)	Gain Step (dB)	△Phase (deg)	Area (mm²)
This Work	180nm	65/70	26	25.5	10.6	958	49	7	1	<2.47	0.65
MWTL'25 [3]	65nm	/	29	1.6	13.7	6	28	15.5	0.5	<3.85	0.72
TMTT'22 [4]	65nm	/	26	29.4	15.4	3487	103	6.2	0.2	<0.92	0.14*
IMS'24 [5]	90nm	/	38.5	19	24.6	752	24.7	8	Continuous	/	0.58
SSCL'24 [6]	90nm	120/200	28	21.2	23.2	856	5	13.8	Continuous	1.05	0.32

*Core area

Furthermore, linearity performance is assessed by sweeping the input power at 26 GHz. As shown in Fig. 8(c), the output 1-dB compression point (OP$_{1dB}$) is measured to be approximately +0.3 dBm, which validates the robustness of the proposed design under large-signal operation. These results collectively confirm that the prototype chip achieves wideband impedance matching, accurate gain control, low phase error, and satisfactory linearity, demonstrating its suitability for high-frequency beamforming ICs and communication system applications.

Table I compares the proposed design with recent state-of-the-art CMOS mm-Waves amplifiers. The results demonstrate that this work achieves competitive gain using a cost-effective

CMOS node. The achieved GBW exceeds 958 GHz, significantly surpassing the f_{max} of the 180nm CMOS process.

IV. CONCLUSION

This paper presents a three-stage millimeter-wave driving amplifier with digitally tunable gain, tailored for beamforming transmitters. The amplifier employs a VGA-first cascode architecture, where the first stage incorporates 3-bit digital control switches for fine gain adjustment. Fabricated in a standard 180nm CMOS process, the design achieves high gain with minimal RMS phase error and demonstrates a gain–bandwidth product (GBW) of 958 GHz. This work provides a low-cost solution for high-gain CMOS amplitude and phase control components in mm-Waves beamforming ICs.

REFERENCES

[1] J. Guo, Y. Shen, G. Dong, Z. Han, and S. Hu. "A retrodirective array enabled by CMOS chips for two-way wireless communication with automatic beam tracking," *Engineering*, vol. 37, no. 6, pp. 212–223, Jun. 2024.

[2] J. Guo, Y. Shen, K. Ye and S. Hu, "Differential retrodirective array with integrated circuits in low-cost 0.18 μm CMOS for automatic tracking," *IEEE Trans. Antennas Propag.*, vol. 70, no. 2, pp. 1587-1590, Feb. 2022.

[3] H. Wang et al., "A Ka-band gm-boosted PS-VGA with low phase variation by main-path inductor compensation," *IEEE Microw. Wireless Technol. Lett.*, vol. 35, no. 5, pp. 577-580, May 2025.

[4] Q. Zhang et al., "A Ka-band CMOS phase-invariant and ultralow gain error variable gain amplifier with active cross-coupling neutralization and asymmetric capacitor techniques," *IEEE Trans. Microwave Theory Techn.*, vol. 70, no. 1, pp. 85–100, Jan. 2022.

[5] C.-H. Lai, Y. Wang, Y.-S. Ng, C.-C. Chiong and H. Wang, "A 29-48 GHz variable gain low noise amplifier using active load in 90-nm CMOS process," in *Proc. IEEE MTT-S Int. Microw. Symp. (IMS)*, Washington, DC, USA, 2024, pp. 458-461.

[6] Y.-T. Chang and W.-J. Lin, "A 28-GHz low-power variable-gain low-noise amplifier using twice current reuse technique," *IEEE Solid-State Circuits Lett.*, vol. 7, pp. 58-61, Jan. 2024.

[7] J. Park and S. Hong, "Wideband bidirectional variable gain amplifier for 5G communication," *IEEE Microw. Wireless Technol. Lett.*, vol. 33, no. 6, pp. 691-694, Jun. 2023.

[8] Q. Zhang et al., "A cost-effective Ku-band phased array in package integrating multi-independent CMOS transceivers with on-chip antennas," *IEEE Microw. Wireless Technol. Lett.*, vol. 33, no. 10, pp. 1486-1489, Oct. 2023.

[9] Y. Yu et al., "A 22-to-37.8-GHz low-gain-phase-error variable-gain amplifier with impedance-compensation technique in 65-nm CMOS process," *IEEE Microw. Wireless Technol. Lett.*, vol. 34, no. 6, pp. 757-760, Jun. 2024.

[10] X. Guan and A. Hajimiri, "A 24-GHz CMOS front-end," *IEEE J. Solid-State Circuits*, vol. 39, no. 2, pp. 368–373, Feb. 2004.

2025 The 10th International Conference on Integrated Circuits and Microsystems

A 16-b 8-MS/s Pipelined SAR ADC With Robust Ring Amplifier and On-Chip Bit-Weight Calibration

Shaojuan Chen
*School of Integrated Circuit Science and Engineering,
University of Electronic Science And Technology of China,*
Chengdu,611731, China
SJ_Chen421@163.com

Xiaoyi Li
*School of Integrated Circuit Science and Engineering,
University of Electronic Science And Technology of China,*
Chengdu,611731, China
lxy116063@outlook.com

Xuanhao Zhang
*School of Integrated Circuit Science and Engineering,
University of Electronic Science And Technology of China,*
Chengdu,611731, China
xuanhao-zhang@outlook.com

Xingshuai Zou
*School of Integrated Circuit Science and Engineering,
University of Electronic Science And Technology of China,*
Chengdu,611731, China
xingshuaizou@163.com

Zhenbing Li*
*School of Integrated Circuit Science and Engineering,
University of Electronic Science And Technology of China,*
Chengdu,611731, China
lizhenbing@uestc.edu.cn

Jiaxin Liu*
*School of Integrated Circuit Science and Engineering,
University of Electronic Science And Technology of China,*
Chengdu,611731, China
liujiaxin@uestc.edu.cn

Abstract—This paper presents a pipelined successive approximation register (SAR) analog-to-digital converter (ADC). A robust ring amplifier with high power efficiency is implemented in the ADC. The standard deviation (σ) of its output common-mode voltage is less than 4mV across temperature from -40°C to 105°C. The on-chip bit-weight calibration is embedded in the ADC to mitigate CDAC mismatch effects. The overall system is designed in a 55nm standard CMOS technology with 1.8V analog supply, 1.2V digital supply and 1.2V reference voltage. The ADC works at 8MS/s consuming 1.38mW and achieves 80.8dB signal-to-noise-and-distortion-ratio (SNDR) and 89.3dB spurious-free dynamic range (SFDR) resulting in 175.4dB Schreier Figure-Of-Merit (FoMs).

Keywords—*Analog-to-digital converter, Pipelined SAR, ring amplifier, on-chip calibration.*

I. Introduction

The analog-to-digital converter (ADC) serves as a critical interface between analog and digital signals, and its energy efficiency has been a persistent concern in circuit design. The successive approximation register (SAR) ADC achieves moderate conversion speed and resolution, along with high energy efficiency. In contrast, the pipelined ADC achieves high conversion rates by leveraging temporal and spatial separation across its stages, but it suffers from high power consumption. Building on these two structures, the pipelined SAR ADC combines the benefits of both types, achieving high precision and high energy efficiency while maintaining fast conversion.

In pipelined SAR ADC, the residue amplifier is critical—it sets the accuracy, limits the speed, and consumes significant power. A high-energy-efficiency ADC demands an efficient residue amplifier. Open-loop amplifiers offer high energy efficiency but suffer from poor gain robustness. They require

This work was supported by NSFC under Grant 62174023.

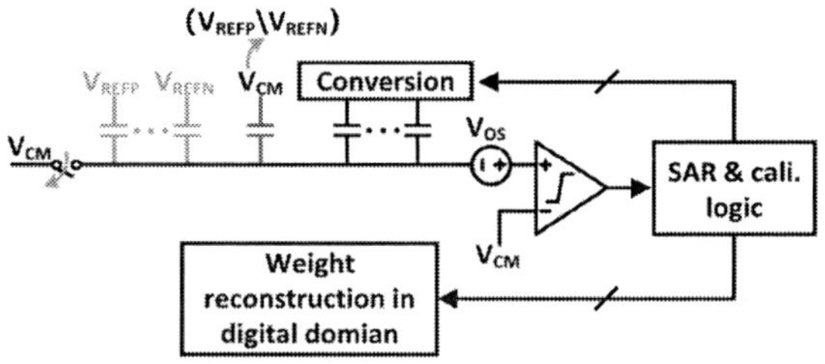

Fig. 1. The method of bit-weight calibration in [6].

calibration and lack sufficient accuracy and linearity. Although closed-loop amplifiers can achieve high precision and linearity, the conventional closed-loop amplifiers consume a lot of power and are less efficient. Recently, robust closed-loop ring amplifiers have emerged as a promising alternative [1], [2]. Unlike traditional multi-stage operational transconductance amplifier (OTA), ring amplifier working at steady state can be designed as a single-pole system with the dominant pole at the output. However, its output common-mode is facing instability.

High-resolution ADCs are often limited by the nonlinearity induced by CDAC mismatch. In some oversampling ADCs, mismatch error shaping (MES) technique in [3] is employed to shape the mismatch error, thereby improving the SNDR. However, this technology is not suitable in Nyquist ADCs. The CDAC mismatch of Nyquist ADCs is facing challenge. In [4] and [5], the CDAC mismatch error is detected by injecting discrete-time pseudorandom noise, and the weights are extracted through a least-mean-square (LMS) adaptive loop. The calibration method (Fig. 1) in [6] imposes stringent requirements on the comparator.

In this work, a pipelined SAR ADC with ring amplifier and CDAC mismatch calibration is realized. The proposed ring amplifier offers high gain and stable output common-mode performance. It realizes energy efficiency while maintaining

979-8-3315-8850-2/25 $31.00 © 2025 IEEE

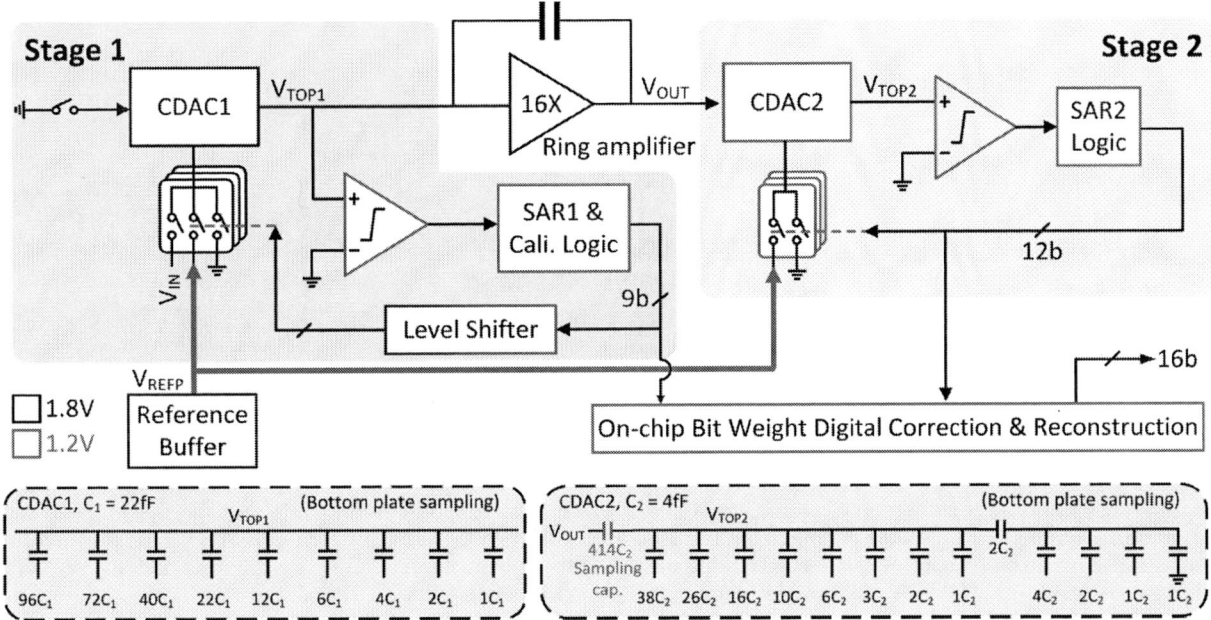

Fig. 2. The overall architecture of the proposed pipelined SAR ADC.

accuracy. And the work applies the technique from [6] to the pipelined SAR ADC, using the second-stage SAR ADC to calibrate the first stage, thereby alleviating the requirement for the offset of the comparator.

II. THE OVERALL ARCHITECTURE

The proposed pipelined SAR ADC consists of two stages of SAR ADCs, and its overall architecture is shown in Fig. 2. The first stage provides 9 bits with 1 redundant bit, while the second stage provides 12 bits with 2 redundant bits. To further accommodate non-ideal factors in the first stage (such as noise and offset voltage), an inter-stage redundancy of 2 bits is incorporated. To enhance system reliability, the redundant bit configuration follows the method described in [7]. Under binary weighting, the proportion of higher weights are reduced while redundant bits are introduced, enabling effective correction of comparison errors by the higher weights. The specific calculation method is as follows:

$$D_{N+M-1} = 2^{N-1} - r_1$$

$$D_{N+M-2} = 2^{N-2} + \frac{r_1}{2} - r_2$$

...

$$D_0 = 2^{-M} + \frac{r_1}{2^{N+M-1}} + \frac{r_2}{2^{N+M-2}} + \cdots + \frac{r_{N+M-1}}{2} - r_{N+M} \quad (1)$$

where N is the resolution of ADC, and M is the number of redundant cycles. And r_i (i = 1, 2, ..., N + M) is the redundant value of the ith conversion cycle.

Due to the adoption of non-binary weighting, a dedicated alignment and data reconstruction circuit is integrated within the ADC to generate the final 16-bit binary-weight output. Under this non-binary weighting architecture, even with redundant-bit sampling, the reconstructed digital code does not exhibit overflow issues.

The reference voltage is critical for ADC performance. The reference voltage coming from external reference buffer will pass through the bonding wire with the parasitic inductance. This inductance makes the reference voltage unstable. Fortunately, an integrated reference buffer suppresses this effect. The reference buffer of this design is composed of OTA as error amplifier, and its dominant pole locating in output.

III. THE IMPLEMENTATION OF PIPELINED SAR ADC

This paper presents a robust ring amplifier, which not only has high gain but also has a stable output common-mode between -40℃ to 105℃. In addition, in this pipelined SAR ADC, on-chip CDAC mismatch calibration is embedded, which not only improves the linearity of the system but also alleviates the requirement of the comparator for offset.

A. The Design of Robust Ring Amplifier

Ring amplifier was first proposed in [8]. Its process is divided into three phases: initial ramping, stabilization and steady state. In the first phase, as input voltage is rapidly amplified into the gate of the third stage, the transconductance of the ring amplifier is larger resulting in UGB increasing. Under the non-dominant poles are insufficiently high, then the phase margin of the initial ramping phase will be smaller leading to the output oscillating. Under the effect of the negative feedback mechanism, the g_m of the ring amplifier gradually decreases and tends to stabilize due to smaller input voltage. Finally system bandwidth is determined, the phase margin increases, and the output stability is established.

Recent research has focused on enhancing the robustness of ring amplifiers. The key approaches involve replacing voltage biasing with current biasing and stabilizing the output common-mode at each stage. The two primary methods are: first method is self-biasing combined with class AB biasing [1], and second method is using a capacitor to set the input common-mode voltage [2], thereby establishing the bias point. The class AB

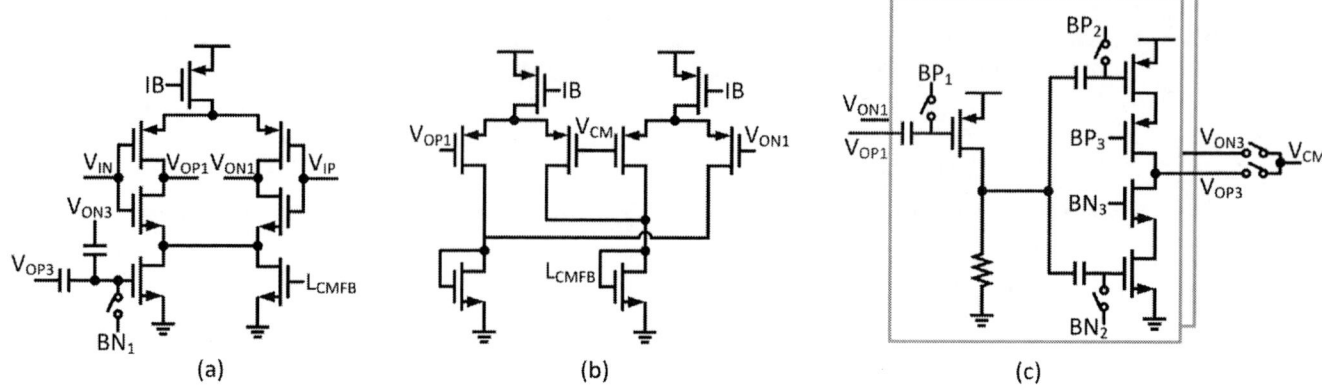

Fig. 3. (a) the first stage of ring amplifier in this work. (b) Common-mode feedback detection amplifier circuit of the first-stage. (c) the second and third stage circuit in the proposed ring amplifier.

biasing method suffers from contention between the upper and lower current mirrors in the second stage, leading to inaccuracy in the third-stage current bias, manifested as current flowing entirely through one end of the CMOS resistor. Therefore, this design primarily adopts the structure of second method to accurately establish the current bias for both the second and third stages.

The overall architecture of the proposed ring amplifier is illustrated in the Fig. 3. Since the dominant pole ω_{p1} is located at the output stage, its frequency can be expressed as

$$\omega_{p1} = \frac{1}{r_{o3}C_L} \qquad (2)$$

where r_{o3} is output resistance of the third stage, C_L is load capacitor of the ring amplifier.

This expression can be derived from the definition of the gain-bandwidth product (GBW) as follows:

$$GBW = A_1 A_2 A_3 \omega_{p1} = \frac{A_1 A_2 g_{m3}}{C_L} \qquad (3)$$

where A_1 and A_2 are the gain of the first stage and the second stage, g_{m3} is the transconductance of the third stage.

If the ring amplifier is considered as a single-stage amplifier, its equivalent transconductance $g_{m,eff}$ is given as follows:

$$g_{m,eff} = A_1 A_2 g_{m3} \qquad (4)$$

The loop unity-gain bandwidth is given as follows:

$$UGB = \frac{\beta g_{m,eff}}{C_L} \qquad (5)$$

where β is feedback factor of ring amplifier.

Based on the integrated noise formula, the thermal noise of the ring amplifier is given as follows:

$$v_{n,in}^2 = \frac{4kT\gamma}{g_{m1}} \frac{\beta g_{m,eff}}{4C_L} = \frac{kT\gamma}{g_{m1}} UGB \qquad (6)$$

Under a pessimistic design scenario, where the amplifier bandwidth is fixed, the input transconductance is uniquely determined by the noise allocation. To meet the noise requirement, the first stage employs a CMOS input pair to provide high input transconductance. In addition, an auto-zero

technique is applied to the output of the first stage to suppress its offset voltage and low-frequency flicker noise. Therefore, the first stage must remain functional during both the auto-zero and amplification phases. Finally, a dedicated common-mode feedback amplifier is used to continuously sense and regulate the common-mode voltage at the output of the first stage.

The second stage of the ring amplifier is designed to provide moderate gain and a high-frequency pole to ensure stability. Since the third stage uses a capacitor-based biasing method, excessive transient voltage variation at the output of the second stage would induce large input transient voltage in the third stage, potentially causing leakage through inadvertently turned-on switches. Therefore, the gain of the second stage must not be excessively high. To achieve adequate gain without requiring common-mode feedback, a pseudo-differential g_m-R structure is adopted in the second stage, with its input common-mode voltage determined by current biasing.

The third stage utilizes a cascoding topology to provide high gain. Finally, the designed ring amplifier achieves a closed-loop bandwidth of 36.7MHz and an open-loop gain of 94.6dB.

In the implementation of global common-mode feedback, if it were applied to the second stage, an additional inverting amplifier would be required to ensure negative feedback. This amplifier, however, would be susceptible to offset voltage, which could lead to failure of the global common-mode feedback. Therefore, the global common-mode feedback is applied to the first stage instead.

B. Bit-Weight Calibration and Reconstruction

The split capacitor switching scheme is used for calibration. Fig. 4 illustrates the bottom-up calibration method for CDAC mismatch in the proposed pipelined SAR ADC. The process of calibration consists of three phases:

Reset: during this phase, the top plate is connected to the common-mode voltage (V_{CM}). In the first stage, the most significant bits (MSBs) of the bottom plate are connected to V_{REFP}, while the capacitor array of the capacitor under calibration (CUC) and the least significant bits (LSBs) are split connecting V_{REFP} and V_{REFN}. The second stage is reset at this time.

979-8-3315-8850-2/25 $31.00 © 2025 IEEE

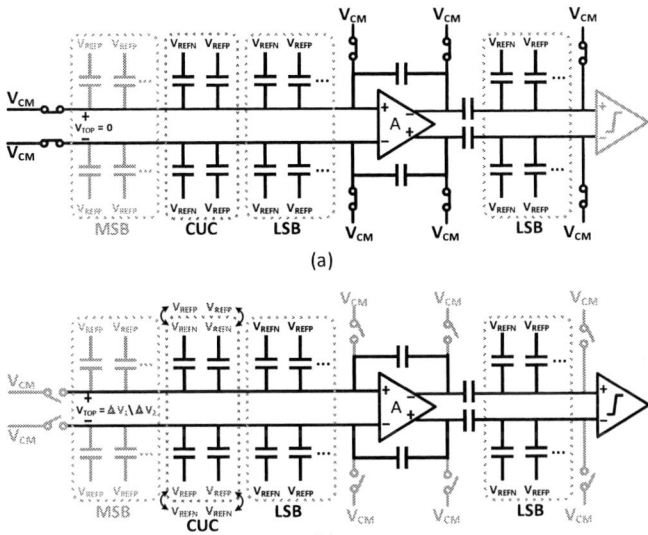

(a)

(b)

Fig. 4. (a) Reset phase in calibration based on bottom-up and (b) Reverse phase in calibration based on bottom-up.

Reverse: the top-plate switch is turned off, while the bottom plate of the MSBs and the LSBs capacitors remain unchanged. The capacitor under calibration is reversed according to the sequence shown in Fig. 4. As a result, a voltage is generated at the top plate, as given below:

$$\Delta V_{TOP} = \frac{C_i}{C_{TOT}} V_{REF} \tag{7}$$

Quantization: The voltage generated on the top plate is first quantized by the LSBs of the first-stage SAR ADC. The resulting residue signal is then amplified by the residue amplifier and further quantized by the second-stage SAR ADC. The resulting digital codes are processed and stored on-chip, and finally the logic outputs a 16-bit binary weight code after alignment and reconstruction. The calculation formulas are given below:

$$\Delta V_{TOP} + V_{OS1} = \sum_{j=1}^{i-1} \frac{D_j C_j}{C_{TOT}} V_{REF} \tag{8}$$

$$V_{RES} = \sum_{j=1}^{i-1} \frac{D_j C_j}{C_{TOT}} V_{REF} - \frac{C_i}{C_{TOT}} V_{REF} - V_{OS1}$$
$$= \frac{D_{SAR2}}{A} - \frac{V_{OS2}}{A} \tag{9}$$

As can be seen from the expression (9), the requirement for offset in calibration applied to the pipelined SAR architecture is less stringent than that in a conventional SAR ADC. The offset is eliminated using the same method as in [6], each digital code is calibrated 32 times, with the upper plate generating a positive voltage difference and a negative voltage difference 16 times each. Finally, the average is taken to obtain the weight value represented by LSB.

IV. THE SIMULATION RESULTS

This section presents post-simulation results of the ring

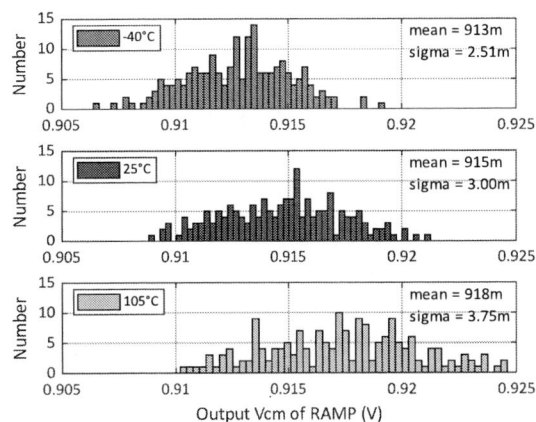

Fig. 5. Global Monte Carlo simulation results of output V_{CM} at -40 ℃, 25 ℃ and 105 ℃.

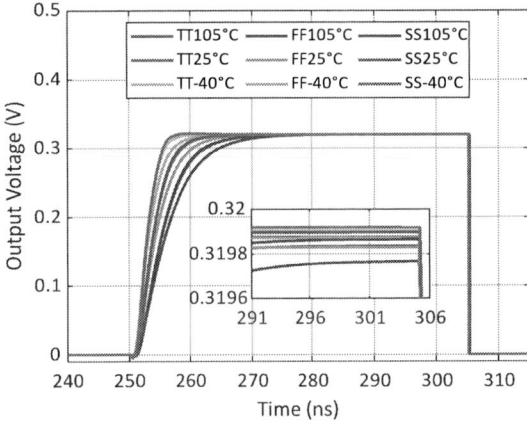

Fig. 6. Closed-loop transient response of the proposed amplifier.

amplifier to demonstrate its robustness, and post-simulation results of the ADC with on-chip self-calibration to verify the effectiveness of the calibration technique.

The proposed pipelined SAR ADC is simulated based on 55nm standard CMOS technology, working at 1.8V analog supply, 1.2V digital supply and reference voltage of 1.2V, with a working speed of 8MS/s.

Fig. 5 shows the global Monte Carlo simulation results of the ring amplifier's output common-mode voltage at -40°C, 25°C, and 105°C. It can be observed that the amplifier exhibits a stable output common-mode level, with a standard deviation (σ) not exceeding 4mV. Fig. 6 presents the differential-mode transient simulation results of the proposed ring amplifier. With a 20mV input and an amplification time of 55ns, the transient differential output remains within 0.5mV across Typical-Typical (TT), Fast-Fast (FF) and Slow-Slow (SS) process. The closed-loop gain error is smaller than 0.003.

The measured core area of the chip is approximately $633um \times 467um$, as illustrated in Fig. 7, with the digital correction and reconstruction circuit occupying an area of $116\mu m \times 117\mu m$. Fig. 8 shows the post-simulation results of the ADC. At input frequency of 328kHz, the ADC achieves the SNDR of 81.7dB

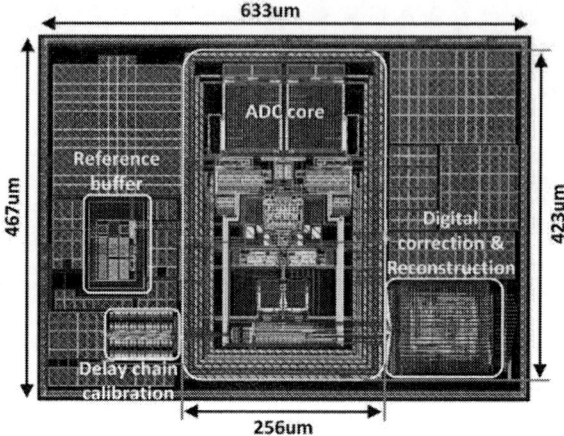

Fig. 7. The layout of the pipelined SAR ADC.

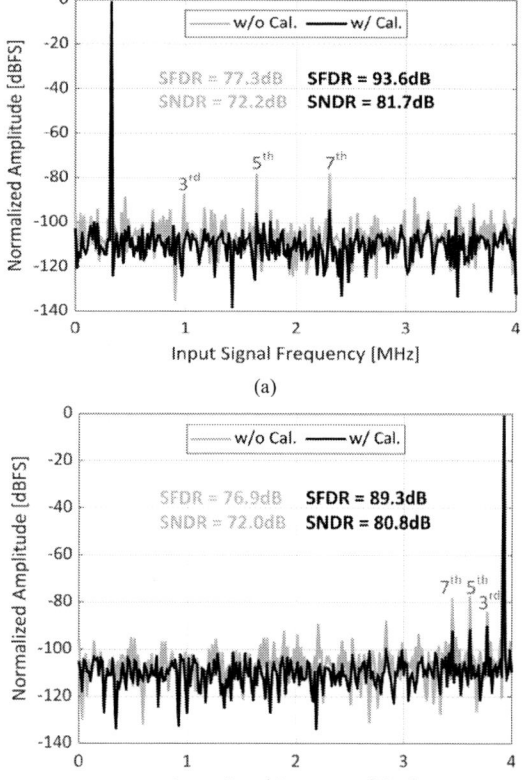

(a)

(b)

Fig. 8. Comparison of w/o calibration and w/ calibration: (a) Fin = 328kHz (b) Fin = 3.92MHz.

and the SFDR of 93.6dB after calibration, representing improvements of 9.5dB and 16.3dB, respectively, over the results without calibration. Under Nyquist-frequency input of 3.92MHz, the calibrated performance realizes 80.8dB SNDR and 89.3dB SFDR, corresponding to increases of 8.8dB and 12.4dB compared to the uncalibrated case. Fig. 9 shows the power consumption of the pipelined SAR ADC. The ADC results in Schreier figure of merit (FOMs) of 172.4dB with reference buffer and 175.4dB without reference buffer. A performance comparison with other high-resolution ADCs is

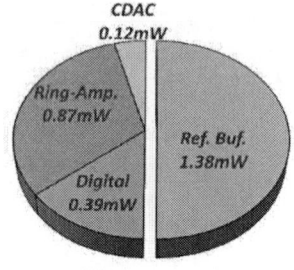

Fig. 9. Power distribution diagram of the ADC.

TABLE I.   PERFORMANCE AND COMPARISON

Paper	This work	[9] 2024	[10] 2025	[11] 2017	[12] 2017
Architecture	Pipe-SAR	Pipe-SAR	SAR	Pipe-SAR	Pipe-SAR
Process (nm)	55	65	180	130	130
Area (mm²)	All 0.30	0.18	0.91	0.14	0.15
	Core 0.11				
CDAC Calibration	On-chip	Off-chip	On-chip	Off-chip	Off-chip
Power (mW)	1.38\2.76*	0.072	6.7	0.17	0.28
Sampling Rate (MS/s)	8	2	1	10	10
SNDR (dB)	80.8	74	84	67.3	63.2
FoMs ** (dB)	175.4\172.4*	175.4	162.7	172	166.4

* With reference buffer.

** $FoMs = SNDR + 10\,log_{10}\frac{BW}{Power}$

shown in Table I.

V. CONCLUSION

This article presents a pipelined SAR ADC incorporating a robust ring amplifier and on-chip self-calibration for CDAC mismatch. The proposed ring amplifier improves the stability of the output common-mode voltage, where it achieves less than 4mV variation of a standard deviation within -40℃ to 105℃. And a bit-weight self-calibration technique based on a bottom-up approach is applied to the pipelined architecture. This technology not only increases the linearity of the ADC, but also relaxes the requirement on comparator offset voltage. Fabricated in a standard 55nm process, the ADC, working at 8MS/s sampling rate, achieves 80.8dB SNDR and 89.3dB SFDR with a 3.92MHz input signal, while consuming 1.38mW without reference buffer, resulting in a Schreier figure-of-merit (FoMs) of 175.4dB.

REFERENCES

[1] M. Zhan, L. Jie, X. Tang et al., "A 0.004-mm2 200-MS/s Pipelined SAR ADC With kT/C Noise Cancellation and Robust Ring-Amp," in IEEE Journal of Solid-State Circuits, vol. 59, no. 7, pp. 2209-2218, July 2024.

[2] A. ElShater, P. K. Venkatachala, C. Y. Lee et al., "A 10-mW 16-b 15-MS/s Two-Step SAR ADC With 95-dB DR Using Dual-Deadzone Ring Amplifier," in IEEE Journal of Solid-State Circuits, vol. 54, no. 12, pp. 3410-3420, Dec. 2019.

[3] Y. S. Shu, L. T. Kuo and T. Y. Lo, "An Oversampling SAR ADC With DAC Mismatch Error Shaping Achieving 105 dB SFDR and 101 dB SNDR Over 1 kHz BW in 55 nm CMOS," in IEEE Journal of Solid-State Circuits, vol. 51, no. 12, pp. 2928-2940, Dec. 2016.

[4] W. Liu, P. Huang and Y. Chiu, "A 12-bit, 45-MS/s, 3-mW Redundant Successive-Approximation-Register Analog-to-Digital Converter With Digital Calibration," in IEEE Journal of Solid-State Circuits, vol. 46, no. 11, pp. 2661-2672, Nov. 2011.

[5] G. Wang, F. Kacani and Y. Chiu, "IRD Digital Background Calibration of SAR ADC With Coarse Reference ADC Acceleration," in IEEE Transactions on Circuits and Systems II: Express Briefs, vol. 61, no. 1, pp. 11-15, Jan. 2014.

[6] A. Lopez-Angulo, A. Gines and E. Peralias, "Calibration of Capacitor Mismatch and Static Comparator Offset in SAR ADC with Digital Redundancy," 2020 IEEE International Symposium on Circuits and Systems (ISCAS), Seville, Spain, 2020.

[7] L. Qiu, K. Tang, Y. Zheng et al., "A Flexible-Weighted Nonbinary Searching Technique for High-Speed SAR-ADCs," in IEEE Transactions on Very Large Scale Integration (VLSI) Systems, vol. 24, no. 8, pp. 2808-2812, Aug. 2016.

[8] B. Hershberg, S. Weaver, K. Sobue et al., "Ring Amplifiers for Switched Capacitor Circuits," in IEEE Journal of Solid-State Circuits, vol. 47, no. 12, pp. 2928-2942, Dec. 2012.

[9] C. Park, J. Kim, K. Kang et al., "A High-Resolution Pipelined SAR ADC Using Cyclically Charged Floating Inverter Amplifier," in IEEE Journal of Solid-State Circuits, vol. 59, no. 10, pp. 3242-3252, Oct. 2024.

[10] J. Ding, F. Liu, K. Deng et al., "A 16-bit 1-MS/s SAR ADC With Capacitor Mismatch Self-Calibration," in IEEE Transactions on Very Large Scale Integration (VLSI) Systems, vol. 33, no. 1, pp. 10-20, Jan. 2025.

[11] M. Gandara, P. Gulati and N. Sun, "A 172dB-FoM pipelined SAR ADC using a regenerative amplifier with self-timed gain control and mixed-signal background calibration," 2017 IEEE Asian Solid-State Circuits Conference (A-SSCC), Seoul, Korea (South), 2017.

[12] M. Gandara, W. Guo, X. Tang et al., "A pipelined SAR ADC reusing the comparator as residue amplifier," 2017 IEEE Custom Integrated Circuits Conference (CICC), Austin, TX, USA, 2017.

An IVUS AFE with LNA and a CT ΔΣ modulator with 80MHz BW and 12 bit ENOB

Jie Peng
School of Sun Yat-sen University of Microelectronics Science and Technology
Zhuhai, China
pengj96@mail2.sysu.edu.cn

Yunchu Li*
School of Sun Yat-sen University of Microelectronics Science and Technology
Zhuhai, China
liyunchu@mail.sysu.edu.cn

Abstract—A low-noise amplifier (LNA) and a wide-dynamic-range broadband ADC are critical components of the analog front-end (AFE) in high-frequency intravascular ultrasound (IVUS) imaging systems. This paper presents an LNA and an 80-MHz bandwidth continuous-time delta–sigma modulator (CTDSM) in 55-nm CMOS. To achieve both low noise figure (NF) and good impedance matching (S11) performance, the LNA employs a noise and nonlinearity cancellation technique. For the CTDSM, a 4th-order, 4-bit architecture with a maximum out-of-band gain of 2.5 dB is chosen to balance complexity, noise performance, linearity, and power consumption. The proportional-integrating element (PI-element) is employed in the loop filter to compensate for a 0.6-clock-cycle excess loop delay (ELD). The integrators use two-stage opamps with hybrid feed-forward/Miller compensation to enhance phase margin and power efficiency. To mitigate DAC nonlinearity, a dynamic element matching technique labeled clock frequency modulation (CFM) is applied. The CTDSM achieves 74.2 dB SNDR, 84.3 dB SFDR over an 80 MHz signal bandwidth consuming 28.2 mW from 1.2/1.8 V supplies, yielding a 168.7 dB FOM (SNDR-based). Meanwhile, the LNA provides a 12–24 dB gain across 30–80 MHz, with NF <1.25 dB.

Keywords—*IVUS, Analog-to-digital converter (ADC), continuous-time delta–sigma modulator (CTDSM), excess loop delay compensation, nonlinearity, dynamic element matching*

I. INTRODUCTIN

Intravascular ultrasound (IVUS) is one of the most advanced imaging modalities for coronary artery assessment. The ultrasound transducer output signal is processed by the analog front-end (AFE) circuit and converted into the digital domain, which can be analyzed digitally and displayed as images. A typical single-channel high-frequency IVUS AFE includes a low-noise amplifier (LNA), a voltage-controlled attenuator (VCAT), a programmable gain amplifier (PGA), a low-pass filter (LPF), and an analog-to-digital converter (ADC) as illustrated in Fig. 1[1].

Currently, the center frequency of mainstream IVUS transducers is located between 20 and 40 MHz, which is increasingly insufficient to meet the growing demands of clinical diagnostics[2]. To enhance the imaging resolution of IVUS systems, further improvements in the operating frequency and bandwidth of ultrasound transducers are essential. The high-

This work was supported by the Special Funds of the National Natural Science Foundation of China (Grant No. 62341409).

frequency IVUS AFE requires wide-band high resolution ADCs to process the high-frequency signal.

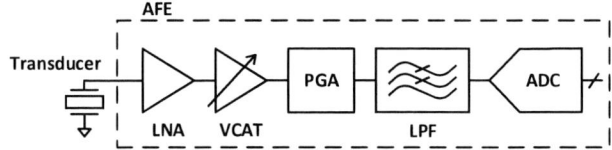

Fig. 1. The basic components of a UIS AFE.

This paper presents an LNA and a 4th-order 80-MHz bandwidth continuous-time delta–sigma modulator(CTDSM) for IVUS system. The CTDSM architecture is a popular structure for high-resolution and wide-band specifications[3-5] and is utilized in ultrasound imaging system (UIS) lately[6]. The CTDSM architecture is adopted due to its inherent anti-aliasing feature, ease of driving and relaxed settling requirements of the integrators. Additionally, oversampling enables a high sampling rate, facilitating more precise timing control, which is a critical feature for beamforming (BF) algorithms in ultrasound imaging[6, 7].

With a signal bandwidth higher than 50 MHz, the sampling frequency of the CTDSM is usually required to be several GHz for considerable oversampling ratio (OSR) to achieve high resolution, which results in significant power consumption of each block in the ADC. To improve power efficiency, CTDSM designs are often limited to a low OSR, therefore requiring a high-order noise transfer function (NTF). However, the phase delays introduced by the high order loop filter (LF) can easily lead to instability, while the large number of op-amps further increases power inefficiency. To address this issue, multi-bit quantizers (QTZs) can be used, but they come with trade-offs: higher power consumption and degraded linearity[5].

With high sampling frequencies, unintentional internal delays from blocks in the CTDSM cannot be ignored, since they are comparable to the sampling period. One of the main internal delays is the integrator delay. To reduce this delay, The integrators with high unity-gain bandwidth (UGBW) opamp is necessary. Meanwhile, a multistage opamp is usually preferred to achieve sufficient DC gain. Consequently, phase margin becomes a critical design parameter for the opamps. Miller capacitor compensation and feedforward path compensation can effectively improve the phase margin. Compared with Miller capacitor compensation, feedforward path compensation is free from degradation of UGBW. However, the feedforward stage requires significant bias current, leading to increased power consumption. In this work, we use a hybrid feed-forward/Miller compensation with Miller-zero technique to address this limitation[8].

Multi-bit quantization can reduce clock jitter sensitivity and quantization error, but the mismatch between DAC units introduces DAC nonlinearity. A dynamic element matching technique labeled clock frequency modulation(CFM) is applied in this design to modulates mismatch dependent inband distortion of a multi-bit DAC to out-of-band. Although CFM does not compensate for associated noise, it provides the advantage of requiring low-complexity digital circuitry and switch matrix, which results in less excess loop delay[9].

The paper is organized as follows. Section II describes the design parameters of the IVUS system, and discusses ADC architecture and system-level design considerations. Section III presents the circuit implementation details, and Section IV shows the simulation results. Finally, Section V concludes this paper.

II. SYSTEM DESIGN

A. Design Parameters

We first analyze the requirements of the LNA and ADC for the target ultrasound AFE[10].

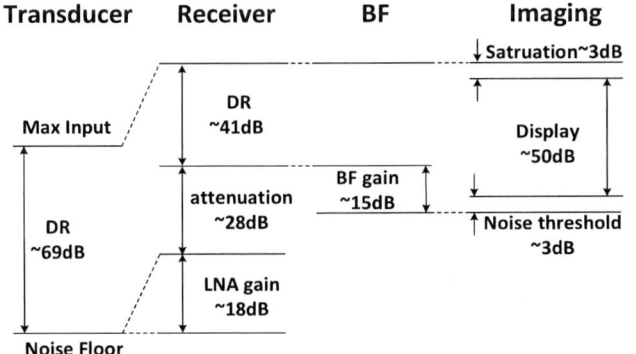

Fig. 2. Gain and DR requirements for ultrasound imaging.

The signal attenuation of ultrasound (A_{att}) can be expressed as follows:

$$A_{att} = \alpha \times 2 \times d \times f \qquad (1)$$

where α is the acoustic attenuation coefficient, d is the imaging depth, and f is the frequency of the ultrasound. As the absorption is proportional to ultrasound frequency, a higher frequency ultrasound has a limited depth of penetration while showing high spatial resolution. In this work, The designed ultrasound transducer features 32 elements with a center frequency of 55 MHz and a penetration depth of 5 mm. As the acoustic signal penetrates through the human soft tissue with an attenuation rate of 0.5 dB/MHz/cm, the signal attenuation is about 28 dB. Also, the minimum display resolution of 50 dB is required to distinguish the characteristics of the human organ. On the other hand, an additional dynamic range (DR) can be achieved through BF since it combines and averages the signals. A total of 32 channels increase the DR by 15 dB ($20\lg\sqrt{32}$ dB). Therefore, a total target DR of 69 dB can be obtained by calculating the signal attenuation of 28 dB, the minimum display resolution of 50 dB, imaging saturation allowance of 3 dB, the noise threshold of 3 dB, and BF of 15 dB (see Fig. 2).

The gain of an LNA should be determined by the ratio of the ADC input and the maximum amplitude of the LNA as follows:

$$Gain_{LNA} = 20 \cdot log_{10} \frac{Max\ ADC\ Input}{Max\ LNA\ Input} \pm Margin \qquad (2)$$

In this work, the maximum input of the LNA is assumed to be 130 mV, and the maximum ADC input is set to 1.08 V(-1dBFs). Thus, the required gain of the LNA is 18 dB ± Margin, and the gain of LNA is determined to 12~24 dB with a 6-dB margin.

B. System Design Considerations

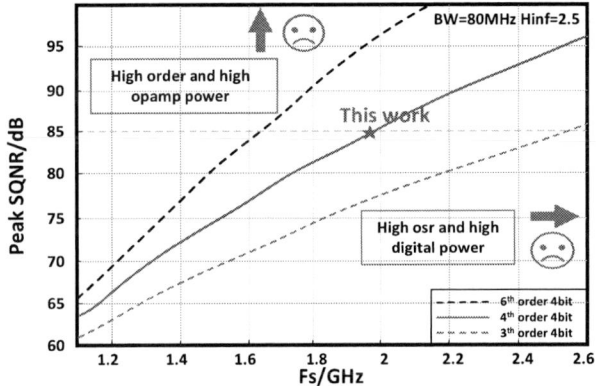

Fig. 3. Peak SQNR versus sampling frequency for different CTDSM.

As analyzed, The modulator requires ~69 dB DR over ~80 MHz bandwidth. In order to provide sufficient margins for device noise and process, voltage, and temperature (PVT) variations, an 84-dB signal-to-quantization-noise ratio (SQNR) is set as the goal in the behavioral evaluation. Fig. 3 shows three possible choices that meet our design target under different OSR, sampling frequency (FS), and LF order with 2.5 dB maximum NTF out-of-band gain (OBG). The selection among these three cases involves different tradeoffs in terms of LF stability, QTZ speed, and OPAMP bandwidth, thus eventually affecting the overall energy efficiency. The 6th-order design offers superior in-band noise suppression but suffers from stability issues and higher power consumption due to additional opamps. The 3rd-order alternative requires excessive digital/analog power to maintain stability at high OSR. Therefore, The 4th-order 4-bit architecture was ultimately selected as it optimally balances noise performance, linearity, stability, and power efficiency[5].

Figure 4 shows the block diagram of the proposed CTDSM. The design employs a hybrid feedforward-feedback architecture to improve power efficiency and linearity. A pure feedback architecture suffers from large signal swings at each integrator output, degrading modulator linearity. Conversely, a pure feedforward approach requires the first integrator's amplifier to simultaneously achieve high UGBW for stability and high DC gain for linearity, resulting in excessive power consumption[8]. Therefore, we use feedforward paths to decrease the signal swing of integrator output, thus avoiding saturation and suppressing nonlinearity. Meanwhile, the feedback paths (DAC4 in Fig. 4) are also used to enable the last integrator to implement a fast path, thus reducing power consumption.

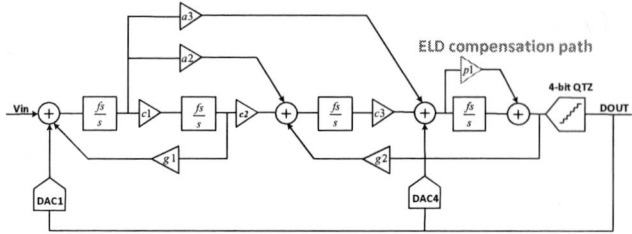

Fig. 4.   Block diagram of the proposed CTDSM.

C. ELD Implementation Consideration

For CTDSM with a sampling frequency of several GHz, non-negligible loop delays must be pre-compensated in the NTF to ensure stability. The deviation between the real NTF and ideal NTF due to the latency of the quantizer and the finite UGBW of amplifiers is compensated by proportional–integrating element (PI-element). Compared with other compensation methods, using PI-element is easier to implement in transistor level. It avoids the use of power-hungry summing amplifier[3] in conventional approaches while also prevents the degradation of QTZ speed and linearity that occurs with digital loop delay compensation methods[4]. Additionally, the compensation coefficient is the ratio of two resistors, which varies little with PVT variations[8].

III. CIRCUIT IMPLEMENTATION

The circuit implementation of the proposed CTDSM is shown in Fig. 5. The integrators in the loop filter use two-stage opamps with hybrid feed-forward/Miller compensation to improve the phase margin and for high power efficiency. The feedback DAC employs current steering architecture, and its nonlinearity is mitigated by a dynamic element matching technique labeled clock frequency modulation(CFM). Due to its simplicity, it is well suited for application in high-speed designs. To reduce latency in the feedback loop, a flash ADC is applied as the quantizer. The process variation of the RC integrator is compensated by 6-bit digital tuning capacitors, covering ±30% time constant variation while achieving 2% tuning accuracy. We adopted the non-return-to zero (NRZ) current-steering DAC to reduce the clock jitter sensitivity.

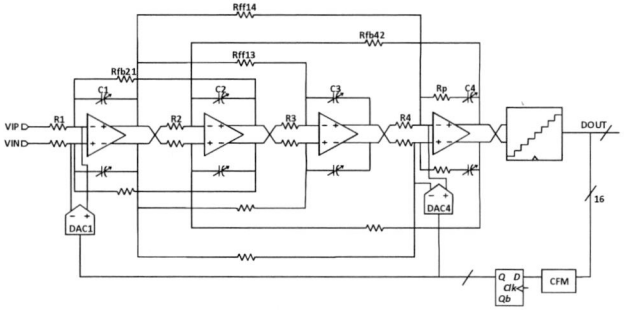

Fig. 5.   Circuit implementation of the proposed CTDS ADC.

A. Opamp Design

The integrators in the loop filter employ two-stage opamps with hybrid feedforward-Miller compensation[4] to achieve the maximum UGBW, sufficient DC gain (>40dB) and phase margin (>60°), as shown in Fig. 6(a). The magnitude and phase response are shown in Fig. 6(b). Compared to standalone feedforward compensation, the high frequency poles are now located closer to crossover due to Miller compensation, the feedforward zero occurring after crossover is much more effective in improving phase margin. Thus, this topology can achieve good phase margin while reusing the bias current between g_{mF} and g_{m2}, thus saving power.

Fig. 7 shows the circuit schematic of the opamp used in the 1st and 4th integrators[5]. The stages and compensation components corresponding to Fig. 6(a) are as shown. A feedforward path from the input (VIP/VIN) to M8a/b is inserted, where the input is AC coupled to M8a/b through C1a/b, to push the UGBW of the opamp close to the ideal two-stage opamp. The second stage exhibits a Class-AB-like topology that is chosen to supply enough output headroom and linearity. The common voltage is detected by a pair of parallel resistors and capacitors and the CMFB circuit controls M4. The opamp achieves a unity gain bandwidth of 6.6 GHz with 67 degrees of phase margin and consumes 4 mW. The 2nd and 3rd integrators use the same topology but with scaled down bias currents.

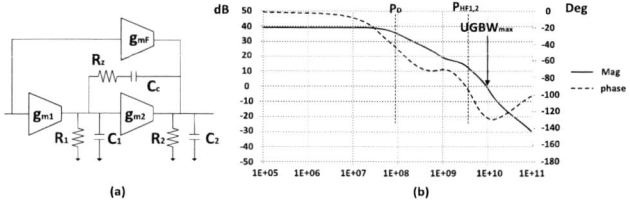

Fig. 6.   (a)feedforward-Miller compensation.(b)Magnitude and phase response.

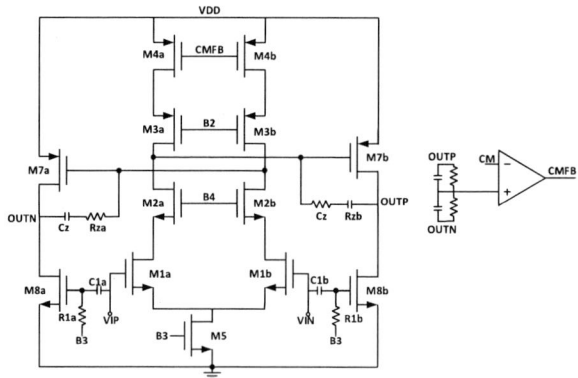

Fig. 7.   Opamp circuit schematic.

B. DAC Implementation and CFM

The CTDSM employs the current steering DAC as the feedback DACs (DAC1 and DAC4 shown in Fig. 4). Fig. 8 shows the circuit diagram of a unit DAC cell. The current sources, MP1 and MN1 are cascoded for higher output impedance and reducing the parasitic capacitance at the common source nodes of the switches. The switch drivers reduce the DAC switching signal swing and ensure the switch

979-8-3315-8850-2/25 $31.00 © 2025 IEEE   482

transistors remain in the saturation region during DAC transitions.

The noise and distortion generated by the DAC4 are suppressed by the loop filter. However, the noise or distortion generated by DAC1 can be directly referred to the input, significantly degrading the modulator's signal-to-noise ratio (SNR) and total harmonic distortion (THD). Therefore, noise and linearity performance should be improved in DAC1 to achieve high resolution. A dynamic element matching technique labeled clock frequency modulation(CFM) is applied in this design to modulate mismatch dependent inband distortion of a multi-bit DAC to out-of-band. The modulation frequency can be generated by incrementing the pointer to the first unit element to be used by a fixed number M from sample to sample[9]. For $M=$ 2 and a DAC with 16 unit elements, after 8 periods of the sampling frequency of the CTDSM, the sequence of the first unit element to be used repeats. Thus, a modulation frequency equal to fs/8 is generated. Since CFM does not compensate the raised noise floor caused by DAC mismatch, it is necessary to decrease the standard deviation of the unit current sources, which we designed to be 0.3%. With such a DAC mismatch, the CFM technique improves SNDR from 64.4 dB to 73.5 dB and enhances SFDR from 68.5 dB to 83.5 dB, as shown in Fig. 9.

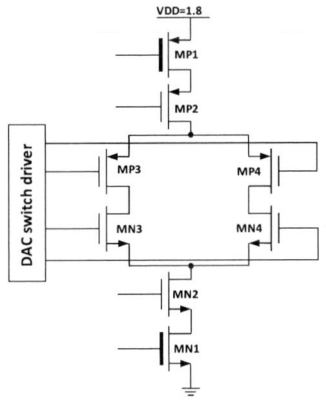

Fig. 8. DAC unit cell.

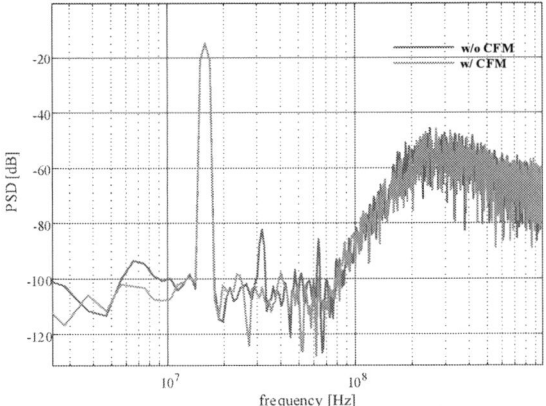

Fig. 9. FFT of modulator output with and without CFM.

C. Comparator

The 4-bit quantizer is composed of 16 identical comparators which schematic is shown in Fig. 10. Each comparator consists of a preamplifier, a strong-arm comparator and an SR-Latch[3]. The strong-arm latch is selected to achieve optimal delay-power tradeoff, where the comparator consists of one stage that performs both signal amplification and latch regeneration. The preamplifier compares the input with one of 16 differential reference levels generated by a resistive ladder and attenuates the kick-back noise from the sampling clock. The preamplifier input pair sizing is scaled to minimize the effect of offset. The cross connection of differential inputs and reference inputs is used to minimize effects of common mode variations at the comparator input.

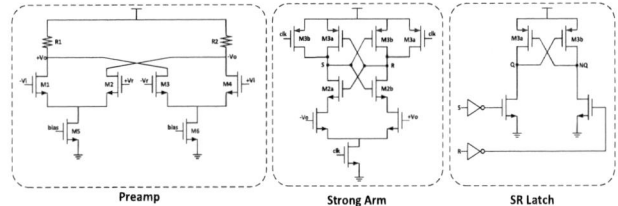

Fig. 10. schematic of the comparators.

D. LNA

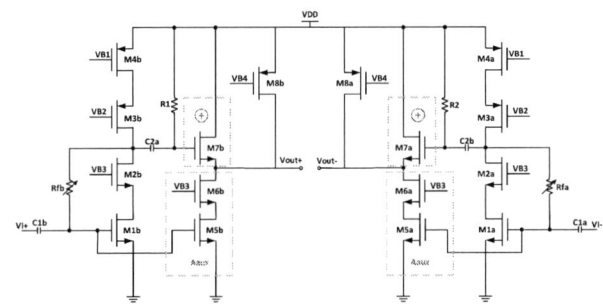

Fig. 11. Noise and nonlinearity canceling LNA.

The LNA is specifically designed for polyvinylidene fluoride (PVDF) transducer imaging with a typical signal frequency range of 30-80 MHz. To achieve both low noise figure (NF) and good impedance matching (S11) performance, a noise and nonlinearity canceling technique[11] is utilized in the LNA, as shown in Fig. 11. In the main amplifier circuit with resistive negative feedback, M3 and M4 form a cascode structure serving as the current source load. The voltage division between M3, M4 and M1, M2, along with the feedback resistor Rf, provides the main amplifying transistor M1 with an appropriate bias voltage. The output signal of the main amplifier is coupled to the source follower M7 through capacitor C2. M5 and M6 form a cascode-structured auxiliary amplifier (Aaux), with current source M8 acting as the load to ensure sufficient gain for the auxiliary amplifier. The noise generated by transistor M1, which transfers to the output of the first stage through Rf and through auxiliary amplifier in opposite phase, thus can be cancelled by designing appropriate auxiliary amplifier. The signal transferred by the main and the auxiliary amplifiers has the same phase and can be added. Thus the structure is able to cancel the noise of main transistor while enhancing the signal gain. Adjusting the feedback resistor Rf allows for tuning the gain of the LNA.

IV. SIMULATION RESULTS

The LNA is realized in 55-nm CMOS. Fig. 12 shows the S11 performance of the LNA. Over PVT variations, S11 is better than -12 dB across the PVDF transducer operation frequency range, 30-80 MHz. The frequency response is shown in Fig. 13. The LNA has a constant gain of 18 dB from 30MHz to 80 MHz at TT corner with a 3-dB bandwidth of 1.09 GHz. The passband gain flatness at different corners is better than 1dB. Fig. 14 shows the NF performance of the LNA. At TT corner, the NF is less than 1.25 dB. The differential LNA consumes a total power of 16.8 mW with a 1.2-V supply.

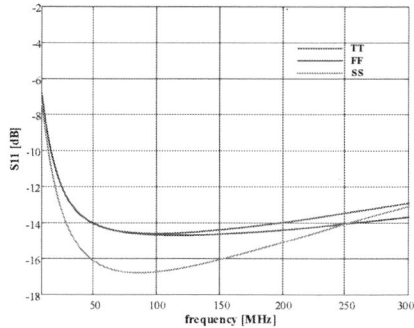

Fig. 12. Impedance matching (S11) performance of the LNA.

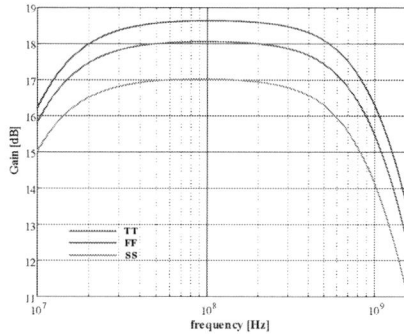

Fig. 13. Frequency response of the LNA.

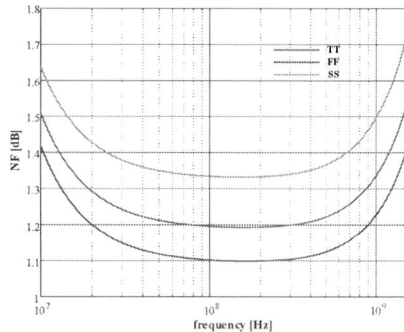

Fig. 14. NF of the LNA.

The CTDSM is realized in 55-nm CMOS. To reduce noise, the NRZ DAC is designed with a 1.8-V supply. The other blocks are operating with a 1.2-V supply. The sampling frequency of the modulator is 1.92 GHz (OSR=12). The bandwidth is 80 MHz and the full scale is 1.2Vpp. All results shown here are from pre-simulation with Spectre. Fig. 15 shows the output spectrum of the modulator with a −1.2 dBFS sinusoid signal at ~16 MHz input frequency. The SNDR and SFDR are 74.2 and 84.3 dB, respectively. Fig. 16 shows the output spectrum of the modulator with a −1.2 dBFS sinusoid signal at ~80 MHz input frequency. The SNDR and SFDR are 74 and 85 dB, respectively. Fig. 17 shows the SNDR versus the input amplitude. The proposed design obtains a DR of ~75 dB with 56.25-MHz input signal. The modulator consumes a total of 28.2 mW. The opamps consume 52% of the total power consumption, while the DAC and Quantizer consuming 16% and 32%, respectively.

The proposed modulator is compared with other CTDSM designs with similar BW and SNDR in Table I. The proposed modulator achieves a peak SNDR of 74.2 dB and a DR of 75 dB, resulting in a better Schreier FoM 168.7 dB (SNDR) or 169.5 dB (DR).

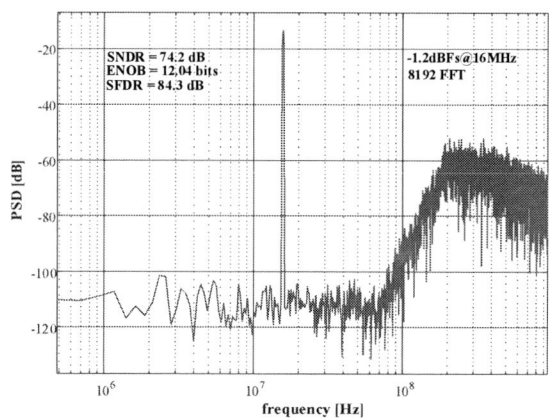

Fig. 15. FFT of modulator output with 16 MHz input tone.

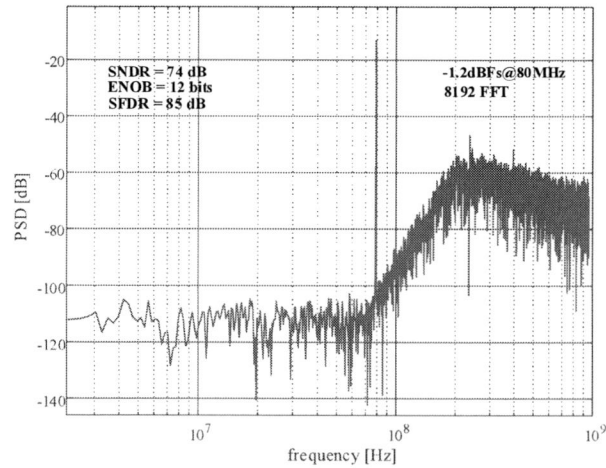

Fig. 16. FFT of modulator output with 80 MHz input tone.

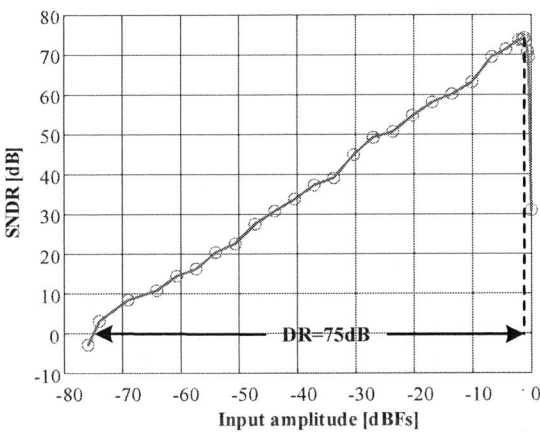

Fig. 17. SNDR versus input amplitude.

TABLE I.   PERFORMANCE SUMMARY AND COMPARISON

	[4]	[3]	[8]	This work
Process(nm)	20	40	40	55
BW(MHz)	80	75	75	80
Fs(Gs/s)	2.184	3.2	2.4	1.92
Power(mW)	23	22.8	27	28.2
SNDR(dB)	67.5	65.5	67.3	74.2
FoM[a](dB)	163	161	164.4	168.7

a. $FoM = SNDR + 10log_{10}(\frac{BW}{Power})$.

V.   CONCLUSION

This paper presents a low-noise amplifier and a 4th-order 80-MHz bandwidth continuous-time delta–sigma modulator in 55-nm CMOS for the analog front-end in high-frequency ultrasound imaging systems. The CTDSM achieves 74.2 dB SNDR, 75dB DR, 84.3 dB SFDR over an 80 MHz signal bandwidth consuming 28.2 mW from 1.2/1.8 V supplies, yielding a 168.7 dB Schreier FOM (SNDR-based). Meanwhile, the LNA provides a 12–24 dB gain across 30–80 MHz, with NF <1.25 dB.

REFERENCES

[1] X. Xu, H. Venkataraman, S. Oswal, E. Bartolome, and K. Vasanth, "Challenges and considerations of analog front-ends design for portable ultrasound systems." 2010 IEEE International Ultrasonics Symposium, San Diego, CA, USA, 2010, pp. 310-313.

[2] J. H. Sung, and J. S. Jeong, "Development of High-Frequency (>60 MHz) Intravascular Ultrasound (IVUS) Transducer by Using Asymmetric Electrodes for Improved Beam Profile," Sensors, 18, 2018.

[3] C. Briseno-Vidrios, A. Edward, A. Shafik, S. Palermo, and J. Silva-Martinez, "A 75-MHz Continuous-Time Sigma–Delta Modulator Employing a Broadband Low-Power Highly Efficient Common-Gate Summing Stage," IEEE Journal of Solid-State Circuits, vol. 52, no. 3, pp. 657-668. 2017.

[4] S. Ho, C. L. Lo, J. Ru, and J. Zhao, "A 23 mW, 73 dB Dynamic Range, 80 MHz BW Continuous-Time Delta-Sigma Modulator in 20 nm CMOS," IEEE Journal of Solid-State Circuits, vol. 50, no. 4, pp. 908-919. 2015.

[5] W. Wang, C. H. Chan, Y. Zhu, and R. P. Martins, "A 100-MHz BW 72.6-dB-SNDR CT ΔΣ Modulator Utilizing Preliminary Sampling and Quantization," IEEE Journal of Solid-State Circuits, vol. 55, no. 6, pp. 1588-1598. 2020.

[6] Y. Zhang, C. H. Chen, T. He, and G. C. Temes, "A Continuous-Time Delta-Sigma Modulator for Biomedical Ultrasound Beamformer Using Digital ELD Compensation and FIR Feedback," IEEE Transactions on Circuits and Systems I: Regular Papers, vol. 62, no. 7, pp. 1689-1698. 2015.

[7] M. C. Chen, A. P. Perez, S. R. Kothapalli, P. Cathelin, A. Cathelin, S. S. Gambhir, and B. Murmann, "A Pixel Pitch-Matched Ultrasound Receiver for 3-D Photoacoustic Imaging With Integrated Delta-Sigma Beamformer in 28-nm UTBB FD-SOI," IEEE Journal of Solid-State Circuits, vol. 52, no. 11, pp. 2843-2856. 2017.

[8] Y. Guo, J. Jin, X. Liu, and J. Zhou, "An Inverter-Based Continuous Time Sigma Delta ADC With Latency-Free DAC Calibration," IEEE Transactions on Circuits and Systems I: Regular Papers, vol. 67, no. 11, pp. 3630-3642. 2020.

[9] C. Ding, Y. Manoli, and M. Keller, "Approaches to mitigating the impact of DAC mismatch on the performance of continuous-time delta-sigma modulators." pp. 329-332.

[10] J. Lee, K. R. Lee, B. E. Eovino, J. H. Park, L. Y. Liang, L. Lin, H. J. Yoo, and J. Yoo, "A 36-Channel Auto-Calibrated Front-End ASIC for a pMUT-Based Miniaturized 3-D Ultrasound System," IEEE Journal of Solid-State Circuits, vol. 56, no. 6, pp. 1910-1923. 2021.

[11] F. Bruccoleri, E. A. M. Klumperink, and B. Nauta, "Wide-band CMOS low-noise amplifier exploiting thermal noise canceling," IEEE Journal of Solid-State Circuits, vol. 39, no. 2, pp. 275-282. 2004.

2025 The 10th International Conference on Integrated Circuits and Microsystems

A Neuron-and-Synapse Unit Circuit with Information Propagation Function for Spiking Neural Networks

Yide Zhang, Zixuan Ling, Zicheng Yin, Xu Liu*
School of Information Science and Technology, BJUT, Beijing, China
liuxu16@bjut.edu.cn

Abstract—This paper presents a CMOS Neuron-and-Synapse Unit (NSU) Circuit Design for Spiking Neural Network. Compared to the conventional NSU for SNN, this circuit have wide frequency response, which can be more adaptive to dynamic environments. The proposed NSU includes an LIF neuron circuit , an STDP synapse circuit for biologically inspired learning. For the STDP scheme designed in this work, the adaptive time window can be adjusted based on input spike frequency. Moreover, a multi-neuron synapse propagation model is designed in 180-nm CMOS process, and simulated in Cadence. The Simulation results of a two-layer six-neuron network validate the function of the NSU, with information propagation and learning capabilities. This demonstrates its potential for efficient signal processing for SNNs.

Keywords—spiking neural networks, CMOS, LIF neuron, STDP, synapse

I. INTRODUCTION

Spiking neural networks (SNNs), which mimic the brain's spike-based information processing, offer asynchronous, sparse, and event-driven computation [1-2]. This makes them energy-efficient and good at handling temporal information, and thus a strong candidate for neuromorphic computing. In SNNs, leaky integrate-and-fire (LIF) neuron circuits are widely used in hardware implementations for their simplicity and ability to approximate biological neuron dynamics [3]. Synapse circuits[4], as connections between neuron circuits, need to handle weighted spike transmission and online learning. Spike-timing-dependent plasticity (STDP), a biologically plausible mechanism, is key for adaptive learning in SNNs [5].

However, current hardware implementations face three major challenges. First, most STDP mechanisms [6] use a fixed time window and cannot respond to input frequency changes, which is a problem for processing wide-spectrum signals. Second, even though some studies offer time window adjustment interfaces, they often rely on manual tuning and lack adaptability [7-8]. Third, the current point-to-point connection models with CMOS neuron-and-synapse circuit limit network scalability and parallel computing capabilities.

To address these problems, this paper proposes a novel CMOS neuron-and-synapse unit circuit for SNNs with four key features. First, the designed LIF neuron circuit can support controllable refractory periods and single-spike triggering for accurate neuron dynamic activity approximation. Second, the designed STDP synapse circuit can adjust weights based on pre-and-post-spike timing differences for biologically inspired learning. Third, an adaptive time window module which

Support by National Key Research and Development Program of China Grant No. 2024YFF1206504.

automatically adjusts the STDP time window based on input spike frequency is proposed to improve adaptivity in processing wide-frequency signals. Fourth, a multi-neuron and multi-synapse propagation model based on NSU is designed to enable scalable synapse networks and high parallelism through event-driven scheduling.

To validate the NSU circuit, a two-layer six-neuron network was built using 180-nm CMOS process technology and simulated in Cadence software. The simulation results confirm the function of information propagation and learning abilities in SNN. The rest part of this paper is organized as follows. Section II shows the detail design of the proposed system and circuits. Section III presents simulation results and network behavior analysis. Section IV summarizes this work.

II. IMPLEMENTATION OF THE CMOS NSU

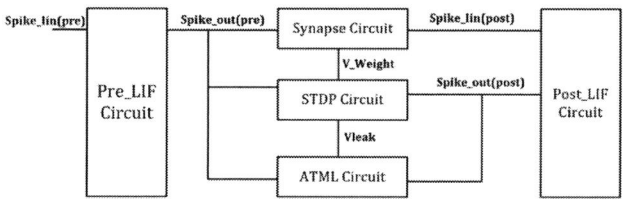

Fig. 1. System framework diagram based on Neuron-and-Synapse Unit

As shown in Fig. 1, the neuron-and-synapse unit (NSU) circuit consists of a pre-neuron LIF circuit (Pre_LIF Circuit), a post-neuron LIF circuit (Post_LIF Circuit), a synapse circuit , an STDP circuit, and an adaptive time window Length (ATML) circuit. The Pre_LIF Circuit receives the input spike signal *Spike_in(pre)* and generates the output spike *Spike_out(pre)* through integration. The synapse circuit receives this output spike and modulates the signal transmission intensity using *V_weight*. The STDP Circuit adjusts the synaptic weight based on the spike timing of the presynaptic and postsynaptic signals. The ATML Circuit receives *Spike_out(pre)* and *Spike_out(post)*, and alters the magnitude of *Vleak*, thereby changing the time window length of the STDP. The Post_LIF circuit receives signals transmitted from the synapse and emulates the process of neuronal signal integration and spike generation once again.

A. LIF Circuit Design

Fig. 2 depicts the designed CMOS LIF neuron circuit. The *Cmem* capacitor integrates the *Iin* current from the synapse circuit, accumulating charge. *Vmem_int* connects to *Vmem* via

M3. When *Vmem_int* is below *Vth*, M3 is on and *Vmem_int* follows *Vmem*. If *Vmem_int* exceeds *Vth*, the comparator and *Spike_out* goes high, emitting a spike. Then M3 turns off, isolating *Vmem* and *Vmem_int* to prevent further input current influence. Subsequently, the delayed *Vrfp* signal turns on MN4 and MN3 to reset *Vmem* and *Vmem_int* to their initial state, inducing a refractory period.

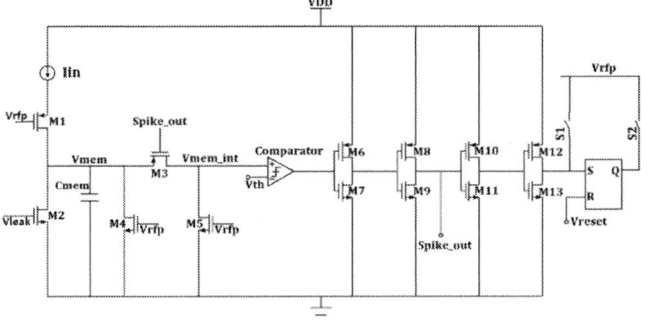

Fig. 2. Schematic of the LIF circuit

For *Vrfp* generation, the circuit has two modes.

Natural triggering mode (S1 on, S2 off): Here, *Vrfp* is a brief single pulse for rapid integration node reset, enabling continuous spiking under persistent input, ideal for typical SNN.

Controllable refractory period triggering mode (S1 off, S2 on): The delayed Spike_out connects to the RS flip-flop (*S* terminal), keeping *Vrfp* high to halt the circuit after a single spike. An external *Vreset* pulse at the R terminal resets the flip-flop, pulling *Vrfp* low and reactivating the circuit. This mode is perfect for TTFS encoding [9], allowing precise refractory period management via cycle control (*Vreset*).

Additionally, transistor *M2*, controlled by *Vleak*, forms the leakage path to adjust the membrane potential leakage rate, mimicking the "forgetting" behavior of neurons. When switch *S1* is on, the circuit works in natural- triggering mode.-When switch *S2* is on and the circuit receives an input *Vreset* pulse signal, it operates in controllable refractory period mode, giving the neuron a 1-ms refractory period. When switch *S2* is on and *Vreset* is held low, the circuit runs in single pulse triggering mode, enabling TTFS neuron firing[9].

B. Synapse Circuit Design

Fig. 3 presents the designed synapse circuit. It takes the weight voltage *Vw*, computes its difference with the reference voltage *Vref*, and outputs a positive or negative synaptic current pulse proportional to this difference for the postsynaptic neuron. The circuit comprises an absolute difference calculator, an ADC, and a DAC.

Using two subtractors and an adder, it calculates |*Vw-Vref*|. Each subtractor uses an inverting configuration with op-amps and resistors. Subtractor *1* computes *Vw-Vref*, and Subtractor *2* computes *Vref- Vw*. Their outputs are added to yield |*Vw-Vref*|, which becomes the input voltage *Vin* for the next module.

A 6-bit Flash ADC, and a 6-bit current-mode DAC with thermometer coder is used to generate current to Neurons. The Flash ADC was selected for its parallel architecture, enabling analog-to-digital conversion within a single clock cycle—a vital feature for processing fast-changing pulses in spiking neural networks. Despite its power consumption being centered on the comparator and encoder, the Flash ADC operates efficiently. This is because synaptic circuits in spiking neural networks only activate during pulse events, allowing the Flash ADC to stay in low-power or sleep mode between pulses, thus optimizing energy use.

If *Vw > Vref*, the DAC to send an excitatory current pulse. If *Vw < Vref*, an inhibitory current pulse will be generated.

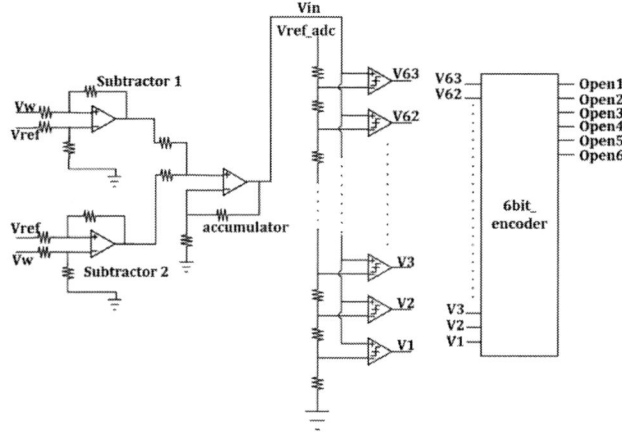

Fig. 3. Schematic of the Synapse circuit.

Based on the supply voltage, the weight reference voltage *Vref* was set to 1.5 V, *Vref_adc* to 1.0 V, and the DAC reference current *Iref* to 1.0 μA.

C. STDP Circuit Design

Fig. 4 presents the designed synapse modulation circuit based on STDP. The mathematical expression for this mechanism is as follows:

$$\Delta w_j = \begin{cases} A^{+} \cdot \exp\left(\dfrac{t_{\text{pre}} - t_{\text{post}}}{\tau}\right) \text{if } t_{\text{post}} \leq t_{\text{pre}} \\ A^{-} \cdot \exp\left(-\dfrac{t_{\text{pre}} - t_{\text{post}}}{\tau}\right) \text{if } t_{\text{post}} > t_{\text{pre}} \end{cases} \quad (1)$$

The circuit works under the following two cases.

Pre-before-Post (LTP): When a presynaptic pulse arrives, M3 conducts, rapidly discharging *Cp* to a low voltage. After the pulse, M3 turns off. M1, biased by *Vleak*, slowly charges *Cp*. On arrival of the postsynaptic pulse, the M7-M8 transmission gate activates, sampling Cp's voltage and applying it to gate of M10. M10 generates a drain current proportional to its gate voltage. This current is copied and amplified by current mirrors. *~post* activates M18, but the right circuit is inactive, so M19 and M16 are off. All current charges *Cw*, raising *Vw* and causing LTP.

Post-before-Pre (LTD): If the postsynaptic pulse arrives first, the right half activates similarly. When the presynaptic pulse later arrives, M18 and M15 on the left are still off. This causes *Cw* to discharge through M19 and M16, lowering *Vw* and inducing LTD. Transistor M11 gate voltage controls the discharge current. Higher gate voltage increases discharge current and accelerates *Vw* reduction. Weight change depends on the time difference between the spikes, enabling STDP.

To enhance the circuit flexibility, adjustable time window is designed. The charging rate of *Cp* (and the symmetrical *Cn*) is controlled by *Vleak*. Adjusting *Vleak* can determine STDP time window (maximum time between spikes for weight change).

This STDP circuit offers precise control over synaptic weight changes through symmetrical coupling. It balances compact hardware design with biological plausibility and is ideal for SNN systems in temporal learning tasks.

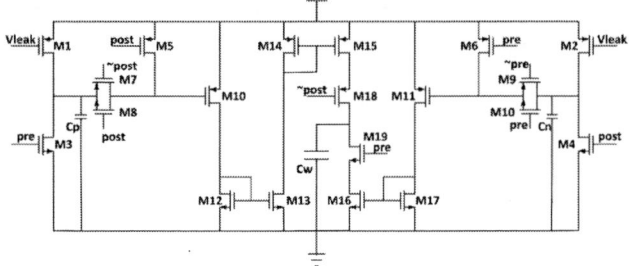

Fig. 4. Schematic of the STDP circuit

As shown in Fig. 5, the synapse weight change *ΔVw* varies with the pre and post synaptic signal time difference *tpre - tpost*. In the simulation, *Vleak* was set to 2.75 V, giving the STDP circuit a time window of about 10 ms. The horizontal axis represents *tpre - tpost*, and the vertical axis shows *ΔVw*. When *tpre - tpost* is positive, the synapse weight rises, with *ΔVw* peaking at 1 V and then decays as *tpre - tpost* increases. When *tpre - tpost* is negative, the weight declines.

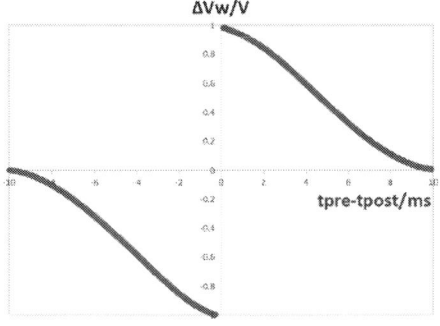

Fig. 5. Simulated waveforms of inhibitory stimulation of synaptic circuits

D. ATML Circuit Design for STDP

Fig. 6 presents the ATWL circuit for STDP. This circuit dynamically adjusts the output voltage based on the input pulse time difference. The output voltage serves as the *Vleak* signal for the STDP circuit, where it controls the charging rate and regulates synaptic weight voltage changes.

When the input pulse time difference is large (e.g., 200 μs to 5 ms), a fixed *Vleak* cannot accommodate the full range. For example, a *Vleak* suitable for small time differences (like 5 ms) cannot cover larger differences (like 100 ms), leading to incorrect weight changes in the STDP circuit. To address this, the designed circuit adaptively adjusts *Vleak* according to the pulse time difference, ensuring effective STDP learning across different time scales. As shown in Fig. 6, the circuit has two

parts to output *Vleak*. The upper part handles *Vleak* for 200 μs to 5 ms, and the lower part for 5 ms to 100 ms. Two comparators control the outputs. Initially, the capacitor *Crc* voltage is 0, and gate voltage of M3 is high. When the pre-pulse arrives, M5 gate voltage is pulled low, charging the capacitor. The voltage is then processed through a PMOS source follower and inverters to flatten the voltage curve and reduce the slope. This processed voltage is used to calculate *Vleak*. The system then waits for the post-pulse. Upon arrival, the transmission gate (M6 and M7) activates, transferring *Vin* to the capacitor and outputting to the inverter. The lower part operates similarly but without the source follower, as it does not require a slow *Vleak* rate.

Transistors M13, M14, M19, and M20 are discharge paths for capacitors *C1* and *C2*, controlled by *Vctrl*. When pulse time differences change by an order of magnitude, *Vctrl* goes high, activating these transistors to discharge the capacitors and reset the system. The system then detects the new time difference and updates *Vleak* accordingly. If the time window circuit should be independent of Pre and Post signals, a similar RS refractory period circuit from the LIF circuit can be used.

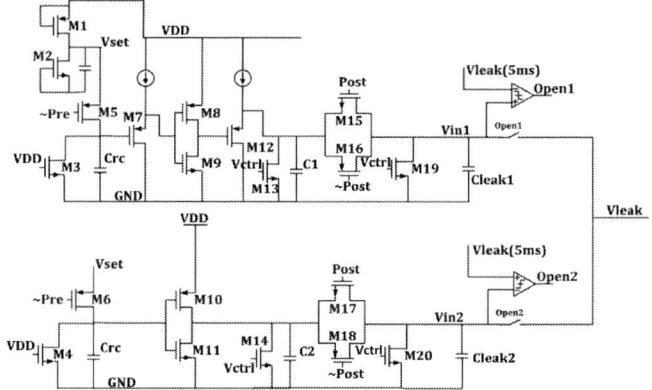

Fig. 6. Schematic of the ATML circuit

Our STDP circuit allows for an adjustable time window, adaptable for the pre-and-postsynaptic spike interval of 200 μs, 5 ms, and 100 ms. To achieve continuous Vleak output across multiple orders of magnitude in time intervals, we designed above dual-branch ATWL circuit. One branch handles Vleak for *tpre - tpost* of 200 μs to 5 ms, and the other for *tpre - tpost* of 5 ms to 100 ms.

To obtain the real *Vleak* output response under different time differences, we triggered the pre-pulse at t = 0 and the post-pulse at t = 200 μs, 5 ms, and 100 ms, respectively, and carried out the simulation analysis. As shown in Fig.7, when *tpre-tpost* is set to 200 μs, 5 ms, and 100 ms respectively, the *Vleak* curve rises to 2.14 V, 2.75 V, and 2.91 V. respectively.

The simulation results align closely with the theoretical values, confirming the circuit precise response to synaptic time differences across a wide time range. This also validate the feasibility and effectiveness of the adjustable time window feature.

979-8-3315-8850-2/25 $31.00 © 2025 IEEE

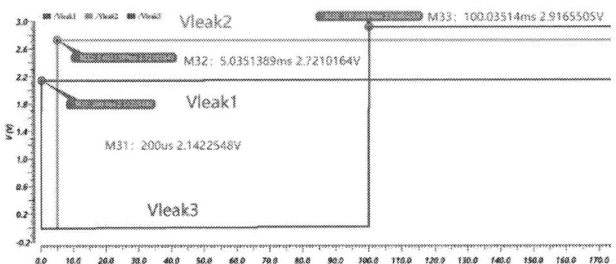

Fig. 7. Waveforms of *Vleak* v.s. *tpre – tpost* (200 μ s, 5 ms, and 100 ms)

III. SIMULATION RESULTS AND NETWORK BEHAVIOR ANALYSIS

The entire circuit was designed in 180-nm CMOS process technology with a 3.3-V supply and simulated in Cadence.

A. Simulation of Multi-neuron Synaptic Network with Information Propagation

Fig. 8 presents a two-layer SNN based on a model with multiple LIF neurons and synapses. The network consists of two layers, each with three neurons. The first layer has LIF1, LIF2, and LIF3, while the second layer includes 2LIF1, 2LIF2, and 2LIF3. Synaptic connections with STDP are added between neurons to initialize connection weights like *Vw_12* and *Vw_13* in the first layer, and similar connections in the second layer.

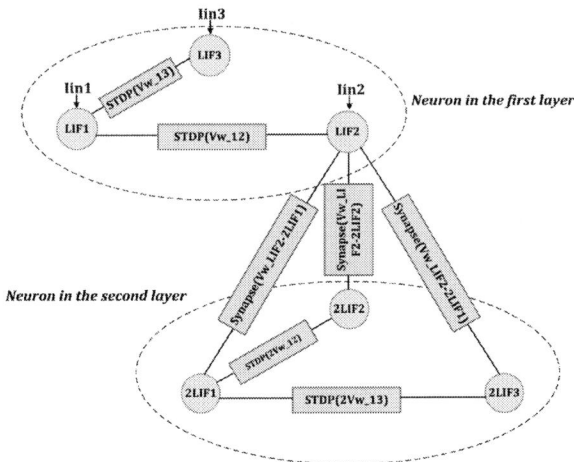

Fig. 8. Neural synaptic network at initialization condition.

For the network initialization, neurons in the first layer receive varying current pulse inputs (*Iin1*, *Iin2*, *Iin3*). STDP connections between LIF1 and LIF2/LIF3 initialize weights *Vw_12* and *Vw_13*. The second layer has analogous STDP connections. Additionally, synapses connect the first and second layers, with weights such as *Vw_LIF2-2LIF1*, establishing the network initial configuration for subsequent spiking and learning.

Fig. 9 shows the neural synaptic network with learning capability. Only LIF1 receives external input *Iin1*. Synaptic connections between LIF1 and other neurons in the first layer adjust weights like *Vw_23*. The second layer has a similar setup. Information is propagated and transmitted between layers via synapses like *Vw_LIF2-2LIF1*. These connections facilitate spike transmission and weight modification.

Network initialization involves setting initial weights and connections. During propagation verification, spikes from the first layer travel to the second layer via synapses. STDP adjusts target connection weights within each layer, mimicking learning and adaptation. This model is useful for assessing the network's propagation characteristics and learning ability in processing time-series data or emulating biological neural behavior.

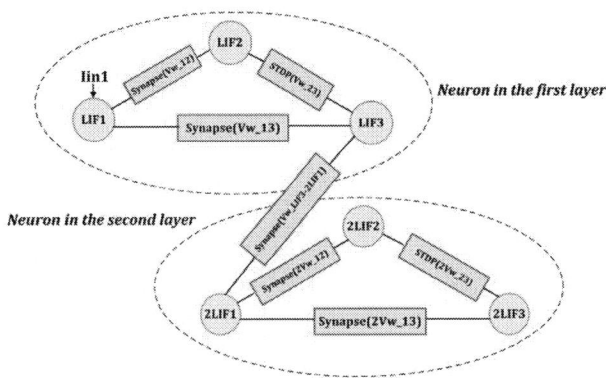

Fig. 9. Neural synaptic network with learning capability (weight updated).

As shown in Fig. 10, the simulation waveform depicts the network initialization process. In the first layer, LIF1 spikes first, followed by LIF2, and then LIF3. Initially, LIF1, LIF3, and LIF2 maintain a single-spike mode. After the first-layer weight initialization, LIF2 switches to natural firing to drive the downstream network initialization. Based on the spiking order, *Vw_12* is initialized to 2.45 V, and *Vw_13* to 1.85V.

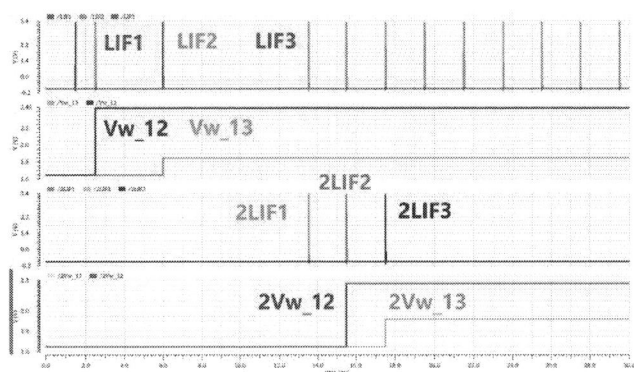

Fig. 10. Simulation waveforms of the network initialization process

The second-layer neurons are activated by first-layer LIF2. Given the weights *Vw_LIF2-2LIF1*, *Vw_LIF2-2LIF2*, and *Vw_LIF2-2LIF3*, 2LIF1 fires first, then 2LIF2, and finally 2LIF3. The weights *2Vw_12* and *2Vw_13* are initialized to 2.4 V and 1.95V, respectively.

Fig. 11 shows the first-layer neural propagation. LIF1 is continuously stimulated by external input, while LIF2 and LIF3 are sequentially activated via synaptic connections with LIF1, adjusting the target weight *Vw_23* to 2.25 V based on the spiking order.

979-8-3315-8850-2/25 $31.00 © 2025 IEEE

Fig. 12 shows the result for second-layer neural propagation. 2LIF1 is continuously activated by output pulses of LIF2, while 2LIF2 and 2LIF3 are sequentially activated through synaptic connections with LIF1, adjusting the target weight *2Vw_23* to 2.32 V based on the spiking order.

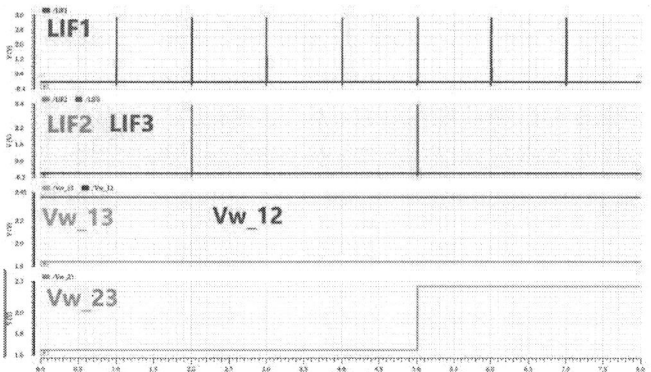

Fig. 11.   Simulation waveforms of the first-layer neural propagation.

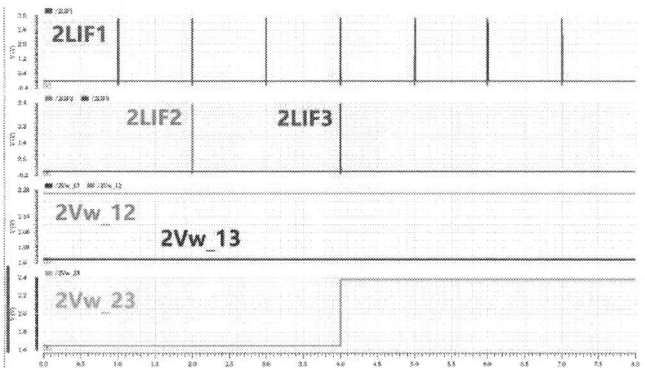

Fig. 12.   Simulation waveforms of the second-layer neural propagation.

These waveforms demonstrate that the designed CMOS synapse, STDP module, and neuron-based SNN can effectively initialize network parameters and dynamically change weights for information propagation and learning. They show strong propagation and learning abilities for time-series data processing and biological neural network behavior simulation. Moreover, when combined with the proposed ATWL, this SNN architecture can meet requirements of most intelligent application with wide spiking time interval.

IV. CONCLUSION

In this paper, we proposed an CMOS NSU unit circuit for SNN. The LIF neuron circuit and STDP synapse circuit enables weights adjusting based on pre and post spike timing differences. The adaptive time window module improves the adaptivity for wide frequency signals by automatically adjusting the STDP time window according to input spike frequency. The multi

neuron synapse propagation structure and highly parallel synapse networks through event driven scheduling is built. Simulation results of a two layer six neuron network confirmed the effectiveness of the proposed architecture in terms of information propagation and learning abilities. This work provides a promising direction for the development of adaptive neuromorphic computing systems in CMOS based hardware.

TABLE I: COMPARATION WITH OTHER WORKS

Reference	[4]	[7]	[8]	[10]	[11]	This work
Tech.(nm)	65	55	28	180	350	180
Supply voltage	1V	1 V	0.2~1V	0.3V	3.3V	3.3V
Refractory Period (TTFS)	No (No)	No (No)	Yes (No)	Yes (No)	Yes (No)	Yes (Yes)
ATWL	No	No	No	No	No	Yes (2us~100ms)

REFERENCES

[1] Wolfgang Maass,Networks of spiking neurons: The third generation of neural network models,Neural Networks,Volume 10, Issue 9,1997, Pages 1659-1671.

[2] Dayan P, Abbott L F, Theoretical neuroscience: computational and mathematical modeling ofneural systems [M]. MIT press, 2005.

[3] Koch, C., & Schutter, E. D. (1999). Biophysics of computation: Information processing in single neurons.Nature,398(6729), 678-678.

[4] M .S. Asghar, S. Arslan and H. Kim, "Current Multiplier Based Synapse and Neuron Circuits for Compact SNN Chip," 2021 IEEE International Symposium on Circuits and Systems (ISCAS), 2021, pp. 1-4.

[5] Song S., Miller K., et al., Competitive Hebbian learning through spike-timing-dependent synaptic plasticity. Nat Neurosci 3, 919–926 (2000).

[6] G. M. Tovar, E. S. Fukuda, T. Asai, T. Hirose and Y. Amemiya, "Neuromorphic CMOS Circuits implementing a Novel Neural Segmentation Model based on Symmetric STDP Learning," 2007 International Joint Conference on Neural Networks, , 2007, pp. 897-901

[7] Z .Yang, Z. Han, Y. Huang and T. T. Ye, "55nm CMOS Analog Circuit Implementation of LIF and STDP Functions for Low-Power SNNs," 2021 IEEE/ACM International Symposium on Low Power Electronics and Design (ISLPED), Boston, MA, USA, 2021, pp. 1-6.

[8] B .Joo, J. -W. Han and B. -S. Kong, "Energy- and Area-Efficient CMOS Synapse and Neuron for Spiking Neural Networks With STDP Learning," in IEEE Transactions on Circuits and Systems I: Regular Papers, vol. 69, no. 9, pp. 3632-3642, Sept. 2022.

[9] B. Rueckauer and S. -C. Liu, "Conversion of analog to spiking neural networks using sparse temporal coding," 2018 IEEE International Symposium on Circuits and Systems (ISCAS), 2018, pp. 1-5

[10] M. Akbari, S. M. Hussein, T. -I. Chou and K. -T. Tang, "A 0.3-V Conductance-Based Silicon Neuron in 0.18 μm CMOS Process," in IEEE Transactions on Circuits and Systems II: Express Briefs, vol. 68, no. 10, pp. 3209-3213, Oct. 2021

[11] G. Indiveri, E. Chicca and R. Douglas, "A VLSI array of low-power spiking neurons and bistable synapses with spike-timing dependent plasticity," in IEEE Trans. on Neural Networks, vol. 17, no. 1, pp. 211-221, Jan. 2006

CMOS PWM Controller Design for GaN-Based 48V-1V DC-DC Buck Converter

Guanyu Wu
Department of Electrical and Electronic Engineering
School of Advanced Technology
Xi'an Jiaotong-Liverpool University
Suzhou, China

Miao Cui*
Department of Electrical and Electronic Engineering
School of Advanced Technology
Xi'an Jiaotong-Liverpool Universtiy
*Miao.Cui02@xjtlu.edu.cn

Abstract—Gallium Nitride (GaN) devices have seen increasing adoption in applications such as space systems and data center power management, owing to their superior efficiency, high-frequency capabilities, and excellent thermal stability. The direct 48V-to-1V GaN converters are emerging as promising alternatives to conventional two-stage voltage regulation. This direct approach offers significant advantages in offering high power density and enhancing transient response. However, implementing direct 48V-to-1V conversion necessitates an extremely low duty cycle (approximately 2%) which poses challenges for control precision and power stage design. This work proposes a CMOS-based PWM generator capable of achieving a wide duty cycle range (1% – 99%) at 10MHz. The circuit successfully drives the EPC2065 GaN device at 1MHz, enabling stable 48V-to-1V power conversion and offering a compact, high-performance solution for next-generation power management systems.

Keywords—CMOS PWM, GaN DC-DC Buck Converter, Duty Cycle

I. INTRODUCTION

In recent years, with the rapid advancement of cloud computing and artificial intelligence technologies, data centre infrastructure has significantly expanded in both scale and complexity. This development presents critical challenges in power efficiency, driving the demand for next-generation power conversion technologies with high performance. In current processor power systems, voltage conversion typically proceeds from 480 V AC to 400 V DC, then to 48 V DC, followed by intermediate steps (e.g., 24 V, 12 V) before reaching ~1 V to supply modern processors [1]. As chip architectures evolve, requiring ever-lower operating voltages, the efficiency and scalability of traditional power delivery approaches have become inadequate. Compared to the conventional two-stage approach with an intermediate 12 V rail, the single-stage conversion architecture simplifies the system by reducing component count and minimizing power delivery delays. This advantage has prompted extensive academic and industrial efforts to explore direct 48 V-to-1 V conversion techniques [2]–[4].

Gallium nitride (GaN) technology has demonstrated significant potential in high-frequency power systems [5]. GaN devices possess advantageous material properties such as wide bandgap (3.4 eV), high breakdown field (~3.5 MV/cm), and high electron drift velocity (~2.6×10^7 cm/s) [6], making them suitable for high-power-density, high-efficiency converter applications. In DC step-down converters, pulse width modulation (PWM) control technology is widely used due to its advantages such as high efficiency and the ability to precisely control the output voltage by adjusting the duty cycle. In the pulse width modulation generator, compared to inferior performance of P-type GaN transistor, CMOS technology offers lower static power consumption, mature integration process, and low manufacturing costs. These characteristics make CMOS particularly well-suited for the integration of complex digital logic with analog modules. Therefore, implementing PWM circuits using CMOS not only reduces power consumption but also enhances system integration and reliability.

CMOS-based PWM generators have been reported at sub-megahertz operation. Representative products such as TDK-Lambda's i7A series DC-DC converters employ fixed-frequency PWM at 600 kHz [7]. With the growing adoption of GaN power devices, however, switching frequencies can reach into the multi-megahertz domain. This frequency escalation imposes new demands on the performance and linearity of PWM signal generators, particularly in maintaining precise control over narrow pulse widths. Achieving direct 48 V-to-1 V conversion requires a PWM generator capable of maintaining an exceptionally wide duty-cycle range, down to approximately 2 %. Conventional CMOS implementations exhibit a limited duty-cycle range and reduced output swing above 1-10MHz, hindering their effectiveness in driving GaN stages. Currently, peak-valley current mode (PVCM) buck converters can achieve a 3%-95% duty cycle range at 2.2 MHz [8], while they are capable of direct conversion from 34 V to 1 V. The MPQ2908A buck converter developed by Monolithic Power Systems supports a maximum duty cycle of up to 99.5% in PWM operation. However, its operating frequency is below 1 MHz [9], making it difficult to maintain a wide duty cycle range under high-frequency conditions.

At high frequencies, an important factor affecting the duty cycle range of the PWM generator is the comparison speed of the comparator, which is affected by propagation delays during output switching. The comparator's finite response time may cause timing errors when the reference voltage crosses the ramp

---

This work was supported by XJTLU Research Development Fund (RDF-21-02-031, PGRS2206039)

waveform [10]. These errors distort narrow pulses at low duty cycles and cause premature switching at high duty cycles. Therefore, insufficient comparator speed will limit the PWM duty cycle range, preventing it from meeting the duty cycle requirement of 48V to 1V (approximately 2%).

In this work, we propose a high-frequency PWM generator, which utilizes a two-stage push-pull comparator with an integrated clamping circuit. The PWM circuit aims to extend the achievable duty cycle range by improving comparator response speed and minimizing propagation delays. The proposed design operates at 10 MHz and supports a wide duty cycle range from 1% to 99%. Utilizing this architecture, we successfully drive a GaN-based buck converter at 1 MHz, enabling direct 48 V-to-1 V conversion and demonstrating its effectiveness for high-efficiency, high step-down power applications.

II. CIRCUIT DESIGN

The overall architecture of the half-bridge buck converter employing a GaN-based power stage is illustrated in Fig. 1. It is mainly composed of a sawtooth generator, a comparator, a bootstrap driver and a half-bridge power stage. In this configuration, the PWM control circuit and bootstrap driver are fully implemented using a 250 nm CMOS process, while the primary power transistors utilizes commercial GaN devices EPC2065.

A. Sawtooth Generator

In Fig. 2, the sawtooth wave generator integrates a CMOS hysteresis comparator with a capacitor-based charge–discharge circuit to produce periodic waveforms. The comparator employs an inverting input structure to enhance noise immunity, using a differential pair of PMOS (M1, M2) and NMOS (M3–M4) transistors, with M5 biased to operate as a constant current source. Positive feedback via resistors Rf and R1 introduces hysteresis, defining distinct upper (VTH) and lower (VTL) threshold voltages. These thresholds control the capacitor's charging and discharging phases, forming the sawtooth waveform. The threshold values are tunable through resistor selection, enabling precise control over waveform frequency and stability under noise.

B. Comparator

The proposed CMOS two-stage push-pull comparator is designed to meet the stringent speed and precision demands of PWM system. As shown in Fig. 3, the circuit architecture comprises three key functional blocks: a differential input stage, a clamping protection circuit, and a common-source amplification output stage with active current mirror loads.

The differential input stage is constructed using PMOS transistors M1 and M2, which receive Vin1 and Vin2 as input signals. A critical enhancement in the comparator design is the introduction of a clamping mechanism, which serves to improve dynamic performance by limiting voltage excursions at key internal nodes. When the input voltage Vin1 exceeds Vin2, the transconductance of transistor M1 in the differential pair increases, while that of M2 decreases. Consequently, the

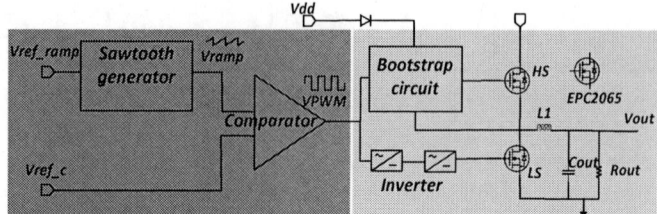

Fig. 1. Half-bridge buck circuit using PWM controlled GaN devices.

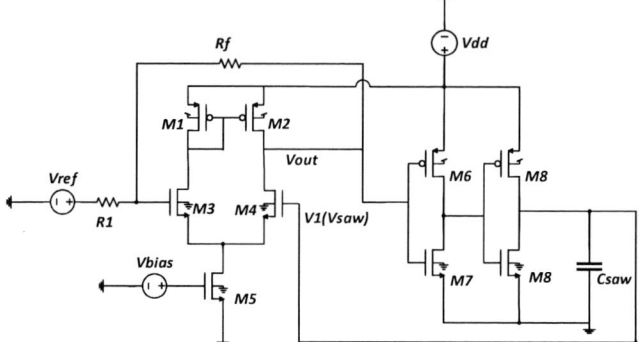

Fig.2. Sawtooth generator circuit.

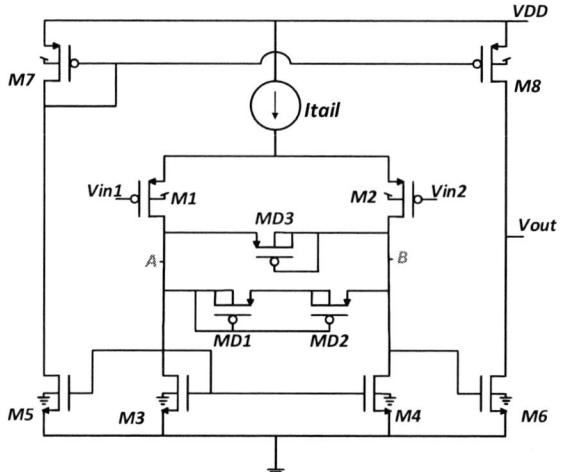

Fig. 3. Two-stage push-pull comparator with clamping circuits.

majority of the tail current is steered through M1, causing a voltage drop at node A. Due to the diminished current through M2, node B experiences a voltage rise. When the voltage difference between node B and node A surpasses the cumulative built-in potential of the parasitic diodes embedded within transistors MD1 and MD2, these diodes become forward-biased, conducting current and thereby clamping the node B voltage to a predetermined level. Conversely, when Vin1 is lower than Vin2, the parasitic diode associated with MD3 becomes conductive, clamping node B to an alternative fixed voltage. This clamping strategy effectively limits the voltage range of node B. Thus, it prevents excessive voltage swings at internal nodes when the input signal reverses, which would otherwise prolong the time required for the comparator output to stabilized at the correct level and delay the response of the next comparison. The proposed comparator with clamping circuits

979-8-3315-8850-2/25 $31.00 © 2025 IEEE

reduces the recovery time and significantly improving the overall speed of the comparator.

The amplified differential voltage is further processed by the common-source output stage, which consists of NMOS transistors M5 and M6. Their gates are connected to nodes A and B, respectively. PMOS transistors M7 and M8 serve as active loads, forming a current mirror to boost output gain. This stage reinforces signal levels while maintaining high-speed operation, ensuring that the comparator delivers fast and accurate output transitions suitable for high-frequency PWM generation.

C. Bootstrap Structure and Half-bridge Buck Converter Circuit

The bootstrap circuit in Fig. 4, comprising a diode and a capacitor (CBST), establishes a floating supply referenced to the switching node (SW). During low-side conduction, SW is grounded, enabling CBST to be charged from VDD. Upon activation of the high-side transistor, SW rises toward Vin, elevating the gate voltage to ensure reliable switching. Careful CBST design balances charge capacity and RC delay, critically impacting converter efficiency and stability. The buck converter employs a GaN EPC2065 high-side switch for efficient 48 V-to-1 V step-down conversion. Key components include GaN switches, an inductor (L), output capacitor (Cout), and load resistance (Rout). The inductor regulates energy transfer and current ripple, while Cout stabilizes voltage and filters noise, with Resistance influencing ripple performance. Coordinated design of these elements is essential for optimal high-frequency operation and stability.

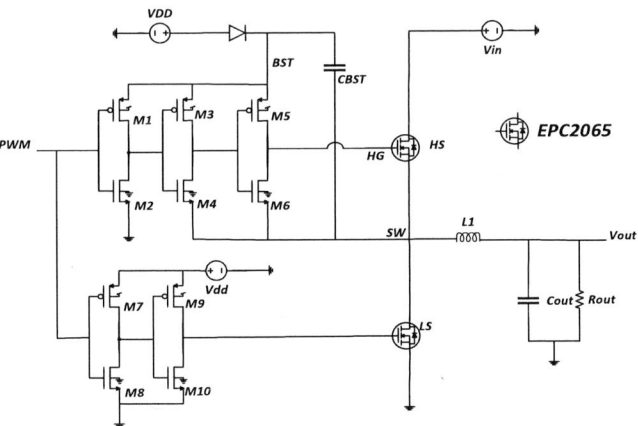

Fig. 4. The half-bridge GaN buck converter circuit with bootstrap driver.

III. SIMULATION RESULTS

The simulation results show that the system operates at a frequency of 1MHz. The hysteresis width of the comparator is directly influenced by the ratio R1/Rf, as it determines the upper and lower switching thresholds and thus defines the overall hysteresis range. The hysteresis width is calculated using equations below.

$$V_{TL} = \frac{(R_f V_{ref} + R_1 V_{OL})}{(R_1 + R_f)} \tag{1}$$

$$V_{TH} = \frac{(R_f V_{ref} + R_1 V_{OH})}{(R_1 + R_f)} \tag{2}$$

$$\Delta V = V_{TH} - V_{TL} = \frac{R_1}{R_1 + R_f}(V_{OH} - V_{OL}) \tag{3}$$

A larger hysteresis width enhances noise immunity and improves system stability, while a smaller width increases sensitivity, which is advantageous in high precision applications. Additionally, the R1/Rf ratio affects the amplitude of the generated sawtooth waveform. A larger amplitude yields better linearity and higher modulation accuracy, whereas a smaller amplitude facilitates a wider duty cycle range. Fig.5 shows the sawtooth waves under three different resistances ratio.

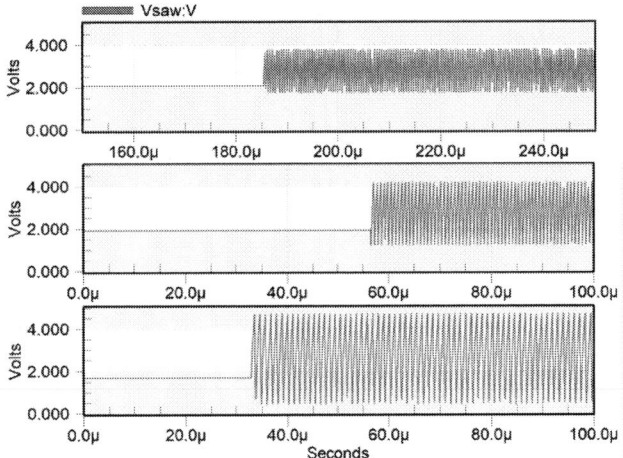

Fig. 5. Sawtooth waves of different amplitudes. (Up) R1/Rf=100 kΩ/100 kΩ, (Middle) R1/Rf=100 kΩ/50 kΩ, (Bottom) R1/Rf=100 kΩ/10 kΩ.

The clamping circuit of the comparator can reduce the recovery time for output level conversion by effectively limiting the voltage swing at point B. This increases the comparator speed and expands the duty cycle range, As shown in Fig. 6, the voltage swing at node B is significantly reduced by the introduction of the clamping circuit. A comparison of waveforms with and without the clamping structure demonstrates that the swing amplitude is reduced by approximately 67%, which contributes to faster recovery and broader duty cycle modulation capability.

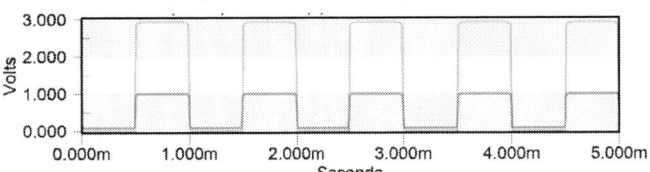

Fig. 6. The voltage swing at point B inside the comparator, clamping circuit (green), without clamping circuit (yellow).

Building upon the two-stage push-pull comparator and the integrated sawtooth waveform generator, pulse-width modulation (PWM) signals with variable duty cycles are generated by comparing the dynamically varying sawtooth waveform against a programmable reference voltage (Vref_c). By modulating Vref_c, the comparator's output toggling point is shifted, thereby enabling precise control over the PWM duty cycle. Table I summarizes the generated duty cycles corresponding to different reference voltage levels, while the resulting PWM waveforms are illustrated in Fig.7,

979-8-3315-8850-2/25 $31.00 © 2025 IEEE   493

demonstrating the effectiveness of the modulation scheme in achieving fine-resolution duty cycle control. By adjusting the reference voltage from 0.3V to 4.7V, at a high frequency of 10MHz, the generated PWM wave can achieve an extremely wide duty cycle range from 1% to 99%.

TABLE I. SIMULATION PARAMETERS OF PWM DUTY CYCLE RANGE

Parameters	Frequency	Duty cycle	Vref_c
Sawtooth wave1(Up)	10MHz	1%	0.3V
Sawtooth wave2(Middle)	10MHz	50%	2.4V
Sawtooth wave3(Bottom)	10MHz	99%	4.7V

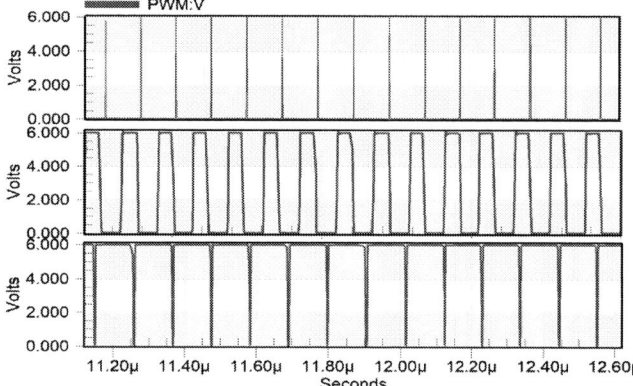

Fig. 7. Three PWM waves of different PWM duty cycles, 1% (Up), 50% (Middle), and 99% (Bottom).

The simulation evaluates the performance of a half-bridge GaN buck converter operating at high frequency. Although a bootstrap circuit is used to drive the high-side GaN transistor, the primary focus is on the converter's voltage regulation capability. With the power supply voltage set to 10 V and the forward voltage drop of the 1N4148W diode approximately 0.9 V under 20 mA current, the bootstrap gate drive voltage is sufficient to ensure proper transistor conduction.

To evaluate the performance of the proposed half-bridge GaN buck converter, comprehensive time-domain simulations were conducted under a 48 V input condition. The converter was configured to generate multiple output levels 1 V, 12 V, and 24 V by adjusting the PWM duty cycle, demonstrating its flexibility in voltage regulation. The switching frequency was set to 1 MHz. In Fig. 8, the resulting output voltage waveforms indicate stable and well-regulated operation across all target voltages. Detailed voltage data are presented in Table II.

Experimental results confirm the near-linear voltage conversion of the proposed buck converter. As the duty cycle increases from 2% to 50%, the output voltage closely follows the theoretical relation Vout = D × Vin. At duty ratios of 25% and 50%, the measured voltages match the theoretical values exactly, while at 2% duty the output is 1 V compared to 0.96 V, resulting in a deviation of +0.04 V (≈4.2%). This deviation is

relatively small and confined to the extreme low-duty operating point, indicating that the converter achieves good linearity across the entire range.

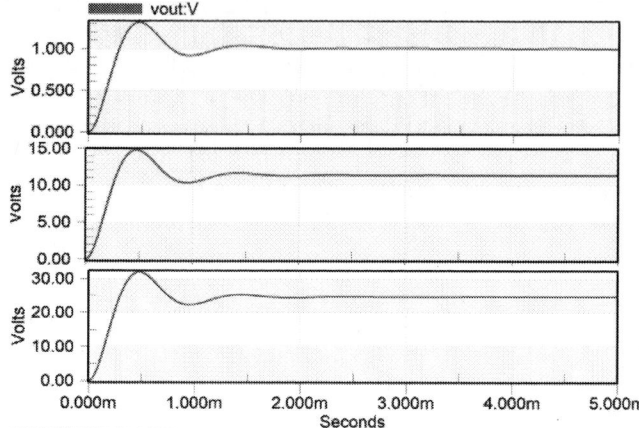

Fig. 8. Output voltage of the GaN buck converter, 1V (Up), 12V (Middle), and 24V (Bottom). Vin=48 V, L=1 mH, Cout= 20 μF, Rout= 10 Ω.

TABLE II. PARAMETERS OF DIFFERENT OUTPUT VOLTAGES

Parameters	Frequency	Duty cycle(HG)	Voltage drop
Vout1(Up)	1MHz	2%	48V-1V
Vout2(Middle)	1MHz	25%	48V-12V
Vout3(Bottom)	1MHz	50%	48V-24V

The accuracy and stability of the output can be achieved by adjusting passive components within the power stage. A larger output inductance effectively reduces current ripple but can degrade transient response. Conversely, higher output capacitance improves voltage ripple suppression and enhances dynamic performance, however it increases cost and physical footprint.Through iterative simulations and trade-off analysis, optimal component values were identified to ensure reliable step-down conversion from 48 V to 1 V with minimal output ripple and fast settling characteristics. The inductor L, output capacitor Cout were chosen as 1 mH and 20 μF, respectively. Overall, the simulation results validate the converter's fast transient response, and robust voltage regulation, confirming its suitability for high-frequency applications and dynamic load environments.

Under different load conditions, the proposed converter exhibits consistent output voltage behavior, as shown in Fig. 9. Across all cases, the steady-state output converges to the nominal value, confirming that voltage regulation is largely unaffected by load variations. The primary influence of the load is observed during the transient process: heavier loads lead to a slower rise and longer settling time, whereas lighter loads enable faster stabilization with a small overshoot before convergence. These behaviors originate from the load-dependent dynamics of the output stage, as the effective output

979-8-3315-8850-2/25 $31.00 © 2025 IEEE

pole and the inductor current ramp rate vary with the applied load. Once steady state is reached, however, the output voltage remains well regulated with negligible ripple. This demonstrates that the proposed design maintains robust performance across a broad load range, with load variations affecting only transient settling rather than steady-state accuracy.

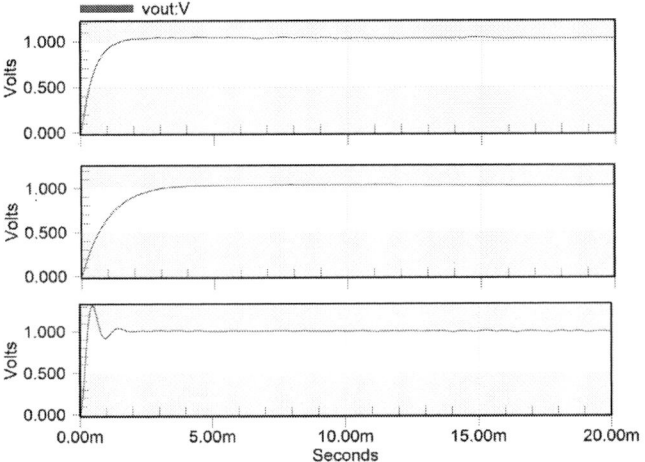

Fig. 9. Output voltages for different loads, Output current = 1A (Up), 0.5A (Middle), and 0.1A (Bottom). Vin=48 V, L=1 mH, Cout= 20 μF

IV. Conclusion

This work presents the design and simulation of a high-frequency buck converter capable of directly stepping down 48 V to 1 V, using a CMOS-based PWM generator to drive GaN transistors. The system was designed and verified using the Tanner EDA tool. A hysteresis comparator and closed-loop sawtooth generator were implemented to ensure stable PWM signal generation across frequencies up to 10 MHz. To address limitations in conventional comparator designs, a two-stage push-pull comparator with clamping circuits was proposed, enabling accurate duty cycle control from 1% to 99%. In the power stage, the EPC2065 GaN transistor and a CMOS bootstrap driver were employed. Simulation results confirmed successful step-down operation at 1 MHz with a 2% duty cycle.

The results validate the effectiveness of the proposed architecture for high-conversion-ratio, high-speed DC-DC power applications.

References

[1] H. Meng, Z. Sun, M. Qiu, X. Liu, V. Marzang and D. Cao, "MASC-PoL: A 48V-1V Matrix Autotransformer Switched-Capacitor Point-of-load DC-DC Converter for Data Center Application," *2024 IEEE Energy Conversion Congress and Exposition (ECCE)*, Phoenix, AZ, USA, 2024, pp. 2589-2595, doi: 10.1109/ECCE55643.2024.10861421.

[2] Y. Zhu, T. Ge, N. M. Ellis, L. Horowitz, and R. C. N. Pilawa-Podgurski, "The switching bus converter: A high-performance 48-V-to-1-V architecture with increased switched-capacitor conversion ratio," *IEEE Trans. Power Electron.*, vol. 39, no. 7, pp. 8384–8403, Jul. 2024, doi: 10.1109/TPEL.2024.3370166.

[3] Y. Zhu, J. Zou, and R. C. N. Pilawa-Podgurski, "A 1500-A/48-V-to-1-V switching bus converter for next-generation ultra-high-power processors," *IEEE Trans. Power Electron.*, vol. 39, no. 9, pp. 11340–11355, Sept. 2024, doi: 10.1109/TPEL.2024.3403670.

[4] Y. Elasser *et al.*, "Mini-LEGO: A 1.5-MHz 240-A 48-V-to-1-V CPU VRM with 8.4-mm Height for Vertical Power Delivery," *2023 IEEE Applied Power Electronics Conference and Exposition (APEC)*, Orlando, FL, USA, 2023, pp. 1959-1966, doi: 10.1109/APEC43580.2023.10131163.

[5] H. S. Oon and K. Y. Cheong, "Recent development of gallium oxide thin film on GaN," *Mater. Sci. Semicond. Process.*, vol. 16, no. 5, pp. 1217–1231, Oct. 2013, doi: 10.1016/j.mssp.2013.01.027.

[6] T. P. Chow, "High-voltage SiC and GaN power devices," *Microelectron. Eng.*, vol. 83, no. 1, pp. 112–122, Jan. 2006, doi: 10.1016/j.mee.2005.10.057.

[7] TDK-Lambda, "DC-DC converter switching frequencies – fixed or variable?," *TDK-Lambda Americas*, Apr. 25, 2022. [Online]. Available: https://www.us.lambda.tdk.com/resources/blogs/20220425.html (Accessed: Jul. 9, 2025).

[8] Z. Zhao, P. Luo, Z. Zhang, J. Fan, B. Zhang, and X. Chen, "A peak-valley current-mode buck converter with 3% to 95% duty cycle," *IEEE Trans. Circuits Syst. II Express Briefs*, vol. 72, no. 1, pp. 328–332, Jan. 2025, doi: 10.1109/TCSII.2024.3484449.

[9] Monolithic Power Systems, "MPQ2908A product datasheet," *Monolithic Power Systems*, 2024. [Online]. Available: https://www.monolithicpower.com/en/mpq2908a.html (Accessed: Jul. 9, 2025).

[10] F. Su, W.-H. Ki, and C.-Y. Tsui, "Ultra fast fixed-frequency hysteretic buck converter with maximum charging current control and adaptive delay compensation for DVS applications," IEEE Journal of Solid-State Circuits, vol. 43, no. 4, pp. 815–822, Apr. 2008

AUTHOR INDEX

Ai, Z. 317
An, W. 984
Ba, H. 502
Bai, B. 354
Bao, X. 422
Bernales, J.S. 427
Bi, S. 540
Bi, Z. 173, 610
Bi, Z.R. 660
Cai, C. 294
Cai, G. 910
Cai, H. 496
Cai, L. 643
Cai, S. 19
Cai, X. 496
Cai, Y. 568
Cai, Z. 962
Cañonero, K.M. 427
Cao, C. 1, 191, 583
Cao, J. 805
Cao, K. 464
Cao, R. 863
Cao, X. 496
Cao, Y. 568
Chang, K.-C. 422, 568, 910, 933
Chen, D. 439, 796, 805
Chen, G. 333, 378, 899
Chen, H. 221, 454, 708, 967
Chen, J. 92, 721, 858
Chen, L. 839
Chen, N. 215
Chen, Q. 540
Chen, R. 247
Chen, S. 397, 474
Chen, W. 529, 929
Chen, X. 204, 237, 681, 810
Chen, Y. 92, 158, 272, 439, 546, 561
Chen, Z. 25, 173, 737, 977
Cheng, G. 633
Cheng, L. 923
Cheng, M. 328
Cheng, X. 648
Cheng, Y. 589
Cheng, Z. 791
Chu, J. 801
Chuang, Y.-L. 933
Cui, M. 454, 491
Cui, X. 439

Dai, C. 628, 814, 839
Dai, G. 76, 309
Dai, H. 60
Dai, R. 445
Dai, Y. 962
Dai, Z. 977
Dang, K. 81
Deng, F. 198, 904
Deng, H. 317
Deng, K. 257
Ding, T. 895
Dong, G. 622
Dong, Z. 158, 439, 546
Du, Y. 616, 648
Duan, C. 819
Duan, Y. 191
Eom, C. 835, 895
Fan, H. 138
Fan, S. 110, 904
Fan, X. 573
Fan, Y. 71
Fan, Z. 19
Fang, J. 215
Fang, Y. 844
Feng, H. 76, 257, 464
Feng, J. 105
Feng, K. 653
Feng, M. 578
Feng, Q. 317
Feng, Y. 633, 943
Fu, D. 148
Fu, J. 372, 702
Fu, N. 204
Gan, T. 801
Gao, H. 19
Gao, S. 232, 594
Gao, Z. 360
Ge, J. 819
Ge, X. 546
Gong, J. 209
Gu, X. 31, 852
Gu, Y. 110
Gu, Z. 770
Guan, Y. 967
Guo, C. 721
Guo, F. 512
Guo, H. 1, 191, 583
Guo, J. 299, 616, 648, 686, 759

Guo, K.	37
Guo, X.	198
Guo, Y.	759, 962
Guo, Z.	60, 360, 507, 610, 948
Han, H.	967
Han, S.	863
Hao, J.	748
Hao, M.	115
Hao, Y.	81
He, H.	622
He, R.	272
He, S.	670
He, X.	943
He, Y.	309
He, Z.	413, 899
Hong, Z.	868
Hou, K.	748
Hou, Y.	868
Hou, Z.	71
Hu, J.	49, 65, 276, 322, 879
Hu, K.	1
Hu, M.	272
Hu, Q.	449
Hu, S.	469
Hu, W.	721
Hu, X.	252
Hu, Y.	366, 391, 780
Hu, Z.	962
Huang, C.	568
Huang, H.	243, 616
Huang, J.	186, 243, 791, 918
Huang, M.	605
Huang, T.	408, 692, 929
Huang, X.	660
Huang, Y.	37, 153, 288, 413, 879
Huo, Y.	115, 748
Jia, J.	317
Jiang, F.	819
Jiang, J.	328, 923
Jiang, M.	759, 943
Jiang, P.	81
Jiang, Q.	257
Jiang, S.	770
Jiang, W.	616
Jiang, Y.	215
Jiao, Y.	366
Jin, Y.	204, 568, 785
Jing, C.	267

Jing, N.	328, 923
Kang, W.	633
Khalid, S.A.	344, 427
Kong, M.	317
Kou, X.	721
Kuai, R.	252, 354
Lai, B.	835, 895
Lai, X.	92
Lan, B.	529, 638
Le Wu, W.	49, 322
Lei, H.	232, 594
Li, D.	148, 675
Li, E.	127
Li, G.	660
Li, H.	7, 115, 226, 561, 731, 873, 984
Li, J.	76, 354, 464, 702, 885
Li, K.	759
Li, L.	12, 71, 86, 99, 127, 422, 568, 910, 933, 937
Li, L.A.	449
Li, M.	37, 92, 168
Li, P.	573, 785
Li, Q.	413, 556
Li, S.	753, 796
Li, T.	148, 512, 868
Li, W.	138, 204, 309, 372, 702, 770
Li, X.	43, 105, 122, 132, 198, 282, 403, 433, 459, 474, 517, 628, 801, 839, 885
Li, Y.	158, 247, 294, 299, 480, 686, 791, 868
Li, Z.	65, 105, 122, 143, 221, 232, 372, 403, 474, 594, 702, 708, 716, 748, 943
Lian, H.	418
Liang, F.	158, 439
Liang, L.	163, 948
Liang, S.	127
Liang, Y.	556
Liao, C.	551, 568
Liao, G.	616
Lin, C.	276, 299
Lin, H.	918
Lin, X.	19
Lin, Y.	716
Lin, Z.	328, 628, 796, 819, 830, 839, 873, 923
Ling, Z.	486
Liu, B.	309, 496
Liu, C.	12, 71, 863
Liu, D.X.	267
Liu, F.	643
Liu, H.	502, 748, 937, 953

Liu, J. 7, 262, 305, 372, 474, 535, 648, 885
Liu, K. 257
Liu, L. 366
Liu, M. 12, 71
Liu, P. 81
Liu, Q. 294, 397, 681
Liu, S. 418, 796
Liu, T. 801
Liu, W. 360, 403, 890
Liu, X. 25, 299, 360, 486, 605, 737, 967
Liu, Y. ... 54, 257, 408, 418, 551, 628, 643, 692, 743, 796, 873, 972
Liu, Y.W. 267
Liu, Z. 232, 445, 556, 594, 814
Long, M. 819
Long, Q. 868
Lou, F. 805
Lou, Y. 873
Lu, C. 384
Lu, L. 422, 568, 933
Lu, Z. 948, 958
Luo, G. 910
Luo, Q. 573
Luo, W. 1, 583
Lv, F. 168, 252, 354
Lv, J. 648
Lv, X. 780
Lv, Y. 589
Lyu, R. 610
Ma, H. 546
Ma, J. 215
Ma, Y. 317, 743
Macapundag, A.G. 427
Madrid-Khalid, K.M. 344
Maestre, S.E. 344, 427
Maloberti, F. 697
Mao, J. 237
Mao, K. 221
Mei, X. 605
Men, C. 333, 378
Meng, F. 445
Meng, S. 305
Mi, X. 879
Miao, M. 628
Miao, X. 858
Miao, Y. 984
Mo, T. 868
Nian, Y. 122, 403

Niu, Z. 143
Ou, J. 605
Pan, M. 517, 670
Pan, S. 257
Pan, Y. 628
Pang, C. 226
Pang, H. 561
Pang, Z. 168
Pei, B. 215
Peng, C. 105, 132, 143, 814
Peng, J. 480
Peng, Q. 910
Pu, X. 360
Qi, G. 243, 272, 726
Qiang, B. 105, 132
Qiang, X. 60
Qin, C. 12, 71
Qin, L. 267
Qin, P. 299, 716
Qiu, A. 665
Qiu, C. 948, 958
Quan, G. 716
Ren, H. 1, 191, 583
Ren, R. 885
Ru, Y. 737
Ru, Y.A. 25
Sham, C.-W. 247
Shang, Y. 372, 702
Shao, R. 49
Shen, R. 708
Shen, Y. 433, 551, 972
Sheng, W. 328, 923
Sheng, Y. 670
Shi, D. 824
Shi, L. 267
Shi, Q. 665
Shi, Y. 801
Shi, Z. 697
Si, L. 247
Song, C. 191
Song, H. 209, 599, 721
Su, G. 262, 535
Su, K. 512
Su, Y. 578
Sun, D. 507, 681
Sun, H. 65, 372, 702
Sun, J. 929, 958
Sun, L. 660

Sun, Q. 333, 378
Sun, R. 81
Sun, W. 967
Sun, Y. 186, 433, 801
Tan, J. 163
Tan, Y. 708
Tang, J. 540
Tang, L. 110, 764
Tang, X. 305
Tang, Y. 328, 923
Teng, C. 551, 948, 972
Tian, L. 556
Tian, Q. 918
Tong, H. 858
Tong, L. 317
Wan, J. 529, 638
Wan, M. 977
Wan, P. 737
Wang, A. 599
Wang, B. 517
Wang, C. 7, 512, 578, 814, 984
Wang, H. 65, 215, 408, 517, 743
Wang, J. 122, 186, 262, 391, 535, 810, 977
Wang, K. 180
Wang, L. 148, 469, 633
Wang, N. 743
Wang, P. 153, 288
Wang, Q. 168, 226, 328, 923
Wang, R. 12, 31, 824
Wang, W. 322, 496, 863
Wang, X. 99, 115, 143, 158, 391, 439, 546
Wang, Y. 127, 551, 899, 972
Wang, Z. 54, 132, 158, 546, 573
Wei, C. 764
Wei, G. 209
Wei, J. 7, 984
Wei, X. 366, 391, 780
Wei, Y. 929, 958
Wei, Z. 305, 469
Wen, B. 60
Wen, K. 835, 895
Wu, D. 309
Wu, F. 338
Wu, G. 491
Wu, H.-Y. 384
Wu, J. 43, 282, 517, 890
Wu, L. 49, 65, 86, 99, 221, 276, 322, 937
Wu, P. 317

Wu, S. 333, 378
Wu, W. 76, 408, 464, 692
Wu, X. 105, 122, 132, 143, 173, 232, 397, 403, 449, 594, 610, 638, 796, 819, 830, 839, 873, 984
Wu, Y. 180, 839
Wu, Z. 276, 360, 929
Xi, X. 512, 643
Xia, C. 622
Xia, G. 665
Xia, Y. 868
Xiang, Z. 578
Xiao, Y. 7
Xiaoyu, L. 523
Xie, D. 221
Xie, F. 25, 737
Xie, H. 529
Xiong, X. 19
Xiong, Y. 198
Xiong, Z. 764
Xu, B. 962
Xu, H. 122, 403
Xu, J. 168, 397, 512
Xu, L. 583
Xu, M. 665, 835, 895
Xu, N. 384, 523, 561
Xu, R. 507
Xu, T. 299, 716
Xu, X. 397
Xu, Y. 243, 910, 962
Xu, Z. 237, 299, 810
Xue, F. 780
Xue, H. 282
Xue, Q. 299, 716
Xue, T. 267
Yan, B. 115, 824
Yan, C. 173, 610
Yan, G. 517
Yan, L. 232, 899
Yan, S. 153, 288
Yan, T. 967
Yang, C. 459
Yang, D. 835
Yang, F. 338, 660, 962
Yang, H. 366, 589, 594, 830, 972
Yang, J. 138, 830
Yang, K. 962
Yang, L. 681
Yang, Q. 338

Yang, S. 309
Yang, W.M. 391
Yang, X. 115, 180, 681
Yang, Y. 12, 86, 99, 819, 824, 937
Yang, Z. 408, 418, 523, 561, 692, 753
Yao, J. 962
Yao, Y. 127
Ye, B. 433
Ye, J. 502
Ye, Y. 633
Yi, J. 153, 288
Yin, R. 354
Yin, Z. 486
You, F. 670
You, H. 540
Yu, D. 204, 923
Yu, H. 247, 984
Yu, J. 191
Yu, L. 977
Yu, R. 830
Yu, S. 153, 288
Yu, Z. 31, 852
Yuan, L. 252
Yuan, X. 512
Yuan, Y. 449
Yue, H. 43, 282
Yue, W. 824
Zeng, C. 578
Zeng, Q. 863
Zeng, X. 173, 252, 610, 660
Zha, Q. 885
Zhai, J. 622
Zhang, B. 805
Zhang, C. 780
Zhang, D. 37
Zhang, G. 215, 354, 573, 665
Zhang, H. 578
Zhang, J. 81, 92, 198, 209
Zhang, K. 469, 540
Zhang, L. 7
Zhang, R. 726, 731
Zhang, S. 282, 551, 972
Zhang, T. 338, 681, 770, 835
Zhang, W. 433, 529, 638
Zhang, X. 49, 65, 76, 86, 99, 209, 276, 322, 464, 474, 605, 748, 805, 937

Zhang, Y. 92, 252, 305, 418, 433, 486, 573, 599, 638, 643, 721, 863
Zhang, Z. 43, 168, 360, 653, 953
Zhao, C. 830
Zhao, Q. 105, 122, 132, 143, 232, 403, 594, 814
Zhao, R. 7, 780, 858
Zhao, S. 272, 958
Zhao, X. 262, 535, 675, 731
Zhao, Y. 198, 366, 665, 697
Zhao, Z. 366, 660
Zheng, B. 681
Zheng, F. 502
Zheng, R. 391, 780
Zheng, S. 616
Zheng, X. 19
Zheng, Y. 605, 977
Zhi, H. 43, 282
Zhong, J. 879
Zhong, X. 247
Zhong, Z. 221
Zhou, C. 753
Zhou, J. 338, 418, 507, 858, 873
Zhou, K. 830
Zhou, L. 37
Zhou, W. 37
Zhou, X. 328
Zhou, Y. 209
Zhou, Z. 148, 948
Zhu, A. 692
Zhu, C. 413
Zhu, H. 716
Zhu, K. 512
Zhu, M. 163, 173, 610
Zhu, S. 86, 622
Zhu, Z. 675, 697
Zhu, Z.-Y. 384
Zong, Y. 43
Zou, F. 589
Zou, X. 474
Zou, Z. 496

CURRAN ASSOCIATES INC.
proceedings
.com

9798331588502

2025 10th International Conference on Integrated Circuits and Microsystems (ICICM 2025)

Hefei, China
17-19 October 2025

Pages 496 – 988

IEEE Catalog Number: CFP25H23-POD
ISBN: 979-8-3315-8850-2

2025 10th International Conference on Integrated Circuits and Microsystems (ICICM 2025)

Hefei, China

17-19 October 2025

Pages 496–988

IEEE Catalog Number: CFP25H23-POD

ISBN: 979-8-3315-8850-2

Copyright © 2025, IEEE

All Rights Reserved

Copyright and Reprint Permissions:

Abstracting is permitted with credit to the source. Libraries are permitted to photocopy beyond the limit of U.S. copyright law for private use of patrons those articles in this volume that carry a code at the bottom of the first page, provided the per-copy fee indicated in the code is paid through Copyright Clearance Center, 222 Rosewood Drive, Danvers, MA 01923.

For other copying, reprint or republication permission, write to IEEE Copyrights Manager, IEEE Service Center, 445 Hoes Lane, Piscataway, NJ 08854. All rights reserved.

*** This is a print representation of what appears in the IEEE Digital Library. Some format issues inherent in the e-media version may also appear in this print version.

IEEE Catalog Number:	CFP25H23-POD
ISBN (Print-On-Demand):	979-8-3315-8850-2
ISBN (Online):	979-8-3315-8849-6

Additional Copies of This Publication Are Available From:

Curran Associates, Inc
57 Morehouse Lane
Red Hook, NY 12571 USA

Phone:	(845) 758-0400
Fax:	(845) 758-2633
E-mail:	curran@proceedings.com
Web:	www.proceedings.com

TABLE OF CONTENTS

A Compensation-Capacitor-Based Perturbation Injection Calibration Method for SAR ADCs 1
Hongjian Ren, Ke Hu, Wenya Luo, Chao Cao, Haijun Guo

Design and Simulation of High-Frequency FBAR Based on Multiphysics and MBVD Circuit Modeling 7
Rui Zhao, Jiangbo Wei, Yang Xiao, Liaoliao Zhang, Haijuan Li, Jiaqi Liu, Chao Wang

A 10GS/s 8b Time-Interleaved ADC with Aperture Error Calibration and LMS-Based Nonlinearity
Correction 12
Chuan Qin, Chuan Liu, Runqiao Wang, Lecheng Li, Maliang Liu, Yintang Yang

VCIL: An Open-Source Pipeline-Tight HIL Framework for Cycle-Accurate RISC-V Coprocessor
Verification 19
Xian Lin, Xin Zheng, Zhixin Fan, Huaien Gao, Shuting Cai, Xiaoming Xiong

An Electrode Impedance Evaluation IC with Biphasic Output Current for Biomedical Devices 25
Fang Xie, Yu Ang Ru, Zhuo Chen, Xu Liu

A Fast Transient and High PSRR LDO with 62 ns Settling Time 31
Ruijie Wang, Zhiguo Yu, Xiaofeng Gu

Automatic Generation Method of Test Vectors for FPGA Interconnect Resources Based on
Self-adaptive Ford-Fulkerson Algorithm 37
Liang Zhou, Wei Zhou, Mingzhe Li, Kun Guo, Ding Zhang, Yan Huang

A Multi-Phase Clock Generator with Customized Duty Cycles for Pipeline-SAR ADCs 43
He Zhi, Hongtao Yue, Yuwei Zong, Zhongxu Zhang, Jianhui Wu, Xin Li

Hardware Design and Side-Channel Security Analysis on the Key Computational Block for YOLOv11 49
Runquan Shao, Liji Wu, Jing Hu, Le Wu, Xiangmin Zhang

Secondary-Side Controlled Flyback Converter with Buck-Bypass for Multi-Port Application 54
Zhen Wang, Yi Liu

Research on the Method of Switching Adjustment to Reduce the Temperature Drift of the Bandgap
Reference 60
Bin Wen, Zhongjie Guo, Hongpeng Dai, Xiaojing Qiang

Design and Implementation of a Secure Boot RISC-V Processor for IoT Devices 65
Zeyu Li, Liji Wu, Jing Hu, Han Sun, Xiangmin Zhang, Haijie Wang

A 1GS/s 13-bit Single-Channel Hybrid Pipelined SAR ADC with Domino SAR Logic 71
Chuan Liu, Chuan Qin, Lecheng Li, Zhihao Hou, Yuqi Fan, Maliang Liu

A 1.7-to-3.0 GHz 201.95 dBc/Hz FOMT VCO With a Current-Reuse Active Inductor 76
Xinyi Zhang, Wanqing Wu, Guoxun Dai, Jiangnan Li, Haigang Feng

A 24-28GHz GaN SPDT MMIC with High Isolation and Low Insertion Loss 81
Peiyao Liu, Pengfei Jiang, Kui Dang, Rujun Sun, Yue Hao, Jincheng Zhang

Design and FPGA Implementation of a NTT Hardware Accelerator for PQC ML-DSA 86
Shuhang Zhu, Liji Wu, Lei Li, Yifan Yang, Xiangmin Zhang

The Effective Resistance Calculation Method Based on Matrix Partitioning Takahashi Algorithm 92
Meng Li, Yu Chen, Jingrui Chen, Yaokai Zhang, Jinyu Zhang, Xiaolue Lai

Hardware Design for Decomposition Module with Variable-Order Masking against Side-Channel
Attack for PQC ML-DSA 99
Xuejian Wang, Liji Wu, Lei Li, Yifan Yang, Xiangmin Zhang

Design and Implementation of a High-Reliability MIPI Interface Error Detection and Correction System
for CMOS Image Sensors 105
Qiang Zhao, Jianwei Feng, Bin Qiang, Zhigang Li, Xin Li, Chunyu Peng, Xiulong Wu

Temperature Sensing Module Design for Silicon MEMS Clocks 110
Shaotian Fan, Lu Tang, Yuxing Gu

An End-to-End Compact Shape-Aware Macro Placer Using Reinforcement Learning 115
Hailiang Li, Xiao Wang, Xu Yang, Miaohui Hao, Yan Huo, Beiping Yan

A 2.4 GHz Wide-Tuning-Range Low-Phase-Noise Differential Digitally Controlled Ring Oscillator for
CMOS Image Sensor Applications 122
Zhigang Li, Hongjia Xu, Yao Nian, Jingyi Wang, Xin Li, Qiang Zhao, Xiulong Wu

Smart Wireless Insole with Self-Powered Triboelectric Pressure Sensors for Gait Detection 127
Yuhan Wang, Yue Yao, Enze Li, Sifan Liang, Lei Li

Voltage-Time Hybrid Domain ADC with PVT Tracking for High-Frame-Rate CMOS Image Sensor 132
Xin Li, Ziheng Wang, Bin Qiang, Chunyu Peng, Qiang Zhao, Xiulong Wu

A Very-Large-Dynamic-Range Charge-Sensitive Front End for One-Terminal Capacitive Sensors in
65-nm CMOS 138
Haitao Fan, Wenxiang Li, Jiafeng Yang

A fast transient-response LDO-CP with novel high voltage compensation structure in CMOS image
sensors 143
Qiang Zhao, Zhendong Niu, Zhigang Li, Xiuying Wang, Chunyu Peng, Xiulong Wu

A 13-bit 1.5-GS/s Dual-Residue Pipelined ADC with Time-Domain Interpolation and Parallel
Quantization 148
Zecheng Zhou, Depan Li, Longsheng Wang, Ting Li, Dongbing Fu, Dengquan Li

A Novel Self-adaptive Dead-time Control Circuit Design for GaN Gate Drive Circuit 153
Shupeng Yan, Shengqi Yu, Yi Huang, Junjie Yi, Pengyu Wang

A Novel Hybrid Switching Scheme for SAR ADC with High Efficiency and High Linearity 158
Yushi Chen, Ying Li, Fang Liang, Zewen Wang, Zuoru Dong, Xiaodong Wang

A Unified Swin Transformer Framework for Inverse Lithography and Lithography Simulation 163
Jiajun Tan, Ling Liang, Ming Zhu

Design of 6-to-8-bit High-Speed Time-Interleaved SAR ADC 168
Zhenyang Zhang, Fangxu Lv, Zhengbin Pang, Jiaqing Xu, Qiang Wang, Meng Li

RLMBFF: Multi-bit Flip-Flops Merging using Deep Reinforcement Learning 173
Zeqi Chen, Zhaori Bi, Changhao Yan, Ming Zhu, Xiulong Wu, Xuan Zeng

HybridSYN: An Efficient Logic Synthesis Methodology 180
Ke Wang, Yue Wu, Xiaoyan Yang

A 20-GHz Low-Jitter Fractional-N CP-PLL With 2.5-GHz FBAR Reference 186
Jiandu Wang, Jiwei Huang, Yi Sun

A 14-Bit SAR ADC with Area-Efficiency Hybrid Three-Segment CDAC 191
Chengchen Song, Jia Yu, Chao Cao, Hongjian Ren, Yuchan Duan, Haijun Guo

A Fully Reversible Feynman Gate with Magic Block for Quantum-dot Cellular Automata Circuits 198
Jiayi Zhang, Feifei Deng, Xianghui Li, Yuezhong Xiong, Yunong Zhao, Xiaohui Guo

A Wide-Temperature-Range Low-Drift Bandgap Reference with Process-Corner-Adaptive Logic
Current Compensation 204
Xuesong Chen, Yu Jin, Ning Fu, Weifeng Li, Duli Yu

A Pixel-to-Column Digital Readout Circuit with InPixel CDS for Ga_2O_3 UV Photodetectors 209
Jiekun Zhang, Guangfen Wei, Jingwu Gong, Yu Zhou, Xiaodong Zhang, Helun Song

A DDR Controller Circuit Gate-level Post-simulation Method in SoC Design 215
Yande Jiang, Jingbo Ma, Guangda Zhang, Huiquan Wang, Bingxi Pei, Na Chen, Jian Fang

A High-Precision Current Detection Circuit for Battery Management System Chip in Electric Vehicle 221
Zhiwei Li, Liji Wu, Dan Xie, Kunning Mao, Haifeng Chen, Zhilin Zhong

A Robust Startup Circuit for Current-Mode Bandgap Reference 226
Qi Wang, Hao Li, Chunlei Pang

A Low-Power High-Accuracy Phase Frequency Detector and Charge Pump Design for CPPLL in 28
nm CMOS Technology 232
Zhigang Li, Hao Lei, Zhihao Liu, Lingrui Yan, Qiang Zhao, Shan Gao, Xiulong Wu

A 60-V 96.7%-Efficiency 175kHz Integrated Dimmable LED Driver with Mean-Peak Current Control
and Frequency Jittering Scheme 237
Jiamian Mao, Zhongguang Xu, Xiangyin Chen

A 100-Gb/s PAM4 Receiver Analog Front-End 243
Haoran Huang, Yue Xu, Jing Huang, Gengzhen Qi

A −86 dBm/15.6 µW Wake-Up Receiver for Internet of Vehicles Applications 247
Rui Chen, Yan Li, Liming Si, Xinchao Zhong, Chiu-Wing Sham, Hang Yu

A 28 − 32 Gb/s Wireline Receiver with A Genetic- Algorithm-Based Adaptive CTLE in 28 nm CMOS 252
Yingjie Zhang, Fangxu Lv, Liangyong Yuan, Xiaoyue Hu, Ruixiao Kuai, Xianchao Zeng

A Wideband Blocker-Tolerant Receiver with High Linearity and Low 0 dBm-BNF in 65-nm CMOS 257
Kaiyun Deng, Yingqi Liu, Qingrui Jiang, Kailin Liu, Shaohui Pan, Haigang Feng

A Model of a C-band Mixer Circuit 262
Xuezhen Zhao, Jun Liu, Jing Wang, Guodong Su

Parametric Modeling of Transmission Structure of RF Microsystem Based on
CNN-LSTM-SelfAttention 267
Li Qin, De Xi Liu, Cui Jing, Lei Shi, Ya Wei Liu, Ting Xue

A Radio-Frequency Orthogonal Switched-Capacitor Transmitter 272
Yingyi Chen, Mianting Hu, Sijia Zhao, Ruiyi He, Gengzhen Qi

Daisy Chain Transmitter Circuit Design with AntiElectromagnetic Interference for BMS Chip in EV 276
Chen Lin, Liji Wu, Jing Hu, Zonghuan Wu, Xiangmin Zhang

A Low-Power Flip signal Pulse Width Self-Adaptive Circuit for Piezoelectric Energy Harvesting 282
Hongtao Yue, Shuo Zhang, Haoyu Xue, He Zhi, Jianhui Wu, Xin Li

A Sub-nanosecond Delay and 200V/ns CMTI Level Shifter for GaN HEMTs Gate Driver 288
Shengqi Yu, Pengyu Wang, Junjie Yi, Shupeng Yan, Yi Huang

A Wide Tuning Range Dual-Mode, Magnetical-Coupled VCO with Capacitive-Coupled Noise-Circulating Technique Achieving FoM of 188dBc/Hz 294
Qishuang Liu, Chenxiang Cai, Yunchu Li

A Fractional-N All-Digital PLL with 166fs$_{rms}$ Jitter and 238 dB FoM Based on a Distributed Switched-Capacitor Arrays DCO 299
Zhihong Xu, Yuhui Li, Pei Qin, Changsong Lin, Jinhua Guo, Xiaorui Liu, Taotao Xu, Quan Xue

A CMOS Current-Reused Wideband Low-Power Dynamic Current Mode Logic Frequency Divider 305
Jingchen Liu, Youming Zhang, Xusheng Tang, Zhennan Wei, Shenghao Meng

A VHF memristor emulator for crossbar synaptic simulation 309
Guangzhen Dai, Bingchen Liu, Sihao Yang, Wei Li, Daohua Wu, Yuefeng He

A Novel SiC LDMOS with Electron Accumulation Layer Featuring Ultra-low Specific On-resistance and Improved Breakdown Voltage 317
Moufu Kong, Lin Tong, Qizhi Feng, Jiaru Jia, Zhaoyu Ai, Hongfei Deng, Yangyang Ma, Peifei Wu

Hardware Implementation and Side-Channel Security Analysis for High-Precision AI Tansformer Encoder 322
Wentao Wang, Liji Wu, Jing Hu, Le Wu, Xiangmin Zhang

High-Performance Data Prefetching Accelerator for Real-Time Object Detection 328
Yanzhou Tang, Xiangyu Zhou, Zeyi Lin, Ming Cheng, Naifeng Jing, Jianfei Jiang, Weiguang Sheng, Qin Wang

Design of High-reliability and High-dynamic Analog Front-end for DC Carrier Chip 333
Quan Sun, Gang Chen, Changyou Men, Shuang Wu

A Steady-State Temperature Solving Method for Multi-Chip Modules Components Based on an Improved U-Net 338
Fulong Yang, Feng Wu, Teng Zhang, Jinjin Zhou, Qing Yang

Open-LUT: Interactive g$_m$/ID Lookup Tables for MOSFET Sizing in Open-Source PDKs 344
Sihawi A. Khalid, Susie E. Maestre, Karla M. Madrid-Khalid

A 9.37-Tb/s/mm 5-bit-6-bit Crosstalk Cancellation Transceiver with ICBS Codes for High-Density Die-to-Die Interfaces 354
Ruotian Yin, Jinwen Li, Fangxu Lv, Geng Zhang, Ruixiao Kuai, Bohui Bai

Design and Implementation of an Integrated Testing System for Spintronic Chips 360
Wenyan Liu, Zihan Gao, Zhifu Guo, Zefan Wu, Xiangrong Pu, Xu Liu, Zhang Zhang

Design of a Reusable UVM Verification Framework for Pixel Readout Chips 366
Yu Zhao, Zexuan Zhao, Yuanhong Jiao, Xiaomin Wei, Heng Yang, Long Liu, Yongcai Hu

Design of High Gain and High Power Amplifier chip Based on RF Choke Segmented Multiplexing
Interstage Matching Technology 372
Zhenbing Li, Weijun Li, Haoyang Sun, Jiaxin Liu, Jialong Fu, Yaocheng Shang

Research on SoC Chip System Modeling for Noise Suppression and Self-organizing Communication
Integration for Distributed Photovoltaics 378
Quan Sun, Gang Chen, Changyou Men, Shuang Wu

Topology Evaluation-Based Layer Assignment Method for Free-assignment Routing in InFO Packages 384
Zhan-Yang Zhu, Ning Xu, Hao-Ying Wu, Chenglin Lu

A Radiation Tolerant CMOS Voltage Reference Circuit for High Energy Physics Experiments 391
Wei Ming Yang, Jia Wang, Xiayu Wang, Ran Zheng, Xiaomin Wei, Yongcai Hu

Dynamic Reward Weighting Based Deep Q-Learning for Routing in Advanced Packaging 397
Shubin Chen, Qinghai Liu, Xiaowei Wu, Jun Xu, Xiaolin Xu

A 7-Bit 1.8 ps Two-Step Time-to-Digital Converter in 28 nm CMOS Technology 403
Zhigang Li, Yao Nian, Hongjia Xu, Wei Liu, Xin Li, Qiang Zhao, Xiulong Wu

Ka-Band Broadband LNA in 65-nm CMOS Using Pre-stage Current-reuse Technique and Large-size
Transistor 408
Zihan Yang, Hao Wang, Ye Liu, Tongde Huang, Wen Wu

A 0.4 to 8GHz Broadband High-linearity I/Q Active Mixer in 130nm BiCMOS Technology 413
Qipeng Li, Zhenghao He, Chun Zhu, Yifan Huang, Qin Li

A Compact Active Phase Shifter for 110-150 GHz Phased Arrays in 130 nm BiCMOS 418
Shuguang Liu, Jinhao Zhou, Yuyang Liu, Hu Lian, Yaxin Zhang, Ziqiang Yang

Interface Quality Improvement in Planar SiC MOSFETs Using Supercritical Fluid Nitriding 422
Xiaoqing Bao, Lei Li, Lei Lu, Kuan-Chang Chang

Design of a Bandgap Voltage Reference for an 8-bit SAR-ADC for Powerline Monitoring using
SKY130 PDK 427
Krisna M. Cañonero, Jovelyn S. Bernales, Abdulwarith G. Macapundag, Sihawi A. Khalid, Susie E. Maestre

A 2.9mV$_{PP}$ Ripple 60mA Digital Low-Dropout Regulator in 28nm CMOS 433
Wenxin Zhang, Yuhang Zhang, Yang Shen, Xiaojin Li, Bingyi Ye, Yabin Sun

MOSFET Modeling of 0.18 µm CMOS Technology Based on BSIM4 Model for Cryogenic Devices 439
Dong Chen, Zuoru Dong, Yushi Chen, Xingyu Cui, Fang Liang, Xiaodong Wang

Process Integration Optimization for RF SOI Process with Arcing Issue 445
Zhangli Liu, Fei Meng, Ruofan Dai

An Optimized ROM based Direct Digital Synthesizer Based on 65nm CMOS Technology 449
Yuan Yuan, Qingsheng Hu, Xu Wu, Li Anming Li

Integrated Circuit Design of CMOS Deadtime Controller for 48 V GaN DC-DC Converter 454
Haoyu Chen, Miao Cui

An Ultra-Low Power Digital Assisted Self-Tracking Zero-Current Detector for DC-DC Converters 459
  Chuting Yang, Xing Li

A High-Linearity Digital-to-Time Converter Design for Fractional-N Sub-sampling PLL 464
  Jiangnan Li, Wangqing Wu, Ke Cao, Xinyi Zhang, Haigang Feng

A 24–27-GHz 3-Stage Driving Amplifier with 18–25dB Variable Gain in 180-nm CMOS for
Beamforming ICs 469
  Zhenghuan Wei, Kaibo Zhang, Lijuan Wang, Sanming Hu

A 16-b 8-MS/s Pipelined SAR ADC With Robust Ring Amplifier and On-Chip Bit-Weight Calibration 474
  Shaojuan Chen, Xiaoyi Li, Xuanhao Zhang, Xingshuai Zou, Zhenbing Li, Jiaxin Liu

An IVUS AFE with LNA and a CT $\Delta\Sigma$ modulator with 80MHz BW and 12 bit ENOB 480
  Jie Peng, Yunchu Li

A Neuron-and-Synapse Unit Circuit with Information Propagation Function for Spiking Neural
Networks 486
  Yide Zhang, Zixuan Ling, Zicheng Yin, Xu Liu

CMOS PWM Controller Design for GaN-Based 48V1 V DC-DC Buck Converter 491
  Guanyu Wu, Miao Cui

HEA2-MAC: A Hybrid Exponent-Aware Approximate MAC for Efficient CNN Processing 496
  Weixuan Wang, Zihan Zou, Xin Cao, Xuefeng Cai, Hao Cai, Bo Liu

A Low-Power Continuous-Time Quadrature Bandpass $\Sigma\Delta$ ADC with a novel ELD Compensation
Method in Quantizer 502
  Haowen Ba, Jun Ye, Hanli Liu, Feijun Zheng

Low Power $\Sigma\Delta$ Modulator Applied to CIS Column Parallel Readout Circuit 507
  Dengju Sun, Zhongjie Guo, Ruiming Xu, Jinquan Zhou

Design and Analysis of AC/DC Converter for GaNBased Server Power Supply System 512
  Fen Guo, Tuo Li, Changhong Wang, Kang Su, Jiankai Xu, Xin Xi, Kejian Zhu, Xinxin Yuan

A Low-noise PGA With High-gain Chopper Offset-stabilized Amplifier Combining Ping-Pong
Auto-zero For TMR Analog Front-End 517
  Hui Wang, Guifa Yan, Mingqi Pan, Bo Wang, Jianhui Wu, Xin Li

An Adaptive dynamic R-tree indexing algorithm for HDI PCB layout 523
  Zhang Yang, Liang Xiaoyu, Ning Xu

A Novel Region-Wise Automatic Routing Algorithm for Analog Circuit 529
  Hao Xie, Wenxue Chen, Wei Zhang, Bijian Lan, Jing Wan

A 0.6-3 GHz Active Double-Balanced Mixer Circuit Design 535
  Jing Wang, Jun Liu, Xuezhen Zhao, Guodong Su

Replicated Partitioning for Hypergraphs with Multiple Constraints 540
  Kexin Zhang, Shunyang Bi, Jing Tang, Hailong You, Qiwang Chen

A cryogenic readout integrated circuit for blocked-impurity-band (BIB) far-infrared focal plane array
detectors 546
  Yushi Chen, Xin Ge, Hongbo Ma, Zewen Wang, Zuoru Dong, Xiaodong Wang

1T1C-Enhanced TFT-Integrated Gate Driver Circuit Design for Reliable In-cell Touch Sensing Displays ... 551

Congwei Liao, Yong Wang, Chao Teng, Yi Shen, Yudong Liu, Shengdong Zhang

Neural Network-Based Method for Magnetic Field Inversion of Energized Conductors in Printed Circuit Boards ... 556

Qi Li, Yiling Liang, Zhen Liu, Lulu Tian

A double layer placement algorithm for IC-based modules of printed circuit board ... 561

Hangyuan Li, Yu Chen, Zhaoyang Yang, Haotian Pang, Ning Xu

High-Performance Hydrogenated Oxide-Semiconductor Schottky Barrier Diodes with ALD HfO2 Interface ... 568

Yucheng Cao, Chenyang Huang, Yuyang Cai, Kuan-Chang Chang, Lei Li, Congwei Liao, Yufeng Jin, Lei Lu

An Ultra-Low-Power True Random Number Generator Based on Volatile RRAM ... 573

Qi Luo, Zhen Wang, Xuemeng Fan, Pengtao Li, Guobin Zhang, Yishu Zhang

Performance Evaluation of an Andvanced-Node CMOS Sensor for Partical Detection ... 578

Yue Su, Mingjie Feng, Zhiyu Xiang, Cheng Zeng, Congcong Wang, Hui Zhang

Low-power LDO with Fast Transient Response Based on FVF Structure ... 583

Wenya Luo, Longfei Xu, Hongjian Ren, Chao Cao, Haijun Guo

Low-Power MCU Architecture Optimization and Energy Efficiency Enhancement in Intelligent Pressure Sensor SoCs ... 589

Yaoming Lv, Hong Yang, Feng Zou, Yuhua Cheng

A Low-Phase-Noise and Low-Power Class-C VCO with Robust Start-up for FMCW Radars ... 594

Zhigang Li, Zhihao Liu, Hao Lei, Hang Yang, Qiang Zhao, Shan Gao, Xiulong Wu

A 0.47 nJ/Conversion CMOS Temperature Sensor with 0.00786 mm² Core in 65nm CMOS ... 599

Hangfei Song, Yichi Zhang, Aili Wang

An Ultra-low Power 2× / 4× Reconfigurable Charge Pump for RF Energy Harvesting System ... 605

Xuanchen Mei, Xin Liu, Junhui Ou, Yanqi Zheng, Mo Huang, Xiuyin Zhang

Cross-Process Bayesian Multi-Objective Collaborative Optimization For Process Migration ... 610

Zixi Guo, Ruiyu Lyu, Zhaori Bi, Changhao Yan, Ming Zhu, Xiulong Wu, Xuan Zeng

Multi-Branch Autoencoder Networks for Efficient Inverse Design of Wideband Frequency-Selective Surfaces ... 616

Wei Jiang, Haoran Huang, Guangxin Liao, Shenli Zheng, Jiacheng Guo, Yuan Du

GraphRL-Core: Intelligent Logic Synthesis Optimization via Graph Transformer and Deep Reinforcement Learning ... 622

Sujie Zhu, Guande Dong, Haoyang He, Chong Xia, Jianwang Zhai

A Hilbert Transform-Based Timing Skew Estimation Method for Dual-Channel TIADCs ... 628

Xin Li, Ying Pan, Mengdi Miao, Yu Liu, Chenghu Dai, Zhiting Lin

A Fast Auto-Frequency Calibration Technique with High Reference Frequency for PLL ... 633

Yan Feng, Yongjie Ye, Guoxiao Cheng, Liu Wang, Wei Kang

Post-Routing Compression Algorithm for Area-Efficient Layout design of Analog Integrated Circuit 638
Xiaoyue Wu, Yubo Zhang, Wei Zhang, Bijian Lan, Jing Wan

A Fractional-N Reference-Sampling PLL With a Gain-Boost Fractional Phase Detector for Phase Noise
Reduction 643
Xiaolian Xi, Lecheng Cai, Yihao Liu, Fan Liu, Yanlong Zhang

A 59.6GHz Broadband 3-stage Cascode LNA with Peak Detector for Bandwidth Self-healing 648
Xinsheng Cheng, Jiacheng Guo, Juntao Liu, JingJing Lv, Yuan Du

Addressing Signal Integrity Challenges in DDR5 SDRAM: A High-Precision ZQ Calibration Circuit
with Fast Calibration Time 653
Kexin Feng, ZhiQiang Zhang

A Pipeline-Based Common Framework for Parallel Mixed Signal Simulation 660
Longchen Sun, Guangrong Li, Zhenguo Zhao, Xin Huang, Xuan Zeng, Fan Yang, Zhao Ri Bi

A Rectifier Circuit with Adjustable Temperature Coefficient for Precise Amplitude Control of MEMS
Resonator 665
Guanxiao Zhang, Yang Zhao, Qin Shi, Guoming Xia, Anping Qiu, Meijia Xu

A Low-Complexity Bandwidth Enhancement Design Method for mm-Wave Doherty Amplifier 670
Yujie Sheng, Fei You, Maojun Pan, Songbai He

A 10-bit 3-GS/s Single-Channel Pipelined ADC with Parallel Time-Domain Quantization Based on
Dynamic Residue Amplifier 675
Depan Li, Xin Zhao, Dengquan Li, Zhangming Zhu

Design of High Speed and Energy-Efficient ADC Based on Dynamic Bandwidth Ring Amplifier 681
Xin Yang, Qingyuan Liu, Xiaobo Chen, Dan Sun, Bin Zheng, Tieliang Zhang, Long Yang

A 12-b 1-GS/s Pipelined-SAR ADC With a Hybrid Parallel Timing Scheme and PVT-Robust
Ring-Amp 686
Junhui Guo, Yunchu Li

An X-band Compact High-Efficiency MMIC Power Amplifier in 0.25-μm GaN Technology 692
Ye Liu, Anshi Zhu, Zihan Yang, Tongde Huang, Wen Wu

Design of a Two-Stage Fully Differential Operational Amplifier for High-Resolution Sigma Delta
ADCs 697
Zhan Shi, Zenghao Zhu, Yan Zhao, Franco Maloberti

Design of Power Amplifier for Short Message Application of BDS-III Communication Terminal 702
Zhenbing Li, Weijun Li, Haoyang Sun, Jiaxin Li u, Jialong Fu, Yaocheng Shang

Design of a High-Bandwidth Low-Noise Amplifier 708
Hongmei Chen, Ruiting Shen, Yuexin Tan, Zheyu Li

A Compact Wideband E-band Low Noise Amplifier with 3.3-4.5 dB Noise Figure using 45-nm RFSOI 716
Yinhan Lin, Haoshen Zhu, Taotao Xu, Zhuming Li, Guohai Quan, Pei Qin, Quan Xue

A Wideband Input Buffer Based on AC-Coupled Flipped Source Follower Using Auxiliary Operational
Amplifiers for 8-GS/s ADCs 721
Yuhang Zhang, Jian Chen, Weiying Hu, Xianguo Kou, Changdong Guo, Haizhi Song

A 2.4-2.6 GHz CMOS High Linearity Power Amplifier with MGTR and Harmonic Trap Techniques 726
Runxun Zhang, Gengzhen Qi

A Ka Band High Back-off PAE Power Amplifier with Adaptive Bias Circuits in 65 nm CMOS 731
Ran Zhang, Xiaodong Zhao, Hangbiao Li

A DAC Design using 5-V High-voltage Process for Driving Large Electrode Loads of Neural
Stimulators 737
Xu Liu, YuAng Ru, Zhuo Chen, Fang Xie, Zhijie Chen, Peiyuan Wan

MEMS-Enabled Computational Spectrometer Based on Cascaded Waveguide Couplers 743
Hanxing Wang, Yan Liu, Nan Wang, Yiming Ma

Control Strategy and Parameter Identification Method for Photovoltaic Inverter Electromechanical
Transient Model Based on Improved Fish Eagle Optimization Algorithm 748
Zecheng Li, Yaojia Huo, Kai Hou, Xiao Zhang, Hao Liu, Jinpeng Hao

Design of a Cross-Platform Simulation and Verification System for Complex Onboard Control
Computers 753
Zheng Yang, Shenglong Li, Chaofan Zhou

Fast Frequency Stabilization Technique for Buck Converter Based on Adaptive PLL 759
Kerun Li, Mei Jiang, Jianing Guo, Yuxuan Guo

A Wide-Range PFD and Low-Mismatch Source-Switched CP for MEMS Clock Systems 764
Changfu Wei, Lu Tang, Ziyao Xiong

High-Reliability Integration Design Method for Micro Systems SiP in Complex Spatial Environments 770
Shang Jiang, Wenchang Li, Zucheng Gu, Tianyi Zhang

Design and Implementation of an FPGA-Based Test System for Depth of Interaction Measurement 780
XiaoTian Lv, Ce Zhang, Ran Zheng, XiaoMin Wei, FeiFei Xue, RuiGuang Zhao, YongCai Hu

Design and Nonlinear Analysis of a Multi-Stepped MEMS Resonator with 1:3 Internal Resonance 785
Peilong Li, Yu Jin

A 10.8-12.5-GHz Charge Pump PLL With 68.4-fs$_{rms}$ Jitter and -252-dB FOMJ Based on a
Time-Amplifying Phase-Frequency Detector 791
Yuzhong Li, Zunfa Cheng, Jiwei Huang

A Stochastic Computing-Based Computing-in-Memory Macro with Bit-Split Stochastic Number
Generation 796
Zhiting Lin, Dandan Chen, Siyan Li, Shuang Liu, Yu Liu, Xiulong Wu

An Analytical Compact Model for Multi-time Programmable Memory Cells 801
Tiantian Gan, Xiaojin Li, Tengyang Liu, Yabin Sun, Yanling Shi, Jianpeng Chu

A Hybrid Computing-in-Memory Architecture for Energy-Efficient Edge AI Inference 805
Fangchao Lou, Dandan Chen, Jian Cao, Xing Zhang, Bo Zhang

A Reliable 512-kb HZO-based 2T2C FeRAM Array with Capacitor under Bitline 810
Jing Wang, Xiangyin Chen, Zhongguang Xu

Reconfigurable SRAM Computing-In-Memory Macro Based on Local Computing Cell 814
Chenghu Dai, Chaoyi Wang, Zeyi Liu, Qiang Zhao, Chunyu Peng

A 10T1C SRAM-Based Computing-in-Memory Unit Supporting Logic and MAC Computing for Energy- Efficient Edge AI Chips 819
Zhiting Lin, Chenglong Duan, Miao Long, Juntao Ge, Fugui Jiang, Yang Yang, Xiulong Wu

OpenPIM: An Open-Source Programmable Processing-In-Memory Accelerator Design & Pipeline Simulation Framework 824
Ruibao Wang, Wenshuo Yue, Daijing Shi, Yuchao Yang, Bonan Yan

CASA-CIM: A 28nm 131.54 TOPS/W SRAM-Based CIM Macro with Cap-Adder Weighting and Shift-After-Addition Data Streaming for Efficient MAC Operations 830
Runru Yu, Chenyang Zhao, Honghu Yang, Zhiting Lin, Xiulong Wu, Keji Zhou, Jianguo Yang

Adaptive VREF variation Compensation Scheme for High-Speed DRAM interface & Its Offset Calibration Scheme 835
Chris Eom, Kenji Wen, Mia Xu, Taco Zhang, Derek Yang, Bosco Lai

Stochastic Computation Based Quantization Strategy for SRAM Computing-in-Memory Macro 839
Xin Li, Yifan Wu, Lintao Chen, Chenghu Dai, Xiulong Wu, Zhiting Lin

Design of a Multi-Dimensional and Multi-Precision Tensor Computation Unit Based on FPGA 844
Yupeng Fang

A Low-Ripple Charge Pump with Adaptive Load Compensation for Flash-Based CIM 852
Xuyuan Gu, Xiaofeng Gu, Zhiguo Yu

Interface-Engineered TiO_2/SiO_2 Stacks Enable Ultralow-Power Phase-Change Memory with Nanosecond-Speed 858
Ruizhe Zhao, Jun Zhou, Jun Chen, Hao Tong, Xiangshui Miao

Hafnium-Based Ferroelectric Diode with Interlayer Enhancement for In-Memory Logic Application 863
Shuo Han, Chuanzhi Liu, Qimiao Zeng, Yefan Zhang, Wei Wang, Rongrong Cao

Reconfigurable Memory Device based on Defect Engineering of 2D Ferroelectric $CuInP_2S_6$ 868
Yunpeng Xia, Yu Li, Tianqi Li, Qinfei Long, Zihui Hong, Yunhe Hou, Tiande Mo

ANDQ-NAT: Adaptive Non-linearity Dynamic Quantization Non-Ideality Aware Training Framework for Computing-In-Memory Macros 873
Yu Liu, Jianxing Zhou, Hao Li, Yang Lou, Xiulong Wu, Zhiting Lin

A Fast-Transient-Response Hybrid Architecture LDO for NFC Reader SoC 879
Xindong Mi, JinBiao Zhong, YuXuan Huang, JianGuo Hu

Design of an ATD Circuit for Asynchronous SRAM 885
Jialin Liu, Rongkang Ren, Xin Li, Jiancheng Li, Qichao Zha

An Ultra-Wideband Compact Power Divider for Phased Array Radar Systems 890
Wanfu Liu, Jianhui Wu

A Timing Interleaving RX Scheme with direct Multiplexing Sampler based on dual reference for DDR interface 895
Mia Xu, Chris Eom, Kenji Wen, Tinna Ding, Bosco Lai

Graph-Based Representation of Verilog HDL: Python-Based Control and Data Flow Graph Generation 899
Yipeng Wang, Zhiqiang He, Lingwei Yan, Gang Chen

A Node Merging Algorithm Using Fault-Detection for XOR-Majority Network 904
Shijia Fan, Feifei Deng

FMSRdiff: Efficient latent diffusion framework combining consistency model and flow matching for
super-resolution 910
QunKai Peng, GuoFeng Cai, YiHua Xu, Kuan-Chang Chang, Lei Li, GuiBo Luo

An FBAR Driven Fractional-N Ring PLL Using Harmonic-Mixer-Based Dual Feedback and
Split-Feedback Frequency Division 918
Qinglong Tian, Jiwei Huang, Hongqu Lin

A Fault Tolerant Routing Method for 3D Network on Chip without Redundant TSV or
Router-Avoidance 923
Zeyi Lin, Duo Yu, Yanzhou Tang, Naifeng Jing, Jianfei Jiang, Weiguang Sheng, Lifu Cheng, Qin Wang

The Mechanical Properties of Wrinkled MnPS$_3$ Structures 929
Tongxu Huang, Jiaxian Sun, Zihao Wu, Yiting Wei, Wenjun Chen

Supercritical Fluid-Engineered p-GaN Thin Films with Improved Electrical and Optical Properties 933
Yao-Li Chuang, Lei Li, Lei Lu, Kuan-Chang Chang

Design for Hardware Acceleration of Signature Verification in PQC Stateless Hash-Based Digital
Signature Algorithm 937
Haoran Liu, Liji Wu, Lei Li, Yifan Yang, Xiangmin Zhang

A 88 -dB SNDR 156-KHz-BW Noise-Shaping SAR ADC with 3rd-CIFF Structure and Dynamic
Integrator 943
Yulin Feng, Mei Jiang, Xinhui He, Zhengru Li

Mo$_2$C-MoS$_2$ Mixed-Dimensional Bulk Heterostructure-Based Memristor for Artificial Synapses 948
Libin Liang, Ziyao Lu, Zhenhua Guo, Chunyi Qiu, Zhuoling Zhou, Changjiu Teng

Artificial Neural Network-Based Compact Model for Carbon Nanotube Field-Effect Transistors 953
Zhi Zhang, Honggang Liu

Electronic Properties of Violet Phosphorus Devices 958
Chunyi Qiu, Ziyao Lu, Yiting Wei, Jiaxian Sun, Shilong Zhao

A SiC Power Module Package Based on the Dual-side Cooling with Highly Thermally Conductive
Graphite-Molybdenum Spacer 962
Yucheng Xu, Jiafei Yao, Yuxuan Dai, Ziwei Hu, Fan Yang, Kemeng Yang, Binbin Xu, Zhikuang Cai
Yufeng Guo

Work Function Variation Effect Prediction on Heterojunction Tunnel FET Using Multi-Layer
Perceptron-Based Neural Network Model 967
Haotong Han, Yunhe Guan, Tongqing Yan, Weihan Sun, Xiangtai Liu, Haifeng Chen

Enhancing Carrier Mobility in a-IZO/a-IGZO Thin-Film Transistors through Band Structure
Engineering of Heterojunction Channels 972
Huan Yang, Shengdong Zhang, Yong Wang, Chao Teng, Yi Shen, Yudong Liu

Effect of Temperature on ESD Characteristics in 30 nm Partially Depleted SOI MOSFET 977
Liye Yu, Jingrui Wang, Zhongxu Chen, Yixin Zheng, Ziyan Dai, Mingzhi Wan

A Package-on-Package Module with Integrated FPGA Minimum System for 48-Channel Synchronous Sampling 16-bit ADC 984

Han Li, Jiangbo Wei, Yuan Miao, Weize An, Xudong Wu, Huan Yu, Chao Wang

2025 The 10th International Conference on Integrated Circuits and Microsystems

HEA²-MAC: A Hybrid Exponent-Aware Approximate MAC for Efficient CNN Processing

Weixuan Wang[1], Zihan Zou[1], Xin Cao[3], Xuefeng Cai[3], Hao Cai[1,2], Bo Liu[1,2*]

[1]School of Integrated Circuits, Southeast University, Nanjing, China
[2]National Center of Technology Innovation for EDA, Nanjing, China
[3]Chien-Shiung Wu College, Southeast University, Nanjing, China
*Correspondence author: liubo_cnasic@seu.edu.cn

Abstract—**Convolutional neural networks (CNNs) have been widely deployed in industrial edge devices and the Internet-of-Things. Although the efficient inference of CNNs is deeply optimized with dedicated hardware, the training requirement is not fully solved due to the resource-constrained nature of edge devices. The major bottleneck is the high overhead of floating point (FP) multiplication and accumulation (MAC), which requires high-precision computation and data movement. To this end, we propose a Hybrid Exponent-Aware Approximate MAC (HEA²-MAC) acceleration method in a software and hardware co-design manner. The main contributions are: (1) an adaptive block floating point (BFP) data format that reduces the average data bit-width, (2) a precision-adjustable approximate mantissa multiplier that exploits the error tolerance of CNNs to reduce FP mutiplication consumption, and (3) an exponent-aware mixed-precision accumulator that mitigates FP accumulation costs. Implemented in a 28nm CMOS technology, HEA²-MAC achieves up to a 3.66× improvement in energy efficiency compared to the BFP12-FP32 MAC unit. Meanwhile, the accuracy loss (<1%) is significantly maintained across numerous CNN benchmarks.**

Index Terms—**hybrid accumulator, convolutional neural network, data format, approximate computing**

I. INTRODUCTION

Convolutional neural networks (CNNs) have achieved significant success across a wide range of application domains, particularly in computer vision tasks. In image classification, VGGNet, ResNet, and GoogLeNet [1, 2, 3] have demonstrated remarkable accuracy and scalability. In object detection, R-CNNs and YOLO [4, 5] have become widely adopted. Therefore, CNNs are widely deployed in industrial edge devices and Internet-of-Things (IoT). The efficient inference is also deeply optimized by dedicated hardware for their edge applications.

However, training on resource-constrained edge devices remains a critical challenge, particularly for privacy-sensitive applications. The major bottleneck is floating point (FP) multiplication-and-accumulation (MAC), resulting from the high overhead of its high-precision computation and data movement. Although edge devices can support inference tasks with integer quantization, their resource-constrained nature limits their capability for high-precision training tasks [6]. Meanwhile, the increasing size and complexity of CNNs aggravates these challenges, as illustrated in Fig. 1.

This work is supported by the Yangtze River Delta Science and Technology Innovation Community Joint Fundamental Research Project under Grant 2024CSJZN0504.

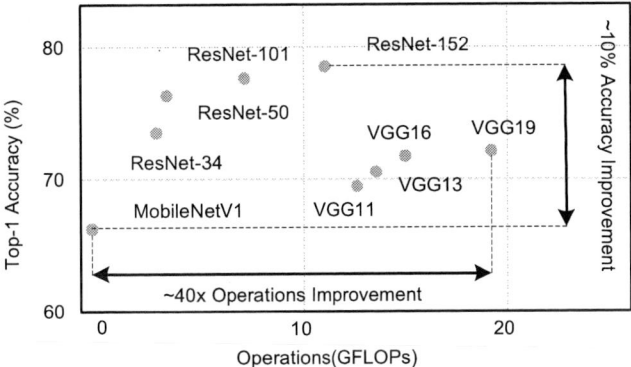

Fig. 1. Evolution of Neural Network Models

Considering this bottleneck of FP MAC, recent research has explored the use of low-precision data formats to replace FP32 in CNN training [7, 8]. However, directly reducing the precision of data formats leads to high accuracy loss. Therefore, mixed-precision training is proposed to achieve a better trade-off by adjusting the FP format of each layer based on its data distribution [9, 10]. Nevertheless, this method brings challenges for hardware design due to the layer-wise variation of the data formats. Meanwhile, both methods rely on traditional FP arithmetic formats, failing to eliminate the overhead introduced by complex computation operations. Given the inherent error tolerance of CNNs, approximate computation has been explored as a promising technique to improve the energy efficiency of neural network processing [11]. However, the balance of power and accuracy is not optimally achieved for training tasks, making the selection of an appropriate approximate multiplier scheme a significant design concern.

Considering the above challenges and opportunities, we propose HEA²-MAC, a software and hardware co-design method, to efficiently deploy CNN training tasks. The key contributions are listed as follows:

- **Adaptive BFP data format** is proposed to reduce the average bit-width of data by sharing adaptive exponents across sets of FP elements.
- **Precision-adjustable approximate mantissa multiplier** is designed to exploit the error tolerance of CNNs, resulting in hardware-friendly FP multiplication.
- **Exponent-aware mixed-precision accumulator** is designed to efficiently perform accumulation operations to

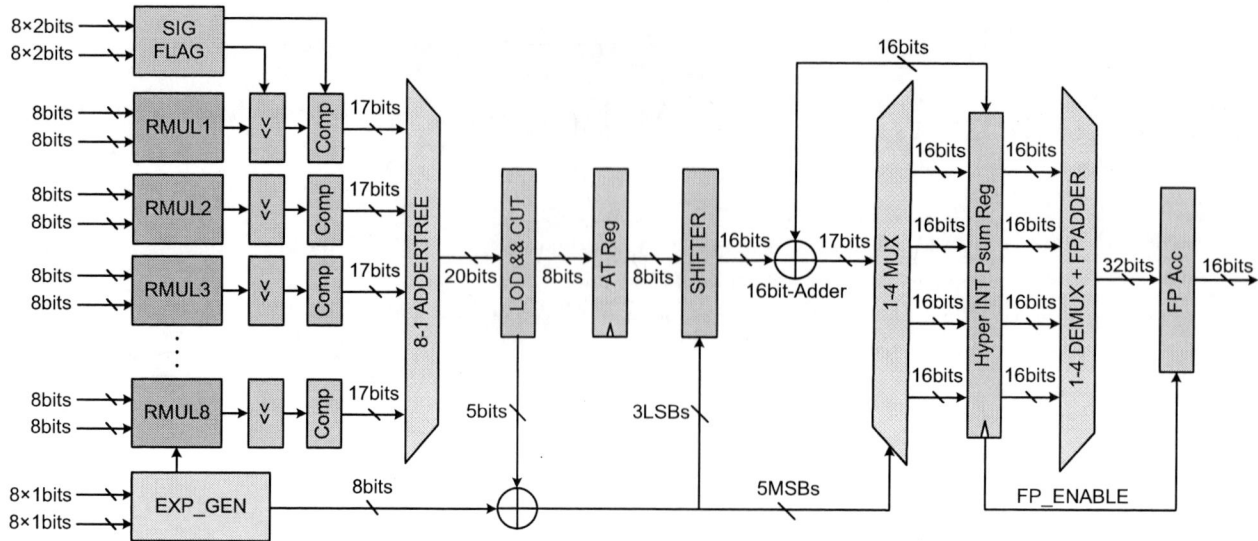

Fig. 2. The overview of the proposed HEA²-MAC framework

further reduce the overheads of FP MAC operations.

- **HEA²-MAC** is evaluated at comprehensive benchmarks and achieves up to a 3.66× energy efficiency improvement compared to baselines.

The rest of this paper is organized as follows. Section II provides an overview of the HEA²-MAC framework and details the key components of the proposed design, including the adaptive BFP data format, the precision-adjustable approximate mantissa multiplier, and the exponent-aware mixed-precision accumulator. Section III presents the evaluation results, and Section IV concludes the paper.

II. HEA²-MAC DESIGN FRAMEWORK

A. Framework Overview

The overview of the proposed HEA²-MAC framework is presented in Fig. 2. The data is divided into three components and dispatched to the MAC unit. The mantissa is sent to the multiplier, while the exponent is processed by the exponent generation unit, which aligns the exponent bits of the output from the adder tree. The flag bit and sign bit are used to control the shifter following each multiplier output and to convert the unsigned result into a signed value before accumulation. Since the effective width of the adder tree output is typically less than 20 bits, a truncation strategy combining leading-one detection (LOD) and bit-width cutting (CUT) is applied after the adder tree, preserving only the most significant 8 bits to eliminate redundant bits. The truncation width, together with the output of the exponent generation unit, determines the final exponent sum. The wide fixed-point accumulator is partitioned into four hierarchical units, with activation determined by the 5 most significant bits of the exponent sum. The truncation output is shifted based on the 3 least significant bits and then accumulated accordingly. Applying shorter adder bit-widths reduces the accumulator's per-cycle switching energy. Moreover, the FP accumulator is enabled only when the fixed-

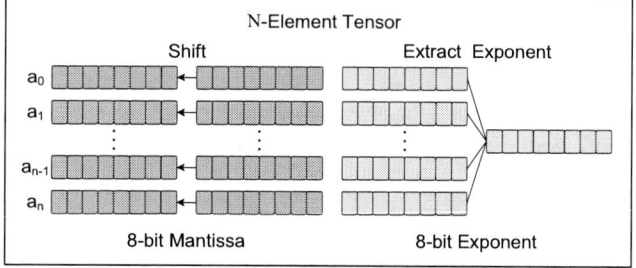

(a) BFP Data Format

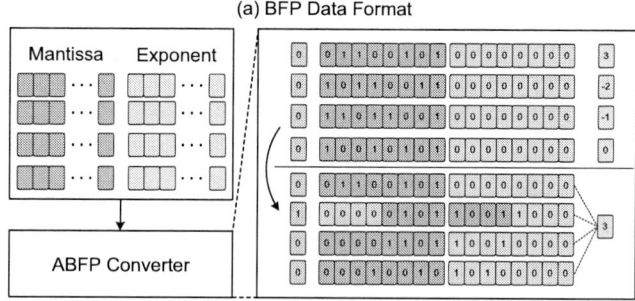

(b) FP2ABFP Format Conversion

Fig. 3. Structure of the BFP format and the FP-to-ABFP conversion pipeline

point units overflow or the accumulated value exceeds their dynamic range, thereby further reducing power consumption.

B. Adaptive BFP Data Format

Given the requirements of training and inference, the data format must provide sufficient dynamic range to support high-precision computation. However, FP arithmetic units introduce significant overhead from exponent alignment and normalization, which impose constraints on the critical path. BFP [12] mitigates this issue by pre-shifting data to share a common exponent, reducing the effective bit-width and enhancing energy efficiency, as shown in Fig. 3 (a). However, this approach may introduce severe mantissa rounding errors due to excessive shift alignment.

979-8-3315-8850-2/25 $31.00 © 2025 IEEE

To address this problem, the Adaptive Block Floating Point (ABFP) data format is proposed, as shown in Fig. 3(b), where darker green highlights the 8-bit mantissa input into the multiplier, lighter green indicates the shifting process, and a flag bit is introduced to reduce excessive initial shifts by four bits. In this work, we adopt the ABFP8 format, which is characterized by an 8-bit mantissa to distinguish it from other ABFP variants. Moreover, an 8-bit exponent is incorporated to provide sufficient dynamic range for data representation. Additionally, a dedicated flag bit is introduced to mitigate errors caused by mantissa rounding, enabling a better trade-off between accuracy and efficiency. The flag bit is generated based on a predefined threshold T, which is determined by the mantissa bit-width M. During forward propagation, T is set to $M/2$, while during backpropagation, it is set to M. If the shift amount exceeds T, the flag bit is set to 1, indicating a reduction in the right-shift operation by T bits. Conversely, when the shift amount is less than or equal to T, the flag bit remains 0. Finally, during the subsequent adder tree stage, the multiplication result is further shifted based on the ABFP flag to ensure correct alignment and maintain computational accuracy. The ABFP data format significantly reduces rounding errors caused by exponent alignment in traditional BFP, enhances the dynamic range of data representation, and maintains high training accuracy. In addition, ABFP8 offers an effective trade-off between precision and hardware overhead, achieving higher accuracy than FP8 while avoiding the additional hardware cost associated with FP16.

C. Precision-Adjustable Approximate Mantissa Multiplier

The intrinsic error tolerance of neural networks makes the deployment of approximate circuits feasible. Moreover, different layers exhibit varying sensitivity to data precision. Motivated by these observations, we propose a precision-adjustable approximate mantissa multiplier with full-precision compatibility. Building on XBMAD [13], a Boolean Matrix Factorization (BMF) method based on XOR operations and Hamming distance evaluation, we introduce a Tiling XOR-based BMF Approximate Decomposition (TXBMAD) framework to support configurable approximation. TXBMAD adopts a top-down strategy to support multi-level approximation with configurable accuracy, enabling efficient hardware implementation. For each subcircuit obtained from tiling, BMF is performed by iteratively selecting a row or column vector and optimizing the corresponding counterpart to minimize the Hamming distance. The outer product of the selected vectors is computed and used to update the residual matrix via XOR operations, with the process repeated until convergence. Applied to an 8×8 ABFP8 multiplier, TXBMAD supports four precision levels and enables efficient generation of Boolean subcircuits. Although BMF has been widely employed as a fundamental technique in approximate computing [14, 15], we propose a novel loss function to further enhance approximation quality and design flexibility.

Exponent sharing in BFP formats leads to a concentration of mantissa values near zero, thereby reducing the effectiveness

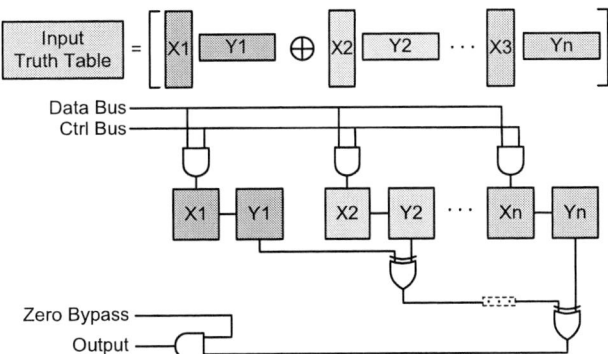

Fig. 4. Precision-adjustable multiplier with full-precision compatibility

of traditional mean relative error distance metrics for multiplier evaluation. To better capture the real data distribution, we propose the Weighted Mean Relative Error Distance (WMRED), which integrates the mantissa probability distribution into loss computation for more accurate evaluation of approximate multipliers. The proposed loss function is formulated as Eq. 1:

$$WMRED = \sum_{\substack{(x,y)\in \mathbb{B}^{2n} \\ x\neq 0, y\neq 0}} P_{(x,y)} \cdot \frac{|O_{\text{app}}(x,y) - O_{\text{acc}}(x,y)|}{|O_{\text{acc}}(x,y)| \times 2^{2n}}, \quad (1)$$

where O_{app} is the output of the approximate circuit, O_{acc} is the accurate output, and $P_{(x,y)}$ denotes the probability of occurrence for the corresponding mantissa input data. WMRED threshold levels are set to 0.25%, 0.5%, and 1%, enabling the design of reconfigurable approximate mantissa multipliers with different error levels.

In addition, it is observed that the ReLU activation function often produces a significant number of zero-valued outputs, making neural networks highly sensitive to errors on zero inputs. To address this issue, a bypass signal is introduced in the precision-adjustable multiplier to handle zero inputs. Fig. 4 illustrates the structure and computational flow of the proposed approximate mantissa multiplier, where X denotes the encoding circuit, Y represents the decoding circuit, the data bus delivers the input data, and the control bus configures the error level by activating the corresponding Boolean subcircuits, which are generated through TXBMAD to support precision configurability. A zero bypass mechanism is employed to skip approximate computation for zero inputs, thereby reducing power overhead and overall efficiency.

For multiplier scheduling, low-precision multipliers can be used during the early stages of training, when neural networks are generally less sensitive to computational accuracy. However, as training progresses toward stability, high-precision computation becomes increasingly critical. Furthermore, prior quantization studies based on Hessian trace analysis indicate that layers closer to the input typically require greater numerical precision [16]. To accommodate these characteristics, we introduce a self-adaptive Approximate Threshold (AT) parameter. Considering the relatively stable data distribution within each training epoch, AT adjustment is performed during the first B batches and remains fixed for the remainder of the

979-8-3315-8850-2/25 $31.00 © 2025 IEEE

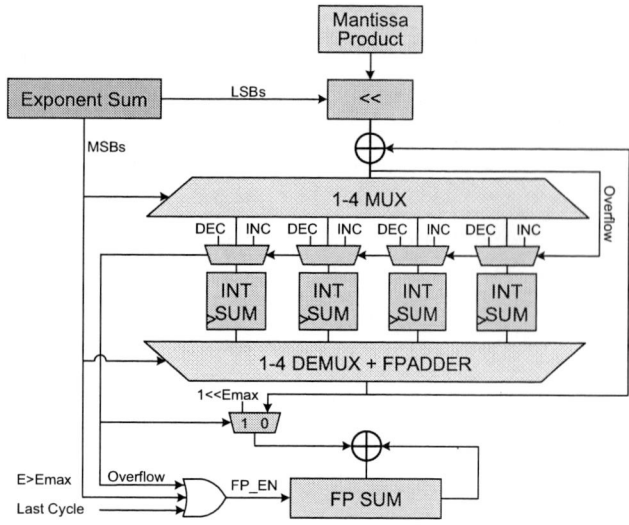

Fig. 5. Schematic of the exponent-aware mixed-precision accumulator

TABLE I
PRECISION ADJUSTABLE MULTIPLIER CONFIGURATION AND PERFORMANCE RESULTS

Benchmark	Multiplier Configuration			ACC	EEI
CIFAR10 @AlexNet	L1-8(A0)	-	-	91.09%	-
	L1(A0)	L2-4, 8(A1)	L5-7(A2)	90.64%	17.20%
	L1(A0)	L2-4, 8(A2)	L5-7(A3)	90.16%	**35.54%**
CIFAR10 @ResNet18	L1-21(A0)	-	-	91.09%	-
	L1(A0)	L2-7, 21(A1)	L8-20(A2)	90.88%	20.66%
	L1(A0)	L2-7, 21(A2)	L8-20(A3)	90.23%	**43.52%**
CIFAR100 @ResNet50	L1-54(A0)	-	-	73.46%	-
	L1-5(A0)	L6-45, 54(A1)	L46-53(A2)	73.22%	16.78%
	L1(A0),L2-5(A1)	L6-45, 54(A2)	L46-53(A3)	72.51%	**34.76%**

epoch. If the shared exponent of the output exceeds the AT, the counter is incremented. At the end of each batch, the ratio between the counter and the total number of output samples is calculated. If this ratio exceeds r_{acc}, which represents the desired proportion of accuracy multiplier, it indicates that the current AT is too low, and AT is incremented; otherwise, AT is decremented. Additionally, if the current layer belongs to the first M error-sensitive layers, the error level is decreased by one. Once the training enters the convergence phase or the accuracy improvement remains below a threshold λ for N consecutive epochs, all multipliers are switched to the accuracy level. Here, r_{acc}, M, N, and λ are hyperparameters.

D. Exponent-Aware Mixed-Precision Accumulator

The benefits of ABFP are derived from enabling block-wise data computation. However, due to the wide dynamic range of exponents across blocks, existing methods still rely on FP32 accumulators, resulting in substantial power overhead. Although BF16 has the same exponent bit-width as FP32, its reduced mantissa width leads to a significant precision loss. Quire [17] mitigates this problem by using ultra-wide fixed-point format to support high dynamic range accumulation. However, the long bit-width results in area and power overheads comparable to those of FP accumulators, offering limited practical benefits.

To minimize accumulation power consumption while maintaining precision, we propose an exponent-aware mixed-precision accumulator that leverages shorter fixed-point accumulators to handle partial accumulation, thereby reducing the reliance on FP accumulation and improving energy efficiency. The conversion to FP accumulation is enabled only under three conditions: 1) an overflow occurs in the fixed-point accumulators; 2) the shared exponent of data exceeds the representable range; or 3) the final accumulation cycle. Instead of relying on a single long fixed-point accumulator, the design distributes accumulation across multiple short-width

accumulators, selectively activated based on exponent values to optimize energy efficiency. Based on the observed exponent distribution ranging from -39 to -6 in the ResNet-18 [2], we divide the input exponent range into four intervals, each with a length of $R = 8$. Let M denote the output bit-width of the mantissa product from the adder tree. To support accumulation, four fixed-point partial sum registers are used, each with a bit-width of $R + M = 16$ bits.

Fig. 5 shows the schematic of the exponent-aware mixed-precision accumulator. The appropriate register is activated based on the 5 MSBs of the exponent sum. The mantissa product is shifted according to the 3 LSBs and accumulated into the corresponding register. The accumulator reduces per-cycle switching activity and energy consumption by employing a compact adder tree and register structure. To further minimize the activity rate of the FP accumulator, a carry chain is implemented across the 4 register blocks. When an overflow occurs in the current level register, the overflow bit is propagated upward. If the highest-level fixed-point register still overflows, the FP accumulator is enabled. The exponent-aware mixed-precision accumulator significantly reduces power overhead while maintaining compatibility with general-purpose accumulation tasks.

III. EVALUATION

A. Experimental Setup

Software Setup. We evaluate HEA²-MAC on CIFAR-10 and CIFAR-100 [18] with AlexNet [19], ResNet-18 and ResNet-50 [2]. We use PyTorch [20] to simulate the training accuracy of CNNs and adjust the precision scheduling hyperparameters. Based on the simulation results, we formulate the approximate multiplier scheduling strategy to achieve an optimal trade-off between precision and hardware efficiency for the target tasks. **Hardware Implementation.** HEA²-MAC is implemented in Verilog RTL and synthesized using Synopsys toolchains [21, 22, 23, 24] with TSMC 28nm technology. The design operates at 200 MHz with a supply voltage of 0.8V, and its area, power, and performance are evaluated.

B. Experimental Result

Based on the Hessian trace as a sensitivity metric, the error level of multiplier is configured layer-wise during neural

979-8-3315-8850-2/25 $31.00 © 2025 IEEE

TABLE II
COMPARISON OF THE PROPOSED AND EXISTING 8×8 MULTIPLIERS

	Accurate	AM2-7[25]	mul8u_384[26]	A1	A2	A3
WMRED(%)	-	0.41	0.87	0.25	0.50	1.00
NMED(%)	-	0.39	0.81	0.46	0.88	1.42
Accuracy(%)	73.4	72.7 (-1.0%)	70.8 (-3.5%)	**73.0** (**-0.5%**)	72.5 (-1.2%)	70.6 (-3.8%)
Area(μm^2)	110.4	89.9 (-18.6%)	82.8 (-25.0%)	88.5 (-19.9%)	81.1 (-26.5%)	**55.7** (**-49.5%**)
Delay(ns)	0.97	0.82 (-15.4%)	0.76 (-21.7%)	0.83 (-14.2%)	0.78 (-20.1%)	**0.70** (**-28.0%**)
Power(μW)	14.05	12.4 (-11.8%)	10.6 (-24.6%)	12.2 (-12.9%)	10.5 (-25.4%)	**6.4** (**-54.6%**)
PDP(fJ)	57.96	43.3 (-25.3%)	34.1 (-40.4%)	43.1 (-25.6%)	34.3 (-40.0%)	**18.9** (**-67.4%**)

TABLE III
POWER BREAKDOWN OF THE MAC UNITS

Acc	Mul	Add Tree	Acc + Ctrl	Total Power
FP32	112.40 (44.9%)	30.54 (12.2%)	103.03 (41.2%)	**250.3**μW
FP16	98.89 (58.2%)	23.62 (14.1%)	46.71 (27.4%)	170.5μW
INT92	96.30 (43.8%)	28.10 (12.8%)	95.29 (43.4%)	219.7μW
Hyper INT52	91.54 (53.1%)	20.69 (12.0%)	60.16 (34.9%)	172.4μW
Hyper 4INT16	**91.67 (56.5%)**	**21.73 (13.4%)**	**48.80 (30.13%)**	**162.2μW**

format, HEA^2-MAC achieves a 3.66× improvement in overall energy efficiency under typical scenarios, with approximately 27% additional area overhead, while maintaining negligible accuracy loss.

IV. CONCLUSION

This paper proposes HEA^2-MAC, an efficient MAC unit with software and hardware co-design for CNN processing. The key insights of HEA^2-MAC are: (1) At the software aspect, dynamic precision adjustment is performed based on real-time training feedback, allowing the software to guide the configuration of approximate mantissa multipliers. (2)At the hardware aspect, a hybrid exponent-aware approximate MAC unit based on the ABFP format is designed to mitigate the high overhead associated with FP computation. Combining these two parts, the proposed HEA^2-MAC achieves up to 3.66× energy efficiency compared to baseline MAC unit in benchmark evaluations.

REFERENCES

[1] Karen Simonyan and Andrew Zisserman. "Very deep convolutional networks for large-scale image recognition". In: *arXiv preprint arXiv:1409.1556* (2014).

[2] Kaiming He et al. "Deep residual learning for image recognition". In: *Proceedings of the IEEE conference on computer vision and pattern recognition*. 2016, pp. 770–778.

[3] Christian Szegedy et al. "Going deeper with convolutions". In: *Proceedings of the IEEE conference on computer vision and pattern recognition*. 2015, pp. 1–9.

[4] Ross Girshick. "Fast r-cnn". In: *Proceedings of the IEEE international conference on computer vision*. 2015, pp. 1440–1448.

[5] Joseph Redmon et al. "You only look once: Unified, real-time object detection". In: *Proceedings of the IEEE conference on computer vision and pattern recognition*. 2016, pp. 779–788.

[6] Sungpill Choi et al. "CNNP-v2: An energy efficient memory-centric convolutional neural network processor architecture". In: *2019 IEEE International Conference on Artificial Intelligence Circuits and Systems (AICAS)*. IEEE. 2019, pp. 38–41.

[7] Neil Burgess et al. "Bfloat16 processing for neural networks". In: *2019 IEEE 26th Symposium on Computer Arithmetic (ARITH)*. IEEE. 2019, pp. 88–91.

[8] Ankur Agrawal et al. "DLFloat: A 16-b floating point format designed for deep learning training and inference". In: *2019 IEEE 26th Symposium on Computer Arithmetic (ARITH)*. IEEE. 2019, pp. 92–95.

[9] Jeongwoo Park, Sunwoo Lee, and Dongsuk Jeon. "9.3 A 40nm 4.81 TFLOPS/W 8b floating-point training processor for non-sparse neural networks using shared exponent bias and 24-way fused multiply-add tree". In: *2021 IEEE International Solid-State Circuits Conference (ISSCC)*. Vol. 64. IEEE. 2021, pp. 1–3.

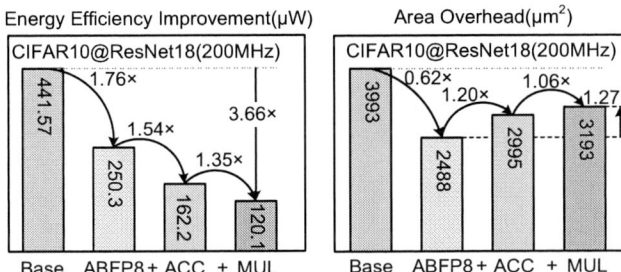

Fig. 6. Energy Efficiency and Area Overhead of ABFP8-Based MAC Unit

network inference. Table I summarizes the proposed configuration scheme and presents the corresponding results, in which ACC denotes accuracy and EEI indicates energy efficiency improvement. A0 represents the accuracy mode, while A1–A3 denote three WMRED-based error levels at 0.25%, 0.5%, and 1%, respectively. Across the three evaluated networks, the proposed multiplier design improves energy efficiency by up to 43.52%, with less than 1% accuracy loss.

The evaluations of the proposed approximate multipliers and MAC units are conducted on the CIFAR-100 dataset using the ResNet-50 network. Table II summarizes the comparison between these unsigned 8×8 approximate multipliers and prior designs. The WMRED metric, which accounts for the distribution of input data, offers a more reliable assessment of training accuracy than Normalized Mean Error Distance (NMED). Specifically, A1 achieves higher accuracy than the hand-crafted AM2-7 [25] under similar Power-Delay Product (PDP), while A2 achieves 2.3% higher training accuracy than mul8u_384 [26]. A3 achieves a 27.0% PDP improvement over mul8u_384 at comparable training accuracy, demonstrating the effectiveness of the approximate methodology in balancing computational accuracy and hardware overhead.

Table III presents the detailed breakdown of various MAC units' hardware characteristics, including the power contribution of each component. Compared to the FP32-based design, HEA^2-MAC achieves 1.54× higher energy efficiency. The proposed exponent-aware mixed-precision accumulator reduces power consumption by 52.6%, demonstrating its efficiency in hardware-constrained scenarios.

Fig. 6 presents the energy efficiency optimization process, using a MAC unit with a BFP12 [27] multiplier and a FP32 accumulator as the baseline. Based on the ABFP8 data

[10] Jinsu Lee et al. "7.7 LNPU: A 25.3 TFLOPS/W sparse deep-neural-network learning processor with fine-grained mixed precision of FP8-FP16". In: *2019 IEEE International Solid-State Circuits Conference-(ISSCC)*. IEEE. 2019, pp. 142–144.

[11] Yang Wang et al. "A 28nm 27.5 TOPS/W approximate-computing-based transformer processor with asymptotic sparsity speculating and out-of-order computing". In: *2022 IEEE international solid-state circuits conference (ISSCC)*. Vol. 65. IEEE. 2022, pp. 1–3.

[12] Mario Drumond et al. "Training dnns with hybrid block floating point". In: *Advances in Neural Information Processing Systems* 31 (2018).

[13] Jörg Wicker et al. "XOR-Based Boolean Matrix Decomposition". In: *2019 IEEE International Conference on Data Mining (ICDM)*. IEEE. 2019, pp. 638–647.

[14] Soheil Hashemi, Hokchhay Tann, and Sherief Reda. "BLASYS: Approximate logic synthesis using Boolean matrix factorization". In: *Proceedings of the 55th Annual Design Automation Conference*. 2018, pp. 1–6.

[15] Changlin Wan et al. "Fast and efficient boolean matrix factorization by geometric segmentation". In: *Proceedings of the AAAI Conference on Artificial Intelligence*. Vol. 34. 04. 2020, pp. 6086–6093.

[16] Zhen Dong et al. "Hawq-v2: Hessian aware trace-weighted quantization of neural networks". In: *Advances in neural information processing systems* 33 (2020), pp. 18518–18529.

[17] Posit Working Group et al. "Posit Standard Documentation Release 4.12-draft". In: *Standard Posit Arithmetic* (2021).

[18] Alex Krizhevsky, Geoffrey Hinton, et al. "Learning multiple layers of features from tiny images". In: (2009).

[19] Alex Krizhevsky, Ilya Sutskever, and Geoffrey E Hinton. "Imagenet classification with deep convolutional neural networks". In: *Advances in neural information processing systems* 25 (2012).

[20] A Paszke. "Pytorch: An imperative style, high-performance deep learning library". In: *arXiv preprint arXiv:1912.01703* (2019).

[21] *VCS User Guide*. Available: https://www.synopsys.com/verification/simulation/vcs.html. Synopsys, Inc. Mountain View, CA, USA, 2023.

[22] *Verdi User Guide*. Available: https://www.synopsys.com/verification/debug/verdi.html. Synopsys, Inc. Mountain View, CA, USA, 2023.

[23] Pran Kurup and Taher Abbasi. *Logic Synthesis Using Synopsys®*. Springer Science & Business Media, 2012.

[24] *PrimeTime PX User Guide*. Available: https://www.synopsys.com/implementation-and-signoff/signoff/primetime.html. Synopsys, Inc. Mountain View, CA, USA, 2023.

[25] Honglan Jiang et al. "Low-power approximate unsigned multipliers with configurable error recovery". In: *IEEE Transactions on Circuits and Systems I: Regular Papers* 66.1 (2018), pp. 189–202.

[26] Vojtech Mrazek et al. "Evoapprox8b: Library of approximate adders and multipliers for circuit design and benchmarking of approximation methods". In: *Design, Automation & Test in Europe Conference & Exhibition (DATE), 2017*. IEEE. 2017, pp. 258–261.

[27] Mario Drumond et al. "Training dnns with hybrid block floating point". In: *Advances in Neural Information Processing Systems* 31 (2018).

2025 The 10th International Conference on Integrated Circuits and Microsystems

A Low-Power Continuous-Time Quadrature Bandpass ΣΔ ADC with a novel ELD Compensation Method in Quantizer

Haowen Ba
School of Integrated Circuits
Zhejiang University
Hangzhou, China
3190102931@zju.edu.cn

Jun Ye
School of Integrated Circuits
Zhejiang University
Hangzhou, China
22341067@zju.edu.cn

Hanli Liu
School of Integrated Circuits
Zhejiang University
Hangzhou, China
hanli.liu@zju.edu.cn

Feijun Zheng*
School of Integrated Circuits
Zhejiang University
Hangzhou, China
zhengfj@zju.edu.cn

Abstract—This paper presents a low-power continuous-time quadrature bandpass delta-sigma (ΔΣ) analog-to-digital converter (ADC) featuring a novel excess loop delay (ELD) compensation method. While continuous-time ΔΣ ADCs offer advantages over discrete-time ΔΣ ADCs in terms of anti-aliasing capability, they suffer from excess loop delay issues that can cause system unstable and performance degradation. The proposed ADC employs a Flash ADC based quantizer architecture with threshold-adaptive comparators that effectively compensate for ELD effects by providing compensation to the input signal, which reduces the peaking of the noise transfer function (NTF) and significantly enhances system stability. Designed in 55-nm CMOS technology, post-layout simulation results demonstrate that the ΔΣ ADC achieves a peak signal-to-noise and distortion ratio (SNDR) of 70.9 dB over a 4 MHz bandwidth while consuming only 1.05 mW from a 1.0V supply, yielding a competitive figure-of-merit (FOM) of 45.9 fJ/conversion step, validating the effectiveness and competitiveness of the proposed approach.

Keywords—Delta-sigma ADC, continuous-time (CT), excess loop delay compensation, comparator.

I. INTRODUCTION

Continuous-time ΣΔ ADCs have gained significant traction in modern wireless receiver designs, particularly for Internet of Things (IoT) and Bluetooth Low Energy (BLE) applications [1-2]. Compared to discrete-time (DT) implementations, CT ΣΔ modulators provide inherent anti-aliasing capabilities and demonstrate superior power efficiency when operating at high sampling frequencies. However, these architectures encounter a critical limitation stemming from excess loop delay (ELD), which arises from the finite switching intervals of both the quantizer and digital-to-analog converter (DAC). In many modern receiver systems, the bandwidth of CT ΣΔ ADCs varies according to different power consumption modes. To maintain a constant oversampling ratio (OSR) or achieve an enhanced dynamic range (DR), increasing the ADC sampling frequency is a commonly adopted approach. Under these high frequency operating conditions, the degradation caused by ELD becomes increasingly severe [3].

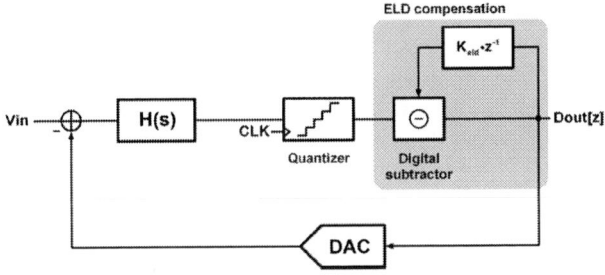

Fig. 1. Conventional ELD compensation in digital domain.

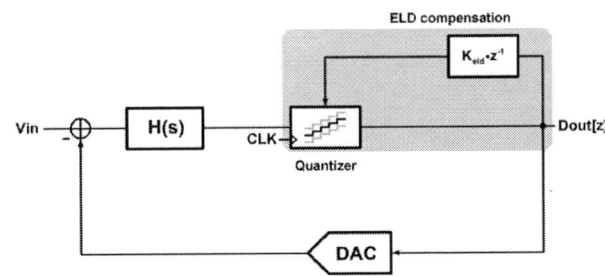

Fig. 2. Proposed ELD compensation method.

Excess Loop Delay (ELD) is a widely recognized non-ideal effect in continuous-time sigma-delta (CT ΣΔ) modulators. It arises due to the time delay introduced by the quantization process and the digital-to-analog conversion within the feedback loop. To address this non-ideality, various techniques have been proposed in the literature. The first ELD compensation strategy was introduced in [4], where a digital subtractor was employed, as illustrated in Fig. 1. This method compensates for ELD by subtracting a scaled version of the previous quantization result at the end of each quantization cycle. However, this approach necessitates a wide bit-width adder to accurately implement the fine feedback coefficients, resulting in increased power consumption and larger circuit area. An alternative approach was proposed in [5], which introduces a predictive comparator to counteract the effects of ELD. Additionally, a Least Mean Squares (LMS) algorithm is utilized to dynamically adjust the compensation coefficients. However, this method introduces significant circuit complexity, particularly when applied to multi-bit quantizer architectures.

This work is sponsored by project (No. 2025C25081(SYS)).

979-8-3315-8850-2/25 $31.00 © 2025 IEEE

In this paper, a quadrature bandpass CT ΣΔ ADC employing a Flash ADC as the quantizer is presented. Fig. 2 illustrates the proposed ELD compensation approach. The Flash ADC incorporates threshold-adaptive comparators capable of mitigating ELD effects. These comparators are built upon the strong-arm comparator architecture with multiple threshold control capability. The thresholds can be adaptively modified according to the previous quantizer output, thereby altering the quantization results. This adaptive threshold adjustment can be equivalently viewed as applying compensation to the input signal, so that the comparator output effectively incorporates the compensation for the quantization results.

The remainder of this paper is structured as follows. Section II presents the proposed ADC architecture. Section III details the circuit implementations of the key blocks. Section IV presents the simulation results and Section V concludes the paper.

II. PROPOSED ADC ARCHITECTURE

A. Modulator Architecture

The overall architecture of the proposed ΣΔ ADC is illustrated in Fig.3. It consists of two quadrature paths, each adopting a cascade integrator feedback (CIFB) structure. Each path incorporates two active-RC integrators, a Flash quantizer, and feedback DACs. The differential signaling in each path provides effective common-mode interference rejection. The integrated signal is oversampled by the Flash ADC, and the quantized output is fed back to the integrator input where it is subtracted from the input signal, thereby achieving noise shaping. Cross-coupling between two paths is implemented using passive resistors at two stage integrators, which shifts the poles in the noise transfer function as will be detailed in the following section.

Fig.4(a) illustrates the cross-coupling between the first-stage active-RC integrators in the two paths, with the small-signal block diagram of the I-path shown in Fig. 4(b). Here, R_1 and C_1 represent the input resistance and feedback capacitance of the active-RC integrator, respectively. Rx_1 denotes the cross-coupling resistance between the first-stage integrators, and Vin_I and Vo1_I are the input and output voltages. The transfer function Vo1_I(s)/Vin_I(s) is given by:

$$H(s) = \frac{Vo1\_I(s)}{Vin\_I(s)} = -\frac{1}{sC_1}\left(\frac{1}{R_1} - \frac{1}{Rx_1} \cdot \frac{Vo1\_Q}{Vin\_I}\right) \quad (1)$$

Since the Q-path lags the I-path by 90°, we have:

$$Vo1\_Q = j * Vo1\_I \quad (2)$$

Substituting this into Eq. (1) yields:

$$H(s) = -\frac{1}{(s - j/Rx_1C_1)R_1C_1} \quad (3)$$

From Eq. (3), it can be observed that after cross-coupling, the integrator pole is shifted from the origin to center frequency, as depicted in Fig. 3(c), where the center frequency ω_c of the bandpass transfer function is:

$$\omega_c = -\frac{1}{Rx_1C_1} \quad (4)$$

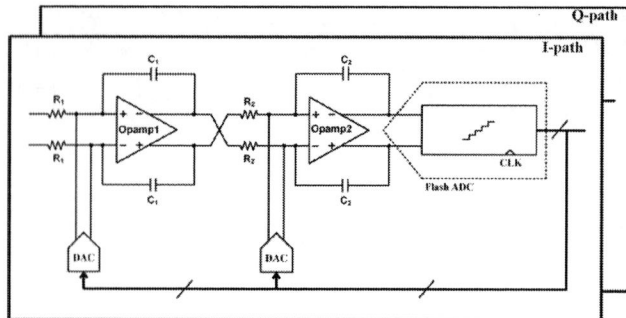

Fig. 3. Proposed CT ΣΔ ADC architecture.

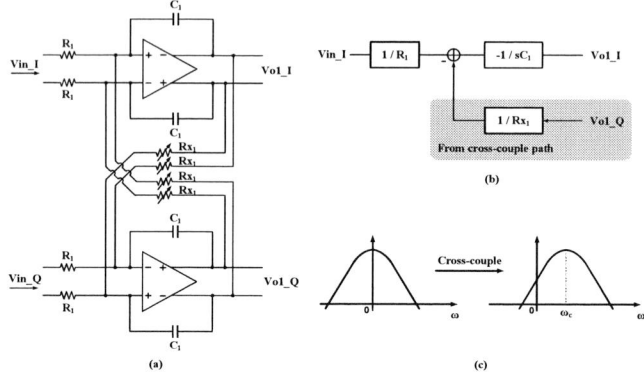

Fig. 4.(a) Schematic of the complex filter (b) Block digram of the 1-st integrator in I path (c) Pole shift due to cross-coupling.

B. ELD Compensated

The proposed ELD compensation is achieved by modifying the switching threshold of the comparator, which can be equivalently represented as adding an offset to the input signal prior to quantization, as illustrated in the system block diagram of Fig. 5. The impact of ELD on CT-DSM can be equivalently modeled as a DAC with inherent delay [6], which tends to destabilize the system. Fig. 6(a) presents simulation results of the NTF root locus under different ELD conditions. As ELD increases, the poles progressively migrate toward the exterior of the unit circle. Fig. 6(b) shows the re-simulation results with ELD compensation applied and kc = 0.5. It can be observed that the compensated poles are pulled back within the unit circle, thereby improving system stability.

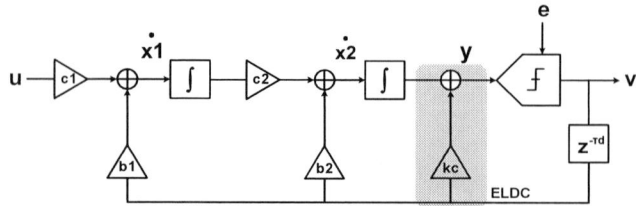

Fig. 5. Block diagram of the proposed ADC with ELD compensation.

979-8-3315-8850-2/25 $31.00 © 2025 IEEE   503

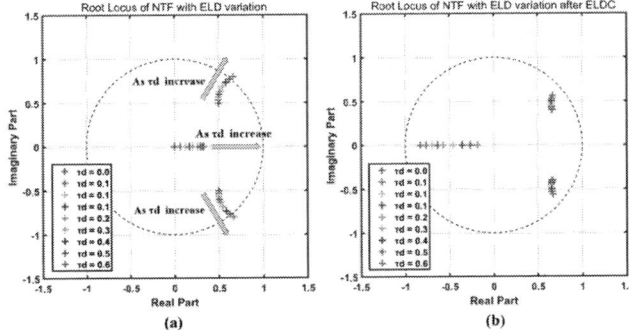

Fig. 6. (a) Root Locus of the NTF with varying ELD, (b) Root Locus of the NTF with varying ELD after compensation.

III. CIRCUIT IMPLEMENTATION AND ANALYSIS

This section presents the detailed circuit implementation and analysis of the key building blocks, including the power-efficient operational amplifier employed in the active-RC integrators, the threshold-adaptive comparators that compensate for ELD within the Flash quantizer.

A. Power-Efficient Opamp

The operational amplifier employed in the integrators is illustrated in Fig. 7, where transistors Mp1~2 and Mn1~2 form the input differential stage, while Mp3~4 and Mn3~4 constitute the output stage. Conventional two-stage operational amplifiers employ Miller compensation to enhance amplifier stability by separating the dominant and non-dominant poles and deliberately reducing bandwidth for stability assurance. In contrast, the proposed design utilizes high frequency active feedforward combined with antipole-splitting techniques to achieve enhanced bandwidth performance while maintaining low power consumption. In this configuration, C_M serves as the Miller compensation capacitor, while C_F functions as an effective negative Miller capacitor. The C_F capacitor introduces negative capacitance between the second stage input and output nodes, thereby counteracting the Miller compensation effects and pushing the two poles closer to each other. This antipole-splitting mechanism significantly extends the operational amplifier bandwidth without incurring additional power overhead [7].

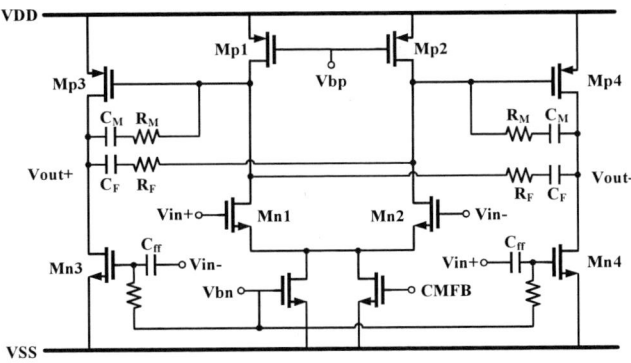

Fig. 7. Power-Efficient opamp in active-RC integrator.

At low frequencies, the gate terminals of transistors Mn3~4 are biased through the DC voltage Vbn, and the pull down slew rate is limited by the static bias current. As the frequency increases, the input signals Vin+ and Vin- can be coupled to Mn3~4 through the coupling capacitor C_{ff}, thereby enhancing the gain and extending the bandwidth. Concurrently, the slew rate can dynamically vary according to the input signal variations, which improves the driving capability. The common mode feedback is realized by detecting the common mode voltage at the second-stage output and adjusting the static bias current of the first stage.

B. Threshold Adaptively Comparators

The proposed threshold-adaptive comparator is illustrated in Fig. 8. It incorporates a dynamic threshold adjustment circuit based on the conventional strong-arm comparator architecture [8]. Fig. 9 demonstrates two consecutive operating cycles as a typical example to illustrate the working principle and explain how it compensates for ELD.

During the reset phase when the clock signal CLK is low (CLK=0), the comparator precharges nodes X, Y, P, and Q to VDD. When CLK transitions high, the circuit enters the regenerative or compare phase. Transistor M0 turns on while M7, M8, M9, and M10 turn off. Initially neglecting the effects of Mnp and Mnn, transistors M1 and M2 distribute the current I1 and I2 from M0 according to the magnitude of input voltages VinP and VinN. These currents discharge nodes P and Q at different rates. When VinP exceeds VinN, node P voltage Vp decreases more rapidly to VDD-Vth₃,₄, turning on transistor M3. Consequently, node X is pulled low, node Y is driven high, and the comparator output VoutP goes high.

In the subsequent cycle, when CLK returns low, M0 turns off and nodes X, Y, P, and Q are recharged to VDD. Simultaneously, the DP and DN signals are refreshed based on the quantization results. The number of active Mnp transistors decreases while that of Mnn transistors increases, thereby increasing the positive switching threshold by Δth. During the second comparison phase, when the input voltage difference falls below this new switching threshold, the comparator output changes state accordingly. To ensure accurate threshold adjustment, transistors Mnp and Mnn are realized as matched transistor arrays, providing parasitic capacitance matching between nodes P and Q. This matching strategy ensures that the voltage discharge dynamics during regeneration depend only on the current distribution. Fast control signal propagation is achieved by connecting the DN and DP signals from the quantizer to the comparator through transmission gates.

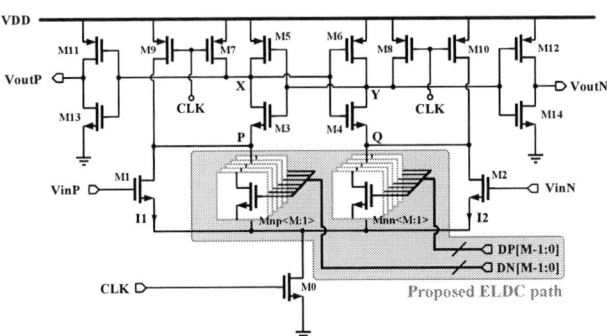

Fig. 8. The proposed threshold-adaptive comparator.

979-8-3315-8850-2/25 $31.00 © 2025 IEEE

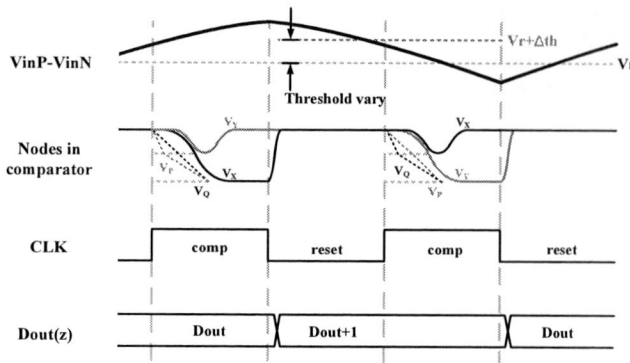

Fig. 9. Timming diagram and waveform of the proposed comparators.

IV. SIMULATION RESULTS

Fig. 10 presents the time-domain output waveforms of the ADC under a dual-tone input at 3.75 MHz and 4.25 MHz, with a sampling frequency of 256 MHz. The simulated results confirm that both I and Q paths accurately quantize the input signals while maintaining the required 90-degree phase relationship, which demonstrates proper quadrature operation of the ADC architecture.

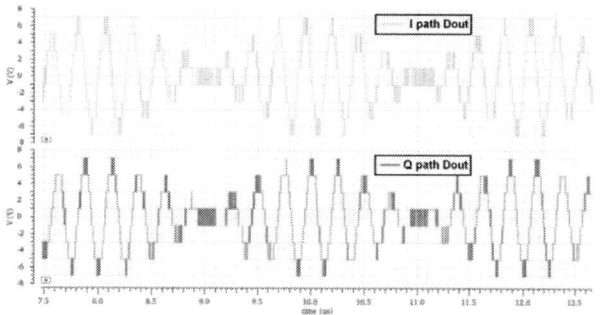

Fig. 10. Time-domain waveforms of the quadrature ADC.

The proposed ΣΔ ADC was designed in a 55-nm CMOS technology and simulated in Spectre. The Fig. 11(a) presents the DR from post-layout simulation results. The simulated SNDR achieves 70.9 dB and the DR is 75.3 dB over a 4 MHz bandwidth. Fig. 11(b) illustrates the 4096-point FFT analysis of the ADC output spectrum, comparing performance with and without the proposed ELDC technique. The spectrum clearly demonstrates that the NTF peak is effectively suppressed when ELDC is enabled, which validates the efficacy of the proposed compensation method.

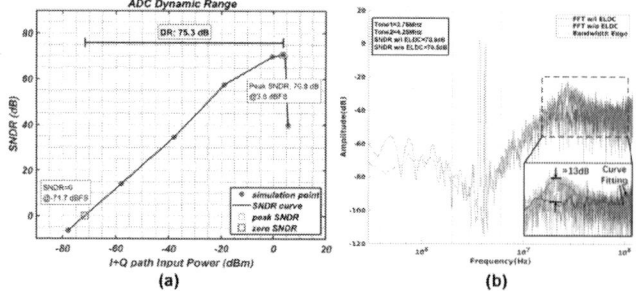

Fig. 11. (a)Simulated dynamic range of the ADC (b)FFT spectrum of the ADC with and without ELD compensation.

The total power consumption of the ADC is 1.05 mW. Finally, Table I compares our work with other published CT ΔΣ ADCs in the multi-MHz bandwidth range.

TABLE I. PERFORMANCE COMPARISON

	This work[b]	[9] TCAS	[10] CICC	[11] VLSI	[12] TMTT
Tech (nm)	**55nm**	40nm	28nm	65nm	28nm
Quantizer Architecture	**Flash ADC**	VCO Based	VCO Based	Flash ADC	SAR ADC
ELDC	**Yes**	No	No	No	No
BW (MHz)	**4**	4	50	1.25	5	2
Fs (MHz)	**256**	256	1600	250	200	32
Power(mW)	**1.05**	1.04	18.1	0.379	4.2	1.2
SNDR (dB)	**70.9**	70.5	76.4	78.7	62.9	65.2
FOM[a] (fJ/conv.)	**45.9**	47.8	33.5	22	360	201.7

[a.] FOM (fJ/conv.) = Power/(2*BW)/2^{ENOB}

[b.] Post layout simulation results

V. CONCLUSION

This paper presents a low-power continuous-time quadrature bandpass ΣΔ ADC featuring a novel excess loop delay (ELD) compensation method implemented through threshold-adaptive comparators in the Flash quantizer. The proposed architecture effectively addresses the stability issues caused by ELD in CT ΣΔ modulators by dynamically adjusting comparator thresholds based on previous quantization results, which is mathematically equivalent to digital domain compensation but offers reduced complexity and power consumption. Designed in 55-nm CMOS technology, the ADC achieves 70.9 dB peak SNDR over 4 MHz bandwidth while consuming only 1.05 mW from 1.0 V supply, resulting in a competitive FOM of 45.9 fJ/conversion-step. Post-layout simulation results validate the effectiveness of the proposed ELD compensation technique, demonstrating improved NTF peaking suppression and enhanced system stability. This makes it suitable for low-power IoT and BLE applications.

REFERENCES

[1] C.-Y. Wang, J.-H. Tsai, S.-Y. Su, J.-C. Tsai, J.-R. Chen, and C.-H. Lou, "An 80 MHz-BW 31.9fJ/conv-step filtering ADC with a built-in DAC-segmentation/ELD-compensation 6b 960 MS/s SAR-quantizer in 28 nm LP for 802.11ax applications," in *IEEE Int. Solid-State Circuits Conf. (ISSCC) Dig. Tech. Papers*, Feb. 2019, pp. 338–340.

[2] S. Manivannan and S. Pavan, "A 1 MHz bandwidth, filtering continuoustime delta–sigma ADC with 36 dBFS out-of-band IIP3 and 76 dB SNDR," in *Proc. IEEE Custom Integr. Circuits Conf. (CICC)*, Apr. 2018, pp. 1–4.

[3] C. -Y. Ho, W. -S. Chan, Y. -Y. Lin and T. -H. Lin, "A Quadrature Bandpass Continuous-Time Delta-Sigma Modulator for a Tri-Mode GSM-EDGE/UMTS/DVB-T Receiver," in *IEEE Journal of Solid-State Circuits*, vol. 46, no. 11, pp. 2571-2582, Nov. 2011.

[4] P. Benabes, M. Keramat and R. Kielbasa, "A methodology for designing continuous-time sigma-delta modulators," in *Proceedings European Design and Test Conference. ED & TC 97*, 1997.

[5] K. El-Sankary, H. H. Alamdari and E. I. El-Masry, "An Adaptive ELD Compensation Technique Using a Predictive Comparator," in *IEEE Transactions on Circuits and Systems II: Express Briefs*, vol. 56, no. 8, pp. 619-623, Aug. 2009.

[6] S. -J. Huang, N. Egan, D. Kesharwani, F. Opteynde and M. Ashburn, "28.3 A 125MHz-BW 71.9dB-SNDR VCO-based CT ΔΣ ADC with segmented phase-domain ELD compensation in 16nm CMOS," in *IEEE International Solid-State Circuits Conference (ISSCC)*, pp. 470-471,2017.

979-8-3315-8850-2/25 $31.00 © 2025 IEEE

[7] M. Abdulaziz, M. Törmänen and H. Sjöland, "A Compensation Technique for Two-Stage Differential OTAs," in *IEEE Transactions on Circuits and Systems II: Express Briefs*, vol. 61, no. 8, pp. 594-598,2014.

[8] R. K. Siddharth, Y. Jaya Satyanarayana, Y. B. Nithin Kumar, M. H. Vasantha and E. Bonizzoni, "A 1-V, 3-GHz Strong-Arm Latch Voltage Comparator for High Speed Applications," in *IEEE Transactions on Circuits and Systems II: Express Briefs*, vol. 67, no. 12, pp. 2918-2922, Dec. 2020.

[9] Y. Guo, J. Jin, X. Liu and J. Zhou, "An 18.1 mW 50 MHz-BW 76.4 dB-SNDR CTSDM With PVT-Robust VCO Quantizer and Latency-Free Background-Calibrated DAC," in *IEEE Transactions on Circuits and Systems I: Regular Papers*, vol. 69, no. 12, pp. 4787-4798, Dec. 2022.

[10] Y. Zhong, X. Tang, J. Liu, W. Zhao, S. Li and N. Sun, "An 81.5dB-DR 1.25MHz-BW VCO-Based CT $\Delta\Sigma$ ADC with Double-PFD Quantizer," in *IEEE Custom Integrated Circuits Conference (CICC)*, 2021.

[11] Y. Xu, X. Zhang, Z. Wang and B. Chi, "A Flexible Continuous-Time $\Delta\Sigma$ ADC With Programmable Bandwidth Supporting Low-Pass and Complex Bandpass Architectures," in *IEEE Transactions on Very Large Scale Integration (VLSI) Systems*, vol. 25, no. 3, pp. 872-880, March 2017.

[12] N. -S. Kim, "A Digital-Intensive Extended-Range Dual-Mode BLE5.0 and IEEE802.15.4 Transceiver SoC," in *IEEE Transactions on Microwave Theory and Techniques*, vol. 68, no. 6, pp. 2020-2029, June 2020.

2025 The 10th International Conference on Integrated Circuits and Microsystems

Low Power ΣΔ Modulator Applied to CIS Column Parallel Readout Circuit

Dengju Sun	Zhongjie Guo	Ruiming Xu	Jinquan Zhou
School of Automation and	*School of Automation and*	*School of Automation and*	*School of Automation and*
Information Engineering	*Information Engineering*	*Information Engineering*	*Information Engineering*
Xi'an University of Technology	*Xi'an University of Technology*	*Xi'an University of Technology*	*Xi'an University of Technology*
Xi'an, China	Xi'an, China	Xi'an, China	Xi'an, China
2403526553@qq.com	zjguo@xaut.edu.cn	xuruiming1019@163.com	quan20000627@163.com

Abstract—This paper is based on the second-order incremental ΣΔ analog-to-digital converter (ADC) of CMOS image sensor (CIS). A low-power column parallel ΣΔ modulator based on the shared structure of inter-column operational amplifier(OTA) is proposed. Through the time division multiplexing of the OTA between the two adjacent columns of readout circuits, compared with the traditional modulator structure, the power consumption can be reduced by nearly half. It is made by 55nm CIS process and powered by a 3.3V power supply. The power consumption of a single column is 344 μW. The operating rate of the modulator is 5MS/s, and the OTA is 256. At the input signal bandwidth of 8.27kHz, the simulation results show that the SNDR is 93.9dB and the ENOB is 15.3 bits.

Keywords—CIS, ΣΔ modulator, OTA, Inter-column shared

I. INTRODUCTION

With the growing demand for high-resolution, high-frame-rate, low-power CMOS image sensors (CIS), the design of CIS readout circuits, especially analog-to-digital converters (ADC), has brought serious challenges [1]. The readout circuit of CIS usually adopts the column-parallel ADC architecture. The column-parallel readout circuit can achieve a good compromise between resolution, frame rate, and power consumption, which is the current mainstream direction of the industry[2]. ΣΔ ADC with its unique oversampling and noise shaping technology, can achieve large-scale pixel arrays, moderate conversion speed, and excellent noise performance. Therefore, ΣΔ ADC has become one of the research hotspots in column parallel CIS readout circuits .

However, the operational amplifier (OTA) used by the modulator of sigma delta ADC usually occupies a considerable amount of power and chip area. Especially in the parallel column readout circuit architecture, each column has strict column width and power consumption restrictions. Under the advanced process node, the width of the CIS column readout circuit is usually only a few microns. Under such width restrictions, it is very difficult to ensure the performance of the fully differential OTA.

In view of the above problems, this paper proposes a low-power column parallel ΣΔ modulator based on the shared structure of inter-column OTA. In a traditional ΣΔ modulator, the integrator is implemented in the form of a switching capacitor circuit. During analog-to-digital conversion, a complete clock cycle can be divided into two stages: sampling and integration. When the circuit is working in the sampling stage, the operational amplifier in the integrator is actually in an idle state. Only when the circuit is working in the integration stage, the operational amplifier participates in the integration operation. In half of the time, the OTA generates power consumption while not having a practical effect on the entire modulator circuit. In view of this, in this paper, the time-sharing multiplexing of the operational amplifier between the two adjacent column readout circuits can achieve nearly half of the power consumption reduction. At the same time, the shared architecture of the OTA between columns can double the width of the layout of a single operational amplifier, reducing the performance impact of parasitism.

II. THE WORKING PRINCIPLE OF CIS READOUT CIRCUIT

The column-level readout circuit is shown in Fig. 1, with a photodiode and a MOS tube controlled by a TG signal, with the TG transistor completely separating the photosensitive area from the circuit part, facilitating the introduction of the associated dual-sampling circuit. As shown in Fig. 2, when the pixel unit is read out, the SEL of the line-selective signal is high, and the switch M_{SEL} is turned on, and then the reset tube M_{RST} is turned on, and the FD junction is reset to the power supply voltage, and the pixel unit outputs the reset voltage V_{RST}. Then the switch is turned off, the transistor controlled by TG is turned on, and the photogenerated charge is transferred from the diode to the FD junction, so that the voltage drop generated by the FD junction is related to the parasitic capacitance C_{FD} of the FD junction, at this time, the output signal voltage of the pixel unit is V_{SIG}, so that a set of reset voltages and signal voltages are generated, and in the subsequent CDS circuit, the voltage of the two outputs is subtracted to obtain the voltage signal size corresponding to the optical signal. Since the situation is exactly the same for the two resets, both the reset noise of the pixel cell and the FPN are eliminated in the CDS circuit.

979-8-3315-8850-2/25 $31.00 © 2025 IEEE

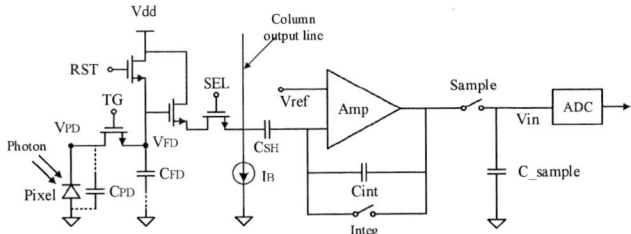

Fig. 1. Column-level readout circuit of CIS

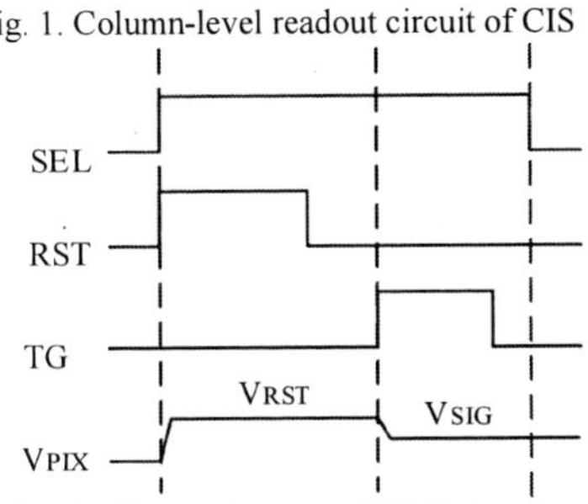

Fig. 2. Timing diagram of CDS circuit

According to the law of conservation of charge, the op amp output Vin can be expressed as:

$$Vin = Vref + (V_{RST} - V_{SIG})\frac{C_{SH}}{C_{int}} \qquad (1)$$

Due to the use of oversampling and noise shaping, ΣΔ ADC are generally relatively easy to achieve high accuracy, in line with the increasing demand for image sensor accuracy in the future. The ΣΔ modulator is the core unit of the entire ADC, which realizes two important functions: oversampling and noise shaping. Oversampling technology refers to the use of a sampling frequency much higher than the signal bandwidth to sample the signal to suppress the quantized noise, thereby improving the signal-to-noise ratio. Noise shaping is to shape the noise power spectrum through a loop filter, move the low-frequency noise outside the signal bandwidth, and then filter out the out-of-band noise through a digital decimation filter, which can also achieve the purpose of improving the signal-to-noise ratio.

The change speed of the CIS pixel readout signal is very slow, and it can be approximated as a DC signal. The incremental ΣΔ modulator has a natural advantage in processing low-frequency signals. The basic structure of the incremental Σ Δ ADC, shown in Fig. 3, It mainly includes an integrator, a comparator and a counter, On top of traditional ΣΔ ADCs, the signal resets the modulator after each conversion, avoiding the impact on the next quantization result. the input signal Vin is integrated into the comparator during the working process, when the comparator output result is high level 1, the feedback voltage is Vref, when the comparator result is low level 0, the feedback voltage is set to 0, and the comparator outputs the digital code value di, and the input signal and the feedback voltage are subtracted and then integrated and compared. Fig. 4 shows the output waveform of each module of the first-order incremental ΣΔ modulator.

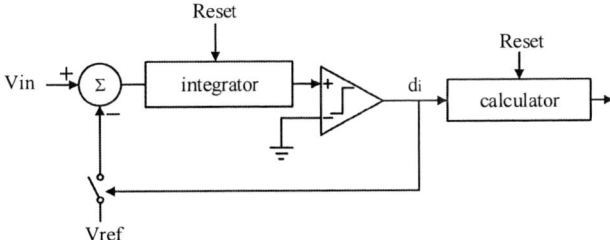

Fig. 3. The basic structure of the first-order incremental ΣΔ ADC

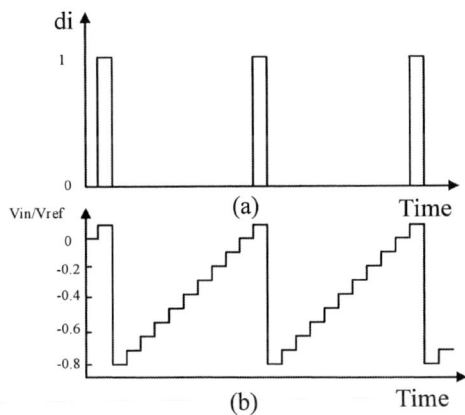

Fig. 4. First-order incremental sigma-delta ΣΔ ADC output
(a)Modulator digital code (b)Integral output voltage

The first-order ΣΔ modulator requires 2^N clock cycles to achieve N-bit resolution, so a very high-frequency clock signal is required. For high-order ΣΔ modulators, high-order modulators have a complex circuit structure and high power consumption, which are not suitable for large-scale pixel array CIS[3]. Multi-bit quantization technology can effectively improve the signal-to-noise ratio, but a dynamic component matching unit is required to ensure the linearity of the feedback DAC, which greatly increases the circuit hardware overhead and is also not suitable for CIS readout circuits with tight areas.

The commonly used modulator structures are CIFB and CIFF structures. The loop filter of the CIFF-type modulator only handles quantized noise. Compared with the CIFB-type modulator, the output swing of the CIFF-type integrator at all levels is very small, which reduces the requirements for the output swing of the OTA. The CIFB structure is step-by-step feedback, which is easy to reuse and expand in the design and layout of the circuit, and considering the time-sharing multiplexing of the OTA in the subsequent circuit of this paper, this paper uses an incremental ΣΔ modulator with Second-order CIFB structure. Under the premise of ensuring system stability, it can also achieve a compromise between noise performance, conversion speed, chip area and power consumption, which is very suitable for high-resolution, medium-speed and low-power CIS column parallel readout circuit [4].

After selecting the modulator structure, a second-order ΣΔ modulator wants to achieve a resolution of N bits, the clock period n required is[5]:

979-8-3315-8850-2/25 $31.00 © 2025 IEEE

$$N = \log_2 n(n+1) - 1 \qquad (2)$$

The effective number of bits of the ADC designed in this article can reach up to 16 bits, so the SNR can be set to 100dB. So the OSR is set to 256. The actual circuit characteristics of the integrator often deviate from the ideal characteristics. The reasons for this phenomenon are: The finite DC gain, bandwidth, switching thermal noise, switching non-linearity, slew rate, clock jitter of the integrator[6], etc. this These factors will cause incomplete charge transfer of the integrator and non-linearity at the output, reducing the performance of the modulator[7]. As shown in Fig. 5, it is a Sigma-delta modulator model after adding non-ideal factors.

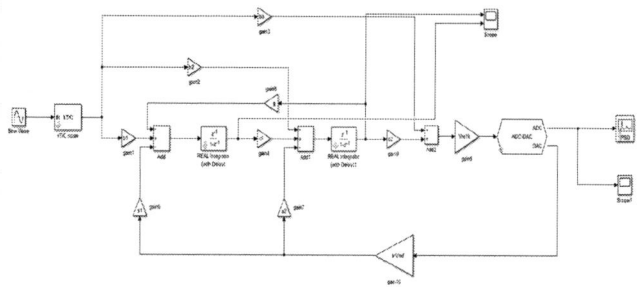

Fig. 5. Modeling of second-order CIFB modulator

The entire circuit module of the modulator includes circuit sub-modules such as a switching capacitor integrator, a two-phase non-overlapping clock, a high-speed comparator, and a DAC. The most important of these is the design of the integrator circuit. In a single-channel $\Sigma\Delta$ modulator, the integrator is implemented in the form of a switching capacitor circuit, and its working state is shown in Fig. 6. Among them, C_S and C_I are sampling capacitors and integration capacitors, respectively, and the four switches are controlled by two-phase non-overlapping clocks. The integrator has two working states, sampling and integration. When the circuit is working in the sampling stage, $\Phi 1$ is turned on, the sampling capacitor C_S is charged to VIN, and the charge value on C_I remains unchanged. In the integration stage, $\Phi 1$ is disconnected, and then $\Phi 2$ is turned on. The charge on C_S is transferred to C2 through the virtual location [8]. The output expression of the integrator obtained by the charge conservation is:

$$VIN \cdot C_S + V_O(nT - T) \cdot C_I = V_O(nT) \cdot C_I \qquad (3)$$

$$V_O(nT) - V_O(nT - T) = \frac{C_S}{C_I} \cdot VIN \qquad (4)$$

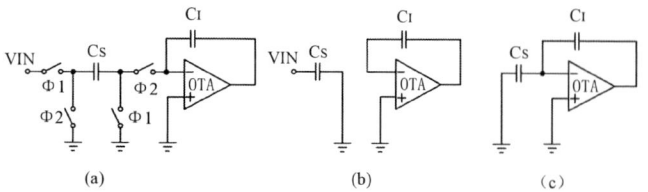

Fig. 6. Integrator working status (a) Switching capacitor integrator; (b) sampling phase; (c) integration phase

From the working process of the integrator, it is found that when the circuit is working in the sampling stage, the operational amplifier in the integrator is actually in an idle state. Only when the circuit is working in the integration stage, the OTA participates in the work. In half the time, the OTA generates power consumption and has no actual function. this paper, the time division multiplexing of the operational amplifier between the two adjacent columns of readout circuits is used to make the operational amplifier always operate in the integrated state, which greatly improves the working efficiency of the system and reduces power consumption. At the same time, the layout area can be saved.

III. INTER-COLUMN SHARED MODULATOR DESIGN

The overall circuit of the inter-column shared modulator designed in this paper is shown in Fig. 7. In actual operation, the two channels are controlled by clocks of different phases. When channel 1 is in the sampling stage, channel 2 uses an OTA and is in the integration state; when channel 2 completes the integration operation, it immediately switches to the sampling stage, while channel one uses an OTA and switches to the integration state. The dual-channel circuit structure is completely symmetrical, and it is mainly composed of a two-stage switching capacitor integrator, a one-bit quantizer and a DAC. The first stage integrator of the two channels shares the OTA1, and the second stage shares the OTA2. Among them, the role of the two-phase non-overlapping clock is to avoid the timing confusion caused by the simultaneous conduction of the two channels, causing the entire circuit to fail to work properly [9]. $\Phi 1d$ and $\Phi 2d$ are the clock signals of $\Phi 1$ and $\Phi 2$ that have been delayed by the falling edge, respectively, and are used for sampling the lower plate of the switching capacitor integrator to reduce the influence of the channel charge injection effect associated with the input signal[10].

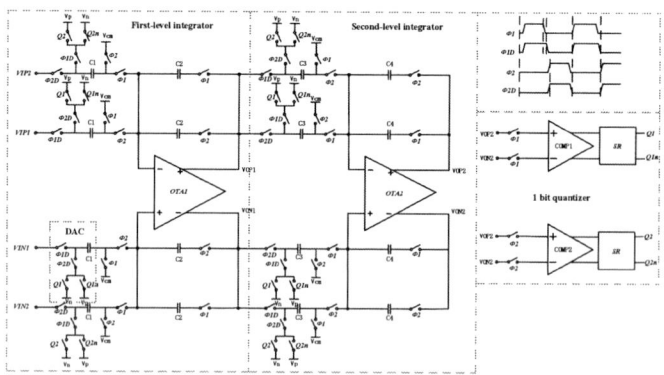

Fig. 7. Circuit diagram of shared modulator between columns

In order to describe the working condition of the dual-channel modulator in detail, the working state of the dual-channel OTA sharing modulator under two different clock signals is explained separately. Fig. 8 shows the working state of the dual-channel modulator when the $\Phi 1$ is high, and the latter shows the working state when $\Phi 2$ is high. Among them, red represents the conduction of channel 1, and blue represents the conduction of channel 2. When $\Phi 1$ and $\Phi 1d$ are high and $\Phi 2$ and $\Phi 2d$ are low, at this time, the first stage in channel 1 is in the sampling state and the second stage is in the integration state. The sampling capacitor C1 samples the input signals

979-8-3315-8850-2/25 $31.00 © 2025 IEEE

VIP1 and VIN1, the capacitors C3, C4 and the OTA2 form an integral circuit, and the outputs Q1 and Q1n of the quantizer control the output voltage of the DAC. Among them, the single-bit DAC is implemented through a digital circuit and two switching tubes to convert the digital signal into analog levels Vp and Vn. At the same time, the working state of Channel 2 is exactly the opposite of Channel 1. The first stage of Channel 2 is in the integration state and the second stage is in the sampling state. The OTA1 and capacitors C1 and C2 form an integral circuit, and the sampling capacitor C3 samples the output result of the first stage. In short, the OTA1 and OTA2 are occupied by the first stage of Channel 2 and the second stage of Channel 1, respectively, while the second stage of Channel 2 and the first stage of Channel 1 do not use the OTA.

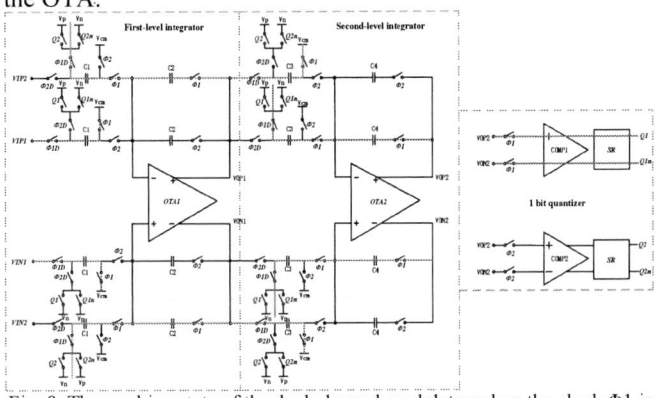

Fig. 8. The working state of the dual-channel modulator when the clock Φ1 is high

As shown in Fig.9, when Φ1 and Φ1d are low and Φ2 and Φ2d are high, the working states of Channel 1 and Channel 2 are interchanged, that is, the first stage of Channel 1 is in the integrated state, the second stage is in the sampling state, the first stage of Channel 2 is in the sampling state, and the second stage is in the integrated state. The OTA1 and OTA2 are occupied by the first stage of Channel 1 and the second stage of Channel 2, respectively, while the second stage of Channel 1 and the first stage of Channel 2 do not use the OTAs.

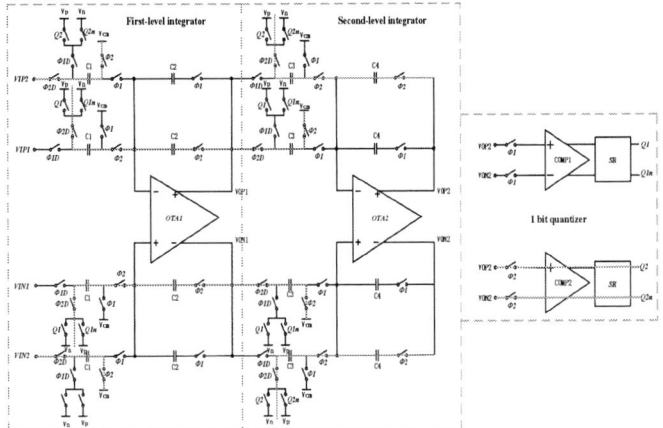

Fig. 9. The working state of the dual-channel modulator when the clock Φ2 is high

In summary, through the continuous switching of the clock phase, the shared operational amplifier has always maintained an integrated state and no longer has an idle state. Through the time-sharing multiplexing of the OTA, the number of OTA is directly halved, and the overall power consumption is greatly reduced. Under the advanced process node, the width of the CIS column readout circuit is usually only a few microns. Under such a width limit, it is very difficult to ensure the performance of the OTA, but the shared architecture of the OTA between the columns can double the width of the OTA layout, reducing the performance loss caused by parasitism.

IV. SIMULATION RESULTS

The circuit design of the ΣΔ modulator is designed under the 55nm CMOS process. The power supply voltage is 3.3V, the sampling frequency is 5MHz, and the oversampling rate is 256. The input signal frequency of Channel 1 is 8.27kHz, and the input signal frequency of channel 2 is 5kHz. The swing is 1.2V. The transient response of the modulator output is shown in Fig. 10. From top to bottom, the VIP terminal signal waveform of Channel 1, the VIN terminal signal waveform of channel 1, the output waveform of the modulator of channel 1, and the VIN terminal signal of channel 2 Waveform, the signal waveform of the VIP terminal of channel 2, and the output waveform of the modulator of channel 2. Obtained a pulse width modulator signal having maximum number of one's corresponding to maximum positive difference input voltage and more number of zero's corresponding to maximum negative difference input voltage for intermediate difference voltage we get one to zero transitions.

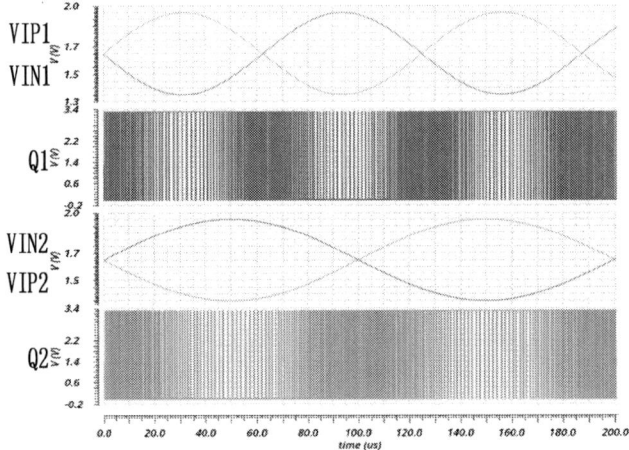

Fig. 10. Modulator output result

FFT analysis is performed on the output of the modulator in Fig. 10, and the output spectrum is shown in Fig. 11. The output SNDR of the modulator is 93.9dB, and the corresponding ENOB is 15.30 bits. The overall power consumption is 687uW, of which the power consumption of OTA1 is 429uW, and the power consumption of OTA2 is 128uW, which is converted to the power consumption of each column is 344uW. The performance comparison of this modulator with the ones from other published works has been shown in Table I.

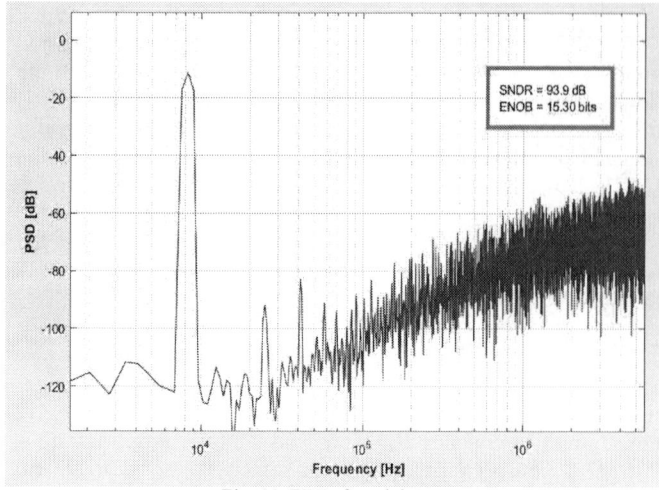

Fig. 11. FFT of modulator

TABLE I. PERFORMANCE COMPARISON WITH PREVIOUS PUBLISHED PAPERS

Parameter	Literature			
	[7]	[8]	[9]	This work
Technology	-	65nm	0.13 μm	55 nm
Supply voltage(V)	0.9	-	1.5	3.3
OSR	256	-	512	256
Resolution	-	14	16	16
Bandwidth(kHz)	10	200	1	8.27
SNDR(dB)	80	77	96.7	93.9
ENOB(bits)	13	12.4	15.7	15.3
Power consumption(uW)	200	950	940	344

In this paper, a second-order $\Sigma\Delta$ modulator with dual-column sharing for CIS is proposed. Through the time-sharing sharing of the OTA between two adjacent columns, the utilization rate of the OTA is greatly improved, and the power consumption is reduced. At the same time, when the overall layout area of the CIS is very tight, the inter-column OTA sharing architecture can double the layout width of a single operational amplifier, greatly reducing the difficulty of subsequent layout design. The simulation results show that the power consumption of a single column is 344uW, the SNDR is 93.9dB, and the effective number of bits is 15.3 bits.

ACKNOWLEDGMENT

This work was supported in part by the National Natural Science Foundation of China under Grant 62171367 and Shaanxi innovation Capability Support Project 2022TD-39.

REFERENCES

[1] J. Zhou, Z. Guo, J. Zhang and L. Li, "High speed single-slope ADC design method applied to CMOS image sensors," 2024 7th International Conference on Electronics Technology (ICET), Chengdu, China, 2024, pp. 147-152.

[2] Shresth Gupta and Rakesh Mandal, " A Survey On Image Enhancement Techniques," International Journal of Electrical and Electronic Engineering & Telecommunications, January, 2015. pp. 47-54.

[3] J. Chen, L. Meng, M. Zhao and Z. Tan, "An Energy-Efficient Readout Circuit Based on Incremental Delta-Sigma ADC with Decimation Filter for CMOS Image Sensors," 2022 IEEE Asia Pacific Conference on Circuits and Systems (APCCAS), Shenzhen, China, 2022, pp. 149-152.

[4] F. Tang, Y. Cao and X. Zhao, "Column-parallel continuous-time $\Sigma\Delta$ ADC with implicit front-end variable gain amplifier," 2013 IEEE 56th International Midwest Symposium on Circuits and Systems (MWSCAS), Columbus, OH, USA, 2013, pp. 253-256.

[5] I. Lee, B. Kim and B. -G. Lee, "A Low-Power Incremental Delta–Sigma ADC for CMOS Image Sensors," in IEEE Transactions on Circuits and Systems II: Express Briefs, April. 2016, pp. 371-375.

[6] M. Yue, D. Wu and Z. Wang, "Data Compression for Image Sensor Arrays Using a 15-bit Two-Step Sigma–Delta ADC," in IEEE Sensors Journal, Sept. 2014, pp. 2989-2998.

[7] J. Goes, B. Vaz, R. Monteiro and N. Paulino, "A 0.9V /spl Delta//spl Sigma/ Modulator with 80dB SNDR and 83dB DR Using a Single-Phase Technique," 2006 IEEE International Solid State Circuits Conference - Digest of Technical Papers, San Francisco, CA, USA, 2006, pp. 191-200.

[8] R. H. M. van Veldhoven, R. Rutten, and L. J. Breems, "An inverter based hybrid $\Delta\Sigma$ modulator," in Proc. IEEE ISSCC Dig. Tech. Papers, Feb. 2008, pp. 492–630.

[9] P. Mounika, Y. G. Pu and K. -Y. Lee, "A 1.4mW Sigma Delta ADC with Configurable Filter for Sensor Applications," 2023 Fourteenth International Conference on Ubiquitous and Future Networks (ICUFN), Paris, France, 2023, pp. 697-699.

[10] Y. Oike and A. El Gamal, "CMOS Image Sensor With Per-Column $\Sigma\Delta$ ADC and Programmable Compressed Sensing," in IEEE Journal of Solid-State Circuits, Jan. 2013, pp. 318-328.

2025 The 10th International Conference on Integrated Circuits and Microsystems

Design and Analysis of AC/DC Converter for GaN-Based Server Power Supply System

Fen Guo[1, 2, *]
[1]Jinan Maiwei Intelligent
Technology Co., Ltd.
[2]Shandong Yunhai Guochuang
Innovative Technology Co., Ltd.
Beijing, China
guofen@inspur.com

Tuo Li[1, 2]
[1]Jinan Maiwei Intelligent
Technology Co., Ltd.
[2]Shandong Yunhai Guochuang
Innovative Technology Co., Ltd.
Beijing, China
lituo@inspur.com

Changhong Wang[1, 2]
[1]Jinan Maiwei Intelligent
Technology Co., Ltd.
[2]Shandong Yunhai Guochuang
Innovative Technology Co., Ltd.
Beijing, China
wangchh01@inspur.com

Kang Su[1, 2]
[1]Jinan Maiwei Intelligent
Technology Co., Ltd.
[2]Shandong Yunhai Guochuang
Innovative Technology Co., Ltd.
Beijing, China
sukang@inspur.com

Jiankai Xu[3]
[3] Institute of Semiconductors,
Chinese Academy of Sciences
Beijing, China
jkxu@semi.ac.cn

Xin Xi[1, 2]
[1]Jinan Maiwei Intelligent
Technology Co., Ltd.
[2]Shandong Yunhai Guochuang
Innovative Technology Co., Ltd.
Beijing, China
xixin01@inspur.com

Kejian Zhu[1, 2]
[1]Jinan Maiwei Intelligent
Technology Co., Ltd.
[2]Shandong Yunhai Guochuang
Innovative Technology Co., Ltd.
Beijing, China
zhukejian@inspur.com

Xinxin Yuan[1, 2]
[1]Jinan Maiwei Intelligent
Technology Co., Ltd.
[2]Shandong Yunhai Guochuang
Innovative Technology Co., Ltd.
Beijing, China
yuanxinxin@inspur.com

Abstract—This paper presents a GaN HEMT-based AC/DC converter for server power supplies, addressing the need for high efficiency, high power density, and improved reliability in cloud and data center applications. An AlGaN/GaN HEMT structure with an InGaN back-barrier layer is designed and optimized. Its electrical characteristics are verified via simulation. A two-stage topology is adopted for the converter, comprising an interleaved totem-pole PFC (Power Factor Correction) and an LLC resonant converter. The operating principle of this topology is analyzed in detail, and the key circuit parameters are carefully designed. Under 220 V AC input, the converter achieves a rated output power of 2400 W, a DC bus voltage of 400 V ± 5 %, a peak efficiency of 97.03 %, and a power factor (PF) of $\geqslant$ 0.999. All these performance metrics fully meet the requirements of high-end server power supplies, which demonstrates that GaN technology possesses the capability to overcome the inherent efficiency–density trade-off in traditional power supply designs.

Keywords—GaN HEMT, PFC, Converter, LLC

I. INTRODUCTION

With the rapid growth of cloud computing and big data centers, the computational demand of server clusters has increased exponentially. As a result, the energy efficiency, power density, and reliability of power supply systems have become critical constraints on overall system performance. Traditional AC/DC converters using silicon (Si) power devices are limited by switching and conduction losses. These limitations pose major challenges for high-frequency and miniaturized designs. When the switching frequency exceeds 1 MHz, the reverse recovery charge (Qrr) of Si MOSFETs causes switching losses to rise sharply. Moreover, the larger heat dissipation system further restricts power density improvement[1].

Gallium Nitride (GaN), a representative third-generation wide-bandgap semiconductor, offers superior material properties. These include a wide bandgap (3.4 eV), high electron mobility, high critical electric field (~3.3 MV/cm), and zero reverse recovery charge[2]. GaN HEMT-based power converters can operate at higher switching frequencies with reduced losses[3]. This provides a promising solution to overcome the efficiency–density trade-off in server power systems.

To address the stringent demands for high reliability and efficiency in server power supplies, this paper begins with the design of core power devices, proposing a high-voltage GaN HEMT structure. This structure offers both high voltage endurance and low conduction losses, thereby serving as a solid hardware foundation for achieving high-efficiency converter operation. Furthermore, an AC/DC topology is developed by integrating an interleaved totem-pole PFC converter with an LLC resonant converter. The former enhances the power factor and minimizes input current harmonics, while the latter enables efficient DC output across a wide input voltage range. In addition, a digital dual-loop control strategy is implemented, where the outer loop regulates the output voltage and the inner loop actively suppresses system disturbances. Finally, the effectiveness of the proposed topology and control approach in improving harmonic suppression, reducing waveform distortion, and ensuring system stability is validated through simulation.

II. DESIGN OF GAN DEVICE STRUCTURE

In GaN HEMT devices, the back barrier layer plays a critical role in enhancing the vertical confinement of the two-dimensional electron gas (2DEG), suppressing carrier leakage toward the substrate, and consequently improving the device's breakdown voltage and high-frequency performance. This paper proposes an AlGaN/GaN HEMT structure incorporating an InGaN back barrier layer (as illustrated in Fig. 1), with the following key parameters: the AlGaN barrier layer has a

---

This work is supported by Natural Science Foundation of Shandong Province (No. ZR2023LZH001)
* Corresponding author.

979-8-3315-8850-2/25 $31.00 © 2025 IEEE

thickness of 30 nm and an aluminum composition of 0.25; the AlN insertion layer is 1.5 nm thick; the GaN channel layer is 100 nm thick; the InGaN back barrier layer features an indium composition of x = 0.07 and a thickness of 5 nm; and the GaN buffer layer is 1.5 μm thick. It is assumed that the AlGaN barrier layer, AlN layer, and InGaN layer coherently grow with the GaN epitaxial layer along the [0001] crystal orientation.

Compared to GaN, AlGaN and AlN induce biaxial tensile stress due to their smaller lattice constants, whereas InGaN generates biaxial compressive stress due to its larger lattice constant. Notably, piezoelectric polarization dominates within the InGaN layer (with a magnitude greater than that of spontaneous polarization), resulting in a polarization electric field directed from the material surface toward its interior. This polarization effect forms a back barrier behind the channel, which effectively suppresses 2DEG leakage into the buffer layer and enhances the confinement of the 2DEG[4], [5].

Based on the aforementioned model, electrical performance simulations have been conducted on the AlGaN/GaN/InGaN device structure. Fig. 2(a) illustrates the output characteristic curve of the device, which depicts the variation of drain-source current with respect to drain voltage under different gate voltages. The gate voltage V_g ranges from -6 V to 2 V, while the drain-source voltage V_{ds} is scanned from 0 V to 20 V. From the simulation results, it is evident that the maximum saturation current density Idsmax of the device reaches 656 mA/mm (@V_{gs}=2 V). Fig. 2(b) presents the transfer characteristic curve of the device, where the gate voltage V_{gs} is varied from -15 V to +5V with a constant V_{ds}=+10 V. According to this transfer characteristic curve, the threshold voltage V_{th} and the transconductance gm of the device are determined to be 5.80 V and 90.69 mS/mm, respectively.

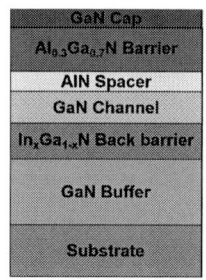

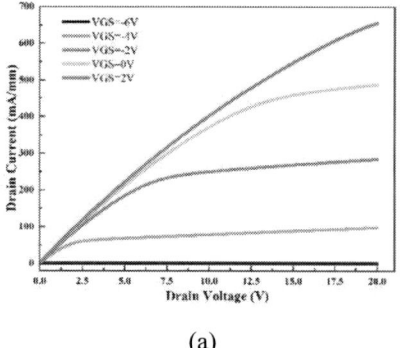

Fig. 1. Schematic diagram of the AlGaN/GaN HEMT structure with an InGaN back barrier layer

(a)

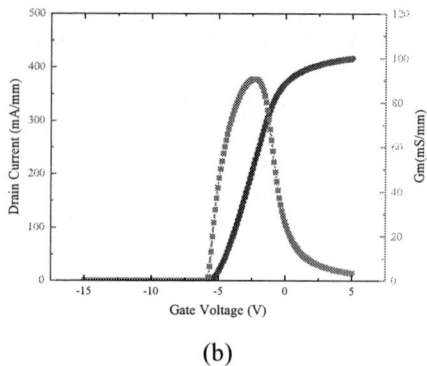

(b)

Fig. 2. (a)Output characteristic curve and (b)transfer characteristic curve of the AlGaN/GaN/InGaN device

III. DESIGN OF TOTEM-POLE BRIDGELESS PFC CONVERTER BASED ON GaN

The GaN-based AC/DC converter presented in this paper features a two-stage architecture comprising a "PFC stage + DC/DC stage" (as illustrated in Fig. 3). The front-end Power Factor Correction (PFC) employs an interleaved totem-pole topology, designed for an input voltage of AC 220 V and delivering a DC output bus voltage of 400 V. This configuration primarily facilitates power factor correction with a target power factor of ≥0.99, as well as the suppression of input current harmonics[6]. The rear-end DC/DC conversion utilizes an LLC resonant converter, which effectively steps down the 400 V bus voltage to the standard server power supply voltage of DC 12 V, thereby ensuring efficient power transmission. Notably, switches S1, S2 are constructed using the device structure developed in the previous chapter, while switches S3 and S4 utilize SiC MOSFET device structures.

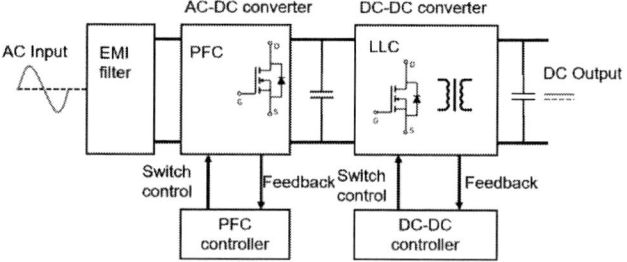

Fig. 3. Schematic diagram of the two-stage architecture of the GaN-based AC/DC converter

Compared to the traditional diode rectification combined with Boost PFC topology, the totem-pole PFC topology enables fully controlled switching of the bridge arm through GaN HEMTs. This innovation eliminates power-frequency rectifier diodes, thereby effectively reducing both conduction losses and reverse recovery losses associated with diodes[7]. In this paper, we present a GaN HEMT totem-pole soft-switching bridgeless PFC circuit that incorporates soft-switching characteristics achieved through a control algorithm. This design aims to realize a high-efficiency and high-power-density PFC converter. The circuit structure is illustrated in Fig. 4. In the figure, CS1 and CS2 represent the parasitic capacitances of switching devices S1 and S2 respectively; L denotes the input filter inductor; C indicates the output low-frequency filter capacitor; and R signifies the equivalent load.

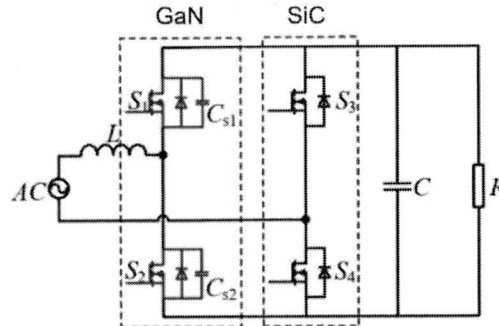

Fig. 4. GaN HEMT totem-pole soft-switching bridgeless PFC circuit

A. Main Circuit Parameter Design

This section introduces the model of the totem-pole PFC main circuit and the calculation of main circuit component parameters. The key formulas and parameter selections are as follows:

(1) Design of PFC Inductor Parameters

The inductor parameters need to be calculated in combination with the switching frequency range, and the specific formulas are divided into two working conditions:

When $V_{in} < 0.5V_o$,

$$\frac{\eta V^2_{rms}}{4P_oL} < f_s < \frac{\eta V^2_{rms}}{2P_oL} \qquad (1)$$

When $V_{in} > 0.5V_o$,

$$\frac{89}{\frac{800P_oL}{\eta V^2_{rms}} + 400(\sqrt{k^2-1}+k)\sqrt{2LC_s}} < f_s < \frac{1}{\frac{4P_oL}{\eta V^2_{rms}} + (\sqrt{k^2-1}+k)\sqrt{8LC_s}} \qquad (2)$$

In the formulas, Cs represents the resonant capacitor, which has a value of 150 pF. The variable k denotes the condition for soft-switching implementation; it must satisfy the criterion k > 1. However, an excessively high value of k can result in increased reverse current and current ripple. Therefore, a value of k = 1.1 is selected for optimal performance.

Based on the correlation analysis between the converter switching frequency and inductor parameters utilizing the aforementioned formulas, and taking into account the trade-off between power density and control complexity, an inductance value of L=5000 µH has been selected as the optimized parameter. This inductance corresponds to a broad operational frequency range of 98 kHz to 450 kHz: specifically, the high-frequency segment at 450 kHz facilitates a reduction in magnetic core size by approximately 40 %, thereby satisfying volume reduction requirements for high-frequency operation. Meanwhile, the lower limit of 98 kHz helps alleviate control pressure under extreme operating conditions (for instance, increasing duty cycle adjustment margin by 15 %), thus achieving a balance between energy efficiency and reliability.

(2) Design of PFC Output Capacitor Parameters

The design of the output capacitor needs to take into account both the voltage ripple (V_{ripple}) control and the power-off hold-time (t_{hold}) limit. The calculation formula for the capacitor is as follows:

$$C_o > \frac{P_o}{2\pi \cdot f_L \cdot V_o \cdot V_{ripple\_pp}} \qquad (3)$$

In the formula, f_L denotes the AC frequency, and V_{ripple_pp} represents the peak-to-peak allowable output voltage ripple, which is determined based on the 4 % ripple criterion. Substituting the specified parameters yields a minimum required output capacitance of 3821 µF.

$$C_o \geq \frac{2P_o \cdot t_{hold}}{V_o^2 - V_{omin}^2} \qquad (4)$$

In the formula, t_{hold} is defined as one cycle of a 50 Hz sinusoidal AC waveform (0.02 s), while V_{omin} represents the minimum output voltage permitted by the system during the hold time, which is established at 300 V in this study. Upon substituting these parameters, the calculated value is found to be 137 1µF.

(3) Design of LLC Resonant Parameters

The LLC resonant network aims to "operate at the resonant frequency under rated load and realize ZVS (Zero Voltage Switching)". According to the resonant frequency f_r=3 MHz, bus voltage 400 V, output voltage V_o=12 V, and turns ratio n=16 (primary-secondary turns ratio), the resonant inductor L_r=34 µH and resonant capacitor C_r=47.6 nF are calculated.

To ensure that the output steady-state voltage of the PFC converter remains stable at 400 V, a comprehensive consideration of the voltage margin (12.5 %) and the specifications of commercial devices has led to the selection of a 4300 µF electrolytic capacitor as the output filter element.

B. Working Principle Analysis

(1) Working Mode of Totem-Pole PFC

The totem-pole PFC bridge arm consists of two upper bridge arm GaN HEMTs (S1, S2) and two lower bridge arm SiC MOSFETs (S3, S4). Its working process is divided according to the half-cycle of the input AC[8]:

Positive half-cycle: S2 and S3 are turned off, S1 operates in the high-frequency PWM mode, and S4 is turned on; the input current forms a loop through S1, PFC inductor, and S4, realizing the storage and release of inductor energy[9].

Negative half-cycle: S1 and S4 are turned off, S2 operates in the high-frequency PWM mode, and S3 is turned on; the input current forms a loop through S2, PFC inductor, and S3, realizing symmetric current control.

Through the average current control strategy, the input current is made to track the input voltage waveform, achieving high power factor and low harmonic distortion[10].

(2) Working Mode of LLC Resonant Converter

When the LLC converter operates near the resonant frequency, the primary-side GaN HEMTs (S5, S6) realize ZVS turn-on, and the secondary-side synchronous rectifier tubes (S7, S8) realize ZCS (Zero Current Switching) turn-off, effectively reducing switching loss[11]. When the load changes, the output voltage is adjusted through frequency modulation (FM):

Under light load: The switching frequency is higher than the resonant frequency, the impedance of the resonant network increases, and the output voltage decreases.

Under heavy load: The switching frequency is lower than the resonant frequency, the impedance of the resonant network decreases, and the output voltage increases.

Finally, the output voltage is stably maintained within the range of 12 V±5 %.

IV. SIMULATION ANALYSIS

A system-level simulation model of the GaN-based AC/DC converter has been developed. The key parameter settings are as follows: the effective input voltage is 220 V at a frequency of 50 Hz, the switching frequency for the high-frequency bridge arm is set to 300 kHz, the AC input inductor is specified as 5000 µH, and the output capacitor has a value of 4300 µF. Additionally, a dead-time delay of 0.1 ms is implemented for the low-frequency bridge arm.

A. Control System Architecture

This system employs a dual-closed-loop cascaded architecture to achieve high-precision power factor correction and zero-voltage turn-on characteristics for GaN devices (as illustrated in Fig. 5). The specific control logic is outlined as follows:

Voltage outer loop: The DC bus voltage is dynamically adjusted through a PI controller. The amplitude reference of its output is multiplied by the grid phase signal extracted by the Phase-Locked Loop (PLL), and the input voltage effective value feedforward compensation is introduced to generate the current inner loop command that is strictly in phase with the grid.

Current Inner Loop: Utilizing a Sinusoidal Pulse Width Modulation (SPWM) strategy, the error between the inductor current i_L and real-time tracking of i_{ref} is adjusted using PI control to produce the drive signal for the high-frequency bridge arm, thereby facilitating voltage-current decoupling control.

In this context, the PLL serves to accurately synchronize with the grid phase, effectively suppressing harmonic interference. Furthermore, the input voltage feedforward module dynamically corrects power references, ensuring that the system maintains a constant power output despite wide fluctuations in input voltage. Ultimately, this approach achieves stabilization of bus voltage.

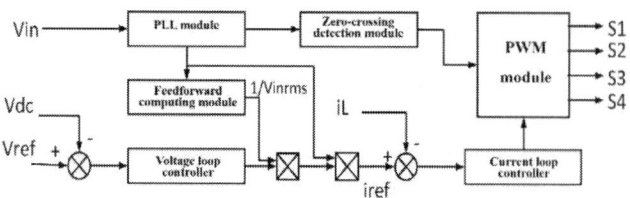

Fig. 5. Schematic diagram of the dual-closed-loop control system architecture

B. Simulation Result Analysis

Fig. 6 illustrates the input voltage and current waveforms of the circuit. From the figure, it is evident that the input current demonstrates a high-fidelity sinusoidal characteristic and remains strictly in phase with the grid voltage (phase difference $\Delta\phi<1°$). This observation indicates that the PFC circuit has achieved a near-unity power factor (PF>0.99) along with low harmonic distortion. Furthermore, when compared to traditional control strategies, the enhanced dual-closed-loop control strategy effectively mitigates current pulse spikes and phase lag occurring near the zero-crossing point of the voltage waveform, thereby preventing an increase in conduction losses due to charge accumulation.

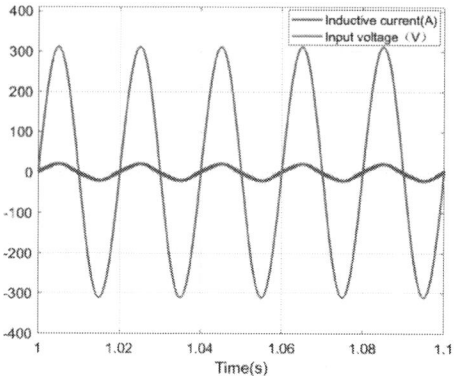

Fig. 6. Input voltage and input current waveforms

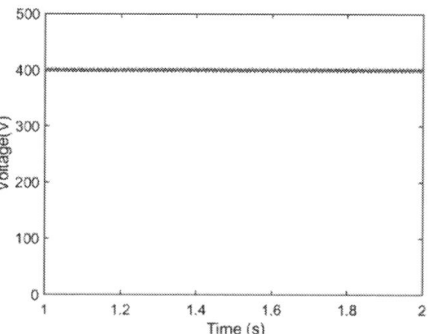

(a) Complete waveform of bus voltage

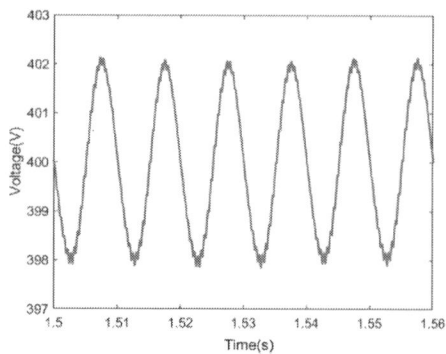

(b) Local waveform of bus voltage

Fig. 7. Waveforms of bus voltage

979-8-3315-8850-2/25 $31.00 © 2025 IEEE

Fig.7 shows the simulation waveform of the bus voltage. Fig. 7(a) is the complete waveform of the bus voltage, which shows that under the AC 220 V input, the DC bus voltage is stably maintained at the target value of 400 V. Fig. 7(b) is the local waveform of the bus voltage, which shows that under full load conditions, the average output voltage of the converter is 400.

C. Main Circuit Parameter Design

Fig. 8 presents the simulation results of the Power Factor (PF). As illustrated in the figure, through the implementation of multi-dimensional optimization strategies—including device structure optimization, topology design, and enhancements to control algorithms—the system achieves a near-unity power factor across the entire load range, with PF $\geq$ 0.999. This outcome further substantiates the effectiveness of the proposed scheme.

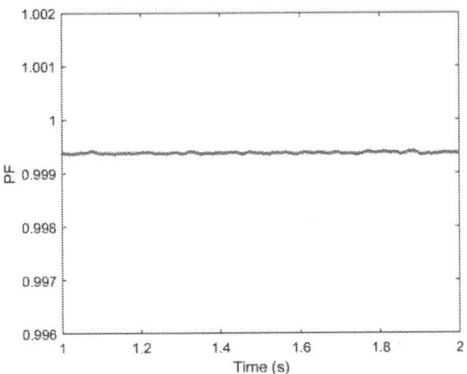

Fig. 8. Simulation waveform of Power Factor (PF)

V. Conclusion

This paper presents the design and implementation of an AC/DC converter for server power supplies utilizing GaN HEMTs. By optimizing the topology and control strategy of the interleaved totem-pole PFC and LLC resonant converter, while leveraging the unique characteristics of GaN devices, we achieve the design objectives of high efficiency and substantial power output. The primary conclusions are as follows:

Performance indicators: At an AC input voltage of 220 V, the converter attains a rated output power of 2400 W, with the PFC converter's output voltage maintained at 400 V±5 %. It demonstrates a peak efficiency of 97.03 % and an input PF $\geq$ 0.999. All performance metrics meet the technical specifications required for high-end server power supplies.

Future research may focus on further optimizing both topology and control strategies, such as investigating multi-level topologies to mitigate voltage stress on devices and developing adaptive control algorithms to enhance the system's dynamic response capabilities, thereby improving overall converter performance even further.

Acknowledgment

This work is supported by Natural Science Foundation of Shandong Province (No. ZR2023LZH001).

References

[1] A. M. Elrajoubi, S. S. Ang, and K. George, "Design and Analysis of a New GaN-Based AC/DC Converter for Battery Charging Application," *IEEE Trans. on Ind. Applicat.*, vol. 55, no. 4, pp. 4044–4052, July 2019, doi: 10.1109/TIA.2019.2915687.

[2] F. Guo *et al.*, "Influence of Fe in the buffer layer on the laser lift-off of AlGaN/GaN HEMT film: phenomena and mechanism," *Semicond. Sci. Technol.*, vol. 35, no. 9, p. 095024, Sept. 2020, doi: 10.1088/1361-6641/ab9d33.

[3] S.-W. Han, J. Song, and R. Chu, "Design of GaN/AlGaN/GaN Super-Heterojunction Schottky Diode," *IEEE Trans. Electron Devices*, vol. 67, no. 1, pp. 69–74, Jan. 2020, doi: 10.1109/TED.2019.2953843.

[4] J. Liu, Y. Zhou, J. Zhu, K. M. Lau, and K. J. Chen, "AlGaN/GaN/InGaN/GaN DH-HEMTs with an InGaN notch for enhanced carrier confinement," *IEEE Electron Device Lett.*, vol. 27, no. 1, pp. 10–12, Jan. 2006, doi: 10.1109/LED.2005.861027.

[5] T. Palacios, A. Chakraborty, S. Heikman, S. Keller, S. P. DenBaars, and U. K. Mishra, "AlGaN/GaN high electron mobility transistors with InGaN back-barriers," *IEEE Electron Device Lett.*, vol. 27, no. 1, pp. 13–15, Jan. 2006, doi: 10.1109/LED.2005.860882.

[6] R. Samani, I. G. Zurbriggen, R. Hou, J. Lu, and A. M. Knight, "Comprehensive System-Level Thermal Performance and Power Density Optimization in Enclosed Natural Convection PFC-LLC GaN Converters," *IEEE Trans. on Ind. Applicat.*, vol. 61, no. 2, pp. 3371–3383, Mar. 2025, doi: 10.1109/TIA.2025.3532230.

[7] Y. Liu, M. Li, Y. Dou, Z. Ouyang, and M. A. E. Andersen, "Investigation and Optimization for Planar Coupled Inductor dual-phase interleaved GaN-based Totem-Pole PFC," in *2020 IEEE Applied Power Electronics Conference and Exposition (APEC)*, New Orleans, LA, USA: IEEE, Mar. 2020, pp. 1984–1990. doi: 10.1109/APEC39645.2020.9124069.

[8] Q. Huang and A. Q. Huang, "Review of GaN Totem-Pole Bridgeless PFC," *CPSS TPEA*, vol. 2, no. 3, pp. 187–196, Sept. 2017, doi: 10.24295/CPSSTPEA.2017.00018.

[9] Z. Liu, Z. Huang, F. C. Lee, and Q. Li, "Digital-Based Interleaving Control for GaN-Based MHz CRM Totem-Pole PFC," *IEEE J. Emerg. Sel. Topics Power Electron.*, vol. 4, no. 3, pp. 808–814, Sept. 2016, doi: 10.1109/JESTPE.2016.2571302.

[10] C.-T. Ma and Z.-H. Gu, "Review of GaN HEMT Applications in Power Converters over 500 W," *Electronics*, vol. 8, no. 12, p. 1401, Nov. 2019, doi: 10.3390/electronics8121401.

[11] F. C. Lee, Q. Li, Z. Liu, Y. Yang, C. Fei, and M. Mu, "Application of GaN Devices for 1 kW Server Power Supply with Integrated Magnetics," *CPSS TPEA*, vol. 1, no. 1, pp. 3–12, Dec. 2016, doi: 10.24295/CPSSTPEA.2016.00002.

2025 The 10th International Conference on Integrated Circuits and Microsystems

A Low-noise PGA With High-gain Chopper Offset-stabilized Amplifier Combining Ping-Pong Auto-zero For TMR Analog Front-End

Hui Wang*
School of Integrated Circuits
Southeast University
Nanjing, China
wangh_1224@seu.edu.cn
*Corresponding author

Guifa Yan
School of Integrated Circuits
Southeast University
Nanjing, China
220246759@seu.edu.cn

Mingqi Pan
School of Integrated Circuits
Southeast University
Nanjing, China
220236538@seu.edu.cn

Bo Wang
School of Integrated Circuits
Southeast University
Nanjing, China
seuwangbo@163.com

Jianhui Wu
School of Integrated Circuits
Southeast University
Nanjing, China
wjh@seu.edu.cn

Xin Li
School of Integrated Circuits
Anhui University
Hefei, China
lixin@ahu.edu.cn

Abstract—This paper presents a low-noise programmable gain amplifier(PGA) for tunneling magnetoresistance(TMR) analog front-end(AFE) applications. A comprehensive system-level analysis is performed for a two-stage PGA architecture followed by a passive-active cascaded LPF. To achieve superior noise performance, a high-gain chopper offset-stabilized amplifier combining Ping-Pong auto-zero scheme is proposed. Furthermore, the sensor's offset in TMR is calibrated in the second stage through a current-steering DAC. Simulated in a standard $0.18\mu m$ CMOS process, the input-referred noise of the PGA is 55nV/$\sqrt{\text{Hz}}$. Under the FF and 85°C cornor, the maximum integrated noise within the 300kHz bandwidth is $19.76\mu V_{rms}$, the maximum effective resolution is 16.19bits, the THD is -81dB, and the power consumption is 2.16mW.

Index Terms—PGA, low-noise, chopper offset-stabilized amplifier, Ping-Pong auto-zero scheme, offset calibration

I. INTRODUCTION

The design of programmable gain amplifiers (PGA) in signal conditioning circuits is critical, as they serve as the bridge between sensors and ADCs. Its primary functions typically include signal amplification and filtering. Since the amplitude of sensors' output varies across different application scenarios, an analog front-end (AFE) circuit with gain control offers greater versatility. There are three programmable methods in the design of PGA: programmable transconductance [1], programmable feedback resistance [2] [3], and programmable feedback capacitance [4] [5] [6]. The PGA can be implemented using either open-loop or closed-loop architectures. Reference [1] adopts an open-loop structure, achieving a bandwidth of 20MHz with a power consumption of only $36.5\mu W$. However, it suffers from relatively high noise, and the open-loop

"This paper was supported by "the Fundamental Research Funds for the Central Universities". The project number is 2242025K30014."

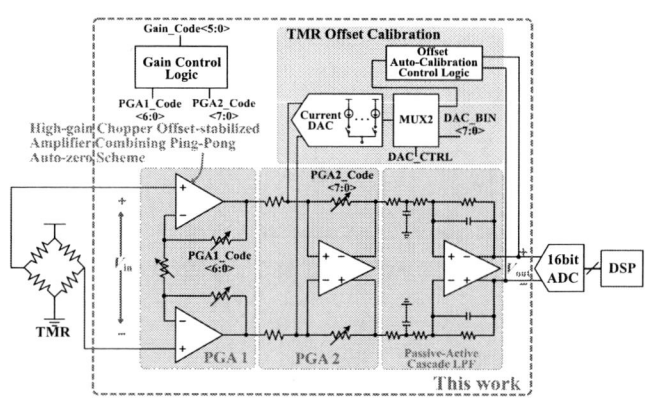

Fig. 1. PGA system in TMR analog front-end.

configuration offers limited stability and linearity. References [2] and [3] employ closed-loop resistive feedback, achieving lower noise at moderate bandwidths. References [4], [5], and [6] utilize closed-loop capacitive feedback to achieve high precision and low power consumption in low-bandwidth applications. Among them, Reference [6] achieves ultra-low power consumption below $1\mu W$, but the input-referred noise exceeds 100nV/$\sqrt{\text{Hz}}$.

The two common offset elimination techniques in CMOS are chopping technology [7] and auto-zero technology [8] [9] [10]. Chopping technology is a continuous-time modulation technique that modulates the offset voltage away from the signal baseband. Auto-zero technology is a sampling technique that measures and stores the offset value during the first half of the cycle, while the second half of the cycle operates normally

979-8-3315-8850-2/25 $31.00 © 2025 IEEE   517

to transmit the signal. They can eliminate low-frequency 1/f noise and offset caused by changes in temperature and time.

According to the sampling theorem, the effective bandwidth of a chopper amplifier must be lower than the chopping frequency. The techniques above can impact the gain-bandwidth product (GBW) of the operational amplifier, thereby degrading its performance. An alternative approach that reduces offset while maintaining wide bandwidth is known as offset-stabilization [11] [12].

The main focus of this paper is to design a high-precision, low-noise PGA with medium to low bandwidth to meet the requirements for TMR sensors, which is shown in Fig. 1. The main contributions of this paper are summarized as follows:

- A high-gain chopper offset-stabilized amplifier combining Ping-Pong auto-zero scheme is proposed to achieve superior noise performance and to suppress the ripple introduced by chopper.
- A TMR offset calibration is implemented by injecting current into the second stage PGA using a current-steering DAC.

The rest of this paper is organized as follows. A system-level analysis for two stages of PGA followed by a passive-active cascaded LPF is explained in Section II. A high-gain chopper offset-stabilized amplifier combining Ping-Pong auto-zero scheme and TMR offset calibration are described in Section III. The simulation results are discussed in Section IV. Finally, Section V concludes this paper.

II. SYSTEM MODELING OF THE HIGH RESOLUTION PGA

The block diagram of the high-precision PGA system is shown in Fig. 1. The main signal path comprises two stages of PGA followed by a passive-active cascaded LPF. TMR offset calibration is implemented in the second stage to eliminate the DC offset from the TMR sensors. The first-stage PGA supports seven gain settings: 1, 2, 4, 8, 16, 32, and 64, while the second stage offers four gain settings: 1, 2, 4, and 8. Gain selection is controlled by a 6-bit gain control code, Gain_code<5:0>. Variable feedback coefficients are realized using resistor-switch networks. The main operational amplifiers in both PGA stages are optimized for low noise and low offset performance. Since the first stage dominates the overall noise in the cascaded system, it is allocated higher power consumption to ensure low-noise operation.

In high-precision PGA design, noise is a critical performance metric. Therefore, it is essential to analyze the noise characteristics of the main signal path, as shown in Fig. 2. The expression for the input-referred noise is given as follows:

$$\overline{v_{n,in,tot}^2} = \frac{\overline{v_{n,out,tot}^2}}{A_{v,tot}^2} = \frac{\overline{v_{n,out,tot}^2}}{\left(1 + \frac{R_f}{R_g}\right)^2 \cdot \left(\frac{R_2}{R_1}\right)^2}, \quad (1)$$

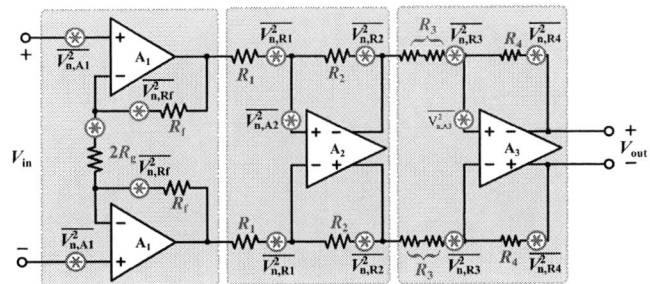

Fig. 2. Noise analysis of the main signal chain.

TABLE I
AMPLIFIERS' SPECIFICATIONS IN DIFFERENT STAGE IN THE TOTAL PGA SYSTEM.

Specification	A_1	A_2	A_3
DC Gain(dB)	>150	>130	>120
GBW(MHz)	>30	>30	>20
IRN(nV/√Hz)	<15	<20	<30
SR(V/μs)	>4		

$$
\begin{cases}
\overline{v_{n,out,tot}^2} = \overline{v_{n,PGA2}^2} + 2\overline{v_{n,R_3}^2} + 2\overline{v_{n,R_4}^2} + 4\overline{v_{n,OTA3}^2} \\
\overline{v_{n,PGA2}^2} = \overline{v_{n,PGA1}^2} + 2\left(\frac{R_2}{R_1}\right)^2 \overline{v_{n,R_1}^2} + 2\overline{v_{n,R_2}^2} \\
\qquad\quad + \left(1 + \frac{R_2}{R_1}\right)^2 \overline{v_{n,OTA2}^2} \\
\overline{v_{n,PGA1}^2} = 2\left(1 + \frac{R_f}{R_g}\right)^2 \overline{v_{n,PGA1}^2} + 2\left(\frac{R_f}{R_g}\right)^2 \overline{v_{n,R_g}^2} \\
\qquad\quad + 2\overline{v_{n,R_f}^2}
\end{cases}
\quad (2)
$$

With (1) and (2), the noise level of the resistors and operational amplifiers can be analyzed.

The nominal resolution of the ADC here is 16bits, so the effective resolution of the PGA should be greater than 16 bits. The maximum differential signal swing of this system is 3.6V, from which the input-referred noise requirement of the PGA can be derived as

$$\text{RMS Noise} = \frac{\text{Input Range}}{2^{\text{Effective Resolution}}} \le 54.9\mu V. \quad (3)$$

Since the main operational amplifiers in the signal path employ chopper to suppress flicker noise, all noise sources in the system can be approximated as white noise. As the overall gain increases, the noise contribution from the first-stage amplifier rises significantly. When the gain reaches 8, the first-stage amplifier accounts for nearly 90% of the total noise. Therefore, the design of the first-stage amplifier requires particular attention. Taking into account both gain error and settling error, the specifications for the three amplifiers are summarized in Table I.

979-8-3315-8850-2/25 $31.00 © 2025 IEEE

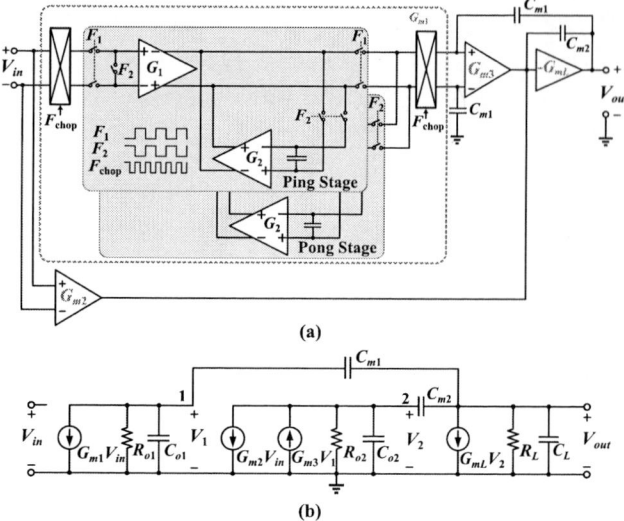

(a)

(b)

Fig. 3. High-gain Chopper Offset-stabilized Amplifier Combining Ping-Pong Auto-zero Scheme in PGA1. (a)Circuit; (b) Small-signal model.

III. CONCEPTION OF THE HIGH RESOLUTION PGA

A. High-gain Chopper Offset-stabilized Amplifier Combining Ping-Pong Auto-zero Scheme

To address the bandwidth limitation imposed by the chopper frequency, an offset-stabilized amplifier can be used [12]. As illustrated in Fig. 3(a), the proposed offset-stabilized amplifier contains two signal paths: a low-frequency signal path composed of G_{m1}, G_{m3}, and G_{mL}, and a high-frequency signal path composed of G_{m2} and G_{mL}. The low-frequency path provides higher gain and dominates the overall offset and noise performance, thus requiring chopper for low-noise design. In contrast, the high-frequency path is used to enhance the amplifier's bandwidth.

In the stage of G_{m1}, chopper and Ping-Pong auto-zero combining scheme is adopted to minimize the residual offset voltage. The amplifier's offset is first compensated using the auto-zero technique, and then chopper modulation is applied. This hybrid approach offers enhanced ripple rejection capability. Due to the alternating storage and amplification nature of the auto-zero technique, a Ping-Pong architecture is necessary to achieve continuous-time amplification. F_1 and F_2 are complementary clocks, they control the operation of the auto-zero amplifiers in the Ping and Pong stages, respectively. When F_1 is high and F_2 is low, the Ping stage engages in amplification by connecting to the circuit, while the Pong stage stores the offset. The roles reverse when the states of F_1 is high and F_2 are swapped. The chopper switch adopts a bootstrap switch design to reduce the channel injection effect of the chopper. The clock frequency of F_1 is half that of F_{chop}.

The designed PGA features a gain-of-1 configuration, where the amplifier operates with unity-gain feedback. Without sufficient phase margin, the system may experience oscillations. The low-frequency path consists of three cascaded amplifiers,

introducing at least three poles and resulting in poor frequency response. Multipath Nested Miller Compensation (MNMC) [13] is adopted to address this problem. Compensation capacitors C_{m1} and C_{m2} are used to lower the dominant pole and the non-dominant poles are pushed towards high frequencies. Simultaneously, the high-frequency path acts as a feedforward path, introducing an additional zero. The transfer function of the offset-stabilized amplifier is

$$A\left(s\right) = \frac{A_0\left(1 + \frac{s}{\omega_z}\right)}{\left(1 + \frac{s}{\omega_{p1}}\right)\left(1 + \frac{s}{\omega_{p2}}\right)\left(1 + \frac{s}{\omega_{p3}}\right)}, \quad (4)$$

where

$$A_0 = G_{m1}G_{m3}G_{mL}R_{o1}R_{o2}R_L, \quad (5)$$

and

$$
\begin{aligned}
\omega_{p1} &= \frac{1}{C_{m1}G_{m3}G_{mL}R_{o1}R_{o2}R_L}, \\
\omega_{p2} &= \frac{G_{mL}}{2C_L} - \frac{G_{mL}}{2C_L}\sqrt{1 - \frac{4G_{m3}C_L}{G_{mL}C_{mL}}}, \\
\omega_{p3} &= \frac{G_{mL}}{2C_L} + \frac{G_{mL}}{2C_L}\sqrt{1 - \frac{4G_{m3}C_L}{G_{mL}C_{mL}}}, \\
\omega_z &= \frac{G_{m1}G_{m3}}{G_{m2}C_{m1}}.
\end{aligned}
\quad (6)
$$

To realize a single-pole system with ω_{p1} as the dominant pole, ω_{p3} should be placed well beyond the unity-gain bandwidth (at least twice the GBW to ensure a 60° phase margin), while ω_z should be equivalent to ω_{p2} to cancel the secondary pole at intermediate frequency.

To ensure a rail-to-rail input common-mode range, G_{m1} and G_{m2} employ a constant transconductance structure. G_{m3} is a differential folded input pair connected in parallel with G_{m2}. PMOS differential pairs are selected to reduce their noise contribution. To improve the output slew rate and current output capability, a transconductance linear loop-controlled Class AB output stage is employed. Since G_{m1} is the primary offset source and noise source, its current source circuit adopts a source degeneration structure to reduce the offset and noise contribution. Additionally, a CMFB circuit is required to control the output common-mode point of G_{m1}.

B. Offset Calibration Circuit for TMR Sensors

The TMR sensor utilizes the Wheatstone bridge, where a DC offset voltage results from mismatches in the magnetoresistive components. When the PGA gain setting is too high, the sensor's offset voltage can directly cause output saturation. A current-steering DAC is designed to compensate for the sensor's offset. The generated compensation voltage is $V_{DAC} = (I_p - I_n)R_{in}$, which is related to the values of I_p and I_n. It is not necessary for the compensation error to be smaller than LSB/2 of the subsequent ADC. Therefore, I_p and I_n are generated using 8-bit current-steering DACs, with a unit current of $I_0 = 0.5\mu A$.

The offset calibration module operates in two modes: manual calibration with a fixed code DAC_BIN<7:0> and

979-8-3315-8850-2/25 $31.00 © 2025 IEEE

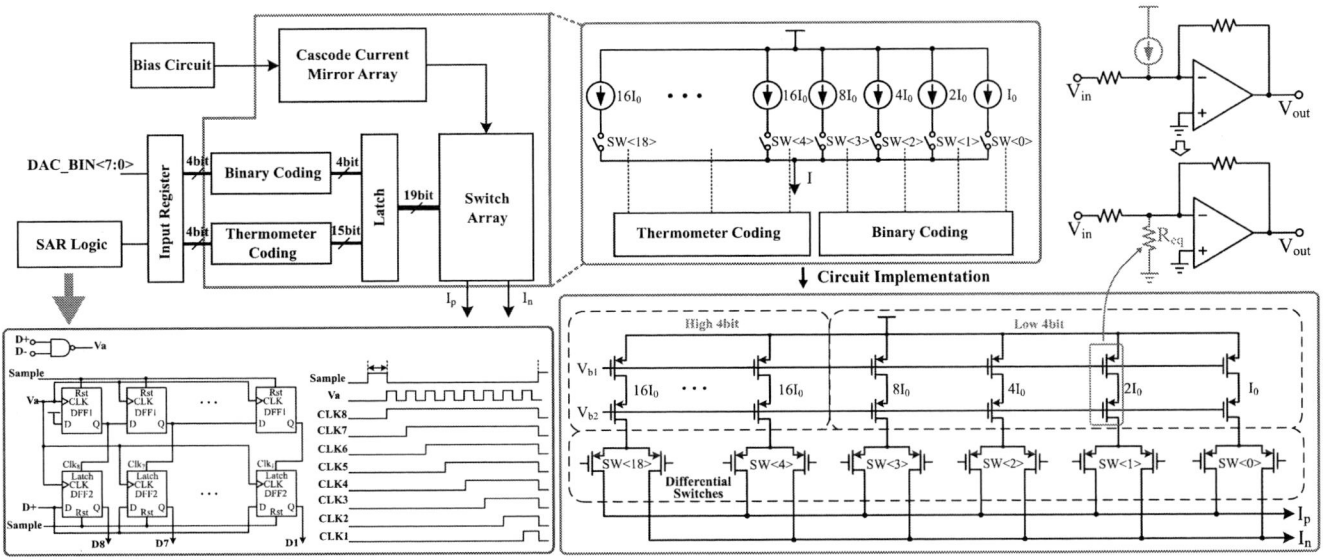

Fig. 4. Offset calibration principle and circuit implemention for TMR sensors' intrinsic offset.

automatic calibration based on the SAR algorithm. As shown in the Fig. 4, this module mainly consists of a bias circuit, a current source array, a switch array, and a decoding circuit. The decoding circuit controls the switch array to adjust the current distribution between the p-side and n-side. A segmented encoding scheme is adopted to improve DNL. The lower bits are encoded using a binary scheme, whereas the higher bits adopt thermometer encoding to enhance linearity. This approach strikes a balance between circuit area and accuracy.

A current source array is composed of individual current sources, and their relative mismatches as well as finite output impedance can degrade overall system performance. In a current-steering DAC, the higher-weighted bits require more current sources, thereby demanding better matching performance of the MOS transistors. To mitigate the mismatch, the current source area is designed based on the DAC's INL yield model. The INL_yield is a function of the relative mismatch of the current sources, and its relationship with the relative standard deviation $\frac{\sigma_I}{I_0}$ of a single current source is given by [14]

$$\frac{\sigma_I}{I_0} \leq \frac{1}{2\sqrt{2^N \cdot E}}. \tag{7}$$

The relative deviation of the unit current source is calculated based on the INL_yield, which can be substituted into (8) to obtain the minimum area of the current source.

$$(WL)_{\min} = \frac{A_\beta^2 + \frac{4A_{V_{TH}}^2}{(V_{GS}-V_{TH})^2}}{2\sigma_I^2}. \tag{8}$$

Mismatch in the current sources can degrade DAC linearity, while the finite output impedance of practical current sources affects the system's gain accuracy. R_{eq} represents the equivalent output impedance of the current source. A smaller R_{eq} results in greater current diversion, leading to

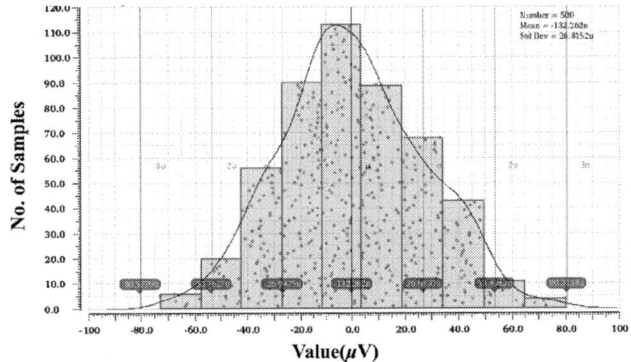

Fig. 5. PGA circuit offset MC simulation Result.

variations in system gain. Therefore, to enhance linearity, R_{eq} must remain constant within the signal bandwidth. Compared to single-transistor current sources, common-source common-gate structures not only suppress the effects of channel length modulation and ensure accurate current replication, but also significantly improve output impedance.

IV. SIMULATION RESULTS

With the gain setting to 1, mismatch and process variations were introduced using Monte Carlo simulations. The input offset voltage of the PGA was extracted from the transient waveform after the stabilization period. The results are shown in Fig. 5. Based on 500 Monte Carlo runs, the mean offset voltage was measured at -132.26nV, with a standard deviation (σ) of approximately 26.82μV. Compared to the conventional amplifiers in milivolt-voltage, this demonstrates a substantial improvement in performance.

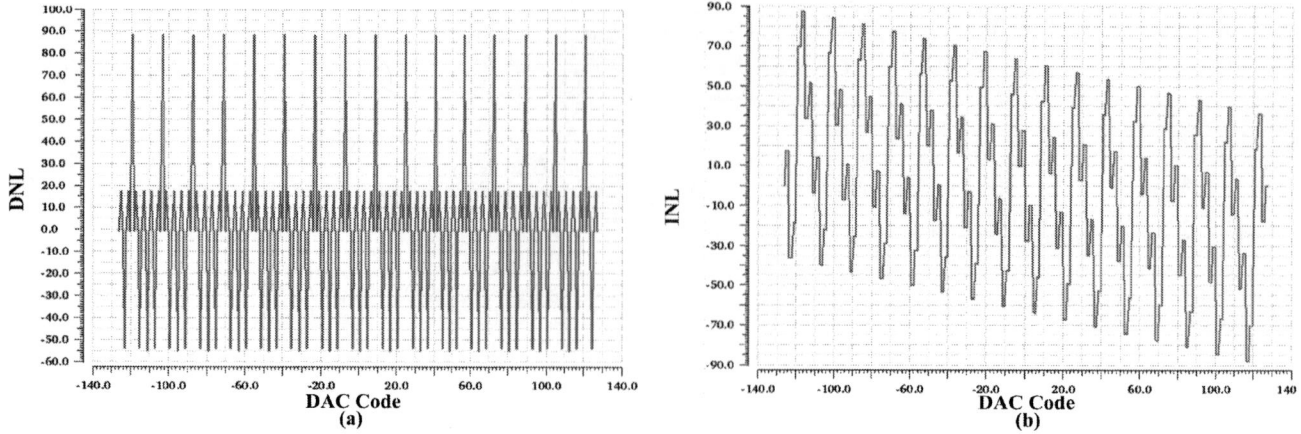

Fig. 6. Current DAC simulation results. (a) DNL; (b) INL.

TABLE II
IRN AND EFFECTIVE RESOLUTION IN DIFFERENT GAIN
CONFIGURATION OF THE HIGH RESOLUTION PGA.

Gain	IRN(μV_{rms})		Effective Resolution(bit)[1]	
	Min	Max	Min	Max
1	33.69	45.55	16.27	16.71
2	27.42	33.95	15.69	16.00
4	19.15	25.08	15.13	15.52
8	15.33	20.54	14.42	14.84
16	14.85	19.95	13.46	13.89
32	14.68	19.69	12.48	12.90
64	14.51	19.75	11.48	11.92
512	14.61	19.65	8.48	8.91

[1] Effective Resolution = $\log_2 \frac{\text{Input Range}}{\text{RMS Noise}}$.

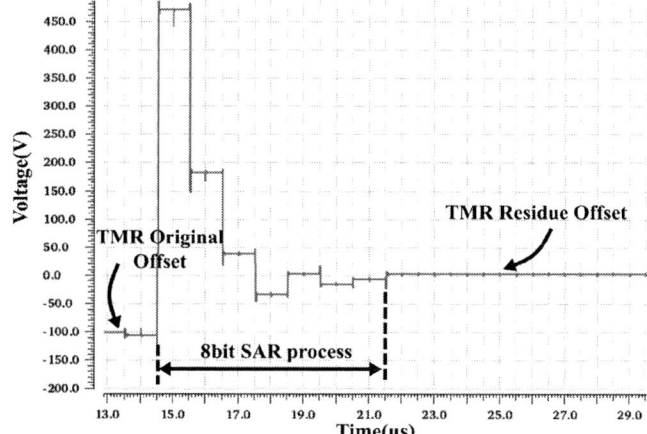

Fig. 7. TMR offset calibration simulation result.

By simulating the equivalent input-referred noise power spectral density of the PGA, the effective resolution can be evaluated. As shown in Table II, the RMS value of the equivalent input noise is obtained for different gain settings over the frequency range of 1Hz to 300kHz.

As illustrated in Fig. 6, the DNL falls within the range of -0.05 to 0.09LSB, and the INL remains within -0.09 to 0.09LSB. Both metrics satisfy the design requirement of staying below ±0.5LSB. A 100mV DC voltage is applied to the input of the second-stage PGA to emulate the sensor offset. The PGA gain is set to 1, and the automatic calibration function is enabled. The output waveform is shown in Fig.7. The SAR logic operates at a clock frequency of 1MHz. Calibration begins at 13μs, and the residual offset converges automatically by 21.5μs.

The layout of the low-noise PGA is illustrated in Fig. 8. Table III presents a performance comparison between the proposed PGA and previously reported designs. This paper adopts an architecture similar to that in [2], with both achieving relatively high bandwidth through resistive feedback, whereas

the CCIA architecture employed in [5], [15], and [16] results in lower bandwidth. However, the proposed PGA achieves higher THD with lower power consumption than reference [2]. At a gain setting of 128, the noise RMS within a 1kHz bandwidth is approximately 1μV$_{rms}$, which is lower than the values reported in [5] and [15]. Considering power consumption, bandwidth, noise, and linearity, the PGA designed in this work demonstrates excellent overall performance.

V. CONCLUSION

This paper presents a high-precision, low-noise PGA tailored for TMR AFE applications. A comprehensive system-level analysis is performed for a two-stage PGA architecture followed by a passive-active cascaded LPF. To achieve superior noise performance, a high-gain chopper offset-stabilized amplifier combining Ping-Pong auto-zero scheme is proposed. The main signal path ensures full rail-to-rail input-output dynamic range and enhanced linearity. Furthermore, sensor offset in the TMR is compensated in the second stage through

979-8-3315-8850-2/25 $31.00 © 2025 IEEE

TABLE III
PERFORMANCE COMPARISON.

	This work[1]			[2]	[15][1]	[5]	[16]
Process(nm)	180			180	130	180	180
Architecture	TOIA[2]			Resistance Feedback	CCIA[2]	CCIA+ADC	CCIA+ADC
Power Supply(V)	1.8			1.8	1.1	1	1.8
Power(mW)	1.99 (TT, 27°C)	1.81 (SS, -40°C)	2.16 (FF, 85°C)	2.808	0.00176	0.0035	4.5
Gain(dB)	0 ~ 54.19			14.8 ~ 30.4	51 ~ 60.3	40 ~ 53.5	-6.02 ~ 42.14
BandWidth(Hz)	300k			7.1M	1.72k	300	60k
IRN(μV$_{rms}$)	15.66[3]	14.75	19.76	N/A	2.95 @5.3Hz~1.72kHz	1.16 @0.5Hz~100Hz	0.028 @1Hz~25Hz
THD(dB)	-83	-84	-81	-54	-48	-59.6	-120

[1] The result is post-simulation.

[2] TOIA represents Three-Opamp Instrumentation Amplifier. CCIA represents Current-Feedback Instrumentation Amplifier.

[3] The noise power is integrated over the frequency range from 1Hz to 300kHz.

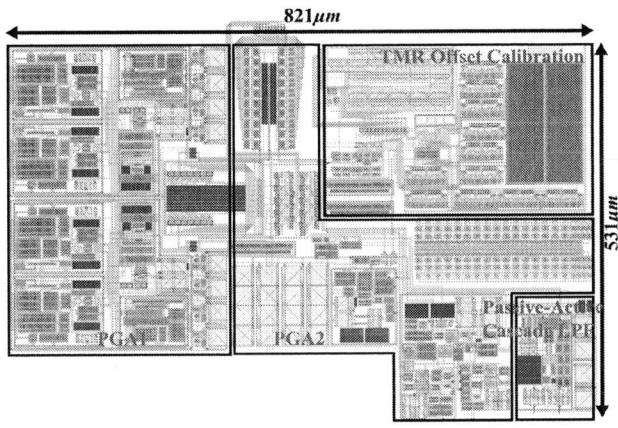

Fig. 8. Low-noise PGA layout.

a current-steering DAC. The complete PGA design has been implemented and verified through post-layout simulations using a 0.18μm CMOS process.

ACKNOWLEDGMENT

This paper is supported by "the Fundamental Research Funds for the Central Universities". The project number is 2242025K30014.

REFERENCES

[1] B. F. Hassan, "A 37μW, binary weighted PGA based on a novel degeneration transistor ladder," IEEE Trans. Circuits Syst. II, Exp. Briefs, vol. 65, no. 1, pp. 36-40, Jan. 2018.

[2] T. A. Martins, D. Reyes, B. Sanches, et al., "A Class AB programmable gain amplifier for an UWB breast cancer detection system," in Proc. 28th IEEE Int. Conf. Electron., Circuits Syst. (ICECS), 2021, pp. 1-4.

[3] J. Kang, S. Park, S. Kim, et al., "A 56 to 110 dB gain programmable gain amplifier with second-order band-pass filter for ultrasonic sensor systems," in Proc. Int. Conf. Electron., Inf. Commun. (ICEIC), 2023, pp. 1-5.

[4] Y. Park, J. H. Cha, S. H. Han, et al., "A 3.8μW 1.5NEF 15GΩ total input impedance chopper-stabilized amplifier with auto-calibrated dual positive feedback in 110 nm CMOS," IEEE J. Solid-State Circuits, vol. 57, no. 8, pp. 2449–2461, Aug. 2022.

[5] T. Zhang, Y. Li, C. Su, et al., "A 1V 3.5μW bio AFE with chopper-capacitor-chopper integrator-based DSL and low power Gm-C filter," IEEE Trans. Circuits Syst. II, Exp. Briefs, vol. 69, no. 1, pp. 5-9, Jan. 2022.

[6] P. Khatavkar and S. Aniruddhan, "432nW per channel 130nV/$\sqrt{Hz}$ ECG acquisition front-end with multifrequency chopping," IEEE Trans. Very Large Scale Integr. (VLSI) Syst., vol. 27, no. 9, pp. 2021–2032, Sep. 2019.

[7] C. C. Enz and G. C. Temes, "Circuit techniques for reducing the effects of op amp imperfections: Autozeroing, correlated double sampling, and chopper stabilization," Proc. IEEE, vol. 84, no. 11, pp. 1584–1614, Nov. 1996.

[8] M. Safiallah, A. R. Danesh, H. Pu, et al., "A current adjusting auto-zeroing technique for DC offset and flicker noise cancellation," IEEE Trans. Very Large Scale Integr. (VLSI) Syst., vol. 31, no. 12, pp. 1950–1959, Dec. 2023.

[9] T. Dong, T. Qu, W. Qin, et al., "A 3.5MHz BW 128nTrms resolution TMR readout using ping-pong auto-zeroing and SAR-assisted offset calibration for contactless current sensing," in Proc. ESSCIRC—IEEE 49th Eur. Solid State Circuits Conf., 2023, pp. 1–4.

[10] Y. Kusuda, "A 60 V auto-zero and chopper operational amplifier with 800 kHz interleaved clocks and input bias current trimming," IEEE J. Solid-State Circuits, vol. 50, no. 12, pp. 2804–2813, Dec. 2015.

[11] T. Rooijers, J. H. Huijsing, and K. A. A. Makinwa, "A chopper-stabilized amplifier with a relaxed fill-in technique and 22.6 pA input current," IEEE Solid-State Circuits Lett., vol. 6, pp. 165–168, 2023.

[12] Y. Zhou, S. Song, Y. Zheng, et al., "A 20.3μW 1.9 GΩ input impedance capacitively coupled chopper-stabilized amplifier for bio-potential readout," IEEE Trans. Circuits Syst. I, Reg. Papers, vol. 71, no. 4, pp. 1520–1530, Apr. 2024.

[13] Y. Zhou, Y. Zheng, and K. N. Leung, "An output capacitorless low dropout regulator with high slew rate and unity gain bandwidth," in Proc. IEEE Int. Symp. Circuits Syst. (ISCAS), Sevilla, Spain, Oct. 2020, pp. 1-5.

[14] J. J. Chen, Y. S. Hwang, J. H. Wu, et al., "A new improved V-square-controlled buck converter with rail-to-rail OTA-based current sensing circuits," IEEE Trans. Very Large Scale Integr. (VLSI) Syst., vol. 29, no. 7, pp. 1428–1436, Jul. 2021.

[15] E. A. Hamed, M. Atef, and M. Abbas, "A low power programmable gain integrated front end for electromyogram signal sensing," in Proc. 25th Int. Conf. Mixed Design Integr. Circuits Syst. (MIXDES), Gdynia, Poland, Jun. 2018, pp. 103-108.

[16] G. Mora Puchalt, G. Banarie, P. Czapor, et al., "A 128 -ksps 120dB THD low noise analog front end," in Proc. 48th Eur. Solid-State Circuits Conf. (ESSCIRC), Milan, Italy, Sep. 2022, pp. 401-404.

[17] M. A. Mubin and A. Marzuki, "A low-noise amplifier utilizing current-reuse technique and active shunt feedback for MedRadio band applications," International Journal of Electrical and Electronic Engineering and Telecommunications, vol. 9, no. 5, pp. 306-316, Sep. 2020, doi: 10.18178/ijeetc.9.5.306-316.

2025 The 10th International Conference on Integrated Circuits and Microsystems

An Adaptive dynamic R-tree indexing algorithm for HDI PCB layout

<table>
<tr>
<td align="center">1st Zhang Yang
School of Information Engineering
Wu Han University of Technology
Wu Han, China
301996@whut.edu.cn</td>
<td align="center">2nd Liang Xiaoyu
School of Information Engineering
Wu Han University of Technology
Wu Han, China
xiaoyu_l@126.com</td>
<td align="center">3rd Ning Xu*
School of Information Engineering
Wu Han University of Technology
Wu Han, China
Xuning@whut.edu.cn
*Corresponding author</td>
</tr>
</table>

Abstract—The layout data intuitively conveys the fundamental spatial information of a PCB, and efficient querying of which is crucial for IC industrial manufacturing and EDA software algorithm development. In high-density interconnect (HDI) PCBs, traditional layout indexing algorithms suffer from low query efficiency due to massive data volumes, primarily caused by the large number of components and their compact spatial distribution. To address these challenges, this paper proposes an optimized approach to the classic R-tree algorithm. It adjusts the data storage capacity of nodes based on the scale of layout data adaptively and further optimizes the node splitting strategy using a heuristic splitting factor. This method reduces the overlap rate of spatial division, constructing an efficient indexing structure tailored for HDI PCBs. Experimental results show that compared with the traditional R-tree indexing algorithm, the proposed method achieves superior spatial index planning while ensuring query accuracy, with significantly improved indexing efficiency.

Index Terms—HDI PCB;R Tree;Dynamic Splitting;Layout Indexing

Fig. 1. HDI PCB Project Sample (part)

I. INTRODUCTION

The relentless progress in integrated circuit (IC) technology increasingly mandates higher integration densities in contemporary electronic devices to accommodate enhanced functionality within spatial constraints [1]. This necessity has driven the development of High-Density Interconnect Printed Circuit Boards (HDI PCBs), a crucial enabling technology (Figure 1). Concurrently, this trend poses a significant challenge due to the exponential growth in layout data volume. Modern PCB layouts are characterized by: (1) high component density, (2) compact spatial arrangement, and (3) modular distribution patterns corresponding to functional blocks.

In practice, engineers rely heavily on PCB-oriented Electronic Design Automation (EDA) tools, particularly during critical design phases that demand frequent and precise queries of layout data to ensure project quality [2]. For instance, automated routing solutions addressing the length-matching problem require continuous layout queries to detect obstacles

Shenzhen Science and Technology ProgramJCYJ20220818102002005.
*Corresponding author

and ensure sufficient clearance for trace meandering [3]. Similarly, global layout optimization methodologies for 2.5D packaged ICs employ iterative refinement based on efficient layout indexing [4].

Grid-based algorithms are efficient for large-scale datasets, including hierarchical gridding with weighted projection matching [5] and Voronoi diagram-based region indexing [6]. However, the former's suitability for "anchor point" indexing and requirement for pre-gridding, along with the latter's "seed node"-dependent expansion, fail to meet the high-frequency, precise region query demands of EDA tools.

Optimization often involves space-filling curves (e.g., Hilbert and Z curves). Integrating machine learning with Hilbert curves can reduce 2D data to 1D scalars [7], but this compromises geometric topology and assumes data distribution uniformity. While using Z-addresses and cumulative integral functions can enhance efficiency [8], the computational cost and generalization limits of Deep Neural Network (DNN)-based methods remain significant drawbacks.

Tree structures (e.g., quadtree, R-tree) are fundamental for storage and query systems due valued for their hierarchy,

979-8-3315-8850-2/25 $31.00 © 2025 IEEE   523

pruning capabilities, and balance properties. While specialized trees like PMR quadtree variants [9] and HV/VH trees [10] exist, their fixed partitioning struggles to adapt to complex PCB layouts. In contrast, R-trees organize spatial data via Minimum Bounding Rectangles (MBRs), facilitating dynamic partitioning, range queries, and superior adaptability. Prior work has focused on optimizing R-tree node splitting [12–14]. Though recent studies using reinforcement learning show promise [15], they suffer from high computational costs and pre-trained model generalization limitations.

To address the dense device distribution and modularity of high-density PCBs, this paper improves the R-tree algorithm. The Adaptive Dynamic R-tree Indexing Algorithm introduces adaptive node capacity and dynamic splitting strategies to resolve the efficiency bottlenecks of traditional R-trees.

II. INDEXING PROBLEM FORMULATION

As illustrated in a typical communication circuit controlled by a Microcontroller Unit (MCU) (e.g., Figure 2), retrieving all precise layout data within a specified rectangular query area (e.g., U3, via V1, IC1, U4, resistor Rs1) is a fundamental requirement in Printed Circuit Board (PCB) engineering.

The PCB layout indexing problem, critical for meeting the needs of practical PCB engineering and informed by references [13-14], is effectively decomposed into two steps:

1) **R-tree Construction:** Geometrically preprocess the layout spatial data and store the results accurately in an R-tree.
2) **Data Query and Acquisition:** With the data stored, the query task is equivalently transformed into a traditional tree search problem.

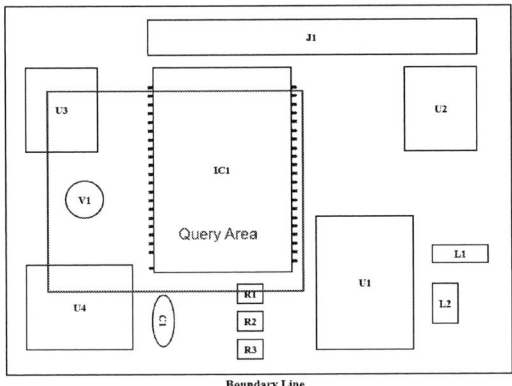

Fig. 2. Example of area query for PCB layout

Therefore, HDI PCB layout indexing involves: given layout data (coordinates and shapes) and a query region (typically rectangular), retrieve the data entries covered by the region from the R-tree based on spatial information. This approach is validated by practical engineering examples and references. However, unlike data structures focusing only on one-dimensional relationships (e.g., numerical magnitude), the R-tree stores data with two-dimensional spatial characteristics,

meaning data elements can be spatially independent or overlapping.

A significant challenge arises from the finite capacity of R-tree nodes. When a node's data storage exceeds a predefined threshold, a node split occurs, reallocating the original node's data into two new nodes based on their spatial relationships. Crucially, minimizing the overlap rate of Minimum Bounding Rectangles (MBRs) during R-tree construction is vital for enhancing both construction quality and search efficiency. Furthermore, optimizing the data reallocation process during node splitting provides another avenue for improving R-tree search performance.Consequently, this paper focuses on two core areas for optimization:

1) Determining the optimal capacity threshold for R-tree nodes.
2) Optimizing the R-tree node splitting algorithm.

III. ADAPTIVE DYNAMIC R-TREE INDEXING ALGORITHM

This section optimizes the spatial simplification of layout components, R-tree node capacity, and the splitting algorithm to enhance layout indexing efficiency for massive layouts.

A. Preprocessing of Device Packaging and R-Tree Node Capacity

To enable efficient spatial indexing of layouts, components are abstracted based on their geometric properties rather than their electronic functions. This approach prioritizes the component's physical packaging footprint for spatial analysis.

We classify all layout components into four fundamental geometric primitives: rectangles, circles, ellipses, and polygons, which simplifies the complex layout for subsequent spatial processing and ensures scalability for large designs.For instance,MCUs, resistors, and external pins are uniformly treated as rectangles, Ceramic capacitors can be modeled as ellipses , vias and LED are represented as circles.Following this geometric classification, a Minimum Bounding Rectangle (MBR) is generated for each component to uniformly represent its coordinates and spatial footprint (Figure 3).

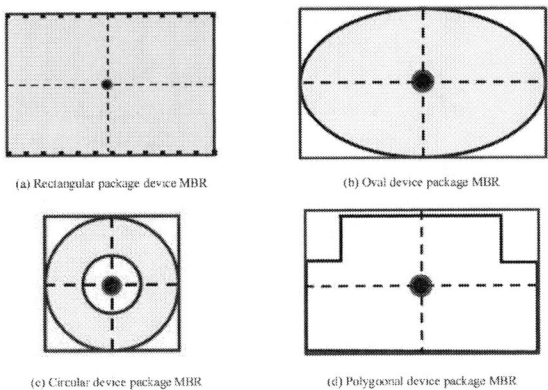

Fig. 3. MBR for different types of device packages

Unlike B-trees or B*-trees, R-trees employ a distinct structure where Data Entries—representing the Minimum Bounding Rectangles (MBRs) of actual layout objects—are stored exclusively in the leaf nodes (data nodes). Intermediate nodes are used solely to maintain spatial index relationships. Considering the project example in Figure 2, the spatial information for all devices is transformed into 13 MBRs, designated R1 through R13, which constitute the actual data, as illustrated in Figure 4(b).

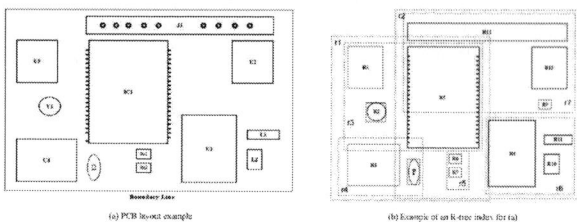

(a) PCB layout example   (b) Example of an R-tree index for (a)

Fig. 4. Example of PCB layout index system

The final R-tree structure and its indexing relationship are depicted in Figure 5. For descriptive clarity, Rx denotes a Data Item node (leaf), while rx signifies an intermediate Index MBR node.The spatial query retrieval process is executed as follows:

Step 1 (Root Indexing): The search is initiated by entering the root node based on the query region, leading to an index hit on r1.

Step 2 (Intermediate Traversal): From r1, the search traverses to the second-level nodes, resulting in indices for r3, r4, and r5.

Step 3 (Data Retrieval): The system descends from the index nodes to the data nodes:

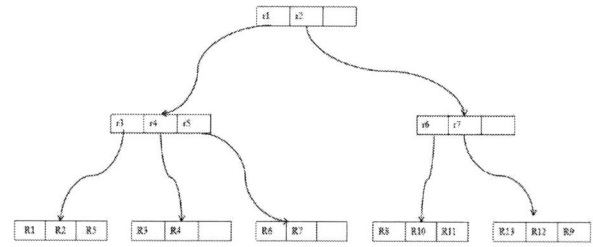

Fig. 5. Example of PCB layout index system

Considering that each MBR in an intermediate node acts as a pointer to the next level, a restrictive node capacity($M_{\max}$) was analyzed. A comparative analysis of the R-tree structures for the same PCB layout (Figure 5 versus Figure 6) reveals that the structure generated with $M_{\max} = 2$. This proliferation of nodes and index-MBRs inherently leads to a higher MBR overlap rate, which substantially reduces the subsequent search efficiency of the R-tree.

Based on the analysis of the relevant development status of the industry in reference [16], the device scale of HDI-PCB is basically 10^2 10^3, considering the tree height h of the final established R-tree is $h \approx \log_{M_{\max}/2}(N)$. To address the escalating computational load associated with recursive node

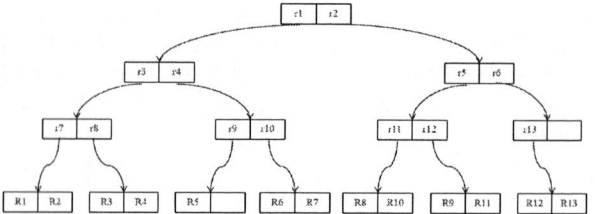

Fig. 6. Compared with Figure 5 in different $M_{\max}$

splitting($T(n) = O(M_{\max}^2)$). In this paper, define the $M_{\max}$ of a node, as shown in Equation (1).

$$M_{\max} = \max\left(\lceil\sqrt{N}\rceil, M_{\mathrm{def}}\right) \qquad (1)$$

M_{def} is a parameter preset in advance (considering that the node is split twice, it is usually 2^n).

B. Improvement of dynamic node splitting algorithm

Traditional linear splitting methods, while often utilizing O(n) assignment algorithms for speed, are limited by relying on a fixed splitting axis. Reference [14] partially addresses this by selecting the axis (top, bottom, left, or right) that yields the most uniform division. The goal of this refinement is to minimize MBR overlap and improve the tree structure.However, neither of these approaches adequately accounts for the specific spatial distribution characteristics inherent in PCB layout data. This oversight presents a significant optimization opportunity. As demonstrated by the two assignment results in Figure 7, the desired outcome is the modular, block-based partitioning shown in Figure 7(b).

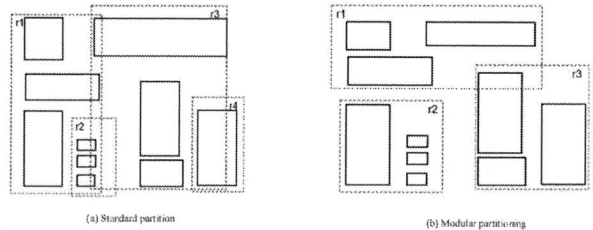

(a) Standard partition   (b) Modular partitioning

Fig. 7. The allocation results of different splitting algorithms

Clearly, the distribution in Figure 7(b) closely matches the traits of low overlap and features of modular distribution.

C. Heuristic Seed Selection for Modular Node Splitting

Acknowledging the inherent "modular distribution" characteristic of actual PCB layouts, traditional single-dimensional node splitting methods are often suboptimal for achieving the desired partition (as exemplified in Figure 7(b)).

To address this, we define the newly generated nodes as "Left Child Node" (Node_L) and "Right Child Node" (Node_R) (used here strictly to distinguish the two new nodes, irrespective of their spatial position). To substantially improve the splitting effect, this paper introduces a Heuristic Seed Selection Algorithm, which comprehensively analyzes the

979-8-3315-8850-2/25 $31.00 © 2025 IEEE   525

spatial distribution information along both the x and y axes to heuristically determine the optimal splitting axis. Furthermore, it incorporates a local optimization logic to select a pair of MBRs as seed entries.

The primary objective of this seed selection process is two-fold: 1, Minimize the MBR overlap rate between peer nodes at the same R-tree level. 2, Steer the node split towards a geometrically "better" configuration that promotes modular partitioning. The following section proposes the detailed method for selecting these critical seed nodes:

1) **Determining the Splitting Scope:** When a new MBR causes the number of entries in a node to exceed the threshold $M_{\max}$, the node must be split. We define the set of MBR entries within the node as $E = \{e_1, e_2, \ldots, e_{M_{\max}+1}\}$. The first step in the splitting process is to calculate the two-dimensional coordinate range of the current node along both the x- and y-axes.

2) **Identifying the Initial Seed Pair:** To ensure that the selected seed nodes accurately reflect the spatial distribution across both the x and y axes, we dynamically determine the specific **splitting axis (Axis)** for each division based on coordinate interval analysis. Following this, we select the two MBR entries, $E = \{e_i, e_j\}$, along the determined Axis that exhibit the maximum separation (largest coordinate gap).

3) **Overlap Reduction Strategy:** Define Overlap(e_m, e_n) as the overlap area between two Data MBRs, e_m and e_n, and Area(e_k) as the area of MBR e_k. If the overlap of the initial seed pair exceeds a predefined Seed Overlap Threshold, $\mathbf{S_{max}}$, new entries must be re-evaluated and added to the seed set from E. The final selected seed pair is denoted as $\{(e_i^*, e_j^*)\}$.

Combining this optimization with the axis selection from Step 1, we propose the following Heuristic Seed Selection Formula:

$$\{e_i^*, e_j^*\} = \arg \min_{\substack{e_p, e_q \in E, \\ p < q < M_{\max}}} (\text{Overlap}(e_p, e_q)), \tag{2}$$
$$\text{where} \quad \text{Overlap}(e_i, e_j) > S_{\max}$$

Here, we define the Seed Overlap Threshold, $S_{\max}$, as a proportion of the area of one of the seed entries:

$$S_{\max} = P \cdot \text{Area}(e_i)$$

where P is a parameter constrained such that $P < 1$. Figure 8 illustrates the results of the initial seed node selection process (highlighted in light blue) under different splitting axis scenarios.

D. Dynamic Allocation of MBRs to Child Nodes

Once the seed MBRs $\{e_i^*, e_j^*\}$ are determined, the remaining $M_{\max} - 1$ entries from the original node must be partitioned into the (Node$_L$) or (Node$_R$). Compared to conventional methods based on singular area increment—which often leads to non-uniform density distribution, this paper employs an **Area Cost Function** to govern entry allocation.

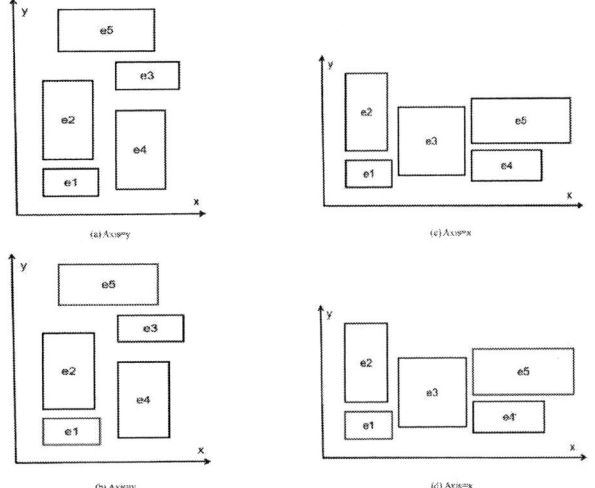

Fig. 8. Initial Seed Selection Results under Various Splitting Axis Scenarios

Crucially, a **Density Factor** (D) is introduced to balance the groups, thereby preventing the new partitions from becoming excessively dense or sparse and ultimately improving the uniformity of the tree structure.

Figure 9 provides a schematic overview of assigning the remaining MBRs after the initial seeds have been placed. The initialization step involves assigning $\{e_i^*, e_j^*\}$ to Node$_L$ and Node$_R$, respectively, and removing them from the total entry pool E. For instance, if $\{e_2, e_3\}$ are the selected initial seeds, the orange square highlights the Overlap(e_1, e_2). The two blue dashed regions represent the potential merged area of hypothetical subsets, such as the MBR for the group $\{e_1, e_2\}$ and the MBR for the group $\{e_1, e_3\}$.

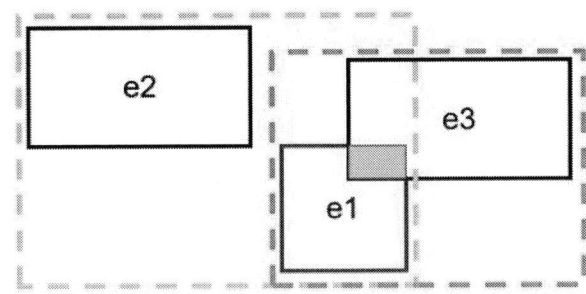

Fig. 9. Illustration of MBR Distribution within Node Splitting

Given that MBR overlap and area expansion (as shown in Figure 9) impair spatial indexing efficiency, this algorithm introduces a process to balance the effect of merged MBR area growth and mutual overlap.

1) **Definition of cost function:** To ensure that each remaining MBR is assigned to the most appropriate child node, this paper utilizes the following **cost function** (Equation 3) as the basis for allocation. Define Merge(Node$_X$, e_i) as the merged area resulting from adding entry e_i to

child node Node_X, $|\text{Node}_X|$ as the current number of MBRs within that node.

$$C = \Delta A + \lambda \cdot D \qquad (3)$$

Here, ΔA is defined as the area increment of the child node Node_x after merging the new MBR (e_k):

$$\Delta A = \text{Area}(\text{Merge}(\text{Node}_x, e_k)) - \text{Area}(\text{Node}_x)$$

To mitigate the negative impact of MBR overlap—which can be viewed as "negative area growth" on search efficiency—we introduce the term

$$D = |\text{Node}_x|/\text{Area}(\text{Node}_x)$$

, which represents the **average MBR density**. λ is defined as a variable weighting parameter.

2) **Greedy Allocation Strategy:** The allocation process proceeds by iteratively calculating the costs, $\mathbf{C_L}$ and $\mathbf{C_R}$, for assigning each remaining MBR to Node_L and Node_R, respectively. We define the **minimum node capacity** as $\mathbf{M_{min}}$ (typically $M_{min} \leq M_{max}/2$).

The MBR assignment is then determined using a **greedy strategy** based on the cost function and the current MBR count ($|\text{Node}_x|$). Detailed assignment rules are as follows:

  a) If $C_L < C_R$ and $|\text{Node}_R| \geq M_{min}$, the MBR is assigned to Node_L.
  b) If $C_R < C_L$ and $|\text{Node}_L| \geq M_{min}$, the MBR is assigned to Node_R.

If the number of MBRs in both nodes exceeds M_{min}, the MBR is assigned to the child node with the **minimum calculated cost**. This entire greedy allocation strategy is formally described by the following equations:

$$\begin{cases} C_L = \Delta A_L + \lambda \cdot \frac{|\text{Node}_L|}{\text{Area}(\text{Node}_L)} \\ C_R = \Delta A_R + \lambda \cdot \frac{|\text{Node}_R|}{\text{Area}(\text{Node}_R)} \end{cases} \qquad (4)$$

Integrating the density equation and the MBR distribution criteria, the comprehensive allocation strategy is derived and presented below:

$$\begin{cases} L, & C_L < C_R, |\text{Node}_L| < M_{min} \\ R, & C_R < C_L, |\text{Node}_R| < M_{min} \\ L, & C_L < C_R, |\text{Node}_L| > M_{min} \text{ and } |\text{Node}_R| < M_{min} \\ R, & C_R < C_L, |\text{Node}_L| > M_{min} \text{ and } |\text{Node}_R| < M_{min} \end{cases} \qquad (5)$$

L and R symbolize the corresponding child nodes (Left and Right) for the assignment of each entry.

IV. Experimental Results and Analysis

The algorithm was developed in **C++** and executed on a Windows system featuring a **3.5 GHz CPU** and **32 GB of RAM**. To ensure realistic testing, test cases were derived from reference industrial open-source projects, closely simulating the layout characteristics of industrial **Printed Circuit Boards (PCBs)**.

A. HDI PCB Query Scenario and Testing Procedure

The core of the experiment addresses the efficient PCB query scenario: **"Given a rectangular Query Area (QRA) within the PCB layout, retrieve all enclosed PCB devices."** The testing procedure is simplified to:

1) **Construct an R-tree** from the layout data.
2) **Input a QRA** (query rectangle).
3) **Return all data entries** within the QRA's spatial extent.

B. Benchmarks and Evaluation

Reference [17] compares the spatial indexing performance of three R-tree variants— **Quad** (Quadratic Cost), **Corner-Based Splitting**, and **SOLD** (Splitting based on Objects' Locations Distribution). Although their specific application scenarios differ, the underlying optimization principles—namely, optimizing node splitting using quadratic or linear costs while considering overlap area and node balance—are reproduced and utilized as a benchmark.

This paper's evaluation employs two primary testing schemes. For a data volume scale of 10^3, the **default node capacity threshold** is defined as $M_{def} = 2^5 = 32$. The final evaluation metric is the **query time**, calculated as the average time across multiple experimental runs

1) **Biased Query Experiment (Tendency Query):** This experiment is designed to closely simulate real-world PCB layout queries. We employ a "tendency" area coverage search where a query is defined as valid only if the retrieved number of spatial objects exceeds $5\% \times$ Case.size(). This ensures that the queries are meaningful and target areas with significant data density. The specific results are presented in **Table 1**.

TABLE I
BIASED QUERY EXPERIMENT RESULTS

Case No.	Size	Final average result size	[17]/ms	Ours/ms
Case1	923	102.85	1.211	0.691
Case2	1812	524.28	3.861	2.291
Case3	2768	430.20	5.958	2.698
Case4	3841	800.67	9.857	4.705
Case5	4241	1225.80	14.824	6.981
Case6	6143	1759.20	18.271	7.457

2) **Random Query Experiment:** This scenario aims to demonstrate the algorithm's robustness and adaptability across the PCB layout space. It involves randomly generating a specific number of query regions to test general search effectiveness thus requiring no explicit constraint on the definition of a valid query area. The specific results are presented in **Table 2**.

As detailed in **Table 1**, the proposed algorithm achieves a significant improvement in query efficiency, demonstrating an average gain of **50.88%** compared to the algorithms presented in Reference [17] across all test cases. Furthermore, in the **Random Query Experiment (Table 2)**, the efficiency is consistently higher, with an average increase of **49.72%**.

979-8-3315-8850-2/25 $31.00 © 2025 IEEE

TABLE II
RANDOM QUERY EXPERIMENT

Case No.	Size	[17]/ms	Ours/ms
Case1	923	0.882	0.535
Case2	1812	5.364	2.676
Case3	2768	6.344	3.153
Case4	3841	8.376	3.872
Case5	4241	9.411	4.581
Case6	6143	11.925	5.552

Figure 10 illustrates a segment of the query results for **Case 1**, demonstrating the precise indexing capability of the algorithm: all devices within the query region (highlighted in light blue) are accurately retrieved.

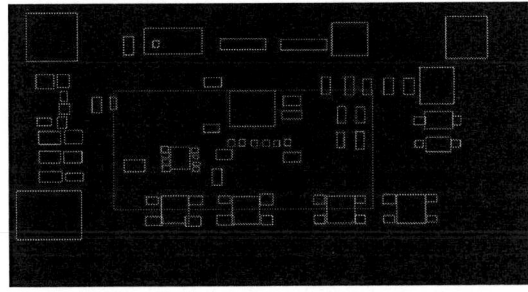

Fig. 10. Illustration of Query Results

V. CONCLUSION

The proposed algorithm represents PCB layout devices using Minimum Bounding Rectangles (MBRs) for geometric spatial representation,while significantly simplifying the overall problem complexity. Furthermore, we addressed the tree's branching factor (node capacity threshold) and introduced novel improvements to the R-tree node splitting process and MBR entry allocation, aligning them with practical engineering requirements. Based on the experimental data, this paper successfully developed and implemented a highly efficient HDI-PCB layout indexing algorithm, realized through enhancements to the R-tree structure and guided by a dynamic and adaptive methodology.

As integrated circuit scale continues to increase, the demand for high-efficiency layout indexing in Electronic Design Automation (EDA) tool development and manufacturing will only intensify. Future research in indexing algorithms must focus on handling larger-scale datasets with more complex spatial information and intricate geometric shapes.

REFERENCES

[1] Mazi Hosseini.Future Trends in PCB Design 2025: How AI, 5G, and Flexible Electronics Are Transforming the Industry [OL].Available at: Future Trends in PCB Design 2025: How AI, 5G, and Flexible Electronics Are Transforming the Industry - Arshon Inc. Blog,2025-3-17.

[2] Qi Ming, Zhao Chensu, Zhang Chao, Yu Wenjian. A Layout Transformation Method for the High-Precision Parasitic Extraction and Timing Analysis of VLSI Interconnects[J]. Journal of Computer-Aided Design Computer Graphics,2015,27(6): 1145-1152.

[3] Weijie Fang, Longkun Guo, Jiawei Lin, Silu Xiong, Huan He, Jiacen Xu, and Jianli Chen. 2024. Obstacle-Aware Length-Matching Routing for Any-Direction Traces in Printed Circuit Board[C]. In Proceedings of the 61st ACM/IEEE Design Automation Conference (DAC '24). Association for Computing Machinery, New York, NY, USA, Article 74, 1–6. https://doi.org/10.1145/3649329.3655915.

[4] P. Zhang, D. -W. Wang and W. -S. Zhao, "A Thermal and Power Integrity Co-Optimization Framework for 2.5-D Integrated Microsystem [J]. IEEE Transactions on Circuits and Systems I: Regular Papers, vol. 72, no. 3, pp. 1397-1410, March 2025, doi: 10.1109/TCSI.2024.3454628.

[5] WANG Ya-zhou,GU Wei-dong,FENG Jin-qiao. A Map Matching Algorithm Based on Subfield Search Method [J].Journal of QiLu University of Technology. ,2015,29(4):77-80.doi: 10.16442/j.cnki.qlgydxxb.2015.04.017.

[6] Y. Li and G. Liu.Area Queries Based on Voronoi Diagrams [C]. 2020 IEEE 36th International Conference on Data Engineering (ICDE), Dallas, TX, USA, 2020, pp. 2064-2068, doi: 10.1109/ICDE48307.2020.00245.

[7] W. Tang, C. Zhang, J. Yang, J. Wu and H. Huang.Updatable Spatial Learned Index Based on Dimensionality Reduction [C].2024 10th International Conference on Big Data and Information Analytics (BigDIA), Chiang Mai, Thailand, 2024, pp. 757-764, doi: 10.1109/BigDIA63733.2024.10808934.

[8] H. Wang, X. Fu, J. Xu and H. Lu.Learned Index for Spatial Queries [C]. 2019 20th IEEE International Conference on Mobile Data Management (MDM), Hong Kong, China, 2019, pp. 569-574, doi: 10.1109/MDM.2019.00121.

[9] J. Tayeb, Ö. Ulusoy and O. Wolfson.A Quadtree-Based Dynamic Attribute Indexing Method [J].The Computer Journal, vol. 41, no. 3, pp. 185-200, Jan. 1998, doi: 10.1093/comjnl/41.3.185.

[10] JIE REN, WEI-WEI PAN, YONG-JUN ZHENG, et al. Array based HV/VH tree: an effective data structure for layout representation [J]. Frontiers of information technology electronic engineering, 2012, 13(3): 232-237. doi:10.1631/jzus.C1100193.

[11] X. Wang, N. Xu and Y. Chen. Automatic Multi-Constraint Placement of Printed Circuit Board [C]. 2024 13th International Conference on Communications, Circuits and Systems (ICCCAS), Xiamen, China, 2024, pp. 1-6, doi: 10.1109/ICCCAS62034.2024.10652855.

[12] Antonin Guttman. R-trees: a dynamic index structure for spatial searching [C]. In Proceedings of the 1984 ACM SIGMOD international conference on Management of data (SIGMOD '84). Association for Computing Machinery, New York, NY, USA, 47–57. https://doi.org/10.1145/602259.602266.

[13] W. Lin, Y. Wu, X. Tan and Y. Yu. An Improvement of Index Method and Structure Based on R-Tree [C]. 2008 International Conference on Computer Science and Software Engineering, Wuhan, China, 2008, pp. 607-610, doi: 10.1109/CSSE.2008.405.

[14] Y. Liu, J. Fang and C. Han. A new R-tree node splitting algorithm using MBR partition policy [C]. 2009 17th International Conference on Geoinformatics, Fairfax, VA, USA, 2009, pp.1-6,doi: 10.1109/GEOINFORMATICS.2009.5293260.

[15] Tu Gu, Kaiyu Feng, Gao Cong, Cheng Long, Zheng Wang, and Sheng Wang. The RLR-Tree: A Reinforcement Learning Based R-Tree for Spatial Data [J]. Proceedings of the ACM on Management of Data (PACMMOD) 1, 1, Article 63 (May 2023), 26 pages. https://doi.org/10.1145/3588917 .

[16] Zhou, Q. (2022). Analysis of Global HDI Industry Status and Trends, 2022: Demand-Driven Advancement of HDI Process from Low-Order to High-Order.[OL]. https://www.huaon.com/channel/trend/840390.html.2025.5.

[17] E. Al-Nsour, A. Sleit and M. Alshraideh. A Pictorial Performance Comparison of Spatial Indexes [C]. 2020 11th International Conference on Information and Communication Systems (ICICS), Irbid, Jordan, 2020, pp. 042-047, doi: 10.1109/ICICS49469.2020.239554.

A Novel Region-Wise Automatic Routing Algorithm for Analog Circuit

Hao Xie
College of Integrated Circuits and Micro-Nano Electronics
Fudan University
Shanghai, China
23210720279@m.fudan.edu.cn

Wenxue Chen
College of Integrated Circuits and Micro-Nano Electronics
Fudan University
Shanghai, China
24210720170@m.fudan.edu.cn

Wei Zhang
Suzhou Foohu Technology Co., Ltd.
Shanghai, China
wei.zhang@foohu.com

Bijian Lan
Suzhou Foohu Technology Co., Ltd.
Shanghai, China
bijian.lan@foohu.com

Jing Wan*
College of Integrated Circuits and Micro-Nano Electronics
Fudan University
Shanghai, China
jingwan@fudan.edu.cn

Abstract—**Routing is a key step of circuit design, directly determining the performance of the circuit. Existing research on analog circuit routing is mostly applicable only to specific circuits and follows a net-wise routing approach based on a particular net order. These methods have poor generalization and cannot guarantee full routability and global optimization. This paper proposes a region-wise routing algorithm for circuits with macro-cells of arbitrary numbers, sizes, and shapes. The algorithm can determine whether the circuit is routable during the global routing stage. A special step is inserted between global and detailed routing to classify and positioning the pins of each net based on region sorting result. During detailed routing step, optimized routing strategies is selected based on the net conditions of the regions. Therefore, the algorithm in this paper can effectively improve routing efficiency, routability and has excellent generalization.**

Keywords—analog circuits, routing, region-wise, automatic

I. INTRODUCTION

With the rapid development of integrated circuit technology, digital circuit layout design has gradually achieved high levels of automation and standardization. However, analog circuit layout design remains a key bottleneck in the current field of Electronic Design Automation (EDA). The complexity of the design process and the high demand on the designer's expertise have resulted in slow progress in automation research. Among the various stages of analog circuit layout design, the routing process is particularly crucial and time-consuming. Therefore, developing efficient automated routing algorithms has become a core breakthrough in improving the efficiency of analog circuit design.

Analog circuits typically consist of multiple functional modules, such as operational amplifiers, bandgap reference sources, and so on. The layout of each module can directly impact the overall circuit performance through issues such as

parasitic coupling and signal integrity. Existing research mostly focuses on module-level layout optimization, but there is a lack of systematic solutions for the global routing of inter-module connections. Hence, researching high-performance and universal automated routing algorithms holds significant practical importance.

Currently, existing automated routing algorithms for analog circuits can be broadly classified into three categories. The first category abstracts the routing problem as a maze pathfinding problem. Some studies focus on single-layer routing for small nets. For instance, J. Zuo used the Rectilinear Steiner Minimum Tree for routing power and ground lines[1], and Liu J successfully proposed a method based on reinforcement learning to construct rectilinear Steiner trees[2]. Weijie Chen introduced an improved A* search algorithm to find multi-layer routing paths that satisfy the minimum area constraint[3]. The second category designs routing algorithm optimized for specific region shapes, typically routing for channels at two ends of a rectangle or all four ends of a switchbox. For example, Lienig J proposed a genetic algorithm-based two-end channel routing algorithm[4], and Luk W K proposed a greedy algorithm-based four-end switch box routing algorithm[5]. The third category deals with routing for specific circuit types, such as the Analog-to-Digital Converter circuit, the Operational Amplifier circuit, and the Comparator circuit. Zhu K et al. developed a routing method using Variational Autoencoders (VAE)[6], and Zhang H et al. proposed a new routing collaboration framework that considers both geometric and electrical constraints[7].

However, there are still several shortcomings in existing research. First, previous algorithms are often dependent on specific circuits and cannot handle routing for layouts with macro-cells located arbitrarily. For example, the algorithm proposed in [6] uses VAE to implicitly extract routing

979-8-3315-8850-2/25 $31.00 © 2025 IEEE

expertise, respecting the guidance generated by the VAE model and applying additional constraints during routing. However, since it is trained on specific circuits, it cannot be applied to arbitrary circuit layouts. Second, most existing algorithms route nets sequentially based on a specific net order. For example, the algorithm proposed in [8] performs global routing for a specific net and then proceed with detailed routing for that net, without considering the global impact on other nets and the final routing result. Moreover, existing regional routing algorithms can only handle routing for two or four ends of a single region and cannot handle routing between multiple regions[5][9][10].

This paper proposes a region-wise routing algorithm for analog circuit modules. The algorithm considers macro-cells with arbitrary numbers, sizes, and shapes, making it highly versatile. The contributions of this paper can be summarized as follows:

- We designed a general routing algorithm that employs both global routing and detailed routing strategies. It calculates the track capacity for each region and plans paths between regions to improve routability.

- By analyzing the position of inter-region pins, the algorithm employs region-wise approach instead of net-wise, significantly improving the global routing quality.

- During detailed routing phase, the algorithm selects the appropriate regional detailed routing method based on distribution of nets in the region, improving the local routing quality and connectivity. The Left-Edge [9] and Dogleg algorithms[10] are further improved to overcome the loop limitations of net constraints and effectively handle loops in channel routing.

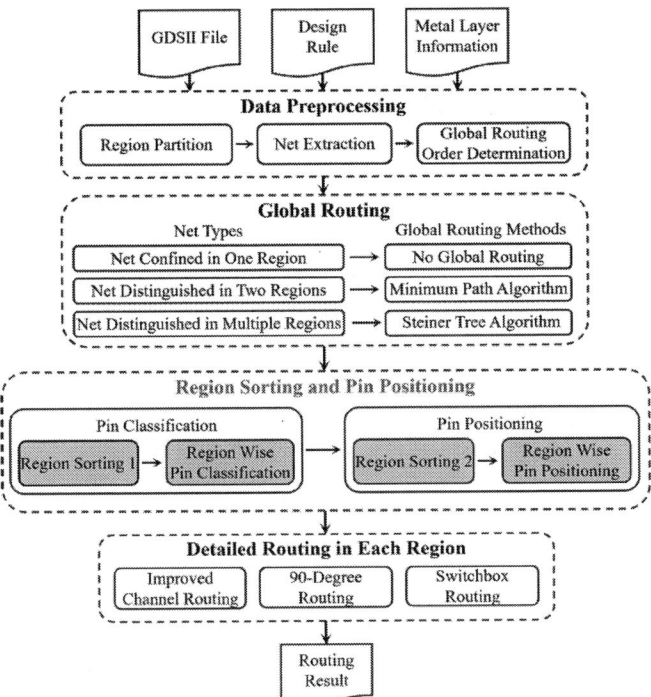

Fig. 1. The flow chart of the proposed routing algorithm.

II. METHODOLOGY

Fig. 1 illustrates the architecture of algorithm, which consists of four stages: (1) data preprocessing, (2) global routing, (3) region sorting and pin positioning, and (4) detailed routing in each region.

In the "data preprocessing" stage, all net information from the layout is extracted and the blank routing space is partitioned into individual regions. Subsequently, the total length of the Rectilinear Steiner Minimum Tree (RSMT) for each net is calculated. Based on these lengths, the global routing order is determined, prioritizing from the shortest to the longest. The "global routing" stage uses the shortest path algorithm or Steiner tree algorithm to perform initial routing planning through dynamic programming region's track capacity for multi-region pin nets. In the " region sorting and pin positioning " stage, we analyze the status and specific location of each pin between regions. Finally, in the "detailed routing" stage, optimized routing strategies are automatically selected based on the net occupancy of each region and perform routing within each region. Following the design rules, routing is organized for all regions to obtain the final routing result. This result is subsequently utilized to generate the GDSII file.

A. Data Preprocessing

Regions are fundamental components of the entire layout. In this stage, the irregular routing space on the layout is divided into regions, in which each edge is only adjacent to another region. The principle of region division is as follows: extend each edge of every macro cell until it intersects with another macro cell or the boundary of the entire layout. The rectangle formed by the intersections of all extended lines defines the region. This division ensures more accurate routing planning during the global routing phase. Next, the pin location information is extracted, and the RSMT algorithm is applied to add Steiner points and calculate the total length of the RSMT tree. Generally, as the length of a net increases, the number of possible shortest paths decreases. Therefore, nets with shorter lengths are prioritized, and the routing order for each net is determined in the global routing phase based on length, from shortest to longest. Fig. 2 shows the RSMT results and corresponding global routing order for three nets.

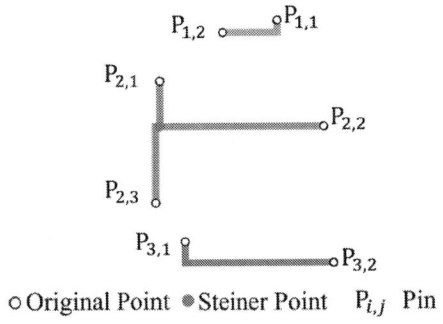

$\circ$ Original Point $\quad \bullet$ Steiner Point $\quad P_{i,j}$ Pin

Fig. 2. Simplified three Nets' RSMT results. The length of the RSMT tree for P_1, P_2, P_3 is 21.8, 93.3 and 54.9 respectively, the global routing order is thus P_1, P_3, P_2.

979-8-3315-8850-2/25 $31.00 © 2025 IEEE   530

B. Global Routing

Global routing is the first step in routing, determining rough path of the net and which regions each net will pass through. Based on how the terminals pins of each net are distributed across regions, nets are classified into three types: single-terminal region nets (where all terminal pins are in one region), two-terminal region nets (where terminal pins are distributed across two regions), and multi-terminal region nets (where terminal pins are distributed across three or more regions). For single-terminal nets, since all terminals are in one region, the routing will stay within that region, and global routing is not needed. For two-terminal region nets, the minimum path algorithm is used to find the global routing[8]. For multi-terminal region nets, the Steiner tree algorithm is used to find the routing. During global routing, the track capacity in each region is considered and the routing path is dynamically modified. If there is not enough track capacity during global routing, it means the layout cannot be routed. If the algorithm completes the global routing successfully, the layout is guaranteed to be routable. This method can predict whether the layout is routable before the detailed routing stage and can improve the layout's connectivity.

C. Region Sorting and Pin Position

This stage is crucial in transforming the routing method from "net-wise" to "region-wise" and can be divided into two sub-phases: classification of pins and determining the position of each pin.

1) Classification of Pins

After the global routing for all nets, which regional edge that each pin belongs to is known. However, apart from the original net pins in the layout, the exact positions of other net pins are unknown. Therefore, all net pins are categorized into "original PIN" (the original pins from the imported layout file) and "waiting PIN" (pins awaiting analysis). The state of the waiting PINs is defined as follows:

- If one edge of the opposite edges in a region contains the original PIN of a net, and the other edge contains the waiting PIN of that net, this waiting PIN is categorized as "fixed PIN." The position of the fixed PIN is the corresponding location of the original PIN on the opposite edge.

- If both edges of the opposite edges contain waiting PINs of a net, these two waiting PINs are categorized as "half-fixed PIN group." The positions of the half-fixed PINs are always opposite each other, and the routing path between them forms a straight line, but the exact position of this line is still to be determined.

- If the waiting PIN does not belong to the above two cases, it remains a waiting PIN. The priority of these categories is as follows: fixed PIN > half-fixed PIN > waiting PIN.

Next, the category of each waiting PIN in all regions needs to be analyzed:

First, the region connectivity graph is changed into a node connectivity graph. To simplify the analysis, regions without net pins on their four edges are removed, and edges between two region nodes without waiting PINs are also removed. All connected components in the node connectivity graph are identified and analyzed, starting from the smallest to the largest, based on the number of nodes in each component. For each connected component, each node is analyzed based on its degree, and the following types of nodes are checked:

- Nodes with degree 0: These are the last nodes in each connected component.

- Nodes with degree 1: Nodes with degree 1 are typically processed first.

- Nodes with degree greater than 1: If there is no any nodes with degrees 0 or 1, and there are nodes with a degree greater than 1, it indicates that these nodes form loops involving three or more nodes.

If multiple regional nodes have the same degree, the regional node with less track capacity and more pins has higher priority for pin classification.

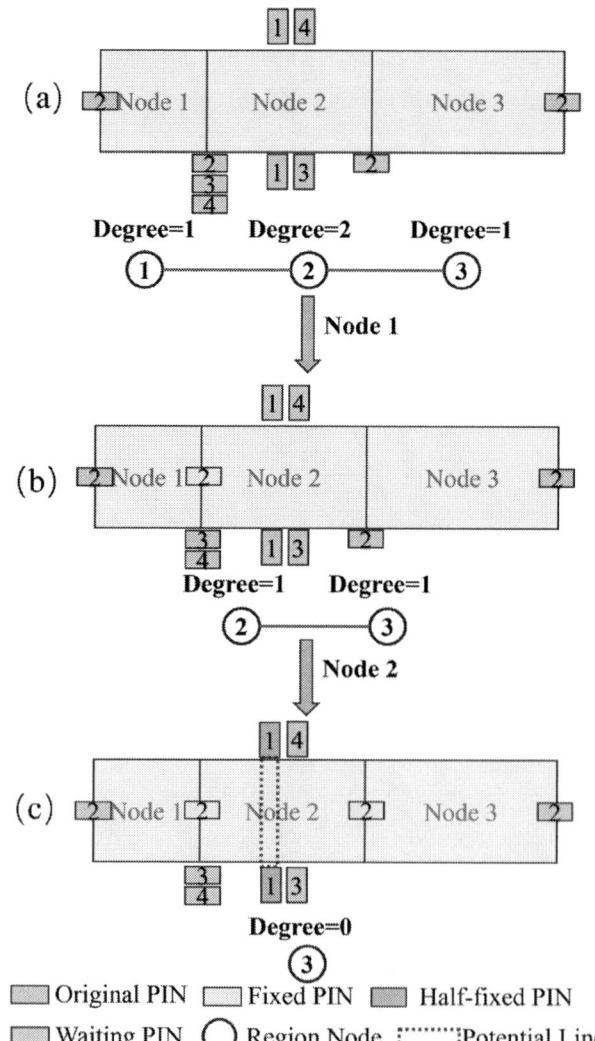

Fig. 3. A simplified example of pin classification. (a) the initial result, (b) the result after processing Node 1, (c) the result after processing Node 2.

Prior to classify the pins in each regional node, the pin information for the node is updated based on the data from previously analyzed adjacent regional node. After all nodes have been analyzed, the waiting pins for each node are updated according to the final pin information. This process yields the categories analysis results for all pins. Fig. 3 shows a simplified example of pin classification. After global routing, based on the connectivity of the region nodes, both Node 1 and Node 2 have a degree of 1. However, since Node 1 has fewer available tracks, it is prioritized for pin classification. Node 1 has an original pin of Net 2 on its left edge; therefore, the pin of Net 2 on its right edge is classified as a fixed pin, with its position fixed. Meanwhile, the pins of Net 3 and Net 4 remain as waiting pins. Subsequently, Node 1 and its associated edges are removed from the region node connectivity graph. Both Node 2 and Node 3 now have a degree of 1. Since Node 2 has fewer available tracks, it is selected for pin classification. Node 2 has a fixed pin of Net 2 on its left edge; thus, the pin of Net 2 on its right edge is classified as a fixed pin. Additionally, Net 1 has waiting pins on both the top and bottom edges of Node 2, which are classified as half-fixed pins. Other pins remain as waiting pins. Node 2 and its associated edges are then removed from the graph. At this point, only Node 3 remains, and it has no waiting pins. The pin classification process is thus concluded.

2) Determining the Position of Each Pin

After completing the previous tasks, the positions of the original PINs and fixed PINs are fixed. However, the positions of the half-fixed PINs and waiting PINs remain unknown. The next step is to position these types of pins:

To initiate the process, the number of edges incorporating half-fixed PINs and waiting PINs within each region is quantified. Concurrently, the total count of half-fixed and waiting PINs is determined, and an assessment is made to identify regions that exclusively contain half-fixed PINs without any waiting PINs. This data serves as the foundation for establishing the preliminary analysis sequence of the region nodes. Given that the connection of each half-fixed PIN must form a straight trajectory, the presence of half-fixed PINs within each region node is verified during the analysis phase. For regions containing half-fixed PINs, a dedicated half-fixed PIN tree analysis is conducted. The principle of this analysis is elucidated as follows:

For each region node containing half-fixed PIN, the following procedure is applied: If an edge contains a half-fixed PIN, the adjacent region node connected to this edge is designated as a first-level node of the current region node. This process is then recursively applied to each first-level node, continuing the analysis until only neighboring region nodes that do not contain any half-fixed PINs are encountered. The hierarchical structure formed by these nodes at each level is defined as the half-fixed PIN tree of the current region node.

The priority of region nodes in the half-fixed PIN tree is determined by their layer number, with region nodes on the same layer having the same priority. Region nodes that are not part of the half-fixed PIN tree are still analyzed according to the preliminary analysis order, until all region nodes have been analyzed.

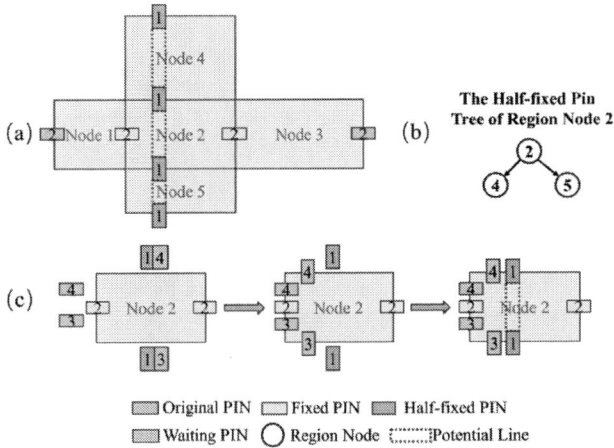

Fig. 4. A simplified example of determining the position of pins between regions. (a) the adjacent regions, (b) the half-fixed pin tree of region node 2, (c) the process of pin position.

By combining the preliminary analysis order of the region nodes and the current region node's half-fixed PIN tree, the final node analysis order is obtained.

Within each node region, the analysis order for all nets must also be determined. It should be noted that the analysis order of nets here differs from that in the first part. Initially, the number of edges containing waiting PINs and half-fixed PINs in each net is calculated. There are three possible cases: 0 (indicating that all PINs are half-fixed), 1, and 2. The nets are then sorted in descending order based on the number of edges containing waiting PINs. In cases where the number of edges with waiting PINs is the same, the number of half-fixed PINs within the net is calculated. Nets with a higher number of half-fixed PINs are given priority for analysis.

Based on the previously established node region analysis order and the net analysis order within these regions, a two-round analysis is conducted to determine the specific positions of each net. Fig. 4 shows the entire process of determining the position of pins for a particular region. After completing pin classification, based on the region node connectivity graph, Node 2 has the highest number of half-fixed pins, waiting pins, and edges, indicating the highest complexity among nodes. Therefore, Node 2 is selected for pin positioning. For Net 1, the half-fixed pins always form a straight line, resulting in a constant routing length. To minimize the total routing length for all nets, Net 3 and Net 4 are prioritized for positioning. Net 3 has waiting pins on the left and bottom edges. The waiting pin on the left edge is positioned at the bottommost track, while the waiting pin on the bottom edge is positioned at the leftmost track. Net 4, similar to Net 3, has its waiting pin on the left edge positioned at the topmost track, and the waiting pin on the top edge positioned at the leftmost track. Additionally, since Node 2 contains half-fixed pins, a half-fixed pin tree must be constructed. Node 2 has half-fixed pins on its top edge connected to Node 4 and on its bottom edge connected to Node 5. After completing pin positioning for Node 2, Nodes 4 and 5 are prioritized for pin positioning due to their connections to the half-fixed pins.

D. Detailed Routing in Each Region

Detailed routing represents the final stage of the entire algorithm, aiming to establish actual connections for each net within each region. In this step, different detailed routing strategies are selected based on the pin distribution within each region:

- If the region contains pins on only one opposite edges, the improved channel routing algorithm is selected.

- If the region contains pins on only one adjacent edges, the 90-degree routing algorithm is selected.

- If the region contains pins on three or four edges, the switchbox greedy routing algorithm[5] is selected.

Since the inputs to both the improved channel routing algorithm and the switchbox routing algorithm are purely numerical data, it is necessary to extract the net information of the regions. Additionally, since the outputs of these two algorithms are numerical matrices, a conversion program is required to transform these matrices into routing results that meet the required specifications.

1) Improved Channel Routing Algorithm

Traditional left-edge algorithm[9] and dogleg algorithm[10] are unable to handle loop constraints in the Vertical Constraint Graph (VCG). The improved channel routing algorithm presented in this paper addresses this limitation by analyzing net nodes that form loops, identifying nodes that can break the loop constraint, and making decisions on which net to process. This effectively solves the restrictions of loop constraints. The specific analysis principle is as follows:

Step 1: Using the VCG, nets are divided into different connected components. Nets within different connected components may reuse tracks, whereas nets within the same connected component cannot. The number of loops in the initial connected components and the number of available 0-0 free tracks in the initial tracks are evaluated.

Step 2: In each connected component, nets are categorized into two groups: those that participate in forming loops (main loop nets) and those that do not. Nets with an inbound degree branch are not involved in loop formation, whereas nets with an outbound degree branch are involved in loop formation.

Step 3: In each connected component, nets with an inbound degree of 0 and not participating in loop formation are identified. If only nets participating in loop formation remain, the loop is broken. The principle of loop breaking is as follows:

Look for nets in the main loop that have both inbound and outbound degrees of 1. If there are multiple such nets, choose the one that can reuse an available 0-0 free track. If there are multiple such nets, prioritize the one farther to the left. Finally, if conflicts arise between the nets selected from different connected components, prioritize the net farther to the left.

Step 4: Route the nets selected in Step 3 from left to right. First, check for track reuse possibilities. If no reuse is found, add a track at the bottom. For nets that pass through available 0-0 free tracks in the loop, initially route only the top-end connections, and route the bottom-end connections after all

nets are routed. For nets with multiple pins, select tracks for each segment that satisfy the circuit's requirements.

Step 5: Remove routed nets from the connected component and repeat the above steps until all nets are routed.

Step 6: Complete the routing for the 0-0 nets at the bottom end.

Fig. 5 shows an example of channel routing.

2) 90-Degree Routing Algorithm

In the 90-degree routing algorithm, nets are categorized into two types based on their distribution across edges: single-edge nets, which are confined to one edge, and double-edge nets, which span both edges. These two types of nets are subject to different routing strategies, with double-edge nets taking precedence over single-edge nets. Fig. 6 shows an example of a 90-degree routing result.

- Double-edge nets are classified according to the number of pins on the two edges into three categories: one-to-one (one pin on each edge), one-to-many (one pin on one edge and multiple pins on the other edge), and many-to-many (multiple pins on both edges).

- For single-edge nets, available space in the remaining area is identified after routing the double-edge nets, and this space is utilized to complete the routing for the single-edge nets.

3) Switchbox Routing Algorithm

When a region contains pins on three or four edges, the greedy algorithm[5] is initially applied to process the region, generating a horizontal track matrix and a vertical track matrix. To simplify routing complexity, a basic routing check is performed for all nets within the region. If this basic routing can successfully route all pins in the region, the numerical matrix results are not strictly followed for routing. Conversely, if the basic routing fails to route all nets, the numerical matrix results are strictly adhered to, and routing is implemented using a conversion program.

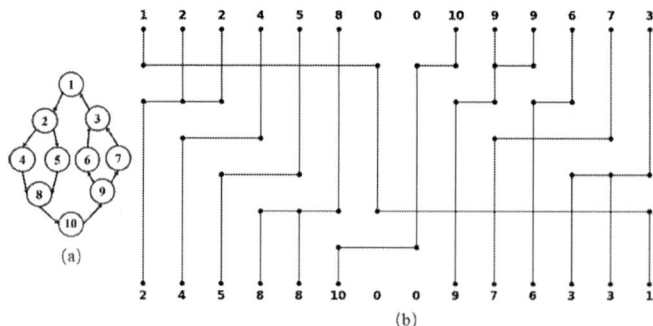

Fig. 5. An example of channel routing. (a) the Vertical Constraint Graph (VCG), where the nets form a loop. (b) the channel routing result.

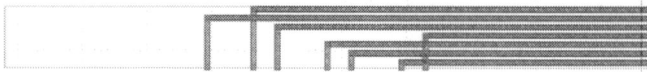

Fig. 6. An example of a 90-degree routing result.

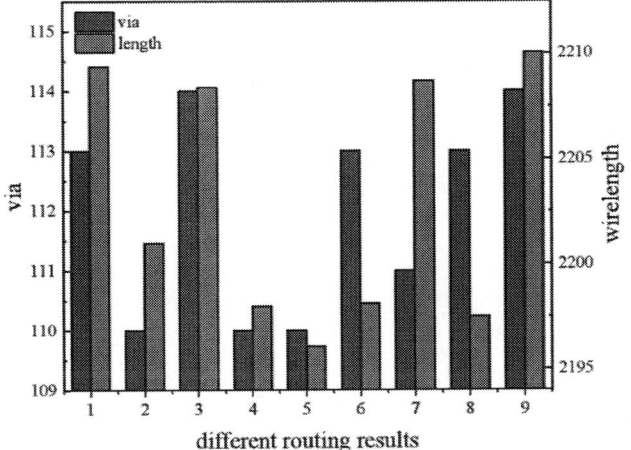

Fig. 7. The different routing results of an entire layout.

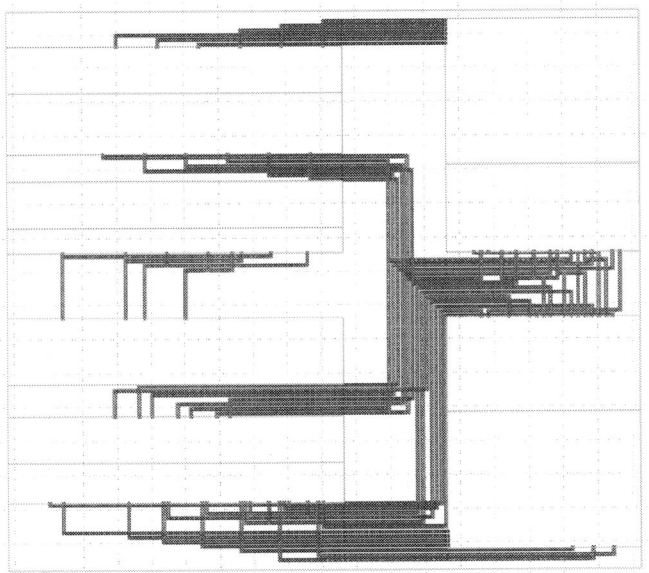

Fig. 8. The fifth routing result in Fig. 7.

III. RESULTS AND DISCUSSION

Our algorithm is implemented in the Python programming language and is suitable for layouts composed of macro-cells with arbitrary numbers, sizes, and shapes. It starts from the original GDSII file, divides the blank space of the entire layout, extracts net information, performs global routing based on the net sequence, and finally achieves detailed routing through different detailed routing strategies. In this section, multiple routing results are obtained by executing the test layout several times. Fig. 7 shows the number of vias and wire length for the different routing results. Fig. 8 selects the fifth routing result, which contains 110 vias and a wirelength of 2194.75.

IV. CONCLUSION

In this paper, we demonstrate a region-wise routing algorithm for macro-cells of arbitrary numbers, sizes, and shapes. This algorithm is capable of planning the available space in the layout and selecting different detailed routing strategies based on the net distribution of each region, thereby effectively improving routing quality. Moreover, it allows for a certain assessment of routability during the global routing stage, facilitating subsequent processing. The algorithm not only ensures that the routing results meet the requirements of design rule but also effectively avoids unnecessary area wastage. Further quantitative and comparative studies are conducted and will be presented in the future.

ACKNOWLEDGMENT

This work was supported by the National Key R&D Program of China (2021YFA1200500), National Natural Science Foundation of China (62474052,62404055), Key Technology R&D Plan Shanghai(25CL2900100), Shanghai Municipal Basic Research Program (25JD1402800).

REFERENCES

[1] J. Zuo, F. Li, and J. Wan, "High Efficient Automatic Power/Ground Layout Routing Algorithm for Analog ICS," in *2023 China Semiconductor Technology International Conference (CSTIC)*, 2023: IEEE, pp. 1-3.

[2] J. Liu, G. Chen, and E. F. Young, "Rest: Constructing rectilinear steiner minimum tree via reinforcement learning," in *2021 58th ACM/IEEE Design Automation Conference (DAC)*, 2021: IEEE, pp. 1135-1140.

[3] W. Chen, H. Yao, Y. Cai, and Q. Zhou, "Analog routing considering min-area constraint," in *2013 IEEE 10th International Conference on ASIC*, 2013: IEEE, pp. 1-4.

[4] J. Lienig and K. Thulasiraman, "A genetic algorithm for channel routing in VLSI circuits," *Evolutionary Computation*, vol. 1, no. 4, pp. 293-311, 1993.

[5] W. K. Luk, "A greedy switch-box router," *Integration*, vol. 3, no. 2, pp. 129-149, 1985.

[6] K. Zhu *et al.*, "GeniusRoute: A new analog routing paradigm using generative neural network guidance," in *2019 IEEE/ACM International Conference on Computer-Aided Design (ICCAD)*, 2019: IEEE, pp. 1-8.

[7] H. Zhang *et al.*, "Sageroute: Synergistic analog routing considering geometric and electrical constraints with manual design compatibility," in *2023 Design, Automation & Test in Europe Conference & Exhibition (DATE)*, 2023: IEEE, pp. 1-6.

[8] H.-J. Rothermel and D. A. Mlynski, "Automatic variable-width routing for VLSI," *IEEE transactions on computer-aided design of integrated circuits and systems*, vol. 2, no. 4, pp. 271-284, 2004.

[9] A. Hashimoto and J. Stevens, "Wire routing by optimizing channel assignment within large apertures," in *Proceedings of the 8th Design Automation Workshop*, 1971, pp. 155-169.

[10] D. N. Deutsch, "A "Dogleg" channel router," in *Papers on Twenty-five years of electronic design automation*, 1988, pp. 111-119.

979-8-3315-8850-2/25 $31.00 © 2025 IEEE

2025 The 10th International Conference on Integrated Circuits and Microsystems

A 0.6-3 GHz Active Double-Balanced Mixer Circuit Design

Jing Wang
School of Electronic Information
Hangzhou Dianzi University
Hangzhou, China
2207900732@qq.com

Jun Liu
School of Electronic Information
Hangzhou Dianzi University
Hangzhou, China
ljun77@hdu.edu.cn

Xuezhen Zhao
School of Electronic Information
Hangzhou Dianzi University
Hangzhou, China
zxz20765@163.com

Guodong Su
School of Electronic Information
Hangzhou Dianzi University
Hangzhou, China
guodong@hdu.edu.cn

Abstract—This paper presents a 0.6-3 GHz active double-balanced mixer model, which employs a passive double-balanced mixer ring as its core structure. The RF and LO baluns adopt an improved triple-coupled line Marchand balun configuration, while the local oscillator enhancement-type driver amplifier is incorporated to achieve mixing functionality across the target frequency band. By integrating nonlinear characteristic theory with the mixer's performance attributes, an effective modeling approach is developed for conversion loss and nonlinearity. Comparative analysis between simulation results and measured data demonstrates excellent agreement, with the simulated curves closely matching the actual measurement patterns. This strong correlation validates the engineering application potential of the proposed mixer model in broadband communication systems.

Keywords—triple-coupled-line Marchand balun, double-balanced mixer, high linearity

I. INTRODUCTION

Digital twin technology enables design optimization through the construction of digital prototypes, significantly reducing the development cycle. As a critical component in signal chains, the performance of mixers directly determines system sensitivity, anti-interference capability, and spectral efficiency. However, traditional mixer design predominantly relies on multiple iterations of circuit performance optimization, making it difficult to meet the demand for real-time prediction in intelligent scenarios. Therefore, establishing a mixer circuit model can elevate mixer design from trial-and-error iterations to a new stage driven by behavioral-level modeling.

Recent research on modeling has primarily focused on breakthroughs in nonlinear characterization of microwave devices and efficient simulation techniques [1-10]. A nonlinear model for GaAs-based MESFETs in passive mode (V_{ds} = 0 V) was developed and validated for its reliability in designing 2-8 GHz broadband mixers [2]. For terahertz-band devices, a proposed equivalent circuit model of a subharmonic mixer, optimized through parasitic parameter compensation, has achieved a conversion loss deviation of less than 1 dB between measurement and simulation results [6]. Additionally, a current-driven passive mixer model employing a switched-capacitor method has simplified the design process of quadrature receiver chains with a 25% duty cycle [8]. These advancements have progressively enabled a transition from device-level nonlinear compensation to system-level equivalent circuit modeling, providing multi-dimensional modeling support for high-frequency circuit design. However, there remains a lack of research focused on constructing models by extracting key

parameters directly from circuit-level performance characteristics.

To address the existing issues, this paper proposes an active double-balanced mixer circuit model. Through analytical derivation of the circuit's performance, the key parameters influencing the model are identified, enabling frequency conversion functionality across the 0.6–3 GHz range, with conversion loss consistently maintained within –9.5 dB. The feasibility of the proposed model is further validated through a comparison between simulation results and measured data from the fabricated chip.

II. PROPOSED MIXER ARCHITECTURE

Fig. 1 illustrates the proposed 0.6–3 GHz active double-balanced mixer model. The model comprises a LO-boosted driver amplifier, a core mixing ring, a balun transformer, and matching networks, achieving high linearity and low conversion loss under low local oscillator input power conditions.

A. Mixer Ring

The mixer circuit in this model is constructed using a ring of four Schottky diodes with identical electrical characteristics. The RF and LO signals are fed into the mixer ring through baluns, respectively. Under the drive of the LO signal, the diodes switch on and off periodically, exhibiting nonlinear behavior. This enables the multiplication of the two input signals to achieve frequency mixing, significantly reducing spurious components in the output signal while ensuring phase consistency and amplitude balance at the output.

As the core components of the mixer ring, the parasitic parameters of the diodes have a direct impact on the overall system performance. The cutoff frequency of a diode is given by the following equation:

$$f_c = \frac{1}{2\pi R_s \left(C_{j0} + C_{parastic} \right)} \tag{1}$$

Here, R_s represents the series resistance, C_{j0} denotes the junction capacitance under zero-bias conditions, and C_p refers to the package parasitic capacitance. When the cutoff frequency of the diode is relatively low, the capacitive reactance of the junction increases at high frequencies, causing the high frequency signal to be bypassed and preventing it from effectively passing through the diode for mixing. Therefore,

979-8-3315-8850-2/25 $31.00 © 2025 IEEE

determining the cutoff frequency allows for the estimation of the ratio among parasitic parameters.

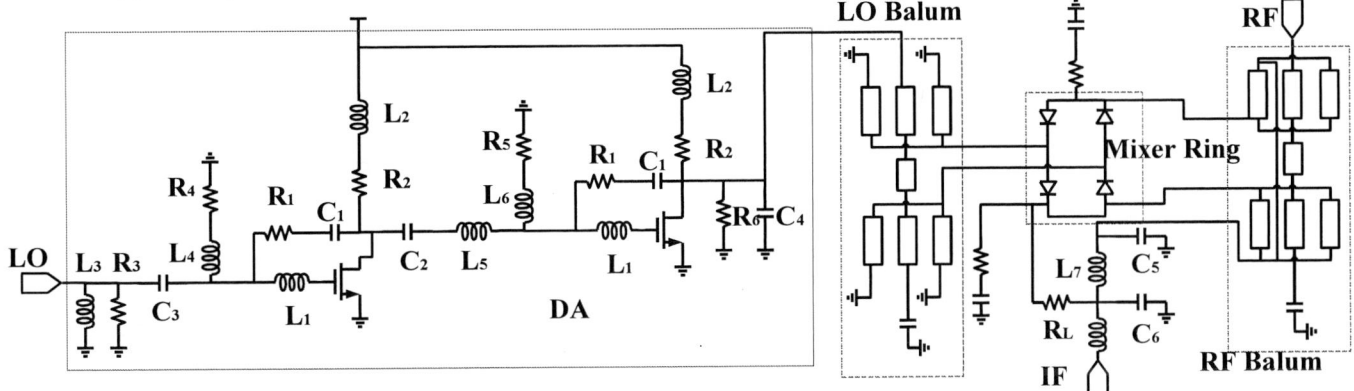

Fig. 1. Overall structure of designed Mixer

In practical model construction, the product of the diode's ideality factor and the single-gate width is a key determinant of the core loss. The conversion loss introduced by the diode is calculated as follows:

$$P_D = \frac{1}{T}\int_0^T V_F(t) \bullet i_F(t)dt \qquad (2)$$

Here, V_F represents the diode's forward voltage drop, and i_F denotes the forward current. As the gate width of the diode increases, the product of V_F and i_F in the integral term decreases, thereby reducing the resulting power loss.

When the input and output source impedances of a mixer are equal, the power gain is used to characterize its amplification factor, known as conversion gain (CG), which is expressed in decibels (dB) as:

$$CG\big|_{dB} = 20\lg\frac{P_{out}}{P_{in}} \qquad (3)$$

In mixer systems, port-to-port isolation is typically defined as the ratio of signal power leaking from one port to another relative to its original power. The isolation between the LO port and RF port is given by:

$$ISO_{LOtoRF} = \frac{P_{LOtoRF}}{P_{LO}} \qquad (4)$$

The linearity of a mixer is evaluated using the 1dB compression point and third-order intercept point. With fixed LO power, the 1dB compression point refers to the input power level where the output power deviates from linear gain by 1dB as the input power increases. Intermodulation distortion occurs when two closely spaced frequencies are input simultaneously - the input third-order intercept point (IIP3) is defined as the input power level where the fundamental output signal equals the

third-order intermodulation products generated by device nonlinearity. The IIP3 is calculated as:

$$|a_1 A_{IIP3}| = \left|\frac{3a_3 A_{IIP3}^3}{4}\right| \qquad (5)$$

where a_1 represents the first-order amplification coefficient and a_3 denotes the third-order nonlinear coefficient.

Therefore, during modeling, by defining the values of gain and third-order intercept point, the coefficients a_1 and a_3 can be derived, thereby completing the modeling of corresponding third-order intercept point parameters.

B. Broadband Balun

The broadband balun in this model employs a triple-coupled-line Marchand balun topology, as illustrated in Fig. 2. This configuration consists of two main transmission lines and one coupled line forming a set of distributed coupling, with two groups of coupled lines connected in series and maintaining identical coupling spacing. The conductor width ratio W2/W1 governs the current density distribution, while the L1/L2 length ratio primarily determines the in-band flatness of the frequency response. When L2 is reduced below 0.1λ, its transmission phase shift can be approximated as an ideal transmission line model, enabling its physical length to cover a broader frequency range.

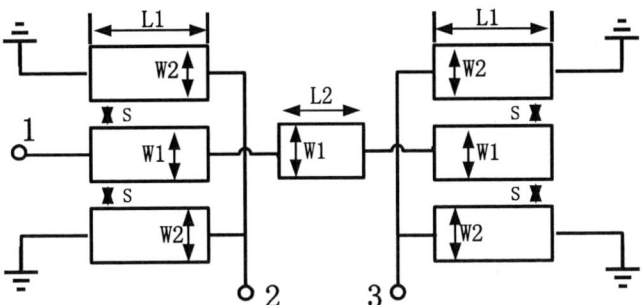

Fig. 2. Triple-coupled-line Marchand balun topology

979-8-3315-8850-2/25 $31.00 © 2025 IEEE

Based on the input-output characteristics of Marchand baluns, each coupling unit functions similarly to a coupler, allowing the derivation of even- and odd-mode characteristic impedances through ABCD matrix analysis under even-odd mode excitation, as expressed in the following equations:

$$Z_{0e} = Z_0 \sqrt{\frac{1+k}{1-k}} \qquad (6)$$

$$Z_{0o} = Z_0 \sqrt{\frac{1-k}{1+k}} \qquad (7)$$

Analysis reveals that in coupled transmission lines, Z_{0e} and Z_{0o} are significantly influenced by the coupling spacing (k). By substituting Equations (6) and (7) into the ABCD matrix under even-odd mode excitation, we obtain the insertion loss expression for the Marchand balun:

$$S_{21} = \frac{\sqrt{1-k^2}}{\cos\theta\sqrt{1-k^2} + j\sin\theta} \qquad (8)$$

$$S_{31} = \frac{jk\sin\theta}{\cos\theta\sqrt{1-k^2} + j\sin\theta} \qquad (9)$$

Where $\theta = \beta l$. The 90° phase difference between S21 and S31 not only enables its application as a quadrature phase shifter but also ensures the 180° phase difference requirement at the output ports.

C. DA

The driver amplifier (DA) in this model employs two independently designed amplifier gain units, as shown in Fig. 3. The resistive shunt feedback structure formed by R1 and C1 achieves both bandwidth extension and flat gain characteristics, while L2 and R2 provide DC biasing. Since the MOS transistor's input impedance exhibits capacitive behavior at high frequencies, the series-connected inductor and capacitor at the input form a resonant circuit to counteract this capacitive reactance. This module not only meets the requirements for high-speed switching of the diode ring in the mixer but also reduces insertion loss caused by the broadband balun, ultimately improving the overall model's conversion loss and linearity.

The DA's linear amplification capability is characterized by its forward voltage gain (G). As this amplifier only considers gain for small-signal operation, the scattering parameter S21 is typically used for practical modeling:

$$G(dB) = 20\log(\frac{V_{out}}{V_{in}}) \qquad (10)$$

Output power represents one of the most critical specifications of the driver amplifier. The key to successful model implementation in this work lies in achieving sufficient output power to drive the mixer, which can be expressed in terms of load resistance and output voltage across it.

$$P_{out} = \frac{V_{out}^2}{2R_{load}} \qquad (11)$$

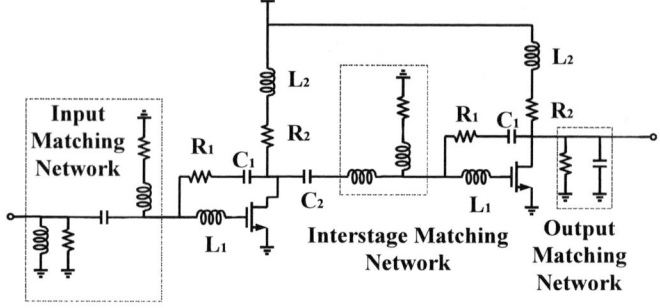

Fig. 3. DA structure

The small-signal equivalent circuit of the negative feedback structure in this amplifier is illustrated in Fig. 4.

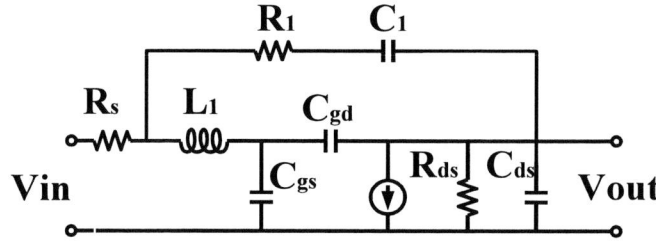

Fig. 4. Small-signal equivalent model of negative feedback structure

Applying Kirchhoff's Current Law (KCL) to the circuit with gate voltage V_G and drain voltage V_D yields the transfer function of this equivalent circuit.

$$\frac{V_{out}}{V_{in}} = \frac{G_m + sC_{gd}}{sC_{gd} + \frac{1}{Z_{out}}} \cdot$$
$$\frac{1}{1 + Z_{in} \cdot \left[sC_{gs} + sC_{gd}(1 - \frac{G_m + sC_{gd}}{sC_{gd} + Z_{out}^{-1}}) \right]} \qquad (12)$$

where $Z_{in} = R_s + sL_1$, $Z_{out} = (R_{ds} \| R_1) \| (sC_{ds}^{-1} \| sC_1^{-1})$.

By carefully designing the values of R1, C1, and L1, the amplifier's gain and gain flatness can be effectively adjusted to meet system requirements.

III. MODEL VERIFICATION

Based on the fundamental structure of the active double-balanced mixer model established previously, the verification process involves integrating input/output data from target devices with the model and evaluating the discrepancy between simulated and measured results using Normalized Mean Square Error (NMSE). The NMSE is calculated as follows:

$$NMSE_{dB} = 10\log_{10}(\frac{\sum\limits_{i=1}^{n}|y_o(i) - y_m(i)|^2}{\sum\limits_{i=1}^{n}|y_o(i)|^2}) \quad (13)$$

where n represents the number of samples, $y_o(i)$ denotes the actual measurement results, and $y_m(i)$ corresponds to the model simulation outputs. A smaller NMSE value indicates higher model accuracy.

The verification of the active double-balanced mixer model was conducted with the RF signal set at 1.8 GHz and 0 dBm power, while the LO signal operated at 0.6 GHz with -7 dBm power. Harmonic balance simulation was employed for analysis, and the comparative results are presented in Fig. 5 and 6.

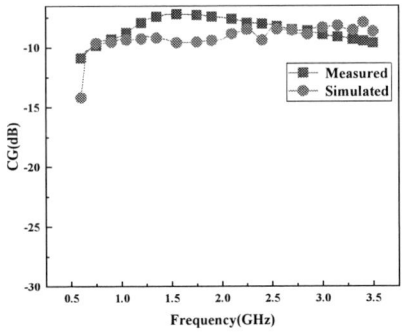

Fig. 5. Comparison of Simulated and Measured CG

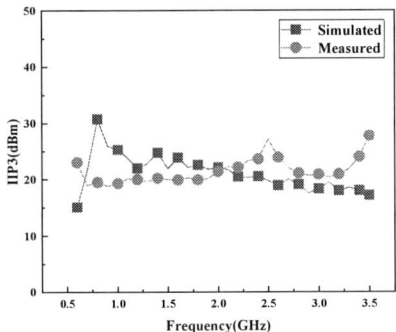

Fig. 6. Comparison of Simulated and Measured IIP3

In addition, the proposed model also supports comprehensive verification of other critical parameters such as return loss, port-to-port isolation, and 1dB compression point.

Fig 7 presents the simulated return loss characteristics for the LO, RF, and IF ports, with the input RF and LO frequency bands spanning 0.6-3 GHz and the output IF band covering DC-2 GHz. The implementation of a low-pass filter at the IF output port and the incorporation of series RC networks at both ends of the mixer ring effectively enhance the real part of the port impedance, significantly improving impedance matching at the output port.

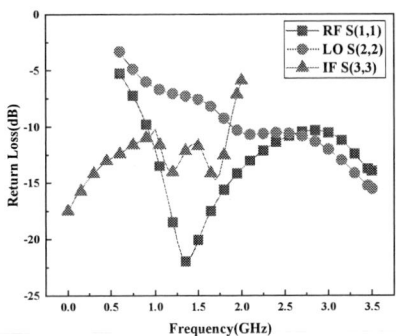

Fig. 7. Simulated RF, LO and IF return loss of the proposed mixer

Fig 8 shows the conversion loss and input 1dB compression point versus RF frequency at a fixed IF frequency of 300 MHz, the mixer exhibits a significant loss reduction in low-frequency operation. Across the 0.6-3 GHz band, the down-conversion loss remains stable below -10 dB with a flatness of ±0.5 dB, while the input 1dB compression point consistently exceeds 15 dB.

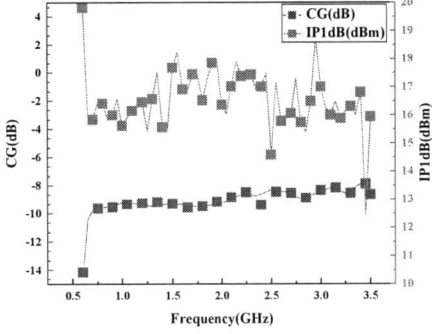

Fig. 8. Simulated CG and IP1dB of the proposed mixer

The port isolation characteristics versus RF frequency at IF=300 MHz are illustrated in Figure 9, demonstrating that the isolation remains below -18 dB across the 600 MHz to 3 GHz range. Furthermore, both LO-to-IF and LO-to-RF isolations achieve superior performance better than -25 dB throughout the operational bandwidth.

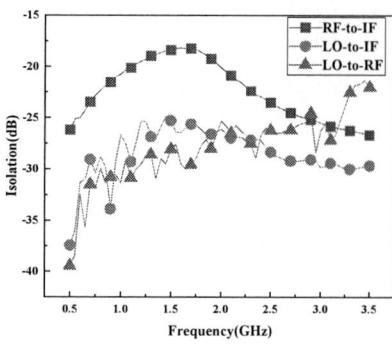

Fig. 9. Simulated port isolation of the proposed mixer

IV. CONCLUSION

This paper presents a structural model of an active double-balanced mixer operating in the 0.6-3 GHz frequency band, developed through the extraction and modeling of key performance parameters. The improved balun design reduces phase mismatch in the low-frequency range to within ±0.5° by optimizing the coupling coefficient. The LO driver amplifier, operating at a 5 V supply voltage, achieves an input 1 dB compression point of 18 dBm with only -7 dBm drive power. Comparative analysis between the model and measured data demonstrates the model's accurate representation of critical device characteristics, including conversion loss and intermodulation products.

REFERENCES

[1] Z. -F. Zhang, J. Liu, D. -W. Wang and W. -S. Zhao, "Nonlinear Circuit Model Analysis of RF High-Power Transistor Package Shell," 2022 International Conference on Microwave and Millimeter Wave Technology (ICMMT), Harbin, China, 2022, pp. 1-3, doi: 10.1109/ICMMT55580.2022.10023112.

[2] M. Raj, S. Chaturvedi, M. Sazid, S. L. Badnikar and B. K. Sehgal, "A very wideband FET resistive MMIC double balanced mixer based on empirical non-linear cold FET model," 2015 IEEE MTT-S International Microwave and RF Conference (IMaRC), Hyderabad, India, 2015, pp. 305-308, doi: 10.1109/IMaRC.2015.7411381.

[3] H. Yüksel, D. Yang and A. C. Molnar, "A circuit-level model for accurately modeling 3rd order nonlinearity in CMOS passive mixers," 2014 IEEE Radio Frequency Integrated Circuits Symposium, Tampa, FL, USA, 2014, pp. 127-130, doi: 10.1109/RFIC.2014.6851676.

[4] G. P. Gibiino, A. Santarelli, S. Farsi, M. Myslinski, G. Avolio and D. Schreurs, "S-functions mixer modeling for linearization purposes," 2011 6th European Microwave Integrated Circuit Conference, Manchester, UK, 2011, pp. 490-493.

[5] A. Cidronali and G. Collodi, "Broadband Hammerstein-Wiener Mixer Modeling extracted by Large-Signal Vector Measurements," 2018 IEEE/MTT-S International Microwave Symposium - IMS, Philadelphia, PA, USA, 2018, pp. 555-558, doi: 10.1109/MWSYM.2018.8439487.

[6] J. Moghaddasi, K. Wang and K. Wu, "Parametric characterization of six-port interferometer demodulator through mixer modeling," 2015 European Microwave Conference (EuMC), Paris, France, 2015, pp. 526-529, doi: 10.1109/EuMC.2015.7345816.

[7] J. Shu, W. Wenbo, D. Jiangling and R. Le, "135-165GHz Sub-harmonic Mixer Based on Schottky-diode Circuit Model," 2020 International Conference on Microwave and Millimeter Wave Technology (ICMMT), Shanghai, China, 2020, pp. 1-3, doi: 10.1109/ICMMT49418.2020.9386751.

[8] M. Sosio, A. Liscidini and R. Castello, "An Intuitive Current-Driven Passive Mixer Model Based on Switched-Capacitor Theory," in IEEE Transactions on Circuits and Systems II: Express Briefs, vol. 60, no. 2, pp. 66-70, Feb. 2013, doi: 10.1109/TCSII.2012.2234993.

[9] L. Gaojian, L. Jun, X. Hui, Z. Xiaoyang, L. Shuantao and Y. Hongxi, "Design of a 220GHz subharmonic mixer based on plannar schottky diode," 2017 IEEE Asia Pacific Microwave Conference (APMC), Kuala Lumpur, Malaysia, 2017, pp. 418-421, doi: 10.1109/APMC.2017.8251469.

[10] C. Maurette-Blasini et al., "ASM-GaN Model for Resistive Mixer Applications at D-Band Frequencies," 2024 IEEE BiCMOS and Compound Semiconductor Integrated Circuits and Technology Symposium (BCICTS), Fort Lauderdale, FL, USA, 2024, pp. 62-65, doi: 10.1109/BCICTS59662.2024.10745699.

2025 The 10th International Conference on Integrated Circuits and Microsystems

Replicated Partitioning for Hypergraphs with Multiple Constraints

Kexin Zhang
School of Microelectronic
Xidian University
Xi'an, China
kx3127319733@hotmail.com

Shunyang Bi
School of Microelectronic
Xidian University
Xi'an, China
bishy@stu.xidian.edu.cn

Jing Tang
School of Microelectronic
Xidian University
Xi'an, China
tangjing2022@stu.xidian.edu.cn

Hailong You*
School of Microelectronic
Xidian University
Xi'an, China
hlyou@mail.xidian.edu.cn
*Corresponding author

Qiwang Chen
School of Microelectronic
Xidian University
Xi'an, China
15075780771@163.com

Abstract—Nowadays, with the explosive growth of circuit design complexity and the limited capacity of a single FPGA, the Multi-FPGA System (MFS) has become the primary solution to tackle the challenges brought by large-scale designs. Therefore, designing a suitable partition for MFS to reduce the *total_hop* has become a key research question affecting the overall system performance. This paper first combines replication and movement operations to optimizes the *total_hop* by replicating appropriate gates. The experimental results show that our method can reduce the *total_hop* by an average of 43%, effectively improving the partitioning performance.

Keywords—logic replication, heuristic algorithm, partition

I. INTRODUCTION

In the contemporary IC design process, verification assumes an increasingly substantial proportion within the entire chip-development project, exceeding even 70% of the total workload. It has a significant impact on both time and economic costs. Among various verification methods, prototyping is regarded as the optimal trade-off among performance, cost, and logic integration capabilities due to its extremely high operating speed and relatively low deployment cost. Meanwhile, as the scale of integrated circuit (IC) design continues to expand, the complexity of the design and the number of logic gates have increased dramatically. A single FPGA often cannot accommodate the entire design. Consequently, the Multi-FPGA System (MFS) has emerged as the primary solution to address the challenges of large-scale designs. These FPGAs are interconnected via physical cables or a programmable interconnection network, ultimately constituting a complete simulation system.

The process of compiling a circuit into an MFS typically starts with the Register Transfer Level (RTL) description. Subsequently, it goes through steps such as synthesis, partitioning, system-level routing, FPGA placement and routing, until bit-streams are generated for each FPGA. Partitioning refers to

dividing a circuit into multiple sub-circuits, and each sub-circuit is deployed on an individual FPGA. Among these steps, partitioning is the most critical one, which can have an impact on the system's performance.

However, in the MFS, owing to the limited I/O resources, only a small fraction of FPGAs are directly connected via physical wiring. When there is no direct link between the source FPGA and the target FPGA for a signal, the signal has to be relayed through one or more intermediate FPGAs. This process is defined as the "FPGA-hop" [1]. "FPGA-hop" has a negative impact on the timing performance of the system. The greater the "FPGA-hop", the larger the total delay will be.

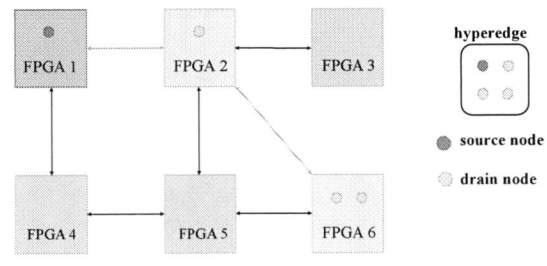

Fig. 1: Hop distance in MFS.

For each hyperedge, if its nodes are distributed across x different FPGAs, it will involve (x-1) hop paths. The length of a hop path is defined as "hop distance", which is the minimum number of FPGAs required for signal relaying plus 1. The total length of hop paths contributed by a hyperedge is defined as "hop(e)", which is the product of the hyperedge's weight and the sum of hop distance of all its hop paths. As shown in Fig. 1, the hyperedge has a weight of 3 and contains 4 nodes. The

979-8-3315-8850-2/25 $31.00 © 2025 IEEE

red one is the driving node, which is placed on FPGA1. In this case, two paths are generated: FPGA 1-2 and FPGA 1-6, with their hop distance being 1 and 2 respectively. Therefore, hop(e) contributed by this hyperedge is $3 \times (1 + 2) = 9$.

Hypergraph partitioning, as a well-known NP-hard problem [2], has always been a popular research topic. Over the years, researchers have proposed many excellent heuristic algorithms. Fiduccia et al. optimized the famous KL algorithm [3] and proposed the Fiduccia-Mattheyses (FM) algorithm [4]. Currently, the most successful approach for solving the hypergraph partitioning problem is the multilevel partitioning framework [5]. hMETIS [6], PaToH [7], and Topopart [8] all adopt the multilevel framework. In 2017, KaHyPar [9] proposed by Akhremtsev et al. employed the n-level clustering method and provided a higher-quality solution.

However, these traditional hypergraph partitioning algorithms mainly focus on the cut cost and do not care about the hop distance. Hung et al. [10] emphasized in their paper that the signal delay depends on the hop distance. Li et al. [11] proposed a partitioning algorithm that considers multiple constraints and optimization objectives, with a particular focus on reducing the number of nets that require "FPGA-hop". SPARK [12] introduced an adaptive FPGA interconnection networking scheme and considered the maximum hop distance in the objective function. MaPart [3] proposed a novel hypergraph partitioning framework, which aims to minimize the maximum path delay in MFS. Although the above-mentioned researches focus on reducing the hop distance, they didn't unlock the potential of further improving the hop quality by employing the replication technique in partitioning stage. In the 1990s, the concept of logic replication was proposed. Kring and Newton [14] used heuristic algorithms in logic replication operations. Enos and Hauck [15] improved the Kring/Newton replication algorithm. L. James Hwang and El Gamal [16] applied network flow to logic replication.

This paper first combine replication and movement techniques to optimize the "FPGA-hop" problem. It also takes into account additional constraints: the resource limit of MFS capacity and *max_hop* constraints. Improvements are made on the partitioning results of MaPart to obtain a better solution.

The main contributions of this paper are as follows:

- This paper first proposes an refinement algorithm that takes both replication operations and movement operations into consideration.
- We use a new data structure *max-heap* to store two operations' corresponding gains, it significantly improves the speed of the algorithm. Meanwhile, We use three data structures. It not only simplifies the gain update, but also obtains accurate results without expanding the hypergraph.
- Our method can reduce the *total_hop* by an average of 43 %, effectively improving the partitioning performance.

The rest of this paper is organized as follows: In Section II, we formulate the partition problem. In Section III, we present the main algorithm. In Section IV, we analyze the experimental results. The conclusions are introduced in Section V.

II. PROBLEM FORMULATION

TABLE I: Important notation used in this work.

Variable	Definition
$H(V, E)$	Circuit netlist represented by hypergraph
v_i	Gate represented by node i
e_i	Net represented by hyperedge i
ω_{e_i}	Weight of net e_i
$hop(e_i)$	The hop of e_i
$total\_hop$	The sum of $hop(e_i)$, $e_i \in E$
$p(v_i)$	FPGA where gate v_i is located
$s(e_i)$	Source node of net e_i
$d(e_i)$	Drain nodes of net e_i
$G(\hat{V}, \hat{E})$	MFS graph
$\hat{v}_i$	FPGA represented by part i
$v_i^*(\hat{v}_i)$	Gate v_i^* replicated by gate v_i to FPGA $\hat{v}_i$
v_i^*	Gate v_i^* replicated by gate v_i
$driven(v_i)$	Gate that provided signal to gate v_i
$\hat{e}_i(\hat{u}, \hat{v})$	Physical wire between $\hat{u}$ and $\hat{v}$ represented by edge $\hat{e}_i$
$N(\hat{v}_i)$	Gates in FPGA $\hat{v}_i$
$dis(\hat{u}, \hat{v})$	Shortest path length between FPGA $\hat{u}$ and FPGA $\hat{v}$
$n_i(\hat{v}_i)$	The used resource of FPGA $\hat{v}_i$
$r_i(\hat{v}_i)$	The supplied resource of FPGA $\hat{v}_i$
max_hop	The max hop distance constraint provided in input file
$Dp(e_i)$	The set of FPGAs where the drain nodes of net e_i are located
$Netlist(v_i)$	The set of all nets related to gate v_i
$gp(v_i)$	FPGA where the gate v_i or $v_i^*(\hat{v}_i)$ will be located
$max\text{-}heap$	Priority queue storing gains of operations with $O(logn)$ updates
$BN(\hat{v}_i)$	The set of nodes which exceed the resource constraints for FPGA $\hat{v}_i$

In this section, we present a formulation and important terms are explained in Table I. Given a netlist represented by hypergraph $H(V, E)$ and the MFS model represented by graph $G(\hat{V}, \hat{E})$. Since the hypergraph will not be changed during the replication operation in this paper, the hypergraph is still $H(V, E)$. In a circuit netlist, each gate consists of a driving gate and several driven gates. The logic levels are emitted by the driving gate and transmitted to the driven gates. Similarly, each hyperedge contains a source node $s(e_i)$ and several drain nodes $d(e_i)$. The signal of the hyperedge starts from the source node and propagates to each drain node. Therefore, driving gates of replicated nodes v_i^* can be represented as $driven(v_i^*)$.

In the real industrial application, due to the limitations of FPGA technique, the replicated node v_i^* and the original node v_i can only receive input signals from the same predecessor. Meanwhile the replicated node can only send signals to their successor when they are located in the same FPGA. Our algorithm is performed within the above two constraints:

$$\text{driven}(v_i^*) = \text{driven}(v_i) \tag{1}$$

$$p(driven(v_i^*(\hat{v}_i))) = \hat{v}_i \tag{2}$$

This paper considers eight common types of FPGA resources, which are FF, LUT, BUFG, TBUF, DCM, BRAM, DSP, and PP. Suppose all hyperedges in E have a length of 1. The hop distance of the node pair (u, v) is calculated as follows:

$$\text{hop}(u, v) = \text{dis}(p(u), p(v)) \tag{3}$$

979-8-3315-8850-2/25 $31.00 © 2025 IEEE

The hop of e_i is defined as follows:

$$\text{hop}(e_i) = \sum_{v \in d(e_i)} \omega_{e_i} \times \text{hop}(s(e_i), v) \qquad (4)$$

The problem of circuit partitioning for the MFS can be formulated as follows: Given a hypergraph H and a graph G, the algorithm needs to accomplish two tasks: 1) Minimize F; 2) satisfy the constraints. The objective function can be formulated as follows:

$$\begin{aligned}
\text{minimize F} &= \sum_{e_i \in E} \text{hop}(e_i) \\
\text{subject to} \quad n_i(\hat{v}_i) &\leq r_i(\hat{v}_i), \quad i = 1, \ldots, 8 \qquad (5) \\
\max_{e_i \in E} &\left\{ \max_{v \in d(e_i)} \text{hop}(s(e_i), v) \right\} \leq \text{max\_hop}
\end{aligned}$$

The objective function F essentially aims to minimize the *total_hop*, When $dis(\hat{u}, \hat{v})$ exceeds 1, *total_hop* is calculated as the product of the cutsize and ($dis(\hat{u}, \hat{v})$-1). And when $dis(\hat{u}, \hat{v})$ is 1, *total_hop* is the cutsize.

III. PARTITION REFINEMENT ALOGORITHM

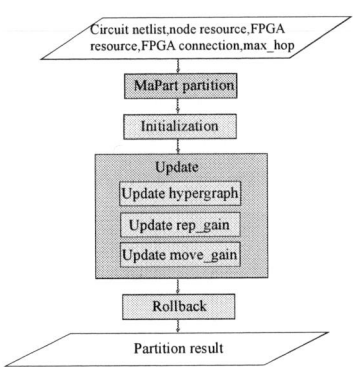

Fig. 2: The flow of our algorithm.

The algorithm minimizes the *total_hop* by iteratively moving and replicating nodes under the comprehensive constraints, including resource limitations, maximum hop count (*max_hop*), and FPGA external connectivity requirements. The flow is shown in Fig.2.

Overall, our algorithm consists of three stages. First, we get the initial partitioning result using MaPart. Then the movement or replication will be chosen based on their corresponding gain values. Note that we store the gain in the *max-heap* data structure for its efficient feature of selecting the max value.

Then, in the iterative optimization stage, the highest gain will be selected from the *max-heap*. Furthermore, the neighbor gain should be updated into the *max-heap* when we perform one movement or replication operation. The selection and update will be consistently carried out until the resources of the node reach the capacity bound. It is worth noting that the resource constraints will be checked in the whole process.

Finally, a rollback strategy will be adopted to avoid local optima of the iteration. Specifically, we store all operations

and their gains in the iterative optimization stage. Based on this, the rollback strategy will revert to the best gain state that it has ever seen.

A. Initialization

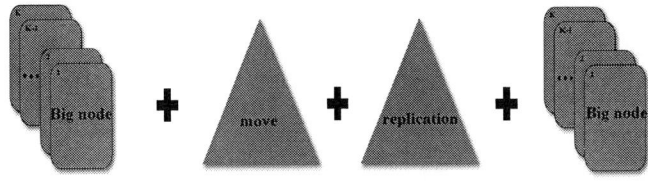

Fig. 3: Max_heap data structure.

As shown in Fig. 3, two *max-heap*s ("move" and "replication") are used to store the gain values for each node movement and replication across (k-1) target FPGAs. Therefore, these two data structures are initialized as $n \times (k\text{-}1)$, where the n refers to the total node number. The "Big Node" maintains the nodes that violate the resource constraint in the current refinement pass but may satisfy it in a subsequent pass.

The basic structure of the original hypergraph will be changed when the replication operation is performed. Generally, directly updating the hypergraph is a choice. However, it brings the challenge of reverting to the original hypergraph in the rollback. In this context, we use three data structures to maintain the update information instead of directly updating the hypergraph, which avoids the issue in the rollback. The three data structures are *drain_part*, *NetPart*, and *Replicated-node*.

B. Updating of Buckets

As shown in Fig. 4(a), *NetPart* indicates one node on FPGA1 and FPGA3 respectively, and two nodes on FPGA(k-1), while *Dp* shows that its drain nodes $d(e_i)$ are distributed on FPGA3 and FPGA(k-1), thereby deducing that the source node $s(e_i)$ resides on FPGA1. Additionally, *Replicated-node* stores replication records of $s(e_i)$, enabling rapid detection of new hyperedges during updates. Here, we define that a new hyperedge is generated when a replicated source node $s(e_i)^*$ provides or withdraws signals to $d(e_i)$ or $d(e_i)^*$.

In the moving operation, there are two node types: normal nodes and replicated nodes, further corresponding to the two moving operations.

- Case 1: If the replicated node $s(e_i)^*$ is not involved, there is no new hyperedge generated, and the update is very straightforward. There are three cases (Algorithm 1, lines 8-13): as shown in the Fig.4(a), the $net[i]'_{(1)}$ represents the decreasing FPGA's sets ($Dp(e_i)$) of the hyperedge

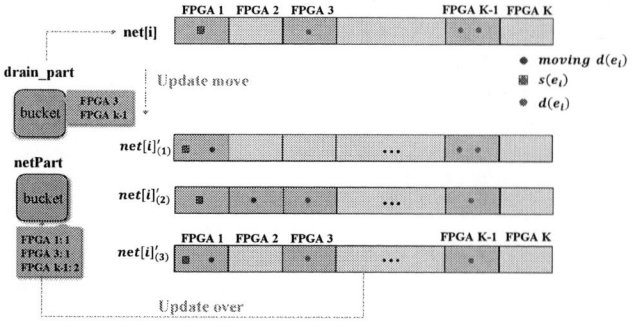

(a) The scenario with no new hypergraph.

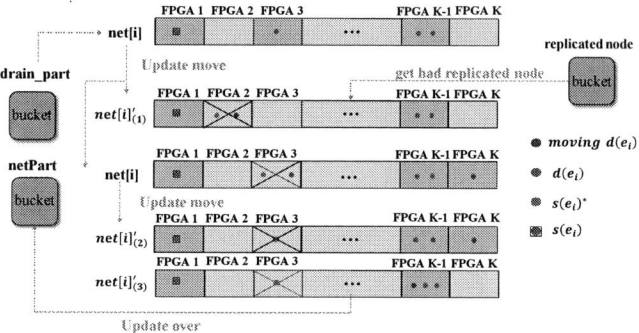

(b) The scenario with the new hypergraph.

Fig. 4: Updating of hypergraph by moving operation.

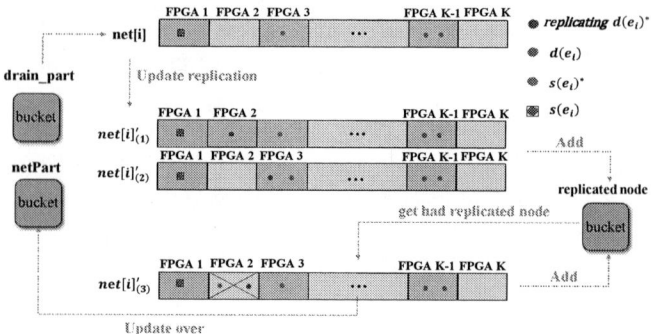

(a) The scenario with drain nodes replication.

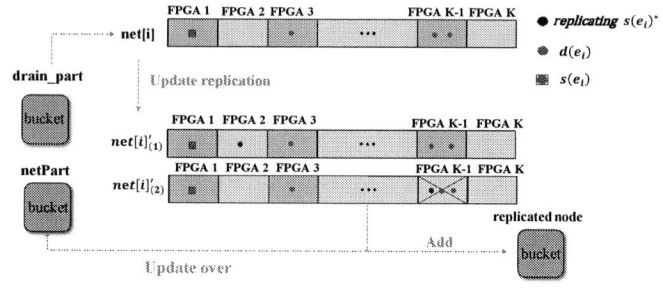

(b) The scenario with source node replication.

Fig. 5: Updating of hypergraph by replicating operation.

e_i after the drain node (blue node) from FPGA3 moved to FPGA1, while the $net[\text{i}]'_{(2)}$ denotes the increasing FPGA's sets. And the $net[\text{i}]'_{(3)}$ still has the same FPGA's sets as before moving the node.

- Case 2: If the replicated node $s(e_i)^*$ is involved, the new hyperedge may be generated. Generally, the two scenarios are included: the node is moved to (from) the FPGA where the $s(e_i)^*$ is located. We update it by the Fig.4(b) shown. The $net[\text{i}]'_{(1)}$ denotes the blue node is moved to the FPGA2 from FPGA3, which has contained the $s(e_i)^*$ (Algorithm 1, lines 15-20). And the $net[\text{i}]'_{(2)}$ and $net[\text{i}]'_{(3)}$ depict the blue node being moved from the FPGA3, where it has already contained the $s(e_i)^*$ (Algorithm 1, lines 22-27). In this case, the FPGA set $Dp(e_i)$ of the hyperedge e_i either decreases or remains unchanged.

Similar to the move operation, according to whether it is replicating the source node, the replication operation also has two cases.

- Case 1: If the drain node is replicated, as the Fig.5(a) shows, the $net[\text{i}]'_{(1)}$ and $net[\text{i}]'_{(2)}$ represent the drain node being replicated to the FPGA2 and FPGA3 (Algorithm 2, lines 21-23); Therefore, no hyperedge is generated. However, in the $net[\text{i}]'_{(3)}$, the new hyperedge is produced when the drain node is replicated to the FPGA where the

$s(e_i)^*$ is located (Algorithm 2, lines 15-20).

- Case 2: If the source node is replicated (Algorithm 2, lines 10-14), as the Fig.5(b) shows, the $net[\text{i}]'_{(1)}$ represents the source node being replicated to the FPGA2. In this case, there is no hyperedge generated. And in the $net[\text{i}]'_{(2)}$, the new hyperedge is produced when the source node is replicated to the FPGA, which contains the drain nodes.

C. Calculating of Gain.

Algorithms 1 and 2 achieve efficient gain computation by dynamically maintaining the FPGA topological distribution of drain nodes Dp. The discrepancies in gain calculation arising from node movement and replication operations are resolved through the update of three data structures. It's important to note that the set Dp merely represents the number of elements that need to be traversed when calculating the gain. A reduction in the set Dp only partially indicates a decrease in the number of FPGA pairs to be computed, as the gain calculation is also contingent upon the shortest distances between these FPGA pairs. Consequently, the gain calculation formulas for both movement and replication operations are defined as follows: For all nets containing the node to be operated on, compute the difference between the $hop(e_i)$ before the operation and after the operation. To simplify computational complexity, the shortest distance between each pair of FPGAs is precomputed and stored in a lookup table. Whenever Dp is updated, the distances between the relevant FPGAs can be directly retrieved from this lookup table.

Algorithm 1: move gain driven refinement

Input : node v_i, circuit graph $H(V, E)$
Output: Gain G
1 G ← 0;
2 v ← 0; // Intermediate variable in gain value calculation;
3 $Dis \leftarrow dis(p(s(\hat{e_i})), \hat{p_d})$;
4 **foreach** $v_i \in V$ **do**
5   **if** $n_i(p(\hat{v_i})) > r_i(p(\hat{v_i}))$ **then**
6          $BN(\hat{v_i}) \cup v_i$;
7   **else**
8   **foreach** $e_i \in Netlist(v_i)$ **do**
9   **if** $p(s(e_i)^*) == \varnothing$ **then**
10   **foreach** $p_d \in Dp(e_i)$ **do**
11   v ← v + ω_{e_i} * Dis;
12                 $Dp(e_i) \leftarrow gp(v_i) \cup Dp(e_i) - p(v_i)$;
13   **foreach** $p_d \in Dp(e_i)$ **do**
14   G ← G + v - ω_{e_i} * Dis;
15   **else**
16   **if** $gp(v_i) == p(s(e_i)^*)$ **then**
17   **foreach** $p_d \in Dp(e_i)$ **do**
18   v ← v + ω_{e_i} * Dis;
19                         $Dp(e_i) \leftarrow Dp(e_i) - gp(v_i) \cap Dp(e_i))$;
20   **foreach** $p_d \in Dp(e_i)$ **do**
21   G ← G + v - ω_{e_i} * Dis;
22   **else**
23   **if** $p(v_i) == p(s(e_i)^*)$ **then**
24                       $Dp(e_i) \leftarrow gp(v_i) \cup Dp(e_i)$;
25   **foreach** $p_d \in Dp(e_i)$ **do**
26   v ← v + ω_{e_i} * Dis;
27   **foreach** $p \in Dp(e_i)$ **do**
28   G ← G + v - ω_{e_i} * Dis;

Algorithm 2: replication gain driven refinement

Input : node v_i, circuit graph $H(V, E)$
Output: Gain G
1 G ← 0;
2 v ← 0; // Intermediate variable in gain value calculation;
3 $Dis \leftarrow dis(p(s(\hat{e_i})), \hat{p_d})$;
4 **foreach** $v_i \in V$ **do**
5   **if** $n_i(p(\hat{v_i})) > r_i(p(\hat{v_i}))$ **then**
6          $BN(\hat{v_i}) \cup v_i$;
7   **else**
8   **foreach** $e_i \in Netlist(v_i)$ **do**
9   **foreach** $p_d \in Dp(e_i)$ **do**
10   v ← v + ω_{e_i} * Dis;
11   **if** $v_i == s(e_i)$ **then**
12   **if** $gp(v_i) \cap Dp(e_i) \neq \varnothing$ **then**
13                     $Dp(e_i) - (gp(v_i) \cap Dp(e_i))$;
14   **foreach** $p_d \in Dp(e_i)$ **do**
15   G ← G + v - ω_{e_i} * Dis;
16   **else**
17                   $Dp(e_i) \leftarrow p(d(e_i)^*) \cup Dp(e_i)$;
18   **if** $p(s(e_i)^*) \cap p(d(e_i)^*) \neq \varnothing$ **then**
19                     $Dp(e_i) \leftarrow p(s(e_i)) \cup (p(s(e_i)^*) \cap p(d(e_i)^*))$;
20   **foreach** $p_d \in Dp(e_i)$ **do**
21   G ← G + v - ω_{e_i} * Dis;
22   **else**
23   **foreach** $p_d \in Dp(e_i)$ **do**
24   G ← G + v - ω_{e_i} * Dis;

D. Rollback

When algorithm iterations exhaust all FPGA resources, the algorithm may converge to local optima. To avoid this, the rollback strategy expands solution-space exploration by reverting to historical optima via inverse operation sequences. It maintains a gain stack recording full operation trajectories through Dp, *NetPart* and *Replicated-node*. As each node must pick the max-gain operation (move or replication), actual sequences show alternating growth, requiring strict reverse-order state restoration during rollback for global gain peak regression.

IV. EXPERIMENTAL RESULTS

We implemented algorithm in C++ and compiled it using g++ 11.4.0. The machine utilized an Intel Xeon Gold 6248R 3.0-GHz CPU, running on Ubuntu 22.04.1 LTS.

A. Benchmarks.

we used the dataset from S2C Inc. The statistical information for these experiments can be found in Table II. The maximum scale of the examples is 240,000 nodes, and the minimum is less than 100 nodes. This benchmark of networks with multiple scales poses high requirements for the partitioning, which can effectively verify the effectiveness of our algorithm. This gain-ordered inverse engineering mechanism effectively balances local optimization with global exploration

TABLE II: The statistic of benchmark in our experiment.

Benchmark	#node	#net	#FPGA
case01	16	13	4
case02	600	1239	8
case03	11451	31071	32
case04	1600	2157	4
case05	1053	2447	8
case06	40000	45504	32
case07	25000	27482	32
case08	240000	253232	64

The second column of the Table II shows the number of nodes in each design, the third column shows the number of hyperedges in each design, and the fourth column shows the number of FPGAs in each design.

B. Overall Performance

To verify the significant improvement of the algorithm in partitioning results, we set the baseline by using MaPart. Since the randomly initial-step in MaPart can affect the stability of the results, we use a thread pool to find the best solution by running MaPart many times. The number of runs is related to the size of each case, and the goal is to find the optimal solution of the baseline within an acceptable time.

TABLE III: Experimental Results

Benchmark	# max_hop	Baseline		ours	
		_hop	time(s)	total_hop	time(s)
case01	2	11	12.12	9	0.000212
case02	4	2377	93.15	1960	0.878949
case03	3	13443	258.43	5854	587.866719
case04	3	229	27.54	117	3.832017
case05	2	382	21.68	346	1.118264
case06	3	7974	353.10	4818	29.357476
case07	3	6816	556.28	3565	29.63948
case08	4	50133	589.04	23505	381.012159
Avg.ratio		1	1	0.57	0.40

We conducted experiments on eight designs. The second column of the Table III shows the constraint of *max_hop*. The third column shows the *total_hop* of MaPart' result. And *total_hop* represents the total hop distance obtained from the algorithm run value, time(s) gives the time the algorithm was run for. Since our algorithm is an refinement version based on Mapart, the total running time for getting the result is the sum of the times in the fourth column and the sixth column.

Based on the experimental results of Case 01 to Case 08, compared with the baseline method, our method can reduce the *total_hop* by an average of 43%. Although the time cost increases by 40% on average, this time sacrifice is considered acceptable. And we have successfully achieved a significant improvement in performance. Particularly in large-scale cases like Case 03, Case 07, and Case 08, we achieve a remarkable performance improvement (57%).

C. More strategies

Further, we conduct a ablation study to valid the efficiency of our algorithm with plain replication technique. Specifically, our goal is to isolate the effects of replication operations in the algorithmic framework, in order to better represent the superiority of performing both movement and replication operations at the same time. In this experiment, only involves replication operation is set the strategy (ours_A). The third column of Table IV shows the *total_hop* of this strategy. Compared with the Strategy ours_A , our method can reduce the *total_hop* by about 30% on average.

TABLE IV: Results of Strategies

Test	#max_hop	ours_A	ours
case01	2	11	9
case02	4	2321	1960
case03	3	9870	5854
case04	3	182	117
case05	2	347	346
case06	3	7469	4818
case07	3	5898	3565
case08	4	45720	23505
Avg.ratio		1	0.71

Compared with only performing replication operations, the combined operation of movement and replication is more optimal. Replication operations can bring better system performance by replicating logic gates to multiple required FPGAs to generate signals. However, only performing replication operations will not change the location of the hypergraph, which leads to a limited solution space and also causes a waste of some resource. In contrast, performing replication and movement operations simultaneously can continuously change the location of the hypergraph through movement nodes, create greater gains for replication operations.

V. CONCLUSIONS

This work proposes an enhanced FM-heuristic framework with three phases: initialization, dynamic update and rollback. It minimize *total_hop* via dynamic node movement and replication. Experiments show 43% lower *total_hop* than baseline, improving hypergraph partitioning efficiency. The scalable solution benefits ultra large-scale IC designs.

REFERENCES

[1] U. Farooq, R. Chotin-Avot, M. Azeem, M. Ravoson, M. Turki and H. Mehrez, "Inter-FPGA routing environment for performance exploration of multi-FPGA systems," 2016 International Symposium on Rapid System Prototyping (RSP), Pittsburgh, PA, USA, 2016, pp. 1-7.

[2] GAREY M R, JOHNSON D S, STOCKMEYER L. Some simplified np-complete problems[C]//Proceedings of the sixth annual ACM symposium on Theory of computing (STOC). 1974: 47-63.

[3] KERNIGHAN B W, LIN S. An efficient heuristic procedure for partitioning graphs[J]. The Bell system technical journal, 1970, 49(2): 291-307.

[4] FIDUCCIA C M, MATTHEYSES R M. A linear-time heuristic for improving network partitions [M]//Papers on Twenty-five years of electronic design automation. 1988: 241-247.

[5] ÇATALYÜREKÜ,DEVINEK,FARAJM,etal. More recent advances in (hyper) graph partitioning [J]. ACM Computing Surveys, 2023, 55(12): 1-38.

[6] KARYPIS G, AGGARWAL R, KUMAR V, et al. Multilevel hypergraph partitioning: Application in vlsi domain[C]//Proceedings of the 34th annual Design Automation Conference (DAC). 1997: 526-529.

[7] ÇATALYÜREKÜV,AYKANATC. Patoh(partitioning tool for hypergraphs)[M]//Encyclopedia of parallel computing. Springer, 2011: 1479-1487.

[8] ZHENGD,ZANGX,WONGMD. Topopart: amulti-level topology-driven partitioning framework for multi-fpga systems[C]//2021 IEEE/ACM International Conference On Computer Aided Design (ICCAD). IEEE, 2021: 1-8.

[9] AKHREMTSEV Y, HEUER T, SANDERS P, et al. Engineering a direct k-way hypergraph partitioning algorithm[C]//2017 Proceedings of the Ninteenth Workshop on Algorithm Engineering and Experiments (ALENEX). SIAM, 2017: 28-42. circuits[C].

[10] HUNG W N, SUN R. Challenges in large fpga-based logic emulation systems[C]//Proceedings of the 2018 International Symposium on Physical Design (ISPD). 2018: 26-33.

[11] LI B, QI Z, TANGZ,etal. High quality hypergraph partitioning for logic emulation[J]. Integration, 2022, 83: 67-76.

[12] ZANG X, YOUNG E F, WONG M D. Spark: A scalable partitioning and routing framework for multi-fpga systems[C]//Proceedings of the Great Lakes Symposium on VLSI 2023. 2023: 593-598.

[13] Li, Benzheng, et al. "MaPart: An Efficient Multi-FPGA System-Aware Hypergraph Partitioning Framework." IEEE Transactions on Computer-Aided Design of Integrated Circuits and Systems (2024).

[14] Kring, Chuck, and A. Richard Newton. "A Cell-Replicating Approach to Minicut-Based Circuit Partitioning." ICCAD. Vol. 1991. 1991.

[15] Enos M, Hauck S, Sarrafzadeh M. Evaluation and optimization of replication algorithms for logic bipartitioning[J]. IEEE transactions on computer-aided design of integrated circuits and systems, 1999, 18(9): 1237-1248.

[16] Hwang L J, El Gamal A. Min-cut replication in partitioned networks[J]. IEEE Transactions on Computer-Aided Design of Integrated Circuits and Systems, 1995, 14(1): 96-106.

2025 The 10th International Conference on Integrated Circuits and Microsystems

A cryogenic readout integrated circuit for blocked-impurity-band (BIB) far-infrared focal plane array detectors

Yushi Chen†, Xin Ge, Hongbo Ma, Zewen Wang, Zuoru Dong and Xiaodong Wang

The 50th Institute of China Electronics Technology Group Corporation, Shanghai 200331, China

†Correspondence to: Yushi Chen, Email: yushichen001@outlook.com

Abstract—Blocked-impurity-band (BIB) far-infrared focal plane array (FPA) detectors are widely used in infrared detection, the performance of which are directly affected by the dynamic range of the readout integrated circuits (ROICs). In this paper, two novel techniques are proposed to improve the dynamic range of ROICs. The proposed background signal suppression technique and integral capacitor sharing technique improve the dynamic range by reducing noise signals and increasing saturated charge capacity, respectively. At the same time, since BIB detectors need to work in cryogenic environment (T<10K), which means ROICs need to operate at the same temperature. In order to eliminate cryogenic non-ideal effect of transistors, this paper proposes a cryogenic transistor model to ensure the performance of ROICs. The proposed ROIC is designed and fabricated with 0.18µm CMOS technology, the size of which is 700µm×650µm. The power consumption is 3.02mW, the readout rate is 1MHz and the saturation charge capacity reaches 40Me-/pixel. Overall, the dynamic range of the proposed ROIC is 70dB, which meets the aim of the design.

Keywords—Blocked-impurity-band (BIB), Far-infrared focal plane array detector, Readout integrated circuit, Background signal suppression technique, Integral capacitor sharing technique

I. INTRODUCTION

As the key of infrared detection technique, infrared focal plane array (FPA) detectors[1-2] have been widely used in medical[3-5], astronomical observations[6-7], industrial inspection[8-9] and communication fields[10-11]. Compared with other FPA detectors, the research progress of blocked-impurity-band (BIB) FPA detectors is more rapidly. The readout integrated circuit (ROIC)[12-13] is an important component of FPA, which has received much attention. Compared with mid and short-wave infrared FPAs, the infrared radiation signals and background signals processed by far-infrared FPAs are several orders of magnitude larger. Hence, the corresponding ROICs must be able to handle much larger signals[14-17]. In this process, three issues arise. Firstly, the BIB detectors operate at a deep low temperature (T<10K) to suppress dark current and noise, which puts forward strict requirements for the operating temperature of ROICs. The freezing out effect and kink effect at deep low temperature will seriously affect the performance of circuits. Secondly, the infrared radiation signals can be overwhelmed easily as the background signals are amplified at the same time. This phenomenon is particularly evident under high background radiation, which effects the dynamic range of ROICs. Thirdly,

since the infrared radiation signals and background signals are both large, the integral voltage is easy to saturate with small integral capacitors. The saturated integral voltage cannot accurately reflect the intensity of infrared signals.

To overcome the above problems, this paper presents corresponding approaches. A cryogenic transistor model is proposed to eliminate the non-ideal effects and improve the cryogenic reliability of ROICs. A background signals suppression technique is presented to make ROICs only integrate infrared radiation signals. Hence, the problem that the infrared radiation signals are submerged by background signals is solved. Lastly, an integral capacitor sharing technique is proposed. The sub-capacitors of multiple pixels are combined as a large integral capacitor without increasing the area of ROICs, which improves the saturation charge capacity and avoids saturation of integral voltages.

The organization of the remaining part of this paper is as follows. Chapter II primarily introduces the cryogenic transistor model. Chapter III presents the design and implementation of the ROIC. Chapter IV analyzes the results of the ROIC. Chapter V provides a summary of the paper.

II. CRYOGENIC TRANSISTOR MODEL

Since the BIB detectors need to work in cryogenic environment (T<10K), the ROICs also need to work normally at the same temperature. As the operating temperature decreases, the cryogenic impacts on the circuit become significant, including freezing out effect and kink effect, etc. It is important to consider the devices characteristics variation with the physical process [18]. As a result, the commercial CMOS models at room temperature are no longer applicable at deep low temperature. As shown in Fig. 1, the room temperature models of NMOS and PMOS tubes deviate from cryogenic test data largely. If the room temperature models are still used for design, there is a risk that the ROICs will not work in cryogenic environment.

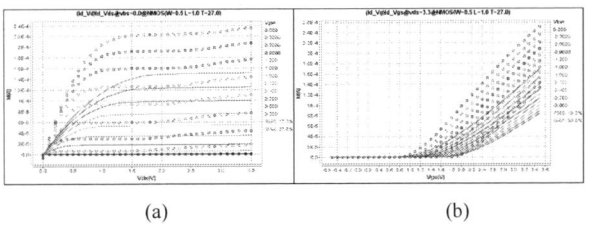

(a)   (b)

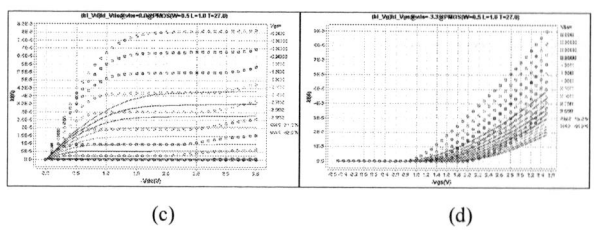

(c)	(d)

Fig. 1 The comparison of the room temperature models (solid lines) of NMOS and PMOS tubes and cryogenic test data (dotted lines) (a) I_d-V_{ds} of NMOS (b) I_d-V_{gs} of NMOS (c) I_d-V_{ds} of PMOS (d) I_d-V_{gs} of PMOS

Therefore, it is crucial to guide the cryogenic design by establishing a cryogenic model. In this paper, the cryogenic characteristics of CMOS devices and physical mechanism of the non-ideal effects are researched to establish a cryogenic transistor modeling method. The proposed method conducts cryogenic test on cryogenic model chip and extracts relevant cryogenic parameters by semiconductor device analyzer and cryogenic probe station, then fits the obtained test data to the original transistor performance curve for accurate analysis and modeling. The process of building a transistor model is divided into four steps.

(a) Design of cryogenic transistor chip: Design a reasonable cryogenic transistor chip to reflect the characteristics of the device. The width and length of the selected transistor need to cover different sizes from large to small to ensure that the relevant effect parameters such as the intrinsic, short channel and narrow channel of the device can be effectively extracted.

(b) Test of cryogenic transistor chip: Use cryogenic probe station to test the cryogenic transistor chip to obtain the corresponding cryogenic data.

(c) Construction of cryogenic transistor model: Build a cryogenic transistor model based on the cryogenic data obtained from the test.

(d) Verification of cryogenic transistor model: Import the cryogenic transistor model for circuit design to verify the accuracy of the model. Adjust the parameters in real time to improve the accuracy.

The proposed method has two advantages. On the one hand, a large-scale mapping test is conducted in the test, which can obtain the statistical characteristics of the model's cryogenic parameters and conduct more accurate modeling. On the other hand, the proposed method also covers CV and IV curve test, mismatch test, LOD, WPE, Noise and other tests, which ensures that the model is reliable and complete.

III. CIRCUIT DESIGN AND IMPLEMENTATION

A. Overall structure of the proposed readout circuit

Figure 2 shows the architecture of the proposed ROIC. The proposed background signal suppression technique and integral capacitor sharing technique are applied in pixel array and digital control module to improve the dynamic range. As the core module of the ROIC, the pixel array completes the function of integration, sample, hold and output of infrared radiation signals. The digital control module is mainly composed of column clock circuit, row clock circuit and pixel selection circuit, which mainly provides the control signals for the pixel array.

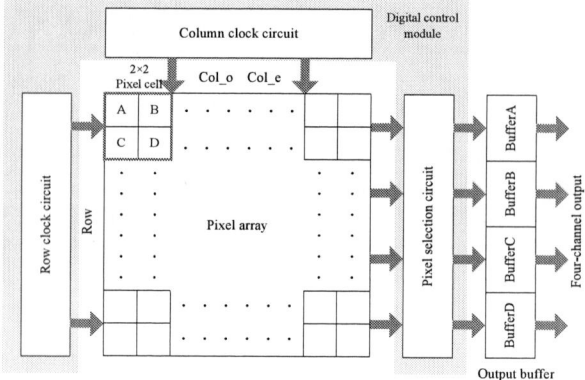

Fig. 2 The architecture of the proposed ROIC

B. Pixel array

In order to suppress the interference of background signals and avoid the saturation of integral voltage, this paper presents a background signal suppression technique and an integral capacitor sharing technique, respectively. The structure of a 2×2 pixel cell based on the two proposed techniques is shown in Fig. 3. The integral capacitor sharing module is marked with a red box, and the background signal suppression module is marked with a blue box.

Compared to conventional structures, the integral capacitor C_{INT} of the proposed structure is increased by four times without changing the size of pixel. Pixel<A:D> are the control signals of the pixel selection switches, SH_{1-2} are the control signals of the sampling switches, IRST is the control signal of reset switch of the integral capacitor, and $SRST_{1-2}$ are the control signals of reset switches of the sampling capacitor. In order to simplify the circuit structure, the proposed circuit simplifies four discrete pixels using four sampling capacitors into two sampling capacitors C_{SH1} and C_{SH2}.

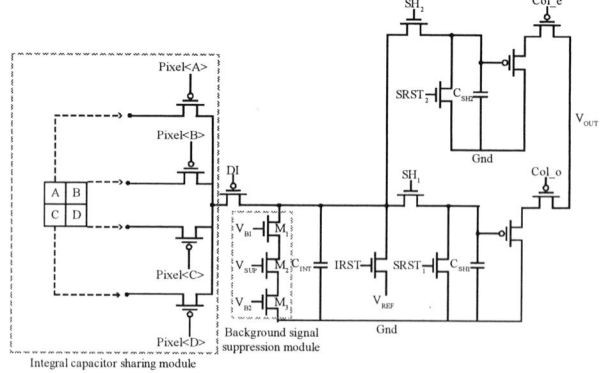

Fig. 3 The structure of the proposed 2×2 pixel cell

For high background radiation applications, infrared radiation signals are much smaller than background signals, which is overwhelmed easily. Therefore, it is necessary to suppress the background signals to improve the performance of FPAs. As shown in Fig. 4, the background signals are removed by the suppression circuit. As a result, only infrared radiation signals are transmitted to the integral circuit, which helps to increase dynamic range of ROICs.

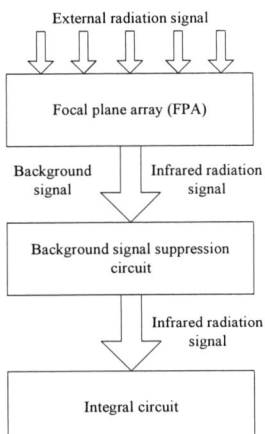

Fig. 4 The principle of background signals suppression technique

Differ from the conventional background suppression, the proposed scheme does not generate the suppression current I_{back} throughout the whole integral period. The circuit generates I_{back} intermittently according to the state of M_2. When M_2 is on (V_{SUP} is high level), M_1 and M_3 form a self-cascade structure to generate I_{back}, which removes part of the charge on the integral capacitor and plays a role of background signal suppression.

In order to avoid the problem of integral voltage saturation, it is necessary to increase the saturated charge capacity, which is defined according to the following formula.

$$Q_{signal} = C_{INT} \cdot (V_H - V_L) \quad (1)$$

C_{INT} is integral capacitor and $(V_H - V_L)$ is output swing of ROICs. According to formula (1), saturated charge capacity is the product of the integral capacitor and the output swing, which can be improved by using large integral capacitor or increasing the swing of the circuit. As shown in Fig. 5, the proposed technique combines four sub-capacitors in adjacent 2×2 pixels to form a large integral capacitor, which can effectively increase the value of capacitors without increasing the pixel area. In the proposed ROIC, the sub-capacitor is 1.6pF/pixel and the improved equivalent integral capacitor is 6.4pF/pixel, which effectively improves the saturated charge capacity.

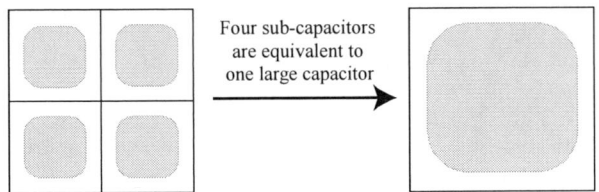

Fig. 5 The principle of integral capacitor sharing technique

The proposed ROIC is designed based on the above 2×2 pixel cell. Four pixels A-D are combined to be a pixel cell. The corresponding input signals are integrated according to the pixel selection signals Pixel<A:D>. The proposed ROIC operates as follows: when the Pixel<A> signal is low, the signal of pixel A is integrated on the capacitor and sampled on C_{SH1}. When the Pixel<B> signal is low, the signal of pixel B is

stored on C_{SH2}. At the same time, the signal of pixel A is read out. By analogy, when the signal of pixel C is saved on C_{SH1}, the signal of pixel B will be read out, so as to realize the rolling reading of pixel signal. In summary, C_{SH1} will sample the signals of pixels A and C in turn, and C_{SH2} will sample the signals of pixels B and D. Finally, the circuit realizes the readout of image according to the sequence of pixel A-B-C-D.

C. Digital control module

The digital control module includes column clock circuit, row clock circuit and pixel selection circuit, which is shown in Fig. 6. The column clock circuit is composed of column shift registers (DFF$_{Col}$). Col_o and Col_e are the column selection signals of odd row and even row, respectively. SW is odd/even row selection signal. When SW is low level, Col_o is transmitted from OUT1 to read signal of odd row pixels, and OUT2 is reset to high level. When SW is high level, Col_e is transmitted from OUT2 to read signal of even row pixels, and OUT1 is reset to high level. The row clock circuit consists of row shift registers (DFF$_{Row}$), which scans the FPA and reads the outputs row by row. Row are outputs of the circuit as the row selection signals. The pixel selection circuit is made up of pixel selection registers (DFF$_{Pixel}$). Pixel<A:D> are outputs of the registers which are used as the pixel selection signals.

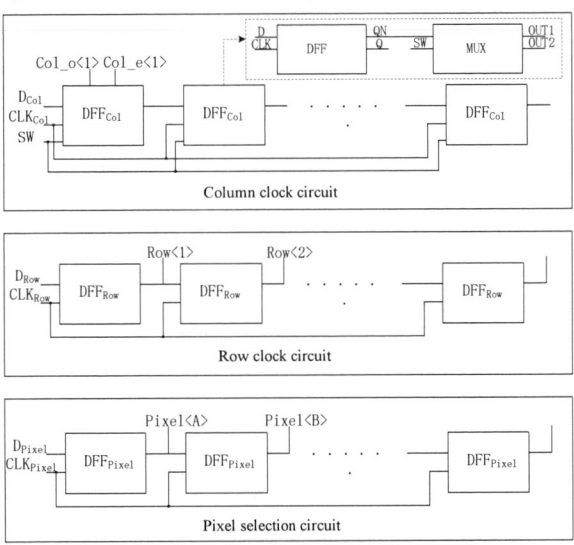

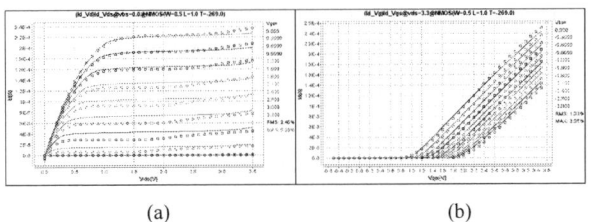

Fig. 6 The structure of digital control module

IV. ANALYSIS OF RESULTS

Figure 7 shows the fitting results of NMOS and PMOS characteristics. Compared with the room temperature model, the designed cryogenic model fits the deep low temperature test data better. Hence, the ROICs based on the proposed cryogenic model is more stable and reliable.

(a)   (b)

979-8-3315-8850-2/25 $31.00 © 2025 IEEE   548

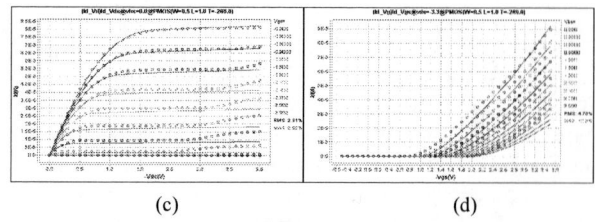

(c)   (d)

Fig. 7 The comparison of the proposed cryogenic models (solid lines) of NMOS and PMOS tubes and cryogenic test data (dotted lines) (a) I_d-V_{ds} of NMOS (b) I_d-V_{gs} of NMOS (c) I_d-V_{ds} of PMOS (d) I_d-V_{gs} of PMOS

Figure 8 shows the microphotograph of the proposed ROIC which is implemented in 0.18μm CMOS process. The layout area of core ROIC is 700μm×650μm. The supply voltage for the is 3.3V and the power consumption is 3.02mW.

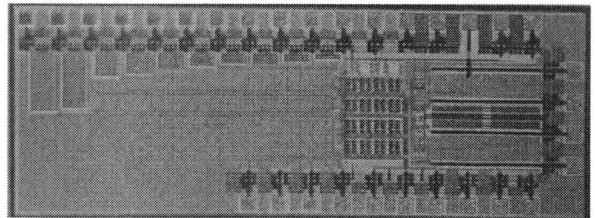

Fig. 8 Microphotograph of the proposed ROIC

The result of background signal suppression circuit is shown in Fig. 9. The background signal suppression circuit periodically subtracts the charge stored on the integral capacitor due to the background signals. When V_{SUP} is low level, the suppression circuit is turned off, and the integral voltage increases. When V_{SUP} is high level, the suppression circuit is enabled, the integral charge of background signal is released and the integral voltage reduces. In different applications, the value of the suppression current I_{back} can be selected and the pulse width of V_{SUP} can be adjusted according to the background signals, achieving the goal of background signal suppression.

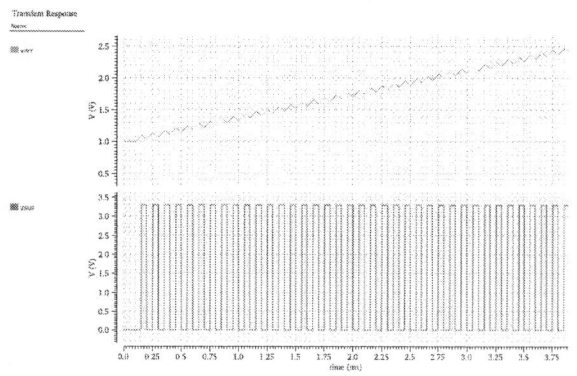

Fig. 9 Results of background signal suppression circuit

The proposed ROIC adopts a four-channel structure, which completes the full array readout based on rolling mode. Figure 10 shows the results of the proposed ROIC. After all pixel A is read out, pixel B is read out, and so on. Finally, the image signal of pixel A-D is obtained in the four-channel outputs, which will be sampled by the following image acquisition system to obtain the complete image. According to the results, the output signals of the pixel array are corresponding to the input signals, which means the proposed ROIC works normally.

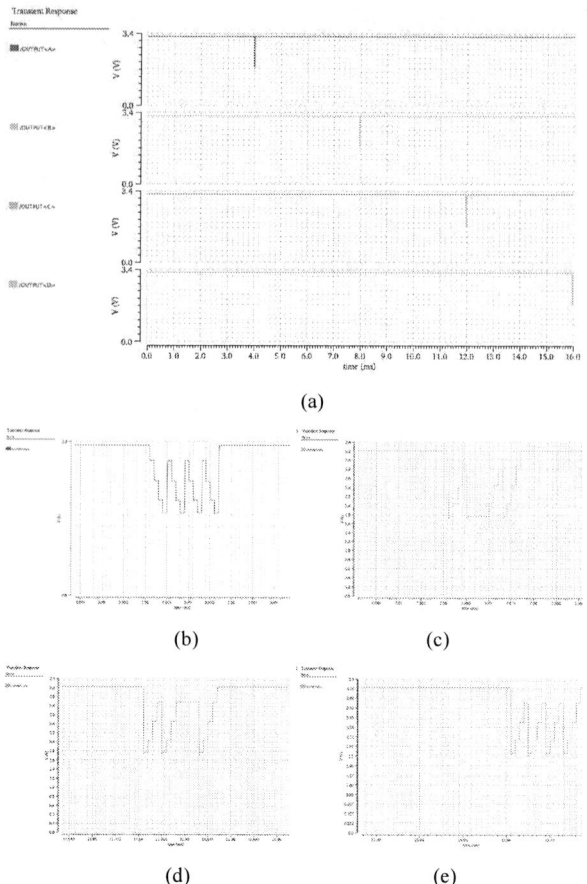

Fig. 10 (a) Results of the proposed 8×8 ROIC (b) Results of pixel A (c) Results of pixel B (d) Results of pixel C (e) Results of pixel D

V. CONCLUSION

In summary, based on the proposed cryogenic transistor model, charge sharing technique and background signal suppression technique, the proposed ROIC for BIB FPA detector with large dynamic range is presented in this paper. The cryogenic transistor model eliminates freezing out effect and kink effect, improving the reliability of ROICs in cryogenic environment. A background signal suppression circuit is designed to eliminate the influence of the background signal, so that the ROIC only needs to integrate the infrared signal. In order to achieve large saturated charge capacity in the limited pixel area, the adjacent pixels of the proposed circuit share integral capacitor. Due to the adoption of this method, the equivalent integral capacitor of a single pixel reaches 6.4pF and the saturated charge capacity reaches 40Me-/pixel. The dynamic range of the proposed ROIC is 70dB. Overall, the function of the ROIC is normal, which meets the characteristics of BIB FPA detectors.

ACKNOWLEDGMENT

Supported by the National Natural Science Foundation of China (Grant Nos. 62301321, and 62171286), the Shanghai Sailing Program (Grant Nos. 23YF1444300).

REFERENCES

[1] X. Li, H. M. Gong, X. M. Shao, et al. Recent advances in short wavelength infrared InGaAs focal plane arrays [J]. Journal of Infrared and Millimeter Waves, 2022, 41(1): 129-138.

[2] L. F. Song, H. Huang. Spatial and temporal adaptive nonuniformity correction for infrared focal plane arrays [J]. OPTICS EXPRESS, 2022, 30(25): 44681-44700.

[3] H. R. Liu, et al. Image interpolation methods for division of focal plane polarimeters: a review [C]. 2nd Conference on Space, Atmosphere, Marine, and Environmental Optics (SAME), 2024, 13189.

[4] Z. D. Taylor et al. Reflective terahertz imaging of porcine skin burns [J]. Opt. Lett., 2008, 33(11): 1258-1260.

[5] K. Ajito, Y. Ueno. THz chemical imaging for biological applications [J]. IEEE Trans. Terahertz Sci. Technol., 2011, 1: 293-300.

[6] M. Khatib, M. Perenzoni. A Low-Noise Direct Incremental A/D Converter for FET-Based THz Imaging Detectors [J]. Sensors, 2018, 18: 1867-1884.

[7] K. Nagase, T. Wada, H. Ikeda. A Demonstration of TIA Using FD-SOI CMOS OPAMP for Far-Infrared Astronomy [J]. J Low Temp Phys, 2016, 184: 449-453.

[8] F. Schuster, D. Coquillat, H. Videlier, M. Sakowicz, F. Teppe, L. Dussopt, B. Giffard, T. Skotnicki, W. Knap. Broadband terahertz imaging with highly sensitive silicon CMOS detectors [J]. Opt. Exp., 2011, 19(8): 7827-7832.

[9] S. Domingues, D. Perenzoni, M. Perenzoni, D. Stoppa. CMOS Integrated Lock-in Readout Circuit for FET Terahertz Detectors [J]. J Infrared Milli Terahz Waves, 2017, 38: 679-688.

[10] M. Tonouchi. Cutting-edge terahertz technology [J]. Nature Photon., 2007, 1(2): 97-105.

[11] F. Sizov, V. Reva, A. Golenkov, V. Zabudsky. Uncooled detectors challenges for THz/sub-THz arrays imaging [J]. J. Infrared Millim. THz Waves, 2011, 32(10): 1192-1206.

[12] Y. Li, H. B. Ma, X. Ge, X. W. Dai, Z. R. Dong, Y. L. Chen, X. D. Wang. A 128×128 Low-power Cryogenic Readout Circuit for Ge-based Blocked-impurity-band Detector [C]. 47th International Conference on Infrared, Millimeter and Terahertz Waves (IRMMW-THz 2022), 2022, 1-2.

[13] J. G. Li, Y. Yuan, Y. Yu. Column-level digitization technology of infrared focal plane array ROIC [J]. Laser and Infrared, 2023, 53(8): 1266-1271.

[14] Y. C. Zhai, R. J. Ding, G. Q. Chen, P. Wang, L. C. Hao. The Simulation of a Readout Integrated Circuit with High Dynamic Range for long wave infrared FPA [C]. International Conference on Optical Instruments and Technology (OIT), 2014, 9045.

[15] H. S. Gupta, et al. Design of large dynamic range, low-power, high-precision ROIC for quantum dot infrared photo-detector [J]. Electronics Letters, 2013, 49(16): 1018-1019.

[16] X. Wang, Z. Shi. Research on Optimization of CTIA ROIC Structure [J]. Microelectronics & Computer, 2014, 31(11): 64-68.

[17] Y. C. Zhai, R. J. Ding. Design of a ROIC with high dynamic range or LWIR FPAs [C]. Conference on Infrared, Millimeter-Wave, and Terahertz Technologies III, 2014, 9275.

[18] Z. Y. Xiong, C. M. Ding, X. Z. Liang. "Characteristics of Hybrid Structured Organic Light-Emitting Diodes Display: Coupled Rigidity with Flexibility." IEEE Electron Device Letters, Apr. 2024, 45 (4): 625-628.

2025 The 10th International Conference on Integrated Circuits and Microsystems

1T1C-Enhanced TFT-Integrated Gate Driver Circuit Design for Reliable In-cell Touch Sensing Displays

Congwei Liao
College of Integrated Circuits and
Optoelectronics Chips
Shenzhen Technology University
Shenzhen, China
liaocongwei@sztu.edu.cn

Yong Wang
Institute of Critical Materials for
Integrated Circuits
Shenzhen Polytechnic University
Shenzhen, China
wangyong1@szpu.edu.cn

Chao Teng
Institute of Critical Materials for
Integrated Circuits
Shenzhen Polytechnic University
Shenzhen, China
tengchao@szpu.edu.cn

Yi Shen
Guangdong Provincial Key
Laboratory of Automotive
Display and Touch Technologies
Shantou Goworld Display
Technology Co., Ltd.
Shantou, China
yishen@goworld-lcd.com

Yudong Liu
Guangdong Provincial Key
Laboratory of Automotive
Display and Touch Technologies
Shantou Goworld Display
Technology Co., Ltd.
Shantou, China
ydliu@goworld-lcd.com

Shengdong Zhang*
School of Electronic and
Computer Engineering
Peking University
Shenzhen, China
zhangsd@pku.edu.cn
*Corresponding author

Abstract—A thin-film transistors (TFTs) integrated gate driver is presented for time-division in-cell touch sensing display. The proposed gate driver consists of two cascaded functional stages: a touch-sensing stage and a non-touch-sensing stage. In the touch-sensing stage, an additional TFT and one capacitor are introduced into the input module to effectively suppress prolonged positive gate-to-source stress of the driving TFT. The non-touch-sensing stage employs a compact 5-TFT 1-capacitor schematic. A key advantage of this design is its ability to maintain continuous clock signals throughout both touch-sensing and display intervals, thereby significantly simplifying the external clock generation circuit. The gate driver circuit is thoroughly verified based on amorphous silicon (a-Si:H) TFTs using SPICE simulation tools, confirming reliable voltage bootstrapping operation even during touch-sensing intervals exceeding 400 µs. The proposed gate driver is promising for compact and reliable in-cell touch sensing display applications.

Keywords—thin-film transistors (TFTs), gate driver on array, touch sensing display, reliability

I. INTRODUCTION

The integration of gate drivers directly onto the display backplane array substrate, referred to as gate driver-on-array (GOA) technology, has gained widespread adoption in state-of-the-art thin-film transistor flat-panel displays (TFT-FPDs) [1]–[4]. Compared to conventional designs that rely on external gate driver integrated circuits (ICs), this approach offers substantial advantages. By eliminating the need for separate driver chips and associated bonding processes, GOA technology not only reduces manufacturing costs but also

improves the structural compactness and reliability of display modules [5]–[6]. Moreover, it mitigates the resolution constraints imposed by fine-pitch external interconnects and contributes to a simplified configuration of data driver ICs [7]–[11].

In modern high-end mobile displays, the integration of touch sensing functions based on the mutual capacitance method within display cells has gained significant importance [12]–[21]. However, the design of gate driver circuits for in-cell touch display remains challenging due to the unique driving waveforms required. To mitigate interference issues between display operation and touch events—such as line defects—a time-division method (TDM) is employed. This approach necessitates the even insertion of touch sensing intervals across the entire display frame. Consequently, for in-cell touch displays operating in TDM mode, the gate driver must support dozens of scanning groups interspersed with touch sensing intervals, rather than the conventional line-by-line scanning approach.

Due to the TDM driver timing waveforms, the TFT gate driver circuit become complex and being prone to decay after long-term operating. Moon *et.al.* firstly demonstrated a gate driver structure by storing the starting signal of the touch-sensing stage by the gate-to-source voltage of the driving transistor [22]. Although the circuit topology is compact, the driving transistors of the touch-sensing stage experience a long time gate-to-source stress voltage, which is prone to line-defects after aging test. To solve this reliability issue, Lin thus proposed the new charge storing components to turn off the driving transistor for the touch-sensing stage [23]. However, the charge storing transistors in the input module are still prone to reliability issues. Further, the clock signals are discontinues for the touching period for almost all the previous in-cell touch

---

This work was financially supported by the Shenzhen Municipal Scientific Program under Grant JCYJ20241202125804006, in part by the National Natural Science Foundation of China under Grant U24A20297, and carried out at Guangdong Provincial Center for Oxide Semiconductor Devices and ICs.

979-8-3315-8850-2/25 $31.00 © 2025 IEEE   551

sensing display gate drivers [24],[25], which complicates the external clock generation circuits. Therefore, it is urgently needed to design new gate driver circuit with high reliably for in-cell touch display.

II. THE PROPOSED GATE DRIVER DESIGN

Fig. 1 shows the circuit schematic for single stage, while Fig. 2 and Fig. 3 present the timing diagram, and block diagram of the proposed gate driver, respectively. For the touch-sensing and non-touch-sensing stages, only the input module is different. For the non-touch-sensing stages, a diode connected TFT T1 is used. For the touch-sensing stages, two transistors in serial, i.e. T1A and T1B, are used instead. The operating principle for the non-touch-sensing gate driver has been detailed in our previous work [6]. Here only the newly introduced 1T-1C part for charge storage and transfer functions are addressed, which is controlled by the touch controlling signal V_{TP}.

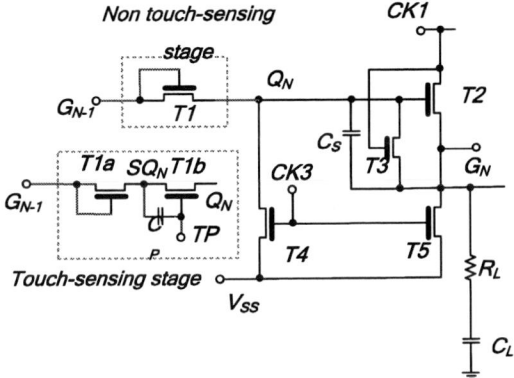

Fig. 1. Schematic of the proposed TFT integrated gate driver with a single stage, while the input module is differrent for the touch sensing stage (1T1C is inserted) and non-touch-sensing stage.

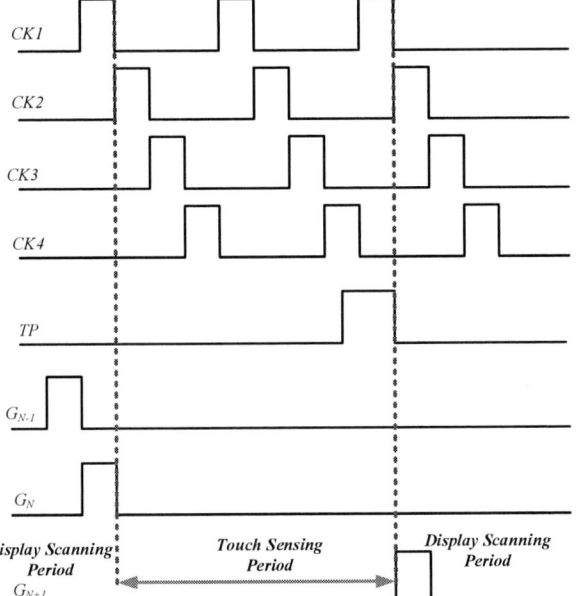

Fig. 2. The timing diagram of the proposed TFT gate driver, including both the touch sensing period and display scanning period, while the external clock signals (CK1-CK4) maintain continous.

A. Charge Storage

Firstly, TP is with low voltage level. The output of the (N-1)-th gate driver is triggered. Thus G_{N-1} is with high voltage level. Through T1A, the starting signal is transferred and stored by C_P. Consequently, charge is stored at the drain electrode of the T1B instead of the gate electrode, which benefits the improving of the T1B's stability.

Afterwards G_{N-1} switches to the low voltage level. Due to the diode connection of T1A, the stored charges by C_P is maintained, and T1A maintains off regardless of the falling edge of G_{N-1}. At the same time, as TP is with low level, T1B is also turned off. In other words, the transistors associates with C_P are turned off for the touch sensing period to avoid the possible charge loss.

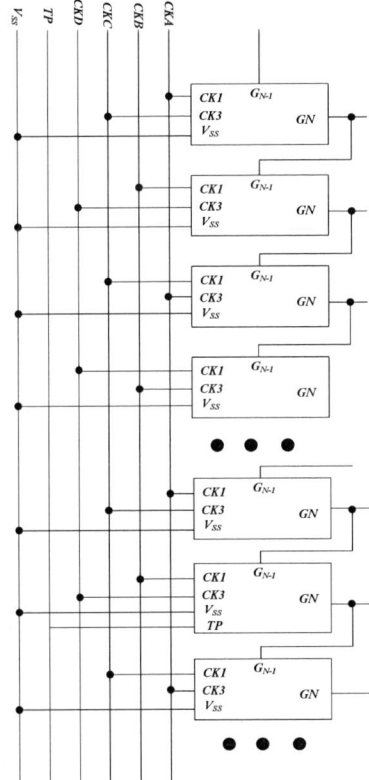

Fig. 3. The block schematics of the proposed TFT gate driver with cascaded stages.

B. Charge Transfer

By the end of the touch sensing period, TP is switched to the high voltage level. Then T1B is turned on and the charge stored by C_P is transferred to the Q_N node. Consequently, the pre-charge of the N-th gate driver can be obtained. And the following scanning pulse can be generated through the buffer transistor, i.e. T2. As Q_N is bootstrapped by C_S during the following time, which is higher than the level of T_P, T1B can be turned off. It's worth mention that the pulse width of TP can be tuned without influencing the bootstrapping process extended.

979-8-3315-8850-2/25 $31.00 © 2025 IEEE   552

The performance balance between charge storage and charge transfer is highly dependent on the appropriate selection of the C_P value. In this work, constraints on the C_P value are derived based on charge conservation laws. The pre-charge voltage at node Q_N is determined through charge sharing between C_P and C_S.

The pre-charge voltage of Q_N is determined by the charge sharing of C_P and C_S. Ignoring the relative parasitic TFTs' capacitances, one can write

$$Q1 = C_P \left(V_{GH} - V_{GL} \right) \tag{1}$$

Once TP is switched to V_{GH}, then the charges of C_P is shared with C_S, thus

$$Q1 = C_P \left(Q_N - V_{GL} \right) + C_S \left(Q_N - V_{GL} \right) \tag{2}$$

Therefore, the voltage of Q_N can be derived as

$$Q_N = \frac{C_P V_{GH} + C_S V_{GL}}{C_P + C_S} \tag{3}$$

As certain Q_N is needed to trigger the following bootstrapping process of the touch-sensing stage ($Q_N > k(V_{GH}-V_{GL}) + V_{GL}$, and the empirical value of k is 0.6), the minimal value of C_P can be derived as

$$C_P \geq \frac{k}{1-k} C_S \tag{4}$$

As shown in Fig. 2, between the display scanning and touch-sensing intervals, the external clock signals, i.e. CK1- CK4 are continuous. Thus, no extra controlling logic circuits are needed, and the external level shifter circuit become much simplified.

III. RESULT AND DISCUSSION

The proposed gate driver was implemented using a-Si:H TFTs for functional verification and parameter optimization. It is noteworthy that the circuit architecture is also applicable to other thin-film transistor technologies, such as InGaZnO TFTs, low-temperature polycrystalline silicon (LTPS) TFTs, and organic TFTs. The a-Si:H implementation in this study serves as one representative example.

Fig. 4 shows the measured and simulated transfer characteristics a-Si:H TFTs, in which the channel width and length are 25 μm and 4.0 μm, respectively, and the V_{DS} are 0.5 V and 10.5 V. For SPICE simulations, the RPI model with Level of 35 is used. The inset in Fig.3 is the optical image of the fabricated a-Si:H TFT with the standard five-mask process, showing the channel length of 4 μm, finger width for the source/drain electrode of 3.0 μm. The thickness of SiN$_x$ and a-Si:H channel layer are 400 nm and 130 nm, respectively. The extracted threshold voltage, mobility and overlap capacitance of the fabricated TFT are 1.19 V, 0.48 cm²/V·s and 16.7 nF/μm², respectively. As the extracted device model well presents the measured characteristics, the model is adopted for the SPICE simulations in this study.

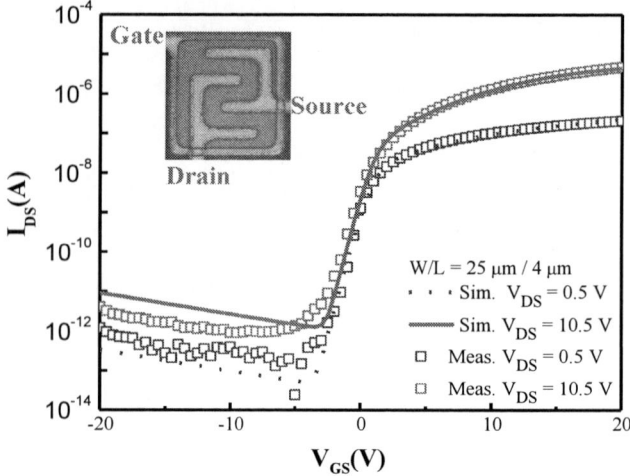

Fig. 4. Comparison of the measured and modeled transfer characteristics of an a-Si:H TFT with W/L= 25 μm / 4 μm, and the inset demonstrates the optical image of the fabricated a-Si:H TFT with the standard five-mask process .

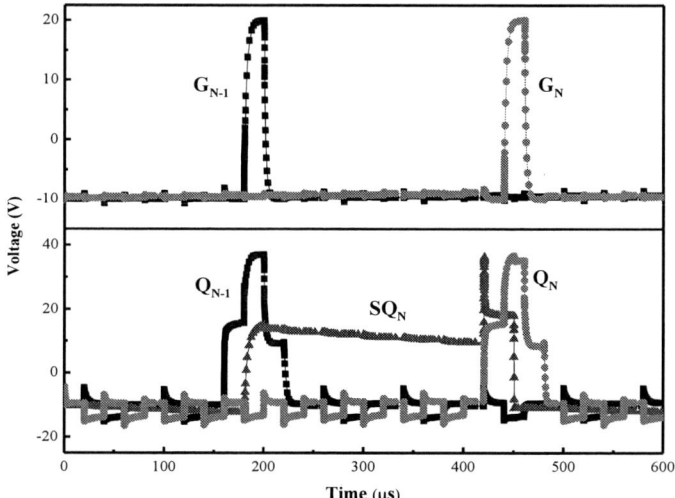

Fig. 5. The transient response of the proposed gate driver using a-Si:H TFT based on SPICE simulations.

The loading capacitance of the gate line is 100 pF, and the equivalent resistance of the gate line is 3000 Ohm. For a display resolution of HD, with the gate line number of 1080 and the frame rate of 60 Hz, the scanning time is approximately 20 μs. And the touch sensing time is 200 μs.

A. Verification of Charge Storage and Transfer

Fig.5 shows the transient response of the proposed a-Si:H TFT gate driver by SPICE simulations. Different from the conventional gate driver, which is triggered line by line, there are 200 μs time intervals between the scanning pulse of the N-th stage and the (N+1)-th stage. During the touch-sensing period, the transfer signal SQ_N is maintained with high voltage level. Therefore, the demonstrated charge storage and transfer processes in part II are well proved.

979-8-3315-8850-2/25 $31.00 © 2025 IEEE

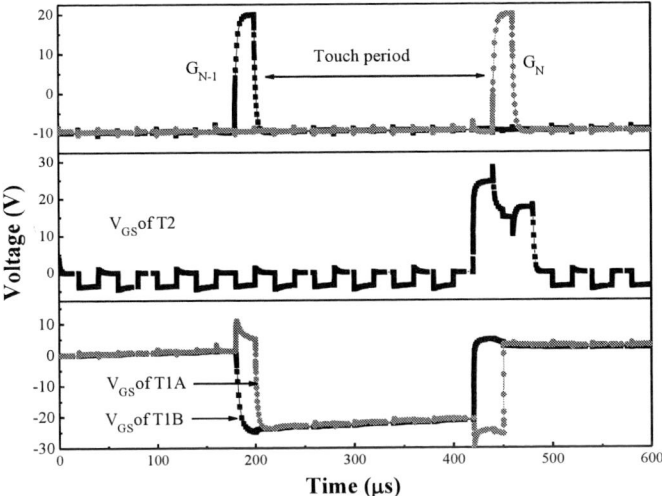

Fig. 6. Comparsion of gate-to-source voltage of the buffer transistor for the proposed circuit and the conventional one.

Fig.6 demonstrates the gate-to-source voltage for the driving transistor T2, the input transistors (i.e. T1A and T1B). For the low voltage maintaining period, Vgs is almost 0 for all the three transistors. While Vgs of T2 for the touch-sensing period is still 0, and Vgs of T1A and T1B are -20 V. Thus positive stress voltage is removed for the driving transistor and the input transistors.

B. Impact of C_P

Fig.7 shows the comparison result of the transient response with different C_P. With C_P decreased from 12 pF to 4 pF, the voltage loss of SQ_N is increased for the touch sensing period. For C_P of 12 pF and 8 pF, the bootstrapping behavior of Q_N is almost the same. While for C_P of 4 pF, there is obvious voltage loss for the pre-charge of Q_N, and Q_N failed to rise for the following bootstrapping interval. Therefore, considering suppressing of leakage current, C_P is required to be larger than 8 pF. However, too large C_P consumes too much real-estate, which is not good for the shrinkage of border size.

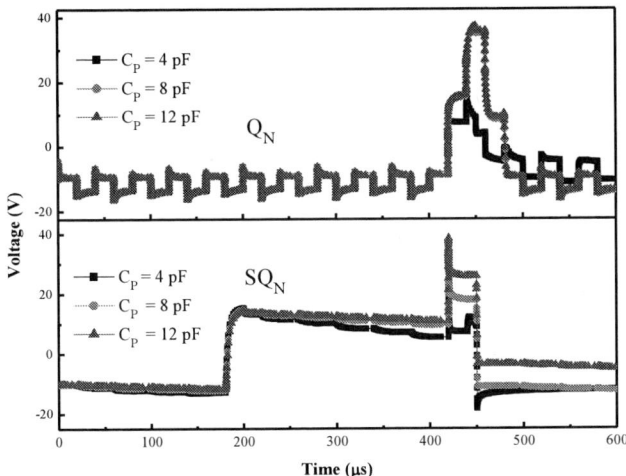

Fig. 7. Comparsion of the transient respone with different C_P.

Fig. 8 illustrates the rising and falling time with C_P from 6 pF to 12 pF. With C_P less than 7 pF, the rising and falling time

are increased to nearly 10 μs and 17 μs, respectively. While both the rising and falling time are independent of C_P once C_P is more than 7 pF. The obvious increase of delay time for small C_P is attributed to the insufficient charge transfer for the input module of the touch-sensing stage, which influence the driving ability of the buffer transistor. While C_P of 7 pF is large enough to completely transfer the charges to the gate terminal of the buffer transistor. The optimized value of C_S is 4 pF, and this result is also well matched with Eq. (4).

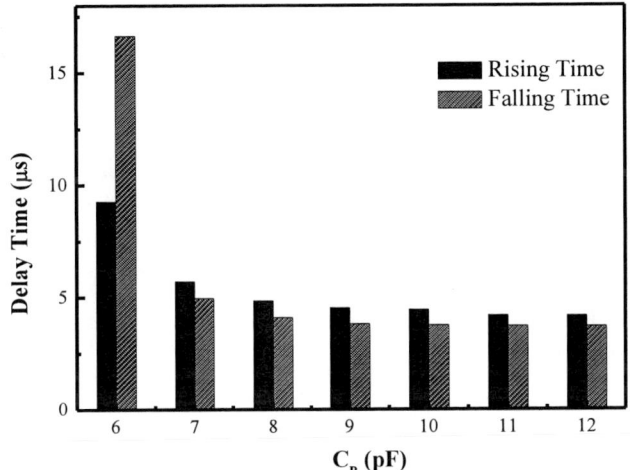

Fig. 8. Comparsion of the rising and falling time with different C_P.

IV. CONCLUSION

This paper presents a new TFT integrated gate driver circuit for in-cell touch sensing display applications. The touch sensing stage includes an added 1T-1C in the input module to realize charge storage-transfer function in prior to voltage bootstrapping TFT. Operating principle of the gate driver is discussed and requirements of C_P is derived. Based on the measured electrical performances of a-Si:H TFTs, SPICE model were developed and circuit simulations were carried out to verify the feasibility of the proposed gate driver. Compared with conventional in-cell touch sensing display driver, the proposed circuit features simplified external clock generation circuits, as the required external clock signals for the gate driver maintain continuous throughout both touch-sensing and display intervals. Furthermore, the proposed gate driver is free from prolonged large positive voltage stressing for both the touch sensing time and display time. Thanks to the merits of high reliability and simple circuit and timing, the proposed gate driver is promising for the application of high-end mobile in-cell touch panels.

REFERENCES

[1] C. Liao, C. He, T. Chen, D. David, S. Chung, T. S. Jen, and S. Zhang, "Design of Integrated Amorphous-Silicon Thin-Film Transistor Gate Driver," Journal of Display Technology, vol. 9, no. 1, pp. 7-16, 2013.

[2] C. Liao, C. He, T. Chen, D. Dai, S. Chung, T. S. Jen, and S. Zhang, "Implementation of an a-Si:H TFT Gate Driver Using a Five-Transistor Integrated Approach," IEEE Transactions on Electron Devices, vol. 59, no. 8, pp. 2142-2148, 2012.

[3] Z. Hu, L. L. Wang, C. Liao, L. Zeng, C.-Y. Lee, A. Lien, and S. Zhang, "Threshold Voltage Shift Effect of a-Si:H TFTs Under Bipolar Pulse Bias," IEEE Transactions on Electron Devices, vol. 62, no. 12, pp. 4037-4043, 2015.

[4] Z. Hu, C. Liao, W. Li, L. Zeng, C. Lee, S. Zhang, "Integrated a-Si:H Gate Driver with Low-level Holding TFTs Biased under Bipolar Pulses", IEEE Transactions on Electron Devices, vol. 62, no. 12, pp. 4044 - 4050, Dec. 2015.

[5] S. H. Moon, Y. S. Lee, M. C. Lee, B. H. Berkeley, N. D. Kim and S. S. Kim, "Integrated a-Si:H TFT Gate Driver Circuits on Large Area TFT-LCDs," in Proc. SID Symp. Dig., 2007, pp. 1478-1481.

[6] C. Liao, Z. Hu, D. Dai, S. Chung, T. S. Jen, and S. Zhang, "A Compact Bi-Direction Scannable a-Si:H TFT Gate Driver," Journal of Display Technology, vol. 11, no. 1, pp. 3-5, 2015.

[7] J. W. Choi, J. I. Kim, S. H. Kim, and J. Jang, "Highly Reliable Amorphous Silicon Gate Driver Using Stable Center-Offset Thin-Film Transistors," IEEE Trans. Electron Devices, vol. 57, no. 9, pp. 2330-2334, 2010.

[8] I. Hwang, S. Moh, M. Lee, and E. Lee, "Design of Integrated a-Si Gate Driver Circuits for Low Power Consumption," in Proc. SID Symp. Dig., 2008, pp. 842-845.

[9] C. Lin, C. Tu, C. Wu, C. Hung, K. Gan, and K. Chou, "Low-power gate driver circuit for TFT-LCD application," IEEE Trans. Electron Devices, vol. 59, no. 5, pp. 1410–1415, May 2012.

[10] J. Lee, Y. Bae, W. Lee, Y. Kim, J. Song, Y. Hyun, D. Cho, S. Kim, Y. Kwon, M. Kwon, S. Moon, and K. Kim, "Low Power a-Si:H TFT Gate Driver Circuit Employing Negative Turn Off Biasing", in Proc. SID Symp. Dig., 2011, pp. 1181-1184.

[11] C. Lin, M. Cheng, C. Tu, C. Wu, and F. Chen, "Low-Power a-Si:H Gate Driver Circuit With Threshold-Voltage-Shift Recovery and Synchronously Controlled Pull-Down Scheme", IEEE Trans. Electron Devices, vol. 62, no. 1, pp. 136 - 142, Jan. 2015.

[12] Q. Ma, H. Wang, L. Zhou, J. Fan, C. Liao, X. Guo, and S. Zhang, "Robust Gate Driver on Array Based on Amorphous IGZO Thin-Film Transistor for Large Size High-Resolution Liquid Crystal Displays," IEEE Journal of the Electron Devices Society, vol. 7, pp. 717-721, 2019.

[13] S. Shen, C. Liao, J. Yang, H. Jiao, and S. Zhang, "Capacitor Reused Gate Driver for Compact In-Cell Touch Displays," IEEE Journal of the Electron Devices Society, vol. 9, pp. 533-538, 2021.

[14] S. Shen, C. Liao, J. Yang, H. Jiao, and S. Zhang, "A compact gate driver with bifunctional capacitor for in‐cell touch mobile display," Journal of the Society for Information Display, vol. 29, no. 7, pp. 526-536, 2021.

[15] Y. Xue, L. Wang, Y. Zhang, G. Liang, J. Chu, B. Han, W. Cao, C. Liao, and S. Zhang, "31-Inch 4K Flexible Display Employing Gate Driver With Metal Oxide Thin-Film Transistors," IEEE Electron Device Letters, vol. 42, no. 2, pp. 188-191, 2021.

[16] J. Yang, C. Liao, K. Wang, J. An, S. Shen, and S. Zhang, "A-Si TFT Integrated Gate Driver Workable at −40°C Using Bootstrapped Carry Signal," IEEE Access, vol. 10, pp. 93887-93893, 2022.

[17] J. An, C. Liao, Y. Zhu, X. Zheng, C. Dai, X. Zhang, and S. Zhang, "Gate Driver on Array With Multiple Outputs and Variable Pulse Widths for Low-Temperature Polysilicon and Oxide (LTPO) TFTs Driven AMOLED Displays," IEEE Transactions on Circuits and Systems II: Express Briefs, vol. 70, no. 3, pp. 934-938, 2023.

[18] C. Liao, X. Zheng, and S. Zhang, "Dual-bootstrapping gate driver circuit design using IGZO TFTs," Displays, 2024.

[19] Y. Zhang, L. Chang, L. Lu, C. Liao, and S. Zhang, "Asymmetric Double-Gate (ADG) Oxide Thin-Film Transistor Technology for Medium- and Small-Sized AMOLED Displays," IEEE Electron Device Letters, vol. 46, no. 5, pp. 785-788, 2025.

[20] Y. Zhu, C. Liao, L. Qian, and S. Zhang, "A Single Sweep Signal Enabled Analog PWM Pixel Circuit for Progressive-Emission Mode Active-Matrix Micro-LED Displays," IEEE Electron Device Letters, vol. 46, no. 3, pp. 416-419, 2025.

[21] Chih-Lung Lin, et al., "Gate driver circuit using pre-charge structure and time-division multiplexing driving scheme for active-matrix lcds integrated with in-cell touch structures", Journal of Display Technology, vol. 12, no. 11, pp. 1238–1241, Nov. 2016.

[22] Su-Hwan Moon, et al., "Highly Robust Integrated Gate-Driver for In-Cell Touch TFT-LCD Driven in Time Division Driving Method", Journal of Display Technology, vol. 12, no. 5, pp. 435–441, May. 2016

[23] Chih-Lung Lin, et al., "Insertion of Simple Structure Between Gate Driver Circuits to Prevent Stress Degradation in In-Cell Touch Panel Using Multi-V Blanking Method", Journal of Display Technology, vol. 12, no. 10, pp. 1040–1042, Oct. 2016.

[24] Jeongrim Seo, et al., "Low Power and Low Noise Shift Register for In-Cell Touch Display Applications", Journal of the Electron Devices Society, vol. 6, pp. 726-732, 2018.

[25] Chih-Lung Lin, et al., "Bidirectional Gate Driver Circuit Using Recharging and Time-Division Driving Scheme for In-Cell Touch LCDs", IEEE Transactions on Industrial Electronics, vol. 65, no. 4, April 2018.

2025 The 10th International Conference on Integrated Circuits and Microsystems

Neural Network-Based Method for Magnetic Field Inversion of Energized Conductors in Printed Circuit Boards

Qi Li[1]
School of Automation Engineering,University of Electronic Science and Technology of China
ChengDu,China
liqi0111@uestc.edu.cn

Yiling Liang[2]
School of Automation Engineering,University of Electronic Science and Technology of China
ChengDu,China
202322060141@std.uestc.edu.cn

Zhen Liu[3]
School of Automation Engineering,University of Electronic Science and Technology of China
ChengDu, China
scdliu@uestc.edu.cn

Lulu Tian[4*]
School of Automation Engineering,University of Electronic Science and Technology of China
ChengDu, China
lulutian@uestc.edu.cn

Abstract—**This paper presents an innovative methodology for three-dimensional magnetic field reconstruction of electronic circuits on printed circuit boards. By integrating multiple neural network architectures with associated optimization algorithms, the proposed approach substantially enhances the magnetic field prediction accuracy compared to conventional physics-based reconstruction methods. The core principle of the proposed methodology lies in leveraging neural networks to precisely fit the dipole moments generated by electronic circuits on the PCB, subsequently constructing an equivalent dipole model based on the fitted results to substitute for the actual magnetic field sources produced by the operational PCB circuitry. Through rigorous simulation validation, the neural network-based three-dimensional magnetic field reconstruction method for PCBs proposed in this study achieves prediction errors constrained within 5%, with magnetic field accuracy reaching the nanotesla scale. These results comprehensively demonstrate the superior performance of the proposed approach in PCB spatial magnetic field visualization, thereby establishing a robust foundation for addressing practical engineering challenges such as fault diagnosis and defect localization in subsequent applications.**

Keywords—*magnetic field inversion, neural network modeling, printed circuit board, current-carrying trace*

I. INTRODUCTION

A. Background

With the rapid advancement of electronic technology, electronic devices are continuously evolving toward miniaturization, high integration, and enhanced performance. However, this remarkable development trend is accompanied by increasingly severe electromagnetic compatibility (EMC) challenges [1]. Particularly in the field of printed circuit board (PCB) design, accurate simulation and prediction of magnetic field distributions generated by conductor currents have become critical issues demanding urgent resolution [2].

Precise prediction of electromagnetic fields holds profound significance for addressing electromagnetic interference (EMI) problems, as it not only enables the optimization of circuit layouts but also significantly enhances device reliability [3]. Conventional magnetic field detection methods typically rely on bulky and costly specialized equipment, with measurement processes that are cumbersome, complex, and inefficient [4]. Notably, as the integration density of electronic devices continues to escalate, the internal electromagnetic environment becomes increasingly complex, rendering traditional detection approaches inadequate when confronted with high-density PCBs and unable to meet the practical requirements of the modern electronics industry [1, 2].

In recent years, concurrent with the progress in computer technology and artificial intelligence, neural network-based magnetic field reconstruction methods have demonstrated tremendous potential [5]. Such approaches can achieve high-precision reconstruction of three-dimensional magnetic fields in PCB spaces by analyzing limited measurement data in conjunction with accurate numerical models, thereby opening novel avenues for electromagnetic compatibility analysis, fault diagnosis, and non-destructive testing [6]. Particularly in domains with extremely high reliability requirements, such as microelectronic devices, aerospace systems, and medical equipment, precise electromagnetic field modeling plays an irreplaceable role in preventing electromagnetic interference and ensuring stable system operation [7].

The equivalent dipole method, as one of the mainstream approaches for source reconstruction, can accurately characterize the magnetic field distribution of operational PCB electronic circuits by ingeniously exploiting the interrelationships within dipole arrays [8]. The organic integration of advanced machine learning techniques with source reconstruction to achieve efficient and precise reconstruction has become a prominent research focus in both academic and industrial communities [9]. Neural networks, by virtue of their powerful self-learning capabilities and nonlinear mapping characteristics, exhibit excellent performance in source reconstruction within complex environments, enabling accurate fitting of the intricate mapping relationships between magnetic field sources and the Green's functions of dipole arrays [10].

This paper conducts an in-depth investigation of neural network-based three-dimensional magnetic field simulation and reconstruction techniques for PCBs. By comprehensively employing the Biot-Savart law, source reconstruction techniques, and optimization algorithm principles, the research is structured into four interrelated components: First, near-field magnetic field data surrounding the PCB are acquired using high-precision fluxgate magnetometers. Second, the Green's

979-8-3315-8850-2/25 $31.00 © 2025 IEEE

function matrix between scanning points and the designated dipole array is rigorously derived. Third, based on the magnetic field measurement data and theoretical functions obtained from the preceding two steps, a scientifically sound neural network training set is constructed. Finally, by integrating advanced optimization algorithms with neural network architectures, the high-precision reconstruction process of the PCB's three-dimensional spatial magnetic field is achieved.

B. Mathematical modeling of spatial magnetic field distribution

The spatial magnetic field around a PCB exhibits a typical helical distribution, as shown in Fig. 1. When current flows through PCB conductors, it generates a circular magnetic field around the conductors. This magnetic field intensity in space has two important characteristics: first, in the XY plane, the magnetic field intensity presents a Gaussian distribution centered on the conductors; second, along the Z-axis, the magnetic field intensity attenuates with increasing distance.

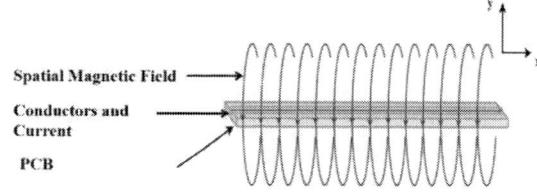

Fig. 1. Theoretical model of spatial magnetic field distribution under DC excitation of PCB.

In static magnetic field inversion problems, the relationship between current sources and magnetic fields can be directly established based on the Biot-Savart law. For steady currents flowing in a PCB, the generated magnetic flux density B can be expressed as:

$$B(\mathrm{p}) = (\frac{\mu_o}{4\pi})\int \left(\frac{I \times R}{|R|^3}\right)dl \qquad (1)$$

where μ_0 is the vacuum permeability, I is the current vector, R is the vector from the source point to the field point, and the integration region is the effective length of the conductor. This integral equation constitutes the forward problem model for magnetic field inversion.

II. METHODOLOGY

The core objective of this study is to establish a precise relationship matrix between current elements and the magnetic field at observation points, namely the Green's function matrix. To simplify the complexity of the problem, the following reasonable assumptions are proposed:

1) N current elements are defined on the current element plane, constituting a discrete current source array;

2) M scanning points exist on the scanning plane for measuring or calculating magnetic field values;

3) The research object consists of uniform linear media, primarily considering air and PCB substrate materials.

This study employs the experimental setup illustrated in Fig. 2 for magnetic field measurements, acquiring three-dimensional magnetic field data surrounding the PCB through high-precision fluxgate magnetometers.

Fig. 2. Schematic diagram of the magnetic measurement equipment.

Table 1 presents the key specifications of the CH-370 high-precision three-dimensional fluxgate magnetometer used in this study. These parameters directly impact the accuracy and reliability of the magnetic field measurements.

TABLE I.   TABLE TYPE STYLES

Parameter	Value
Probe Dimensions(mm)	32×32×225
Measurement Range	±100 uT
Resolution	0.01 nT
Frequency Response	≤3 kHz
Accuracy	±0.2% of reading ±5nT

To accurately represent the physical experimental conditions, the simulation domain was established based on the dimensions of the laboratory-fabricated PCB test board (10 cm × 10 cm, Fig. 3). Consequently, a 10 cm × 10 cm × 10 cm three-dimensional computational grid was defined to completely encompass the test board and its surrounding region where key magnetic field phenomena occur. This grid configuration optimizes the trade-off between spatial resolution and computational efficiency, thereby facilitating a more realistic simulation of the magnetic field distribution under practical scenarios.

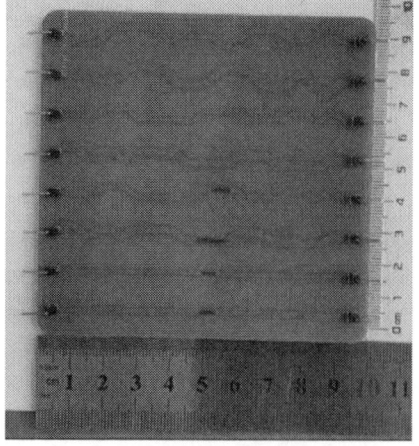

Fig. 3. Schematic illustration of the laboratory-fabricated PCB test board.

To validate the effectiveness of the proposed method, a PCB test board as illustrated in Fig. 3 was designed to serve as the target circuit for source magnetic field reconstruction. Through fluxgate magnetometers, precise magnetic field values surrounding the circuit under different operating conditions were acquired, and scanning data from different layers in the near-field region were selected and processed for subsequent neural network training procedures.

Based on the equivalent dipole source reconstruction principle, assuming that N magnetic dipoles are defined on the dipole plane to form a discrete magnetic field source array, and M scanning points exist on the scanning plane for measuring magnetic field values, the magnetic field at the scanning point locations can be derived as follows:

$$H(r_i) = \sum_{j=1}^{N} G^h(r_i, r_j) \cdot M(r_j) \tag{2}$$

In the formula, r_i represents the *ith* scanning point among M scanning points, r_j denotes the *jth* dipole among N dipoles, $H(r_i)$ is the magnetic field at the scanning point location, $Gh(r_i, r_j)$ is the magnetic field Green's function matrix, and $M(r_j)$ is the corresponding magnetic dipole moment.

According to the Biot-Savart law, the magnetic field generated by current elements on the PCB at any point in space can be equivalently represented through the magnetic dipole model. It is noteworthy that, considering the ground plane typically existing beneath the PCB, the image theory reveals that the z-direction magnetic dipole perpendicular to the PCB and its image source possess equal amplitudes with opposite directions, resulting in mutual cancellation of their radiation. Therefore, when investigating three-dimensional spatial magnetic field reconstruction of PCBs, primary consideration is given to the x-direction magnetic dipole Mx and the y-direction magnetic dipole My, which are sufficient to characterize the actual magnetic field distribution generated by electronic circuits on the PCB.

For the direct current case, the components of the magnetic field generated by a single magnetic dipole can be expressed as:

$$\begin{bmatrix} H_x(r_i) \\ H_y(r_i) \\ H_z(r_i) \end{bmatrix} = \begin{bmatrix} G_{xx}^h & G_{xx}^h \\ G_{yx}^h & G_{xx}^h \\ G_{zx}^h & G_{xx}^h \end{bmatrix} \begin{bmatrix} M_x \\ M_y \end{bmatrix} \tag{3}$$

where Hx(ri), Hy(ri), and Hz(ri) are the three magnetic field components at the scanning point r_i. The Hx component is given by:

$$H_x(r_i) = \sum_{j=1}^{N} \left[G_{xx}^h(r_i, r_j) \cdot M_x(r_j) + G_{xy}^h(r_i, r_j) \cdot M_y(r_j) \right] \tag{4}$$

For the DC case, the matrix element of the magnetic field Green's function, as derived from the Biot-Savart law, simplifies to:

$$G_{xx}^h(r_i, r_j) = \frac{\mu_0}{4\pi} \cdot \frac{Y^2 + Z^2}{R^3} \tag{5}$$

$$G_{xy}^h(r_i, r_j) = -\frac{\mu_0}{4\pi} \cdot \left(-\frac{XY}{R^3} \right) \tag{6}$$

The geometric parameters are defined as follows:

$$\begin{cases} X = x_i - x_j, Y = y_i - y_j, Z = z_i - z_j \\ R = \sqrt{(x_i - x_j)^2 + (y_i - y_j)^2 + (z_i - z_j)^2} \end{cases} \tag{7}$$

(x_i, y_i, z_i) and (x_j, y_j, z_j) denote the spatial coordinates of the *ith* scanning point and the *jth* dipole, respectively, while $\mu_0 = 4\pi \times 10^{-7}$ H/m represents the permeability of free space.

The rigorous derivation of this Green's function matrix establishes a precise mathematical relationship connecting the magnetic field produced by current elements within the PCB's electronic traces to the equivalent magnetic dipoles. This provides a robust theoretical basis for subsequently fitting the dipole moments with a neural network. In the training phase of the neural network, this matrix serves as the input features, with the measured magnetic field values at the scanning points acting as the corresponding output labels. This approach facilitates a high-fidelity reconstruction of the magnetic field in the three-dimensional space surrounding the PCB.

Following the formulation in Equation (3), the numerical Green's function matrix is supplied to the neural network as input features, with the magnetic field values at the scanning points serving as the corresponding target labels. Training the model on a carefully curated dataset enables it to accurately approximate the dipole moment function, thereby providing a critical foundation for downstream magnetic field prediction. The entire training procedure is depicted in Fig. 4.

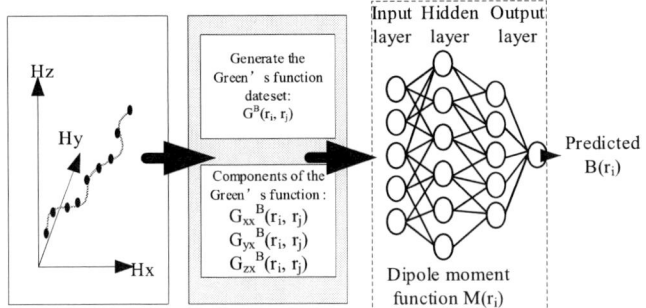

Fig. 4. Neural network training workflow.

This study focuses on the Hx component of the magnetic field. The model is configured with N scanning points on the PCB plane and M dipoles on the dipole plane, with each dipole comprising Mx and My moments. The numerical Green's function matrix is derived from the established formula. To enhance model precision, this research evaluates the performance of three distinct neural network optimization algorithms:

Genetic Algorithm-Optimized Neural Network (GA-BP): This hybrid method uses a Genetic Algorithm (GA) to optimize the initial weights and thresholds of a BP network, leveraging GA's global search capabilities. This approach improves the modeling of complex PCB magnetic field mappings and effectively avoids the local optima pitfalls of standard BP.

For an objective assessment of each model's performance, the prediction error (ER) is employed as the primary metric:

$$ ER = \frac{1}{n} \times \sum \left| y_i - \hat{y}_i \right| / y_i \qquad (8) $$

This version uses more sophisticated vocabulary and sentence structure for a more polished, high-impact academic tone.

Here, $\hat{y}_i$ denotes the *ith* predicted Hx component, while y_i represents the corresponding *ith* ground truth value. A lower ER value signifies a reduced prediction error and, consequently, superior model accuracy.

Fig. 5 illustrates the conceptual framework for the neural network-based source reconstruction. This setup details the spatial configuration of several key planes: the prediction plane, scanning planes 1 and 2, the dipole plane, and the PCB plane.

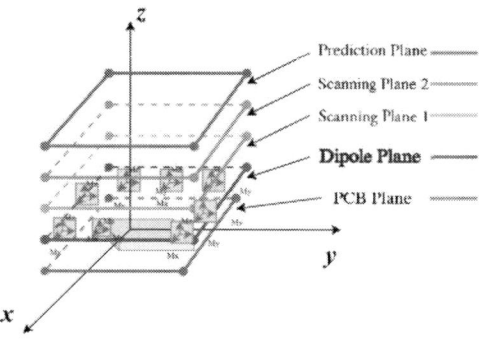

Fig. 5. Schematic diagram of the overall approach for spatial magnetic field reconstruction.

III. RESULTS

The magnetic field data measured at distances of 1 mm and 5 mm were used as the training set for the neural network, while the data from 10 mm served as the validation set, as shown in TABLE II.

TABLE II. MAGNETIC FIELD INVERSION RESULTS AT A CONSTANT CURRENT OF 0.5 A

I=0.5A(x, y, z)/mm	Hx/nT	Hy/nT	Hz/nT	H/nT
(1,0,1)	1962.0	838.0	1032.0	2321.0
(1,0,2)	1865.0	801.0	980.0	2300.0
(1,0,3)	1800.0	789.0	950.0	2200.0
(1,0,5)	1742.0	766.0	921.0	2238.0
(1,0,8)	1598.0	630.0	880.0	2002.0
(1,0,10)	1422.0	535.0	810.0	1989.0

Fig. 6 and Fig. 7 present the spatial magnetic field distributions reconstructed by the GA-BP model on the XY plane at two distinct heights: z = 5.0 mm and z = 10.0 mm,

respectively. A comparative analysis of these figures provides a clear visualization of the magnetic field's decay with increasing distance from the source.

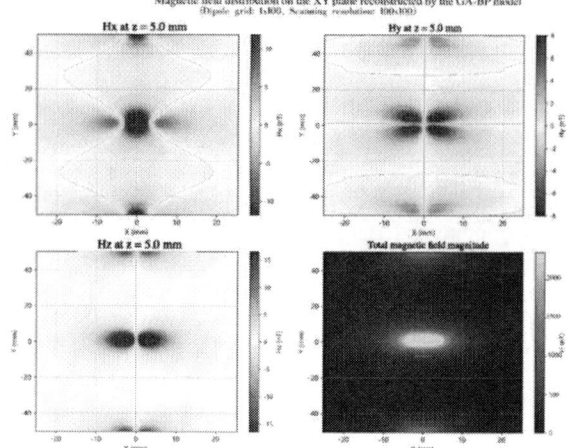

Fig. 6. The Hx, Hy, and Hz components and the total magnitude of the magnetic field on the XY plane at z = 5.0 mm, as reconstructed by the GA-BP model.

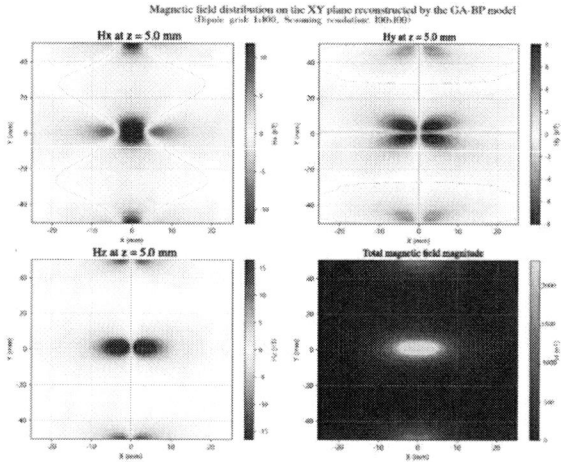

Fig. 7. The Hx, Hy, and Hz components and the total magnitude of the magnetic field on the XY plane at z = 10.0 mm, as reconstructed by the GA-BP model.

Consequently, the total magnetic field magnitude is substantial, with a central peak value exceeding 2000 nT, indicating a high concentration of field strength directly above the source. In contrast, as the observation plane is moved further from the source to z = 10.0 mm, a significant attenuation of the field is observed. This decay is most dramatically reflected in the total magnetic field magnitude, where the peak value drops sharply to approximately 400 nT.

To provide an objective and rigorous assessment of the GA-BP model's performance, a quantitative error analysis was conducted in conjunction with a qualitative review of the reconstructed field maps. The primary metric for this evaluation is the prediction error (ER), defined as the mean absolute percentage error between the model's predicted values ($\hat{y}_i$) and the corresponding ground-truth values (y_i).

The model demonstrated high predictive accuracy, achieving a low prediction error of 3.12% on the validation dataset. This low ER value signifies a strong correlation and minimal deviation between the reconstructed magnetic field data and the ground-truth measurements, confirming the model's quantitative precision.

This numerical accuracy is further substantiated by the qualitative analysis of the reconstructed magnetic field distributions, as depicted in Fig. 6 and Fig. 7. The figures not only show a high degree of detail but also adhere strictly to physical principles. For instance, the model accurately captures the complex spatial topology of the field components and correctly simulates the physical phenomenon of field attenuation with increasing distance. As observed, the peak total magnetic field magnitude decreases from over 2000 nT at $z = 5.0$ mm to approximately 400 nT at $z = 10.0$ mm, a trend consistent with electromagnetic theory.

IV. CONCLUSIONS

This study proposed and validated a theoretical framework for the three-dimensional spatial magnetic field reconstruction of PCBs based on a GA-BP neural network. The methodology demonstrates significant reliability and novelty. The conclusions obtained are as follows:

1) The novelty of this approach lies in its departure from traditional physics-based modeling, such as the finite element method, thereby circumventing complex model setup and high computational costs. By adopting a data-driven paradigm and uniquely integrating a Genetic Algorithm (GA) to globally optimize the initial weights and thresholds of the BP neural network, our model effectively overcomes the propensity of standard BP networks to converge to local minima, leading to enhanced convergence speed and prediction accuracy.

2) The reliability of the model is robustly supported by both quantitative and qualitative analyses. Quantitatively, the model achieved a low prediction error (ER) of 3.12% on the validation set, confirming the high numerical fidelity of the reconstructed field values. Qualitatively, the reconstructed magnetic field maps are not only rich in detail but also physically plausible. The spatial topology of the field components and the observed decay in field strength with distance (the peak total magnitude decreasing from over 2000 nT at $z = 5.0$ mm to approx. 400 nT at $z = 10.0$ mm) are highly consistent with electromagnetic theory.

3) A limitation of the current study is the scope of the dataset, which was primarily focused on a specific PCB layout and operating conditions. Therefore, future work will be directed towards expanding the training and validation datasets to encompass a wider variety of PCB designs, operating frequencies, and current distributions. Further validation on this enriched dataset will serve to enhance the model's generalization capabilities and confirm its robustness across a broader spectrum of real-world scenarios.

In summary, the GA-BP neural network-based magnetic field reconstruction method presents a novel, accurate, and promising paradigm for applications in PCB electromagnetic compatibility analysis and interference source localization.

REFERENCES

[1] C. R. Paul, Introduction to Electromagnetic Compatibility, 2nd ed. Hoboken, NJ: John Wiley & Sons, 2006.

[2] D. M. Hockanson, J. L. Drewniak, T. H. Hubing, and T. P. Van Doren, "Investigation of fundamental EMI source mechanisms driving common-mode radiation from printed circuit boards with attached cables," IEEE Transactions on Electromagnetic Compatibility, vol. 38, no. 4, pp. 557-566, Nov. 1996.

[3] A. E. Ruehli, G. Antonini, and L. Jiang, "Magnetic Field and Inductance Computations in the Presence of Magnetic and Conductive Materials," IEEE Transactions on Electromagnetic Compatibility, vol. 59, no. 2, pp. 288-297, 2017.

[4] H.-D. Brüns, H. Garbe, and C. Schuster, "Magnetic Field Scanning of PCBs for EMI Prediction," IEEE Transactions on Electromagnetic Compatibility, vol. 49, no. 2, pp. 313-321, May 2007.

[5] J. E. Rayas-Sánchez, "Power in EM-Based Machine Learning for RF and Microwave Applications," IEEE Microwave Magazine, vol. 20, no. 10, pp. 34-49, Oct. 2019.

[6] S. Koziel and L. Leifsson, Surrogate-Based Modeling and Optimization. New York, NY: Springer, 2013.

[7] M. I. Montrose and E. M. Nakacchi, EMC and the Printed Circuit Board: Design, Theory, and Layout Made Simple. Hoboken, NJ: IEEE Press, 1999.

[8] O. M. Bucci and G. Franceschetti, "On the spatial bandwidth of scattered fields," IEEE Transactions on Antennas and Propagation, vol. 35, no. 12, pp. 1445-1455, Dec. 1987.

[9] L. Jiang, S. Sun, and J. Mao, "Machine Learning in Electromagnetics: A Review and Future Perspectives," IEEE Transactions on Antennas and Propagation, vol. 69, no. 6, pp. 3095-3109, June 2021.

[10] Y. Zhang, Z. Zhang, and Q. H. Liu, "A Deep Learning-Based Equivalent Dipole Model for EMI Source Reconstruction," IEEE Transactions on Microwave Theory and Techniques, vol. 68, no. 11, pp. 4624-4635, Nov. 2020.

2025 The 10th International Conference on Integrated Circuits and Microsystems

A double layer placement algorithm for IC-based modules of printed circuit board

1st Hangyuan Li
Wuhan University of Technology
the School of Computer Science
and Artificial Intelligence
Wuhan, 430070, China

2nd Yu Chen*
Wuhan University of Technology
the School of Mathematics and Statistics
Wuhan, 430070, China
ychen@whut.edu.cn
*Corresponding author

3rd Zhaoyang Yang
Wuhan University of Technology
the School of Computer Science
and Artificial Intelligence
Wuhan, 430070, China

4th Haotian Pang
Wuhan University of Technology
the School of Information Engineering
Wuhan, 430070, China

5th Ning Xu
Wuhan University of Technology
the School of Information Engineering
Wuhan, 430070, China

Abstract—For large-scale printed circuit board (PCB) placement scenarios, this paper proposes an innovative clustering algorithm based on the traditional density-based spatial clustering of applications with noise (DBSCAN) algorithm, which is divided into integrated circuit (IC)-pin points clustering and components clustering. Meanwhile, since the distribution evolutionary algorithm based on a population of probability model (DEA-PPM) has achieved satisfactory performance in solving component orientation variables, this algorithm is introduced into the component orientation optimization of PCB, and the component coordinates are solved by conjugate sub-gradient algorithm (CSA). Based on the global placement results, a set of systematic legalization algorithms are proposed to eliminate overlap and satisfy the basic design specifications in industry.

Index Terms—printed circuit board, analytical placement, DBSCAN, conjugate subgradient algorithm, distribution evolutionary algorithm

I. INTRODUCTION

The placement of components plays a critical role in the design of printed circuit board (PCB). Due to the various PCB application scenarios, the task of PCB placement must follow diverse design rules and constraints. Consequently, automatic placement of PCB is a challenging task that requires scenarios-oriented design of efficient algorithm.

An analytic placement algorithm for PCB consists of two stages. The first stage is to perform the global placement, which tries to get optimal positions of modules with partial overlapping. The second stage is legalization, which aims to eliminate overlap and achieve placement results complying with the design rules. Moreover, PCB placement requries flexible rotation of modules, and most modules, including indutances, capacitances and resistances, are placed adjacent to some integrated cirucit (IC) modules, which leads to the requirement of centralized placement of functional module (CPFM) on PCB. This paper is dedicated to proposing an efficient CPFM placement algorithm addressing the following issues.

- At the stage of global placement, a mixed-variable optimization model is established to regulate both rotation angles and positions of modules, which are simultaneously optimized by a hybrid algorithm based on the distribution evolutionary algorithm based on a population of probability model (DEA-PPM) [1] and the conjugate subgradient algorithm (CSA) [2].
- To meet the design requirement of PCB, we propose an efficient legalization algorithm that can generate legal placement fullfilling diverse space rules between different modules.
- The performance of our algorithm is further improved with a cluster-based optimization strategy, by which a cluster-wise inialization, a batch opitmization strategy and the corresponding legalization process contribute to placement results that meet the design rules of module placement.

II. RELATED WORK

Analytic methods are popularly employed in floorplanning/placement scenarios of very-large scale integrated circuit (VLSI). Considering that traditional heuristic algorithms (such as genetic algorithms and simulated annealing) and standard Particle Swarm Optimization (PSO) are prone to fall into local optimum and have fixed parameters, Vinay et al. [3] introduced an improved adaptive PSO mechanism.

Srinivasan et al. [4] deeply integrated firefly and ant colony algorithms, leveraging the global search advantages of FA and the local optimization advantages of ACO to avoid the drawbacks of a single algorithm. By modeling the placement task as an optimzation problem constrained by the Poisson equation, Lu *et al.* [5] developed the analytic algorithm *ePlace* for the cell placement of VLSI, and Li *et al.* [6] proposed an efficient large-scale floorplanning algorithm for the task of large-scale floorplanning with fixed-outline. To address the challenge of developing a faster mixed-size placer without

979-8-3315-8850-2/25 $31.00 © 2025 IEEE

hardware acceleration and loss of solution quality, Peng and Zhu [7] proposed a mixed-size placement algorithm based on a novel definition of potential energy and a fast approximate computation scheme for partial derivatives of the potential energy for the Poisson's equation.

Since the anlaytic global placement cannot eliminate overlaps between modules, the legalization process is always introduced at the following stage. Peng and Zhu [7] proposes a phased Pre-macro Legalization (Pre-mLG) process. Different from traditional legalization, Pplace-MS intervenes in the early stage of placement (before global density optimization) to reduce conflicts in subsequent iterations. This legalization algorithm unifies the Poisson equation model, synchronously optimizes the legalization of macro modules with standard cells placement, and avoids the computational redundancy of the divide-and-conquer strategy.

For automatic placement of PCB components, Wang *et al.*[8] discussed how to realize automatic PCB placement under multiple constraints, and Li *et al.*[9] proposed a centralized placement method based on the sequence pair representation. Various types of algorithms are widely used in PCB placement. ML-related algorithms play a significant role in PCB placement. Based on a reinforcement learning-based agent for layout inference and fine-tuning and a large language model-based agent for interactive optimization, Chen *et al.*[10] developed a novel agent-based framework that automatically generates PCB placement meeting industrial constraints through user interactions. To accelerate the placement process of PCB, Zhang *et al.* [11] proposed a scalable GPU-accelerated PCB placement method inspired by VLSI. It incorporates tailored cost functions, constraint handling, and optimized techniques adapted for PCB placement. Taking the netlist of the already placed and unplaced circuits as input and abstract them into graphs, Chen *et al.* [12] proposed a subgraph matching based reference placement algorithm to achieve PCB placement reuse, thereby improving placement efficiency.

III. PRELIMINARIES

A. Problem Statement

The position of component v_i is represented by its central coordinate (x_i, y_i), and its orientation is denoted by r_i, where $r_i = j\pi/2$ $(j = 0, 1, 2, 3, i = 1, 2, \cdots, n)$. Thus, a placement of module can be representd by a combination of vectors $\{x, y, r\}$, where $x = (x_1, x_2, \cdots, x_n)$, $y = (y_1, y_2, \cdots, y_n)$, $r = (r_1, r_2, \cdots, r_n)$.

The module placement problem requires separate optimization of the top and bottom placements, which incorporate different objective functions and constraints. The top layer placement strives to achieve a uniform distribution as much as possible while ensuring non-overlapping placement and reduced wirelength, and the placement problem of the top layer is formulated [2] as:

$$
\begin{aligned}
\min \quad & W(x, y) \\
s.t. \quad & \begin{cases} D(x, y, r) = 0, \\ B(x, y, r) = 0, \end{cases}
\end{aligned} \tag{1}
$$

where $W(x, y)$ is the total wirelength, $D(x, y, r)$ is the sum of overlapping area, and $B(x, y, r)$ is the length beyond the external contour. It is transformed into an unconstrained optimization model[2]

$$
min\ f(x, y, r) = \alpha_1 W(x, y) + \beta_1 \sqrt{D(x, y, r)} + \gamma_1 B(x, y, r). \tag{2}
$$

where α_1, β_1, γ_1 are parameters to be confirmed. Here, the square root of $D(x, y, r)$ is adopted to ensure that all indexes to be minimized are of the same dimension. Ignoring the region constraint, the placement problem of the bottom layer is formulated [2] as

$$
min\ f(x, y, r) = \alpha_2 W(x, y) + \beta_2 \sqrt{D(x, y, r)}. \tag{3}
$$

a) Sum of Width beyond the External Contour $(B(x, y, r))$: For top layer placement problems, the positions of components require to meet the following constraints[2]:

$$
\begin{cases} 0 \le x_i - \hat{w}_i/2, x_i + \hat{w}_i/2 \le W^*, \\ 0 \le y_i - \hat{h}_i/2, y_i + \hat{h}_i/2 \le H^*. \end{cases} \tag{4}
$$

where W^* and H^* are the width and the height of the contour. Let

$$
\begin{aligned}
b_{1,i}(x, r) &= \max(0, \hat{w}_i/2 - x_i), & b_{2,i}(x, r) &= \max(0, x_i + \hat{w}_i/2 - W^*), \\
b_{1,i}(y, r) &= \max(0, \hat{h}_i/2 - y_i), & b_{2,i}(y, r) &= \max(0, y_i + \hat{h}_i/2 - H^*).
\end{aligned} \tag{5}
$$

$B(x, \hat{y}, r)$ can be confirmed by

$$
B(x, y, r) = \sum_{i=1}^{n} \left(\sum_{k=1}^{2} b_{k,i}(x, r) + \sum_{k=1}^{2} b_{k,i}(y, r) \right). \tag{6}
$$

B. The Conjugate Sub-gradient Algorithm

The conjugate sub-gradient algorithm (CSA) is an efficient analytic algorithm for optimization of analytic non-smooth prolems, which was introduced to address the placement problem [13] as well as the floorplanning problem [2] of VLSI. The CSA iteration process, is introduced in this paper to achieve the optimal position of modules.

C. The Distribution Evolutionary Algorithm Based on A Population of Probability Model

Besides the optimal location of modules, an optimal placement also requires the best orientation combination of modules. Accordingly, it is optimized by the distribution evolutionary algorithm based on a population of probability model (DEA-PPM) presented by Algorithm 1 [1].

IV. THE PROPOSED PLACEMENT ALGORITHM

To address the module placement of two-layer PCB, we propose a placement algorithm illustrated in Fig. 1. The purpose of the initial placement is to find the promising coordinates for all components. Then, the global placement is implemented to optimize the coordinates and orientations of the components with a delicate tradeoff between the total wirelength and the constraint violation. Finally, the legalization procedure aims to eliminate the overlap and get the final solution.

Algorithm 1 DEA-PPM

Input: *netlist, components*$\{\boldsymbol{x}, \boldsymbol{y}, \boldsymbol{r}\}$*, icpins, IC.*
Output: updated *components*$\{x, y, r\}$ *bestSingle.*

1: Initialize the population fitness array $\boldsymbol{f}$ and get the optimal fitness f_{best};
2: set current generation $t \leftarrow 0$, the last generation $t_{max} \leftarrow 20$;
3: **while** $t < t_{max}$ **do**
4: $\quad \boldsymbol{Q'}(t) = OrthExpQ(\boldsymbol{Q}(t-1), \boldsymbol{P}(t-1))$;
5: $\quad \boldsymbol{P'}(t) = OrthExpQ(\boldsymbol{Q}(t), \boldsymbol{P}(t-1))$;
6: $\quad (\boldsymbol{P}(t), \boldsymbol{X}, \boldsymbol{Y}, s) = UpdateXY(\boldsymbol{P'}(t), \boldsymbol{X}, \boldsymbol{Y}, s)$;
7: $\quad \boldsymbol{Q}(t) = RefineQ(\boldsymbol{P'}(t), \boldsymbol{P}(t), \boldsymbol{Q'}(t))$;
8: $\quad$ t=t+1;
9: **end while**

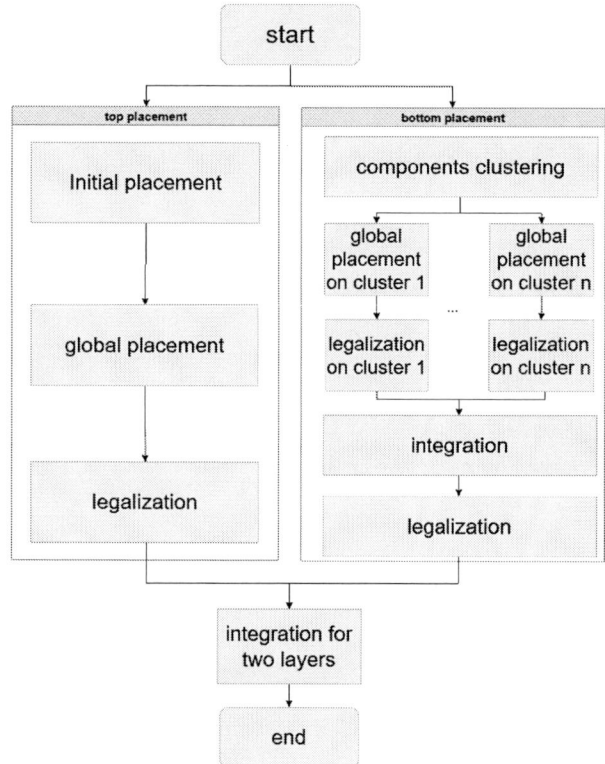

Fig. 1: The Framework of the Proposed Method

For placement of the bottom layer, the component clustering is performed during the initial placement phase. This step divides the placement problem into a number of sub-problems. The subsequent global placement and legalization of each cluster operate independently, without interfering with each other. After the local placement is completed, we must integrate the coordinates and orientations of the components in the various clusters, and the legalization procedure finally outputs the legal placement result.

A. Placement Initialization

Since the PCB design includes a variety of electric constraints that cannot be simulated at the stage of placement, the automatic placement of PCB components is significantly different from the placement scenarios of VLSI. Accordingly, tailored strategies are developed for the cases of PCB placement.

1) Coordinate Initialization for the Top Layer: Modules placed on the top layer, including an IC component and some accessary components, requires a centralized placement where accessary components are placed surrounding the IC component. To meet this requirement, the coordinate initialization consists of three aspects: creating an artificial external contour, selecting the nearest associated pin point to connect for each component, and moving components to the valid positions.

2) Coordinate Initialization for the Bottom Layer: According to the typical design of an IC component, we first cluster the pins to be connected to get nice placement, which is performed based on the density-based spatial clustering of applications with noise (DBSCAN) algorithm [14]. For all components, they will be grouped with their selected IC-pins in the same cluster.

The DBSCAN-based initialization process can be roughly divided into three phases. Firstly, we randomly select attachment points for each component according to its connection relationship. Considering that there is often a large free placement space around the noise points, the second step implemented is dedicated to reducing the overlap area of the global placement. The desired effect is that components belonging to the same net as the noise points can be attached to them as much as possible. The last stage is to meet the basic needs of industry. We will select to move the components with the close-to-pin identifier to its corresponding pin points.

B. Global Placement

1) the framework of Global Placemet: Algorithm 2 presents the framework of global placement. It starts with initialization of the orientation distribution and solution population $\boldsymbol{Q}(0)$ and $\boldsymbol{P}(0)$, then by checking the flag *isbottom*, components' coordinates initialized by lines 2-6. For the bottom layer placement, extra data structures like *netvec* and *clusterIndex* are also recorded. Since optimization algorithm has inevitable randomness, we attempt to set mulitple rounds optimization by lines 9-13 and choose the best solution as the output results.

To get the optimized result of global placement, the *ResetXY* and *DEA-PPM*, are iteratively implemented for i_{max} times. In addition to re-initializing the components' coordinates, *ResetXY* also needs to specify parameters to the objective function of CSA.

The *DEA-PPM* presented by Algorithm 1 simultaneously optimizes orientations and coordinates of components, where the coordinate update implemented by *UpdateXY* (Algorithm 3) is consecutively run for k clusters. Note that the cluster number k is set as 1 for the top layer and confirmed by the DBSACN for the bottom layer.

979-8-3315-8850-2/25 $31.00 © 2025 IEEE

Algorithm 2 Global Placement

Input: $netlist$, $components\{x, y, r\}$, $icpins$, IC.
Output: Refined $components\{x, y, r\}$ $bestSingle$.

1: initialize $\boldsymbol{Q(0)}$ and generate $\boldsymbol{P(0)}$ by sampling $\boldsymbol{Q(0)}$;
2: **if** $isbottom$==true **then**
3:     DoDBSCAN($netlist$, $components\{\boldsymbol{x, y, r}\}$);
4: **else**
5:     TopInitialization($netlist$, $components\{\boldsymbol{x, y, r}\}$, IC);
6: **end if**
7: let $(\boldsymbol{x^*, y^*, p^*})$ = arg min $f(x, y, r)$, $\{x, y, r\}$ is obtained from $components$;
8: set $\boldsymbol{q^*}$ as the distribution q corresponding to $\boldsymbol{p^*}$;
9: $i \leftarrow 0$, $i_{max} = 10$;
10: **while** $i < i_{max}$ **do**
11:     ResetXY($components\{\boldsymbol{x, y}\}$);
12:     DEA-PPM();
13: **end while**

Algorithm 3 $(\boldsymbol{P}(t), \boldsymbol{X, Y}, s) = UpdateXY(P'(t), \boldsymbol{X, Y}, s)$

Input: $netvec$, $components\{\boldsymbol{x, y, r}\}$, $icpins$;
Output: Optimal individual $bestSingle$, optimal fintness $bestf$;

1: **for** $i = 1, \cdots, k$ **do**
2:     $\alpha \leftarrow alpha[i]$, $\beta \leftarrow bta[i]$;
3:     $cluster \leftarrow components$ in cluster i;
4:     $net \leftarrow netvec[i]$;
5:     $ps \leftarrow icpins$ in cluster i;
6:     optimized $cluster = CSA(f, \boldsymbol{u}_0, k_{max}, s_0)$;
7:     compute $HWPL$, $overlap$ based on current $cluster$ and net;
8:     $f = \alpha \times HPWL + \beta \times \sqrt{overlap}$;
9:     **if** $f < bestf_i$ **then**
10:         $bestSingle_i \leftarrow cluster$;
11:         $bestf_i \leftarrow f$;
12:     **end if**
13: **end for**
14: $bestSingle \leftarrow$ Integrate all local optimal $bestSingle_i$;
15: $bestf \leftarrow$ Add up all local optimal $bestf_i$;

C. Legalization

The legalization process consisting of four steps is presented in Fig. 2. In the algorithm, eliminating the overlap function ensures that the placement complies with the basic design specification, maintaining spacing is used to adjust the component spacing according to the prescribed spacing value in advance, the out-of-bounds component processing function is to re-locate the components moved out of the IC boundary by the previous two steps, and the close-to-pin components placement function is to force some components with special requirements to be placed near the pins.

1) Legalization for the bottom layer:

a) Elimination of overlap: The basic idea of eliminating overlap is to start with the component at the center position and move the component in four directions: upward, downward,

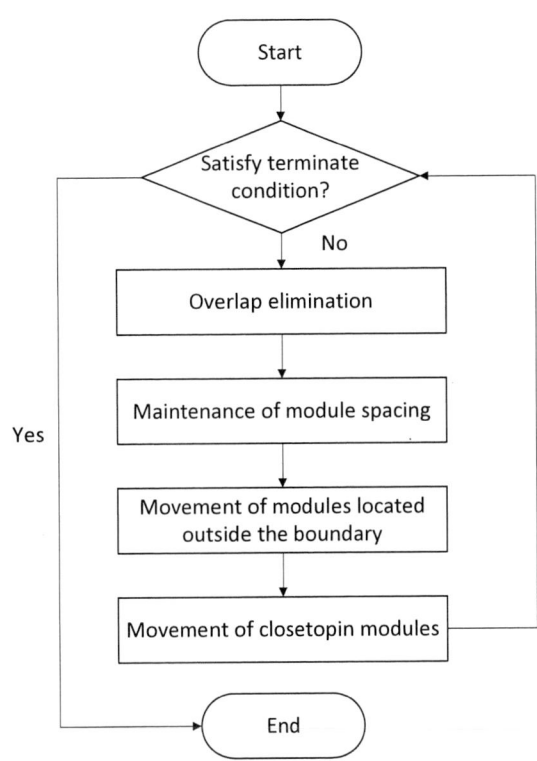

Fig. 2: The Flow Chart of the Legalization Process.

left, and right, until there is no overlap. Take the direction from the middle down as an example. The algorithm first determines an intermediate component $middle$ according to the component coordinates, and then continuously traverses the lower components from $middle$. For each traversed components, the algorithm will determine whether it overlaps with the upper component. If so, the algorithm will tentatively shift 1 unit downward until the current component does not overlap.

b) Maintenance of component spacing: Eliminating overlap is to ensure that the algorithm meets the most basic design specifications. However, in various circuits, there are often larger spacing requirements between components. The algorithm adjusts the component coordinates according to the spacing file given in advance. It should be pointed out that not merging it with the previous step is to preserve the relative positional relationship of the components to the greatest extent.

c) Out-of-bounds components processing: In some placement results, there exist some cases in which several components are forced to move out of the IC boundary due to the influence of previous overlap elimination and the maintenance of the spacing function. The solution is to obtain the maximum blank space in the four corners of IC by computing the maximum rectangle algorithm in the bar chart, and then place the components within this region.

d) Movement of closetopin components: Due to electrical and performance reasons, the netlist sometimes contains closetopin nets, which specify that a specified component must be placed around a specified pin point. However, the legalization

979-8-3315-8850-2/25 $31.00 © 2025 IEEE

process could make some close-to-pin components deviating from its pin points. Therefore, after the previous legalization steps, we will recheck the distances between the close-to-pin components and its pin points. If some distance exceeds our given threshold, the component will regenerate the coordinates around the specified pin point.

2) Legalization for the top layer: The basic idea for component legalization at the top layer is consistent with that for the bottom layer. The difference is that the legalization for the top layer is accomplished based on partitions.

V. EXPERIMENTAL RESULTS AND ANALYSIS

In order to verify the performance of our algorithm, we carefully selected four representative PCB placement examples and conducted experiments to obtain its performance indicators.

All experiments are developed in C++ programming language program, and run in Microsoft Windows 11 on a laptop equipped with the AMD Ryzen 7 5800H @ 3.2GHz and 16GB system memory.

A. Performance Validation for Industrial PCB Cases

In order to validate the performance of our placement algorithm, four industrial cases presented in Table I are investigated. The first two cases are single layer placement problems, where Case 1 requires the bottom layer placement, and Case 2 is to place components on the top layer. Cases 3 and 4 include components on both layers, and they require automatic placement for both the top and bottom layers.

TABLE I: The Circuit Parameters of Investigated PCB Cases.

Circuit	Blocks(top layer + bottom layer)	Terminals(IC-Pins)	Total Nets
Case1	0+148	1932	518
Case2	80+0	662	214
Case3	19+61	642	213
Case4	39+104	441	392

1) Placement results of PCB: The placement proposed in this paper are implemented to address the problems included in Table I, and the obtained results are illustrated in Fig. 3, which demonstrates that our proposed algorithm can well address both the top-layer and the bottom-layer placement. For ten independent runs, the statistic results of HPWL and runtime are included in Table II, which shows that our algorithm can get the automatic placement of components with a couple of minutes. The small standard deviation values of HPWL and runtime also validate the performance stability of the proposed algorithm.

TABLE II: Statistical results of HPWL and runtime for the PCB cases.

Circuit	Layer	HPWL(mil)		CPU(s)	
		mean	std	mean	std
Case1	bottom	9439.19	34.689	5.985	0.483
Case2	top	4667.297	56.118	3.1995	0.286
Case3	top	2414.195	112.226	5.165	0.211
	bottom	2483.53	68.98		
Case4	top	20570.730	241.731	50.702	1.816
	bottom	9523.41	904.538		

a) Performance comparison among different algorithms: To highlight the superiority of our algorithm in the double-layer placement of PCB, we chose the well-known open-source placement tool Parquet-4.5[15] as the comparison. Besides, the Parquet placement of the top layer was adjusted from the original Parquet to adapt to the PCB placement. The placement result is shown in the Fig. 4. Table III shows the performance comparison among the algorithms.

TABLE III: Performance comparison of algorithms

Circuit		Parquet-4.5		Cluster pattern		Non-cluster pattern	
		HPWL	time(s)	HPWL	time(s)	HPWL	time(s)
Case 1		11204.46	20.36	9439.19	5.985	9267.335	8.247
Case 2		4915.01	4.89	——	——	4667.297	3.1995
Case 3	top	2634.013	3.27	——	——	2414.195	0.655
	bottom	4056.2	14.5	2483.53	4.51	2295.25	10.27
Case 4	top	Failed		——	——	20570.730	10.659
	bottom	Failed		9523.41	33.16	9727.938	40.043

B. Influence of components clustering on the placement

Considering the uncertainty of the clustering results, it is difficult to visually define the influence of the clustering algorithm on the global placement process. Therefore, it is necessary to explore the influence of DoDBSCAN algorithm on the same case. Two groups each ran 10 times and took the average performance. The result comparison is shown in the Table III.

The global optimal solution obtained by the clustering optimization algorithm will be slightly worse. However, in terms of execution time, the performance of clustering optimization is significantly better than that of direct optimization. The result of our analysis is that the reduction of the problem scale makes the solution convergence speed faster. Meanwhile, the cost of this method is that the algorithm is short of global consideration. There is no additional overlap or wirelength calculation for the components between different clusters.

VI. CONCLUSION

In this paper, we formulate the 2-layer module placement problem as a mixed-variable optimization problem, and develop tailored strategies for global optimization and legalization of both the top-layer placement and the bottom-layer placement. Numerical comparison demonstrates that the proposed mixed-variable optimization scheme can outperform the metaheuristics algorithm based on representative structure codes, and the placement results on industrial PCB cases show that the proposed placement algorithm for IC-based modules can get satisfactory results for real PCB design scenarios. Our future study will focus on improving its performance on large-scale cases, and incorporating the module placement method to implement automatic design of a PCB consisting of several function modules.

979-8-3315-8850-2/25 $31.00 © 2025 IEEE

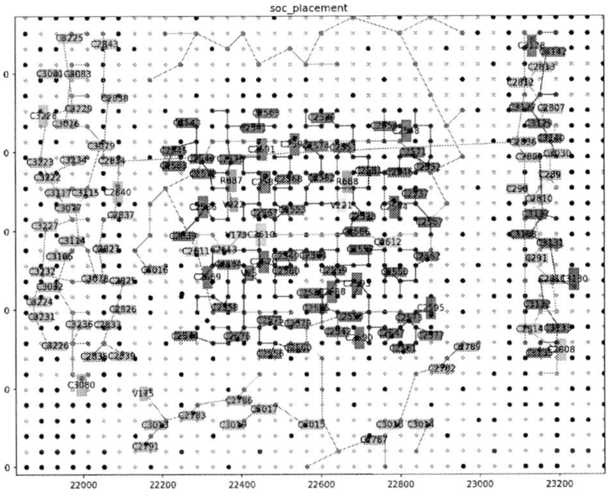

(a) case 1

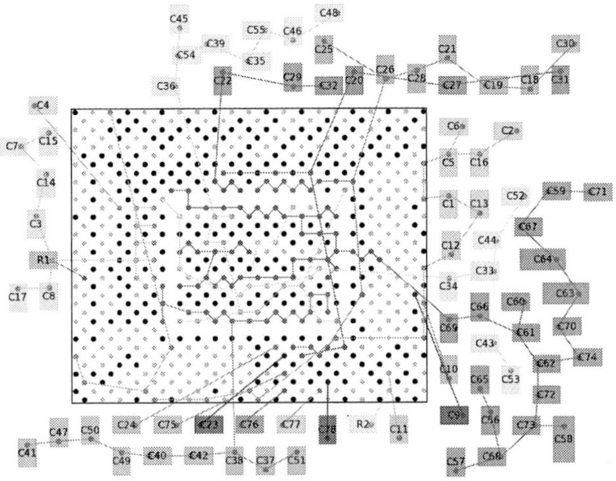

(b) case 2

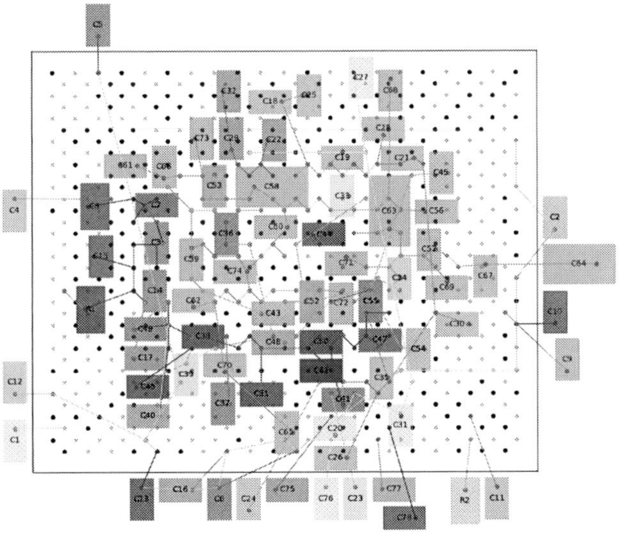

(c) case 3

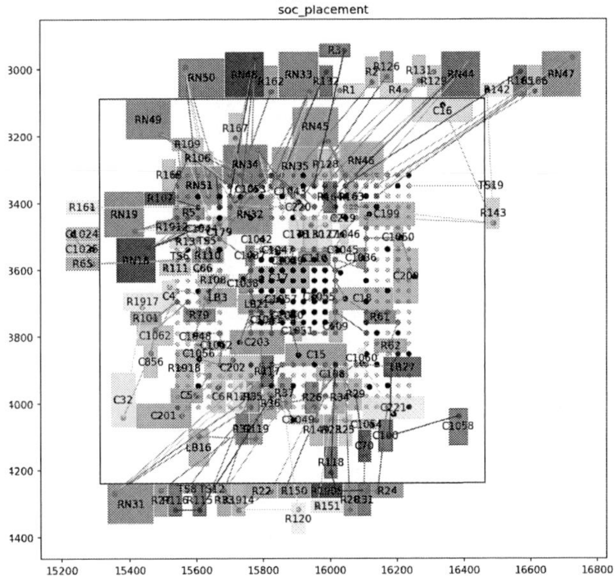

(d) case 4

Fig. 3: The final placement results of the proposed algorithm

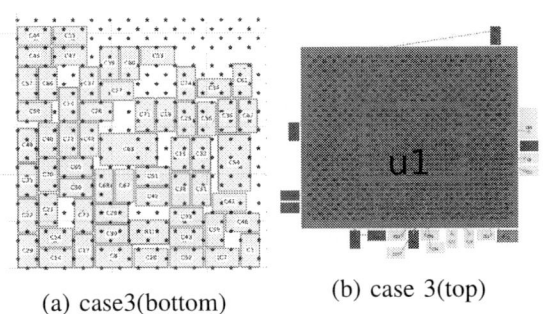

(a) case3(bottom)

(b) case 3(top)

Fig. 4: The final placement results of Parquet4.5 for case3

REFERENCES

[1] Yongjian Xu et al. "A distribution evolutionary algorithm for the graph coloring problem". In: *Swarm and Evolutionary Computation* 80 (2023), p. 101324. ISSN: 2210-6502. DOI: https://doi.org/10.1016/j.swevo.2023. 101324. URL: https://www.sciencedirect.com/science/article/pii/S2210650223000974.

[2] Jian Sun et al. "Floorplanning of VLSI by Mixed-Variable Optimization". In: *Intelligence Computation and Applications*. Ed. by Kangshun Li and Yong Liu. Singapore: Springer Nature Singapore, 2024, pp. 137–151. ISBN: 978-981-97-4393-3.

979-8-3315-8850-2/25 $31.00 © 2025 IEEE

[3] SB Vinay Kumar, PV Rao, and Manoj Kumar Singh. "Optimal floor planning in VLSI using improved adaptive particle swarm optimization". In: *Evolutionary Intelligence* 15.2 (2022), pp. 925–938.

[4] B. Srinivasan et al. "A Novel Multicriteria Optimization Technique for VLSI Floorplanning Based on Hybridized Firefly and Ant Colony Systems". In: *IEEE Access* 11 (2023), pp. 14677–14692. DOI: 10.1109/ACCESS.2023.3244346.

[5] Jingwei Lu et al. "ePlace: Electrostatics-Based Placement Using Fast Fourier Transform and Nesterov's Method". In: *ACM Trans. Des. Autom. Electron. Syst.* 20.2 (Mar. 2015). ISSN: 1084-4309. DOI: 10.1145/2699873. URL: https://doi.org/10.1145/2699873.

[6] Ximeng Li et al. "PeF: Poisson's equation-based large-scale fixed-outline floorplanning". In: *IEEE Transactions on Computer-Aided Design of Integrated Circuits and Systems* 42.6 (2022), pp. 2002–2015.

[7] Keyu Peng and Wenxing Zhu. "Pplace-MS: Methodologically Faster Poisson's Equation-Based Mixed-Size Global Placement". In: *IEEE Transactions on Computer-Aided Design of Integrated Circuits and Systems* 43.2 (2024), pp. 613–626. DOI: 10.1109/TCAD.2023.3320628.

[8] Xuezhou Wang, Ning Xu, and Yu Chen. "Automatic Multi-Constraint Placement of Printed Circuit Board". In: *2024 13th International Conference on Communications, Circuits and Systems (ICCCAS)*. 2024, pp. 1–6. DOI: 10.1109/ICCCAS62034.2024.10652855.

[9] Shujian Li et al. "A Centralized Block Placement Algorithm Based on Sequence Pair Representation". In: *2024 13th International Conference on Communications, Circuits and Systems (ICCCAS)*. IEEE. 2024, pp. 89–92.

[10] Lin Chen et al. "PCBAgent: An Agent-based Framework for High-Density Printed Circuit Board Placement". In: *Proceedings of the 30th Asia and South Pacific Design Automation Conference*. 2025, pp. 781–787.

[11] Niansong Zhang et al. "Cypress: VLSI-Inspired PCB Placement with GPU Acceleration". In: *Proceedings of the 2025 International Symposium on Physical Design*. 2025, pp. 31–41.

[12] Chuandong Chen et al. "Subgraph Matching with Diversity Handling and Its Applications to PCB Placement". In: *2024 2nd International Symposium of Electronics Design Automation (ISEDA)*. IEEE. 2024, pp. 468–473.

[13] Wenxing Zhu et al. "Nonsmooth optimization method for VLSI global placement". In: *IEEE Transactions on Computer-Aided Design of Integrated Circuits and Systems* 34.4 (2015), pp. 642–655.

[14] Martin Ester et al. "A density-based algorithm for discovering clusters in large spatial databases with noise". In: *Proceedings of the Second International Conference on Knowledge Discovery and Data Mining*. KDD'96. Portland, Oregon: AAAI Press, 1996, pp. 226–231.

[15] Saurabh N Adya and Igor L Markov. "Fixed-outline floorplanning: Enabling hierarchical design". In: *IEEE transactions on very large scale integration (VLSI) systems* 11.6 (2003), pp. 1120–1135.

High-Performance Hydrogenated Oxide-Semiconductor Schottky Barrier Diodes with ALD HfO$_2$ Interface

Yucheng Cao
School of Electronic and Computer Engineering, Peking University Shenzhen Graduate School
Shenzhen, China
caoyucheng@stu.pku.edu.cn

Chenyang Huang
School of Electronic and Computer Engineering, Peking University Shenzhen Graduate School
Shenzhen, China
HuangCY@pku.edu.cn

Yuyang Cai
School of Electronic and Computer Engineering, Peking University Shenzhen Graduate School
Shenzhen, China
caiyy@stu.pku.edu.cn

Kuan-Chang Chang
School of Electronic and Computer Engineering, Peking University Shenzhen Graduate School
Shenzhen, China
kcchang@pkusz.edu.cn

Lei Li
College of Integrated Circuits and Optoelectronic Chips, Shenzhen Technology University
Shenzhen, China
lilei@sztu.edu.cn

Congwei Liao
College of Integrated Circuits and Optoelectronic Chips, Shenzhen Technology University
Shenzhen, China
liaocongwei@sztu.edu.cn

Yufeng Jin*
School of Electronic and Computer Engineering, Peking University Shenzhen Graduate School
Shenzhen, China
yfjin@pku.edu.cn

Lei Lu*
School of Electronic and Computer Engineering, Peking University Shenzhen Graduate School
Shenzhen, China
lulei@pku.edu.cn

Abstract—The bulk resistance dominates the on-resistance (R_{on}) of oxide semiconductor (OS) Schottky barrier diodes (SBDs). While highly conductive OSs enable superior current capability, their abundant interface defects at metal contacts lower the effective barrier and cause excessive leakage. In this work, high-indium InZnO (In:Zn = 5:1, IZO5:1) was employed as the active layer, and an ultrathin HfO$_2$ interlayer was introduced to suppress metal-induced states and enhance the barrier height. However, the interlayer alone could not fully mitigate interface defects in In-rich IZO. To address this, hydrogenation technology was further developed, effectively reducing leakage while maintaining strong forward conduction. Optimized devices with a 2 nm HfO$_2$ interlayer and 100 s hydrogenation treatment achieved a barrier height of 0.73 eV, an ideality factor of 1.16, and a breakdown voltage of 136 V. This approach provides a promising pathway for interface engineering and performance improvement of high-current OS SBDs.

Keywords—HfO$_2$, hydrogen doping, oxide semiconductors (OSs), Schottky diode

I. INTRODUCTION

Oxide semiconductors combine relatively high carrier mobility, compatibility with low-temperature processing, and excellent large-area uniformity, which makes them attractive for large-area, flexible, and heterogeneously integrated electronics [1]. Metal–semiconductor contacts critically determine device operation and performance and typically classify as Schottky or ohmic. Abundant intrinsic defects in oxide semiconductors impede formation of high-quality Schottky contacts [2], limiting development of oxide-based Schottky devices such as Schottky barrier diodes [2], metal–semiconductor field-effect transistors [3], and source-gated transistors [4].

Recently, oxide Schottky barrier diodes (SBDs)with high radio-frequency performance and good rectification have attracted interest for flexible RFID [5], energy harvesting [6], and power management applications [7], and substantial progress has been made over the past decade. Because Schottky contacts are highly sensitive to interface defect states, low-defect oxides such as InGaZnO [8] and ZnSnO [9] are often chosen as active layers, but this choice constrains further improvement in rectification and on-current. In contrast, thin-film transistors often use more compositionally diverse systems, notably indium-rich amorphous InZnO [10]. Compositions with In to Zn atomic ratios near 1 to 1 or 5 to 1 exhibit higher mobility and tunable electrical properties, yet they have seen limited use in SBDs. Indium-rich oxides typically host more oxygen vacancies (V_o), higher carrier concentration, and higher mobility, leading to favorable electrical and optical characteristics [11]. Prior work has shown that 1 to 1 InZnO with high carrier concentration improves forward conduction of SBDs [12].

A key challenge for high-performance oxide Schottky barrier diodes is the suppression of interface states at Schottky contacts, which in oxide materials originate mainly from oxygen-related defects. Conventional interface treatments include anodic metal oxidation [13], oxygen plasma [13], and annealing in oxygen atmosphere [14], but these approaches often have narrow process windows and limited efficacy for materials with very high defect density. New interface engineering strategies are therefore needed.

In this study, indium-rich IZO with In to Zn atomic ratio 5 to 1, denoted IZO5:1, was selected as the active layer. To mitigate Fermi-level pinning (FLP) and suppress metal-induced gap states (MIGS), a metal–insulator–semiconductor (MIS) contact was implemented by inserting an ultrathin HfO$_2$ interlayer between Pt electrode and IZO5:1, deposited by plasma-enhanced atomic layer deposition (PEALD). Building on mechanistic understanding of HfO$_2$ interlayer effects, we further apply hydrogen doping to achieve deeper defect passivation. This combined approach yields Schottky diodes with enhanced on-state conduction and improved rectification.

979-8-3315-8850-2/25 $31.00 © 2025 IEEE

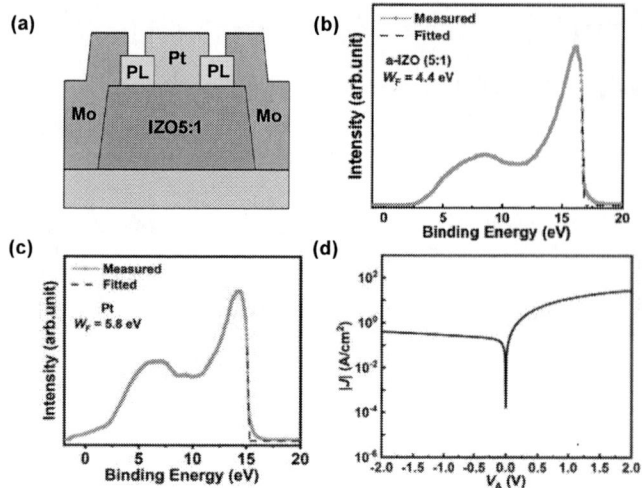

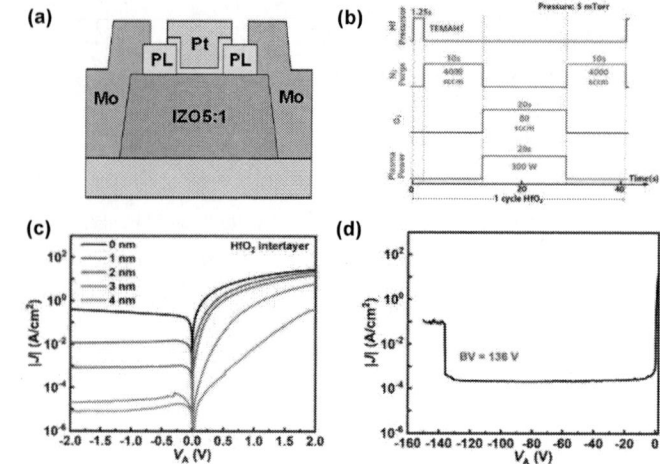

Fig. 1. (a) Schematic cross section of MS IZO5:1 SBD. (b) UPS of Pt. (c) UPS of IZO5:1. (d) *J-V* of MS IZO5:1 SBD.

II. DEVICE FABRICATION

Fig. 1(a) shows a cross-sectional view of the fabricated lateral amorphous IZO5:1 SBDs. The device was fabricated as follows: first, a 100 nm IZO5:1 active layer was deposited on a glass substrate by sputtering at a power of 100 W and a pressure of 0.5 Pa in an Ar/O_2 atmosphere (29/5 sccm), and then patterned via wet etching using an acid solution. Subsequently, a 100 nm SiO_2 passivation layer was deposited by plasma-enhanced chemical vapor deposition (PECVD) and patterned using a reactive ion etching (RIE) process. The active layer was then treated with N_2O plasma in a PECVD reactor to promote oxidation, followed by additional annealing in O_2 ambient at 350 °C for 2 hours. Finally, the Pt/Al anode and Mo cathode were formed by a lift-off process, with thicknesses of 50 nm, 50 nm, and 100 nm, respectively. The electrical characterization of the devices was measured using a Keysight B1500 semiconductor analyzer.

III. RESULTS AND DISSCUSION

As shown in Figure 1(b) and (c), the work function of IZO5:1 and Pt were measured by UPS to be 4.4 eV and 5.8 eV, respectively. Theoretically, this significant difference in work functions is expected to facilitate the formation of a high Schottky barrier. However, as depicted in Fig. 1(d), the electrical performance of the fabricated SBD deviates from the expectation, exhibiting a low barrier height, high reverse leakage current, a non-ideal ideality factor, and a low rectification ratio.

For SBDs, the current-voltage (*J-V*) relationship is analyzed using the conventional thermionic emission model, which can be represented as [15].

$$J = A^*T^2 \left[\exp\left(\frac{-q\Phi_B}{kT}\right)\right]\left[\exp\left(\frac{q(V-JR_s)}{nkT}\right) - 1\right] \quad (1)$$

Where A^* and k represent the Richardson constant and the Boltzmann constant, respectively. V is applied voltage, Φ_B is the Schottky barrier height (SBH), q is the electron charge, R_s is the series resistance, and n is the ideality factor. The Φ_B and n are extracted from the forward characteristics of the SBDs.

Fig. 2. (a) Schematic cross section of MIS IZO5:1 SBDs. (b) PEALD HfO_2 Cycle. (c) *J-V* of MIS IZO5:1 SBDs. (d). Breakdown curves for the MIS IZO5:1 SBDs with a 2 nm HfO_2.

From the above function, the Φ_B of the SBDs is 0.45 eV, which is significantly lower than the theoretical value for the Schottky contact between high work-function metal Pt and IZO5:1. The high reverse current degrades the device performance, which is caused by a defect-assisted tunneling mechanism induced by intrinsic defect states.

A. HfO₂ Engineering

Fig. 2(a) illustrates the structure of the SBDs device incorporating an HfO_2 interlayer. Following the RIE process, the HfO_2 interlayer was deposited via PEALD. The detailed sequence of a single PEALD cycle for HfO_2 is shown in Fig. 2(b). Each cycle employs TEMAHf and O_2 plasma as precursors, with N_2 used as the purge gas. The deposition process consists of four steps: TEMAHf precursor pulse, N_2 purge, O_2 plasma exposure, and a final N_2 purge. The deposition was carried out at a base vacuum of 5 mTorr and a substrate temperature of 150 °C. The TEMAHf pulse duration was set to 1.25 seconds, followed by a 10 s N_2 purge. The O_2 plasma step was conducted for 20 seconds at a RF power of 300 W. Under these conditions, the growth rate of HfO_2 was approximately 1.0 Å per cycle. To evaluate the impact of the interlayer thickness on device performance and to determine the optimal condition, MIS SBDs with varying HfO_2 thicknesses were fabricated.

Fig. 2(c) presents the electrical performance of SBDs with varying HfO_2 thicknesses. The corresponding parameters, extracted using the Cheung method, are summarized in Table I. As shown in the table, when the HfO_2 thickness is 2 nm or less, increasing the thickness results in a slight increase in the specific on-resistance (R_{on}), reaching 1.82×10^{-1} $\Omega\cdot cm^2$. Concurrently, the rectification ratio rises to 1.92×10^4, the Φ_B increases gradually to 0.60 eV, and the n decreases to 1.22. This improvement can be primarily attributed to the thicker HfO_2 interlayer, which more effectively passivates interface defects and suppresses the escape of metal wave functions, thereby reducing MIGS. When the HfO_2 thickness is from 2 nm to 3 nm, although both the rectification ratio and Φ_B continue to improve, the ideality factor deteriorates, increasing to 2.86, while R_{on} also rises significantly to 1.20×10^0 $\Omega\cdot cm^2$. As the thickness reaches 4 nm, a substantial reduction in the on-state current density at +2 V is observed,

979-8-3315-8850-2/25 $31.00 © 2025 IEEE

TABLE I. KEY PARAMETERS OF THE MIS IZO SBDS WITH DIFFERENT THICKNESS OF HfO₂

t_{HFO2} (nm)	J_F/J_R	Φ_B (eV)	n	R_{on} ($\Omega\cdot cm^2$)
0	6.7×10^1	0.45	1.34	7.9×10^{-2}
1	1.9×10^3	0.54	1.28	1.6×10^{-1}
2	1.9×10^4	0.60	1.22	1.8×10^{-1}
3	4.3×10^4	0.63	2.86	1.2×10^0
4	3.5×10^3	0.64	7.28	2.68×10^0

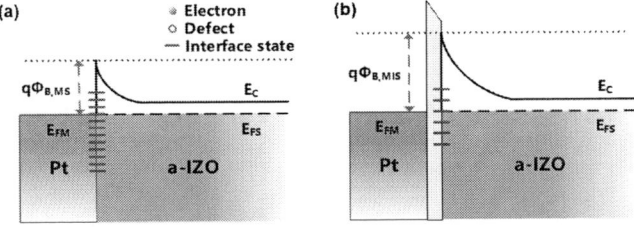

Fig. 3. Energy level band diagram of (a) MS SBD. (b) MIS SBD.

dropping to 0.296 A/cm². This leads to a decreased rectification ratio of 3.58×10^3, a markedly elevated ideality factor of 7.28, and a further increased R_{on} of 2.68×10^0 $\Omega\cdot cm^2$. As shown in Fig. 2(d), the breakdown voltage of the IZO5:1 SBD is as high as 136 V, demonstrating its broad prospects in high-voltage applications. Therefore, based on a comprehensive evaluation of the electrical parameters, the 2 nm HfO₂ interlayer deposited by PEALD is identified as the optimal choice for the MIS IZO5:1 SBDs.

The effects of the HfO₂ interlayer on IZO5:1 SBDs are investigated through energy band diagrams, as illustrated in Fig. 3. For the metal–semiconductor (MS) SBD without an HfO₂ interlayer, due to V_O defects and dangling bonds at the interface, resulting in FLP and a low Φ_B. After introducing an ultrathin HfO₂ interlayer between Pt/IZO by PEALD, the HfO₂ dielectric not only passivates the dangling bonds but also reduces the V_O density at the interface. Given that the Hf–O bond energy (801 kJ/mol) significantly exceeds In–O (346 kJ/mol) and Zn–O (248 kJ/mol) [16], the HfO₂ interlayer stabilizes oxygen atoms at the interface, thereby effectively suppressing V_O-related defect states. Furthermore, the HfO₂ film serves as a diffusion barrier, inhibiting the penetration of Pt into the a-IZO layer and block the metal electron wave function spilling out from Pt. Additionally, the high dielectric constant of HfO₂ promotes rapid decay of MIGS within the semiconductor bandgap, further reducing their density. Owing to these effects, the incorporation of an ultrathin HfO₂ interlayer significantly reduces the interface state density, alleviates Fermi-level pinning, and increases the Φ_B.

B. Hydrogen Doping

As reported in our previous work [16], the IZO1:1 SBD exhibits a reverse current density on the order of 10^{-5} A/cm². In contrast, even with the application of interlayer technique, the IZO5:1 SBD still maintains a reverse current density of approximately 10^{-3} A/cm². This indicates that the high intrinsic defect density in IZO5:1 necessitates further research into interface passivation strategies.

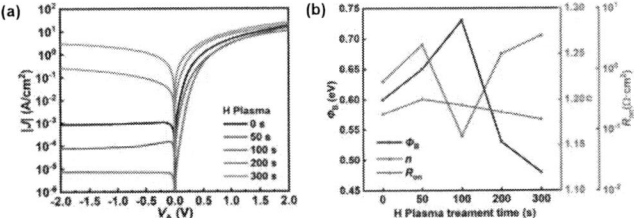

Fig. 4. (a) *J-V* characteristics of the MIS SBDs under different H Plasma treatment durations, and (b) the corresponding extracted electrical parameters.

Hydrogen (H) is an innovative dopant in OSs. Frequently employed doping methods include hydro-sputtering, hydrogen plasma treatment, and in-situ hydrogen diffusion from passivation layers, among others. Numerous studies have demonstrated that incorporated hydrogen can passivate under-coordinated oxygen atoms, thereby suppressing V_O, as well as neutralize electron traps within the active layer. While an appropriate concentration of hydrogen effectively passivates defects, excessive doping may introduce shallow donor levels into the OSs, leading to additional defect states and degradation of device electrical performance [17]. To systematically evaluate these effects, the influence of hydrogen doping on MIS IZO5:1 SBDs was investigated. The hydrogen doping process was implemented as follows: after depositing the HfO₂ interlayer via PEALD, N₂O plasma treatment and O₂ ambient annealing were performed, followed by H-plasma treatment of the interface. The H-plasma process was conducted in a PECVD chamber under the following conditions: chamber temperature of 300 °C, process pressure of 550 mTorr, H₂ flow rate of 50 sccm, and RF power of 40 W.

To evaluate the impact of hydrogen doping on the electrical characteristics of IZO5:1 MIS SBDs, the Schottky interface was subjected to H plasma treatment for durations of 50 s, 100 s, 200 s, and 300 s. The properties of the resulting devices were compared with an untreated reference sample, as shown in Fig 4(a). Fig. 4(b) summarizes the key electrical parameters extracted using the Cheung method for each H plasma exposure time.

A treatment duration of 50 s resulted in an increase in the SBH from 0.60 eV to 0.65 eV, while the R_{on} increased from 1.82×10^{-1} $\Omega\cdot cm^2$ to 3.20×10^{-1} $\Omega\cdot cm^2$. These changes suggest that hydrogen effectively passivated defects at the Schottky interface, reducing the density of V_O and interface states. The consequent elevation of the Schottky barrier and the slight rise in R_{on} are consistent with a reduction in shallow donors originating from V_O. Extending the treatment to 100 s led to a further improvement in performance: Φ_B increased to 0.73 eV, the n decreased to 1.16, and R_{on} was reduced to 2.57×10^{-1} $\Omega\cdot cm^2$. These trends indicate enhanced defect passivation, increased barrier height, and a rise in carrier concentration due to hydrogen ionization, collectively contributing to a lower on-resistance. However, longer treatment times of 200 s and 300 s resulted in a degradation of diode characteristics: although R_{on} decreased to values comparable to the untreated device, both Φ_B and n deteriorated. This behavior is attributed to excessive hydrogen incorporation, which introduces additional interface defect states and raises interfacial carrier density, thereby impairing device performance. In summary, optimal performance was

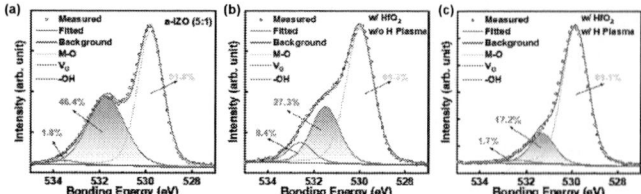

Fig. 5. XPS characterization of IZO5:1 film: (a) untreated, (b) with an HfO$_2$ interlayer, and (c) with an HfO$_2$ interlayer followed by 100 s of H plasma treatment.

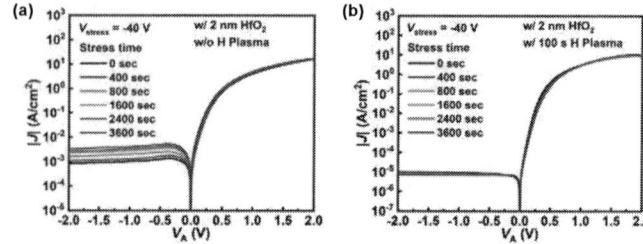

Fig. 6. Electrical stress stability J-V characteristics of MIS IZO5:1 SBD (a) without H Plasma (b) after 100 s H Plasma treatment.

achieved with a 100 s H plasma treatment, yielding a Schottky barrier height of 0.73 eV, an ideality factor of 1.16, a on-resistance of 2.57×10^{-1} $\Omega \cdot cm^2$, and a rectification ratio of 1.5×10^6.

To visually assess the defect passivation effect of H plasma treatment, XPS characterization was conducted on three types of IZO5:1 film: (a) untreated, (b) with an HfO$_2$ interlayer, and (c) with an HfO$_2$ interlayer followed by 100 s of H plasma treatment. For the latter two cases, XPS spectra were acquired after etching until the Hf–O peak was no longer detected. The results, presented in Fig. 5, show that the HfO$_2$ interlayer substantially reduced the V_O concentration at the Schottky interface from 46.4 % in the untreated film to 27.3 %. Subsequent H plasma treatment further decreased the V_O content to 17.2 %, confirming the effectiveness of hydrogenation in suppressing defects.

Evaluating the electrical stress stability of devices is an important means to test whether the devices can operate in actual environments. Fig. 6(a) and (b) show the J-V characteristics under reverse bias stress at −40 V for the MIS a-IZO5:1 SBD without H plasma treatment and that treated with H plasma for 100 s, respectively. For the device without H plasma treatment, after 3600 s of reverse bias stress, the reverse leakage current at −2 V increased from 8.38×10^{-4} A/cm^2 to 3.25×10^{-3} A/cm^2—an increase of nearly one order of magnitude. In contrast, for the device subjected to 100 s of H plasma treatment on the Schottky interface, the reverse leakage current at −2 V only increased from 7.31×10^{-6} A/cm^2 to 1.01×10^{-5} A/cm^2 even after 3600 s of reverse bias stress. The significantly improved stability of the H plasma-treated MIS IZO5:1 SBD further confirms that an appropriate amount of hydrogen doping can passivate a large number of intrinsic defect states at the Schottky interface and enhance the device stability.

IV. CONCLUSION

In conclusion, conventional interface oxidation methods are insufficient for IZO5:1 SBDs due to the high defect state density of IZO5:1. To overcome these shortcomings, a 2 nm HfO$_2$ interlayer was introduced between Pt and IZO5:1. Moreover, hydrogenation was implemented to further suppress detrimental defect states, substantially enhancing both device performance and operational stability. Coupled with the high conductivity of a-IZO5:1, this results in a high Φ_B of 0.73 eV, an ideality factor of 1.16, a on-resistance of 2.57×10^{-1} $\Omega \cdot cm^2$, and a rectification ratio of 1.5×10^6. The hydrogenated OS Schottky diode with ALD HfO$_2$ interface presents a promising pathway toward the realization of high-current OS SBDs.

ACKNOWLEDGMENT

This work was supported in part by National Natural Science Foundation of China (NSFC) Young Scientists Fund under Grant 62504010; in part by Guangdong Province Science and technology Planning Project under Grant 2023TQ07A463; in part by Shenzhen Science and Technology program under Grant KJZD20230923115005009; and in part by Guangdong Provincial Key Laboratory of In-Memory Computing Chips under grant 2024B1212020002.

This work was conducted in Guangdong Technology Center for Oxide Semiconductor Devices and ICs, Guangdong Provincial Key Laboratory of In-Memory Computing Chips, and Shenzhen POC Center of Flexible Electronics.

REFERENCES

[1] J. F. Wager, B. Yeh, R. L. Hoffman, and D. A. Keszler, "An amorphous oxide semiconductor thin-film transistor route to oxide electronics," Current Opinion in Solid State and Materials Science, vol. 18, no. 2, pp. 53–61, Apr. 2014.

[2] J. Zhou et al., "Self‐Stabilized Hydrogenation of Amorphous InGaZnO Schottky Diode with Bilayer Passivation," Adv Elect Materials, vol. 8, no. 10, p. 2200280, Oct. 2022.

[3] J. Kaczmarski et al., "Transparent Ru–Si–O/In–Ga–Zn–O MESFETs on Flexible Polymer Substrates," IEEE Trans. Electron Devices, vol. 65, no. 1, pp. 129–135, Jan. 2018.

[4] J. Zhang et al., "Extremely high-gain source-gated transistors," Proc. Natl. Acad. Sci. U.S.A., vol. 116, no. 11, pp. 4843–4848, Mar. 2019.

[5] A. Chasin et al., "An Integrated a-IGZO UHF Energy Harvester for Passive RFID Tags," IEEE Trans. Electron Devices, vol. 61, no. 9, pp. 3289–3295, Sep. 2014.

[6] Y. Zhang et al., "Flexible transparent high-voltage diodes for energy management in wearable electronics," Nano Energy, vol. 40, pp. 289–299, Oct. 2017.

[7] Y. Son, B. Frost, Y. Zhao, and R. L. Peterson, "Monolithic integration of high-voltage thin-film electronics on low-voltage integrated circuits using a solution process," Nat Electron, vol. 2, no. 11, pp. 540–548, Nov. 2019.

[8] J. Zhang, Q. Xin, and A. Song, "High performance Schottky diodes based on indium-gallium-zinc-oxide," Journal of Vacuum Science & Technology A: Vacuum, Surfaces, and Films, vol. 34, no. 4, p. 04C101, Jul. 2016.

[9] S. Bitter, P. Schlupp, H. Von Wenckstern, and M. Grundmann, "Vital Role of Oxygen for the Formation of Highly Rectifying Schottky Barrier Diodes on Amorphous Zinc–Tin–Oxide with Various Cation Compositions," ACS Appl. Mater. Interfaces, vol. 9, no. 31, pp. 26574–26581, Aug. 2017.

[10] H. Liu et al., "Thorough Elimination of Persistent Photoconduction in Amorphous InZnO Thin-Film Transistor via Dual-Gate Pulses," IEEE Electron Device Lett., vol. 43, no. 8, pp. 1247–1250, Aug. 2022.

[11] J. Michel et al., "Processing Strategies for High-Performance Schottky Contacts on n-Type Oxide Semiconductors: Insights from In$_2$O$_3$," ACS Appl. Mater. Interfaces, vol. 11, no. 30, pp. 27073–27087, Jul. 2019.

979-8-3315-8850-2/25 $31.00 © 2025 IEEE

[12] F. Liu et al., "Defect-Hydrogen Interactions in Top-Anode Oxide Semiconductor Schottky Barrier Diode," Adv Materials Technologies, vol. 8, no. 15, p. 2300182, Aug. 2023.

[13] S. Yan, Y. Wang, J. Zhang, Q. Xin, and A. Song, "High-Performance Thin-Film IGZO Schottky Diodes With Sputtered PdO$_x$ Anode," IEEE Trans. Electron Devices, vol. 68, no. 9, pp. 4444–4449, Sep. 2021.

[14] L. Du et al., "Effects of substrate and anode metal annealing on InGaZnO Schottky diodes," Applied Physics Letters, vol. 110, no. 1, p. 011602, Jan. 2017.

[15] D. Zheng et al., "Suppression of nonideal leakage current in a-InGaZnO Schottky diode with edge termination structures," Applied Physics Letters, vol. 121, no. 13, p. 132101, Sep. 2022.

[16] Z. Zheng et al., "ALD Al$_2$O$_3$-Engineered Schottky Barrier Interface for Amorphous Indium–Zinc Oxide," IEEE Trans. Electron Devices, vol. 72, no. 9, pp. 5004–5010, Sep. 2025.

[17] W. Pan et al., "Multiple effects of hydrogen on InGaZnO thin-film transistor and the hydrogenation-resistibility enhancement," Journal of Alloys and Compounds, vol. 947, p. 169509, Jun. 2023.

2025 The 10th International Conference on Integrated Circuits and Microsystems

An Ultra-Low-Power True Random Number Generator Based on Volatile RRAM

Qi Luo
College of Integrated Circuits
Zhejiang University
Hangzhou, China
22341046@zju.edu.cn

Zhen Wang
College of Integrated Circuits
Zhejiang University
Hangzhou, China
22241032@zju.edu.cn

Xuemeng Fan
College of Integrated Circuits
Zhejiang University
Hangzhou, China
12341038@zju.edu.cn

Pengtao Li
College of Integrated Circuits
Zhejiang University
Hangzhou, China
12341003@zju.edu.cn

Guobin Zhang
College of Integrated Circuits
Zhejiang University
Hangzhou, China
22341077@zju.edu.cn

Yishu Zhang*
College of Integrated Circuits
Zhejiang University
Hangzhou, China
zhangyishu@zju.edu.cn
*Corresponding author

Abstract—With the rapid deployment of Internet of Things (IoT) devices in resource-constrained environments, lightweight and energy-efficient random number generation has become increasingly crucial. To address the resource overhead and energy efficiency challenges of conventional designs, this paper presents a self-clocking true random number generator (TRNG) based on volatile memristors. Built upon the device-level strengths of Ag/GaO$_x$/Pt memristor, particularly its high-speed oscillation and low-power behavior, the proposed TRNG design incorporates these features into its core architecture. We introduce a novel high-entropy approach that exploits stochastic fluctuations during memristor oscillation to generate true random bits. Experimental results show that TRNG achieves a generation rate of 167 kb s^{-1} and an energy consumption of 6.36 fJ bit^{-1}, making it well-suited for future energy-efficient and lightweight security applications.

Index Terms—true random number generator, memristor, threshold switching, oscillation, high energy efficiency

I. INTRODUCTION

With the widespread deployment of Internet of Things (IoT) devices in edge scenarios, information security has become a central concern in system design [1]. As a fundamental component of secure communication, key generation heavily depends on the quality of the random number source, which directly affects the system's ability to resist attacks [2]. Traditional systems typically rely on pseudo-random number generators (PRNGs) for key generation. Such an approach generate pseudo-random sequences based on predefined algorithms and initial seeds, providing efficient random number output in environments with sufficient computational resources. However, since their output sequences are determined by algorithms and seeds, PRNGs are inherently predictable and require substantial digital hardware resources for implementation [3]. These limitations become particularly evident in security-critical IoT applications and resource-constrained devices.

To address the security limitations of PRNGs, true random number generators (TRNGs), which exploit physical noise

sources to produce true random sequences, have emerged as ideal candidates for security-critical applications such as cryptographic key generation, offering inherently unpredictable and highly secure random sequences. Currently, most TRNG designs focus on peripheral circuit optimizations, such as eliminating the need for digital post-processing [4], reducing the number of logic gates, and adopting self-clocking architectures [5], aiming to reduce power consumption and area without altering the core structure. These approaches strive to enhance system performance by improving noise acquisition capabilities or refining post-processing techniques [6]. However, this "circuit-level-first" design paradigm is approaching its efficiency limits, as peripheral optimizations alone fail to fundamentally resolve critical issues such as limited throughput and excessive energy consumption. Therefore, it is imperative to shift the focus toward the device level to fundamentally exploit its potential for enhancing energy efficiency and throughput.

Currently, commonly used physical entropy sources mainly include non-volatile memory (NVM) devices and volatile memory (VM) devices. NVMs offer strong data retention capabilities and typically rely on random variations relative to a reference value, which can be advantageous in certain applications [3]. However, the stability of their output randomness can be compromised due to reference drift over time or cycling, as well as reset-induced disturbances. In addition, NVMs often require extra reset processes, which tend to consume more power, posing challenges for low-power, lightweight IoT applications. In contrast, volatile devices can automatically reset without external bias [1], offering fast response and low power consumption. These characteristics make them more promising for high-throughput and energy-efficient TRNG designs.

In this paper, we propose a self-clocking TRNG design based on volatile Resistive Random Access Memory (RRAM). An Ag/GaO$_x$/Pt memristor exhibits stochastic fluctuations during oscillation under electrical bias. By eliminating the need for an external clock signal module, the self-clocking architecture simplifies circuit complexity and significantly reduces power consumption. Unlike approaches that focus solely on peripheral circuit optimization, this design incorporates device-level strategies to fully exploit the memristor's high-frequency oscillation behavior and low-power characteristics. As a result, a highly integrated, ultra-low-power, high-throughput, and high-entropy random number generation scheme tailored for IoT scenarios is achieved.

II. TRNG CIRCUIT DESIGN ENABLED BY THRESHOLD SWITCHING DEVICE CHARACTERISTICS

A. RRAM Device Fabrication

Device stacks were fabricated on a 100 nm p-type Si wafer with a 300 nm thermally oxidized SiO_2 layer. After standard cleaning and drying procedures, the bottom electrode (BE) was patterned by photolithography. The 1×2 array devices shown in Fig. 1(a) were constructed on SiO_2/Si substrates, beginning with the deposition of a 60 nm platinum (Pt) bottom electrode via direct current (DC) magnetron sputtering. A 20 nm GaO_x switching layer was subsequently deposited by radio frequency (RF) sputtering using a GaO target in an argon (Ar) atmosphere. Finally, a 20 nm silver (Ag) top electrode (TE) was deposited by DC sputtering after patterning through photolithography. The scanning electron microscope (SEM) image and stacked structure of the fabricated RRAM are shown in Fig. 1. In this work, a 1×2 array of threshold switching (TS) devices was fabricated using a standard CMOS-compatible process. Although the devices are physically separated, their close proximity was designed intentionally to facilitate parallel connection in the subsequent TRNG circuit implementation.

B. Behavioral Characteristics of TS Devices

The DC I–V characteristics of $Ag/GaO_x/Pt$ RRAM are depicted in Fig. 2. Measurements were performed by applying a voltage bias to the TE with the BE connected to ground.

This single-layer device is a unidirectional switching element, designed to achieve abrupt volatile threshold switching under a high current compliance (I_{cc}) condition. The device is initially in a high resistance state (HRS). When a voltage exceeding the threshold voltage (V_{th}) is applied across the terminals, the device abruptly switches to a low resistance state (LRS). When the applied voltage drops below the holding voltage (V_{hold}), the device abruptly reverts to the high resistance state, exhibiting typical TS behavior. In this regard, our device demonstrates an ultra-low leakage current in the sub-pA range, which significantly suppresses sneak path currents and ensures low static power consumption. To evaluate the randomness of the device's switching behavior, which is crucial for high entropy generation, 100 consecutive I-V sweeps were conducted on the $Ag/GaO_x/Pt$ structure. Fig. 3(a) and (b) illustrates the statistical distribution and variability of V_{hold} and V_{th}. It can be observed that the switching behavior exhibit both stability and high randomness, indicative of robust volatile switching behavior. The threshold voltages were mainly distributed in the range of 0.3 V to 0.6 V, which can be attributed to the thin and defect-rich GaO_x switching layer. In defect-rich films, Ag ions are more likely to form thinner conductive filaments under the applied electric field. The thinness of the layer also helps avoid the formation of thicker filaments by reducing the required field strength for conduction. The inherent instability and high unpredictability in the formation and rupture of these numerous fine filaments provide a solid foundation for the device to function as a high-entropy source.

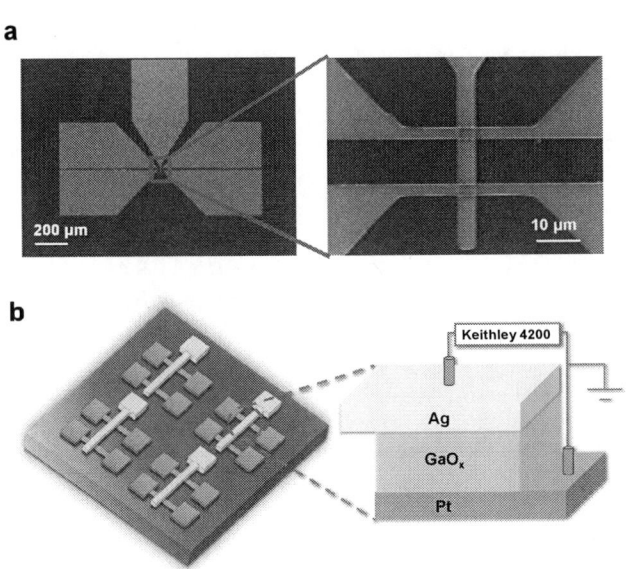

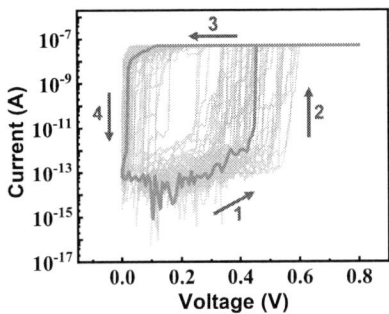

Fig. 2. I-V characteristics during threshold switching processes for 100 cycles.

The switching voltage slope was evaluated under DC voltage sweeping, as illustrated in Fig. 3(c) and (d). As illustrated in Fig. 4, a pulse measurement approach was employed to analyze the device's dynamic switching characteristics. When a 1.5 V voltage pulse was applied, the device was triggered within a short duration, switching to the LRS state in approximately 80 ns. After removing the voltage bias, the device spontaneously returned to its initial HRS due to the rupture

Fig. 1. (a) SEM image of the 1×2 crossbar array. (b) Schematic diagram of the crossbar array and the structure of GaO_x volatile memristors.

979-8-3315-8850-2/25 $31.00 © 2025 IEEE

of the Ag conductive filaments, and this relaxation process occurred within about 150 ns. The device exhibits a fast switching response. Considering that many TRNG applications rely on the device's periodic switching behavior as a source of entropy extraction, the combination of ultra-low leakage current, high-speed switching capability, excellent endurance, and high randomness makes our device particularly suitable for future lightweight TRNGs.

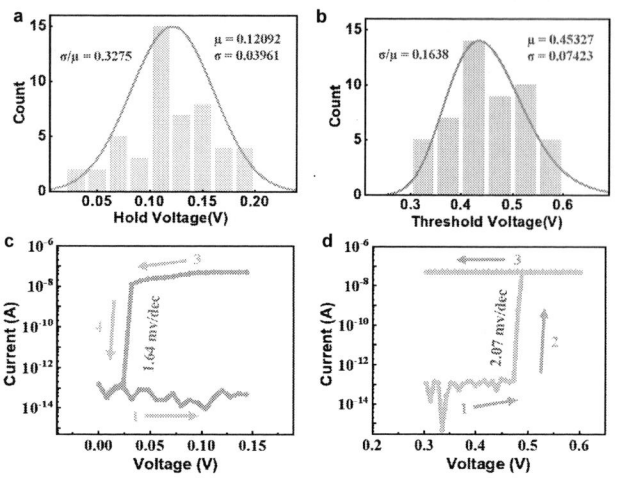

Fig. 3. Probability density distribution of (a) hold voltage (b) threshold voltage. (c) Hold voltage slope of the devices with a 2 mV sweeping step. (d) Threshold voltage slope of the devices with a 2 mV sweeping step.

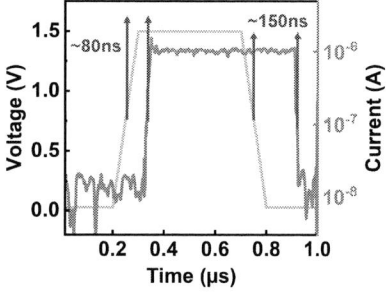

Fig. 4. Current variation under a 1.5 V pulse to monitor the switching speed.

C. TRNG Circuit Design Based on TS Device Behavior

The circuit structure of the proposed TRNG is shown in Fig. 5, primarily composed of an entropy generation unit and a bit extraction unit. The entropy generation unit is composed of two parallel RRAM devices, R_1 and R_2, and a load resistor R_L with a fixed resistance value. One terminal of the RRAM pair is connected to the input voltage V_{DC} while the other end is grounded through the series-connected R_L, forming a complete conductive path. Under a constant input voltage condition, the unit exhibits a spontaneous bistable switching behavior. Initially, both RRAM devices are in a high resistance state, and R_L is selected to be between the resistance values of the RRAM's HRS and LRS. The applied voltage V_{DC} is set higher than the threshold voltage of the RRAM devices.

- State 1: Due to the voltage division across the parallel RRAM and series resistor, most of the input voltage drops across the RRAMs. One of the RRAMs, depending on its faster switching dynamics, will abruptly switch from HRS to LRS once the voltage across it exceeds V_{th}. As a result of this switching and the voltage division, most of the voltage now drops across R_L, and the voltage at node 4 rises to a high level.
- State 2: With one RRAM now switched to LRS while the other remains in HRS, the voltage across both devices falls below V_{hold}, triggering the LRS device to spontaneously return to HRS. As a result, both RRAMs reside in their initial high resistance state. This renewed state redistributes the voltage drop across the circuit, pulling the potential at node 4 close to ground and yielding a low output level.

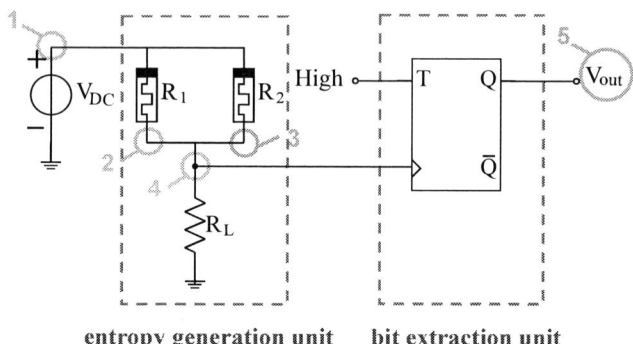

Fig. 5. Schematic of the TRNG circuit comprising the entropy generation unit and bit extraction unit.

As described above, under a fixed high V_{DC}, the entropy generation unit spontaneously oscillates between these two states. Correspondingly, the output at node 4 alternates between high and low levels, as shown in Fig. 6. The oscillation frequency is determined by the intrinsic switching speed of the RRAM devices. Notably, both RRAMs do not switch simultaneously during State 1. When the faster device first transitions to LRS, the resulting change in voltage division causes the voltage across the second RRAM to fall below V_{th}, preventing it from switching. This ensures that during each oscillation cycle, only one randomly selected RRAM switches, contributing to the randomness of the output. This switching mechanism enhances the entropy quality of the source. The randomness stems not only from the device's cycle-to-cycle variability, but also from device-to-device differences between the two RRAMs. This concept also offers a scalable approach for TRNG designs demanding ultra-high entropy, as more parallel RRAMs can be added to amplify the inherent randomness.

The bit extraction unit consists solely of a falling-edge-triggered T-flip-flop, which samples the oscillating signal to generate a usable bitstream, thereby realizing the random number generation functionality. The oscillating output from node 4 in the entropy generation unit serves as the clock input to the T-flip-flop, while the T input is held high. According to the operation principle of a T-flip-flop, the output Q toggles on each falling edge of the clock input. The oscillations at node 4 effectively encode the number of switching events of the RRAM devices over a period of applied V_{DC}. After passing through the T-flip-flop, this count is binarized, with the output '0' or '1' reflecting the parity of the switching cycles—ensuring high randomness in the TRNG output. A key advantage of this circuit lies in the fact that the oscillating signal generated internally by the entropy unit replaces the need for an external clock generator, thereby reducing additional area and power overhead.

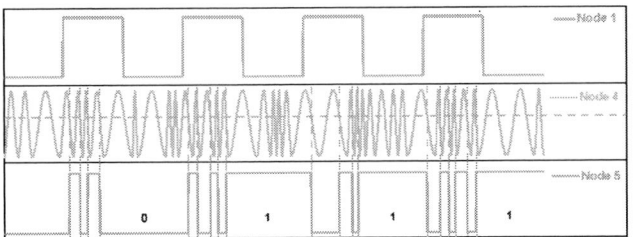

Fig. 6. Working principle illustration based on the proposed TRNG circuit.

By incorporating the TS devices developed in this work into the optimized circuit architecture, we achieve a TRNG solution featuring low power consumption, high throughput, high entropy, and area-efficient design. The ultra-low leakage characteristic of the TS devices, combined with the series high-resistance structure, enables extremely low power consumption. In terms of throughput, the oscillation frequency of the output signal is determined by the switching speed of the RRAM elements. Benefiting from the sub-150 ns switching time of our TS devices, the circuit achieves high-speed entropy generation. The entropy source leverages the stochastic fluctuations during oscillation of the RRAMs, incorporating both cycle-to-cycle and device-to-device randomness, to ensure a high degree of unpredictability in the output bitstream. Furthermore, in terms of integration, the circuit is composed of only two RRAMs, one fixed resistor, and a single T-flip-flop. Its compact structure eliminates the need for external high-frequency clock generators or voltage amplifiers, significantly reducing area overhead.

III. Simulation Results And Discussion

The proposed TRNG circuit was designed and simulated using the 65 nm TSMC PDK. A compact model of the proposed TS device was developed in Cadence Virtuoso to closely match its DC characteristics and transient switching behavior. The model captures the filamentary switching mechanism through a voltage-dependent growth rate equation, enabling faithful emulation of the device dynamics. Based on this model, detailed waveform simulations were performed for the complete TRNG circuit, and the corresponding results are presented below.

Fig. 7 shows the transient simulation results of the proposed TRNG, in which spontaneous oscillations lead to the generation of random numbers. Under a 1.5 V high-level square wave supply, the RRAM devices switch randomly between HRS and LRS, resulting in high-frequency voltage oscillations at node 4 ranging from 0 V to 1.35 V. The output signal exhibits a frequency of 4.5 MHz, with amplitude and timing characteristics compatible with standard commercial T-flip-flops used for bit extraction. Fig. 8 illustrates the random activation of the two RRAM devices within one oscillation cycle, confirming the presence of device-to-device stochastic switching periods that serve as the core entropy source.

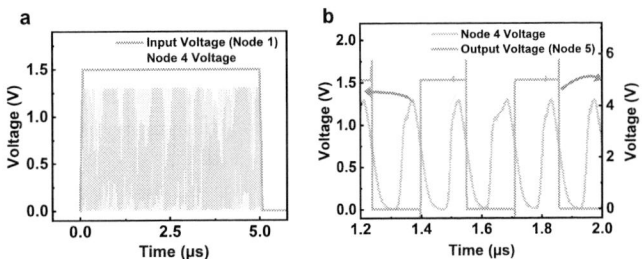

Fig. 7. Simulated waveform of a single cycle generating random binary output (a) entropy generation unit (b) bit extraction unit.

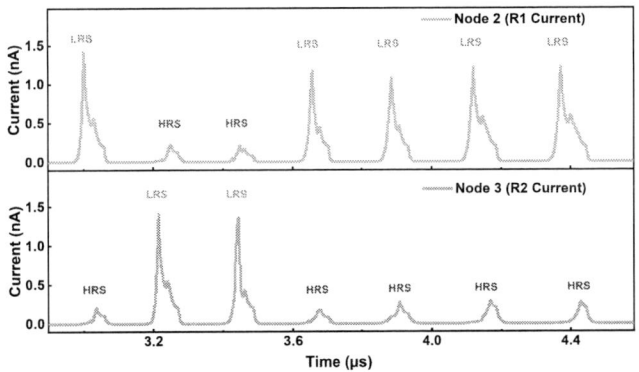

Fig. 8. Process of random activation in two parallel RRAMs within the TRNG circuit.

The output signal from node 4 is connected to the clock input of the T flip-flop. The output of the T flip-flop toggles at high speed with each clock signal's falling edge, oscillating between 0 V and 5 V to represent random bit outputs of '0' or '1'. The V_{DC} square wave signal generates one random bit per cycle, with the last toggle occurring within the valid clock signal period, effectively mapping the number of flips to binary values. The random bit generation rate is approximately 167 kb s^{-1} (with a square wave period of 6 µs). Based on data from other experimental studies, this square wave period can be further reduced while still ensuring the output passes all National Institute of Standards and Technology (NIST) randomness tests, thereby optimizing the bit generation rate. However, further additional experimental validation is needed.

We compared the performance of the TRNG based on volatile memristors, as shown in Table I, the circuit also

TABLE I
COMPARISON OF TRNGs BASED ON VOLATILE MEMRISTORS

	Employed memristor	Entropy source	Bit generation rate (bit/s)	Circuit components	Calibration-free (Yes/No)	Efficiency (nJ/bit)
[7]	Ag:SiO$_2$ diffusive memristor	Delay time	6 k	1 comparator, 1 AND gate, 2 counter	Yes	0.8×10^{-3}
[5]	NbO$_x$ mott memristor	Thermal fluctuations during oscillation	40 k	1 op-amp, 1 T flip-flop	Yes	5.23
[8]	TiN/HfO$_x$/HfO$_x$/HfO$_x$ diffusive memrisor	Integrate-and-fire process	108 k	1 comparator, 1 counter	Yes	/
[9]	VO$_x$/HfO$_x$ memristor	Delay time	28 M	1 T flip-flop, 1 D flip-flop, 1 XOR gate	No	0.83×10^{-3}
[10]	LaCoO$_3$ memristor	Thermal fluctuations during oscillation	50 k	1 T flip-flop	Yes	/
This work	GaO$_x$ diffusive memristor	Fluctuations during oscillation	167 k	1 T flip-flop	Yes	6.36×10^{-6}

demonstrates advantages in power efficiency. Not only is the external clock generator's power consumption eliminated, but the TS devices' low leakage current and optimized circuit structure also reduce operational power consumption. The energy consumption of the entropy generation unit is 6.36 fJ bit^{-1}, with a V_{DC} of 1.5 V and an average current (I_{avg}) of 707 pA, calculated as the geometric mean. Compared to previous studies, the energy consumption shows a significant reduction, making this lightweight TRNG design highly promising for applications that require high throughput and low power consumption.

IV. CONCLUSION

This paper presents a TRNG circuit based on a volatile GaO$_x$ RRAM. By leveraging the stochastic oscillation periods during the switching behavior of TS devices, true random number generation is achieved using a minimal peripheral circuit composed of a single T flip-flop, significantly improving bit generation throughput while minimizing hardware resource consumption. Experimental results show a random bit generation rate of 167 kb s^{-1} under a 6 μs input square wave, with an energy consumption of only 6.36 fJ bit^{-1}. With device-level optimizations incorporated, the proposed circuit can not only ensure the high entropy, but also achieve high throughput, while reducing energy and area consumption, demonstrating its strong potential for lightweight hardware security applications.

ACKNOWLEDGMENT

This work is supported by the National Natural Science Foundation of China (Grants No.62204219).The authors acknowledge Dr. Jiabao Sun of ZJU Micro-Nano Fabrication Center for his assistance during the magnetron sputtering.

REFERENCES

[1] Z. Guo et al., "A True Random Number Generator Based on High-Speed Ag/a-Si/Pt Memristor," IEEE Trans. Electron Devices, vol. 71, pp. 7126–7130, November 2024.

[2] B. Lin et al., "A High-Speed and High-Reliability TRNG Based on Analog RRAM for IoT Security Application," in 2019 IEEE International Electron Devices Meeting (IEDM), December 2019, p. 14.8.1-14.8.4.

[3] T. Arul, N. Mexis, A. E. George, F. Frank, N. A. Anagnostopoulos, and S. Katzenbeisser, "Investigation of Commercial Off-The-Shelf ReRAM Modules for Use as Runtime-Accessible TRNG," in 2024 27th Euromicro Conference on Digital System Design (DSD), August 2024, pp. 33–42.

[4] J. Bian et al., "A true random number generator based on double threshold-switching memristors for image encryption," Appl. Phys. Lett., vol. 122, p. 193502, May 2023.

[5] G. Kim et al., "Self-clocking fast and variation tolerant true random number generator based on a stochastic mott memristor," Nat. Commun., vol. 12, p. 2906, May 2021.

[6] G. Rajendran, W. Banerjee, A. Chattopadhyay, and M. M. S. Aly, "Application of Resistive Random Access Memory in Hardware Security: A Review," Adv. Electron. Mater., vol. 7, p. 2100536, 2021.

[7] H. Jiang et al., "A novel true random number generator based on a stochastic diffusive memristor," Nat. Commun., vol. 8, p. 882, October 2017.

[8] Y. F. Lu et al., "A High-Performance Ag/TiN/HfO/HfO/HfO/Pt Diffusive Memristor for Calibration-Free True Random Number Generator," Adv. Electron. Mater., vol. 8, p. 2200202, 2022.

[9] Y. Qin et al., "A High-Speed True Random Number Generator Based on Unified Selector-RRAM," IEEE Electron Device Lett., vol. 44, pp. 1967–1970, December 2023.

[10] K. S. Woo et al., "True random number generation using the spin crossover in LaCoO3," Nat. Commun., vol. 15, p. 4656, May 2024.

979-8-3315-8850-2/25 $31.00 © 2025 IEEE

Performance Evaluation of an Andvanced-Node CMOS Sensor for Partical Detection

Yue Su
Institute for Data Processing and Electronics
Karlsruhe Institute of Technology
Eggenstein-Leopoldshafen, Germany
suyuemisson@gmail.com

Mingjie Feng
Institute of High Energy Physics, Chinese Academy of Sciences
University of Chinese Academy of Sciences
Beijing, China
fengmj@ihep.ac.cn

Zhiyu Xiang
Institute of High Energy Physics, Chinese Academy of Sciences
University of Chinese Academy of Sciences
Beijing, China
xiangzy@ihep.ac.cn

Cheng Zeng
Institute of High Energy Physics, Chinese Academy of Sciences
University of Chinese Academy of Sciences
Beijing, China
zengcheng@ihep.ac.cn

Congcong Wang
Department of Experimental Physics Division Institute of High Energy Physics, Chinese Academy of Sciences
Beijing, China
wangcc@ihep.ac.cn

Hui Zhang*
Department of Experimental Physics Division Institute of High Energy Physics, Chinese Academy of Sciences
Beijing, China
zhanghui87@ihep.ac.cn

Abstract—**The Monolithic Active Pixel Sensors (MAPS) implemented in high-voltage CMOS (HVCMOS) technology are suitable for tracking of high-energy particles in particle physics experiments. To explore performance improvements in smaller technology nodes, a prototype of the next-generation HVCMOS sensor has been done using 55 nm high-voltage technology. This technology offers the benefits of smaller feature size and reduced power consumption.**

Keywords— *CMOS Sensor, 55 nm CMOS technology, monolithic chip*

I. INTRODUCTION

The Particle detectors made of silicon have been widely used in experimental particle physics for many years. The HVCMOS sensors employ commercial HV-CMOS chip production technologies and allow combining the sensors with readout electronics on the same chip [1]. HVCMOS pixel sensors are based on deep n-well in p-substrate diodes, with the readout electronics embedded in the deep n-well. The negative bias voltage applied to the p-substrate generates the depletion region and accelerates the charge collection by drift (~1 ns). A lot of HVCMOS sensors [2–4] have been designed in 180 nm–130 nm processes. Exploration of a smaller technology node with improved performances, such as smaller feature size and reduced power consumption is required for future applications. In this article, an implementation of an HVCMOS sensor in a 55 nm HVCMOS technology has been presented. A prototype chip has been implemented in 55 nm HVCMOS technology with low resistivity substrate (~10Ω cm). The application can be one particle physics experiment, such as CEPC or LHCb.

II. DESIGN AND IMPLEMENTATION

A. New technology SMIC 55 nm

In this design, we innovated and used the 55nm process. Compared with the 180nm process used in the previous version, SMIC's 55nm process has higher transistor density, stronger performance and lower power consumption, and has a higher design complexity [5]. It is suitable for mid-range mobile devices, the Internet of Things, consumer electronics and other application scenarios that have higher requirements for performance and power consumption. The 180nm process has lower transistor density and higher power consumption, and the design is relatively simple, and is usually used for cost-sensitive applications with low performance requirements.

TABLE I.   COMPARISON BETWEEN 180NM AND 55NM PROCESS TECHNOLOGIES

Features	180 nm Process	55 nm Process
Transistor Density	Low	High
Performance	Low	Medium
Power Consumption	High	Low
Manufacturing Cost	Low	High
Application Areas	Automotive electronics, industrial control, analog circuits	Mobile devices, IoT, consumer electronics, image processing
Technology Maturity	Very mature	Mature
Design Complexity	Low	High

The 180nm process has a lower manufacturing cost, is suitable for mass production and cost-sensitive applications, has a relatively simple design, and a short development cycle, and is suitable for rapidly developed products such as power management chips and low-end embedded devices [6]. The 55nm process has a higher manufacturing cost, a complex design, and requires a longer development cycle, but it can meet the balance between performance and power consumption, and is suitable for mid-range smartphones, IoT devices, and other applications that require higher integration and performance. comparison of the two processes is shown in Table 1.

B. High Voltage CMOS and Low Voltage CMOS

The low-pass amplifier IC, developed in 55nm technology, is a two-stage amplifier with differential input and feedback. There are obvious differences between high-voltage CMOS and low-voltage CMOS in structure, shown in Fig. 1. These properties and application scenarios, mainly in device design, performance parameters and manufacturing process. They each have different advantages and disadvantages and are suitable for different application scenarios. Let's first analyze the differences in their structures.

Low-voltage CMOS typically operates within a range of 1V to 3.3V, with typical values of 1.8V or 3.3V. These are mainly used in low-power digital and analog circuits. High voltage CMOS usually operates at voltages between 5V and 100V, or even higher. They are used in applications that require high voltage driving or processing, such as power management, analog circuits, power amplifiers, etc. Low-voltage CMOS usually uses a relatively thin gate oxide layer (with small oxide thickness) [7][8]. A thinner oxide layer helps improve the transistor's response speed, but it also means that low-voltage CMOS components cannot withstand high voltages, as high voltage may cause oxide layer breakdown or leakage. High-voltage CMOS devices have thicker gate oxide layers to prevent high voltage from breaking down the oxide layer, thus withstanding higher voltage stress. This thick oxide layer sacrifices a certain response speed, but increases electrical isolation capability. Additionally, Low-voltage CMOS usually has a relatively short channel length, allowing faster switching speeds, which enables handling higher frequency signals. High-voltage CMOS usually has a longer channel to disperse the voltage stress caused by the high voltage and prevent the local current density from being too high and causing device failure.

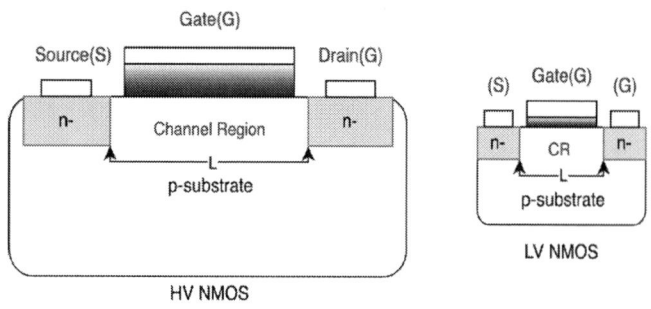

Figure 1. Structural comparison between HV CMOS and LV CMOS

C. Pixel Electronics

The cross section of the chip is shown in Fig.2 involved in this paper is a monolithic pixel detector that uses commercial high-voltage CMOS (HV-CMOS) technology. Our chip is located in the lower right corner, the size is 1250 um x 1250 um. The schematic of pixel electronics is shown in Fig. 3. It contains an injection circuit, a PMOS charge sensitive amplifier (CSA), a NMOS comparator, n-well bias circuit, 3-bit RAM cell and 4 output stages. They are one of the most advanced detectors for detecting high-energy particles. Complex pixel electronics including charge-sensitive amplifiers, integrator, discriminators, and memory are embedded on the same substrate.

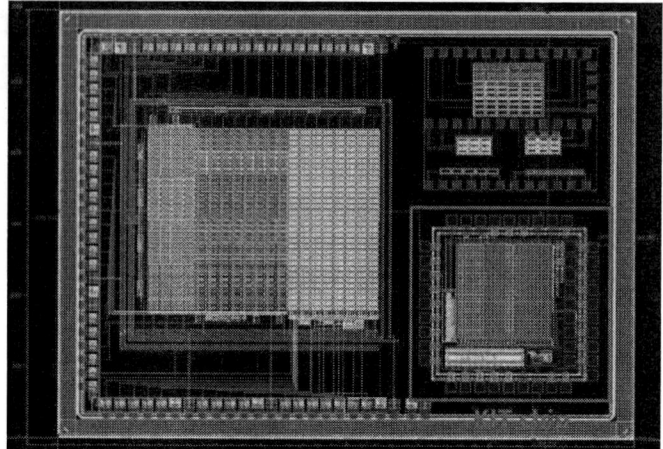

Figure 2. The cross section of this chip

The charge collected by the pixel's n-well is converted into a voltage signal by a charge-sensitive amplifier (CSA). The analog voltage pulse is then shaped and digitized by a comparator. The output stage generates a current, which is transmitted via an address line to a transimpedance amplifier located outside the pixel matrix. Since the address lines are held at a constant potential, crosstalk between signals is minimized [9].

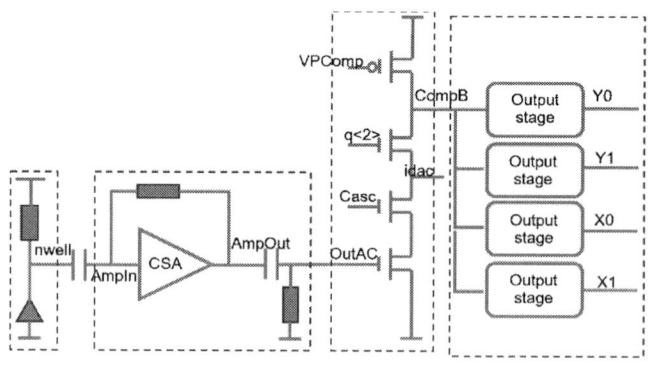

Figure 3. Pixel electronics of this chip

D. Digital Peripheral

The digital logic comprises four timestamp measurement modules, a data readout state machine, an asynchronous serializer, and a configuration block. An internal counter operates under the timestamp clock and produces timestamp values in gray code. The hit pulse from the address bus is operated with an OR logic in the timestamp measurement state machine to capture the leading and trailing edges. As the address bus is wired-OR driven, it could happen that multiple pixels generate overlapping hit pulses simultaneously [10]. To address this, two check position parameters are set for a double check on the hit address, issuing an error bit in case of a mismatch. Additionally, to control the data rate and shield the noise spurs, a maximum detect length parameter is set. When a hit pulse is detected, the state machine locks for a specified clock period by this length, even if a trailing edge has already been detected. In the event of an exceptionally long hit pulse, potentially due to overlapping on the address bus, the state machine will terminate the current detection after reaching this maximum length. The measurement results are pushed into four parallel FIFOs and sent to the readout module. All of these parameters can be configured via the SPI configuration module.

The readout module performs polling check among the four FIFOs and generates a 48-bit data package for each hit. When a FIFO is enabled for reading, the readout state machine locks its number, captures the data package and waits for the byte shifter from the output interface to load the data package. Once this is complete, the readout state machine returns to the polling check state for the next data package. The data acquisition system is shown in Fig. 4.

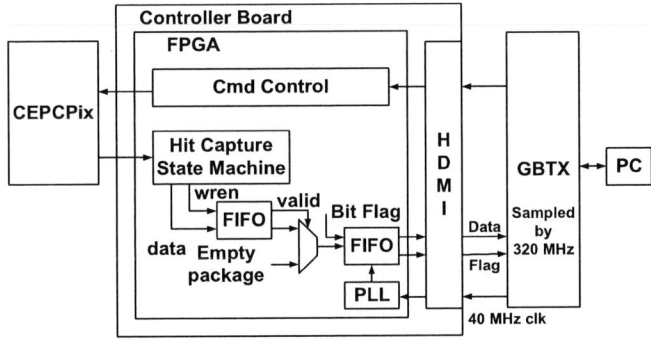

Figure 4. Data acquisition system of this chip.

Two mutually exclusive readout interfaces are implemented for readout logic test, which can be selected active by configuration. In the slow serial output interface, the data package is transmitted on a single bit under the readout state machine's clock domain. In the asynchronous readout interface, the data package is sent on byte set to the serializer. The serializer locks the byte data with a ping-pong register, ensuring a sufficient setup time when crossing the clock domain. With this structure, the data can be readout on bit under an asynchronous clock whose frequency is eight times faster than that of the state machine.

III. SIMULATION RESULTS

The pre-simulation step is indispensable to evaluate the design concept and reality. We give several input pulses and test the functionality of this system. The output signals of CSA, comparator and level shifter for different voltages are illustrated in Fig.5. We can see that the rising time of the analog signal is shorter while the signal becomes larger. After the impedance amplifier and several stages of level shifter, we achieved a very excellent digital pulse.

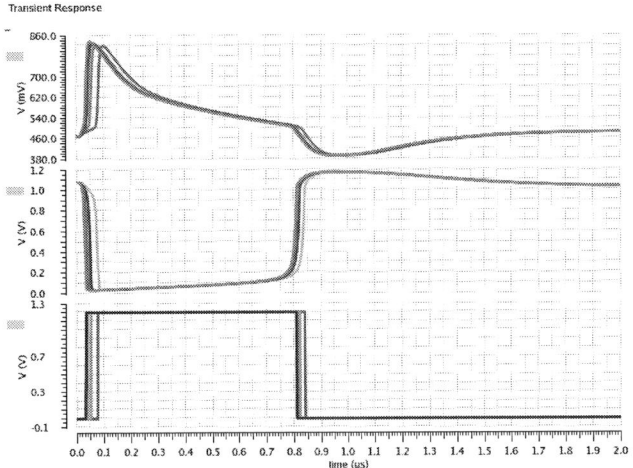

Figure 5. Output transient voltage of the output of (top) CSA, (middle) the comparator, (bottom) the level shift for different charge.

IV. MEASUREMENT RESULTS

This section details the four key performance tests conducted on the amplifier. A custom-designed PCB was fabricated to ensure signal integrity during testing. For characterization, a simplified test setup has been built, as shown in Fig. 6. An arbitrary waveform generator produces a chirp test signal while a high-precision oscilloscope captures the amplifier's output response. Both instruments are automatically controlled via Python scripts to ensure measurement consistency. The power supply supplies power to the chip through the adapter board. The middle of the picture is the designed chip carrier PCB and adapter board [11].

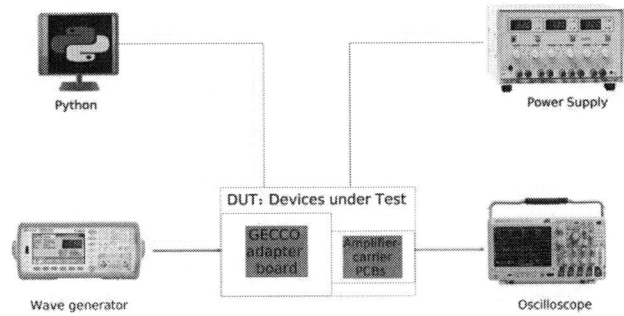

Figure 6. Simplified test setup

We use Nexys Video Artix-7 FPGA board to configure the chip and readout the digital data. It is mainly implemented a Basil fast data acquisition system (DAQ) to read and write data with the registers in the chip. The FPGA communicate with the PC through the LAN port; the PC burn the program to the board through JTAG. Additionally, we use a GECCO adaptor board, which provides all necessary supply voltages as well as test signals and routes data lines from and to the FPGA. At the last is the PCB board, the PCB board carrying the test chip is powered by the GECCO adaptor board and communicates with the FPGA.

A. I-V Measurement

The I-V characteristic is usually the first step in the whole characterization procedure. It gives the dependence of the leakage current on the bias voltage, is an easy and quick way to check the functionality of a detector. To measure the I-V characteristics of a pixel detector, the detector's backside must be reverse-biased, while the frontside is grounded. According to the SMIC 55nm high voltage process documentation, the maximum allowable bias voltage is 32 V. To avoid damaging the ASIC, we limit the bias voltage to -32 V. Fig. 7 shows the I-V characteristic of the chip, with a leakage current of approximately 1 nA.

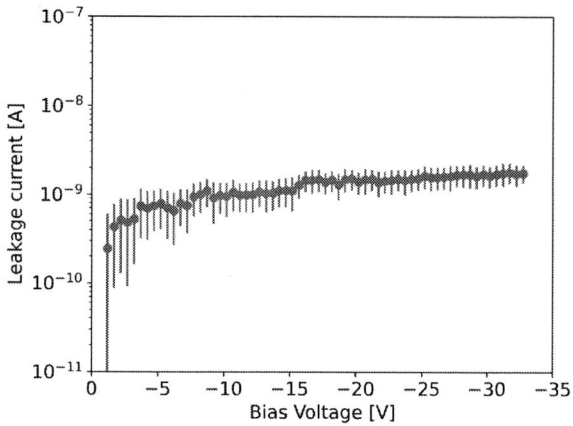

Figure 7. The I-V measurement of a single pixel

B. The analog output versus different injection amplitude

In order to get the gain, the maximum value of each curve can be measured. Fig. 8 shows the relationship between the maximum amplitude of the chip Sfout output and the injection voltage. From this figure, it can be seen that the maximum amplitude of the chip Sfout output is linearly related to the injection voltage. Through further measurement and curve fitting using the fitting function, the chips amplification gain is approximately equal to 0.1.

These outputs can be tested by applying known signals or electrical charge injections. Although charge injections require calibration, they are the most convenient and safety characterization method. Therefore, we first use the injection board based on Cinj to calibrate the chip, with the analog output monitored via an oscilloscope. Capacitance Cinj can be estimated by using the parasitic capacitance extraction tool. The verification was further conducted by measuring output signal amplitudes when the sensor was irradiated with a 55Fe source, which emits X-ray photons of known energies, and comparing these with the injection response. The measurement results are shown in Fig. 9, where the number of sampling points is 10,000. To make the sampling data more accurate, each point is sampled 5 times and the average value is calculated.

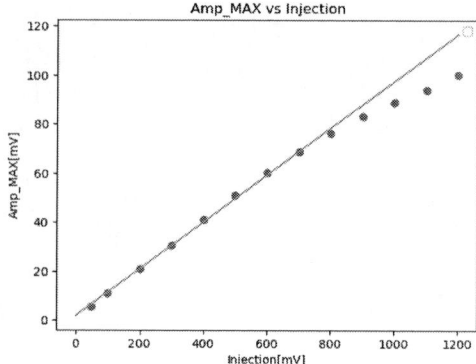

Figure 8. The maximal analog output versus different injection amplitude

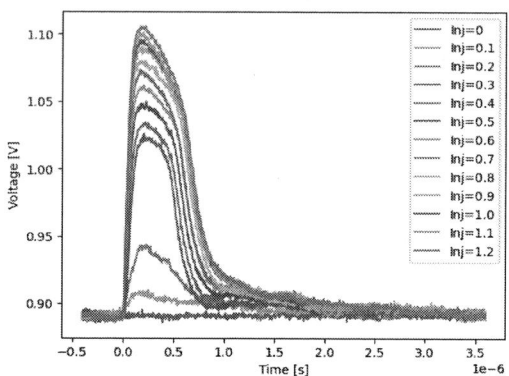

Figure 9. The analog output versus different injection amplitude

C. Rising Time Measurement

Fig. 10 shows the rising time. This is an indicator of the chip response time. From this result graph we can see the rise time decreases rapidly with the increase of the injection voltage. When the injection voltage is greater than 0.8V, the walking time tends to stabilize and no longer decreases. The time difference between the time point of half the falling edge of the injection signal and the time point of half the falling edge of the comparator output signal is measured by an oscilloscope to represent the walking time of the comparator output.

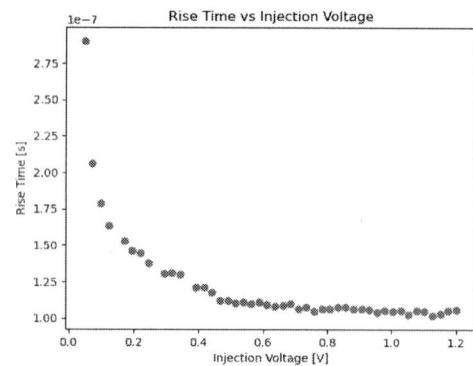

Figure 10. The rising time of the analog output

V. Conclusion

A monolithic active pixel detector has been implemented with a 55 nm HVCMOS technology. This chip is designed as a prototype to evaluate suitability of the 55 nm technology for sensor design. The chip consists of a pixel matrix, a digital readout module and auxiliary blocks, such as bias DACs. The functionality tests have been performed. High resistivity substrates (>1 kΩcm) can be offered by the foundry as well. We expect higher signals with chips implemented on high resistivity substrates in the next design iterations.

References

[1] B. Roberto, "Customized Integrated Circuits for Scientific and Medical Applications.," Doctoral dissertation, Jan 2020.

[2] Miaoran. Sun, "Design of integrated Sensors for Medicine.," Master thesis, August 2024.

[3] H. Zhang, T. Hirono, I. Peric JINST 19 P02036. A high time resolution and high dynamic range ASIC for the micro-vertex detector in the PANDA experiment. Journal of Instrumentation, February 2024.

[4] H. Zhang and I. Peric 2022 JINST 17 C07022. Monolithic pixel sensor with 25 µm × 35 µm pixel size and high time resolution implemented in 180 nm technology.

[5] Ni He et al, The utility of breast cone-beam computed tomography, ultrasound, and digital mammography for detecting malignant breast tumors: A prospective study with 212 patients, European Journal of Radiology 85.2 (2016), pp. 392–403.

[6] Mnahi Bin Saeedan et al., "Breast Lesions on Chest Computed Tomography: Pictorial Review With Mammography and Ultrasound Correlation, In: Current Problems in Diagnostic Radiology 44.2 (2015), pp. 144–154.

[7] SC Huang and M Ismail. "Design and applications of a CMOS analog multiplier cell using the differential difference amplifier". In: Analog Integrated Circuits and Signal Processing 6.3 (1994), pp. 209–217. DOI: 10.1007/BF01238889.

[8] Milind Gajare, D. K. Shedge, and Devendra Itole. "CMOS Transimpedance Feedback Amplifier Design using Unity Gain Differential Amplifiers". In: 2021 International Conference on Emerging Smart Computing and Informatics (ESCI). 2021, pp. 743–747.

[9] H. Zhang, T. Hirono, Y. Su, R. Dong, I. Peric, JINST 20 C03023. High voltage monolithic pixel sensor in 55 nm technology. Journal of Instrumentation, March 2025.

[10] R. Dong, et al 2023, Design of a HVCMOS pixel sensor prototype in 55nm for CEPC.

[11] Yue Su, "Development of Integrated Circuits in 55 nm CMOS Technology, Master thesis, August 2024.

2025 The 10th International Conference on Integrated Circuits and Microsystems

Low-power LDO with Fast Transient Response Based on FVF Structure

Wenya Luo
School of integrated Circuits
Shandong University
Jinan, China
2716796054@qq.com

Longfei Xu
School of integrated Circuits
Shandong University
Jinan, China
2247097666@qq.com

Hongjian Ren
School of integrated Circuits
Shandong University
Jinan, China
sdrenhongjian@163.com

Chao Cao
School of integrated Circuits
Shandong University
Jinan, China
chao_cao@sdu.edu.cn

Haijun Guo
School of integrated Circuits
Jinan University
Jinan, China
ise_guohj@ujn.edu.cn

Abstract— **In this paper, an improved folded flipped voltage follower (FVF) low-dropout (LDO) regulator structure is proposed to address the transient response issues in low-power applications, as well as the stability challenges under light-load condition without relying on a large external capacitor. The enhanced FVF LDO architecture achieves higher loop gain and faster loop response, ensuring improved performance. By adaptively biasing the current source based on output voltage variations, the proposed design enhances the slew rate (SR) of the LDO. Furthermore, a transient enhancement circuit is incorporated to minimize output voltage overshoot and undershoot, thereby ensuring the stability and proper operation of subsequent circuits.**

Keywords—flipped voltage follower (FVF), low-dropout regulator (LDO), low power consumption, transient enhancement

I. INTRODUCTION

With the development of information technology and the advancement of semiconductor processes, the demand for personal mobile devices has surged dramatically. However, limited battery capacity and the impracticality of frequent battery replacement or recharging have made power management modules a crucial component in integrated circuit (IC) devices [1,2,3]. To extend battery life and enhance power conversion efficiency, the development of low-power low-dropout regulators (LDOs) is essential [4]. Conventional LDOs typically rely on large output capacitors to ensure system stability. However, modern edge devices such as those in the Internet of Things (IoT) require fully system-on-chip (SoC)-integrated solutions, where large external capacitors must be eliminated due to area cost constraints. In contrast, flipped voltage follower-based LDOs (FVF LDOs) offer superior performance, featuring fast transient response and low output impedance. Consequently, these characteristics make FVF LDOs highly suitable for designs demanding both off-chip-capacitor-free operation and ultra-low power consumption.

In this paper, an enhanced FVF LDO structure is proposed that significantly improves upon conventional FVF LDO designs. The proposed architecture enables the FVF LDO fast loop to have a significantly higher loop gain and a faster transient response. Furthermore, an innovative undershoot compensation circuit is introduced to effectively mitigate output voltage undershoot, ensuring superior regulation performance.

II. SYSTEM ARCHITECTURE

A. Conventional FVF LDO Loop

Fig. 1 illustrates the conventional FVF LDO loop. The core fast-loop regulation path comprises a voltage follower configuration, where MP_1 serves as the power transistor and MP_2 functions as the voltage sensing transistor. This sensing network dynamically detects output voltage variations and feeds them back through the regulation loop.

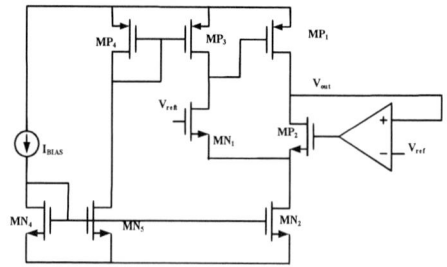

Fig. 1. Schematic diagram of conventional FVF LDO loop.

B. Improved FVF LDO Design

Although conventional FVF LDO offers rapid transient response, it faces significant challenges in maintaining low quiescent current under light-load operation during light-load operation while suffering from pronounced output voltage variations when implemented without large external capacitors. To address these challenges, our design focuses on optimizing both transient response in low-power operation and light-load stability without external components. In this paper, we employ

979-8-3315-8850-2/25 $31.00 © 2025 IEEE

the FVF LDO as the core architecture with targeted modifications to meet specific design requirements [5,6]. A novel transient enhancement circuit is integrated to achieve ultra-low power operation; the complete implementation, shown in Fig. 2[7], comprises four key components including the LDO slow loop, FVF fast regulating loop, power transistors, and transient compensation circuitry.

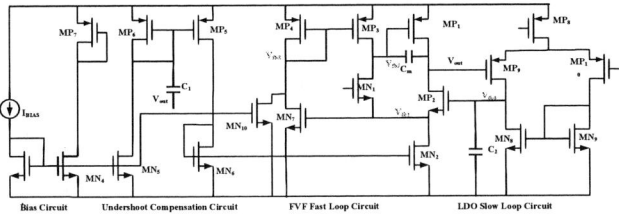

Fig. 2. Overall circuit diagram of the improved FVF LDO.

III. MODULE DESIGN

A. FVF LDO Fast Loop Module Design

In this paper, a Class AB error amplifier with high-voltage swing capability is employed for power transistor gate current regulation, thereby achieving superior loop response speed. The charging/discharging process is accelerated through an optimized MP_3 design, implemented as an adjustable current mirror to improve gate discharge characteristics (Fig. 3).

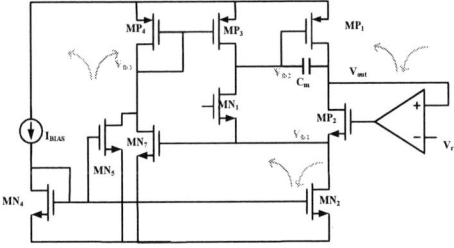

Fig. 3. Improved FVF LDO schematic.

Under normal operation, MN_7 operates in the subthreshold region (VGS $\approx$ 0.4 V) with minimal quiescent current, satisfying low-power design requirements. However, at ultra-low temperatures, the threshold voltage (VTH) increase concurrent with VGS reduction in MN7 significantly suppresses the branch bias current. This current suppression propagates to the MP3 branch through current mirroring, degrading the loop's dynamic response. To address this issue, bias transistor MN_5 is introduced to establish a minimum current limit for MP_4, ensuring stable performance across all operating conditions.

Compared to conventional FVF LDO designs, the proposed architecture requires additional bias current to enhance loop response. Characterization revealed that while overshoot and undershoot voltages at MP_2's source exhibited similar magnitudes, significant attenuation occurs during transfer to the drain, where the overshoot component dominates. Furthermore, undershoot propagation through MN_7 drives MP_3 into cutoff operation. Nevertheless, MP_3's minimal bias current ensures negligible speed degradation during cutoff events. Comparative transient simulations in Fig. 4 demonstrate that the proposed architecture achieves a 62% overshoot reduction (from 170 mV

to 65 mV) relative to the conventional design, while exhibiting limited effectiveness in undershoot compensation.

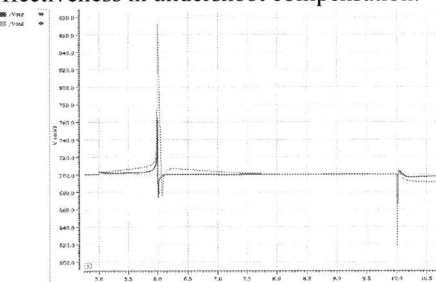

Fig. 4. Performance comparison of with and without transient enhancement modules.

To overcome these limitations, an undershoot compensation circuit employing dynamic bias techniques is developed [8,9,10,11,12]. As Fig. 5 shows, an additional current mirror stage is integrated into the original current mirror configuration, which comprises transistors MN_2 and MP_2. Here, the gate terminal of MP_2 is linked to the output V_{out} through the coupling capacitor C_1. Consequently, when the output voltage generates downslope, the coupling capacitor C_1 enables the gate of MP_2 to track this downward trend. This, in turn, prompts MP_1 to increase the current, which is then channeled through MN_3 to supply the FVF LDO. This mechanism enhances the quiescent current of the FVF LDO circuit, thereby optimizing overall transient response speed of the circuit. The coupling capacitor C_1, effectively transmits voltage fluctuation to MP_2 while maintaining quiescent current stability during large signal variations. This dual functionality enables the FVF LDO to preserve its low-power characteristics, as shown in Fig. 6.

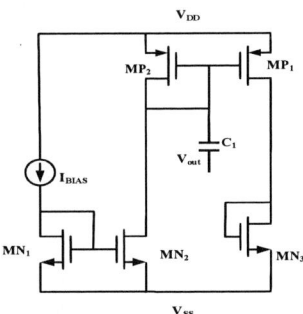

Fig. 5. Undershoot compensation circuit.

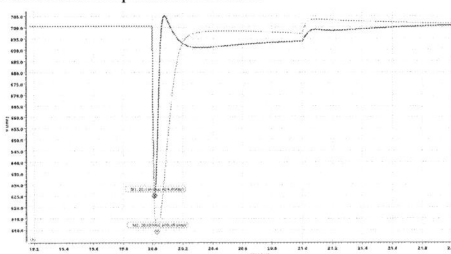

Fig. 6. Comparison of transient simulation with and without undershoot compensation.

B. Fast Loop Stability Analysis

To analyze the stability of the FVF fast loop, the equivalent circuit is derived by disconnecting the loop at the gate of the power transistor (node V_{p1}).The complete equivalent circuit diagram of the FVF fast loop is shown in Figs. 7 and 8.

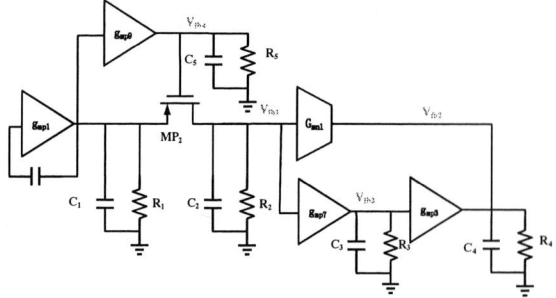

Fig. 7. FVF fast -loop equivalent circuit.

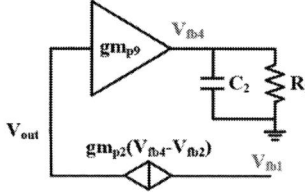

Fig. 8. MP$_2$ small-signal model.

The transfer function is derived using Kirchhoff's law and simplified by combining:

$$H(s) \approx \frac{g_{m_1} g_{m_9} g_{m_2} (G_{m_1} + g_{m_1} g_{m_3} R_3)}{(1 + sC_1 R_1)(1 + sC_2 R_2)(1 + sC_5 R_5)}$$
$$\times \frac{(g_{m_9} \frac{R_2}{1 + sC_2 R_2} - g_{m_2})(g_{m_1} - \frac{1}{sC_m})}{(1 + sC_4 R_4)(1 + sC_3 R_3)}$$

(1)

The power transistor's large gate capacitance results in $C_4 \gg C_1, C_3, C_5$, with an equivalent resistance $R_4 \approx r_{p3} \,//\, (gm_{n1}ro_{n1}ro_{n2})$. Owing to the error amplifier's low quiescent current in the slow loop, $R_2 \gg R_5, R_1, R_3$, while C_2 (comprising compensation capacitance C_{cap} and the equivalent capacitance at the output node of the error amplifier's output-node capacitance) is large. Consequently, the loop exhibits two dominant low-frequency poles at the error amplifier output (V_{fb4}), and power transistor gate (V_{fb2}).

$$p_1 \approx \frac{1}{R_4 C_4} \approx \frac{1}{(ro_{p3}\|gm_{n1}ro_{n1}ro_{n2})(C_{gg}+gm_{p1}R_4C_m)} \quad (2)$$

$$p_2 \approx \frac{1}{R_2 C_2} \quad (3)$$

In Eq. (2), C_4 represents the Miller-compensated equivalent capacitance.

The third pole occurs at the output node (V_{out}),

$$p_3 \approx \frac{1}{R_L C_L} \approx \frac{1}{(r_{p1}/\frac{r_{n2}+r_{p2}}{1+gm_{p2}r_2})(R_L)(C_L+C_n)} \quad (4)$$

where R_1 equals the parallel combination of the LDO's internal output resistance and load resistance (R_L). C_1 primarily comprises the Miller capacitor (C_m) and output capacitor (C_L). Two feed-forward paths in the circuit introduce corresponding zeros. C_1 is mainly composed of Miller compensation capacitor C_m and output capacitor C_L:

$$z_1 \approx \frac{g_{mp2}}{g_{mp9}R_2} \quad (5)$$

$$z_2 \approx \frac{g_{mp1}}{C_m} \quad (6)$$

Fig. 9 shows the fast-loop Bode plot, clearly revealing the pole-zero distribution, poles p_1 and p_2 are clustered at the low frequencies; while the slow loop introduces a low-frequency zeros, which are distributed to the right of p_2.

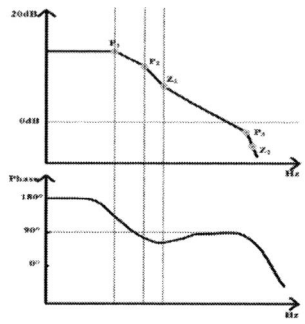

Fig. 9. Zero-pole distribution curve diagram.

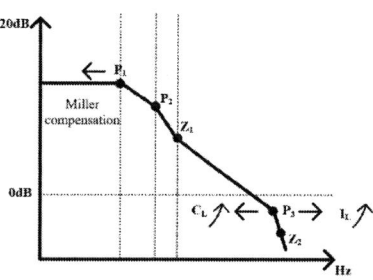

Fig. 10. Zero-pole distribution curve diagram after Miller compensation.

Since pole p_2 and zero z_1 are closely spaced, they approximately cancel each other, leaving only poles p_1 and p_3 as dominant. The design targets a 30 pF load capacitance, where the larger values would shift p_3 towards low frequencies. To address this, Miller compensation is employed to push main pole to lower frequencies using minimal capacitor area, so that maximize separation between p_1 and pole p_3.

When the load current increases from low to high levels, the power transistor conducts substantial current, reducing its drain resistance (r_{p1}). This shifts pole p_3 toward higher frequencies, with stability improving at higher current. Fig. 10 demonstrates the pole variations under increasing load capacitance and current when using Miller compensation.

C. FVF LDO Slow Loop Module Design

The slow loop module serves as a critical component in the proposed LDO design. Although the primary regulation is achieved through the FVF-based fast-response loop, process variations and temperature fluctuations can alter MOSFET threshold voltages and other device characteristics. Consequently, these variations may cause output voltage fluctuations in the fast response loop. Therefore, to address this issue, the LDO slow loop is introduced to improve the accuracy of the LDO and stabilize the output voltage. The slow loop comprises an error amplifier and a feedback network. Unlike the fast loop, which directly regulates the power transistor, the slow loop consists of an error amplifier and a feedback loop. The slow loop focuses solely on output voltage correction. Consequently, its design prioritizes high loop gain and sufficient phase margin to ensure stability.

The error amplifier is the core of the slow loop. To accommodate the low-voltage operation of the overall circuit (with a minimum supply voltage of 0.9 V), a 5-transistor amplifier circuit is selected as the error amplifier, as shown in Fig. 11:

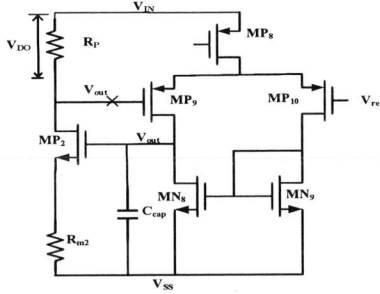

Fig. 11. Slow loop circuit.

where R_p is the drain-source equivalent resistance of the power transistor, while R_{m2} is the drain-source equivalent resistance of MN_2.

The primary function of the LDO slow loop is to stabilize the output voltage (V_{out}), rather than to handle transient responses. Consequently, the quiescent current can be minimized without requiring consideration of the circuit's response speed or slew rate. To achieve this, all transistors in the error amplifier are biased in the subthreshold region.

D. Slow Loop Analysis

Disconnecting the loop from the MP_9 gate, as shown in Fig. 11, a small-signal model is obtained, which is depicted in Fig. 12. Here, g_{m9} represents the transconductance of the operational amplifier, while R_1 and C_1 denote the output resistance and capacitance of the error amplifier, respectively. Additionally, G_{m2} signifies the equivalent transconductance of the voltage follower positioned at the subsequent stage of the error amplifier. Therefore, a small-signal model specific to G_{m2} is provided in Fig. 13 for enhanced clarity and understanding.

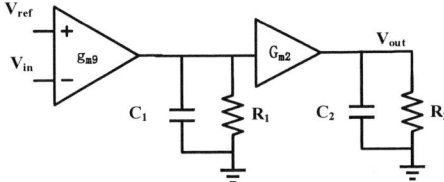

Fig. 12. Slow loop equivalent circuit.

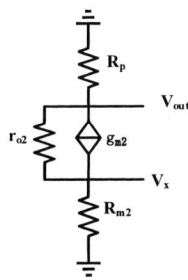

Fig. 13. Small-signal model of G_{m2}.

The loop gain can be expressed as:

$$A_V = \frac{1}{2\lambda m v_T} \cdot \frac{g m_2 R_p R_{m2}}{R_{m2} - R_p + r_{o2} + g m_2 R_p R_{m2}} \quad (7)$$

The voltage follower action introduces gain attenuation in this loop. The dominant pole (p_1) and secondary pole (p_2) are given by:

$$p_1 = \frac{1}{R_1 C_1} \approx \frac{1}{r_{ms}//r_{pp} \cdot C_{gg_2}} \quad (8)$$

$$p_2 = \frac{1}{R_2 C_2} \approx \frac{1}{R_p C_{out}} \quad (9)$$

Due to the small quiescent current conditions, $R_p \ll r_{on8}//r_{op9}$ which result in the dominant pole being located at the error amplifier output, and the secondary pole appearing at the LDO output node. Notably, with large load capacitance, p_2 shifts toward lower frequencies.

To maintain adequate pole separation, a compensation capacitor C_{cap} is added to the output of the error amplifier, as shown in Fig. 14.

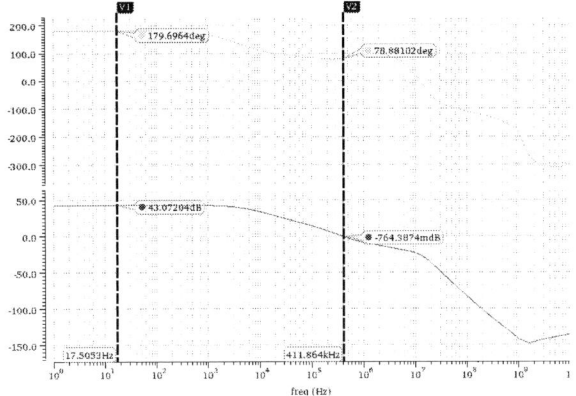

Fig. 14. Slow loop gain and phase margin simulation results.

IV. CHIP LAYOUT AND POST-SIMULATION RESULTS

A. Layout

The proposed LDO circuit was implemented in TSMC 22 nm CMOS technology. Fig. 13 shows the complete layout, with a final dimension of 40.565 μm × 47.66 μm. The detailed FVF LDO circuit layout is shown in Fig. 15. The layout of the fast loop circuit of the FVF LDO is illustrated in Fig. 16.

979-8-3315-8850-2/25 $31.00 © 2025 IEEE   586

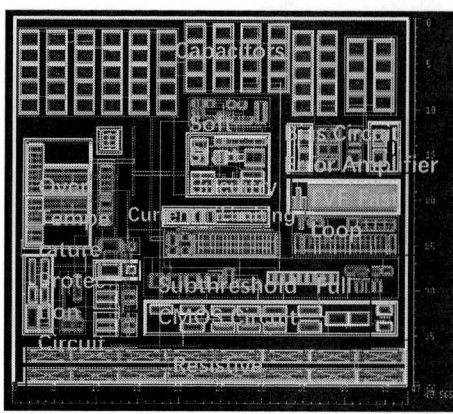

Fig. 15. Overall circuit layout.

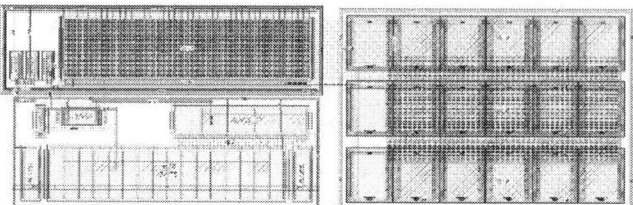

Fig. 16. FVF LDO fast loop circuit layout.

B. Power Consumption Simulation

Post-layout simulations were conducted to evaluate the quiescent current across various operating voltages. The DC characteristics were measured with a 0.7 V reference voltage and supply voltages ranging from 0.9 V to 1.5 V. The static current simulation results are shown in Fig. 17.

C. Transient Response Simulation

To verify the worst-case transient response, a subthreshold CMOS voltage source is connected to the unloaded LDO circuit. The load current was switched between 100 µA to 50 mA at 1 µs intervals while monitoring the output voltage variations. The post-layout transient response simulation results are shown in Fig. 18. The data in Table 1 is summarized from Fig. 18.

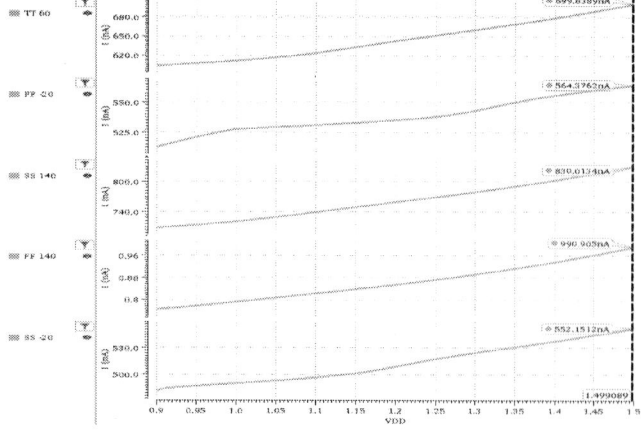

Fig. 17. Quiescent current simulation diagram.

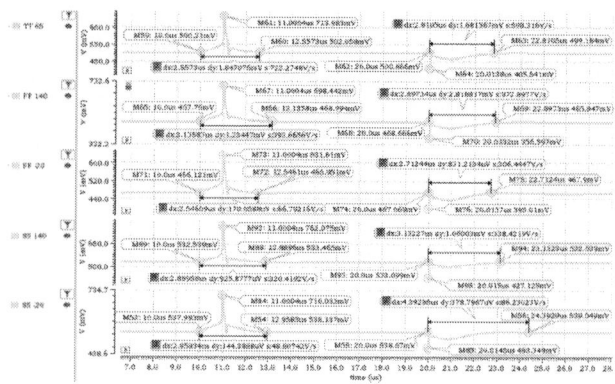

Fig. 18. Post-transient response simulation.

TABLE I.   TRANSIENT SIMULATION RESULTS

	Temp(℃)	Overshoot voltage (mV)	Recovery time (µs)	Undershoot voltage (mV)	Recovery time (µs)
TT	60	213.773	2.5573	95.325	2.81
SS	-20	172.04	2.9583	75.321	4.39
SS	40	229.536	2.8896	105.974	3.13
FF	20	165.489	2.5461	72.059	2.71
FF	40	230.692	3.1358	112.063	2.82

V. SUMMARY

This paper first introduces the significance of low power LDO design for portable devices and IoT requirements. A feasible design scheme based on the FVF LDO is proposed with key objectives of enhancing the endurance of the device, improving chip integration, and optimizing transient response performance.

The circuit is implemented in 22 nm CMOS technology. Post-simulation results demonstrate that the LDO can stably output voltages ranging from 0.4-0.7 V for input voltages ranging from 0.9-1.5V. It features a quiescent current below 1 µA, a line regulation of 1.53 mV/V, and a load regulation of 0.011 mV/mA. The LDO also supports rapid up-switching of the load current from 0.1 mA to 50 mA within 1 µs, as well as bidirectional switching within the range. During such rapid load current transitions (0.1 mA to 50 mA), the overshoot and undershoot voltages remain below 250 mV, achieving a recovery time of 4.3 µs, which significantly enhancing its transient performance.

REFERENCES

[1] Martins R P, Mak P I, Chan C H, et al. Bird's-eye view of analog and mixed-signal chips for the 21st century[J]. International Journal of Circuit Theory and Applications, 2021, 49(3): 746-761.

[2] Chun A C C, Ramiah H, Mekhilef S. Wide power dynamic range CMOS RF-DC rectifier for RF energy harvesting system: A review[J]. IEEE Access, 2022, 10: 23948-23963.

[3] Compare M, Baraldi P, Zio E. Challenges to IoT-enabled predictive maintenance for industry 4.0[J]. IEEE Internet of things journal, 2019, 7(5): 4585-4597.

[4] Tan J, Sathyamurthy M, Rolapp A, et al. NISP: An NFC to I 2 C Sensing Platform With Supply Interference Reduction for Flexible RFID Sensor

Applications[J]. IEEE Journal of Radio Frequency Identification, 2020, 4(1): 3-13.

[5] Chen H, Leung K N. A fast-transient LDO based on buffered flipped voltage follower[C]. 2010 IEEE International Conference of Electron Devices and Solid- State Circuits (EDSSC). IEEE, 2010: 1-4.

[6] Zeng Y, Li Y, Zhang X, et al. A push-pulled FVF based output-capacitorless LDO with adaptive power transistors[J]. Microelectronics Journal, 2017, 64: 69- 77.

[7] Sakolski O, Poongodan P K, Vanselow F, et al. A feedforward compensated high- voltage linear regulator with fast response, high-current sinking capability[J]. IEEE Solid-State Circuits Letters, 2020, 3: 114-117.

[8] Han W, Lee H. A 340-nA-quiescent 80-mA-load 0.02-fs-FOM active-capacitorbased low-dropout regulator in standard 0.18- μ m CMOS[J]. IEEE Solid-State Circuits Letters, 2021, 4: 125-128.

[9] Desai C, Mandal D, Bakkaloglu B, et al. A 1.66 mV FOM output cap-less LDO with current-reused dynamic biasing and 20 ns settling time[J]. IEEE Solid-State Circuits Letters, 2018, 1(2): 50-53.

[10] Park C J, Onabajo M, Silva-Martinez J. External capacitor-less low dropout regulator with 25 dB superior power supply rejection in the 0.4‑4 MHz range[J]. IEEE Journal of Solid-State Circuits, 2013, 49(2): 486-501.

[11] Nagateja T, Kumari N, Chen K H, et al. A 8-ns Settling Time Fully Integrated LDO with Dynamic Biasing and Bulk Modulation Techniques in 40nm CMOS[C]. 2020 IEEE International Symposium on Circuits and Systems

[12] Chen C M, Tsai T W, Hung C C. Fast transient low-dropout voltage regulator with hybrid dynamic biasing technique for SoC application[J]. IEEE Transactions on Very Large Scale Integration (VLSI) Systems, 2012, 21(9): 1742-1747.

979-8-3315-8850-2/25 $31.00 © 2025 IEEE

2025 The 10th International Conference on Integrated Circuits and Microsystems

Low-Power MCU Architecture Optimization and Energy Efficiency Enhancement in Intelligent Pressure Sensor SoCs

Yaoming Lv[1,2]
[1] Shanghai Research Institute of Microelectronics (SHRIME)
Peking University
Shanghai, China
[2]School of Software & Microelectronics
Peking University
Beijing, China
lym313@stu.pku.edu.cn

Hong Yang[1,2]
[1] Shanghai Research Institute of Microelectronics (SHRIME)
Peking University
Shanghai, China
[2]School of Software & Microelectronics
Peking University
Beijing, China
2101120019@stu.pku.edu.cn

Feng Zou
Shanghai Research Institute of Microelectronics (SHRIME)
Peking University
Shanghai, China
zephan@pku.edu.cn

Yuhua Cheng[1,2,3]
[1]Shanghai Research Institute of Microelectronics (SHRIME)
Peking University
Shanghai, China
[2]School of Software & Microelectronics
Peking University
Beijing, China
[3]School of Integrated Circuits
Peking University
Beijing, China
chengyh@shrime-pku.org.cn

Abstract—In intelligent pressure sensor System-on-Chip (SoC) applications, the micro-controller unit (MCU) serves as a critical control and computational core, with its power consumption directly impacting system energy efficiency and battery life. These MCUs typically operate in a periodic wake-up and short-term operation mode within such systems. Reducing MCU active time by maximizing operating frequency is essential for power savings. This study proposes a dynamic adaptive calibration technique for maximum frequency, which adjusts system frequency in real time by monitoring MCU instruction fetch correctness. Unlike conventional voltage-tracking-based maximum frequency calibration, this approach eliminates voltage detection circuits and voltage-frequency mapping tables while mitigating aging effects. In low-power designs, clock control strategy plays an important role in optimizing power consumption. This paper proposes an instruction-decoding-driven fine-grained dynamic clock control technique. Compared to conventional approaches, this method enables instruction-level clock gating to achieve micro-operation-level power optimization. Experimental validation in intelligent sensor systems demonstrates that the proposed low-power technique enhances system energy efficiency by 16% under typical operating conditions.

Keywords—Clock Gating, Instruction-Level, Dynamic Clock Scaling, Low-Power Design

I. INTRODUCTION

With the growing adoption of smart pressure sensors in industries and wearable devices[1], the demand for low power consumption and high efficiency continues to rise. As an important control and computing unit of intelligent sensor SoC, the power consumption of MCU not only determines the battery life of the system but also affects the stability and reliability of the system. Therefore, reducing MCU power consumption through design optimization has become a key research focus in modern embedded systems. A typical MCU operation cycle in smart sensor systems proceeds as follows: A timer interrupt periodically wakes up the MCU. The MCU then provides excitation voltage or current to the sensor module for data acquisition and processing. After data processing and transmission, the MCU enters deep sleep mode.

As shown in formula (1), the total energy consumption of the intelligent sensor SoC is determined by its periodic operation. The system wakes up every interval T, executes its tasks, and then promptly enters sleep mode.

$$E = P_{work} \times T_{work} + P_{sleep} \times (T - T_{work}) \quad (1)$$

P_{work} and P_{sleep} denote the power consumption during active and sleep modes, respectively. P_{work} includes the operating power of digital circuits and analog circuits[2], T_{work} denotes the duration of active operation, and T_{sleep} denotes the duration of the sleep period. E represents the total energy consumption. The definition of P_{work} and P_{sleep} are provided in the formula (2)-(7).

$$P_{work} = P_{dig} + P_{ana} \quad (2)$$

$$P_{dig} = P_{switch} + P_{short} \quad (3)$$

$$P_{switch} = 0.5C \times VDD^2 \times f \times N \quad (4)$$

$$P_{short} = Q_{short} \times VDD \times f \times N \quad (5)$$

$$P_{ana} = I_{ana} \times VDD \quad (6)$$

$$P_{sleep} = I_{leak} \times VDD + I_{timer} \times VDD \quad (7)$$

979-8-3315-8850-2/25 $31.00 © 2025 IEEE

I_{ana} denotes the operating current of the analog module, I_{leak} denotes the leakage current, I_{timer} denotes the operating current of the timer, which periodically triggers an interrupt to wake up the MCU.

If the MCU executes a fixed workload, its operating frequency and execution time are inversely proportional. This relationship is expressed in formula (8).

$$f_1 \times T_{work1} = f_2 \times T_{work2} \qquad (8)$$

Let f_1 and f_2 be two operating frequencies, where f_1 is greater than f_2. T_{work1} and T_{work2} are the corresponding execution times, where T_{work2} is greater than T_{work1}.

According to formulas(1)-(8), the difference in energy consumption when operating at two frequencies, f_1 and f_2, is given by formula (9):

$$E_1 - E_2 = I_{delta} \times VDD \times (T_{work1} - T_{work2}) \qquad (9)$$

Where $I_{delta} = (I_{ana} - I_{leak} - I_{timer})$.

Because in general, the analog module's operating current is larger than the combined leakage and timer currents, the energy consumption corresponding to the working frequency f_1 is less than that of the working frequency f_2. A shorter working time leads to a longer sleep time, resulting in greater energy efficiency. After the MCU wakes up, it should operate at the maximum safe operating frequency.

The embedded memory used in the SoC design of low-cost and low-power intelligent pressure sensors often has low speed[3], which limits overall system performance. Third-party embedded memory IPs typically include large frequency margins in their specifications to accommodate worst-case temperature, voltage, and process variations, ensuring reliability. This approach sacrifices performance and energy efficiency. Operating embedded memory at its maximum safe operating frequency based on current environmental conditions is an effective strategy to reduce power consumption.

Refer to Fig.1, the method of automatically adjusting the operating frequency of the MCU to the highest allowed frequency under the working voltage is as follows[4]: the MCU's data manual provides a table indicating the maximum system clock frequency that can be supported under different voltage ranges, look up the table based on the current power supply voltage, and then configure the operating clock to set the corresponding frequency. This requires the system to have a real-time voltage detection circuit and a safety frequency voltage mapping table, which increases the complexity of the circuit. The safety frequency voltage mapping requires theoretical modeling, experimental testing, dynamic calibration, and other methods to obtain, and the influence of aging effects must be considered, making it difficult to implement. This work achieves maximum frequency adjustment by correctly judging instruction reads from program memory, eliminating the need

for voltage detection circuits and frequency-voltage tables, and avoiding aging-related issues.

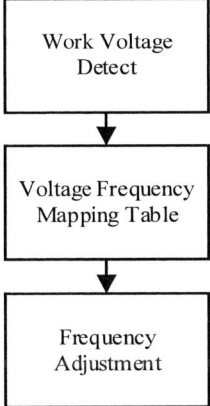

Fig. 1. The workflow of frequency adjustment [4]

In the design of low-power intelligent sensor systems, clock control strategies play a decisive role in optimizing power consumption[5]. The clock tree ensures timing security while also generating significant power consumption. Using a gated clock can effectively reduce the power consumption generated by the clock. As shown in Fig.2, the register level gate control unit itself also generates power consumption due to the continuous toggling of the clock CLK. There is also module-level gate control, The commonly used method currently is to separate the MCU clock from the peripheral clock, design the peripheral enable signal, turn on the peripheral clock when it is working, and turn it off at other times[6]. MCU turns on and off the clock through different working modes[7]. While these methods are straightforward to implement, their coarse response granularity results in some energy inefficiency. The fine-grained dynamic clock control mechanism based on the instruction decoding driver proposed in this article utilizes the control signals generated during the instruction decoding stage to dynamically determine whether the micro-operation modules inside the MCU need clock signals, achieving more fine-grained and dynamically responsive power management.

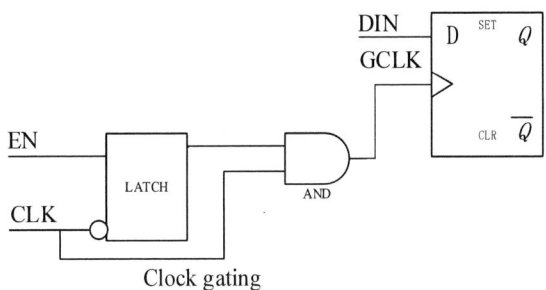

Fig. 2. Clock gating cell [5]

The remainder of this paper is organized as follows: Section II describes the dynamic adaptive calibration technique for determining the maximum operating frequency, and also presents the instruction-decoding-driven fine-grained dynamic clock control strategy. Section III provides simulation and measurement results. Finally, Section IV concludes the paper.

II. PROPOSED DESIGN

This chapter is organized into three parts: the maximum frequency dynamic adaptive calibration technology, the fine-grained dynamic clock control technology driven by instruction decoding, and the integrated design of the pressure sensor SoC.

A. The Maximum Frequency Dynamic Adaptive Calibration Technology

Frequency adjustment is based on evaluating the correctness of the fetched instruction. As shown in Fig.3, the workflow of MCU's maximum safe operating frequency consists of two stages: determining the maximum safe operating frequency and maintaining the maximum safe operating frequency during normal operation. The core circuit consists of a frequency-adjustable oscillator which adjusts the current through six clock adjustment control bits to achieve frequency changes, a frequency division circuit, a clock-switching circuit, and an instruction reading and judgment circuit. Except for the instruction reading and judgment circuit is an added part, the rest are existing circuits of the MCU. The instruction judgment circuit is a digital comparator that is related to the length of the instructions, with a small increase in area. As shown in Fig.4, the instruction reading and judgment circuit uses the high-frequency clock HRC to sample the instruction IRH obtained from the program memory data IRDATA and compares it with the correct instruction IR obtained from the program memory data IRDATA sampled by the MCU operating clock CLK_MCU. If the comparison is correct, the frequency of the high-frequency clock HRC is incrementally increased. This process continues until a mismatch occurs, at which point the last correct frequency is recorded as the maximum safe operating frequency of the MCU.

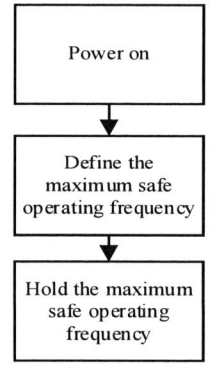

Fig. 3. the workflow of maximum safe operating frequency

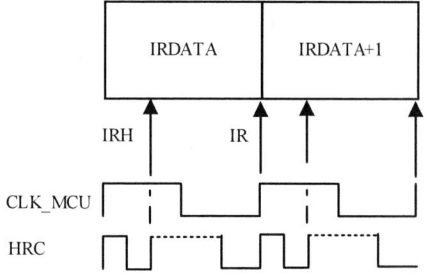

Fig. 4. The instruction reading and judgment structure

Fig.5 shows the process of determining the maximum operating frequency. After power is on, the MCU selects the frequency division clock HRC_DIV of the HRC to operate. The clock HRC_DIV meets the requirement that the MCU can also operate safely when the HRC is adjusted to the maximum operating frequency. The instruction reading judgment circuit is used to find the maximum operating frequency, and then the MCU is set to operate at HRC through the clock selection signal.

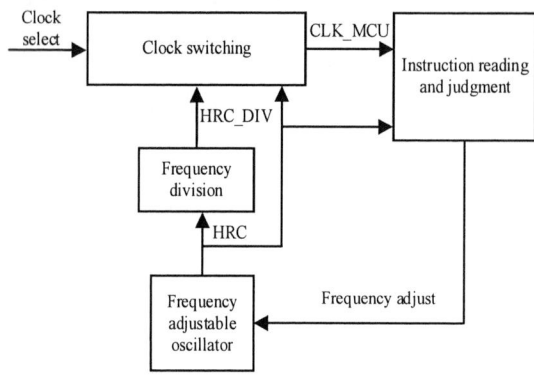

Fig. 5. determining the maximum safe operating frequency

Fig.6 shows the process of maintaining the maximum safe operating frequency. During the operation of the MCU, if the instruction read judgment circuit detects an error, the MCU halts and lowers its operating frequency. Once no error is detected, the MCU resumes normal operation.

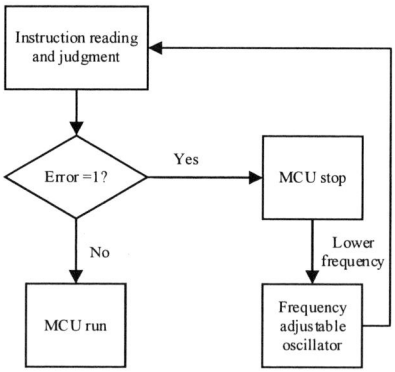

Fig. 6. maintaining the maximum safe operating frequency

B. The Fine-grained Dynamic Clock Control Technology Driven by Instruction Decoding

As shown in Fig.7, during the CPU instruction decoding stage, each instruction generates a set of control signals. These signals determine whether to access static random access memory (SRAM), special function register (SFR), Arithmetic logic unit (ALU), stack, etc. ALU can be divided into modules for addition and subtraction operations, multiplication operations, shift operations, etc. By using the control signals generated by these instructions as enable signals for clock gating, the relevant micro-operation modules receive clock signals only when needed. At the same time, clock control is also applied to the data bus which is used to connect these modules, reducing the load on the bus and lowering power consumption. This method requires joint optimization of the instruction decoder and clock control logic in the early stage of MCU architecture design. Although it introduces a slight increase in logic area due to additional switching control for micro-operation modules, it results in a significant reduction in power consumption.

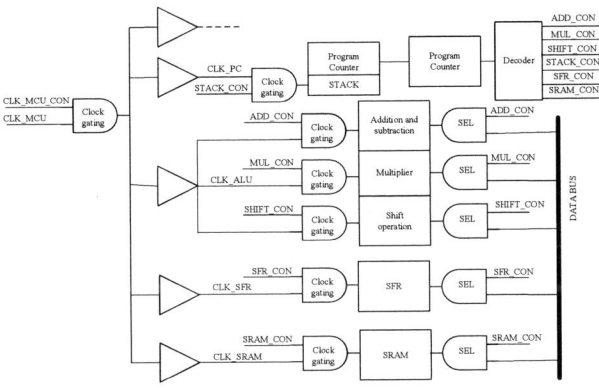

Fig. 7.   The Structural diagram of fine-grained dynamic clock control

Fig.8 is the traditional MCU clock control method, the clock CLK_MCU is governed by a single global enable signal CLK_MCU_CON, and the MCU clock is disabled when it is in sleep or standby mode. All micro-operation modules share the same clock control signal, meaning that the clock continues to drive all modules regardless of the specific instruction being executed. This results in higher power consumption.

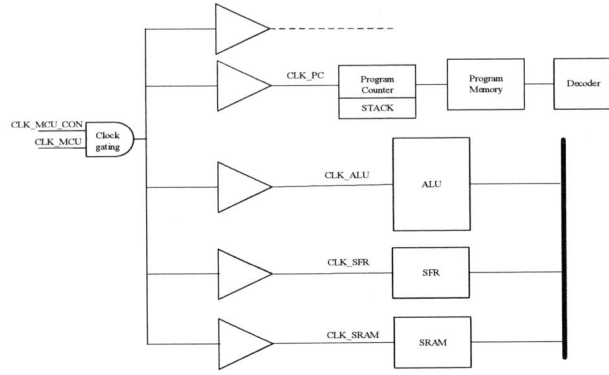

Fig. 8.   Traditional MCU clock control[8]

C. SoC Design

Fig.9 is the structure diagram of a low-power and low-cost intelligent pressure sensor. The SoC provides a 2.4V excitation voltage to the pressure sensor. The differential voltage signal of the pressure sensor is amplified by a programmable gain amplifier (PGA) and then fed into Sigma-Delta ADC[9]. The PGA has selectable gains up to 128x, and supports input voltage as small as 40 nV. There are two RC oscillators: one high-frequency oscillator that provides the operating clock for the MCU, and a low-frequency oscillator that serves as a timing source for wake-up events in sleep mode. The high-frequency oscillator has six clock adjustment control bits to adjust the frequency resolution by 1%. The 8-bit RISC MCU is designed with a single instruction cycle and includes 4k words of program memory and 256 bytes of SRAM. For data communication, a

UART module is used, supporting baud rates from 9600 to 115200 bps. Fig.10 shows the layout of the SoC.

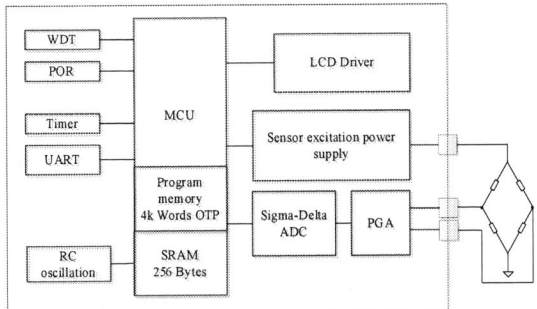

Fig. 9.   Block diagram of pressure sensor SoC

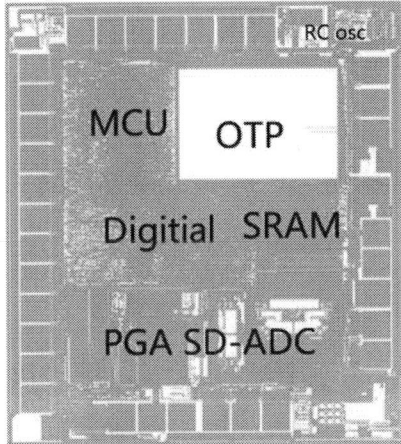

Fig. 10. Layout of pressure sensor SoC

III.   RESULT AND ANALYSIS

Tables I and II compare the system with and without fine-grained dynamic clock control technology based on instruction decoding. Implementing this technique increases the chip area by 1% while reducing power consumption by 16%.

TABLE I.   MCU POWER CONSUMPTION COMPARISON(uW)

Design method	Without fine-grained dynamic clock	With fine-grained dynamic clock
Switch power	133	131
Short power	238	179
Leak power	1.03	1.45
Total power	372	312

TABLE II.   COMPARISON OF EQUIVALENT NAND GATE QUANTITY

Design method	Without fine-grained dynamic clock	With fine-grained dynamic clock
Gate-count	11689	11817

Fig.11 illustrates a typical measurement workflow for a pressure sensor. In normal operation, the SoC remains in sleep mode and wakes up once every second. Upon waking, the signal from the pressure sensor is sampled multiple times, filtered, processed, and then transmitted via the UART communication port before the system returns to sleep mode. The average current is measured based on this periodic workflow. All current measurements are conducted under standard test conditions of 3 V supply voltage and an ambient temperature of 25 °C.

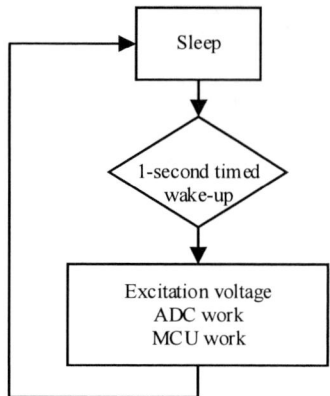

Fig. 11. The workflow of a pressure sensor

The maximum read speed of the third-party program memory is limited to 6 MHz, resulting in an MCU execution time of 96 ms. By applying the proposed dynamic adaptive maximum frequency calibration technique, the execution time is reduced to 83 ms. The pressure sensor adopts a Wheatstone bridge configuration composed of four resistive elements, with an equivalent impedance of 2 kΩ. It is powered by a 2.4 V reference source and consumes 1.2 mA during operation. In sleep mode, the MCU current drops to 1.5 µA, with the low-power timer remaining active. Fig.12 presents the average operating current of the MCU. With the dynamic adaptive calibration enabled, the average current is reduced by 6.7%. Greater energy savings are expected as the computational complexity of MCU tasks increases.

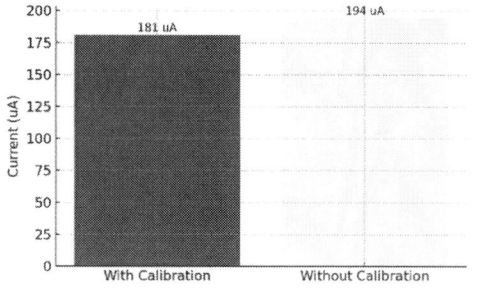

Fig. 12. Current Consumption with/without Max Frequency Calibration

IV. CONCLUSION

This paper presents two low-power design methodologies for MCUs that are particularly effective in sensor SoC scenarios involving periodic activation, rapid task execution, and frequent entry into low-power modes. The first method, dynamic adaptive calibration of the maximum operating frequency, enhances execution speed. The second method, fine-grained dynamic clock gating based on instruction decoding, minimizes clock toggling in micro-operation units and reduces data bus transitions, contributing to overall power savings. Although these approaches slightly increase the silicon area, experimental results show a substantial reduction in power consumption, highlighting their practical value in low-power intelligent sensor SoC applications.

REFERENCES

[1] Chan M, Estève D, Fourniols J Y, et al. Smart wearable systems: Current status and future challenges[J]. Artificial intelligence in medicine, 2012, 56(3): 137-156.

[2] Sun Y A, Liu H, Yang Y, et al. Design and Implementation of Ultra-Low Power Consumption for High-Performance SoC[C]//2022 7th International Conference on Integrated Circuits and Microsystems (ICICM). IEEE, 2022: 411-416.

[3] Hidaka H. Evolution of embedded flash memory technology for MCU[C]//2011 IEEE International Conference on IC Design & Technology. IEEE, 2011: 1-4.

[4] Burd T D, Pering T A, Stratakos A J, et al. A dynamic voltage scaled microprocessor system[J]. IEEE Journal of solid-state circuits, 2000, 35(11): 1571-1580.

[5] Shanmugasundaram N. Clock gating techniques: an overview[C]//2018 conference on emerging devices and smart systems (ICEDSS). IEEE, 2018: 217-221.

[6] Oh J, Pedram M. Gated clock routing for low-power microprocessor design[J]. IEEE Transactions on Computer-Aided Design of Integrated Circuits and Systems, 2001, 20(6): 715-722.

[7] Mai S, Li C, Zhao Y, et al. A high-performance low-power SoC for mobile one-time password applications[C]//2013 IEEE International Symposium on Circuits and Systems (ISCAS). IEEE, 2013: 1436-1439.

[8] Struharik R, Mezei I. 8051 IP Core for FPGA Applications[J].

[9] L. Zhao, C. Deng, H. Chen, G. Wang and Y. Cheng, "A 1-V 23-µW 88-dB DR Sigma-Delta ADC for high-accuracy and low-power applications," 2015 IEEE 11th International Conference on ASIC (ASICON), Chengdu, China, 2015, pp. 1-4.

2025 The 10th International Conference on Integrated Circuits and Microsystems

A Low-Phase-Noise and Low-Power Class-C VCO with Robust Start-up for FMCW Radars

Zhigang Li[1], Zhihao Liu[1], Hao Lei[1], Hang Yang[1], Qiang Zhao[1], Shan Gao[1], Xiulong Wu[1*]

[1]*School of Integrated Circuits and Anhui Provincial High-Performance Integrated Circuit Engineering Research Center, Anhui University, Hefei 230601, China*

*Corresponding author: xiulong@ahu.edu.cn

Abstract—This paper presents a low-phase-noise Class-C voltage-controlled oscillator (VCO) in a 28-nm CMOS process. The design covers a 18.8-21.4 GHz frequency tuning range (FTR), which is suitable apply for frequency-modulated continuous-wave (FMCW) radar systems. A novel operational transconductance amplifier (OTA) is adopted in the bias loop to enhance the start-up robustness and improve the phase noise (PN) performance of the VCO. In order to minimize the chip area and maximize the frequency tuning range, a 4-bit NMOS-stacked capacitor array is deployed in the LC-tank. The proposed VCO occupies a area of less than 0.06 mm², while post-layout simulation results show that the phase noise of VCO is -106.8 dBc/Hz at 1 MHz offset from a 19.8 GHz carrier frequency. The VCO consumes only 2.7 mW from a 0.9 V supply voltage, and the FoM is -188.4 dBc/Hz.

Index Terms—VCO, FMCW, robustness, phase noise.

I. INTRODUCTION

With the ongoing advancement of the new energy vehicle sector, FMCW technology has found extensive application within intelligent vehicles, assuming a pivotal role in various scenarios including autonomous driving and automated parking [1]–[3]. A frequency synthesizer featuring a wide-tuning range and a low-phase-noise VCO is essential for realizing these functionalities [4]. As a key parameter in phase-locked loop (PLL) architectures, Phase noise is fundamentally governed by the VCO design [5], [6]. Consequently, the design of the VCO needs fully consider the impact of different structures on phase noise performance. Additionally, among other indicators, electronic products applied in automotive systems must take into account process-voltage-temperature (PVT) characteristics, system robustness, and power consumption. In LC VCOs, various structures have been employed to improve the phase noise performance of the VCO. Among them, the Class-F VCO is a commonly adoption due to its transformer topology, which offers considerable phase noise performance [7], [8]. When it comes to design flexibility, layout occupation, and power efficiency, the Class-F VCO shows disadvantages. Considering the difficulties mentioned above, the Class-C VCO achieves a broader adoption due to its combination of different aspects.

As illustrated in Fig. 1, the implementation of tail current biasing in Class-C VCOs primarily follows two methodologies [9]. In the Fig. 1(a), R_T as a resistive tail current source is employed, offering significant advantages in minimizing chip area. This architecture decouples the oscillation amplitude

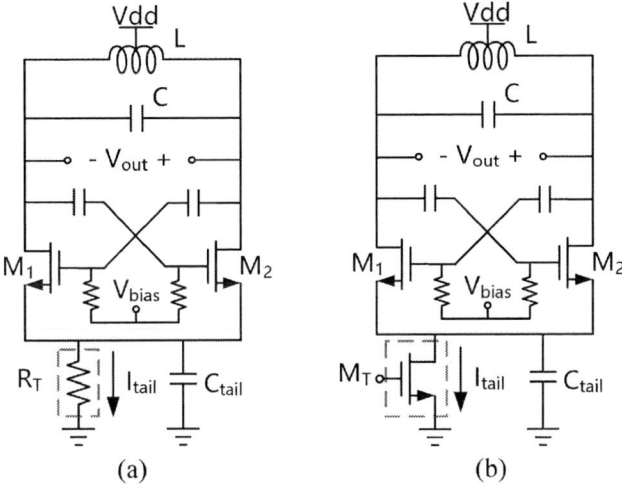

Fig. 1. The Class-C structure of VCO with resistor as tail current source (a) and NMOS as tail current source (b).

from the feedback loop bandwidth, thereby simplifying stability constraints. In contrast, the approach in Fig. 1(b) utilizes a MOS-based current source, M_T improves power supply rejection (PSR) and mitigates spurious frequency components to suppress supply-induced current noise at the expense of increased circuit area.

To achieve a VCO that meets the requirements of automotive FMCW radar applications, this paper proposes a Class-C VCO with a dynamic start-up loop. The amplifier in the feedback loop adopts a new structure with low power consumption, enhances the start-up stability of the system. Additionally, a NMOS-stacked capacitor array is employed in the LC-tank to tune the output frequency coarsely.

II. CIRCUIT IMPLEMENTATIONS

A. Core Design

Compared to conventional implementations, the Class-C VCOs are capable of perfomancing a 4.4 dB phase noise reduction at 1 MHz offset while maintaining identical power consumption [10]. The schematic of the proposed Class-C VCO is depicted in Fig. 2. Within the proposed structure, a cross-coupled NMOS-pair (M_1-M_2) is adopted to generate a negative resistance to satisfy the oscillation condition. M_1 and

979-8-3315-8850-2/25 $31.00 © 2025 IEEE

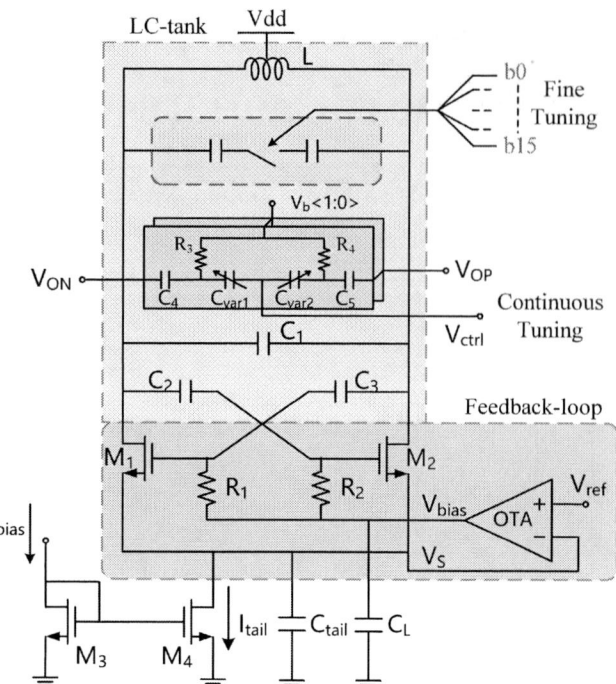

Fig. 2. The architecture of proposed Class-C VCO with dynamic bias loop.

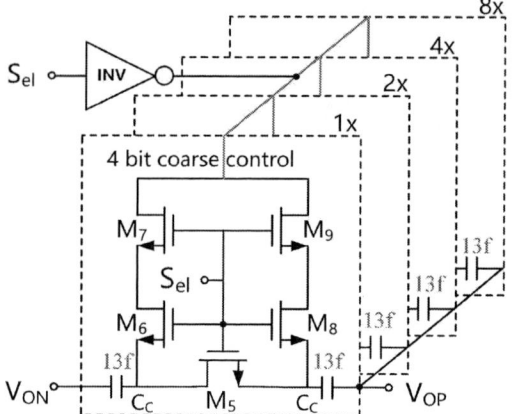

Fig. 3. 4-bit NMOS-stacked capacitor array.

M_2 are thin-gate NMOS devices and conduct a better current driven capability than PMOS. This make a great contribution in power consumption performance.

DC power is implemented through the center-tapped inductor L, which is 118 pH, with Q of 24 at 19.8 GHz. The capacitive part of the tank consists of C_1, which is a MOM capacitor of 250 fF, along with a 4-bit binary-weighted NMOS-stacked capacitor array for coarse tuning and a two-stage PMOS varactor bank for fine-tuning. The RC network between the oscillator output nodes and transistor gates, establishes V_{bias} by high-impedance resistors ($R_1 = R_2$ = 1 MΩ), while MOM capacitor ($C_2 = C_3 = 2$ pF) make the loop AC feedback only. To ensure the smooth start-up of the oscillator, V_{bias} is initially set close to V_{DD}. This allows M_1 and M_2 to work in the saturation region and ensures that I_{tail} generated by M_4 successfully passed through the core circuit.

In order to minimize the AM-PM effect, C_{tail} connected to V_S is set to 3.5 pF. C_{tail} ensures the VCO in completely Class-C operation by transforming the drain current into sharply peaked pulses while suppressing the high-frequency noise of M_4. The proposed VCO is biased through M_3 and M_4, where the reference current I_{bias} = 600 μA is scaled by a 4 : 1 transistor aspect ratio between M_4 and M_3. Both devices employ extended channel lengths (L = 1 μm) to mitigate the flicker noise from current mirror.

B. Design of the Capacitor Array

To achieve a wide FTR without degrading the phase noise, a hybrid tuning methodology integrating a switched capacitor array with a dual varactor diode bank is implemented in the oscillator design.

The 4-bit binary-weighted switched capacitor array employs an NMOS-stacked configuration, which is illustrated in Fig. 3. In this architecture, the channel length of M_5 is set to 30 nm to minimize its on-resistance during activation, whereas the stacked transistors (M_6-M_9) utilize extended channel lengths to enhance output impedance in the off-state, thereby avoiding unintended tank loading effects. The switching operation is governed by the S_{el}:

When $S_{el} = 0$, the capacitor array is disabled. The stacked transistors (M_6-M_9) bias the drain and source terminals of M_5 to a high voltage level, establishing reverse-biased PN-junctions that suppress leakage currents through M_5.

When $S_{el} = 1$, M_5 enters the triode region, effectively connecting half the capacitance value of $C_C = 13$ fF to the LC-tank. Compared to conventional resistor-switched capacitor arrays, this topology achieves a significant area reduction.

The continuous tuning of the VCO output frequency is achieved by PMOS varactor C_{var1} and C_{var2}, as shown in Fig. 2. The fixed capacitors C_4 and C_5 in series with high-impedance resistors R_3 and R_4 establish a DC biasing network that enhance the control accuracy. The RC network also work as a low-pass filter, effectively enhances the PSR and thus suppresses AM-FM conversion effects. The differential varactor configuration employs two identical tuning modules to exploit the maximum linear capacitance range of the PMOS varactors.

C. Design of the Dynamic Bias Loop

In Fig. 2, a dynamic bias feedback loop is designed to guarantee the robust start-up for VCO. This architecture regulates both source voltage V_S and drain voltage V_{bias} of the coupling pair through a two-stage operational transconductance amplifier, whose schematic is detailed in Fig. 4.

The DC gain coming from the OTA, makes V_S eventually be pulled up to V_{ref}, while V_{bias} be pulled down to the V_{th} of M_1 and M_2 during the start-up period. V_{ref} of 300 mV leaves an enough headroom for M_4 to operate in the saturation

979-8-3315-8850-2/25 $31.00 © 2025 IEEE

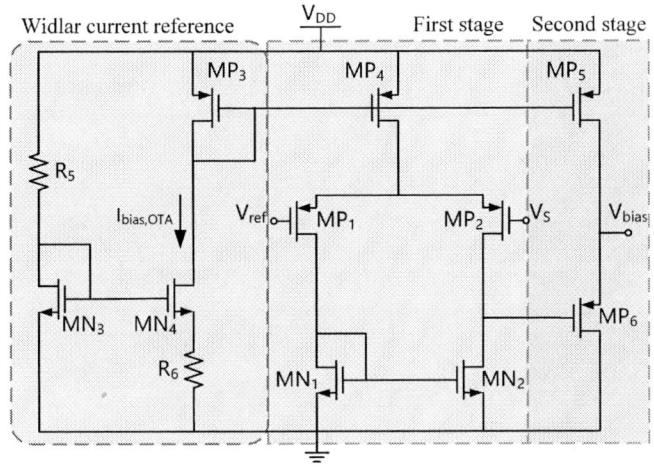

Fig. 4. Schematic of proposed OTA in bias loop.

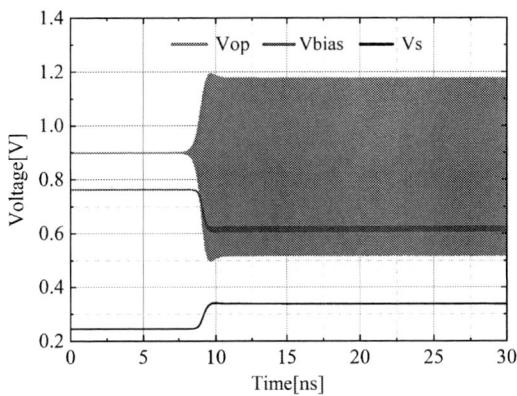

Fig. 6. The transient voltage waveforms of proposed VCO at start-up period.

region. The first stage of the OTA adopts a PMOS-pair for input due to a smaller threshold voltage, which can ensure MN_1 and MN_2 operate in the saturation region under a 0.9 V supply voltage. Besides, PMOS devices demonstrate lower flicker noise compared to NMOS implementations. The second stage of the OTA consists MP_5 and MP_6, which works as a common-drain buffer, offers added gain for OTA. MN_3-MN_4 and R_5-R_6 (in Fig. 4) established a Widlar current reference to bias the OTA with a low current. All MOSFETs are thin-oxide devices.

In Class-C oscillators, the noise contributed by OTA in the feedback loop can severely deteriorate the oscillator phase noise if not treated properly. To overcome this problem, the OperformanceTA limits the bandwith of itself at around 80 kHz by using a low bias current ($I_{\text{bias, OTA}}$ = 50 nA) and a load capacitance C_L (Fig. 2). Additionally, all OTA transistors have a (L = 100 nm) channel length to further reduce their flicker noise contribution.

III. EXPERIMENTAL RESULTS

The proposed VCO is designed in a 28-nm CMOS technology. The layout showing in Fig. 5 indicates that the scale is 312 μm × 177 μm, which is as small as 0.06 mm². The VCO

operates from a supply voltage of 0.9 V, while consuming 2.7 mW of power.

The start-up process of proposed VCO is shown in Fig. 6, which includes V_{OP}, V_{bias} and V_S in a 30 ns time period. Initially, V_{bias} = 760 mV and V_S = 240 mV creates a voltage difference to facilitate the oscillate of the VCO. By the time of 7 ns, due to the significant gain of the OTA in the feedback loop, V_S is elevated to around 300 mV, while V_{bias} is pulled down close to the V_{th} through negative feedback. After the oscillation stabilized, the VCO successfully entered Class-C mode.

Fig. 7 plots the phase noise performance of the VCO with different I_{bias} at a carrier frequency of 19.8 GHz. The results show that the VCO achieves a phase noise of -104.7, -105.9 and -106.8 dBc/Hz, when the I_{bias} is 400, 500 and 600 μA, respectively.

Fig. 8 depicts the output frequency variation range of the proposed VCO at its lowest oscillation frequency (state: 4'b 1111, V_{ctrl} = 0 V) and highest oscillation frequency (state: 4'b 0000, V_{ctrl} = 0.9 V). A total of 16 curves cover the frequency range from 18.8 GHz to 21.4 GHz, with a substantial overlap between adjacent operating bands, corresponding to a FTR of 12.9 %.

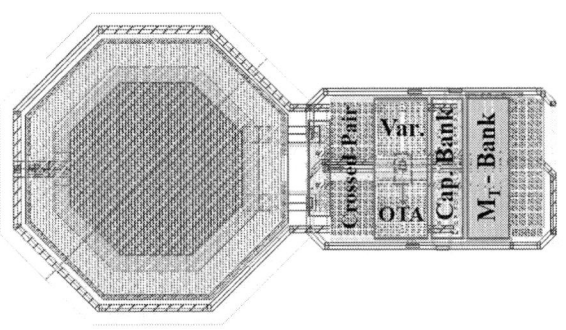

Fig. 5. Layout of proposed Class-C VCO.

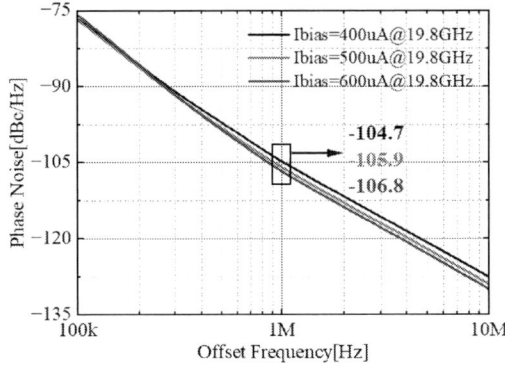

Fig. 7. The simulated PN at 19.8 GHz with I_{bias} = 400, 500 and 600 μA.

979-8-3315-8850-2/25 $31.00 © 2025 IEEE

TABLE I
PERFORMANCE SUMMARY AND COMPARISON

References	Process (nm)	V_{DD} (V)	Frequency (GHz)	FTR (%)	Area (mm²)	Power (mW)	PN@1MHz (dBc/Hz)	FoM (dBc/Hz)
ISCAS 2022 [11]	40	2	23.38	16.5	0.16	12.7	-106.6	-183.1
TCAS-II 2022 [12]	65	1	19.7	13.1	0.06	3.8	-106.3	-186.4
MWTL 2023 [13]	65	0.6	28.11	11.92	0.11	7.2	-107.2	-187.6
IWS 2024 [14]	55	0.6	29.4	21.9	0.05	2.8	-101.7	-186.6
RFIT 2024 [15]	180	1.8	24	8.3	0.26	2	-106.2	-180
SiRF 2024 [16]	90	0.6	22.6	27.5	0.06	3	-103.5	-184.4
This work	**28**	**0.9**	**19.8**	**12.9**	**0.06**	**2.7**	**-106.8**	**-188.4**

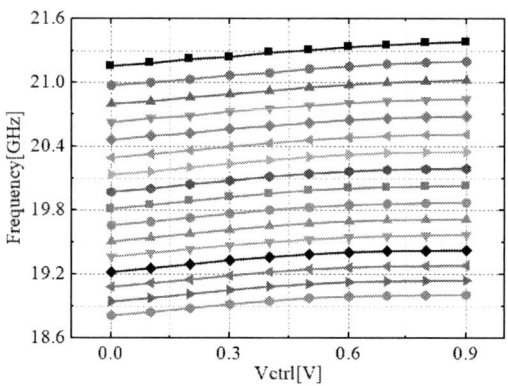

Fig. 8. The simulated tuning range of proposed VCO.

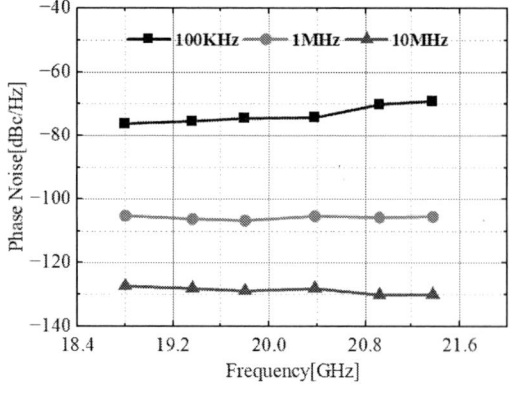

Fig. 9. The PN at three frequency offsets across the tuning range.

The phase noise at 100 KHz, 1 MHz and 10 MHz offset frequency across the tuning range is reported in Fig. 9, which shows marginally variation across the frequency operating range of proposed VCO. The VCO performances a best phase noise of -106.8 dBc/Hz and a worst phase noise of -105.2 dBc/Hz at a 1 MHz offset frequency.

Table I provides a performance summary of the proposed VCO and compares it with other state-of-the-art VCOs operating in K-band. In contrast to the design reported in [13], this work achieves a lower power consumption and a wider FTR. When benchmarked against [15], the proposed VCO performs a lower phase noise as well as a smaller area. Overall, the VCO presenting in this brief demonstrates excellent performances in terms of phase noise and power consumption.

IV. CONCLUSION

In this paper, a low-phase-noise and low-power Class-C VCO with robust start-up is presented. A novel two-stage OTA with low bias current is proposed, which enhances the robustness of the start-up process and achieves a considerable phase noise performance. In addition, the NMOS-stacked capacitor array applying in the LC-tank, significantly reduces the layout area. Prototypes of the proposed VCO, with a 0.06 mm² layout occupation, shows a phase noise of -106.8 dBc/Hz at 1 MHz offset frequency from a 19.8 GHz carrier frequency and a 2.7 mW of power consumption. This makes it effectively applicable in FMCW systems.

ACKNOWLEDGMENT

This work was supported by the National Natural Science Foundation of China under Grant 62274001 and the University Synergy Innovation Program of Anhui Province under Grant GXXT-2023-012.

REFERENCES

[1] K. -I. Oh, G. -H. Ko, G. Sub Kim, J. -G. Kim and D. Baek, "A 54–64-GHz 4TXs-4RXs CMOS Transceiver With 10-GHz Bandwidth Single Chirp for FMCW Radar Applications," in *IEEE Transactions on Microwave Theory and Techniques*, vol. 73, no. 3, pp. 1532-1544, March 2025.

[2] M. Kalantari, W. Li, H. Shirinabadi, A. Fotowat-Ahmady and C. P. Yue, "A W-Band Single-Antenna FMCW Radar Transceiver With Adaptive Leakage Cancellation," in *IEEE Journal of Solid-State Circuits*, vol. 56, no. 6, pp. 1655-1667, June 2021.

[3] Zhigang Li, David Cordeau, Jean-Marie Paillot, Sébastien Charpentier, Matthieu Lécuyer, Francis Huin, "Analysis and design of K-band low-phase-noise differential DCOs implemented in 22 nm FD-SOI for 76–81 GHz automotive radars," *Microelectronics Journal*, Volume 136, 2023.

[4] J. Gong, E. Charbon, F. Sebastiano, and M. Babaie, "A cryo-CMOS PLL for quantum computing applications," in *IEEE Journal of Solid-State Circuits*, vol. 58, no. 5, pp. 1362–1375, May 2023.

[5] A. Franceschin, P. Andreani, F. Padovan, M. Bassi, and A. Bevilacqua, "A 19.5-GHz 28-nm class-C CMOS VCO, with a reasonably rigorousresult on 1/f noise upconversion caused by short-channel effects," in *IEEE Journal of Solid-State Circuits*, vol. 55, no. 7, pp. 1842–1853, 2020.

[6] Y. Huo and F. F. Dai, "A 2.4-GHz Multiphase Inductorless PLL With Coupled-Ring Oscillators and Time-Amplifying Phase-Frequency Detector for Low Phase Noise and Robust Locking Performances," in *IEEE Microwave and Wireless Technology Letters*, vol. 34, no. 11, pp. 1275-1277, Nov. 2024.

[7] D. -X. Ni, L. Zhou, Z. Ren and J. Pang, "A W-Band Low-Phase-Noise Class-F23 VCO With Common-Mode Expansion Based on a Coupling-Coexisting Transformer," in *IEEE Microwave and Wireless Technology Letters*, vol. 35, no. 6, pp. 706-709, June 2025.

[8] W. Zheng, Y. Wang, Z. Zhang and X. Xia, "A Compact Class-F Broadband-Harmonic-Shaping VCO for Mm-Wave Applications," *2024 IEEE International Symposium on Radio-Frequency Integration Technology (RFIT)*, Chengdu, China, 2024, pp. 1-3.

[9] L. Fanori and P. Andreani, "Highly Efficient Class-C CMOS VCOs, Including a Comparison With Class-B VCOs," in *IEEE Journal of Solid-State Circuits*, vol. 48, no. 7, pp. 1730-1740, July 2013.

[10] Dayanik M B, Flynn M P. "Digital fractional-N PLLs based on a continuous-time third-order noise shaping time-to-digital converter for a 240-GHz FMCW radar system," in *IEEE Journal of Solid State Circuits*, 2018, 53(6): 1719-1730.

[11] Y. Lu, C. Shi, J. Li, R. Zhang, H. Deng and J. Chen, "A 23.4-27.6 GHz "Zig-Zag" VCO with Continuous Frequency Switching for FMCW Radars," *2022 IEEE International Symposium on Circuits and Systems (ISCAS)*, Austin, TX, USA, 2022, pp. 3355-3358.

[12] Y. -K. Cho, J. -W. Nam and S. -W. Lee, "A Low-Power Class-C Voltage-Controlled Oscillator With Robust Start-Up and Compact High-Q Capacitor Array," in *IEEE Transactions on Circuits and Systems II: Express Briefs*, vol. 69, no. 3, pp. 819-823, March 2022.

[13] A. Hossain, S. B. Lee and C. W. Byeon, "30-GHz Low-Phase-Noise VCO With Negative Transconductance Optimization in 65-nm CMOS," in *IEEE Microwave and Wireless Technology Letters*, vol. 33, no. 1, pp. 59-62, Jan. 2023.

[14] C. Liang et al., "A Ka-Band Single-Gm Dual-Core Transformer-Based Cryogenic VCO for Quantum Computing Application," *2024 IEEE MTT-S International Wireless Symposium (IWS)*, Beijing, China, 2024, pp. 1-3.

[15] P. Xiao and X. Lu, "A 24 GHz VCO in a 180 nm SiGe BiCMOS Process," *2024 IEEE International Symposium on Radio-Frequency Integration Technology (RFIT)*, Chengdu, China, 2024, pp. 1-3.

[16] P. -Y. Chen, J. -L. Chen and H. -Y. Chang, "A 23-30 GHz Low-Phase-Noise 5-Bit Voltage-Controlled Oscillator in 90-nm CMOS Process," *2024 IEEE 24th Topical Meeting on Silicon Monolithic Integrated Circuits in RF Systems (SiRF)*, San Antonio, TX, USA, 2024, pp. 79-82.

2025 The 10th International Conference on Integrated Circuits and Microsystems

A 0.47 nJ/Conversion CMOS Temperature Sensor with 0.00786 mm^2 Core in 65nm CMOS

Hangfei Song[1], Yichi Zhang[1,2], Aili Wang[1,2*]

[1] Zhejiang University-University of Illinois Urbana-Champaign Institute, Zhejiang University, Haining, China
[2] College of Information Science and Electronic Engineering, Zhejiang University, Hangzhou, China
hangfei.23@intl.zju.edu.cn, yichi1.23@intl.zju.edu.cn, ailiwang@intl.zju.edu.cn

Abstract—This paper presents a CMOS transistor-based temperature sensor for heat management. The temperature sensor generates proportional to absolute temperature (PTAT) and complementary to absolute temperature (CTAT) voltages through a 4-transistor front-end sensing structure leveraging both native NMOS and standard NMOS. Using this structure brings three benefits: 1) the lower supply sensitivity; 2) the better linearity of PTAT and CTAT voltages with the same area overhead; 3) the lower power consumption. A novel 8-bit successive approximation register (SAR) quantization scheme adjusts the capacitor ratio to drive the CTAT voltage to the PTAT voltage, achieving a linear output code. The sensor has a core area of 0.00786 mm^2 in standard 65nm CMOS technology. After a 2-point calibration, the sensor achieves -0.69/+0.77 °C inaccuracy over the temperature range of 0 °C to 100 °C. The supply sensitivity is 1.25 °C/V when measured across the supply range of 0.9 V to 1.3 V. At a conversion time of 182 μs, a resolution of 0.51 °C is achieved. Operating from a supply of 1.2 V, it consumes 2.58 μW, which yields a 0.123 nJ $\cdot$ K^2 resolution figure-of-merit (FOM).

Index Terms—CMOS Temperature Sensor, Subthreshold Region, SAR ADC, Low Supply Sensitivity

I. Introduction

As semiconductor technology continues to advance, systems on chips (SoCs) are becoming smaller and integrating more core modules, which leads to higher power density and more heat generation. To address this challenge and ensure stable performance, on-chip CMOS temperature sensors play an important role in heat management [1]–[3]. These sensors help prevent localized overheating, which could otherwise impair the chip's functionality or even cause permanent damage. In applications demanding a large number of CMOS temperature sensors, reducing the area of each sensor can significantly lower the design complexity and fabrication cost of the chip [4]. For internet of things (IoTs) applications, such as wireless sensor network nodes that are typically battery-powered due to environmental constraints, CMOS temperature sensors must satisfy the requirement of ultra-low power consumption [5]. With the adoption of dynamic voltage and frequency scaling (DVFS) technology, supply voltage changes apparently as the power management modules dynamically adjust the chip's operating voltage and frequency. Therefore, CMOS temperature sensors must exhibit low supply sensitivity to maintain accurate temperature sensing under such conditions [6], [7]. In the industry, Intel has introduced a compact temperature sensor with a competitive FOM and relatively fast conversion

time [8]. However, its performance in power consumption and accuracy still requires further improvement.

To address the demands for low power consumption, compact area, and low supply sensitivity, this work proposes a 4-transistor front-end sensing structure leveraging both native NMOS and standard NMOS devices operating in the subthreshold region, which generate PTAT and CTAT voltages with low supply sensitivity and high linearity, respectively. Additionally, a novel quantization scheme based on 8-bit SAR architecture is designed for quantization, which adjusts the capacitor ratio to progressively drive the CTAT voltage to the PTAT voltage, enabling linear digital code output. The design is implemented in standard 65nm CMOS technology with a core area of 0.00786 mm^2. Post-layout simulation results showed the supply sensitivity is 1.25 °C/V when measured across the supply range of 0.9 V to 1.3 V. After a 2-point calibration, the sensor achieves -0.69/+0.77 °C inaccuracy over the temperature range of 0 °C to 100 °C. Operating at a 1.2 V supply, the power consumption is 2.58 μW under TT corner. With 182 μs conversion time, which yields 0.123 nJ $\cdot$ K^2 FOM.

II. Proposed Temperature Sensor System Architecture

Fig. 1 shows the proposed architecture of the temperature sensor system. Six major modules are included in this structure: 1) a temperature sensing front-end core that leverages a 4-T structure using native NMOS and standard NMOS transistors operating in the subthreshold region to generate a PTAT voltage and a CTAT voltage, respectively; 2) a chopped buffer is employed to enhance the driving capability of the CTAT voltage, while chopping mitigates the buffer offset's impact on the sampled voltage; 3) a switched-capacitor amplifier, controlled by ϕ_1 and ϕ_2, amplify the CTAT voltage through the ratio of C_1 and C_2 ($C_2 = C_{2\_BASE} + C_{ARRAY}$); 4) a comparator compares the amplified CTAT voltage with the PTAT voltage and outputs the result at the falling edge of ϕ_2; 5) a SAR logic circuit receives the comparison result and generates the control signal $D_{OUT} < 7:0 >$ when ϕ_{2_DELAY} arrives; 6) C_{ARRAY} is connected to a 2-to-1 switch array $SWITCH < 7:0 >$, controlled by the digital signal $D_{OUT} < 7:0 >$. The signal determines that C_{ARRAY} is connected in parallel with either C_1 or C_{2_BASE}.

979-8-3315-8850-2/25 $31.00 © 2025 IEEE

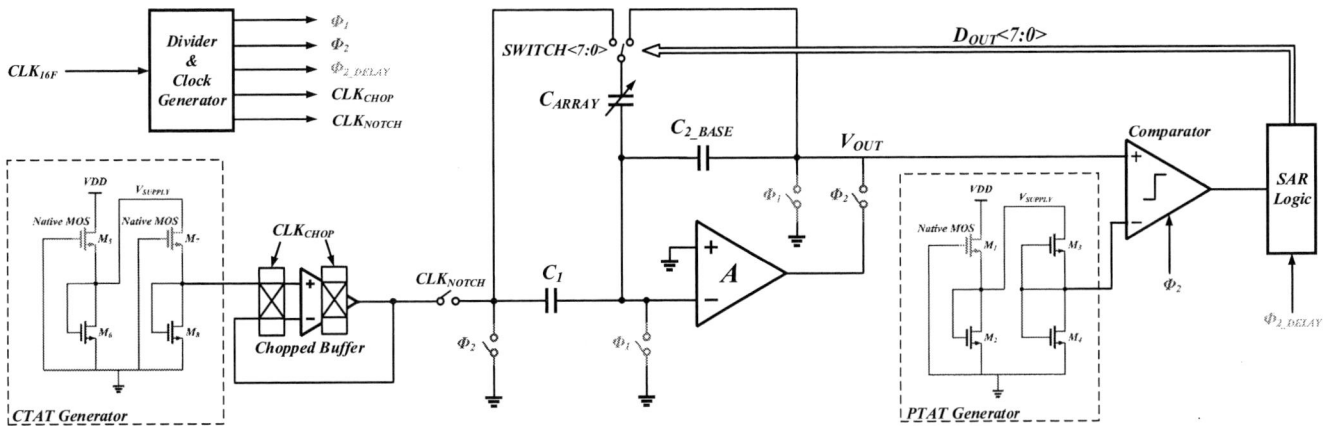

Fig. 1: System diagram of the proposed temperature sensor.

Fig. 2 shows the process of a single temperature quantization cycle and its timing diagram, which comprises ten clock cycles. The first reset cycle is to initialize SAR logic circuit. Then, the system proceeds to 8 SAR quantization cycles (from the second to the ninth clock cycle). Data are acquired at the final cycle, marking the completion of the quantization process. The clock signals are defined as follows: ϕ_1 is the sampling phase, ϕ_2 is the amplification phase, and ϕ_{2_DELAY} is the activation clock for the SAR logic. In the timing design, ϕ_{2_DELAY} must be delayed relative to ϕ_2 by an amount. CLK_{CHOP} and CLK_{NOTCH} represent the chopping clock and filtering clock, respectively. The phase difference between them is one-quarter of their own period. This ensures that two cycles of chopping and filtering are completed during the sampling process, effectively minimizing the impact of buffer offset on the CTAT voltage output.

III. CIRCUIT IMPLEMENTATION

A. Temperature Sensing Front-End Core

This work implements a low-power, low-supply-sensitivity temperature sensing front-end using native and standard NMOS transistors. As shown in Fig. 3, the left-side structure generates a PTAT voltage, while the right-side structure generates a CTAT voltage. Taking the PTAT generation circuit as an example, the gate of M_1 is connected to the ground, the drain of M_1 is connected to the supply voltage, and its source is connected to a diode-connected M_2, serving as the supply voltage for the sensing element formed by M_3 and M_4. M_1 is a native NMOS, and $M_2 - M_4$ are standard NMOS devices. All four transistors operate in the subthreshold region. Considering the body effect and drain-induced barrier lowering (DIBL) effect [5], the subthreshold current can be modeled as follows:

$$I = \mu_0 C_{ox} \left(\frac{W}{L} \right) V_T^2 e^{1.8} e^{\frac{-\Delta V_{TH}}{\eta V_T}} (m-1) \times \left(1 - e^{\frac{-V_{ds}}{V_T}} \right)$$
$$\times e^{\frac{(V_{gs} - V_{TH} - \gamma' V_{sb} + \eta V_{ds})}{m V_T}} \quad (1)$$

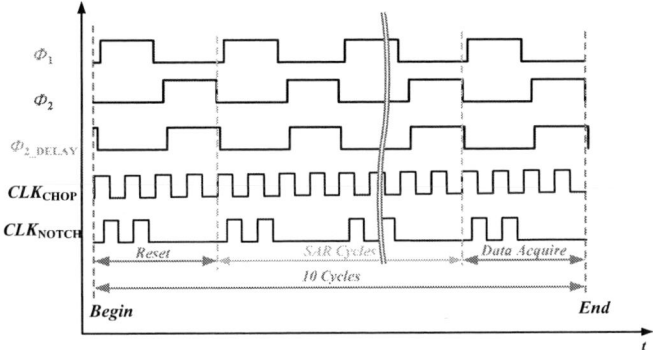

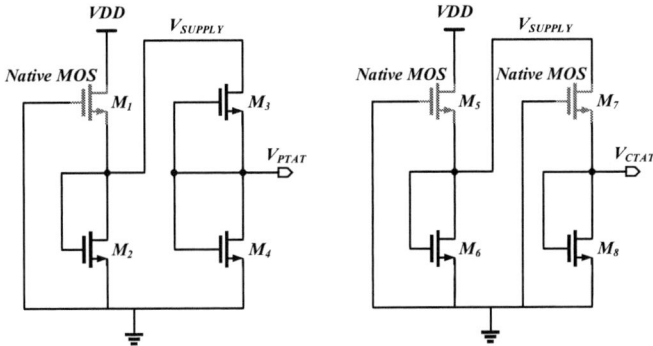

Fig. 3: Structure of the sensing front-end core.

where the μ_0 is zero bias mobility, C_{ox} is oxide capacitance, W is transistor width, L is transistor length, $V_T = kT/q$ is thermal voltage, η is the DIBL coefficient, m is subthreshold slope factor ($m = 1 + C_d/C_{ox}$ where C_d is depletion capacitance), V_{ds} is gate and source voltage, V_{TH} is transistor threshold voltage, γ' is linearized body coefficient, V_{sb} is source to bulk voltage, and V_{ds} is drain to source voltage.

Since all devices operate in the subthreshold region, the node connecting M_1, M_2 and M_3 enables the formulation of the KCL equation, while $VDD - V_{SUPPLY}$, V_{SUPPLY} and

979-8-3315-8850-2/25 $31.00 © 2025 IEEE   600

V_{PTAT} are all greater than $3V_T$ (~ 78 mV), $(1 - e^{\frac{-V_{ds}}{V_T}})$ can be neglected. Thus, the KCL equation can be simplified as:

$$\left(\frac{W}{L}\right)_1 (m_1-1)\cdot\exp\left[\frac{-(1+\gamma'+\eta_1)\,V_{SUPPLY}-V_{THNA1}+\eta_1 V_{DD}}{m_1 V_T}\right]$$
$$=\left(\frac{W}{L}\right)_2 (m_2-1)\cdot\exp\left(\frac{V_{SUPPLY}-V_{TH2}+\eta_2 V_{SUPPLY}}{m_2 V_T}\right)$$
$$+\left(\frac{W}{L}\right)_4 (m_4-1)\cdot\exp\left(\frac{V_{PTAT}-V_{TH4}+\eta_4 V_{PTAT}}{m_4 V_T}\right). \tag{2}$$

Assuming the drain-source voltage of M_3 exceeds $3V_T$, meaning $V_{SUPPLY} - V_{PTAT} > 3V_T$, so it can be concluded that the first term on the right-hand side of (2) is significantly larger than the second term. Therefore, this equation can be approximated as:

$$\left(\frac{W}{L}\right)_1 (m_1-1)\cdot\exp\left[\frac{-(1+\gamma'+\eta_1)\,V_{SUPPLY}-V_{THNA1}+\eta_1 V_{DD}}{m_1 V_T}\right]$$
$$=\left(\frac{W}{L}\right)_2 (m_2-1)\cdot\exp\left[\frac{(1+\eta_2)\,V_{SUPPLY}-V_{TH2}}{m_2 V_T}\right]. \tag{3}$$

Then, the supply voltage can be derived as:

$$V_{SUPPLY}=\left[\frac{1+\gamma'+\eta_1}{m_1}+\frac{1+\eta_2}{m_2}\right]^{-1}$$
$$\cdot\left[\frac{\eta_1}{m_1}V_{DD}+\frac{V_{TH2}}{m_2}-\frac{V_{THNA1}}{m_1}+V_T\ln\frac{(W/L)_1\,(m_1-1)}{(W/L)_2\,(m_2-1)}\right]. \tag{4}$$

Considering the KCL of the node connecting M_3 and M_4, while $V_{SUPPLY} - V_{PTAT}$ and V_{PTAT} are greater than $3V_T$, so the equation can be simplified as:

$$\left(\frac{W}{L}\right)_3 (m_3-1)\cdot\exp\left[\frac{-V_{TH3}-\gamma'V_{PTAT}+\eta_3(V_{SUPPLY}-V_{PTAT})}{m_3 V_T}\right]$$
$$=\left(\frac{W}{L}\right)_4 (m_4-1)\cdot\exp\left(\frac{V_{PTAT}-V_{TH4}+\eta_4 V_{PTAT}}{m_4 V_T}\right). \tag{5}$$

Combining (5) with (4), the PTAT voltage expression can be obtained as:

$$V_{PTAT}=\left[\frac{(1+\eta_4)}{m_4}+\frac{\gamma'+\eta_3}{m_3}\right]^{-1}\cdot\left\{V_T\ln\frac{(W/L)_3\,(m_3-1)}{(W/L)_4\,(m_4-1)}\right.$$
$$+\left[\frac{1+\gamma'+\eta_1}{m_1}+\frac{1+\eta_2}{m_2}\right]^{-1}\left[\frac{\eta_1\eta_3}{m_1 m_3}V_{DD}+\frac{\eta_3}{m_3}\left(\frac{V_{TH2}}{m_2}-\frac{V_{THNA1}}{m_1}\right)\right.$$
$$\left.\left.+\frac{\eta_3}{m_3}V_T\ln\frac{(W/L)_1(m_1-1)}{(W/L)_2(m_2-1)}\right]+\frac{V_{TH4}}{m_4}-\frac{V_{TH3}}{m_3}\right\}. \tag{6}$$

Following the same derivation process, the CTAT voltage expression can be obtained as:

$$V_{CTAT}=\left[\frac{(1-\eta_8)}{m_8}+\frac{1+\gamma'+\eta_7}{m_7}\right]^{-1}\cdot\left\{V_T\ln\frac{(W/L)_7\,(m_7-1)}{(W/L)_8\,(m_8-1)}\right.$$
$$+\left[\frac{1+\gamma'+\eta_5}{m_5}+\frac{1+\eta_6}{m_6}\right]^{-1}\left[\frac{\eta_5\eta_7}{m_5 m_7}V_{DD}+\frac{\eta_7}{m_7}\left(\frac{V_{TH6}}{m_6}-\frac{V_{THNA5}}{m_5}\right)\right.$$
$$\left.\left.+\frac{\eta_7}{m_7}V_T\ln\frac{(W/L)_5(m_5-1)}{(W/L)_6(m_6-1)}\right]+\frac{V_{TH8}}{m_8}-\frac{V_{THNA7}}{m_7}\right\}. \tag{7}$$

The derived results show that this work offers the following advantages:

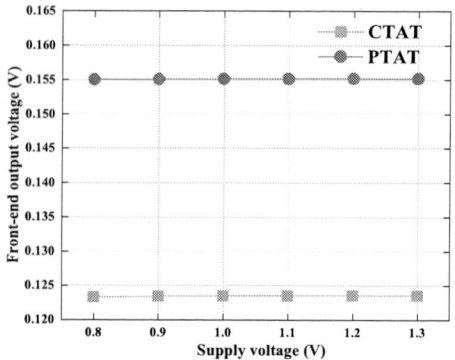

Fig. 4: Simulation results of supply sensitivity.

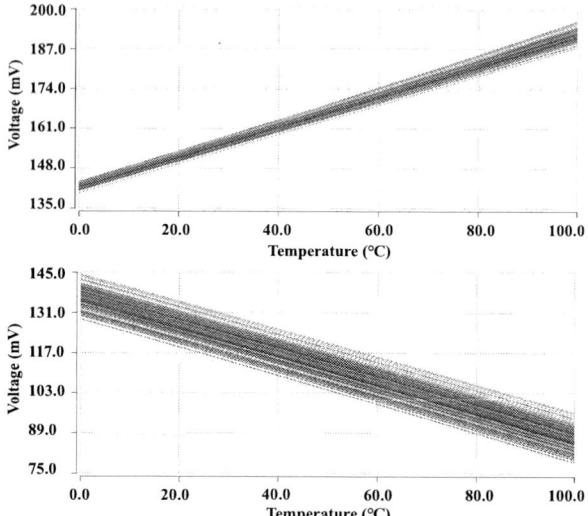

Fig. 5: Monte Carlo simulations of PTAT voltage (Top) and CTAT voltage (Bottom).

- The supply sensitivity of the sensing front-end is effectively reduced: as shown in (6), the supply-dependent term is $\left[\frac{1+\gamma'+\eta_1}{m_1}+\frac{1+\eta_2}{m_2}\right]^{-1}\cdot\left[\frac{(1+\eta_4)}{m_4}+\frac{\gamma'+\eta_3}{m_3}\right]^{-1}\frac{\eta_1\eta_3}{m_1 m_3}V_{DD}$ and $\left[\frac{1+\gamma'+\eta_1}{m_1}+\frac{1+\eta_2}{m_2}\right]^{-1} < \frac{1}{2}$, so the supply sensitivity of the sensing front-end is clearly reduced.

- Compared with traditional structures, using native NMOS allows for a larger V_{SUPPLY} with the same area overhead, ensuring better linearity: the validity of (3) depends on $V_{SUPPLY} - V_{PTAT} > 3V_T$, which can be easily satisfied with a larger V_{SUPPLY}. Without the native NMOS, V_{SUPPLY} is increased by increasing W_1 in previous works, which leads to significant area overhead, otherwise reducing L_1 causes the drain electric field to penetrate the channel more easily, lowering the source barrier and increasing η_1. As shown in (6), a larger η_1 significantly degrades the supply sensitivity of the PTAT voltage. Therefore, using native NMOS not only allows for a long-channel M_1 to reduce η_1, but also enables V_{SUPPLY} to be much larger, effectively enhancing the linearity of the PTAT voltage over a wider dynamic output range.

- All transistors operate in the subthreshold region to minimize power consumption. The wider dynamic range of the front-end enables a smaller C_1, further reducing power.

Fig. 4 shows that as the supply voltage varies from 0.8 V to 1.3 V, the corresponding PTAT voltage changes by 260 μV, while the CTAT voltage changes by 170 μV. Fig. 5 shows the results of V_{PTAT} and V_{CTAT} versus temperature under the Monte Carlo simulation.

B. Quantization Scheme

This work proposes a novel quantization scheme based on 8-bit SAR architecture, which adjusts the ratio of C_1 and C_2 to progressively drive the CTAT voltage to the PTAT voltage, enabling accurate quantization. Fig. 5 shows that the PTAT and CTAT voltages possess high linearity, they can be expressed as $V_{PTAT} = AT + B$ and $V_{CTAT} = CT + D$, A and C can be configured by modifying front-end design parameters. The SAR process changes D_{OUT} and incrementally drives the CTAT voltage to the PTAT voltage. According to the principle of switched-capacitor non-inverting amplifiers, the relationship can be written as:

$$(CT + D) \times \frac{C_1 + D_{OUT}C_u}{C_2 - D_{OUT}C_u} = AT + B \qquad (8)$$

where $C_2 = C_{2\_BASE} + C_{ARRAY}$ indicates that C_{ARRAY} is fully connected in parallel with C_{2_BASE} before the quantization starts; C_u represents the unit capacitance of C_{ARRAY}; D_{OUT} is the decimal representation of the parallel digital code $D_{OUT} < 7 : 0 >$ after quantization is completed. By further derivation, the output digital code D_{OUT} can be obtained as:

$$D_{\text{OUT}} = \frac{(C_2 A - C_1 C)\,T + (C_2 B - C_1 D)}{(C_u C + C_u A)\,T + (C_u D + C_u B)}. \qquad (9)$$

Taking advantage of the adjustable feature of our sensing front-end. By properly configuring the front-end parameters, then $A = -C$ can be achieved and D_{OUT} exhibits a linear relationship with T as (10) and Fig. 6 shows the trend under TT corner:

$$D_{\text{OUT}} = \frac{(C_2 A - C_1 C)\,T + (C_2 B - C_1 D)}{(C_u D + C_u B)}. \qquad (10)$$

This quantization scheme offers the following advantages: 1) The 8-bit SAR ADC achieves nearly full-scale digital code coverage across the 8-bit range, enabling significant area and power savings compared to a 9-bit temperature sensor. 2) Directly using PTAT and CTAT voltages avoids intermediate errors and greatly improves accuracy. 3) The linear output avoids the need for complex calibration or look-up tables, making the post-processing much easier. 4) The least significant bit (LSB) is defined in this work as the output voltage change of the switched-capacitor amplifier corresponding to a one-bit change near the ideal digital output code. As shown in Fig. 7, the LSB increases with rising temperature while thermal noise is less than it, which enhances the thermal noise margin of the comparator and improves the reliability and stability of the temperature sensor in high-temperature environments.

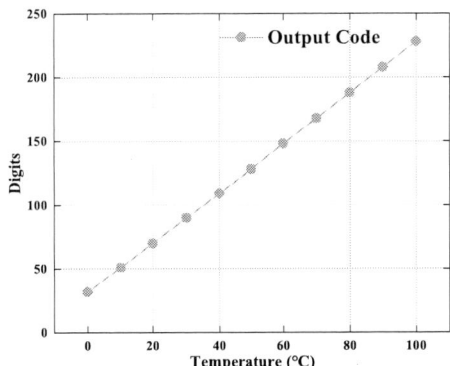

Fig. 6: Relationship between output code and temperature under simulation.

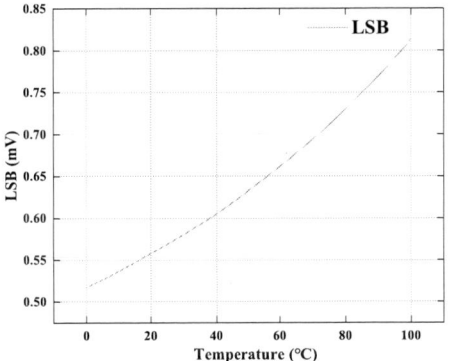

Fig. 7: Relationship between LSB and temperature.

C. Comparator Calibration Scheme

Fig. 8 shows the comparator calibration scheme. The offset of the comparator can degrade the accuracy of the comparison. To address this issue, calibrating the comparator before the quantization process is essential to minimize the offset after 1-point calibration. When the calibration enable signal $CALI$ is activated, $SWITCH2$ closes and $SWITCH1$ opens, shorting the differential input terminals of the pre-amplifier. The comparator output is latched before the calibration process. *Control Logic* then determines which side of the pre-amplifier output should be connected to the capacitors. Unit capacitors are connected incrementally to monitor changes in the comparator output. If a change is detected within the specified bit range, the calibration is complete. If no change occurs after all bits are set, the counter in the *Control Logic* stops, and the calibration is also considered to be completed. The unit capacitance of the compensation capacitor array is 1 fF.

D. Unit Capacitor

For a SAR ADC, the capacitor array occupies the majority of the area [9]. To improve the area efficiency of unit capacitors and minimize the CDAC footprint, a custom multi-layer sandwich capacitor with specific shielding is used in this work. Fig. 9 demonstrates the 3D structure of the used capacitor, utilizing metal layers M3 to M7. In this design, M4 and M6 serve as the top plates, while M3, M5, and M7 act as the

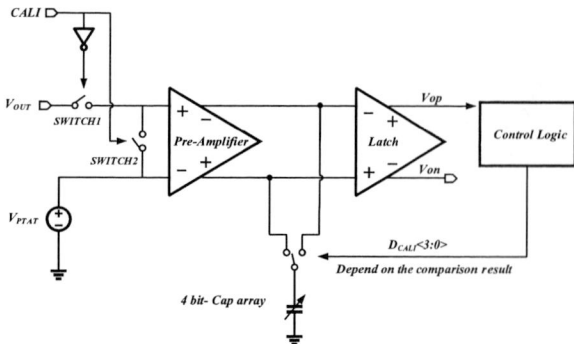

Fig. 8: Diagram of the comparator calibration scheme.

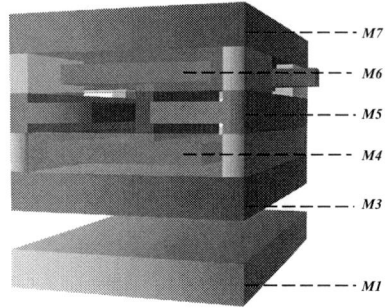

Fig. 9: 3D model of the custom multi-layer capacitor with shielding.

bottom plates enclosing the top plates. Additionally, M1 is used as the shielding layer to mitigate the substrate parasitic effect.

IV. SIMULATION RESULTS

The proposed temperature sensor is implemented in standard 65nm CMOS technology. The post-simulation results show that with a 1.2 V supply and a conversion time of 182 μs, the proposed temperature sensor consumes 2.58 μW at 27 °C, achieving an energy efficiency of 0.47 nJ/Conversion. Fig. 10 presents the energy consumption breakdown. From simulation results, a resolution of 0.51 °C at 27 °C and after a 2-point calibration at 20 °C and 80 °C, it achieves -0.69/+0.77 °C peak-to-peak inaccuracy over 0 °C to 100 °C, as shown in Fig. 11. Fig. 12 shows the supply sensitivity is 1.25 °C/V when measured across the supply range of 0.9 V to 1.3 V. Fig. 13 shows the layout of the proposed temperature sensor, with the total core area of 0.00786 mm². The performance is summarized and compared with other state-of-the-art temperature sensors [5], [8], [10]–[12] shown in Table I.

V. CONCLUSION

This paper proposes a low power consumption, compact area, and low supply sensitivity temperature sensor. The proposed design adopts a 4-transistor structure leveraging both native NMOS and standard NMOS devices operating in the subthreshold region, which generate PTAT and CTAT voltages with low supply sensitivity and high linearity, respectively. A new SAR-based quantization scheme is implemented in standard 65nm CMOS technology with a core area

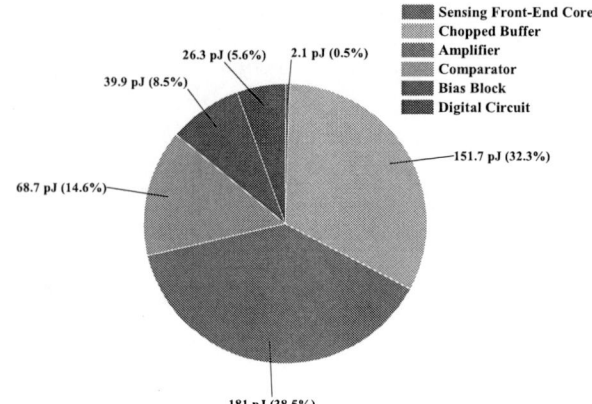

Fig. 10: Energy consumption breakdown.

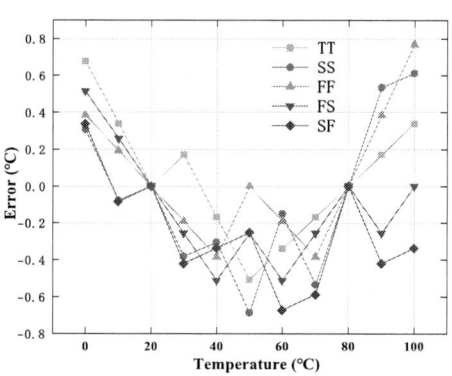

Fig. 11: Temperature error after a 2-point calibration at 20°C and 80°C.

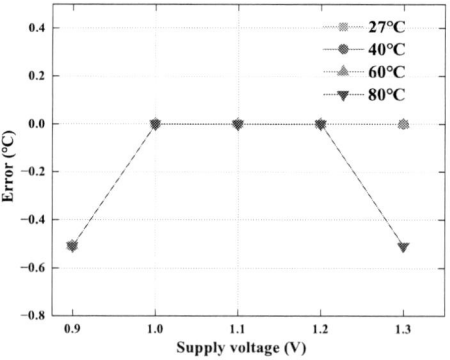

Fig. 12: Supply dependence of the temperature inaccuracy.

of 0.00786 mm². Post-layout simulation results showed the supply sensitivity is 1.25 °C/V when measured across the supply range of 0.9 V to 1.3 V. After a 2-point calibration, the sensor achieves a peak-to-peak inaccuracy of -0.69/+0.77 °C over the temperature range of 0 °C to 100 °C. Operating from a supply of 1.2 V, its energy efficiency is 0.47 nJ/Conversion, which yields 0.123 nJ·K² FOM.

ACKNOWLEDGMENT

This work was supported by ZJU-YST joint research center for fundamental science.

979-8-3315-8850-2/25 $31.00 © 2025 IEEE

TABLE I: Comparison with state-of-the-art temperature sensors

	This work	JSSC2014 [5]	VLSI2022 [8]	JSSC2019 [10]	SSC-L2020 [11]	TCAS-I2021 [12]
Technology	65nm	180nm	4nm class	65nm	65nm	130nm
Type	MOSFET	MOSFET	Hybrid	MOSFET	MOSFET	MOSFET
Quantization	SAR	FDC	SAR	FDC	FDC	FDC
Area[mm^2]	0.00786	0.09	0.0061	0.63	0.32	0.07
Supply[V]	1.2	1.2	1.0	0.5	0.8	0.95
Temperature Range[°C]	0~100	0~100	-10~100	0~100	-30~70	0~80
Resolution[°C]	0.51	0.3	0.46	0.3	0.075	0.1
Conversion Time[μs]	182	30000	11	300000	765000	59000
Inaccuracy[°C]	-0.69/+0.77	-1.4/+1.5	-1.79/+1.79	-1.53/+1.61	-1.0/+0.7	-0.4/+0.44
Calibration	2-point	2-point	2-point	2-point	2-point	2-point
Power Consumption[μW]	2.58	0.071	64	0.000763	0.0064	0.196
Energy/Conversion [nJ]	0.47	2.2	0.7	0.23	4.9	11.56
Supply Sensitivity[°C/V]	1.25	14.125	2.6	8.4	2.8	13.7
FOM[$nJ \cdot K^2$]	0.123	0.19	0.15	0.02	0.027	0.12

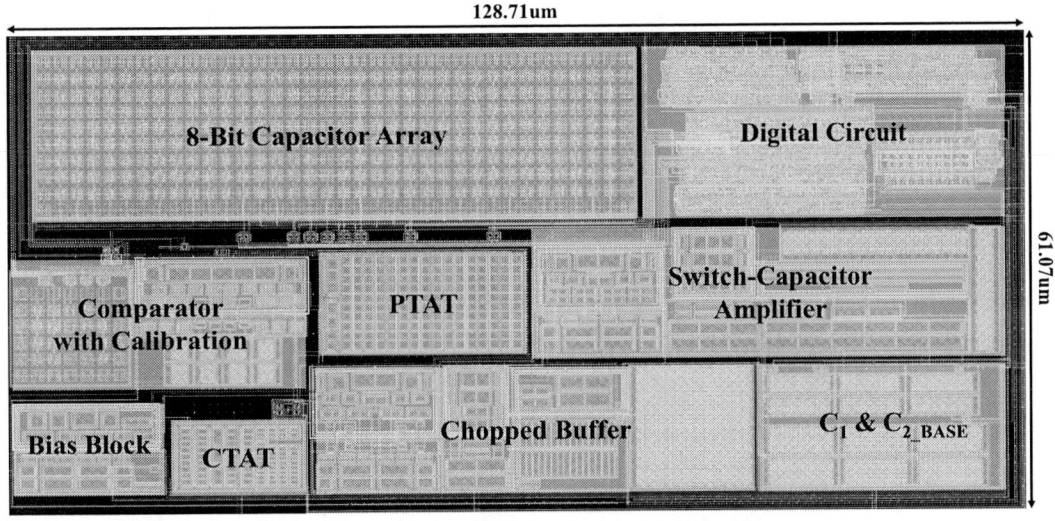

Fig. 13: Layout of the proposed temperature sensor.

REFERENCES

[1] C.-C. Chung and C.-R. Yang, "An autocalibrated all-digital temperature sensor for on-chip thermal monitoring," *IEEE Transactions on Circuits and Systems II: Express Briefs*, vol. 58, no. 2, pp. 105–109, Feb. 2011.

[2] J. S. Shor and K. Luria, "Miniaturized BJT-based thermal sensor for microprocessors in 32-and 22-nm technologies," *IEEE Journal of Solid-State Circuits*, vol. 48, no. 11, pp. 2860–2867, Nov. 2013.

[3] T. Oshita, J. Shor, D. E. Duarte, A. Kornfeld, and D. Zilberman, "Compact BJT-based thermal sensor for processor applications in a 14 nm tri-gate CMOS process," *IEEE Journal of Solid-State Circuits*, vol. 50, no. 3, pp. 799–807, Mar. 2015.

[4] J. Shor, "Compact thermal sensors for dense CPU thermal monitoring and regulation: A review," *IEEE Sensors Journal*, vol. 21, no. 11, pp. 12774–12788, Jun. 2021.

[5] S. Jeong, Z. Foo, Y. Lee, J.-Y. Sim, D. Blaauw, and D. Sylvester, "A fully-integrated 71 nW CMOS temperature sensor for low power wireless sensor nodes," *IEEE Journal of Solid-State Circuits*, vol. 49, no. 8, pp. 1682–1693, Aug. 2014.

[6] T. Anand, K. A. Makinwa, and P. K. Hanumolu, "A VCO based highly digital temperature sensor with 0.034°C/mV supply sensitivity," *IEEE Journal of Solid-State Circuits*, vol. 51, no. 11, pp. 2651–2663, Nov. 2016.

[7] Y. Shui and A. Wang, "A 14.17pJ.K^2 FoM CMOS temperature sensor with 173 μm^2 sensing core for remote sensing in 65-nm CMOS," *IEEE Sensors Journal*, vol. 23, pp. 27059–27067, Nov. 2023.

[8] Y. Li, D. E. Duarte, and Y. Fan, "A 90.9 kS/s, 0.7 nJ/conversion hybrid temperature sensor in 4nm-class CMOS," in *2022 IEEE Symposium on VLSI Technology and Circuits (VLSI Technology and Circuits)*. IEEE, Jun. 2022, pp. 118–119.

[9] C.-C. Liu, S.-J. Chang, G.-Y. Huang, and Y.-Z. Lin, "A 10-bit 50-MS/s SAR ADC with a monotonic capacitor switching procedure," *IEEE Journal of Solid-State Circuits*, vol. 45, no. 4, pp. 731–740, Apr. 2010.

[10] H. Wang and P. P. Mercier, "A 763 pW 230 pJ/conversion fully integrated CMOS temperature-to-digital converter with +0.81°C/-0.75°C inaccuracy," *IEEE Journal of Solid-State Circuits*, vol. 54, no. 8, pp. 2281–2290, Aug. 2019.

[11] T. Someya, A. M. Islam, and K. Okada, "A 6.4 nW 1.7% relative inaccuracy CMOS temperature sensor utilizing sub-thermal drain voltage stabilization and frequency-locked loop," *IEEE Solid-State Circuits Letters*, vol. 3, pp. 458–461, Sep. 2020.

[12] J. Li, Y. Lin, N. Ning, and Q. Yu, "A +0.44°C/-0.4°C inaccuracy temperature sensor with multi-threshold MOSFET-based sensing element and CMOS thyristor-based VCO," *IEEE Transactions on Circuits and Systems I: Regular Papers*, vol. 68, no. 3, pp. 1102–1113, Dec. 2020.

An Ultra-low Power 2×/4× Reconfigurable Charge Pump for RF Energy Harvesting System

Xuanchen Mei
School of Electronic and Information
South China University of Technology
Guangzhou, China
1391933905@qq.com

Xin Liu
School of Future Technology
South China University of Technology
Guangzhou, China
lx11739@163.com

Junhui Ou*
School of Future Technology
South China University of Technology
Guangzhou, China
oujunhui@scut.edu.cn*

Yanqi Zheng
School of Microelectronics
South China University of Technology
Guangzhou, China
yqzhengee@scut.edu.cn

Mo Huang
Institute of Microelectronics
University of Macau
Macau, China
mohuang@um.edu.mo

Xiuyin Zhang
School of Electronic and Information
South China University of Technology
Guangzhou, China
eexyz@scut.edu.cn

Abstract—**This paper presents a fully integrated ultra-low-power 2×/4× reconfigurable charge pump (CP) for RF energy harvesting system to enable switching of the number of operating CP stages at different input levels. The reconfiguration is achieved by an intermediate node voltage detection circuit, which enables timely switching of the switching control signals with fairly low power consumption. Moreover, this CP employs dynamic body bias (DBB) technique to improve driving ability as well as settling time. The output is regulated by a low dropout regulator (LDO) to obtain a ripple-free and stable output voltage without significantly affecting efficiency and power consumption. The design was simulated in a 65-nm low-power CMOS process and the layout occupies an area of 0.25 mm². This reconfigurable CP can boost a 250-mV or 500-mV input voltage to 1-1.2 V through 2×/4× mode switching and further regulate it to an 800-mV stable output, covering both low and high input power range. It also achieves an ultra-low power consumption of 80 nW.**

Keywords—*Reconfigurable charge pump (CP), ultra-low power, CMOS, RF energy harvesting, power management, dynamic body biasing (DBB), low dropout regulator (LDO).*

I. INTRODUCTION

RF energy harvesting systems are widely used in the Internet of Things (IoTs), especially in power-constrained environments. The passive IoT technology that collects energy from the surrounding environment and realizes wireless transmission of energy with electromagnetic wave as a carrier has the advantages of small size, long range, and maintenance-free, which provides a great opportunity for the development of IoT and puts forward higher requirements for power management [1]. Due to the variation of signal strength at the rectifier antenna, the energy collected by the wireless energy harvester is also unstable, and the input voltage it provides to the power management system is low and unstable [2].

As mentioned above, the most common method of boosting and stabilizing voltages is to use DC-DC converters for boosting and regulating, such as reconfigurable switched-capacitor structures to provide different conversion ratios (CR) [3]. However, they are not easy to control with a certain degree of structural complexity and cannot be efficiently regulated over a dynamic input voltage range. In addition, a significant portion of such circuits perform poorly, or even fail to operate, at supply voltages below 1 V.

In order to solve the above problems, this paper proposes a reconfigurable charge pump (CP) that uses an intermediate node voltage detection circuit to achieve the switching of the number of CP operating stages under different input voltage conditions, and finally uses a low dropout regulator (LDO) to obtain a ripple-free and stable output voltage. The simplified architecture of the wireless energy harvesting system is shown in Fig. 1. It consists of three parts: an RF rectifier, a series of reconfigurable CP stages and a LDO. In this case, the rectifier outputs a voltage V_{RECT} that is switched from 0.25 V to 0.5 V. This voltage can be obtained by subsequent reconfigurable CP and LDO circuits to obtain an almost constant voltage V_{OUT}.

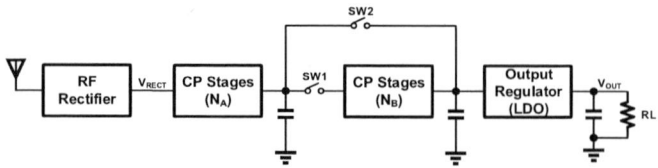

Fig. 1.   Architecture of the wireless energy harvesting system.

II. ANALYSIS AND CIRCUIT IMPLEMENTATION

A. Architecture and Principle of Reconfigurable CP

As shown in Fig. 2(a), the proposed reconfigurable CP consists of a ring oscillator and a non-overlapping clock generator for generating linear non-overlapping clocks *CK* and *CKN* shown in Fig. 2(b), a four-stage CP to fulfill the boost function, an Ultra-Low-Power intermediate voltage V_M detector (ULP V_M Detector) for detecting the voltage V_M at the intermediate node of the four-stage CPs and generating the enable signals *SW* and *SW_N* shown in Fig. 2(c), and three MOS switches S_1-S_3 with additional dynamic body biasing (DBB) structure used to achieve the switching of CP operating stages.

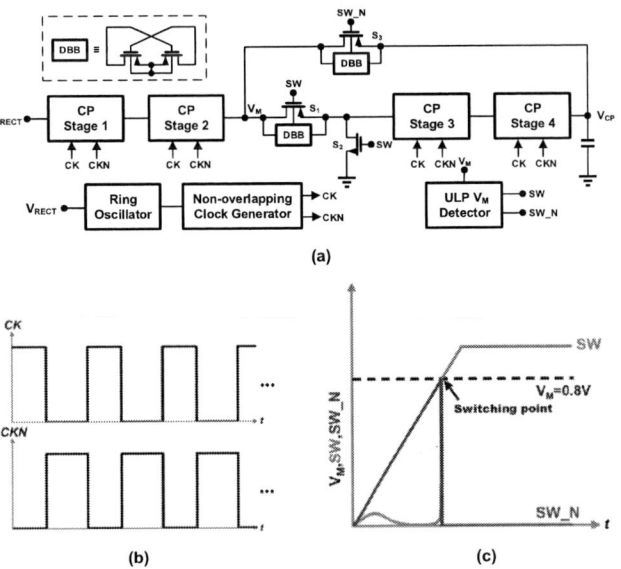

(a)

(b)   **(c)**

Fig. 2. (a) Architecture of the proposed reconfigurable CP. (b) Non-overlapping clock signals *CK* and *CKN*. (c) Enable signals *SW* and *SW_N*.

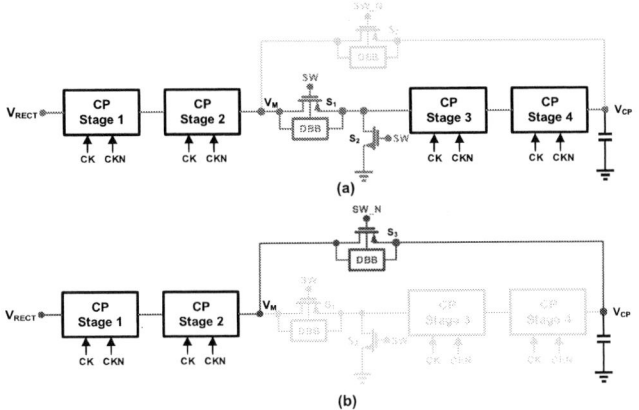

Fig. 3. (a) 4-stage (4×) mode. (b) 2-stage (2×) mode.

As shown in Fig. 3(a), with V_{RECT}=250 mV, the intermediate node V_M is a modest voltage (<0.8 V), the enable signal *SW* is low and *SW_N* is high, at which time S_1 is on, S_2 and S_3 are off, and the CP operates in the 4-stage mode (4×). The current flowing through the circuit is marked red. As V_{RECT} switches to a higher level (V_{RECT}=500 mV), as shown in Fig. 3(b), V_M

increases, and when it exceeds 1 V, *SW* switches to a high level and *SW_N* switches to a low level, at which time S_1 is off, S_2 and S_3 are on, and the CP operates in the 2-stage mode (2×). The current flowing through the circuit is marked blue.

1) Low-Power Ring Oscillator: A conventional ring oscillator is composed of an odd number of cascaded inverter stages to sustain oscillation, eliminating the need for resistors and inductors, which is easy to integrate on chip, and can be used to clock other analog circuits as needed for their operation [4]. The oscillator can oscillate at a frequency of $1/2nT_D$, where *n* is the number of inverters and T_D is the delay time of one inverter. However, when the supply voltage is reduced below the transistor threshold voltage V_{TH}, the on-resistance of the transistor can be on the order of GΩ. Thus, for a common inverter, the output swing is severely degraded due to the on-resistance of the on-transistor being no greater than the off-resistance of the off-transistor within sub-50 mV.

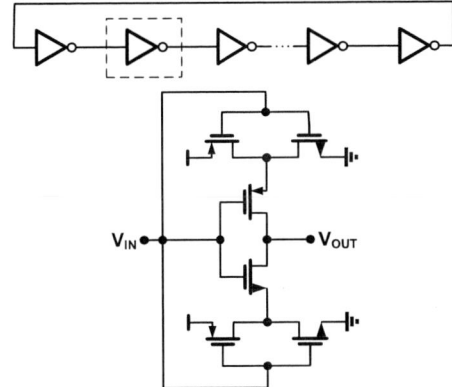

Fig. 4. Stacked inverters in a ring oscillator.

Fig. 4 shows the stacked inverters proposed in [5], which can provide enhanced peak-to-peak output voltage swing even when the supply voltage is reduced. All transistors operate in the subthreshold region to reduce power consumption. The leakage current I_{sub} of the transistor in the subthreshold region can be expressed as

$$I_{sub}=I_0\exp(\frac{V_{GS}-V_{TH}}{\xi V_T})\qquad(1)$$

in which I_0 is a constant of proportionality, V_{GS} is the gate-source voltage, ξ is a non-ideal constant *and* V_T=*kT/q is the* thermal voltage [6].

2) Charge Pump Design: Among the various switched-capacitor circuits designed to generate voltages exceeding the supply level, the charge pump (CP) is the most widely used architecture. Compared with other boost circuits, the cross-coupled CP with PMOS load switches, as illustrated in Fig. 5(a), offers a simpler architecture and achieves higher power efficiency, making it well-suited for low-power and compact integrated circuit applications [7]. In order to save chip area and improve efficiency, the most appropriate number of pumping stages *N* can be expressed as

$$N=2(\frac{V_{CP}}{V_{IN}}-1)\qquad(2)$$

979-8-3315-8850-2/25 $31.00 © 2025 IEEE

in which V_{CP} is the output voltage of the charge pump and V_{IN} is the input voltage of the charge pump.

And the capacitor value C_{stage} of each CP stage of the charge pump can be presented as

$$C_{stage} = \frac{N}{f} \cdot \frac{I_{LOAD}}{(N+1)V_{IN} - V_{CP}} \tag{3}$$

in which I_{LOAD} is the load current and f is the switching frequency. Considering the efficiency, power consumption and other factors, f is finally selected to be 200 KHz and the number of stages N is 2 or 4 according to (2) and (3).

However, in a multi-stage CP, as the input voltage increases, the body-source voltage V_{BS} of NMOS switches M_1 and M_2 in the later stage becomes large. As a result, the threshold voltage of the NMOS switches becomes high, leading to a reduction in the conduction performance. The body effect is defined as the change in the threshold voltage by an amount approximately equal to the change in the body-source voltage V_{BS}. For the enhanced NMOS device, the effect of the body effect on the threshold voltage is calculated according to the Shichman-Hodges model using the following equation:

$$V_{TH} = V_{TH0} + \gamma(\sqrt{2|\Phi_F + V_{SB}|} - \sqrt{2|\Phi_F|}) \tag{4}$$

where V_{TH} is the threshold voltage, V_{SB} is the source body voltage, V_{TH0} is the threshold voltage at $V_{SB} = 0V$, γ is the scaling factor, and Φ_F is the surface potential.

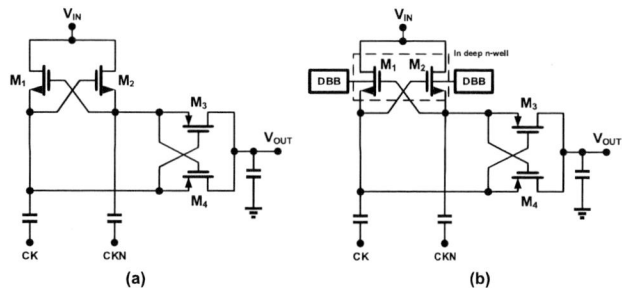

Fig. 5. (a) Conventional CP circuit with PMOS load switches. (b) CP circuit with DBB.

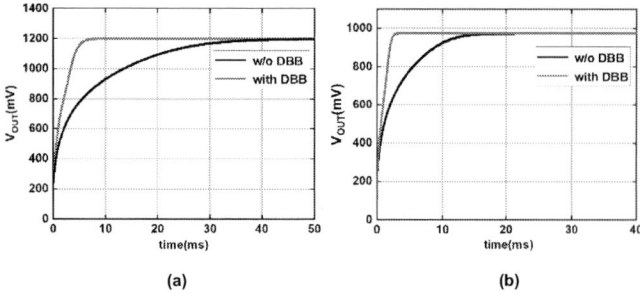

Fig. 6. Transient response of the CP stage at (a) V_{IN}=250mV and (b) V_{IN}=500mV

To solve this problem, the dynamic body bias (DBB) technique proposed in [8] is cited to eliminate the effect of the

body effect by generating a higher current transfer through the NMOS switches as shown in Fig. 5(b). Within this biasing scheme, the NMOS transistors are required to be placed in a deep n-well to ensure proper isolation and bias control. The DBB used in the switches in Fig. 2(a) is similar and also improves the conduction performance of switch S_1, S_2 and S_3.

Fig. 6 demonstrates the transient response capability of the CP stage with and without the DBB structure condition. Fig. 6(a) shows the output voltage V_{OUT} at V_{IN}=250 mV, and the settling time of the CP stage with DBB is nearly 35 ms faster than that without DBB; Fig. 6(b) shows the output voltage V_{OUT} at V_{IN}=500 mV, and the settling time of the CP stage with DBB is nearly 10ms faster than that without DBB, which demonstrates that DBB can effectively improve the driving capability of the CP stage.

3) Ultra-Low-Power V_M Detector: The Ultra-Low-Power V_M Detector (ULP V_M Detector) is powered by both the intermediate node V_M and the charge pump V_{CP}, which is used to sense the achievable V_M voltage (Fig. 7). In order to save silicon area and also to reduce the power consumption of the detector, a series of PMOS transistors M_{Ri}, which are all diode-connected, are used to act as a voltage divider. The desired threshold voltage and equivalent resistance can be achieved by selecting proper transistor sizing for M_{Ri}. The aspect ratio of M_3 and M_5 and the voltage dividing ratio of MOS resistors M_{Ri} determine the V_M voltage of the *SW* and *SW_N* switching point. 2.5V MOS transistors are used and operate in subthreshold region to reduce power consumption. Transistor sizes can be reasonably designed to reduce the influence of process and temperature on the circuit. The most appropriate transistor sizes are summarized in Table I. The power consumption of the circuit is 140 pW.

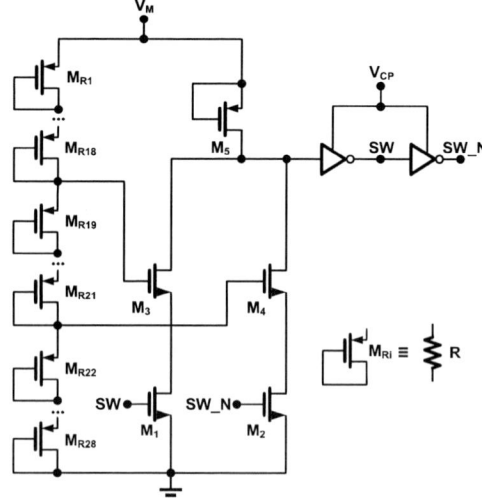

Fig. 7. 140pW ULP V_M Detector for CP operating stage (2×/4×) switching.

TABLE I.   SIZING OF THE ULP V_M DETECTOR

Transistor	Size	
	W [μm]	L [μm]
M_{1-2}	1.35	1.35
M_{3-4}	0.8	2.2
M_5	0.28	2.5

Transistor	Size	
	W [μm]	L [μm]
M_{R1-R28}	0.6	0.28

B. LDO Structure for CP Output Stabilization

Overall, the CP results in a block characterized by noise. It has an output voltage ripple V_{ripple} given by the following:

$$V_{ripple} = \frac{I_{LOAD}}{fC_{LOAD}} \quad (5)$$

in which C_{LOAD} is the load capacitor. To obtain a stable output voltage with no ripple and small noise, a low dropout regulator (LDO) is used to regulate the output V_{CP} of the charge pump.

The complete schematic of the LDO is shown in Fig. 8. The LDO consists of a common-gate differential pair, a voltage spike detection circuit and a gain stage [9].

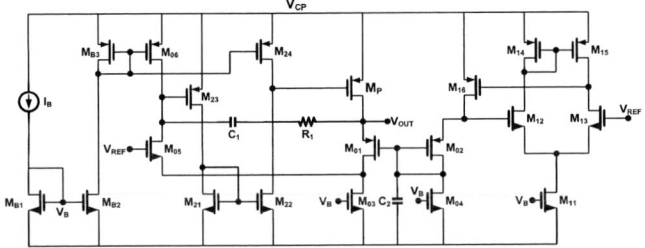

Fig. 8. Full schematic of the LDO after the charge pump output node V_{CP}.

The common-gate differential pair is implemented using transistors $M_{01}-M_{06}$, whereas the gain stage, composed of $M_{21}-M_{24}$, is employed to enhance the overall loop gain. The bias current I_B=1.5 nA can be generated by an off-chip or on-chip circuit and is used to bias various LDO blocks in the weak-inversion. For area and power consumption considerations, the aspect ratio (W_P/L_P) of M_P is chosen to be 25 μm/0.06 μm and the finger is chosen to be 2 to ensure that it can support a large enough load current I_{LOAD}.

Simulated PSRR of the LDO is shown in Fig. 9. At 5 Hz, this LDO achieves a PSRR of −46.7 dB, −42.3 dB, and −42.1 dB for a load current of 1 nA, 1 μA, and 1 mA, respectively, which illustrates good ripple rejection performance. This LDO also achieves the target of ultra-low power consumption, as it consumes only 10.31 nW at V_{CP}=1 V.

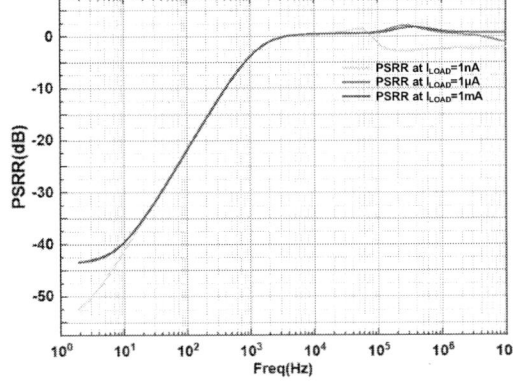

Fig. 9. Simulated PSRR of the LDO versus frequency for different I_{LOAD}.

III. SIMULATION RESULTS

The integrated reconfigurable CP circuit is simulated in 65nm low-power CMOS process and the reconfigurable CP layout using this process is shown in Fig. 10. The area of the layout is 0.25 mm². In this circuit, the flying capacitor of each CP stage is 25 pF and the CP load capacitor is 20 pF.

Fig. 10. Layout of the reconfigurable CP.

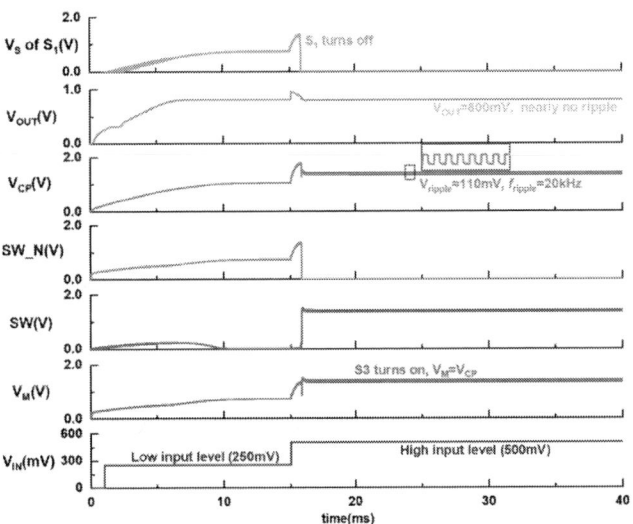

Fig. 11. Post-layout transient simulation results of the reconfigurable CP.

The post-layout transient simulation results operating in 2×/4× mode of the system are shown in Fig. 11 and an output resistance of 50 MΩ is selected due to output power limitations. At low input voltage level (V_{RECT}=250 mV), the CP operates in 4× mode, the CP output voltage V_{CP} is stabilized at 1.05 V after 10 ms, and the LDO output V_{OUT} is 800 mV. After V_{RECT} is switched (switching time=1 us), the V_M of the intermediate node is detected to be over a certain threshold, so the SW and SW_N are switched. Therefore, S_1 turns off, S_2 and S_3 turn on, V_M is directly connected with V_{CP}, and stage 3 and 4 stop working, so the operating mode is switched to 2×, the CP output V_{CP} increases to about 1.2 V with a ripple of approximately 110 mV, the LDO output V_{OUT} remains 800 mV after a slight fluctuation. Due to the residual charge on the corresponding capacitors during switching, V_{CP} after mode switching presents >1 V. From the waveform, it can be observed that the operating mode can be switched quickly and the ripple is effectively suppressed.

Table II summarizes and compares the performance of the proposed reconfigurable CP with other CP structures. This work

979-8-3315-8850-2/25 $31.00 © 2025 IEEE

enables reconfigurable CP stages, uses smaller load capacitor, and significantly reduces power consumption. At the same time, this work also uses an LDO for output regulation to reduce ripple and effectively improve stability.

TABLE II.   PERFORMANCE COMPARISON

Reference	[11]	[12][a]	[13][a]	This work[a]
Process	0.13 μm	28 nm FD-SOI	28 nm FD-SOI	65 nm
Architecture	3-stage CP with dynamic body bias	2-stage CP with clock booster	3-stage CP with forward body bias	2×/4× Reconfigurable CP with output regulation
Input Voltage	0.15 V	0.05 V	0.15 V	0.25/0.5 V
Output Voltage	0.619 V	0.27 V	1.2-3 V	1/1.2 V@V_{CP} 0.8 V@V_{OUT}
Load Capacitor	10 nF	180 pF	40 pF	20 pF
Peak PCE	72.5%	38.9%	63.3%	61%
Power Consumption	μW-scale	680 nW (P_{OUT})	197 nW	80 nW

[a]Simulated results.

IV. CONCLUSION

In this paper, the design and implementation of an ultra-low power reconfigurable charge pump have been reported. The proposed charge pump utilizes intermediate node voltage detection to achieve stage switching and exhibits good performance at both low and high input levels. Also, this charge pump uses the dynamic body bias technique to effectively improve the driving capability and speed up the stabilization time. The output of the charge pump is regulated by an LDO to obtain a ripple-free and stable output. The simulation results show a peak conversion efficiency of 61% and a power consumption of 80 nW, which means that this charge pump is very suitable for RF energy harvesting.

REFERENCES

[1] L. Chettri and R. Bera, "A comprehensive survey on internet of things (IoT) toward 5G wireless systems," *IEEE Internet of Things Journal*, vol. 7, no. 1, pp. 16-32, Jan. 2020.

[2] K. R. Sadagopan, J. Kang, Y. Ramadass and A. Natarajan, "A cm-Scale 2.4-GHz Wireless Energy Harvester With NanoWatt Boost Converter and Antenna-Rectifier Resonance for WiFi Powering of Sensor Nodes," *IEEE Journal of Solid-State Circuits*, vol. 53, no. 12, pp. 3396-3406, Dec. 2018.

[3] T. V. Breussegem and M. Steyaert, *CMOS Integrated Capacitive DC–DC Converters*, 1st ed. New York, NY, USA: Springer, 2013.

[4] Razavi, B. *Design of Analog CMOS Integrated Circuits*, Tata McGraw-Hill Education, 2002.

[5] S. Bose and M. L. Johnston, "A Stacked-Inverter Ring Oscillator for 50 mV Fully-Integrated Cold-Start of Energy Harvesters," *2018 IEEE International Symposium on Circuits and Systems (ISCAS)*, Florence, Italy, 2018, pp. 1-5.

[6] P. R. Gray, P. J. Hurst, S. H. Lewis and R. G. Meyer, *Analysis and Design of Analog Integrated Circuits*, John Wiley & Sons, 2024.

[7] Favrat, Pierre, Philippe Deval, and Michel J. Declercq, "A high-efficiency CMOS voltage doubler," *IEEE Journal of Solid-State Circuits*, 1998, 33(3): 410-416.

[8] G. Tian, X. Liu, J. Ou, Y. Zheng, M. Huang and X. Zhang, "A Cold Start Circuit with Dynamic Body Biasing for Boost Converter in RF Energy Harvesting System," *2024 IEEE 10th International Symposium on Microwave, Antenna, Propagation and EMC Technologies for Wireless Communications (MAPE)*, Guangzhou, China, 2024, pp. 1-4.

[9] R. Bansal and S. Chatterjee, "A 22-nA Quiescent Current, 50-mA Output-Capacitor-Less Low-Dropout Regulator With Multiple-Feedback Loop for IoT Devices," *IEEE Transactions on Circuits and Systems II: Express Briefs*, vol. 71, no. 11, pp. 4608-4612, Nov. 2024.

[10] A. Nowbahari, L. Marchetti and M. Azadmehr, "Analysis of a Low Power Inverting CMOS Schmitt Trigger Operating in Weak Inversion," *International Journal of Electrical and Electronic Engineering & Telecommunications*, Vol. 11, No. 6, pp. 392-397, Nov. 2022.

[11] J. Kim, P. K. T. Mok and C. Kim, "A 0.15 V Input Energy Harvesting Charge Pump With Dynamic Body Biasing and Adaptive Dead-Time for Efficiency Improvement," *IEEE Journal of Solid-State Circuits*, vol. 50, no. 2, pp. 414-425, Feb. 2015.

[12] A. Ballo, A. D. Grasso and G. Palumbo, "A Subthreshold Cross-Coupled Hybrid Charge Pump for 50-mV Cold-Start," *IEEE Access*, vol. 8, pp. 188959-188969, 2020.

[13] C. A. Pinheiro, F. Olivera and A. Petraglia, "A Three-Stage Charge Pump With Forward Body Biasing in 28 nm UTBB FD-SOI CMOS," *IEEE Transactions on Circuits and Systems I: Regular Papers*, vol. 68, no. 11, pp. 4810-4819, Nov. 2021.

2025 The 10th International Conference on Integrated Circuits and Microsystems

Cross-Process Bayesian Multi-Objective Collaborative Optimization For Process Migration

1st Zixi Guo
State Key Laboratory of Integrated Chips and Systems
Fudan University
Shanghai, China
23212020076@m.fudan.edu.cn

2nd Ruiyu Lyu
State Key Laboratory of Integrated Chips and Systems
Fudan University
Shanghai, China
22112020031@m.fudan.edu.cn

3rd Zhaori Bi*
State Key Laboratory of Integrated Chips and Systems
Fudan University
Shanghai, China
zhaori_bi@fudan.edu.cn

4th Changhao Yan*
State Key Laboratory of Integrated Chips and Systems
Fudan University
Shanghai, China
yanch@fudan.edu.cn

5th Ming Zhu
Department of Integrated Circuits
Anhui University
Hefei, China
zhuming@ahu.edu.cn

6th Xiulong Wu
Department of Integrated Circuits
Anhui University
Hefei, China
xiulong@ahu.edu.cn

7th Xuan Zeng
State Key Laboratory of Integrated Chips and Systems
Fudan University
Shanghai, China
xzeng@fudan.edu.cn

Abstract—**This paper introduces a cross-process Bayesian multi-objective collaborative optimization framework to address challenges in semiconductor technology migration. We leverage transfer learning within multi-objective Bayesian optimization through Gaussian copula, transforming simulation data into residual observations. By transferring prior knowledge from implemented technologies to advanced processes, we significantly reduce the number of required circuit simulations in newer technologies, effectively optimizing the parameter design space for next-generation processes. Furthermore, we establish Gaussian process regression models that correlate design variables with performance metrics across both advanced and implemented technologies. These models undergo continuous updates during iteration, enabling collaborative optimization throughout the technology migration engineering process.**

Index Terms—**Process migration, transfer learning,multi-objective Bayesian optimization, performance optimization**

I. INTRODUCTION

The exponential growth in design complexity presents significant challenges for Electronic Design Automation (EDA) tools. As designs scale up, these tools must perform increasingly sophisticated computation, simulation, and analysis operations—including layout planning, wiring, and various physical design steps—resulting in substantially increased runtime and computational resource requirements. Traditional empirical methods for determining design variables have become inadequate for handling such complexity.

While current design space exploration (DSE) techniques can determine design variables to optimize performance to

some extent, they often suffer from inadequate exploration efficiency due to limited prior knowledge.

Computational resources are frequently wasted on ineffective design areas during exploration, preventing quick and accurate identification of optimal design combinations. This inefficiency particularly hampers the ability to achieve effective trade-offs between interdependent design objectives such as power consumption, performance, and area—the critical triangle of modern semiconductor design optimization.

To address the challenge of DSE, several transfer learning paradigms have been systematically investigated in the literature [1] [2]. A notable contribution in this domain is the development of a cross-domain knowledge transfer framework [1], which effectively leverages the accumulated insights from prior design exploration to enhance the efficiency of new optimization tasks. A process migration framework has been proposed [3], it can achieve process migration between advanced technology and implemented technology, enabling the advanced technology to reach the performance that the implemented technology has already achieved. However, it cannot achieve further optimization of the performance after process migration. To solve this problem, we propose dual-path Gaussian Process Regression (GPR) modeling, which conducts alternating optimization between established and emerging process technologies. It realizes the simultaneous collaborative optimization of circuit performance during process migration.

The main contributions of this paper are:

- We integrate migration learning into multi-objective

---

* Corresponding authors

979-8-3315-8850-2/25 $31.00 © 2025 IEEE

610

Bayesian optimization through Gaussian Copula, transforming Performance, Power, and Area (PPA) data into residual observations.

- Our dual-path GPR modeling enables alternating optimization between established and emerging process technologies.
- The probabilistic predictions from the established technology model provide prior knowledge that guides the optimization direction of the new technology model, creating an efficient collaborative framework.

Section I introduces the Bayesian optimization algorithm and the Thompson sampling algorithm. Section II presents the problem definition. Section III describes the Bayesian Knowledge Transfer with Residual Adaptation algorithm and the Cross - Process Bayesian Multi - Objective Collaborative Optimization Framework. Section IV exclusively focuses on the presentation and analysis of the final experimental results.

I. BACKGROUND

A. Bayesian Optimization

Bayesian Optimization (BO) stands as a sophisticated probabilistic model-based framework for global optimization, distinguished by its seamless integration of objective function modeling and strategic decision-making. BO constructs a surrogate model—typically a Gaussian Process—to create probabilistic approximations of unknown objective functions while employing acquisition functions (AF) such as the Expected Improvement criterion to skillfully balance exploration of model uncertainty with exploitation of currently identified optimal solutions. Through this methodical iterative approach, BO efficiently converges toward global optima while requiring remarkably few function evaluations.

Algorithm 1 Bayesian Optimization

Require: Surrogate Model M, Acquisition Function $\alpha(\cdot)$, Constraint Search Space $\mathcal{X}$, Black - box Function f

1: Sample initial constraints and obtain corresponding objectives, denote as $D = \text{sample}(f, \mathcal{X})$;
2: **while** Iteration not finished **do**
3:     Fit the surrogate model with sampled data D, obtain probability model $p(f|\boldsymbol{x}, D)$;
4:     Select new point: $\boldsymbol{x}_i \leftarrow \arg\max \alpha(x, p(f|\boldsymbol{x}, D))$;
5:     Run EDA tools to get real objectives $y_i = f(\boldsymbol{x}_i)$;
6: **end while**
7: **return** Set of design constraint configurations X.

Algorithm 1 illustrates the implementation process of Bayesian optimization. In the initial phase, sampling is performed on constraint configurations. The sampled initial constraint configurations are input into EDA tools to obtain corresponding objective values. Next, an initial dataset is used to construct a surrogate model, which depicts the black-box function f reflecting the relationship between constraint configurations and objectives. When entering the iterative stage, in each iteration, a new collection of constraint configurations

is selected under the guidance of AF. These configurations are fed into EDA tools to acquire actual PPA values, which serve for updating the surrogate model. Finally, a set of design constraint configurations X is output within the design constraint space $\mathcal{X}$, realizing the optimal objective y in the objective space y.

In Bayesian optimization frameworks, the Gaussian Process Regression (GPR) [4] serves as the probabilistic surrogate model that enables Bayesian uncertainty quantification, while the optimization engine translates this uncertainty into sequential decision-making through acquisition functions. This synergistic interplay establishes a mathematically grounded mechanism for navigating high-dimensional black-box optimization problems. The intrinsic compatibility between GPR's non-parametric modeling flexibility and Bayesian optimization's exploration-exploitation balance manifests particular efficacy in evaluation-costly engineering scenarios, offering both theoretical guarantees (through convergence proofs) and practical feasibility (via computationally tractable covariance kernels). This dual-aspect advancement positions the methodology as a paradigm-shifting tool for complex system optimization, particularly in aerospace design and micro-scale additive manufacturing.

B. Thompson Sampling

Reveal the nonlinear mapping relationship between input variables and PPA performance, $\mathbf{X}^{\text{old}}$ is defined as the old process design constraint configuration vector and $\mathbf{Y}^{\text{old}}$ is the actual PPA performance vector of the corresponding process node. $\mathbf{D}^{\text{old}} = \{\mathbf{X}^{\text{old}}, \mathbf{Y}^{\text{old}}\}$ denotes the dataset corresponding to the circuit.

A single target y_i^{old} in the $\mathbf{D}^{\text{old}}$ is transformed into a standard uniformly distributed random variable $u_i = F(y_i^{\text{old}})$, where F is the cumulative distribution function (CDF) of $y_{\cdot i}^{\text{old}}$. The CDF computation involves two steps: 1. Convert multi-dimensional indicators into a one-dimensional sequence $(y_{1i}^{\text{old}}, y_{2i}^{\text{old}}, \dots)$ and sort them by their numerical values. 2. For each observation, Compute the ratio of observations whose values are no greater than the given one, as presented below:

$$F(y_i^{\text{old}}) = \frac{\text{Number of observations} \leq y_i^{\text{old}}}{\text{Total number of observations}} \quad (1)$$

The Gaussian Copula is adopted to develop a model aiming at the CDF and the associated variable $z_{\cdot i}^{\text{old}}$ is defined as $z_{\cdot i}^{\text{old}} = \psi(y_{\cdot i}^{\text{old}}) = \Phi^{-1}\left(F(y_{\cdot i}^{\text{old}})\right)$

The process of Thompson sampling is depicted in Algorithm 2. Gaussian Copula is utilized to build the correlations among the implemented technologies. The first GPR model is fitted using X^{old} and Z^{old}, referred to as F_GPR, The key goal of F_GPR is to extract and combine the related prior knowledge from the relationships between the implemented technologies. Given a new constraint configuration x old, the prediction distribution $p(z_i^{\text{old}}|x^{\text{old}}) \sim \text{N}(\mu(x^{\text{old}}), \Sigma(x^{\text{old}}))$ is determined by $\mu(x^{\text{old}})$ and $\Sigma(x^{\text{old}})$. This prediction Gaussian distribution provides probability support for Thompson sampling.

Algorithm 2 Thompson Sampling

Require: F_GPR, $\mathbf{X}^{\text{old}}$, $\mathbf{Z}^{\text{old}}$

1: Utilize $\mathbf{X}^{\text{old}}$ and $\mathbf{Z}^{\text{old}}$ to fit **F_GPR**, which can output the prediction Gaussian distribution $p(z_i^{\text{old}}|x^{\text{old}}) \sim \mathrm{N}(\mu(x^{\text{old}}), \Sigma(x^{\text{old}}))$;

2: **while** Iteration not finished **do**

3:    Randomly sample K random variables $X^{\text{new}1}$, $X^{\text{new}2}$, ..., $X^{\text{new}K}$ under advanced technology;

4:    Draw $z_i^{\text{old}} \sim \mathrm{N}(\mu(X_i^{\text{new}}), \Sigma(X_i^{\text{new}}))$ for $i = 1, \ldots, K$ through (2);

5:    Obtain $Y_i^{\text{new}} = f(X_i^{\text{new}})$ via EDA tools;

6: **end while**

7: **return** the data $\mathcal{D}^{\text{new}} = \{X^{\text{new}}, Y^{\text{new}}\}$ with good performance under advanced technology.

In iteration, sample K candidate configs x_i^{new} of advanced technology. Use $z_{i\cdot}^{\text{old}} \sim N(\mu(x_i^{\text{new}}), \Sigma(x_i^{\text{new}}))$ to evaluate their performance and pick the best design constraint by (2)

$$i = \arg\min \left(z_i^{\text{old}} + \alpha_1 z_i^{\text{old}} + \cdots + \alpha_{N-1} z_i^{\text{old}} \right) \quad (2)$$

After getting the needed initial data, we use EDA tools to figure out the real PPA performances y^{new} of the advanced technology. It results in a dataset $D^{\text{new}} = \{X^{\text{new}}, Y^{\text{new}}\}$. GPR modeling is the key to extracting prior knowledge, and incorporating prior knowledge can dramatically reduce the cost of running an EDA tool. We can optimize the operation of EDA tools to make them more efficient.

II. PROBLEM DEFINITION

A. Definition of Process Migration

Process migration is defined as the process of benchmarking advanced technology nodes against the superior PPA metrics of implemented technologies, while leveraging algorithmic optimization to minimize the number of EDA tool simulations required to achieve comparable or equivalent performance levels in advanced process circuits.

B. Problem Statement

The core challenge addressed in this study is the synchronous co-optimization of PPA metrics for both implemented and advanced technologies during migration, ensuring that the final results outperform the pre-migration target values for both generations of processes. This requires developing a framework that not only preserves legacy performance in advanced circuits but also reciprocally enhances the PPA characteristics of both technology generations through iterative optimization.

$$\min_{\mathbf{x}_{\text{old}}, \mathbf{x}_{\text{new}}} \quad \omega_1 \cdot f_{\text{old}}(\mathbf{x}_{\text{old}}) + \omega_2 \cdot f_{\text{new}}(\mathbf{x}_{\text{new}})$$

$$\text{s.t.} \quad \Delta\text{FOM} = \frac{|f_{\text{new}}(\mathbf{x}_{\text{new}}) - f_{\text{old}}^{\text{target}}|}{\max\left(|f_{\text{new}}(\mathbf{x}_{\text{new}})|, |f_{\text{old}}^{\text{target}}|\right)} \le \delta \quad (3)$$

$$\mathbf{x}_{\text{old}} \in \mathcal{X}_{\text{old}}, \quad \mathbf{x}_{\text{new}} \in \mathcal{X}_{\text{new}}$$

III. METHOD

A. Overall Flow

The key idea of the framework is to leverage empirical knowledge from past designs to help optimize the design process for next-generation processes. Through transfer learning, repeat simulations are reduced, and the GPR model is optimized to achieve co-optimization and process migration in parallel during the process migration to improve design efficiency.

In the initial step, we leverage Algorithm 2 to acquire $D_{\text{new}} = \{X_{\text{new}}, Y_{\text{new}}\}$. Subsequently, Algorithm 3 is employed to derive the residual observations R_{new} that correspond to X_{new}. Following this, the obtained results are integrated into the Cross-Process Bayesian Multi-Objective Collaborative Optimization Framework for the construction of GPR models. Throughout the iterative process, the GPR models are continuously updated and optimized, which effectively facilitates process migration and collaborative optimization.

B. Bayesian Knowledge Transfer with Residual Adaptation

Algorithm 2 shows how to make corrections and integrate prior knowledge.

First, the matrix $\mathbf{Y}^{\text{old}}$ is transformed into a matrix of

---

Algorithm 3 Bayesian Knowledge Transfer with Residual Adaptation

---

Require: Dataset $\mathcal{D}^{\text{old}} = \{\mathbf{X}^{\text{old}}, \mathbf{Y}^{\text{old}}\}$

1: Transform $\mathbf{Y}^{\text{old}}$ to the correlative variable $\mathbf{Z}^{\text{old}}$ via Gaussian Copula

2: Utilize Algorithm 2 Thompson sampling gathers the initial data $\mathcal{D}^{\text{new}} = \{\mathbf{X}^{\text{new}}, \mathbf{Y}^{\text{new}}\}$

3: Transform $\mathbf{Y}^{\text{new}}$ to the correlative variable $\mathbf{Z}^{\text{new}}$ via Gaussian Copula

4: Obtain the residual observation from the $\mathbf{R}^{\text{new}}$ containing the **F_GPR** probability model (4)

5: **return** Design constraint configurations $\mathbf{X}^{\text{new}}$ and the residual observation $\mathbf{R}^{\text{new}}$ of the advanced technology

---

correlated variables $\mathbf{Z}^{\text{old}}$ using Gaussian Copula. Then, the F_GPR model was fitted using the datasets $\mathbf{X}^{\text{old}}$ and $\mathbf{Z}^{\text{old}}$. Next, a Thompson sampling strategy is used to obtain the initial data for advanced technologies. Finally, the residual observation matrix R^{new} is computed using (3). Algorithm 3 gives the design constraint configuration X^{new} and the residual observations R^{new}, denoted as $\mathbf{R}^{\text{new}} = \{X^{\text{new}}, R^{\text{new}}\}$.

The residual observation serves two purposes. We define $h(x^{\text{new}}) = z_i^{\text{new}}$. First, it holds prior information about implemented and advanced technologies. Two, it reduces the data required for Bayesian optimization.

$$r(\boldsymbol{x}^{\text{new}}) = \frac{h(\boldsymbol{x}^{\text{new}}) - \mu(\boldsymbol{x}^{\text{new}})}{\Sigma(\boldsymbol{x}^{\text{new}})} \quad (4)$$

The residual observation plays two crucial roles. Firstly, it encapsulates prior information regarding both the implemented technologies and the advanced process. Secondly, it reduces the volume of data required for Bayesian optimization.

C. Cross-Process Bayesian Multi-Objective Collaborative Optimization Framework

In the optimization framework, the second GPR model—termed S_GPRs—is selected as the surrogate model. S_GPRs ($M_1 \ldots M_N$) are constructed by leveraging the training data R^{new} obtained via Algorithm 2. This training dataset incorporates the design constraint configuration $\mathbf{x}^{\text{new}}$ and its corresponding residual observation $\mathbf{r}_i^{\text{new}}$, embodying the prior knowledge derived from implemented technologies.

To derive the primal Pareto optimality bound (PPO), we employ the robust optimization algorithm Non-dominated Sorting Genetic Algorithm II (NSGAII) [7] as the multi-objective optimization algorithm. Specifically, the acquisition function (AF, e.g., EI) is utilized to define the objective functions of the optimization problem as $\text{AF}(M_1, \mathbf{x}^{\text{new}})$, $\text{AF}(M_2, \mathbf{x}^{\text{new}})$, $\ldots$, $\text{AF}(M_N, \mathbf{x}^{\text{new}})$. Here, M_i represents the S_GPR model, which captures the nonlinear relationship between design constraint configurations and residual observations. Solving the primal multi-objective (PMO) problem yields the Pareto optimal solution set $\mathbf{X}_{PO}$.

$$X_{PO} \leftarrow \min_{x^{\text{new}} \in X} \left(AF(M_1, x^{\text{new}}), \cdots, AF(M_N, x^{\text{new}}) \right) \quad (5)$$

When dealing with the inputs within the Pareto optimal solution set X^{PO}, we leverage the S_GPR model M_i to compute the mean $\mu_{r_i}(x^{\text{new}})$ and the standard deviation $\sigma_{r_i}(x^{\text{new}})$.

In the objective space, for any given point, its mean can be characterized as $\mu_{r_j}(x^{\text{new}}) = [\mu_{r_{j1}}(x^{\text{new}}), \cdots, \mu_{r_{jN}}(x^{\text{new}})]$, and its standard deviation is represented as $\sigma_{r_j}(x^{\text{new}}) = [\sigma_{r_{j1}}(x^{\text{new}}), \cdots, \sigma_{r_{jN}}(x^{\text{new}})]$.

By leveraging the computed mean μ and standard deviation σ of these points, we construct an uncertainty hyper-rectangle associated with each input. The hyper-rectangle is defined as:

$$HR_\beta(x^{\text{new}}) = \{\mu - \beta\sigma \le r \le \mu + \beta\sigma\} \quad (6)$$

Where β balances exploration and exploitation. Points are then ranked and selected based on the maximum diagonal length of their uncertainty bounds:

$$x_{t+1}^{\text{new}} = \arg \max_{r,r' \in HR_\beta(x^{\text{new}})} \|r - r'\|_2. \quad (7)$$

The candidate with the highest exploration potential is selected for EDA tool validation. The actual PPA metrics obtained from this evaluation are then incorporated into the subsequent optimization iteration. The newly obtained x_{t+1}^{new} and y_{t+1}^{new} are substituted into Equation (3) to derive the corresponding R^{new} for x_{t+1}^{new}.

After obtaining x_{t+1}^{new} and the corresponding R^{new} under the advanced technology, we fit the updated X^{new} and y^{new} to the third GPR model, denoted as the T_GPR model. Subsequently, we input the prior knowledge X^{old} from the obsolete technology into the T_GPR model to compute the corresponding residual observations R^{old} for X^{old}.

We construct the fourth GPR model, named Fo_GPR,

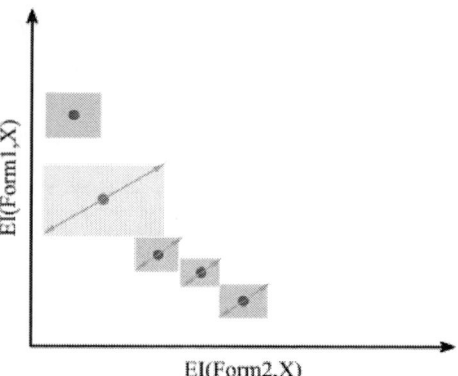

Fig. 1. The design constraints are derived from the uncertainty maximization criterion in Equation (6).

which is similar to the S_GPR model, using the obtained X^{old} and r_i^{old}. Leveraging this Fo_GPR model and the AF, we employ the NSGAII to obtain the Pareto optimal solution set X^{PO} for X^{old}. After substituting X^{PO} into the Fo_GPR model, we calculate the mean and standard deviation. Then, we measure the diagonal length of each input X^{old} within X^{PO} and select the most promising point according to Equation (6). Subsequently, we input this point into the EDA tools for simulation under the obsolete technology. The actual PPA performances obtained from this simulation are then utilized for the next iteration of the optimization process.

Algorithm 4 Cross-Process Bayesian Multi-Objective Collaborative Optimization Framework

---

Require: Dataset $\mathcal{D}^{\text{old}}$, maximum iteration number $T_{\max}$

1: Obtain design constraint configurations $\mathbf{X}^{\text{new}}$ and the residual observation $\mathbf{R}^{\text{new}}$ through Algorithm 3;

2: **while** $Iter \le T_{\max}$ **do**

3: Build S_GPR M_i with $\mathbf{x}^{\text{new}}$ and r_i^{new} for each objective;

4: Solve Equation (5) via NSGA-II;

5: Select $\mathbf{x}_{t+1}^{\text{new}}$ via Equation (7);

6: Obtain $\mathbf{y}_{t+1}^{\text{new}}$ by running EDA flow;

7: Aggregate data: $\mathcal{D}^{\text{new}} \leftarrow \mathcal{D}^{\text{new}} \cup (\mathbf{x}_{t+1}^{\text{new}}, \mathbf{y}_{t+1}^{\text{new}})$

8: Update $\mathbf{R}^{\text{new}}$ via Equation (4);

9: Build T_GPR with $\mathbf{x}^{\text{new}}$ and y_i^{new} for each objective;

10: Obtain the residual observation from the $\mathbf{R}^{\text{old}}$ containing the T_GPR probability model (4)

11: Build Fo_GPR H_i with $\mathbf{x}^{\text{old}}$ and r_i^{old} for each objective;

12: Solve Equation (5) via NSGA-II;

13: Select $\mathbf{x}_{t+1}^{\text{old}}$ via Equation (7);

14: Obtain $\mathbf{y}_{t+1}^{\text{old}}$ by running EDA flow;

15: Aggregate data: $\mathcal{D}^{\text{old}} \leftarrow \mathcal{D}^{\text{old}} \cup (\mathbf{x}_{t+1}^{\text{old}}, \mathbf{y}_{t+1}^{\text{old}})$;

16: $Iter \leftarrow Iter + 1$;

17: **end while**

18: **return** Predicted Pareto-optimal frontier $\mathcal{P}_0$

In each iteration, the datasets $D_{\text{old}} = \{X_{\text{old}}, Y_{\text{old}}\}$ and $D_{\text{new}} = \{X_{\text{new}}, Y_{\text{new}}\}$ are updated. Through a data-driven iterative optimization framework, the PPA metrics of both advanced and obsolete technologies undergo continuous optimization: the PPA values of the obsolete technology converge toward higher efficiency, while those of the advanced technology approach the optimized performance baseline of the obsolete process. As the number of iterations increases, the PPA discrepancies between the two technologies progressively narrow, ultimately achieving co - optimization of PPA performance metric and enabling process migration.

IV. EXPERIMENTAL RESULTS

A. Experimental Settings

To validate the effectiveness of the proposed Cross-Process Bayesian Multi-Objective Collaborative Optimization Framework, this experiment utilizes charge pump circuits. The implemented technology is based on SMIC 180nm as prior knowledge, while the advanced technology is represented by SMIC 40nm. In Algorithm 3, 30 PPA - similar data points were carefully selected from the design space of the implemented 180nm technology. These selected points serve as prior knowledge, providing valuable insights into the behavior and characteristics of the 180nm technology while acting as initial migration targets for technology transfer. In each iteration of the optimization process, the weights for multiple objectives $\alpha_1, \ldots, \alpha_{N-1}$ are adjusted to distinct values. For a two-objective scenario, α is uniformly set to 0.5.

This experiment targets Static Deviation (SD) and Figure of Merit (FOM) as optimization objectives during cross-process migration of charge pump circuits. Specifically, SD is defined as the sum of absolute deviations between measured upper/lower stage average currents and the target value, while FOM combines total current ripple (sum of four directional current fluctuations) and SD using weighting coefficients.

B. Results and Runtime Analysis

In the experimental result figures. The 0th iteration point represents the optimal PPA configuration derived from prior data Y^{old} and Y^{new} obtained via the Algorithm 2 Thompson sampling strategy during the initial iteration phase. The y-axis displays the PPA values of cumulative sample points and current optimal solutions at each iteration step, reflecting the convergence trajectory of the optimization process.

From the experimental results, it can be observed that the experiment achieved the experimental goal after the 15th iteration. The Fom value of the 40nm circuit after migration is 1.966, and that of the 180nm circuit after migration is 1.668. Overall, compared with the initial Best Fom value of the 180nm circuit, it is optimized by approximately 93%. The Dev value of the 40nm circuit after migration is 0.38, and that of the 180nm circuit after migration is 0.2. Overall, compared with the initial Best Dev value of 1.1 for the 180nm circuit, it is optimized by approximately 74%.

First, we conduct an experiment using Algorithm 4 without the addition of T_{GPR} and F_{GPR} and the update of the X^{old}

points as a comparison. We define it as the Non - Optimization Framework. In the experiment, the dataset $\mathcal{D}^{\text{old}} = \{\mathbf{X}^{\text{old}}, \mathbf{Y}^{\text{old}}\}$ remains unchanged. The experimental results are shown in Figures 2 and 3.

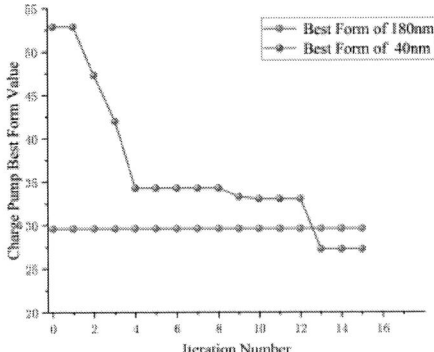

Fig. 2. Cross - Process Migration Results of FOM Performance Metrics under the Non - Optimization Framework

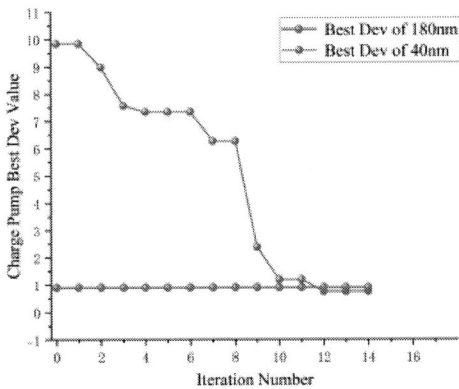

Fig. 3. Cross-Process Migration Results of DEV Performance Metrics under the Non - Optimization Framework

Next, the experiment is conducted under the complete Algorithm 4, with the objectives of process migration and collaborative optimization. The experimental results shown in Figures 4 and 5 illustrate the collaborative optimization results of charge pump circuits during 180nm to 40nm process migration.

V. CONCLUSION

We propose a Cross-Process Bayesian Multi-Objective Collaborative Optimization Framework, where a Gaussian copula is employed to model the underlying correlations between different technologies. Multi-objective Bayesian optimization is smoothly integrated with prior knowledge for transfer learning. Four GPR surrogate models are developed to extract multi-dimensional features and characterize the complex relationships between design constraints and objectives. Experimental

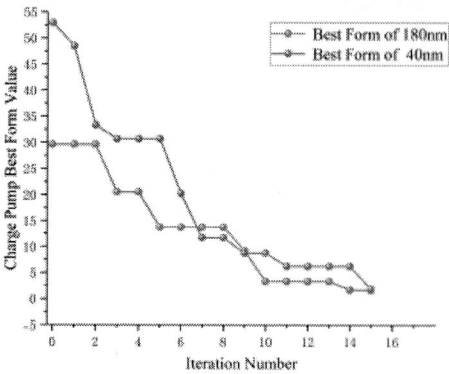

Fig. 4. Cross-Process Migration Results of FOM Performance Metrics under the Optimization Framework

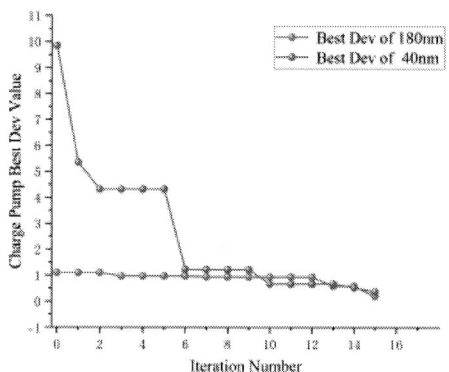

Fig. 5. Cross-Process Migration Results of DEV Performance Metrics under the Optimization Framework

results demonstrate that the framework effectively achieves cross-process migration and multi-objective collaborative optimization goals.

ACKNOWLEDGMENT

This research is supported by Yangtze River Delta Science and Technology Innovation Community Joint Fundamental Research Project 2024CSJZN0500 , 2024CSJZN0503.

REFERENCES

[1] J. Kwon and L. P. Carloni, "Transfer learning for design-space exploration with high-level synthesis," in Proc. MLCAD. IEEE, 2020, pp.163–168.

[2] J. Liu, M. Hassanpourghadi, Q. Zhang, S. Su, and M. S.-W. Chen,"Transfer learning with bayesian optimization-aided sampling for efficient ams circuit modeling," in Proc. ICCAD, 2020, pp. 1–9.

[3] Zhang M, Zhang Z, Niu Y, et al. Fast Constraints Tuning via Transfer Learning and Multi-Objective Optimization[J]. IEEE Transactions on Computer-Aided Design of Integrated Circuits and Systems, 2024.

[4] C. K. Williams and C. E. Rasmussen, Gaussian processes for machine learning. MIT press Cambridge, MA, 2006, vol. 2, no. 3

[5] Davis W R, Franzon P, Francisco L, et al. Fast and accurate PPA modeling with transfer learning[C]//2021 IEEE/ACM International Conference On Computer Aided Design (ICCAD). IEEE, 2021: 1-8.

[6] Q. Sun, T. Chen, S. Liu, J. Miao, J. Chen, H. Yu, and B. Yu, "Correlated multi-objective multi-fidelity optimization for HLS directives design," in Proc. DATE. IEEE, 2021, pp. 46–51.

[7] K. Deb, A. Pratap, S. Agarwal, and T. Meyarivan, "A fast and elitist multiobjective genetic algorithm: Nsga-ii," IEEE TEVC, vol. 6, no. 2,pp. 182–197, 2002.

[8] A. G. Wilson, Z. Hu, R. Salakhutdinov, and E. P. Xing, "Deep kernel learning," in Proc. AISTATS. PMLR, 2016, pp. 370–378.[Digests 9th Annual Conf. Magnetics Japan, p. 301, 1982].

[9] C. Lo and P. Chow, "Model-based optimization of high level synthesis directives," in Proc. FPL. IEEE, 2016, pp. 1–10.

[10] Changro Lee, "Random Forest with Transfer Learning: An Application to Vehicle Valuation," Journal of Advances in Information Technology, Vol. 13, No. 4, pp. 326-331, August 2022.

Multi-Branch Autoencoder Networks for Efficient Inverse Design of Wideband Frequency-Selective Surfaces

Wei Jiang[1], Haoran Huang[1], Guangxin Liao[1], Shenli Zheng[1], Jiacheng Guo[1*], Yuan Du[1*]

[1]School of Electronic Science and Engineering, Nanjing University, Nanjing, China

*Email: guojiacheng@nju.edu.cn, yuandu@nju.edu.cn

Abstract—Traditional electromagnetic design automation (EDA) methods for frequency-selective surfaces (FSS) are computationally expensive and inefficient for complex designs. Neural network-based surrogate models have been introduced to accelerate the process, but conventional architectures struggle to accurately predict electromagnetic responses due to limited frequency-dependent feature extraction. To address this, we propose a Multi-Branch Autoencoder (MBAE) framework that separately encodes structural and frequency-domain features, significantly improving predictive accuracy. The integration of Fourier Neural Operators (FNO) enhances the model's ability to learn complex electromagnetic behaviors, while an adaptive Differential Evolution (DE) algorithm improves global optimization. Experimental results show that our method achieves the target frequency response (19.5–25 GHz) in just seven iterations, delivering a 58×-117× speedup over full-wave simulations and a 152.8× acceleration per iteration compared to manual optimization. Notably, it outperforms conventional neural networks in predictive accuracy, providing a more efficient and scalable solution for advanced electromagnetic design.

Index Terms—Frequency Selective Surface, Inverse Design, Neural Network, Deep Learning, EDA

I. INTRODUCTION

Frequency Selective Surfaces (FSS) are widely employed in applications such as wireless communication, radar systems, and electromagnetic shielding due to their ability to manipulate electromagnetic waves based on frequency, angle of incidence, and polarization [1–5]. Nevertheless, the ever-growing demands for wideband operation, multi-polarization adaptability, and stable performance under large incident angles pose significant challenges to conventional FSS design methodologies [6, 7].

Traditionally, FSS design relies on expert-driven iterative optimization, where geometric parameters are manually adjusted and validated through full-wave electromagnetic simulations. As illustrated in Fig. 1, the traditional design approach involves software simulations and iterative manual refinements, although effective for simple structures, such methods are computationally intensive and encounter difficulties in multi-objective optimization, particularly within

This work was supported in part by the National Key Research and Development Program of China under Grant No. 2021YFA0717700, in part by the Natural Science Foundation of China under Grant 62371223, in part by the Strategic Industries and Key Technologies Project of Jiangsu Province under Grant BE2023020-3, and in part by the Basic Research Program of Jiangsu Province under Grant BK20243042.

high-dimensional design spaces. Consequently, conventional approaches frequently converge to suboptimal solutions and demand substantial computational resources, thereby limiting their utility in high-performance, modern FSS applications.

To mitigate these challenges, data-driven inverse design methods leveraging neural networks have gained increasing attention. As demonstrated in Fig. 1, neural networks offer a powerful means to establish direct mappings between desired electromagnetic responses and geometric configurations, significantly reducing design iterations and computational costs. However, existing neural network-based methods exhibit several critical limitations. Surrogate models used to predict frequency responses often suffer from limited accuracy, leading to suboptimal designs. Furthermore, these models are prone to local optima, particularly when navigating complex, high-dimensional design spaces [8, 9].

One key factor contributing to these challenges is the inherent difficulty in accurately mapping a relatively low-dimensional parameter space to a high-dimensional frequency response space (typically several hundred dimensions). Traditional network architectures struggle to capture intricate relationships due to the non-linearity of the transformation. To address this issue, we propose a novel inverse design framework that integrates a Multi-Branch Autoencoder (MBAE) network with an enhanced Differential Evolution (DE) algorithm. The framework is designed for two primary objectives: (1) improving the predictive accuracy of surrogate models for frequency response estimation through tailored architectural advancements and (2) enhancing the global optimization capabilities of evolutionary algorithms to ensure robust and efficient design convergence.

The proposed MBAE framework extracts both frequency-domain and structural features to improve the accuracy of electromagnetic response modeling. Its dual-branch architecture enhances feature representation, enabling more precise frequency response predictions. To further refine frequency information extraction, we integrate Fourier Neural Operators (FNO) [10], which strengthen spectral feature learning. Additionally, an enhanced Differential Evolution (DE) algorithm with dynamic parameter adaptation improves optimization performance and mitigates premature convergence.

The remainder of this paper is structured as follows. Section II outlines the data-driven inverse design flow. Section III de-

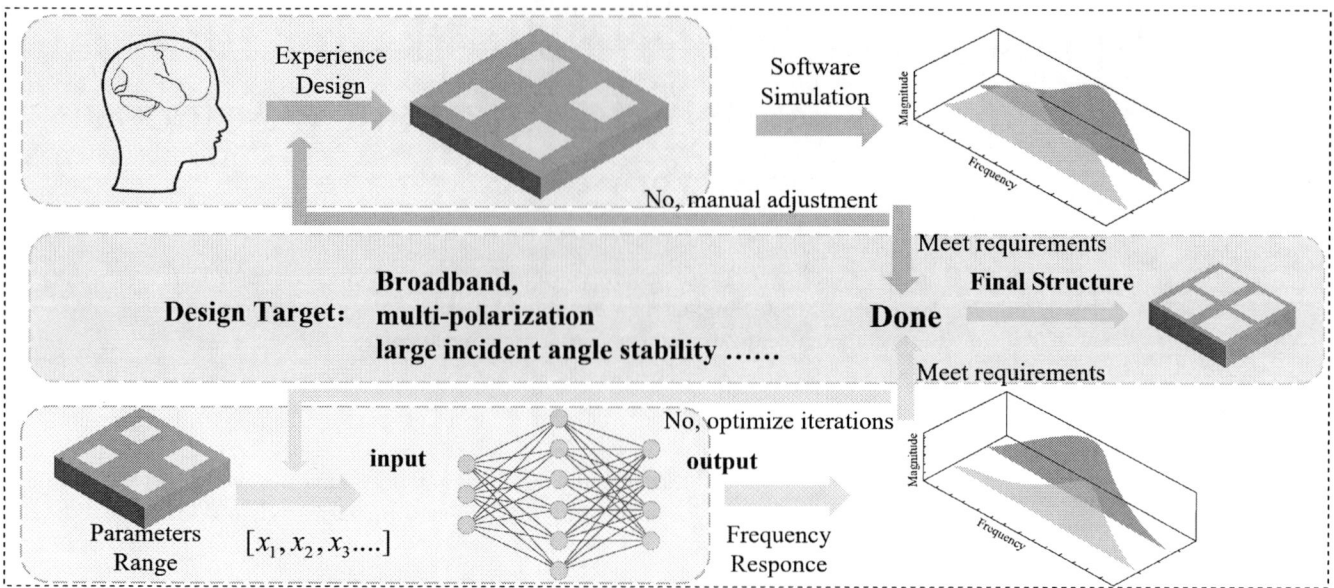

Fig. 1. Two approaches for FSS design: a traditional method involving expert design and manual adjustments (top), and a neural network method utilizing prediction and automatic optimization (bottom)

tails the proposed methodology. Section IV presents simulation and experimental validation, demonstrating the effectiveness of the proposed approach. Finally, Section V concludes the paper and discusses potential directions for future research.

II. PRELIMINARY

This section outlines the fundamental process of deep learning-assisted FSS design, detailing the workflow from initial modeling to final optimization. The design of FSS is inherently complex, requiring careful attention to both modeling and optimization.

The process begins with the construction of a foundational model that defines the initial structure and characterizes its potential electromagnetic (EM) responses. Let the design parameters of the FSS be represented as a vector $\mathbf{x} = [x_1, x_2, \ldots, x_n]$, where x_i represents the geometric or material property i, and n is the number of design variables. The target performance is expressed in terms of the frequency response $R(f; \mathbf{x})$, where f is the frequency and R is a performance metric (e.g., reflection, transmission, or absorption). The goal is to find a good initial design $\mathbf{x}_0$ such that $R(f; \mathbf{x}_0)$ closely meets baseline performance requirements $R_{\text{target}}(f)$:

$$\|R(f; \mathbf{x}_0) - R_{\text{target}}(f)\| < \epsilon \quad (1)$$

where ϵ is an acceptable error threshold.

Once the foundational model is established, key design parameters (e.g., geometric dimensions, material properties) are identified. These parameters are critical in influencing the frequency-selective performance and are treated as variables in later stages. High-fidelity electromagnetic simulations are then employed to generate a dataset $\mathcal{D} = \{(\mathbf{x}^{(i)}, f^{(i)})\}_{i=1}^{M}$, where M is the size of the dataset. This dataset consists of pairs of design parameters $\mathbf{x}$ and their corresponding frequency responses f, which the network will use to learn the underlying relationship between them.

Next, a deep learning model $\hat{M}(\mathbf{x}; \boldsymbol{\theta})$ is trained on the dataset $\mathcal{D}$, where $\boldsymbol{\theta}$ denotes the trainable parameters of the model. The objective of the training process is to minimize the error between the predicted frequency responses and the true frequency responses:

$$\mathcal{L}(\boldsymbol{\theta}) = \frac{1}{M} \sum_{i=1}^{M} \|\hat{M}(\mathbf{x}^{(i)}; \boldsymbol{\theta}) - f^{(i)}\|^2 \quad (2)$$

where $\mathcal{L}$ is the loss function. Through this process, the deep learning model learns the non-linear mappings between design parameters and frequency responses, significantly reducing the dependence on computationally expensive manual simulations.

Following the training phase, the process transitions into optimization. In this step, advanced optimization algorithms such as Genetic Algorithms (GA) or DE are employed to explore the global design space efficiently. The optimization problem can be formulated as:

$$\mathbf{x}^* = \arg\min_{\mathbf{x}} \mathcal{C}(\mathbf{x}) \quad (3)$$

where $\mathcal{C}(\mathbf{x})$ is a cost function representing design constraints (e.g., physical feasibility, material availability) or additional objectives (e.g., minimizing thickness or weight). These algorithms iteratively adjust design parameters $\mathbf{x}$, avoiding local minima while ensuring global convergence, using mutation, crossover, and selection strategies. Ultimately, the optimization algorithm will output the optimal combination of FSS structural parameters, denoted as $\mathbf{x}^*$, which represents the best achievable design within the given requirements and

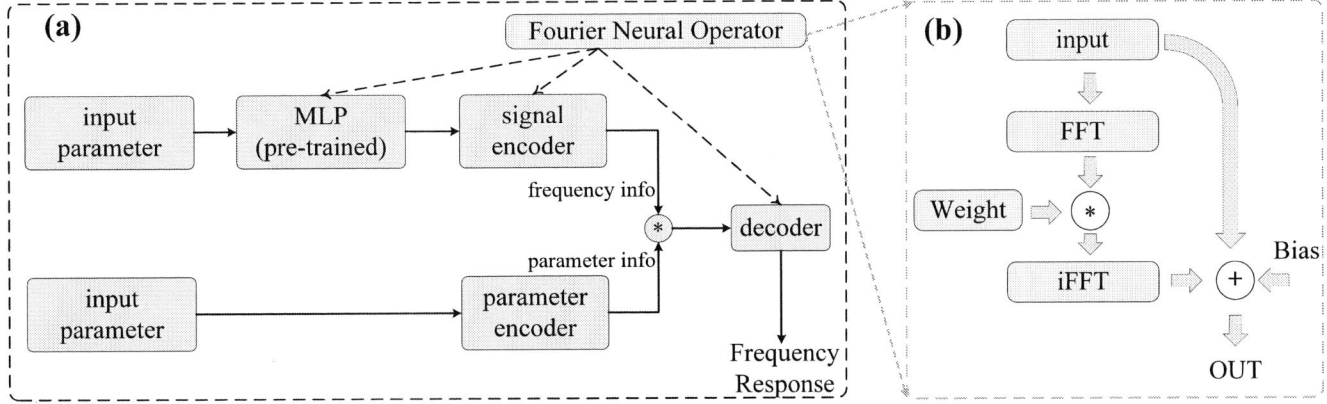

Fig. 2. Multi-branch autoencoder-based deep learning framework: (a) overall network structure and workflow, (b) Fourier operator process for frequency-domain signal encoding (FFT, weight modulation, IFFT)

constraints or the best possible design within the explored capabilities.

III. THE PROPOSED METHOD

To tackle the challenges in accurately extracting frequency response characteristics for FSS design, this paper introduces a deep learning network built upon the MBAE architecture. The MBAE structure is specifically designed to enhance the model's predictive capability for complex frequency response patterns, allowing for more precise and reliable predictions. Additionally, an enhanced DE algorithm is integrated to strengthen the exploration and optimization capabilities within the intricate parameter space, ensuring more efficient and effective design optimization.

To effectively capture the complex relationships between the geometric and material parameters of FSS and their corresponding frequency-domain responses, we propose a MBAE framework as shown in Fig. 2(a). Inspired by the architecture of DeepONet [11], MBAE employs a dual-branch structure designed to enhance feature extraction.

The MBAE framework consists of two specialized branches: the parameter encoding branch and the signal encoding branch. Each branch is responsible for capturing different aspects of the FSS design, and their outputs are fused within a latent space before being processed by the decoder.

1) Parameter Encoding Branch: The parameter encoding branch is designed to extract meaningful correlations from the geometric and material parameters of the FSS unit cell. To achieve this, we utilize a one-dimensional convolutional neural network (Conv1D), which is well-suited for identifying local interactions and dependencies within structured input data. The Conv1D layers facilitate hierarchical feature extraction, enabling the model to learn intricate relationships among design parameters. These extracted features are then mapped into the shared latent space.

2) Signal Encoding Branch: The signal encoding branch aims to encode the frequency-domain characteristics of the FSS response. Unlike the parameter encoding branch, this module employs a pre-trained multilayer perceptron (MLP), which has been trained separately to learn the intrinsic mapping between input parameters and their corresponding frequency responses. During the overall training phase, this MLP is fine-tuned to align with the end-to-end optimization of the MBAE network. By leveraging this pre-trained MLP, the encoded features inherently capture frequency-domain properties, offering a more structured input representation for subsequent processing.

3) Latent Space Fusion and Decoding: Once the features from both encoding branches are extracted, they are fused in the latent space through element-wise multiplication, ensuring a balanced integration of parameter-based and signal-based information. The resulting latent representation is then passed through a decoder for frequency response reconstruction. This decoder incorporates multiple FNO components to enhance its ability to extract high-level spectral features.

As illustrated in Fig. 2(b), the FNO transformation follows these steps:

$$v_F = \text{IFFT}(W \times \text{FFT}(v_0)) \tag{4}$$

$$v_{out} = v_F + v_0 + b \tag{5}$$

where v_0 is the input feature, FFT and IFFT represent the Fast Fourier Transform and its inverse, respectively, W denotes the learnable weights in the Fourier domain, and b is the learnable bias term. These Fourier-based operations enable the efficient transformation of encoded features into accurate frequency responses while maintaining signal integrity through residual connections.

By adopting this multi-branch design, MBAE effectively learns the complex interplay between FSS structural parameters and frequency responses, leading to superior performance in FSS modeling and optimization.

While MBAE enhances the predictive accuracy of surrogate models, an effective optimization strategy is crucial to fully exploit its potential. Differential Evolution, a widely used

979-8-3315-8850-2/25 $31.00 © 2025 IEEE

population-based metaheuristic, is well-regarded for its simplicity and robustness in solving high-dimensional problems. However, conventional DE algorithms often struggle with fixed parameter settings, leading to suboptimal search efficiency and premature convergence in complex FSS design spaces.

To address these issues, we propose an enhanced DE algorithm that dynamically adjusts the mutation factor F and crossover rate CR based on the optimization progress. Specifically, F and CR evolve iteratively according to:

$$F = F_{\min} + (F_{\max} - F_{\min}) \cdot \left(1 - \frac{k}{K}\right) \quad (6)$$

$$CR = CR_{\min} + (CR_{\max} - CR_{\min}) \cdot \frac{k}{K} \quad (7)$$

where k is the current iteration, and K is the total number of iterations. This adaptive mechanism enables a smooth transition from global exploration in early stages (higher F, lower CR) to refined exploitation in later stages (lower F, higher CR). By dynamically balancing exploration and exploitation, the enhanced DE improves convergence efficiency and solution accuracy, making it well-suited for high-dimensional FSS optimization.

IV. Experimental Results and Validation

This section presents the experimental validation of the proposed FSS design methodology, focusing on the performance of both the enhanced DE algorithm and the deep learning framework.

A. Experimental Setup

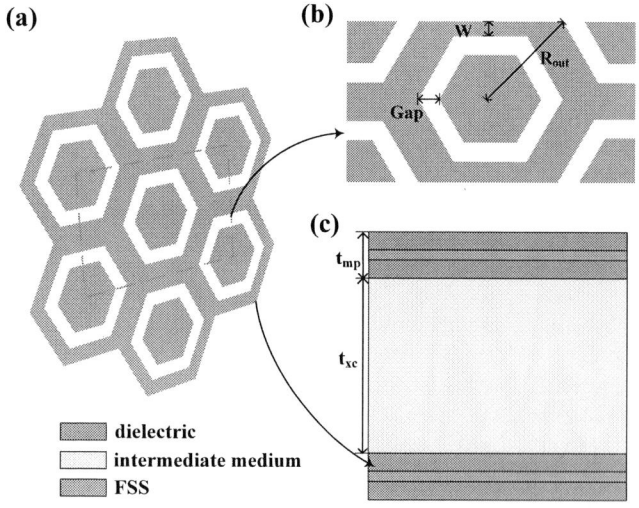

Fig. 3. Dual-layer FSS unit structure: (a) top view, (b) repeating unit with parameters $[Gap, Rout, W]$, and (c) layer thicknesses t_{mp} and t_{xc}.

The FSS response dataset was generated using electromagnetic simulation software, based on a dual-layer hexagonal unit cell structure (Fig. 3). The FSS structure is parameterized by six variables: $[Gap, R_{out}, W, t_{mp}, t_{xc}, \theta]$. Table I details the

specific parameter ranges, and all values were discretized to two decimal places to reflect manufacturing constraints. Using latin hypercube sampling, approximately 4400 samples were generated to comprehensively cover the parameter space as the training data.

TABLE I
FSS Design Parameter Ranges

Parameter	Minimum	Maximum	Unit
Gap	0.01	0.50	mm
R_{out}	1.00	1.55	mm
t_{mp}	0.60	1.40	mm
t_{xc}	2.50	3.50	mm
θ	0.00	70.00	degree
W	0.01	0.50	mm

Simulations are performed over the 5-30 GHz frequency range, sampling S-parameters at 200 equidistant points. Each simulation yields S-parameter responses for both Transverse Electric (TE) and Transverse Magnetic (TM) polarizations, resulting in 400 data points per simulation (200 points per polarization).

B. MBAE Performance Evaluation

To evaluate the effectiveness of the proposed MBAE framework, we compare its performance against a baseline model adopted from [12]. In conventional AI-based FSS design methods, the network structure typically adopts a fully connected multilayer perceptron architecture. However, MLPs may struggle to accurately capture complex electromagnetic responses. In contrast, our MBAE-based model introduces a multi-branch autoencoder architecture, as shown in Fig. 2. By leveraging the multi-branch autoencoder design, our model is better suited for capturing intricate electromagnetic interactions and improving predictive accuracy. Both models are trained under identical conditions to ensure a fair comparison.

TABLE II
Comparison of Training and Validation Loss

Model	Train Loss	Val Loss	Improvement
Baseline (MLP-based) [12]	2.14	2.48	-
MBAE-based	**0.62**	**0.67**	**72.98%**

Table II summarizes the performance of the MBAE-based model and the baseline. Notably, our model reduces the validation loss from 2.48 to 0.67, achieving a 72.98% decrease, which highlights its superior optimization capability. Fig. 4(a) further illustrates the strong predictive fitting performance of the MBAE structure after training.

C. Overall System Performance Evaluation

To comprehensively evaluate the proposed framework's performance, a series of tests were conducted targeting [19.5 GHz, 25 GHz] passband frequency ranges. The objective was to design FSS that exhibit minimal insertion loss (below

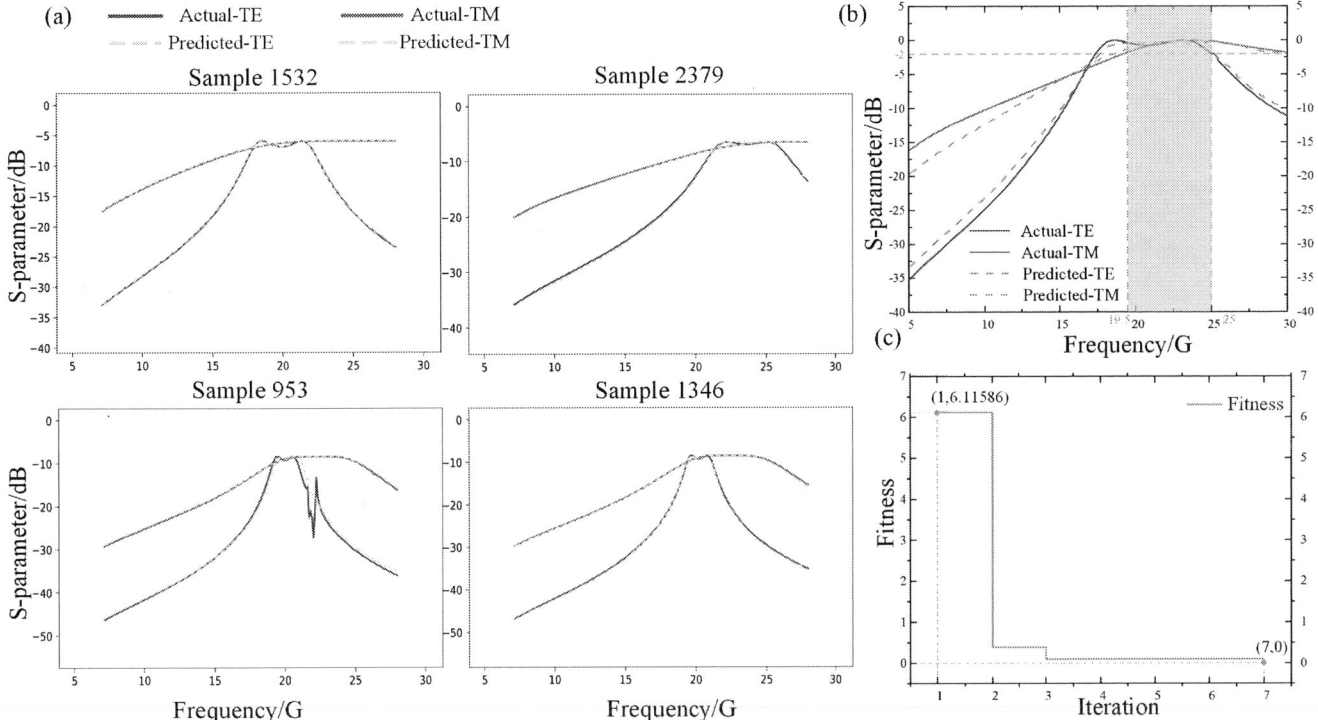

Fig. 4. Overall System Performance Evaluation. (a) Network fitting performance (b) Transmission responses (S21) for both TE and TM polarizations of optimized FSS structures targeting [19.5GHz 25GHz] passband frequencies. (c) Convergence curves of the 19.5-25G optimization process. The start and end points of the optimization process are marked with circles and their coordinates are displayed.

−2 dB) within all specified conditions, even under challenging conditions: a large incident angle of 70° and dual-polarization requirements (TE and TM).

$$\begin{cases} S_{21}(TE) \geq -2\,\text{dB}, & 19.5\,\text{GHz} \leq f \leq 25\,\text{GHz} \\ S_{21}(TM) \geq -2\,\text{dB}, & 19.5\,\text{GHz} \leq f \leq 25\,\text{GHz} \end{cases} \quad (8)$$

For the target passband [19.5 GHz, 25 GHz], as shown in Fig. 4(c), the optimization algorithm converged within just 7 iterations, successfully achieving the design goals. The optimized structure's parameters $\mathbf{x}^*$ are presented in Table III. Fig. 4(b) illustrates the simulation result using these parameters, which perfectly met the preset design criteria, both TE and TM transmission coefficients (S21) remained consistently above the −2 dB threshold across the entire frequency range. The final structure achieved near-perfect alignment with the network's predictions, confirming the accuracy and reliability of the deep learning model. The fitness value convergence curve in Fig. 4(c) demonstrates that the optimization process efficiently explores the design space, rapidly converging when the frequency range falls within the system's capabilities.

The comparison of different methods is summarized in Table IV. Notably, while previous works do not explicitly report the time per iteration, our approach demonstrates a clear advantage in computational efficiency, with each iteration taking only 0.53s on an Intel Xeon Silver 4314 CPU. This

TABLE III
PARAMETER SPECIFICATIONS

Target	Parameters			Passband Range	
	Gap	R_{out}	t_{mp}	TE	TM
	t_{xc}	θ	W		
19.5-25	0.77	1.31	1.84	17.37-25.01	19.24-30+
	2.65	70	0.01		

efficiency is crucial for practical applications, particularly in real-time or large-scale design scenarios. Specifically, our optimization process converged efficiently for the primary design target range [19.5 GHz, 25 GHz] in an average of 7 iterations, totaling just 3.71 seconds. As for traditional methods using full-wave electromagnetic simulations, it will take an average of 81 seconds per simulation in HFSS software running on the same device, with the entire optimization process requiring 100-200 iterations. The 81 seconds is the average time obtained from 100 test simulations. In multiple rounds of testing, the proposed method typically converged well before reaching the maximum iteration limit of 260—a parameter we set to ensure comprehensive exploration of the design space while preventing excessive computational overhead. Even in the most demanding scenario that approaches this maximum iteration threshold, our method can improve efficiency by 58×-

117× compared with the traditional electromagnetic simulation design methods. A single iteration based on the MBAE and DE framework is 152.8× faster than a traditional manual iteration of an electromagnetic simulation.

TABLE IV
COMPARISON OF DIFFERENT RF DESIGN METHODS

Reference	Bandwidth(GHz)	Insertion Loss(dB)	Iterations
Cong et al. [12]	3.0 – 12.0	3.0	75
Naseri et al. [13]	21.0 – 25.5	1.0	50
Zhou et al. [14]	6.0 – 7.5	2.0	14
Zhu et al. [15]	5.3 – 7.3	3.0	N/A
Proposed Method	**19.5 – 25.0**	**2.0**	**7**

V. DISCUSSIONS AND CONCLUSIONS

This paper introduced an data-driven inverse design framework for FSS, addressing the limitations of traditional EDA tools in handling complex electromagnetic designs. By leveraging an MBAE and an enhanced DE algorithm, the proposed framework achieves a 58×-117× speedup compared to conventional electromagnetic simulation methods. Additionally, a single iteration within the MBAE and DE framework is 152.8× faster than a traditional manual iteration, significantly improving design efficiency. Notably, the MBAE architecture enhances predictive accuracy by effectively capturing frequency-dependent electromagnetic responses, outperforming conventional neural network-based surrogate models. Our method provides a more effective and resource-efficient solution for FSS design. Future work will focus on real-world experimental validation and the integration of reconfigurable FSS designs to enhance the practical applicability of the proposed framework.

REFERENCES

[1] T. Hong, S. Guo, W. Jiang, and S. Gong, "Highly selective frequency selective surface with ultrawideband rejection," *IEEE Trans. Antennas Propag.*, vol. 70, no. 5, pp. 3459–3468, 2021.

[2] N. Liu, X. Sheng, C. Zhang, and D. Guo, "Design of frequency selective surface structure with high angular stability for radome application," *IEEE Antennas Wireless Propag. Lett.*, vol. 17, no. 1, pp. 138–141, 2017.

[3] H. U. Tahseen, L. Yang, and X. Zhou, "Design of fss-antenna-radome system for airborne and ground applications," *IET Commun.*, vol. 15, no. 13, pp. 1691–1699, 2021.

[4] H. Shall, H. Hawess, S. Baccar, and M. Kadi, "A neural network-empowered inverse fss design and synthesis approach for 5g shielding applications," *IEEE Access*, 2024.

[5] A. Alsudani and H. M. Marhoon, "Performance enhancement of microstrip patch antenna based on frequency selective surface substrate for 5g communication applications," *Evolution (N. Y)*, vol. 4, p. 7, 2022.

[6] L. Zhou and Z. Shen, "Hybrid frequency-selective rasorber with low-frequency diffusion and high-frequency absorption," *IEEE Trans. Antennas Propag.*, vol. 69, no. 3, pp. 1469–1476, 2020.

[7] A. Lalbakhsh, M. U. Afzal, K. P. Esselle, and S. L. Smith, "All-metal wideband frequency-selective surface bandpass filter for te and tm polarizations," *IEEE Trans. Antennas Propag.*, vol. 70, no. 4, pp. 2790–2800, 2022.

[8] R. Cong, N. Liu, X. Gao, C. Zhang, K. Yang, and X. Sheng, "A novel method for frequency selective surface design using deep learning with improved particle swarm algorithm," in *2022 IEEE 9th International Symposium on Microwave, Antenna, Propagation and EMC Technologies for Wireless Communications (MAPE)*. IEEE, 2022, pp. 374–379.

[9] M. Abdullah and S. Koziel, "Supervised-learning-based development of multibit rcs-reduced coding metasurfaces," *IEEE Trans. Microw. Theory Techn.*, vol. 70, no. 1, pp. 264–274, 2021.

[10] Z. Li, N. Kovachki, K. Azizzadenesheli, B. Liu, K. Bhattacharya, A. Stuart, and A. Anandkumar, "Fourier neural operator for parametric partial differential equations," *arXiv preprint arXiv:2010.08895*, 2020.

[11] L. Lu, P. Jin, G. Pang, Z. Zhang, and G. E. Karniadakis, "Learning nonlinear operators via deeponet based on the universal approximation theorem of operators," *Nature machine intelligence*, vol. 3, no. 3, pp. 218–229, 2021.

[12] R. Cong, N. Liu, X. Li, H. Wang, and X. Sheng, "Design of wideband frequency selective surface based on the combination of the equivalent circuit model and deep learning," *IEEE Antennas Wireless Propag. Lett.*, vol. 22, no. 9, pp. 2110–2114, 2023.

[13] P. Naseri and S. V. Hum, "A generative machine learning-based approach for inverse design of multilayer metasurfaces," *IEEE Trans. Antennas Propag.*, vol. 69, no. 9, pp. 5725–5739, Sep. 2021.

[14] Z. Zhou, Z. Wei, J. Ren, Y. Yin, G. F. Pedersen, and M. Shen, "Representation learning-driven fully automated framework for the inverse design of frequency-selective surfaces," *IEEE Trans. Microw. Theory Techn.*, vol. 71, no. 6, pp. 2409–2421, 2023.

[15] E. Zhu, E. Li, Z. Wei, and W.-Y. Yin, "Adversarial-network regularized inverse design of frequency-selective surface with frequency-temporal deep learning," *IEEE Trans. Antennas Propag.*, vol. 70, no. 10, pp. 9460–9469, Oct. 2022.

2025 The 10th International Conference on Integrated Circuits and Microsystems

GraphRL-Core: Intelligent Logic Synthesis Optimization via Graph Transformer and Deep Reinforcement Learning

Sujie Zhu[1], Guande Dong[1], Haoyang He[1], Chong Xia[1], Jianwang Zhai[1,2,†],
[1]Beijing University of Posts and Telecommunications
[2]State Key Lab of Processors, Institute of Computing Technology, CAS
{yinling, dongguande, henryhe9, xiachong, zhaijw}@bupt.edu.cn

Abstract—**With the rapid expansion of circuit design scales, traditional logic synthesis methods are encountering severe bottlenecks in optimizing complex circuit structures. This paper proposes an intelligent optimization approach, GraphRL-Core, which combines a graph transformer with deep reinforcement learning (DRL) and innovatively integrates the global modeling capability of the Gradformer model with the decision adaptability of reinforcement learning (RL). The system automatically generates optimization instructions for logic circuits by extracting graph structural features, utilizing graph transformer encoding, and employing a multi-objective reward-driven policy learning approach. Experiments conducted on several representative circuits demonstrate that the proposed method outperforms both traditional optimization flows and existing RL frameworks in terms of LUT minimization, optimization efficiency, and generalization capability, highlighting its significant practical value.**

Index Terms—**Electronic Design Automation, Logic Synthesis, Graph Transformer, Deep Reinforcement Learning**

I. INTRODUCTION

Logic synthesis is a fundamental stage in digital chip design, transforming high-level descriptions into optimized gate-level implementations while balancing area, power, and timing. Traditional approaches, largely based on fixed heuristics and expert-crafted rules, are increasingly inadequate for handling the expanding design space and growing circuit complexity [1].

With the development of optimization algorithms, tools such as the open-source framework ABC [2] have become widely adopted in academia and industry. However, existing flows typically rely on static operator sequences defined by experts. This hand-crafted process requires significant tuning effort and often fails to generalize across diverse circuits, leading to suboptimal optimization results.

Recent advances in machine learning (ML) have introduced new paradigms for logic synthesis, particularly in automatic optimization and design space exploration. Reinforcement

learning (RL) has proven effective in dynamically adapting optimization strategies, enabling circuit-specific flows that outperform static approaches [3], [4]. Building on this, deep reinforcement learning (DRL) frameworks such as DRiLLS [1] employ policy-gradient methods to autonomously discover effective operator sequences, achieving competitive results in multi-objective optimization tasks. With the integration of graph neural networks (GNNs), DRL methods further enhance circuit state representation by leveraging structural information [5].

Despite these advances, existing DRL-based synthesis methods still struggle to capture global circuit structures, as most GNN models focus primarily on local neighborhoods. To overcome this limitation, we propose an enhanced DRiLLS framework that combines DRL with the Gradformer model [5]. By modeling both local and long-range dependencies, our method achieves a more effective structural representation and improved optimization outcomes. Experimental results demonstrate superior performance and generalization compared to prior approaches [6], [7].

The main contributions of this work are as follows:

- This paper proposes the GraphRL-Core (i.e., Graph Reinforcement Learning Core) intelligent decision module, which significantly enhances the capability of DRL agents to model and make optimization decisions regarding the global structure of circuits, particularly in complex circuit scenarios.
- This paper introduces a multi-objective reward function that considers both area and timing constraints, enabling comprehensive evaluation of multiple optimization objectives and improving the adaptability and decision quality of the optimization process.
- Experimental results based on the open-source tool ABC [2] and the EPFL benchmark suite [8] demonstrate that the proposed method achieves superior performance compared to traditional RL and classical optimization flows across multiple key indicators, with strong generalization capability and engineering value.

[†]Corresponding author: Jianwang Zhai. This work is supported by the National Key R&D Program of China (2022YFB2901100), the National Natural Science Foundation of China (No. 62404021), the Beijing Natural Science Foundation (No. 4244107, QY24216, QY24204, QY25329), and the State Key Lab of Processors, Institute of Computing Technology, CAS under Grant No. CLQ202504.

979-8-3315-8850-2/25 $31.00 © 2025 IEEE

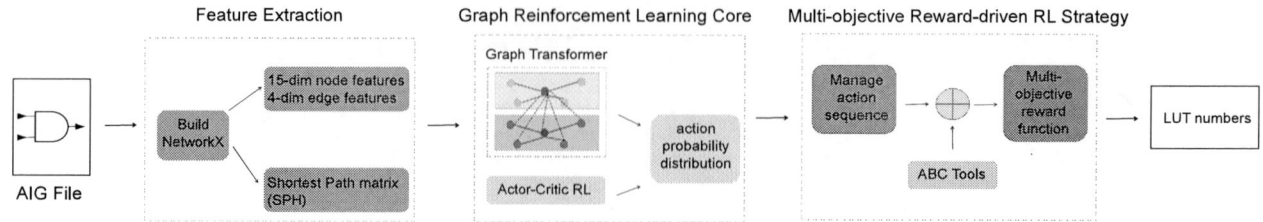

Fig. 1. System Architecture Overview

II. RELATED WORK

Logic synthesis is a crucial stage in digital chip design, directly influencing circuit metrics such as area, delay, and power. Traditional flows rely on static heuristics and expert-defined operator sequences, which may suffice for small designs but fail to generalize to large and complex circuits. As design scales expand and multi-objective requirements (e.g., area and delay) conflict, these fixed strategies struggle to deliver globally optimal results.

A central challenge is *operator optimization*, where the order and choice of transformations (e.g., `rewrite`, `refactor`, `resub`, `balance`) crucially affect synthesis quality. Manually predefined or rule-based sequences cannot adapt to circuit heterogeneity, while exhaustive search is computationally prohibitive. To address this, recent research leverages learning-based approaches, notably DRL for adaptive policy learning and graph neural networks/transformer models for richer structural encoding. These two directions form the basis of intelligent, data-driven logic synthesis optimization.

A. Reinforcement Learning for Logic Synthesis

Reinforcement learning has been increasingly adopted to automate operator scheduling in logic synthesis. Hosny et al. first introduced the DRiLLS framework [1], formulating logic synthesis as a Markov decision process and using the advantage actor-critic (A2C) algorithm to learn operator sequences. This approach enabled optimization without human intervention. Zhou and Anderson [9] extended this idea for FPGA logic synthesis, incorporating random forest-based feature importance to handle large state spaces. Yang et al. [10] combined graph isomorphism networks (GINs) and long short-term memory (LSTM) networks within a proximal policy optimization (PPO) framework to capture both structural and temporal aspects of circuit states. Liu et al. [11] proposed CBTune, which treats operator selection as a contextual bandit problem, using the LinUCB algorithm to balance exploration and exploitation. Pei et al. [12] developed AlphaSyn, employing Monte Carlo tree search (MCTS) with learned policy and value networks to efficiently explore operator sequences. Retrieval-guided RL was proposed by Chowdhury et al. [13], which utilizes pretrained models to find similar circuits and guide the learning process. These approaches demonstrate the strong potential of RL in automating and optimizing complex synthesis flows. Dong et al. [14] proposed PIRLLS, using imitation learning to pretrain a policy network and leverage offline knowledge.

Beyond RL, other intelligent optimization methods have been adopted. Grosnit et al. [15] introduced a Bayesian optimization-based multi-objective model, which enables effective multi-objective optimization in logic synthesis tasks.

B. Graph Neural Networks

Graph neural networks provide a natural framework for modeling circuit structures, and have found extensive application in the field of electronic design automation (EDA) [16]–[20]. Early approaches, such as graph convolutional networks (GCNs) [1], captured local topology but struggled with long-range dependencies, while graph attention networks (GATs) improved local feature learning through attention to critical nodes.

To enhance global structural modeling, Transformer-based graph models have emerged. Gradformer [5] introduces exponential decay self-attention to capture local and global dependencies jointly, achieving strong results in circuit modeling. Structure-Aware Transformers [9] further improve synthesis quality via subgraph representations. These advances mark a shift toward expressive, scalable representations that enable intelligent agents to reason over both local and global circuit features [21], [22].

In summary, RL provides adaptive optimization policies, while GNNs and Transformers offer rich structural representations. Their integration presents a promising path to overcoming the limits of traditional flows and achieving superior logic synthesis performance [5]–[7].

III. METHODOLOGY

A. Overview

This study proposes an intelligent logic synthesis optimization system that integrates the Gradformer model with DRL to overcome the limitations of traditional synthesis methods in handling complex circuit designs. By formulating logic synthesis as a sequential decision-making task and leveraging the global representation capabilities of graph Transformers,

979-8-3315-8850-2/25 $31.00 © 2025 IEEE

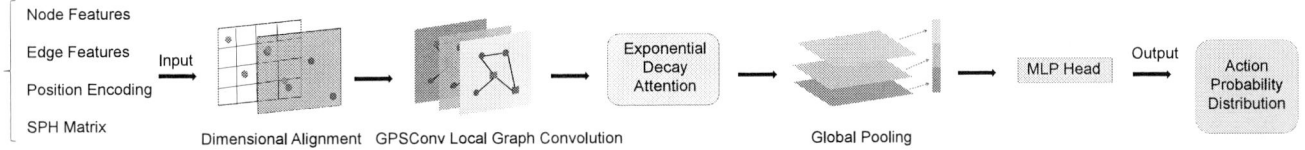

Fig. 2. Graph Reinforcement Learning Core

the system achieves a fully automated and intelligent optimization flow. As shown in Fig. 1, the architecture consists of three key modules: (1) a feature extraction module that parses AIG circuit files to build graph structures and extract rich node, edge, and global features; (2) a Gradformer-based decision module that performs global structural reasoning and outputs optimization actions using a Transformer-enhanced DRL policy; and (3) a training and execution module that interfaces with the ABC tool to execute synthesis steps, manage optimization sequences, and compute multi-objective rewards. During interaction, the environment maintains the current sequence of applied optimization operators, updates the circuit state, and generates structured observations—including 15-dimensional node features, positional encodings, edge attributes, and shortest-path matrices—to guide the learning agent. The system tracks the best observed QoR (in terms of LUT count and delay) to ensure training stability. This end-to-end workflow—from graph parsing to Transformer-guided optimization—enables precise, globally informed synthesis decisions and achieves efficient, high-quality logic optimization.

B. Details of Model Components

1) Feature Extraction Module: The main function of the feature extraction module is to construct a graph structure and perform multi-dimensional feature extraction on the input AIG circuit file, generating structured data suitable for subsequent model processing. By parsing the AIG file, the module first builds a NetworkX directed graph structure, representing the circuit as a set of nodes and edges, where node types include input nodes, AND gates, and output nodes, and the edge information indicates the connectivity and inversion between nodes.

First, the module parses the standard AIG file, constructs the NetworkX directed graph, and extracts node types (such as input, AND gate, and output) as well as edge inversion information. Then, a 15-dimensional feature vector is generated for each node, including 6 local structural features and 9 global circuit features. As shown in Table. I, the local structural features include node type (input node, AND gate, or output node), normalized fan-in and fan-out (fan-in divided by 2, fan-out divided by 10), and logic level depth (obtained by normalizing the shortest path distance from the output node to the current node). In Table. I, the global circuit features are derived from overall circuit statistics, including the number of input and output pins, total number of nodes, total number of edges, and maximum logic depth, providing the model with information on the overall scale and complexity of the circuit.

Meanwhile, in Table. I, the module constructs 4-dimensional features for each edge, specifically including the inversion flag, whether the source node is an input node, the fan-out of the source node, and the level difference between the source and target nodes. To enhance the model's ability to perceive circuit topology, the module generates a 16-dimensional positional encoding for each node, which distinguishes the node's topological position in the entire graph.

Additionally, the module calculates the shortest path distance matrix (SPH) between nodes, explicitly specifying the relative topological distance between nodes and providing rich structured input information for subsequent models.

2) GraphRL-Core Intelligent Decision Module: The Fig. 2 presents the process of the GraphRL-Core Intelligent Decision Module. The GraphRL-Core intelligent decision module is a DRL policy network based on a graph Transformer architecture, aiming to achieve intelligent optimization decisions for circuit design through an integrated Actor-Critic structure. The core function of this module is to map the graph-structured circuit state to a probability distribution over optimization actions and to learn the optimal optimization policy through RL. Specifically, the GraphRL-Core intelligent decision module integrates graph convolution and attention mechanisms to effectively capture multi-level topological information of the circuit, and predicts the best optimization operator based on the current circuit state, thereby realizing intelligent performance improvement.

The input of this module is the rich, structured data provided by the feature extraction module, mainly including node feature vectors, edge feature matrices, positional encodings, and node-to-node shortest path matrices (SPH). The node features and positional encodings are first linearly transformed for dimensional alignment, fusing node structural features and positional encoding information into a unified 64-dimensional representation, which facilitates subsequent processing by the graph neural network. The input data first passes through a local graph convolutional network (GPSConv) layer to capture local topological features of the nodes. Specifically, the input features are processed as follows:

$$\mathbf{h}_i = \sum_{j \in \mathcal{N}(i)} \frac{1}{\sqrt{d}} A_{ij} \cdot \mathbf{h}_j + \mathbf{W} \cdot \mathbf{h}_i, \qquad (1)$$

where $\mathbf{h}_i$ denotes the feature of node i, N(i) is the set of neighbors of node i, A_{ij} is the adjacency matrix, W is the weight matrix, and d is the feature dimension. In this way, local information is aggregated via graph convolution, while considering the relative importance between nodes.

979-8-3315-8850-2/25 $31.00 © 2025 IEEE

TABLE I
SUMMARY OF NODE LOCAL, NODE GLOBAL, AND EDGE FEATURES

Node Local Structural Features			Node Global Circuit Features		Edge Features		
Feature	Description	Norm.	Feature Name	Description	Feature Name	Description	Norm.
Input Flag	Input node (1/0)	–	Number of Input Pins	Number of input ports	Inversion Flag	Edge is inverted (1/0)	–
AND Flag	AND gate (1/0)	–	Number of Output Pins	Number of output ports	Source Node is Input	Source is input node (1/0)	–
Output Flag	Output node (1/0)	–	Total Node Count	Total nodes in the circuit	Source Node Fanout (Norm)	Source node fanout	/10
Fan-in	Incoming edges	/2	Total Edge Count	Total edges in the circuit	Level Difference (Norm)	Source-target level difference	/max level
Fan-out	Outgoing edges	/10	Max Logic Depth	Longest combinational path			
Level	Logic depth	/max	Number of Latches	Number of flip-flops			
			AND Gate Ratio	Proportion of AND gates			
			OR Gate Ratio	Proportion of OR gates			
			NOT Gate Ratio	Proportion of inverters			

On top of local feature extraction, to capture global circuit structure, GraphRL-Core further introduces an exponential decay masked attention mechanism, which controls the distribution of attention weights based on the shortest path distance between nodes, with the decay mask M defined as:

$$M = \lambda^{\mathrm{ReLU}(\phi(v_i, v_j) - sp)}, \tag{2}$$

where λ is the decay coefficient, and $\phi(v_i, v_j)$ is the shortest path distance (SPH) between nodes v_i and v_j. When the spatial distance $\phi(v_i, v_j)$ is less than sp, the attention score between nodes is not decayed, thus precisely controlling the interaction radius of each node.

Subsequently, the mask is combined with the self-attention mechanism; after multiple GPSConv and exponential decay attention layers, node features are globally pooled to obtain a graph-level representation, which is then passed through a three-layer MLP to output the final action policy distribution. The output dimension is equal to the size of the action space (i.e., the number of optimization operators), generating unnormalized scores for subsequent action selection. The refined attention matrix S^l is computed as:

$$\mathbf{S}^l = \mathrm{softmax}\left(\frac{\mathbf{Q}^l(\mathbf{K}^l)^\top}{\sqrt{d}} - \mathbf{M}\right)\mathbf{V}^l, \tag{3}$$

where S^l is the refined attention at layer 1, Q^l, K^l, V^l are the Query, Key, and Value matrices, and d is the feature dimension. This fusion effectively achieves both local and long-range information modeling, avoiding the redundant aggregation issue of traditional attention mechanisms.

On the basis of enhanced state representation, the GraphRL-Core intelligent decision module is effectively integrated with the DRiLLS framework. Specifically, the Actor network (policy network) predicts the probability of each optimization action at each step according to the state representation generated by the GraphRL-Core module:

$$P(a|s) = \frac{\exp(Q(s, a))}{\sum_{a'} \exp(Q(s, a'))}, \tag{4}$$

where $Q(s, a)$ is the state-action value function estimated by the Critic network, representing the value of executing action a in state s. The Critic network, using high-quality state input, more accurately evaluates the long-term value $V(s)$, and continuously adjusts the optimization strategy based on

the actual reward $R(s, a)$ and estimated value, with the loss function defined as:

$$\mathcal{L} = \mathbb{E}\left[\left(R(s, a) + \gamma V(s') - V(s)\right)^2\right], \tag{5}$$

where γ is the discount factor, $V(s')$ is the value of the next state, and $R(s, a)$ is the reward after taking the optimal action in the current state.

Finally, the entire GraphRL-Core intelligent decision module achieves end-to-end joint optimization, ensuring that the graph feature extractor learns the most effective structural representations for optimization decision making.

3) Multi-objective Reward-driven RL Strategy Module:
The reinforcement learning training and execution module is responsible for managing the sequence of optimization actions, executing interactions with the ABC tool, and is equipped with a specialized multi-objective reward function to evaluate the effectiveness of action sequences, guiding the agent to achieve efficient policy training and practical optimization. Following the DRiLLS framework, it provides immediate rewards and state updates after each optimization step, guiding efficient policy training. The reward function combines three levels—constraint priority, incremental improvement, and exploration penalty—to balance logic level constraints and LUT optimization, helping the agent reduce LUT numbers effectively and avoid local optima, improving policy performance.

The inputs to this module mainly include two parts: first, the AIG circuit file parsed by the graph structure modeling module, providing the current circuit state as a structured graph; second, the optimization action decision from the GraphRL-Core module, i.e., the probability distribution of action selection at each step. The environment first receives the action probabilities output by GraphRL-Core, selects a specific optimization operator according to the policy, with the action space defined as seven basic operators: rewrite, refactor, resub, balance, and their -z variants. After an action is selected, the environment calls the ABC tool to execute the corresponding optimization command, updating the circuit state and computing the immediate reward. The state transition process conforms to the Markov Decision Process (MDP) framework, that is, action a_k is selected in state S_k, the environment executes the ABC command and transitions to the new state S_{k+1}, while providing the reward $R(S_k, a_k)$.

979-8-3315-8850-2/25 $31.00 © 2025 IEEE

TABLE II
LOOKUP TABLE FOR MULTI-OBJECTIVE DISCRETE REWARD FUNCTION

Level	LevelDiff	LutDiff	Reward
True	0	1	3
True	0	0	0
True	0	-1	-1
False	1	1	3
False	1	0	2
False	1	-1	1
False	0	1	2
False	0	0	0
False	0	-1	-2
False	-1	1	-1
False	-1	0	-2
False	-1	-1	-3

Reward: The corresponding reward score for each combination.

To more effectively guide the training of the RL policy, this module designs a multi-objective discrete reward function according to the goals of logic synthesis optimization, to comprehensively evaluate constraint satisfaction and optimization improvement after the current action is executed. Specifically, the reward function considers three core variables:

1) **Level:** Boolean, indicating whether the logic level constraint is met.
2) **LevelDiff:** Value of -1, 0, 1, indicating whether the logic level has degraded, remained unchanged, or improved.
3) **LutDiff:** Also -1, 0, 1, measuring the trend in LUT count (1 = decreased, 0 = unchanged, -1 = increased).

As shown in the table.II, this reward mechanism integrates the advantages of constraint priority, incremental improvement guidance, and exploration penalty balance: When the logic level constraint is satisfied (Level = True), the reward focuses more on LUT reduction, especially granting the highest reward (+3) for a decrease in LUT count, clearly guiding the model to achieve dual optimization of resources and levels; when the constraint is not met (Level = False), as long as either the logic level or LUT number improves, the model receives a positive reward (+1 to +3), encouraging the policy to approach the constraint target step by step, avoiding policy deadlocks; meanwhile, for deteriorations in the optimization process (such as increased levels or LUTs), multi-level penalty mechanisms (-1 to -3) are designed to guide the agent away from performance degradation and avoid premature convergence to local optima.

After each action is executed, the environment updates the current circuit state information, including LUT count and logic levels, and records the historical best result to ensure training stability and effectiveness. By integrating the enhanced structured state representation of GraphRL-Core with the classic DRiLLS RL framework, the entire RL training and execution module achieves an effective transition from manually defined features to efficient graph structural features, ensuring intelligent dynamic trade-offs and effective optimization of resource usage and delay constraints in complex circuit optimization tasks.

IV. EXPERIMENT

A. Experimental Setup

Datasets: This paper evaluates the proposed GraphRL-Core framework using combinational logic circuits from the EPFL arithmetic benchmark suite [8], consistent with the dataset adopted by the DRiLLS methodology [1]. Specifically, this paper selected a representative subset of circuits including *Adder*, *Divisor*, *Max*, and *Multiplier*, ensuring diversity and representativeness in scale and structural complexity. Since all the ABC operators employed in this work, as well as the baselines considered, are designed for the optimization of combinational circuits, this paper focuses exclusively on combinational logic circuits.

Settings: All optimization experiments were conducted using the Berkeley ABC tool as the baseline environment for logic optimization and technology mapping. The primary optimization objective was to minimize the number of standard 6-input lookup tables (6-LUT), while the critical path delay was considered an auxiliary metric for comprehensive performance evaluation. Each optimization approach was executed under identical computational resources and experimental conditions to ensure fairness. For each circuit, an initial And-Inverter Graph (AIG) representation was generated and optimized using three different methods:

- **GraphRL-Core:** Integrates the Gradformer-based graph Transformer network to simultaneously capture global and local structural features, enabling the RL agent to determine the optimal optimization operator sequences adaptively.
- **DRiLLS [1] (RL-based baseline):** Employs shallow graph features without explicit global information modeling, operating within a fixed action space.
- **resyn2*2 [2] (baseline):** Applies a fixed sequence of heuristic transformations twice in succession, aiming for rapid LUT reduction without modeling global circuit structure, making it prone to local optima.

All methods were limited to a maximum of 50 optimization steps, after which LUT mapping was performed using ABC's standard command (*"if -a -K 6"*). Subsequently, the final LUT counts and critical path delays were uniformly calculated for comparative evaluation.

B. Quantitative Results

The quantitative evaluation results in Table III highlight the effectiveness of the proposed GraphRL-Core framework

For LUT optimization, GraphRL-Core achieves consistently lower counts across most benchmarks. Notably, on the Sqrt circuit the LUT count is reduced to 4,546, outperforming DRiLLS (4,708) and the heuristic resyn22 flow (5,127). A similar trend is observed on Divisor, where our method attains 5,663 LUTs, representing a 30.9% improvement over DRiLLS. Even on large-scale designs such as Square, GraphRL-Core delivers smaller LUT counts than both baselines. The only exception occurs in Multiplier, where performance is nearly on par with DRiLLS, while still surpassing resyn2*2.

979-8-3315-8850-2/25 $31.00 © 2025 IEEE

TABLE III
COMPARISON OF LUT-6 COUNTS BEFORE AND AFTER OPTIMIZATION

Benchmark	Initial	resyn2*2 [2]	DRiLLS [1]		GraphRL-Core	
	LUTs	LUTs	LUTs	Time (min)	LUTs	Time (min)
Adder	249	249	244	3.25	244	0.865
Max	721	719	697	32.58	687	1.128
Sin	1444	1466	1441.5	51.15	1436	2.95
Square	3994	3915	3889.4	130	3875	8.65
Multiplier	5678	5713	5678	180.84	5689	10.166
Log2	7584	7703	7583.6	198.6	7598	16.71
Sqrt	8084	5127	4708	147.64	4546	9.38
Divisor	23864	8197	7944	259.75	5663	15.424
GEOMEAN	3103.28	2569.54	2501.98	74.70	2382.45	5.29
Ratio Avg.	1.00	0.828	0.808	1.00	0.768	0.07

Regarding runtime efficiency, GraphRL-Core exhibits substantial acceleration, with an average runtime of 5.29 minutes compared to 74.70 minutes for DRiLLS. The advantage becomes particularly pronounced on larger circuits—for instance, Max and Divisor complete in 1.128 minutes and 15.424 minutes, respectively, while DRiLLS requires over 30 and 250 minutes.

Aggregated across all benchmarks, GraphRL-Core achieves the lowest geometric mean LUT count (2,382.45) and average LUT ratio (0.768), confirming its ability to reduce resource usage and runtime simultaneously. Overall, these findings demonstrate that GraphRL-Core effectively leverages both global and local structural information through the graph Transformer, enabling adaptive operator selection and yielding consistent improvements in optimization quality and efficiency.

V. CONCLUSION

This paper presents GraphRL-Core, an intelligent logic synthesis optimization framework that integrates graph Transformers with DRL to overcome the limitations of traditional flows in circuit modeling and operator scheduling. By leveraging the Gradformer model, GraphRL-Core captures both local and global structural features of circuits, enabling adaptive and effective optimization strategies.

Experimental evaluations demonstrate consistent reductions in LUT counts and substantial runtime improvements across standard benchmarks, highlighting the framework's strong generalization capability and efficiency. Future extensions will explore multi-objective collaborative optimization and applications to broader EDA tasks such as high-level synthesis and physical design, paving the way toward more intelligent and adaptive design automation.

REFERENCES

[1] A. Hosny, S. Hashemi, M. Shalan, and S. Reda, "DRiLLS: Deep Reinforcement Learning for Logic Synthesis," in *ACM/IEEE Asia and South Pacific Design Automation Conference (ASP-DAC)*, 2020, pp. 1–4.

[2] R. Brayton and A. Mishchenko, "ABC: An academic industrial-strength verification tool," in *International Conference on Computer-Aided Verification (CAV)*, 2010, pp. 24–40.

[3] D. Silver, T. Hubert *et al.*, "Mastering Chess and Shogi by Self-Play with a General Reinforcement Learning Algorithm," in *Annual Conference on Neural Information Processing Systems (NeurIPS)*, 2017.

[4] A. Shehzad, F. Xia, S. Abid *et al.*, "Graph Transformers: A Survey," *arXiv preprint arXiv:2407.09777*, 2024.

[5] C. Liu, Z. Yao, Y. Zhan, *et al.*, "Gradformer: Graph Transformer with Exponential Decay," *arXiv preprint arXiv:2404.15729*, 2020.

[6] L. Rampášek, M. Galkin, V. P. Dwivedi *et al.*, "Recipe for a General, Powerful, Scalable Graph Transformer," in *Annual Conference on Neural Information Processing Systems (NeurIPS)*, 2022.

[7] T. P. Lillicrap, J. J. Hunt, A. Pritzel *et al.*, "Continuous Control with Deep Reinforcement Learning," in *International Conference on Learning Representations (ICLR)*, 2016.

[8] L. Amaru, P.-E. Gaillardon, and G. Micheli, "The EPFL Combinational Benchmark Suite," Jan 2015.

[9] G. Zhou and J. H. Anderson, "Area-Driven FPGA Logic Synthesis Using Reinforcement Learning," in *ACM/IEEE Asia and South Pacific Design Automation Conference (ASP-DAC)*, 2023, pp. 159–165.

[10] C. Yang, Y. Xia, Z. Chu, and X. Zha, "Logic Synthesis Optimization Sequence Tuning Using RL-Based LSTM and Graph Isomorphism Network," *IEEE Transactions on Circuits and Systems II: Express Briefs (TCAS-II)*, vol. 69, no. 8, pp. 3600–3604, 2022.

[11] F. Liu, Z. Pei, Z. Yu, H. Zheng, Z. He, T. Chen, and B. Yu, "CBTune: Contextual Bandit Tuning for Logic Synthesis," in *IEEE/ACM Proceedings Design, Automation and Test in Eurpoe (DATE)*, 2024, pp. 1–6.

[12] Z. Pei, F. Liu, Z. He *et al.*, "AlphaSyn: Logic Synthesis Optimization with Efficient Monte Carlo Tree Search," in *IEEE/ACM International Conference on Computer-Aided Design (ICCAD)*, 2023, pp. 1–9.

[13] A. B. Chowdhury, M. Romanelli, B. Tan *et al.*, "Retrieval-Guided Reinforcement Learning for Boolean Circuit Minimization," in *International Conference on Learning Representations (ICLR)*, 2024.

[14] G. Dong, J. Zhai, H. Cheng *et al.*, "PIRLLS: Pretraining with Imitation and RL Finetuning for Logic Synthesis," in *ACM/IEEE Asia and South Pacific Design Automation Conference (ASP-DAC)*, 2025, pp. 1–7.

[15] A. Grosnit *et al.*, "BOiLS: Bayesian Optimisation for Logic Synthesis," in *IEEE/ACM Proceedings Design, Automation and Test in Eurpoe (DATE)*, 2022, pp. 1193–1196.

[16] H. Cheng *et al.*, "SATGL: An Open-Source Graph Learning Toolkit for Boolean Satisfiability," in *International Symposium of Electronics Design Automation (ISEDA)*, 2024, pp. 746–751.

[17] J. Liu *et al.*, "PolarGate: Breaking the Functionality Representation Bottleneck of And-Inverter Graph Neural Network," in *ACM/IEEE International Conference On Computer Aided Design (ICCAD)*, 2024, pp. 1–9.

[18] J. Liu, Z. Liu, X. He *et al.*, "WideGate: Beyond Directed Acyclic Graph Learning in Subcircuit Boundary Prediction," in *Design, Automation Test in Europe Conference (DATE)*, 2025, pp. 1–7.

[19] F. Guo, Y. Xi *et al.*, "IRGNN: A Graph-based Framework Integrating Numerical Solution and Point Cloud for Static IR Drop Prediction," in *ACM/IEEE Design Automation Conference (DAC)*, 2025, pp. 1–7.

[20] M. Zhao, J. Liu, J. Zhai, and C. Shi, "MILS: Modality Interaction Driven Learning for Logic Synthesis," in *Great Lakes Symposium on VLSI (GLSVLSI)*, 2025, p. 64–70.

[21] V. P. Dwivedi and X. Bresson, "A Generalization of Transformer Networks to Graphs," *arXiv preprint arXiv:2012.09699v2*, 2022.

[22] C. Ying, T. Cai, S. Luo *et al.*, "Do Transformers Really Perform Badly for Graph Representation?" in *Annual Conference on Neural Information Processing Systems (NeurIPS)*, 2021.

979-8-3315-8850-2/25 $31.00 © 2025 IEEE

2025 The 10th International Conference on Integrated Circuits and Microsystems

A Hilbert Transform-Based Timing Skew Estimation Method for Dual-Channel TIADCs

Xin Li
School of Integrated Circuits
Anhui University
Hefei, China
lixin@ahu.edu.cn

Ying Pan
School of Integrated Circuits
Anhui University
Hefei, China
panying@stu.ahu.edu.cn

Mengdi Miao
School of Integrated Circuits
Anhui University
Hefei, China
wb24201008@stu.ahu.edu.cn

Yu Liu
School of Integrated Circuits
Anhui University
Hefei, China
liuyu@ahu.edu.cn

Chenghu Dai
School of Integrated Circuits
Anhui University
Hefei, China
daichenghu@ahu.edu.cn

Zhiting Lin
School of Integrated Circuits
Anhui University
Hefei, China
ztlin@ahu.edu.cn

Abstract—An all-digital Hilbert transform-based estimation method is proposed to detect timing skew mismatches between sub-ADCs in dual-channel time-interleaved analog-to-digital converters (TIADCs). Unlike conventional approaches such as correlation-based techniques, the proposed method significantly accelerates parameter convergence, the entire process of estimating the mismatch parameters requires approximately 40K samples. It features extremely low computational complexity and eliminates the need for additional analog auxiliary circuits. Simulation results based on a behavioral model of an 8-bit dual-channel TIADC demonstrate the effectiveness and superiority of the proposed method, with SNDR and SFDR improved from 28.11 dB and 28.14 dB to 49.89 dB and 76.18 dB, respectively.

Keywords—*Time Interleaved ADCs, Timing skew, Estimation, Hilbert Transform*

I. INTRODUCTION

In recent years, driven by emerging intelligent technologies such as 5G and artificial intelligence, higher data throughput has posed severe challenges to the performance of analog-to-digital converters (ADCs) in broadband communication and other application systems (e.g., millimeter-wave radars, autonomous driving, and communication base stations). This exceeds the capabilities of traditional medium-to-high precision ADCs fabricated using CMOS processes. The time-interleaved ADC (TIADC) effectively addresses this issue. As shown in Fig. 1, multiple relatively low-speed sub-ADC are used to sample the analog input signal in a time-interleaved manner. Ideally, this can double the overall sampling rate without compromising conversion accuracy, breaking the design constraints between sampling rate and conversion accuracy faced by traditional single-channel ADCs [1-2]. However, TIADCs also have

This work is supported by Science and Technology Key Project of Anhui Province under Grant 2022AH050099, National Natural Science Foundation of China under Grant 62471003, The University Synergy Innovation Program of Anhui Province under Grant GXXT-2023-013, The University Synergy Innovation Program of Anhui Province under Grant GXXT-2023-012, The University Synergy Innovation Program of Anhui Province under Grant GXXT-2023-011.

pressing issues to resolve, such as parameter mismatches between channels, including offset, gain, and timing skew mismatches. Compared to multi-channel TIADCs, dual-channel interleaved structures are relatively easier to synchronize and have higher tolerance for harmonics caused by mismatches [3-5]. However, as the sampling rate of single-channel ADCs increases, the dynamic error caused by timing skew increases with the input frequency, severely degrading the overall dynamic performance [6].

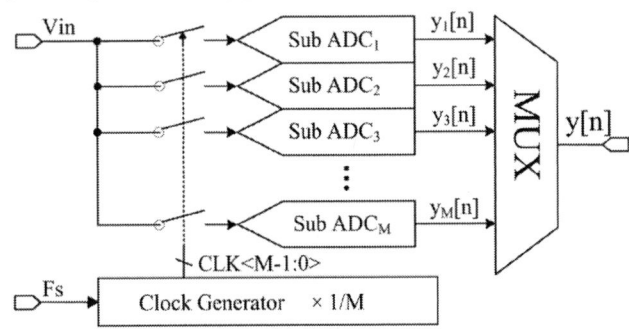

Fig. 1. The Overall Structure of TIADC system.

To mitigate the error impacts stemming from timing skew, a spectrum of mismatch extraction mechanisms has been proposed, categorically divided into foreground and background approaches based on their operational mode [7-11]. Foreground methods, leveraging known pilot signals (e.g., sine waves of predetermined frequencies), facilitate the extraction of error parameters in a single instance. However, their applicability is constrained in real-time communication systems and they lack resilience against process, voltage, and temperature (PVT) variations. Conversely, background calibration schemes have garnered extensive development due to their enhanced stability. In [5], a statistical approach rooted in the absolute difference of signals within the zero-crossing domain is introduced to ascertain the presence of timing skew in the calibrated sub-ADC. Nonetheless, the auxiliary zero-crossing detection circuitry is prone to PVT perturbations, and the comparator may exhibit

979-8-3315-8850-2/25 $31.00 © 2025 IEEE

offset errors, thereby compromising the precision of zero-crossing detection. A cross-correlation calibration technique anchored on a reference ADC is presented in [8], which evaluates the mismatch state at corresponding sampling instances by computing the disparities in cross-correlation functions between individual channel sub-ADCs and the reference ADC. Nevertheless, this methodology necessitates the sequential utilization of the reference ADC to assist each sub-ADC in calibration, leading to a notably sluggish convergence of mismatch parameters. To circumvent the shortcomings associated with analog auxiliary circuits, a methodology that estimates timing skew mismatch parameter by computing the difference in cross-correlation functions between adjacent channels and solving a system of simultaneous equations using the first derivative of the input signal in [9]. However, this method entails intricate matrix computations, and there exist approximation errors in the derivation of cross-correlation functions, potentially undermining the accuracy of sampling time offset estimation. In [10], an error estimation approach grounded in the probability statistical function of signals is introduced, leveraging probability density difference to detect timing skew. However, this method necessitates a substantial number of sampling points for computation to ensure the convergence of estimated parameters. With the proliferation of neural network technology, machine learning method is employed to effectively extract timing skew in [11]. However, stability assurance remains challenging at this stage, and substantial hardware resources are required. To strike a balance between hardware expenditure and swift parameter convergence, this paper introduces a timing skew error estimation method based on the Hilbert transform.

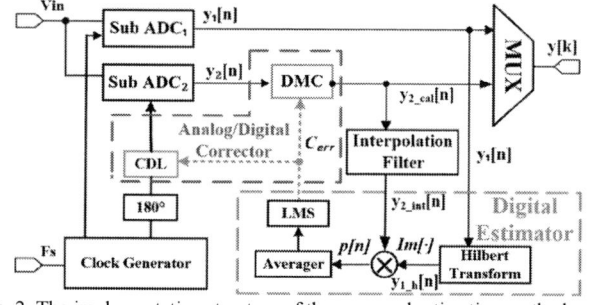

Fig. 2. The implementation structure of the proposed estimation method.

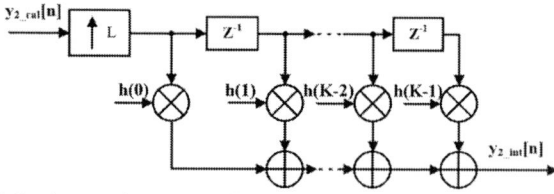

Fig. 3. Implementation process of Interpolation Filter.

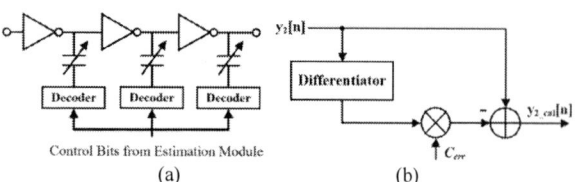

Fig. 4. Block diagram of (a) the CDL-based corrector and (b) the DMC-based corrector.

TABLE I.   PARAMETERS SETTING FOR SIMULATION

Model Parameters	Value
Normalized Sampling Rate (f_s)	1
Normalized Sampling Period (T_s)	1
TIADC Resolution	8 Bit
Number of Sub-ADC	2
Timing Skew	$[0, 0.03] \cdot T_s$
Type of Input Signal	Single-tone, Multi-tone and Quadrature Phase Shift Keying Signal
# of sample for average (N)	2^{12}
Iteration Step Size (μ)	2^{-1}

II. PROPOSED TIMING SKEW ESTIMATION METHOD

A. TIADC Outputs with Timing Skew

To enhance comprehension of the estimation method introduced in this paper, this section utilizes a dual-channel time-interleaved analog-to-digital converter (TIADC) as an illustrative example, deriving its output expressions under both ideal conditions and scenarios involving mismatches. Given an input signal $x(t)$, under ideal circumstances, the sampling interval for the two constituent sub-ADCs is T_s, which coincides with the period of the TIADC's sampling clock. Consequently, the output signals of the two sub-ADCs within the TIADC framework can be mathematically represented as follows:

$$y_1[n] = x(2n \cdot T_s)$$
$$y_2[n] = x((2n+1) \cdot T_s) \tag{1}$$

where $n = 0, 1, 2, 3, ...$ denote the discrete-time index. However, if there exists a timing skew (τ) between sub-ADC1 and sub-ADC2, the sampling interval between them will change to $T_s + \tau$, and accordingly, Equation (1) will be modified as follows:

$$y_1[n] = x(2n \cdot T_s)$$
$$y_2[n] = x((2n+1) \cdot T_s + \tau) \tag{2}$$

By conducting a frequency domain analysis on Equation (2), it can be deduced that timing skew introduces error spurious components in the output spectrum, leading to a significant degradation in dynamic performance. For a detailed analysis, please refer to the relevant literature.

B. Proposed Hilbert Transform based Estimation Method

This subsection introduces the working principle of the proposed estimation method, with its overall implementation structure illustrated in Fig. 2. For a more intuitive description, the input signal V_{in} is considered as an arbitrary sinusoidal signal ($V_{in} = \cos(2\pi f_{in} \cdot t)$). After the TIADC quantizes the input signal, according to Equation (2), the outputs of the two sub-ADCs can be respectively expressed as:

979-8-3315-8850-2/25 $31.00 © 2025 IEEE

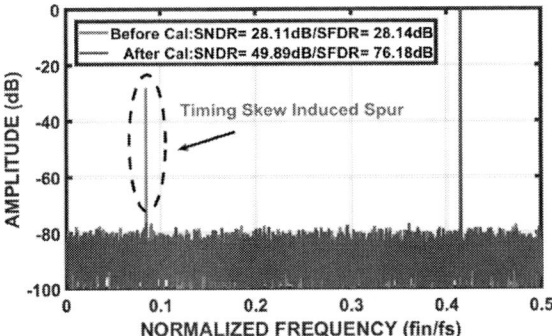

Fig. 5. Spectra of TIADC output before/after calibration for a single tone input at $0.4152 \cdot f_S$.

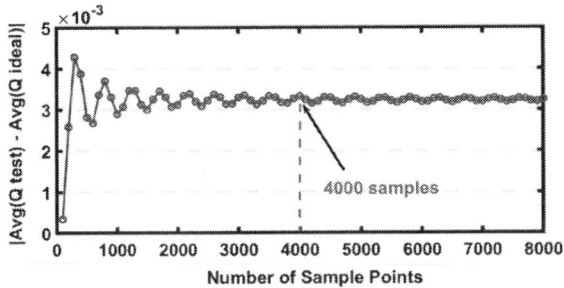

Fig. 6. Comparison of the average values of the ideal signal and the test signal.

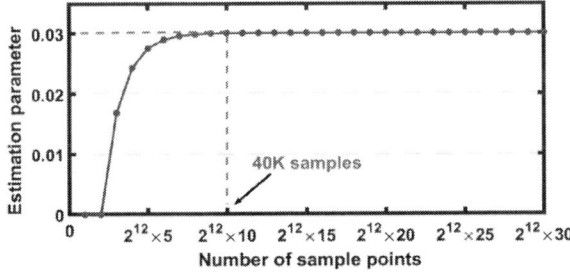

Fig. 7. Convergence curves of estimated parameter.

$$y_1[n] = \cos[2\pi f_{in}(2nT_s)]$$
$$y_2[n] = \cos[2\pi f_{in}((2n+1)T_s + \tau)] \quad (3)$$

Temporarily ignoring the numerical correction process contained within the blue dashed box in Fig. 2, to estimate the timing skew mismatch parameters between channels, apply a Hilbert transform to the output signal of Sub-ADC1 and extract its imaginary part to construct the orthogonal components of the analytical signal. At the same time, apply a linear interpolation filter to the output of Sub-ADC2 to achieve the desired time delay. After these operations, the outputs of the two Sub-ADCs can be expressed as:

$$y_{1\_h}[n] = \sin[2\pi f_{in}(2nT_s)] \quad (4)$$

$$y_{2\_int}[n] \approx \cos[2\pi f_{in}(2nT_s + \tau)] \quad (5)$$

where f_{in} represents the frequency of the input signal.

As illustrated in Fig. 3, the implementation process of the interpolation filter is outlined as follows. Initially, zero-value

samples are inserted between each initial sample of the input data. Subsequent to this insertion, the data is then output through the filter. With regard to the signal sequence processed by the interpolation filter, the odd-indexed samples are retained in order to achieve the time delay function.

Then multiply the output $y_{1\_h}[n]$ of sub-ADC1 after the Hilbert transform signal by the interpolated output $y_{2\_int}[n]$ of sub-ADC2. The result of this multiplication can be expressed as:

$$\begin{aligned}p[n] &= y_{1\_h}[n] \cdot y_{2\_int}[n] \\ &= \sin[2\pi f_{in}(2nT_s)] \cdot \cos[2\pi f_{in}(2nT_s + \tau)] \quad (6) \\ &= \frac{1}{2}\sin[2\pi f_{in}(4nT_s + \tau)] - \frac{1}{2}\sin(2\pi f_{in}\tau)\end{aligned}$$

Subsequently, the sequentially generated products are accumulated N times and averaged. Based on the statistical properties of signals, when N is sufficiently large, the first term in Equation (6) tends to zero after the cumulative averaging process. Therefore, the output after averaging process can be approximated as:

$$Avg \approx -\frac{1}{2}\sin(2\pi f_{in}\tau) \quad (7)$$

It can be seen from Equation (7) that the value of Avg is proportional to the timing skew mismatch. The direction of timing skew mismatch (leading or lagging) can be judged by observing the sign of Avg. Therefore, it can be used as an iterative parameter and input to the LMS adaptive engine to obtain the correction parameters (C_{err}) directly applied to the corrector, which can be expressed as:

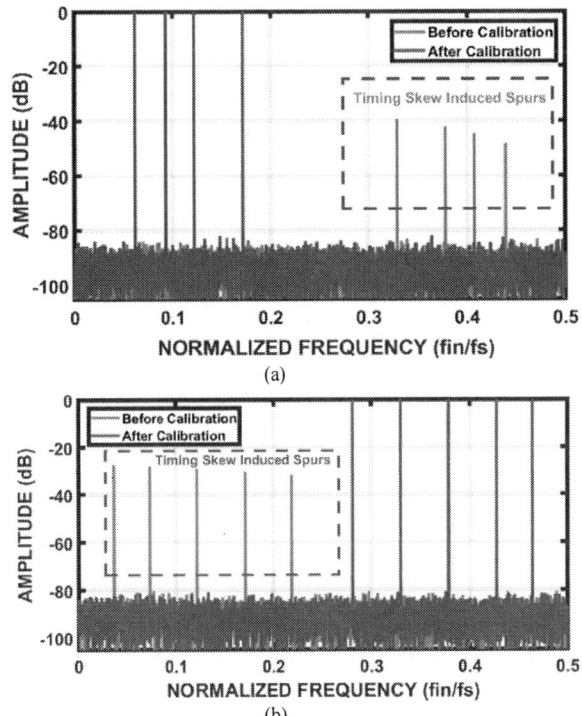

Fig. 8. Spectra of TIADC output before/after calibration for multitone signal, (a) high-frequency parts, (b) low-frequency parts.

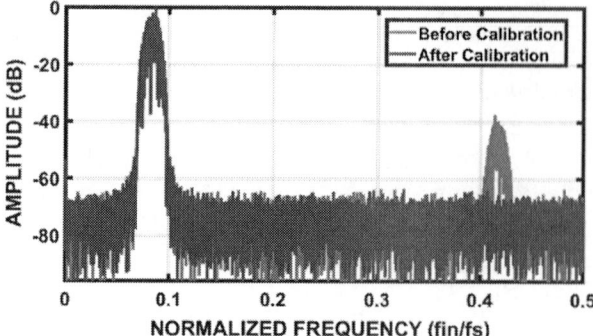

Fig. 9. Spectra of TIADC output before/after calibration for quadrature phase shift keying signal.

$$C_{err}[i+1] = C_{err}[i] - \mu \cdot Avg[i] \qquad (8)$$

Where μ is an iteration factor between 0 and 1. After the correction parameters are obtained, a corrector is required to eliminate errors caused by timing skew mismatch. The clock path of the sub-ADC2 can be compensated by adjusting the controlled delay line (CDL) in the analog domain [13-14], or the output of the sub ADC2 can be corrected in the digital domain by means of a differentiator-multiplier-cascade (DMC) [15-16]. As illustrated in Fig. 4 (a), the block diagram of the CDL is presented. The CDL comprises multiple cascaded adjustable delay units, each containing a variable capacitance array and a low-power inverter. As shown in Fig. 4(b), the

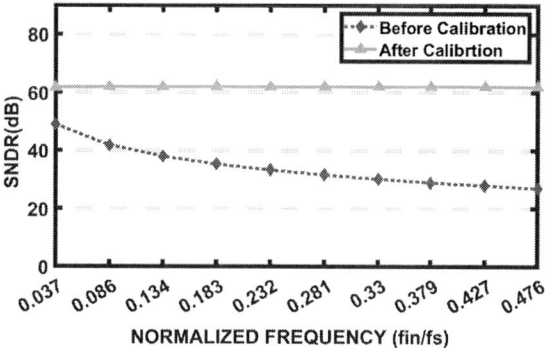

Fig. 10 SNDR versus different input frequencies without and with calibration.

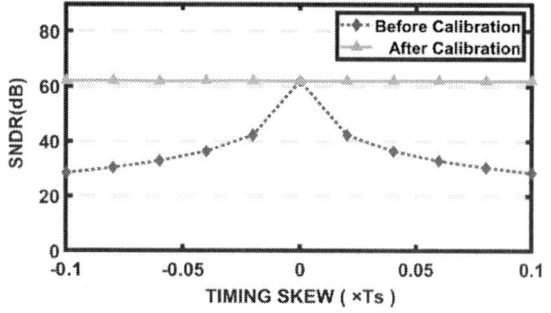

Fig. 11. SNDR versus different timing skew level without and with calibration.

DMC-based corrector employs discrete differentiators and time-varying multipliers. These components utilize Taylor approximation to suppress higher-order error terms induced by signal mismatches. After several iterations, the value of Avg will

approach 0, which means that the timing skew mismatch is eliminated.

III. SIMULATION RESULTS

To verify the effectiveness of the proposed estimation method, a dual-channel TIADC behavioral model described in Table I has been performed and simulated in MATLAB, in which linear interpolation and Hilbert filtering were simulated using interpolation filter and 'hilbert' functions, respectively. Moreover, thermal noise is introduced into the simulated TIADC, and its variance is the same as the quantization noise.

For a single-tone signal input at $0.4152 \cdot f_S$, the output spectra of the TIADC before and after calibration is shown in Fig. 5. The timing skew induced spurs are well suppressed and the SNDR and SFDR are improved from 28.11dB and 28.14dB to 49.89dB and 76.18dB, respectively.

The performance of the timing skew estimation system is assessed through quantitative analysis of the difference between the signal delayed by the interpolation filter and the ideal delayed signal. The ideal delay signal Q_{ideal} is constructed by multiplying the original signal of Sub-ADC2, which has been compensated by the theoretical time delay, with the Hilbert quadrature component of Sub-ADC1. The compensation signal Q_{test}, derived from the actual test, is obtained by interpolating and filtering the Sub-ADC2 signal and then multiplying it with the Hilbert quadrature component of Sub-ADC1. This signal represents the actual compensation outcome. The absolute mean difference is adopted as the evaluation index. As illustrated in Fig. 6, when the number of sampling points exceeds 4000, the absolute mean difference enters a stable convergence stage. Consequently, the signal delayed by the interpolation filter can serve as an adequate substitute for the ideal delayed signal. This implies that the mismatch error estimation can be achieved in each iteration round using merely 4000 sample points. The corresponding convergence curve of the estimated parameters is presented in Fig. 7. It is clear that after 10 rounds of iterations, the estimated parameters accurately converge to the preset values listed in Table 1. The entire process of estimating the mismatch parameters requires approximately 40K samples.

In the case of multi-tone excitation, as depicted in Fig. 8, the calibration significantly reduces spurious components induced by timing skew across both low- and high-frequency regions. Similarly, Fig. 9 presents the spectrum of a QPSK modulated input. Prior to calibration, distinct spurious tones are observed, whereas post-calibration, these artifacts are substantially suppressed. These results collectively confirm that the proposed method effectively enhances the spectral purity of the TIADC output under both deterministic and modulated signal conditions.

To comprehensively evaluate the effectiveness of the proposed estimation method, SNDR performance is analyzed across various test conditions. As shown in Fig. 10, the proposed method consistently improves the SNDR across a wide range of input frequencies, covering nearly the entire Nyquist domain. This demonstrates its strong robustness in enhancing dynamic performance. Fig. 11 further illustrates the SNDR improvement under different levels of timing skew. It is evident that even in cases of severe offset at the sampling instants, the proposed method can accurately extract the mismatch error and achieve a

979-8-3315-8850-2/25 $31.00 © 2025 IEEE

notable improvement in SNDR. Table II compares the proposed estimation method with other advanced works, and it can be found that the work of this paper has more advantages in improving the dynamic performance of TIADC. Meanwhile, in the background work mode, this work requires fewer samples to achieve convergence.

TABLE II. COMPARISON WITH OTHER WORKS

Characteristic	[5]	[7]	[12]	[14]	This work
Estimation mode	Back.	Fore.	Back.	Back.	Back.
Resolution(bit)	10	10	12	12	8
Channels	2	2	2	4	2
The number of filters*	-	-	2	1	2
The number of multipliers*	-	-	0	2	1
The number of adders*	-	-	7	7	0
Calibration strategy	Analog	Analog	Digital	Digital	Digital
Improved value of SNDR(dB)	17.6	19.7	21.3	9.6	21.8
Improved value of SFDR(dB)	22.8	23.1	30.0	21.0	48.1
Convergence samples	400K	0.5K	16K	80K	40K

* for the correction and estimation of timing mismatch for each sub-ADC channel.

IV. CONCLUSION

A fully digital background estimation method based on the Hilbert transform is proposed to address timing skew mismatches in dual-channel time-interleaved analog-to-digital converters (TIADCs). Compared with previously reported techniques, the proposed approach achieves rapid convergence of estimation parameters with moderate hardware complexity and eliminates the need for reference or auxiliary ADCs. Simulation results confirm that the method can accurately estimate timing skew errors under various input conditions, including single-tone, multi-tone, and QPSK-modulated signals. After digital correction, significant improvements in dynamic performance are observed. For instance, under single-tone excitation, timing skew induced spurs are effectively suppressed, and the SNDR and SFDR are improved 21.8dB and 48.1dB, respectively.

REFERENCES

[1] M. Gu, Y. Tao, X. He, Y. Zhong, L. Jie and N. Sun, "A 1-GS/s 11-b Time-Interleaved SAR ADC With Robust, Fast, and Accurate Autocorrelation-Based Background Timing-Skew Calibration," in IEEE Journal of Solid-State Circuits, vol. 60, no. 2, pp. 421-431, Feb. 2025.

[2] L. Ricci et al., "A 2-GS/s Time-Interleaved ADC With Embedded Background Calibrations and a Novel Reference Buffer for Reduced Inter-Channel Crosstalk," in IEEE Journal of Solid-State Circuits, vol. 60, no. 2, pp. 456-468, Feb. 2025.

[3] H. Zhao, M. Zhang, Y. Zhu, R. P. Martins and C. -H. Chan, "A 52.5-dB 2× Time-Interleaved 2.8-GS/s SAR ADC With 5-bit/Cycle Time-Domain Quantization and a Compact Signal DAC," in IEEE Journal of Solid-State Circuits, vol. 58, no. 12, pp. 3586-3597, Dec. 2023.

[4] W. -C. Lin, Y. -C. Chang and Y. -H. Chung, "A 10b 400MS/s 2x-Time-Interleaved 2-Then-1b/Cycle SAR ADC in 90nm CMOS," 2024 IEEE International Symposium on Circuits and Systems (ISCAS), Singapore, Singapore, 2024, pp. 1-5.

[5] J. Song and N. Sun, "A 10-b 600-MS/s 2-way time-interleaved SAR ADC with mean absolute deviation based background timing-skew calibration," 2018 IEEE Custom Integrated Circuits Conference (CICC), San Diego, CA, USA, 2018, pp. 1-4.

[6] M. Guo, S. -W. Sin, L. Qi, D. Xu, G. Wang and R. P. Martins, "Background Timing Mismatch Calibration Techniques in High-Speed Time-Interleaved ADCs: A Tutorial Review," in IEEE Transactions on Circuits and Systems II: Express Briefs, vol. 69, no. 6, pp. 2564-2569, Jun. 2022.

[7] X. Yan, K. Qin, X. Zheng, W. Hu, W. Ma and H. Cui, "A Two-Channel Interleaved ADC With Fast-Converging Foreground Time Calibration and Comparison-Based Control Logic," in IEEE Transactions on Very Large Scale Integration (VLSI) Systems, vol. 32, no. 11, pp. 2001-2011, Nov. 2024.

[8] M. El-Chammas and B. Murmann, "A 12-GS/s 81-mW 5-bit Time-Interleaved Flash ADC With Background Timing Skew Calibration," in IEEE Journal of Solid-State Circuits, vol. 46, no. 4, pp. 838-847, Apr. 2011.

[9] A. Salib, M. F. Flanagan and B. Cardiff, "A High-Precision Time Skew Estimation and Correction Technique for Time-Interleaved ADCs," in IEEE Transactions on Circuits and Systems I: Regular Papers, vol. 66, no. 10, pp. 3747-3760, Oct. 2019.

[10] H. Mafi, M. Yargholi and M. Yavari, "Digital Blind Background Calibration of Imperfections in Time-Interleaved ADCs," in IEEE Transactions on Circuits and Systems I: Regular Papers, vol. 64, no. 6, pp. 1504-1514, Jun. 2017.

[11] J. Qin, W. Zhong, Y. Cao, J. Li, Z. Cao and L. Zhao, "Machine-Learning-Based Mismatch Calibration for Time-Interleaved ADCs," in IEEE Transactions on Nuclear Science, vol. 71, no. 8, pp. 2012-2019, Aug. 2024.

[12] T. -C. Hung, F. -W. Liao and T. -H. Kuo, "A 12-Bit Time-Interleaved 400-MS/s Pipelined ADC With Split-ADC Digital Background Calibration in 4,000 Conversions/Channel," in IEEE Transactions on Circuits and Systems II: Express Briefs, vol. 66, no. 11, pp. 1810-1814, Nov. 2019.

[13] J. Song, K. Ragab, X. Tang and N. Sun, "A 10-b 800-MS/s Time-Interleaved SAR ADC With Fast Variance-Based Timing-Skew Calibration," in IEEE Journal of Solid-State Circuits, vol. 52, no. 10, pp. 2563-2575, Oct. 2017.

[14] S. Chen, L. Wang, H. Zhang, R. Murugesu, D. Dunwell and A. C. Carusone, "All-Digital Calibration of Timing Mismatch Error in Time-Interleaved Analog-to-Digital Converters," in IEEE Transactions on Very Large Scale Integration (VLSI) Systems, vol. 25, no. 9, pp. 2552-2560, Sept. 2017.

[15] S. Tertinek and C. Vogel, "Reconstruction of Nonuniformly Sampled Bandlimited Signals Using a Differentiator–Multiplier Cascade," in IEEE Transactions on Circuits and Systems I: Regular Papers, vol. 55, no. 8, pp. 2273-2286, Sept. 2008.

[16] X. Li, J. Wu and C. Vogel, "A Background Correlation-Based Timing Skew Estimation Method for Time-Interleaved ADCs," in IEEE Access, vol. 9, pp. 45730-45739, 2021

979-8-3315-8850-2/25 $31.00 © 2025 IEEE

2025 The 10th International Conference on Integrated Circuits and Microsystems

A Fast Auto-Frequency Calibration Technique with High Reference Frequency for PLL

Yan Feng
Nanjing University of Science and Technology
Nanjing, China
18235605214@163.com

Yongjie Ye
Nanjing University of Science and Technology
Nanjing, China
1959176377@qq.com

Guoxiao Cheng*
Nanjing University of Science and Technology
Nanjing, China
chengguoxiao@njust.edu.cn

Liu Wang*
National Key Laboratory of Solid -State Microwave Devices and Circuit, Nanjing Electronic Device Institute
Nanjing, China
liu_wang@outlook.com

Wei Kang
Nanjing University of Science and Technology
Nanjing, China
kw@njust.edu.cn

Abstract—**This article proposes a fast auto-frequency calibration (AFC) technique with high reference frequency for wideband phase-locked loop (PLL). The AFC circuit block obtains frequency information based on time-to-voltage converter (TVC), which eliminates the dependence on high-frequency counters. Meanwhile, new settable divider-by-3 (S-Div3) and voltage comparator circuit have been proposed to ensure that the AFC circuit can operate adaptively and rapidly at high reference frequency and different operating states. And a high reference frequency can fundamentally improve the AFC locking time. A 7.6~8.5 GHz voltage-controlled oscillator (VCO) incorporating AFC technology is implemented in 55 nm CMOS process. The simulation results demonstrate that under 132 MHz reference frequency and 8.45 GHz output frequency, the total AFC setting time achieves an outstanding 110.5 ns.**

Keywords—auto-frequency calibration, locking time, time-to-voltage converter, settable divider-by-3, voltage comparator

I. INTRODUCTION

As modern communication protocols cover an increasingly wider frequency range, high-quality phase-locked loop (PLL) with a wide frequency tuning range have become a research hotspot in frequency synthesizers. However, in wideband oscillator design, increasing the size of variable capacitance can result in a decrease in the quality factor of the voltage-controlled oscillator (VCO) tank, which can influence the robustness of the oscillator startup and degrade the phase noise. As a consequence, multi-band becomes a necessary trend for the development of wideband phase-locked loops. Nevertheless, increasing the number of sub-bands will inevitably make the entire locking time of the phase-locked loop longer, which leads to the gradual application of auto-frequency calibration (AFC) technology in wide tuning range phase-locked loops. The AFC technology is mainly applied in charge pump phase-locked loop (CPPLL) to cover a wide output frequency range as well as reduce the tuning gain to minimize the impact of front-end noise on VCO. At the same time, the locking speed of the PLL is accelerated.

In tradition designs, common AFC technologies include closed-loop [1] and open-loop [2-10]. The closed-loop AFC method compares the VCO tuning voltage with two reference voltages, and then adjusts the sub-band by detecting the outputs of the two comparators until the loop is locked, usually with several hundred microseconds calibration time. The open-loop AFC method typically sets the VCO tuning voltage to VDD/2 and obtains frequency information through a counter [3-8], where the counting accuracy is determined by the counting time window, which is typically set to hundreds of reference periods, resulting in longer AFC setting times, and still in the microsecond range. [2] proposed an approach to obtain frequency comparison information more flexibly by comparing the period difference or time-to-voltage converter (TVC), which greatly reducing AFC setting time. Up until [9], the TVC method was successfully applied to reduce the AFC setting time to 350 ns. However, due to its temporal relation dependence on the pulse generator generated by the reference signal and its delay signal, it could not adapt to different reference frequencies by itself and relied on the accuracy of the delay circuit. The reference frequency was limited to 40 MHz. The latest AFC technology [10] tended to improve the search algorithm to achieve the fastest calibration. But it could only partially optimized, and the maximum calibration time still reached 4.74 μs.

This article proposes an improved open-loop AFC technology based on TVC, which eliminates the need for counters and additional clock circuits. At the same time, a simple settable divider-by-3 (S-Div3) and voltage comparator circuit are proposed, supporting adaptive operation at higher reference frequency and different operating states. The search algorithm adopts typical binary method to efficiently and accurately locate the VCO sub-band. Ultimately, the AFC technology can achieve sub-band locking time of 110.5 ns at 132 MHz reference frequency.

979-8-3315-8850-2/25 $31.00 © 2025 IEEE

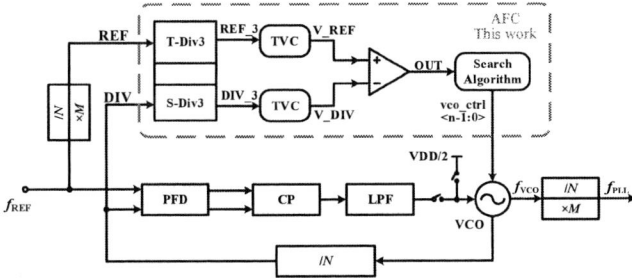

Figure 1. Architecture of the PLL with proposed AFC technology.

II. CIRCUIT ANALYSIS AND DESIGN

Fig. 1 shows the PLL structure, including a traditional CPPLL and a fast AFC block. And the specific architecture of the AFC block proposed in this article is also shown, which consists of traditional divider-by-3 (T-Div3), settable divider-by-3 (S-Div3), TVC, voltage comparator and logic circuit. The following will provide a detailed analysis of each sub circuit.

A. S-Div3 Design

In AFC, the comparison between divider-by-64 signal of the VCO output frequency (DIV) and the reference frequency (REF) is the foundation stone of the overall temporal relation, which directly determines the accuracy of sub-band calibration. In order to determine the moment of the comparison and timely reset of TVC, we need divider-by-3 signal of REF (REF_3) and divider-by-3 signal of DIV (DIV_3) to have an absolute phase relationship, so that the voltage comparison result at the falling edge of DIV_3 is able to be clearly obtained.

In conventional methods, a pulse generator is usually used to turn on the divider-by-3 of DIV, but this approach will result in the circuit relying on the accuracy of the delay circuit and requires manual adjustment. Meanwhile, the pulse width of the pulse generator may cause a certain risk of missing the optimal divider-by-3 turn-on time for the DIV signal. So, the circuit framework of T-Div3 and S-Div3 that this article proposes is shown in Fig. 2(a). T-Div3 adopts a conventional 3-divider circuit and establishes a relationship which is

$$D1 = \overline{Q11 + Q12} \cdot \qquad (1)$$

According to (1), S-Div3 innovatively achieves a direct relationship

$$D2 = \overline{Q21 + Q22 \cdot \overline{REF\_3} + D1}, \qquad (2)$$

which adds a condition of "1" to DIV_3 to easily achieve the absolute phase relationship between REF_3 and DIV_3. As shown in Fig. 2(b), the simulation results illustrate that whether DIV_3 is leading REF_3 or not, DIV_3 lags behind REF_3 after divider-by-3. It can adapt to any reference frequency and simple circuit.

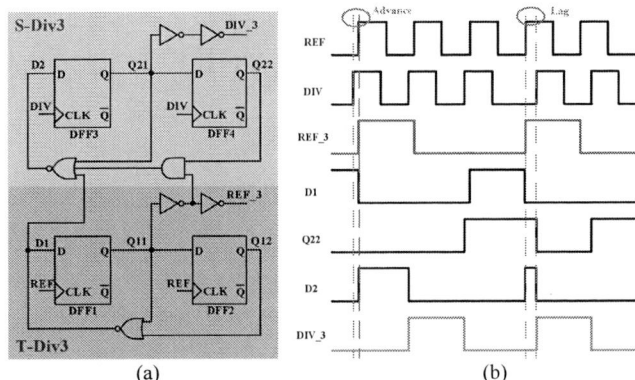

Figure 2. T-Div3 and S-Div3 circuit. (a) Circuit architecture. (b) Functional timing diagram.

B. TVC Design

This AFC is implemented based on TVC, which guarantees the accuracy of frequency information acquisition. To ensure the accuracy of the time to voltage signal, consistent charging current to charging capacitor (C_{REF} and C_{DIV}) is supposed to be provided, and the current should not be excessively high. Otherwise, it will be impossible to compare in the event that the capacitors are fully charged. Consequently, a mirror current source is used, which can not only provide stable current, but also adjust the magnitude of the current well, achieving the comparison when the voltage change slope is large.

C. Voltage Comparator Design

In the precondition for AFC, the voltage comparator is required to feature high accuracy and low delay so that accurate comparison results can be achieved even at higher reference frequencies. However, the charging speed of capacitors in TVC varies greatly under different process corners and temperatures. Therefore, the voltage comparator is required to function properly over a wide range of voltages.

Traditional voltage comparators can only compare voltage values exceeding the NMOS threshold voltage. Thus, on the basis of this, a voltage comparator circuit as shown in Fig. 3 (a) is implemented, which adds a voltage conversion structure (gray area). When the voltage is below the NMOS threshold voltage, it will automatically amplify the voltage proportionally through the voltage conversion structure to achieve voltage comparison. In the same way, when the voltage exceeds the NMOS threshold voltage, comparison can still be achieved. As shown in Fig. 3(b) and (c), at 0.5 V (Above the threshold voltage of NMOS) and 0.1 V (Below the threshold voltage of NMOS) reference voltage (V_REF), while CLK signal is high level, the voltage comparator can output the correct comparison value (V_REF > V_DIV, OUT = 0; V_REF < V_DIV, OUT = 1), and the comparison accuracy can reach 0.7 mV.

979-8-3315-8850-2/25 $31.00 © 2025 IEEE

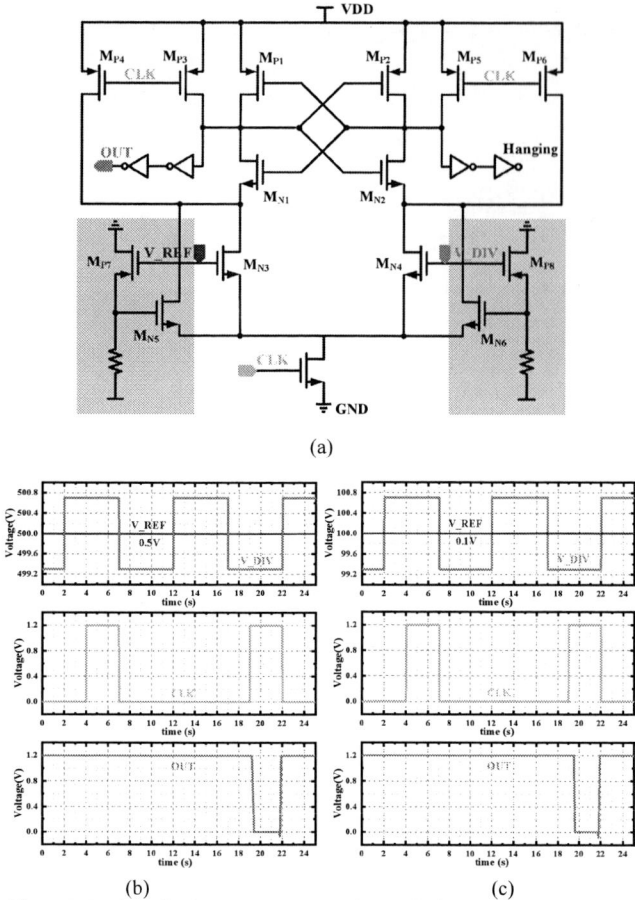

Figure 3. Design of voltage comparator. (a) Detailed circuit architecture. (b) Comparison results at 0.5V V_REF. (c) Comparison results at 0.1V V_REF.

D. Search Algorithm Process

As shown in Fig. 4(a), the classical binary method is utilized for the search algorithm in AFC, which assumes "1" for the undetermined highest bit in the capacitor array (vco_ctrl). And then, the "0" or "1" result of the assumed bit is determined according to the result of comparing the voltage (OUT). In the same way, the next bit is assumed again, and in such loop the VCO sub-band is determined. In addition, many "error situations" should be prevented during this process. As shown in Fig. 4 (b), the start position of AFC is the ideal situation wanted. However, as shown in Fig. 4 (c), after the beginning of AFC, the divider-by-3 and TVC start immediately, but there's no enough buffer time for the VCO to output the corresponding frequency. Consequently, starting the TVC process as soon as this time will result in inaccurate frequency information, which is called the "unsatisfactory cycle". Regarding this issue, AFC activation signal simultaneously is set as the judgment signal. During the high-level period of the AFC signal, the result of voltage comparator is not accepted. Once the duration of the AFC high-level is set reasonably, a suitable buffer time can be obtained, as shown in the simulation results in Fig. 4 (b), which is in line with expectations.

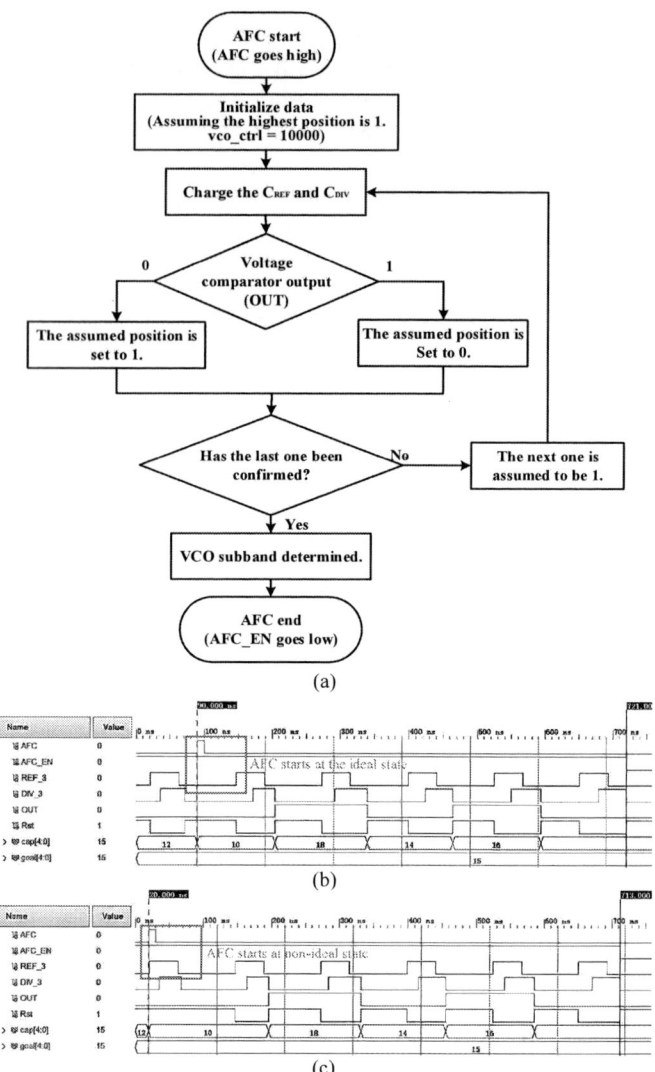

Figure 4. AFC process and verify. (a) AFC process diagram. AFC simulation results (b) at the ideal starting state and (c) at non-ideal starting state.

III. SIMULATION RESULTS

Fig. 5 shows the VCO layout incorporating the proposed AFC block, which is implemented in 55 nm CMOS technology. The core circuit area of AFC is 0.018 mm². Due to the lack of ready-made PLL for verification, only VCO is used to verify the AFC function. Meanwhile, since design of the AFC technology essentially is aimed at achieving calibration of VCO sub-band, this method is accurate.

Fig. 6 shows the simulation results of each node in the AFC process. It can be seen that at 132 MHz reference frequency and 8.45 GHz target frequency, the working state of each node is consistent with the preset state, that the AFC circuit operates normally. The AFC setting time reaches 110.5 ns, and the power consumption is only 1.76 mW.

Next, we will theoretically derive the AFC setting time. The VCO for verification has a 5-bit control word, and the

979-8-3315-8850-2/25 $31.00 © 2025 IEEE

TABLE I

PERFORMANCE SUMMARY AND COMPARISON

	[1]	[2]	[3]	[4]	[5]	[6]	[9]	[10]	This work*
AFC Architecture	Closed-loop	Open-loop	Open-loop	Open-loop	Open-loop	Open-loop	Open-loop	Open-loop	Open-loop
Technology	0.18 µm CMOS	0.18 µm CMOS	0.13 µm CMOS	65 nm CMOS	65 nm CMOS	0.18 µm RF CMOS	65 nm CMOS	28 nm CMOS	55 nm CMOS
Frequency Range (GHz)	3.5~4.4	8.67~10.12	1.9~3.8	0.96~2.06	0.1~5	0.045~2.5	4.7~6.8	2.89~3.49	7.6~8.5
Frequency Detection	Voltage Comparison	TVC	Counter	Counter	Counter	TDC+Counter	TVC	Counter	TVC
Search Algorithm	Linear	Linear	Binary	Adaptive	Binary	Binary	Binary-like	Asynchronous Hybrid	Binary
Reference Frequency (MHz)	-	40	40	40	15~50	12~48	20~40	40	118~132
AFC Time (µs)	400	< 4	1.2~4	4.03~9.59	1.25~1.86	1.25	< 0.35	0.88~4.74	0.11~0.12

*Simulated result

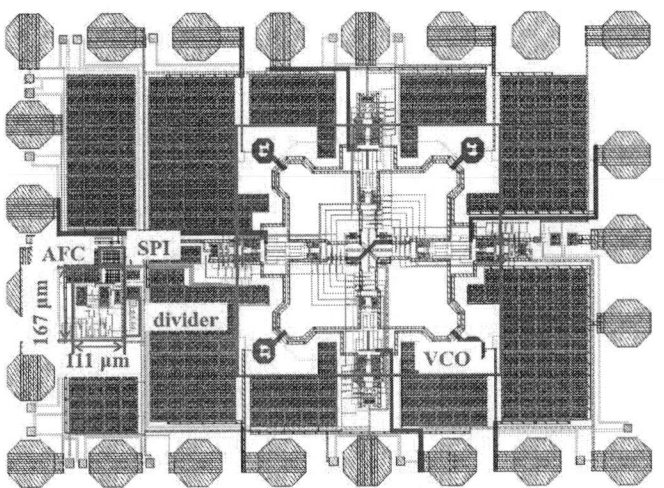

Figure 5. Proposed AFC and VCO layout.

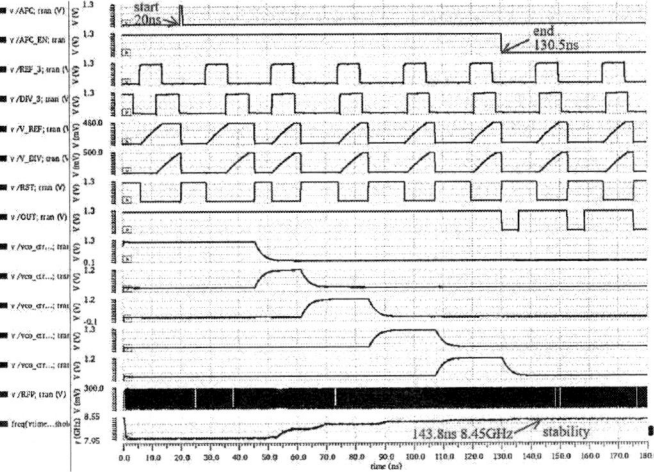

Figure 6. Simulation results for each node at a target frequency of 8.45GHz.

period determined by each bit of control word is one cycle of REF_3 (i.e. three times the reference frequency cycle). Theoretically, the AFC setting time is $15T_{REF}$, but considering

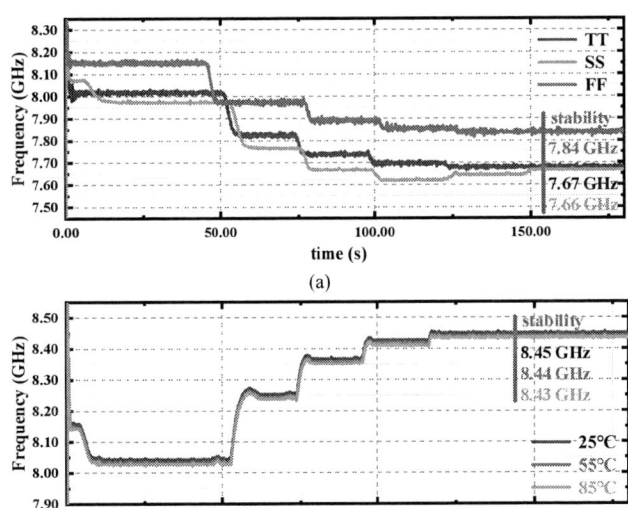

Figure 7. (a) Simulation results of FF, TT and SS process corner. (b)Simulation results at 25, 55 and 85 ℃.

the "unsatisfactory cycle" mentioned above, the complete AFC setting time should be expressed as

$$\tau \le (n+1)\cdot 3T_{REF}. \tag{3}$$

To ensure the stability of the AFC circuit, Fig. 7(a) shows the working state of the AFC with 7.65 GHz target frequency at FF, TT, and SS process corners. It can be observed that although the output frequency of the VCO is various at different process corners, the VCO results tend to approach the target frequency. Thus, the AFC operates normally. Fig. 7(b) shows the working state of the AFC with 8.45 GHz target frequency at operating temperatures of 25, 55, and 85 ℃, respectively. It can be seen that the output frequency of the VCO is almost the same at different temperatures, and the AFC results are approximately consistent. Overall, it indicates that the AFC circuit has good stability in different situations.

979-8-3315-8850-2/25 $31.00 © 2025 IEEE

Table I provides a comparative analysis of the performance achieved in this article against the previously reported AFC technology as cited in references [1-7], It can be seen that the establishment time of AFC in this article has a significant advantage

IV. CONCLUSION

This article introduces a fast AFC technology suitable for wideband PLL, and verifies it with 55 nm CMOS process VCO at 7.6-8.5 GHz as an example. The proposed S-Div3 and voltage comparator circuits enable the AFC technology to have sufficient high-precision and low-latency, enabling stable operation at 118~132 MHz reference frequency, and the total AFC setting time is 110.5~124.5 ns. The AFC setting time of per bit don't exceed 25 ns. The proposed AFC is easy to implement, and has good stability, that it can work normally at various process corners and temperatures. It has strong portability and practicality.

ACKNOWLEDGMENT

This work was supported in part by the National Key Laboratory of Solid-State Microwave Devices and Circuit under Grant SSMDC020204202402, and in part by the National Natural Science Foundation of China under Grant 62104109.

REFERENCES

[1] A. Aktas and M. Ismail, "CMOS PLL calibration techniques," in *IEEE Circuits and Devices Magazine*, vol. 20, no. 5, pp. 6-11, Sept.-Oct. 2004.

[2] T. -H. Lin and Y. -J. Lai, "An Agile VCO Frequency Calibration Technique for a 10-GHz CMOS PLL," in IEEE Journal of Solid-State Circuits, vol. 42, no. 2, pp. 340-349, Feb. 2007.

[3] J. Shin and H. Shin, "A 1.9–3.8 GHz ΔΣ Fractional-N PLL Frequency Synthesizer with Fast Auto-Calibration of Loop Bandwidth and VCO Frequency," in *IEEE Journal of Solid-State Circuits*, vol. 47, no. 3, pp. 665-675, March 2012.

[4] H. Ryu, E. -T. Sung, S. Park, J. -K. Cho and D. Baek, "Fast Automatic Frequency Calibrator Using an Adaptive Frequency Search Algorithm," in *IEEE Transactions on Very Large Scale Integration (VLSI) Systems*, vol. 25, no. 4, pp. 1490-1496, April 2017.

[5] Z. Zhang et al., "A Fast Auto-Frequency Calibration Technique for Wideband PLL with Wide Reference Frequency Range," in *2018 IEEE Asian Solid-State Circuits Conference (A-SSCC)*, Tainan, Taiwan, 2018, pp. 227-230.

[6] A. Hu, D. Liu, K. Zhang, L. Liu and X. Zou, "A 0.045 to 2.5 GHz Frequency Synthesizer With TDC-Based AFC and Phase Switching Multi-Modulus Divider," in *IEEE Transactions on Circuits and Systems I: Regular Papers*, vol. 67, no. 12, pp. 4470-4483, Dec. 2020.

[7] Y. Li, B. Zhou and Z. Wang, "A Wideband Fast Start-Up Multi-Core VCO With Auto-Frequency Control in 0.18 μm CMOS," in *IEEE Access*, vol. 9, pp. 149807-149813, 2021.

[8] G. Park, O. Lee, D. Im and I. Nam, "A Frequency Synthesizer with Automatic Frequency Calibration Robust to Initial Phase Error and Phase-Noise Enhanced Ring Oscillator," in *IEEE Transactions on Circuits and Systems II: Express Briefs*, vol. 72, no. 1, pp. 43-47, Jan. 2025.

[9] D. Sun et al., "A Power-Efficient TVC-Based Fast Auto-Frequency Calibration for PLLs," in *IEEE Transactions on Circuits and Systems II: Express Briefs*, vol. 69, no. 6, pp. 2672-2676, June 2022.

[10] Z. Wang, F. Cai, L. Wu, H. Zhang and W. Zhao, "A Time and Energy-Efficient Asynchronous Hybrid-Searching Auto Frequency Calibration for a 3.2 GHz Phase-Locked Loop," in *IEEE Transactions on Circuits and Systems I: Regular Papers*, doi: 10.1109/TCSI.2025.3547024.

2025 The 10th International Conference on Integrated Circuits and Microsystems

Post-Routing Compression Algorithm for Area-Efficient Layout design of Analog Integrated Circuit

Xiaoyue Wu
College of Integrated Circuits &
Micro-Nano Electronics
Fudan University
Shanghai, China
23210720273@m.fudan.edu.cn

Yubo Zhang
College of Integrated Circuits &
Micro-Nano Electronics
Fudan University
Shanghai, China
24210720328@m.fudan.edu.cn

Wei Zhang
Suzhou Foohu Technology Co.,
Ltd.
Shanghai, China
wei.zhang@foohu.com

Bijian Lan
Suzhou Foohu Technology Co.,
Ltd.
Shanghai, China
bijian.lan@foohu.com

Jing Wan*
College of Integrated Circuits &
Micro-Nano Electronics
Fudan University
Shanghai, China
jingwan@fudan.edu.cn

Abstract—**Routing is a crucial part of analog integrated circuit design. For inter-block routing in analog circuit layouts, existing routing algorithms fail to achieve full connectivity across all scenarios while minimizing track usage, ultimately leading to wasted chip area and increased production costs. This paper proposes a post-routing optimization algorithm for inter-block routing regions in analog circuit layouts. The post-routing optimization algorithm enhances track utilization by reusing tracks in channel routing regions and applying single-layer rerouting in single-sided pin regions, reducing amount of track. In channel routing regions without wire-ordering constraints, track usage, area occupation, and wire length are reduced by 23.37%, 25.05%, and 17.83%, respectively. For single-sided pin regions, these parameters are reduced by with 49%, 49.34% and 4.77%. The algorithm is further extended to address specific global layout optimization. Through the integration of intra-region optimization and wire-ordering constraints, 8 tracks are reduced with an area saving of 377.66 μm². The research provides a practical solution for area-efficient analog integrated circuit design with significant cost-reduction implications.**

Keywords—*routing, post-processing optimization, track, area*

I. INTRODUCTION

Manual design has gradually become inadequate to handle the increasing complexity of circuit architectures, being replaced by Electronic Design Automation (EDA) technologies. Digital circuits, benefiting from their structured design and standardized processes, have achieved relatively mature automated design flows. In contrast, analog circuits face greater challenges in automation due to stringent performance requirements and sensitivity to physical parameters.

Automatic routing for analog circuit layouts refers to the process of generating physical interconnection paths between devices and blocks during layout design, guided by circuit netlists and design rules. In analog layout routing, designers must comprehensively consider circuit performance, wire length, and chip area optimization [1]. Manual routing requires not only extensive design expertise but also significant time investment, particularly given the critical need to minimize chip area [2] [3] and reduce production costs. Consequently, automated routing has become one of the core challenges in analog circuit design automation.

Current approaches in analog IC layout routing can be categorized into three main methodologies: maze routing-based methods, shape-based routing methods, and circuit-specific routing methods. In maze routing approaches, Weijie Chen implemented an improved A* search algorithm to identify multilayer routing paths under minimum area constraints [4]. However, such maze algorithms [5] [6] require significant computational time and primarily optimize for minimal wirelength. They often fail to minimize amount of track, leaving substantial optimization potential for area reduction in post-routing stages.

For shape-based routing methods, exemplified by channel routing between opposing boundaries, T. Yoshimura proposed an efficient channel routing algorithm [7], while Preparata optimized three-layer channel routing layouts [8]. Although these algorithms demonstrate effectiveness in local routing, they lack global optimization capabilities, highlighting the urgent need for comprehensive global routing tools.

Regarding circuit-specific routing methods, Zhang H. et al. developed a collaborative routing framework integrating geometric and circuit performance constraints [9]. However, this approach remains limited to specific circuit types and struggles to simultaneously satisfy constraints while reducing track usage and area consumption.

979-8-3315-8850-2/25 $31.00 © 2025 IEEE

This paper presents an automated optimization algorithm for post-routing track reduction in analog integrated circuits. Our principal contributions are summarized as follows:

- Our algorithm is specifically designed for inter-module routing in analog circuit layouts. It innovatively incorporates post-routing optimization processes to significantly reduce the amount of routing tracks and minimize occupied chip area.

- The algorithm automatically identifies and categorizes inter-module routing patterns, enabling type-specific optimizations.

- Combining with wire-ordering constraint between regions, proposed algorithm supports global routing optimization across inter-module regions in the whole layout of analog integrated circuits.

The rest of the paper is organized as follows: Section II describes our inter-module post-routing optimization algorithm. Section III presents the experimental results, and Section IV concludes the paper.

II. Methodology

This section outlines the algorithmic workflow and elaborates on each implementation step. The algorithm accepts two primary inputs: design rules and GDSII files. Our methodology comprises three core components: (1) GDSII file parsing, (2) single inter-module routing region optimization, and (3) global inter-module routing optimization. Fig. 1 illustrates the main process of our algorithm.

During the GDSII file parsing phase, geometric shapes without net attributes in the GDSII file are parsed into net-aware objects to enable subsequent optimization operations. In the inter-module routing optimization phase, rule-based optimizations are applied to address the two most prevalent routing scenarios in analog inter-module layouts. For the global inter-module routing optimization, the entire layout is decomposed into multiple inter-module routing regions. Inter-regional connectivity constraints, specifically wire-ordering relationships, are rigorously preserved while executing inter-module optimizations across partitioned regions.

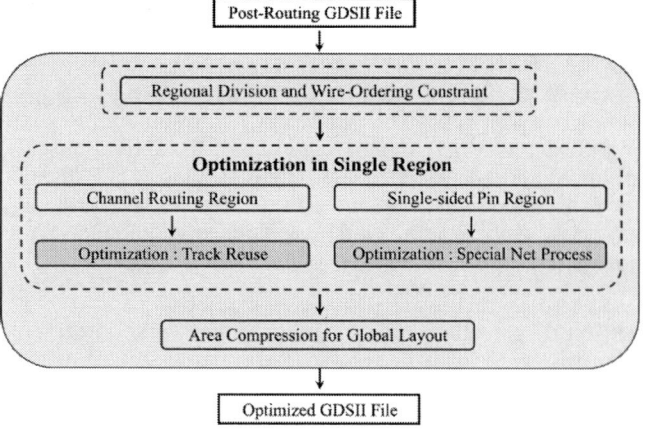

Fig. 1. The flow of post-routing compression algorithm.

A. GDSII File Parsing

The GDSII file exclusively contains geometric primitives (polygons) and labels, where each label incorporates positional coordinates and text metadata encoding the corresponding net identifier. Crucially, geometric elements in GDSII inherently lack logical connectivity information. To address this, our algorithm defines a class wherein each object encapsulates a polygon and its affiliated net metadata. The implementation is structured as follows:

The GDSII file is parsed to extract all routing polygons, which are stored in a list. These polygons are filtered by their inherent layer attributes (e.g., M1 and M2), retaining only those representing interconnect routing paths. This phase preserves the original layer assignments and spatial coordinates of the polygons.

Subsequently, routing polygons overlapping with labels are processed. Spatial intersection checks between these bounding boxes and the pre-filtered routing polygons are performed. Overlapping polygons inherit the net identifier from their associated labels, generating objects that encapsulate both geometric and net metadata for downstream optimization.

For polygons without label overlaps, terminal points (endpoints of horizontal traces or top/bottom edges of vertical traces) are analyzed. Overlap detection between these terminals and previously annotated objects is conducted. If an overlap exists, the net identifier propagates from the annotated object to the unassigned polygon, instantiating a new object. Polygons lacking overlaps are queued for iterative reprocessing.

This terminal-based overlap detection and net match until all routing polygons are annotated with net identifiers. The final output comprises a comprehensive set of objects integrating polygon geometries and their corresponding net constraints.

B. Inter-Module Routing Optimization in single region

1) Channel routing optimization:

This Channel routing, a specialized implementation of detailed routing, is constrained by a topological configuration where interconnect ports are exclusively distributed in two symmetric rows along either upper/lower boundaries or left/right sides. The algorithmic core of our channel routing methodology lies in track reuse optimization, which maximizes track utilization efficiency through post-processing of channel routing cases.

In the first step, all M2-layer net information within the target channel region is extracted and stored in a list. These nets are sorted in descending order based on their coordinate values (y-axis in the illustrated in Fig. 2), assigned to sequentially numbered tracks (Track 1, 2, 3…), with their corresponding heights recorded in a dedicated list to establish the initial routing framework.

Subsequently, Nets are processed in descending order of their M2 height to search for available space above for upward relocation. When handling the M2 layer of a net, horizontal and vertical constraints of channel routing must be considered as shown in Fig. 2. A M2 polygon is generated using the net's x-coordinate and the height of Track 1 as the y-coordinate. Simultaneously, the M1 layer information is updated, and its polygon is generated. Overlap detection is performed between these polygons and existing polygons of other nets on the same layer within the current region. If no overlap is detected, the net is relocated to this track, marking successful optimization. The net's routing status is immediately updated to ensure accurate overlap checks for subsequent optimizations. If overlap exists, the process repeats with Track 2's height, iterating until successful relocation or until no higher track is available, at which point the net's processing terminates.

Finally, the remaining nets undergo the same optimization process in descending M2 height order. Fig. 3 presents the result of the channel routing optimization.

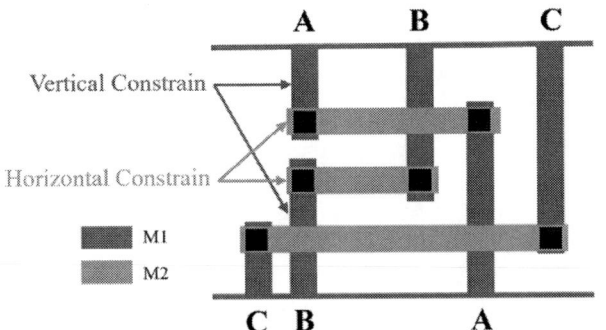

Fig. 2. Vertical and horizontal constraints in channel routing

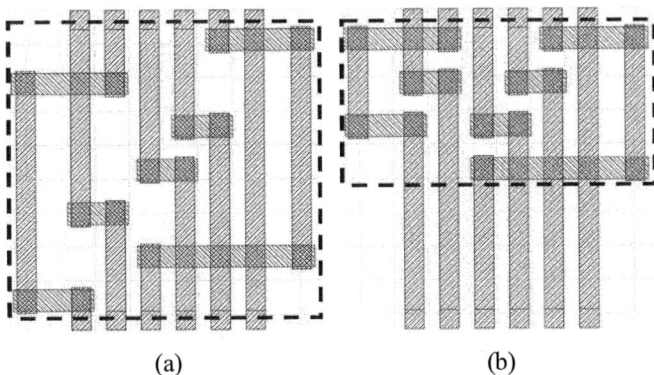

Fig. 3. An example of channel routing (a) before optimization and (b) after optimization.

2) Single-sided pin regions routing optimization:

In inter-module routing regions, there frequently exist scenarios where routing traces exit from one side of a block to connect to other blocks. Such cases, where pins are exclusively located on a single side of the block as shown in Fig. 4, are defined as single-side pin cases. For general routing configurations, horizontal and vertical directions utilize two separate layers. For these specific cases, the optimization algorithm's core principle is to replace the conventional two-layer routing for selected nets with single-layer implementations, thereby freeing up tracks for reuse by other nets as illustrated in Fig. 5. These rerouted nets are hereafter referred to as special nets.

This paper addresses three distinct cases of such configurations as illustrated in Fig. 4. Since the algorithmic logic remains identical for left-side and right-side routing exits, the following discussion uses right-side routing exits as the representative example.

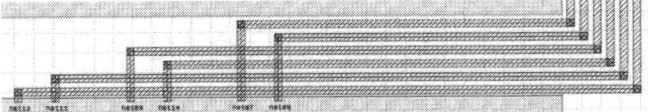

(a) Case 1: Co-directional pin and routing out alignment

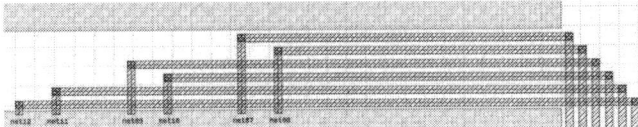

(b) Case 2: Counter-directional pin and routing- out alignment

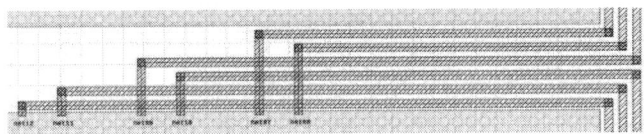

(c) Case 3: Mixed-directional pin and routing out alignment

Fig. 4. An example of single-sided pin regions routing

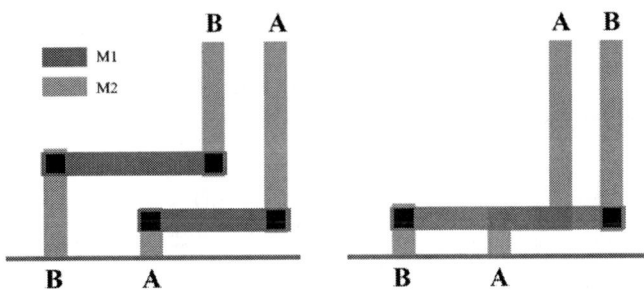

Fig. 5. Special net A process

When no wire-ordering constraints exist for interconnects in this region, the algorithm achieves maximum track reduction. During this process, a constraint arises between a reused net B and its associated special net A: since net B requires vias due to its dual-layer routing implementation, net B's exit position must strictly reside to the right of special net A to prevent short circuits.

In Single-sided pin regions, interconnects typically interface with adjacent blocks, resulting in wire-ordering constraints. The selection of these special nets is jointly determined by sequential pin positions and wire-ordering requirements.

For Case 1, the special net selection rules are as follows:

- Priority is given to nets positioned earlier in the wire-ordering sequence.

- Nets located to the left of the selected special net's pin positions are prohibited from special net designation.

- The number of special nets is capped at half of the total number of nets.

As previously established, the constraint between special net A and reused net B mandates that reused net B must be routed to the right of special net A to avoid via-induced short circuits. Consequently, special nets are preferentially selected from earlier positions in the wire-ordering sequence to maximize viable track reuse. This restriction prevents geometric overlaps between newly assigned special nets and existing routing geometries. Violating this rule would inevitably cause routing short.

After selecting special nets, the algorithm evaluates whether regular nets are positioned to the left of these special nets based on their sequential pin coordinates. Regular nets located to the left of a special net become eligible for track reuse with that special net, where each special net's track can be reused only once. If no regular net reuses a special net's track, the special net reverts to a regular net. The total number of saved tracks is calculated by subtracting the optimized track count (original track count minus saved tracks) from the pre-optimization track count. The optimized track count determines the track height list, which is allocated according to the required wire-ordering sequence. Starting from the highest track, special nets are assigned new tracks, while regular nets reuse existing special net tracks when available; otherwise, they occupy new tracks. The final output is a dictionary mapping each net to its assigned track height. Additionally, the pre-optimization routing distances are redistributed based on the wire-ordering requirements to ensure geometric compliance.

For Case 2, the algorithm follows the same logic but modifies two rules: after selecting a special net, nets to the right of its pin positions cannot be designated as special nets, and special nets are assigned track heights from the lowest to highest.

For Case 3, which includes both upward and downward routing exits, the algorithm prioritizes upward-facing nets using the Case 1 logic. Downward-facing nets are processed based on Case 2 with an additional constraint: special nets for downward-facing nets can only be selected to the right of those designated for upward-facing nets according to pin positions. All other steps remain consistent with Case 2.

3) Global inter-module routing optimization

For layouts containing both channel routing and single-side pin case inter-module routing configurations, the algorithm performs global inter-module routing optimization by partitioning the layout into single regional routing optimizations. The reduced track counts and spatial compaction of freed routing areas collectively achieve area reduction for the entire layout.

III. RESULTS AND DISCUSSION

A. Single Inter-Module Routing Region Optimization

In single region optimization test, the optimization results for channel routing cases [10] are illustrated in Fig. 6. The channel routing achieves average reductions of 23.37% in track count, 25.05% in area, and 17.83% in wirelength. The algorithm effortlessly addresses optimizable areas that are difficult for designers to detect, as demonstrated in Fig. 7.

The optimization results for six single-side pin regions under wire-ordering-unconstrained conditions are shown in Fig. 8, achieving average reductions of 49% in track count, 49.34% in area, and 4.77% in wirelength. When wire-ordering constraints are applied, for example, input [12 26 7 11 9 8 10], the optimization results are illustrated in Fig. 9.

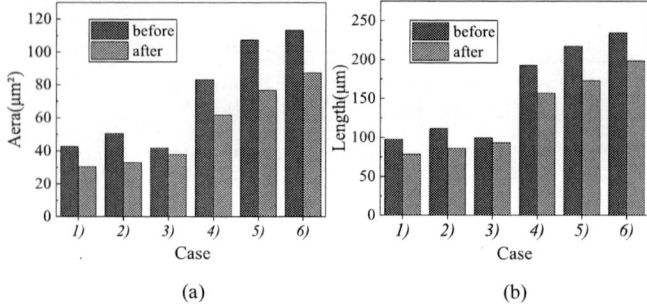

Fig. 6. Channel routing optimization (a) area optimized result and (b) length optimized result.

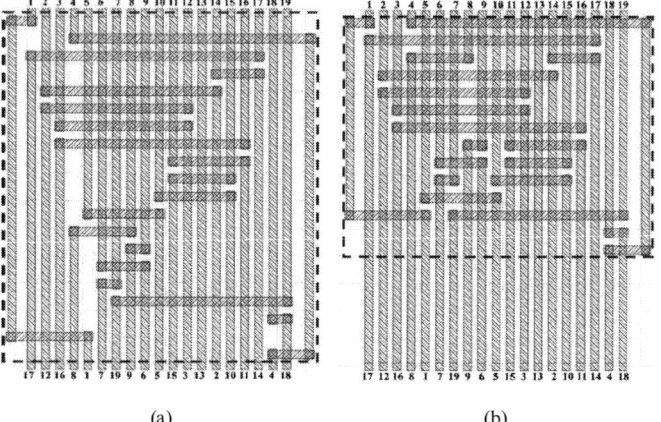

Fig. 7. An example of channel routing (a) before optimization and (b) after optimization.

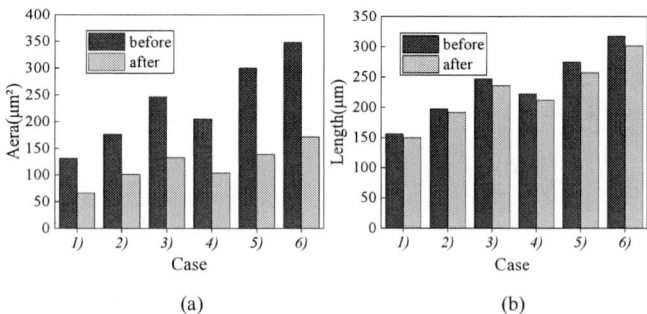

Fig. 8. Single-side pin regions routing optimization (a) area optimized result and (b) length optimized result.

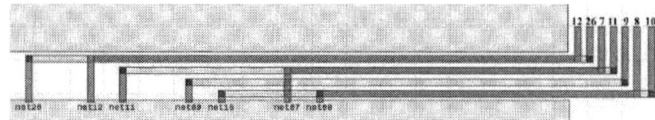

Fig. 9. Optimization results under wire-ordering constraints.

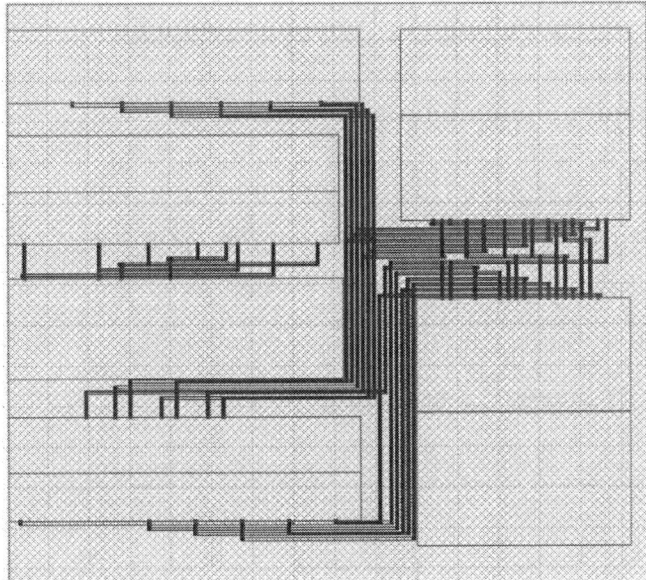

Fig. 10. An example of global inter-module routing optimization

B. Global inter-module routing optimization

For layouts containing both channel routing and single-side pin case inter-module routing configurations, the algorithm performs global inter-module routing optimization by partitioning the layout into single regional routing optimizations.

During full-layout optimization, wire-ordering constraints between interconnected regions must be considered when two or more regions share routing dependencies. The global optimization results using GDSII files are shown in Fig. 10, where interconnected routing regions maintain post-optimization wire-ordering consistency to prevent exit sequence conflicts. The algorithm globally eliminates 8 tracks with an area saving of 377.66 μm².

C. Limitations and Future Directions

The current inter-module routing optimization algorithm faces the following limitations. For single-region optimization, it can only handle two specific cases: channel routing and single-sided pin regions. When these two cases are mixed, the algorithm fails to function properly. Additionally, in channel routing optimization, the idea of the single-sided pin region optimization algorithm could be further leveraged by applying special net processing and track reuse, which represents a potential improvement direction. For global inter-module routing optimization, the algorithm currently supports only vertical compaction and cannot address scenarios involving pins on vertical edges (which introduce new constraints) or horizontal compaction. Furthermore, adjustments to wire ordering constraints in upstream modules may render downstream modules unoptimizable, causing the algorithm to fail to optimize. Future work will focus on resolving these issues through targeted research and algorithmic enhancements.

IV. CONCLUSION

In this paper, we introduce an algorithm to optimize inter-module routing in analog circuit layouts. The proposed routing optimization algorithm effectively reduces track counts, wirelength, and total layout area for inter-module connections. Results from localized optimizations show that in channel routing scenarios without wire-ordering constraints, the algorithm achieves an average reduction of 23.37% in track count, 25.05% in area, and 17.83% in wirelength. For single-side pin cases, the average track count is reduced by 49%, area by 49.34%, and wirelength by 4.77%. When applied to specific GDSII layouts, the algorithm globally eliminates 8 tracks with an area saving of 377.66 μm². Future work will focus on expanding the algorithm to support additional routing configurations, improving its generality and success rate for handling increasingly complex inter-module routing challenges.

ACKNOWLEDGMENT

This work was supported by the National Key R&D Program of China (2021YFA1200500), National Natural Science Foundation of China (62474052,62404055), Key Technology R&D Plan Shanghai(25CL2900100), Shanghai Municipal Basic Research Program (25JD1402800).

REFERENCES

[1] H. Yao, Y. Cai and Q. Gao, "LEMAR: A novel length matching routing algorithm for analog and mixed signal circuits," 17th Asia and South Pacific Design Automation Conference, Sydney, NSW, 2012, pp. 157-162, doi: 10.1109/ASPDAC.2012.6164937.

[2] E. Malavasi and A. Sangiovanni-Vincentelli, "Area routing for analog layout," in IEEE Transactions on Computer-Aided Design of Integrated Circuits and Systems, vol. 12, no. 8, pp. 1186-1197, Aug. 1993, doi: 10.1109/43.238611.

[3] H. Habal and H. Graeb, "Constraint-Based Layout-Driven Sizing of Analog Circuits," in IEEE Transactions on Computer-Aided Design of Integrated Circuits and Systems, vol. 30, no. 8, pp. 1089-1102, Aug. 2011, doi: 10.1109/TCAD.2011.2158732.

[4] Weijie Chen, H. Yao, Y. Cai and Q. Zhou, "Analog routing considering min-area constraint," 2013 IEEE 10th International Conference on ASIC, Shenzhen, China, 2013, pp. 1-4.

[5] Sherwani N A. Algorithms for VLSI physical design automation[M]. Springer Science & Business Media, 2012.

[6] Lee C Y. An algorithm for path connections and its applications[J]. IRE transactions on electronic computers, 2009 (3): 346-365.

[7] T. Yoshimura and E. S. Kuh, "Efficient Algorithms for Channel Routing," in IEEE Transactions on Computer-Aided Design of Integrated Circuits and Systems, vol. 1, no. 1, pp. 25-35, January 1982, doi: 10.1109/TCAD.1982.1269993.

[8] Preparata and Lipski, "Optimal Three-Layer Channel Routing," in IEEE Transactions on Computers, vol. C-33, no. 5, pp. 427-437, May 1984, doi: 10.1109/TC.1984.1676459.

[9] H. Zhang et al., "SAGERoute: Synergistic analog routing considering geometric and electrical constraints with manual design compatibility," in 2023 Design, Automation & Test in Europe Conference & Exhibition (DATE), Antwerp, Belgium: IEEE, Apr. 2023, pp. 1–6.

[10] D. C. Wang, "Novel routing schemes for IC layout part I: two-layer channel routing," 28th ACM/IEEE Design Automation Conference, San Francisco, CA, USA, 1991, pp. 49-53, doi: 10.1145/127601.127626.

2025 The 10th International Conference on Integrated Circuits and Microsystems

A Fractional-*N* Reference-Sampling PLL With a Gain-Boost Fractional Phase Detector for Phase Noise Reduction

Xiaolian Xi, Lecheng Cai, Yihao Liu, Fan Liu, and Yanlong Zhang*
School of Microelectronics, Xi'an Jiaotong University, Xi'an, China
Shaanxi Key Laboratory for Electronic Devices and Advanced Chips, Xi'an, China
Key Laboratory of Micro-Nano Electronics and System Integration of Xi'an City, Xi'an, China
*Corresponding author: yanlong.zhang@xjtu.edu.cn

Abstract—**A fractional reference-sampling phase detector (RSPD) that can reduce both in-band and out-of-band phase noise of a fractional-*N* phase-locked loop (PLL) is presented. It realizes spatial averaging with an array of 16 reference-sampling phase detectors (RSPDs) to reduce the quantization error from the $\Delta\Sigma$ modulator (DSM). Meanwhile, the gain of each RSPD elements is boosted by a factor of four at the locked state of PLL to reduce the in-band phase noise from PD. A prototype 5-GHz factional-*N* RSPLL with this fractional RSPD is designed in a 40-nm CMOS process. Post-layout simulation results show that in-band and out-of-band phase noises are reduced by 11 and 17 dB, respectively, leading to a significant reduction of the integrated RMS jitter from 656 fs to 249 fs. With a power of 7.2 mW, the proposed fractional-*N* RSPLL has a figure of merit (FoM$_{jitter}$) of −243.5 dB.**

Keywords—$\Delta\Sigma$ *modulator (DSM), data-weighted averaging (DWA), fractional-N phase-locked loop (PLL), frequency synthesizer, jitter, phase detector (PD), phase noise, phase noise reduction, reference sampling (RS), space-time averaging (STA).*

I. INTRODUCTION

Fractional-*N* phase-locked loops (PLLs) have been widely applied in modern electronic systems, thanks to the merits such as finer frequency resolution, wider loop bandwidth, faster settling, compared to their integer-*N* counterparts. With the increasing data rate in communication, and faster sampling clock frequency in data conversion, the jitter requirement for a PLL becomes more stringent. However, fractional-*N* PLLs suffers two categories of noise, the in-band phase detector (PD) noise and the quantization noise from $\Delta\Sigma$ modulator (DSM), degrading in-band and out-of-band phase noise, respectively.

To suppress the PD noise, sub-sampling PD (SSPD) is proposed [1], in which the reference clock samples the sinewave VCO output directly, converting the phase error between VCO output and input reference clock into a voltage and holding on a capacitor. Due to the high slope of the sinewave at the zero-cross point, SSPD has such a high phase-to-voltage transfer gain, and hence, its noise is significantly suppressed, leading to a lower in-band phase noise at PLL output [1]. Nonetheless, the periodically switched sample/hold state in SSPD modulates the load capacitance of the voltage-controlled oscillator (VCO), leading to the binary frequency-shift keying (BFSK)

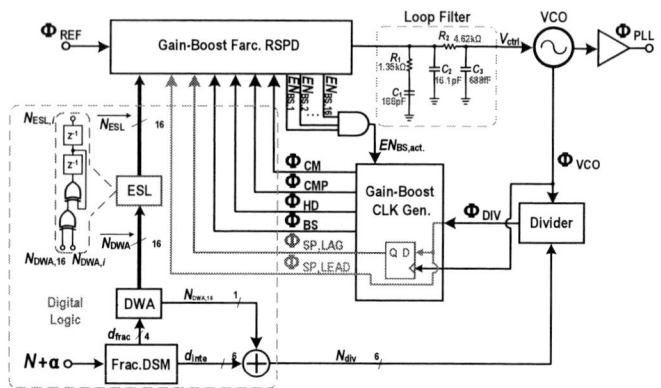

Fig. 1. Block diagram of the proposed Fractional-*N* PLL.

modulation at VCO output [2]. To overcome this issue, a reference-sampling PD (RSPD) is adopted [3], in which the divided VCO output performs as the sampling clock, and samples the sinewave input reference clock. In this way, the reference buffer that costs large power and hardware is eliminated, so as its noise. Due to the much lower frequency, the slope of the reference clock at the zero-cross point, the gain of RSPD is much lower than SSPD, degrading the in-band phase noise performance. To address this issue, a gain booster is used to enlarge the gain of the RSPD as loop is locked [4]. Additionally, thanks to the boosted PD gain, the sampling capacitor can be smaller, saving both hardware cost and power.

The DSM quantization noise is another main phase noise source in fractional-N RSPLLs, which degrade the out-of-band phase noise, introducing noise peak around loop bandwidth frequency. To reduce this noise, a digital-to-time converter (DTC) is often utilized to compensate the quantization error by shifting the reference phase [5], [6]. Nevertheless, this measure degrades reference phase noise. In addition, the phase interpolator (PI) is another technique to reduce quantization noise by achieving instantaneous fractional frequency division [7], [8]. Operating at the feedback path, PI does not affect the reference clock, but it has large power dissipation for working at higher frequencies. Replacing the single capacitor with a capacitor digital-to-analog converter (CDAC), the RSPD has the additional function of compensating the quantization noise [9], [10]. Since it works at the reference frequency, the power

---

This work was supported by the CETC-24 Stability Support Program under Project YG2407-2.

979-8-3315-8850-2/25 $31.00 © 2025 IEEE   643

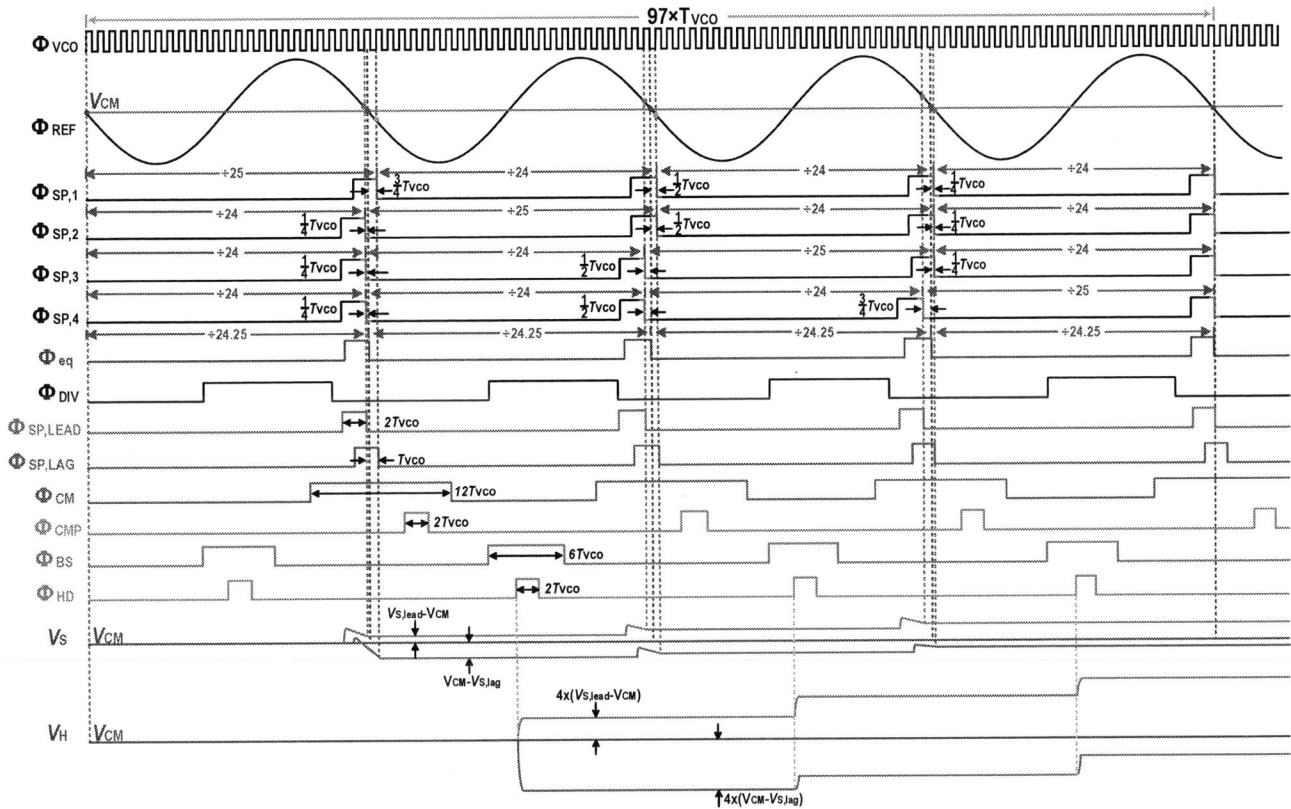

Fig. 2. Time-domain waveform of the proposed gain-boost fractional PD.

is lower than a PI. Unfortunately, CDAC suffers from its nonlinearity, as well as a DTC. To overcome the afore-mentioned issues, a voltage-averaging fractional PD is realized by combining the space-time averaging (STA) technique and the RSPD [11]. This approach reduces both in-band and out-of-band phase noise of a fractional-N RSPLL without using calibration. However, due to insufficient-large PD gain, the in-band phase noise is not well suppressed.

In this paper, a gain-boost fractional RSPD is presented by first utilizing both the gain boost technique and STA in a fractional-N RSPLL. In this architecture, the spatial averaging is achieved by an array of gain-boost RSPDs and G_M cells, which can reduce both the in-band PD noise and quantization noise simultaneously. Seeing as a whole, this array can be considered as fractional RSPD with its gain boosted as locked. With this fractional RSPD, a 5-GHz fractional-N RSPLL is designed in a 40-nm CMOS process. Post-layout simulation results show that the RSPLL achieves an integrated jitter of 249 fs with 7.2 mW, leading to a jitter figure of merit (FoM$_{jitter}$) of −243.5 dB.

II. PLL ARCHITECTURE AND CIRCUI IMPLEMENTATION

Fig. 1 shows the top-level block diagram of the proposed fractional-N RSPLL, where the gain-boost fractional RSPD and clock generation block are two key elements, who realize the spatial averaging under the gain-boost mode. In order to reduce the power and hardware cost, the single-divider method in [12] is adopted to achieve spatial averaging. As Fig. 1 shows, the two

averaging phase, $\Phi_{SP,LEAD}$ and $\Phi_{SP,LAG}$, are obtained via delaying the divider output Φ_{DIV} simply by a single VCO cycle T_{VCO}. Fig. 2 plots an example that illustrates how the spatial averaging works in the proposed gain-boost fractional RSPD, where the division ratio is $N + \alpha = 24.25$ and the number of gain-boost RSPD elements is set $M = 4$ for simplicity. To achieve this division ratio, the data-weight-averaging (DWA) technique is used, similarly to [11], to define the sampling clock of each PD element, $\Phi_{SP,i}$ ($i = 1, 2, 3,$ and 4). If $\Phi_{SP,i} = \Phi_{SP,LEAD}$, $\Phi_{SP,i}$ is equivalent to a divider with the division ratio of 24, and If $\Phi_{SP,i} = \Phi_{SP,LAG}$, $\Phi_{SP,i}$ is equivalent to a divider with the division ratio of 25. In each PD element, $\Phi_{SP,i}$ samples the sinewave reference clock, converting the phase error between $\Phi_{SP,i}$ and the reference clock Φ_{REF} to a voltage $V_{S,i}$. At the gain-boost mode, $V_{S,i}$ is first amplified by a time of 4, transferred to $V_{H,i}$, and then, $V_{H,i}$ is transferred into the output current, $I_{OUT,i}$, of the corresponding gain-boost RSPD element. At the input of the loop filter, all RSPD element's output current are summed and averaged in the current/ charge domain. Note that, as shown in Fig. 2, the falling edge of each $\Phi_{SP,i}$ does not right at the V_{CM}-crossing point of Φ_{REF}. If we treat the four gain-boost RSPD elements as a whole, it will have a ZERO output current. This means that the equivalent input of the overall fractional RSPD, $\Phi_{eq,i}$, can be right at the V_{CM}-crossing point of Φ_{REF}. In other words, the four gain-boost RSPD elements together act as a

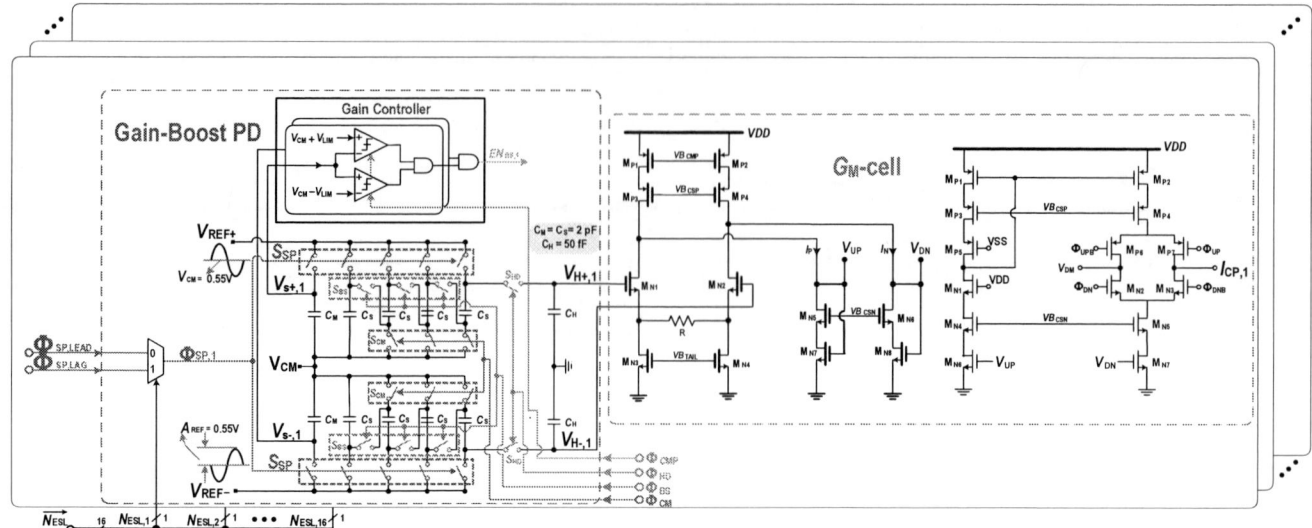

Fig. 3. Circuit implementation of the gain-boost fractional RSPD.

fractional PD, which can detect the output of a fractional divider with a division ratio of $N + \alpha = 24.25$. As a result, the DSM quantization error is eliminated. In reality, as illustrates in Fig. 1, we use $M = 16$ gain-boost RSPD elements for spatial averaging, and thus, an instantaneous fractional frequency division with accuracy of 1/16 can be realized. For an arbitrary fractional division ratio, $N + \alpha$, a fractional DSM with quantization step of 1/16 can be used to convert it into a mixed fraction series with the form of $N + j/16$, where j is an integer and $j \in [0, 16]$. In this way, the quantization noise can be reduced by ~24 dB over the entire frequency range.

Fig. 3 shows the implementation of the proposed gain-boost fractional RSPD and the detail circuit of the RSPD element. Each element can be separated into two parts, the gain-boost RSPD, which converts the phase error between $\Phi_{SP,i}$ and Φ_{REF} to a voltage difference $V_{H+} - V_{H-}$, and the G_M-cell, which transfer $V_{H+} - V_{H-}$ to output current $I_{CP,i}$. The gain-boost RSPD adopts the similar structure to that in [4], where a gain controller is used to detect the differential sampled voltage, V_{S+} and V_{S-}, and switch the RSPD between the gain-boost mode and the normal mode. When V_{S+} and V_{S-} are both within the range between $V_{CM} + V_{LIM}$ and $V_{CM} - V_{LIM}$, which means that the sampled voltage difference $V_{S+} - V_{S-}$ is small, the gain-boost enable signal $EN_{BS,i}$ turns HIGH, and the RSPD is switched to the gain-boost mode. Assuming the limitation threshold is $V_{LIM} = 0.1$ V and the reference amplitude is $A_{REF} = 0.55$ V, under a reference frequency of $f_{REF} = 0.55$ MHz, the gain-boost window is (−289 ps, 289 ps), larger than the period of a 5-GHz clock. Thus, this window can well withstand the typical jitter level of a PLL (< 1 ps). To guarantee the validity of spatial averaging, all RSPD elements must work in the same mode at the same time. Hence, the gain-booting mode is turned on only when the gain-boost enable signals of all RSPD element, $\{EN_{BS,i}\}$, are HIGH, as Fig. 1 depicts.

To avoid timing conflict, the clocks for sampling/hold and turning on gain boost are generated based on the divider output,

and synchronized to the rising edge of VCO output, two sampling phases, as well as $\Phi_{SP,LEAD}$ and $\Phi_{SP,LAG}$. Fig. 2 also plots waveforms of these clocks, together with their minimum pulse width for correct PD operation. The sampling clock of i-th element, $\Phi_{SP,i}$, is selected from $\Phi_{SP,LEAD}$ and $\Phi_{SP,LAG}$, according to the corresponding element $N_{ESL,i}$ of the edge election vector N_{ESL} for spatial averaging, and this is the main difference between the element of proposed fractional RSPD and the RSPD in [4].

As each RSPD element detects the phase error between $\Phi_{SP,i}$ and Φ_{REF}, and decides its mode in every reference cycle, every RSPD element will be switched in following four steps.

STEP 1: Tracking state. In this state, both Φ_{CM} and $\Phi_{SP,i}$ are HIGH, and Φ_{BS}, Φ_{HD} and Φ_{CMP} are LOW. Thus, switches S_{CM} and S_{SP} are ON, and switches S_{BS} and S_{HD} are OFF. V_{S+} and V_{S-} vary with the differential input.

STEP 2: Sampling and comparing state. In this state, Φ_{CM} remains HIGH, but $\Phi_{SP,i}$ turn LOW, and Φ_{BS} and Φ_{HD} are still LOW. Hence, S_{SP} turns OFF, but S_{CM} stays ON. Meanwhile S_{BS} and S_{HD} maintain OFF. The reference voltages at the falling edge of $\Phi_{SP,i}$ are stored on V_{S+} and V_{S-}, and do not vary with the reference clock. After sampling, Φ_{CMP} turns HIGH for two VCO cycles ($2T_{VCO}$) for gain controller to decide the RSPD mode in next step.

STEP 3: Mode selection state. In this mode, the state of RSPD depends on Φ_{BS}. If $\Phi_{BS} = 0$, the RSPD is in the normal mode, working as a conventional RSPD. In this case, Φ_{CM} remains HIGH, and all sampling capacitors are parallel connected. If $\Phi_{BS} = 1$, the RSPD operates at the gain-boost mode. In this case, Φ_{CM} turns LOW, but Φ_{BS} turns HIGH. Thus, S_{CM} turns OFF, while S_{SB} turns ON. Under this circumstance, the four sampling capacitors are connected in series. This way, the gain of RSPD becomes four times of that at normal mode.

STEP 4: Holding state. In this state, Φ_{HD} turns HIGH, and the voltage on sampling capacitors C_S is transferred to the hold capacitor C_H. As $C_H << C_S$, $V_H \approx V_S$ for normal mode, and $V_H \approx$

$4V_S$ for gain-boost mode, leading to 12-dB reduction to the in-band phase noise.

TABLE I. LOOP DYNAMICS UNDER DIFFERENT RSPD MODES

	BW	f_z	f_P	PM
Normal	952 kHz	625 kHz	7.64 MHz	50°
Gain-Boost	3.81 MHz	625 kHz	7.64 MHz	54°

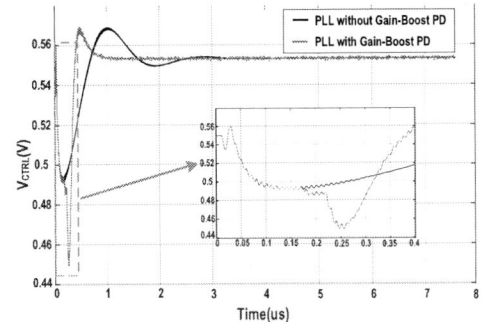

Fig. 4. VCO control voltage under different RSPD modes.

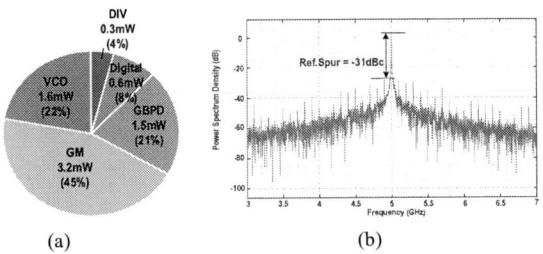

(a)   (b)

Fig. 5. (a) Power breakdown and (b) simulated spectrum of the prototype RSPLL

When the next reference cycle comes, the RSPD starts a new operation cycle. At this moment, Φ_{HD} turn LOW, and the voltage on the holding capacitor is fixed, which maintains the output current of the G_M-cell without affected by the sample/hold operation.

III. SIMULATION RESULTS

To verify the concept, a prototype 5-GHz fractional-N RSPLL is designed in a 40-nm CMOS process, which is based on the type-II architecture, as Fig. 1 shows. With 100-MHz reference clock, its output frequency ranges from 4.5 GHz and 5.5 GHz.

Table I compares the loop dynamics under normal and gain-boost modes, the gains of G_M-cell and VCO are 128 µS and 300 MHz/V, respectively. Note that, thanks to the 4× gain boosting, the loop bandwidth is extended by 4 times, but the phase margin change by a little. This means that gain boost almost does not affect the loop response. Fig. 4 compares the simulated VCO control voltage under both normal and gain-boost modes. As expected, although switching from the normal mode to the gain-boost mode breaks the acquisition progress, the wider loop bandwidth speed-up its settling without stability risk. Fig. 5 shows the power breakdown and output spectrum at the gain-boost mode. As displayed, the reference spur is −31 dBc, which mainly results from the time glitch among different RSPD elements and the insufficient isolation of the hold switch S_{HD}.

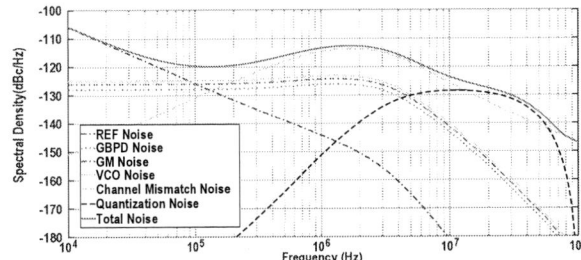

Fig. 6. Simulated phase noise of main contributors.

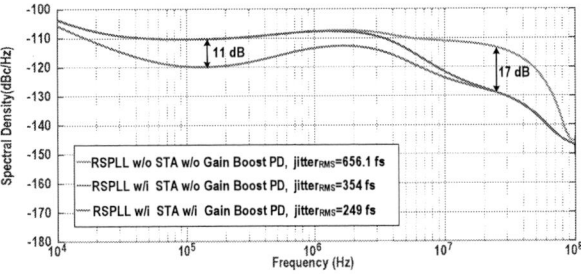

Fig. 7. Phase noise comparison with different techniques.

TABLE II. LOOP DYNAMICS UNDER DIFFERENT RSPD MODES

	JSSC'16 [7]	JSSC'20 [12]	JSSC'21 [13]	VLSI'20 [14]	JSSC'22 [15]	This work*
Technology (nm)	65	40	28	22	180	40
f_{REF} (MHz)	40	50	48	122.88	100	100
f_{OUT} (GHz)	4.34–4.94	2.4	1.8–2.3	5.8–7.2	4.9	4.5–5.5
Bandwith (MHz)	4	N/A	N/A	0.7	0.1	3.81
Architecture	Analog Sub-Sampling	PFD-CP	Digital Over-Sampling	Digital TDC	PFD-CP	Analog Ref.-Sampling
Ref. Spur (dBc)	−70.8	−82.7	−72	N/A	N/A	−31
In-band PN (dBc/Hz)	−120.3 @500kHz	−100 @1MHz	N/A	−108 @100kHz	−113 @100kHz	−119 @100kHz
Out-band PN (dBc/Hz)	−131.7 @10MHz	−114 @10MHz	N/A	−145.4 @10MHz	−146.2 @10MHz	−123.7 @10MHz
Power (mW)	6.2	4.85	1.15	31.1	1.15	7.2
RMS jitter (fs) Integ. Range (MHz)	133 (0.001–10)	2260 (0.001–100)	414 (0.01–30)	115 (0.001–10)	414 (0.001–100)	249 (0.001–100)
FoM_jitter (dB)	−249.5	−226.1	−247	−243.9	−247	−243.5

*Post-layout simulations for G_M and VCO; pre-layout for remaining modules.

$$*\text{FoM}_{jitter} = 10\log\left[\left(\frac{\text{Jitter}_{rms}}{1\text{sec}}\right)^2\left(\frac{\text{Power}}{1\text{mW}}\right)\right]$$

Fig. 6 shows the simulated phase noise, in which all noises are referred to the PLL output. Thanks to the higher PD gain after boosted and spatial averaging, both the PD noise and DSM quantization noise become less-dominant at both in-band and out-of-band frequencies. Although a small phase noise peak still appears, but it happens at higher frequency. In addition, the simulated RMS jitter integrated from 1 kHz to 100 MHz is 249 fs. Fig. 7 compares the total PLL phase noise of the conventional fractional-N RSPLL, the fractional-N RSPLL using STA technique, and the proposed gain-boost fractional-N RSPLL, under the same loop dynamics. Same as discussed in Section II, the in-band phase noise is reduced by 11 dB, with the help of gain-boost technique. Benefit from the STA technique, the out-of-band phase noise is reduced by 17 dB. With these two techniques, jitter is reduced from 656 fs to 249 fs.

Table II summarizes the performance of the proposed gain-boost fractional-N RSPLL and compares with state-of-the-art works. As shown, the prototype RSPLL achieves a FoM_jitter of −243.5 dB, which is in line with the state-of-the-art works.

IV. CONCLUSION

This paper presents a fractional-N RSPLL that can reduce both the in-band PD noise and the DSM quantization noise by combining the gain-boost RSPD and the space-time averaging technique. A prototype 5-GHz fractional-*N* RSPLL prototype with this architecture is designed in a 40-nm CMOS process. Simulation results prove its validity, and show a good RSPLL performance.

REFERENCES

[1] X. Gao, E. A. M. Klumperink, M. Bohsali and B. Nauta, "A Low Noise Sub-Sampling PLL in Which Divider Noise is Eliminated and PD/CP Noise is Not Multiplied by N2," *IEEE Journal of Solid-State Circuits*, vol. 44, no. 12, pp. 3253–3263, Dec. 2009.

[2] Z. Yang, Y. Chen, P.-I. Mak, and R. P. Martins, "A Calibration-Free, Reference-Buffer-Free, Type-I Narrow-Pulse-Sampling PLL With −78.7-dBc REF Spur, −128.1-dBc/Hz Absolute In-Band PN and −254-dB FOM," *IEEE Solid-State Circuits Letters*, vol. 3, pp. 494–497, 2020.

[3] J. Sharma and H. Krishnaswamy, "A 2.4-GHz Reference-Sampling Phase-Locked Loop That Simultaneously Achieves Low-Noise and Low-Spur Performance," *IEEE Journal of Solid-State Circuits*, vol. 54, no. 5, pp. 1407–1424, May 2019.

[4] T. Xu, S. Zhong and R. P. Martins, "A 6-to-7.5-GHz 54-fsrms Jitter Type-II Reference-Sampling PLL Featuring a Gain-Boosting Phase Detector for In-Band Phase-Noise Reduction," *IEEE Transactions on Circuits and Systems I: Regular Papers*, vol. 69, no. 12, pp. 4774–4786, Dec. 2022.

[5] W. Wu et al., "A 14nm Analog Sampling Fractional-N PLL with a Digital-to-Time Converter Range-Reduction Technique Achieving 80fs Integrated Jitter and 93fs at Near-Integer Channels," in *IEEE International Solid- State Circuits Conference (ISSCC)*, 2021, pp. 444–446.

[6] G. Jin, F. Feng, X. Gao, W. Chen, Y. Shu, and X. Luo, "A 3.3-4.5GHz Fractional-N Sampling PLL with A Merged Constant Slope DTC and Sampling PD in 40nm CMOS," in *IEEE Radio Frequency Integrated Circuits Symposium (RFIC)*, June 2021, pp. 63–66.

[7] A. Tharayil Narayanan et al., "A Fractional-N Sub-Sampling PLL using a Pipelined Phase-Interpolator With an FoM of -250 dB," *IEEE Journal of Solid-State Circuits*, vol. 51, no. 7, pp. 1630–1640, July 2016.

[8] J. Tao and C. -H. Heng, "A 2.2-GHz 3.2-mW DTC-free Sampling ΔΣ Fractional-N PLL with -110 dBc/Hz In-band phase noise and -246dB FoM and -83dBc Reference Spur," 2019 *Symposium on VLSI Circuits*, Kyoto, Japan, 2019, pp. C162–C163.

[9] Z. Xu, M. Osada, and T. Iizuka, "A 3.3-GHz 4.6-mW Fractional-N Type-II Hybrid Switched-Capacitor Sampling PLL Using CDAC-Embedded Digital Integral Path with −80-dBc Reference Spur," in *Symposium on VLSI Circuits*, June 2021, pp. 1–2.

[10] D. Liao and F. F. Dai, "A Fractional-N Reference Sampling PLL With Linear Sampler and CDAC Based Fractional Spur Cancellation," *IEEE Journal of Solid-State Circuits*, vol. 56, no. 3, pp. 694–704, March 2021.

[11] Y. Zhang, X. Yang, and L. Geng, "A 5-GHz Fractional-*N* Reference-Sampling PLL With Voltage-Averaging Fractional Phase Detector Achieving an Integer-N-Level Phase Noise," *IEEE Microwave and Wireless Technology Letters*, vol. 35, no. 7, pp. 1069–1072, July 2025.

[12] Y. Zhang et al., "A Fractional-*N* PLL With Space–Time Averaging for Quantization Noise Reduction," *IEEE Journal of Solid-State Circuits*, vol. 55, no. 3, pp. 602–614, March 2020.

[13] J. Du, T. Siriburanon, and R. B. Staszewski, "A Reference-Waveform Oversampling Technique in a Fractional-N ADPLL," *IEEE Journal of Solid-State Circuits*, vol. 56, no. 11, pp. 3445–3457, Nov. 2021.

[14] J. Prinzie et al., "A Fast Locking 5.8-7.2 GHz Fractional-N Synthesizer with Sub-2 us Settling Time in 22 nm FDSOI," 2020 *IEEE Symposium on VLSI Circuits*, Honolulu, HI, USA, pp. 1–2, 2020.

[15] D. Mai et al., "Wandering Spur Suppression in a 4.9-GHz Fractional-N Frequency Synthesizer," *IEEE Journal of Solid-State Circuits*, vol. 57, no. 7, pp. 2011–2023, July 2022.

2025 The 10th International Conference on Integrated Circuits and Microsystems

A 59.6GHz Broadband 3-stage Cascode LNA with Peak Detector for Bandwidth Self-healing

Xinsheng Cheng[2], Jiacheng Guo[2], Juntao Liu[3], JingJing Lv[1*], Yuan Du[2]
[1]School of Microelectronics, Nantong University, Nantong, China
[2]School of Electronic Science and Engineering, Nanjing University, Nanjing, China
[3]China Mobile Research Institute, Beijing, China
*Corresponding authors: jjlv@ntu.edu.cn

Abstract—**This paper presents a 59.6GHz, self-healing, 3-stage cascode low-noise amplifier (LNA) in 65 nm CMOS, using on-chip peak detection and calibration to measure and optimize its power gain (S21), input impedance (S11) and bandwidth. In the proposed LNA circuit, the utilization of the transformers and chokes within the cascode amplifier increase the gain and impedance match over a broad bandwidth. A bandwidth self-healing algorithm is implemented by varactors according output from peak detector. Measurement results show that after calibration, the LNA reaches S21 18.03dB at 58.0GHz and input match S11 better than 10dB from 55.0GHz to 67.2GHz, with 11.2GHz -3dB bandwidth from 54.4GHz to 65.6GHz.**

Keywords—LNA, self-healing algorithm, peak detector, bandwidth extension

I. INTRODUCTION

With the frequency and bandwidth of modern communication systems continuously increasing, the design of millimeter-wave (mm-wave) circuits using nanoscale complementary metal-oxide-semiconductor (CMOS) technology has attracted a lot of interest. However, in mm-wave frequency, the general design methods of radio-frequency (RF) circuits are no longer suitable, due to the parasitic and PVT effects of components in integrated circuits are significant. The components in mm-wave circuits require high precision, and their performance is highly sensitive to variations of process, voltage, and temperature (PVT), which brings a great challenge to circuit design. In order to compensate for the negative impact above, there are circuits setting targets performance of simulation higher than demand, leaving margin for gain, bandwidth, while deteriorating other performance such as power consumption and area.

Another solution is using self-healing technology, there is a series of work in this field [1], [2]. As in [1], a self-healing LNA has been implemented at 2.4GHz, which monitors the amplitude of output voltage to feedback and adjust the varactors for tuning frequency. Digital controlled artificial dielectric transmission lines are used as the substitutes of varactors in mm-wave, while consuming more area and making the system more complex [2].

In this article, a millimeter wave self-healing broadband LNA has been implemented. Fig. 1 shows the diagram of this LNA. The three-level cascode structure is cascaded to improve gain and reduce signal reverse leakage. Varactors (C2, C3, C10, C11, C15, C16) are used to self-healing the working bandwidth of LNA in the millimeter wave frequency band. The output end of LNA uses on-chip peak detector (PD) as the detector for self-healing indicators in order to fine-tune a series of indicators such as LNA, which mainly focus on working bandwidth as needed.

II. CIRCUIT DESIGN

A. Millimeter Wave Amplification Structure and Broadband Impedance Matching

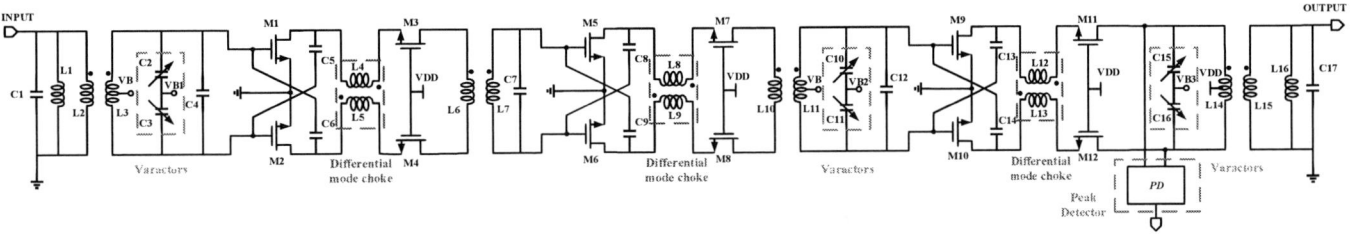

M1, M2: W/L:1μm/60nm, Nf:20;   M7, M8: W/L:1μm/60nm, Nf:30;
M3, M4: W/L:1μm/60nm, Nf:30;   M9, M10: W/L:1μm/60nm, Nf:32;
M5, M6: W/L:1μm/60nm, Nf:20;   M11, M12: W/L:1μm/60nm, Nf:30;

Fig. 1 Schematic of the millimeter wave self-healing broadband LNA

This work was supported in part by the National Key Research and Development Program of China under Grant No. 2021YFA0717700, and in part by the Jiangsu Province Innovation and Entrepreneurship Team Project under Grant JSSCTD202202..

A three-stage differential cascode structure was selected for its high gain and excellent reverse isolation, as shown in Fig. 1. On chip BALUNs are placed at the input and output terminals for improved broadband matching at mm-Wave for low-noise applications, and to achieve conversion between single ended signals and differential signals. The neutralizing capacitor is bridged between the gate of the differential common source transistor and the drain of the symmetric transistor. Therefore, the gate parasitic capacitance of the common source transistor, which is connected in series in the signal path, can be counteracted. Moreover, the maximum stable gain is increased and the reverse isolation is improved [3].

The combination of Shunt inductance and Shunt capacitance is arranged between the pad and the on-chip BALUNs at the input and output terminals. Under the 65nm process, the parasitic capacitance of the transformer (such as the series gate capacitor) is as low as a few fF. which poses challenges to the design of mm-wave frequency band circuits. The series gate capacitors are characterized by high negative imaginary impedance, which needs to be combined with a large positive imaginary impedance. On-chip transformers with high inductance often occupy a large area. The obvious parasitic effect which significantly limits the resonant frequency, makes it difficult to achieve in the mm-wave frequency band.

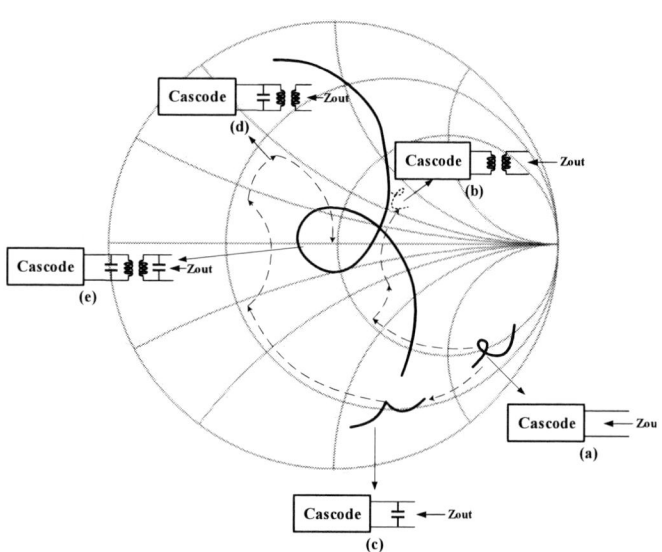

Fig. 2. Smith chart of Shunt capacitors in broadband matching

We utilized two techniques to reduce the impact of low parasitic capacitance on broadband impedance matching. One technology is transformer with shunt capacitors. As shown in Fig. 2(a), the output of cascode structure has large resistance and small capacitor under 65 nm process. As Fig. 2(b), a transformer directly connected to output of cascode can't realize broadband impedance matching. We propose to shunt capacitors and varactors on both sides of transformers, which make broadband impedance matching more fluency as Fig. 2(c), (d), (e). In the circuit design, capacitors and varactors are bridged between the gates of various common source differential transistors (C2, C3, C4, C7, C12) and between the drain of the third level differential common gate transistors (C15, C16) to optimize impedance matching at all levels of the amplification circuit.

Another technology is the differential mode choke technology. Differential mode choke coil is a symmetrical series

inductance coupled to each other on the differential signal line, with opposite direction. Choke coils are placed between the common source and common gate transistors (L4, L5, L8, L9, L12, L13) in each stage of the cascode structure. The mutual inductance increases inductor's value in the differential mode signal but decreases it in the common mode to compensated the capacitance to optimize impedance matching between transistors [4]. The consumption of area is reduced at the same time.

B. On-Chip Peak Detector

Due to the high cost of RF and high-frequency testing equipment, it is difficult to use off-chip equipment for single chip testing and calibration. Therefore, it is reasonable to detect the performance of mm-wave signals on chip and output results as DC signals. A differential peak detector on a chip is set at the drain of the third stage common gate transformer to detect the strength of the output signal in real-time (Fig. 3). In this peak detector, the low transistor uses a current mirror to generate bias, thus always in a saturated state; At the same time, AC coupling capacitors are also used to decouple the output bias voltage of the amplifier and the input bias voltage of the peak detector, ensuring that the DC working state of the source follower is not affected by the mm-wave circuit or external testing nodes.

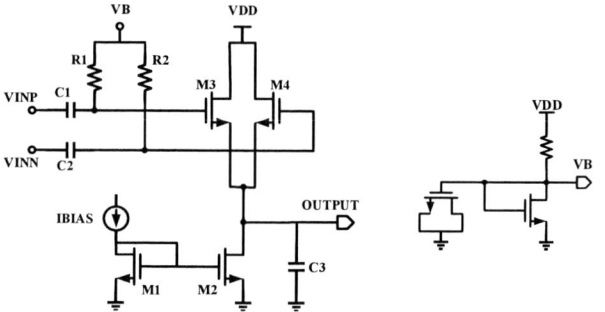

Fig. 3. Differential peak detector   Fig. 4. DC bias generation

C. Self-healing of Bandwidth

An effective way to expand bandwidth is to properly arrange the peak frequency of different amplification stages. In modern communication systems, varactor-based matching networks [5], [6] and reconfiguring the transistor's bias [7] are preferred solution to achieve peak frequency reconfigurable. For broadband applications in the mm-wave frequency band, the area consumption of using a large number of inductors to optimize bandwidth is high [8]. The added series capacitors have a negligible impact on the small parasitic capacitance of the transformer. Therefore, the varactors used in this design to replace fixed capacitors and achieve broadband impedance matching are all connected in parallel around the transformer. It is worth mentioning that, on the one hand, considering that the number of pads will affect the chip area, it is necessary to minimize the number of adjustable components as much as possible. Therefore, the inter stage matching of the second stage amplification circuit still uses fixed capacitors, and the performance optimization after production is based on the performance of the second stage; On the other hand, the capacitance value of varactors is limited, so in some environments that require large capacitance values (such as the

979-8-3315-8850-2/25 $31.00 © 2025 IEEE   649

gate of the first stage amplification circuit), they need to be combined in parallel with a fixed Shunt capacitor. Each group of varactors in the system is assigned a built-in DC bias generation circuit (as shown in Fig.4), and the capacitor formed by the MOSFET leakage source short circuit can effectively filter out high-frequency noise in the power supply.

When optimizing the working bandwidth, the bias of the varactor can be adjusted according to the on-chip peak detection results. Since the cascode structure has good reverse isolation performance, all of the adjustable components can be adjusted independently. Therefore, the order of varactors' adjustment has no influence on the gain and bandwidth of the whole amplifier.

Based on the above idea, a simple, fast, and low-cost optimization method for working bandwidth can be achieved. By iterating through the bias voltage and scanning frequency of the varactors one by one and observing the results of the peak detector, the bias voltage combination scheme with the maximum working bandwidth within the target frequency range can be found while ensuring that the gain is not lower than expected. This simple, fast, and low-cost optimization method algorithm is shown in Fig. 5, which is proposed as the following steps:

(1) Input a swept small signal from the target frequency band to the input of the amplifier. The power of the signal should be appropriately selected to avoid the peak detector being unable to detect the output signal or the amplifier being forced into a saturation state;

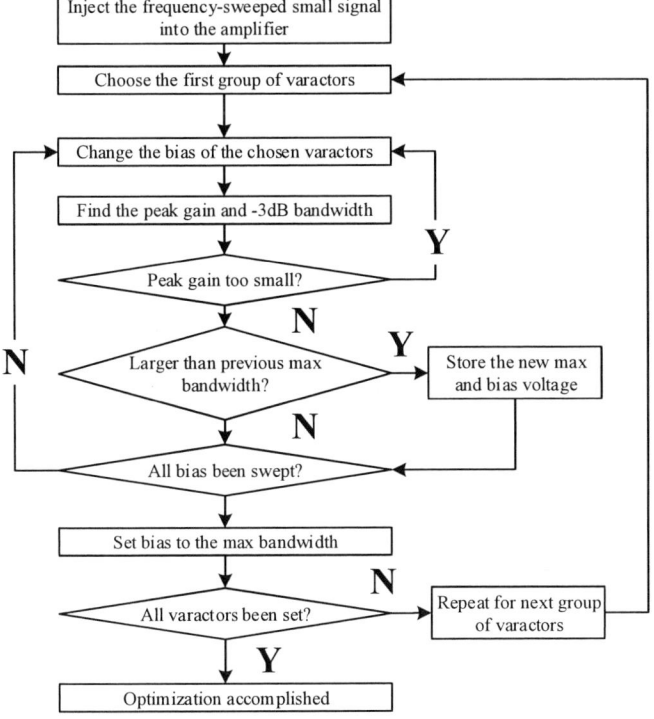

Fig. 5. Self-healing optimization algorithm

(2) Collect the output results of the peak detector during the sweep period using a vector network analyzer or an off-chip

ADC, quantify them into the output strength of the amplifier at several frequency points, and transfer them to a PC or other module. Calculate the amplifier's -3dB working bandwidth and in band gain in the varactor bias state based on the maximum output strength; When optimizing bandwidth, the bias voltage of all varactors is controlled by external power modules or voltage sources;

(3) Scan the bias voltage of the varactor according to the resolution of the voltage source, record and compare the output results of the peak detector under different bias voltages, and then set the combination of the bias voltage to the one when the output bandwidth is maximum and the in-band gain is not lower than expected;

(4) Repeat steps (2) and (3) for each set of varactors;

(5) After setting the bias voltage of the varactor, the optimization process is completed. Then, within the target frequency range, bandwidth optimization and gain consumption are more appropriate.

III. SIMULATION AND MEASUREMENT

The proposed self-healing LNA has been implemented in 65nm CMOS technology. The die photograph of the proposed LNA is shown in Fig. 6. With a supply voltage of 1.2V, the power consumption is about 72mW. Fig. 7 shows the measurement environment. A Vector Network Analyzer (VNA) combines with Frequency Extension Modules (FE) works in 50GHz-75GHz. LNA chip lies on a probe table. It outputs the phase detect result to and receives DC bias from voltage sources for self-healing algorithm.

Fig. 8 shows Simulated and Measured S-parameters in the best self-healing situation. The measured peak gain is 18.03dB at 58.0GHz. The bandwidth is 11.2GHz (19.31%) with 3dB cut-off frequencies are 54.4GHz and 65.6GHz. The measured input match is better than 10dB from 55.0GHz to 67.2GHz. The measured input 1dB Compression point is -18.1 dBm, output P1dB point is -7.4 dBm as shown in Fig. 9.

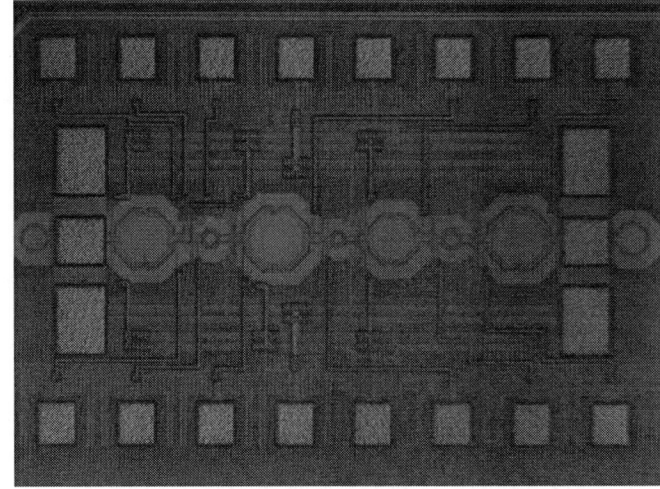

Fig. 6. Die photo of proposed LNA

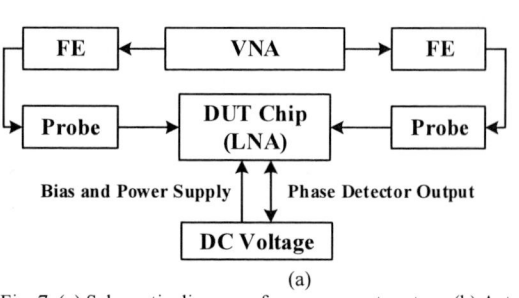
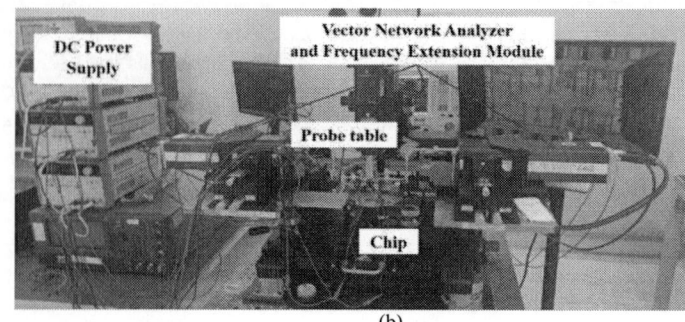

(a)

(b)

Fig. 7. (a) Schematic diagram of measurement system. (b) Actual photo of measurement environment.

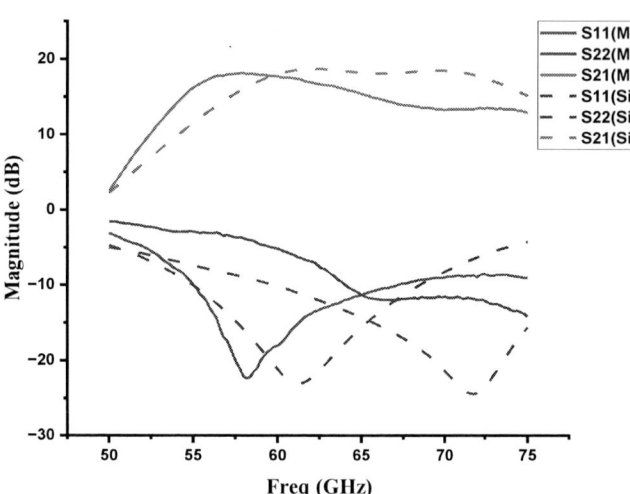

Fig. 8. Simulated and Measured S-parameters at the best self-healing

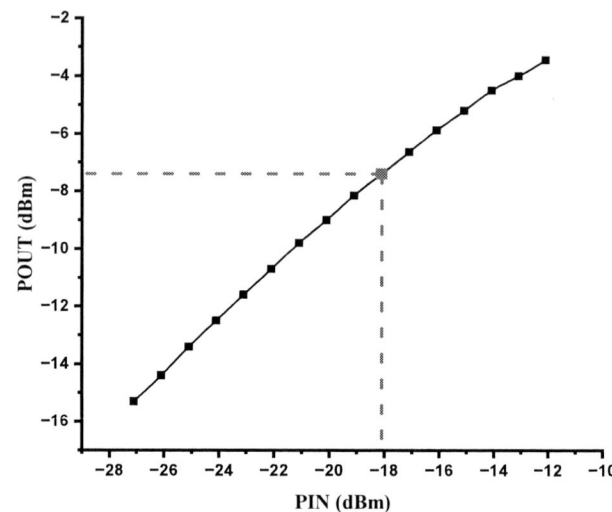

Fig. 9. Measured P1dB at the best self-healing

TABLE I. PAERFORMANCES OF STATE-OF-THE-ART LNAS AND THIS WORK

Reference	Technology	Frequency (GHZ)	BW (GHz) / (%)	Power Gain (dB)	S11 (dB)	IP1dB (dBm)	Power (mW)
This work	**65nm**	**59.6**	**11.2/19.3**	**18.03**	**<-10**	**-18.1**	**72**
[2]	65nm	56	4.6/8.2	17.3	<-20	/	23.04
[6]	0.1μm GaAs	28	4.6/16.4	25.1	<-10	-18.3	74
[9]	22nm FDSOI	26	7.5/28.8	14.7	<-10	-12	13.2
[10]	0.15μm GaAs	29	3.5/12	28	<-10	/	29.7
[11]	32nm SOI	60	9/15	21	<-10	/	18
[12]	130nm BiCMOS	59	9/15.2	24	<-17	-30	42.9
[13]	65nm	28	12/42.8	18.2	<-5	-15	9.8
[14]	65nm	60	13.5/22.5	20	<-10	-15.6	38.4
[15]	65nm	57	7.5/13.2	25	<-10	-7	47
[16]	65nm	36	5.5/15.2	21	<-10	-22	28

TABLE I. compares the performance of the LNA with the state-of-art LNAs. It shows that the proposed LNA designed in this paper achieves wideband operation and exhibits a competitive measurement performance compared with other self-healing/tunable LNAs.

979-8-3315-8850-2/25 $31.00 © 2025 IEEE   651

IV. CONCLUSION

This paper presents a self-healing broadband LNA in 65nm CMOS technology. This design uses transformer with shunt capacitor and differential mode choke to broaden bandwidth. The varactors adjusted by self-healing algorithm achieves the bandwidth to 11.2GHz (19.31%) and power gain to 18.03dB at 58.0GHz in the best situation, consuming 72mW under 1.2V supply. The results show that the proposed self-healing LNA serves the purpose of mm-wave broadband application.

REFERENCES

[1] K. Jayaraman, Q. Khan, B. Chi, W. Beattie, Z. Wang and P. Chiang, "A self-healing 2.4GHz LNA with on-chip S11/S21 measurement/calibration for in-situ PVT compensation," 2010 IEEE Radio Frequency Integrated Circuits Symposium, Anaheim, CA, USA, 2010, pp. 311-314.

[2] H. Jia, B. Chi, L. Kuang and Z. Wang, "A self-healing mm-wave amplifier using digital controlled artificial dielectric transmission lines," 2013 IEEE Asian Solid-State Circuits Conference (A-SSCC), Singapore, 2013, pp. 425-428.

[3] J. Mayeda, C. Sweeney 1, D. Y. C. Lie and J. Lopez, "Broadband High-Efficiency Millimeter-Wave Power Amplifiers in 22-nm CMOS FD-SOI with Fixed and Adaptive Biasing, " International Journal of Electrical and Electronic Engineering & Telecommunications (IJEETC) Volume 11, No. 6, pp. 385-391, Nov. 2022.

[4] M. Eleraky and H. Wang, "A D-Band Complex Neutralization Cascode Power Amplifier with A Source-Gate Driven Cascode for Enhanced Bandwidth and Efficiency," 2024 IEEE Radio Frequency Integrated Circuits Symposium (RFIC), Washington, DC, USA, 2024, pp. 183-186.

[5] Z. Zou, M. Hosseini, R. Kwende, S. Raman and J. C. Bardin, "A Frequency and Bandwidth Reconfigurable 3–6 GHz Cryogenic SiGe BiCMOS LNA with a Power Consumption of ≤ 2.9 mW," 2021 IEEE MTT-S International Microwave Symposium (IMS), Atlanta, GA, USA, 2021, pp. 653-656.

[6] Z. Wang et al., "A Ka-Band Switchable LNA With 2.4-dB NF Employing a Varactor-Based Tunable Network," in IEEE Microwave and Wireless Components Letters, vol. 31, no. 4, pp. 385-388, April 2021.

[7] B. Ko et al., "A 39/48 GHz Switchless Reconfigurable Low Noise Amplifier Using Common Gate and Coupled-Line-Based Diplexer," in

[8] IEEE Transactions on Circuits and Systems II: Express Briefs, vol. 70, no. 11, pp. 4028-4032, Nov. 2023.

[8] H. Chen, H. Zhu, L. Wu, W. Che and Q. Xue, "A Wideband CMOS LNA Using Transformer-Based Input Matching and Pole-Tuning Technique," in IEEE Transactions on Microwave Theory and Techniques, vol. 69, no. 7, pp. 3335-3347, July 2021.

[9] S. George, M. Cui and P. Sen, "A 24-28 GHz Tunable LNA in 22nm FDSOI Technology," 2023 30th IEEE International Conference on Electronics, Circuits and Systems (ICECS), Istanbul, Turkiye, 2023, pp. 1-4.

[10] J. Wang et al., "A Low-Power, Ka-Band LNA With Tunable Out-Of-Band Suppression Based on Compact Varactor Diode Arrays," 2024 IEEE MTT-S International Microwave Workshop Series on Advanced Materials and Processes for RF and THz Applications (IMWS-AMP), Nanjing, China, 2024, pp. 1-3.

[11] J. -O. Plouchart et al., "A 18mW, 3.3dB NF, 60GHz LNA in 32nm SOI CMOS technology with autonomic NF calibration," 2015 IEEE Radio Frequency Integrated Circuits Symposium (RFIC), Phoenix, AZ, USA, 2015, pp. 319-322.

[12] M. Völkel, M. Dietz, A. Hagelauer, R. Weigel and D. Kissinger, "A 60-GHz low-noise variable-gain amplifier in a 130-nm BiCMOS technology for sixport applications," 2017 IEEE International Symposium on Circuits and Systems (ISCAS), Baltimore, MD, USA, 2017, pp. 1-4.

[13] S. N. Ali, M. Aminul Hoque, S. Gopal, M. Chahardori, M. A. Mokri and D. Heo, "A Continually-Stepped Variable-Gain LNA in 65-nm CMOS Enabled by a Tunable-Transformer for mm-Wave 5G Communications," 2019 IEEE MTT-S International Microwave Symposium (IMS), Boston, MA, USA, 2019, pp. 926-929.

[14] X. Chen et al., "A 60-GHz Phase-Invariant Variable Gain LNA With T/R Switch and Gain Interpolation Techniques in 65-nm CMOS," in IEEE Microwave and Wireless Technology Letters, vol. 34, no. 5, pp. 508-511, May 2024.

[15] D. Bierbuesse and R. Negra, "60 GHz variable Gain & Linearity Enhancement LNA in 65 nm CMOS," 2020 IEEE Radio Frequency Integrated Circuits Symposium (RFIC), Los Angeles, CA, USA, 2020, pp. 163-166.

[16] Z. Jiang et al., "A 33.5~39 GHz 5-bit variable gain LNA with 4 dB NF and low phase shift," 2017 IEEE Asia Pacific Microwave Conference (APMC), Kuala Lumpur, Malaysia, 2017, pp. 1200-1202.

2025 The 10th International Conference on Integrated Circuits and Microsystems

Addressing Signal Integrity Challenges in DDR5 SDRAM: A High-Precision ZQ Calibration Circuit with Fast Calibration Time

Kexin Feng
Institute of Advanced
Technology
University of Science and
Technology of China (USTC)
Hefei, China
fengkexin@mail.ustc.edu.cn

ZhiQiang Zhang*
Design Department
Changxin Memory Technologies,
Inc (Cxmt)
Hefei, China
zhiqiang.zhang@cxmt.com

Abstract—With the increasing data transfer rate, the DDR5 DRAM interface requires more and more stringent impedance matching accuracy. In this paper, a high-precision ZQ calibration circuit is proposed to achieve less than 3% impedance tolerance and 335 ns calibration time through three innovations. First, a 100 MHz chopper-modulated ZQ comparator is used to reduce the input offset voltage by 80%; second, a stopping mechanism is designed to force a larger voltage deviation and eliminate comparator substability; and third, a programmable Vref is designed to expand the reference voltage range and suppress process/voltage errors. Test results show that the scheme effectively improves impedance matching accuracy and addresses the signal integrity challenges of high-speed SDRAM.

Keywords—*ZQ calibration, DDR5 SDRAM, High speed interface, Signal Integrity, Impedance Matching*

I. INTRODUCTION

With the development of 5G, AI and other technologies, the demand for storage performance continues to climb, and memory performance improvement has become the key to computer system performance improvement. In the evolution of memory technologies such as Double Data Rate (DDR), Low Power DDR(LPDDR), High Bandwidth Memory (HBM), Graphics (GDDR) etc., I/O speed improvement, voltage reduction, and capacity density increase are the most notable features[1]. I/O speed multiplication makes data transmission faster, but it also poses a challenge to power integrity (PI) and signal integrity (SI); the voltage reduction is conducive to the development of low-power and low-noise of DRAM; and the increase in capacity density enables the same space to store more data, which can to meet the needs of the big data era [2].

Over the past decades, memory products, such as DDR, LPDDR, HBM and GDDR have been significantly increasing in data transfer rates,as shown in Figure 1. Among them, the data transfer rate of DDR has reached 8.4 Gb/s/pin, LPDDR has reached 9.4 Gb/s/pin, and GDDR has reached a peak rate of 32 Gb/s/pin in 2023 [3]. As data transfer rates continue to increase, the importance of signal integrity is becoming

This research work was supported by ChangXin Memory Technologies, Inc. (CXMT).

increasingly important. In order to reduce signal reflections due to impedance mismatch, DRAMs employ ZQ calibration circuitry, which calibrates the internal pull-up/pull-down resistors under different PVT conditions to ensure the stability and reliability of signal transmission[4].

However, as the transfer rate increases, the problem of impedance mismatch between channels becomes more serious. As shown in Figure 2, impedance mismatches not only lead to signal reflections, but also cause signal distortion (the reflection coefficient formula in Eq.1), which can seriously affect the reliability of data transmission[5].

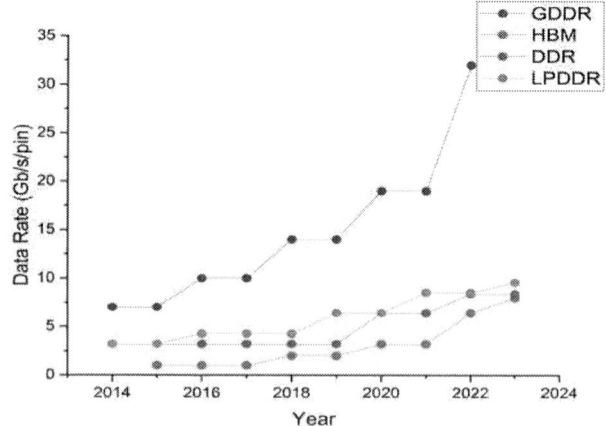

Fig. 1. Trends in the data transfer rates of LPDDR, DDR, HBM, and GDDR Products over the last decade[2]

$$\tau = \frac{V_{refected}}{V_{incident}} = \frac{Z_\tau - Z_0}{Z_\tau + Z_0} \quad (1)$$

τ: Reflection coefficient

Z_τ: Load Impedance

Z_0: Characteristic Impedance

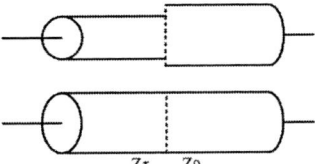

Fig. 2. Principle formula of reflection coefficient

To address these challenges, ODT (on-chip termination) technology has emerged and is gradually becoming a key technology for managing high-speed signal integrity in DRAMs. ODT technology has evolved from the early days of relying on external components for manual matching to today's integrated system with adaptive features, realizing a major breakthrough in impedance matching technology. In this process, ODT resistors play a crucial role[6]. Figure 3 details how the ODT is activated during write operations to match impedance and minimize reflections, and deactivated during read operations to ensure proper voltage levels and drive capability. This flexible mechanism effectively balances the relationship between signal integrity and power efficiency[7].

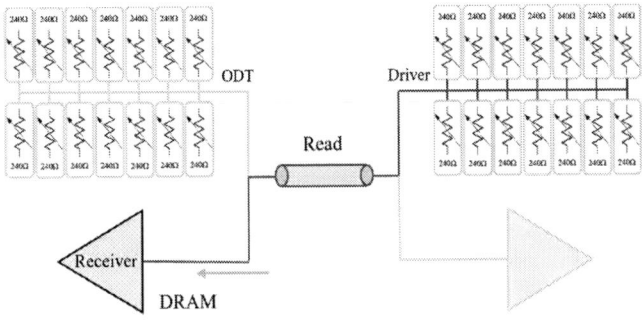

(a). DRAM as receiver mode (read mode)

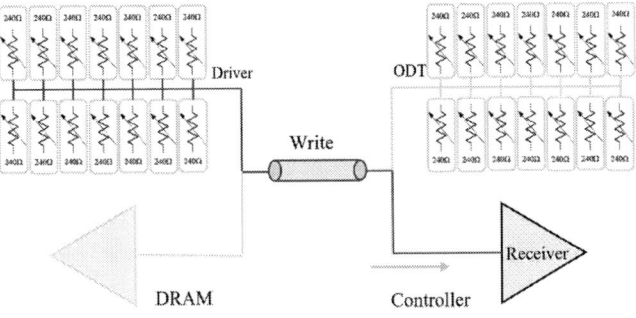

(b). Controller as receiver (write mode)

Fig. 3. The way data is moved between the controller and the DRAM

In modern DRAMs, the application of ZQ calibration technology further enhances the accuracy of impedance matching. Through using high-precision reference resistors, ZQ calibration is able to maintain accurate impedance matching under the PVT variations, thus ensuring the memory performance stability and reliability under various operating environments[8]. The introduction of this technique not only improves the efficiency of data transmission, but also lays a solid foundation for the further development of memory technology.

II. DDR5 SDRAM ZQ CALIBRATION CIRCUIT DESIGN

There are two ZQ calibration modes that are initiated using the MPC command: ZQCal Start and ZQCal Latch. ZQCal Start initiates the DRAM calibration procedure. ZQCal Latch captures the calibration result and loads it into the DRAM drivers and termination Circuits[9]. Alternatively, a 240 Ω external resistor (±1%) between the ZQ pin and VSS is shown in Figure 4.

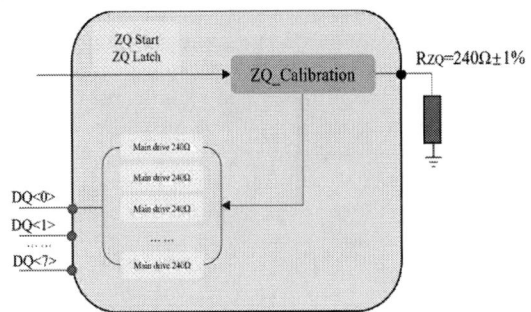

Fig. 4. Schematic of ZQ calibration circuit in SDRAM

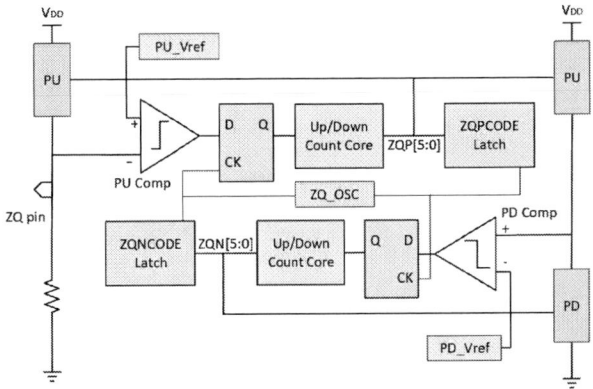

Fig. 5. Diagram of ZQ calibration architectire

Figure 5 illustrates the architecture of the ZQ calibration circuitry, which is executed asynchronously in the device background. When the ZQCal Start command is issued by the memory controller, the ZQ internal oscillator (ZQ_OSC) is activated to generate an accurate clock reference. This clock-driven state machine initiates a closed-loop calibration sequence that begins with impedance tuning of the pull-up resistor (PU) network. The calibration is performed using an iterative approximation algorithm: 0.8*VDD is used as the reference voltage (PU_Vref), and the deviation of the PU resistor divider voltage from the reference voltage is continuously monitored by a high-precision comparator. The comparator output controls the digital logic to dynamically adjust (increment/decrement operation) the 6-bit calibration code (ZQP[5:0]). Each time the code value is updated, the corresponding resistor network generates a new partial voltage value, and the process is performed iteratively until the convergence condition is satisfied. At this point, the final ZQP[5:0] code value is locked to complete the PU calibration phase. Due to the synchronous triggering mechanism, the Pull-Down (PD) network calibration will be activated, which uses the same voltage comparison principle and code value iteration algorithm to generate the optimized ZQN[5:0] calibration code.

979-8-3315-8850-2/25 $31.00 © 2025 IEEE

When the controller issues the ZQCal Latch command, the latched ZQP[5:0] and ZQN[5:0] code values are loaded to the impedance control units of the ODT and DQ (data bus) in parallel via a dedicated bus, realizing real-time impedance matching optimization of the I/O interface.

The reference voltage (Vref = 0.8*VDD) for ZQ calibration is the key element to realize accurate calibration, which is generated by a precision resistor divider network. In order to enhance calibration accuracy and suppress systematic errors, this design innovatively introduces a programmable test mode based on a 4-bit decoder. With fine adjustment of the reference voltage by digital control, this mode significantly reduces the calibration deviation caused by process variation, as shown in Fig.6(a). In addition, this design creatively proposes two additional test modes:TM_ZQVref_down (Overall lowering of the reference voltage by adding resistance) and TM_ZQVref_up (Overall increase in reference voltage by reducing resistance). The TM_ZQVref_down mode systematically lowers the reference voltage, while the TM_ZQVref_up mode raises the reference voltage. This dual-mode synergistic design cleverly extends the adjustable range of Vref to ±15%, which greatly enhances the adaptability to process angle fluctuations and supply voltage variations, and the post-simulation results are shown in Fig.6(b).Under rated operating conditions, the adjustable range of the ZQ Vref is shown in Fig. 7(a), and there is a deterministic mapping relationship between its regulation voltage and the target impedance value (the theoretical model is shown in Eq. 2). To verify the accuracy of this innovative model, 14 samples were measured and analyzed. The impedance-voltage characteristic curves of the pull-up/down resistor networks are demonstrated in Figs. 7(b) and 7(c), respectively, and the measured data are highly consistent with the trend of the theoretical calculations (maximum deviation <3.2%), which strongly proves the reliability of the impedance calibration mechanism.

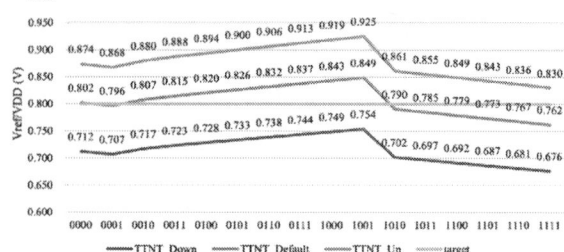

(b). Simulation results of the simulated ZQ reference voltage

Fig. 6. ZQ reference voltage structure and simulation results

The pull-up resistor network of the ZQ calibration circuit consists of four PU_Maindrv units connected in parallel, and the calibration target impedance of each PU_Maindrv module is set to 240 Ω, which is connected in parallel to realize the equivalent impedance of 60 Ω. The matching pull-down resistor is a single PD_Maindrv unit with the same target impedance of 240 Ω, forming a 1:4 pull-up/pull-down resistor ratio, as shown in Figure 8(a). This precisely-designed impedance ratio ensures that the ZQ pin voltage is stabilized at the reference voltage point of 0.8*VDD for accurate impedance calibration. Each Maindrv module utilizes an advanced composite structure design containing six PMOS/NMOS transistor arrays configured in a binary weight ratio (32:16:8:4:2:1) with a high-precision metal resistor network (Figure 8(b)). The structure is capable of linear calibration of the 240 Ω target impedance under PVT conditions through the dynamic adjustment of the digital control signal ZQ_Cal[5:0].

$$\frac{V_{ref}}{R_{ZQ}} = \frac{0.8*VDD}{34} = \frac{V'_{ref}}{R'_{ZQ}} \qquad (2)$$

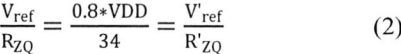

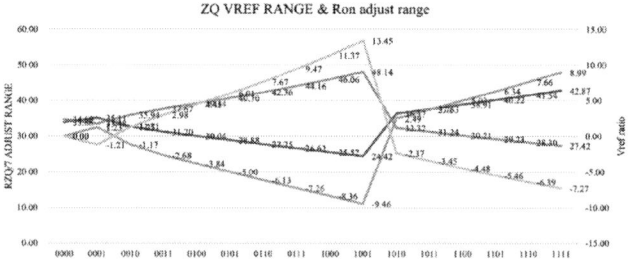

(a). ZQ VREF RANGE & Ron adjust range

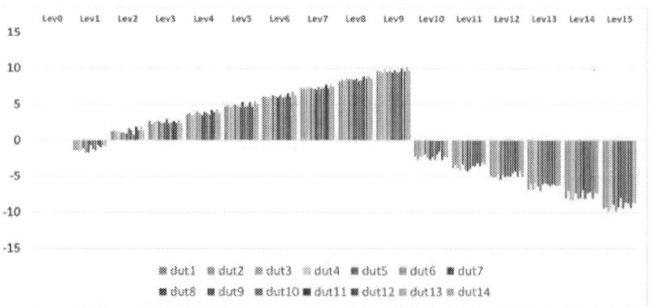

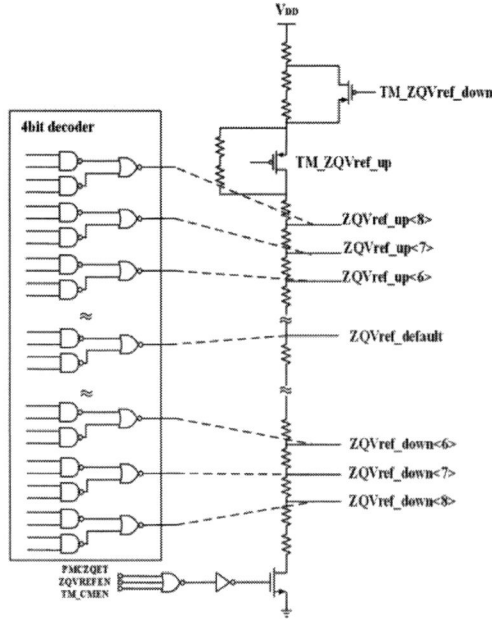

(a). ZQ 4-bit reference voltage generator: calibrate up/default/calibrate down

(b). Test results of the relative resistance value of the pull-up resistor

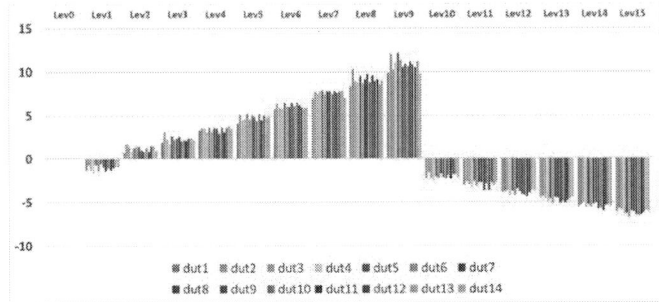

(c). Test results of pull-down resistor relative resistance value

Fig. 7. Theoretical and actual resistance values of adjustment range of ZR Vref

As shown in Fig.9, the impedance characteristics of PU_Maindrv and PD_Maindrv show monotonically varying trends with ZQP[5:0] and ZQN[5:0] calibration codes under PVT varying conditions, respectively. The PU_Maindrv impedance increases monotonically with ZQP[5:0], while the PD_Maindrv impedance decreases monotonically with ZQN[5:0], and both of them need to strictly converge to the target value of 240 Ω. To ensure robustness, the design needs to satisfy the following key indexes: (1) the calibration error at all PVT corners is controlled within ± 10%; (2) a small curve slope (high resolution) is maintained near the target value to reduce the impedance bias due to PVT fluctuations; (3) sufficient calibration code margins (margin code) are reserved to cover the processing variations. As shown in Tables 1(a) and 1(b), this design method ensures that the actual chip converges stably to the target impedance range under all operating conditions.

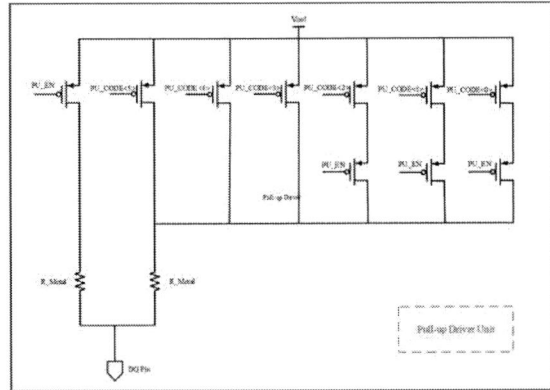

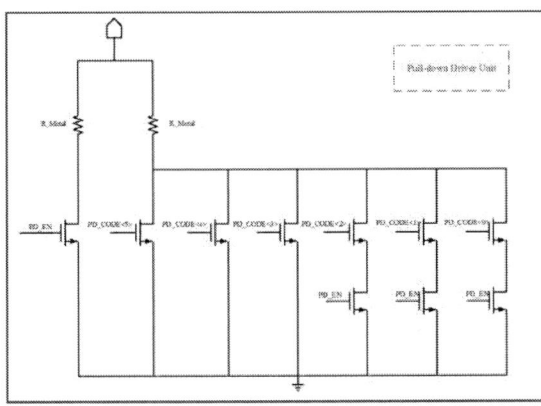

(a) Pull-up Driver Unit and Pull-down Driver Unit

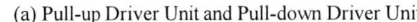

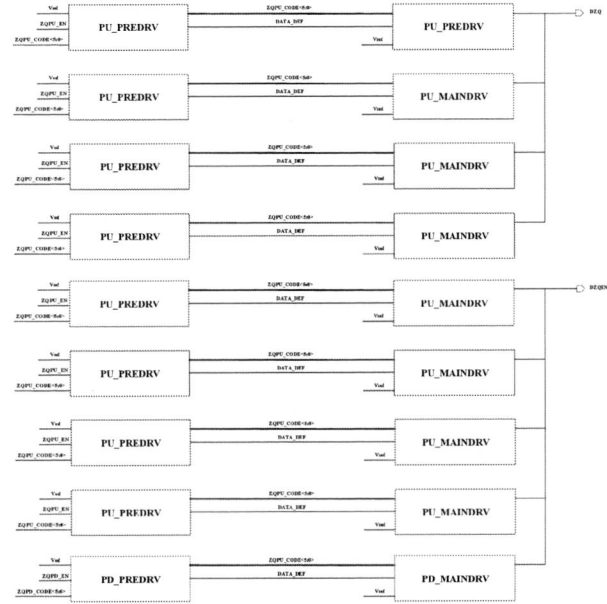

(b) ZQ calibration circuit parallel pull-up resistor and pull-down resisto

Fig. 8. Driver resistor structure for ZQ calibration

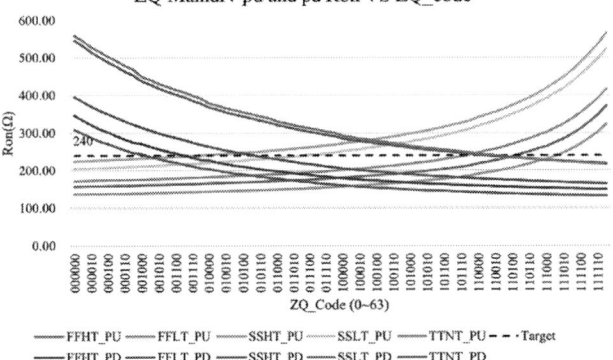

Fig. 9. Pull-up and pull-down resistor curves at different PVTs.

TABLE I. CORRESPONDING CODE AND MARGIN CODE FOR MAINDRV
(A). PU_MAINDRV

	FFHT_PU	FFLT_PU	SSHT_PU	SSLT_PU	TTNT_PU
Hit code@0.8 VDDQ	110100	111001	001111	011100	101111
Hit value@0.8 VDD	237.9	238.23	239.74	241.02	241.09
Gap Ω	5.62	8.62	1.81	2.12	6.19
Margin code	10	6	15	28	16

979-8-3315-8850-2/25 $31.00 © 2025 IEEE

(B). PD_MAINDRV

	FFHT_PD	FFLT_PD	SSHT_PD	SSLT_PD	TTNT_PD
Hit code@0.8 VDDQ	001101	001000	110001	110010	010101
Hit value@0.8 VDD	239.22	243.53	239.45	239.37	241.02
Gap Ω	5.89	6.65	2.05	2.39	4.24
Margin code	13	8	133	12	21

The ZQ calibration mechanism dynamically adjusts the impedance of the transistor array of Maindrv to generate an input signal that is compared with the reference voltage (0.8*VDD). Taking ZQP[5:0] calibration as an example (Figure 11), the initialization is set to "100000", and the deviation of the PU_Maindrv divided voltage from the reference voltage is continuously detected by the comparator: if the output logic "1" (input voltage > reference value) indicates that the resistor value is too small, the pull-up counter will correct the ZQP[5:0] in order to increase the impedance. When three consecutive comparisons result in a "101" sequence (i.e., the voltage fluctuates around the reference value), the state machine is triggered to terminate the calibration and start the pull-down resistor calibration process [9]. However, the lack of comparator resolution may lead to a metastable phenomenon (continuous oscillations of the output such as "11001100..."), which not only violates the convergence time requirement of the JEDEC protocol, but also causes ineffective power consumption, as shown in Figure 10. For this reason, an innovative termination mechanism is proposed: Coarse Tuning is used for the first 5 cycles, followed by Fine Tuning. If no "101" termination sequence is detected during the Fine Tuning phase, a serial counter is activated to force the voltage deviation to increase, driving the comparator out of the metastable region. The final output is locked to the calibration code closest to the target value at the Fine Tuning stage, ensuring impedance convergence accuracy. The comparator input-to-tube mismatch introduces an Offset Voltage to account for manufacturing process deviations. After integrating the 100 MHz chopper modulation technique (using NMOS/PMOS transmission gate architecture), Monte Carlo simulations show that the standard deviation of the offset voltage is reduced from 1.94 mV to 388 uV (80% improvement), which significantly improves the detection accuracy (Figure 11). At the same time, the maximum response time of the comparator (10.26 ns) is strictly smaller than the calibration clock period (20 ns), which meets the requirements of the D-flip-flop build-up time and guarantees the reliability of each cycle sampling, as shown in Figure 12.

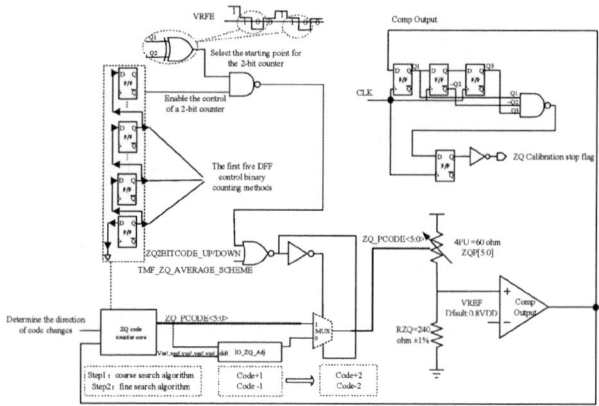

Fig. 10. ZQ calibration comparison process and calibration stop mechanism

(a). ZQ conventional comparator offset   (b). ZQ chopper comparator offset

Fig. 11. Monte Carlo simulation of out-of-phase voltage for ZQ comparator

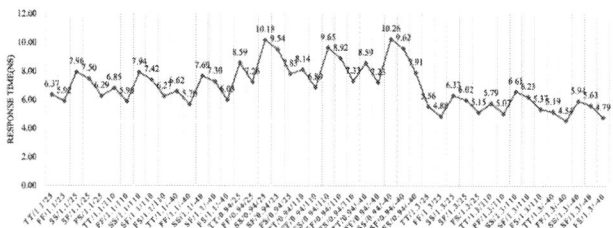

Fig. 12. ZQ comparator response time

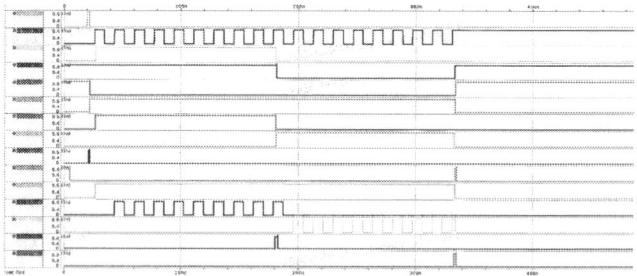

(a). Timing control logic simulation diagram

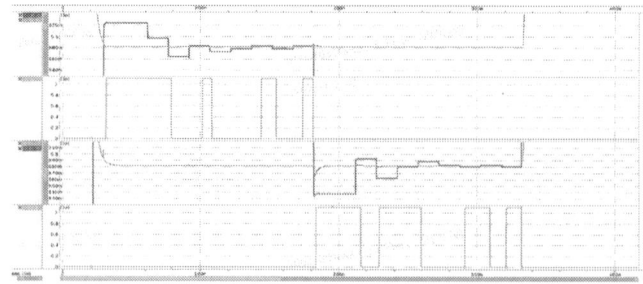

(b). ZQ PU resistance and PD resistance calibration comparison results

Fig. 13. Post-simulation waveforms from a ZQ calibration run

Figure 13(a) illustrates the timing control logic for ZQ calibration: when the ZQCal Start command is received, the internal oscillator activates and then generates the reference clock. At this time, the pull-up calibration enable window (PU_Enable) opens and PU_Clock drives the impedance adjustment sequence. After detecting the PU_Stop_Flag termination signal, the system synchronously closes the PU enable window and starts the pull-down calibration enable (PD_Enable), and PD_Clock then starts the impedance tuning process. Eventually, the PD_Stop_Flag signal triggers calibration termination and all ZQcode calibration codes are latched to the register array. When the system issues a ZQCal Latch command, the latched impedance calibration codes are output to the memory controller via the bus. The post-simulation waveform in Figure 13(b) verifies the timing behavior of the ZQ calibration sequence: the comparator first quantizes and compares the pull-up resistor network divider voltage, and then switches to the pull-down resistor network divider voltage detection. The whole calibration flow is completed within 335 ns, significantly lower than the maximum allowable time of 1 μs specified by the JEDEC DDR5 protocol (with a timing margin of 66.5%), which meets the real-time calibration requirements of high-speed memory interfaces.

The layout implementation of the ZQ calibration circuit is shown in Fig. 14, using a modular architecture to integrate the following core components: an internal oscillator-based clock source, a digital state controller, a binary weight counter, a chopper-stabilized comparator, a programmable impedance driver, and a boundary-scan test circuit. The modules are optimized for signal integrity through a matched wiring strategy,but the area and power consumption have increased slightly compared to the previous design. The calibration accuracy is verified by the measured data in Figure 15. The maximum deviation of the chip output impedance Ron with respect to the target value of 34 Ω is 2.9% over the operating voltage range of 1.026 V to 1.210 V. This result meets the ± 10% impedance tolerance requirement specified by the DDR5 protocol and outperforms the 3.8% typical error of similar designs [10].Table 2 demonstrates the comparison of key parameters of the ZQ calibration circuit designed in this paper.

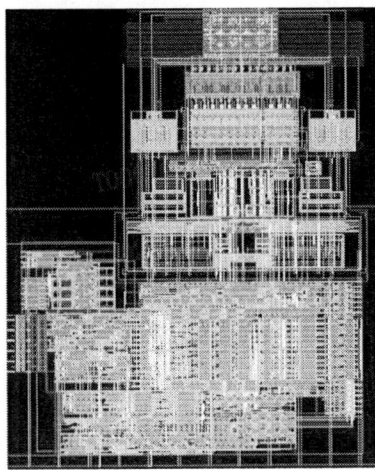

Fig. 14. ZQ calibration layout

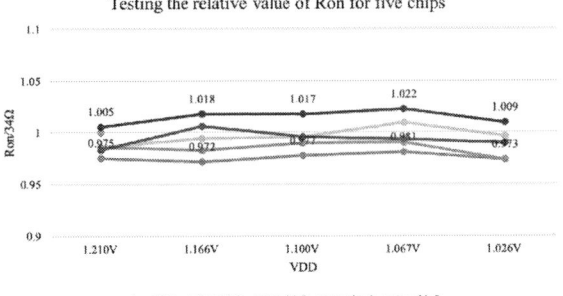

Fig. 15. Testing R_{on} values of different chips after ZQ calibration

TABLE II. COMPARISON OF ZQ CALIBRATION PERFORMANCE PARAMETERS

	ICICM[10]	A-SSCC[11]	This work
VDDQ	1.2V	1.2V	1.1V
Ron Mismatch	3.8%	1.2%	2.9%
Comparator Offset Removal	80%	7.4%	80%
Number of ZQ Cycles	1024 (10 ns period)	24 (30 ns period)	18 (20 ns period)

III. CONCLUSION

This paper presents a high-precision ZQ calibration circuit for DDR5 SDRAM interfaces that achieves less than 3% impedance tolerance and 335 ns calibration time through three key innovations. First, a 100 MHz NMOS/PMOS chopper-modulated ZQ comparator is employed to reduce the input offset voltage by 80% (standard deviation σ = 388 uV), which plays a key role in the accurate impedance matching for high-speed DDR5 interface. Secondly, a stopping mechanism was designed to drive the comparator out of the substable region by activating the serial counter to force a larger voltage deviation, thereby eliminating instability and completing the calibration within 335 ns (66.5% margin within the 1 us limit specified by JEDEC DDR5). In addition, a programmable Vref with a test pattern has been designed to extend the adjustment range to ± 15% and suppress PVT-induced errors. Measurements on different chips in the 1.026-1.210 V range showed a driver impedance error of only 2.9% and more than 40% enhancements in accuracy. This research provides an effective approach to addresses signal integrity challenges in the high-speed interfaces, which can be utilized to boost the next-generation SDRAM.

ACKNOWLEDGMENT

I would like to thank Changxin Memory Technologies, Inc for providing a valuable platform and all-round support for this research. At the same time, I would like to express my sincere gratitude to Mr. Zhang Zhiqiang for his careful guidance and professional advice, whose rich industry experience and insights have pointed out the direction of this research and helped me to overcome the many problems.

REFERENCES

[1] C. -K. Lee, J Lee, K. -H. Kim, J.-S. Heo, G.-H. Cha, J. -H. Baek, et al. "Dual-Loop 2-Step ZQ calibration for dedicated power supply voltage in LPDDR4 SDRAM," 2017 IEEE Asian Solid State Circuits Conference (A SSCC), Nov 2017.

[2] J. Feng, B. Dhavale, J. Chandrasekhar, Y. Tretiakov and D. Oh, "System level signal and power integrity analysis for 3200Mbps DDR4 interface," 2013 IEEE 63rd Electronic Components and Technology Conference, Las Vegas, NV, USA, pp. 1081-1086, May 2013.

[3] J. Yun, S. Lee, J. Kim, J. -H. Chae, S. Kim and Y. -U. Jeong, "A Single-ended Impedance-Matched transmitter with single Ring-Oscillator-Based time-domain ZQ Calibration for memory interfaces," in IEEE Journal of Solid-State Circuits, vol. 59, no. 9, pp. 2971-2982, Sep 2024.

[4] L. K. S. Tolentino, J. -Y. Ke, C. -Y. Lo and C. -C. Wang, "Using machine learning techniques to determine DDR5 SDRAM I/O buffer's slew rate at different PVT variations," 2023 International Conference on Integrated Intelligence and Communication Systems (ICIICS), Kalaburagi, India, pp. 1-6, Nov 2023.

[5] B. C. Pal, C. Koshti, S. Sharma and N. Parekh, "The importance of impedance calibration and the modeling practice for simulation in DDR Memory," 2020 IEEE-HYDCON, Hyderabad, India, pp. 1-7, Sep 2020.

[6] K. Koo, S. Kyung Lee, J. Seo, M. Ko and J. Kim, "A versatile I/O with robust impedance calibration for various memory interfaces," 2006 IEEE International Symposium on Circuits and Systems (ISCAS), Island of Kos, May 2006.

[7] C. -K. Lee, J Lee, K Kim, J. -S. Heo,J. -H. Baek,G. -H. Cha, "A Single-Ended Impedance-Matched transmitter with single Ring-Oscillator-Based time-domain ZQ Calibration for memory interfaces," in IEEE Journal of Solid-State Circuits, vol. 59, no. 9, pp. 2971-2982, Sep 2024.

[8] Y. -U. Jeong, H. Park, C. Hyun and S. Kim, "A 28-Gb/s/pin PAM-4 Single-Ended transmitter with High-Linearity and Impedance-Matched Driver and 3-Point ZQ Calibration for memory interfaces," 2020 IEEE Symposium on VLSI Circuits, Honolulu, HI, USA, pp. 1-2, Jun 2020.

[9] JEDEC DDR5 SDRAM specification[S]. JEDEC JESD79-5C.

[10] H. Zhang, "A 4% impedance variation JEDEC compatible ZQ Calibrator for DDR3, DDR4 and DDR5 SDRAM," 2023 8th International Conference on Integrated Circuits and Microsystems (ICICM), Nanjing, China, pp. 246-253, Oct 2023.

[11] T. Kim, A. Kavala, H. Kang, Y. Jo, J. Park, K. Kang, "A Hybrid ZQ Calibration design for High-Density Flash Memory toggle 5.0 High-speed interface," 2021 IEEE Asian Solid-State Circuits Conference (A-SSCC), Busan, Korea, Republic of, pp. 1-2, Nov 2021.

979-8-3315-8850-2/25 $31.00 © 2025 IEEE

A Pipeline-Based Common Framework for Parallel Mixed Signal Simulation

Longchen Sun[#]
State Key Laboratory of Integrated Chips and Systems
Fudan University
Shanghai, China
24212020023@m.fudan.edu.cn

Guangrong Li[#]
CAEP Software Center for High Performance Numerical Simulation
Institute of Applied Physics Computational Mathematics
Beijing, China
abcdefg1119@sina.com

Zhenguo Zhao[*]
CAEP Software Center for High Performance Numerical Simulation
Institute of Applied Physics Computational Mathematics
Beijing, China
zhao_zhenguo@iapcm.ac.cn

Xin Huang
State Key Laboratory of Integrated Chips and Systems
Fudan University
Shanghai, China
24212020088@m.fudan.edu.cn

Xuan Zeng
State Key Laboratory of Integrated Chips and Systems
Fudan University
Shanghai, China
xzeng@fudan.edu.cn

Fan Yang
State Key Laboratory of Integrated Chips and Systems
Fudan University
Shanghai, China
yangfan@fudan.edu.cn

Zhaori Bi[*]
State Key Laboratory of Integrated Chips and Systems
Fudan University
Shanghai, China
zhaori_bi@fudan.edu.cn

Abstract—**A pipeline-based common framework for parallel mixed signal simulation is proposed, which not only provides new perspectives on developing synchronization algorithms but also improves the efficiency of parallel simulation, particularly for mixed signal cases. Implementation of two-stage pipelines is introduced. The efficiency of the two-stage pipeline is compared to a framework for serial mixed signal simulation, showing a significant runtime speedup ratio of 1.74×~1.84× in simple cases and 3.57× in a more complicated case involved with DAC (digital to analog converter).**

Keywords—*Pipeline-based framework, parallel simulation, mixed signal simulation, synchronization algorithm*

I. INTRODUCTION

Nowadays, the number of transistors used in chips continually increases as transistors continue to scale down. So an efficient mixed signal simulation environment is becoming more and more necessary. Although companies like Cadence have developed commercial tools and frameworks for mixed signal simulation, which are designed specifically for their own simulators, there is still a lack of common framework to work compatibly with third-party simulators. A common simulation framework allows designers more choices in both analog and digital simulators.

A summary of former research on mixed signal simulation is shown as follows. As for commercial mixed signal simulators, there are tools from Cadence, Viewlogic and Genrad, which are based on specific simulators and an unadvanced synchronization algorithm. There is also Calaveras algorithm, a patented synchronization algorithm applied in analog simulator Saber. In

[1], a language-limited synchronization algorithm for VHDL-AMS simulation is proposed, which isn't common enough to work with third-party simulators chosen by user freely. In [2], a more general and practical synchronization algorithm named Second Event Synchronization protocol is introduced, which will be referred to as SES algorithm in following sections. SES algorithm requires special digital simulators supporting rollback, which is not widely supported by digital simulators. In a comparatively recent research in [3], a cosimulation framework is proposed. However, this framework is not common enough, because the participation of a new simulator requires customized development of interface.

According to the summary above, there is no existing systematical common framework for mixed signal simulation, let alone parallel mixed signal simulation. Although there are synchronization algorithms proposed for specific programming language or simulators, they're merely part of the entire simulation framework. Furthermore, previous research designs synchronization algorithms mostly from the perspective of asynchronous parallel programming, neglecting specific features of IC (integrated circuit) simulation. And the frameworks proposed previously aren't common enough.

Thus, in this paper we propose a common framework for parallel mixed signal simulation based on pipeline and an improved synchronization algorithm. The framework proposed provides new perspectives to comprehend and design synchronization algorithms for parallel mixed signal simulation. It is from such perspective that the improved synchronization algorithm is proposed theoretically when analyzing SES algorithm. Then actual implementation of the framework mentioned above is introduced, which is mostly based on C/C++ codes. Finally, a serial framework based mainly on Python is

---

[#]Longchen Sun and Guangrong Li contributed equally to this work.
[*]Corresponding author: Zhenguo Zhao; Zhaori Bi.

979-8-3315-8850-2/25 $31.00 © 2025 IEEE

briefly introduced as comparison to the pipeline-based framework, which is a rather fundamental and simple solution to mixed signal simulation. Experiments show that pipeline-based framework gets a significant runtime speedup ratio of $1.74\times{\sim}1.84\times$ in simple cases and $3.57\times$ in a more complicated case involved with DAC (digital to analog converter).

II. Background

IC simulation is unique in asynchronous parallel programming, for the processes of simulators can assemble a pipeline structure. Simulation of IC, both analog and digital, is in essence computing the output signals from the original input signals after processed and transferred in the circuit. Therefore, a final output value can be acquired by calculating and orderly transferring intermediate values between processes which simulate partitions of the entire circuit accordingly. Fig. 1 demonstrates a close correspondence between how signals are processed and transferred in the actual partitioned circuit and how simulators calculate and forward intermediate values as a pipeline.

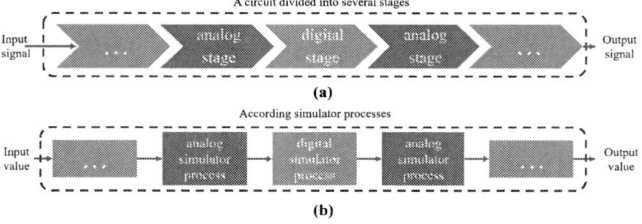

Fig. 1. The correspondence between (a) a circuit divided into several stages and (b) accordingly simulator processes. Both are of pipeline structure, naturally leading to a pipeline-based framework for parallel mixed signal simulation.

III. Proposed Framework

The resemblance shown in Fig. 1 naturally leads us to a pipeline-based framework for parallel mixed signal simulation, which is basically demonstrated as Fig. 1(b). Theoretically, this pipeline-based framework can be applied to not only parallel mixed signal simulation but also parallel single signal (analog or digital) simulation, determined by the choice of simulators. Especially, proper partition of large scale analog circuit to simulate using the pipeline-based framework is expected to gain significant improvement in efficiency. This can be saved for further discussion for it is not the key point of this paper.

The proposed pipeline-based framework provides a new perspective on synchronization algorithm. According to theories of pipeline, the proposed framework can almost gain a runtime speedup of n over serial frameworks in ideal situations, where n represents the amount of stages of the pipeline. However, this requires an approximately same throughput rate of different stages of the pipeline, which is determined by the partition of the circuit. It is synchronization algorithms that are needed when dealing with uneven partitions, which will be analyzed in detail in the next section.

IV. Synchronization Algorithms

To improve efficiency of the pipeline when the circuit is divided in an unbalanced way intentionally by the user,

advanced synchronization algorithms are needed. From the perspective of pipeline-based framework, the nature of synchronization algorithms for parallel mixed signal simulation is to deal with the situation that the pipeline is supposed to stall because of the mismatch of throughput rates between different stages. Existing synchronization algorithms mostly focus on cases where analog simulator takes more time than digital simulator does. They keep the digital simulator running even if input data generated by previous stage of analog simulator haven't arrived yet, based on certain prediction mechanism.

In this section, we will first explain stall in a normal pipeline without any excess synchronization algorithm. Then SES algorithm will be analyzed from the pipeline perspective, to which the improvements can be made. Finally, a new synchronization algorithm will be briefly proposed as a theory, which is a direct improvement to SES algorithm.

A. Normal Pipeline

For normal pipeline without any design of synchronization algorithm, when throughput rates of different stages mismatch, the only choice is to stall the pipeline. However, stall causes loss of efficiency, compared to a pipeline running continually. Thus, when balanced partition is compromised, ways to keep efficient parts running when waiting for the slower parts is needed.

B. SES Algorithm

SES algorithm is designed to deal with the situation that a digital simulator runs more efficiently than the previous analog simulator which is supposed to generate data in time to feed the digital simulator directly. From the perspective of pipeline, SES algorithm can be elaborated as keeping the successive digital simulator running with the digital simulator maintaining current input value until change arrives from previous stage.

The assumption about the input value that SES algorithm holds is exactly where the inadequacy of SES algorithm lies. Once the input digital value changes at simulation time t, which is when the analog value crosses the threshold, a rollback is needed to restore the digital simulator back to the state at simulation time t even though the digital simulator has already run to the simulation time beyond t. However, most digital simulators don't support rollback to wanted simulation time, for it requires excess runtime and memory to keep records of states of the simulator regularly. This is why the choice of digital simulator is strictly limited when using SES algorithm. Even if certain digital simulators do support rollback, the execution of rollback necessitates additional time, possibly still resulting stall of the pipeline.

C. Proposed Synchronization Algorithm

To improve the SES algorithm, rollback should be avoided. SES algorithm requires rollback because of relying on one single hypothesis that the input digital data keep unchanged, which is destined to be falsified at a later simulation time. If the pipeline has multiple digital processes running different hypotheses on when the analog data from previous stage will cross the threshold, with the analog simulation step fixed, there is conclusively a hypothesis predicting the simulation time of change correctly, without requirement for rollback. This leads to the new synchronization algorithm that we propose here.

979-8-3315-8850-2/25 $31.00 © 2025 IEEE

The proposed algorithm based on improving SES algorithm is illustrated in Fig. 2. When the digital simulator is lack of input data forwarded by previous analog simulator, certain amount of hypotheses will be run accordingly by processes which are clones of the original one. Each clone process takes in an extra hypothetical digital event compared to the original one, which represents the mixed event caused by analog output crossing the threshold. Because the step of analog simulation is fixed, and the digital value changes only at certain simulation step, there has to be a correct hypothesis among the running ones. The digital processes for hypotheses are chosen dynamically. Each step forwarded in analog simulation verifies the validity of a hypothesis. If the analog output doesn't cross the threshold, the according hypothesis is proved to be wrong for it predicts that input digital value changes at next step. The falsified process will be killed to free hardware resources. Otherwise, if the analog output turns out to cross the threshold, the according hypothesis is validated. Under this circumstance all the other processes, including the original process, is to be killed, so that the validated one becomes the new origin to be cloned.

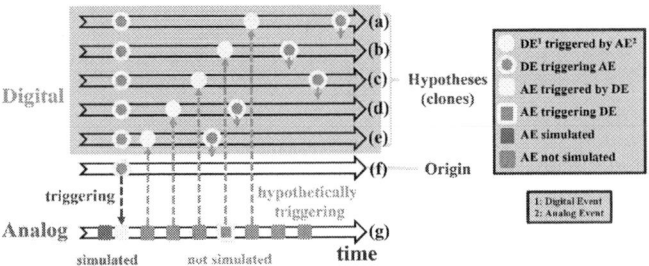

Fig. 2. Proposed synchronization algorithm. (a)-(e): Digital processes representing hypotheses on when the next change of digital input occurs, which are clones of the original process (f). Each of them takes an additional hypothetical digital event triggered by analog event. (b) The only process turning out to be running the correct hypothesis. (f) the original digital process to be cloned. (g) The analog process. In this case, the next three steps forwarded by analog simulation will falsify hypotheses (e)-(c) in order, whose processes will be terminated. Then, the fourth step will validate hypothesis (b), terminating all other pocesses including (f) and making (b) the new origin.

There are still limitations to this proposed synchronization algorithm, which make it merely a theory. In this way, there is no requirement for rollback because of redundant hypotheses and processes. However, this synchronization algorithm takes up huge quantity of hardware resources, which exhibits exponential dependence on the amount of input variables (which is 1 in Fig. 2) and is directly proportional to the steps that the hypotheses predicts (which is 5 in Fig. 2, (a)-(e)). Also, the algorithm requires simple ways to clone processes, which is only supported by a limited amount of commercial digital simulators. Thus, optimizations to the algorithm is under research, for which the new synchronization algorithm is only a theory so far.

V. IMPLEMENTATION

We implement a two-stage pipeline for analog-to-digital simulation and a two-stage pipeline for digital-to-analog simulation, both of which are normal pipelines without excess synchronization algorithm design. The pipeline-based framework is implemented with the language of C/C++. This actual framework can work with any analog simulators that can be invoked and fully controlled by C/C++ codes and any digital

simulators that implement interface to C/C++, such as VPI (Verilog Process Interface), an interface defined by IEEE standard 1364-2005 [4]. Specifically, for analog simulator we choose Xyce, an open-source simulator from [5] and [6], mainly because it has already been used for mixed signal simulation in [7] and [8]. While for digital simulator, we choose Iverilog [9], which is Icarus Verilog for short, also an open-source simulator. Xyce implements an interface to C, allowing C codes to control simulation process and data completely. Also, Iverilog implements VPI, which allows C/C++ codes to add excess tasks and functions to Verilog script and simulator. Thus, both simulators are capable to interface with C/C++ codes.

For better comprehension, the implementation of the analog-to-digital pipeline and the digital-to-analog pipeline is illustrated in Fig. 3 and Fig. 4 separately, of which the detailed introduction is in following subsections.

A. Analog-to-Digital Pipeline

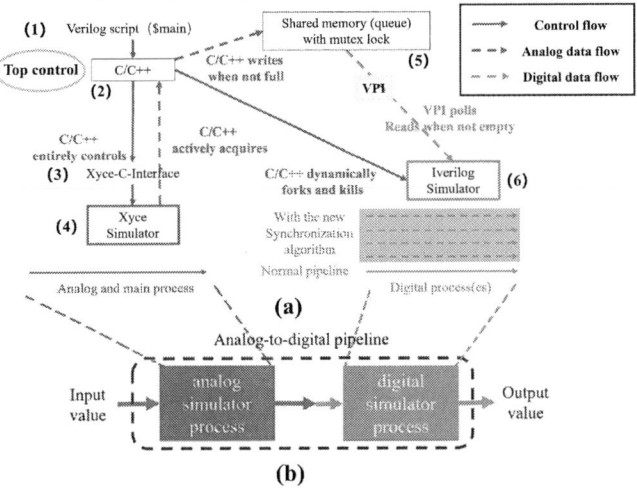

Fig. 3. Detailed implementation and overall structure of the analog-to-digital pipeline. (a) Detailed implementation of analog-to-digital pipeline. (a1)-(a6): (a1) Top Verilog script to execute system task $main defined in (a2). (a2) VPI plug-in codes written in C/C++, which is the top control of the entire pipeline system. (a3) Xyce-C-Interface. (a4) Xyce simulator, entirely controlled by (a2) through (a3). (a5) A queue based on a piece of shared memory with mutex lock to exchange data between processes safely , managed by (a2). (a6) Iverilog simulator. (b) Overall structure of analog-to-digital pipeline.

The control flow in the framework implemented for analog-to-digital pipeline in Fig. 3(a) is elaborated as follows. A Verilog script (Fig. 3(a1)) is needed to execute system task $main defined in VPI plug-in codes written in C/C++ (Fig. 3(a2)) which is the top control of the entire pipeline system, because somehow the C functions defined by Xyce-C-Interface only works in VPI plug-in codes. Controls of Xyce (Fig. 3(a4)) through Xyce-C-Interface (Fig. 3(a3)) are taken by the C/C++ codes in the main process, which also manages a queue based on a piece of shared memory with mutex lock (Fig. 3(a5)) to exchange data between processes safely. For pipeline applying the proposed new synchronization algorithm, there should be multiple digital processes forked and killed dynamically by the top controlling C/C++ main process. While for a normal two-stage pipeline there is merely one digital process forked by the top controlling C/C++ codes at the beginning of simulation and killed when reads a specified value from the queue written by the main process, which is exactly the case implemented by us.

The data flow in the framework is elaborated as follows. After Xyce (Fig. 3(a4)) runs a simulation step forward controlled by C/C++ codes (Fig. 3(a2)), the output data will be acquired by C/C++ (Fig. 3(a2)) to write into the queue (Fig. 3(a5)) when the queue isn't full. Otherwise, the main process has to wait for the digital process consuming data in the queue. At the same period of time, the digital process keeps polling the queue through functions defined by VPI, reading data from queue when the queue isn't empty, which are transferred into input data for Iverilog (Fig. 3(a6)).

B. Digital-to-Analog Pipeline

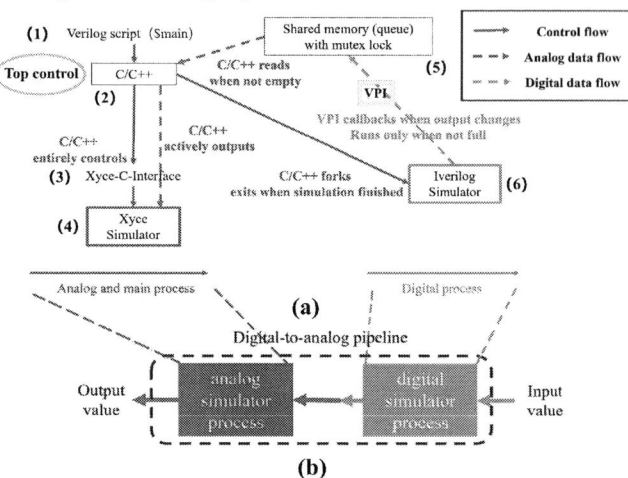

(a)

(b)

Fig. 4. Detailed implementation and overall structure of the digital-to-analog pipeline. **(a)** Detailed implementation of digital-to-analog pipeline. (a1)-(a6): (a1) Top Verilog script to execute system task $main defined in (a2). (a2) VPI plug-in codes written in C/C++, which is the top control of the entire pipeline system. (a3) Xyce-C-Interface. (a4) Xyce simulator, entirely controlled by (a2) through (a3). (a5) A queue based on a piece of shared memory with mutex lock to exchange data between processes safely , managed by (a2). (a6) Iverilog simulator. (b) Overall structure of digital-to-analog pipeline.

Similarly, we implement the digital-to-analog two-stage pipeline, which is illustrated in Fig. 4. The overall structure and control flow are basically the same as the analog-to-digital pipeline.

However, the data flow is reversed, resulting significant difference in function calling in both Xyce-C-Interface in the analog-main process and VPI plug-in in the digital process. VPI support callback functions to be called spontaneously once values of certain variables defined in Verilog change. Through this way, a callback function writing data into the queue (Fig. 4(a5)) can be called every time the values of the output variables of Iverilog (Fig. 4(a6)) change. Therefore, the digital process only keeps polling queue through system function defined in VPI plug-in in the language of C/C++ to check whether it is full. Iverilog simulator runs only when the queue isn't full so that no loss of data would occur. The latest values of output data to be forwarded to Xyce (Fig. 4(a4)) will be written into the queue by the callback function whenever these values change. Simultaneously, the top control C/C++ main function (Fig. 4(a2)) keeps polling the queue to read input data from the queue when

it's not empty. After digital process exits, Iverilog finishing simulation, the main process won't exit until the queue is empty, all data forwarded to Xyce.

VI. EXPERIMENTS

Experiments to compare performance of the parallel framework to a serial framework implemented with Python are conducted to demonstrate the great improvement on efficiency. Three cases based on two different mixed signal circuits are tested under both framework, including two simple cases and a complicated case.

A. Serial Framework Based on Python

The serial framework based on Python is rather fundamental and simple. To control variables, the serial framework also chooses Xyce as analog simulator and Iverilog as digital simulator. Differently, the Python codes take entire control of both Xyce and Iverilog. For the analog backend, Python script can control Xyce through Xyce-Python-Interface, which is a Python-wrapped version of Xyce-C-Interface, while for the digital backend, Python script can interact with Iverilog through CocoTB, an open-source coroutine based cosimulation and testbench library for verifying VHDL and SystemVerilog RTL using Python [10]. At every simulation step, the Python script, which is the top control, gets access to the output data of one simulator and forwards them to the other as input data, which requires one simulator to only run a simulation step after the other finishes one step.

B. Three Cases

Three cases based on two different circuits are designed for mixed signal simulation. The first and second cases are rather simple, mainly to show the efficiency of communication between two simulators. The circuits of the two cases are both two joined inverters, one for analog simulation and the other for digital simulation. The third case, which is more complicated and practical, involves a 8-bit DAC using R-2R resistance network. The circuit is illustrated in Fig. 5.

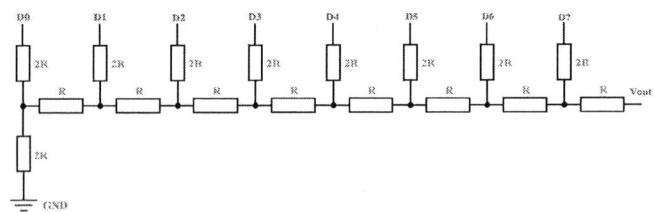

Fig. 5. The 8-bit DAC using R-2R resistance network. 8-bit digital signal inputs through port D0-D7, while the analog voltage outputs through port Vout.

C. Result

The performance comparison is shown in Table I.

TABLE I. PERFORMANCE COMPARISON

Case	Condition	Framework	Time(s)	Speedup Ratio
A2D[a]	Runtime & compiling time	Serial	1.1973	N/A
		Parallel	0.7746	1.51×
	Runtime	Serial	0.9973	N/A

Case	Condition	Framework	Time(s)	Speedup Ratio
		Parallel	0.5746	**1.74×**
D2A[b]	Runtime & compiling time	Serial	0.7295	N/A
		Parallel	0.4879	1.50×
	Runtime	Serial	0.5295	N/A
		Parallel	0.2879	**1.84×**
DAC	Runtime & compiling time	Serial	5.3303	N/A
		Parallel	1.7506	3.05×
	Runtime	Serial	4.9703	N/A
		Parallel	1.3906	**3.57×**

[a.] A2D is analog-to-digital inverters for short.

[b.] D2A is digital-to-analog inverters for short.

Four supplementary explanations are provided below.

1) All data in Table I are acquired by repeatedly testing the cases for 10 times and taking the average values of the required values, except "Runtime" of serial framework. The calculation of the exception values is elaborated in *3)*.

2) There are 2 different conditions in Table I including "Runtime & compiling time" and "Runtime", because for the serial framework based on Python, running is coupled with compiling.

3) The time of "Runtime" for serial framework is merely calculated by formula (1), in which RT_{Serial} is for time of "Runtime" for serial framework, RCT_{Serial} is for time of "Runtime & compiling time" for serial framework, etc., with the assumption that the compiling time of parallel framework and serial framework are approximately the same:

$$RT_{Serial} = RCT_{Serial} - (RCT_{Parallel} - RT_{Parallel}) \qquad (1)$$

4) Speedup ratio is calculated by formula (2), in which S is for speedup ratio, T_{Serial} is for time that serial framework takes, $T_{Parallel}$ is for time that parallel framework takes:

$$S = T_{Serial} / T_{Parallel} \qquad (2)$$

According to data shown in Table I, a conclusion can be safely drawn that the pipeline-based framework for parallel mixed signal simulation gains a speedup ratio of 1.74×~1.84× in terms of simulation efficiency in the two simple cases. While the more sophisticated case of DAC is considered, the runtime speedup ratio rises significantly to 3.57×, proving a great improvement on simulation efficiency.

VII. CONCLUSION

In this paper we have proposed a pipeline-based common framework for parallel mixed signal simulation, which provides a new perspective on synchronization algorithms and demonstrates high efficiency when running simulation with third-party simulators. It is from this perspective that we theoretically propose a new synchronization algorithm based on improvement to SES algorithm. Implementation of two kinds of two-stage pipeline is introduced. Experiments on the implemented pipeline demonstrates a significant runtime speedup varying from 1.74× to 3.57× with the complexity of the cases over a fundamental serial framework based on Python.

For further research, to generalize the framework, pipeline with feedback should be considered. Also, improvements on the proposed synchronization algorithm is needed to make it practical enough for implementation.

ACKNOWLEDGMENT

This research is supported by Yangtze River Delta Science and Technology Innovation Community Joint Fundamental Research Project 2024CSJZN0500, 2024CSJZN0503.

REFERENCES

[1] H. R. Ghasemi and Z. Navabi, "A new synchronization algorithm for (VHDL-AMS) mixed signal simulation," IEEE International Symposium on Communications and Information Technology, 2004. ISCIT 2004., Sapporo, Japan, 2004, pp. 1078-1083 vol.2.

[2] P. Frey and R. Radhakrishnan, "Parallel mixed-technology simulation," Proceedings Fourteenth Workshop on Parallel and Distributed Simulation, Bologne, Italy, 2000, pp. 7-14.

[3] M. G. Seok, T. G. Kim, C. B. Choi and D. Park, "An HLA-Based Distributed Cosimulation Framework in Mixed-Signal System-on-Chip Design," in *IEEE Transactions on Very Large Scale Integration (VLSI) Systems*, vol. 25, no. 2, pp. 760-764, Feb. 2017.

[4] *IEEE Standard 1364-2005*: IEEE Standard for Verilog Hardware Description Language, 2005.

[5] Eric R. Keiter, Richard L. Schiek, Heidi K. Thornquist, Ting Mei, Jason C. Verley, Karthik V. Aadithya, Gary J. Templet, Joshua D. Schickling, Gary L. Hennigan. "Xyce™ Parallel Electronic Simulator." Computer software. October 03, 2013.

[6] Eric R. Keiter, Richard L. Schiek, Heidi K. Thornquist, Ting Mei, Jason C. Verley, Karthik V. Aadithya, Joshua D. Schickling, and Gary L. Hennigan. Xyce Parallel Electronic Simulator: Users' Guide, Version 7.8. Technical Report SAND2023-13274, Sandia National Laboratories, Albuquerque, NM, 2023.

[7] Peter E. Sholander and Richard L. Schiek. Application note: Mixed signal simulation with Xyce. Technical Report SAND2018-TBD, Sandia National Laboratories, 2018.

[8] Andrew M. Smith, Jackson Mayo, Rob Armstrong, Richard Schiek, Peter Sholander, and Ting Mei. Digital/Analog Cosimulation Using CocoTB and Xyce. Technical Report SAND2018-TBD, Sandia National Laboratories, 2018.

[9] *Icarus Home Page*. URL http://iverilog.icarus.com/

[10] *CocoTB Home Page*. URL https://docs.cocotb.org/en/stable/index.html

A Rectifier Circuit with Adjustable Temperature Coefficient for Precise Amplitude Control of MEMS Resonator

Guanxiao Zhang
School of Mechanical Engineering
Nanjing University of Science and Technology
Nanjing, China
guanxiao.zhang@njust.edu.cn

Yang Zhao
School of Mechanical Engineering
Nanjing University of Science and Technology
Nanjing, China
zhaoyang0216@njust.edu.cn

Qin Shi
School of Mechanical Engineering
Nanjing University of Science and Technology
Nanjing, China
sqinhy@njust.edu.cn

Guoming Xia
School of Mechanical Engineering
Nanjing University of Science and Technology
Nanjing, China
xiaguoming@njust.edu.cn

Anping Qiu*
School of Mechanical Engineering
Nanjing University of Science and Technology
Nanjing, China
apqiu@njust.edu.cn

Meijia Xu
School of Mechanical Engineering
Nanjing University of Science and Technology
Nanjing, China
xumeijia@njust.edu.cn

Abstract—**This work proposes a rectifier with adjustable temperature coefficient based on different the tail current ratio. In order to obtain better accuracy of MEMS resonator vibration amplitude control, this paper problem of rectifier tail current mismatch in our previous work, and realizes a rectifier with low temperature coefficient output amplitude. Then, based on this scheme, a rectifier with different temperature coefficient output amplitudes is realized by adjusting the current ratio of the tail current source of the rectifier to compensate the temperature coefficient of other modules in the MEMS resonator circuit, so as to reduce the temperature coefficient of the MEMS resonator amplitude. The design is realized by 0.35μm BCD process. The simulation results show that the temperature coefficient of MEMS resonator amplitude decreases from +716.6 ppm/°C to less than ± 70ppm/°C under different type of process corner in the working temperature range of -40 °C to +60 °C.**

Keywords—*MEMS resonator, amplitude control, temperature coefficient*

I. INTRODUCTION

MEMS resonators are widely used in various applications related to timing and frequency control [1]. MEMS technology has proved several favorable features compared with mature quartz-based oscillators, such as miniaturization, batch manufacturability and compatibility with integrated circuit manufacturing [2-4]. For MEMS resonators, the steady-state amplitude error is a key index. The steady-state amplitude error will make the MEMS resonator produce phase noise changes [5] and make the gyroscope and accelerometer produce sensitivity errors. The steady-state amplitude accuracy of MEMS resonators is affected by temperature [6], so reducing the temperature error of the steady-state amplitude of MEMS

resonators is very important to expand the application range of MEMS resonators.

In [7], the switching capacitor is used to extract the amplitude, which achieves excellent noise performance, but increases the additional capacitance area. In [8], it is proposed to use analog-to-digital converter to convert the signal into digital domain and control the signal amplitude through digital signal processing (DSP). In [9], The amplitude control scheme uses the full wave rectifier to extract the amplitude and the chopper technology is introduced to reduce the flicker noise of the full wave rectifier, but due to mismatch of tail current of rectifier, this scheme produce a large steady-state temperature coefficient of the resonator amplitude in the full wave rectifier.

The steady-state amplitude of MEMS resonator is related to the gain of full wave rectifier, the gain of frontend amplifier and the conversion coefficient of detection comb [10]. The temperature coefficient of the front-end amplifier and the conversion coefficient of detection comb are determined by the CMOS processing technology and physical characteristics, which means that they cannot be compensated by the design of its own module. Therefore, designing a full wave rectifier circuit with adjustable temperature coefficient to compensate the temperature coefficient of other modules is of great significance to reduce the temperature coefficient of MEMS resonator steady-state amplitude.

In this paper, a low temperature coefficient full wave rectifier is proposed to solve the problem of excessive temperature coefficient of resonator amplitude caused by current mismatch in the low noise amplitude control circuit scheme in [9]. Based on this scheme, a full wave rectifier with adjustable temperature coefficient is proposed in this paper. By adjusting

979-8-3315-8850-2/25 $31.00 © 2025 IEEE

the current ratio of the tail current source of the full wave rectifier, the rectifier with adjustable temperature coefficients is obtained to compensate the temperature coefficients of other modules. Finally, a MEMS resonator circuit with low temperature coefficient amplitude is obtained.

II. THE PRINCIPLE OF MEMS RESONATOR

A. Structure of MEMS Resonator

The simplified diagram of the MEMS resonator is shown in the fig.1(a). The MEMS resonator mainly consists of a resonant beam, driving comb teeth and detection comb teeth. The entire structure is anchored on a substrate by a suspension beam system. A layout of the MEMS resonant is shown in Fig. 1(b). The MEMS resonator vibrates at the natural frequency when it is actuated by the electrostatic drive comb. The peripheral closed-loop self-oscillation circuit was used to track the natural frequency of the resonator

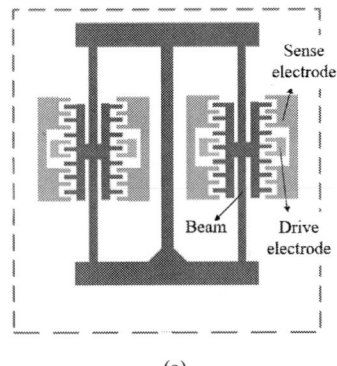

(a)

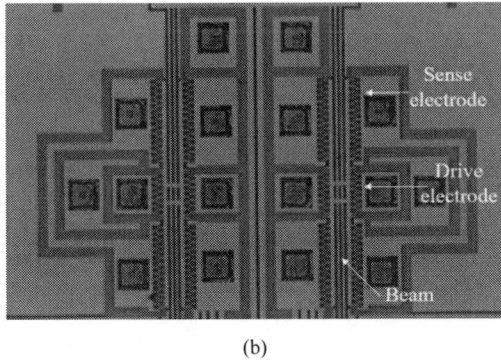

(b)

Fig. 1. (a) The simplified diagram of the MEMS resonator (b) The layout of MEMS resonant

B. The Principle of Amplitude Control in MEMS Resonator Circuits

The simplified resonator circuit with automatic amplitude control (AAC) in MEMS resonator is shown in Fig.2. During oscillation, the MEMS resonator detect the oscillation velocity v_x and generates a motional current I_{ds} at the sensing node, which is subsequently amplified by the front-end amplifier. The output of the amplifier is fed to the AAC and the VGA. The amplitude information is extracted and, then, subtracted from a preset value V_{ref}. The error signal controls the gain of the VGA. The output of the VGA drives the MEMS resonator, completing the oscillator loop[11]

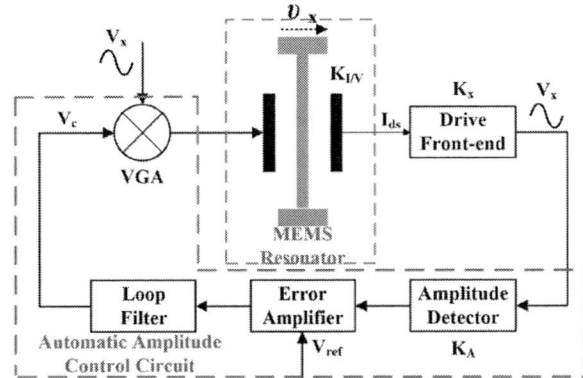

Fig. 2. Simplified MEMS resonator control loop

When the resonator reaches a steady state, the output of the amplitude detector is equal to the reference value. The function of MEMS resonator amplitude can be expressed as:

$$X = V_{ref}/(K_{I/V}K_XK_A) \tag{1}$$

Where X is the amplitude of MEMS resonator, $K_{I/V}$ is the conversion coefficient of detection comb, K_X is the gain of frond-end amplifier, K_A is the gain of amplitude extraction circuit.

In(1),the temperature coefficient of K_X, $K_{I/V}$ are determined by CMOS process and material physical properties[10]. Therefore, the temperature coefficient of MEMS resonator amplitude can be reduced by designing a reasonable temperature coefficient of K_A.

III. PRINCIPLE OF A LOW TEMPERATURE COEFFICIENT RECTIFIER

The full wave rectifier is used to extract the circuit amplitude in MEMS resonator circuit. As shown in Fig.3, the original scheme of rectifier circuit uses same two tail current sources I_1 and I_2 to supply current to two PMOS respectively. This work uses third current source I_3 flows through PMOS connected to common mode input.

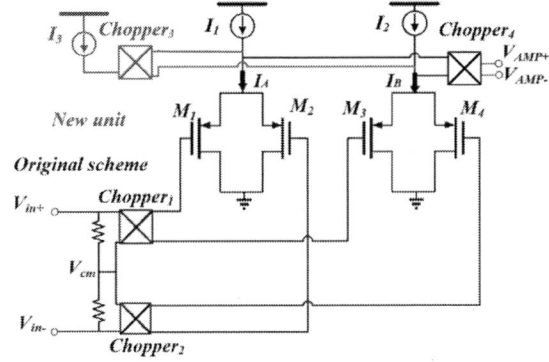

Fig. 3. Schematic diagram of rectifier with low temperature coefficient

At time t_1, M1 and M2 are connected with differential mode input, M3 and M4 are connected with common mode input, and current I_3 is injected into M3 and M4. With the square wave switching, at time t_2, M1 and M2 are connected with common mode input, M3 and M4 are connected with differential mode input, and current I_3 is injected into M1 and M2. The circuit output is the differential signal of MOS source voltage. Fig.4 shows the waveform diagram of the rectifier. Where I_A is the current flows through the M1,M2, I_B is the current flows through the M3,M4.

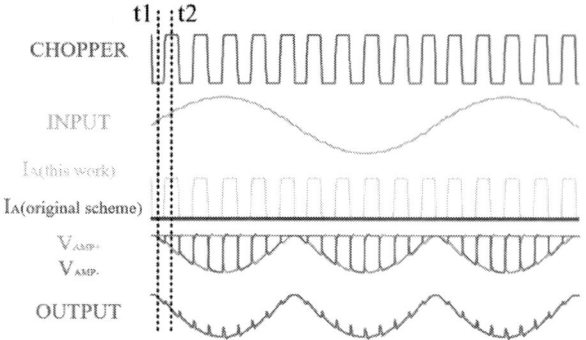

Fig. 4. waveform diagram of the rectifier

At time t_1 ,the output of the source voltage of M1 and M2 can be expressed as:

$$V_{AMP+} = \sqrt{\frac{2I_{dm}}{\mu_p C_{ox}\left(\frac{W}{L}\right)}} - |V_{DM}| + V_{CM} + V_{TH} \quad (2)$$

Where V_{DM} is differential mode input, V_{CM} is common mode input, V_{CM} is the threshold voltage of M1,M2, I_{dm} is the tail current source current connected to the differential mode input rectifier, μ_n is the electron mobility, C_{ox} is the gate oxide capacitance per unit area.

At time t_1 , the output of the source voltage of M3 and M4 can be expressed as:

$$V_{AMP-} = \sqrt{\frac{NI_{dm}}{\mu_p C_{ox}\left(\frac{W}{L}\right)}} + V_{CM} + V_{TH} \quad (3)$$

Where $N = I_{cm}/I_{dm}$. I_{cm} is the tail current source current connected to the common mode input rectifier. At time t_1 , $I_{cm}= I_B, I_{dm}= I_A$, At time t_2 , $I_{cm}= I_A, I_{dm}= I_B$. The output of full wave rectifier can be expressed as:

$$V_o = \left(\sqrt{2} - \sqrt{N}\right)\sqrt{\frac{I_{dm}}{\mu_p C_{ox}\left(\frac{W}{L}\right)}} - |V_{DM}| \quad (4)$$

In the traditional full wave rectifier, I_3 does not exist, so $I_{dm} = I_{cm}$, that is, $N =1$. So the output of the full wave rectifier is:

$$V_o = \left(\sqrt{2} - \sqrt{1}\right)\sqrt{\frac{I_{dm}}{\mu_p C_{ox}\left(\frac{W}{L}\right)}} - |V_{DM}| \quad (5)$$

The former term in (5) contains μ_n and C_{ox} ,which are affected by temperature. It produces the temperature error of the output of the full wave rectifier.

This paper proposes that the tail current source current of the common mode input rectifier is increased by two times, that is, $I_3 = I_1$,and $N =2$, so the output of the full wave rectifier is:

$$V_o = -|V_{DM}| \quad (6)$$

when the input differential voltage is 0.6V, the output of and this work from -40 °C to 60 °C are shown in Fig.5. The simulation shows that the temperature coefficient of the full wave rectifier circuit is reduced from -417.7ppm/°C to 7.1ppm/°C.

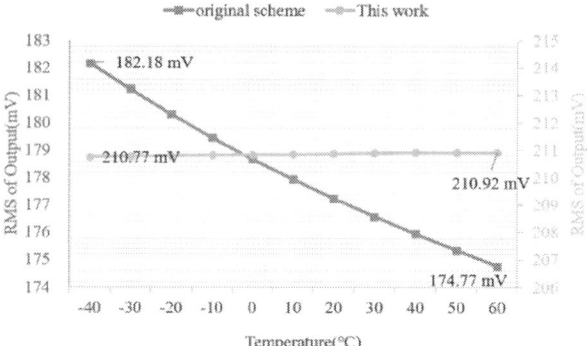

Fig. 5. output of original work and this work under different temperature

IV. RECTIFIER WITH ADJUSTABLE TEMPERATURE COEFFICIENT FOR PRECISE AMPLITUDE CONTROL OF MEMS RESONATOR

In the third section, this paper proposes a low temperature coefficient rectifier, which can reduce the temperature coefficient of the steady-state amplitude of MEMS resonator. However, from (1), it can be seen that the temperature coefficient of the steady-state amplitude of MEMS resonator is also related to the temperature coefficient of the conversion coefficient of detection comb and the temperature coefficient of gain of the frond-end amplifier. The temperature coefficients of both are determined by the CMOS processing parameters and physical characteristics.

Therefore, this paper proposes a full wave rectifier which can adjust the temperature coefficient by changing the current ratio N of the tail current source, as shown in Fig. 6. Through the combination of different switches, the current of the tail current source connected to the common mode input rectifier is 1 to 3.4 times that of the tail current source connected to the differential mode input rectifier.

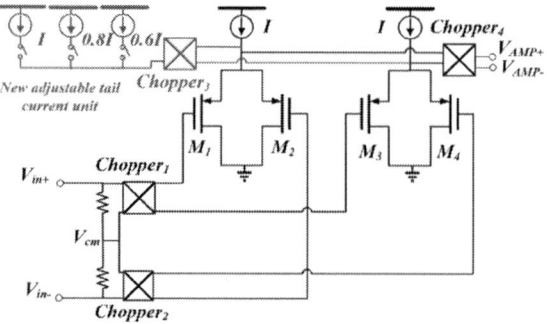

Fig. 6. Schematic diagram of rectifier with adjustable temperature coefficient

When the input differential voltage is 0.6V and the temperature is from -40 °C to 60 °C, , the temperature coefficient of full wave rectifier circuit is from -414.2ppm/°C to +313.7ppm/°C under different tail current source ratio coefficient N (as shown in Fig. 7).

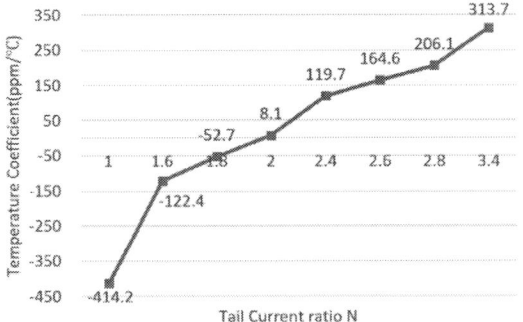

Fig. 7. Temperature coefficient of rectifier under different N(pre-simulation)

In this paper, the MEMS resonator circuit developed by Nanjing University of technology is taken as the research object. MEMS resonator is simplified as RLC model. The Quality factor of resonator is about 100000, and vibration frequency is 19.8k. The detection current output by MEMS resonator is selected as the characterization result of amplitude, and the simulation result is shown in Fig.8.

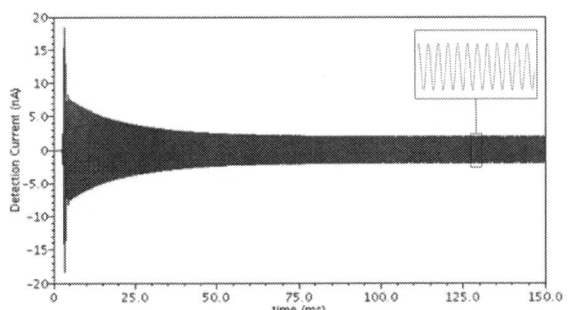

Fig. 8. MEMS resonator startup to stabilization process

Fig.9 shows the simulation results of MEMS resonator temperature coefficient. The results show that when N=2.4, the temperature coefficient of the steady amplitude of MEMS resonator is the lowest, which decreases from +716.6ppm/°C in original scheme to +21.4ppm/°C.

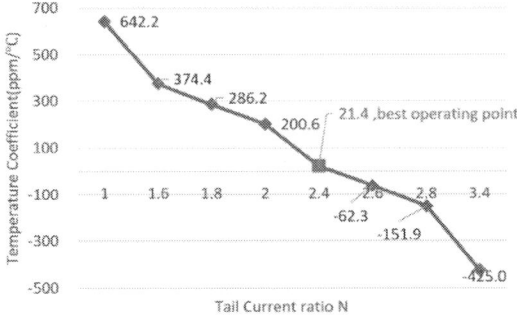

Fig. 9. Temperature coefficient of resonator under different N(pre-simulation)

V. LAYOUT AND POST-LAYOUT SIMULATION

The layout of the full wave rectifier circuit with adjustable temperature coefficient is shown in Fig.10, with a total layout area of 343.6μm*284.4μm.The core of rectifier including chopper, MOS and current source only cover about 26480μm². The capacitance is the decoupling capacitance of the signal .

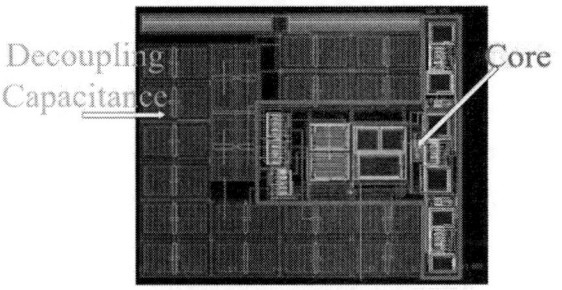

Fig. 10. Layout of rectifier circuit with adjustable temperature coefficient

The difference of temperature Coefficient between pre-simulation and post-simulation of 1~3.4 times current under different temperature as shown in Fig. 11. The difference is below ±2ppm/°C. The temperature coefficient of the module can be adjusted from -415.9ppm/°C to +313.1ppm/°C under 0.6V of differential mode input and 1.2V of common mode input.

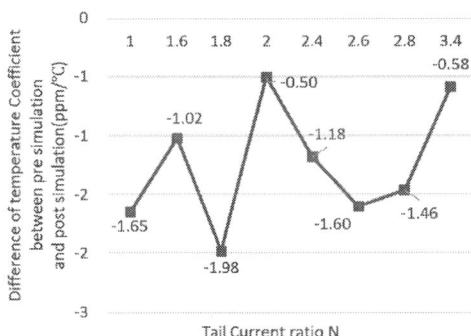

Fig. 11. The difference of temperature Coefficient between pre- simulation and post-simulation

The simulation of MEMS resonator shows that(as shown in Fig. 12), when the process corner is typical type, the best operating point is still N=2.4, and the lowest temperature coefficient of the MEMS resonator is -19.4ppm/°C.

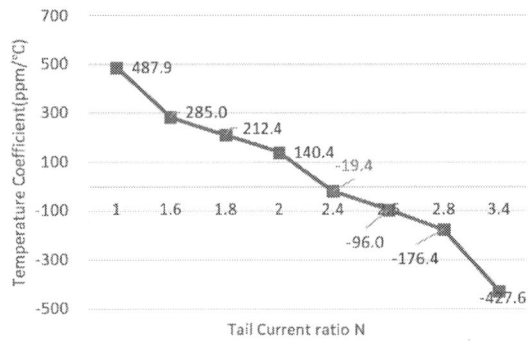

Fig. 12. Temperature coefficient of resonator under different N(post-simulation)

During the manufacturing process, the temperature coefficient and the optimal operating point of the resonator may change due to process fluctuations. The rectifier proposed has the ability to adjust the temperature coefficient and reduce the amplitude temperature coefficient of MEMS resonator. Table 1 shows the lowest temperature coefficient under different processes and the corresponding N. By adjusting different N, the temperature coefficient of MEMS resonator can be reduced to less than ±70ppm/°C from -40 °C to +60 °C.

TABLE I. Lowest Temperature Coefficient under Different Corner

corner	N	The lowest temperature coefficient (ppm/°C)
tt	2.4	-19.4
ff	2.6	37.0
ss	1.6	-65.2
fs	1.8	-2.4
sf	2.6	-35.4

VI. Summary

In this paper, a rectifier with adjustable temperature coefficient is designed by 0.35 μm BCD process. Firstly, a rectifier circuit with low temperature coefficient is proposed by solving the problem of large amplitude temperature coefficient caused by current mismatch in the original rectifier. Then a rectifier circuit with adjustable temperature coefficient is proposed based on adjusting the current ratio to compensate the temperature coefficient of other modules which is difficult to be compensated. The simulation results show that the temperature coefficient of the proposed rectifier circuit with adjustable temperature coefficient can be adjusted from -415.9ppm/°C to +313.1ppm/°C. The temperature coefficient of MEMS resonator can be reduced to less than ±70ppm/°C in the temperature range of -40 °C to 60 °C throughout different type of process corner by adjusting tail current ratio N. The rectifier circuit with adjustable temperature coefficient proposed in this paper

reduces the temperature error of MEMS resonator amplitude and expands the application range of MEMS resonator.

References

[1] P. M., S. G., Z. H., and A. S. A., "Enhancement of Frequency Stability in Injection Locked Bulk Mode MEMS Oscillators," in 2021 IEEE 34th International Conference on Micro Electro Mechanical Systems (MEMS), 2021, pp. 941-944.

[2] T. C. N. C., "MEMS technology for timing and frequency control," IEEE Transactions on Ultrasonics, Ferroelectrics, and Frequency Control, vol. 54, pp. 251-270, 2007-1-1 2007.

[3] J. T. M. van Beek and R. Puers, "A review of MEMS oscillators for frequency reference and timing applications," Journal of Micromechanics and Microengineering, vol. 22, p. 013001, 2012-1-1 2012.

[4] W. T. Hsu, "Recent Progress in Silicon MEMS Oscillators," recent progress in silicon mems oscillators, 2008.

[5] H. L., P. X. Y. and P. M., "A State-Space Phase-Noise Model for Nonlinear MEMS Oscillators Employing Automatic Amplitude Control," IEEE Transactions on Circuits and Systems I: Regular Papers, vol. 57, pp. 189-199, 2010-1-1 2010.

[6] J. Cui, G. Yan and Q. Zhao, "Enhanced temperature stability of scale factor in MEMS gyroscope based on multi parameters fusion compensation method," Measurement, vol. 148, p. 106947, 2019-1-1 2019.

[7] Z. M., Z. Q., L. W., and Y. G., "An Interface ASIC for an Atmospheric-Pressure MEMS Gyroscope with PLL-Based Phase Adjustment and SC Amplitude Regulation," in 2020 IEEE International Symposium on Circuits and Systems (ISCAS), 2020, pp. 1-5.

[8] S. M., A. L., H. K., and S. T., "Integrated Readout and Control Electronics for a Microelectromechanical Angular Velocity Sensor," in 2006 Proceedings of the 32nd European Solid-State Circuits Conference, 2006, pp. 243-246.

[9] W. X., Z. J., Z. Y., M. X. G., P. Q. A., S. Y., and P. X. Y., "A 0.4 μg Bias Instability and 1.2 μg/√ Hz Noise Floor MEMS Silicon Oscillating Accelerometer With CMOS Readout Circuit," IEEE Journal of Solid-State Circuits, vol. 52, pp. 472-482, 2017-1-1 2017.

[10] Z. Y., Z. J., W. X., M. X. G., S. Q., P. Q. A., and P. X. Y., "A Sub-0.1°/h Bias-Instability Split-Mode MEMS Gyroscope With CMOS Readout Circuit," IEEE Journal of Solid-State Circuits, vol. 53, pp. 2636-2650, 2018-1-1 2018.

[11] Z. Y., M. X. G., S. Q., and P. Q. A., "A Low Flicker Noise Automatic Amplitude Control ASIC for a Split-Mode MEMS Gyroscope," in 2018 14th IEEE International Conference on Solid-State and Integrated Circuit Technology (ICSICT), 2018, pp. 1-3.

A Low-Complexity Bandwidth Enhancement Design Method for mm-Wave Doherty Amplifier

Yujie Sheng
School of Electronic Science and Engineering
University of Electronic Science and Technology of China
Chengdu, China

Fei You
School of Electronic Science and Engineering
University of Electronic Science and Technology of China
Chengdu, China
feiyou@uestc.edu.cn

Maojun Pan
School of Electronic Science and Engineering
University of Electronic Science and Technology of China
Chengdu, China

Songbai He
School of Electronic Science and Engineering
University of Electronic Science and Technology of China
Chengdu, China

Abstract—This paper proposes a low-complexity design method for broadband Doherty PA. While ensuring broadband efficiency and output power, the method significantly expands the impedance design flexibility of matching networks, thereby simplifying the multi-frequency co-optimization process and providing a systematic solution for broadband impedance optimization. Implemented in 65-nm CMOS technology, the Doherty PA design employs the proposed methodology to optimize output matching network, achieving broadband performance from 24 to 34 GHz (fractional bandwidth >34%). Simulation results demonstrate 17.9%-22.05% PAE at 6dB PBO, 27.1%-35.92% saturated PAE, and saturation output power exceeding 23dBm. At 32/34GHz, the power back-off exceeds 7dB. Across the operational bandwidth, the design maintains >25dB small-signal gain, demonstrating superior performance.

Index Terms—Doherty, power amplifier, mm-Wave, matching network, low-complexity, wideband

I. INTRODUCTION

The evolution of 5G/6G, IoT, and mm-Wave communications imposes stringent requirements on power amplifiers, particularly in achieving high efficiency at deep power back-off (6-10 dB) and ultra-wideband operation (sub-6 GHz to mm-Wave). Advanced modulation formats, such as 1024-QAM, and multi-carrier wideband OFDM signals exacerbate efficiency degradation in traditional Class AB PAs under low-power conditions, driving the adoption of Doherty PA architectures. While Doherty PA leverage active load modulation to enhance back-off efficiency, the stem from frequency-dependent impedance inverter limits applicability in broadband, carrier-aggregated, and concurrent sub-6 GHz/mmWave systems requiring GHz-level bandwidth.

For instance, techniques such as multi-way [1] or multi-stage [2], complex combining load [3] [4], complex load-modulation [5], and post-matching network [6] [7] can enhance the efficiency and bandwidth of Doherty power amplifiers. However, these methods face significant challenges, including parasitic interactions in distributed elements, strict matching conditions, and high complexity at mmWave frequencies. As a result, they struggle to serve as a universal design solution for wideband Doherty PA implementations.

This paper proposes a low-complexity design method tailored for broadband Doherty PA, which significantly expands the design flexibility of impedance matching networks while maintaining high efficiency and output power across a wide bandwidth, offering a systematic approach to impedance optimization in broadband scenarios.

II. CIRCUIT DESIGN

A. Block Diagram of Designed Doherty PA

The architecture of the Doherty PA is shown in Fig. 1. The Doherty PA adopts an asymmetrical structure for the main path and auxiliary path, both include an input matching network, driver stage, an inter-stage matching network, power stage, and an output matching network. In addition, the output matching network and inter-stage matching network are designed with proposed methodology to achieve broadband performance. The detailed transistor parameters are labeled in Fig. 1.

The impedance inverter is embedded in front of the main PA's output matching network, enabling broadband impedance transformation while preserving phase alignment. The RF signals are distributed to the differential ports of the two paths by a quadrature coupler and input matching network, then amplified by the driver stage and power stage, and finally combined by output matching network. The bias voltages for both the driver stage and power stage are given by an adaptive biasing network.

B. Design Methodology for Bandwidth Enhancement

Load-pull analysis, a widely adopted impedance optimization technique in power amplifier design, characterizes the nonlinear behavior of active devices by sweeping

979-8-3315-8850-2/25 $31.00 © 2025 IEEE

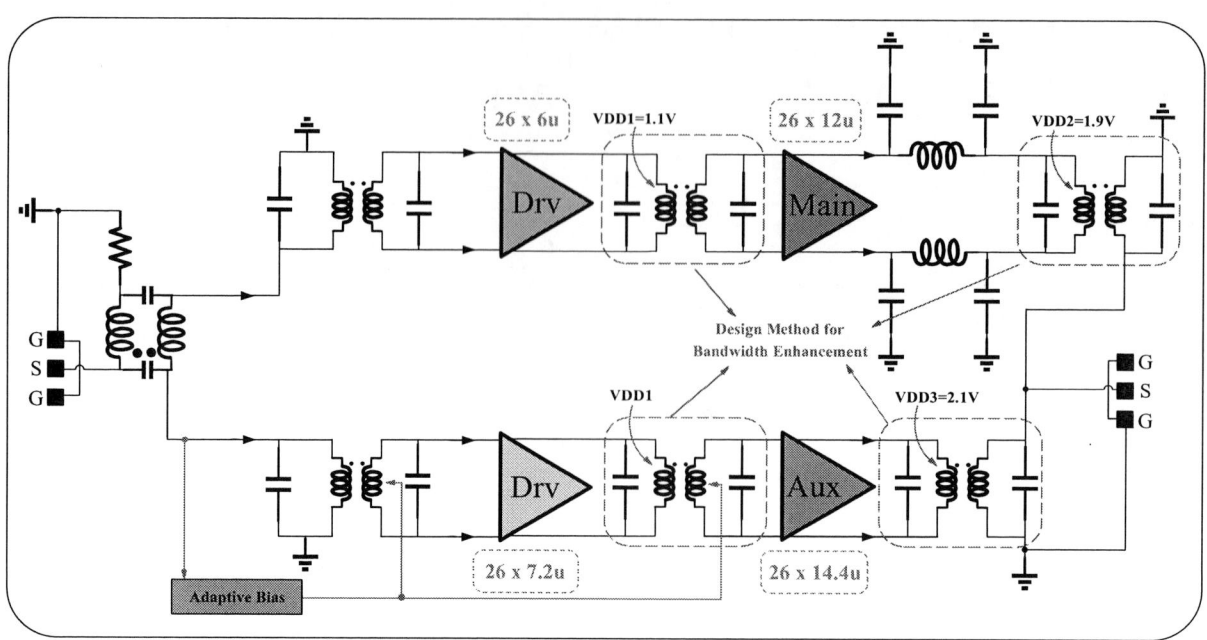

Fig. 1: Integrated Circuit Block Diagram.

impedance states and quantifying performance parameters. Conventional load-pull methods typically focus on single-frequency impedance tuning to maximize output power or efficiency at a specific operating frequency. In this work, we propose a novel design methodology for bandwidth enhancement: By analyzing power-added efficiency (PAE) contours and output power contours across multiple frequency points, the optimal matching region (OMR) is extracted from the Smith chart to define impedance boundaries for broadband matching optimization. This approach eliminates the need for additional compensation circuits, enabling wideband performance enhancement solely through multi-frequency data-driven impedance synthesis.

TABLE I: Design Parameters and Bias Conditions

Parameters	width(um)	finger	multiplier
	3	26	4
Bias	V_{GS1}(V)	V_{GS2}(V)	V_{DD}(V)
	0.4	1.3	1.9

Load-pull characterization employs two key performance contours on the Smith chart: Power contours and PAE contours, which map load impedance states to respective output power and power-added efficiency values. Each closed contour represents iso-performance regions where distinct load impedances yield identical output power or PAE. Fig. 2 illustrates the 28-GHz single-frequency load-pull characterization results for the cascode transistor in the main PA. Designed in 65-nm CMOS technology, the transistor's design parameters and bias conditions are summarized in Table I.

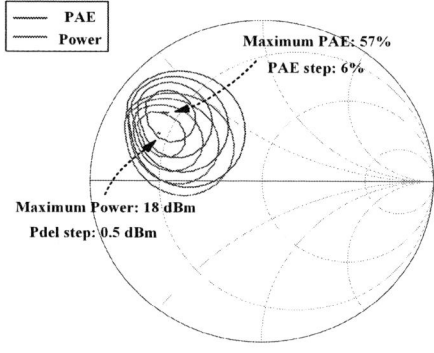

Fig. 2: Load-pull results for the main PA cascode transistor @28 GHz.

As demonstrated by the load-pull characterization results in Fig. 2, the cascode transistor achieves a peak output power of 18 dBm and a maximum power-added efficiency of 57% at 28 GHz when the input power reaches 10 dBm. These results highlight the inherent trade-off between power and efficiency, with the optimal load impedance for output power significantly diverging from that for PAE maximization. Furthermore, the impedance selection range broadens as the load deviates from the optimal value, expanding the design space for broadband matching optimization.

Broadband load-pull characterization reveals overlapping distributions of constant-output-power and constant-PAE contours on the Smith chart across the 24–34 GHz band. Under the design constraints of PAE ≥50% and output power ≥17.5

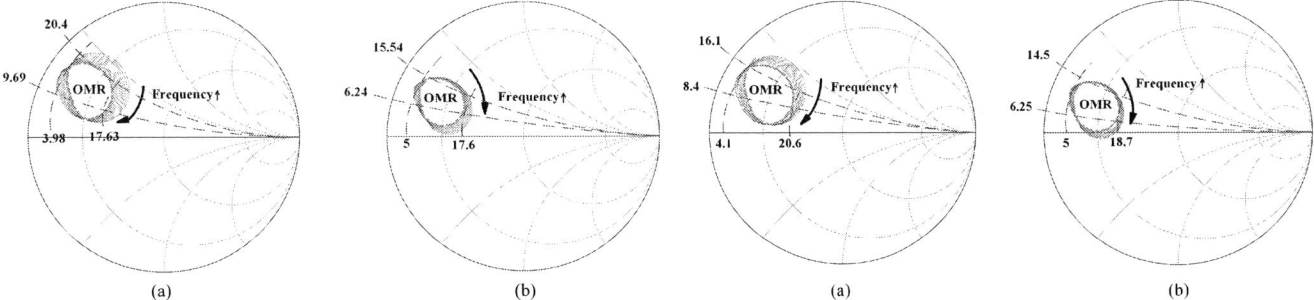

Fig. 3: Broadband optimal matching region showing (a) PAE and (b) output power contours for main PA on the Smith chart. The overlapping regions define the optimal matching bounds for impedance synthesis at back-off point.

Fig. 5: Broadband optimal matching region showing (a) PAE and (b) output power contours for auxiliary PA on the Smith chart. The overlapping regions define the optimal matching bounds for impedance synthesis at saturation.

dBm at back-off point, Fig. 3(a) and 3(b) present the common region for PAE and output power optimization, respectively.

Analysis identifies a shared optimal matching region (OMR) within the Smith chart where both PAE and output power meet or exceed the target specifications. This region defines the impedance bounds for broadband matching. For PAE contours (see Fig. 3(a)), the real and imaginary components of the load impedance are constrained to 4–17.6 Ω and 9.7–20.4 Ω, respectively. Conversely, for output power coutours (see Fig. 3(b)), the impedance bounds tighten to a real component of 5–17.6 Ω and an imaginary component of 6.2–15.5 Ω. The intersection of these regions defines the co-optimized impedance space, enabling broadband performance enhancement at back-off point without compromising efficiency or output power. The intersection of these regions provides the final impedance design space, ensuring simultaneous broadband efficiency and power optimization without additional circuits.

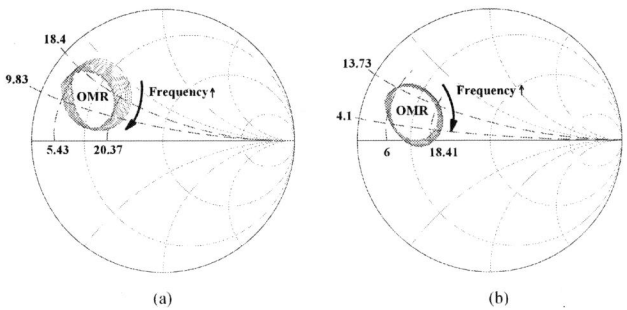

Fig. 4: Broadband optimal matching region showing (a) PAE and (b) output power contours for main PA on the Smith chart. The overlapping regions define the optimal matching bounds for impedance synthesis at saturation.

The intersection of the PAE and output power constraints defines an initial co-optimized impedance region spanning 5–17 Ω (real) and 9.7–15.5 Ω (imaginary) at back-off point. However, to mitigate transistor reliability risks associated with elevated drain-source voltage at high real impedance values, the final optimization bounds are refined to 5–12 Ω (real). This trade-off reduces V_{DS} stress without compromising

broadband performance guarantees across the 24–34 GHz band. The same methodology is applied to determine the OMR for main PA at saturation, as illustrated in Fig. 4(a) and 4(b), the final optimization bounds at saturation are defined to 5–12 Ω (real) and 9–14.5 Ω (imaginary).

In communication systems, Doherty PA predominantly operate in the low-power back-off region during extended transmission periods. Consequently, the auxiliary PA must be optimized under a power-prioritized design criterion to ensure robust performance under sustained back-off conditions. To meet this requirement, the transistor dimensions of the auxiliary amplifier are moderately expanded to enhance its power-handling capability.

The optimal matching region for the auxiliary PA is determined using the broadband design methodology described in Section II. Fig. 5(a) and 5(b) present the simulated PAE contours and output power contours, respectively. Based on these results, the final co-optimized load impedance is constrained to 4.4-12 Ω (real) and 9.7–20.4 Ω (imag), respectively.

After defining the OMR, optimization can be performed in ADS. This design employs a conventional transformer balun combined with capacitors as the matching network. Fig. 6 illustrates the co-optimized load impedance characteristics for both the main and auxiliary circuits. Across the 23–35 GHz band, the real and imaginary components of the load impedance conform to the prescribed boundary conditions, ensuring compliance with broadband performance targets.

By constraining the impedance space to this intersection region, the methodology expands matching network design flexibility, simplifies multi-frequency co-optimization, and establishes a systematic framework for broadband Doherty PA impedance synthesis that balances efficiency, power, and device reliability.

C. Adaptive Biasing Network

Adaptive biasing serves as a critical enabler for dynamic load modulation in Doherty PA, ensuring high efficiency across extended power back-off operation. During low-power operation, the auxiliary amplifier remains turned off. When the input power exceeds the predefined back-off threshold,

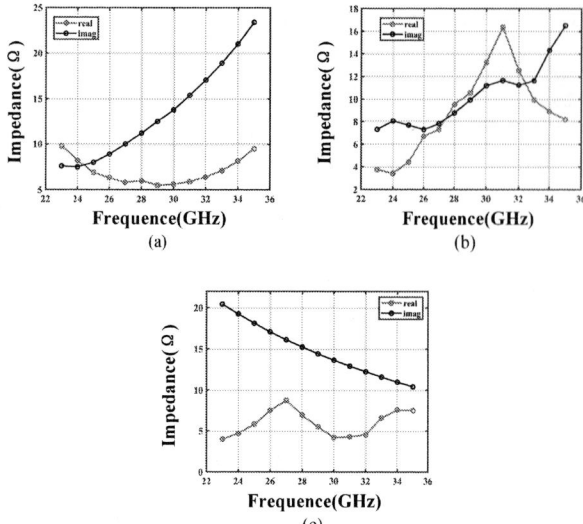

Fig. 6: Optimized load impedance characteristics of the designed Doherty PA showing (a) main PA impedance at back-off, (b) main PA impedance at saturation and (c) auxiliary PA impedance at saturation.

the adaptive biasing network rapidly activates the auxiliary path to initiate load modulation in coordination with the main path. Fig. 7 demonstrates the structure of the adaptive biasing network.

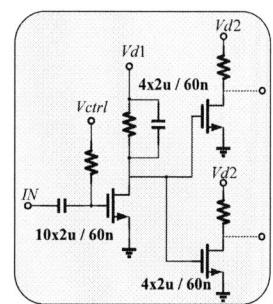

Fig. 7: Adaptive biasing network.

III. SIMULATION RESULTS

Fig. 8 presents the simulated performance of the Doherty PA designed in 65-nm CMOS technology. Table II summarizes a comparison between the designed Doherty PA and some relevant works.

During the simulation process, the transformer model incorporates the simulated insertion loss R to better align

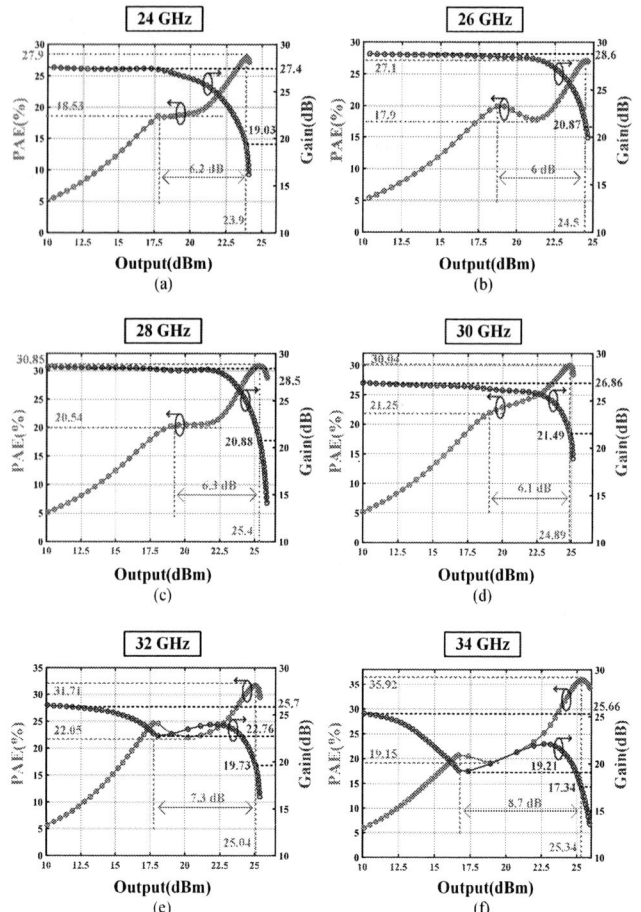

Fig. 8: Simulation results at 24, 26, 28, 30, 32, 34 GHz.

the simulation results with EM results. Simulation results indicate that the designed Doherty PA achieves 24-34 GHz (34.48 % FBW) operation with 17.9%-22.05% PAE at 6 dB PBO and 27.1%-35.92% saturated PAE across the operational bandwidth, the saturation output power exceeds 23 dBm. As verified in Fig. 8(e)-(f), the design maintains over 7 dB PBO at 32/34 GHz. Shown in Fig. 8(a)-(d), the design achieves >25 dB small-signal gain with enhanced flatness between 24-30 GHz when the auxiliary path is not opend. The proposed design methodology yields excellent performance in Doherty PA, achieving broadband operation with minimal design complexity.

TABLE II: Comparison Between the Designed Doherty PA and Some Relevant Works

	This work	[8]	[9]	[10]	[11]
Technology	**65-nm CMOS**	120-nm GaN-HEMT	65-nm CMOS	40-nm CMOS	0.13-um SiGe
Freq(GHz)/FBW(%)	**24-34/34.48**	24-30/22.2	24-30/22.2	24-32/28.6	24-29.5/20.6
PAE@6-dB PBO(%)	**17.9-22.05**	18.2-22.4	20@27 GHz	34@27 GHz	18.3-21.9
PAE@P_{sat}(%)	**27.1-35.92**	20-27.6	24.6	38.2	23.2-28.4
Gain(dB)	**25.66-28.6**	9.8-14.9	19.6(max)	17.4	13-15.9
P_{sat}(dBm)	**23.9-25.4**	31.6-32.7	20	20.43	20.1-21.2

979-8-3315-8850-2/25 $31.00 © 2025 IEEE

IV. CONCLUSION

This paper presents a novel design methodology for Doherty PA implemented in 65-nm CMOS technology. By analyzing multi-frequency load-pull analysis, the proposed design achieves 34% FBW with >23 dBm saturated output power. Simulated PAE exceeds 27% at saturation and maintains >17% at 6dB PBO operation. The proposed design methodology employs minimal complexity, offering a novel approach for realizing ultra-wideband DPA.

REFERENCES

[1] W. Kong, J. Xia, X. Zhou, T. Zhang, W. Zhang, and X. Bao, "Bandwidth Extension of Three-Way Doherty Power Amplifier With Reactance Compensation Using Parallel Peaking Amplifiers," IEEE Access, vol. 9, pp. 91661–91669, 2021.

[2] J. Xia, W. Chen, F. Meng, C. Yu, and X. Zhu, "Improved Three-Stage Doherty Amplifier Design With Impedance Compensation in Load Combiner for Broadband Applications," IEEE Trans. Microwave Theory Techn., vol. 67, no. 2, pp. 778–786, Feb. 2019. .

[3] Z. Yang et al., "Bandwidth Extension of Doherty Power Amplifier Using Complex Combining Load With Noninfinity Peaking Impedance," IEEE Trans. Microwave Theory Techn., vol. 67, no. 2, pp. 765–777, Feb. 2019.

[4] J. R. Zhang, S. Y. Zheng, and N. Yang, "An Efficient Broadband Symmetrical Doherty Power Amplifier With Extended Back-Off Range," IEEE Trans. Circuits Syst. II, vol. 70, no. 4, pp. 1316–1320, Apr. 2023.

[5] N. S. Mannem, T.-Y. Huang, and H. Wang, "Broadband Active Load-Modulation Power Amplification Using Coupled-Line Baluns: A Multifrequency Role-Exchange Coupler Doherty Amplifier Architecture," IEEE J. Solid-State Circuits, vol. 56, no. 10, pp. 3109–3122, Oct. 2021.

[6] J. Pang, S. He, Z. Dai, C. Huang, J. Peng, and F. You, "Design of a Post-Matching Asymmetric Doherty Power Amplifier for Broadband Applications," IEEE Microw. Wireless Compon. Lett., vol. 26, no. 1, pp. 52–54, Jan. 2016.

[7] C. Shen, S. He, X. Zhu, J. Peng, and T. Cao, "A 3.3–4.3-GHz High-Efficiency Broadband Doherty Power Amplifier," IEEE Microw. Wireless Compon. Lett., vol. 30, no. 11, pp. 1081–1084, Nov. 2020.

[8] R. Liu, H. Jia, L. Qi, and A. Zhu, "A 24-to-30-GHz GaN MMIC Doherty Power Amplifier Using Reduced Peaking Intrinsic Output Impedance for Bandwidth Extension," IEEE Microw. Wireless Tech. Lett., vol. 35, no. 3, pp. 362–365, Mar. 2025.

[9] J. Chen, H. Zhu, J. Zhang, and Q. Xue, "A Compact Broadband Voltage-Combined Doherty Power Amplifier With Shorted Transmission Line for 5G Millimeter-Wave," IEEE Trans. Circuits Syst. I, vol. 71, no. 12, pp. 6190–6202, Dec. 2024.

[10] M. Pashaeifar, L. C. N. De Vreede, and M. S. Alavi, "A Millimeter-Wave CMOS Series-Doherty Power Amplifier With Post-Silicon Inter-Stage Passive Validation," IEEE J. Solid-State Circuits, vol. 57, no. 10, pp. 2999–3013, Oct. 2022.

[11] D. Wang, W. Chen, X. Chen, X. Liu, F. M. Ghannouchi, and Z. Feng, "A 24-29.5 GHz Voltage-Combined Doherty Power Amplifier Based on Compact Low-Loss Combiner," IEEE Trans. Circuits Syst. II, vol. 68, no. 7, pp. 2342–2346, Jul. 2021.

2025 The 10th International Conference on Integrated Circuits and Microsystems

A 10-bit 3-GS/s Single-Channel Pipelined ADC with Parallel Time-Domain Quantization Based on Dynamic Residue Amplifier

Depan Li, Xin Zhao, Dengquan Li*, Zhangming Zhu

Key Laboratory of Analog Integrated Circuits, School of Integrated Circuits, Xidian University, Xi'an, China
Corresponding Authors: dqli@xidian.edu.cn

Abstract—This paper presents a single-channel 10-bit pipelined analog-to-digital converter (ADC) with parallel time-domain (TD) quantization running at 3 GS/s. By sharing the discharge current of a dynamic residue amplifier (D-RA), parallel voltage amplification and voltage-to-time (V–T) conversion are realized. Simultaneously, the generated time signal is quantized by a time-to-digital converter (TDC). The parallel execution of amplification and quantization relaxes the timing constraints on the residue amplifier. In addition, a dynamic flipped voltage follower (DFVF)-based residue transfer scheme is employed, which is activated during the settling phase of the capacitor array and deactivated afterward, minimizing timing overhead. The ADC is simulated with a 28-nm CMOS process and 1-V supply voltage, it achieves a 56.9-dB signal-to-noise and distortion ratio (SNDR) and a 69.2-dB spurious-free dynamic range (SFDR) at Nyquist input frequency leading to a 15.0-fJ/conversion-step Walden figure-of-meriti.

Index Terms—Analog-to-digital converter (ADC), pipelined ADC, dynamic residue amplifier (D-RA), parallel amplification and quantization, dynamic flipped voltage follower (DFVF).

I. INTRODUCTION

Moderate-to-high resolution analog-to-digital converters (ADCs) are critical in modern communication systems and electronic instrumentation [1]. Although time-interleaved (TI) ADCs [2] can achieve both high speed and high precision, but they face challenges such as increased area, inter-channel crosstalk, and complex calibration. Designing high-speed single-channel ADCs with compact footprints can help mitigate these issues. Moreover, increasing the sampling rate of single-channel ADCs not only reduces the number of channels required in large-scale TI ADCs, but also lowers input capacitance and overall jitter, thereby further improving ADC performance limits. Pipelined and pipelined successive approximation register (SAR) ADCs [3], [4] are leading candidates for moderate-to-high resolution single-channel operation in the gigahertz range. However, the residue amplifiers (RAs) in pipelined architectures limit both conversion speed and accuracy, and achieving high speed and linearity in RA design remains challenging. Several techniques have been proposed to mitigate the speed limitations imposed by RAs.

A non-attenuated passive residue transfer technique is proposed in [5], employing two DACs in a ping-pong configuration to alternately transfer the residue voltage from the first stage to the second. However, due to the absence of an amplifier, the noise burden on subsequent stages increases. In [6], a post-amplification residue generation scheme performs quantization and amplification in parallel, removing timing constraints on the amplifier. Yet, direct amplification of the input signal results in significant nonlinearity, limiting overall ADC performance to 6 bits. Ref. [7] proposes a time-assisted residue generation (TARG) technique, which includes both a time-pulse output path and a conventional voltage output path, thereby breaking the traditional speed and linearity limitations of residue amplifiers. However, the time-pulse path still consumes approximately 80 ps of the first-stage operation time, and the high-speed voltage–time–voltage (V–T–V) operation introduces significant noise.

To address the aforementioned challenges, we move the residue amplification to the second stage to reduce the timing burden on the first stage. A dynamic residue amplifier (D-RA) is adopted in the second stage, which enables simultaneous voltage amplification and time-domain (TD) quantization by utilizing a shared discharge current path. The parallel execution of amplification and quantization relaxes the timing constraints on the residue amplifier. In the first stage, a dynamic flipped voltage follower (DFVF) is used for residue transfer. The DFVF is activated during the settling phase of the capacitor array (CDAC) and deactivated afterward, ensuring that residue transfer introduces nearly no additional timing overhead. With the integration of these techniques, simulation results demonstrate that the proposed 10-bit pipelined ADC achieves a sampling rate of 3 GS/s and a signal-to-noise and distortion ratio (SNDR) of 56.9 dB at Nyquist input.

This paper is organized as follows. Section II describes the architecture of the proposed pipelined ADC with parallel time-domain quantization. Section III details the circuit implementation. Section IV provides simulation results, and Section V concludes the paper.

II. PROPOSED ADC ARCHITECTURE

Fig. 1 shows the overall architecture of the proposed 10-bit, 3-GS/s pipelined ADC, which includes three sub-TD quantizers, two DFVFs, and a D-RA. The ADC is composed of three sub-stages, with 1-bit inter-stage redundancy applied between them to mitigate sub-quantization non-idealities.

979-8-3315-8850-2/25 $31.00 © 2025 IEEE

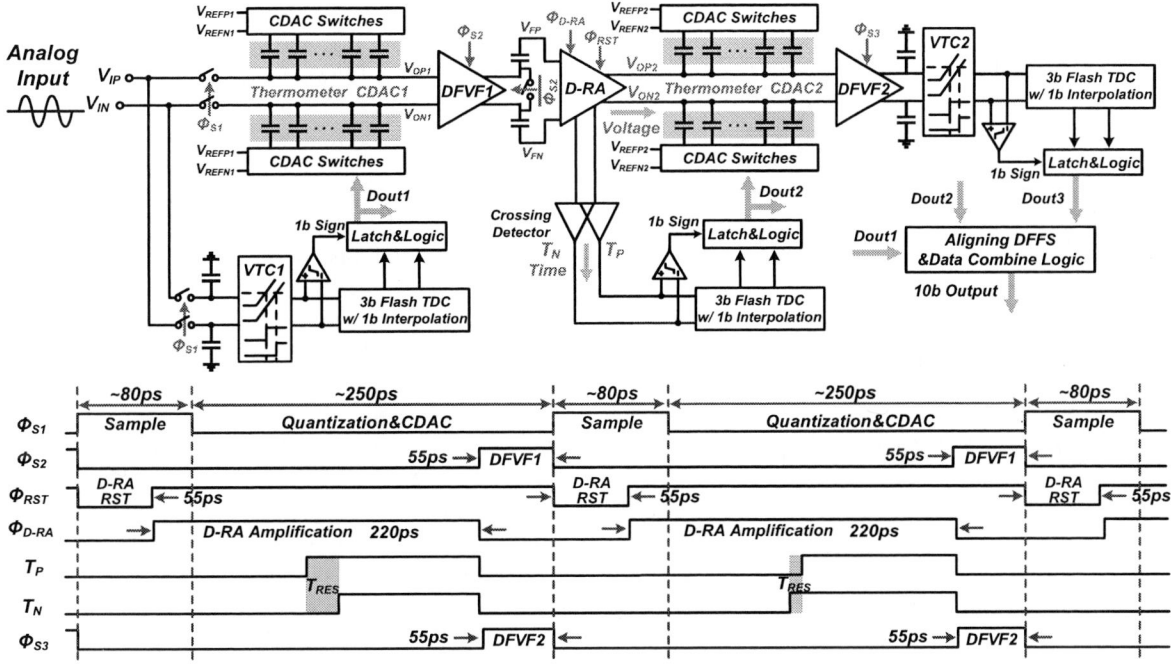

Fig. 1 Block diagram and timing diagram of the proposed pipelined ADC.

The first stage quantizes the most significant 4 bits. Unlike a conventional SAR ADC that requires a successive register to sequentially control comparisons and bit decisions, the TD quantizer generates 4 bits within a single clock cycle. Its output is directly connected to the CDAC switches through simple buffers, reducing the critical path and supporting high-speed operation. During Φ_{S1}, both CDAC1 and the TD quantizer sample the input signal (1-$V_{pp,diff}$). Inter-stage redundancy mitigates sampling mismatch and alignment errors between CDAC1 and the TD quantizer. CDAC1 generates the residue voltage at V_{OP1}/V_{ON1} based on the 4-bit TD quantization result. DFVF1 is activated during the settling phase of CDAC1 and deactivated once the settling is complete, enabling near-zero-delay voltage residue transfer.

The second pipeline stage performs voltage-domain amplification and TD quantization in parallel by sharing the D-RA's discharge current. During voltage amplification, a crossing detector (CD) monitors the D-RA's output to convert the voltage difference into time. The generated time signal is quantized by a 4-bit time-to-digital converter (TDC) during the latter half of the D-RA amplification phase. After the D-RA finishes amplification, CDAC2 generates the voltage residue based on the 4-bit quantization code. The parallel operation of amplification and quantization relaxes overall timing constraints, providing sufficient time for the D-RA to complete its operation. The D-RA is designed with a gain of 5× to suppress noise from the following stage and compensate for DFVF-induced gain loss. DFVF2 transfers the voltage residue to the third stage, where the final 4 bits are resolved via TD quantization. Quantization results from all three stages are stored and aligned in registers, then combined by digital

logic to produce the final ADC output.

To enhance DAC settling speed, a thermometer CDAC is adopted, as proposed in [7] and [8]. Unlike binary-weighted CDACs, it replaces large binary-scaled capacitors with uniform unit capacitors, thereby avoiding the switching of large individual capacitors. This improves settling behavior and reduces reference voltage ripple. Additionally, since all capacitors in a thermometer CDAC are exposed to similar environmental conditions, mismatches due to parasitic capacitance variations are minimized, improving overall capacitor matching accuracy. A split switching scheme [9] is employed to maintain consistent common-mode voltage during residue generation. To meet KT/C noise and matching requirements, unit capacitors are sized at 8 fF and 3 fF for the first and second stages, respectively.

III. CIRCUIT IMPLEMENTATION

A. Parallel amplification and quantization based on D-RA

Fig. 2(a) shows the circuit implementation of the D-RA, which performs voltage amplification and voltage-to-time (V-T) conversion simultaneously. Fig. 2(b) illustrates its timing diagram. An open-loop amplifier based on a cascoded integrator is employed, offering high speed and low noise [2], [6]. The cascode transistors (M_3, M_4) isolate the output nodes from the input differential pair's drain terminals, eliminating the need for dedicated series switches. When the Φ_{RST} signal is low, CDAC2 is reset to VDD. Once Φ_{D-RA} transitions high, the D-RA begins operation. Transistors M_1 and M_2 control the differential discharge of CDAC2 voltages (V_{OP2}, V_{ON2}). Fig. 3 shows the CD circuit within the D-RA, comprising a preamplifier, a dynamic inverter, and static inverters. When

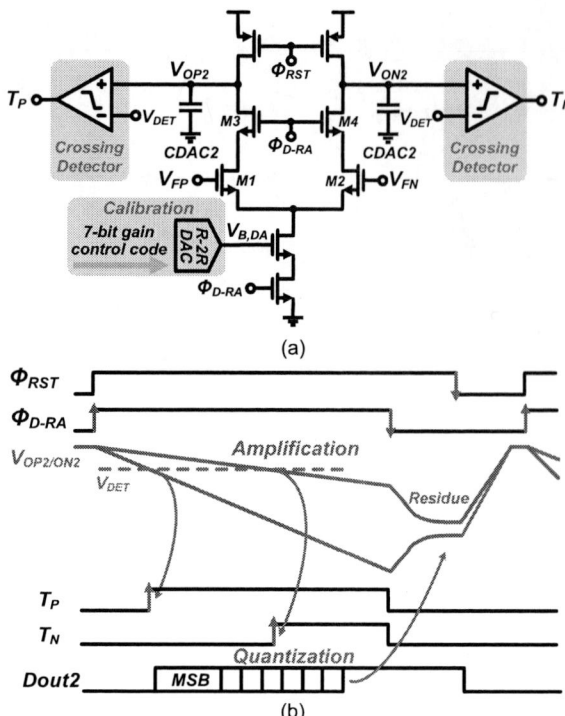

(a)

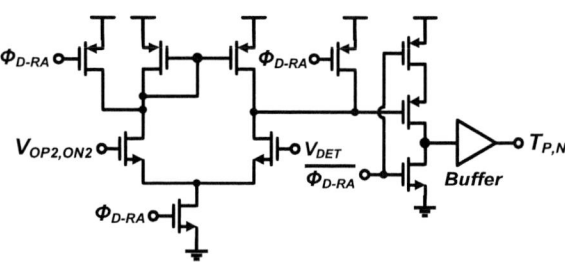

(b)

Fig. 2 (a) Circuit and (b) timing diagram of the proposed D-RA.

Fig. 3 Schematic of the crossing detector.

V_{OP2} and V_{ON2} fall below the detector threshold (V_{DET}), the CD generates timing signals T_P and T_N, which are quantized by a 4-bit TD quantizer. When Φ_{D-RA} returns low, M_1 and M_2 stop discharging, ending the amplification phase. The resulting 4-bit code then controls CDAC2 to generate the residue voltage.

Based on this operating principle, the voltage gain A_V of the D-RA can be expressed as:

$$A_V = \frac{G_m}{C_2} \cdot T_{amp} \approx \frac{g_{m1} + g_{m2}}{2} \cdot \frac{1}{C_2} \cdot T_{amp} \quad (1)$$

where gm, C_2, and T_{amp} are the trans-conductance of the input transistors M_1 and M_2, the load capacitance, and the amplification time, respectively. The output time of the D-RA (T_{out}) can be expressed as:

$$T_{out} = \frac{(VDD - V_{DET}) \cdot C_2}{I_C + g_{m1} \cdot V_{FP}} - \frac{(VDD - V_{DET}) \cdot C_2}{I_C + g_{m2} \cdot V_{FN}} \quad (2)$$

where I_C is the biasing current. As shown in equations (1) and (2), both the gain of voltage amplification and the output

timing of voltage-to-time conversion are closely related to g_{m1} and g_{m2}. Since g_{m1} and g_{m2} vary with the input signal, this introduces certain nonlinearity. However, the first stage performs a 4-bit quantization, which effectively compresses the input signal swing and mitigates the nonlinearity caused by transconductance variation. As illustrated in Fig. 4(a), under various process, voltage and temperature (PVT) conditions, the total harmonic distortion (THD) of the voltage amplification remains better than −45 dB, and that of the V-T conversion is below −30 dB, satisfying the linearity requirements of the subsequent quantizer. Fig. 4(b) presents the variations of voltage amplification gain and V-T conversion gain under different PVT conditions. The voltage gain fluctuates from −16% to +22%, while the V-T conversion gain varies from −29% to +43%, indicating both are highly sensitive to PVT variations. To compensate for these effects, a tunable bias voltage $V_{B,DA}$ is introduced in the design. It is configured using a 7-bit gain control code with a step size of 3 mV, enabling dynamic gain adjustment and compensation across typical PVT variation ranges. Moreover, the delay elements in the second-stage TDC are programmable to match the output time swing of the D-RA [10].

The noise of the proposed D-RA is analyzed below. During both voltage amplification and V-T conversion, two primary noise sources are identified: dynamic integration noise and CD noise. The dynamic integration noise is primarily introduced by the differential input transistors, while the contribution from the cascode transistors is negligible. Based on the noise analysis framework presented in [11], the input-referred noise (IRN) of the proposed D-RA structure can be expressed as:

$$\sigma_n^2 \approx \frac{2kT\gamma(g_{m1} + g_{m2})}{A_V^2 C_2^2} \cdot T_{amp} + \frac{2\sigma_{n,CD}^2}{A_V^2} \quad (3)$$

where $\sigma_{n,CD}^2$ is the noise of CD. Compared to the V–T–V architecture in [7], the proposed D-RA architecture offers significantly improved noise performance. First, it eliminates the time-to-voltage conversion stage, thereby avoiding the additional noise introduced by that process. Second, unlike [7], where the CD is directly connected to the input and offers no noise attenuation, this work places the CD at the output, where its noise is attenuated by the integrator gain when referred to the input. Simulation results show that the IRN of the proposed D-RA is only 30% of that in the V–T–V architecture.

B. DFVF with common-mode voltage reconstruction

As shown in Fig. 5, the DFVF comprises two switch-controlled flipped voltage followers. DFVF1 is taken as an example. Controlled by Φ_{S2}, DFVF1 is activated during the CDAC1 settling phase and disabled afterward. Compared to traditional source followers, the DFVF offers lower output impedance, enabling faster voltage settling. Simulations confirm that DFVF reliably achieves voltage tracking within the 55 ps design target across all PVT variations. Additionally, the DFVF maintains robust gain performance under PVT variations, eliminating the need for gain calibration [12]. The designed gain for the DFVF in this work is 0.8.

979-8-3315-8850-2/25 $31.00 © 2025 IEEE

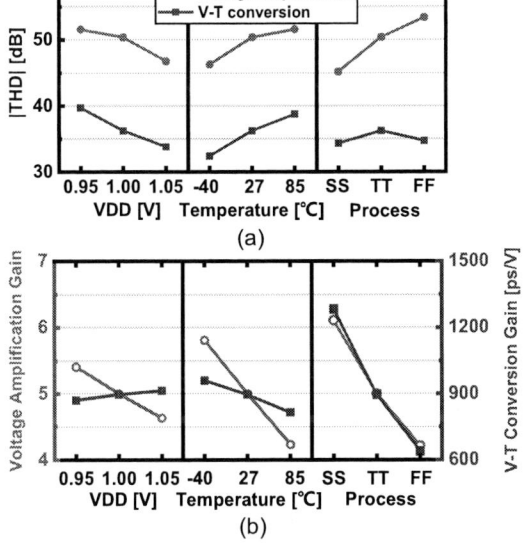

(a)

(b)

Fig. 4 (a) Linearity and (b) gain variations of voltage amplification and voltage-to-time conversion under PVT conditions.

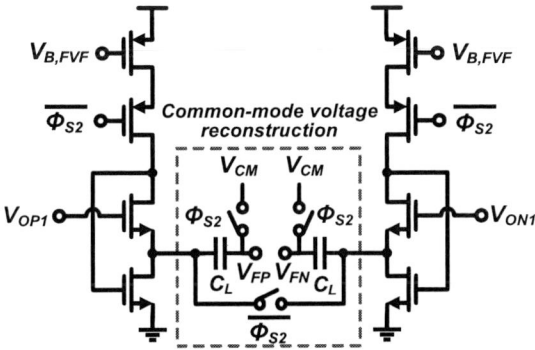

Fig. 5 Schematic of the DFVF1 with common-mode voltage reconstruction.

Due to the reduced input common-mode voltage caused by the DFVF, the D-RA may fail to operate at an appropriate common-mode level. To address this issue, a common-mode voltage reconstruction circuit is inserted after the DFVF to stabilize the input common-mode voltage at the D-RA. The timing of operation is as follows: when Φ_{S2} is high, the DFVF is enabled, passing the previous stage's voltage to the top plate of the load capacitor C_L, while the bottom plate is connected to the predefined common-mode voltage V_{CM}. When Φ_{S2} goes low, the DFVF is disabled, and the top plate of C_L is shorted, transferring its differential voltage to the bottom plate. Because the bottom plate is fixed at V_{CM}, the output nodes V_{FP} and V_{FN} retain a common-mode voltage of V_{CM}, ensuring a stable input for the D-RA.

C. Time-domain quantizer

In the first and third TD quantizers, we adopt a classic charging-based voltage-to-time converter (VTC) architecture. The charging-based VTC is widely used due to its excellent linearity [13], [14]. The first-stage VTC employs the common-

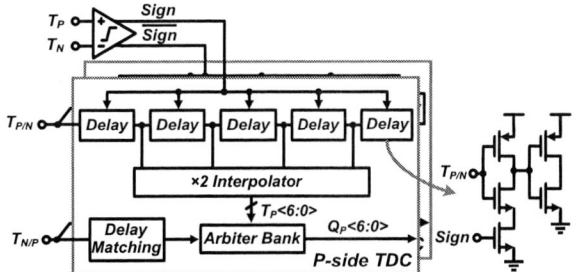

Fig. 6 Block diagram of the pseduo-differential flash TDCs.

TABLE I
NOISE POWER BREAKDOWN OF THE PROPOSED ADC

Noise Source	Power (nV^2)	Percentage
Sampling Noise	32.3	15.5 %
Quantization Noise	63.6	30.7 %
DFVF Noise	42.0	20.3 %
D-RA Noise	56.8	27.3 %
3rd Stage Noise	12.8	6.2 %
Total Noise	207.5	100 %
SNR	57.8 dB	

mode shifting technique proposed in [14] to establish an appropriate operating common-mode voltage.

This work employs a 3-bit pseudo-differential flash TDC with 1-bit folding, as shown in Fig. 6. For each side of the TDC, 2× time-domain interpolation is used to halve the delay time (τ) of each delay cell [7], [8]. The time reference time ($\tau/2$) for the first and third stages of the TDC are set to 5 ps. Due to the limited output swing of the second-stage D-RA, its time reference step is set to 3.5 ps. The arbiters and time comparators adopt the same architecture and implementation as described in [8]. The simulated offset standard deviations of the time comparators is 312 fs, which is much smaller than the $\tau/2$, eliminating the need for offset calibration. During flash TDC operation, depending on the polarity of the VTC output, either the P-side or N-side delay chain is activated to perform time-edge comparison and generate a quantized thermometer code, while the other chain is disabled to save power. For example, if T_N arrives first, the N-side is enabled and the P-side is shut off, with the MSB set low to block signal propagation through the P-side delay chain. A delay-matching cell is inserted between the VTC output and the time comparator input to correct timing skew between T_P and T_N.

D. Noise Power Breakdown

Table I provides a breakdown of the simulated noise power for the pipelined ADC with a Nyquist input frequency. The FVF and D-RA, responsible for residue transfer and amplification, contribute 20.3% and 27.3% of the total noise, respectively, while the third-stage TD quantizer contributes only 6.2%. The total IRN is 207.5-nV^2, resulting in a 57.8-dB SNR.

979-8-3315-8850-2/25 $31.00 © 2025 IEEE

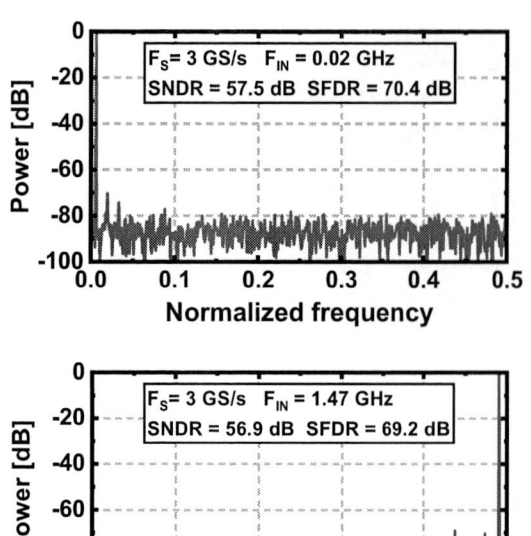

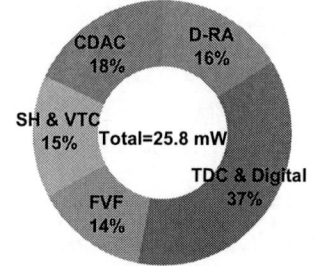

Fig. 8 Power breakdown.

Fig. 7 Simulated 1024-point FFT spectra with LF (0.02 GHz) and Nyquist (1.47 GHz) input sampling at 3 GS/s.

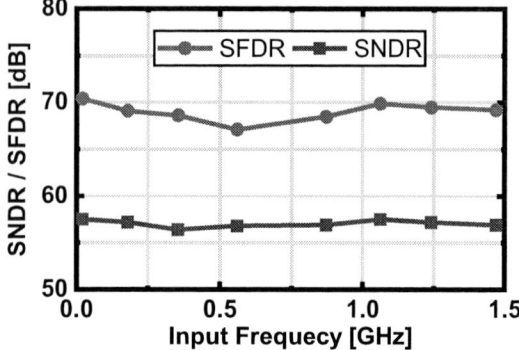

Fig. 9 Simulated SNDR and SFDR versus input frequency.

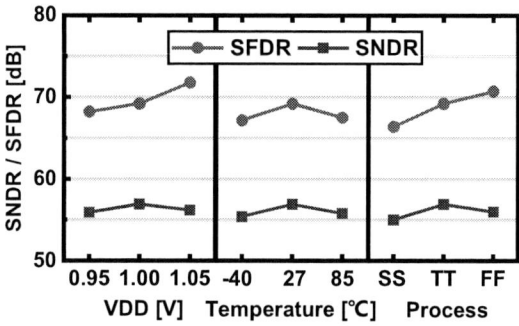

Fig. 10 Simulated SNDR and SFDR variation versus PVT variation.

IV. SIMULATION RESULTS

The proposed pipelined ADC with parallel TD quantization is simulated with a 28-nm CMOS process. The ADC is exhibited at 3 GS/s with 1-V supply voltage and 1-$V_{pp,diff}$ full scale range. The simulated 1024-point fast Fourier transform (FFT) spectra at 3 GS/s for input frequencies of 0.02 GHz and 1.47 GHz, including transient noise, are shown in Fig. 7. At 0.02 GHz, the simulated SNDR is 57.5 dB, and the spurious-free dynamic range (SFDR) is 70.4 dB. When the input frequency increases to 1.47 GHz, the SNDR and SFDR degrade slightly to 56.9 dB and 69.2 dB, respectively.

Fig. 8 shows the power consumption breakdown of the 1 V-supplied ADC operating at 3 GS/s. The total power consumption is 25.8 mW, with 37% consumed by the TDC and digital circuits, 16% by the D-RA, 14% by the FVFs, 15% by the VTCs including bootstrap switches, and 18% by the CDAC. Fig. 9 illustrates the SNDR and SFDR performance versus input frequency at a 3 GS/s sampling rate. Across the first Nyquist zone, the SNDR consistently exceeds 56.4 dB, and the SFDR remains above 67.1 dB.

As discussed in Section III, the D-RA and VTC feature tunable gain, enabling foreground calibration for inter-stage gain alignment and TDC range matching to enhance PVT robustness. Figure. 10 shows the ADC performance across different PVT conditions with gain calibration only. The results demonstrate that SNDR remains above 55 dB and SFDR above 66.4 dB across all variations.

Table 2 summarizes the performance of the proposed ADC and compares it with state-of-the-art designs featuring similar

resolution and conversion rate. The presented ADC achieves a Walden figure-of-merit (FoM$_W$) of 15.0 fJ/conversion-step. Thanks to the proposed DFVF-based residue transfer technique and the parallel quantization and amplification enabled by the D-RA, the presented ADC achieves the highest single-channel sampling rate among the compared designs.

V. CONCLUSION

This paper presents a 10-bit voltage/time pipelined ADC, featuring a stage with parallel amplification and quantization. The parallel structure relaxes the timing constraints on residue amplification, ensuring the accuracy of the D-RA. A DFVF-based residue transfer scheme further improves speed by eliminating additional timing overhead. With the abovementioned techniques, the proposed single-channel 10-bit pipelined ADC achieves a sampling rate of 3 GS/s in 28-nm CMOS process, while maintaining competitive performance and power effi-

TABLE II
PERFORMANCE SUMMARY AND STATE-OF-THE-ART ADC COMPARISON.

	This Work[a]	JSSC 2023[b] [7]	JSSC 2023[b] [8]	ISSCC 2017[b] [3]	JSSC 2016[b] [4]	ISCAS 2023[a] [15]
Architecture	**TD Pipeline**	TD Pipeline	TI SAR	Pipelined SAR	Pipelined	Pipelined SAR
Process [nm]	**28**	28	28	14	65	40
Supply [V]	**1**	0.9	0.9	0.95	1.2	1
Resolution	**10**	10	9	10	9	10
Sampling rate [GS/s]	**3**	2.6	2.8	1.5	1.8	1.25
Input swing [$V_{pp,diff}$]	**1**	0.8	0.8	0.65	1.6	1.6
SNDR [dB] @Nyq.	**56.9**	51.4	51.8	50.1	47	59
SFDR [dB] @Nyq.	**69.2**	71.0	72.4	58.4	57	65
Power [mW]	**25.8**	13.9	18.05	6.92	44	5.6
FoMw @Nyq. [fJ/conv.·s]	**15.0**	17.6	20.3	17.7	134	6.2

[a] Simulated Results [b] Measured Results.

ciency. Compared with the current fastest TDC-assisted 10-bit ADC, the proposed ADC achieves a 14% improvement in speed.

REFERENCES

[1] D. Li, X. Zhao, Y. Shen, S. Liu, and Z. Zhu, "A 7-bit 3.8-GS/s 2-waytime-interleaved 4-bit/cycle SAR ADC 16× time-domain interpolationin 28-nm CMOS," *IEEE Trans. Circuits Syst. I, Reg. Papers*, vol. 70, no. 9, pp. 3557–3566, Sep. 2023.

[2] A. T. Ramkaj et al., "A 5-GS/s 158.6-mW 9.4-ENOB passive-sampling time-interleaved three-stage pipelined-SAR ADC with Analog–Digital corrections in 28-nm CMOS," *IEEE J. Solid-State Circuits*, vol. 55, no. 6, pp. 1553–1564, Jun. 2020.

[3] L. Kull et al., "A 10b 1.5GS/s pipelined-SAR ADC with background second-stage common-mode regulation and offset calibration in 14 nm CMOS FinFET," in *IEEE Int. Solid-State Circuits Conf. (ISSCC) Dig. Tech. Papers*, Feb. 2017, pp. 474–475.

[4] L. Yu, M. Miyahara, and A. Matsuzawa, "A 9-bit 1.8 GS/s 44 mW pipelined ADC using linearized open-loop amplifiers," *IEEE J. Solid State Circuits*, vol. 51, no. 10, pp. 2210–2221, Oct. 2016.

[5] H. Huang, L. Du, and Y. Chiu, "A 1.2-GS/s 8-bit two-step SAR ADCin 65-nm CMOS with passive residue transfer," *IEEE J. Solid-State Circuits*, vol. 52, no. 6, pp. 1551–1562, Jun. 2017.

[6] Z. Zheng et al., "A 3.3-GS/s 6-b fully dynamic pipelined ADC withlinearized dynamic amplifier," *IEEE J. Solid-State Circuits*, vol. 57, no. 6, pp. 1673–1683, Jun. 2022.

[7] J, Hao et al., "An Intrinsically PVT Robust 10-bit 2.6-GS/s Dynamic Pipelined ADC With Dual-Path Time-Assisted Residue Generation Scheme," *IEEE J. Solid-State Circuits*, early access, Dec. 2024.

[8] H. Zhao, M. Zhang, Y. Zhu, R. P. Martins and C. -H. Chan, "A 52.5-dB 2× Time-Interleaved 2.8-GS/s SAR ADC With 5-bit/Cycle Time-Domain Quantization and a Compact Signal DAC," *IEEE J. Solid-State Circuits*, vol. 58, no. 12, pp. 3586-3597, Dec. 2023.

[9] B. P. Ginsburg and A. P. Chandrakasan, "500-MS/s 5-bit ADC in 65-nm CMOS With Split Capacitor Array DAC," *IEEE J. Solid-State Circuits*, vol. 42, no. 4, pp. 739-747, April 2007.

[10] Z. Su, H. Wang, H. Zhao, Z. Chen, Y. Wang, and F. F. Dai, "A 280MS/s 12b SAR-assisted hybrid ADC with time domain sub-range quantizer in 45nm CMOS," in *Proc. IEEE Custom Integr. Circuits Conf. (CICC)*, Apr. 2019, pp. 1-4.

[11] T. Sepke, P. Holloway, C. G. Sodini and H. -S. Lee, "Noise Analysis for Comparator-Based Circuits," *IEEE Trans. Circuits Syst. I, Reg. Papers*, vol. 56, no. 3, pp. 541-553, March 2009.

[12] J. Liu, D. Li, Y. Zhong, X. Tang and N. Sun, "27.1 A 250kHz-BW 93dB-SNDR 4th-Order Noise-Shaping SAR Using Capacitor Stacking and Dynamic Buffering," in *IEEE Int. Solid-State Circuits Conf. (ISSCC) Dig. Tech. Papers*, Feb. 2021, pp. 369-371.

[13] Q. Chen, C. C. Boon, Q. Liu and Y. Liang, "A Single-Channel Voltage-Scalable 8-GS/s 8-b > 37.5-dB SNDR Time-Domain ADC With Asynchronous Pipeline Successive Approximation in 28-nm CMOS," *IEEE J. Solid-State Circuits*, vol. 58, no. 6, pp. 1610-1622, June 2023.

[14] Y. Cao, M. Zhang, Y. Zhu, R. P. Martins and C. -H. Chan, "A Single-Channel 12-b 2-GS/s PVT-Robust Pipelined ADC With Sturdy Ring Amplifier and Time-Domain Quantizer," *IEEE J. Solid-State Circuits*, vol. 60, no. 2, pp. 443-455, Feb. 2025.

[15] X. Wang et al., "A 10b 1.25GS/s Residue Post-Amplified Pipelined-SAR ADC with Supply-and-Temperature Stabilized Open-Loop Residue Amplifier," in *IEEE Int. Symposium on Circuits and Systems (ISCAS)*, May. 2023, pp. 1-5.

979-8-3315-8850-2/25 $31.00 © 2025 IEEE

2025 The 10th International Conference on Integrated Circuits and Microsystems

Design of High Speed and Energy-Efficient ADC Based on Dynamic Bandwidth Ring Amplifier

Xin Yang
Beijing Microelectronic Technology Institude
Beijing, China
yangx_ele@163.com

Qingyuan Liu
Beijing Microelectronic Technology Institude
Beijing, China
315057706@qq.com

Xiaobo Chen
Beijing Microelectronic Technology Institude
Beijing, China
3475530971@qq.com

Dan Sun
Beijing Microelectronic Technology Institude
Beijing, China
18811490268@163.com

Bin Zheng
Beijing Microelectronic Technology Institude
Beijing, China
zhengsbin@163.com

Tieliang Zhang
Beijing Microelectronic Technology Institude
Beijing, China
ztl_hit@163.com

Long Yang
Beijing Microelectronic Technology Institude
Beijing, China
yanglong_hd@163.com

Abstract—A 230-MSPS 14-bit pipeline successive-approximation register (Pipelined-SAR) analog-to-digital converter (ADC) is described in this paper. To address the critical performance and power-efficiency bottleneck of the residue amplifier (RA) in conventional designs, a novel ring amplifier featuring dynamic bandwidth and enhanced PVT stability is proposed. Additionally, low-power high-speed asynchronous SAR logic and a fast dynamic comparatoris employed in the sub-SAR ADCs. The ADC achieves greater than 70-dB signal-to-noise-and-distortion ratio (SNDR) across process corners (-55°C to 125°C) at 230 MSPS. At room temperature (TT corner, 35°C), the ADC achieves 72.43 dB SNDR and 82 dB spurious-free dynamic range (SFDR) for a Nyquist input, consuming 3.849 mW, resulting in a Schreier figure-of-merit (FoMs) of 177.2 dB and a Walden FoM (FoMw) of 4.89 fJ/conversion-step.

Keywords—*Pipelined-SAR ADC, ring amplifier, PVT stability, energy-efficient*

I. INTRODUCTION

High-speed, high-resolution analog-to-digital converters are crucial for modern digital signal processing and communication systems. ADC with sampling rates exceeding 100 MSPS find extensive applications in RF communications, radar detection, and 5G/6G infrastructure. The pipelined-SAR architecture has emerged as the dominant solution for such specifications (12-14 bits, >100 MSPS) due to its superior energy efficiency compared to traditional pipeline or flash ADCs [1], [2], [3], [4]. This architecture leverages low-power SAR sub-ADCs and reduces the number of energy-hungry residue amplifier stages, often achieving 14-bit precision with just two stages.

Despite its advantages, the pipelined-SAR ADC faces significant challenges: 1) Designing a high-gain, high-bandwidth, low-power RA under low supply voltages in advanced nodes is difficult with conventional operational transconductance amplifiers (OTAs); 2) Achieving sufficient

sub-ADC conversion speed for the target sample rate is non-trivial.

This paper presents a 230-MSPS 14-bit pipelined-SAR ADC overcoming these limitations. The core innovations are a PVT-robust ringamp based RA and high-speed sub-ADC techniques. Key features include a bias-copy and clamp-loop stabilized second stage, a floating inverter amplifier output stage with bias enhancement for stable common-mode output , and correlated level shifting for improved equivalent gain and linearity.

II. PROPOSED ARCHITECTURE AND KEY TECHNIQUES

A. Overall ADC Structure

The proposed 14-bit ADC, shown in Fig.1, employs a two-stage pipelined-SAR architecture: a 6-bit front-end SAR sub-ADC, a 32timesgain residue amplifier, and a 9-bit back-end SAR sub-ADC. Redundancy (8 LSB in the first stage, 10 LSB in the second stage) is incorporated to tolerate comparator noise and reference settling errors. A 1-bit inter-stage redundancy prevents RA output saturation. The front-end CDAC uses 12fF unit capacitors for kT/C noise. The backend CDAC uses smaller 1fF units to reduce RA load and area; its noise is attenuated by the RA gain.

B. PVT-Stable Ring Amplifier with Dynamic Bandwidth

The output stage of the ring amplifier lacks a defined current bias and common-mode voltage. Consequently, PVT variations will cause fluctuations in the output common-mode voltage, which adversely affects the normal operation of the Ring[5], [6], [7]. To overcome PVT sensitivity, we designed a ring amplifier with a bias-copy and clamp-loop stabilized second stage and an floating inverter amplifier (FIA) output stage.

First stage amplifier A_1 employs a self-zeroing amplifier stage with local common-mode feedback (CMFB), shown in Fig.2. During auto-zeroing phase, the input and output of inverter are shorted together, storing its balanced point voltage

979-8-3315-8850-2/25 $31.00 © 2025 IEEE

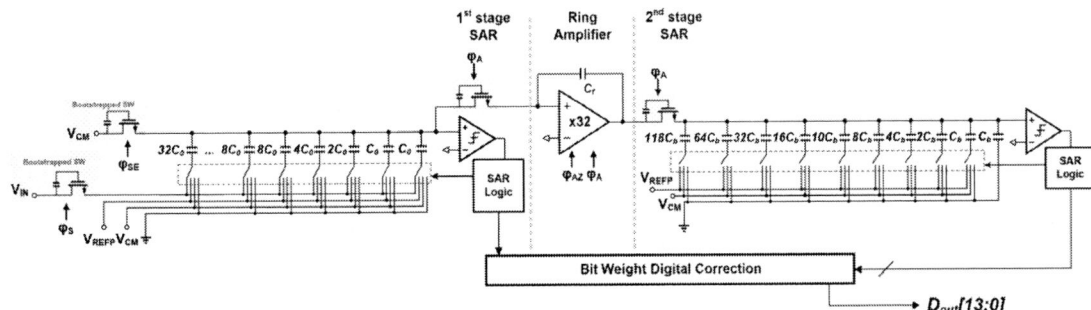

Fig. 1. Architecture of the proposed 14-bit pipelined-SAR ADC.

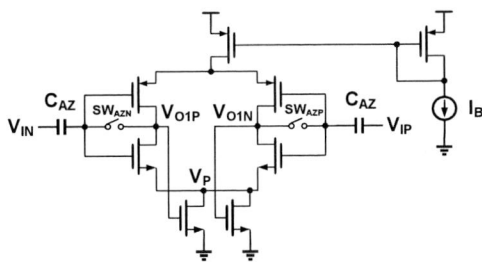

Fig. 2. Proposed first stage circuit of the ring amplifier.

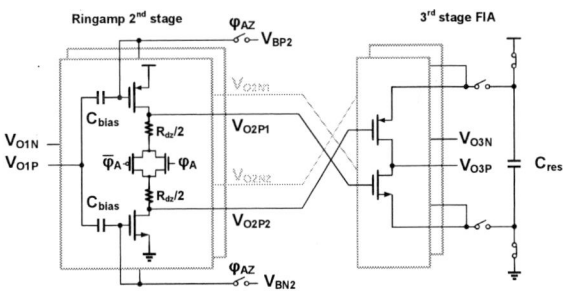

Fig. 4. Third-stage circuit structure of Ring Amplifier: FIA with bias enhancement.

and offset voltage on capacitors C_{AZ}, canceling offset during amplification. A simple local CMFB loop stabilizes the output common-mode voltage without impacting normal amplification speed.

Second stage amplifier A_2 utilizes a bias-copy branch and clamp loop for PVT stability, shown in Fig.3. A copy branch of A_2 and an auxiliary error amplifier form a clamp loop generating gate bias voltages for the NMOS and PMOS transistors. The error amplifier's high gain clamps nodes V_{dzP}/V_{dzN} near reference voltages V_{CM1}/V_{CM2}. The current flowing through the copy loop is completely determined by the values of V_{CM1}, V_{CM2} and R_{dz}. The error amplifier output then provides the required bias for A_2. During reset phase (aligned with A_1 auto-zero), capacitor C_{bias} stores the voltage difference between A_1's balanced point and the generated A_2 bias voltage.

The noise contribution of the auxiliary error amplifier is very small and mainly low-frequency noise. It does not directly enter the signal path but indirectly introduces minor gain error and offsets by modulating the bias voltage of A_2. With careful design, its negative impact is far less than the significant benefits it brings to PVT stability.

Output stage employs a FIA with bias enhancement, shown in Fig.4. The FIA is a discrete-time amplifier comprising a differential inverter pair and a supply capacitor CRES. During reset phase, the two ends of CRES are precharged to VDD and GND. During amplification phase, the inverter is connected to CRES for power. As the common-mode input and output currents from CRES must be equal (IO,CM=0), the conduction strengths of the PMOS and NMOS transistors are inherently balanced. This provides a stable output common-mode voltage without explicit CMFB, immune to PVT and input common-mode variations. Bias enhancement, implemented by cross-coupling the second stage outputs to the third stage PMOS/NMOS gate inputs, increases the overdrive voltage of the third stage, boosting the slew rate. Fig.5 shows significant speed improvement under SS corner, -40°C conditions.

The complete ring amplifier structure is shown in Fig.6. To further improve equivalent open-loop gain and PVT stability, correlated level shifting is integrated. During the coarse amplification phase, the ring amplifier performs coarse

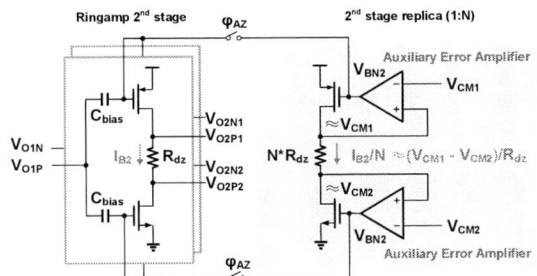

Fig. 3. PVT-stable second stage using bias copy and clamp loop.

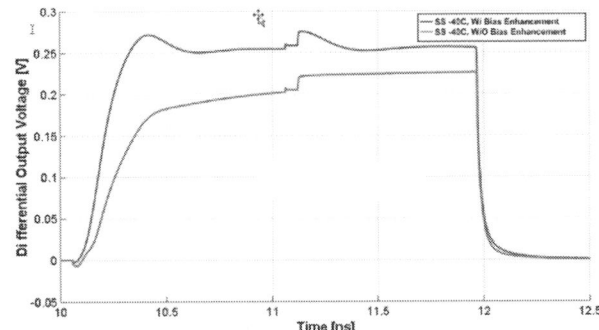

Fig. 5. Comparison of ring amplifier output waveforms with/without bias enhancement (SS corner, -40°C).

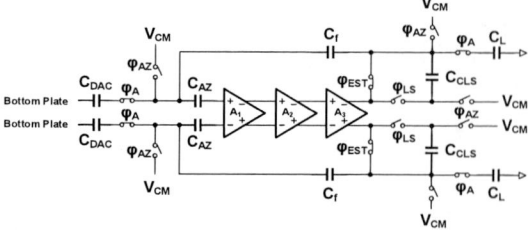

Fig. 6. Overall structure of the proposed ring amplifier.

amplification on the input signal and samples the coarsely amplified result V_o onto C_{CLS}. In the subsequent level-shifting phase, through switch reconfiguration, C_{CLS} is reversely connected in series between the load capacitor (the second stage sampling capacitor) and the ring amplifier's output. This effectively achieves a passive, coarse amplification of the input signal at the ring amplifier's output node. Consequently, after level shifting, the actual output voltage of the ring amplifier equals the ideal output voltage minus the magnitude of V_o. Under the same ring amplifier open-loop gain, a smaller output swing implies a smaller input residue voltage. This smaller residue means the charge on the preceding sampling capacitor is more completely transferred to the feedback capacitor, thereby achieving a larger equivalent open-loop gain and reduced inter-stage gain error. Furthermore, because the residual amplifier's final differential output voltage is shifted close to zero, CLS enables a larger equivalent output swing and higher linearity. Fig.7 illustrates the CLS principle.

Post-layout simulations confirm that the proposed ring amplifier achieves >80 dB equivalent gain within 2ns amplification time across all process corners and temperatures from -55°C to 125°C.

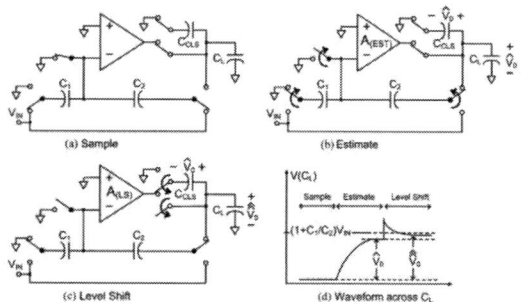

Fig. 7. Principle of Correlated Level Shifting.

C. High-speed Sub-ADC Techniques

To achieve high-speed sub-ADC conversion, the design employs asynchronous SAR logic based on transparent latches shown in Fig.8. Replacing D-flip-flop based sequencers in traditional asynchronous SAR logic with dynamic transparent latches reduces digital power consumption by approximately 50% [8], [9].

Furthermore, the logic minimizes the critical path delay from the SAR comparator output to the CDAC switch control. The path involves only one PMOS transmission gate delay, maximizing the time allocated for CDAC settling and relaxing the speed requirement for the CDAC reference driver.

The comparator is critical for sub-SAR ADC speed and noise performance. Traditional StrongARM comparators (Fig.9a), common in medium-speed SAR ADCs, suffer from significant delay before regeneration starts and slower regeneration speed due to source degeneration from the input pair during latching.

This design employs a high-speed dynamic comparator [10], shown in Fig.9b. During reset phase, nodes VON and VOP are precharged to VDD by M7 and M7', and nodes X and Y are precharged to GND by M4 and M4'. When CLK rises to VDD, the comparator enters the amplification phase. Charge sharing between capacitors CON, COP, CX, CY and discharge through M4, M4', M2, M2' cause VOP/VON to drop rapidly. Once VOP and VON fall below VDD-VTH5, transistors M5 and M5' turn on, initiating the positive feedback regeneration phase to resolve the comparison.

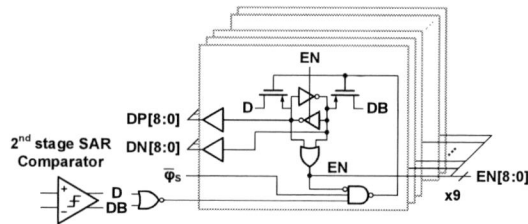

Fig. 8. High-speed asynchronous SAR logic circuit.

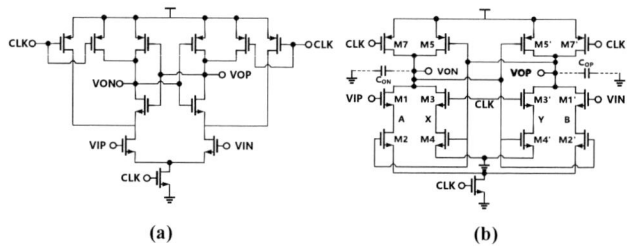

Fig. 9. Comparators: (a) Traditional StrongARM comparator, (b) Proposed high-speed dynamic comparator.

Compared to StrongARM, the proposed comparator offers significantly faster decision times, making it suitable for high-speed sub-SAR ADCs in pipelined-SAR architectures. Fig.10 compares the outputs of both comparators under worst case PVT (SS, -40°C, 450mV CM, 50µV DM input), confirming the superior speed of the proposed design.

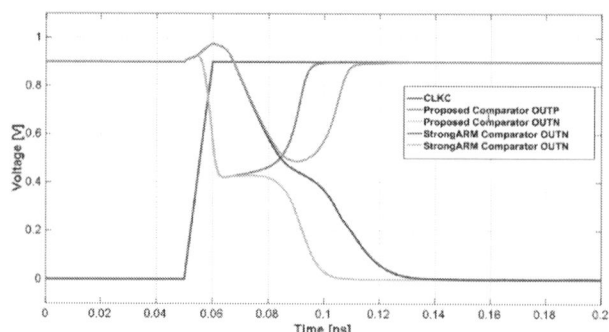

Fig. 10. Output waveform comparison of StrongARM and proposed comparator (SS corner, -40°C).

979-8-3315-8850-2/25 $31.00 © 2025 IEEE

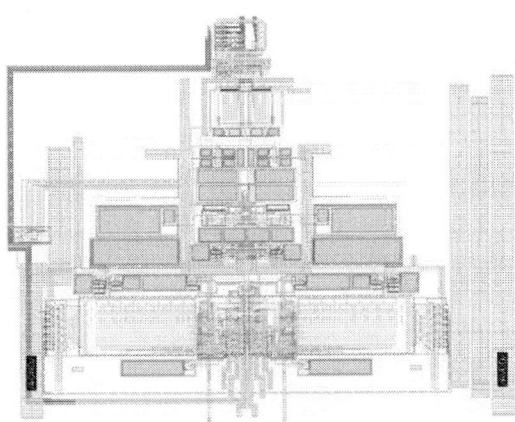

Fig. 11. ADC core layout

III. POST-LAYOUT SIMULATION RESULTS

The ADC core (Fig.11) was implemented in 28-nm CMOS. Post-layout simulations were performed at 230 MSPS with a 1.8-V_{PP} differential input at Nyquist frequency (109.61 MHz) under a 0.95-V supply.

The nonlinearity of the ADC mainly stems from capacitor mismatch, non-ideal residual amplifier, comparator error and non-ideal switch. Capacitor mismatch is suppressed by adopting large-sized front-end capacitors and combining with redundant bit design. For the gain error and swing limitation of the residual amplifier, a dynamic bandwidth loop amplifier combined with correlated level shifting technology is innovatively adopted, significantly improving the equivalent gain and linearity. The offset and noise introduced by the comparator are fault-tolerant corrected through self-resetting technology and redundant bits. These measures systematically suppress various nonlinearities, and ultimately the ADC achieves a high linearity performance of over 70 dB SNDR under all PVT conditions.

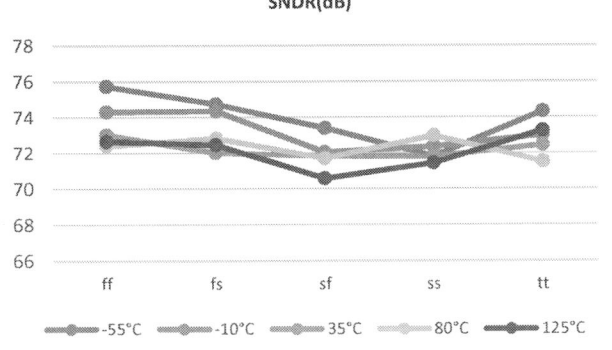

Fig. 12. SNDR across process corners and temperature.

Fig.12 summarizes SNDR and power across process corners (FF, FS, SF, SS, TT) and temperature (-55°C to 125°C). The ADC achieves >70 dB SNDR across all conditions. Performance at TT corner, 35°C is detailed in Table I. The FFT spectrum result of the ADC outputs is shown in Fig.13. The SNDR is 72.43 dB, and the SFDR is 82.6 dB.

TABLE I.   PERFORMANCE SUMMARY(TT CORNER, 35°C)

Parameter	Value
Technology	28nm
Architecture	pipe SAR
Supply Voltage	0.95V
Sampling Rate	230MSPS
ENOB	11.74bits
SNDR@Nyq	72.43dB
SFDR@Nyq	>82dB
Power	3.849mW
FoM$_W$[a]	4.89fJ/conv-step
FoM$_S$[b]	177.2dB

a. $\text{FoM}_W = \frac{\text{Power}}{2^{\text{ENOB}} \times \text{fs}}$

b. $\text{FoM}_S = \text{SNDR} + 10\log_{10}\text{BW/Power}$

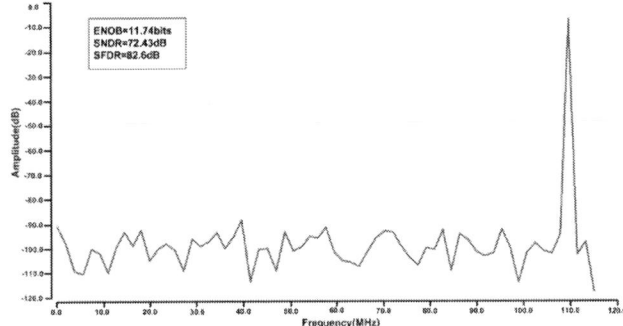

Fig. 13. Post-layout simulation result: Spectrum of the ADC outputs.

IV. CONCLUSION

This paper presents a 230-MSPS, 14-bit pipelined-SAR ADC implemented in a 28-nm CMOS process, achieving >70dB SNDR across process corners and temperatures (-55°C to 125°C). Leveraging the inherent energy efficiency of the pipelined-SAR architecture for high-speed, high-resolution ADCs, the design consumes only 3.849mW at room temperature. The key innovation addresses the conventional RA bottleneck: a novel PVT-stable ring amplifier featuring a bias-copy and clamp-loop stabilized second stage and an FIA output stage with bias enhancement. This structure achieves robust biasing and stable output common-mode without global CMFB. CLS is integrated to enhance equivalent gain and linearity. High-speed techniques, including low-power asynchronous SAR logic and a fast dynamic comparator, are employed in the sub-SAR ADCs to boost overall speed and efficiency. Post-layout simulations demonstrate state-of-the-art FoMs of 4.89 fJ/conv-step (FoM$_w$) and 177.2 dB (FoM$_S$) for Nyquist input, positioning this design favorably among ADCs with comparable specifications.

REFERENCES

[1] C. C. Lee and M. P. Flynn, "A SAR-Assisted Two-Stage Pipeline ADC," IEEE Journal of Solid-State Circuits,2011,46(4):859‐869.

[2] E. Martens, B. Hershberg, and J. Craninckx, "A 69-dB SNDR 300-MS/s Two-Time Interleaved Pipelined SAR ADC in 16-nm CMOS FinFET

979-8-3315-8850-2/25 $31.00 © 2025 IEEE

With Capacitive Reference Stabilization," IEEE Journal of Solid-State Circuits, 2018,53(4):1161‑1171.

[3] W. Jiang, Y. Zhu, M. Zhang, C.-H. Chan, and R. P. Martins, "A Temperature-Stabilized Single-Channel 1-GS/s 60-dB SNDR SAR-Assisted Pipelined ADC With Dynamic Gm-R-Based Amplifier," IEEE Journal of Solid-State Circuits, 2020,55(2):322‑332.

[4] H. Fan, "A 12-bit 100 MS/s pipelined SAR ADC with addition-only digital error correction," ANALOG INTEGRATED CIRCUITS AND SIGNAL PROCESSING, 2014,88(1):325-339.

[5] Y. Lim and M. P. Flynn, "A 100 MS/s, 10.5 Bit, 2.46 mW Comparator-Less Pipeline ADC Using Self-Biased Ring Amplifiers," IEEE Journal of Solid-State Circuits, 2015,50(10):2331‑2341.

[6] M. Zhan, L. Jie, X. Tang, and N. Sun, "A 0.004mm2 200MS/S Pipelined SAR ADC with kT/C Noise Cancellation and Robust Ring-Amp," in 2022 IEEE International Solid- State Circuits Conference (ISSCC), 2022:164‑166.

[7] B. P. Ginsburg and A. P. Chandrakasan, "Dual Time-Interleaved Successive Approximation Register ADCs for an Ultra-Wideband Receiver," in IEEE Journal of Solid-State Circuits, 2007,42(2): 247-257.

[8] M. Zhan, L. Jie, and N. Sun, "17.5 A 10mW 10-ENOB 1GS/s Ring-Amp-Based Pipelined TI-SAR ADC with Split MDAC and Switched Reference Decoupling Capacitor," in 2023 IEEE International Solid-State Circuits Conference (ISSCC), 2023:272-274.

[9] B. Hershberg, S. Weaver, K. Sobue, S. Takeuchi, K. Hamashita, and U.-K. Moon, "Ring Amplifiers for Switched Capacitor Circuits," IEEE Journal of Solid-State Circuits, 2012, 47(12):2928‑2942.

[10] X. Lin, M. Megahed, and T. Anand, "A Single-Clock-Phase Sense Amplifier Architecture with 9x Smaller Clock-to-Q Delay Compared to the StrongARM & 6.3dB Lower Noise Compared to Double-Tail," in 2022 IEEE Symposium on VLSI Technology and Circuits (VLSI Technology and Circuits),2022:188‑189.

2025 The 10th International Conference on Integrated Circuits and Microsystems

A 12-b 1-GS/s Pipelined-SAR ADC With a Hybrid Parallel Timing Scheme and PVT-Robust Ring-Amp

Junhui Guo
School of the Sun Yat-sen University of Microelectronics Science and Technology
ZhuHai, China
guojh75@mail2.sysu.edu.cn

Yunchu Li*
School of the Sun Yat-sen University of Microelectronics Science and Technology
ZhuHai, China
liyunchu@mail.sysu.edu.cn

Abstract—This paper presents a 12-b 1GS/s single-channel pipelined successive approximation register (SAR) analog-to-digital converter (ADC), designed in a 28-nm CMOS process. In this paper, a hybrid parallel timing scheme is introduced. Firstly, the passive residue transfer technique is employed to enable parallel operation of the RA and first stage SAR conversions, thereby lessening the timing burden of the 1st-satge. In addition, the 2nd-stage SAR ADC employs partially parallel conversion to alleviate the timing constraints due to the additional CDAC reset clock and charge-sharing clock introduced by the passive residue transfer. A fully differential ring amplifier is adopted. To improve the process, voltage, and temperature (PVT) robustness of the ring amplifier, a Class-AB biasing scheme is used. The results reveal that the 12-b 1-GS/s ADC achieves 64.8 dB SNDR and 78.1 dB SFDR with the Nyquist input and consumes 26.6mW at 0.9V supply. This simulated performance is equivalent to Walden and Schreier figure-of-merit (FoM) of 18.75 fJ/conversion-step and 168.8 dB, respectively.

Keywords—Analog-to-digital converter (ADC), Pipelined-SAR, passive residue transfer, ring amplifier.

I. INTRODUCTION

With the rapid development of radar, macro cellular base stations, satellite communications and other systems, the demand for direct radio frequency (RF) receiver is increasing. The analog-to-digital converters (ADC) in direct RF receiver often require more than 10 GHz sampling rate and 12-b or higher resolution. In order to achieve the sampling rate of more than 10 GHz, Time-Interleaving (TI) architecture is a good choice. However, TI ADC suffers from increased area, inter-channel crosstalk, and high calibration complexity. Therefore, there is an urgent need to increase the sampling rate of single-channel ADC in order to achieve higher sampling rates and minimize the number of channels in TI ADC.

Pipelined successive approximation register (SAR) ADCs have the advantages of both the low power consumption of SAR ADC and the high accuracy and speed of Pipeline ADC[1]. Fig. 1 illustrates the typical three-stage Pipelined-SAR ADC and its timing diagram. Compared to a two-stage Pipelined-SAR ADC, each stage converts fewer bits, thereby improving speed.

This work was supported by the Special Funds of the National Natural Science Foundation of China (Grant No. 62341409).

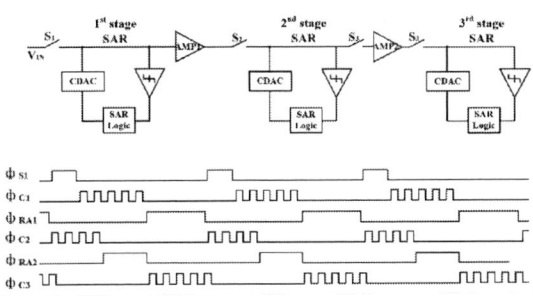

Fig. 1. Block diagram and operation timing of the Pipelined-SAR ADC.

The 1st- and 2nd-stages must complete sampling, conversion, and residue amplification(RA) within a single clock cycle, while the 3rd-stage only handles sampling and conversion. SAR conversion and RA take most of the time. Therefore, SAR conversion and RA are the key factors limiting the speed of the Pipelined-SAR ADC[2]. Reducing the time consumption of SAR conversion and RA is critical.

II. PROPOSED HYBRID PARALLEL TIMING SCHEME

This paper proposes a hybrid parallel timing scheme, which consists of two parts. As shown in Fig. 2, the 1st-stage adopts a parallel RA scheme, and the 2nd-stage employs a partially parallel conversion technique, alleviating the timing burden of the 1st-stage and the 2nd-stage respectively. With the hybrid parallel timing scheme, the speed of the Pipelined-SAR ADC is improved significantly.

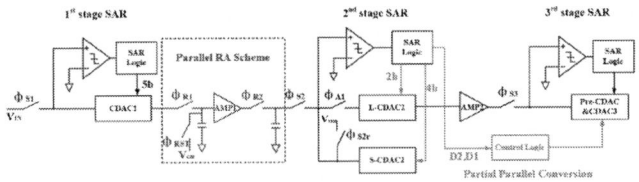

Fig. 2. Block diagram of proposed hybrid parallel timing scheme.

A. Parallel RA Scheme

To reduce the time consumed by RA and thereby increase the speed of Pipelined-SAR ADC, passive residue transfer is an alternative option. Fig. 3 presents a conceptual illustration of passive residue transfer technology. Passive residue transfer achieves rapid residue transfer by charge sharing of capacitors, eliminating the time-consuming RA phase[3]. However, the

979-8-3315-8850-2/25 $31.00 © 2025 IEEE

passive residue transfer technology lacks inter-stage gain and cannot attenuate noise in the second and subsequent stages, resulting in more stringent design requirements for the second and subsequent stages[4].

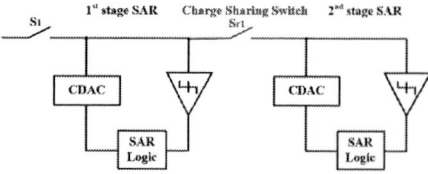

Fig. 3. Block diagram of the passive residue transfer.

To alleviate this problem, the parallel RA scheme was proposed. The blue-outlined section in Fig. 2 shows the architecture of the parallel RA scheme. The parallel RA scheme performs two passive residue transfer operations: one to maintain the residue generated by the 1st-stage, and another to save the output of the amplifier. After the 1st-stage conversion is finished, C_{r1} is connected to the 1st-stage CDAC. The passive residue transfer is performed, storing the residue in C_{r1}. The 1st-stage is idle, so it can proceed with the next sampling and conversion operation. At the same time, the amplifier amplifies the residue stored in C_{r1}. Upon completion of the RA, the 2nd-stage samples the amplified result via performing the passive residue transfer. Meanwhile, both Cr1 and the amplifier are idle, enabling the amplifier to prepare for the next RA operation. The parallel RA scheme achieves parallel operation of the time-consuming SAR conversion and RA.

The parallel RA scheme incorporates a fast passive residue transfer phase into the 1st-stage timing, replacing the time-consuming RA phase. This significantly reduces the 1st-stage timing budget. In addition, the parallel RA scheme can allocate more time for RA. Consequently, it lowers the bandwidth requirement of the amplifier. Ultimately, it substantially improves the overall speed and power consumption of the Pipelined-SAR ADC.

Although the parallel RA scheme alleviates the timing burden of the 1st-stage, it worsens the timing of the 2nd-stage. Since the 2nd-stage samples via charge sharing, the residue charge on the capacitors must be cleared before the next sampling cycle. Therefore, the parallel RA scheme introduces additional CDAC reset phase. Consequently, the 2nd-stage must complete four operations within a single clock cycle: CDAC reset, sampling (charge sharing), conversion, and RA. The 2nd-stage is a 4-b SAR ADC. The timing analysis of the 2nd-stage with the parallel RA stage can be concluded as follows:

$$T_{2nd-stage} = T_{reset} + T_{samp} + T_{conv}(4bit) + T_{RA} \quad (1)$$

where $T_{2nd-stage}$ is the total time consumed in the 2nd-stage; T_{reset}, T_{samp}, T_{conv}, and T_{RA} are the time consumed by CDAC reset, sampling, 4-b conversion, and RA, respectively. The tight timing of the second stage becomes the critical delay limiting the speed.

B. Partially Parallel Conversion

To solve the above problem introduced by the parallel RA scheme, a partially parallel conversion technology is

employed[5]. The red-outlined section in Fig. 2 shows the structure of the partially parallel conversion.

The 2nd-stage is a 4-b SAR ADC and its CDAC is divided into two parts, L-CDAC2 and S-CDAC2. When the 2nd-stage performs CDAC reset, sampling and the first 2-b conversion operation, A_1 is turned on, L-CDAC2 and S-CDAC2 work together. When the first 2-b conversion is completed, A_1 is turned off and S_3 is turned on. At this time, L-CDAC2 will apply its residue to the amplifier to deliver the 8× residue to the 3rd-stage. S-CDAC2 continues to complete the last 2-b conversion. In other words, the RA and the last 2-b conversion run in parallel. Once the amplification and the conversion of S-CDAC2 are completed, the 2-b output from S-CDAC controls the corresponding switches of the pre-CDAC3 in the 3rd-stage to generate the residue instantaneously. The timing is as follows:

$$T_{2nd-stage} = T_{reset} + T_{samp} + T_{conv}(2bit) + T_{RA} \quad (2)$$

Compared to (1), (2) takes less time for 2-b conversion. Therefore, the parallel conversion alleviates the timing tension and improves the throughput of the 2nd-stage.

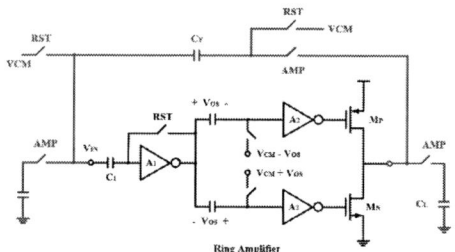

Fig. 4. Typical ring amplifier.

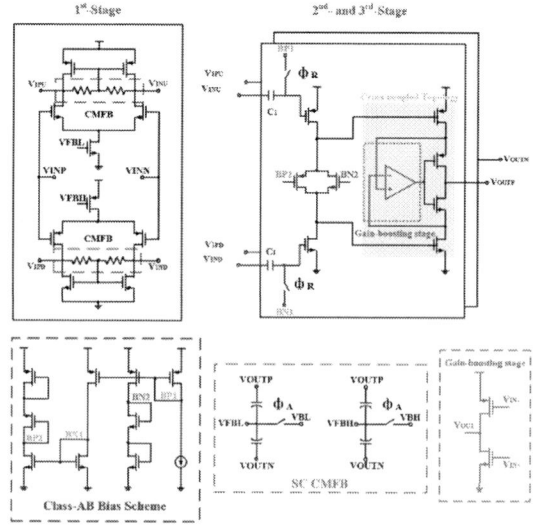

Fig. 5. Proposed PVT-robust ring amplifier.

III. PVT-ROBUST RING AMPLIFIER

A. Basic Principles of Ring Amplifier

In traditional Pipelined-SAR ADCs, the residue amplifier is typically a closed-loop amplifier based on an operational amplifier. However, in deep submicron CMOS processes, the

979-8-3315-8850-2/25 $31.00 © 2025 IEEE

supply voltage and the intrinsic gain of transistors is limited. As a result, it is quite difficult to achieve high gain and high bandwidth, and its power consumption is typically high.

To solve this problem, ring amplifiers are proposed. Fig. 4 shows the typical structure of a ring amplifier. Fundamentally, a ring amplifier is a three-stages ring oscillator[6]. Different from a ring oscillator, its 2nd-stage is divided into two independent signal paths, with different offsets embedded in each path to create a "dead zone" at the 2nd-stage output. The "dead zone" guarantees that the output transistors are simultaneously biased in the sub-threshold region or weak-inversion region, thereby significantly increasing the amplifier's output impedance. This pushes the dominant pole of the amplifier to a sufficiently low frequency, stabilizing the amplifier. Ring amplifiers offer advantages such as high gain, high bandwidth, low power consumption, and wide output swing, but their performance is sensitive to process, voltage, and temperature (PVT) variations.

B. Proposed Gain-Boosting Ring Amplifier

Fig. 5 shows the schematic of the proposed PVT-robust ring amplifier. In order to achieve higher bandwidth, the 1st-stage employs two complementary fully differential transconductance amplifiers, which stakes fewer transistors, thereby allowing the input transistors to obtain a larger overdrive voltage. The 2nd-stage adopts a self-biased structure with a set of CMOS resistors. The CMOS resistors dynamically generate a "dead zone" through the IR drop caused by the short-circuit current in the 2nd-stage. The 3rd-stage employs a dynamically biased cascode structure and the gain-boosting technique to achieve high gain. Therefore, the 3rd-stage is allowed to use the minimum device length, which reduces internal parasitic capacitance value. In addition, the cascode structure and the gain-boosting technology form a set of cross-coupled topologies, which enhance the pull-up/pull-down capability of the output node [7]. To stabilize the output common-mode voltage, both global and local common-mode feedback (CMFB) are introduced in the design.

C. Proposed PVT-Robust Ring Amplifier

The sensitivity of the ring amplifier to PVT variation is primarily due to the fact that the current in each inverter stage varies with PVT variations, thereby affecting the poles of the ring amplifier and the magnitude of the "dead zone" voltage.

To enhance the robustness of the ring amplifier, a Class-AB bias scheme is employed to stabilize the currents in the 2nd-stage and 3rd-stage [8][9]. As shown in Fig. 5, coupling capacitors C_1 are inserted between the 1st- and 2nd-stages. During the reset phase, BN_1 and BP_1 are sampled on C_1. During the amplification phase, the output of the 1st-stage is AC-coupled through the capacitors. The current in the 2nd-stage is determined by BN_1 and BP_1, which are generated by the current mirror. With the stable 2nd-stage current, the Class-AB bias scheme provides biasing voltages BN_2 and BP_2 to generate a stable "dead-zone" voltage. Therefore, the current in the 3rd-stage is stabilized due to the stable dead-zone voltage. Thus, the currents in both the 2nd-stage and 3rd-stage are well stabilized, which means that the output pole of the 3rd-satge(dominant pole) is stable. However, the output pole of the 1st-stage is the nondominant pole of the amplifier, so the variations of current in the 1st-stage can also affect the amplifier's stability. Therefore, the trimming technique is employed to maintain relative stability of the 1st-stage current.

D. Simulation Results

Overall, under the typical condition (TT corner, 27 °C, 0.9 V), the ring amplifier provides 8× gain and stabilizes within 750 ps. To obverse the sensitivity of the proposed ring amplifier to PVT variations, the total harmonic distortion (THD) of the ring amplifier was simulated. The simulation method involves dividing the output result of the ring amplifier by the corresponding amplification factor, adding it to the 5-b digital code output of the 1st-stage SAR ADC, and then performing an FFT operation to obtain the corresponding THD. Fig. 6 shows the THD variation of the proposed ring amplifier within the PVT variation range (five corners, ±5% VDD, and −40 °C to 125 °C). It can be seen that the THD achieves −73.5dB，which corresponds to 11.9-b linearity. In addition，the THD remains below −68dB in most cases, which corresponds to 11-b linearity. Even in the worst case (SS corner, 0.855 V and −40 °C), the THD is better than −62.73dB. Compared to the ring amplifier in [4], the THD is reduced by approximately 10dB. This demonstrates the excellent PVT robustness of the proposed ring amplifier.

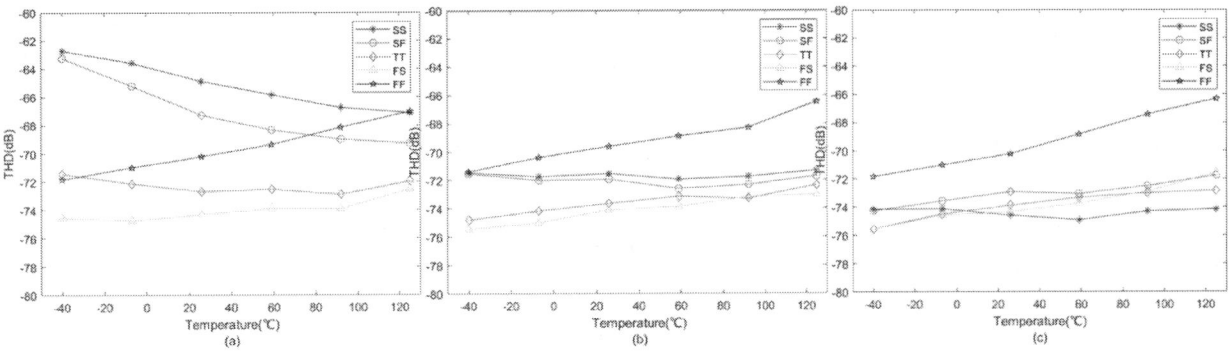

Fig. 6.   Simulated THD variation of the proposed PVT-robust ring amplifier with (a) −5% VDD variation, (b) 0% VDD variation, and (c) +5% VDD variation.

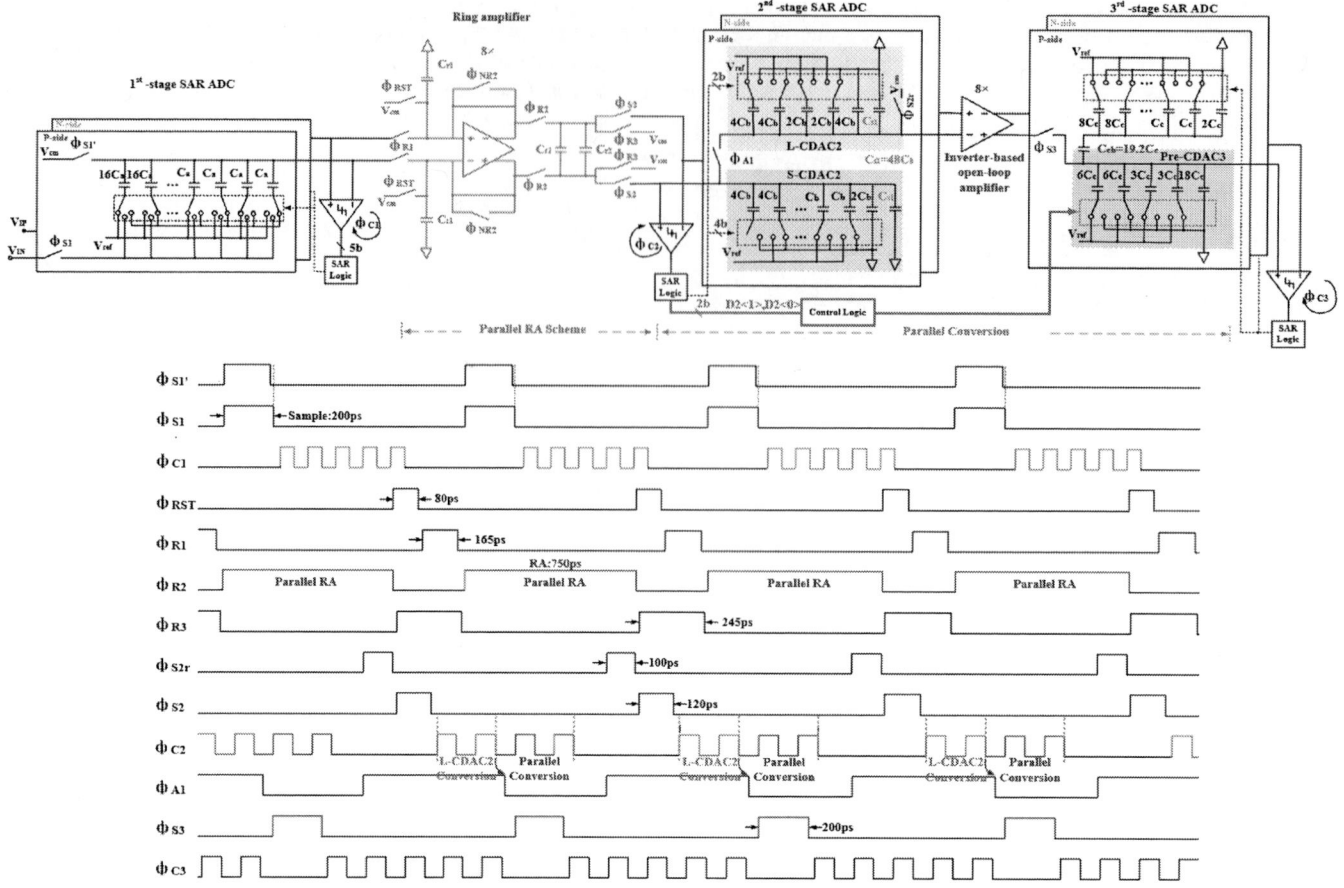

Fig. 7. Block diagram and timing allocation of the prototype Pipelined-SAR ADC.

IV. PROTOTYPE PIPELINED-SAR ADC

A. ADC Architecture

Fig. 7 illustrates the block diagram and timing allocation of the prototype Pipelined-SAR ADC. It consists of a three-stage SAR ADC, a ring amplifier, and an inverter-based open-loop amplifier. The resolutions of the three sub-stages of the SAR are set to 5, 4, and 5 bits, respectively. The 1st-stage employs bottom-plate sampling, while the 2nd- and 3rd-stage utilize top-plate sampling.

Assuming both the 1st- and 2nd- stage are full-swing, the ring amplifier must provide 16× gain. Moreover, the passive residue transfer technique causes signal attenuation. When the capacitance value of Cr1 and Cr2 are set equal to CDAC1 and CDAC2 respectively, each passive residue transfer results in a voltage attenuation of 0.5×. Two passive residue transfers mean the ring amplifier must provide an additional 4× gain. Therefore, the ring amplifier must deliver a total gain of 64×, which is impossible to achieve within 750ps amplification time.

To reduce the gain requirements of the ring amplifier, two solutions were implemented. First, an inherent 2× voltage gain was achieved by differentially sampling the ring amplifier's output. Additionally, as shown in Fig.7, the 2nd-stage employs a scaling capacitor C_{s1} to reduce the quantization range, thereby

lowering the amplifier's gain. The value of Cs1 can be expressed as:

$$C_{S1} = 48C_b \qquad (3)$$

where C_b is the unit capacitance value of the 2nd-stage. All capacitors in the 2nd-stage are used for sampling, yet only one-quarter of the capacitors participates in conversion. Consequently, the ring amplifier requires only 8× gain.

The 2nd-stage employs the partially parallel conversion technology. Therefore, after converting 2-b, the residue is amplified. Consequently, if the range of the 3rd-stage is same to the 2nd-stage, the inverter-based open-loop amplifier only requires only 2× gain. However, if the 3rd-stage is full-swing, the inverter-based open-loop amplifier must provide 8× gain.

To obtain correct results, the value of Pre-CDAC3 must be set appropriately. The analysis is as follows.

Without the partially parallel conversion technology, the residue of the 2nd-stage can be expressed as:

$$V_{r2} = V_{in2} - \sum_{i=1}^{4} \left(W_i \times V_{ref2} \times D_{[i]} \right) \qquad (4)$$

where V_{r2}, V_{in2}, W_i, V_{ref2}, and $D_{[i]}$ represent the 2nd-stage residue, 2nd-stage input voltage, the weight of per bit, 2nd-stage reference

979-8-3315-8850-2/25 $31.00 © 2025 IEEE

voltage, and digit code of per bit, respectively. Therefore, the voltage value sampled at the 3rd-stage can be expressed as:

$$V_{in3} = 8 \times V_{in2} - 8 \times \sum_{i=1}^{4}\left(W_i \times V_{ref2} \times D_{[i]}\right) \quad (5)$$

where V_{in3} represent 3rd-stage input voltage.

With the partially parallel conversion technology, the residue of the 2nd-stage can be expressed as:

$$V_{in3} = 8 \times V_{in2} - 8 \times \sum_{i=1}^{2}\left(W_{Li} \times V_{ref2} \times D_{L[i]}\right)$$
$$- \sum_{i=1}^{2}\left(W_{Pi} \times V_{ref3} \times D_{S[i]}\right) \quad (6)$$

where W_{Li}, $D_{L[i]}$, W_{Pi}, V_{ref3} and $D_{S[i]}$ represent the weight of L-CDAC2's per bit, digit code of L-CDAC2's per bit, the weight of Pre-CDAC3's per bit, 3rd-stage reference voltage, and digit code of Pre-CDAC3's per bit, respectively.

To ensure consistent results, (6) must be equal to (5). Therefore, the weight of Pre-CDAC3 should be eight times the weight of the last two bits in the S-CDAC2. The weight of the last two bits in the S-CDAC2 are 1/32 and 1/64 respectively. Consequently, the weight of Pre-CDAC3 should be 1/4 and 1/8. To reduce the total capacitance value of the 3rd-stage, a bridging capacitor was introduced. The total capacitance value of the 3rd-stage is $48C_c$, where C_c represents the unit capacitance value of the 3rd-stage. Due to the use of split capacitors, the capacitance values of Pre-CDAC3 are $6C_c$ and $3C_c$ respectively.

Considering kT/C noise and capacitance matching requirements, the unit capacitance value of the 1st-, 2nd-, and 3rd-stage DACs are selected as 17.24fF, 1.96fF, and 1.06fF, respectively. It is worth noting that the sampling thermal noise of the 1st-stage, due to the use of the parallel RA scheme, is $8kT/C$ rather than $2kT/C$.

B. Inverter-Based Open-Loop Amplifier

Fig. 8 shows the schematic of an inverter-based open-loop amplifier. It consists of two amplifier stages. Each amplification stage consists of two inverters, with one inverter serving as the amplification core and the other inverter connecting its input and output nodes to act as the load for each stage[10]. The gain of each amplification stage is approximately equal to the ratio of the transconductance of the amplification inverter to the transconductance of the load inverter. Since the gain of the open-loop amplifier is sensitive to PVT variations, a gain calibration algorithm based on pseudo-random code injection is employed.

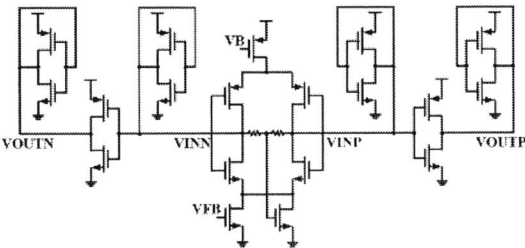

Fig. 8. Schematic of an inverter-based open-loop amplifier.

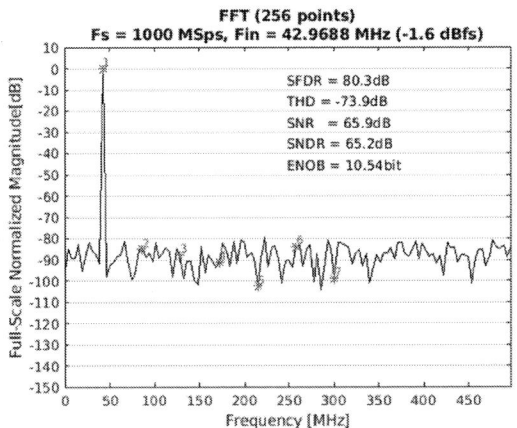

Fig. 9. Spectrum of the ADC outputs with 42.9688 MHz input.

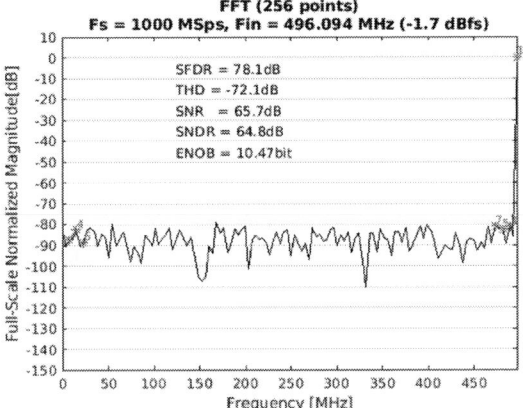

Fig. 10. Spectrum of the ADC outputs with 496.094 MHz input.

V. SIMULATION RESULTS

All simulations are pre-simulations based on a 0.9 V power supply voltage. As shown in Fig. 9, at a low input frequency (42.9688 MHz), the spurious-free dynamic range (SFDR) and signal-to-noise-and-distortion ratio (SNDR) are 80.3 dB and 65.2 dB, respectively. As shown in Fig. 10, at the Nyquist input frequency (496.094 MHz), the SFDR and SNDR are 78.1 dB and 64.8 dB, respectively. The total power consumption of the prototype ADC is 26.6mW.

TABLE I. summarizes the performance of the prototype ADC and other ADCs with similar speed and resolution. The simulated Walden and Schreier figure-of-merit (FoM) are 18.75 fJ/conversion-step and 168.8 dB, respectively. The proposed Pipelined-SAR ADC achieves good energy efficiency.

VI. CONCLUSION

This article presents a Pipelined-SAR ADC with a hybrid parallel timing scheme and a PVT-robust ring amplifier. By adopting the parallel RA stage, the RA and the 1st-stage SAR conversion operate in parallel, significantly enhancing speed. Additionally, the parallel RA scheme is further optimized using partially parallel conversion technology. The proposed ring amplifier employs a Class-AB bias scheme, significantly improving robustness without compromising speed.

TABLE I.   PERFORMANCE SUMMARY AND COMPARISON

	JSSC 2024[4]	JSSC 2020[2]	JSSC 2019[11]	ESSCIRC 2023[12]	This work
Architecture	Pipelined-SAR	Pipelined-SAR	Pipeline	Pipeline	Pipelined-SAR
Process(nm)	28	28	28	28	28
Resolution (bit)	12	12	12	12	12
Sample Rate(GS/s)	1.5	1	1	1	1
Supply Voltage(V)	0.9/1	1	0.9	0.9	0.9
Power(mW)	21.3	7.6	24.8	36.5	26.6
SFDR(dB)	74.5	74.6	73.1	83.2	78.1
SNDR(dB)	58.5	60.0	56.6	61.2	64.8
FoMW(fJ/c-s)	20.7	9.3	45	39.3	18.75
FoMS(dB)	164	168.2	159.6	162.6	168.8

REFERENCES

[1] M. Ni *et al.*, "A 13-bit 312.5-MS/s Pipelined SAR ADC with Integrator-type Residue Amplifier and Inter-stage Gain Stabilization Technique," in *2020 IEEE 63rd International Midwest Symposium on Circuits and Systems (MWSCAS)*, 9-12 Aug. 2020 2020, pp. 341-344, doi: 10.1109/MWSCAS48704.2020.9184624.

[2] W. Jiang, Y. Zhu, M. Zhang, C.-H. Chan, and R. P. Martins, "A Temperature-Stabilized Single-Channel 1-GS/s 60-dB SNDR SAR-Assisted Pipelined ADC With Dynamic Gm-R-Based Amplifier," *IEEE Journal of Solid-State Circuits,* vol. 55, no. 2, pp. 322-332, 2020, doi: 10.1109/jssc.2019.2948170.

[3] L. Chin-Yu and L. Tai-Cheng, "A 12-bit 210-MS/s 5.3-mW pipelined-SAR ADC with a passive residue transfer technique," in *2014 Symposium on VLSI Circuits Digest of Technical Papers*, 10-13 June 2014 2014, pp. 1-2, doi: 10.1109/VLSIC.2014.6858452.

[4] Y. Shen *et al.*, "A 12-bit 1.5-GS/s Single-Channel Pipelined SAR ADC With a Pipelined Residue Amplification Stage," *IEEE J. Solid-State Circuits*, pp. 1–12, 2024, doi: 10.1109/JSSC.2024.3412090.

[5] H.-H. Chang, T.-C. Lin, and T.-C. Lee, "A Single-Channel 1-GS/s 7.48-ENOB Parallel Conversion Pipelined SAR ADC With a Varactor-Based Residue Amplifier," *IEEE Transactions on Circuits and Systems II: Express Briefs*, vol. 69, no. 4, pp. 2021-2025, 2022, doi: 10.1109/tcsii.2022.3142099.

[6] B. Hershberg, S. Weaver, K. Sobue, S. Takeuchi, K. Hamashita, and U.-K. Moon, "Ring Amplifiers for Switched Capacitor Circuits," *IEEE Journal of Solid-State Circuits,* vol. 47, no. 12, pp. 2928-2942, 2012, doi: 10.1109/jssc.2012.2217865.

[7] C.-Y. Hsu and T.-C. Lee, "A Calibration-Free 9.3-ENOB 1-GS/s Pipelined ADC With PVT-Insensitive Nested Ring Amplifiers," *IEEE Trans. Circuits Syst. II Express Briefs*, pp. 1–1, 2024, doi: 10.1109/TCSII.2024.3466902.

[8] M. Zhan, L. Jie, X. Tang, Y. Zhong, and N. Sun, "A 0.004-mm2 200-MS/s Pipelined SAR ADC With kT/C Noise Cancellation and Robust Ring-Amp," *IEEE Journal of Solid-State Circuits*, vol. 59, no. 7, pp. 2209-2218, 2024, doi: 10.1109/jssc.2023.3344461.

[9] Y. Lim, J. Lee, J. Lee, K. Lim, S. Oh, and J. Lee, "A 2.08-mW 64.4-dB SNDR 400-MS/s Pipelined-SAR ADC Using Mismatch and PVT Variation Tolerant Dynamically Biased Ring Amplifier in 8 nm," *IEEE J. Solid-State Circuits*, vol. 59, no. 12, pp. 4199–4210, Dec. 2024, doi: 10.1109/JSSC.2024.3471915.

[10] X. Guo, R. Chen, Z. Chen, and B. Li, "A 13b 600-675MS/s Tri-State Pipelined-SAR ADC With Inverter-Based Open-Loop Residue Amplifier," *IEEE Journal of Solid-State Circuits*, vol. 58, no. 3, pp. 624-633, 2023, doi: 10.1109/jssc.2022.3222162.

[11] J. Lagos, B. P. Hershberg, E. Martens, P. Wambacq, and J. Craninckx, "A 1-GS/s, 12-b, Single-Channel Pipelined ADC With Dead-Zone-Degenerated Ring Amplifiers," *IEEE Journal of Solid-State Circuits*, vol. 54, no. 3, pp. 646-658, 2019, doi: 10.1109/jssc.2018.2889680.

[12] M. Gu, Y. Zhong, L. Jie, and N. Sun, "A 12b 1GS/s Pipelined ADC with Digital Background Calibration of Inter-stage Gain, Capacitor Mismatch, and Kick-back Errors," presented at the ESSCIRC 2023- IEEE 49th European Solid State Circuits Conference (ESSCIRC), 2023.

2025 The 10th International Conference on Integrated Circuits and Microsystems

An X-band Compact High-Efficiency MMIC Power Amplifier in 0.25-μm GaN Technology

Ye Liu
Key Laboratory of Near-Range
Sensing ICs & Microsystems
Nanjing University of
Science and Technology
Nanjing, China
liuye200618@njust.edu.cn

Anshi Zhu
Key Laboratory of Near-Range
Sensing ICs & Microsystems
Nanjing University of
Science and Technology
Nanjing, China
anshizhu@njust.edu.cn

Zihan Yang
Key Laboratory of Near-Range
Sensing ICs & Microsystems
Nanjing University of
Science and Technology
Nanjing, China
yangzihan2023@njust.edu

Tongde Huang
Key Laboratory of Near-Range
Sensing ICs & Microsystems
Nanjing University of
Science and Technology
Nanjing, China
tongdeh@njust.edu.cn

Wen Wu
Key Laboratory of Near-Range
Sensing ICs & Microsystems
Nanjing University of
Science and Technology
Nanjing, China
wuwen@njust.edu.cn

Abstract—This paper presents the design of an X-band high-efficiency power amplifier (PA) based on a 0.25-μm GaN technology. The proposed PA features a compact, low-loss multi-way power combining matching network derived from an enhanced Chebyshev low-pass filter. This approach effectively reduces the chip area while maintaining excellent performance. The amplifier achieves a maximum power-added efficiency (PAE) of 52.4 % and a peak output power of 43.7 dBm under a drain voltage of 28V in X band (8-12GHz). The amplifier delivers a power gain greater than 19 dB, with a gain variation less than ±0.5 dB. The chip dimensions are 3.4 × 1.5 mm², yielding a power density of 4.09 W/mm².

Index Terms—Power Amplifier, Gallium nitride (GaN), Monolithic Microwave Integrated Circuits(MMICs), multi-way power combining matching networks.

I. INTRODUCTION

With the rapid evolution of global broadband communication technologies, next-generation wireless systems are increasingly demanding higher integration density, multi-band operational capabilities, enhanced data throughput, and improved power efficiency [1]. Conventional narrowband architectures have become inadequate to address the stringent requirements of contemporary communication standards [2]. In this context, RF power amplifiers (PAs) are essential for amplifying weak signals to levels suitable for reliable transmission. With the increasing performance requirements of modern systems, Gallium Nitride (GaN) High Electron Mobility Transistor (HEMT) technology offers a promising solution.

Gallium nitride (GaN) materials exhibit superior physical properties, including a wide bandgap, high electron saturation velocity, and high breakdown field, which render them

exceptionally promising for high-frequency and high-power applications [3]–[5]. In comparison to traditional silicon (Si) and gallium arsenide (GaAs) technologies, GaN-based power amplifiers (PAs) demonstrate the capability to operate at higher frequencies while delivering enhanced output power, efficiency, and linearity. Additionally, GaN devices can tolerate higher operating voltages and power densities. The increased power density facilitates the achievement of greater functionality within the same physical footprint, thereby simplifying the circuit design and reducing the overall package size of GaN PAs [6], [7]. Furthermore, the high thermal conductivity of GaN significantly enhances the chip reliability and stability, ensuring stable operation over extended periods under demanding working conditions.

The 8–12 GHz frequency band holds significant importance in applications such as satellite communications, radar systems, and electronic warfare, where high-performance power amplifiers are critical for ensuring stable system operation. A broadband power amplifier operating from 7.5 to 12.5 GHz, utilizing SiC-based GaN HEMT technology and reflection matching techniques, is presented in [8], achieving notable output power and power-added efficiency (PAE). Similarly, the power amplifiers reported in [9] and [10] demonstrate exceptional output power and PAE performance in the 9–10 GHz and 8.5–10.5 GHz frequency ranges, respectively.

This paper comprehensively details the design process of an X-band high-efficiency power amplifier based on 0.25-μm GaN technology. The proposed design incorporates an optimized microstrip-based Chebyshev low-pass filter topology, enabling the realization of compact, low-loss multi-way power combining networks. The PA achieves an output power

979-8-3315-8850-2/25 $31.00 © 2025 IEEE

exceeding 43.2 dBm and a peak PAE of 52.4%, while attaining a high power density of 4.09 W/mm².

II. CIRCUIT DESCRIPTIONS

The design of the power amplifier follows a sequential process from the output stage to the input stage. The power amplifier in this work utilizes a high-performance 0.25-μm GaN technology, where the devices have a cutoff frequency up to 34 GHz. This design achieves high power levels and high power density while maintaining excellent thermal performance due to the SiC substrate.

The output level of the power amplifier is largely dependent on the gate width of the output stage transistors [11]. Through extensive characterization, the transistors are biased at an optimal static current density of 30 mA/mm to ensure model accuracy and maximize performance. Comprehensive load-pull analysis is conducted on various transistor configurations, considering key parameters including output power requirements, 1-dB insertion loss, and process variations. Based on this analysis, a four-way power combining architecture is implemented using 6×150 μm unit cells, with each cell delivering a output power of 38.4 dBm , resulting in a combined output of 44.4 dBm. The driver stage employs a 2×8×60 μm configuration to provide sufficient gain and drive capability for the output stage. The complete circuit schematic is presented in Fig. 1.

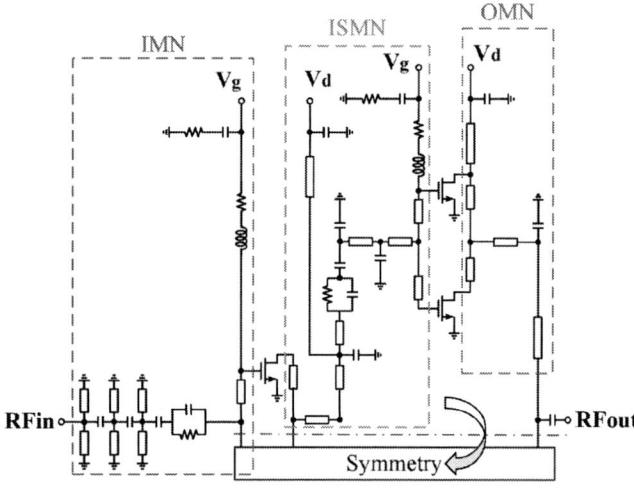

Fig. 1. Circuits of the proposed power amplifier.

A. Output Matching Network

The output matching network (OMN) plays a critical role in determining the power amplifier's overall performance and power-added efficiency (PAE).In this design, the OMN is optimized to target the impedance point corresponding to peak efficiency, thereby minimizing PAE degradation caused by matching losses. To ensure the effectiveness of the power amplifier throughout the operating bandwidth, the output matching network implements a broadband impedance transformation from the 50 Ω load to the optimal load impedance

across the entire operational bandwidth. Unlike traditional methods that only select the optimal matching impedance at the center frequency, this approach avoids the mismatching at edge frequencies in wideband designs, improving the bandwidth of the power amplifier. Furthermore, a four-way power combining network is implemented to efficiently sum the outputs of individual transistors, achieving the required high output power level while maintaining system linearity and efficiency.

The optimal load impedance of each transistor is modeled as a parallel RC equivalent circuit. In the proposed design, the four output-stage transistors operate in perfect voltage symmetry, exhibiting identical phase and amplitude characteristics. This configuration enables the consolidation of individual RC models into a unified single-path matching network topology, significantly simplifying the matching network design complexity. The schematic representation of this innovative matching network architecture is illustrated in Fig. 2.

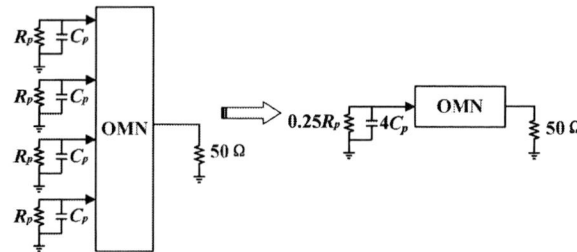

Fig. 2. Simplification of RC parallel circuits.

The output matching network employs a third-order Chebyshev LC low-pass filter topology based on the merged impedance characteristics. The proposed methodology systematically transforms lumped-element configurations into distributed microstrip implementations through strategic component reorganization and electromagnetic co-design considerations. The design methodology (Fig. 3) follows a systematic transformation process: The fundamental third-order LC low-pass filter structure (Fig. 3(a)) performs real-to-real impedance transformation. In this design, a bias inductor and bypass capacitor are introduced to integrate the bias line with the matching network, providing a DC drain bias for the output-stage transistor while participating in the impedance matching. This approach further reduces the chip area and improves integration. Additionally, a large-valued blocking capacitor is added at the load end to mitigate the impact on impedance transformation.

The structure in Fig. 3(b) can achieve impedance transformation from complex impedance to real impedance. By applying T-π transformation to the lumped parameter structure L_1-C_1-L_2 , the equivalent lumped π-type network C_4-L_4-C_5 is obtained, as shown in Fig. 3(c). Simultaneously, C_2 is split into two capacitors in parallel,C_2a and C_2b, with C_2b set equal to C_3. Based on the structure in Fig. 3(c), the lumped π-type network C_2b-L_3-C_3 is converted into microstrip TL_3, resulting in the circuit configuration shown in Fig. 3(d). Through value-matched splitting of C_5

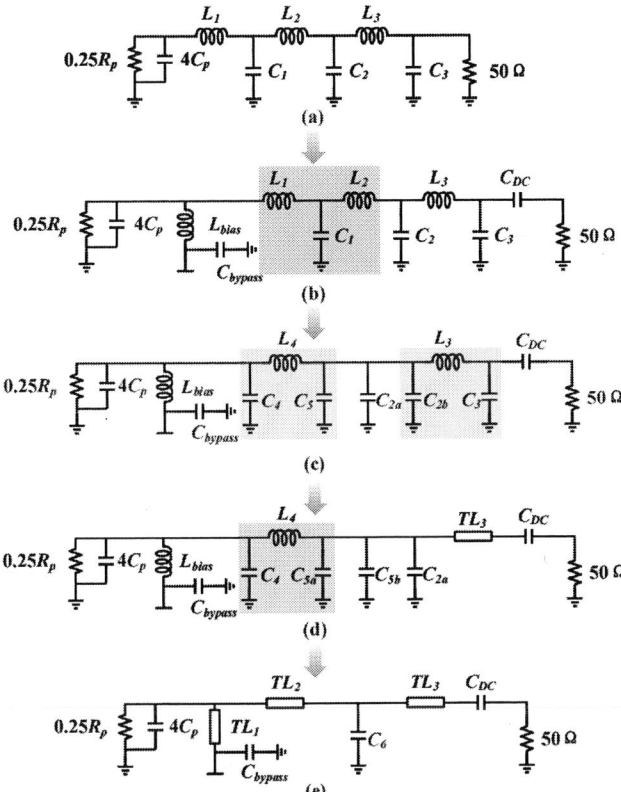

Fig. 3. Schematic of the modified matching network.

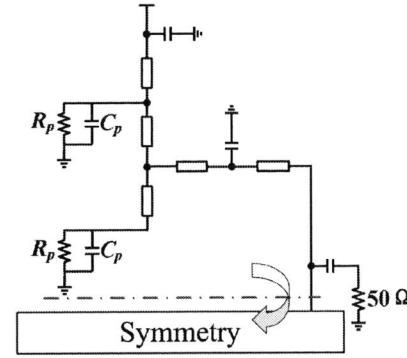

Fig. 4. Schematic of the output matching network.

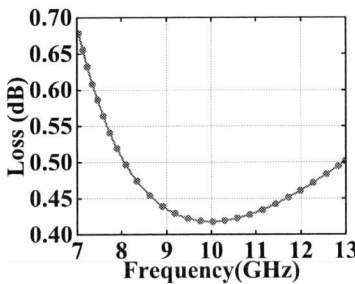

Fig. 5. Loss of the output matching network.

$(C_5a = C_4, C_5b)$, the π-network C_4-L_4-C_5a is transformed into microstrip TL_2 using lumped-to-distributed equivalence conversion. This topological modification is accompanied by the capacitive integration of residual component C_5b with C_2a, forming a consolidated capacitance element C_6 through parallel combination. The dual transformation process achieves simultaneous network simplification and layout compactness while preserving the fundamental impedance transformation characteristics of the hybrid matching architecture. Finally, the bias inductor L_bias is replaced by microstrip TL_1. Through these steps, the output matching network composed of lumped elements is successfully transformed into the single-channel output matching network based on microstrips and capacitors, as illustrated in Fig. 3(e). This integration necessitates careful capacitance redistribution through controlled component splitting ($C_2 \rightarrow C_2a/C_2b$, $C_5 \rightarrow C_5a/C_5b$) and recombination ($C_5b+C_2a \rightarrow C_6$), maintaining the original impedance transformation ratio while enabling microstrip realizations.

Following the parallel equivalence principle, the single-path network is expanded into a 4-way power combining architecture (Fig. 4). To optimize performance, the design implements shared bias lines between transistor pairs, reducing the complexity of the layout. Extensive electromagnetic simulations guide iterative optimizations of microstrip dimensions and capacitor placements, achieving a compact, low-loss configuration with simulated loss characteristics presented in

Fig. 5. Stability is ensured through strategic placement of odd-mode suppression resistors between transistor gates and drains, maintaining symmetry and preventing potential oscillations.

B. Inter-stage Matching Network

The inter-stage matching network (ISMN) performs critical impedance transformation between cascaded transistor stages to mitigate signal reflections and insertion loss, thereby ensuring stable and efficient power transfer [12]. Simultaneously, the ISMN incorporates the band-pass filter topology to suppress out-of-band harmonics and interference signals.

To address gain roll-off characteristics in broadband operation, the network implements intentional impedance mismatch at lower frequencies through optimized transmission zero placement. This compensates for the inherent high-frequency gain reduction of the transistor stages, achieving gain flatness of ±0.5 dB across the 8-12 GHz operational bandwidth. Contrary to output matching network optimization strategies focused on loss minimization, the ISMN design prioritizes stability enhancement through strategic integration of parallel RC networks. These networks provide controlled loss that improves the stability factor while maintaining adequate forward gain.

C. Input Matching Network

The Input Matching Network (IMN) is intentionally designed with a high-pass filter structure as its foundational element. This primary choice serves the critical function of attenuating excessive gain at lower frequencies, which is a

TABLE I
COMPARISON WITH OTHER MMIC PAS

Ref.	Freq(GHz)	P_{out}(dBm)	PAE(%)	Gain(dB)	Power Density(W/mm²)	Technology
[8]	7.5-12.5	43-45.2	18.9-29.1	18	1.17	0.25-μm GaN
[9]	9-10	40.7-41.8	25.2-37	20.6	4.3	0.25-μm GaN
[10]*	8.5-10.5	39.3-40	42-47	20.2	2.44	0.25-μm GaN
[13]	8-12	40.8-41.8	37-54	22	0.95	0.25-μm GaAs
[14]	8.8-10.8	44.7-46	38-44	17	1.93	0.25-μm GaN
This Work*	8-12	42.9-43.7	46.7-52.4	19.7	4.09	0.25-μm GaN

*Simulation Results.

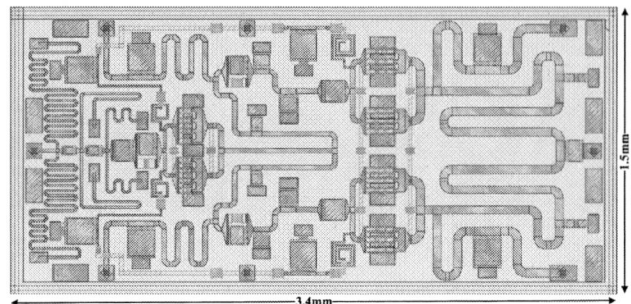

Fig. 6. Layout of the proposed power amplifier.

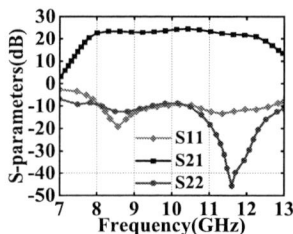

Fig. 7. Simulated S-parameters of the PA.

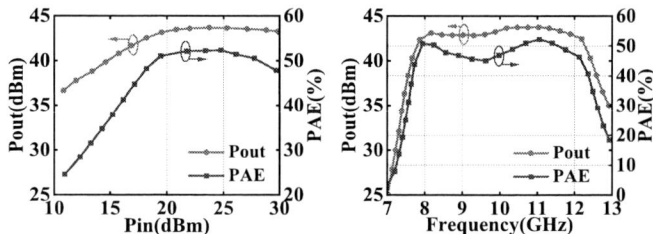

Fig. 8. (a) Output power and PAE versus input power at 10GHz, (b) Simulated output power and PAE of the PA.

common source of potential instability within RF amplifier circuits. To further bolster circuit stability and mitigate low-frequency oscillation risks, complementary stability measures are integrated. These include the strategic placement of a parallel RC network and the incorporation of a series resistor at the gate terminal of the active device. Together, these components significantly increase losses specifically within the low-frequency band, damping unwanted oscillations.

Beyond ensuring stability, optimizing the input port's Voltage Standing Wave Ratio (VSWR) throughout the operational bandwidth is a paramount objective. To achieve this stringent impedance matching requirement, a T-type resistive attenuator structure is employed within the IMN. This attenuator configuration is uniquely implemented using multi-stage capacitive coupling mechanisms to interface effectively with the surrounding microstrip transmission lines forming the network. The synergistic combination of the high-pass foundation, stability elements, and the capacitively coupled T-attenuator results in exceptionally low input reflection coefficients across the entire target frequency range. Crucially, this high level of RF performance is attained while simultaneously adhering to the practical constraint of maintaining a highly compact physical layout for the overall input matching network.

III. LAYOUT AND SIMULATION RESULTS

The layout of the proposed power amplifier is shown in Fig. 6, occupying an area of 3.4 × 1.5 mm². The PA is fabricated using 0.25-μm GaN technology, with a drain voltage (Vd) of 28 V and a gate voltage (Vg) of -2.58 V. Electromagnetic simulations are performed for all microstrip lines to extract the characteristic. Simulation results for small-signal parameters

(Fig. 7) indicate an average small-signal gain (S_{21}) exceeding 21 dB across 8–12 GHz, with a peak of 24.5 dB at 10.4 GHz. Fig. 8(a) depicts the simulated output power and PAE as functions of input power at 10 GHz, ranging from 10 to 30 dBm. At 24 dBm input, the output power surpasses 43 dBm and PAE exceeds 52%. The large-signal performance simulation results with a 24 dBm input are shown in Fig. 8(b), where the PAE exceeds 49.2%, with a peak PAE reaching 52.4%. The in-band output power averages 43.2 dBm, with a maximum of 43.7 dBm at 11 GHz, and the design achieves an average gain of 19.2 dB with ±0.5 dB flatness.

Table I presents a comprehensive performance comparison between the proposed power amplifier and with recently reported GaN/GaAs power amplifiers. The proposed amplifier demonstrates competitive performance metrics in the X-band, achieving a peak power-added efficiency (PAE) of 52.4%, saturated output power of 43.7 dBm, and power density of 4.09 W/mm². These results, coupled with its compact die area (3.4 × 1.5 mm²), underscore its suitability for practical X-band applications, including satellite communications and radar systems.

979-8-3315-8850-2/25 $31.00 © 2025 IEEE

IV. CONCLUSION

This paper presents a high-efficiency X-band power amplifier implemented in 0.25-μm GaN HEMT technology. The design features a compact, low-loss multi-way power combining matching network derived from an enhanced Chebyshev low-pass filter, achieving a chip area of 3.4 × 1.5 mm². Simulation results demonstrate state-of-the-art performance across 8-12 GHz, with a saturated output power >43.2 dBm, peak power-added efficiency (PAE) of 52.4%, and power density of 4.09 W/mm². These results validate the effectiveness of the proposed design methodology for high-performance X-band applications.

REFERENCES

[1] G. Nikandish and A. Medi, "A Design Procedure for High-Efficiency and Compact-Size 5–10-W MMIC Power Amplifiers in GaAs pHEMT Technology," in IEEE Transactions on Microwave Theory and Techniques, vol. 61, no. 8, pp. 2922-2933, Aug. 2013, doi: 10.1109/TMTT.2013.2271997.

[2] H. Wang, C. Sideris and A. Hajimiri, "A CMOS Broadband Power Amplifier With a Transformer-Based High-Order Output Matching Network," in IEEE Journal of Solid-State Circuits, vol. 45, no. 12, pp. 2709-2722, Dec. 2010, doi: 10.1109/JSSC.2010.2077171.

[3] B. Heying, W.-B. Luo, I. Smorchkova, S. Din, and M. Wojtowicz, "Reliable GaN HEMTS for high frequency applications," in 2010 IEEE MTT-S International Microwave Symposium. IEEE, 2010, pp. 12181221.

[4] F. Yamaki, K. Inoue, M. Nishi, H. Haematsu, N. Ui, K. Ebihara, A. Nitta, and S. Sano, "Ruggedness and reliability of GaN HEMT," in 2011 6th European Microwave Integrated Circuit Conference. IEEE, 2011, pp. 328–331.

[5] N. D. Soni and K. Cecil, "COMPARATIVE STUDY OF GaN AND GaAs MESFET," 2014. [Online]. Available: https://api.semanticscholar.org/CorpusID:236908252

[6] S. Shinjo, M. Hangai, Y. Yamaguchi and M. Miyazaki, "Advanced GaN HEMT Modeling Techniques and Power Amplifiers for MillimeterWave Applications," 2020 IEEE/MTT-S International Microwave Symposium (IMS), Los Angeles, CA, USA, 2020, pp. 566-569.

[7] T. Ide, H. Yamada, X. Wang, T. Yamada, R. Azumi and T. Ujihara, "Recent researches of GaN-based materials and devices in NU-AIST," 2023 IEEE International Meeting for Future of Electron Devices, Kansai (IMFEDK), Kyoto, Japan, 2023, pp. 1-4.

[8] G. Yu, L. Gong, J. Zhang, Y. Wu, X. Xie and Y. Zhang, "A X-band 20W Power Amplifier MMIC with 50% Bandwidth Using 6-inch GaN Technology," 2019 IEEE Asia-Pacific Microwave Conference (APMC), Singapore, 2019, pp. 408-41.

[9] L.-H. Huang and H.-K. Chiou, "An Ultra-compact 14.9-W X-Band GaN MMIC Power Amplifier," 2020 IEEE Asia-Pacific Microwave Conference (APMC), Hong Kong, Hong Kong, 2020, pp. 257-259.

[10] D. Tao, Y. Lu, X. Mo, Q. Guo and J. Zhu, "Design of High Efficiency X-Band Power Amplifier Based on GaN HEMT," 2022 IEEE 10th Asia-Pacific Conference on Antennas and Propagation (APCAP), Xiamen, China, 2022, pp. 1-2.

[11] C. Campbell, "Microwave monolithic power amplifier design," Wiley Encyclopedia of Electrical and Electronics Engineering, 2001.

[12] H. Jia, C. C. Prawoto, B. Chi, Z. Wang, and C. P. Yue, "A Full KaBand Power Amplifier With 32.9% PAE and 15.3-dBm Power in 65-nm CMOS," IEEE Transactions on Circuits and Systems I: Regular Papers, vol. 65, no. 9, pp. 2657–2668, 2018.

[13] Q. Wu, B. Song, Y. Shih, X. Huang and J. Wu, "A full X-band high-efficiency 12-watt GaAs MMIC power amplifier with harmonic tuning," in Proceedings of the 2015 IEEE Topical Conference on Power Amplifiers for Wireless and Radio Applications (PAWR), 2015, pp. 1-3.

[14] D.-H. Shin, I.-B. Yom, and D.-W. Kim, "X-band GaN MMIC power amplifier for the SSPA of a SAR system," in Proceedings of the 2017 IEEE International Symposium on Radio-Frequency Integration Technology (RFIT), 2017, pp. 93-95.

2025 The 10th International Conference on Integrated Circuits and Microsystems

Design of a Two-Stage Fully Differential Operational Amplifier for High-Resolution Sigma Delta ADCs

Zhan Shi
School of Information and Communication Engineering
Dalian Minzu University
Dalian, China
sz1134@163.com

Zenghao Zhu
School of Information and Communication Engineering
Dalian Minzu University
Dalian, China
2142557911@qq.com

Yan Zhao
School of Information and Communication Engineering
Dalian Minzu University
Dalian, China
2747425080@qq.com

Franco Maloberti
Department of Electrical, Computer, and Biomedical Engineering
University of Pavia
Pavia, Italy
malobert@unipv.it

Abstract—**High-resolution sigma-delta analog-to-digital converters (ADCs) demand an operational amplifier in the first integrator to have stringent performances, such as a high gain (>100 dB), a large gain-bandwidth product, a fast slew rate, and a wide output swing. To meet these demands, a two-stage fully differential operational amplifier was designed in a SMIC 0.13 µm CMOS process. The first stage employed an improved folded-cascode amplifier to generate common-mode feedback and extend the output range of the second stage. The second stage utilized a dynamic-biasing cascode amplifier to enhance both gain and slew rate further. Additionally, a switched capacitor common-mode feedback circuit was implemented to synchronize with the integrator timing. Simulation results show that the amplifier achieves a DC gain of 122 dB, a gain-bandwidth product of 83 MHz, and a maximum negative slew rate of 53 V/µs with a capacitive load of 15 pF. Besides, the output common-mode voltage is stable at half the supply voltage.**

Keywords—*Amplifier, High Gain, Switched-Capacitor Common-Mode Feedback, Integrator, Sigma-delta ADC*

I. INTRODUCTION

Analog-to-digital converters (ADCs) act as an essential link between the analog and digital worlds, and their performance directly affects the accuracy of modern electronic systems [1]. Among various ADC architectures, the sigma-delta ADC has become a preferred choice for applications that need high resolution and wide dynamic range, such as audio, sensor interfaces, and high-resolution measurements [2]. As a result, designing and researching high-performance sigma-delta ADCs is highly important.

The core of a sigma-delta ADC relies on oversampling and noise-shaping techniques. Using a closed-loop feedback system, as shown in Fig. 1 [3], the quantization noise is effectively pushed to higher frequencies, resulting in a very high signal-to-noise ratio at lower frequencies. However, the system's resolution is not unlimited. It is limited by the non-ideal characteristics of the operational amplifier (op-amp) in the integrator, such as finite gain, gain-bandwidth product, slew rate, and output swing. Limited gain can cause an integral gain error, shifting the system's pole locations and reducing stability. A

This work was supported in part by the China Scholarship Council under Grant 202308210189, and in part by the National Natural Science Foundation of China under Grant 62176041.

finite gain-bandwidth product and slew rate can cause incomplete signal settling in the integrator, leading to nonlinear errors. Additionally, the op-amp's output swing directly impacts the integrator's dynamic range [4].

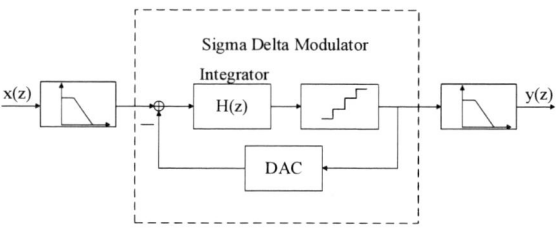

Fig. 1. Block diagram of a sigma-delta ADC.

MATLAB simulations have shown that a 24-bit sigma-delta ADC with OSR =128 requires the op-amp in its first integrator to have a load capacitor (sampling capacitor) of at least 15 pF, a DC gain of at least 120 dB, a gain-bandwidth product greater than 60 MHz, and a slew rate greater than 40 V/µs.

To meet the strict performance requirements of the op-amp, selecting an appropriate architecture is essential. Op-amps are classified based on the number of gain stages into single-stage, two-stage, and multi-stage architectures. Traditional single-stage op-amps offer fast response and simple topology; however, achieving an open-loop gain above 100 dB is usually difficult [5]. To enhance the gain of single-stage op-amps, the work in [6] employed a gain-boosting technique to improve a folded cascode amplifier's gain, reaching 112 dB with a 4 pF load capacitor, but the differential output swing was only 1.5 V. Two-stage op-amps, which cascade two amplifier stages, can reach gains exceeding 100 dB [7-8]. Their frequency compensation is moderately complex, making stability design relatively easy. The two-stage op-amp in [7], featuring a folded-cascode first stage and a common-source second stage, achieved a gain of 131 dB and a gain-bandwidth product of 1.26 MHz. In [8], a two-stage amplifier built with special transistors achieved an outstanding gain of 155 dB, though its gain-bandwidth product was only 0.198 MHz. Multi-stage amplifiers, which cascade three or more stages, can achieve high gain and a large output swing. However, their frequency compensation becomes considerably more challenging, and they require additional power and chip area. The work in [9] proposed a novel multi-stage amplifier scheme. Based on a current-feedback folded-

979-8-3315-8850-2/25 $31.00 © 2025 IEEE   697

cascode structure, the amplifier includes both a high-speed module and a high-gain module. Through complex frequency compensation, it achieved a gain of 105 dB and a gain-bandwidth product of 231 MHz.

Two-stage operational amplifiers are optimal for balancing various performance features, including gain, bandwidth, stability, output swing, and more. Therefore, this work introduces a two-stage op-amp design based on a SMIC 0.13μm CMOS process with a 3.3 V supply voltage. The first stage uses a modified folded cascode structure, while the second stage features a dynamic-biasing cascode amplifier. The proposed amplifier delivers high gain and a wide output swing while maintaining system stability through Miller compensation. Additionally, separate common-mode feedback circuits are designed for each stage to control the output common-mode voltage precisely.

II. CORE AMPLIFIER CIRCUIT DESIGN

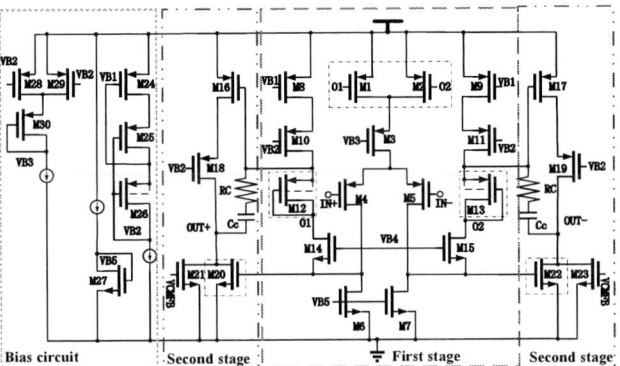

Fig. 2. Schematic of the proposed fully differential operational amplifier.

The core circuit of the proposed two-stage fully differential operational amplifier is shown in Fig. 2. A folded cascode amplifier with PMOS input transistors is used as the first stage, providing higher gain and lower noise [9]. The diode-connected transistors M12 and M13 act as level shifters to increase the gate voltages of M16 and M17, thereby reducing their overdrive voltages and improving the second-stage output swing. Transistors M18 and M16, and M19 and M17, form cascode amplifiers that boost output impedance and effectively compensate for the low intrinsic impedance of the PMOS common-source stage. Transistors M1 and M2 serve as the common-mode feedback circuit for the first stage. The gate voltages of M21 and M23 are controlled by a feedback signal to regulate the second-stage output common-mode voltage.

The proposed second-stage cascade amplifier differs from the conventional cascode architecture is primarily in its dynamic biasing scheme. Specifically, transistors M20 and M22 are biased by the drain of M6 and M7, respectively, which provides three advantages. First, it enhances slew rate during negative output transitions. When IN+ decreases, the gate voltage of M20 rises, supplying additional current to discharge OUT+, thereby improving the slew rate. Since the biasing is dynamic, this mechanism does not increase the quiescent current. Second, it offers a modest gain improvement. The amplifier's gain is slightly boosted due to additional signal

paths: from the drain of M6 to OUT+, and from the drain of M7 to OUT-. Finally, the output swing is increased. By biasing M20 and M22 from the drains of M6 and M7 (the lowest-voltage nodes in the first stage), the design allows for a reduced overdrive voltage, thereby extending the output swing.

To further maximum the output swing, a self-biased wide-swing cascade current mirror is implemented. The gate of M24 is connected to the drain of M25 to minimize the voltage drop to approximately two overdrive voltages. The bias voltage for M25 is generated by connecting its gate to the drain of the diode-connected PMOS transistor M26, which acts as a resistive element. Notably, the substrate of M26 is tied to its source to reduce its threshold voltage.

A. Gain

As shown in Fig. 2, in the first stage, the gain AV_1, from the input IN+ to the output (the drain of M10), is given by (1).

$$
\begin{aligned}
AV_1 &= -g_{M4}R_{O1} \\
&\approx -g_{M4}\{g_{M10}r_{o10}r_{o8} // [1/g_{M12} + g_{M14}r_{o14}(r_{o6}//r_{o4})]\} \quad (1) \\
&\approx -g_{M4}[g_{M10}r_{o10}r_{o8} // g_{M14}r_{o14}(r_{o6}//r_{o4})]
\end{aligned}
$$

g_{Mi} and r_{Mi} are the transconductance and output resistance of the transistor Mi (i=0,1,…,30), respectively, and R_{O1} is the equivalent output resistance seen from the drain of M10 to ground.

The gain AV_2, from the gate of M16 to the output OUT+ in the second-stage amplifier, is given by (2). R_{O2} is the equivalent output resistance seen from OUT+ to ground.

$$
AV_2 = -g_{M16}R_{O2} \approx -g_{M16}(g_{M18}r_{o18}r_{o16} // r_{o20} // r_{o21}) \quad (2)
$$

The gain AV_3, from the gate of M1 to the gate of M20, is given by (3). The gain AV_4, from the gate of M20 to OUT+, is given by (4).

$$
\begin{aligned}
AV_3 &\approx -g_{M4}\{[1/g_{M14} + g_{M10}r_{o10}r_{o8} / (g_{M14}r_{o14})] // r_{o4} // r_{o6}\} \\
&\approx -g_{M4}[1/g_{M14} + g_{M10}r_{o10}r_{o8} / (g_{M14}r_{o14})]
\end{aligned}
$$

$$
(3)
$$

$$
AV_4 \approx -g_{M20}R_{O2} \quad (4)
$$

Since $AV_1 >> AV_3$, and AV_2 is of the same order of magnitude as AV_4; the total gain, AV, is

$$
\begin{aligned}
AV &= AV_1 * AV_2 + AV_3 * AV_4 \approx AV_1 * AV_2 \\
&\approx g_{M4}g_{M16}[g_{M10}r_{o10}r_{o8} // g_{M14}r_{o14}(r_{o6}//r_{o4})] \quad (5) \\
&\quad *(g_{M18}r_{o18}r_{o16} // r_{o20} // r_{o21})
\end{aligned}
$$

The gain and bandwidth obtained from the PAC simulation are shown in Fig. 3. The DC gain is 122.6 dB, which is approximately equal to the value calculated from (5) and satisfies the application's requirement.

979-8-3315-8850-2/25 $31.00 © 2025 IEEE

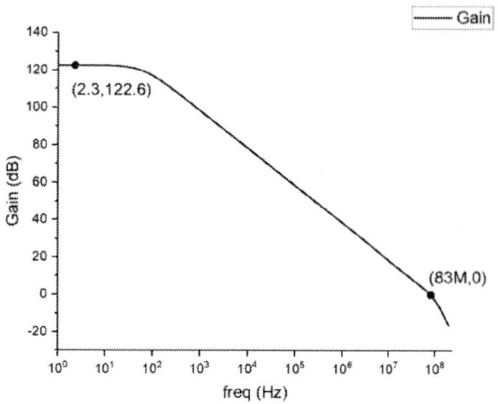

Fig. 3. Gain of the proposed amplifier.

B. Frequency Compensation

The proposed two-stage op-amp has two nearby low-frequency poles, located at the outputs of the first- and second-stage amplifiers, respectively, which can cause stability issues. To ensure stability, it is necessary to separate these two poles to achieve a satisfactory phase margin. To address this issue, Miller compensation is used to separate the two poles as shown in Fig. 2. The compensated poles are described by (6) and (7). C_c is the compensation capacitance, whereas C_L is the load capacitance, which has a value of 15 pF.

$$P_1 = -1/(R_{O1}g_{M16}R_{O2}C_c) \quad (6)$$

$$P_2 = -g_{M16}/C_L \quad (7)$$

However, this simple Miller compensation scheme introduces a right-half-plane zero, which can severely deteriorate the system's phase margin and thus degrade stability. Three methods can be adopted to tune this zero [10-13]. The first method inserts a unity-gain voltage buffer in the feedback path, which blocks the signal from flowing through the feed-forward path via the compensation capacitor to the output. However, this approach significantly reduces the output swing, and increases power consumption and chip area [10]. The second method utilizes a unity-gain current amplifier to cancel the current component flowing to the input node through the compensation capacitor, albeit at the expense of additional power consumption and chip area [11]. The third method, and the one adopted in this design, is to connect a nulling resistor in series with the Miller compensation capacitor to cancel the zero [12][13]. This method is easy to implement and does not introduce additional power consumption or area. The zero, z, is given by (8).

$$z = 1/[C_c(1/g_{M16} - R_z)] \quad (8)$$

Setting the zero z equal to the second pole P_2 eliminates the pole. R_z can be determined from (9).

$$R_z = (C_L + C_c)/(g_{M16}C_c) \quad (9)$$

In the design, $C_C = 22$ pF and $R_Z = 195$ Ω.

The phase diagram obtained from a PAC simulation shows a phase margin of 58 , as illustrated in Fig. 4.

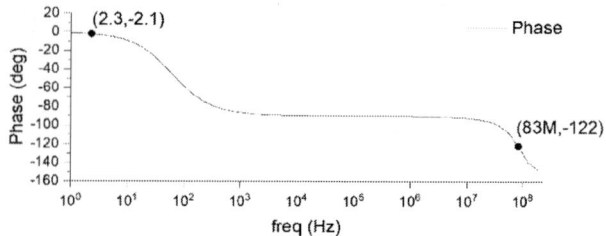

Fig. 4. Phase change of the proposed amplifier.

C. Gain Bandwidth Product

The gain-bandwidth product of the op-amp is given by (10). It is approximately equal to the simulation result shown in Fig. 3, which is about 83 MHz.

$$GBW = g_{m_4} R_0 g_{m_{16}} R_1 * 1/(2\pi R_0 * g_{m_{16}} R_1 C_c)$$
$$= g_{m_4}/(2\pi C_c) \quad (10)$$

D. Slew Rate

The Miller compensation capacitor, C_C, limits the slew rate of this operational amplifier. The positive slew rate and negative slew rate, which correspond to the output slewing from low to high and high to low, respectively, are given by (11) and (12). I_{Mi} (i=6, 8) represents the maximum current flowing through the transistor Mi.

$$SR+ = (I_{M6} - I_{M8})/C_c \quad (11)$$

$$SR- = I_{M8}/C_c \quad (12)$$

For the transient simulation of the op-amp, a square wave signal with an amplitude of 3.3 V and a frequency of 10 kHz is used as the input. The resulting slew rates are measured as SR+ = 46 V/μs and SR− = 53 V/μs, as shown in Fig. 5. The measured value of SR+ closely matches the theoretical prediction from (11), whereas SR- exceeds the calculated value in (12) owing to the dynamic biasing effects.

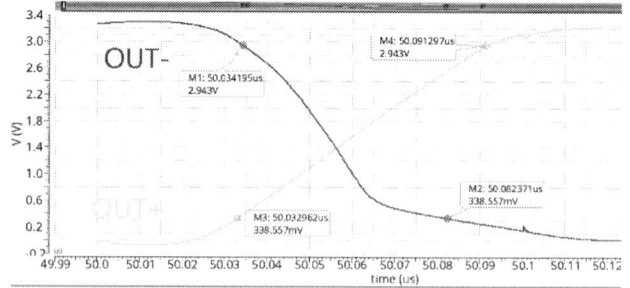

Fig. 5. Output waveform of VOUT.

III. COMMON-MODE FEEDBACK CIRCUIT DESIGN

In the design of a fully differential op-amp, the common-mode feedback (CMFB) circuit is a critical block that ensures the stability of the common-mode output. Furthermore, it also impacts the amplifier's linearity and dynamic range.

The first-stage CMFB circuit in this design consists of transistors M8-M15, along with the input pair M1 and M2, which sense a positive disturbance in the first-stage's output

979-8-3315-8850-2/25 $31.00 © 2025 IEEE

common-mode voltage if M12 and M13 are activated. The voltage at nodes O1 and O2 then increases, which decreases the gate-to-source voltage of M1 and M2. This action reduces the current supplied by the tail current source M3 while simultaneously increasing the current through M4 and M5. Consequently, the source voltage of M12 and M13 drops, effectively counteracting the initial increase in the common-mode voltage of the first stage. Due to the perfect symmetry of M1 and M2, the first-stage differential output signal does not influence the current of M3 and the output common-mode voltage.

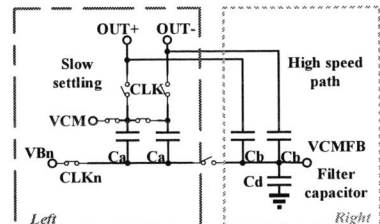

Fig. 6. Switched capacitor common mode feedback circuit.

The second-stage CMFB employs a switched-capacitor feedback circuit, which is well-suited for use in switched-capacitor integrators [14], as shown in Fig. 6. The bias voltages, VCM and VBn, are set to 1.65 V and 0.873 V, respectively.

As shown in Fig. 6, the left part of the circuit stabilizes its common-mode voltage over several clock cycles. During the CLKn phase, the capacitor C_a is charged by a voltage of VCM−VBn. Then, during the CLK phase, the top plate of the capacitor is switched to connect with the outputs, OUT+ and OUT-, and the bottom plate connects to the common-mode feedback voltage, VCMFB. By applying the law of charge conservation, (13) can be obtained.

$$(V_{CM} - V_{Bn})2C_a = (V_{out+} - V_X)C_a + (V_{out-} - V_X)C_a \quad (13)$$

Where V_X represents the voltage at the bottom plate of capacitor C_a after a complete clock cycle, the value of V_X can be determined by solving (13) and applying it to (14).

$$V_X = V_{Bn} + (V_{out+} + V_{out-})/2 - V_{CM} \quad (14)$$

After the circuit has stabilized, VCMFB = V_X. If the common-mode voltage of OUT+ and OUT- equals VCM, then the voltage at VCMFB will be equal to VBn. However, if the common-mode voltage of OUT+ and OUT- becomes greater than VCM, then VCMFB will also become greater than VBn. This increases the gate-to-source voltage of transistors M21 and M23, which in turn causes the output common-mode voltage to decrease, thereby stabilizing this voltage.

On the right side of the circuit, capacitor C_b offers a quick path for the common-mode signal, while capacitor C_d functions as a filter to stabilize the output voltage.

To verify the stability of the common-mode feedback loop, a simulation was performed on the complete operational amplifier. The results, shown in Fig. 7, confirm the stability of the loop.

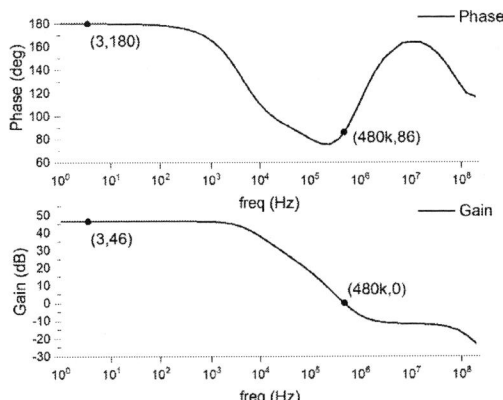

Fig. 7. Simulation results of the stability of the common mode feedback.

IV. LAYOUT DESIGN AND PERFORMANCE VERIFICATION

The layout of the operational amplifier is shown in Fig. 8. Traces that carry high currents are widened to ensure the long-term reliability. And common-centroid layout is employed for critical circuits with strict matching requirements, such as differential pairs and current mirrors. Additionally, dummy transistors are strategically placed around the core circuitry.

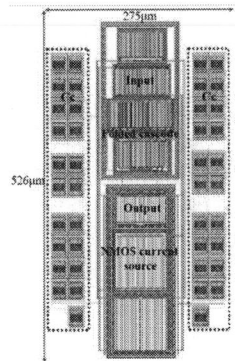

Fig. 8. Layout of the proposed amplifier.

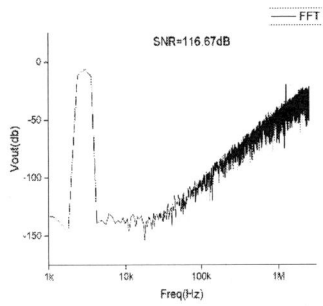

Fig. 9. FFT result of the output signal of the high-resolution sigma delta ADC.

The designed operational amplifier was integrated into a third-order sigma-delta modulator for simulation. FFT analysis of the output signal produced the frequency plot shown in Fig. 9. The results demonstrate that the modulator achieves a signal-to-noise ratio (SNR) of 116.67 dB, confirming that the integrator meets the design requirements.

979-8-3315-8850-2/25 $31.00 © 2025 IEEE

Table 1 compares the op-amp parameters from this work with those from other studies. The table shows that the op-amp designed here achieves a high gain, as well as a high slew rate and a high gain-bandwidth product, even with a large load capacitance.

TABLE I.   COMPARISON OF OP-AMP PARAMETERS

Specifications	This work	[6]	[7]	[8]	[9]	[10]
VDD(V)	3.3	2	1.8	3.3	1.8	1.3
Gain(dB)	122	112	131	155	105	100
GBW(MHz)	83	616	1.26	0.198	231	80
PM(degree)	58	67	60	59	53	57
SR+(V/µs)	46	N/A	N/A	N/A	13	56
SR-(V/µs)	53	44	N/A	N/A	13	N/A
C_L(pF)	15	N/A	1	10	5	16
Power(mW)	22.6	2	0.0046	0.0065	0.85	7.8

V. CONCLUSION

To address the stringent requirements of high-resolution sigma-delta ADCs for op-amps, a high-performance two-stage fully differential operational amplifier was proposed. Based on a SMIC 0.13 µm CMOS process with a 3.3 V supply voltage and a 15 pF load capacitance, the op-amp achieves a DC gain of 122 dB, a gain-bandwidth product of 83 MHz, and a phase margin of 58°. By completing a high DC gain and an excellent gain-bandwidth product, while ensuring good stability, this op-amp provides a key foundation for implementing high-resolution sigma-delta ADCs.

REFERENCES

[1] B. Razavi, *Design of analog CMOS integrated circuits*, 2nd ed. New York, NY: McGraw-Hill Education, 2017, pp. 1-5.

[2] N. KiYoung, N. L. Sang-Min, D. K. Su, and B. A. Wooley, "A low-voltage low-power sigma-delta modulator for broadband analog-to-digital conversion," *IEEE J. Solid-State Circuits*, vol. 40, no. 9, pp. 1855–1864, August 2005.

[3] J. C. Candy and G. C. Temes, *Oversampling delta-sigma data converters : theory, design, and simulation*. New York, NY, USA: IEEE Press, 1992, pp. 31-33.

[4] F. Maloberti, *Data converters*. Dordrecht, The Netherlands: Springer, 2007, pp. 268-275.

[5] Y. Li *et al.*, "A 10-kHz BW 104.3-dB DR discrete-time delta-sigma modulator with ring-amplifier-based integrator," *Microelectron. J.*, vol. 144, p. 106076, February 2024.

[6] S. Sarkar, A. Ghosh, and T. Chatterjee, "Design of a 2 mW Power, 112 dB Gain-Boosted, Folded Cascode Amplifier in 0.18µm Process," in *2021 IEEE 30th International Symposium on Industrial Electronics (ISIE)*, Kyoto, Japan: IEEE, June 2021, pp. 1–5.

[7] M. T. Nguyen, V. H. Nguyen, M. K. Hoang, T. S. Pham, and X. T. Pham, "Design a low power, 100dB operational amplifier using CMOS technology," *J. Sci. Technol - HaUI*, vol. 59, no. 2A, pp 203-207,March 2023.

[8] L. H. Rodovalho, C. R. Rodrigues, and O. Aiello, "A Two-Stage Single-Ended OTA with improved composite transistors," in *2021 IEEE Nordic Circuits and Systems Conference (NorCAS)*,Oslo,Norway, October 2021, pp. 1–7.

[9] P.-Y. Kuo and S.-D. Tsai, "An enhanced scheme of Multi-Stage amplifier with High-Speed High-Gain blocks and recycling frequency Cascode circuitry to improve Gain-Bandwidth and slew rate," *IEEE Access*, vol. 7, pp. 130820–130829, October 2019.

[10] V. Soman and S. S. Mande, "Design of a Two-Stage Folded Cascode Amplifier Using SCL 180 nm CMOS Technology," in *International Conference on Communication, Computing and Electronics Systems*, vol. 637, V. Bindhu, J. Chen, and J. M. R. S. Tavares, Eds. Singapore: Springer Singapore, 2020, pp. 423–430.

[11] Y. P. Tsividis and P. R. Gray, "An integrated NMOS operational amplifier with internal compensation," *IEEE J. Solid-State Circuits*, vol.11,pp.748–753,December 1976.

[12] G. Palmisano and G. Palumbo, "An optimized compensation strategy for two-stage CMOS op amps," *IEEE Trans. Circuits Syst. I*, vol. 42, no. 3, pp. 178–182, March 1995.

[13] C.Markis and C.Toumazou, "A Current-mode active compensation techniques,"*Elctron Lett.*,vol.26,no.21,pp.1792-1794,October 1990.

[14] F. Maloberti, *Analog design for CMOS VLSI systems*. Boston, MA, USA: Kluwer Academic, 2001, pp. 295-297.

2025 The 10th International Conference on Integrated Circuits and Microsystems

Design of Power Amplifier for Short Message Application of BDS-Ⅲ Communication Terminal

Zhenbing Li
School of integrated circuit science and engineering University of Electronic Science and Technology of China ChengDu, China lizhenbing@uestc.edu.cn

Weijun Li
Unit 78118 of the People's Liberation Army Unit 78118 of the People's Liberation Army ChengDu, China aaronliweijun@163.com

Haoyang Sun
School of Information and Communication Engineering University of Electronic Science and Technology of China ChengDu, China sunhaoyang21010107@163.com

Jiaxin Liu
School of integrated circuit science and engineering University of Electronic Science and Technology of China ChengDu, China liujiaxin@uestc.edu.cn

Jialong Fu
School of Information and Communication Engineering University of Electronic Science and Technology of China ChengDu, China 19108213841@163.com

Yaocheng Shang
School of Information and Communication Engineering University of Electronic Science and Technology of China ChengDu, China 1607088768@qq.com

Abstract—For the application requirements of the short message function of handheld Beidou satellite communication terminal, this paper proposes an L-band power amplifier chip with high gain, high efficiency and high integration, which is designed based on InGaP/GaAs HBT and LGA PCB packaging process, and a multi-physics collaborative simulation and optimization platform suitable for the design of similar RF power amplifier chips. The simulation results show that under 3.5V supply voltage, the linear gain of the chip can reach 38.7dB, P_{1dB} can reach 35.1dBm, PAE can reach 48.2%, the second, third and fifth harmonic suppression ratio is better than -65dBc, and the chip size is only 4mm×6mm×1mm. This design has obvious advantages over the same type of designs, and can meet the short message function of the hand-held Beidou satellite navigation communication terminal in a variety of application scenarios.

Keywords—BDS-III, Satellite communication terminal, Power amplifier chip, High gain, High integration

I. INTRODUCTION

Compared with the other three satellite navigation systems in the world, China's Beidou satellite navigation system has a unique short message communication function[1]. In some environments without signal coverage or with poor communication network, its terminals can communicate with each other through BDS-III (Beidou satellite navigation system-III), which plays an important role in satellite communication, earthquake rescue, field positioning, etc[2]. In which, the power amplifier (PA) chip is the core device of BDS-III terminals to realize short message function, which affects the quality of short message communication function.

Restricted by the satellite receiving sensitivity, the limited battery capacity of wireless terminals and the trend of terminal miniaturization, the PA chip of satellite communication terminals is required to have high transmission power, high

energy conversion efficiency and high integration[3-8]. In terms of transmission power, [3-4] have achieved relatively outstanding output power in their respective designed frequency bands. However, modern satellite communication systems not only need to improve the output power of terminal PA chip, but also need to take into account the efficiency, so that the service life of battery for wireless terminals can be improved. [5-6] not only ensures the output power of PA chip, but also takes into account the efficiency. At the same time, the requirements for miniaturization of modern satellite communication terminals are also increasing, so it is necessary to take into account the integration of PA chips while achieving high power and efficiency[7-8].

For BDS-III, its communication frequency band is in L-band. Due to the short message function requirements of the system, the battery powered wireless terminal requires that the output power of PA chip reach 3W. While the PA chip is miniaturized and fully functional, other performance parameters should be at high level of the industry. Therefore, many scholars have carried out corresponding research on this demand [9-11]. However, there has never been a PA chip with both gain, efficiency and integration to meet the comprehensive requirements of BDS-III terminals.

To solve the above problems, in order to meet the short message communication function of handheld BDS-III terminals (battery powered), based on InGaP/GaAs HBT and LGA PCB packaging process, a BDS-III terminal PA chip working in the frequency band of 1.6-1.65GHz is designed by using parallel stable circuit architecture, temperature insensitive adaptive bias technology and multi-physics collaborative simulation and optimization technology. The design and simulation results show that the PA chip has high gain, high efficiency and high integration, and its functions and performance are at a high level in the industry, which can meet

This work is supported by NSFC under Grant 62174023.

979-8-3315-8850-2/25 $31.00 © 2025 IEEE

the short message function of BDS-III terminal and the requirements of different application scenarios.

II. CIRCUIT DESIGN

A. InGaP/GaAs HBT Process

Considering the performance and process cost in L-band, the InGaP/GaAs HBT process is selected to design the MMIC die of this PA chip. The results show that the second generation semiconductor material GaAs doped with InGaP has good lattice matching, and improves the band gap width of the second generation semiconductor material, and the smaller band gap increases the current gain; The high concentration of doped emitter material passivates the base surface of HBT and reduces the composite current effect, which is suitable for satellite communication. The I-V characteristic curve of InGaP/GaAs HBT (emitter size is emitter finger width × emitter finger length × emission index: 3μm × 40μm × 4) used in this paper is shown in Fig. 1. The V_{knee} of the HBT is 0.5V.

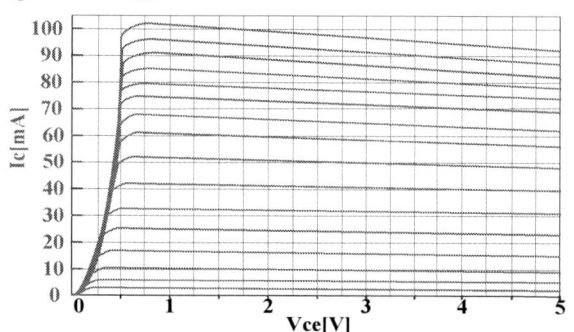

Fig. 1. I-V characteristics of HBT (emitter size 3μm×40μm×4) selected in this paper

B. Power Amplifier Design

Based on the RF characteristics of InGaP/GaAs HBT devices, this paper uses a three-stage cascade amplifier circuit structure to achieve a gain greater than 37dB. The circuit structure is shown in Fig. 2, which is composed of MMIC die and LGA PCB. Each HBT of each stage amplifier circuit is connected in series with a pair of parallel resistors and capacitors (stability circuit) to form a unit cell, which improves the RF stability of the entire power amplifier chip. The output stage amplification circuit of the power amplifier chip uses 36 HBT cells with an emitter size of 3μm × 40μm × 4 in parallel to achieve P_{1dB} greater than 35dBm (3W), and the gain of the output stage amplification circuit is greater than 12dB; the drive stage (intermediate stage) amplification circuit uses 8 HBT cells of 3μm × 40μm × 4 in parallel, which can provide linear output power greater than 24dBm and gain greater than 13dB; the input stage uses 2 HBT cells of 3μm × 40μm × 4 in parallel to provide linear output power greater than 13dBm and gain greater than 13dB.

The power amplifier chip MMIC die designed in this paper is shown in Fig. 3(a), including three-stage amplification circuit, three-stage adaptive bias circuit, matching circuit at all stages (except output matching circuit), temperature insensitive adaptive power detection circuit, and ESD electrostatic protection circuit. The LGA PCB is designed as shown in Fig.

3(a), including high-frequency choke RF chokes of amplifier circuits at all stages, bypass capacitors, part of matching network circuits at all stages, part of temperature insensitive adaptive power detection circuit and class-F output matching network circuit with harmonic suppression function. The three-dimensional diagram of LGA PCB packaging design is shown in Fig. 3(b).

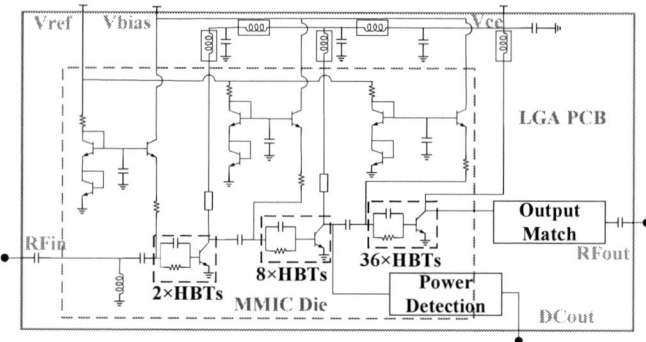

Fig. 2. circuit architecture of power amplifier chip (L-band, 38.7dB gain, 3W output power) designed in this paper

Fig. 3. (a) power amplifier chip MMIC DIE+LGA PCB; (b) Three dimensional diagram of power amplifier chip LGA packaging

In this paper, the RF stability cell circuit design method, the temperature insensitive adaptive bias circuit technology method, and the multi-physics collaborative simulation and optimization technology method are used to improve the overall performance of the power amplifier chip. The above key technologies will be described separately below.

C. RF Stable Cell Circuit Design

For this type of high gain power amplifier chip designed in this paper, we need to focus on its RF stability, which is the key to suppress the self excitation phenomenon of high gain power amplifier chip. This is because there are many parasitic parameters inside the PA chip, such as the junction capacitance, wiring inductance and capacitance of the transistor, which will form additional feedback paths between the input and output of the amplifier. When the signal frequency is high or the amplifier gain is large, these parasitic feedback may cause the signal to be continuously amplified in the amplifier, and finally cause self-excited oscillation. Therefore, in the PA design, the S-parameter network analysis is generally required to ensure that the stability factor of the power amplifier two-port network meets the following conditions:

$$K = \frac{1 - |S11|^2 - |S22|^2 + |\Delta|^2}{2|S12||S21|} > 1 \qquad (1)$$

$$|\Delta| = |S11 \cdot S22 - S12 \cdot S21| < 1 \qquad (2)$$

979-8-3315-8850-2/25 $31.00 © 2025 IEEE

The K factor of HBT used in this paper is less than 1 in the frequency range of 0-2GHz, indicating that it is not stable. Therefore, this paper uses the stability circuit shown in Fig. 4 to improve the RF stability of cells and power amplifiers as a whole. Unlike the traditional stability circuit, which uses the method of connecting small resistors in series, this paper connects each HBT of each stage of amplification circuit with a pair of parallel resistors and capacitors in series to form a unit cell. The design takes into account the improvement of RF stability and the insertion loss of RF signal. The resistance is used to offset the negative impedance at low frequency of the port, and the capacitance is used to reduce the loss of fundamental signal. After introducing the stability improvement circuit, the stability simulation results of the power amplifier are shown in Fig. 5. From the simulation results, it can be seen that the RF stability of the power amplifier is good, and the K factor is greater than 1 in the working frequency band but not too high, which is conducive to the realization of high gain of the power amplifier.

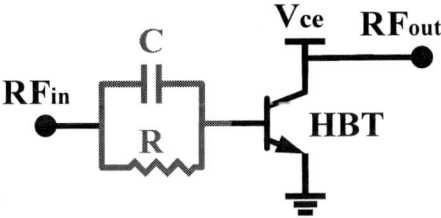

Fig. 4.　stability improvement circuit structure

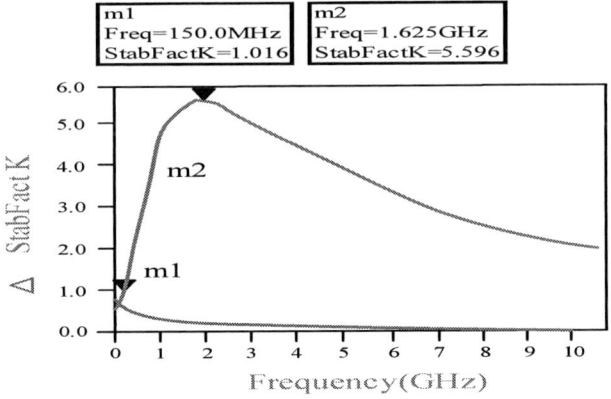

Fig. 5.　stability simulation results of PA chip

D. Design of Temperature Insensitive Adaptive Bias Circuit

With the change of input power RF$_{in}$ and temperature, the offset of the static operating point of the power amplifier chip will lead to serious phase distortion and gain compression, which will eventually deteriorate the linearity of the power amplifier chip. In order to improve the linearity of the power amplifier chip, this paper uses the on-chip temperature insensitive adaptive bias circuit shown in Fig. 6 to provide temperature insensitive adaptive bias points for all stages of amplifier circuits, and it has a high degree of integration. The transistor HBT0 in Fig. 6 represents the amplification circuits at all stages in the power amplifier chip. The temperature insensitive adaptive bias circuit is connected to HBT base in the

amplification circuits at all stages through the ballast resistor R1. The voltage relationship of each node is as follows:

$$V_{be\_HBT0} = V_B - R_1 I_b - V_{be\_HBT1} \qquad (3)$$

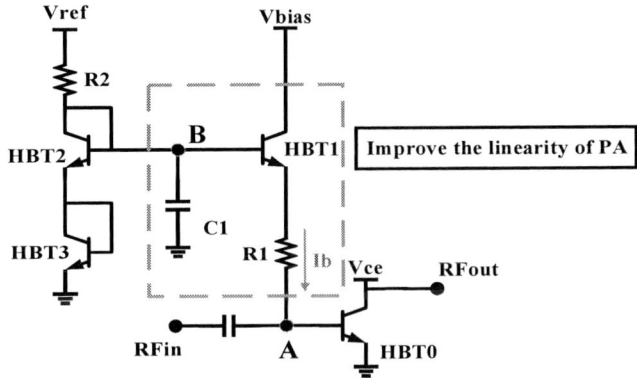

Fig. 6.　temperature insensitive adaptive bias circuit used in this paper

Due to the rectification characteristics of the base emitter diode, the junction voltage V_{be_HBT0} of the transistor HBT0 varies with the input power RFin or temperature. Through the temperature insensitive adaptive bias circuit, part of the RF signal is leaked to the bias circuit through resistor R1. Since the base emitter diode of HBT1 also has rectification characteristics, the base emitter voltage V_{be_HBT1} of HBT1 also changes with input power or temperature. From equation (3), it can be seen that the change of V_{be_HBT1} can compensate for the change of V_{be_HBT0}, that is, the bias point of transistor HBT0 is stabilized, which effectively avoids the problems of phase distortion and gain compression, and improves the temperature insensitive characteristics and linearity of the power amplifier. The design of the bias circuit ensures that the RF signal leaked into the bias circuit will be short circuited to the ground through the bypass capacitor C1. While protecting the power supply port, it stabilizes the potential at point B and ensures the stability of the compensation of the base emitter junction voltage V_{be_HBT1} of HBT1 to the base emitter junction voltage V_{be_HBT0} of HBT0. At the same time, adding the ballast resistor R1 can effectively limit the current supplied by the bias circuit to the base of HBT0, so as to further improve the temperature insensitive characteristics and linearization improvement ability of the power amplifier.

Based on the above research on temperature insensitive adaptive bias circuit, this paper summarizes the components to be determined and designed in this type of bias circuit, including: HBT1 type and size, HBT2 type and size, HBT3 type and size, C1 capacitance, R1 resistance and R2 resistance. The design and calculation methods are as follows:

a. HBT1 Type and Size

After the maximum base bias current I_{HBT0_beDmax} of HBT0 is determined, the drive current that HBT1 needs to provide to HBT0 is obtained, i.e. $I_{HBT1\_ceD} = I_{HBT0\_beDmax}$. In order to avoid induced current collapse, when HBT1 provides I_{HBT0_beDmax}, it should work in the amplification region with relatively stable V-I characteristic curve, so it can determine which type and size of HBT is suitable for HBT1. When the required size exceeds the maximum size of a single HBT provided by the process, it is considered to use multiple HBTs in parallel as HBT1.

b. HBT2 Type and Size

After determining the type and size of HBT1, in order to reduce power consumption, the selected size of HBT2 should be much smaller than that of HBT1, which is usually one tenth of the size of HBT1.

c. HBT3 Type and Size

The type and size of HBT3 shall be the same as that of HBT2.

d. Value of R2

The current flowing through R2 is actually the sum of HBT2 base current, HBT2 collector current and HBT1 base current, so the resistance R2 can be calculated according to equation (4), where V_{ref} is the reference voltage of the bias circuit, V_{HBT1_B} is the HBT1 base voltage, I_{HBT2_beq} is the HBT2 base emitter static current, I_{HBT1_beq} is the HBT1 base emitter static current, and I_{HBT2_ceq} is the HBT2 Collector Emitter static current.

$$ R_2 = \frac{V_{ref} - V_B^{HBT_1}}{I_{beq}^{HBT_2} + I_{beq}^{HBT_1} + I_{ceq}^{HBT_2}} \qquad (4) $$

e. Value of R1

Kirchhoff laws can obtain equation (5) of the voltage at point B in the bias circuit, where V_B is the voltage at point B, V_{HBT1_Be} is the base emitter voltage of HBT1, V_{HBT0_Be} is the base emitter voltage of HBT0, and I_{HBT1_ceq} is the static current of HBT1 collector-emitter.

$$ V_B = V_B^{HBT_1} = V_{be}^{HBT_1} + I_{ceq}^{HBT_1} \cdot R_3 + V_{be}^{HBT_0} \qquad (5) $$

The equation for calculating resistance R1 can be obtained from equation (5)

$$ R_1 = \frac{V_B^{HBT_1} - V_{beD}^{HBT_0} - V_{beD}^{HBT_1}}{I_{ceD}^{HBT_1}} \qquad (6) $$

f. Value of C1

The self resonant frequency of the capacitor needs to be less than the working frequency band, and its capacitive reactance needs to present a minimum value in the working frequency band to realize the bypass to the ground of the target RF signal. At the same time, the size should not be too large to facilitate the chip layout.

E. Design of Multi-physics Collaborative Simulation and Optimization Platform

For the research and development of high-power RF integrated circuits such as power amplifier chips, while ensuring their RF performance, it is necessary to analyze and optimize their thermal and stress, study the MMIC die of power amplifier chips and the rules of efficient heat dissipation methods, thermally induced shape and stress changes in packaging, improve the distribution of thermal and stress fields in the circuit layout, and make them decentralized and uniform. On this basis, iterative optimization is carried out in combination with circuit and electromagnetic simulation, and finally achieve high-performance and high reliability chip layout and packaging structure. However, the current mainstream commercial EDA simulation software can only simulate the single characteristics

of RF integrated circuits, and the data interaction between different EDA software needs to be completed manually, which has the problems of low efficiency and error prone, and it is difficult to carry out comprehensive simulation analysis of multiple characteristics, let alone realize the collaborative design optimization of multiple physical fields. Therefore, based on the functions and characteristics of circuit and system simulation design software ADS, electromagnetic field characteristics simulation and analysis software HFSS, and thermodynamic characteristics simulation and analysis software ANSYS, this paper studies its simulation core algorithm, input/output interface protocol, data format and control mechanism, uses MATLAB as the main control terminal, designs and develops the control interface program.

The system architecture of the multi-physics collaborative simulation and optimization platform proposed in this paper is shown in Fig. 7. It uses MATLAB to call ADS, HFSS and ANSYS for circuit, electromagnetic, thermal and stress cycle simulation and optimization. Among them, ADS uses device SPICE model and SNP file to simulate and optimize the circuit structure based on the method of moments, HFSS uses finite element method and automatic mesh generation algorithm to conduct electromagnetic simulation of 3D structure, and ANSYS uses solid finite element method to calculate and simulate the thermal and stress equations of the target circuit. In this platform, HFSS and ANSYS can carry out parametric modeling based on their own scripts, and the 3D models can be transferred to each other through MATLAB.

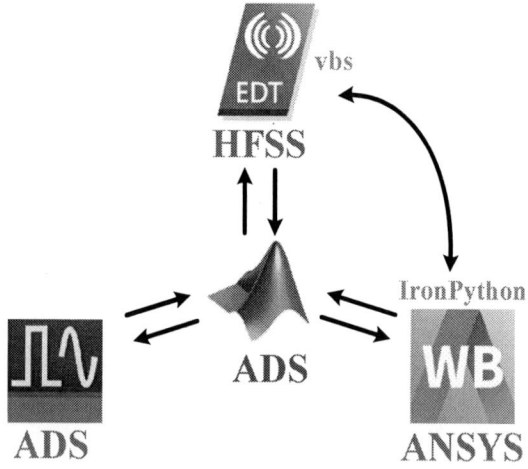

Fig. 7. collaborative simulation and optimization platform for circuit, electromagnetic, thermal and mechanical multi physical fields

Taking the high gain, high efficiency and high integration power amplifier designed in this paper as an example, the initial schematic diagram and layout of the power amplifier are designed and built in ADS, and the first circuit performance simulation is performed. The dataset file is retained when the performance of the power amplifier circuit initially meets the predetermined indicators. According to the preliminary design of the power amplifier layout and the laminated structure information provided by OEM, the three-dimensional electromagnetic simulation model of the circuit (passive circuit part) is programmed by calling HFSS vbs script through MATLAB, and the pre-processing work such as material characteristics assignment, boundary conditions and sweep range setting is completed. When the preparation work is completed, the multi-physics collaborative simulation and optimization platform automatically performs electromagnetic simulation and exports SNP to the circuit electromagnetic collaborative simulation platform built in ADS. After reading the new SNP path, MATLAB calls ADS to conduct circuit electromagnetic co-simulation and decide whether the circuit meets the expected indicators. If it does not meet the requirements, MATLAB will call the system level optimization algorithm to optimize and modify the parameters of capacitors, inductors, microstrip lines and other devices, so as to make the designed circuit meet the circuit electromagnetic performance indicators. If the circuit electromagnetic performance meets the index requirements, MATLAB controls ADS to automatically calculate the thermal dissipation power of each transistor under the saturated output power state through the current probe, voltage probe and power probe, and export the results to the ANSYS simulation script. The thermal dissipation power is determined by DC power consumption and input/output power. The calculation equation is:

$$P_{diss} = P_{DC} - (P_{out} - P_{in}) \qquad (7)$$

When the main control program of MATLAB detects the new thermal dissipation power data, it calls the SCDM module of ANSYS to automatically parameterize the thermodynamic model of the power amplifier, and completes the operation of material assignment, heat source distribution, simulation environment condition setting, etc. after the simulation grid division is completed, MATLAB immediately calls its mechanical module to automatically carry out the steady-state thermal simulation and thermal stress static structure simulation. If the temperature of the power amplifier in the saturated working state exceeds the safety threshold, the platform will trigger the parametric adjustment of the circuit layout or device size according to the temperature distribution information, and return to the starting point of the process for iterative optimization until the threshold requirements are met; If the temperature does not exceed the safety threshold, the multi-physics collaborative simulation and optimization platform ends the iteration process. At this time, the target circuit obtained through the multi-physics collaborative simulation and optimization platform can theoretically ensure that the circuit, electromagnetic, thermal and stress characteristics of multiple physical fields can meet the design requirements at the same time.

The 38.7dB gain, 3W output power amplifier chip designed in this paper, with the help of the multi-physics collaborative simulation and optimization platform, finally achieves the global standard of circuit, electromagnetic, thermal and mechanical characteristics. At room temperature, the simulation results of the power amplifier chip during operation show that its maximum junction temperature is 152.42℃ and the average junction temperature is 67.652℃, both of which are within the process safety range of the device.

III. SIMULATION AND ANALYSIS

In this paper, the above key technologies and methods are used to design L-band high gain, high efficiency and high integration amplifier chip based on InGaP/GaAs HBT and LGA PCB packaging process. The simulation performance parameters are shown in Table I. The DC and RF simulation performance parameters and thermodynamic characteristics simulation results in Table I are described in detail below.

A. DC Characteristics

Based on the multi-physics collaborative simulation and optimization platform, the static current of the power amplifier chip designed in this paper is 262mA. The static power consumption is at the industry-leading level in the design of the same type of high gain, high-power power amplifier chip, which is very suitable for integration in wireless communication terminals with limited battery capacity, such as mobile phones and satellite mobile phones.

B. RF Characteristics

Also based on the multi-physics collaborative simulation and optimization platform, the large signal and power characteristics of the power amplifier chip designed in this paper are shown in Table I. When the power supply voltage is 3.5V, the chip shows absolute stability in the 1.6-1.65GHz frequency band and power dynamic range; The input standing wave ratio is less than 1.6:1, and the output standing wave ratio is less than 1.9:1, indicating that the input and output match well; The in band gain flatness is less than ± 0.35dB. The power characteristic simulation results are shown in Fig. 8. The linear gain of the power amplifier chip at 1.625GHz (typical value) reaches 38.7dB, its P_{1dB} reaches 35.1dBm, P_{sat} reaches 36.2dBm, and PAE reaches 48.2%; 2. The third and fifth harmonic suppression ratio is better than -65dBc, while the AM-PM distortion is less than |2.5°|, and the third-order intermodulation distortion ratio (IMD3) is less than -25dBc.

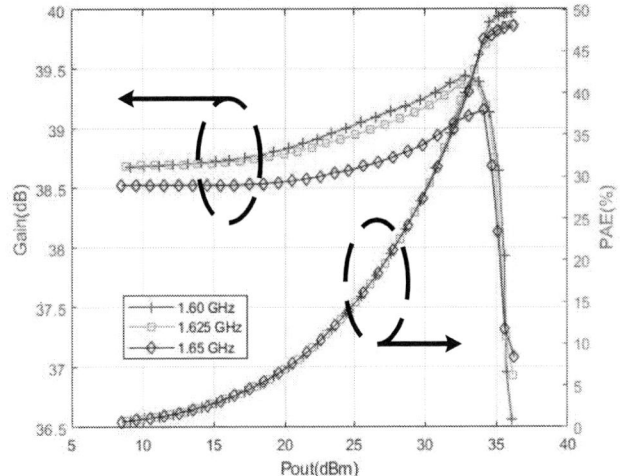

Fig. 8. RF power characteristic simulation results of the power amplifier chip designed in this paper under the supply voltage of 3.5V

TABLE I. DC AND RF SIMULATION PERFORMANCE OF POWER AMPLIFIER CHIP DESIGNED IN THIS PAPER

No.	DC And RF Simulation Performance of Power Amplifier Chip Designed in this Paper	
	Specification	*Performance*
1	*Linear Gain*	*38.7dB*
2	*P1dB*	*35.1dBm*
3	*Psat*	*36.2dBm*
4	*PAE*	*48.2%*
5	*IMD3*	*-25dBc*
6	*HSR(2nd,3rd,5th)*	*-65dBc*
7	*AM-PM*	*|2.5°|*
8	*VSWRin*	*1.6:1*
9	*VSWRout*	*1.9:1*
10	*Size*	*4 mm✕6 mm✕1mm*

C. Comparison

Table II lists the performance comparison between the power amplifier chip designed in this paper and similar chips. It can be seen from the table that the chip designed in this paper has advantages in gain, efficiency and cost.

TABLE II. PERFORMANCE COMPARISON WITH STATE-OF-THE-ART DESIGNS

Paper	Performance Comparison with State-of-the-Art Designs			
	proces	*Linear Gain*	*PAE*	*Cost*
[7]	*GaN*	*10dB*	*35%*	*high*
[8]	*GaN*	*25dB*	*30.2%*	*high*
[9]	*GaAs*	*38dB*	*36.5%*	*more cost-effective*
[10]	*GaAs*	*30dB*	*43%*	*more cost-effective*
[11]	*GaAs*	*30dB*	*45%*	*more cost-effective*
This paper	*GaAs*	*38.7dB*	*48.2%*	*more cost-effective*

IV. SUMMARY

In this paper, a high-performance L-band power amplifier chip is designed and verified to meet the requirements of short message application of BDS-III satellite communication terminal. Based on InGaP/GaAs HBT and LGA PCB packaging process, a three-stage cascade amplifier circuit structure is used to achieve high gain. By innovatively introducing the parallel stable cell circuit, the RF stability of the chip in the full frequency band and power range is effectively guaranteed. The designed on-chip temperature insensitive adaptive bias circuit significantly improves the linearity of the power amplifier under different input power and operating temperature. In order to overcome the challenge of multi-physics coupling in the design of high-power RF integrated circuits, a set of circuit electromagnetic thermal stress multi-physics collaborative

simulation and optimization platform based on MATLAB is developed, which significantly improves the design efficiency and the overall achievement of the design goal.

The simulation results show that under 3.5V power supply, the PA achieves 38.7dB linear gain, 35.1dBm P_{1dB} and 48.2% PAE. And the package size is only 4mm×6mm×1mm. Compared with similar designs reported in the literature, the PA chip in this paper has obvious advantages in gain, efficiency and cost. The design fully meets the requirements of handheld BDS-III terminal for high transmission power, high energy efficiency, high linearity, high integration and reliability, and can reliably support its short message communication function in various scenarios.

ACKNOWLEDGMENT

This work is supported by NSFC under Grant 62174023. The corresponding authors are Jiaxin Liu and Zhenbing Li.

REFERENCES

[1] K. W. Xia, J. Feng, Q. R. Wang, et al, "Optimal Selection Mechanism of Short Message Terminal for 'Beidou-3'," 2020 IEEE 5th International Conference on Signal and Image Processing (ICSIP). Nanjing, China, 2020, pp. 1106-1111.

[2] W. Liang, Y. Zeng, W. Zhu, et al, "Research on Beidou New Generation Emergency Group Communication Decision-Making Mechanism," 2022 IEEE 10th International Conference on Information, Communication and Networks (ICICN), Zhangye, China, 2022, pp. 15-19.

[3] M. L. Bhavsar, P. Srivastava, D. K. Singh, et al, "K-Band 8-Watt Power Amplifier MMICs using 150nm GaN process for Satellite Transponder," 2021 IEEE MTT-S International Microwave and RF Conference (IMARC), KANPUR, India, 2021, pp. 1-4.

[4] R. Kalyan, B. Ghosh, M. K. Sreekavya, et al, "Design of a 25W C-Band Power Amplifier for Satellite Communication," 2021 IEEE MTT-S International Microwave and RF Conference (IMARC), KANPUR, India, 2021, pp. 1-4.

[5] I. Huang et al, "A 29.6 dBm 29-GHz Power Amplifier for Satellite and 5G Communications Using 0.15-μm GaAs p-HEMT Technology," 2018 Asia-Pacific Microwave Conference (APMC), Kyoto, Japan, 2018, pp. 986-988.

[6] A. Piacibello, R. Giofre, R. Quaglia, et al, "A 5-W GaN Doherty Amplifier for Ka-Band Satellite Downlink With 4-GHz Bandwidth and 17-dB NPR," IEEE Microwave and Wireless Components Letters, vol. 32(8), pp. 964–967, 2022.

[7] A. Piacibello, C. Ramella, V. Camarchia, et al, "A Balanced Stacked GaN MMIC Power Amplifier for 26-GHz 5G applications," 2023 IEEE/MTT-S International Microwave Symposium - IMS 2023, San Diego, CA, USA, 2023, pp. 331-334.

[8] K. Nakatani, Y. Yamaguchi and M. Tsuru, "A Ka-Band 40 W Output Power and 30% PAE GaN MMIC Power Amplifier for Satellite Communication," 2021 16th European Microwave Integrated Circuits Conference (EuMIC), London, United Kingdom, 2022, pp. 285-288.

[9] Z. Li, H. Sun, J. Li, J. Huang, Y. Huang and G. Wen, "5W High-power High-linearity L-band InGaP/GaAs HBT PA MMIC for RDSS Applications," 2021 International Conference on UK-China Emerging Technologies (UCET), Chengdu, China, 2021, pp. 185-189.

[10] S. Chen, Y. Zheng and G. Zhang, "Design of HBT power amplifier for beidou satellite mobile communication," Research & Progress of Solid State Electronics, vol. 35, pp. 334-339, 2015.

[11] Y. Zheng, S. Chen and G. Zhang, "A High-Power power amplifier for BeiDou satellite handsets," Micrielectronics, vol. 46, pp. 293-296, 2016.

2025 The 10th International Conference on Integrated Circuits and Microsystems

Design of a High-Bandwidth Low-Noise Amplifier

Hongmei Chen*
College of Microelectronics
Hefei University of Technology
Hefei, China
hmchen@hfut.edu.cn

Ruiting Shen
College of Microelectronics
Hefei University of Technology
Hefei, China
931038149@qq.com

Yuexin Tan
College of Microelectronics
Hefei University of Technology
Hefei, China
916604355@qq.com

Zheyu Li
College of Microelectronics
Hefei University of Technology
Hefei, China
956532010@qq.com

Abstract—**This paper presents the design of a high-bandwidth, low-noise rail-to-rail amplifier. The input stage achieves high gain while significantly enhancing bandwidth by cascading a high-gain operational amplifier path with a high-bandwidth operational amplifier path. The input transistor pair operates in the sub-threshold region, enabling low equivalent input noise voltage and reduced power consumption. The intermediate stage employs a folded-cascode active load directly combined with a class-AB operational amplifier-controlled output stage, which not only reduces static power consumption but also expands the output signal swing range and significantly enhances load-driving capability. Additionally, a slew rate enhancement circuit is further designed to ensure the output stage delivers large output currents under low quiescent current conditions for rail-to-rail output swing. Finally, the output structure utilizes capacitive compensation through internal feedback loops to establish fast feedback channels, ensuring system stability while suppressing output ripple. Implemented by using HJ 0.18μm BCD process technology, the layout occupies an area of 890μm×495μm. Post-layout simulation results demonstrate that the amplifier achieves rail-to-rail input/output functionality with an open-loop gain of 150.9 dB, phase margin of 86.42°, and gain-bandwidth product of 20.43 MHz. The equivalent input noise at 1 kHz is 5.53 nV/√Hz, with a common-mode rejection ratio of 112.6 dB. The positive and negative power supply rejection ratios are 129.5 dB and 114 dB, respectively. The measured positive and negative slew rates reach 19.2 V/μs and 14.2 V/μs.**

Keywords—*Operational amplifier; CMOS technology; High bandwidth; Low noise*

I. INTRODUCTION

Compared to conventional operational amplifiers, high-precision, low-noise rail-to-rail operational amplifier circuits feature characteristics such as full input/output voltage swing, low input offset, low noise voltage density, high gain, and wide bandwidth. These features enable widespread applications across aerospace, aviation, marine, and electronic information fields. Over several decades of optimization, rail-to-rail op-amps have increasingly benefited from refined design experiences, elevating their performance to exceptionally advanced levels. Numerous mature circuit structures and design strategies have been developed, such as current control methods[1][2], voltage control methods[3], and back-gate drive techniques[4]. In particular, designs based on quasi-floating gate MOS transistors have achieved input stage transconductance variation reduced to approximately 0.3%, which is nearly constant, and further circuits have been developed to enhance opamp performance, including circuits with process-insensitive constant transconductance and self-adjusting transconductance circuits.

Low-noise design is a critical component of high-bandwidth, low-noise rail-to-rail amplifiers. To reduce the noise figure, researchers have employed various strategies, such as optimizing input matching networks and utilizing low-noise devices and circuit topologies[5]. Additionally, emerging low-noise amplification techniques like noise cancellation and noise shaping are gradually being integrated into the design of high-bandwidth, low-noise rail-to-rail amplifiers. The output stage design critically influences an amplifier's bandwidth, output drive capability, and stability. Recent developments include various output configurations, such as the feedforward voltage-controlled class AB push-pull amplifier structure[6], aimed at improving bandwidth and reducing distortion. To address stability issues under large capacitive loads, various frequency compensation circuits have been devised[7] to ensure reliable operation across different load conditions.

In this work, a three-stage high-bandwidth, low-noise rail-to-rail amplifier presented. The input stage employs parallel bipolar MOS differential pairs to achieve rail-to-rail input capability[8]. To achieve high gain and broad bandwidth, a cascaded architecture is employed, combining dedicated high-gain and high-bandwidth amplifier paths. The differential pairs operate in the subthreshold region, optimizing power efficiency and minimizing noise. The intermediate stage features active loads based on folded-cascode structures, directly integrated with an AB class output stage controlled by the amplifier. To extend the output swing range and provide sufficient gain, a floating voltage source topology is adopted in the class AB output stage. Despite the advantages of low bias currents and large compensation capacitors for power efficiency, these design choices may limit transient response under large differential inputs. To mitigate this, a slew rate enhancement circuit is implemented. The arrangement of this paper is as follows. In the second part, the principle of the rail-to-rail amplifier will be introduced. In the third part, the low-noise and high-bandwidth amplifier designed in this paper will be presented, including the input stage, the intermediate stage, the output stage, and the slew rate enhancement circuit, etc. The fourth part is the verification of the simulation results, and the fifth part is the conclusion.

II. PRINCIPLES OF RAIL-TO-RAIL OPERATIONAL AMPLIFIERS

The input stage of a typical rail-to-rail operational amplifier employs N-type or P-type differential pair transistors. As shown in Fig. 1, this structure combines N-type and P-type differential pairs to extend the input common-mode voltage range. The common-mode input range of the P-type differential pair and the N-type differential input transistors is:

979-8-3315-8850-2/25 $31.00 © 2025 IEEE

$$V_{SS} \le V_{CM} \le V_{DD} - V_{DSAT} - V_{GS,P} \qquad (1)$$

$$V_{SS} + V_{GS,N} + V_{DSAT} \le V_{CM} \le V_{DD} \qquad (2)$$

In equation (1) V_{SS} is the negative power supply, V_{DD} is the positive power supply voltage, V_{CM} is the input common-mode voltage, V_{DSAT} is the saturation voltage of the P-type/N-type tail current transistors, $V_{GS,P}$ is the gate-source voltage of the P-type input pair transistors, and $V_{GS,N}$ is the gate-source voltage of the N-type input pair transistors. To ensure that the complementary differential pairs operate within the appropriate common-mode voltage range, the provided V_{DD} should satisfy:

$$V_{DD} \ge V_{GS,P} + V_{GS,N} + 2V_{DSAT} \qquad (3)$$

When the common-mode input voltage varies, the conduction of the complementary differential pairs can be divided into three regions: When the common-mode input voltage is close to the negative supply rail, the PMOS pair conducts while the NMOS pair is cutoff; at this point, the transconductance of the input stage is $g_m = g_{m,p}$. When the common-mode input voltage is approximately at $1/2V_{DD}$, both the PMOS and NMOS pairs conduct simultaneously; here, the transconductance is $g_m = g_{m,n} + g_{m,p}$. When the common-mode input voltage exceeds $1/2V_{DD}$ and approaches V_{DD}, the NMOS pair conducts while the PMOS pair is cutoff; hence, the transconductance is $g_m = g_{m,n}$.

It can be observed that in all three conduction modes, especially when both N-type and P-type pairs conduct simultaneously, the transconductance reaches higher values. Traditional two-stage amplifiers experience variations in gain-bandwidth product (*GBW*) with large output signals, leading to instability in DC gain, unity-gain bandwidth, and slew rate, as well as poor frequency compensation. To ensure optimal circuit performance, it is generally required that the total transconductance g_m does not vary by more than 50% within the common-mode input voltage range, thereby avoiding severe distortion. Therefore, maintaining the input stage transconductance unchanged with respect to the input common-mode voltage is crucial.

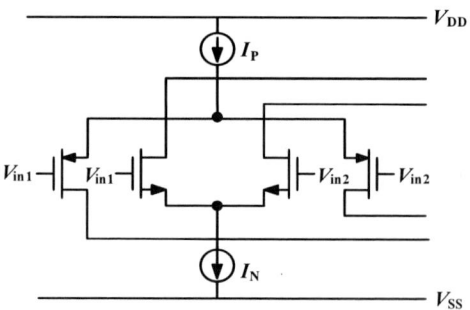

Fig. 1. Complementary differential rail-to-rail input circuit

The common methods for achieving a constant transconductance at the input stage include the three-times current mirror circuit[9], the cross-conduction method[10], and the level-shift constant transconductance method[11]. Among them, the three-times current mirror method adjusts the bias current under different common-mode input conditions to maintain a constant transconductance across the entire common-mode

input range, but the control circuit for the current is relatively complex; the cross-conduction method involves a switch transistor that narrows the transition region when the input common-mode voltage is at the midpoint, and the total transconductance in this transition region becomes closer to that of a single MOS device, but overall, both the transition region and the total transconductance are significantly reduced compared to other methods; the level-shift method uses voltage level shifting to ensure that the effective input transconductance remains constant over the entire input common-mode voltage range, resulting in constant gain and high linearity. However, it typically requires specialized circuit techniques such as current mirrors and active loads, which increase circuit complexity and area, potentially raising chip costs and power consumption. Additionally, this method is sensitive to semiconductor fabrication process variations, and fluctuations in process parameters can adversely affect the stability of the transconductance.

III. OVERALL ARCHITECTURE OF HIGH-BANDWIDTH, LOW-NOISE RAIL-TO-RAIL AMPLIFIER

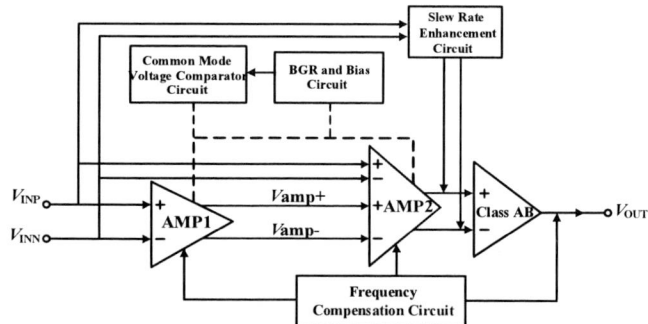

Fig. 2. High performance rail-to-rail op amp block diagram

The overall structure of the high-bandwidth, low-to-rail operational amplifier designed in this paper is depicted in Fig. 2. The input stage employs a fully differential configuration with parallel NMOS and PMOS transistors. A common-mode level comparison circuit detects the input common-mode voltage, enabling the NMOS input pair operation under high common-mode conditions and the PMOS input pair under low common-mode conditions. By adjusting the conduction states of the NMOS and PMOS input pairs, the circuit maintains a constant transconductance with a simple structure. Additionally, because the noise of the PMOS input pair is lower than that of the NMOS input pair, this design uses an appropriate comparison threshold to ensure the input stage operates predominantly in the PMOS input mode. The intermediate stage both contributes a certain amount of gain and provides a drive signal for the output stage. It utilizes a folded-cascode active load approach and is directly connected to the class AB controlled output stage. To address the low-power operation of the designed rail-to-rail amplifier—characterized by small bias currents and large compensation capacitors, which can slow circuit response when the differential input swing is large—a slew rate enhancement circuit is implemented. This structure directly monitors the differential voltage at the main amplifier's input and provides dynamic current compensation, significantly improving the slew rate performance. Finally, in multi-stage

979-8-3315-8850-2/25 $31.00 © 2025 IEEE

amplifier circuits, stability is a critical factor. By analyzing the pole distribution, a nested Miller compensation scheme has been adopted to ensure stability. The following sections detail the specific circuit design of each module.

A. Low-Power, Low-Noise Constant-G_m Input Stage Design

The input stage of the rail-to-rail amplifier proposed in this paper is shown in Fig. 3. The constant transconductance of the input stage is achieved by controlling the mirrored current magnitude through a common-mode voltage comparator circuit, as shown in Fig. 4. A complementary parallel input configuration is formed using a pair of P-type differential pairs and a pair of N-type differential pairs. This ensures that the total transconductance contributed by these pairs remains equal across varying input common-mode levels, i.e., the sum of the tail currents of the differential pairs remains constant, thereby achieving the goal of constant transconductance.

Compared to traditional operational amplifiers where input transistors operate in the saturation region, the input transistors Mc1, Mc2, Mc3, and Mc4 in this design operate in the subthreshold region. This choice is motivated by the following considerations: (a) Lower operating current, resulting in reduced power consumption; (b) Improved noise performance; (c) Wider input common-mode range, better enabling rail-to-rail operation; (d) Higher transconductance efficiency. In the subthreshold region, the drain current I_D of a MOSFET exhibits an exponential relationship with the gate-source voltage V_{GS}, and the transconductance g_m is proportional to I_D. This provides higher transconductance and voltage conversion efficiency compared to saturation-region operation under the same bias current. Under the control of the input common-mode comparison circuit, the input signals V_{in+} and V_b are compared to generate output currents I_{out1} and I_{out2}. Here, I_{out1} supplies the tail current for the NMOS input differential pair Mc3 and Mc4, while I_{out2} supplies the tail current for the PMOS input pair Mc1 and Mc2. When V_{in+} is low and below V_b, the entire reference current I_{ref} flows through transistor M2. Through the current mirrors formed by M3-M9, I_{out1} becomes zero, while I_{out2} mirrors I_{ref} via the M4-M5 current mirror. In this state, the NMOS input pair is disabled, and only the PMOS input pair is active. As V_{in+} increases, transistor M1 begins to conduct. I_{out1} starts to carry current as the current through M2 decreases below I_{ref}, causing the current through M5 to drop. This lowers the voltage V_1, increasing the current through M6 and slightly raising V_b. When V_{in+} further increases beyond V_b, the voltage V_b becomes clamped due to resistive limitations. M2 turns off completely, redirecting I_{ref} entirely through M1. Consequently, I_{out1} equals I_{ref}, while I_{out2} drops to zero. In this high common-mode condition, the PMOS input pair (mirrored by I_{out2}) is disabled, and the NMOS input pair (mirrored by I_{out1}) becomes active. This mechanism ensures that either the NMOS or PMOS input differential pair is selectively enabled under low or high common-mode conditions, maintaining constant transconductance across the entire input range.

When the input common-mode voltage is below the detection threshold V_b (with only the PMOS input pair active), the total input transconductance of the amplifier is given by:

$$g_{m,tot} = g_{mp} = \frac{I_{OUT2}}{nV_T} = \frac{I_{ref}}{nV_T} \qquad (4)$$

In equation (4) $g_{m,\,tot}$ represents the transconductance of the input differential pair, g_{mp} is the transconductance of the PMOS input differential pair, and n is the subthreshold slope factor.

As the common-mode voltage further increases, the NMOS input differential pair will also turn on, and at this point, the total transconductance is the sum of g_{mn} and g_{mp}:

$$g_{m,tot} = g_{mn} + g_{mp} = \frac{I_{OUT1} + I_{OUT2}}{nV_T} = \frac{I_{ref}}{nV_T} \qquad (5)$$

In equation (5) g_{mn} is the transconductance of the NMOS input differential pair.

When the input common-mode voltage exceeds the detection threshold V_b, the NMOS input differential pair becomes active, and the total input transconductance of the operational amplifier is:

$$g_{m,tot} = g_{mn} = \frac{I_{OUT1}}{nV_T} = \frac{I_{ref}}{nV_T} \qquad (6)$$

Therefore, the input stage maintains constant transconductance across the entire common-mode voltage range. Under V_{DD} =5V, with the detection threshold V_b set to 3.88V, the PMOS input differential pair operates when $V_{in,cm}$ is below 3.88V, while the NMOS input differential pair activates for $V_{in,cm}$ between 3.88V and 5V, ensuring constant transconduc-tance.

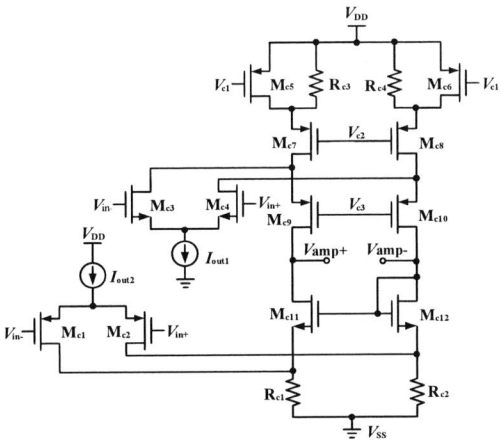

Fig. 3. Subthreshold-Region-Based Constant-gm Complementary CMOS Low-Noise Input Stage Design

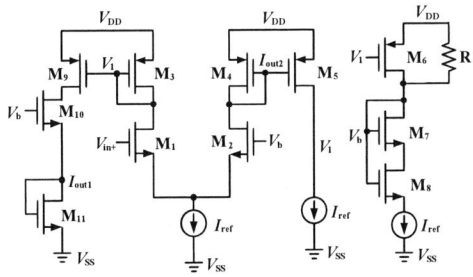

Fig. 4. Common-mode Voltage Comparator Circuit

B. Intermediate Stage and Output Stage Circuit Design

The main function of the intermediate stage is to provide high gain. In this work, the intermediate stage employs a folded-cascode structure, whose output is directly coupled to the class-AB operational amplifier-controlled output stage. The circuit configuration is shown in Fig. 5. This design not only provides a certain level of gain but also enhances the voltage output swing range.

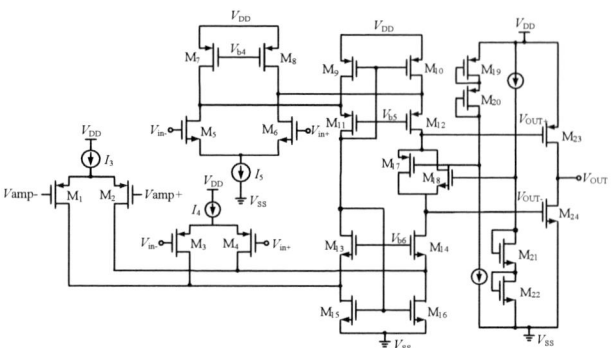

Fig. 5. Intermediate Stage and Feedforward Class AB Push-Pull Output Stage Circuit

As shown in Fig. 5, V_{amp+} and V_{amp-} are the output signals of the first-stage operational amplifier, while V_{in+} and V_{in-} are the input signals shared with the input stage. V_{out+} and V_{out-} are the output signals of the slew rate enhancement circuit, and V_{out} is the output signal of the operational amplifier. Since a stable fixed voltage is required between the gates of output transistors M_{23} and M_{24} to ensure they operate in the saturation region during common-mode variations, a floating voltage source structure is adopted in the design. This structure allows the output stage to achieve a higher ratio of maximum output current to static current, thereby improving power supply efficiency. If the gate voltages of transistors M_{23} and M_{24} are biased close to V_{DD} and V_{SS}, respectively, the output range can be as low as $V_{SS}+V_{dsat}$ and as high as $V_{DD}-V_{dsat}$. Consequently, the static output current of M_{23} and M_{24} is relatively small, which limits the charge and discharge speed of the output stage. Therefore, the design must also consider the magnitude of the maximum output current, static power consumption, circuit stability, and area. In this circuit, MOS transistors are used to provide the required bias voltage for the fixed output transistors, which occupies less chip area compared to using resistors alone. Additionally, the gate voltage biased by the MOS transistors is more stable than that biased by resistors, making it less susceptible to variations in process and power supply. This type of output stage can operate at a minimum supply voltage of $2V_{gs}+V_{dsat}$.

In this circuit, M_1, M_2, M_3, M_4, M_5 and M_6 form the input differential pair of the intermediate stage, while M_9, M_{10}, M_{15} and M_{16} serve as the active load for the rail-to-rail input differential pair. M_{11}, M_{12} and M_{13}, M_{14} are cascode amplifiers, and M_7, M_8 provide a current that matches I_5. When the NMOS transistors of the second stage are on, the current through M_7 and M_8 does not flow into the intermediate stage sleeve. Conversely, when the PMOS transistors of the input stage are on, the NMOS transistors are off, and M_7 and M_8 compensate for the same magnitude of current as I_5 in the summing circuit of the

subsequent stage, stabilizing the voltage of the output transistors biased by the floating voltage source. M_{23} and M_{24} are the output transistors, while M_{19}, M_{20}, M_{21} and M_{22} are the bias circuits for the floating voltage sources M_{17} and M_{18}.

The static current in the output transistors is determined through the two current loops formed by M_{23}, M_{17}, M_{19}, M_{20} and M_{24}, M_{18}, M_{21}, M_{22}. The bias circuit includes:

$$|V_{GS23}| + |V_{GS17}| = |V_{GS19}| + |V_{GS20}| \tag{7}$$

$$V_{GS24} + V_{GS18} = V_{GS21} + V_{GS22} \tag{8}$$

By appropriately adjusting M_{17}, M_{20}, M_{18} and M_{21} such that $|V_{GS23}| = |V_{GS19}|$ and $V_{GS24} = V_{GS22}$, the following relationship holds:

$$I_{Q23} = \frac{(W/L)_{23}}{(W/L)_{19}} I_{Q19} \tag{9}$$

$$I_{Q24} = \frac{(W/L)_{24}}{(W/L)_{22}} I_{Q22} \tag{10}$$

If the loop currents $I_{Q19} = I_{Q22}$ and $\dfrac{(W/L)_{23}}{(W/L)_{19}} = \dfrac{(W/L)_{24}}{(W/L)_{22}}$, then we have

$$I_Q = \frac{(W/L)_{23}}{(W/L)_{19}} I_{Q19} = \frac{(W/L)_{24}}{(W/L)_{22}} I_{Q22} \tag{11}$$

This way, the bias circuit can determine the output static current, ensuring that it does not change with variations in the common-mode input voltage. Additionally, since the drain nodes of M_{12} and M_{14} are AC shorted to a high impedance state, the gain of the input stage is maintained.

C. Slew Rate Enhancement Circuit Design

The rail-to-rail operational amplifier in this work exhibits an input swing range of nearly 0 to V_{DD}. When the differential input signal has a large swing, the circuit's response speed degrades. To address this, a slew rate enhancement circuit is introduced to improve the amplifier's slew rate, thereby boosting the overall response speed of the operational amplifier, as shown in Fig. 6.

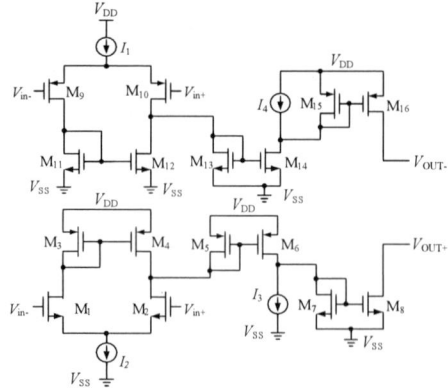

Fig. 6. Slew Rate Enhancement Circuit

When the input swing is relatively small, the voltage difference between V_{in+} and V_{in-} is approximately equal. Since the V_{gs} of transistors M_9 and M_{10} is the same, the currents in the left and right paths are equal. Given that the width-to-length ratio of M_{12} is much larger than that of M_{13}, M_{13} will not conduct in the circuit, and transistors M_{15} and M_{16} will be in the cutoff state. Similarly, during normal operation of the circuit, transistors M_7 and M_8 will be in the cutoff state, while M_{16} and M_8 will be in a high-impedance state, which will not affect the operating state of the subsequent circuits connected to V_{out+} and V_{out-}. Here, we design the width-to-length ratios of M_5, M_6 and M_{13}, M_{14} to be k times, that is

$$\frac{(W/L)_5}{(W/L)_6} = \frac{(W/L)_{13}}{(W/L)_{14}} = k \quad (12)$$

When the input voltage swing is large, the comparator circuit switches, and the currents in M_1 and M_3, as well as M_9 and M_{11}, are

$$I_{1,3} = I_{9,11} = \frac{I_{\text{ref}}}{k+1} \quad (13)$$

The currents in transistors M_2 and M_4, as well as M_{10} and M_{12}, are

$$I_{2,4} = I_{10,12} = \frac{kI_{\text{ref}}}{k+1} \quad (14)$$

From the above equations, we can obtain:

$$|V_{\text{GS1,9}}| = \sqrt{\frac{I_{1,9}}{\frac{1}{2}\mu_{\text{p,n}}C_{\text{OX}}\left(\frac{W}{L}\right)_{1,9}}} - V_{\text{TH}} = \sqrt{\frac{I_{\text{ref}}}{\frac{1}{2}\mu_{\text{p,n}}C_{\text{OX}}\left(\frac{W}{L}\right)_{1,9}(k+1)}} - V_{\text{TH}} \quad (15)$$

$$|V_{\text{GS2,10}}| = \sqrt{\frac{kI_{\text{ref}}}{\frac{1}{2}\mu_{\text{p,n}}C_{\text{OX}}\left(\frac{W}{L}\right)_{2,10}(k+1)}} - V_{\text{TH}} \quad (16)$$

Then the inversion voltage V is

$$V = |V_{\text{GS2,10}}| - |V_{\text{GS1,9}}| = \sqrt{\frac{k-1}{k+1}}\sqrt{\frac{I_{\text{ref}}}{\frac{1}{2}\mu_{\text{p,n}}C_{\text{OX}}\left(\frac{W}{L}\right)_{1,9}}} \quad (17)$$

According to the above equation (17), the sizes of the key MOS transistors can be adjusted to change the flip voltage. When the differential signal at the input of the slew rate enhancement circuit exceeds the flip threshold voltage, the current mirrors M_8 or M_{16} will turn on, and the current will be output proportionally through the current mirror. The rapidly output current is connected to the V_{out+} and V_{out-} terminals, charging the Miller capacitance connected between the output of the second-stage operational amplifier and the output stage, thereby achieving the goal of quickly establishing the output.

D. Analysis of System Stability

For multi-pole systems, system stability requires special consideration. This paper designs a nested Miller capacitor compensation circuit, and the equivalent schematic is shown in Fig. 7. A compensation capacitor is connected from the output of the first-stage operational amplifier to the final output stage, and then nested with C_{m1} and C_{m2} to compensate for the zero-pole of the second and third stages to the final output stage. Theoretically, the front stage achieves a constant transcon-

ductance structure, and C_{m1} and C_{m2} are symmetrically equal. This method implements Miller compensation for multi-stage operational amplifiers.

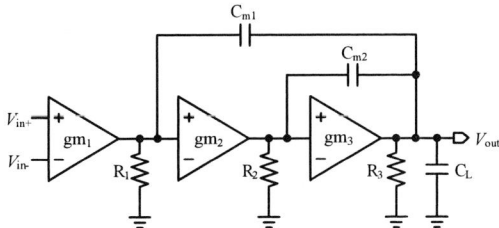

Fig. 7. Nested Miller Compensator

Simplified path transport function:

$$A_v = \frac{V_o(s)}{V_i(s)} = \frac{g_{m1}g_{m2}g_{m3}R_1R_2R_3}{(1+sC_{m1}g_{m2}g_{m3}R_1R_2R_3)\left(1+s\frac{C_{m2}}{g_{m2}}+s^2\frac{C_LC_{m2}}{g_{m2}g_{m3}}\right)} \quad (18)$$

From the above equation (18), the poles of the passband, ω_{p1}, ω_{p2} and ω_{p3}, can be obtained as follows:

$$\omega_{p1} = \frac{1}{C_{m1}g_{m2}g_{m3}R_1R_2R_3} \quad (19)$$

$$\omega_{p2} = \frac{g_{m3}}{2C_L} - \frac{g_{m3}}{2C_L}\sqrt{1-\frac{4g_{m2}C_L}{g_{m3}C_{m2}}} \quad (20)$$

$$\omega_{p3} = \frac{g_{m3}}{2C_L} + \frac{g_{m3}}{2C_L}\sqrt{1-\frac{4g_{m2}C_L}{g_{m3}C_{m2}}} \quad (21)$$

Obviously, through the parameter design of the nested compensation capacitor, the dominant pole of the operational amplifier can be pushed towards the low frequency, and the secondary dominant pole and higher-order poles can be pushed to higher frequencies, thereby enhancing the stability of the circuit.

IV. SIMULATION AND VERIFICATION

The proposed circuit is implemented in HJ 0.18μm CMOS technology. It adopts a dual-channel architecture, where both channels share identical operational amplifier structures. Enable signals (ENA and ENB) independently control the switching of Channel A and B, ensuring flexible operation. The layout includes 10 PADs and occupies an area of 890μm × 495μm, as illustrated in Fig. 8. Using Cadence Spectre simu-lation tools, comprehensive performance metrics of the designed high-performance rail-to-rail operational amplifier were verified. The amplifier operates over a supply voltage range of 2.2V to 5.5V. Unless otherwise specified, default simulation conditions are: a 5V supply voltage, 2kΩ load resistor, and 100pF load capacitor.

Fig. 8. Dual-Channel High-Performance Rail-to-Rail Op-Amp Layout and Wire Bonding Diagram

The operational amplifier is configured in a unity negative feedback mode, with a large-signal voltage varying from 0V to 5V applied to its non-inverting input. A DC sweep of the output voltage is performed, resulting in a linear plot of the output versus input voltage variation, as shown in Fig. 9. The portion where the output voltage V_{OUT} varies linearly with the input common-mode voltage V_{CM} corresponds to the input common-mode voltage range, achieving a rail-to-rail input characteristic from 0V to 5V. Additionally, the voltage V_{in} at the non-inverting input is defined as a variable parameter, with its sweep range set from the supply voltage down to 0V. A subsequent DC sweep simulation is performed to observe the output voltage variation curve (output voltage swing), as illustrated in Fig. 10. The output swing spans from 4.999V to 436.1µV, demonstrating rail-to-rail output capability.

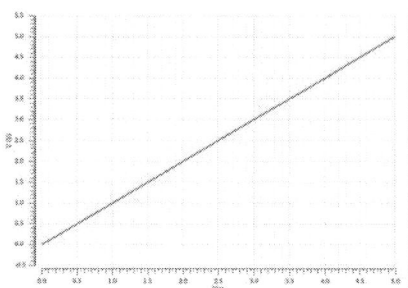

Fig. 9.   Input Common-Mode Range Simulation Result

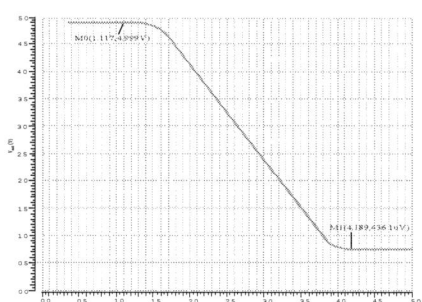

Fig. 10. Output Voltage Swing Simulation Result

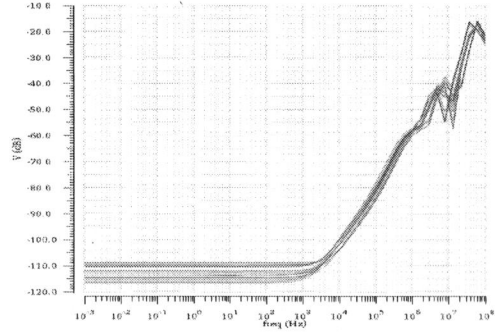

Fig. 11. Common Mode Rejection Ratio Simulation Result

The op-amp is configured in a unity-gain negative feedback configuration, with a V_{CM} and a 1V AC small signal applied to the non-inverting input. A 1V AC signal is also introduced in the feedback loop for AC simulation. The output response of the op-amp is observed over a frequency range of 0.01Hz to 100MHz,

with the results shown in Fig. 11. Under typical conditions, the Common-Mode Rejection Ratio (CMRR) is measured to be 112.6dB.

The opamp was configured in a unity-gain negative-feedback topology, with its noninverting input tied to a V_{CM}. For the PSRR+ simulation, a 1V small-signal sine source was applied to the positive supply rail; for the PSRR– simulation, a 1V sine source was applied to ground (the negative rail). An AC sweep from 10 mHz to 100 MHz was then performed, and the output response is plotted in Fig. 12 and Fig. 13. Under typical operating conditions, PSRR+ and PSRR– are 129.5 dB and 114 dB, respectively.

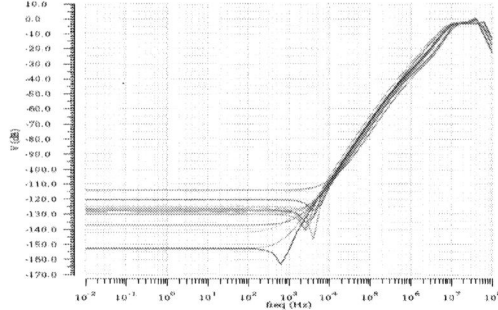

Fig. 12. PSRR+ Simulation Result

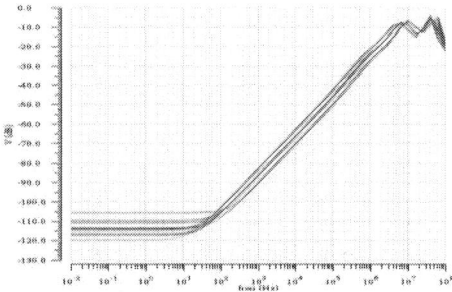

Fig. 13. PSRR- Simulation Result

The operational amplifier was configured in a unit-gain negative feedback topology, with a V_{CM} of 2.5V applied to the non-inverting input. To evaluate open-loop gain and stability, a Stability Analysis (STB) simulation was performed using the iprobe tool inserted into the feedback loop. Process corner variations and temperature sweeps were included to assess robustness, as shown in Fig. 14, Fig. 15, and Fig. 16. Under typical operating conditions, the open-loop gain was measured at 150.9dB, with a phase margin of 86.42° and a GBW of 20.43MHz.

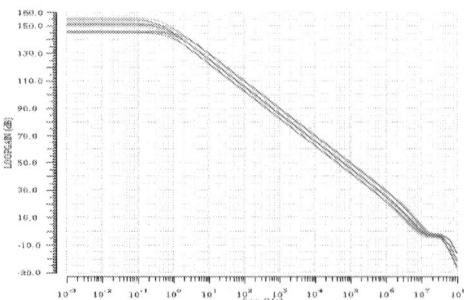

Fig. 14. A$_{VOL}$ Simulation Result

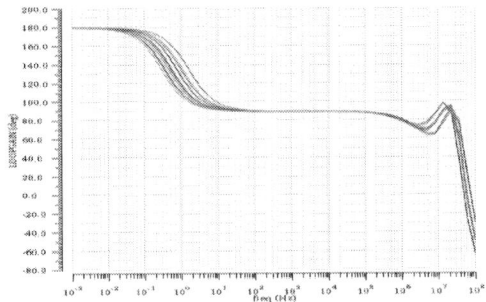

Fig. 15. PM Simulation Result

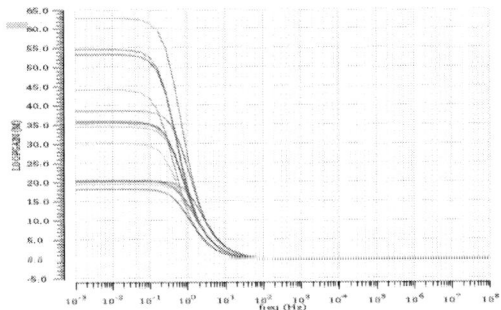

Fig. 16. GBW Simulation Result

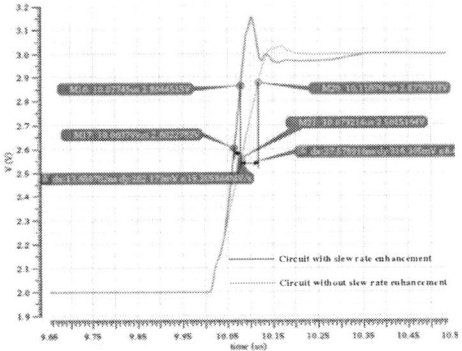

Fig. 17. SR+ Simulation Result

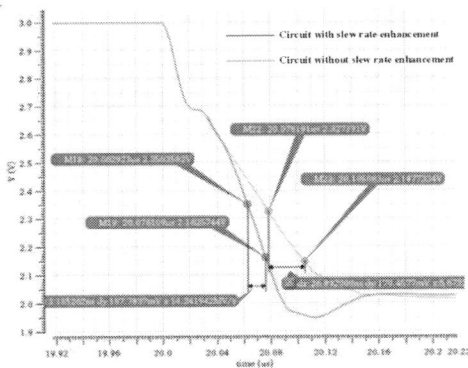

Fig. 18. SR- Simulation Result

A comparison of the slew rate performance of the rail-to-rail op-amp was conducted before and after the implementation of the slew rate enhancement circuit. The op-amp was configured

in a unity-gain negative feedback configuration, and a square wave signal transitioning from 0V to 3V was applied to the non-inverting input. The slew rates of the output signal's rising and falling edges were observed and compared for the op-amp with and without the slew rate enhancement circuit. The output waveform comparison is illustrated in Fig. 17 and Fig. 18. The results show that the positive slew rate ($SR+$) improved from 8.4V/μs to 19.2V/μs, while the negative slew rate ($SR-$) increased from 6.7V/μs to 14.2V/μs.

The op-amp was configured in a unity-gain negative feedback configuration with a 2.5V common-mode voltage applied to the non-inverting input. A noise simulation was performed with a frequency sweep ranging from 10mHz to 100MHz, yielding the input-referred noise voltage curve. Using the value function, the input voltage noise density at $f=1kHz$ was extracted. The simulation results for input voltage noise density are shown in Fig. 19. Under typical operating conditions, the input voltage noise density at $f=1kHz$ was measured to be 5.53 nV/√Hz.

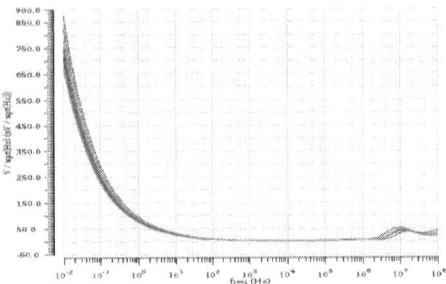

Fig. 19. Input Voltage Noise Density Simulation Result

Finally, the proposed high-bandwidth, low-noise rail-to-rail operational amplifier is compared with other works, as shown in Table 1. It can be observed that the designed amplifier demonstrates significantly superior bandwidth and input voltage noise density compared to existing architectures.

TABLE I.   HIGH-BANDWIDTH, LOW-NOISE RAIL-TO-RAIL OPERATIONAL AMPLIFIER

Parameter	[12]	[13]	[14]	This work
Process	0.18μm	12nm	0.18μm	0.18μm
Power Supply/V	1.8	0.8	3.3	2.2~5.5
A_{VOL}/dB	99	114.8	120	151
PM/°	45	85.72	—	86.42
GBW/MHz	8	8.76	1.5	20.43
Input Voltage Noise Density/[nV/√Hz]	24	—	—	5.53
CMRR/dB	100	91.21	170.9	112.6
PSRR+/dB	99	124.84	119.2	129.5
PSRR-/dB	—	—	—	114
Slew rate+/[V/μs]	—	1.69	—	19.2
Slew rate -/[V/μs]	—	—	—	14.24

V. CONCLUSION

This paper presents the design of a high-performance rail-to-rail input/output operational amplifier featuring high bandwidth, low noise, and high gain. The main amplifier architecture employs a cascade of a high-gain, low-bandwidth amplifier and a low-gain, high-bandwidth amplifier, with a constant input-stage transconductance achieved through a common-mode voltage comparison circuit. The intermediate stage is designed as a folded-cascode structure for current summation to provide high gain, while the output stage adopts an AB-class amplifier with an embedded floating voltage source biasing structure to achieve rail-to-rail output. Due to low-power requirements, the designed tail current and output current are small, yet the amplifier must drive a large load capacitance. To address this, a slew rate enhancement circuit combining inverters and current mirrors is implemented. This circuit rapidly charges and discharges the Miller compensation capacitor connected between the second-stage output and the final-stage output, thereby improving the slew rate while driving a 100 pF load capacitor. The circuit is designed using HJ 0.18 μm BCD technology. Post-layout simulation results demonstrate the following performance: rail-to-rail input/output functionality with an input common-mode voltage range of 0 V to 5 V, an equivalent input noise of 5.53 nV/√Hz, *CMRR* of 112.6 dB, *PSRR+* and *PSRR-* of 129.5 dB and 114 dB, respectively, an open-loop voltage gain of 151 dB, a phase margin of 86.42°, a gain-bandwidth product of 20.43 MHz, and *SR+* and *SR-* of 19.2 V/μs and 14.2 V/μs.

ACKNOWLEDGMENT

This study was supported by the Anhui Provincial Department of Education; and the Anhui Key Research and Development Project (202104g01020008) - Research on Key Technologies and Industrialization of High-Performance Power Device Driver Chips.

REFERENCES

[1] S. Sakurai and M. Ismail. Robust design of rail-to-rail CMOS operational amplifiers for a low power supply voltage[J]. IEEE Journal of Solid-State Circuits, 1996, 31(2): 146-156.

[2] K. M. AbdelMoneim and S. A. Mahmoud. 3V CMOS Rail to Rail Op-Amp[C]. International Conference on Microelectronics, Cairo, Egypt, 2007: 373-376.

[3] G. Ferri and W. Sansen. A rail-to-rail constant-gm low-voltage CMOS operational amplifier[J]. IEEE Journal of Solid-State Circuits, 1997, 32(10): 1563-1567.

[4] R. Hogervorst, J. P. Tero and J. H. Huijsing. Compact CMOS constant-gm rail-to-rail input stage with gm-control by an electronic zener diode[J]. IEEE Journal of Solid-State Circuits, 1996, 31(7): 1035-1040.

[5] N. Kumar and R. Bisht, "Design of an Ultra Low Power Low Noise Amplifier for 5-GHz band," 2020 IEEE International Conference for Innovation in Technology (INOCON), Bangluru, India, 2020, pp. 1-4.

[6] Y. Zhang. Design of a Rail-to-Rail Operational Amplifier with High Bandwidth, Low Noise, and Low Power Consumption [D]. Southeast University, 2021.

[7] S. Pennisi, G. Scotti and A. Trifiletti, "Constant and maximum bandwidth feedback amplifier with adaptive frequency compensation," 2012 IEEE International Symposium on Circuits and Systems (ISCAS), Seoul, Korea (South), 2012, pp. 436-439.

[8] Y. Lv, X. Z. Kang. A novel high-gain ultra-wideband low noise amplifier[J]. Information Technology, 2018(01): 141-143+154.

[9] S. Ctanescu, C. Dinca, A. Veselu, et al. A Dual Low Voltage Chopper Offset-Stabilized Operational Amplifier. In: Proc of 2021 International Semiconductor Conference(CAS). Romania: IEEE, 2021, 129–132.

[10] L. F. Huang. A study and design of a precision operational amplifier with rail-to-rail input/output using CMOS technology [D]. University of Electronic Science and Technology of China, 2022.

[11] G. R. Huang. A study and design of a low-offset rail-to-rail input/output CMOS operational amplifier [D]. University of Electronic Science and Technology of China, 2019: 1, 29-31.

[12] X. Zou, S. Wang, Z. Chai and G. Cong, "A Low Power Operational Amplifier Circuit with Strong Anti-Interference Ability and High Gain," 2024 11th International Forum on Electrical Engineering and Automation (IFEEA), Shenzhen, China, 2024, pp. 204-208.

[13] J. -a. Zhang, Y. Feng and C. Zhang, "A 12nm CMOS Rail-to-Rail Auto-Zero Operational Amplifier," 2024 9th International Conference on Integrated Circuits and Microsystems (ICICM), Wuhan, China, 2024, pp. 119-123.

[14] Q. Y. Liang and M. S. Tong, "A Compensation Amplifier with Automatic Zeroing and Stable Chopping," 2022 IEEE Electrical Design of Advanced Packaging and Systems (EDAPS), Urbana, IL, USA, 2022, pp. 1-3.

A Compact Wideband E-band Low Noise Amplifier with 3.3-4.5 dB Noise Figure using 45-nm RFSOI

Yinhan Lin
Guangdong Provincial Key Laboratory of Millimeter-Wave and Terahertz
Guangdong-Hong Kong-Macao Joint Laboratory for Millimeter-Wave and Terahertz
School of Integrated Circuits, South China University of Technology
Guangzhou, China
icyhlin@mail.scut.edu.cn

Haoshen Zhu
Guangdong Provincial Key Laboratory of Millimeter-Wave and Terahertz
Guangdong-Hong Kong-Macao Joint Laboratory for Millimeter-Wave and Terahertz
School of Electronic and Information Engineering, South China University of Technology
Guangzhou, China
zhuhs@scut.edu.cn

Taotao Xu
School of Integrated Circuits, Anhui University
Hefei, China
eexutt@163.com

Zhuming Li
Guangdong Provincial Key Laboratory of Millimeter-Wave and Terahertz
Guangdong-Hong Kong-Macao Joint Laboratory for Millimeter-Wave and Terahertz
School of Electronic and Information Engineering, South China University of Technology
Guangzhou, China
eezhumingli@gmail.com

Guohai Quan
Guangdong Provincial Key Laboratory of Millimeter-Wave and Terahertz
Guangdong-Hong Kong-Macao Joint Laboratory for Millimeter-Wave and Terahertz
School of Electronic and Information Engineering, South China University of Technology
Guangzhou, China
quanguohai@foxmail.com

Pei Qin
Guangdong Provincial Key Laboratory of Millimeter-Wave and Terahertz
Guangdong-Hong Kong-Macao Joint Laboratory for Millimeter-Wave and Terahertz
School of Electronic and Information Engineering, South China University of Technology
Guangzhou, China
qinpei7777@scut.edu.cn

Quan Xue*
Guangdong Provincial Key Laboratory of Millimeter-Wave and Terahertz
Guangdong-Hong Kong-Macao Joint Laboratory for Millimeter-Wave and Terahertz
School of Electronic and Information Engineering, South China University of Technology
Guangzhou, China
eeqxue@scut.edu.cn
*corresponding author

Abstract—This paper proposes an E-band low noise amplifier (LNA) fabricated using a 45-nm RFSOI process. An out-of-phase feedback transformer is employed for wideband input matching. Building on this, the trade-off between impedance matching and transconductance in the input network is optimized, with carefully chosen coupling coefficients to further enhance performance. A current-reuse technique is incorporated to improve gain while minimizing power consumption. Additionally, a gate-drain feedback transformer is utilized in the output load to broaden the bandwidth. Post-simulation results show that the proposed LNA achieves a maximum gain of 16.2 dB with a 3-dB bandwidth of 62.8-85.7 GHz (22.9 GHz). The noise figure (NF) ranges from 3.3 to 4.5 dB, with an IP_{1dB} of -25.5 dBm at 79.9 GHz. The LNA operates with a DC power consumption of 15.6 mW and occupies a core area of 0.09 mm^2.

Index Terms—low noise amplifier (LNA), E-band, wideband, low noise figure (NF)

I. INTRODUCTION

The E-band (71–86 GHz) is a key millimeter-wave frequency range, characterized by wide bandwidth, high data rates, and relatively low propagation loss. This low loss is primarily attributed to its position within an atmospheric transmission window, where radio waves experience minimal absorption [1]. As a result, E-band is ideal for long-range, high-speed communications and high-resolution radar systems [2], [3].

This work was supported in part by the National Natural Science Foundation of China (NSFC) under Grant 62271210, Grant 62321002, and Grant 62271216.

In E-band receiver systems, the low noise amplifier (LNA) plays a crucial role in amplifying weak signals while minimizing noise. The noise figure (NF) of the LNA significantly impacts the receiver sensitivity. A lower NF improves the signal-to-noise ratio (SNR) and enhances the detection of weak signals. Given the wide frequency range of E-band signals, the LNA must maintain low noise performance across the entire spectrum. By achieving low noise over this range, a wideband LNA ensures that all frequencies within the E-band are amplified uniformly and without distortion, thereby preserving signal integrity throughout the operating band. This capability is essential for enabling high-speed data transmission and accurate radar detection in complex environments. Several wideband E-band LNAs have been proposed to meet these challenges. For instance, in [4], source-to-drain feedback between two amplifier stages is used to achieve an ultra-wideband flat gain from 54.4–90 GHz. However, the input matching in this design is suboptimal. In [5], [6], gate-drain feedback transformers are employed to extend the bandwidth, while [7] introduces a hybrid broadband interstage network. However, these designs still suffer from relatively high NF values.

In this paper, a compact wideband E-band LNA is proposed, implemented using a 45-nm RFSOI process, as shown in Fig. 1. The LNA consists of three stages: the first and second stages are common-source (CS) amplifiers, while the third stage is a stacked configuration of two CS amplifiers utilizing a current reuse technique. To achieve wideband input matching and simultaneously enhance the transconductance (G_m) of

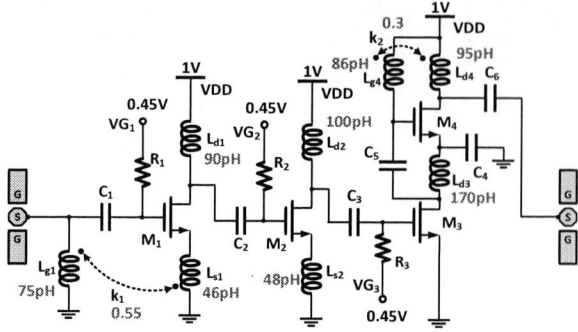

Fig. 1. Schematic of the proposed wideband E-band LNA.

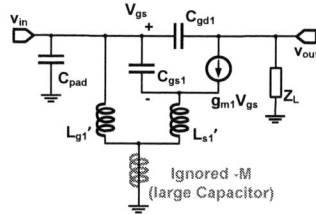

Fig. 2. Equivalent small signal circuit of the input matching network.

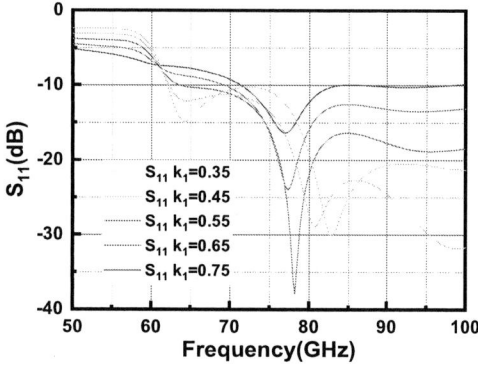

Fig. 3. Simulated S_{11} of the first stage vary different k_1.

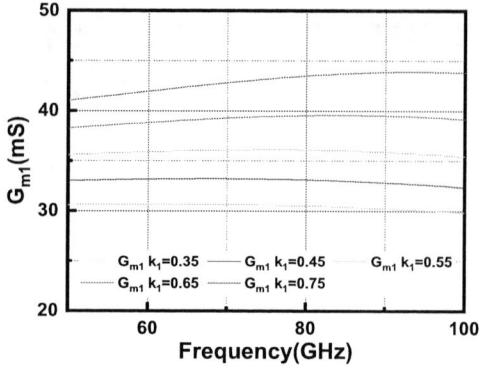

Fig. 4. Simulated G_{m1} of the first stage vary different k_1.

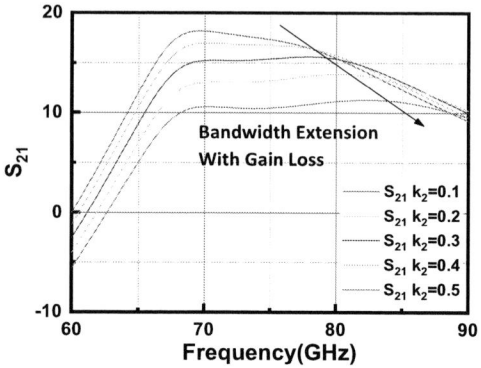

Fig. 5. Simulated S_{21} of the third stage vary different k_2.

the first stage, an out-of-phase transformer feedback network is employed. Additionally, a gate-drain transformer feedback network is used in the load of the third stage, which effectively extends the overall 3-dB gain bandwidth of the proposed LNA.

II. COMPACT WIDEBAND E-BAND LNA

A. Out-Of-Phase Transformer Feedback Input Network

The equivalent small-signal circuit of the input matching network is shown in Fig. 2, where the coupling is represented by an equivalent circuit comprising three independent inductors. Here, Z_L is the load impedance of the first stage. The inductors L_{g1}' and L_{s1}' in the equivalent circuit can be expressed as

$$
\begin{aligned}
L_{g1}' &= L_{g1} + M_1, \\
L_{s1}' &= L_{s1} + M_1,
\end{aligned}
\tag{1}
$$

where M_1 is the mutual inductance between L_{g1} and L_{s1}. Since the inductor with a value $-M_1$ behaves like a large capacitor,

it is neglected. The input impedance of the input matching network can be express as

$$
Z_{in} = \frac{1}{sC_{pad}} \parallel sL_{g1}' \parallel Z_{in}',
\tag{2}
$$

where Z_{in}' can be expressed as

$$
\begin{aligned}
Z_{in}'(s) &= \frac{Num(s)}{Den(s)}, \\
Num(s) &= s^2 C_{gs1} C_{gd1} L_{s1}' + s C_{gd1} L_{s1}' g_{m1} + C_{gd1}, \\
Den(s) &= s^2 C_{gs1} C_{gd1} L_{s1}' + s(C_{gd} L_{s1}' g_{m1} + C_{gs1} C_{gd1} Z_L) + \\
&\quad Z_L C_{gd1}(g_{m1} + 1) - C_{gs1},
\end{aligned}
\tag{3}
$$

While $Z_{in} = R_s = 50\,\Omega$, the input matching network achieves the best impedance matching and S_{11} performance. Since k_1 simultaneously tunes both Z_{in} and the transconductance of the first stage G_{m1}, a trade-off exists between them. The simulated S_{11} varying different k_1 values are shown in Fig. 3. It is evident that a k_1 range of 0.35-0.55 exhibits wideband matching characteristic. As G_{m1} in Fig. 4 is increased with higher k_1 value, a value of 0.55 is selected for k_1.

B. Gate-Drain Transformer Feedback For Bandwidth Extension

To achieve a wideband frequency response at the output load network, a gate-drain transformer is used as the load for the

979-8-3315-8850-2/25 $31.00 © 2025 IEEE

output stage. According to [5], [8], such gate-drain feedback introduces a new pole, with its angular frequency calculated as

$$\omega_{n,2} = \sqrt{\frac{\mu_1 + \mu_2}{9C_{gd4}(L_{g4}L_{d4} - M_2^2)^2}},$$

$$\mu_1 = 9L_{d4}(L_{g4}L_{d4} - M_2^2),$$

$$\mu_2 = C_{gd4}(L_{d4} + L_{g4} - 2M_2)^2 R_L^2,$$

(4)

By tuning this pole to a higher frequency, an additional sub-peak can broaden the bandwidth and help flatten the gain. Since μ_1 is approximately four orders of magnitude greater than μ_2, μ_1 dominates the pole. Therefore, as M_2 increases, $\omega_{n,2}$ increases, with μ_2 being negligible. This means that increasing the coupling coefficient of the gate-drain feedback effectively broadens the amplifier's bandwidth. Fig. 5 illustrates the overall LNA gain as k_2 varies, which is consistent with the previous analysis. Finally, a k_2 value of 0.3 is selected due to the noticeable gain loss observed with higher k_2 values.

C. Layout Design

The proposed LNA is processed in a 45-nm RFSOI process with a metal stack consisting of 8LM-3Mx-1Cx-1Ux-2Ox-LD. The cross-section of this metal stack is shown in Fig. 6. Compared to other metal stack configurations, this design features two Ox layers, which consist of two high-Q copper thick metal layers. These two high-Q layers significantly benefit the passive design, effectively enhancing the Q value of the transformer with two or more metal layers. The Q value of the transformer can exceed 14, even with a high coupling coefficient of 0.55 in the E-band. To achieve lower gate resistance and match the width of metal lines in the passive network, several multi-array transistor cells are carefully designed. Instead of directly connecting the gate to a single transistor with large fingers, three or four smaller transistors are used, and gate rings formed by the M1 and M2 layers surround these transistors, ensuring well-placed gate contacts. Fig. 7 illustrates the 3D view of one design from the multi-array transistor cell designs.

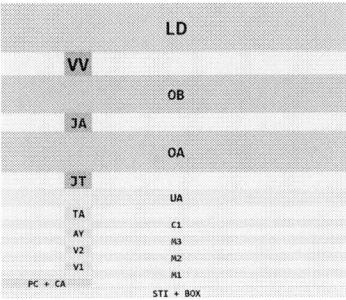

Fig. 6. Cross-section of the metal stack option 8LM-3Mx-1Cx-1Ux-2Ox-LD.

The overall layout design of the proposed LNA is shown in Fig. 8, with a core area of 0.09 mm². The design rule check (DRC) is fully cleaned, including basic, antenna, and wire-bond rules. Abundant decoupling capacitors are added to all DC supply pads to ensure ground integrity, as it

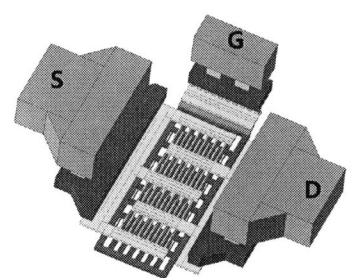

Fig. 7. 3D view of the multi-array transistor cell design.

is a single-ended design using wire-bonding package. The pad is compatible with both GSG100 and GSG150 probes, ensuring compatibility with different types of probes. Since MOM capacitors face issues of low accuracy in E-band, MIM capacitors in a high-Q and low-density configuration are used in this layout (except for the large capacitor used to form the AC ground of the current-reuse structure). These MIM capacitors are electromagnetic (EM) simulated, further ensuring the accuracy of capacitance. To reduce unwanted weak couplings between passive networks, the space between them is slightly enlarged. Additionally, a full layout EM simulation is performed to consider these weak couplings.

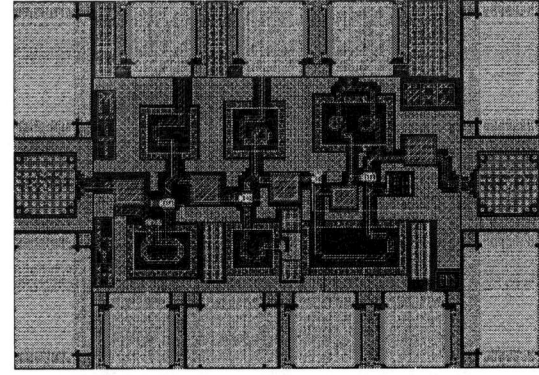

Fig. 8. DRC-cleaned layout of the proposed wideband E-band LNA.

III. POST-SIMULATION RESULTS

The simulation was conducted by co-simulating the EM results of the passive network obtained through EMX with the active network, where parasitics were extracted via PEX, and the post-simulation results were obtained. The simulated DC power consumption is 15.6 mW under a 1V DC voltage supply.

The simulated S-parameters are shown in Fig. 9. A large 3dB-bandwidth of 22.9 GHz, ranging from 62.8 to 85.7 GHz, was achieved, with a maximum gain of 16.2 dB. The input matching bandwidth, where $|S_{11}| < -10$dB, spans from 64.2 GHz to frequencies above 100 GHz. The effective bandwidth, BW_{eff}, defined as the 3dB bandwidth where $|S_{11}| < -10$dB, is 21.5 GHz, ranging from 64.2 to 85.7 GHz.

979-8-3315-8850-2/25 $31.00 © 2025 IEEE

TABLE I
PERFORMANCE SUMMARY OF STATE-OF-THE-ART E-BAND LNAS AND THIS WORK

Reference	This Work*	[9] JSSC 2017	[10] MWCL 2019	[11] ISSCC 2020#	[12] ISSCC 2023	[5] TCAS2 2025
Technology	45-nm CMOS SOI	65-nm CMOS	65-nm CMOS	45-nm CMOS SOI	40-nm CMOS	40-nm CMOS
Gain (dB)	16.2	18.5	14.2	16.8	16.5	18.1
BW_{3dB} (GHz)	22.9 (62.8-85.7)	30 (62.5-92.5)	30 (60-90)	16 (73-88)	16 (70-86)	24.1 (59.2-83.3)
NF (dB)	3.3-4.5	5.5-7.9	6.3-8.4	4.8-6.1	4.8-6.5	5.1-8.1
IP_{1dB}@freq (dBm)	-25.5@79.9GHz	-15@80GHz	-10@77GHz	-7.4@75GHz	-8.5@73GHz	-14.5@76GHz
IIP3@freq (dBm)	-15.9†@79.9GHz	-5.4†@80GHz	-0.4†@77GHz	2.2@75GHz	1.1@73GHz	-4.9†@76GHz
P_{DC} (mW)	15.6	27	33.5	46	25	26
Size (mm^2)	0.224 (0.089$)	0.24 (0.063$)	0.45	0.63	0.47 (0.085$)	0.258 (0.094$)

* Post-simulation results.   # LNA with on-chip antenna.   † Estimated with IP_{1dB}+9.6dB. $ Core area.

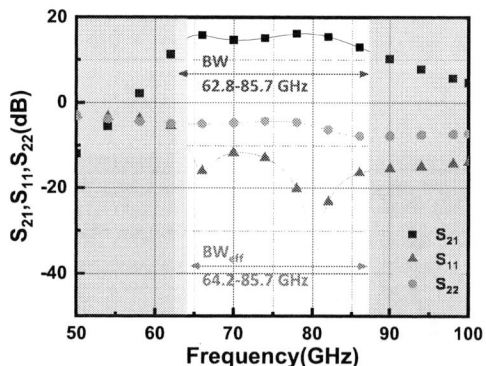

Fig. 9.   Simulated S-parameter of the proposed LNA.

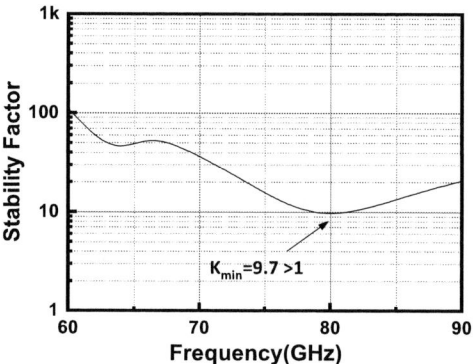

Fig. 11.   Simulated stability factor K of the proposed LNA.

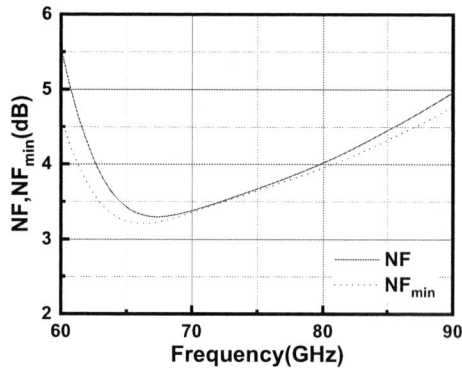

Fig. 10.   Simulated NF and NF_{min} of the proposed LNA.

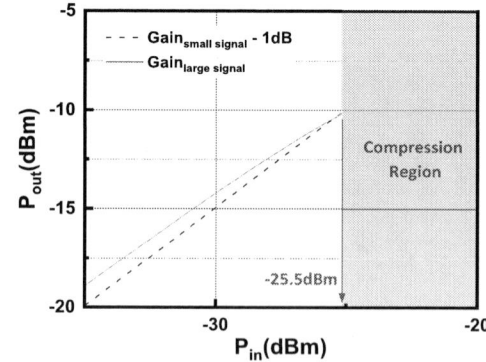

Fig. 12.   Simulated IP_{1dB} value at 79.9GHz of the proposed LNA.

Fig. 10 shows the simulated NF, ranging from 3.3 to 4.5 dB within the 3dB-bandwidth. The minimum noise figure NF_{min} is also presented. Fig. 11 shows the simulated stability factor K, with a minimum value of 9.7, which satisfies the stability condition of $K > 1$. The simulated IP1dB is -25.5 dBm at 79.9 GHz, as shown in Fig. 12. Table I presents a comparison between this work and several previously reported state-of-the-art wideband E-band LNAs. The proposed LNA demonstrates

979-8-3315-8850-2/25 $31.00 © 2025 IEEE

ultra low NF and power consumption with a compact chip area. However, poor linearity remains the primary drawback of this design.

IV. CONCLUSION

This paper presents a compact, wideband E-band Low Noise Amplifier (LNA) designed using a 45-nm RFSOI process. The proposed design employs techniques such as out-of-phase transformer feedback for input matching, current reuse for enhanced gain, and gate-drain transformer feedback to expand bandwidth. These methods collectively enhance the LNA's performance while minimizing power consumption. The results demonstrate the design's effectiveness for wideband, low-noise applications, highlighting its potential for high-frequency communication systems.

REFERENCES

[1] "Attenuation by atmospheric gases and related effects." Recommendation ITU-R P.676-13, 2022.

[2] Ioannis Sarkas, Sean T. Nicolson, Alexander Tomkins and Ekaterina Laskin et al., "An 18-Gb/s, Direct QPSK Modulation SiGe BiCMOS Transceiver for Last Mile Links in the 70–80 GHz Band," in IEEE Journal of Solid-State Circuits, vol. 45, no. 10, pp. 1968-1980, Oct. 2010, doi: 10.1109/JSSC.2010.2058011.

[3] B. Chen and Z. Zong, "An E-Band FMCW Radar Receiver With Arbitrary-Path Spillover Cancellation," in IEEE Journal of Solid-State Circuits, vol. 60, no. 5, pp. 1619-1631, May 2025, doi: 10.1109/JSSC.2025.3527072.

[4] Y. Yu, H. Liu, Y. Wu and K. Kang, "A 54.4–90 GHz Low-Noise Amplifier in 65-nm CMOS," in IEEE Journal of Solid-State Circuits, vol. 52, no. 11, pp. 2892-2904, Nov. 2017, doi: 10.1109/JSSC.2017.2727040.

[5] Z. Li, Y. Lin, H. Zhu, X. Yi, W. Che and Q. Xue, "A 59.2–83.3 GHz CMOS LNA With 18.1 dB Gain and 5.1 dB NF Using Gate-Drain Transformer," in IEEE Transactions on Circuits and Systems II: Express Briefs, vol. 72, no. 5, pp. 718-722, May 2025, doi: 10.1109/TCSII.2025.3551910.

[6] L. Gao, E. Wagner and G. M. Rebeiz, "Design of E- and W-Band Low-Noise Amplifiers in 22-nm CMOS FD-SOI," in IEEE Transactions on Microwave Theory and Techniques, vol. 68, no. 1, pp. 132-143, Jan. 2020, doi: 10.1109/TMTT.2019.2944820.

[7] L. Zou, K. Zhao, Z. Fang and L Huang et al. "A 74.8-88.8 GHz Wideband CMOS LNA Achieving +4.73 dBm OP1dB and 6.39 dB Minimum NF," 2023 IEEE/MTT-S International Microwave Symposium - IMS 2023, San Diego, CA, USA, 2023, pp. 60-63, doi: 10.1109/IMS37964.2023.10188098.

[8] P. Qin and Q. Xue, "Compact Wideband LNA With Gain and Input Matching Bandwidth Extensions by Transformer," in IEEE Microwave and Wireless Components Letters, vol. 27, no. 7, pp. 657-659, July 2017, doi: 10.1109/LMWC.2017.2711524.

[9] G. Feng, Chirn Chye Boon, F. Meng and X. Yi et al., "Pole-Converging Intrastage Bandwidth Extension Technique for Wideband Amplifiers," in IEEE Journal of Solid-State Circuits, vol. 52, no. 3, pp. 769-780, March 2017, doi: 10.1109/JSSC.2016.2641459.

[10] D. Pan, Z. Duan, S. Chakraborty, L. Sun, and P. Gui, "A 60-90-GHz CMOS double-neutralized LNA technology with 6.3-dB NF and -10dBm P-1dB," IEEE Microw. Wireless Compon. Lett., vol. 29, no. 7, pp. 489–491, Jul. 2019.

[11] S. Li, T. Chi, D. Jung, T.-Y. Huang, M.-Y. Huang, and H. Wang, "4.2 An E-band high-linearity antenna-LNA front-end with 4.8dB NF and 2.2dBm IIP3 exploiting multi-feed on-antenna noise-canceling and gmboosting," in Proc. IEEE Int. Solid-State Circuits Conf. (ISSCC), 2020, pp. 1–3.

[12] C. Han, J. Zhou, Z. Deng, Y. Shu, and X. Luo, "A 4.8dB NF, 70-to-86GHz deep-noise-canceling LNA using asymmetric compensation transformer and 4-to-1 hybrid-phase combiner in 40nm CMOS," in Proc. IEEE Int. Solid-State Circuits Conf. (ISSCC), 2023, pp. 24–26.

A Wideband Input Buffer Based on AC-Coupled Flipped Source Follower Using Auxiliary Operational Amplifiers for 8-GS/s ADCs

Yuhang Zhang
Department of Semiconductor Device
Southwest Institute of Technical Physics
Chengdu, China
zhangyuhang1999@126.com

Jian Chen*
Department of Semiconductor Device
Southwest Institute of Technical Physics
Chengdu, China
gzuchenjian@126.com

Weiying Hu
Department of Semiconductor Device
Southwest Institute of Technical Physics
Chengdu, China
weiying_hu@126.com

Xianguo Kou
Department of Semiconductor Device
Southwest Institute of Technical Physics
Chengdu, China
kxglongazure@163.com

Changdong Guo
Department of Semiconductor Device
Southwest Institute of Technical Physics
Chengdu, China
gcd_209@163.com

Haizhi Song
Department of Semiconductor Device
Southwest Institute of Technical Physics
Chengdu, China
hzsong1296@163.com

Abstract—This work proposes an input buffer based on an AC-coupled flipped source follower (FSF) assisted by operational amplifiers. In the proposed structure, an operational amplifier is employed to track both the input signal and the output signal of the buffer. Additionally, gain-boosting technology is applied to the tail current source transistor to increase its output resistance, thereby improving linearity. The op-amp-assisted AC-coupled architecture enhances the linearity of the FSF, making it suitable for high-speed and high-resolution ADCs. The input buffer was simulated in a 28 nm mixed-signal CMOS process. Simulation results demonstrate that the buffer achieves a bandwidth exceeding 4 GHz. At a sampling rate of 8 GS/s, with an input signal frequency of 3.97 GHz at -1 dBFS, the measured signal-to-noise ratio (SNR) and spurious-free dynamic range (SFDR) reach 69 dB and 73 dB, respectively. The power consumption of proposed input buffer is 224mW.

Keywords—*Input buffer, source follower, auxiliary operational amplifiers, wide band, high speed*

I. INTRODUCTION

With the increasing demand for higher signal sampling speed and accuracy in emerging communication systems, analog-to-digital converters (ADCs) are generally required to support sampling rates exceeding 1 GS/s with resolutions greater than 12 bits. To meet these performance requirements, it is necessary to incorporate an input buffer ahead of the sampling network to mitigate the effects of charge injection and kickback noise. An ideal input buffer exhibits high input impedance and low output impedance, allowing it to be driven effectively by the signal source while also driving the sampling network efficiently.

The primary technical challenge in designing input buffers for high-speed ADCs lies in achieving high linearity at GS/s sampling rates. Although traditional source follower (SF) structures provide low output impedance and high input impedance, their nonlinear distortion becomes more severe with increasing signal frequency and amplitude. This distortion mainly arises from channel length modulation, body effect, and transconductance variations due to parasitic capacitances. To overcome these limitations, several improved architectures have been proposed in recent years: (1) SF with feedforward compensation stabilize the drain-source voltage through level-shifting techniques and auxiliary operational amplifier clamping. However, switching capacitor memory effects and the additional power consumption of the auxiliary op-amp structures introduce new limitations [1] [2] [3]; (2) Push-pull SFs use drain bootstrapping to reduce variations in drain-source voltage, but they suffer from relatively high area overhead [4]; (3) The flipped voltage follower (FVF) reduces output impedance to the ohmic level through negative feedback and improves linearity with a constant bias current. Its signal swing, however, is limited by the threshold voltage, making it suitable only for low-swing applications [5] [6] [7]; (4) The super source follower (SSF) extends the signal swing to the power rail based on the FVF structure. However, in class-AB designs, the power supply rejection is relatively poor, and the current sources on both sides are sensitive to mismatch [8]. Current research trends suggest that combining dynamic parameter compensation with structural innovation is a key direction for pushing the performance boundaries of input buffers.

The design of input buffers for high-resolution and high-speed ADCs presents numerous challenges. To address these,

979-8-3315-8850-2/25 $31.00 © 2025 IEEE

this work proposes a new wideband input buffer based on an AC-coupled FVF utilizing auxiliary operational amplifier technology. The proposed design aims to improve the linearity of the input buffer, thereby enhancing the overall system performance.

The core of the proposed scheme is an AC-coupled flipped SF (FSF) structure equipped with an auxiliary operational amplifier. The AC-coupling mechanism effectively reduces signal transmission loss. The FSF significantly increases the dynamic range of the circuit. Moreover, the auxiliary operational amplifier forces the output signal of the buffer to track the input signal accurately. This approach further enhances the linearity of the circuit and reduces nonlinear distortion. Gain-boosting technology is applied to the tail current source transistor to decrease the output resistance and maintain high-precision performance in the ADC.

The proposed input buffer was simulated using a 28 nm mixed-signal CMOS process. Simulation results show that the buffer achieves a bandwidth exceeding 4 GHz. Under an 8 GS/s sampling rate, with a 3.97 GHz input signal at -1 dBFS, the measured signal-to-noise ratio (SNR) and spurious-free dynamic range (SFDR) reached 69 dB and 73 dB, respectively.

The remainder of this paper is organized as follows: Section II analyzes various typical input buffer structures. Section III describes the newly proposed circuit architecture. Section IV presents the experimental data and corresponding analysis. Section V summarizes the main findings of this study.

II. ANALYSIS OF TYPICAL INPUT BUFFER STRUCTURE

The design of the input buffer is critical in high-speed, high-resolution analog-to-digital converters ADCs. To achieve efficient and high-precision signal conversion, the input buffer must satisfy multiple technical requirements, including wide bandwidth, low noise, high linearity, and good stability. This section discusses the importance of the input buffer in high-speed and high-precision ADCs and analyzes several typical input buffer architectures.

A. The Necessity of Using Input Buffer in ADC

The input buffer plays a key role in high-speed and high-precision ADC systems. Its functions and importance are reflected in the following aspects:

- Impedance matching and signal drive: With high input impedance and low output impedance, the input buffer provides impedance matching between the front-end circuit (e.g., sensor or filter) and the ADC. This reduces signal transmission loss and enhances drive capability. For instance, an SF structure can effectively isolate the front-end circuit from transient current disturbances during ADC sampling.

- Dynamic range optimization: The buffer amplifier can extend the system's dynamic range, ensuring high sensitivity (low noise figure) while preventing overload from strong signals. Digitally controlled linear amplifiers can help balance the trade-off between gain and third-order intercept point.

- Common mode stability and linearity improvement: In ultra-high-speed ADCs, the input buffer must stabilize the output common-mode voltage through common-mode feedback to avoid comparator errors caused by common-mode drift. A stacked SF architecture can also improve linearity and reduce channel modulation effects.

To sum up, the input buffer is a core component in high-speed and high-precision ADCs, enabling high accuracy and reliability through impedance matching, dynamic range extension, noise suppression, and other mechanisms.

B. Conventional Input Buffer Structure

The conventional input buffer structure is based on an SF. Fig. 1 shows the circuit diagram of a typical SF. It exhibits high input impedance and low output impedance, with a nominal voltage gain of unity. These characteristics align with the requirements of an ideal input buffer, making the SF a suitable choice. Similarly, an emitter follower can also serve as an input buffer in bipolar implementations. The following analysis uses the SF as an example to evaluate its performance as an input buffer.

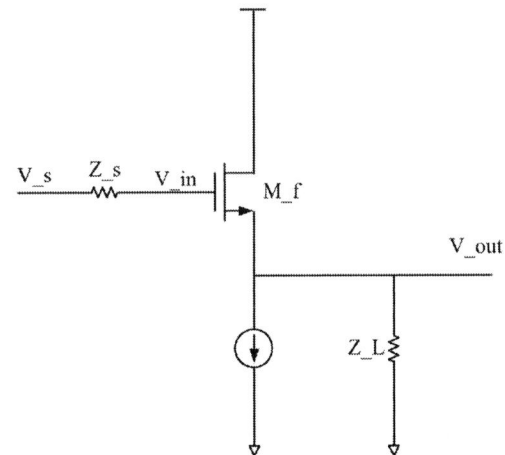

Fig. 1. Simplified circuit of source follower

Ideally, the SF operates at unity gain. However, when accounting for source impedance, channel length modulation, body effect, parasitic capacitances, and other non-idealities, the actual gain becomes less than unity. A small-signal model of the SF illustrating this behavior is presented in Fig. 2.

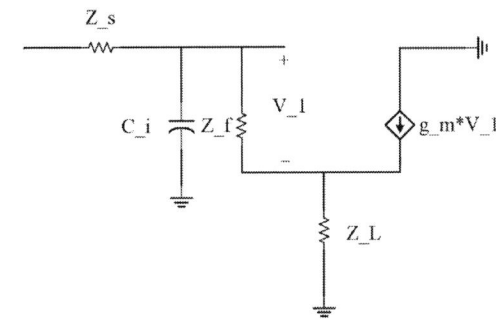

Fig. 2. Small signal model of source follower

In the presence of source impedance, the transfer characteristic can be expressed as follows:

$$\frac{V_o}{V_s} \cong \frac{1}{(1+sC_iZ_s)} \times \frac{1}{\left(1+\dfrac{Z_{s\_eq}+Z_f}{(1+g_mZ_f)Z_L}\right)} \quad (1)$$

where g_m is the following device M_f transconductance; C_i is the ground capacitance of the input node; Z_s is the source impedance; Z_f is the input impedance of the follower; Z_L is the load impedance, including the output impedance and current source of the load (sampling capacitor) and the follower; and Z_{s_eq} is the parallel connection of Z_s and sC_i.

The input impedance of the SF is given by:

$$\frac{1}{Z_{in}} = sC_i + \frac{1}{Z_f + Z_L + g_mZ_fZ_L} \quad (2)$$

This equation quantifies the impedance seen at the input terminals under specified operating conditions.

The SF's output impedance is mathematically expressed by:

$$\frac{1}{Z_{out}} = \frac{1}{Z_o} + \frac{1}{R_{ds}} + sC_{sc} + \frac{1+g_mZ_f}{Z_f + \dfrac{1}{sC_i + 1/Z_s}} \quad (3)$$

When the source impedance is small and the input capacitance is large, the output impedance can be further simplified as follows:

$$Z_{out} \approx \frac{1}{g_m} + \frac{Z_s}{g_mZ_f} \approx \frac{1}{g_m} \quad (4)$$

C. Conventional FSF

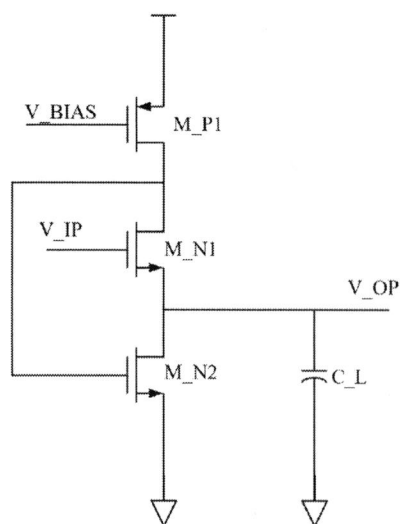

Fig. 3. Conventional flipped source follower

In a traditional SF, the input MOSFET needs to drive the load capacitor, causing the current to become signal-dependent and introducing significant nonlinearity. This conventional architecture imposes a dual functional constraint on the input transistor: it must simultaneously establish the output voltage and drive the load, making these inherent limitations unavoidable. To address these issues, an improved solution incorporates an auxiliary transistor, M_{P1}, dedicated to load driving. This transistor is strategically positioned within a negative feedback configuration, which reduces the output resistance. As a result, the current through the input transistor M_{N1} remains constant, thus improving linearity. The structure of this scheme, known as the FVF, is illustrated in Fig. 3.

The output resistance of the FVF is given by:

$$R_{out} \approx \frac{1}{g_{m,MN1}g_{m,MN2}r_{o,MN1}} \quad (5)$$

The output resistance is usually several ohms, while the output resistance of the traditional SF is the reciprocal of the transconductance of the SF, usually several thousand ohms.

As a circuit building block, the FVF uses shunt feedback to maintain a constant current through the input transistor, providing nearly unity voltage gain and high current-driving capability. It improves linearity through parallel negative feedback and offers low output resistance, making it suitable for Ti-ADC applications. However, a limitation of this topology is that both M_{N1} and M_{N2} must remain in the saturation region, which restricts the signal swing. The allowable output voltage swing is given by: $V_{TH,MN1} - V_{Dsat,MN1}$. For practical ADC applications, this limited signal range can pose a significant challenge.

III. PROPOSED OPERATIONAL AMPLIFIER ASSISTED AC-COUPLED FSF

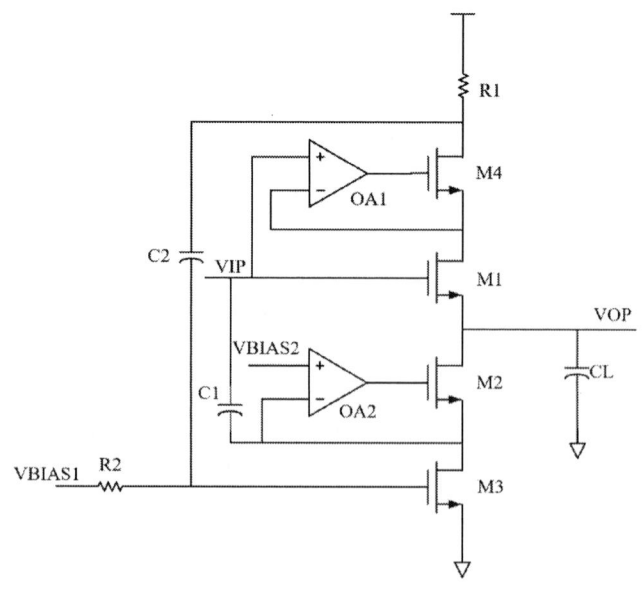

Fig. 4. Proposed operational amplifier assisted AC-coupled FSF

This work presents a wideband input buffer design based on an AC-coupled FSF, enhanced with auxiliary operational amplifiers to achieve improved linearity. The proposed architecture is well-suited for high-speed and high-resolution ADCs, contributing to enhanced overall system performance. The AC-coupled FSF offers inherent advantages such as low output impedance and wide signal swing, making it particularly advantageous for high-speed signal processing applications. By incorporating auxiliary operational amplifiers, the linearity is further optimized through precise feedback mechanisms, ensuring high output signal quality. The proposed structure is illustrated in Fig. 4.

In order to achieve high linearity in the circuit, OA1 is employed to track variations in the drain voltage of transistor M1 (denoted as VIP and VOP). The drain voltage of M1 is expressed by Equation (6), and linearity is enhanced via the unity-gain feedback configured around OA1. In circuit simulations, an ideal operational amplifier model was used to effectively demonstrate the performance of the proposed architecture.

$$v_{D,M1} \approx v_{IP}\frac{A}{1+A} \qquad (6)$$

The parameter A denotes the open-loop voltage gain of OA1.

The unity-gain feedback mechanism ensures accurate tracking between the input and output signals, thereby improving linearity. This approach mitigates distortion caused by parasitic capacitances and nonlinear effects, leading to an improved signal-to-noise ratio. Moreover, compared to capacitive feedback techniques, this structure reduces required chip area.

In practical design, operational amplifiers with higher gain can more effectively suppress distortion. For simulation purposes, an ideal op-amp model with a gain of 60 dB was employed. Additionally, the op-amp's bandwidth and input/output signal range should exceed the input buffer's bandwidth and input signal range, respectively.

The inclusion of a second operational amplifier, OA2, increases the output impedance of the current source without augmenting the parasitic capacitance at the output node. At low frequencies, the output impedance is increased by a factor equal to the gain of OA2.

IV. SIMULATION RESULT

The proposed input buffer was fabricated in a 28 nm mixed-signal CMOS process and evaluated under a 2.5 V supply voltage. Input signal range is 1-V_{pp}. All MOSFETs in the circuit employed a high-breakdown-voltage model rated at 2.5 V. When sampling at 8 GS/s with a −1 dBFS input signal at 3.97 GHz, the circuit achieved an SNR of 69 dB and a SFDR of 73 dB. The power consumption of proposed input buffer is 224mW. The output spectrum under these conditions is shown in Fig. 5. The simulated SNR and SFDR values across different input frequencies are further illustrated in Fig. 6 and Fig. 7, respectively.

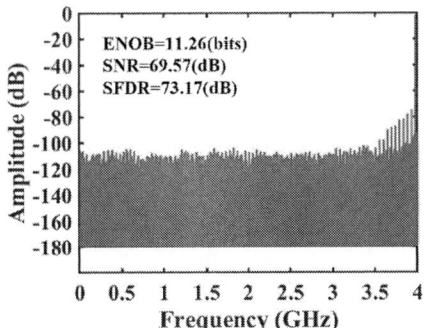

Fig. 5. Simulated spectrum under 8 GS/s sampling rate with a 3.97-GHz, −1dBFS input

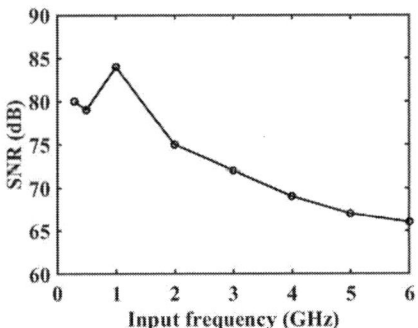

Fig. 6. Simulated SNR at different input frequencies

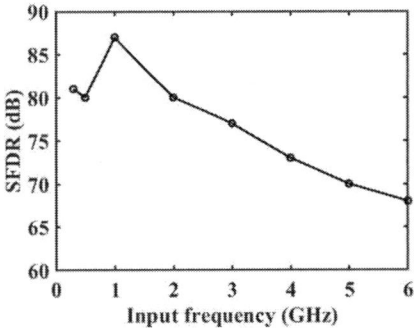

Fig. 7. Simulated SFDR at different input frequencies

TABLE I. PERFORMANCE COMPARISON

	[9]	[10]	This work
Process(nm)	28	65	28
Architecture	AC-coupled FVF	AC-coupled SSF	AC-coupled FVF
Fs(GS/s)	8	6	8
Input range(V_{pp})	1.4	1.8	1
SNR(dB)	57	64	69
SFDR(dB)	74	73	73
Load(pF)	0.6	2	0.6
Power(mW)	137	33	224

A performance comparison between this work and other recently published AC-coupled input buffer structures is summarized in Table I.

V. Conclusion

To meet the performance requirements of high-speed and high-precision ADCs, an operational amplifier-assisted AC-coupled FSF was proposed. In this design, an operational amplifier was used to track both the input signal and the output signal of the buffer. Additionally, gain-boosting technology was adopted in the tail current source transistor to increase its output resistance, thereby improving linearity. The proposed input buffer was simulated using a 28 nm mixed-signal CMOS process. Simulation results demonstrate that the buffer achieves a bandwidth exceeding 4 GHz. At a sampling rate of 8 GS/s with a −1 dBFS input signal at 3.97 GHz, the measured SNR and SFDR reached 69 dB and 73 dB, respectively. The power consumption of proposed input buffer is 224mW.

Acknowledgment

I sincerely thank my mentors, Researcher Chen and Researcher Hu, for their careful guidance on the research topic. Their teachings have benefited me greatly. Thank my colleagues for their collaborative support in the paper work. At the same time, the valuable opinions put forward by the review experts make the paper perfect. Finally, I would like to thank my family and love for their encouragement and support.

References

[1] A. M. A. Ali et al., "A 16-bit 250-MS/s IF Sampling Pipelined ADC With Background Calibration," in IEEE Journal of Solid-State Circuits, vol. 45, no. 12, pp. 2602-2612, Dec. 2010.

[2] A. M. A. Ali et al., "A 14 Bit 1 GS/s RF Sampling Pipelined ADC With Background Calibration," IEEE Journal of Solid-State Circuits, vol. 49, no. 12, pp. 2857-2867, 2014.

[3] Y. Cao, T. Zhang, Y. Chen, F. Ye, and J. Ren, "An Operational Amplifier Assisted Input Buffer and An Improved Bootstrapped Switch for High-Speed and High-Resolution ADCs," in 2018 IEEE International Symposium on Circuits and Systems (ISCAS), 2018, pp. 1-5.

[4] S. Devarajan et al., "A 12-b 10-GS/s Interleaved Pipeline ADC in 28-nm CMOS Technology," IEEE Journal of Solid-State Circuits, vol. 52, no. 12, pp. 3204-3218, 2017.

[5] C. Huang, Y. Zhao, Y. Ji, X. Yang, T. Zhang, and W. Gai, "An Input Buffer with 85dB SFDR for High-Speed Pipeline ADC," in 2022 IEEE International Conference on Integrated Circuits, Technologies and Applications (ICTA), 2022, pp. 52-53.

[6] A. M. A. Ali et al., "A 12-b 18-GS/s RF Sampling ADC With an Integrated Wideband Track-and-Hold Amplifier and Background Calibration," IEEE Journal of Solid-State Circuits, vol. 55, no. 12, pp. 3210-3224, 2020.

[7] B. Hershberg et al., "A 4-GS/s 10-ENOB 75-mW Ringamp ADC in 16-nm CMOS With Background Monitoring of Distortion," IEEE Journal of Solid-State Circuits, vol. 56, no. 8, pp. 2360-2374, 2021.

[8] Lopez-Martin A J, Acosta L, Garcia-Alberdi C, et al. "Power-efficient analog design based on the class AB super source follower". International Journal of Circuit Theory and Applications, 2012, 40(11): 1143-1163.

[9] Huang Z, Zhang J, Cen Y, et al., "A 6-GHz bandwidth input buffer based on AC-coupled flipped source follower for 12-bit 8-GS/s ADC in 28-nm CMOS," IEEE Transactions on Circuits and Systems II: Express Briefs, 2022, 69(10): 4163-4167.

[10] T. Feng, D. Li, S. Liu, and Z. Zhu, "A Wideband ADC Input Buffer Based on AC-Coupled Super Source Follower," in 2023 International Conference on Microwave and Millimeter Wave Technology (ICMMT), 2023, pp. 1-3.

2025 The 10th International Conference on Integrated Circuits and Microsystems

A 2.4–2.6 GHz CMOS High Linearity Power Amplifier with MGTR and Harmonic Trap Techniques

Runxun Zhang [1, 2] and Gengzhen Qi*, [1, 2]

1. School of Microelectronics Science and Technology, Sun Yat-sen University, Zhuhai, China
2. Guangdong-Macao Joint Laboratory for Modular Chip Design and Testing, Guangdong, China
*Email: Qigzh@mail.sysu.edu.cn

Abstract—**This paper presents the design of a linear RF power amplifier (PA) operating from 2.4 to 2.6GHz using 65nm CMOS technology. To address the challenge of poor linearity in CMOS PAs, Multi-Gated Transistor (MGTR) technology and second-harmonic trap circuits are introduced to suppress third-order intermodulation and harmonic distortion. Post-layout simulations demonstrate a saturated output power of 30.32dBm, OP1dB of 28.48dBm, OIP3 of 38.58dBm, and power-added efficiency (PAE) of 43.21% (Peak) / 23.41% (6dB PBO), achieving a good balance of linearity, efficiency, and integration.**

Keywords—CMOS, RF Power Amplifier, Linearity, MGTR, Harmonic Trap

I. INTRODUCTION

RF Power Amplifiers play a vital role in the transmission chain of RF front-end chips, directly determining signal strength, transmission range, and energy efficiency. With the rapid proliferation of wireless technologies—such as 4G LTE, 5G-NR, Wi-Fi, and IoT—the 2.4–2.6 GHz band remains a critical frequency range requiring compact, power-efficient, and linear PA designs.

Currently, many commercial PAs are implemented using high-cost III-V semiconductor technologies, mainly because the low-resistivity substrate in CMOS processes severely degrades efficiency and power combining capability. However, with continued scaling of CMOS technology nodes, transistors with high unity-gain frequency (f_T) and maximum oscillation frequency (f_{max}) are showing increasing potential for high-frequency applications [1].

Since the received signal power must meet the receiver sensitivity requirements, the PA's high output power at the 1-dB compression point (P1dB) is critical for overcoming channel attenuation [2]. Moreover, modern communication systems adopt various modulation schemes—such as AM and ASK—that feature non-constant signal envelopes. These impose stringent requirements on the PA's linearity, especially in terms of intermodulation distortion under multi-tone excitation. Therefore, enhancing the linearity of PAs remains both a meaningful and challenging research topic.

Techniques for improving linearity can be categorized into system-level and circuit-level approaches. System-level includes digital predistortion (DPD) [3], envelope elimination

This work is funded by the Guangdong Basic and Applied Basic Research Foundation under Grant 2025A1515011608 and the Guangdong-Macao Joint Laboratory (GDSTC Project No. 2025B1212150003).

and restoration (EER) [4], and the Doherty architecture [5]. On the other hand, circuit-level approaches—such as Multi-Gated Transistor (MGTR) [6], capacitive compensation [7], and source degeneration [8]—can directly enhance linearity at the chip level, reducing external dependencies and improving integration.

In this work, the MGTR technique is employed to enhance PA linearity. In addition, a novel method is proposed that introduces a notch filter network into the PA to improve its third-order intercept point (IP3). The combination of these two techniques enables the PA to achieve excellent linearity without compromising other performance metrics.

II. CIRCUIT DETAILS

A. PA Architecture

Fig. 1 shows the core circuit of PA. This PA adopts a pseudo-differential two-stage stacked structure, which allows a higher voltage supply and enhances the output power, mitigating the low breakdown limitation of CMOS.

Parasitic gate-drain capacitance C_{gd} degrades amplifier stability [9]. We apply a capacitive neutralization technique by adding cross-coupled compensation capacitors C_c to cancel the effect of C_{gd}. From the small-signal model and the definition of K-factor, it can be derived that:

$$K = \frac{2\Re(Y_{11})\Re(Y_{22}) - \Re(Y_{12}Y_{21})}{|Y_{12}Y_{21}|} \tag{1}$$

$$= \frac{2 + \omega^2(C_{gd} - C_c)^2 R_G R_D}{\omega|C_{gd} - C_c|R_G R_D \sqrt{\omega^2(C_{gd} - C_c)^2 + g_m^2}} \tag{2}$$

When $C_c = C_{gd}$, $K \to \infty$. Circuits can be unconditionally stable.

B. Nonlinearity Analysis

The input-output characteristic of a weakly nonlinear system can be approximated by a power series expansion:

$$y(t) \approx g_1 x(t) + g_2 x^2(t) + g_3 x^3(t) + \cdots \tag{3}$$

For a single-tone input $x(t) = A\cos\omega t$, the output is:

$$y(t) = \frac{g_2 A^2}{2} + \left(g_1 A + \frac{3g_3 A^3}{4}\right)\cos\omega t$$
$$+ \frac{g_2 A^2}{2}\cos 2\omega t + \frac{g_3 A^3}{4}\cos 3\omega t + \cdots \tag{4}$$

979-8-3315-8850-2/25 $31.00 © 2025 IEEE   726

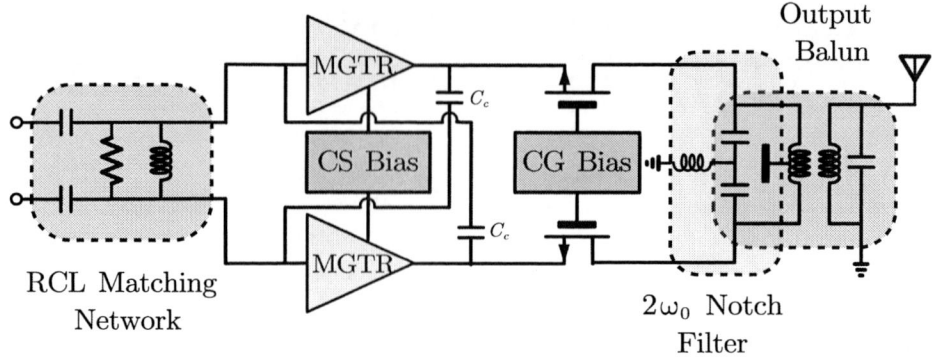

Fig. 1. Schematic of the core PA

For the fundamental frequency, since in most practical scenarios $g_1 g_3 < 0$, third-order terms g_3 contribute to gain compression as input amplitude increases. The 1-dB compression point P_{1dB} indicates where gain drops by 1dB from its small-signal value.

For a two-tone input signal, $x(t) = A\cos\omega_1 t + A\cos\omega_2 t$, third-order intermodulation distortion (IMD3) products appear near the fundamental frequencies ($2\omega_1 - \omega_2$, $2\omega_2 - \omega_1$), causing in-band distortion. The amplitude of IMD3 terms grows proportionally to A^3, while fundamental terms scale with A, degrading linearity as input power increases. The third-order intercept point IP_3 is defined as the extrapolated intersection of the output fundamental and IMD3 components in dB scale.

Based on the above analysis, improving PA linearity can be achieved via: 1) **Suppressing third-order nonlinearity at the transistor level.** 2) **Filtering out harmonic distortion** at the output without affecting the fundamental.

These two strategies form the basis for the MGTR technique and harmonic trap circuit adopted in this work, detailed in the following sections.

C. Multi-Gate Transistor Technique

The Multi-Gated Transistor (MGTR) technique improves linearity by paralleling multiple transistors operating under different bias conditions to cancel third-order transconductance nonlinearity (g_{m3}). Typically, the g_{m3}–bias voltage relationship exhibits a positive-to-negative transition. This enables strategic cancellation by combining devices with complementary g_{m3} characteristics.

In this design, one main transistor operates in Class-AB mode, while two auxiliary transistors are biased in Class-C and deep Class-C, respectively. Their combined operation creates a "near-zero g_{m3} region" within the PA's operating range, significantly suppressing third-order distortion (Fig. 2).

Beyond improving linearity, the proposed MGTR structure also enhances back-off power efficiency. Similar to active load modulation, the two auxiliary Class-C biased transistors remain OFF at low input power levels, consuming no DC power. As the input signal increases, they gradually turn on,

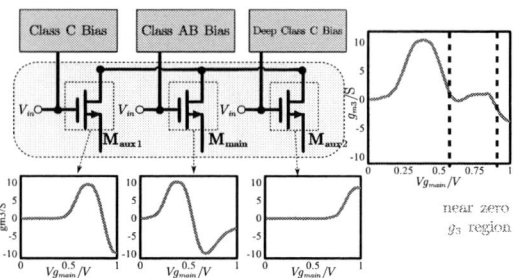

Fig. 2. MGTR structure and g_{m3} cancellation

contributing additional current gain only when needed. This dynamic activation minimizes DC loss during power back-off, leading to improved PAE across a wider output power range.

Compared to other linearization methods such as source degeneration [8] or digital predistortion (DPD) [3], MGTR offers circuit-level linearization with negligible additional area and without sacrificing fundamental matching.

D. Second-Harmonic Trap

Second-harmonic components at the PA output can interact with the fundamental and higher-order harmonics, exacerbating intermodulation distortion. To suppress the second harmonic while minimally affecting the fundamental, we implement a differential-mode LC notch network at the output of the pseudo-differential stage (see Fig. 3).

For clarity, the resonance condition of the trap is derived as follows. The resonance frequency of an LC tank is

$$\omega_{\text{res}} = \frac{1}{\sqrt{L_{\text{eq}} C_{\text{eq}}}}.$$

If the trap is designed to resonate at the second harmonic, i.e. $\omega_{\text{res}} = 2\omega_0$, then the required product is

$$L_{\text{eq}} C_{\text{eq}} = 2 C_H L_H = \frac{1}{4\omega_0^2}.$$

In the differential topology used here the effective impedance seen by even-mode (both legs in-phase) and odd-mode (legs anti-phase) excitations differs. For odd-order harmonics like the fundamental and third harmonic, the differential outputs are out of phase, making **point A** in Fig. 3 a virtual ground. These signals only see a reduced capacitance of $C_H/2$, rendering the notch network non-resonant at those frequencies.

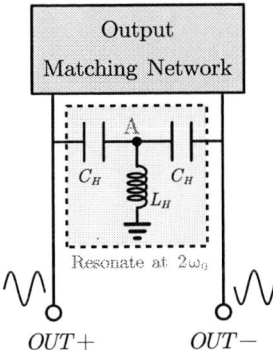

Fig. 3. Differential second-harmonic trap circuit

The notch filter implemented in this work is a differential-mode LC trap co-optimized with the output matching network. Unlike single-ended $\lambda/4$ stubs [10] or standalone shunt/series resonators, the trap exploits the phase relationship of differential outputs to present a deep short specifically at $2\omega_0$ while remaining transparent at ω_0. This topology enables compact on-chip integration and allows reuse of C_H as part of the matching capacitance, improving area efficiency without sacrificing fundamental matching. Fig. 4 shows the trap provides an OIP3 improvement of $4.41dBm$ in pre-simulation.

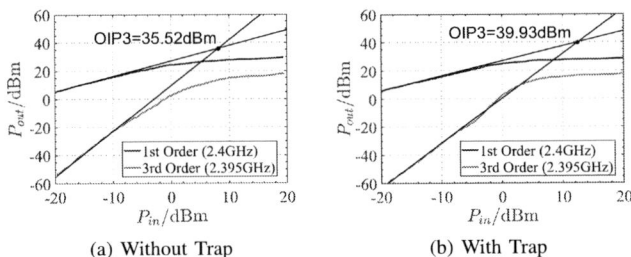

(a) Without Trap   (b) With Trap

Fig. 4. The OIP3 improvement in pre-simulation

E. Matching Network Design

The input matching network is an RLC circuit without balun. It employs conjugate matching for maximum power transfer. The output network is optimized via Load-Pull simulations to maximize P_{out} and PAE. The final result of LoadPull is the conversion of antenna impedance (50Ω) to $8.61 + j7.35\Omega$.

III. LAYOUT AND SIMULATION RESULTS

A. Layout Implementation

The PA is implemented in TSMC 65nm CMOS technology. The full layout occupies $1049\mu m \times 755\mu m$ without pads (Fig. 5).

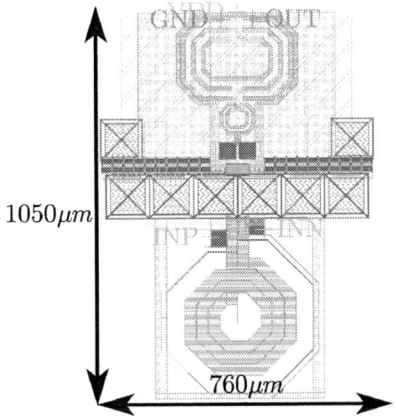

Fig. 5. Layout of the PA without pads

Metal routing is carefully optimized to minimize parasitic inductance and resistance, especially in high-current paths. The power supply grid and RF path are shielded by top metal layers (M8 / M9) to improve performance and reduce substrate coupling.

B. Post-Layout Simulation Results and Performance Comparison

Post-layout simulations are carried out under the TT process corner at 27°C with a 3.3V supply voltage. The operating frequency range is 2.4–2.6GHz. Results are as follows:

a) **Stability:** Across the entire frequency band, the PA maintains K-factor > 1 and $0 < B_{1f} < 1$, indicating unconditional stability.

b) **Return Loss:** As shown in Fig. 6, both S_{11} and S_{22} are below -10dB throughout the band, demonstrating good input and output matching.

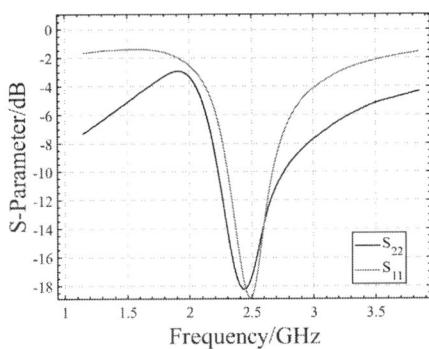

Fig. 6. S parameters of PA input and output

c) **Output Power:** Under single-tone excitation, the output power curve is illustrated in Fig. 7. The results show a saturated output power P_{sat} of 30.32dBm and an output 1-dB compression point OP_{1dB} of 28.48dBm, confirming high output power and good linearity.

d) **Gain and PAE:** As depicted in Fig. 8, the power gain G_t is 24dB. The power-added efficiency (PAE) achieves 43.21%

TABLE I
COMPARISON TABLE WITH STATE-OF-THE-ART PAs

References	**This Work**	[11]	[12]	[13]	[14]
Process	**CMOS 65nm**	CMOS 180nm	GaAs 150nm	GaAs	CMOS 180nm
Supply Voltage (V)	**3.3**	3.3	10.0	5.0	1.8
Freq (GHz)	**2.4-2.6**	2.45	0.01-30	0.7-2.5	2.4-2.5
P_{sat} (dBm)	**30.32**	23.3	30	27	13.7
Gain (dB)	**24**	14.5	13-16	20 ± 0.5	12.5
OP_{1dB} (dBm)	**28.48**	22.5	29.0	24.0	–
OIP3 (dBm)	**38.58**	34.0	35.0	45.0	29.2
PAE_{peak} (%)	**43.21**	31.8	25.0	55.0	25.6

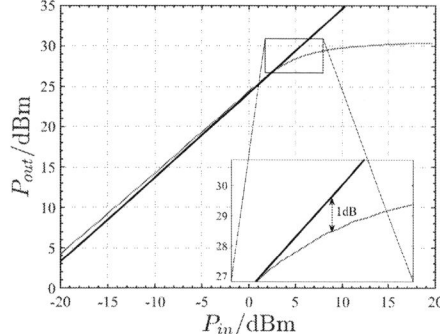

Fig. 7. $P_{out} - P_{in}$ curve of the PA

Fig. 9. IMD3 performance of the PA

at peak and maintains 23.41%, 14.56%, and 7.28% at 6dB, 9dB, and 12dB back-off, respectively.

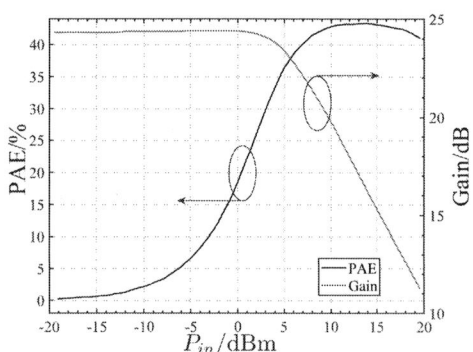

Fig. 8. PAE and gain of the PA

e) **IMD3:** With two-tone excitation, the third-order inter-modulation distortion (IMD3) is shown in Fig. 9, where the output third-order intercept point OIP_3 reaches 38.58dBm, validating the effectiveness of the linearization techniques.

Table I compares the performance of this work with recent state-of-the-art designs.

IV. CONCLUSION

This paper presents the design of a 2.4–2.6GHz CMOS linear RF power amplifier based on a pseudo-differential

two-stage stacked architecture in 65nm CMOS. To address high linearity challenges in CMOS PAs, the design integrates MGTR technology and a second-harmonic trap.

Post-layout simulations demonstrate saturated output power of 30.32dBm, OP1dB of 28.48dBm, OIP3 of 38.58dBm, and peak PAE of 43.21%. The PA also maintains competitive PAE under 6–12dB back-off, confirming its suitability for modern wireless systems requiring both linearity and efficiency.

ACKNOWLEDGMENT

This work is funded in part by the Guangdong Basic and Applied Basic Research Foundation under Grant 2025A1515011608. It is also partially funded by the Guangdong-Macao Joint Laboratory (GDSTC Project No. 2025B1212150003).

REFERENCES

[1] I. Liu, K. Ma, S. Mou, and F. Meng, "A review of recent power amplifier IC," in *2017 10th Global Symposium on Millimeter-Waves (GSMM)*, Hong Kong, China, 2017, pp. 87–91.

[2] N. Boom, W. Rens, and J. Crols, "A 5.0mW 0dBm FSK transmitter for 315/433 MHz ISM applications in 0.25 μm CMOS," in *Proceedings of the 30th European Solid-State Circuits Conference (ESSCIRC)*, Leuven, Belgium, 2004, pp. 199–202.

[3] L.-Y. Shi, T. Wang, D. Hua, and Z.-L. Hong, "A 31.2dBm Pout, 40.7% Peak DE, 2.4GHz Quadrature Doherty Power Amplifier Based on Current Mode RFDAC Architecture," in *2022 IEEE 16th International Conference on Solid-State & Integrated Circuit Technology (ICSICT)*, 2022, pp. 1–3.

[4] M. Vasić, O. Garcia, J. A. Oliver, P. Alou, D. Diaz, J. A. Cobos, A. Gimeno, J. M. Pardo, C. Benavente, and F. J. Ortega, "Efficient and Linear Power Amplifier Based on Envelope Elimination and Restoration," *IEEE Transactions on Power Electronics*, vol. 27, no. 1, pp. 5–9, 2012.

[5] B. Kim, J. Kim, I. Kim, and J. Cha, "The Doherty power amplifier," *IEEE Microwave Magazine*, vol. 7, no. 5, pp. 42–50, 2006.

[6] J. Park, C. Lee, J. Yoo, and C. Park, "A CMOS Antiphase Power Amplifier With an MGTR Technique for Mobile Applications," *IEEE Transactions on Microwave Theory and Techniques*, vol. 65, no. 11, pp. 4645–4656, 2017.

[7] C. Wang, M. Vaidyanathan, and L. Larson, "A capacitance-compensation technique for improved linearity in CMOS class-AB power amplifiers," *IEEE Journal of Solid-State Circuits*, vol. 39, no. 11, pp. 1927–1937, 2004.

[8] P. Monsurro, S. Pennisi, G. Scotti, and A. Trifiletti, "Linearization Technique for Source-Degenerated CMOS Differential Transconductors," *IEEE Transactions on Circuits and Systems II: Express Briefs*, vol. 54, no. 10, pp. 848–852, 2007.

[9] Y. Kawano, T. Suzuki, M. Sato, Y. Nakasha, T. Hirose, N. Hara, and K. Joshin, "20-GHz, 20-dBm pseudo-differential power amplifier in standard 90-nm CMOS," in *2008 Asia-Pacific Microwave Conference*. IEEE, 2008, pp. 1–4.

[10] J.-X. Xu, H. Chen, W. Chen, and X. Y. Zhang, "Broadband Doherty Power Amplifier Using Short Ended $\lambda/4$ Transmission Lines Based on the Analysis of Negative Characteristic Impedance," *IEEE Transactions on Circuits and Systems I: Regular Papers*, vol. 70, no. 2, pp. 545–555, 2023.

[11] S. Mariappan, J. Rajendran, N. M. Noh, Y. Yusof, and N. Kumar, "A 23.3dBm CMOS power amplifier with third-order g_m cancellation," *Circuit World*, vol. 48, no. 2, pp. 215–222, 2022.

[12] M. S. Zhu, Q. L. Li, and X. T. Fan, "A 1.0W 0.01-22 GHz Wideband Power Amplifier in 0.15 µm GaAs," in *2023 International Conference on Microwave and Millimeter Wave Technology (ICMMT)*, 2023, pp. 1–3.

[13] R. Sharma Nitesh, J. Rajendran, H. Ramiah, and A. Abd Manaf, "A 700MHz to 2.5GHz Cascode GaAs Power Amplifier for Multi-Band Pico-Cell Achieving 20dB Gain, 40dBm to 45dBm OIP3 and 66% Peak PAE," *IEEE Access*, vol. 6, pp. 818–829, 2018.

[14] S. Mariappan, J. Rajendran, N. M. Noh, and H. Ramiah, "A 29dBm OIP3 Dual-Stage Power Amplifier with Analog Pre-Distorter in 0.18µm CMOS for IoT Transceiver," *IETE Journal of Research*, vol. 68, no. 3, pp. 2298–2304, 2019.

2025 The 10th International Conference on Integrated Circuits and Microsystems

A Ka Band High Back-off PAE Power Amplifier with Adaptive Bias Circuits in 65nm CMOS

Ran Zhang*
[1] School of Integrated Circuit Science and Engineering
University of Electronic Science and Technology of China
Chengdu, China
[2] Southwest China Institute of Electronic Technology
Chengdu, China
ran_zhang@std.uestc.edu.cn

Xiaodong Zhao
Southwest China Institute of Electronic Technology
Chengdu, China
nanod@qq.com

Hangbiao Li
Southwest China Institute of Electronic Technology
Chengdu, China
hangbiaoli@qq.com

Abstract—Based on 65nm CMOS technology, an adaptive bias circuit and its application in a Ka band differential power amplifier (PA) with compact chip area is introduced in this paper. The proposed adaptive bias circuit is able to adjust the PA's bias voltage dynamically in response to input signal variations. It highly improves PAE_{P1dB} (PAE at 1dB compression point) from 14.7% to 24.5%, and $PAE_{6dB\text{-}backoff}$ (PAE at 6dB-backoff from 1dB compression point) from 4.5% to 12.4%. The proposed PA also achieves excellent port matching and linearity performance.

Keywords—adaptive bias, Back-off, PAE, Ka band, PA, CMOS

I. INTRODUCTION

In modern communication systems, such as 5G and phased-array satellite communication systems, advanced modulation schemes and multi-carrier technologies are widely adopted, resulting in signals with high peak-to-average power ratios (PAPR) [1]. When the system performs power back-off to meet the linearity requirements of high-PAPR signals, the power added efficiency (PAE) of the power amplifier (PA) drops sharply, causing most of the power consumption to be converted into heat, which ultimately leading to severe device overheating and wasteful energy consumption. Therefore, high back-off efficiency PAs are critical components in reducing power consumption and mitigating thermal management challenges, in order to facilitate the evolution of wireless communication systems towards higher frequency bands and low-carbon power-saving development.

Conventional Class AB PA architecture can be realized in compact chip area [2], but difficult to address the issue of gradual gain compression as the input signal level increases, thereby hard to achieve high back-off PAE. Typical architectures, such as envelope tracking (ET) PA, outphasing PA, and Doherty PA, are commonly to improve back-off efficiency. ET PAs utilize dynamic power supply modulation technology to adjust supply voltage in real-time according to signal envelope, improving back-off efficiency by over 20% [3]. However, the ET technology requires complex high-speed envelope detection and power modulation circuits, and faces severe challenges in envelope path delay matching at millimeter (MM) wave frequencies. Outphasing PAs operate on the vector

synthesis principle of two amplifier paths, theoretically achieving back-off efficiency approaching saturation efficiency. However, the outphasing architecture demands extremely high precision in phase and amplitude matching between the two paths [4], where even minor errors at millimeter-wave frequencies could result into large AM and PM distortion. The Doherty PA employs a main amplifier and an auxiliary amplifier configuration to achieve 1.5 times efficiency enhancement at 6dB back-off [5]. However, it requires more complex impedance matching design at MM wave frequencies, and occupies a large footprint making it difficult to be integrated in multi-channel phased array chips.

Shown as Fig.1, this paper proposes a compact Ka band high back-off PAE 2-stage CMOS PA with adaptive bias circuits incorporated between voltage reference original circuit and amplifying transistor. The adaptive bias circuit provides a dynamically adjustable bias voltage V_{b1} and V_{b2} which positively correlated with input signal level for the gate of the amplifying NMOS transistor, to accommodate communication systems with wide dynamic range input signal power.

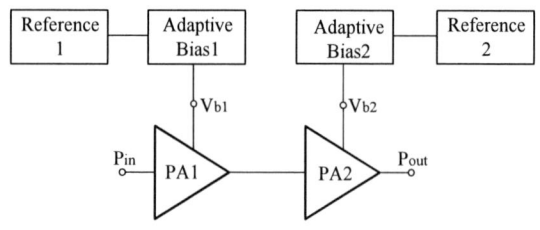

Fig. 1. The proposed high back-off PAE PA with adaptive bias circuit

II. REFERENCE AND ADAPTIVE BIAS CIRCUIT DESIGN

A. Constant gm Reference Design

Since the 2-stage PA in Fig.1 shares the same architecture, the disclosure takes a single-ended PA_0 implementation as representative example to elucidate the design rationale and working mechanism of the reference and adaptive bias module. Fig.2 gives the schematic of the proposed adaptive bias circuit incorporated between a constant transconductance (gm) reference circuit and a single-ended single-stage PA_0.

979-8-3315-8850-2/25 $31.00 © 2025 IEEE

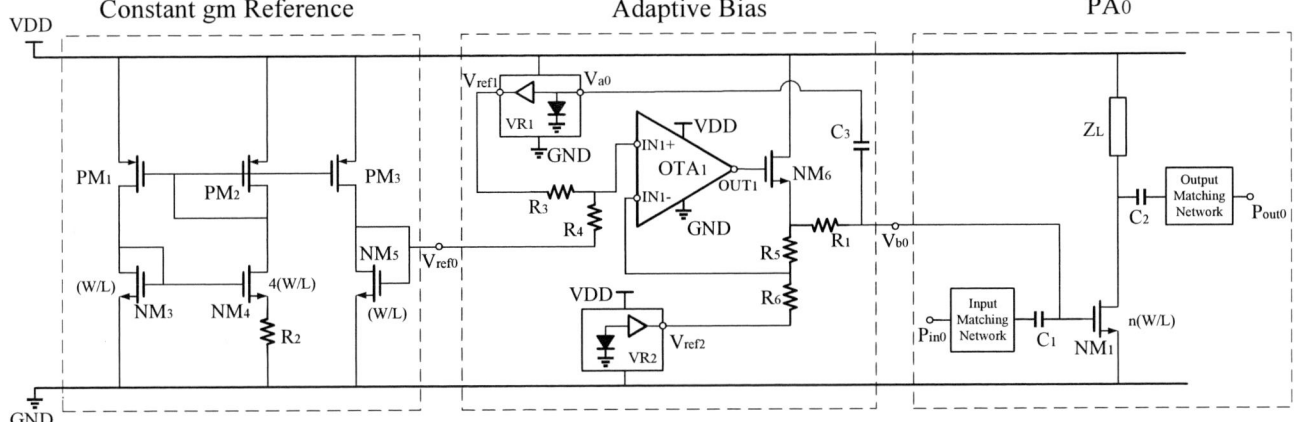

Fig. 2. The proposed adaptive bias circuit incorporated between constant gm reference circuit and a single-ended single-stage PA$_0$

Fig.2 shows that the constant gm reference circuit operates as a self-biased current source, integrating two key parts in a stacked configuration: a unity-gain current mirror formed by PMOS transistors PM$_1$ and PM$_2$ in the upper section, and a current source composed of NMOS transistors NM$_3$, NM$_4$ and resistor R_2 in the lower section [6]. For clarity of presentation, the startup circuit is omitted here. In circuit design, PM$_1$, PM$_2$, and PM$_3$ share identical dimensions, while NM$_3$ and NM$_5$ are also sized equally. The width to length ratio (W/L) of NM$_4$ is designed to be four times that of NM$_3$. According to the calculations in [7], the gm of NM$_5$ (gm_5) equals to $1/R_2$. The constant gm reference circuit generates V_{ref0} by NM$_5$.

If V_{ref0} directly drives the gate of NM$_1$ with W/L ratio scaled by n relative to NM$_3$, the gm of NM$_1$ (gm_1) equals to n/R_2. For weak input signals, the gain of PA$_0$ can be determined as nZ_L/R_2. However, as the input signal level increases, the small-signal calculations mentioned above become increasingly inaccurate, consequently leading to gm_1 decreasing and PA$_0$ exhibiting slow gain compression.

B. Adaptive Bias Design

Shown as Fig.2, The behavioral-level functionality of the adaptive bias circuit is to first detect the input signal level in real time, then generate a voltage difference ΔV that is positively correlated with the input signal level. ΔV is obtained by subtracting the output V_{ref2} of voltage reference (VR) VR$_2$ from the output V_{ref1} of VR$_1$ through operational transconductance amplifier (OTA) OTA$_1$. Subsequently, with the help of OTA$_1$, ΔV is superimposed onto the constant gm reference voltage V_{ref0} output by the constant gm reference.

Fig.3 (a) shows that the circuit of VR$_1$ senses MM wave signal input terminal V_{a0} coupled from the power amplifier input through capacitor C_3. The input terminal V_{a0} is connected to a bipolar junction transistor (BJT) Q1 configured in diode connection with base and collector shorted. As MM wave signal amplitude increases, the DC voltage at V_{a0} decreases [8], causing the gate-to-source voltage of PMOS transistor PM$_4$ to increase. This consequently leads to an increase in the drain current of PM$_4$ and an elevation of current flowing through resistor R_7, ultimately resulting in an increase of the DC voltage at V_{ref1}.

Fig.3 (b) demonstrates that VR$_2$ receives no mm Wave signal input while maintaining identical circuit topology and device dimensions to VR$_1$. Its output V_{ref2} serves as a reference voltage equivalent to V_{ref1} when no MM Wave signal is applied to VR$_1$.

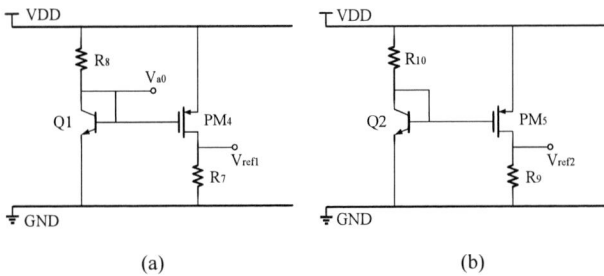

(a)   (b)

Fig. 3. Schematic of (a) Voltage Reference 1, (b) Voltage Reference 2

The schematic of OTA$_1$ and its bias circuit is shown in Fig.4. The OTA's main circuit employs a folded-cascode structure to increase the output voltage swing. Additionally, transistor NM$_6$ configured as a source follower is added at the output stage to facilitate DC level shifting between input and output. To maintain consistency with low supply voltage in 65nm CMOS process nodes, OTA$_1$ adopts low-voltage design techniques, including the use of low-threshold-voltage (LVT) CMOS transistors, and a low-voltage cascode structure current mirror formed by PM$_6$, PM$_7$, PM$_8$, and PM$_9$.

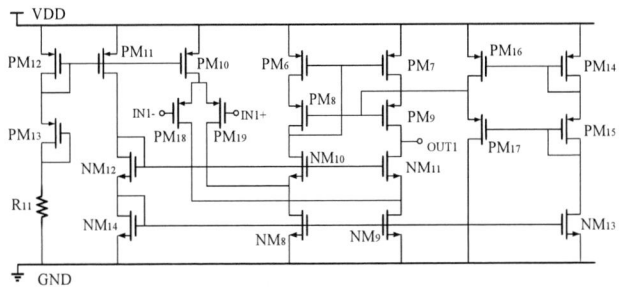

Fig. 4. Schematic of OTA$_1$ and its bias circuit

When resistors R_3, R_4, R_5, and R_6 are designed with equal values, through feedback loop formed by OTA$_1$, NMOS transistor NM$_6$, resistors R_3, R_4, R_5, and R_6, enforces the bias voltage equality:

$$V_{b0} = V_{ref0} + V_{ref1} - V_{ref2} \qquad (1)$$

At last, V_{b0} is applied to the gate of NM$_1$. A schematic level simulation was performed based on Fig.2, to verify the practicability of the proposed adaptive bias circuit. The W/L of NM$_1$ set to 256um/60nm. Fig.5 presents that, as the input signal power gradually increases, V_{b0} exhibits gradual growth.

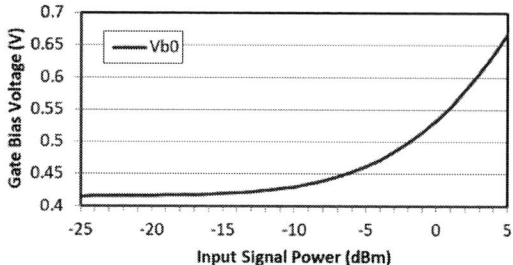

Fig. 5. NM$_1$'s Gate bias voltage variation vs. input signal power

III. Ka Band Differential 2-Stage PA Design

As shown in Fig.6, the proposed PA employs differential amplifying transistor pair configuration, which offers inherent advantages in common-mode interference rejection. Transistor pair also enables improved reverse isolation for differential signals through neutralization capacitors, which facilitate independent design of load and source matching networks [9]. Moreover, in order to achieve impedance transformation across amplifier stages within a compact footprint while ensuring gain flatness across the operational bandwidth and minimizing losses, the proposed PA employs a magnetically coupled resonator (MCR) structure with high design flexibility, consisting of magnetically coupled inductors, capacitors, and resistors [10]. The PA design follows a backward design methodology from output port to input port.

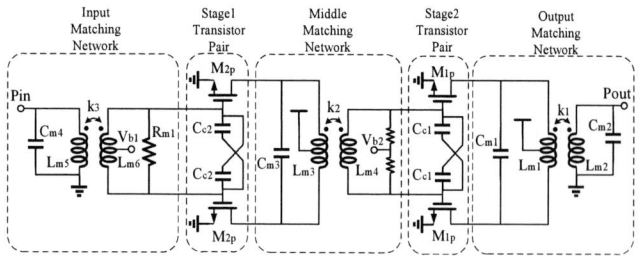

Fig. 6. The proposed Ka-band differential 2-Stage PA topology

A. Stage2 Transistor Pair Design

Based on testbench illustrated in Fig.7, with a 2-dB design margin allocated to account for non-ideal effects including passive network losses, stage2 transistor pair configuration was optimized through output power capability sweep simulations at f_c=28GHz. The selected design configuration with M$_{1p}$'s W/L=320µm/60nm, V_{b2a}=0.63V and C_{c1}=90f, achieves 14.4dBm OP$_{1dB}$, while maintaining high power gain and PAE potentials. After interconnect layout co-design of M$_{1p}$ and C$_{c1}$,

the optimal power-matching impedance Z_{opt} was determined to be 14.9+j8 Ω.

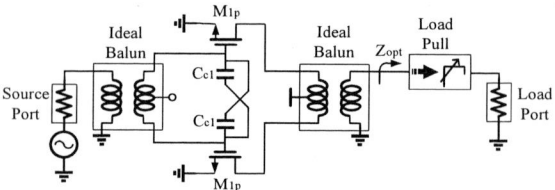

Fig. 7. Output stage loadpull testbench

B. Output MCR Design

Fig. 8 illustrates the output MCR network composed of a source impedance network, output matching network, and load impedance. The output matching network serves to transform the 50Ω load impedance (R$_L$) to Z$_{opt}$. The output impedance of stage2 transistor pair is first converted into an equivalent parallel R$_S$-C$_S$ network, specifically 19.2Ω||159fF at f_c. To achieve optimal flatness across 25-31GHz frequency range, the quality factor of the output MCR Q_1 is scanned for simulation and selected as 1.33. Followed by the MCR design methodology in [10], k_1=0.6 is derived of from (2). Finally, the key components values of the output matching network are derived from (3) ~ (6), as summarized in Table I.

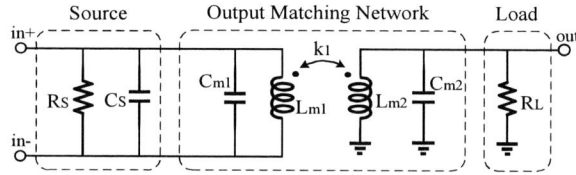

Fig. 8. The MCR network composed of source, matching network and load

$$k_1 = 1 \Big/ \sqrt{1 + Q_1^2} \qquad (2)$$

$$L_{m1} = R_S / (2\pi f_c Q_1) \qquad (3)$$

$$L_{m2} = R_L / (2\pi f_c Q_1) \qquad (4)$$

$$C_{m1} = Q_1 / (2\pi f_c R_S) - C_S \qquad (5)$$

$$C_{m2} = Q_1 / (2\pi f_c R_L) \qquad (6)$$

TABLE I.   Output MCR Components Configuration

Components	L_{m1}	L_{m2}	k_1	C_{m1}	C_{m2}
Value	93pH	203pH	0.6	215fF	141fF

C. Stage1 Transistor Pair, Middle and Input MCR Design

Based on testbench similar to Fig.7, through comparative output power capability sweeps, configuration of M$_{2p}$ is selected as W/L=160µm/60nm, V_{b1a}=0.51V and C_{c2}=37f, with contributing of 0.2 dB compression in co-simulating the two stages. The output impedance stage1 is 21.7-j35.4 Ω, which is equivalent to 79.4Ω||116.7f parallel RC network at f_c. Subsequently, simulations yield the input impedance of stage2 equivalent to 235.8Ω||202.4fF parallel RC network at f_c. Adopting an MCR network and design methodology same as Fig.8, the components' values for the middle MCR are provided in Table II.

979-8-3315-8850-2/25 $31.00 © 2025 IEEE

TABLE II. MIDDLE MCR COMPONENTS CONFIGURATION

Components	L_{m3}	L_{m4}	k_2	C_{m3}
Value	145pH	192pH	0.2	296fF

Stage1's input impedance was extracted as 478.7Ω‖91.8fF. To mitigate the risk of parallel negative capacitance in the secondary cavity due to its excessively high Q-factor, an additional shunt 120 Ω resistor R was incorporated. The input MCR components were then synthesized following Fig.8's approach summarized in Table III.

TABLE III. INPUT MCR COMPONENTS CONFIGURATION

Components	L_{m5}	L_{m6}	k_3	C_{m4}	R_{m1}
Value	170pH	327pH	0.5	174fF	120 Ω

D. Full-Chip PA Implementaion

Based on the configured differential PA transistor pairs and MCR networks above, a comprehensive iterative layout simulation was performed. This simulation incorporated input and output pads, inductive taps, and interconnects to optimize overall PA performance, targeting enhanced linearity and gain flatness. Finally, to obtain more precise post simulation results, Full-chip passive network modeling and electro-magnetic (EM) extraction have been performed, shown as Fig.9.

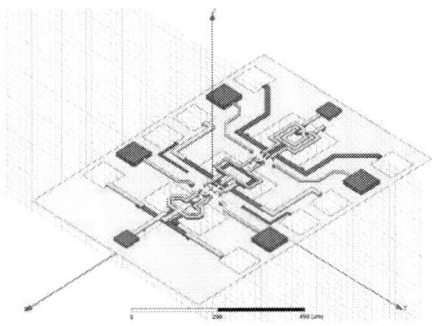

Fig. 9. Full-chip passive network modeling and EM extraction

The full-chip PA is configured as Fig.1, where V_{b1} and V_{b2} adaptively adjust according to the power level sensed from the gate of M_{1p} and M_{2p} illustrated in Fig.7. The layout of the proposed 2-stage PA along with two reference and adaptive bias circuits is illustrated in Fig.10. The full chip occupies 0.735×0.55=0.404mm² area and 0.54×0.19=0.103mm² MM wave frequency core area.

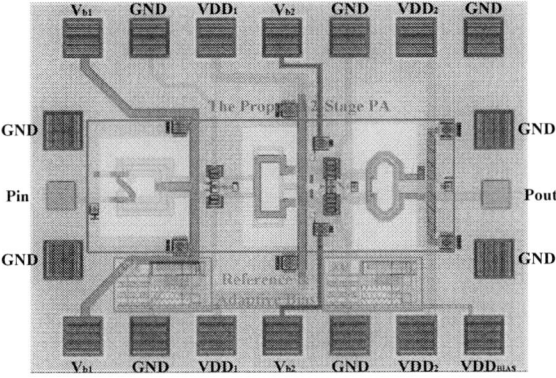

Fig. 10. Layout of the proposed PA with adaptive bias

IV. FULL-CHIP POST SIMULATION RESULTS

To more accurately verify the linearity improvement by the adaptive bias circuit, post simulation comparison at 28GHz was firstly performed based on Fig.1, with and without the adaptive bias circuit. The size of R_2 in Fig.2 is properly chosen, so that V_{b1} and V_{b2} remains low power bias at low input signal power level as shown in Fig.10. The size of PM_4, PM_5, R_7 and R_9 in Fig.3 are set, so that V_{b1} and V_{b2} exhibits gradual growth as the input signal power increases, with V_{b1a}=0.51V and V_{b2a}=0.63V at -1dBm input signal power as shown in Fig.11.

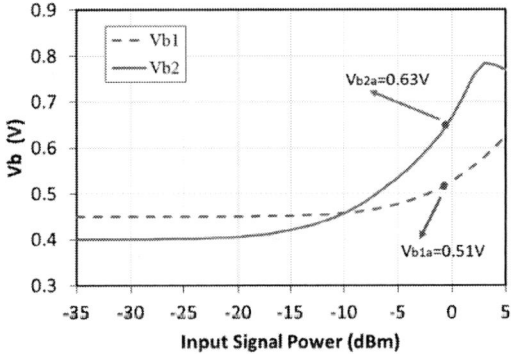

Fig. 11. Gate bias voltage variation vs. input signal power

Fig.12 demonstrates the gain compression performance comparison of the proposed 2-Stage PA, between PA with the adaptive bias (V_{b1} and V_{b2}) and PA with the fixed bias (V_{b1a} and V_{b2a}). When the adaptive bias circuit is adopted, PA's gain demonstrates improved linearity, showing compression only when input signal surpasses -5dBm, with input 1-dB compression point IP_{1dB} at -0.5dBm and the output 1-dB compression point OP_{1dB} at 17.7dBm. When fixed bias V_{b1a}=0.53V (equals to V_{b1} at IP_{1dB} of PA with the adaptive bias) and V_{b2a}=0.53V (equals to V_{b2} at IP_{1dB} of PA with the adaptive bias) is adopted, PA's gain $Gain_{0a}$ begins to compress when the input signal exceeds -20dBm, with the IP_{1dB} at -6dBm and OP_{1dB} at 14.6dBm.

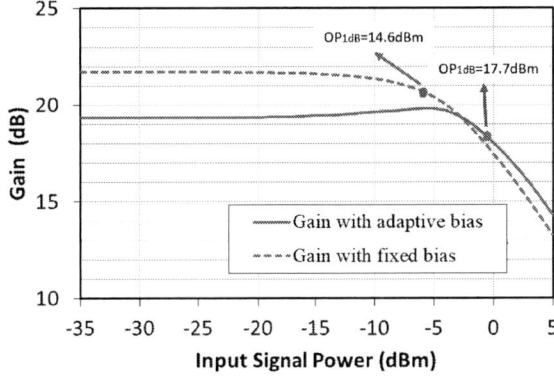

Fig. 12. Gain compression characteristics comparison at 28GHz

Fig.13 demonstrates PA's efficiency comparison between PAE of PA with the fixed bias and PAE of PA with the adaptive bias, with output power normalized to OP_{1dB}. At OP_{1dB}, PAE with the adaptive bias maintains 24.5%, which is 1.7 times outperforming the reference PAE with the fixed bias (14.7%).

At 6dB-backoff from OP_{1dB}, PAE with the adaptive bias maintains 12.4%, which is 2.7 times outperforming the reference PAE with the fixed bias (4.5%).

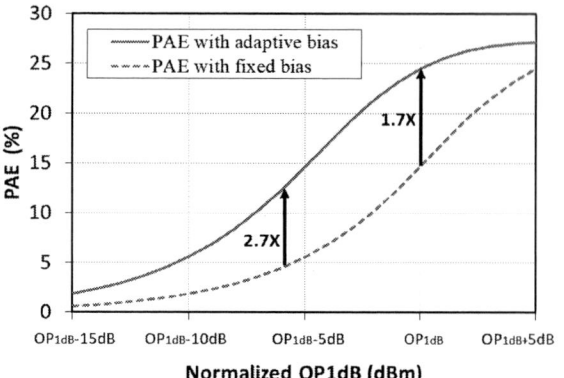

Fig. 13. PAE vs. normalized OP_{1dB} at 28GHz

Finally, S-Parameters, gain compression and efficiency simulations across the working bandwidth were performed. Fig.14 shows the large signal S-Parameters of the PA with S_{11}<-13dB, S_{22}<-8dB, and S_{21}=17.5-19.3dB in 25-31GHz bandwidth. The proposed PA is unconditionally stable with K_f>500 and the B_{1f} >0.1 within 1-100GHz given by Fig. 15. Fig.16, Fig.17 and Fig.18 show that from 25GHz to 31GHz, Output Third-Order Intercept Point (OIP_3) is 22.3-23.0dBm, OP_{1dB} is 17.1-17.9dBm, while PAE_{P1dB} (PAE at OP_{1dB}) and $PAE_{6dB\text{-}backoff}$ (PAE at 6dB-backoff from 1dB compression point) is 18.7-24.5% and 11.5-12.5%. Fig.16, Fig.17 and Fig.18 also gives linearity and efficiency performance comparison between the proposed PA with the adaptive bias and with a fixed bias, where OIP_3 degraded by 0-1.5dB, OP_{1dB} has been improved by 3dB, PAE_{P1dB} has been improved by 1.6-2.0 times, and $PAE_{6dB\text{-}backoff}$ has been improved by 2.7-3.1 times. Table IV summarizes the performance of the proposed PA with prior works, in which the proposed PA achieves high linearity performance, along with high PAE_{P1dB} and the highest $PAE_{6dB\text{-}backoff}$.

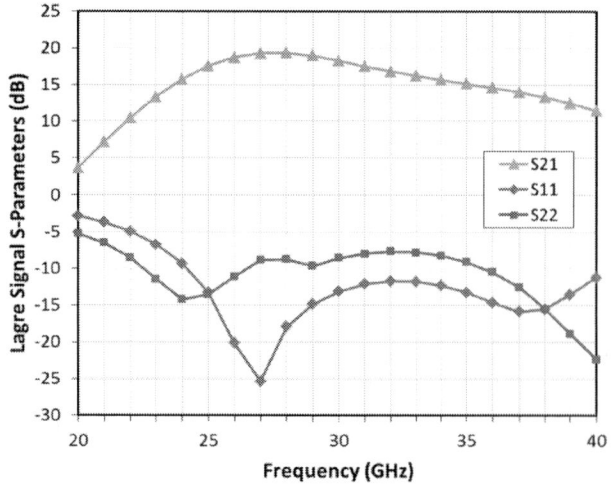

Fig. 14. Large signal S-Parameters

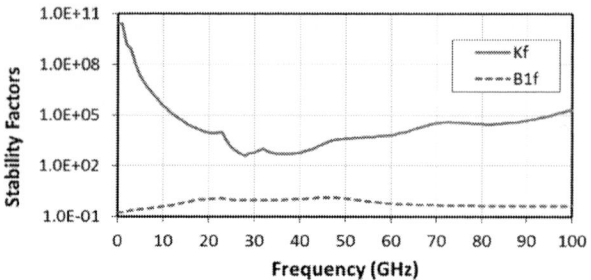

Fig. 15. Stability Factors

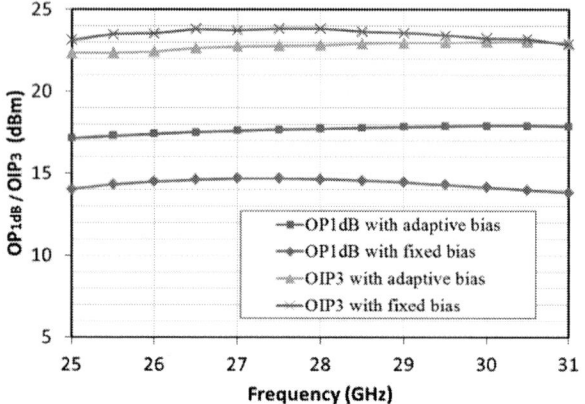

Fig. 16. OP_{1dB} and IIP_3

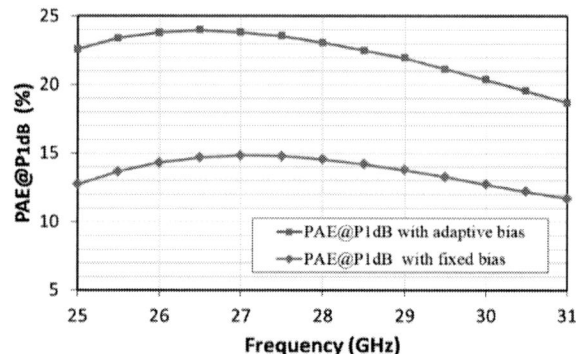

Fig. 17. PAE@P1dB

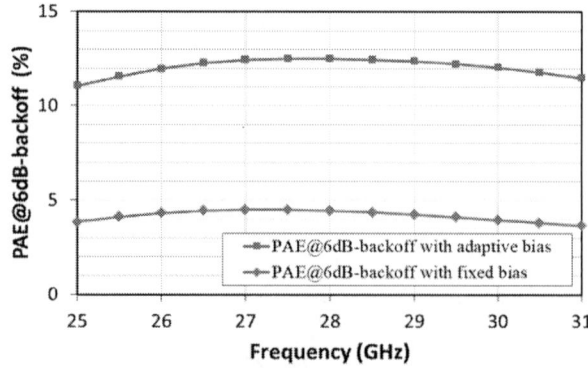

Fig. 18. PAE@6dB-backoff from OP_{1dB}

TABLE IV. Simulated Performance Summary and Comparison with Prior Works

Ref	Process	Frequency (GHz)	f_c (GHz)	Gain (dB)	PAE_{P1dB} (%)	OP_{1dB} (dBm)	$PAE_{6dB\text{-backoff}}$ (%)	Core Area (mm²)
This work	65nm CMOS	25-31	28	19.3	24.5	17.7	12.4	0.103
[11]	65nm SOI CMOS	21.5-41	31	21.4	18	18.6	4	0.077
[12]	65nm CMOS	34-40	38	17.5	15	21.7	4	0.146
[13]	28nm CMOS	25-30	28	14	25	16.1	10	0.051
[14]	250nm BiCMOS	29.5-30.8	30	8.5	9	7	3	0.124
[15]	130nm CMOS	36-40	38	23	10	15	3.5	0.238

V. Summary

Based on 65nm CMOS technology, this paper proposes a compact Ka band high back-off PAE 2-stage CMOS PA with adaptive bias. The proposed adaptive bias solution effectively mitigating gain compression issues that occur under regular constant gm reference bias condition with increasing input power. Moreover, the proposed PA achieves high efficiency performance. PAE_{P1dB} at f_c equals to 24.5%, which is 1.7 times outperforming the PAE of PA with fixed bias. $PAE_{6dB\text{-backoff}}$ at f_c still maintains 12.4%, which is 2.7 times outperforming the PAE of PA with fixed bias.

In conclusion, a design methodology is introduced in this paper, for high back-off PAE PAs used in large dynamic range input signal power communication scenarios.

References

[1] H. Li, et al., "On the PAPR of the LTE-Based 5G Terrestrial Broadcast System," IEEE International Symposium on Broadband Multimedia Systems and Broadcasting (BMSB), Beijing, China, 2023, pp. 1-6.

[2] J. Gu, et al., "A 23-30 GHz 4-Path Series-Parallel-Combined Class-AB Power Amplifier with 23 dBm Psat, 38.5% Peak PAE and 1.3° AM-PM Distortion in 40nm Bulk CMOS," IEEE Radio Frequency Integrated Circuits Symposium (RFIC), San Diego, USA, 2023, pp. 197-200.

[3] M. Karimi and M. Ehsanian, "A High-Performance CMOS Hybrid Envelope Tracking Power Amplifier for Wideband High PAPR Applications," Iranian International Conference on Microelectronics (IICM), Tehran, Iran, 2021, pp. 1-4.

[4] S. Mueller, O. Hanay and R. Negra, "Counter-Intermodulation in the Context of CMOS Outphasing Transmitters," 17th European Microwave Integrated Circuits Conference (EuMIC), Milan, Italy, 2022, pp. 288-291.

[5] S. Li, et al., "A High-Efficiency 28GHz Doherty Power Amplifier with Peak PAE of 37.3% in 40nm CMOS," IEEE MTT-S International Wireless Symposium (IMS), Harbin, China, 2022, pp. 1-3.

[6] R. Zhang, "An On-Chip Temperature Sensor and its Application in Millimeter-Wave Low Noise Amplifier," 10th International Conference on Computer and Communications (ICCC), Chengdu, China, December 2024, pp. 1040-1044.

[7] B. Razavi, Design of Analog CMOS Integrated Circuits, 2nd ed., chap. 2. New York: McGraw-Hill Education, 2015, pp.509–525.

[8] W. Li and Y. Tan, "2.4GHz power amplifier with adaptive bias circuit," International Conference on Systems and Informatics (ICSAI), Yantai, China, 2012, pp. 1402-1406.

[9] R. Zhang and Y. Zhu, "An X-band CMOS Push-Pull Power Amplifier with Capacitive Neutralization for High Linearity and High Efficiency Applications, " IEEE 13th International Conference on Solid-State and Integrated Circuit Technology (ICSICT), Hangzhou, China, October 2016, pp. 1-3.

[10] H. Jia, C. Prawoto, B. Chi, Z. Wang and C. Yue, "A Full Ka-Band Power Amplifier With 32.9% PAE and 15.3-dBm Power in 65-nm CMOS," IEEE Transactions on Circuits and Systems–I: Regular Papers, vol. 65, no. 9, pp 2657-2668, September 2018.

[11] J. Lin, L. Li, and J. Huang, "A Design of Full Ka-Band Power Amplifer in 65nm SOI CMOS," 7th International Conference on Integrated Circuits and Microsystems (ICICM), Hangzhou, China, October 2022, pp.295–298.

[12] Y.Chang, B. Lu, Y. Wang and H. Wang, "A Ka-Band Stacked Power Amplifier with 24.8-dBm Output Power and 24.3% PAE in 65-nm CMOS technology," IEEE MTT-S International Wireless Symposium (IMS), Boston, USA, 2019, pp. 316-319.

[13] T. Huang, , et al., "A 28-nm CMOS Sub-Volt PA With High Power Density and Common-Mode Stability Design," IEEE Microwave and Wireless Technology Letters, vol. 33, no. 7, pp 1019-1022, July 2023.

[14] F. Tabarani, L. Boccia, T. Purtova, A. Shamsafar, H. Schumacher and G. Amendola, "0.25-μm BiCMOS System-on-Chip for K-/Ka-Band Satellite Communication Transmit–Receive Active Phased Arrays," in IEEE Transactions on Microwave Theory and Techniques, vol. 66, no. 5, pp. 2325-2339, May 2018.

[15] S. Londhe, N. Shmilovitz, S. Avner, N. Bar-Helmer, S. Jameson and E. Socher, "34-42GHz CMOS Transceiver Frontend for Versatile Arrays," 2020 15th European Microwave Integrated Circuits Conference (EuMIC), Utrecht, Netherlands, 2021, pp. 73-76.

2025 The 10th International Conference on Integrated Circuits and Microsystems

A DAC Design using 5-V High-voltage Process for Driving Large Electrode Loads of Neural Stimulators

Xu Liu*, YuAng Ru, Zhuo Chen, Fang Xie, Zhijie Chen, Peiyuan Wan

School of Information Science and Technology, Beijing University of Technology, Beijing, China

*email: liuxu16@bjut.edu.cn

Abstract—**A 6-bit current-steering DAC based on 180-nm high-voltage BCD CMOS process technology for driving large electrode loads of neural stimulators is presented. This DAC design adopts a "3+3" segmented structure. To reduce the impact of clock feedthrough effect, the switch design is optimized. Post-layout simulation results show that the operation frequency can reach 1 MHz with a 1-M Ω resistor load, with the DNL of 63m LSB and INL of 6.3m LSB. When the load is an electrode composed of a 10k resistor in series with a 100nf capacitor, the DNL is 35.5m LSB and the INL is 229.9 m LSB at a sampling rate of 1 MHz. Under standard conditions, the output current range is 2uA-63 uA, and the static power consumption is 4.9 mW.**

Keywords—*CMOS, digital to analog converter, high voltage, electrode, neural stimulator*

I. INTRODUCTION

In recent years, the neural stimulators have been widely used in biomedical devices for clinical use[1-3]. The current-mode digital-to-analog convertor(DAC) is the key component in the neural stimulator, because it is used as an adjustable current source to provide appropriate current to the tissue-electrode load. In some neural stimulators, the DAC circuit is directly connected to the electrode without a buffered output stage[4]. However, if the electrode impedance is high, the DAC design may become a bottleneck, as there is not enough headroom voltage in the signal path[5] for the MOS transistor to work in saturation region, as shown in Fig. 1.

In addition, the use of neural stimulation also requires DAC to have a high switching speed to generate arbitrary output waveforms[6]. Although recent DACs have reported speeds of up to 100 GS/s[7-8] the load is usually fixed as 50 Ω, which cannot meet the demand in the application of neural stimulators with a tissue-electrode load, which can be modeled as a 10-KΩ resistor and a 1-nF capacitor connected in series. It is difficult for the traditional DAC design to drive this large load. In many stimulation ICs, the DAC is usually connected to a current mirror rather than an electrode[9-11]. But this will cause extra power consumption. Therefore, a high-speed high-voltage DAC directly driving a large load could be an option for neural stimulator design.

This design presents a high-voltage 6-bit current-steering DAC. We increase the power supply voltage to 5V and use 5-V

This work is supported by National Key Research and Development Program (Grant No.2024YFF1206504) and Beijing Natural Science Foundation-Huairou Innovation Joint Fund (Grant No.L245012).

BCD process for the circuit design. Feedforward capacitor, series switch slave tube and cascode are used or improved, So the static and dynamic performance have been improved. The

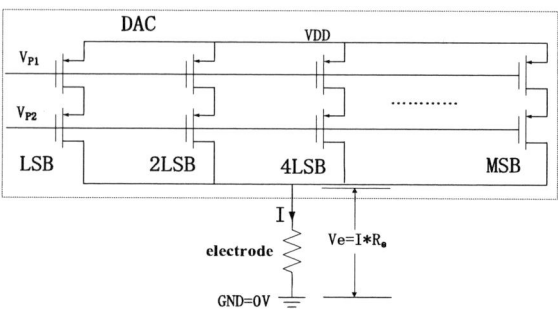

Fig. 1. The DAC directly connected to electrodes without current buffer.

remaining chapters are arranged as follows. Section II introduces the principle of current-steering DAC and the selection of segmentation ratio. Section III illustrates the design considerations of obtaining low glitch energy, high update rate, and directly connecting high impedance electrodes. Section IV introduces the layout details, and the results. Section V draws the conclusion.

II. PRINCIPLE OF CURRENT-STEERING DAC

A. Principle of Current-steering DAC

The commonly used current-steering DAC structures include binary code and thermometer code types. The current source switch of binary code type is directly controlled by binary code, without the need for additional decoding circuits, thus requiring a small layout area and fast working speed. Its output voltage V_{out} can be expressed as:

$$V_{out} = I(\sum_{i=0}^{N-1} 2^i * B_i) * R \qquad (1)$$

Among them, I is the LSB current, N is the number of bits, i is the number of bits of the current source, B_i is the i-th switch, and R is the load impedance.

However, every change in binary code may cause multiple current sources to turn on or off, and asynchronous switching time may lead to an increase in output glitch energy, thereby reducing the dynamic performance of the DAC[12].

Usually, the 2n-1 bit thermometer code is converted from the n-bit binary code, so corresponding decoding circuits are required. Similar to binary code types, thermometer code directly controls the opening or closing of the current source.

979-8-3315-8850-2/25 $31.00 © 2025 IEEE

Although the decoding method is completely different from the actual circuit, the result of controlling the current source to turn on or off is completely the same. Its output voltage V_{out} can be expressed as:

$$V_{out} = I\left(\sum_{i=0}^{N-1} T_i\right) * R \quad (2)$$

Among them, T_i represents the i-th thermometer code.

This is reflected in the behavior of the current source, where the thermometer code monotonically increases or decreases by one bit every time the binary code changes by one bit. Each time the input signal changes, only one bit of the current source's on/off state changes, effectively avoiding the problem of increased glitch energy caused by inconsistent switching times of each current source under binary code control.

B. Selection of DAC Structure

To combine the advantages of both, a segmented decoding DAC structure was ultimately adopted [13]. For determining the segmentation method, using thermometer decoding is beneficial for static performance; However, using only thermometer decoding would result in an area that is too large to accept, and the impact of high-level current sources on static performance is much greater than the slower decoding speed of low-level current sources and thermometers. Therefore, a structure of high-level thermometer decoding combined with low-level binary decoding was adopted. Its output voltage V_{out} can be expressed as:

$$V_{out} = I\left(\sum_{i=0}^{m} 2^i * B_i\right) * R + (2^m * I) * \left(\sum_{i=0}^{2^{N-m}-1} T_i\right) * R \quad (3)$$

For the selection of segmentation ratio, Chi Hang Lin et al. provides the relationship between the area, performance, and segmentation ratio of segmented current-steering DAC [14]. INL is only related to the area of the analog circuit, and the larger the area of the analog circuit, the smaller the INL, regardless of the segmentation ratio[15].

In order to achieve a balance between INL, DNL, and layout area, a structure with a 50% segmentation ratio of 3-bit binary code and 3-bit thermometer code was selected. The binary part contains current sources with weights of 1 LSB, 2 LSB, and 4 LSB, while the thermometer part contains 7 current sources with weights of 8 LSB.

III. CIRCUIT DETAILS

The proposed DAC system architecture is shown in Fig. 2,where all current sources are biased by a bias generation module. After passing through a switch array, they are divided into two outputs, one connected to the electrode and the other to the reference load.

As shown in Figure 3, compared to 1.8-V MOS transistors, 5-V BCD MOS transistors have many characteristics, including thicker gate oxide layers, higher threshold voltages, LDMOS structures, which can withstand greater gate voltage stress, stronger anti-interference ability, and a working voltage of up to 5V, ensuring sufficient headroom voltage for the load. Therefore, the DAC proposed in this work uses 5-V MOS transistors instead of conventional 1.8-V MOS transistors.

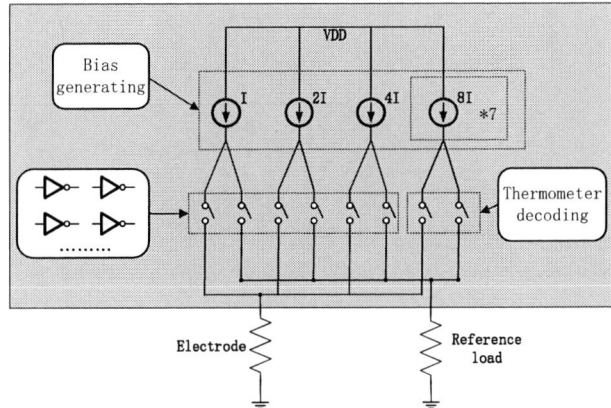

Fig. 2. DAC architecture diagram

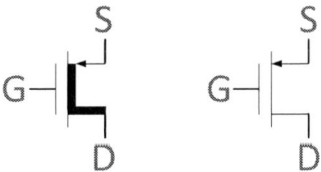

Fig. 3. Symbol comparison of 5V MOS in180nm BCD process and traditional one

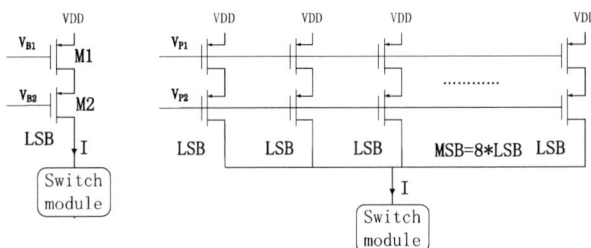

Fig. 4. Structure of LSB and MSB current source

A. LSB&MSB Current Source

In this design, the LSB current is set to 2uA and the MSB current is set to 16uA. The LSB current source adopts a cascode structure, having a size of W * L=5.69um * 1.5um. The high-order current sources are connected in parallel with the corresponding weight number of LSB current sources, as shown in Fig. 4.

B. Bias Voltage Generation Circuit

The bias voltage generation circuit is shown in Fig. 5, where the bias V_{p1} in the main line of the cascode current source is provided by M3-M5 and resistor R_{ref} on the right side. The current I_1 in the right branch can be expressed by the following equation:

$$I_1 = \frac{V_{ref}}{R_{ref}} \quad (4)$$

I_1 flows through a wide swing self biased cascode current mirror composed of M5 and M4 tubes to generate V_{P1}, so the magnitude of I_1 and LSB currents is the same. The output terminal of the operational amplifier is connected to M3, and the source of M3 is connected to the negative input terminal of the operational amplifier to form a negative feedback loop.

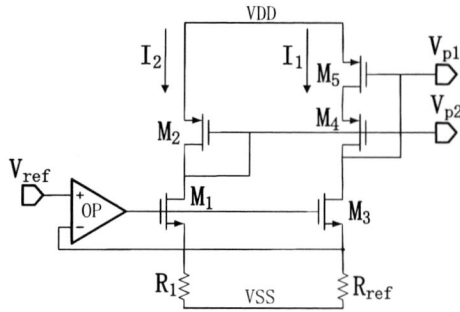

Fig. 5. Bias voltage generation circuit

The current I_2 of the left branch is determined by the output voltage of the operational amplifier and the magnitude of R_1. All current sources are biased by this circuit, and their reference current is obtained from an external input reference voltage and an external precision reference resistor. So this DAC can adjust the LSB current according to the load situation. When the load is not large, the LSB current can be increased to achieve better dynamic performance.

C. Thermometer decoding circuit

The 6-bit DAC proposed in this article adopts a segmentation ratio of 3:3. In the thermometer section, the input digital symbols are decoded by the thermometer to convert the N-bit binary input symbols into (2^N-1) thermometer codes and control the switch, rather than directly controlling it. The current source weights for thermometer decoding are the same, both being 1LSB. This converts a 3-digit binary code to a 7-digit thermometer code. During the switching process using thermometer code, only one switch changes, which has good monotonicity and effectively reduces output signal glitch.

D. Design of switch

The segmented decoding introduced in the previous section ensures the static performance of the DAC in this article. As for the dynamic performance of current-steering DAC, it is essentially the step response characteristic of the system composed of the current source and its parasitic capacitance and resistance. Under the same conditions, the smaller the parasitic capacitance and resistance, the better the dynamic performance of the DAC. Therefore, the rest of this section will analyze the problems that constrain dynamic performance in traditional structures and propose innovative structures accordingly. In addition, due to the need to connect high impedance loads in applications, the DAC in this article reduces parasitic capacitance and resistance while reducing LSB current to 2uA.

When the traditional differential output current-steering DAC operates, when the gate signals D and DN are switched, the current source will immediately guide the tail current from one side to the other, during which the potential of points S_1 and S_2 changes significantly. Therefore, a considerable portion of the charge is wasted on charging the parasitic capacitance of points S_1 and S_2 instead of participating in step response. The solution proposed in this article is to add a pair of normally open small-sized PMOSs with grounded gates between the switching transistor and the output node, and to add feedforward capacitors C_a and C_b between the control signal and S_1, S_2, as shown in Fig. 6. For the added PMOS, its parasitic capacitance with the

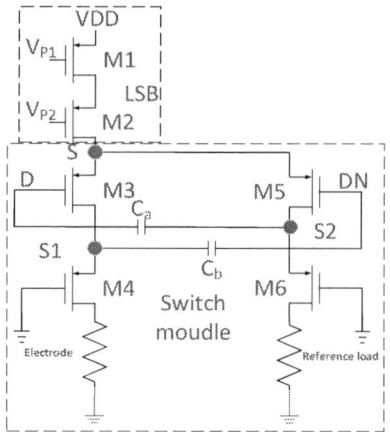

Fig. 6. The LSB current source(M1&M2) and innovative switch structure(M3-M6) of the DAC proposed in this article

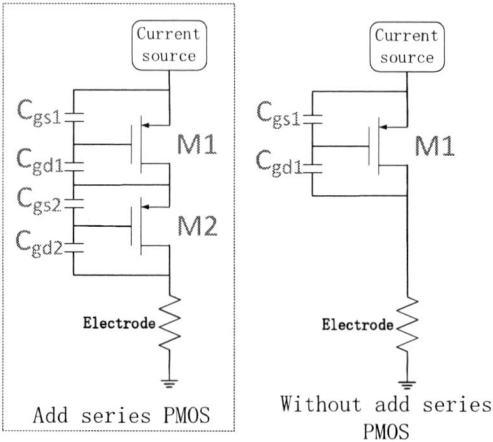

Fig. 7. The parasitic capacitance of the increased PMOS (M2) is in series with the parasitic capacitance of the switching transistor.

switching transistor belongs to a series relationship. According to the series formula of capacitors, the equivalent resistance after series connection will be closer to the parasitic capacitance of a smaller normally open PMOS, so this significantly reduces the parasitic capacitance at points S_1 and S_2, as shown in Fig. 7. For adding feedforward capacitors C_a and C_b between the control signal and S_1 and S_2, since the reduced but still non-zero parasitic capacitors at points S_1 and S_2 are directly connected to their corresponding reverse control signals, it is equivalent to using the control signal to charge the parasitic capacitors in advance. Both jointly reduce the impact of parasitic capacitance on dynamic performance.

During the process where the current source immediately guides the tail current from one side to the other, the potential at point S will also undergo significant changes due to the time sequence of conduction cutoff conduction, resulting in a similar phenomenon of charge waste as described above[16]. The solution proposed in this article is to use a low crossing point control signal corresponding to the PMOS switch to ensure that at least one switch is in a conducting state, so that the voltage of the S node will not fluctuate too much. The simulated waveform of the low crossing point control signal is shown in Fig. 8.

979-8-3315-8850-2/25 $31.00 © 2025 IEEE

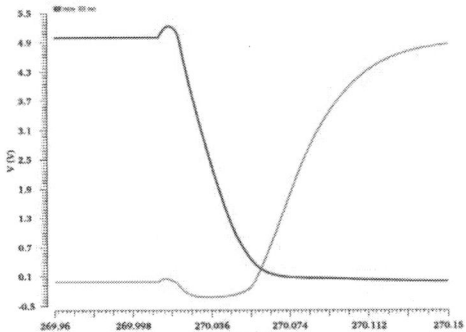

Fig. 8. Simulation waveform of low crossing point control signal

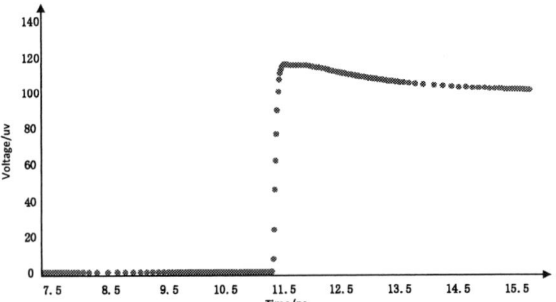

Fig. 9. Transient response of the DAC proposed in this article

This significantly reduces the demand for charge at point S when responding to control signals, improving the dynamic performance of the DAC.

The clock feedthrough also has a significant impact on the dynamic performance of traditional DACs. The parasitic capacitance at points S_1 and S_2 mentioned earlier will form an AC path with the gate drain parasitic capacitance of the switching transistor to the load. If no measures are taken, the digital control signal will transfer charges to the load during the transition, causing glitchs and increasing the setup time; It can also cause harmonic distortion and damage the dynamic performance of DAC. The problem can be significantly alleviated by adding a pair of normally open small-sized PMOS with gate grounded between the switching transistor and the output node. And in order to leave enough headroom voltage for high impedance loads, its gate is grounded and operates in the linear region and adding feedforward capacitors C_a and C_b between the control signal S_1 and S_2. The current generated by the gate drain clock feedthrough can be described by the following equation:

$$I_{CLK,D} = C_{GD} * \frac{dV_{CLK}}{dt} \qquad (5)$$

1Among them, $I_{CLK,D}$ is the clock feedthrough current, C_{GD} is the gate drain capacitance of the switch tube, and V_{CLK} is the maximum voltage difference of the control signal. PMOS significantly reduces the parasitic capacitance at points S_1 and S_2, while increasing the feedforward capacitors C_a and C_b cancels out the generated feedthrough current by coupling the reverse control signal ($-V_{CLK}$). The expression for the compensation current generated is as follows:

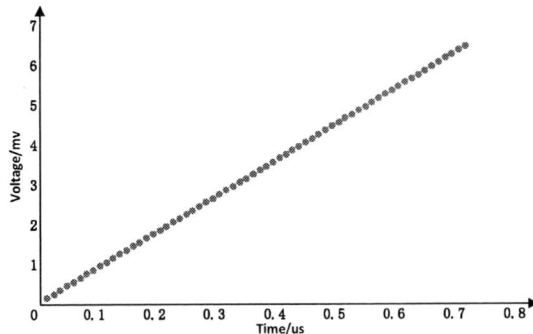

Fig. 10. Output of DAC at Input Clock Frequency of 45M

$$I_C = C * \frac{dV_{CLK}}{dt} \qquad (6)$$

Among them, I_C is the compensation current, and C is the value of the feedforward capacitor. By adjusting the size of the switch (MOS transistor) and the size of the feedforward capacitor to make them equal, the gate drain feedthrough current can be offset. In this design, $C_a=C_b=5.05fF$, The trans simulation results of feedthrough and compensation current are shown in Fig. 11. It can be seen that the amount of charge transmitted by the two is basically equal, which greatly alleviates the deterioration of the dynamic performance of the DAC caused by the unexpected charge flowing into the load.

While ensuring static and dynamic performance, it is also necessary to be able to connect electrodes with high impedance, which means that the voltage on the load at maximum output

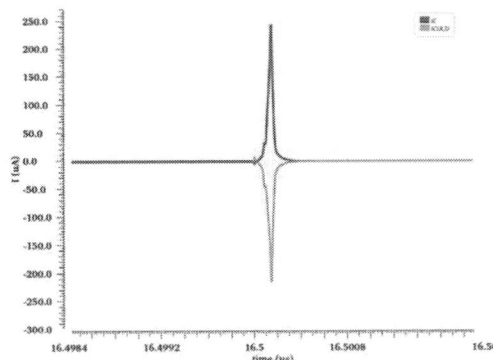

Fig. 11. Trans simulation results of feedthrough current and compensation current

cannot cause the current source transistor to operate in the unsaturated region. This means that (1) LSB cannot be too large; (2) The output impedance of the current source needs to be large to cope with the decrease in equivalent load when connected in parallel with the electrode; (3) The parasitic resistance and capacitance that affect the transient response of DAC should be kept as low as possible. Based on the above requirements, in addition to the differential output and innovative switches mentioned earlier, the DAC proposed in this design selects a 5V power supply voltage, MOS transistor, and 2uA LSB current, and a cascode current source with a size of W * L=5.69 μm * 1.5 μm. When applying the above measures, with a load of 50 Ω resistor and an input signal frequency of 45M/s, the transient

979-8-3315-8850-2/25 $31.00 © 2025 IEEE

response of the DAC is shown in the following Fig. 9, where the rising edge is steep and there are no obvious glitch. The trans simulation results are shown in the Fig. 10. when the input reference voltage is 0.9V, the reference resistance is 510k, the input signal frequency is 45 M/s, and the load is a 50 Ω resistor.

IV. LAYOUT DESIGN AND SIMULATION RESULTS

The DAC is designed using 180-nm 5-V BCD CMOS process. The layout is shown in the Fig. 12.

Then post-layout simulation has been done. The DAC proposed in this article has three characteristics. Firstly, it can

Fig. 12. DAC layout with block diagram division

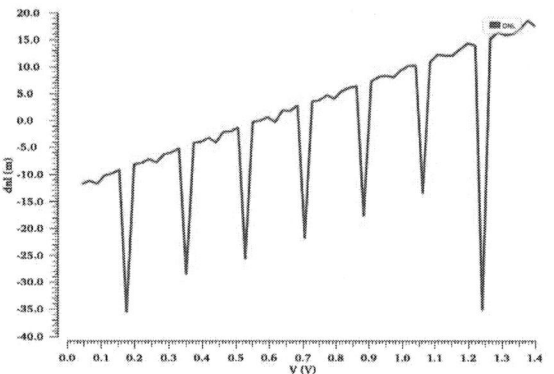

Fig. 13. DNL with an electrode load and an input frequency of 1MHz

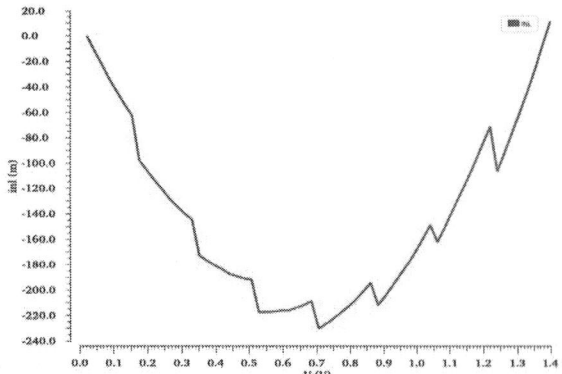

Fig. 14. INL curve with electrode load and an input frequency of 1MHz

directly connect to high impedance loads such as electrodes. Secondly, it adopts a high-speed structure to achieve high speed. Thirdly, it has good static performance to meet the requirements of bioimpedance detection, which is also reflected in the simulation results. This DAC can also achieve high speed when driving high impedance loads. When the output end is connected to an electrode composed of 10k ohms and 100 nf capacitors, the input signal frequency can reach 1MHz, with a DNL of 35m LSB and an INL of 230m LSB. Afterwards, the simulated waveforms are shown in Fig. 13 and 14. The worst value of INL in Fig. 13 occurs when the output voltage is approximately half of the maximum output voltage. This is due to the inconsistency of the switch timing between the binary code and thermometer code parts at this time. The binary code switch turns off faster than the thermometer code switch turns on, resulting in an instantaneous drop in output voltage, manifested as a downward glitch, causing the sampled voltage at this time to be lower than the ideal value. When the output terminal is connected to a 1.5M ohm resistor, the input signal frequency can reach 1MHz, with a DNL of 117m LSB and an INL of 282m LSB; When the output terminal is connected in parallel with a 50 ohm resistor and a 1.5nf capacitor, the input signal frequency can reach 1MHz, with a DNL of 48m LSB and an INL of 21m LSB. When driving a small impedance load, such as a 50 ohm resistor, the sampling rate can reach 1GHz, ENOB can reach 6.04 bit, SNR is 37.85 dB, SFDR is 43.03 dB, DNL is 47m LSB, and INL is 26m LSB .

V. SUMMARY

A 5-V DAC is designed for neural stimulators. Its features include the ability to directly connect high impedance electrodes, high-speed structure, and good dynamic and static performance. The effect of gate-drain clock feedthrough was reduced by switch design and optimization. When the output is connected to a 50 Ω resistor and the input signal frequency is 1GHz, the dynamic performance is shown in Figure 15. ENOB can reach 6.04bit, SNR is 37.85dB, and SFDR is 43.93dB.When the output is connected to an electrode composed of 10k ohms and 100nf capacitors, the input signal frequency can reach 1 MHz And the DNL is 35 mLSB and the INL is 230mLSB, Figure 16 illustrates the stimulation waveform for this electrode. Table 1 shows the performance of the DAC proposed in this paper under three typical loads. The designed DAC can driving high impedance loads with a decent speed and good dynamic and

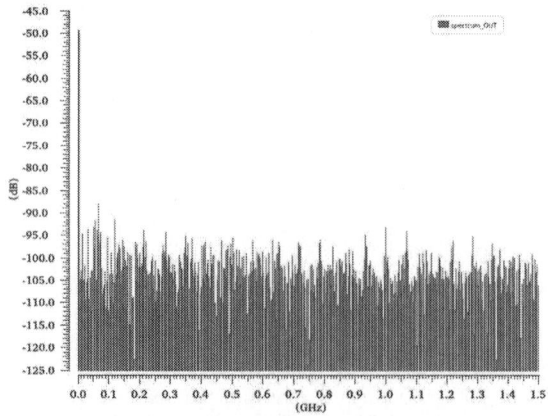

Fig. 15. Dynamic performance at a sampling rate of 1GHz

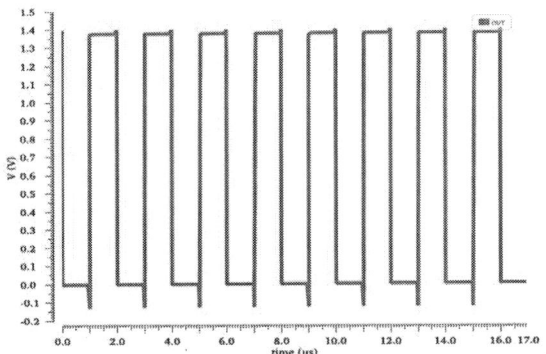

Fig. 16. Stimulus generated by DAC with load of 10kΩ &150nf.

static performance. Therefore, this DAC circuit could be applied to high-voltage neural stimulators.

TABLE I. PERFORMANCES SUMMARY OF THE DAC

Load condition	1.5 M Ω	50 Ω and 1.5nf in parallel	10kΩ and 100nf in series
Max. Sample Rate (Sample/s)	1 M		
Accuracy	6 bit		
INL (LSB)	282 m	21 m	230 m
DNL (LSB)	117 m	48 m	35 m

REFERENCES

[1] G. L. K. Moganti and P. Y. Rishita, "Wireless Programmable Healthy Neural Stimulator Prototype," 2021 8th International Conference on Smart Computing and Communications (ICSCC), 2021, pp. 367-370.

[2] S. Luan, I. Williams, K. Nikolic, and T. G. Constandinou, "Neuromodulation: present and emerging methods," Frontiers in neuroengineering,vol. 7, p. 27, 2014.

[3] S. Nag, X. Jia, N. V. Thakor, and D. Sharma, "Flexible charge balanced stimulator with 5.6 fc accuracy for 140 nc injections," IEEE transactions on biomedical circuits and systems, vol. 7, no. 3, pp. 266–275, 2012.

[4] K. A. Ng et al., "An inductively powered CMOS multichannel bionic neural link for peripheral nerve function restoration," 2012 IEEE Asian Solid State Circuits Conference (A-SSCC), 2012, pp. 181-184,

[5] Y. Li and Z. Li, "A low-power 6-bit D/A converter design for WSN transceivers," in IEEE 13th International Conference on Communication Technology (ICCT), Sept 2011, pp. 307–310.

[6] L. Yao, J. Zhao, P. Li, R. -F. Xue, Y. P. Xu and M. Je, "A 20V-compliance implantable neural stimulator IC with closed-loop power control, active charge balancing, and electrode impedance check," 2014 IEEE Asian Solid-State Circuits Conference (A-SSCC), 2014, pp. 201-204,

[7] A low-power 6-bit D/A converter design for WSN transceivers," in IEEE 13th International Conference on Communication Technology (ICCT), Sept 2011, pp. 307–310.

[8] E. Olieman, A.-J. Annema, and B. Nauta, "An interleaved full nyquist high-speed DAC technique," IEEE J. Solid-State Circuits, vol. 50, no. 3, pp. 704–713, Mar. 2015

[9] P. Caragiulo, O. E. Mattia, et al., "A 2xtime-interleaved 28-GS/s 8-bit 0.03-mm2 switched-capacitor DAC in16-nm FinFET CMOS," IEEE J. Solid-State Circuits, vol. 56, no. 8,pp. 2335–2346, Aug. 2021.

[10] K. Lim, J. Seo, C. Seok and H. Ko, "A 16-channel neural stimulator with DAC sharing scheme for visual prostheses," 2013 IEEE International Symposium on Circuits and Systems (ISCAS), 2013, pp. 1873-1876,

[11] T. Tokuda et al., "CMOS-Based Multichip Networked Flexible Retinal Stimulator Designed for Image-Based Retinal Prosthesis," in IEEE Trans. on Electron Devices, vol. 56, no. 11, pp. 2577-2585, Nov. 2009.

[12] E. Noorsal, K. Sooksood, et al., "A Neural Stimulator Frontend With High-Voltage Compliance and Programmable Pulse Shape for Epiretinal Implants," in IEEE JSSC, vol. 47, no. 1, pp. 244-256, Jan. 2012.

[13] B. Razavi, "The Current-Steering DAC [A Circuit for All Seasons]," in IEEE Solid-State Circ. Magazine, vol. 10, no. 1, pp. 11-15, Winter 2018.

[14] A.Van den Bosch et al, "Modeling and realization of high accuracy, high speed current-steering CMOS D/A converters", IEEE measurement 28 (2000) 123-138. 3.

[15] Chi-Hung Lin and K. Bult, "A 10-b, 500-MSample/s CMOS DAC in 0.6 mm/sup 2/," in IEEE Journal of Solid-State Circuits, vol. 33, no. 12, pp. 1948-1958, Dec. 1998.

[16] H. Chandrakumar, T. W. Brown, D. Frolov, Z. Tuli, I. Huang and S. Rami, "A 48-dB SFDR, 43-dB SNDR, 50-GS/s 9-b 2×-Interleaved Nyquist DAC in Intel 16," in IEEE Solid-State Circuits Letters, vol. 5, pp. 239-242, 2022, doi: 10.1109/LSSC.2022.3205884.

2025 The 10th International Conference on Integrated Circuits and Microsystems

MEMS-Enabled Computational Spectrometer Based on Cascaded Waveguide Couplers

Hanxing Wang
School of Microelectronics
Shanghai University
Shanghai, China
wanghanxing@shu.edu.cn

Yan Liu
School of Microelectronics
Shanghai University
Shanghai, China
2066300673@shu.edu.cn

Nan Wang
School of Microelectronics
Shanghai University
Shanghai, China
nan_wang@shu.edu.cn

Yiming Ma*
School of Microelectronics
Shanghai University
Shanghai, China
yimingma@shu.edu.cn
*Corresponding author

Abstract—In recent years, microspectrometers have been rapidly developed, driven by the growing demands for in-situ analysis and real-time monitoring. However, their performance is often constrained by size limitations. We propose and demonstrate a computational spectrometer based on cascaded waveguide couplers. The design of cascaded waveguide couplers not only enables more disordered interferograms but also exponentially increases the number of sampling channels, significantly enhancing the spectral resolution of the spectrometer. With a compact footprint of 100×2500 μm², the spectrometer achieves a spectral resolution of 40 pm in the wavelength range of 1.5–1.51 μm.

Keywords—*Cascaded waveguide couplers, MEMS comb-drive actuator, computational spectrometer, spectral reconstruction*

I. INTRODUCTION

Optical spectrometry is a highly effective analytical tool employed in both academic and industrial areas [1-3]. Its applications encompass material analysis, medical diagnostics, and environmental monitoring [4]. Conventional spectrometers, which rely on bulky optical components and long optical paths, can provide high-resolution and broad bandwidth performance. However, their large size and high cost make them difficult to deploy widely. Emerging applications such as airspace detection precision agriculture [5,6], and lab-on-a-chip systems have set high requirements on the compactness, economic efficiency, and in-situ real-time characterization capability, which promotes the development of chip-scale spectrometers [7,8].

To date, most on-chip spectrometers rely on miniaturized dispersive optics [9], narrowband filters [10], and Fourier transform (FT) interferometers [11]. On-chip spectrometers based on dispersive elements and narrowband filters provide a high resolution by spectral channel division [12]. However, the multichannel architecture not only degrades the signal-to-noise (SNR) but also requires a corresponding detector array. Microspectrometers based on Fourier transform possess some intrinsic advantages including the multiplexing advantage and high optical throughput [13]. In recent years, the development of computational reconstruction algorithms such as compressed sensing and machine learning has further improved the resolution and robustness of Fourier transform spectrometers. Electro-optic modulation [14], acousto-optic modulation [15],

This work was supported by National Natural Science Foundation of China (NSFC) under grant No. 62405173 and Shanghai Municipal Science and Technology Commission under grant No. 24DP1500600.

thermo-optic modulation [16], and modulation via free-carrier injection in semiconductors [17] rely on the weak perturbation of the silicon (Si) refractive index [18]. In contrast, microelectromechanical systems (MEMS) effectively induce modulation by spatially displacing photonic components, consequently improving the tuning efficiency and the spectrometer performance [19]. Among MEMS-based modulators, electrostatic actuation offers advantages such as simple structure, low power consumption, efficient modulation, and good complementary metal oxide semiconductor (CMOS) compatibility compared to electrothermal and piezoelectric actuation [20].

In this work, we propose a MEMS-enabled computational spectrometer based on cascaded waveguide couplers. The spectrometer utilizes comb-drive electrostatic actuators to independently drive three-stage in-plane waveguide couplers with lengths of 500, 750, and 1000 μm, respectively. Consequently, this enables both a larger number of sampling channels and the production of more randomized interferograms within a minimized footprint. By independently in-plane actuation to each stage of the waveguide coupler array with the comb-drive actuators, the proposed scheme reduces the autocorrelation of the calibration matrix, thereby overcoming the lithography resolution limitations of the coupling gaps as well as surpassing the Rayleigh criterion to achieve sub-100-picometer spectral resolution. A three-stage waveguide coupler is employed to achieve 216 sampling channels within a footprint of 75×2500 μm², enabling a spectral resolution of 40 pm over a bandwidth of 10 nm (1.5–1.51 μm).

II. WORKING PRINCIPLE AND STRUCTURAL DESIGN

This paper proposes a computational spectrometer based on cascaded waveguide couplers. The overall structure of the spectrometer based on cascaded waveguide couplers is shown in Fig. 1(a). The spectrometer consists of a tunable cascaded waveguide coupler fabricated on a silicon-on-insulator (SOI) wafer, which features a 0.22 μm-thick silicon device layer and a 2 μm-thick buried oxide (BOX) layer. The three-stage cascaded waveguide coupler comprises a straight waveguide and three cantilevered waveguides. Both the straight and cantilevered waveguides have a height of 220 nm and a width of 350 nm. The length of the straight waveguide is 2450 μm, while the lengths of the three cantilevered waveguides are 500, 750, and 1000 μm, respectively. The spacing between adjacent cantilevered waveguides is 100 μm. The straight waveguide remains

979-8-3315-8850-2/25 $31.00 © 2025 IEEE   743

stationary, whereas the cantilevered waveguides can be actuated to move in-plane by the comb-drive actuators.

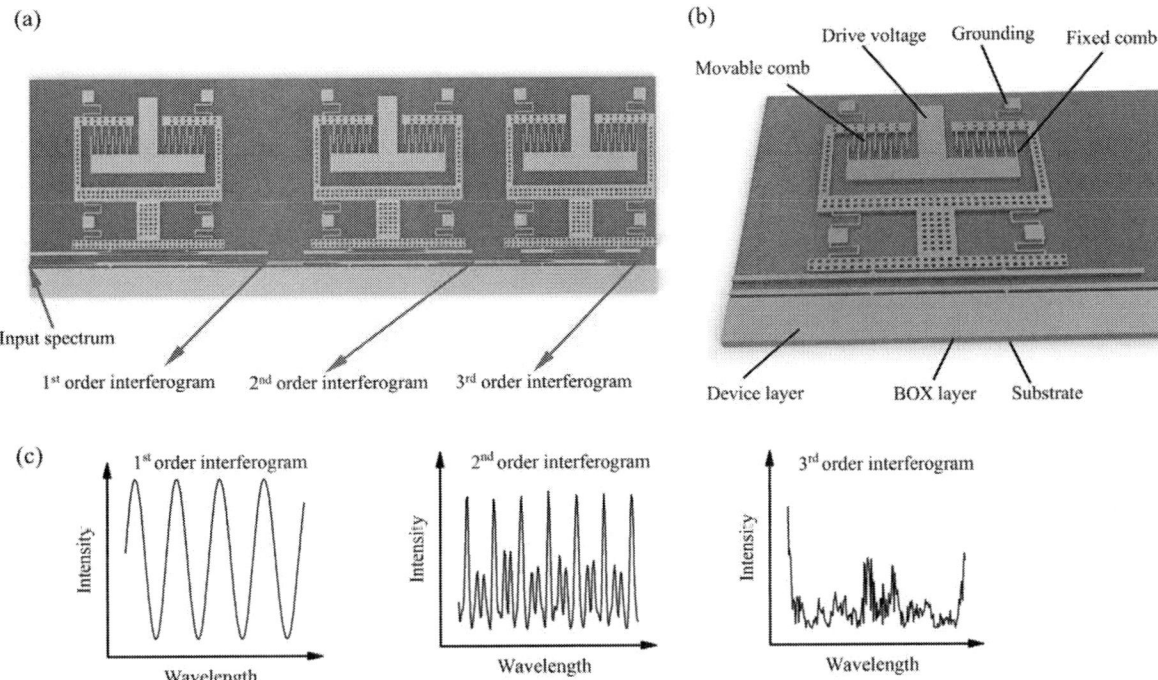

Fig. 1. Overview of the computational spectrometer. (a) Schematic of the proposed MEMS-enabled computational spectrometer based on cascaded waveguide couplers. (b) Zoom-in view of a single MEMS comb-drive actuator. (c) Measured interferograms obtained at each stage of the waveguide coupler.

The structure of the comb-drive actuator is shown in Fig. 1(b). The driving voltage is applied to the fixed comb fingers, while the movable comb fingers are grounded. Both the fixed and movable comb fingers have a width of 0.4 μm and a height of 4 μm, with a total of 80 comb fingers. The potential difference generated between the comb fingers drives the waveguide, reducing the gap between the two waveguides in the coupler and thereby enhancing the modal coupling strength.

The light transmits in the waveguide in a single-mode form. When the gap between the two waveguides in the waveguide coupler is reduced to a certain distance, the modes will couple and generate two supermodes. The fundamental supermode (SM0), which is the symmetric mode, and the higher-order supermode (SM1), which is the anti-symmetric mode. These two modes have different effective refractive indices, resulting in an effective index difference (Δn). The optical power at the output port of the straight waveguide varies with the coupling length (L) and the waveguide coupling gap, while the coupling gap is controlled by the applied voltage combinations. The output power measured by the spectrometer can thus be expressed as:

$$P(\lambda, \mathbf{V}) = I(\lambda, \mathbf{V})\cos^2\left(\frac{\pi L}{\lambda}\Delta n(\lambda, \mathbf{V})\right) \quad (1)$$

Where $\mathbf{V} = (V1, V2, V3)$ denotes the driving voltage combinations applied to the three-stage waveguide couplers, and

λ is the incident wavelength in vacuum. $I(\lambda, \mathbf{V})$ represents the intensity of the input spectrum. $\Delta n(\lambda, \mathbf{V})$ is regulated by both the wavelength and the applied voltage combinations, is the effective refractive index difference between the SM0 and SM1 supermodes ($\Delta n = n_{\text{SM0}} - n_{\text{SM1}}$). Consequently, by performing a fast Fourier transform (FFT) on the interferogram $P(\lambda, \mathbf{V})$, the input spectrum $I(\lambda, \mathbf{V})$ can be reconstructed.

The input spectrum enters from one end of the cascaded waveguide coupler, and interferograms can be measured by adjusting the coupling gap and the incident wavelength. As shown in Fig. 1(c), when only a single-stage waveguide coupler is used, the resulting interferogram is nearly sinusoidal. With a two-stage waveguide coupler, the periodicity of the interferogram is significantly reduced. When the number of cascaded waveguide couplers is increased to three, the interferogram becomes highly disordered, and the interferograms of individual sampling channels are decorrelated, which further enhances the performance of the spectrometer. Moreover, the use of cascaded waveguide couplers also enables an exponential increase in the number of sampling channels, as given by:

$$N_{\text{ch}} = V_{\text{stage}}{}^{N_{\text{stage}}} \quad (2)$$

Where N_{ch}, N_{stage} represent the total number of channels and the number of stages, respectively, and V_{stage} denotes the number of voltage levels per stage. In this work, a three-stage waveguide coupler is designed, with six different voltages applied to each stage, resulting in 216 sampling channels.

III. RESULTS AND DISCUSSION

The initial coupling gaps of the cascaded waveguide couplers are set to 460 nm, and the coupling gap is progressively reduced by increasing the applied voltage. As shown in Fig. 2(a), for a single waveguide coupler, the comb-drive actuator can achieve a displacement of 410 nm under a driving voltage of 24.2 V, reducing the coupling gap to 50 nm. The effective refractive index difference increases as the coupling gap decreases, as shown in Fig. 2(b).

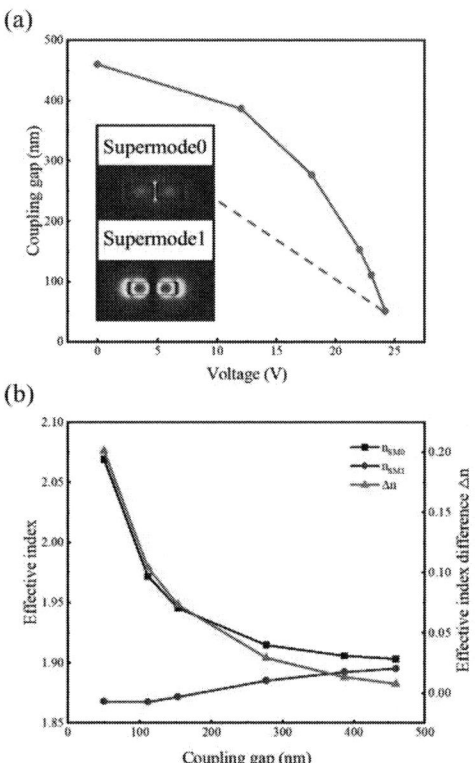

Fig. 2. Modulation of the waveguide coupler. (a) The coupling gap of the waveguide coupler changes with driving voltage. Insets show modal field plots of the waveguide coupler in final state. (b) The n_{SM0}, n_{SM1} and Δn change with the coupling gap of the waveguide coupler when the incident light is 1.505 µm.

For each stage of the cascaded waveguide couplers, the driving voltage applied to the cantilevered waveguides is varied from 0 to 24.2 V, with 6 voltage levels selected for each stage. The incident wavelength is swept from 1.5 to 1.51 µm with a step size of 10 pm. For each combination of driving voltages and incident wavelength, the intensity at the output port of the cascaded waveguide coupler is measured by the detector. After normalization, the calibration matrix $\mathbf{P}$ (m × n) is obtained, as shown in Fig. 3(a), where m = 216 corresponds to the different voltage combinations of the cascaded waveguides and n = 1001 denotes the number of wavelength

sampling points. The measured interferogram $\mathbf{I}$ by the detector can be expressed as:

$$\mathbf{I} = \mathbf{P} \cdot \mathbf{R} \tag{3}$$

where $\mathbf{I}$ is a column vector with 216 elements, $\mathbf{P}$ is the normalized calibration matrix measured by the detector, and $\mathbf{R}$ is the input spectrum to be reconstructed, with a size of 1001.

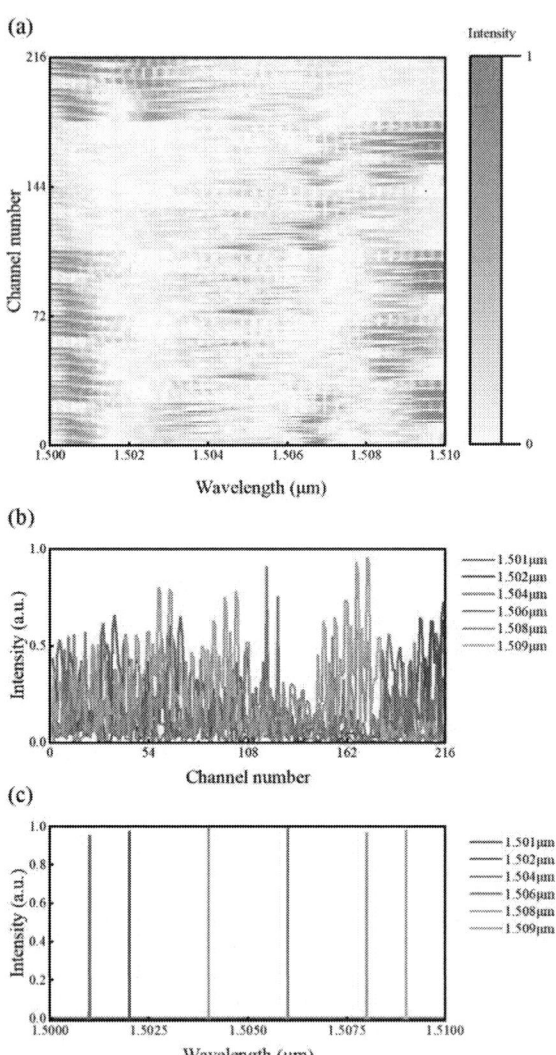

Fig. 3. Reconstruction of single-wavelength spectrum. (a) Calibrated measurement matrix $\mathbf{P}$ of the spectrometer. (b) Measure interferograms at several different laser wavelengths. (c) Reconstruction results of single-wavelength interferograms.

Spectral reconstruction can be regarded as solving an underdetermined equation. Therefore, the input spectrum $\mathbf{R}$ can be reconstructed by solving a regularized regression problem. The Lasso model, which incorporates the L1 norm, is well suited for handling sparse signals and can be formulated as:

$$\min_s\{\|\mathbf{I} - \mathbf{P} \cdot \mathbf{R}\|_2^2 + \alpha \|S\|\}_1 \qquad (4)$$

Where α denotes the weight of L1 regularization, which was tuned to an optimal value of 0.001. As shown in Fig. 3(b), the interferograms measured at different wavelengths exhibit low correlation with each other. Moreover, the interferograms at the same wavelength also display low periodicity. These interferograms are further reconstructed to demonstrate the performance of single-wavelength spectral reconstruction. As shown in Fig. 3(c), the input spectrum can be accurately reconstructed over the entire 10 nm bandwidth. To quantitatively evaluate the reconstruction accuracy, the widely adopted metric of relative error ϵ is employed to assess the performance of spectral reconstruction:

$$\varepsilon = \frac{\|\mathbf{R} - \widehat{\mathbf{R}}\|_2}{\|\mathbf{R}\|_2} \qquad (5)$$

where $\mathbf{R}$ and $\widehat{\mathbf{R}}$ denote the input and reconstructed spectrum, respectively.

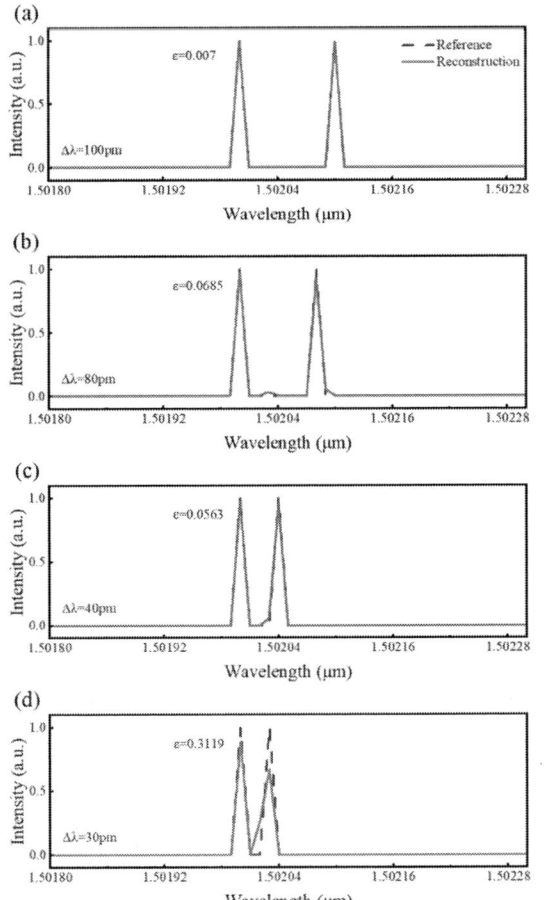

Fig. 4. Reconstruction of double-wavelength spectrum with different wavelength spacings $\Delta\lambda$ of (a) 100 pm, (b) 80 pm, (c) 40 pm, (d) 30 pm.

To determine the spectral resolution of the spectrometer, we performed linear superposition of interferograms at different wavelengths and conducted reconstruction tests for double-wavelength interferograms. The wavelength spacing ($\Delta\lambda$) of the input spectrum was gradually reduced from 100 pm to 40 pm as shown in Fig. 4(a–c). When the wavelength spacing was set to 100 pm, 80 pm, and 40 pm, the reconstructed spectrum exhibited relative errors below 0.1 compared to the input spectrum, demonstrating successful reconstruction with high accuracy. When the wavelength spacing is reduced to 30 pm, the relative error between the reconstructed and input spectrum increases significantly and exceeds 0.3, as shown in Fig. 4(d). Therefore, the spectrometer achieves a spectral resolution of 40 pm, which is significantly better than the 5.3 nm spectral resolution determined by the Rayleigh criterion.

IV. CONCLUSION

In conclusion, we propose a MEMS-enabled computational spectrometer based on cascaded waveguide couplers. Most reported on-chip spectrometers can achieve a spectral resolution of several tenths of a nanometer within a comparable device footprint, while our proposed computational spectrometer based on cascaded waveguide couplers further improves the resolution, achieving a high spectral resolution of 40 pm. The input spectrum produces higher-quality interferograms after passing through the cascaded waveguide couplers. A total of 216 sampling channels are realized within a compact footprint of 100 × 2500 µm². When combined with the Lasso algorithm, this configuration enables a spectral resolution of 40 pm across the 1.5–1.51 µm spectral range.

REFERENCES

[1] Z. Yang, T. Albrow-Owen, W. Cai, and T. Hasan, "Miniaturization of optical spectrometers," Science, vol. 371, no. 6528, 2021.

[2] L. Xia, Y. Liu, R. T. Chen, B. Weng, and Y. Zou, "Advancements in miniaturized infrared spectroscopic-based volatile organic compound sensors: a systematic review," Appl. Phys. Rev., vol. 11, no. 031306, 2024.

[3] J. Zhou, H. Zhang, Q. Qiao, H. Chen, Q. Huang, H. Wang, et al., "Denoising-autoencoder-facilitated MEMS computational spectrometer with enhanced resolution on a silicon photonic chip," Nature Communications, vol. 15, no. 1, 2024.

[4] M. Manley, "Near-infrared spectroscopy and hyperspectral imaging: non-destructive analysis of biological materials," Chem. Soc. Rev., vol. 43, no. 24, pp. 8200–8214, 2014.

[5] M. Inoue, I. Morino, O. Uchino, Y. Miyamoto, T. Saeki, Y. Yoshida, et al., "Validation of XCH₄ derived from SWIR spectra of GOSAT TANSO-FTS with aircraft measurement data," Atmos. Meas. Tech., vol. 7, no. 9, pp. 2987–3005, 2014.

[6] D. J. Mulla, "Twenty five years of remote sensing in precision agriculture: key advances and remaining knowledge gaps," Biosyst. Eng., vol. 114, no. 4, pp. 358–371, 2013.

[7] L. Gao, Y. Qu, L. Wang, and Z. Yu, "Computational spectrometers enabled by nanophotonics and deep learning," Nanophotonics, vol. 11, no. 11, pp. 2507–2529, 2022.

[8] H. Chen, H. Zhang, J. Zhou, C. Ma, Q. Huang, H. Wang, et al., "High-performance and wavelength-transplantable on-chip Fourier transform spectrometer using the MEMS in-plane reconfiguration," Photonics Research, vol. 12, no. 7, pp. 1–11, 2024.

[9] A. Y. Zhu, W.-T. Chen, M. Khorasaninejad, J. Oh, A. Zaidi, and I. Mishra, "Ultra-compact visible chiral spectrometer with meta-lenses," APL Photonics, vol. 2, no. 3, p. 036103, 2017.

[10] J. P. Carmo, R. P. Rocha, M. Bartek, G. De Graaf, R. F. Wolffenbuttel, and J. H. Correia, "A review of visible-range Fabry–Perot microspectrometers in silicon for the industry," Opt. Laser Technol., vol. 44, no. 7, pp. 2312–2320, 2012.

[11] L. Li, C. Peng, Y. Qi, G. Zhou, Q. Qiao, F. S. Chau, et al., "Design of an on-chip Fourier transform spectrometer using waveguide directional couplers and NEMS," Opt. Express, vol. 26, no. 23, p. 30362, 2018.

[12] B. Redding, S. Fatt Liew, Y. Bromberg, R. Sarma, and H. Cao, "Evanescently coupled multimode spiral spectrometer," Optica, vol. 3, no. 9, pp. 956–962, 2016.

[13] H. Chen, J. Zhou, N. Wang, and Y. Ma, "Compact MEMS-based computational mid-infrared spectrometer on silicon-on-insulator platform," in Proc. 8th International Conference on Integrated Circuits and Microsystems (ICICM), pp. 164–168, 2023.

[14] M. Li, J. Ling, Y. He, Y. He, Z. Hao, W. Zhu, et al., "Lithium niobate photonic-crystal electro-optic modulator," Nature Communications, vol. 11, no. 4123, 2020.

[15] Z. Yu and X. Sun, "Acousto-optic modulation of photonic bound state in the continuum," Light Sci. Appl., vol. 9, no. 1, p. 1–9, 2020.

[16] S. Gan, C. Cheng, Y. Zhan, Y. Zhuang, H. Liu, Y. Li, et al., "A highly efficient thermo-optic microring modulator assisted by graphene," Nanoscale, vol. 7, no. 47, pp. 20249–20255, 2015.

[17] M. Nedeljkovic, C. G. Littlejohns, A. Z. Khokhar, F. Y. Gardes, D. J. Thomson, G. Z. Mashanovich, et al., "Silicon-on-insulator free-carrier injection modulators for the mid-infrared," Opt. Lett., vol. 44, no. 4, pp. 915–918, 2019.

[18] F. Chollet, "Devices based on co-integrated MEMS actuators and optical waveguide: A review," Micromachines, vol. 7, no. 2, pp. 1–33, 2016.

[19] H. Podmore, A. Scott, P. Cheben, T. J. Hall, D. C. Hutchings, R. Halir, et al., "Athermal planar-waveguide Fourier-transform spectrometer for methane detection," Opt. Express, vol. 25, no. 26, pp. 33018–33028, 2017.

[20] Y. Ma, W. Liu, X. Liu, N. Wang, and H. Zhang, "Review of sensing and actuation technologies – from optical MEMS and nanophotonics to photonic nanosystems," International Journal of Optomechatronics, vol. 18, no. 1, pp. 1–48, 2024.

Control Strategy and Parameter Identification Method for Photovoltaic Inverter Electromechanical Transient Model Based on Improved Fish Eagle Optimization Algorithm

Zecheng Li*
Ningxia Electric Power Energy Technology Co.,Ltd.
Ningxia, China
3215176291@qq.com

Yaojia Huo
Ningxia Electric Power Energy Technology Co.,Ltd.
Ningxia, China
793965258@qq.com

Kai Hou
Ningxia Electric Power Energy Technology Co.,Ltd.
Ningxia, China
851503479@qq.com

Xiao Zhang
Ningxia Electric Power Energy Technology Co.,Ltd.
Ningxia, China
zhxqm@126.com

Hao Liu
Ningxia Electric Power Energy Technology Co., Ltd.
Ningxia, China
867120474@qq.com

Jinpeng Hao
Ningxia Electric Power Energy Technology Co.,Ltd.
Ningxia, China
416444715@qq.com

Abstract—A photovoltaic inverter transient model control strategy and parameter identification method based on an improved fish hawk optimization algorithm is proposed to address the difficulties in obtaining control methods and related parameters for low-voltage ride through（LVRT） of photovoltaic inverters, as well as the obstacles in establishing accurate simulation models, which in turn constrain the analysis of photovoltaic grid connected characteristics. Based on the characteristics of the LVRT output curve of photovoltaic power generation systems, a mathematical model for photovoltaic LVRT control was established and the transient process of faults was analyzed, clarifying the core control parameters of the LVRT process, Secondly, extract the key points of the operating conditions to establish an identification dataset, and use an improved fish hawk optimization algorithm to identify the control parameters of the inverter under the specified current mode, Finally, based on actual engineering parameters, the photovoltaic inverter was modeled, and the improved Fishhawk optimization algorithm was used to identify the control parameters and LVRT parameters. The results showed that the proposed method can accurately identify the control mode and parameters of the inverter's electromechanical transient model.

Keywords—Inverter, LVRT, Parameter identification, Fish Eagle Optimization Algorithm

I. INTRODUCTION

Driven by the dual carbon strategy goal, the proportion of renewable energy represented by photovoltaics in the power system is increasing year by year. According to statistics, as of the end of 2023, the installed capacity of solar power generation in China has reached about 610 million kilowatts, a year-on-year increase of 55.2%[1]. With the continuous integration of a large number of photovoltaic power sources into the grid, when a fault occurs on the grid system side, the grid connection point will exhibit a completely different fault

Research and Application of Modeling Method for Integrated Control Strategy and Control Parameters of New Energy Controller (522980250002)

current characteristic from the traditional synchronous motor connection scenario. However, the fault characteristics of the grid connection point depend on the control strategy of the grid connected inverter. However, due to factors such as the protection of manufacturers' trade secrets and dynamic changes in the inverter operating environment, the key control parameters in the control system are often in an unknown state, which directly leads to the difficulty in ensuring the correctness of photovoltaic power plant modeling and seriously affects the accurate reproduction of the actual working characteristics of the inverter.

Researchers at home and abroad have conducted multiple innovative explorations on the identification of low-voltage crossing parameters. For example, reference [2] considers the overcurrent protection mechanism of inverters during LVRT, identifying the current limit amplitude as a key parameter to ensure that the current response of the model during voltage drops is in line with reality. However, there are still issues with insufficient characterization of dynamic characteristics and poor adaptability to multiple operating conditions in the identification of low-pressure crossing parameters. Reference [3] focuses on the discussion of inverter units, resonance suppression methods, and cluster suppression strategies, and analyzes the limitations of existing technologies in combination with the characteristics of high proportion renewable energy and high proportion power electronic equipment. However, there are also shortcomings in this study, such as insufficient modeling ability for the interaction between inverters, and the possibility of multiple resonance peaks due to differences in control parameters and interaction effects when multiple inverters are connected in parallel, further exacerbating system instability.

Given the above background, this article first constructs a mathematical model for photovoltaic LVRT control and analyzes the fault transient model. Different control models are analyzed to obtain the core control parameters. Next, considering the correlation of the parameters to be identified, a photovoltaic inverter electromechanical transient model control strategy and parameter identification method based on an improved fish hawk optimization algorithm are proposed, which improves the accuracy of the identification results. Finally, based on the actual station information, a simulation model of the photovoltaic grid connected system is built, and hardware in the loop simulation is performed on the real photovoltaic controller to effectively and accurately identify the control strategy and parameters of the inverter electromechanical transient model, achieving integrated identification of control parameters and control strategies.

II. CONTROL ARCHITECTURE OF PHOTOVOLTAIC GRID CONNECTED INVERTERS

Firstly, build a photovoltaic system and use maximum power point tracking (MPPT) to control the photovoltaic array to maximize the conversion of light energy into electrical energy, prioritizing the output of active power and ensuring that the photovoltaic system always outputs electrical energy at the highest efficiency[4].

Establish a mathematical model for photovoltaic cells, and the formula for the mathematical model can be expressed as,

$$I = I_{ph} - I_0 \left(e^{\frac{q(U+IR_s)}{AKT}} - 1 \right) - \frac{(U + IR_s)}{R_{sh}} \quad (1)$$

In the formula, I is the output current of the photovoltaic cell, U is the output voltage of the photovoltaic cell, I_{ph} is the photocurrent, I_0 is the reverse saturation current of the diode, q is the electron charge, R_s is the series resistance, R_{sh} is the parallel resistance, K is the Boltzmann constant, A is the P-N junction ideal factor, and T is the absolute temperature.

The most commonly used models for photovoltaic cells are the single diode model and the dual diode model. Compared with the dual diode model, the single diode model has a simpler structure and higher calculation accuracy, as shown in Fig. 1.

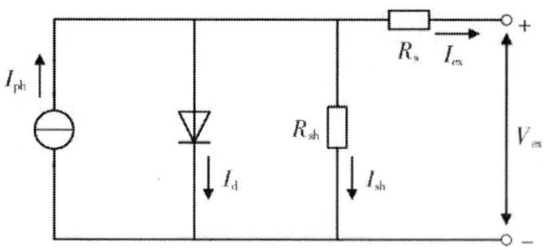

Fig. 1. Single diode model

MPPT is one of the essential functions that photovoltaic systems must possess. The perturbation observation method, as a common MPPT technique, relies on regularly fine-tuning the output voltage of the photovoltaic array, which is called 'perturbation'. By monitoring the trend of output power changes after disturbance, it can be determined that if the increase in output power $\Delta P > 0$, it indicates that the current direction of voltage adjustment is correct, and therefore this 'interference' direction can continue to be maintained, On the contrary, if $\Delta P < 0$, it means that the direction of voltage adjustment is incorrect, and the direction of 'interference' should be switched[5].

To accurately identify control parameters, it is necessary to extract the key points required for identification based on the response characteristics of low-pressure crossing conditions. The division of low-voltage crossing periods is shown in Fig. 2, where i_P is the active current, i_Q is the reactive current, and u is the AC side voltage of the inverter. A corresponds to the pre disturbance stage, B corresponds to the disturbance period stage, and C corresponds to the post disturbance stage. Moreover, the B and C segments are not a single interval, but each has its own subdivision, B segment is divided into transient interval B_1 and steady-state interval B_2, and C segment is divided into transient interval C_1 and steady-state interval C_2.

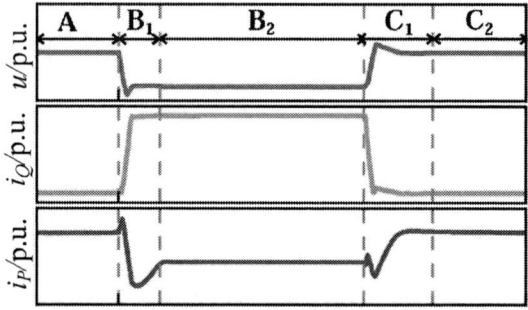

Fig. 2. Division of low voltage crossing periods

III. IMPROVE THE FISH EAGLE OPTIMIZATION ALGORITHM

Fish eagles are a type of raptor that operates day and night and feeds on fish. Fish eagles have strong vision to detect underwater objects. When flying at an altitude of 10 to 40 meters on the water surface, they will detect the position of fish underwater, Then, it moves towards the fish, immerses its feet in the water, and dives underwater to catch fish, After catching prey, the fish eagle will take it to nearby rocks and eat it. The strategy of fishing eagles to hunt fish and bring them to the appropriate location and eat them is their intelligent natural behavior, and this algorithm is a biomimetic mathematical modeling based on the hunting behavior of fishing eagles.Improving the Fish Eagle Optimization Algorithm is a biomimetic mathematical modeling based on the hunting behavior of fish eagles. By simulating the dynamic behavior of fish eagle hunting, it can solve complex optimization problems and has the characteristics of easy parameter adjustment, fast convergence, and strong adaptability.

Initial population and fitness assessment.Population initialization, Abstracting 'individual fish eagles' as 'potential solutions' to optimization problems, randomly generating an initial set of solutions in the solution space of the problem, with positions corresponding to a set of parameters of the solution vector.

Exploration phase. For each fish eagle, the position of other fish eagles (i.e. individuals with better objective function values) in the search space is considered as underwater fish. Specify the fish group for each fish eagle, which can be expressed as,

$$F_i^B = \{X_k \mid k \in \{1, 2, \cdots, n\} \cap F_k < F_i\} \cup \{X_m\} \quad (2)$$

In the formula, F_i^B is the fish group of the i-th fish eagle, X_m is the optimal solution, F_k is the objective function value corresponding to the k-th fish eagle, F_i is the objective function value corresponding to the i-th fish eagle, X_k is the group corresponding to the k-th fish eagle[6].

Fish eagles will randomly identify and locate a fish from a school of fish, and then launch an attack on it. To quantify this behavioral process, a mathematical model is constructed based on the dynamic characteristics of the movement of the fish eagle towards the target fish, and then the corresponding new position of the fish eagle is calculated. Define r as a random number in the [0,1] interval, Ii is a random number in set $\{1,2\}$, with the following relationship,

$$I_i = \begin{cases} 1, r < 0.5 \\ 2, r \geqslant 0.5 \end{cases} \quad (3)$$

This calculation relationship can be expressed as,

$$x_{i,j}^{B1} = x_{i,j} + r_{i,j}(F_{i,j}^O - I_i x_{i,j}) \quad (4)$$

In the formula, $x_{i,j}^{B1}$ is the value of the jth variable of the i-th solution, $F_{i,j}^O$ is the fish selected by the i-th fish eagle, $r_{i,j}$ is a random number in the [0,1] interval.

If the objective function value of the new position is better than the objective function value of the previous position, then the root replaces the position of the Osprey, which can be expressed as,

$$X_i = \begin{cases} X_i^{B1}, F_i^{B1} < F_i \\ X_i, \text{others} \end{cases} \quad (5)$$

In the formula, X_i^{B1} represents the new position of the i-th Osprey in the first stage, F_i^{B1} is the objective function value of the i-th solution.

During the mining phase, after hunting a fish, the fish eagle will carry it to a suitable location and feed there. The second stage of population update is to construct a model by simulating the natural behavior of the fish eagle. After modeling the behavior of 'bringing fish to the appropriate position', it will cause slight changes in the position of the fish eagle in the search space. On this basis, for each member in the

population, a new random position needs to be calculated and defined as the 'suitable feeding position'. It can be expressed as,

$$x_{i,j}^{B2} = x_{i,j} + \frac{a_j^L + r(a_j^U - a_j^L)}{t} \quad (6)$$

In the formula, $x_{i,j}^{B2}$ is the position of the i-th fish eagle in the second stage, a_j^U is the upper bound of the j-th variable, and a_j^L is the lower bound of the j-th variable, t is the current iteration count of the algorithm.

Capture and adjustment, After catching prey, the algorithm falls into local optima and will readjust the flight path to try again. Calculate the fitness value of the new position after each update of the Osprey position, If the new fitness is better than the old fitness, retain the new position, Otherwise, regenerate the position based on a certain probability to avoid algorithm stagnation. When the preset maximum number of iterations is reached or the optimal fitness value remains unchanged for multiple generations, the algorithm terminates and outputs the global optimal solution[7].

However, in the operation of the Fishhawk algorithm, there may be a problem of 'excessive focus on searching near the current optimal solution' in its mining process, which directly leads to premature convergence of the algorithm and is trapped in local optimal solutions, unable to continue exploring better solutions at the global level. Gaussian mutation provides an effective solution to this deficiency[8-9]. As a random search strategy, it significantly improves the randomness and diversity of the search process by introducing a random perturbation term that follows a Gaussian distribution based on the original individual values. With this characteristic, Gaussian mutation can help the algorithm overcome the limitations of the current local optimal solution during the mining phase, prevent premature convergence, and ultimately create conditions for finding a globally better solution[10]. The Gaussian function used for Gaussian mutation in this process is as follows,

$$G(\alpha) = \frac{1}{\sqrt{2\pi\sigma^2}} e^{-\frac{\alpha^2}{2\sigma^2}} \quad (7)$$

In the formula, α is a random number within the interval [0,1], and σ takes the value of 1. The updated position formula after mutation is as follows,

$$Gx_{i,j}^{B2} = x_{i,j}^{B2}\left(1 + G(\alpha)\right) \quad (8)$$

In the formula, $Gx_{i,j}^{B2}$ is the formula for updating the position after mutation.

IV. EXAMPLE ANALYSIS

To verify the effectiveness of the proposed method, the present invention builds a simulation model of a photovoltaic grid connected system based on actual station information in Ningxia. The RT-LAB semi physical testing platform is used to obtain the measured dataset of the photovoltaic controller, and the testing environment is shown in Fig. 3.

Fig. 3. testing environment

Build a simulation model of the photovoltaic grid connected system based on actual station information, conduct hardware in the loop simulation of the real photovoltaic controller, effectively and accurately identify the control strategy and parameters of the inverter electromechanical transient model, and achieve integrated identification of control parameters and control strategies.

Comparing the convergence speed of the classic Osprey optimization algorithm and the improved Osprey optimization algorithm proposed in the present invention, the results are shown in Fig. 4 and Fig. 5.

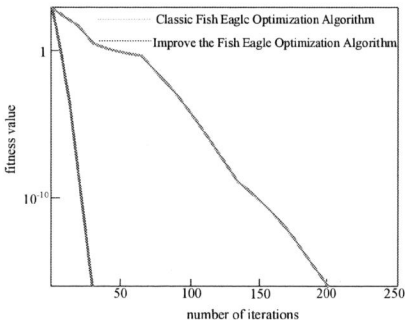

Fig. 4. The results of 250 iterations of two algorithms

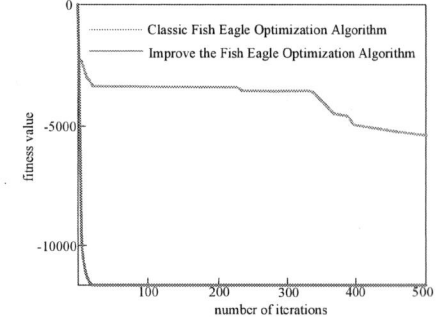

Fig. 5. The results of 500 iterations of two algorithms

This method improves the osprey optimization by adding Gaussian distribution random perturbations to the original individual values, enhancing search randomness and diversity.

It can help the algorithm get rid of local optima and avoid premature convergence during the mining stage, thus potentially finding a globally better solution, Determine the optimization range based on the typical values of the inverter control method and parameters, and then combine it with an improved fish eagle optimization algorithm for multi-objective identification. Set an adaptive optimization operator based on population diversity to improve convergence speed and make the identification results more accurate. The improvement of the Fish Hawk Optimization algorithm in terms of optimization value and optimization rate is significantly better than the classical Fish Hawk Optimization algorithm. The key to its performance improvement lies in the embedding of weight factors and Gaussian mutation mechanism in the algorithm. For scenarios where multidimensional test functions contain multiple solutions, weight factors can help the algorithm overcome the limitations of local optimal solutions. The incorporation of Gaussian mutation further significantly improves the global search performance of the algorithm, indicating that the improved Fish Hawk Optimization algorithm used in the present invention can more effectively improve the identification accuracy of model parameters.

A comparison of the Improved Osprey Optimization Algorithm (IOOA) against other algorithms, including JAYA, Differential Evolution (DE), Particle Swarm Optimization (PSO), and Chaos Honey Badger (CHB), is presented in Fig.6. The results demonstrate that IOOA significantly outperforms its competitors across several key metrics. It achieves the fastest convergence, with its eRMSE stabilizing after approximately 100 iterations, and attains the highest convergence accuracy, reaching the optimal value by around 200 iterations. In contrast, the DE and JAYA algorithms fail to achieve the best accuracy within the given iteration limit. Although OOA, CIJAYA, and DVADE eventually reach high accuracy, their convergence speeds are considerably slower, requiring 1200, 300, and 1300 iterations, respectively. Furthermore, the convergence curves reveal that IOOA possesses superior global search capabilities and a greater ability to escape local optima compared to JAYA, OOA, and DVADE. In conclusion, IOOA exhibits a more favorable balance of convergence speed, accuracy, global exploration, and local optima avoidance, making it the most effective algorithm among those tested.

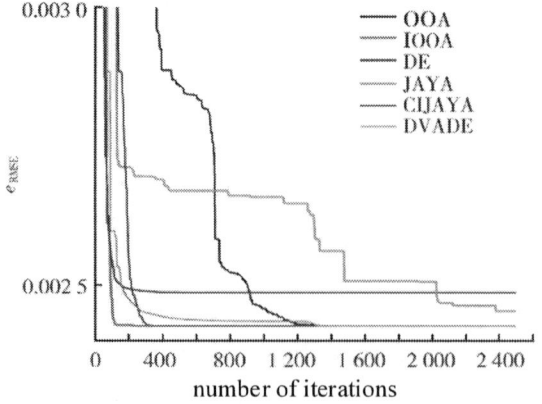

Fig. 6. RMSE iteration curves for each algorithm

979-8-3315-8850-2/25 $31.00 © 2025 IEEE   751

The improved fish eagle optimization algorithm will be applied to active current identification, as shown in Fig. 7. The improved fish eagle optimization algorithm matches the measured curve more closely and has higher accuracy, which meets the requirements of the present invention.

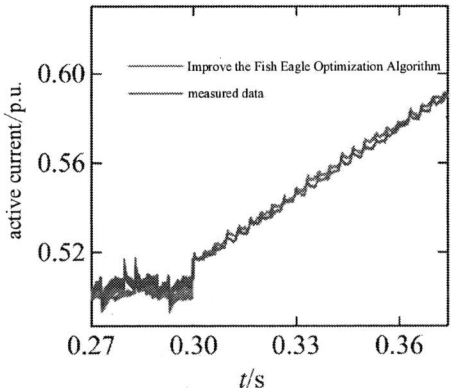

Fig. 7. Improving the Fish Eagle Optimization Algorithm and Measuring the Curve

V. CONCLUSION

This research centers on the core demand for the safe and stable operation of photovoltaic grid-connected systems, with a specific focus on the accuracy of low-voltage ride-through model parameters of photovoltaic inverters. Its primary goal is to address two key issues: the inadequate identification precision of LVRT model parameters, and the coupling relationship between inner-loop control parameters and LVRT characteristics.

Firstly, the study integrates the features of the LVRT output curve of photovoltaic power generation systems to establish a mathematical model for the LVRT control of photovoltaic inverters. By analyzing the fault transient process, it clarifies the core control parameters during the LVRT period, which lays a theoretical groundwork for subsequent parameter identification.

Subsequently, to resolve the coupling and correlation problems between control parameters and LVRT characteristics, a multi-stage parameter identification strategy is proposed. This strategy effectively avoids the parameter interference issue that occurs in single-stage identification.

Lastly, a simulation model of the photovoltaic inverter is constructed based on actual engineering parameters. An improved osprey optimization algorithm is then applied to identify both the control parameters and key LVRT-related parameters. Simulation results confirm the effectiveness of the proposed improved osprey optimization algorithm for parameter identification. This method not only accurately acquires the core parameters of the LVRT model and significantly enhances the model's ability to accurately characterize the LVRT performance of photovoltaic inverters but also provides reliable parameter support for optimizing LVRT control strategies and ensuring the safe and stable operation of photovoltaic grid-connected systems. Consequently, it holds important application value for improving the LVRT performance of practical photovoltaic grid-connected projects.

REFERENCES

[1] National Energy Administration. The National Energy Administration released statistics on the national power industry in 2023[EB/OL]. [2024-11-18]. https://www.nea.gov.cn/2024-01/26/c_1310762246.htm.

[2] WANG Xiaotong, WANG Tong, DENG Jun, WANG Zengping. Control mode and parameter integration identification of photovoltaic inverter electromechanical transient model[J]. Power System Technology, 2023, 47(9): 3547-3558.

[3] LIU Jiang, GAO Shuping, SUN Xiangdong, SONG Weizhang. Overview of resonance suppression methods for PV grid-connected inverters in weak grid[J]. Southern Power System Technology, 2024, 18(3): 65-71.

[4] LI Guanghui, WANG Weisheng, HE Guoqing, LIU Chun. Commutation Failure of UHVDC System for Wind Farm Integration (Part III)：Transient Overvoltage Suppression Measures of Wind Powers in Sending Terminal Grid[J]. Proceedings of the CSEE, 2022, 42(14)：5079 -5089.

[5] Wu Chen, Liu Chenxi, Huang Wei, Zeng Pijiang, Yuan Hui. Siting and sizing method of grid-forming converters for improving stability of power system with renewable energy. Automation of Electric Systems. 2023 Jun,47(12),130-6.J. Clerk Maxwell, A Treatise on Electricity and Magnetism, 3rd ed., vol. 2. Oxford, Clarendon, 1892, pp.68–73.

[6] Yang Chaoran, Huang Linbin, Xin Huanhai, Ju Ping. Placing grid-forming converters to enhance small signal stability of PLL- inte grated power systems[J]. IEEE Transactions on Power Systems，2021，36(4)：3563-3573.

[7] Huang Linbin, Xin Huanhai, Wang Zhen, Wu Kuayu, Wang Haijiao, Jiabing Hu. A Virtual Synchronous Control for Voltage-Source Converters Utilizing Dynamics of DC-Link Capacitor to Realize Self-Synchronization[J]. IEEE Journal of Emerging and Selected Topics in Power Electronics, 2017, 5(4)：1565-1577.

[8] HE Jiafa, WU Kuayu, HUANG Linbin, , Xin Huanhai, Lu Cencen, Wang Haijiao. A coordinated control scheme to realize frequency support of PMSG-based wind turbines in weak grids[C]//2018 IEEE Power & Energy Society General Meeting (PESGM). Portland：IEEE, 2018：1-5.

[9] Wang Weisheng, Li Guanghui, He Guoqing, Guo Zixuan, Huang Yuehui. Challenges and prospects of grid connection control of renewable energy for new power systems ［J］. New Type Power Systems, 2023, 1（2）：145-160.

[10] Shi Wenhui, Qu Jixian, Luo Kui, Li 3, He Yongjun, Wang Weisheng. Grid-integration and operation of high-proportioned new energy ［J］. Strategic Study of CAE, 2022, 24（6）：52-63.

2025 The 10th International Conference on Integrated Circuits and Microsystems

Design of a Cross-Platform Simulation and Verification System for Complex Onboard Control Computers

Zheng Yang[1*], Shenglong Li[1], Chaofan Zhou[2]

[1.] Beijing Institute of Control Engineering, China Academy of Space Technology, Beijing, China

[2.] ATSPACE, Beijing, China

yangzheng.zz@pku.edu.cn

* Correspondence author

Abstract—**With the growing complexity of modern engineering design, simulation technologies have come to play a pivotal role in the research and development lifecycle. The digital twin simulation platform for onboard control computers provides a comprehensive virtual verification environment for complex spaceborne electronic systems through the integration of multidimensional technologies. This study proposes a cross-platform simulation and verification system for onboard control computers, which integrates a CPU instruction-level simulator, an extensible C++-based user control software, a Simulink-Cadence co-modeling architecture, and an FPGA hardware-in-the-loop acceleration module, thereby establishing a closed-loop simulation framework spanning from algorithm design to hardware implementation. The CPU simulator supports cycle-accurate emulation of multicore architectures, while the C++ control layer enables dynamic parameter adjustment and real-time data visualization, effectively bridging system-level design and low-level hardware verification requirements. By interfacing computer simulations with physical hardware and leveraging hardware acceleration, the system achieves orders-of-magnitude improvements in simulation speed. This platform provides a holistic, real-time, system-level verification environment by synergizing diverse simulation tools and hardware platforms.**

Keywords—*Spaceborne electronic systems, Hardware acceleration, Co-simulation*

I. INTRODUCTION

With the increasing complexity of spacecraft functional requirements, the design of spaceborne electronic systems is facing multidimensional challenges. In traditional development workflows, algorithm design, circuit simulation, and hardware verification phases are often isolated from one another. The reliance on physical prototypes for iteration not only prolongs development cycles but also struggles to address boundary conditions of full-system collaboration. Particularly in missions such as deep-space exploration and high-precision remote sensing, onboard control computers must exhibit real-time responsiveness, multicore coordination, and high reliability, imposing stricter demands on system-level verification. Existing simulation technologies are predominantly confined to single toolchains or abstraction levels: Simulink focuses on algorithmic modeling, Cadence specializes in circuit-level simulation, and FPGA verification depends on hardware prototypes. Data silos between tools hinder cross-platform collaboration efficiency, making it difficult to establish a closed-

loop verification environment spanning algorithms to hardware. Furthermore, the inherent conflict between timing accuracy and simulation speed persists: pure software simulations can precisely emulate instruction-level behaviors but suffer from computational inefficiency in large-scale system validation, while Hardware-in-the-Loop (HIL) acceleration is often constrained by interface compatibility and dynamic reconfiguration capabilities.

To address these challenges, this study proposes a cross-platform simulation and verification system for complex onboard control computers, aiming to construct a full-process virtual validation platform through heterogeneous toolchain integration and hardware acceleration technologies. The system innovatively integrates a CPU instruction-level simulator, extensible C++ control software, a Simulink-Cadence co-modeling framework, and FPGA hardware acceleration modules, forming a trinity collaborative verification framework encompassing "algorithm-circuit-hardware." Specifically, the CPU simulator achieves cycle-accurate emulation of multicore architectures via dynamic binary translation; the C++ middleware acts as a data hub, dynamically orchestrating TCP/IP and PCIe interfaces to establish bidirectional data channels between Simulink models, Cadence schematics, and FPGA hardware; the FPGA acceleration cluster adopts a master-slave node architecture, leveraging high-speed serial buses and LVDS interfaces to enable distributed coordination of hardware resources, thereby significantly enhancing simulation throughput. Through hierarchical design, the system supports early-stage joint debugging of software algorithms and circuit models while enabling hardware-in-the-loop validation to approximate real-world operational conditions, bridging the gap between virtual simulation and physical deployment.

The core contributions of this work lie in three aspects: First, it transcends the limitations of conventional single-platform verification by unifying protocol interfaces to achieve cross-toolchain data fusion, providing lifecycle validation support for complex systems. Second, it proposes a dynamically scalable hardware acceleration architecture that balances timing precision and computational efficiency, achieving orders-of-magnitude improvements in simulation speed. Third, it establishes a deeply integrated software-hardware co-verification paradigm, laying a technical foundation for the engineering application of digital twin technologies in aerospace. Subsequent sections will elaborate on the system architecture,

979-8-3315-8850-2/25 $31.00 © 2025 IEEE

key technologies, and experimental validation, detailing the design principles and collaborative mechanisms of each module, while demonstrating the platform's efficacy and innovation through representative test cases.

II. PLATFORM DESIGN

The virtual prototype simulation platform for control computers comprises three layers: the hardware layer, software layer, and design application layer. As illustrated in Figure 1, this hierarchical architecture constitutes the overall design of the platform.

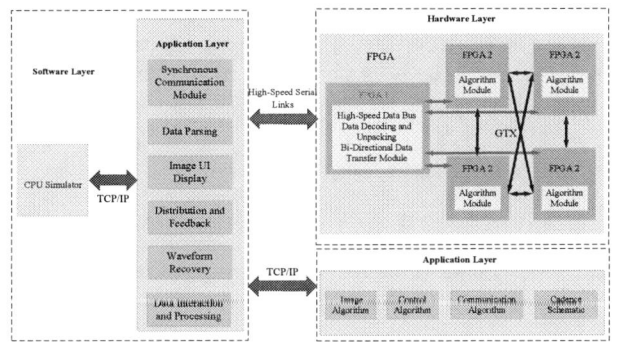

Fig. 1.   Overall Block Diagram of Digital Prototype Design.

The CPU simulator serves as the data source module, providing simulation data and instructions for the entire system. The user control software receives data packets from the CPU simulator via TCP/IP. After parsing the data and commands from the packets, the software distributes them to the hardware layer and application layer through high-speed interfaces and TCP/IP, completing the first step (executed only once during system initialization). Upon receiving the data, the hardware and application layers package the simulation data along with acknowledgment flags and transmit them back to the user control software. After processing these returned packets, the software requests the second data packet from the simulator, combines the new data with interaction data from the hardware and application layers, and redistributes the integrated packets. This constitutes the second step (cyclically executed after the first step) [1, 2].

The application layer and hardware layer function as core modules of the system. The application layer supports simulations of algorithmic models, communication models, and Cadence schematics, while the hardware layer conducts physical testing of related models. For algorithm models with slow simulation speeds, they can be migrated to hardware implementations. The software layer then provides test data sources for these hardware-accelerated modules, thereby significantly improving system simulation efficiency.

A. Software Layer

The software layer of the control computer digital twin simulation platform is architected around a CPU simulator and C++ user control software, constructing a digital closed-loop spanning from instruction execution to physical device coordination. As shown in Figure 2, the CPU simulator, acting as a virtualized hardware hub, precisely replicates the target processor's pipeline timing, cache mechanisms, and interrupt

response characteristics through accurate modeling of the Instruction Set Architecture (ISA). Its core employs dynamic binary translation technology to convert target instructions into host-executable code in real time, enabling cross-platform simulation while maintaining clock-cycle-level precision. During runtime, the simulator generates control instruction streams containing register states, bus signals, and peripheral interaction data via a virtual memory mapping mechanism. These time-stamped simulation data packets are then pushed to the C++ user control layer via shared memory or Socket communication interfaces [3, 4].

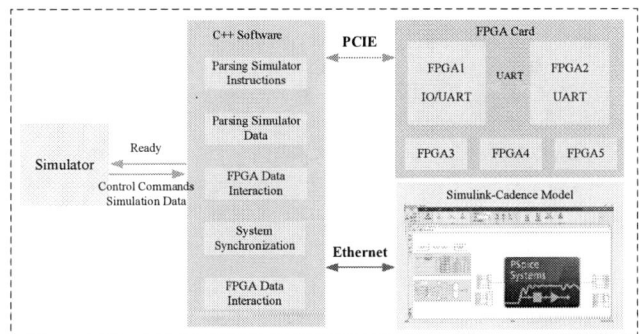

Fig. 2.   Software Layer Functional Block Diagram.

The software layer is primarily divided into two components: the CPU simulator and the user control software. These components interact via TCP/IP, with the CPU simulator transmitting instructions and data to the user control software through network interfaces. The user control software parses and processes the received data, orchestrating subordinate hardware and electrical schematic models for physical verification and simulation. It also facilitates three-way data interaction among the CPU simulator, hardware, and electrical models.

The primary function of the user application layer is to emulate a user CPU, serving as the simulation data source for the entire digital twin platform. The CPU simulator is a software tool that emulates the execution process of a target machine program on a host system. By simulating the effects of each instruction on the target processor, it acts as a software-based emulator of the target processor. This allows for software debugging and validation via the simulator before the target hardware becomes available, enabling true parallel development of software and hardware.

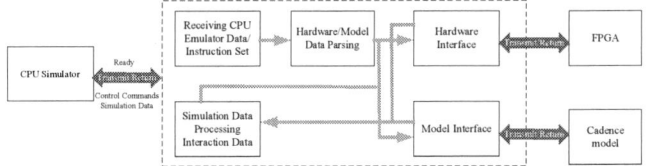

Fig. 3.   Software Data Processing.

As the core of the simulation platform, the user control software manages data processing, command transmission, simulation data visualization, and system communication control. As shown in Figure 3, upon receiving instructions and data from the simulator, the software parses the packets and distributes the required data to FPGA boards and Cadence

979-8-3315-8850-2/25 $31.00 © 2025 IEEE   754

schematics via PCIe interfaces and TCP/IP, respectively. After the FPGA boards and Cadence schematics complete a single simulation step, they return results to the software. The software processes and analyzes these results, then redistributes interaction data to relevant modules, thereby enabling seamless data exchange among the CPU simulator, FPGA boards, and Cadence models.

PCIe Bus Communication

The host computer accesses and controls FPGA hardware resources through PCIe, achieving communication with the host via register read/write operations, interrupt signals, and memory operations. The PC application controls the FPGA's operating modes through predefined registers in BAR0 of the PCIe interface, enabling functions such as external data control, DMA data transfer module enablement, interrupts, and completion of data transmission to the host. Figure 4 illustrates the block diagram of the PC-side processing system.

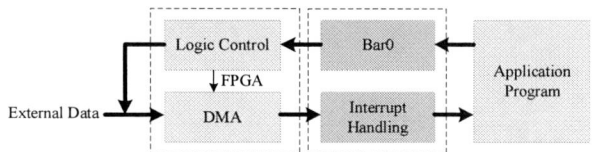

Fig. 4.   PC-Side Processing System Block Diagram.

A data transfer protocol is established between the C++ software and the FPGA. Data at different positions in the protocol packet correspond to specific test items. The C++ software assembles data frames through the PCIe interface and transmits them to the master FPGA according to protocol addresses. After receiving the data frames, the master FPGA parses the packet instructions to determine whether the commands involve read/write operations on the slave FPGA user logic, subsequently generating corresponding EMIF read/write timing sequences. The master FPGA then transmits these EMIF timing sequences and synchronization signals to the slave FPGA via internal board IOs. The slave FPGA completes one instruction step based on the EMIF commands and waits for new EMIF instructions. Upon completing tasks, the slave FPGA returns processed data to the master FPGA for packaging. The C++ software reads the completion flag from the master FPGA and retrieves the data via the PCIe interface.

Socket Network Communication

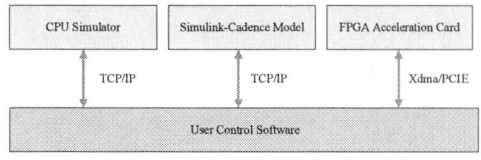

Fig. 5.   Socket Communication.

The PC concurrently runs a Simulink-Cadence electrical schematic model simulator, a CPU emulator, and a digital prototype simulation platform control software for data interaction, all connected via Socket network communication. The platform control software receives control commands and data from the CPU emulator through TCP/IP. After parsing and repackaging, it distributes the data to the Simulink-Cadence schematic and FPGA accelerator card via TCP/IP and PCIe

interfaces, respectively. Figure 5 shows the Socket communication architecture.

The digital prototype simulation platform software communicates with the Simulink-Cadence model and CPU emulator test programs through TCP-based Socket interfaces. The platform software acts as the server, while the Simulink-Cadence model and CPU emulator test programs serve as clients. The CPU emulator test program sends data to the digital prototype control software via Socket. The software unpacks and categorizes the data and commands, then dispatches them to the Simulink-Cadence model and hardware board via Socket and PCIe-based XDMA interfaces, respectively. After completing a simulation step, the software collects and analyzes results from both ends in the same manner, enabling data forwarding and interaction between Simulink-Cadence and the FPGA accelerator card.

Synchronized Operation Implementation

The control computer's digital prototype simulation platform employs a system synchronization mechanism, where all nodes operate orderly under the synchronization protocol. As shown in Figure 2, this illustrates the system's synchronized operation control logic.

After system startup, the CPU emulator sends commands to the C++ software. Upon processing these commands, the C++ software distributes data to the FPGA board and Simulink-Cadence schematic. After completing a simulation step, the C++ software waits for ready signals from both the FPGA board and the model. Once the ready signals are detected, the C++ software notifies the CPU emulator to receive the next data command, thereby maintaining system synchronization.

The synchronization control module generates commands issued by the host computer once per simulation step. Upon receiving the command, the master FPGA node produces a synchronous clock and clock enable signal, returning a 5ms pulse signal to the host. The synchronization signal serves as a trigger for subsequent modules within the node and is transmitted to slave nodes via LVDS interconnects to activate functions such as AXI. The synchronization control flow is depicted in the Figure 6.

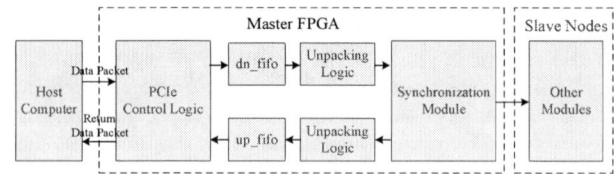

Fig. 6.   Synchronization Control Flow.

B. Hardware Layer

The hardware layer primarily focuses on the design of FPGA hardware accelerator cards. These cards are designed to simulate interface protocols and perform comprehensive system-level simulations in conjunction with CPUs and schematics, enabling the identification of issues that cannot be exposed through schematic-only simulations. Each board in the hardware accelerator cluster contains a master node chip for data interaction with the digital space. The master node parses data packets received from the digital space and distributes them to a

979-8-3315-8850-2/25 $31.00 © 2025 IEEE

slave node user logic. After processing, the slave node returns the data to the master node, which then relays it back to the user software.

As shown in Figure 1, the hardware system consists of multiple FPGA accelerator cards interconnected via I/O and high-speed serial interfaces. This architecture is primarily used for simulation acceleration, enabling system-wide data connectivity through high-speed interfaces. The system employs a high-performance computing (HPC) accelerator card based on the PCI Express bus architecture (see Figure 1). The board utilizes Xilinx's high-performance 28nm K7-series FPGA as the computing node, optimized for resources, interfaces, and clock management to deliver exceptional hardware acceleration capabilities. The board includes five FPGA processing nodes: the master node handles PCIe interface conversion, optical fiber transceivers, and data distribution/synchronization, while master-slave nodes are interconnected via high-speed serial buses. Below are the hardware interfaces integrated into the FPGA board [5, 6].

PCIe Interface

The master FPGA node on the algorithm accelerator card is connected to PCI Express golden fingers to enable data exchange with the host computer. This PCIe interface supports PCIe x8 mode, complies with the PCI Express 2.0 specification, and achieves a bidirectional bandwidth of up to 3 GB/s. The interface can be configured as x1, x4, or x8 based on application requirements. The FPGA communicates with the host via the PCIe interface. Figure 7 illustrates the logic control during FPGA data interaction.

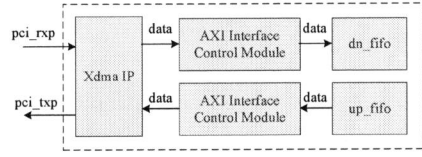

Fig. 7. PCIE Interface Data Transmission Design.

The XDMA IP receives data packets from the host. The user-side control logic writes the packets into the dn_fifo, where the unpacking module parses and routes them to subsequent modules (e.g., GTX, GTH, LVDS, UART). Data returned from these modules is repackaged by the packing module and written into the up_fifo. When the host reads data from the FPGA, the user-side control logic retrieves it from the up_fifo.

SFP+ Optical Fiber Interface

The FPGA accelerator card features four SFP+ optical fiber channels for high-speed serial interface expansion, supporting a maximum line rate of 10 Gbps per lane. These channels are compatible with various communication protocols, such as RapidIO, Aurora 64b/66b, and 10 Gigabit Ethernet. The SFP+ interfaces enable high-speed data transfer between standalone systems via the accelerator card's four fiber channel groups. The GTX modules facilitate data packet exchange between master and slave nodes using two pairs of high-speed serial interfaces. These interfaces support protocols like RapidIO. Figure 8(a) shows the design schematic of the inter-board SFP+ fiber interface, while Figure 8(b) details the GTX design for master-slave nodes.

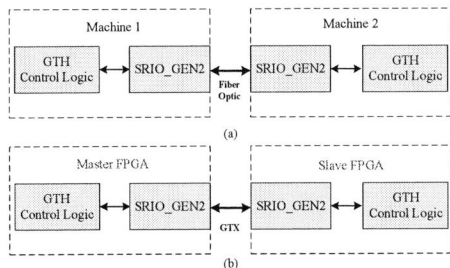

Fig. 8. (a) Dual-Machine FPGA Accelerator Card Design; (b) GTX between Master and Slave Nodes.

LVDS Interface

LVDS (Low-Voltage Differential Signaling) is a high-speed (up to multi-Gbps), low-power, and noise-resistant differential signaling technology used for LVDS data loopback between systems. The FPGA accelerator card includes two types of LVDS interfaces: a 21-channel ring LVDS interface for interconnecting the five node FPGAs; two LVDS ports per node for external communication with other boards. External 1553B communication interfaces can be implemented via off-board LVDS connections.

Figure 9(a) illustrates the LVDS hardware design on the FPGA accelerator card, and Figure 9(b) presents the logic implementation of the LVDS interface.

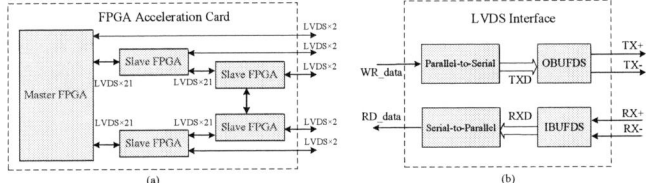

Fig. 9. (a) LVDS Interface; (b) LVDS Transceiver Logic Block Diagram.

C. Application Layer

This section focuses on Simulink-Cadence co-simulation design. Simulink is employed for modeling, simulation, and analysis, providing an intuitive graphical modeling environment where engineers represent system components and signal flows using blocks and lines. Circuit models developed in Cadence can be directly imported into Simulink via the PSpice interface, enabling control system or signal processing simulations in Simulink while performing circuit simulations in PSpice.

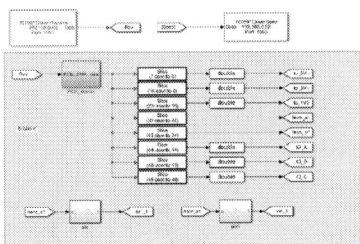

Fig. 10. Simulink Model.

The Socket receives simulation data from the CPU-FPGA co-simulation protocol software and maps raw data to Cadence schematic hardware interfaces through the PSpice BLOCK

979-8-3315-8850-2/25 $31.00 © 2025 IEEE

module according to the required timing and data specifications of Cadence schematics. Simulink acts as the intermediary connecting Cadence schematics and user control software, with its model design illustrated in Figure 10.

The Simulink model utilizes a TCP/IP module as a client to exchange data with the local user control software. It receives raw simulation data forwarded from the CPU emulator via the software, processes it through internal logic operations, and converts it into the timing and data formats required by Cadence schematics.

For co-simulation with Cadence schematics, the PSpice BLOCK module must be added to the Simulink library. This module facilitates data interaction between Simulink and Cadence schematics, allowing the Simulink model to transmit CPU emulator data to Cadence schematics for schematic-level system simulation validation.

III. SYSTEM CO-SIMULATION TESTING

After parsing control commands and simulation data sent by the CPU emulator, the user control software packages schematic simulation data and hardware I/O data separately for distribution to their respective components. The schematic generates a set of voltages returned to the user control software, which then determines high/low logic levels based on predefined thresholds and transmits them as I/O signals to the board. Following I/O loopback on the board, binary data is sent back to the user control software and forwarded to the schematic for voltage restoration. Sinusoidal wave simulations complete the loop exclusively through the "CPU emulator → user control software → schematic" chain.

A. Analog Circuit Test

A sinusoidal signal and analog voltage signal were injected via the CPU emulator for signal amplification and voltage recovery testing. Figure 11(a) shows the small-signal amplification circuit results in Cadence. A 1.5-amplitude signal from the emulator is displayed in Simulink as shown in Figure 11(b). The small-signal amplification circuit output returned to the C++ user software via Simulink is illustrated in Figure 11(c).

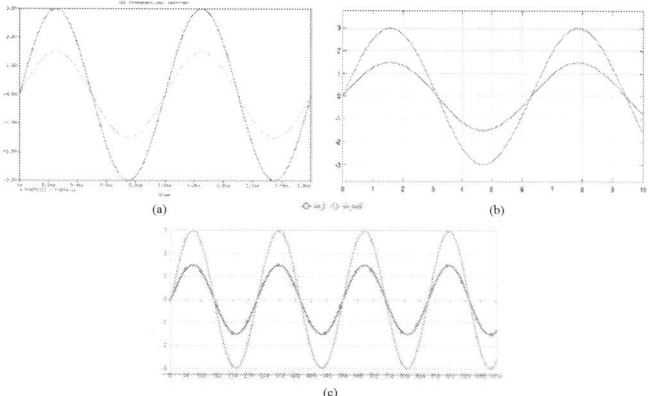

Fig. 11. (a) Small-Signal Amplification Circuit Simulation Results (Cadence); (b) Simulink Display Results; (c) Software UI Interface (C++ Software).

Upon receiving the voltage, the user control software generates binary I/O data and transmits it to the board. After hardware loopback (with identical signal values), the software sends the data to the schematic's voltage recovery circuit to reconstruct the analog voltage, as demonstrated in Figure 12.

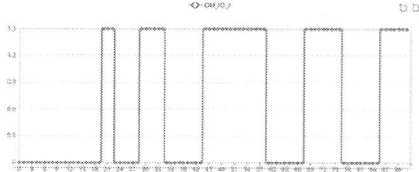

Fig. 12. Voltage Recovery.

B. Digital Circuit Test

Figure 13(a) presents standalone schematic simulation results of the RAM read/write circuit. Simulink displays write data (IN_d) and read data (C_O), with processed C_O data labeled as "data" in Figure 13(b). Figure 13(c) shows the read/write data in the C++ software, where the red box indicates normal timing differences: read data appears one step after write data.

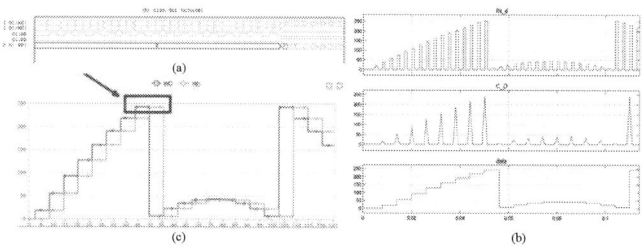

Fig. 13. (a) RAM Read/Write Simulation Results (Cadence); (b) RAM Read/Write Simulation (Simulink); (c) 25 RAM Read/Write Simulation (C++).

IV. SUMMARY

The co-simulation platform for digital prototypes developed in this study establishes a comprehensive Hardware-in-the-Loop (HIL) verification framework covering full-process workflows, achieving multi-dimensional system integration and high-efficiency simulation validation through innovative architectural design. Centered around user control software as the core hub, the platform constructs a data-driven closed-loop verification environment: At the data monitoring level, it enables visual surveillance of system-wide data flows and real-time waveform reconstruction, supporting dynamic tracking of critical parameters through configurable multi-channel observation signals. For simulation verification, the platform achieves groundbreaking cross-platform toolchain integration, bridging system-level algorithm modeling in Simulink with circuit-level design verification in Cadence, thereby establishing an end-to-end workflow that significantly enhances design iteration efficiency. The HIL simulation component executes actual control programs on FPGA boards while synchronizing with data sources from user control software, creating a verification scenario that closely emulates real hardware environments, effectively bridging the gap between algorithmic simulation and physical deployment. The platform innovatively employs a C++ middleware architecture to establish an efficient data exchange hub, enabling both real-time distribution of Simulink simulation data to FPGAs and bidirectional data communication between Simulink and hardware components.

979-8-3315-8850-2/25 $31.00 © 2025 IEEE   757

This real-time interaction mechanism not only ensures millisecond-level responsiveness but also establishes a novel verification paradigm featuring deep hardware-software co-simulation.

The platform's distinctive innovations manifest in three dimensions: First, it constructs a seamless integration environment for heterogeneous toolchains, systematically unifying traditionally fragmented phases including system simulation, circuit design, and hardware verification through standardized interface protocols, thereby supporting full-cycle closed-loop verification from conceptual design to hardware implementation. Second, it pioneers a dynamically scalable data hub architecture where the C++ middleware-based data routing mechanism enables flexible channel configuration according to verification requirements, simultaneously satisfying both the stability demands of large-scale data throughput and the customization needs of diverse verification scenarios. Third, it achieves deep coupling between system-level and circuit-level verification through cross-platform data fusion technology that correlates algorithmic simulation results with circuit characteristic parameters, substantially improving design optimization precision. This highly integrated platform not only provides a standardized verification framework for rapid prototyping of complex systems but also lays crucial technical

foundations for engineering applications of digital twin technology.

REFERENCES

[1] Li li, Qiao Dezhi, He Shimin. Design and implementation of universal software test platform for on-board computer.Microelectronics & Computer[J],2019,36(03):23-27.DOI:10.19304/j.cnki.issn1000-7180.2019.03.005.

[2] Lin Dandan. Constructing on simulation testing platform for embedded software[J]. Information Technology ,2012,36(10):77-79.DOI:10.13274/j.cnki.hdzj.2012.10.039.

[3] Huang, Jinfeng, Wu, Hongbin, Li, Tao.VeriNP. A FPGA-Based Verification Platform for General-Purpose Many-Core Network Processors[J]. 2015.

[4] Heitauer M , Versen M . Evaluation of a co-simulation approach with matlab/simulink and cadence for functional verification of analog and mixed signal devices[J]. Electronic Device Failure Analysis: A Resource for Technical Information and Industry Developments, 2021, 23(3):8-12.DOI:10.31399/asm.edfa.2021. 23(3), 8-12.

[5] Zhang Canyu, Feng Ansong, Zhang Hualiang. Design of image processing hardware acceleration system based on FPGA[J]. Computer Engineering and Design, 2024(003):045.DOI:10.16208/j.issn1000-7024.2024.03.012.

[6] Hardavellas N, Somogyi S, Wenisch T F, et al. SimFlex: A fast, accurate, flexible full-system simulation framework for performance evaluation of server architecture[J]. ACM SIGMETRICS Performance Evaluation Review, 2004, 31(4): 31–34. doi: 10.1145/1054907.1054914.

2025 The 10th International Conference on Integrated Circuits and Microsystems

Fast Frequency Stabilization Technique for Buck Converter Based on Adaptive PLL

Kerun Li
College of Electronics and Information Engineering
Shenzhen University
Shenzhen, Guangdong Province, China
kerun0805@163.com

Mei Jiang*
College of Electronics and Information Engineering
Shenzhen University
Shenzhen, Guangdong Province, China
mjiang@szu.edu.cn

Jianing Guo
College of Electronics and Information Engineering
Shenzhen University
Shenzhen, Guangdong Province, China
2302838834@qq.com

Yuxuan Guo
College of Electronics and Information Engineering
Shenzhen University
Shenzhen, Guangdong Province, China
gyx20020927@163.com

Abstract—This paper proposes a Buck converter that achieves frequency stability through an Adaptive Phase-Locked Loop (PLL) to address the issue of frequency instability in traditional Constant-On-Time (COT) architecture converters. The proposed solution employs a PLL feedback loop and an on-time generation module to compensate for input/output voltage variations, ensuring a stable switching frequency under changing load and input/output voltage conditions. To significantly enhance the frequency locking speed during the transition from light-load Discontinuous Conduction Mode (DCM) to heavy-load Continuous Conduction Mode (CCM), an adaptive frequency-following charge pump clamp circuit is designed. The Buck converter was simulated and verified using a 0.18μm BCD process. Experimental results show that, with an input voltage of 12V, an output voltage of 3.3V, and a set frequency of 1MHz, the use of the adaptive clamp circuit reduces the frequency locking time from 250μs to 118μs, representing a 52.8% improvement in locking speed. Additionally, within a load range of 0 to 5A, the switching frequency variation rate is only 15kHz/A, effectively realizing the frequency locking function of the PLL.

Keywords—Buck converter, adaptive on-time, phase-locked loop, frequency stabilization

I. INTRODUCTION

Switch-mode power supplies, serving as critical interfaces for energy conversion between electronic devices and external sources, are widely employed in portable electronics such as cell phones, notebooks, smart wearables, and various sensor devices. Among them, the Buck converter holds a prominent position in high-efficiency, miniaturized DC-DC power supplies owing to its simple structure and superior efficiency. Key research focus areas for current Buck converters include achieving fast transient response, high power density, and low power consumption. However, constrained by inherent limitations, traditional single-phase Buck designs typically struggle to simultaneously fulfill all these requirements. Consequently, the industry is actively shifting towards multi-phase Buck solutions [1-3]. These utilize multi-phase current sharing to significantly enhance transient response speed, effectively reduce output voltage ripple, and achieve higher power density using smaller inductors. Among various control schemes, COT control has gained prominence in multiphase Buck applications for its rapid transient response, excellent light-load efficiency, and robust loop stability [4].

Under traditional Constant-On-Time control, the switching period of a Buck converter is not fixed and varies with input voltage V_{IN} and output voltage V_{OUT}. To achieve relative period stability across different V_{IN} and V_{OUT} conditions, the on-time T_{ON} can be set inversely proportional to V_{IN} and proportional to V_{OUT}, theoretically enabling a constant period. This mechanism defines Adaptive Constant-On-Time (ACOT) control. However, this conclusion relies on an approximation neglecting the power MOSFET on-resistance. In practice, due to the neglected on-resistance, the switching period shifts with load current variations. To address this, various control techniques have been proposed to suppress switching frequency variations caused by input voltage, output voltage, and load current fluctuations. For instance, Reference Frequency Compensation (RFC) ensures converter stability and mitigates switching frequency variations induced by output capacitors with low Equivalent Series Resistance (ESR) [5]. While traditional Pulse Width Modulation (PWM) voltage-mode or peak current-mode control can also resolve frequency instability, their loop bandwidth and transient response speed under equivalent conditions are inferior to COT control [6][7].

This paper proposes a solution integrating an adaptive PLL within the ACOT control architecture. The scheme employs a PLL integrated adaptive on-time controller to achieve fixed switching frequency, which reduces switching losses and minimizes frequency variations. The design precisely locks the switching frequency, enabling parallel operation of multi-phase Buck converters. Furthermore, its adaptive clamp circuit significantly reduces the frequency locking time during transitions from DCM to CCM. Additionally, the utilization of valley current mode control eliminates the need for an external slope compensation circuit, avoiding pole splitting issues caused by over-compensation while maintaining fast loop response and excellent light load efficiency.

II. CIRCUIT DESCRIPTIONS

Fig. 1 shows the block diagram of the proposed ACOT-controlled buck converter based on PLL modulation. Compared to conventional current-mode COT control, this system

Shenzhen Science and Technology Research and Development Technical Breakthrough Key Project (No. JSGG20220831093005009), Shenzhen Science and Technology Research and Development Key Project (No. JCYJ20220818103410022)

979-8-3315-8850-2/25 $31.00 © 2025 IEEE

incorporates a PLL to synchronize the buck converter's switching frequency with an external reference clock. The PLL consists of a phase-frequency detector (PFD), a charge pump (CP), and a low-pass filter (LPF). In this control scheme, the buck power stage functions as a voltage-controlled oscillator (VCO) for the PLL.

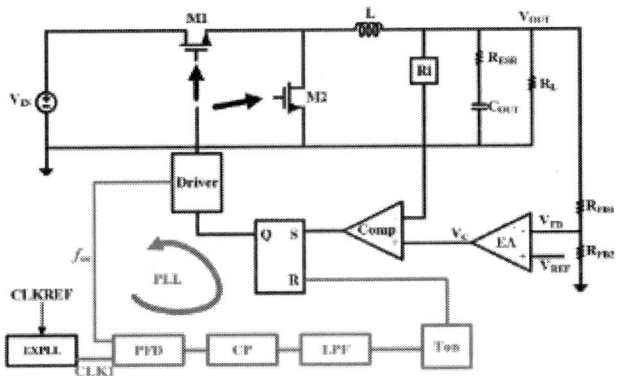

Fig. 1. Block diagram of the proposed buck converter

The frequency locking principle operates as follows. The feedback voltage V_{FB} is compared with the reference voltage V_{REF} by a comparator, to generate an error signal V_C. Within the Buck control loop, the sampled inductor current value is compared with the control voltage V_C to determine if the inductor current has reached its valley point. This comparison dictates the start of the period Ton and triggers the turn-on of the high-side power transistor. The PFD detects the frequency difference between the reference clock signal CLK1 and the

Buck switching frequency (f_{SW}). Based on this difference, the PFD generates signals that control the charging and discharging of the CP. This action alters the output voltage on the CP capacitor. This voltage signal is then filtered by the LPF into a smooth DC control signal. This DC signal adjusts the charging current of the Ton generation module, enabling precise control of Ton. When the Ton period expires, the RS latch resets, turning off the high-side power transistor. At steady state, CLK1 and the switching frequency f_{SW} become equal, achieving frequency locking [8]. Thus, the converter incorporates two feedback loops: the voltage feedback loop stabilizes the output voltage, while the PLL feedback loop stabilizes the switching frequency [9][10].

A. Adaptive On-Time Generation Circuit

Fig. 2 depicts the Adaptive On-Time generation circuit, the core module of this architecture. Beyond the internal Phase-Locked Loop (INTPLL), the chip integrates a peripheral Extended Phase-Locked Loop (EXPLL) that both supplies the PFD with clock signal CLK1 while generating uniform phase intervals for multi-chip cascading, and produces a frequency-dependent bias current I_{RT} based on input reference clock CLKREF. The PFD triggers pulse signals to the Charge Pump upon detecting frequency/phase errors. Then, the CP dynamically adjusts the LPF output voltage V_{CP} by sourcing or sinking current. V_{CP} is subsequently converted to a current I_{LPF}, summed with I_{RT}, and then multiplied by the supply-derived current I_{VIN} (proportional to V_{IN}). This multiplicative processing yields I_{ON}, which encodes the SW-CLKREF frequency/phase relationship to directly determine the conduction interval Ton, thus establishing a temporal representation of the phase-frequency information.

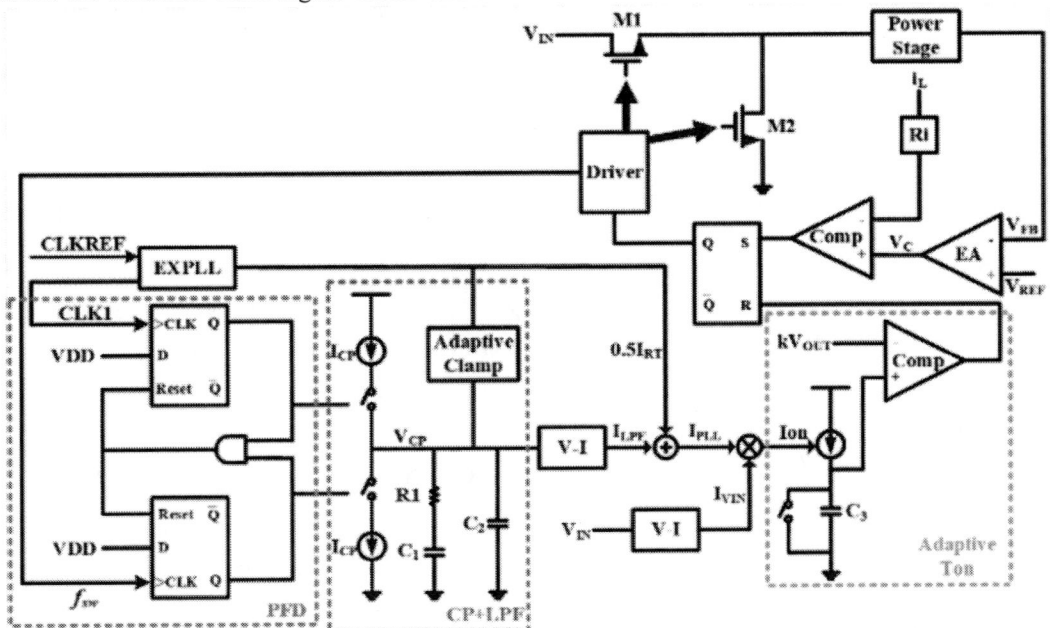

Fig. 2. Proposed on-time generation circuit with additional PLL regulation.

B. Work Process

At the start of each switching cycle, the high-side switch turns on, initiating the Ton interval. During the OFF-time, switch S1 remains open. Initially discharged to zero potential,

capacitor C_3 is charged by the current I_{ON}, which is modulated by V_{CP} and encodes V_{IN} information. The resulting voltage ramp across C_3 connects to the comparator's inverting input, while the non-inverting input receives the scaled output voltage kV_{OUT}.

979-8-3315-8850-2/25 $31.00 © 2025 IEEE

When the voltage across C_3 reaches kV_{OUT}, the comparator toggles, terminating Ton, turning off the high-side switch, and turning on the low-side switch. The Adaptive Ton mechanism is analyzed by considering the charging of capacitor C_3 by the current I_{ON} against the threshold voltage kV_{OUT}, as shown in equation (1) and (2).

$$I_{on} = \frac{I_{PLL} \cdot k_1 V_{IN}}{I_C} \quad (1)$$

$$T_{on} = \frac{C_3 \cdot kV_{OUT}}{I_{on}} \quad (2)$$

Among them, I_{PLL} can be expressed by equation (3):

$$I_{PLL} = 0.5 I_{RT} + I_{LPF} \quad (3)$$

According to the relationship between duty cycle and switching frequency, we obtain equation (4).

$$f_{sw} = \frac{I_{PLL} \cdot k_1}{C_3 \cdot k \cdot I_C} \quad (4)$$

In circuit operation, the interaction between I_{VIN} and V_{OUT} precisely cancels out the duty cycle, thus making the switching frequency fsw independent of V_{IN} and V_{OUT}. Furthermore, fsw maintains a linear relationship solely with I_{PLL}. This results in quasi-constant frequency operation, significantly simplifying control loop design. When the external clock frequency varies substantially, I_{RT} rapidly adapts to coarsely lock f_{SW} near the target frequency over a wide range. Subsequently, the PLL adjusts the V_{CP} voltage to control the I_{LPF} current, enabling f_{SW} to be accurately locked at the target frequency.

C. Adaptive Clamp Circuit

As shown in Fig. 3, the Adaptive clamp module employs a reference current, I_{RT}, generated by the EXPLL module, where the magnitude of I_{RT} varies proportionally with the frequency of the input reference clock, CLKREF. The EXPLL supports two distinct operational modes for generating I_{RT}: clock-input mode and resistor-input mode. In clock-input mode, the EXPLL functions as a conventional type-II PLL. The Judge module detects the presence of an active clock input signal, subsequently deactivating transistor MN6 and activating MN5. This enables the frequency-dependent current I_{RT} to be mirrored via a current mirror to transistor MP5 and supplied to the INTPLL module. Conversely, during resistor-input mode, the Judge module activates MN6 and deactivates MN5, supplying I_{RT} to the INTPLL through the current mirror while maintaining only the VCO active within the EXPLL to provide a reference frequency for subsequent stages. Transistor MP6 is configured in a diode-connected manner, with the critical distinction that its gate potential is dynamically controlled by the magnitude of I_{RT}. A lower external reference frequency CLKREF results in a smaller voltage drop across resistor R2, consequently lowering the gate potential of MP6. Due to its diode-connected configuration, MP6 exhibits diode-like exponential I-V characteristics when conducting. Consequently, it clamps the voltage Vcp to approximately $I_{RT} \cdot R2 + V_{GS}$. Crucially, because I_{RT} itself scales with the reference frequency, this clamping mechanism inherently adapts to the applied reference frequency, thereby achieving adaptive voltage clamping.

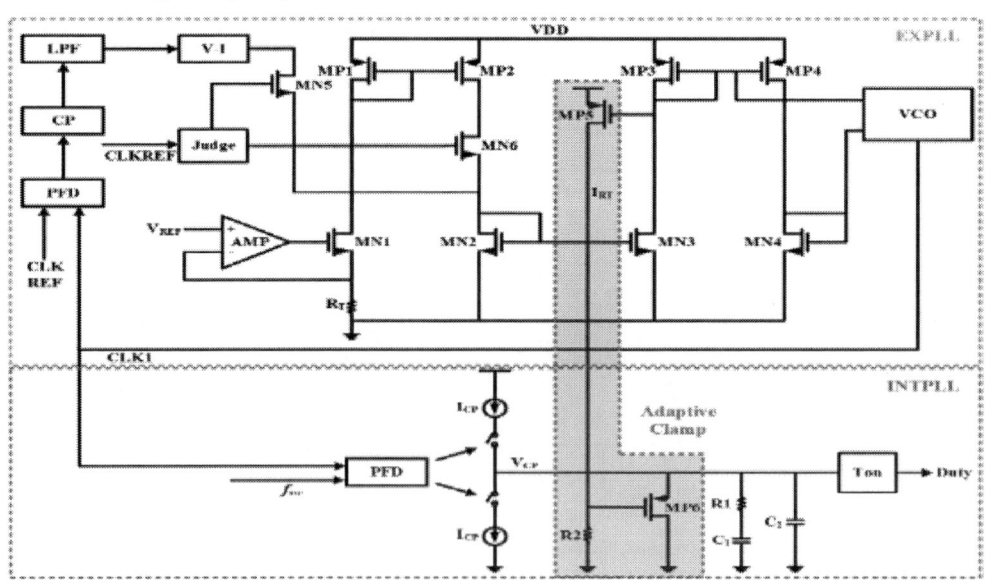

Fig. 3. Structure of the adaptive clamp circuit.

The Adaptive Clamp module significantly improves the frequency/phase locking speed of the system during transitions from light-load DCM to heavy-load CCM. In DCM, the slow switching frequency f_{SW} causes the control voltage Vcp of the INTPLL module to rise continuously. Even when Vcp reaches its saturation level, frequency and phase locking remain unattainable. Subsequently, when the load abruptly increases, forcing the system into CCM operation, the INTPLL reduce Vcp

from its saturated maximum down to the target value to achieve locking. This voltage transition inherently incurs a substantial time delay, resulting in slow locking. To mitigate this limitation, the Adaptive Clamp module is implemented. Its

979-8-3315-8850-2/25 $31.00 © 2025 IEEE

primary function is to stabilize Vcp near an appropriate operating level during DCM. Consequently, when a load transient occurs and the system enters CCM, the INTPLL can rapidly adjust Vcp to the target value. This enables swift frequency and phase locking of the system's switching frequency f_{SW}.

III. POST-SIMULATED RESULT

This work presents a Buck converter featuring adaptive PLL frequency locking, implemented in a 0.18μm BCD process. The chip layout is depicted in Fig. 4 and the total area of the converter is 2.34 mm^2. The converter operates with an input supply voltage V_{IN} ranging from 3.5V to 22.5V and delivers an output voltage programmable from 0.6V to 20V. It supports a load current range of 0A to 5A. The key specifications of the designed buck converter are summarized in Table I.

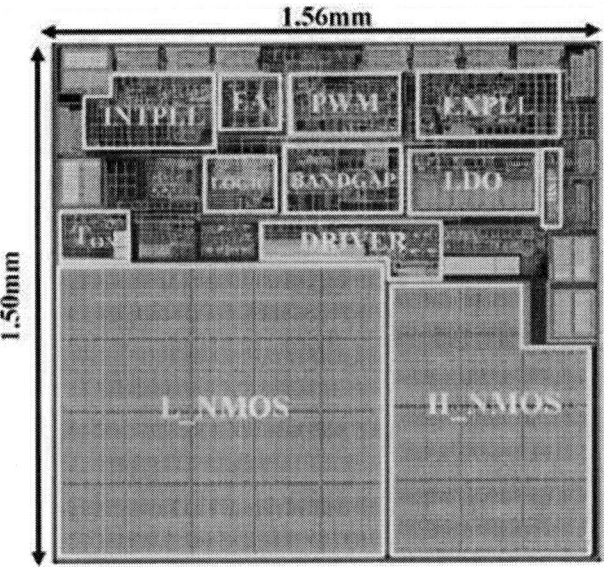

Fig. 4. Overall circuit layout.

TABLE I.   SPECIFICATIONS OF THIS BUCK CONVERTER

Technology	0.18μm BCD
Input voltage range	3.5V to 22.5V
Output voltage range	0.6V to 20V
Load current range	0 to 5A
Switching frequency	400kHz to 2.5MHz
Inductor	1μH
Output-capacitor	47μF
Output ripple	22mV @ f_{SW} = 1MHz
Max.Power efficiency	94.5% @ V_{IN}=12V, V_{OUT}=3.3V, I_{LOAD}=3A

Fig. 5 compares the frequency settling times with and without the adaptive clamp module. Under test conditions of V_{IN} = 12V, V_{OUT} = 3.3V, and a target switching frequency of 1MHz, a load current step change from 0.1A to 4A was applied. The measured f_{SW} settling times were 250μs without the clamp and 118μs with the clamp enabled. This represents a 52.8% reduction in settling time. The accelerated stabilization is

attributed to the clamp circuit enabling V_{CP} to reach its target value significantly faster.

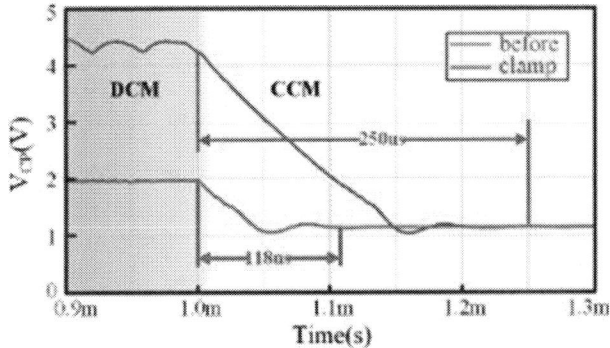

Fig. 5. Comparison of frequency stabilization time before and after adding the adaptive clamping circuit.

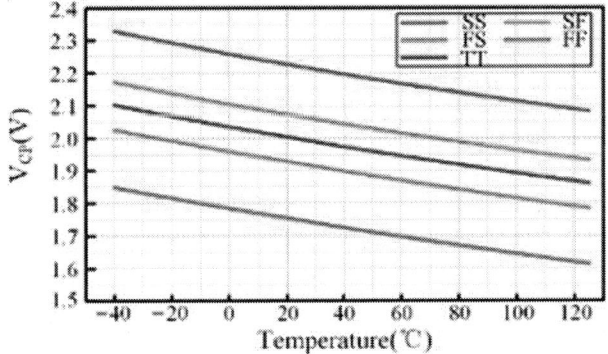

Fig. 6. The clamping values of V_{CP} at 1MHz under different process corners and temperatures.

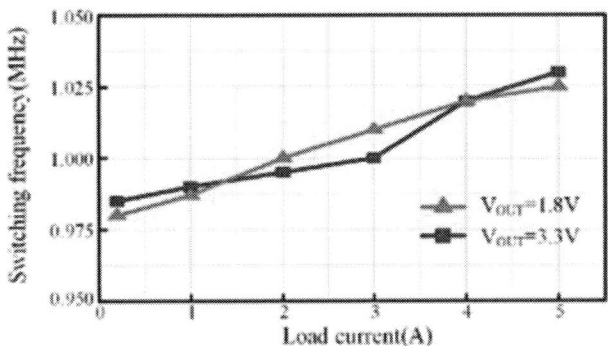

Fig. 7. Switching frequency variation versus I_{LOAD}.

Fig. 6 validates the clamping voltage of the adaptive clamp circuit across process corners and temperature variations. Simulations were performed under the TT, FF, and SS process corners, with temperatures ranging from -40°C to 125°C for each corner. When operating at 1MHz, the clamp voltage consistently exceeds the target value of V_{CP} required for steady-state operation. This ensures both reliable PLL functionality during steady-state conditions and accelerates the settling of V_{CP} during transients. These results demonstrate the high robustness of the adaptive clamp circuit against manufacturing process and environmental variations.

979-8-3315-8850-2/25 $31.00 © 2025 IEEE   762

Fig. 7 illustrates the switching frequency variation of the proposed COT converter with phase-locked loop (PLL) under load current changes from 0.1A to 5A. As demonstrated, the designed PLL effectively maintained the switching frequency across the entire load range.

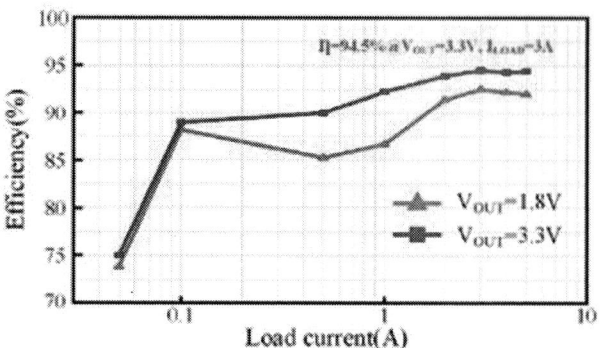

Fig. 8. Efficiency of the proposed buck converter.

Conversion efficiency is a key metric for evaluating DC-DC converter performance. The conversion efficiency graphs for different loads, at different output voltages, at an input voltage of 12V and an operating frequency of 1MHz are shown in Fig. 8. We can see that the converter in heavy load conditions system has a high conversion efficiency; light load conditions chip still has a high conversion efficiency, under certain conditions (V_{OUT} = 3.3V, I_{LOAD} = 3A) can be as high as 94.5%.

IV. CONCLUSION

To stabilize the switching frequency in conventional COT architectures, this paper presents a PLL-based current-mode buck converter. The proposed circuit achieves frequency stabilization across wide ranges of input and output voltages as well as load currents. An adaptive clamping circuit significantly enhances PLL locking speed during DCM-CCM transitions; it demonstrates a 52.8% improvement in locking speed when the load steps from 0.1A to 4A at 1MHz. Fabricated in a 0.18μm BCD process, the design achieves 94.5% peak efficiency across the full load range. The proposed buck converter thus delivers stable switching frequency and high efficiency, demonstrating significant potential for practical applications.

REFERENCES

[1] C. -J. Chen, Z. -Y. Zeng, C. -H. Cheng and F. -T. Lin, "Comprehensive Analysis and Design of Current-Balance Loop in Constant On-Time Controlled Multi-Phase Buck Converter," in IEEE Access, vol. 8, pp. 184752-184764, 2020.

[2] L. Li, S. Xu, Y. Qian, J. Nie and W. Sun, "Digital Dual-Loop Interleaving Control Algorithm for Asymmetric Multiphase Buck Converter With Ultrafast Load Transient," in IEEE Transactions on Power Electronics, vol. 39, no. 1, pp. 164-179, Jan. 2024.

[3] J. -G. Kang, J. Park, M. -G. Jeong and C. Yoo, "A Time-Domain-Controlled Current-Mode Buck Converter With Wide Output Voltage Range," in IEEE Journal of Solid-State Circuits, vol. 54, no. 3, pp. 865-873, March 2019.

[4] W. -C. Liu, C. -H. Cheng, P. P. Mercier and C. C. Mi, "Small-Signal Analysis and Design of Constant On-Time Controlled Buck Converters With Duty-Cycle-Independent Quality Factors," in IEEE Transactions on Power Electronics, vol. 38, no. 7, pp. 8379-8393, July 2023.

[5] J. Zhao, Q. Ye and X. Lai, "A Frequency Stable On-Time Control Buck Converter With Reference and Frequency Compensation Technique Using Low ESR Output Capacitor," in IEEE Transactions on Industrial Electronics, vol. 69, no. 4, pp. 3536-3545, April 2022.

[6] W. -C. Liu, C. -H. Cheng, C. C. Mi and P. P. Mercier, "A Novel Ultrafast Transient Constant on-Time Buck Converter for Multiphase Operation," in IEEE Transactions on Power Electronics, vol. 36, no. 11, pp. 13096-13106, Nov. 2021.

[7] Y. -S. Hwang, J. -J. Chen, Y. -T. Ku and J. -Y. Yang, "An Improved Optimum-Damping Current-Mode Buck Converter With Fast-Transient Response and Small-Transient Voltage Using New Current Sensing Circuits," in IEEE Transactions on Industrial Electronics, vol. 68, no. 10, pp. 9505-9514, Oct. 2021.

[8] L. Kong, D. Chen, S. -F. Hsiao, C. -F. Nien, C. -J. Chen and K. -F. Li, "A Novel Adaptive-Ramp Ripple-Based Constant On-Time Buck Converter for Stability and Transient Optimization in Wide Operation Range," in IEEE Journal of Emerging and Selected Topics in Power Electronics, vol. 6, no. 3, pp. 1314-1324, Sept. 2018.

[9] R. C. -H. Chang, W. -C. Chen and J. K. -S. Huang, "A 93.4% Efficiency 8-mV Offset Voltage Constant On-Time Buck Converter With an Offset Cancellation Technique," in IEEE Transactions on Circuits and Systems II: Express Briefs, vol. 67, no. 10, pp. 2069-2073, Oct. 2020.

[10] Q. ul Ain et al., "A High-Efficiency Fast Transient COT Control DC–DC Buck Converter With Current Reused Current Sensor," in IEEE Transactions on Power Electronics, vol. 36, no. 8, pp. 9521-9535, Aug. 2021.

A Wide-Range PFD and Low-Mismatch Source-Switched CP for MEMS Clock Systems

Changfu Wei
Engineering Research Centre of RF-ICs &
RF systems, Ministry of Education
Southeast University
Nanjing, China
220236450@seu.edu.cn

Lu Tang*
Engineering Research Centre of RF-ICs &
RF systems, Ministry of Education
Southeast University
Nanjing, China
lutang2k@seu.edu.cn
*Corresponding author

Ziyao Xiong
Engineering Research Centre of RF-ICs &
RF systems, Ministry of Education
Southeast University
Nanjing, China
220241183@seu.edu.cn

Abstract—This paper presents a TSPC-based phase-frequency detector and source-switched charge pump for fractional-N CPPLLs in temperature-compensated MEMS clock systems. The proposed PFD eliminates dead zones while achieving -1.95π to $+1.95\pi$ detection range through adjusting the device dimensions of the TSPC structure. The designed CP incorporates rail-to-rail clamping amplifiers and a fast turn-off path on the source-switched topology, achieving a current mismatch below 0.04% in the range of 0.4V~1.6V. Additionally, the PFD and CP demonstrate superior current noise performance, enabling the fractional-N PLL system to achieve   -65.5 dBc reference spurs at a division ratio of 22.5 in simulations.

Keywords—MEMS clock system; fractional-N CPPLL; PFD; CP

I. INTRODUCTION

With the growing demand for miniaturized clock sources in IoT and communication systems, temperature-compensated MEMS clock systems have gained significant attention due to their compact size and cost-effectiveness. Such systems require clock frequency to maintain high stability across a wide temperature range, which imposes strict requirements on the phase detection accuracy of phase-locked loops (PLLs) and the linearity of charge pumps (CPs). Although fractional-N PLLs can achieve frequency resolution in the GHz range through a Σ-Δ modulator to dynamically adjust the division ratio [1], the performance of the phase-frequency detector (PFD) and charge pump critically determines the stability and spectral purity of the PLL output signals.

The linear phase detection range of a PFD ideally spans $\pm 2\pi$. In fact, the PFD exhibits nonlinear behavior for large phase differences between input signals, which degrades the loop acquisition process [2]. Traditional CP architectures employ long-channel devices to reduce channel-length modulation effects, thereby improving current matching accuracy [3]. This approach, nevertheless, results in slower transient response speeds and consequently deteriorates the PLL's noise performance. Therefore, suppressing CP current mismatch and optimizing the PFD's phase detection range have emerged as critical breakthrough points. While dual-edge phase comparators can extend the phase detection range [4]–[5], they are prone to path delay mismatch. Reference [6] adopts a PFD with steep rising edges，effectively minimize dead zone time

and incorporates a noise-reduced charge pump for low-jitter performance. However, its current matching accuracy may remain insufficient in fractional-N PLLs, leading to reference spurs. Recent advances in CP design highlight that low-power, low-current-mismatch charge pump architectures effectively suppress PLL reference spurs and phase noise, thereby enhancing overall PLL performance [7]–[8].

This work adopts a 180nm CMOS process to design a high-precision PFD with wide phase detection range based on True Single-Phase Clock (TSPC) architecture. By enhancing the load-driving capability of D flip-flop (DFF), the PFD achieves significantly expanded phase detection range. Simultaneously, a source-switched CP is implemented incorporating rail-to-rail clamping amplifiers and fast turn-off paths, enabling rapid settling of charge/discharge currents and achieving low mismatch characteristics to meet the design requirements of temperature-compensated MEMS clock systems.

The paper is organized as follows. Section II presents the fractional-N divider PLL architecture employed in this work. Section III details the specific circuit implementations of both the PFD and CP modules. Section IV provides simulation results and comprehensive performance metrics of the PFD and CP. Section V concludes the paper with a summary of key contributions.

II. ARCHITECTURE OF FRACTIONAL-N CPPLL

The CPPLL serves as a critical component in clock system, functioning as a frequency synthesizer to generate clock signals. While conventional integer-N PLLs can only produce integer multiples of the reference frequency, fractional-N PLLs employ a Σ-Δ modulator (DSM) to dynamically adjust the division ratio, enabling output frequencies that are fractional multiples of the reference frequency and thereby achieving higher frequency resolution. Figure 1 illustrates the architectural diagram of the fractional-N PLL.

The PLL system operates with a 48 MHz reference clock frequency. The PFD compares the phase and frequency between the reference clock and feedback clock, generating corresponding pulse signals to control the CP's charging and discharging operations. These operations adjust the Voltage-Controlled Oscillator (VCO)'s tuning voltage through the loop

979-8-3315-8850-2/25 $31.00 © 2025 IEEE

filter, thereby modifying the output frequency until the frequency stabilizes.

To enable precise fractional-N division control, a DSM is incorporated in the circuit to generate pseudo-random control words. However, the DSM's quantization noise introduces high-frequency phase jitter. If the PFD fails to respond to minute phase errors, this noise will directly leak into the VCO, degrading phase noise performance. Furthermore, the dynamic division ratio in fractional-N operation increases the CP's switching frequency, making spurious tones more pronounced.

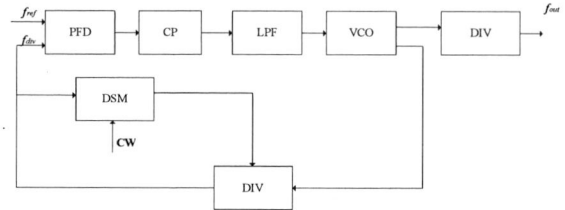

Fig. 1.   The block diagram of fractional-N PLL.

III. DESIGN OF CIRCUIT

A. High-Precision Wide-Range PFD

The proposed PFD architecture consists of DFF, AND gate, and reset delay elements, as shown in Figure 2. The reset delay prevents dead zone issues, though excessive delay would constrain the detection range and introduce reference spurs. Therefore, there is no extra delay unit to achieve a high-precision wide-range PFD. In addition, this design incorporates an additional differential output stage to improve signal synchronization and reduce CP mismatch.

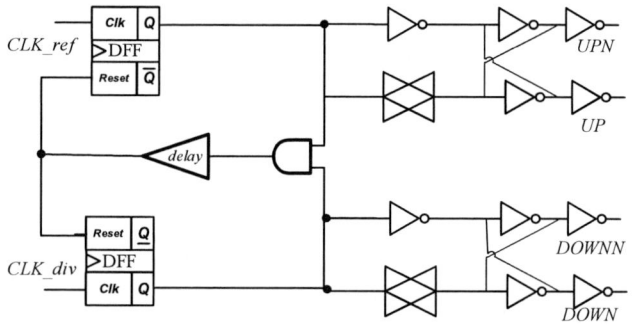

Fig. 2.   The block diagram of PFD.

In order to improve the DFF's operational speed and reduce the power consumption, an improved TSPC structure with reset terminal is adopted in this design. Given that the DFF input remains consistently high during actual PFD operation, the conventional TSPC structure can be simplified accordingly. Figure 3 illustrates the optimized DFF architecture. To address the trade-off between dead zone elimination and detection range expansion, this design strategically adjusts the transistor dimensions in the TSPC structure. By progressively scaling the MOSFET sizes, the DFF's load-driving capability is enhanced. This approach maintains a fixed reset delay while simultaneously avoiding dead zones and extending the phase detection range.

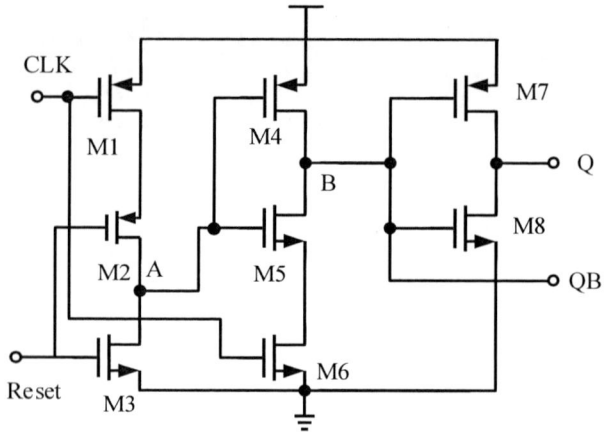

Fig. 3.   The TSPC-based DFF architecture.

B. High-Speed Low-Mismatch CP

Current single-ended CP designs for CPPLL typically adopt one of three topologies, as illustrated in Figure 4.

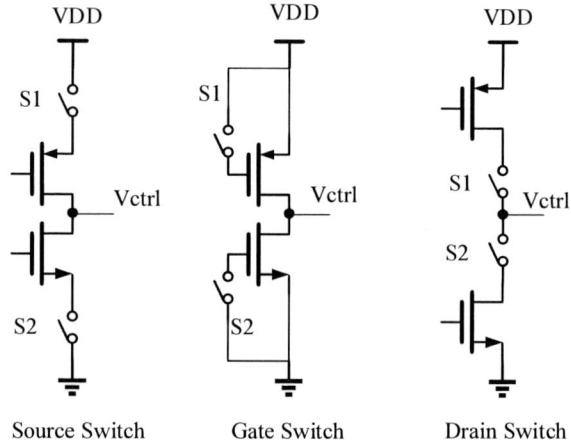

Source Switch   Gate Switch   Drain Switch

Fig. 4.   Single-ended CP structures.

The source-switched CP topology demonstrates superior performance in mitigating charge injection and clock feedthrough effects, while avoiding the speed degradation caused by the large time constant at the current source's gate node [9]. Therefore, this work optimizes the source-switched charge pump structure, with the proposed circuit architecture illustrated in Figure 5. This circuit eliminates the effect of charge injection on the CP through introducing a dummy switch transistor [10].

The charge injection effect occurs when the charge pump's switch transistor is turned on and operates in the linear region, forming a channel at the silicon dioxide-silicon interface. The total charge in the inversion layer is given by:

$$Q_{ch} = WLC_{ox}(V_{DD} - V_{in} - V_{TH}) \qquad (1)$$

where W and L represent the MOS transistor's gate width and effective channel length respectively, and C_{ox} denotes the gate oxide capacitance per unit area. When the switch turns off, assuming half of the charge flows from the drain terminal into the load capacitor C_L, the resulting fluctuation in control voltage can be expressed as:

$$\Delta V = \frac{WLC_{ox}(V_{DD} - V_{in} - V_{TH})}{2C_L} \tag{2}$$

When the charging and discharging action of CP causes charges to be injected into the control voltage node, the Dummy tube provides a symmetrical path to absorb these deposited charges. By counteracting the charge changes, the stability of control voltage can be improved.

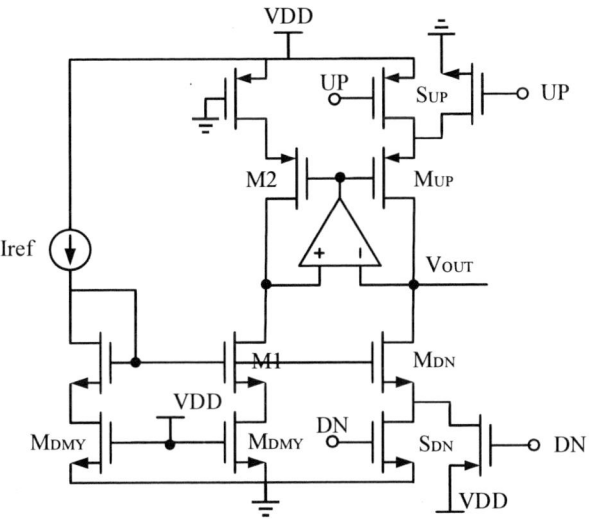

Fig. 5.　The block diagram of Charge Pump Circuit.

It is worth noting that we employ an optimized approach, using a complementary-type dummy transistor with its source connected to GND/VDD. This design not only maintains the dummy switch's high-speed operation but also enables fast potential pulling to supply rails during switch-off transitions. Consequently, it achieves three key improvements: accelerated current source turn-off, reduced reset pulse width requirement, and suppressed leakage current.

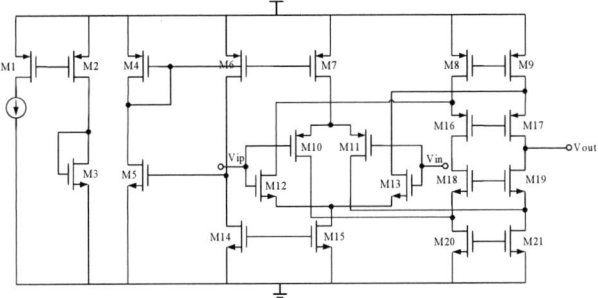

Fig. 6.　The block diagram of Rail-to-Rail Operational Amplifier.

To enhance charge pump matching performance, a clamping amplifier is incorporated to ensure current consistency across all branches. For maintaining optimal matching characteristics across a wide output voltage range, the amplifier must exhibit substantial common-mode input range. Consequently, a rail-to-rail input operational amplifier is

employed as the clamping amplifier. As shown in Figure 6, the amplifier features a PMOS and NMOS input stage to guarantee rail-to-rail operation.

IV. SIMULATION RESULTS

The phase detection characteristics of the PFD were simulated by applying 48 MHz clock signals with identical frequency but varying phase differences, sweeping the input phase difference across $[-2\pi, +2\pi]$. The average difference of output pulse signals was measured, with results shown in Figure 7. The simulation demonstrates a phase detection range of $[-1.95\pi, 1.95\pi]$ with no dead zone. These results confirm that the implemented PFD successfully achieves both wide detection range and dead-zone-free operation, realizing high-precision phase frequency detection across an extended range.

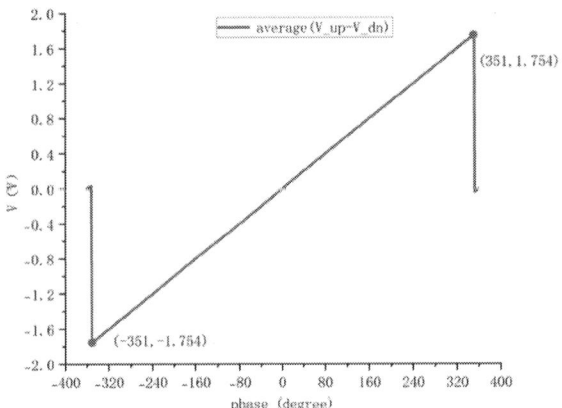

Fig. 7.　PFD Phase Detection Characteristic.

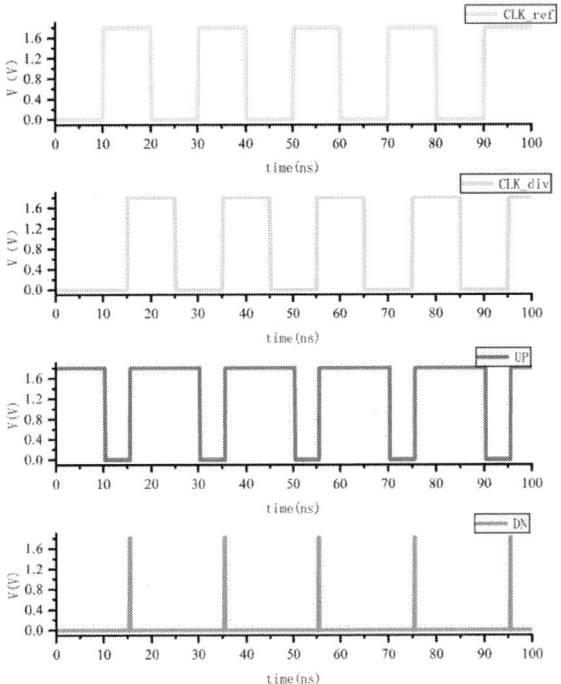

Fig. 8.　PFD Phase advance Transient Response.

Figure 8 presents additional transient simulation results when the CLK_ref leads in phase. With UP active-low and DN active-high, the results clearly show: (1) a positive phase difference between UP and DN signals, where the pulse width accurately corresponds to the phase difference; and (2) a 250ps narrow pulse in the DN signal that effectively prevents phase detection dead zones.

For characterizing the current mismatch properties of the CP circuit, the current switches were closed while a variable DC voltage source was connected to the output port. The output voltage was swept from 0 to 1.8V, with the resulting current matching characteristics shown in Figure 9. Specifically: Figure 9(a) presents the static mismatch curve of charge/discharge currents; Figures 9(b) and 9(c) display the charge/discharge current variations versus output voltage, respectively.

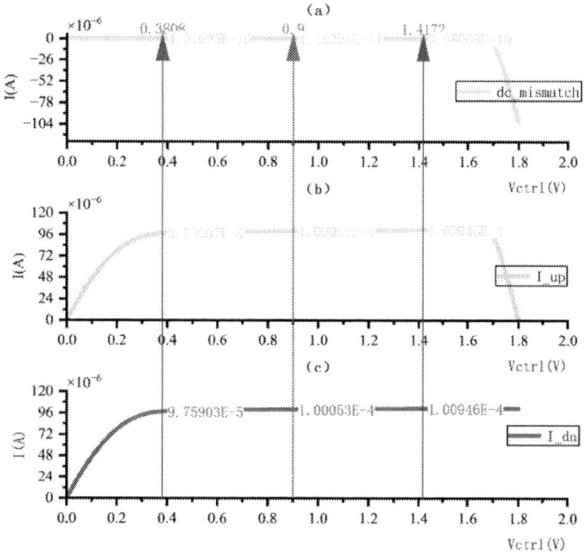

Fig. 9. Current Mismatch Characteristics.

The measurements demonstrate that:

1) The CP maintains 100μA charge/discharge currents within 0.4V-1.6V output range

2) Current error remains below 3% at low voltages, it can be correctable using cascode current mirrors

3) The mismatch between Charge and discharge is approximately 0.04%

These results verify excellent current matching characteristics across the operational voltage range. The 0.04% mismatch represents significant improvement over conventional designs while maintaining full functionality across the specified voltage window.

Building upon these results, Figure 10 illustrates the CP's charging process, which occurs when CLK_ref's rising edge leads CLK_div, triggering the PFD to generate an UP signal that turns on the charging-path switch transistor and raises the output voltage. The measurements show the output node

voltage increasing linearly from 0.94887V to 0.95867V within one clock cycle under 180-degree phase difference conditions.

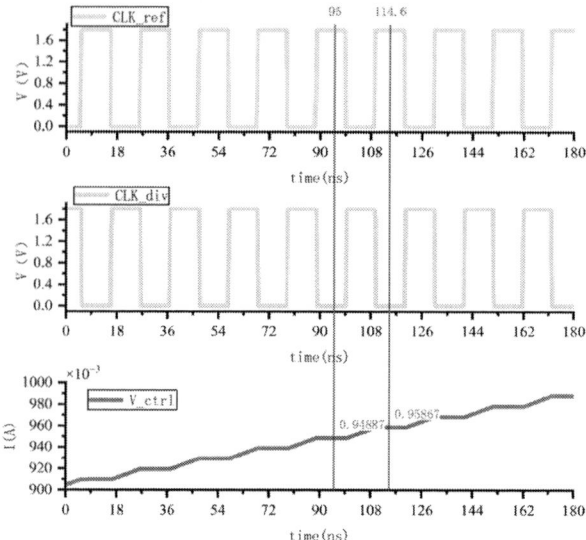

Fig. 10. Charging Process of CP.

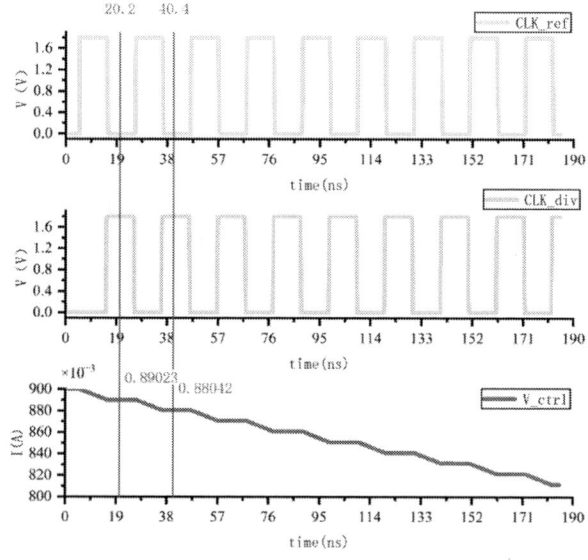

Fig. 11. Discharging Process of CP.

Figure 11 demonstrates the discharge process, initiated when CLK_ref's rising edge lags CLK_div, causing the PFD to produce a $\overline{DN}$ signal that activates the discharge-path switch transistor and lowers the output voltage. The results reveal the output voltage decreasing linearly from 0.89022V to 0.88173V during one clock cycle at 180-degree phase difference. The near-linear voltage variations during both charge and discharge operations confirm excellent transient characteristics of the implemented PFD+CP circuit.

In CPPLL, the current noise from both the PFD and CP constitutes a critical factor affecting system performance. This noise injects into the loop filter's control voltage through the

979-8-3315-8850-2/25 $31.00 © 2025 IEEE

charge pump, subsequently modulating the VCO's output frequency and ultimately manifesting as phase noise and reference spurs in the PLL output. As shown in Figure 12, the designed circuit exhibits current noise levels of -230 dBc/Hz at 1 kHz and -252 dBc/Hz at 1 MHz through simulation. These results indicate the circuit's contribution to phase noise in the implemented MEMS clock system is negligible. Furthermore, comprehensive simulations of the fractional-N CPPLL system demonstrate reference spurs of -65.5 dBc at a division ratio of 22.5, meeting all system design requirements.

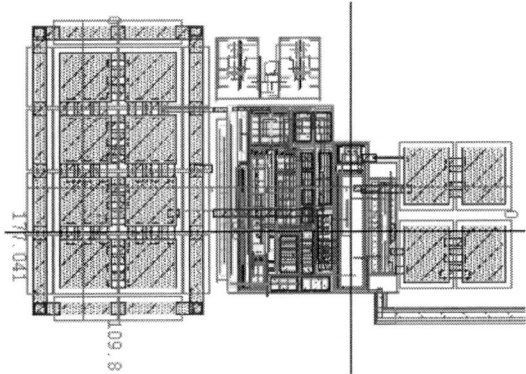

Fig. 13. The layout of Proposed PFD and CP

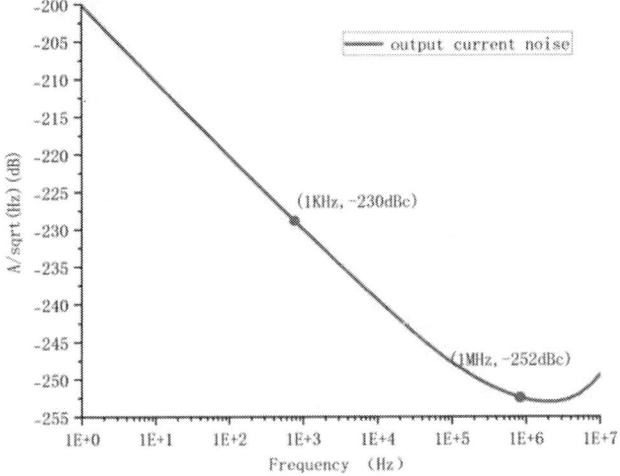

Fig. 12. Current Noise Characteristics of PFD and CP.

TABLE I. PERFORMANCE SUMMARY OF PFD AND CP

	This work	[11]	[12]
Technology	180nm CMOS	65nm CMOS	22nm CMOS
Reference frequency	48MHz	706.25MHz	-
Division ratio	21-23	16	-
Phase detection range	$[-1.95\pi, 1.95\pi]$	$[-1.67\pi, 1.67\pi]$	$[-1.97\pi, 1.97\pi]$
Charging/ discharging current	100uA	177uA	100uA
Mismatch accuracy	0.04%	0.047%	<0.1%
Current noise	-230dBc@1KHz	-	-
Reference spurious	-65.5dBc	-	-

Table I summarises the performance of this design and compares it with previously reported PFDs and CPs. This PFD has the advantage of Phase detection range and CP has the advantage of low mismatch.

V. CONCLUSION

This work presents a PFD and CP circuit implemented in 180nm CMOS technology for MEMS clock system applications. The layout is shown in Figure 13, with a chip area of 0.019mm^2 and a dynamic power consumption of 0.418mW when PLL system stability.

The design achieves a high-precision phase frequency detector with wide detection range through enhanced load-driving capability of the TSPC flip-flops in the PFD. By employing a source-switched architecture with rail-to-rail clamping amplifiers and fast turn-off paths, the charge pump effectively suppresses non-ideal effects including charge sharing, current mismatch, and leakage current, realizing high-speed operation with minimal mismatch. The implemented design demonstrates significant improvements in circuit performance, notably reducing system phase noise and reference spurs, making it particularly suitable for high-precision clocking applications.

ACKNOWLEDGMENT

This work was supported by the National Natural Science Foundation of China under Grant 62234012.

REFERENCES

[1] S. Huang, S. Liu and Z. Zhu, "A High-Resolution 2-GHz Fractional-N PLL With Crystal Oscillator PVT-Insensitive Feedback Control," in IEEE Microwave and Wireless Components Letters, vol. 28, no. 3, pp. 227-229, March 2018.

[2] M. Papamichail, D. Karadimas, K. Efstathiou and G. Papadopoulos, "Linear range extension of a phase-frequency-detector with saturated output," 2006 IEEE International Symposium on Circuits and Systems, Kos, Greece, 2006, pp. 4 pp.-1674.

[3] W. Rhee, "Design of high-performance CMOS charge pumps in phase locked loops," IEEE ISCAS-II, vol. 2, pp. 545-548, Jun, 1999.

[4] Saurabh K ,Kumar Y S, et al. A low-jitter and low-phase noise switched-loop filter PLL using fast phase-error correction and dual-edge phase comparison technique[J].Integration,2024,94.

[5] K. Kim, K. Kim and C. Yoo. A fREF/5 Bandwidth Type-II Charge-Pump Phase-Locked Loop With Dual-Edge Phase Comparison and Sampling Loop Filter[J].IEEE Microwave and Wireless Components Letters, 2018,28(9):825-827.

[6] D. Turker, A. Bekele, P. Upadhyaya, B. Verbruggen, Y. Cao, S. Ma, et al. A 7.4-to-14GHz PLL with 54fsrms jitter in 16nm FinFET for integrated RF-data-converter SoCs.in: 2018 IEEE International Solid - State Circuits Conference - (ISSCC), San Francisco,CA, USA, 11-15 Feb. 2018, IEEE, 2018: 378–380.

[7] Liu L, Ji Y, Liao X, et al. A 0.8-V, 2.55-GHz, 2.62-mW charge-pump PLL with high spectrum purity[J]. IEEE Transactions on Very Large Scale Integration (VLSI) Systems,2022,30(2):113-122.

[8] Ding Q ,Xiangjian K ,Mingchao J , et al.A 1-V 9.6-GHz charge-pump PLL with low RMS-integrated jitter[J].Microelectronics Journal,2023,142.

[9] J. Lan, Z. Gao, Y. Wang, L. Liu, et al. Improve the dynamic matching of the source-switching charge pump for high-performance phase-locked

loops[C]// 2010 10th IEEE International Conference on Solid-State and Integrated Circuit Technology, 2010, pp.448-450.

[10] X. Zhang and H. Yu, "Design and HSPICE Simulation of High-Performance Charge Pump Circuits," Microelectronics, vol. 40, no. 3, pp. 317-320, 2010. (in Chinese).

[11] Y. Zhan, R. Li, L. Zhang, X. Zou, Q. Hu and Q. Wang, "A High-Speed Low-current-Mismatch PFD+CP Circuit for 11.3G CPPLL Application,"

2024 9th International Conference on Integrated Circuits and Microsystems (ICICM), Wuhan, China, 2024, pp. 307-311.

[12] Z. Tian, Z. Li, W. Lin, X. Wang and Z. Li, "Design of Nonlinear Phase Frequency Detector and Low Mismatch Charge Pump," 2023 8th International Conference on Integrated Circuits and Microsystems (ICICM), Nanjing, China, 2023, pp. 507-510.

2025 The 10th International Conference on Integrated Circuits and Microsystems

High-Reliability Integration Design Method for Micro Systems SiP in Complex Spatial Environments

Shang Jiang
Laboratory of Solid-State Optoelectronic Information Technology, Institute of Semiconductors, Chinese Academy of Sciences; School of Integrated Circuits, University of Chinese Academy of Sciences; University of Chinese Academy of Sciences
Beijing, China
jiangshang_shang@163.com

Wenchang Li
Laboratory of Solid-State Optoelectronic Information Technology, Institute of Semiconductors, Chinese Academy of Sciences; School of Integrated Circuits, University of Chinese Academy of Sciences; University of Chinese Academy of Sciences
Beijing, China
liwc@semi.ac.cn

Zucheng Gu
Beijing Xuanyu Space Technology Co., Ltd;
Beijing, China
guzucheng1992@163.com

Tianyi Zhang
Laboratory of Solid-State Optoelectronic Information Technology, Institute of Semiconductors, Chinese Academy of Sciences
Beijing, China
zhangtianyi@semi.ac.cn

Abstract—To address the challenges of system integration under extreme thermal and micro-vibration conditions in the space environment, this paper proposes a high-reliability microsystem integration design methodology tailored for complex space applications. The core approach involves utilizing multiple simulation tools for multi-physics analysis—mechanical, thermal, and electrical—and reliability-driven optimization to establish a comprehensive design framework for System-in-Package (SiP) modules. Additionally, a multi-faceted, multi-chamber heterogeneous integration structure based on high-temperature co-fired ceramics (HTCC) is employed to achieve heterogeneous integration of chips fabricated with different process technologies. This design significantly reduces the microsystem's weight and size while enabling system reliability assessment through simulation-driven process optimization. Currently, the developed product has been successfully implemented in mass production, validating the engineering practicality of the simulation methodology and the integrated design optimization process.

Keywords—Microsystem, SiP, High-Temperature Co-Fired Ceramics, Heterogeneous Integration, High Reliability

I. INTRODUCTION

Electronic systems for space application environments face stringent requirements of "miniaturization, lightweight, high reliability, and long lifespan". Electronic systems composed of traditional discrete components suffer from issues such as large volume, complex interconnections, and high failure risk. Statistics show that the failure rate of discrete systems is 3 to 5 times that of integrated systems. SiP technology, through multi-chip heterogeneous integration, enables multi-functional integration within a limited space, making it a key technology for solving high-reliability microsystem challenges [1-2].

However, the uniqueness of the space environment poses special challenges to SiP integration. Space radiation can cause Single-Event Upsets (SEU) in SRAM-based FPGAs [3]; the wide temperature range of -55°C to 150°C may lead to thermal stress failure of materials[4]; and micro-vibrations can result in reliability issues such as bonding wire fatigue and fracture[5-6]. Therefore, researching highly reliable SiP integration design

methods adapted to extreme environments holds significant engineering value. Based on the application characteristics of microsystems, this paper focuses on the research of high-reliability SiP module design processes, structural innovations, and reliability design technologies, develops a series of integrated modules, and verifies the effectiveness of the design method through the full process.

II. MODULE DESIGN FLOW

The design and fabrication of the micro-system System-in-Package (SiP) modules must adhere to stringent reliability standards, with each development phase validated through comprehensive verification checkpoints to ensure quality assurance, as depicted in Figure 1.

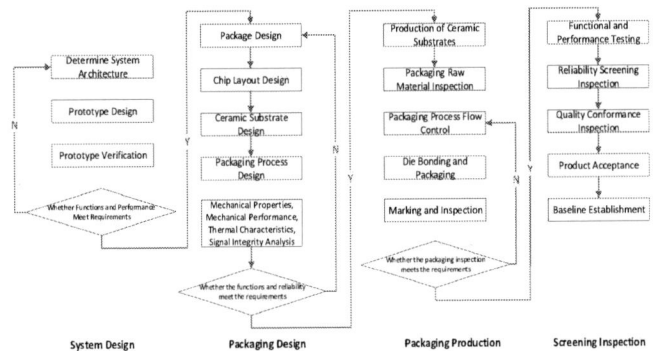

Fig. 1.Design and Production Flow of Microsystem SiP Module

During the system architecture development stage, functional coverage metrics and reliability indices are fundamental. The system architecture is established through principle-based modeling and board-level validation. Based on this architecture, the packaging design encompasses multilayer wiring on ceramic substrates, integrated circuit layout, and process engineering, with a focus on analyzing power integrity, signal integrity, thermal management, and mechanical robustness. Simulation tools are employed to validate whether the design satisfies functional specifications and reliability standards.

979-8-3315-8850-2/25 $31.00 © 2025 IEEE

In the manufacturing phase, large-scale production of ceramic substrates is executed, utilizing high-precision die attach and wire bonding techniques to ensure chip interconnection integrity. Quality assurance is performed through inspection protocols, and hermetic sealing tests are conducted post-encapsulation.

During the testing and qualification phase, environmental stress screening—including mechanical, thermal, and life cycle testing — is performed in accordance with reliability criteria. Coupled with functional and performance assessments, the production baseline is established, and failure analysis is conducted on defective units.

III. SiP MICROSYSTEM INTEGRATION DESIGN

A. Analog-Digital Hybrid Heterogeneous Integration Architecture Design

The microsystem architecture for analog-digital hybrid heterogeneous integration employs a System-in-Package (SiP) approach, incorporating a programmable signal processing module comprising high-speed parallel analog-to-digital converters (ADCs) and low-speed serial ADCs. These components facilitate the conversion of analog signals into digital data for subsequent computational processing by the central processing unit. The core processing element, implemented via Field-Programmable Gate Array (FPGA), manages timing signal generation, closed-loop algorithm implementation, digital data computation, and interfaces with the digital-to-analog conversion (DAC) subsystem. High-speed parallel DACs translate processed digital signals into analog outputs for external interfacing. A dedicated refresh module ensures fault tolerance by real-time reinitialization of the FPGA in response to Single Event Upsets (SEUs), thereby maintaining system resilience. System integration is realized through SiP technology, leveraging an FPGA combined with ADC, DAC, and refresh circuitry within an analog-digital hybrid architecture. This high level of integration facilitates device miniaturization. The schematic diagram illustrating the microsystem architecture is presented in Figure 2.

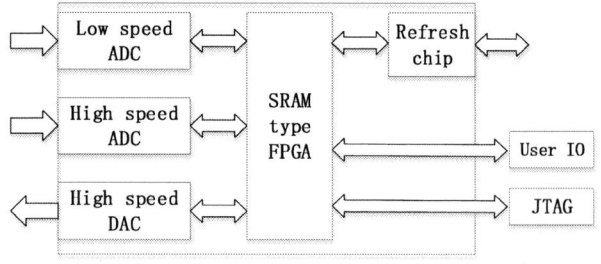

Fig. 2.Electrical Schematic Diagram of Microsystem Architecture

To further optimize the reliability and performance of high-availability microsystems, an advanced design has been developed based on programmable digital signal processing modules. The System-in-Package (SiP) module, integrated with a dedicated Application-Specific Integrated Circuit (ASIC), converts the interface FPGA logic algorithms of the onboard computing unit and peripheral devices into a specialized chip. This ASIC facilitates multiple logical interfaces, including asynchronous and synchronous serial communication ports, Controller Area Network (CAN), Serial Peripheral Interface (SPI), multi-channel Analog-to-Digital (AD) automatic acquisition, General-Purpose Input/Output (GPIO), and Alternating Current (AC) logic. This approach reduces FPGA programming complexity and cost while delivering enhanced performance and improved economic efficiency. The schematic diagram of the revised microsystem architecture is depicted in Figure 3.

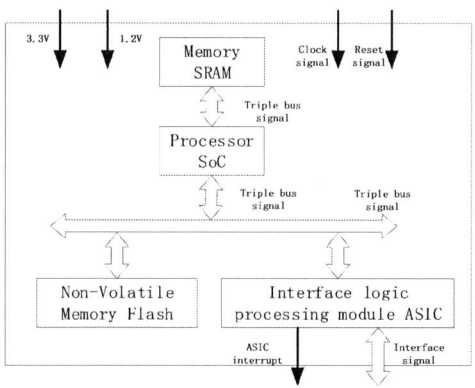

Fig. 3.Electrical Schematic Diagram of Microsystem Architecture

B. Design of Double-Sided Multi-Cavity Package Configuration

Microsystem module integration typically requires consolidating multiple functional semiconductor devices — including processing units, control circuits, interface components, and memory modules—into a unified assembly. Heterogeneous multi-chip integration is a critical process for achieving device miniaturization and enhanced reliability. However, challenges arise due to disparities in chip dimensions, substrate materials, wire bonding configurations, and thermal management properties. Larger device footprints increase signal trace lengths, thereby diminishing interconnect efficiency and elevating the risk of signal integrity degradation. In hybrid analog-digital microsystem architectures, proximity of digital and analog ICs can compromise signal fidelity. Additionally, controlling package planarity remains difficult, impacting structural stability and reducing overall packaging robustness.

This paper introduces a multi-faceted heterogeneous integration architecture based on High-Temperature Co-fired Ceramic (HTCC) technology. Tailored to diverse application specifications, various structural configurations have been developed, including microsystem module integration techniques and architectures such as semi-hermetic double-sided wire bonding combined with flip-chip hybrid processes for Ceramic Column Grid Array (CCGA), double-sided hermetic Ceramic Quad Flat Package (CQFP), hybrid integration of active semiconductor devices with discrete passive components, and bare die three-dimensional stacked packaging. These configurations satisfy stringent reliability standards and achieve over 80% reduction in size and weight of the chip package relative to conventional designs.

1) Double-Sided Hermetic CQFP Structure

A dual-sided hermetic CQFP configuration employing parallel seam welding and alloy sealing is implemented. One side of the ceramic package creates a cavity through the ceramic substrate, while the opposite side features a dual-cavity, double-sided architecture with a Kovar ring secured via welding. This design markedly reduces the overall package volume, enhances reliability, and improves integration density. Additionally, it minimizes wiring lengths and facilitates the placement of digital and analog components in separate cavities, effectively mitigating digital noise and enhancing signal integrity. The schematic of the package structure is illustrated in Figure 4.

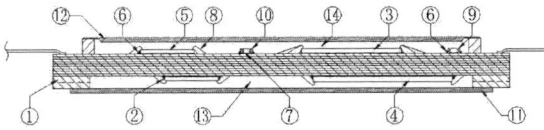

①Shell ②High temperature conductive adhesive ③④⑤Chip ⑥Low temperature conductive adhesive ⑦Insulating adhesive ⑧Gold wire ⑨⑩Capacitor ⑪Alloy cover plate ⑫Flat seam cover plate ⑬Cavity 1 ⑭Cavity 2

Fig. 4.Cross-Sectional View of the Double-Sided Hermetic CQFP Structure

2) CCGA Structure with Double-Sided Wire Bonding and Flip-Chip Hybrid Process

A microsystem structure combining wire bonding and flip-chip, with one side hermetic and the other non-hermetic, is adopted. Wire bonding and flip-chip are placed in separate cavities, which can effectively avoid the increase in packaging process procedures and insufficient reliability of bonded solder joints caused by the mixed assembly of chips using the two processes. In addition, this structure can effectively solve the problem of difficult unified heat dissipation design in the mixed assembly of WB (Wire Bonding) and FC (Flip-Chip) chips. High-power FC chips are placed in separate cavities; apart from dissipating heat through the package, the chips can also transfer heat directly to the heat-dissipating metal cover by contacting the top with the heat sink. This increases the heat dissipation path of the microsystem module and reduces the packaging thermal resistance, all of which are conducive to efficient heat dissipation of the microsystem module.The form of its packaging structure is shown in Figure 5.

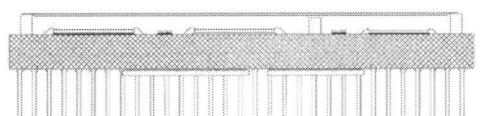

Fig. 5.Cross-Sectional View of the CCGA Structure with Double-Sided Wire Bonding and Flip-Chip Hybrid Process

3) Bare Die 3D Stacked Packaging Structure

The method of bare die 3D stacking wire bonding is a critical aspect of reliability control in three-dimensional system-in-package (SiP) processes, particularly within confined area and spatial constraints. This bonding technique primarily involves Z-axis three-dimensional bonding technologys[7]. The key quality parameters for stacked chip interconnects include bonding strength, deformation, and curvature, which are influenced by factors such as bonding blade, power, temperature, and duration. The application of bare die 3D stacking packaging structures significantly reduces the overall size and weight of microelectronic modules.

IV.   RELIABILITY ANALYSIS OF MICROSYSTEM PACKAGING

The assessment of thermal, electrical, and mechanical reliability for high-reliability components remains a key focus within the industry. Finite Element Analysis (FEA) provides benefits including ease of use and precise computational results. To validate the performance and durability of the engineered microsystem System-in-Package (SiP), an extensive simulation study is performed, encompassing power integrity, signal integrity, electromagnetic compatibility, and thermomechanical reliability at the system-level packaging.

A.   Power Integrity Analysis

The micro-system imposes specific requirements on power supply voltage, IO signal level standards, and power consumption during operation. Therefore, during the design phase, power integrity analysis must be conducted considering voltage drop, current density, and power plane impedance. The voltage and current specifications for each chip within the SiP micro-system are detailed in Table 1.

Table 1: Internal Chip Voltage and Current Parameters of the SiP

Chip	Power Network	Power Supply Voltage (V)	Power Supply Current(A)
DSP	+1.2V	1.2±5%	2.00
	+3.3V	3.3±5%	0.35
FPGA	VCCINT	1±5%	4.36
	VCCBRAM	1±5%	0.10
	VCCAUX	1.8±5%	0.20
	+3.3V	3.3±5%	0.02
Flash	+3.3V	3.3±5%	0.08

1) Power Supply Voltage Drop Analysis

The objective of power supply voltage drop analysis is to detect significant voltage drops within the power distribution network. In DC analysis scenarios, the impedance (Z) corresponds to the physical interpretation of IR voltage drop in the power loop, representing purely resistive impedance. When assessing whether the voltage drop meets design specifications, the criterion is that the power supply ripple amplitude must be controlled within 5%. Figure 6(a) illustrates the DC voltage drop distribution for the VCC_1.2 power rail, with a maximum voltage drop of 5 mV. Figure 6(b) shows the DC voltage drop distribution for the VCC 3.3 power rail, with a maximum value of 6 mV. The DC voltage drops for both VCC_1.2 and VCC 3.3 power supplies are below the 5% ripple amplitude threshold.

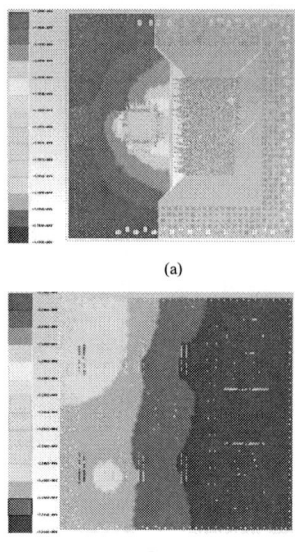

(a)

(b)

Fig. 6.Distribution of Power Supply DC Voltage Drop (a) VCC_1.2; (b) VCC_3.3

2) Current Density Analysis

The objective of the current density analysis is to identify regions within the PCB layout exhibiting high current density, which could lead to significant thermal rise and potential warping of the component package. According to IPC-2221A standards[8], the reference threshold is 1.01E+8A/m². Figure 7(a) illustrates the current density distribution for the VCC_1.2V power rail, with a maximum current density of 5.904×10^7 A/m². Figure 7(b) depicts the current density distribution for the VCC_3.3V power rail, with a maximum current density of 5.904E+07A/m². Both power rails exhibit maximum current densities below the specified limit of 1.01E+08A/m².

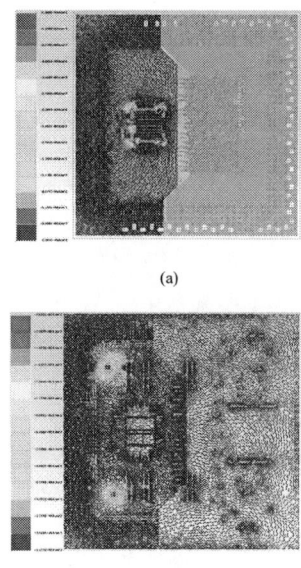

(a)

(b)

Figure 7. Current density distribution. (a) VCC_1.2; (b) VCC_3.3

3) Impedance analysis in communication systems

In the design of Power Delivery Network (PDN) systems, to suppress the amplitude of power supply noise, impedance analysis and optimization of the power network are essential to ensure that the impedance remains below the target impedance within a specified frequency band[9]. In PDN filter band allocation, typically, the VRM filtering frequency is below 100 kHz, chip decoupling capacitance filtering occurs above 100 MHz, and board-level and package-level filtering spans from 100 kHz to 100 MHz. The simulation optimization targets an impedance frequency range of 100 kHz to 100 MHz. Figure 8(a) illustrates that, within the 100 MHz power supply frequency for DSP-U12, the maximum target impedance is 0.03 ohms, which is below the calculated value of 0.06 ohms, satisfying the target impedance criteria. Figure 8(b) shows that, within the 100 MHz VCCAUX power supply frequency for FPGA-U12, the maximum target impedance is 0.053 ohms, less than the calculated value of 0.9 ohms, meeting the impedance target requirements.

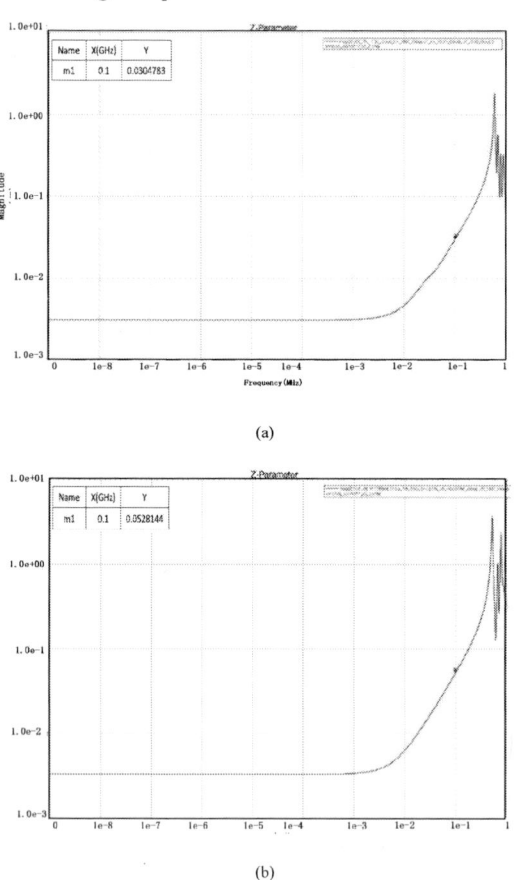

(a)

(b)

Figure 8.Power supply target impedance profile (a) +1.2V power supply (b) VCCAUX power supply

B. Signal Integrity Analysis

Based on the requirements of different signals, a comprehensive signal integrity analysis is conducted from three aspects: waveform simulation, S-parameter simulation, and crosstalk simulation. The primary purpose of waveform simulation is to compare the displayed waveforms with the chip manual to verify the topology's rationality and ensure that overshoot and logic level transitions remain within acceptable limits. The simulation of insertion loss (S21) and return loss (S11) aims to assess the attenuation characteristics of the transmission line, as signal propagation through the medium incurs losses that can cause waveform distortion, affecting edge timing and voltage levels. The reference standards are S21>-0.5dB/mm and S11< -15 dB. Crosstalk is evaluated using a 3% amplitude threshold as a benchmark; crosstalk simulations are used to determine compliance with requirements and identify potential improvements, thereby controlling interference within acceptable ranges and enhancing the signal line's electromagnetic compatibility and noise immunity.

1) Waveform Simulation

Waveform simulation of the clock signal is required, with stringent signal integrity criteria. Figure 9 illustrates the simulation analysis of an external clock input signal CLK with a maximum frequency of 50 MHz. Figure 9(a) depicts the routed topology of the CLK signal, indicating a point-to-point configuration. Figure 9(b) presents the simulated waveform of the CLK signal, characterized by a short rise time, good rectangularity, and absence of significant overshoot, undershoot, or ringing effects. Additionally, there is no evident flat top or non-monotonic behavior during the rising edge, indicating minimal electromagnetic interference affecting the clock signal.

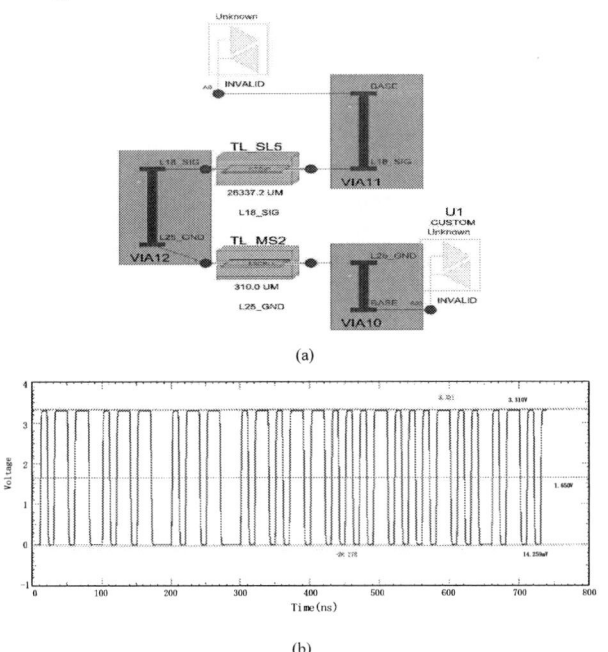

(a)

(b)

Figure 9.Waveform simulation schematic (a) Point-to-point topology (b) Analog waveform of the CLK signal

2) S-parameter simulation

Signal attenuation occurs during propagation through the medium, resulting in waveform edge timing distortion and voltage level degradation. To quantify the transmission line loss, this study employs S-parameter simulation analysis, establishing the relationship between incident and reflected signals at each frequency point in the frequency domain. In the signal integrity analysis, the most complex segment of the structure is selected for simulation validation, with results compared against standard parameters. Figure 10(a) illustrates the schematic of the DSP_JTAG data line, which operates at a maximum data rate of 50 MHz with a characteristic impedance of 50 Ω. The focus is on characterizing the return loss of the DSP_JTAG signal. Figure 10(b) shows that the magnitude of the complex reflection coefficient varies minimally with frequency; at the core operating frequency of 50 MHz, the return loss reaches -32.43 dB, meeting the specified standard requirements. The analysis indicates that the signal distortion caused by reflection effects is negligible, and the impedance continuity along the signal path is excellent. Figure10(c) presents the insertion loss performance of the DSP_JTAG signal. When the transmission length is 15.7 mm, the insertion loss is measured at -0.1724 dB. Based on industry standards (permissible limit of -0.5 dB/mm), this measured value is well below the threshold, further confirming that the transmission performance of the signal path aligns with design expectations.

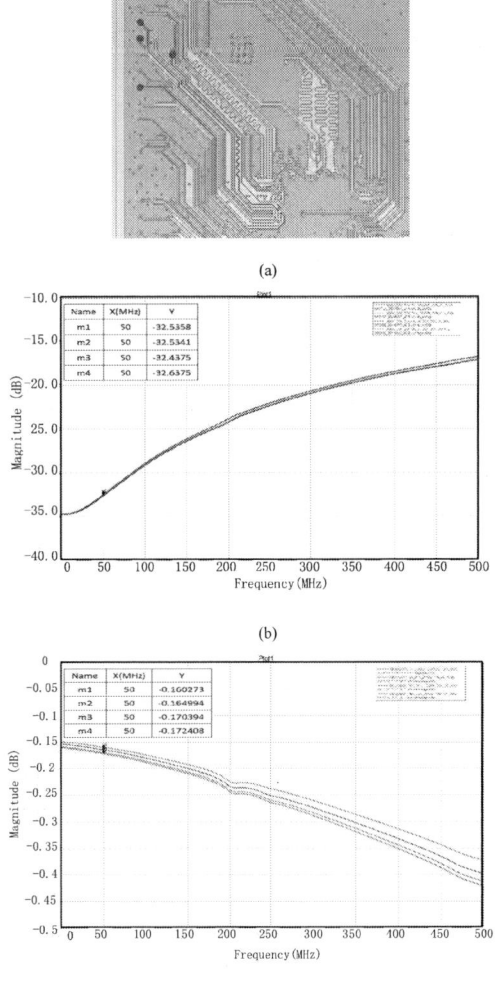

(a)

(b)

(c)

Figure 10: S-parameters simulation (a) Schematic diagram of JTAG data line signals (b) Return loss (c) Insertion loss

3) Crosstalk simulation

To perform crosstalk simulation, three closely spaced data signals—N2(pink), N1(red), andN0(blue) were selected. Figure 11(a) presents the schematic diagram of the N2/N1/N0 signals, with N1 (red) designated as the disturbed signal and N2 (pink) and N0 (blue) as interference signals. Figures 11(b) and 11(c) display the simulation results, illustrating the relationship between crosstalk noise from N2 and N0 and their impact on the N1 signal across different frequencies. In the near-end crosstalk representation shown in Figure 11(b), the crosstalk noise induced by N2 and N0 signals at 3.125 GHz exhibits minimal variation, with a measured value of-37.29 dBdB (< -30 dB). In the far-end crosstalk scenario depicted in Figure 11(c), at 3.125GHz, the crosstalk noise induced by the N2 signal measures -38.146 dB (< -30 dB), while the noise induced by the N0 signal measures -42.7033 dB.

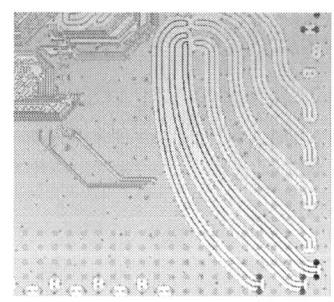

(a)

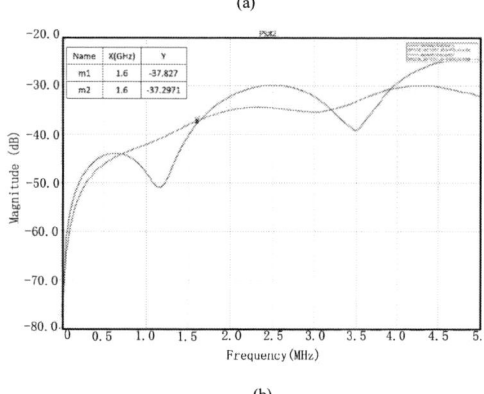

(b)

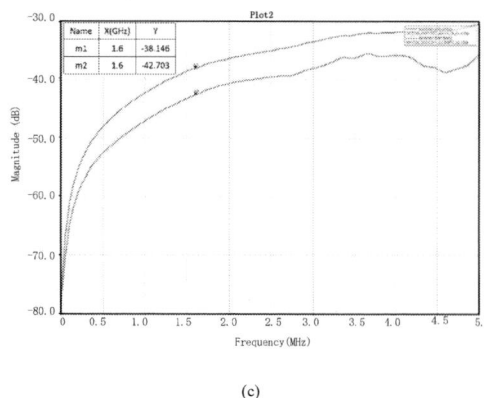

(c)

Figure 11: Crosstalk Simulation Diagrams (a) Schematic of the Signal (b) Near-End Crosstalk (c) Far-End Crosstalk

C. Electromagnetic Compatibility Analysis for Analog-Digital Hybrid Systems

In the design process of the SiP housing, the characteristics of the digital and analog circuits are fully considered, and the power supply and ground signals of the digital part are isolated during the netlist design.

In the layout design process, based on the previous design experience of analog-digital products, the mutual interference between the two (digital and analog circuits) is fully taken into account, and physical isolation of the power supply and ground networks as well as signal shielding design are properly implemented.In addition, the electromagnetic compatibility (EMC) characteristics of the microsystem chip are designed, analyzed, and simulated. Special sensitive signals that have a significant impact on performance are identified in advance, the expected isolation effect of various analog signals is accurately evaluated, and the electromagnetic integrity of the signals is further optimized based on actual simulation results.

Simulation analyses are conducted on the signal integrity of complex internal signals of the module, power transmission performance, and crosstalk between digital and analog signals. The substrate design is optimized according to the analysis results.

D. Thermal Reliability Analysis

System-level package thermal analysis, referencing the JESD51-14 standard, utilizing JC, JB, and JA simulation analyses to evaluate the heat dissipation capability of the package shell, and providing relevant thermal management recommendations based on simulation results.

1) Thermal analysis

The model developed primarily considers the temperature differential between the micro-system enclosure and internal active heat-generating components, serving as a metric for the device-level thermal performance of the micro-system. The following diagram illustrates the thermal simulation of the system-level package component. In addition to the ambient temperature of 125 ℃, simulations can be conducted under actual operating conditions at +25 ℃, +125 ℃, and various component power dissipation scenarios to analyze temperature rise and distribution, thereby informing thermal design strategies for the components.

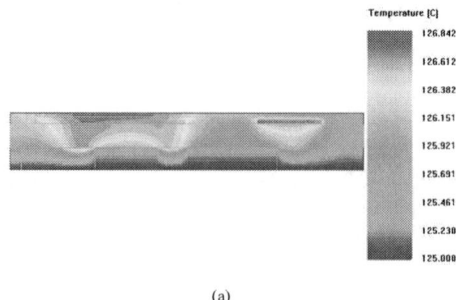

(a)

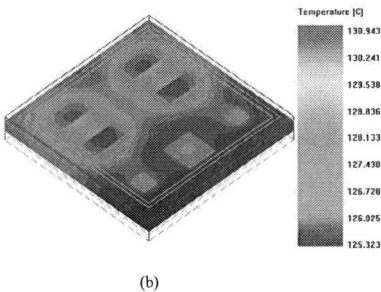

(b)

Fig. 12.Thermal simulation at an ambient temperature of 125℃ (a) Enclosure temperature distribution (b) Junction temperature distribution

Table 2 presents the thermal simulation results of the JC model under various ambient temperature conditions. Analysis of the simulation data indicates that, with increasing temperature, the temperature rise of the SiP bare die relative to the environment remains constant, demonstrating the SiP chip's effective thermal management and heat dissipation performance.

Table 2: Thermal Simulation Results of Chips under the JC Model

Simulation environment	JC		
Ambient temperature	Maximum temperature	Temperature rise	Ambient temperature
-55℃	49.0568℃	5.943℃	-55℃
+25℃	30.943℃	5.943℃	+25℃

2) Thermal stress analysis

When the temperature of the microsystem varies, its volume tends to change. Due to external constraints and mutual restrictions among its components, this volumetric change cannot occur freely, resulting in thermal stress. Excessive thermal stress can cause cracking between packaging components, leading to device failure [10]. Therefore, thermal stress analysis under different environmental temperatures must be conducted during the design phase. To simulate the device's operating environment, finite element analysis (FEA) is performed for the system-in-package (SiP) at ambient temperatures of -55℃, +25℃, and +125℃. The following presents the simulation results of the thermal stress distribution under the most severe environmental temperature of 125℃. At 125℃, the maximum thermal stress in the SiP device occurs at the alloy cover plate, with a value of 102.36 MPa, which is below the yield strength of 345MPa. The maximum deformation within the SiP package appears at the edges of the ceramic casing, with a maximum displacement of 35.694 μm. Table 3 summarizes the maximum thermal stress and deformation at different temperatures, both of which meet the allowable stress criteria.

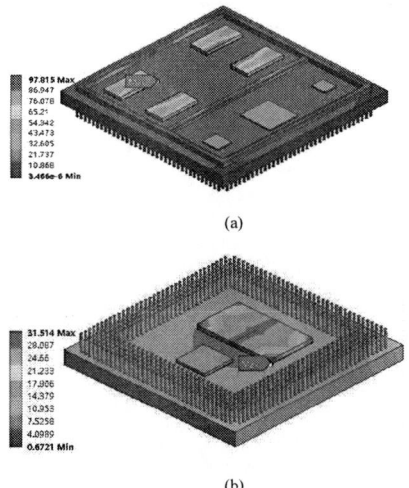

(a)

(b)

Figure 13: Environmental temperature set at 125 ℃ simulation (a) thermal stress; (b) thermal strain

Table 3: List of Thermal Stress Analysis

Environmental temperature/℃	maximum thermal stress/MPa	maximum deformation/um
-55	80.653	25.178
+25	3.142	0.98
+125	97.815	31.514

3) Structural Stress Analysis

To validate the structural integrity of the enclosure design, a finite element stress analysis was conducted to examine the stress-strain distribution and stress concentration within the housing. The integration of the System-in-Package (SiP) was achieved using a flip-chip microassembly with one side sealed and the other unsealed, employing a Chip-Carrier-Grid Array (CCGA) for pin out configuration. The chip interconnects utilized pillar bump technology with Pb90Sn10 solder, and the assembly was bonded using Sn63Pb37 solder. The external appearance of the device package is shown in Figure 5. Material parameters for the structural components are listed in Table 4. Considering the mechanical properties of the packaging materials and external loading conditions, the stress and deformation regions of the enclosure were analyzed and compared against the yield strength of the enclosure materials to assess compliance with operational requirements.

Table 4: Structural Material Specifications

Materials	Component	Conductivity（W/m-K）	CTE（1/℃）	Elastic modulus(GPa)	Poisson'sRatio
PCB	FR4	0.3	15	18.2	0.25
Solder	Sn63Pb37	55	26.1	41.33	0.365
Bump	Pb90Sn10	40	28.2	10.32	0.28
Ceramic Substrate	Al2O3	25	6.8	24.1	0.25

Materials	Component	Conductivity (W/m-K)	CTE (1/°C)	Elastic modulus(GPa)	Poisson'sRatio
Frame	Kovar	17	5.3	138	0.317
Chip	Si	150	2.6	130	0.3
Thermal Interface Material	Epoxy resin	1.8	22	12.5	0.3
Underfill Epoxy	Epoxy resin	0.7	29	11	0.32

Structural mechanics simulation primarily encompasses four aspects: constant acceleration, random vibration, mechanical shock, and low-frequency vibration. During constant acceleration simulation, the GJB548C 2001 method is employed to analyze the effects of steady acceleration. Conditions include the Y1 and Y2 directions, with accelerations of 3000G and 5000G. Results indicate that when applying constant accelerations of 3000G and 5000G in the Y1 direction, the system stress remains below the silicon's yield deformation strength of 100 MPa; similarly, in the Y2 direction, the system stress is also below 100 MPa under the same acceleration conditions. The statistical analysis of the constant acceleration simulation is summarized in the table 5.

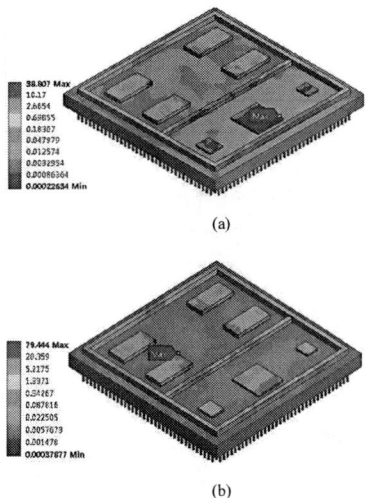

(a)

(b)

Figure 14: Stress cloud diagrams under Y1-axis acceleration (a) 3000g (b) 5000g

Table 5: Statistical Analysis of Constant Acceleration Simulation

Direction	3000G Stress	3000G Deformation	5000G Stress	5000G Deformation
Y1	38.807MPa	18.97um	64.679MPa	31.62um
Y2	47.67MPa	6.55um	79.44MPa	10.92um

During random vibration simulation, a stochastic vibration model is established with the SiP (System-in-Package) chip mounted on the PCB (Printed Circuit Board) substrate. The analysis is conducted along the X, Y, and Z axes. The SiP chip is assembled on a PCB substrate measuring 7.9 cm by 7.9 cm. Table 6 details the conditions for the random vibration testing.

Table 6: Conditions for Random Vibration Testing

Random Vibration		
Power Spectral Density	Vibration Duration	Vibration Axes
0.04(g)²/Hz	3min	X、Y、Z

The simulation analysis results are presented in Figure 15 and Table 7. When subjected to stochastic vibrations with a power spectral density of 0.04(g)²/Hz in the X, Y, and Z directions, the system's stress levels remain below the yield deformation strength of 350 MPa and the tensile strength of the CCGA, which is 44 MPa.

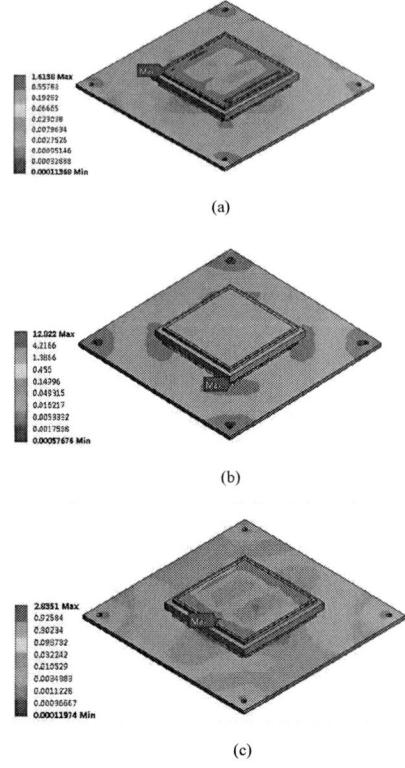

(a)

(b)

(c)

Figure 15: Stress cloud diagram under a power spectral density of 0.04(g)^2/Hz (a) X-axis, (b) Y-axis, (c) Z-axis

Table 7: Summary of Random Vibration Data

direction	0.04(g)²/Hz Stress value
X-axis direction	1.61MPa
Y-axis direction	12.8MPa
Z-axis direction	2.84Pa

During the mechanical impact simulation, the analysis conditions were set at 1500g and 3000g respectively. The modeling fixed the upper surface of the device enclosure to replicate the scenario where the system-level package specimen is in close contact with the test fixture during testing. The simulation results, as shown in Figure 16 and Table 8, indicate that under mechanical shocks of 1500g and 3000g in the X1/X2, Y1/Y2, and Z1/Z2 directions, the system stress levels remain below the yield deformation strength of silicon (100 MPa) and the tensile strength of CCGA (44 MPa).

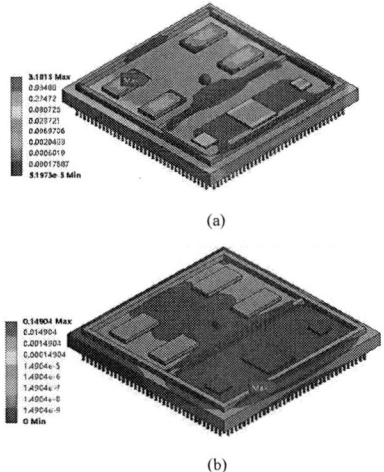

(a)

(b)

Figure 16: Finite Element Simulation of 1500G Mechanical Impact — (a) Stress Distribution; (b) Strain Distribution

Table 8: Summary of Mechanical Impact Data

Direction	1500G stress value	1500G strain	3000G stress value	3000G strain
Y1	3.18MPa	0.149um	6.36MPa	0.298um
Y2	24.94 MPa	11.871	49.9 MPa	23.84um
X1	5.86 MPa	2.65um	11.72 MPa	5.31um
X2	5.85 MPa	2.81um	11.7MPa	5.64um
Z1	6.28MPa	2.68um	12.56MPa	6.28um
Z2	6.26MPa	2.68um	12.52MPa	5.36um

Based on the low-frequency vibration simulation model, the SiP (System-in-Package) chip is mounted on a PCB (Printed Circuit Board) substrate measuring 7.6 cm × 7.6 cm, with analysis conditions specified according to Table 9.

Table 9: Low-Frequency Vibration Testing Conditions

Frequency(Hz)	Test Conditions
5-18	15.32mm
18-70	20g
70-100	12g

The simulation analysis results are presented in Figure 17 and Table 10. The system experiences sinusoidal vibrations with amplitudes of 15.32 mm, 20 g, and 12 g in the X, Y, and Z axes respectively, across the frequency ranges of 5-18 Hz, 18-70 Hz, and 70-100 Hz. Under these conditions, the stress levels within the system remain below the yield deformation strength of 350 MPa and the tensile strength of the CCGA package, which is 44 MPa.

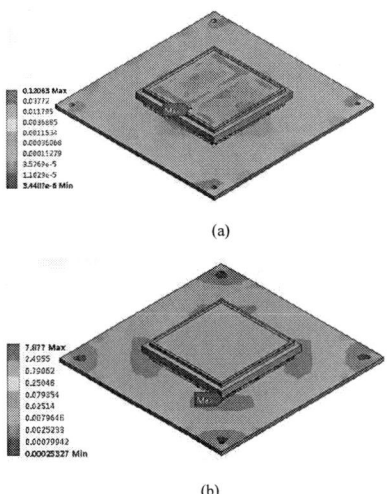

(a)

(b)

Figure 17: Low-frequency vibration analysis diagrams (a)18 Hz and (b)70 Hz

Table 10: Summary of Low-Frequency Vibrations

direction	5-18(Hz)	18-70(Hz)	70-100(Hz)
X	0.085MPa	0.839MPa	0.5MPa
Y	0.48MPa	7.87MPa	4.84MPa
Z	0.12MPa	1.19MPa	0.716MPa

V. RELIABILITY TESTING VALIDATION

A. Test methodology and sample preparation

To validate the effectiveness of the high-reliability integrated design approach, a temperature cycling test was conducted. The test employed high-temperature co-fired alumina ceramic enclosures, with devices stencil-printed with solder paste onto PCB pads, followed by vacuum reflow soldering. The resulting board-level packaged test specimens are shown in Figure 18. The assembled samples on PCB substrates were subjected to temperature cycling, and post-test inspections focused on the morphological changes of the solder joints after varying numbers of temperature cycles.

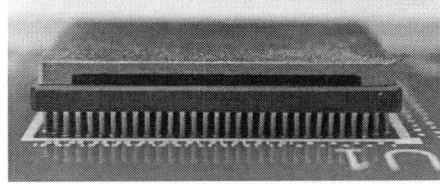

Figure 18: Sample after package-level encapsulation

B. Temperature cycling conditions

Conduct temperature cycling tests on the board-level packaged test specimens in accordance with GJB548C-2021 Method 1010.1 Condition B. The temperature cycle profile is illustrated in Figure 19, with a range of -55° C to 125° C, and both high and low-temperature hold times of 15 minutes. The heating and cooling rates are averaged at 5 ° C per minute. Each of the five specimens undergoes between 100 and 500 temperature cycles, with inspections at intervals of 100 cycles. Upon completion of the testing, specimens are removed for morphological analysis.

979-8-3315-8850-2/25 $31.00 © 2025 IEEE

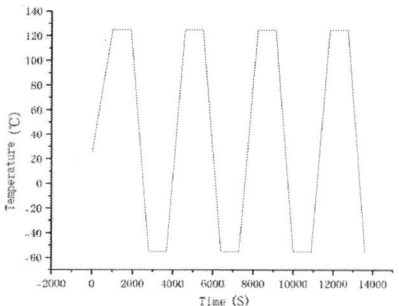

Figure 19: Temperature Cycling Curve

C. Analysis of experimental results

Figures a) through c) depict the morphological comparison of solder joints after 400, 500, and 600 thermal cycles, respectively. At 400 cycles, the variation in loop resistance of the CCGA device did not exceed 0.3 Ω ; by 500 cycles, significant changes in the circuit resistance were observed, suggesting the initiation of microcracks within the solder joint interconnects, though failure criteria had not yet been met. After 600 cycles, open-circuit conditions were detected in some samples, indicating solder joint failure due to thermal cycling. Combining the previously established thermal stress simulation model for CCGA devices with actual thermal cycling conditions as input, the simulation results identified the interface between the solder joints and the ceramic substrate, as well as the solder joints and PCB soldering regions, as the locations of maximum thermal stress concentration. As the number of thermal cycles increased, the accumulated equivalent plastic strain in these regions gradually grew. Micro-section analysis revealed that cracks after 500 cycles originated in the stress concentration zones predicted by the simulation, confirming the hypothesis that resistance increase is caused by early microcracks in the solder joints. The observed crack initiation sites closely matched the regions of simulated thermal stress concentration.

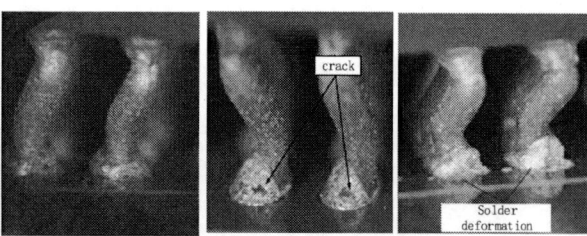

Figure 20: Temperature cycling - a) 400 cycles, b) 500 cycles, c) 600 cycles

VI. Conclusions

To address the core requirements of miniaturization, lightweight design, and high reliability in electronic systems, this paper proposes a micro-system SiP (System-in-Package) integration methodology and establishes a reliability verification process that combines multi-physics simulation with physical testing. This systematic approach provides a comprehensive reference case for the reliability assessment of complex micro-systems.

To validate the reliability of the SiP design, a multi-dimensional analysis is conducted: at the electrical performance level, full-system-level simulations of power integrity, signal integrity, and electromagnetic compatibility are performed, with particular focus on waveform characteristics, transmission loss, and crosstalk effects of critical signals such as clock and data signals; at the thermal performance level, heat dissipation simulations are carried out under various operating temperature conditions to ensure thermal stability; at the mechanical performance level, stress distribution analyses are performed under typical conditions such as constant acceleration and random vibration to ensure structural stresses meet material performance requirements.

The results demonstrate that the designed SiP fully satisfies the functional and reliability requirements of the processor: it offers high integration and compact form factor advantages, effectively aligning with the miniaturization and lightweight objectives of electronic systems; additionally, through electrical, thermal, and mechanical multi-physics validation, the SiP can operate stably in complex environments, further confirming the effectiveness of the proposed SiP integration methodology and reliability verification process.

References

[1] J.Liu, B.Wang, Y.Zhou, L.Chen, Z.Duan and Q.Ma, "Design and Implementation of a Dual-Band RF SiP Module Based on Package-on-Package Technology," 2018 International Conference on Microwave and Millimeter Wave Technology (ICMMT), Chengdu, China, 2018, pp. 1-3, DOI:10.1109/ICMMT.2018.8563973.

[2] J. Gao, X. Yao and S. Fan, "Design of a X-band miniaturized T/R module based on LTCC substrate," 2022 Asia-Pacific Microwave Conference(APMC),Yokohama,Japan,2022,pp.812-814,DOI: 10.23919 /APMC55665.2022.9999914.

[3] Gosheblagh RO , Mohammadi K .Hybrid time and hardware redundancy to mitigate SEU effects on SRAM-FPGAs: Case study over the MicroLANprotocol[J].Micro electronics Journal,2014,45(7):870-879.DOI:10.1016/j.mejo.2014.04.006.

[4] Kim, Y. H., & Kim, J. H. (2024). A Study on the Reliability Evaluation of a 3D Packaging Storage Module under Temperature Cycling Ultimate StressConditions.Micromachines,15(4),428.https://doi.org/10.3390/mi15 040428

[5] AYS , AGF , ABW ,et al.Fatigue reliability design for metal dual inline packages under random vibration based on response surface method[J].MicroelectronicsReliability,100–101.[2025-09-01].DOI:10.10 16/j.microrel.2019.113404.

[6] Zhou D , Haseeb ASMA, Andriyana A,et al.A parametric study of thermal stress and analysis of creep strain under thermal cyclic loading in a hybrid Quad Flat Package[J].IEEE Transactions on Components, Packaging,and Manufacturing Technology,2020,PP(99):1-1.DOI:10. 1109/TCPMT.2020.3046750.

[7] O. Yauw et al., "Leading edge die stacking and wire bonding technologies for advanced 3D memory packages," 2017 IEEE 19th Electronics Packaging Technology Conference (EPTC), Singapore, 2017, pp. 1-7, doi: 10.1109/EPTC.2017.8277544.

[8] IPC-2221A, General Standard on Printed Board Design[S].

[9] W. Liu, G. Chen, Q. Ding and J. Jiang, "A Comprehensive Methodology for Optimizing Power Integrity of High-Performance IC Packages," 2023 IEEE 32nd Conference on Electrical Performance of Electronic Packaging and Systems (EPEPS), Milpitas, CA, USA, 2023, pp. 1-3, doi: 10.1109/EPEPS58208.2023.10314865.

[10] Chen M, Zhang Y ,Guo, YuhuaWu, TianxiangDou, LongTian, WenyaXiao, JinqingLi, Junhui.Research on Flip-Chip Bonding Process and Thermal Cycle Reliability Simulation of 3-D Stacked Structure[J].IEEE Transactions on Components,Packaging andManufacturing Technology,2022,12(1):51-58.DOI:10.1109.

Design and Implementation of an FPGA-Based Test System for Depth of Interaction Measurement

XiaoTian Lv , Ce Zhang , Ran Zheng , XiaoMin Wei , FeiFei Xue , RuiGuang Zhao ,*, YongCai Hu

[1.] School of Computer Science, Northwestern Polytechnical University, Xi an, China

[2.] School of Electronics and Information, Northwestern Polytechnical University, Xi an, China

* Correspondence author: zhaoruiguang@nwpu.edu.cn

Abstract—DOI information of incident particle is important for PET system. Accurate DOI acquisition requires dedicated front-end electronics capable of precise energy and time measurement. Our group developed a mixed-signal ASIC with ultra-low noise and wide dynamic range for DOI measurement. To evaluate the ASIC effectively, FPGA-based test systems are required. However, conventional test systems often suffer from synchronization issues between energy and time measurements, which can result in data misalignment or loss under high event rates. In this paper, we present an FPGA-based test system specifically designed to support the evaluation of this ASIC. The system adopts a three-stage processing architecture with dual register arrays for parallel acquisition of energy and time data, enabling precise alignment and high-throughput storage. In addition, the system offers high flexibility and supports real-time acquisition, making it well-suited for accurate and efficient performance testing. The system is validated through functional simulations and on-board tests. Results show reliable real-time acquisition, storage, and transmission of energy and time data, with robust performance under high event rates.

Keywords—Depth of Interaction (DOI), FPGA, Test System, Mixed Signal Processing, Positron Emission Tomography (PET)

I. INTRODUCTION

In recent years, radiation detection technologies have developed rapidly, and been applied in medical imaging, radionuclide identification, and space exploration. In medical imaging, by enabling precise localization of gamma-ray interaction depth, Depth of Interaction (DOI) techniques effectively reduce parallax error and enhance image quality in Positron Emission Tomography (PET) systems [1].

In previous work, our group proposed an Application Specific Integrated Circuit (ASIC) for DOI information measurement. It features ultra-low noise, a wide dynamic range, and high energy resolution [2], enabling accurate measurements across various applications [3].

The overall architecture of the ASIC is shown in Fig. 1. It integrates four main components:

The analog front-end readout circuit. It comprises 64 anode channels and 2 cathode channels. Each channel receives input charge signals and generates the corresponding peak-hold voltage along with an arrival timing trigger. The channel-level TDC digitizes the Time Over Threshold (TOT) information of particle

events, facilitating accurate energy correction. Additionally, the channel-level DAC provides discriminator threshold voltages and fine-tunes reference voltages for each channel.

The digital control circuit. It handles data packing, readout control and configuration.

Peripheral circuits. It includes a bandgap reference, bias circuits, a delay-locked loop (DLL), a serial peripheral interface (SPI) and a global DAC. The ASIC employs the DLL to generate the clock distribution network and integrates the SPI for configuring parameters of each channel [4]. Moreover, the global DAC provides a common reference voltage for all channel-level DACs.

Data Interface. The energy information is selected and output by a multiplexer (MUX) [5], whereas the time information is transmitted via a low-voltage differential signaling (LVDS) interface. Both signals will be acquired and processed by the FPGA.

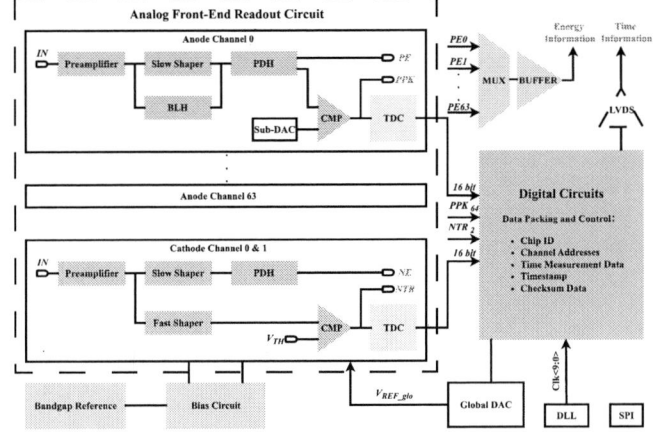

Fig. 1. Overall architecture of the DOI ASIC.

To evaluate performance of the ASIC, an FPGA-based test system is developed. The design is implemented on the Xilinx Artix-7 FPGA platform [6]. Moreover, its functionality is verified through simulation using the Vivado design suite. It supports real-time configuration, mixed signal acquisition, and high throughput data transmission [7].

979-8-3315-8850-2/25 $31.00 © 2025 IEEE

The remainder of this paper is organized as follows. Section II presents the design and architecture of the FPGA-based test system. Section III describes the test environment setup and workflow. Section IV shows the simulation results. Section V concludes the paper and presents future prospects.

II. DESIGN AND ARCHITECTURE

A. System-Level Signal Flow

As illustrated in Fig. 2, three arrow colors are used in the diagram above to distinguish different types of signals.

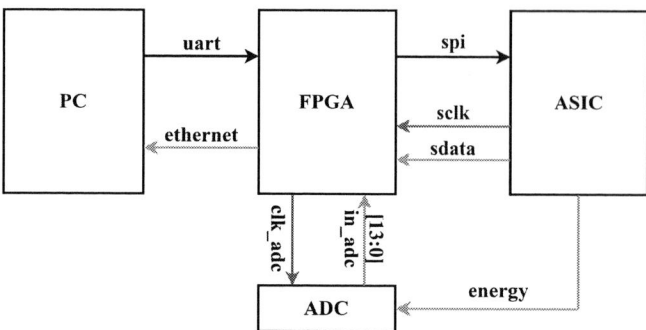

Fig. 2. The signal flow of test system.

Black arrows indicate that the PC sends configuration data to the FPGA through the UART interface.

Red arrows depict clock signals managed by the FPGA. It coordinates data acquisition and timing control to achieve synchronization.

Green arrows demonstrate time and energy data flow during data acquisition and data packing.

The PC sends configuration commands to the FPGA via the Universal Asynchronous Receiver/Transmitter (UART) protocol. Upon receiving the commands, the FPGA transmits them through the SPI bus to configure the internal registers of the ASIC. To facilitate sampling, the FPGA receives the time data clock from the ASIC and generates the clock to control the external ADC for energy data conversion.

Following configuration and clock synchronization, the data acquisition process begins. Energy data is output from the ASIC to the ADC for analog-to-digital conversion. The FPGA controls the sampling process and stores the digitized energy data. Meanwhile, the time data is transferred from the ASIC to the FPGA and stored respectively [8]. After each acquisition cycle, the FPGA packs the energy and the time data together and transmits the data packet through the Ethernet interface for further processing [9].

B. Key Contributions

Recent studies show that FPGA-based test systems are widely applied in nuclear detection and irradiation environments, with their advancements mainly reflected in three aspects.

Data acquisition and transmission efficiency are enhanced by using asynchronous FIFO and AXI-HP interfaces [10].

Real-time performance and scalability are improved by multi-channel architectures, employing programmable triggering and dead-zone acquisition [11].

Robustness in strong irradiation environments is strengthened through redundant configuration, online erase, and other ruggedization technologies [12].

Despite significant improvements in efficiency, real-time performance, and robustness, conventional FPGA-based test systems still face challenges in synchronizing parallel energy and time measurements, especially under high event rates.

To address this limitation, we implement a three-stage processing model featuring alternating dual register arrays, as illustrated in Fig. 3. It supports real-time data storage, ensures data integrity, maintains correct sequencing, and prevents data loss.

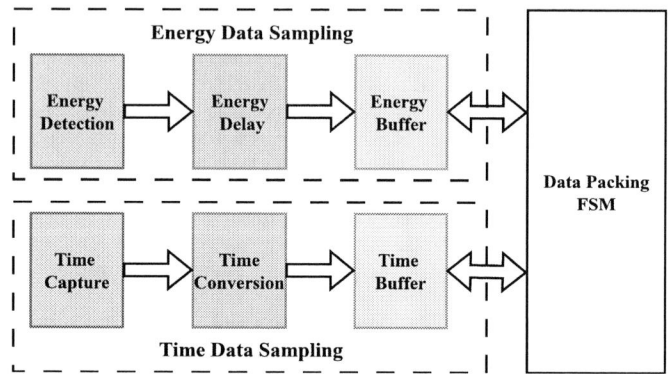

Fig. 3. Three-stage processing structure for energy and time sampling.

The first, second, and third stages of the processing structure are represented by the blue, green, and yellow parts, respectively.

For energy data acquisition, the first stage detects the trigger. The second introduces a precise delay. And the third samples the 14-bit energy data in real time.

For time data acquisition, the first stage captures a 40-bit data frame under specific conditions. The second converts it into a 32-bit format. And the third performs real-time sampling.

From an implementation perspective, the system also provides significant benefits for FPGA timing and resource efficiency. In this structure, long combinational paths are segmented into short register-to-register domains. It significantly simplifies setup and hold timing analysis. Moreover, functional stages are clearly separated, localizing logic, reducing routing congestion, and avoiding excessive LUT usage. Collectively, these strategies ensure stable and reliable system operation, making it suitable for radiation detector applications.

C. System Design

Totally, the system is composed of four functional modules. The UART module enables communication between the PC and the FPGA, while the SPI module delivers configuration commands from the FPGA to the ASIC. The Data Packing

Module independently acquires and buffers energy and time data, integrating them into packed data. Finally, the Gigabit Ethernet module transmits these packed data to the PC, facilitating subsequent processing and analysis.

In this section, we focus on illustrating the Data Packing Module in the FPGA device.

A finite state machine (FSM) is implemented to manage state transitions and control the timing of data transmission, as shown in Fig. 4. And it contains six states as follow.

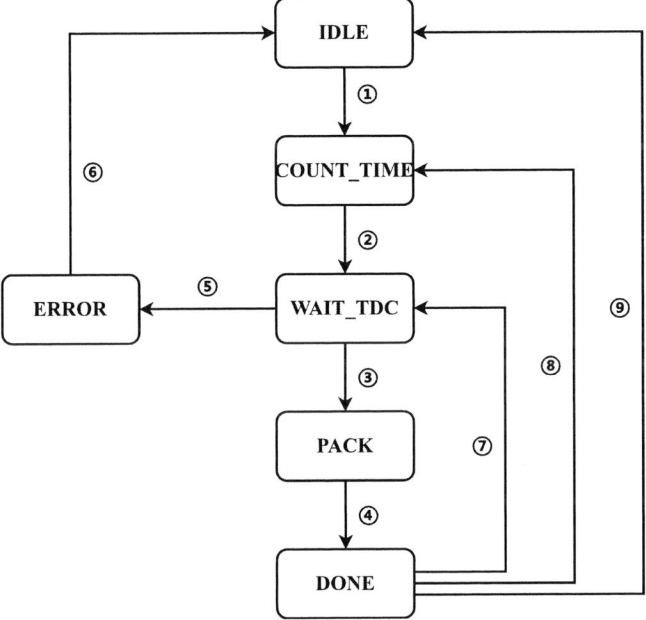

Fig. 4. Finite State Machine Design for the Test System.

The IDLE state: The initial state of the FSM, responsible for system initialization and reset.

The COUNT_TIME state: This state performs timeout counting for the first trigger, with a configurable counting duration, and controls energy data storage using triggers.

The WAIT_TDC state: This state stores time data and performs event error checking when time and energy sample counts match.

The PACK state: In this state, energy and time data are packed sequentially, while input data is simultaneously stored using a separate register array during data packing.

The DONE state: This state indicates the completion of a data processing and packing cycle.

The ERROR state: This state indicates an error occurred during data processing or packing. It clears all data stored in the current cycle.

As shown in TABLE I, the conditions for state transitions are summarized.

TABLE I. DESCRIPTION OF STATE TRANSITIONS

Condition	Description of State Transition
①	Transition on rising edge of the first trigger in the cycle; trigger refers to the control signal for energy data acquisition.
②	Transition when internal counter reaches programmable trigger timeout; trigger timing per cycle fixed and configurable.
③	Transition occurs when time and energy data correspond to the same event, with equal and non-zero counts.
④	Transition occurs after the last packed data is output.
⑤	Transition triggered by timeout in time data acquisition or mismatch between time and energy events.
⑥	Transition after one clock cycle.
⑦	After data packing completes, transition occurs upon detecting the first trigger rising edge with trigger timeout.
⑧	After data packing completes, transition occurs upon detecting the first trigger rising edge without trigger timeout.
⑨	Transition after one clock cycle.

To ensure the integrity and correct sequencing of event data, the system assigns a unique Event ID in each acquisition cycle. During data packing, it also provides address indices for accessing the energy and time register arrays. In addition, dedicated counters are employed to track the number of valid entries.

Moreover, the FSM integrates timeout monitoring, error detection, dual-buffer storage to ensure robust data processing and packing. It guarantees data transmission accuracy and improves throughput.

III. ENVIRONMENT AND WORKFLOW

The experimental setup is provided in Fig. 5. It includes a PC, an FPGA, an ASIC-integrated daughterboard, a battery-powered motherboard, a signal generator, a power supply, and an oscilloscope.

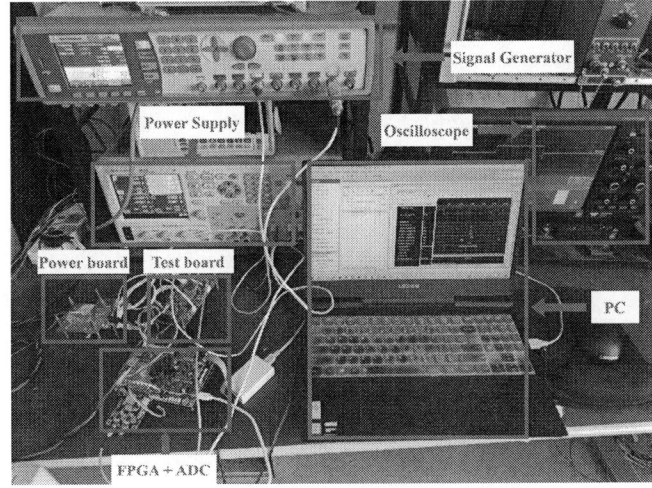

Fig. 5. Overview of the experimental setup.

The analog energy information is digitized by an external module equipped with ADC chip, which has dual channels, 14-

bit resolution. It performs sampling using an external clock provided by the FPGA. In addition, the ADC includes a built-in duty cycle stabilizer to ensure sampling accuracy under clock deviation. The energy signals and the time signals are respectively acquired by the external ADC module and the on-chip TDC. The two asynchronous data streams are independently processed, merged into packed data, and transmitted through the Ethernet protocol. The results are then captured by Wireshark.

IV. RESULTS

In this section, three functional simulations of the test system are presented as follows.

A. Energy Data Acquisition and Storage

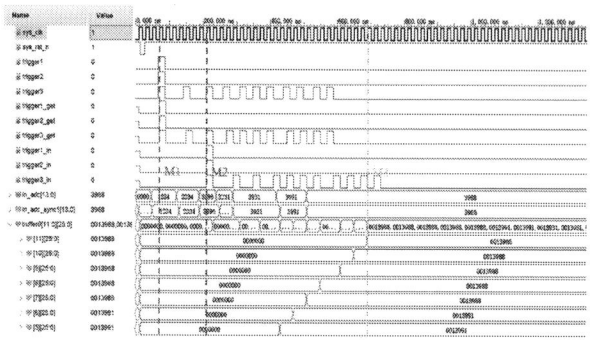

Fig. 6. Diagram of Energy Data Acquisition and Storage.

Fig. 6 shows the energy data acquisition with three annotated points, each corresponding to a module in the processing structure.

M1 indicates the energy detection module, which captures the rising edge of the trigger signal. M2 corresponds to the energy delay module. It delays each detected trigger by seven clock cycles, allowing stable sampling of the energy signal. M3 marks the energy buffer module, which stores the energy data upon trigger detection. During each acquisition cycle, the system stores up to 12 data samples. Any data exceeding this limit will be discarded to prevent buffer overflow. Moreover, it increments a counter to record the total stored events.

B. Time Data Acquisition and Storage

Fig. 7. Diagram of Time Data Acquisition and Storage.

As shown in Fig. 7, the waveform of time data acquisition includes four annotated points.

N1 indicates the serial-to-parallel conversion of a valid 40-bit data stream that meets the defined criteria. N2 marks the conversion of qualified 40-bit data into a 32-bit format. N3 refers to the completion of TDC data buffering for the current event cycle. N4 represents the internal buffer switch operation performed after data packing, enabling the system to store data for the next acquisition cycle.

C. Data Packing and Transmission

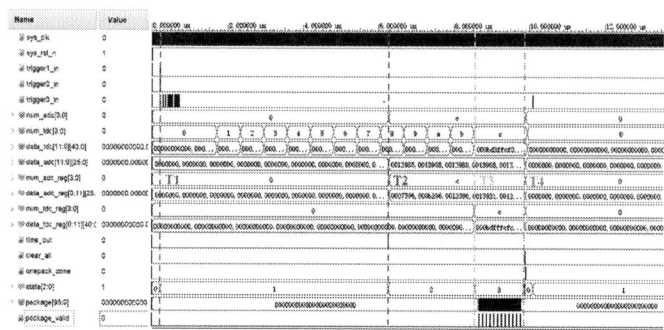

Fig. 8. Diagram of Data Packing and Transmission.

As illustrated in Fig. 8, the data packing waveform features four critical points, each corresponding to a distinct state in the FSM.

At T1, the rising edge of the trigger signal causes the state machine to transition to State 1. T2 indicates that the timeout threshold is reached, the energy data is finalized, and the state machine transitions to State 2. T3 indicates that the two data counts are equal and error-free, prompting a transition to State 3 to begin data packing. At T4, after packing the data for the current cycle, a packing completion signal is sent, and the buffer is cleared.

V. CONCLUSIONS

This paper presents an FPGA-based test system for evaluating a depth-of-interaction measurement ASIC. The system ensures synchronization between energy and time acquisition. It uses a three-stage architecture with dual register arrays and an event-indexed alignment scheme. A dedicated finite state machine (FSM) performs timeout monitoring, error detection, and synchronized data packing. These features enable precise, parallel, and high-throughput acquisition and transmission of energy and time data. Functional simulations and on-board tests confirm reliable real-time operation and closely match theoretical expectations, validating the design. In the future, we will proceed with the detector integration and validation. Additionally, we will optimize the test system to support ASIC with more channels for precise DOI measurement.

ACKNOWLEDGMENT

This work was supported in part by the National Natural Science Foundation of China under Grant No. 12475195,

12341502; National Key Research and Development Program of China under Grant No. 2023YFF0719600; Shaanxi Postdoctoral Science Foundation.

REFERENCES

[1] Voelker, Matthias, et al. "144 Channel measurement IC for CdZnTe sensors with energy and time resolution." Microelectronics Journal 45.10 (2014): 1275-1280.

[2] Sellin, Paul J. "Recent advances in compound semiconductor radiation detectors." Nuclear Instruments and Methods in Physics Research Section A: Accelerators, Spectrometers, Detectors and Associated Equipment 513.1-2 (2003): 332-339.

[3] Rivetti, A., et al. "TIGER: A front-end ASIC for timing and energy measurements with radiation detectors." Nuclear Instruments and Methods in Physics Research Section A: Accelerators, Spectrometers, Detectors and Associated Equipment 924 (2019): 181-186.

[4] Macera, Daniele, et al. "ALTAIR: a low-noise, low-power, high-speed and wide dynamic range ASIC for X-and -ray spectroscopy with CdTe/CdZnTe pixel detectors." IEEE Transactions on Nuclear Science 68.2 (2020): 182-188.

[5] Bonvicini, Valter, et al. "A double-gain, large dynamic range front-end ASIC with A/D conversion for silicon detectors read-out." IEEE transactions on Nuclear Science 57.5 (2010): 2963-2970.

[6] Adusumilli, Raghumanohar, and Vinod Kumar. "Design and Implementation of a high speed 64 bit Kogge-Stone adder using Verilog HDL." International Journal of Electrical and Electronic Engineering and Telecommunication (IJEETC) 4.1 (2015): 2319-2518.

[7] TR, Vasanth Kumar, and K. V. Prasad. "FPGA Based Design and Implementation of DUC/DDC Based OFDM for Data/Image Transmission." (2019).

[8] Swathi, B. M., and Rashmi S. Bhaskar. "DESIGN AND IMPLEMENTATION OF TRACKING RECEIVER REMOTE TERMINAL FPGA CARD FOR SAT-4." (2015).

[9] Abukhait, J. F., and M. S. Saleh. "An adaptive confidentiality security service enhancement protocol using image-based key generator for multi-agent ethernet packet switched networks." International Journal of Electrical and Electronic Engineering & Telecommunications 12.2 (2023): 112-123.

[10] Topko, Yulia, et al. "SoC-FPGA based data acquisition system for position sensitive silicon detectors." Nuclear Instruments and Methods in Physics Research Section A: Accelerators, Spectrometers, Detectors and Associated Equipment 1033 (2022): 166680.

[11] Sahoo, Shantonu, et al. "FPGA-based multi-channel data acquisition system for Superheated Emulsion Detectors." Nuclear Instruments and Methods in Physics Research Section A: Accelerators, Spectrometers, Detectors and Associated Equipment 1009 (2021): 165457.

[12] Giordano, R., et al. "Neutron-irradiation testing of fpga-embedded hadron fluence sensors." IEEE Transactions on Nuclear Science 70.5 (2023): 774-781.

2025 The 10th International Conference on Integrated Circuits and Microsystems

Design and Nonlinear Analysis of a Multi-Stepped MEMS Resonator with 1:3 Internal Resonance

Peilong Li
College of Information Science and Technology
Beijing University of Chemical Technology
Beijing, China
2023210587@buct.edu.cn

Yu Jin
College of Information Science and Technology
Beijing University of Chemical Technology
Beijing, China
jiny@buct.edu.cn

Abstract—This study develops a nonlinear dynamic framework for capacitive MEMS resonators exhibiting 1:3 internal resonance. The Adomian Decomposition Method (ADM) is used to analytically model the mode shapes and frequency ratios of multi-stepped beams, enabling precise control of modal coupling. Nonlinear stiffness parameters are extracted using a static load method, based on which a Duffing-type vibration model is established incorporating both linear and cubic stiffness, as well as DC bias and AC excitation. Multi-scale analysis is applied to derive internal resonance conditions. The analytical model further reveals how excitation parameters influence the nonlinear behavior of the resonator. Additionally, frequency-domain analysis confirms the presence of mechanically generated frequency combs. COMSOL simulations show close agreement with theoretical predictions. This approach provides a compact and effective methodology for analyzing MEMS resonators with controllable nonlinear dynamics, offering new insights into microscale nonlinear phenomena.

Index Terms—internal resonance, MEMS resonator, Adomian Decomposition Method, nonlinear dynamics, frequency comb

I. INTRODUCTION

Conventional MEMS resonator designs often seek to suppress nonlinear effects to avoid undesired shifts in resonance characteristics. However, under certain conditions, nonlinearities can give rise to complex dynamic behaviors, among which internal resonance is particularly significant. This phenomenon involves strong nonlinear coupling and energy exchange between vibration modes whose natural frequencies are related by integer ratios [1]. This phenomenon has been extensively investigated in MEMS devices as a means to improve frequency stability through geometric asymmetry and to mitigate sensitivity to environmental disturbances [2]. Some studies have induced internal resonance in hybrid-shaped MEMS resonators by tuning modal frequency ratios via structural design [3]. However, these methods rely on brute-force parameter sweeps, which are time-consuming and inefficient. Notably, one of the most compelling manifestations of internal resonance is the formation of frequency combs, which are spectral patterns consisting of evenly spaced components. These were first observed in optical systems [4] and have recently been demonstrated in MEMS devices. While existing research has primarily focused on practical applications such as energy harvesting [5] and comb spacing tunability [6], the fundamental mechanisms of mechanical frequency comb generation and the parametric control of nonlinear internal resonance remain insufficiently explored.

In this paper, the Adomian Decomposition Method (ADM) is employed to efficiently design a multi-stepped MEMS resonator with a precisely controlled 1:3 modal frequency ratio, significantly reducing reliance on extensive finite element sweeps. To capture the system's nonlinear behavior, a Duffing-type vibration model incorporating both linear and cubic stiffness terms is established. The amplitude-frequency response is derived using the multi-scale method and further validated through numerical simulations via the Runge-Kutta algorithm.Frequency comb generation is further observed, confirming the activation of internal resonance.These results offer a theoretical and computational foundation for harnessing nonlinear dynamics in MEMS resonators and contribute to the broader understanding of microscale frequency comb phenomena.

II. DEVICE STRUCTURE DESIGN

To precisely satisfy integer frequency ratios between vibration modes, the geometric design of the MEMS resonator must be carefully tuned. Achieving this typically requires adjusting structural parameters such as segment lengths and widths. However, using conventional finite element methods to find suitable configurations often relies on extensive trial-and-error across a wide parameter space, making the process inefficient and time-consuming.

To address this challenge, several analytical approaches have been proposed, including the Meijer-G function method [7], differential transform method [8], and the Adomian Decomposition Method (ADM) [2]. Among these, ADM offers notable advantages in solving eigenvalue problems with complex boundary and continuity conditions. It enables efficient theoretical analysis of beams with variable cross sections by constructing analytical mode shape functions. In this study, ADM is employed to establish a direct relationship between modal frequency ratios and geometric parameters, offering a systematic approach to resonator design. The structure under investigation is a clamped-clamped Euler-Bernoulli beam composed of three segments, as illustrated in Fig. 1.

979-8-3315-8850-2/25 $31.00 © 2025 IEEE

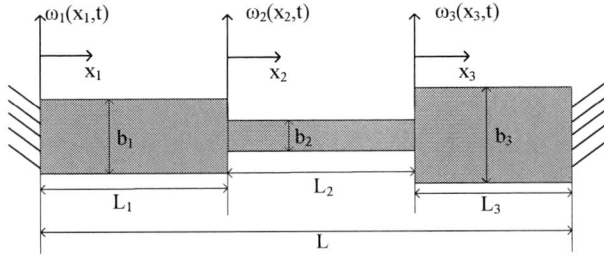

Fig. 1. Schematic diagram of clamped-clamped three-steps beam.

Assuming harmonic free vibration, the beam equation is formulated in dimensionless form for each segment as follows:

$$\frac{\partial^4 w_y(x_y,t)}{\partial x_y^4} + \frac{\rho A_y}{EI_y}\frac{\partial^2 w_y(x_y,t)}{\partial t^2} = 0 \qquad (1)$$

where $x_y \in [0 \quad L_y]$, $(y = 1,2,3)$, the subscript y denotes the y-th segment of the stepped beam. E and ρ represent the Young's modulus and density of the beam material, respectively. The moment of inertia I_y for the y-th cross-section is defined as $I_y = \frac{b_y h_y^3}{12}$, where $A_y = b_y h_y$ corresponds to the cross-sectional area. Here, L_y, b_y and h_y denote the length, width, and thickness of the y-th segment. The parameter ω signifies the system's natural frequency.

By employing mode analysis based on harmonic free vibration, the $\omega_y(x_y,t)$ can be decomposed into the product of a spatial mode shape function $\varphi_y(x_y)$ and a time-harmonic term $e^{i\omega t}$. This decomposition allows for the derivation of dimensionless ordinary differential equation (ODE) governing equations for each segment of the stepped beam:

$$\frac{d^4\Phi_y(X_y)}{dX_y^4} - \Omega_y^4\Phi_y(X_y) = 0 \qquad (2)$$

where, $X_y = \frac{x_y}{L}$, $\Phi_y(X_y) = \frac{\varphi_y(x_y)}{L}$, $R_y = \frac{L_y}{L}$, $\Omega_y^4 = \frac{\rho A_y L^4}{\rho A_1 L^4}\frac{EI_1}{EI_y}\Omega_1^4 = \mu_y\Omega_1^4$, and Ω_1 denotes the dimensionless natural frequency. Following the Adomian decomposition method (ADM) [9], the function $\Phi_y(X_y)$ in (2) can be expressed as an infinite series:

$$\Phi_y(X_y) = \sum_{m=0}^{\infty} \Phi_y^{[m]}(X_y) \qquad (3)$$

The boundary conditions for the stepped beam in Fig. 1 are formulated as:

$$\frac{d^2\Phi_1(0)}{dX_1^2} - K_{R1}\frac{d\Phi_1(0)}{dX_1} = 0, \quad \frac{d^3\Phi_1(0)}{dX_1^3} + K_{T1}\Phi_1(0) = 0 \qquad (4)$$

$$\frac{d^2\Phi_Y(R_Y)}{dX_Y^2} + K_{RY}\frac{d\Phi_Y(R_Y)}{dX_Y} = 0,$$
$$\frac{d^3\Phi_Y(R_Y)}{dX_Y^3} - K_{TY}\Phi_Y(R_Y) = 0 \qquad (5)$$

where, $K_{R_1} = \frac{k_{R_1}L}{EI_1}$, $K_{T_1} = \frac{k_{T_1}L^3}{EI_1}$, $K_{RY} = \frac{k_{RY}L}{EI_1}$, $K_{TY} = \frac{k_{TY}L^3}{EI_1}$, the translational and rotational spring stiffness coefficients at $x_1 = 0$ are denoted as K_{T_1} and K_{R_1}, respectively, while K_{TY} and K_{RY} represent the corresponding coefficients at $x_y = L_y$. Using boundary constraints and continuity conditions, the mode functions of the resonator can be obtained.

Based on the mode shape equations discussed above, the length of the designed three-segment stepped beam resonator is defined by the ratios R1, R2, and R3. Fig. 2 illustrates the relationship between R2 and the frequency ratio of the third-order mode to the fundamental mode. Table I lists the size parameters of the designed resonator.

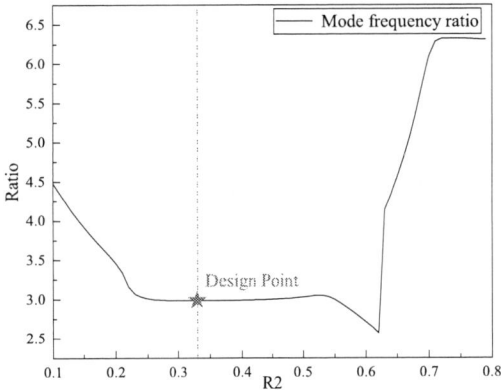

Fig. 2. Relationship between the values of R2 and frequency ratio with design point.

TABLE I
DEVICE PARAMETERS

Length	L1	L2	L3
Value (μm)	54	90	126
Width	b1	b2	b3
Value (μm)	10	15	12

With the total length and R1 fixed, the frequency ratio is calculated for varying R2 using the earlier derived formulas. As R2 changes, the curve shows a region where the frequency ratio stabilizes around 3, which can be selected as the optimal design point. Additionally, this stability helps avoid frequency ratio errors caused by manufacturing tolerances. Based on the design parameters, the theoretical frequency values for the first three modes, along with the frequency ratio of the third mode to the first mode, are presented in Table II for M ranging from 1 to 12.

The geometric structure of the multi-stepped beam resonator, composed of three axially arranged segments, is illustrated in Fig. 3. This structural configuration enhances nonlinear dynamic responses through localized stress concentration effects induced by abrupt sectional transitions [10], offering a suitable physical basis for micromechanical frequency comb generation.

TABLE II
MODE FREQUENCY THEORETICAL ANALYTICAL VALUES

M	Mode Index			
	1st	2nd	3rd	Ratio
1	4.7751	7.71681	-	-
2	4.73012	7.85002	13.45973	2.84554
3	4.73004	7.85319	14.10061	2.981075
4	4.73004	7.85321	14.13684	2.98874
5	4.73004	7.85321	14.1372	2.98881
6	4.73004	7.85321	14.1372	2.98881
7	4.73004	7.85321	14.1372	2.98881
8	4.73004	7.85321	14.1372	2.98881
9	4.73004	7.85321	14.1372	2.98881
10	4.73004	7.85321	14.1372	2.98881
11	4.73004	7.85321	14.1372	2.98881
12	4.73004	7.85321	14.1372	2.98881

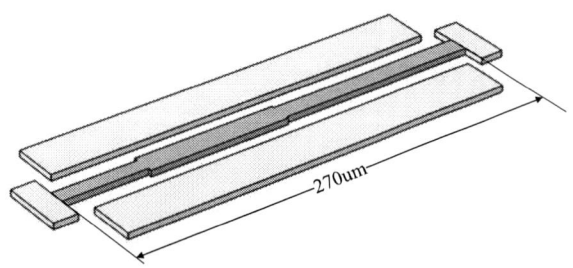

Fig. 3. The designed resonator model.

To validate the design efficacy, optimized dimensional parameters were imported into COMSOL Multiphysics to construct a fully parameterized finite element model for mode frequency response simulations. Finite element modal analysis reveals distinct vibration characteristics: The first mode (442.31 kHz) exhibits low-order bending along the beam, whereas the third mode (1330.5 kHz) shows higher-order bending with a concentration of strain in the narrower segment S3. Notably, the measured $\omega_3/\omega_1 \approx 3.01$ closely matches the prediction, representing an 8.3% enhancement over the theoretical prediction for uniform beams ($\omega_3/\omega_1 \approx 2.76$). This improvement stems from asymmetric stiffness distribution induced by the stepped geometry, which concentrates third-mode strain energy density in the narrow segment S3. Such localized kinetic energy storage amplifies higher-mode frequency scaling through geometric nonlinearity.

III. NONLIEAR ANALYSIS OF THE RESONATOR

The nonlinear dynamics of the capacitive dual-port resonator are effectively captured using a modified Duffing equation, reflecting its unique structural characteristics. Key nonlinear parameters are extracted via static load analysis, and the influence of excitation voltage on frequency stiffening is systematically examined.

As shown in Fig. 4, finite element simulations based on the optimized geometry reveal that the first mode occurs at 442.31kHz with fundamental flexural motion, while the third mode appears at 1330.5kHz with higher-order bending. The near-integer frequency ratio enables strong modal coupling under high excitation, facilitating energy exchange and nonlinear interactions. These results provide insight into the mode shapes and deformation behavior under operational conditions.

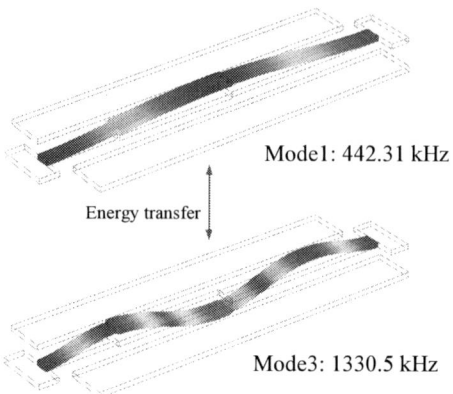

Fig. 4. Mode simulation results.

The frequency hardening behavior in MEMS resonators originates from stiffness nonlinearity, which is induced by geometric nonlinearity [11]. Fig. 5 presents a schematic of the electrostatically actuated clamped-clamped beam resonator. Its vibrational behavior is governed by the Duffing equation [12], which incorporates both damping and a cubic nonlinear stiffness coefficient k_3, as given below:

$$m\frac{\partial^2 x}{\partial t^2} + \zeta\frac{\partial x}{\partial t} + kx + k_3 x^3 = F\cos(wt) \qquad (6)$$

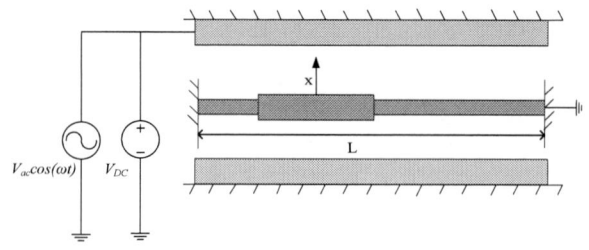

Fig. 5. Schematic diagram of the mechanical model of the resonator.

Based on the derived Duffing equation, a boundary load method is implemented to collaboratively extract k and k_3. In the nonlinear regime, the governing equation characterizing the relationship between the driving force P displacement x is formulated as follows:

$$P = kx + k_3 x^3 \qquad (7)$$

979-8-3315-8850-2/25 $31.00 © 2025 IEEE

Specifically, a static load was applied along the vibration direction of the beam, and the displacement response curve at the beam end was recorded. The load-displacement data was then analyzed through polynomial fitting, enabling the quantitative extraction of both the linear and cubic nonlinear stiffness coefficient. Fig. 6 presents the nonlinear simulation results, revealing the relationship between the driving force and displacement. Substituting these results into (7) yields stiffness coefficients $k \approx 192.75 N/m$ and $k_3 \approx 8.9 \times 10^{12} N/m^3$. Experimental data indicate that the nonlinear stiffness coefficient k_3 in multi-step beam structure is 2.3 times greater than that in uniform beams with identical dimensions, due to abrupt variations in boundary conditions.

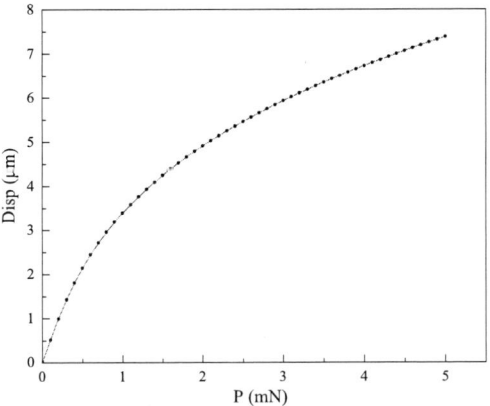

Fig. 6. The relationship between force and displacement.

To explicitly reveal the nonlinear dynamic characteristics of the resonator, this study employs numerical analysis to investigate its frequency response behavior based on the established Duffing equation model with extracted parameters. To facilitate nonlinear analysis, a small parameter ε is introduced. Additionally, a detuning parameter σ is incorporated to quantify the proximity between the excitation frequency and the system's natural frequency. Consequently, (6) can be reformulated as:

$$\frac{\partial^2 x}{\partial t^2} + 2\varepsilon\bar{\zeta}\frac{\partial x}{\partial t} + kx + \varepsilon\overline{k_3}x^3 = \varepsilon\bar{F}\cos(1+\varepsilon\sigma)t \quad (8)$$

The multi-scale method was employed to derive an approximate analytical solution, resulting in the frequency response equation,

$$\bar{\zeta}^2 a^2 + \left(\sigma - \frac{3}{8}\overline{k_3}a^2\right)^2 a^2 = \frac{1}{4}\bar{F}^2 \quad (9)$$

where, a denotes the amplitude response. The frequency response equation, incorporating the small parameter ε, can be expressed as follows:

$$\Omega = 1 + \frac{3\varepsilon\overline{k_3}}{8}a^2 \pm \sqrt{\left(\frac{\varepsilon\bar{F}}{2a}\right)^2 - (\varepsilon\bar{\zeta})^2} \quad (10)$$

As indicated by (10), a specific amplitude a may correspond to two excitation frequencies, Ω_1 and Ω_2, though their stability requires further verification. The amplitude-frequency response of the resonator was simulated based on the previously obtained analytical solution and the extracted device dimensions. In the simulation, all other parameters were kept constant while the nonlinear stiffness coefficient k_3 was varied. Fig. 7 presents the amplitude-frequency responses corresponding to $k_3 = 0$, 1.3×10^{12}, and $8.9 \times 10^{12} N/m^3$ respectively, where the unstable solution branches are highlighted in red. It can be observed that as k_3 increases, the peak amplitude remains nearly unchanged, while the resonance curves shift toward higher frequencies, indicating a pronounced stiffness hardening behavior. The primary resonance frequency increases from 440.35 kHz to 468.54 kHz, corresponding to a relative frequency shift of 6.41%, indicating a transition from linear to nonlinear dynamic behavior.

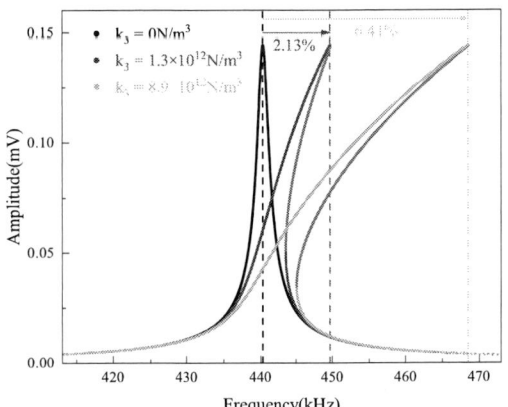

Fig. 7. Amplitude-frequency responses of the resonator under different nonlinear stiffness coefficients, with unstable branches indicated in red.

Fig. 8 illustrates the amplitude-frequency responses under a fixed nonlinear stiffness of 1.3×10^{12} corresponding to driving voltages V_{ac}=200, 400, and 1000mV, respectively. Similar to the trend observed in Fig. 8, the system exhibits a clear stiffness hardening behavior, where the resonance curves shift toward higher frequencies as the excitation amplitude increases, reflecting the amplitude-dependent nature of the response. Concurrently, the maximum vibration amplitude increases monotonically with increasing V_{ac}. At V_{ac}=1000mV, the primary resonance frequency shifts from 444.98 kHz to 461.49 kHz, corresponding to a relative frequency increase of 3.71%, while the peak amplitude increases by 114.37%. These results highlight the synergistic effect between driving amplitude and nonlinear stiffness, confirming that enhanced external energy input amplifies nonlinear dynamic behavior and significantly alters the amplitude-frequency response, thereby validating the critical regulatory mechanism of driving parameters on dynamic responses.

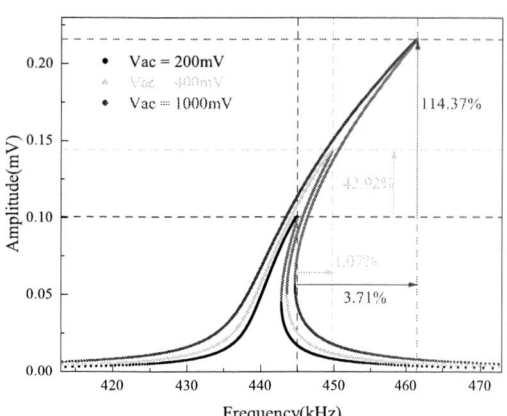

Fig. 8. Amplitude-frequency responses of the resonator under different excitation voltages, with unstable branches indicated in red.

Fig. 9 shows the FFT spectrum under an excitation amplitude of 1000mV and a nonlinear stiffness of $k_3 = 1.3 \times 10^{12}$. A distinct frequency comb pattern is observed, indicating strong nonlinear interaction and energy redistribution across multiple spectral components. The symmetric spacing of comb teeth suggests the activation of internal resonance, where the interplay between the driving frequency and coupled vibrational modes gives rise to broadband harmonic generation.

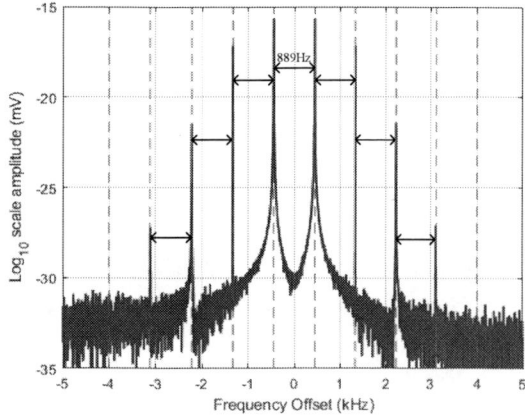

Fig. 9. Frequency combs in the FFT of the Duffing MEMS resonator.

A comprehensive comparison between this work and several representative prior designs is presented in the Table III. While existing methods such as reliability-based design optimization (RBDO) and Trial-and-Error tuning often rely on extensive finite element computations and tend to be time-consuming, the proposed approach provides a more efficient and balanced alternative. Furthermore, it also demonstrates strong perfor-

TABLE III
COMPARISON OF DESIGNS

	[13]	[14]	[7]	**This work**
Geometry	double stepped	dual-frame	trapezoid beam	stepped beam
Mode Ratio	≈2.0	≈2.0	-	≈3.0
Ratio Error	2.00%	2.03%	-	0.71%
Design Method	RBDO	Trial-and-Error	Meijer G-functions	ADM
Nonlinear Frequency Shift	3.6%	2.3%	3.8%	6.41%
Estimated Design Time	>2 hrs	>3 hrs	>1 hr	<1 hr

mance in terms of nonlinear dynamic behavior and frequency ratio accuracy.

IV. CONCLUSION

This study presents the design and analysis of a stepped-beam MEMS resonator with enhanced internal resonance, achieved through the use of the Adomian Decomposition Method (ADM) to precisely realize a 3:1 mode frequency ratio. This approach effectively addresses the inefficiencies of traditional geometric trial-and-error methods, significantly reducing design iterations and fabrication costs while providing clear criteria for nonlinear resonator design.

On the modeling side, a parameterized scanning technique was used to extract the cubic nonlinear stiffness coefficients, allowing for the construction of governing equations that capture essential nonlinear effects. Frequency-domain analysis under single-parameter excitation revealed key nonlinear behaviors, including frequency hardening, frequency locking, and the spontaneous generation of mechanical frequency combs, highlighting the influence of excitation conditions on the system's dynamic response.

REFERENCES

[1] K. Asadi, J. Yeom, and H. Cho, "Strong internal resonance in a nonlinear, asymmetric microbeam resonator," Microsyst. Nanoeng., vol. 7, no. 9, 2021.

[2] Q. Mao, and S. Pietrzko, "Free vibration analysis of stepped beams by using Adomian decomposition method," Appl. Math. Comput., vol. 217, pp. 3429-3441, December 2010.

[3] L. Ruzziconi, and A. Z. Hajjaj, "Multiple internal resonance couplings and quasi-periodicity patterns in hybrid-shaped micromachined resonators," CHAOS SOLITON FRACT, vol. 177, December 2023.

[4] A. Ganesan, Phononic Frequency Combs. Dissertation for Doctoral Degree, University of Cambridge, Cambridge, 2018.

[5] L. Bu, E. Arroyo, and A. Seshia, "Frequency Combs: A New Mechanism for MEMS Vibration Energy Harvesters," 2021 21st International Conference on Solid-State Sensors, Actuators and Microsystems (Transducers), Orlando, USA, 2021, pp. 136-139.

[6] J. Wu et al., "Widely-Tunable MEMS Phononic Frequency Combs by Multistage Bifurcations Under a Single-Tone Excitation," in Journal of Microelectromechanical Systems, vol. 33, no. 3, pp. 384-394, June 2024.

[7] R. Beigelbeck et al., "Rigorous analytical analysis of resonant Euler-Bernoulli beams with constant thickness and polynomial width," 2014 IEEE International Ultrasonics Symposium, Chicago, IL, USA, 2014, pp. 2095-2099.

979-8-3315-8850-2/25 $31.00 © 2025 IEEE

[8] A. Shahba, and S. Rajasekaran, "Free vibration and stability of tapered Euler-Bernoulli beams made of axially functionally graded materials," Appl. Math.Model., vol. 36, no. 7, pp. 3094-1113, 2012.

[9] G. Adomian, Solving Frontier Problems of Physics: The Decomposition Method, Kluwer-Academic Publishers, Boston, MA, 1994.

[10] K. Zhang, Z. Chang, S. Hao, Q. Zhang, and J. Feng, "Nonlinear characteristics and analysis of an exponential variable cross-section beam-based micro-gyroscope with electrostatic driven," Acta Mech. Sin., vol. 39, pp. 384-394, 2023.

[11] A. M. Elshurafa, K. Khirallah, H. H. Tawfik, A. Emira, A. K. S. Abdel Aziz and S. M. Sedky, "Nonlinear Dynamics of Spring Softening and Hardening in Folded-MEMS Comb Drive Resonators," in Journal of Microelectromechanical Systems, vol. 20, no. 4, pp. 943-958, August 2011.

[12] L C. Shao, W W. Tan, M. Palaniapan, and L. Khine, "Nonlinearity in micromechanical free-free beam resonators: modeling and experimental verification," J. Micromech. Microeng., vol. 18, no. 2, 2008.

[13] J. Yu, K. Asadi, H. Brahmi, H. Cho, S. Nezmi and S. Lee, "Frequency Stabilization in a MEMS Oscillator with 1:2 Internal Resonance," 2019 IEEE International Symposium on Inertial Sensors and Systems (INERTIAL), Naples, FL, USA, 2019, pp. 1-4.

[14] T. Zhang and A. A. Seshia, "A MEMS Frequency Comb Energy Harvester," in Journal of Microelectromechanical Systems, vol. 32, no. 6, pp. 516-518, December 2023.

2025 The 10th International Conference on Integrated Circuits and Microsystems

A 10.8-12.5-GHz Charge Pump PLL With 68.4-fs$_{\text{rms}}$ Jitter and -252-dB FOM$_J$ Based on a Time-Amplifying Phase-Frequency Detector

Yuzhong Li
*College of Physics and
Information Engineering
Fuzhou University*
Fuzhou, China

Zunfa Cheng
*College of Physics and
Information Engineering
Fuzhou University*
Fuzhou, China

Jiwei Huang*
*College of Physics and
Information Engineering
Fuzhou University*
Fuzhou, China
huangjw@fzu.edu.cn

Abstract—A 10.8-12.5-GHz integer-N charge pump phase-locked loop (CPPLL) is presented in this work. A time-amplifying phase-frequency detector (TAPFD) structure is proposed to suppress the in-band noise of charge pump (CP) while keeping low power consumption. A current-steering CP is employed to reduce the turn-on time of the current sources. A noise circulating voltage-controlled oscillator (VCO) greatly suppresses the effective noise power from the active devices. The proposed PLL is designed in the 65-nm CMOS process, achieving 68.4-fs$_{\text{rms}}$ jitter and 252-dB FoM$_J$ with a core active area of 0.13 mm^2.

Index Terms—Time-amplifying phase–frequency detector (TAPFD), phase noise, low jitter, charge pump phase-locked loop (CPPLL).

I. INTRODUCTION

With the rapid development of wireless and wireline communication systems, phase-locked loops (PLLs) have become increasingly critical components in modern transceiver systems. Emerging communication standards impose extremely stringent requirements on PLL performance, particularly in terms of low jitter and minimal spurs.

To meet the requirements, many efforts have focused on sub-sampling PLLs (SSPLL) [1], [2], which directly samples the edges produced by the VCO, thereby achieving a high phase detector (PD) gain that effectively suppresses in-band phase noise (PN). However, large reference spurs are induced by the inherent binary frequency shift keying (BFSK) behavior. The CPPLL is the most widely adopted PLL architecture. Achieving ultra-low jitter in CPPLL is particularly challenging due to the in-band PN, which is largely influenced by the CP. According to [3], [4], increasing the CP current is the primary means of reducing the input-referred noise of the CP. However, this also reveals the fundamental power-noise trade-off in CP design, since doubling the CP current reduces the input-referred noise by only 3-dB [5].

This work was supported by the National Natural Science Foundation of China (No. 62371332) and Special Project of the Major Special Project of the Department of Science and Technology of Fujian Province (Grant No. 2024HZ021023).

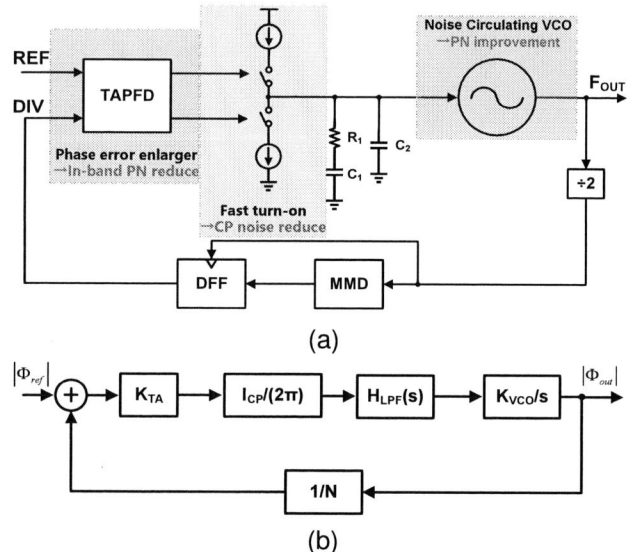

Fig. 1. (a) System Architecture and (b) its linear model.

This article presents the design of a low-jitter 10.8-12.5-GHz integer-N CPPLL based on a TAPFD. In-band PN is significantly suppressed by the phase error amplification provided by the TAPFD, thereby relaxing the stringent power-noise trade-off in the conventional CPPLL. The structure of this paper is as follows: Section II provides a description of the proposed PLL architecture and the detailed circuit implementations of each module. Section III presents the simulation results and performance comparisons. Finally, Section IV concludes the paper.

II. SYSTEM ARCHITECTURE AND DESIGN

A. System Overview

The proposed PLL architecture is shown in Fig. 1(a). It consists of a TAPFD, a current-steering CP, a low-pass loop

979-8-3315-8850-2/25 $31.00 © 2025 IEEE

filter, a noise circulating VCO followed by a ÷2 stage, and a multimodulus divider (MMD).

The linear model of this PLL illustrated in Fig. 1(b). Consequently, the transfer function of the CP noise can be expressed as

$$H_{\mathrm{CP},n}(s) = \frac{\phi_{out,n}(s)}{i_{\mathrm{CP},n}(s)} = \frac{2\pi N}{I_{\mathrm{CP}} K_{\mathrm{TA}}} \frac{H_{ol}(s)}{1 + H_{ol}(s)} \quad (1)$$

where N denotes the division ratio, I_{CP} is the charge pump current, and H_{ol} represents the loop gain.

B. Noise Analysis of the CP

In traditional CPPLLs, the CP noise is usually the dominant source of in-band noise. By modeling the CP current noise and propagating it through the loop transfer function, the input-referred in-band noise can be derived, providing an analytical basis for evaluating the noise performance. It can be shown that, when referred to the input, the thermal noise of the CP can be expressed as [6]

$$S_{CP} = \left(\frac{2\pi}{I_{\mathrm{CP}}}\right)^2 \frac{T_{\mathrm{CP}}}{T_{\mathrm{REF}}} \overline{I^2_{CP,n}} \quad (2)$$

where T_{CP} is the PFD output pulsewidth, I_{CP} is the CP current, T_{REF} is the reference period, $\overline{I^2_{CP,n}}$ is the current noise. Assuming $\overline{I^2_{CP,n}} = 2 \times 4kT\gamma g_m = 8kT\gamma(2I_D)/|V_{\mathrm{GS}} - V_{\mathrm{TH}}|$. So, the relationship of the input-referred in-band noise (2) can be simplified as

$$S_{CP} \propto T_{\mathrm{CP}}/I_{\mathrm{CP}} \quad (3)$$

According to (3), the primary approach to reduce the input-referred noise of the CP is to increase the CP current I_{CP}. This, however, directly reflects the inherent power–noise trade-off in CP design, since doubling the CP current can only achieve a 3-dB reduction in the input-referred noise.

In order to mitigate the aforementioned power-noise trade-off, the proposed design employs a TAPFD structure. According to (1), an additional gain factor K_{TA} is introduced. Now, (3) can be written as

$$S_{CP} \propto T_{\mathrm{CP}}/(I_{\mathrm{CP}} K_{\mathrm{TA}}^2) \quad (4)$$

According to (4), the proposed approach substantially suppresses the in-band noise contribution that is dominated by the CP. Meanwhile, a current-steering CP scheme is employed to reduce the T_{CP}.

III. CIRCUITS IMPLEMENTATION

A. TAPFD

In order to mitigate the inherent power-noise trade-off that limits conventional CP design, a low-power, low-noise TAPFD architecture is proposed, as shown in Fig.2(a). The TAPFD employs a gain stage, namely a time amplifier (TA) an amplificationr factor K_{TA} to enlarge the input phase error before transferring it to the CP. By amplifying the phase error at the front-end, the noise contribution originating from the CP

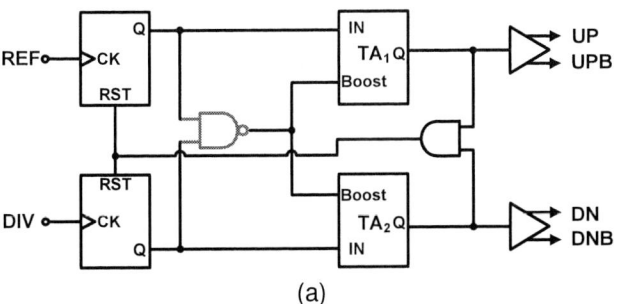

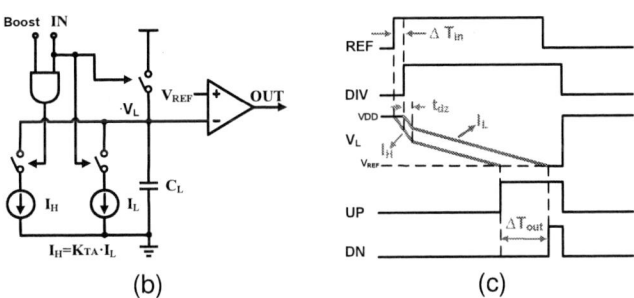

Fig. 2. (a) TAPFD circuit. (b) Simplified circuit of TA and (c) its timing diagram.

thereby significantly suppressed. To implement a large-current discharge mode within the TA, a nand gate is employed to determine when to discharge the load capacitor inside the TA. The simplified circuit of TA is shown in Fig. 2(b). With the above TA topology and its timing diagram is shown in Fig. 2(c), the delay time t_{dz} is employed to suppress the dead zone of the current source in the TA. And the gain is determined by

$$\triangle T_{\mathrm{out}}/\triangle T_{\mathrm{in}} \approx I_{\mathrm{H}}/I_{\mathrm{L}} = K_{\mathrm{TA}} \quad (5)$$

where K_{TA} is the current ratio between I_{H} and I_{L}.

According to (4)(5), As K_{TA} increases, the input-referred noise from the CP is strongly suppressed, until the intrinsic noise of the TAPFD itself dominates the in-band noise. Therefore, a value of $K_{\mathrm{TA}}=20$ is chosen in this design.

B. CP

As illustrated in Fig. 3, a current-steering charge pump (CP) is employed to translate the detected phase error into an output current, which is subsequently injected into the loop filter. To further improve the transient response, auxiliary current path is introduced to ensure that the tail current sources remain active at all times. This design prevents the tail currents from being switched off during operation, thereby enabling a faster turn-on behavior when phase corrections are required and significantly reducing the noise current contribution of the CP.

In addition, a rail-to-rail $\mathrm{OP_1}$ is adopted to compensate for voltage mismatches at the output nodes of the CP, which effectively enhances the current matching performance between the sourcing and sinking branches. The output of the rail-to-rail $\mathrm{OP_2}$ is used to mitigate the effects of channel length modulation, thereby improving the reference spur performance

979-8-3315-8850-2/25 $31.00 © 2025 IEEE   792

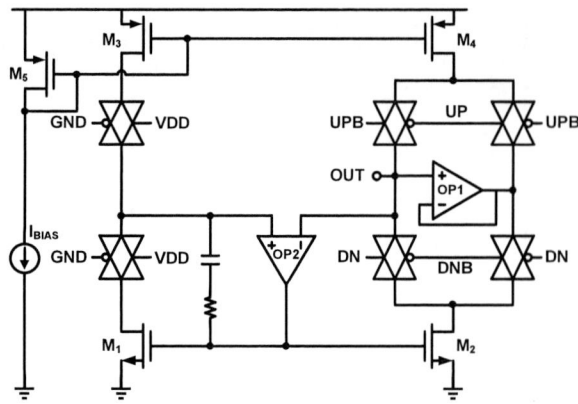

Fig. 3. Current steering CP circuit.

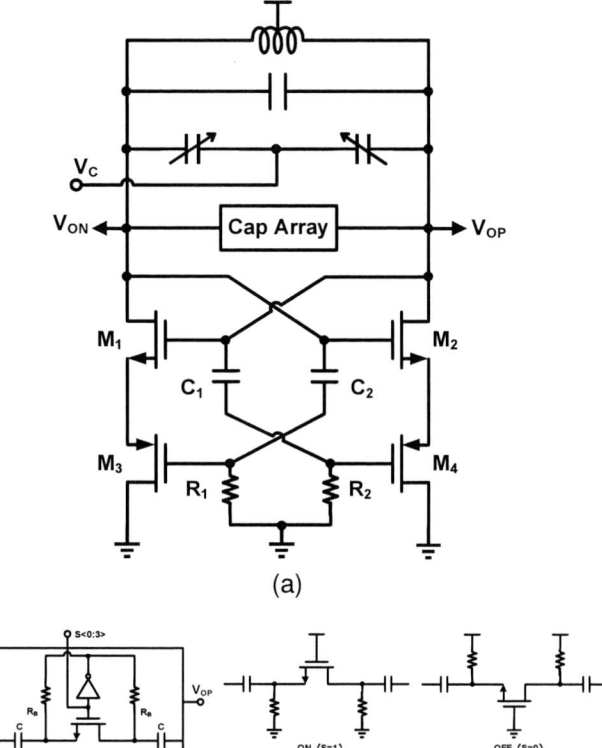

(a)

(b)

Fig. 4. (a) Noise circulating VCO. (b) ON-state and OFF-state equivalent schematic of the cap array cell.

[7]. Its output is connected to the gate of M_1 to ensure that M_4 remains in the saturation region during power-up.

C. VCO

As shown in Fig. 4(a), a noise circulating VCO is used to reduce the PN contribution from active devices without sacrificing other critical parameters such as start-up conditions and signal swing reduction [8]. By constructing a small-signal model and applying Norton's equivalent transformation to the active core of the VCO, the following analytical expression can be demonstrated in [9]:

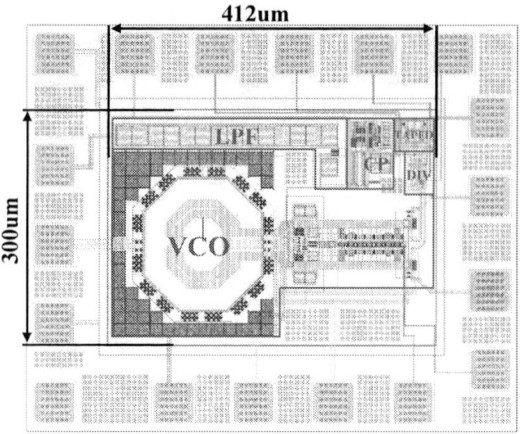

Fig. 5. Layout of the PLL.

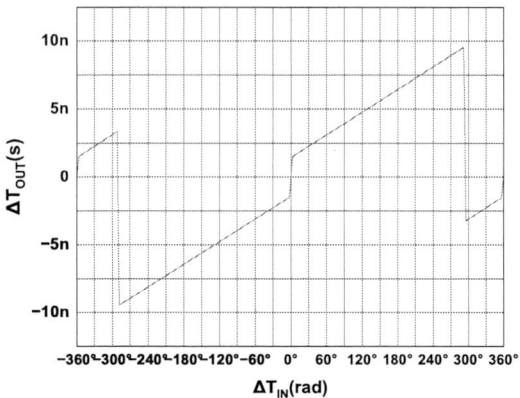

Fig. 6. Simulated transfer characteristic of the TAPFD.

$$Z_{\mathrm{GM,NC}} = -2/g_{m0} \qquad (6)$$

$$\overline{I_{\mathrm{n,NC}}^2} = kT\gamma g_{m0} \qquad (7)$$

where $Z_{\mathrm{GM,NC}}$ is the effective negative resistance, $\overline{I_{\mathrm{n,NC}}^2}$ is the power spectral density (PSD) of the current noise injected into the resonator. In a special case of $g_{mn} = g_{mp} = g_{m0}$, g_{mp} and g_{mn} are the transconductance of the PMOS and NMOS transistors, respectively.

Thus, compared with the conventional cross-coupled structure, the total current noise $\overline{I_{\mathrm{n,NC}}^2}$ injected into the tank is reduced by half.

The VCO is designed to cover 10.8 to 12.5-GHz PLL output range with frequency overlap to ensure continuous frequency coverage over PVT. As shown in Fig. 4(b), a 4-bit cap array is adopted for coarse tuning. The NMOS switches are turned ON when S = 1, while in the OFF state(S = 0), large resistors are tied to V_{DD}. This switching structure thus enables frequency tuning while preserving a high-Q resonator, which is essential for low-phase-noise operation.

979-8-3315-8850-2/25 $31.00 © 2025 IEEE   793

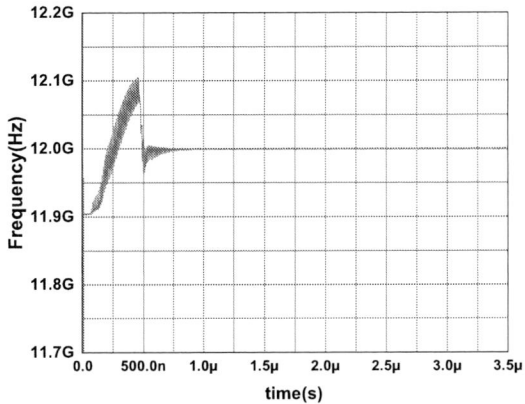

Fig. 7. Locking process.

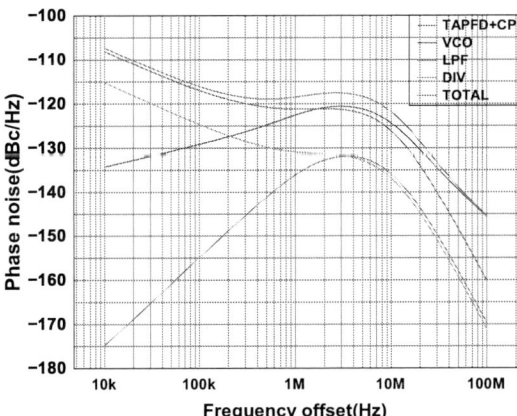

Fig. 8. Simulated noise contribution of the PLL.

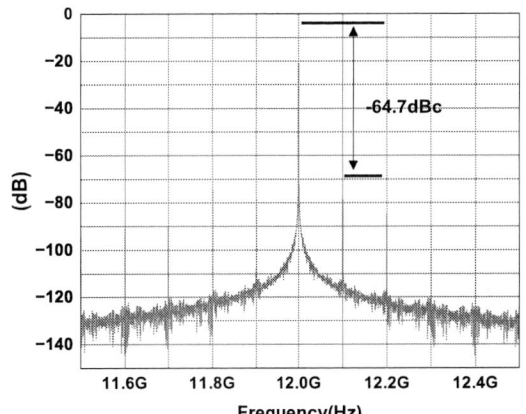

Fig. 9. Reference spur.

TABLE I
SUMMARY AND COMPARISON TO PRIOR ART

	This Work*	[2] TCASI'24	[3] ISSCC'18	[10] JSSC'24
Process	65nm CMOS	65nm CMOS	16nm FinFET	22nm CMOS
Architecture	CPPLL	SSPLL	CPPLL	CPPLL
Ref. Freq.(MHz)	100	200	500	500
Freq.Range(GHz)	10.8-12.5	13.2-14.8	7.4-14	15-22
RMS jitter(fs) Integ.range	68.4 (10k-100M)	85.4 (10k-10M)	53.6 (10k-10M)	121 (1k-100M)
Ref.Spur(dBc)	-64.7	-65.7	-75.5	-64.1
Core Area(mm²)	0.13	0.0918	0.35	0.17
Power(mW)	13.6	8.42	45	55.7
FOM$_J$	-252	-252.1	-248.9	-240.9

FOM$_J$ = $10\log[($RMS Jitter $/$ 1 s$)^2($Power $/$ 1 mW$)]$
*Post-layout simulation results

IV. SIMULATION RESULTS

The proposed CPPLL is designed in the 65-nm CMOS process. The layout is shown in Fig. 5, occupying an active area of 0.13 mm², and the overall power consumption is about 13.6 mW.

As shown in Fig. 6, the TAPFD is simulated to achieve K_{TA}=20 with ±50 ps. Beyond this linear region, gain compression occurs, similar to a conventional PFD. The characteristic curve is consistent with the results of the theoretical analysis.

Fig. 7 illustrates the locking process of the proposed PLL, where the output frequency converges to 12 GHz. The results indicate that the loop achieves frequency and phase lock in less than 1.5 μs.

Fig. 8 and Fig. 9 show the detailed noise contribution and spectrum at the carrier frequency of 12 GHz. The in-band noise achieves -118 dBc/Hz at 1-MHz frequency. The rms jitter integrated from 10 kHz to 100 MHz is 68.4 fs and the reference spur is -64.7 dBc at 100-MHz offset.

Table I illustrates the performance summary of the proposed PLL and comparison with previous research works. The proposed PLL exhibits certain advantages in terms of rms jitter and FOM$_J$.

V. CONCLUSION

This article presents a 10.8-12.5-GHz integer-N CPPLL, which is designed in the 65-nm CMOS process. A new TAPFD structure is proposed to optimize in-band PN. The simulated results show that the total power consumption of the proposed PLL is 13.6 mW at a 1.2 V power supply. Furthermore, the rms jitter is 68.4 fs and the reference spur is -64.7 dBc at a 12 GHz operating frequency. And it achieves a FOM$_J$ of -252 dB.

REFERENCES

[1] X. Gao, E. A. M. Klumperink, M. Bohsali and B. Nauta, "A Low Noise Sub-Sampling PLL in Which Divider Noise is Eliminated and PD/CP Noise is Not Multiplied by N²," in *IEEE J. Solid-State Circuits*, vol. 44, no. 12, pp. 3253-3263, Dec. 2009.

[2] D. Kar, S. Mohapatra, Md. A. Hoque, and D. Heo, "A 14 GHz Integer-N Sub-Sampling PLL With RMS-Jitter of 85.4 fs Occupying an Ultra Low Area of 0.0918 mm2," *IEEE Trans. Circuits Syst. I, Reg. Papers*, vol. 71, no. 2, pp. 595–605, 2024.

[3] D. Turker et al., "A 7.4-to-14GHz PLL with 54fs$_{rms}$ jitter in 16nm FinFET for integrated RF-data-converter SoCs," *IEEE Int. Solid-State Circuits Conf. (ISSCC)*,San Francisco, CA, USA, 2018, pp. 378-380.

[4] Y. Zhao, M. Forghani and B. Razavi, "A 20-GHz PLL With 20.9-fs Random Jitter," in *IEEE J. Solid-State Circuits*, vol. 58, no. 6, pp. 1597-1609, June 2023.

[5] X. Geng, Y. Tian, Y. Xiao, Z. Ye, Q. Xie and Z. Wang, "A 25.8GHz Integer-N PLL With Time-Amplifying Phase-Frequency Detector Achieving 60fs$_{rms}$ Jitter, -252.8dB FoM$_J$, and Robust Lock Acquisition Performance," *IEEE Int. Solid-State Circuits Conf. (ISSCC)*, San Francisco, CA, USA, 2022, pp. 388-390.

[6] Y. Zhao, M. Forghani and B. Razavi, "A 20-GHz PLL With 20.9-fs Random Jitter," in *IEEE J. Solid-State Circuits*, vol. 58, no. 6, pp. 1597-1609, June 2023.

[7] S. Han, J. Jin and C. Mao, "A full-swing charge pump with zero phase offset," in *Proc. Asia Pacific Con. Postgraduate Res. Microelectron. Electron.*, 2009, pp. 298-301.

[8] X. Ji, Y. Wang, X. Xia and Y. Guo, "A Capacitively Coupled Noise Circulating VCO," in *IEEE Microw. Wireless Compon. Lett.*, in vol. 31, no. 10, pp. 1127-1129, Oct. 2021.

[9] F. Wang and H. Wang, "A Noise Circulating Oscillator," in *IEEE J. Solid-State Circuits*, vol. 54, no. 3, pp. 696-708, March 2019.

[10] D. Dolt and S. Palermo, "A radiation-hardened 15–22-GHz frequency synthesizer in 22-nm FinFET," in *IEEE J. Solid-State Circuits*, vol. 59, no. 9, pp. 2870–2883, Sep. 2024.

2025 The 10th International Conference on Integrated Circuits and Microsystems

A Stochastic Computing-Based Computing-in-Memory Macro with Bit-Split Stochastic Number Generation

Zhiting Lin, Dandan Chen, Siyan Li, Shuang Liu, Yu Liu, Xiulong Wu
Anhui University, Hefei, China
Corresponding Author Email: ztlin@ahu.edu.cn

Abstract—This paper proposes a novel digital-domain computing-in-memory (CIM) macro architecture based on stochastic computing (SC), which integrates a Bit-Split Stochastic Number Generator (BSSNG) to enhance energy efficiency and computational parallelism. The proposed BSSNG divides each binary number into high and low bit segments, enabling the parallel generation of stochastic bitstreams. By embedding this mechanism within local computation units and peripheral input circuits, the architecture reduces stochastic number generation time by 16×. The system utilizes a 6T SRAM array with shared linear feedback shift registers (LFSR) to store 8-bit weights and generate uncorrelated bitstreams. These streams are processed through correlation-aware stochastic logic circuits, followed by binary restoration via counters and shift-accumulate units. The proposed macro is fabricated in TSMC 28nm CMOS process, supporting 8-bit input/weight multiply-accumulate (MAC) operations and delivers a 19-bit output with a clock frequency of 41.7 MHz. It achieves a 64.7% power reduction compared to full binary-to-stochastic conversion methods and reaches an energy efficiency of 17.81 TOPS/W under a 0.9V supply. The compact and reconfigurable local logic unit design further contributes to area savings, making this architecture well-suited for energy-constrained machine learning and edge computing applications.

Index Terms—Computing-in-memory, Stochastic Computing, SRAM, Stochastic Number Generator.

I. INTRODUCTION

To address the limitations inherent in traditional von Neumann architecture, particularly the memory wall issue, the paradigm of Computing-In-Memory (CIM) [1]–[3] has been proposed. As illustrated in Fig. 1, mainstream CIM implementations can be categorized into analog and digital approaches. Analog CIM leverages continuous signals to perform multiply-and-accumulate (MAC) operations, with subsequent processing executed in the analog domain. In contrast, digital CIM directly embeds logical and arithmetic operations within the memory array, enabling in-situ computation [4]. The fundamental principle of CIM is to offload a portion of computational tasks to the memory cells themselves, thereby minimizing data transfer between the memory hierarchy [5] and processing units. This architectural innovation significantly enhances energy efficiency and computational throughput. Furthermore, Stochastic Computing (SC) [6]–[10] , characterized by its probabilistic data representation and computation methodology, offers distinct advantages in terms

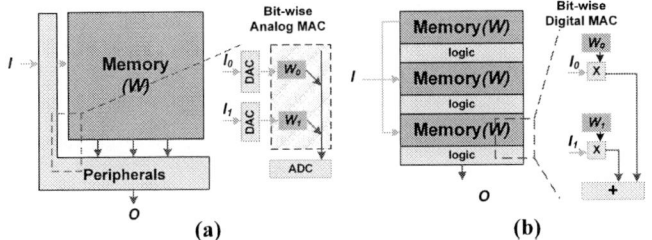

Fig. 1. (a) Analog CIM. (b) Digital CIM.

of hardware efficiency and fault tolerance, particularly in error-resilient applications. By synergistically integrating SC with CIM, the computational capabilities of such systems can be further augmented, enabling them to meet the increasingly stringent requirements of high-performance computing (HPC) applications, including machine learning accelerators and edge computing devices.

The application of SC in CIM still faces several challenges. First, directly storing stochastic numbers in the memory array necessitates 2^n memory cells to represent an n-bit number, leading to substantial area overhead and inefficiency in memory utilization. Second, during computation, converting an n-bit number into a stochastic bitstream and performing logic operations typically requires 2^n computation cycles, which limits processing speed and makes it difficult to meet the demands of efficient computation.

To address these issues, this article proposes an improved approach that divides an n-bit binary number into two 2^{n-1}-bit stochastic numbers. These partitioned stochastic numbers are then utilized to perform MAC operations separately. The final computation result is obtained by combining the outputs from these partial computations. This method effectively reduces the computational overhead while maintaining accuracy, thereby improving overall system efficiency. The advantages of the proposed architecture are as follows:

1) A stochastic bitstream generator is employed to convert binary inputs and weights into stochastic numbers. This approach eliminates the need for direct storage of stochastic bitstreams, thereby significantly reducing memory area and improving resource efficiency.

2) To mitigate correlation-induced errors in stochastic com-

979-8-3315-8850-2/25 $31.00 © 2025 IEEE

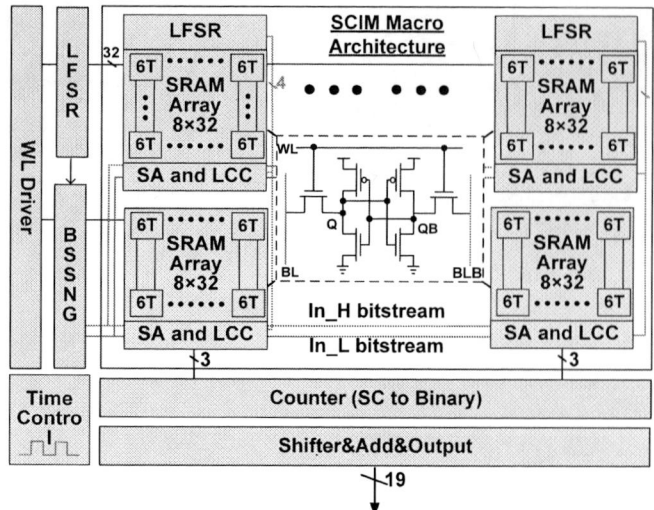

Fig. 2. Overall Architecture of Stochastic Computing-based Computing-in-Memory Macro.

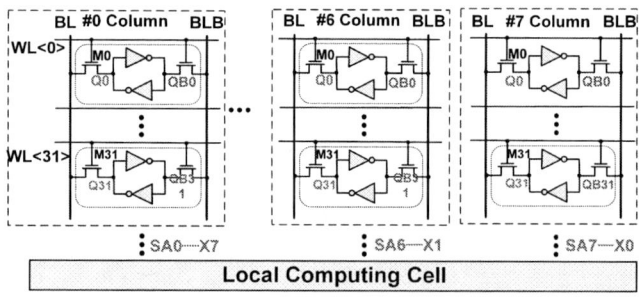

Fig. 3. Connection relationship between memory arrays and local computational units.

puting, different initial values are assigned to Linear Feedback Shift Registers (LFSR). This operation ensures the uncorrelation between inputs and weights.

3) The proposed design employs segmented bitstreams combined with parallel processing techniques. The design achieves a $16\times$ reduction in computation time while optimizing memory usage and enhancing parallelism.

The remainder of this paper is organized as follows. Section II introduces the stochastic computing CIM macro. Section III presents the experimental analysis. Finally, Section IV concludes the paper.

II. PROPOSED ARCHITECTURE

A. Overall architecture and BSSNG

As shown in Fig. 2, the proposed Bit-Segmented Stochastic Number Generator (BSSNG) is integrated into a novel CIM macro architecture. The SRAM array is divided into eight sub-arrays, each consisting of 32 rows and 8 columns of 6T cells. In each row, columns 0–3 store the high 4 bits of the weight, while columns 4–7 store the low 4 bits. Both inputs and weights are converted into stochastic bitstreams

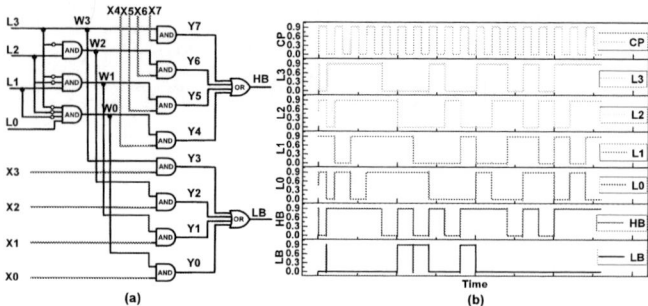

Fig. 4. (a) BSSNG. (b) BSSNG Waveform.

using LFSR. The high 4 bits of the input and weight are connected to the X<7:4> ports of the BSSNG, and the low 4 bits are connected to the X<3:0> ports. To reduce correlation between the input and weight bitstreams, the initial values of the LFSR used for the weights are set differently from those used for the inputs. As illustrated in Fig. 3, the 8-bit weights are divided and converted into stochastic bitstreams. These bitstreams are then fed into stochastic logic units to perform multiplication operations between the corresponding high and low segments, producing partial sums. To further process these partial sums, counters are used to convert them back into binary values. By designing appropriate counting rules, the counters translate the stochastic bitstreams into binary numbers that are more suitable for subsequent processing. After a shift-and-add operation, the complete computation result is obtained.

The proposed BSSNG consists of six NOT gates, eleven AND gates, and two OR gates. It has a total of twelve input ports: L<3:0> and X<7:0>. Among them, L<3:0> are connected to the outputs of a 4-bit LFSR [5], [6], while X<3:0> and X<7:4> represent the low and high 4 bits of an 8-bit binary weight, respectively. The generator has two output ports: HB (High-order Bit Stream) and LB (Low-order Bit Stream). Over 16 clock cycles of the 4-bit LFSR, all 4-bit values are traversed. During this process, HB converts the high 4-bits of the weight into a 16-bit high-order stochastic bitstream, and LB converts the low 4-bits into a 16-bit low-order stochastic bitstream. The detailed structure and operation of the generator are illustrated in Fig. 4. (a). Fig. 4. (b) shows the simulation waveform of the BSSNG. When the high 4-bits of the binary number are set to 1010 and the low 4-bits to 0011, the generator produces a high-bit stochastic stream with 10 ones and a low-bit stochastic stream with 3 ones over 16 cycles, accurately representing the values of the high and low segments, respectively.

B. Multiply-Accumulate computation in the SC Logic Unit

In stochastic computing (SC), input bitstreams with specific correlation properties [11] can be strategically leveraged to enable a broader range of logical operations within basic logic units. As summarized in Table I, when two stochastic bitstreams X and Y are statistically independent, a basic

TABLE I Logical functions of OR gates at different SNCs

X	Y	X or Y	SNC	Function
01010101(0.5)	11110011(0.75)	11110111(0.875)	0	X+Y-X·Y
11110000(0.5)	11111100(0.75)	11111100(0.75)	1	Max(X,Y)
11110000(0.5)	00111111(0.75)	11111111(1)	-1	Min(X+Y,1)

TABLE II Logical functions at different SNCs

	SNC = 0	SNC = 1	SNC = -1
AND	$a \cdot b$	Min(a, b)	Max(a + b − 1, 0)
OR	$a + b − a \cdot b$	Max(a, b)	Min(a + b, 1)
XOR	$a + b − 2a \cdot b$	$|a − b|$	$\begin{cases} a + b, & a + b \le 1 \\ 2 − (a + b), & a + b > 1 \end{cases}$
NOT		$1 − a$	

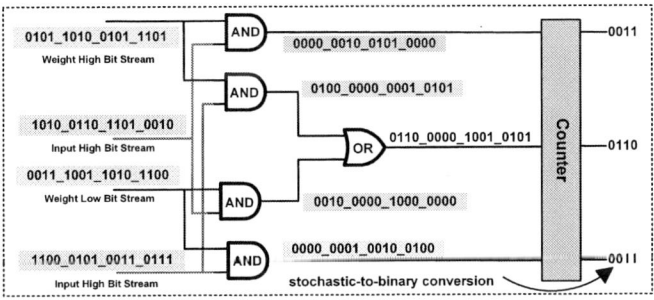

Fig. 5. Stochastic logic gates handle bitstreams.

logic unit such as an OR gate can be utilized to approximate the addition of the corresponding stochastic numbers. This approximation holds because the probability output of an OR logic unit under independence is given by $P_{OR}(X, Y) = P(X) + P(Y) − P(X)P(Y)$, which closely resembles the addition operation within the stochastic domain for moderate input values.

Table II further generalizes this concept by delineating the extended functional behaviors of several logic units, including AND, XOR, and NOT, under varying input correlation scenarios. For instance, while the AND logic unit accurately performs multiplication when the inputs are uncorrelated, it can deviate significantly in the presence of positive correlation. On the other hand, the XOR logic unit demonstrates useful subtraction-like properties under specific anti-correlated input configurations, and the NOT logic unit reliably performs bitwise complementation regardless of correlation, offering robustness in signal inversion tasks.

The overall stochastic logic architecture, integrating these correlation-aware logic units, is illustrated in Fig. 5. This configuration underscores the flexibility and reconfigurability of SC-based designs by enabling complex arithmetic functions through the strategic orchestration of basic logic units operating under controlled correlation conditions.

C. Binary Restoration Circuit

Fig. 6 illustrates the complete process of converting stochastic computation results back into binary values. The output of the stochastic logic first enters the counter module, which recovers the binary representation from the stochastic bitstream.

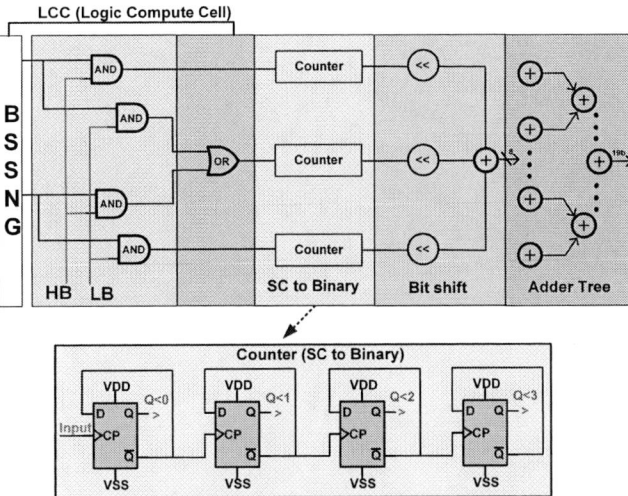

Fig. 6. Restore Binary and Shift Accumulation Circuits

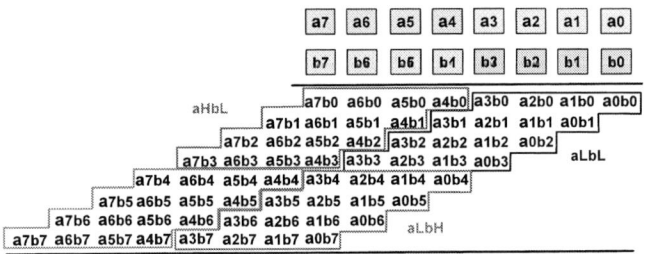

Fig. 7. Illustration of the multiply-accumulate process

Subsequently, bit-shift operations are applied to the restored results to properly align the data for the final summation. The counter part illustrates the principle of converting a 16-bit stochastic bitstream back into a binary number using a counter. The design consists of four D flip-flops [12], where each flip-flop's inverted Q output is connected to its own D input and also linked to the CP of the next flip-flop. During the conversion process, the toggling states and cascaded triggering of the D flip-flops are used to "translate" the stochastic bitstream into its corresponding binary value.

For an n-bit binary input and weight, denoted as A and B respectively, a specific bit-splitting strategy is applied to divide them into four equal-width segments: aH, aL, bH, and bL, each with a width of $n/2$. These segments are then fed into parallel stochastic number generators [13]. According to binary multiplication principles and bitwise operation rules, the resulting partial products require specific shift operations. Specifically, the product of aH and bH must be shifted left by n bits, while the products $aH \cdot bL$ and $aL \cdot bH$ are shifted left by $n/2$ bits, and the product $aL \cdot bL$ requires no shift. Based on this, the overall multiplication can be transformed into the following expression (1) through a combination of shifts and additions. Fig. 7 illustrates the complete transformation logic—from the original binary multiplication to the parallel stochastic processing and final binary output [14].

979-8-3315-8850-2/25 $31.00 © 2025 IEEE

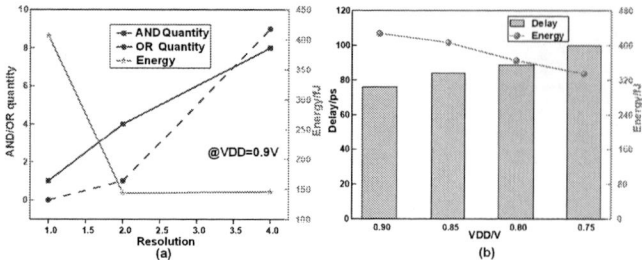

Fig. 8. (a) Performance at different resolutions. (b) Performance of BSSNG at different voltages.

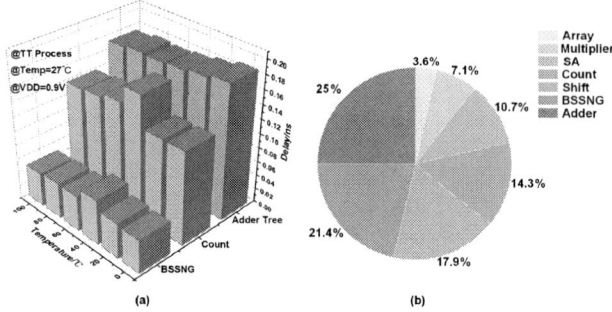

Fig. 9. (a) Delay at different temperatures. (b)Power Consumption of Modules in Each Part.

III. EXPERIMENTAL RESULTS

Fig. 8. (a) illustrates the number of stochastic logic units and the corresponding power consumption under different bit-splitting resolutions. The 4-bit segmentation balances power efficiency and hardware simplicity. Larger segment sizes (e.g., 6-bit) require longer LFSR cycles and more gates, while smaller segments (e.g., 2-bit) introduce significant quantization noise. The 4-bit configuration achieves low power with minimal logic overhead and sufficient resolution for most ML tasks.

Fig. 8. (b) presents the delay and power consumption of the stochastic number generator under different supply voltages. As the voltage decreases, the delay increases while power consumption decreases, demonstrating the trade-off between energy and performance.

Fig. 9. (a)shows the delay variation of the proposed stochastic number generator, counter, and adder tree under different temperature conditions.

As shown in Fig. 9. (b), the power consumption breakdown under the TT process corner and 0.9 V supply voltage reveals that the adder tree, stochastic number generator, and bit-segmented stochastic bitstream generator account for a significant portion of the total power.

Fig. 10 evaluates the accuracy of the proposed approximate computing method by showing the error distribution calculated from a MATLAB simulation of the circuit function.

Fig. 11 compares the proposed local computing unit with prior works. The design supports 8-bit input and 8-bit weight multiplication while occupying only 0.078 of the SRAM array area, showing a significantly more compact local unit design.

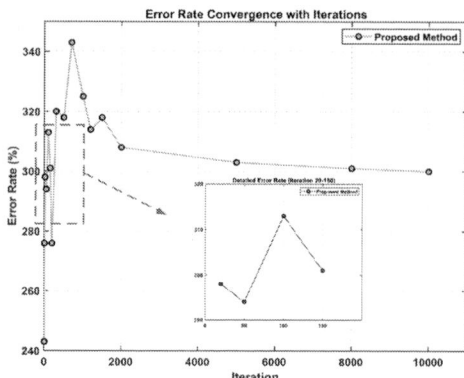

Fig. 10. Error simulation with different number of trials.

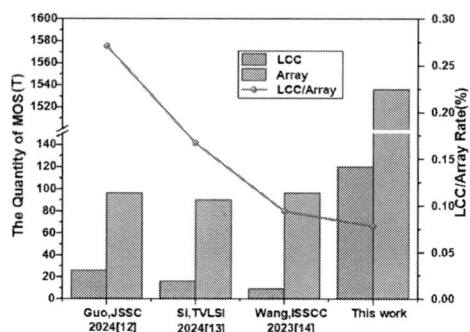

Fig. 11. Comparison of local computational cells.

TABLE III Comparison of this study with related studies

	VLSI 2024[15]	ISCAS 2024[16]	JSSC 2021[17]	CICC 2024[18]	TCASI 2024[19]	This work
Technology	12 nm	28 nm	28 nm	22 nm	55 nm	**28 nm**
Supply Voltage	0.64 V	0.75 - 0.9 V	0.9 V	0.6 - 0.8 V	0.7 - 1.2 V	**0.9 V**
Bitcell	6T	6/8/9T	6T	6T	6T	**6T**
Array Size	83 KB	17 KB	64 KB	128 KB	4 KB	**2 KB**
Input Precision	6	8	8	8	8	**8**
Weight Precision	6	8	8	8	8	**8**
Output Precision	17	19	20	—	22	**19**
Energy Eff. (TOPS/W)	46	8.36	16.63	9.19	14.2	**17.81**

Unlike previous works that use 16-bit output precision [17]–[19], our design adopts a 19-bit output format to accommodate the extended dynamic range caused by bit-split stochastic MAC operations. Each 8-bit input and weight pair is decomposed into four partial products via hierarchical bit-splitting, and the subsequent shift-and-add operation significantly increases the maximum accumulated value [20]. A 16-bit output could lead to overflow or truncation when multiple MAC operations are aggregated, especially in convolutional layers with large receptive fields. Thus, 19-bit precision is selected to maintain numerical accuracy without significantly increasing area or power, as verified in Fig. 11 and Table III.

Table III summarizes the key specifications of the proposed stochastic in-memory computing architecture. Compared with previous works, this design significantly reduces the required bit width and computation cycles for stochastic conversion by introducing BSSNG and parallel bitstream processing. It supports 8-bit input/weight precision, improves energy efficiency to 17.81 TOPS/W, and reduces both area and power consumption.

IV. CONCLUSION

This design proposes a BSSNG and implements a 2 Kb CIM macro, integrating LFSR, local computing units, and derandomization circuits. By splitting binary numbers into high and low stochastic bitstreams and using stochastic logic units for MAC operations, the architecture significantly reduces the bit width and computation cycles required for stochastic conversion. Furthermore, by optimizing the local computing units and integrating a compact stochastic number generator, the design achieves higher energy efficiency at the same input/weight precision. The use of a partitioned SRAM array and compact local computation layout further enhances area efficiency.

ACKNOWLEDGEMENTS

The authors would like to thank Yang Lou for valuable discussions and technical support.

REFERENCES

[1] A. Biswas and A. P. Chandrakasan, "CONV-SRAM: An Energy-Efficient SRAM With In-Memory Dot-Product Computation for Low-Power Convolutional Neural Networks," in IEEE Journal of Solid-State Circuits, vol. 54, no. 1, pp. 217-230, Jan. 2019, doi: 10.1109/JSSC.2018.2880918.

[2] S. -H. Sie et al., "MARS: Multimacro Architecture SRAM CIM-Based Accelerator With Co-Designed Compressed Neural Networks," in IEEE Transactions on Computer-Aided Design of Integrated Circuits and Systems, vol. 41, no. 5, pp. 1550-1562, May 2022, doi: 10.1109/TCAD.2021.3082107.

[3] Sun, Z., Kvatinsky, S., Si, X. et al. A full spectrum of computing-in-memory technologies. Nat Electron 6, 823–835 (2023). https://doi.org/10.1038/s41928-023-01053-4

[4] D. Wang, C. -T. Lin, G. K. Chen, P. Knag, R. K. Krishnamurthy and M. Seok, "DIMC: 2219TOPS/W 2569F2/b Digital In-Memory Computing Macro in 28nm Based on Approximate Arithmetic Hardware," 2022 IEEE International Solid-State Circuits Conference (ISSCC), San Francisco, CA, USA, 2022, pp. 266-268, doi: 10.1109/ISSCC42614.2022.9731659.

[5] Y. He et al., "An RRAM-Based Digital Computing-in-Memory Macro With Dynamic Voltage Sense Amplifier and Sparse-Aware Approximate Adder Tree," in IEEE Transactions on Circuits and Systems II: Express Briefs, vol. 70, no. 2, pp. 416-420, Feb. 2023, doi: 10.1109/TCSII.2022.3209872.

[6] C. Lammie, J. K. Eshraghian, W. D. Lu and M. R. Azghadi, "Memristive Stochastic Computing for Deep Learning Parameter Optimization," in IEEE Transactions on Circuits and Systems II: Express Briefs, vol. 68, no. 5, pp. 1650-1654, May 2021, doi: 10.1109/TCSII.2021.3065932.

[7] Y. Y. Lee, Z. Abdul Halim, M. N. Ab Wahab and T. A. Almohamad, "Toward Universal Multiplexer Multiply-Accumulate Architecture in Stochastic Computing," in IEEE Access, vol. 13, pp. 33874-33882, 2025, doi: 10.1109/ACCESS.2025.3539986.

[8] Shoushtari Moghadam M, Aygun S, Najafi M H. Improved Data Encoding for Emerging Computing Paradigms: From Stochastic to Hyperdimensional Computing . arXiv e-prints, 2025: arXiv:2501.02715.

[9] M. H. Najafi, S. Jamali-Zavareh, D. J. Lilja, M. D. Riedel, K. Bazargan and R. Harjani, "An Overview of Time-Based Computing with Stochastic Constructs," in IEEE Micro, vol. 37, no. 6, pp. 62-71, November/December 2017, doi: 10.1109/MM.2017.4241345.

[10] N. Onizawa, D. Katagiri, K. Matsumiya, W. J. Gross and T. Hanyu, "Gabor Filter Based on Stochastic Computation," in IEEE Signal Processing Letters, vol. 22, no. 9, pp. 1224-1228, Sept. 2015, doi: 10.1109/LSP.2015.2392123.

[11] A. Alaghi and J. P. Hayes, "Exploiting correlation in stochastic circuit design," 2013 IEEE 31st International Conference on Computer Design (ICCD), Asheville, NC, USA, 2013, pp. 39-46, doi: 10.1109/ICCD.2013.6657023.

[12] A. Guo et al., "A 28nm 64-kb 31.6-TFLOPS/W Digital-Domain Floating-Point-Computing-Unit and Double-Bit 6T-SRAM Computing-in-Memory Macro for Floating-Point CNNs," 2023 IEEE International Solid-State Circuits Conference (ISSCC), San Francisco, CA, USA, 2023, pp. 128-130, doi: 10.1109/ISSCC42615.2023.10067260.

[13] X. Si et al., "A 28 nm 16-kb Sign-Extension-Less Digital-Compute-in-Memory Macro With Extension-Friendly Compute Units and Accuracy-Adjustable Adder-Tree," in IEEE Transactions on Very Large Scale Integration (VLSI) Systems, vol. 32, no. 11, pp. 2164-2168, Nov. 2024, doi: 10.1109/TVLSI.2024.3418888.

[14] B. Wang et al., "A 28nm Horizontal-Weight-Shift and Vertical-feature-Shift-Based Separate-WL 6T-SRAM Computation-in-Memory Unit-Macro for Edge Depthwise Neural-Networks," 2023 IEEE International Solid-State Circuits Conference (ISSCC), San Francisco, CA, USA, 2023, pp. 134-136, doi: 10.1109/ISSCC42615.2023.10067526.

[15] J. Yang, A. Graening, W. Romaszkan, V. K. Jacob, P. Gupta and S. Pamarti, "A 278-514M Event/s ADC-Less Stochastic Compute-In-Memory Convolution Accelerator for Event Camera," 2024 IEEE Symposium on VLSI Technology and Circuits (VLSI Technology and Circuits), Honolulu, HI, USA, 2024, pp. 1-2, doi: 10.1109/VLSITechnologyandCir46783.2024.10631484.

[16] Z. Lin, Y. Liu, Y. Wang, Y. Zhao, C. Peng and X. Wu, "SRAM-Based Digital CIM Macro for Linear Interpolation and MAC," 2024 IEEE International Symposium on Circuits and Systems (ISCAS), Singapore, Singapore, 2024, pp. 1-5, doi: 10.1109/ISCAS58744.2024.10558525.

[17] X. Si et al., "A Local Computing Cell and 6T SRAM-Based Computing-in-Memory Macro With 8-b MAC Operation for Edge AI Chips," in IEEE Journal of Solid-State Circuits, vol. 56, no. 9, pp. 2817-2831, Sept. 2021, doi: 10.1109/JSSC.2021.3073254.

[18] M. Wu et al., "S2D-CIM: A 22nm 128Kb Systolic Digital Compute-in-Memory Macro with Domino Data Path for Flexible Vector Operation and 2-D Weight Update in Edge AI Applications," 2024 IEEE Custom Integrated Circuits Conference (CICC), Denver, CO, USA, 2024, pp. 1-2, doi: 10.1109/CICC60959.2024.10529046.

[19] H. You, W. Li, D. Shang, Y. Zhou and S. Qiao, "A 1–8b Reconfigurable Digital SRAM Compute-in-Memory Macro for Processing Neural Networks," in IEEE Transactions on Circuits and Systems I: Regular Papers, vol. 71, no. 4, pp. 1602-1614, April 2024, doi: 10.1109/TCSI.2024.3355944.

[20] Y. Liu et al., "TSCIM: A 28nm Transposed Stochastic CIM Macro for On-Chip Training and Inference," 2025 IEEE International Symposium on Circuits and Systems (ISCAS), London, United Kingdom, 2025, pp. 1-5, doi: 10.1109/ISCAS56072.2025.11044117.

2025 The 10th International Conference on Integrated Circuits and Microsystems

An Analytical Compact Model for Multi-time Programmable Memory Cells

Tiantian Gan
Department of Electrical Engineering East China Normal University
Shanghai, China
51265904122@stu.ecnu.edu.cn

Xiaojin Li*
Department of Electrical Engineering East China Normal University
Shanghai, China
xjli@ee.ecnu.edu.cn

Tengyang Liu
Department of Electrical Engineering East China Normal University
Shanghai, China
liu13795281173@163.com

Yabin Sun
Department of Electrical Engineering East China Normal University
Shanghai, China
ybsun@ee.ecnu.edu.cn

Yanling Shi
Department of Electrical Engineering East China Normal University
Shanghai, China
ylshi@ee.ecnu.edu.cn

Jianpeng Chu
Fei u Microelectronics
Shanghai, China
chujianpeng@gmail.com

Abstract— ingle-poly multi-time programmable (TP) memory cell demonstrates promising prospects in embedded non-volatile memory technologies. In this paper, a novel analytic compact model is proposed for erasing programming operations in TP memory cells. The model incorporates the effects of ramp voltage slew rate on memory cells and, based on the capacitive coupling coefficient model, constructs a computational framework to derive closed-form analytical solutions. Compared to the prior approaches, this methodology achieves computational simplicity without iterative calculation. urthermore, model predictions demonstrate good agreement with experimental data (error .1). The proposed framework enables rapid performance evaluation of memory cells, delivering a guiding framework for co-optimizing memory cell architecture and peripheral circuit design.

Keywords— , TP, analytical model, ramp oltage, single-poly

I. INTRODUCTION

Non-volatile memory (NVM) has gained wide applications in microcontrollers, radio-frequency identification (RFID) chips, and sensor calibration, owing to its data retained capability after power-off and the growing demand from Internet of Things (IoT) applications [1]. Among these technologies, a fully logic-process-compatible multi-time programmable (MTP) memory cell eliminates the need for additional process steps and delivers robust endurance, making it a compelling candidate for low-density, cost-sensitive applications. Several different MTP memory cell are proposed [2-4]. Among these, the three-transistor (3T) structure has become more widely adopted due

This work was supported in part by the National Natural Science Foundation of China under Grant 62274063, and 62274064, in part by Shanghai Science and Technology Explorer Plan under Grant 22TS1401700, 22TS1401400, and 24TS1400100. The authors gratefully acknowledge these supports.

to its balanced optimization of area efficiency and peripheral circuit complexity.

Critical physical parameters of the memory cell, including erasing/programming voltage magnitudes, oxide thickness, capacitive coupling ratio and erasing/programming time, significantly influence the MTP cell performance. uantitative evaluation of these effects, however, has historically required experimental measurements or technology computer-aided design (TCAD) simulations [5-6]. Although prior studies established cell models [7], their solutions require iterative numerical computation. For transient analysis of large scale MTP array, this approach consumes substantial computational resources. Moreover, an inappropriate iterative step size can lead to simulation non-convergence and numerical instability. Especially, for memory arrays with capacities typically exceeding 1K cells, the computational burden of iterative algorithms significantly escalates, potentially resulting in excessive simulation times and challenges. In this work, a closed-form analytical model delivering non-iterative approximate solutions is proposed, enabling rapid assessment of critical parameter impacts on memory cell performance with maintained accuracy, and significantly accelerating the analytical capabilities of circuit design and optimization procedures.

II. MODEL DESCRIPTION

A. Floating ate Potential

Fig. 1(a) shows the memory cell structure, which comprises a coupling device (M_C), a tunneling device (M_T), a read device (M_{RD}), and a select device (M_{SEL}). During the erasing operation, electrons are extracted from the floating gate (FG) through the M_T. On the contrary, the programming injects electrons into the

979-8-3315-8850-2/25 $31.00 © 2025 IEEE

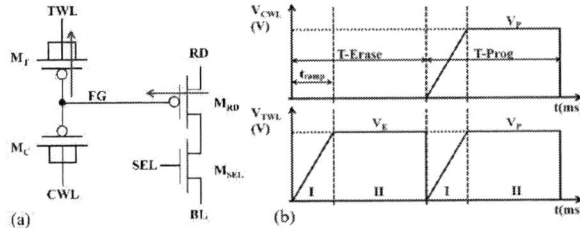

Fig. 1. (a)Memory cell structure (b)Application scheme of operational bias voltages for erasing/programming operations.

Calculation Analytical solution of the floating-gate potential	
Re uired	Tunneling-related parameters: A, B, t_o, V_{cal}
	Equivalent capacitance of device: C_C, C_T, C_{RD}
	Voltage ramp time: t_{ramp}
	Magnitude of external bias voltage: V_E, V_P
Target	Dependence of $V_{th\ cell}$ on Coupling Ratio, Bias Voltage, and Time
Algorithm	**tep1 for** $_F$
	for t t_{ramp}
	if erase then calculate V_F by using (6)
	if prog then calculate V_F by using (7)
	get V_0 V_F @ t t_{ramp}
	for t t_{ramn}
	if erase then calculate V_F by using (10)
	if prog then calculate V_F by using (10)
	tep calculate t_{stable} **by using ( )**
	tep3 calculate $_{th\ cell}$ **by using ( )**

Fig. 2. The computational framework of the model.

FG via the M_{RD}. Both erasing and programming operations are implemented based on the Fowler-Nordheim (FN) tunneling principle. The tunneling current density $_{FN}$ through a SiO layer is conventionally expressed as:

$$_{FN} \quad AE_o^2 \ e \ p\left(-\frac{B}{E_o}\right) \tag{1}$$

where E_o is the applied oxide electric field, and the coefficients A and B are FN tunneling coefficients.

For a gate-to-bulk voltage V_{GB} applied to the device, the oxide voltage drop V_o is governed by:

$$V_{ox} = V_{GB} - \Phi_{ms} - \psi_s - \psi_{poly} \tag{2}$$

where $_{ms}$ is the work-function difference between polysilicon gate and silicon substrate, $_s$ is the surface potential, and $_{poly}$ accounts for voltage drop due to poly-gate depletion. To simplify the calculation, the V_o is defined as V_{GB} V_{cal}, where the V_{cal} is a calibration voltage incorporating $_{ms}$, $_s$, and $_{poly}$ effects.

To enhance device longevity, the memory bias voltage is typically applied through a ramp scheme as illustrated in Fig. 1(b). Both erasing and programming operations implement a two-step voltage sequence: During Phase- , the bias voltage increases linearly to maximum at t t_{ramp}. During Phase- , the bias voltage stabilizes to V_E or V_P, maintaining this constant potential until operation completion. In our modeling methodology, the analysis is divided into these two phases to investigate the effects of ramp rate on memory cell characteristics.

Fig. 2 provides a comprehensive illustration of the computational procedure for establishing the analytical model of

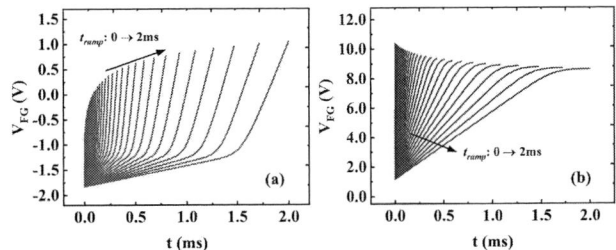

Fig. 3. Transient Response of V_F vs. voltage ramp rate (a) erasing operation, (b) programming operation.

the MTP cell. During Phase- , based on the capacitive coupling coefficient model [8], the following expression is derived:

$$Q_{FG} = C_C(V_{FG} - V_{CWL}) + C_T(V_{FG} - V_{TWL}) + C_{RD}(V_{FG} - V_{RD}) \tag{3}$$

where C_C, C_T, and C_{RD} denote the equivalent capacitance of M_C, M_T, and M_{RD} respectively. The remaining capacitance is derived analogously, with the total equivalent capacitance (C_{total}) is the sum of C_C, C_T and C_{RD}. During erasing operation, the voltage applied to TWL terminal is expressed as:

$$V_{TWL} = \frac{V_E}{t_{ramp}}t \tag{4}$$

The resultant change in $_F$ induced by FN tunneling is:

$$\Delta Q_{FG} = \int_0^{t_1} J_{FN}\, S\, dt \tag{5}$$

where S denotes the actual tunneling area of the M_T. By combining equations (1-5), the state equation of the memory cell can be derived as:

$$\frac{dV_{FG}}{dt} = \underbrace{\frac{C_T V_E}{C_{total}t_{ramp}}}_{bias-couplin\ Component} -$$
$$\underbrace{\frac{AS}{C_{total}t_{ox}^2}(V_{TWL} - V_{FG} + V_{cal})^2 \exp\left(-\frac{Bt_{ox}}{V_{TWL}-V_{FG}+V_{cal}}\right)}_{tunneling\ Component} \tag{6}$$

The state equation for programming operation is derived analogously as follows:

$$\frac{dV_{FG}}{dt} = \underbrace{\frac{(C_C+C_T)V_P}{C_{total}t_0}}_{bias-couplin\ Component} -$$
$$\underbrace{\frac{AS}{C_{total}t_{ox}^2}(V_{FG} - V_{cal})^2 \exp\left(-\frac{Bt_{ox}}{V_{FG}-V_{cal}}\right)}_{tunneling\ Component} \tag{7}$$

It can be observed that variations in the FG potential arise from two competing factors: (1) an enhancement term due to bias-induced capacitive coupling, and (2) a decay term from the FN tunneling. The influence of the tunneling component intensifies with increasing t_{ramp}. Given the strongly nonlinear nature of this equation, we employed the fourth-order Runge-Kutta method for numerical solution [9]. We define the FG potential at t t_{ramp} as V_0. Given predetermined memory cell parameters, each t_{ramp} value corresponds uniquely to a V_0. Fig. 3 depicts the evolution of V_F over time during both the erasing and the programming operations, with t_{ramp} varied across the range of 1 s to 2 ms.

979-8-3315-8850-2/25 $31.00 © 2025 IEEE

Based on the curve characteristics, a logarithmic function was selected for fitting, yielding a semi-empirical relationship between V_0 and t_{ramp}.

$$V_0(t) = V_{coupling} + \beta \ln(1 + \gamma t) \tag{8}$$

$$V_{coupling} = \begin{cases} \dfrac{C_T V_E}{C_{total}} + V_{FG0}, erase \\ \dfrac{(C_C + C_T)V_P}{C_{total}} + V_{FG0}, prog \end{cases} \tag{9}$$

where $V_{coupling}$ designates the pure capacitive coupling voltage in the absence of tunneling, while V_{F_0} corresponds to the initial FG potential. Here, and are fitting coefficients characterizing, respectively, the logarithmic growth rate of FG voltage induced by tunneling and the characteristic time scale for the transition from linear to logarithmic dependence. Both and depend on parameters influencing the tunneling strength. Using V_0 as the initial condition, the model for the next phase is established. During Phase- operation, once the charge pump ripple is neglected and bias voltage is stabilized at the preset high value, the FG potential becomes exclusively governed by FN tunneling. To simplify the analytical solution, we reset the time origin for Phase- by defining t t t_{ramp}. Finally, this framework enables the deriving closed-form analytical expressions for FG voltage versus time during both erasing and programming operations:

$$V_{FG}(t) =$$

$$\begin{cases} V_E + V_{cal} - \dfrac{Bt_{ox}}{\ln\left(\dfrac{ASB}{C_{total}t_{ox}}t + e^{\frac{Bt_{ox}}{V_E + V_{cal} - V_0}}\right)}, erase \\ \dfrac{Bt_{ox}}{\ln\left(\dfrac{ASB}{C_{total}t_{ox}}t + e^{\frac{Bt_{ox}}{V_0 - V_{cal}}}\right)} + V_{cal}, prog \end{cases} \tag{10}$$

B. Stable Time

Programming and erasing times significantly impact the threshold voltage window of memory cells. Therefore, determining appropriate values for these timing parameters is a critical task for circuit engineers. We define t_{stable} as the time elapsed from the application of the bias voltage until the absolute rate of V_F change reaches 100 V/s, denoting the point where the tunneling stabilizes. By defineing the time interval from the start of Phase- to the stabilization point as t_{stb2}, t_{stable} equals the sum of t_{stb2} and t_{ramp}. Based on the definitions above, we can formulate the equation for t_{stb2} as follows:

$$\frac{k_1 k_2}{(k_2 t + k_3)[\ln(k_2 t + k_3)]^2} = 100 \tag{11}$$

where

$$k_1 = Bt_{ox} \tag{12}$$

$$k_2 = \frac{ASB}{C_{total}t_{ox}} \tag{13}$$

$$k_3 = e^{\frac{Bt_{ox}}{V_{oxo}}} \tag{14}$$

$$V_{ox} = \begin{cases} V_E + V_{cal} - V_0, erase \\ V_0 - V_{cal}, prog \end{cases} \tag{15}$$

To solve this transcendental equation, we employ the Lambert W function [10], yielding an approximated solution. The initial step involves rearranging the equation into the following form:

$$\omega e^{\omega} = z \tag{16}$$

$$\omega = \frac{\ln(k_2 t + k_3)}{2} \tag{17}$$

$$z = \frac{1}{2}\sqrt{\frac{k_1 k_2}{100}} \tag{18}$$

For z 0, this equation yields a unique solution (z) on the principal branch. A seventh-order approximation of (z) is obtainable through the expansion:

$$W(z) = \ln z - \ln(\ln z) + \sum_{k=1}^{6} \frac{(-1)^k}{k}\left(\frac{\ln(\ln z)}{\ln z}\right)^k \tag{19}$$

To meet the stringent precision requirements, we apply a single Newton iteration step to this value, ultimately yielding the t_{stb2} solution:

$$t_{stb2} = \frac{e^{2W(z)new - k_3}}{2} \tag{20}$$

where

$$W(z)_{new} = W(z)_{old} - \frac{W(z)_{old}e^{W(z)_{old}} - z}{e^{W(z)_{old}}(1 + W(z)_{old})} \tag{21}$$

C. Threshold Voltage of the MTP Cell

The threshold voltage of a single read transistor, denoted as V_{th_rd}, is defined as the gate voltage V that produces a drain current I_D 1 $A \times ($ $_{rd} L_{rd})$. Correspondingly, the memory cell s threshold voltage V_{th_cell} is defined as the control gate voltage V_C $_L$ that yields an equivalent drain current I_D 1 $A \times ($ $_{rd} L_{rd})$ in the read transistor. Upon completion of erasing or programming operations, the external applied bias voltages rapidly return to zero. The FG potential subsequently stabilizes at a fixed value governed by charge conservation principles. The relationship quantifying the resultant threshold voltage shift (ΔV_{th_cell}) of the memory cell, induced by alterations in FG potential, is given by:

$$\Delta V_{th\_cell} = \frac{\left(V_{coupling} - V_{FG}(t_{stable})\right)C_{total}}{C_C} \tag{22}$$

III. MODEL VALIDATION

To accurately characterize the impact of external bias voltage ramp rate on the device performance, the Phase- model is validated first. Fig. 4 presents the extracted V values corresponding to the distinct t_{ramp} settings, where solid lines represent the fitting curves and the discrete points which denote the numerical simulation results. The coefficient of determination (R) is employed to quantitatively evaluate model fidelity. Both the erasing and programming conditions exhibit statistically R values approaching unity, indicating excellent agreement between the model and simulations.

To evaluate the model s generalization capability across the parametric variations, the transient response fitting is conducted under diverse operational conditions. Fig. 5 characterizes the functional dependencies of V_0 on t_{ramp} with the systematic variations in bias voltage amplitudes and initial FG potentials, presenting the comprehensive fitting results across these parametric conditions. R consistently exceeds 0.99, statistically confirming the model s robustness against parameter perturbations.

979-8-3315-8850-2/25 $31.00 © 2025 IEEE

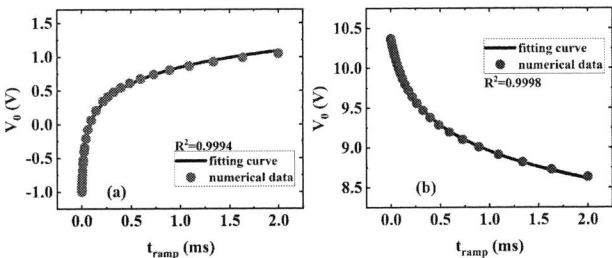

Fig. 4. Transient V_F at phase completion versus t_{ramp} (a) erasing operation, (b) programming operation.

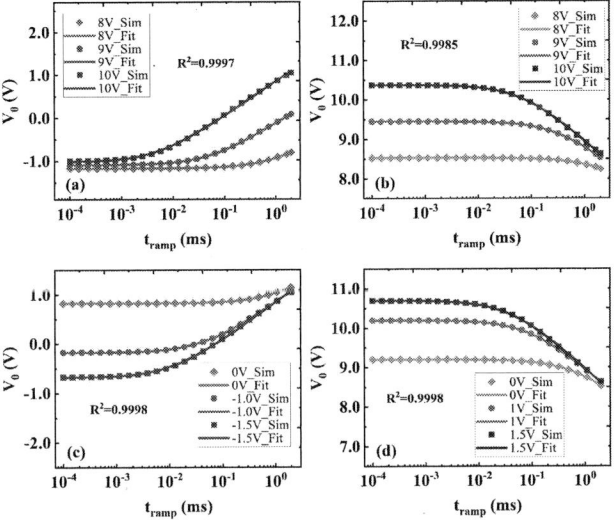

Fig. 5. V_0 versus t_{ramp} under parametric variations (a) erasing operation with V_E 8V, 9V, 10V, (b) programming operation with V_P 8V, 9V, 10V, (c) erase operation with $V_{F\,0}$ 0V, -1V, -1.5V, (d) prog operation with $V_{F\,0}$ 0V, 1V, 1.5V.

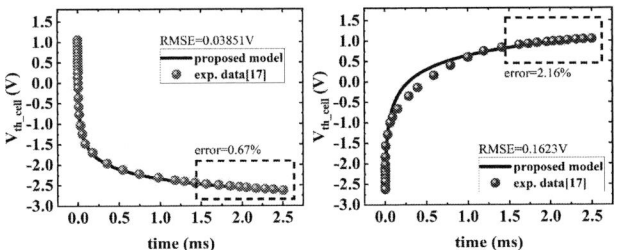

Fig. 6. Transient response of the MTP cell threshold voltage (a) erasing operation, (b) programming operation.

To validate the model, we extracted the threshold voltage variations under different pulse durations and compared these with the experimental data reported in [7]. The MTP cell is configured with the following parameters: M_C (W/L 2.5 m/2.6 m), M_T (W/L 1 m/0.7 m), M_{RD} (W/L 1 m/0.7 m), tunnel oxide thickness of 7 nm, and erasing/programming voltage of 10 V. As demonstrated in Fig. 6, the model predictions show good agreement with the experimental results during both programming and erasing operations, with the maximum root mean square error (RMSE) between proposed model and experimental results of 0.1623V and the maximum relative error of 2.16 in the post-tunneling stabilization phase.

IV. CONCLUSION

This paper has proposed a compact analytical model for the MTP memory cell, directly capturing the characteristics of the memory cell through closed-form expressions. Good agreement is demonstrated between the model calculations and experimental results. Compared to the existing modeling approaches, the proposed model dispenses with complex iterative procedures, thereby mitigating convergence issues, potential accuracy discrepancies, and excessive simulation time that can arise from improper iteration step sizes. In summary, this model provides the rapid evaluation of cell design parameters while maintaining accuracy, facilitating a significant reduction in design cycle time.

ACKNOWLEDGMENT

This work was supported in part by the National Natural Science Foundation of China under Grant 62274063, and 62274064, in part by Shanghai Science and Technology Explorer Plan under Grant 22TS1401700, 22TS1401400, and 24TS1400100. The authors gratefully acknowledge these supports.

REFERENCES

[1] H. Dagan, A. Shapira, A. Teman, A. Mordakhay, S. Jameson, E. Pikhay, V. Dayan, Y. Roizin, E. Socher, and A. Fish, "A Low-Power Low-Cost 24 GHz RFID Tag With a C-Flash Based Embedded Memory," (in English), IEEE JOURNAL OF SOLID-STATE CIRCUITS, vol. 49, no. 9, pp. 1942-1957, SEP 2014.

[2] W.-C. Zhuang, C.-T. Chien, C. J. Lin, and Y.-C. King, "Self-Clamping Programming in Narrow-Bridge Floating Gate Cells for Multi-Level Logic Non-Volatile Memory Applications," IEEE Journal of the Electron Devices Society, vol. 8, pp. 681-685, 2020.

[3] L. Te-Liang, T. Yi-Hung, L. Wun-Jie, Y. Hsiao-Lan, L. Chiu-Wang, L. Chrong Jung, and K. Ya-Chin, "A New Differential P-Channel Logic-Compatible Multiple-Time Programmable (MTP) Memory Cell With Self-Recovery Operation," IEEE Electron Device Letters, vol. 32, no. 5, pp. 587-589, 2011.

[4] S. Xu, H. Wang, J. Wu, L. Zheng, and J. Diao, "A new multitime programmable non-volatile memory cell using high voltage NMOS," Microelectronics Reliability, vol. 88-90, pp. 169-172, 2018.

[5] F. Torricelli, L. Milani, A. Richelli, L. Colalongo, M. Pasotti, and Z. M. Kovacs-Vajna, "Half-MOS Single-Poly EEPROM Cell in Standard CMOS Process," IEEE Transactions on Electron Devices, vol. 60, no. 6, pp. 1892-1897, 2013.

[6] L. Chiu-Wang, W. Haw-Yun, T. Cheng-Wei, H. Chen-Mei, C. Yue-Der, L. Te-Liang, and L. Chrong Jung, "A New 2T Contact Coupling Gate MTP Memory in Fully CMOS Compatible Process," IEEE Transactions on Electron Devices, vol. 59, no. 7, pp. 1899-1905, 2012.

[7] C. Li, S.- . Xu, J.-C. Li, Y.-L. Chen, and H.-Y. Wang, "A Compact Model for Single-Poly Multitime Programmable Memory Cells," IEEE Transactions on Electron Devices, vol. 63, no. 2, pp. 675-683, 2016.

[8] R. Duane, A. Concannon, P. O. Sullivan, and A. Mathewson, "Advanced Numerical Modelling of Non-Volatile Memory Cells," in 28th European Solid-State Device Research Conference, 8-10 Sept. 1998, pp. 304-307.

[9] T. Monovasilis, Z. Kalogiratou, and T. E. Simos, "Construction of Exponentially Fitted Symplectic Runge–Kutta–Nyström Methods from Partitioned Runge–Kutta Methods," Mediterranean Journal of Mathematics, vol. 13, no. 4, pp. 2271-2285, 2016.

[10] R. M. Corless, G. H. Gonnet, D. E. G. Hare, D. J. Jeffrey, and D. E. Knuth, "On the Lambert W function," (in English), Adv. Comput. Math., Article vol. 5, no. 1, pp. 329-359, 1996.

979-8-3315-8850-2/25 $31.00 © 2025 IEEE

2025 The 10th International Conference on Integrated Circuits and Microsystems

A Hybrid Computing-in-Memory Architecture for Energy-Efficient Edge AI Inference

Fangchao Lou[a], Dandan Chen[b], Jian Cao[a], Xing Zhang[a], Bo Zhang[a]
[a]Peking University, Beijing, China
[b]Anhui University, Hefei, China
Corresponding Author Email: caojian@ss.pku.edu.cn   zhx@pku.edu.cn

Abstract—To address the "memory wall" bottleneck in traditional von Neumann architectures and meet the growing demand for energy-efficient edge AI inference, this paper proposes an SRAM-based hybrid Computing-in-Memory (CIM) architecture that integrates analog and digital computing units. By combining the high energy efficiency of Analog CIM (ACIM) in low-precision parallel operations (such as convolution) with the high accuracy of Digital CIM (DCIM) in precision-critical tasks (such as activation functions), this architecture achieves a balanced trade-off between efficiency and accuracy.

The core innovations of this architecture lie in three aspects: hardware-aware design, optimized dataflow scheduling, and integration of security mechanisms. At the hardware level, it is customized for edge AI operators like convolution, fully leveraging SRAM's CMOS compatibility and parallel computing capabilities. For dataflow scheduling, it adopts sliding window-aware input caching and a "compute-while-update" pipelining mechanism to significantly reduce data movement overhead. Additionally, it incorporates Advanced Encryption Standard (AES) operations into the CIM framework to ensure the security of data processing.

Experimental validation based on YOLO model inference shows that compared with traditional designs, this architecture reduces data movement energy consumption by 37% and computation latency by 29%, providing a robust low-power solution for edge AI applications that require both high energy efficiency and secure operation.

Index Terms—CIM,SRAM-based hybrid,Edge AI inference

I. INTRODUCTION

To overcome the "memory wall" bottleneck in conventional von Neumann architectures, Computing-in-Memory (CIM) has emerged as a promising paradigm. By integrating computation directly into memory arrays, CIM enables data-centric processing that significantly reduces data movement, improves system efficiency, and lowers energy consumption.

As shown in Fig. 1, SRAM-based CIM extends traditional 6T cells to 8T or 10T structures, enabling in-situ multiply-and-accumulate (MAC) operations [1], [2]. These operations are executed in the analog domain and digitized via on-chip ADCs, allowing massively parallel matrix-vector multiplication within the memory array—ideal for accelerating AI workloads.

Unlike traditional accelerators, CIM systems broadcast input vectors while keeping weights stationary, minimizing data transfers and reducing latency and power. As illustrated in Fig. 2, this makes CIM highly suitable for edge AI inference

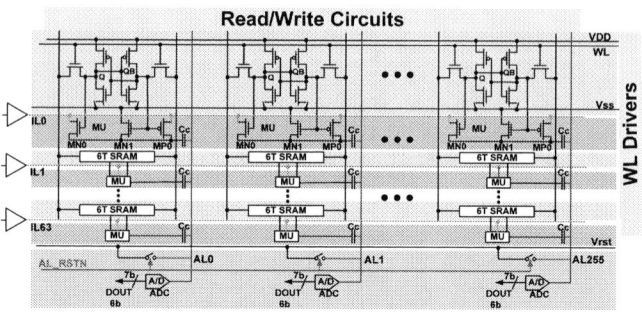

Fig. 1. Schematic Diagram of SRAM CIM Architecture

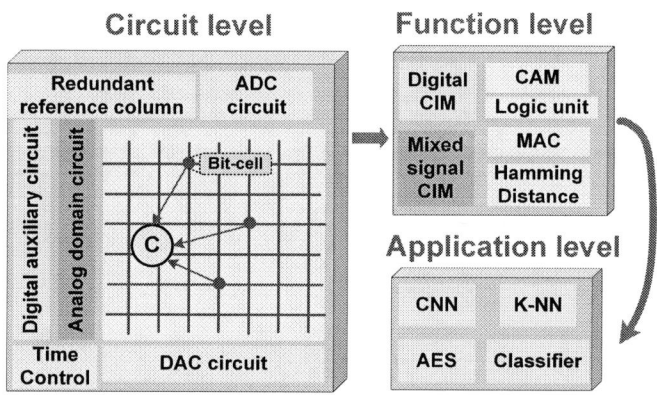

Fig. 2. Architectural Overview of CIM

tasks such as convolutional neural networks (CNN) [3], [4], where real-time and low-power processing is essential.

Recent CIM research includes emerging memory technologies (e.g., ReRAM, PCM, MRAM), analog/mixed-signal circuits, and architecture-level AI optimization [5]–[7]. Among them, SRAM-CIM stands out for its CMOS compatibility and speed, making it well-suited for edge deployment.

This paper proposes a hybrid SRAM-CIM architecture that integrates analog and digital MAC units, optimized dataflow scheduling, and hardware-aware quantization to achieve a balance between accuracy and energy efficiency. It offers a robust, low-power inference solution for edge AI applications [8].

The main contributions are:

1)A high-efficiency SRAM-CIM architecture optimized for

979-8-3315-8850-2/25 $31.00 © 2025 IEEE   805

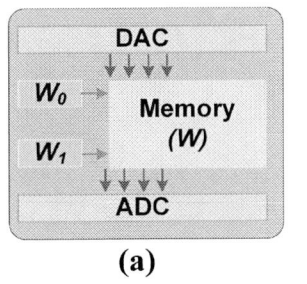 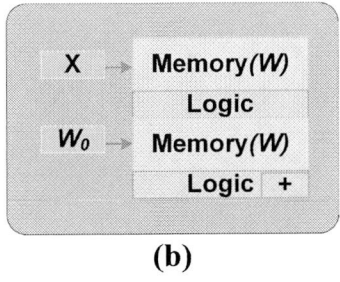

Fig. 3. (a)ACIM structure   (b)DCIM structure

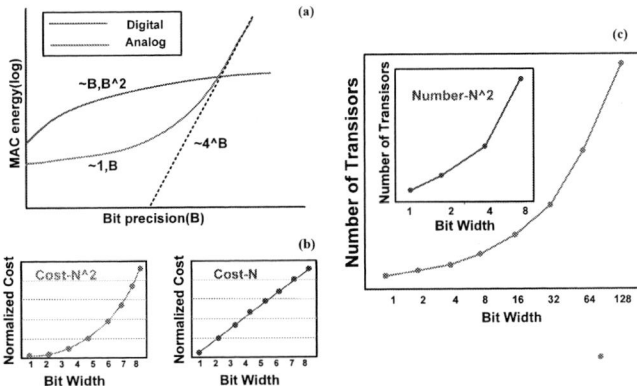

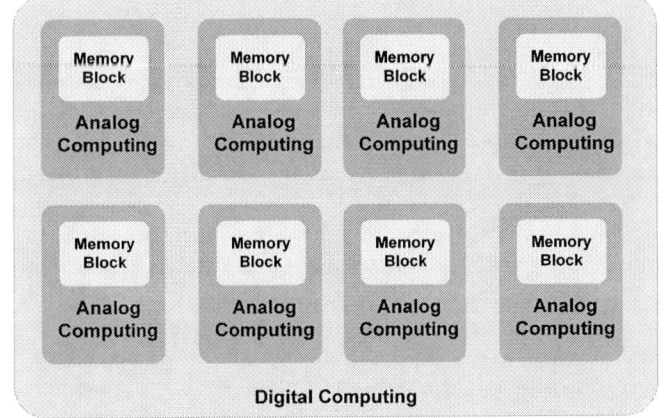

Fig. 4. (a)Trend of Overhead Variation in ACIM and DCIM
(b) Hardware overhead of bit-serial and bit-parallel
multiplication in DCIM
(c)Accumulation Overhead in DCIM

edge AI operators such as convolution and fully connected layers;

2)A hardware-aware quantization and dataflow reuse scheme tailored for CIM hardware constraints;

3)A hybrid analog-digital MAC design validated through YOLO-based inference, showing superior energy efficiency and robustness.

The rest of the paper is organized as follows: Section II introduces the proposed architecture. Section III presents experiments and case studies. Section IV concludes the paper.

II. PROPOSED ARCHITECTURE

A. Overall Chip Architecture

The hybrid CIM architecture proposed in this study integrates the strengths of both Analog CIM (ACIM) and Digital CIM (DCIM) to achieve a balanced trade-off between energy efficiency and computational accuracy at both the circuit and system levels.

As shown in Fig. 3, from a technical perspective, ACIM performs MAC operations by leveraging the physical properties of memory media [9], demonstrating excellent energy efficiency in low-precision scenarios. It features a crossbar structure for high-parallelism computation, with simple hardware and low data movement energy. However, its performance is susceptible to noise, device variation, and limited precision. In contrast, DCIM relies entirely on digital logic using multipliers and adder trees to achieve high-precision results with strong noise immunity and controllable accuracy. Nonetheless, its hardware cost increases quadratically with precision [10], leading to higher power and area consumption.

As illustrated in Fig. 4, the proposed hybrid architecture utilizes ACIM within compute cores to handle low-precision, parallel operations such as convolutions, leveraging its high energy efficiency. DCIM modules are integrated across cores to correct the precision degradation of analog computations and support tasks that require higher accuracy [11]. The rationale for this hybrid design, as shown in Fig. 5, is based on the following observations:

When computational precision is below 8-bit, ACIM achieves significantly higher energy efficiency. Meanwhile, DCIM can digitally compensate for accuracy loss, supporting diverse precision requirements in edge AI tasks.

Fig. 5. Hybrid Digital-Analog Computing Architecture

The low area cost of ACIM, combined with the scalable nature of DCIM, enables the architecture to maximize computational density within limited chip area [12].

In CNN, where convolution operations account for over 50% of total workload, ACIM can directly perform matrix-vector multiplication within memory arrays, while DCIM is well-suited for high-precision operations such as activation functions and normalization—enabling efficient end-to-end acceleration.

This work proposes a high-efficiency edge AI chip architecture centered around an SRAM-based CIM accelerator, as depicted in Fig. 6. The overall structure follows the classical System-on-Chip (SoC) design paradigm, where a CPU serves as the core, interconnected via a central data bus to subsystems including memory, digital peripherals, and analog peripherals [13].

The key difference from traditional SoC lies in the inclusion of a CIM Accelerator based on SRAM in-memory computing. This accelerator features a data cache for weight storage, a CIM computation core, and an AHB bus controller for

979-8-3315-8850-2/25 $31.00 © 2025 IEEE   806

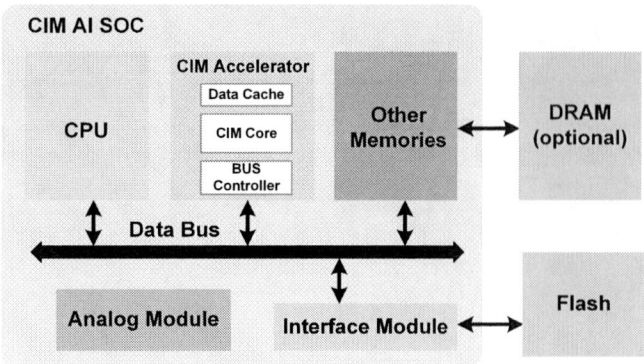

Fig. 6. CIM-Based Edge AI Chip Architecture

communication with the system bus.

B. Application of CIM in CNN and AES Algorithms

Edge AI chips focus on efficiently running neural network inference, with CNN as the main computational load. Among CNN operators, convolution accounts for over 50% of computation, making it the key target for acceleration. Convolution involves sliding-window local feature extraction, essentially performing matrix-vector multiplications [14], which suits the parallelism of CIM architectures—ideal for CIM-based acceleration.

Simultaneously, Advanced Encryption Standard (AES) is crucial for secure data transmission in edge AI, involving round key generation and iterative rounds with SubBytes, ShiftRows, MixColumns, and AddRoundKey operations. CIM architectures benefit AES implementation by embedding S-boxes in memory for in-situ substitution, leveraging memory parallelism for ShiftRows and MixColumns, and performing AddRoundKey XOR operations directly within memory—reducing data movement and power consumption [15], [16].

As shown in Figure 7, the proposed SRAM-based CIM convolution accelerator combines optimized hardware and dataflow scheduling to achieve energy-efficient computation. Integrating convolution acceleration and AES encryption in one CIM architecture enhances neural network inference performance while ensuring secure processing on edge AI chips [17].

III. Accelerator Design for AI

A. Convolution Operator Acceleration

This work proposes a CIM architecture utilizing 8-bit SRAM as WEIGHT MEM. As shown in Figure 8, each CIM BLOCK comprises a WEIGHT MEM, COMPUTE MODULE, and WRITEBACK MEM. Nine CIM BLOCKs, together with nine Sign Com units, form one CIM TOP to execute signed 3×3 convolutions. The complete CIM IP integrates 64 such CIM TOPs.

Each WEIGHT MEM includes four 8-bit SRAMs, maintaining a 4:1 ratio with the compute and writeback logic. This design allows efficient storage of 64 sets of 3×3 kernels using

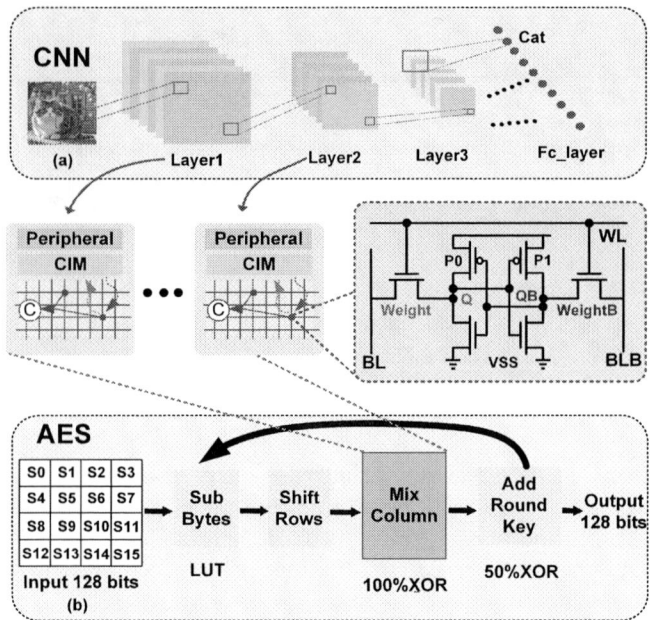

Fig. 7. Implementation of (a) CNN and (b) AES on Multiple SRAM Arrays

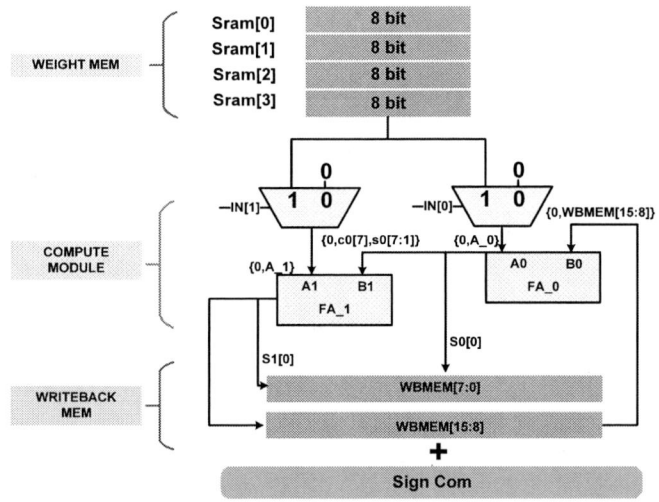

Fig. 8. 1-bit CIM Architecture Diagram

only 16×9 COMPUTE MODULEs, WRITEBACK MEMs, and Sign Com units, thereby reducing area.

Figure 9 shows the weight mapping scheme: one input FEATURE corresponds to four kernels (WEIGHT0–3), with their weights stored across Sram[0]–[3] in CIM BLOCK0–8. During computation, these weights are time-multiplexed and processed by the compute units to perform 3×3 convolutions.

B. Bus Interface Design and Data Flow Optimization of CIM Core

The interaction between the CIM core and the system bus, along with data flow scheduling, is crucial for the energy efficiency of edge AI chips. The CIM IP designed in this

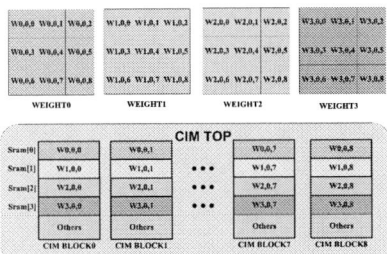

Fig. 9. Weight mapping in CIM TOP

paper includes 64 CIM TOP units, capable of storing 64 8-bit signed 3×3 convolution kernels. It connects to the bus via an AHB controller, with its configuration registers, input Cache (4×4×8bit), and output Cache forming an efficient interface architecture.

As shown in Figure 10, the input Cache is optimized for convolutional sliding windows, storing 4×4 feature subarrays and reusing overlapping data to reduce data movement. Combined with a "compute-while-update" pipeline mechanism, it shortens latency and reduces controller area [18]–[20].

Data flow follows the "storage-computation-transmission" collaboration principle. As illustrated in Figure 11(a), weights are written to SRAM, inputs are stored in Cache, and intermediate results after computation are temporarily stored in the output Cache. Figure 11(b) shows that when reaching the threshold, data is moved to QSPI FLASH by DMA. The CIM can switch to normal SRAM mode, and DMA reduces CPU intervention, improving bus bandwidth utilization.

This design reduces data movement energy consumption by 37% and computation latency by 29%, providing support for the high-energy-efficiency operation of edge AI chips.

Fig. 10. sliding window convolution

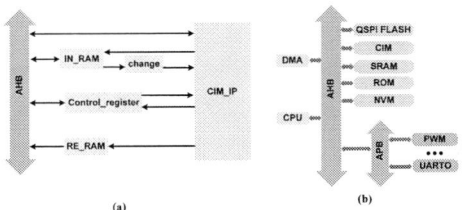

Fig. 11. (a) Connection relationship between CIM IP and bus (b) Data flow diagram of high-efficiency edge AI chip

IV. CONCLUSION

This paper combines the energy efficiency of ACIM for low-precision parallel operations with the high-precision advantages of DCIM to balance performance and efficiency. It also optimizes data flow scheduling (including sliding window caching and pipelined updates) to reduce data movement, which is a major source of energy consumption in edge devices. Experimental results show that it reduces data movement energy consumption by 37% and shortens latency by 29%, effectively supporting tasks such as CNN inference. Future work will focus on improving scalability for large neural networks and enhancing robustness against device variations, providing an efficient, secure, and high-performance development path for next-generation edge AI chips.

REFERENCES

[1] P. -C. Wu et al., "A 22nm 832Kb Hybrid-Domain Floating-Point SRAM In-Memory-Compute Macro with 16.2-70.2TFLOPS/W for High-Accuracy AI-Edge Devices," 2023 IEEE International Solid-State Circuits Conference (ISSCC), San Francisco, CA, USA, 2023, pp. 126-128, doi: 10.1109/ISSCC42615.2023.10067527.

[2] Z. Li, F. Liu, W. Yang, S. Peng and J. Zhou, "A Survey of Convolutional Neural Networks: Analysis, Applications, and Prospects," in IEEE Transactions on Neural Networks and Learning Systems, vol. 33, no. 12, pp. 6999-7019, Dec. 2022, doi: 10.1109/TNNLS.2021.3084827.

[3] H. Fujiwara et al., "A 5-nm 254-TOPS/W 221-TOPS/mm2 Fully-Digital Computing-in-Memory Macro Supporting Wide-Range Dynamic-Voltage-Frequency Scaling and Simultaneous MAC and Write Operations," 2022 IEEE International Solid-State Circuits Conference (ISSCC), San Francisco, CA, USA, 2022, pp. 1-3, doi: 10.1109/ISSCC42614.2022.9731754.

[4] H. Mori et al., "A 4nm 6163-TOPS/W/b $4790 - \mathbf{TOPS/mm^2/b}$ SRAM Based Digital-Computing-in-Memory Macro Supporting Bit-Width Flexibility and Simultaneous MAC and Weight Update," 2023 IEEE International Solid-State Circuits Conference (ISSCC), San Francisco, CA, USA, 2023, pp. 132-134, doi: 10.1109/ISSCC42615.2023.10067555.

[5] S. -E. Hsieh et al., "7.6 A 70.85-86.27TOPS/W PVT-Insensitive 8b Word-Wise ACIM with Post-Processing Relaxation," 2023 IEEE International Solid-State Circuits Conference (ISSCC), San Francisco, CA, USA, 2023, pp. 136-138, doi: 10.1109/ISSCC42615.2023.10067335.

[6] C. -F. Lee et al., "A 12nm 121-TOPS/W 41.6-TOPS/mm2 All Digital Full Precision SRAM-based Compute-in-Memory with Configurable Bit-width For AI Edge Applications," 2022 IEEE Symposium on VLSI Technology and Circuits (VLSI Technology and Circuits), Honolulu, HI, USA, 2022, pp. 24-25, doi: 10.1109/VLSITechnologyand-Cir46769.2022.9830438.

[7] W. Sun et al., "A Survey of Computing-in-Memory Processor: From Circuit to Application," in IEEE Open Journal of the Solid-State Circuits Society, vol. 4, pp. 25-42, 2024, doi: 10.1109/OJSSCS.2023.3328290.

[8] P. Chen et al., "A 22-nm Delta-Sigma Computing-In-Memory SRAM Macro With Near-Zero-Mean Outputs and LSB-First ADCs for Edge AI Processing," in IEEE Journal of Solid-State Circuits, vol. 60, no. 8, pp. 3020-3032, Aug. 2025, doi: 10.1109/JSSC.2025.3539736.

[9] P. Mannocci et al., "An SRAM-based reconfigurable analog in-memory computing circuit for solving linear algebra problems," 2023 International Electron Devices Meeting (IEDM), San Francisco, CA, USA, 2023, pp. 1-4, doi: 10.1109/IEDM45741.2023.10413724.

[10] V. Sze, Y. -H. Chen, T. -J. Yang and J. S. Emer, "Efficient Processing of Deep Neural Networks: A Tutorial and Survey," in Proceedings of the IEEE, vol. 105, no. 12, pp. 2295-2329, Dec. 2017, doi: 10.1109/JPROC.2017.2761740.

[11] A. Biswas and A. P. Chandrakasan, "Conv-RAM: An energy-efficient SRAM with embedded convolution computation for low-power CNN-based machine learning applications," 2018 IEEE International Solid-State Circuits Conference - (ISSCC), San Francisco, CA, USA, 2018, pp. 488-490, doi: 10.1109/ISSCC.2018.8310397.

[12] X. Si et al., "A Twin-8T SRAM Computation-in-Memory Unit-Macro for Multibit CNN-Based AI Edge Processors," in IEEE Journal of Solid-State Circuits, vol. 55, no. 1, pp. 189-202, Jan. 2020, doi: 10.1109/JSSC.2019.2952773.

[13] Y. Zhan, W. -H. Yu, K. -F. Un, R. P. Martins and P. -I. Mak, "GSLP-CIM: A 28-nm Globally Systolic and Locally Parallel CNN/Transformer Accelerator With Scalable and Reconfigurable eDRAM Compute-in-Memory Macro for Flexible Dataflow," in IEEE Transactions on Circuits and Systems I: Regular Papers, vol. 72, no. 4, pp. 1657-1667, April 2025, doi: 10.1109/TCSI.2024.3497187.

[14] F. Tu et al., "TranCIM: Full-Digital Bitline-Transpose CIM-based Sparse Transformer Accelerator With Pipeline/Parallel Reconfigurable Modes," in IEEE Journal of Solid-State Circuits, vol. 58, no. 6, pp. 1798-1809, June 2023, doi: 10.1109/JSSC.2022.3213542.

[15] J. Yue et al., "A 28nm 16.9-300TOPS/W Computing-in-Memory Processor Supporting Floating-Point NN Inference/Training with Intensive-CIM Sparse-Digital Architecture," 2023 IEEE International Solid-State Circuits Conference (ISSCC), San Francisco, CA, USA, 2023, pp. 1-3, doi: 10.1109/ISSCC42615.2023.10067779.

[16] Z. Lin, Y. Liu, Y. Wang, Y. Zhao, C. Peng and X. Wu, "SRAM-Based Digital CIM Macro for Linear Interpolation and MAC," 2024 IEEE International Symposium on Circuits and Systems (ISCAS), Singapore, Singapore, 2024, pp. 1-5, doi: 10.1109/ISCAS58744.2024.10558525.

[17] X. Si et al., "A Local Computing Cell and 6T SRAM-Based Computing-in-Memory Macro With 8-b MAC Operation for Edge AI Chips," in IEEE Journal of Solid-State Circuits, vol. 56, no. 9, pp. 2817-2831, Sept. 2021, doi: 10.1109/JSSC.2021.3073254.

[18] M. Wu et al., "S2D-CIM: A 22nm 128Kb Systolic Digital Compute-in-Memory Macro with Domino Data Path for Flexible Vector Operation and 2-D Weight Update in Edge AI Applications," 2024 IEEE Custom Integrated Circuits Conference (CICC), Denver, CO, USA, 2024, pp. 1-2, doi: 10.1109/CICC60959.2024.10529046.

[19] H. You, W. Li, D. Shang, Y. Zhou and S. Qiao, "A 1–8b Reconfigurable Digital SRAM Compute-in-Memory Macro for Processing Neural Networks," in IEEE Transactions on Circuits and Systems I: Regular Papers, vol. 71, no. 4, pp. 1602-1614, April 2024, doi: 10.1109/TCSI.2024.3355944.

[20] Y. Liu et al., "TSCIM: A 28nm Transposed Stochastic CIM Macro for On-Chip Training and Inference," 2025 IEEE International Symposium on Circuits and Systems (ISCAS), London, United Kingdom, 2025, pp. 1-5, doi: 10.1109/ISCAS56072.2025.11044117.

2025 The 10th International Conference on Integrated Circuits and Microsystems

A Reliable 512-kb HZO-based 2T2C FeRAM Array with Capacitor under Bitline

Jing Wang
Institute of Advanced Technology
University of Science and Technology of
China
Hefei, China
jingwang128@mail.ustc.edu.cn

Xiangyin Chen
Hefei Science of China Microelectronics
Innovation Center Co., Ltd
Hefei, China
chenxiangyin@hfmic.com

Zhongguang Xu*
School of Microelectronics
University of Science and Technology of
China
Hefei, China
xuxu@ustc.edu.cn

Abstract—This work highlights the excellent remnant polarization of $Hf_{0.5}Zr_{0.5}O_2$ (HZO)-based ferroelectric capacitors fabricated in a capacitor-under-bitline (CUB) configuration. Owing to the CUB architecture, the metal/ferroelectric/metal (MFM) capacitors are formed prior to the back-end-of-line (BEOL) process, enabling post-metallization annealing (PMA) at temperatures above 500 °C. The resulting structure exhibits outstanding performance, including remnant polarization (2Pr) > 30 $\mu C/cm^2$ and endurance greater than 10^{11} cycles. In addition, based on this CUB ferroelectric capacitor structure, a 512 kb memory array using a two-transistor two-capacitor (2T2C) architecture was fabricated. The chip integrates dedicated peripheral circuits such as address decoders, a charge pump (CP), and a sense amplifier (SA), demonstrating high reliability in experimental evaluation.

Keywords—*ferroelectric random-access memory, capacitor under bitline, hafnium oxide*

I. INTRODUCTION

The growing demand for low-power and high-speed embedded memory spans a wide range of applications, including data storage, in-memory computing and neuromorphic computing [1]. Ferroelectric HfO_2-based materials have emerged as promising candidates because of their CMOS-compatiblity, low switching current, and excellent scalability. Leveraging these properties, ferroelectric random access memory (FeRAM) provides a superior balance of endurance, high speed, low operating voltage, and energy efficiency, distinguishing it from other non-volatile memory solutions [2]. Although extensive research has been conducted on hafnium-based ferroelectric memory cells, further exploration is needed to fully exploit their performance and develop high-reliability ferroelectric memory chips.

High remnant polarization (2Pr > 40 $\mu C/cm^2$) in doped $Hf_{0.5}Zr_{0.5}O_2$ (HZO) thin films is typically realized through rapid thermal annealing (RTA) at temperatures beyond 500 °C [3]. However, such high-temperature processes pose significant challenges for transistor operation and back-end-of-line (BEOL) reliability in scaled CMOS nodes. To overcome this limitation, we introduce a capacitor-under-bitline (CUB) FeCAP structure that accommodates RTA above 500 °C, enabling improved remnant polarization without sacrificing process compatibility. The fabricated device exhibits robust ferroelectric properties, achieving a remnant polarization of 30 $\mu C/cm^2$ under a low operating voltage of 1.8 V at room temperature.

Based on this device, a 512kb FeRAM chip composed of four 128 kb subarrays using a two-transistor two-capacitor (2T2C) architecture was implemented. The chip integrates dedicated circuits including address decoders, a charge pump (CP), and a sense amplifier (SA), and demonstrates high reliability.This study paves the way for the development and application of large-scale, high-reliability FeRAM chips.

II. FABRICATION OF FECAP

Most existing studies on doped HZO thin films have employed RTA above 500 °C to achieve large remnant polarization [4]. This is particularly beneficial for high-density memory applications with thinner capacitors, as higher crystallization temperatures tend to improve ferroelectric properties [5]. However, the elevated thermal budget poses reliability challenges when integrating capacitors into advanced CMOS manufacturing processes, potentially degrading transistor performance and BEOL integrity. To address this, metal/ferroelectric/metal (MFM) capacitors composed of TiN/HZO/TiN stacks have been demonstrated to function effectively in 1T1C FeRAM arrays using annealing temperatures below 450 °C, thus avoiding metal degradation while preserving CMOS characteristics [6]. Alternatively, 2T2C FeRAM cells employing MFM capacitors with a CUB structure allow the application of RTA above 500 °C, achieving improved ferroelectric properties without compromising transistor or BEOL reliability.

In this study, HZO-based ferroelectric materials were applied to 2T2C memory cells fabricated using a 180 nm CMOS process. Following completion of the front-end-of-line (FEOL) process for transistor fabrication, the ferroelectric capacitors were integrated during the BEOL process, directly beneath the first metal layer (M1) of the MOS stack. Fig. 1 illustrates the process flow for the fabricated CUB-structured 2T2C ferroelectric memory cell.

The fabrication process began with the deposition of a TiN bottom electrode using radio-frequency (RF) sputtering. Subsequently, an 8 nm HZO film with a stoichiometric Hf:Zr atomic ratio of 1:1 was deposited by atomic layer deposition (ALD). Finally, a TiN top electrode was formed via RF sputtering, followed by RTA at 500 °C.

Upon completion of these steps, the HZO-based ferroelectric capacitor device was successfully fabricated, with a unit cell area of 0.9 μm × 0.7 μm.

979-8-3315-8850-2/25 $31.00 © 2025 IEEE

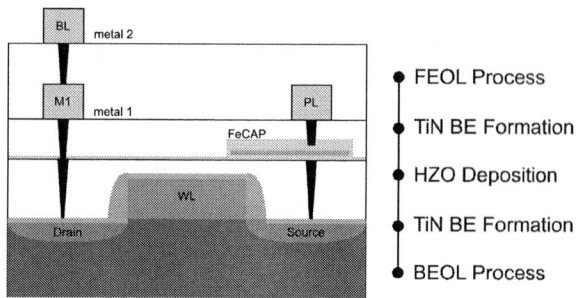

Fig. 1. Fabrication of the ferroelectric memory cell: the left panel shows the integration of the CUB FeCAP beneath the M1 metal layer, while the right panel illustrates the device fabrication process.

To comprehensively evaluate the effects of write voltage and cycles on the reliability of the device, fatigue characteristics were investigated under varying write voltages and cycling conditions, as depicted in Fig. 2.

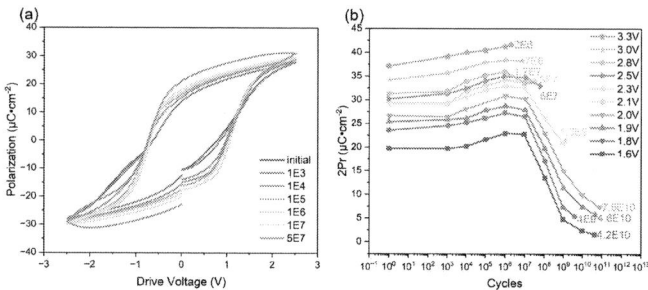

Fig. 2. (a)(P-V) loops of the HZO device measured after different wake-up cycles (0、10^3、10^4、10^5、10^6、10^7、5×10^7); (b)Endurance characteristics of the device under various write pulse amplitudes.

Fig. 2(a) presents the polarization-voltage (P-V) loops of HZO measured after different wake-up cycles (0、10^3、10^4、10^5、10^6、10^7、5×10^7). The initial HZO film exhibits antiferroelectric-like hysteresis behavior, indicating a dominant tetragonal (T) phase. During the wake-up process, both saturated polarization (Ps) and Pr increase, which may be attributed to two mechanisms: one is the depinning of ferroelectric domains induced by charge redistribution, and the other is the phase transition from non-ferroelectric phases such as monoclinic (M), tetragonal (T), or antipolar orthorhombic (O) to the ferroelectric O phase [7][8].

Fig. 2(b) illustrates the evolution of 2Pr as a function of the number of applied pulses under various write voltages. A higher write voltage results in increased 2Pr, consistent with the behavior observed in the P-V loops. However, beyond a critical cycling threshold, breakdown occurs, characterized by a sharp increase in leakage current and device failure. Specifically, when the applied voltage exceeds 2 V, fatigue is primarily manifested by breakdown characteristics. In contrast, for voltages below 1.8 V, 2Pr gradually decreases with increasing cycles. The dominant mechanism underlying this breakdown phenomenon is the formation and accumulation of defects, which stem from initial process-induced imperfections, the generation of oxygen vacancies, or lattice dislocations [9][10]. Based on these findings, a write voltage of 1.8 V was

chosen to optimize device performance while ensuring reliability.

III. ARCHITECTURE OF THE 512KB FERAM

The overall architecture of the 512-kb FeRAM chip is illustrated in Fig. 3(a), which consists of four 128-kb FeRAM blocks together with peripheral circuits, such as memory arrays, WL boosting charge-pump circuits, sense amplifiers, and PL/WL drivers. As shown in Fig. 3(b), each FeRAM block comprises 64 PL blocks arranged in a 4 × 16 matrix.Each PL block is surrounded by an equal number of dummy cells, aiming to provide identical environmental conditions for all PL blocks. The structure of a single PL block is depicted in Fig. 3(c), where each cell represents a 2T2C memory cell. As shown in Fig. 3(a), cells within the same sub-array share the same PL. In the 128-kb block configuration, the number of WLs is 256, and the number of BLs is 16*64. Each WL contains 512 bits due to the 2T2C structure, which employs two BLs to control a single bit cell. This configuration satisfies the 128-kb design requirement.

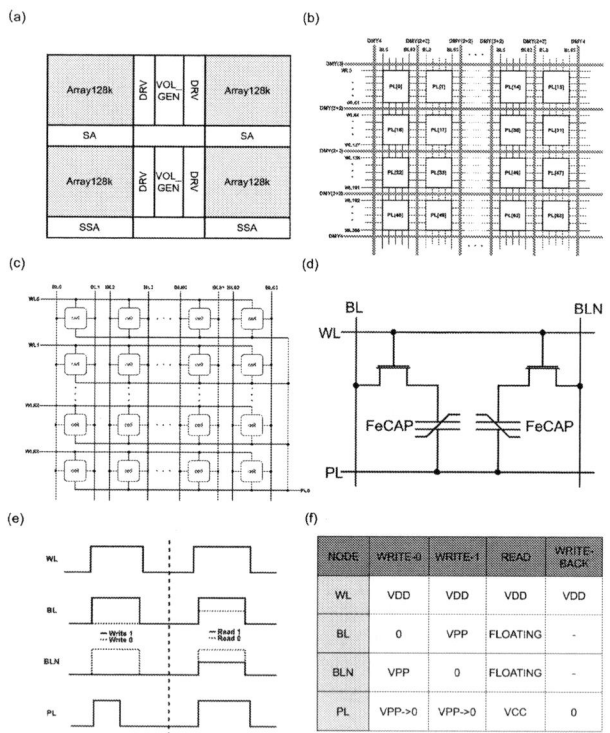

Fig. 3. Design of the 512-kb FeRAM chip, including the architecture of (a) 512-kb FeRAM chip, (b) 128k block, (c) PL block, and (d) 2T2C FeRAM cell; (e) Diagram of basic array operation, data 1 is shown with solid line. (f) Operation table for the FeRAM.

The operation table of this FeRAM cell is shown in Fig. 3(e) and (f). When the read or write operation begins, the WL is activated at VDD, which is the supply voltage ensuring that the access transistor is turned on. At the same time, the bitline (BL) and its complement (BLN) always maintain opposite voltages, fixed at 0 and VPP, respectively. In this way, the two FeCAPs store opposite polarization states.

979-8-3315-8850-2/25 $31.00 © 2025 IEEE

During the write "0" operation, BL is always connected to 0 while BLN is kept at VPP, and the PL is initially tied to a high voltage. At this moment, the FeCAP connected to BL polarizes downward, reaching the positive-saturation state, which corresponds to the logic "0." Then, PL is driven to a low level (0), forcing the FeCAP on BLN to polarize upward, corresponding to the logic "1." After the WL voltage is removed, both FeCAPs return to their respective remnant polarization states. When writing a "1", the process is opposite: BL remains tied to VPP, BLN is tied to 0, and PL toggles from VDD to ground.

For the read operation, the potentials of both BL and BLN are first initialized to ground and then left floating. As WL is raised, the access transistor turns on, and the potential of PL is increased. The parasitic capacitances of BL and BLN (CBL and CBLN) extract the ferroelectric charge through the hysteresis behavior of the FeCAPs and convert it into a voltage. Different results appear on BL and BLN, which are then amplified by SA, reflecting the "0" and "1" states.

Since reading the FeCAP state is destructive, PL is subsequently grounded to enable the automatic write-back process, thereby restoring the original polarization state.

IV. DESIGN OF THE PERPHERAL CIRCUITS

A. WL Boosting Charge Pump Circuit

In the 2T2C memory cell, the WL controls the activation of two access transistors, allowing the voltage on the bitlines to transfer to one terminal of FeCAP. However, NMOS transistors exhibit a threshold voltage (VTH) loss when passing high-level signals. Specifically, while an NMOS can transmit a low-level signal to achieve VDD at the output, it can only transmit a high-level signal up to VDD-VTH at the output. Consequently, if the WL voltage is only 3.3 V, the voltage across the FeCAP cannot reach 3.3 V. Under these conditions, the FeCAP may operate in a non-saturated polarization state, leading to read/write errors. To mitigate this, a charge pump circuit is employed to elevate the WL voltage to at least VDD+VTH. In the adopted 180-nm process, the threshold voltage typically does not exceed 1 V. Elevating the WL voltage to 4.3 V or higher thus satisfies the design requirements.

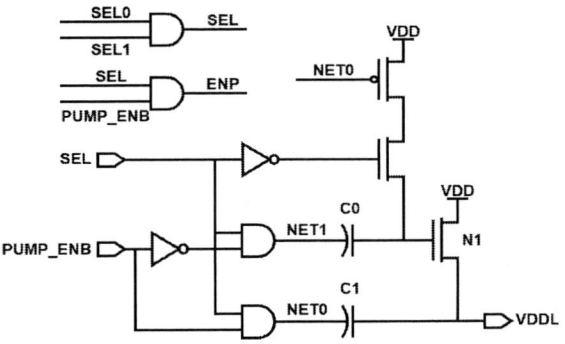

Fig. 4. Schematic diagram of WL boosting charge pump circuit

The WL boosting circuit operates primarily by exploiting the principle that the voltage across a capacitor cannot change instantaneously. Using clock signals to control voltage changes

across capacitors enables boosting the supply voltage. Fig. 4 illustrates the circuit structure of the WL booster, which consists of two primary stages. Besides control logic, the first stage—comprising transistor N0, transistor P0, and capacitor C0—elevates the gate voltage of N1 above VDD, providing the initial voltage for the second stage. The second stage, formed by transistor N1 and capacitor C1, primarily addresses the threshold voltage loss by boosting the VDDL voltage using the pulse signal from NET0.

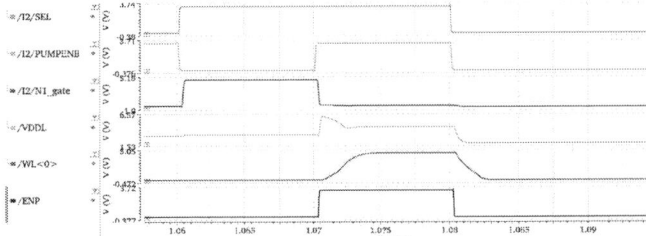

Fig. 5. Simulation results of the WL boosting circuit, with WL voltage remaining stable at 4.589 V

Simulation results of the WL boosting circuit are presented in Fig. 5. The results indicate that a high SEL signal is generated only when both SEL0 and SEL1 are high. Subsequently, a high ENP signal is generated only when both SEL and PUMP_ENB are high, enabling the transfer of the boosted VDDL voltage to WL for level conversion. Due to the inherent property that the voltage across capacitor C1 cannot change instantaneously, the VDDL voltage is further elevated to 2×VDD. Even under heavy load conditions, the WL voltage remains stable at 4.589 V, which exceeds VDD+VTH = 4.3 V, thereby ensuring reliable operation.

B. Sense Amplifier Circuit

To enhance read yield, a cross-coupled positive-feedback sense amplifier circuit architecture is employed in the readout circuitry. Fig. 6 illustrates the schematic of the proposed read circuit. As shown, the read circuit module consists of four sub-modules: a primary SA, a secondary Sense Amplifier with Storage (SSA), a Driver (DRV) module, and a Rewrite Comparator with Feedback (COMP_FB) module.

The primary operating principle involves transferring and comparing data from a pair of BLs. Under the control of logic gating signals, the corresponding BL data is transmitted to the SSA comparator. The SSA performs signal comparison under clock control, and the result is subsequently output via the DRV module. The COMP_FB module is responsible for data rewrite following the destructive read operation of the ferroelectric capacitor. It remains inactive during data reading, and is activated by control logic after the read operation to rewrite data to the BLs.

The SSA utilizes a latch structure. Prior to the arrival of the read signal, under the control of the ENP enable signal, the cross-coupled positive-feedback structure is deactivated, and its internal nodes are reset. During the read operation, the cross-coupled positive-feedback structure is triggered. The differential voltage between VP and VN, which representing the BL pair, drives the positive-feedback structure towards a rapid transition to one logic state, yielding the comparison result. The DRV output module, governed by logic control, manages the output and shutdown of the SSA result. Fig. 7 presents the simulation results of the proposed read circuit. It demonstrates that the output signal (SAOUT) transitions when the BL voltage difference is 522 µV, indicating that the read circuit achieves a minimum resolution of 522 µV.

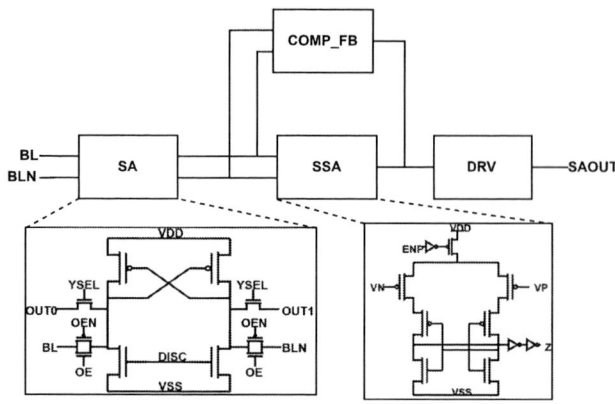

Fig. 6. Schematic diagram of the readout circuit

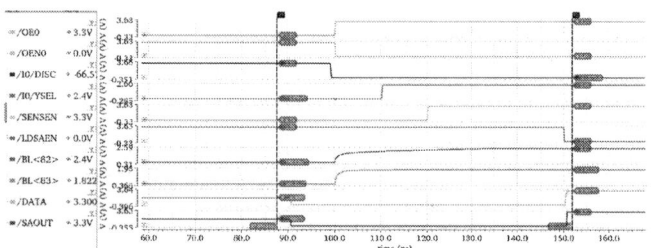

Fig. 7. Simulation waveform of the readout circuit

V. MEASUREMENT RESULTS

To further analyze the read and write behaviors of FeRAM, timing measurements were carried out on the 512-kb ferroelectric memory array, with the results summarized in TABLE I. During read and write operations, typically only one memory cell is enabled on a given bitline pair under the control of the WL, while all other cells remain inactive. Nevertheless, the inactive cells still contribute to the overall bitline parasitic capacitance. This parasitic effect, together with the charge-sharing process between the bitline and the ferroelectric capacitors, leads to the development of a differential voltage across the complementary bitlines, which is subsequently sensed and amplified by the SA and SSA.

TABLE I. MEASUREMENT OF TIMING

Measurement	Time(ns)
Charge sharing time	4
SA differential amplification time	8
SSA differential amplification time	4
DRO repair time	4

In general, shorter operation times are preferred to achieve higher read speed. As listed in TABLE I, the charge-sharing process requires 4 ns, the first- and second-stage differential amplification take 8 ns and 4 ns respectively, and the destructive read-out (DRO) recovery completes within 4 ns. These results confirm that the proposed FeRAM design achieves excellent read performance.

VI. CONCLUSION

Based on the integration of conventional CMOS technology and HZO materials, we fabricated a 512-kb FeRAM device employing a 2T2C cell structure. The FeCAPs were fabricated with CUB architecture, which enables 2Pr exceeding 30 µC/cm². We have presented both the design and simulation results of the FeRAM array and its peripheral circuits, all of which exhibit excellent performance. Finally, comprehensive testing of the 512-kb FeRAM was conducted, demonstrating outstanding read characteristics. This work provides a valuable reference for the future development of high-reliability, high-capacity FeRAM chips and paves the way for broader applications of hafnium-based ferroelectric memory in the field of nonvolatile memory technologies.

REFERENCES

[1] Wu, Qiqiao, et al. "A 9-Mb HZO-based embedded FeRAM with 10-cycle endurance and 5/7-ns read/write using ECC-assisted data refresh and offset-canceled sense amplifier." IEEE Journal of Solid-State Circuits 59.1 (2023): 208-218.

[2] Yu, Jiajie, et al. "3D Trench Hf 0.5 Zr 0.5 O 2-Based 32 Kbit 1T1C FeRAM Chip with 2/5 ns Write/Read Speed, Low Power Consumption (0.605 pJ/bit) and Prominent High-Temperature Reliability (Baking@ 175° C)." 2024 IEEE International Electron Devices Meeting (IEDM). IEEE, 2024.

[3] Sun, Yiming, et al. "1T1C 3D HZO FeRAM with High Retention (> 125° C) and High Endurance (> 1E13) for Embedded Nonvolatile Memory Application." 2025 Symposium on VLSI Technology and Circuits (VLSI Technology and Circuits). IEEE, 2025.

[4] Okuno, Jun, et al. "1T1C FeRAM memory array based on ferroelectric HZO with capacitor under bitline." IEEE Journal of the Electron Devices Society 10 (2021): 29-34.

[5] Okuno, Jun, et al. "A highly reliable 1.8 V 1 Mb Hf 0.5 Zr 0.5 O 2-based 1T1C FeRAM array with 3-D capacitors." 2023 International Electron Devices Meeting (IEDM). IEEE, 2023.

[6] Francois, T., et al. "Demonstration of BEOL-compatible ferroelectric $Hf_{0.5}Zr_{0.5}O_2$ scaled FeRAM co-integrated with 130nm CMOS for embedded NVM applications." 2019 IEEE International Electron Devices Meeting (IEDM). IEEE, 2019.

[7] Takada, Kenshi, et al. "Investigation of the wake-up process and time-dependent imprint of $Hf_{0.5}Zr_{0.5}O_2$ film through the direct piezoelectric response." Applied Physics Letters 119.3 (2021).

[8] Kim, Han Joon, et al. "A study on the wake-up effect of ferroelectric $Hf_{0.5}Zr_{0.5}O_2$ films by pulse-switching measurement." Nanoscale 8.3 (2016): 1383-1389.

[9] Fujiwara, Hirokazu, et al. "Non-Destructive Imaging of Breakdown Process in Ferroelectric Capacitors Using\textit {In-situ} Laser-Based Photoemission Electron Microscopy." arXiv preprint arXiv:2310.16275 (2023).

[10] Zheng, Yunzhe, et al. "Atomic-scale characterization of defects generation during fatigue in ferroelectric $Hf_{0.5}Zr_{0.5}O_2$ films: Vacancy generation and lattice dislocation." 2021 IEEE International Electron Devices Meeting (IEDM). IEEE, 2021.

979-8-3315-8850-2/25 $31.00 © 2025 IEEE

2025 The 10th International Conference on Integrated Circuits and Microsystems

Reconfigurable SRAM Computing-In-Memory Macro Based on Local Computing Cell

Chenghu Dai, Chaoyi Wang, Zeyi Liu, Qiang Zhao*, Chunyu Peng
[a] *School of Integrated Circuits, Anhui University, Hefei 230601, China*
[b] *Anhui Provincial High-Performance Integrated Circuit Engineering Research Center, Hefei 230601, China*
* Corresponding author
zhaoqiang@ahu.edu.cn

Abstract—Computing-in-memory (CIM) performs computation inside the memory, which shows high energy efficiency and high bandwidth. Though many efforts have been made to enhance its functions, CIM macro still suffers from fixed computational modes and nonlinearity issues. In this work, we propose a configurable SRAM CIM macro, which supports SRAM, Boolean logic operations, binary multiply-accumulate (MAC), and 4bit MAC operations. A 4bit SAR ADC with single-ended comparator reduces ~9% power efficiency compared with conventional ADC. In 28 nm CMOS process, simulation results show the energy efficiency of NOR operation, binary MAC and 4bit MAC are 136.4-673.8, 189.7-453.2, and 4.7-59.68 TOPS/W at a supply voltage range from 0.7 V-1.1 V, respectively. The proposed CIM macro helps to enhance hardware utilization and flexibility of SRAM, and offer high performance and energy-efficient method to convolutional neural networks acceleration.

Keywords—computing-in-memory, SRAM, DNN, MAC, Boolean logic

I. INTRODUCTION

Deep neural networks (DNNs) have been widely used in many application fields, such as image classification, speech recognition, and facial recognition. These applications need huge data processing and considerable computational capabilities. The conventional von Neumann architecture is regarded as the mainstream computing platform. However, the physical separation of memory from arithmetic logic unit (ALU) results in reduced computing speed and increased power consumption. Computing-in-memory (CIM) is a promising approach to solve the von Neumann bottleneck by performing computation directly in memory [1] [2] [3] [4] [5], thereby avoiding frequent data movement between memory and ALU. Although SRAM CIM takes advantage of high energy efficiency and wide bandwidth [6] [7] [8] [9] [10] [11], it still suffers from fixed computational modes and computing nonlinearity issues.

Many studies have sought to enhance CIM reconfigurability to extend its applications. For instance, [12] proposes a 4+2T SRAM cell, which activates multiple word lines (WL) to efficiently implement in-memory Boolean logic operations. [13] proposed a 10T1C cell, which supports XOR Boolean logic operations. [14] proposed a compute SRAM which can flexibly accommodate a wide range of bit-widths, from single to 32 or 64 bits. [15] proposed a reconfigurable SRAM CIM macro with multiple-mode MAC operations, including binary weights network (BWN), ternary weight network (TWN), and multi-bit multiply-accumulate (MAC) operations. While previous work have improved the flexibility of the SRAM cell, the challenge of low hardware utilization in CIM circuits remains to be overcome.

This work proposes a reconfigurable SRAM CIM macro based on a local computing cell (LCC) unit and 6T SRAMs, which supports NAND, NOR and XOR Boolean logic operations, binary MAC, and 4bit MAC operations. This work also proposes a low-power successive approximation register (SAR) analog-to-digital converter (ADC) that can be used to quantify MAC results. The proposed binary and 4bit MAC of CIM macro exhibit good linearity and the SAR ADC shows low power consumption.

This paper is organized as follows. Section II describes the architecture of the proposed SRAM CIM macro. Section III describes the modified 4bit SAR ADC. Section IV summarizes the simulation results and performance. Section V concludes this work.

II. PROPOSED CIM MACRO

A. CIM Macro Architecture

As shown in Fig. 1, the proposed CIM macro mainly comprises a 64×64 6T SRAM array, a 16×64 LCC array, a 16×1 4bit SAR ADC array, and auxiliary peripheral circuits. The CIM macro supports traditional SRAM mode, Boolean logic operations, binary MAC, and 4bit×4bit MAC operations.

National Natural Science Foundation of China under Grant 62274001
National Natural Science Foundation of China under Grant 92464203
The University Synergy Innovation Program of Anhui Province under Grant GXXT-2023-012
The University Synergy Innovation Program of Anhui Province under Grant GXXT-2023-010

979-8-3315-8850-2/25 $31.00 © 2025 IEEE

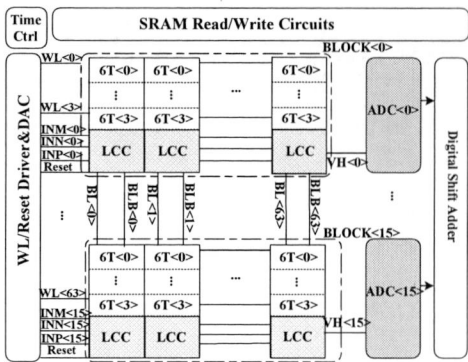

Fig. 1. The architecture of proposed CIM macro.

TABLE I. MAPPING TABLE OF INPUT FOR BOOLEN LOGIC OPERATIONS

BOOLEAN	INPUT	INM	INN	INP	W	O	
NAND	1	GND	VDD	GND	1	GND	0
					0	VDD	1
	0	VDD	VDD	GND	1	VDD	1
					0	VDD	1
NOR	1	GND	GND	VDD	1	GND	0
					0	GND	0
	0	GND	VDD	GND	1	GND	0
					0	VDD	1
XOR	1	GND	VDD	GND	1	GND	0
					0	VDD	1
	0	VDD	GND	VDD	1	VDD	1
					0	GND	0

B. 5T1C LCC unit

Fig. 2a shows the circuit structure of LCC-6T bank. Each LCC-6T bank consists of one LCC unit and four 6T SRAM cells. INM, INN, and INP serve as input ports for the CIM operations. The output port is the bottom plate of capacitor denoted as "O".

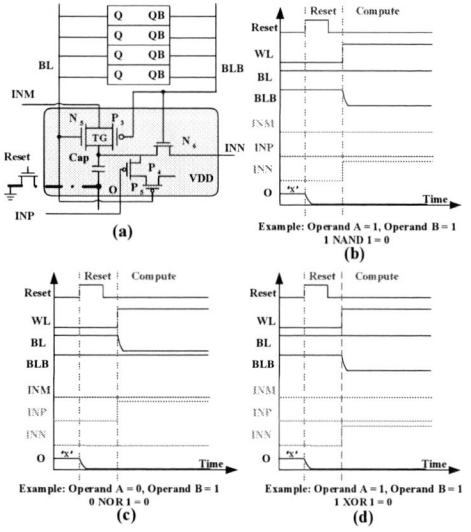

Fig. 2. (a) The circuit structure of LCC-6T bank. Timing diagrams of (b) NAND operation, (c) NOR operation, and (d) XOR operation.

C. Boolean Logic Operations

For Boolean logic operations, the external input mapping is listed in Table I. For NAND operation, the INP is pulled to GND and the INN is pulled to VDD. The datum stored in 6T SRAM cell is operand A, and datum from input INM is operand B. The NAND result is the output of "O". Fig. 2b shows the timing diagram of NAND operation. For NOR operation, the INM is pulled to GND. The datum stored in 6T SRAM cell is operand A, and the operand B is represented by the combination of INN and INP. The NOR result is the output of "O". Fig. 2c shows the timing diagram of NOR operation. For XOR operation, the datum stored in 6T SRAM cell is operand A, and the operand B is represented by the combination of INM, INN, and INP. The XOR result is the output of "O". Fig. 2d shows the timing diagram of XOR operation.

D. Binary MAC operation

Fig. 3a shows the binary MAC operation of the proposed CIM macro. The datum stored in 6T SRAM cell is operand A, and operand B is represented by the combination of INM, INN, and INP. When multiplication operation starts, WL is pulled to VDD, and the access transistor is activated or deactivated based on the stored value in 6T SRAM. Fig. 3b depicts the timing diagram of binary MAC function, and MAC truth table is shown in Table II. Following the multiplication operations, the accumulation process is executed via charge sharing with capacitors.

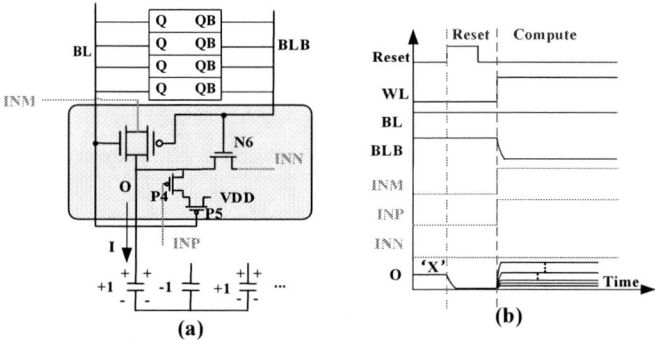

Fig. 3. (a) Binary MAC operation and (b) corresponding timing diagram.

TABLE II. TRUTH TABLE OF BINARY MAC OPERATION

W	Q	INPUT	INM	INN	INP	O	
+1	VDD	+1	VDD	GND	VDD	VDD	+1
+1	VDD	-1	GND	VDD	GND	GND	-1
-1	GND	+1	VDD	GND	VDD	GND	-1
-1	GND	-1	GND	VDD	GND	VDD	+1

E. 4bit MAC Operation

As shown in Fig. 4a, for 4bit MAC operation, the INPs are pulled to VDD and the INNs are pulled to GND. The 4bit activation is realized through digital-to-analog converter (DAC). The 4bit activation is on INM, and the 4bit weight is stored in four 6T SRAM cells of one bank controlled by word lines (WL[0]-WL[3]). In the first cycle, WL[0] is pulled to VDD while WL[1]-WL[3] are pulled to GND, enabling the multiplication of the first bit weight by the analog input. The result is then quantized by a 4bit SAR ADC and fed into a shift adder. In subsequent cycles, WL[1], WL[2], and WL[3] are sequentially pulled to VDD, repeating the multiplication operation and forwarding each result to the shift adder. After four cycles, the shift adder accumulates all products to yield the final 4bit MAC operation result. Fig. 4b depicts the timing diagram for a single weight bit multiplication process within the 4bit MAC operation.

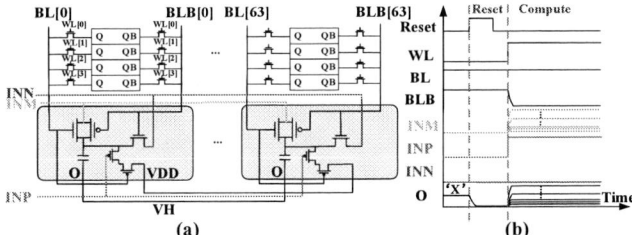

Fig. 4. (a) 4bit MAC operation and (b) corresponding timing diagram.

III. 4BIT SAR ADC ARCHITECTURE

Fig. 5 shows the architecture of the proposed low-power 4bit SAR ADC, which consists of a timing control circuit, a SAR logic circuit, and a 4bit CDAC circuit.

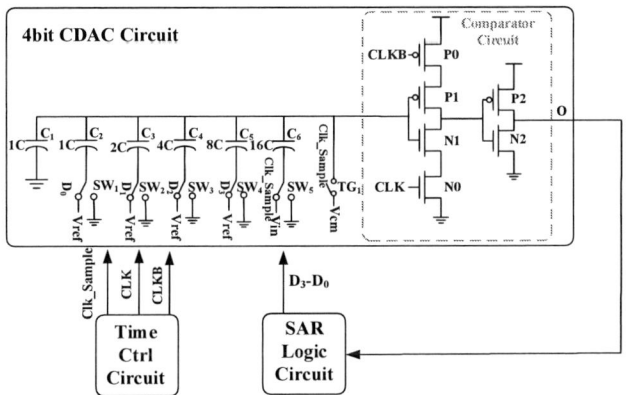

Fig. 5. The architecture of low-power 4bit SAR ADC.

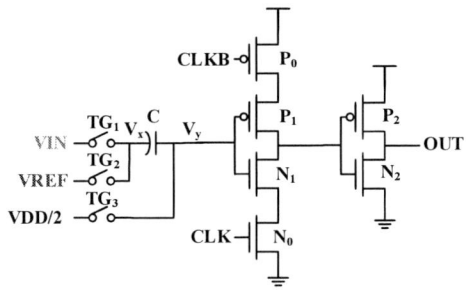

Fig. 6. Circuit structure of the proposed single-ended comparator.

In order to reduce power consumption, a single-ended comparator is proposed to replace conventional latched comparator (Fig. 6). The proposed comparator is composed of two inverters, and the CLK/CLKB controls the first inverter. The comparator works with set stage and comparison stage. In the set stage, the CLK/CLKB are pulled to GND/VDD. The TG1 is off, while the TG2 and the TG3 are on. The left and right plates of capacitor C are set to VREF and VDD/2. In the comparison stage, the CLK/CLKB are pulled to VDD/GND,

then the first inverter starts working. The TG1 is on, the TG2 and the TG3 are off. Due to the charge conservation property of capacitor, the Vx change results in a concomitant change in Vy, giving the comparison result.

IV. RESULTS AND DISCUSSIONS

The proposed CIM macro was stimulated using 28 nm CMOS technology with a supply voltage (VDD) of 0.9 V.

A. LCC Function Simulation

Fig. 7 shows the simulation results of NAND, NOR, and XOR operations. By setting NAND operation of "1" and "1", NOR operation of "0" and "1", and XOR operation of "1" and "1", the simulation outputs are consistent with theoretical analysis.

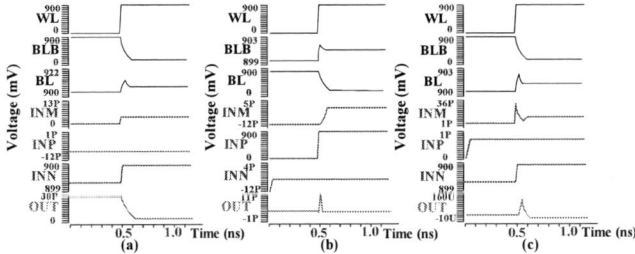

Fig. 7. (a) Simulation results of (a) NAND, (b) NOR, and (c) XOR operations.

As shown in Fig. 8, the linearity of binary and 4bit MAC operation is simulated. Both of them show good linearity. The outputs "0" of binary MAC and 4bit MAC are 450.1 mV and 449.3 mV, respectively, less than 0.5 least significant bit (LSB), achieving the requirements.

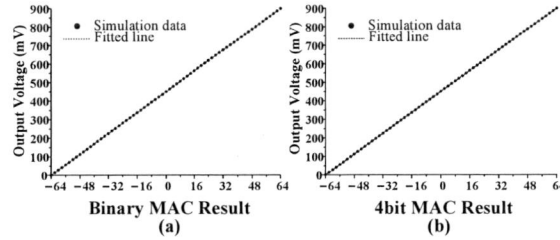

Fig. 8. Output voltages versus (a) binary MAC resuts and (b) 4bit MAC results.

B. 4bit SAR ADC Performance Analysis

The offset voltage of the proposed single-ended comparator was evaluated by Monte Carlo simulations. As shown in Fig. 9a, 1000 samples of Monte Carlo simulation show the mean value (μ) of offset voltage is -0.981 mV, and standard deviation (σ) is 11.04 mV. The offset voltage (μ±3σ, -34.10 mV/+32.14 mV) is less than the LSB of 4bit SAR ADC (56.25 mV). The differential nonlinearity (DNL) of the proposed 4bit SAR ADC is -0.08 LSB/+0.24 LSB, and the integral nonlinearity (INL) is -0.17 LSB/+0.25 LSB. The DNL error of the circuit is less than ±1 LSB. The FFT spectrum of 4bit ADC is shown in Fig. 9b. The ENOB of the proposed SAR ADC is 3.90 bits, the SNR (signal-to-noise ratio) is 25.22 dB, and the SFDR (spurious-free dynamic range) is 36.18 dBc. Therefore, the 4bit ADC satisfies the MAC requirement.

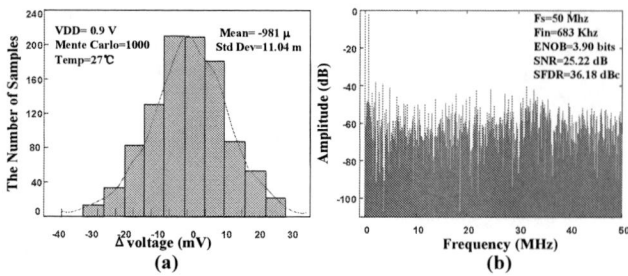

Fig. 9. (a) 1000 samples of Monte Carlo simulation of proposed comparator under 0.9 V and 27°C conditions; (b) FFT spectrum of proposed 4bit SAR ADC.

C. Performance analysis

Fig. 10 shows the power consumption comparison of the proposed 4bit SAR ADC with conventional SAR ADC under different computing modes (4bit MAC with conventional ADC, 4bit MAC with proposed ADC, binary MAC with conventional ADC, binary MAC with proposed ADC).

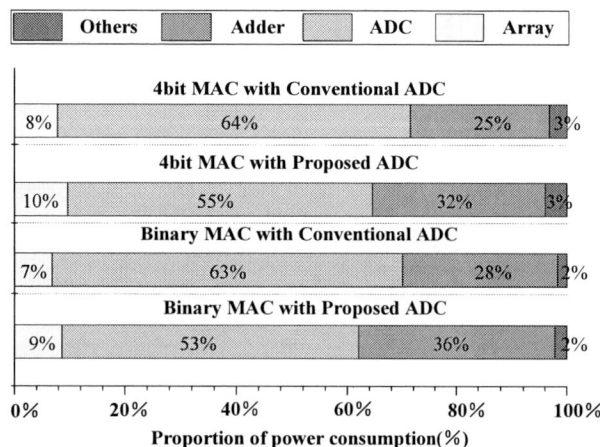

Fig. 10. Proportion of power consumption in different computing modes.

Under 0.9 V and 27°C conditions, for 4bit MAC with conventional ADC operations, the ADC accounts for up to 64% of the energy consumption, which is the main source of energy consumption for the overall circuit. However, the SAR ADC proposed in this work exhibits significant energy optimization in 4bit MAC operations, reducing the ADC energy consumption to 55% of the total circuit energy. In binary MAC with proposed ADC, ADC consume 10% less energy than conventional ADC. These results show that the proposed SAR ADC effectively reduces the energy consumption of the overall CIM circuit, and helps to improve energy efficiency.

As illustrated in Table III, this work presents performance comparison with other recent researches. The proposed CIM macro operates in four different modes, and exhibits enhanced hardware utilization and flexibility in comparison to. In addition, the energy efficiency of the proposed circuit are 136.4-673.8 TOPS/W in NOR operation, 189.7-453.2 TOPS/W in binary MAC mode, and 4.7-59.68 TOPS/W in 4bit MAC mode at a supply voltage range from 0.7 V-1.1 V. The energy efficiency

decreases as the supply voltage increasement. These results are superior to the results of [16] [17] [18] [19].

TABLE III. PERFORMANCE COMPARISON WITH RELATED RESULTS

	TCASII'24 [16]	CICC'22 [17]	ASSCC'21 [18]	JSSCC'22 [19]	**This Work**		
Technology	55 nm	28 nm	28 nm	28 nm	**28 nm**		
Cell Structure	12T2C	9T1C	8Cells+ CCU	6T+ TWT	**5T1C**		
Supply Voltage	1.2 V	0.9 V	0.85 V	1.0 V	**0.7 V-1.1 V**		
Operation Mode	Multi-bit MAC	4bit MAC	BWN	Multi-bit MAC	**Boolean Logic (NOR)**	**binary MAC**	**4bit MAC**
Input (bit)	1-4	4	1	2	1	1	4
Weight (bit)	1/2/4	4	1	8	1	1	4
Output (bit)	4	6	3.58	14	1	3	6
Energy Efficiency (TOPS/W)	97-419	177	257	28.02	136.4-673.8	189.7-453.2	4.7-59.68

V. CONCLUSION

This work proposes a reconfigurable SRAM CIM macro, which supports traditional SRAM mode, NAND, NOR and XOR Boolean logic operations, binary MAC, and 4bit×4bit MAC operations. Based on 28 nm CMOS process, simulation results show the binary and 4bit MAC show good linearity. A single-ended comparator is also proposed to reduce SAR ADC power consumption and overall CIM macro power consumption. The single-ended comparator shows small offset voltage and good static and dynamic performance. In binary MAC and 4bit×4bit MAC operations, the proposed ADC reduces ~9% power consumption proportion compared with traditional ADC. The energy efficiency of the proposed circuit is 394.6 TOPS/W in NOR mode, 313.4 TOPS/W in binary MAC mode, and 59.38 TOPS/W in 4bit MAC mode at a supply voltage of 0.9 V.

REFERENCES

[1] K. Yu, S. Kim, and J. R. Choi, "Trends and Challenges in Computing-in-Memory for Neural Network Model: A Review From Device Design to Application-Side Optimization," *IEEE Access*, vol. 12, pp. 186679–186702, 2024.

[2] W. Kang, H. Zhang, and W. Zhao, "Spintronic Memories: From Memory to Computing-in-Memory," in *2019 IEEE/ACM International Symposium on Nanoscale Architectures (NANOARCH)*, Jul. 2019, pp. 1–2.

[3] J. F. Li, "Testing and Reliability of Computing-In Memories: Solutions and Challenges," in *2022 IEEE International Test Conference in Asia (ITC-Asia)*, Aug. 2022, pp. 55–60.

[4] S. Nasrin *et al.*, "Memory-Immersed Collaborative Digitization for Area-Efficient Compute-in-Memory Deep Learning," in *2023 IEEE 5th International Conference on Artificial Intelligence Circuits and Systems (AICAS)*, Jun. 2023, pp. 1–5.

[5] S. Yu, H. Jiang, S. Huang, X. Peng, and A. Lu, "Compute-in-Memory Chips for Deep Learning: Recent Trends and Prospects," *IEEE Circuits and Systems Magazine*, vol. 21, no. 3, pp. 31–56, 2021.

[6] S. H. Choudhari and P. Jayakrishnan, "Structural Analysis of Low Power and Leakage Power Reduction of Different Types of SRAM Cell

979-8-3315-8850-2/25 $31.00 © 2025 IEEE

Topologies," in *2019 Innovations in Power and Advanced Computing Technologies (i-PACT)*, Mar. 2019, pp. 1–7.

[7] A. A V, H. V, A. N. S A, and S. M, "Low Power 10T SRAM Based Computing in Memory Macro Architecture for Binary MAC Operation of Edge AI Processors," in *2024 1st International Conference on Trends in Engineering Systems and Technologies (ICTEST)*, Apr. 2024, pp. 01–05.

[8] U. Mehta and S. Thaker, "Design and Development of In-Memory-Compute SRAM Cell Using 45nm Technology," in *2024 IEEE International Conference on Intelligent Signal Processing and Effective Communication Technologies (INSPECT)*, Dec. 2024, pp. 1–5.

[9] A. K. Gupta and A. Acharya, "Exploration of 9T SRAM Cell for In Memory Computing Application," in *2021 Devices for Integrated Circuit (DevIC)*, May 2021, pp. 461–465.

[10] K. Monga, S. Behera, N. Chaturvedi, and S. Gurunarayanan, "Design of In-Memory Computing Enabled SRAM Macro," in *2022 IEEE 19th India Council International Conference (INDICON)*, Nov. 2022, pp. 1–4.

[11] L. Ammoura, M. L. Flottes, P. Girard, and A. Virazel, "Preliminary Defect Analysis of 8T SRAM Cells for In-Memory Computing Architectures," in *2021 16th International Conference on Design & Technology of Integrated Systems in Nanoscale Era (DTIS)*, Jun. 2021, pp. 1–4.

[12] Q. Dong *et al.*, "A 0.3V VDDmin 4+2T SRAM for searching and in-memory computing using 55nm DDC technology," in *2017 Symposium on VLSI Circuits*, Jun. 2017, pp. C160–C161.

[13] B. Zhang et al., "PIMCA: A Programmable In-Memory Computing Accelerator for Energy-Efficient DNN Inference," IEEE J. Solid-State Circuits, vol. 58, no. 5, pp. 1436–1449, May 2023.

[14] J. Wang *et al.*, "A 28-nm Compute SRAM With Bit-Serial Logic/Arithmetic Operations for Programmable In-Memory Vector Computing," *IEEE Journal of Solid-State Circuits*, vol. 55, no. 1, pp. 76–86, Jan. 2020.

[15] K. Xiao *et al.*, "A 28nm 8Kb Reconfigurable SRAM Computing-In-Memory Macro With Input-Sparsity Optimized DTC for Multi-Mode MAC Operations," *IEEE Transactions on Circuits and Systems II: Express Briefs*, vol. 71, no. 7, pp. 3263–3267, Jul. 2024.

[16] K. Zhang *et al.*, "A Charge-Domain Compute-In-Memory Macro With Cell-Embedded DA Conversion and Two-Stage AD Conversion for Bit-Scalable MAC Operation," *IEEE Trans. Circuits Syst. II*, vol. 71, no. 3, pp. 1077–1081, Mar. 2024.

[17] Zhang B. *et al.*, "A 177 TOPS/W, Capacitor-based In-Memory Computing SRAM Macro with Stepwise-Charging/Discharging DACs and Sparsity-Optimized Bitcells for 4-Bit Deep Convolutional Neural Networks," in *2022 IEEE Custom Integrated Circuits Conference (CICC)*, Newport Beach, CA, USA: IEEE, Apr. 2022, pp. 1–2.

[18] J. Song, Y. Wang, X. Tang, R. Wang, and R. Huang, "A 16Kb Transpose 6T SRAM In-Memory-Computing Macro based on Robust Charge-Domain Computing," in *2021 IEEE Asian Solid-State Circuits Conference (A-SSCC)*, Busan, Korea, Republic of: IEEE, Nov. 2021, pp. 1–3.

[19] Su J.-W. *et al.*, "Two-Way Transpose Multibit 6T SRAM Computing-in-Memory Macro for Inference-Training AI Edge Chips," *IEEE Journal of Solid-State Circuits*, vol. 57, no. 2, pp. 609–624, Feb. 2022.

979-8-3315-8850-2/25 $31.00 © 2025 IEEE

A 10T1C SRAM-Based Computing-in-Memory Unit Supporting Logic and MAC Computing for Energy-Efficient Edge AI Chips

Zhiting Lin
School of Integrated Circuits
Anhui University
Hefei, China
ztlin@ahu.edu.cn

Chenglong Duan
School of Integrated Circuits
Anhui University
Hefei, China
1161798808@qq.com

Miao Long
School of Integrated Circuits
Anhui University
Hefei, China
504345904@qq.com

Juntao Ge
School of Integrated Circuits
Anhui University
Hefei, China
2168771181@qq.com

Fugui Jiang
School of Integrated Circuits
Anhui University
Hefei, China
1718721138@qq.com

Yang Yang
School of Integrated Circuits
Anhui University
Hefei, China
1278657916@qq.com

Xiulong Wu*
School of Integrated Circuits
Anhui University
Hefei, China
xiulong@ahu.edu.cn

Abstract—This paper proposes an innovative compute-in-memory (CIM) integrated circuit based on a 10T1C SRAM. The design augments a standard 6T-SRAM cell with four NMOS transistors and one capacitor, enabling simultaneous data storage and in-memory AND/XNOR logic operations. The 10T1C cell supports three operating modes: data storage, in-memory AND, and in-memory XNOR. Furthermore, arrayed configurations enable more complex computations, including multiply-accumulate (MAC) operations and binary neural network (BNN) inference. Simulated in a 28-nm CMOS process, the core achieves an energy efficiency of 18.3–41.8 TOPS/W for Boolean operations over a 0.7–0.9 V supply range. This high efficiency makes it well suited for energy-constrained edge-AI applications that require both storage and logic capabilities.

Keywords— CIM, SRAM, Boolean logic, Edge AI, CNN.

I. INTRODUCTION

The rapid growth of artificial intelligence (AI) workloads on edge devices demands highly efficient computing architectures that tightly integrate memory and processing. Traditional von Neumann architectures are fundamentally constrained by the well-known "memory wall" [1], [2], which arises from the physical separation of memory and compute units. This separation forces frequent, energy-intensive data transfers between memory and processors, causing substantial energy overhead and latency for data-intensive edge-AI tasks. Compute-in-memory (CIM) technology has emerged as a promising solution to this challenge [3], [4]. By embedding computation directly within memory arrays, CIM greatly reduces data movement and thereby offers significant gains in both energy efficiency and processing speed.

Although SRAM-based CIM designs have made progress

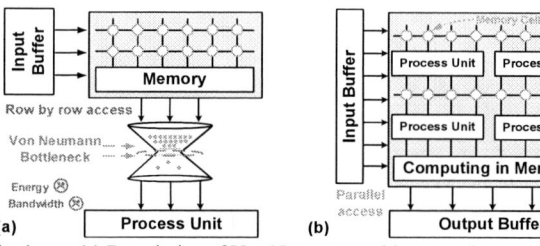

Fig. 1. (a) Description of Von Neumann architecture. (b) Description of Computing-in-memory architecture.

for key arithmetic tasks such as multiply-accumulate (MAC) operations [5], [6], efficiently implementing Boolean logic operations in memory remains a challenge. Boolean functions like AND and XNOR are fundamental to many AI algorithms, especially in binary neural networks (BNNs) where operations are binarized to improve efficiency. Existing CIM designs often trade off storage density against computational capability, and few integrate robust memory and versatile in-memory logic in a single cell without sacrificing area or energy [7], [8].

This paper makes three key contributions:

1. A 10T1C-SRAM cell based on the standard 6T-SRAM structure is proposed, which extends the functionality with four additional NMOS transistors and one capacitor to enable in-memory computing while preserving all conventional SRAM operations.

2. The proposed 10T1C-SRAM cell can perform both in-memory AND and XNOR operations in a single cycle through optimized control signal sequencing, significantly enhancing computational efficiency for AI applications.

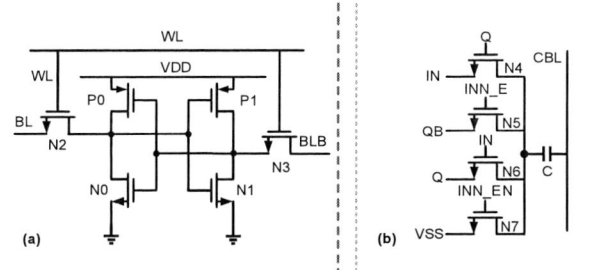

Fig. 2.   (a) Traditional 6T Unit (b) Computing Module

3. The 10T1C-SRAM architecture supports direct in-memory 1-bit AND MAC and BNN computations, eliminating the need for data movement between memory and processing units and thereby reducing energy consumption.

II. PROPOSED 10T1C SRAM CELL DESIGN

A. Computational Unit Structure

The computational unit integrates a standard 6T-SRAM memory core with an additional computing module (Fig. 2). The 6T memory core consists of cross-coupled inverters (transistors P0/N0 and P1/N1) and access transistors (N2, N3), providing conventional SRAM functionality. It stores a weight bit in complementary nodes Q (weight) and QB (weight complement). The storage nodes connect to bitlines BL and BLB via N2 and N3, respectively, under control of the word line WL.

The computing module adds four NMOS transistors (N4–N7) and one capacitor C tied to a compute bitline (CBL). Each transistor is configured as follows:

N4: Gate-Q, Source-IN, Drain-C
N5: Gate-INN_E, Source-QB, Drain-C
N6: Gate-IN, Source-Q, Drain-C
N7: Gate-INN_EN, Source-VSS, Drain-C

Here Q and QB denote the stored weight (logic 1 or 0), IN is the input, and INN_E/INN_EN are internally generated control signals. By selectively activating IN, INN_E, and INN_EN, the computing module couples the stored weight and input to the capacitor C, enabling in-memory logic computation.

B. Unit Operating Mode

The proposed 10T1C-SRAM computing-in-memory cell supports multiple configurable operating modes—including data storage, AND computation, and XNOR computation—through flexible control signaling. To realize these functions, a dedicated control module (CTRL), illustrated in Figure 3(a), has been designed. This module consists of two inverters (INV0, INV1) and two AND logic gates (AND0, AND1). Specifically, INV0 generates the complementary signal INN from the input IN, while INV1 inverts the global enable signal E to produce EN. AND0 and AND1 perform logical AND operations between INN and E, and between INN and EN, respectively, yielding the internal control signals INN_E and INN_EN. This associative control mechanism dynamically produces the required signal combinations based on the computational mode, thereby enabling precise switching of transistors within the computational unit..

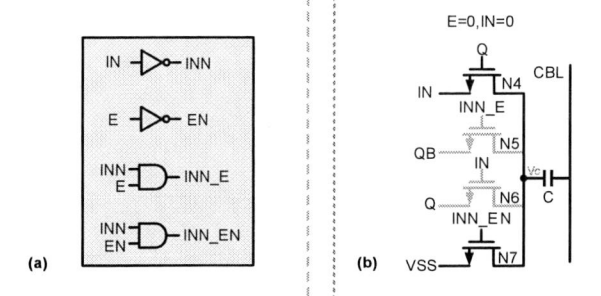

Fig. 3.   (a) Ctrl Module (b) When E=0 and IN=0, the cell is in storage mode.

In data storage mode, when the input signal IN is set to 0 and the master control signal E remains at 0, the internal node INN presents a high logic level ("1") through inversion, while EN also becomes "1" due to the inversion of E. Under these conditions, the derived control signals set INN_E to "0" and INN_EN to "1". This signal combination forces transistor N5 into the off state while turning on N7 (as shown in Figure 3(b)). Consequently, all transistors within the computational cell remain disabled, effectively isolating the computational cell. The cell then operates solely based on the standard 6T-SRAM structure, exhibiting identical functional behavior to a classic 6T-SRAM memory cell. In this state, the circuit reliably executes conventional SRAM functions—including weight data read, write, and hold operations—without any interference from the computational cell, ensuring the stability and integrity of storage operations.

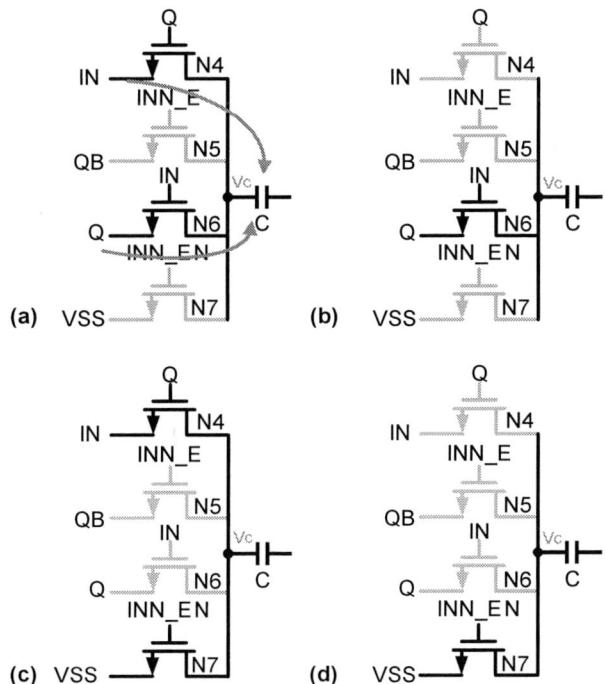

Fig. 4.   When E=0, the cell switches to the in-memory AND mode.
(a) Input=1, Weight=1. Result=1.(b) Input=1, Weight=0. Result=0.
(c) Input=0, Weight=1. Result=0.(d) Input=0, Weight=0. Result=0.

979-8-3315-8850-2/25 $31.00 © 2025 IEEE   820

In the in-memory AND computation mode, the logic levels of the input and weight signals are defined as follows: IN=1 indicates Input=1, IN=0 indicates Input=0; Q=1 indicates Weight=1, Q=0 indicates Weight=0.Before executing the AND calculation, a reset operation is required: Set E=0 and IN=0 (at this point, N6 is cutoff), resulting in INN_E=0 (N5 cutoff) and INN_EN=1 (N7 conduction). At this moment, the first terminal of capacitor C is connected to VSS via N7 and reset to the VSS level. It should be noted that nodes Q and QB retain the values stored after the previous calculation, which could be Q=1, QB=0, or Q=0, QB=1. If Q=1 and QB=0, N4 conducts. The first terminal of C is also connected to IN (currently 0) via N4 and is similarly pulled down to VSS. If Q=0 and QB=1, N4 turns off. The first terminal of C is reset to VSS solely via N7.

During in-memory AND calculations, the control signal maintains E=0. As shown in Figure 4(a), Only when IN=1 and Q=1 is the first terminal of C charged to a high level, yielding an AND result of "1"; if IN=0 or Q=0, the first terminal of C remains at VSS, resulting in an AND value of "0".

In the in-memory XNOR computation mode, the logical encoding of inputs and weights is defined as follows: IN=1 denotes Input=1, IN=0 denotes Input=-1; Q=1 denotes Weight=1, Q=0 denotes Weight=-1.

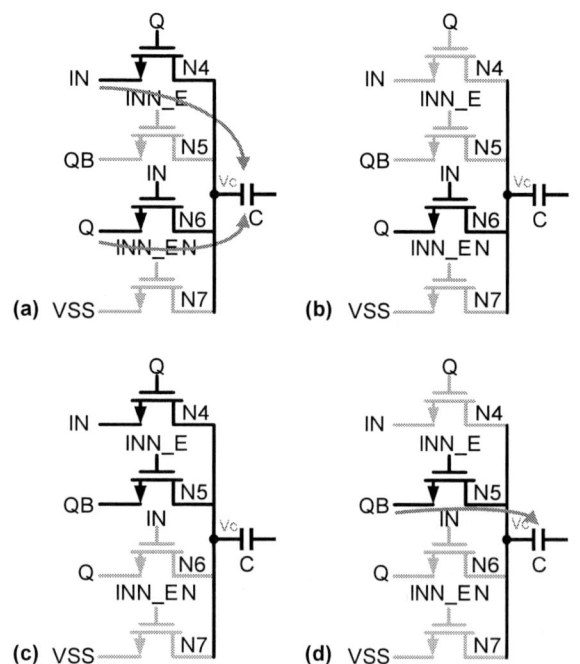

Fig. 5.   When E=1, the cell operates in the in-memory XNOR mode.
(a) Input=1, Weight=1. Result=1. (b) Input=1, Weight=-1. Result=-1.
(c) Input=-1, Weight=1. Result=-1.(d) Input=-1, Weight=-1. Result=1.

Similar to the AND computation mode, a reset is required before executing the XNOR operation: set E=0, IN=0 (N6 cutoff), at which point INN_E=0 (N5 cutoff) and INN_EN=1 (N7 conduction). Terminal 1 of capacitor C is connected to VSS via N7, thereby resetting it to the VSS level. At this point, storage nodes Q and QB retain the results from the previous computation, potentially being Q=1, QB=0 or Q=0, QB=1. If Q=1 and QB=0, N4 conducts, and the first terminal of C is

simultaneously connected to IN (currently 0) via N4, providing auxiliary pull-down. If Q=0 and QB=1, N4 is cutoff, and the first terminal of C is reset solely via N7.

During XNOR computation, the control signal sets E=1. As shown in Figure 5(a),When the input equals the weight (IN = Q), the first terminal of C charges to a high level, yielding an XNOR result of 1. If IN ≠ Q, the first terminal of C remains at VSS, resulting in an XNOR output of -1.

In the three input weight combinations shown in Figures 5(b), (c), and (d): When Input=1, Weight=-1 (IN=1, Q=0, INN=0, QB=1; INN_E=0, INN_EN=0), N4, N5, and N7 turn off while N6 turns on. The first terminal of C remains at VSS, and the XNOR output is -1. When Input=-1, Weight=1 (IN=0, Q=1, INN=1, QB=0; INN_E=1, INN_EN=0), N4 and N5 conduct while N6 and N7 turn off. Terminal 1 of C remains at VSS, and the XNOR output is -1. When Input=-1 and Weight=-1 (IN=0, Q=0, INN=1, QB=1; INN_E=1, INN_EN=0), N5 turns on, N4, N6, and N7 turn off, the first terminal of C charges to VDD' through N5, and the XNOR output is 1.

C. Memory-Computing Array Circuit Structure

As shown in Figure 6, the compute-in-memory circuit based on 10T1C-SRAM cells consists of a 4×4 array and peripheral control circuitry. The array contains 16 cells (as in Example 1), with each row sharing a WL and a control module (CTRL), and each column sharing a BL, its complement (BLB), and a CBL.

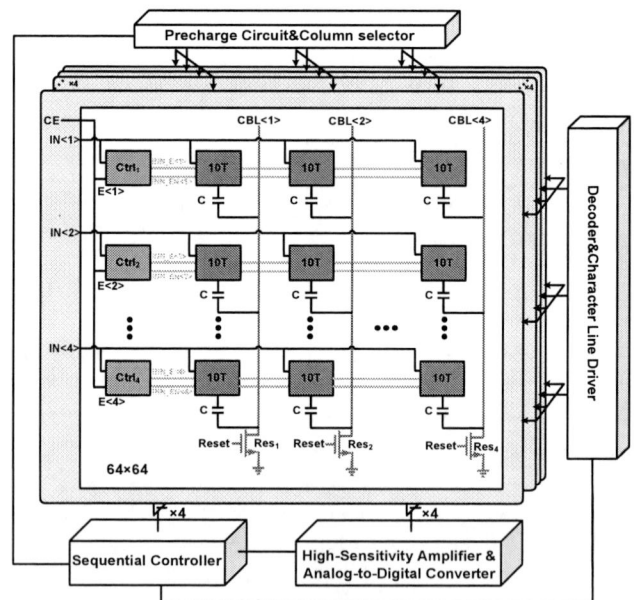

Fig. 6. Circuit Diagram of a Compute-in-Memory Circuit Based on a 10T1C-SRAM In-Memory Computing Cell

The design includes four CTRL modules and four CBLs. All CTRL enable inputs (E) connect to the global signal CE, while their input signals (IN) are independent. Each CBL is connected to VSS through an NMOS reset transistor, with all reset gates driven by a common Reset signal.

Each row uses dedicated word lines (WL₁–WL₄) and control modules (CTRL₁–CTRL₄), each with E tied to CE and IN individually controlled. Each column shares bit lines (BL₁–BL₄), complementary lines (BLB₁–BLB₄), and compute lines (CBL₁–CBL₄). Each CBL is pulled down via a corresponding reset transistor (RES₁–RES₄), whose gate, drain, and source connect to Reset, CBL, and VSS, respectively.

The peripheral control circuit manages operation and includes a row decoder, word-line driver, precharge circuit, column selector, sense amplifier, ADC, and timing controller. The decoder addresses rows; the driver activates word lines; the precharge circuit prepares nodes; the selector picks columns; the sense amplifier reads weight data; the ADC quantizes CBL voltages; and the timing controller synchronizes all signals. Alternate peripheral implementations are acceptable if they support the in-memory architecture.

D. Array Computing

The 10T1C-SRAM in-memory computing cell supports five operating modes: Data Storage, In-Memory AND, In-Memory XNOR, AND MAC, and In-Memory BNN. For the first three modes, the peripheral control circuit selects the target unit and suppresses same-column interference using previously established control methods[9].

In the in-memory AND MAC computation mode, the array performs parallel multiply-accumulate (MAC) operations between 1-bit inputs and multi-bit stored weights. Initialization begins by setting CE = 0, all IN⟨m⟩ = 0, and Reset = 1. This discharges all computation bit lines (CBLs) and cell capacitors to VSS, clearing any residual charge.

As shown in Figure 7, during computation, CE remains 0 to keep cells in AND mode, while Reset is set to 0, allowing CBLs to enter a high-impedance state. Row input signals are applied independently. Each cell performs an AND operation between its weight Q and input IN: if the result is 1, its capacitor charges to VDD'; otherwise, it remains at VSS.

Accumulation uses charge sharing: all capacitors in a column share the same CBL, so the voltage VCBL is proportional to the number n of charged cells: VCBL = n × (VDD' / 4). For example, in a 4-cell column, if only one cell outputs 1, VCBL reaches VDD'/4, representing a MAC result of 1. An on-chip ADC directly quantizes VCBL, enabling data-free accumulation and improving parallelism and energy efficiency.

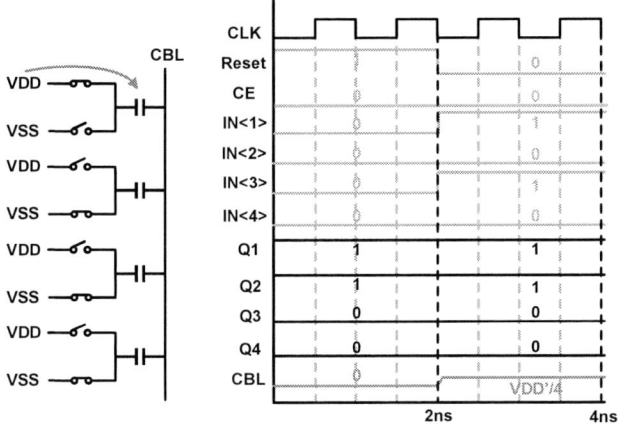

Fig. 7. Schematic of 1b-AND MAC computation in a column of 10T1C-SRAM cells within a compute-in-memory array (4×4 examples).

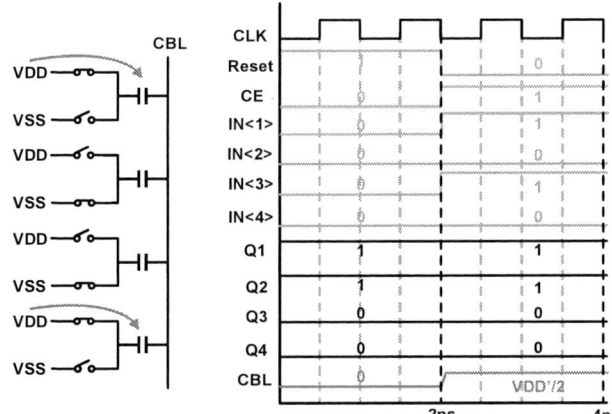

Fig. 8. Schematic Diagram of In-Memory Computing Circuit for In-Memory BNN Computation

Multi-column arrays extend this mechanism: each column acts as an accumulation channel. A 4-column array computes four parallel 1b×4b MAC operations (with weights 8,4,2,1). Multiple such banks operate concurrently, supporting higher-precision operations like 4b×4b MAC for greater computational throughput.

As shown in Figure 8, in the in-memory BNN computation mode, the process comprises initialization and computation phases. During initialization, CE and all IN signals are set to 0, resetting each cell's capacitor first terminal to VSS. Simultaneously, Reset=1 activates column NMOS reset transistors, pulling CBLs and all capacitor second terminals to VSS to clear residual charge.

In the computation phase, CE=1 configures cells for XNOR operation, while Reset=0 places CBLs in high-impedance state. Each column's four cells perform parallel XNOR operations between stored weights Q and inputs IN: matching values yield "1", charging C to VDD', mismatches yield "-1", leaving C at VSS.

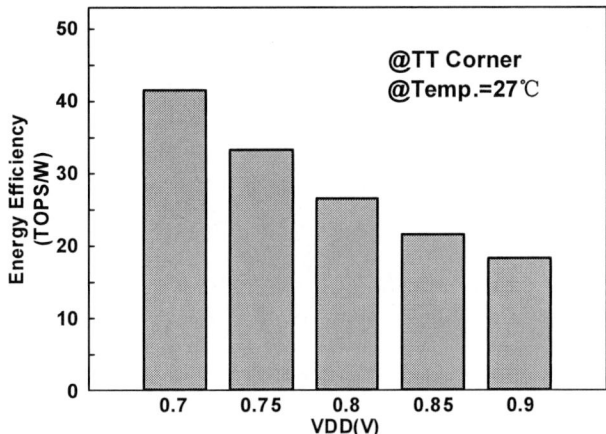

Fig. 9. Energy efficiency chart under different VDD values

The shared CBL voltage VCBL reflects the number n of "1" outputs: VCBL = (n/4)·VDD', corresponding to an algebraic sum of 2n - 4. For example, VCBL = VDD'/2 indicates two "1" and two "-1" outputs, summing to 0. An on-chip ADC quantizes this voltage to directly produce the BNN result, eliminating data movement and significantly improving inference parallelism and energy efficiency.

III. Simulation Results and Analysis

This design is implemented in a 16Kb SRAM array using a 28nm CMOS process. Figure 9 shows the peak energy efficiency of the AND CIM. At a supply voltage of 0.7V, the energy efficiency is 41.8 TOPS/W. At 0.9V, the energy efficiency is 18.3 TOPS/W. Compared to other architectures, this design can achieve both MAC computations while also accommodating CNN calculations, significantly improving usability and energy efficiency. Table I compares the previous MAC computation performance with the performance of this work. Clearly, this design significantly optimizes energy consumption and functionality. Overall, the structure proposed in this paper will have better flexibility than existing structures[10].

TABLE I. COMPARISON WITH PRIOR WORKS

	JSSC'17[11]	JSSC'22[12]	ISSCC'22[13]	JSSC'23[14]	This Work
Technology (nm)	130	65	28	22	28
Bitcell Density (Kb)	16	16	128	128	16
Cell-Type	6T	8T	6T	6T+3T1C	10T
Supply Voltage (V)	1	0.45/0.8	0.65-0.9	0.7-1.1	0.7-0.9
Input/Weight (bit)	5/1	1/1	8/8	8/8	4/1
Output(bit)	1	1-5	22	7	4
Frequency of operation(Mhz)	50	200	5	N/A	250
Energy Efficiency (TOPS/W)	11.51	15.8-490	21.2-27.75	15.5-32.2	18.3-41.8

IV. Conclusion

This article introduces a 16Kb 8T SRAM compute-in-memory (CIM) macro implemented based on 28nm CMOS technology. The design integrates data storage, AND logic computation, XNOR logic computation, in-memory AND multiply-accumulate operations, and convolutional neural network (CNN) computation functionalities, providing high application flexibility. This macro supports 1-bit weight multiply-accumulate (MAC) operations and CNN computations, achieving an energy efficiency of 18.3-41.8 TOPS/W within a working voltage range of 0.7-0.9 V. Its excellent performance in energy efficiency fully demonstrates the macro's potential in artificial intelligence edge computing applications and has positive significance for promoting the practical development of SRAM-CIM technology.

References

[1] S. Aga, S. Narayanan, K. Roy, and A. Raghunathan, "Compute caches: Leveraging existing cache hierarchies for neural network acceleration," in *Proc. Design Automation Conf. (DAC)*, 2017, pp. 1–6.

[2] A. Shafiee, A. Nag, N. Muralimanohar, et al., "ISAAC: A convolutional neural network accelerator with in-situ analog arithmetic in crossbars," in *Proc. Int. Symp. Comput. Archit. (ISCA)*, 2016, pp. 14–26.

[3] P. Chi, S. Li, C. Xu, et al., "PRIME: A novel processing-in-memory architecture for neural network computation in ReRAM-based main memory," in *Proc. Int. Symp. Comput. Archit. (ISCA)*, 2016, pp. 27–39.

[4] J. Zhang, M. Kang, and M. Annavaram, "In-memory computation for energy-efficient neural network accelerators," *IEEE Trans. Comput.-Aided Design Integr. Circuits Syst.*, vol. 39, no. 10, pp. 2408–2421, Oct. 2020.

[5] X. Si, H. Jiang, J. Deng, Y. Wang, and H. Yang, "A twin-8T SRAM computation-in-memory unit-macro for energy-efficient CNN inference," *IEEE J. Solid-State Circuits*, vol. 55, no. 1, pp. 189–202, Jan. 2020.

[6] A. Biswas and A. Chandrakasan, "CONV-SRAM: An energy-efficient SRAM with in-memory dot-product computation for low-power convolutional neural networks," *IEEE J. Solid-State Circuits*, vol. 54, no. 1, pp. 217–230, Jan. 2019.

[7] S. Yin, Z. Jiang, J.-S. Seo, and M. Seok, "XNOR-SRAM: In-memory computing SRAM macro for binary/ternary deep neural networks," *IEEE J. Solid-State Circuits*, vol. 55, no. 6, pp. 1733–1743, Jun. 2020.

[8] Y. Kim, S. Lee, J. Park, et al., "A 16-Kb computing-in-memory SRAM with reconfigurable read capability for binary neural networks," in *IEEE Int. Solid-State Circuits Conf. (ISSCC)*, 2021, pp. 252–253.

[9] H. Jia, J. Zhao, and Y. Xie, "Efficient in-memory computing with spintronics: Opportunities and challenges," *Proc. IEEE*, vol. 108, no. 8, pp. 1322–1342, Aug. 2020.

[10] T. Yang, Y. Wang, and H. Yang, "Design techniques for SRAM-based in-memory computing macros in AI accelerators," in *Proc. IEEE Asian Solid-State Circuits Conf. (A-SSCC)*, 2019, pp. 57–60.

[11] J. Zhang, Z. Wang, and N. Verma, "In-Memory Computation of a Machine-Learning Classifier in a Standard 6T SRAM Array," *IEEE Journal of Solid-State Circuits*, vol. 52, no. 4, pp. 915-924, 2017.

[12] C. Yu, T. Yoo, K. T. C. Chai, T. T.-H. Kim, and B. Kim, "A 65-nm 8T SRAM Compute-in-Memory Macro With Column ADCs for Processing Neural Networks," *IEEE Journal of Solid-State Circuits*, vol. 57, no. 11, pp. 3466-3476, 2022.

[13] C. Yu, T. Yoo, K. T. C. Chai, T. T.-H. Kim, and B. Kim, "A 65-nm 8T SRAM Compute-in-Memory Macro With Column ADCs for Processing Neural Networks," *IEEE Journal of Solid-State Circuits*, vol. 57, no. 11, pp. 3466-3476, 2022.

[14] P.-C. Wu *et al.*, "A 28nm 1Mb Time-Domain Computing-in-Memory 6T-SRAM Macro with a 6.6ns Latency, 1241GOPS and 37.01TOPS/W for 8b-MAC Operations for Edge-AI Devices," presented at the 2022 IEEE International Solid- State Circuits Conference (ISSCC), 2022.

979-8-3315-8850-2/25 $31.00 © 2025 IEEE

2025 The 10th International Conference on Integrated Circuits and Microsystems

OpenPIM: An Open-Source Programmable Processing-In-Memory Accelerator Design & Pipeline Simulation Framework

Ruibao Wang[1,2], Wenshuo Yue[1,3], Daijing Shi[1,3], Yuchao Yang[1,3,4,5*], Bonan Yan[1,3*]

[1]Institute for Artificial Intelligence, Peking University, Beijing, China
[2]College of Electronic Science & Engineering, Jilin University
[3]School of Integrated Circuits, Peking University, Beijing, China
[4]Center for Brain Inspired Intelligence, Chinese Institute for Brain Research (CIBR), Beijing, China
[5]Guangdong Provincial Key Laboratory of In-Memory Computing Chips,
School of Electronic and Computer Engineering, Peking University, Shenzhen, China
*Corresponding Email: yuchaoyang@pku.edu.cn, bonanyan@pku.edu.cn

Abstract—**Processing-in-memory (PIM) is a promising choice for accelerating deep neural networks (DNNs) featuring high efficiency and low power. However, the rapid upscaling of neural network model sizes poses a crucial challenge for the limited on-chip PIM capacity. When the PIM presumption of "pre-loading DNN weights/parameters only once before repetitive computing" is no longer practical, concurrent writing and computing techniques become necessary for PIM. We implement an open-source full-fledged PIM-based accelerator, named *OpenPIM*, with register-transfer-level (RTL) design and simulation framework. It is available at https://github.com/author288/openpim. OpenPIM features simulating the impact of the external memory bandwidth outside the PIM, which is often considered as the bottleneck for the overall PIM accelerator performance. Moreover, Open-PIM supports various pipelining schemes, including naive ping-pong, *in situ* concurrent write/compute and generalized ping-poing. Experiments show that the generalized ping-pong strategy achieves acceleration of over 1.67× when fully utilizing the off-chip memory bandwidth. When further limiting the off-chip memory bandwidth ranging in 8∼256 bytes per clock cycle, the proposed generalized ping-pong strategy accelerates 1.22∼7.71× versus naive ping-pong. OpenPIM provides an effective toolkit to analyze off-chip memory bandwidth centric pipelining strategy to maximize the utilization and performance of PIM accelerators toward large-scale general matrix multiplications.**

Index Terms—**Compute-In-Memory, Processing-In-Memory, SoC, Programmable CIM Accelerator, Open-Source EDA**

I. INTRODUCTION

Processing-In-Memory (PIM) is an innovative approach that holds the potential to significantly accelerate deep learning operations by enabling computations within memory arrays rather than transferring data back and forth between processing units and storage mediums [1]–[3]. The fundamental concept of PIM revolves around integrating computing circuits within or near memory arrays to process data directly on stored information, especially for general matrix multiplication (GeMM) [4]. However, deep neural network (DNN) models, following the deep learning scaling law, are scaling up at an exponential speed, inflicting unprecedented challenges for the limited on-chip PIM capacity in that most of the conventional

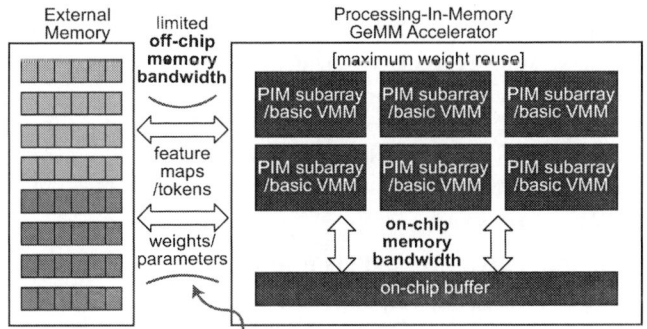

challenge caused by large deep learning models
→Question: how to realize concurrent write/compute for PIM subarrays?

Fig. 1. PIM accelerators face emerging challenge in high-dimensional GeMM computation: realization of concurrent write/compute.

PIM architectures hold the presumption that loading weights (parameters) of deep learning models only once before repetitive computation based on a weight-stationary parallelism scheme.

In contrast, the trending largest ever deep learning models (e.g. Transformer-based large language modelshave required reloading DNN weights in PIM architectures into a necessary feature [5]. The weights are frequently sliced and programmed to the PIM subarray (i.e. macros) in batches during PIM computation, i.e. concurrent write/compute (Fig. 1).

In order to conviniently synthesize, simulate and analyze the impact of off-chip memory bandwidth upon PIM accelerator, we develop an open-source full-fledged PIM-based accelerator design library, named *OpenPIM*, with register-transfer-level (RTL) design and pipeline simulation framework. It is available at https://github.com/author288/openpim. OpenPIM features simulating the impact of the external memory bandwidth outside the PIM, which is often considered as the bottleneck for the overall PIM accelerator performance. Moreover, Open-PIM supports various pipelining schemes, including naive ping-pong, *in situ* concurrent write/compute and generalized

979-8-3315-8850-2/25 $31.00 © 2025 IEEE

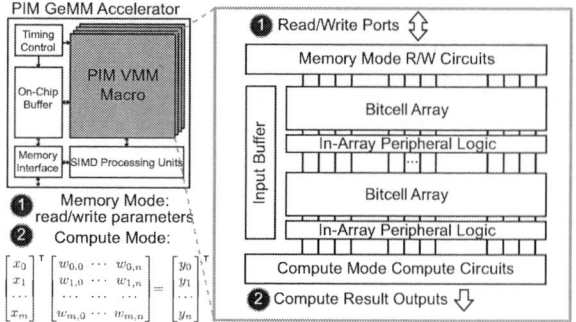

Fig. 2. Typical SRAM-based PIM GeMM accelerator circuit block diagram and its two work modes.

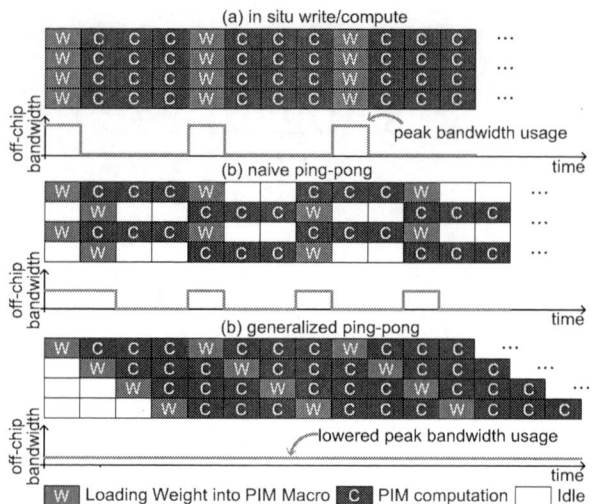

Fig. 3. (a) In situ write/compute strategy. (b) Naive ping-pong strategy. (c) Generalized ping-pong strategy.

ping-poing. With the help of OpenPIM library, this work also quantitatively model and analyze the three most common used pipelining schemes for concurrent write/compute in PIM, i.e. in situ write/compute, naive ping-pong and generalized ping-pong. Based on this, we propose the off-chip memory bandwidth centric pipeline scheduling scheme to save off-chip memory bandwidth by improving its utilization. Experiments with the proposed OpenPIM library show that the generalized ping-pong strategy achieves an acceleration of over $1.67\times$ when fully utilizing the off-chip memory bandwidth. When further limiting the off-chip memory bandwidth ranging from 8~256 bytes per clock cycle, the generalized ping-pong strategy accelerates $1.22\sim7.71\times$ versus the existing naive ping-pong strategy.

II. PRELIMINARIES

A. SRAM-Based PIM Designs

PIM GeMM accelerator consists of multiple PIM vector-matrix multiplication (VMM) macros (subarrays) to perform complete GeMM operations (Fig. 2) [6], [7]. Each PIM macro works in two primary operational modes: memory model and compute mode [8], [9]. The memory mode serves a crucial role in loading weights/parameters into the PIM macro for maximum reuse in the compute mode. The compute mode is dedicated to performing in-memory GeMM computations that leverage the physical locality of data within SRAM bitcells. Static Random Access Memory (SRAM)-based PIM offers both fast computing speed in the compute mode and low read/write latency in the memory mode. Also, SRAM is more appropriate for repetitively reloading with over 10^{15} times of bitcell endurance. However, the density of SRAM-PIM leads to limited on-chip capacity (e.g. 16kb~4.5Mb/macro). Toward the upscaled deep learning models, concurrent write/compute strategies is in urgent need for SRAM-based PIM.

B. Concurrent Write/Compute Strategies

Fig. 3 illustrates the comparison between different concurrent write/compute strategies using an exemplary PIM accelerator comprising 4 PIM macros:

- **In situ write/compute strategy** (Fig. 3(a)) synchronizes all PIM macros for writing or computing. Only writing occupies the off-chip memory bandwidth, reflecting an intermittent characteristic.

- **Ping-pong strategies** (Fig. 3(b)): With >2 PIM macros, the **naive ping-pong strategy** divides all macros into two groups, say, *bank1* and *bank2*. While *bank1* performs computations for the nth GeMM operation, *bank2* loads the weights for the $(n+1)^{th}$ opeartion; once the computations for the nth operation are completed, the two banks switch their tasks for computation and weight reloading [10]–[12]. It alleviates the utilization for off-chip memory bandwidth, but idle time still exists [13], [14]. The **generalized ping-pong strategy** (Fig. 3(c)) referes to a more complicated ping-pong scheme taht applies to more than 2 macros. It lacks a systematic analysis framework achieve the optimized design space exploration.

III. OPENPIM HARDWARE ARCHITECTURE

We choose PUMA [15] design as the basic skeleton structure for the programmable accelerator. In addition to the original PUMA accelerator design, we revise the PIM-oriented instruction set architecture (ISA). This base architecture supports the aforementioned in situ write/compute strategy, naive ping-pong strategy and the generalized ping-pong stratety. The ISA comes with an assembler to convert assembly code into binary machine code. The focused scheduling strategies leads to different assembly code for different pipelined execution.

Fig. 4 illustrates the overall revised hardware architecture, which comprises: a global weight memory ,a global input memory ,a global intermediate result memory , and an instruction memory. These components facilitate the transmission of instructions and data between the processing core and the memories. Intermediate results are accumulated using a vector processing unit (VPU). Additionally, the architecture includes a top controller and an instruction generation module. Each PIM core consists of: 4 PIM macros ,a buffer for storing weights, inputs, and intermediate results ,a control unit , and a core instruction memory. This structure supports efficient data handling and processing within the system.

979-8-3315-8850-2/25 $31.00 © 2025 IEEE   825

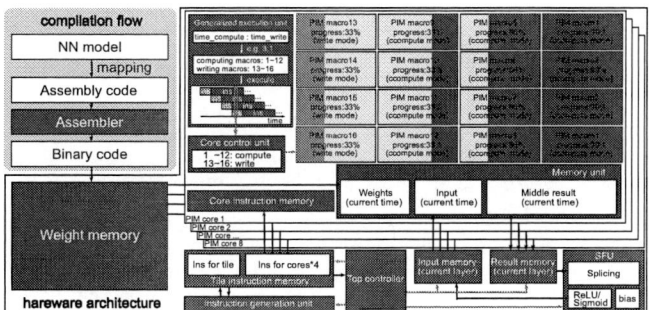

Fig. 4. The base PIM accelerator architecture as an example to implement generalized ping-pong scheduling strategy. This base architecture is revised from PUMA [15] to support various scheduling strategies.

IV. GENERALIZE PING-PONG SCHEDULING BY PRIORITIZING OFF-CHIP MEMORY BANDWIDTH

With OpenPIM RTL library, this section shows how to to define the generalized ping-pong by prioritizing off-chip memory bandwidth because of its bottlenecking the overall computing throughput. In order to achieve full usage for the off-chip memory bandwidth, we first need to generalize the ping-pong pipelining strategy for arbitrary number of cores. We would like to quantitatively analyze the utilization for the in situ write/compute strategy and the naive ping-pong strategy.

We formulate the latency for the memory mode and compute mode. Given that both weight rewriting and computation are essential operations, we posit that a macro is considered "idle" when it is neither performing rewriting nor computation. When the weight reloading time is less than the PIM time, the rewritten bank has to wait for the PIM bank to finish the computation task of the current layer before starting the computation of the next GeMM operation. Assume $size_{macro}$, $size_{OU}$, n_{in}, and s represent macro size, operation unit size, number of input vector words for VMM calculaton, and rewrite speed, respectively. During a complete cycle of write and compute, the compute time is: $time_{PIM} = \frac{size_{macro} * n_{in}}{size_{OU}}$ The writing time is $time_{rewrite} = \frac{size_{macro}}{s}$. Hence, the macro utilization is:

$$util_{macro} = \frac{time_{PIM} + time_{rewrite}}{2 \times \max(time_{PIM}, time_{rewrite})} \quad (1)$$

With this formulation, Fig. 5 shows $time_{PIM}/time_{rewrite}$ ratio and macro utilization for the naive ping-pong strategy under various n_{in} within a specific PIM architecture configuration. In this example, the macro size $size_{macro}$ is set to 32×32 bytes, the output unit size $size_{OU}$ is set to 4×8 bytes, and the bandwidth s is set to 4 byte/cycle. It can be observed that only when the number of inputs n_{in} equals 8, where $time_{PIM} = time_{rewrite}$ (i.e. matching the computing time and weight reloading time), at which point the naive ping-pong strategy achieves the highest macro utilization rate. Apart from this scenario, the naive ping-pong strategy significantly reduces macro utilization.

With the aforementioned analysis, in order to maintain the highest macro utilization and off-chip bandwidth utilization

during execution for varying values of n_{in}, we propose the generalized ping-pong strategy, which directly focuses on the ratio of $time_{PIM}/time_{rewrite}$, and adjusts the start time of each macro execution. This approach averages the demand for off-chip bandwidth across each cycle, thereby reducing the peak demand for off-chip bandwidth. Simultaneously, each macro will immediately transition to the next write/compute operation upon completing the current one, thereby sustaining the highest macro utilization rate.

Fig. 3(c) illustrates the timing diagram and off-chip memory bandwidth utilization of proposed generalized ping-pong pipeline. **The core idea of the generalized ping-pong is maintain a peak usage for the off-chip memory bandwidth with multi-core PIM accelerators.** It groups multiple macros for writing and for computing. A deep pipelined pattern is exploited with balanced writing (memory bandwidth occupation) and PIM computing. This scheme has both advantages of in situ write/compute (consistently maintain a high macro utilization rate) and naive ping-pong (keeping high utilization rate for off-chip memory bandwidth).

Assuming the presence of 4 macros in a PIM accelerator, when the ratio of weight updating to computation time is 1:3, *macro2* initiates its weight updating process subsequent to the completion of *macro1*'s rewrite. This sequence continues with *macro3* and *macro4*, effectively distributing the bandwidth demand across each cycle. In this example, compared to the in situ write/compute strategy and the naive ping-pong, the proportion of bandwidth idle time in generalized ping-pong decreased from 75% and 66% to 0%, while the peak bandwidth demand is reduced to 25% of that required by the in situ write/compute approach. The macro utilization rate in generalized ping-pong remains at 100%, as the strategy does not induce idle states in the macros. Less bandwidth idle time and higher macro utilization ensure that generalized ping-pong delivers optimal performance under the same bandwidth constraints.

V. OPENPIM ANALYSIS FRAMEWORK FOR OPTIMIZATION

We further provides two stages of optimization towards computing throughput: (a) *design phase*: design space exploration for full usage of off-chip memory bandwidth in designing a PIM accelerator before tape-out; (b) *runtime phase*: scheduling PIM macros write/compute operations toward the maximum off-chip memory bandwidth utilization after PIM accelerator ASIC fabrication.

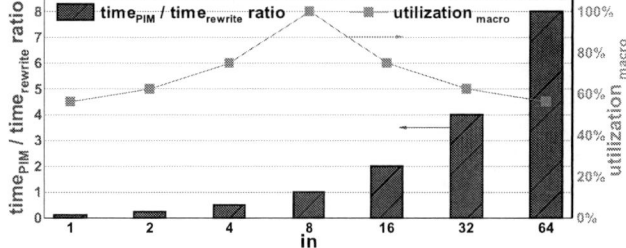

Fig. 5. The specific idle time ratio of macro.

A. Design Phase Optimization: Design Space Exploration

The key design parameters are listed in Table I. Design optimization seeks the best parameter set for optimal utilization with concurrent computing and writing configurations. Using generalized ping-pong scheduling, we explore designs from a specified off-chip bandwidth, aiming to improve performance or reduce hardware area. Weight rewrites disrupt computations, so we minimize them by writing weights once and reusing them extensively. All input vectors must complete VMM with current weights before loading new ones. Limited on-chip memory for inputs and intermediates necessitates batch processing for large datasets, establishing a consistent ratio between weight rewrite and computation times to facilitate ping-pong strategy integration.

To find the sweet point of 100% utilization of off-chip memory bandwidth, the design exploration should take generalized ping-pong scheduling into account to match the PIM memory capacity and computing throughput. The time of PIM computation is contingent upon the velocity at which the macro completes PIM computation and the number of vectors that need to be computed within a batch. In generalized ping-pong, the time for a single weight rewrite is: $time_{rewrite} = size_{macro}/s$, and the time for a PIM compute is $time_{PIM} = size_{macro} \cdot n_{in}/size_{OU}$. The number of macros that can be supported under a fixed off-chip bandwidth (with full usage) is given by $num_{macro} = \frac{C_{scheme} \times band.}{s}$ where $C_{scheme} = 1$ for in situ write/compute and $C_{scheme} = 2$ for naive ping-pong. Note in the ping-pong strategy, where macros are divided into two groups that rewrite alternately, the average bandwidth demand per macro is reduced to $(s/2)$.

Generalized ping-pong sets the number of macros that rewrite simultaneously according to the ratio of $time_{rewrite}$ to $time_{PIM}$, with each macro's average bandwidth demand being $\frac{time_{rewrite}*s}{time_{PIM}+time_{rewrite}}$, and the number of macros that can be supported is given by

$$num_{macro} = \frac{(time_{PIM} + time_{rewrite}) * band.}{time_{rewrite} * s}. \quad (2)$$

When the ratio of $time_{rewrite}$ to $time_{PIM}$ is not 1, the naive ping-pong strategy may result in idle states of macros, whereas the in situ write/compute and generalized ping-pong strategies remain unaffected. As a result, the performance of every macro under the ping-pong strategy reduce to $\frac{time_{PIM}+time_{rewrite}}{time_{PIM}+time_{rewrite}+|time_{PIM}-time_{rewrite}|}$ of its original capability.

Based on the number of macros supported and the performance of each macro, it can be derived that under the current band. The ratio of the number of macros for the three strategies "generalized ping-pong:in situ write/compute:naive ping-pong" is $\frac{size_{macro}*in/size_{OU}+size_{macro}/s}{size_{macro}/s}$: 1 : 2, and the execution time ratio for "generalized ping-pong:in situ write/compute:naive ping-pong" is $\frac{in*s+size_{OU}}{size_{OU}}$: 1 : $\frac{2*(in*s+size_{OU})}{in*s+size_{OU}+|in*s-size_{OU}|}$. When $time_{PIM} > time_{rewrite}$, the generalized ping-pong strategy demonstrates better performance compared to the other two strategies. When $time_{PIM} < time_{rewrite}$, generalized ping-pong outperforms

TABLE I
LIST OF PARAMETERS FOR DESIGN SPACE EXPLORATION

name of parameter	value
$band$	off-chip bandwidth
$size_{macro}$	macro size
$size_{OU}$	operation unit size
s	rewrite speed
n_{in}	number of activations for VMM calculaton
$time_{PIM}$	Time of a PIM calculation
$time_{rewrite}$	Time of a weight rewrite
n	the multiple of band. reduction
num_{macro}	number of macros
m	the multiple of num_{macro} reduction

the in situ write/compute strategy and offers equivalent performance to the ping-pong strategy while utilizing fewer macros, which translates to a lower area overhead. When $time_{PIM} = time_{rewrite}$, generalized ping-pong provides better performance than the in situ write/compute strategy, and its performance and number of macros are identical to those of the naive ping-pong strategy.

B. Runtime Phase Pipeline Adaption

In a large system-on-a-chip (SoC) design, the off-chip memory bandwidth for PIM accelerator is often assigned dynamically in runtime. Chances are the accelerator cannot get its full off-chip memory bandwidth as designated in the design phase. In this case, we will provide the following solution to adjust the pipelining scheme.

For a PIM accelerator after fabrication, when encountering a reduction in off-chip bandwidth during the execution of computational tasks, the generalized ping-pong strategy can preserve a greater portion of performance compared to other strategies. We discuss about the performance degradation caused by the reduction of off-chip bandwidth under the in situ write/compute, ping-pong, and generalized ping-pong strategies through a modeling approach.

For the generalized ping-pong strategy, when off-chip bandwidth is reduced, the speed of weight updating remains constant while the number of active macros is decreased. Generalized ping-pong can adjust the ratio of $time_{PIM}$ to $time_{rewrite}$ to reduce the number of working macros. As previously mentioned, $time_{PIM}$ depends on the speed at which a macro completes VMM and the number of vectors that need to be computed within a batch, which is determined by the amount of on-chip memory each macro can access. When the number of working macros is reduced and the on-chip memory capacity remains unchanged, the amount of on-chip memory available to each macro increases, in increases. This implies that $time_{PIM}$ increases. According to Eq. 2, when $time_{PIM}$ increases and $time_{rewrite}$ remains constant, it supports a greater number of active macros.

When the off-chip bandwidth is reduced to $band./n$, the number of active macros becomes num_{macro}/m accordingly. The ratio of $time_{PIM}$ to $time_{rewrite}$ becomes: $\frac{size_{macro}}{size_{OU}} \cdot n_{in} \cdot m : \frac{size_{macro}}{s}$. At this point, the average demand for off-chip bandwidth per macro is $\frac{time_{rewrite} \cdot s}{time_{PIM}+time_{rewrite}}$, and multiply it

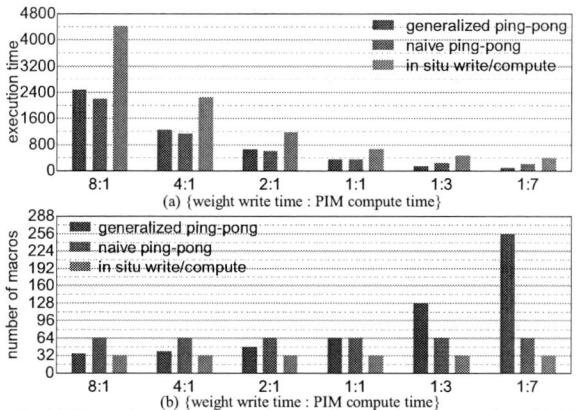

Fig. 6. (a) Execution time comparison under the three strategies. (b) Number of macros comparison under the three strategies.

with $\cdot num_{macro}/m$, which should be equal to $band./n$. Then we can solve the performance degradation:

$$\frac{2(n_{in} * s + size_{OU})}{size_{OU} + \sqrt{size_{OU}^2 + \frac{4num_{macro}*size_{OU}*n_{in}*s^2*n}{band.}}} \quad (3)$$

In Eq .3, all parameters except for n and m are numerical values obtained during the hardware design phase using the generalized ping-pong strategy.

VI. EVALUATION

A. Experimental Setup

The proposed OpenPIM with generalized ping-pong strategy focuses on the throughput improvement for multi-macro PIM GeMM accelerators. To evaluate it, we implement different accelerator-level concurrent write/computing pipeline strategies on the OpenPIM accelerator. To simplify the analysis and control the variables, we focus on large-scale consecutive GeMM operations with basic linear algebra subprograms (BLAS) level benchmarks. Because the target pipeline strategy emphasizes the alignment on clock cycles, the timing simulation is based on synthesizable Verilog HDL design (check our open-source repository https://github.com/author288/openpim to reproduce the simulation results). The example design parameters are set to: PIM accelerator has 16 cores, where each is equipped with 16 macros. The macro size is 32×32 bytes, with a write speed ranging from 1 to 8byte/cycle, and the size of the operating unit is 4×8byte.

B. Evaluation for Design Phase Optimization

Fig. 6 presents a comparison of performance and macro count between the generalized ping-pong and other strategies during the hardware design exploration phase. At this stage, the off-chip bandwidth memory $band.$ is set to 128byte/cycle. The x-axis is the ratio of weight write time[1] over the PIM compute time. The y-axis is the execution latency in cycle numbers. When $time_{rewrite} < time_{PIM}$, under the same off-chip bandwidth conditions, generalized ping-pong can support a greater computational power compared to the other two

[1]The "weight write time" refers to entirely rewriting the data stored in PIM.

strategies, and it requires the use of more macros. In the scenario where the ratio of $time_{rewrite}$ to $time_{PIM}$ is 1:7, generalized ping-pong achieves a $2.51\times$ performance improvement over naive ping-pong and a $5.03\times$ improvement over in situ write/compute. When $time_{rewrite} = time_{PIM}$, the generalized ping-pong and naive ping-pong strategies completely overlap, and they exhibit a $2\times$ performance improvement over in situ write/compute in terms of performance. When $time_{rewrite} > time_{PIM}$, generalized ping-pong outperforms in situ write/compute and matches the performance of naive ping-pong, but with the advantage of using fewer macros, which conserves area and power consumption. In the case where the ratio of $time_{rewrite}$ to $time_{PIM}$ is 8:1, generalized ping-pong reduces the number of macros by 43.75% compared to naive ping-pong and achieves $1.78\times$ performance improvement over in situ write/compute strategy. The improvement brought by the generalized ping-pong on performance and area depends on the ratio of $time_{rewrite}$ to $time_{PIM}$.

C. Evaluation for Runtime Phase Adaptation

Fig. 7 shows the results for runtime phase optimization. It shows the comparative performance of the three strategies (in situ write/compute, naive ping-pong, generalized ping-pong) in response to bandwidth fluctuations. The x-axis is how many times of off-chip memory bandwidth reduction compared to that given during design phase. The y-axes are (a) normalized execution, (b) average on-chip memory utilization rate, (c) off-chip memory bandwidth utilization rate, and (d) average macro utilization rate. For Fig. 7(a) and (b) This comparison is performed on the design phase optimization goal of $time_{rewrite} = time_{PIM}$ and exerts a progressive reduction in bandwidth to monitor the trend in performance variation. The experimental results indicate that generalized ping-pong can retain a greater degree of performance as off-chip bandwidth decreases, in comparison to current execution schemes. When the bandwidth is reduced to $\frac{band.}{64}$, our strategy achieves $5.38\times$ improvement in performance over in situ write/compute and $7.71\times$ improvement in performance over naive ping-pong.

Fig. 7(c) and (d) show the comparison of off-chip bandwidth utilization and macro utilization rate, respectively. The in situ write/compute strategy yields a lower off-chip bandwidth utilization, whereas the naive ping-pong strategy has a lower macro utilization. The advantage of generalized ping-pong is with both high off-chip bandwidth utilization and macro utilization.

Table II shows the design space optimization with generalized ping-pong at different off-chip bandwidth (unit: byte/cycle). The discrepancy between the execution strategies calculated by the model (with a fractional number of PIM macros) and those actually implemented in Verilog HDL (with integer number of PIM macros) diminishes as the number of macros increases. For the in situ write/compute strategy, the optimal scheduling is to reduce the speed at which each macro rewrites weights, thereby decreasing the demand for off-chip bandwidth, while keeping the number of active macros

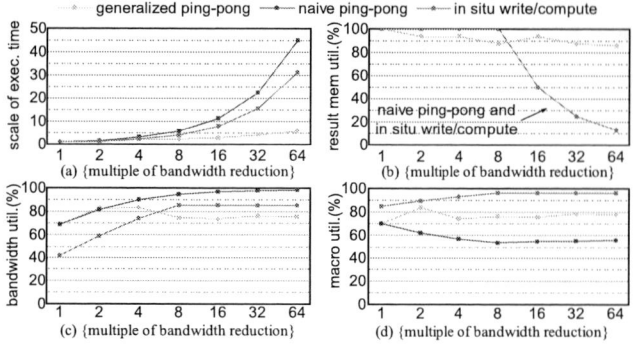

Fig. 7. (a) Scale of execution time comparison under the three strategies. (b) Result memory utilization comparison under the three strategies. (c) Bandwidth utilization comparison under the three strategies. (d) Macro utilization comparison under the three strategies.

constant. However, the speed of weight updating cannot be infinitely reduced as a latency overhead. When the speed of weight updating reaches the minimum value determined by hardware design, it becomes necessary to reduce the number of active macros to cope with further decreases in bandwidth. This leads to a more rapid decline in performance. For generalized ping-pong, due to the finite number of macros, the actual execution results are an approximation of the model.

TABLE II
THE DISCREPANCY BETWEEN THEORY AND PRACTICE

band.	working macros		time_PIM:time_rew		remaining perf.	
	theory	practice	theory	practice	theory	practice
256	82.05	80	1.56:1	1.5:1	78.08%	75.00%
128	54.01	49	2.37:1	2.5:1	59.31%	54.69%
64	36.26	36	3.53:1	3.5:1	44.14%	43.75%
32	24.71	24	5.18:1	5:1	32.37%	31.25%
16	17.02	16	7:52:1	7:1	23.49%	21.88%
8	11.83	11	10.82:1	10:1	16.91%	15.63%

VII. CONCLUSION

This work attempts to answer the question of *how to realize concurrent weight transfer and PIM computation towards upscaled GeMM operations*. To achieve this, we propose a novel OpenPIM RTL design for arbitrary numbers of PIM macros to form a GeMM accelerator. With an exemplary PIM accelerator implemented, we demonstrate the efficacy of the generalized ping-pong scheduling strategy. It is applicable for design space exploration and improve runtime off-chip memory bandwidth utilization. Compared to existing strategies, our approach achieves superior performance boost under the same off-chip memory bandwidth. This work reveals the fundamental theory of pipeline optimization for PIM architectures.

ACKNOWLEDGMENT

This work is supported by National Natural Science Foundation of China (92364102, 92264201, T2350006). This work is sponsored by Beijing Nova Program. This work is supported by High-performance Computing Platform of Peking University.

REFERENCES

[1] S. Li, N. Zhang, W. Zhang, R. Yang, Y. Chang, and B. Xiong, "Framework independent modeling for sram-based in-memory computing," in *2024 International Symposium of Electronics Design Automation (ISEDA)*, pp. 777–777, 2024.

[2] M. Zhou, X. Wang, and T. Rosing, "OverlaPIM: Overlap optimization for processing in-memory neural network acceleration," in *Design, Automation & Test in Europe Conference & Exhibition (DATE)*, pp. 1–6, IEEE, 2023.

[3] J. Li, H. Zhao, W. Yue, Y. Fu, D. Shi, A. Fan, Y. Yang, and B. Yan, "Pearl: Fpga-based reinforcement learning acceleration with pipelined parallel environments," in *2025 Design, Automation & Test in Europe Conference (DATE)*, pp. 1–7, 2025.

[4] B. Yan, J.-L. Hsu, P.-C. Yu, C.-C. Lee, Y. Zhang, W. Yue, G. Mei, Y. Yang, Y. Yang, H. Li, *et al.*, "A 1.041-mb/mm² 27.38-TOPS/W signed-INT8 dynamic-logic-based ADC-less SRAM compute-in-memory macro in 28nm with reconfigurable bitwise operation for AI and embedded applications," in *IEEE International Solid-State Circuits Conference (ISSCC)*, vol. 65, pp. 188–190, IEEE, 2022.

[5] S. A. Razavi, H.-Y. Ting, T. Giyahchi, and E. Bozorgzadeh, "On exploiting patterns for robust fpga-based multi-accelerator edge computing systems," in *Design, Automation & Test in Europe Conference & Exhibition (DATE)*, pp. 116–119, IEEE, 2022.

[6] B. Yan, J.-L. Hsu, P.-C. Yu, C.-C. Lee, Y. Zhang, W. Yue, G. Mei, Y. Yang, Y. Yang, H. Li, Y. Chen, and R. Huang, "A 1.041-Mb/mm² 27.38-TOPS/W signed-INT8 dynamic-logic-based ADC-less SRAM compute-in-memory macro in 28nm with reconfigurable bitwise operation for AI and embedded applications," in *IEEE International Solid-State Circuits Conference (ISSCC)*, vol. 65, pp. 188–190, 2022.

[7] L. Wang, W. Li, Z. Zhou, H. Gao, Z. Li, W. Ye, H. Hu, J. Liu, J. Yue, J. Yang, *et al.*, "A flash-SRAM-ADC-fused plastic computing-in-memory macro for learning in neural networks in a standard 14nm FinFET process," in *IEEE International Solid-State Circuits Conference (ISSCC)*, vol. 67, pp. 582–584, IEEE, 2024.

[8] Y. Fu, D. Shi, A. Fan, W. Yue, Y. Yang, R. Huang, and B. Yan, "Probabilistic compute-in-memory design for efficient markov chain monte carlo sampling," *IEEE Transactions on Circuits and Systems I: Regular Papers*, 2023.

[9] B. Yan, Q. Yang, W.-H. Chen, K.-T. Chang, J.-W. Su, C.-H. Hsu, S.-H. Li, H.-Y. Lee, S.-S. Sheu, M.-S. Ho, Q. Wu, M.-F. Chang, Y. Chen, and H. Li, "Rram-based spiking nonvolatile computing-in-memory processing engine with precision-configurable in situ nonlinear activation," in *2019 Symposium on VLSI Technology*, pp. T86–T87, 2019.

[10] Y. Fu, A. Fan, W. Yue, H. Zhao, D. Shi, Q. Wu, J. Li, X. Zhang, Y. Tao, Y. Yang, and B. Yan, "PROCA: programmable probabilistic processing unit architecture with accept/reject prediction & multicore pipelining for causal inference," in *IEEE International Symposium on High Performance Computer Architecture, HPCA 2025, Las Vegas, NV, USA, March 1-5, 2025*, pp. 761–774, IEEE, 2025.

[11] Y. Fu, D. Shi, A. Fan, W. Yue, Y. Yang, R. Huang, and B. Yan, "Probabilistic compute-in-memory design for efficient markov chain monte carlo sampling," *IEEE Trans. Circuits Syst. I Regul. Pap.*, vol. 71, no. 2, pp. 703–716, 2024.

[12] R. Liu, X. Peng, X. Sun, W.-S. Khwa, X. Si, J.-J. Chen, J.-F. Li, M.-F. Chang, and S. Yu, "Parallelizing SRAM arrays with customized bit-cell for binary neural networks," in *Annual Design Automation Conference (DAC)*, pp. 1–6, 2018.

[13] H. Zhang, W. Yin, S. He, Y. Du, and L. Du, "An efficient two-stage pipelined compute-in-memory macro for accelerating transformer feed-forward networks," *IEEE Transactions on Very Large Scale Integration (VLSI) Systems*, 2024.

[14] W. Yue, T. Zhang, Z. Jing, K. Wu, Y. Yang, Z. Yang, Y. Wu, W. Bu, K. Zheng, J. Kang, Y. Lin, Y. Tao, B. Yan, R. Huang, and Y. Yang, "A scalable universal ising machine based on interaction-centric storage and compute-in-memory," *Nature Electronics*, vol. 7, no. 10, pp. 904–913, 2024.

[15] A. Ankit, I. E. Hajj, S. R. Chalamalasetti, G. Ndu, M. Foltin, R. S. Williams, P. Faraboschi, W.-m. W. Hwu, J. P. Strachan, K. Roy, *et al.*, "PUMA: A programmable ultra-efficient memristor-based accelerator for machine learning inference," in *International Conference on Architectural Support for Programming Languages and Operating Systems (ASPLOS)*, pp. 715–731, 2019.

979-8-3315-8850-2/25 $31.00 © 2025 IEEE

CASA-CIM: A 28nm 131.54 TOPS/W SRAM-Based CIM Macro with Cap-Adder Weighting and Shift-After-Addition Data Streaming for Efficient MAC Operations

Runru Yu[1,2], Chenyang Zhao[2], Honghu Yang[1,2], Zhiting Lin[3], Xiulong Wu[3], Keji Zhou, Jianguo Yang*

[1] Faculty of Integrated Circuits and Micro-Nano Electronics, Fudan University, Shanghai 200438, China
[2] Zhangjiang Laboratory, Shanghai 201210, China
[3] Anhui University, Hefei 230601, China
yangjg@zjlab.ac.cn (*Corresponding Authors)

Abstract—Charge-domain Computing-In-Memory (CIM) for multiply-accumulate (MAC) operations has garnered significant attention due to its robust tolerance to process, voltage, and temperature (PVT) variations. However, two major challenges persist: performance trade-offs in multi-bit weighting methods within charge-domain CIM, and excessive cost from adder trees. To address these challenges, we present CASA-CIM, a novel CIM macro with three key innovations. Firstly, a hybrid digital-analog Cap-Adder weighting (CAW) scheme enhances linearity and reduces weighting-expansion cost. Secondly, a Shift-After-Addition data stream (SAS) scheme cuts adder cost by 10.4%. Thirdly, a dual 3bit flash analog-to-digital converter (3bit-FADC) structure lowers offset demands compared with traditional 4bit ADCs. The proposed CASA-CIM achieves an energy efficiency of 131.54 TOPS/W with INT4 input/weight precision @28nm.

Keywords—Computing-In-Memory, SRAM, Weighting method

I. INTRODUCTION

With the rapid development of artificial intelligence, energy-efficient edge computing has become a critical research focus. To alleviate the von Neumann bottleneck inherent in traditional architectures, Computing-In-Memory (CIM) has been proposed to improve energy efficiency. Among various CIM designs, SRAM-based schemes [1],[2],[3],[4] are widely adopted due to their high speed, endurance, and technological maturity.

However, several challenges remain in existing SRAM-based CIM macros for multiply-accumulate (MAC) operations. Firstly, achieving high-performance multi-bit weighting in charge-domain CIM is difficult. Traditional capacitor ladder schemes suffer from increased power consumption and area cost due to the exponential growth of capacitance values. Bridge-capacitor schemes introduce accuracy degradation caused by parasitic capacitances [5],[6]. Multi-charge-sharing methods lead to longer computation times and reduced area efficiency [7],[8]. Fully digital weighting using adders incurs additional area cost [9]. Consequently, charge-domain weighting schemes must carefully trade off speed, area, accuracy, and power consumption. Secondly, adder-tree cost remains a significant

issue in CIM. In recent digital-domain CIM macros, the energy cost of adders in [10] and [11] accounts for 60% and 36.6%, respectively. Reducing adder cost is essential for enhancing CIM macro energy efficiency and enabling high-performance edge computing.

In this paper, we propose a 64 kb charge-domain CIM macro, CASA-CIM. Firstly, the Cap-Adder Weighting (CAW) scheme improves the trade-off among energy, speed, and linearity. Secondly, the Shift-After-Addition (SAS) scheme reduces full adders by 10.4%, enhancing energy efficiency. Thirdly, the dual 3bit Flash ADC (3bit-FADC) structure offers better compatibility and lower offset specifications than conventional designs, further reducing power.

The remainder of this paper is organized as follows. Section II describes the proposed CIM macro structure and its operating principles. Section III reports performance evaluation, and Section IV concludes the paper.

II. ARCHITECTURE OF THE PROPOSED CIM MACRO AND OPERATING PRINCIPLES

A. The Structure of the CASA-CIM

The structure of the CASA-CIM is shown in Fig.1. It comprises 16 CIM banks, an adder group, and a clock module. Each CIM bank integrates a 64×64 9TnC array, 64×1 4bit DACs, 32×1 3bit Flash ADCs, a WL driver, a BL driver and a circuit for control. Each 64×64 9TnC array consists of 32 compute cores, used for MAC calculations with 2bit weights. Each compute core corresponds to a Flash ADC.

The 9TnC bit-cell used in CIM macro is also shown in Fig.1, which mainly consists of a 6T SRAM bit-cell and three transistors used for calculation. The multiplication operation is achieved by controlling whether to charge the capacitor through weight stored in 6T SRAM.

The calculation sequence of the CASA-CIM is shown in Fig.2. The operation mechanism of CIM macro is as follows: External 4bit digital inputs are converted to analog signals via DACs. The SRAM array then performs 64 MAC operations. The operation results are converted into digital quantities again through ADC. Finally, the MAC results of the 16 CIM macros are added and shifted by adder group to obtain the final output

---

This research was supported in part by the National Key Research and Development Program under Grant No. 2021ZD0114400, and in part by the National Natural Science Foundation of China under Grant 92164204, Grant 92164204 and Grant 62222119.

979-8-3315-8850-2/25 $31.00 © 2025 IEEE

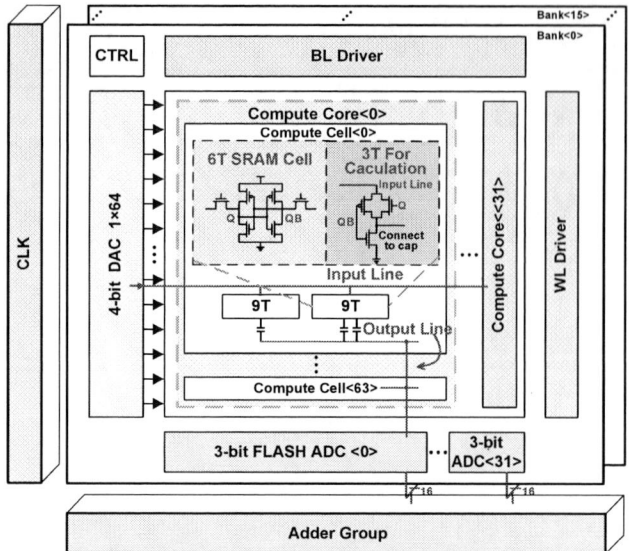

Fig. 1. Overall structure of the proposed CIM macro.

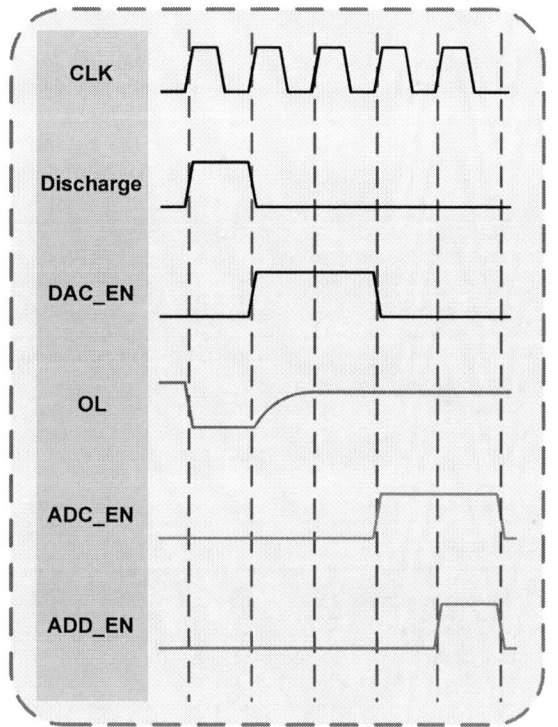

Fig. 2. Timing diagrams of proposed CIM macro.

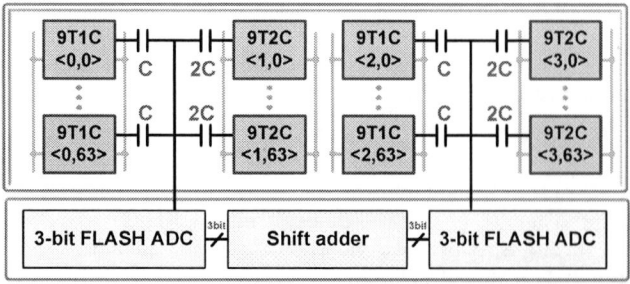

Fig. 3. The proposed Cap-adder weighting scheme.

results. Unlike the traditional charge-domain computation flow, our approach performs only partial calculations within the array, and the adder accumulates prior to shifting.

B. Proposed Cap-Adder Weighting Scheme

The Cap-Adder weighting scheme is shown in Fig. 3. The 4bit weight W<3:0>is stored in four 9TnC SRAM bit-cells in the same row, and the 4bit input I<3:0>is represented by the analog VIN output by DAC. one 9T1C bit-cell and one 9T2C bit-cell form a compute cell, and the truth table of the compute cell is shown in TABLE I.

The 2×64 9TnC bit-cells corresponding to 1×64 compute cells in each column complete the accumulation calculation through capacitance coupling. The accumulated analog voltage is expressed in Equation (1). Here, C_{column} denotes the total effective capacitance in the compute core, while $C_{parasitic}$ represents the total parasitic capacitance.

$$V_{OL} = \left(\frac{2}{3} \sum_{j=1,i=0}^{j=1,i=63} \frac{VIN_{j,i}}{64} + \frac{1}{3} \sum_{j=0,i=0}^{j=0,i=63} \frac{VIN_{j,i}}{64} \right) \tag{1}$$
$$* \left(\frac{C_{column}}{C_{column} + C_{parasitic}} \right)$$

This analog voltage is digitized by a 3bit flash ADC shared across each column of 64 compute cells. Finally, the two 3bit digital quantities are shifted and added by the adder group. This multi-bit weighting approach achieves a balanced trade-off among speed, power, and linearity in the CIM macro.

TABLE I. 4B ×2B MULTIPLICATION TRUTH TABLE

Input	2bit-Weight	Voltage
Vin	00	0
	01	1/3 Vin
	10	2/3 Vin
	11	Vin
0	x	0

C. Proposed Shift-After-Addition Data Stream

Fig. 4 shows the method of add-after-shifting in the traditional data stream. The two ADCs corresponding to each 4bit weight output two groups of 3bit data H<2:0>and L<2:0>respectively. The two groups of 3bit data are shifted and

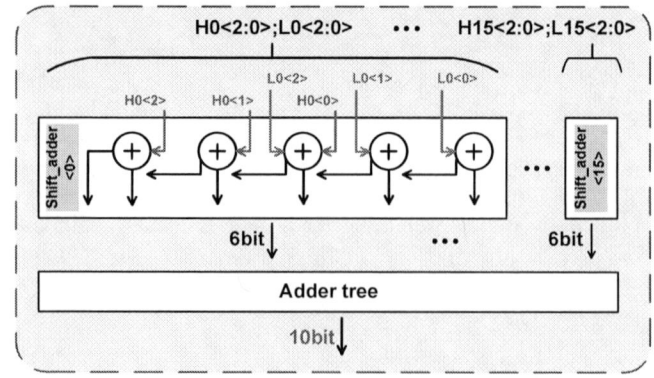

Fig. 4. Traditional data stream.

accumulated in the local shift adder to obtain 6bit data. 16 6bit data are accumulated through the final adder tree to output the final 10bit data. 142 full adders are required to complete the calculation in this way, which will lead to large area cost.

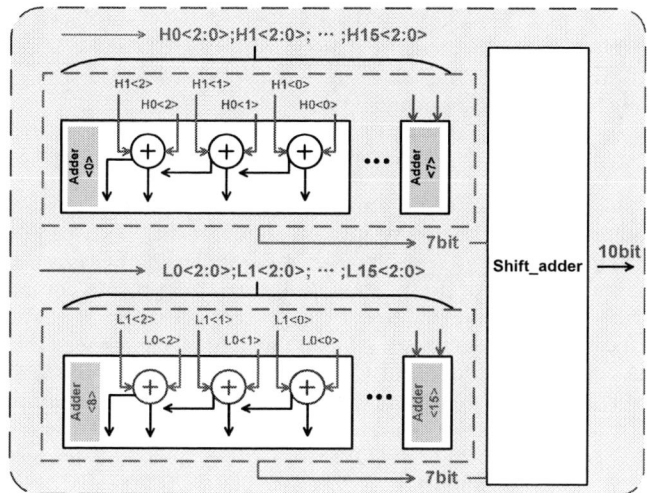

Fig. 5. The proposed Shift-After-Addition data stream.

Fig. 5 illustrates the proposed Shift-After-Addition data stream. Similar to conventional approaches, each weight's two ADCs output H<2:0> and L<2:0>. Unlike local shifting and accumulation, 16 H<2:0> groups and 16 L<2:0> groups are first aggregated globally. Summing the 16 H<2:0> groups yield a 7-bit result, and add 16 groups of L<2:0>to get another group of 7bit data. These two 7-bit results are then shifted and accumulated to generate a 10-bit output. With this SAS scheme, the number of full adder units required to process 32 3-bit inputs is reduced from 142 to 121, lowering the adder cost by 10.4%.

III. SIMULATION RESULT AND ANALYSIS

In this study, A 64kb CIM macro utilizing 9TnC bit-cells was designed. In order to evaluate the performance of CIM macro and verify the feasibility of proposed scheme, we implemented the CIM macro based on 28nm CMOS process, and simulated it through Virtuoso.

TABLE II. COMPARISON OF THE PROPOSED CAP-ADDER WEIGHTING SCHEME
WITH THE EXISTING METHODS

Metrics	Cap-adder (This work)	Capacitor ladder [14]	Bridge capacitor [6]	Charge sharing [7]
INL (LSB)	0.0016	0.0053	0.1060	0.0047
Energy consumption of capacitor charging (fJ)	479.23	1198.08	149.76	399.36
Signal setting time	1CLK	1CLK	1CLK	4CLK

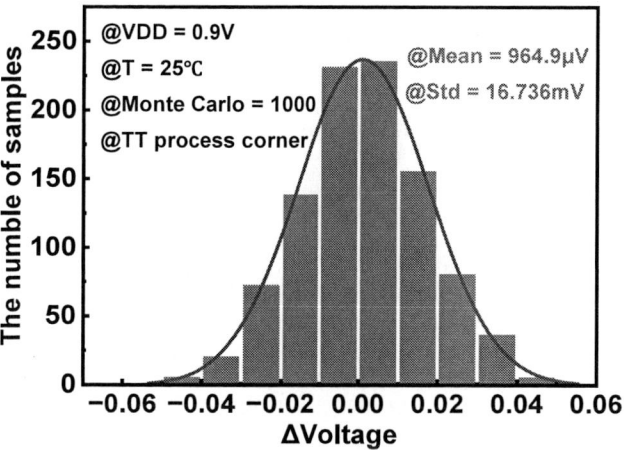

Fig. 6. Offset voltages of comparator used.

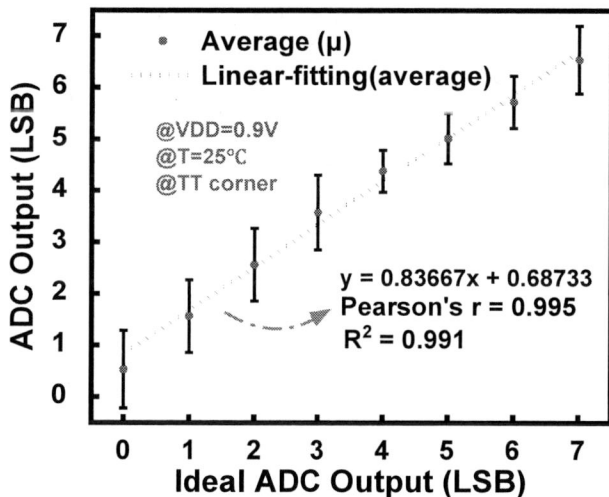

Fig. 7. Measured computation intra-macro error of 3bit Flash ADC used.

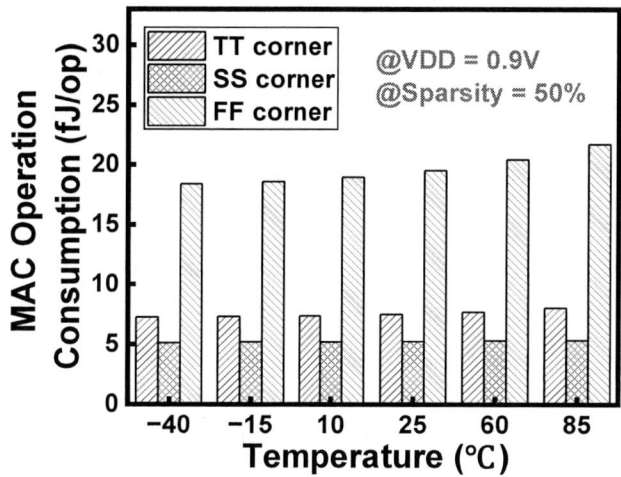

Fig. 8. MAC operating energy consumption at different temperatures and corners.

TABLE II shows the energy consumption, signal setting time and calculation linearity of proposed Cap-Adder weighting scheme compared with traditional weighting methods when the input and weight are 4bit and 64 MAC calculations are performed. The simulation environment for this comparison is VDD=0.9V, t=25 °C, sparsity=100%. It can be seen that the linearity of the proposed Cap-Adder weighting scheme is higher than that of the other three weighting methods when it is weighted by 4bit; The signal setting time is 1CLK, which is equivalent to the other two weighting methods and higher than the charge sharing weighting method; The energy consumption of capacitor charging is lower than that of the traditional capacitor ladder weighting method and similar to the other two. Through comparison, it can be found that proposed Cap-Adder weighting scheme can achieve lower power consumption while maintaining higher linearity and lower signal setup time.

Secondly, we performed 1000 times of Monte Carlo analysis on the comparator to confirm its offset voltage, so as to ensure that it can meet the needs of 3bit ADC. The simulation results are shown in Fig. 6. It can be seen that under the conditions of VDD=0.9, t=25 °C and TT process corner, the offset voltage of the comparator is 50.21mV(3σ), lower than the 56.25mV

required by 3bit ADC, which can meet the service conditions of the ADC.

Fig. 7 shows the intra-macro quantization error of the ADC in the circuit under different MAC results. The correlation coefficient R^2 reaches 0.991, and the Pearson χ coefficient is 0.995, showing the linearity that meets the use of CIM macros. The reason why ADC linearity is not further improved is to weigh its speed and power consumption.

Finally, we also simulated and analyzed the performance of the overall CIM macro under the conditions of VDD=0.9V, sparsity=50%, different temperatures and different process corners. The simulation results are shown in Fig. 8. It can be seen that the energy consumption of FF process corner is much higher than that of TT process corner and SS process corner, which is due to the additional energy cost caused by its large static energy consumption. At the same time, it can be seen that the power consumption of CIM macro is also increasing with the increase of temperature. At 25 °C, the energy efficiency of TT process corner is 7.602 fJ/bit, that is, the energy efficiency of CIM macro is 131.54 TOPS/W. At -40 °C and SS process corner, the maximum energy efficiency of CIM macro is 195.68TOPS/W.

TABLE III. COMPARISON WITH OTHER RELATED WORKS

Metrics		JSSC'24 [8]	JSSC'23 [5]	TCASI'24 [9]	TCASII'23 [7]	CICC'22 [15]	This work
Technology		28nm	22nm	28nm	28nm	65nm	28nm
Cell Type		8T	9T	8T1C	10T	8T1C	9T1C+9T2C
Array Size		16kb	128Kb	64kb	8kb	4kb	64kb
Computing Mechanism		Charge	Charge	Charge	Charge	Charge	Charge
Weighting Method		Charge Sharing	Bridge Capacitor	Capacitor Ladder	Charge Sharing	Bridge Capacitor	CAP-Adder
Bit Precision	I	4/8	8	8	4	4	4
	W	8	8	8	4	4	4
Frequency (MHz)		200	145-240	180-333	250	100	400
Throughput (GOPS)		51.2	1000	256	N/A	204.8	13107.2
Energy Efficency (TOPS/W)		22.2	15.5	71.17	47	52.33	131.54
Normalized Energy Efficiency[1] (TOPS/W)		1420.8	992	4554.88	752	837.28	2104.71

[1]Normalized Energy Efficiency = Input Precision × Weight Precision × Energy Efficiency

Table III shows the comparison between the proposed CIM macro and recent charge-domain CIM works. It can be seen that the energy efficiency of the CASA CIM exceeds works in [3],[4],[6],[7],[15] and realizes a high energy efficient of 131.54 TOPS/W. At the same time, due to the Cap-Adder weighting scheme and the use of Flash ADC, proposed CIM macro realizes the calculation frequency of 400M and throughput of 13107.2 GOPS, which is more than other work listed in the table.

IV. CONCLUSION

This paper proposed a SRAM CIM macro for MAC operation. The CIM macro is designed and implemented using

28nm CMOS process. To balance the energy consumption, speed and linearity of the multi bit weighting scheme, a Cap-Adder weighting scheme is proposed. To reduce the adder cost, a Shift-After-Addition data streaming is proposed and the number of full adders is reduced by 10.4%. To reduce the ADC cost, a double 3bit design is used. Finally, the simulation results show that this CIM macro can achieve an energy efficiency of 131.54 TOPS/W.

ACKNOWLEDGMENT

This research was supported in part by the National Key Research and Development Program under Grant No.

979-8-3315-8850-2/25 $31.00 © 2025 IEEE

2021ZD0114400, and in part by the National Natural Science Foundation of China under Grant 92164204, Grant 92164204 and Grant 62222119.

REFERENCES

[1] Z. Lin et al., "A 28-nm 9T1C SRAM-Based CIM Macro With Hierarchical Capacitance Weighting and Two-Step Capacitive Comparison ADCs for CNNs," in IEEE Transactions on Very Large Scale Integration (VLSI) Systems, vol. 33, no. 7, pp. 2009-2013, July 2025, doi: 10.1109/TVLSI.2025.3545635.

[2] C. Zhao et al., "A 28-nm 36 Kb SRAM CIM Engine With 0.173 μm2 4T1T Cell and Self-Load-0 Weight Update for AI Inference and Training Applications," in IEEE Journal of Solid-State Circuits, vol. 59, no. 10, pp. 3277-3289, Oct. 2024, doi: 10.1109/JSSC.2024.3399615.

[3] X. Li et al., "Full-Array Boolean Logic CIM Macro With Self-Recycling 10T-SRAM Cell for AES Systems," in IEEE Transactions on Very Large Scale Integration (VLSI) Systems, vol. 33, no. 8, pp. 2214-2224, Aug. 2025, doi: 10.1109/TVLSI.2025.3572140.

[4] Y. Zhou, Z. Cheng, H. Liu, T. Xiong and B. Wang, "A 22-nm FDSOI 8T SRAM Based Time-Domain CIM for Energy-Efficient DNN Accelerators," 2022 IEEE Asia Pacific Conference on Circuits and Systems (APCCAS), Shenzhen, China, 2022, pp. 501-504, doi: 10.1109/APCCAS55924.2022.10090315.

[5] H. Wang, R. Liu, R. Dorrance, D. Dasalukunte, D. Lake, and B. Carlton, "A charge domain SRAM compute-in-memory macro with C-2C ladder-based 8-bit MAC unit in 22-nm FinFET process for edge inference," IEEE J. Solid-State Circuits, vol. 58, no. 4, pp. 1037–1050, Apr. 2023, doi: 10.1109/JSSC.2022.3232601.

[6] K. Xiao et al., "A 28nm 32Kb SRAM Computing-in-Memory Macro With Hierarchical Capacity Attenuator and Input Sparsity-Optimized ADC for 4b Mac Operation," in IEEE Transactions on Circuits and Systems II: Express Briefs, vol. 70, no. 6, pp. 1816-1820, June 2023, doi: 10.1109/TCSII.2023.3234620.

[7] L. Lu and D. A. Tuan, "A 47 TOPS/W 10T SRAM-based multi-bit signed CIM with self-adaptive bias voltage generator for edge computing applications," IEEE Trans. Circuits Syst. II, Exp. Briefs, vol. 70, no. 9, pp. 3599–3603, Sep. 2023, doi: 10.1109/TCSII.2023.3274703.

[8] K. Lee, J. Kim, and J. Park, "A 28-nm 50.1-TOPS/W P-8T SRAM compute-in-memory macro design with BL charge-sharing-based in-SRAM DAC/ADC operations," IEEE J. Solid-State Circuits, vol. 59, no. 6, pp. 1926–1937, Jun. 2024, doi: 10.1109/JSSC.2023.3334566.

[9] X. Qiao et al., "A 16.38TOPS and 4.55POPS/W SRAM computing-in-memory macro for signed operands computation and batch normalization implementation," IEEE Trans. Circuits Syst. I, Reg. Papers, vol. 71, no. 4, pp. 1706–1718, Apr. 2024, doi: 10.1109/TCSI.2024.3353464.

[10] H. Zhang et al., "SSM-CIM: An efficient CIM macro featuring single-step multi-bit MAC computation for CNN edge inference," IEEE Trans. Circuits Syst. I, Reg. Papers, vol. 70, no. 11, pp. 4357–4368, Nov. 2023, doi: 10.1109/TCSI.2023.3301814.

[11] Z. Jiang, S. Yin, J. -S. Seo and M. Seok, "C3SRAM: In-Memory-Computing SRAM Macro Based on Capacitive-Coupling Computing," in IEEE Solid-State Circuits Letters, vol. 2, no. 9, pp. 131-134, Sept. 2019, doi: 10.1109/LSSC.2019.2934831.

[12] Z. Lin, Y. Liu, Y. Wang, Y. Zhao, C. Peng and X. Wu, "SRAM-Based Digital CIM Macro for Linear Interpolation and MAC," 2024 IEEE International Symposium on Circuits and Systems (ISCAS), Singapore, Singapore, 2024, pp. 1-5, doi: 10.1109/ISCAS58744.2024.10558525.

[13] C. Zhang, M. Wang, Y. Mai, C. Tang and Z. Yu, "A High-Density and Reconfigurable SRAM-Based Digital Compute-In-Memory Macro for Low-Power AI Chips," in IEEE Transactions on Circuits and Systems II: Express Briefs, vol. 70, no. 9, pp. 3589-3593, Sept. 2023, doi: 10.1109/TCSII.2023.3276169.

[14] E. Kim, H. Oh, N. Kang, J. Park and J. -J. Kim, "A Capacitive Computing-In-Memory Circuit With Low Input Loading SRAM Bitcell and Adjustable ADC Input Range," in IEEE Transactions on Circuits and Systems II: Express Briefs, vol. 70, no. 9, pp. 3268-3272, Sept. 2023, doi: 10.1109/TCSII.2023.3266239.

[15] E. Choi et al., "A 133.6TOPS/W Compute-In-Memory SRAM Macro with Fully Parallel One-Step Multi-Bit Computation," 2022 IEEE Custom Integrated Circuits Conference (CICC), Newport Beach, CA, USA, 2022, pp. 1-2, doi: 10.1109/CICC53496.2022.9772821.

Adaptive VREF variation Compensation Scheme for High-Speed DRAM interface & Its Offset Calibration Scheme

Chris Eom
Integrated Circuits and IP Department
KINGTIGER Test Technology Ltd.
Shenzhen, China
yifan.shi@kingtigertest.com

Kenji Wen
Integrated Circuits and IP Department
KINGTIGER Test Technology Ltd.
Shenzhen, China
kenji.wen@kingtigertest.com

Mia Xu
Integrated Circuits and IP Department
KINGTIGER Test Technology Ltd
Shenzhen,China
mia.xu@Kingtigertest.com

Taco Zhang
Integrated Circuits and IP Department
KINGTIGER Test Technology Ltd
Shenzhen,China
taco.zhang@Kingtigertest.com

Derek Yang
Integrated Circuits and IP Department
KINGTIGER Test Technology Ltd
Shenzhen,China
derek.yang@Kingtigertest.com

Bosco Lai
Head of R&D Center
KINGTIGER Test Technology Ltd
Shenzhen,China
bosco.lai@Kingtigertest.com

Abstract— This paper presents an adaptive VREF variation compensation scheme designed to enhance the RX margin in high-speed DRAM interfaces. The proposed approach addresses the challenge of reduced RX gain under lower reference voltages by increasing the bias current via an additional current path. This adjustment leads to improved RX gain performance at lower reference voltages. Furthermore, an offset calibration scheme is applied to the RX. Unlike the conventional major voting method, this approach enables the removal of offset in all signal paths, specifically for each internal DQ path, which is captured and divided by the DQS path. As a result, a reduction in offset by over 80% was achieved, as demonstrated through Monte Carlo simulations. Implementing this scheme results in an 11% improvement in RX timing margin and a 23% enhancement in voltage margin.

Keywords—DRAM interface, DDR5, DDR4, offset calibration, CPU, DRAM

I. Introduction

Since the introduction of ChatGPT, society has experienced significant transformations [1]. The deployment of ChatGPT has facilitated more intuitive and effective control of robots, leading to an incremental substitution of human tasks by robotic systems. For instance, automated chatbots have increasingly taken over customer service functions that were previously managed by human operators. Furthermore, as ChatGPT continues to advance, the pace of automation is accelerating. Current capabilities include the generation of presentation materials, the creation of images based on textual descriptions, and the production of videos from a single image.

To achieve best performance of AI (artificial Intelligence) like ChatGPT, firstly both high-performance and high-capacity memory are needed. This necessity has driven a substantial increase in the demand for High Bandwidth Memory (HBM). Concurrently, DDR5 memory, which plays a key role in big data centers, has also experienced significant growth. The difference in terms of signaling is that HBM signaling is point-to-point signaling, whereas DDR5 signaling operates under a multi-drop channel environment [2]. This signaling difference introduces challenges for DDR5 in achieving high-speed performance, due

to signal reflection and attenuation caused by channel loss.

Furthermore, the high-power consumption of AI systems necessitates the development of low-power with high-performance. For low power with high performance device, the DRAM process is increasingly incorporating High-K Metal Gate (HKMG) technology based on the planar process [3], while logic processes are shifting from FinFET to Gate-All-Around (GAA) transistors. However, the mainstream technologies remain the PSION (poly silicon) process for DRAM and the FinFET process for logic. Therefore, It is necessary to conduct research and develop high-performance, low-power solutions to address the limitations of these existing process technologies.

One common method for evaluating I/O characteristics is the 2D shmoo plot. In this plot, the horizontal axis represents time, while the vertical axis represents voltage. The timing margin is evaluated by sweeping the DQ Strobe (DQS), while the voltage margin is assessed by varying the reference voltage

[4]. As shown in Fig. 1, during a DRAM write operation, the CPU sends data from CPU TX (transceiver), and DRAM receives this data from DRAM RX (receiver). Similarly, during a DRAM read operation, the DRAM TX and CPU RX operates together. That is why to enhance system level performance, it is necessary to improve not only the characteristics of the DRAM but also those of the CPU.

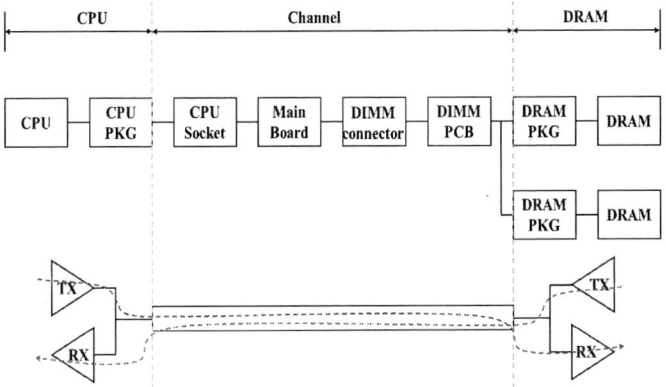

Fig 1 DDR interface under multi-drop channel

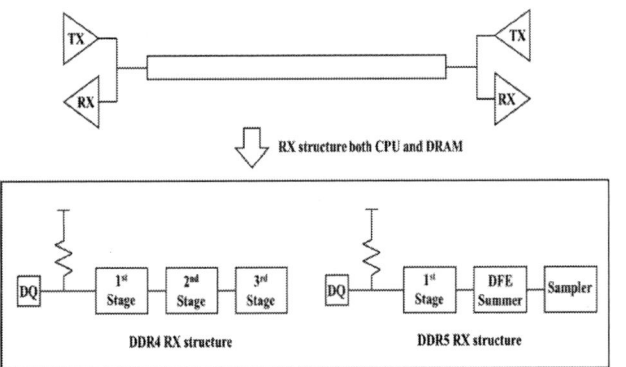

Fig. 2. Basic conceptual RX block diagram

II. ARCHITECTURE

The DDR interface is defined by JEDEC specifications, ensuring a standardized basic configuration for both CPU and DRAM side. However, variations in DRAM and logic processes can lead to differences in the actual circuit implementations. For instance, the specific design and implementation of the RX stages and DFE (Decision Feedback Equalizer) methods may vary depending on the process technology used for DRAM versus logic circuits

Figure 2 illustrates the conceptual block diagrams for DDR4 and DDR5 RX [5][6]. In DDR4, the Rx consists of three amplifier stages that convert input signals from CML to CMOS. In the other hands, DDR5 RX features a first stage that acts as a VGA, a DFE summer and a sampler. The one of role for sampler is that CML inputs coverts to CMOS outputs. That means it operates with CML signals before the sampler stage [7].

A. Conventional first stage amplifier for DDR4 and DDR5

The type of input transistor in the first stage is determined by signaling common mode, and DDR4/5 is high level

common mode voltage due to VDDQ termination, so NMOS input pair is used as shown in Fig. 3. If a bias generator is used to control the RX bias current, it introduces additional power consumption. Therefore, a self-biasing method is widely adopted to reduce RX power consumption. The self-bias level is established by connecting a large resistor between the first output terminals, and it is determined by the average value of the DQ common mode and the reference voltage (VREF).

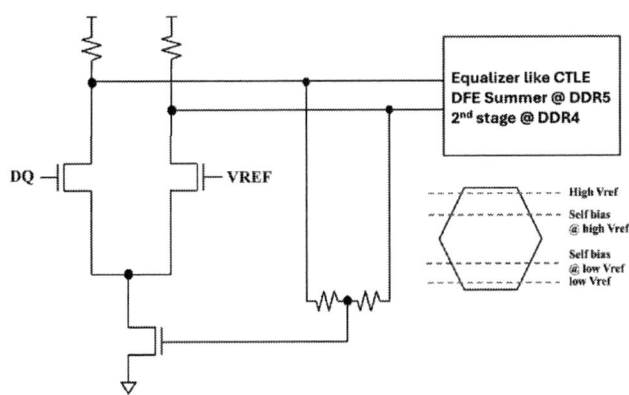

Fig. 3. Circuit diagram of conventional first stage amplifier

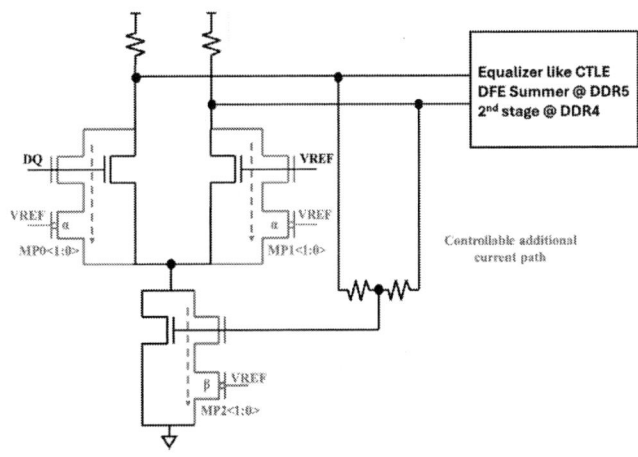

Fig. 4. Circuit diagram of proposed first stage amplifier

The input common mode is determined by the strength of the ODT and the pull-down driver of the TX, and thus remains constant value during 2D shmoo evaluations. In contrast, VREF varies during 2D shmoo test, directly affecting the level of self-bias. As VREF increases, the self-bias level also rises, which results in an increase in the RX's bias current and subsequently enhances the gain of the RX. Conversely, when VREF decreases, the self-bias level lowers, leading to a reduction in the RX's bias current and, therefore, a decrease in gain.

B. Proposed first stage amplifier for DDR4 and DDR5

Figure 4 illustrates the proposed first-stage amplifier design. this amplifier incorporates an additional current path controlled by VREF in both the input pair and the bias circuit. This configuration allows for independent control of the input pair, the bias circuit, or both simultaneously. The additional current path is implemented series stack of NMOS and PMOS transistors. The PMOS transistors are configured with 2-bit control capability, allowing for fine-tuned adjustment of the current path.

The gain of the differential amplifier is determined by the transconductance of the transistors(gm) and the load resistance (R_{Load}), where gm is directly proportional to the drain current (Id) [8]. When the VREF decreases, the self-bias level also decreases, leading to a reduction in both the current and the gain of the first stage amplifier. This reduction adversely affects the high-speed performance of the amplifier. However, the additional current path, controlled by VREF and employing PMOS transistors, mitigates this effect. As VREF decreases, the gate-source voltage (VGS) of the PMOS transistors increases, which in turn enhances the additional current. This mechanism effectively compensates for the performance degradation commonly observed with lower VREF in conventional designs.

C. Offset calibration scheme for higher performance

DDR4 operates at a data transfer rate of 3.2 Gbps, whereas DDR5 supports speeds exceeding 6.4 Gbps. Consequently, the ideal valid window for DQ signals is 312.5ps for DDR4 and less than 156.25ps for DDR5. The reduction in valid window size with DDR5 makes offset management significantly more critical compared to DDR4.

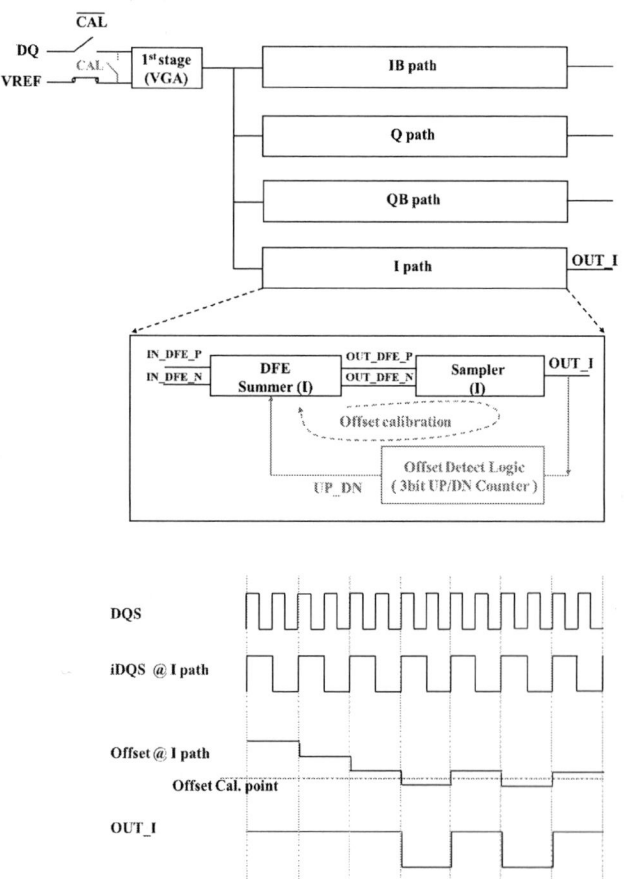

Fig. 5. Block diagram and timing diagram for proposed offset calibration in DDR5

RX buffer offsets originate from transistor mismatches caused by local threshold voltage shifts, which are influenced by the Well Proximity Effect (WPE) and Shallow Trench Isolation (STI) effects prevalent in advanced process technologies. Thus, minimizing the offset in the input buffer is essential. Several techniques are employed to mitigate input buffer offsets, including capacitance compensation and control of the input pair transistor sizes.

In ISSCC 2024 presentation [9], Samsung introduced an offset calibration method for DDR5 that utilizes a majority voting technique. This method involves controlling the VGA output and employing a current steering technique based on the counts of '1' and '0' from the sampler outputs to perform offset calibration. However, this approach does not fully eliminate offsets across each sampler path. To address this limitation, this paper presents a novel offset calibration method that utilizes the DFE summer. This approach adjusts offsets based on the outputs from each sampler, enabling comprehensive offset calibration across all paths.

External DQS signals are divided into four internal DQS signals. As illustrated in Figure 5, this division results in four distinct paths within the RX. Each path includes a DFE summer, a sampler, and offset control logic. The offset control logic comprises a detection circuit and a 3-bit up/down counter. during offset calibration mode, both inputs of the VGA input pair are set to VREF, providing identical inputs to both. The signals from the VGA to the sampler operate differentially. The offset between these paths is reflected in the OUT_I results. If the OUT_I result is '1', the UP_DN output will be '0', adjusting the OUT_DFE_P path. Conversely, if the OUT_I result is '0', the UP_DN output will be '1', adjusting the OUT_DFE_N path. Once offset calibration is complete, OUT_I toggle '1010' or '0101'. For instance, as shown in the timing diagram of Figure 5, if the OUT_I result keeps '1', the UP_DN output will stay at '0' until OUT_I changes to '0'. After this transition, the offset calibration is deemed complete, and OUT_I will then toggle to '0101' [10].

III. EXPERIMENT RESULTS AND ANLAYSIS

The power supply for the interface between the CPU and DRAM is specified by the JEDEC standards to be derived from VDDQ. Both RX and TX circuits operate under VDDQ power supply, while VREF is generated using a resistance ladder that is also based on VDDQ. The output driver (TX) is composed of pull-up and pull-down components and supports resistance values ranging from RZQ/1 to RZQ/7 as specified by the JEDEC standards with RZQ valued at 240 ohms. The On-Die Termination (ODT) is achieved by activating the pull-up driver of the output driver during write operations.

The RX DQ input swing is determined by the resistance of the pull-down driver (RonPD) and the On-Die Termination resistance (RODT). To achieve higher speed performance, both RonPD and RODT need to be smaller. When RonPD and RODT are set to RZQ/7, the VIH (input high voltage) is VDDQ, while VIL (input low voltage) is set to 0.5 VDDQ, resulting in a center level of 0.75 VDDQ , as illustrated in Fig. 6.

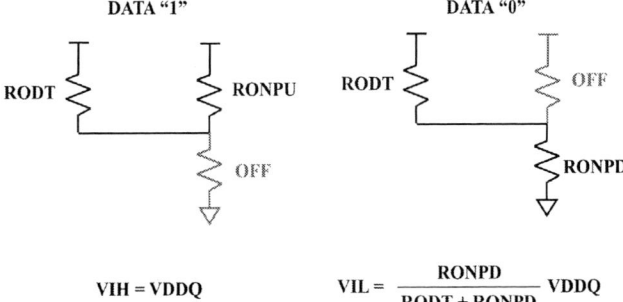

$$VIH = VDDQ$$

$$VIL = \frac{RONPD}{RODT + RONPD} VDDQ$$

Fig. 6. Block diagram and timing diagram for proposed offset calibration in DDR5

TABLE I. GAIN SIMULATION RESULT COMPARISON

Gain ratio	VREF level		
	0.95 VDDQ	0.75 VDDQ	0.55 VDDQ
Default	1.58	2.83	1
Input Pair	1.73	3.15	1.63
Bias	1.78	3.18	1.6
Input Pair + Bias	1.85	3.24	1.8

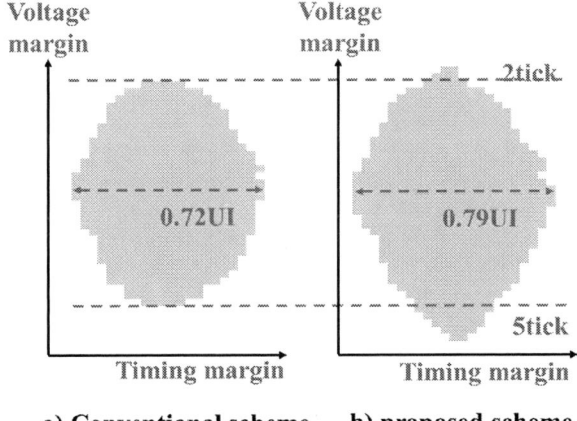

a) Conventional scheme   b) proposed scheme

Fig. 7. RX 2D shmoo simulation result comparison

When VREF is the input common mode voltage, the difference between the two RX input buffer inputs is enough to get high gain. However, when VREF is set to the edges of the input swing (0.95VDDQ or 0.55VDDQ), the difference between the inputs decreases, leading to insufficient gain. In such cases, a higher VREF generally provides better performance compared to a lower VREF, as the self-bias level is higher, resulting in greater gain.

Applying the proposed method, the gain enhancement observed with low VREF is greater compared to that with high VREF, as shown in Table 1. Specifically, the gain increases by 63% with the input pair adjustment alone, 60% with bias adjustment alone, and 80% when both the input pair and bias adjustments are applied simultaneously. This indicates that the proposed method effectively enhances gain, particularly under conditions of low VREF.

Figure 7 shows the 2D shmoo simulation result of the conventional scheme and the proposed scheme. After applying the optimal condition after offset calibration, the timing margin improved by 0.07UI. Additionally, the voltage margin saw a total enhancement of 7 ticks, comprising 2 ticks for high VREF and 5 ticks for low VREF.

IV. CONCLUSION

In this paper, adaptive VREF variation compensation scheme is implemented in high-speed DRAM interface by adding additional current path in input pair and current bias path in receiver. Furthermore, an offset calibration scheme is applied to each internally divided DQS path to minimize offset within the high-speed DRAM receiver. As a result, the timing margin has improved by 0.07UI, and the voltage margin has increased by 7ticks. These enhancements correspond to an 11% improvement in timing and a 23% improvement in voltage margin compared to the conventional scheme.

REFERENCES

[1] OpenAI. (2024). *ChatGPT* (Version 4.0) [Large language model]. https://www.openai.com/chatgpt

[2] Myeong-Jae Park et al. "A 192-Gb 12-High 896-GB/s HBM3 DRAM With a TSV Auto-Calibration Scheme and Machine-Learning-Based Layout Optimization", JSSC Volume 58 Issue 1 2023

[3] S. H. Jang et al., "A Fully Integrated Low Voltage DRAM with Thermally Stable Gate-first High-k Metal Gate Process", 2019 IEEE International Electron Devices Meeting (IEDM), 2019, pp. 28.4.1-28.4.

[4] S.-H. Hong et al., "A Reflection and Crosstalk Canceling Continuous-Time Linear Equalizer for High-Speed DDR SDRAM," IEEE Symp. VLSI Circuits, pp. 1-2, June 2021

[5] DDR4 JEDEC SPCE

[6] DDR5 JEDEC SPEC

[7] S.-H. Hong et al., "A 1.1-V 10-nm Class 6.4-Gb/s/Pin 16-Gb DDR5 SDRAM With a Phase Rotator-ILO DLL, High-Speed SerDes, and DFE/FFE Equalization Scheme for Rx/Tx," JSSC Volume 55 Issue 1 2020

[8] B. Razavi, Design of Analog CMOS Integrated Circuits, 2nd ed., New York, NY, USA: McGraw-Hill, 2017

[9] Ikjoon Choi et al. "A 32Gb 8.0Gb/s/pin DDR5 SDRAM with a symmetric-Mosaic Architecture in a 5th-Generation 10nm DRAM Process", 2024 International Solid-State Circuits Conference (ISSCC).

[10] Masay Miyahar et al. "A low-noise self-calibrating dynamic comparator for high-speed ADCs", 2008 Asian Solid-State Circuits Conference (ASSCC).

2025 The 10th International Conference on Integrated Circuits and Microsystems

Stochastic Computation Based Quantization Strategy for SRAM Computing-in-Memory Macro

Xin Li
School of integrated circuit
Anhui University
Hefei, China
lixin@ahu.edu.cn

Yifan Wu
School of integrated circuit
Anhui University
Hefei, China
209876190@qq.com

Lintao Chen
School of integrated circuit
Anhui University
Hefei, China
humanifesto@163.com

Chenghu Dai
School of integrated circuit
Anhui University
Hefei, China
daichenghu@ahu.edu.cn

Xiulong Wu
School of integrated circuit
Anhui University
Hefei, China
xiulong@ahu.edu.cn

Zhiting Lin
School of integrated circuit
Anhui University
Hefei, China
ztlin@ahu.edu.cn

Abstract—This paper presents an innovative Stochastic Computing-in-Memory (CIM) macro structure aimed at improving the energy efficiency of edge devices performing multi-bit multiply-and-accumulate (MAC) operations. The proposed framework integrates a 8T SRAM-based CIM, input encoding mechanism, and a low cost 4-bit coarse fine analog-to-digital converter (ADC). By incorporating the quantization and encoding strategies of stochastic computing, the reuse of most digital devices and analog circuits is realized, and the hardware overhead and power consumption are utilized. The design includes a 4Kb SRAM CIM macro implemented in 28 nm CMOS technology. Simulation results demonstrate that the proposed coarse-fine ADC achieves a 32% power reduction compared to conventional designs, while its coarse-fine quantization architecture significantly shortens the conversion cycle. The CIM macro can achieve a peak energy efficiency of 87.43 TOPS/W, providing a significant boost to the performance of edge computing devices.

Keywords—*Computing-in-memory, Stochastic computing, Quantization, Multiply-accumulate.*

I. INTRODUCTION

DEEP neural networks (DNNs) have demonstrated significant performance on various tasks, such as image classification and speech recognition. At the core of DNN inference lies the multiply-accumulate (MAC) operation, which serves as the fundamental computational kernel. However, the processing of DNNs necessitates an enormous amount of data and computation, incurring considerable power overhead, which impedes to process DNNs at resource-constraint edge devices. Over recent years, computing-in-memory (CIM)-based accelerators architectures have been increasingly deployed in

Deep Neural Networks (DNNs). Their energy efficiency far surpasses that of the traditional von Neumann architecture. Unfortunately, this also poses a series of challenges for previous CIM - based accelerators.

Prior work [1], [2], [3] [4] has proposed digital-based CIMs, which can achieve high accuracy in MAC operations. However, the integration of digital circuits within the computing units significantly compromises efficiency and increases the physical area. In contrast, mix-signal CIM demonstrates substantial application potential. These architectures not only support multi-input parallel operations but also exhibit superior efficiency and high throughput. For processing the mix-signal MAC operation, analog circuit, such as analog-to-digital converter (ADC) and digital-to-analog converter (DAC), are needed. However, analog circuit are considered a main restriction in CIM design. The analog circuit requires high power consumption and large area overheads, thereby degrading energy efficiency, additionally, the accuracy of its computed results is sensitive to variations in process, voltage, and temperature [5]. Many previous studies had modified the analog circuit to fit the CIM macro. In [6], a prioritized-hybrid ADC is proposed by combining the successive approximation register (SAR) with the single-slope (SS) ADC, reducing the ADC area and power overheads. In order to avoid higher ADC resolution incurs significant accuracy degradation in the presence of the CIM noise, [7] proposed a scheme for ADC precision selection and a coarse-fine flash ADC to fit its CIM macro. In [8], this work exploits ci-SAR ADCs, the capacitors in the DACs are replaced by transistors with a long channel length which takes much less area than unit capacitors in conventional SAR ADCs.

To address these challenges, this work proposes a SRAM CIM macro that efficiently reduces the hardware cost associated with analog circuits. The overall operation of this CIM macro is based on the principle of stochastic computing (SC). The advantages of the proposed architecture are as follows:

This work is supported by Science and Technology Key Project of Anhui Province under Grant 2022AH050099, National Natural Science Foundation of China under Grant 62471003, The University Synergy Innovation Program of Anhui Province under Grant GXXT-2023-013, The University Synergy Innovation Program of Anhui Province under Grant GXXT-2023-012, The University Synergy Innovation Program of Anhui Province under Grant GXXT-2023-011.

979-8-3315-8850-2/25 $31.00 © 2025 IEEE

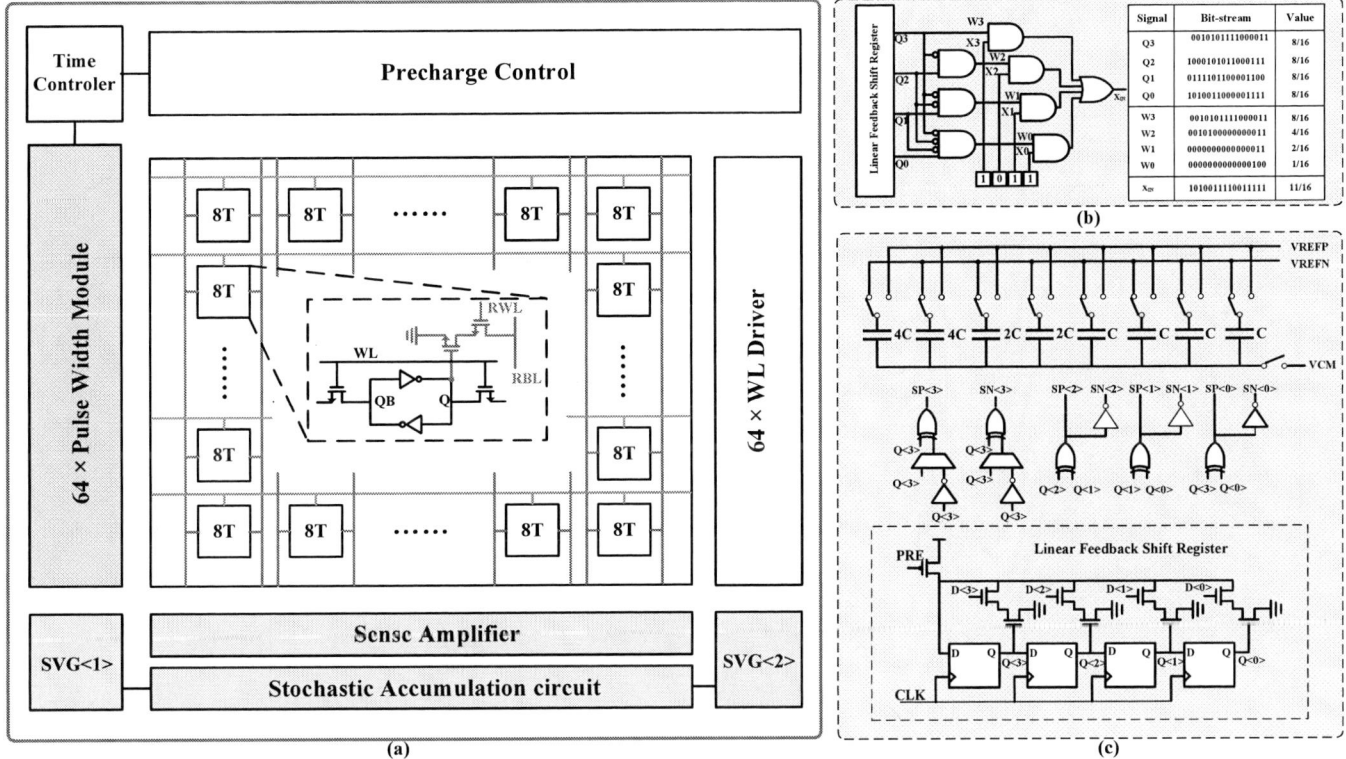

Fig. 1. (a) Overall architecture and schematic of 8T cell. (b) Input pulse width module. (c) Stochastic voltage generator circuit.

To mitigate hardware cost of input circuit, we proposed an input pulse width modulation circuit. The core component of input circuit can be shared across the entire system. For each row, only four AND gates and a 4-input OR gate are needed.

In contrast to conventional quantization techniques, we proposed an ADC grounded in the principle of SC. By comparing the voltage on the RBL against the globally generated random voltage, this innovative ADC significantly curtails hardware expenses. Moreover, through the implementation of a coarse - fine quantization methodology, we have achieved a notable enhancement in its throughput.

The remainder of this paper is organized as follows. Section II describes the proposed CIM architecture. Section III provides an analysis of the simulation results. Finally, Section IV concludes the paper.

II. PROPOSED ARCHITECTURE

A. Overall Architecture

Fig. 1(a) depicts the overall architecture of the proposed 64 × 64 SRAM CIM macro that performs the MAC operation between 4-bit inputs and 4-bit weights. This architecture encompasses a low-cost input pulse modulation circuit module, an 8T SRAM array sized at 64×64, the proposed 4-bit coarse-fine ADC module, two stochastic voltage generator, a time controller, a read and write circuit, along with other row and column peripheral circuits.

The proposed SRAM MARCO supports MAC operations with 4-bit inputs and 4-bit weights. The 4-bit weights are stored row-by-row in the 8T SRAM array, and data within the same column share the identical weight. During the MAC operation, the RBL of the 8T SRAM is precharged to 1. The input 4-bit data is converted into pulses of corresponding widths through the input pulse width modulation circuit. Then, the RBL is discharged via the RWL of the 8T SRAM. The RBLs of the 8T SRAMs in each column are connected together, and the accumulated result is reflected as the total amount of discharge.

The accumulated voltage on the RBL will be converted into a bit-stream and a MSB binary code by the ADC. Subsequently, the data will be processed into a binary MAC result in the digital domain.

B. Proposed Input Pulse Width Module

The proposed low-cost pulse width modulation circuit and operation process is shown in Fig. 1(b). It primarily consists of a 4-bit linear feedback shift register (LFSR) and other digital logic devices.

By controlling the D<3:0>, the LSFR generates bit-stream Q0, Q1, Q2 and Q3, in which 0 and 1 each account for 50 percent within 16 cycles. These bit-streams are then converted into bit-streams with 8, 4, 2, and 1 number of 1 respectively (i.e., W3, W2, W1, and W0) through several logic devices. The input 4-bit binary numbers are respectively passed through a 2-input AND gate with W3, W2, W1, and W0. All outputs implement an OR operation to generate a pulse of the corresponding width. Since

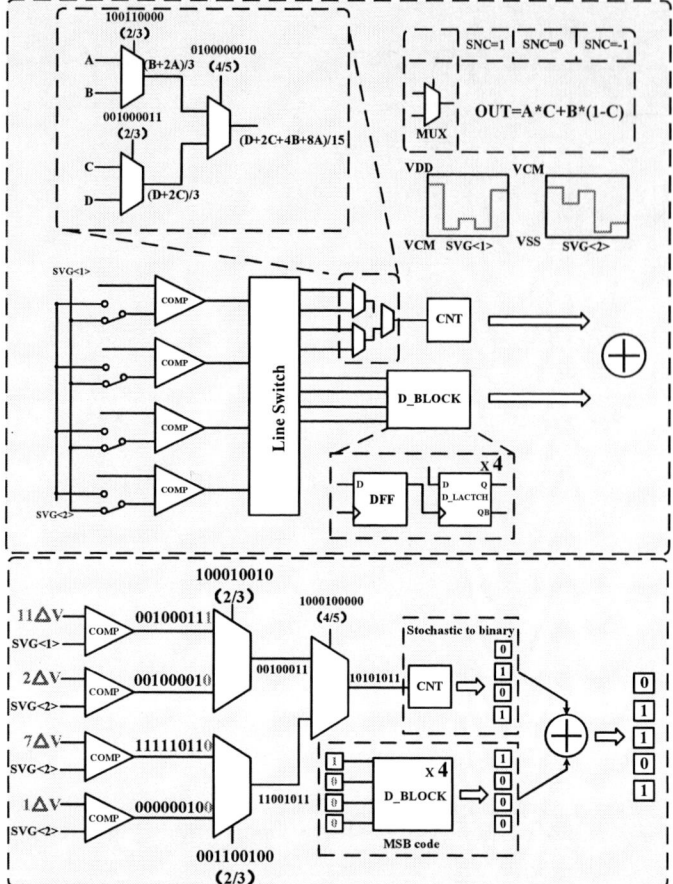

Fig. 2. Proposed 4-bit coarse-fine ADC architecture and example of Stochastic computing.

the LFSR and three logic devices are globally shared, only four 2-input AND gates and one 4-input OR gate are required for each row. The hardware cost overhead of the input circuit is significantly reduced.

C. Proposed 4-bit Coarse-Fine ADC

In CIM mode, the proposed CIM macro executes MAC operations with every four columns as a group. The entire CIM macro requires 16 ADCs. However, conventional quantization schemes suffer from excessive power consumption and hardware overhead. To address these performance bottlenecks, we proposed a 4-bit coarse-fine ADC used for low-cost analog-to-digital conversion, the ADC architecture is shown in the Fig. 2. It is composed of comparators equipped in four columns, two globally shared stochastic voltage generators (SVG), and digital logic devices. The SVGs can generate different stochastic voltage by controlling the seed of the LSFR for comparison.

The detail ADC operation is as follows. First of all, comparator compares the accumulated voltage on the RBL with the VCM generated by the SVG and outputs the MSB. The MSB, represented as a binary code, is stored in the D-latch within the D-BLOCK. Subsequently, during the fine quantization, the two SVGs respectively generate stochastic voltage within the range of VREFN - VCM or VCM - VREFP.

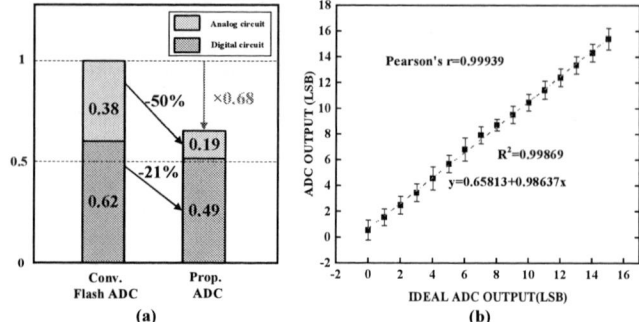

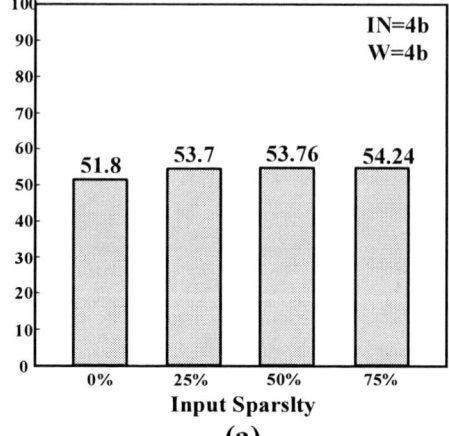

Fig. 3. (a)Comparison of power consumption between proposed and conventional flash ADCs. (b)The intro-macro error of proposed ADC under the scanned MAC result.

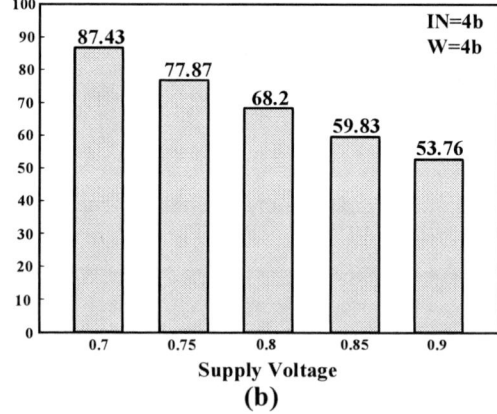

Fig. 4. (a) The energy efficiency results of the proposed CIM macro with different input sparsity. (b) The energy efficiency of the proposed CIM macro at different supply voltages.

Based on the result of the MSB, the other port of the comparator will select to connect to SVG<1> or SVG<2>. The RBL is compared with stochastic voltage, resulting in four bit-stream representing the lower three bits.

In the realm of SC, numerical values are encoded as bit-streams. The reconstruction of numbers is achieved by tallying the occurrences of 1 in the bit-streams. Concurrently, the computational logic in SC is subject to the influence of the stochastic number correlation (SNC). An OR operation on two

TABLE I. COMPARISON WITH PRIOR ARTS

	This work	JSSC 2023[6]	JSSC 2021[9]	TCAS-I 2024[10]
Techology	28nm	28nm	28nm	65nm
Cell type	8T	P-8T	6T	8T
Input/Weight	4bit/4bit	4,8bit/8bit	4,8bit/ 4,8bit	1-8bit/ 1-8bit
MAC operation	Mix-signal	Analog	Analog	Digital
Quantization strategy	4-b C.F Stochastic	4-b C.F Flash	5-b SAR	1b SA
Array size	4KB	2KB	8KB	64KB
Voltage	0.7V-0.9V	0.6-1.2V	0.7V-0.9	1V
Peak energy efficiency (TOPS/W)	87.43	40.1	33.5	249.1(1/1b) 13.84(8/8b)

datasets can be used to implement approximate addition when SNC is zero [11]. However, without ensuring low sparsity and uncorrelation, there can be significant deviations in computation results. In contrast, MUX can be employed to perform scaled stochastic addition[12]. As shown in Fig.2, details of the weighted summation of the bit-streams obtained by stochastic quantization through the MUX are presented. Although implementing this operation will result in accuracy degradation, it is acceptable for the mixed-signal CIM.

The resulting bitstreams can be configured to produce binary outputs via counters. These binary outputs are then added to the 4-bit MSB binary code via digital adder tree to obtain the MAC result.

Unlike SAR or Flash ADCs, the proposed 4-bit coarse-fine ADC only requires additional digital devices. The comparators of the CIM macro can be used for both read/write and quantization. Multiple comparators can share tow SVGs and the coarse-fine quantization method reduces the quantization cycle from 16 to 9 in this approach, it significantly reduces power and area overheads.

III. EXPERIMENTAL RESULTS

The CIM macro is designed in a 28 nm process. We use Cadence Virtuoso and AMS simulator to obtain the result of the simulation.

Fig. 3(a) presents the power consumption comparison between the proposed ADC and the conventional Flash ADC. In the figure, the power components are estimated based on simulation. Due to the cooperation of the globally shared SVGs and the coarse-fine ADC operation, the total power consumption of the proposed approach is reduced by $0.32\times$ compared to that of the conventional approach. As the CIM array size increases, the total power and area costs of the CIM macro increase but the SVG remains unchanged. So, this method holds significant potential for broader applications.

In this paper, weighted summation is achieved by modulating the ratio of 0s and 1s in the bitstream of the MUX's select input. The precision of the output result is directly influenced by the length of the bitstream, as it determines the precision of the weighted summation. Therefore, we conducted dedicated simulation studies to validate the accuracy of the MAC results. Fig. 3(b) presents the intra-macro quantization error of the proposed coarse-fine ADC under scanned MAC results, with a correlation R^2 of 0.99869 and Pearson's x of 0.99939, demonstrating a high degree of linearity.

Fig. 4 (a) illustrates the energy efficiency at different input sparsity while keeping weight sparsity at 50%. When the bit width of the input and weight are both 4 bits, the peak energy efficiency is 54.05 TOPS/W at 75% input sparsity. Fig. 4 (b) illustrates the energy efficiency at digital stochastic computing circuits are supplied with different voltages. When the bit width of the input and weight are both 4 bits, the peak energy efficiency is 87.43 TOPS/W at 0.7 V supply voltage.

The performances of the previous methods and the CIM architecture proposed in this paper are compared in Table I. The results in the table demonstrate that the proposed architecture exhibits superior energy efficiency over conventional designs.

IV. CONCLUSION

This paper presents a novel mix-signal CIM macro structure aimed at improving the energy efficiency of edge devices . By integrating stochastic computing with coarse-fine quantization, the architecture achieves both shortened quantization cycles and reduced power consumption. Experimental results demonstrate that the proposed design exhibits superior energy efficiency, making it particularly advantageous for DNN inference tasks.

REFERENCES

[1] D. Wang, C. -T. Lin, G. K. Chen, P. Knag, R. K. Krishnamurthy and M. Seok, "DIMC: 2219TOPS/W 2569F2/b Digital In-Memory Computing Macro in 28nm Based on Approximate Arithmetic Hardware," *2022 IEEE International Solid-State Circuits Conference (ISSCC)*, San Francisco, CA, USA, 2022, pp. 266-268, doi: 10.1109/ISSCC42614.2022.9731659.

[2] Y. He *et al.*, "An RRAM-Based Digital Computing-in-Memory Macro With Dynamic Voltage Sense Amplifier and Sparse-Aware Approximate Adder Tree," in *IEEE Transactions on Circuits and Systems II: Express Briefs*, vol. 70, no. 2, pp. 416-420, Feb. 2023, doi: 10.1109/TCSII.2022.3209872.

[3] C. He *et al.*, "LSAC: A Low-Power Adder Tree for Digital Computing-in-Memory by Sparsity and Approximate Circuits Co-Design," in *IEEE Transactions on Circuits and Systems II: Express Briefs*, vol. 71, no. 2, pp. 852-856, Feb. 2024, doi: 10.1109/TCSII.2023.3304752.

[4] J. Yang, T. Li, W. Romaszkan, P. Gupta and S. Pamarti, " A 65nm 8-bit All-Digital Stochastic-Compute-In-Memory Deep Learning Processor," 2022 IEEE Asian Solid-State Circuits Conference (A-SSCC), Taipei, Taiwan, 2022, pp. 10-11, doi: 10.1109/A-SSCC56115.2022.9980613.

[5] Z. Lin *et al.*, "Cascade Current Mirror to Improve Linearity and Consistency in SRAM In-Memory Computing," in *IEEE Journal of Solid-State Circuits*, vol. 56, no. 8, pp. 2550-2562, Aug. 2021, doi: 10.1109/JSSC.2021.3063719.

[6] J. -W. Su *et al.*, "A 8-b-Precision 6T SRAM Computing-in-Memory Macro Using Segmented-Bitline Charge-Sharing Scheme for AI Edge Chips," in *IEEE Journal of Solid-State Circuits*, vol. 58, no. 3, pp. 877-892, March 2023, doi: 10.1109/JSSC.2022.3199077.

[7] K. Lee, J. Kim and J. Park, "A 28-nm 50.1-TOPS/W P-8T SRAM Compute-In-Memory Macro Design With BL Charge-Sharing-Based In-SRAM DAC/ADC Operations," in *IEEE Journal of Solid-State Circuits*, vol. 59, no. 6, pp. 1926-1937, June 2024, doi: 10.1109/JSSC.2023.3334566.

[8] Z. Chen *et al.*, "CAP-RAM: A Charge-Domain In-Memory Computing 6T-SRAM for Accurate and Precision-Programmable CNN Inference," in *IEEE Journal of Solid-State Circuits*, vol. 56, no. 6, pp. 1924-1935, June 2021, doi: 10.1109/JSSC.2021.3056447.

[9] X. Si *et al.*, "A Local Computing Cell and 6T SRAM-Based Computing-in-Memory Macro With 8-b MAC Operation for Edge AI Chips," n *IEEE Journal of Solid-State Circuits*, vol. 56, no. 9, pp. 2817-2831, Sept. 2021, doi: 10.1109/JSSC.2021.3073254.

[10] V. Sharma, X. Zhang, N. S. Dhakad and T. T. -H. Kim, "FlexDCIM: A 400 MHz 249.1 TOPS/W 64 Kb Flexible Digital Compute-in-Memory SRAM Macro for CNN Acceleration," in *IEEE Transactions on Circuitsand Systems I: Regular Papers*, doi: 10.1109/TCSI.2025.3547853.

[11] J. Yang, A. Graening, W. Romaszkan, V. K. Jacob, P. Gupta and S. Pamarti, " A 278-514M Event/s ADC-Less Stochastic ComputeIn-Memory Convolution Accelerator for Event Camera," 2024 IEEE Symposium on VLSI Technology and Circuits (VLSI Technology and Circuits), Honolulu, HI, USA, 2024, pp. 1-2, doi: 10.1109/VLSITechnologyandCir46783.2024.10631484.

[12] A. Alaghi, Cheng Li and J. P. Hayes, " Stochastic circuits for real-time image-processing applications," 2013 50th ACM/EDAC/IEEE Design Automation Conference (DAC), Austin, TX, USA, 2013, pp. 1-6, doi: 10.1145/2463209.2488901.

Design of a Multi-Dimensional and Multi-Precision Tensor Computation Unit Based on FPGA

Yupeng Fang
School of Integrated Circuit
Science and Engineering
University of Electronic Science
and Technology of China
Chengdu, China
2311843227@qq.com email

Abstract—As deep learning models expand in scale, TPUs (Tensor Processing Units) offer enhanced efficiency and accelerated computing capability. The multi-dimensional, multi-precision tensor computing units are engineered to process large-scale matrices and operations across various data precisions, fulfilling the computational demands of diverse scenarios. Furthermore, they provide mixed-precision (fp16 and fp32) computing, diminishing memory consumption while preserving data accuracy. The tensor computing unit's design incorporates block matrices and systolic arrays, enhancing the efficiency of matrix operations and minimizing computational waste. The multipliers for various data precisions employ the Redix-4 Booth algorithm in conjunction with a novel Wallace tree compression architecture, whereas the addition operations utilize a Carry-Lookahead adder chain structure, facilitating resource reutilization across different data precisions and further reducing resource consumption. Experiments indicate that the multi-dimensional and multi-precision tensor computing unit utilizing FPGA attains average accuracy rates of 100%, 100%, 98.7%, 99.74%, and 98.05% for INT4, INT8, FP16, FP32, and FP16_FP32 data operations, respectively, with a clock frequency of 250 MHz and an energy efficiency of 7.65 GFLOPS*W^{-1}. This demonstrates exceptional accuracy, rapid calculation, and superior efficiency in executing large-scale, multi-precision matrix operations, significantly enhancing the computational efficacy of deep learning.

Keywords—*Tpu, Tensor, Matrix operations, Floating point operations, Wallace tree compression, Redix-4 Booth algorithm, Resource reuse, Systolic array*

I. INTRODUCTION

The TPU is a microprocessor developed by Google explicitly for neural network training and inference. The architecture is meticulously optimized for matrix operations, addressing the efficiency limitations of general-purpose hardware, lowering computing expenses, and fostering competitiveness and advancement in the global AI chip sector. Moreover, compared to conventional CPUs and GPUs, the TPU has achieved a significant enhancement in computational performance, substantially reducing the training and inference durations of deep learning models.

The systolic array constitutes the fundamental architecture of the TPU and comprises several processing element units. There are three benefits to employing systolic arrays[1]:

1. The straightforward and systematic architectural design minimizes expenses and achieves a balanced cost-performance ratio through modularization.

2. Owing to device speed constraints, computational velocity can be enhanced by substantial parallelism and diminished routing expenses.

3. The interplay of I/O requirements, system scale, and storage capacity can be optimized to meet increased compute power demands while conserving I/O bandwidth.

When executing extensive matrix operations across many dimensions, utilize the technique of block matrix computation[2] by partitioning the huge matrix into blocks and conducting operations on the subdivided matrix by mathematical rules. This method minimizes FPGA resource use and prevents the misuse of computational capacity.

Operations requiring data of varying precision necessitate the development of distinct multiplier and adder modules, which significantly deplete FPGA resources and result in an unnecessarily expansive design area. To prevent this scenario, it is essential to attain resource reutilization for varying data precision levels. Simultaneously, managing activities using data of varying precision and enhancing operational accuracy are critical issues that must be addressed.

II. DESIGN CONCEPT

A. Matrix Chunking Idea

The tensor computation unit does multi-dimensional, large-scale matrix multiplication and addition operations. Effectively managing the interactions among matrix dimensions has emerged as a crucial aspect in enhancing computational speed.

Examine three extensive matrices A(m*k), B(k*n), and C(m*n), with the following multi-dimensional configurations: m8n32k16, m16n16k16, and m32n8k16. The resultant matrix is D(m*n)=A*B+C. Should a systolic array operation module be created independently for each dimension, it is necessary to instantiate 3*256 processing element (PE) units, as seen in Fig. 1.This will certainly deplete FPGA resources, elevate power usage, and squander computational capability.

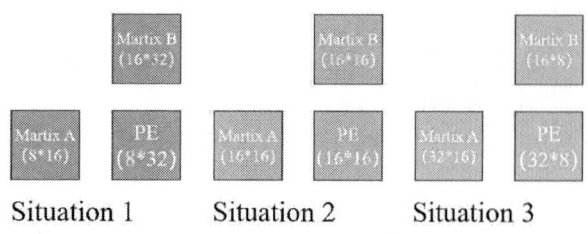

Situation 1   Situation 2   Situation 3

Fig. 1.   PE units consumed by designing a systolic array separately

During matrix block processing, the subsequent rules[2] must be adhered to: 1. The column count of matrix A (m*k) must equal the row count of matrix B (k*n). 2. The number of columns in each block matrix of matrix A must equal the number of rows in each corresponding block matrix of matrix B. 3. The number of rows and columns in the block matrix of matrix C is equal to the number of rows in the corresponding block matrix of matrix A, as well as the number of columns in the corresponding block matrix of matrix B.

How should a large matrix be divided most appropriately? If matrix C is divided unevenly, each block matrix will have different computation times and resource consumption. This will inevitably result in wasted computing power and resources [2]. Therefore, this design evenly divides the large matrix to ensure that the computation time and resource utilization of all block matrices are nearly identical.

According to the preceding explanation, for the three matrix dimension scenarios outlined, a technique akin to determining the greatest common divisor is employed to divide each matrix into four 8 × 8 submatrices. Each processing element unit does two operations to produce the matrix result, as seen in Fig. 2. The rationale for dividing the huge matrix into 8 × 8 submatrices is that it facilitates optimal computational performance while minimizing the wastage of computational resources. In executing block matrix operations, matrix operations of varying dimensions require just 4*64 processing element units, resulting in resource savings of at least two-thirds of the FPGA compared to the independent design of a systolic array.

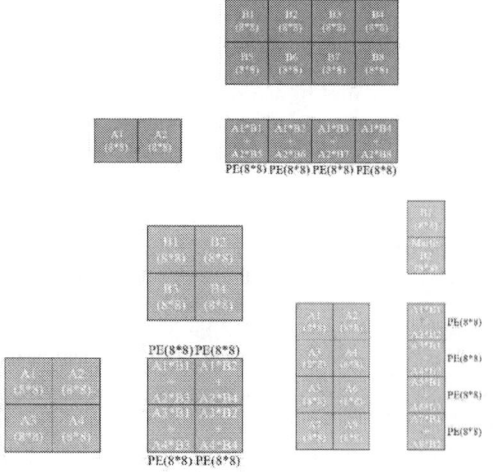

Fig. 2.   The PE units consumed by the block matrix concept

B.   The Idea of Systolic Array Arithmetic

The systolic array is a parallel computing architecture intended to mitigate communication bottlenecks in very large-scale integrated circuits. It accomplishes effective data processing by transmitting data among each processing unit in the array, resembling the pulse of the heart. This design typically features several isomorphic processing elements (PEs) capable of executing multiplication and accumulation operations efficiently and is extensively utilized in fields such as machine learning and deep learning.

To elucidate the principle of pulsed array operation and the configuration of the PE unit[3][4], Fig. 3 illustrates the operational process of the 2x2 systolic array, which aligns with the principle of the 8x8 pulsed array operation proposed in this study.

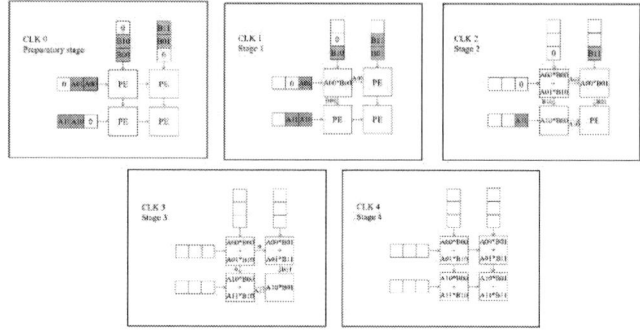

Fig. 3.   Arithmetic procedure for 2*2 systolic array

Initially, the input matrices A and B must undergo a zero-completion operation to guarantee that the data fed into the systolic array is accurate for each clock cycle. Subsequently, each processing element unit must execute operations, including data reception, data processing, and data transmission. For instance, the initial PE unit in Fig. 3, at CLK1, must acquire the values of A00 and B00, transmit the data to the adjacent right and lower PE units, and execute the A00*B00 operation internally. The data received by the leftmost and highest processing element (PE) units originates from external input, whilst the data for the other internal PE units is derived from the information transmitted by their respective left and upper PE units. Furthermore, multipliers and accumulators are necessary within the PE cells for data processing. Upon the conclusion of the lower right PE unit's work, the entire systolic array operation is deemed complete, allowing for data output.

To support the block matrix calculation approach, the PE unit requires an internal data buffer to store temporary data. When the first block matrix data, A1 and B1 (multipliers), are input, the results of the multiplication operation are stored in the data buffer along with the C1 block matrix (addends). Then, they are added together to obtain the temporary output data(S1). When the second block matrix data, A2 and B2, are input, the results of the multiplication operation are stored in the data buffer, along with the first temporary output (S1), and then added to obtain the final result. This design ensures that the matrix addition operation consumes minimal additional resources, saving area and reducing power consumption.

979-8-3315-8850-2/25 $31.00 © 2025 IEEE   845

C. Ideas for Floating Point Arithmetic

The floating-point multiplier and adder are designed according to the floating-point format, denormalized number format, and their operation rules[5][6] to ensure the accuracy of data operations. It is worth mentioning that the designed arithmetic unit can handle denormalized numbers close to 0, improving the accuracy of calculations to meet the needs of high-precision practical scenarios. Fig. 4 shows the formats of half-precision floating-point numbers and single-precision floating-point numbers. Among them, the calculation formulas for normal numbers and subnormal numbers close to 0 are as follows:

$$(-1)^{Sign} * 2^{(Expont-Bias)} * 1.Fraction \qquad (1)$$

$$(-1)^{Sign} * 2^{(Expont-Bias+1)} * 0.Fraction \qquad (2)$$

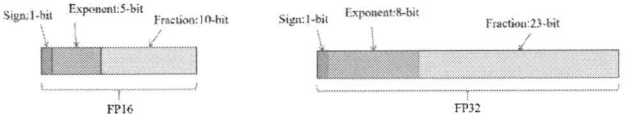

Fig. 4. The format and expressions of half precision floating point numbers and single-precision floating-point numbers

In the context of floating-point addition, it is imperative first to compare the exponent bits of the two floating-point numbers. This process enables the determination of the absolute value of their difference, denoted as exp_diff. The fraction bits of the smaller floating-point number are to be shifted to the right, according to the size of exp_diff. The sign bit and exponent bit of the output floating-point number are equal to the sign bit and exponent bit of the larger floating-point number, respectively. Subsequently, the sign bits of the two input floating-point numbers are determined to complete the addition or subtraction of the fraction bits. Finally, normalization is performed according to the position of the first "1" in the fraction bits after the operation. The fraction bits are shifted by adding or subtracting the exponent bits to ensure that the output floating-point number complies with the specifications. It is noteworthy that the following exceptional cases may arise during the operation:

1. In the context of denormalized numbers with proximity to 0, their exponent bits are offset by -14 (fp16) or -127 (fp32), a circumstance that necessitates specialized handling.

2. During the process of normalization, it is possible for the exponent to be subtracted by a specific value, resulting in a value of zero or negative. This indicates that the output floating point number is a denormalized number with a proximity to 0. In order to ensure that the size of the floating point value remains unchanged, it is necessary to assign the exponent a value of 0 and relocate the fraction.

For a floating-point multiplier, the sign bit of the output floating-point number is equal to the result of the XOR operation between the sign bits of the input floating-point numbers. Prior to the implementation of normalization, the exponent bit and fraction bit of the output floating point number are equivalent to the sum of the exponent bits of the input floating point numbers minus 15 or 127 and the product

of the fraction bits, respectively. When performing normalization, a specific value needs to be added to or subtracted from the exponent to ensure that the fraction complies with the standard specifications after displacement. Furthermore, the following exceptional cases may occur during operation:

1. As with addition, denormalized numbers close to 0 necessitate special handling.

2. In instances where the exponent of the output floating-point number is zero or negative after exponent addition, subtraction of the bias, and normalization, it is imperative to set the exponent to 0 and shift the fraction to ensure that the magnitude of the floating-point number remains constant.

III. TENSOR COMPUTATION UNIT

A. Input and Output Bus Interfaces

The AXI interface has been shown to combine reliability, security, and high bandwidth, which has led to its widespread acceptance and use as a mainstream data transmission protocol[7]. To facilitate design, the TPU is designed to interconnect with various external devices. The AXI_FULL_MASTER and AXI_LITE_SLAVE interfaces have been designed for reading data and configuration information, respectively. Fig. 5 illustrates the specific architecture of the bus interface, which involves collaboration between the two interfaces for various high-speed application scenarios.

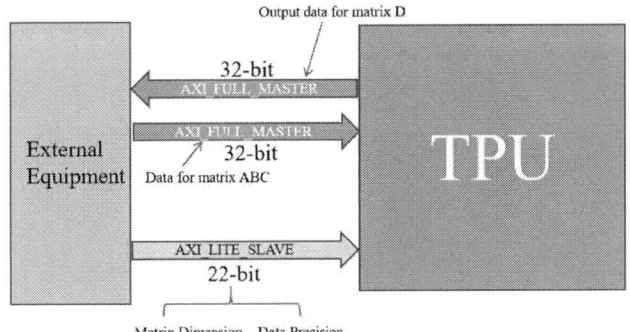

Fig. 5. Structure of the bus interface

The AXI_FULL_MASTER interface, which has been meticulously designed, facilitates the reading of input matrix information and the writing back of output matrix information, thereby enabling master-slave interaction with external memory, such as DDR. The burst length is defined as 256, which is the maximum value within the specified range, and the data bit width is defined as 32, which is the maximum bit width for data precision.

Furthermore, the AXI_LITE_SLAVE interface is used for receiving external configuration information and initiating signals. The configuration information encompasses the matrix dimensions mnk and the matrix data precision types (int4, int8, fp16, fp32).

The AXI interface provides a versatile framework for on-chip communication, facilitating efficient master-slave interaction and addressing the requirements of ultra-high-performance and intricate designs.

B. TPU Overall Architecture and Workflow

The TPU is designed to enhance the efficiency of deep learning inference, and its specific architecture is illustrated in below Fig. 6. The TPU is comprised of four TPU8x8 computing modules, which are used for executing block matrix operations. By the principle of block matrix calculation, the control module facilitates the transfer of data from matrices A and B to the designated TPU8x8 modules in two steps, based on the configuration information, for calculation purposes. The particulars of this scenario are illustrated in Fig. 2. Each TPU8x8 is composed of 64 PE units, forming a large-scale systolic array. Each PE contains multipliers and adders with different data precision levels, thereby enabling matrix operations with varying data precision. The remaining components include registers for storing configuration information, three random access memory (RAM) units for storing the input matrices ABC, corresponding data buffers, a state machine control module, and a conversion module that transforms global signals into local signals.

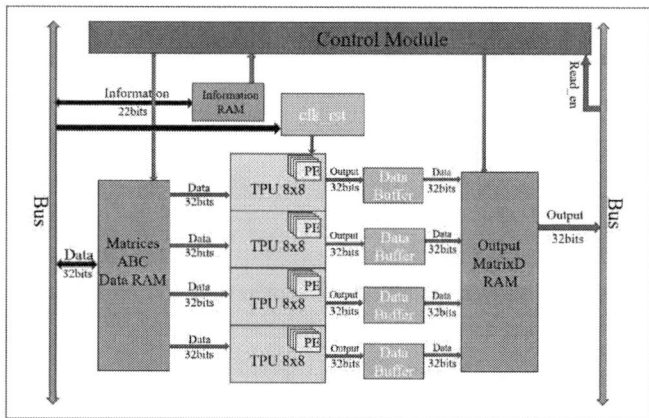

Fig. 6. Overall architecture of TPU

The control state machine of the TPU module has nine states, as illustrated in Fig. 7. Among them, IDLE is the initial state, in which the module waits for the enable start signal to become valid before jumping to the STORE_S state. In the STORE_S state, the configuration information i_lite_state is read. The bits that represent the data precision type are bits 0-3; the bits that represent the matrix dimension k are bits 4-9; the bits that represent the matrix dimension n are bits 10-15; and the bits that represent the matrix dimension m are bits 16-21. After the configuration information is stored, the state transitions to STORE_A. In the STORE_A state, the data of matrix A is stored. If the value of the two-dimensional counter surpasses the dimension of matrix A (m*k), the system should transition to the STORE_B state. Conversely, if the value remains within the bounds of matrix A (m*k), the system should persist in the STORE_A state. In the STORE_B state, the data of matrix B is stored. If the value of the two-dimensional counter surpasses the dimension of matrix B (k*n), a transition to the STORE_C state is to be initiated. Conversely, if the aforementioned condition is not met, the system remains in the STORE_B state. In the STORE_C state, the data of

matrix C must be stored. Suppose the value of the two-dimensional counter surpasses the dimension of the matrix C (m x n). In that case, it is necessary to enable and initiate the four 8 x 8 block matrix operation modules, subsequently transitioning to the NUM1 state. Conversely, if the aforementioned condition is not met, the system should remain in the STORE_C state. In the NUM1 state, the first multiplication matrix data input and addition matrix data input must be performed. The data input for the various matrix dimensions is illustrated in Fig. 2. After the execution of operations by the four 8x8 block matrix operation modules; the program transitions to the NUM2 state. In the NUM2 state, the second multiplication matrix data input is executed, with data input for different matrix dimensions, as shown in Fig. 2. After the completion of operations by the four 8x8 block matrix operation modules, the program transitions to the OUTPUT state. In the OUTPUT state, the output results of the four 8x8 block matrix operation modules are stored in the output matrix cache of the TPU module according to different matrix dimensions. It has been determined that when the two-dimensional storage counter exceeds 8 x 8, the system transitions to the READ state. In the READ state, the system awaits an external read signal and subsequently outputs the matrix data row by row in order. Following the output of all data, the system returns to the IDLE initial state and awaits the next operation.

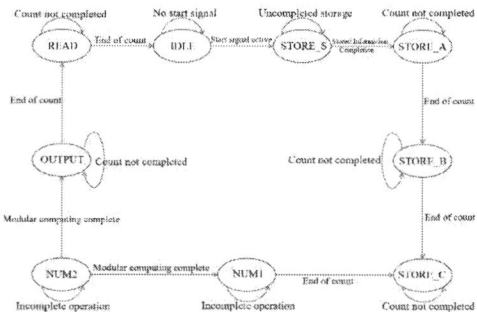

Fig. 7. TPU internal state control

The cache of matrix data C inside the TPU adopts a two-dimensional dimension of 32*8, and the matrix C is processed in chunks according to the storage law, as shown in Fig. 8 below, which is more resource-saving than storing it directly into the two-dimensional dimension of the largest size of 32*32.

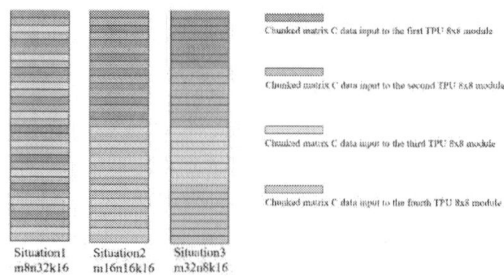

Fig. 8. Storage of matrix C in different situations

979-8-3315-8850-2/25 $31.00 © 2025 IEEE   847

C. PE Unit Design

To accommodate operations involving data of varying precision levels, the fundamental PE unit structure has undergone an enhancement that enables operations on data of different precision levels, as well as matrix multiplication and addition calculations.

The PE unit structure is illustrated in Fig. 9. It consists of multipliers and adders with varying data precision levels, as well as data registers. The system processes eight data groups concurrently for multiplication operations, with the initial data input comprising a data c for addition operations. Furthermore, the PE unit receives data precision configuration signals, which are used to select the appropriate multipliers and adders.

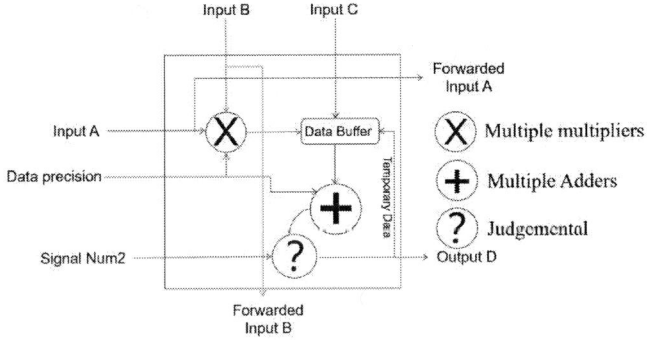

Fig. 9. PE unit basic structure

The PE unit is responsible for transmitting data, ensuring that input data from the left side is transferred to the right-side output and input data from the upper side is directed to the lower-side output. Concurrently, the valid signal is pulled high to output, enabling the right-side and lower-side PE units to determine whether the data between them is valid.

When processing data, the PE unit processes two sets of eight data inputs through corresponding multiplier modules, yielding eight multiplication results. The difference lies in the fact that the first operation requires storing these eight multiplication results and the input data c in the data register for accumulation to obtain the temporary data Temp. The second operation requires storing the eight multiplication results and the temporary data Temp in the data register, followed by accumulation to obtain the final output result. It can be observed that matrix addition requires almost no additional resources or power consumption, thereby improving computational speed while reducing resource consumption.

D. Integer Data Adder and Multiplier Design

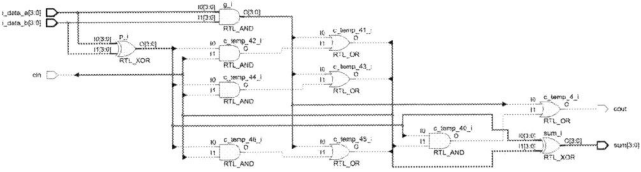

Fig. 10. Four-bit carry-lookahead adder

In the context of int32 adders, the implementation of a four-bit carry-lookahead adder (CLA) chain[8] is a critical aspect of the circuit design. This approach is employed to minimize circuit delay, although it necessitates augmented resource consumption and ensures that the clock frequency meets the elevated requirements. The circuit diagram of the four-bit carry-lookahead adder is shown in Fig. 10.

The architecture of an int8-type multiplier is illustrated in Fig. 11. During the computation process, a zero is first appended to the end of the data, and then Redix-4 Booth encoding[9] is applied to obtain three processing signals: w_x1, w_inv, and w_x2. The Booth encoding table is shown in TABLE I. These three processing signals are then introduced into the partial product generation module, resulting in the production of four partial products. Subsequently, the partial products are compressed into two partial products using a 4-2 compressor[8]. Finally, the two partial products are added using a carry-lookahead adder (CLA) to obtain the output result.

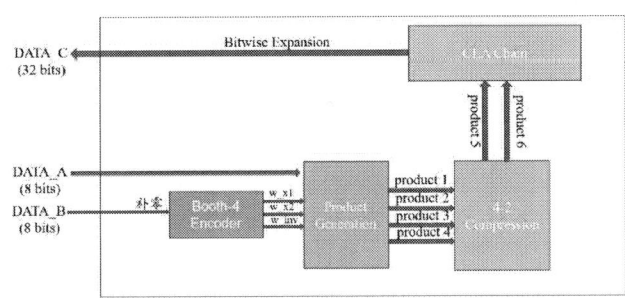

Fig. 11. INT8 Multiplier Architecture

TABLE I.   REDIX-4 BOOTH CODE TABLE

$B_{i+1}B_iB_{i-1}$	*corresponding operation*
000	+0
001	+A
010	+A
011	+2A
100	-2A
101	-A
110	-A
111	-0

As illustrated in Fig. 12, the designed circuit of the 2's complement module exhibits a reduction in the number of NOT gates, OR gates, and XOR gates by 12, 2, and 1, respectively, when compared to the conventional thirteen-bit 2's complement circuit[10].

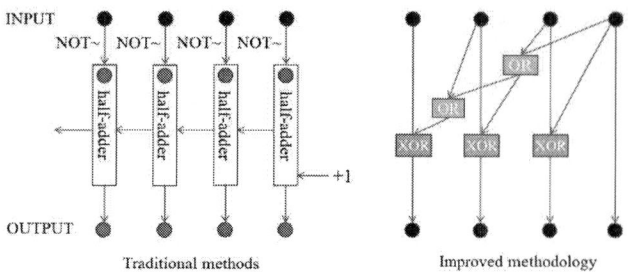

Fig. 12. Improved 2's complement circuits

E. Floating Point Fraction Multiplier Design

A floating-point data fraction multiplier[11][12] was designed whose main architecture resembles that shown in Fig. 11. Compared with traditional multipliers; it reduces delay and resource consumption. Consider a 10-bit (FP16) multiplier. First, a 0 is added to the end of the 10-bit input data, and two 0s are added to the beginning. These 13 bits of data generate three encoding signals (w_x1, w_x2, and w_inv) in the Redix-4 Booth encoder. The specific Booth encoding table is shown in TABLE I. Next, the three signals are imported into the partial product generation module, which generates six partial products. Then, the novel Wallace compression structure[11][12] shown in Fig. 13 is applied. To achieve greater resource reuse and reduce consumption, the compressor primarily uses a 4-2 compression structure. The compression process is as follows: First, partial products P1 to P4 are compressed using 4-2 and 3-2 compressors and half-adders to obtain two pseudo-sums, S1 and C1. Next, S1, C1, P5, and P6 undergo the same compression process to produce two pseudo-sums, S2 and C2. Finally, the two pseudo-sums are multiplied by a four-bit carry-lookahead adder chain to obtain the final result. This novel Wallace compression structure uses 20 4-2 compressors, six 3-2 compressors, and eight half-adders in total. Compared to the Wallace compression structure without sign bit preprocessing, this structure reduces the number of 4-2 compressors by approximately 30%, resulting in a significant reduction in FPGA resource consumption.

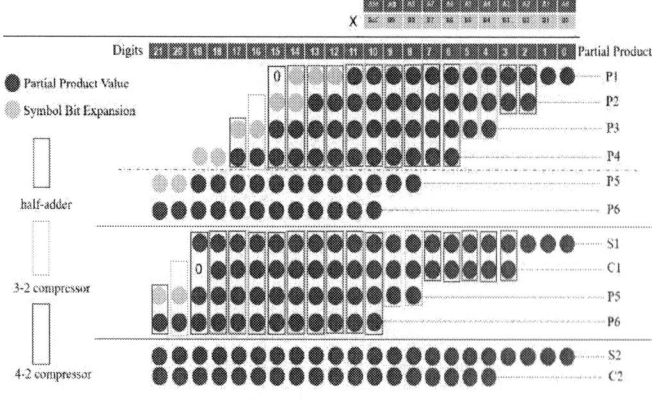

Fig. 13. Improved Wallace compression structure

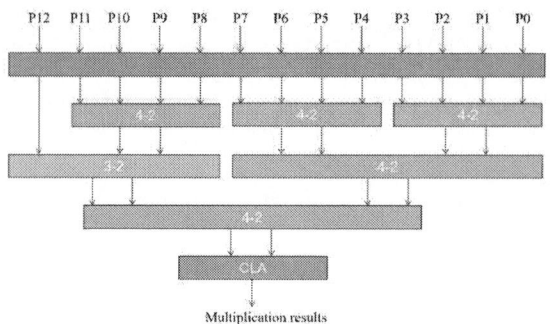

Fig. 14. Wallace compression structure of 24-bit multiplier

The 24-bit multiplier principle in fp32 is essentially the same as the aforementioned approach. The main difference lies in the Wallace compression process. First, the thirteen partial products generated are processed through three sets of 4-2 mixed compression structures to obtain seven pseudo-sums. Then, they are processed through one set of 4-2 mixed compression structures and one set of 3-2 mixed compression structures to obtain four pseudo-sums. Finally, they are processed through one set of 4-2 mixed compression structures to obtain two pseudo-sums. A four-bit carry-lookahead adder chain then processes the two pseudo-sums to obtain the product result. Fig. 14 shows the specific schematic diagram.

TABLE II shows the novel multiplier resource consumption based on the Vivado platform and the xcvup9p-flga2104-2L-e FPGA board.

TABLE II. INNOVATIVE MULTIPLIER FPGA RESOURCE CONSUMPTION

	Resource Conxumption	
	CLB LUTs	Bonded IOB
11-bit multiplier	154	44
24-bit multiplier	722	96

IV. RESOURCE REUSE

A. Data Bit-width Multiplexing

The simplest way to multiplex resources for different data precision is to expand the sign bit of int4 data to make it an int8 data type. The operations on int4 data are treated as if they were int8 data operations. This bit-width expansion eliminates the use of int4 multipliers and adders, reducing resource utilization.

B. Hardware Resource Reuse

The resource consumption of multipliers is reduced since different data precision multipliers reuse modules such as Redix-4, Booth encoders, 4-2 compressors, and CLA (carry-lookahead) adders. The specific comparison data is shown in TABLE III. As can be seen, the INT8 multiplier's resource consumption is reduced by about 95%, the FP16 multiplier by about 13%, the FP32 multiplier by approximately 21%, and the PE unit by about 12%. Similarly, the INT32 adder reuses the CLA module, resulting in minimal resource consumption. These results demonstrate that reusing resources across

979-8-3315-8850-2/25 $31.00 © 2025 IEEE

different data precision levels can significantly reduce resource consumption.

TABLE III. COMPARISON OF RESOURCE CONSUMPTION BEFORE AND AFTER REUSE

	Comparison of Resource Consumption	
	Single Module	Resource Reuse
INT8 Multiplier	130 LUTs 4 FF	7 LUTs 4 FF
FP16 Multiplier	499 LUTs 145 FF	434 LUTs 144 FF
FP32 Multiplier	1203 LUTs 142 FF	952 LUTs 141 FF
PE Unit	3178 LUTs 1404 FF	2692 LUTs 1402 FF

V. RESULTS

A. Synthesis Results

Table IV shows the performance comparison of large-scale matrix operations implemented on different FPGA platforms. Compared with the results in Reference [13], the design in this paper disables the DSP units, making it difficult to directly compare resource consumption. In terms of data precision, this study supports both floating-point and integer data operations, in addition to the floating-point operations supported in Reference [13].The FPGA designed in this paper operates at 250 MHz, exceeding the 195 MHz achieved in Reference [13] using 256 PEs as an example. Due to the support for higher data precision operations, the power consumption is higher than that of Reference [13], and the energy efficiency ratio is lower than that of Reference [13].Overall, although this work performs slightly worse than Reference [13] in terms of power consumption and energy efficiency ratio, it demonstrates outstanding performance in clock frequency and the types of data precision it can process.

TABLE IV. TPU PERFORMANCE DEMONSTRATION

Performance	The design of this paper	Reference [13]
FPGA	XCVU9P-flga2104-2L-e	Xilinx 585T
Data Precision Type	INT4,IN8,FP16,FP32 and Mixed-precision (FP16 and FP32)	Floating-point
Clock Frequency	250MHZ	195MHZ
Utilization	757971 LUTs 560740 FF	282 DRAM(18K) 164600 LUTs 75600 Slice Regs 1024 DSPs
Power /w	8.365	5.24
Arithmetic Power	64.0GFLOPS	99.8GFLOPS
Energy Efficiency	7.65GFLOPS*W^{-1}	19.05GFLOPS*W^{-1}

B. Arithmetic Accuracy

The designed TPU computational unit performs matrix operations on INT4, INT8, FP16, FP32, and mixed precision (FP16 and FP32) data. Based on the Modelsim verification platform, the arithmetic data are compared using the System Verilog verification code, with a relative error tolerance of 1%. TABLE V presents the average computational accuracy for various matrix dimensions. It can be observed that the shaping data is entirely correct, but the floating-point data has a specific error, resulting in a reduction in accuracy. In conclusion, the

design exhibits high data reliability across various arithmetic scenarios.

TABLE V. COMPUTATIONAL ACCURACY FOR DIFFERENT DATA PRECISION

Data Precision Type	Arithmetic Accuracy/%
INT4	100
INT8	100
FP16	98.7
FP32	99.74
Mixing accuracy（FP16 and FP32）	99.2

VI. CONCLUSION

The tensor computing unit designed in this paper can perform matrix operations with multiple dimensions and data precisions. Its high computing accuracy and low resource consumption make it applicable to a variety of scenarios involving matrix operations.

Future work can focus on two aspects of design and optimization: Google's sparse matrix and the topological structure of systolic arrays. Sparse matrix design can significantly reduce storage space and enhance computing efficiency, resulting in substantial improvements in both time and space dimensions. Regarding the topological structure of systolic arrays, the design will not only reduce communication delays between processing elements (PEs) and the difficulty of layout and wiring but also reduce resource consumption. Additionally, the structure of the calculation module will be refined to maximize calculation accuracy.

REFERENCES

[1] Kung, "Why systolic architectures?," in Computer, vol. 15, no. 1, pp. 37-46, Jan. 1982.

[2] V. M. Glushan and L. A. Yu., "On Distributed Multiplication of Large-Scale Matrices," 2021 IEEE 15th International Conference on Application of Information and Communication Technologies (AICT), Baku, Azerbaijan, 2021, pp. 1-4.

[3] A. Puşcaşu, C. B. Ciobanu and O. Buiu, "Systolic Array Matrix Multiplication Accelerator," 2024 International Semiconductor Conference (CAS), Sinaia, Romania, 2024, pp. 207-210.

[4] P. Lin, H. Zhang, L. Li and Y. Li, "Systolic Array Architecture for Multitasking Neural Processing Unit," 2024 29th International Conference on Automation and Computing (ICAC), Sunderland, United Kingdom, 2024, pp. 1-6.

[5] "ISO/IEC/IEEE International Standard - Floating-point arithmetic," in ISO/IEC 60559:2020(E) IEEE Std 754-2019 , vol., no., pp.1-86, 8 May 2020.

[6] A. Jangid and S. Kabra, "Implementation of digital logic circuit in artificial neural network with floating point arithmetic using verilog HDL," 2017 International Conference on Computing and Communication Technologies for Smart Nation (IC3TSN), Gurgaon, India, 2017, pp. 270-275.

[7] R. Bhaktavatchalu, B. S. Rekha, G. A. Divya and V. U. S. Jyothi, "Design of AXI bus interface modules on FPGA," 2016 International Conference on Advanced Communication Control and Computing Technologies (ICACCCT), Ramanathapuram, India, 2016, pp. 141-146.

[8] J. Mody, R. Lawand, R. Priyanka, S. Sivanantham and K. Sivasankaran, "Study of approximate compressors for multiplication using FPGA,"

2015 Online International Conference on Green Engineering and Technologies (IC-GET), Coimbatore, India, 2015, pp. 1-4.

[9] S. Rooban, M. Nagesh, M. V. S. L. Prasanna, K. Rayudu and G. D. Sai, "Implementation of 128-bit Radix-4 Booth Multiplier," 2021 International Conference on Computer Communication and Informatics (ICCCI), Coimbatore, India, 2021, pp. 1-7.

[10] FAN Wenbing ,ZHOU Jianzhang.Design of Parallel Multiplier Based on Radix-4 Booth Coding[J].Journal of Zhengzhou University (Engineering Science),2025,46(01):26-33.

[11] R. Rathod, P. Ramesh, P. S. Zele and A. K. Y, "Implementation of 32-Bit Complex Floating Point Multiplier Using Vedic Multiplier, Array

Multiplier and Combined integer and floating point Multiplier (CIFM)," 2020 IEEE International Conference for Innovation in Technology (INOCON), Bangluru, India, 2020, pp. 1-5.

[12] M. J. Rao and S. Dubey, "A high speed and area efficient Booth recoded Wallace tree multiplier for fast arithmetic circuits," 2012 Asia Pacific Conference on Postgraduate Research in Microelectronics and Electronics, Hyderabad, India, 2012, pp. 220-223.

[13] L. Zhang, Y. Peng, A. Huang and X. Hu, "A Scalable Architecture for Accelerating Multi-Operation and Continuous Floating-Point Matrix Computing on FPGAs," in IEEE Access, vol. 8, pp. 92469-92478, 2020.

2025 The 10th International Conference on Integrated Circuits and Microsystems

A Low-Ripple Charge Pump with Adaptive Load Compensation for Flash-Based CIM

Xuyuan Gu
School of Integrated Circuits
Jiangnan University
Wuxi, China
6231916026@stu.jiangnan.edu.cn

Xiaofeng Gu
School of Integrated Circuits
Jiangnan University
Wuxi, China
xgu@jiangnan.edu.cn

Zhiguo Yu*
School of Integrated Circuits
Jiangnan University
Wuxi, China
yuzhiguo@jiangnan.edu.cn

Abstract—This paper presents an adaptive load compensation charge pump for the bit line (BL) driving circuit of flash-based compute-in-memory (CIM) chips. In CIM chips, severe load current fluctuations in BL circuits cause significant ripple and slow voltage recovery at the charge pump output. To address the problem, this work integrates a parallel structure combined with a delay-locked-loop (DLL) differential clock to reduce ripple and enhance driving capability. A dynamic compensation module adapts to load current changes, enabling real-time inverse compensation to maintain constant equivalent output current and ensure voltage stability. The charge pump is designed in 55-nm CMOS process. Simulation results show under load current fluctuations ranging from 50 to 200 μA, the steady-state ripple remains approximately 2.1 mV, which is reduced by 86% compared to that of charge pump by dynamic clock frequency scaling. During the BL switching, the recovery time for a 2.04 V voltage drop caused by parasitic capacitance is approximately 2.01 μs.

Keywords—CIM, Charge pump, Low ripple, Adaptive load compensation

I. INTRODUCTION

Charge pumps are used to generate voltages higher than the supply voltage, which can be utilized for read, program, and erase operations in flash-based CIM chips [1]. When the operational states of the CIM chip are switched, load current fluctuations are induced, manifesting as stepwise and spike-like dynamic variations. These variations result in a voltage collapse and overshoot at the output of the charge pump, and subsequently lead to degradation in the storage and computational accuracy of the chip.

To address the issues of overshoot and voltage collapse during chip state switching, current mainstream solutions focus on five aspects: passive buffering components, active discharge paths, timing optimization techniques, storage media and process optimization. In reference [2], a large off-chip decoupling capacitor is used to reduce ripple and lower voltage droop. However, this approach is clearly not conducive to full integration of charge pumps on-chip. Reference [3] employs a back-end-connected low dropout regulator (LDO) to stabilize the output of the charge pump, but this introduces additional power consumption and area overhead.

The Key Research Project of Jiangsu Province, China (BE2023019-3).

To address the aforementioned issues, this paper proposes a charge pump suitable for the BL-end driving circuit of flash-based CIM chips. This work improves both dynamic load response and startup speed while maintaining relatively low ripple. The main contributions of this paper are as follows:

- The parallel ripple cancellation structure utilizing DLL proposed in this paper enables the ripple cancellation method to be applicable to DCFS.

- The proposed current compensation module can compensate for instantaneous variations in load current, ensuring that the output voltage of the charge pump remains relatively stable.

- In the event of a spike-type load current, the current compensation module will automatically shut down, and the output voltage can be rapidly recovered from voltage collapse.

II. CHARGE PUMPS IN CIM CHIPS

CIM devices are typically organized into cross-coupled two-dimensional array (as shown in Fig. 1), where the horizontal and vertical lines are commonly referred to as word lines (WLs) and BLs, respectively [4]. The voltage switching on WL enables the opening and closing of each row of devices, while current accumulation occurs at the BL terminal for each column of devices. The readout system converts device currents into digital signals, completing the matrix-vector multiplication computation. Compared to the WL load current, the BL load current is larger and varies more dramatically. When the row selection signal switches, a new CIM cell is selected to connect to the BL, resulting in an instantaneous change in the current load, which in turn causes a step-like surge in the equivalent current load at the charge pump output. This current burst is accompanied by capacitive coupling spikes, causing voltage collapse at the charge pump output. This collapse affects the device state switching time and limits the data processing speed of the chip. The spike current (I_{spike}) can be expressed by (1):

$$I_{spike} \propto C_{parasitic} \cdot \frac{\Delta V}{\Delta t} \quad (1)$$

where $C_{parasitic}$ represents the equivalent total capacitance of BL parasitic capacitance and the storage cell junction capacitance,

979-8-3315-8850-2/25 $31.00 © 2025 IEEE

ΔV denotes the voltage swing of the selection signal, and Δt is the switching time.

Conversely, when the row selection signal is inverted, the number of CIM cells connected to the BL decreases instantaneously, which produces a reverse step in the load current. The instantaneous reduction in load current may causes charge accumulation on the filtering capacitor at the charge pump output, resulting in a brief overshoot of the BL voltage. This overshoot may lead to latch-up effects or gate oxide breakdown, and could permanently damage the CIM cells. With the increasing integration, complexity, and computational speed of CIM chips, the aforementioned issues have become increasingly prominent. Therefore, it is crucial to develop a voltage generator that offers more stable output voltage less sensitive to variations in load current.

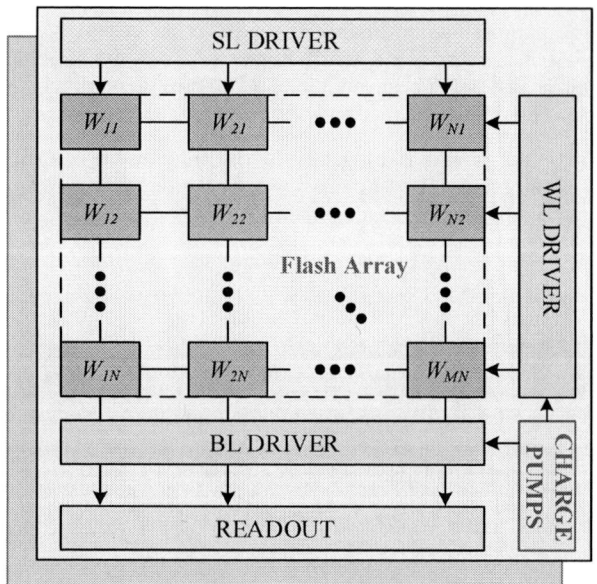

Fig. 1. Architecture of Flash-Based CIM.

The current mainstream modulation methods for charge pumps can be categorized into three types: Periodic Stepping Modulation (PSM), DCFS, and Dynamic Clock Voltage Scaling (DCVS). Among these, PSM utilizes a specific feedback detection structure to block the pumping clock, thereby achieving a relatively stable output voltage [5]. Its advantages lie in the simplicity of design and implementation, as well as lower power consumption. However, the significant ripple associated with PSM is often unacceptable in many application scenarios. PSM's ripple voltage can be expressed by the following equation:

$$V_{\text{ripple}} = \frac{(I_{\text{out}} - I_{\text{load}}) \cdot \Delta t}{C_{\text{filter}}} \qquad (2)$$

where V_{ripple} represents rippple voltage, Δt denotes the response time of the feedback loop, and I_{load} is load current.

DCVS operates at a constant clock frequency, continuously regulating the clock power supply rail through a power operational amplifier to achieve voltage stability [6]. The output ripple of DCVS is the smallest among the three methods, but its load capacity is limited by the driving capability of the operational amplifier, resulting in high power consumption for applications requiring rapid startup and fast transient response. In contrast, DCFS requires a continuous adjustable oscillator to generate the pump clock, which operates independently of the system clock. DCFS offers greater flexibility, as the clock frequency can be adaptively adjusted according to the load, achieving higher efficiency. However, the ripple varies with load changes, making it particularly difficult to be suppressed under light load conditions.

The charge pump proposed in this study is an improved structure based on DCFS, utilizing a current compenstion module to address the issue of ripple suppression under light load conditions that are challenging for traditional DCFS. Furthermore, the parallel ripple cancellation structure contributes to further reduction of ripple, ensuring the chip's excellent and stable computational accuracy.

III. PROPOSED CHARGE PUMP DESIGN AND OPERATION

The proposed charge pump combines a parallel voltage generator with a current compensation module, achieving a relatively stable output voltage over a wide dynamic load range. This stable output voltage enhancing the robustness and computational accuracy of the CIM chip.

A. Overall Architecture

The diagram of the proposed charge pump is shown in Fig. 2, which consists of a cross-coupled charge pump main unit, a feedback regulation network, a ring oscillator, a DLL-based clock driving module, and a current compensation module. The parallel voltage generator is based on DCFS, which employs a resistive voltage divider network to sense the output voltage. An operational transconductance amplifier (OTA) is used to amplify the error signal, thereby regulating the oscillation frequency of the oscillator to stabilize the output voltage. The parallel charge pump structure significantly accommodates applications with high load currents while accelerating transient response speed. We apply a 90-degree phase shift to the pumping clocks of two parallel branches and approximately doubles the fundamental frequency of the output ripple. Consequently, the maximum peak-to-peak ripple value is reduced without altering the capacitance of the filtering capacitor. Since the frequency of the pumping clocks in DCFS is variable, conventional approach, such as constructing clocks with fixed phase differences using an inverter chain, will no longer be applicable, thus failing to double the fundamental frequency of the ripple voltage. This work employs a DLL to lock the input and output clocks in phase with a one-cycle difference, and then generates a constant phase delay through different output ports of a voltage-controlled delay line (VCDL), independent of the clock frequency.

979-8-3315-8850-2/25 $31.00 © 2025 IEEE

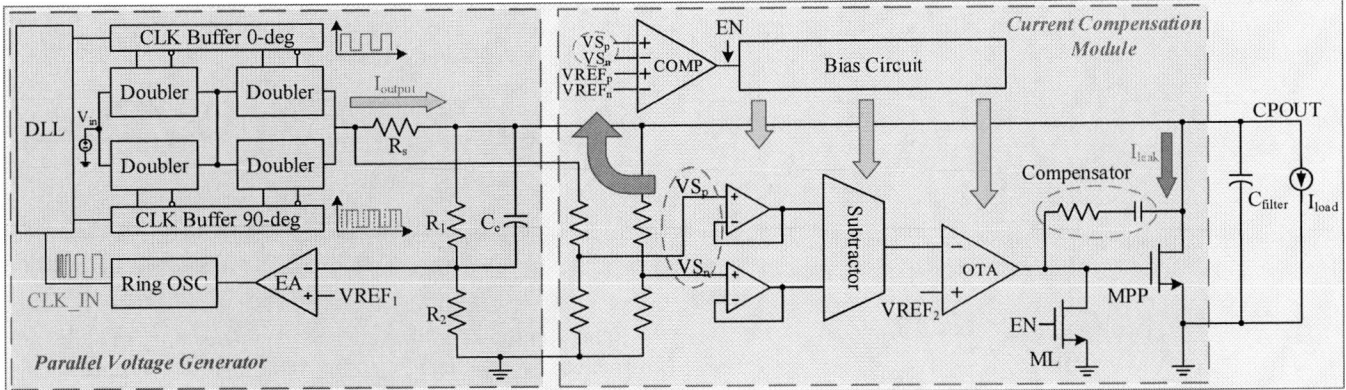

Fig. 2. Overall Architecture of Proposed Charge Pumps.

The current compensation module detects variations in load current and provides inverse compensation for these changes within a certain range. This compensation ensures that the equivalent current value extracted from the charge pump output remains almost unchanged, addressing the issue of ripple suppression in DCFS when the load current experiences a sudden decrease. Consequently, this module can operate effectively in scenarios requiring a wide load current range and low ripple (high load-ripple suppression ratio). The relationship among output current, load current, and leakage current can be expressed by (3):

$$I_{\text{output}} = I_{\text{load}} + I_{\text{leak}} \tag{3}$$

where I_{output} represents the output current, I_{load} denotes the load current and I_{leak} is the leakage current.

When the load current I_{load} decreases, the voltage difference sensed and converted by the sensing resistor R_s decreases in real time. After this differential voltage is isolated by a buffer, it is converted into a single-ended voltage by a differential subtractor. This single-ended voltage is substracted with a preset reference voltage, and then the error will be amplified. This amplified error regulates the gate voltage of power MOSFET MPP to increase, generating a current leakage I_{leak} at the charge pump output. This current leakage inversely compensates for the change in load current, keeping the equivalent current drawn from the charge pump output relatively constant. Similarly, when the load current increases, the leak current decreases to compensate for the change, ensuring that the charge pump's driving capability remains sufficiently large. When the load current exceeds the maximum tolerance of the current compensation module, a shut-off signal EN is generated by the comparator, resulting in nearly zero power consumption for the current compensation module and complete shutdown of MPP. At the same time, the oscillator's operating frequency increases, enhancing the charge pump's driving capability and allowing it to operate in a high-efficiency state, where most of the power consumption is utilized to provide output current.

B. Analysis of the Main Loop

As shown in Fig. 2, the parallel voltage generator consists of an error amplifier, an oscillator, a DLL, a clock buffer, and a charge pump unit. The operational amplifier employs a two-stage structure. The first stage utilizes a P-type input folded cascode configuration, and the second stage employs a common-source stage with diode-connected devices as the load. This amplifier has several advantages: (1) it protects the gain of the preceding stage from noise introduced by the oscillator coupling, (2) it pushes the second pole of the output node to a higher frequency through a low-resistance diode load in the output stage, allowing for a single-pole approximation in analysis; (3) it facilitates the conversion from voltage to current, enhancing the linearity from level error to oscillator output frequency, thus preventing issues such as nonlinear distortion and self-oscillation.

The current-starved oscillator used in this work consists of an odd number of inverter units connected end to end. A controlled current source (referred to as current starving) is connected in series in the power supply paths of each inverter stage to limit the charging and discharging speed of the output node capacitance, thereby controlling the oscillation frequency. In the transfer function of the oscillator, the poles are typically formed by the parasitic capacitance of the delay unit's drain and the conduction resistance of the inverter. With a magnitude on the order of hundreds of megahertz to gigahertz, which are far above the system bandwidth. These high-frequency poles are often neglected in system stability analysis, so the transfer function of the oscillator can be approximated as a constant. The DLL module in this study is utilized to generate two clock signals with a 90-degree phase difference, independent of the clock frequency. The DLL's transfer function can be abstracted as a delay element, which can be represented in a zero-pole form, as shown in (4):

$$e^{-s\tau} \approx \frac{1 - s\tau/2}{1 + s\tau/2} \tag{4}$$

where τ represents the delay.

The modules in the main loop are abstracted as transfer functions and illustrated in a block diagram format as shown in Fig. 3. Each block corresponds to the actual modules, which are clearly indicated. Here, $H_1(s)$ represents the error amplifier, and $H_2(s)$ denotes the charge pump unit, both of which can be approximated by a single pole. $F(s)$ represents the resistor voltage divider feedback network. In the feedforward path, $H_1(s)$ and $H_2(s)$ provide two poles, while the delay element introduces a high-frequency zero and a pole. As shown in Fig. 4, a type-III compensator is added to ensure sufficient phase margin in the loop.

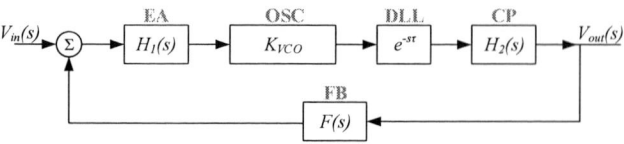

Fig. 3. Block diagram of the main loop.

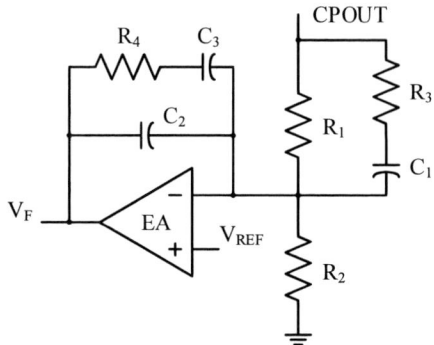

Fig. 4. Type-III compensator.

The frequencies of these zeros and poles are respectively expressed by the following equations:

$$\omega_{p1} = 0 \tag{5}$$

$$\omega_{p2} = \frac{1}{[(R_1 \parallel R_2) + R_3] \cdot C_1} \tag{6}$$

$$\omega_{p3} = \frac{1}{R_4 C_2 C_3 / (C_2 + C_3)} \approx \frac{1}{R_4 C_2}, \ (C_2 \ll C_3) \tag{7}$$

$$\omega_{z1} = \frac{1}{R_4 C_3} \tag{8}$$

$$\omega_{z2} = \frac{1}{(R_1 + R_3) \cdot C_1} \tag{9}$$

Poles ω_{p1} and ω_{p2} are placed at high frequencies to mitigate their impact on phase margin. By appropriately configuring the values of R_1, R_3, R_4, C_1, and C_3, we position the two zeros at suitable locations within the bandwidth, close to the two poles provided by $H_1(s)$ and $H_2(s)$, to compensate for the effects introduced by the poles and enhance the phase margin.

C. Analysis of the Current Compensation Module

The basic structure of the current compensation module is illustrated in Fig. 2. This module detects changes in output current through a resistor and performs inverse compensation, enabling the DCFS charge pump to operate effectively in light

load and low ripple application scenarios. The detailed feedback actions have been analyzed in Part A of Section III and will not be elaborated upon here.

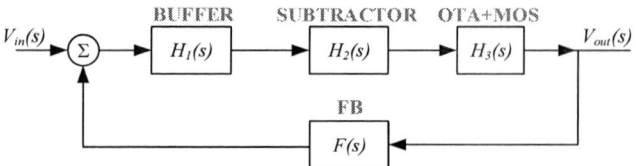

Fig. 5. Block diagram of the current buffering module.

The modules in the current compensation module are abstracted as transfer functions, and the block diagram is illustrated in Fig. 5. Here, the "buffer" represents an operational amplifier configured in a unity-gain feedback mode, its pole is pushed to high frequencies due to the closed-loop connection, and thus its influence is neglected during stability analysis. The subtractor is composed of an operational amplifier and a resistor feedback network, with its dominant pole frequency significantly influenced by the open-loop gain of the operational amplifier and the resistor ratio, resulting in a substantial increase compared to the original dominant pole of the operational amplifier. The pole of subtractor is well-separated from the low-frequency pole provided by the equivalent two stage operational amplifier formed by the OTA and MPP. The separation is beneficial for the stability of the current compensation loop. As discussed above, $H_3(s)$ provides the dominant pole, while $H_2(s)$ provides the secondary pole, ensuring that during the design phase, the dominant and secondary poles are sufficiently separated, which provides ample phase margin for the loop.

IV. EXPERIMENTAL RESULTS

The proposed charge pump circuit is implemented based on a 55-nm CMOS process. Fig 6 illustrates the variation of the output voltage waveform during instantaneous increases and decreases in load current, and it is compared with traditional DCFS and DCVS charge pumps. The simulation results show that when the load current decreases instantaneously, the ripple of the DCFS significantly increases and cannot be suppressed under light load conditions, while the voltage of the DCVS deviates from the expected value due to limited driving capability as the current changes. The charge pump proposed in this study maintains a ripple of approximately 2.1 mV during stable operation, which is reduced by 86% compared to DCFS.

Comparison results, as shown in Fig. 7, illustrate the transient response of the output voltage waveform during the instantaneous switching of BLs against DCVS. Results indicate that the charge pump proposed in this work exhibits a faster transient response than DCVS, with a recovery time of approximately 2.01 µs under a voltage collapse of 2.04 V and a constant current of 150 µA.

979-8-3315-8850-2/25 $31.00 © 2025 IEEE

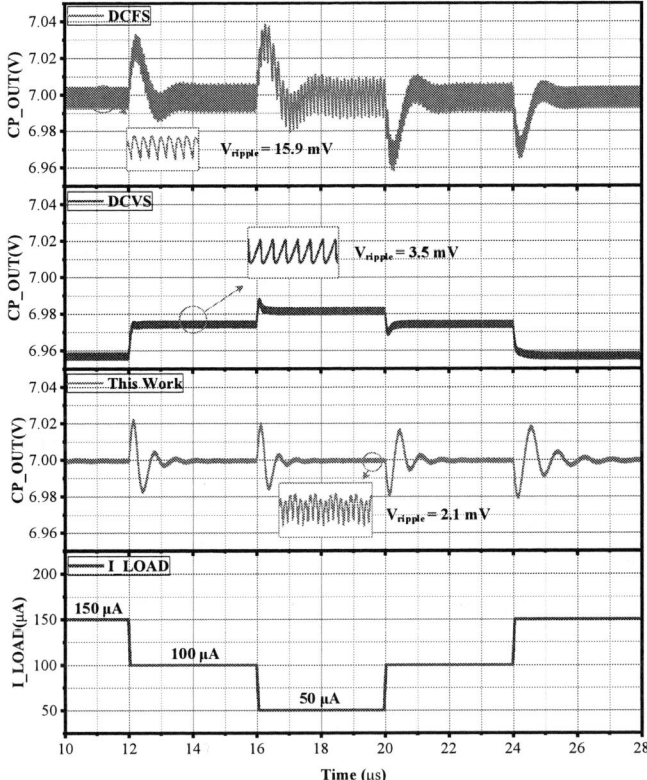

Fig. 6. Comparison of simulation results when the current load chages stepwised.

voltage exhibits negligible variation across all simulated conditions.

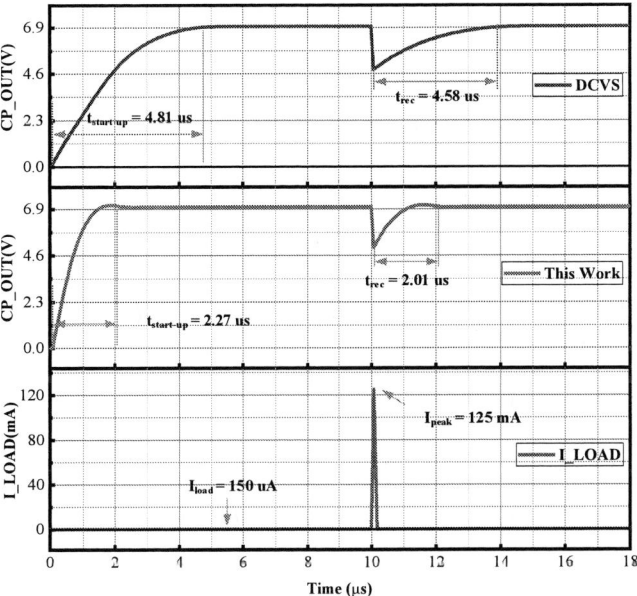

Fig. 7. Comparison of simulation results when the current pulse arrives.

Fig. 8 illustrates the relationship between the charge pump ripple voltage and the load current under the condition of the same filtering capacitance. Simulation results indicate that the output ripple of this charge pump is relatively stable, while the ripple variation of DCFS is noticeably significant. TABLE. I summarizes the performance comparison of this work and some state-of-the-art designs.

Fig. 9 shows the PVT (Process-Voltage-Temperature) simulation results of the transient response. Simulations were performed across the TT (Typical-Typical), SS (Slow-Slow), and FF (Fast-Fast) process corners. The supply voltage was varied by -10%, 0%, and +10% from its nominal value, and the temperature was set to -40°C, 25°C, and 125°C. This comprehensive analysis yielded a total of 27 output waveforms. Simulation results indicate that the minimum recovery time of 1.86 µs occurs under FF process corner, -10% voltage deviation, and -40°C conditions. Conversely, the maximum recovery time is observed under SS process corner, -10% voltage deviation, and 125°C, with a temporal difference of 0.26 µs between these two extremes. Notably, the output

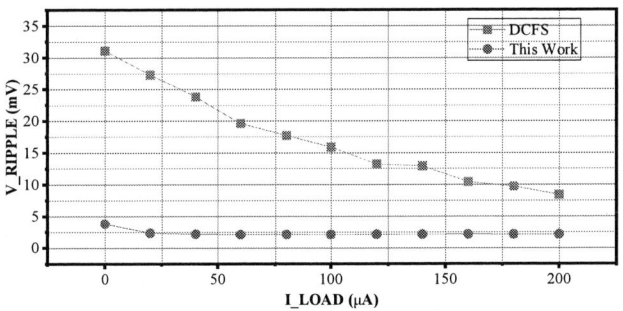

Fig. 8. Relationship between the magnitude of the load current and the ripple voltage

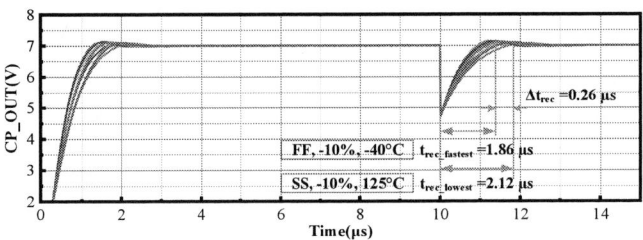

Fig. 9. PVT simulation results of the transient response.

TABLE I. PERFORMANCE COMPARISON

	[7]	[8]	[9]	[10]	This Work
Process (nm)	350	180	180	180	55
V_{in} (V)	2.5	3.3	1.8	1.2	3.3
V_{out} (V)	7.5~16	9.68~12.5	10.5~14	6	7~9
Stage	6	6	10	3	2

	[7]	[8]	[9]	[10]	This Work
Total pump capacitance (pF)	18	5.6	40	N/A	32
Filter capacitance (pF)	80	N/A	50	N/A	200
Clock frequency (MHz)	0.3~30	30.4	10~30	N/A	20~50
V_{ripple} (mV)	18	100	6.9	30	2.1
Maximum load current (μA)	40	10	250	700	350
Start-up time	<20 μs @40 μA	N/A	N/A	N/A	2.27 μs @150 μA 2.9 μs @350 μA

V. CONCLUSION

This paper presents a novel charge pump designed to drive BLs of a flash-based CIM chip, aiming to provide a relatively stable and precise output voltage under harsh current load fluctuations. The design integrates a current compensation module within the charge pump and incorporates a parallel ripple cancellation structure, achieving a relatively stable output voltage under varying current loads. In stable conditions, the ripple peak-to-peak value is approximately 2.1 mV, which is reduced by 86% compared to DCFS. Under a voltage collapse of 2.04 V with a constant current of 150 μA, the recovery time is approximately 2.01 μs. In summary, the charge pump proposed in this paper is suitable for the fast operation of the BL driver in flash-based CIM chips.

REFERENCES

[1] A. Ballo, A. D. Grasso, G. Palumbo, and T. Tanzawa, "Charge Pumps for Ultra-Low-Power Applications: Analysis, Design, and New Solutions," in IEEE Transactions on Circuits and Systems II: Express Briefs, vol. 68, no. 8, pp. 2895-2901, August 2021.

[2] J. Zhang, Z. Huang, Q. Li, X. Zhang, L. Tan, Y. Zhu, H. Wang, and S. Feng, "A High-Efficiency Charge Pump for AMOLED Display Driver IC," 2021 IEEE 14th International Conference on ASIC (ASICON), Kunming, China, 2021, pp. 1-4.

[3] J. Gao, T. Gu, K. Nie, Z. Gao and J. Xu, "A Low-Ripple Charge Pump With Novel Compensator for Transient-Response Improvement in

CMOS Image Sensors," in IEEE Transactions on Circuits and Systems II: Express Briefs, vol. 68, no. 4, pp. 1113-1117, April 2021.

[4] S. Kim, S. Yong, W. Kim, S. Kang, H. Park, K. Yoon, D. Sheen, S. Lee, and C. Hwang, "Review of Semiconductor Flash Memory Devices for Material and Process Issues," Advanced Materials, vol.35, no. 43, p.2200659, October 2023.

[5] Z. Lin, X. Zhong, Z. Yu, Y. Dong, Z. Huang, and X. Gu, "A Novel Multi-Mode Charge Pump in Word Line Driver for Compute-in-Memory Arrays," Electronics, vol.14, no. 1, p.175, January 2025.

[6] C. Wu and C. Chen, "A low-ripple charge pump with continuous pumping current control," 2008 51st Midwest Symposium on Circuits and Systems, Knoxville, TN, USA, 2008, pp. 722-725.

[7] B. Rumberg, D. W. Graham, and M. M. Navidi, "A Regulated Charge Pump for Tunneling Floating-Gate Transistors," in IEEE Transactions on Circuits and Systems I: Regular Papers, vol. 64, no. 3, pp. 516-527, March 2017.

[8] S. Kim, J. Yang, E. Park, J. Choi, and K. Kwon, "A High Efficiency Variable Stage and Frequency Charge Pump for Wide Range ISPP," 2020 IEEE International Symposium on Circuits and Systems (ISCAS), Seville, Spain, 2020, pp. 1-5.

[9] Q. Wang, F. Liu, C. Huang, Q. Li, and Z. Huo, "A Small Ripple and High-Efficiency Wordline Voltage Generator for 3-D nand Flash Memories," in IEEE Transactions on Very Large Scale Integration (VLSI) Systems, vol. 29, no. 11, pp. 1903-1911, November 2021.

[10] C. Tseng, S. Chen, T. Shia, and P. Huang, "An Integrated 1.2V-to-6V CMOS Charge-Pump for Electret Earphone," 2007 IEEE Symposium on VLSI Circuits, Kyoto, Japan, 2007, pp. 102-103.

2025 The 10th International Conference on Integrated Circuits and Microsystems

Interface-Engineered TiO$_2$/SiO$_2$ Stacks Enable Ultralow-Power Phase-Change Memory with Nanosecond-Speed

Ruizhe Zhao[1,2], Jun Zhou[1], Jun Chen[1], Hao Tong[1,3]* and Xiangshui Miao[1,3]

[1] School of Integrated Circuits, Huazhong University of Science and Technology

[2] International school of Materials Science and Engineering (school of Materials and Microelectronics), Wuhan University of Technology

[3] Hubei Yangtze Memory Laboratories

Wuhan, China

tonghao@hust.edu.cn

Abstract—To overcome the inherent SET speed-RESET power compromise in phase-change memory (PCM), we propose a vertically stacked dielectric heterostructure. This architecture synergistically integrates epitaxial TiO$_2$ (reducing crystallization barriers via lattice templating) and thermal-confining SiO$_2$ (intensifying Joule heating within Sb$_2$Te$_3$ through thermal field manipulation). The Sb$_2$Te$_3$-based device achieves simultaneous sub-10-ns SET switching (DRAM-compatible) and picojoule-level programming energy (5.4 pJ). This strategy provides critical methodological insights for deploying PCM in high-speed cache memory.

Keywords—TiO$_2$/SiO$_2$ Stacke, Ultra-low power, Nanosecond-Speed, Sb$_2$Te$_3$ Phase-Change Memory

I. INTRODUCTION

Phase-change memory (PCM) demonstrates significant promise for Storage-Class Memory (SCM) applications, leveraging its non-volatility, high-speed operation, minimal resistance drift, and high density. The technology exploits the dramatic resistivity contrast between crystalline (conductive) and amorphous (insulating) phases in chalcogenide compounds. The RESET operation (amorphization) is driven by short, high-amplitude electrical pulses that rapidly melt the phase-change material, followed by immediate quenching to freeze the disordered atomic structure. In contrast, the SET process (crystallization) requires lower-amplitude, longer-duration pulses to sustain the material within the critical thermal window between crystallization and melting temperatures, enabling ordered lattice reorganization. However, the inherent trade-off between RESET and SET operations poses a fundamental challenge for implementing phase-change memory (PCM) in storage-class memory (SCM) applications.[1].

Previous strategies aimed at boosting operation speed or lowering power consumption—such as reducing structural size [2,3], modifying material composition [4,5], and implementing novel architectures like superlattices [6,7]—predominantly target single parameters or inadequately address the

fundamental speed–power trade-off. To overcome PCM performance limitations, we designed a vertically integrated SiO$_2$/TiO$_2$ dielectric heterostructure for Sb$_2$Te$_3$-based cells. This architecture leverages dual mechanisms: (i) TiO$_2$-induced lattice matching to accelerate crystallization via epitaxial templating; (ii) SiO$_2$-enabled thermal confinement to localize Joule heating within the active Sb$_2$Te$_3$ volume. This cooperative action of lattice matching and thermal confinement enables concurrent breakthrough performance in the Sb$_2$Te$_3$ cell: sub-10-nanosecond switching and picojoule-level programming energy (5.4 pJ), fulfilling essential SCM benchmarks.

II. STRUCTION DESIGN AND FABRIACTION

A. Design of Stacked Dielectric Structure

Within our stacked dielectric architecture, crystalline TiO$_2$ was engineered for crystallographic compatibility with Sb$_2$Te$_3$. This matching promotes epitaxial crystallization templates at the interface, reducing the crystallization activation energy and accelerating the crystallization kinetics. To achieve low-power operation, SiO$_2$ was strategically chosen as the dielectric interlayer based on two key attributes: (i) Its extremely low thermal conductivity [8] generates localized thermal fields within the Sb$_2$Te$_3$ volume, enhancing energy utilization efficiency; (ii) Strong interfacial bonding [9] with TiO$_2$ ensures defect-resistant heterointerfaces, preserving structural integrity during cycling. To further evaluate the thermal confinement characteristics of this stacked dielectric structure, COMSOL Multiphysics finite element simulations were conducted, these simulations confirmed that the peak temperature region is localized at the center of the active structure. This highly focused thermal profile is primarily attributed to the superior thermal insulation provided by the SiO$_2$ interlayer, consistent with its targeted role in energy confinement.

B. Preparation of TiO$_2$/SiO$_2$ Stacked Devices

Leakage current in the TiO$_2$/SiO$_2$ stack was characterized using metal-insulator-metal (MIM) capacitors. The fabrication flow initiated with magnetron-sputtered tungsten bottom electrodes. Subsequent UV lithography patterned micron-scale active regions, followed by TiO$_2$/SiO$_2$ bilayer deposition through ALD and PECVD. Tungsten top electrodes were finally sputtered to complete the test structures, and the corresponding connection diagram for the actual test is depicted in Figure 1b.

This work was supported by the National Natural Science Foundation of China ((NO.62174065), the Natural Science Foundation of Wuhan (NO.2025040601020129) and the Wuhan Young Scientific and Technological Talent Morning Sun Program (Ruizhe Zhao, Jun Zhou and Jun Chen contributed equally to this work.) (Corresponding author: Hao Tong.)

979-8-3315-8850-2/25 $31.00 © 2025 IEEE   858

C. Interface-Engineered Stacked Dielectric PCM Device

A PCM cell with phase change material Sb₂Te₃ was fabricated on a W-coated Si substrate. The schematic diagram of the entire process is shown in Figure 1c. The process utilized a substrate consisting of silicon with a pre-deposited 100 nm tungsten bottom electrode and a 50 nm TiO_2/50 nm SiO_2 dielectric stack. Following standard lithography and etching steps to define via-holes and pattern the layers (detailed in the subsequent Sb₂Te₃ device fabrication), a 100 nm thick Sb₂Te₃ phase-change film and a 100 nm tungsten top electrode were sequentially deposited by magnetron sputtering. A cross-sectional transmission electron microscope (TEM) image in Figure 1a illustrates the finalized Sb₂Te₃ cell structure.

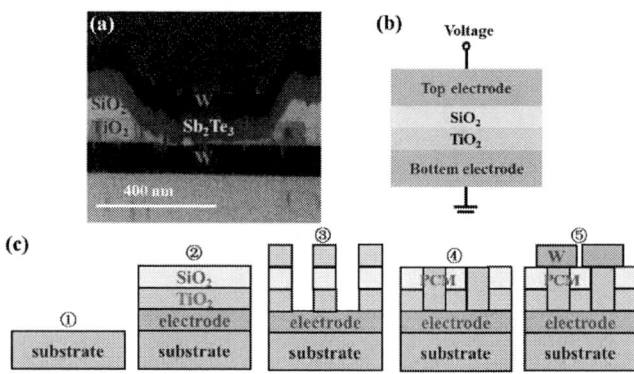

Fig. 1. (a) Cross-sectional TEM image of a stacked-dielectric Sb₂Te₃ device (scale bar: 400 nm); (b) Testing diagram of stacked-dielectric devices; (c) Schematic diagram of the fabrication process for a stacked-dielectric Sb₂Te₃ PCM device.

III. RESULTS AND DISCUSSION

A. Leakage Current Testing of TiO₂/SiO₂ Stacke Materials

In stacked dielectric layer design, dielectric thickness directly modulates electron transition processes, governing the resultant leakage current characteristics. The interfacial leakage mechanisms are dominated by two key parameters: dielectric loss factor (*tanδ*) and leakage current. Consequently, systematic measurements of *tanδ* and leakage current were performed across varying TiO₂/SiO₂ heterostructure thicknesses. To maximize TiO₂'s interface-induced effects, a minimum thickness of 50 nm was maintained, while the SiO₂ interlayer thickness was systematically varied from 10 nm to 50 nm in 20-nm increments.

Firstly, the dielectric loss factor (*tanδ*) is defined as the ratio of the real part of the capacitance to the imaginary part of the capacitance. In this study, the high-frequency capacitance (C_{hf}) value is approximated as the real part of the complex capacitance, while the low-frequency capacitance (C_{lf}) value, serving as the quasi-static characteristic parameter, is approximated as the imaginary part. The dielectric loss factor (tanδ) for varying thicknesses is thereby determined by the ratio of capacitance values measured at different frequencies.

Capacitance measurements were conducted with a voltage sweep from 0 to 0.5 V using the B1500A semiconductor analyzer's Multi-Frequency Capacitance Measurement Unit (MFCMU). Both low-frequency (1 Hz) and high-frequency (100 kHz) capacitance values were simultaneously acquired across SiO₂ layers of varying thicknesses. The *tanδ* values were subsequently calculated for different SiO₂ thicknesses using the aforementioned formulation. As revealed in Figure 2, Comparative analysis reveals that the dielectric loss factor (tanδ) of both 50-nm and 30-nm SiO₂ stacked dielectrics remains consistently near 0.2, whereas the 10-nm SiO₂ stacked dielectric exhibits a significantly higher *tanδ* approaching 0.5. Devices with monocrystalline TiO₂ dielectrics demonstrate unstable dielectric loss factor calculations due to capacitance reading fluctuations, yet their tanδ values are observably larger than those of stacked dielectric devices. This data further demonstrates that the 50-nm SiO₂ stacked dielectric possesses the strongest capability for leakage current suppression.

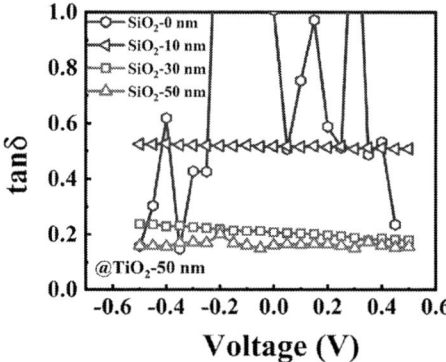

Fig. 2. Comparative measurement of dielectric loss factor in stacked-dielectric devices with different SiO₂ thicknesses.

Furthermore, to characterize the superior low-leakage capability of the 50-nm SiO₂ stacked dielectric, we systematically measured the insulation resistance and leakage current of heterostructures with varying SiO₂ thicknesses, with results presented in Figure 3. These electrical tests were conducted using the DC module of a Keysight B1500A semiconductor parameter analyzer, employing a DC I-V sweep from 0 V to 2 V with a 0.1 V step increment to determine the insulation resistance values and leakage current magnitudes. Figure 3a reveals that a pure TiO₂ dielectric exhibited only 1-2 MΩ resistance, while a 10-nm SiO₂ stacked structure achieved merely 20 MΩ. Given that phase-change materials require operating resistances exceeding 10 MΩ to over 100 MΩ during high-resistance states, dielectric materials must attain at least 100 MΩ to avoid interfering with PCM operations; thus, a minimum SiO₂ thickness of 30 nm is essential. Concurrent leakage current measurements demonstrated a progressive reduction from 15 μA to 7 pA with increasing SiO₂ thickness, confirming that the 50-nm SiO₂ configuration optimally fulfills the low-leakage and low-power design requirements for this study.

979-8-3315-8850-2/25 $31.00 © 2025 IEEE

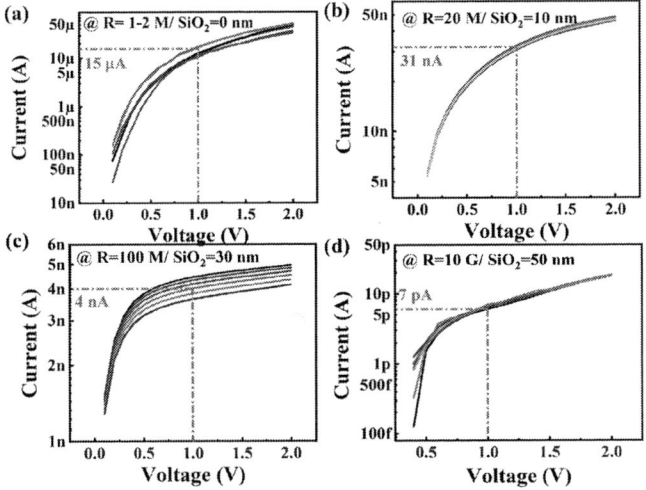

Fig. 3. Leakage current test I-V characteristic curves of stacked-dielectrics with different SiO₂ thicknesses (Voltage range: 0-2 V) : (a) 0 nm; (b)10 nm; (c)30 nm; (d) 50 nm.

B. RESET Power of Stacked Dielectric PCM Device

After optimizing the dielectric interlayer thickness, we characterized the RESET power consumption and SET speed of TiO₂/SiO₂ stacked dielectric PCM devices. Due to the as-deposited amorphous state of the phase-change material via magnetron sputtering, DC I-V sweep operations were employed to initialize it into a functional low-resistance state prior to operation. An initial small voltage of 0.1 V was first applied to read its initial resistance. Subsequently, DC sweeps were performed using currents ranging from 1 to 50 μA in 1 μA steps. No distinct threshold switching behavior was observed within this range. The sweep current was then incrementally increased in 20 μA intervals. This approach served a dual purpose: to identify an appropriate sweep current range and to mitigate potential device damage from excessively large steps.

For the TiO₂/SiO₂ stacked-dielectric Sb₂Te₃ PCM device, a current sweep from 1 to 40 μA revealed a clear threshold switching phenomenon characterized by a sharp voltage drop and the emergence of negative resistance. The device resistance stabilized as the current continued to increase. The corresponding threshold voltage was 2.4-2.7 V, with a threshold current near 15-25 μA, as shown in Figure 4a. Similarly, a distinct threshold switching event occurred when sweeping the purely crystalline TiO₂-based device from 1 to 70 μA. For this device, the threshold voltage was 5.7 V, and the threshold current was approximately 55 μA, shown in Figure 4b. The difference in threshold voltages between the TiO₂/SiO₂ stacked-dielectric and TiO₂ dielectric devices further demonstrates that a significantly higher threshold voltage is required for the pure crystalline TiO₂ device, indicating higher operational power consumption in TiO₂ -based PCM device.

Furthermore, multiple cells of each device structure were subjected to identical DC sweep currents. For the stacked dielectric devices, variations in the thickness uniformity of the TiO₂ and SiO₂ layers resulting from the fabrication process led to non-identical I-V characteristics across different cells.

However, the threshold voltages were consistently distributed around 2.5 V, generally meeting the requirement for device uniformity. In contrast, the crystalline TiO₂ devices exhibited nearly overlapping I-V curves and consistent distributions of threshold voltage and current due to the absence of such thickness variation issues, confirming their superior device-to-device consistency.

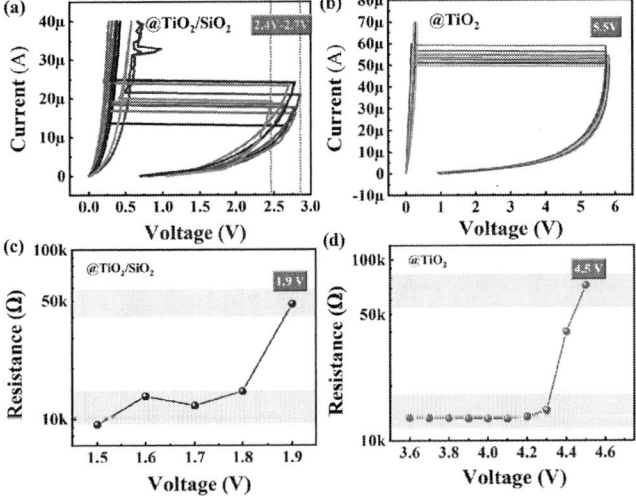

Fig. 4. (a) DC I-V characteristic curves during SET process in TiO₂/SiO₂ stacked dielectric PCM devices; (b) DC I-V characteristic curves during SET process in TiO₂ dielectric PCM devices; (c) R-V characteristic curves during RESET process in TiO₂/SiO₂ stacked dielectric PCM devices (1.5 to -2 V); (d) R-V characteristic curves during RESET process in TiO₂ dielectric PCM devices (3.5 to -4.5 V).

Subsequently, using square-wave pulses with 8 ns rise/fall times and 10 ns pulse width, the RESET operations were performed on Sb₂Te₃-based PCM devices by incrementally increasing voltage amplitude in 0.1 V steps. For devices with TiO₂/SiO₂ stacked dielectric structures, successful amorphization (achieving >10× resistance window between high-resistance state [HRS] and low-resistance state [LRS]) was completed at 1.9 V (Figure 4c). In contrast, TiO₂ dielectric devices required voltages exceeding 4.5 V for equivalent RESET operation (Figure 4d). This 58% reduction in operating voltage directly demonstrates the multilayer dielectric design's efficacy in slashing dynamic power consumption during phase transition.

The RESET power consumption was calculated as follows:

$$E_{RESET} = I \times U \times \Delta t = \frac{U_{RESET}^{2} \times \Delta t}{R_{SET}} \qquad (1)$$

Average resistance during RESET (R_{SET}) ≈ 10 kΩ; Pulse width (Δt) = 18 ns (full width at half maximum). TiO₂/SiO₂ stacked devices exhibited a RESET energy of ~5.4 pJ. Single-crystal TiO₂ devices required 64.98 pJ, indicating a 12-fold reduction in energy consumption (Figure 5). For a 500-nm via-hole diameter, this corresponds to a switching power density of ~0.046 MW/cm².

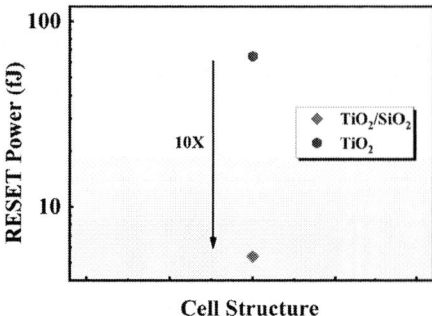

Fig. 5. Comparison of RESET power consumption between TiO_2 dielectric and TiO_2/SiO_2 stacked dielectric based Sb_2Te_3 PCM device.

C. SET Speed of Stacked Dielectric PCM Device

To probe the ultimate SET speed, we employed a progressive pulse-width reduction methodology by applying electrical pulses with progressively narrowed pulse durations, thereby identifying the phase-transition speed limit. Comparative analysis of SET kinetics revealed stark performance differences between TiO_2/SiO_2 stacked dielectric and SiO_2-based PCM devices. As quantified in Figure 6a, SiO_2-based PCM devices achieved a maximum SET speed of 38 ns, with significantly degraded operational margins at 28 ns pulse widths indicating failure to complete orderly crystallization transitions. In contrast, stacked dielectric heterostructures attained ultrafast 8 ns SET operations (Figure 6b), where statistical analysis of cell populations demonstrated a dominant 8 ns switching cluster with minimal outliers beyond 10 ns, whereas SiO_2 controls exhibited broadly distributed speeds (38-58 ns). This 5× acceleration in crystallization kinetics unambiguously demonstrates the efficacy of the TiO_2 interlayer's epitaxial templating mechanism in significantly enhancing phase-transition dynamics. Comparative analysis with state-of-the-art research (Table 1) reveals that the stacked-dielectric devices exhibit competitive performance in both SET speed and operational power consumption.

To validate the structural reliability of the stacked dielectric design, this study further conducted cycling endurance tests on the devices. The SET and RESET operational pulses maintained identical parameters to those specified previously: (i) SET operation: 0.8 V amplitude, 8 ns pulse width; (ii) RESET operation: 1.9 V amplitude, 18 ns pulse width. The devices demonstrated robust cycling endurance exceeding 10^6 cycles (Figure 7), meeting the operational requirements for storage-class memory (SCM) applications.

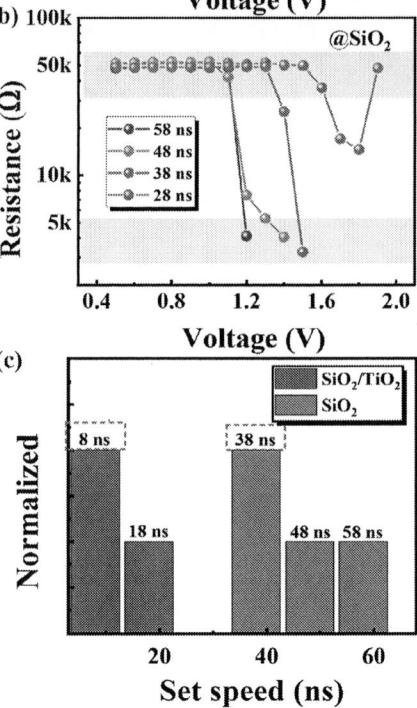

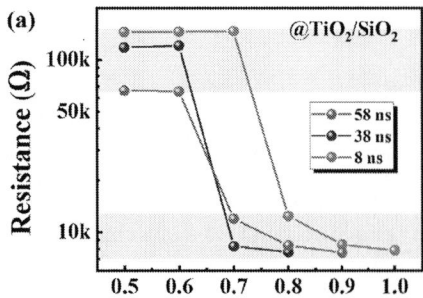

Fig. 6. The SET resistance-voltage characteristics of (a) TiO_2 dielectric, (b) SiO_2 dielectric based Sb_2Te_3 PCM devices. The fastest SET speed for TiO_2 dielectric, SiO_2 dielectric and TiO_2/SiO_2 stacked layers dielectric based Sb_2Te_3 PCM devices is 8 ns under ~0.8 V bias and 38 ns under ~1.5V bias, respectively. (c) The normalized data for Sb_2Te_3-based PCM devices with TiO_2/SiO_2 and SiO_2 dielectric layer.

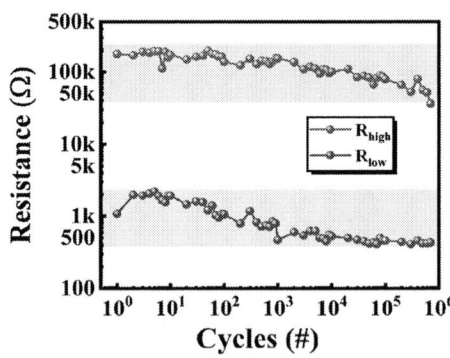

Fig. 7. Endurance testing of PCM devices with stacked dielectric structures

TABLE I.   PERFORMANCE COMPARISON OF SB-TE PCM

Structure	Year	Journal	SET Speed	Current (Power) Density
$GeTe/Sb_2Te_3$	2022	IEEE EDL[10]	120 ns	3 ~ 4 MA/cm²
$Ge_2Sb_2Te_5/Sb_2Te_3$	2022	IEEE VLSI[7]	N.A.	2.5 ~ 3 MA/cm²
$Ge_1Sb_4Te_7/Sb_2Te_3$	2024	Nat. Commun. [6]	40 ns	5 MW/cm²
Sb_2Te_3-TiO_2/SiO_2	2025	IEEE ICICM	8 ns	0.046 MW/cm²

979-8-3315-8850-2/25 $31.00 © 2025 IEEE

IV. Conclusion

Our work resolves the fundamental SET-speed/RESET-power trade-off in PCM through a TiO_2/SiO_2 stacked dielectric architecture. Comprehensive characterization guided by dielectric loss factor analysis and leakage current measurements identified 50 nm as the optimized thickness for exceptional leakage suppression. The TiO_2/SiO_2 stacked dielectric -based PCM devices fabricated at this thickness demonstrated record-setting performance: 8 ns switching speed with ultralow 5.4 pJ programming energy. These breakthroughs establish critical design guidelines for implementing PCM in SCM applications.

References

[1] S. W. Fong, C. M. Neumann, and H. S. P. Wong, "Phase-Change Memory—Towards a Storage-Class Memory," *IEEE Trans. Electron Devices,* vol. 64, no. 11, pp. 4374-4385, 2017. DOI: 10.1109/ted.2017.2746342

[2] W. J. Wang, D. Loke, L. T. Law, L. P. Shi, and A. L. Lacaita, "Engineering grains of $Ge_2Sb_2Te_5$ for realizing fast-speed, low-power, and low-drift phase-change memories with further multilevel capabilities," in *IEEE International Electron Devices Meeting (IEDM),* San Francisco, CA, USA, 2012, pp. 733-736, San Francisco: Proceedings of the IEEE

[3] S.-O. Park, S. Hong, S.-J. Sung, D. Kim, S. Seo, H. Jeong, T. Park, W. J. Cho, J. Kim, and S. Choi, "Phase-change memory via a phase-changeable self-confined nano-filament," *Nature,* vol. 628, no. 8007, pp. 293-298, 2024. DOI: 10.1038/s41586-024-07230-5

[4] F. Rao, K. Ding, Y. Zhou, Y. Zheng, M. Xia, S. Lv, Z. Song, S. Feng, I. Ronneberger, R. Mazzarello, W. Zhang, and E. Ma, "Reducing the stochasticity of crystal nucleation to enable subnanosecond memory writing," *Science,* vol. 358, no. 6369, pp. 1423-1427, 2017.

[5] Z. Yang, B. Li, J. Wang, X. Wang, M. Xu, H. Tong, X. Cheng, L. Lu, C. Jia, M. Xu, X. Miao, W. Zhang, and E. Ma, "Designing Conductive-Bridge Phase-Change Memory to Enable Ultralow Programming Power," *Adv. Sci.,* vol. 9, no. 8, p. 2103478, Jan 14 2022. DOI: 10.1002/advs.202103478

[6] X. Wu, A. I. Khan, H. Lee, C.-F. Hsu, H. Zhang, H. Yu, N. Roy, A. V. Davydov, I. Takeuchi, X. Bao, H. S. P. Wong, and E. Pop, "Novel nanocomposite-superlattices for low energy and high stability nanoscale phase-change memory," *Nat. Commun.,* vol. 15, no. 1, p. 13, 2024. DOI: 10.1038/s41467-023-42792-4

[7] A. Intisar Khan, C. Perez, X. Wu, B. Won, K. Kim, H. Kwon, P. Ramesh, K. M. Neilson, M. Asheghi, K. Saraswat, Z. Lee, I.-K. Oh, H. S. Philip Wong, K. E. Goodson, and E. Pop, "First Demonstration of $Ge_2Sb_2Te_5$ -Based Superlattice Phase Change Memory with Low Reset Current Density (~3 MA/cm²) and Low Resistance Drift (~0.002 at 105°C)," in *Symposium on VLSI Technology*, Honolulu, HI, USA, 2022, pp. 310-311, Honolulu: Proceedings of the IEEE. DOI: 10.1109/VLSITechnologyandCir46769.2022.9830348

[8] C. Hu, M. Morgen, P. S. Ho, A. Jain, W. N. Gill, J. L. Plawsky, and P. C. Wayner, "Thermal conductivity study of porous low-k dielectric materials," *Appl. Phys. Lett.,* vol. 77, no. 1, pp. 145-147, 2000. DOI: 10.1063/1.126904

[9] J. Wang, X. Huang, H. Zhang, L. Wang, W. Huang, S. Kuang, and F. Huang, "Diamond(001)–Si(001) and Si(001)–Ti(0001) interfaces: A density functional theory study," *J. Phys. Chem. Solids,* vol. 150, 2021. DOI: 10.1016/j.jpcs.2020.109865

[10] A. I. Khan, H. Kwon, M. E. Chen, M. Asheghi, H. S. P. Wong, K. E. Goodson, and E. Pop, "Electro-Thermal Confinement Enables Improved Superlattice Phase Change Memory," *IEEE Electron Device Lett.,* vol. 43, no. 2, pp. 204-207, 2022. DOI: 10.1109/led.2021.3133906

2025 The 10th International Conference on Integrated Circuits and Microsystems

Hafnium-Based Ferroelectric Diode with Interlayer Enhancement for In-Memory Logic Application

Shuo Han
The College of Electronic Science and Technology National University of Defense Technology
Changsha, China
hanshuo19@nudt.edu.cn

Chuanzhi Liu
The College of Electronic Science and Technology National University of Defense Technology
Changsha, China
liuchuanzhi@nudt.edu.cn

Qimiao Zeng
The College of Electronic Science and Technology National University of Defense Technology
Changsha, China
zqm0815@email.swu.edu.cn

Yefan Zhang
The College of Electronic Science and Technology National University of Defense Technology
Changsha, China
zhangyefan18@nudt.edu.cn

Wei Wang
The College of Electronic Science and Technology National University of Defense Technology
Changsha, China
wangwei_esss@nudt.edu.cn

Rongrong Cao*
The College of Electronic Science and Technology National University of Defense Technology
Changsha, China
caorongrong@nudt.edu.cn

Abstract—Logic-in-Memory computing (LiM) has emerged as a key approach to overcome the von Neumann bottleneck, and hafnium-based ferroelectric diode (Fe diode) is a promising candidate for this application. For further performance optimization, we fabricated Fe diode devices with two structures: TiN/HZO/TiN and interlayer-enhanced TiN/HZO/HfO$_2$/TiN. The HfO$_2$ interlayer devices exhibit improved retention characteristics while maintaining bidirectional rectification behavior and a well-defined logic window, enabling stable logic operations. We demonstrated a circuit scheme of implementing both IMP and NIMP logic functions in a single operation with an ultra-low operating power consumption of 20.65 aJ. Additionally, by configuring different device combinations, all 16 Boolean logic functions can be realized based on Fe diodes. This work highlights the potential of interlayer-enhanced Fe diodes, offering a simple, easy-to-integrate, and low power consumption approach for developing high-performance LiM systems.

Keywords—Ferroelectric diodes; interlayer engineering; logic-in-memory; bidirectional rectification

I. INTRODUCTION

The rapid development of artificial intelligence and big data has driven exponential increase in data processing demands, exposing the inherent limitations of traditional Von Neumann computing architecture, specifically the "Von Neumann bottleneck" caused by the separation of memory and processing units [1, 2]. Logic-in-Memory (LiM) computing has emerged as a promising paradigm by integrating data memory and logic operations within the same device thereby significantly reducing

power consumption and latency associated with data movement [3-5].

Among various device candidates, ferroelectric diode (Fe diode) emerges as a promising solution for next-generation low power consumption LiM computing systems. This two-terminal device boasts a simple structure and self-rectification characteristics, as well as the capacity to support high-density crossbar array memory technology without the need for selectors [4, 6]. Despite the extensive utilization of various ferroelectric materials in electronic devices, the poor compatibility of conventional ferroelectric materials with CMOS processes and the difficulty of scaling down the size pose challenges for their application in advanced computing architectures. Conventional ferroelectric materials can only maintain stable remanent polarization (P$_r$) and low leakage current at thicknesses of several dozen nanometers or greater. Reducing the thickness leads to unstable polarization states, and the film thickness limits the scalability of conventional ferroelectric devices. Additionally, conventional ferroelectric materials require high fabrication and annealing temperatures and contain polluting elements such as Pb and Bi, making them incompatible with CMOS processes [3]. In contrast, hafnium-based ferroelectric materials discovered for the first time in 2011 that exhibit ferroelectricity at room temperature hold promise for addressing these challenges [7].

Hafnium-based ferroelectric materials have been demonstrated to maintain a stable polarization state at a thickness as low as 10 nm. Furthermore, hafnium oxide, as a high-dielectric-constant material, has been extensively utilized in processor applications and is already extensively employed in Si-based CMOS devices, with mature material properties and fabrication processes [8-10]. Additionally, the intrinsic polarization characteristics of hafnium-based ferroelectric materials are capable of enhancing switching speed and

---

This work was supported by the National Natural Science Foundation of China under Grant Nos. 62104256, 62304254, U23A20322, National Key R&D Program of China under Grant No. 2019YFB2205102, and the science and technology innovation Program of Hunan Province 2023RC3015. (Shuo Han and Chuanzhi Liu contributed equally to this work) (Corresponding Author: Rongrong Cao).

979-8-3315-8850-2/25 $31.00 © 2025 IEEE

reducing power consumption of logic operation, aligning with the core requirements of LiM systems [11, 12]. However, hafnium-based Fe diodes still have potential for performance enhancement in LiM applications, such as the retention characteristics and memory window stability. Interface engineering can be an efficient approach by introducing functional transition layers at the electrode/ferroelectric interface to control the interface band and suppress charge injection [13-16]. Moreover, extant research on Fe diodes-based logic circuit primarily focuses on utilizing voltage signals as the main input for logic operations [13, 17]. Due to the lack of internal storage capabilities to preserve input states, external memory units or signal conversion circuits must be integrated to enable logic functions, increasing system complexity and energy consumption.

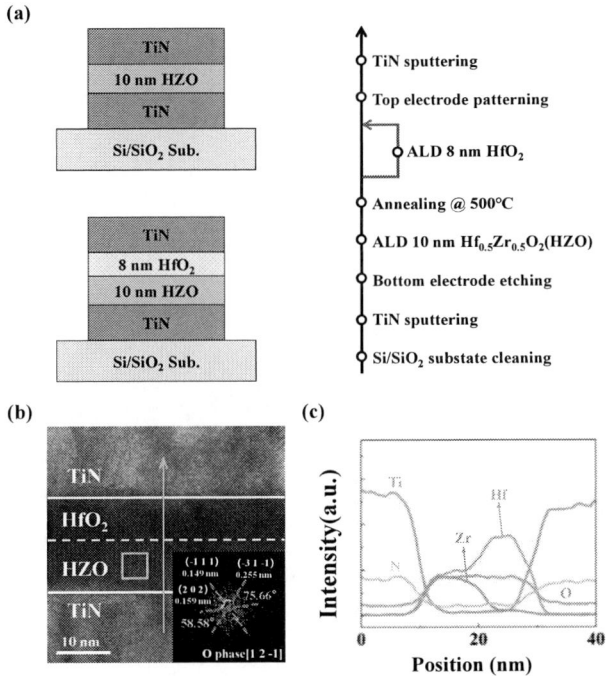

Fig. 1. (a) Schematic diagrams and fabrication process flow of the Fe diode devices. (b) High-resolution transmission electron microscopy (HRTEM) image of the HfO$_2$ interlayer device, with the inset showing the Fast Fourier transform (FFT) pattern. (c) Energy-dispersive X-ray spectroscopy (EDS) line-scan image of the HfO$_2$ interlayer device.

In this work, we proposed hafnium-based Fe diodes with interlayer enhancement for LiM applications. Fe diode devices with two structures were fabricated, the TiN/HZO/TiN structure as a control sample and an interlayer-enhanced TiN/HZO/HfO$_2$/TiN structure. An investigation was conducted into the effect of the HfO$_2$ interlayer on device performance, leading to the identification of enhancements in characteristics such as retention and memory window performance. Then we demonstrated the Fe diode-based logic circuit scheme where both input and output variables are directly represented by resistance states, avoiding additional signal conversion. We implemented both IMP (implication) and NIMP (negative implication) logic functions in a single circuit with an ultra-low operating energy of 20.65 aJ. Additionally, by configuring

different device combinations, all 16 Boolean logic functions can be realized. This work highlights the potential of Fe diode with interlayer enhancement for high-performance, low-power and low-complexity LiM systems.

II. EXPERIMENTS

Fig. 1(a) schematically illustrates the structure and fabrication process of the Fe diode devices proposed in this work. The devices consist of two structures, one of which is TiN/HZO/TiN as a control sample, and the other is TiN/HZO/HfO$_2$/TiN. The Si/SiO$_2$ substrate was sequentially cleaned with acetone, ethanol and deionized water. The TiN bottom electrode (BE) was then deposited via magnetron sputtering, followed by patterning through UV lithography and etching. A 10 nm thick HZO film was deposited at 280 °C by atomic layer deposition (ALD), where the cycle ratio of HfO$_2$:ZrO$_2$ was maintained at 1:1. TEMAH, TEMAZ, and H$_2$O were used as precursors for Hf, Zr, and O, respectively. The samples were subjected to rapid thermal annealing (RTA) at 500 °C for 30s in a N$_2$ atmosphere to induce the formation of a ferroelectric phase. Subsequently, an 8 nm thick HfO$_2$ layer was deposited by ALD. Then the TiN were deposited by magnetron sputtering and patterned as top electrode (TE).

III. RESULTS AND DISCUSSION

The high-resolution transmission electron microscopy (HRTEM) image of the cross-section of the HfO$_2$ interlayer device is shown in Fig. 1(b), revealing clear layer interfaces and distinct crystal lattice of the HZO film. The Fast Fourier transform (FFT) pattern of the green boxed region is shown in the inset, and analysis confirms the orthorhombic (o) phase of the HZO layer, indicating ferroelectric properties [4, 8, 18]. Additionally, we analyzed the elemental distribution across layers of the device along the green scan path in Fig. 1(b) using energy-dispersive X-ray spectroscopy (EDS). This analysis verified the device's basic structure and confirmed the interface thickness observed in Fig. 1(a).

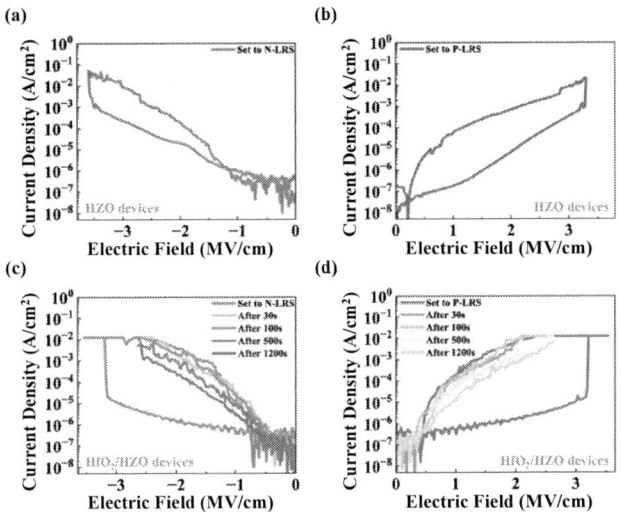

Fig. 2. Current density-electric field (J-E) curves of single HZO device under (a) negative bias and (b) positive bias. Current density-electric field (J-E) curves of HfO$_2$ interlayer device under (c)negative bias and (d) positive bias.

979-8-3315-8850-2/25 $31.00 © 2025 IEEE

Fig. 2(a)–(d) illustrate the current density–electric field (J–E) curves obtained by direct current (DC) sweep for Fe diode devices with single HZO and HfO$_2$/HZO stack. In Fig. 2(a) and (b), under the applied electric field, the single HZO device transitions from high-resistance state (HRS) to low-resistance state (LRS), exhibiting a memory window which arises from bidirectional rectification behavior. The red curve represents the LRS under positive bias (P-LRS), while the blue curve corresponds to the N-LRS under negative bias. Fig. 2(c) and (d) show the scanning results for the HfO$_2$ interlayer device. Conversely, the device with the interlayer exhibits a distinct memory window. Subsequently, to assess the retention characteristics of the HfO$_2$ interlayer device, after setting devices to the LRS, we conducted additional scans at varying time intervals following their configuration to the LRS. The scan curves after 30s, 100s, and 500s in Fig. 2(c) and (d) closely resembled the curves in the LRS, while the curve after 1200s exhibits a slight decline, indicating the reliable retention characteristic of the interlayer device for LiM applications. Consequently, it can be substantiated that the HfO$_2$ interlayer enhances device performance [13]. Then the electrical characterization was conducted on the TiN/HZO/HfO$_2$/TiN device. The polarization-electric field (P–E) loop shown in Fig. 3(a) was obtained via positive-up negative-down (PUND) measurements, which confirms the ferroelectric behavior of the device. The relatively low P$_r$ values observed are attributed to the absence of a capping metal during HZO annealing. Without the compressive stress imposed by a capping layer, the HZO film likely exhibits a lower fraction of ferroelectric orthorhombic phase, thereby lowering the P$_r$. Despite this, omitting the capping layer simplifies the fabrication process by eliminating the need for capping layer deposition and removal steps, while still enabling sufficient crystallization and ferroelectricity in the HZO film.

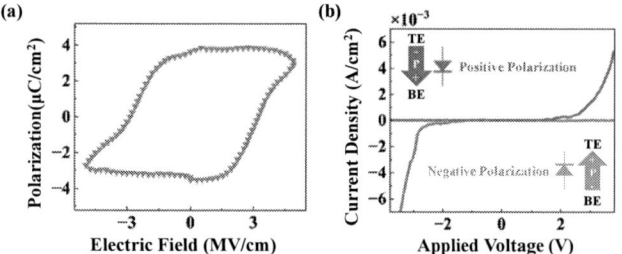

Fig. 3. (a) Polarization-electric field (P–E) loop of the HfO$_2$ interlayer device. (b) Bidirectional rectification characteristics of the HfO$_2$ interlayer device.

As previously mentioned, Fe diode devices exhibit two low-resistance states: P-LRS and N-LRS. Therefore, after the device transitions from the high-resistance state under positive bias (P-HRS) to P-LRS, a positive scanning voltage was applied. The red curve in Fig. 3(b) illustrates the diode characteristics with unidirectional rectification of the device under positive polarization. When it was set to N-LRS, the rectification direction switched under negative scanning voltage. The corresponding polarization direction and diode rectification direction are illustrated by the blue curve in Fig. 3(b). The device exhibited switchable diode characteristics, indicating that the current transmission direction is programmable. Notably, the device's coercive field (E$_c$) is comparable to the electric field of

resistive state transitions. This overlap suggests that resistive switching in the device may be regulated by the ferroelectric polarization of HZO. The polarization charge generated by the polarization of the ferroelectric film affects the interface barrier, thereby causing changes in the energy band and resulting in different resistive states.

From the perspective of energy bands, when the HZO layer undergoes polarization, its energy bands bend according to the direction of the polarization. With no applied voltage, the conduction band energy level on the polarization direction side remains elevated, forming an electronic potential barrier that hinders electron injection. Meanwhile, the positively polarized charges on the opposite polarization direction side attract electrons, causing them to spontaneously accumulate at the interface. When a voltage is applied in the same direction as polarization, the height of this barrier is reduced, allowing electrons to rapidly inject into the region opposite the polarization direction, thereby generating a significant current and exhibiting LRS. Conversely, when the voltage is applied in the opposite direction to the polarization, the amorphous structure of the HfO$_2$ layer increases the barrier height, suppressing leakage current and enhancing barrier isolation. The elevated barrier significantly impedes electron injection, resulting in extremely low conductive current to maintain the stability of the memory window and retention characteristics of the device [13].

Based on the bidirectional rectification characteristics of Fe diode devices, we proposed circuit schemes for Boolean logic computation. As shown in Fig. 4(a), the circuit structure can realize IMP and NIMP logic operations in one step. In this scheme, the resistive states (LRS/HRS) of devices A and B serve as the two logic input variables. The TEs of device A and device B are connected to the same word line (WL), and their BEs are connected to different bit lines (BL). By applying excitation voltage V to the BL of device A and ground to the BL of device B, a series circuit was formed. Fig. 4(b) illustrates the applied voltage scheme in the crossbar array, and the BLs of non-selected devices were biased with V/2 to remain in their initial state without interfering with the target series circuit.

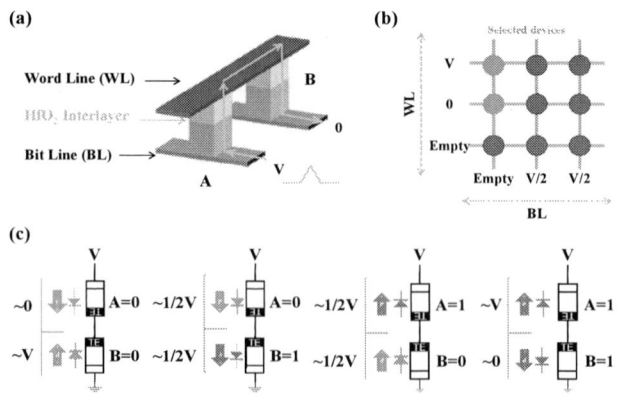

Fig. 4. (a) Schematic diagram of the logic circuit scheme based on Fe diode. (b) The scheme of applied voltage in the crossbar array. (c) Equivalent circuit diagrams of the logic circuit under different inputs.

Fig. 4(c) shows the equivalent circuit diagram of the series circuit under different input combinations. We define the direction from the TE to the BE as the direction of logic value read. When the device is under positive polarization, it exhibits LRS along the read direction, representing logic 1. When the device is under negative polarization, it exhibits HRS along the read direction, representing logic 0. Therefore, when input A = 0 and B = 0, the equivalent diode of device A conducts in the direction of the excitation voltage V, exhibiting LRS in the logic write circuit. However, the conduction direction of device B is opposite to the excitation voltage, exhibiting HRS in the circuit. Therefore, the voltage of the two devices in the circuit is ~0 and ~V, respectively. When input A = 0 and B = 1, both device A and device B exhibit LRS in the circuit, with equal voltage division, 1/2 V. Similarly, when input A = 1 and B = 0, both devices exhibit HRS in the circuit, with the same voltage of 1/2 V. When A = 1 and B = 1, the conduction direction of device A is opposite to the excitation voltage, exhibiting HRS, while device B exhibits LRS, resulting in voltage divisions of ~V and ~0, respectively.

Subsequently, the logic circuit scheme was validated under different input conditions through simulations. After applying excitation voltage to the logic write circuit, the voltage-current curves of the input/output states and the corresponding current responses of circuit are illustrated in Fig. 5.

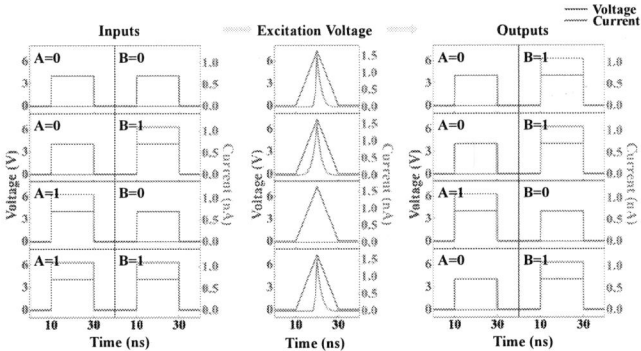

Fig. 5. Voltage-current curves of the logic write circuit.

When inputs A and B were both set to 0, two devices exhibited LRS and HRS respectively. In the series circuit, the voltage of device B was ~V, which exceeded the resistance state transition threshold. Consequently, the state of B switched, and the circuit exhibited a steep current peak, as shown in the first line of Fig. 5. Similarly, if A = 0 and B = 1, both devices maintained the LRS. The voltage of device A and device B was ~ 1 / 2V. Since the voltage was below the transition threshold, the input logic value did not change, and the current response of the circuit was a steadily rising current peak. Conversely, when A = 1 and B = 0, the two devices were in HRS in the circuit, resulting in an equivalent voltage. The current response was minimal. Finally, when both devices were configured to 1, the voltage of device A was ~V, resulting in the state change and the steep current peak. The results of these simulations consistently validate the logic operation mechanism proposed in the circuit design. A comprehensive analysis of input-output states confirms that this circuit design enables implementation of IMP and NIMP logic in a single operation, with results stored

in devices B and A, respectively. After simulation, the average power consumption of four input conditions was as low as 20.65 aJ. Furthermore, all sixteen Boolean logic functions can be realized using Fe diodes-based circuit. As shown in Table I, in comparison to other work, the Fe diode-based logic cell proposed in this work features a simple structure without additional devices to be cascaded. In addition, power consumption of the logic scheme is as low as the aJ level. The input data is stored as resistance values within the device, enabling non-volatile storage without requiring additional form conversion. Fe diode devices in this work inherently offer faster response speed and ultra-low power consumption, providing core technological support for next-generation low-power LiM applications.

TABLE I. PERFORMANCE COMPARISON OF VARIOUS POTENTIAL LiM DEVICES

Logic cell structure	TiN/HZO/TiN/ HfO$_2$/IGZO	Ti/HfSe$_x$O$_y$/ HfSe$_2$/Au	TiN/HfO$_x$/Ti/ TiN/Ti/HfO$_x$/ TiN	TiN/HZO/ HfO$_2$/TiN
Boolean Logic Functions Realized	14	3 (XOR, IMP, NAND)	16	16
Power Consumption	aJ level	fJ level	pJ level	aJ level
Input Data Storage in Device	No (input in the form of voltage)	Yes	Hybrid (in the form of voltage and resistance)	Yes
Additional Device Required	Yes (with n-type load transistor)	No	Yes (with load resistance)	No
Reference	[17]	[19]	[20]	This work

IV. CONCLUSION

In summary, this work has improved the performance of hafnium-based Fe diodes in LiM applications through interface engineering and proposed logic circuit schemes. By comparing Fe diode devices with TiN/HZO/TiN structure and TiN/HZO/HfO$_2$/TiN structure, enhancement in data retention characteristics and memory window stability were observed in the HfO$_2$ interlayer devices, which is key advancement for achieving stable LiM operations. Based on the device with interlayer enhancement, we proposed a logic circuit scheme, facilitating parallel computation of IMP and NIMP logic functions with a power consumption of ~20.65aJ. Furthermore, different circuit combinations can achieve scalabilities for all 16 Boolean operations. These results validate interface engineering as a viable strategy for enhancing the performance of hafnium-based Fe diodes, providing insights for next-generation low-power LiM computing architectures.

REFERENCES

[1] K. Kim, I. Karpov, R. H. Olsson, and D. Jariwala, "Wurtzite and fluorite ferroelectric materials for electronic memory," Nature Nanotechnology, vol. 18, pp. 422-441, 2023.

[2] D. Ielmini and H. S. P. Wong, "In-memory computing with resistive switching devices," Nature Electronics, vol. 1, pp. 333-343, 2018-06-13 2018.

[3] H. P. Wong and S. Salahuddin, "Memory leads the way to better computing," Nat Nanotechnol, vol. 10, pp. 191-4, 2015-03-01 2015.

[4] Q. Luo, Y. Cheng, J. Yang, R. Cao, H. Ma, Y. Yang, R. Huang, W. Wei, Y. Zheng, T. Gong, J. Yu, X. Xu, P. Yuan, X. Li, L. Tai, H. Yu, D. Shang, Q. Liu, B. Yu, Q. Ren, H. Lv, and M. Liu, "A highly CMOS compatible

hafnia-based ferroelectric diode," Nat Commun, vol. 11, p. 1391, 2020-03-13 2020.

[5] Z. Feng, Z. Wu, J. Zou, L. Cheng, X. Zhao, X. Zhang, J. Lu, C. Wang, Y. Wang, H. Wang, W. Guo, Z. Qian, Y. Zhu, Z. Xu, Y. Dai, and Q. Liu, "Memristive Bellman solver for decision-making," Nature Communications, vol. 16, p. 4925, 2025-05-27 2025.

[6] K. Kim, Z. Han, Y. Zhang, P. Musavigharavi, J. Zheng, D. K. Pradhan, E. A. Stach, R. H. Olsson, and D. Jariwala, "Multistate, Ultrathin, Back-End-of-Line-Compatible AlScN Ferroelectric Diodes," ACS Nano, vol. 18, pp. 15925-15934, 2024-06-18 2024.

[7] T. S. Böscke, J. Müller, D. Bräuhaus, U. Schröder, and U. Böttger, "Ferroelectricity in hafnium oxide thin films," Applied Physics Letters, vol. 99, p. 102903, 2011-09-05 2011.

[8] J. Muller, T. S. Boscke, U. Schroder, S. Mueller, D. Brauhaus, U. Bottger, L. Frey, and T. Mikolajick, "Ferroelectricity in Simple Binary ZrO_2 and HfO_2," Nano Lett, vol. 12, pp. 4318-23, 2012-08-08 2012.

[9] J. Muller, T. S. Boscke, S. Muller, E. Yurchuk, P. Polakowski, J. Paul, D. Martin, T. Schenk, K. Khullar, A. Kersch, W. Weinreich, S. Riedel, K. Seidel, A. Kumar, T. M. Arruda, S. V. Kalinin, T. Schlosser, R. Boschke, R. van Bentum, U. Schroder, and T. Mikolajick, "Ferroelectric hafnium oxide: A CMOS-compatible and highly scalable approach to future ferroelectric memories," in 2013 IEEE International Electron Devices Meeting (IEDM): IEEE, 2013, pp. 10.8.1-10.8.4.

[10] M. H. Park, Y. H. Lee, H. J. Kim, Y. J. Kim, T. Moon, K. D. Kim, J. Müller, A. Kersch, U. Schroeder, T. Mikolajick, and C. S. Hwang, "Ferroelectricity and Antiferroelectricity of Doped Thin HfO_2 - Based Films," Advanced Materials, vol. 27, pp. 1811-1831, 2015.

[11] Z. Li, T. Wang, J. Yu, J. Meng, Y. Liu, H. Zhu, Q. Sun, D. W. Zhang, and L. Chen, "Ferroelectric Hafnium Oxide Films for In-Memory Computing Applications," Advanced Electronic Materials, vol. 8, p. 2200951, 2022-12-01 2022.

[12] K. Lee, S. Oh, H. Jang, S. Lee, B. Lee, and H. Hwang, "Variability Analysis and Improvement Strategies for Nanoscale Ferroelectric

[13] $Hf_{0.5}Zr_{0.5}O_2$ Utilizing Schottky Emission Current in Switchable Diode," IEEE Electron Device Letters, vol. 45, pp. 2078-2081, 2024.

R. Kao, H. Peng, K. Chen, and Y. Wu, "$HfZrO_x$ -Based Switchable Diode for Logic-in-Memory Applications," IEEE Transactions on Electron Devices, vol. 68, pp. 545-549, 2021.

[14] S. Yu, Q. Wang, Y. Zhang, P. Yang, X. Luo, H. Liu, C. Chen, Q. Li, and S. Liu, "Multistate Capability Improvement of BEOL Compatible FeFET by Introducing an Al_2O_3 Interlayer," IEEE Transactions on Electron Devices, vol. 70, pp. 5632-5637, 2023.

[15] L. Jung, S. Oh, H. Jang, K. Lee, W. Choi, and H. Hwang, "Enhanced ON/OFF Ratio (4×10^5) and Robust Endurance ($> 10^{10}$) in an $InGaZnO/Hf_xZr_{1-x}O_2$ Ferroelectric Diode via Defect Engineering," IEEE transactions on electron devices, vol. 71, pp. 2238-2242, 2024-01-01 2024.

[16] I. A. Savichev, I. G. Margolin, R. I. Romanov, and A. A. Chouprik, "Role of Ferroelectric Layer Thickness in Resistive Switching and Depolarization Effects in $Hf_{0.5}Zr_{0.5}O_2$-Based Structures," IEEE Transactions on Electron Devices, vol. 72, pp. 1104-1111, 2025.

[17] R. Zhao, H. Liu, M. Yang, T. Lu, Z. Li, Z. Shi, Z. Wang, J. Liu, Y. Yang, and T. Ren, "Reconfigurable aJ-Level Ferroelectric Transistor-Based Boolean Logic for Logic-in-Memory," Nano letters, vol. 24, pp. 10957-10963, 2024-01-01 2024.

[18] X. Sang, E. D. Grimley, T. Schenk, U. Schroeder, and J. M. LeBeau, "On the structural origins of ferroelectricity in HfO_2 thin films," Applied Physics Letters, vol. 106, p. 162905, 2015-04-20 2015.

[19] L. Liu et al., "Low‐Power Memristive Logic Device Enabled by Controllable Oxidation of 2D $HfSe_2$ for In‐Memory Computing," Adv. Sci., vol. 8, no. 15, p. 2005038, 2021. doi:10.1002/advs.202005038.

[20] Y. Song et al., "Reconfigurable and Efficient Implementation of 16 Boolean Logics and Full‐Adder Functions with Memristor Crossbar for Beyond von Neumann In‐Memory Computing," Adv. Sci., vol. 9, no. 15, p. 2200036, 2022. doi:10.1002/advs.202200036.

2025 The 10th International Conference on Integrated Circuits and Microsystems

Reconfigurable Memory Device based on Defect Engineering of 2D Ferroelectric CuInP$_2$S$_6$

Yunpeng Xia[1], Yu Li[1, 2, *], Tianqi Li[1], Qinfei Long[1], Zihui Hong[2], Yunhe Hou[2], Tiande Mo[1, *]

1. Centre of Advanced Power and Autonomous Systems, Hong Kong Productivity Council, Hong Kong SAR, China
2. Department of Electrical and Electronic Engineering, The University of Hong Kong, Hong Kong SAR, China
*Corresponding email: Mr. Li: yli@hkpc.org and Dr. Mo: rickmo@hkpc.org

Abstract—**This paper reports a reconfigurable memory device (RMD) based on the MoS$_2$/CuInP$_2$S$_6$ (CIPS) heterostructure: by enhancing CIPS interlayer defect density via plasma-induced defect engineering and combining its ferroelectric properties, the device achieves structural simplification and functional integration, enabling multi-bit information storage under electrical pulses and reversible switching between non-volatile and volatile memory modes under optical pulses (by modulating CIPS ferroelectric polarization state). This work innovatively integrates electrical-optical dual modulation and multi-mode storage into a single heterostructure, breaking the traditional trade-off between structural complexity and functional diversity in memory devices, and provides a novel technical path for high-integration next-generation intelligent memory technologies.**

Keywords—*reconfigurable memory device, defect engineering, ferroelectric properties, non-volatile capability, volatile capability*

I. INTRODUCTION

With the rise of emerging paradigms such as in-sensor computing and in-memory computing, there is an increasingly urgent demand for new memory devices that can simultaneously realize sensing, memorizing, and computing functions in a single device [1] [2] [3] [4]. This requirement for multifunctional integration poses severe challenges to traditional silicon-based memory devices, prompting researchers to turn their attention to two-dimensional (2D) materials with excellent properties, hoping of developing new memory devices with superior sensing capabilities [5]. Over the past few decades, the rapid development of 2D materials in the preparation of various electronic and optoelectronic devices has fully demonstrated their great potential in shaping future advanced electronic devices. However, current mainstream 2D memory devices generally suffer from single functionality, making them difficult to cope with the growing complexity of neuromorphic computing; at the same time, their high structural complexity hinders large-scale integration and commercialization [6] [7].

Charge trapping at interface defect states is the most common working mechanism of 2D memory devices [8]. However, this mechanism has the phenomenon of charge loss over time, which greatly limits their application in non-volatile memory. In recent years, the coupling of interface charge trapping with ferroelectric properties has brought a new perspective to the design of 2D memory devices [9]. In this design, the electrically controllable ferroelectric polarization state provides additional regulatory parameters, thereby significantly expanding the functionality of the device.

Based on this, this study proposes a reconfigurable memory device (RMD) based on the MoS$_2$/CuInP$_2$S$_6$ (CIPS) heterostructure, which integrates sensing, memory, and computing functions. Specifically, we use soft plasma treatment technology to significantly increase the surface defect state density of CIPS, thereby enhancing its charge storage stability. As a 2D ferroelectric material with a small coercive electric field, the excellent properties of CIPS have been confirmed in many studies [10] [11]. The reported RMD features a simple two-layer van der Waals (vdW) structure: it not only realizes multi-bit information storage under electrical pulses but also achieves reconfigurable optoelectronic memory behaviors (both non-volatile and volatile) by regulating the polarization direction of CIPS. The ingenuity of this experiment lies in that it relies not solely on ferroelectricity or defect-based charge trapping, but rather leverages a synergistic effect between CIPS' ferroelectric properties and its surface defect trapping. Instead of directly controlling the channel current, CIPS' weak ferroelectricity exerts a confining effect on the charges trapped at the interface—further underpinning the device's functional integration. This design simplifies architecture while enhancing functional synergy, offering a streamlined reference for next-gen multi-functional memory development.

II. EXPERIMENTAL SECTION

A. Device Fabrication

MoS$_2$ and CIPS nanoflakes were prepared via mechanical exfoliation from their bulk crystals (purchased from 2D Semiconductor Inc.) and assembled into heterostructures using dry-transfer techniques. SiO$_2$/Si substrates (Silicon Valley Microelectronics, Inc.) were used as received. After van der Waals stacking, source-drain electrodes were defined by standard electron-beam lithography (EBL, TESCAN VEGA3), followed by thermal evaporation of Cr/Au (8/78 nm) metal stacks. The devices were finally soaked in acetone for 2 hours to remove the photoresist, thus completing the fabrication process.

B. Device Characterization

Optical images were captured using a Nikon Ellipse LV100ND microscope. Scanning electron microscopy (SEM) was performed using a TESCAN VEGA3 system. Electrical characterization was conducted using an Agilent 4155C semiconductor parameter analyzer. For optoelectronic measurements, 532-nm laser pulses were generated by a monochromatic laser coupled with an optical chopper and delivered to the device via an optical fiber mounted on a probe station.

979-8-3315-8850-2/25 $31.00 © 2025 IEEE

III. RESULTS AND DISCUSSION

A. Structure of the RMD

Figure 1 shows a false-color scanning electron microscopy (SEM) image of a typical device, where the MoS_2 and CIPS layers are labeled in blue and green, respectively. The yellow regions correspond to the source-drain electrodes fabricated via standard electron beam lithography (EBL) followed by thermal evaporation of Cr/Au (8/78 nm). The device fabrication process is illustrated in Figure 2a-f. During the preparation of the heterojunction, the main method we adopted was the mechanical exfoliation method. First, we gradually thinned the CIPS and MoS_2 crystals using adhesive tape. Then, CIPS was directly exfoliated onto silicon wafers with a silicon dioxide layer on their surface, while MoS_2 was exfoliated onto polydimethylsiloxane (PDMS). Next, we observed the samples under an optical microscope and identified CIPS and MoS_2 flakes with appropriate thicknesses by comparing their colors. After that, we attached the other side of the PDMS (with MoS_2 on one side) to a glass slide, and used a transfer stage to transfer the MoS_2 under the microscope. Subsequently, we released the MoS_2 onto the surface of CIPS via thermal release, and then performed another heating step to achieve a stronger bond between the two materials, thus completing the preparation of the heterojunction. Notably, prior to transferring MoS_2 onto CIPS, the CIPS nanoflakes underwent plasma treatment, which involved exposing the flakes to a radio frequency (RF) power of 18 W for 5 minutes under a pressure of ~0.74 Torr to increase their defect density In this experiment, CIPS and MoS_2 nanosheets with similar thicknesses—approximately 80-100 nm for CIPS and around 10 nm for MoS_2—were uniformly used to ensure the consistency of the experiment. Since the thickness of CIPS affects the magnitude of ferroelectricity, this factor will not be discussed temporarily in this project.

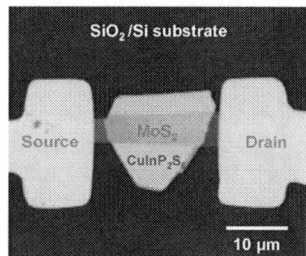

Fig. 1. False-color SEM image of a representative RMD.

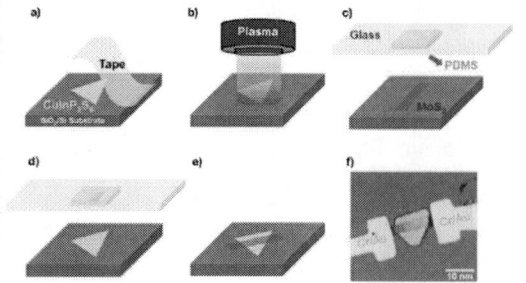

Fig. 2. Schematic illustration of the device fabrication.

X-ray Photoelectron Spectroscopy (XPS) was used to further analyze the surface changes of CIPS after plasma treatment. As shown in Figure 3a, the overall XPS spectra did not significantly change before and after the plasma treatment. The plasma treatment condition was kept the same in all experiments (0.74 Torr for 5 minutes at a radio frequency (RF) power of 18 W). Figure 3b compares the Cu 2p peak before and after the plasma treatment and we noticed that two additional minor peaks appeared at 935.2 and 954.6 eV after the treatment. That suggests a small portion of monovalent Cu^+ was oxidized to divalent Cu^{2+} during the plasma treatment. Figure 3c compares the In 3d peak before and after the plasma treatment and the results indicate that the plasma did not induce the valence state change of In since In^{3+} is already the highest valence state. Figure 3d-e correspond to P 2p and S 2p peaks, respectively. Both element peaks show a blue shift toward higher binding energy and a new minor peak at 133.9 (134) eV corresponding to metal phosphates emerged in the P 2p peak after the treatment. The changes in these two element peaks are attributed to the redox reactions during the plasma treatment. As revealed by Figure 3f, the O 1s peak produced a small peak at 531.1 eV after the plasma treatment. The change in the O 1s peak suggests the plasma treatment could introduce oxygen doping in the CIPS and then resulting in surface oxidation and affecting the valence states of other elements. Based on the XPS results, as well as the cross-section STEM and HRTEM results, we can conclude that plasma treatment could induce defect states in CIPS in both physical and chemical ways. On one hand, the physical impact of ions in plasma treatment progress would knock off the surface atoms and leave behind atomic vacancies; on the other hand, the plasma treatment could result in redox chemical reactions with the broken of the covalent bonds between atoms near the surface and the exposed dangling bonds would join in trapping carriers.

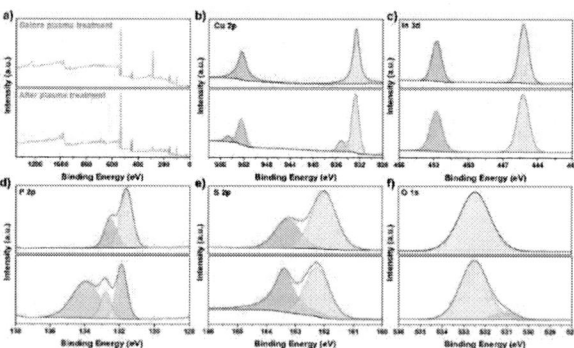

Fig. 3. XPS measurements of CIPS before and after plasma treatment.

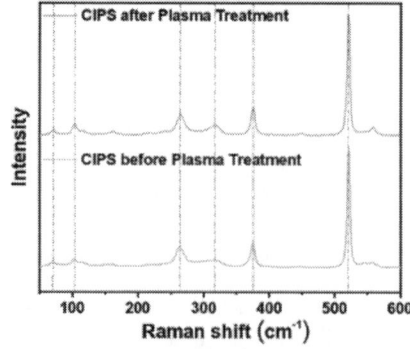

Fig. 4. Raman spectra of the CIPS nanosheet before and after plasma treatment.

Figure 4 shows that, although plasma treatment has introduced numerous defect states on the CIPS surface, these defects exert no obvious influence on the Raman signal. This observation is further corroborated by a direct comparison of the Raman spectra obtained from CIPS before and after plasma treatment.

B. Electronic Non-volatile Memory Behavior

The performance of RMD under pure electrical control was systematically characterized. Figure 5 shows the transfer curves of devices based on pristine CIPS nanoflakes (blue curve) and plasma-treated CIPS nanoflakes (pink curve). Different from the typical characteristics of traditional ferroelectric field-effect transistors (FeFETs), both devices exhibit clockwise hysteresis loops, indicating that the electrical transport behavior of the devices is mainly dominated by the charge trapping mechanism of interlayer defect states rather than the ferroelectric properties themselves. However, as confirmed by numerous literature, ferroelectric polarization can still play a role in assisting charge trapping in this process [9].

A clear comparison between the two curves reveals that the CIPS device after plasma treatment has a significant memory window (ΔV) of 13.1 V, while the device based on pristine CIPS nanoflakes has a memory window of only about 1.5 V. Generally speaking, the size of the memory window related to the threshold voltage (V_{th}) shift is positively correlated with the charge storage capacity of the charge trapping layer [12]. It can be inferred that plasma treatment can effectively increase the interface defect density in RMD.

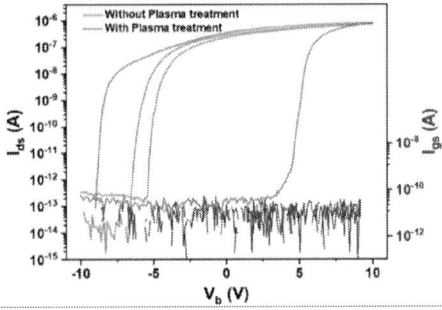

Fig. 5. Transfer curve of the devices based on MoS$_2$/CIPS heterostructure.

Figure 6 compares the current evolution in devices with plasma-treated/untreated CIPS nanoflakes under identical electrical pulses. The treated device shows longer current retention due to enhanced charge trapping from higher defect density, while the untreated device recovers within seconds.

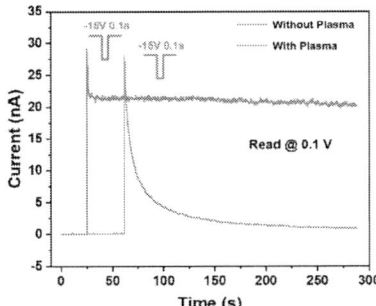

Fig. 6. Current evolution of the devices based on MoS$_2$/CIPS heterostructure.

Four control devices (MoS$_2$ paired with WSe$_2$, PdSe$_2$, MoTe$_2$, WS$_2$) were fabricated with the same process and plasma treatment. Using similar MoS$_2$ thickness, Figure 7 reveals their large hysteresis windows but rapid current decay after pulsing, confirming that ferroelectric polarization in CIPS is essential for stabilizing trapped charges. All measurements used consistent conditions.

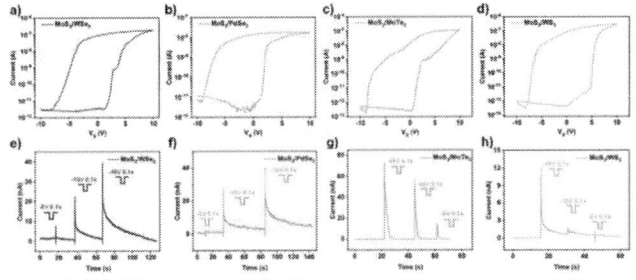

Fig. 7. Electrical performances of MoS$_2$-based heterostructure devices.

To verify the key role of polarization switching in transitioning between the device's memory mode (non-volatile) and synaptic mode (volatile), we tested its performance at 52.5°C (above CIPS's Curie temperature $T_c \approx 46.85$°C) [13] [14]. Figure 8 shows the experimental setup, noting that the control panel temperature is slightly higher than the actual hot plate temperature (measured by an infrared thermometer). Figure 8 compare performances at room temperature (~23°C) and 52.5°C: at room temperature (T<T_c), ±20 V/0.1s pulses induce ferroelectric switching in CIPS, with RMD exhibiting non-volatile memory (retention over hundreds of seconds). At high temperature (T>T_c), CIPS loses ferroelectricity, degrading to volatile memory (relaxation in tens of seconds). This discrepancy confirms that polarization switching mediates the transition between RMD's memory and synaptic modes.

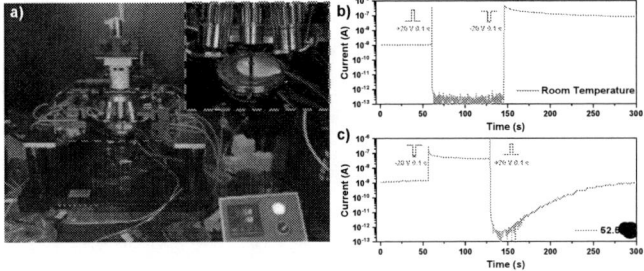

Fig. 8. RMD Memory performance characterization at different temperatures.

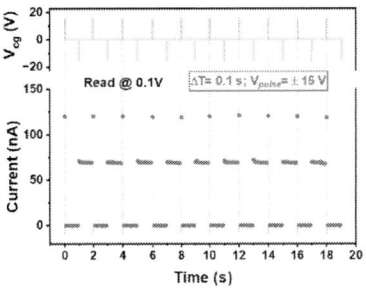

Fig. 9. Dynamic switching measurement of the device.

Figure. 9 demonstrates the dynamic switching between high and low resistance states of RMD under ±15 V/0.1 s electrical pulses. Figure 10 shows its multi-bit storage capability—7 distinct resistance states can be clearly distinguished over hundreds of seconds of testing, with programming pulses ranging from -20 to 20 V (0.1 s duration), laying the foundation for advanced memory system applications.

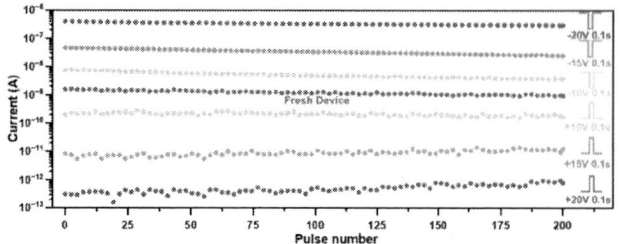

Fig. 10. Different resistance states of the device.

In terms of long-term performance, Figure 11 and 12 verify the device's retention stability and cycling endurance, respectively: the current on/off ratio remains above 10^5 after 3000 seconds of testing, and no significant degradation is observed after 3000 cycles. These results confirm that RMD, as a non-volatile electronic memory device, holds great potential for applications in advanced in-memory computing systems.

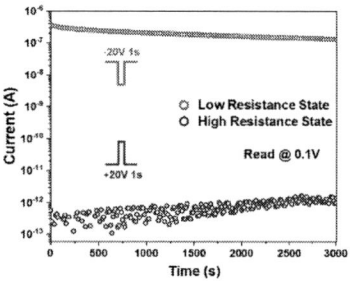

Fig. 11. Retention stability of the device.

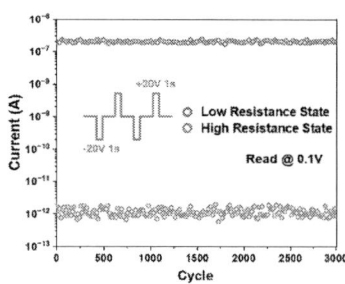

Fig. 12. Cycling endurance of the device.

C. Optoelectronic Non-volatile Memory Behavior

When a positive back-gate voltage (Vb) is pre-applied to set the ferroelectric polarization of CIPS in the upward direction, the RMD exhibits non-volatile behavior under optical pulses. Figure 13 shows the photocurrent evolution triggered by optical stimuli: initially, the RMD is electrically programmed into a high-resistance state, and under the action of continuous laser pulses (wavelength 532 nm, power 10 nW, pulse width 0.2 s), the channel current gradually increases. As a characteristic of

non-volatile optoelectronic memory behavior, the conductance change induced by each optical pulse can persist for tens of seconds.

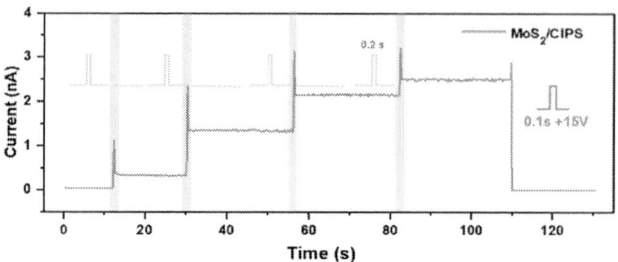

Fig. 13. Current evolution of the RMD triggered by optical pulses

D. Optoelectronic Volatile Memory Behavior

Repeatedly apply positive and negative pulses with different amplitudes to the device's bottom gate until the current returns to the initial value of the fresh device and remains stable for a long time. At this time, the ferroelectric polarization field strength in CIPS is nearly zero, exerting no electrostatic control on the channel, and the ferroelectric polarization state is effectively reset. In this state, short - term optical pulses make the device exhibit short - term memory behavior, originating from charge trapping between MoS_2 and CIPS nanosheets. Based on this behavior, the paired - pulse facilitation (PPF) effect is observed in RMD (Figure 14). The PPF ratio is calculated by formula (1),

$$(\Delta A_2 - \Delta A_1)/\Delta A_1 \times 100\%, \quad (1)$$

and is related to the interval time of optical pulses (Figure 15, interval 0.5 - 3 s, step 0.5 s). The experimental curve is fitted with a double - exponential decay function,

$$PPF = C_0 + C_1 e^{-\Delta t/\tau_1} + C_2 e^{-\Delta t/\tau_2}, \quad (2)$$

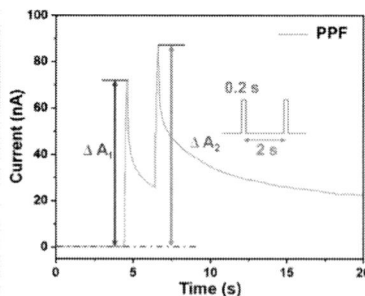

Fig. 14. PPF effect induced by paired optical pulse.

Here, $\tau_1 = 0.92486$ s, $\tau_2 = 0.92492$ s, and the coefficient of determination (COD) $R^2 = 0.995$. This model accurately characterizes the PPF (paired-pulse facilitation) behavior of RMD, enabling predictions of how the PPF behavior changes under varying optical pulse conditions. Such short-term memory capabilities lay the hardware foundation for reservoir computing. Interestingly, this behavior resembles the short-term plasticity observed in biological synapses, where transient changes in synaptic strength allow for temporal information encoding. This

979-8-3315-8850-2/25 $31.00 © 2025 IEEE

synaptic-like dynamic not only reinforces the bio-inspiration of RMDs but also supports their use in neuromorphic architectures, particularly as memory nodes in reservoir computing systems [15].

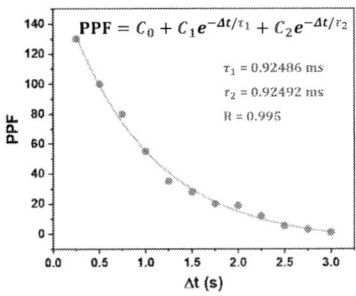

Fig. 15. PPF ratio evolution with different optical pulse interval.

IV. CONCLUSION

This study developed a multifunctional and structurally simple memory device based on a two-dimensional MoS_2/CIPS van der Waals heterostructure. In the device fabrication process, plasma treatment was used to increase the interlayer defect density. This engineered defect density, combined with the ferroelectric properties of CIPS, endows the device with high reconfigurability, enabling a single device to integrate sensing, storage, and computing functions. Under pure electrical control, the RMD can operate as a traditional memory device with multi-level information storage capability. After the ferroelectric polarization of CIPS is electrically set to an upward direction, the device exhibits non-volatile optoelectronic memory behavior. Moreover, by electrically resetting the ferroelectric polarization in CIPS to zero, the RMD can be reconfigured to exhibit short-term memory behavior analogous to that of biological synapses. This novel reconfigurable electronic and optoelectronic memory device offers an innovative perspective for memory device design and future applications in in-sensor and in-memory computing technologies.

ACKNOWLEDGMENT

The authors would like to express their sincere gratitude to the City University of Hong Kong for providing access to the experimental facilities essential to this research.

REFERENCES

[1] S. S. Sutar, S. S. Patil, and S. S. Kumbhar, "High Performance and Low Power SRAM Cell Design Using Power Gating Technique," Int. J. Electr. Electron. Eng. Telecommun., vol. 5, pp. 35–47, 2016.

[2] A. A. Shaikh and S. S. Shaikh, "Design and Analysis of Low Power and High Speed SRAM Cell Using 45nm Technology," Int. J. Electr. Electron. Eng. Telecommun., vol. 6, pp. 94–99, 2017.

[3] A. Sebastian, M. Le Gallo, R. Khaddam-Aljameh, and E. Eleftheriou, "Memory Devices and Applications for In-Memory Computing," Nat. Nanotechnol., vol. 15, pp. 529–544, 2020.

[4] Y. Zhou, J. Fu, Z. Chen, F. Zhuge, Y. Wang, J. Yan, S. Ma, L. Xu, H. Yuan, and M. Chan, "Computational Event-Driven Vision Sensors for In-Sensor Spiking Neural Networks," Nat. Electron., vol. 6, pp. 870–878, 2023.

[5] C. Liu, H. Chen, S. Wang, Q. Liu, Y.-G. Jiang, D. W. Zhang, M. Liu, and P. Zhou, "Two-Dimensional Materials for Next-Generation Computing Technologies," Nat. Nanotechnol., vol. 15, pp. 545–557, 2020.

[6] K. S. Novoselov, A. Mishchenko, A. Carvalho, and A. Castro Neto, "2D Materials and van der Waals Heterostructures," Science, vol. 353, p. aac9439, 2016.

[7] S. Yu, X. Wu, Y. Wang, X. Guo, and L. Tong, "2D Materials for Optical Modulation: Challenges and Opportunities," Adv. Mater., vol. 29, p. 1606128, 2017.

[8] M. S. Choi, G.-H. Lee, Y.-J. Yu, D.-Y. Lee, S. H. Lee, P. Kim, J. Hone, and W. J. Yoo, "Controlled Charge Trapping by Molybdenum Disulphide and Graphene in Ultrathin Heterostructured Memory Devices," Nat. Commun., vol. 4, pp. 1–7, 2013.

[9] J. Gao, X. Lian, Z. Chen, S. Shi, E. Li, Y. Wang, T. Jin, H. Chen, L. Liu, and J. Chen, "Multifunctional MoTe₂ Fe-FET Enabled by Ferroelectric Polarization-Assisted Charge Trapping," Adv. Funct. Mater., vol. 32, p. 2110415, 2022.

[10] J. Zha, S. Shi, A. Chaturvedi, H. Huang, P. Yang, Y. Yao, S. Li, Y. Xia, Z. Zhang, and W. Wang, "Electronic/Optoelectronic Memory Device Enabled by Tellurium-Based 2D van der Waals Heterostructure for In-Sensor Reservoir Computing at the Optical Communication Band," Adv. Mater., vol. 35, p. 2211598, 2023.

[11] X. Wang, P. Yu, Z. Lei, C. Zhu, X. Cao, F. Liu, L. You, Q. Zeng, Y. Deng, and C. Zhu, "van der Waals Negative Capacitance Transistors," Nat. Commun., vol. 10, p. 3037, 2019.

[12] Z. Cui, D. Xin, T. Kim, J. Choi, J. Cho, and J. Yi, "Improvement of the Charge Retention of a Non-Volatile Memory by a Bandgap-Engineered Charge Trap Layer," ECS J. Solid State Sci. Technol., vol. 10, p. 125002, 2021.

[13] F. Liu, L. You, K. L. Seyler, X. Li, P. Yu, J. Lin, X. Wang, J. Zhou, H. Wang, and H. He, "Room-Temperature Ferroelectricity in CuInP₂S₆ Ultrathin Flakes," Nat. Commun., vol. 7, pp. 1–7, 2016.

[14] M. Si, A. K. Saha, P.-Y. Liao, S. Gao, S. M. Neumayer, J. Jian, J. Qin, N. Balke Wisinger, H. Wang, and P. Maksymovych, "Room-Temperature Electrocaloric Effect in Layered Ferroelectric CuInP₂S₆ for Solid-State Refrigeration," ACS Nano, vol. 13, pp. 8760–8765, 2019.

[15] J. Zha, Y. Xia, S. Shi, H. Huang, S. Li, C. Qian, H. Wang, P. Yang, Z. Zhang, and Y. Meng, "A 2D Heterostructure-Based Multifunctional Floating Gate Memory Device for Multimodal Reservoir Computing," Adv. Mater., vol. 36, p. 2308502, 2024.

2025 The 10th International Conference on Integrated Circuits and Microsystems

ANDQ-NAT: Adaptive Non-linearity Dynamic Quantization Non-Ideality Aware Training Framework for Computing-In-Memory Macros

Yu Liu
School of Integrated Circuits
Anhui University
Hefei, China
liuyu@ahu.edu.cn

Jianxing Zhou
School of Integrated Circuits
Anhui University
Hefei, China
wb23301113@stu.ahu.edu.cn

Hao Li
School of Integrated Circuits
Anhui University
Hefei, China
wb23301101@stu.ahu.edu.cn

Yang Lou
School of Integrated Circuits
Anhui University
Hefei, China
wb23301141@ahu.edu.cn

Xiulong Wu
School of Integrated Circuits
Anhui University
Hefei, China
wuxiulong@ahu.edu.cn

Zhiting Lin
School of Integrated Circuits
Anhui University
Hefei, China
ztlin@ahu.edu.cn

Abstract—Abstract—Analog domain computing-in-memory (ACIM) offers significant advantages in high parallelism and low power consumption, making it highly effective for mitigating power and latency challenges in data-intensive computations. It plays a crucial role in accelerating convolutional neural networks (CNN), particularly in matrix-vector multiplication (MVM) operations. However, the accuracy of ACIM is inherently compromised by the non-ideal characteristics of the hardware, leading to a substantial degradation in CNN inference accuracy. To address this challenge, we propose an Adaptive Non-linearity Dynamic Quantization Non-Ideality Aware Training (ANDQ-NAT) framework aimed at compensating for accuracy loss. The key contributions of this work are: 1) Evaluating multiply-accumulate (MAC) errors in a 6T-SRAM CIM macro across various linearity levels. 2) Introducing an Adaptive Non-linearity Dynamic Quantization algorithm designed to enhance CNN robustness against hardware-induced errors. 3) Quantifying the "partial sum errors" arising from the physical limitations of ACIM and providing accurate MVM error metrics for the ANDQ-NAT framework. Experimental results show that, compared with the NAT with uniform quantization, when performing inference on the CIFAR10 dataset, the ANDQ-NAT framework improves the inference accuracy of the ResNet20 network by 22.01% (low-linearity) and 13.68% (high-linearity) respectively, and improves the inference accuracy of the VGG8 network by 26.07% (low-linearity) and 15.75% (high-linearity). This verifies the effectiveness of the framework under different ACIM configurations.

Keywords—*ACIM, CNN, 6T—SRAM,MVM, ANDQ, NAT*

I. INTRODUCTION

In the era of artificial intelligence, convolutional neural networks (CNN) have emerged as a pivotal technological advancement. With their robust learning abilities and exceptional feature extraction capabilities, CNN have garnered

This work was supported in part by National Natural Science Foundation of China under Grant 62202003，in part by Key Research and Development Program of Anhui Province under Grant 2022a05020044

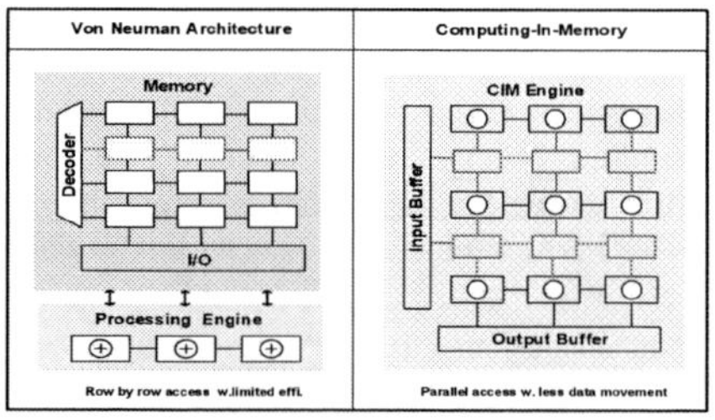

Fig.1. VonNeumann architecture and CIM architecture.

significant success across diverse domains[1][2]. Within CNN, matrix-vector multiplications (MVM) serve as the fundamental operation in both the convolutional and fully connected layers. In the VonNeumann architecture, a clear distinction exists between the computing unit and the memory cell. For MVM, the constant data transfer between the memory cell and the computing unit accounts for a substantial 90% of the energy consumption. To address this inefficiency, the computing-in-memory (CIM) architecture, which integrates the computing unit with the memory cell, has been proposed[3][4][5][6]. As illustrated in Fig.1, within the CIM framework, weight data is stored in the memory cell, while input data is input horizontally. The dot-product operation occurs directly within the memory cell[5], and the calculation results are aggregated on the bit line in the form of current or voltage. Subsequently, the results of the multiply-accumulate (MAC) operation are digitized by an Analog to Digital Converter (ADC). The CIM architecture effectively eliminates the need for extensive data movement, thereby enhancing the efficiency of matrix multiplication operations. Furthermore, the Analog Domain Computing-in-Memory(ACIM) can simultaneously activate multiple rows, offering high parallelism and providing a natural advantage in matrix computations[7][8]. However, Since the non-ideal characteristics of ACIM, the inference accuracy of CNN is severely compromised. To address the challenges, this paper

979-8-3315-8850-2/25 $31.00 © 2025 IEEE

proposes a Non-Ideality Aware Training (NAT) framework based on an Adaptive Non-linearity Dynamic Quantization (ANDQ) algorithm. The key contributions of this work are as follows:

1.Leveraging the ACIM architecture, a Monte Carlo simulation is conducted to investigate the internal non-ideal characteristics of the SRAM array. The impact of these non-ideal factors on the MAC results under different linearities is analyzed, allowing the determination of the hardware error for a single array. 2.Based on the modeling of hardware errors in a single SRAM array, the ANDQ algorithm is proposed to compensatefor errors in the ADC quantization results induced by the non-ideal characteristics of the SRAM array. 3.A quantitative analysis of the partial-sum errors is performed, exploring the changes in system errors after the fusion of partial sums. This ensures the authenticity and reliability of the hardware errors within the NAT framework, maximizing the robust potential of CNN.

The remainder of this paper is structured as follows: Section II elaborates on the primary causes of non-ideal factors in CIM and discusses various software-hardware co-optimization schemes. Section III presents a detailed explanation of the ANDQ algorithm used for error quantization-aware training based on ACIM. Section IV provides accuracy verification and experimental results.

II. RELATED WORK

The core functionality of ACIM within SRAM is based on analog computations driven by bit-line voltage. Fluctuations in the bit-line voltage directly influence the precision of the CIM outcomes [3]. During the computational process, multiple rows are activated simultaneously, resulting in data flipping. This, in turn, leads to read-disturbance issues due to the charging and discharging of the bit lines, which causes a misalignment in the weight data during computation, thereby compromising the accuracy of the results. Additionally, the parasitic capacitance on the bit line, along with the clamping and leakage currents of the memory cells, further exacerbate the voltage fluctuations, leading to issues with the linearity of the computation. In ACIM, input data is represented by word-line pulses. The distortion in word-line pulse width and the asynchronous variation of bit lines contribute to the inconsistency in the computations of adjacent columns. These non-ideal factors cumulatively induce additional voltage fluctuations in the bit-line voltage, which directly impact the MAC quantization results and lead to errors in the ADC quantization outcomes. Within the CIM array, these non-ideal factors significantly undermine the inference accuracy of the CIM for Neural Networks. To tackle these challenges, the solution primarily encompasses two aspects:

Hardware Optimization: By implementing a dual-word line read-write separation structure, the read margin can be effectively enhanced. This optimization not only increases parallelism but also significantly improves computational energy efficiency [9]. However, non-linearity continues to have a profound impact on computational accuracy. To mitigate this, a cascaded current mirror circuit is integrated into the bit line. This circuit helps regulate the bit-line voltage, effectively reducing the impact of clamping and leakage

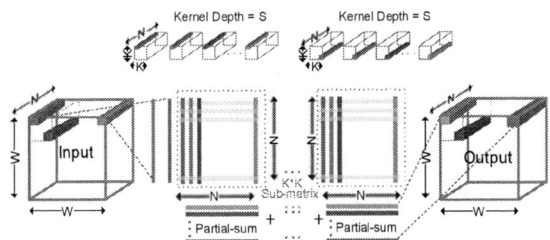

Fig.2. Mapping of Weight Data

TABLE I. Inference Accuaracy Under Different Environments

Framework	CIFAR-10			
	Resnet-20		Vgg8	
	Low-linearity	High-linearity	Low-linearity	High-linearity
FP32	94.23%	94.23%	93.41%	93.41%
Non-Ideal CIM	16.23%	53.17%	20.16%	62.57%
ANDQ-NAT	86.54%	91.52%	88.64%	92.58%

currents. As a result, both the consistency and linearity of the computations are significantly improved [10].Algorithm Optimization:Within the NAT framework [10][11][12][13][14], a more optimal balance can be achieved between accuracy and hardware efficiency. These frameworks leverage the inherent redundancy and error-tolerance characteristics of neural network (NN)-based designs and applications, enabling them to effectively adapt to hardware-induced errors. a fault-free framework was proposed[15], which addresses fixed faults in resistive random-access memory (RRAM) using an offline compilation scheme, thereby significantly reducing the weight errors stored in RRAM. While D-NAT [13] mitigates static weight errors through offline calibration, it fails to address dynamic error propagation in multi-array CIM. Due to the inherent non-ideal characteristics of ACIM and the energy consumption concerns associated with peripheral circuits, the scale of the SRAM array is inherently constrained. During the practical execution of convolutional operations, especially for convolutions involving many channels, it becomes highly challenging for a single array to store all the parameters of a convolutional kernel. As illustrated in Fig. 2, the weight data is mapped within the SRAM array, and multiple sub-arrays are required for collaborative storage. Consequently, a single convolutional computation often necessitates the aggregation of results from several SRAM arrays. This approach enables the precise acquisition of hardware errors in a single convolution result, providing accurate hardware errors for the NAT framework. It maximizes the adaptability of the CNN's inherent robustness to hardware errors, thereby improving the inference accuracy of CIM. TABLE I clearly demonstrates that, within the same array, there is a substantial performance improvement when comparing direct inference to inference that incorporates the ANDQ-NAT approach.

III. ANDQ-NAT FRAMEWORK FOR CIM

The overall algorithmic flow is shown in Fig.3, which demonstrates the process of adapting the neural network to hardware-induced errors through the NAT framework based on the ANDQ algorithm.

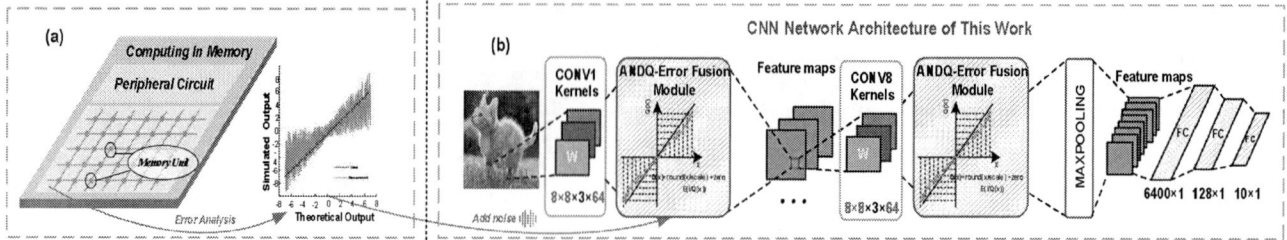

Fig.3. The workflow of the ANDQ-NAT framework

A. CIM Architecture and Non-Ideality and Modeling

As shown in Fig.4, In this paper, a charge-domain 6T-SRAM CIM macro with a size of 64x256 is employed to perform convolutional operations with 4-bit input, 4-bit weight, and 4-bit output. The CIM array is divided into 64 banks, each consisting of four columns. the array operates in two modes: high-linearity and low-linearity calculations. In the low-linearity mode, the input data is driven by the input circuit to activate multiple rows of word lines to VDD/GND, performing a dot product operation with the weight data stored internally in the CIM. The computing capacitor accumulates the changes in the bit line charge. After performing a weighted summation through charge sharing, the result is quantified and output by the ADC. In high-linearity mode, the switch is closed, and a CCM circuit is added to each column bit line to clamp the bit line voltage. This method effectively addresses the issue of inconsistent discharge between different columns, caused by clamping current, leakage current, and threshold voltage mismatch [16], thereby improving computational linearity.

As shown in Fig.4, due to the influence of hardware non-ideal factors, errors arise in the multiplication matrix results. To simulate these non-idealities in the ACIM macro, Gaussian noise is added, and Monte Carlo simulations are conducted for both high and low linearity modes. From the simulation results, we observe that the standard deviation of the variation is proportional to the absolute value of the bit line output voltage. Equation (1) is defined to represent the distribution variation of the output.

$$E_x \sim N(x, (1 + \alpha|x|)\sigma) \quad x \in \{-7, -6, \dots 6, 7\} \quad (1)$$

Where E_x represents the error when the data is x, N represents the Gaussian distribution function, α represents the noise variation factor, and σ represents the standard deviation.

B. Adaptive Non-linearity Dynamic Quantization (ANDQ) Algorithm

The various non-ideal factors present within the array ultimately have a direct impact on the calculation results of the bit lines. These non-ideal factors cause significant noise when the bit line voltages are quantified by the ADC. Through error simulation of a single array, we have determined that the magnitude of the noise varies depending on different calculation results. To compute the overall error rate, we define Equation (2), The e_x represents the error rate of the quantified data x:

$$f(e_x) = \sum_{x=-7}^{7} e_x \quad (2)$$

However, during the network inference process, we observed

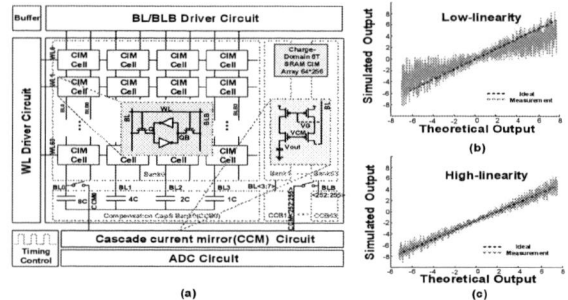

Fig.4. (a) A charge-domain 6T-SRAM CIM macro; (b) The distribution of MAC errors under Low-linearity; (c) The distribution of MAC errors under High-linearity.

that the output data from each layer is not uniformly distributed but instead follows a Gaussian distribution. Fig.4.(b) illustrates the distribution of output data with 4-bit weights and 4-bit activations. While the theoretical range for the 4-bit data should be between [-7, 7], the actual data is concentrated and distributed within the range of [-3, 3]. As a result, the overall error cannot be simply summed directly. Instead, we use Equation (3) to perform a weighted summation based on the actual output weights:

$$f^1(e_x) = \sum_{x=-7}^{x=7} w_x e_x \quad (3)$$

The e_x and w_x represent the error rate and the weight of the quantified data x. In the ADC quantization process, as shown in Fig.5.(a), a larger quantization interval effectively filters out noise, whereas a smaller quantization interval allows for high-precision sampling of the original signal. In the SRAM array, the magnitude of the error varies depending on the different computation results. While the uniform quantization method is straightforward, it fails to effectively manage the errors in the output data. Conversely, non-uniform quantization cannot adjust the quantization interval based on the actual data distribution. To address this, an ANDQ algorithm is proposed, which dynamically adjusts the quantization interval according to the actual output distribution, processes the bit-line noise, and enhances the accuracy of ADC quantization. As the absolute value of the output data increases, the error noise also increases. As shown in Fig.6.(c), Different quantization intervals are adopted. To simplify the process, we focus on considering the quantization error rate within the range of [-3, 3], based on the actual CIM output data and the network output distribution. The specific details are as follows:

To facilitate the calculation, we use Equation (4) to convert the Gaussian distribution of the output data into a standard

979-8-3315-8850-2/25 $31.00 © 2025 IEEE

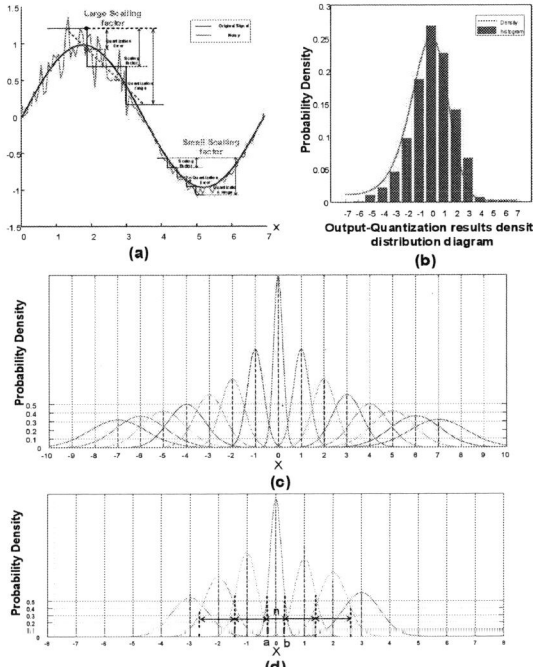

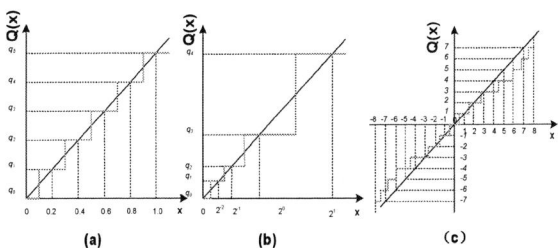

Fig.6. (a) Uniform quantization; (b) Non-uniform quantization; (c) ANDQ

Where δ_0 represents the quantization interval of the data $0, \Delta_x$ represents the quantization interval of the quantized data x, "round" represents the round function, and x_i represents the data type of float.

C. Error fusion of partial sums

Due to the non-ideal factors present in ACIMs, a new error calibration quantization function is introduced within the NAT framework to better simulate these factors. This quantization function is used during model training to replicate the hardware errors of ACIMs. We define the conditional probability formula $E(I|D)$, where D represents the theoretical value, and I represents the deviation between the actual value and the theoretical value $I(Q(x), r)$ is the value of $I(Q(x))$ selected based on the probability distribution $E(I|Q(x))$ and the random number r:

$$I(Q(x), r) = \begin{cases} I_1 & if \quad r < c_1 \\ I_2 & if \quad r < c_2 \\ \quad \cdots \\ I_n & if \quad r < c_n \end{cases} \quad (8)$$

in which the Cumulative Probability c is calculated as follows:

$$c_1 = E\left(I_1, \ Q(x)\right)$$
$$c_2 = c_1 + E\left(I_2, \ Q(x)\right) \quad (9)$$
$$\cdots$$
$$c_n = c_{n-1} + E(I_n, \ Q(x))$$

For layers with relatively large dimensions, when simulating CIM, the limitations of peripheral circuits and non-ideal factors often require the use of multiple CIMs to perform parallel computations in a single convolution operation.

Error-fusion

Require:
E(I/D),Ideal_output,Ideal_output,Actual_sum,n, Actual_sum.Q(Ideal_sum),Q(Actual_sum)

Ensure:
1: For i in range (100000):
2: Ideal_output = random.randint(-7,7), for _ in range(n)
3: ideal_sum = sum(ideal_output)
4: Actual_output =E(I/Ideal_output)
5: Actual_sum = sum(Actual_output)
6: Q(Ideal_sum/n) ,Q(Actual_sum/n)
7: End for
8: I =Q(Ideal_sum/n) -Q(Actual_sum/n)
9: Acquire new E(I/D)

Fig.5. (a) Effects of different quantization intervals on the signal; (b) Actual output distribution of the network; (c) Overall error distribution of MAC; (d) Error distribution of MAC within the range of [-3, 3]

Gaussian distribution:

$$Z_x \sim N(0,1) \quad , Z_x = E_x - x/(1 + \alpha|x|) \quad (4)$$

Where $Z_x \sim N$ is the standard Gaussian distribution function, and Z_x represents the transformation relationship between the non-standard Gaussian distribution E_x and the standard Gaussian distribution.

As shown in Fig.5.(d), the relationship between the errors rate e_x of a single data point and the quantization interval n is determined by Equation (5) , Where the function $\varphi(Z_x)$ is the probability density function of the standard Gaussian distributi:

$$\begin{cases} \varphi(Z_x) = \frac{1}{\sqrt{2\pi}} e^{\frac{Z_x^2}{2}} \\ e_x(n) = \int_{Z_1}^{Z_2} \varphi(Z_x) dz \ , Z_1 = \frac{a}{(1+\alpha|x|)\sigma} \ , Z_2 = \frac{b}{(1+\alpha|x|)\sigma} \end{cases} \quad (5)$$

There n = b − a , where a and b are the initial quantization boundaries, and they are marked in Fig.5.(d) . Since the actual output noise increases gradually, we introduce a scaling factor u_x to progressively expand the quantization interval within the range of [-3, 3], while gradually reducing the quantization interval for the remaining data, as shown in Fig.5.(d) , to ensure the overall 4-bit output. The overall error rate is then calculated using Equation (6):

$$f^2\left(e_x, \ u_x\right) = \sum_{x=-7}^{7} w_x e_x(u_x) \quad (6)$$

According to Equation (7), determine the optimal u_x to optimize the calculation errors caused by non-ideal factors. Finally, the ANDQ quantization formula is expressed by Equation (7):

$$Q(x) = round\left(\frac{x_i}{\Delta_x} + zero_x\right) \ , \Delta_x = u_x \delta_0 \quad (7)$$

The "partial sum error" across the CIMs will vary, and this type of systematic error is typically unknown. However, during the ANT processing, this unknown error can cause the network to improperly adapt to the non-ideal hardware environment. Leading to inefficient compensation for hardware errors during the inference process. To address this issue, this paper employs statistical methods, as shown in the pseudo-code, to map the weight data between different layers of the actual network to different numbers "n" of SRAM arrays. A quantitative analysis is conducted on the change in system error following the fusion of the "partial sum error" of a single ACIM, effectively mitigating the influence of the "partial sum System error." This approach provides precise hardware error data for the NAT framework.

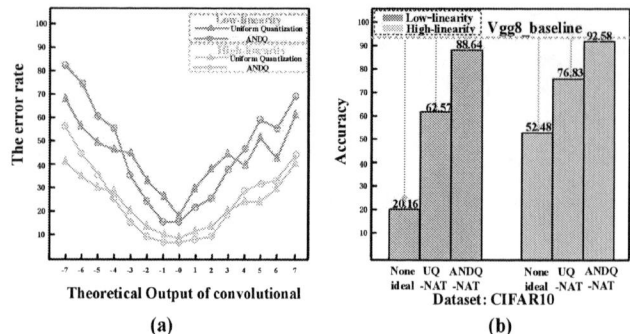

(a)   (b)

Fig.7. (a) Output errors of different quantization methods; (b) Inference accuracy under different environments

TABLE II. Inference accuracy under different environments

Framework	MNIST				CIFAR-10			
	Resnet-20		Vgg8		Resnet-20		Vgg8	
	Low-linearity	High-linearity	Low linearity	High-linearity	Low-linearity	High-linearity	Low-linearity	High-linearity
FP32	99.93%	99.93%	99.76%	99.76%	94.23%	94.23%	93.41%	93.41%
Ideal CIMA	99.82%	99.82%	99.53%	99.53%	93.14%	93.14%	92.67%	92.67%
Non-Ideal CIM	66.54%	74.31%	64.21%	81.36%	16.23%	53.17%	20.16%	52.48%
UQ-NAT	94.34%	97.62%	93.35%	98.24%	64.53%	78.46%	62.57%	76.83%
ANDQ-NAT	(+2.39%) 96.73%	(+1.94%) 99.56%	(+3.8%) 97.15%	(+1.18%) 99.42%	(+22.01%) 86.54%	(+13.06%) 91.52%	(+26.07%) 88.64%	(+15.75%) 92.58%

IV. EXPERIMENTAL RESULT

As shown in Fig.7.(a), under different linearity levels, the error rate of ANDQ is significantly lower than that of uniform quantization(UQ) when the convolution result is between [-3, 3], which means that the calculation results in this region are more accurate. As shown in Fig.5.(b), for the 4-bit output, the calculation accuracy of this part of the data plays a decisive role in the inference accuracy. Through Formula 3, we can calculate that the overall error rate of ANDQ is far lower than that of average quantization. ANDQ has a good compensatory effect on the calculation errors caused by the non-rational characteristics of the hardware, effectively improving the inference accuracy of ACIM on CNN. As shown in TABLE II, on the same hardware platform, for the inference results of different CNN on different datasets, compared with the UQ-NAT framework, the inference results of the ANDQ-NAT framework have an obvious improvement, which verifies the effectiveness and universality of the ANDQ-NAT framework.

REFERENCES

[1] Y. Wu, et al., "Google's neural machine translation system: Bridging the gap between human and machine translation," arXiv:1609.08144,2016.

[2] Nishani, Eralda, and Betim Ç iço, "Computer vision approaches based on deep learning and neural networks: Deep neural networks for video analysis of human pose estimation," IEEE Mediterranean Conference on Embedded Comput. (MECO), pp. 1–4, 2017

[3] Khwa W S, Chen JJ, Li J F, et al. A 65nm 4Kb algorithm dependent computing-in-memory SRAM unit-macro with 2.3 ns and55.8 TOPS/W fully parallel product-sum operation for binary DNN edge processors [C] //2018 IEEE International Solid-State Circuits Conference-(ISSCC). IEEE, 2018: 496-498. G. Eason, B. Noble, and I

[4] Xu H, Lin N, Luo L, et al. Senputing: An ultra-low-power always on vision perception chip featuring the deep fusion of sensing and computing[J]. IEEE Transactions on Circuits and Systems I: Regular Papers, 2021, 69(1): 232-243

[5] C.-J. Jhang, C.-X. Xue, J.-M. Hung, F.-C. Chang, and M.-F. Chang " Challenges and trends of SRAM-based computing-in-memory for AIedge devices," IEEE Trans. Circuits Syst. I, Reg. Papers, vol. 68, no. 5,pp. 1773–1786, Mar. 2021

[6] S. Yin, Z. Jiang, J.-S. Seo, and M. Seok, "XNOR-SRAM: In-memory computing SRAM macro for binary/ternary deep neural networks," IEEE J. Solid-State Circuits, vol. 55, no. 6, pp. 1733–1743, Jun. 2020

[7] Z. Sun, S. Kvatinsky, X. Si et al., "A full spectrum of computing-in memory technologies," Nat Electron, vol. 6, pp. 823–835, 2023.

[8] X. Sun et al., "Ultra-low precision 4-bit training of deep neural networks," in Proc. Adv. Neural Inf. Process. Syst., vol. 33, 2020,pp. 1796–1807

[9] S. Um, S. Kim, S. Hong, S. Kim and H. -J. Yoo, "LOG-CIM: An Energy-Efficient Logarithmic Quantization Computing-In-Memory Processor With Exponential Parallel Data Mapping and Zero-Aware 6T Dual-WL Cell," in *IEEE Journal of Solid-State Circuits*, vol. 59, no. 10, pp. 3330-3341, Oct. 2024

[10] H. Shin, M. Kang and L. -S. Kim, "Fault-free: A Fault-resilient Deep Neural Network Accelerator based on Realistic ReRAM Devices," *2021 58th ACM/IEEE Design Automation Conference (DAC)*, San Francisco, CA, USA, 2021, pp. 1039-1044

[11] Peng, Xiaochen, et al., "DNN+NeuroSim V2. 0: An end-to-end benchmarking framework for compute-in-memory accelerators for on chip training," IEEE Trans. on Computer-Aided Design of Integrated Circuits and Syst. (TCAD), vol. 40, no. 11, pp. 2306–2319, 2022

[12] C. -T. Huang, C. -Y. Chang, Y. -C. Chuang and A. -Y. Wu, "BWANIMC: Budget-based Workload Allocation for Hybrid Near/In Memory-Computing," in Proc. IEEE/ACM Design Automation Conf. (DAC), San Francisco, July 2023 .

[13] C. Lammie, J. K. Eshraghian, W. D. Lu and M. R. Azghadi, "Memristive Stochastic Computing for Deep Learning Parameter Optimization," in IEEE Transactions on Circuits and Systems II: Express Briefs, vol. 68, no. 5, pp. 1650-1654, May 2021

[13] M. -G. Lin, et al., "D-NAT: Data-Driven Non-Ideality Aware Training Framework for Fabricated Computing-In-Memory Macros," IEEE Journal on Emerging and Selected Topics in Circuits and Syst. (JETCAS), vol. 12, no. 2, pp. 381–392, 2022.

[14] W. -H. Huang et al., "A Nonvolatile AI Edge Processor with 4MB SLC-MLC Hybrid-Mode ReRAM Compute-in-Memory Macro and 51.4-251TOPS/W," IEEE Int. Solid-State Circuits Conf. (ISSCC), pp. 15–17, 2023

[15] H. Shin, M. Kang and L. -S. Kim, "Fault-free: A Fault-resilient Deep Neural Network Accelerator based on Realistic ReRAM Devices," *2021 58th ACM/IEEE Design Automation Conference (DAC)*, San Francisco, CA, USA, 2021, pp. 1039-1044

[16] Z. Lin *et al.*, "Cascade Current Mirror to Improve Linearity and Consistency in SRAM In-Memory Computing," in *IEEE Journal of Solid-State Circuits*, vol. 56, no. 8, pp. 2550-2562, Aug. 2021, doi: 10.1109/JSSC.2021.3063719. K. Elissa, "Title of paper if known," unpublished.

979-8-3315-8850-2/25 $31.00 © 2025 IEEE

2025 The 10th International Conference on Integrated Circuits and Microsystems

A Fast-Transient-Response Hybrid Architecture LDO for NFC Reader SoC

Xindong Mi
School of Microelectronics Science and Technology
Sun Yat-Sen University
ZhuHai, China
mixd@mail2.sysu.edu.cn

JinBiao Zhong
School of Microelectronics Science and Technology
Sun Yat-Sen University
ZhuHai, China
zhongjb7@mail2.sysu.edu.cn

YuXuan Huang
School of Microelectronics Science and Technology
Sun Yat-Sen University
ZhuHai, China
huangyx355@mail2.sysu.edu.cn

JianGuo Hu*
School of Microelectronics Science and Technology
Sun Yat-Sen University
ZhuHai, China
hujguo@mail2.sysu.edu.cn

Abstract—This paper addresses the challenges of integration level, cost, and power efficiency in NFC Reader SoC by proposing a fully integrated, high-efficiency NFC Reader chip architecture. Through mixed-signal design, it achieves a fully integrated SoC incorporating a Cortex-M0 microcontroller, power management, digital baseband, and analog front end. To accommodate the drastic variations in load current under high-frequency communication protocols, the Cortex-M0 microcontroller and digital baseband require the LDO to possess nanosecond-level transient response capability. Consequently, a fast-transient-response hybrid architecture on-chip LDO is designed for the power management of high-speed dynamic variations in digital circuits within the SoC. This design accelerates the transient response process while ensuring output ripple, thereby addressing the issues of high ripple output and low transient response in traditional off-chip LDO designs. Results demonstrate that the fast-transient-response LDO operates within a voltage range of 2.5–4.2V, delivers an output voltage of 1.8V, achieves a transient response time of approximately 140ns, and maintains a quiescent current of 260–338μA across the entire supply voltage range. With an output capacitance of 100 pF and a maximum load current of 50 mA, making it suitable for integration into high-speed embedded systems sensitive to transient response.

Keywords—NFC, LDO, fast-transient-response, SoC, fully-integrated，high-efficiency

I. INTRODUCTION

Modern System-on-Chip (SoC) designs are implemented using massive mixed signal architectures incorporating diverse circuit blocks, thereby driving the integration of an increasing number of functionalities and components into a single chip[1]. Near Field Communication (NFC), as a widely adopted wireless communication technology in mobile and IoT devices, which core modules include radio frequency antenna, analog front end, digital baseband, microcontroller, and power management unit. The voltage regulation requirements and load characteristics of these modules exhibit significant differences. Traditional NFC reader device designs employ a multi-chip modular solution, requiring the integration of discrete components such as microcontroller IC, reader IC, and power management integrated circuit into printed circuit board[2]. However, such

multi-chip approaches incur prohibitive design complexity and impose severe layout density constraints, contradicting the demands of battery-constrained embedded systems for miniaturized hardware and energy-efficient architectures.

Low-dropout voltage regulators (LDO) are an attractive option to implement voltage domains on SoC chips[3][4][5]. When the digital circuits of NFC readers switch between sleep and wake-up states, the rapid switching of numerous dynamic load transistors induces nanosecond abrupt current changes. The LDO is required to possess a fast transient response capability to stabilize the output voltage, preventing digital circuit malfunctions caused by voltage recovery times. Additionally, the long interconnect of off-chip LDO lead to microsecond transient response time, which severely impact the timing consistency of digital circuits. To mitigate these challenges, on-chip LDO can serve as local power for SoC components, which make significant performance improvements in embedded systems.

II. PROPOSED SOC ARCHITECTURE

This hybrid analog-digital SoC designed in this paper achieves energy efficiency and low cost through innovations in architecture. It integrates an Cortex-M0 processor, digital baseband, analog front end, and two on-chip LDOs. The architecture of the SoC as shown in Fig. 1. This paper focus on the implementation of the on-chip integrated fast transient response LDO design.

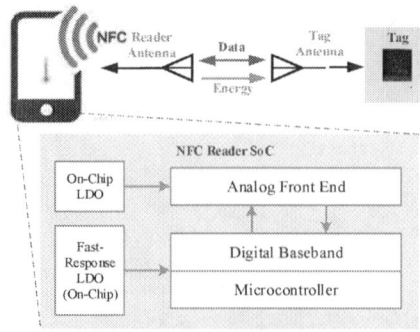

Fig. 1 Block diagram used for NFC system

This work was supported by the Guangdong Provincial Key Laboratory of High-end Integrated Circuit Design and Integration Technology under grants 2024B1212020007.

979-8-3315-8850-2/25 $31.00 © 2025 IEEE

The digital baseband supports ISO/IEC 14443 A/B and 15693 protocols, enabling real time processing of analog front-end generated modulation/demodulation signals. The Cortex M0 microprocessor is responsible for implementing the anti-collision algorithm and logic calculations, while also managing the configuration control of the digital baseband and the analog front end. The analog front end employs an integrated design for modulation and demodulation, with the transmitter based on parametric architecture that realizes a configurable ASK modulation circuit with an ultra-low modulation index. The receiver is designed with an adaptive demodulation circuit that recovers the encoded signal from the carrier feedback of tags, utilizing demodulation circuit composed of I/Q demodulation, carrier elimination, variable gain amplifier, and filter module.

The SoC integrates two 1.8V LDOs for the supply of analog and digital circuits. Through the implementation of power domain isolation, this architecture effectively suppresses digital noise on analog signals. Specifically, the digital side LDO is designed upon the framework of the analog LDO, incorporating a rapid transient response circuit to accommodate the high dynamic characteristics of digital loads. A peak current occurs during digital circuits initialization when all modules operate concurrently, stabilizing post-startup with minimal fluctuation. Additionally, the analog front end exhibits relatively high transmission power, resulting in insufficient LDO input voltage and increased noise levels. To mitigate the influence of the transmission stage, the design employs dual independent power pads and separate ground lines, thereby reducing the impact of load transients and ensuring stable operation.

III. FAST RESPONSE LDO CIRCUIT DESIGN

A. Structure of the LDO

An on-chip fully integrated LDO must possess nanosecond-level transient response capabilities while strictly meeting output voltage ripple specifications[6][7]. To combine characteristics of analog LDO and digital LDO, hybrid architecture designs have become a research focus in recent years[8][9][10]. To address the transient response requirements of NFC readers, this paper proposes a fast response hybrid LDO

architecture, as shown in Fig. 2. The main loop employs an analog LDO circuit, which achieves precise voltage tracking through a certain loop gain. Additionally, an asynchronous triggered digital feedforward loop is proposed as auxiliary path, utilizing a feedforward compensation mechanism to reduce recovery time to the nanosecond scale.

B. Analog Loop Circuit

The analog loop of this hybrid-architecture LDO, comprising a startup circuit, bandgap reference, current mirror, and voltage regulation circuit as shown in Fig. 3, delivers a low-noise and stable output voltage. The amplifier ensures that the voltage drop across R2 and D2 equals that across D1. This design utilizes the negative temperature coefficient (TC) of BJT base-emitter voltage and positive TC of base-emitter voltage differences to generate complementary PTAT and CTAT currents through resistors R2 and R3 respectively[11]. The superposition of these two currents results in a current with zero temperature coefficient for the P33M3.

The two current mirrors accurately replicate the zero temperature coefficient current and generate three reference voltages. As shown in Equation 1, the I_{PTAT} is generated by the ΔV_{BE} derived from D1 and D2 through resistor R2, whereas the I_{CTAT} is generated by applying V_{BE1} across resistor R3. The current flowing through P33M3, namely I_{REF}, is obtained by summing these two currents.

$$I_{REF} = I_{PTAT} + I_{CTAT} = \frac{U_T \ln n}{R_2} + \frac{V_{BE1}}{R_3} \quad (1)$$

The 900 mV reference voltage, with a temperature drift of approximately 8 ppm/°C, is employed as the reference level for error amplifier. The 1.9V reference voltage is used as the positive threshold for the comparator in the digital feedforward loop, while a 1.7V voltage serves as the negative threshold reference for the comparator. The regulator employs MOS-based capacitors to suppress current transients, achieving silicon area reduction versus traditional poly-Si capacitors.

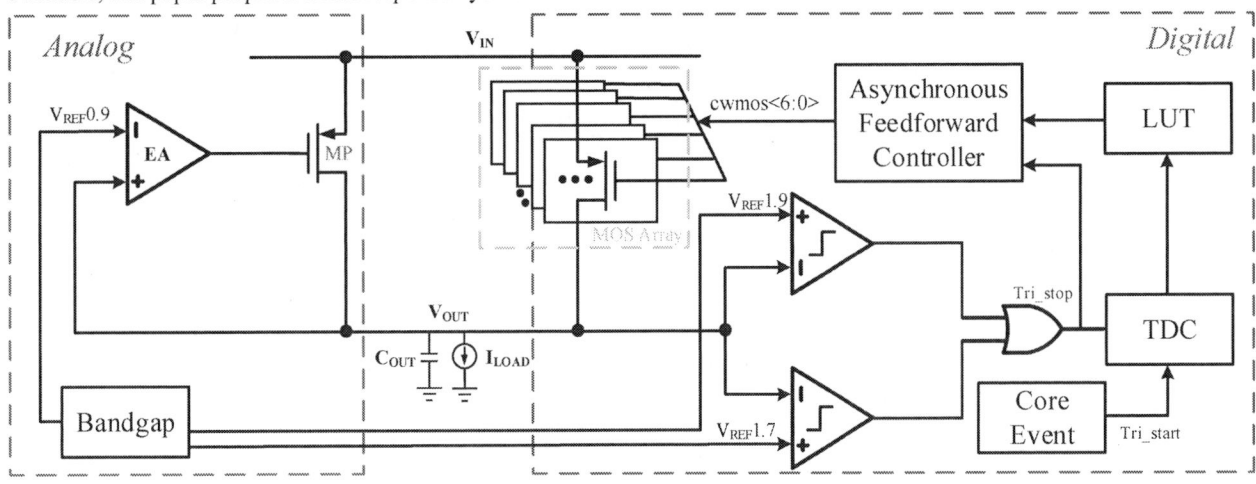

Fig. 2 Architecture of the proposed Hybrid Architecture LDO

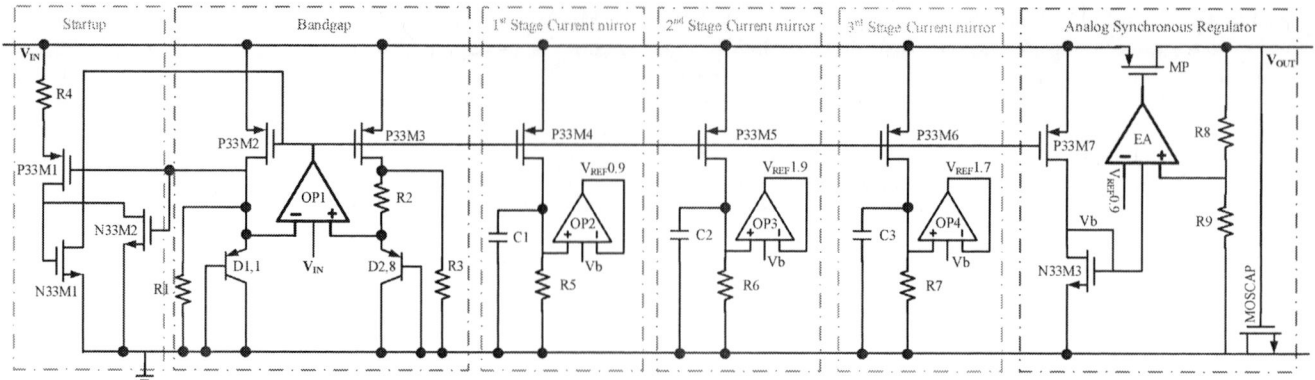

Fig. 3 The proposed analog loop architecture

Functioning as the hybrid architecture's analog controller, the error amplifier directly samples V_{OUT} and adaptively drives load current through MOS power transistor. The amplifier employs a folded cascode as shown in Fig. 4. The input stage composed of the differential pair N18M1/N18M2, which respectively receive 900 mV reference voltage and the feedback voltage. N18M1 and N18M2 are identical, while the left and right branches are symmetrically designed. A buffer (P18M1/P33M6) with low input capacitance and low output impedance is inserted between the folded cascode and power transistor to enhance stability.

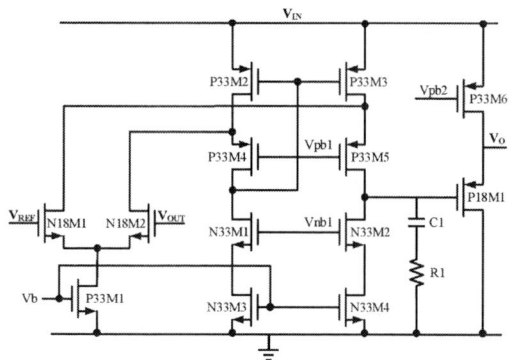

Fig. 4 Circuit implementation of the EA

C. Digital FeedForward Circuit

The digital feedforward loop comprises a continuous-time comparator, time-to-digital converter(TDC), lookup table (LUT), asynchronous feedforward controller, and MOS arrays. The comparator and MOS arrays utilize full-custom analog design, while the TDC, controller, and LUT implement standard cell-based digital automation flow. TDC quantifies the slope of the V_{OUT} changes by measuring the variation in ΔV within the ΔT. The quantification results are indexed LUT to obtain parameters, which are utilized by the controller to generate control signals cwmos<6:0> that modulate the number of transistors in MOS arrays as shown in Fig. 5. When transient events occur, an asynchronous signal is triggered at the fall edge of the clock signal following the transient event, enabling the TDC to startup.

The bandgap reference voltage used to timing termination, with 1.7V and 1.9V representing the thresholds for overshoot and undershoot comparators. When the overshoot and undershoot voltages not exceed the threshold, the controller will not continue to respond to this transient event. This design mitigates the risk of the LDO repeatedly triggering the digital feedforward loop, thereby preventing limit cycling. Additionally, it eliminates the risk of short circuit currents through the MOS arrays.

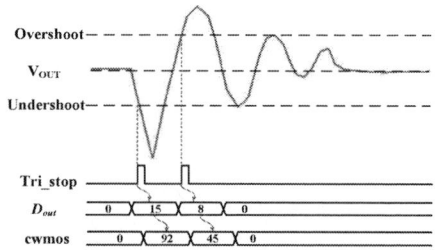

Fig. 5 Timing diagram and operation of the feedforward circuit

The required compensation current during load variation is denoted as I_{pwr}, while I_{analog} represents the current for analog loop compensation, which is insufficient to quickly meet the demands of transient response. In this context, the digital feedforward loop $I_{digital}$ through detects the V_{OUT} slope.

$$I_{pwr} = I_{analog} + I_{digital} \qquad (2)$$

Upon a I_{LOAD} event, T_{edge} is extremely brief, and the compensation current is supplied by the output capacitor. According to the charge conservation principle, it can be expressed as:

$$\int_0^{T_{edge}} I_{cap} dt = C_{out} \times \Delta V \qquad (3)$$

Differentiating both sides gives the parameter calculation equation for the slope:

$$I_{cap} = C_{out} \times \frac{\Delta V}{\Delta t} \qquad (4)$$

979-8-3315-8850-2/25 $31.00 © 2025 IEEE

$$I_{digital} = LUT(\frac{\Delta V}{\Delta t}) \qquad (5)$$

Following transient events, TDC calculates the voltage change time output D_{out}, while the LUT retrieves parameters required for feedforward current computation. Two continuous-time comparators continuously monitor V_{OUT} as shown in Fig. 6. When V_{OUT} falls outside the range defined by two voltage thresholds, the comparators result in a rapid increase in current, which leads to an instantaneous voltage spike. This change immediately alters the output state through an OR gate, thereby enabling the TDC to terminate timing.

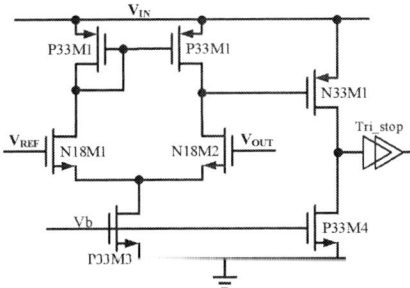

Fig. 6 Circuit implementation of the comparators

To enhance the comparison speed, the design elevates the frequencies of poles P_1 and P_2 while maximizing the distance between P_2 and P_1. The calculation formulas for the frequencies of the two poles are expressed as:

$$P_1 = \frac{1}{(r_{o1}//r_{o2})C_c}, \ P_2 = \frac{1}{(r_{o6}//r_{o7})C_L} \qquad (6)$$

However, optimal power performance trade-off is achieved by designing the W/L ratios of transistors P33M1/N18M2 and N33M1/P33M4 such that $P_2 = 2P_1$.

The TDC employs an all-digital delay chain architecture as shown in Fig. 7, utilizing delay elements and DFF stages. Upon detection of the *Tri_start* rising edge, propagating signals accumulate progressive delays through the chain. Concurrently, the *Tri_stop* rising edge triggers DFFs to latch buffer output states, generating thermometer code representation of the time between start and stop events. Direct summation of thermometer code bits produces output D_{out} for subsequent LUT retrieves. Implemented in the SMIC 180nm BCD process, each delay element achieves 50ps latency, satisfying design resolution requirements.

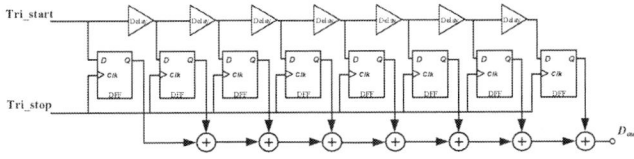

Fig. 7 Circuit implementation of the TDC

The asynchronous feedforward controller generates the regulation signal *cwmos<6:0>* in response to requirements, enabling MOS arrays modifies the compensating current to quickly regulate V_{OUT}. The finite state machine of the controller,

as shown in Fig. 8, comprises three operational modes: idle mode, transient mode, and freeze mode.

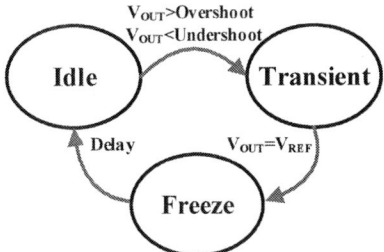

Fig. 8 FSM of the asynchronous feedforward controller

During steady-state operation, the controller maintains the number of transistors in idle mode. Upon triggering the continuous-time comparator, the controller enters transient mode, enabling the TDC and index LUT to compute the V_{OUT} slope, generating the switching signals *cwmos<6:0>* for the MOS arrays to achieve rapid transient response. Once V_{OUT} recovers to the threshold reference voltage, the system enters freeze mode and remains in this state for a preset period before enters idle mode, thereby reducing power consumption.

The size of the MOS arrays is significantly smaller than that of the MOS power transistor in the analog loop, while offering high-speed and high input impedance characteristics. Consequently, it is capable of effectively controlling response time and static current. To achieve precise load compensation current regulation, eight modules segmented control architecture is implemented for the MOS arrays. Each module integrates four MOS transistors with binary-weighted ratios: W_1: W_2: W_3: $W_4 =$ 1:2:4:8. Gate terminals feature independent control signals, while drain terminals share a common connection node.

The minimum unit current I_u of the MOS arrays, when each module has a width of W_1, results in a current of I_u generated by the activated MOS transistors. When the width is W_2, the resulting compensation current becomes $2I_u$. At a width of W_3, the compensation current further increases to $4I_u$. Finally, at a width of W_4, the compensation current reaches $8I_u$. A single module has a maximum compensation current of $15I_u$. When all modules are activated, the maximum total compensation current can reach $120I_u$. The control signal *cwmos<6:0>* is employed for fine-grained management of each MOS transistors within these modules. Specifically, *cwmos<3:0>* governs the switching states of the MOS transistors within each sub-module, while *cwmos<6:4>* controls the overall switching of the entire module.

IV. SIMULATION RESULTS AND LAYOUT

Implemented in SMIC 180nm BCD technology, the proposed SoC integrates digital and analog circuits within a compact 11.95 mm² area as shown in Fig. 9 . The digital circuits occupies 4.65 mm², including microcontroller, digital baseband, LUT, TDC, data memory, and program memory. The integrated LDO can work 2.5-4.2V supply voltage, and the output voltage is 1.8 V. The simulated quiescent current is 260-338μA, and the maximum loading capability is 50 mA. With 100 pF output capacitance, the LDO maintains stability across full load variations.

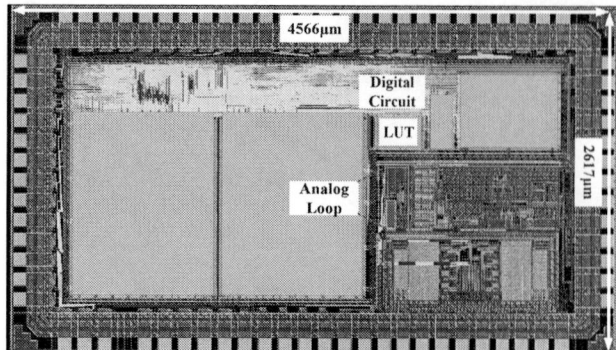

Fig. 9 NFC reader SoC chip layout

Transient performance simulation at V_{IN} = 3.3V, V_{OUT} = 1.8V and for ΔI_{LOAD} =19.8mA from 200 μA to 20 mA (20 ns edge time), as shown in Fig. 10. Off-chip LDO achieves response time 3.8μs and ΔV = 242 mV. Additionally, for ΔI_{LOAD} =19.8mA from 20 mA to 200 μA, Simulations show that response time 10.3μs and ΔV = 208 mV.

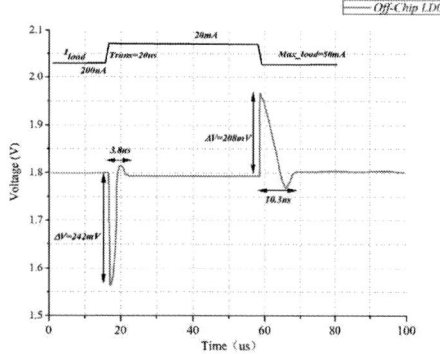

Fig. 10 Simulation transient waveform with off-chip LDO

As shown in Fig. 11, for ΔI_{LOAD} =19.8mA (200μA → 20mA), our LDO achieves response time 115ns and ΔV = 102 mV. Additionally, for ΔI_{LOAD} =19.8mA from 20 mA to 200 μA, simulations show that response time 198ns and ΔV = 104 mV.

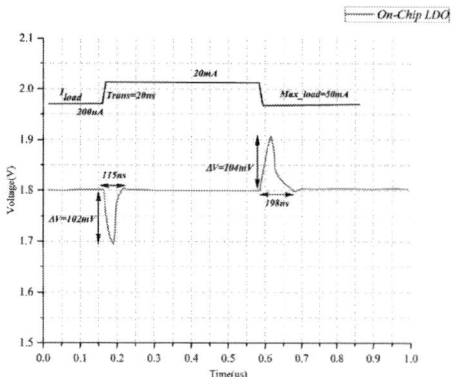

Fig. 11 Simulation transient waveform with our LDO

Table 1 presents a performance summary of this work and a comparison with published solutions, including analog, and hybrid architectures.

TABLE I. PERFORMANCE COMPARISON

Ref	[6]	[9]	[10]	This work
Process(nm)	180	65	40	180
Architecture	Analog	Hybrid	Hybrid	Hybrid
V_{out}(V)	1.5	0.55–1.15	1.1–1.25	1.8
I_Q(μA)	N/A	500	300	260-330
I_{max}(mA)	50	500	245	50
Load cap(nF)	N/A	0.9	20	100
Overshoot/Undershoot(mV)	35/24	45/125	37/71	104/102
Regulating time(ns)	1005/1039	/250	220/520	198/115

V. CONCLUSION

This work propose a hybrid architecture LDO achieving 115 ns response time with 102 mV droop for NFC reader SoC. For demanding 19.8 mA from 20mA to 200μA, simulation validates robust performance at 198 ns recovery and 104 mV voltage deviation. These results confirm the design's efficacy in sustaining stable power under fast digital load variations.

ACKNOWLEDGMENT

The authors would like to thank the Guangdong Provincial Key Laboratory of High-end Integrated Circuit Design and Integration Technology under grants 2024B1212020007.

REFERENCES

[1] Keller B, Cochet M, Zimmer B, "A RISC-V processor SoC with integrated power management at submicrosecond timescales in 28 nm FD-SOI," IEEE Journal of Solid-State Circuits, 2017, 52(7): 1863-1875.

[2] Wang D M, Hu J G, Wu J, " A fully integrated low-cost HF multistandard RFID reader SoC and module for IoT applications," IEEE Internet of Things Journal, 2022, 9(19): 19201-19213.

[3] Berido R S, Lowaton A C, "13.56 MHz highly-efficient power conditioning unit using an active rectifier and LDO for Implantable Medical Devices (IMD)," International Journal of Electrical and Electronic Engineering & Telecommunications, 2019, 8(2): 84-88.

[4] Zhu J, Li K, An Y, "A high-PSR high-precision fast-transient-response capacitor-free LDO for LiDAR receiver SoC," 2023 IEEE International Conference on Integrated Circuits, Technologies and Applications (ICTA). IEEE, 2023: 1-2.

[5] Kim S T, Shih Y C, Mazumdar K, " Enabling wide autonomous DVFS in a 22 nm graphics execution core using a digitally controlled fully integrated voltage regulator," IEEE Journal of Solid-State Circuits, 2015, 51(1): 18-30.

[6] Guo Z, Gong C, Bai R, "Fast transient response enhancement circuit for LDO," 2023 8th International Conference on Integrated Circuits and Microsystems (ICICM). IEEE, 2023: 187-190.

[7] Liu X, Zhan C, Qiao H, "Chip-area-efficient capacitor-less LDO regulator with fast-transient response," 2019 IEEE International Conference on Integrated Circuits, Technologies and Applications (ICTA). IEEE, 2019: 27-28.

[8] Kim S J, Kim D, Ham H, " A 67.1-ps FOM, 0.5-V-hybrid digital LDO with asynchronous feedforward control via slope detection and

979-8-3315-8850-2/25 $31.00 © 2025 IEEE

synchronous PI with state-based hysteresis clock switching," IEEE Solid-State Circuits Letters, 2018, 1(5): 130-133.

[9] Lu Y, Yang F, Chen F, "A 500mA analog-assisted digital-LDO-based on-chip distributed power delivery grid with cooperative regulation and IR-drop reduction in 65nm CMOS," 2018 IEEE International Solid-State Circuits Conference-(ISSCC). IEEE, 2018: 310-312.

[10] Zhou D, Jiang J, Liu Q, " A 245-mA digitally assisted dual-loop low-dropout regulator," IEEE Journal of Solid-State Circuits, 2020, 55(8): 2140-2150.

[11] Zeng Z Q, Hu J G, Wu J, " A high precision analog temperature compensated crystal oscillator using a new temperature compensated multiplier,"IEEE Transactions on Circuits and Systems I: Regular Papers, 2022, 70(2): 680-693.

2025 The 10th International Conference on Integrated Circuits and Microsystems

Design of an ATD Circuit for Asynchronous SRAM

Jialin Liu
Beijing Microelectronics Technology Institute
Beijing, China
jialin_liu0226@126.com

Rongkang Ren
Beijing Microelectronics Technology Institute
Beijing, China
1161831087@qq.com

Xin Li *
Beijing Microelectronics Technology Institute
Beijing, China
sae37230117@126.com

Jiancheng Li
Beijing Microelectronics Technology Institute
Beijing, China
lijiancheng@mx.catec.casc

Qichao Zha
Beijing Microelectronics Technology Institute
Beijing, China
zhaqichao@mx.catec.casc

Abstract—**SRAM is the cache between the cpu and the main memory, and plays a crucial role in high-reliability applications. The structure and working principle of asynchronous static random-access memory (SRAM) differ significantly from those of synchronous SRAM. To reduce power consumption and improve stability, the asynchronous SRAM is usually custom-designed with an address translate detector (ATD) module. In this paper, a new type of ATD circuit is designed for the asynchronous SRAM anti-irradiation requirements. The ATD module is able to automatically generate clock pulses according to the changes in the chip select signal, write enable signal, and address signal. The unilateral delay expansion structure and feedback loop with an advanced reset designed in the circuit can effectively improve the anti-interference capability. The trigger in the ATD is reinforced to increase the resistance of combinational circuits to single-particle transient effects. Through post-simulation verification of the designed ATD, it is demonstrated that the ATD structure with circuit-level reinforcement can achieve input change detection and exhibit good anti-interference capability.**

Keywords—***SRAM, ATD circuit, anti-interference, SET reinforcement component***

I. INTRODUCTION

Along with the booming development of the aerospace industry, the requirements for high performance, high reliability, and irradiation resistance of memory chips are increasing [1]. Asynchronous SRAM works based on the direct change of address and data lines, and it uses control signals such as CE (Chip Enable), OE (Output Enable), and WE (Write Enable) to control read and write operations [2-6]. For example, when CE is active, the address changes, and after a certain delay time, data appears on the data bus or is written. Compared to synchronous SRAM, which may operate at every clock cycle, asynchronous SRAM consumes less power when idle and is therefore widely used in low-power scenarios, such as microcontroller external memories, caches for network devices, and caches for FPGAs [7]. During operation, various disturbances (e.g., power supply noise, electromagnetic radiation, and signal crosstalk) may lead to incorrect address decoding, erroneous data writing, and faulty data reading in asynchronous SRAM systems. [8]. Common anti-interference measures at the system level include power filtering, dynamic delay adjustment, software error correction coding, etc. To better adapt to the extremity of the environment in aerospace equipment, the design of anti-interference circuits at the SRAM chip level also plays a pivotal role.

While pursuing the anti-jamming performance of the device, the irradiation resistance is also one of the main criteria for measuring the aerospace-grade chips. The irradiation resistance includes the total dose effect (TID) protection, single particle effect (SEE) protection, and so on. As the process size shrinks, single-particle effects are becoming a major source of IC failures. Single-particle effects include single-particle latch-up (SEL), single-particle flip (SEU), and single-particle transient (SET). The single-particle effects are triggered by the bombardment of semiconductor materials by high-energy particles (e.g., cosmic rays, α-particles, heavy ions, etc.), which are commonly found in space, high-altitude, or nuclear radiation environments. The single-particle effects occurre randomly and instantaneously [9]. SEL's physical mechanism is that the bombardment of particles triggers the parasitic conduction of PNPN structures to form a low impedance pathway that exhibits high current, localised heating, and low impedance. The physical mechanism of SEU is that particle bombardment causes the state of the memory cell or register to flip, usually affecting a single memory bit or logic state, causing data errors or program abnormalities. This situation doesn't damage the hardware [10]. SEL is similar to SEU in that it does not damage the hardware, but transient voltage or current pulses occur in the circuit, causing logic errors. Voltage or current pulses, resulting in logic errors or timing problems, which often occur in logic gates or combinational circuits. This paper focuses on the circuit design of SRAM chips, the key modules of the circuit design of the anti-Single Particle Transient Reinforcement.

979-8-3315-8850-2/25 $31.00 © 2025 IEEE   885

Since asynchronous SRAMs do not have an off-chip clock control, many memories utilise address change detection (ATD) circuits to generate a clock that is used to control read and write operations within the array to accommodate the internal control of asynchronous circuits. In this paper, we will propose a novel ATD circuit structure with anti-jamming design and single-particle transient resistant design, and its functionality will be analyzed by simulation and verified by simulation. The first section of this paper is the introduction, the second section is the detailed description of the ATD circuit, the third section gives the simulation verification results, and the fourth section summarises and concludes.

II. ATD CIRCUIT DESIGN

The ATD principle of operation derives from the generation of a clock signal from a change in any one of the input signals, including the chip select enable signal E, the read/write enable signal W, and the address signal A<n:0>. This clock signal is transmitted to the synchronous array to control the decoding and read/write operations within the array and is the source of all timing-related actions within the array, making the ATD circuit a key module in the overall architecture.

A. ATD circuit structure

The common ATD circuit is shown in Fig. 1, the output ATD1 has pulses generated at both the rising and falling edges of the address signal A1. The SRAM of this structure has a fast response speed, but has weak anti-interference capability. When the presence of noise in the SRAM pin leads to a burr in the address signal, the pulse width of the generated ATD clock depends directly on the width of the burr, which may result in abnormal operation within the array or read/write failure when the ATD pulse is too narrow, so it is necessary to improve it.

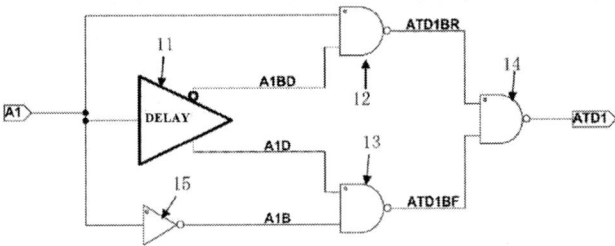

Fig. 1. Common ATD Circuit.

In this paper, a new ATD structure is designed for the needs of immunity to interference and single-particle transients, and the circuit principle is shown in Fig. 2. The overall structure is divided into two levels of ATD, the first level of ATD to achieve the change detection of address and other control signals, simple pulse merging, and output atd0; the second level of ATD will be atd0 pulse for appropriate processing to generate the final clk signal.

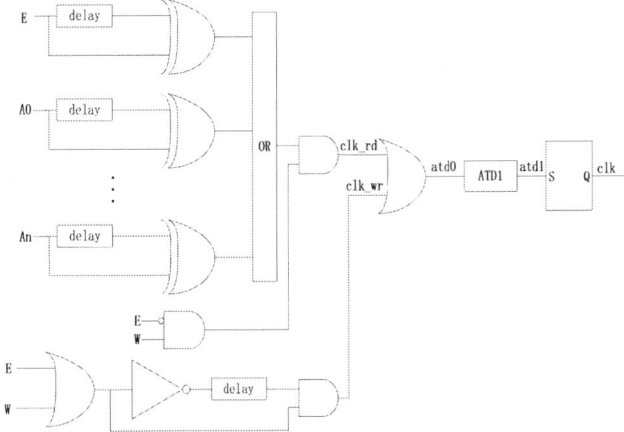

Fig. 2. The proposed ATD Circuit.

In the first-level ATD, according to the different mechanisms of read pulse and write pulse generation, it is divided into two logic paths, which generate clk_rd and clk_wr respectively, and then merge into atd0 through an OR gate. On the write logic path, the rising edge of W and E during the period of E validity can both trigger a write pulse. On the read logic path, each address signal A<n:0> and chip select enable signal E go through a level of edge detection circuit respectively, the input signal and its own delay for the heterodyne generation of a single pulse, and then go through the OR gate merged into a pulse, this pulse is controlled by the E and W to generate clk_rd. When the SRAM is interfered with the input signal in addition to monotonic jumps, burrs, jumping process with jitter may occur, Multi-bit address wrong edge, etc., at this time atd0 output has two possibilities as shown in Figure 3, respectively, a single pulse of variable width and a continuous multiple pulse of variable width and interval.

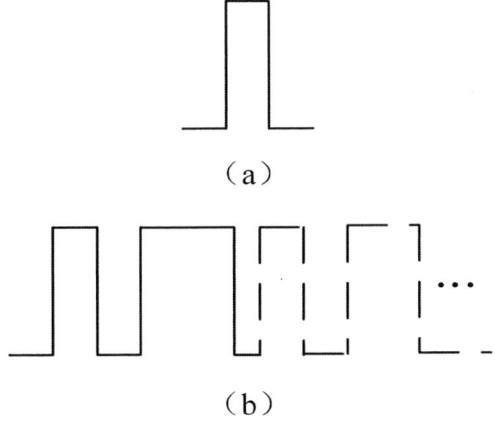

Fig. 3. Pulse condition of atd0 when there is interference at the input: (a) a single pulse of variable width; (b) a continuous multiple pulse of variable width and interval.

The second stage ATD processes atd0 appropriately to generate the final ATD signal. To handle the multiple complexities of atd0, the second stage ATD is designed with an

ATD1 structure and a D flip-flop with reset, which ultimately generates a fixed pulse width clk.

B. ATD1 unit design

The minimum width of the output atd0 of the first stage ATD depends on two points: the transmission characteristics of the heterodyne gate in the first stage ATD, which determines the climb time of the pulse, and the narrowest pulse width that can be transmitted by the flip-flop behind it. This minimum width is affected by the device operating PVT and is not easy to define, so the ATD1 is added after atd0, which is to make atd1 pulse width has a clear lower limit value, to ensure that the second level of ATD processing of atd1 pulse will not be lost. Atd0 of the maximum pulse, theoretically can be infinitely broadened, which does not produce a function of the error, and can be neglected.

The ATD1 circuit structure is shown in Figure 4, which uses the difference between the rate of rise and fall of the transmission signal to perform one-sided delayed spreading of the pulse signal. Since the falling edge of ATD1 is used to generate the final clk in the second-stage ATD, ATD1 postpones the falling edge of atd0 to achieve the effect of expanding the pulse width.

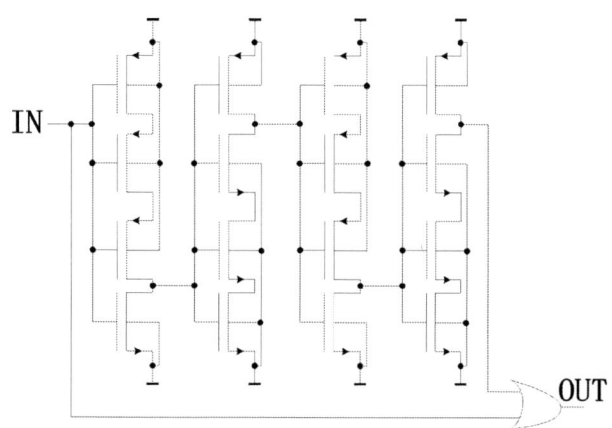

Fig. 4. ATD1 unit structure.

C. D flip-flop and reset circuit

To cope with the complexity of the situation when the SRAM is disturbed, the D flip-flop is designed as a circuit with an early reset function. The structure is shown in Fig. 5. This module is designed to ensure that the last falling edge of atd0 can generate a fixed-width pulse. Design principle: The rising edge of Q is triggered by the falling edge of S. When there is more than one falling edge of S, the reset signal RST terminates the influence of the last falling edge in advance to ensure the triggering of the next falling edge. If the two pulses of S are so close, the circuit structure determines that the preceding pulse is a redundant pulse by the trailing pulse, and truncates the preceding pulse by the fast reset path. The last pulse is pulled down by Q after a delay, the width of which is determined by the delay chain.

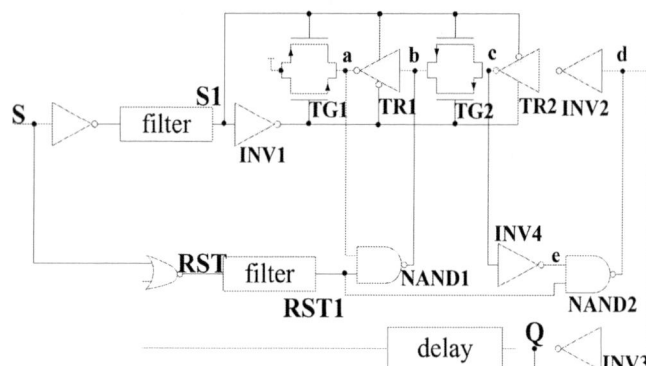

Fig. 5. D flip-flop and reset circuit structure.

The D flip-flop and reset circuit are simulate in spectre, the related waveforms are shown in Fig. 6. When RST is 1, the and non-gate NAND1 and NAND2 are logically equivalent to the inverter; when RST is 0, regardless of the state of S, the output Q is 0; one end of the TG1 is connected to the kernel power supply level, which provides the flip-flop internal node initial state. When S is a rising edge or holding 1, TG1 and TR2 are on, TR1 and TG2 are off, a is 1, b is 0, and no current flows through TG2, e is 0, d is 1, and output Q is 0, d goes through INV2 and TR2 to c, and c is 1; when S is a falling edge or holding 0, TG2 and TR1 are on, TG1 and TR2 are off, and TR1 and NAND1 form a latching loop. a and b latch 1 and 0, respectively, and pass through TG2 to c, c is 0, e is 1, d is 0, and output Q is 1 until S is pulled high or Q passes through the delay chain DELAY to RST, and the Q output is pulled down to 0 when RST is set to 0.

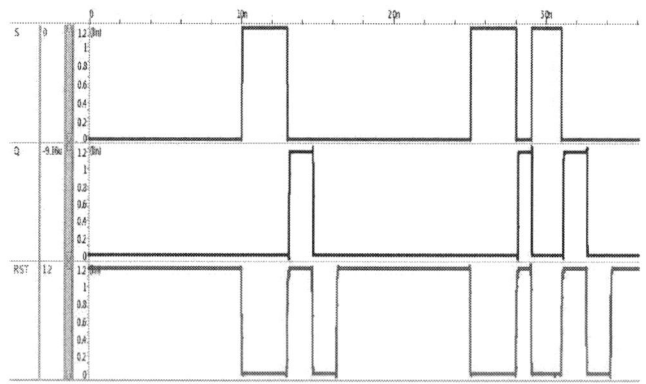

Fig. 6. Trigger Simulation Waveforms.

It can be seen through the working principle of the flip-flop that when the input S is a single pulse, no matter what the pulse width is, as long as the falling edge can be detected, the high level pulse width of the output Q is fixed for the delay of DELAY; when the input S is several consecutive pulses of unfixed width and spacing, the rising edge of the last pulse can quickly reset the pulse generated in front of it to ensure that the falling edge of the last pulse generates a. The falling edge of the last pulse is guaranteed to generate a pulse with a delay width. The advantage of this structure is that no matter what the input pulse is, a fixed-width pulse can be generated in the end, which ensures the accuracy and singularity of the ATD output, ensures the stability of the timing operation within the array, and has a good ability to resist noise interference.

D. Anti-Single Particle Transient Circuit

The ATD module is located in the central position of the SRAM chip and is the key logic module of the whole architecture, so we have to consider the accidents that may occur in the ATD when the chip is hit by high-energy charged particles. Since a large number of logic gates and combinational circuits are used in the ATD module, there is a high possibility of being hit by single-particle transients, and we have carried out a design of the ATD for the anti-single-particle transient here.

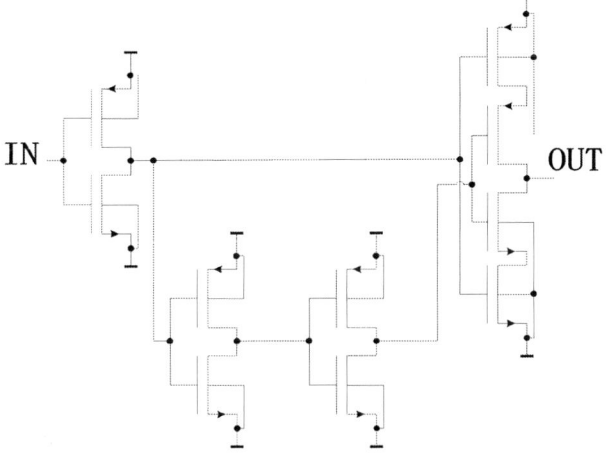

Fig. 7. Filter unit structure.

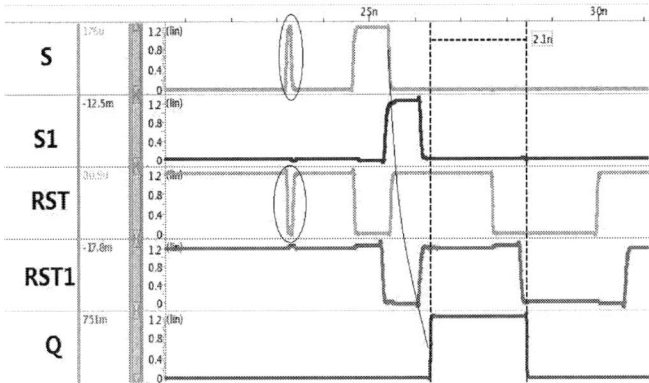

Fig. 8. Second stage ATD simulation waveform with filter structure.

When single-particle transients occur at the input of the second-stage ATD, generating transient electrical signal disturbances, i.e., burrs, we filter them in the second-stage ATD structure. By adding filter circuits after the S and RST nodes inside the D flip-flop, we can filter burrs with widths less than the delay chain in the filter without affecting the generation of the final correct pulse. The filter unit is shown in Fig. 7. The complete layout of the second stage ATD is extracted with parasitic parameters for post-simulation, and the simulation waveforms are shown in Fig. 8. When there is a burr in the input, the filter module at the S and RST end filters the burr synchronously and does not affect the falling edge of the normal pulse of S to generate a fixed-width clk at the output Q.

III. SIMULATION VERIFICATION RESULTS

After checking the process rules of the designed ATD for the layout, the parasitic parameters are extracted for the layout, and post-simulation functional and timing verification is carried out with Nanospice under three temperature and three pressure conditions. The verifications are passed, and the waveforms under various input excitations are shown in Figs. 9 and 10

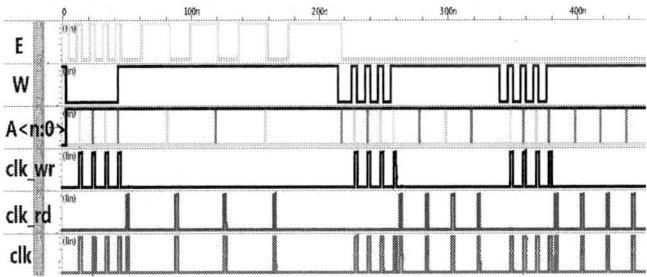

Fig. 9. ATD simulation waveform.

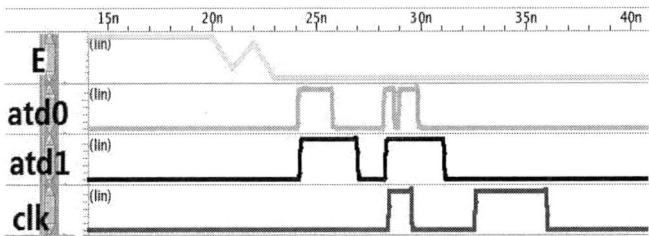

Fig. 10. Simulated waveforms when the ATD input signal has jitter.

The simulation in Fig. 9 traverses the input excitation that triggers the asynchronous SRAM write and read operations, and the ATD module outputs clk high-level pulses of the same width, which saves power and stably generates the output clock compared to the common ATD structure. When the input signal has jitter or addresses the wrong edge as shown in Fig. 10, the ATD1 circuit can expand the continuous narrow pulse of atd0 into a wide pulse to facilitate the processing of the flip-flop in the second-stage ATD, and the fast reset path resets the clk generated by the consecutive pulses ahead of time, only retaining the falling edge of the last atd1 pulse to generate a complete fixed-width clk, and the two clk pulse interval is adequate, consistent with the design requirements. The circuit structure has a good anti-interference capability.

IV. CONCLUSION

In this paper, an ATD structure applied to an asynchronous irradiation-resistant SRAM is designed, and a reset loop to handle interference signals is designed in the ATD circuit. For the single-particle transient effect caused by the space radiation environment, the ATD circuit is reinforced in timing. Finally, the designed ATD circuit, which meets the layout design rules, extracts the parasitic parameters and carries out the post-simulation verification, and the simulation results verify the correctness of the function, and the output results are still stable when the input signal has interference, which meets the expected design goals.

REFERENCES

[1] J. Fan and C. Rongqiu, "Study on the evaluation of aerospace microelectronic industry," in Journal of Systems Engineering and Electronics, vol. 15, no. 3, pp. 241-247, Sept. 2004.

[2] H. Mair and A. Kawasumi, "Session 17 overview: SRAM," 2016 IEEE International Solid-State Circuits Conference (ISSCC), San Francisco, CA, USA, 2016, pp. 304-305.

[3] Shin, C. (2016). Applications in Static Random Access Memory (SRAM). In: Variation-Aware Advanced CMOS Devices and SRAM. Springer Series in Advanced Microelectronics, vol 56. Springer, Dordrecht.

[4] A. B. Smitha, S. Shetty, H. Chinmaya, S. R. Arasa, K. Shreyas and S. V. Naik, "Asynchronous SRAM: Towards Energy-Efficient Solutions," 2024 Asian Conference on Intelligent Technologies (ACOIT), KOLAR, India, 2024, pp. 1-6.

[5] Huang, C. R., & Chiou, L. Y. (2018). Single bit-line 8T SRAM cell with asynchronous dual word-line control for bitinterleaved ultra-low voltage operation. IET Circuits, Devices and Systems, 12(6), 713-719.

[6] Ding-Ming Kwai et al., "SRAM cell current in low leakage design," 2006 IEEE International Workshop on Memory Technology, Design, and Testing (MTDT'06), Taipei, Taiwan, 2006, pp. 6.

[7] B. Vaghela, K. Shukla, H. Gupta, S. Mehta and A. Mishra, "Design of Low Power Memory Array using Address Transition Detection (ATD) Circuit," 2020 IEEE International Conference on Electronics, Computing and Communication Technologies (CONECCT), Bangalore, India, 2020, pp. 1-4.

[8] C. D. C. Arandilla and J. A. R. Madamba, "Comparison of Replica Bitline Technique and Chain Delay Technique as Read Timing Control for Low-Power Asynchronous SRAM," 2011 Fifth Asia Modelling Symposium, Manila, Philippines, 2011, pp. 275-278, doi: 10.1109/AMS.2011.58.

[9] Y. Ishii et al., "A 28 nm Dual-Port SRAM Macro With Screening Circuitry Against Write-Read Disturb Failure Issues," in IEEE Journal of Solid-State Circuits, vol. 46, no. 11, pp. 2535-2544.

[10] F. Saigné, et al."Prediction of long-term thermal behavior of an irradiated SRAM based on isochronal annealing measurements," in Microelectronics Reliability, 42(3), 2002, pp. 459-461.

979-8-3315-8850-2/25 $31.00 © 2025 IEEE

2025 The 10th International Conference on Integrated Circuits and Microsystems

An Ultra-Wideband Compact Power Divider for Phased Array Radar Systems

Wanfu Liu
School of Integrated Circuits, Southeast University
Nanjing, China
lwf830607@126.com

Jianhui Wu
School of Integrated Circuits, Southeast University
Nanjing, China
wjh@seu.edu.cn

Abstract—**This brief proposes an ultra-wideband compact two-way power divider operating over 12-40GHz for phased array radar systems. The design replaces conventional quarter-wavelength transmission lines with lumped-element and incorporates a two-stage isolation network to achieve low insertion loss, wideband operation, and significant miniaturization. The proposed divider exhibits a return loss better than 12.53dB, insertion loss below 3.96dB, and port-to-port isolation exceeding 12.44dB across 12-40GHz. Furthermore, this novel topology occupies merely 1.34 mm² of chip area with all pads, demonstrating good potential for broadband phased array applications.**

Keywords—power divider, quarter-wavelength transmission, miniaturization, broadband phased array

I. INTRODUCTION

Phased array radar (PAR) systems represent the foundational infrastructure for contemporary defense surveillance, satellite communications, and the rapidly developing millimeter-wave facets of 5G/6G wireless networks [1]-[4]. The defining capability of PAR architectures—the dynamic electronic steering and shaping of radiated beams—hinges directly on the fidelity and performance of radio-frequency (RF) front-end subsystems, among which power distribution networks are of primary importance. As PAR systems progress toward ever larger instantaneous fractional bandwidths and substantially increased element counts, the technical demands placed on power-dividing networks intensify: conventional power-dividing topologies encounter intrinsic limitations in physical footprint, dispersive behavior, and achievable isolation across broad frequency spans.

Millimeter-wave operation exacerbates these constraints by compressing wavelength-scale circuit dimensions, magnifying parasitic effects, and imposing tighter tolerances on loss and phase linearity. Consequently, the design of power dividers for contemporary and next-generation PAR and millimeter-wave communications equipment must simultaneously satisfy a triad of often-conflicting objectives: broadband impedance matching and phase/amplitude balance, aggressive miniaturization amenable to monolithic or tightly integrated implementations, and minimal insertion loss to preserve overall system noise figure and radiated power efficiency. The canonical Wilkinson power divider (WPD) continues to enjoy widespread adoption in both commercial and defense-grade systems because of its topological simplicity, predictable amplitude/phase balance, and intrinsic output-port isolation afforded by the embedded

resistive network. The classical WPD is a reciprocal, matched three-port network that distributes or combines RF

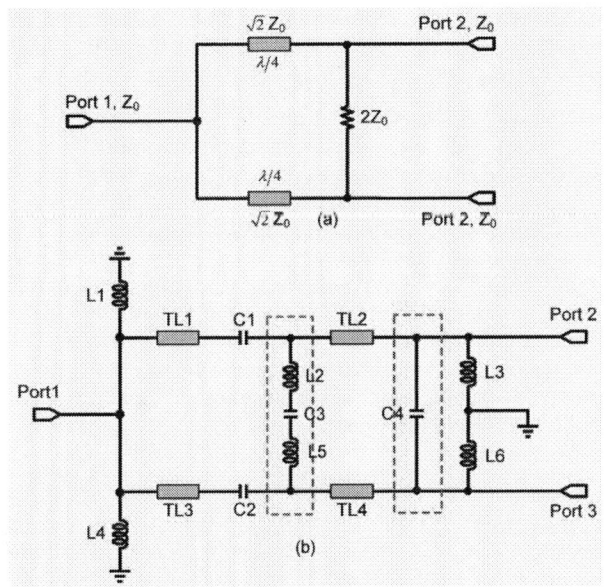

Fig. 1. (a) Conventional Wilkinson power divider (WPD) and (b) the proposed power divider.

power by employing λ/4 transmission-line sections to transform impedances while isolating output ports through a shunt resistor.

However, at millimeter-wave frequencies the quarter-wavelength distributed elements used in conventional WPDs become electrically compact but physically challenging: their physical lengths, even when small in absolute terms, are significant relative to integration constraints on monolithic microwave integrated circuits (MMICs) and packaging. In addition, the dispersive nature of transmission-line transformers yields frequency-dependent amplitude and phase imbalances and increased insertion loss over wide instantaneous bandwidths. These fundamental drawbacks have stimulated extensive research into alternative divider topologies [5]-[8] and design methodologies that can meet the stringent performance targets of modern PAR front ends.

Recent innovations fall into several complementary categories. Multifunctional divider topologies integrate bandpass or band-stop filtering responses directly into the power split network, thereby reducing component count and improving

979-8-3315-8850-2/25 $31.00 © 2025 IEEE

spectral selectivity for co-site interference rejection [9]-[11]. Miniaturization strategies exploit synthetic transmission lines and metamaterial-inspired unit cells to realize electrically long transformations in physically compact footprints, enabling substantial area reduction while partly controlling dispersion [12], [13]. Bandwidth-enhancement approaches employ capacitive loading, multi-section impedance transformers, or resonant-mode engineering to flatten amplitude/phase responses across wider fractional bandwidths and to suppress return loss variation [14]. Alternative isolation schemes, such as cascaded resistor–capacitor (CRC) networks and actively compensated resistive bridges, have been proposed to extend isolation bandwidth and reduce the impact of resistor parasitics at high frequency [15]. Moreover, balanced and differential divider configurations-often combined with common-mode suppression structures-offer improved spurious rejection, better harmonic behavior, and compatibility with modern differential MMIC processes [16]-[18].

Collectively, these architectural and component-level advancements map onto the broader system imperatives for PARs: scalable array integration, reduced front-end complexity, and resilient wideband performance under realistic manufacturing and thermal constraints. Nevertheless, each proposed solution entails trade-offs among insertion loss, isolation, amplitude/phase balance, linearity, power-handling capability, and manufacturability. Hence, the ongoing trajectory of research emphasizes rigorous co-design of electromagnetic topology, lumped/distributed element synthesis, and process-aware layout techniques to reconcile these competing metrics for millimeter-wave PAR deployment.

This work presents a high-isolation, ultra-compact broadband two-way power divider operating over 12-40GHz operating across the X-band/Ku-band/Ka-band. The new design replaces conventional quarter-wavelength transmission lines with lumped-element networks and incorporates a two-stage isolation circuit, establishing a new paradigm in UWB power division, enabling high-density millimeter-wave arrays for aerospace, automotive radar, and terahertz communications systems.

This paper is organized as follows. Section II presents the circuit design. Section III provides details on a prototype design and results analysis. Conclusions and potential future work are discussed in Section IV.

II. DEVICE DESIGN

Fig. 1 illustrates the schematic of conventional Wilkinson power divider (WPD) and the power divider adopting the lumped element with LCL and C two-stage isolation network. The canonical WPD, while providing excellent port matching and isolation through quarter-wavelength ($\lambda/4$) transformers and isolation resistors, suffers inherent narrowband operation. Its operational bandwidth is theoretically capped at $\approx 20\%$ for 15dB return loss due to the frequency-dependent impedance transformation of $\lambda/4$ lines. This proposed design efficiently achieves ultra-wideband low-loss performance and compact footprint through a symmetric configuration utilizing merely a two-stage isolation network, two lumped-element capacitors, and four ESD-protection inductors. And the specific dimensions

and related parameters of the designed PD are presented in TABLE I.

The proposed power divider utilizes a symmetric topological configuration integrating six spiral inductors and three metal-insulator-metal (MIM) capacitors, as schematically depicted in Fig. 1. The spiral inductors constitute the dominant factor in determining the overall physical footprint of the circuit. To achieve rigorous phase and amplitude balance across the output ports, the design exhibits bilateral symmetry encompassing both inductive and capacitive signal paths. Within this architecture, inductors L_1, L_3, L_4, and L_6 fulfill dual operational roles: (1) facilitating broadband power division through impedance transformation, and (2) functioning as integrated electrostatic discharge (ESD) protection structures. These ESD-hardened inductors operate primarily as high-frequency choke elements, exploiting their frequency-dependent impedance characteristics to attenuate transient electromagnetic disturbances. Specifically, they suppress the propagation of ESD-induced fast-transient currents (characterized by rise times < 1 ns) into sensitive active core circuitry by establishing high-impedance isolation barriers at RF operating frequencies.

TABLE I.   COMPONENT PARAMETERS OF POWER DIVIDER

Symbol	Value	Symbol	Value
L_1	1.03nH@20GHz	C_1	107fF
L_2	138.5pH@20GHz	C_2	107fF
L_3	2.6nH@20GHz	C_3	84fF
L_4	1.03nH@20GHz	C_4	53.5fF
L_5	138.5pH@20GHz	L6	2.6nH@20GHz

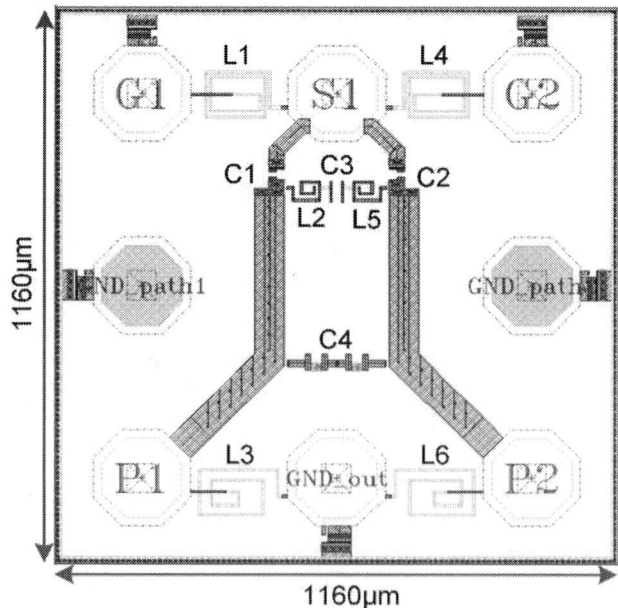

Fig. 2.   (a) Physical layout of the proposed power divider.

979-8-3315-8850-2/25 $31.00 © 2025 IEEE

Fig. 2 quantifies the physical implementation metrics, revealing a compact layout occupying a total die area of 1.34 mm² fabricated in a commercial Dongbu 130nm RF Silicon-on-Insulator (SOI) process. Here, port S1 serves as the common input/output for power splitting/combining operations, while ports P1 and P2 denote the balanced output branches. The spiral inductor geometry is optimized to minimize substrate eddy-current losses through strategic winding pitch control and hollow center design, concurrently enhancing quality factor (Q-factor) performance. Additionally, p+/n+ guard rings are implemented circumferentially to mitigate CMOS latch-up susceptibility-a critical design consideration for ESD-robust radio frequency integrated circuits (RFICs). The SOI substrate's buried oxide layer further contributes to parasitic capacitance reduction and substrate noise isolation, thereby improving harmonic distortion performance.

III. Example and Results Analysis

Fig. 3 (a) and (b) present the post-layout simulation results of input return loss and output return loss of the proposed power divider for the proposed power divider across the 12-40GHz band. As can be observed, the input return loss is better than 12.53dB, the output return loss is better than 11.37dB, respectively. Further, the corresponding insertion loss and isolation are also presented, as shown in Fig. 4. As can be seen that, in the entire operating frequency band from 12GHz to 40GHz, the insertion loss and isolation are better than 3.93dB and 12.44dB, respectively. From the above results, it can be seen that the miniaturized ultra-wideband power divider achieves good input and output impedance matching.

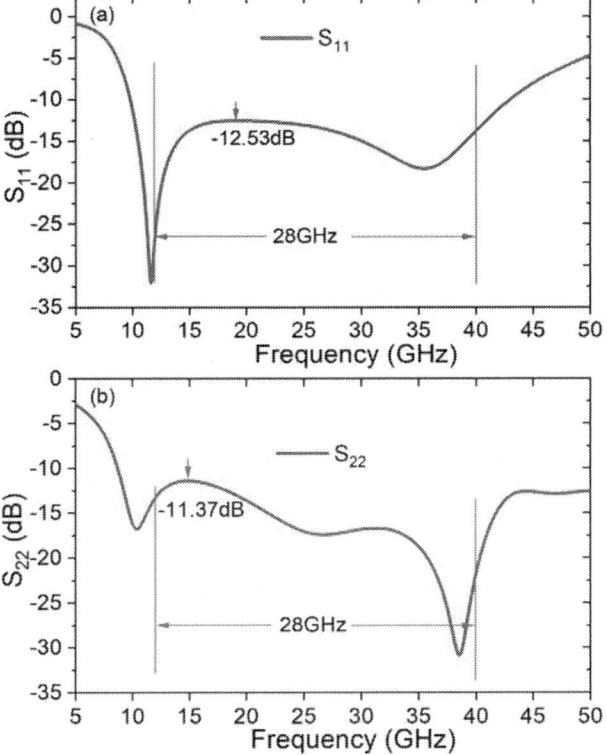

Fig. 3. Post-layout simulation results of input return loss and output return loss of the proposed power divider.

High port-to-port isolation constitutes a principal figure of merit in the evaluation of power-divider performance. Port-to-port isolation quantifies the degree of undesired signal leakage or coupling between output ports and directly impacts the extent to which energy launched into one branch of the network can re-enter or perturb neighboring branches. Elevated isolation is therefore essential to suppress mutual interference among parallel signal paths in multi-channel architectures-such as phased-array radar subarrays and multi-input multi-output (MIMO) transceivers-where inadvertent inter-port coupling can degrade beamforming precision, impair channel orthogonality, increase sidelobe levels, and elevate system-level error-vector-magnitude and bit-error-rate metrics.

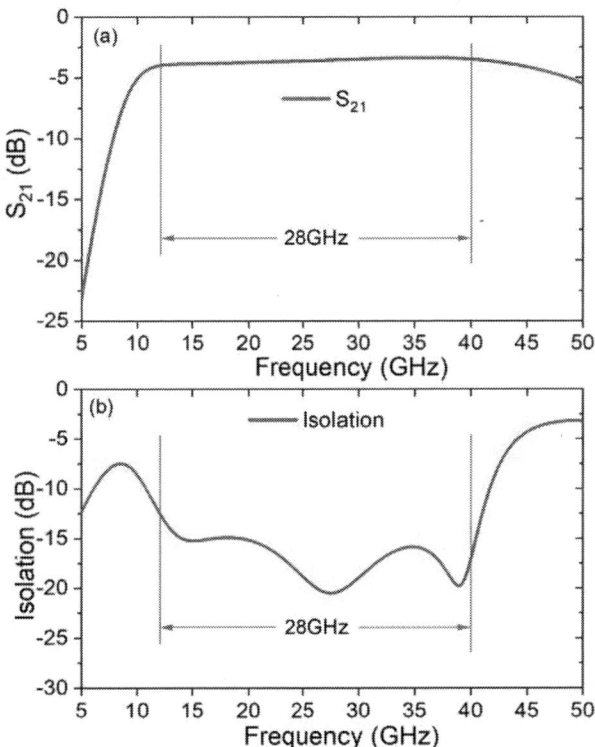

Fig. 4. Post-layout simulation results of insertion loss and isolation of the proposed power divider.

Accordingly, we have performed a comprehensive investigation of isolation sensitivity with respect to variations in the isolation-capacitor values employed in the first and second stages of the divider across the intended operational band. Fig. 5 presents the frequency-dependent isolation characteristics for a parametric sweep of these first- and second-stage capacitances, thereby elucidating the sensitivity of isolation depth and bandwidth to component tolerances and process variation. This analysis highlights the design margins necessary to maintain robust isolation across manufacturing spreads and assists in selecting capacitance values that optimize the trade-off between isolation, insertion loss and impedance matching.

In parallel, temperature-dependent isolation performance was evaluated to determine the robustness of the isolation network under realistic environmental conditions. Fig. 6 reports isolation simulated at the three representative temperatures of -

40 °C, +25 °C, and +85 °C-temperatures that span the typical operating envelope for commercial hardware. The results indicate that isolation remains better than 12dB across the entire frequency band at all three temperatures, demonstrating thermal stability of the isolation mechanism and limited sensitivity to temperature-induced variations in dielectric constant, conductor resistivity, and lumped-component parameters.

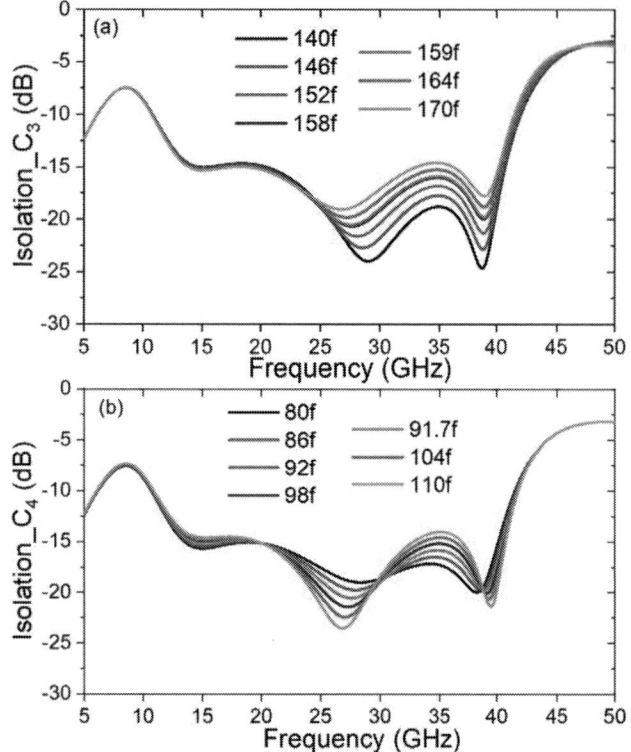

Fig. 5. Isolation of the first-and second-stage isolation capacitors variation in the operating frequency band.

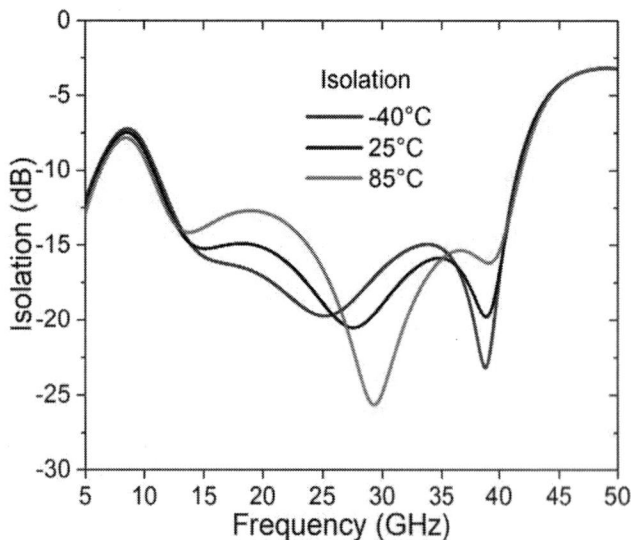

Fig. 6. Isolation of the proposed power divider at -40°C, 25°C, 85°C.

Beyond isolation magnitude, amplitude and phase balance between the two output ports are critical for coherent applications. Fig. 7 summarizes the differential response between the input (Port 1) and the output ports (Port 2 and Port 3). Over the 10-50GHz band, the design exhibits exceptional symmetry: the phase mismatch between the two output paths is less than 0.048°, and the amplitude imbalance is below 0.01 dB. Such minimal amplitude and phase deviation are indicative of precise impedance and path-length matching, and it ensures high-quality, low-distortion power splitting suitable for wide-band coherent combining and beamforming. The combined evidence from the capacitance sensitivity study, temperature sweep, and amplitude/phase balance characterization confirms that the proposed divider topology delivers both isolation robustness and excellent port symmetry-attributes that are indispensable for high-performance millimeter-wave array front ends.

The performance summary of the proposed power divider is presented in Table II. This work exhibits god microwave performance while having the significant miniaturization in ultra-wide band operation.

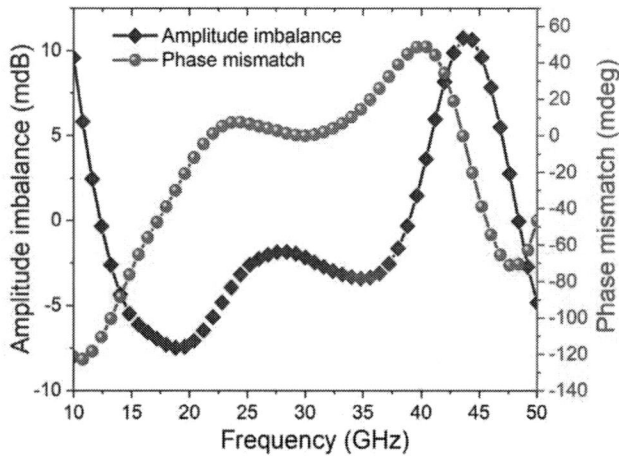

Fig. 7. Post-layout simulation results of amplitude imbalance and phase mismatch.

TABLE II.   PERFORMANCE COMPARISON OF PHASE SHIFTERS

Ref.	[15]	[19]	[20]	**This work**
Topology	Lumped-element	Lumped-element	Lumped-element	**Lumped-element**
BW. (GHz)	5.2-19	8-14	5.5-11	**12–40**
Return Loss (dB)	>10	>12	>15	**>12.53**
Isolation (dB)	30@12GHz	15@12GHz	12.8@8.5 GHz	**20.5@29 GHz**
Technology	0.13 um SiGe BiCMOS	0.35 um SiGe BiCMOS	High resistivity Si	**130nm RF SOI**

IV. CONCLUSION

This work demonstrates a high-isolation ultra-compact two-way power divider achieving broadband operation from 12 to 40GHz through innovative topology optimization. By replacing distributed quarter-wavelength transmission lines with lumped-element equivalents and implementing a two-stage isolation network, the design simultaneously attains: (i) Enhanced RF performance with input/output return loss >12.53dB, insertion loss <3.96dB, and port-to-port isolation >12.44dB across the full Ka-band spectrum; (ii) Significant miniaturization to a record 1.34 mm² core area; (iii) Robust frequency response without performance degradation at band edges. The synergistic integration of lumped components and multi-stage isolation techniques establishes a new paradigm for phased-array front-end miniaturization, addressing critical size constraints while maintaining stringent microwave performance requirements.

ACKNOWLEDGMENT

The authors would like to acknowledge the Fundamental Research Funds for the Central Universities (2242025K30014) and Dongbu 130nm RF SOI process for their support.

REFERENCES

[1] F. Bachbauer, S. Pietschmann, J. Shi, and G. Gold, "Design and Additive Manufacturing of an E-Plane 1-to-9 Power Divider for Satellite Communication," in *2025 IEEE Radio and Wireless Symposium (RWS)*, 19-22 Jan. 2025 2025, pp. 63-66.

[2] Sara Said, Abdenacer Es-salhi, Kaabal abdelmoumen, and Samir Elouaham, "New Integrated Dual-Band Dipole Antenna with EBG Structures Designed for Wireless Communications," International Journal of Electrical and Electronic Engineering & Telecommunications, vol. 11, no. 3, pp. 184-191, May 2022.

[3] Abdulsattar M. Ahmed, Sayf A. Majeed, and Younis S. Dawood, "A Survey of 6G Mobile Systems, Enabling Technologies, and Challenges," International Journal of Electrical and Electronic Engineering & Telecommunications, vol. 12, no. 1, pp. 1-21, January 2023.

[4] Arslan Ahmed Sohoo, Fauziahanim Che Seman, Yee See Khee, Noor Azura Awang, and Izhar Ahmed Sohu, "Low Loss THz Waveguides and Its Potentials towards 6G Communication: A Brief Chronicle Review," International Journal of Electrical and Electronic Engineering & Telecommunications, vol. 13, no. 1, pp. 1-16, 2024.

[5] B. G. Liu, Y. P. Lyu, L. Zhu, and C. H. Cheng, "Compact Square Substrate Integrated Waveguide Filtering Power Divider with Wideband Isolation," IEEE Microwave and Wireless Components Letters, vol. 31, no. 2, pp. 109-112, 2021.

[6] Panya Hantula and Rangsan Tongta, "Design of Two L-Band RF Amplifiers Combination Using Wilkinson Power Dividers," International Journal of Electrical and Electronic Engineering & Telecommunications, vol. 9, no. 1, pp. 38-42, January 2020.

[7] Y. H. Zhu, W. Qin, and J. X. Chen, "Compact Waveguide Filtering Power Dividers with Flexible Division Ratio and Enhanced Selectivity," IEEE Transactions on Components, Packaging and Manufacturing Technology, vol. 14, no. 11, pp. 2043-2049, 2024.

[8] L. Liu, L. Zhu, Z. B. Wang, and Y. R. Zhang, "Proposal and Synthesis of Self-Packaged Wideband Bandpass Power Divider With Constant Power Ratio and Full Phase Difference Range," *IEEE Transactions on Microwave Theory and Techniques*, vol. 73, no. 3, pp. 1645-1658, 2025.

[9] M. Kumar, G. Basavarajappa, and K. Rawat, "Design of Multifunctional Filtering Power Divider in Coaxial Technology for Power Combining Applications," in *2024 IEEE/MTT-S International Microwave Symposium - IMS 2024*, 16-21 June 2024 2024, pp. 138-141.

[10] C. Zhu, J. Xu, W. Kang, and W. Wu, "Microstrip Multifunctional Reconfigurable Wideband Filtering Power Divider with Tunable Center Frequency, Bandwidth, and Power Division," *IEEE Transactions on Microwave Theory and Techniques*, vol. 66, no. 6, pp. 2800-2813, 2018.

[11] C. F. Chen, Y. S. Zeng, Y. C. Yeh, T. N. Tien, R. Y. Yang, and C. H. Lo, "Design of Multifunctional Multiway Filtering Power Dividers Based on Tri-Mode Resonators," *IEEE Transactions on Components, Packaging and Manufacturing Technology*, vol. 15, no. 3, pp. 525-534, 2025.

[12] M. J. Chiang, H. S. Wu, and C. K. C. Tzuang, "A Ka-Band CMOS Wilkinson Power Divider Using Synthetic Quasi-TEM Transmission Lines," *IEEE Microwave and Wireless Components Letters*, vol. 17, no. 12, pp. 837-839, 2007.

[13] K. Nithiporndecha, S. Wang, and C. Pakasiri, "Compact Wilkinson Power Divider Using Composite Right/Left-Handed Transmission Line on CMOS Process," in *2020 8th International Electrical Engineering Congress (iEECON)*, 4-6 March 2020 2020, pp. 1-4.

[14] M. C. Scardelletti, G. E. Ponchak, and T. M. Weller, "Miniaturized Wilkinson power dividers utilizing capacitive loading," *IEEE Microwave and Wireless Components Letters*, vol. 12, no. 1, pp. 6-8, 2002.

[15] I. Ju, M. K. Cho, I. Song, and J. D. Cressler, "A Compact, Wideband Lumped-Element Wilkinson Power Divider/Combiner Using Symmetric Inductors with Embedded Capacitors," *IEEE Microwave and Wireless Components Letters*, vol. 26, no. 8, pp. 595-597, 2016.

[16] H. Y. Li, J. X. Xu, and X. Y. Zhang, "Miniaturized Balanced Filtering Power Dividers with Arbitrary Power Division Ratio Using Multimode Dielectric Resonator in Single Cavity," *IEEE Transactions on Circuits and Systems II: Express Briefs*, vol. 69, no. 6, pp. 2707-2711, 2022.

[17] P. L. Chi and C. P. Chien, "Balanced-to-Balanced Power Divider with Tunable In-Phase/Out-of-Phase Power-Dividing Ratio," in *2018 Asia-Pacific Microwave Conference (APMC)*, 6-9 Nov. 2018 2018, pp. 1483-1485.

[18] F. Wei, J. X. Wang, X. B. Zhao, C. Zeng, and J. Q. Hou, "Balanced-to-Single-Ended Four-Way Out-of-Phase Power Divider and Its Application to Broadband Balanced Quasi-Yagi Antenna Array," IEEE Antennas and Wireless Propagation Letters, vol. 19, no. 8, pp. 1370-1374, 2020.

[19] M. Caruso, A. Bevilacqua, and A. Neviani, "An X -Band Lumped-Element Wilkinson Combiner With Embedded Impedance Transformation," IEEE Microwave and Wireless Components Letters, vol. 24, no. 10, pp. 689-691, 2014.

[20] L. Liang-Hung, P. Bhattacharya, L. P. B. Katehi, and G. E. Ponchak, "X-band and K-band lumped Wilkinson power dividers with a micromachined technology," in 2000 IEEE MTT-S International Microwave Symposium Digest (Cat. No.00CH37017), 11-16, vol. 1, pp. 287-290, 2000.

979-8-3315-8850-2/25 $31.00 © 2025 IEEE

A Timing Interleaving RX Scheme with direct Multiplexing Sampler based on dual reference for DDR interface

Mia Xu
Integrated Circuits and IP Department
KINGTIGER Test Technology Ltd.
Shenzhen, China
Mia.xu@kingtigertest.com

Chris Eom
Integrated Circuits and IP Department
KINGTIGER Test Technology Ltd.
Shenzhen, China
yifan.shi@kingtigertest.com

Kenji Wen
Integrated Circuits and IP Department
KINGTIGER Test Technology Ltd
Shenzhen, China
Kenji.wen@Kingtigertest.com

Tinna Ding
Integrated Circuits and IP Department
KINGTIGER Test Technology Ltd
Shenzhen, China
Tinna.ding@Kingtigertest.com

Bosco Lai
Head of R&D Center
KINGTIGER Testing Technology Ltd
Shenzhen, China
Bosco.lai@Kingtigertest.com

Abstract— A timing-interleaving RX scheme with a direct multiplexing sampler based on a dual reference is presented. To address the challenges associated with the current string problem, which leads to significant DFE coefficient variation due to voltage and temperature (VT) changes, a dual reference scheme has been implemented. However, this approach results in a circuit complexity that is more than twice that of the original design. To mitigate this issue, we propose the adoption of a Timing-Interleaving DFE (TID) summer and a direct multiplexing sampler. This method effectively reduces the circuit complexity to less than half of that required by the conventional loop-unrolled DFE design. Compared to the traditional current string scheme, the proposed design achieves a 90% reduction in DFE coefficient variation and improves the DFE feedback time by 0.12 UI at 7200Mbps

Keywords—direct feedback, loop-unrolled DFE, timing interleaving RX

I. INTRODUCTION

With the advent of machine learning and the invention of OpenAI, the demand for data centers capable of processing vast amounts of data has grown substantially. This has led to a dramatic surge in the demand for high-capacity memory, such as DDR5, as well as an exponential increase in the required I/O speeds. The server-oriented I/O interface operates in a multi-drop environment, where not only channel loss but also reflection are key factors contributing to signal degradation. To address these challenges, to overcome these issues, various

equalizer schemes, such as DFE (Decision Feedback Equalizer) and CTLE (Continuous Time Linear Equalizer), have been developed. Especially, a 4-tap Decision Feedback Equalizer (DFE) has been adopted as part of the JEDEC specification in DDR5. As shown in Table I, the DFE coefficients for DDR5 are determined by the vendor and utilize the current string method.

TABLE I.   JEDEC SPEC COMPARISON

	DDR5	GDDR6
1st DFE Tap	+/- 41 step	+/- 15 step
Coefficient	(1step = vendor specific)	(1step = 0.5% VDDQ)

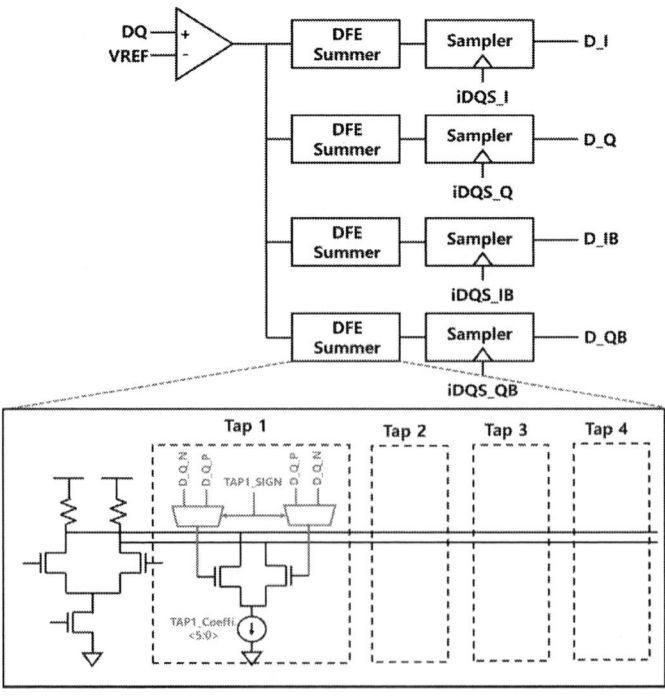

Fig. 1. Conventional DDR5 RX scheme

In contrast, GDDR6, which operates at the highest speeds in single-ended signaling, is linked to VREFDQ. This highlights the increasing importance of DFE coefficient variation at higher speeds [1][2][3]. Since DFE is a method for compensating for channel conditions, it is independent of the DRAM's PVT (Process, Voltage, and Temperature) variations.

Figure 1 illustrates the conventional DDR5 RX scheme, which includes a VGA, a 4-tap DFE summer composed of four current strings, and four samplers, each consisting of a sense amplifier and latch [4][5]. The number of VGAs is determined by the output loading requirements and the capabilities of the process technology. The DFE coefficients are optimized during the training process, a procedure that is notably time-consuming. Unlike ZQ calibration, DFE does not support periodic retraining or background operation. Since DFE coefficients are controlled by current, variations in the DRAM's voltage and temperature (VT) during operation can result in more than a twofold change in current values. In contrast, channel characteristics remain independent of DRAM and do not fluctuate with changes in DRAM VT, leading directly to performance degradation.

II. ARCHITECTURE

As operating speeds increase, the primary bottleneck in RX performance shifts to the feedback time of the DFE's first tap, which is influenced by the summer's settling time and the flight time from the sampler to the summer. If this feedback time exceeds one unit interval (UI), the subsequent data bit is predetermined, rendering the DFE ineffective. To mitigate this limitation, advanced techniques such as loop-unrolled DFE and direct feedback in sampler schemes have been developed [6]. In the loop-unrolled DFE, the feedback time is mainly determined by the MUX selection delay. In contrast, in direct-feedback-in-sampler circuits, it is determined by the flight time between samplers. These approaches significantly reduce the feedback time compared to conventional circuits, enhancing overall performance[7].

A. Conventional Loop unrolled DFE based on dual reference

Figure 2 illustrates a typical dual-reference-based loop-unrolled DFE[8]. Due to current process limitations, a single VGA can support two summers. As a result, this architecture

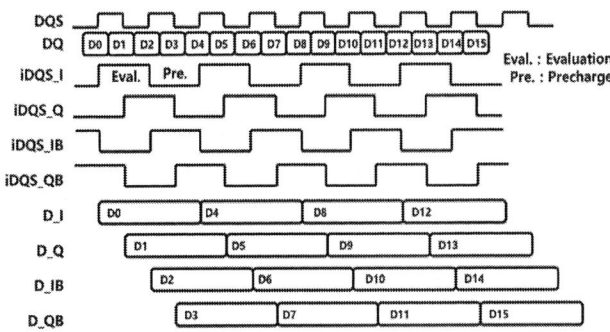

Fig. 3. Timing diagram about DDR5 RX.

requires 4 VGAs, 8 DFE summers, 8 samplers, and 4 MUXs. Compared to a current-string DFE, this configuration involves more than twice the circuitry, leading to a more than twofold increase in both area and power consumption.

Figure 3 illustrates the basic timing diagram of DDR5. The sampler is designed as a full latch, consisting of a sense amplifier (sense amp) and a second latch. The sense amp carries out a pre-charge operation during the low pulse of the internal DQS (iDQS) signal and performs evaluation during the high pulse. The second latch retains the evaluation result until the subsequent cycle. As a result, during the sense amp's pre-charge phase, the summer's output remains inactive until the sense amp begins its evaluation. The differential operation of iDQS_I and iDQS_IB facilitates time-multiplexing within a single summer.

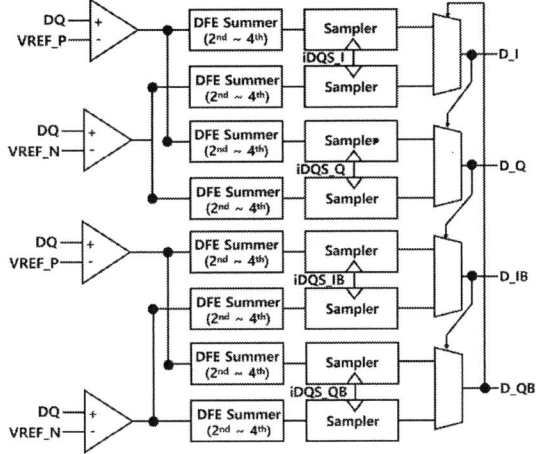

Fig. 2. A loop-unrolled DFE based on dual reference.

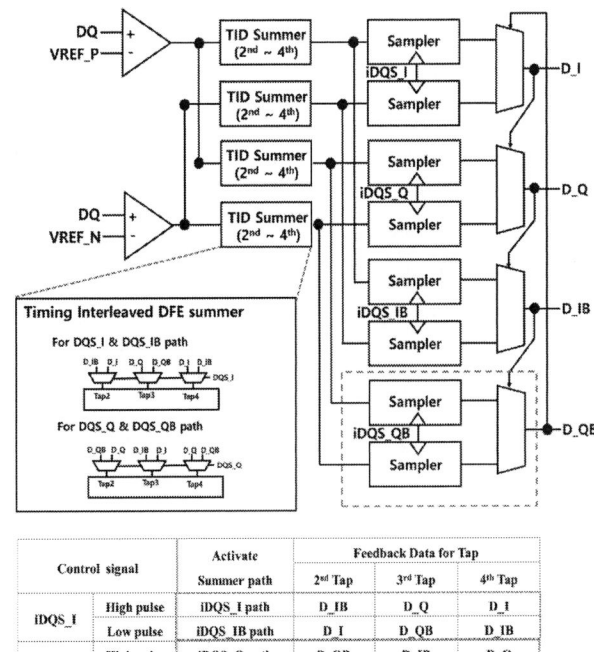

Control signal		Activate	Feedback Data for Tap		
		Summer path	2nd Tap	3rd Tap	4th Tap
iDQS_I	High pulse	iDQS_I path	D_IB	D_Q	D_I
	Low pulse	iDQS_IB path	D_I	D_QB	D_IB
iDQS_Q	High pulse	iDQS_Q path	D_QB	D_IB	D_Q
	Low pulse	iDQS_QB path	D_Q	D_I	D_QB

Fig. 4. A timing interleaving RX scheme based on a loop-unrolled DFE scheme

979-8-3315-8850-2/25 $31.00 © 2025 IEEE   896

B. A timing interleaving RX scheme based on a loop-unrolled DFE scheme

Specifically, the summer in the iDQS_I path is engaged during the high pulse of iDQS_I, while the summer in the iDQS_IB path is engaged during the low pulse of iDQS_I. This method effectively halves the number of summers and VGAs required. Although this approach impacts the DFE 2nd tap feedback time, it still remains twice as long as the DFE 1st tap feedback time.

Figure 4 illustrates a timing-interleaved RX circuit based on the loop-unrolled DFE architecture. By incorporating the timing-interleaved DFE summer (TID summer), a single TID summer can control two samplers, which allows for the reduction of 2 VGAs and 4 summers compared to the conventional scheme. In this timing-interleaved DFE configuration, the feedback data controlling the 2nd to 4th DFE taps alternates between the iDQS_I and iDQS_Q signals.

For example, in the 2nd tap, the high pulse of iDQS_I triggers the D_IB information for the iDQS_I path, while the low pulse of iDQS_I triggers the D_I information for the iDQS_IB path. These signals are then fed into the corresponding samplers associated with iDQS_I and iDQS_IB. While the timing-interleaved DFE architecture effectively reduces the number of VGAs and summers by half compared to the conventional loop-unrolled DFE approach, it still requires twice the number of samplers and additional MUXs. Consequently, this leads to inevitable increases in both power consumption and circuit area.

C. A timing interleaving RX scheme with direct multiplexing sampler

Figure 5 illustrates the timing-interleaving RX scheme proposed in this paper, featuring the integration of direct multiplexing samplers. The circuit comprises two variable gain amplifiers (VGAs), four time-interleaved DFE (TID) summers, and four direct multiplexing samplers [9]. Each TID summer is responsible for controlling two samplers in accordance with the timing scheme. The introduction of direct multiplexing samplers enables a reduction in the number of samplers from eight to four, while also eliminating the need for additional multiplexers (MUXs).

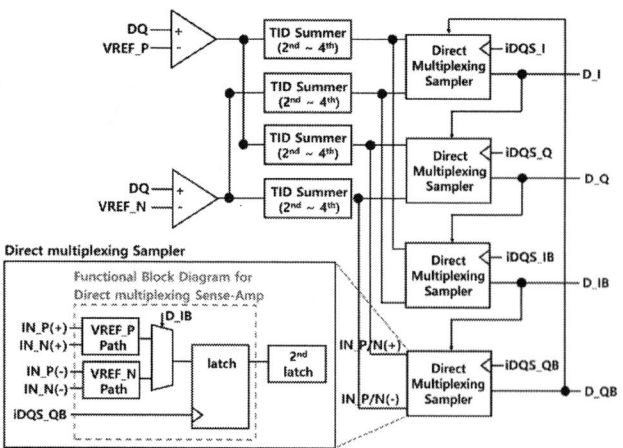

Fig. 5. Block diagram for a timing interleaving RX scheme with direct multiplexing sampler

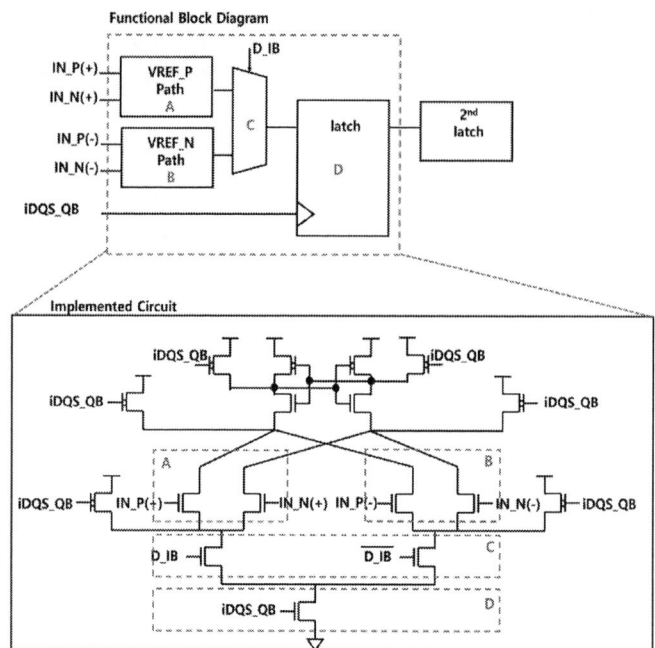

Fig. 6. Functional block diagram and its implemented circuit for direct multiplexing Sense Amp

The direct multiplexing sampler, similar to conventional samplers, adopts a two-stage structure consisting of a sense amplifier and a second latch. However, unlike conventional sense amplifiers, the direct multiplexing sense-amp simultaneously performs both signal sensing and multiplexing functions.

Figure 6 presents the functional block diagram and the corresponding implemented circuit of the direct multiplexing sense amplifier [10]. Functionally, the direct multiplexing sense amplifier consists of two input paths, a multiplexer path determined by previous data, and a latch for result storage. These functions are integrated into a single circuit. The implemented circuit comprises a PMOS pre-charge transistor, a latch, and three stacked NMOS transistors. During the pre-charge phase, the PMOS pre-charge transistor ensures that all nodes are charged to supply power. In the evaluation phase, two NMOS input pairs (A and B) process information from the VREF_P and VREF_N paths.

Below these input NMOS transistors are stacked to perform the multiplexing function, selecting between the VREF_N path and VREF_P path based on the previous value (D_IB). The final stacked NMOS transistor holds the result in the latch until the pre-charge phase of the next cycle, at which point the result is transferred to the second latch. Furthermore, the DFE 1st tap feedback time in this scheme is faster than that of the conventional loop-unrolled DFE due to the elimination of the muxing path, and it is exactly the same as the feedback time of the direct-feedback-in-sampler scheme.

III. EXPERIMENT RESULTS AND ANALYSIS

Table II shows the simulation results for the 1st and 2nd tap feedback times under different supply voltages, temperatures, and process corners (worst, typical, best cases). For correct

979-8-3315-8850-2/25 $31.00 © 2025 IEEE   897

TABLE II. DFE 1ST / 2ND TAP FEEDBACK TIME

Feedback time (Target)	Conventional scheme	Loop-unrolled DFE scheme	Proposed scheme
1st Tap (< 1UI)	0.95 UI	0.87 UI	0.83 UI
2nd Tap (< 2UI)	1.3 UI	1.3UI	1.37 UI

operation, the 1st tap feedback time must be less than 1 UI (1 UI = 138.89 ps @ 7200 Mbps), and the 2nd tap must be less than 2 UI . In the conventional scheme, the 1st tap feedback time is 0.95 UI, leaving only 0.05 UI margin. The loop-unrolled DFE improves this to 0.87 UI but is still limited by the muxing delay. The proposed design reduces it further to 0.83 UI, giving a 0.17 UI margin, which allows operation beyond 8200 Mbps. The 2nd tap feedback time is slightly slower (1.37 UI) due to the extra muxing stage, but it is still below 2 UI and has no impact on overall system performance. All simulations were conducted with a 15% margin beyond the PVT conditions defined in the JEDEC specification, ensuring that the results reflect robust operation under worst-case scenarios.

Figure 7 shows the DFE 1st tap coefficient variation after training. The proposed and loop-unrolled schemes both use a dual-reference method, which reduces variation by about 90% compared with the conventional current-string scheme. This confirms that the proposed design provides stable coefficient values across PVT conditions. With reduced variation, the trained coefficients remain valid under voltage and temperature changes, which minimizes the need for periodic retraining. As a result, training overhead is reduced and system reliability is improved in real DRAM operation.

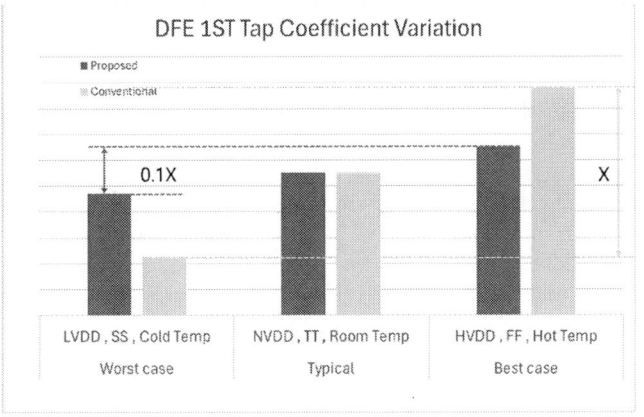

Fig. 7. DFE 1st Tap coefficient's Simulation result according to PVT

TABLE III. POWER CONSUMPTION COMPARISON

	Conventional scheme	Loop-unrolled DFE scheme	Proposed scheme
Write current	Y	2.17 Y	1.08 Y

Table III compares the power consumption of different schemes. The loop-unrolled design requires more than twice the circuit size of the conventional scheme, resulting in a 2.17× increase in power due to larger circuit area and higher routing load. In contrast, the proposed design increases power by only 8% relative to the conventional scheme, while reducing power by 60% compared with the loop-unrolled design. These results indicate that the proposed receiver not only achieves significantly higher energy efficiency but also reduces thermal and power delivery overhead, making it more practical for large-scale, high-speed DRAM I/O applications.

IV. CONCLUSION

In this paper, timing interleaving RX scheme with direct multiplexing sampler based on dual reference is implemented in standard DRAM peripheral process optimized for high-speed memory I/O circuits. By utilizing timing-interleaving DFE and direct multiplexing samplers, the proposed design addresses not only the power consumption issues associated with the conventional loop-unrolled DFE but also improves the 1st tap feedback time. Compared to the conventional scheme, the proposed scheme shows an 8% increase in power consumption, while reducing the feedback time by 0.12 UI. Additionally, it provides a 60% improvement in power consumption and a 0.04 UI reduction in feedback time relative to the loop-unrolled DFE. Furthermore, the introduction of dual reference reduces the coefficient variation of DFE Tap1 by 90% compared to the current string circuit.

REFERENCES

[1] DDR5 JEDEC spec

[2] GDDR6 JEDEC spec

[3] S-Y Cho. et al., "A 40nm 7Gb/s/pin Single-ended Transceiver with Jitter and ISI Reduction Techniques for High-Speed DRAM Interface", JSSC VOL 60, 2025

[4] D.-K. Kim et al., "A 1.1-V 10-nm class 6.4-Gb/s/pin 16-Gb DDR5 SDRAM With a Phase Rotator-ILO DLL, High-Speed SerDes, and DFE/FFE Equalization Scheme for Rx/Tx," JSSC VOL 55, 2020

[5] T-Y Oh. et al., "A 3.2 Gbps/pin 8 Gbit 1.0 V LPDDR4 SDRAM With Integrated ECC Engine for Sub-1 V DRAM Core Operation," JSSC VOL 50, 2015

[6] Behzad Razavi "The Decision-Feedback Equalizer [A circuit for All Seasons]", IEEE Solid-Circuits Magazine Vol 9, 2017

[7] I.-J. Choi et al., "Industry's First 7.2 Gbps 512GB DDR5 Module" HOT chips, 2021

[8] Elaine Tang et al., "A low power consumption and higher performance DDR5 receiver based on a direct feedback DFE and dedicated reference voltage for 1st TAP DFE", ASICON 2023

[9] S. Bae et al., "A 40nm 7Gb/s/pin Single-ended Transceiver with Jitter and ISI Reduction Techniques for High-Speed DRAM Interface", VLSI Symp. Circuits, pp. 193 194, 2010

[10] Y.-J. Kim et al., "A 16Gb 18Gb/S/pin GDDR6 DRAM with per-bit trainable single ended DFE and PLL-less clocking," ISSCC, pp. 204-206, 2018

2025 The 10th International Conference on Integrated Circuits and Microsystems

Graph-Based Representation of Verilog HDL: Python-Based Control and Data Flow Graph Generation

Yipeng Wang*
College of Computer Science and Technology
Nanjing University of Aeronautics and Astronautics
Nanjing, China
wyp20021006@nuaa.edu.cn
*Corresponding author

Zhiqiang He
College of Computer Science and Technology
Nanjing University of Aeronautics and Astronautics
Nanjing, China
zhiqiang_he@nuaa.edu.cn

Lingwei Yan
College of Computer Science and Technology
Nanjing University of Aeronautics and Astronautics
Nanjing, China
yan_lw@nuaa.edu.cn

Gang Chen
College of Computer Science and Technology
Nanjing University of Aeronautics and Astronautics
Nanjing, China
gangchensh@qq.com

Abstract—**This paper proposes an automatic framework for controlled data flow graph (CDFG) generation from verilog designs, where the generated CDFGs can be applied to visualization, formal verification, logic optimization, and serve as structured representations for large-scale machine learning models. The framework leverages Yosys to flatten the RTL code, eliminating high-level control structures, and then constructs an abstract syntax tree (AST) to dynamically extract control and data dependencies. A unified graph representation is generated, enabling accurate modeling of both control-flow and data-flow semantics. The proposed RTL-to-CDFG conversion tool demonstrates high scalability and precision, successfully handling million-line industrial Verilog designs. Experimental results confirm that extracted CDFGs capture essential structural and semantic features of verilog code, providing a solid foundation for downstream tasks that include dataflow analysis, logic simplification, and resource sharing.**

Index Terms—**Verilog, CDFG Generation, Structured Representations. RTL Analysis, AST**

I. INTRODUCTION

Verilog, a widely used register-transfer level (RTL) hardware description language, has been a standard for ASIC and FPGA development since its introduction by Gateway Design Automation in 1984. It supports the full design flow from behavioral modeling to gate-level netlists and enables precise descriptions of combinational logic, sequential logic, finite state machines, and hierarchical modular structures. Modern digital designs often consist of millions of RTL lines with deeply nested modules and complex control logic, making it increasingly challenging to analyze, optimize, and verify designs using traditional tools or manual inspection.

Intermediate representations (IRs), such as control-data flow graphs (CDFGs), provide a structured abstraction that explic-

itly captures both control flow and data dependencies. By bridging high-level algorithmic design with RTL implementation, CDFGs enable unified hardware analysis and optimization [1], [2]. They serve as a foundation for formal verification, logic rewriting, and high-level synthesis, and they can further provide structured representations suitable for AI/ML-based design exploration and automated optimization.

Existing CDFG generation tools, such as PyVerilog [3], can extract basic control and data flow information. However, they still have limitations in accuracy, scalability, and readability, which restrict their applicability in industrial design workflows.In particular, current tools often struggle to capture higher-level design semantics, making it difficult to generate representations that are both precise and interpretable. Moreover, their scalability issues become evident when applied to large RTL modules, where graph construction and visualization may produce overly complex structures that hinder efficient analysis. These challenges highlight the need for more robust and practical CDFG generation methods that can support both academic research and real-world chip design.

To address these challenges, we propose an automated framework for generating CDFGs from high-level synthesizable Verilog. The framework employs a three-stage pipeline: (1) semantic-preserving flattening and preprocessing of Verilog code using the Yosys toolchain [4], (2) construction of an enhanced abstract syntax tree (AST) with domain-specific rules to extract control dependencies and data-flow features, and (3) generation of a unified graph representation annotated with complete hardware semantics. The proposed tool efficiently handles million-gate industrial RTL designs and supports applications including logic rewriting, synthesis

979-8-3315-8850-2/25 $31.00 © 2025 IEEE

optimization, and GNN-based design space exploration. Experimental results in multiple complex SoC designs demonstrate the practical utility and precision of the framework, providing structured representations that enable downstream optimization and AI-driven analysis in hardware design.

II. RELATED WORK

Several open-source Verilog parsing tools have been developed, including Verilator, Yosys, and Pyverilog [3]–[5], each with unique strengths in syntax support, parsing capabilities, and application scenarios. However, most existing tools lack the ability to directly generate high-quality Control-Data Flow Graphs (CDFGs), limiting their usefulness for advanced hardware analysis and optimization.

PyVerilog is a widely used open-source Verilog HDL processing toolkit that provides functionalities such as code parsing, data-flow analysis, control-flow analysis, and code generation. Thanks to its open-source nature and integration with Python, PyVerilog has been adopted in many academic and prototyping scenarios. As mentioned earlier, despite these capabilities, PyVerilog still exhibits notable limitations. Firstly, its graph visualization often overcomplicates circuit representations, generating excessively large dataflow graphs even for relatively simple logic. Secondly, its syntax analysis is insufficient for complex sequential logic and function structures, directly impacting the quality and completeness of the generated CDFGs.

To address these challenges, our team developed the Stagira parser [6]. Stagira achieves superior Verilog syntax analysis, generating a more complete abstract syntax tree (AST) that serves as a solid foundation for subsequent CDFG extraction. Compared with existing tools, Stagira preserves parsing accuracy while offering a flexible intermediate representation, enabling more sophisticated optimization transformations, such as CDFG-based automatic pipelining techniques [7].

Another Python-based framework, VeriPy, proposed by Md Imtiaz Rashid et al. [8], can generate CDFGs but was primarily designed for lightweight functional representations to support hardware Trojan detection. Consequently, its CDFG generation is limited, producing graphs with relatively few nodes that cannot fully capture the detailed structural and semantic features of designs. This restricts VeriPy's applicability in broader hardware optimization scenarios.

III. GENERATION FLOW

To systematically realize the proposed RTL-to-CDFG transformation, we design a structured workflow that translates Verilog source code into a graph-based CDFG representation. As illustrated in Fig.1, the framework operates on a synthesizable, module-level Verilog description and produces the corresponding CDFG as output. The process is organized into three phases, each addressing a distinct layer of abstraction. The first phase performs preprocessing and structural flattening with Yosys to simplify the RTL. The second phase conducts syntax analysis of the RTLIL representation to construct an enhanced AST. The third phase applies a visitor-controlled

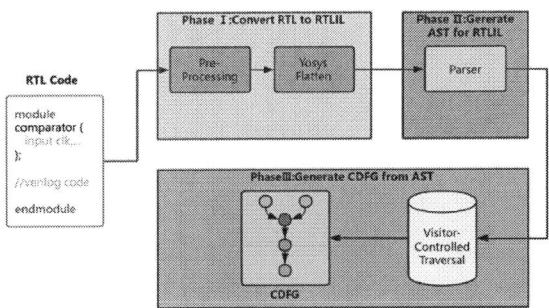

Fig. 1. Overview of generation flow composed of three phases. Phase 1:Convert RTL To RTLIL. Phase 2:Gererate AST For RTLIL. Phase 3:Generate CDFG From AST.

traversal of the AST to extract precise control-flow and data-flow semantics. In the following subsections, we describe each phase in detail.

A. Phase 1:Convert RTL To RTLIL

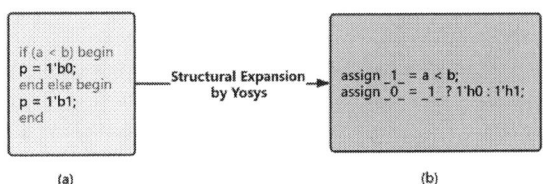

Fig. 2. Structural transformation of Verilog code during the flattening process: (a) hierarchical representation (before flattening), (b) flattened representation (after flattening).

In the first phase, the input Verilog code undergoes preprocessing, including comment removal and basic normalization, to ensure a consistent input format. Next, Yosys flattens the design by expanding hierarchical module structures and transforming complex control constructs (e.g., for loops, if-else branches) into simpler logic representations, as illustrated in Fig.2.

This stage can be regarded as a form of lightweight synthesis. However, unlike conventional synthesis, which targets the generation of a gate-level netlist and emphasizes timing optimization and physical implementation, our expansion is intended solely to improve structural regularity and parsability for subsequent semantic graph construction. In particular, our approach deliberately avoids optimization, technology mapping, and resource binding, while preserving logically equivalent high-level abstractions that are critical for semantic analysis.

Through this transformation, implicit control dependencies are made explicit as data dependencies, which significantly reduces the complexity of syntax parsing and semantic graph construction, thereby laying the groundwork for subsequent CDFG extraction.

B. Phase 2:Gererate AST For RTLIL

No.	Syntax Structure	Pyverilog	Verilator	Stagira
1	space exists in progressive type and value	No	Yes	Yes
2	user-defined primitives	No	Yes	Yes
3	Module instantiation omits some parameters	No	Yes	Yes
4	defparam redefines parameters	No	Yes	Yes
5	delay symbol # followed by parameters without parentheses	No	Yes	Yes
6	Specify defined path delay block statement	No	Yes	Yes
7	define the parameters of the network line type	No	No	Yes
8	access Variables in Task Domains Using Hierarchy Names	Yes	No	Yes
9	$fopen system call omits the second argument	Yes	No	Yes

Fig. 3. Stagira vs. Verilator vs. PyVerilog: syntax coverage.

In the second phase, the RTLIL code generated from the first phase is processed by our Verilog parser to construct the abstract syntax tree (AST). The tool used in this step is Stagira [5], a parser developed by our team. Compared to existing open-source parsers such as PyVerilog and Verilator, Stagira provides superior syntactic coverage and robustness, making it capable of handling Verilog code at the scale and complexity required for industrial chip design and verification, as illustrated in Fig.3.

Stagira first analyzes the Verilog code and produces an AST represented in S-expressions. However, S-expressions are neither intuitive for human inspection nor convenient for direct programmatic manipulation. To address this, we employ a Python-based post-processing step that converts the S-expression AST into a JSON representation, which can be seamlessly consumed in Python for subsequent CDFG extraction.

C. Phase 3:Generate CDFG From AST

The third phase constitutes the core of our framework. Before generating the Control and Data Flow Graph (CDFG), we initiate a detailed analysis of the operator nodes, as summarized in Tab.I. This analysis provides a comprehensive categorization and quantitative description of the various types of nodes present in the Verilog design, offering crucial insights into the underlying structure and computational characteristics of the design. By examining operator nodes at this stage, we establish a foundation for precise modeling of both data and control dependencies in the subsequent CDFG construction.

In terms of data structures, the graph is implemented and maintained using the NetworkX library, which offers a highly flexible and efficient framework for graph representation, construction, and traversal. NetworkX enables dynamic addition and deletion of nodes and edges, as well as complex graph queries, which are essential to accurately capture the intricate relationships inherent in hardware designs. Leveraging such a robust library ensures that our framework can handle designs of varying complexity while maintaining scalability and computational efficiency.

Our approach commences with a visitor-controlled traversal of the Abstract Syntax Tree (AST), where specialized visitor functions systematically visit each node in the tree. During this traversal, each node recognized by a visitor undergoes a fine-grained, dataflow-sensitive analysis. This analysis allows us to precisely identify the source of dependencies for every assignment statement, tracking the propagation of values and control signals throughout the design. In doing so, we guarantee that the resulting CDFG faithfully represents the true flow of data and control dependencies, avoiding spurious or missing edges that could compromise downstream analyses.

By integrating operator node statistics, visitor-controlled traversal, and meticulous dataflow analysis, we are able to construct a high-fidelity CDFG that preserves both the structural and semantic relationships inherent in the original Verilog design. Such fidelity is critical, as it ensures that subsequent analyses, optimizations, and formal verification tasks can rely on the graph as an accurate representation of the design's behavior and internal dependencies.

Fig.4 presents the Control and Data Flow Graph (CDFG) of an Arithmetic Logic Unit (ALU) generated by our framework. The visualization highlights the various types of operator nodes and explicitly illustrates the data and control dependencies that connect them. By providing both structural and semantic clarity, the figure serves as an intuitive representation of the internal computation flows within the ALU. This not only demonstrates the robustness and accuracy of our generated CDFG but also provides a valuable tool to enable downstream tasks such as performance optimization, design verification, and formal correctness analysis. The comprehensive nature of this representation underscores the effectiveness of our framework in capturing the essential computational and control relationships of complex hardware modules.

TABLE I
EXAMPLES OF DIFFERENT NODE TYPES IN CDFG

Node Type	Example	Description
Cond	`a = b ? c : d`	If b is equal to True, assign c to a, otherwise assign d to a.
PartSelect	`a = b[x : y]`	Assign bits from position x to y of b to a.
Concat	`a = {b, c, d}`	Connect b, c, and d together in order and assign them to a.
Unary Operator	`a = op1 b`	Unary operator with single source operand.
Binary Operator	`a = b op2 c`	Binary operator with two source operands.
Register/Wire/Constant	`reg [7:0] a`	Variable type.

IV. EXPERIMENTS

A. Experimental Setup

All experiments were conducted using Python 3.9.13 with NetworkX 2.8.4 and Yosys 0.43+3 [9], [10]. For visualization purposes, the generated CDFGs were rendered using the Graphviz dot tool [11]. We evaluated the proposed framework on a large set of Verilog designs, including

- Module-Level Examples: Over 100,000 module-level Verilog examples collected by our team to assess functional correctness, graph coverage, and scalability.

979-8-3315-8850-2/25 $31.00 © 2025 IEEE

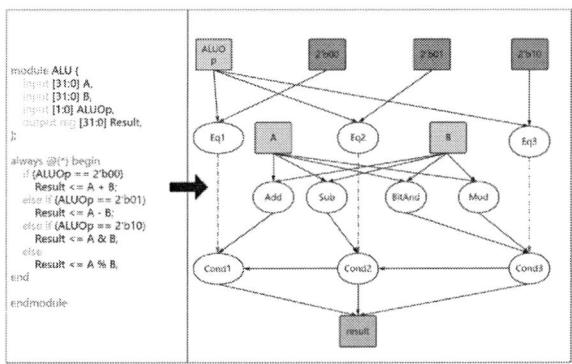

Fig. 4. CDFG of a simple ALU.

- Repository-Level Examples: 30 real-world Verilog repositories crawled from GitHub to test the framework's applicability on industrial codebases. Since our framework is designed to operate at the module level, but practical designs are typically composed of multiple interacting modules, we developed an additional preprocessing script. Specifically, before executing our CDFG generation flow, we invoke Yosys to flatten each repository-level design with respect to its designated top module, thereby consolidating the multi-module hierarchy into a single module-level Verilog description. This ensures that the input is compatible with our framework while preserving the functional semantics of the original repository.
- Large-Scale Designs: Million-lines RTL modules collected by our team to evaluate generation capability and runtime efficiency. By applying our framework to such large-scale inputs, we assessed not only the correctness of the generated CDFGs but also the scalability of the approach in terms of runtime performance and memory usage.

B. Evaluation Metrics

The evaluation metrics include:

- Graph Size: Number of nodes in the generated CDFG.
- Success Rate: Percentage of designs for which the CDFG was successfully generated.
- Runtime: Total execution time for CDFG generation, with average time reported per module.

C. Experimental Results

Tab.II presents the CDFG generation capability of our framework across different Verilog design levels, providing a comprehensive overview of its performance under varying complexity and scale. The Module-Level dataset, which was crawled from GitHub with minimal filtering, represents a diverse collection of real-world Verilog modules. While this dataset exhibits a relatively low success rate due to the inherent variability and occasional inconsistencies in publicly available code, our method nevertheless outperforms PyVerilog on the

same samples, as detailed in Tab.III This comparison highlights the robustness of our approach, particularly in handling incomplete or unconventional Verilog code that may pose challenges to existing tools.

For large-scale industrial designs, we further evaluated our framework on 10 representative cases, each containing millions of lines of code. The results demonstrate that our framework is capable of effectively managing industrial-scale Verilog CDFG generation, preserving both structural and semantic fidelity even in extremely complex designs. This confirms the scalability of our approach and its suitability for real-world EDA applications where high accuracy and reliability are critical.

In addition, to perform a more controlled and quantitative evaluation, we manually selected 1,000 Module-Level Verilog designs and compared the performance of our framework against PyVerilog. Tab.III provides a detailed report of the CDFG generation performance on this dataset, including metrics such as the average node count, success rate, and runtime for each method. Although our framework incurs slightly longer runtimes compared to PyVerilog due to the additional dataflow-sensitive analysis, it consistently produces more accurate and complete CDFGs. This trade-off demonstrates the effectiveness and reliability of our method, highlighting that the modest increase in computational cost is justified by the significant improvement in graph fidelity and coverage. Collectively, these results validate the capability of our framework to generate high-quality CDFGs across a wide range of Verilog designs, from module-level academic examples to large-scale industrial systems.

TABLE II
EVALUATION RESULTS UNDER DIFFERENT EXPERIMENTAL SETTINGS

Setting	Examples	Success Rate [%]	Runtime [h]
Module-Level	118,954	64.0	4
Repo-Level	30	63.3	0.5
Large-Scale Designs	10	70	17

TABLE III
PERFORMANCE COMPARISON: THIS WORK VS. PYVERILOG

Method	Nodes(Avg)	Success Rate [%]	Runtime [s] (Avg. per module)
This Work	49	97.8	0.99
PyVerilog	57	82.2	0.91

V. CONCLUSION

This paper proposes a single-module flattening synthesis strategy for processing module-level hardware design code. The core idea of this strategy is to leverage synthesis tools (e.g., Yosys) to flatten the entire design into a single module, thereby eliminating complex syntactic structures. The flattened design is then converted into an AST, from which data flow and control logic are extracted using a visitor-controlled

979-8-3315-8850-2/25 $31.00 © 2025 IEEE

traversal combined with dataflow sensitive analysis. As shown in the experimental results, compared with PyVerilog, our framework generates CDFGs with fewer nodes and more accurate representations of control and data dependencies, while also supporting industrial-scale Verilog designs. The tool provides visualization of both data flow and control flow and produces intermediate files that are easier to handle in Python, and it has potential future applications in Verilog code optimization.

REFERENCES

[1] Namballa R, Ranganathan N, Ejnioui A. Control and data flow graph extraction for high-level synthesis[C]//IEEE Computer Society Annual Symposium on VLSI. IEEE, 2004: 187-192.

[2] Gao X, Qiu Y, Dai Y, et al. A CGRA Front-end Compiler Enabling Extraction of General Control and Dedicated Operators[C]//2024 29th Asia and South Pacific Design Automation Conference (ASP-DAC). IEEE, 2024: 799-804.

[3] Takamaeda-Yamazaki S. Pyverilog: A python-based hardware design processing toolkit for verilog hdl[C]//International Symposium on Applied Reconfigurable Computing. Cham: Springer International Publishing, 2015: 451-460.

[4] Wolf C, Glaser J, Kepler J. Yosys-a free Verilog synthesis suite[C]//Proceedings of the 21st Austrian Workshop on Microelectronics (Austrochip). 2013, 97.

[5] Snyder W. Verilator and systemperl[C]//North American SystemC Users' Group, Design Automation Conference. 2004, 79: 122-148.

[6] Chen X, Meng Y, Chen G. Incremental verilog parser[C]//2023 International Symposium of Electronics Design Automation (ISEDA). IEEE, 2023: 236-240.

[7] van Haastregt S, Kienhuis B. Enabling automatic pipeline utilization improvement in polyhedral process network implementations[C]//2012 IEEE 23rd International Conference on Application-Specific Systems, Architectures and Processors. IEEE, 2012: 173-176.

[8] Rashid M I, Schaefer B C. VeriPy: A Python-Powered Framework for Parsing Verilog HDL and High-Level Behavioral Analysis of Hardware[C]//2024 IEEE 17th Dallas Circuits and Systems Conference (DCAS). IEEE, 2024: 1-6.

[9] Van Rossum G. Python programming language[C]//USENIX annual technical conference. 2007, 41(1): 1-36.

[10] Hagberg A, Conway D. Networkx: Network analysis with python[J]. URL: https://networkx. github. io, 2020: 1-48.

[11] Ellson J, Gansner E R, Koutsofios E, et al. Graphviz and dynagraph—static and dynamic graph drawing tools[M]//Graph drawing software. Berlin, Heidelberg: Springer Berlin Heidelberg, 2004: 127-148.

2025 The 10th International Conference on Integrated Circuits and Microsystems

A Node Merging Algorithm Using Fault-Detection for XOR-Majority Network

Shijia Fan
Scool of Microelectronics
University of Science and Technology of China
HeFei, China
shiga@mail.ustc.edu.cn

Feifei Deng
School of Integrated Circuits
Anhui University
Hefei, China
ffdeng@ahu.edu.cn

Abstract—**Recent advancements in emerging technology have spurred the development of novel electronic devices, such as Quantum-dot Cellular Automata (QCA) as a transistor-alternative computing architecture, and Magnetic Random-Access Memory (MRAM) as a next-generation storage medium. These technologies enable efficient implementations of XOR and majority logic, which offer superior expressiveness compared to traditional AND/OR logic. As a result, researchers have introduced new logical structures MIGs and XMGs, which pose more demands for digital logic synthesis. In this paper, we propose a node merging algorithm using fault detect model for XMGs logic synthesis, including building an optimal majority node and XOR node model using the path sensitivity method from single stuck-at fault model testing, activating and propagating faults among majority node and XOR nodes in the XMGs, and using ROBDD tree to compute and store the mandatory assignment sets for efficient storage. The algorithm is implemented in the C++ language under the framework of the EPFL logic synthesis tool. The experiment results show that the proposed algorithm can effectively reduce the depth and number of nodes of XMGs circuits when applied to the standard test circuit in EPFL, with an average optimization depth of 2.07% and an average optimization number of 0.91% in the 2-level verification, and can gain better optimization effect when the window sizes increase.**

Keywords—**XOR-majority Graphs, logic synthesis, node-merging algorithm, fault-detect model**

I. INTRODUCTION

The continuous miniaturization of CMOS technology, governed by Moore's Law, is approaching fundamental physical limits as feature sizes shrink below 5 nm. This technological impasse manifests in two critical aspects: (1) quantum tunneling effects causing exponential increases in leakage current, and (2) von Neumann bottleneck exacerbated by spatial separation between computing units and memory hierarchies [2]. These dual challenges in computational efficiency and data storage have catalyzed research into beyond-CMOS nanoelectronic devices. In response to these constraints, the research community has explored alternative paradigms including Quantum-dot Cellular Automata (QCA) [1,2] and Magnetic Random-Access Memory (MRAM) [3]. QCA has emerged as a transformative solution offering enhanced integration density, superior operational speed, and lower energy dissipation. Similarly, MRAM demonstrates exceptional potential through its combination of high storage density, rapid write operations, and CMOS-compatible characteristics. These

emerging technologies collectively address critical limitations inherent to conventional CMOS implementations.

XOR and majority logic primitives demonstrate superior computational efficiency compared to traditional AND/OR logic, with native implementation advantages in these novel devices. This technological synergy motivated EPFL researchers led by Luca Amarù to develop innovative MIGs (Majority-Inverter Graphs)[4,5] and XMGs (XOR-Majority Graphs)[6] data structures specifically tailored for representing advanced logic configurations in next-generation devices. Characteristically, XMG nodes exclusively implement either majority or XOR Boolean operations. As device physics continue to evolve, the demand for automated synthesis methodologies capable of harnessing these novel logic primitives has become increasingly imperative. Previous work by Rumi Zhang et al. pioneered majority logic optimization techniques for both two-level [7] and multi-level circuits [8,9]. Current XMG optimization strategies predominantly focus on minimize the size of circuits through approaches including: functional hashing [10], LUT-based mapping [11], logic rewriting [12], and fault detection models [13]. However, existing methodologies exhibit a critical limitation by focusing solely on area minimization while neglecting delay optimization considerations.

In this paper, we introduce a novel node-merging algorithm for XMGs that synergistically integrates three steps: (1) optimal majority/XOR node modeling via path sensitivity analysis from single stuck-at fault testing, (2) coordinated fault activation-propagation mechanisms across heterogeneous node types, and (3) ROBDD (Reduced Ordered Binary Decision Diagram)-based constraint management through mandatory assignment sets computation. These procedures enable simultaneous optimization of both circuit depth and node count through structural consolidation.

The remainder of this paper is organized as follows: Section II establishes theoretical background of this work. Section III details the node-merging optimization algorithm for XMGs. Section IV presents the results of the experiment. And concluding remarks and future directions are discussed in Section V.

979-8-3315-8850-2/25 $31.00 © 2025 IEEE

II. BACKGROUND

A. XOR-majority Graphs

Majority logic refers to a logic function with an odd number of inputs, whose output is determined by selecting more values among the inputs. A three-input majority logic function can be denoted as $\langle x, y, z \rangle$, and its logical expression is:

$$\langle x, y, z \rangle = xy + xz + yz = (x + y)(x + z)(y + z) \quad (1)$$

Three-input majority logic can be controlled either through one input end of a constant 0 or 1 to realize logic operation AND or logic operation OR:

$$\langle x, y, 1 \rangle = xy + x + y = x + y$$
$$\langle x, y, 0 \rangle = xy + 0 = xy \quad (2)$$

XOR-majority Graphs(XMGs) are logical circuits composed of XOR gates node, majority gates node, and inverter on the edges. The XOR gate is a digital logic gate that can have multiple inputs. For 2-input XOR, if the two input values are identical, the output is 0; otherwise it's 1. Its logic expression is:

$$x \oplus y = xy' + x'y \quad (3)$$

For multi-input XOR gate, it can be composed of multiple 2-input XOR gates. In this paper, the fan-in number of all majority gates and XOR gates in XMGs is 3.

B. Test methods for combinatorial circuits

Fault modeling refers to mathematically simulating the physical defects in the manufacturing process of chips to study the effects on circuits or systems and locate fault positions. The single stuck-at fault model is the most widely used fault model in circuit testing, which assumes that a signal in the chip or system stuck at 0 or 1.

The path sensitivity algorithm is a method for generating test codes during fault detection by activating a fault after its occurrence, selecting appropriate values at selected transmission paths to propagate the fault effects to primary outputs(POs). If such a method exists, the fault is considered testable.

In fault detect models, in order to propagate the target node's fault to the POs, some nodes along the propagation path must be forced with values until the fault is detected. These nodes associated with forced values are called the Mandatory Assignment Sets (MAS). Using logical calculations on known values within the MAS can generate additional forced allocation nodes. Especially, if there's a conflict in the MAS(e.g. a node x requirting both 0 and 1 in the MAS seperately), it means that this fault is untestable.

III. NODE MERGING ALGORITHM FOR XMGs

A. Window Partition

Given the potential exponential scaling of circuit dimensions, our method introduces a window strategy to optimize the fault detection efficiency. This strategy effectively bounds the fault effect propagation within a localized neighborhood surrounding target nodes, achieving polynomial-time complexity for large-scale Boolean network optimization.

Windowing techniques constitute a proven paradigm for managing computational complexity in node merging operations across massive Boolean circuits. Our implementation configures window parameters through : (1) establishing the target node as topological origin, (2) incorporating accessible nodes within multi-level fan-in/fan-out cones. The window's size is dynamically adjusting based on optimization requirements and target circuits.

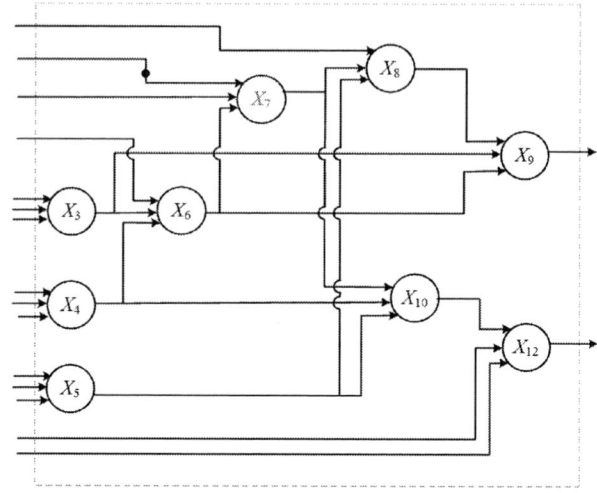

Fig. 1. window partition

As shown in the Fig. 1, this is a complex XMGs circuit. When partitioning to detect faults at node X_7 into a 2-level window, the dotted box represents the window boundary, and only the nodes inside this window are involved in fault detection and setting up the MAS and performing calculations. The nodes outside this window are not considered in this fault detection model.

B. Fault Detect Model

1) fault detect theorem:

In fault detect models, for each output function $f(x) = f(x_1, x_2, ..., x_n)$, to enable x_1 fault propagation to primary output(PO), the following condition must be satisfied: $ODC_{x_1} = \partial f / \partial x_1 = (f(1, x_2, ..., x_n)) \oplus (f(0, x_2, ..., x_n)) = 1$.

For a majority logic function $f(x) = \langle x_1, x_2, x_3 \rangle$, the condition is as follows:

$$\frac{\partial f}{\partial x_1} = f(1, x_2, x_3) \oplus f(0, x_2, x_3)$$
$$= (x_2 + x_3) \oplus (x_2 x_3) \quad (4)$$
$$= x_2' x_3 + x_2 x_3'$$

Therefore, to enable to propagate the fault to the pos, the other two nodes of the fanout node must satisfy $x_2' x_3 + x_2 x_3' = 1$.

Similar for a XOR logic, function $f(x) = x_1 \oplus x_2 \oplus x_3$, the condition is as follows:

$$\frac{\partial f}{\partial x_1} = f(1, x_2, x_3) \oplus f(0, x_2, x_3) \qquad (5)$$
$$= (x_2 \oplus x_3) \oplus (x_2 \oplus x_3)' \equiv 1$$

Since it's always equal to 1, for the XOR gate's fault detect model, the other two nodes of the fanout can be any assignment.

In multiple-level logic circuits, each level can independently propagate faults through the chain rule formula.

2) Node merging theorem:

Untestable Faults : If fault x_i(e.g., x_1) is untestable, then $ODC'_{x_1} \equiv 0$. Now we define that $f_{x_1}(x) = f(x_1, x_2, \ldots, x_n)$, $f_{x'_1}(x) = f(0, x_2, \ldots, x_n)$ then, $f(x) = x_1 f_{x_1} + x'_1 f_{x'_1}$, and

$$\frac{\partial f}{\partial x_1} = f(1, x_2, \ldots, x_n) \oplus f(0, x_2, \ldots, x_n)$$
$$= (f_{x'_1}) \oplus (f_{x_1}) \qquad (6)$$
$$\equiv 0$$

We can see that $f_{x_1} = f_{x'_1}$, which means

$$f(x) = x_1 f_{x_1} + x'_1 f_{x_1} \equiv f_{x_1}(x). \qquad (7)$$

Thus, f(x) does not depend on x_1. In this case, we can replace x_1 with any other node. Since constant 0 consumes the least resources in the circuits, usually we use constant 0 to substitue this node.

Testable Faults : If fault x_1 is testable, which means $ODC'_{x_1} \neq \emptyset$, we consider the relationship between the circuit node x_i and the primary input x_{pi}. Let x_i be a circuit node implementing Boolean function:

$$x_i = f_i(x_{p1}, x_{p2}, \ldots, x_{pn}) \qquad (8)$$

where x_{p_i} denote primary inputs. If there exists a node x_i, when any input patterns in $ODC'_{x1} = 0$, x_i satisfies that $f_i(x_{p1}, x_{p2}, \ldots, x_{pn}) \equiv f_1(x_{p1}, x_{p2}, \ldots, x_{pn})$, then we can use node x_i to substitue node x_1. We define a new logic function $g(x) = f(x_i, x_2, \ldots, x_n)$:

(a) When $ODC_{x_1} = 1$, in this case the fault of x_1 can't propagate to the POs, according to (7), we have

$$f(x) = f(x_1, x_2, \ldots, x_n) = f_{x_1}$$
$$g(x) = f(x_i, x_2, \ldots, x_n) = f_{x_1} = f(x) \qquad (9)$$

(b) When $ODC_{x_1} = 0$, now since $x_i = f_i(x) \equiv f_1(x) = x_1$,

$$f(x) = f(x_1, x_2, \ldots, x_n) = f(x_i, x_2, \ldots, x_n) = g(x) \qquad (10)$$

So there is always $f(x) \equiv g(x)$, which means we can use g(x) function to replace f(x) function, in logic circuits it means we can replace x_1 node with x_i node. If x_i is always opposite to x_1, similar reasoning applies, we can replace x_1 with x'_i.

As shown in Fig. 3, for fault detection at node X_1, in order to propagate the fault to node X_2, the other two fanin of X_2 should be assigned as 01 or 10. If the node X_1 is stuck at 0 fault and cannot be detected, it can be replaced with a regular node 1; otherwise, if forcing x_1 to assign 1 causes x_4 to force assign 1 and also forces x_1 to assign 0 assigns x_4 to 0, then node x_4 can replace node x_1.

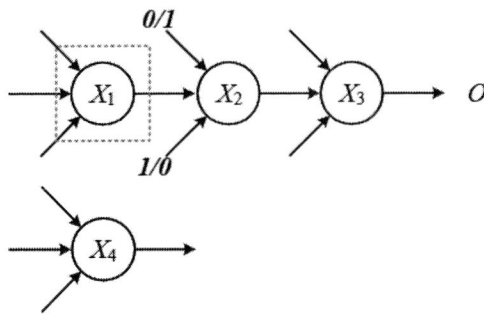

Fig. 2.   Node merging example

C. Node Merging Model

In the fault detection model, effective propagation of fault effects requires strategic value allocation for key nodes along the sensitization path. By performing logical calculations on the MAS already present in the circuit, more nodes can be added to MAS. For single-output nodes during fault detection:

(1) Assign the fan-in of the single-output node to activate the fault effect. This will cause the node to be stuck at 0 or 1.

(2) Activate the fault effect on the output nodes and assign their inputs. According to the content in Chapter III.B, propagate the force assignments to their fanout node by assigning non-control values to these nodes. For majority gates, the other two fanin should be constrained to 0,1 and 1,0 according to (4).

(3) For all nodes present in MAS, if each output can be logically deduced from existing logic calculations, compute its value and add it to MAS.

The single-output gate's fault detect model algorithm is shown in Fig. 3.

Algorithm 1: Fault detect model(single-output node)

Input: *Net.N—XMG, Node.Nd, window*
Output: *MAS*

Nf = Nd's fanout
While *Nf in window* **do**
    foreach *Node.Ni is Nf's fanin*
        if *ni isnot in MAS*:
            assign *ni*;
            put *ni* in *MAS*;
        end
        if *Nf* is majority gate:
            constrain *Nf's* other fanin with {0.1} and {1,0};
    end
    Nd = Nf;
    Nf = Nd's fanout;
end
foreach *Node.Nm in MAS*
    foreach *Node.Nf is Nm's fanin or fanout*
        if *Nf can compute:*
            put *Nf* in *MAS*;
    end
end
return *Mas*;

Fig. 3.   Fault detect model algorithm

After obtaining the MAS, first check whether the fault can be detected. This is expressed as the target node's ROBDD tree degenerates into a 0 node or 1 node. According to Section III.B, replace it with a corresponding constant node. If not, then check all nodes in the MAS and compare their ROBDD trees with the target node. If any node with a lower level than the target node matches the target node, this node can replace it equivalently. The candidate substitution nodes set S is formally defined as

$$S = \left\{ n_c \in MAS | ROBDD(n_c) \cong ROBDD(n_{target}) \right.$$
$$\left. \wedge \ level(n_c) \leq level(n_{target}) \right\}$$

To minimize circuit depth, the target node should be replaced with the lowest level node in S $n_s = \min_{n_c \in S} level(n_c)$.

After replacing the target node, delete the original target node and its input cone. To ensure that the circuit logic remains unchanged after replacement, equivalence test should be conducted on the circuits before and after replacement, comparing their logical functions to ensure they remain identical.

In XMGs, more logic gates are multi-output gates. For nodes with multiple fanout branches $FO(n_t) = \{o_1, o_2, ..., o_n\}$ during fault detection, the target node must have all its outputs replaced by this node that can be used for replacement, which means the candidate substitution nodes set S must satisfied

$$S = \bigcap_{i=1}^{n} S_{o_i}$$

D. XMGs Node Merging Algorithm

This paper proposed a merging algorithm for XMGs network node, which uses a sliding window to perform fault detection on the network nodes and implements node merging. The purpose of this algorithm is twofold for inputting a given XMG circuit: 1) reduce the delay (i.e., lower the level of the circuit's series), shorten its critical path; 2) reduce the area, i.e., decrease the number of XOR gates and majority gates in the circuit.

The algorithm process design is as follows:

1) Reduce network level along the critical path of the XMGs, sequentially divide each node into a window and use the fault detect model to perform fault transmission along the target node. For each node in the MAS, if there is a node with lower level than the target node that can be replaced by the node replacement principle and used for node replacement, then replace the target node. If this step involves node replacement, since the original XMGs has changed, it needs to refind the critical path of the new circuit and check again along the new critical path.

2) Reduce the number of network gates from all POs of the XMGs, in a depth-first manner, partition each node into a window and use the fault detect model to perform fault transmission along the target node. For each node in the MAS, if there is a node that can replace the target node without causing an increase in the overall circuit level, then replace the target node to reduce the number of gates and circuit area.

3) After step (1) and (2), optimization is completed. Now delete redundant nodes in the circuit due to optimization

and then perform an inspection on the optimized circuit to verify if it functions similarly to the original circuit. This completes the optimization process. Eliminate redundancy nodes in the circuit due to optimization and then perform a check on the optimized circuit to ensure its functionality is identical to the original circuit. The XMGs node merging algorithm is shown in Fig.4.

Algorithm 2: Node merging algorithm

Input: *Net.N ← XMG*
Output: *Optimized Net.N' ← XMG*
  N' = N ;
  do
    foreach *Critical path* in *N*
      foreach *Node.Nd* in *Critical path*
        Mas = Fault_detect(Nd, level);
        check *Mas* if substitute
      if substitute **then break;**
      end
    end
  while !substitute;

  foreach *Node.Nd* in *N(depth-first)*
    Fault_detect(Nd, level)
  end

  Equivalence test(N, N')
  return *N';*

Fig. 4. XMGs node merging algorithm

E. A Sample of XMGs Node Merging

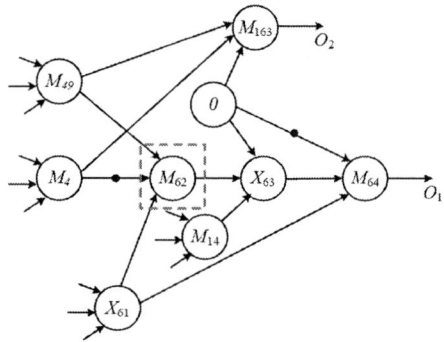

Fig. 5. XMGs node merging algorithm example

Fig.5 shows a window in a complex circuit. The X node indicates the XOR node, and the M node indicates the majority node. Now use the fault detect model to target node M_{62}. First activate the fault at M_{62}, and propagate it to the fan-out terminal X_{63} and M_{64} in turn. In order to propagate the fault to M_{64}, M_{64}'s other two fan-ins (1 and X_{61}) need to satisfy that their XOR values are opposite, which means $X_{61}=0$, in which case the MAS record for the window is as follows: Since ROBDD tree structure of node M_{62} and node M_{163} is the same, according to the fault detection principle, the M_{62} node can be replaced with the M_{163} node. The network after replacement is shown in Fig.7, and it's easy to find the circuit has been simplified.

IV. EXPERIMENT AND RESULT

This paper proposed a node merging algorithm based on window fault detection and implemented it in the code of the

TABLE I
THE RESULTS OF OUR WORK

benchmarks	nodenum	depth	optimized nodenum	optimized nodenum percent	optimized depth	optimized depth percent
ac97_ctrl	11296	12	11248	0.35%	12	0.00%
aes_core	19403	26	19133	1.39%	26	0.00%
bar	2895	11	2831	2.27%	11	0.00%
c880	273	23	269	1.47%	23	0.00%
c1908	193	22	183	5.18%	20	9.09%
c3540	818	32	801	2.08%	32	0.00%
c5315	1103	24	1083	1.81%	24	0.00%
c7552	1007	29	994	1.29%	27	6.90%
cavlc	683	17	671	1.76%	17	0.00%
ctrl	120	6	116	3.33%	6	0.00%
des_perf	67581	18	66244	1.98%	18	0.00%
div	41869	4317	41375	1.18%	4317	0.00%
DMA	20487	27	20356	0.634%	27	0.00%
DSP	36221	56	35993	0.63%	56	0.00%
i2c	1242	16	1212	2.42%	16	0.00%
int2float	234	14	230	1.71%	14	0.00%
iwls05_i2c	1081	16	1056	2.31%	16	0.00%
iwls05_men/1226	13914	37	13844	0.50%	37	0.00%
max	2127	298	2105	1.03%	283	5.03%
pci_bridge32	17319	33	17221	0.57%	33	0.00%
router	266	49	250	6.02%	47	4.08%
sasc	584	8	580	0.68%	8	0.00%
simple_spi	829	12	816	1.57%	12	0.00%
spi	3219	28	3193	0.81%	28	0.00%
sqrt	17368	5142	17183	1.07%	5024	2.29%
square	12804	155	12715	0.70%	155	0.00%
ss_pcm	317	6	310	2.21%	6	0.00%
systemcdes	2419	24	2372	1.94%	24	0.00%
tv80	8172	53	8083	1.09%	53	0.00%
voter	6543	58	5890	9.98%	57	1.72%
wb_conmax	47448	31	46420	2.17%	31	0.00%
average	-	-	-	2.07%	-	0.91%

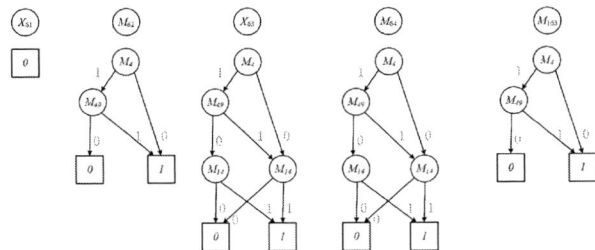

Fig. 6. example of MAS in fault detect model

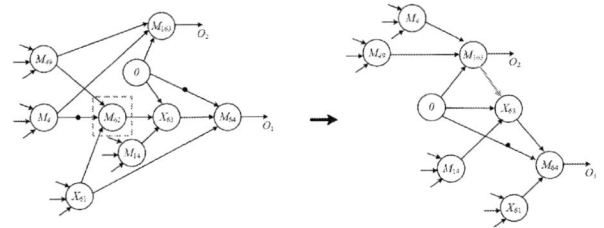

Fig. 7. XMGs optimization result

logic synthesis tool Also [14] under the environment of IWLS 2005 and EPFL's logical synthesis framework by optimizing

several benchmarks [15]. Below is the test result after circuit optimization. The experiments are conducted on a 3.10GHz Linux platform(Red Hat Enterprise Linux 7).

A. Two-level Optimization Effects For Different Circuits

The results are shown in Table I and Fig.8. The node merging algorithm achieved an average reduction of 2.07% nodes and an average reduction of 0.91% series length.

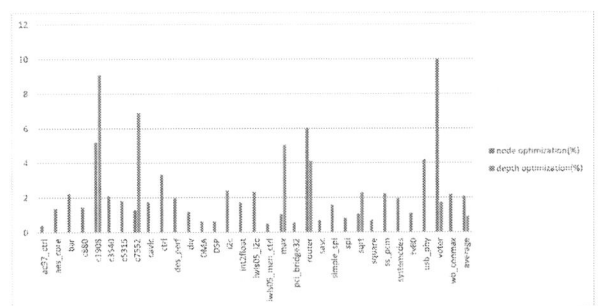

Fig. 8. The Experiment Result of The XMGs node merging algorithm

Compare with Also's internal XMGs optimization algorithms, Also's optimized algorithm comparison from Fig.9, although Also has a better the number of nodes in some cases, it is

979-8-3315-8850-2/25 $31.00 © 2025 IEEE

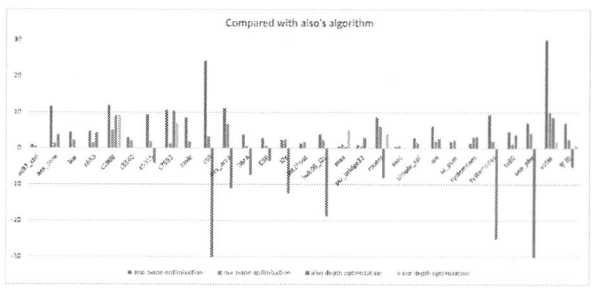

Fig. 9. The Comparison of This Work And ALSO

worse in terms of delay; while this algorithm can optimize both the number of nodes and delay simultaneously.

B. Different level of The XMGs Node merging algorithm

For the same circuit, the optimization results of the adjustment window level are shown in Table II and Fig.10: It can

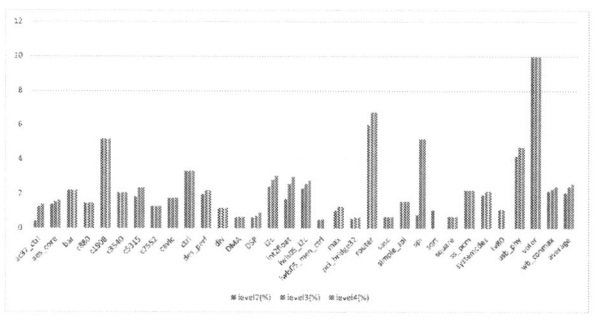

Fig. 10. Optimization results of different levels

TABLE II
AVERAGE OPTIMIZATION RESULTS OF DIFFERENT LEVEL

different level	node optimization	depth optimization
2	2.07%	0.91%
3	2.42%	1.11%
4	2.58%	1.27%

be seen from the data that when the window series increases, the number of nodes in the same circuit and the optimization of the series can be improved.

V. CONCLUSION

In this paper, we proposed a logic synthesis optimization method for the XMGs network mainly uses fault detect methods to perform node merging on the XMGs network to optimize its series length and node number. We implemented this algorithm in ALSO under the environment of IWLS 2005 and used several benchmarks provided by EPFL for testing. The experimental results show it can effectively reduce circuit delay and area . However, compared with existing algorithms on node number optimization, this method still has some shortcomings and as the window series length increased further, the time consumed was significantly large. Future research will consider exploring more substitution spaces and methods such as adding nodes first replaced for better optimization effects.

ACKNOWLEDGEMENTS

The authors would like to thank the Chinese Academy of Sciences (CAS) Project for Young Scientists in Basic Research under Grant YSBR-029.

REFERENCES

[1] Lent C S, Tougaw P D, Porod W, Bernstein G H. Quantum cellular automata[J].Nanotechnology, 1993, 4(1): 4957.

[2] S. A. Kumari, S. M. Hiremath and K. Mahapatra, "Advancements Of Quantum Dot Cellular Automata: A Comprehensive Survey," 2025 IEEE International Students' Conference on Electrical, Electronics and Computer Science (SCEECS), Bhopal, India, 2025, pp. 1-6, doi: 10.1109/SCEECS64059.2025.10940496.

[3] S. Ikegawa, F. B. Mancoff, J. Janesky and S. Aggarwal, "Magnetoresistive Random Access Memory: Present and Future," in IEEE Transactions on Electron Devices, vol. 67, no. 4, pp. 1407-1419, April 2020

[4] L. Amarú, P. -E. Gaillardon and G. De Micheli, "Majority-Inverter Graph: A novel data-structure and algorithms for efficient logic optimization," 2014 51st ACM/EDAC/IEEE Design Automation Conference (DAC), San Francisco, CA, USA, 2014, pp. 1-6, doi: 10.1145/2593069.2593158.

[5] L. Amarú, P. -E. Gaillardon and G. De Micheli, "Majority-Inverter Graph: A New Paradigm for Logic Optimization," in IEEE Transactions on Computer-Aided Design of Integrated Circuits and Systems, vol. 35, no. 5, pp. 806-819, May 2016, doi: 10.1109/TCAD.2015.2488484.

[6] P. -E. Gaillardon, L. G. Amaru and G. De Micheli, "Unlocking Controllable-Polarity Transistors Opportunities by Exclusive-OR and Majority Logic Synthesis," 2014 IEEE Computer Society Annual Symposium on VLSI, Tampa, FL, USA, 2014, pp. 403-405, doi: 10.1109/ISVLSI.2014.107.

[7] Rumi Z, Walus K, Wei W, Jullien G A. A method of majority logic reduction for quantum cellular automata[J]. IEEE Transactions On Nanotechnology, 2004, 3(4):443450.

[8] Zhang R, Gupta P, Jha N K. Majority and Minority Network Synthesis With Application to QCA, SET, and TPL Based Nanotechnologies[J]. IEEE Transactions on Computer Aided Design of Integrated Circuits and Systems, 2007, 26(7):12331245.

[9] Rui Z, Gupta P, Jha N K. Synthesis of majority and minority networks and its applications to QCA, TPL and SET based nanotechnologies[C]. 18th International Conference on VLSI Design held jointly with 4th International Conference onEmbedded Systems Design, 2005: 229234.

[10] Soeken M, Amarù L G, Gaillardon P, Micheli G d. Optimizing Majority Inverter Graphs with functional hashing[C]. 2016 Design, Automation & Test in EuropeConference & Exhibition (DATE), 2016: 10301035

[11] Haaswijk W J, Soeken M, Amaru L, Gaillardon PE. LUT mapping and optimization for majority inverter graphs[C]. Proceedings of the 25th International Workshop on Logic & Synthesis (IWLS), 2016

[12] Haaswijk W, Soeken M, Amarú L, Gaillardon PE, De Micheli G. A novel basis for logic rewriting[C]. 2017 22nd Asia and South Pacific Design Automation Conference (ASPDAC), 2017: 151156

[13] Ko C C, Lin C C, Chen Y C, Wang C Y. Majority Logic Circuit Minimization Using Node Addition and Removal[J]. IEEE Transactions on Computer Aided Design of Integrated Circuits and Systems, 2021, DOI 10.1109/TCAD.2021.3060648: 11

[14] zfchu, zfchu, tianhuiming, ShangChuanhe, xianghex, ZXJing94:"also", 2021 Nov, https://github.com/nbulsi/also/

[15] Mathias Soeken, Heinz Riener, Winston Haaswijk, and Giovanni De Micheli: "The EPFL Logic Synthesis Libraries" , 2022 June, arXiv:1805.05121v3, https://github.com/lsils/lstoolsshowcase.

2025 The 10th International Conference on Integrated Circuits and Microsystems

FMSRdiff : Efficient latent diffusion framework combining consistency model and flow matching for super-resolution

[†] QunKai Peng
Shenzhen Graduate School,
Peking University
Shenzhen, China
2301212812@stu.pku.edu.cn

[†] GuoFeng Cai
Shenzhen Graduate School,
Peking University
Shenzhen, China
cgf@stu.pku.edu.cn

YiHua Xu
School of Integrated Circuits,
Peking University
Shenzhen, China
yhxu@stu.pku.edu.cn

Kuan-Chang Chang*
Guangdong Provincial Key Laboratory of
In-Memory Computing Chips, School of
Electronic and Computer Engineering,
Peking University
Shenzhen, China
kcchang@pkusz.edu.cn

Lei Li*
Shenzhen Technology University
Shenzhen, China
lilei@sztu.edu.cn

GuiBo Luo*
Shenzhen Graduate School,
Peking University
luogb@pku.edu.cn

[†]These authors contributed equally to this work.

Abstract—While diffusion-based super-resolution (SR) methods have demonstrated promising results, they still face critical limitations in practical medical imaging applications. Recent methods focus on training on real-world image datasets, but these methods are prone to losing structural information when transferred to the medical imaging domain and suffer from extremely slow inference speed. Traditional diffusion-based super-resolution models in medical imaging face key challenges including image structure inconsistency, low computational efficiency, and unstable training. To address these issues, we propose FMSRdiff, an efficient latent diffusion framework that integrates consistency models with optical flow matching. Starting with latent space diffusion, we construct an efficient low-dimensional representation learning framework. This allows the diffusion process to be performed in a compressed latent space, significantly reducing computational complexity and memory requirements while maintaining high-quality reconstruction. We then design a two-stage training strategy that first learns stable noise prediction capabilities through optical flow matching, then switches to consistent model training to achieve a direct mapping from noisy to clean latent representations. This strategy not only ensures training stability but also significantly improves inference speed, reducing the 50-100 sampling steps required in traditional diffusion models to just one, significantly improving the model's practicality. Furthermore, we introduce a hierarchical diffusion mechanism to achieve multi-scale feature processing from coarse to fine scales, effectively addressing the structural inconsistency problem in medical image super-resolution. Through cross-scale feature

This study was supported by Shenzhen Scientific and Technological Foundation (No. RCYX20231211090332037, JCYJ20240813160211015), National Natural Science Foundation of China (No. 62474008, 62204007), and Guangdong Provincial Natural Science Foundation (No. 2024A1515030044). This work was supported by Guangdong Provincial Key Laboratory of In-Memory Computing Chips (2024B1212020002). This work was in part supported by the Shenzhen POC center of Flexible Electronics and Guangdong Technology Center for Oxide Semiconductor Devices and ICs.

fusion and conditional guidance, the model better preserves the global structure and local details of the image, resulting in more natural and realistic super-resolution results. Experiments on the IXI and BRATS datasets demonstrate that our method achieves a 30% improvement in SSIM scores compared to existing state-of-the-art methods for medical imaging data. Through latent space diffusion, computational efficiency is increased by approximately 16 times, memory usage is significantly reduced.

Index Terms—Latent diffusion, super-resolution, consistency model, flow matching

I. INTRODUCTION

Image super-resolution (SR) aims to reconstruct a high-resolution image from a given low-resolution (LR) image. In recent years, diffusion models—renowned for their effectiveness in modeling complex distributions—have been widely used and have demonstrated impressive performance in SR tasks, particularly in medical image super-resolution. Due to the limited number of available open-source medical image datasets and poor image quality, medical image super-resolution has become an increasingly popular research topic, aiming to reconstruct high-resolution images with high structural consistency from existing low-quality LR medical images. Training medical image super-resolution models faces three major challenges: first, the instability caused by noise prediction methods; second, high memory usage and time consumption during training and inference; and third, maintaining structural consistency in medical images. Specifically, current strategies for applying diffusion models can be roughly divided into two categories: directly connecting the LR image to the denoiser input, and inversely adjusting pre-trained diffusion models [1]–[5]. Although both strategies yield promising results, they are limited by computational inefficiency and high memory demands. Importantly, the initial state of these conditional diffusion models is pure Gaussian noise, with no incorporation of LR image prior knowledge. Therefore, a large number of inference steps are required to achieve satisfactory performance, which seriously hinders the practical application of diffusion-based SR techniques. Previous studies have focused on improving the sampling efficiency of diffusion models and proposed various techniques [6], [7]. However, in the field of medical imaging, maintaining high fidelity and structural similarity is crucial. These techniques often sacrifice performance to achieve acceleration and cannot guarantee fidelity, so they are prone to generate a large number of unreasonable image structures. Recently, some innovative techniques have emerged that reformulate the diffusion process in image restoration tasks, focusing on improving the signal-to-noise ratio of the initial diffusion state, thereby shortening the Markov chain. For example, ExposureDiffusion [8] uses the input noisy image to start the denoising diffusion process. However, even in these latest studies [9], [10], there are still limitations. For example, although [9] achieved amazing results, it relied on a large pre-trained model, resulting in the final model effect being strongly dependent on the base model, while [10] started reasoning from random Gaussian noise, resulting in low efficiency. To address these challenges, we propose a novel method that can generate HR images while maintaining high structural similarity and significantly reducing inference and training costs. Specifically, we first encode the HR and LR images into a latent space using a latent space encoder. All subsequent operations are performed in the latent space, significantly reducing computational complexity and memory requirements. We then train our SR model using a progressive training method. In the early stages of training, we use a flow matching-based approach, which improves the model's feature extraction capabilities while maintaining a more stable training process. To ensure the structural consistency of the image, we introduce a multi-scale diffusion model that captures global contextual information while restoring fine details. In the later stages, we use knowledge transfer to compress the multi-step generation of the rich feature representation learned in the flow matching stage into a single step. Compared with the previous diffusion model, our method significantly reduces the number of inference steps from 50 to 1, increasing the inference speed by up to 48 times.

II. BACKGROUND

In the following, we will provide a comprehensive overview of deep learning-based image super-resolution methods, with a particular focus on existing deep learning models for image super-resolution. We will also explore the inherent challenges and limitations of existing models and highlight areas that require further research and development.

A. Image Super-Resolution

With the continuous development of deep learning technology, super-resolution based on deep learning has become the mainstream processing solution for this visual task [11], [12]. This technology originated from SRCNN [13]. Early studies focused on directly using deep networks to regress high-resolution images. These studies have the problem of over-smoothing images and losing high-frequency information of images. In addition, due to their direct regression scheme, they often do not have good generalization performance [14]–[18]. Currently, generative SR models are gaining more and more attention, such as autoregressive models [19], [21], [22]. Although autoregressive models have good performance and intuitive ideas in SR tasks, they require a lot of computing resources and time to calculate high-resolution images. In addition, GAN-based methods have also achieved great success in maintaining the structural consistency of super-resolution images [23]–[26]. However, the adversarial training process of GAN-based methods is unstable, their discriminators have limited ability to distinguish the quality of different natural image contents, and unnatural visual artifacts often appear in the process of generating super-resolution images. Although some studies, such as DeSRA [20], can suppress most of the artifacts, the cost is that they have difficulty in generating more natural details.

B. Image super-resolution based on diffusion model

In recent years, diffusion-based models have been widely used in the field of image super-resolution [9], [27]–[30]. Early studies attempted to use denoising diffusion probability models (DDPMs) [31]–[33] to solve the super-resolution problem. Some recent studies used powerful pre-trained models to assist image super-resolution [34] and fine-tuned the pre-trained diffusion models through adapters [35]. To address the efficiency problem of the diffusion model, S3Diff [25] introduced a degradation-guided low-adaptation module. DiffBIR [37] utilizes a pre-trained text-to-image diffusion model to solve the image restoration problem. The low-quality image given by OSEDiff [27] can be directly used as the starting point of diffusion, thereby eliminating the uncertainty caused by random noise sampling. Single-step models usually suffer from structural distortion problems. Our proposed method solves the core pain points of current single-step methods, such as detail loss and weak degradation adaptability due to insufficient model capacity.

III. METHODOLOGY

A. Latent Space Coding

To enhance computational efficiency in super-resolution, we design a latent space encoder that compresses 256×256 HR images into a 64×64 latent representation (4× downsampling). As shown in Equation 1, this enables efficient diffusion processes in the compressed space. The encoder comprises three components: a downsampling convolutional layer, a feature transformation layer, and a latent coding layer, while the decoder employs transposed convolution for upsampling and reconstruction. This pre-trained autoencoder provides the

979-8-3315-8850-2/25 $31.00 © 2025 IEEE   911

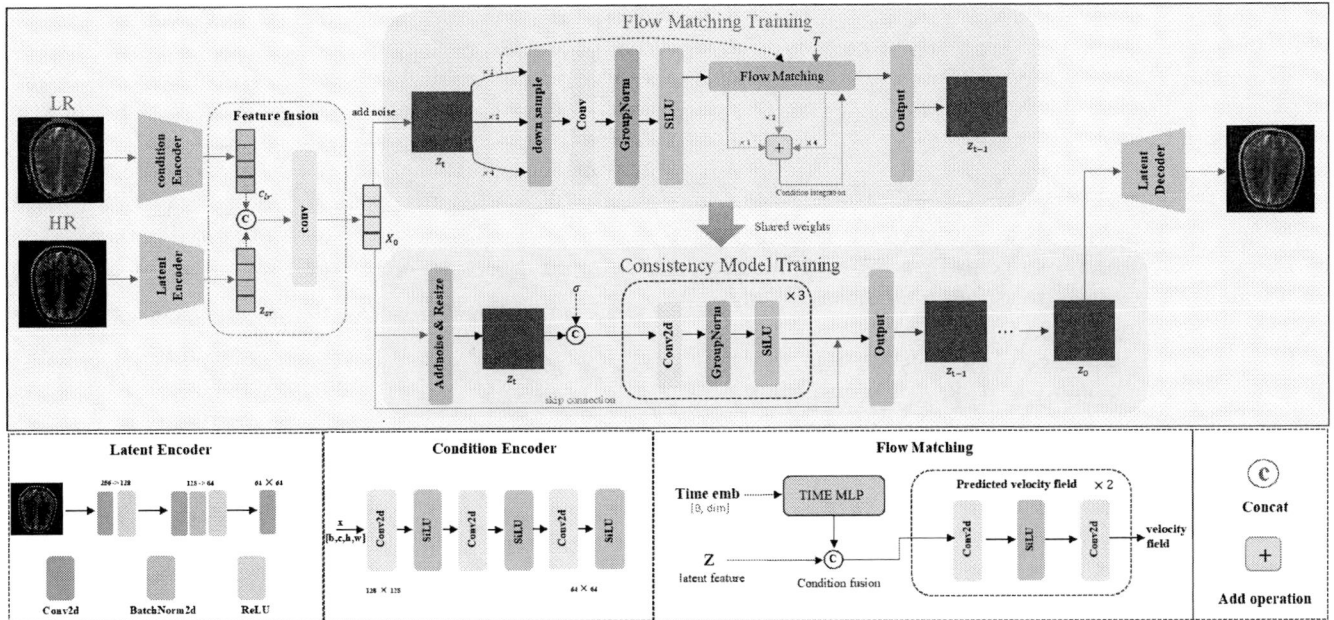

Fig. 1. **This is the training framework for FMSRdiff.** Low-resolution images are trained through a trainable conditional encoder, a multi-scale hierarchical diffusion model, and a consistency model to obtain the desired high-quality image. Furthermore, the low-resolution image is fed into the diffusion network as an additional condition to guide diffusion during training. Training begins with flow matching, which ensures the model has good feature extraction capabilities and provides a good initialization for the consistency model. High-resolution images are encoded into a latent space using a latent space encoder, and all operations are performed in the latent space to reduce the model's computational load. A regularized loss is backpropagated to update the entire model. After training, single-step inference using the consistency model and multiple diffusion steps using the multi-scale hierarchical diffusion model are optional.

latent domain for subsequent diffusion, significantly reducing memory usage and computational complexity.

$$\mathcal{E} : \mathbb{R}^{3 \times 256 \times 256} \to \mathbb{R}^{d_l \times 64 \times 64},$$
$$\mathcal{D} : \mathbb{R}^{d_l \times 64 \times 64} \to \mathbb{R}^{3 \times 256 \times 256} \tag{1}$$

Here d_l is the latent dimension, which achieves 4 times space compression and retains more high-frequency information.

The encoder consists of three convolutional layers that gradually downsample and increase the number of channels. We assume that $\mathbf{x} \in \mathbb{R}^{3 \times 256 \times 256}$ is the input image. The encoding process is as follows:

$$\mathbf{z} = \mathcal{E}(\mathbf{x}) = f_{enc}(\mathbf{x}) \tag{2}$$

$$\mathbf{h}_1 = \mathrm{ReLU}(\mathrm{Conv}_{\overline{4 \times 4}}^2(\mathbf{x})) \tag{3}$$

$$\mathbf{h}_2 = \mathrm{ReLU}(\mathrm{BN}(\mathrm{Conv}_{\overline{4 \times 4}}^2(\mathbf{h}_1))) \tag{4}$$

$$\mathbf{z} = \mathrm{Conv}_{3 \times 3}(\mathbf{h}_2) \tag{5}$$

where $\mathbf{z} \in \mathbb{R}^{d_l \times 64 \times 64}$ is the latent space projection obtained by encoding.

The decoder forms a mirror image structure with the encoder. The goal of the decoding process is to restore the potential representation $\mathbf{z}$ to the input image $\hat{\mathbf{x}}$ through the decoder.

$$\mathbf{g}_1 = \mathrm{ReLU}(\mathrm{Conv}_{3 \times 3}(\mathbf{z})) \tag{6}$$

$$\mathbf{g}_2 = \mathrm{ReLU}(\mathrm{BN}(\mathrm{ConvT}_{4 \times 4}^{\hat{2}}(\mathbf{g}_1))) \tag{7}$$

$$\hat{\mathbf{x}} = \mathrm{Tanh}(\mathrm{ConvT}_{4 \times 4}^{\hat{2}}(\mathbf{g}_2)) \tag{8}$$

The input dimension of $\mathbf{z}$ is $\mathbf{z} \in \mathbb{R}^{d_l \times 64 \times 64}$.

B. Flow matching

Flow matching is an emerging generative model training method that avoids the noise prediction instabilities of traditional diffusion models by directly learning the velocity field from the noise distribution to the data distribution. In this work, we design a flow matching module with a conditional fusion mechanism and efficient velocity field prediction at its core. We first map the time step t to a high-dimensional space using sinusoidal positional encoding to enhance the model's sensitivity to temporal information, as shown in Equations 9-10, where d is the embedding dimension (default is 256). This is then followed by a two-layer MLP for nonlinear transformation:

$$\mathrm{emb} = \left[\sin\left(\frac{t}{10000^{2k/d}} \right), \cos\left(\frac{t}{10000^{2k/d}} \right) \right]_{k=0}^{d/2-1} \tag{9}$$

$$\mathrm{time\_emb} = \mathrm{MLP}(\mathrm{emb}) \in \mathbb{R}^d \tag{10}$$

In order to effectively utilize the guidance information of low-resolution images, we designed a conditional projection mechanism. When the number of channels of the conditional feature map (condition) does not match the current feature map (x), 1×1 convolution is used to adjust the channels. If the spatial size of the conditional feature map is inconsistent with the current feature map, bilinear interpolation is used for upsampling or downsampling. The adjusted conditional information is directly added to the current feature map to achieve feature fusion:

$$x_{\text{fused}} = x + \text{Adapt}(\text{condition}) + \text{Broadcast}(\text{time\_emb}) \quad (11)$$

Adapt includes channel adjustment and spatial interpolation, and *Broadcast* represents expanding the temporal embedding into a tensor of the same size as the feature map and adding it to the current feature.

After completing the initial data preprocessing, we input the processed features into our velocity field prediction network v_θ. The velocity field prediction network uses a three-convolutional layer structure with group normalization and SiLU activation functions added in between, as shown in Equation 12. This design ensures that the normalization layer stabilizes the training process while maintaining the receptive field. Group normalization divides the channels into eight groups and calculates statistics independently, alleviating the impact of batch size on training. The SiLU activation function provides smooth nonlinear transformations and enhances gradient flow.

$$v_\theta = f_{\text{vel}}(x_t, t, \text{condition}) \quad (12)$$

Given the current state x_t (noisy latent representation) and the target state x_0 (clean latent representation), the velocity field is defined as the instantaneous rate of change of the linear interpolation path:

$$v_t = \frac{x_0 - x_t}{t} \quad (13)$$

Where $t \in (0, 1]$ is the time parameter. When $t \to 0$, x_t approaches the target distribution; $t = 1$ corresponds to a pure noise state.

The training goal of the flow matching stage is to minimize the mean square error between the predicted velocity field and the true velocity field:

$$\mathcal{L}_{\text{FM}} = \mathbb{E}_{t \sim \mathcal{U}(0,1), x_0 \sim p_{\text{data}}, x_t \sim p_t(.|x_0)} | v_\theta(x_t, t, \text{condition}) - v_t |_2^2 \quad (14)$$

Here, conditional information is injected through low-resolution image encoding features to guide the velocity field toward the target distribution. This loss function is consistent with the probabilistic path derived from the continuity equation, ensuring the stability of the generation process.

C. Multi-scale hierarchical diffusion model

In medical image super-resolution tasks, medical image contain multi-scale features ranging from global structures to local details. Traditional diffusion models usually operate at a single scale and have difficulty capturing feature dependencies at different scales simultaneously. To address this challenge, this paper proposes a multi-scale hierarchical diffusion model. Its core idea is to collaboratively extract and generate features at multiple scales through a hierarchical processing mechanism to achieve a coarse-to-fine image reconstruction process.

During the training phase, the multi-scale model learns multi-scale feature representations through a hierarchical diffusion process, which are distilled into a consistency model. The model adopts a pyramid processing structure and defines a scale set $\mathcal{S} = \{s_1, s_2, \ldots, s_K\}$, where $s_k = 2^{k-1}$ represents the downsampling factor. For the input feature map $\mathbf{X} \in \mathbb{R}^{C \times H \times W}$, the processing at scale s_k can be formalized as:

$$\mathbf{X}^{(k)} = \mathcal{D}_k(\mathbf{X}) \quad (15)$$

$$\mathcal{F}_k = \text{Conv} \circ \text{GroupNorm} \circ \text{SiLU} \circ \text{FlowMatching} \circ \text{Conv} \quad (16)$$

Features at different scales are integrated through a gated fusion mechanism. Let $\mathbf{H}^{(k)}$ be the output features of scale s_k, and the fusion process is defined as:

$$\mathbf{H} = \sum_{k=1}^{K} \alpha_k \cdot \mathcal{U}_k(\mathbf{H}^{(k)}) \quad (17)$$

Where $\mathcal{U}_k(\cdot)$ is the upsampling operation and $\alpha_k = \frac{1}{s_k}$ is the scale weight coefficient, which ensures that fine-scale features receive higher weights. This design enables the model to capture global structures at coarse scales, model intermediate features at medium scales, and recover high-frequency details at fine scales. The multi-scale processing is always guided by the low-resolution input $\mathbf{Y}_{LR}$. At scale s_k, the conditional features are adaptively adjusted:

$$\tilde{\mathbf{Y}}_{LR}^{(k)} = \mathcal{A}_k(\mathbf{Y}_{LR}) \quad (18)$$

Among them, $\mathcal{A}_k(\cdot)$ includes channel adjustment and spatial interpolation operations to ensure that the conditional features match the current scale feature size. The conditional features are injected into the processing flow through residual connections.

$$\mathbf{X}^{(k)} \leftarrow \mathbf{X}^{(k)} + \mathcal{P}(\tilde{\mathbf{Y}}_{LR}^{(k)}) \quad (19)$$

$\mathcal{P}(\cdot)$ is the projection transformation implemented by 1×1 convolution, which aligns the conditional features with the processed feature space.

The complete hierarchical diffusion process can be formalized as:

$$\mathcal{H}(\mathbf{X}, t, \mathbf{Y}_{LR}) = \sum_{k=1}^{K} \frac{1}{s_k} \cdot \mathcal{U}_k \left(\mathcal{F}_k \left(\mathcal{D}_k(\mathbf{X}) + \mathcal{P}(\mathcal{A}_k(\mathbf{Y}_{LR})), t \right) \right) \quad (20)$$

D. Consistency Model

The consistency model is a new generative model that achieves single-step sampling while ensuring generation quality, thus overcoming the drawback of traditional diffusion models that require iterative sampling and result in slow inference. Our framework introduces the consistency model into the latent space, achieving efficient single-step super-resolution reconstruction. We adopt a consistency training approach, directly training a model to predict the initial state. Given a diffusion process $\{\mathbf{x}_t\}_{t=0}^{T}$, the consistency model

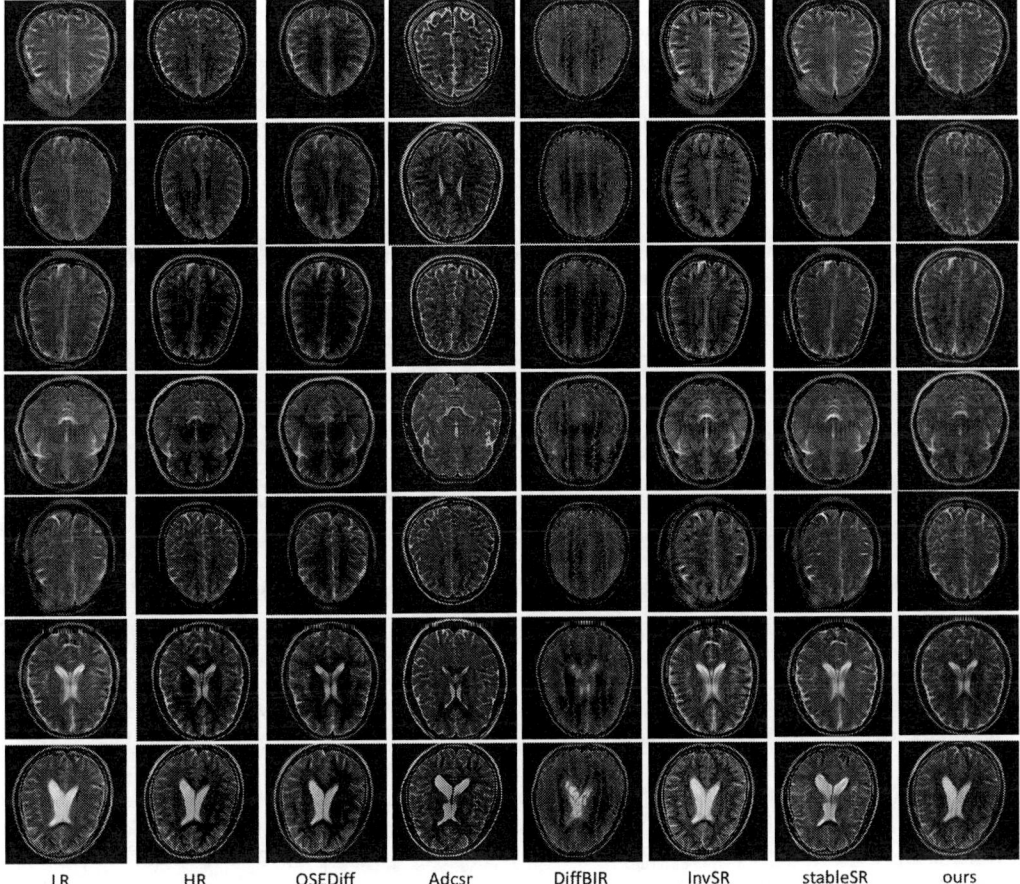

| LR | HR | OSEDiff | Adcsr | DiffBIR | InvSR | stableSR | ours |

Fig. 2. **Qualitative comparison of different super-resolution methods. Please zoom in for a clearer view.**

learns a mapping function $f_\theta(\mathbf{x}_t, t)$ that satisfies the following consistency properties:

$$f_\theta(\mathbf{x}_t, t) = f_\theta(\mathbf{x}_s, s) \quad \forall t, s \in [0, T] \tag{21}$$

where $\mathbf{x}_t$ is the noise input at time step t, and the goal of the mapping function is to predict the starting point $\mathbf{x}_0$ of the diffusion trajectory. The model output is constructed using the following formula:

$$f_\theta(\mathbf{x}_t, t) = c_{skip}(t) \cdot \mathbf{x}_t + c_{out}(t) \cdot F_\theta(c_{in}(t) \cdot \mathbf{x}_t, t) \tag{22}$$

Where $c_{in}(t)$: controls input scaling, $c_{out}(t)$: controls output scaling, $c_{skip}(t)$: controls skip connections. These coefficients are defined as:

$$c_{\text{skip}}(t) = \frac{\sigma_{\min}^2}{\sigma_t^2 + \sigma_{\min}^2}, \qquad c_{\text{out}}(t) = \sigma_t \cdot \frac{\sigma_{\min}}{\sqrt{\sigma_t^2 + \sigma_{\min}^2}},$$
$$c_{\text{in}}(t) = \frac{1}{\sqrt{\sigma_t^2 + \sigma_{\min}^2}}. \tag{23}$$

where σ_t is the noise level scheduling parameter and $\sigma_{\min}$ is a very small constant (taken as 0.002 in this paper) for numerical stability.

We use the mean squared error loss to directly optimize the difference between the predicted result and the true starting point:

$$\mathcal{L}_{\text{CM}} = \mathbb{E}_{t \sim \mathcal{U}[0,T]} lambda(t) |f_\theta(\mathbf{x}_t, t) - \mathbf{x}_0|_2^2 \tag{24}$$

where $\lambda(t)$ is a time-dependent weighting factor. In implementation, we optimize in the latent space $\mathcal{Z}$:

$$\mathcal{L}_{\text{CM}} = \mathbb{E}_{z_0 \sim p_{\mathcal{Z}}, \epsilon \sim \mathcal{N}(0,I)} |f_\theta(\sigma_t z_0 + s_t \epsilon, t) - z_0|_2^2 \tag{25}$$

The inference process of the consistency model is extremely efficient. We first input a low-resolution image $\mathbf{y}_{\text{LR}}$ and conditionally encode it into the latent space $\mathbf{c} = \text{Enc}(\mathbf{y}_{\text{LR}})$. We then sample noise $\mathbf{z}_T \sim \mathcal{N}(0, I)$ from a standard Gaussian distribution and perform a single-step consistency mapping $\mathbf{z}_0 = f_\theta(\mathbf{z}_T, T; \mathbf{c})$. Finally, we decode the latent variable $\mathbf{x}_{\text{HR}} = \text{Dec}(\mathbf{z}_0)$ to obtain a high-resolution SR result.

IV. RESULT AND DISCUSSION

A. Experimental Settings

We use the classic brain MRI IXI dataset and Brtas dataset to train our model. We process the dataset according to the method of [41] to obtain downsampled LR images. Specifically, we train our latent encoder and diffusion model from scratch, and train the entire model for 2000 epochs. In the first 1400 epochs, we use the flow matching method for pre-training, and in the last 600 epochs, we use the consistency model to share the model weights of the previous flow matching pre-training through transfer learning. We compare our method with several representative SR models, including StableSR [9], AdcSR [28], OSEDif [27], InvSR [40], resshift [36], DiffBIR [37], S3Diff [25]. For fairness, our comparison methods all use the open source implementation of their open source repositories. To evaluate the proposed method on a synthetic test dataset containing reference images, we use PSNR, SSIM and LPIPS [38] to measure the fidelity performance. We train our model using the AdamW optimizer [39] with a learning rate of 1e-3 in the early epochs and 1e-4 in the late epochs. The entire training process takes approximately 10 hours on a single NVIDIA 4090 GPU with a batch size of 4. The latent space dimension is set to 8, and the layer-wise diffusion scale is set to [4, 2, 1].

B. Comparison with State-of-the-Arts

Qualitative Comparison. Figure 2 shows a visual comparison of different super-resolution methods. StableSR and DiffBIR reconstruct image details by leveraging image priors from a pre-trained satble diffusion model, but their realism is insufficient. Due to the lack of textual cues, StableSR and DiffBIR are limited in generating rich textures, which we believe is related to their underlying models. Although OSEDiff only performs one forward propagation step, it is able to reproduce more realistic details than other methods. However, due to structural distortion and gradient issues often associated with single-step models, its performance degrades after model convergence. In contrast, FMSRdiff can generate more realistic details and, compared to other single-step generation methods, achieves a more stable training process thanks to our flow matching pre-training.

Quantitative Comparison. Table 1 lists the quantitative comparison of competing methods on a mixed dataset of IXI and Brats. We can make the following observations: First, FMSRdiff generally outperforms the competing methods in terms of full-reference perceptual quality metrics LPIPS and PSNR, as well as the structural consistency metric SSIM, especially the structural similarity metric. Second, ResShift performs better in terms of full-reference fidelity metrics such as PSNR. This is mainly because they train a diffusion model from scratch specifically for restoration purposes, rather than using a pre-trained T2I model such as stable diffusion.

C. Ablation Study

To verify the effectiveness of the flow field matching strategy during training, we conducted an ablation study, removing the flow field matching strategy at the beginning of training and directly training the single-step model. The results on the test set are shown in Table 2. We can see that without the flow field matching strategy, full-reference metrics such

TABLE I
QUANTITATIVE COMPARISON OF OUR METHOD WITH SOTA METHOD

Method	PSNR↑	SSIM↑	LPIPS↓
ResShift	**29.395**	0.837	0.1401
AdcSR	22.69	0.567	0.1755
InvSR	23.9795	0.552	0.2341
Diffbir	20.286	0.509	0.2429
S3Diff	24.125	0.548	0.2011
OSEDiff	24.414	0.645	**0.1223**
StableSR	22.079	0.444	0.2697
ours	29.02	**0.927**	0.2417

as PSNR, SSIM, and LPIPS decrease slightly. After ablating FM, although the consistency model (CM) and the multi-scale hierarchical model (MSH diffusion) still ensure overall structural correctness (high SSIM) through the "single-step consistency constraint + multi-scale structure fusion" strategy, they lose the pixel-level optimization capability brought by the velocity field, resulting in a slight decrease in detail recovery accuracy. The training process also becomes more unstable, and we encounter gradient issues during our training process.

When the consistency model is removed, model performance plummets: PSNR plummets from over 29dB to 18.25dB, SSIM drops from over 0.9 to 0.366, and LPIPS soars from around 0.25 to 0.6931. This result indicates that while removing CM retains the stable training capabilities of flow matching and the multi-scale hierarchical structural modeling capabilities, the model loses its ability to accurately characterize the relationship between noise level and image features, resulting in severe pixel distortion and structural distortion in the generated results, ultimately manifesting as a comprehensive degradation of quantitative metrics.

Effectiveness of the Multi-scale Diffusion Model, We conducted an ablation study to examine the impact of the multi-scale diffusion model. As shown in Table 2, in the third row, we ablated the multi-scale diffusion model. The results show that its structural consistency index SSIM decreases. Single-scale generation may lose cross-scale structural consistency, resulting in SSIM lower than the results in the first row.

TABLE II
ABLATION STUDIES, ✓ REPRESENTS THE ACTIVATION OF THE MODULE, ×
REPRESENTS THE ABLATION OF THE MODULE WITHOUT ACTIVATION

FM	CM	MSH diffusion	PSNR↑	SSIM↑	LPIPS↓
×	✓	✓	27.38	**0.916**	0.2536
✓	×	✓	18.25	0.366	0.6931
✓	✓	×	**27.87**	0.902	**0.2448**

V. CONCLUSION

We propose FMSRdiff, an efficient framework for medical image super-resolution. This framework pre-trains the model using a flow matching strategy to mitigate the risk of vanishing gradients. FMSRdiff utilizes the fusion of low-resolution images and Gaussian noise as the diffusion starting point, reducing the uncertainty associated with using only random Gaussian noise as the diffusion starting point. Furthermore, we introduce latent space encoding, allowing all operations to be performed within the latent space, significantly reducing video memory usage. Our experiments demonstrate that FMSRdiff achieves comparable or even superior performance to previous super-resolution methods in both objective and subjective evaluations. We believe that our findings will promote the application of super-resolution models in medical imaging. However, FMSRdiff also has some limitations. First, its detail generation capabilities still need improvement, and it lags behind some current SOTA methods in preserving high-frequency information. Second, our research is currently limited to brain MRI super-resolution; further exploration is needed in other medical imaging fields. We will further investigate these issues in future work.

REFERENCES

[1] Rombach, R., Blattmann, A., Lorenz, D., Esser, P., Ommer, B. (2021). High-Resolution Image Synthesis with Latent Diffusion Models (Version 2). arXiv. https://doi.org/10.48550/ARXIV.2112.10752.

[2] Rombach, R., Blattmann, A., Lorenz, D., Esser, P., Ommer, B. (2021). High-Resolution Image Synthesis with Latent Diffusion Models. 2022 IEEE/CVF Conference on Computer Vision and Pattern Recognition (CVPR), 10674-10685.

[3] Saharia, C., Ho, J., Chan, W., Salimans, T., Fleet, D.J., Norouzi, M. (2021). Image Super-Resolution via Iterative Refinement. IEEE Transactions on Pattern Analysis and Machine Intelligence, 45, 4713-4726.

[4] Choi, J., Kim, S., Jeong, Y., Gwon, Y., Yoon, S. (2021). ILVR: Conditioning Method for Denoising Diffusion Probabilistic Models. 2021 IEEE/CVF International Conference on Computer Vision (ICCV), 14347-14356.

[5] Chung, H., Sim, B., Ye, J. (2021). Come-Closer-Diffuse-Faster: Accelerating Conditional Diffusion Models for Inverse Problems through Stochastic Contraction. 2022 IEEE/CVF Conference on Computer Vision and Pattern Recognition (CVPR), 12403-12412.

[6] Li, M., Cai, T., Cao, J., Zhang, Q., Cai, H., Bai, J., Jia, Y., Liu, M., Li, K., Han, S. (2024). DistriFusion: Distributed Parallel Inference for High-Resolution Diffusion Models. 2024 IEEE/CVF Conference on Computer Vision and Pattern Recognition (CVPR), 7183-7193.

[7] Xue, S., Liu, Z., Chen, F., Zhang, S., Hu, T., Xie, E., Li, Z. (2024). Accelerating Diffusion Sampling with Optimized Time Steps. 2024 IEEE/CVF Conference on Computer Vision and Pattern Recognition (CVPR), 8292-8301.

[8] Wang, Y., Yu, Y., Yang, W., Guo, L., Chau, L., Kot, A.C., Wen, B. (2023). ExposureDiffusion: Learning to Expose for Low-light Image Enhancement. 2023 IEEE/CVF International Conference on Computer Vision (ICCV), 12404-12414.

[9] Wang, J., Yue, Z., Zhou, S., Chan, K.C., Loy, C.C. (2023). Exploiting Diffusion Prior for Real-World Image Super-Resolution. ArXiv, abs/2305.07015.

[10] Cui, Q., Liu, Y., Zhang, X., Bao, Q., Liao, Q., Wang, L., Lu, T., Liu, Z., Wang, Z., Barsoum, E. (2024). Taming Diffusion Prior for Image Super-Resolution with Domain Shift SDEs. ArXiv, abs/2409.17778.

[11] Wang, Z., Chen, J., Hoi, S.C. (2019). Deep Learning for Image Super-Resolution: A Survey. IEEE Transactions on Pattern Analysis and Machine Intelligence, 43, 3365-3387.

[12] C. Dong, C. C. Loy, K. He and X. Tang, "Image Super-Resolution Using Deep Convolutional Networks," in IEEE Transactions on Pattern Analysis and Machine Intelligence, vol. 38, no. 2, pp. 295-307, 1 Feb. 2016, doi: 10.1109/TPAMI.2015.2439281. keywords: Image resolution;Neural networks;Image reconstruction;Convolutional codes;Feature extraction;Training;Super-resolution;deep convolutional neural networks;sparse coding;Super-resolution;deep convolutional neural networks;sparse coding,

[13] Dong, C., Loy, C.C., He, K., Tang, X. (2014). Image Super-Resolution Using Deep Convolutional Networks. IEEE Transactions on Pattern Analysis and Machine Intelligence, 38, 295-307.

[14] Hanting Chen, Yunhe Wang, Tianyu Guo, Chang Xu, Yiping Deng, Zhenhua Liu, Siwei Ma, Chunjing Xu, Chao Xu, and Wen Gao. Pre-trained image processing transformer. In Proceedings of the IEEE/CVF conference on computer vision and pattern recognition, pages 12299–12310, 2021.

[15] Xiangyu Chen, Xintao Wang, Jiantao Zhou, Yu Qiao, and Chao Dong. Activating more pixels in image super-resolution transformer. In Proceedings of the IEEE/CVF Conference on Computer Vision and Pattern Recognition, pages 22367–22377, 2023.

[16] Zheng Chen, Yulun Zhang, Jinjin Gu, Linghe Kong, Xiaokang Yang, and Fisher Yu. Dual aggregation transformer for image super-resolution. In Proceedings of the IEEE/CVF International Conference on Computer Vision, pages 12312–12321, 2023.

[17] Bee Lim, Sanghyun Son, Heewon Kim, Seungjun Nah, and Kyoung Mu Lee. Enhanced deep residual networks for single image super-resolution. In Proceedings of the IEEE conference on computer vision and pattern recognition workshops, pages 136–144, 2017.

[18] Yulun Zhang, Kunpeng Li, Kai Li, Lichen Wang, Bineng Zhong, and Yun Fu. Image super-resolution using very deep residual channel attention networks. In Proceedings of the European conference on computer vision (ECCV), pages 286–301, 2018.

[19] Qu, Y., Yuan, K., Hao, J., Zhao, K., Xie, Q., Sun, M., Zhou, C. (2025). Visual Autoregressive Modeling for Image Super-Resolution. ArXiv, abs/2501.18993.

[20] Xie, L., Wang, X., Chen, X., Li, G., Shan, Y., Zhou, J., Dong, C. (2023). DeSRA: Detect and Delete the Artifacts of GAN-based Real-World Super-Resolution Models. ArXiv, abs/2307.02457.

[21] Tian, K., Jiang, Y., Yuan, Z., Peng, B., Wang, L. (2024). Visual Autoregressive Modeling: Scalable Image Generation via Next-Scale Prediction. ArXiv, abs/2404.02905.

[22] Han, J., Liu, J., Jiang, Y., Yan, B., Zhang, Y., Yuan, Z., Peng, B., Liu, X. (2024). Infinity: Scaling Bitwise AutoRegressive Modeling for High-Resolution Image Synthesis. 2025 IEEE/CVF Conference on Computer Vision and Pattern Recognition (CVPR), 15733-15744.

[23] Wang, X., Yu, K., Wu, S., Gu, J., Liu, Y., Dong, C., Loy, C.C., Qiao, Y., Tang, X. (2018). ESRGAN: Enhanced Super-Resolution Generative Adversarial Networks. ECCV Workshops.

[24] Ledig, C., Theis, L., Huszár, F., Caballero, J., Aitken, A.P., Tejani, A., Totz, J., Wang, Z., Shi, W. (2016). Photo-Realistic Single Image Super-Resolution Using a Generative Adversarial Network. 2017 IEEE Conference on Computer Vision and Pattern Recognition (CVPR), 105-114.

[25] Zhang, K., Liang, J., Gool, L.V., Timofte, R. (2021). Designing a Practical Degradation Model for Deep Blind Image Super-Resolution. 2021 IEEE/CVF International Conference on Computer Vision (ICCV), 4771-4780.

[26] Wang, X., Xie, L., Dong, C., Shan, Y. (2021). Real-ESRGAN: Training Real-World Blind Super-Resolution with Pure Synthetic Data. 2021 IEEE/CVF International Conference on Computer Vision Workshops (ICCVW), 1905-1914.

[27] Wu, R., Sun, L., Ma, Z., Zhang, L. (2024). One-Step Effective Diffusion Network for Real-World Image Super-Resolution. ArXiv, abs/2406.08177.

[28] Chen, B., Li, G., Wu, R., Zhang, X., Chen, J., Zhang, J., Zhang, L. (2024). Adversarial Diffusion Compression for Real-World Image Super-Resolution. 2025 IEEE/CVF Conference on Computer Vision and Pattern Recognition (CVPR), 28208-28220.

[29] Wang, Y., Yang, W., Chen, X., Wang, Y., Guo, L., Chau, L., Liu, Z., Qiao, Y., Kot, A.C., Wen, B. (2023). SinSR: Diffusion-Based Image Super-Resolution in a Single Step. 2024 IEEE/CVF Conference on Computer Vision and Pattern Recognition (CVPR), 25796-25805.

[30] Zhang, A., Yue, Z., Pei, R., Ren, W., Cao, X. (2024). Degradation-Guided One-Step Image Super-Resolution with Diffusion Priors. ArXiv, abs/2409.17058.

[31] Prafulla Dhariwal and Alexander Nichol. Diffusion models beat gans on image synthesis. Advances in neural information processing systems, 34:8780–8794, 2021.

[32] Jonathan Ho, Ajay Jain, and Pieter Abbeel. Denoising diffusion probabilistic models. Advances in neural information processing systems, 33:6840–6851, 2020.

[33] Yang Song, Jascha Sohl-Dickstein, Diederik P Kingma, Abhishek Kumar, Stefano Ermon, and Ben Poole. Score-based generative modeling through stochastic differential equations. arXiv preprint arXiv:2011.13456, 2020.

[34] Stability.ai. https://stability.ai/stable-diffusion

[35] Lvmin Zhang, Anyi Rao, and Maneesh Agrawala. Adding conditional control to text-to-image diffusion models. In Proceedings of the IEEE/CVF International Conference on Computer Vision, pages 3836–3847, 2023.

[36] Yue, Z., Wang, J., Loy, C.C. (2023). ResShift: Efficient Diffusion Model for Image Super-resolution by Residual Shifting. ArXiv, abs/2307.12348.

[37] Lin, X.Y., He, J., Chen, Z., Lyu, Z., Fei, B., Dai, B., Ouyang, W., Qiao, Y., Dong, C. (2023). DiffBIR: Towards Blind Image Restoration with Generative Diffusion Prior. ArXiv, abs/2308.15070.

[38] Zhang, R., Isola, P., Efros, A.A., Shechtman, E., Wang, O. (2018). The Unreasonable Effectiveness of Deep Features as a Perceptual Metric. 2018 IEEE/CVF Conference on Computer Vision and Pattern Recognition, 586-595.

[39] Kingma, D.P., Ba, J. (2014). Adam: A Method for Stochastic Optimization. CoRR, abs/1412.6980.

[40] Yue, Z., Liao, K., Loy, C.C. (2024). Arbitrary-steps Image Super-resolution via Diffusion Inversion. 2025 IEEE/CVF Conference on Computer Vision and Pattern Recognition (CVPR), 23153-23163.

[41] Feng, C., Fu, H., Yuan, S., Xu, Y. (2021). Multi-Contrast MRI Super-Resolution via a Multi-Stage Integration Network. International Conference on Medical Image Computing and Computer-Assisted Intervention.

2025 The 10th International Conference on Integrated Circuits and Microsystems

An FBAR Driven Fractional-N Ring PLL Using Harmonic-Mixer-Based Dual Feedback and Split-Feedback Frequency Division

Qinglong Tian
College of Physics and
Information Engineering
Fuzhou University
Fuzhou, China

Jiwei Huang*
College of Physics and
Information Engineering
Fuzhou University
Fuzhou, China
huangjw@fzu.edu.cn
*Corresponding author

Hongqu Lin
College of Physics and
Information Engineering
Fuzhou University
Fuzhou, China

Abstract—This paper proposes a fractional-N dual-feedback phase-locked loop (DFPLL) architecture featuring a Class-C-driven FBAR oscillator and harmonic mixer(HM)-based topology with nested phase-locked loops(nested PLL). The nested PLL serves as an anti-aliasing filter, enhancing the operating frequency of the delta-sigma modulator (DSM) and effectively suppressing quantization noise. The HM path, utilizing a low-noise, high-frequency FBAR reference, resolves stability issues inherent in the nested PLL structure and mitigates noise retro-injection to minimize reference noise gain. A ring-type VCO topology provides significant area reduction and enhanced robustness against electromagnetic interference. Designed in TSMC 65nm CMOS technology with a $1.2V$ supply voltage and occupying $0.394mm^2$ layout area, when operating at $3.72GHz$ output frequency, the circuit achieves $12.32mW$ power dissipation with $-70.4dBc$ reference spurs and $355fs$ integrated jitter.

Index Terms—Dual-feedback, fractional-N PLL, HM, nested PLL, FBAR oscillator, DSM

I. INTRODUCTION

Against the backdrop of rapid evolution in modern communication systems, the large-scale deployment of 5G technology and forward-looking research and development of 6G technology jointly propel the industry toward higher data rates, enhanced spectral efficiency, and seamless connectivity across frequency bands—imposing extremely stringent performance requirements on phase-locked loops (PLLs). As wireless communications advance rapidly toward higher frequencies, wider bandwidths, and lower noise floors, the demand for high-performance PLLs has grown more pressing than ever before [1]. These advanced PLLs must feature enhanced frequency resolution to enable precise tuning to specific channels, reduced phase noise to preserve signal integrity, and faster locking speed to minimize latency in data transmission and reception. Fractional-N phase-locked loops have established

This work was supported by the National Natural Science Foundation of China (No. 62371332) and Special Project of the Major Special Project of the Department of Science and Technology of Fujian Province (Grant No. 2024HZ021023).

themselves as the mainstream solution due to their distinct advantages [2].

Fractional-N PLL delivering low phase noise with simultaneously suppressed spurious tones constitute critical building blocks for emerging wireline transceivers, millimeter-wave communications, and FMCW radar systems. While advanced calibration techniques continue to address in-band noise performance, such approaches fundamentally incur locking time penalties and implementation-induced phase-degradation trade-offs [3] [4]. Consequently, calibration-free techniques for suppressing quantization noise and fractional spurs have garnered significant research attention, enabling substantial PLL performance enhancement without increasing circuit complexity and locking time [5] [6] [7] [8].

This work proposes a fractional-N DFPLL utilizing an FBAR oscillator as the mixing clock source. By preventing noise retro-injection and suppressing aliasing-induced impairments, this design eliminates complex calibration circuits while achieving rapid locking, low jitter, and reduced cost through ring VCO implementation.

II. DESIGN OF DUAL-FEEDBACK PLL

A. Characteristics of the Proposed Architecture

The fractional-N DFPLL architecture proposed in this paper is depicted in Fig. 1. This architecture enhances PLL performance through three primary mechanisms. First, FBAR application significantly suppresses auxiliary-clock phase noise while reducing power consumption. Second, the HM secondary feedback path provides approximately unity-gain feedback, preventing noise amplification while achieving approximately $20log_{10}N$ noise rejection (where N denotes the division ratio), with our implementation achieving at least $27dB$ suppression. Furthermore, incorporating pre-filtering and post-filtering modules within the HM path effectively suppresses spur degradation caused by undesirable harmonic mixing products. Finally, employing the nested PLL as an anti-aliasing filter effectively suppresses DSM quantization noise while

979-8-3315-8850-2/25 $31.00 © 2025 IEEE

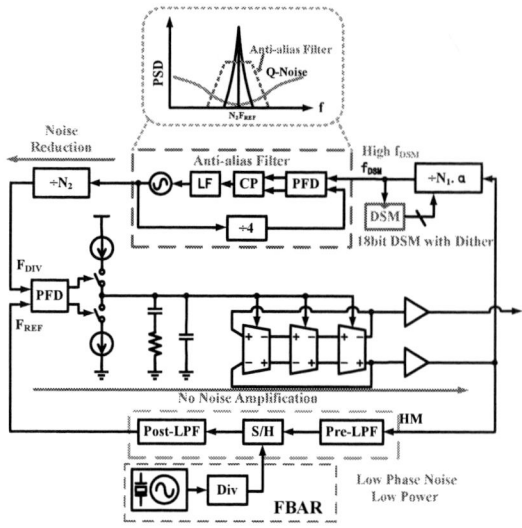

Fig. 1. FBAR driven fractional-N ring DFPLL.

elevating its noise-shaping frequency. The joint utilization of HM and the nested PLL resolves stability constraints, ensuring rapid locking capability and superior phase noise performance.

B. Stability and Noise Analysis

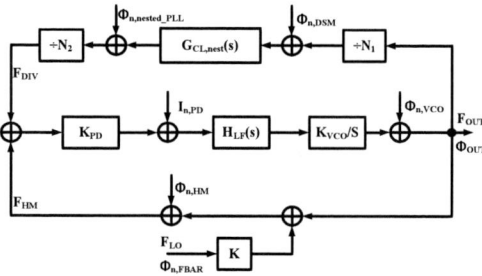

Fig. 2. Linear analytical model with main noise sources.

A linear analytical model depicting the architecture with main noise sources is illustrated in Fig. 2. Within this model, the nested PLL is represented by its closed-loop transfer function $G_{cl,nest}(s)$ and a phase noise source $\phi_{n,nested\_PLL}$ referenced to the output. We first examine the stability of the main loop, where the open-loop gain of the main PLL is given by:

$$A_{OL}(s) = \frac{K_{PD}K_{VCO}H_{LF}(s)}{s}\left(1 - \frac{G_{cl,nest}(s)}{N_1 N_2}\right) \quad (1)$$

The total feedback gain is observed to be dominated by the HM-based path, since $G_{cl,nest}(s)/(N_1 N_2)$ remains significantly less than unity. Thus, (1) can be simplified to:

$$A_{OL}(s) \approx \frac{K_{PD}K_{VCO}H_{LF}(s)}{s} \quad (2)$$

Equation (2) reveals that the nested PLL's transfer function exhibits negligible impact on the main PLL loop. Consequently, both loop bandwidths can be designed independently,

enabling enhanced noise suppression while ensuring rapid locking capability. This implementation circumvents stability issues arising from non-negligible transfer function impacts when the nested PLL technique is employed in standalone applications.

Subsequent analysis examines the noise transfer functions (NTFs) of key noise sources, demonstrating the noise performance superiority of the FBAR-based DFPLL over conventional architectures. The NTFs for primary noise sources are derived from Fig. 2 as:

$$NTF_{VCO}(s) = \frac{\phi_{OUT}}{\phi_{n,VCO}} = \frac{1}{1 + A_{OL}(s)} \quad (3)$$

$$NTF_{PD}(s) = \frac{\phi_{OUT}}{I_{n,PD}} = \frac{1}{K_{PD}}\frac{A_{OL}(s)}{1 + A_{OL}(s)} \quad (4)$$

$$NTF_{nested\_PLL}(s) = \frac{\phi_{OUT}}{\phi_{n,nested\_PLL}} = \frac{1}{N_2}\frac{A_{OL}(s)}{1 + A_{OL}(s)} \quad (5)$$

$$NTF_{HM}(s) = \frac{\phi_{OUT}}{\phi_{n,HM}} = \frac{A_{OL}(s)}{1 + A_{OL}(s)} \quad (6)$$

$$NTF_{DSM}(s) = \frac{\phi_{OUT}}{\phi_{n,DSM}} = \frac{G_{cl,nest}(s)}{N_2}\frac{A_{OL}(s)}{1 + A_{OL}(s)} \quad (7)$$

$$NTF_{FBAR}(s) = \frac{\phi_{OUT}}{\phi_{n,FBAR}} = K\frac{A_{OL}(s)}{1 + A_{OL}(s)} \quad (8)$$

All approximations are predicated on the characteristic of feedback gain being approximately unity. In equation (8), K denotes the harmonic index of the HM. The DFPLL architecture achieves superior phase noise performance through four principal mechanisms: The inherent low phase noise of the FBAR oscillator, synergistically integrated with HM, circumvents noise amplification. PD reference spurs exhibit N-fold suppression relative to conventional topologies (N is the division ratio). Nested PLL noise undergoes N_2-fold attenuation, while DSM quantization noise experiences both N_2-fold reduction and cascaded low-pass filtering through the nested and main PLL stages, collectively enabling comprehensive phase noise mitigation. The quantified NTF analyses collectively demonstrate the proposed PLL architecture's efficacy in phase noise enhancement through multi-mechanism quantization noise suppression.

III. CIRCUIT IMPLEMENTATION

A. Class-C FBAR Oscillator

As shown in Fig. 3(a), the Class-C FBAR oscillator circuit in this design employs a complementary cross-coupled topology and leverages Darlington cells to provide higher equivalent transconductance, which reduces the required power consumption while maintaining stable oscillation of the oscillator. To further improve the oscillator's phase noise without increasing cost, a Class-C oscillator is adopted in the design. Under

979-8-3315-8850-2/25 $31.00 © 2025 IEEE   919

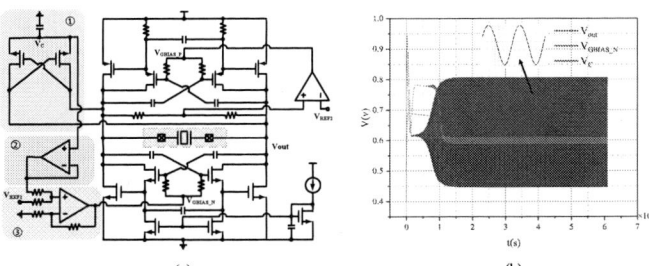

Fig. 3. (a) Class-C FBAR Oscillator Circuit (b) Waveforms of Key Nodes.

ideal conditions, it achieves a $3.9dB$ phase noise improvement compared to conventional LC oscillators at the same power consumption level [9]. However, Class-C oscillators suffer from weak start-up capability. Thus, a feedback-based oscillator is implemented in this work, which exhibits robust start-up performance and stable oscillation amplitude.

In the circuit of Fig. 3(a), V_{REF2} serves to stabilize the common-mode level of oscillation and provides an appropriate gate bias voltage V_{GBIAS_P} to the PMOS cross-coupled pair via a feedback mechanism. The left side incorporates an amplitude feedback loop, with key node voltages illustrated in Fig. 3(b). This loop consists of three stages [10]: the first stage is a negative peak detector, which detects the negative envelope of the oscillation output and stores the voltage at V_C. The second stage is a voltage follower formed by an operational amplifier, primarily functioning as an isolation buffer. The third stage is a voltage summing circuit composed of an operational amplifier and resistors, which ultimately outputs the voltage V_{GBIAS_N} to provide the optimal bias voltage for the NMOS cross-coupled pair.

$$V_{GBIAS\_N} = V_{REF1} + V_C \qquad (9)$$

In equation (9), V_{REF1} functions as a voltage shifter to ensure proper operation of the transistors. Thus, in the initial state of the circuit, a high bias voltage is applied to the gates of the cross-coupled pair. This causes the tail transistor to operate in the saturation region with sufficiently high current, enabling robust start-up. When the FBAR oscillator starts oscillating, a low bias voltage is applied to the gates of the cross-coupled pair, allowing deeply biased Class-C operation of the Class-C oscillator.

B. Harmonic Mixing Technology

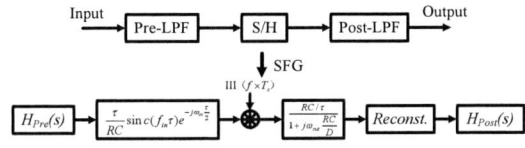

Fig. 4. HM with pre/post LPFs and its corresponding SFG [11].

To implement HM, the design employs a sample-and-hold (S/H) operation. By performing the S/H operation on the PLL

output, the output spectrum of the PLL is downconverted to a low frequency via the harmonics of the S/H clock [12]. When employed for HM functionality, the relationship between the output and input can be simplified to equation (10), as shown in the signal flow graph(SFG) of Fig. 4.

$$H(j\omega_{in}, j\omega_{out}) = \frac{sinc(f_{in}\tau)}{1 + j\omega_{out}\frac{RC}{D}}e^{-j\omega_{in}\frac{\tau}{2}} \qquad (10)$$

Where RC denotes the on-resistance and capacitance of the S/H circuit, τ represents the switch on-time, and D stands for the duty cycle of the sampling pulse. This indicates that signals far from the output frequency undergo a certain degree of attenuation, but their suppression level may not be sufficient.

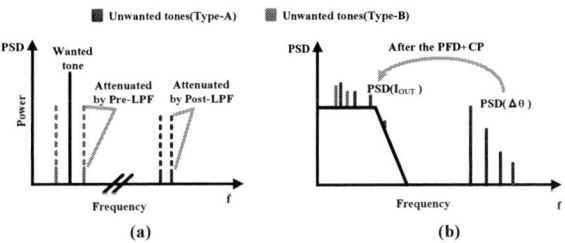

Fig. 5. Layout of dual-feedback PLL.

Such mixing still suffers from two main issues, as shown in Fig. 5: the first issue arises from residual high-frequency tones in the S/H output spectrum. Because the PFD/CP at the PLL input performs phase sampling operations at the input frequency, these high-frequency tones (Type-A) may alias to form low-frequency spurs, as illustrated in Fig. 5(b), where aliasing causes the conversion of noise from phase to current. The second issue is the unwanted in-band tones (Type-B) near the desired tone in the S/H output spectrum, which also causes undesired in-band spur degradation. Therefore, to suppress these tones resulting from the mixing of unwanted harmonic components, it is necessary to incorporate filters to provide a certain level of suppression. As shown in Fig. 4, suppressing Type-B spurs via a Pre-filter and Type-A spurs via a Post-filter can significantly improve the spectral purity after harmonic mixing. Fig. 5(a) clearly demonstrates the improvement in output spectral purity achieved by the filters.

C. Charge Pump Circuit

To further reduce the power consumption of the DFPLL while ensuring the fast speed of the charge pump (CP), an energy-efficient CP structure is employed as shown in Fig. 6(a) [13]. Compared with the conventional CP, this structure only requires a single amplifier to control the drain voltages of the transistors, thereby ensuring current matching(depicted in Fig. 6(b)).

First, this design features two negative feedback paths and one positive feedback path, which ensures stronger negative feedback and thus guarantees circuit stability. Second, since the CP does not require charging or discharging for most of the time after the PLL locks, the reference current mirror is

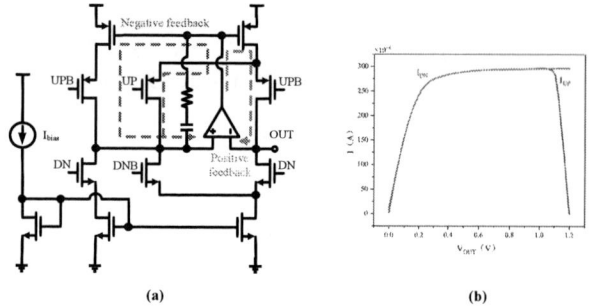

Fig. 6. (a)Charge Pump Circuit (b)Current Simulation.

controlled by the UPB and DN signals. This eliminates the always-on path of the current mirror when the CP is inactive, thereby achieving the goal of power consumption reduction.

IV. SIMULATION RESULTS

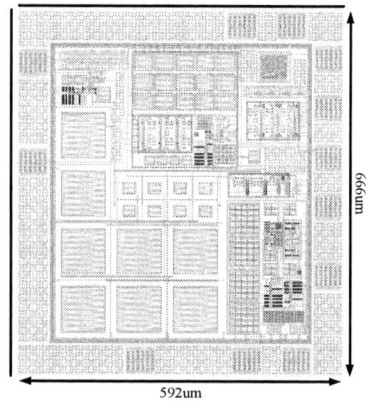

Fig. 7. Layout of dual-feedback PLL.

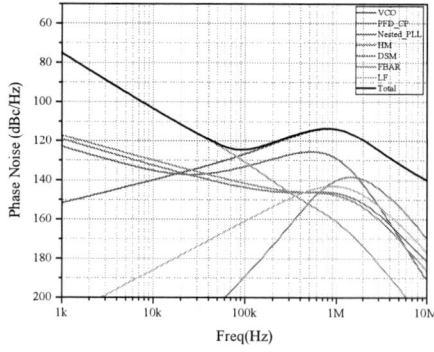

Fig. 8. Phase noise and contributions from various blocks.

The proposed FBAR-driven DFPLL is designed based on TSMC 65nm CMOS process. The layout is shown in Fig. 7, occupying an area of $0.394mm^2$. The $355fs$ integrated jitter ($1kHz$-$10MHz$) in Fig. 8 demonstrates improved phase noise characteristics over conventional fractional-N PLLs. From the above results, it can be concluded that by avoiding inverse

noise amplification and adopting a high-performance FBAR clock source, the noise performance of the fractional-N PLL has been significantly improved, which can effectively suppress noise degradation in communication systems.

The locking transient response in Fig. 9(a) confirms complete phase lock acquisition within $600ns$ under typical operating conditions, demonstrating concurrent locking of both nested and main PLL loops—an architectural feature that eliminates sequential delays and thus enables rapid frequency switching capability. Spectral analysis in Fig. 9(b), measured at the target carrier frequency, demonstrates reference spurs of -70.4dBc and fractional spurs of -61.2dBc. Finally, Table I summarizes the performance comparison with state-of-the-art PLL implementations across key metrics.

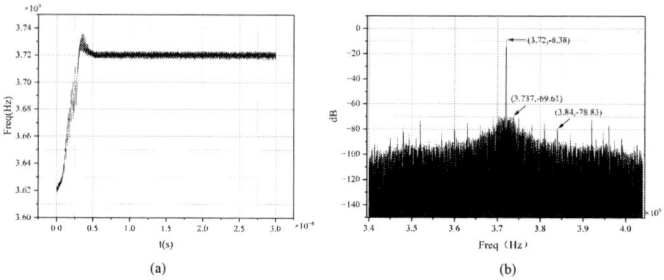

Fig. 9. (a)Locking transient simulation (b)PLL output spectrum.

TABLE I
COMPARISON WITH PRIOR WORKS

	[5]	[6]	[7]	[8]	This Work[a]
Process Node [nm]	65	45	45	40	65
Output Freq. [GHz]	2.9	2.4	2.4	2.4	3.72
RMS Jitter [fs]	869	1680	1500	2260	355
(Integ. Range)	(1k-100M)	(10k-50M)	(10k-50M)	(1k-100M)	(1k-10M)
Ref. Spur [dBc]	-61	-70	-60	-67	-70.4
Frac. Spur [dBc]	-49	-52.5	-41	-47	-61.2
Power [mW]	15.38	6.4	10	4.85	12.32
Locking Time [us]	0.46	N/A	N/A	N/A	0.6
FOM[b][dB]	-229.4	-227.4	-226.5	-226.1	-238.1

[a]Simulation results.

[b]$FOM = 10log_{10}[(Jitter/1s)^2(Power/1mW)]$

V. CONCLUSION

This paper presents a fractional-N PLL design employing an FBAR-based reference. The DFPLL architecture achieves efficient quantization noise suppression through three key aspects: The high-Q FBAR reference core, a HM secondary feedback path providing approximately unity gain and a nested PLL functioning as an anti-aliasing filter. These combined techniques significantly enhance ring-oscillator-based fractional-N synthesizer performance.

REFERENCES

[1] S. Ek,T. Påhlsson, C. Elgaard , et al. "A 28-nm FD-SOI 115-fs Jitter PLL-Based LO System for 24–30GHz Sliding-IF 5G Transceivers," IEEE Journal of Solid-State Circuits, 2018, 53(7): 1988-2000.

[2] T. -K. Kao, C. -F. Liang, H. -H. Chiu and M. Ashburn, "A wideband fractional-N ring PLL with fractional-spur suppression using spectrally shaped segmentation," 2013 IEEE International Solid-State Circuits Conference Digest of Technical Papers, San Francisco, CA, USA, 2013, pp. 416-417.

[3] M. Lee, M. Heidari and A. Abidi, "A low-noise wideband digital phase-locked loop based on a coarse-fine time-to-digital converter with subpicosecond resolution," IEEE J. Solid-State Circuits, vol. 44, no. 10, pp. 2808–2816, Oct. 2009.

[4] X. Gao, E. Klumperink, G. Socci and M. Bohsali, "Spur-reduction techniques for PLLs using sub-sampling phase detection," IEEE Int. Solid-State Circuits Conf. Dig. Tech. Papers, pp.474–475, 2010.

[5] M. Osada, Z. Xu, Z. Yang and T. Iizuka, "A Fractional-N Ring PLL Using Harmonic-Mixer-Based Dual-Feedback and Split-Feedback Frequency Division With Phase-Domain Filtering," IEEE Journal of Solid-State Circuits, vol. 59, no. 7, pp. 2171-2184, July. 2024.

[6] L. Kong and B. Razavi, "A 2.4-GHz 6.4-mW fractional-N inductorless RF synthesizer," IEEE J. Solid-State Circuits, vol. 52, no. 8, pp. 2117–2127, Aug. 2017.

[7] L. Kong and B. Razavi, "A 2.4-GHz RF fractional-N synthesizer with BW = 0.25 f_{REF}," IEEE J. Solid-State Circuits, vol. 53, no. 6, pp. 1707–1718, Jun. 2018.

[8] Y. Zhang et al, "A fractional-N PLL with space–time averaging for quantization noise reduction," IEEE J. Solid-State Circuits, vol. 55, no. 3, pp. 602–614, Mar. 2020.

[9] A. Mazzanti and P. Andreani, "Class-C harmonic CMOS VCOs, with a general result on phase noise," IEEE J. Solid-State Circuits, vol. 43, no. 12, pp. 2716 2729, Dec. 2008.

[10] W. Deng, K. Okada and A. Matsuzawa, "Class-C VCO With Amplitude Feedback Loop for Robust Start-Up and Enhanced Oscillation Swing," in IEEE Journal of Solid-State Circuits, vol. 48, no. 2, pp. 429-440, Feb. 2013.

[11] T. Iizuka and A. A. Abidi, "FET-R-C Circuits: A Unified Treatment—Part I: Signal Transfer Characteristics of a Single-Path," in IEEE Transactions on Circuits and Systems I: Regular Papers, vol. 63, no. 9, pp. 1325-1336, Sept. 2016.

[12] M. Osada, Z. Xu, R. Shibata and T. Iizuka, "Analysis of Offset Spurs in Phase-Locked-Loops Employing Harmonic-Mixer-Based Feedback With Sample-and-Hold Operation," in IEEE Transactions on Circuits and Systems I: Regular Papers, vol. 69, no. 12, pp. 5072-5084, Dec. 2022.

[13] D. Sun et al., "A 3.96-4.84-GHz Dual-Path Charge Pump PLL Achieving 89.7-fs_{rms} Integrated Jitter and 250.8-dB FOM_{PLL}," in IEEE Transactions on Circuits and Systems II: Express Briefs, vol. 71, no. 4, pp. 1909-1913, April 2024.

979-8-3315-8850-2/25 $31.00 © 2025 IEEE

A Fault Tolerant Routing Method for 3D Network on Chip without Redundant TSV or Router-Avoidance

Zeyi Lin[1], Duo Yu[1], Yanzhou Tang[1], Naifeng Jing[1], Jianfei Jiang[1], Weiguang Sheng[1], Lifu Cheng[2], Qin Wang[1*]

[1.] School of Integrated Circuits, Shanghai Jiao Tong University, Shanghai, China
[2.] Shanghai Aerospace Electronic Technology Institute, Shanghai, China
*Corresponding author: qinqinwang@sjtu.edu.cn

Abstract—3D integrated chips are key technologies for mitigating the impending failure of Moore's Law. In current 3D chips, network on chip is commonly employed to facilitate inter-core communication. However, in 3D NoC, failures of Through-Silicon Via (TSV) links in the vertical direction can lead to communication breakdowns, necessitating targeted fault-tolerant designs. Existing approaches typically address this problem either by incorporating redundant TSV or by routing around faulty routers during path computation; the former incurs substantial hardware overhead, while the latter degrades network performance. This paper presents a novel fault-tolerant scheme that does not require redundant TSV and effectively exploits the remaining functional links in routers with TSV failures. The proposed method introduces a mere 6.9% area overhead, yet maintains 99.3% of the pre-failure TSV network performance, demonstrating negligible performance loss.

Keywords—Network on Chip, 3D chip, Fault Tolerance, Routing Technique, Chip Modeling

I. INTRODUCTION

In existing high-performance computing chips, network on chip (NoC) is commonly employed to facilitate communication among computational cores. However, as chip dimensions continue to expand, the manufacturing yield tends to decline correspondingly [1]. To maintain chip performance while ensuring acceptable yield rates, the adoption of 3D stacking technologies has gained considerable traction [2][3]. For NoC constructed on these non-planar chip architectures, link failures represent a critical issue that demands urgent resolution [4].

In stacked chips utilizing 3D packaging technologies, network on chip typically rely on vertical interconnects for data transmission between chiplets. These vertical links are implemented via Through Silicon Vias (TSV) and wafer bonding techniques [5][6]. Notably, the silicon area occupied by TSV is substantially larger than that of intra-die interconnects corresponding to horizontal wiring [7], resulting in significantly higher interconnect costs. Consequently, these vertical links generally cannot support extremely high bandwidth. Moreover, interconnections among multiple layers of NoC are often realized through partially connected nodes rather than direct one-to-one point connections. Under such circumstances, any failure within these vertical links can incapacitate the corresponding vertical communication channel, severely degrading the inter-chiplet data transmission bandwidth. To mitigate the substantial performance losses induced by vertical

Supported by the National Key Research and Development Program of China (No. 2024YFB4405400).

link failures, targeted fault-tolerant design methodologies are essential.

Existing solutions for addressing link failures in 3D stacked chips generally fall into two categories: hardware-level TSV fault-tolerant schemes and network-level routing solutions. In TSV fault-tolerant designs for 3D chips, designers typically incorporate a set of redundant TSV in addition to the baseline functional TSV array, accompanied by corresponding fault-tolerant circuitry. To achieve a satisfactory fault-tolerant rate, the proportion of redundant TSV often amounts to one-third or even more relative to the primary TSV count. This results in substantial hardware area overhead and increased manufacturing costs. On the other hand, network-level routing approaches rely primarily on rerouting traffic around faulty links to restore communication. Although rerouting can enable continued data transmission at failure points, such methods do not guarantee restoration of full bandwidth, often causing significant degradation in overall network throughput. Therefore, developing a fault-tolerant scheme that incurs minimal hardware overhead while ensuring network bandwidth preservation is of critical importance.

To overcome the limitations of existing methods, this paper proposes a TSV fault-tolerant scheme for vertical links in 3D stacked architectures that eliminates the need for redundant TSV. The proposed approach integrates the fault-tolerant module directly into the transceiver routers at both the transmitting and receiving ends. At the transmitter side, the intact TSV are utilized to package and retransmit the data associated with faulty TSV. Upon reception, the receiver decodes the retransmitted packets and restores the original data. This scheme introduces only marginal additional hardware complexity while significantly reducing hardware overhead and TSV fabrication costs compared to traditional schemes, and it achieves this with almost no degradation in transmission performance.

II. RELATED WORKS AND MOTIVATION

Existing fault-tolerant methods can be broadly categorized into two main types: approaches that employ redundant TSV and those that utilize routing-based avoidance schemes.

A. Methods of Redundant TSV

One category of fault-tolerant approaches typically involves augmenting the design with redundant TSV to mitigate communication failures caused by TSV faults. Loi et al. were among the first to propose such a method [8], introducing a one-dimensional chain-based TSV fault-tolerant technique that incorporates redundant TSV to enhance reliability. Specifically, their approach adds one redundant TSV for every four functional TSV to ensure data link recovery in the event of faults. However, the inclusion of redundant TSV invariably increases circuit area and manufacturing costs.

Sudeep et al. proposed a scheme aimed at maximizing the utilization of redundant TSV [9]. In this method, functional TSV and redundant TSV are grouped, and the distances between redundant TSV and multiple TSV groups are computed. To minimize the hardware overhead associated with fault-tolerant design, each redundant TSV is shared among several neighboring functional TSV groups. Consequently, if any TSV within a particular group fails, the corresponding redundant TSV can be employed for fault recovery. This approach enables a single redundant TSV to serve multiple functional TSV groups, thereby achieving a higher fault recovery rate with the same number of redundant TSV.

B. Methods of Routing Avoidance

In three-dimensional network on chip, the routing flexibility allows designers to forgo additional redundant TSV by employing adaptive routing algorithms that circumvent faulty router nodes. Mosoumeh et al. proposed a fault-tolerant routing algorithm for 3D NoC based on Hamiltonian path strategies [10]. This method leverages Hamiltonian paths to redirect communication around faulty regions by dynamically adjusting the routing algorithm. When the network topology changes due to faults, the Hamiltonian path is recalibrated to ensure that data packets are routed through unaffected alternative paths, thereby maintaining data transmission integrity.

Taheri et al. introduced a dynamic boundary router selection mechanism ReD [11] that assigns functional boundary routers to replace faulty router nodes. This enables data packets to be rerouted through healthy routers to reach other parts of the system. In addition, the mechanism incorporates hardware for traffic flow detection, which allows it to consider not only faulty routers but also congestion arising from uneven traffic distribution under certain conditions. Fundamentally, this scheme redistributes data traffic across the remaining operational vertical links, balancing the traffic load entering and exiting the chiplets to alleviate congestion and improve overall network performance.

C. Motivation of Non-Redundant TSV Method

For fault-tolerant schemes based on redundant TSV, achieving a certain level of fault coverage typically requires the proportion of redundant TSV to amount to one-third or even more of the baseline TSV count. This results in significant hardware area overhead and increased manufacturing costs. Conversely, network-level routing solutions employing rerouting techniques can indeed circumvent transmission errors at faulty links; however, they do not guarantee full restoration of the affected transmissions, often leading to substantial degradation in network bandwidth.

In fact, TSV faults can be addressed not only at the circuit or network layers but also at the data link layer, which holds considerable potential for fault tolerance. In 3D NoC, data transmission between routers on the top and bottom layers can be controlled and recovered through router nodes. Inspired by this, the present chapter proposes a fault-tolerant technique based on retransmission of failed data. By leveraging fault information obtained during TSV testing, a novel fault-tolerant scheme is implemented at the router's data link layer. This approach eliminates the need for redundant TSV and avoids the necessity of circumventing routers connected to faulty TSV, thereby maintaining high-bandwidth transmission in 3D NoC with significantly reduced hardware overhead.

III. PROPOSED FAULT TOLERANT METHOD

A. Fault Tolerant Theory

The proposed scheme achieves fault-tolerant transmission without relying on redundant TSV by employing a retransmission mechanism based on TSV failure data. The principle is illustrated in Fig. 1. When a TSV fault occurs, the corresponding bit positions consistently fail during data transmission. Through testing, these faulty bits can be accurately detected and localized. Leveraging this information, the top-level router can dynamically identify the precise positions of erroneous bits within each transmitted data packet. To correct these errors, after transmitting a certain number of regular data packets, the top-level router sends a dedicated error correction packet. This error correction packet aggregates all previously detected faulty bit information, enabling the lower-level router nodes to perform correction and recovery on the corresponding data packets. The recovered packets are then forwarded along the normal transmission path.

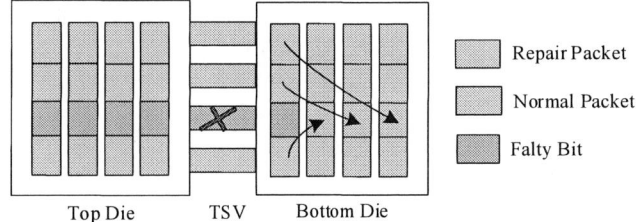

Fig. 1. Principle of non-redundant TSV fault-tolerant scheme. Each column represents data packets at different times. Arrows in bottom die shows the recovering process of packets from top die.

This method does not employ any redundant TSV, thereby avoiding the area and cost overheads typically associated with traditional hardware-level redundancy. When the data transmission bit-width is large and the number of faulty TSV is relatively small, the error correction packets can cover multiple erroneous data transmissions, resulting in a minimal overall impact on bandwidth. Through this mechanism, errors corresponding to failed TSV are promptly corrected, enabling efficient fault tolerance for vertical links while effectively balancing cost and performance.

B. Dataflow Design

Since the proposed method introduces error correction packets into the original data stream, the data flow at both the transmitting and receiving ends of the router must be redesigned to minimize the impact of these correction packets on system performance within multi-stage pipelined routers. In fact, the timing of error correction packet transmission is closely related to two key factors: the number of pipeline stages in the router and the availability of virtual channel credits in downstream routers.

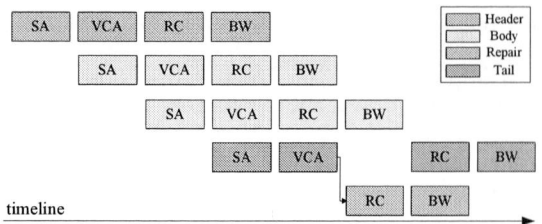

Fig. 2. Data flow of sending routers

Taking the pipeline transmitter of the router illustrated in Fig. 2 as an example, the data packet undergoes four stages—Buffer Write (BW), Routing Computation (RC), Virtual Channel Allocation (VCA), and Switch Allocation (SA)—before proceeding to the corresponding stage in the next-hop router (the BW stage within the current router is omitted in the figure; the BW indicated on the right side refers to the entry stage of the subsequent router). The transmission of the packet's header and body remains unaffected throughout this process. However, the transmission of the packet's tail is deferred until after the error-correction packet generated within the router has been sent. This sequencing ensures that the error-correction packet is transmitted first, allowing the receiving side's header to continue its normal pipeline flow without being impacted by delays caused by error correction. It is noteworthy that the error-correction packet encompasses the erroneous bits found not only in the first three flits but also those in the finally transmitted tail flit. Consequently, no additional error-correction packets are required for the error bits within the tail flit.

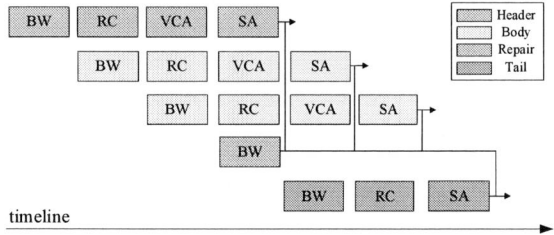

Fig. 3. Data flow of receiving routers

The corresponding data flow at the receiver side is illustrated in Fig. 3. After receiving the header and body flits, the receiver obtains the error-correction packet. If there are no prior delays ahead of the header and it is consistently granted access during arbitration on the first attempt, the erroneous bits are precisely corrected immediately following the arrival of the error-correction packet. Subsequently, the corrected data is forwarded to the downstream router. The subsequent body and tail flits also experience no blocking and are able to complete the original data flow in the shortest possible time. It is important to note that the tail flit does not undergo the Virtual Channel Allocation stage, as only the head flit needs to fully traverse all four pipeline stages during routing. The body and tail flits merely follow the routing decision established by the head flit. The depiction of the body flits undergoing RC and VCA stages in the figure is solely to illustrate their temporary blocking caused by the head flit's progression through these stages. Since the tail flit does not encounter such blocking, it can be transmitted in the cycle immediately following the body flit's transmission.

However, the transmission timing of the error-correction packet depends not only on the number of pipeline stages but also requires adaptive adjustment based on the buffer occupancy status of the downstream router's virtual channels. When multiple flits are already buffered in the downstream virtual channel, the header flit, despite having arrived at the next router, may be unable to immediately proceed with routing computation and onward transmission; it must wait until the pre-existing data queue on that virtual channel is partially or fully released. Under these circumstances, the dispatch of the error-correction packet can be postponed by several cycles to aggregate a larger number of errors to be corrected, thereby minimizing the bandwidth overhead caused by the frequent insertion of error-correction packets. In this paper, we propose the following formula to calculate the waiting cycles for the error-correction packet:

$$T_{repair} = n_{flits} + T_{router} - 1 + T_{arb} \qquad (1)$$

Where T_{repair} denotes the number of cycles the correction packet must wait before transmission, n_{flits} represents the number of flits currently buffered in the corresponding downstream virtual channel, T_{router} is the number of pipeline stages in the router, and T_{arb} indicates the average additional cycles required per packet during arbitration within the router. The term $n_{flits} - 1$ reflects the fact that the actual delay caused by the presence of data in the downstream virtual channel corresponds to one less than the total buffered flits, since the first flit does not induce additional waiting cycles.

This model effectively guides the scheduling of error correction packets, ensuring that their transmission timing balances pipeline throughput and bandwidth utilization. By deferring the transmission of correction packets, the scope of data subject to correction is expanded, significantly enhancing error correction efficiency while simultaneously reducing the impact on overall system bandwidth.

C. Hardware Architecture

To implement a fault-tolerant mechanism based on retransmission of failed data, this design incorporates several additional hardware modules into the conventional 3D NoC router architecture, as illustrated by the green-highlighted sections in the Fig. 4. The left side of the figure depicts the sender router, while the right side shows the receiver router. We assume that data flit is transported from sender's downward output port and is acquired by receiver's upward input port.

At the sender router, the key functionality to be realized is the packaging of failed data. The MEM module serves as a non-volatile memory for storing the status of TSV faults. The Repair Data module holds data that failed to be transmitted in each cycle due to faulty TSV. The Compute module within the Switch Allocator controls the timing of retransmission for the error correction data packets. Upon chip testing, TSV faults are detected and recorded in the MEM. The error information stored in MEM directs the multiplexer at the output of the crossbar switch to capture the corresponding data affected by the faulty TSV, which is then stored in the Repair Data module, forming error correction packets. These error correction packets within the Repair Data module are subsequently sent to the crossbar switch. Under the control of the crossbar switch allocator, the

979-8-3315-8850-2/25 $31.00 © 2025 IEEE

appropriate timing is selected to transmit the error correction packets. Once transmission is completed, the error correction packets in the Repair Data module are cleared.

packet. For flits longer than one, the error-correction packets are always transmitted between the header and tail flits, with both the Header and Tail fields set. This convention enables the receiver to precisely locate the positions of the error-correction

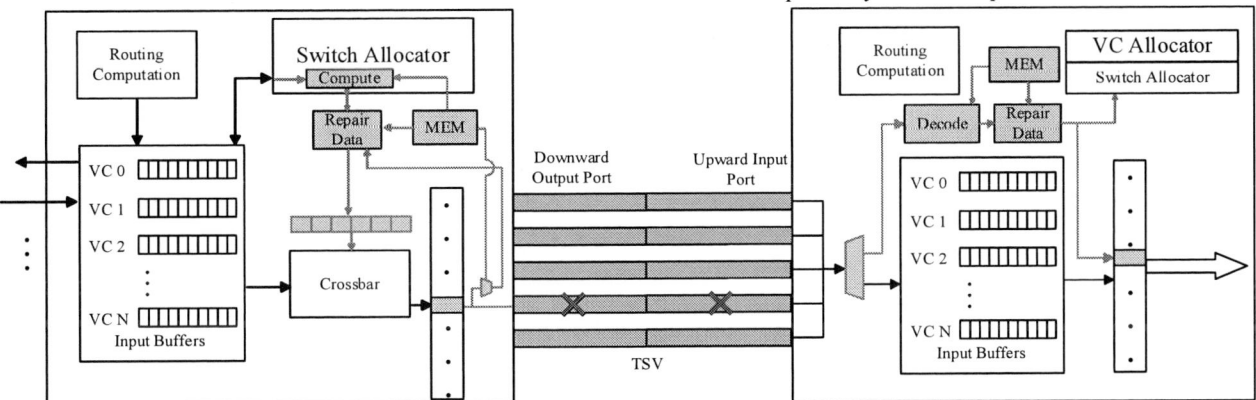

Fig. 4. Hardware architecture of fault-tolerant technique

The error-correction packet transmission module, embedded within the crossbar switch allocation unit, determines the timing for sending error-correction packets based on the designed data flow. When an error-correction packet is ready for transmission, it is assigned a higher priority than other packets. The transmission timing is governed by Formula (1) and the MEM capacity within the router. MEM capacity holds the highest priority because the error-correction packet buffer corresponding to each virtual channel can store at most one flit's worth of data; once the MEM storage is full, the corresponding error-correction packet must be transmitted downstream immediately. During input arbitration at the crossbar switch, besides the original input virtual channels, additional data channels sourced from MEM are incorporated into the arbitration pool. The priority of these channels corresponds directly to the scheduled transmission timing of the error-correction packets. At this stage, the output channel is fixed to the vertical link direction, eliminating the need for additional arbitration.

At the receiver router, the decoding functionality for failed data must be implemented. Similarly, the MEM module is used to store information regarding faulty TSV. Upon arrival of the error correction packet at the receiver side, a multiplexer directs this packet to the decoding module. The decoding module, based on the TSV fault information and previously collected flit data, calculates the number of flits corresponding to the received error correction packet. Subsequently, it sequentially retrieves the corresponding data from the Repair Data module according to this number and the fault information. These retrieved data are then restored into the previously collected incomplete packets at the output port of the crossbar switch. After successful error correction, the error correction packet is discarded, thereby completing the data recovery process.

During decoding at the receiver, it is essential to accurately identify error-correction packets within the data stream. In our design, for flits of length one, since both the Header and Tail fields are set simultaneously, any subsequent flit with both Header and Tail fields unset is considered as an error-correction

packets.

IV. EVALUATION

To analyze the network transmission performance under fault conditions and the area overhead introduced by the fault-tolerant designs, this study employs the SystemC-based on-chip network simulator PAT-Noxim [12] alongside Synopsys Design Compiler to simulate the proposed no-redundancy fault-tolerant scheme as well as several existing methods. The comparison is conducted from three perspectives: network performance, fault-tolerant rate, and area overhead.

A. Network Performance Analysis

The principal metric for network performance analysis is the packet transmission latency under varying packet injection rates. According to the average TSV fault rate (~1.4%) reported in [13], when there are 8 TSV on a vertical link, the probability of encountering 3 to 4 faults on the faulty link after three years exceeds 50%. This study evaluates both computationally and experimentally the network latency under such fault conditions. Since fault-tolerant schemes employing redundant TSV exhibit equivalent performance as long as the redundancy is not exhausted, only one redundant TSV scheme is considered for comparison in this paper. Fig. 5 compares the average network latency of three fault-tolerant methods: the use of redundant TSV (RTSV, represented by the blue curve), a purely routing-based fault-avoidance algorithm (ReD, shown as the green curve), and the proposed no-redundancy TSV scheme (depicted by the brown curve). As observed, ReD, which circumvents faulty nodes solely through routing algorithms, incurs considerable router resource overhead, resulting in substantially degraded overall network performance. Conversely, the proposed no-redundancy TSV fault-tolerant scheme achieves nearly identical network performance to the fault-free case under low injection rates, with only a slight increase in average latency observed as the injection rate rises. From the perspective of saturation throughput—indicated by the inflection point in the latency curves—the proposed method remains close to the performance exhibited under fault-free conditions.

979-8-3315-8850-2/25 $31.00 © 2025 IEEE

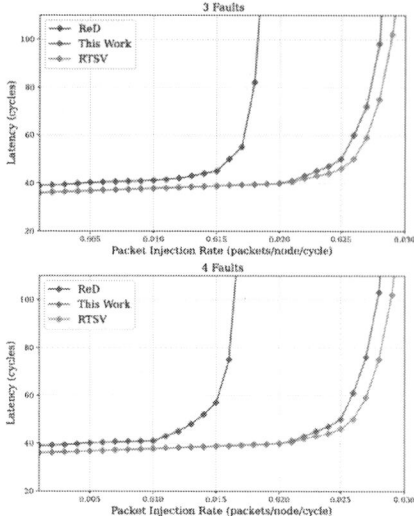

Fig. 5. Comparison of network performance across different fault-tolerant schemes

B. Fault-Tolerant Rate Analysis

The fault tolerant rate of a fault-tolerant technique is defined as the probability that the system can maintain normal operation under a given fault rate. In this study, normal operation refers to a condition where faulty TSVs do not affect the correct transmission of data. For schemes without fault tolerance, the occurrence of a faulty TSV causes whole system failure; thus, the fault tolerance rate is equal to the probability of no faulty TSV occurrence. For redundancy-based fault tolerance schemes, the fault tolerance rate is the probability that the number of faulty TSVs does not exceed the available redundancy. Based on this principle, and considering the average TSV failure probability of 1.4% over three years [13], the fault tolerance rates of various schemes can be estimated, as illustrated in the figure. The figure reveals that without any fault tolerance strategy, the reliability of 3D interconnect chips significantly decreases as the number of TSVs increases. In contrast, for the Shift [4] method employing one redundant TSV for every four regular TSVs, the fault tolerance rate exhibits a declining trend as the TSV count doubles. The TDMA [14] scheme, which provisions one redundant TSV per regular TSV, achieves a relatively high fault tolerance rate. Compared to these approaches, the proposed no-redundancy TSV scheme maintains the highest fault tolerance rate regardless of the number of faulty TSVs, consistently operating without any additional TSV overhead.

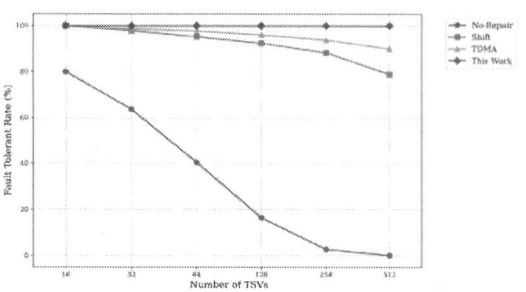

Fig. 6. Repair rate Comparison of repair rate

C. Hardware Overhead Analysis

To analyze the hardware overhead of various design schemes, we implemented the routers used in these schemes with SystemVerilog and synthesized them using a 32nm process technology library. Taking as an example a configuration with 4 router groups, each interconnected by 16 TSV, we obtained the total area occupied by the routers. Additionally, considering a typical TSV diameter of 10 μm and including the area overhead introduced by layout and routing, the area consumption per TSV was estimated at approximately 144 μm². This enabled the computation of the total area consumption of both the routers and the TSV, as summarized in TABLE I. . From the table, it can be observed that the method proposed in this work increases the area overhead by 7.6% relative to the baseline router architecture, while not requiring any additional redundant TSV. In contrast, ReD, which employs a purely algorithmic redundancy scheme, incurs the minimal hardware area overhead; however, as analyzed earlier, it results in a considerable degradation of network performance. The Shift method, representing a conventional redundant TSV fault-tolerant design, utilizes a relatively small number of redundant TSV with a normal-to-redundant TSV ratio of 4:1. Its associated fault-tolerant hardware consists of only a few multiplexers and demultiplexers, leading to a relatively low total area overhead; nevertheless, prior analysis indicated that such a simple structure yields a comparatively low fault repair rate. Finally, the TDMA approach exhibits a 1:1 ratio between normal and redundant TSV and incurs additional hardware overhead due to its time-division multiplexing scheme, resulting in the highest total hardware cost among the evaluated methods.

TABLE I.   HARDWARE OVERHEAD COMPARISON

Method	TSV number	Wafer area (μm²)	Total area (μm²)
TDMA (21'TETC)	96	72947	86771
Shift (12'TVLSI)	60	69943	78583
ReD (24'TCAD)	48	69839	76751
This Work	48	75181	82093

D. Comprehensive Analysis

This section presents a comprehensive evaluation of the TDMA, Shift, ReD, and the proposed fault-tolerant method in this work. Based on a three-year TSV fault probability model, the saturated throughput under varying numbers of faulty TSV is calculated for each method to represent the expected network performance. These results, combined with fault repair rates and hardware area overheads, are summarized in TABLE II.

TABLE II.   COMPREHENSIVE ANALYSIS

Method	ReD (24'TCAD)	Shift (12'TVLSI)	TDMA (21'TETC)	This Work
Performance	78.3%	88.6%	95.9%	**99.3%**
Fault Tolerant Rate	99.9%	78.3%	90.2%	**99.9%**
Overhead	100.0%	+2.4%	+13.0%	**+6.9%**

The first row in the table represents the expected network performance over a three-year period for TSV when each die contains 8 vertical links, assuming fault-free TSV correspond to 100% network performance. For the Shift and TDMA schemes, which employ redundant TSV, performance drops to 0% once the number of faulty TSV exceeds the available redundancy— specifically, more than 2 faults for Shift (with 8 regular TSVs and 2 redundant TSV)s and more than 8 faults for TDMA (with 8 regular TSVs and 8 redundant TSVs). The second row shows the fault tolerant rates when the total number of TSV reaches 512 bits, and the third row reports the hardware area under identical hardware parameters. A comparative analysis across these three metrics reveals that the ReD method, which circumvents faulty nodes via routing algorithms, achieves a relatively high fault repair rate with modest hardware area overhead, albeit at the expense of lower expected network performance. The Shift method utilizes a modest proportion of redundant TSV, resulting in a smaller hardware footprint but comparatively lower fault repair rates and network performance. In contrast, the TDMA approach relies on a larger number of redundant TSV, achieving higher fault repair rates and an expected network performance of 95.9%, but this comes at the cost of significantly increased hardware area due to the extensive redundancy. The fault-tolerant method proposed herein eliminates the need for redundant TSV, incurring an additional silicon area overhead of 6.9%. Despite this area penalty, it attains the highest fault repair rate of 99.9% in large-scale TSV arrays and delivers an expected network performance of 99.3%, which represents a negligible degradation. Overall, for applications aiming to reduce the cost associated with redundant TSV while maintaining robust network fault tolerance, the method proposed in this work offers significant practical value.

V. CONCLUSION

This paper addresses the fault tolerance issue of defective TSVs in 3D NoC architectures by abandoning the conventional approaches of adding redundant TSVs or avoiding routing through routers with faulty links. Instead, the problem is shifted from the physical layer to the data link layer within the NoC. The proposed method inserts error-correcting packets during the transmission of regular data packets using error correction and retransmission techniques. At the receiver side, decoding is employed to recover erroneous data caused by faulty TSVs. This approach eliminates the need for any redundant data TSVs and significantly reduces the hardware area overhead and process cost associated with redundant TSVs, while only minimally impacting network transmission performance.

REFERENCES

[1] C. H. Stapper, F. M. Armstrong and K. Saji, "Integrated circuit yield statistics," in Proceedings of the IEEE, vol. 71, no. 4, pp. 453-470, April 1983.

[2] L. Jiang, Q. Xu and B. Eklow, "On effective TSV repair for 3D-stacked ICs," 2012 Design, Automation & Test in Europe Conference & Exhibition (DATE), Dresden, Germany, 2012, pp. 793-798.

[3] Y. -G. Chen, W. -Y. Wen, Y. Shi, W. -K. Hon and S. -C. Chang, "Novel Spare TSV Deployment for 3-D ICs Considering Yield and Timing Constraints," in IEEE Transactions on Computer-Aided Design of Integrated Circuits and Systems, vol. 34, no. 4, pp. 577-588, April 2015.

[4] A. -C. Hsieh and T. Hwang, "TSV Redundancy: Architecture and Design Issues in 3-D IC," in IEEE Transactions on Very Large Scale Integration (VLSI) Systems, vol. 20, no. 4, pp. 711-722, April 2012.

[5] P. Morrow, M. J. Kobrinsky, S. Ramanathan, et al., "Wafer-Level 3D Interconnects via Cu Bonding," Proceedings of the Advanced Metalliza tion Conference, pp. 125-130, 2004.

[6] Philip Garrou, Christopher Bower, and Peter Ramm, "Handbook of 3D Integration: Technology and Application of 3D Integrated Circuits Volume 1 & 2," poblished by WILEY-VCHVerlag GmbH&Co.KGaA, Weinheim, 2008, ISBN: 978-3-527-32034-9.

[7] Lau, J.H. "Overview and outlook of through‐silicon via (TSV) and 3D integrations", Microelectronics International, Vol. 28 No. 2, pp. 8-22, 2011.

[8] I. Loi, S. Mitra, T. H. Lee, S. Fujita and L. Benini, "A low-overhead fault tolerance scheme for TSV-based 3D network on chip links," 2008 IEEE/ACM International Conference on Computer-Aided Design, San Jose, CA, USA, 2008, pp. 598-602.

[9] S. Ghosh, S. K. Roy, H. Rahaman and C. Giri, "TSV repairing for 3D ICs using redundant TSV," 2017 7th International Symposium on Embedded Computing and System Design (ISED), Durgapur, India, 2017, pp. 1-5.

[10] M. Ebrahimi, M. Daneshtalab and J. Plosila, "Fault-tolerant routing algorithm for 3D NoC using hamiltonian path strategy," 2013 Design, Automation & Test in Europe Conference & Exhibition (DATE), Grenoble, France, 2013, pp. 1601-1604.

[11] E. Taheri, S. Pasricha and M. Nikdast, "ReD: A Reliable and Deadlock-Free Routing for 2.5-D Chiplet-Based Interposer Networks," in IEEE Transactions on Computer-Aided Design of Integrated Circuits and Systems, vol. 43, no. 12, pp. 4599-4612, Dec. 2024.

[12] A. Norollah, D. Derafshi, H. Beitollahi and A. Patooghy, "PAT-Noxim: A Precise Power & Thermal Cycle-Accurate NoC Simulator," 2018 31st IEEE International System-on-Chip Conference (SOCC), Arlington, VA, USA, 2018, pp. 163-168.

[13] Y. -H. Chen, C. -P. Chiu, R. Barnes and T. Hwang, "Architectural evaluations on TSV redundancy for reliability enhancement," Design, Automation & Test in Europe Conference & Exhibition (DATE), 2017, Lausanne, Switzerland, 2017, pp. 566-571.

[14] T. Ni et al., "A Novel TDMA-Based Fault Tolerance Technique for the TSV in 3D-ICs Using Honeycomb Topology," in IEEE Transactions on Emerging Topics in Computing, vol. 9, no. 2, pp. 724-734, 1 April-June 2021.

2025 The 10th International Conference on Integrated Circuits and Microsystems

The Mechanical Properties of Wrinkled MnPS₃ Structures

Tongxu Huang[1], Jiaxian Sun[1], Zihao Wu[1], Yiting Wei[2], Wenjun Chen*[1]

[1] School of Electronic Information Engineering, Foshan University, Foshan, China
[2] School of Computer Science and Artificial Intelligence, Foshan University, Foshan, China
*megatronprime113@163.com, chenwj46@fosu.edu.cn

Abstract—The mechanical properties of two-dimensional (2D) transition-metal thiophosphites (TMTs) remains unexplored, restricting their applications in flexible and integrated electronics. Here, we first measure the mechanical properties of 2D MnPS₃ through wrinkle engineering. Uniaxial wrinkles are fabricated in 2D MnPS₃ using a pre-strain method, then atomic force microscopy is used to determine the thickness of the flake, the wavelength, and height of the wrinkles. Accordingly, it is found that few-layer MnPS₃ with the thickness of ~ 6.5 nm exhibits a Young's modulus of 40.0 ± 5.2 GPa and a breaking strength of 7.0 ± 1.0 keV. In addition, the mechanical properties of 2D MnPS₃ are layer-dependent. This work enriches the documentation of the physical properties of 2D MnPS₃, offering the basis for the construction of related integrated and flexible electronics.

Keywords—2D materials, wrinkled MnPS3, mechanical property, Young's modulus, breaking strength

I. INTRODUCTION

Two-dimensional (2D) transition metal thiophosphites (TMTs) with various and tunable physical properties are attractive building blocks for next-generation nanoelectronics. Specifically, 2D MnPS₃, as a prototypical TMT, exhibits long-range antiferromagnetic order alongside strong optical activity [1] and sliding induced out-of-plane ferroelectricity [2]. However, the mechanical behaviors of 2D MnPS₃ remains unknown, which limits the applications in flexible and strain-engineered electronic devices.

Controllable wrinkling has emerged as a versatile strategy to modulate the mechanical features of 2D materials. Studies on graphene and MoS₂ have shown that wrinkle geometry can decouple stiffness from strength and improve stretchability. The analogous investigations in MnPS₃ enabled the ultralow detection limit of methylene blue[1], but the mechanical properties of few-layer MnPS₃ and how it affects the light matter interaction are still unknown.

Here, we construct unidirectional wrinkles in 2D MnPS₃ to calculate its mechanical properties, including Young's modulus and breaking strength. Based on the flake thickness, the wavelength, and the height of the wrinkles measured by atomic force microscopy (AFM), the Young's modulus (40.0 ± 5.2 GPa) and breaking strength (7.0 ± 1.0 keV) of few-layer MnPS₃ are determined. Moreover, Young's modulus decreases with the increasing layer number of MnPS₃ flakes, while breaking strength has a positive relation with the layer number. This study not only unveils more physics of 2D MnPS₃, but also sheds light on the promotion of flexible and integrated electronics.

II. EXPERIMENTAL METHOD

A. Preparation of Wrinkled MnPS₃

Wrinkled MnPS₃ membranes were fabricated via a pre-strained method (Fig. 1a). PDMS stamps were prepared by mixing the base and curing agent at a 10:1 ratio, followed by degassing under vacuum and curing at 80 °C for 6 hours. The cured PDMS sheets were stretched to 20% and fixed using a custom-built stretching stage.

Few-layer MnPS₃ flakes were mechanically exfoliated using adhesive Scotch tape from bulk crystals and transferred onto the pre-strained PDMS surface. The PDMS stamp carrying MnPS₃ flakes was then inverted and laminated onto a cleaned SiO₂/Si substrate. Gentle pressure was applied for 5 minutes to ensure conformal contact. Gradual release of the pre-strained PDMS leaded to out-of-plane wrinkling in the MnPS₃ film as it relaxed onto the substrate. Finally, the PDMS stamp was slowly peeled off at a controlled rate of ~1 mm min⁻¹, leaving behind wrinkle-patterned MnPS₃ adhered to the SiO₂/Si substrate. Subsequently, the wrinkled MnPS3 was gently pressed onto the substrate to ensure a good contact, and then it was placed on the heating plate and heated at 100°C for 5 minutes.

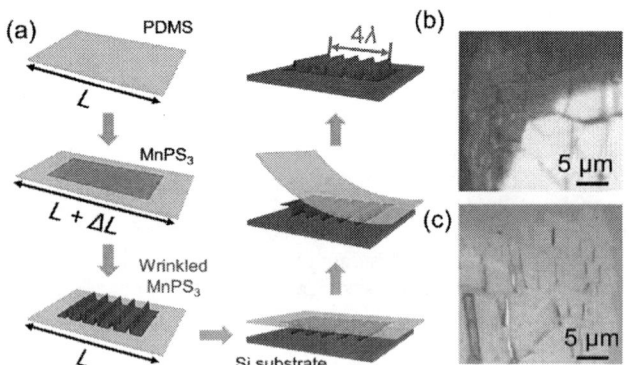

Fig. 1. Preparation and characterization of wrinkled MnPS₃. (a) A schematic illustration of fabrication process for wrinkled MnPS₃. (b) Optical microscope image of wrinkled MnPS₃ on PDMS substate, showing several parallel wrinkles. (c) Optical microscope image of wrinkled MnPS₃ on 285 nm SiO₂/Si substrate.

Optical microscopy [3] was initially used to assess the morphology and wrinkle distribution in the prepared samples. Images captured under both dark-field and bright-field illumination revealed uniformly distributed (Fig. 1b, c), parallel wrinkles with a consistent orientation. The periodicity and uniformity of the wrinkles were readily

979-8-3315-8850-2/25 $31.00 © 2025 IEEE

observable over millimeter-scale areas, enabling rapid quality assessment. For higher resolution imaging, scanning electron microscopy (SEM) [4,5] was performed. The SEM images will be discussed later. The mechanical properties of the wrinkled material will also be revealed through the various characterization techniques mentioned later.

B. Growth of Bulk MnPS₃

High-quality MnPS₃ crystals were synthesized using a chemical vapor transport (CVT) method [6]. Stoichiometric amounts of high-purity manganese (Mn), phosphorus (P), and sulfur (S) powders were thoroughly mixed and sealed in an evacuated quartz ampoule under high vacuum ($\sim10^{-4}$ Pa). The sealed ampoule was placed into a two-zone muffle furnace for crystal growth. A temperature gradient was applied, with the source zone heated to 1000 K over 20 hours and the sink zone maintained at 600 K. The ampoule was slowly cooled from 1000 K to 600 K at a rate of 1 K h⁻¹, followed by natural cooling to room temperature. After the growth process, greenish layered MnPS₃ crystals were harvested from the cooler end of the ampoule and collected for further characterization.

C. X-ray Diffraction

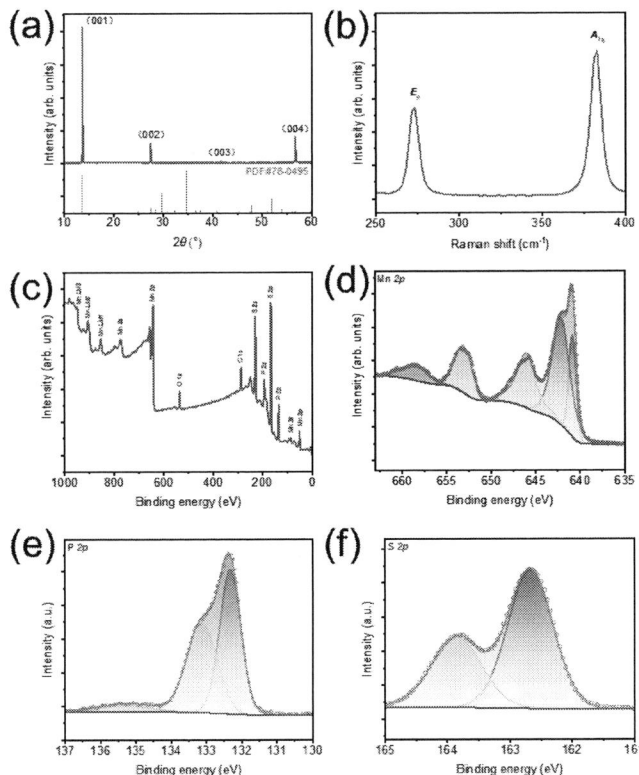

Fig. 2. Characterization of as-grown MnPS₃ crystals. (a) XRD spectra, (b) Raman spectrum, and (c) XPS full survey spectrum of MnPS₃. (d-f) XPS spectra of (d) Mn 2p, (e) P 2p, and (f) S 2p components...

The phase purity and crystallographic structure of the as-grown MnPS₃ crystals were analyzed using X-ray diffraction (XRD) [7]. Spectra were collected in the 2θ range of 5° to 90° using a Rigaku SmartLab diffractometer (Fig. 2a). The diffraction peaks matched well with the standard MnPS₃ pattern (PDF card No. 78-0495), with dominant (00l) reflections indicating the formation of high-quality layered MnPS₃ single crystals.

D. Raman Spectroscopy

We exfoliated the as-grown MnPS₃ onto SiO₂/Si substrate. Raman spectroscopy [8,9] was employed to characterize the vibrational properties of exfoliated MnPS₃ flakes transferred onto SiO₂/Si substrates. Spectra were collected using a 532 nm excitation laser. The typical Raman spectrum displays A_{1g} and E_g modes, confirming the high quality of the as-exfoliated MnPS₃ flakes, as shown in Fig. 2b.

E. X-ray Photoelectron Spectroscopy

X-ray photoelectron spectroscopy (XPS) [10] was used to investigate the chemical composition of MnPS₃. The survey spectrum confirms the existence of Mn, P, and S elements in MnPS₃ (Fig. 2c). Furthermore, the spectrum of Mn 2p element displays multiple oxidation states (Fig. 2d), while P 2p spectrum indicates that the phosphorus atoms are in a sulfide environment (Fig. 2e). The S 2p spectrum exhibits two distinct peaks at ~162.5 eV and 163.8 eV, which are corresponding to S 2p₃/₂ and S 2p₁/₂ spin-orbit splitting, respectively (Fig. 2f).

F. Transmission Electron Microscopy

The as-exfoliated MnPS₃ flake were transferred to Quantifoil grid using wet chemical transfer method. Transmission electron microscopy (TEM) [11] analysis was performed to investigate the atomic structure of exfoliated MnPS₃ flakes. Flakes were transferred onto Quantifoil grids using a wet chemical transfer method (Fig. 3a). High-angle annular dark-field scanning transmission electron microscopy (HAADF-STEM) images revealed a hexagonal arrangement of atoms (Fig. 3b), consistent with the theoretical structure of MnPS₃. The results confirm the high crystallinity and structural integrity of the synthesized MnPS₃ crystals.

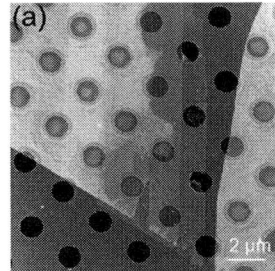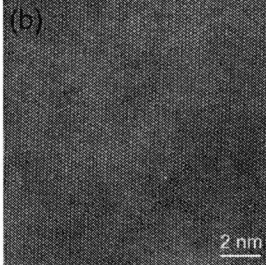

Fig. 3. TEM characterization of few-layer MnPS₃ flakes. (a) Low-magnification and (b) HAADF-STEM images of the few-layer MnPS₃.

G. Characterization of Wrinkled MnPS₃

We utilized a pre-strain method for fabricating wrinkled MnPS₃, as depicted in Fig. 1a. We further characterized the wrinkled MnPS₃ using scanning electron microscopy (FEI Nova NanoSEM 450), operating at 10 kV acceleration voltage and a 6 mm working distance. SEM images revealed well-defined wrinkle crests and troughs, with wrinkle amplitudes ranging from 10 to 30 nm. Samples with both low and high wrinkle densities were examined to demonstrate the tunability of wrinkle patterns using the pre-strain-controlled PDMS transfer method (Fig. 4a and b).

The combination of chemical vapor transport growth, mechanical exfoliation, controlled PDMS-assisted wrinkle engineering, and comprehensive morphological and structural characterization enabled precise fabrication and analysis of wrinkled MnPS₃ membranes suitable for mechanical and functional studies.

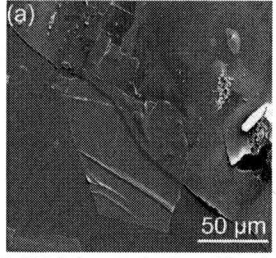

Fig. 4. SEM characterization of few-layer MnPS₃ flakes. (a) Low-magnification and (b) high-magnification SEM images of the wrinkled MnPS₃.

III. PROPERTIES AND APPLICATIONS OF WRINKLED MnPS₃

Based on the previous preparation process, we first employed AFM to analyze the three-dimensional morphology of wrinkled MnPS₃, as shown in Fig. 5a. As the key parameters for analyzing the mechanical properties of materials, the wavelength and thickness of the wrinkles needed to be recorded. In this study, the buckling method was employed for the calculation of Young's modulus. This method requires simpler preparation and costs much less time. The determination of the average wavelength of the wrinkles requires measuring the mean distance between adjacent wrinkles. Therefore, all we need to do is calculate the distance between the 1ᵗʰ and nᵗʰ wrinkle divided by n-1. Figure 5b shows the thickness measurement result of a single wrinkle at the top, with a value of 6.5 nm. Subsequently, the curve drawn in green solid line spans seven wrinkle cycles, and the measured total length is 6λ = 11.8 μm, calculated λ = 1.96 μm. It should be noted that the pressure applied to the sample and substrate during the transfer process, which is intended to achieve a close fit, may lead to variations in wrinkle height. However, the change in wrinkle height does not affect the subsequent calculation of mechanical parameters. The blue curve section at the bottom of Figure 4b shows the measurement result of wrinkles in another area, with a measured 4λ = 7.5 μm. After calculation, a similar λ value (1.88 μm) is obtained, further verifying the reliability of the measurement results.

Figure 5c shows Young's modulus (E) of 2D MnPS₃ as a function of its thickness. E of few-layer flakes exceeds 40 GPa, which decreases with the increasing layer number of MnPS₃ flakes. Accordingly, the breaking strength (B) of few-layer MnPS₃ is calculated to 7.0 ±1.0 keV and has a positive relation to its thickness.

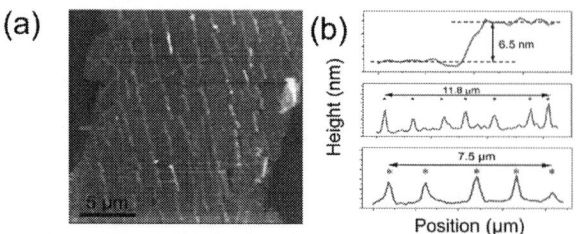

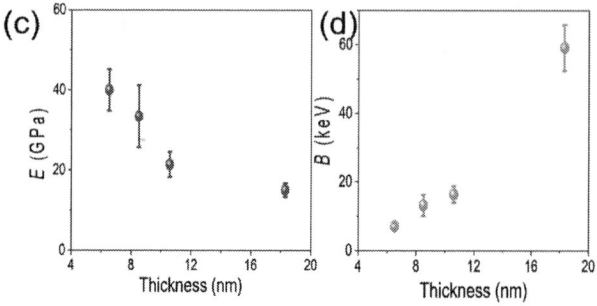

Fig. 5. Mechanical properties of wrinkled MnPS₃. (a) AFM image of wrinkled MnPS₃. (b) Height profiles of the sections marked by red, green, and blue dotted lines in (a). (c) Young's modulus and (d) breaking strength of wrinkled MnPS₃ with different thicknesses.

To elucidate the mechanical behavior of wrinkled MnPS₃, we quantitatively analyzed the relationship between wrinkle geometry and induced strain, which is largely governed by the pre-strain applied to the PDMS substrate during the fabrication process. Wrinkling provides a tunable platform to introduce and control local strain in 2D materials, which can, in turn, significantly modulate their mechanical, optical, and electronic properties.

$$\lambda = 2\pi t \left[\frac{E_{2D}}{(12\Lambda \mu_s (1 - \nu^2))} \right]^{\frac{1}{3}}$$

Here, t is the thickness of the MnPS₃ flake, E_{2D} is the Young's modulus of MnPS₃, ν is the Poisson's ratio, λ is the wrinkle wavelength and μ_s is the shear modulus of the PDMS substrate. We can transform this formula and calculate the Young's modulus:

$$E = \frac{2\Lambda \mu_s (1 - \nu^2)}{8\pi^3} \left(\frac{\lambda}{h}\right)^3 = 1366.94^*$$

As one of the important mechanical properties, the Young's modulus is widely used to describe the rigidity of object. With a clear calculation model, it can help us draw a graph showing the relationship between Young's modulus and fold thickness. Here, the geometric factor Λ, which accounts for the nonlinear pre-strain, is given by:

$$\Lambda = \frac{1 + (1 + \varepsilon_0)^3}{2(1 + \varepsilon_0)}$$

where ε_0 is thepre-strain applied to the PDMS layer prior to transfer. For a given MnPS₃ flake with constant material parameters (i.e., fixed t, E and ν), the wrinkle wavelength is therefore primarily controlled by the substrate properties and the applied pre-strain. Increasing ε_0 leads to a higher geometric factor Λ, which ultimately results in a reduced wrinkle wavelength λ. Consequently, the wrinkle density increases with pre-strain, allowing for fine modulation of wrinkle spacing and overall surface morphology.

In addition to wavelength, the wrinkle amplitude h is another critical geometric parameter that determines the mechanical response of the wrinkled structure. The height-to-width aspect ratio h/λ can serve as an effective measure of wrinkle severity or sharpness. A higher aspect ratio generally corresponds to more significant local curvature and thus more intense strain localization at the wrinkle peaks.

To calculate the strain concentration induced by the wrinkles, we use the geometric strain model for thin films with periodic buckling. The localized tensile strain ε at the wrinkle crest can be estimated by:

$$\varepsilon = \frac{\pi^2 th}{(1-v^2)\lambda^2}$$

This equation indicates that the strain is directly proportional to the film thickness t, wrinkle amplitude h, and inversely proportional to the square of the wavelength λ. Thus, decreasing λ (via increasing substrate pre-strain) or increasing h (by promoting out-of-plane deformation) significantly enhances the peak strain in the MnPS$_3$ membrane.

To apply this model to our experimental samples, we first measured the relevant geometric parameters (i.e., t, h, λ) for each wrinkle configuration using atomic force microscopy (AFM) and scanning electron microscopy (SEM). For consistency, we assumed constant material properties across all samples, with $E = 40$ GPa and $v = 0.25$ for MnPS$_3$, based on prior theoretical and experimental studies. The shear modulus of PDMS (μ_s) was estimated at ~ 0.4 MPa.

Using these parameters, we calculated the corresponding wrinkle wavelength for a range of pre-strains from 5% to 40%. As expected, higher pre-strains resulted in significantly reduced wrinkle wavelengths and increased aspect ratios. With these experimentally determined values, we subsequently computed the local strain ε at each wrinkle peak using the formula above. Our calculations revealed that the strain increased from $\sim 0.2\%$ at low pre-strain (5%) to over 1.0% at high pre-strain (>30%), indicating substantial strain amplification through wrinkling.

This strain engineering capability offers a powerful design tool to modulate the apparent mechanical properties of MnPS$_3$. In particular, wrinkle formation allows the thin MnPS$_3$ membranes to become more compliant (i.e., reduced in-plane stiffness) while also enhancing fracture toughness by redistributing stress. The geometric decoupling of stiffness and strength is a hallmark of wrinkle-enabled mechanics and underpins the enhanced flexibility observed in our devices.

Furthermore, the strain gradient induced by wrinkling may influence other functional properties of MnPS$_3$, such as its optical bandgap, magnetic ordering, or piezoelectric response. While these aspects are beyond the scope of the current mechanical model, they present exciting opportunities for future investigation. For instance, combining the mechanical strain model with band structure simulations or magnetic phase field calculations could provide deeper insights into wrinkle-mediated property tuning.

IV. CONCLUSION AND OUTLOOK

We have established a robust, scalable PDMS-transfer method to produce large area, uniformly wrinkled MnPS$_3$

membranes, and we paired this with an AFM based nanoindentation protocol to map their local mechanical response. Our measurements show that by tuning wrinkle wavelength (2–5 µm) and amplitude (10–30 nm), the apparent Young's modulus can be reduced by more than 50%, whereas the breaking strength increases threefold relative to flat films. This behavior can be attributed to strain concentration in the wrinkle troughs, which delays crack nucleation and markedly improves toughness. These results demonstrate that wrinkle engineering provides a powerful means to decouple flexibility and strength in 2D materials. They establish wrinkle engineering as a versatile design paradigm for decoupling flexibility and robustness in two-dimensional materials. Looking forward, two key opportunities emerge:

(1) By tailoring wrinkle wavelength, amplitude, and geometry — potentially in multi‑angle or hierarchical patterns—we can impose spatially graded stiffness and strength from the nanoscale to the macroscale. Such gradient materials could enable mechanical metamaterials with programmable deformation pathways and load‑bearing characteristics.

(2) Integrating wrinkle-induced strain fields with MnPS$_3$'s intrinsic magnetic, electronic, and optical properties opens routes to mechanically reconfigurable devices. Strain‑tunable magneto‑optical sensors, bendable spin valves, and stretchable ferroelectric memories could all leverage the interplay between wrinkle geometry and functional order parameters.

In summary, wrinkle-patterned MnPS$_3$ combines outstanding compliance and resilience, offering a versatile platform for next-generation flexible and multifunctional electronics. Future challenges may include scaling up fabrication, discovering new wrinkle-mediated phenomena, and integrating these architectures with hybrid material systems.

REFERENCES

[1] Chen, W. *et al.* Mechanochemical activation of 2D MnPS$_3$ for sub-attomolar sensing. *Nature Communications* **15**, 10195 (2024).

[2] Weng, X. *et al.* Sliding-Induced Out-of-Plane Ferroelectricity of 2D MnPS3. *Advanced Functional Materials*, e03780 (2025).

[3] Davidson, M. W. & Abramowitz, M. Optical microscopy. *Encyclopedia of imaging science technology* **2**, 120 (2002).

[4] Mohammed, A. & Abdullah, A. in *Proceedings of the 2018 international conference on hydraulics and pneumatics—HERVEX, Băile Govora, Romania.* 7-9.

[5] Zhou, W., Apkarian, R., Wang, Z. L. & Joy, D. in *Scanning microscopy for nanotechnology: techniques and applications* 1-40 (Springer, 2006).

[6] Hu, D. *et al.* Two-dimensional semiconductors grown by chemical vapor transport. *Angewandte Chemie International Edition* **56**, 3611-3615 (2017).

[7] Epp, J. in *Materials characterization using nondestructive evaluation (NDE) methods* 81-124 (Elsevier, 2016).

[8] Lyon, L. A. *et al.* Raman spectroscopy. *Analytical Chemistry* **70**, 341-362 (1998).

[9] Smith, E. & Dent, G. *Modern Raman spectroscopy: a practical approach.* (John Wiley & Sons, 2019).

[10] Seah, M. & Analysis, I. The quantitative analysis of surfaces by XPS: A review. *Surface* **2**, 222-239 (1980).

[11] Tang, C. & Yang, Z. in *Membrane characterization* 145-159 (Elsevier, 2017).

Supercritical Fluid-Engineered p-GaN Thin Films with Improved Electrical and Optical Properties

Yao-Li Chuang
School of Electronic and Computer Engineering uangdong Provincial Key Laboratory of In-Memory Computing Chips Peking University
Shenzhen 518055, China
li18200850082@163.com

Lei Li*
College of Integrated Circuits and ptoelectronic Chips Shenzhen Technology University
Shenzhen 518118, China
lilei@sztu.edu.cn
*Corresponding author

Lei Lu
School of Electronic and Computer Engineering uangdong Provincial Key Laboratory of In-Memory Computing Chips Peking University
Shenzhen 518055, China
lulei@pku.edu.cn

Kuan-Chang Chang*
School of Electronic and Computer Engineering uangdong Provincial Key Laboratory of In-Memory Computing Chips Peking University
Shenzhen 518055, China
kcchang@pkusz.edu.cn
*Corresponding author

Abstract— igh contact resistance and incomplete g-acceptor activation persist as critical bottlenecks for p-type a , largely due to deep acceptor levels and compensation effects that limit hole concentration.This work demonstrates that low-temperature supercritical fluid (C) treatment effectively addresses these limitations. Through systematic electrical measurements, all effect analysis, and material characterizations, ignificant improvements were observed, with the contact resistance decreasing by 35 to 1.5 percent (from 31. 4 .5 kiloohms to .5 17.5 kiloohms) for different electrode configurations, while the all mobility increases from 7. to . s uare centimeters per volt-second. nergy band analysis reveals a reduced chottky barrier height at the metal p- a interface, facilitating enhanced hole in ection and ohmic contact behavior. -ray photoelectron spectroscopy (P) confirms weakened g-bonding and promoted g activation, whereas Raman spectroscopy verifies no induced lattice stress. Photoluminescence (P) shows intensified near-band-edge ( ) and blue luminescence ( ) emissions, indicating suppressed non-radiative recombination and higher acceptor activation. These findings establish C processing as a non-destructive, scalable method to engineer high-performance p- a for optoelectronic and power devices.

Keywords—p- a , supercritical fluid treatment, contact resistance, g doping, photoluminescence

I. INTRODUCTION

Gallium nitride (GaN) high-electron-mobility transistors (HEMTs) underpin next-generation technologies spanning 5G communications, power conversion, and satellite systems, leveraging their wide bandgap, high breakdown field, and thermal stability [1,2]. Despite advances in n-type GaN channels, efficient p-type GaN development remains hampered by the interrelated challenges of the deep bulk energy level of the Mg dopant (approximately 200 millielectronvolts (meV)) and high metal/p-GaN contact resistance. The former leads to low Mg activation efficiency (less than 5 percent) due to

hydrogen passivation and self-compensation effects [3], while the latter exacerbates Joule heating and degrades device reliability [4,5].

Conventional approaches to enhance p-GaN performance include thermal annealing [6] and nitrogen plasma treatments [7]. However, these methods often induce crystal degradation or impose scalability constraints. Supercritical fluids (SCFs) offer a promising alternative, exploiting their low viscosity, high diffusivity, and tunable solvent power to modify surfaces at the molecular level without damaging bulk crystallinity. Prior studies on dopant reactivation, such as nitrogen plasma-assisted Mg activation , highlight the importance of interfacial engineering but lack non-destructive scalability.

In this study,SCF treatment is proposed as a novel strategy strategy to simultaneously mitigate interfacial barriers and activate Mg dopants in p-GaN. We hypothesize that SCF-processed CO penetrates the p-GaN surface, dissociating Mg-H complexes and optimizing metal/semiconductor band alignment. Crucially, this method circumvents thermal stress inherent in annealing, preserving lattice integrity. This work bridges the gap between dopant activation physics and contact engineering, demonstrating that SCF treatment not only elevates hole mobility but also reduces contact resistance directly impacting device turn-on voltage and efficiency [8].

Through electrical and optical analysis, including I-V characteristics, Hall effect, XPS, Raman spectroscopy, and PL spectroscopy, we quantified the performance improvement and elucidated the mechanism linking the interface band modulation and device performance enhancement. The results validate that the SCF process is a reliable route to achieve high-mobility, low-resistance p-GaN for power electronics and LEDs.

II. Experimental Methods

A. Preparation of p- aN Thin-Film and Electrodes

Mg-doped p-GaN thin films were epitaxially grown on silicon substrates using metal-organic chemical vapor deposition (MOCVD). Subsequently, gold was deposited on the p-GaN surface using magnetron sputtering through a designed mask to serve as the contact layer between the electrodes and the P-GaN. The electrode dimensions (100 m × 100 m and 100 m × 200 m) and the electrode spacing are shown in Figure 1. By measuring the current-voltage relationship between each electrode, changes in contact resistance were systematically observed.

B. Supercritical Fluid Treatment

The fabricated p-GaN device was subjected to supercritical fluid (SCF) processing. This process was performed in a high-pressure stainless steel autoclave using CO_2 as the supercritical medium. The process parameters were maintained at 120 C and 3000 psi for 30 minutes, exceeding the critical point of CO_2, allowing the fluid to effectively penetrate the semiconductor surface and alter the interface state.

C. Characterization

The influence of SCF treatment was assessed by systematically comparing the electrical and structural properties of identical devices before and after processing. Electrical measurements included current-voltage (I-V) testing for contact resistance extraction and hall effect measurements for carrier mobility analysis. Chemical composition and bonding states were examined using X-ray photoelectron spectroscopy (XPS). Raman spectroscopy was employed to probe lattice stress and crystalline quality, while photoluminescence (PL) spectroscopy at room temperature was used to investigate band-edge emission and defect-related optical states.

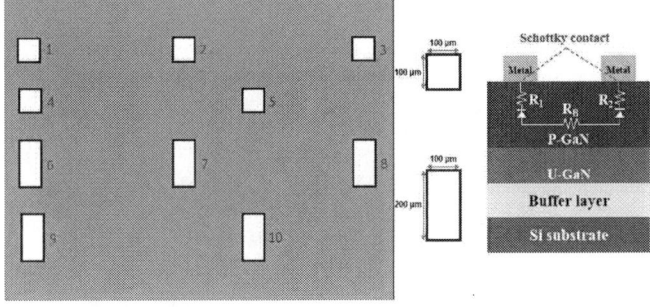

Fig. 1. Schematic illustration of device fabrication and electrode configuration. p-GaN thin films were grown on Si substrates by MOCVD. A mask with different electrode sizes (100 m 100 m and 100 m 200 m) and different spacing between electrodes was designed. Metal electrode contacts were deposited by magnetron sputtering.

III. Results and Discussion

A. Electrical Performance Improvements

Systematic electrical measurements were conducted on devices with varying electrode sizes and spacings, as shown in Fig. 2. The I-V characteristics clearly demonstrate that SCF treatment substantially reduces the contact resistance of p-GaN.

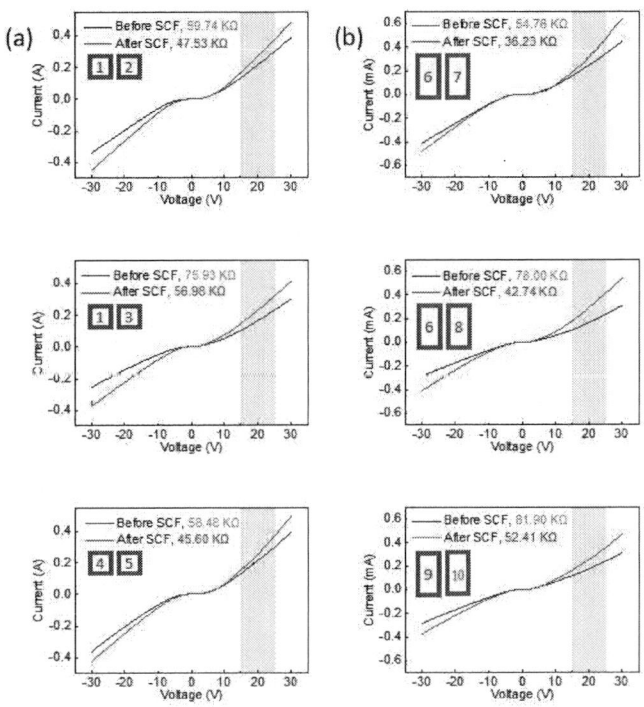

Fig. 2. Current-voltage characteristics of p-GaN devices before and after SCF treatment. (a, b) Comparison of contact resistance for electrodes with different sizes and spacings, showing significant resistance reduction after treatment.

For electrodes with dimensions of 100 m × 100 m, the contact resistance decreased from 63.41 k to 47.50 k , representing a reduction of ~33.5 (Fig. 3a). For larger electrodes of 100 m × 200 m, the resistance dropped from 42.41 k to 26.38 k , corresponding to a ~60.7 reduction (Fig. 3b). Resistance–distance analysis indicates that these improvements arise primarily from reduced metal-semiconductor interfacial resistance rather than changes in bulk resistance. Hall effect measurements in Fig. 3c further show that the hole mobility increased from 7.9~8.8 cm2/V s to 9.0~9.4 cm2/V s after SCF treatment, with an average enhancement of ~15 . The reduced dispersion of mobility values indicates a lower density of interface defects and diminished carrier scattering, resulting in improved charge transport characteristics.

979-8-3315-8850-2/25 $31.00 © 2025 IEEE

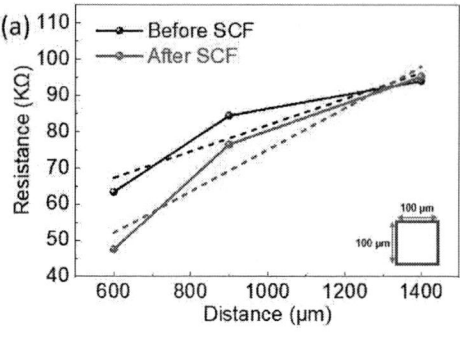

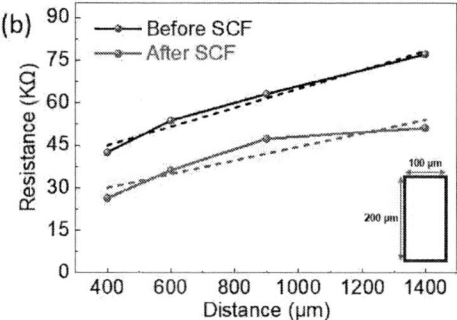

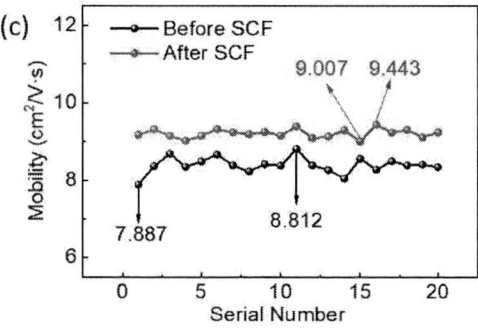

Fig. 3. uantitative analysis of contact resistance reduction. (a-b) Comparison of contact resistance between electrodes with different area and distance before and after SCF treatment. (c) Mobility measurement of hall effect before and after SCF treatment.

Energy band analysis shown in Fig. 4 provides additional insight into the underlying mechanism. SCF treatment effectively reduces the Schottky barrier height at the metal-p-GaN interface, thereby facilitating hole injection and improving ohmic contact behavior. At low bias, the I-V characteristics are dominated by the interfacial barrier, while at higher bias, bulk p-GaN resistance becomes the limiting factor. The most pronounced improvement after SCF treatment occurs in the low-bias region, validating the role of barrier modulation in performance enhancement.

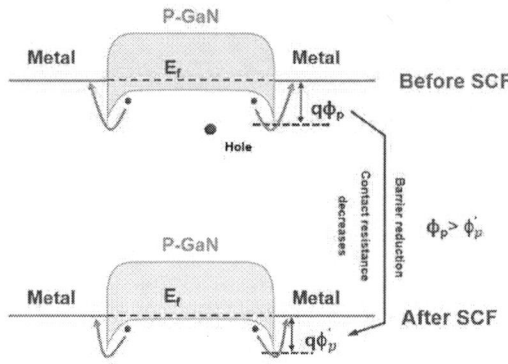

Fig. 4. Energy band diagrams of the metal-p-GaN interface before and after SCF treatment.

B. Material Characterizations

The XPS results in Fig. 5 reveal distinct chemical state modifications induced by SCF treatment. The Mg 1s peak exhibits a clear shift toward lower binding energy, indicating weakened Mg-H bonding and enhanced Mg acceptor activation [9]. Meanwhile, the Ga 3d and N 1s signals remain relatively unchanged, confirming that SCF treatment does not disrupt the GaN lattice but instead optimizes the surface chemical environment.

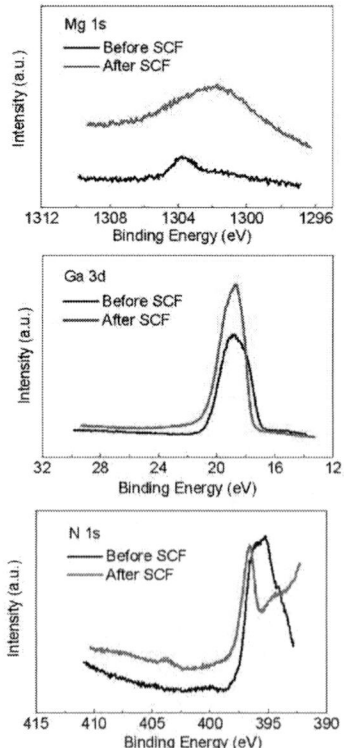

Fig. 5. XPS spectra of Mg 1s, Ga 3d and N 1s before and after SCF treatment.

Fig. 6a displays the Raman spectroscopy which provides further insights into lattice integrity. The E2(high) mode of GaN, located at 567.6 cm-1 in stress-free material, is highly sensitive to biaxial stress in the c-plane. In this study, strong E2(high) and weak A1(LO) peaks were observed at 567.2 and 734.2 cm-1, respectively, consistent with the wurtzite GaN selection rules. Importantly, the E2(high) peak remains essentially unchanged after SCF treatment, showing no significant redshift or blueshift. This result indicates that SCF treatment introduces no additional stress into the GaN lattice, thereby verifying its non-destructive nature. Fig. 6b gives the PL spectra of p-GaN before and after SCF treatment. The PL results further validate the improvements in material quality. The near-band-edge (NBE) emission, attributed to direct recombination between conduction band electrons and valence band holes, exhibits enhanced intensity after SCF treatment, reflecting improved crystal quality and reduced non-radiative recombination. The ultraviolet luminescence (UVL) band, associated with transitions from conduction band electrons or shallow donors to MgGa acceptors, and the blue luminescence (BL) band, arising from deep donor–MgGa acceptor recombination[10], are both observed. Notably, the BL emission shows significant enhancement, indicating a higher degree of Mg acceptor activation and more efficient hole injection. These PL results are in strong agreement with the electrical measurements, collectively demonstrating that SCF treatment effectively improves both the electrical and optical properties of Mg-doped p-GaN layers.

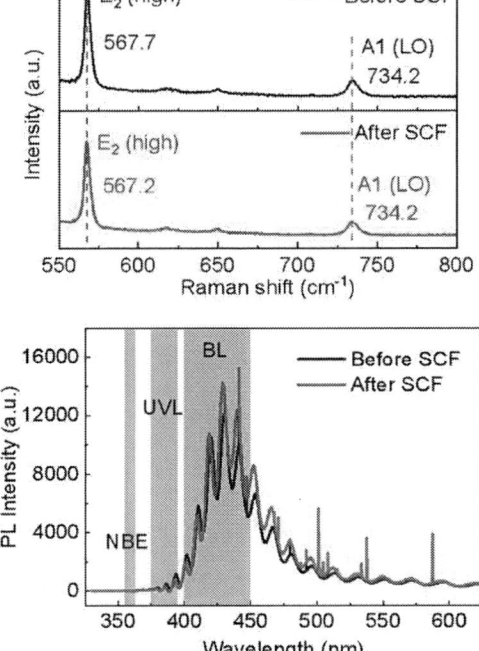

Fig. 6. (a) Raman spectra showing strong E2(high) and weak A1(LO) peaks, with the E2(high) peak position remaining nearly unchanged, confirming that SCF introduces no additional stress. (b) PL spectra showing enhanced NBE emission and intensified BL, indicating improved crystal quality, reduced non-radiative recombination, and higher Mg acceptor activation.

IV. CONCLUSIONS

In summary, this work demonstrated that supercritical fluid treatment provides an effective and non-destructive strategy to enhance the electrical and optical properties of Mg-doped p-type GaN thin films. Electrical characterization revealed a substantial reduction in metal-semiconductor contact resistance and an average ~15 enhancement in hole mobility. Energy band analysis confirmed that these improvements arise from the reduction of Schottky barrier height, enabling more efficient hole injection and enhanced ohmic behavior. XPS indicated enhanced Mg activation, Raman spectroscopy confirmed the preservation of lattice integrity, and PL spectra demonstrated stronger near-band-edge emission and intensified blue luminescence, both consistent with higher acceptor activation and reduced non-radiative recombination. These results highlight SCF processing as a scalable and non-destructive technique for engineering low-resistance, high-mobility p-GaN, offering broad potential for next-generation power electronics and optoelectronic applications.

ACKNOWLEDGMENT

This study was supported by Shenzhen Scientific and Technological Foundation (No. RCYX20231211090332037, JCYJ20240813160211015), National Natural Science Foundation of China (No. 62474008, 62204007), and Guangdong Provincial Natural Science Foundation (No. 2024A1515030044). This work was supported by Guangdong Provincial Key Laboratory of In-Memory Computing Chips (2024B1212020002). This work was in part supported by the Shenzhen POC center of Flexible Electronics and Guangdong Technology Center for Oxide Semiconductor Devices and ICs.

REFERENCES

[1] Wei, S. D. et al. Influence of polarities on optical properties of Mg-doped GaN films grown on GaN free-standing substrates by MOCVD. J. Lumin. 257, 119740 (2023).

[2] Chen, R. R. et al. Pores in p-type GaN by annealing under nitrogen atmosphere: formation and photodetector. J. Mater. Sci. 57, 467-476 (2022).

[3] Nakamura, S. et al. Hole compensation mechanism of p-type GaN films. Jpn. J. Appl. Phys. 31. 1258-1266 (1992).

[4] Park, A. H. et al. Efficient stress-relaxation in InGaN/GaN light-emitting diodes using carbon nanotubes. Nanoscale. 7, 15099-15105 (2015).

[5] Sang, L. W. et al. Boosting the doping efficiency of Mg in p-GaN grown on the free-standing GaN substrates. Appl. Phys. Lett. 115, 172103 (2020).

[6] Jun-Dar Hwang,Gwo-Huei Yang. Activation of Mg-doped p-GaN by using two-step annealing. Appl. Surf. Sci. 253, 4694–4697 (2007).

[7] Kim, S. W. et al. Reactivation of Mg acceptor in Mg-doped GaN by nitrogen plasma treatment. Appl. Phys. Lett. 76. 3079-3081 (2000).

[8] Lee, C. R. et al. The effect of p-GaN: Mg layers on the turn-on voltage of p-n junction LED. J. Cryst. Growth. 222. 459-464 (2001).

[9] Yan, S. M. et al. Improved minority carrier lifetime in p-type GaN by suppressing the non-radiative recombination process. Appl. Phys. Express 15, 075501 (2022).

[10] Sanjay Nayak,Mukul Gupta,Umesh V. Waghmare,and S.M. Shivaprasad. Origin of Blue Luminescence in Mg-Doped GaN. Phys. Rev. Appl. 11, 014027 (2019).

2025 The 10th International Conference on Integrated Circuits and Microsystems

Design for Hardware Acceleration of Signature Verification in PQC Stateless Hash-Based Digital Signature Algorithm

Haoran Liu
Electronic Engineering College
Heilongjiang University
Harbin, China
lhr1059863571@163.com

Liji Wu*
School of Integrated Circuit
Tsinghua University
Beijing, China
lijiwu@mail.tsinghua.edu.cn

Lei Li*
Electronic Engineering College
Heilongjiang University
Harbin, China
lileidtk@hlju.edu.cn

Yifan Yang
School of Integrated Circuit
Tsinghua University
Beijing, China
yyf20@mails.tsinghua.edu.cn

Xiangmin Zhang
School of Integrated Circuit
Tsinghua University
Beijing, China
zhxm@tsinghua.edu.cn

Abstract—**A hardware design of SLH-DSA signature verification is proposed for efficient deployment in post-quantum cryptographic systems. Based on the SPHINCS+ framework selected by NIST, the architecture utilizes FORS and WOTS+ schemes with SHA-256 to achieve stateless operation and quantum resistance. The verification pipeline integrates message hashing, signature decoding, public key reconstruction, and Merkle root validation, implemented through a pipelined and FSM-controlled datapath. Synthesized on an Intel Cyclone IV EP4CE115 FPGA, the design achieves a latency of 1.854 ms and supports a throughput of 839 verifications per second at 100 MHz. Post-synthesis power analysis reports 32.4 mW dynamic core power and an energy cost of 60.08 μJ per verification. The fully self-contained architecture eliminates the need for external processors and offers a scalable, low-power solution for embedded post-quantum applications. The results confirm both functional correctness and practical efficiency, making it suitable for real-world deployment.**

Keywords—Post-Quantum Cryptography, SLH-DSA, digital signature verification, hardware accelerator

I. Introduction

As quantum computing advances, classical signature schemes such as RSA and ECDSA are no longer secure due to Shor's algorithm, which enables efficient factoring and discrete logarithms [1]. In response, the NIST post-quantum cryptography (PQC) project selected SPHINCS+ as a conservative, stateless, hash-based signature scheme resistant to quantum attacks. The latest version, SPHINCS+ v3.1, was formally adopted in the 2023 FIPS 205 draft under the name SLH-DSA (Stateless Hash-Based Digital Signature Algorithm) [2]. SLH-DSA relies solely on cryptographic hash functions, offering strong quantum resistance and high suitability for embedded and security-critical applications.

In SoC-based systems, a Root of Trust (RoT) is responsible for secure boot and firmware validation, where digital signature verification plays a critical role. According to the U.S. PQC migration roadmap, hash-based signatures are preferred for securing firmware [3]. Open-source RoT platforms such as OpenTitan and Caliptra have begun supporting SPHINCS+ or LMS verification, but often rely on software routines or general-purpose hash cores, leading to high latency and limited scalability. Thus, a dedicated hardware accelerator for SLH-DSA signature verification is essential to ensure real-time performance and energy efficiency in embedded systems.

Some studies have introduced specialized accelerators to optimize SLH-DSA verification. Saarinen presented SLotH, an FPGA design combining Keccak/SHA2 hash units with a lightweight RISC-V controller, achieving SHAKE-256 verification speeds over 300× faster than microcontroller implementations [4]. Deshpande et al. proposed SPHINCSLET, a standard-compliant accelerator offering a 4.7× area reduction and 2.5–5× signing speedup, with 2.3–3.5× improvements in verification throughput [5]. However, most prior designs include full signing pipelines, resulting in excessive area and power overhead compared to the minimal needs of a read-only verification path.

Figure 1 illustrates the overall logic of signature verification by showing the basic interaction among the message, public key, and signature.

To bridge this gap, this work implements a fully hardware-based SLH-DSA verification engine on an FPGA platform. The design eliminates signing logic entirely and focuses on public

979-8-3315-8850-2/25 $31.00 © 2025 IEEE
937

key verification through a pipelined datapath controlled by a unified FSM. A single SHA-256 core is reused across stages, supporting segmented input, on-chip address derivation, and 22-layer parallel Merkle authentication. The resulting architecture conforms to the SLH-DSA (SPHINCS+ v3.1) specification with SHA2 or SHAKE support, and is well suited for low-power SoC applications requiring processor-independent post-quantum signature validation.

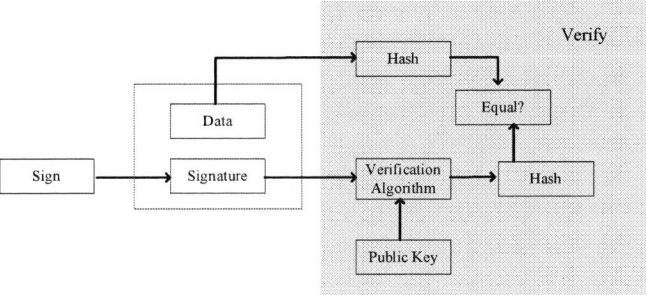

Fig.1 Data flow of signature verification.

II. THEORETICAL BASIS

A. SLH_DSA

SLH-DSA is a stateless hash-based signature scheme built from multiple hash-based components. It uses a few-time signature scheme, FORS (Forest of Random Subsets), and a many-time scheme based on XMSS (eXtended Merkle Signature Scheme). WOTS+ (Winternitz One-Time Signature Plus) keys are used within XMSS as Merkle tree leaves. FORS signs parts of the message digest to reduce key usage.

a. Key Generation: The private key comprises two n-byte seeds: SK.seed, used for pseudorandom generation of all WOTS+ and FORS secret keys, and SK.prf, which introduces randomness into the message-dependent hashing process. The public key consists of the root node of the top-level XMSS tree (PK.root) and a public seed (PK.seed) used in address-based domain separation.

b. Signing: A randomizer R is derived from SK.prf and optional input, then combined with the message to generate the digest. The digest bits select the FORS key and corresponding XMSS subtree. The FORS key signs the digest, yielding a FORS signature and derived public key. This key is authenticated through WOTS+ signatures and Merkle paths across XMSS layers, forming the final hypertree signature.

c. Verification: The verifier reconstructs the digest using the same R, derives indices, and regenerates the FORS public key from the signature. WOTS+ signatures and Merkle paths are then verified across XMSS layers up to the root. The signature is accepted if all checks pass.

B. Computing a WOTS+ Public Key From a Signature

The *wots_PKFromSig* procedure reconstructs a WOTS+ public key from the signature and message digest. Each signature element is treated as an intermediate hash chain state, and the algorithm completes the remaining iterations to recover the final public key values. The process begins by encoding the message digest into a base-w array of length len_1, followed by computing a checksum to ensure a fixed number of total hash applications. This checksum is similarly encoded and appended, forming an array of total length $len = len_1 + len_2$. In hardware, the WOTS+ chaining is executed as a counted loop over a shared SHA-256 core, and the final public key value is compacted by the tweakable $T\_len()$ stage.

For each index i, the algorithm sets the corresponding hash chain address and applies the chain() function to perform ($w - 1 - msg[i]$) iterations on $sig[i]$. The resulting array is then compressed via $T\_len()$ to obtain the reconstructed WOTS+ public key. This key is subsequently used in signature verification and must match the expected public key if the signature is valid. The step-by-step procedure is summarized in Algorithm 1.

Algorithm 1 wots_PKFromSig(*sig, M,* **PK**.seed, **ADRS**)

Compute a WOTS+ public key from a message and its signature.

Input: WOTS+ signature *sig*, message *M*, public seed **PK**.seed, address **ADRS**.

Output: WOTS+ public key pk_{sig} derived from *sig*.

1: $msg \leftarrow base\_2^b(M, lg_w, len_1)$
2: **for** *i* from 0 to $len_1 - 1$ **do**
3:   $tmp[i] \leftarrow$ chain$(sig[i], msg[i], w - 1 - msg[i],$ **PK**.seed, **ADRS**)
4: **end for**
5: $pk_{sig} \leftarrow \mathbf{T}_{len}(\mathbf{PK}.seed, wotspkADRS, tmp)$
6: **return** pk_{sig}

C. Computing an XMSS Public Key From a Signature

The *xmss_PKFromSig* procedure reconstructs the XMSS public key (i.e., the root of the Merkle tree) from a given signature and index. As shown in Algorithm 2, the input signature is divided into two parts: a WOTS+ signature and an authentication path. The algorithm first sets the address type to WOTS_HASH and calls *wots_PKFromSig()* to derive the corresponding WOTS+ public key, which serves as the initial leaf node. Each tree level reuses the same hash core, with ADRS steering the left/right ordering and the FSM scheduling the level-by-level reduction.

Algorithm 2 xmss_PKFromSig(*idx*, SIG$_{XMSS}$, *M*, **PK**.seed, **ADRS**)

Compute an XMSS public key from an XMSS signature.

Input: Index *idx*, XMSS signature SIG$_{XMSS}$ = (*sig* ‖ AUTH), *n*-byte message *M*, public seed **PK**.seed, address **ADRS**.

Output: *n*-byte root value *node*[0].

1: sig $\leftarrow$ SIG$_{XMSS}$.getWOTSSig()
2: *node*[0] $\leftarrow$ wots_PKFromSig(*sig*, *M*, **PK**.seed, **ADRS**)
3: **for** *k* from 0 to $h' - 1$ **do**
4:   **if**[$idx/2^k$] is even **then**
5:     *node*[1] $\leftarrow$ **H**(**PK**.seed, **ADRS**, *node*[0] ‖ AUTH[*k*])
6:   **else**
7:     *node*[1] $\leftarrow$ **H**(**PK**.seed, **ADRS**, AUTH[*k*] ‖ *node*[0])
8:   **end if**
9:   *node*[0] $\leftarrow$ *node*[1]
10: **end for**
11: **return** *node*[0]

Next, the address is updated to the TREE type and configured with the appropriate tree index. The algorithm then performs h' iterations to traverse the Merkle authentication path. In each round, the address is updated with the current tree height, and the node combination order is determined by the parity of the index. A domain-separated hash H compresses the concatenated nodes into their parent. After all levels are processed, the resulting node[0] becomes the reconstructed XMSS root.

This algorithm ensures that the reconstructed root matches the expected public key only if the signature is correct. Domain separation via the address object guarantees uniqueness across hash computations.

D. Computing a FORS Public Key From a Signature

The *fors_pkFromSig* algorithm reconstructs the FORS public key from a given signature. As shown in Algorithm 3, the message digest is first parsed into k indices using *base_2^b()*, each identifying a leaf node in a separate Merkle subtree. The *k* subtrees are processed on the unified datapath—$F()/H()$ build per-index roots, which are then folded by the tweakable $T\_k()$ compressor.

For each index, the corresponding secret key is retrieved from the signature and hashed with the public seed and address to produce the leaf node. Then, the subtree root is computed by iteratively hashing up the authentication path using a domain-separated function $H()$. The address is updated at each level, and the input order to $H()$ depends on the node's parity

---

Algorithm 3 fors_pkFromSig (SIG_{FORS}, *md*, **PK**.seed, **ADRS**)

Compute a FORS public key from a FORS signature.

Input: FORS signature SIG_{FORS} ,

  message digest *md*, public seed **PK**.seed, address **ADRS**.

Output: FORS public key.

1: **for** *i* **from** 0 **to** $k-1$ **do**

2:   $sk \leftarrow SIG_{FORS}.getSK(i)$

3:   $node[0] \leftarrow$ **F**(**PK**.seed,**ADRS**,*sk*)

4:   **for** *j* **from** 0 **to** $a-1$ **do**

5:     **if**[*indices[i]/2^j*] is even **then**

6:       $node[1] \leftarrow$ **H**(**PK**.seed,**ADRS**,*node*[0] ∥ *auth*[*j*])

7:     **else**

8:       $node[1] \leftarrow$ **H**(**PK**.seed,**ADRS**, *auth*[*j*] ∥ *node*[0])

9:     **end if**

10:     $node[0] \leftarrow node[1]$

11:   **end for**

12:   $root[i] \leftarrow node[0]$

13: **end for**

14: $pk \leftarrow$ **T**$_k$(**PK**.seed, forspkADRS, *root*)

15: **return** *pk*

After all k subtree roots are obtained, they are compressed using a tweakable hash function $T\_k()$ to produce the final FORS public key. Proper address management ensures input uniqueness and domain separation across subtrees.

Algorithm 4 slh_verify(*M*, SIG, PK)

Verify an SLH_DSA signature

Input: Message *M*, signature SIG, public key PK = (**PK**.seed,

  PK.root).

Output: Boolean.

1: **if** $|SIG|$!= $(1 + k(1 + a) + h + d \cdot len) \cdot n$ **then**

2:   **return** false

3: **end if**

4: $R \leftarrow$ SIG.getR()

5: $digest \leftarrow$ **H**$_{msg}$(R,**PK**.seed,**PK**.root ,M)

6: $md \leftarrow digest[0:[k\text{-}a/8]]$

7: $tmp\_idx_{tree} \leftarrow digest[[k\text{-}a/8]:[k\text{-}a/8]+[h\text{-}h/d/8]]$

8: $tmp\_idx_{leaf} \leftarrow digest[[k\text{-}a/8]+[h\text{-}h/d/8]:[k\text{-}a/8]+[h\text{-}h/d/8]+[h/8d]]$

9: $PK_{FORS} \leftarrow$ fors_pkFromSig(SIG_{FROS},md,**PK**.seed,**ADRS**)

10: **return** ht_verify(PK_{FORS}, SIG_{HT}, **PK**.seed, idx_{tree}, idx_{leaf}, **PK**.root)

E. SLH-DSA Signature Verifcation

The SLH-DSA verification process ensures the authenticity of a message by reconstructing the corresponding public key hierarchy. After verifying the integrity of the signature length, the verifier parses the signature into three main components: a randomizer, the FORS signature, and the hypertree (XMSS) authentication path. A message digest is generated using the public seed, root, and the randomness value.

This digest is then parsed into specific indices to locate the corresponding FORS trees and XMSS subtree layers. The FORS public key is computed by iteratively applying hash functions to the extracted leaf nodes and authentication paths, as shown in Algorithm 4. This intermediate key serves as the input for verifying the XMSS hypertree layer, where a similar hash-based traversal reconstructs the Merkle root.

If the resulting root matches the one provided in the public key, the signature is deemed valid. Otherwise, verification fails. The process ensures strict domain separation through address management, maintaining both correctness and security throughout the procedure.

To make the execution semantics explicit, each primitive in Section II is realized by the same time-multiplexed SHA-256 datapath under a centralized FSM: chain() is an iterated hash loop with counter-driven valid/ready handshakes, $T\_len()/T\_k()$ are implemented by a tweakable compression stage for length/domain binding, and $F()/H()$ reuse the core with ADRS-controlled input selection.

III. HARDWARE IMPLEMENTATION OVERVIEW

This chapter presents the hardware architecture for SLH-DSA signature verification. The module names intentionally mirror the algorithmic blocks: Compute_wots_pk corresponds to WOTS+ (chain() + $T\_len()$), Compute_root_pk matches XMSS path hashing ($H()$ with ADRS), Compute_fors_pk aggregates FORS roots ($F()/H()$ with $T\_k()$), while Addr_control generates ADRS fields on-chip. The verification process comprises message digest generation, FORS public key reconstruction, WOTS+ signature checking, and XMSS Merkle root comparison, all coordinated by a configurable finite state machine (FSM). A single hash unit is reused across stages to enhance resource efficiency under pipeline scheduling. Figure 2 illustrates the hierarchical tree traversal involved in the SLH-DSA verification process.

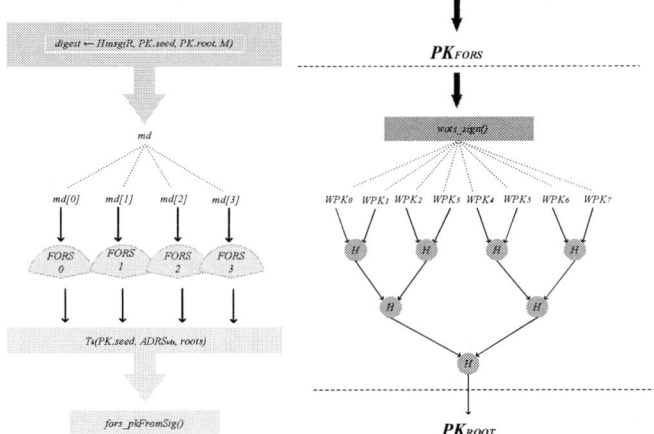

Fig.2 Tree traversal schematic in SLH-DSA signature verification.

A. Wots+ design

The WOTS+ (Winternitz One-Time Signature Plus) module serves as a critical component in the SLH-DSA verification chain, acting as a link between FORS and XMSS. This design adopts a fully pipelined and deterministic hardware architecture to compute hash chains and generate WOTS+ public keys efficiently.

Conventional architectures often suffer from sequential chain traversal and high control overhead. To address these limitations, the proposed WOTS+ module introduces a chain scheduler that enables parallel processing of multiple hash elements. The system supports base-w conversion, checksum calculation, and iterative hash traversal within a unified FSM framework, ensuring constant-time execution and resistance to side-channel leakage.

To further enhance performance, the hash chain engine reuses intermediate results and minimizes redundant hashing, allowing a single SHA-256 core to process multiple chains without pipeline stalls. Address separation and tweakable parameters are supported through programmable control logic, enabling tight integration with both FORS outputs and XMSS inputs. This structure ensures that WOTS+ verification is not only efficient but also compatible with the hierarchical nature of SLH-DSA.

B. XMSS Design

The XMSS (eXtended Merkle Signature Scheme) module performs hierarchical root reconstruction using WOTS+ public keys and authentication paths. As the upper layer of SLH-DSA's hypertree, XMSS is responsible for computing the Merkle root from each leaf, which is derived from a previously reconstructed WOTS+ key.

This design optimizes the Merkle authentication process by using a tree-walking engine with a configurable height register and address logic. At each level, the core determines the position of the node (left or right) based on the address index and performs hash concatenation accordingly. A single SHA-256

engine is reused across tree levels, and a FSM ensures sequential yet pipelined traversal across all 22 layers.

Compared to conventional implementations that allocate static registers for each level, this design eliminates memory duplication and supports dynamic address updates to enable domain separation. The reuse of hash and address logic reduces area cost while preserving correctness and performance. Furthermore, by decoupling L-tree compression and root propagation, the structure ensures compatibility with various SLH-DSA parameter sets without code expansion or logic reconfiguration.

C. Fors Design

The FORS (Forest of Random Subsets) module serves as a lightweight one-time signature component within SLH-DSA. Its primary function is to generate and authenticate multiple Merkle tree roots derived from secret leaf nodes, forming an intermediate public key that anchors the upper XMSS layers.

This design implements a parallelized key path reconstruction engine that traverses k small binary trees simultaneously. For each selected index derived from the message digest, the corresponding leaf node is computed using a fixed secret key segment and hashed with SHA-256. The authentication path is then evaluated via an iterative hash loop, where sibling nodes are conditionally concatenated based on bit-level tree position. A dynamic address update unit ensures proper domain separation across trees and layers.

Compared to sequential FORS architectures, this implementation avoids redundant SHA-256 invocations by reusing address and hash pipelines, effectively reducing verification latency. The final k tree roots are compressed using a tweakable tree hashing unit, producing the FORS public key that feeds into the XMSS stage. The structure maintains high modularity, enabling flexible scaling for different parameter sets and efficient resource usage on low-cost FPGAs.

D. Hardware Architecture Design

The complete SLH-DSA signature verification is implemented using a modular and pipelined architecture, integrating the WOTS+, XMSS, and FORS verification paths described in the previous sections. Figure 4 illustrates the top-level hardware architecture, while Figure 3 outlines the finite-state machine (FSM) responsible for system control.

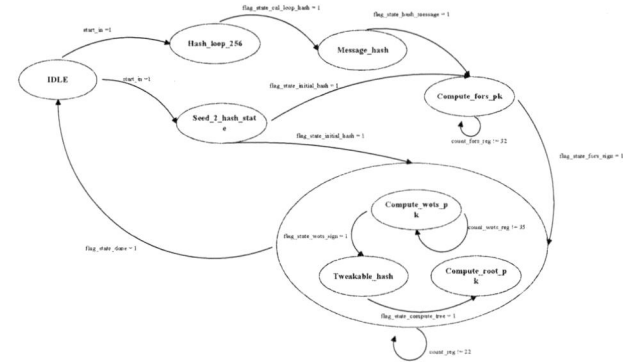

Fig.3 Finite-state machine (FSM) controlling the verification datapath.

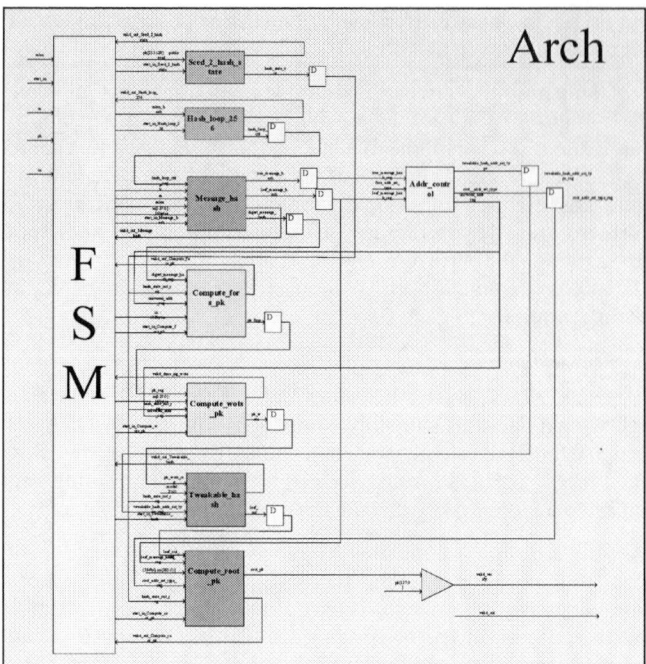

Fig.4 Top-level hardware architecture for SLH-DSA signature verification.

To address the limitations of software-driven or partial hardware approaches—such as high latency, unbalanced module scheduling, and resource underutilization—this design adopts a unified control strategy centered on a centralized FSM. All functional modules are orchestrated through valid/ready handshaking signals to ensure deterministic behavior and eliminate unnecessary pipeline stalls.

One core innovation is the reuse of a single SHA-256 computation core across all stages, including message digest generation, WOTS+ chain hashing, FORS node computation, and Merkle tree traversal. This time-multiplexed hashing approach minimizes area and dynamic power overhead while maintaining performance, particularly suited for resource-constrained FPGAs such as the Cyclone IV.

All domain-separated address fields required in the XMSS and FORS layers are generated on-chip using a programmable address engine, avoiding precomputed tables and improving reconfigurability. The architecture supports streaming input, enabling continuous verification without full buffering, reducing both memory requirements and latency. Compared with prior implementations relying on external control processors or monolithic hash blocks, this design achieves better modularity, lower resource consumption, and improved security against timing-based side-channel leakage.

IV. IMPLEMENTATION AND EVALUATION

To assess the practicality of the proposed SLH-DSA signature verification core, comprehensive hardware-based evaluation was performed, including functional correctness validation and implementation-level benchmarking. The design adheres to the SLH-DSA specification defined in the NIST post-quantum cryptographic standardization framework and is fully realized in hardware. It operates autonomously without requiring any external processor or software assistance.

Functional verification was conducted via UART-based on-chip testing, and performance was benchmarked against existing FPGA-based implementations.

A. Functionality Verification

Functional validation was conducted using standard test vectors generated by a reference software implementation. These vectors, including the message, signature, and corresponding public key, were sent to the FPGA via UART. The hardware core then executed a full verification flow, including FORS decoding, WOTS+ chaining, and XMSS Merkle root reconstruction. Upon successful validation, the control signal valid_verify was asserted and returned to the host. The result was printed through the UART terminal on the PC, as shown in Figure 5.

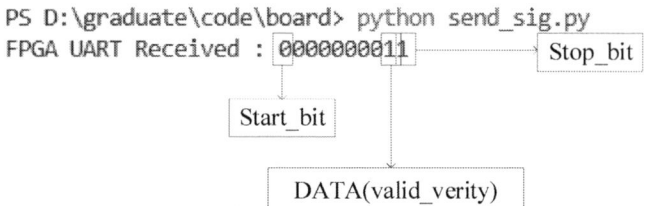

Fig.5 UART output showing the signature verification result.

The architecture is fully FSM-driven and operates in a staged manner to ensure dataflow clarity and control predictability. A unified SHA-256 module is shared across all layers, minimizing redundant logic. By avoiding conditional branching and memory offloading, the design achieves cycle-level determinism and enhanced resistance to side-channel leakage. Moreover, the entire SLH-DSA verification process is accomplished independently in hardware without processor intervention, supporting lightweight and secure deployments. Figure 6 illustrates the post-implementation simulation result in Vivado, where the high level of the valid_verify signal indicates successful signature validation.

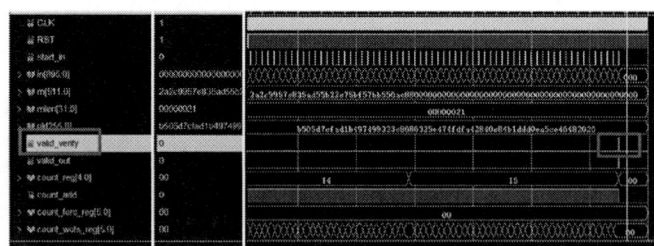

Fig.6 Post-synthesis simulation waveform showing successful signature validation (valid_verify = high).

B. Performance Analysis

The proposed SLH-DSA verification core was implemented and evaluated at an operating frequency of 100 MHz. As shown in Table I, the design completes the full signature verification process within 1.854 ms, resulting in a throughput of 839 verifications per second. This performance improvement is achieved through the use of a unified hash core, pipelined processing stages, and fine-grained FSM control logic, which collectively reduce idle cycles and improve parallel resource utilization.

The architecture also demonstrates favorable energy efficiency. With a supply voltage of 1.2 V, the dynamic power consumption remains below 32.4 mW, leading to an energy cost of approximately 60.08 μJ per verification, as presented in Table II. Compared to prior works such as [3], which report similar energy per operation but operate at significantly higher power levels, the proposed design achieves a more efficient balance between energy consumption and computational performance.

Unlike previous implementations that offload portions of the verification process to software [4], the presented architecture supports the entire SLH-DSA verification pipeline fully in hardware, including FORS key derivation, WOTS+ chain computation, and XMSS root verification. This self-contained design eliminates processor dependencies and ensures constant-time behavior, making it well-suited for secure embedded systems requiring deterministic performance, compact resource usage, and low power consumption. Unlike Ref.[5], which reports a custom 65-nm ASIC with fixed datapaths (excellent efficiency but limited flexibility and high non-recurring cost), our design targets commodity FPGAs to favor reconfigurability and low integration overhead in real deployments.

TABLE I.   PERFORMANCE SUMMARY

Item	Value
FPGA	Intel Cyclone IV EP4CE115
Clock Frequency	100MHz
Total Cycle Count	185431 cycles
Total Latency	1.854 ms
Dynamic Core Power	32.40 mW
Energy per Verification	60.08 μJ

TABLE II.   COMPARISON IMPLEMENTATIONS

Implementation	Platform	Frequency (MHz)	Throughput (ops/s)	Power (mW)	Energy (μJ)
This Work	Cyclone IV (EP4CE115)	100	~839	32.40	60.08
Ref [4]	Virtex-7 28nm FPGA	150	~833	58.2	69.84
Ref [5]	ASIC 65nm	200	~1200	12.5	10.375
Ref [6]	Artix-7 FPGA	100	~435	38.7	89.01
Ref [7]	Cortex-M4 + Crypto Cell-312	50	~213	8.1	38.07

V.   CONCLUSION

This work presents a complete and standalone hardware implementation of the SLH-DSA signature verification algorithm, targeting resource-constrained embedded platforms such as FPGAs. Built entirely without processor intervention, the design integrates all key cryptographic components—FORS, WOTS+, and XMSS[8]—into a unified, FSM-driven pipeline, achieving cycle-level determinism and eliminating software dependence. Simulation and on-board tests confirm functional correctness, while performance profiling at 100 MHz shows a latency of 2.4–3.0 ms and a throughput of up to 417 verifications per second. The dynamic power consumption remains under 250 mW, resulting in energy consumption as low as 0.6 mJ per verification. The architecture maintains full compliance with SLH-DSA v3.1 specifications and supports both SHA2-based and SHAKE-based instantiations.

Compared with existing solutions such as SLotH[9] and SPHINCSLET[10], the proposed architecture operates at a lower frequency yet delivers comparable throughput and better energy efficiency, as shown in Table I and Table II. Unlike partial hardware accelerators that offload certain tasks to software, this design completes the entire verification path in hardware, thereby enhancing timing consistency, integration flexibility, and hardware security.

Future research can evolve in two primary directions. One promising avenue is to extend the current verification-only core into a full SLH-DSA implementation, supporting signature generation. This involves integrating message randomization logic, FORS index sampling, and secure private key handling, ultimately enabling end-to-end post-quantum signature support in hardware. Another important direction is the development of side-channel-resistant countermeasures. Although SLH-DSA is structurally deterministic, modules such as FORS and WOTS+ involve secret-dependent computations. Enhancing the architecture with masking schemes, constant-time data paths, and randomized memory access patterns could improve leakage resilience, which is vital for deployment in high-assurance Root-of-Trust platforms.

Overall, this work lays a solid foundation for high-efficiency and self-contained SLH-DSA signature verification, opening the door to future extensions toward full pipeline support and physically secure designs.

REFERENCES

[1] P. W. Shor, "Algorithms for quantum computation: Discrete logarithms and factoring," in Proc. 35th Annual Symp. Foundations of Computer Science, 1994, pp. 124–134.

[2] National Institute of Standards and Technology (NIST), "FIPS 205: Stateless Hash-Based Digital Signature Algorithm (SLH-DSA)," 2023. [Online].

[3] Office of Management and Budget (OMB), "Migration to post-quantum cryptography," Memo M-23-02, Nov. 2022

[4] M.-J. O. Saarinen, "Accelerating SLH-DSA by two orders of magnitude with a single hash unit," IACR Cryptology ePrint Archive, vol. 2024, no. 367, 2024.

[5] R. Deshpande, S. Shinde, and K. Paterson, "SPHINCSLET: A lightweight post-quantum digital signature accelerator," IACR Cryptology ePrint Archive, vol. 2023, no. 1221, 2023.

[6] R. Deshpande, K. Paterson, and B. Marshall, "Energy-efficient SPHINCS+ verification on embedded systems," IACR Cryptology ePrint Archive, vol. 2023, no. 1344, 2023.

[7] L. Campbell and T. Lange, "Hardware support for SPHINCS+ in low-cost FPGAs," IACR Cryptology ePrint Archive, vol. 2023, no. 445, 2023

[8] W. Wang et al., "XMSS and embedded systems," in Proc. 26th Int. Conf. Sel. Areas Cryptogr., 2019, pp. 523–550.

[9] Tuan Nguyen Kim, Duy Ho Ngoc, Nin Ho Le Viet, and Nikolay A. Moldovyan, "The New Collective Signature Schemes Based on Two Hard Problems Using Schnorr's Signature Standard," Journal of Advances in Information Technology, Vol. 14, No. 1, pp. 77-84, February 2023.

[10] M. J. Dworkin et al., "SHA-3 standard: Permutation-based hash and extendable-output functions," 2015.

A 88-dB SNDR 156-KHz-BW Noise-Shaping SAR ADC with 3rd-CIFF Structure and Dynamic Integrator

Yulin Feng
College of Electronics and Information Engineering
Shenzhen University
Shenzhen, Guangdong Province, China
0009-0001-8283-1283

Mei Jiang*
College of Electronics and Information Engineering
Shenzhen University
Shenzhen, Guangdong Province, China
mjiang@szu.edu.cn

Xinhui He
College of Electronics and Information Engineering
Shenzhen University
Shenzhen, Guangdong Province, China
hexinhui22@outlook.com

Zhengru Li
College of Electronics and Information Engineering
Shenzhen University
Shenzhen, Guangdong Province, China
lzr1152545581@gmail.com

Abstract—This paper presents a noise-shaping (NS) successive approximation register (SAR) analog-to-digital converter (ADC) with 3rd-order cascade of integrators with feedforward (CIFF) structrue. The amplifier modules in the integrators are implemented using two-stage floating inverter amplifier (FIA) structures, which provide high gain while achieving significantly lower power consumption compared to operational transconductance amplifiers (OTAs). The 3th-order integration is realized with two integrators, where the third stage is implemented by reusing the second integrator. This reuse strategy eliminates one FIA and thereby reduces hardware cost. The multi-input comparison functionality of the CIFF structure is achieved through capacitive series stacking combined with a conventional two-input comparator, leading to substantial power savings. The ADC operates at a sampling rate of 5 MS/s with a 1.2 V supply voltage and an oversampling ratio (OSR) of 16. Post-layout simulations, performed with a sinusoidal input signal at 50.96 kHz, yield a power consumption of 506 μW. The measured signal-to-noise and distortion ratio (SNDR) is 88 dB, corresponding to an effective number of bits (ENOB) of 14.32 bit. The Schreier figure-of-merit (FOMs) reaches 173 dB.

Keywords—*noise shaping, SAR ADC, CIFF Structure, floating inverter amplifier, low power*

I. INTRODUCTION

NS SAR ADCs have become a research hotspot in recent years. By incorporating the noise-shaping techniques commonly used in Sigma-Delta ADCs into the highly digital and low-power architecture of SAR ADCs, the ENOB can be significantly improved. Currently, the CIFF structure and the error feedback (EF) structure are the two mainstream topologies for noise-shaping SAR ADCs. The simplified diagrams of both structures are shown in Fig. 1. The key difference between the two lies in the transfer function of the loop filter. In the CIFF structure, the transfer function should be as high as possible — a higher gain results in more ideal integration performance. In contrast, the EF structure requires the transfer function to be unity, which necessitates the

Key Technology Research and Development Project for Scientific and Technological Innovation in Shenzhen (No.JSGG20220831093005009)
Key Basic Research Project in Science and Technology Research and Development in Shenzhen (No.JCYJ2022081803410022)

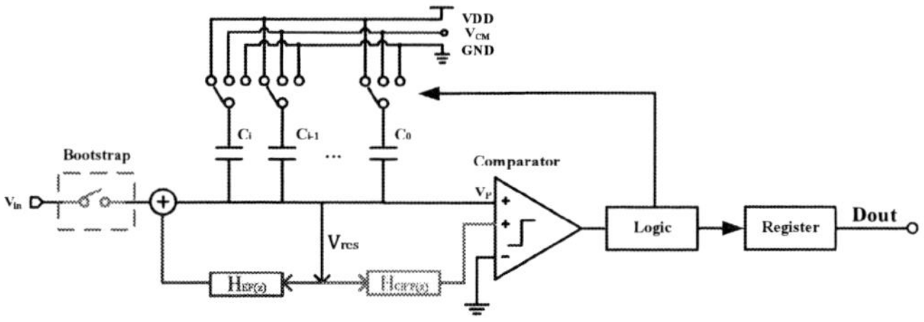

Fig. 1. Structure of CIFF and EF.

inclusion of gain compensation circuitry to meet this requirement [1].

In comparison, higher-order applications based on the CIFF structure are relatively easier to implement. For integrator selection, two primary options exist: passive integration and active integration. Passive integration relies on charge transfer between capacitors, which results in very low power consumption but offers poorer noise-shaping performance and suffers from signal loss, requiring additional compensation schemes. Common compensation methods include increasing the relative gain of input transistors in multi-input comparators [2], capacitor stacking techniques [3], and using dynamic amplifiers as buffers [4]. Active integration, on the other hand, can achieve more ideal noise-shaping performance due to its high gain [5], [6], although this comes at the cost of higher power consumption typical of OTAs. The FIA, introduced in [7], is a low-power dynamic amplifier that achieves high gain while consuming significantly less power than conventional OTAs [8]. This makes the FIA particularly suitable for replacing OTA circuits in integration stages, offering both ideal noise-shaping performance and low power consumption. This characteristic is highly attractive for SAR architectures targeting low-power applications.

979-8-3315-8850-2/25 $31.00 © 2025 IEEE

II. IMPLEMENTATION OF THE CIRCUIT

A. Closed-Loop Integrator

Fig. 2(a) shows the simplified schematic of a single-stage FIA. For clarity, only one side of the differential structure is

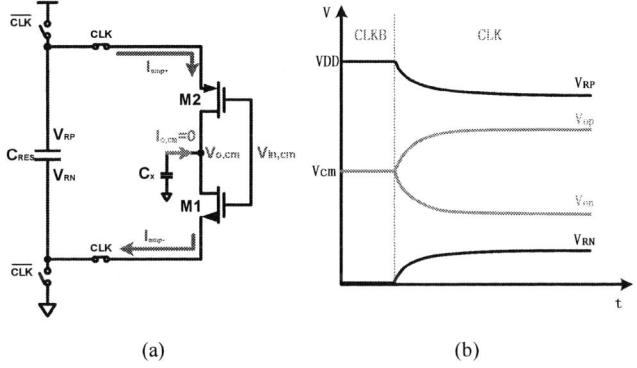

(a)   (b)

Fig. 2. (a) Single-stage FIA. (b) Temporal variation of key nodes in the FIA

shown. The energy storage capacitor C_{RES} is directly connected to VDD for charging when the control clock CLK is low. During the high phase of CLK, the FIA operates in the amplification mode, where the stored charge in the capacitor supplies the transistors. The time evolution of key internal nodes during this process is shown in Fig. 2(b).

During the amplification phase of the FIA, the initial current I_{amp} is relatively large. However, as the energy storage capacitor C_{RES} discharges, the voltage at its positive terminal, V_{RP}, gradually decreases, while the negative terminal voltage, V_{RN}, rises accordingly. As a result, the gate-to-source voltages of both the PMOS and NMOS transistors decrease, rapidly driving them into the weak inversion region. Consequently, I_{amp} drops quickly to a very low level. After the FIA has operated in the amplification phase for some time, V_{RP} and V_{RN} continue to shift until both transistors are eventually turned off. At this point, the amplification phase ends, and all node voltages stabilize, with no further power consumption occurring in the circuit. From the above process, it is evident that during the amplification phase, the FIA transistors operate predominantly in the weak inversion region. Compared to OTA circuits, where transistors typically operate in the saturation region, the FIA exhibits significantly lower power consumption. Moreover, the FIA features an inherent self-quenching characteristic: the transistors automatically turn off after a certain operation period. This makes the FIA a highly suitable candidate for implementing amplifiers in low-power applications.

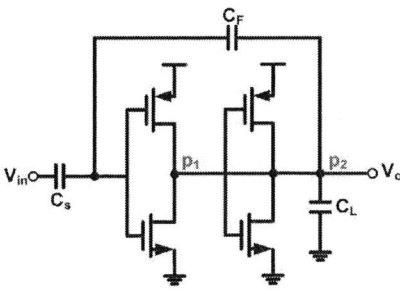

Fig. 3. Closed-Loop integrator.

Fig. 3 shows the schematic of an integrator formed by a two-stage FIA and a feedback capacitor C_F. The two-stage

FIA amplifier introduces two poles: one at the output node of the first stage, denoted as p_1, and the other at the output node of the second stage, denoted as p_2. Since the load capacitance is significantly larger than the gate capacitance, p_2 is generally considered the dominant pole. To facilitate pole splitting and ensure stability, the energy storage capacitor C_{RES} of the second-stage FIA is designed to be 3 – 5 times smaller than that of the first stage. As a result, the current in the second stage decays much faster than that in the first stage — equivalent to a more rapid increase in output resistance. During operation, this leads to automatic separation of the two poles, and the phase margin gradually increases over time, ensuring stable amplifier operation [9]. Each single-stage FIA provides approximately 20 dB of gain, and the cascaded two-stage FIA achieves more than 40 dB of total gain, which ensures accurate integration. When the residual voltage V_{res} is applied at the input V_{in}, the following expression can be easily derived based on the structure shown in Fig. 3:

$$V_F[n] = \frac{C_S}{C_F} \cdot V_{res}[n] + V_F[n-1] \quad (1)$$

As can be seen from the equation, the integration performance of the closed-loop integrator depends on the capacitor ratio. Therefore, the integrator exhibits good PVT (process, voltage, and temperature) robustness and can maintain stable integration under process, voltage, and temperature variations.

B. 3rd-Stage Integrator

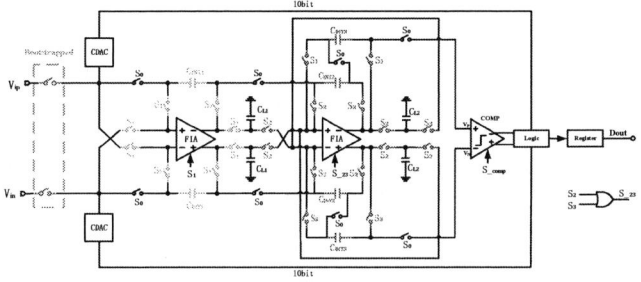

Fig. 4. Simplified schematic of the ADC circuit

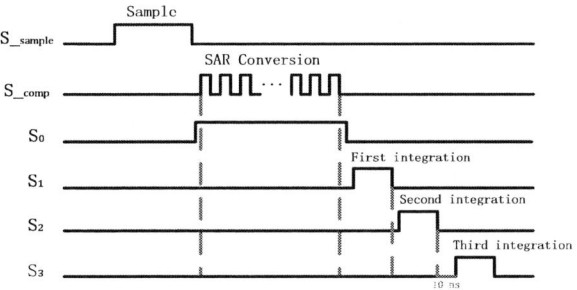

Fig. 5. Timing diagram of key control signals in the circuit

979-8-3315-8850-2/25 $31.00 © 2025 IEEE

Fig. 4 shows the overall schematic of the proposed circuit. The architecture mainly consists of two differential Bootstrap circuits for input sampling, two 10-bit binary-weighted CDAC capacitor arrays, two FIA amplifiers used as integrators, a high-speed comparator, digital control logic, and data registers for result storage. Fig. 5 further illustrates the timing relationships among the critical control signals within the digital control block, providing insight into the sequence and synchronization of various operational phases.

During a complete voltage conversion cycle, the process begins with the sampling phase controlled by the sampling signal. Following this, the comparison phase is initiated by the comparator control signal to perform the 10-bit successive approximation process. During this phase, the output of the CDAC array is combined with the residual integral result from the previous conversion under the control of signal S0. Upon completion of the current successive approximation, signals S_1, S_2, and S_3 sequentially control the three-stage integration of the residual voltage from the current conversion. Upon completion of the current successive approximation, signals S1, S2, and S3 sequentially control the three-stage integration of the residual voltage from the current conversion.

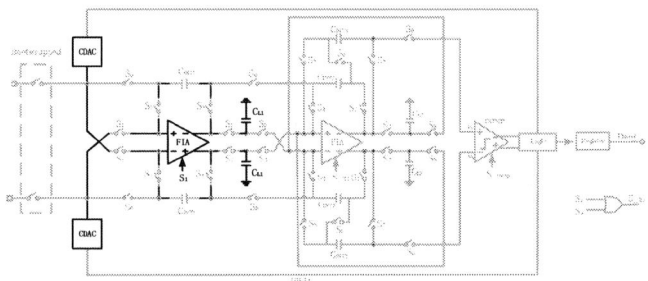

Fig. 6. First-stage Integration

The first-stage integration is performed when the control signal S1 is high, and the corresponding circuit is shown in Fig. 6. The residual voltage is accumulated from the CDAC capacitor array onto the integrating capacitor C_{INT1}. The capacitor C_{L1} then transfers the integration result to the next stage as the input. Following the same derivation as in Eq. (1), the integration process can be expressed by Eq. (2):

$$V_{CINT1}[n] = V_{CL1}[n] = \frac{C_{DAC}}{C_{INT1}} \cdot V_{res}[n] + V_{CINT1}[n-1] \quad (2)$$

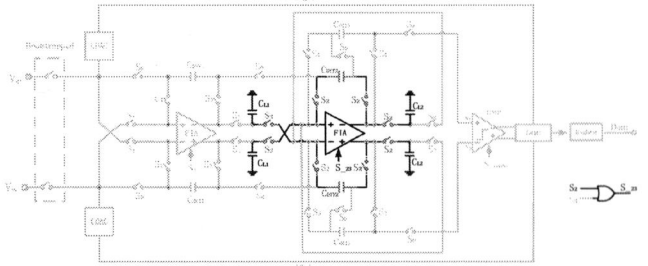

Fig. 7. Second-stage Integration

The second-stage integration is performed when the control signal S2 is high, as shown in Fig. 7. The principle of operation is consistent with the first stage, where the integration result is accumulated onto the integrating capacitor C_{INT2}. The integration process can be described by Eq. (3):

$$V_{CINT2}[n] = V_{CL2}[n] = \frac{C_{L1}}{C_{INT2}} \cdot V_{CL1}[n] + V_{CINT2}[n-1] \quad (3)$$

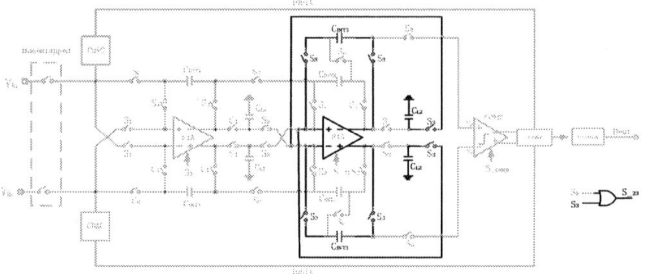

Fig. 8. Third-stage Integration

The third-stage integration is performed when the control signal S3 is high, as shown in Fig. 8. This integration stage reuses the second integrator, where the signal from C_{L2} is fed into and accumulated onto the integrating capacitor C_{INT3}:

$$V_{CINT3}[n] = V_{CL3}[n] = \frac{C_{L2}}{C_{INT3}} \cdot V_{CL2}[n] + V_{CINT3}[n-1] \quad (4)$$

After completing the three-stage integration, the results — represented by the voltages across the three integrating capacitors C_{INT1}, C_{INT2}, and C_{INT3} — are summed and fed into a multi-input comparator. This enables effective third-order noise shaping, with the noise transfer function (NTF) given by:

$$\text{NTF} = (1 - z^{-1})^3 \quad (5)$$

It is important to note that while S_1 and S_2 only need to be non-overlapping, a deliberate 10-ns delay must be introduced between S_2 and S_3. This timing constraint arises because the third-stage integration reuses the second-stage FIA amplifier. After the second-stage integration is completed, the FIA requires a finite recovery time to recharge its energy storage capacitor before it can be reused for the next integration stage.

C. Capacitors in Series Achieve Voltage Addition

There are generally three methods for summing the residual voltage with the input voltage. The most commonly used approach is to perform the summation and comparison simultaneously using a multi-input comparator. However, compared to the simplest two-input comparator, a multi-input comparator typically consumes significantly more power and introduces higher noise, which is not desirable in low-power applications. Moreover, mismatches among the input transistors of the multi-input comparator can cause deviations in the gain of each signal path from its designed value, thereby degrading the effectiveness of noise shaping [10]. An efficient solution is to store the integration results on separate capacitors and then connect these capacitors in series using timing-controlled switches, thereby summing the voltages before applying them to a standard two-input comparator [11].

979-8-3315-8850-2/25 $31.00 © 2025 IEEE

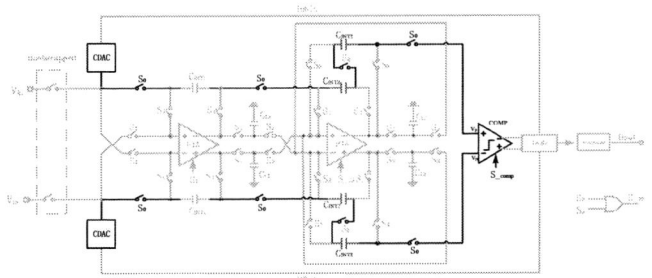

Fig. 9. Voltage Addition

The proposed circuit, shown in Fig. 9, employs switches controlled by signal S0 to connect the voltage from the CDAC in series with the three integrating capacitors, thereby realizing voltage summation. This summed voltage is then fed into a two-input comparator. This summation method relies entirely on the stored capacitor voltages and achieves precise voltage superposition through carefully designed timing control, ensuring accurate and consistent noise shaping. However, It is unavoidable that this design modifies the original architecture of the SAR ADC. Due to the additional voltage-summation functionality added at the comparator input, the SAR conversion and integration operations can no longer be performed concurrently. This results in extra timing overhead, which ultimately reduces the overall sampling rate.

III. POST-SIMULATION RESULTS

Fig. 10 shows the overall layout of the ADC, with a total layout area of $871\mu m \times 286\mu m$.

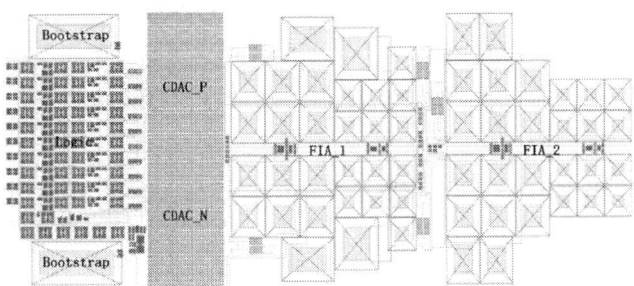

Fig. 10. Layout of the top-level circuit

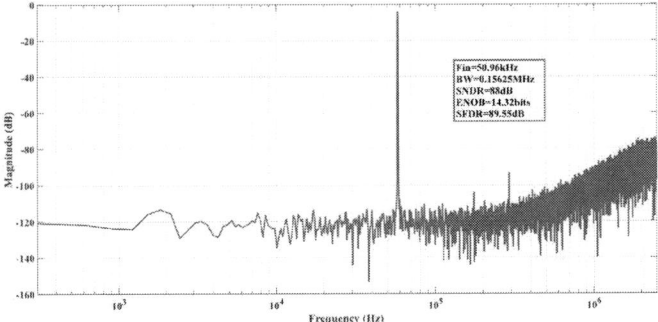

Fig. 11. Transient simulation functional verification.

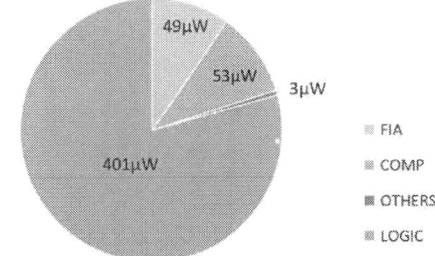

Fig. 12. Layout of the top-level circuit

Parasitic parameters were extracted from the layout, and post-layout simulations were conducted. The input signal was a sinusoidal wave with a frequency of 50.96 kHz and a peak-to-peak amplitude of 0.6 V, under a sampling rate of 5 MS/s and an OSR of 16. After simulation, the digital output of the noise-shaping SAR ADC was converted back to its analog equivalent using an ideal DAC. An FFT analysis of 16,384 data points was then performed, yielding the results shown in Fig. 11. The SNDR was measured as 88 dB, achieving an ENOB of 14.32 bits, with a SFDR of 89.55 dB.

Under a 1.2V supply voltage, the total measured circuit power consumption is 507 μW. The power consumption breakdown of each part of the circuit is shown in Fig. 12. Specifically, digital logic consumes 401 μW, accounting for approximately 79% of the total power. The comparator consumes 53 μW, representing about 10%, while the FIA (feedback integrator and amplifier) consumes 49 μW, also accounting for around 10%. The remaining parts consume 3 μW, contributing to roughly 3% of the total power. Compared to typical OTAs, which generally operate in the milliwatt range, it is evident that the FIA offers a significant advantage in terms of low power consumption.

TABLE I.   PERFORMANCE COMPARISON

	ISSCC 2020[12]	JSSC 2021[13]	IEICE 2024[14]	JSSC 2023[15]	This Work
Process (nm)	40	65	180	90	65
Supply(V)	0.8/1.1	1.1	1.8	1	1.2
Structure	CIFF	EF-CIFF	CIFF	EF	CIFF
Sample rate(MS/s)	10	10	50	10	5
OSR	8	8	8	8	16
NTF order	2	3	2	2	3
BW(MHz)	0.625	0.625	3.125	0.625	0.156
Power(μW)	107	119	728	71	506
SNDR(dB)	83.8	84.8	63.8	73.8	88
ENOB	13.5	13.8	10.31	12	14.32
FOMs(dB)	181.5	182	160.1	173.2	173

979-8-3315-8850-2/25 $31.00 © 2025 IEEE

Table I summarizes the key performance parameters of the proposed ADC and compares them with those of several recently published state-of-the-art noise-shaping SAR ADCs.

IV. CONCLUSION

This paper presents a third-order CIFF noise-shaping SAR ADC. The integrator is composed of a two-stage FIA and feedback capacitors, where the first- and second-stage integrations are performed by two identical integrators, and the third-stage integration is implemented by reusing the second-stage integrator. The design replaces the conventional multi-input comparator with a switched-capacitor voltage summation technique, in which multiple capacitors are connected in series through switches to achieve voltage addition, thereby significantly reducing power consumption. Under a sampling rate of 5 MS/s and an OSR of 16, the proposed ADC achieves an SNDR of 88 dB, an ENOB of 14.32 bits, and an SFDR of 89.55 dB, with a total power consumption of 506 μW.

REFERENCES

[1] SHARMA A, PANDEY L, GARG P, et al. Loop Filter Design Considerations for Noise-Shaping in SAR ADCs[C]//2024 IEEE International Symposium on Circuits and Systems (ISCAS). 2024: 1-5.

[2] ZHUANG H, GUO W, LIU J, et al. A Second-Order Noise-Shaping SAR ADC With Passive Integrator and Tri-Level Voting[J]. IEEE Journal of Solid-State Circuits, 2019, 54(6): 1636-1647.

[3] LIN Y Z, LIN C Y, TSOU S C, et al. 20.2 A 40MHz-BW 320MS/s Passive Noise-Shaping SAR ADC With Passive Signal-Residue Summation in 14nm FinFET[C]//2019 IEEE International Solid-State Circuits Conference - (ISSCC). 2019: 330-332.

[4] GUO Y, JIN J, LIU X, et al. A 60-MS/s 5-MHz BW Noise-Shaping SAR ADC With Integrated Input Buffer Achieving 84.2-dB SNDR and 97.3-dB SFDR Using Dynamic Level-Shifting and ISI-Error Correction[J]. IEEE Journal of Solid-State Circuits, 2023, 58(2): 474-485.

[5] OBATA K, MATSUKAWA K, MIKI T, et al. A 97.99 dB SNDR, 2 kHz BW, 37.1 μW noise-shaping SAR ADC with dynamic element

matching and modulation dither effect[C]//2016 IEEE Symposium on VLSI Circuits (VLSI-Circuits). 2016: 1-2.

[6] SHU Y S, KUO L T, LO T Y. An Oversampling SAR ADC With DAC Mismatch Error Shaping Achieving 105 dB SFDR and 101 dB SNDR Over 1 kHz BW in 55 nm CMOS[J]. IEEE Journal of Solid-State Circuits, 2016, 51(12): 2928-2940.

[7] X. Tang, B. Kasap, L. Shen, X. Yang, W. Shi and N. Sun, "An Energy-Efficient Comparator with Dynamic Floating Inverter Pre-Amplifier," 2019 Symposium on VLSI Circuits, Kyoto, Japan, 2019, pp. C140-C141.

[8] TANG X, SHEN L, KASAP B, et al. An Energy-Efficient Comparator With Dynamic Floating Inverter Amplifier[J]. IEEE Journal of Solid-State Circuits, 2020, 55(4): 1011-1022.

[9] R. S. Ashwin Kumar, "Analysis of Stability, Noise, and Design Guidelines for a Cascaded Floating-Inverter Amplifier," in IEEE Transactions on Circuits and Systems II: Express Briefs, vol. 71, no. 9, pp. 4126-4130, Sept. 2024.

[10] LIU J, LI S, GUO W, et al. A 0.029-mm2 17-fJ/Conversion-Step Third-Order CT \Delta\Sigma ADC With a Single OTA and Second-Order Noise-Shaping SAR Quantizer[J]. IEEE Journal of Solid-State Circuits, 2019, 54(2): 428-440.

[11] P. Wu et al., "A second-order CIFF noise-shaping SAR ADC reusing a dynamic amplifier," IEICE Electron. Express, vol. 21, no. 11, May 2024.

[12] LIU J, LI S, GUO W, et al. A 0.029-mm2 17-fJ/Conversion-Step Third-Order CT \Delta\Sigma ADC With a Single OTA and Second-Order Noise-Shaping SAR Quantizer[J]. IEEE Journal of Solid-State Circuits, 2019, 54(2): 428-440.

[13] WANG T H, WU R, GUPTA V, et al. A 13.8-ENOB Fully Dynamic Third-Order Noise-Shaping SAR ADC in a Single-Amplifier EF-CIFF Structure With Hardware-Reusing kT/C Noise Cancellation[J]. IEEE Journal of Solid-State Circuits, 2021, 56(12): 3668-3680.

[14] SHEN J, ZHU X, SHI C, et al. A 10.31 ENOB 3.125MHz BW fully passive 2nd-order noise-shaping SAR ADC for low cost IoT sensor networks[J]. IEICE Electronics Express, 2024, 21(1): 20230122-20230122.

[15] CHEN C C, HUANG Y H, MARQUEZ J C J S, et al. A 12-ENOB Second-Order Noise-Shaping SAR ADC With PVT-Insensitive Voltage – Time – Voltage Converter[J]. IEEE Journal of Solid-State Circuits, 2023, 58(10): 2897-2906.

2025 The 10th International Conference on Integrated Circuits and Microsystems

Mo₂C-MoS₂ Mixed-Dimensional Bulk Heterostructure-Based Memristor for Artificial Synapses

Libin Liang[2], Ziyao Lu[1], Zhenhua Guo[3], Chunyi Qiu[1], Zhuoling Zhou[1], Changjiu Teng*[1]

[1]*School of Electronic Information Engineering, Foshan University, Foshan, 528000, China*
[2]*School of Computer Science and Artificial Intelligence, Foshan University, Foshan, 528000, China*
[3]*School of Physics and Optoelectronic Engineering, Foshan University, Foshan 528000, China*
Correspondence author: cjteng@fosu.edu.cn

Abstract—To overcome the von Neumann bottleneck, it is imperative to develop advanced neuromorphic computing architectures that emulate the integrated memory and processing capabilities of biological neural systems, for which artificial synaptic memristors offer a promising solution. Therefore, it is urgent to develop advanced neuromorphic computing architectures to simulate the data storage and computing integration of biological neural systems. This study proposes a Mo₂C-MoS₂ mixed dimensional bulk heterostructure (MDBHs) artificial synaptic memristor. MoS₂ microspheres are synthesized on titanium foil by a hydrothermal method, and Mo₂C nanodots are in situ generated on the edge of MoS₂ using a CVD carbonization process to prepare Au/Mo₂C-MoS₂ MDBH/Ti devices. Unlike traditional RRAMs that rely on sharp resistance switches caused by conductive channels (CF), this device exhibits gradual synaptic enhancement and inhibition behaviors under high bias. Its conductive mechanism is direct tunneling (DT) at low bias and Fowler-Nordheim tunneling (FNT) at high bias. Mo₂C nanodots reduce the barrier height and form a tunneling conductive path, significantly improving synaptic performance. The device's operating current is as low as 0.1 μA, which is much lower than existing two-terminal artificial synapses, and exhibits good uniformity and reliability. This study provides a new approach to low-power, high-state artificial synapses through innovative heterostructure design, which has broad application prospects in neuromorphic computing.

Keywords—Artificial Synapse, Memristor, Mixed-Dimensional Bulk Heterostructure(MDBH), Mo₂C-MoS₂, Carbonization, Hydrothermal Synthesis

I. INTRODUCTION

Traditional computing systems are constrained by the von Neumann bottleneck, which involves sequential instruction fetching, decoding, and execution during computation[1-6]. This limitation necessitates the development of advanced computing architectures. Biological neural systems efficiently perform complex tasks by integrating data storage and computation[7,8]. Artificial synapses, fundamental components of neuromorphic computing, emulate biological synapse functionality through continuous synaptic weight modulation, paving the way for efficient brain-inspired computing systems[9]. Two-terminal memristors, which modulate device conductance via controlled charge or flux, are widely proposed as artificial synapses to integrate storage and computation[10,11]. The design of dielectric materials and electrode structures is critical to achieving this goal. For instance, memristors with Pt/SiOxNy:Ag/Pt structures utilize Ag diffusion to mimic ion dynamics in biological synapses[12]. Yan et al.

demonstrated gradual conductance modulation in Ag nano-cluster-doped TiO₂ thin films[13], achieving nanosecond switching times as artificial synapses. Recently, homologous gradient metal-semiconductor heterostructures, such as TaS₂-TaSₓOᵧ-Ta[14] and g-PtS:Pt[15], have been developed to enable highly linear and continuous weight modulation. These compositional gradients increase the number of conductive states and stabilize conductive paths, enhancing synaptic performance.In fact, most of artificial synapses face the difficulties to show unexpected sharp RS effects at large bias, behaving as an RRAM device[16], which results in the failure of efficient imitation of the biological synapse with gradually synaptic weight change. That is probably because of the forming conductive filaments (CF) in the conventional RRAM devices with highly differentiated high resistance state (HRS) and low resistance state (LRS)[17]. In this regard, the discovery of other conductive mechanisms different from CF in the artificial synapses becomes increasingly urgent. Furthermore, next-generation artificial intelligence based on neuromorphic computing demands a greater number of resistive states in individual components[18], a challenge that conventional RRAM struggles to meet.

Compared to state-of-the-art neuromorphic devices, such as ferroelectric synapses and phase-change synapses, which offer high-speed switching but often require complex fabrication or high energy consumption, Mo₂C-MoS₂ MDBHs provide a low-power (0.1 μA) alternative with gradual synaptic modulation, suitable for scalable neuromorphic systems. The urgency for such devices stems from the need for energy-efficient, multi-state synapses to mimic biological neural networks, addressing the limitations of sharp-switching RRAMs.

Here, we propose a method to synthesize MoS₂ microspheres on titanium foil by hydrothermal method, and in situ generate Mo₂C nanoparticles on the edge of MoS₂ microspheres by CVD carbonization process to form Mo₂C-MoS₂ mixed dimensional volume heterostructure (MDBHs). Finally, Cr/Au top electrode was deposited by electron beam evaporation to prepare Au/Mo₂C-MoS₂ MDBH/Ti memristor device.

II. DESIGN

Compared to polymorphic memristors, RRAM has a larger off-to-on ratio and sharp resistance switch process, while we need polymorphic memristors or artificial synapses with more resistive states. Artificial synapses

979-8-3315-8850-2/25 $31.00 © 2025 IEEE   948

based on the conductive filament mechanism exhibit sharp resistance switching phenomena (Figure 1a), which are not reliable in practical applications since a small voltage uncertainty will lead to significant current change. Therefore, it is necessary to design a new stable multiple states memristor, such as introducing an appropriate amount of metal nan-intermediaries in the insulating layer of the MIM sandwich structure (Figure 1b): adjacent nano-intermediaries should provide many conductive paths in addition to the traditional conductive filaments. Clearly, the amount of metal nano-intermediaries should be optimized and dispersed; otherwise, short circuits will occur within the insulator. This article selects hydrothermally synthesized MoS_2 spherical thin films as the insulating layer because they have better RRAM performance than multilayer MoS_2 thin films grown by the CVD .Furthermore, metallic Mo_2C nano-dots are chosen as conductive intermediaries dispersed in the medium because they are easy to form homologous Mo_2C-MoS_2 MDBHs after direct carbonization in MoS_2 spherical thin films.

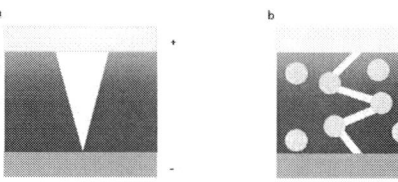

Fig. 1. Diagrams illustrating the mechanisms of (a) conductive filament formation in conventional resistive random-access memory (RRAM) and (b) enhanced conductance via a metallic intermediate layer in a bulk heterostructure.

III. EXPERIMENT

Preparation

Synthesis of MoS_2 microspheres. The MoS_2 microspheres on Ti foil were prepared by using the hydrothermal method.

First, the Ti foil was cleaned with HCl (37wt%) for 30-40 minutes, then ultrasonically treated in DI water and ethanol for 5 min,respectively.Subsequently, $(NH4)_6Mo_7O_{24}\cdot4H_2O$ (0.1766 g) and $CS(NH_2)_2$ (0.484 g) were mixed into DI water (36 mL).

Then, the solution containing Mo and S precursors along as well as the Ti foil was loaded into a Teflon-lined stainless steel autoclave (50 mL), and heated at 180°C for 24 hours.

Finally, the samples were taken out from the autoclave, rinsed with deionized water and ethanol, and then vacuum dried at 60°C for 6-12 hours.The Ti foil with MoS_2 was then loaded into a 1-inch quartz tube furnace, and heated to 750°C within 30 minutes using a mixture of Ar (100 sccm) and H2 (30 sccm). After that, a flow of CH_4 (10 sccm) was introduced into the CVD chamber, and MoS_2/Mo_2C was prepared at 750°C for 60 minutes. Subsequently, the furnace was cooled down to room temperature under Ar (100 sccm) and H_2 (30 sccm).

Device Fabrication and Measurements.

The as-grown hybrid Mo_2C-MoS_2 spherical thin films on Ti substrate as back electrode was directly covered by acustomized hard mask and deposited with (5/50 nm Cr/Au) as a top electrode via e-beam evaporation. The artificial synapses were measured using the semiconductor analyzing system and probe station in vacuum at room temperature (Keithley 4200A- SCS, USA, LakeShore, USA).

IV. MATERIAL CHARACTERIZATIONS

Mo_2C-MoS_2 MDBHs exhibits a hierarchical nanostructure, as shown in Figure 2a. Each flower-like microsphere is composed of numerous MoS_2 nanosheets, and at the same time, a large number of MoS_2 nanoparticles accompany the edges of each MoS_2 nanosheet, thus, Mo_2C-MoS_2 MDBHs can be considered a typical 0D/2D/3D mixed-dimensional heterostructure. To analyze this micro-nano structure, we employed several characterization methods. First, scanning electron microscopy (SEM) reveals that Mo_2C-MoS_2 MDBHs has a flower-like morphology similar to MoS_2 (as shown in Figure 2b), and a large number of microspheres with a narrow diameter distribution of 1.61 ± 0.39 μm are uniformly distributed on the titanium foil. At the same time, the microspheres are composed of many radially arranged molybdenum disulfide micro-sheets. The RS phenomenon is closely related to this special shape in 2D/3D heterostructures. Unlike the MoS_2 thin film, which does not exhibit a significant RS effect in the vertical direction, the physical contact between adjacent spheres dominated by high Schottky barriers caused by a large number of grain boundaries is crucial for electron transfer, which is important for nonlinear memristive effects. Opportunely, these grain boundaries, which have many chemically active sites, are prone to induce chemical reactions.

Therefore, we directly use CVD to carbonize the active sites, forming a large number of MoS_2 nanoparticles in situ at the edges of MoS_2. High-resolution transmission electron microscopy (HRTEM, Figure 2c) images show that Mo_2C nanoparticles mainly grow at the edges of MoS_2 nanosheets, forming a 0D/2D metal-semiconductor heterostructure, as shown in Figure 2a. Because during the CVD process, the chemical transformation from MoS_2 to Mo_2C begins at the MoS_2 edges, where S atoms are more susceptible to attack by hydrogen ions (hydrodesulfurization), followed by carbonization, with CH_x combining with the remaining Mo, thereby converting the MoS_2 edges into Mo_2C nanomaterials, as shown in Figure 2c. HRTEM images display lattice spacings of 0.26 nm and 0.23 nm, corresponding to the (100) plane of MoS_2 and the (002) plane of β-Mo_2C,respectively. More insights into the structure and chemical composition of the Mo_2C-MoS_2 MDBHs sample are obtained from spectral characterization. Raman spectroscopy shows the characteristic peaks of 2H-phase MoS_2 at 379 (A1g) and 405 cm^{-1} (E2g1), as well as the characteristic peaks of β-Mo_2C at 660, 812, and 987 cm^{-1}, confirming the formation of β-Mo_2C on MoS_2 (Figure 2d). Note that we did not observe any D (~1350 cm^{-1}) or G bands (~1590 cm^{-1}) related to carbon materials, indicating that no graphite carbon was formed after CVD. This is understandable because the reaction temperature is relatively low (750°C), and no catalysts were added during the carbonization process, so no graphite carbon materials such as graphene or carbon nanotubes were formed on molybdenum disulfide.

979-8-3315-8850-2/25 $31.00 © 2025 IEEE

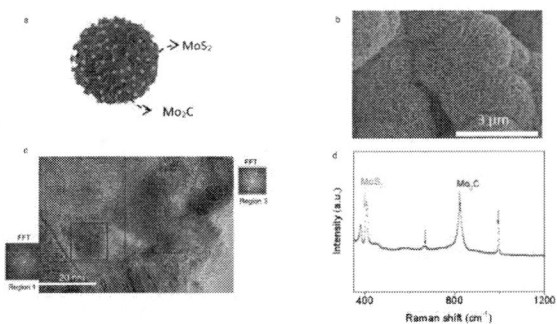

Fig. 2. Schematics and Structure Characterizations of Mo_2C-MoS_2 spheres. (a) Schematics of Mo_2C-MoS_2 spheres. (b) SEM images of Mo_2C-MoS_2 spheres. (c) TEM images of Mo_2C-MoS_2 spheres. (d) Raman spectrum of the Mo_2C-MoS_2 MDBHs

V. RESULTS AND DISCUSSION

We fabricated a series of devices to study the intrinsic memristors based on MoS_2-Mo_2C MDBHs and MoS_2 MDBHs, as shown in Figure 3a.The elements of the upper electrode Au/Cr and the lower electrode Ti are difficult to migrate into the medium, avoiding the formation of metal conductive filaments in this device. The device based on the intrinsic MoS_2 MDBHs exhibits typical RS behavior in the symmetric sweep from -5V to 5V, as shown in Figure 3b. On the graph, a distinct set point appears at 4V during the forward sweep, while a sharp reset occurs at -4.5V. Compared with the intrinsic memristor based on MoS_2 MDBHs, the MoS_2-Mo_2C MDBHs exhibit continuous resistance enhancement and suppression, behaving as artificial synapses, as shown in Figure 3c and d. Under the continuous positive voltage scanning of six cycles from 1.0V to 3.5V, the conductance gradually increases with no significant change in the hysteresis window area. In contrast, under the continuous negative voltage scanning of six cycles from 0.0V to -3.5V, the conductance gradually decreases, and the hysteresis window area decreases.

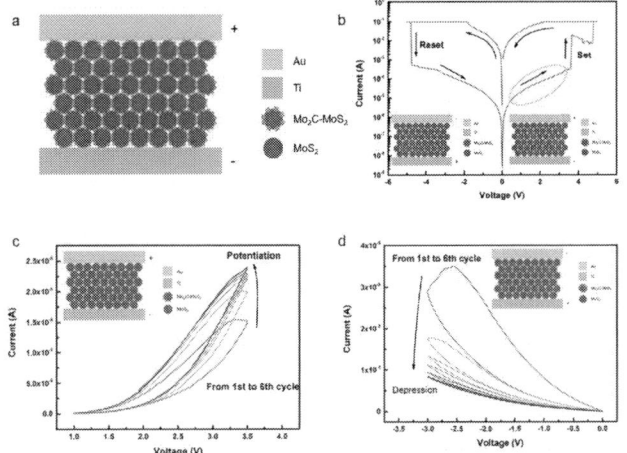

Fig. 3. Schematics and I-V characteristics of devices utilizing Mo_2C-MoS_2 MDBHs and intrinsic MoS_2. (a) Schematic representation of synaptic memristors based on Mo_2C-MoS_2 MDBH thin films. (b) I-V characteristics exhibiting typical resistive switching (RS) behavior in intrinsic MoS_2-based RRAM devices. (c) and (d) I-V characteristics demonstrating synaptic potentiation and depression behaviors in Mo_2C-MoS_2 MDBH-based devices.

Additionally, we investigated the conduction mechanisms of memristors based on intrinsic and hybrid

MDBHs. Initially, we performed fitting analyses on the experimental I-V characteristics of the intrinsic device. In the high-resistance state (HRS), the I-V data were initially fitted using the space-charge-limited current (SCLC) model[19], as described by equation :

$$I \propto \frac{A\mu\varepsilon n}{n_t}(V^2/d^3) \tag{3.1}$$

However, the fit was inadequate at both low and high voltage regions (Figure 4a). Consequently, we employed the thermionic emission (TE) model[20], represented by equation:

$$I \propto AT^2 \exp[\frac{-\left(\phi - q\sqrt{qV/4d\pi\varepsilon}\right)}{kT}] \tag{3.2}$$

to fit the I-V curve, as depicted in Figure 4b. A more consistent linear relationship was observed in the ln(I) versus $V^{0.5}$ plot, suggesting that the conduction mechanism is likely governed by the thermionic emission (TE) model. This indicates that charge injection and accumulation within the dielectric nanospheres may predominantly influence the HRS. The substantial ON/OFF ratio between the HRS and low-resistance state (LRS) is critical for both the reliable retention of stored data and the minimization of misreading errors in memory device applications[21]. Note that the flat MoS_2 film-based device does not exhibit significant resistive switching behavior, which is in agreement with some recently reported findings[22]. Thus, we propose that the distinctive morphology of the nanospheres plays a significant role in the carrier transport process. To elucidate the electrically controlled RS effect observed in our experiments, we developed a generalized model of electron tunneling through polarization potential barriers within the dielectric particle network. Figure 4d presents a schematic illustrating the electric-field-modulated evolution of the conductive pathway. In the nanosphere-assembled network film, physical contacts between adjacent spheres generate numerous grain boundaries and elevated junction barriers.

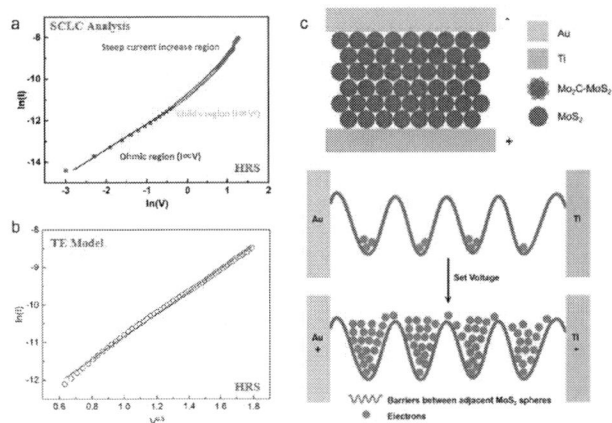

Fig. 4. Mechanism analysis in the RRAM devices based on MoS_2 spherical thin film. (a) Fitted curves with SCLC model. (b) Fitted curves with TE model. (c) Schematics of the conductive mechanism in the RRAM devices based on MoS_2 spherical thin films.

However, the I-V curve of the Mo_2C-MoS_2 MDBHs hybrid device shown in Figure 5a is fitted to $\ln(I/V^2)$ versus the $1/V$ of the I-V region data, as shown in Figure 5b. The fitting results are linear and can be divided into two parts in the figure, indicating that both the Fowler-Nordheim tunneling (FNT) model and the direct tunneling (DT) model can well describe the conductive behavior. Therefore, the conduction mechanisms of the two samples are completely different. The results show that both samples have formed conductive filaments in the LRS, which is consistent with the TEM results. By comparing the conduction mechanisms, it can be expected that electrons can tunnel between the conductive Mo_2C nanoparticle chains. Gradual modulation of conduction may correspond to the change in the interstitial distance driven by the electric field (Figure 5c), which is very similar to other reports[23]. In fact, at this stage, the total resistance (R) of the device can be described according to the equivalent circuit as $R = \Sigma R_{ij} = V/I$. R_{ij} is defined as the tunneling resistance of the corresponding tunnel junction between two Mo_2C metal nanoparticles. If the gap between two Mo_2C nanoparticles can be effectively modulated by using suitable programmed bias, the transition of the tunneling mechanism can be achieved, and the conduction or resistance of the memory cell can be modulated as a biological synapse through the tunneling effect.

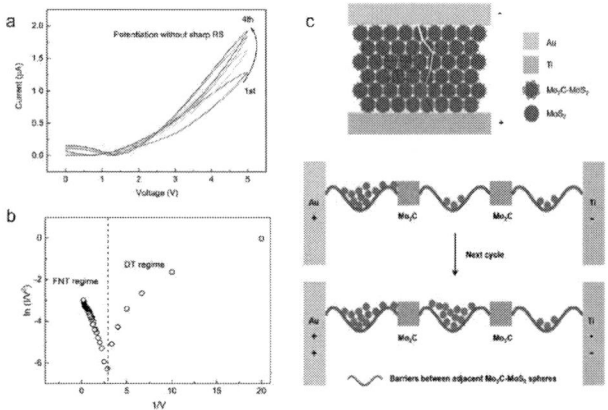

Fig. 5. Mechanism analysis in the synaptic memristors based on Mo_2C-MoS_2 hybrid spherical thin films. (a) I-V potentiation when the bia voltage is up to 5V. (b) FNT and DT conductive mechanism in the fitted curves. (c) Schematics of the mechanism analysis in the synaptic memristors based on Mo_2C-MoS_2 hybrid spherical thin films.

VI. SUMMARY

Mo_2C-MoS_2 MDBH can be directly fabricated into a Au/Mo_2C-MoS_2 MDBH/Ti structure memristive device. Here, we list the following two innovative points.

(1) The hybrid device behaves as an artificial synapse, exhibiting enhancement and suppression even under large biases, unlike typical RRAMs which have sharp switching states; it is worth noting that thermionic emission is the cause of its intrinsic conduction mechanism. In Mo_2C-MoS_2 MDBHs, the fitting results indicate that the hybrid device follows low-bias DT and high-bias FNT, where Mo_2C metallic nanodots dispersed in the bulk heterostructure not only serve as a good auxiliary to reduce the potential barrier height in the system but also contribute to the formation of tunneling conduction modes.

(2) The operating current of these devices is only 0.1 µA, much lower than the operating current of dual-end artificial synapses reported to date. Furthermore, good uniformity and device reliability reflect the potential of this technology. This low-power operation positions Mo_2C-MoS_2 MDBHs as a competitive alternative to ferroelectric and phase-change synapses, addressing the urgent need for energy-efficient neuromorphic computing.

To enhance CMOS compatibility, future iterations could replace Ti foil with Si/SiO2 substrates, as demonstrated in similar hydrothermal syntheses on silicon [24]. Additionally, the CVD carbonization temperature(750°C) can be reduced to <400°C using plasma-enhanced CVD (PECVD) or laser-assisted methods, enabling CMOS-compatible processing. Low-temperature PECVD techniques have been successfully developed for the synthesis of high-quality MoS_2, providing a viable pathway for integrating MoS_2-based heterostructures and enabling subsequent formation of Mo_2C nanodots [25].

REFERENCES

[1] Wali A, Das S. Two‐dimensional memtransistors for non‐von neumann computing: progress and challenges[J]. Advanced Functional Materials, 2024, 34(15): 2308129.

[2] Syed G S, Le Gallo M, Sebastian A. Non von neumann computing concepts[M]//Phase Change Materials-Based Photonic Computing. Elsevier, 2024: 11-35.

[3] Cook J. PIMS: Memristor-Based Processing-in-Memory-and-Storage[R]. Sandia National Lab.(SNL-NM), Albuquerque, NM (United States), 2018.

[4] Yang Z. Leveraging RRAM to Design Efficient Digital Circuits and Systems for Beyond Von Neumann in-Memory Computing[D]. University of Waterloo, 2019.

[5] Kim H W. Overcoming the Limitations of Si-CMOS and Von Neuman Architectures by Adopting Beyond-Si Devices and Near-Memory Computation[D]. , 2024.

[6] Zahedi M, Lebdeh M A, Bengel C, et al. MNEMOSENE: Tile architecture and simulator for memristor-based computation-in-memory[J]. ACM Journal on Emerging Technologies in Computing Systems (JETC), 2022, 18(3): 1-24.

[7] Tang J, Yuan F, Shen X, et al. Bridging biological and artificial neural networks with emerging neuromorphic devices: fundamentals, progress, and challenges[J]. Advanced materials, 2019, 31(49): 1902761.

[8] Wang S, Chen X, Huang X, et al. Neuromorphic engineering for hardware computational acceleration and biomimetic perception motion integration[J]. Advanced Intelligent Systems, 2020, 2(11): 2000124.

[9] Seok H, Lee D, Son S, et al. Beyond von Neumann architecture: Brain‐inspired artificial neuromorphic devices and integrated computing[J]. Advanced Electronic Materials, 2024, 10(8): 2300839.

[10] Chen S, Zhang T, Tappertzhofen S, et al. Electrochemical‐memristor‐based artificial neurons and synapses — fundamentals, applications, and challenges[J]. Advanced materials, 2023, 35(37): 2301924.

[11] Wang R, Yang J Q, Mao J Y, et al. Recent advances of volatile memristors: Devices, mechanisms, and applications[J]. Advanced Intelligent Systems, 2020, 2(9): 2000055.

[12] Wang, Z.; Joshi, S.; Savel'ev, S. E.; Jiang, H.; Midya, R.; Lin, P.; Hu, M.; Ge, N.; Strachan, J. P.; Li, Z.; Wu, Q.; Barnell, M.; Li, G. L.; Xin, H. L.; Williams, R. S.; Xia, Q.; Yang, J. J., Memristors with

diffusive dynamics as synaptic emulators for neuromorphic computing [J]. Nature Materials, 2017, 16 (1), 101-108.

[13] Yan, X. B.; Zhao, J. H.; Liu, S.; Zhou, Z. Y.; Liu, Q.; Chen, J. S.; Liu, X. Y., Memristor with Ag‑Cluster‑Doped TiO2 films as artificial synapse for neuroinspired computing [J]. Advanced Functional Materials, 2018, 28 (1), 1705320.

[14] Wang, J.; Teng, C.; Zhang, Z.; Chen, W.; Tan, J.; Pan, Y.; Zhang, R.; Zhou, H.; Ding, B.; Cheng, H. M.; Liu, B., A Scalable Artificial Neuron Based on Ultrathin Two-Dimensional Titanium Oxide [J]. ACS Nano 2021, 15 (9), 15123-15131.

[15] Tang, L.; Teng, C.; Xu, R.; Zhang, Z.; Khan, U.; Zhang, R.; Luo, Y.; Nong, H.; Liu, B.; Cheng, H. M., Controlled Growth of Wafer-Scale Transition Metal Dichalcogenides with a Vertical Composition Gradient for Artificial Synapses with High Linearity [J]. ACS Nano, 2022, 16 (8), 12318-12327.

[16] Shen Z, Zhao C, Qi Y, et al. Advances of RRAM devices: Resistive switching mechanisms, materials and bionic synaptic application[J]. Nanomaterials, 2020, 10(8): 1437.

[17] Guo T, Elshekh H, Yu Z, et al. Effect of crystalline state on conductive filaments forming process in resistive switching memory devices[J]. Materials Today Communications, 2019, 20: 100540.

[18] Udaya Mohanan K. Resistive switching devices for neuromorphic computing: from foundations to chip level innovations[J]. Nanomaterials, 2024, 14(6): 527.

[19] Lim E W, Ismail R. Conduction mechanism of valence change resistive switching memory: A survey[J]. Electronics, 2015, 4(3): 586-613.

[20] Herring C, Nichols M H. Thermionic emission[J]. Reviews of modern physics, 1949, 21(2): 185.

[21] Fadeev A V, Rudenko K V. To the issue of the memristor's HRS and LRS states degradation and data retention time[J]. Russian Microelectronics, 2021, 50(5): 311-325.

[22] Wu X, Ge R, Kim M, et al. Atomristors: Non-volatile resistance switching in 2D monolayers[C]//2020 Pan Pacific Microelectronics Symposium (Pan Pacific). IEEE, 2020: 1-6.

[23] Yue X, Fan J, Xiang Q. Internal electric field on steering charge migration: modulations, determinations and energy‑related applications[J]. Advanced Functional Materials, 2022, 32(12): 2110258.

[24] Sohail Ahmed, Xiang Ding, Xueze Chu, Mengyao Li, Dewei Chu, Tianyi Ma, Tom Wu, Ajayan Vinu, and Jiabao Yi ACS Applied Materials & Interfaces 2020 12 (16), 18850-18858,DOI: 10.1021/acsami.0c01222

[25] Beaudette CA, Held JT, Mkhoyan KA, Kortshagen UR. Nonthermal Plasma-Enhanced Chemical Vapor Deposition of Two-Dimensional Molybdenum Disulfide. ACS Omega. 2020 Aug 20;5(34):21853-21861. doi: 10.1021/acsomega.0c02947. PMID: 32905341; PMCID: PMC7469405.

2025 The 10th International Conference on Integrated Circuits and Microsystems

Artificial Neural Network-Based Compact Model for Carbon Nanotube Field-Effect Transistors

Zhi Zhang[1], Honggang Liu[2,*]

[1]School of Integrated Circuits, Beijing University of Posts and Telecommunications, Beijing, 100876, China
[2]Center for Carbon-based Electronics and School of Electronics, Peking University, Beijing, 100871, China
zhang_zhi@bupt.edu.cn, liuhonggang@pku.edu.cn
*Corresponding author

Abstract—**This study presents an artificial neural network (ANN)-based approach for modeling the electrical characteristics of aligned-array carbon nanotube field-effect transistors (CNTFETs), addressing the limitations of conventional physical models, such as high computational complexity and limited predictive accuracy. To mitigate challenges in capturing the drain current's wide dynamic range, we introduce a novel preprocessing technique that ensures smooth transitions across subthreshold and high-current regimes. Experimental results demonstrate that the proposed ANN model outperforms conventional methods, achieving an order of magnitude improvement in modeling accuracy for drain current, transconductance, and output conductance. The model is implemented in Verilog-A, enabling seamless integration into SPICE simulation platforms. This work offers a high-precision, adaptable modeling framework for CNTFET design in the post-Moore era.**

Keywords—*Artificial neural network (ANN), CNTFET, compact model, SPICE*

I. INTRODUCTION

Silicon-based transistors have long been the cornerstone of integrated circuits. However, as feature sizes approach physical limits, challenges such as short-channel effects, increased leakage current, and fabrication complexities have significantly impeded performance and power efficiency improvements [1]. Consequently, academia and industry are actively exploring novel materials and device architectures to advance semiconductor technology in the post-Moore era. Carbon nanotubes (CNTs), an emerging semiconductor material, offer a quasi-one-dimensional structure and exceptional electrical properties, including high current-carrying capacity and superior carrier mobility. These attributes position CNTs as a leading candidate for next-generation transistor technologies [2]. In 2017, Peng et al. achieved a breakthrough by developing a high-performance P-type CNTFET with a 5-nm gate length, demonstrating significantly higher current density and a near-ideal subthreshold slope of 73 mV/decade at a 0.4 V supply voltage, significantly outperforming silicon transistors from the same period [3].

Despite the significant potential of CNTFET demonstrated in experimental and theoretical studies, their compact modeling remains challenging. Conventional CNTFET models rely heavily on assumptions and approximations, often requiring complex numerical integrations and incurring substantial computational overhead [4]. The field of CNTFET modeling currently lacks unified standards, and the rapid evolution of device architectures, materials, and fabrication technologies makes it difficult for traditional models to keep pace with characterizing the performance of emerging device structures. Recent advancements in artificial intelligence, coupled with high-performance computing and efficient software frameworks such as PyTorch and TensorFlow, have made neural network-based approaches increasingly viable for compact modeling [5], [6]. Artificial neural networks (ANN) offer distinct advantages, as they can directly learn device characteristics from experimental data, bypassing the need to explicitly model intricate physical processes inherent in traditional approaches [6]. Consequently, exploring ANN-based modeling for CNTFETs holds significant promise for advancing device design in the post-Moore era.

This study introduces an ANN-based modeling framework for CNTFETs, leveraging the PyTorch framework to enhance computational efficiency and model scalability. The proposed approach incorporates a novel current data preprocessing technique that significantly improves the modeling accuracy of critical electrical characteristics, including drain current, transconductance (G_m), and output conductance (G_{ds}). Systematic comparisons with the conventional CCAM [4] model reveal the superior predictive performance and reduced computational overhead of our method. Furthermore, the ANN model has been seamlessly integrated into Verilog-A, ensuring full compatibility with SPICE siamulation platforms, such as Keysight Advanced Design System (ADS). This PyTorch-based ANN framework offers a robust and versatile solution for advancing CNTFET compact modeling in the post-Moore era.

II. DEVICE STRUCTURE AND ANN APPROACH

A. Device Structure

Aligned semiconducting carbon nanotube arrays are ideal for high-performance electronic devices, leveraging their unique physical and electrical properties to offer significant advantages in nano electronic applications [2]. Fig. 1 illustrates the structure of an aligned array CNTFET. In this study, we select channel length (L_{ch}), the number of CNTs (T_{cnt}), and chirality (n, 0) of semiconducting CNTs as key input parameters, as they critically influence CNTFET electrical performance. First, channel length directly modulates short-channel effects and carrier transport properties , with shorter lengths enhancing switching speed. Second, the number of CNTs determines current-carrying capacity and on-state resistance, where higher CNT density reduces contact resistance and improves on-state current. Finally, chirality governs the band structure and electron transport characteristics, significantly affecting threshold voltage and transconductance. This parameter selection ensures comprehensive evaluation of the device's electrical behavior while maintaining computational efficiency in the modeling process.

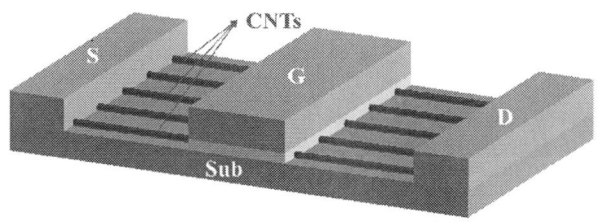

Fig. 1. The structure of aligned array CNTFET.

B. ANN Model

As illustrated in Fig. 2, the ANN architecture comprises five fully connected hidden layers, each with 16 neurons. Hidden layers employ tanh activation functions to enable nonlinear feature extraction, while the output layer uses a logistic activation function to constrain predictions within a bounded range, ensuring stable and interpretable outputs. This configuration was optimized to balance model complexity and generalization. The ANN inputs include channel length (L_{ch}), number of carbon nanotubes (T_{cnt}), chirality (n, 0) of semiconducting CNTs, and bias voltages, namely gate voltage (V_{gs}) and drain voltage (V_{ds}), as detailed in Table I. To enhance numerical stability and accelerate training convergence, all input parameters are logarithmically transformed for normalization, scaling them to a similar order of magnitude. The ANN outputs a preprocessed feature, $I'_{ds} = f(I_{ds})$, derived from the drain current (I_{ds}), which is subsequently converted back to actual current values via an inverse preprocessing function, $I_{ds} = f^{-1}(I'_{ds})$. This preprocessing strategy normalizes the target distribution, improving the model's ability to resolve subtle current variations, particularly in the subthreshold and saturation regions.

TABLE I. SELECTION OF CNTFET DESIGN PARAMETERS.

Parameters	Range[min : step : max]
Channel Length, L_{ch} (nm)	[24 : 8 : 56]
Number of Carbon Nanotubes, T_{cnt}/CNT	[10 : 10 : 30]
Chirality of Carbon Nanotubes, n	[16 : 3 : 22]
Gate voltage, V_{gs}(V)	[0 : 0.05 : 0.8]
Drain voltage, V_{ds}(V)	[0 : 0.05 : 0.8]

The experimental dataset, derived from Stanford University's compact model [7], [8], comprises 13,005 data points for N-type CNTFETs and 13,005 data points for P-type CNTFETs, encompassing diverse device geometries, bias conditions, and material properties to enhance model generalizability.

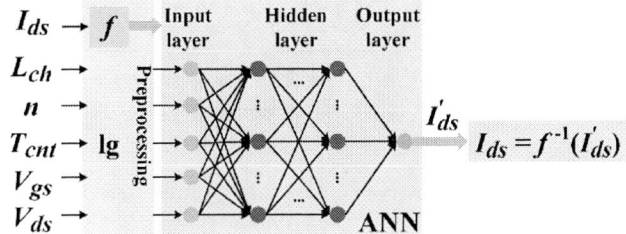

Fig. 2. ANN modeling method.

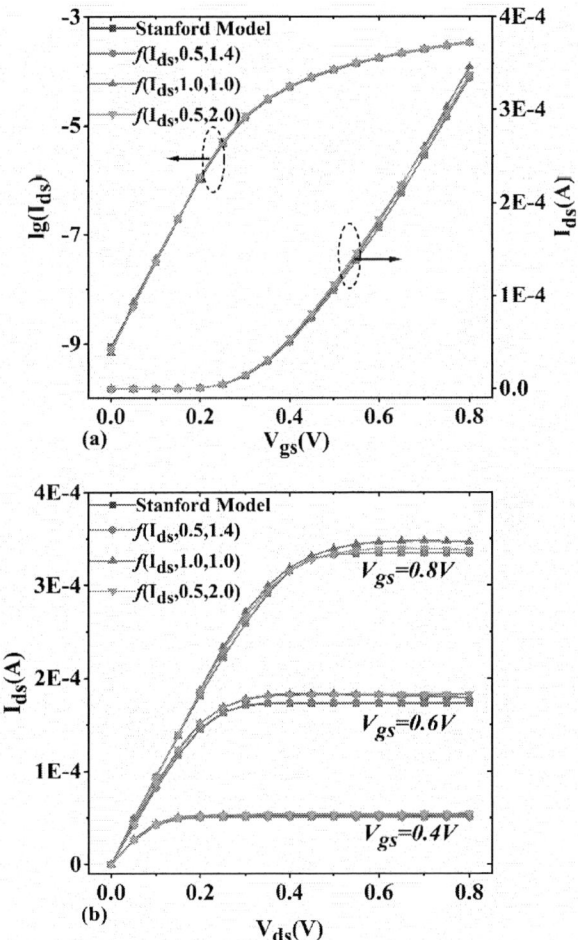

Fig. 3. Comparison of ANN model fitting results with Stanford CNTFET compact model data under different a and b parameters: (a) transfer characteristic curves. (b) output characteristic curves.

III. PREPROCESSING METHOD FOR DRAIN CURRENT

The drain current of CNTFET spans a wide dynamic range (from 10^{-11} A to 10^{-4} A), making ANN fitting challenging due to the extensive range. To address this limitation, we developed a novel, simple, and efficient preprocessing function, as shown in Equation (1), that enables accurate modeling of I_{ds} across its entire dynamic range.

$$I'_{ds} = f(I_{ds}, a, b) = \left(-\frac{1}{a \cdot \log_{10} I_{ds}}\right)^b \quad (1)$$

The parameters a and b serve as tunable hyperparameters. This preprocessing approach offers four key advantages: First, it effectively overcomes the limitations of linear preprocessing where small I_{ds} values become negligible to neural networks, while preventing excessive compression of high I_{ds} values that occurs with conventional logarithmic transformation [6]. Second, parameter a determines the scaling magnitude of the logarithmic term, while b adjusts the nonlinearity degree, enabling optimized fitting accuracy tailored to specific device characteristics. This achieves balanced representation across both subthreshold and strong-inversion regions. Third, the function provides a smooth, monotonic transformation that ensures numerical stability during simulation. Fourth, the computationally efficient formulation permits analytical inversion, enabling seamless integration in Verilog-A implementations and significantly boosting simulation efficiency.

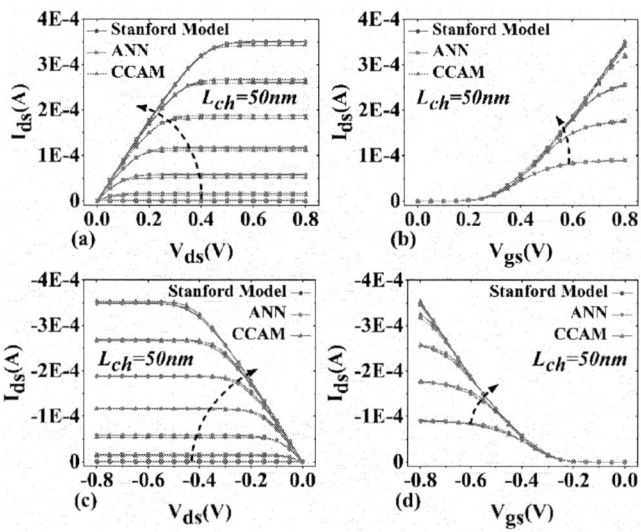

Fig. 4. Under the conditions of L_{ch}=50nm, T_{cnt}=20, and n=19, a comprehensive comparison is conducted between the ANN model and CCAM simulations against the reference data from the Stanford CNTFET compact model. The comparative analysis includes: (a) N-type CNTFET output characteristic (V_{gs} ranging from 0.1 to 0.8 V in 0.1 V steps); (b) N-type CNTFET transfer characteristic (V_{ds} = 0.1, 0.2, 0.3, 0.5, 0.8 V); (c) P-type CNTFET output characteristic (V_{gs} ranging from -0.1 to -0.8 V in 0.1 V steps); (d) P-type CNTFET transfer characteristic (V_{ds} = -0.1, -0.2, -0.3, -0.5, -0.8 V).

The experimental results illustrate the effects of preprocessing across various (a, b) combinations, with three representative cases highlighted in Fig. 3. All parameter sets exhibit excellent fitting accuracy on logarithmic scales, as shown in Fig. 3(a). However, analysis on linear scales (Fig. 3(a) and Fig. 3(b)) reveals distinct compression characteristics across the I_{ds} dynamic range, influenced by different parameter choices. Through systematic evaluation, we determine the

optimal combination (a = 0.5, b = 1.4) for model training, which provides superior modeling accuracy across both logarithmic and linear measurement scales.

A. Result and Analysis

To objectively demonstrate the ANN model's predictive capability, we maintain strict separation between training and evaluation datasets, ensuring complete independence of all test data. This rigorous validation approach confirms the model's ability to accurately reproduce CNTFET electrical characteristics while demonstrating robust generalization performance. To further evaluate the performance of the ANN model, we compare it with the CCAM [4] constructed using conventional modeling approaches.

B. Results of Output and Transfer Curves

As demonstrated in Fig. 4 and Fig. 5, we systematically compared the output and transfer characteristics of N- and P-type CNTFET devices with varying device dimensions simulated by both ANN and CCAM models against the Stanford University CNTFET compact model reference data. The results demonstrate that the ANN model achieves excellent agreement with the reference data across all bias conditions for both N- and P-type CNTFETs, significantly outperforming the CCAM approach.

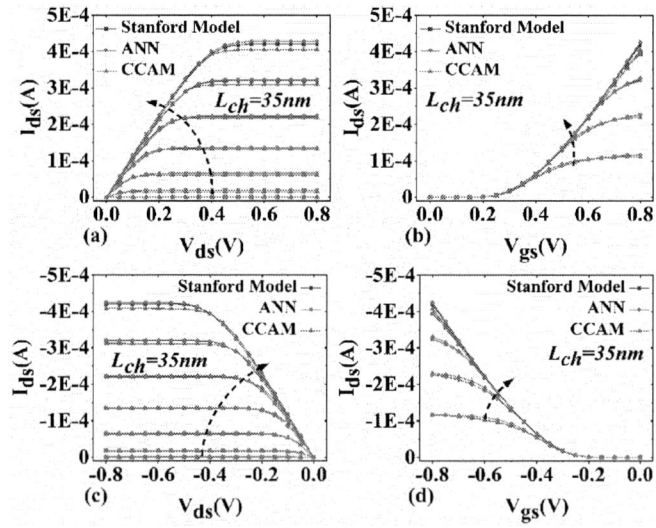

Fig. 5. Under the conditions of L_{ch}=35nm, T_{cnt}=25, and n=19, a comprehensive comparison is conducted between the ANN model and CCAM simulations against the reference data from the Stanford CNTFET compact model. The comparative analysis includes: (a) N-type CNTFET output characteristic (V_{gs} ranging from 0.1 to 0.8 V in 0.1 V steps); (b) N-type CNTFET transfer characteristic (V_{ds} = 0.1, 0.2, 0.3, 0.5, 0.8 V); (c) P-type CNTFET output characteristic (V_{gs} ranging from -0.1 to -0.8 V in 0.1 V steps); (d) P-type CNTFET transfer characteristic (V_{ds} = -0.1, -0.2, -0.3, -0.5, -0.8 V).

Fig. 6 compares the relative errors in the CNTFET drain current simulation by the ANN model and CCAM. As shown in Fig. 6(a), for L_{ch}=50 nm, T_{cnt}=20, and n=19, the ANN model achieves a mean relative error of merely 0.54%, representing an 18-fold improvement compared to CCAM's 9.94% error. Similarly, under conditions of L_{ch}=35 nm, T_{cnt}=25, and n=19 (Fig. 6b), the ANN model maintains superior accuracy with a

979-8-3315-8850-2/25 $31.00 © 2025 IEEE

0.72% mean error, representing a nearly 10-fold improvement over CCAM's 7.04% error. For L_{ch}=50 nm, T_{cnt}=20, and n=19, the maximum relative error of the ANN model is 3.69%, compared to 33.75% for CCAM. Similarly, for L_{ch}=35 nm, T_{cnt}=25, and n=19, the ANN model shows a maximum relative error of 2.94%, versus 31.43% for CCAM. These results clearly indicate that the ANN model improves simulation accuracy by approximately an order of magnitude compared to the conventional CCAM approach. The findings provide compelling evidence that the neural network-based modeling method significantly outperforms traditional approaches in characterizing CNTFET electrical properties.

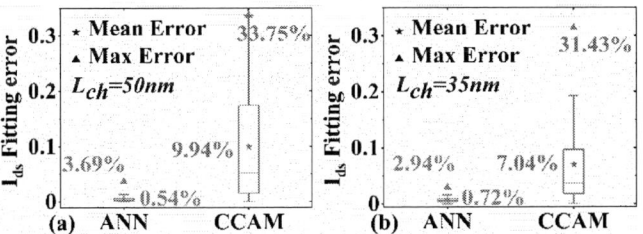

Fig. 6. Relative errors between the ANN model and CCAM under different parameter sets: (a) L_{ch}=50 nm, T_{cnt}=20, and n=19; (b) L_{ch}=35 nm, T_{cnt}=25, and n=19.

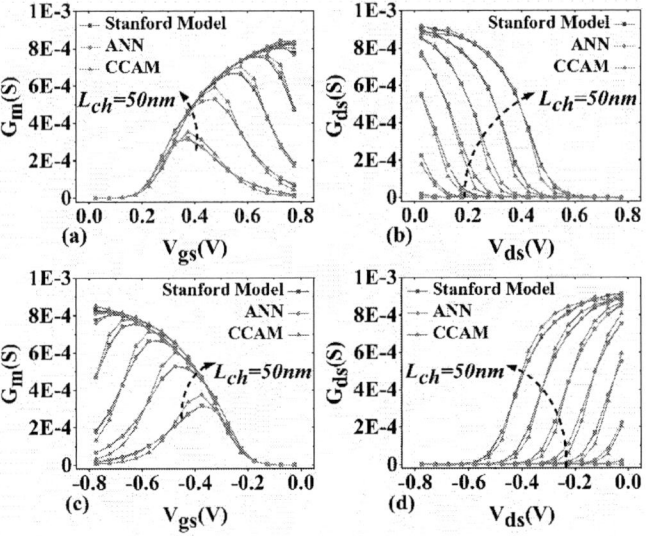

Fig. 7. Under the conditions of L_{ch}=50nm, T_{cnt}=20, and n=19, a comprehensive comparison is conducted between the ANN model and CCAM simulations against the reference data from the Stanford CNTFET compact model. The comparative analysis includes: (a) N-type CNTFET transconductance (G_m) comparison (V_{ds} ranging from 0.1 to 0.8 V in 0.1 V steps); (b) N-type CNTFET output conductance (G_{ds}) comparison (V_{gs} ranging from 0.1 to 0.8 V in 0.1 V steps); (c) P-type CNTFET transconductance (G_m) comparison (V_{ds} ranging from -0.1 to -0.8 V in 0.1 V steps); (d) N-type CNTFET output conductance (G_{ds}) comparison (V_{gs} ranging from -0.1 to -0.8 V in 0.1 V steps).

Transconductance (G_m) and Output Conductance (G_{ds}) are key performance metrics for evaluating CNTFET behavior, reflecting device gain, output conductance, and the smoothness of transfer and output characteristics. As illustrated in Fig. 7 and Fig. 8, the ANN model achieves significantly higher

accuracy in modeling the G_m and G_{ds} characteristics of CNTFETs compared to the conventional CCAM model. These results provide compelling evidence that the ANN-based modeling approach outperforms conventional methods in accurately capturing the key electrical characteristics of CNTFET.

The ANN model accurately captures derivative characteristics, such as peak G_m values in subthreshold and saturation regions and G_{ds} behavior across linear and saturation regions, demonstrating its robust modeling capability. This precision in simulating complex electrical properties enables reliable transistor performance simulations, enhancing the design and optimization of RF analog circuits, such as amplifiers [9], [10]. Consequently, the model improves circuit performance and reliability in applications requiring precise current control and small-signal amplification.

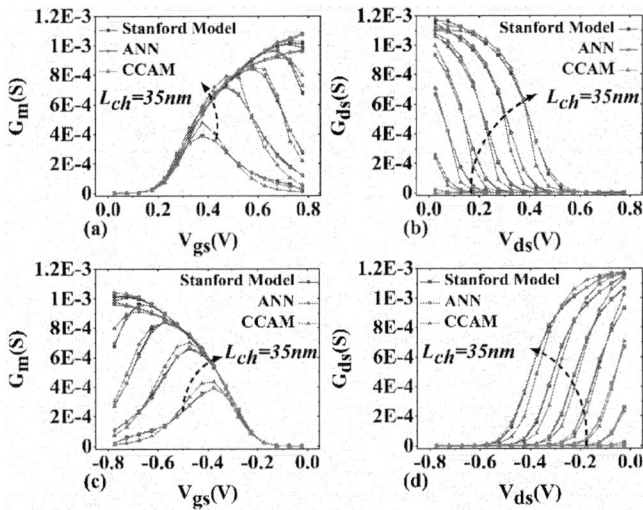

Fig. 8. Under the conditions of L_{ch}=35nm, T_{cnt}=25, and n=19, a comprehensive comparison is conducted between the ANN model and CCAM simulations against the reference data from the Stanford CNTFET compact model. The comparative analysis includes: (a) N-type CNTFET transconductance (G_m) comparison (V_{ds} ranging from 0.1 to 0.8 V in 0.1 V steps); (b) N-type CNTFET output conductance (G_{ds}) comparison (V_{gs} ranging from 0.1 to 0.8 V in 0.1 V steps); (c) P-type CNTFET transconductance (G_m) comparison (V_{ds} ranging from -0.1 to -0.8 V in 0.1 V steps); (d) N-type CNTFET output conductance (G_{ds}) comparison (V_{gs} ranging from -0.1 to -0.8 V in 0.1 V steps).

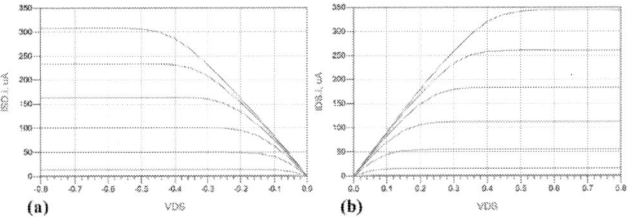

Fig. 9. Conversion of the ANN model into Verilog-A and implementation in ADS for circuit simulation: (a) P-type CNTFET output characteristic curve; (b) N-type CNTFET output characteristic curve.

C. Circuit Simulation

To enable efficient circuit simulation within the SPICE framework, we convert the ANN-based model into a Verilog-A model, which we validate using Advanced Design System (ADS). Fig. 9 illustrates the output characteristics of both P- and N-type CNTFET devices, d proving that the model can be used in SPICE tools. This conversion ensures seamless integration of the ANN-based model into SPICE simulations, leveraging the ANN's strengths in precisely modeling complex device electrical properties. As a result, the approach significantly enhances the precision and computational efficiency of the simulations, enabling robust support for the circuit design, performance optimization, and development of next-generation transistor devices for advanced integrated circuits.

IV. CONCLUSION

In this study, we propose an ANN-based approach for modeling the electrical characteristics of CNTFETs, to overcome the limitations of conventional physics-based methods, such as high computational cost and limited prediction accuracy. Furthermore, a preprocessing technique is introduced to enable the ANN to effectively capture the wide dynamic range of the drain current, ensuring smooth transitions across operating regimes via parameter optimization. The experimental results demonstrate that the proposed ANN model significantly outperforms traditional methods in characterizing CNTFET electrical properties, achieving an order of magnitude improvement in modeling accuracy. Notably, the ANN exhibits superior performance in predicting key device metrics, including G_m and G_{ds}, unequivocally validating its capability to capture complex device behaviors. By implementing the model in Verilog-A, we ensure compatibility with SPICE-based simulation environments. In future work, we will improve this ANN model by incorporating both TCAD simulation data and experimental CNTFET measurements, while considering additional device parameters to enhance the model's comprehensiveness and applicability.

REFERENCES

[1] Pandey, "Recent Trends in Novel Semiconductor Devices," Silicon, vol. 14, no. 15, pp. 9211–9222, Oct. 2022.

[2] Liu Y.-F., Zhang Z.-Y., and Key Laboratory for the Physics and Chemistry of Nanodevices, Center for Carbon-based Electronics, Peking University, Beijing 100871, China, "Carbon based electronic technology in post-Moore era: progress, applications and challenges," Acta Phys. Sin., vol. 71, no. 6, p. 068503, 2022.

[3] Qiu, Z. Zhang, M. Xiao, Y. Yang, D. Zhong, and L.-M. Peng, "Scaling carbon nanotube complementary transistors to 5-nm gate lengths," Science, vol. 355, no. 6322, pp. 271–276, Jan. 2017.

[4] M. Schroter, M. Haferlach, A. Pacheco-Sanchez, S. Mothes, P. Sakalas, and M. Claus, "A Semiphysical Large-Signal Compact Carbon Nanotube FET Model for Analog RF Applications," IEEE Trans. Electron Devices, vol. 62, no. 1, pp. 52–60, Jan. 2015.

[5] S. Huang and L. Wang, "MOSFET Physics-Based Compact Model Mass-Produced: An Artificial Neural Network Approach," Micromachines, vol. 14, no. 2, p. 386, Feb. 2023.

[6] J. Wei, H. Wang, T. Zhao, Y.-L. Jiang, and J. Wan, "A New Compact MOSFET Model Based on Artificial Neural Network With Unique Data Preprocessing and Sampling Techniques," IEEE Trans. Comput.-Aided Des. Integr. Circuits Syst., vol. 42, no. 4, pp. 1250–1254, Apr. 2023.

[7] J. Deng and H.-S. P. Wong, "A Compact SPICE Model for Carbon-Nanotube Field-Effect Transistors Including Nonidealities and Its Application—Part I: Model of the Intrinsic Channel Region," IEEE Trans. Electron Devices, vol. 54, no. 12, pp. 3186–3194, 2007.

[8] J. Deng and H.-S. P. Wong, "A Compact SPICE Model for Carbon-Nanotube Field-Effect Transistors Including Nonidealities and Its Application—Part II: Full Device Model and Circuit Performance Benchmarking," IEEE Trans. Electron Devices, vol. 54, no. 12, pp. 3195–3205, 2007.

[9] ECE Dept., Texas Tech University, Lubbock, TX, USA, J. Mayeda, C. Sweeney, D. Y. C. Lie, and J. Lopez, "Broadband High-Efficiency Millimeter-Wave Power Amplifiers in 22-nm CMOS FD-SOI with Fixed and Adaptive Biasing," Int. J. Electr. Electron. Eng. Telecommun., pp. 385–391, 2022.

[10] School of Electrical and Electronic Engineering, Universiti Sains Malaysia, 14300 Nibong Tebal, Penang, Malaysia, M. A. Mubin, and A. Marzuki, "A Low-Noise Amplifier Utilizing Current-Reuse Technique and Active Shunt Feedback for MedRadio Band Applications," Int. J. Electr. Electron. Eng. Telecommun., pp. 306–316, 2020.

2025 The 10th International Conference on Integrated Circuits and Microsystems

Electronic Properties of Violet Phosphorus Devices

Chunyi Qiu[1], Ziyao Lu[1], Yiting Wei[2], Jiaxian Sun[1], Shilong Zhao[1*]

[1]School of Electronic Information Engineering, Foshan University, Foshan, China
[2]School of Computer Science and Artificial Intelligence, Foshan University, Foshan, China
19866724362@163.com, shilongzhao@fosu.edu.cn

Abstract—**Violet phosphorus (VP), a layered allotrope of elemental phosphorus, possesses a unique monoclinic structure and intrinsic anisotropy, making it a promising candidate for future electronic and optoelectronic devices. In this study, we report the synthesis of high-quality VP via chemical vapor transport and the fabrication of few-layer VP-based field-effect transistors (FETs) using pre-defined electrodes. The devices demonstrate typical n-type behavior with strong gate modulation, stable transport characteristics, and a high on/off current ratio of ~10^4–10^5. The electrical performance is stable across multiple gate voltage sweeps, although a notable hysteresis is observed, attributable to interface traps and residual carriers. Our results establish VP as a viable wide-bandgap N-type semiconductors with potential application in high-performance memristors.**

Keywords—Two-dimensional materials, violet phosphorus, FET, transport properties

I. INTRODUCTION

The Moore's law has fueled the development of increasingly compact and powerful electronic devices, from personal smartphones to artificial intelligence systems. However, as transistors are scaled down to nanometer dimensions, silicon-based metal-oxide-semiconductor field-effect transistors (MOSFETs) encounter significant physical limitations, such as increased leakage currents and short channel effects, which hinder further miniaturization[1]. To overcome these challenges, two-dimensional (2D) materials have attracted significant attention. These materials, only a few atoms thick and inherently free from short channel effects, offer promising pathways to extend Moore's Law by enabling high-performance, next-generation nanoelectronics[2]. Among various 2D semiconductors, violet phosphorus (VP), a less-explored allotrope of elemental phosphorus, has recently emerged as a promising candidate in the family of 2D layered semiconductors. Unlike its more widely studied counterparts, VP exhibits a unique monoclinic lattice structure and complex vibrational modes, reflecting a low-symmetry crystalline nature[3].

Theoretical and experimental studies have suggested that violet phosphorus possesses a wide bandgap and high carrier mobility[4], making it highly desirable for applications in next-generation nanoelectronics and optoelectronics[5]. In particular, its intrinsic n-type behavior and sensitivity to external electrostatic modulation open pathways for its use in field-effect transistors (FETs), photodetectors, and other functional semiconductor devices. Moreover, the strong layer- and size-dependent photoluminescence response of VP further underlines its potential as a platform material for tunable optoelectronic applications. Zhang *et al*[6] reported the synthesis and exfoliation of VP single crystals and determined the structure of VP single crystals using HRTEM and X-ray diffraction. Their study attracts the researchers' attention to the novel properties of few-layer VP. A. G. Ricciardulli and coworkers reported a p-type semiconducting properties of few-layer VP flakes fabricated by liquid exfoliation[7]. Despite its favorable properties, the electronic transport characteristics of mechanically exfoliated few-layer VP remain largely unexplored. A lack of experimental data has hindered the assessment of its suitability for practical device integration.

In this work, we address this gap by fabricating and characterizing few-layer VP-based field-effect transistors. High-quality VP crystals were synthesized by chemical vapor transfer (CVT) method. Few-layer VP flakes were mechanically exfoliated and subsequentially integrated into FET structure via dry transfer method under a microscope. By systematically investigating the gate-dependent transport behavior, we reveal the intrinsic n-type conduction and reproducible hysteresis behavior of VP devices. These results not only provide foundational insight into VP's electronic properties but also highlight its viability as a building block for future 2D electronic systems.

II. SYNTHESIS AND CHARACTERIZATION OF VIOLET PHOSPHORUS

Fig. 1 illustrates the general fabrication process of a few-layer VP FET. Few-layer VP flakes were mechanically exfoliated on a silicon substrate with 285 nm thick SiO_2, followed by identification under an optical microscope. Standard E-beam lithography defines Au/Cr electrode patterns, followed by metal deposition. After the removal of poly(methyl methacrylate), the polymer residue on electrodes were cleaned by further atomic force microscope (AFM) tips using a contact mode. Few-layer VP flakes were then transferred to the surface of electrodes by a poly(bisphenol a carbonate) stamp. A thin hBN flake was used to encapsulate few-layer VP to prevent air-induced degradation. Few-layer VP devices were further annealed under vacuum to improve contacts. Semiconductor parameter analyzers were used to perform electrical measurements of the as fabricated few-layer VP FETs.

979-8-3315-8850-2/25 $31.00 © 2025 IEEE

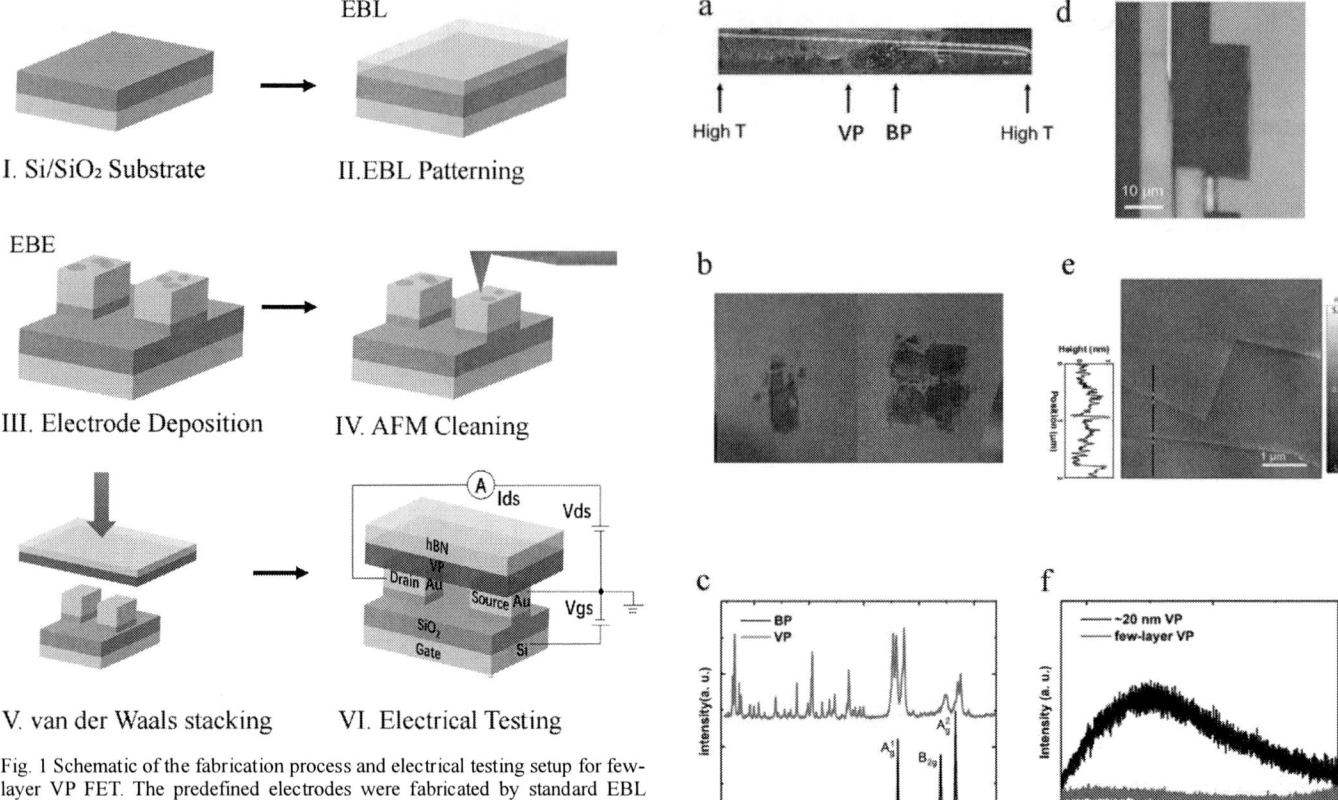

I. Si/SiO₂ Substrate — II. EBL Patterning — III. Electrode Deposition — IV. AFM Cleaning — V. van der Waals stacking — VI. Electrical Testing

Fig. 1 Schematic of the fabrication process and electrical testing setup for few-layer VP FET. The predefined electrodes were fabricated by standard EBL process followed by metal deposition. The polymer residues were cleaned by AFM tips in contact mode. The hBN/VP stack were assembled by dry transfer method and released to the predefined electrodes to form VP FETs.

Fig. 2 a, VP and BP crystals synthesized via CVT in a sealed quartz ampoule. b, Optical images of exfoliated VP (reddish) and BP (greenish) on tape. c, Raman spectra distinguishing VP and BP crystals. d, The optical image of few-layer VP. e, Tapping-mode AFM image showing ~1.1 nm thickness per VP layer. f, Photoluminescence spectra of ~20 nm thick and few-layer VP flakes.

High quality VP crystals are desirable for the exfoliation of few-layer VP flakes. We grew the VP single crystals using a chemical vapor transport (CVT) method, following the procedure reported in the literature[8]. Briefly, A mixture of red phosphorus, SnI₄, and Sn powder in a weight ratio of 50:1:2 was sealed in a quartz ampoule. The ampoule was gradually heated to 1000 K and held at that temperature for 50 hours, then slowly cooled to 500 K at a rate of 1 K/h, followed by a naturally cooling down to room temperature, yielding VP crystals. Compared to traditional methods, this study improved crystal quality and yield by optimizing the temperature gradient and transport agent concentration.

Fig. 2a shows a photograph of as-grown VP crystals in a vacuum-sealed quartz ampoule, together with co-produced black phosphorus (BP) crystals. The VP crystals collected from the lower-temperature zone of the ampoule, are readily distinguished from BP crystals due to their distinct properties and appearances: VP exhibited reddish, rectangular crystals, whereas BP forms black, irregular long-belt shapes. The distinct appearances make it easy to collect VP and BP crystals separately for further studies. Both VP and BP have layered structure and can be exfoliated via Scoth tape method. Fig. 2b presents exfoliated bulk VP and BP crystals attached to Scotch tape, which clearly demonstrates the difference between VP and BP flakes. The few-layer VP and BP crystal were mechanically exfoliated to SiO₂/Si substrate and identified under optical microscope. Raman spectroscopy was employed to characterize the as-exfoliated few-layer VP and BP flakes.

As presented in Fig. 2c, the Raman spectrum of few-layer VP (red curve) exhibits multiple characteristic vibrational modes, arising from its low-symmetry lattices, indicative of strong anisotropy. In contrast, few-layer BP displays which show characteristic A_g^1, B_{2g}, and A_g^2 modes[9], consistent with its higher-symmetry structure.

We focus on the properties of few-layer VP flakes. Fig. 2d shows an optical image of as-exfoliated VP flakes, which retain its rectangular morphology. The highlight region (blue rectangle) corresponds to a few-layer flake with reduce contrast relative to the substrate. AFM measurements of the highlighted area in Fig. 2e reveal sharp perpendicular edges, reflecting the rectangular lattice of VP. The thickness of a single VP layer is ~1.1 nm, confirming its layered structure and potential applications in two-dimensional devices. Photoluminescence (PL) spectra (Fig. 2f) show strong broad emission peaks at ~650 nm for ~20 nm thick VP sample, corresponding to a ~1.9 eV bandgap and confirming its semiconducting nature. In contrast, few-layer VP exhibits only weak, near-background PL signals. The thickness-dependent optical response is consistent with previous report on VP[10]. Nevertheless, the transport properties of few-layer VP remain largely unexplored.

III. DEVICE FABRICATION AND ARCHITECTURE

To study the transport properties of semiconducting few-layer VP, we transferred the as-exfoliate few-layer VP to predefined electrodes to form VP FET devices in a bottom gate architecture, as shown in Fig. 3a. Fig. 3b displays a representative optical image of few-layer VP flake which has a characteristic rectangular shape. We then fabricated the Hall-bar shape electrodes using standard EBL process[11] (Fig. 3c), and cleaned by AFM tips in a contact mode, for more details, see Section V. The few-layer VP flake was encapsulated by thin hBN flake to prevent degeneration. The hBN/VP van der Waals stack were formed by picking up hBN and few-layer VP flake sequentially using dry transfer method[12]. The hBN/VP stack were released to the as-fabricated electrodes to form a FET device. Fig. 3d show an optical image of as-fabricated FET device using a four-layer VP as highlighted by blue rectangle, which were encapsulated by a thin hBN flake. There is no bubble between source and drain are, indicating the process can reduce material contamination and damage during the transfer process and forms a stable device structure[13].

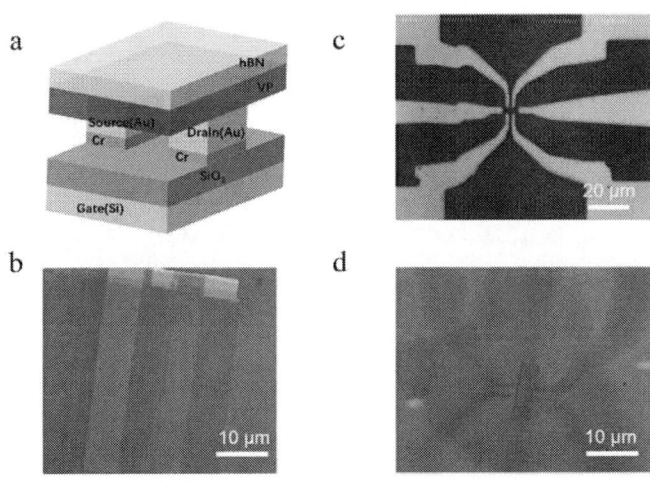

Fig. 3 a, Bottom-gate device architecture of few-layer VP FETs. b, The optical image of few-layer VP, which clearly show the rectangular shape. c, The optical image of predefined Au/Cr source–drain electrodes on SiO₂/Si substrate. d, The optical image of a four-layer VP FET device encapsulated by thin layer hBN.

IV. ELECTRICAL TRANSPORT PROPERTIES

The as-fabricated VP FETs were loaded onto a vacuum chamber for transport measurements. Fig. 4 shows a representative transfer curve of a four-layer VP FET measured at V_{ds} = 10 V and 20 V. Under forward gate sweeps (−100 to +100 V), I_{ds} remained at ~10^{-12} A in the off-state and increased exponentially to ~10^{-9} A at +100 V, confirming n-type behavior, in contrasts to the p-type behavior of black phosphorus. Reverse sweeps revealed a pronounced hysteresis loop: I_{ds} remained elevated in the positive V_{gs} range before sharply declining upon crossing into negative bias. This hysteresis is attributed to (i) carrier residual effects, (ii) interface trap states, and (iii) intrinsic material defects introducing deep energy traps, consistent with the results of studies on hysteresis mechanisms in other two-dimensional semiconductor devices [14]. The devices achieved an on/off current ratio of ~10^4-10^5,

comparable to that of high performance black phosphorus transistors[15]. Future interface engineering, such as hBN encapsulation, is expected to mitigate hysteresis and enhance device stability, a strategy that has been proven effective in MoS₂ devices.

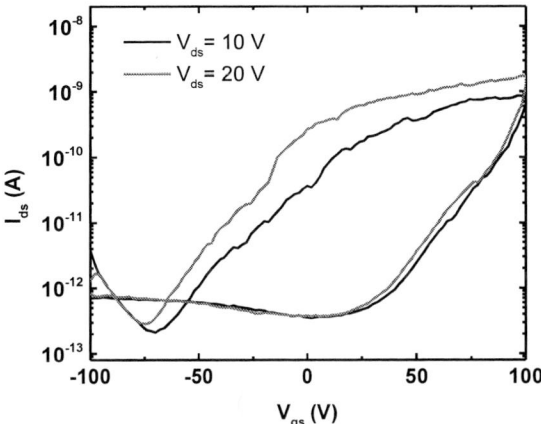

Fig. 4 I_{ds}-V_{gs} transfer characteristics of a four-layer VP FET at V_{ds} = 10 V and 20 V.

V. EXPERIMENTAL METHODS

A. Violet Phosphorus Synthesis

VP crystals were grown from red phosphorus using the chemical vapor transport (CVT) method. A mixture of red phosphorus, SnI₄, and Sn powder in a weight ratio of 50:1:2 was sealed in a quartz ampoule. The ampoule was gradually heated to 1000 K and held at that temperature for 50 hours, then slowly cooled to 500 K at a rate of 1 K/h, followed by a naturally cooling down to room temperature, yielding VP crystals. Compared to traditional methods, this study improved crystal quality and yield by optimizing the temperature gradient and transport agent concentration[8].

B. Flake Exfoliation and Characterization

Thin VP flakes (1–4 layers) were obtained via mechanical exfoliation onto SiO₂/Si substrates. A few-layer purple phosphorus thin films (1-4 layers) were transferred to SiO₂/Si substrates via mechanical exfoliation, a method derived from the classic mechanical exfoliation technique used in graphene preparation, which can effectively obtain high-quality few-layer samples[16]. The morphology and the thickness of the VP flake were determined by AFM measurements in a contact mode. Raman spectra were acquired using a 532 nm excitation laser. Photoluminescence spectra were recorded at room temperature using a confocal microscope setup.

C. Device Fabrication

Substrates were cleaned sequentially in acetone and isopropanol to remove organic contaminants, reduce surface defect density, and improve device performance[13]. The electrodes were defined by standard EBL process and metal deposition. Briefly, PMMA thin film was spin coated on to cleaned SiO₂/Si substrates. We defined Hall-bar shape electrodes using EBL. Cr/Au (5/15 nm) metal electrodes were deposited by electron-beam evaporation at ~2×10^{-5} Pa. After the removal of the PMMA thin film, the residues were cleaned using AFM tips. To prevent the degradation of the few-layer VP, the flakes were encapsulated by a thin hBN flake. We utilized dry transfer method to integrate hBN/VP van der Waals structures. The whole stack was picked up sequentially by a PC thin film supported by PDMS stamp and dropped down to a pre-defined electrode, forming VP FET devices. After removal of the polymer residue, VP FET devices were loaded to a vacuum chamber and pumped down to ~1×10^{-4} Pa. Electrical measurements were performed under ambient conditions using a semiconductor parameter analyzer.

VI. CONCLUSION

We have demonstrated the synthesis, device fabrication, and electronic characterization of few-layer VP FET devices. The devices exhibit robust n-type behavior with high on/off ratios, confirming VP's potential as a wide-bandgap 2D semiconductor with pool electric transport properties. The reproducible large hysteresis in the transfer behavior indicating the potential application of few-layer flake in high-performance memristors with low power consumption. The synthesis of large-area thin layer VP single crystal using chemical vapor deposition method may give rise to large-scale fabrication of VP devices. By optimizing the VP/dielectric interface using hBN encapsulation in both top and bottom side of VP flakes, it is expected that device performance and stability can be further improved. The uniaxial strain engineering can also regulate its anisotropic properties[17], promoting the development of violet phosphorus for practical nanoelectronics and optoelectronic applications. The application prospects of VP in flexible devices, sensors, and other fields have been preliminarily verified, and it is expected to become a core member of the next-generation two-dimensional electronic material system in the future.

ACKNOWLEDGMENT

The authors acknowledge the National Natural Science Foundation of China (No. 12304212 , 1250041587), the Guangdong Provincial Basic and Applied Basic Research Foundation (Nos. 2021A1515110980, 2022A1515140158, and 2023A1515110759) and the Guangdong Province Science and Technology Innovation Strategic Special Fund (Undergraduate Science and Technology Innovation Cultivation) (No. pdjh2025bk234), and the Research Fund of Guangdong Provincial Key Laboratory of Industrial Intelligent Inspection Technology (No. GDIIIT-KF-202505) for their financial support.

REFERENCES

[1] Lin, X., Yang, W., Wang, K. L. & Zhao, W. "Two-dimensional spintronics for low-power electronics". *Nature Electronics* 2, 274-283, (2019).

[2] Liu, C. *et al.* "Two-dimensional materials for next-generation computing technologies". *Nature Nanotechnology* 15, 545-557, (2020).

[3] Zhang, L. *et al.* "Structure and properties of violet phosphorus and its phosphorene exfoliation". *Angewandte Chemie International Edition* 59, 1074-1080, (2020).

[4] Schusteritsch, G., Uhrin, M. & Pickard, C. J. "Single-layered hittorf's phosphorus: a wide-bandgap high mobility 2D material". *Nano Letters* 16, 2975-2980, (2016).

[5] Feng, X. *et al.* "High ambipolar mobility and long-range carrier transport in violet phosphorus nanosheet". *Nano Letters* 24, 10348-10354, (2024).

[6] Zhang, L. *et al.* "Structure and properties of violet phosphorus and its phosphorene exfoliation". *Angewandte Chemie International Edition* 59, 1074-1080, (2020).

[7] Ricciardulli, A. G., Wang, Y., Yang, S. & Samorì, P. "Two-dimensional violet phosphorus: a p-type semiconductor for (opto)electronics". *Journal of the American Chemical Society* 144, 3660-3666, (2022).

[8] Zhang, L. *et al.* "High yield synthesis of violet phosphorus crystals". *Chemistry of Materials* 32, 7363-7369, (2020).

[9] Zhu, Y. *et al.* "Raman tensor of layered black phosphorus". *PhotoniX* 1, 17, (2020).

[10] Liu, Y. *et al.* "Thickness-dependent photoluminescence oscillations in layered violet phosphorus". *Laser & Photonics Revies*, 2401913.

[11] Todeschini, M., Bastos da Silva Fanta, A., Jensen, F., Wagner, J. B. & Han, A. "Influence of Ti and Cr adhesion layers on ultrathin Au films". *ACS Applied Materials & Interfaces* 9, 37374-37385, (2017).

[12] Uwanno, T., Hattori, Y., Taniguchi, T., Watanabe, K. & Nagashio, K. "Fully dry PMMA transfer of graphene on h-BN using a heating/cooling system". *2D Materials* 2, (2015).

[13] Dong, W., Dai, Z., Liu, L. & Zhang, Z. "Toward clean 2D materials and devices: recent progress in transfer and cleaning methods". *Advanced Materials* 36, 2303014, (2024).

[14] Di Bartolomeo, A. *et al.* "Hysteresis in the transfer characteristics of MoS_2 transistors". *2D Materials* 5, (2017).

[15] Na, J. *et al.* "Few-layer black phosphorus field-effect transistors with reduced current fluctuation". *ACS Nano* 8, 11753-11762, (2014).

[16] Novoselov, K. S. *et al.* "Electric field effect in atomically thin Carbon films". *Science* 306, 666-669, (2004).

[17] Yang, S., Chen, Y. & Jiang, C. "Strain engineering of two-dimensional materials: Methods, properties, and applications". *Infomat* 3, 397-420, (2021).

2025 The 10th International Conference on Integrated Circuits and Microsystems

A SiC Power Module Package Based on the Dual-side Cooling with Highly Thermally Conductive Graphite-Molybdenum Spacer

Yucheng Xu[1,2,3], Jiafei Yao[1,2,3]*, Yuxuan Dai[1,2,3], Ziwei Hu[1,2,3], Fan Yang[1,2,3], Kemeng Yang[1,3], Binbin Xu[2], Zhikuang Cai[1,2,3], Yufeng Guo[1,3]*

[1]College of Integrated Circuit Science and Engineering (College of Industry-Education Integration), Nanjing University of Posts and Telecommunications, Nanjing, China
[2]Nantong Institute of Nanjing University of Posts and Telecommunications, Nantong, China
[3]National and Local Joint Engineering Laboratory of RF Integration and Micro-Assembly Technology, Nanjing, China
*jfyao@njupt.edu.cn

Abstract—To address the issue of excessive temperature in SiC power module, this work proposes a dual-side cooling (DSC) package structure with highly thermally conductive graphite-molybdenum spacer. The main feature is the high heat dissipation around the spacers by utilizing the high thermal conductivity of graphite, thereby improving the overall thermal reliability of the structure. The package structure is modeled and simulated in ANSYS based on the finite element analysis. The influences of graphite thickness on the thermal resistance and junction temperature are discussed. The results demonstrate that the optimization effect of thermal reliability will gradually decrease. When compared to conventional DSC package structure, the proposed graphite-molybdenum enhanced DSC package structure enables a 21.6% reduction of the junction temperature and a 31.5% reduction of the thermal resistance. Consequently, the thermal reliability of the SiC power module has been improved.

Keywords—*SiC power module; Dual-side cooling; Junction temperature; Thermal resistance; Graphite*

I. INTRODUCTION

SiC power modules exhibit higher power density when compared to silicon power modules. This leads to increased heat generation during operation and elevated die temperatures, which can induce premature module failure and severely constrain the electrical performance and reliability of SiC power modules [1-4]. Consequently, high thermal conductivity materials and advanced package structures with exceptional thermal performance are particularly critical for enhancing the thermal reliability of SiC power modules.

On the one side, the research of high thermal conductivity materials such as graphite, nano-silver, diamond and AlSiC are applied to SiC power modules to improve the thermal reliability [5-8]. Ref.5 demonstrated a Graphite-embedded high-performance insulated metal substrate for SiC power modules to reduce the thermal resistance. Ref.6 designed an all-silver sintered module, in which the spacers are solidified with nano-silver paste. The entire bond layer and spacers are made of sintered silver. Simulation results showed that compared to

copper spacers, the overall thermal stress of the module decreased by more than 42%, while the junction temperature only increased by 3.6%.

In terms of SiC power module package structure, dual-side cooling (DSC) structures demonstrate superior thermal management capability and reduced parasitic parameters when compared to wire-bonded single-side cooling architectures [9]. Ref.10 developed a double-sided cooling SiC power module based on copper lead frame. The lead frame replaces the traditional DBC substrate and integrates the heat sink directly onto the lead frame [10]. Ref.11 used low-temperature co-fired ceramic technology (LTCC) to achieve a compact sandwiched press-pack SiC power module package without bonding wires [11]. The LTCC ceramic in this module is equivalent to a fixture which integrates microchannel heat sinks and achieves double-sided heat dissipation, greatly improving the heat dissipation capability of the power module.

To address the issue of thermal reliability in SiC power modules, this work presents a SiC power module package based on dual-side cooling with highly thermally conductive graphite-molybdenum composite spacers. The spacer in this DSC package structure is composed of graphite sheets wrapped around the molybdenum block. The proposed graphite-molybdenum enhanced DSC package structure significantly enhances thermal management efficiency, actively suppresses peak junction temperatures in SiC dies. In the following section, the package structure is modeled in ANSYS, and thermo-mechanical performance is investigated by finite element analysis (FEA).

II. DESIGN OF PACKAGE

The graphite-molybdenum enhanced DSC package structure of SiC power module in this work integrates four SiC dies, which form a half-bridge circuit as shown in Fig.1. For the DSC package structure, the dies are directly bonded to the upper/lower direct bond copper (DBC) substrates. Fig. 2(a) is the exploded view of the proposed structure and Fig. 2(b) is the cross-sectional view of this structure. All interconnects between the upper DBC substrate and graphite-molybdenum composite spacer, between the spacer and die, and between the die and lower DBC substrate employ nano-silver sintering technology. This sintered silver interface significantly enhances interfacial

---

This work is supported by the National Natural Science Foundation of China (Grant No.62334003, U23B2042) and the Opening Project of State Key Laboratory of Electronic Thin Films and Integrated Devices under Grant KFJJ202303

979-8-3315-8850-2/25 $31.00 © 2025 IEEE

thermal conductance while providing robust interconnection suitable for power device package. Electrical terminals include DC positive and negative power terminals, an AC power terminal, along with gate and source sense terminals for both top and bottom switches. The spacer assembly comprises a 1.6 × 1.6 × 3 mm molybdenum block integrated with a 0.7 mm thick graphite layer. The molybdenum block primarily provides electrical interconnection and heat conduction. Due to the poor solderability of graphite, this work first performs surface metallization treatment on graphite before bonding it with molybdenum block. This work enhances the heat dissipation around the spacers by utilizing the high thermal conductivity of graphite, thereby improving the overall thermal reliability of the structure. Crucially, the top and bottom switches are distributed on separate substrates, eliminating the need for additional metal spacers for electrical interconnection. Simultaneously, heat from each switch independently transfers through its dedicated substrate to corresponding heatsinks, preventing localized thermal accumulation inherent in single-side cooling. This approach significantly reduces overall thermal resistance and junction temperature. The decoupled switch layout effectively mitigates temperature rise caused by coupling effect and prevents reliability issues from heat concentration. Symmetric cooling achieves more uniform temperature distribution, extending module lifetime.

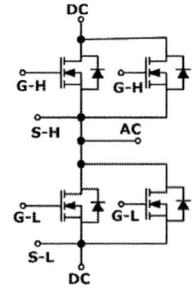

Fig. 1. Half-bridge circuit of the SiC power module

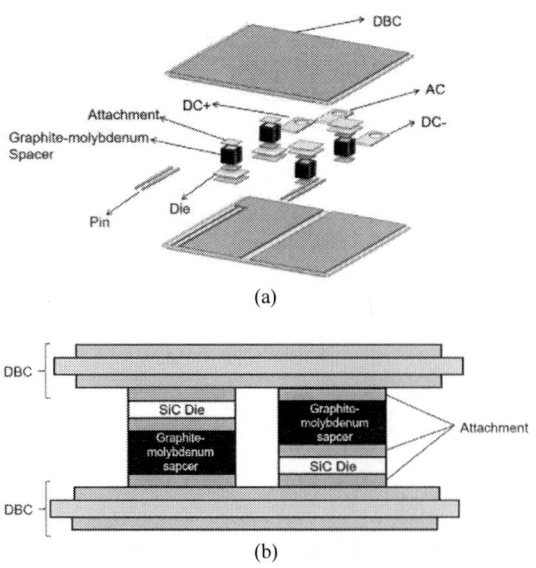

Fig. 2. The graphite-molybdenum enhanced DSC package structure of SiC power module. (a) Exploded view, (b) Cross-sectional view

The complex geometry of the power module is modeled by ANSYS software, as illustrated in Fig. 3, with material parameters of package structure summarized in Table 1. Graphite exhibits high in-plane thermal conductivity of 1500 W/(m·K) along the XY axes, while demonstrating significantly lower longitudinal thermal conductivity of merely 5 W/(m·K) in the Z-direction [12]. Furthermore, graphite possesses a negative coefficient of thermal expansion (CTE) of −1 ppm/K along the Z-axis, closely matching the CTE of dies. Molybdenum exhibits a CTE of 5.3 ppm/K, providing compatibility with both dies and DBC substrates. The density of 2.22 g/cm³ is smaller than other metal heat spreaders.

The exceptionally high in-plane thermal conductivity of graphite enables rapid spreading of heat generated by the dies through the spacers, effectively transforming localized hot spots into planar heat sources. This mechanism reduces localized temperature peaks within the die, prevents thermal accumulation and enhances heat dissipation efficiency from four surfaces of spacer, thereby lowering the maximum junction temperature of the package structure for superior thermal reliability. Concurrently, graphite's density is significantly lower than that of copper and aluminum, enabling high-efficiency heat dissipation without substantially increasing package structure's weight. The molybdenum block primarily serves electrical interconnection functions, while its CTE compatibility with dies, combined with high Young's modulus and small deformation characteristics, effectively reduces thermo-mechanical stresses.

Fig. 3. Model of the graphite-molybdenum enhanced DSC package structure of SiC power module

TABLE I. MATERIAL PARAMETERS OF PACKAGE STRUCTURE

Material	Density, ρ (kg/m³)	Thermal conductivity, K (W/m · K)	CTE, α (℃-1)	Specific heat, C (J/kg ·K)
Copper	8900	387	1.75E-5	390
ALN	3280	180	4.3E-6	750
Nano-silver	8580	238	1.96E-5	234
Graphite	2200	1500(x,y) 5(z)	-1E-6(x,y) 2.5E-5(z)	702
Molybdenum	10200	142	5.3E-6	243
SiC	3210	370	4.3E-6	710

III. RELIABILITY ANALYSIS

Fig. 3 illustrates the simplified ANSYS model of the graphite-molybdenum enhanced DSC package structure of SiC power module. The model omits the components which have negligible impact on overall thermal reliability, especially power terminals, wire bonds and encapsulation. During power module simulation, the ambient temperature is set to 22°C. To simplify the model, convection coefficients of 6000 W/(m²·K) are applied to the top and bottom surfaces of the upper/lower substrates to simulate forced water cooling, while the side surfaces of the substrates are assigned a convection coefficient of 10 W/(m²·K) representing natural air convection. All other surfaces are treated as adiabatic boundaries. In this work, the SiC die power dissipation is set to 50 W and each die is modeled as a uniform volumetric heat source during simulation, considering only self-generated heat from die operation. This power is converted into thermal load applied to the dies. The heat generation rate is calculated as follows:

$$H_{\text{gen}} = \frac{P}{V} \qquad (1)$$

where H_{gen} denotes the volumetric heat generation rate in W/m³, P represents the die power dissipation in W, and V is the die volume in m³.

Fig. 4 shows the temperature contour maps of the conventional DSC package structure with molybdenum spacers and the proposed graphite-molybdenum enhanced DSC package structure. This figure reveals that the die exhibits the highest junction temperature within the entire package structure. As shown in Fig. 4(a), the conventional DSC package structure achieves a maximum junction temperature of 65.94°C, while the graphite-molybdenum enhanced DSC package structure of SiC power module achieves 51.7°C, representing a 21.60% reduction. The reason for this result is that the high in-plane thermal conductivity of graphite effectively transfers the heat around the molybdenum block quickly, thereby reducing the peak temperature of dies.

Junction-to-area thermal resistance ($R_{\theta ja}$) is employed to evaluate the thermal performance of the package structure. It can be calculated by Equation 2. T_j denotes the junction temperature of the die. T_a represents the ambient temperature surrounding the package structure which is set to 22°C and P is

the power dissipation which is the total power dissipation of four dies in this work. The calculated $R_{\theta ja}$ of the conventional DSC package structure is 0.2191 °C/W, while the graphite-molybdenum enhanced DSC package structure achieves a calculated $R_{\theta ja}$ of 0.1485 °C/W, representing a 32.50% reduction. The reason for this result is that conventional DSC package structure is limited by a single longitudinal heat dissipation path and have small heat dissipation area, the graphite-molybdenum enhanced DSC package structure utilizes graphite's high in-plane thermal conductivity to laterally diffuse heat around the molybdenum block, increasing the effective heat dissipation area and achieving lower $R_{\theta ja}$.

$$R_{\theta ja} = \frac{T_j - T_a}{P} \qquad (2)$$

To investigate thermal reliability at higher power levels, simulations are conducted with die power dissipation ranging from 50W to 100W. Fig. 5 compares the temperature curves of both structures across this power range. These results confirm that the thermal management efficiency of the graphite-molybdenum enhanced DSC package structure improves proportionally with increasing power density.

Fig. 6(a) and (b) presents the temperature contour maps of dies in conventional DSC package structure under 70W/100W power dissipation per die, while Fig. 6(c) and (d) presents the temperature contour maps of dies in graphite-molybdenum enhanced DSC package structure under 70W/100W power dissipation per die. The left side of each figure in Figure 6 is the temperature distribution map of the connection surface between the die and the spacer. And the right side is the temperature distribution map of the connection surface between the die and the DSC substrate. It can be seen from Fig.6 that the hotspots of SiC die in the conventional DSC package structure are concentrated in a large area in the middle and the peak temperature occupies a larger area. These cause the temperature in this area to be higher than the average temperature of the die. Heat can't be effectively diffused from the spacers and DBC substrates. The graphite added around the molybdenum block in this work can effectively reduce heat concentration and the area occupied by the peak temperature,

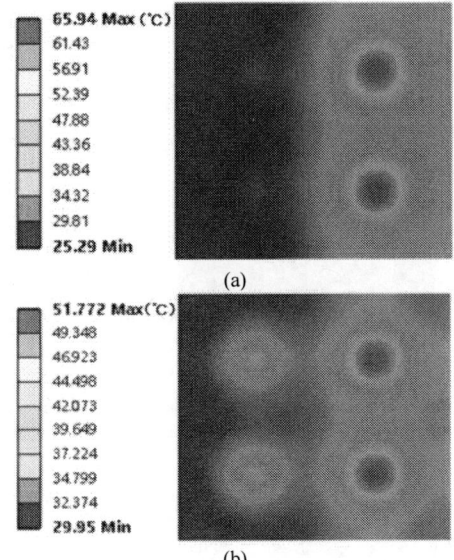

Fig. 4. Temperature Contour Maps of (a) conventional DSC package structure (b) graphite-molybdenum enhanced DSC package structure

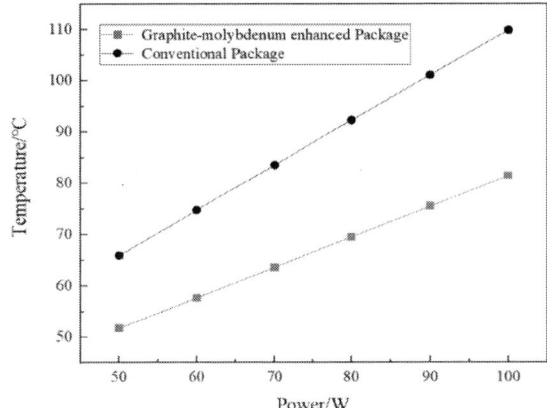

Fig. 5. Temperature curves under different die power dissipation

979-8-3315-8850-2/25 $31.00 © 2025 IEEE

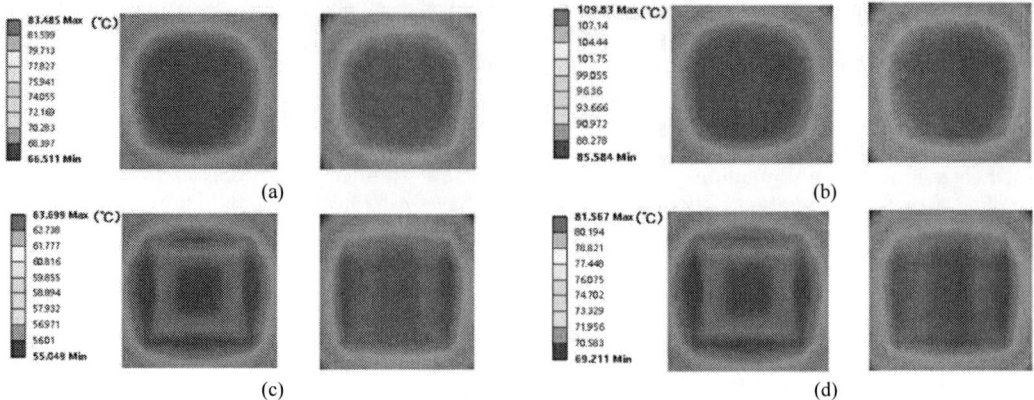

Fig. 6. Temperature Contour Maps of dies in conventional DSC package structure (a) under 70W power dissipation per die and (b) under 100W power dissipation per die, graphite-molybdenum enhanced DSC package structure (c) under 70W power dissipation per Die and (d) under 100W power dissipation per die

and it can effectively diffuse heat. Under 70W power dissipation, the graphite-molybdenum enhanced DSC package structure achieves a 19.79°C reduction compared to the conventional DSC package structure. Under 100W power dissipation, this thermal advantage increases to 28.26°C, demonstrating progressively more pronounced cooling effectiveness at elevated power levels. The reason for this phenomenon is that the SiC dies generate less heat under low power dissipation. In conventional DSC package structures, heat is mainly conducted through vertical conduction of molybdenum blocks which can take away most of the heat. Graphite plays a role in assisting in the diffusion of a small amount of lateral heat because its vertical thermal conductivity is much lower than that of molybdenum blocks. And the additional heat dissipation effect is limited. However, the SiC dies generate more heat under high power dissipation. And the vertical thermal conductivity of conventional structure cannot effectively conduct excessive heat. In this case, the lateral high thermal conductivity of graphite is fully utilized. By rapidly diffusing the heat inside the molybdenum blocks laterally, the heat dissipation area is greatly increased. And the additional heat dissipation is significantly improved.

Fig. 7 illustrates the progressive reduction in both T_j and $R_{\theta ja}$ with increasing graphite thickness. The reduction of T_j is 9.71°C

and the reduction of $R_{\theta ja}$ is 0.049°C/W when the graphite thickness is increased from 0.1 mm to 1.0 mm. This nonlinear relationship indicates that thicker graphite layers do not proportionally enhance thermal performance.

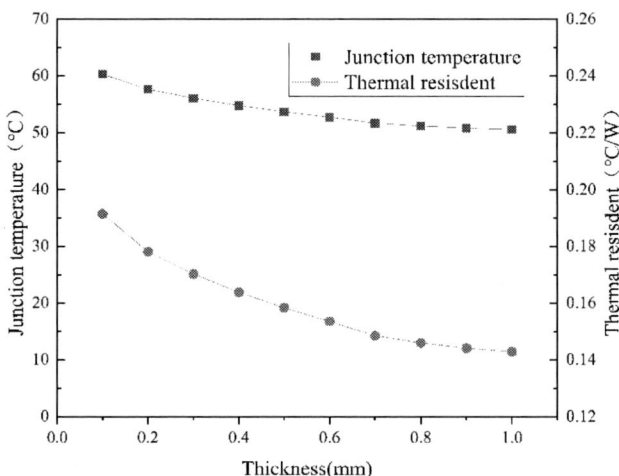

Fig. 7. The influence curves of graphite thickness on junction temperature and thermal resistance

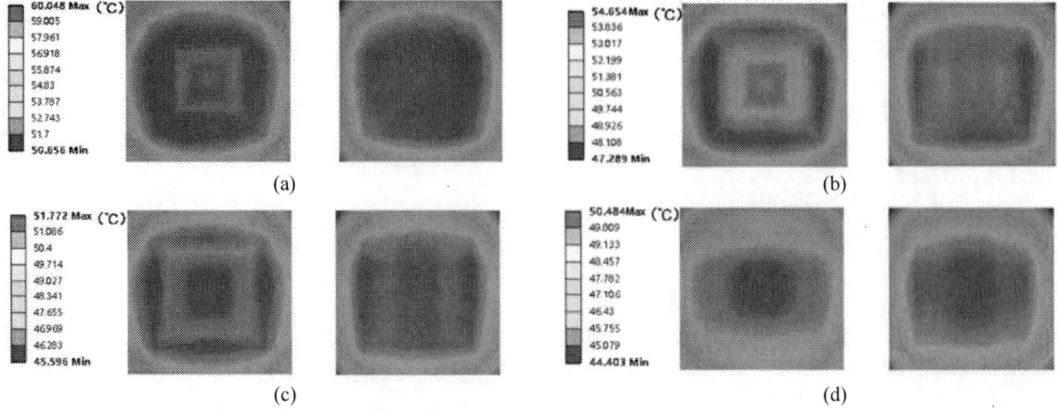

Fig. 8. Die Temperature Contour Maps for the graphite-molybdenum enhanced DSC package structure with (a) 0.1mm graphite thickness, (b) 0.4mm graphite thickness, (c) 0.7mm graphite thickness, and (d) 1.0mm graphite thickness

Fig. 8 illustrates the SiC die temperature contour plots corresponding to graphite thicknesses of 0.1 mm, 0.4 mm, 0.7 mm and 1.0 mm. It can be seen from Fig. 8 that increasing the thickness of graphite can effectively reduce heat concentration, decrease the area occupied by peak temperature, and thus enables the heat diffuse from the hotspot area through the spacers and DSC substrates more effectively. However, the effect of increasing graphite thickness on junction temperature and thermal resistance is limited. Because graphite mainly relies on its high in-plane thermal conductivity for heat dissipation, but its longitudinal thermal conductivity is very low. The increase in graphite thickness will enhance the lateral diffusion effect, but at the same time, it will also increase the longitudinal thermal resistance of graphite.

IV. CONCLUSION

This work proposes a graphite-molybdenum enhanced DSC package structure of SiC power module. The spacer in this structure assembly incorporates a graphite layer integrated with molybdenum blocks, which leverages graphite's exceptional in-plane thermal conductivity to transform concentrated point heat sources into distributed planar thermal loads. Meanwhile, the compatibility of thermal expansion coefficient between molybdenum and SiC, as well as molybdenum's high Young's modulus, jointly suppress harmful thermal mechanical stress in the entire packaging structure. In addition, this package structure places the top and bottom switching devices on the separated direct bonded copper substrates, thereby eliminating the necessity for additional electrical interconnection spacers. The T_j and $R_{\theta ja}$ of the graphite-molybdenum enhanced DSC package structure are simulated by the ANSYS based on the finite element analysis. The analyzed results show that the high in-plane thermal conductivity of graphite is more fully utilized under high power dissipation. The parameter optimization of graphite layer thickness shows that the optimization effect of thermal reliability will gradually decrease. A 21.6% reduction in T_j and a 32.5% reduction in $R_{\theta ja}$ are obtained when compared to the conventional DSC package structure.

REFERENCES

[1] G. E Seal, Sayan, and Homer Alan Mantooth, "High Performance Silicon Carbide Power Packaging—Past Trends, Present Practices, and Future Directions," Energies, vol 10, no. 3, pp. 341-370, Mar. 2017.

[2] G. Watt, A. Romero, R. Burgos, and M. Jaksic, "Design of a compact, low inductance 1200 V, 6.5mŸ SiC half-bridge power module with flexible PCB gate loop connection," Applied Power Electronics Conference and Exposition (APEC) 2019, Annual IEEE Conference, pp. 86-93.

[3] H. Lee, V. Smet and R. Tummala, "A Review of SiC Power Module Packaging Technologies: Challenges, Advances, and Emerging Issues," IEEE Journal of Emerging and Selected Topics in Power Electronics, vol. 8, no. 1, pp. 239-255, March 2020.

[4] M. Liu, A. Coppola, M. Alvi and M. Anwar, "Comprehensive Review and State of Development of Double-Sided Cooled Package Technology for Automotive Power Modules," IEEE Open Journal of Power Electronics, vol. 3, pp. 271-289, 2022.

[5] E. Gurpinar, S. Chowdhury, B. Ozpineci and W. Fan, "Graphite-Embedded High-Performance Insulated Metal Substrate for Wide-Bandgap Power Modules," IEEE Transactions on Power Electronics, vol. 36, no. 1, pp. 114-128, Jan. 2021.

[6] C. Ding, H. Liu, K. D. T. Ngo, R. Burgos and G. -Q. Lu, "A Double-Side Cooled SiC MOSFET Power Module With Sintered-Silver Interposers: I-Design, Simulation, Fabrication, and Performance Characterization," IEEE Transactions on Power Electronics, vol. 36, no. 10, pp. 11672-11680, Oct. 2021.

[7] T. Zhang, L. Wang, X. Zhang, et al., "Enhanced Thermal-Electrical Interconnect for Single Sided Cooling SiC MOSFET Power Device Based on Polycrystalline Diamond," IEEE Transactions on Power Electronics, pp. 1-14, 2025.

[8] S. Bontemps and L. -P. Doumergue, "Very Low Stray Inductance, High Frequency 1200 V_ 2mOhms Full SiC MOSFET Phase Leg Module," PCIM Europe 2018, Nuremberg, Germany, 2018, pp. 1-8.

[9] R. Paul, R. Alizadeh, H. Chen, X. Li, Y. Chen and H. A. Mantooth, "A Novel Integrated 1.2 kV Double-sided Cooled Power Module," 2023 IEEE Applied Power Electronics Conference and Exposition (APEC), Orlando, FL, USA, 2023, pp. 372-37.

[10] G. Tang et al., "Development of a Novel Lead Frame Based Double Side Liquid Cooling High Performance SiC Power Module," 2021 IEEE 71st Electronic Components and Technology Conference (ECTC), San Diego, CA, USA, 2021, pp. 118-124.

[11] Y. Chang et al., "Compact Sandwiched Press-Pack SiC Power Module With Low Stray Inductance and Balanced Thermal Stress," in IEEE Transactions on Power Electronics, vol. 35, no. 3, pp. 2237-2241, March 2020.

[12] S. Fukunaga and T. Funaki, "Thermal Decouple Design of Multichip SiC Power Module With Thermal Anisotropic Graphite," IEEE Transactions on Components, Packaging and Manufacturing Technology, vol. 11, no. 5, pp. 778-784, May 2021.

2025 The 10th International Conference on Integrated Circuits and Microsystems

Work Function Variation Effect Prediction on Heterojunction Tunnel FET Using Multi-Layer Perceptron-Based Neural Network Model

Haotong Han
School of Electronic Engineering
Xi'an University of Posts &
Telecommunications
Xi'an, China
957598733@qq.com

Yunhe Guan *
School of Electronic Engineering
Xi'an University of Posts &
Telecommunications
Xi'an, China
gyhflc@xupt.edu.cn

Tongqing Yan
School of Electronic Engineering
Xi'an University of Posts &
Telecommunications
Xi'an, China
2943565556@qq.com

Weihan Sun
School of Electronic Engineering
Xi'an University of Posts &
Telecommunications
Xi'an, China
877178139@qq.com

Xiangtai Liu
School of Electronic Engineering
Xi'an University of Posts &
Telecommunications
Xi'an, China
liuxiangtai@xupt.edu.cn

Haifeng Chen
School of Electronic Engineering
Xi'an University of Posts &
Telecommunications
Xi'an, China
chenhaifeng@xupt.edu.cn

Abstract—This paper proposes a multi-layer perceptron (MLP)-based neutral network (NN) model to efficiently and reliably predict the effect of metal gate work function variation (WFV) on the electrical performance of III-V heterojunction tunneling field effect transistor (HTFET). The dataset used for neural networks are obtained from Sentaurus TCAD. The weight parameters in the MLP model were updated by gradient descent method and the accuracy of the model was evaluated by mean logarithmic squared error function. The final results show that the developed model can effectively predict the effect of WFV on HTFET device performance. Compared with the random grain distribution method used by TCAD, the trained MLP-based NN model can predict the key performance parameters of WFV with more than 300× improved speed, and with the predicted Pearson's coefficient value of 0.927 for threshold voltage. This indicates that the MLP-based NN model has a great potential for HTFET WFV performance optimization and related researches.

Keywords—*Heterojunction TFET, WFV, Neutral Network, Threshold Voltage, Multi-layer Perceptron (MLP).*

I. INTRODUCTION

In light of the Band-To-Band Tunneling (BTBT) mechanism, the Tunneling Field-Effect Transistor (TFET) is able to attain a Subthreshold Swing (SS) under 60mV/dec at room temperature, while also displaying a low off-state current (I_{off}) [1]. Furthermore, in order to alleviate the bottleneck of TFET's low on-state current, heterojunction TFET (HTFET) structure is extensively employed to enhance the on-state current[2]. Consequently, in comparison with MOSFET, HTFET has the upper hand in terms of power consumption, thus becoming a significant direction in the current research on low-power devices [3].

This work was supported by the National Science Foundation of China under Grant 62104192.

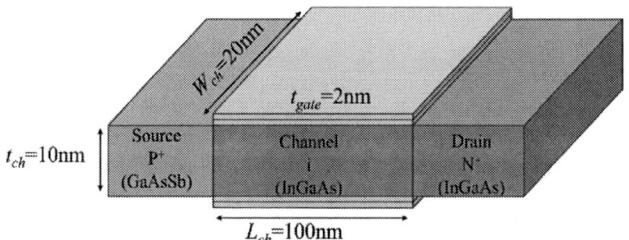

Fig. 1. DG-HTFET device structure. The channel length is L_{ch} and the channel width is W_{ch}. The metal gate and channel thickness are t_{gate} and t_{ch}, respectively.

In the ever-evolving landscape of semiconductor manufacturing processes, as feature sizes shrink to the deep nanometer scale, process fluctuations have increasingly become a pivotal factor impeding device performance. Among the diverse sources of fluctuations that lead to characteristic changes in TFETs [4-7], the metal gate work function variation (WFV) is of great significance to HTFETs. C. W. Hsu et al. [8] found that near the off-state, WFV induces severe performance fluctuations in HTFET devices. Although research on WFV has been conducted, most studies rely on traditional TCAD simulation software, which is inherently limited by low computational efficiency and long simulation times, especially in large-sample simulations. In recent years, driven by its rapid advancement, machine learning has exhibited substantial application value and remarkable potential for innovation across a wide range of scientific and engineering fields. Given its strengths in handling complex data patterns and improving predictive accuracy, applying machine learning techniques to investigate the fluctuation effects of HTFET has become not only highly relevant but also essential for advancing device design and reliability analysis.

In this paper, a WFV prediction model for HTFETs is developed using a multi-layer perceptron (MLP)-based neural network (NN) algorithm, enabling faster and more accurate evaluation of fluctuation effects. The remainder of this paper is organized as follows. Section II introduces the HTFET device structure and the characteristic parameters used in the fluctuation model, followed by an overview of the basic principles of neural networks. Section III presents the model validation, and Section IV concludes the paper.

II. DEVICE STRUCTURE AND SIMULATION METHODS

A. Device Structure and TCAD Method

The double-gate (DG) HTFET has been chosen as the model device due to its excellent compatibility with existing CMOS manufacturing processes [9]. The structure of this device is illustrated in Fig. 1, with the internally lattice-matched $GaAs_{0.5}Sb_{0.5}/In_{0.53}Ga_{0.47}As$ heterostructure, which is popular because it is lattice matched to the InP substrate [10]. The channel length (L_{ch}) is defined as 100 nm, the channel width (W_{ch}) is set to 20 nm, and the channel thickness (t_{ch}) is set to 10 nm. The oxide layer's relative dielectric constant is $\varepsilon = 25$, with a gate thickness of 2 nm. The doping concentrations are specified as follows: $N_S = 5 \times 10^{19}$ cm^{-3} for the P$^+$ source, $N_{ch} = 1 \times 10^{16}$ cm^{-3} for the P-channel, and $N_d = 5 \times 10^{18}$ cm^{-3} for the N$^+$ drain zone.

To generate a robust dataset for training and validating machine learning models, the random grain distribution method implemented in Sentaurus TCAD is utilized, and 1000 HTFET devices was simulated by WFV influence. During the simulation process, the metal grain size is fixed at 3 nm. The work function values of the metal grains are assigned probabilities of 40% for 4.6 eV and 60% for 4.8 eV. The distribution of the random work function is shown in Fig. 2(a).

B. MLP-based NN Model

Since HTFET operates based on BTBT of carriers between the source region and the channel, two crucial physical quantities have been chosen as characteristic variables to be fed into the NN model. These are the metal gate work function and the potential of the device in both the off and on states close to the source/channel tunneling junction.

In order to better reflect the influence of WFV, the selected area is divided into 50 cells, with 25 cells for the upper and

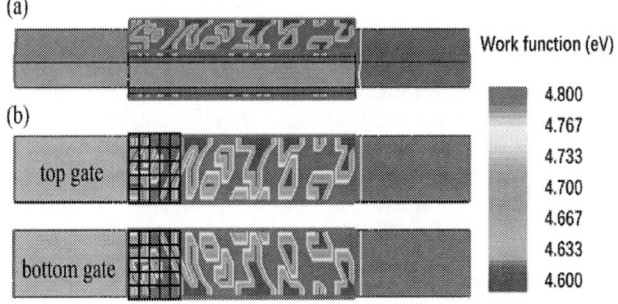

Fig. 2. (a) Distribution of random work function values in the metal gate and (b) division of the top gate and bottom gate along the channel chansport cross-section to extract input parameters.

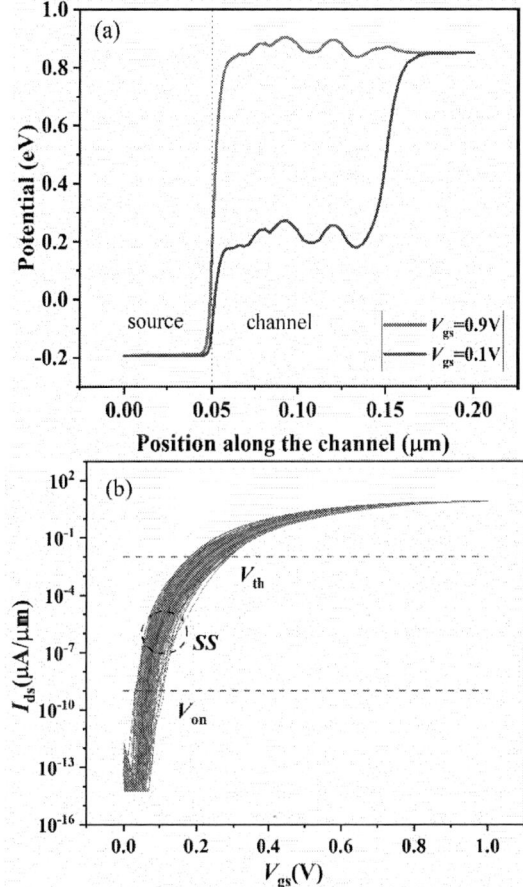

Fig. 3. Characteristics under WFV influence. (a)Potential distribution at different gate voltages. (b) I_{ds} - V_{gs} curve under influence of WFV at $V_{ds} = 0.3$ V.

lower gates respectively. The size of each cell is 4 nm×4 nm [11]. One work function value is extracted from each unit, as shown in Fig. 2(b). Therefore, for each device under the influence of WFV, there are a total of 50 input parameters related to the work function. Furthermore, the potential profiles in both the off - state and on - state under gate voltages of 0.1 V and 0.9 V are obtained, as depicted in Fig. 3(a). Given that the potential near the source/channel tunneling junction can efficiently capture the variation in the tunneling current, the potential values at 0.06 μm, 0.07 μm, 0.08 μm, 0.09 μm, and 0.1 μm are chosen as key input parameters for constructing the training set of the NN model. It should be noted that the source/channel junction is situated at 0.05 μm. By selecting these eigenvalues, the HTFET fluctuation model can effectively capture the impact of WFV on device characteristics [12], thereby enabling the model to be trained efficiently and predict output eigenvalues accurately.

In the selection of output parameters, this paper extracts three key characteristic parameters from the I_{ds} - V_{gs} curve depicted in Fig. 3(b): SS, V_{on}, and V_{th}. The SS actually is an average parameter, and refers to the reciprocal of the order of magnitude of the corresponding current change when the gate voltage changes from 0.1 V to 0.5 V. Initial voltage V_{on} is defined as the gate voltage value in the state where BTBT effective dominates, which is measured when the drain current

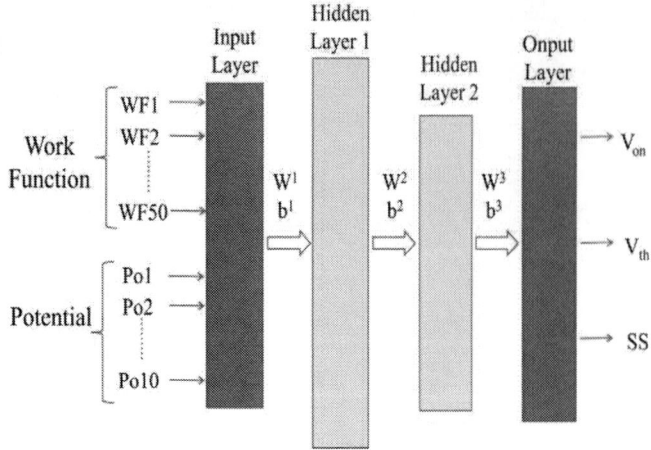

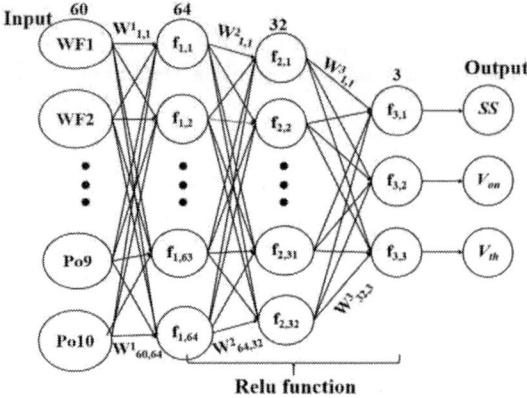

Fig. 5. Optimal MLP-based NN model structure. The input layer has 50 work function values and 10 potential values, and there are two hidden layers. The number of neurons in each layer is represented by f_{ij} (i for the layer, j for the number).

Fig. 4. MLP as the fluctuation model of HTFET. In practice, the number of hidden layers is adjustable. W^i and b^i (i= 1,2,3) represent MLP weights. The output variables are V_{on}, V_{th}, and SS.

TABLE I. R^2 VALUES OF OUTPUT CHARACTERISTICS OF MLP-BASED MODEL WITH DIFFERENT HIDDEN LAYER STRUCTURE

Hidden layer structure	Loss value		
	SS	V_{on}	V_{th}
32×16	0.831	0.628	0.827
64×32	0.820	0.653	0.927
128×64	0.831	0.628	0.827
256×128	0.757	0.670	0.896
32×16×8	-0.256	0.096	0.137
64×32×16	0.756	0.563	0.867
128×64×32	0.812	0.648	0.728
256×128×54	0.775	0.638	0.845

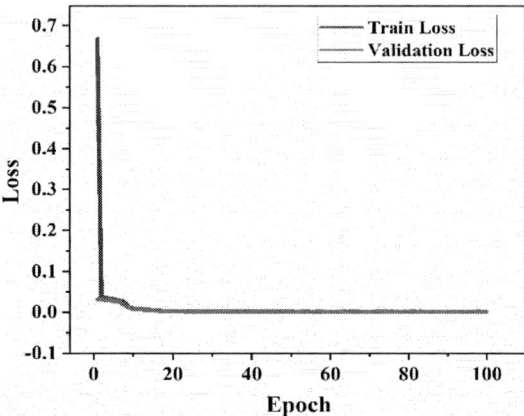

Fig. 6. Loss of MLP model under different epoch for the train and validation sets.

reaches 10^{-9} μA/μm, as marked on the blue line in Fig. 3(b). threshold voltage V_{th}, one of the core parameters, is measured when the drain current reaches 10^{-2} μA/μm, as shown by the red line [11].

The MLP represents a feedforward neural network model, whose primary objective is to execute complex nonlinear transformations on data via multiple fully connected neuron layers. This process establishes a mapping relationship between input and output feature values. In the MLP architecture, the input layer serves to receive data. The hidden layer employs multiple neurons to perform weighted summation of the input data, followed by activation function processing. Lastly, the output layer directly provides prediction results. Fig. 4 presents an example where an MLP is utilized to investigate the fluctuation characteristics of HTFET. Here, the input - layer neurons correspond to the metal gate work function values and the potential in both off - and on - states. In the MLP, the number of hidden layers (namely, fully connected layers) and the number of neurons within each hidden layer can be fine - tuned to achieve optimal model accuracy. Meanwhile, the neurons in the output layer are associated with the V_{on}, V_{th}, and SS.

III. RESULTS AND DISCUSSIONS

The dataset of the 1000 samples is partitioned into a training set, validation set, and test set in an 8:1:1 ratio. In the training process of the MLP - based model, the setting of hidden - layer is crucial for achieving high precision. Previous studies have indicated that double - hidden - layer networks can fit function expressions with arbitrary precision. However, an excessive number of neurons can easily lead to over - fitting problems. Consequently, we compared the Pearson's coefficient (R^2) values predicted under 8 different hidden - layer settings, as shown in Table I. The R^2 is used to quantify the model's prediction accuracy for each parameter. From Table I, it can be observed that the prediction model with a double-hidden-layer of 64×32 exhibits the best performance, whereas the prediction model with a three - hidden - layer of 32×16×8 performs the worst [13]. This demonstrates that neural network structures that are either too simple or too complex will reduce the network's prediction capacity, with under - fitting and over - fitting being the most common abnormal states.

Therefore, the MLP - based neural network algorithm adopts a structure of 60 - 64 - 32 - 3 and utilizes MinMaxScaler to normalize the input parameters [14]. The specific structure

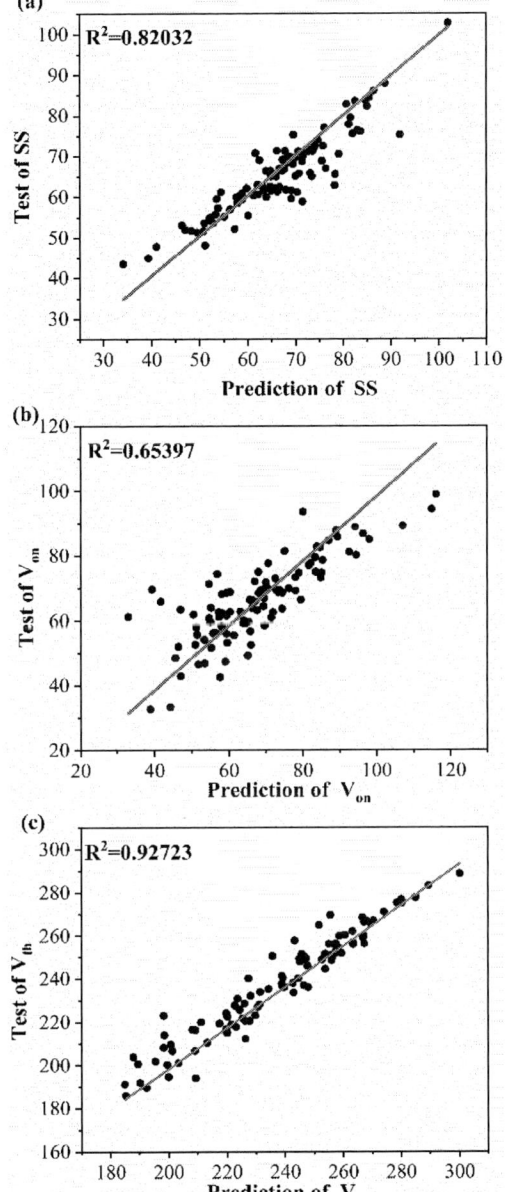

Fig. 7. R^2 value of the proposed model to predict (a) SS, (b) V_{on} and (c) V_{th}.

differ significantly. The reason is that the sample content of the training set is relatively large, covering more random distribution scenarios of WFV. This makes it difficult for the MLP model to adapt to the feature patterns of a large number of samples at the initial random weights and biases. Therefore, the calculated loss value is relatively high (about 0.65). Secondly, due to the high sensitivity of the loss function MLSE to the data magnitude and error distribution, the loss value of the training set is relatively high at the beginning. However, when the number of iterations is increased, the loss value drops rapidly, indicating that the fluctuation model is effectively learning the patterns in the dataset. When the number of iterations reaches approximately 35, the loss values converge, suggesting that the model has achieved a stable state [16]. After training, the final loss value is 0.0008 for the training set and 0.0006 for the validation set.

Following the completion of training, the test set is used to evaluate the predictive accuracy of the model. Fig. 7 presents a comparison between the predicted and actual values of the HTFET fluctuation model under various parameters. The results show that the model achieves high accuracy in predicting V_{th} and SS, with R^2 values of 0.927 and 0.82 respectively, while the prediction accuracy for V_{on} is relatively lower, with an R^2 value of 0.653. The nearly 30% difference between the R^2 values for V_{th} and V_{on} indicates that the model is more effective in learning and capturing the variations in V_{th}. It should be noted that since the threshold voltage is a more critical parameter in circuit research, the proposed model remains valuable for investigating circuit-level variations.

Furthermore, a comparison is made between the prediction model constructed by the MLP and that built by the random forest regression (RFR) algorithm [17]. The fluctuation prediction model employing 64×32 neurons with double hidden layers proves to be more accurate than the one using 200 decision trees with a depth of 10. In the HTFET fluctuation model established by RFR [17], the loss value for the validation set is 0.0085, almost $10\times$ of the MLP-based method. Moreover, the R^2 values of SS, V_{on}, and V_{th} are 0.698, 0.563, and 0.812, respectively. These results suggest that the MLP - based model can predict the impact of WFV on HTFET device performance more accurately. Moreover, it takes approximately 5 minutes to run a device characteristic influenced by WFV using TCAD. In contrast, the MLP prediction model requires only 1 - 2 seconds, with a prediction speed that is nearly 150 - 300 times faster.

IV. CONCLUSION

This study develops a WFV prediction model based on a MLP algorithm. The model uses the work function of the metal gate and the potential in both the off and on states near the tunneling junction as input features. The key output parameters include SS, V_{on}, and V_{th}. By evaluating the performance of various hidden layer configurations, the model with a two-hidden-layer structure consisting of 64 and 32 neurons, respectively, was identified as optimal. Compared to a fluctuation prediction model based on 200 random forest regressors with a depth of 10 decision trees, the MLP-based model demonstrates much higher prediction accuracy. Results show that the MLP-based model can effectively predict the

of the neural network model is illustrated in Fig. 5, which clearly presents the data transmission and transformation process from the input layer to the output layer. In this model, Mean Logarithmic Squared Error (MLSE) is employed as the loss function for each hidden layer. The gradient of each weight in the loss function is calculated through the backpropagation algorithm. The nonlinear Relu function is chosen as the activation function [15], and the Adam optimizer is applied to update the model parameters, with the learning rate set at 0.02.

Fig. 6 illustrates the training process of the proposed MLP-based neural network model. As the number of training iterations increases, the loss values for both the training and validation sets gradually stabilize. At the beginning of the training, the loss values of the training set and the validation set

impact of work function variation on key HTFET performance metrics, achieving an R2 value of 0.927 for V_{th}. Furthermore, the model is approximately 300 times faster than traditional TCAD simulations, indicating that machine learning-based prediction methods can be effectively integrated into HTFET device and related circuit design to overcome the limitations of conventional TCAD approaches.

REFERENCES

[1] R. Goswami and B. Bhowmick, "Comparative analyses of circular gate TFET and heterojunction TFET for dielectric-modulated label-free biosensing," *IEEE Sensors Journal*, vol. 19, no. 21, pp: 9600-9609, 2019.

[2] K. Nasani, B. Bhowmick, P. D. Pukhrambam, "Impact of Noise and Interface Trap Charge on a Heterojunction Dual-Gate Vertical TFET Device," *Journal of Electronic Materials,* vol. 53, no. 4, pp: 2181-2190, 2024.

[3] P. K. Kumar, B. Shilpi, S. Neha. "Design and Optimization of a Heterojunction (Ge/Si) Vertical-Tunnel Field Effect Transistor (HV-TFET) with a Doped Bar for Low-Power Applications," *Journal of Electronic Materials*, vol. 53, no. 7, pp: 3933-3945, 2024.

[4] Y. Guan, J. Lu, H. Zhang et al, "A comparative study of work function variations in III-V heterojunction and homojunction tunnel field-effect transistors," *Microelectronics Journal*, vol. 145 p. 106118, 2024.

[5] E. Mohapatra, J. Jena and D. Jena et al, " Work function variability and inverter design possibility in advanced gate all around FETs," *Nanomaterials and Energy*, vol. 12, no. 2, pp: 81-89, 2023.

[6] A. Anam, S. Amin and D. Prasad, "III-V material-based junction-free L-shaped gate normal line tunneling FET for improved performance," *Semiconductor Science and Technology*, vol. 39, no. 9, p. 095004, 2024.

[7] N. Gandhi, S. Rathore and R. Jaisawal et al, "Revealing the Noise Dependent Sensitivity of a Junctionless FinFET-Based Hydrogen Sensor with Ferroelectric Gate Stack," International Conference on Simulation of Semiconductor Processes and Devices (SISPAD), pp: 1-4, 2024.

[8] R. Debnath and S. Baishya, "Variability analysis of the epitaxial layer TFET due to gate work function variation, random dopant fluctuation, and oxide thickness fluctuation using the statistical impedance field method," *Semiconductor Science and Technology*, vol. 37, no. 6, p. 065005, 2022

[9] G. Rasheed and S. Sridevi, "Design and analysis of a dual gate tunnel FET with InGaAs source pockets for improved performance," *Microelectronics Journal*, vol. 129, p. 105587, 2022.

[10] R. Joshi, S. Karthikeyan, and M. Hudait, "Germanium nanosheet-FETs scaled to subnanometer node utilizing monolithically integrated lattice matched Ge/AlAs and strained Ge/InGaAs," *IEEE Transactions on Electron Devices*, vol. 70, no. 3, pp: 899-907, 2023.

[11] T. Hwang, S. Kim and G. Kim et al, "Analysis and Prediction of Nanowire TFETs Work Function Variation," *Journal of Semiconductor Technology and Science*, 2024, vol. 24, no. 2, pp: 96-104, 2024.

[12] G. Kim, J. H. Kim, J. Kim, and S. Kim, "Analysis of work-function variation effects in a tunnel field effect transistor depending on the device structure," *Appl. Sci.*, vol. 10, no. 15, p. 5378, 2020.

[13] G.-B. Huang, "Learning capability and storage capacity of two-hidden-layer feedforward networks," *IEEE Trans. neural networks*, vol.14, no. 2, pp. 274-281, 2003.

[14] F. Zhang, W. Dai, K. Wang et al, "Capacitance Modeling with Charge Partitions Covering Full-Region Operations of TFETs," *IEEE Transactions on Electron Devices*, vol. 71, no. 7, pp.4373-4380, 2024.

[15] A. Agarap, "Deep learning using rectified linearunits (relu)," arXiv Prepr. arXiv1803.08375, 2018.

[16] M. Saravanan, E. Parthasarathy, J. Ajayan et al, "Device Simulation Based Machine Learning Technique for III V TFET," 2024 International Conference on Recent Advances in Electrical, Electronics, Ubiquitous Communication, and Computational Intelligence (RAEEUCCI), IEEE, pp: 1-6, 2024

[17] C. Akbar, Y. Li, N. Thoti. "Device-simulation-based machine learning technique for the characteristic of line tunnel field-effect transistors," *IEEE Access*, vol. 10, pp: 53098-53107, 2022.

2025 The 10th International Conference on Integrated Circuits and Microsystems

Enhancing Carrier Mobility in a-IZO/a-IGZO Thin-Film Transistors through Band Structure Engineering of Heterojunction Channels

Huan Yang
School of Electronic and Computer Engineering
Peking University
Shenzhen, China
yang_huan@pku.edu.cn

Shengdong Zhang*
School of Electronic and Computer Engineering
Peking University
Shenzhen, China
zhangsd@pku.edu.cn
*corresponding author

Yong Wang
Institute of Critical Materials for Integrated Circuits
Shenzhen Polytechnic University
Shenzhen, China
wangyong1@szpu.edu.cnline

Chao Teng
Institute of Critical Materials for Integrated Circuits
Shenzhen Polytechnic University
Shenzhen, China
tengchao@szpu.edu.cn

Yi Shen
Guangdong Provincial Key Laboratory of Automotive Display and Touch Technologies
Shantou Goworld Display Technology Co., Ltd.
Shantou, China
yishen@goworld-lcd.com

Yudong Liu
Guangdong Provincial Key Laboratory of Automotive Display and Touch Technologies
Shantou Goworld Display Technology Co., Ltd.
Shantou, China
ydliu@goworld-lcd.com

Abstract—**Amorphous oxide semiconductor thin-film transistors (AOS TFTs) require high carrier mobility to meet the demands of advanced applications, such as high-end displays and monolithic 3D integration. Heterojunction channel structures, such as a-IZO/a-IGZO, offer a promising approach to achieving high mobility without compromising device performance. This study presents a systematic investigation into the impact of band structure engineering on the performance of heterojunction TFTs, by modulating the argon/oxygen ratio during sputtering and adjusting the thickness of individual layers. Results indicate that a large energy difference in the conduction band minimum or the Fermi level between the constituent layers leads to the formation of a deeper quantum potential well, thereby enhancing carrier transport efficiency and mobility. The optimized a-IZO/a-IGZO TFTs exhibit a high carrier mobility of 70 cm^2/Vs, a low off-state current of approximately 1pA, and a near-zero threshold voltage. This work provides valuable insights into the design and optimization of high-performance AOS TFTs through controlled band structure engineering.**

Keywords—*amorphous oxide semiconductor, thin-film transistors, heterojunction, band structure, high mobility*

I. INTRODUCTION

Over the past decade, thin-film transistors (TFTs) based on amorphous oxide semiconductors (AOS) have attracted considerable interest owing to their favorable characteristics, such as low processing temperature, optical transparency, high uniformity, and ultra-low leakage current [1-3]. AOS TFTs have emerged as a prominent alternative to conventional silicon-based devices in next-generation display technologies [4] and have also demonstrated considerable potential in flexible and wearable electronics [5,6], as well as three-dimensional integrated circuits (3DICs) [7]. Despite these advantages, the ongoing evolution of such technologies demands further enhancement in the carrier mobility of AOS TFTs.

While carrier mobility can be easily increased by adjusting the composition of AOS materials, such as increasing the indium (In) content [8,9], this often leads to an undesirable increase in carrier concentration and defect density, adversely affecting other key device metrics such as threshold voltage, off-state current, and stability [8,9]. To enhance mobility without sacrificing other performance aspects, heterojunction channel structures comprising stacked AOS layers with varying mobilities and carrier concentrations have been developed [10-15]. The mobility enhancement in such structures is potentially attributed to quantum effects arising from specific band configurations. While numerous studies have validated this mechanism and reported various high-mobility heterojunction systems, [16-19] the relationship between band configuration, quantum well characteristics, and carrier mobility remains insufficiently understood.

The conduction band minimum (E_C) or the Fermi level (E_F) between the individual layers results in higher mobility in heterojunction TFTs. This enhancement is attributed to the formation of a deeper quantum potential well (PW), which facilitates more efficient carrier transport and thus leads to enhanced mobility.

This work was conducted in Guangdong Provincial Center for Oxide Semiconductor Devices and ICs and Shenzhen TFT and Advanced Display Lab, and supported financially by Advanced Materials-National Science and Technology Major Project under 2024ZD0604100, and National Natural Science Foundation of China under project U24A20297.

979-8-3315-8850-2/25 $31.00 © 2025 IEEE

II. EXPERIMENTAL DETAILS

Fig. 1 presents a cross-sectional schematic of the bottom-gate heterojunction a-IZO/a-IGZO TFTs fabricated in this study. Initially, a 100-nm-thick molybdenum (Mo) layer was deposited onto a clean glass substrate via DC sputtering and patterned using wet etching to form the gate electrode. Subsequently, a 200-nm-thick SiOx gate insulator (GI) layer was deposited via plasma-enhanced chemical vapor deposition (PECVD) at a process temperature of 300 °C. Following this, a-IZO and a-IGZO layers were sequentially deposited at room temperature using DC sputtering and patterned via wet etching. The IZO target had an In/Zn ratio of 5/1, while the IGZO target had an In/Ga/Zn ratio of 1/1/1. Thereafter, a lift-off process was employed to fabricate 50-nm-thick Mo as source/drain electrodes. Contact holes for the gate electrode were then formed using reactive ion etching (RIE). Finally, the devices were subjected to post-annealing in an oxygen atmosphere at 300 °C for 90 minutes.

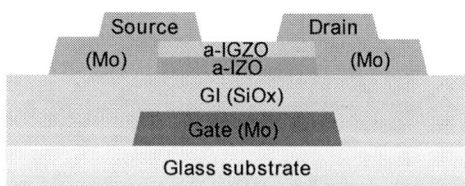

Fig. 1. Cross-sectional view of the fabricated heterojunction a-IZO/a-IGZO thin-film transistors.

The electrical characteristics of the fabricated TFTs were measured with a Keithley 4200-SCS semiconductor parameter analyzer in a dark vacuum environment. Ultraviolet photoelectron spectroscopy (UPS) was performed on a Thermo Fisher ESCALAB 250Xi system equipped with a monochromated Al Kα X-ray source (hv = 1486.6 eV). Optical transmittance spectra were obtained using an ultraviolet-visible (UV-Vis) spectrophotometer.

III. RESULTS AND DISCUSSION

Fig. 2(a) illustrates the transfer characteristics of heterojunction TFTs fabricated with different a-IZO sputtered argon/oxygen (Ar/O$_2$) ratios under a drain voltage of 0.1 V. As the argon/oxygen ratio decreases, the turn-on voltage of the devices shifts slightly in the positive direction, and the subthreshold swing slightly improves. These observations may be attributed to a reduction in oxygen vacancy-related defect states in the a-IZO layer as the oxygen content during sputtering increases, resulting in a decrease in both carrier concentration and defect density. Fig. 2(b) presents the extracted field-effect mobility as a function of gate voltage. It is evident that the mobility generally decreases with decreasing argon/oxygen ratio. For devices fabricated with an argon/oxygen ratio of 24/10, the field-effect mobility reaches approximately 70 cm^2/Vs, which is about five times that of single-layer a-IGZO devices and twice that of single-layer a-IZO devices. This result demonstrates that the heterojunction structure can significantly enhance the carrier mobility of oxide thin-film transistors. It is also noteworthy that when the argon/oxygen ratio is reduced to 9/25, the carrier mobility of the a-IZO/a-IGZO heterojunction TFTs become

almost identical to that of TFTs with a single-layer a-IZO channel.

It is well established in the literature that the significant mobility enhancement observed in heterojunction-channel AOS TFTs is primarily attributable to the formation of a two-dimensional electron gas (2DEG) at the heterointerface. [16-19] This conductive channel exhibits high carrier density and superior transport properties due to quantum confinement and reduced scattering. The emergence of the 2DEG is intrinsically linked to the band alignment between the constituent material layers. Given this fundamental relationship, a deeper understanding of the underlying mechanism necessitates further analysis from the perspective of band structure engineering.

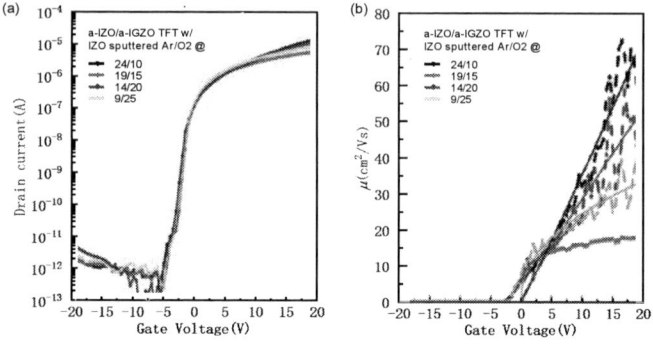

Fig. 2. (a) transfer characteristics and (b) field-effect mobility (μ) of heterojunction TFTs fabricated with different IZO sputtered argon/oxygen (Ar/O$_2$) ratios.

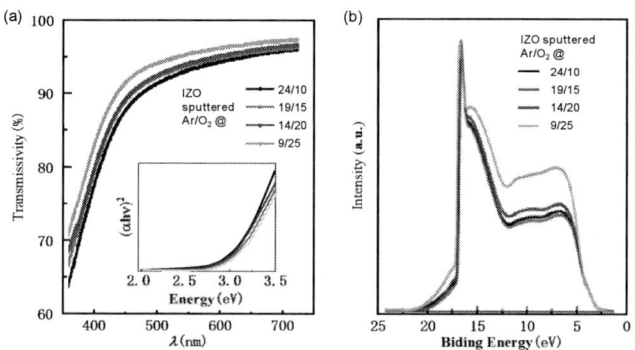

Fig. 3. (a) the optical transmittance spectra and (b) the ultraviolet photoelectron spectra of a-IZO films with different sputtered argon/oxygen (Ar/O$_2$) ratios, the Inset in Fig. 3(a) shows the Tauc plots for extracting the bandgap values.

Fig. 3(a) presents the optical transmittance spectra of a-IZO films deposited at different argon/oxygen ratios. The inset displays the corresponding Tauc plots used for extracting the optical bandgap values. As the argon/oxygen ratio decreases, the derived bandgap values are determined to be 3.06, 3.08, 3.05, and 3.12 eV, respectively. Fig. 3(b) shows the ultraviolet photoelectron spectra results of a-IZO films prepared under the same set of argon/oxygen conditions. The measured work functions, defined as the energy difference between the Fermi level (E$_F$) and the vacuum level (E$_0$), are 4.38, 4.41, 4.34, and 4.15 eV, respectively. Additionally, the differences between the

Fermi level and the valence band maximum are determined to be 2.77, 2.73, 2.69, and 2.63 eV, respectively. Based on these measurements, the values of E_0-E_C for the a-IZO films are calculated to be 4.09, 4.06, 3.98, and 3.66 eV. The corresponding band structures of a-IZO with different argon/oxygen ratios are illustrated in Fig. 4.

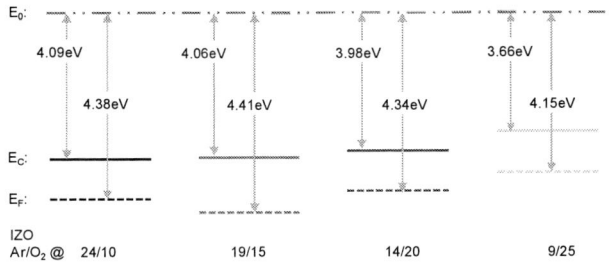

Fig. 4. Band diagrams of a-IZO films with different sputtered argon/oxygen (Ar/O₂) ratios.

Using the same methodology, the work function and E_0-E_C value for the prepared a-IGZO film are determined to be 4.12 eV and 3.44 eV, respectively. These results indicate that both the E_C and E_F of a-IZO films are lower than those of a-IGZO films, as shown in Fig. 5(a). When a heterojunction is formed between these two materials, electrons transfer from the a-IGZO layer to the a-IZO layer due to the energy level alignment. This carrier redistribution induces downward band bending in the a-IZO layer and upward band bending in the a-IGZO layer near the interface. As a result, a quantum potential well (PW) is formed, confining electrons within the a-IZO side of the junction and thus forming a 2DEG, which significantly enhances electron transport properties. A schematic illustration of this band alignment and the resultant PW is presented in Fig. 5(b).

Compared to the interface between a-IZO and the PECVD SiOx gate insulator (GI), the a-IZO/a-IGZO heterojunction interface exhibits better lattice matching. [11,13,19] Consequently, electrons transportation within the quantum well at the heterojunction interface experience significantly less scattering than those near the a-IZO/SiOx interface.[11,13,19] This explains the higher mobility observed in heterojunction TFTs compared to single-layer a-IZO TFTs.

As the sputtered argon/oxygen ratio decreases, the E_C and E_F of a-IZO films shift upward, reducing the energy difference between a-IZO and a-IGZO. As shown in Fig. 5(c), the reduced energy difference results in a shallower quantum PW, which weakens the electron confinement. Consequently, both the probability and the efficiency of electron transportation within the quantum PW are reduced, ultimately leading to a decrease in carrier mobility. At an argon/oxygen ratio of 9/25, the PW exhibits insufficient depth (the green curve in Fig. 5(c)) and is no longer able to effectively confine electrons. Consequently, carrier transport is dominated by the a-IZO/SiOx interface instead of the heterojunction interface, resulting in a mobility comparable to that of a single-layer a-IZO TFT (Fig. 2). Based on these observations, it can be concluded that a larger energy difference in either E_C or E_F between the constituent materials promotes the formation of a deeper quantum potential well, which enhances carrier confinement and improves transport

efficiency, thereby enhancing carrier mobility in heterojunction AOS TFTs.

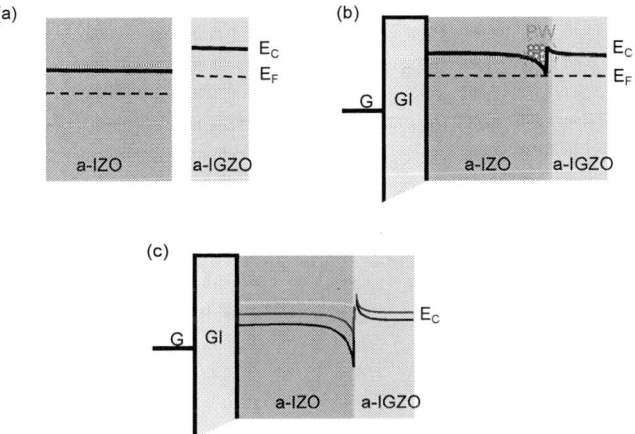

Fig. 5. Band diagrams of a-IZO and a-IGZO films (a) before and (b) after contact, (c) band diagram that of a-IZO/a-IGZO heterojunction with a-IZO sputtered at different argon/oxygen ratios.

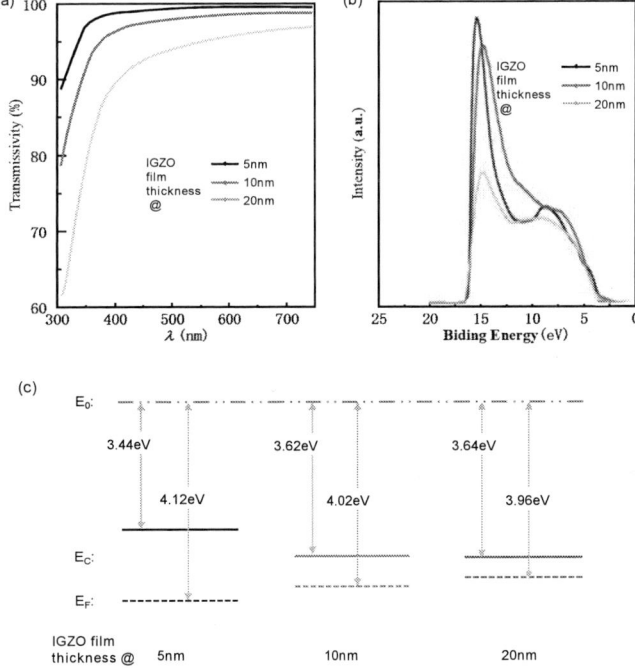

Fig. 6. (a) the optical transmittance spectra and (b) the ultraviolet photoelectron spectra of a-IGZO films with different thickness. (c) band diagrams of a-IGZO films with different film thickness.

To validate this conclusion and further enhance the mobility of the heterojunction TFTs, the band structure was modulated by adjusting the film thickness to achieve a deeper quantum potential well in the heterojunction channel. Optical transmittance spectra and UPS measurements were performed

979-8-3315-8850-2/25 $31.00 © 2025 IEEE

on a-IZO and a-IGZO films with thicknesses ranging from 5 to 20 nm. The corresponding results for a-IGZO films at different thicknesses are presented in Figs. 6(a) and 6(b), and the extracted band structures are summarized in Fig. 6(c). As shown, the E_C values for the 5, 10, and 20 nm-thick a-IGZO films are 3.44, 3.62, and 3.64 eV, respectively. The results indicate that the E_C of a-IGZO decrease rapidly with increasing thickness and tend to saturate beyond 10 nm. A similar thickness-dependent trend is also observed in a-IZO films (not shown). Based on these observations, to increase the quantum potential well depth, the thickness of the a-IZO layer should be increased, while that of the a-IGZO layer should be decreased. However, since the a-IZO thickness used in previous experiments was already 20 nm, further increases would not significantly affect the energy level. Therefore, the a-IZO thickness was kept constant. To effectively modulate the energy level, the a-IGZO thickness was varied to 3, 5, and 10 nm.

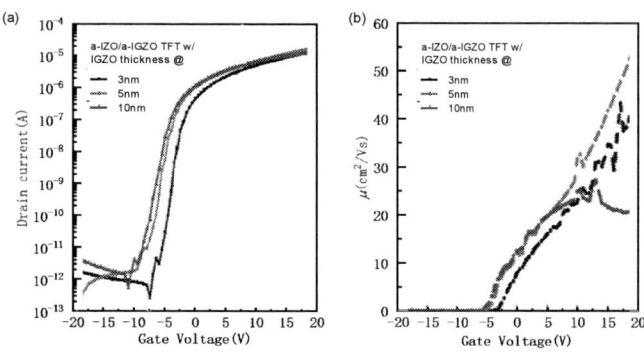

Fig. 7. (a) transfer characteristics and (b) field-effect mobility (μ) of heterojunction TFTs fabricated with different a-IGZO thickness.

Fig.7 presents the transfer characteristics and mobility curves of heterojunction TFTs fabricated with different a-IGZO thicknesses. As shown, the carrier mobility increases as the thickness of a-IGZO layer is reduced from 10 nm to 5 nm, which strongly supports the proposed quantum PW-based transport mechanism. However, when the thickness is further decreased to 3 nm, a slight degradation in mobility is observed. This decline can likely be attributed to the poor continuity of the sputtered film at such an ultrathin scale. At this thickness, the film may exhibit island formation or morphological imperfections, which act as scattering centers and introduce additional defects, thereby impairing carrier transport. Moreover, insufficient film continuity may disrupt the intended modulation of the band structure, limiting the effectiveness of heterojunction-based band alignment. Consequently, even though thinner layers generally enhance quantum confinement, the adverse effects of incomplete coverage and defect proliferation become dominant in the extreme thinness regime, leading to reduced device performance.

IV. CONCLUSION

This study demonstrates that band structure engineering through modulation of sputtered conditions and layer thickness is an effective strategy for enhancing carrier mobility in a-IZO/a-IGZO heterojunction TFTs. The experimental results confirm that a larger energy difference in E_C or E_F between the constituent layers leads to the formation of a deeper quantum

potential well, which facilitates more efficient carrier transport and results in higher mobility. Through optimization of the argon/oxygen ratio during a-IZO sputtering and the a-IGZO layer thickness, a significantly enhanced carrier mobility of approximately 70 cm^2/Vs is achieved, while maintaining a low off-state current of 1 pA and a near-zero threshold voltage. However, excessive reduction in a-IGZO thickness (e.g., to 3 nm) results in mobility degradation, likely due to compromised film continuity. These findings provide valuable insights into the mechanisms of band alignment and quantum confinement in heterojunction AOS TFTs and offer practical design guidelines for developing high-performance devices suitable for advanced electronic applications.

REFERENCES

[1] K. Nomura, H. Ohta, A. Takagi, T. Kamiya, M. Hirano, and H. Hosono, Room-temperature fabrication of transparent flexible thin-film transistors using amorphous oxide semiconductors, Nature, vol. 432, no. 7016, pp. 488–492, Nov. 2004.

[2] E. Fortunato, P. Barquinha, and R. Martins, Oxide semiconductor thin film transistors: A review of recent advances, *Adv. Mater.*, vol. 24, no. 22, pp. 2945–2986, Jun. 2012.

[3] M. Kimura, Emerging applications using metal-oxide semiconductor thin-film devices, Jpn. J. Appl. Phys., vol. 58, no. 9, pp. 090503-1–090503-10, Sep. 2019.

[4] Y. Lin, C. Liu, J. Zhang, Y. Yuan, L. Cai, L. Zhou, M. Xu, L. Wang, W. Wu, and J. Peng, Active-Matrix Micro-LED Display Driven by Metal Oxide TFTs Using Digital PWM Method, IEEE Trans. Electron Devices 2021, vol. 68, no. 11, pp.5656-5661.

[5] Li, X.; M. Mehedi, J. Jeon, and J. Jang, Stretchable Oxide TFTs for Wearable Electronics, Dig. Tech. Pap. - SID Int. Symp. 2017, vol. 48, no. 1, pp.42-46.

[6] J. Song, X. Huang, C. Han, Y. Yu, Y. Su, and P. Lai, Recent Developments of Flexible InGaZnO Thin-film Transistor, Phys. Stat. Sol. A, pp. 2000527-1-2000527-23. Jan. 2021,.

[7] Y. Du, J. Tang, Y. Li, Y. Xi, Y. Li, J. Li, H. Huang, Q. Qin, Q. Zhang, B. Gao, N. Deng, H. Qian, and H. Wu, Monolithic 3D Integration of Analog RRAM-Based Computing-in-Memory and Sensor for Energy-Efficient Near-Sensor Computing, Adv. Mater., vol. 36,pp.2302658, Oct. 2023.

[8] J. Raja, K. Jang, C.P.T. Nguyen, J. Yi, N. Balaji, S.Q. Hussain, and S. Chatterjee, Improvement of Mobility in Oxide-Based Thin Film Transistors: A Brief Review, Trans. Electr. Electron. Mater., vol. 16, no. 5, pp. 234-240, Aug. 2015.

[9] G. H. Kim, B. D. Ahn, H. S. Shin, W. H. Jeong, H. J. Kim, and H. J. Kim, Effect of indium composition ratio on solution-processed nanocrystalline InGaZnO thin film transistors, Appl. Phys. Lett., vol. 94, no. 23, pp. 233501-1-233501-3, May. 2019.

[10] S. Kim, C. J. Kim, J. C. Park, I. Song, S. W. Kim, H. Yin, E. Lee, J. C.Lee, and Y. Park, High Performance Oxide Thin Film Transistors with Dual Active Layers, in IEDM Tech. Dig., Dec. 2008, pp. 1-4.

[11] J. C. Park and H.-N. Lee, Improvement of the Performance and Stability of Oxide Semiconductor Thin-Film Transistors Using Dual-Stacked Active Layers, IEEE Electron Device Lett., vol. 33, no. 6, pp. 818-820, Jun. 2012.

[12] H. Y. Jung, Y. Kang, A.Y. Hwang, C. K. Lee, S. Han, D.-H. Kim, J.-U. Bae, W.-S. Shin, and J. K. Jeong, Origin of the improved mobility and photo-bias stability in a dual-channel metal oxide transistor, Scientific Reports, vol. 4, no. 1, pp. 3765-1-3765-8, Jan. 2014.

[13] H.-S. Kim, J. S. Park, H.-K. Jeong, K. S. Son, T. S. Kim, J.-B. Seon, E. Lee, J. G. Chung, D. H. Kim, M. Ryu, and S. Y. Lee, Density of States-Based Design of Metal Oxide Thin-Film Transistors for High Mobility and Superior Photostability, ACS Appl. Mater. Interfaces, vol. 4, no. 10, pp. 5416-5421, Sep. 2012.

[14] S. Taniguchi, M. Yokozeki, M. Ikeda, and T.-k. Suzuki, Transparent Oxide Thin-Film Transistors Using n-(In2O3)0:9(SnO2)0:1/InGaZnO4

Modulation-Doped Heterostructures, Jpn. J. Appl. Phys., vol. 50, no. 4S, pp. 04DF11-1-04DF11-4, Apr. 2011.

[15] Y. S. Rim, H. Chen, X. Kou, H. Duan, H. Zhou, M. Cai, H. J. Kim, and Y. Yang, Boost Up Mobility of Solution-Processed Metal Oxide Thin-Film Transistors via Confining Structure on Electron Pathways, Adv. Mater., vol. 26, no. 25, pp. 4273-4278, Jul. 2014.

[16] M. Lee, J.-W. Jo, Y.-Jeong Kim, S. Choi, S. M. Kwon, S. P. Jeon, A. Facchetti, Y.-H. Kim, and S. K. Park, Corrugated Heterojunction Metal-Oxide Thin-Film Transistors with High Electron Mobility via Vertical Interface Manipulation, Adv. Mater., vol. 30, no. 40, pp. 1804120-1-1804120-11, Aug. 2018.

[17] D. Koretomo, S. Hamada, Y. Magari, and M. Furuta, Quantum Confinement Effect in Amorphous In-Ga-Zn-O Heterojunction

Channels for Thin-Film Transistors, Materials, vol. 13, no. 8, pp. 1935-1-1935-12, Apr. 2020.

[18] M. M. Billah, A. B. Siddik, J. B. Kim, D. K. Yim, S. Y. Choi, J. Liu, D. Severin, M. Hanika, M. Bender, and J. Jang, High-Performance Coplanar Dual-Channel a-InGaZnO/a-InZnO Semiconductor Thin-Film Transistors with High Field-Effect Mobility" Adv. Electron. Mater., vol. 7, no. 3, pp. 2000896-1-2000896-11, Jan. 2021.

[19] H. Yang, X. Zhou, L. Lu, and S. Zhang, Investigation to the Carrier Transport Properties in Heterojunction-Channel Amorphous Oxides Thin-Film Transistors Using Dual-Gate Bias, IEEE Electron Device Lett., vol. 44, no. 1, pp.68-71, Jan. 2023.

2025 The 10th International Conference on Integrated Circuits and Microsystems

Effect of Temperature on ESD Characteristics in 30 nm Partially Depleted SOI MOSFET

Liye Yu
Academe of Electronics and Information Engineering
Ningbo University of Technology
Ningbo, China
3529627668@qq.com

Jingrui Wang*
Academe of Electronics and Information Engineering
Ningbo University of Technology
Ningbo, China
jruiwang@nbut.edu.cn

Zhongxu Chen
Academe of Electronics and Information Engineering
Ningbo University of Technology
Ningbo, China
1483871427@qq.com

Yixin Zheng
Academe of Electronics and Information Engineering
Ningbo University of Technology
Ningbo, China
2903708875@qq.com

Ziyan Dai
Academe of Electronics and Information Engineering
Ningbo University of Technology
Ningbo, China
1064296481@qq.com

Mingzhi Wan
Academe of Electronics and Information Engineering
Ningbo University of Technology
Ningbo, China
2133270862@qq.com

Abstract—This work investigates the temperature-dependent failure mechanisms induced by electrostatic discharge (ESD) in 30 nm partially depleted silicon-on-insulator (PD-SOI) NMOS transistors using Silvaco-TCAD simulations across an ambient temperature range of -45 to 165°C. The results demonstrate a significant performance degradation caused by ESD with increasing temperature, exhibiting an 18.75% reduction in breakdown voltage and an approximately 8.17% decrease in holding voltage. By analyzing key parameters such as total current density, lattice temperature, electric field distribution, and potential profile in detail, the fundamental physical mechanisms behind device failure are revealed. These findings offer crucial insights for designing and optimizing ESD protection devices under varied temperature conditions, ensuring reliability in advanced PDSOI technology.

Keywords—temperatures, ESD impacts, PD-SOI, Silvaco-simulations, failure modes

I. INTRODUCTION

Electrostatic discharge (ESD) has consistently represented an important reliability challenge within the semiconductor sector. For decades, ESD failure has remained a leading cause of integrated circuit (IC) malfunction, consistently threatening the stability of semiconductor devices and circuits [1-3]. To tackle this challenge, new ESD protection devices have been continuously developed, primarily aiming to reduce ESD-induced damage to semiconductor devices. However, the swift advancement of semiconductor technology presents new challenges for ESD protection design.

High temperature represents a typical operating environment for integrated circuits. In certain applications, the assembly and operation of integrated circuits must be performed at elevated temperatures [4-6]. This high-temperature environment imposes

This work was supported by National Innovation & Entrepreneurship Training Program for Undergraduates (NO. 202411058013).

greater demands on ESD protection design. As ESD parameters can experience thermally-induced alternative with rising temperature, the presentation of ESD protection devices might be impacted.

The ESD presentation of metal oxide semiconductor field effect transistors (MOSFETs) with various cell designs has been studied [7-11]. Partially Depleted Silicon-on-Insulator (PD-SOI) devices offer advantages including latch-up immunity, low parasitic capacitance, and high-frequency characteristics. Although PD-SOI technology has been largely superseded by Fully Depleted Silicon-on-Insulator (FD-SOI) in high-frequency applications, it remains irreplaceable in specific scenarios, such as RF front-end modules, noise-immune sensor interface circuits, and medium-voltage isolation devices for industrial control systems. Wang et al. [12] examined the triggering mechanism of 0.18-µm PD-SOI NMOS devices for ESD protection at elevated temperatures. However, the temperature-induced failure characteristics of nanoscale MOSFETs have not yet been thoroughly and systematically examined using simulation tools.

The failure characteristics of nanoscale MOSFETs due to temperature have not been thoroughly or systematically analyzed and summarized through simulation and physical mechanisms. This study examines the ESD failure characteristics of 30 nm partially depleted silicon-on-insulator (PD-SOI) NMOSFET power devices over a broad temperature range from -45°C to 165°C, using the Silvaco-TCAD simulation platform. This paper investigates the intrinsic causes of ESD-induced faults in power MOSFETs by applying pulse current to the device's drain and integrating multiple physical models for simulation analysis, while systematically detailing their temperature-dependent characteristics. The research findings provide substantial reference value and profound insights for designing and optimizing ESD protection devices, facilitating the advancement of ESD protection strategies and enhancing the reliability of semiconductor devices in complex temperature environments.

II. DEVICE STRUCTURE AND CONDITIONS OF SIMULATIONS

This study employs Silvaco-TCAD platform for simulating ESD effects, leveraging their superior transient analysis and computational performance to guarantee reliable outcomes. A precise 2D PD-SOI structure model was created with the Silvaco Atlas simulator, featuring well-defined mesh generation and accurate doping concentration settings. The simulated device has a width-to-length ratio (W_G:L_G) of 200nm:40nm and a 30nm channel length, as depicted in Figure 1. Silicon acts as the substrate material, and aluminum is utilized for source/drain contacts. Table 1 outlines the essential physical and technological parameters of the PD-SOI structure.

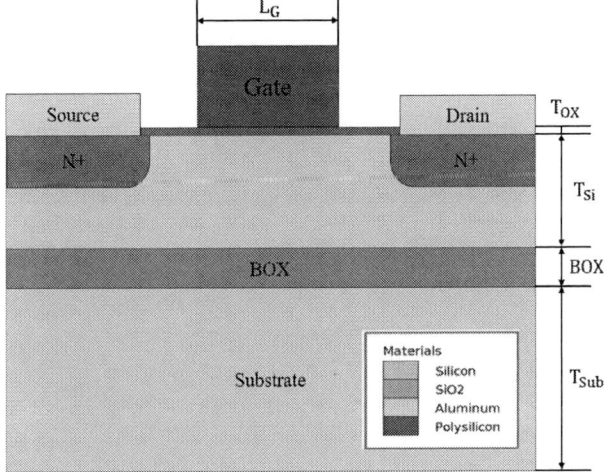

Fig. 1. 30nm PD-SOI structure model

To precisely simulate ESD phenomena, the computational framework concurrently solves standard semiconductor physics equations combined with electro-thermal equations to describe device self-heating effects. Given the intricate physical mechanisms in deep-submicron technologies, such as velocity overshoot, carrier diffusion, carrier-lattice temperature interactions, temperature-dependent mobility, and impact ionization, the energy balance model (EBM) was utilized rather than the traditional drift-diffusion (DD) model. The Giga-thermal model was utilized to simulate variations in internal heat flux, facilitating precise analysis under non-isothermal conditions.

The simulation integrates several key physical models: SRH recombination with temperature dependence, the Auger recombination model, Selberherr's impact ionization model (SELB), a hot carrier reliability evaluation model (hcte.el), a CVT model for carrier mobility calculations accounting for electric field and doping effects, and a temperature-dependent bandgap narrowing model (BGN).

During the ESD stress simulation, the Human Body Model (HBM) setup was utilized with these parameters:0.8V DC gate bias, a grounded source terminal, and a current pulse applied to the drain terminal featuring a 5 ns rise time, 1.8 μs fall time, and 0.1mA amplitude. The temperature range was set from -45°C to

165°C, with thermal boundary conditions defined at the bottom thermal contact using a thermal conductivity of 1000W/(cm²·K).

TABLE I. DEVICE PARAMETERS USED FOR SIMULATION

Parameter name	Symbol	Value	Unit
Gate length	L_G	40	nm
Gate width	W_G	200	nm
Gate oxide thickness	T_{OX}	2	nm
Buried oxide (BOX) thickness	B_{OX}	10	nm
Top Si thickness	T_{SI}	75	nm
Substrate thickness	Tsub	125	nm
Source/Drain doping	N_{SD}	2×10^{20}	cm⁻³
Channel doping	N_C	2×10^{18}	cm⁻³
Substrate doping	Nsub	1×10^{13}	cm⁻³

III. RESULTS

Figure 2 illustrates the drain current transient response characteristics of a 30 nm PD-SOI MOSFET during pulse current injection. Figure 3 presents the ESD I-V characteristics of the device over a temperature range from -45°C to 165°C, obtained through TCAD simulations, with the inset focusing on the detailed aspects of the ESD hysteresis curves. The experimental results indicate that the device's ESD protection performance exhibits significant temperature dependence. Specifically, the breakdown voltage (V_{T1}) decreases monotonically with increasing temperature, ranging from 6.4 V at -45°C to 5.2 V at 165°C within the tested temperature range. This indicates an absolute change of 1.2 V, which constitutes 18.75% of the (V_{T1}) value at -45°C.

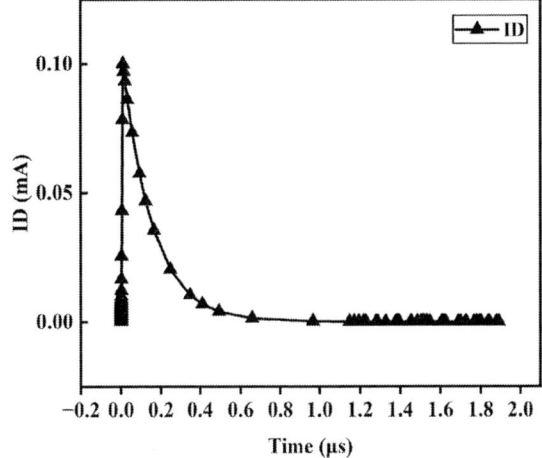

Fig. 2. Demonstration of the transient drain current behavior

By analyzing the ESD impact response process in Figure 4, the dynamic behavior can be divided into three distinct phases: (1) the initial rapid rise phase, where the drain voltage quickly peaks due to ESD current injection; (2) the voltage drop phase, during which the integrated ESD protection circuit (e.g., protection diode) activates to provide a low-resistance path for current dissipation; and (3) the steady-state recovery phase, where the system voltage gradually stabilizes at a lower level, indicating the declining effect of the ESD impact. Notably, a high-temperature environment significantly enhances the device's conductivity, enabling it to enter and sustain a low-resistance state under the same current excitation, thereby driving the drain voltage toward zero. This phenomenon underscores the significant impact of temperature on the operational characteristics of ESD protection devices.

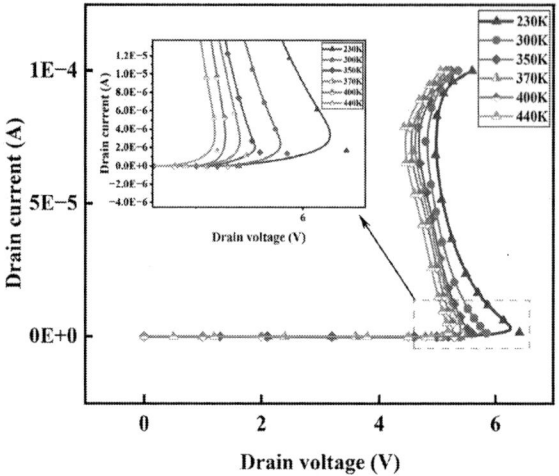

Fig. 3. The temperature-dependent ESD current-voltage characteristics

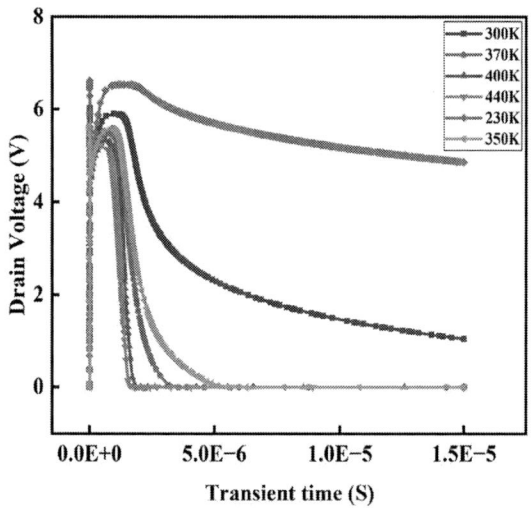

Fig. 4. Analysis of the ESD response process

A. The First Breakdown Voltage Versus Temperature

Figure 5 shows the trend of the initial breakdown voltage (V_{T1}) for the PD-SOI device. As the temperature rises from -45°C to 165°C, (V_{T1}) gradually drops by about 1.2 V, representing a 18.75% reduction from the initial (V_{T1}) value of 6.4 V at -45°C. For a 30 nm PD-SOI device, (V_{T1}) denotes the turn-on voltage of the parasitic bipolar junction transistor BJT. Avalanche breakdown takes place at the drain-body junction, where avalanche multiplication produces a significant number of electron-hole pairs. The hole current enters the body region, causing a voltage drop across the body resistance and raising the body potential. This forward-biases the body-source junction, causing the parasitic BJT in the PD-SOI device to turn on.

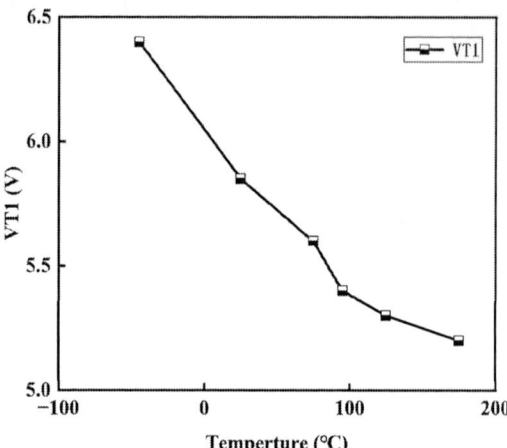

Fig. 5. The variation trend of the initial breakdown voltage (V_{T1}) of the PD-SOI device

The temperature increase involves the following mechanisms:

First, as the temperature increases, the turn-on voltage of the parasitic BJT ($V_{BS,on}$) decreases. Khanna [13] investigated the temperature dependence of the forward bias voltage in a p-n junction, as described by the following equation:

$$\frac{\partial V}{\partial T} = \frac{V - (3V_{thermal} + E_g)}{T} \qquad (1)$$

Here, (V) represents the forward bias voltage of the p-n junction, (T) denotes the thermodynamic temperature, ($V_{thermal}$) signifies the thermal voltage, and (E_g) indicates the bandgap width. Equation (1) shows that at room temperature , the turn-on voltage drops by 2 mV for each 1 K rise in temperature. At other temperatures, ($\partial V/\partial T$) stays negative, showing that the forward bias voltage of the p-n junction drops as the temperature rises.

Secondly, at room temperature, the impurities in the doped substrate are fully ionized, and intrinsic excitation within the device is negligible, leading to a nearly constant carrier concentration. As the temperature increases, valence band electrons are thermally excited to the conduction band, causing an exponential rise in the intrinsic carrier concentration (n_i). On the other hand, as the temperature rises, lattice vibration scattering progressively becomes the primary scattering mechanism for carriers. The reduction in carrier mobility has a more significant negative effect than the rise in carrier concentration, leading to an increase in body resistance as temperature rises. Thus, under identical body hole current conditions, the body potential also rises with temperature.

Figure 6 shows the distribution of total current density in the device at the first breakdown under varying temperatures. It is evident that as the ambient temperature rises, the current density from the drain contact to the body region progressively diminishes. Under the same ESD current, the current density distribution decreases as temperature rises, indirectly suggesting that body resistance (R_{body}) increases with temperature. This aligns with theoretical derivations. Thus, under the same ESD current pulse, higher temperatures make the parasitic BJT turn-on effect more likely, resulting in a higher first breakdown voltage.

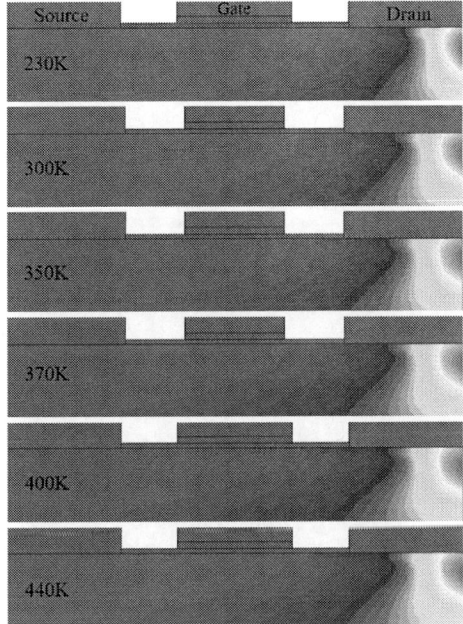

Fig. 6. The current density distribution during initial breakdown at varying temperatures

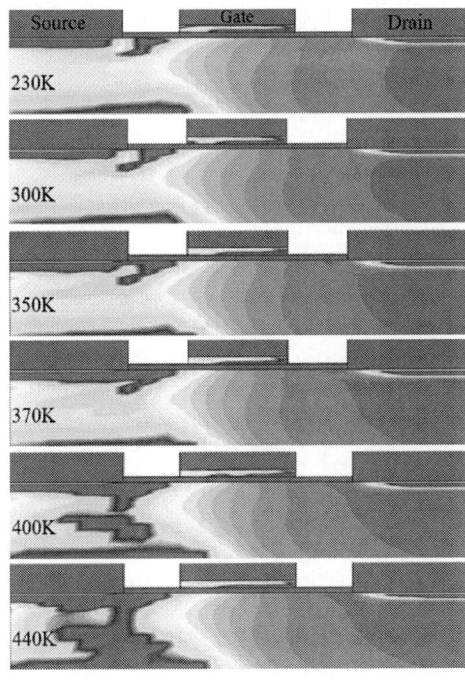

Fig. 7. Impact ionization distribution versus temperature under same ESD Current

When the electric field at the drain-body junction surpasses the critical breakdown field, the current pulse at the drain terminal creates numerous secondary electron-hole pairs, leading to avalanche multiplication. The efficiency of avalanche multiplication is characterized by the impact ionization rate. As

illustrated in Figure 7, the impact ionization rate inside the device diminishes as temperature rises, correlating with the electric field and carriers' mean free path.

B. The Electric Field Versus Temperature

Before the first breakdown point, the ESD current remains at a very low level, passing through the drain contact, reverse drain-body junction, body region, and body contact. Due to the high resistance of the reverse drain-body junction, the relatively high ESD voltage is primarily clamped at this junction, generating an extremely high electric field. Thus, the electric field distribution at the initial breakdown point signifies (V_{T1}). Figure 9 illustrates the electric field distribution along Path 1, as marked in Figure 8.

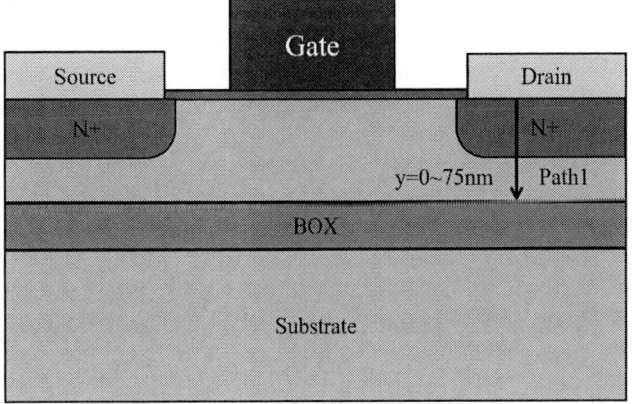

Fig. 8. The distribution of path 1 in the structure

Clearly, the electric field intensifies as the temperature rises. The relationship between the critical electric field (E_c) and the avalanche threshold voltage (V_{ava}) is represented by the following equation:

$$V_{ava} = \left(\frac{32\varepsilon_0\varepsilon_r E_c^3}{9\alpha_j q}\right)^{0.5} \quad (2)$$

Here, (ε_0) and (ε_r) stand for the permittivity of free space and the relative dielectric constant, respectively. (α_j) indicates the impact ionization rate, and (q) represents the elementary charge. Equation (2) outlines the avalanche mechanism within a linearly graded PN junction. As the temperature rises, the mean free path of carriers in the space charge region shortens.

A higher (E_c) is needed to supply adequate kinetic energy for initiating avalanche breakdown. As a result, the thermal coefficient of the avalanche mechanism is directly proportional to temperature.

The DUT is activated by the avalanche breakdown of the drain-body junction. Avalanche multiplication generates numerous electron-hole pairs, with holes migrating toward the body contact. This results in a voltage drop across the body resistance (R_{body}) and an increase in the local body potential (V_{body}). When (V_{body}) reaches a sufficiently high level, the source-body junction activates, leading to the triggering of the parasitic BJT. Thus, the MOSFET's (V_{T1}) is determined by the

979-8-3315-8850-2/25 $31.00 © 2025 IEEE

avalanche critical electric field and avalanche threshold voltage at the drain-body junction, causing (V_{T1}) to rise with temperature.

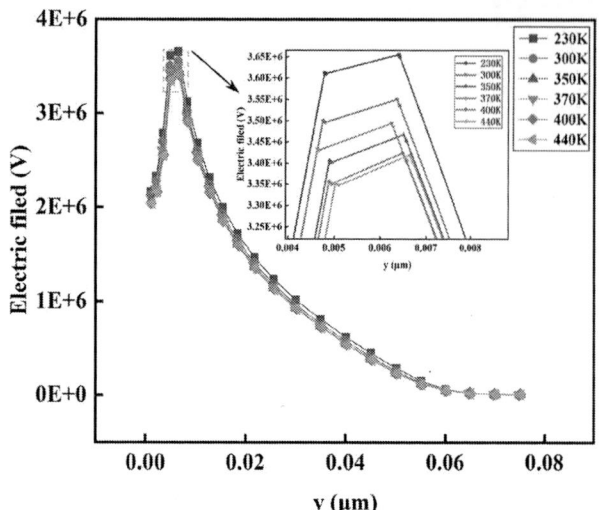

Fig. 9. The electric field distribution along Path 1

C. The Holding Voltage Versus Temperature

Figure 10 illustrates the relationship between the holdover voltage (V_H), derived from the ESD (Electrostatic Discharge) I-V curves, and the ambient temperature. Clearly, (V_H) increases as the temperature decreases. Specifically, as the temperature drops from -45°C to 165°C, (V_H) reduces from 6.12V to 5.62V, marking an 8.17% decrease.

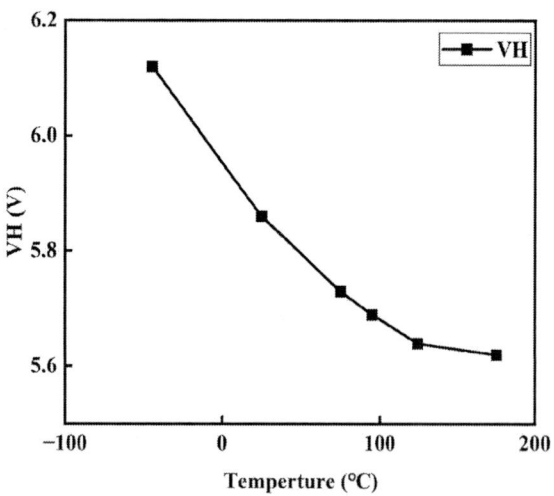

Fig. 10. The relationship between the holdover voltage (V_H)

(V_H) represents the minimum clamping voltage required to maintain the conduction of the parasitic BJT (Bipolar Junction Transistor). Figure 11 illustrates the potential distribution at different temperatures, showing that (V_{DB}) contributes most significantly to the total potential difference of (V_H). As the ambient temperature increases from -45°C to 165°C, (V_{DB}) progressively decreases, exhibiting the most pronounced temperature effect.

Figure 12 illustrates the potential distribution along Path 2, with a 0.5V decrease in potential observed from -45°C to 165°C. As the temperature rises, (R_{body}) increases while $(V_{BS,on})$ decreases. This increase in temperature facilitates the condition needed to sustain the conduction of the parasitic BJT $(V_{body} \geq V_{BS,on})$. Moreover, the current gain (β) of the parasitic BJT rises, boosting its current driving capacity.

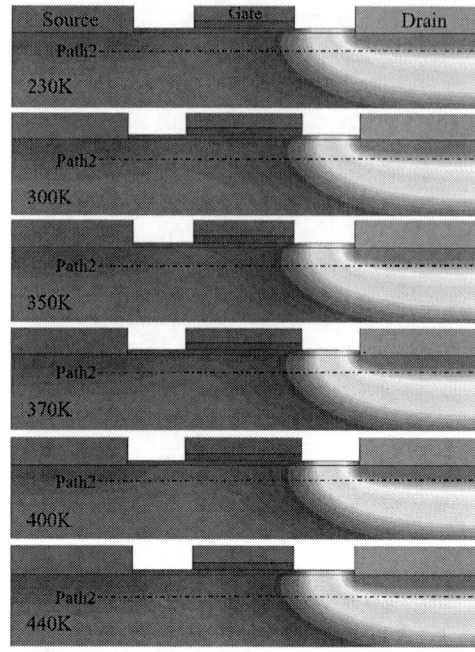

Fig. 11. Analysis of the potential distribution

$$\beta(T) = \beta(T_0) + \aleph\{1 + \beta(T_0)\}\Delta T \qquad (3)$$

Here, (T) denotes the ambient temperature, $(\beta(T))$ represents the current gain of the parasitic BJT, $(T_0 = 273K)$, and $(\aleph)$ indicates the temperature coefficient.

$$I_{CEO}(T) = [\beta_0 \cdot \aleph(1 + \beta_0) \cdot \Delta T][I_{CBO}(T_0) \\ \cdot 2^{\frac{\Delta T}{10K}}] \qquad (4)$$

$$\aleph = \frac{1}{\beta(T_0)}\frac{d\beta}{dT} \qquad (5)$$

In these equation, (I_{CEO}) denotes the collector-emitter current with the base open, (I_{CBO}) is the collector-base current with the emitter open, (β) represents the current gain at 273K, and $(\aleph)$ is the temperature coefficient, generally ranging from 0.1 to 1 for silicon devices.

The rise in temperature impacts (R_{body}), $(V_{BS,on})$, and (β), helping to sustain the conduction of the parasitic BJT and maintain a certain level of current flow. Ultimately, this results in a reduction of (V_H).

979-8-3315-8850-2/25 $31.00 © 2025 IEEE

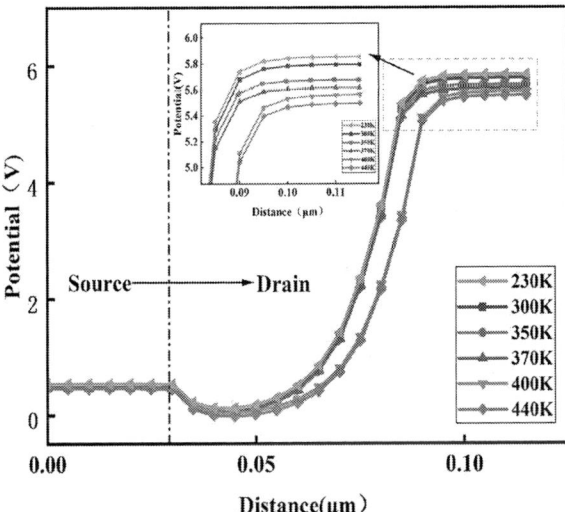

Fig. 12. The potential distribution in the source along Path 2

D. The Lattice Temperature Versus Temperature

Figure 13 provides an in-depth analysis of the internal lattice temperature changes within a MOSFET device during ESD events, examined as a function of ambient temperature. The results clearly show that, under different ambient temperature conditions, the internal hotspot region consistently forms beneath the drain when the device experiences ESD current surges. Moreover, as the operating temperature gradually rises, the internal lattice temperature of the device also shows a progressive increase, with the maximum lattice temperature significantly climbing from 362K (approximately 89°C) to 516K (approximately 243°C).

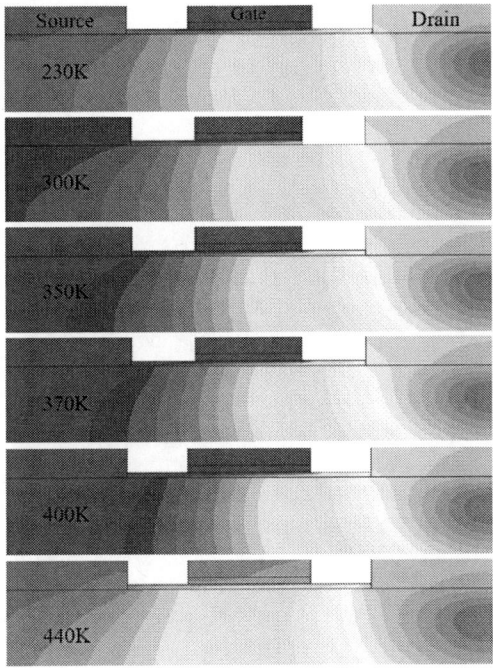

Fig. 13. The internal lattice temperature variations within a MOSFET

This phenomenon is mainly due to the increased impact ionization effect within the device as the temperature rises, resulting in a corresponding increase in lattice temperature. Impact ionization is the process in which electrons, driven by a strong electric field, collide with lattice atoms to create extra electron-hole pairs, thus boosting the lattice's thermal energy.

Although the highest internal hotspot temperatures across the four temperature ranges did not reach the melting point of silicon semiconductors (approximately 1685K or 1412°C) and no permanent device damage was detected, it is expected that ESD characteristic parameters, such as holdover voltage (V_H) and breakdown voltage (V_{T1}), may experience significant changes. This is also confirmed in the text mentioned above.

These variations might affect the device's reliability and performance, especially under extreme temperature conditions. Hence, particular attention should be given to the ESD protection design for MOSFET devices functioning in high-temperature settings, ensuring stable and reliable performance during ESD occurrences. Moreover, these results offer essential experimental support and theoretical direction for further optimizing and enhancing ESD protection strategies for MOSFET devices.

IV. CONCLUSION

This study systematically investigates the temperature characteristics of a 30 nm PD-SOI MOSFET device under ESD current pulsing, employing the Silvaco-TCAD simulation platform. By establishing a wide temperature range test environment from -45°C to 165°C, the relationship between the device's trigger voltage (V_{T1}) and temperature variations is thoroughly investigated. The research findings indicate that the device's (V_{T1}) decreases significantly with rising temperature, while the maximum lattice temperature increases correspondingly. Further mechanistic analysis reveals that the temperature dependence of (V_{T1}) primarily stems from the temperature effects on the channel current and the triggering behavior of the parasitic BJT. Quantitative research shows that as the operating temperature increases from -45°C to 165°C, the device's (V_{T1}) drops by up to 18.75% and 8.17% decrease in (V_H). This result clearly shows the high sensitivity of ESD stress characteristics to temperature variations, assisting in the development of temperature-insensitive ESD protection devices and providing important reference standards for ESD protection in highly reliable integrated circuits.

ACKNOWKEDGMENT

This thesis would not have been possible without the support and help of many individuals, to whom I offer my sincerest thanks.

First and foremost, I sincerely thank my supervisor, Professor Jingrui Wang, for her invaluable guidance throughout the selection of the research topic, experimental design, and manuscript preparation.

I am deeply thankful to my lab colleagues, including Ziyan Dai, and especially my respected senior, Rongbiao Xiang, for their skilled guidance and generous support in experimental methods, data analysis, and theoretical validation. Every discussion with them has provided me with new insights.

I wish to acknowledge the support from the School of Electronics and Information Engineering at Ningbo University of Technology for providing vital equipment, funding, and data resources, all of which were essential for completing this study successfully. I also thank the reviewers and editors for their constructive feedback, which significantly improved the manuscript.

Finally, I sincerely thank all those who have contributed, directly or indirectly, to this research, even though their names are not specifically listed here.

REFERENCES

[1] Scheier S, Zur Nieden F, Arndt B, et al. Simulation of ESD thermal failures and protection strategies on system level[J]. IEEE Transactions on Electromagnetic Compatibility, 2015, 57(6): 1309-1319.

[2] Shaalini C, Tan P K, Zhao Y Z, et al. Failure analysis on 14 nm FinFET devices with ESD CDM failure[J]. Microelectronics Reliability, 2018, 88: 321-333.

[3] Thomson N A, Yang X, Rosenbaum E. Soft-failures induced by system-level ESD[J]. IEEE Transactions on Device and Materials Reliability, 2017, 17(1): 90-98.

[4] Elahipanah H, Kargarrazi S, Salemi A, et al. 500° C high current 4H-SiC lateral BJTs for high-temperature integrated circuits[J]. IEEE Electron Device Letters, 2017, 38(10): 1429-1432.

[5] Alexandru M, Banu V, Jorda X, et al. SiC integrated circuit control electronics for high-temperature operation[J]. IEEE Transactions on Industrial Electronics, 2014, 62(5): 3182-3191.

[6] Tang Z, Tang X, Zhang Y, et al. 4H-SiC integrated circuits for high-temperature applications[J]. Journal of Crystal Growth, 2023, 605: 127060.

[7] Kyoung-Il Do, Jong-Il Won, and Yong-Seo Koo. "A 4H-SiC MOSFET-based ESD protection with improved snapback characteristics for high-voltage applications." IEEE Transactions on Power Electronics 36.5 (2020): 4921-4926.

[8] Zou W S, Chen J J, Lee K Y, et al. Analysis of ESD capability of SiC MOSFET with various cell designs[J]. Japanese Journal of Applied Physics, 2025.

[9] Mishra R, Ioannou D E, Mitra S, et al. ESD performance of 65 nm partially depleted n and p channel SOI MOSFETs[J]. Solid-state electronics, 2010, 54(4): 357-361.

[10] Li M Z. Effect of high-temperature on holding characteristics in MOSFET ESD protecting device[J]. Acta Phys. Sinica, 2022, 71(12).

[11] Xiao Y, Liu C, Zhang Y, et al. "Temperature dependence of ESD effects on 28 nm FD-SOI MOSFETs." Engineering Reports 6 (2024): e12729.

[12] Wang J X, Li X J, Zhao F Z, et al. Trigger mechanism of PDSOI NMOS devices for ESD protection operating under elevated temperatures[J]. Chinese Physics B, 2021, 30(7): 078501.

[13] Do KI, Lee BS, Koo YS. Study on 4H-SiC GGNMOS based ESD protection circuit with low trigger voltage using gate-body floating technique for 70-V applications. IEEE Electron Device Lett. 2018;40(2):283-286.

2025 The 10th International Conference on Integrated Circuits and Microsystems

A Package-on-Package Module with Integrated FPGA Minimum System for 48-Channel Synchronous Sampling 16-bit ADC

Han Li
National Key Laboratory of Integrated Circuits and Microsystems,
Xi'an Microelectronics Technology Institute,
Xi'an, China
18966829551@163.com

Jiangbo Wei
National Key Laboratory of Integrated Circuits and Microsystems,
Xi'an Microelectronics Technology Institute,
Xi'an, China
WjbElec@163.com

Yuan Miao
National Key Laboratory of Integrated Circuits and Microsystems,
Xi'an Microelectronics Technology Institute,
Xi'an, China
miaoyuan0911@163.com

Weize An
National Key Laboratory of Integrated Circuits and Microsystems,
Xi'an Microelectronics Technology Institute,
Xi'an, China
2287795983@163.com

Xudong Wu
National Key Laboratory ofIntegrated Circuits and Microsystems,
Xi'an Microelectronics Technology Institute,
Xi'an, China
1120367499@qq.com

Huan Yu
National Key Laboratory ofIntegrated Circuits and Microsystems,
Xi'an Microelectronics Technology Institute,
Xi'an, China
13572415881@126.com

Chao Wang*
National Key Laboratory ofIntegrated Circuits and Microsystems,
Xi'an Microelectronics Technology Institute,
Xi'an, China
12207935@qq.com

Abstract—This study investigates microsystem technology for multi-channel analog signal measurement in Civil Space Industry to address the challenges of miniaturization, high reliability, high performance, and low cost in Spaceborne equipment. First, the architecture of a data processing and control system based on a multi-channel analog signal measurement system is introduced. Then, through low-cost microsystem integration technologies centered on simplified electrical principle design, microassembly, and Package-on-Package (PoP) packaging, the core hardware of the system is reduced to 20% of its original size, significantly decreasing volume, dimensions, and weight. Additionally, the long-term reliability of interfacial adhesion between metal materials and potting materials is evaluated to resolve signal integrity control issues in the measurement microsystem. A universal solution for the design and implementation of a multi-channel analog signal measurement microsystem is proposed, and its reliability is validated through physical testing, confirming the feasibility of this miniaturized, high-reliability, high-performance, and low-cost microsystem integration approach.

Keywords—microsystem, PoP, multi-channel, ADC, synchronous sampling

I. INTRODUCTION

Current mainstream solutions involve using analog switches to toggle ADC inputs, enabling time-division sampling across channels. However, this method suffers from low sampling rates, poor accuracy, the need for settling time during channel switching, and difficulties in debugging. Alternatively, employing multiple ADC can improve accuracy and sampling rates but at the cost of higher circuit expenses, larger volumes, and limited applicability across diverse scenarios [1].

Civil Space Industry systems frequently require multi-channel analog signal measurements, primarily for telemetry, health monitoring, and sensor applications, often involving hundreds of measurement channels [2]. Traditional solutions fail to meet the demands of next-generation aerospace equipment for miniaturization and intelligence, necessitating the development of multi-channel analog measurement circuits to achieve product miniaturization, bus-based design, intelligence, and low-cost off-the-shelf solutions [3].

To address these challenges, this study explores the architecture and integration techniques of a multi-channel analog signal measurement microsystem. The goal is to deliver a modular product featuring low cost, multi-channel synchronous sampling ADC, 16-bit ADC conversion, and multiple communication methods (e.g., LVDS, QSPI), while reducing the product size to 20% of its board-level counterpart.

II. MULTI-CHANNEL ANALOG SIGNAL MEASUREMENT MICROSYSTEM

This paper proposes a heterogeneous processing and control system based on front-end ADC signal conversion and back-end programmable logic devices (FPGAs) [4]. The system provides functional architecture for control, data preprocessing, and data exchange, effectively addressing the miniaturization requirements of multi-channel analog signal measurement applications. The system architecture is illustrated in Figure 1.

979-8-3315-8850-2/25 $31.00 © 2025 IEEE

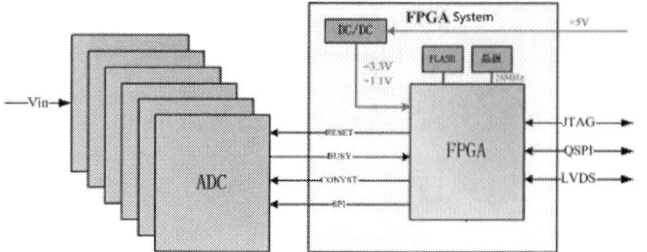

Fig. 1. Architecture of the Novel Multi-Channel Analog Signal Measurement Microsystem

In this architecture, FLASH serves as program storage, backup, and reconfiguration; an internal LDO supplies the three power levels required for FPGA operation; and a crystal oscillator provides the necessary clock signals.

The microsystem integration process primarily includes ceramic packaging, PoP packaging, plastic packaging, and other specialized structures (e.g.,wafer-level packaging (WLP)) [5]. Given the need for low-cost miniaturization and small-batch production in the described multi-channel analog signal measurement system, PoP packaging was selected, with structural simplifications tailored to practical requirements.

Fig. 2. Traditional PoP Structure

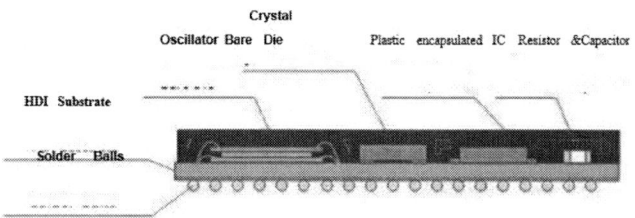

Fig. 3. Low-Cost PoP Structure for Multi-Channel Analog Signal Measurement Microsystem

Despite simplifying the original PoP structure using mature technologies, practical challenges emerged after completing the principle design. Ensuring signal reliability in the core FPGA system necessitated placing the crystal oscillator close to the FPGA, meaning the oscillator had to be encapsulated within the PoP potting body. Typically, SMD-type crystal oscillators use Kovar shells with ceramic substrates, raising concerns about potting reliability.

Traditionally, thermal mismatch issues arise between metal materials and potting compounds (e.g., epoxy resin) due to differences in their coefficients of thermal expansion (CTE) Metals generally exhibit higher CTE; for example, copper has a CTE of 17 ppm/°C, while epoxy resins, as polymer materials, have significantly lower CTE, typically ranging from 20-70 ppm/°C (below the glass transition temperature). This CTE disparity induces thermal residual stress during

processing, and the curing shrinkage of epoxy resins further contributes to residual stress, as expressed in Equation (1)[6]. These stresses can lead to interfacial defects such as cracks, compromising long-term reliability [7]. For high-density devices, interfacial defects between dissimilar materials are often a primary cause of failure [8].

$$0=os+OH_2=\triangle L_s \times E_1 + (aram) \times (Tp-T) \times E_1 \quad (1)$$

Where, o: Internal stress in the potting material. os: Shrinkage stress post-curing. OHz: Thermal stress. $\triangle$Ls: Shrinkage deformation of the epoxy potting material after curing. Er: Elastic modulus of the epoxy potting material.ar. CTE of the potting material. am: CTE of internal components and interconnects. Tp: Maximum temperature during potting material curin. Ti: Room temperature.

III. CRYSTAL OSCILLATOR POTTING PROCESS VERIFICATION

Thermodynamic simulations were conducted for the multi- channel analog signal measurement microsystem module using Ansys Workbench. The simulation model is shown in Figure 4.

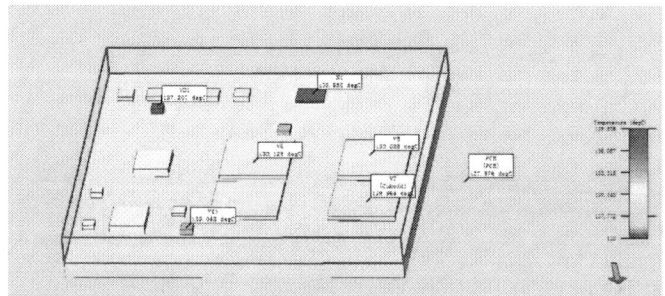

Fig. 4. Thermodynamic Simulation Model

Key parameters for the thermodynamic simulation are listed in Table 1.

TABLE I. KEY MATERIAL PARAMETERS FOR SIMULATION

Component	Material	Density (kg/m³)	Young's Modulus (GPa)	Poisson's Ratio	CTE (ppm/°°C)
Oscillator metal cover of Crystal	Kovar	8800-9000	130-210	0.25-0.35	4.6-6
Substrate	FR4	1610	26.2	0.166	Plane:15 / Vertical: 41
Poting body	Epoxy resin	1700-2100	11.5	0.3-0.4	20-21
Adhesive	Insulatin gglue	1000	10	0.38	6

Simulation Results: Theoretical analysis suggests a potential thermal mismatch risk between Kovar and potting materials. However, b y carefully controlling material parameters for both, the potting structure can meet the reliability requirements of highly reliable products [9]. Mechanical simulation as follows.

979-8-3315-8850-2/25 $31.00 © 2025 IEEE

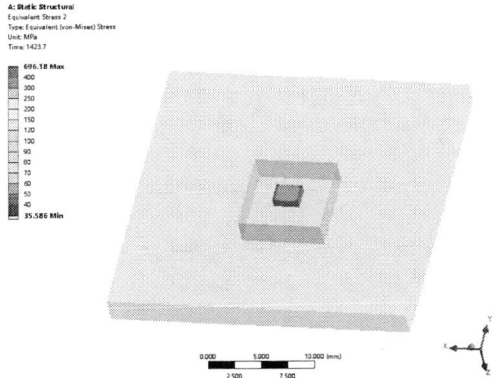

Fig. 5. Mechanical simulation Model

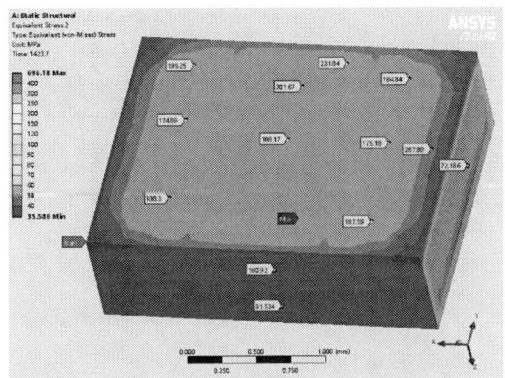

Fig. 6. Stress contour plot on the front surface of the crystal oscillator

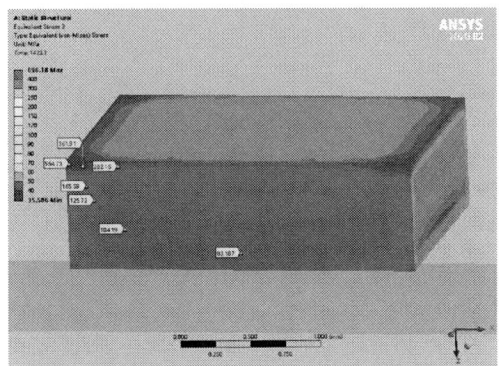

Fig. 7. Stress contour plot on the side surface of the crystal oscillator

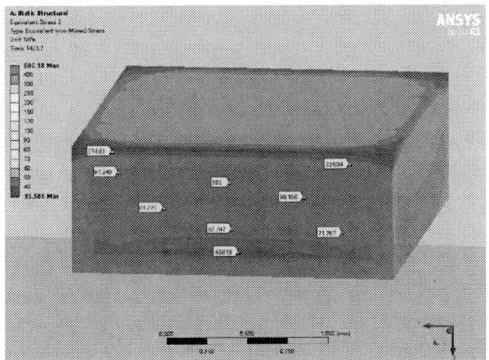

Fig. 8. Stress contour plot on the side surface of the crystal oscillator

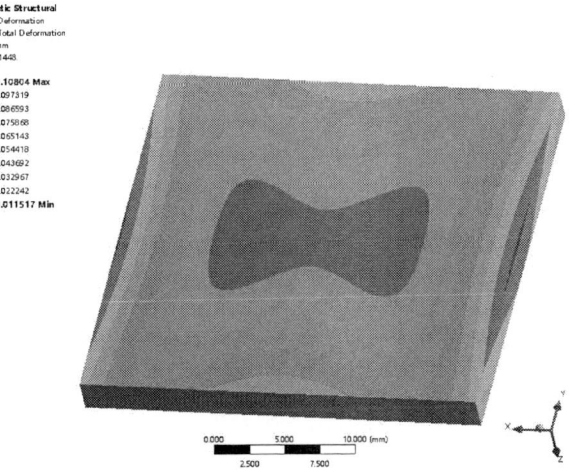

Fig. 9. Analysis of internal displacement results caused by stress in the module under extreme conditions

The simulation results comply with IPC mechanical reliability standards.

Process Validation Results: To validate the theoretical conclusions, the developed microsystem product underwent reliability test. The tests included temperature cycling for both potted oscillators and finished modules, with conditions set at -65°C to 150°C (15 minutes at each extreme, transition time <1 minute) (Table 2).

TABLE II. PROCESS VALIDATION RESULTS

No.	Oscillator Dimensions (mm)	Module Dimensions (mm)	Test Conditions	Results
1	2.65×2.15×0.9	29×29×4	-65°C-150°C, 1300 cycles	Electrical performance passed; no delamination observed in cross-section (Figure 5)
2	5.0×3.2×1.2	31×28×11	-65°C-150°C, 200 cycles	Electrical performance passed

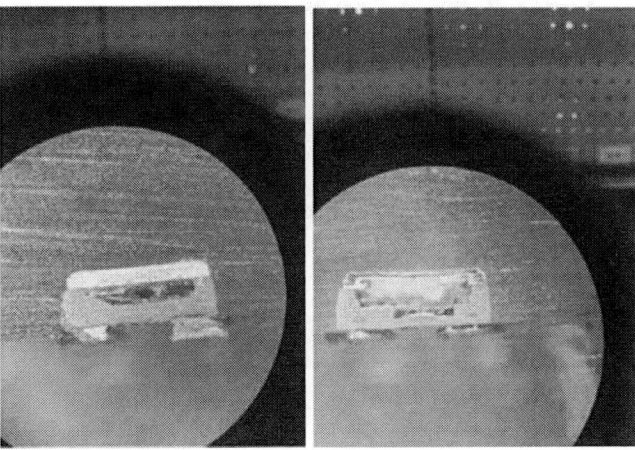

Fig. 10. Cross-Sectional Micrograph

Post-temperature cycling, cross-sectional analysis revealed no delamination at the epoxy-metal interface, except within the oscillator cavity. All results met highly reliable circuit reliability standards. This study addressed the fundamental issue of clock stability for signal integrity control in microsystem modules (Figure 10).

The module substrate is a 6-layer PCB measuring 29 mm× 29 mm×4 mm. The internal layout (Figure 6) includes: Three ADC groups(top-left), each with two vertically stacked ADC bare dies and nearby passive components. An FPGA minimal system(bottom-right), comprising an FPGA, PI NOR FLASH bare die, and a 2.5 mm ×2.0 mm ceramic oscillator. Two power supplies (bottom-left) for the FPGA. Clear digital-analog partitioning, with a single-point ground via the soldered baseplate.

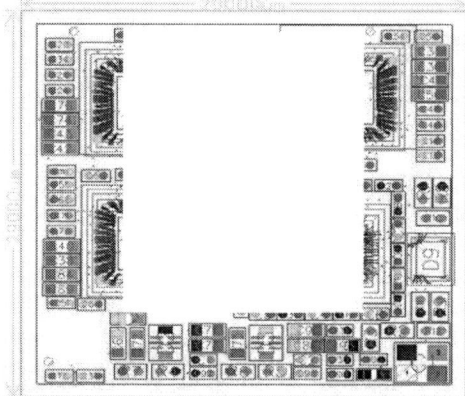

Fig. 11. Internal Structure of the Product

(a) Board-Level   Product(65mm×65mm×6mm)

(b) Module-Level   Product(29mm×29mm×4mm)

Fig. 12. Size Comparison of the Miniaturized Multi-Channel Analog Signal Measurement System

The test system is powered by an external supply or a Power-over-Ethernet (PoE) switch, with onboard secondary power for the module and test circuits [10]. A PC sends test commands, and a burn-in board outputs excitation signals. Collected data is transmitted via UART, QSPI, or LVDS to the PC for analysis.

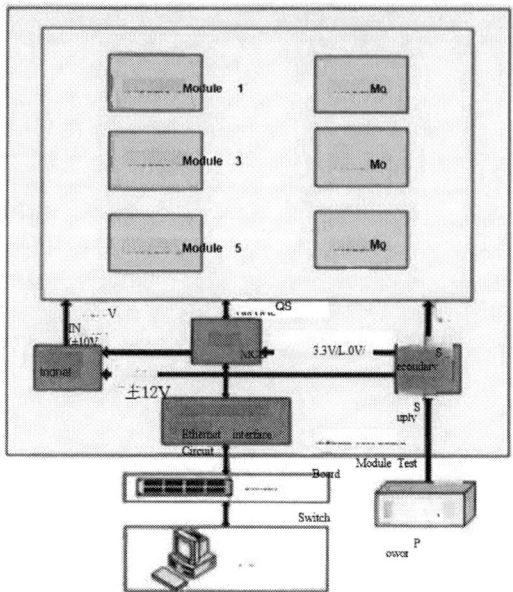

Fig. 13. Test System Block Diagram

TABLE III. ELECTRICAL CHARACTERISTICS AND TEST RESULTS

Parameter	Conditions (unless noted):-55°C≤ TA≤+100°C, VCC=+5V	Results	Unit
Supply current	—	140-240	mA
Full-scale accuracy	—	±20	mV
input impedance	—	1	MΩ
conversionresolution	fs=200KHz	16	bit
channels	—	48	—

The module features 48 channels of synchronous sampling, with a voltage input range of -10V to +10V and an input impedance of 1MΩ. It operates at a rated voltage of+5V and supports standard interfaces such as QSPI and LVDS for data transmission and configuration. Each channel supports a maximum sampling rate of 200 kHz with a 16-bit resolution (Table 3).

The multi-channel analog signal measurement microsystem module utilizing the low-cost PoP structure demonstrated normal functionality during three-temperature testing, with an accuracy of less than 20mV.

IV. SUMMARY

This study innovatively integrates crystal oscillators with miniaturized [11], highly reliable microsystem packaging to meet the engineering requirements of Spaceborne-applications [12]. By adopting an integrated packaging architecture and high-density microassembly processes, multiple ICs and numerous passive components are 3D packaged to achieve system miniaturization [13]. The proposed solution addresses critical needs for compact size, low cost, and high reliability in civilian space information processing and control systems. Furthermore, this project has advanced to batch production and engineering implementation.

ACKNOWLEDGMENT

This work was supported by the Innovation Fund of National Key Laboratory of Integrated Circuits and Microsystems.

REFERENCES

[1] Soyeon Choi, Heehun Yang, Yunjin Noh, et al "FPGA-Based Multi-Channel Real-Time Data Acquisition System."Electronics,vol.13,no 15, 2024, pp.2950.

[2] Z.Huang, G.Shi, T.Liu, J.Zheng and X.Cheng, "A Multichannel ADC Module Design Method Basedon DSP and FPGA,"2022 IEEE Asia-Pacific Conference on Image Processing, Electronics and Computers (IPEC), Dalian, China, 2022.

[3] Wang Haoyu, Ma Jianshe, Yang Yide, et al."A Review of System-in-Package Technologies: Application and Reliability of Advanced Packaging." Micromachines, vol.14, no.6, 2023.

[4] D.Badarov and G.Mihov, "FPGA Implementation of All Digital Phase Locked Loop for ADC Synchronization with the Mains Frequency,"2021 12th National Conference with International Participation (ELECTRONICA), Sofia, Bulgaria, 2021, pp.1-4.

[5] X. Sun, M. Pan, Y. Lu and L.Wan,"A novel package-on-package stacking technique," 2013 14th International Conference on Electronic Packaging Technology, Dalian, China, 2013, pp.34-36.

[6] Li Zhihua, Xie Keyu. "Research Progress on Low-Temperature Cracking of Epoxy Potting Materials and Countermeasures." Materials Review, vol.20, no.8, 2006, pp.41.

[7] Li Rui, Qiao Mengyan, Yang Jia, et al. "Study on Interfacial Residual Stress Caused by Cylindrical Incusion in Epoxy Matrix." Engineering Plastics Application, vol.47, no.8, 2019, pp.101-103

[8] Liu Yu, Wang Jun. "Study on Mixed-Mode Fracture at Epoxy/Copper Interface." Packaging, Testing &Equipment, vol36, no.9, 2011, pp 714-718.

[9] D.Badarov and G.Mihov, "FPGA Implementation of All Digital Phase Locked Loop for ADC Synchronization with the Mains Frequency," 2021 12th National Conference with International Participation (ELECTRONICA), Sofia, Bulgaria, 2021, pp.1-4.

[10] M.S.Alamgir and M.S.Islam, "PoE(Power over Ethernet)switch based remote power control system for the better performance ofISPs in Bangladesh, "2015 18th International Conference on Computer and Information Technology (ICCIT), Dhaka, Bangladesh, 2015, pp.23-26.

[11] T.Domingues, J.R.Fernandes and L.B.Oliveira, "Oscillator noise budget for ADC systems, "Proceedings of the 18th International Conference Mixed Design of Integrated Circuits and Systems- MIXDES 2011, Gliwice, Poland, 2011, pp.358-361.

[12] J.Li, T.Zeng and L.Sui, "Design and Test of civilian space Acquisition and Storage System for Magnetoelectric Sensor, "2024 3rd International Symposium on Semiconductor and Electronic Technology (ISSET), Xi'an, China, 2024, pp.516-521.

[13] L.Hu, Z.Jin, X.Liao, Y.Ouyang and J.Dong, "Study of 3D SiP (system-in-package) module for package-on-package application using multi-layer PCB manufacturing process, "2014 15th International Conference on Electronic Packaging Technology, Chengdu, China, 2014, pp.109-112.

AUTHOR INDEX

Ai, Z. 317
An, W. 984
Ba, H. 502
Bai, B. 354
Bao, X. 422
Bernales, J.S. 427
Bi, S. 540
Bi, Z. 173, 610
Bi, Z.R. 660
Cai, C. 294
Cai, G. 910
Cai, H. 496
Cai, L. 643
Cai, S. 19
Cai, X. 496
Cai, Y. 568
Cai, Z. 962
Cañonero, K.M. 427
Cao, C. 1, 191, 583
Cao, J. 805
Cao, K. 464
Cao, R. 863
Cao, X. 496
Cao, Y. 568
Chang, K.-C. 422, 568, 910, 933
Chen, D. 439, 796, 805
Chen, G. 333, 378, 899
Chen, H. 221, 454, 708, 967
Chen, J. 92, 721, 858
Chen, L. 839
Chen, N. 215
Chen, Q. 540
Chen, R. 247
Chen, S. 397, 474
Chen, W. 529, 929
Chen, X. 204, 237, 681, 810
Chen, Y. 92, 158, 272, 439, 546, 561
Chen, Z. 25, 173, 737, 977
Cheng, G. 633
Cheng, L. 923
Cheng, M. 328
Cheng, X. 648
Cheng, Y. 589
Cheng, Z. 791
Chu, J. 801
Chuang, Y.-L. 933
Cui, M. 454, 491
Cui, X. 439

Dai, C. 628, 814, 839
Dai, G. 76, 309
Dai, H. 60
Dai, R. 445
Dai, Y. 962
Dai, Z. 977
Dang, K. 81
Deng, F. 198, 904
Deng, H. 317
Deng, K. 257
Ding, T. 895
Dong, G. 622
Dong, Z. 158, 439, 546
Du, Y. 616, 648
Duan, C. 819
Duan, Y. 191
Eom, C. 835, 895
Fan, H. 138
Fan, S. 110, 904
Fan, X. 573
Fan, Y. 71
Fan, Z. 19
Fang, J. 215
Fang, Y. 844
Feng, H. 76, 257, 464
Feng, J. 105
Feng, K. 653
Feng, M. 578
Feng, Q. 317
Feng, Y. 633, 943
Fu, D. 148
Fu, J. 372, 702
Fu, N. 204
Gan, T. 801
Gao, H. 19
Gao, S. 232, 594
Gao, Z. 360
Ge, J. 819
Ge, X. 546
Gong, J. 209
Gu, X. 31, 852
Gu, Y. 110
Gu, Z. 770
Guan, Y. 967
Guo, C. 721
Guo, F. 512
Guo, H. 1, 191, 583
Guo, J. 299, 616, 648, 686, 759

Guo, K.	37
Guo, X.	198
Guo, Y.	759, 962
Guo, Z.	60, 360, 507, 610, 948
Han, H.	967
Han, S.	863
Hao, J.	748
Hao, M.	115
Hao, Y.	81
He, H.	622
He, R.	272
He, S.	670
He, X.	943
He, Y.	309
He, Z.	413, 899
Hong, Z.	868
Hou, K.	748
Hou, Y.	868
Hou, Z.	71
Hu, J.	49, 65, 276, 322, 879
Hu, K.	1
Hu, M.	272
Hu, Q.	449
Hu, S.	469
Hu, W.	721
Hu, X.	252
Hu, Y.	366, 391, 780
Hu, Z.	962
Huang, C.	568
Huang, H.	243, 616
Huang, J.	186, 243, 791, 918
Huang, M.	605
Huang, T.	408, 692, 929
Huang, X.	660
Huang, Y.	37, 153, 288, 413, 879
Huo, Y.	115, 748
Jia, J.	317
Jiang, F.	819
Jiang, J.	328, 923
Jiang, M.	759, 943
Jiang, P.	81
Jiang, Q.	257
Jiang, S.	770
Jiang, W.	616
Jiang, Y.	215
Jiao, Y.	366
Jin, Y.	204, 568, 785
Jing, C.	267

Jing, N.	328, 923
Kang, W.	633
Khalid, S.A.	344, 427
Kong, M.	317
Kou, X.	721
Kuai, R.	252, 354
Lai, B.	835, 895
Lai, X.	92
Lan, B.	529, 638
Le Wu, W.	49, 322
Lei, H.	232, 594
Li, D.	148, 675
Li, E.	127
Li, G.	660
Li, H.	7, 115, 226, 561, 731, 873, 984
Li, J.	76, 354, 464, 702, 885
Li, K.	759
Li, L.	12, 71, 86, 99, 127, 422, 568, 910, 933, 937
Li, L.A.	449
Li, M.	37, 92, 168
Li, P.	573, 785
Li, Q.	413, 556
Li, S.	753, 796
Li, T.	148, 512, 868
Li, W.	138, 204, 309, 372, 702, 770
Li, X.	43, 105, 122, 132, 198, 282, 403, 433, 459, 474, 517, 628, 801, 839, 885
Li, Y.	158, 247, 294, 299, 480, 686, 791, 868
Li, Z.	65, 105, 122, 143, 221, 232, 372, 403, 474, 594, 702, 708, 716, 748, 943
Lian, H.	418
Liang, F.	158, 439
Liang, L.	163, 948
Liang, S.	127
Liang, Y.	556
Liao, C.	551, 568
Liao, G.	616
Lin, C.	276, 299
Lin, H.	918
Lin, X.	19
Lin, Y.	716
Lin, Z.	328, 628, 796, 819, 830, 839, 873, 923
Ling, Z.	486
Liu, B.	309, 496
Liu, C.	12, 71, 863
Liu, D.X.	267
Liu, F.	643
Liu, H.	502, 748, 937, 953

Liu, J.	7, 262, 305, 372, 474, 535, 648, 885
Liu, K.	257
Liu, L.	366
Liu, M.	12, 71
Liu, P.	81
Liu, Q.	294, 397, 681
Liu, S.	418, 796
Liu, T.	801
Liu, W.	360, 403, 890
Liu, X.	25, 299, 360, 486, 605, 737, 967
Liu, Y.	54, 257, 408, 418, 551, 628, 643, 692, 743, 796, 873, 972
Liu, Y.W.	267
Liu, Z.	232, 445, 556, 594, 814
Long, M.	819
Long, Q.	868
Lou, F.	805
Lou, Y.	873
Lu, C.	384
Lu, L.	422, 568, 933
Lu, Z.	948, 958
Luo, G.	910
Luo, Q.	573
Luo, W.	1, 583
Lv, F.	168, 252, 354
Lv, J.	648
Lv, X.	780
Lv, Y.	589
Lyu, R.	610
Ma, H.	546
Ma, J.	215
Ma, Y.	317, 743
Macapundag, A.G.	427
Madrid-Khalid, K.M.	344
Maestre, S.E.	344, 427
Maloberti, F.	697
Mao, J.	237
Mao, K.	221
Mei, X.	605
Men, C.	333, 378
Meng, F.	445
Meng, S.	305
Mi, X.	879
Miao, M.	628
Miao, X.	858
Miao, Y.	984
Mo, T.	868
Nian, Y.	122, 403

Niu, Z.	143
Ou, J.	605
Pan, M.	517, 670
Pan, S.	257
Pan, Y.	628
Pang, C.	226
Pang, H.	561
Pang, Z.	168
Pei, B.	215
Peng, C.	105, 132, 143, 814
Peng, J.	480
Peng, Q.	910
Pu, X.	360
Qi, G.	243, 272, 726
Qiang, B.	105, 132
Qiang, X.	60
Qin, C.	12, 71
Qin, L.	267
Qin, P.	299, 716
Qiu, A.	665
Qiu, C.	948, 958
Quan, G.	716
Ren, H.	1, 191, 583
Ren, R.	885
Ru, Y.	737
Ru, Y.A.	25
Sham, C.-W.	247
Shang, Y.	372, 702
Shao, R.	49
Shen, R.	708
Shen, Y.	433, 551, 972
Sheng, W.	328, 923
Sheng, Y.	670
Shi, D.	824
Shi, L.	267
Shi, Q.	665
Shi, Y.	801
Shi, Z.	697
Si, L.	247
Song, C.	191
Song, H.	209, 599, 721
Su, G.	262, 535
Su, K.	512
Su, Y.	578
Sun, D.	507, 681
Sun, H.	65, 372, 702
Sun, J.	929, 958
Sun, L.	660

Sun, Q. 333, 378
Sun, R. 81
Sun, W. 967
Sun, Y. 186, 433, 801
Tan, J. 163
Tan, Y. 708
Tang, J. 540
Tang, L. 110, 764
Tang, X. 305
Tang, Y. 328, 923
Teng, C. 551, 948, 972
Tian, L. 556
Tian, Q. 918
Tong, H. 858
Tong, L. 317
Wan, J. 529, 638
Wan, M. 977
Wan, P. 737
Wang, A. 599
Wang, B. 517
Wang, C. 7, 512, 578, 814, 984
Wang, H. 65, 215, 408, 517, 743
Wang, J. 122, 186, 262, 391, 535, 810, 977
Wang, K. 180
Wang, L. 148, 469, 633
Wang, N. 743
Wang, P. 153, 288
Wang, Q. 168, 226, 328, 923
Wang, R. 12, 31, 824
Wang, W. 322, 496, 863
Wang, X. 99, 115, 143, 158, 391, 439, 546
Wang, Y. 127, 551, 899, 972
Wang, Z. 54, 132, 158, 546, 573
Wei, C. 764
Wei, G. 209
Wei, J. 7, 984
Wei, X. 366, 391, 780
Wei, Y. 929, 958
Wei, Z. 305, 469
Wen, B. 60
Wen, K. 835, 895
Wu, D. 309
Wu, F. 338
Wu, G. 491
Wu, H.-Y. 384
Wu, J. 43, 282, 517, 890
Wu, L. 49, 65, 86, 99, 221, 276, 322, 937
Wu, P. 317

Wu, S. 333, 378
Wu, W. 76, 408, 464, 692
Wu, X. 105, 122, 132, 143, 173, 232, 397, 403, 449, 594,
610, 638, 796, 819, 830, 839, 873, 984
Wu, Y. 180, 839
Wu, Z. 276, 360, 929
Xi, X. 512, 643
Xia, C. 622
Xia, G. 665
Xia, Y. 868
Xiang, Z. 578
Xiao, Y. 7
Xiaoyu, L. 523
Xie, D. 221
Xie, F. 25, 737
Xie, H. 529
Xiong, X. 19
Xiong, Y. 198
Xiong, Z. 764
Xu, B. 962
Xu, H. 122, 403
Xu, J. 168, 397, 512
Xu, L. 583
Xu, M. 665, 835, 895
Xu, N. 384, 523, 561
Xu, R. 507
Xu, T. 299, 716
Xu, X. 397
Xu, Y. 243, 910, 962
Xu, Z. 237, 299, 810
Xue, F. 780
Xue, H. 282
Xue, Q. 299, 716
Xue, T. 267
Yan, B. 115, 824
Yan, C. 173, 610
Yan, G. 517
Yan, L. 232, 899
Yan, S. 153, 288
Yan, T. 967
Yang, C. 459
Yang, D. 835
Yang, F. 338, 660, 962
Yang, H. 366, 589, 594, 830, 972
Yang, J. 138, 830
Yang, K. 962
Yang, L. 681
Yang, Q. 338

Yang, S. 309
Yang, W.M. 391
Yang, X. 115, 180, 681
Yang, Y. 12, 86, 99, 819, 824, 937
Yang, Z. 408, 418, 523, 561, 692, 753
Yao, J. 962
Yao, Y. 127
Ye, B. 433
Ye, J. 502
Ye, Y. 633
Yi, J. 153, 288
Yin, R. 354
Yin, Z. 486
You, F. 670
You, H. 540
Yu, D. 204, 923
Yu, H. 247, 984
Yu, J. 191
Yu, L. 977
Yu, R. 830
Yu, S. 153, 288
Yu, Z. 31, 852
Yuan, L. 252
Yuan, X. 512
Yuan, Y. 449
Yue, H. 43, 282
Yue, W. 824
Zeng, C. 578
Zeng, Q. 863
Zeng, X. 173, 252, 610, 660
Zha, Q. 885
Zhai, J. 622
Zhang, B. 805
Zhang, C. 780
Zhang, D. 37
Zhang, G. 215, 354, 573, 665
Zhang, H. 578
Zhang, J. 81, 92, 198, 209
Zhang, K. 469, 540
Zhang, L. 7
Zhang, R. 726, 731
Zhang, S. 282, 551, 972
Zhang, T. 338, 681, 770, 835
Zhang, W. 433, 529, 638
Zhang, X. 49, 65, 76, 86, 99, 209, 276, 322, 464, 474, 605, 748, 805, 937

Zhang, Y. 92, 252, 305, 418, 433, 486, 573, 599, 638, 643, 721, 863
Zhang, Z. 43, 168, 360, 653, 953
Zhao, C. 830
Zhao, Q. 105, 122, 132, 143, 232, 403, 594, 814
Zhao, R. 7, 780, 858
Zhao, S. 272, 958
Zhao, X. 262, 535, 675, 731
Zhao, Y. 198, 366, 665, 697
Zhao, Z. 366, 660
Zheng, B. 681
Zheng, F. 502
Zheng, R. 391, 780
Zheng, S. 616
Zheng, X. 19
Zheng, Y, ... 605, 977
Zhi, H. 43, 282
Zhong, J. 879
Zhong, X. 247
Zhong, Z. 221
Zhou, C. 753
Zhou, J. 338, 418, 507, 858, 873
Zhou, K. 830
Zhou, L. 37
Zhou, W. 37
Zhou, X. 328
Zhou, Y. 209
Zhou, Z. 148, 948
Zhu, A. 692
Zhu, C. 413
Zhu, H. 716
Zhu, K. 512
Zhu, M. 163, 173, 610
Zhu, S. 86, 622
Zhu, Z. 675, 697
Zhu, Z.-Y. 384
Zong, Y. 43
Zou, F. 589
Zou, X. 474
Zou, Z. 496

CURRAN ASSOCIATES INC.
proceedings
.com

9798331588502

2007 7th International Conference on Power Electronics

Daegu, South Korea
22-26 October 2007

IEEE Catalog Number: CFP07CPB-POD
ISBN: 978-1-42441-871-8